Human Anatomy and Physiology

Elaine N. Marieb, R.N., Ph.D.
Holyoke Community College

The Benjamin/Cummings Publishing Company, Inc.

Redwood City, California • Fort Collins, Colorado
Menlo Park, California • Reading, Massachusetts • New York
Don Mills, Ontario • Wokingham, U.K. • Amsterdam • Bonn
Sydney • Singapore • Tokyo • Madrid • San Juan

Sponsoring Editor: Connie J. Spatz

Senior Developmental Editor: Patricia S. Burner

Developmental Editor: Robin Fox

Production Director: Laura Argento

Production and Art Coordination: Pat Waldo and Deborah Gale/Partners in Publishing

Production Assistants: Linda E. Seabright-Medina and Jo Jackson

Text and Cover Design: Gary Head

Dummy Artist: Gary Head

Photo Editor: Darcy Lanham

Photo Researcher: Stuart Kenter

Artists: Martha Blake, Raychel Ciemma, Barbara Cousins, Charles W. Hoffman, Georg Klatt, Jeanne Koelling, Stephanie McCann, Linda McVay, Ken Miller, Carla Simmons, and Nadine Sokol

Copyeditors: Janet Greenblatt and Pearl Vapnek

Proofreader: Steve Sorensen

Indexer: Steve Sorensen

Editorial Assistant: Lisa Donohoe

Compositor: Graphic Typesetting Service

Film Preparation: Color Response, Inc.

Figure acknowledgments begin on page C-1.

About the cover: Photograph by Annie Leibovitz, featuring Olympic medalist Evelyn Ashford. © Annie Leibovitz/Contact Press Images.

Library of Congress Cataloging-in-Publication Data
Marieb, Elaine Nicpon, 1936-
Human anatomy and physiology/Elaine N. Marieb.
p. cm.—(The Benjamin/Cummings series in the life sciences)
Includes bibliographies and index.
ISBN 0-8053-0122-4
1. Human physiology. 2. Anatomy, Human. I. Title. II. Series.
[DNLM: 1. Anatomy. 2. Physiology. QS 4 M334h]
QP31.2.M36 1989
612
DNLM/DLC 88-39523
for Library of Congress CIP

CDEFGHIJK-RN-8932109

The Benjamin/Cummings Publishing Company, Inc.
390 Bridge Parkway
Redwood City, California 94065

About the Author

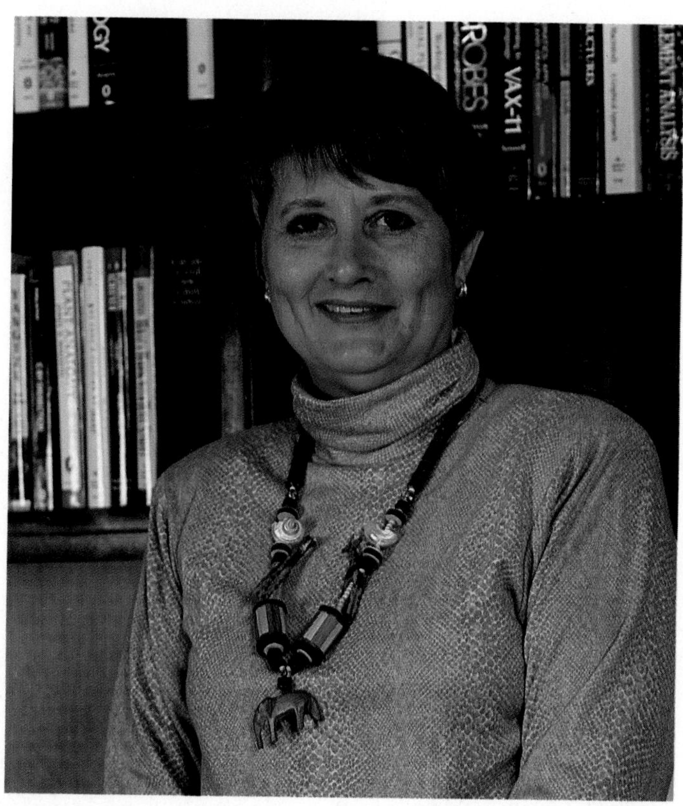

When Elaine Marieb walks into her classroom, she faces the same challenge all anatomy and physiology instructors do. She must communicate a vast amount of complex information to students in a way that stimulates their interest, not dampens it. Twenty years of teaching in both the laboratory and the classroom has enhanced her sensibilities about pedagogy and presentation. The insights gained from this experience are distilled in *Human Anatomy and Physiology*—a book that truly addresses the problems students encounter in this course.

Dr. Marieb started her teaching career at Springfield College, where she taught anatomy and physiology to physical education majors. She then joined the faculty of the Biological Science Division of Holyoke Community College in 1969 after receiving her Ph.D. in Zoology from the University of Massachusetts at Amherst. As a result of her contact with students at Holyoke Community College, many of whom were pursuing degrees in nursing, she developed a desire to better understand the relationship of the scientific study of the human body and the clinical aspects of nursing practice. While continuing to teach full-time, Dr. Marieb earned an associate degree in nursing in 1980 from Holyoke Community College. She continued her nursing education, receiving a bachelor's degree in nursing from Fitchburg State College and then a master's degree in gerontological nursing from the University of Massachusetts in 1985. Dr. Marieb is currently teaching at Holyoke Community College, where she was honored with the "Outstanding Educator Award" for excellence in classroom instruction.

Dr. Marieb is an accomplished author, with several successful textbooks and laboratory manuals to her credit. These publications are listed on the page facing the title page. This text—*Human Anatomy and Physiology*—is the culmination of her career as an educator and author.

Preface to the Instructor

As a teacher, you know that teaching is not just the presentation of facts. You must provide information in a framework that encourages genuine understanding, devise new presentations to get students over conceptual hurdles, and help students apply what they have learned to new situations. All the while you hope that you can help them exceed expectations—both theirs and yours—and that you can inspire in them a real love of the subject.

After many years of teaching human anatomy and physiology to nursing students, my curiosity about the clinical aspects of anatomy and physiology led me to the classroom—not as a teacher, but as a nursing student. When I sat on the other side of the desk among allied health students of all ages, it was easy to see how my explanations of anatomy and physiology could be improved. I became convinced that new approaches to many topics could excite and challenge students' natural curiosity, and I decided to write this book.

I was very fortunate in collaborating with Benjamin/Cummings, a publisher that shared my goal: to set new standards for pedagogical and visual effectiveness in an anatomy and physiology text.

Unifying Themes

The story of anatomy and physiology would not be coherent or logical without the development of unifying themes. Three integrating themes, presented in Chapter 1 and expanded upon throughout the text, form the common thread that unifies, organizes, and sets the tone for the entire book.

Interrelationships of body organ systems. The fact that nearly all regulatory mechanisms require interaction of several organ systems is continually emphasized. For example, Chapter 6, which describes the growth and maintenance of bone tissue, emphasizes the importance of muscle pull on bones to maintain bone strength; Chapter 10, which deals with the structure and function of muscle tissue, discusses the importance of bones as levers that muscles use to promote body movements. Unique summary figures called Homeostatic Interrelationship Diagrams show the relationship of each body system to all other systems. This approach should help students think of the body as a dynamic community of interdependent parts rather than as a number of isolated structural units. The wall chart entitled "Physiological Interactions" will also help students achieve an appreciation for the interrelationships of systems.

Homeostasis. The normal and most desirable condition of body functioning is homeostasis. Its loss or destruction always leads to some type of pathology—temporary or permanent. Pathological conditions are introduced and integrated with the text material as appropriate. However, clinical examples are used only to clarify and illuminate normal functioning, not as an end in and of themselves. For example, Chapter 20, which deals with the structure and function of blood vessels, explains how the ability of healthy arteries to expand and recoil ensures continuous blood flow and proper circulation. Also discussed in this chapter, however, are the effects on homeostasis when arteries lose their elasticity: high blood pressure and all of its attendant problems. These homeostatic imbalances are indicated visually by a red seesaw symbol that has been tipped off balance. Whenever students see the unbalanced seesaw in art or text, the concept of disease as a loss of homeostasis is reinforced.

Complementarity of structure and function. Students are encouraged to develop a critical understanding of the structure of an organ, tissue, or cell as a prerequisite to comprehending its function. The fundamental concepts of physiology are carefully explained and related to structural characteristics that promote or allow the various functions to occur. For example, the heart can act as a double blood pump because its chambers are interconnected by muscle bundles that coil like figure eights with no beginning and no end.

Special Features

Art and Photo Program. Art plays a critical role in helping students visualize anatomical and physiological concepts. Writing this book was a process of translating my knowledge about anatomy and physiology into words and illustrations. The developmental editors and I met with each artist to brainstorm on each piece of art. Our team of medical illustrators read the manuscript, did additional research when necessary, and then provided realistic, highly accurate figures to depict the anatomy and devised innovative approaches to illustrate the physiological concepts. Each illustration was carefully reviewed at every stage of its development to ensure accuracy.

The art program is full color throughout, and color is used not only in an aesthetic sense but in a functional sense as well. For example, the color protocol for the parts and organelles of the cell established in Chapter 3 is used throughout the book whenever cellular structures are shown. Thus, the nucleus is always illustrated in shades of purple and the mitochondria in orange. Likewise, ATP is signalled by a bright yellow sunburst wherever it appears.

Consistency of this nature encourages automatic learning on the part of the student and continually reinforces their memory. In many figures, the body location or general structure of a body part is first illustrated in a simple diagram for orientation, and then the structures of interest are shown in enlarged, detailed view. Not only are light and scanning electron micrographs used abundantly, but images produced by modern medical scanning techniques (CAT, PET, and MRI) are inserted where appropriate to enhance the student's grasp of structure. Additionally, the highly acclaimed Bassett photographs of dissected cadavers appear throughout the book and a collection of them is provided in Appendix D.

Illustrated Tables. The use of illustrations within tables is a highly effective and efficient method of helping students learn important information. Thus most chapters have at least one illustrated table. These tables run the gamut in level of complexity from the exceptionally complete tissue tables in Chapter 4 to the simple, but highly effective table of orientation and directional terms in Chapter 1. In particular, the illustrated gross muscle anatomy tables in Chapter 10 are unique, because, in addition to doing what all such tables do (summarizing the location, attachments, and general functions of the skeletal muscles), these tables include an essay overview of the functions of each group of muscles and describe how they interact and play against one another. This information provides an added dimension not seen in other textbooks at this level.

Color Coding. Each body system has been assigned a specific color. For example, all the chapters concerned with the cardiovascular system, Chapters 18–20, are identified by a red color tab at the top of the page. Thus, if you wish to quickly flip to the chapters on this system, you would simply open the book to the section identified by the red-edged pages. Likewise, all Homeostatic Interrelationship Diagrams, which summarize the interactions of the body's organ systems, and several tables and figures identify the organ systems by these same color codes. The brief table of contents on p. xiv provides a guide to interpreting the color coding scheme.

Physical Education Applications. Because most students majoring in physical education must take a course in human anatomy and physiology, this text describes physiological changes or demands during exercise. The text also explains the "why" of common sports injuries, and discusses other concerns of athletes. These physical education applications are signified by a bicycle symbol preceding the sections.

"A Closer Look" Boxes. Many chapters of this book have special topic boxes. Some of these boxes deal with timely topics such as AIDS (pp. 700–701), cocaine (p. 461), and steroid use by athletes (p. 281). Other boxes describe recent medical advances such as the new technique of using skeletal muscle to repair a decompensated heart (pp. 614–615) or in vitro fertilization (pp. 964–965). Some boxes are expanded discussions of important topics in the text such as neurotransmitters (p. 366) or the role of the pineal gland in mediating the effect of light (and jet lag) on the body (p. 560).

Life Span Approach. Most chapters close with a "Developmental Aspects" section that summarizes the embryonic development of organs of the system and then examines how they change throughout the life span. Diseases particularly common at certain periods of life are pointed out, and problems encountered by the elderly are emphasized.

Content and Organization

The material in this book is presented in five units encompassing a total of 30 chapters. Each unit is self-contained and its individual chapters can be assigned in any sequence without loss of continuity. Following is an overview of the book's organization.

Unit 1: Organization of the Body (Chapters 1–4). Chapter 1 introduces the student to the basic organization of the body, defines homeostasis and explains the basis of control systems, and introduces the terminology that will be used thereafter in dealing with anatomical concepts. Chapter 2 provides students with an unintimidating treatment of chemistry, complete with all the background they need to comprehend physiological principles. Chapter 3 provides an up-to-date treatment of the cell and its organelles, including the cytoskeleton, which is so important in cellular division and intracellular transport processes. The chapter also introduces the concept of membrane potential, which must be understood before the activity of excitable cells (muscle and nerve cells) makes sense. The tabular figures in Chapter 4 are a complete histology learning tool—a combination of photomicrographs, line art, and descriptions of structure, function, and location of body tissues.

Unit 2: Covering, Support, and Movement of the Body (Chapters 5–10). This unit first considers skin and then the skeletal and muscular systems and their interactions in promoting body support, protection, and mobility. A thorough discussion of bone remodeling and its controls is found in Chapter 6. Chapter 9 provides a current and understandable presentation of muscle cell anatomy and physiology and includes illustrations not found in other textbooks. For instance, Figure 9.11 (p. 253) on excitation-contraction coupling is a model of simplicity that accurately and understandably explains some very difficult concepts.

Unit 3: Regulation and Integration of the Body (Chapters 11–17). Unit 3 provides comprehensive treatment of the two great regulatory systems of the body, the nervous system and the endocrine systems. Chapter 11 sets the stage by focusing on the relationship between neuron anatomy and physiology. The illustrated tables on cranial nerves in Chapter 13 are superb learning tools. Chapter 15, on neural integration, explores the various levels of motor and sensory integration, command neurons, and current theories about the operation of higher mental functions such as intelligence and memory. Chapter 16, "The Special Senses," features illustrations not seen elsewhere. For example, Figure 16.11 (p. 499) is an exquisitely beautiful depiction of rods and cones.

Unit 4: Maintenance of the Body (Chapters 18–27). Unit 4 presents body systems and mechanisms that maintain homeostasis on a day-to-day basis—cardiovascular, lymphatic, immune, respiratory, digestive, and urinary systems and water, electrolyte, and acid-base balance of body fluids. The discussion of autoregulation of blood flow in Chapter 20 is a highly understandable explanation of a difficult topic. Chapter 22, on the immune system, is a thorough treatment of an increasingly important and fast-moving field.

Unit 5: Continuity (Chapters 28–30). The final unit considers the reproductive system, sexual development, pregnancy, embryonic development, and basic concepts of heredity. The discussion in Chapter 28 of the hormonal regulation of the female cycles has been praised by reviewers as the clearest explanation of any text. Chapter 30 provides a substantial mini-exposition of the basic understandings of human genetics.

In-Text Learning Aids

A number of pedagogical devices have been incorporated into *Human Anatomy and Physiology* to enhance its utility as a learning tool.

Combined Chapter Outline and Objectives. Each chapter begins with an outline of the major topics of the chapter with specific objectives for each topic. Page references allow students to easily refer to pertinent text discussions.

Preview of Selected Key Terms. A list of important terms and their phonetic pronunciations is presented at the beginning of each chapter. Important terms within the chapter text are also boldfaced and phonetic spellings are provided.

Related Clinical Terms. At the end of each chapter, a list of clinical terms that are relevant to the contents of the chapter is provided.

Chapter Summary. Thorough chapter summaries in a study outline format with page references will help students review material they have just read.

Review Questions. A variety of question formats—multiple choice, matching, short answer essay, and clinical application questions—are found at the end of each chapter. This set of questions provides an important tool for students to test their knowledge of the material. Answers to the multiple choice and matching questions can be found in Appendix B.

Appendices. The appendices offer several useful references for the student. They include Appendix A: The Metric System; Appendix B: Answers to the Multiple Choice and Matching Questions; Appendix C: Words, Roots, Prefixes, Suffixes, and Combining Forms; Appendix D: Cadaver Photos (featuring color photographs of dissected cadavers from the Bassett photo collection); and Appendix E: Periodic Table.

Glossary. At the end of this book, students will find an extensive glossary that includes both the definitions and phonetic pronunciations of key terms.

Supplements

The ancillary package that accompanies *Human Anatomy and Physiology* has been carefully developed to assist instructors and students in using and deriving the greatest benefit from this text.

Instructor's Manual. The Instructor's Manual, prepared by Jerri Lindsey and William Matthai of Tarrant County Junior College, provides detailed annotated lecture outlines, suggested class activities/discussion topics, a list of drugs and related disorders for each chapter, and answers to all essay/short answer questions in the text.

Test Bank. Jerri Lindsey and William Matthai helped to organize and prepare a comprehensive test bank that is keyed to the text. They were careful to offer the right mix of true/false, matching, multiple choice, essay, and clinical questions. This class-tested test bank is available on Microtest, a microcomputer test generation program for the IBM PC, AT XT, the Apple II family, and the Macintosh computers. This software is available for qualified adopters of *Human Anatomy and Physiology* by contacting the publisher or your local representative.

Overhead Acetate Transparencies. Over 140 full-color transparencies are available to qualified adopters of this text. These acetate transparencies feature the most important illustrations from the text. The type has been enlarged for easy viewing in large lecture halls.

Photo Atlas. Authored by Robert A. Chase, M.D., Professor of Anatomy at Stanford University, *The Bassett Atlas of Human Anatomy* offers an extraordinary collection of full-color dissection photos. An inexpensive alternative to the costly collections now available, it will be extremely useful to your students in their laboratories.

Human Dissection Slides. Qualified adopters of *The Bassett Atlas of Human Anatomy* or the *Human Anatomy and Physiology* text or laboratory manuals, will be eligible for a set of 85 complimentary dissection slides taken from *The Bassett Atlas of Human Anatomy.*

Laboratory Manuals. Elaine N. Marieb has authored three very successful lab manuals that are complementary to this textbook: *Human Anatomy and Physiology Lab Manual: Brief Version*, second edition (1987); *Human Anatomy and Physiology Lab Manual: Cat Version*, third edition (1989); and *Human Anatomy and Physiology Lab Manual: Fetal Pig Version*, third edition (1989). One of these will meet the needs of your laboratory course, depending on the length of the course and the animal of dissection.

Student Study Guide. This study guide/workbook by Elaine N. Marieb will help students study concepts from laboratory and lecture through a variety of exercises: identification, matching, labeling, completion, true/false, definitions, and over 100 coloring exercises. With all answers at the end of the book, it is an ideal way for students to review the basic concepts of human anatomy and physiology.

"Physiological Interactions" Wall Chart. This unique wall chart reinforces the principal theme of the book—the interrelationship of body systems. This poster will serve as a road map for students throughout the course.

* * *

The proof of the effectiveness of this text as a teaching tool will be its usefulness in the classroom. I would appreciate hearing from you about the ways this book might be improved in future editions.

Elaine N. Marieb
Holyoke Community College
Department of Biology
303 Homestead Avenue
Holyoke, Massachusetts 01040

Acknowledgments

Although the writing of a textbook is an exercise in lonely perseverence for the author, bringing the book to its final printed form requires the collaborative effort of many dedicated and skilled people. My publisher, the Benjamin/Cummings Publishing Company, spared no effort in its commitment to publish an accurate, instructive, and beautiful book. Over 300 reviews were commissioned enlisting comments and suggestions from both generalist academicians and specialists in various niches of anatomy and physiology. These manuscript reviewers are named on p. xii.

From beginning to end, reviewers' contributions have been of inestimable value to the development of this project. During the initial phases, instructors generously provided us with their syllabi and responded to questionnaires probing what they would expect of an ideal anatomy and physiology textbook. During intermediate stages, focus group participants met and provided constructive criticism to author and editorial staff. The review process continued throughout all drafts of the work and included both the art sketches and final inked art. Although an error-free manuscript is a near impossibility in a work of this magnitude, there will surely be very few because of the careful work and expertise of our reviewers. For this I am extremely grateful. A very special debt of gratitude is owed to Jon Mallatt of Washington State University who reviewed the entire manuscript—some chapters more than once. His sensitivity to possible areas of student confusion and his care in checking the accuracy of the anatomical art resulted in his becoming an important member of the "team," and someone whom I depended upon greatly as my second pair of eyes.

I want to thank my many students and colleagues who were generous with their time and thoughts. They endured endless questions and provided valuable feedback and insights on the clarity and appropriateness of my writing style. They did not always tell me what I wanted to hear, but assured of the sincerity of their criticism, I always listened.

I owe an immense debt of gratitude to our team of artists, which included Martha Blake, Raychel Ciemma, Barbara Cousins, Charles W. Hoffman, Georg Klatt, Jeanne Koelling, Stephanie McCann, Linda McVay, Ken Miller, Carla Simmons, and Nadine Sokol. Their superb illustrations made the body come alive and will surely contribute to the success of this book. Although most of the art was newly drawn, some outstanding illustrations were borrowed from *Biology*, published by Benjamin/Cummings. I wish to thank the author, Neil Campbell, for permission to use the selected pieces.

The marketing department, in particular Anne Manly, went the extra mile to produce the wall chart, "Physiological Interactions." This beautiful poster will serve as an invaluable study aid for students in this course.

The photo program supporting the text is also exceptional, a credit to Darcy Lanham who headed the photo research program. Her perception of exactly what was needed to support a particular concept or show a specialized bit of anatomy was matched by her tireless searches to find that "just right" photograph or micrograph. A special thanks is due Lucile Bassett who gave us the rights to use the dissection photographs produced by her husband Dr. David Bassett, and to Dr. Robert A. Chase, of Stanford University, who searched through some 1600 photographs to find those that would best show the desired body structures.

Gary Head, who has received awards for the excellence of his book designs, outdid himself in creating the design for this book. He provided a crisp, efficient design that effectively showcased the art and special pedagogical features of the book and that would be inviting rather than intimidating to the student. The design chosen for the cover is particularly striking and conveys the feeling of vigor that permeates the physiological discussions in the text.

Editors are editors are editors, but mine were special. Andy Crowley, my first sponsoring editor, got the ball rolling and infused the project with enthusiasm and vision. Pat Burner, my developmental editor at Benjamin/Cummings, surely and carefully guided the manuscript's progress until it was ready to be delivered to the production team. The format to be taken by many of the illustrated tables was Pat's brainchild, and her editorial wisdom has left its earmarks throughout the text. Robin Fox did the developmental edit on most of the manuscript, and there is little doubt that the book is better for her careful work.

My sponsoring editor, Connie Spatz, came to Benjamin/Cummings as Life Sciences editor after this project had begun. Her consistently cheerful spirit and encouragement became my sole mainstay at many points. I respect Connie for her honesty, her exquisite perceptiveness of human mood, her incredible drive to publish the best book possible, and her seemingly endless energy directed toward that end. She is a publishing professional par excellence.

Pat Waldo and Deborah Gale of Partners in Publishing, the company that handled the production of this book, also deserve a hearty round of applause. They managed to orchestrate the publication of the book under a tight deadline without compromising its integrity. Complementing and coordinating the production process in-house at Benjamin/Cummings was Laura Argento. Although her contributions were many, perhaps Laura's greatest contribution was her ability to foresee possible problems before they happened, allowing us to take circumventing measures.

When I finally made the decision to write this book, my largest and most ambitious work, I chose Benjamin/Cummings for several reasons. I know from experience that Benjamin/Cummings is committed to publishing high-quality texts that are valuable learning tools for students, and is willing to place the resources of the company behind this commitment. When I wrote my first book for Benjamin/Cummings in the late seventies, I became acquainted with Jim Behnke who was then Life Sciences Editor. Now, 10 years later, Jim is the president of the company. Even so, he reviewed and commented on nearly all phases of the development of this book. I trust him implicitly and at least partly because of this trust, I have found the relationship between publisher and author to be a satisfying and rewarding one.

Last, but certainly not least, I want to thank my friends who still remembered that I was their friend during these four years of social denial.

This book is dedicated to my students, whose enthusiasm inspired me to write about the subject I love so much; to Curtis Smith, who gave me the courage to try; and to my beloved husband "Zeb," who suffered with me during the many months of unrelenting work and rejoiced with me in its completion.

Focus Group Participants

Robert Agnew, *Northlake College*
Joe Anders, *North Harris County College*
Jennifer Breckler, *San Francisco State University*
Jean Cons, *College of San Mateo*
Carl Knight, *Eastfield College*
Sylvia Lianides, *West Valley College*
Jerri Lindsey, *Tarrant County Junior College*
Gordon Locklear, *Chabot College*
Gail Matson, *American River College*
Robert Nabors, *Tarrant County Junior College*
David Smith, *San Antonio College*
Greg Smith, *St. Mary's College*
Richard Symmons, *California State University at Hayward*

Manuscript and Art Reviewers

Thomas Adams, *Michigan State University*
Robert Agnew, *Northlake College*
Joe Anders, *North Harris County College*
Robert Anthony, *Triton College*
Delon Barfuss, *Georgia State University*
Wayne Becker, *University of Wisconsin at Madison*
John Benson, *Ohio University*
Frank Binder, *Marshall University*
Lanier Byrd, *Saint Philips College*
Gene Block, *University of Virginia*
Franklyn Bolander, *University of South Carolina*
Robert Boley, *University of Texas at Arlington*
James Boyson, *San Antonio College*
Michael Bowes, *Humboldt State University*
Jennifer Breckler, *San Francisco State University*
Alan Brush, *University of Connecticut*
Warren Burggren, *University of Massachusetts at Amherst*
Michael Burke, *University of Kansas*
Ray Burkett, *Shelby State College*
Kenneth Bynum, *University of North Carolina at Chapel Hill*
Colin Campbell, *Pima Community College*
Wayne Carley, *Lamar University*
Tom Cole, *Phillips College*
Richard Connett, *University of Rochester*
Darrell Davies, *Kalamazoo Valley Community College*
Scott Dunham, *Illinois Central College*
Klaus Elgert, *Virginia Technological Institute*
Victor Eroschenko, *University of Idaho*
Jim Ewig, *Towson State University*
Douglas Fonner, *Ferris State College*
John Frehn, *Illinois State University*
Anne Funkhauser, *University of the Pacific*
Gregory Garman, *Centralia College*
Stephen George, *Amherst College*
Carol Gerding, *Cuyahoga Community College*
Todd Gleeson, *University of Colorado at Boulder*
Judy Goodenough, *University of Massachusetts at Amherst*
Ann Hagan, *American University*
Ann Harmer, *Orange Coast College*
Deborah Hettinger, *Texas Lutheran College*
C. J. Hubbard, *Northern Illinois University*
Margaret Hudson, *Seattle University*
Karen Jones, RN, *Cardiovascular Surgery, Inc.*, Evansville, Indiana
Donald Kisiel, *Suffolk Community College*
William Kleinelp, *Middlesex County College*
Carl Knight, *Eastfield College*
Robert Laird, *University of Central Florida*
Jerri Lindsey, *Tarrant County Junior College*
Alan Liss, *State University of New York at Binghamton*
Mark Lomolino, *University of Arizona*
Jon Mallatt, *Washington State University*
William Matthai, *Tarrant County Junior College*
Gary Matthews, *State University of New York at Stony Brook*
Esmail Meisami, *University of Illinois at Champagne-Urbana*
Doug Merrill, *Rochester Institute of Technology*
Lew Milner, *North Central Technical College*
John Minnich, *University of Wisconsin at Milwaukee*
Robert Nabors, *Tarrant County Junior College*
Evan Oyakawa, *California State University at Los Angeles*
Harry Painter, *North Virginia Community College*
S. Arthur Reed, *University of Hawaii*
Curtis Smith, *Mount Holyoke College*
David Smith, *San Antonio College*
Wendy Smith, *Northeastern University*
Phil Sokolove, *University of Maryland*
Richard Symmons, *California State University at Hayward*
Sunny Tripp, *University of Notre Dame*
Guy Van Cleave, *Modesto Junior College*
Edward Wallen, *University of Wisconsin at Parkside*
James Waters, *Humboldt State University*
Richard Welton, *Southern Oregon State College*
Glenn Yoshida, *Los Angeles Southwest College*
Yameen Zubari, *Catonsville Community College*

Preface to the Student

This book is written for you. In a way, it is written by my students, since it incorporates their suggestions, the answers to questions they most often ask, and explanatory approaches that have produced the greatest success in helping them learn about the human body. Human anatomy and physiology is more than just interesting—it is fascinating. To help get you involved in the study of this exciting subject, a number of special features are incorporated throughout the book.

The tone of this book is intentionally informal and unintimidating. There is absolutely no reason that you can't enjoy your learning task. This book is meant to be a guide to the understanding of your own body, not an encyclopedia of human anatomy and physiology. I have tried to be selective about the information to be included and have chosen facts that stress essential concepts. Physiological concepts are explained thoroughly; abundant analogies are used and examples are taken from familiar events whenever possible.

The illustrations and tables are designed with your learning needs in mind. The tables are summaries of important information in the text and should be valuable resources for reviewing for an exam. In all cases, the figures are referenced where their viewing would be most advantageous to help in understanding the text. Those depicting physiological mechanisms often use the flow-chart format so that you can see where you've been and where you're going. Special topic boxes, called "A Closer Look," alert you to advances in medicine or present scientific information that can be applied to your daily life.

Each chapter begins with an outline of its major topics (and their page references), with specific learning objectives listed for each topic. Also at the beginning of each chapter is a list of selected key terms and their phonetic pronunciations that provides a preview to some of the important terms included in the chapter. Important terms within the chapter text are highlighted by being bold-faced (in dark type), and phonetic spellings are provided for terms that are likely to be unfamiliar to you. To use these phonetic aids, you will need to remember the following rules:

1. Syllables are separated by dashes.
2. Accent marks follow stressed syllables. The primary stress is shown by ′, and the second stress by ″.
3. Assume that vowels are short unless a bar, indicating a long vowel sound, appears above it.

For example, the phonetic spelling of "thrombophlebitis" is throm″-bō-fleh-bī′-tis. The fourth syllable (bī′) receives the greatest stress, and the first syllable (throm″) gets the secondary stress. The "ō" in the second syllable and the "ī" in the fourth syllable are long. A more complete explanation of the pronunciation system can be found at the beginning of the glossary.

Any exam causes anxiety. To help you better prepare for an exam or comprehend the material you have just read, thorough summaries complete with page references are found at the end of each chapter as are review questions that use a combination of testing techniques (multiple choice, matching, short answer essay, and clinical applications).

I hope that you enjoy *Human Anatomy and Physiology* and that this book makes learning about the body's structures and functions an exciting and rewarding process. Perhaps the best bit of advice I can give you is that memory depends on understanding. Thus, if you strive to achieve understanding instead of rote memorization, your memory will not fail you very often.

I would appreciate hearing from you about your experiences with this textbook or suggestions for improvements in future editions.

Elaine N. Marieb

Elaine N. Marieb
Holyoke Community College
Department of Biology
303 Homestead Avenue
Holyoke, Massachusetts 01040

Brief Contents

UNIT 4

MAINTENANCE OF THE BODY 566

UNIT 5

CONTINUITY 912

APPENDICES

Detailed Contents

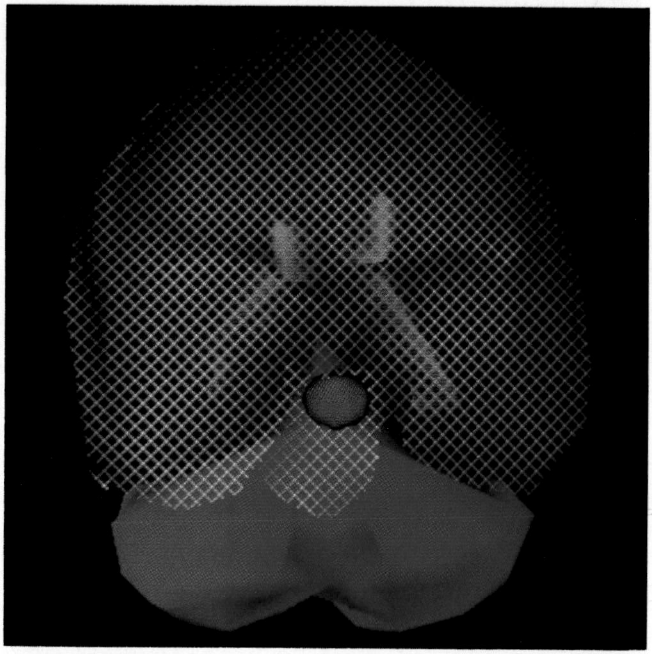

CAT scan of the brain.

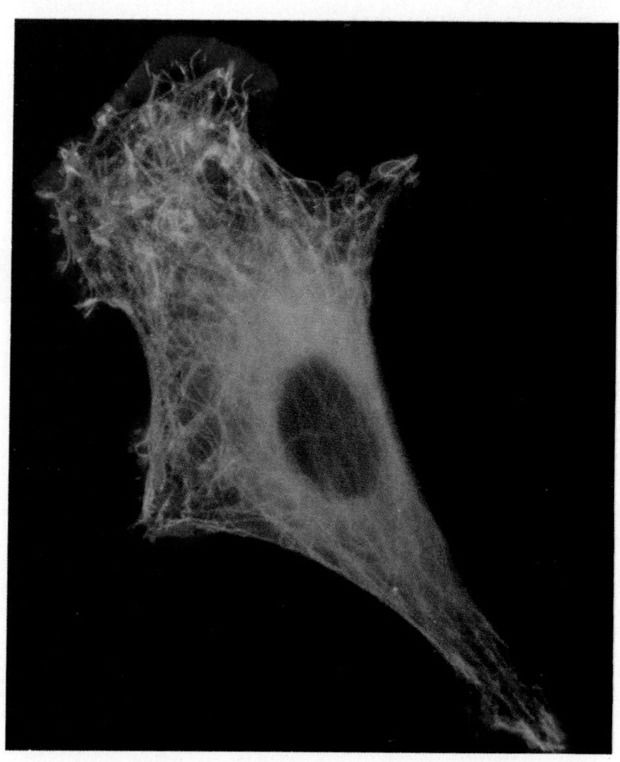

Micrograph of the cytoskeleton (1200x).

4 Tissues: The Living Fabric 100

UNIT 2

COVERING, SUPPORT, AND MOVEMENT OF THE BODY 132

5 The Integumentary System 134

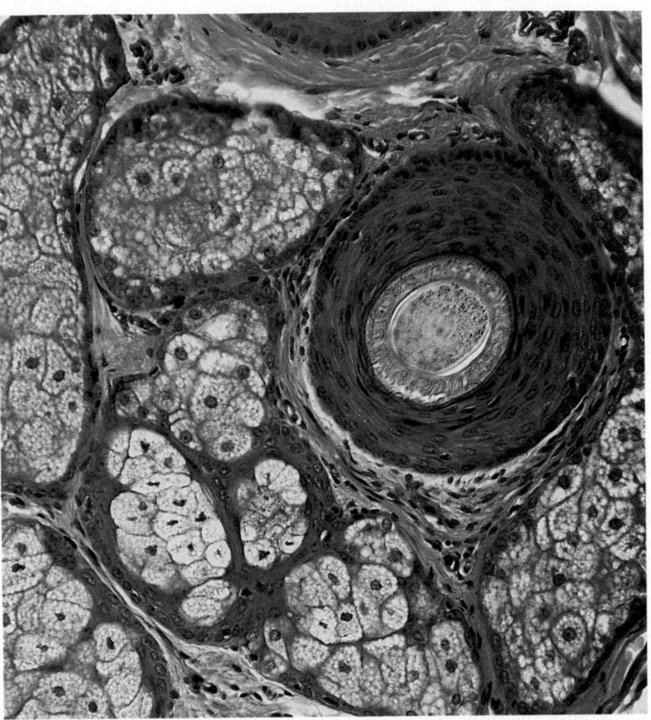

Light micrograph of sebaceous glands surrounding a hair follicle (200x).

6 Bones and Bone Tissue 152

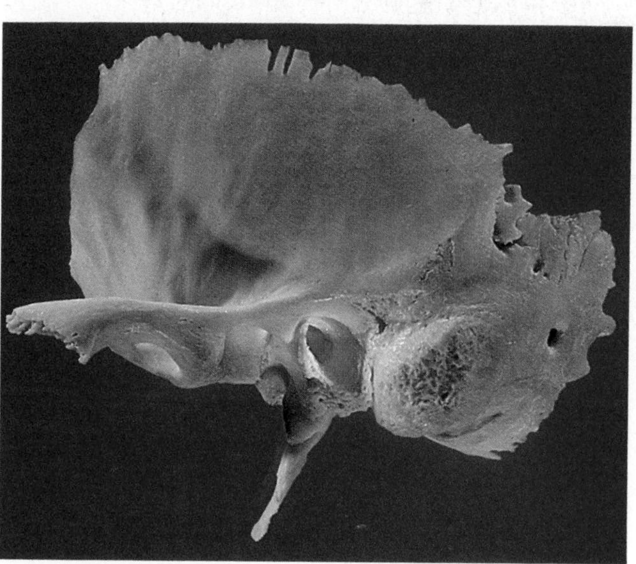

Temporal bone.

8 Joints 215

9 Muscles and Muscle Tissue 239

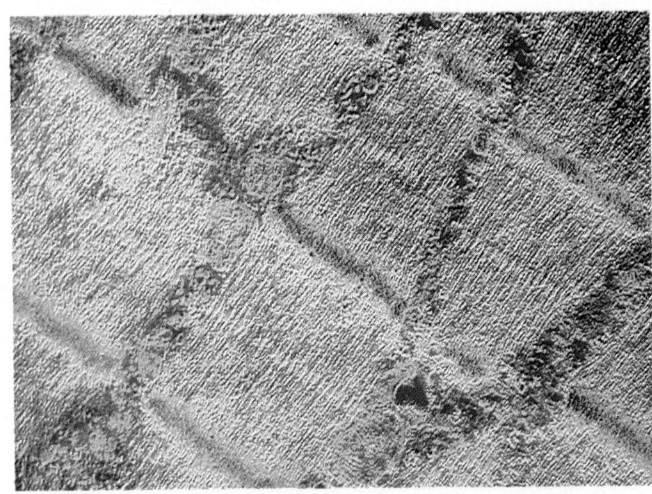

Transmission electron micrograph of myofibrils in striated skeletal muscle (12,800x).

12 The Central Nervous System 375

13 The Peripheral Nervous System and Reflex Activity 420

14 The Autonomic Nervous System 447

15 Neural Integration 464

16 The Special Senses 486

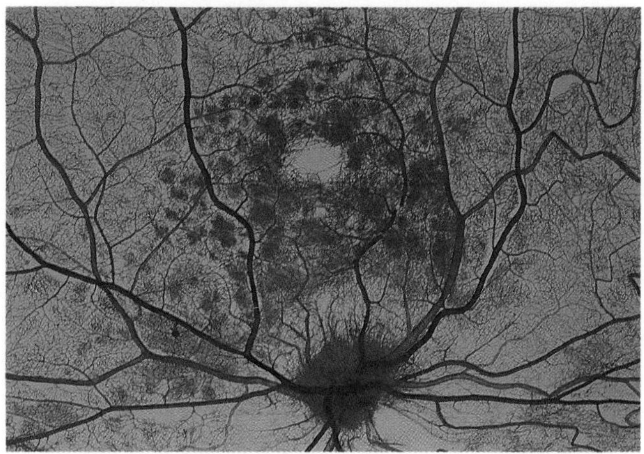

Human retina.

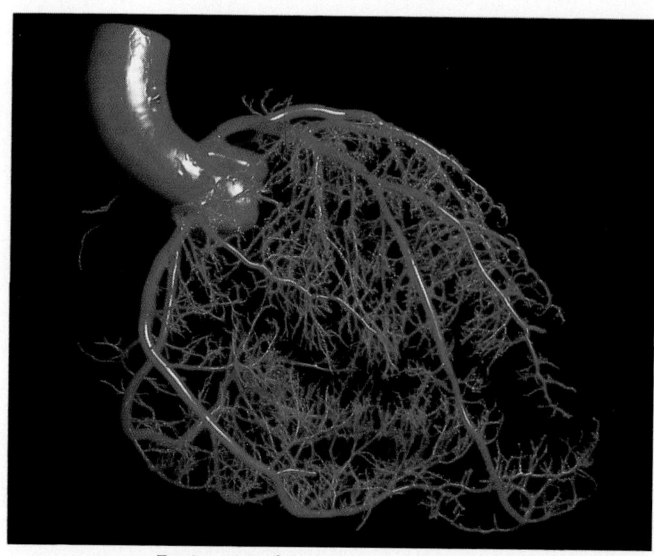

Resin cast of coronary arteries supplying the heart.

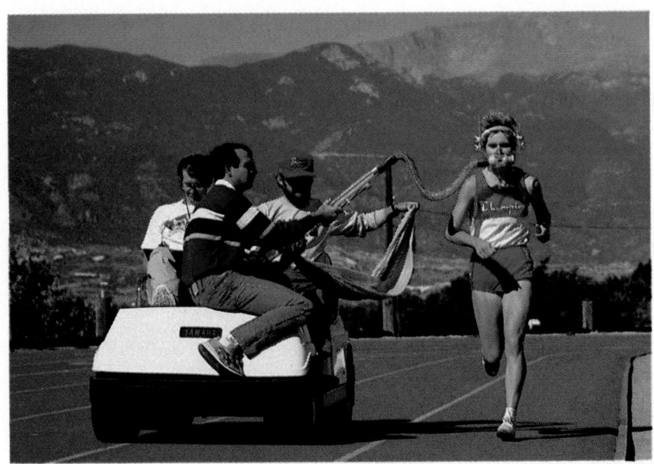

Scientists at U.S. Olympic Training Center measuring oxygen and carbon dioxide levels during exercise.

25 Nutrition, Metabolism, and Body Temperature Regulation 804

26 The Urinary System 855

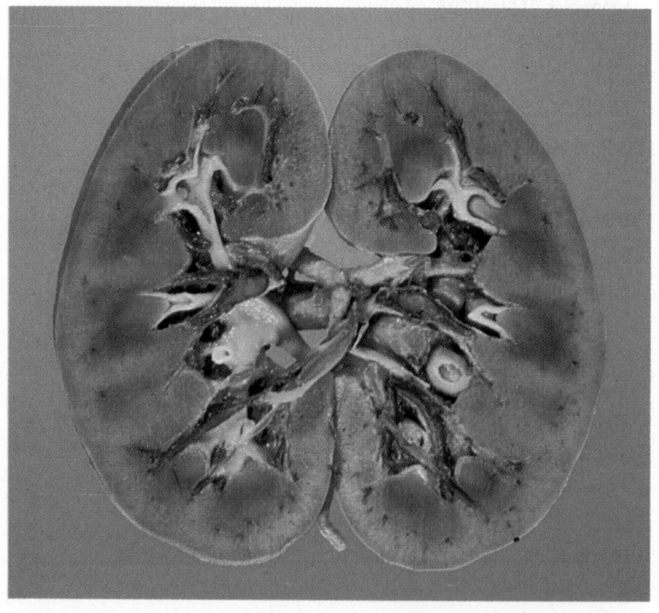

Coronally sectioned kidney.

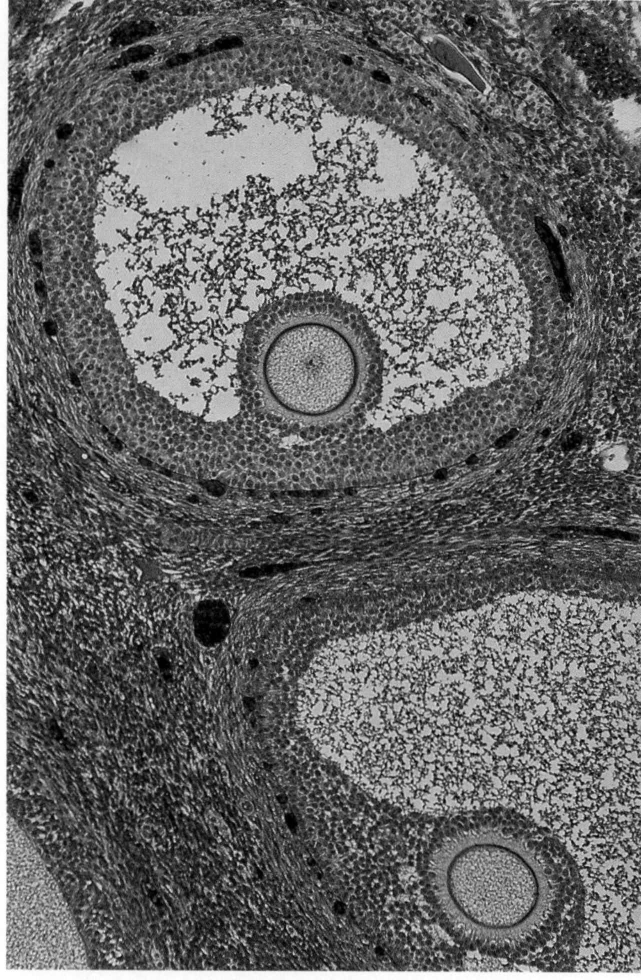

Light micrograph of an ovarian follicle (375x).

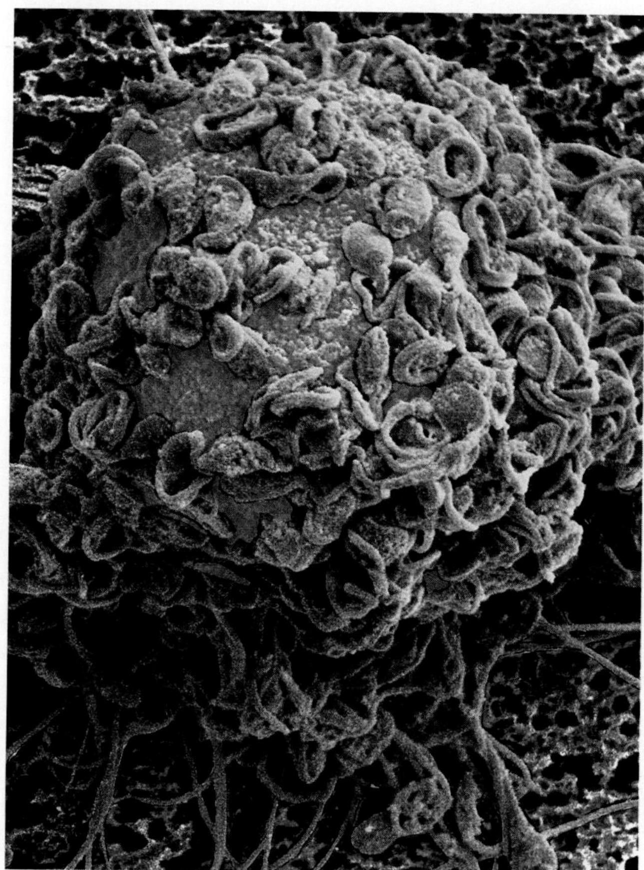

Scanning electron micrograph of human sperm on an egg (1400x).

The four chapters of this initial unit introduce you to the sciences of anatomy and physiology and provide the framework for your subsequent studies of the human body and its workings. Chapter 1 first defines anatomy and physiology and then describes the basic organization and principal functions of the living body. It completes its orientation task by presenting anatomical terminology that will be used throughout the text. Chapters 2–4 begin to ascend the hierarchy of body structure in more detail, first describing its chemical units, then cells, and finally tissues. Subsequent units in the book examine the organ systems of the body.

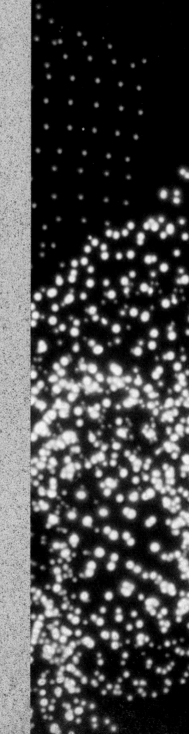

Computer graphic showing the helical shape of DNA.

UNIT 1

ORGANIZATION OF THE BODY

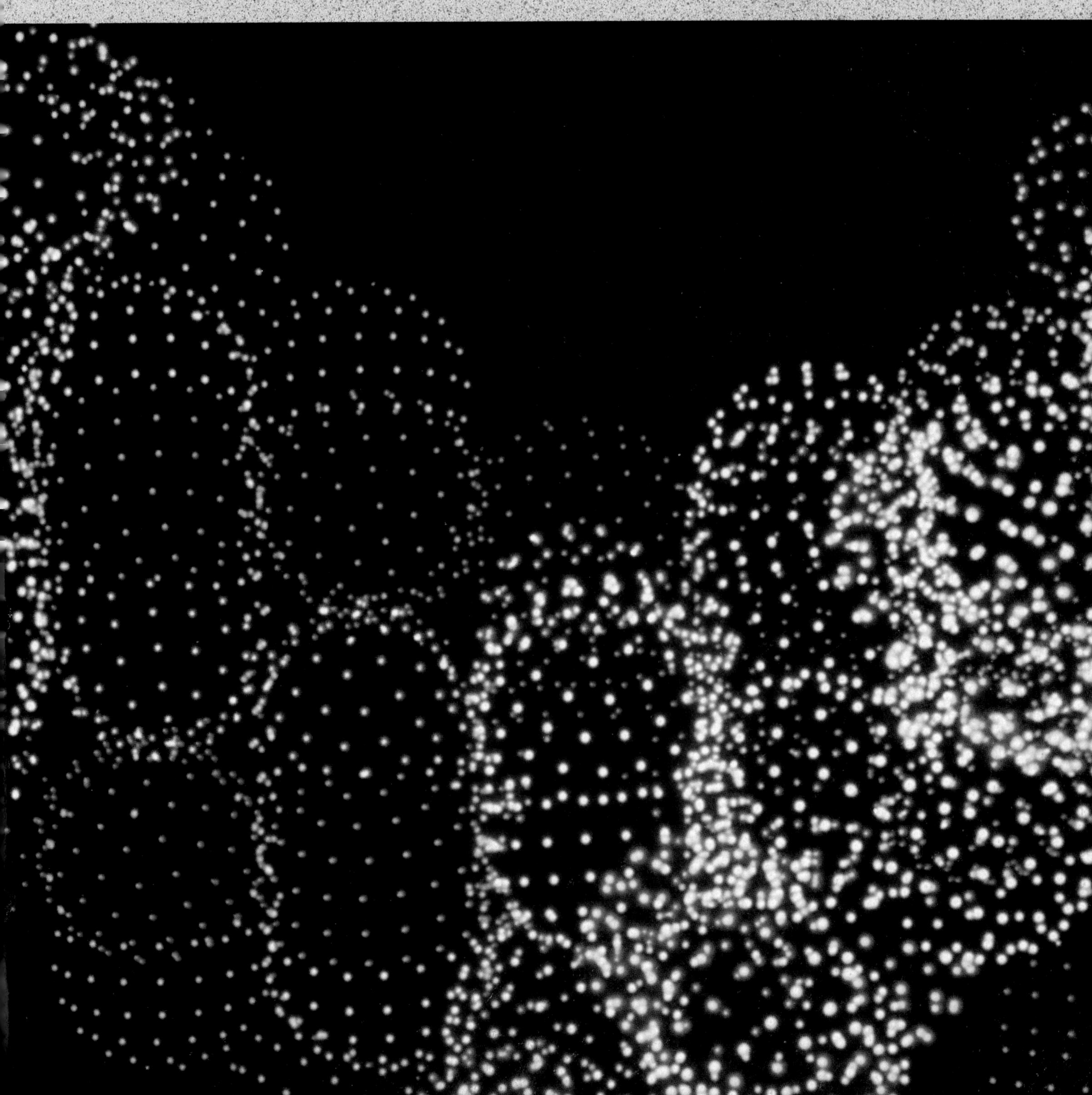

CHAPTER 1

The Human Body: An Orientation

Chapter Outline and Student Objectives

An Overview of Anatomy and Physiology (pp. 5–6)

1. Define anatomy and physiology and describe the subdivisions of anatomy.
2. Explain the principle of complementarity.

The Hierarchy of Structural Organization (pp. 6–7)

3. Name (in order of increasing complexity) the different levels of structural organization that make up the human body, and explain their relationships.
4. List the 12 organ systems of the body and briefly explain the major function(s) of each system.

Maintenance of Life (pp. 7–12)

5. List the functional characteristics common to humans (and other organisms) and explain the importance of each to maintaining life.
6. List the survival needs of the body.

Homeostasis (pp. 12–15)

7. Define homeostasis and explain its importance.
8. Define negative feedback and describe its role in maintaining body homeostasis.
9. Define positive feedback and explain why it usually causes homeostatic imbalance. Also note specific situations in which it contributes to homeostasis, or normal body function.

The Language of Anatomy (pp. 15–22)

10. Define the anatomical position.
11. Use correct anatomical terminology to describe body directions, regions, and body planes or sections.
12. Locate the major body cavities, name their subdivisions, and list the major organs in each cavity or subdivision.
13. Name the specific serous membranes and note their common function.
14. Name the nine regions or four quadrants of the abdominopelvic cavity and list the organs they contain.

Preview of Selected Key Terms

Anatomy (*ana* = apart; *tomy* = to cut) The science of the structure of living organisms.

Physiology (*physio* = nature; *ology* = the study of) The science of the functioning of living organisms.

Complementarity of structure and function The close relationship that exists between a structure and its function; i.e., structure determines functions.

Cell The unit of structure and function in living organisms.

Tissue A group of functionally similar cells forming a distinct structure.

Organ A part of the body formed of two or more tissues and adapted to carry out a specific function; e.g., the stomach.

Organ system A group of organs that work together to perform a vital body function; e.g., the nervous system.

Organism The living animal (or plant), which represents the sum total of all of its organ systems working together to maintain life.

Metabolism (*meta* = a change) The chemical changes that occur within the body.

Homeostasis (hō″-mē-ō-stā′-sis) (*homeo* = the same; *stasis* = standing still) A state of body equilibrium, or the maintenance of a stable internal environment of the body.

Section A cut through the body (or an organ) that is made along a particular plane; a thin slice of tissue prepared for microscopic study.

Parietal (puh-rī′-eh-tul) (*parie* = wall) Pertaining to the walls of a cavity.

Visceral (vih′-ser-ul) (*viscus* = an organ in a body cavity) Pertaining to an internal organ of the body or the inner part of a structure.

As you make your way through this book, you will be learning about one of the most fascinating subjects possible—your own body. Such a study is not only interesting and highly personal, but timely as well.

Currently, an information blizzard is in progress, with news of some medical advance appearing almost daily. To appreciate emerging discoveries in genetic engineering, understand new techniques for detecting and treating disease, and make use of published facts on how to stay healthy, it is increasingly important to learn about the workings of your body. For those of you preparing for a career in the health sciences, the study of anatomy and physiology has added rewards because it provides the strong foundation needed to support your clinical experiences.

This chapter first defines and contrasts anatomy and physiology and discusses the levels of organizational complexity in the human body. Then it reviews needs and functional processes common to all living organisms. Three essential concepts—the complementarity of structure and function, hierarchy of structural organization, and homeostasis—will form the bedrock for your study of the human body and will provide threads that unify the topics discussed in this book. The final section of the chapter deals with the language of anatomy—the terminology "tools" that anatomists use whenever they describe the body or its parts.

An Overview of Anatomy and Physiology

Although many branches of science contribute to an understanding of the human body, the most critical concepts are embodied in two subdivisions of biology: anatomy and physiology. Anatomy and physiology are complementary sciences. **Anatomy** is the study of the form, or structure, of body parts and of how these parts relate to one another, whereas **physiology** concerns the functioning of the body's structural machinery, that is, how the parts of the body work and carry out their life-sustaining activities.

Topics of Anatomy

Anatomy is a broad field with many subdivisions, each providing enough information to be a course in itself. It is most conveniently studied by using preserved specimens or fixed tissue sections.

Gross anatomy is the study of large body structures such as the heart, lungs, and kidneys, which can be seen easily and examined without any type of magnifying instruments. The term *anatomy* (derived from the Greek words meaning "to cut apart") is related most closely to gross anatomical studies, since in such studies preserved animals or their organs are dissected (cut up) to be examined. Studies of gross anatomy can be approached in different ways. In *regional anatomy*, all of the various structures (muscles, bones, blood vessels, nerves, etc.) in one particular region of the body, such as the abdomen or leg, are examined at the same time. In *systemic (sis-teh'-mik) anatomy*, the gross anatomy of the body is studied system by system. For example, when studying the cardiovascular system, you would examine the heart and the blood vessels of the entire body.

Microscopic anatomy concerns structures too small to be seen without the aid of a microscope. For most such studies, exceedingly thin slices, or sections, of body tissues are stained and mounted on slides for examination under the microscope. Subdivisions of microscopic anatomy include *cellular anatomy*, or *cytology* (sī-tah'-luh-jē), which studies the cells of the body, and *histology* (his-tah'-luh-jē), the study of tissues.

Developmental anatomy concerns the structural changes in an individual from conception through old age. *Embryology* (em"-brē-ah'-luh-jē), a subdivision of developmental anatomy, concerns only the developmental changes that occur before birth.

There are also a number of highly specialized branches of anatomy used primarily for medical diagnosis and scientific research. For example, in *pathological anatomy*, the structural changes in body cells, tissues, and organs caused by disease are studied on both gross and microscopic levels, and in *molecular biology*, the structure of molecules (chemical substances) necessary for body structure and function are investigated. Molecular biology is actually a separate branch of biology, but it falls under the anatomy "umbrella" when we push anatomical studies into the subcellular level, where molecules provide the fundamental links between structure and function. *Radiographic anatomy* is the study of anatomy by means of X-ray images. X-ray technicians are trained in this area, and radiology is valuable to clinicians evaluating patients for certain bone disorders, tumors, and other conditions that cause anatomical changes. As you can see, subjects of interest to anatomists range from easily seen structures down to the smallest molecule, and their study provides us with a static image of the architecture of the body and its parts.

As you will quickly learn, the most important tools for studying anatomy are observation, manipulation, and a mastery of anatomical terminology. A simple example should serve to illustrate how these tools work together in an anatomical study. Let's assume that your topic is freely movable joints of the body. In the laboratory, you will be able to *observe* an animal joint, noting how its parts fit together. You can work the joint (*manipulate* it) to determine its range of motion. Then, using *anatomical terminology*, you can name its parts and describe their relationships so that other students (and your instructor) will have no trouble understanding you. Although most of your observations will be made with the naked eye or with the help of a microscope, medical technology has developed a number of sophisticated tools that can peer into the body without disrupting it. These include CAT and MRI scans and a number of other exciting medical imaging techniques discussed in the box in Chapter 2, pp. 32–33.

Topics of Physiology

Physiology is also subdivided into several specialized areas. Common subdivisions of physiology consider the operation of specific organ systems. Thus, *renal physiology* concerns urine production and kidney function; *neurophysiology* explains the workings of the nervous system; *cardiac physiology* examines the operation of the heart; and so forth. While anatomy provides the static image, physiology reveals the dynamic nature of the workings of the living body.

Physiology often focuses on the cellular or molecular level because what the body can do depends on the operation of its individual cells, and what cells can do ultimately depends on the chemical reactions that go on within them. In addition, an in-depth understanding of physiology also rests on the principles of physics, which help to explain electric currents, blood pressure, and the way muscles use bones to cause body movements. Body functions such as nervous system activity, muscular contraction, and digestion are incomprehensible without an understanding of the underlying chemistry and physics. Thus, basic chemical and physical principles will be presented in Chapter 2 and throughout the book as needed to explain physiological topics.

Complementarity of Structure and Function

Although it is possible to study anatomy and physiology in isolation from one another, they are truly inseparable sciences because function always reflects structure; that is, what a structure is capable of doing depends critically on its specific architecture. This is called the **principle of complementarity of structure and function.** For example, bones can provide support and protection to body organs because they contain hard mineral deposits; blood flows in one direction through the heart because the heart has valves that prevent backflow; and the lungs can serve as a site for gas exchange because the walls of their air sacs are extremely thin. This textbook consistently stresses the intimacy of this relationship to make learning meaningful. In all cases, a description of the anatomy of a structure is accompanied by an explanation of its function, emphasizing the structural characteristics contributing to function.

The Hierarchy of Structural Organization

The human body incorporates many levels of structural complexity (Figure 1.1). The simplest level of the structural hierarchy is the *chemical level,* which we will study in Chapter 2. At this level, **atoms,** tiny building blocks of matter, combine to form molecules such as water, sugar, and proteins. Molecules, in turn, associate in very specific ways to form microscopic **cells,** the living structural and functional units of an organism. The *cellular level* is examined in Chapter 3. Individual cells vary widely in size and shape, a reflection of their unique functions in the body. All cells can utilize nutrients and maintain their boundaries, but only certain cell types can form the transparent lens of the eye, secrete mucus, or conduct nerve impulses.

The simplest organisms are composed of single cells, but in complex organisms, such as human beings, the hierarchy continues to the *tissue level.* **Tissues** consist of groups of similar cells that have a common function. The four basic tissue types in the human body are epithelium, muscle, connective tissue, and nervous tissue. Each tissue type has a characteristic role in the body, and we will discuss those roles in detail in Chapter 4. Briefly, epithelium covers the body surface and lines its cavities; muscle causes movement; connective tissue supports the body and protects its organs; and nervous tissue provides a means of rapid internal communication by transmitting electrical impulses.

At the *organ level* of organization, extremely complex physiological processes become possible. An **organ** is a discrete structure composed of at least two tissue types, and four is more commonplace. The stomach is a good example of an organ: Its lining is an epithelium that produces digestive juices; the bulk of its wall is composed of muscle that acts to churn and mix stomach contents (food); its connective tissue helps to reinforce the soft muscular walls; and its nerve fibers increase digestive activity by stimulating the muscle to contract more vigorously and the glands to secrete more digestive juices. You can think of each organ of the body as a highly specialized functional center responsible for a necessary activity that no other type of organ can perform.

The next level of organization is the *organ system level.* Organs that cooperate and work closely with one another to accomplish a common purpose are said

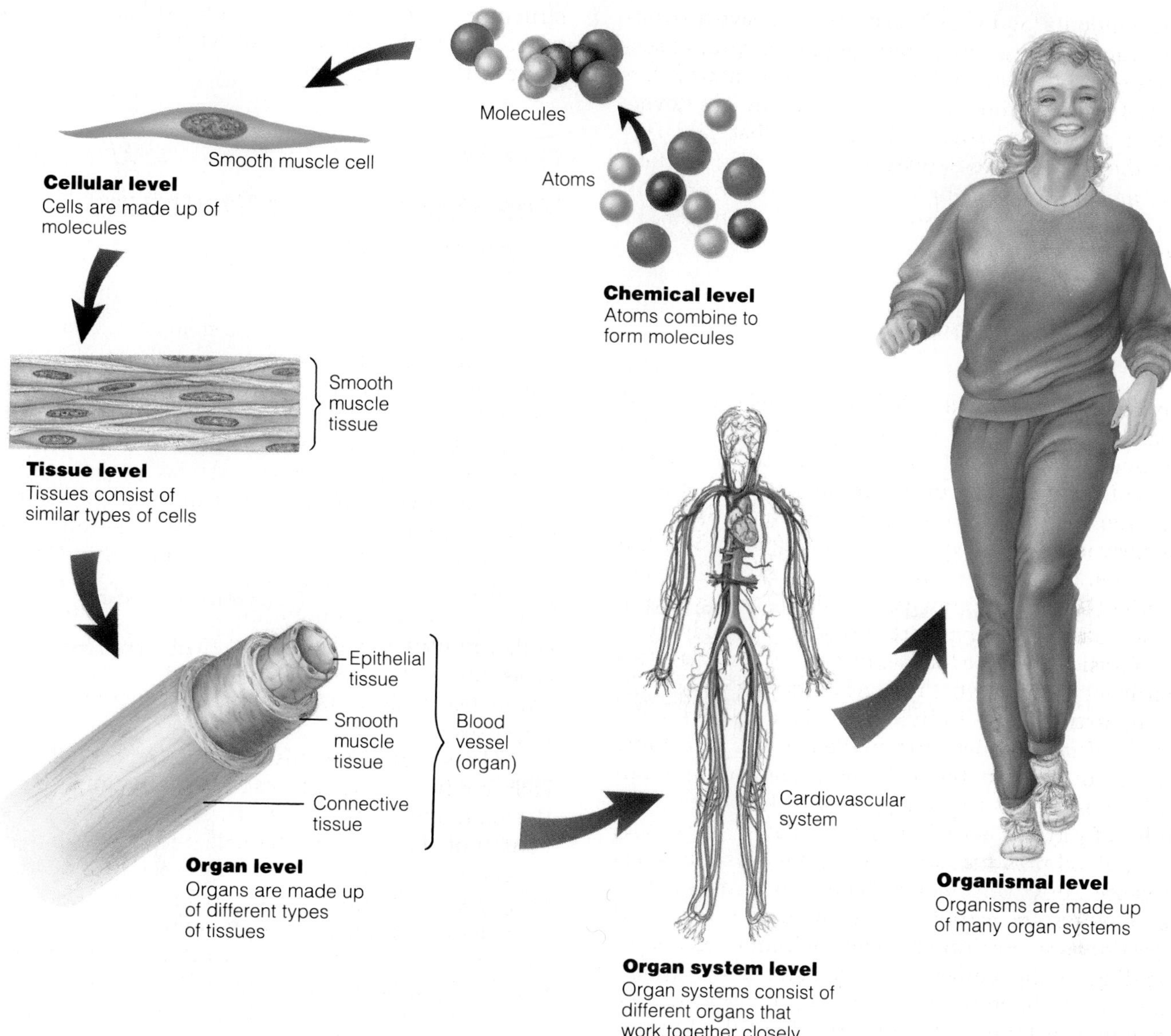

Figure 1.1 Levels of structural complexity. In this diagram, components of the cardiovascular system are used to illustrate the various levels of complexity in a human being.

to be part of a particular **organ system.** For example, organs of the respiratory system—the lungs, bronchi, trachea, and others—ensure that air flows into and out of the lungs continuously so that oxygen can enter the blood and carbon dioxide can be removed. Organs making up the digestive system—the mouth, esophagus, stomach, intestine, and so forth—break down ingested food so that the nutrients can be absorbed into the blood. The digestive system also removes indigestible food residues from the body. Besides the respiratory and digestive systems, the other systems of the body are the integumentary, skeletal, muscular, nervous, endocrine, cardiovascular, lymphatic, immune, urinary, and reproductive systems. Figure 1.2 provides a brief overview of these organ systems, which we will study in more detail in Units 2–5. The highest level of organization is the **organism,** exemplified by a living human being. The *organismal level* represents the sum total of all levels of complexity working continuously and in unison to promote life.

Maintenance of Life

Functional Characteristics

Now that we have introduced the structural levels composing the human body, the question that naturally follows is, What does this highly organized human

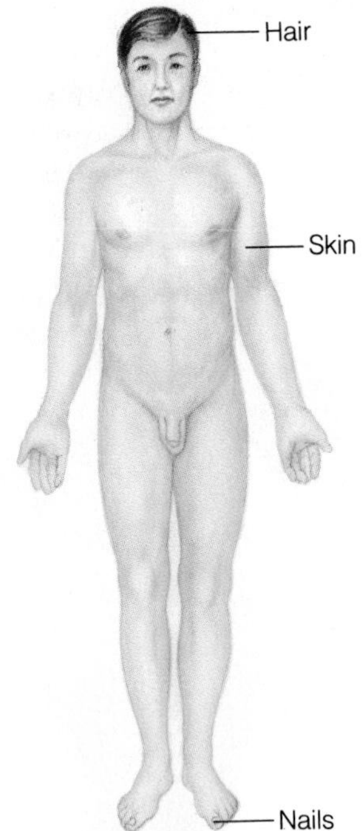

(a) Integumentary system

Forms the external body covering; protects deeper tissues from injury; synthesizes vitamin D; location of cutaneous (pain, pressure, and temperature) receptors.

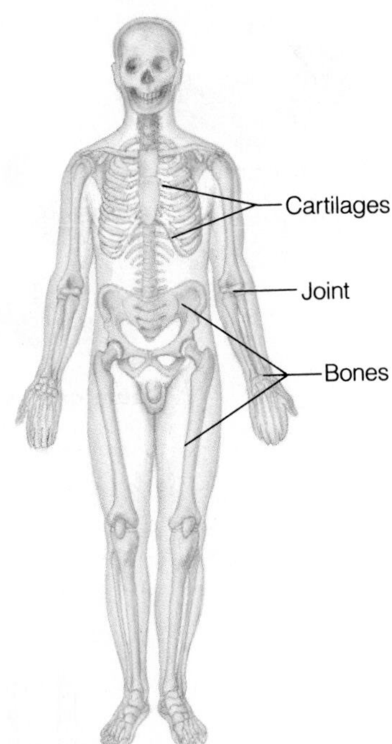

(b) Skeletal system

Protects and supports body organs; provides a framework the muscles use to cause movement; blood cells are formed within bones.

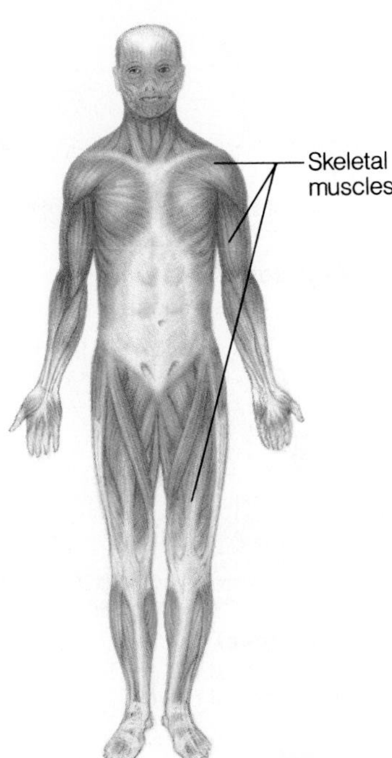

(c) Muscular system

Allows manipulation of the environment, locomotion, and facial expression.

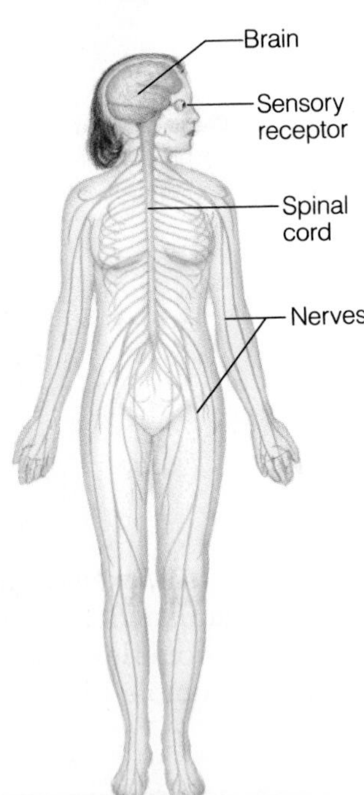

(d) Nervous system

Fast-acting control system of the body; responds to internal and external changes by activating appropriate muscles and glands.

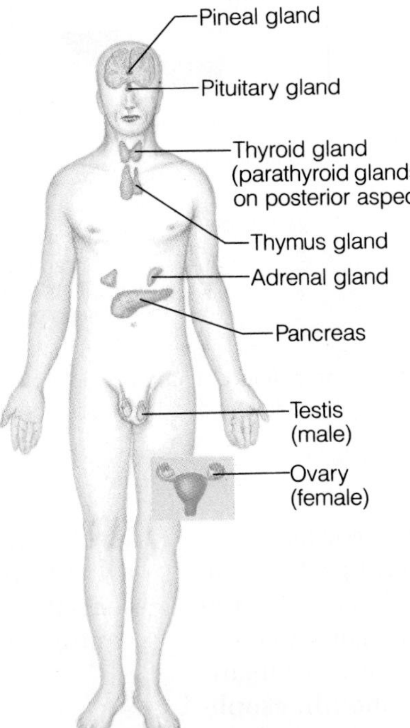

(e) Endocrine system

Glands secrete hormones that regulate processes such as growth, reproduction, and nutrient use (metabolism) by body cells.

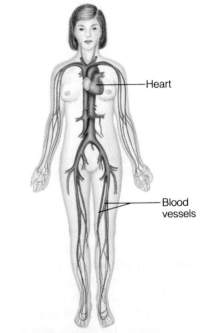

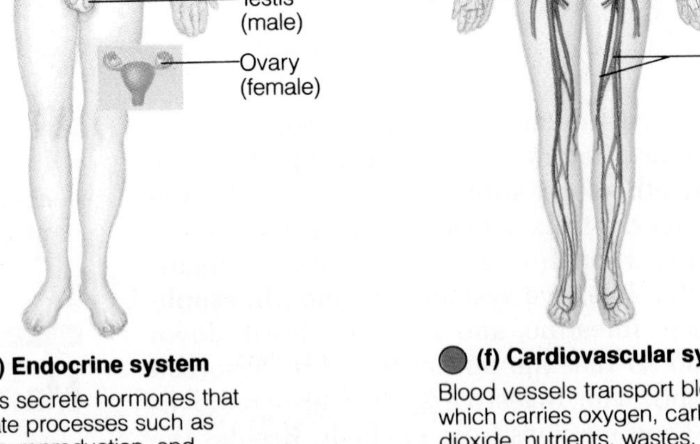

(f) Cardiovascular system

Blood vessels transport blood which carries oxygen, carbon dioxide, nutrients, wastes, etc.; the heart pumps blood.

Figure 1.2 Summary of the body's organ systems. The structural components of each organ system are illustrated in the diagrammatic view. The major functions of the organ system are listed beneath each illustration. Note that the immune system (**h**) is a functional system rather than an organ system in the true sense.

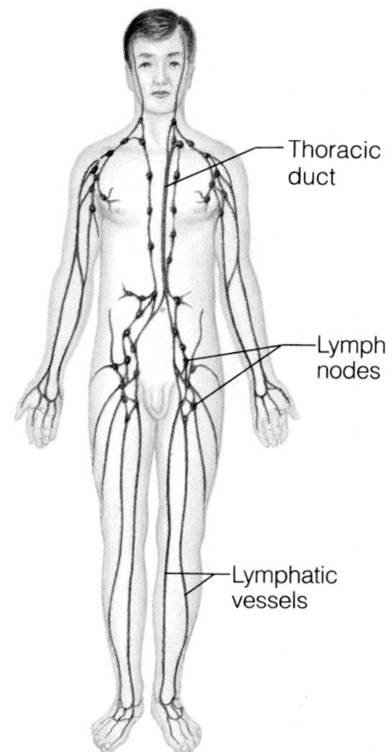

(g) Lymphatic system

Picks up fluid leaked from blood vessels and returns it to blood; disposes of debris in the lymphatic stream; houses white blood cells involved in immunity.

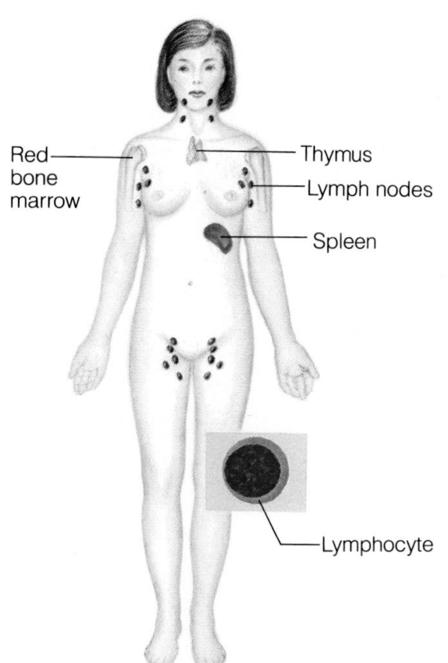

(h) Immune system

A functional system that protects the body via the immune response, in which foreign substances are attacked by lymphocytes and/ or antibodies.

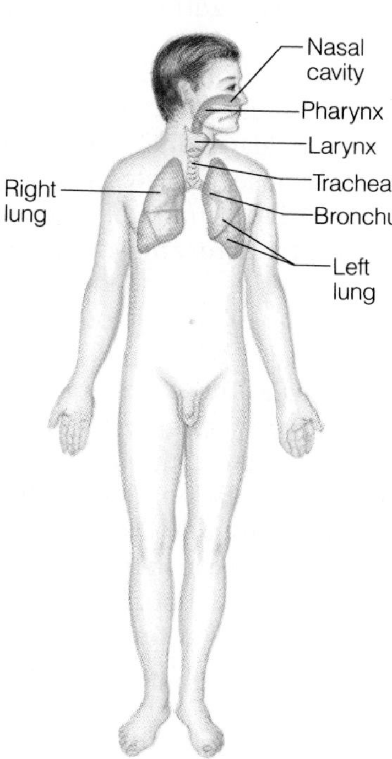

(i) Respiratory system

Keeps blood constantly supplied with oxygen and removes carbon dioxide; the gaseous exchanges occur through the walls of the air sacs of the lungs.

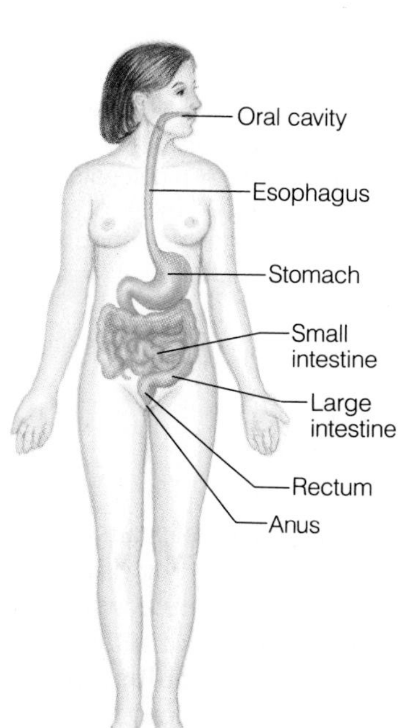

(j) Digestive system

Breaks down food into absorbable units that enter the blood for distribution to body cells; indigestible foodstuffs are eliminated as feces.

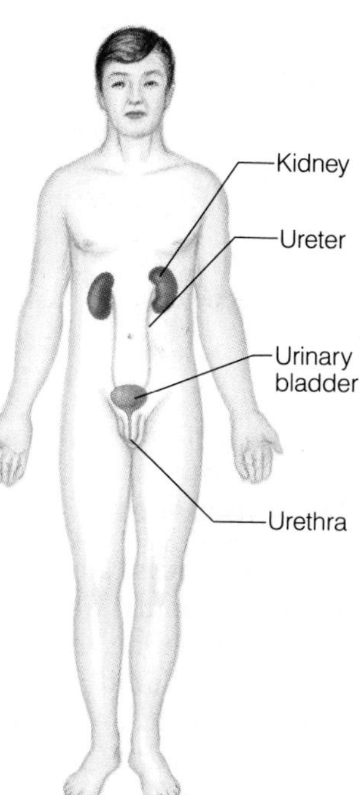

(k) Urinary system

Eliminates nitrogenous wastes from the body; regulates water, electrolyte, and acid-base balance of the blood.

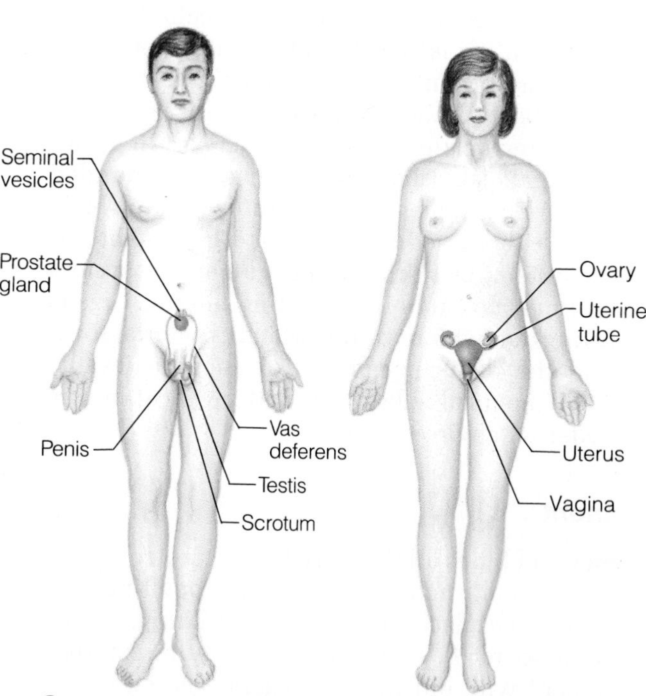

(l) Male reproductive system **(m) Female reproductive system**

Overall function of the reproductive system is production of offspring. Testes produce sperm and male sex hormone; ducts and glands aid in delivery of viable sperm to the female reproductive tract. Ovaries produce eggs and female sex hormones; remaining structures serve as sites for fertilization and development of the fetus.

body do? Like all complex animals, human beings maintain their boundaries, move, respond to environmental changes, take in and digest nutrients, carry out metabolism, dispose of wastes, reproduce themselves, and grow. We will discuss each of these necessary life functions briefly here and in more detail in later chapters.

It cannot be emphasized too strongly that the multicellular state and the parceling out of vital body functions to several different organ systems result in an interdependence among all body cells. So, too, just as no man is an island, no man's (or woman's) organ systems work in isolation. Rather, they work together cooperatively to promote the well-being of the entire body (Figure 1.3). Because this theme will be emphasized throughout this book, it seems appropriate to identify the most important organ systems contributing to each of the functional processes. As you read through this material, you may want to refer back to the more detailed descriptions of the organ systems provided in Figure 1.2.

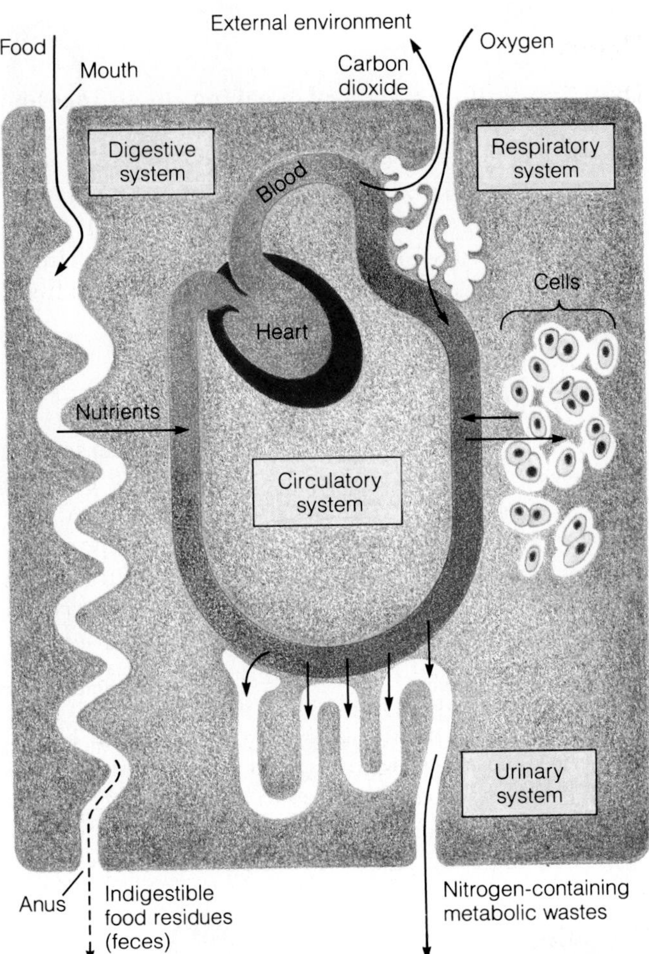

Figure 1.3 Examples of selected interrelationships among body organ systems. The digestive and respiratory systems, in contact with the external environment, take in nutrients and oxygen, respectively, which are then distributed by the blood to all body cells. Elimination from the body of metabolic wastes is accomplished by the urinary and respiratory systems.

Maintenance of Boundaries

An important characteristic of any living organism is its ability to **maintain its boundaries** so that its internal environment remains distinct from the external environment surrounding it. In single-celled organisms, the boundary is an external limiting membrane that functions to contain its contents and to allow needed substances in while restricting entry of potentially damaging or unnecessary substances. In a similar manner, all the cells of our body are surrounded by a semipermeable membrane. Additionally, the body as a whole is enclosed and protected by the integumentary system, or skin. The integumentary system plays an important role in protecting internal organs from drying out (which would be fatal), bacterial invasion, and the damaging effects of an unbelievable number of chemical substances and physical factors in the external environment.

Movement

Movement includes all the activities promoted by the muscular system, such as voluntarily propelling ourselves from one place to another by walking, swimming, and so forth, and manipulating the external environment with our nimble fingers. The muscular system is aided by the skeletal system, which provides the bony framework that the muscles pull on as they work. Movement also occurs as substances such as blood, foodstuffs, and urine are propelled, respectively, through internal organs of the cardiovascular, digestive, and urinary systems.

Responsiveness

Responsiveness, or *irritability,* consists of sensing changes (stimuli) in the environment and then reacting, or responding, to these changes. For example, when you cut your hand on broken glass, you involuntarily pull your hand away from the painful stimulus (the broken glass). It is not necessary to think about it—it just happens! Likewise, when carbon dioxide in your blood rises to dangerously high levels, chemical sensors respond by sending messages to brain centers controlling respiration, and your breathing rate is accelerated.

Because nerve cells are highly irritable and can communicate rapidly with each other by conducting electrical impulses, the nervous system bears the major responsibility for responsiveness. However, all body cells exhibit irritability to some extent.

Digestion

Digestion is the process of breaking down ingested foodstuffs to simple molecules that can then be absorbed into the blood for distribution to all body cells by the cardiovascular system. In a simple, one-celled organism like an amoeba, the cell itself is the "digestion factory," but in the complex multicellular human body, the digestive system performs this function for the entire community of body cells.

Metabolism

Metabolism is a broad term that encompasses all chemical reactions that occur within body cells. It includes breaking down complex substances into their simpler building blocks, or chemical components, synthesizing more complex cellular structures from simpler substances, and using nutrients and oxygen to generate (via *cellular respiration*) the high-energy ATP molecules that power cellular activities. Metabolism depends on the digestive and respiratory systems to make nutrients and oxygen available to the blood and on the cardiovascular system to see to it that these needed substances are distributed throughout the body. Metabolism is regulated chiefly by hormones secreted by glands of the endocrine system.

Excretion

Excretion is the process of removing *excreta* (ek-skrē′-tuh), or wastes, from the body. If the body is to continue to operate as we expect it to, it must get rid of the nonuseful products of digestion and metabolism. Several organ systems participate in excretion. Indigestible food residues leave the body via the terminal digestive system organs. The urinary system disposes of nitrogen-containing metabolic wastes, such as urea and uric acid, which are flushed out in urine. Carbon dioxide, a by-product of cellular respiration, is carried in the blood to the lungs, where it leaves the body in exhaled air.

Reproduction

Reproduction, the formation of offspring, can occur on the cellular or organismal level. Cellular reproduction involves cell division, whereby the original cell produces two identical daughter cells that may then be used for body growth or repair. Reproduction of the human organism, or the making of a whole new person, is the major task of the reproductive system. It involves the formation and union of reproductive cells, sperm and egg, to form a fertilized egg, which then develops into a bouncing baby within the mother's body. The reproductive system is directly responsible for producing offspring, but its function is exquisitely regulated by hormones of the endocrine system.

The reproductive organs of males and females are quite different (see Figure 1.2, l and m). This reflects a division of labor in the reproductive process. The male's testes produce sperm, and his accessory reproductive organs deliver the sperm into the female's reproductive tract, where the remaining events take place. The female's ovaries produce eggs, and her accessory structures provide the site for fertilization, then protect and nurture the developing fetus, and finally participate in the birth process.

Growth

Growth is an increase in size and is usually accomplished through an increase in the number of cells via cell division. However, individual cells also increase in size when not dividing. For true growth to occur, constructive activities must occur at a faster rate than cell-destroying activities.

Survival Needs

The ultimate goal of nearly all body systems is to maintain life. However, life is extraordinarily fragile and requires several factors acting together for its persistence. These factors, or survival needs, include food, oxygen, water, appropriate temperature, and atmospheric pressure.

Nutrients, taken in via the diet, contain the chemical substances used for energy and cell building. Most plant-derived foods are rich in carbohydrates, vitamins, and minerals, whereas most animal foods are rich in proteins and fats. Carbohydrates are the major energy fuel for body cells. Proteins and to a lesser extent fats are essential for building cell structures. Fats also cushion body organs, form insulating layers, and constitute a reserve of energy-rich fuel. Selected minerals and vitamins are required for the chemical reactions that go on in cells and for oxygen transport in the blood. Calcium helps to make bones hard and is required for blood clotting.

All the nutrients in the world are useless unless **oxygen** is also available, because the chemical reactions that release energy from foods are *oxidative* reactions, requiring oxygen. Without this vital gas, cells can survive for only a few minutes. Approximately 20% of the air we breathe is oxygen. It is made available to the blood and body cells by the cooperative efforts of the respiratory and cardiovascular systems.

Water accounts for 60% to 80% of body weight and is the single most abundant chemical substance in the body. It provides the liquid environment necessary for chemical reactions and the fluid base for body secretions and excretions. Water is obtained chiefly from ingested foods or liquids and is lost from the body by evaporation from the lungs and skin and in body excretions.

If chemical reactions are to proceed at life-sustaining rates, **body temperature** must be maintained at around 37°C (98°F). As body temperature drops below this point, physiological reactions become slower and slower, and finally stop. Excessively high body temperature causes chemical reactions to proceed too rapidly, and body proteins begin to break down. At either extreme, death occurs. Most body heat is generated by the activity of the skeletal muscles.

The force exerted on the surface of the body as a result of the weight of air is referred to as **atmospheric pressure.** Breathing and the exchange of oxygen and carbon dioxide in the lungs depend on appropriate atmospheric pressure. At high altitudes, where the air is thin and the atmospheric pressure is lower, gas exchange may be inadequate to support cellular metabolism. However, the bone marrow eventually compensates by forming more red blood cells.

The mere presence of these survival factors is not sufficient to maintain life. They must be present in appropriate amounts as well; excesses and deficits may be equally harmful. For example, while oxygen is essential, extremely high levels of oxygen are toxic to body cells. Similarly, the food ingested must be of high quality and in proper amounts; otherwise, nutritional disease, obesity, or starvation is a likely outcome. Also, while the listed needs are the ones most crucial to survival, they do not even begin to encompass all of the body's needs. For example, we can, if we must, live without gravity, but the quality of life suffers.

Homeostasis

When you really think about the fact that your body contains trillions of cells, that thousands of physiological processes go on in it each second, and that remarkably little usually goes wrong with it, you begin to appreciate what a marvelous machine your body really is. Walter Cannon, an American physiologist of the early twentieth century, spoke of the "wisdom of the body," and he coined the word **homeostasis** (hō″-mē-ō-stā′-sis) to describe its ability to maintain relatively stable internal conditions even in the face of continuous change in the outside world. Although the literal translation of homeostasis is "unchanging," the term does not really designate a static, or unchanging, state. Rather, it indicates a *dynamic* state of equilibrium, or a balance, in which internal conditions change and vary, but always within relatively narrow ranges.

In general, the body is said to be in homeostasis when its cells' needs are adequately met and functional activities are occurring smoothly. But maintaining homeostasis is much more complex than it may appear at first glance. Virtually every organ system has a role to play in maintaining constancy of the internal environment. Not only must adequate blood levels of vital nutrients be continuously present, but heart activity and blood pressure must be constantly monitored and adjusted so that the blood is propelled with adequate force to reach all body tissues. Additionally, wastes must not be allowed to accumulate, and body temperature must be precisely controlled to ensure proper conditions for metabolism. An unbelievable variety of chemical, thermal, and neural factors act and interact in complex ways—sometimes bolstering and sometimes impeding the ability of the body to maintain its "steady rudder."

General Characteristics of Control Mechanisms

Communication within the body is essential for homeostasis. Communication is accomplished chiefly by the nervous and endocrine systems, which use electrical impulses delivered by nerves or blood-borne hormones as information carriers. The details of how these two great regulating systems operate are the subjects of later chapters, but the fundamental characteristics of control systems that promote homeostasis will be explained here.

Regardless of the variable being regulated, all homeostatic control mechanisms have a minimum of three interdependent components (Figure 1.4). The first component is the *control center,* which determines the *set point* at which a variable is to be maintained, analyzes the input it receives, and then determines the appropriate response.

The second element is a *receptor.* Essentially, it is some type of sensor that monitors the environment and responds to changes, called *stimuli,* by sending information (input) to the control center. The flow of information from the receptor to the control center occurs along the so-called *afferent pathway.*

The third component is the *effector,* which provides the means by which the control center can cause a response (output) to the stimulus. Information flows from the control center to the effector along the *efferent pathway.* The results of the response then *feed back* to influence the stimulus, either depressing it (negative feedback) so that the whole control mechanism is shut off or enhancing it (positive feedback) so that the reaction is continued at an even more vigorous rate.

Now that we have constructed the framework that describes the basic flow of information in control systems, we are ready to explain how negative and positive feedback mechanisms help maintain body homeostasis.

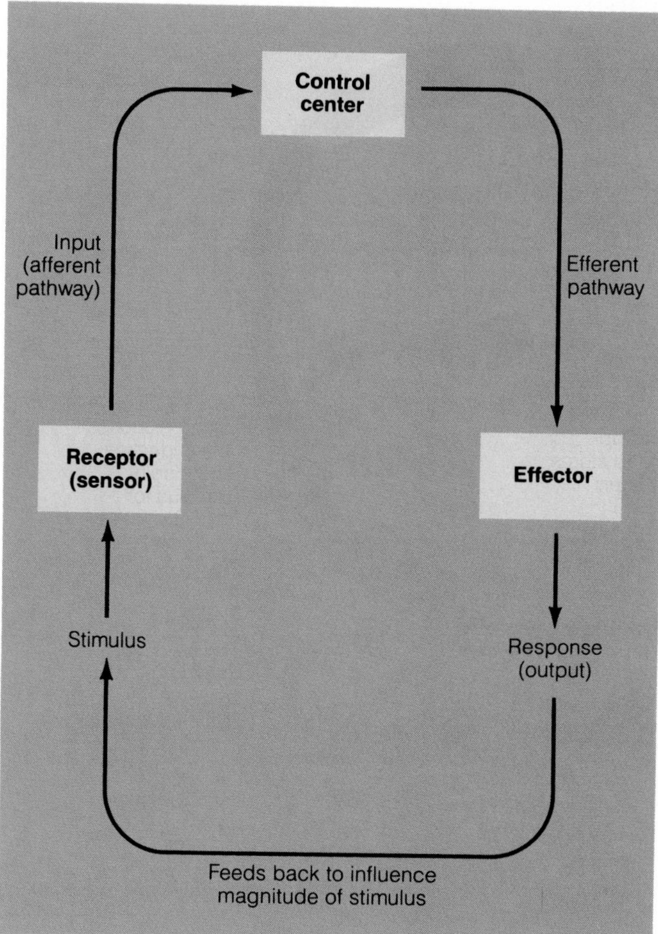

Figure 1.4 The elements of a control system. Communication between the receptor, control center, and effector is essential for normal operation of the system.

Negative Feedback Mechanisms

Most homeostatic control mechanisms are **negative feedback mechanisms.** In such systems, an increase in the output of the system feeds back and decreases the input into the system. The net effect is to decrease the original stimulus or reduce its effects, slowing the activity or shutting it off entirely. These mechanisms cause the variable to change in a direction *opposite* to that of the initial change; thus the name "negative" feedback mechanisms. A frequently cited example of a nonbiological negative feedback system is a home heating system connected to a temperature-sensing thermostat. In this situation, the thermostat houses both the receptor and the control center. If the thermostat is set at 20°C (68°F), the heating system (effector) will be triggered ON when the house temperature drops below that setting. As the furnace produces heat and the air is warmed, the temperature begins to rise, and when it reaches 20°C or slightly higher, the thermostat sends a signal to shut off the furnace. This process results in a cycling of "furnace-ON" and "furnace-OFF" so that the temperature in the house stays very near the desired temperature of 20°C. Your body "thermostat," located in a part of your brain called the hypothalamus, operates in a similar fashion.

Regulation of body temperature by the hypothalamus illustrates one of the many ways the nervous system maintains the constancy of the internal environment. The endocrine system is equally important in these endeavors, however, and a good example of a hormonally controlled negative feedback mechanism is the control of blood glucose levels by pancreatic hormones (Figure 1.5).

If body cells are to carry out normal metabolism, they must have continuous access to glucose, their major fuel for producing cellular energy, or ATP. Blood sugar levels are normally maintained around 90 milligrams (mg) of glucose per 100 milliliters (ml) of blood.* Let's assume that you have lost your willpower and have downed four jelly doughnuts. The doughnuts are quickly broken down to sugars by your digestive system, and as the sugars flood into the bloodstream, blood sugar levels spike upward, disrupting homeostasis. The rising glucose levels stimulate the insulin-producing cells of the pancreas, which respond by secreting insulin into the blood. Insulin accelerates the uptake of glucose by most body cells and encourages the storage of excess glucose as glycogen in the liver and muscles, so that a reserve supply of glucose is "put into the larder," so to speak. Consequently, blood sugar levels ebb back toward the normal set point, and the stimulus for insulin release diminishes.

Glucagon, the other pancreatic hormone, has the opposite effect. Its release is triggered as blood sugar levels decline below the set point. Suppose, for example, that you have skipped lunch and it is now about 2 P.M. Since your blood sugar level is running low, glucagon secretion is stimulated. Glucagon targets the liver, causing it to release its glucose reserves into the blood. Consequently, blood sugar levels increase back into the homeostatic range.

The body's ability to regulate its internal environment is fundamental. Body temperature and blood glucose levels are only two of the variables that need to be regulated. There are hundreds! Other negative feedback mechanisms regulate heart rate, blood pressure, the rate and depth of breathing, and blood levels of oxygen, carbon dioxide, and minerals. We will discuss many of these mechanisms when we consider the different organ systems of the body.

Not all homeostatic mechanisms operate in the same way. The positive feedback mechanisms described next are fundamentally different from negative feedback mechanisms. However, all homeostatic mechanisms have the same goal: prevention of sudden severe changes within the body.

*The metric system is described in Appendix A.

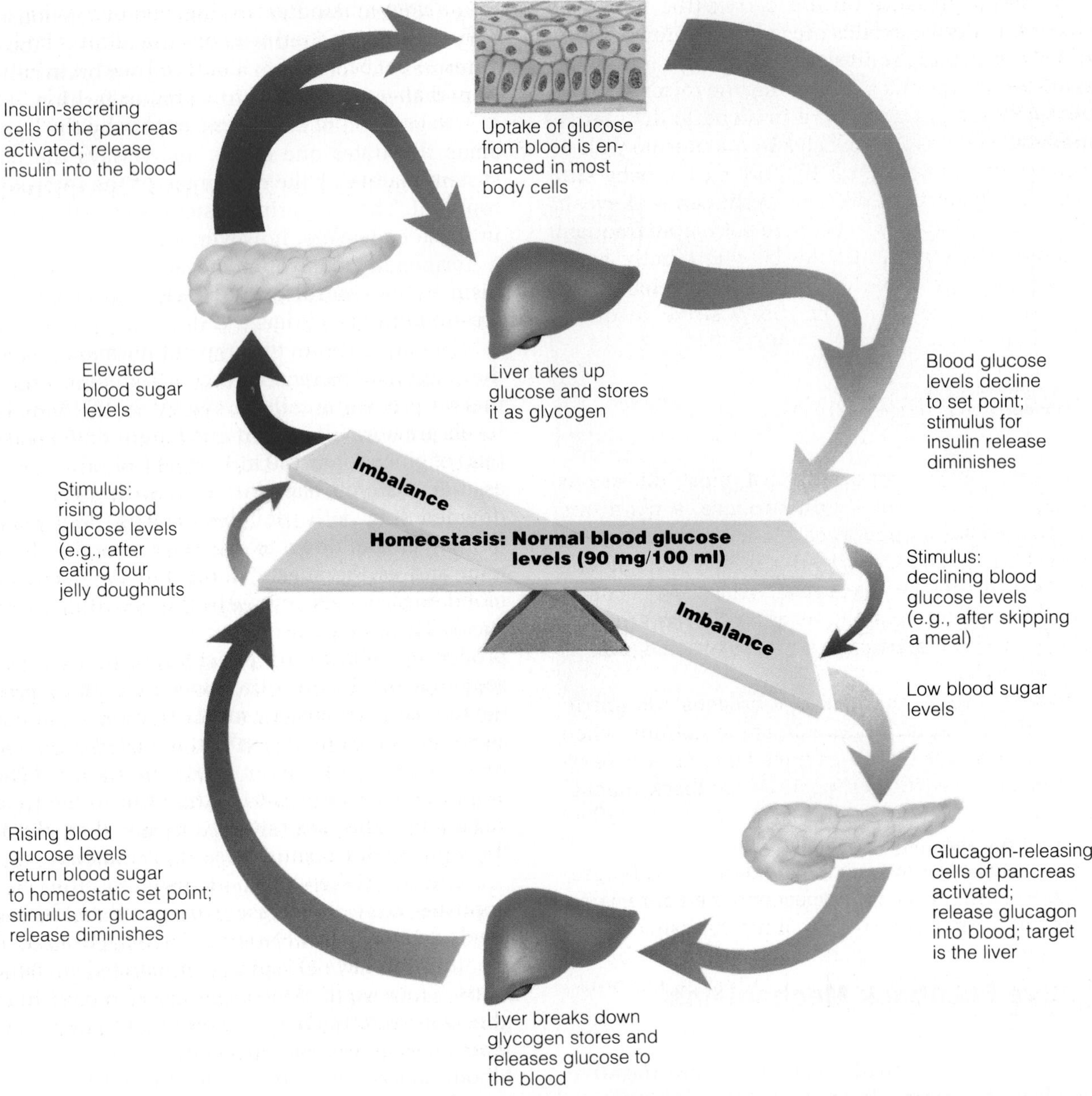

Figure 1.5 Regulation of blood glucose levels by a negative feedback mechanism involving pancreatic hormones.

Positive Feedback Mechanisms

In **positive feedback mechanisms** the result or response of the mechanism enhances the original stimulus, so that the activity (output) is accelerated. This feedback mechanism is said to be "positive," because the change that occurs proceeds in the *same* direction as the initial disturbance. In contrast to negative feedback controls, which maintain some physiological function or keep blood chemicals within narrow ranges, positive feedback mechanisms usually control episodic (infrequent) events that do not require continuous adjustments. Typically, they set off a train of events that may be self-perpetuating and quite explosive. Because of these characteristics, positive feedback mechanisms are often referred to as *cascades* (from the French word meaning "to fall"). Because positive feedback mechanisms are likely to race out of control, they are rarely used to promote the moment-to-moment well-being of the body. However, there are a few good examples of their use as homeostatic mechanisms, such as in the control of blood clotting and the enhancement of labor contractions.

The hormone oxytocin enhances labor contractions during the birth of a baby in the following way. As the baby moves down into its mother's birth canal, the increased pressure on her cervix (the muscular outlet of the uterus) excites pressure receptors located there. The receptors send rapid nerve impulses to the brain, which responds by triggering the release of oxytocin. Oxytocin is transported in blood to the uterus, where it stimulates the muscles in the uterine wall to contract even more vigorously, forcing the baby still further into the birth canal. This cyclic series of events causes the contractions to become both more frequent and more powerful, until the baby is finally born. At this point, the stimulus for oxytocin release (the pressure) ends, shutting off the positive feedback mechanism.

Homeostatic Imbalance

Homeostasis is so important that most disease is regarded as a result of its disturbance, a condition called **homeostatic imbalance.** With age, body organs become less efficient, as do the body's control systems. As a result, the internal environment becomes less and less stable, placing us at an ever greater risk for illness and producing the changes we recognize as the aging process.

Another important source of homeostatic imbalance occurs in certain pathological situations when the usual negative feedback mechanisms are overwhelmed and destructive positive feedback mechanisms take over. Some instances of heart failure reflect this phenomenon.

Examples of homeostatic imbalance will be provided throughout this book to accentuate your understanding of normal physiological mechanisms. These homeostatic imbalance sections are preceded by the symbol ⚠ to alert you to the fact that an abnormal condition is being described.

The Language of Anatomy

Although most of us are naturally curious about our bodies, this curiosity is sometimes dampened when we are confronted with the terminology unique to the study of anatomy and physiology. Let's face it. You can't just pick up an anatomy and physiology book and read it as though it were a novel. Unfortunately, confusion is inevitable without such specialized terminology. For example, if you are looking at a ball, "above" always means the area over the top of the ball. Other directional terms can also be used consistently because the ball is a totally symmetrical object. All sides and surfaces are equivalent. The human body, of course, has many protrusions, bends, and unique landmarks; thus, the question becomes, Above what? To prevent misunderstanding, anatomists have a universally accepted terminology that allows body structures to be located and identified with a minimum of words and a high degree of precision. This language of anatomy is presented and explained next.

Anatomical Position and Directional Terms

To accurately describe body parts and position, it is essential to have an initial reference point and indications of direction. The anatomical reference point is a standard body position called the **anatomical position.** In the anatomical position, the body is erect and the arms are hanging at the sides of the body with the palms forward and the thumbs pointed away from the body. You can see the anatomical position in Figure 1.6. It is essential to understand and remember the anatomical position because most of the directional terminology used in this book refers to an individual's body *as if it were in this position, regardless of its actual position.*

A number of **directional terms** are used by medical personnel and anatomists to explain precisely where one body structure is in relation to another. For example, we could describe the relationship between the ears and the nose informally by stating, "The ears are located on each side of the head to the right and left of the nose and slightly higher than the nose." Using anatomic terminology, this condenses to, "The ears are lateral and superior to the nose." Clearly, using anatomical terminology saves a good deal of description and is less ambiguous. Commonly used directional terms are defined and illustrated in Table 1.1. Although most of these terms are also used in everyday conversation, keep in mind that their anatomical meanings are very precise.

Regional Terms

Regional terms used to designate specific areas of the body are indicated in Figure 1.6. The common term for each of these body regions is also provided to orient you to the region being referred to.

Body Planes and Sections

The study of anatomy often involves dissection, in which the body or its organs are *sectioned* (cut) along an imaginary line called a *plane.* The most frequently used planes are the sagittal, frontal, and transverse planes, which lie at right angles to one another. Notice that a section bears the same name as the plane along

Table 1.1 Orientation and Directional Terms

Term	Definition	Illustration	Example
Superior (cranial)	Toward the head end or upper part of a structure or the body; above		The forehead is superior to the nose
Inferior (caudal)	Away from the head end or toward the lower part of a structure or the body; below		The navel is inferior to the breastbone
Anterior (ventral)*	Toward or at the front of the body; in front of		The breastbone is anterior to the spine
Posterior (dorsal)*	Toward or at the back of the body; behind		The heart is posterior to the breastbone
Medial	Toward or at the midline of the body; on the inner side of		The groin is medial to the thigh
Lateral	Away from the midline of the body; on the outer side of		The eye is lateral to the bridge of the nose
Intermediate	Between a more medial and a more lateral structure		The collarbone is intermediate between the breastbone and shoulder
Proximal	Closer to the origin of the body part or the point of attachment of a limb to the body trunk		The elbow is proximal to the wrist
Distal	Farther from the origin of a body part or the point of attachment of a limb to the body trunk		The knee is distal to the thigh
Superficial	Toward or at the body surface		The skin is superficial to the skeleton
Deep	Away from the body surface; more internal		The pelvis is deep to the buttock muscles

*Whereas the terms *ventral* and *anterior* are synonymous in humans, this is not the case in four-legged animals. *Ventral* specifically refers to the "belly" of a vertebrate animal and thus is the inferior surface of four-legged animals. Likewise, although the dorsal and posterior surfaces are the same in humans, the term *dorsal* specifically refers to an animal's back. Thus, the dorsal surface of four-legged animals is their superior surface.

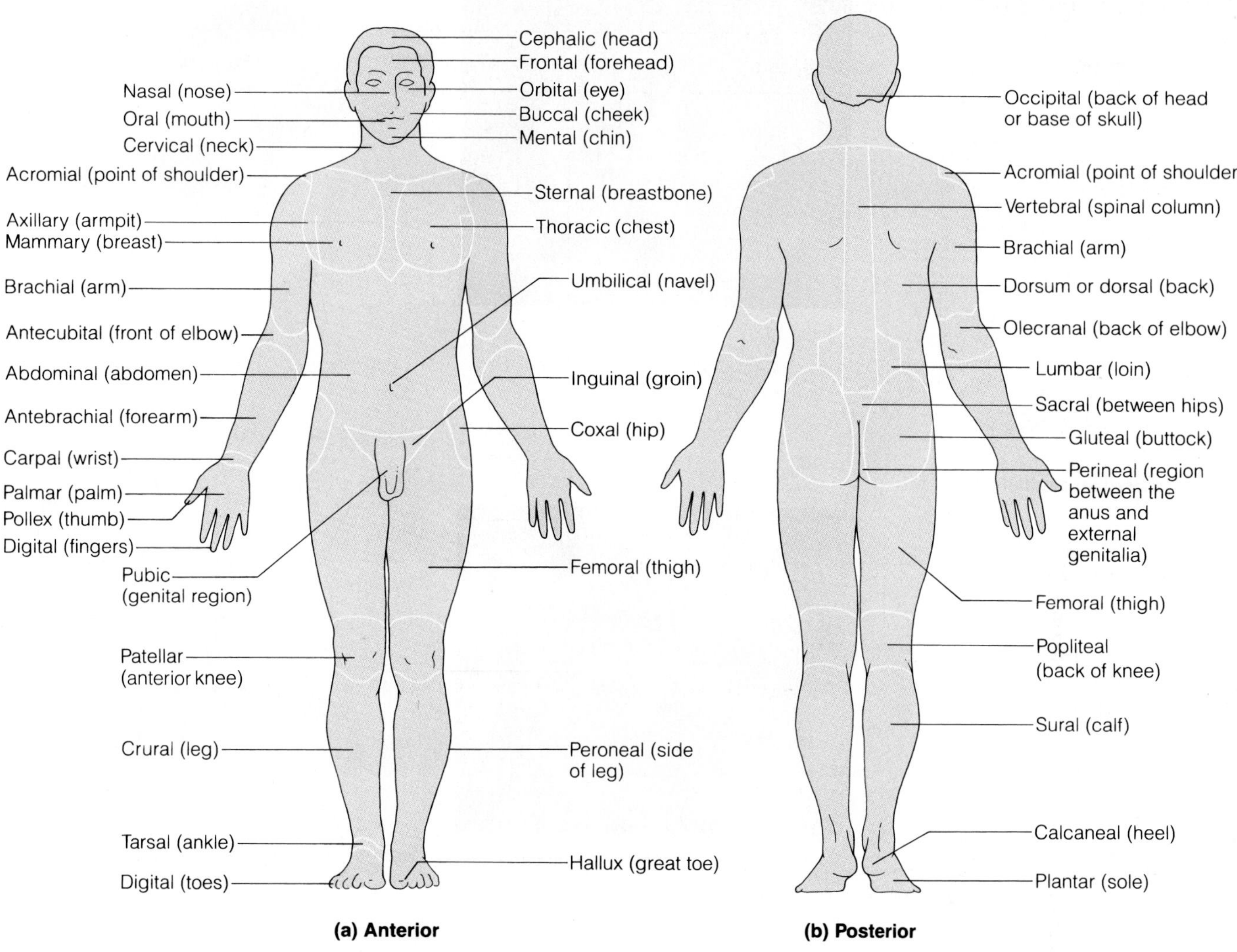

Figure 1.6 Regional terms used to designate specific body areas. The body is depicted in anatomical position.

which the cut is made; thus, a cut along a sagittal plane produces a sagittal section.

A **sagittal** (sa′-jih-tul) **plane** runs longitudinally and divides the body or organ into right and left portions. If the sagittal plane is exactly midline and the parts are symmetrical and equal, the plane is more precisely called a **midsagittal** or **median sagittal plane** (Figure 1.7a). All other sagittal planes are referred to more specifically as **parasagittal planes.**

A **frontal plane,** like the sagittal plane, runs longitudinally. However, in this case, the body or organ is divided into anterior and posterior portions. A frontal plane is also called a **coronal** (kuh-rō′-nul) **plane** (Figure 1.7b).

A **transverse plane** runs horizontally across the long axis of the body or organ, dividing it into superior and inferior parts (Figure 1.7c). When organs are sectioned along the transverse plane for microscopic examination, the tissue sections or slices are commonly referred to as **cross sections.** Cuts made along a plane intermediate between a horizontal and longitudinal plane are said to be made on an **oblique plane.**

Sectioning a body or organ along different planes often results in very different views. For example, a transverse section of the body trunk at the level of the kidneys would show kidney structure in cross section very nicely; a frontal section of the body trunk would show a different view of kidney anatomy; and a midsagittal section would miss the kidneys completely.

Body Cavities and Membranes

Within the axial (vertical axis) portion of the body are two major closed body cavities that contain internal organs.

Dorsal Body Cavity

The **dorsal body cavity,** located nearer to the dorsal, or posterior, surface of the body, is subdivided into a

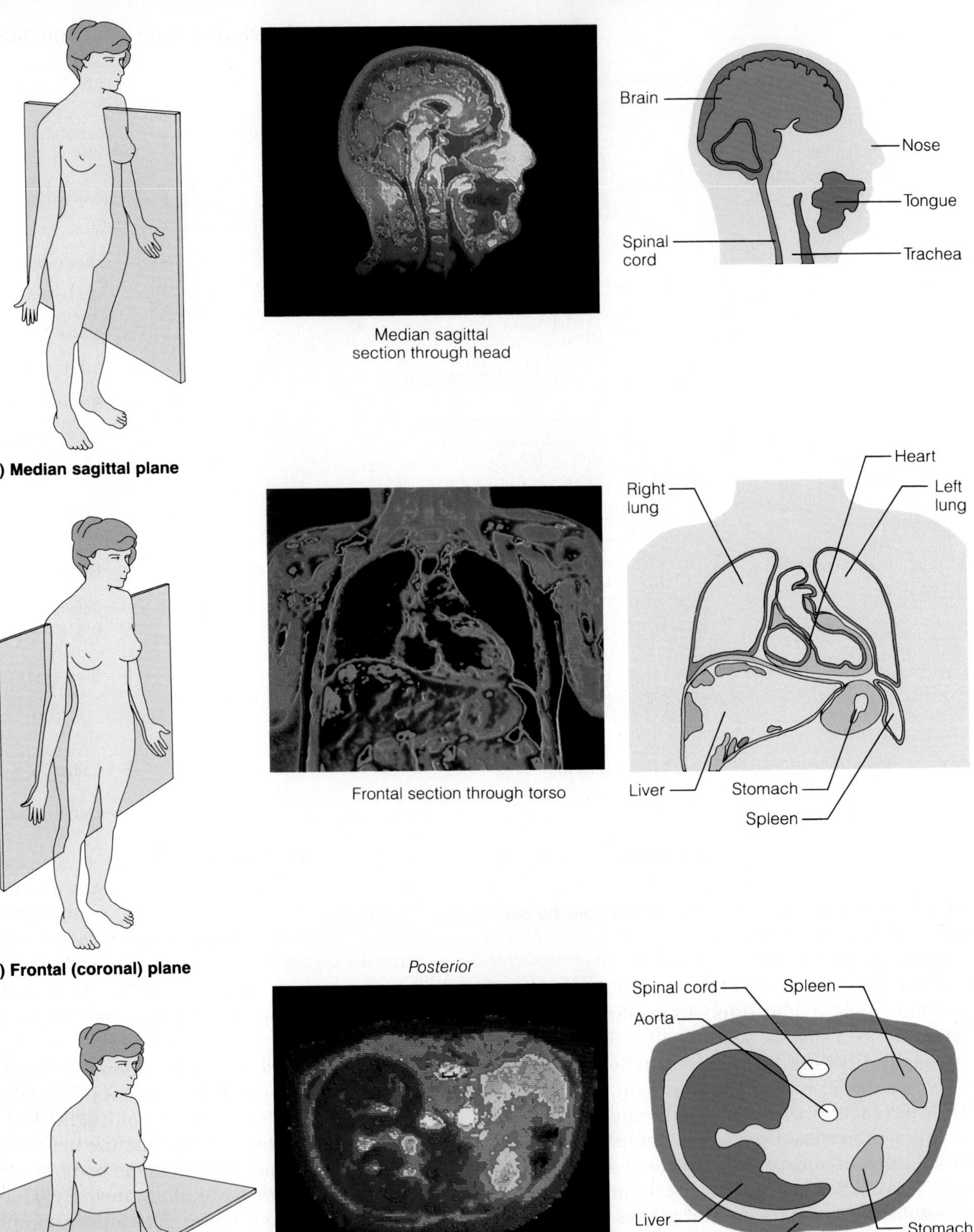

Figure 1.7 Planes of the body. The left side of the figure illustrates three major planes of space (sagittal, frontal, and transverse) relative to humans in the anatomical position. Selected areas of the body, visualized using magnetic resonance imaging (MRI) scans of the body taken at corresponding planes, are illustrated in the center of the figure. Diagrams identifying body organs seen in the MRI scans are shown on the right.

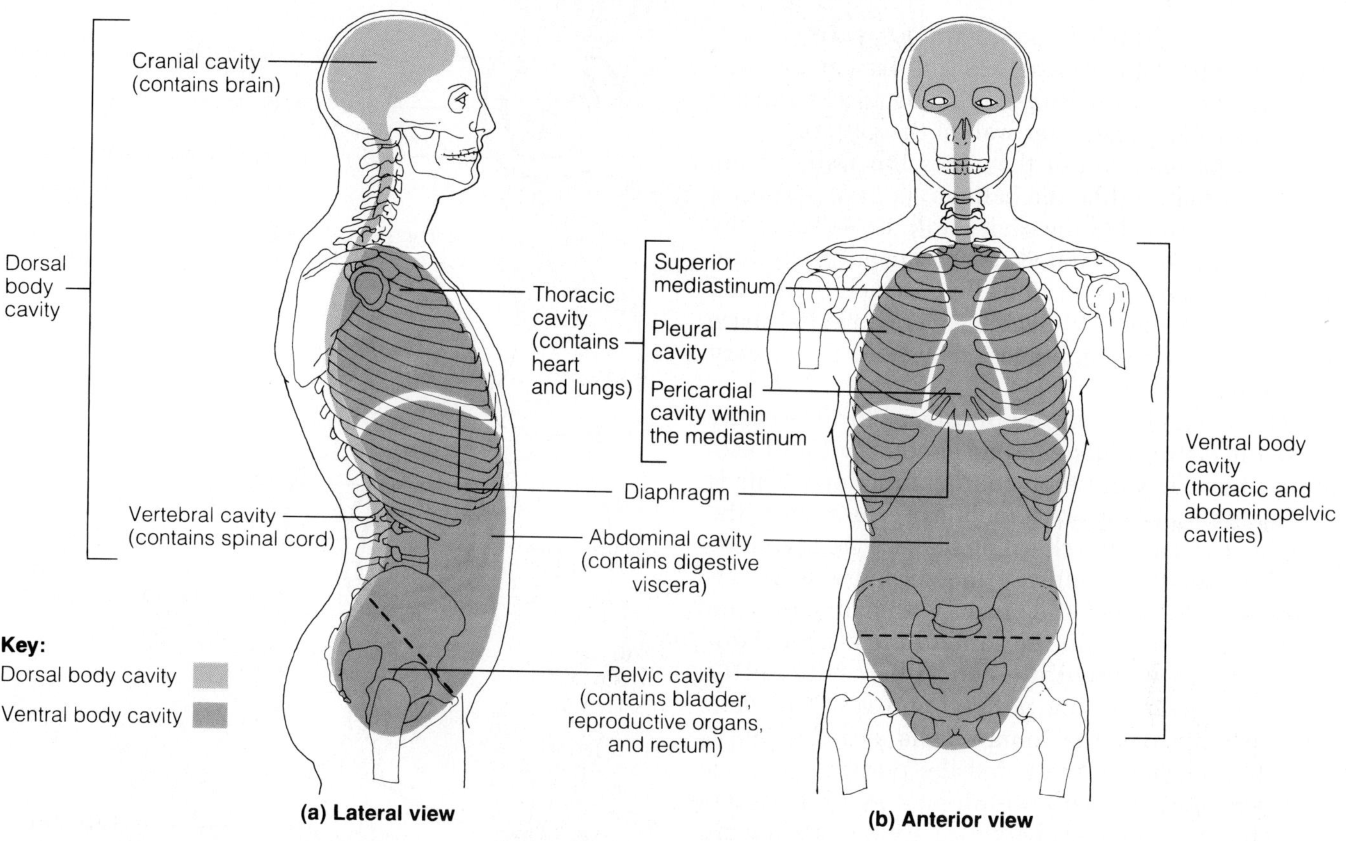

Figure 1.8 Dorsal and ventral body cavities and their subdivisions.

cranial cavity, within which the brain is encased by the skull, and a **vertebral cavity,** or **spinal cavity,** which forms a bony enclosure around the delicate spinal cord (Figure 1.8). Since the spinal cord issues from and is essentially a continuation of the brain, the cranial and spinal cavities are continuous with one another. The vital and extremely fragile organs housed by the dorsal body cavity are afforded excellent protection by the hard bony walls of the dorsal cavity.

Ventral Body Cavity

The more anterior and larger of the closed body cavities is the **ventral body cavity** (see Figure 1.8). Like the dorsal cavity, it has two major subdivisions, in this case housing a group of internal organs collectively called the *viscera* (vih′-ser-uh). The more superior **thoracic** (thor-a′-sik) **cavity,** surrounded by the ribs and muscles of the chest, is further subdivided into lateral **pleural** (plur′-ul) **cavities,** each housing a lung, and the medial **mediastinum** (mē″-dē-uh-stī′-num). The mediastinum contains the **pericardial** (payr″-ih-kar′-dē-ul) **cavity,** which encloses the heart, and the remaining thoracic organs (esophagus, trachea, and others).

The thoracic cavity is separated from the more inferior **abdominopelvic** (ab-dah″-mih-nō-pel′-vik) **cavity** by the diaphragm, a large, dome-shaped muscle. The abdominopelvic cavity, as its name suggests, has two parts. However, these regions are not physically separated by a muscular or membrane wall. Its superior portion, the **abdominal cavity,** contains the stomach, intestines, spleen, liver, and other organs. The inferior part, the **pelvic cavity,** contains the bladder, some reproductive organs, and the rectum. As Figure 1.8a reveals, the abdominal and pelvic cavities are not aligned with each other; rather, the bowl-shaped pelvis is tipped away from the perpendicular.

When the body is subjected to physical trauma (as often happens in an automobile accident), the most vulnerable abdominopelvic organs are those within the abdominal cavity, because the cavity walls of that portion are formed only of trunk muscles and are not reinforced by bone. The pelvic organs receive a somewhat greater degree of protection from the bony pelvis in which they reside. ■

The walls of the ventral body cavity and the outer surfaces of the organs it contains are covered with an exceedingly thin, double-layered membrane, the **serosa** (sur-ō′-suh), or **serous membrane.** The part of the membrane lining the cavity walls is generally called the **parietal** (puh-rī′-eh-tul) **serosa.** It continues on to form the **visceral serosa,** covering the organs in the cavity. You can visualize this relationship by pushing your fist into a limp balloon (Figure 1.9a). The part of the balloon that clings closely to your fist can be compared to the visceral serosa clinging to the organ's external surface. The outer wall of the balloon represents the parietal serosa that lines the walls of the cavity and that, unlike the balloon, is never exposed but is always fused to the cavity wall. In the body, the serous layers are separated not by air but by a thin, lubricating fluid, called *serous fluid,* which is secreted by both membranes. Although there is a potential space between the two membranes, they tend to lie very close to each other.

The serous fluid allows the organs to slide easily across the cavity walls and one another without friction as they carry out their routine functions. This is extremely important when mobile organs such as the pumping heart and the churning stomach are involved.

The specific naming of the serous membranes depends on the cavity and organs with which they are associated. For example, the *parietal* and *visceral pericardia* are associated with the pericardial cavity and the heart (Figure 1.9b). The *parietal* and *visceral pleurae* (plur′-ē) cover the thorax walls and the lungs, respectively (Figure 1.9c), and the *parietal* and *visceral peritoneums* (payr″-ih-tuh-nē′-ums) are associated with the abdominopelvic cavity and its organs (Figure 1.9d).

Inflammation of the serous membranes, accompanied by a deficit of lubricating fluid, leads to excruciating pain as the organs stick together and drag across one another, as anyone who has experienced *pleurisy* (inflammation of the pleurae) or *peritonitis* (inflammation of the peritoneum) can attest. ■

Other Body Cavities

In addition to the major closed body cavities, there are a number of other smaller body cavities, many of which are located in the head and most of which are open to the body exterior:

1. Oral cavity. This is the cavity commonly referred to as the mouth, which contains the teeth and tongue. The oral cavity is continuous with the cavity of the digestive organs, which also opens to the exterior at the anus.

2. Nasal cavity. Located within the nose, the medially divided nasal cavity is part of and continuous with the passages of the respiratory system.

3. Orbital cavities. The orbital cavities house the eyes and present them in an anterior position.

4. Middle ear cavities. The middle ear cavities carved into the temporal bone of the skull contain tiny bones associated with the transmission of sound to the organ of hearing in the inner ear.

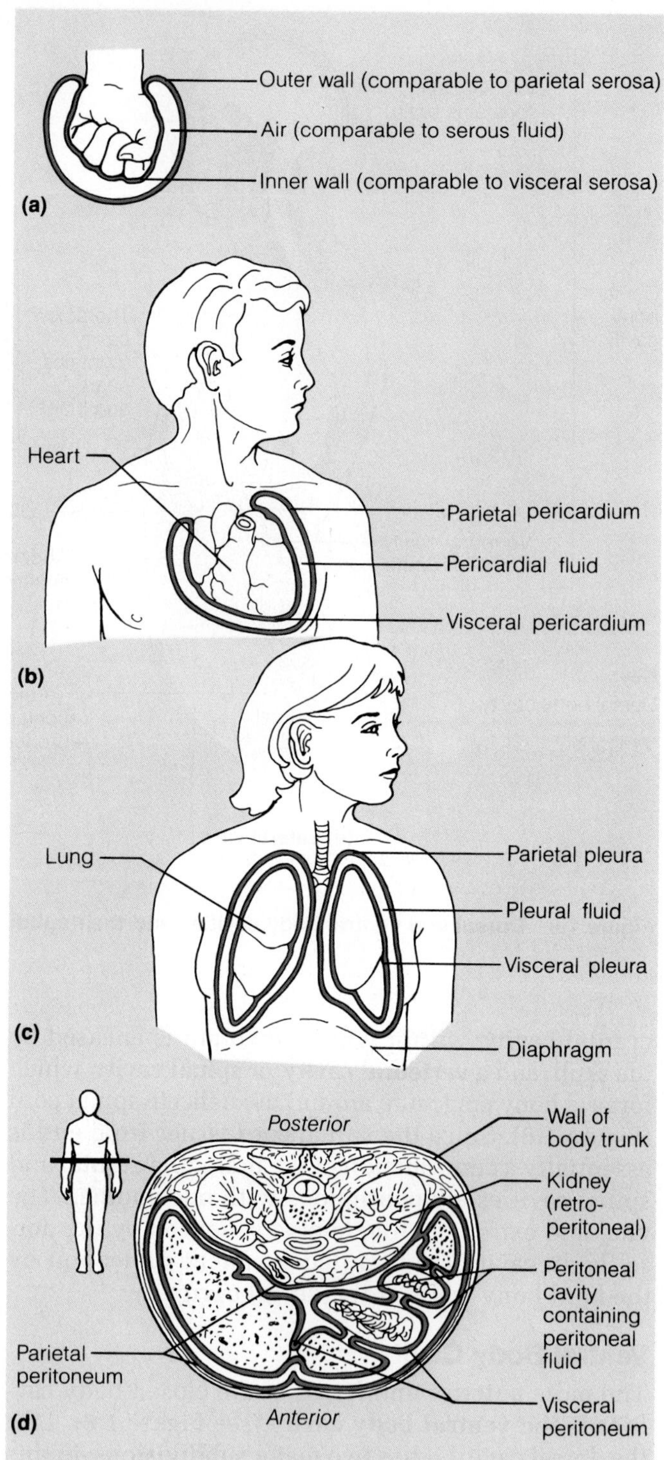

Figure 1.9 Serous membrane relationships. (**a**) A fist is thrust into a flaccid balloon to demonstrate the relationship between the parietal and visceral serous membrane layers. (**b**) The parietal pericardium is the outer layer lining the pericardial cavity; the visceral pericardium clings to the external heart surface. (**c**) The visceral pleura surrounding each lung is continuous with the parietal pleura that adheres to the walls of the thoracic cavity. (**d**) Transverse section through the body trunk at the level of the liver and kidneys. The parietal peritoneum lines the wall of the abdominopelvic cavity; the visceral peritoneum covers the external surfaces of most organs within that cavity. However, some abdominal organs are retroperitoneal (behind the peritoneum). This is true of the kidneys.

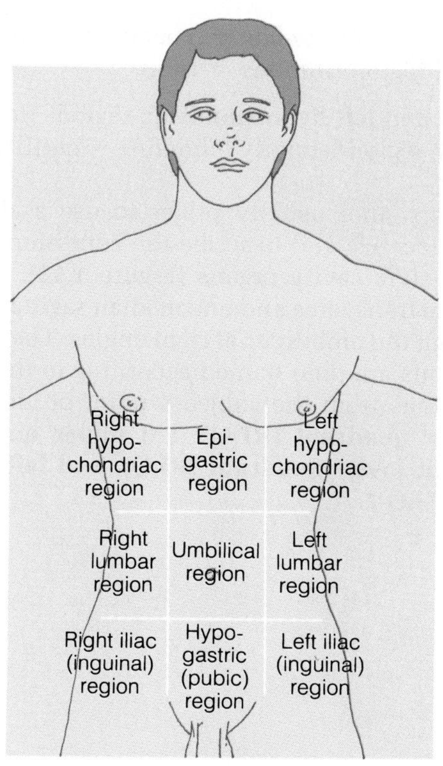

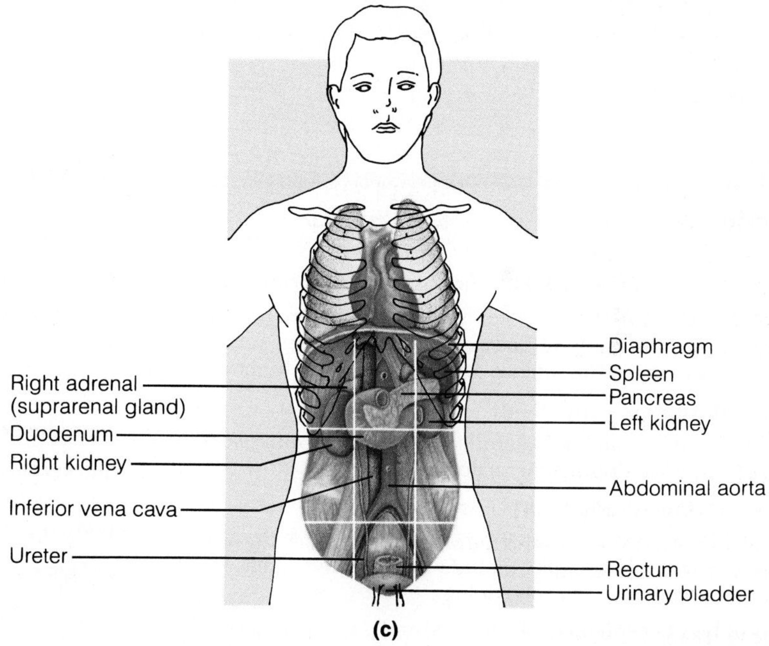

Figure 1.10 The nine abdominopelvic regions. (**a**) Division of the abdominopelvic cavity into nine regions delineated by four planes. The superior transverse plane is just inferior to the ribs; the inferior transverse plane is just superior to the hip bones; and the parasagittal planes lie just medial to the nipples. (**b**) Anterior view of the abdominopelvic cavity showing the superficial organs. (**c**) Anterior view of the abdominopelvic cavity showing the deep organs; the more superficial liver, stomach, and intestines do not appear.

5. Synovial (sih-nō′-vē-ul) **cavities.** Synovial cavities are enclosed within fibrous capsules that surround the freely movable joints of the body. Like the serous membranes of the ventral body cavity, the membranes lining the synovial cavities secrete a lubricating fluid that reduces friction as the bones move across one another.

Abdominopelvic Regions and Quadrants

Since the abdominopelvic cavity is fairly large and contains many organs, it is often helpful to divide it into smaller areas for study. One division method, used primarily by anatomists, uses two transverse and two parasagittal planes to divide the cavity into nine regions (Figure 1.10):

- The **umbilical region** is the centermost region deep to and surrounding the umbilicus (navel).
- The **epigastric region** is located superior to the umbilical region (*epi* = upon, above; *gastric* = stomach).
- The **hypogastric (pubic) region** is located inferior to the umbilical region (*hypo* = below).
- The **right** and **left iliac,** or **inguinal** (in′-gwih-nul), **regions** are located laterally to the hypogastric region (*iliac* = superior part of the hip bone).

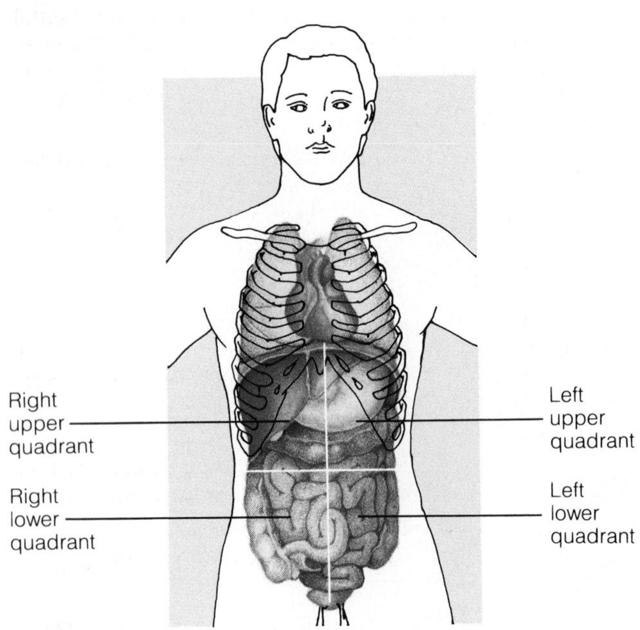

Figure 1.11 The four abdominopelvic quadrants. In this scheme, the abdominopelvic cavity is divided into four quadrants by two planes. The figure shows superficial organs within each quadrant.

- The **right** and **left lumbar regions** lie laterally to the umbilical region (*lumbus* = loin).
- The **right** and **left hypochondriac regions** flank the epigastric region laterally (*chondro* = cartilage).

Medical personnel usually prefer to use a simpler scheme to localize and describe the condition of the abdominopelvic cavity organs (Figure 1.11). In this scheme, one transverse and one median sagittal plane pass through the umbilicus at right angles. The resulting *quadrants* are then named according to their relative positions from the subject's point of view: the **right upper quadrant (RUQ), left upper quadrant (LUQ), right lower quadrant (RLQ),** and **left lower quadrant (LLQ).**

Chapter Summary

AN OVERVIEW OF ANATOMY AND PHYSIOLOGY (pp. 5–6)

1. Anatomy is the study of body structures and their relationships; physiology is the science of how body parts function.

Topics of Anatomy (pp. 5–6)

2. Subdivisions of anatomy include gross anatomy, microscopic anatomy, and developmental anatomy.

Topics of Physiology (p. 6)

3. Typically, physiology studies the functioning of specific organs or organ systems. Examples include cardiac physiology, renal physiology, and muscular physiology.

4. Physiology is explained by chemical and physical principles.

Complementarity of Structure and Function (p. 6)

5. Anatomy and physiology are inseparable: What a body can do depends on the unique architecture of its parts. This is called the complementarity of structure and function.

THE HIERARCHY OF STRUCTURAL ORGANIZATION (pp. 6–7)

1. The levels of structural organization of the body, from simplest to most complex, are: chemical, cellular, tissue, organ, organ system, and organismal.

2. The 12 organ systems of the body are the integumentary, skeletal, muscular, nervous, endocrine, cardiovascular, lymphatic, immune, respiratory, digestive, urinary, and reproductive systems.

MAINTENANCE OF LIFE (pp. 7–12)

Functional Characteristics (pp. 7–11)

1. All living organisms carry out certain vital functional activities necessary for life, including maintenance of boundaries, movement, responsiveness, digestion, metabolism, excretion, reproduction, and growth.

Survival Needs (pp. 11–12)

2. Survival needs include nutrients, water, oxygen, appropriate body temperature, and atmospheric pressure.

HOMEOSTASIS (pp. 12–15)

1. Homeostasis is a dynamic equilibrium of the internal environment. All body systems contribute to homeostasis, but the nervous and endocrine systems are most important.

General Characteristics of Control Mechanisms (p. 12)

2. Control systems of the body contain at least three elements: control center, receptor(s), and effector(s).

Negative Feedback Mechanisms (p. 13)

3. Negative feedback mechanisms turn off or reduce the original stimulus. Body temperature, heart rate, breathing rate and depth, and blood levels of glucose and certain ions are regulated by negative feedback mechanisms.

Positive Feedback Mechanisms (pp. 14–15)

4. Positive feedback mechanisms enhance the initial stimulus, leading to an enhancement of the response. Positive feedback mechanisms rarely contribute to homeostasis, but blood clotting and labor contractions are regulated by such mechanisms.

Homeostatic Imbalance (p. 15)

5. With age, the efficiency of negative feedback mechanisms declines, and positive feedback mechanisms occur more frequently. These changes underlie certain disease conditions.

THE LANGUAGE OF ANATOMY (pp. 15–22)

Anatomical Position and Directional Terms (p. 15)

1. In the anatomical position, the body is erect, facing forward, arms at sides with palms forward.

2. Directional terms allow the position of the body parts to be precisely described. Terms commonly used to describe body directions and orientation include: superior/inferior; anterior/posterior; ventral/dorsal; medial/lateral; proximal/distal; and superficial/deep.

Regional Terms (p. 15)

3. Regional terms are used to designate specific areas of the body (see Figure 1.6).

Body Planes and Sections (pp. 15–17)

4. The body or its organs may be cut along planes, or imaginary lines, to produce different types of sections. Frequently used planes are sagittal, frontal, and transverse.

Body Cavities and Membranes (pp. 17–21)

5. The body contains two major closed cavities: the dorsal cavity, subdivided into the cranial and spinal cavities; and the ventral cavity, subdivided into the superior thoracic cavity and the inferior abdominopelvic cavity.

6. The walls of the ventral cavity and the surfaces of the organs it contains are covered with thin membranes, the parietal and visceral serosae, respectively. The serosae produce a thin lubricating fluid that decreases friction during organ functioning.

7. There are several smaller body cavities. Most of these are in the head and open to the exterior.

Abdominopelvic Regions and Quadrants (pp. 21–22)

8. The abdominopelvic cavity may be divided by four planes into nine abdominal regions (epigastric, umbilical, hypogastric, right and left iliac, right and left lumbar, and right and left hypochondriac), or by two planes into four quadrants.

Review Questions

Multiple Choice/Matching

1. The correct sequence of levels forming the structural hierarchy is (a) organ, organ system, cellular, chemical, tissue, organismal; (b) chemical, cellular, tissue, organismal, organ, organ system; (c) chemical, cellular, tissue, organ, organ system, organismal; (d) organismal, organ system, organ, tissue, cellular, chemical.

2. The structural and functional unit of life is (a) a cell, (b) an organ, (c) the organism, (d) a molecule.

3. Which of the following is a *major* functional characteristic of all organisms? (a) movement, (b) growth, (c) metabolism, (d) responsiveness, (e) all of these.

4. Two of the following organ systems bear the major responsibility for ensuring homeostasis of the internal environment. Which two? (a) nervous system, (b) digestive system, (c) circulatory system, (d) endocrine system, (e) reproductive system.

5. Choose the structure or organ that matches the given directional term.
(a) distal: the elbow/the wrist
(b) lateral: the hipbone/the umbilicus
(c) superior: the nose/the chin
(d) anterior: the toes/the heel
(e) superficial: the scalp/the skull

6. Assume that the body has been divided by a median sagittal plane, a frontal plane, and a transverse plane made at the level of each of the organs listed below. Which organs would not be visible in all three sections? (a) urinary bladder, (b) brain, (c) lungs, (d) kidneys, (e) small intestine, (f) heart.

7. Relate each of the following conditions or statements to either the dorsal body cavity or the ventral body cavity.
(a) surrounded by the bony skull and the vertebral column
(b) includes the thoracic and abdominopelvic cavities
(c) contains the brain and spinal cord
(d) located more anteriorly
(e) contains the heart, lungs, and digestive organs

8. Which of the following relationships is *incorrect?* (a) visceral peritoneum/outer surface of small intestine, (b) parietal pericardium/outer surface of heart, (c) parietal pleura/wall of thoracic cavity.

9. Which ventral cavity subdivision has no bony protection? (a) thoracic cavity, (b) abdominal cavity, (c) pelvic cavity.

Short Answer Review Questions

10. According to the principle of complementarity, how does anatomy relate to physiology?

11. Construct a table that lists the 12 organ systems of the body, name two organs of each system, and describe the overall or major function of each system.

12. List and describe briefly five external factors that must be present or provided to sustain life.

13. Define homeostasis.

14. Compare and contrast the operation of negative and positive feedback mechanisms in maintaining homeostasis. Provide two examples of variables controlled by negative feedback mechanisms and one example of a process regulated by a positive feedback mechanism.

15. Describe and assume the anatomical position. Why is an understanding of this position important? What is the importance of directional terms?

16. Define plane and section.

17. Provide the anatomical term that correctly names each of the following body regions: (a) arm, (b) thigh, (c) chest, (d) fingers and toes, (e) anterior aspect of the knee.

18. (a) Make a diagram showing the nine abdominopelvic regions, and name each region. Name two organs (or parts of organs) that could be located in each of the named regions. (b) Make a similar sketch illustrating how the abdominopelvic cavity may be divided into quadrants, and name each quadrant.

Clinical Application Questions

19. John has been suffering agonizing pain with each breath and has been informed by the physician that he has pleurisy. (a) Specifically, what membranes are involved in this condition? (b) What is their usual role in the body? (c) Explain why John's condition is so painful.

20. How is the concept of homeostasis (or its loss) related to disease and aging? Provide examples to support your reasoning.

21. When at the clinic, Harry was told that blood would be drawn from his anticubital region. What body part was Harry asked to hold out? Later, the nurse came in and give Harry a shot of penicillin in his gluteal region. Did Harry take off his shirt or drop his pants to receive the injection? Before Harry left, the nurse noticed that Harry had a nasty bruise on his cervical region. What part of his body was black and blue?

CHAPTER 2

Chemistry Comes Alive

Chapter Outline and Student Objectives

PART I: BASIC CHEMISTRY

Definition of Concepts: Matter and Energy (pp. 25–26)

1. Differentiate clearly between matter and energy and between potential energy and kinetic energy.
2. Describe the major energy forms.

Composition of Matter: Atoms and Elements (pp. 26–30)

3. Define chemical element and list the four elements that form the bulk of body matter.
4. List the subatomic particles; describe their relative masses, charges, and positions in the atom.
5. Define atom, atomic weight, isotope, and radioisotope.

How Matter Is Combined: Molecules and Mixtures (pp. 30–33)

6. Distinguish between a compound and a mixture. Define molecule.
7. Compare solutions, colloids, and suspensions.

Chemical Bonds (pp. 34–38)

8. Differentiate between ionic and covalent bonds. Contrast these bonds with hydrogen bonds.
9. Compare and contrast polar and nonpolar compounds.

Chemical Reactions (pp. 38–41)

10. Contrast synthesis, decomposition, and exchange reactions. Explain why chemical reactions in the body are often irreversible.
11. Describe factors that affect chemical reaction rates.

PART II: BIOCHEMISTRY: THE COMPOSITION AND REACTIONS OF LIVING MATTER

Inorganic Compounds (pp. 41–44)

12. Explain the importance of water and salts to body homeostasis.
13. Define acid and base, and explain the concept of pH.

Organic Compounds (pp. 44–57)

14. Compare the building blocks, general structures, and biological functions of organic molecules in the body.
15. Explain the role of dehydration synthesis and hydrolysis in the formation and breakdown of organic molecules.
16. Compare the functional roles of neutral fats, phospholipids, and steroids in the body.
17. Contrast structural and functional proteins.
18. Describe the mechanism of enzyme activity.
19. Compare and contrast DNA and RNA.
20. Explain the role of ATP in cell metabolism.

Preview of Selected Key Terms

Matter The substance composing physical objects; exists as gases, liquids, and solids.

Energy (*ergon* = work) The capacity to do work; may be stored (potential energy) or in action (kinetic energy).

Element One of a limited number of unique varieties of matter that composes substances of all kinds; e.g., carbon, hydrogen, oxygen.

Atom (*atomos* = indivisible) The smallest particle of an elemental substance that exhibits the properties of that element; composed of protons, neutrons, and electrons.

Chemical reaction Chemical process in which molecules are formed, changed, or broken down.

Molecule A particle consisting of two or more atoms joined by chemical bonds.

Mixture A substance consisting of particles dispersed together and in which the particles retain their unique properties; e.g., a solution, colloid, or suspension.

Ionic bond (ī-ah'-nik) A chemical bond formed by electron transfer between atoms.

Covalent bond (*co* = with; *valent* = having power) A chemical bond created by electron sharing between atoms.

Enzyme (en'-zīm) (*zyme* = causing to ferment) A protein that acts as a biological catalyst to speed up chemical reactions.

Organic Pertaining to carbon-containing substances, such as proteins, fats, and carbohydrates.

Inorganic Pertaining to chemical substances that do not contain carbon, such as water and salts.

pH (pē-āch) The measure of the relative acidity or alkalinity of a solution; hydrogen ion concentration in moles per liter.

Adenosine triphosphate (uh-den′-uh-sēn trī″-fos′-fāt) **(ATP).** The organic molecule that stores and releases chemical energy for cellular use in the body.

Why study chemistry in an anatomy and physiology course? The answer is very simple. The food you eat and the medicines you take when you are ill are composed of chemical substances. Indeed, your entire body is made up of chemicals, thousands of them, continuously interacting with one another at an incredible pace. Although it is possible to study anatomy without much reference to chemistry, chemical reactions underlie all physiological processes—movement, digestion, the pumping of your heart, and even your thoughts. This chapter presents the basics of chemistry and biochemistry (the chemistry of living material), providing the background you will need to understand body functions.

PART 1: BASIC CHEMISTRY

Definition of Concepts: Matter and Energy

Matter is the "stuff"—the gases, liquids, and solids—of the universe. With some exceptions, it can be seen, smelled, and felt. More precisely, matter is anything that occupies space and has mass. For all practical purposes, we can consider mass to be the same as weight, and we will use these terms interchangeably. However, this usage is not quite accurate. *Mass* is the amount of matter (or chemical substance) that an object contains, and an object has a constant mass, regardless of where it is—at sea level, on a mountaintop, or in outer space. However, *weight* reflects and varies with gravity; the greater the gravitational pull, the greater the weight. Astronauts are said to be weightless when in outer space, but they still have the same mass as they do on earth. The science of chemistry studies the nature of matter, especially how its building blocks are put together and interact with one another.

Compared with matter, **energy** is less tangible. Energy has no mass, does not take up space, and can only be measured by its effects on matter. Energy is defined as the capacity to do work, or to put matter into motion. The greater the work done, the more energy expended doing it. A baseball player who has just hit the ball over the fence has used much more energy than a batter who just bunts the ball!

Actually, energy is a topic of physics, but matter and energy are inseparable. Matter is the substance, and energy is the mover of the substance. All living things are composed of matter and, to grow and function, they all require energy. It is the release and use of energy by living systems that gives us the elusive quality called "life." Thus, it is worth taking a brief detour to introduce the kinds of energy used by the body as it does its work.

Potential and Kinetic Energy

The total energy of any object consists of two components: potential energy and kinetic energy. **Potential energy** is the energy an object has because of its position (or internal structure) in relation to other objects. It is stored, or inactive, energy that has the *potential*, or capability, to do work, but is not presently doing so. The batteries in an unused toy have potential energy, as does water behind a dam.

Kinetic (kih-neh′-tik) **energy** is energy associated with a moving object; it is energy in action, the form that is actually doing work. Moving objects do work by transferring their motion to other objects; for example, the push you give a swinging door sets it into motion. Kinetic energy is seen in the constant movement of the tiniest particles of matter (atoms and molecules), as well as in large objects. The kinetic energy of individual molecules is expressed as heat, or thermal energy, which is continuously transferred from warmer to cooler objects. The hotter an object is, the faster its molecules move. As you will see shortly, the thermal energy of chemical substances is an important factor in chemical reactions.

Potential energy becomes kinetic energy and is capable of doing work when it is released. For example, a stick of dynamite causes an explosive effect on matter after its fuse is lit. Similarly, dammed water is transformed into a rushing torrent when the dam is opened, whereupon it can move a turbine at a hydroelectric plant, or charge a battery. Energy is constantly being changed from potential to kinetic and back again.

Forms of Energy

The body uses energy in several forms, all of which can exist as potential or kinetic energy. **Chemical energy** is the form stored in the bonds of chemical substances. When the bonds are broken, the potential energy is released for use and becomes kinetic energy. For example, as foods are broken down in the body, some of the released energy is converted into the kinetic energy of your moving arm. Chemical energy is the most fundamental form of energy in living organisms, because it is used to run all functional processes.

Electrical energy reflects the movement of charged particles. In your home, electrical energy exists in the flow of electrons along the household wiring. In your body, electrical currents are generated as charged particles called ions move across cell membranes. The nervous system uses electrical currents, called nerve impulses, to transmit messages from one part of the body to another.

Mechanical energy is *directly* involved in moving matter. When you ride a bicycle, your legs provide the mechanical energy that moves the pedals.

Radiant energy, or **electromagnetic** (ih-lek″-trō-mag-neh′-tik) **energy,** is energy that travels in waves. These waves, which vary in length, are collectively called the *electromagnetic spectrum*. Besides the wavelengths that we can see, called light, the electromagnetic spectrum includes infrared waves, radio waves, ultraviolet waves, and X-rays. Light is important to our ability to see. When it strikes the retinas of our eyes, it sets off a chain of reactions that eventually results in vision. Ultraviolet waves are responsible for the sunburn we get at the beach, but also stimulate our body to make vitamin D. X-rays have little to do with normal body functioning, but they are important in the study of anatomy and in medical diagnosis.

With few exceptions, one form of energy can be converted to another. For example, chemical energy (gasoline) that powers the motor of a speedboat is converted into the mechanical energy of the whirling propeller that allows the boat to skim across the water. All energy conversions are relatively inefficient; some of the initial energy supply is always "lost" to the environment as heat. (It is not really lost, however, because energy cannot be created or destroyed; but that part given off as heat is at least partly *unusable*.) You can easily demonstrate this principle. Electrical energy is converted into light energy in a light bulb, but if you touch a lit bulb, you will soon discover that some of the electrical energy is producing heat instead. All energy conversions that occur in the body liberate heat, which helps to maintain our body temperature and enables other chemical reactions to occur at life-sustaining rates. Some of the liberated heat always escapes from the body; consequently, if we are to maintain normal body temperature, we must continually stoke our internal "furnace" by ingesting food.

Composition of Matter: Atoms and Elements

All matter is composed of fundamental substances called **elements,** unique substances that cannot be broken down into simpler substances by ordinary chemical methods. Among the well-known elements are oxygen, carbon, gold, silver, copper, and iron. At present, 108 elements are known, of which 92 are naturally occurring; the rest are made artificially in accelerator devices. Four elements—carbon, oxygen, hydrogen, and nitrogen—make up about 96% of body weight; 20 others are present in the body in lesser, or trace, amounts. The elements contributing to body mass and their importance in the body are given in Table 2.1.

Each element is composed of more or less identical particles or building blocks, called **atoms.** Atoms are unimaginably small. The smallest are less than 0.1 nanometer (nm) in diameter, and the largest only about five times as large. (1 nm = 0.00000001 [or 10^{-8}] centimeter (cm), or four-billionths of an inch!) Each element's atoms differ from those of all other elements and give the element its unique physical and chemical properties. Physical properties are those we can detect with our senses (such as color and texture) or measure (such as boiling point and freezing point). Chemical properties pertain to the way atoms interact with other atoms (bonding behavior) and account for the fact that iron rusts, gasoline burns in air, animals can digest their food, and so on.

Each element is designated by a one- or two-letter chemical shorthand called an **atomic symbol,** usually the first letter(s) of the element's name. For example, C stands for carbon, O for oxygen, and Ca for calcium. In a few cases, the atomic symbol is taken from the Latin name for the element; for example, sodium is indicated by Na, from the Latin word *natrium*.

Atomic Structure

The word *atom* comes from the Greek word meaning "indivisible," and until the present century, atomic indivisibility was accepted as a scientific truth. We now know that atoms are clusters of even smaller particles called protons, neutrons, and electrons and that even those subatomic particles can be split into smaller components with high-technology tools. Even so, this does not disprove the old idea of atom indivisibility

Table 2.1 Common Elements Composing the Human Body

Element	Atomic symbol	% Body mass	Functions
Oxygen	O	65.0	A major component of both organic (carbon-containing) and inorganic (non-carbon-containing) molecules; as a gas, it is necessary for the production of cellular energy (ATP)
Carbon	C	18.5	A primary component of all organic molecules, which include carbohydrates, lipids (fats), proteins, and nucleic acids
Hydrogen	H	9.5	A component of all organic molecules; as an ion (proton), it influences the pH of body fluids
Nitrogen	N	3.2	A component of proteins and nucleic acids (genetic material)
Calcium	Ca	1.5	Found as a salt in bones and teeth; its ionic form is required for muscle contraction, conduction of nerve impulses, and blood clotting
Phosphorus	P	1.0	Part of calcium phosphate salts in bones and teeth; also present in nucleic acids; part of ATP
Potassium	K	0.4	Its ion (K^+) is the major positive ion (cation) in cells; necessary for conduction of nerve impulses and muscle contraction
Sulfur	S	0.3	Component of proteins, particularly muscle proteins
Sodium	Na	0.2	As an ion (Na^+), sodium is the major positive ion found in extracellular fluids (outside of cells); important for water balance, conduction of nerve impulses, and muscle contraction
Chlorine	Cl	0.2	Ionic chlorine (Cl^-) is the most abundant negative ion (anion) in extracellular fluids
Magnesium	Mg	0.1	Present in bone; also an important cofactor in a number of metabolic reactions
Iodine	I	0.1	Needed to make functional thyroid hormones
Iron	Fe	0.1	Component of hemoglobin (which transports oxygen within red blood cells) and some enzymes
Chromium Cobalt Copper Fluorine Manganese Molybdenum Selenium Silicon Tin Vanadium Zinc	Cr Co Cu F Mn Mo Se Si Sn V Zn		These elements are referred to as *trace elements* because they are required in very minute amounts; many are found as part of enzymes or are required for enzyme activation

because an atom loses the unique properties of its element when it is split into its subatomic particles.

An atom's subatomic particles differ in mass, electrical charge, and position in the atom. An atom has a central nucleus containing protons and neutrons tightly bound together. The *nucleus*, in turn, is surrounded by orbiting electrons (Figure 2.1a). **Protons** (p^+) bear a positive charge, while **neutrons** (n^0) are neutral; thus, the nucleus is positively charged. Protons and neutrons are heavy subatomic particles and have approximately the same mass, which is arbitrarily designated as 1 atomic mass unit (1 amu). Since all of the heavy subatomic particles are concentrated in the nucleus, the nucleus is fantastically dense and accounts for nearly the entire mass (99.9%) of the atom. The tiny **electrons** (e^-) bear a negative charge equal in strength to the positive charge of the proton. However, an electron has only about 1/1800 the mass of a proton, and the mass of an electron is usually designated as 0 amu. All atoms are electrically neutral overall, which indicates that every atom has equal numbers of positively charged protons and negatively charged electrons. For example, iron has 26 protons and 26 electrons. Choose any atom and the same relationship between the numbers of protons and electrons holds.

The *planetary model*, illustrated in Figure 2.1a, is a simplified (and now outdated) model of atomic structure. As you can see, it depicts electrons moving around the nucleus in fixed, generally circular orbits, like planets around the sun. But we can never determine the exact location of electrons at a particular time because they jump around following unknown trajectories. So, instead of speaking of discrete orbits, chemists talk about *orbitals*—regions around the nucleus in which a given electron or electron pair is likely to be found most of the time. This more modern model of atomic structure, called the *orbital model*, has proved to be more useful for predicting the chemical behavior of atoms. As illustrated in Figure 2.1b, the orbital model depicts *probable* regions of greatest

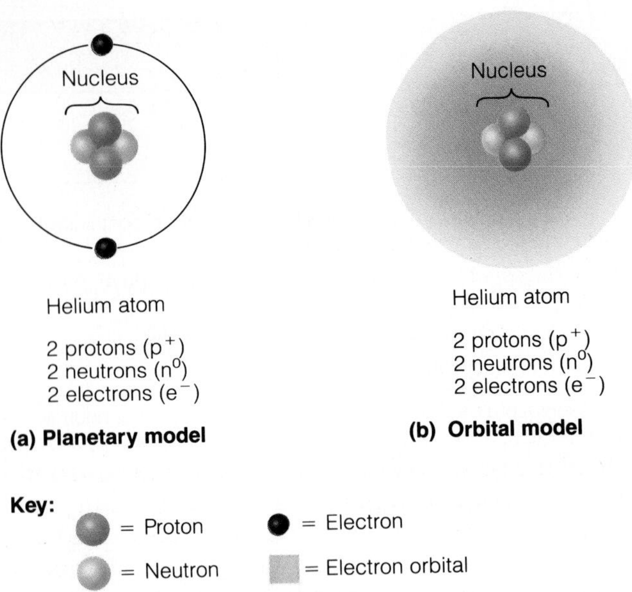

Figure 2.1 The structure of an atom. The dense central nucleus contains the protons and neutrons. (**a**) In the planetary model of atomic structure, the electrons move around the nucleus in fixed orbits. (**b**) The orbital model recognizes that we never know exactly where electrons are; therefore electrons are shown as a cloud of negative charge.

electron density by denser shading and abandons the orbit lines used by the planetary model. However, because of the simplicity of the planetary model, most of the descriptions of atomic structure in this text will use that model.

Hydrogen is the simplest atom, with just one proton and one electron. You can visualize the spatial relationships within the hydrogen atom by imagining it enlarged with a diameter equal to the length of a football field. In that case, the nucleus could be represented by a lead ball the size of a gumdrop in the exact center of the sphere and the lone electron pictured as a fly buzzing about unpredictably within the sphere. This mental picture should serve as a reminder that most of the volume of an atom is empty space, and nearly all of the mass is concentrated in its central nucleus.

Identifying Elements

All protons are alike, regardless of the atom being considered. The same is true of all neutrons and all electrons. So what determines the unique properties of each element? The answer resides in the fact that atoms of different elements are composed of *different numbers* of protons, neutrons, and electrons.

The simplest and smallest atom, hydrogen, has one proton, one electron, and no neutrons (Figure 2.2). Next is the helium atom, with two protons, two neutrons, and two orbiting electrons; lithium follows with three protons, four neutrons, and three electrons. If we were to continue this step-by-step progression, we would get a graded series of atoms containing from 1 to 108 protons, an equal number of electrons, and a slightly larger number of neutrons at each step.

Although such a time-consuming exercise might help us see the regular progression from the atoms of one element to those of the next, there are easier ways of becoming informed about the various elements. All we really need to know to identify a particular element is its atomic number, mass number, and atomic weight. Taken together, these indicators provide a fairly complete profile of each element.

Atomic Number

The **atomic number** of any atom is equal to the number of protons in the nucleus and is written as a subscript to the left of its atomic symbol. Hydrogen, with one proton, has an atomic number of 1 ($_1H$); helium, with two protons, has an atomic number of 2 ($_2He$); and so on. Since the number of protons is always equal to the number of electrons in an atom, the atomic number indirectly tells us the number of electrons in the atom

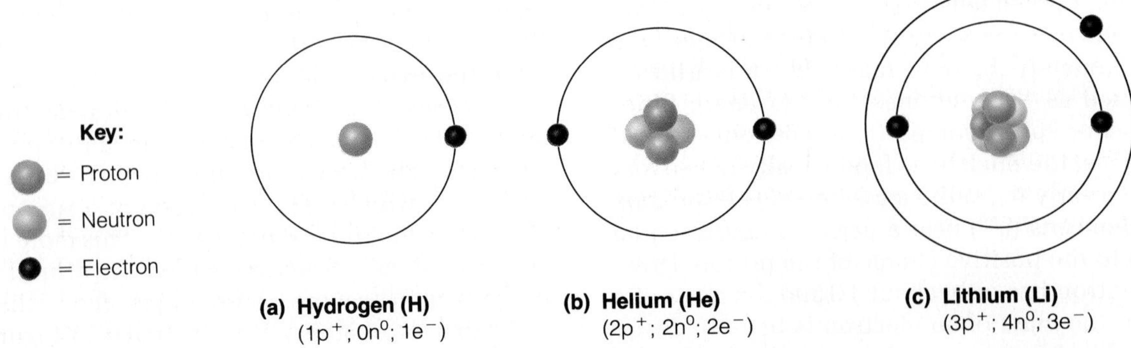

Figure 2.2 Atomic structure of the three smallest atoms.

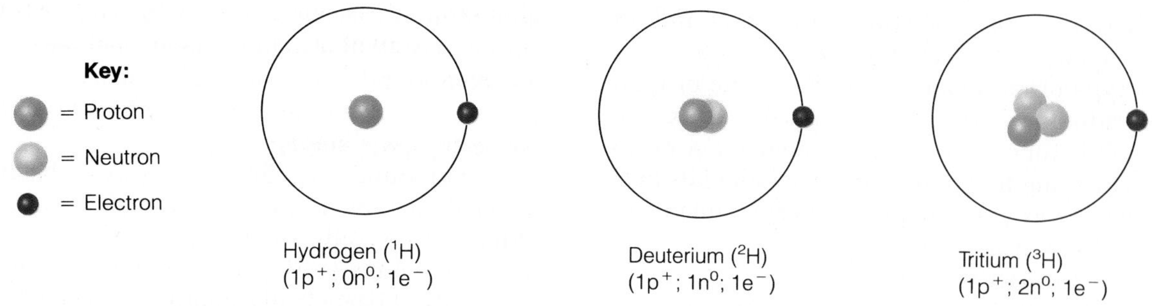

Figure 2.3 Isotopes of hydrogen.

as well. As will be explained shortly, this is important information indeed, because electrons determine the chemical behavior of atoms.

Mass Number

The **mass number** of an atom is the sum of its protons and neutrons. (The mass of the electrons is so small that it is ignored.) Hydrogen has only one proton in its nucleus, so its atomic and mass numbers are the same: 1. Helium, with two protons and two neutrons, has a mass number of 4. The mass number is usually indicated by a superscript to the left of the atomic symbol. Thus, helium can be described by the notation ^4_2He. This simple notation allows us to deduce the total number and kinds of subatomic particles in any atom because it indicates the number of protons (the atomic number), the number of electrons (the atomic number), and the number of neutrons (mass number minus atomic number).

From what we have said so far, it may appear as if each element has one, and only one, type of atom representing it. This is not really the case. Nearly all known elements have two or more atomic variants called **isotopes** (ī′-suh-tōps), which have the same atomic number, but vary in their mass numbers (and therefore their weights). Another way of saying this is that all the isotopes of an element have the same number of protons (and electrons), but vary in the number of neutrons they contain. Earlier, when we said that hydrogen has a mass number of 1, we were speaking of ^1H, its most abundant isotope; some hydrogen atoms have a mass of 2 or 3 amu, which means that they have one proton and, respectively, one or two neutrons (Figure 2.3). Similarly, carbon exists in three isotopic forms: ^{12}C, ^{13}C, and ^{14}C. Each of the carbon isotopes has six protons (otherwise it would not be carbon), but the first has six neutrons, the second has seven, and the third has eight.

Atomic Weight

It would seem that atomic weight should be the same as atomic mass, and this would be so if atomic weight referred to the weight of a single atom. However, **atomic weight** is an average of the relative weights (mass numbers) of all the isotopes of an element, taking into account their relative abundance in nature. As a rule, the atomic weight of an element is approximately equal to the mass number of its most abundant isotope. For example, the atomic weight of hydrogen is 1.008, which reveals that its lightest isotope (^1H) is present in much greater amounts than its ^2H or ^3H forms.

Radioisotopes

The heavier isotopes of many elements are unstable, and their atoms decompose spontaneously into more stable forms by emitting particles, or energy, from their nuclei. This process of atomic decay is called radioactivity, and isotopes that exhibit this behavior are called **radioisotopes** (rā″-dē-ō-ī′-suh-tōps). The why of this event is complex, and you only need to know that the dense nuclear particles are composed of even smaller particles called *quarks* that associate in one way to form protons and in another way to form neutrons. Apparently, the nuclear "glue" that holds these quarks together is less effective in the heavier isotopes.

The disintegration of a radioactive nucleus may be compared to a tiny explosion. In some cases, the atom is changed to an atom of a different element, a transformation called *transmutation* (tranz″-myū-tā′-shun). Types of radioactive emission include *alpha* (α) and *beta* (β) *particles* and *gamma* (γ) *ray emission*. The α emission has the lowest penetrating power and is least damaging to living tissue; γ emission has the greatest penetrating power.

In α particle emission, heavy α particles, each consisting of two protons and two neutrons, are ejected from the nucleus; radium-226 (^{226}Ra) decays by α particle emission. In β particle emission, β particles, essentially massless particles with either a positive or negative charge, are ejected from the nucleus; β particles are emitted by radioisotopes of carbon, hydrogen, and iodine. Isotopes that decay by γ emission emit electromagnetic energy (γ rays) that is massless and uncharged; γ rays are similar to X-rays and are extremely damaging to living tissue.

Each type of radioisotope produces one or more forms of radiation and tends to lose its radioactive behavior with time. The time required for a radioisotope to lose one-half of its activity is called its *half-life.* The half-lives of radioisotopes vary dramatically from hours to thousands of years.

Because radioactivity can be detected with scanners and other types of detectors, radioisotopes are valuable tools for biological research and medicine. Most radioisotopes used in the clinical setting are used for diagnosis, that is, to localize damaged or cancerous tissues. For example, iodine-131 is used to test the activity of the thyroid gland, to determine its size, and to detect thyroid cancer. The sophisticated PET scans described in the box on p. 32 use radioisotopes to probe the workings of molecules deep within our bodies. Because all types of radioactivity, regardless of the purpose for which they are used, damage living tissue, the radioisotopes selected for diagnostic procedures are those with the shortest half-lives. Radium-226, cobalt-60, and certain other radioisotopes are used to destroy localized cancers. For example, radium is sometimes used to destroy uterine cancer in women. ■

How Matter Is Combined: Molecules and Mixtures

Molecules and Compounds

Most atoms do not exist in the free state, but instead are chemically combined with other atoms. Such a combination of two or more atoms held together by chemical bonds is called a **molecule.** (The nature of the chemical bonds that unite atoms will be described shortly.)

If two or more atoms of the *same* element combine, the resulting substance is called a *molecule* of that element. When two hydrogen atoms bond, the product is a molecule of hydrogen gas and is written as H_2. Similarly, when two oxygen atoms combine, a molecule of oxygen gas (O_2) is formed. Not all molecules of elements are diatomic (two-atom) molecules; sulfur atoms, for instance, commonly combine to form molecules containing eight sulfur atoms (S_8).

When two or more *different* kinds of atoms bind together, they form molecules of a **compound.** Two hydrogen atoms can combine with one oxygen atom to form the compound water (H_2O), and four hydrogen atoms can combine with one carbon atom to form the compound methane (CH_4). At the risk of being repetitious, notice that molecules of methane and water are compounds, but molecules of hydrogen gas are not, since compounds always contain atoms of at least two different elements.

Compounds are chemically pure substances, and all of their molecules are identical. Thus, just as an atom is the smallest particle of an element that still exhibits the properties of the element, a molecule is the smallest particle of a compound that still displays the specific characteristics of the compound. This is an important concept because the properties of compounds are usually very different from those of the atoms they contain.

Mixtures

Mixtures are substances composed of two or more components physically intermixed together. Although most matter in nature exists in the form of mixtures, there are only three basic types: solutions, colloids, and suspensions. Mixtures differ from compounds in several ways:

1. None of the properties of atoms or molecules are changed when they become part of a mixture. The fact that no chemical bonding occurs between the components of a mixture is the chief difference between mixtures and compounds.

2. The substances making up a mixture can be separated by physical means—straining, filtering, evaporation, and so on, depending on the precise nature of the mixture. Compounds, by contrast, can be separated into their constituent atoms only by chemical means (breaking bonds).

3. Some mixtures are homogeneous, whereas others are heterogeneous. To say that a substance is *homogeneous* means that a sample taken from any part of the substance will have exactly the same composition (in terms of the atoms or molecules it contains) as any other sample. A bar of 100% pure (elemental) iron is homogeneous, as are all compounds. *Heterogeneous* substances vary in their makeup from place to place. For example, iron ore is a heterogeneous substance that contains iron as well as many other elements.

Solutions

Solutions are homogeneous mixtures of two or more components that may be gases, liquids, or solids. Examples include the air we breathe (a mixture of gases), salt water (a mixture of salt, which is a solid, and water), and rubbing alcohol (a mixture of two liquids, alcohol and water). The substance present in the greatest quantity is called the **solvent** (or dissolving medium); solvents are usually liquids or gases. Substances present in smaller amounts are called **solutes.**

Water is the body's chief solvent. Most solutions in the body are *true solutions* containing gases, liquids, or solids dissolved in water. True solutions are usually transparent. Salt water (saline solution) and a mixture of glucose and water are true solutions. The solutes of true solutions are very minute, usually in the form of individual atoms and molecules. Consequently, they are not visible to the naked eye, do not settle out, and do not scatter light. Thus, if a beam of light is passed through a true solution, you will not see the path of light.

Expressing the Concentration of Solutions. True solutions are often described in terms of their *concentration,* which may be specified in various ways. Frequently, solutions used in a college laboratory or in a hospital are described in terms of the *percent* (parts per 100 parts) of the solute in the solution. The percentage designation always refers to the solute percentage, and unless otherwise noted, water is assumed to be the solvent.

Another way of expressing the concentration of a solution is in terms of its **molarity** (mō-layr'-ih-tē), or moles per liter. This method is more complicated, but is much more useful. To understand molarity, you must first understand what a mole is. A **mole** of any element or compound is the amount of that substance, weighed out in grams, that is equal to its atomic weight or *molecular weight* (sum of the atomic weights). This is easier than it seems. An example will illustrate this concept.

Glucose is $C_6H_{12}O_6$, which indicates that it has 6 carbon atoms, 12 hydrogen atoms, and 6 oxygen atoms. To compute the molecular weight of glucose, you would look up the atomic weight of each of its atoms and compute its molecular weight as follows:

Atom	Number of atoms		Atomic weight		Total atomic weight
C	6	×	12.011	=	72.066
H	12	×	1.008	=	12.096
O	6	×	15.999	=	95.994
					180.156

Then, to make a one-*molar* solution of glucose, you would weigh out 180.156 grams (g), one gram molecular weight, of glucose and add enough water to make 1 liter (L) of solution. Thus, a one-molar solution (abbreviated 1.0 *M*) of a chemical substance is one gram molecular weight of the substance (or one gram atomic weight in the case of elemental substances) in 1 L (1000 ml) of solution.

The beauty of using the mole as the basis of preparing solutions is the precision of this measurement. One mole of any substance always contains exactly the same number of solute particles, that is, 6.02×10^{23}.* This number is called *Avogadro's* (ah"-vuh-gah'-drōz) *number.* So whether we weigh out 1 mole of glucose (180 g) or 1 mole of water (18 g) or 1 mole of methane (16 g), in each case we will have 6.02×10^{23} molecules of that substance. This allows almost mind-boggling precision to be achieved.

Colloids

Colloids (kah'-loydz) are heterogeneous mixtures that often appear translucent or milky. Although the solute particles are larger than those in true solutions, they still do not settle out. However, they do scatter light, and the path of a light beam shining through a colloidal mixture is visible.

Colloids have many unique properties, including the ability of some to undergo *sol-gel transformations,* that is, to change reversibly from a fluid (sol) state to a more solid (gel) state. Jell-O is a familiar example of a nonliving colloid that changes from a sol to a gel when refrigerated (and which will liquefy again if placed in the sun). The semifluid material within living cells is also considered a colloid, and its sol-gel changes underlie many important cell functions, such as cell division.

Suspensions

Suspensions are heterogeneous mixtures that have large, often visible solutes. Perhaps the best example of a suspension is a mixture of sand in water. In the body, a good example is blood, in which the living blood corpuscles are suspended in the fluid portion of blood, or blood plasma. If left to stand, the suspended particles will settle out unless some means—mixing, shaking, or, in the case of blood, circulation—is used to keep them in suspension.

As you can see, all three types of mixtures are prominent in both living and nonliving systems. In fact, living material is the most complex mixture of all, since it contains all three kinds of mixtures superimposed upon one another in the same area.

*The important exception to this rule concerns molecules that ionize and break up into charged particles (ions) in water, such as salts, acids, and bases (see p. 42). For example, simple table salt (sodium chloride) breaks up into two types of charged particles; therefore, in a 1.0 *M* solution of sodium chloride, there are actually *2 moles* of solute particles in solution.

A CLOSER LOOK Medical Imaging: Illuminating the Body

New scanning techniques that make pictures of our insides have reached a level of sophistication that is difficult to comprehend. By bombarding the body with energy, these imaging techniques not only reveal the structure of our internal organs, but also wring out information about the private and, until now, secret workings of their molecules. These new breeds of imaging techniques are changing the face of medical diagnosis.

Until about 30 years ago, the magical but murky X-ray was the only means of extracting information from within a living body. What X-rays did and still do best was visualize hard, bony structures and peer at the chest to locate abnormally dense structures (tumors, TB nodules) in the lungs. The 1950s saw the advent of nuclear medicine, which uses radioisotopes to scan the body, and ultrasound techniques. In the 1970s, CAT, PET, and MRI scanning techniques were introduced.

The best known of the new imaging devices is *computerized axial tomography (CAT)*, a refined version of X-ray equipment. Instead of compressing a three-dimensional structure into a flat, two-dimensional image, a CAT scanner confines its beam to a thin slice of the body and ends the confusion resulting from images of overlapping structures seen in conventional X-rays. CAT's clarity has all but eliminated exploratory surgery. As the patient is slowly moved through the doughnut-shaped CAT machine, its X-ray tube rotates around the body and sends beams from all directions to a specific level of the patient's body. Different tissues absorb the radiation in varying amounts, depending on their relative density. The device's computer translates this information into a detailed, cross-sectional picture of the body region scanned. CAT scans are at the forefront for evaluating most problems that affect the brain, as well as abdominal problems and certain skeletal problems, particularly those of the spine.

Just as the X-ray spawned a "big brother," so too did nuclear medicine in the guise of *positron-emission tomography (PET)*. The special advantage of PET is that its images send messages about *metabolic processes*. The introduction of PET was exciting news because a central premise of medicine is that every pathology reflects some underlying biochemical defect. Presently, PET is most used as a research tool, but its greatest value has been its ability to provide insights into brain activity in those affected by mental illness, Alzheimer's disease, and epilepsy. In preparation for a PET scan of the brain, the patient is given an injection of exceptionally short-lived radioisotopes that have been tagged to biological molecules (such as glucose) and then positioned in the PET scanner. As the radioisotopes are absorbed by metabolically active brain cells, high-energy γ rays are produced. The computer analyzes the γ ray emission and produces a picture of the brain's biochemical activity in vivid blues and yellows. Despite the excitement generated by PET, it seems destined to very restricted use because of the almost prohibitive cost of the equipment.

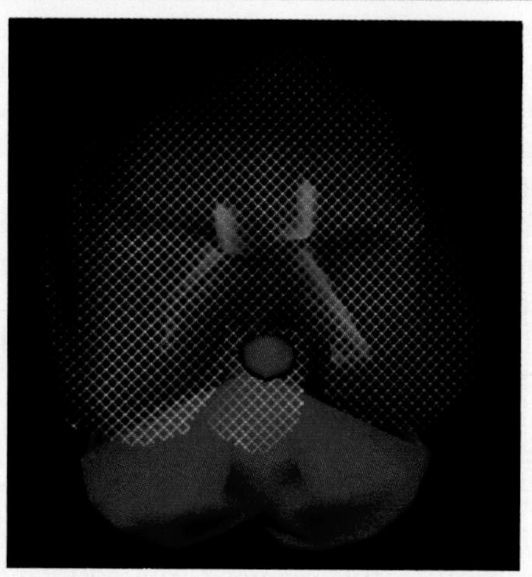

(a) CAT scan of the brain. The red spherical object is a pituitary tumor.

Sonography, or *ultrasound imaging*, has distinct advantages over the approaches examined so far. First, the equipment is inexpensive, and second, it employs very high-frequency sound waves (ultrasound) as its energy source. Ultrasound, unlike ionizing forms of radiation, has no adverse effects on living tissues (as far as we know). The body's tissues are probed with pulses of sound waves (emitted by an electrically excited crystal). The sound waves, reflected and scattered to different extents by body tissues, cause echoes that are analyzed to construct visual images of the outlines of body organs of interest. Basically, this method uses sound echos to detect objects in the same manner as do bats and the sonar devices used by industry and the military forces. Because of its safety, ultrasound is the imaging technique of choice for obstetrics (i.e., for determining fetal age and position and locating the placenta). Since sound waves have very low penetrating power and are rapidly dissipated in air, sonography is of little value for looking at air-filled structures (the lungs) or those protected by bone (the brain and spinal cord).

Another technique that depends on nonionizing radiation is *magnetic resonance imaging (MRI)*, a technique that uses magnetic fields 3,000 to 40,000 times stronger than the earth's to pry information from the body's molecules. The patient is inserted into a chamber within a huge magnet. Hydrogen molecules spin like tops in the magnetic field, and their energy is further enhanced by radio

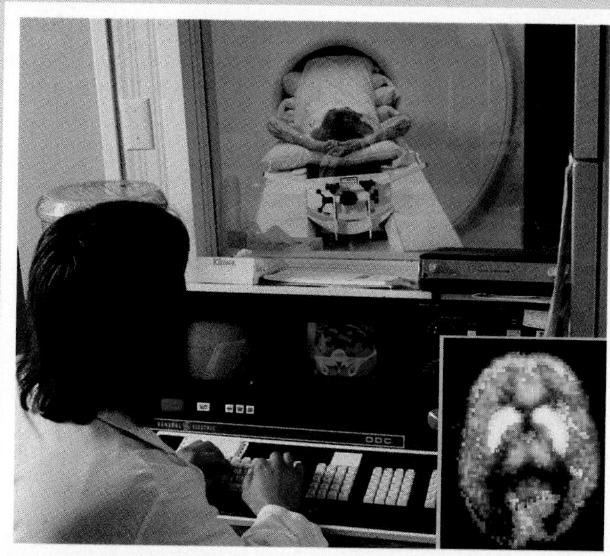

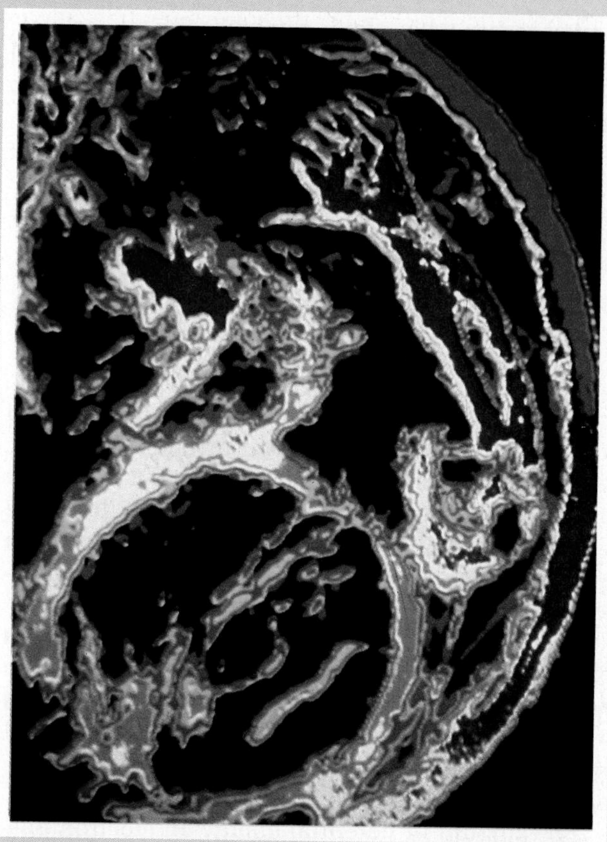

(b) A patient is shown being moved into a PET scanner. The inset shows a PET scan of the brain. (c) Ultrasound image of a fetus in utero; the head and arm are clearly visible.

waves. When the radio waves are turned off, the energy is released and translated by the computer into a visual image. MRI is immensely popular because of its ability to do many things a CAT scan cannot. Since it focuses on the behavior of hydrogen atoms of water molecules, it is much better at visualizing soft tissues of the body. For example, it can differentiate between the fatty white matter and the more watery gray matter of the brain (see Figure 1.7) and enable physicians to see the delicate nerve fibers in the spinal cord. Since dense structures do not show up at all in MRI, the bones of the skull and/or vertebral column do not impair the view. MRI is also particularly good at detecting degenerative disease of various kinds; multiple sclerosis plaques do not show up well in CAT scans, but are dazzlingly clear in MRI scans. MRI can also be used to tune in on metabolic reactions, such as processes that generate energy-rich ATP molecules. Despite its advantages, the powerful magnetic field of MRI presents some thorny problems. For example, it can "suck" metal objects, such as implanted pacemakers, loose tooth fillings, and embedded shrapnel, from the body, and in a lighter vein, erase the magnetic tape on your charge cards. MRI is extremely costly and requires special quarters to insulate it from all external radio waves. But perhaps its biggest impediment is that there is no convincing evidence that such huge, variable magnetic fields can be used without risk to the body, and studies that indicate that harmful changes may occur at the molecular level are beginning to trickle in.

As you can see, modern medical science has many remarkable new diagnostic tools at its disposal, and there is little question that they have made their mark. CAT scans, along with nuclear medicine, now account for about 25% of imaging. Ultrasonography, because of its safety and low cost, has become the most widespread of the new techniques. Even so, let us not forget conventional X-rays, because they are still the workhorse of diagnostic imaging techniques and account for better than half of all imaging currently done.

Chemical Bonds

As noted earlier, when atoms combine with other atoms, they are held together by **chemical bonds.** A chemical bond is not a physical structure like a pair of handcuffs linking two people together, but is instead an energy relationship that occurs between electrons of the reacting atoms.

The Role of Electrons in Chemical Bonding

Electrons forming the electron cloud around the nucleus of an atom occupy regions of space called **electron shells** that consecutively surround the atomic nucleus. The maximum number of electron shells any atom may have is seven (numbered 1 to 7 from the nucleus outward), but the actual number depends on the number of electrons the atom possesses. Each electron shell contains one or more orbitals.

It is important to understand that each electron shell represents a different **energy level,** because this prompts you to think of electrons as particles with a certain amount of potential energy. In general, the terms *electron shell* and *energy level* are used interchangeably.

The amount of potential energy an electron possesses depends first and foremost on the energy level it occupies because the attractive force between the positively charged nucleus and negatively charged electrons is greatest close to the nucleus and falls off with increasing distance. There is a good deal of information in this statement, because if it is really understood, it explains why electrons farthest from the nucleus (1) have the greatest potential energy—it takes more energy to overcome the nuclear attraction and reach the more distant energy levels, and (2) are most likely to become involved in chemical interactions with other atoms—they are least tightly held by their own atomic nucleus and most easily influenced by external forces (other atoms and molecules).

Each electron shell has a maximum number of electrons that it can hold. Shell 1, the first shell immediately surrounding the nucleus, can accommodate only 2 electrons; shell 2 holds a maximum of 8; shell 3 can accommodate 18 electrons. Subsequent shells hold larger and larger numbers of electrons. The shells tend to be filled with electrons consecutively. For example, shell 1 is completely filled before any electrons appear in shell 2.

In the bonding of atoms, the only important electrons are those found in the atom's outermost energy level. Inner electrons usually do not take part in bonding because they are more tightly held by the atomic nucleus. When the outermost energy level of an atom either is filled to capacity or contains eight electrons, a stable state is reached. Such atoms are chemically *inert* (ih-nert'); that is, they shun interaction with other atoms and are unreactive. A group of elements called the *inert gases,* which include helium and neon, have their outermost energy shells filled to capacity (Figure 2.4a). However, atoms in which the outermost energy

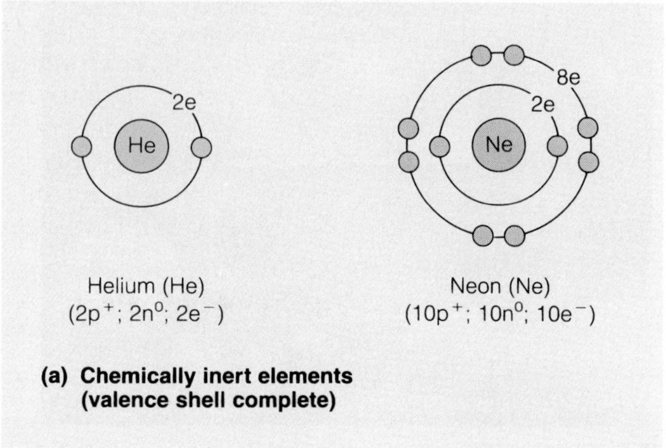

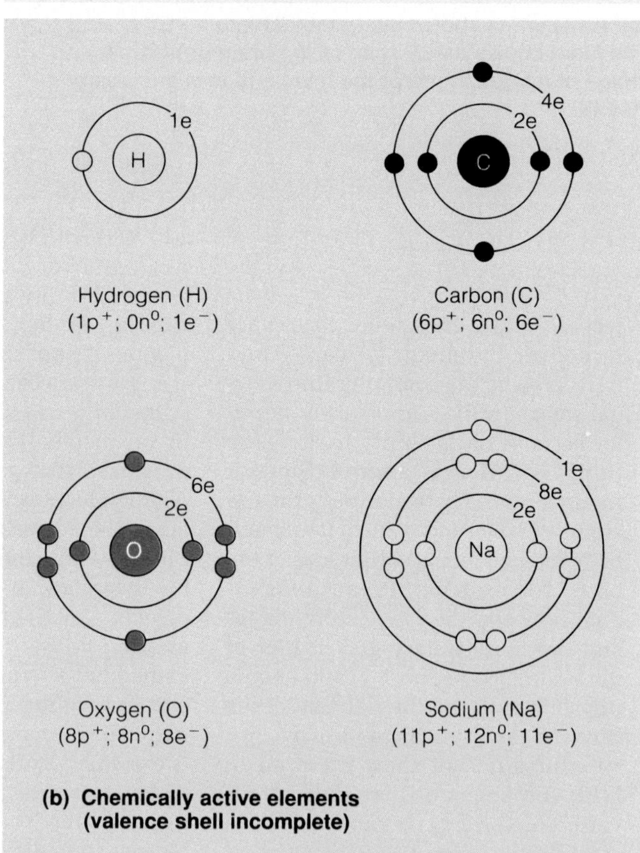

Figure 2.4 Chemically inert and reactive elements. (a) Helium and neon are chemically inert because in each case the outermost energy level (valence shell) is fully occupied by electrons. (**b**) Elements in which the valence shell is incomplete are chemically reactive. Such atoms tend to interact with other atoms to gain, lose, or share electrons to fill their valence shells. (*Note:* To simplify the diagrams, each atomic nucleus is shown as a circle with the atom's symbol in it; individual protons and neutrons are not shown.)

level is not filled, or which contain fewer or more than eight electrons in that energy level (Figure 2.4b), tend to gain, lose, or share electrons with other atoms to achieve the stable configuration. A possible point of confusion must be addressed here. Regardless of their size and regardless of the number of electrons that can occupy some of the higher energy levels, all atoms are restricted to a total of eight electrons in their outermost shell. It is only these electrons that take part in bonding behavior. For this reason, the term *valence* (vā′-lents) *shell* is used to specifically indicate an atom's outermost energy level containing electrons that are chemically reactive. Hence, the key to chemical reactivity is the **octet** (ok-tet′) **rule**, or *rule of eights*. Except for shell 1, which is filled with two electrons, atoms tend to interact in such a way that they will have eight electrons in their valence shell.

Types of Chemical Bonds

Three major types of chemical bonds—ionic, covalent, and hydrogen bonds—result from the attractive forces between atoms.

Ionic Bonds

Atoms are electrically neutral, but when electrons are completely transferred from one atom to another, the precise balance of + and − charges is lost, and charged particles called **ions** are formed. The result is an **ionic** (ī-ah′-nik) **bond.** The atom that gains one or more electrons acquires a net negative charge and is called an *anion* (an′-ī″-un); the atom that loses electrons acquires a net positive charge and is called a *cation* (kat′-ī″-un). (It might help to think of the "t" in "cation" as a plus (+) sign.) Both anions and cations are formed whenever electron transfer between atoms occurs. Since opposite charges attract, these ions tend to stay close together.

Perhaps the best example of ionic bonding is the formation of sodium chloride (NaCl), or common table salt, by the interaction of sodium and chlorine atoms (Figure 2.5). Sodium, atomic number 11, has one electron in its valence shell, and it would be very difficult to attempt to fill this shell by adding seven more. However, if this single electron is lost, shell 2 with eight electrons becomes the outermost energy level containing electrons and is full. Thus, by losing the lone electron in its third energy level, sodium achieves stability and becomes a cation (Na^+). On the other hand, chlorine, atomic number 17, needs only one electron to fill its valence shell; thus, by accepting an electron, chlorine becomes an anion and achieves the stable state. The ideal situation is for sodium to donate an electron to chlorine, and that is exactly what happens in the interaction between these two atoms. The ions created in this exchange attract each other, forming sodium chloride. Ionic bonds are commonly formed between atoms with one or two valence shell electrons (the metallic elements, such as sodium, calcium, and potassium) and atoms with seven valence shell electrons (including chlorine, fluorine, and iodine). Most ionic compounds fall in the chemical category called *salts*.

In the absence of water, ionic compounds such as sodium chloride do not exist as individual molecules. Instead, they exist in the form of *crystals*, large arrays of cations and anions held together by ionic bonds (see Figure 2.11).

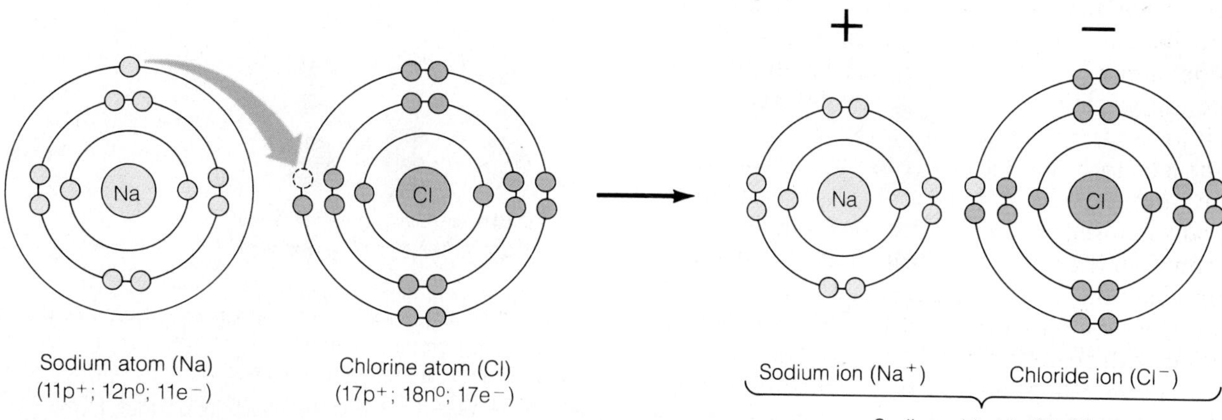

Figure 2.5 Formation of an ionic bond. Both sodium and chlorine atoms are chemically reactive because their valence shells are incompletely filled. Sodium gains stability by losing one electron, whereas chlorine becomes stable by gaining one electron. After electron transfer, sodium becomes a sodium ion (Na^+), and chlorine becomes a chloride ion (Cl^-). The oppositely charged ions attract each other.

Sodium chloride is an excellent example to illustrate the difference in properties between a compound and its constituent atoms. Sodium is a silvery white metal, and chlorine in its molecular state is a poisonous green gas (and the basis of bleach). However, sodium chloride is a white crystalline solid that we sprinkle on our food.

Covalent Bonds

Electrons do not have to be completely transferred (lost or gained) for atoms to achieve stability. Instead, they may be *shared* so that each atom is able to fill its outer electron shell for at least part of the time. Electron sharing produces molecules in which the shared electrons constitute **covalent** (kō-vā′-lent) **bonds.**

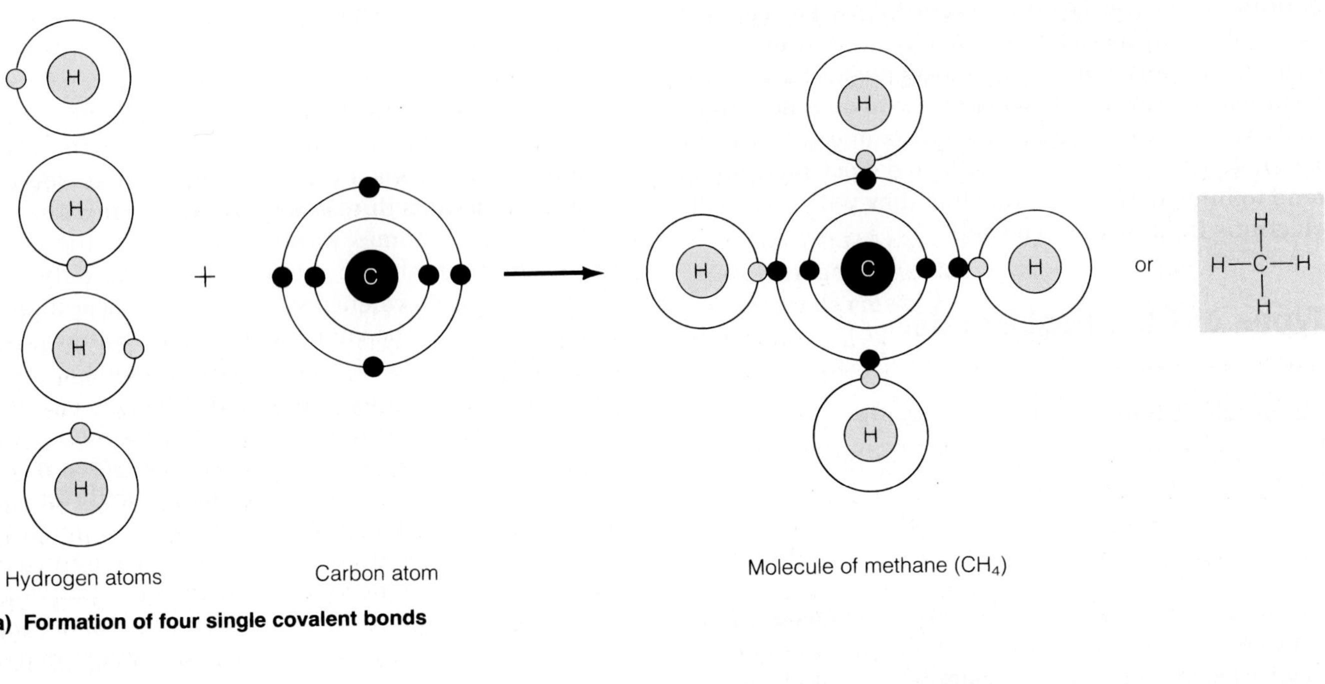

(a) Formation of four single covalent bonds

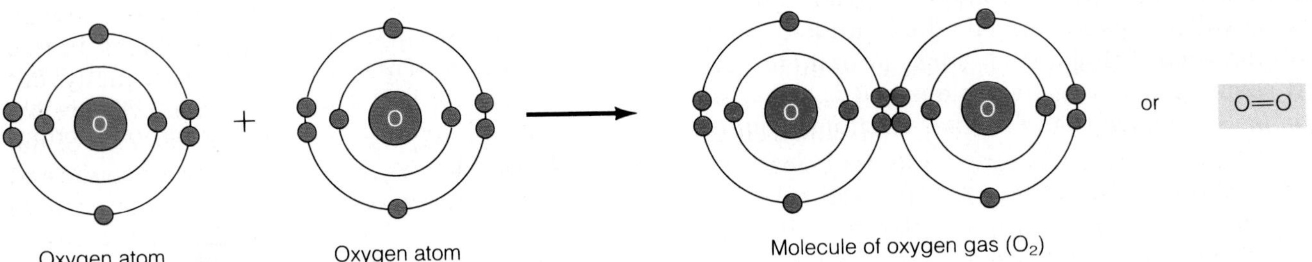

(b) Formation of a double covalent bond

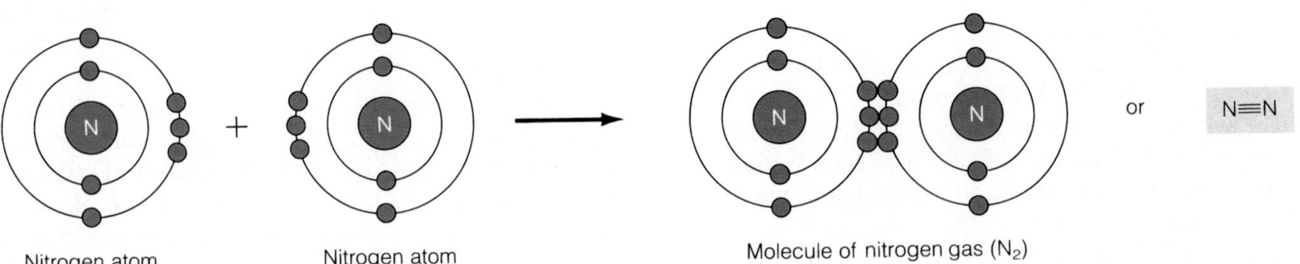

(c) Formation of a triple covalent bond

Figure 2.6 Formation of covalent bonds. In each example, the electrons of interacting atoms are shown in different colors to make the electron-sharing pattern easier to follow. In the shaded boxes to the right, each covalent bond is indicated by a single line between the electron-pair-sharing atoms. (**a**) Formation of a molecule of methane; carbon shares four electron pairs with four hydrogen atoms. (**b**) Formation of a molecule of oxygen gas; each oxygen atom shares two electron pairs with its partner, forming a double covalent bond. (**c**) When a molecule of nitrogen gas is formed, the two nitrogen atoms share three electron pairs, forming a triple covalent bond.

Hydrogen with its single electron can fill its only shell (shell 1) by sharing a pair of electrons with another atom. When it shares with another hydrogen, a molecule of hydrogen gas is formed. The shared electron pair orbits around each atom (and the molecule as a whole), satisfying the stability needs of each. Hydrogen can also share an electron pair with different kinds of atoms to form a compound (Figure 2.6a). Carbon has four electrons in its outermost shell, but needs eight to achieve stability, whereas hydrogen has one electron, but needs two. When a methane molecule (CH_4) is formed, carbon shares four pairs of electrons with four hydrogen atoms (one pair with each hydrogen). Again, the shared electrons orbit and "belong to" the whole molecule, ensuring the stability of each atom.

One pair of shared electrons forms a *single covalent bond* (indicated by a single line connecting the atoms, such as H–H). But in some cases, atoms share two or three electron pairs (Figure 2.6b and c), resulting in *double* or *triple covalent bonds* (indicated by double or triple lines such as O=O or N≡N).

Polar Molecules. In the covalent bonds discussed thus far, the electrons have been shared equally between the atoms of the molecule. The molecules formed are electrically balanced and are more specifically called *nonpolar molecules*. This is not always the case. When two or more atoms form covalent bonds, the resulting molecule always has a definite three-dimensional shape, with bonds formed at definite angles. A molecule's shape plays a major role in determining what other molecules or atoms it can interact with; it may also result in unequal electron pair sharing and *polar molecules*. This is especially true in nonsymmetrical molecules that contain atoms with different electron-attracting abilities. The ability of atoms to attract electrons is related to the number of valence shell electrons and to atomic size. In general, *small* atoms with six or seven valence shell electrons, such as oxygen, nitrogen, and chlorine, attract electrons very strongly. This capability of electron-hungry atoms is called *electronegativity*. Conversely, most atoms with only one or two valence shell electrons tend to be *electropositive*; that is, their electron-attracting ability is so low that they usually lose their valence shell electrons to other atoms. Potassium and sodium, each with one valence shell electron, are good examples of electropositive atoms.

Carbon dioxide and water illustrate the importance of molecular shape and the relative electron-attracting abilities of atoms in determining whether a covalently bonded molecule is nonpolar or polar in nature. Carbon dioxide (CO_2) is formed when carbon shares four electron pairs with two oxygen atoms (two pairs are shared with each oxygen). Oxygen is highly electronegative and attracts the shared electrons much more strongly than does carbon. However, since the carbon dioxide molecule is linear (Figure 2.7a), the electron-pulling ability of one oxygen atom is offset by that of the other, like a tug-of-war at a standoff. As a result, the electrons are shared equally by all the atoms, and carbon dioxide is a nonpolar compound.

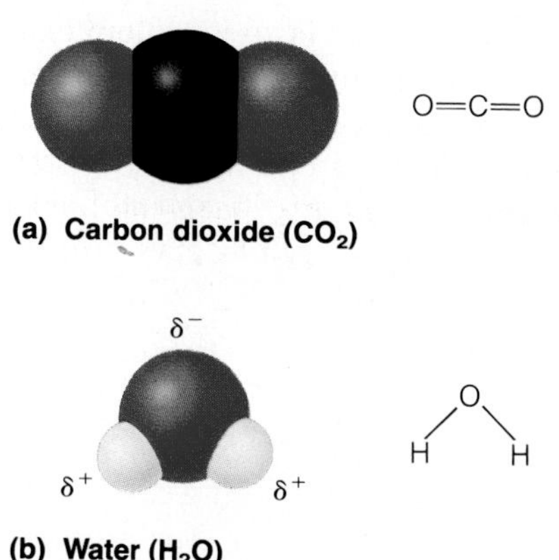

Figure 2.7 Molecular models illustrating the three-dimensional structure of carbon dioxide and water molecules.

On the other hand, a water molecule (H_2O) is V-shaped (Figure 2.7b). The two hydrogen atoms are located at one end of the molecule, and the oxygen atom is at the opposite end. This arrangement allows the oxygen atom to pull the shared electrons toward itself and away from the two hydrogen atoms. In this case, the electron pairs are *not* shared equally, but spend more time in the vicinity of oxygen. Since electrons are negatively charged, the oxygen end of the molecule becomes slightly more negative (indicated by δ^-) and the hydrogen end slightly more positive (indicated by δ^+). Because water has two poles of charge, it is a *polar molecule*, or *dipole* (dī′-pōl). Polar molecules can orient toward other dipoles or toward charged particles (such as ions and proteins), and they play an essential role in the chemical reactions that go on in body cells. The polarity of water is particularly significant, as you will see later in this chapter.

Different types of molecules exhibit different degrees of polarity, and we can see a gradual change from ionic to nonpolar covalent bonding as summarized in Figure 2.8. Ionic bonds (complete electron transfer) and nonpolar covalent bonds (equal electron sharing) are just the extremes of a continuum, with various degrees of unequal sharing in between.

Hydrogen Bonds

Hydrogen bonds are weak bonds that form when a hydrogen atom, already covalently linked to an electronegative atom (usually nitrogen or oxygen), is

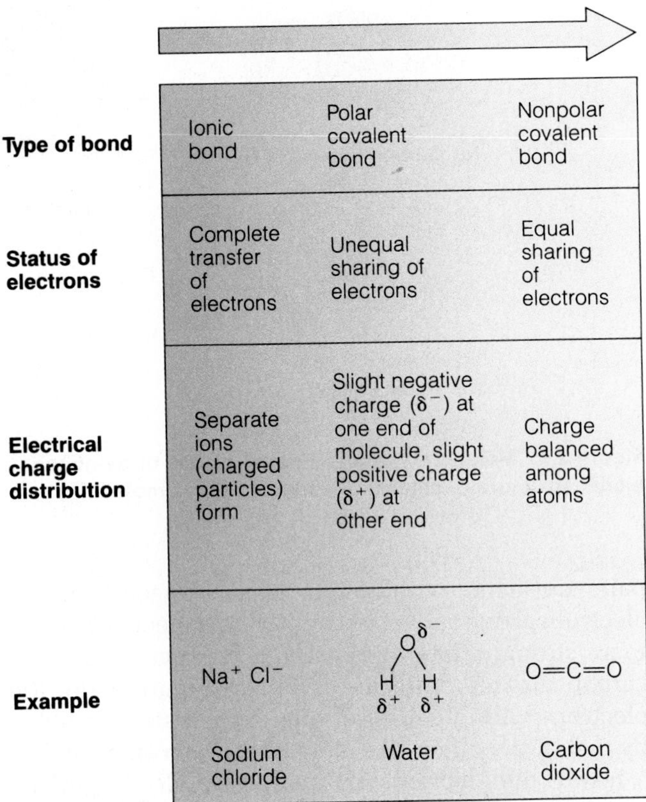

Type of bond
Ionic bond
Polar covalent bond
Nonpolar covalent bond
Status of electrons
Complete transfer of electrons
Unequal sharing of electrons
Equal sharing of electrons
Electrical charge distribution
Separate ions (charged particles) form
Slight negative charge (δ⁻) at one end of molecule, slight positive charge (δ⁺) at other end
Charge balanced among atoms
Example
Na⁺ Cl⁻
Sodium chloride
Water
O=C=O
Carbon dioxide

Figure 2.8 Comparison of ionic, polar covalent, and nonpolar covalent bonds. This display compares the status of electrons involved in bonding and the electrical charge distribution in the molecules formed.

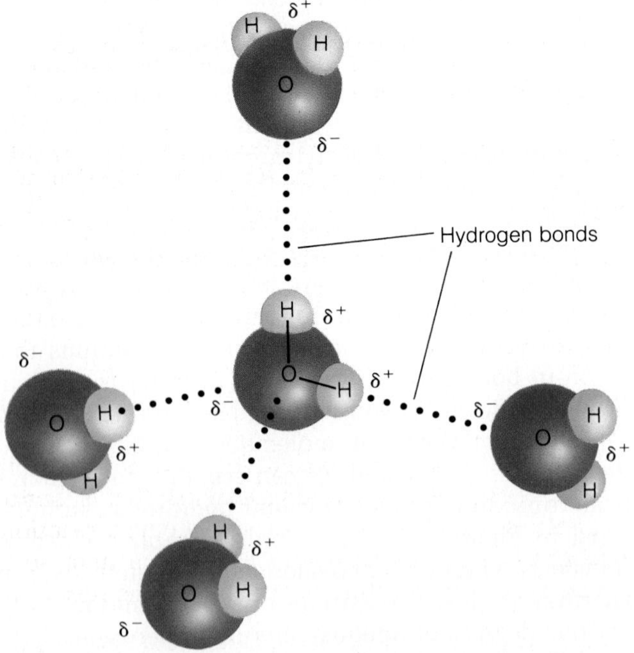

Figure 2.9 Hydrogen bonding between polar water molecules. The slightly positive ends (indicated by δ^+) of the water molecules become aligned with the slightly negative ends (indicated by δ^-) of other water molecules.

attracted by another electronegative atom. Hydrogen bonding is common between dipoles such as water molecules, because the slightly positive hydrogens of one molecule are attracted to the slightly negative oxygen atoms of other molecules (Figure 2.9). The tendency of water molecules to cling together and form films, referred to as *surface tension*, helps explain why water beads up into spheres when it sits on a hard surface.

Hydrogen bonds can also be important *intramolecular bonds*, binding different parts of a single large molecule together into a specific three-dimensional shape. Biological molecules containing large numbers of atoms, such as proteins and DNA, have numerous hydrogen bonds that help maintain and stabilize their structures.

Chemical Reactions

As noted earlier, all particles of matter are in constant motion because of their kinetic energy. The movement of atoms or molecules in a solid is usually limited to vibratory movements because the particles are united by fairly rigid bonds. But in a liquid or gas, the particles dart about randomly, sometimes colliding with one another and interacting to undergo chemical reactions. A **chemical reaction** occurs whenever chemical bonds are formed, rearranged, or broken. Since all chemical bonds represent stored chemical energy, all chemical reactions involve the absorption or release of energy.

Chemical Equations

Earlier, we discussed the joining of two hydrogen atoms to form hydrogen gas and the combining of four atoms of hydrogen and one atom of carbon to form methane. These reactions can be written in symbolic form as **chemical equations**:

$$H + H \rightarrow H_2 \text{ (hydrogen gas)}$$

and

$$4H + C \rightarrow CH_4 \text{ (methane)}$$

Notice that a number written as a subscript indicates that the atoms are joined by chemical bonds. But a number written as a prefix denotes the number of unjoined atoms or molecules. Hence, CH_4 reveals that four hydrogen atoms are bonded together with carbon to form the methane molecule, but 4H signifies four unjoined hydrogen atoms.

A chemical equation is like a sentence describing what happens in a reaction. It contains the following information: the number and kinds of reacting substances, or *reactants;* the chemical composition of the *product(s);* and, in balanced equations, the relative proportion of each reactant and product. In the preceding examples, reactants are atoms, as indicated by their atomic symbols (H, C). The product in each case is a molecule, as indicated by its *molecular formula* (H_2, CH_4). The equation for the formation of methane may be read *as either* "four atoms of hydrogen plus one atom of carbon yield one molecule of methane" or "four moles of hydrogen atoms plus one mole of carbon yields one mole of methane." Generally speaking, interpretation as moles is more practical because it is impossible to measure out one atom or one molecule of anything!

Patterns of Chemical Reactions

Most chemical reactions exhibit one of three patterns: they are either synthesis, decomposition, or exchange reactions.

When atoms or molecules combine to form a larger, more complex molecule, the process is called a **synthesis reaction.** A synthesis reaction, which always involves bond formation, can be represented as

$$A + B \rightarrow AB$$

Energy is required to make the bonds of most large molecules found in the body; therefore, synthesis reactions are energy-absorbing reactions. They are the basis of constructive *(anabolic)* activities in body cells, such as the joining of small molecules called amino acids to form large protein molecules (Figure 2.10a). Synthesis reactions are especially prominent in rapidly growing tissues.

A **decomposition reaction** occurs when a molecule is broken down into smaller molecules or its constituent atoms:

$$AB \rightarrow A + B$$

Essentially, decomposition reactions are reverse synthesis reactions; bonds are broken and energy is released. Decomposition reactions underlie all degradative *(catabolic)* processes that occur in body cells. For example, the bonds of large glycogen molecules are broken to release simpler molecules of glucose sugar (Figure 2.10b). The breakdown of food by the organs of our digestive system also involves decomposition reactions.

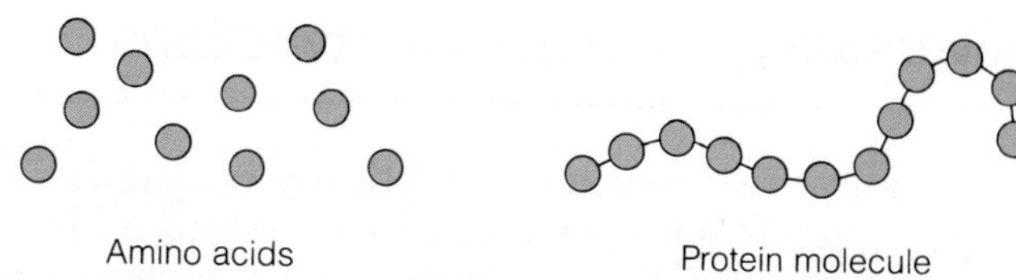

(a) Example of a synthesis reaction: amino acids are joined to form a protein molecule

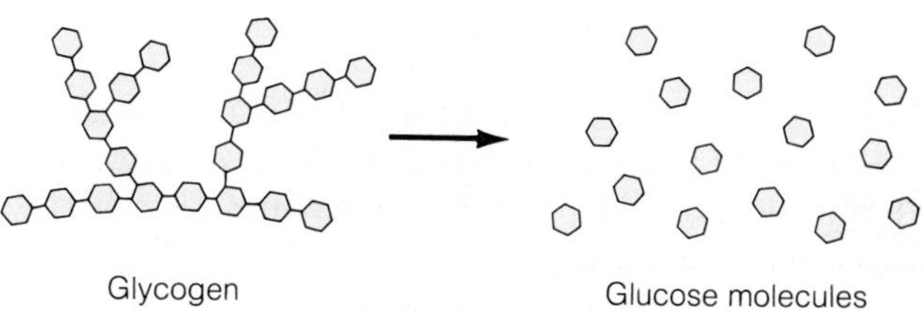

(b) Example of a decomposition reaction: breakdown of glycogen to release glucose units

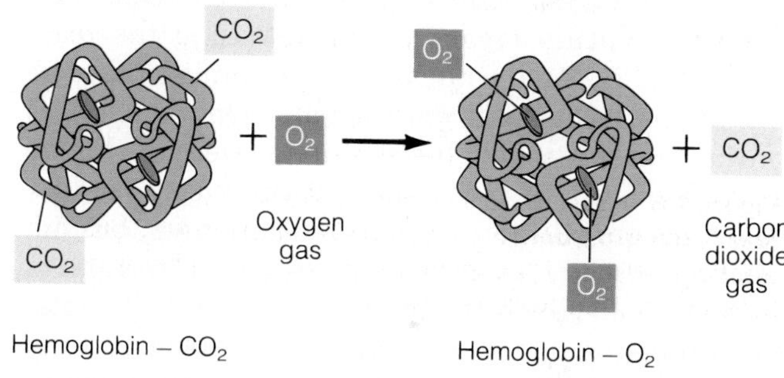

(c) Example of an exchange reaction: hemoglobin unloads carbon dioxide gas (CO_2) and picks up oxygen gas (O_2) as blood flows through the lungs

Figure 2.10 Patterns of chemical reactions. (**a**) In synthesis reactions, smaller particles (atoms, ions, or molecules) are bonded together to form larger, more complex molecules. (**b**) Decomposition reactions involve bond breaking. (**c**) In exchange, or displacement, reactions, bonds are both made and broken.

Exchange, or **displacement, reactions** involve both synthesis and decomposition reactions; that is, bonds are both made and broken. In an exchange reaction, portions of reactant molecules change partners, so to speak, resulting in different product molecules:

$$AB + C \rightarrow AC + B \quad \text{and} \quad AB + CD \rightarrow AD + CB$$

An exchange reaction occurs when hemoglobin, the iron-containing pigment in our red blood cells, unloads carbon dioxide and picks up oxygen in the lungs (Figure 2.10c).

Reversibility of Chemical Reactions

All chemical reactions are theoretically reversible. If chemical bonds can be made, they can be broken, and vice versa. Reversibility is indicated by a double arrow. When the arrows differ in length, the longer arrow indicates the direction in which the reaction proceeds more rapidly:

$$A + B \rightleftharpoons AB$$

This example indicates that the product (AB) is made more rapidly than it is broken down. Over time, the product accumulates and the reactants (A and B), decrease in amount.

When the arrows are of equal length, as in

$$A + B \rightleftharpoons AB$$

the reaction goes in both directions at the same rate; that is, for each molecule of product (AB) formed, one product molecule breaks down, releasing the reactants A and B. Such a chemical reaction is said to be in a state of **chemical equilibrium.** Once chemical equilibrium is reached, there is no further *net change* in the amounts of reactants and products. Product molecules are still formed and broken down, but the situation already established when equilibrium is reached (such as greater numbers of product molecules) remains unchanged. This is analogous to the admission scheme used by many large museums in which tickets are sold according to time of entry. If 300 tickets are issued for the 9 A.M. admission, 300 people will be admitted when the doors open. Thereafter, when 6 people leave, 6 are admitted, and when another 15 people leave, 15 more are allowed in. Although there is a continual turnover, the museum always contains about 300 art lovers throughout the day.

Although all chemical reactions are reversible in theory, many show so little tendency to go in the reverse direction that they are irreversible for all practical purposes. This is true of many biological reactions. Chemical reactions that release energy when going in one direction will not go in the opposite direction unless energy is absorbed or put back into the system. For example, our cells break down glucose via a series of energy-releasing reactions that yield carbon dioxide and water, and some of the energy released is trapped in the bonds of ATP. Since the cells need the energy of ATP to power various cell functions (and more glucose will be along with the next meal), this particular reaction is never reversed in our cells. Furthermore, if a product of a reaction is continuously removed from the reaction site, it is unavailable to take part in the reverse reaction. The carbon dioxide that is released as glucose is degraded, leaves the cell, enters the blood, and is eventually removed from the body by the lungs.

Factors Influencing the Rate of Chemical Reactions

For atoms and molecules to react chemically, they must *collide* with enough force to overcome the repulsion between their electrons so that interactions between valence shell electrons become possible. Changes in electronic configuration—the basis of bond making and breaking—cannot occur long distance, and the force of collision depends on the velocity of particle movement. Solid, forceful collisions between rapidly moving particles with high kinetic energy are much more likely to result in interactions than are those in which the particles merely graze each other or touch lightly.

Particle Size

Smaller particles move faster than larger ones (at the same temperature) and therefore tend to collide both more frequently and more forcefully. Hence, the smaller the reacting particles, the faster the rate of a chemical reaction at a given temperature and concentration.

Temperature

Because increasing the temperature of a substance increases the kinetic energy of its particles and the force of their collisions, chemical reactions proceed more quickly at higher temperatures. When the temperature drops, the speed of particle movement declines, and reactions occur more and more slowly.

Concentration

Chemical reactions proceed most rapidly when the reacting particles are present in high concentration, since the larger the number of randomly moving particles in a given space, the greater the likelihood of successful collisions. As the concentration of the reactants declines, chemical equilibrium is eventually reached unless additional reactants are added or products are removed from the reaction site.

Catalysts

Whereas many chemical reactions that occur in nonliving systems can be speeded up simply by adding heat, drastic increases in body temperature are life threatening because important biological molecules are destroyed. Still, at normal body temperatures, most chemical reactions would proceed very slowly—far too slowly to maintain life—were it not for the presence of catalysts. **Catalysts** (ka′-tuh-lists) are substances that increase the rate of chemical reactions without themselves becoming chemically changed or part of the product. Protein molecules called **enzymes** (en′-zīmz) are biological catalysts; the presence of enzymes is the single most important factor determining the rate of chemical reactions in living systems. Enzymes and their mode of action are described later in this chapter.

PART 2: BIOCHEMISTRY: THE COMPOSITION AND REACTIONS OF LIVING MATTER

Compounds contributing to body structure and function fall into one of two major classes: inorganic and organic compounds. **Organic compounds** contain carbon. All organic compounds are covalently bonded molecules, and many are large.

All other chemicals in the body are considered **inorganic compounds.** These include water, salts, and many acids and bases. Organic and inorganic compounds are equally essential for life. Trying to decide which is more valuable is like trying to decide whether the ignition system or the engine is more essential to the operation of a car!

Inorganic Compounds

Water

Water is the most abundant and important inorganic compound found in living material, making up from 60% to 80% of the volume of most living cells. The versatility of this vital fluid is incredible, as reflected in its various properties.

1. High heat capacity. Water has a *high heat capacity;* that is, it absorbs and releases large amounts of heat before changing appreciably in temperature itself. Its predominence in living matter prevents sudden changes in body temperature from external factors, such as sun or wind exposure, or from internal conditions that cause rapid heat liberation, such as vigorous muscle activity. As part of blood, water redistributes heat among body tissues, ensuring temperature homeostasis.

2. High heat of vaporization. When water evaporates, or vaporizes, it changes from a liquid to a gas (water vapor). This transformation requires large amounts of heat energy to disrupt the hydrogen bonds that hold water molecules together. This property is extremely beneficial when we sweat. As perspiration (mostly water) evaporates from our skin, large amounts of heat are removed from the body and lost to the external environment, providing an efficient cooling mechanism.

3. Polarity/solvent properties. Water is an unparalleled solvent and suspension medium for both inorganic and organic molecules. Indeed, water is often called the *universal solvent*. Biochemistry is "wet chemistry"; biological molecules do not react chemically unless they are in solution, and virtually all chemical reactions occurring in the body depend on water's solvent properties.

Because of their polar nature, water molecules orient themselves with their slightly negative ends toward the positive ends of the solutes, and vice versa, first attracting them, then surrounding them. This enclosure of particles by *hydration layers* (layers of water molecules) shields the particles from the effects of other charged substances in the vicinity. This explains why ionic compounds and other small reactive molecules (such as acids and bases) dissociate in water, their ions separating from each other and becoming evenly scattered in the water, forming true solutions (Figure 2.11). Water also forms hydration layers around large charged molecules such as proteins, preventing them from settling out of solution. Such protein-water mixtures are biological colloids.

Because of its solvent properties, water serves as the body's major transport medium. Nutrients, respiratory gases, and metabolic wastes can be carried throughout the body because they dissolve in blood plasma, and many metabolic wastes are excreted from the body in urine, another watery fluid.

Specialized molecules that act as body lubricants also use water as their dissolving medium. Such lubricating fluids include mucus, which helps to ease feces through the bowel, and the serous fluids that reduce friction and help prevent adhesions between visceral organs.

4. Reactivity. Water is an important *reactant* in many chemical reactions. For example, foods are digested to their building blocks by the addition of a water molecule to each bond to be broken. Such decomposition reactions are more specifically called *hydration*, or *hydrolysis* (hī-drah′-lih-sis), *reactions*. Conversely, when large carbohydrate or protein molecules

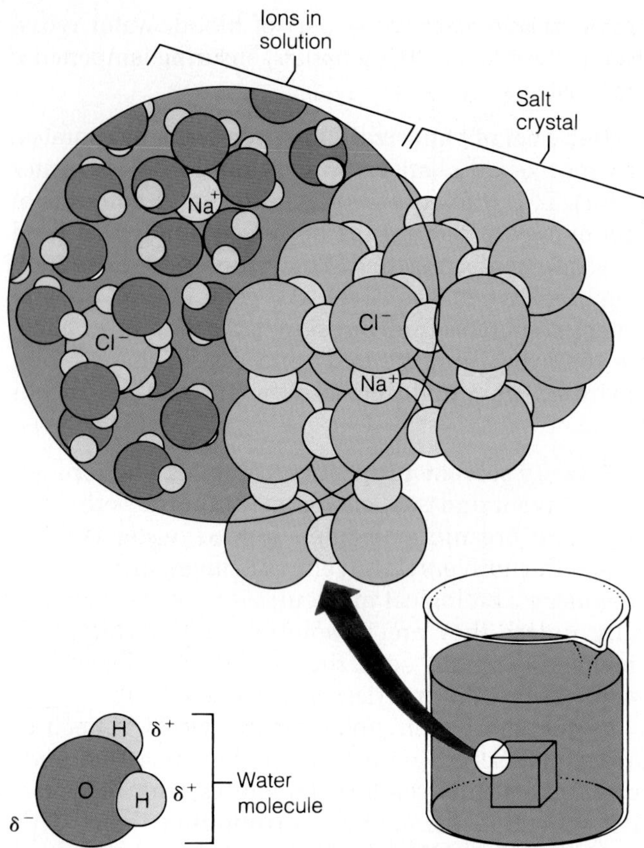

Figure 2.11 Dissociation of a salt in water. The slightly negative ends of the water molecules (δ^-) are attracted to Na^+, whereas the slightly positive ends of water molecules (δ^+) orient toward Cl^-, causing the ions to be pulled off the crystal lattice.

are synthesized from smaller molecules, a water molecule is removed for every bond formed, in a reaction called *dehydration synthesis.*

5. Cushioning. Finally, by forming a resilient cushion around certain body organs, water helps protect them from physical trauma. The cerebrospinal fluid surrounding the brain reflects water's *cushioning* role.

Salts

A **salt** is an ionic compound consisting of cations other than H^+ and anions other than the hydroxyl ion (OH^-). As already noted, when salts are dissolved in water, they dissociate into their component ions (Figure 2.11). For example, sodium sulfate (Na_2SO_4) dissociates into the ions $2Na^+$ and SO_4^{2-}. This process occurs rather easily, because the ions are already formed; all that remains is for water to overcome the attraction between the oppositely charged ions. All ions are *electrolytes* (ih-lek′-trō-līts), substances that conduct an electrical current in solution. (Note that groups of atoms that bear an overall charge, such as sulfate, are called *polyatomic ions.*)

Salts of many metal elements, such as NaCl, Ca_2CO_3 (calcium carbonate), and KCl (potassium chloride), are common in the body. However, the most plentiful salts are the calcium phosphates, which contribute to the hardness of bones and teeth. In their ionized form, salts play vital roles in body function. For instance, the electrolyte properties of ionic sodium and potassium are essential for the transmission of nerve impulses and the contraction of muscles. Moreover, ionic iron forms part of the hemoglobin molecules that transport oxygen within red blood cells, and ionic zinc and copper are important to the activity of some enzymes. Other important functions of the elements found in body salts are summarized in Table 2.1.

Maintaining proper ionic balance in our body fluids is one of the most critical homeostatic roles of the kidneys. When this balance is severely disturbed, virtually nothing in the body works. All the physiological activities listed above and thousands of others are disrupted and ultimately grind to a stop. ■

Acids and Bases

Like salts, acids and bases are electrolytes; that is, they ionize and dissociate in water and can then conduct an electrical current. However, unlike salts, acids and some bases are covalently bonded molecules. Therefore, the ions must be formed before dissociation can occur. In high concentrations, acids and bases can be extremely damaging to living tissue.

Acids

Acids have a sour taste and can react with many metals or "burn" a hole in your rug; but the most useful definition of an acid for our purposes is that it is a substance that releases hydrogen ions (H^+) in detectable amounts. Since a hydrogen ion is just a hydrogen nucleus, or "naked" proton, acids are also defined as *proton donors.*

When acids are dissolved in water, they release hydrogen ions (protons) and anions. The anions have little or no effect on acidity; it is the concentration of protons that determines the acidity of a solution. For example, hydrochloric acid (HCl), an acid produced by stomach cells that aids digestion, dissociates into a proton and a chloride ion:

$$\underset{}{HCl} \longrightarrow \underset{\text{proton}}{H^+} + \underset{\text{anion}}{Cl^-}$$

Other acids found or produced in the body include acetic acid ($HC_2H_3O_2$, commonly abbreviated as HAc), or vinegar, sulfuric acid (H_2SO_4), and carbonic acid (H_2CO_3). The molecular formula for an acid is easy to recognize because the hydrogen is always written first.

Bases

Bases have a bitter taste, feel slippery, and are *proton acceptors.* Common inorganic bases include the *hydroxides* (hī-drox′-īdz), such as magnesium hydroxide (milk of magnesia) and sodium hydroxide (lye). Like acids, hydroxides dissociate when dissolved in water, but in this case, hydroxyl ions (OH^-) and cations are liberated. For example, ionization of sodium hydroxide (NaOH) produces a hydroxyl ion and a sodium ion; the hydroxyl ion then binds to (accepts) a proton present in the solution. This produces water and simultaneously reduces the acidity (hydrogen ion concentration) of the solution:

$$\underset{}{NaOH} \longrightarrow \underset{\text{cation}}{Na^+} + \underset{\text{hydroxyl ion}}{OH^-}$$

and then

$$OH^- + H^+ \longrightarrow \underset{\text{water}}{H_2O}$$

Bicarbonate ion (HCO_3^-), an important base in the body, is particularly abundant in blood, where it constitutes the so-called alkaline reserve of the body. Ammonia (NH_3), a common waste product of protein breakdown in the body, is also a base. It has one pair of unshared electrons that strongly attracts protons. By accepting a proton, ammonia becomes an ammonium ion:

$$NH_3 + H^+ \longrightarrow \underset{\text{ammonium ion}}{NH_4^+}$$

pH: Acid-Base Concentration

The greater the number of hydrogen ions in solution, the more acidic the solution is said to be; conversely, the greater the concentration of hydroxyl ions (the lower the concentration of H^+), the more basic, or *alkaline* (al′-kuh-lin or al′-kuh-līn), the solution becomes. The relative concentration of hydrogen ions in various body fluids is measured in concentration units called **pH** (pē-āch) **units.**

The idea for a pH scale was devised by a Danish biochemist and part-time beer brewer named Sören Sörensen in 1909. He was searching for a convenient means of checking the acidity of his alcoholic product to prevent its spoilage by bacterial action. (Many bacteria are inhibited by acidic conditions.) The pH scale that resulted is based on the concentration of hydrogen ions in a solution, expressed in terms of moles per liter, or molarity. The pH scale runs from 0 to 14 and is logarithmic; that is, each successive change of one pH unit represents a tenfold change in hydrogen ion concentration (Figure 2.12).

At a pH of 7, the number of hydrogen ions exactly equals the number of hydroxyl ions, and the solution

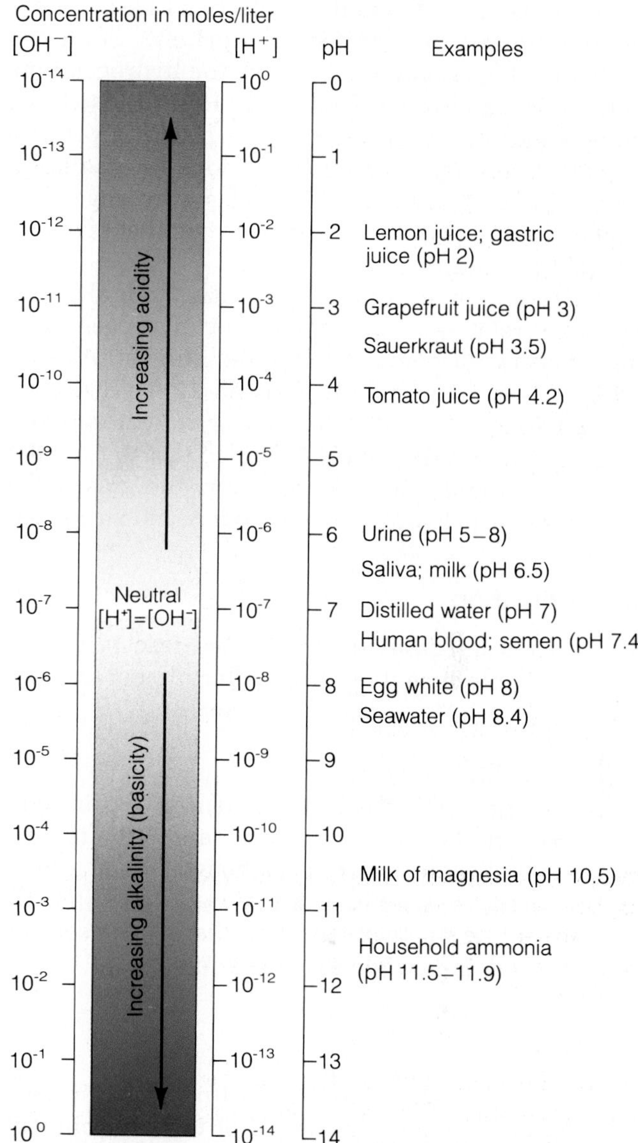

Figure 2.12 The pH scale and pH values of representative substances. The pH scale is based on the number of hydrogen ions in solution. The actual concentration of hydrogen ions ($[H^+]$) (expressed in moles per liter) and the corresponding hydroxyl concentration ($[OH^-]$) are indicated for each pH value noted. At a pH of 7, the concentrations of hydrogen and hydroxyl ions are equal and the solution is neutral.

is said to be *neutral*—neither acidic nor basic. Absolutely pure (distilled) water has a pH of 7. Solutions with a pH lower than 7 are acidic; the hydrogen ions outnumber the hydroxyl ions. The lower the pH, the more acidic the solution. A solution with a pH of 6 has ten times as many hydrogen ions as a solution with a pH of 7; a pH of 3 indicates a hydrogen ion concentration 10,000 times greater than that of a neutral (pH 7) solution.

Solutions with a pH higher than 7 are alkaline, and the relative concentration of hydrogen ions decreases by a factor of 10 with each higher pH unit. Thus, solutions with a pH of 8 and 12 have, respectively, 1/10 and 1/100,000 the number of hydrogen ions present in a solution of pH 7. Notice that the smaller the concentration of hydrogen ions, the greater the concentration of hydroxyl ions becomes, and vice versa.

Neutralization

When acids and bases are mixed, they react with each other in displacement reactions to form water and salt:

$$\underset{\text{acid}}{HCl} + \underset{\text{base}}{NaOH} \rightarrow \underset{\text{water}}{H_2O} + \underset{\text{salt}}{NaCl}$$

This type of reaction is called a *neutralization reaction*, because the combining of H^+ and OH^- to form water neutralizes the solution. Although the salt formed in this reaction is written in molecular form (NaCl), remember that it actually exists as dissociated sodium and chloride ions when dissolved in water.

Buffers

Biochemical reactions are extraordinarily sensitive to even minute changes in the pH of the environment. Homeostasis of acid-base balance is carefully regulated by the kidneys and lungs and by chemical systems (proteins and other types of molecules) called *buffers*. Buffers resist abrupt and large swings in the pH of body fluids by releasing H^+ (acting as acids) when the pH begins to rise and by binding with protons (acting as bases) when the pH drops. Since blood comes into close contact with nearly every body cell, its pH is particularly critical. Normally, blood pH is slightly alkaline and varies within a very narrow range (7.35–7.45). If it exceeds these limits by more than a few tenths of a pH unit, death becomes a distinct possibility. The approximate pH of several body fluids and of a number of common substances appears in Figure 2.12. Acid-base balance and buffers are described in more detail in Chapter 27.

Organic Compounds

The molecules unique to living systems—proteins, carbohydrates, fats, and nucleic acids—all contain carbon and hence are **organic compounds.** Although organic compounds are distinguished by the fact that they contain carbon and inorganic compounds are defined as compounds that lack carbon, there are a few irrational exceptions to this generalization that you should be aware of. Carbon dioxide, carbon monoxide, and carbides, for example, all contain carbon but are considered inorganic compounds.

What makes carbon so special that "living" chemistry depends on its presence? To begin with, no other *small* atom is so precisely *electroneutral*. That is, carbon neither loses nor gains electrons, but instead always shares them. Furthermore, with four valence shell electrons, carbon can, and often does, form up to four covalent bonds with other elements, as well as with other carbon atoms. As a result, carbon is found in long, chainlike molecules (common in fats), ring structures (typical of carbohydrates and steroids), and many other structures uniquely suited for their specific roles in the body.

Carbohydrates

Carbohydrates, a group of molecules that includes sugars and starches, represent 1% to 2% of cell mass. Carbohydrates contain carbon, hydrogen, and oxygen, and, with slight variations, the hydrogen and oxygen atoms occur in the same 2:1 ratio as in water. This ratio is reflected in the word *carbohydrate* (meaning "hydrated carbon").

A carbohydrate can be classified according to size and solubility as a monosaccharide ("one sugar"), disaccharide ("two sugars"), or polysaccharide ("many sugars"). Monosaccharides are the structural units, or building blocks, of the other carbohydrates. In general, the larger the carbohydrate molecule, the less soluble it is in water.

Monosaccharides

Monosaccharides (mah″-nō-sa′-kuh-rīdz), or *simple sugars*, are single-chain or single-ring structures containing from three to seven carbon atoms (Figure 2.13a). Usually, the carbon, hydrogen, and oxygen atoms occur in the ratio 1:2:1, so a general formula for a monosaccharide is $(CH_2O)_n$, where *n* is equal to the number of carbons in the sugar. Glucose, for example, has six carbon atoms, and its molecular formula is $C_6H_{12}O_6$; and ribose, with five carbons, is $C_5H_{10}O_5$.

Monosaccharides are named generically according to the number of carbon atoms they contain. The most important monosaccharides in the body are pentose (five-carbon) and hexose (six-carbon) sugars. For example, *deoxyribose* (dē-ok″-sē-rī′-bōs″), a pentose, is a structural component of DNA, and *glucose*, a hexose, is blood sugar. Two other hexoses, galactose and fructose, are *isomers* (ī′-suh-merz) of glucose. That is, they have the same molecular formula ($C_6H_{12}O_6$), but their atoms are arranged differently, giving them different chemical properties (see Figure 2.13a). When ingested in the diet, galactose and fructose are converted to glucose in the liver for use by the body cells.

Disaccharides

A **disaccharide** (dī-sa′-kuh-rīd″), or *double sugar*, is formed when two monosaccharides are joined together by **dehydration synthesis** (Figure 2.13b). In this synthesis reaction, a water molecule is lost as the bond is formed, as illustrated by the synthesis of sucrose (sū′-krōs):

$$\underset{\text{glucose + fructose}}{2C_6H_{12}O_6} \rightarrow \underset{\text{sucrose}}{C_{12}H_{22}O_{11}} + \underset{\text{water}}{H_2O}$$

Notice that the molecular formula for sucrose contains two hydrogen atoms and one oxygen atom less than the total number of hydrogen and oxygen atoms in the glucose and fructose, because a water molecule is released during bond formation.

Important disaccharides in the diet are *sucrose* (glucose + fructose), which is cane or table sugar; *lactose* (glucose + galactose), found in milk; and *maltose* (glucose + glucose), also called malt sugar (see Figure 2.13b). Since disaccharides are too large to pass through cell membranes, they must be digested to their simple sugar units to be absorbed from the digestive tract into the blood. This decomposition process is called **hydration** or **hydrolysis** and is essentially the reverse of dehydration synthesis. A water molecule is added to each bond, breaking the bonds and releasing the simple sugar units (see Figure 2.13b).

Polysaccharides

Polysaccharides (pah″-lē-sa′-kuh-rīdz″) are long chains of simple sugars linked together by dehydration synthesis (Figure 2.13c). Such a long, chainlike molecule made up of many similar units is called a *polymer*. Polysaccharides are large, fairly insoluble molecules and thus are ideal forms for storing carbohydrates. Another consequence of their large size is that they lack the sweetness of the simple and double sugars. Only two polysaccharides are of major importance to the body: starch and glycogen. Both are polymers of glucose. Only their bond angles differ.

Starch is the storage carbohydrate form found in plants. The actual number of glucose units composing a starch molecule is very high and variable. When we eat starchy foods such as grain products and potatoes, the starch must be digested to its glucose units to be absorbed.

Glycogen (glī′-kō-gin), the storage carbohydrate of animal tissues, is stored primarily in skeletal muscle and liver cells. Like starch, it is highly branched and is a very large molecule (see Figure 2.13c). When blood sugar levels drop precipitously, liver cells break down glycogen and release its glucose units to the blood. Since there are many branch endings from which glucose can be released simultaneously, body cells have almost instant access to glucose fuel.

Carbohydrate Functions

The major function of carbohydrates in the body is to provide a ready, easily used source of cellular fuel. Most cells can use only a limited number of simple sugars, and glucose is at the top of the "cellular menu." Within cells, glucose is broken down and oxidized in a series of reactions collectively called *cellular respiration*. The final products of cellular respiration are carbon dioxide and water. The bond energy released during glucose breakdown is trapped in the bonds of other molecules called adenosine triphosphate, or ATP. ATP temporarily stores the energy and ultimately releases it for use in the energy-consuming reactions in the cell as *its* bonds are broken. The overall process of cellular respiration can be represented as

$$\text{glucose} + \text{oxygen} \rightarrow \text{carbon dioxide} + \text{water} + \text{ATP}$$

When not immediately needed to generate bond energy of ATP molecules, dietary carbohydrates are converted to glycogen or fat and stored. Those of us who "get fat" from eating too many carbohydrate-rich snacks have a first-hand, personal awareness of this conversion process!

Carbohydrates are also used to a small extent for structural purposes. For example, some sugars are attached to the external surfaces of cell membranes, where they act as "road signs" for cell-to-cell interaction. When protein intake is inadequate, some sugars may be converted by the liver to the building blocks needed to produce proteins.

Lipids

Lipids are organic compounds that are insoluble in water but dissolve readily in other lipids and in organic solvents such as alcohol, chloroform, and ether. Like carbohydrates, all lipids contain carbon, hydrogen, and oxygen, but the proportion of oxygen in lipids is much lower. In addition, phosphorus is found in some of

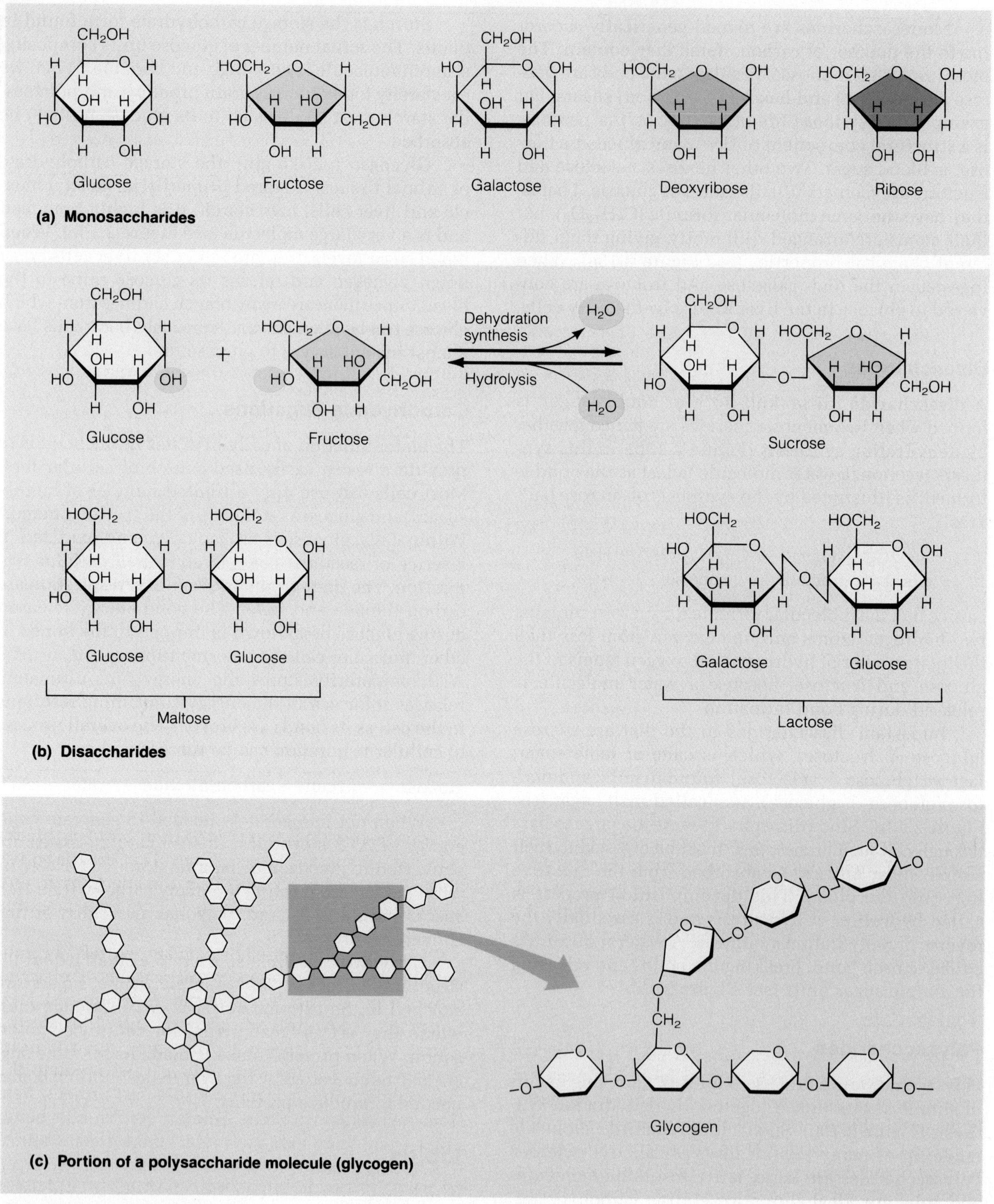

Glucose Fructose Galactose Deoxyribose Ribose

(a) Monosaccharides

Glucose + Fructose

Dehydration synthesis

Hydrolysis

H_2O

Sucrose

Glucose Glucose

Maltose

Galactose Glucose

Lactose

(b) Disaccharides

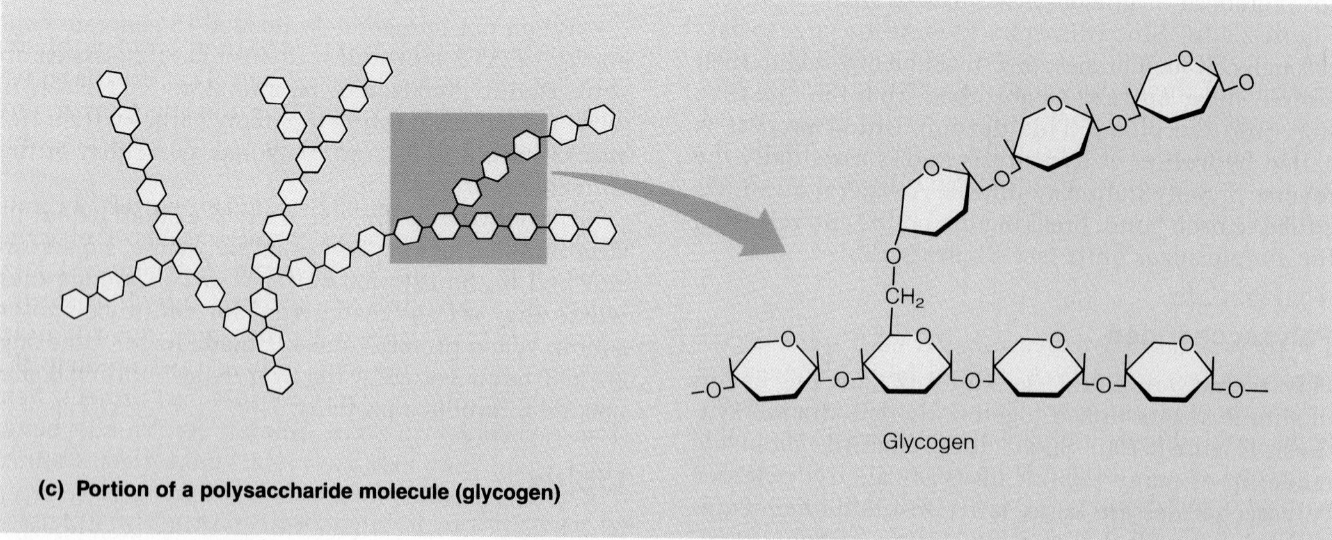

(c) Portion of a polysaccharide molecule (glycogen)

Figure 2.13 Carbohydrate molecules. (**a**) Monosaccharides important to the body. The hexose sugars glucose, fructose, and galactose are isomers: Their molecular formulas are all the same ($C_6H_{12}O_6$), but the arrangement of their atoms differs, as shown in this diagram. Deoxyribose and ribose are pentose sugars. (**b**) Common disaccharides. Disaccharides are two monosaccharides linked together by dehydration synthesis, a process involving the removal of a water molecule at the bond site. The formation of sucrose from glucose and fructose molecules is illustrated. In the reverse reaction (hydrolysis), sucrose is broken down to glucose and fructose by the addition of a water molecule to the bond. Other important disaccharides are maltose and lactose.(**c**) Simplified representation of part of a glycogen molecule. Glycogen is a polysaccharide formed from linked glucose units.

Table 2.2 Representative Lipids Found in the Body

Lipid type	Location/function
Neutral fat (triglyceride)	In fat deposits (subcutaneous tissue and around organs); protects and insulates body organs; the major source of *stored* energy in the body
Phospholipid (phosphatidyl choline; cephalin; others)	Chief component of cell membranes; may participate in the transport of lipids in plasma; prevalent in nervous tissue
Steroids:	
Cholesterol	The basis for manufacture of all body steroids
Bile salts	Breakdown products of cholesterol; released by the liver into the digestive tract, where they aid fat digestion and absorption
Vitamin D	A fat-soluble vitamin produced in the skin on exposure to UV radiation; necessary for normal bone growth and function
Sex hormones	Estrogen and progesterone (female hormones) and testosterone (a male hormone) are produced in the gonads and are necessary for normal reproductive function
Adrenal cortical hormones	Cortisol, a glucocorticoid, is a metabolic hormone necessary for maintaining normal blood glucose levels; aldosterone helps to regulate salt and water balance of the body by targeting the kidneys
Other lipoid substances:	
Fat-soluble vitamins:	
A	Found in orange-pigmented vegetables and fruits; converted in the retina to retinal, a part of the photoreceptor pigment involved in vision
E	Found in plant products such as wheat germ and green leafy vegetables; claims have been made (but not proved in humans) that it promotes wound healing and contributes to fertility; may help to neutralize highly reactive particles called free radicals believed to be involved in triggering some types of cancer
K	Made available to humans largely by the action of intestinal bacteria; also prevalent in a wide variety of foods; necessary for proper clotting of blood
Prostaglandins	Group of fatty acid derivatives found in all cell membranes; diverse effects, including stimulation of uterine contractions, regulation of blood pressure, and control of gastrointestinal tract motility; involved in inflammation
Lipoproteins	Lipoid and protein-based substances that transport fatty acids and cholesterol in the bloodstream; major varieties are high-density lipoproteins (HDLs) and low-density lipoproteins (LDLs)

the more complex lipids. Lipids are usually classified in terms of their solubility and include *neutral fats, phospholipids* (fos″-fō-lip′-idz), *steroids* (stayr′-oydz), and a number of other lipoid substances. Table 2.2 gives the locations and functions of some representative lipids found in the body.

Neutral Fats

The neutral fats, or *triglycerides* (trī″-glih′-ser-īdz), are known as fats and oils. A neutral fat is composed of two types of building blocks, *fatty acids* and *glycerol* (glih′-ser-ol) (Figure 2.14a). Fatty acids are linear chains of carbon and hydrogen atoms (hydrocarbon chains) with an organic acid group (–COOH) at one end; glycerol is a modified simple sugar (a sugar alcohol). Three fatty acid chains are attached to a single glycerol molecule by dehydration synthesis, resulting in an E-shaped molecule. The glycerol molecule is constant in all neutral fats, but the fatty acid chains vary, resulting in different types of neutral fat molecules. The neutral fats tend to be large molecules, often consisting of hundreds of atoms. Consequently, ingested fats and oils must be broken down to their building blocks before they can be absorbed. Neutral fats are the body's most concentrated source of usable energy fuel, and when they are broken down, they yield large amounts of energy.

The hydrocarbon chains make neutral fats nonpolar molecules. Since polar and nonpolar molecules do not interact, oil (or fats) and water do not mix. Consequently, neutral fats are well suited for storing energy fuel in the body and are found primarily in fat deposits beneath the skin, where they insulate the deeper body tissues from heat loss and protect them from mechanical trauma. Women are usually better English channel swimmers than men; this is attributed partly to their thicker subcutaneous fatty layer, which helps insulate them from the bitterly cold water of the channel.

Neutral fats may be solid (fats) or liquid (oils). How solid a neutral fat is at a given temperature depends on two factors: the length of its fatty acid chains and their degree of saturation. Carbon-containing compounds with only single covalent bonds are

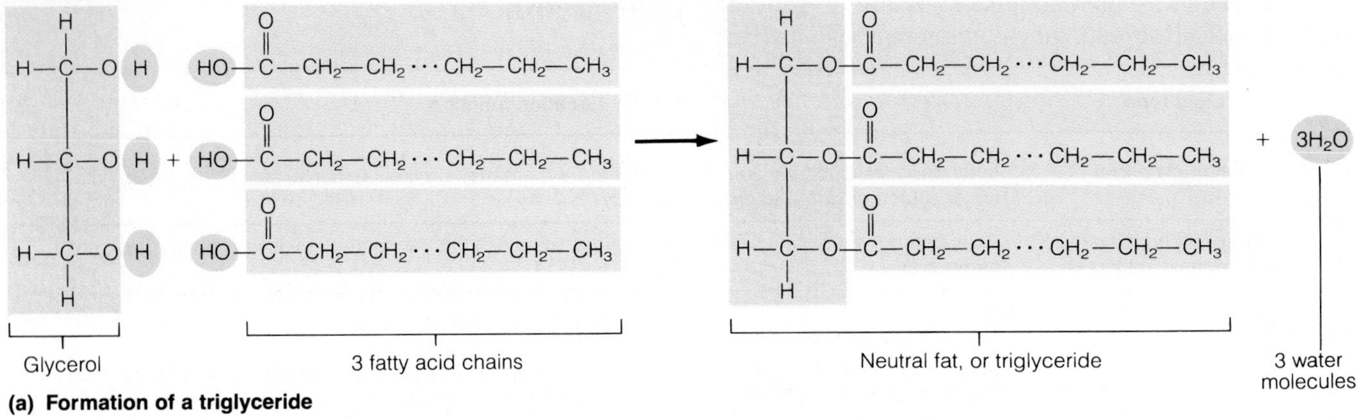

Glycerol

3 fatty acid chains

Neutral fat, or triglyceride

3 water molecules

(a) Formation of a triglyceride

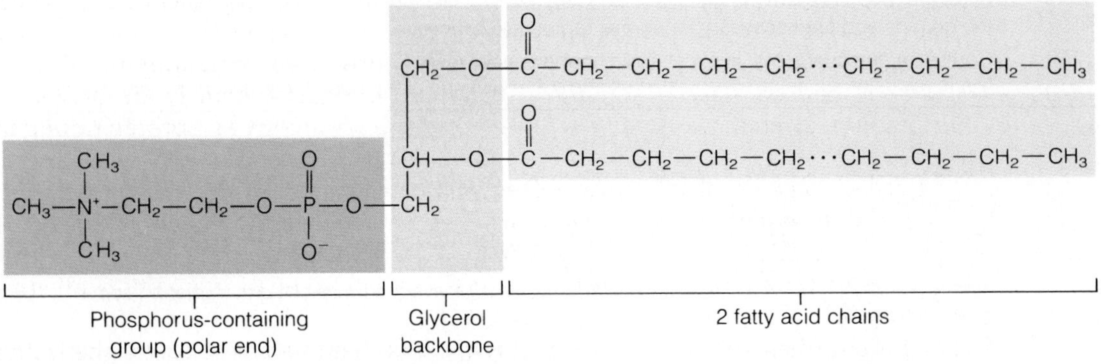

Phosphorus-containing group (polar end)

Glycerol backbone

2 fatty acid chains

(b) Phospholipid molecule (phosphatidyl choline)

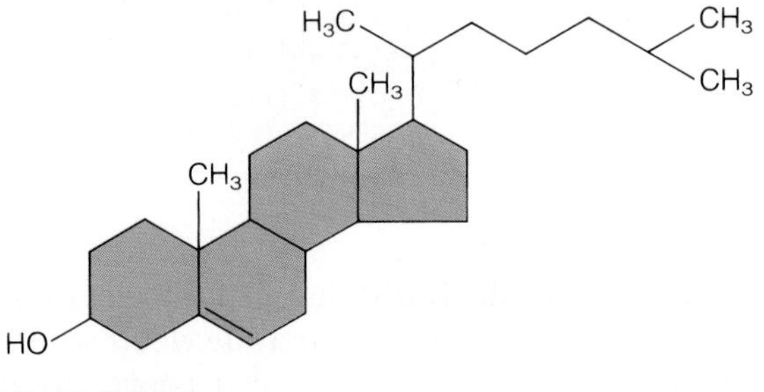

(c) Cholesterol

Figure 2.14 Lipids. (**a**) The neutral fats, or triglycerides, are synthesized by dehydration synthesis. In this process, three fatty acid chains are attached to a single glycerol molecule, and a water molecule is lost at each bond site. (**b**) Structure of a typical phospholipid molecule. Two fatty acid chains and a phosphorus-containing group are attached to the glycerol backbone. (**c**) The generalized structure of cholesterol. Cholesterol is the basis for all steroids formed in the body. The steroid nucleus is shaded.

referred to as *saturated molecules*. Organic molecules that contain one or more double or triple bonds are said to be *unsaturated* or *polyunsaturated*. Neutral fats with short fatty acid chains and/or unsaturated fatty acids are liquid at room temperature and are typical of plant lipids. We are familiar with these unsaturated fats as oils used for cooking, such as olive, peanut, corn, and safflower oils. Longer fatty acid chains and/or more saturated fatty acids are common in animal fats such as butter fat and the fat of meats, which are solid at room temperature.

Saturated fats, along with cholesterol, have been implicated as substances that encourage the deposit of fatty substances on artery walls and eventually arteriosclerosis (hardening of the arteries). As a result, margarine made from polyunsaturated fats has been promoted as a product that allows us to "have our cake and eat it too"—a good-tasting spread that (unlike butter) does not damage our arteries. The reasoning is understandable; however, recent studies have indicated that certain natural fish oils may have even greater protective effects than the highly advertised polyunsaturated plant oils. ■

Phospholipids

Phospholipids, or *complex lipids,* are modified triglycerides with a phosphorus-containing group and two, rather than three, fatty acid chains (Figure 2.14b). The phosphorus-containing group gives phospholipids their distinctive chemical properties. Although the hydrocarbon portion of the molecule is nonpolar and interacts only with nonpolar molecules, the phosphorus-containing part is polar and attracts other polar or charged particles, such as water or ions. As you will see in Chapter 3, cells use this unique characteristic of phospholipids in the construction of their limiting membranes. A number of biologically important phospholipids and their functions are listed in Table 2.2.

Steroids

Steroids (Figure 2.14c) are structurally quite different from neutral fats and phospholipids. Instead of hydrocarbon chains, steroids have hydrocarbon rings. However, like neutral fats, steroids are fat-soluble and contain little oxygen. The single most important steroid is *cholesterol* (kuh-les′-ter-ol″). We ingest cholesterol in animal products such as eggs, meat, and cheese, and our liver produces a certain amount of cholesterol, regardless of our dietary intake.

Cholesterol has earned a "bad press" because of its role in arteriosclerosis, but it is absolutely essential for human life. Cholesterol is found in cell membranes and is the raw material of vitamin D, steroid hormones, and bile salts. Although steroid hormones are present in the body in only small quantities, they are vital to homeostasis. Without sex hormones, reproduction would be impossible, and when we are deprived of cortisol and aldosterone produced by the adrenal gland, death results.

Proteins

Protein composes 10% to 30% of cell mass and is the basic structural material of the body. However, not all proteins are construction materials; many proteins play active and vital roles in ensuring normal cell function. Proteins have the most varied functions of any molecules in the body. All proteins contain carbon, oxygen, hydrogen, and nitrogen, and many contain sulfur and phosphorus as well.

Amino Acids and Peptide Bonds

The building blocks of proteins are molecules called **amino acids,** of which there are 20 common types. All amino acids have two important functional groups in common: a basic group called an *amine* (uh′-mēn) *group* (–NH_2), and an organic *acid group* (–COOH) that allows the amino acid to act as a proton donor. In fact, all amino acids have identical structures except for a third functional group called the *R group.* Differences in the numbers and arrangement of the atoms making up the R group make each amino acid unique in its bonding behavior and relative acidity or alkalinity (Figure 2.15).

Proteins are long chains of amino acids. The amino acids are joined together by dehydration synthesis, with the amine end of one amino acid linked to the acid end of the next. The resulting bond produces a characteristic arrangement of linked atoms and is called

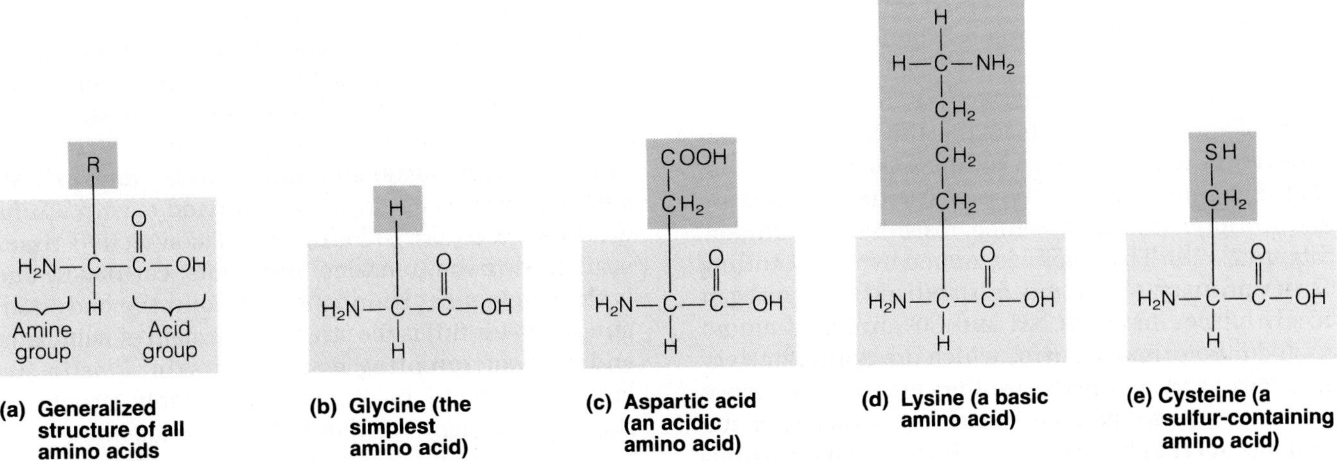

Figure 2.15 Amino acid structures. (**a**) Generalized structure of amino acids. All amino acids have both an amine (–NH_2) group and an acid (–COOH) group; they differ only in the atomic makeup of their R groups (green). (**b**)-(**e**) Specific structures of four amino acids. The simplest (glycine) has an R group consisting of a single hydrogen atom. An acid group in the R group makes the amino acid more acidic. An amine group in the R group makes it more basic. The presence of a sulfhydryl (–SH) group in the R group hints that this is an amino acid likely to participate in intramolecular bonding.

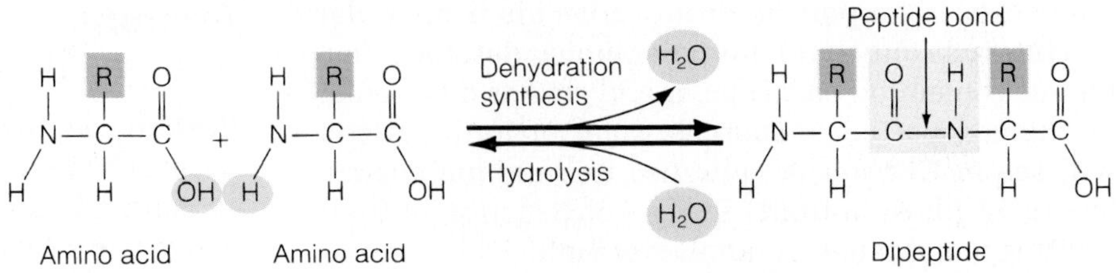

Figure 2.16 Amino acids are linked together by dehydration synthesis. The acid group of one amino acid is bonded to the amine group of the next, with the loss of a water molecule. The bond formed is called a peptide bond. Peptide bonds are broken when water is added to the bond (i.e., during hydrolysis).

a *peptide bond* (Figure 2.16). Two united amino acids form a *dipeptide,* three a *tripeptide,* and ten or more a *polypeptide.* Polypeptides containing more than 50 amino acids are called proteins; however, most proteins are large, complex molecules *(macromolecules)* containing from 100 to over 1000 amino acids.

The types of amino acids and the sequence in which they occur give each protein unique properties. We can think of the 20 amino acids as a 20-letter "alphabet" used in specific combinations to form "words" (proteins). Just as a change in one letter can produce a word with an entirely different meaning (flour → floor), a change in the kind or position of an amino acid can produce a protein with a different function. Of course, some letter changes will produce nonsense words (flour → fllur); likewise, many changes in amino acid combinations will result in nonfunctional proteins. Nevertheless, there are thousands of different proteins in the body, each with distinct functional properties, and all of them are constructed from different combinations of the same 20 amino acids.

Structural Levels of Proteins

Proteins can be described in terms of four structural levels. The linear sequence of amino acids composing the polypeptide chain is called the *primary structure* of a protein. This structure, which resembles a strand of amino acid "beads," is the backbone of the protein molecule (Figure 2.17a).

However, few proteins exist as a simple, linear chain of amino acids; instead, most twist or fold upon themselves to form a more complex *secondary structure.* The most common type of secondary structure is the *alpha* (α) *helix,* which resembles a "Slinky" (Figure 2.17b). The α helix is formed by a tight coiling of the primary chain and is stabilized by hydrogen bonds formed between NH and CO groups of amino acids in the primary chain, which are approximately four amino acids apart. Another type of secondary structure is the *beta* (β)*-pleated sheet,* in which the primary polypeptide chains do not coil, but are linked side by side by hydrogen bonds to form a pleated, ladderlike structure (Figure 2.17c). Notice that in this structure the hydrogen bonds link together *different* polypeptide chains, whereas those in α helices link different parts of the *same* chain together.

Many proteins have *tertiary* (ter'-shē-ayr"-ē) *structure,* the next higher level of complexity, which is superimposed on secondary structure, usually the α helix. For example, the α helix becomes twisted and folded upon itself to form a roughly spherical molecule or *globular protein* (Figure 2.17d). This highly specific structure is maintained by both covalent and hydrogen bonds. When two or more polypeptide chains aggregate to form a complex protein, we say that the protein has *quaternary* (kwah'-ter-nuh"-rē) *structure* (Figure 2.17e). Hemoglobin exhibits this structural level, but relatively few proteins have quaternary structure.

In all cases, the ultimate overall structure of any protein is specified by its primary structure. This important concept reflects the fact that the types and relative positions of amino acids in the protein backbone determine where bonds can be formed to produce coiled or folded structures that achieve higher levels of complexity.

Structural and Functional Proteins

The different structural levels confer distinct properties on proteins, allowing them to be classified as structural or functional proteins. As a rule, proteins reaching only the secondary level (α helix, β-pleated sheet, or a combination of the two) are linear, insoluble, and very stable. Most **structural proteins** fall into this category. Structural proteins, also called *fibrous proteins* because of their extended, strandlike appearance, form an integral part of body structures. They provide tensile strength to tissue and include such proteins as collagen, keratin, and elastin. *Collagen* (kah'-luh-jin) is found in all connective tissue structures (bone, tendons, ligaments, cartilages) and is the single most abundant protein in the body. *Keratin* (kayr'-uh-tin) is the structural protein of hair, nails, and the waterproofing material of skin. *Elastin* (ih-las'-tin) is found in durable but flexible structures, such as the ligaments that bind bones together.

Proteins that possess tertiary or quaternary structure are most often **functional proteins.** Functional proteins are water-soluble, mobile, chemically active molecules, and they play crucial roles in virtually all biological processes. Some (antibodies) help to provide immunity, others (protein-based hormones) help to regulate growth and development, and still others

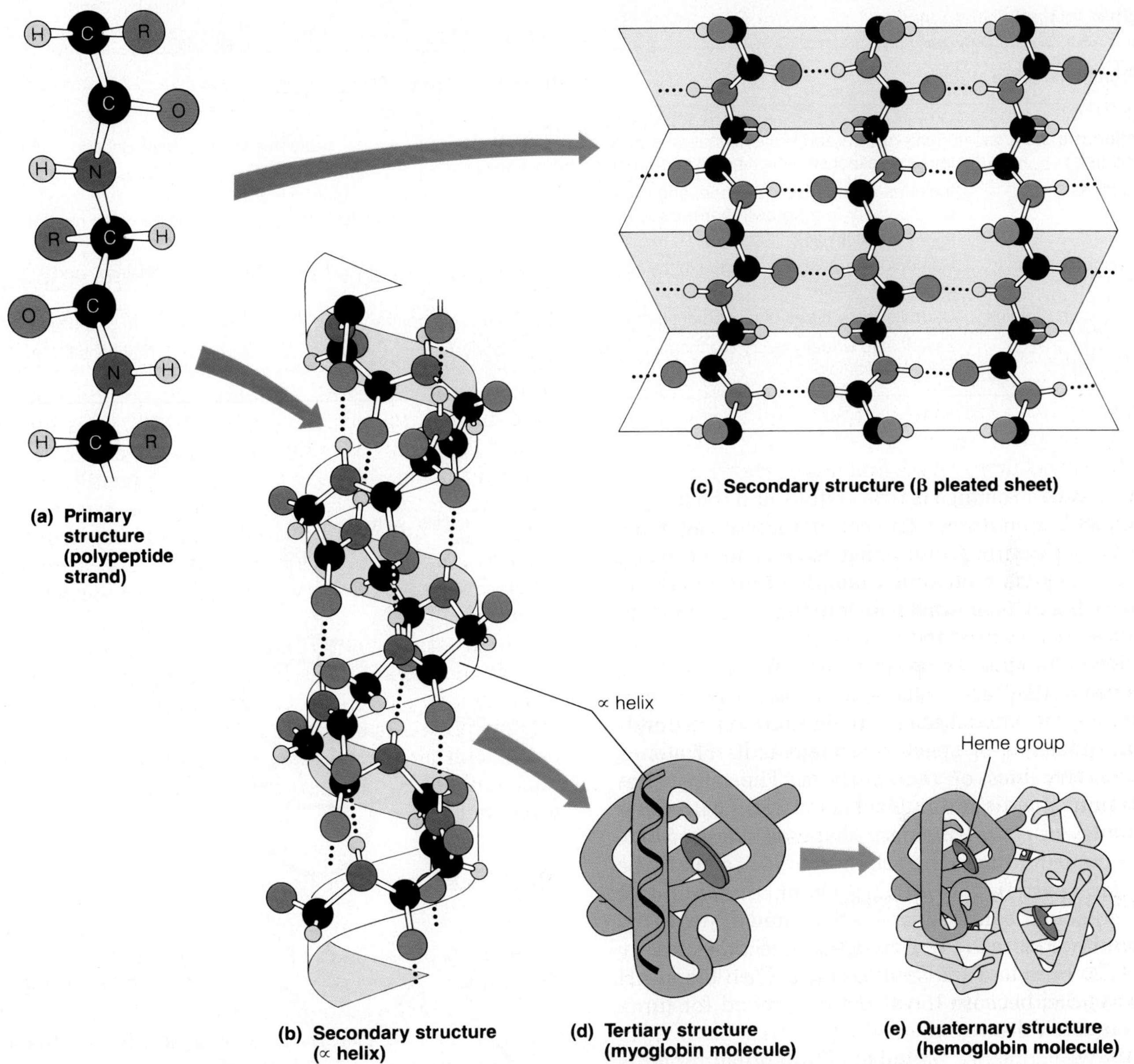

(a) Primary structure (polypeptide strand)

(b) Secondary structure (∝ helix)

(c) Secondary structure (β pleated sheet)

(d) Tertiary structure (myoglobin molecule)

(e) Quaternary structure (hemoglobin molecule)

Figure 2.17 Levels of protein structure. (**a**) Primary structure. Amino acids are linked together to form a polypeptide strand. (**b**) Secondary structure—the α helix. The primary chain is coiled upon itself to form a spiral structure that is maintained by hydrogen bonds (indicated by broken lines). (**c**) Secondary structure—the β-pleated sheet. Two or more primary chains are linked side by side by hydrogen bonds to form an undulating ribbonlike structure. (**d**) Tertiary structure. The α helix coils and folds upon itself to form a basically spherical molecule that is maintained by intramolecular bonds. The molecule shown is myoglobin. (**e**) Quaternary structure of hemoglobin. Hemoglobin is formed from four polypeptide chains that are linked together in a specific manner.

(enzymes) are catalysts that oversee just about every chemical reaction that goes on within the body. The roles of these and other functional proteins are summarized in Table 2.3.

Protein Denaturation

Structural proteins are exceptionally stable, but functional proteins are quite the opposite. The activity of a functional protein depends on its specific three-dimensional structure, and intramolecular bonds, particularly hydrogen bonds, are critically important in maintaining that structure. However, hydrogen bonds are fragile and easily disrupted by many chemical and physical factors, such as excessive acidity or heat. At a pH of 7.5, virtually all hydrogen bonds are broken, but if normal pH (about 7.4) is restored, the bonds reform. Extreme heat also ruptures hydrogen bonds, but

Table 2.3 Representative Classes of Functional Proteins

Functional class	Roles in the body
Antibodies (gamma globulins)	Highly specialized proteins that recognize and inactivate bacteria, toxins, and some viruses; they function in the immune response, which helps to protect the body from invading foreign substances
Hormones	Includes acid-derived, peptide, and protein hormones, which help to regulate metabolic activity, growth, and development (e.g., growth hormone is an anabolic hormone necessary for optimal growth; insulin helps regulate blood sugar levels)
Transport proteins	Hemoglobin transports oxygen in the blood; other transport proteins in the blood carry iron, cholesterol, and other substances
Contractile proteins	Underlie many types of movement in the body, such as muscle contraction, cell division, and sperm propulsion
Catalysts (enzymes)	Essential to virtually every biochemical reaction in the body, they increase the rates of chemical reactions by at least a millionfold

in this case, the change is irreversible, and the protein is said to be *denatured*. The coagulation of egg white (primarily albumin protein) that occurs when you boil or fry an egg is an obvious example of protein denaturation. There is no way to restore the white, rubbery protein to its original translucent form.

When functional proteins within the body are denatured, they are incapable of performing their physiological roles, because their function depends on the presence of specific arrangements of atoms, called *active sites*, on their surfaces. The active sites are regions that fit and interact chemically with other molecules of complementary shape or charge. Since atoms contributing to the active sites may actually be very far apart in the primary chain, disruption of intramolecular bonds separates them and destroys the active site (Figure 2.18). Hemoglobin becomes totally unable to bind and transport oxygen when blood pH is too acidic, because the structure needed for function has been destroyed.

We will consider a number of functional proteins in conjunction with the organ systems or functional processes to which they are closely related. However, since enzymes are required for the normal functioning of all body cells and systems, we will consider these incredibly complex molecules here.

Enzymes and Enzyme Activity

Enzymes are functional proteins that act as biological catalysts; that is, they regulate and accelerate the rate of biochemical reactions, but are not used up or changed in those reactions. Enzymes cannot force chemical reactions to occur between molecules that would not otherwise react; they can only increase the speed of reaction. We might think of enzymes as molecular "bellows" used to fan a sluggish fire into flaming activity, because in the absence of enzymes, biochemical reactions proceed so slowly that for all practical

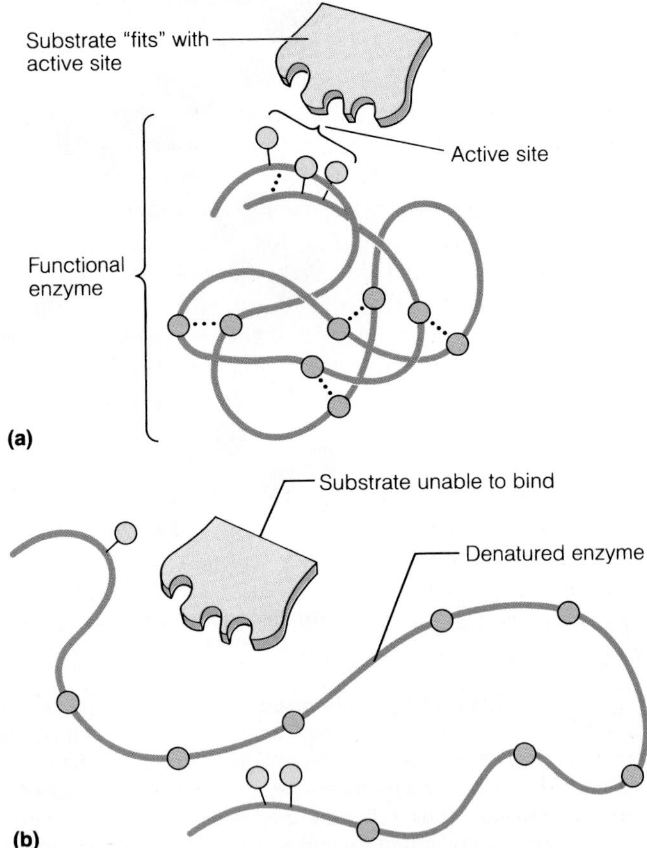

Figure 2.18 Denaturation of a functional molecule such as an enzyme. (**a**) The molecule's globular structure is maintained by intramolecular bonds. Atoms composing the active site of the enzyme are shown as stalked particles. The substrate, or interacting molecule, has a corresponding binding site, and the two sites fit together precisely. (**b**) Breaking the intramolecular bonds that maintain the secondary and tertiary structure of the enzyme now results in a linear molecule, with atoms of the former active site widely separated. Enzyme-substrate binding can no longer occur.

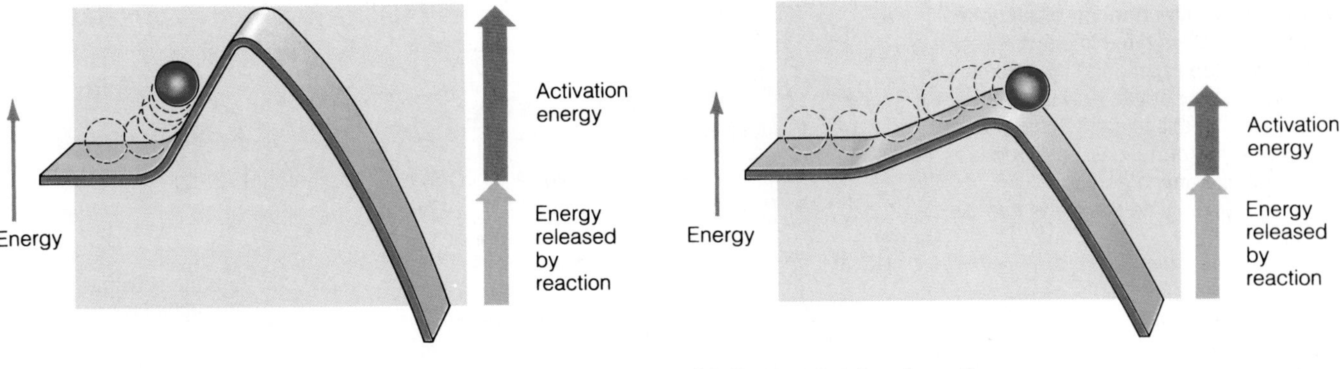

Figure 2.19 Comparison of the energy barrier in a noncatalyzed reaction and in the same reaction catalyzed by an enzyme. In each case, the reacting particles, indicated simply by a ball, must achieve a certain energy level before they can interact. The amount of energy that must be absorbed to reach that state and overcome the energy barrier, represented by the peak, is called activation energy. In (**a**), the noncatalyzed reaction, the amount of activation energy needed is much higher than that needed in (**b**), the enzyme-catalyzed reaction.

purposes they do not occur at all. Enzymes increase the rates of these reactions by a factor of about 1 million!

Some enzymes are purely protein in nature. In other cases, the enzyme consists of two parts—a protein portion and a *cofactor.* Depending on the enzyme, the cofactor may be an ion of a metal element such as copper or iron (in which case it is called a *prosthetic group*), or an organic molecule needed to assist the reaction in some particular way. Most often, the organic "helpers" are derived from vitamins (especially the B complex vitamins) and are more precisely called *coenzymes*.

Each enzyme is highly specific; that is, it controls only a single chemical reaction or a small group of related reactions. Thus, the presence of specific enzymes determines not only which reactions will be speeded up, but also which reactions will occur (no enzyme, no reaction). Most enzymes are named according to the type of reaction they catalyze. Thus, there are *hydrolases* (hī′-druh-lā-siz), which add water during hydrolysis reactions; *oxidases* (ok′-sih-dā-siz), which add oxygen; and so on. Most enzyme names can be recognized by the suffix *-ase*.

Some enzymes are produced in an inactive form that must be activated in some way before they can function. For example, digestive enzymes produced in the pancreas are not activated until they reach the small intestine, where they actually do their work. If they were produced in active form, the pancreas would digest itself. Sometimes, enzymes are inactivated immediately after they have performed their catalytic function. This is true of enzymes that promote blood clot formation when the wall of a blood vessel has been damaged. Once the clotting process has been triggered, those enzymes are inactivated; otherwise, you would have blood vessels full of solid blood instead of one protective blood clot!

How do enzymes perform their catalytic role? As we said earlier, a chemical reaction cannot occur unless the interacting molecules reach a certain energy level, reflected in their speed of movement. More precisely, every reaction requires that a certain amount of energy, called *activation energy,* be absorbed to "prime" the reaction and push the reactants to an energy level whereby their random collisions will be forceful enough to ensure interaction. This is true regardless of whether the overall reaction is ultimately energy absorbing or energy releasing.

One obvious way of increasing molecular energy is to increase the temperature, but in living systems, this would result in protein denaturation. (This is why a high fever can be a serious event.) The role of an enzyme is to ensure that a reaction can occur without abnormal increases in body temperature; it does this by decreasing the amount of activation energy required (Figure 2.19). The exact means by which enzymes accomplish this remarkable feat is not fully understood. However, we know that they decrease the randomness of reactions by binding the reacting molecules temporarily to the enzyme surface and presenting them to each other in the proper position for chemical interaction to occur.

Three basic steps appear to be involved in the mechanism of enzyme action (Figure 2.20).

1. The enzyme forms a bond with the substance(s) on which it acts. These substances are called the *substrates* of the enzyme. Substrate binding occurs at the enzyme's active site, a cleft or crevice on the enzyme's surface. Substrate binding causes the active site to undergo a structural change that allows the substrate and the active site to fit together very precisely. This dynamic process of enzyme recognition for the proper substrate is called *induced fit*.

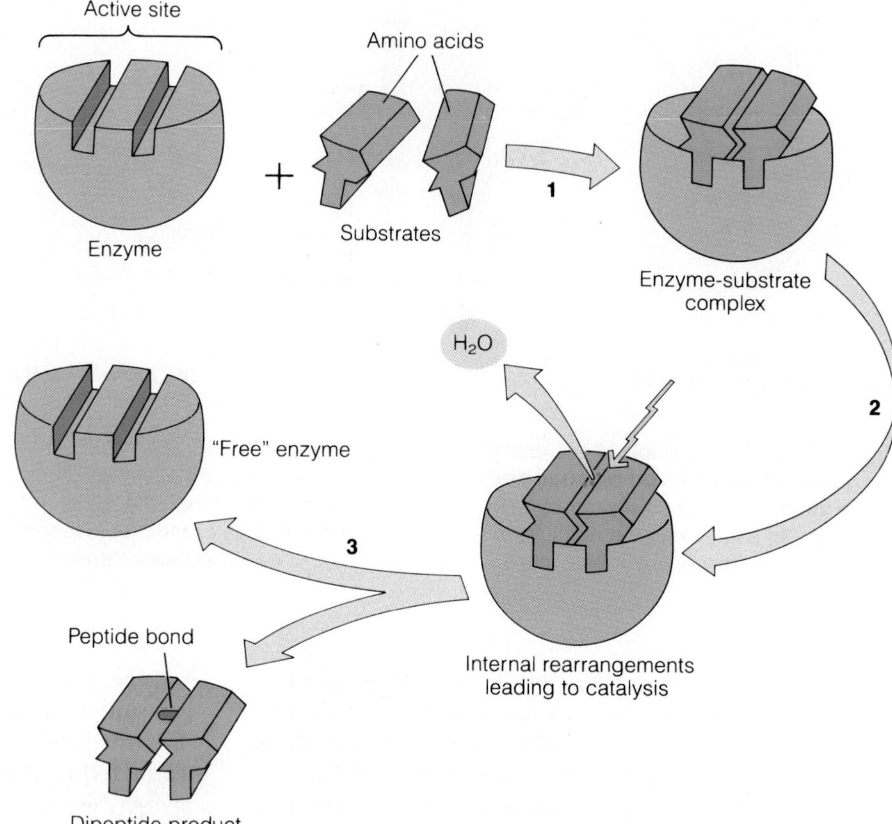

Figure 2.20 Mechanism of enzyme action. Each enzyme is highly specific in terms of the reaction(s) it can catalyze and bonds properly to only one or a few substrates. In this example, the enzyme catalyzes the formation of a dipeptide from specific amino acids.
Step 1: The enzyme-substrate complex is formed.

Step 2: Internal rearrangements occur. In this case, energy is absorbed (indicated by the yellow arrow) as a hydroxyl ion is removed from one amino acid and a hydrogen ion from the other. As the water molecule is ejected, the peptide bond is formed.
Step 3: The enzyme releases the product of the reaction, the dipeptide.

2. The enzyme-substrate complex undergoes an internal rearrangement that forms the product.

3. The enzyme releases the product of the reaction. This step shows the catalytic role of the enzyme: If the enzyme became part of the product, it would be a reactant and not a catalyst.

Since the unaltered enzymes can act again and again, cells need only small amounts of each enzyme. Usually, catalysis occurs with incredible speed; most enzymes can catalyze millions of reactions per minute.

Nucleic Acids (DNA and RNA)

The **nucleic** (noo-klē′-ik) **acids** include two major classes of molecules, **deoxyribonucleic** (dē-ok′-sē-rī″-bō-noo-klē″-ik) **acid (DNA)** and **ribonucleic acid (RNA).** Nucleic acids, composed of carbon, oxygen, hydrogen, nitrogen, and phosphorus, are the largest molecules in the body, and their structural units, **nucleotides,** are quite complex. Each nucleotide consists of three components joined together by dehydration synthesis (Figure 2.21a): a nitrogen-containing base, a pentose sugar, and a phosphate group. There are five varieties of nitrogen-containing bases that can contribute to nucleotide structure: *adenine* (a′-dih-nēn), abbreviated A; guanine (gwah′-nēn), G; cytosine (sī′-tō-sēn), C; thymine (thī′-mēn), T; and *uracil* (yer′-uh-sil), U. Adenine and guanine are large, two-ring bases, whereas cytosine, thymine, and uracil are smaller, single-ring bases.

Though DNA and RNA are both composed of nucleotides, they differ in many respects. DNA is found in the nucleus (control center) of the cell, where it constitutes the genetic material, or genes. DNA has two fundamental roles: It replicates itself before a cell divides, ensuring that the genetic information in the descendent cells is identical, and it provides instructions for building every protein in the body. By providing the directions for protein synthesis, DNA determines what type of an organism you will be—frog, human, oak tree—and directs your growth and development according to the information it contains. Although we have said that enzymes govern all chemical reactions, remember that enzymes, too, are proteins formed at the direction of DNA. RNA is located chiefly outside the nucleus and can be considered a "molecular slave" of DNA. That is, RNA carries out the orders for protein synthesis issued by DNA.

DNA is a long, double-stranded polymer—a double chain of nucleotides (Figure 2.21b and c). The bases in DNA are A, G, C, and T, and its pentose sugar is *deoxyribose* (as reflected in its name). The two nucleotide chains are held together by hydrogen bonds

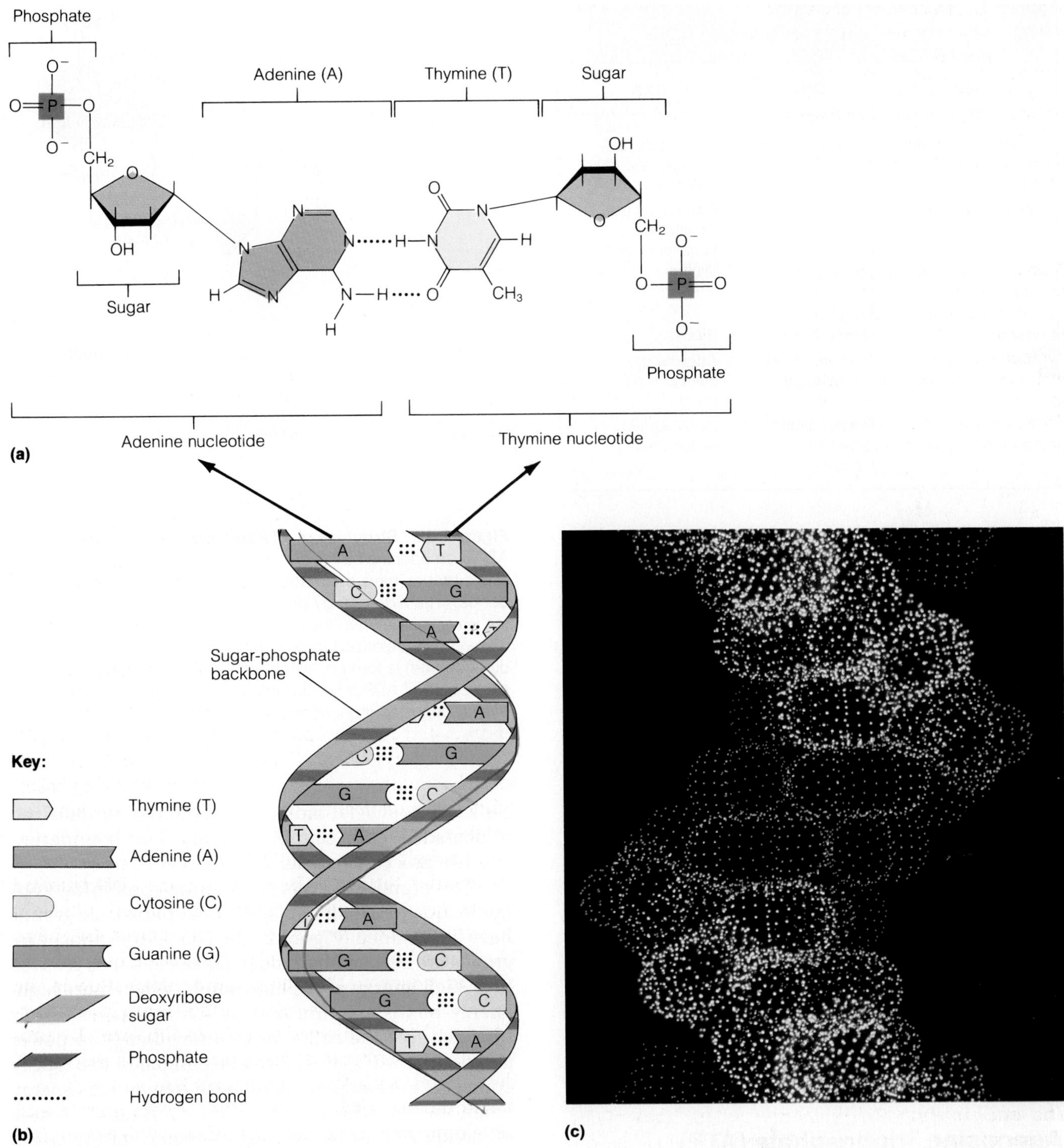

Figure 2.21 Structure of DNA. (**a**) The unit of DNA is the nucleotide, which is composed of a deoxyribose sugar molecule linked to a phosphate group, with a base attached to the sugar. Two linked nucleotides are illustrated. (**b**) DNA is a coiled double polymer of nucleotides (a double helix). The backbones of the ladderlike molecule are formed by alternating sugar and phosphate units. The rungs are formed by the binding together of complementary bases (A-T and G-C) by hydrogen bonds (shown by broken lines). (**c**) Computer-generated image of a DNA molecule.

Table 2.4 Comparison of DNA and RNA

Characteristic	DNA	RNA
Major cellular site	Nucleus	Cytoplasm (cell area outside the nucleus)
Major function	Is the genetic material; directs protein synthesis; replicates itself before cell division	Carries out the genetic instructions for protein synthesis
Sugar	Deoxyribose	Ribose
Bases	Adenine, guanine, cytosine, thymine	Adenine, guanine, cytosine, uracil
Structure	Double strand coiled into a double helix	Single straight or folded strand

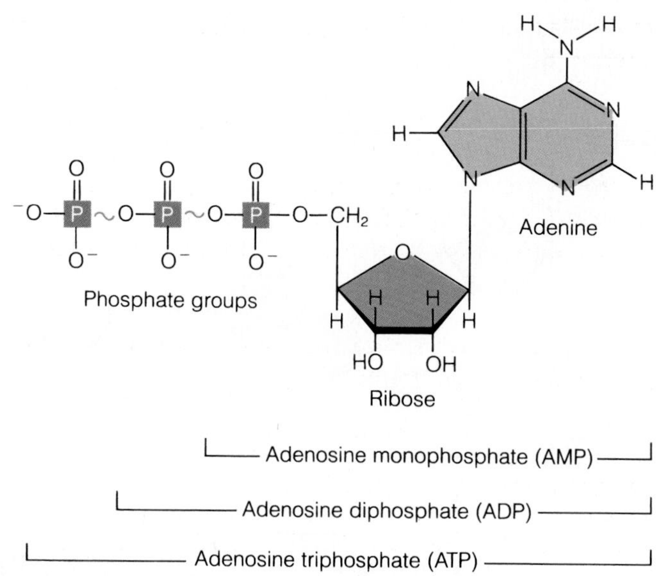

Figure 2.22 Structure of ATP (adenosine triphosphate). ATP is an adenine nucleotide to which two additional phosphate groups have been attached by high-energy phosphate bonds. (The high-energy phosphate bonds are indicated by the wavy lines.) When the terminal phosphate group is cleaved off, energy is released to do useful work and ADP (adenosine diphosphate) is formed. When the terminal phosphate group is cleaved off of ADP, a similar amount of energy is released and AMP (adenosine monophosphate) is formed.

between the bases, so that a ladderlike molecule is formed. The alternating sugar and phosphate molecules of each chain are called the *backbone;* the two backbones form the "uprights" of the "ladder," while the joined bases form the "rungs." Bonding of the bases is very specific: A always bonds to T, and G always bonds to C. A and T are therefore called *complementary bases,* as are C and G. A specific base sequence (say, ATGA) on one nucleotide strand would necessarily be bonded to a complementary base sequence (TACT) on the other strand. The whole molecule is coiled into a spiral, staircase-like structure called a *double helix.*

RNA molecules are single strands of nucleotides. RNA bases include A, G, C, and U (U replaces the T found in DNA), and its sugar is *ribose* instead of deoxyribose. Three varieties of RNA exist, distinguished by their relative size and shape, and each has a highly specific role to play in carrying out DNA's instructions. We will discuss DNA replication and the relative roles of DNA and RNA in protein synthesis in Chapter 3. Table 2.4 provides a comparison of DNA and RNA.

Adenosine Triphosphate (ATP)

Although glucose is the most important cellular fuel, none of the chemical energy contained in its bonds can be used directly to power chemical reactions. The energy released during glucose breakdown is captured and stored in small "packets" of energy—the bonds of **adenosine triphosphate** (uh-den′-uh-sēn trī″-fos′-fāt), or **ATP.** The synthesis of ATP is an all-important cellular function because it provides a form of chemical energy that is universally usable by all body cells. Without ATP, molecules cannot be synthesized or degraded, cells cannot maintain their boundaries, and life processes cease.

Structurally, ATP is an adenine-containing RNA nucleotide to which two additional phosphate groups have been added (Figure 2.22). The "extra" phosphate groups are attached by unique chemical bonds referred to as **high-energy phosphate bonds.** When these high-energy bonds are ruptured by hydrolysis, energy is liberated that the cell can use immediately to power a particular activity. ATP can be compared to a tightly coiled spring ready to uncoil with tremendous energy when the catch is released. An important advantage of having ATP as the "energy currency" of the cell is that the amount of energy released by breaking its high-energy bonds corresponds closely to that needed to drive most biochemical reactions. As a result, cells are protected from excessive energy release that might be damaging, and energy squandering is kept to a minimum.

Cleavage of the terminal phosphate bond of ATP yields a molecule with two phosphate groups—*adenosine diphosphate (ADP)* and an inorganic phosphate group, indicated by Pi—accompanied by a release of energy (E):

$$ATP \rightleftharpoons ADP + Pi + E$$

As ATP's bonds are broken to provide energy for cellular needs, ADP accumulates. Sometimes, the second high-energy phosphate bond is also cleaved, liberating a similar amount of energy and producing adenosine monophosphate (AMP). ATP supplies are replenished as glucose and other fuel molecules are decomposed and their bond energy is released. The same amount of energy that is liberated when ATP's terminal phosphates are cleaved off must be captured and used to drive the reverse reaction to reattach phosphates and re-form the high-energy bonds.

Related Clinical Terms

Acidosis (a″-sih-dō′-sis) (*acid* = sour, sharp) A condition of acidity or low pH (below 7.35) of the blood; high hydrogen ion concentration.

Alkalosis (al″-kuh-lō′-sis) A condition of basicity or high pH (above 7.45) of the blood; low hydrogen ion concentration.

Heavy metals Metals with toxic effects on the body, including arsenic, mercury, and lead; iron, also included in this group, is toxic in high concentrations.

Ionizing radiation Radiation that causes atoms to ionize; e.g., radioisotopes and X-rays.

Ketosis (kē-tō′-sis) A form of acidosis resulting from excessive ketones (breakdown products of fats) in the blood; common during starvation and acute attacks of diabetes mellitus.

Radiation sickness Disease resulting from exposure of the body to radioactivity; digestive system organs are most affected.

Chapter Summary

PART 1: BASIC CHEMISTRY

DEFINITION OF CONCEPTS: MATTER AND ENERGY (pp. 25–26)

1. Matter is anything that takes up space and has mass. Energy is the capacity to do work or put matter into motion.

Potential and Kinetic Energy (p. 25)

2. Energy exists as potential energy (stored energy or energy of position) and kinetic energy (active or working energy).

Forms of Energy (p. 26)

3. Forms of energy involved in body functioning are chemical, electrical, radiant, and mechanical. Of these, chemical (bond) energy is most important.

4. Energy may be converted from one form to another, but some energy is always unusable (lost as heat) in such transformations.

COMPOSITION OF MATTER: ATOMS AND ELEMENTS (pp. 26–30)

1. Elements are unique substances that cannot be decomposed into simpler substances by ordinary chemical methods. Four elements (carbon, hydrogen, oxygen, and nitrogen) make up 96% of body weight.

Atomic Structure (pp. 26–28)

2. The building blocks of elements are atoms.

3. Atoms are composed of positively charged protons, negatively charged electrons, and uncharged neutrons. Protons and neutrons are located in the atomic nucleus; electrons are outside the nucleus in areas of space called orbitals, within the electron shells. In any atom, the number of electrons equals the number of protons.

Identifying Elements (pp. 28–29)

4. Atoms may be identified by their atomic number (p^+) and mass number ($p^+ + n^0$). The chemical notation ^4_2He means that helium (He) has an atomic number of 2 and a mass number of 4.

5. Isotopes of an element differ in the number of neutrons they contain. The atomic weight of any element is approximately equal to the mass number of its most abundant isotope.

Radioisotopes (pp. 29–30)

6. Many heavy isotopes are unstable (radioactive) and decompose to more stable forms by emitting α or β particles or γ rays. Radioisotopes are useful in medical diagnosis and biochemical research.

HOW MATTER IS COMBINED: MOLECULES AND MIXTURES (pp. 30–33)

Molecules and Compounds (p. 30)

1. A molecule is the smallest unit resulting from the chemical bonding of two or more atoms. If the atoms are different, they form a molecule of a compound.

Mixtures (pp. 30–31)

2. Mixtures are physical combinations of solutes in a solvent. The components of a mixture retain their individual properties.

3. The types of mixtures, in order of increasing solute size, are solutions, colloids, and suspensions.

4. Solution concentrations are typically designated in terms of percent or molarity.

CHEMICAL BONDS (pp. 34–38)

The Role of Electrons in Chemical Bonding (pp. 34–35)

1. Electrons of an atom occupy areas of space called electron shells or energy levels. Electrons in the shell farthest from the nucleus (valence shell) are most energetic.

2. Chemical bonds are energy relationships between valence shell electrons of the reacting atoms. Atoms with a full valence shell or eight valence shell electrons are chemically unreactive (inert); those with an incomplete valence shell interact with other atoms to achieve a stable electronic configuration.

Types of Chemical Bonds (pp. 35–38)

3. Ionic bonds are formed when valence shell electrons are completely transferred from one atom to another.

4. Covalent bonds are formed when atoms share electron pairs. If the electron pairs are shared equally, the molecule is nonpolar; if they are shared unequally, it is a polar molecule, or a dipole.

5. Hydrogen bonds are weak bonds between hydrogen and nitrogen or oxygen. They bind together different molecules (e.g., water molecules) or different parts of the same molecule.

CHEMICAL REACTIONS (pp. 38–41)

Chemical Equations (pp. 38–39)

1. Chemical reactions involve the formation, breaking, or rearrangement of chemical bonds.

Patterns of Chemical Reactions (p. 39)

2. Chemical reactions include synthesis, decomposition, and exchange reactions. All chemical reactions involve energy exchanges.

Reversibility of Chemical Reactions (p. 40)

3. If reaction conditions remain unchanged, all chemical reactions eventually reach a state of chemical equilibrium in which the reaction proceeds in both directions at the same rate.

4. All chemical reactions are theoretically reversible, but many biological reactions go only in one direction because of energy requirements and/or the removal of reaction products.

Factors Influencing the Rate of Chemical Reactions (pp. 40–41)

5. Chemical reactions occur only when particles collide and valence shell electrons interact.

6. The smaller the reacting particles, the greater their kinetic energy and hence the reaction rate. Increasing the temperature or concentration of the reactants, as well as the presence of catalysts, increases chemical reaction rates.

PART 2: BIOCHEMISTRY: THE COMPOSITION AND REACTIONS OF LIVING MATTER

INORGANIC COMPOUNDS (pp. 41–44)

1. Most inorganic compounds do not contain carbon. Those found in the body include water, salts, and inorganic acids and bases.

Water (pp. 41–42)

2. Water is the single most abundant compound in the body. It absorbs and releases heat slowly, acts as a universal solvent, participates in chemical reactions, and cushions body organs.

Salts (pp. 42)

3. Salts are ionic compounds that dissolve in water and act as electrolytes. Calcium and phosphorus salts contribute to the hardness of bones and teeth. Ions of salts are involved in many physiological processes.

Acids and Bases (pp. 42–44)

4. Acids are proton donors; in water, they ionize and dissociate, releasing hydrogen ions (which account for their properties) and anions.

5. Bases are proton acceptors. The most important inorganic bases are the hydroxides; bicarbonate ion and ammonia are important bases in the body.

6. pH is a measure of hydrogen ion concentration of a solution (in moles per liter). A pH of 7 is neutral; a higher pH is alkaline, and a lower pH is acidic. Normal blood pH is 7.35–7.45. Buffers help to prevent excessive changes in the pH of body fluids.

ORGANIC COMPOUNDS (pp. 44–57)

1. Organic compounds contain carbon. Those found in the body include carbohydrates, lipids, proteins, and nucleic acids, all of which are synthesized by dehydration synthesis and digested by hydrolysis. All of these biological molecules contain C, H, and O. Proteins and nucleic acids also contain N.

Carbohydrates (pp. 44–45)

2. Carbohydrate building blocks are monosaccharides, the most important of which are hexoses (glucose, fructose, galactose) and pentoses (ribose, deoxyribose).

3. Disaccharides (sucrose, lactose, maltose) and polysaccharides (starch, glycogen) are composed of linked monosaccharide units.

4. Carbohydrates, particularly glucose, are the major energy fuel for forming ATP. Excess carbohydrates are stored as glycogen or converted to fat for storage.

Lipids (pp. 45–49)

5. Lipids dissolve in fats or organic solvents, but not in water.

6. Neutral fats are found chiefly in fatty tissue; they serve as insulation and reserve body fuel.

7. Phospholipids are modified neutral fats that have polar and nonpolar portions. They are found in all plasma membranes.

8. The steroid cholesterol is found in cell membranes and is the basis of steroid hormones, bile, and vitamin D.

Proteins (pp. 49–54)

9. The unit of proteins is the amino acid; 20 common amino acids are found in the body.

10. Many amino acids joined by peptide bonds form a polypeptide. A protein (one or more polypeptides) is distinguished by the number and sequence of amino acids in its chain(s) and by the complexity of its three-dimensional structure.

11. Structural proteins, such as keratin, collagen, and elastin, have secondary structure (α helix or β-pleated sheet).

12. Functional proteins (enzymes, some hormones, antibodies, hemoglobin, and others) achieve tertiary or quaternary structure and are globular, soluble molecules. Functional proteins are subject to denaturation by extremes of temperature or pH.

13. Enzymes, biological catalysts, increase the rate of chemical reactions by decreasing the amount of activation energy needed. They do this by combining with the reactants and holding them in the proper position to interact. Many enzymes require cofactors to function.

Nucleic Acids (DNA and RNA) (pp. 54–56)

14. Nucleic acids include deoxyribonucleic acid (DNA) and ribonucleic acid (RNA). The structural unit of nucleic acids is the nucleotide, which consists of a nitrogenous base (adenine, guanine, cytosine, thymine, or uracil), a sugar (ribose or deoxyribose), and a phosphate group.

15. DNA is a double-stranded helix: it contains deoxyribose and the bases A, G, C, and T. DNA specifies protein structure and replicates itself exactly before cell division.

16. RNA is single-stranded; it contains ribose and the bases A, G, C, and U. RNA is involved in carrying out DNA's instructions for protein synthesis.

Adenosine Triphosphate (ATP) (pp. 56–57)

17. ATP is the universal energy compound of body cells. Some of the energy liberated by the breakdown of glucose and other food fuels is captured in the bonds of ATP molecules and transferred to energy-consuming reactions.

Review Questions

Multiple Choice/Matching

1. Which of the following forms of energy are in use during vision? (a) chemical, (b) electrical, (c) mechanical, (d) radiant.

2. All of the following are examples of the four major elements contributing to body mass except (a) hydrogen, (b) carbon, (c) nitrogen, (d) sodium, (e) oxygen.

3. The mass number of an atom is (a) equal to the number of protons it contains, (b) the sum of its protons and neutrons, (c) the sum of all of its subatomic particles, (d) the average of the mass numbers of all of its isotopes.

4. A deficiency in this element can be expected to reduce the hemoglobin content of blood: (a) Fe, (b) I, (c) F, (d) Ca, (e) K.

5. Which set of terms best describes a proton? (a) negative charge, massless, in the orbital; (b) positive charge, 1 amu, in the nucleus; (c) uncharged, 1 amu, in the nucleus.

6. The subatomic particles responsible for the chemical behavior of atoms are (a) electrons, (b) ions, (c) neutrons, (d) protons.

7. Which of the following are molecules of a compound? (a) N_2, (b) C, (c) $C_6H_{12}O_6$, (d) NaOH, (e) S_8.

8. Which of the following does *not* describe a mixture? (a) Properties of its components are retained, (b) chemical bonds are formed, (c) components can be separated physically, (d) includes both heterogeneous and homogenous examples.

9. A mixture that is homogeneous and transparent and does not scatter light is (a) a true solution, (b) a colloid, (c) a compound, (d) a suspension.

10. When a pair of electrons are shared between two atoms, the bond formed is called (a) a single covalent bond, (b) a double covalent bond, (c) a triple covalent bond, (d) an ionic bond.

11. Molecules formed when electrons are shared unequally are (a) salts, (b) polar molecules, (c) nonpolar molecules.

12. Which of the following covalently bonded molecules are polar?

```
                        H                 H
                        |                 |
(a) H—Cl   (b) H— C —H   (c) Cl— C —Cl   (d) O=O
                        |                 |
                        H                 Cl
```

13. Identify each reaction as (a) a synthesis reaction, (b) a decomposition reaction, or (c) an exchange reaction.

(1) 2 Hg + $O_2 \rightarrow$ 2 HgO
(2) HCL + NaOH $\rightarrow$ NaCl + H_2O

14. Factors that accelerate the rate of chemical reactions include all but (a) the presence of catalysts, (b) increasing the temperature, (c) decreasing the temperature, (d) increasing the concentration of the reactants.

15. Which of the following molecules is an inorganic molecule? (a) sucrose, (b) cholesterol, (c) collagen, (d) sodium chloride.

16. Water's importance to living systems reflects (a) its polarity and solvent properties, (b) its high heat capacity, (c) its high heat of vaporization, (d) its chemical reactivity, (e) all of these.

17. Acids (a) release hydroxyl ions when dissolved in water, (b) are proton acceptors, (c) cause the pH of a solution to rise, (d) release protons when dissolved in water.

18. A chemist, during the course of an analysis, runs across a chemical that he determines to be composed of carbon, hydrogen, and oxygen in the proportion 1:2:1 and whose molecular shape is six-sided. It is probably (a) a pentose, (b) an amino acid, (c) a fatty acid, (d) a monosaccharide, (e) a nucleic acid.

19. A neutral fat consists of (a) glycerol plus up to three fatty acids, (b) a sugar-phosphate backbone to which two amino groups are attached, (c) two to several hexoses, (d) amino acids that have been thoroughly saturated with hydrogen.

20. A chemical has an amine group and an organic acid group. It does not, however, have any peptide bonds. It is (a) a monosaccharide, (b) an amino acid, (c) a protein, (d) a fat.

21. The lipid used as the basis of vitamin D, sex hormones, and bile is (a) neutral fats, (b) cholesterol, (c) phospholipids, (c) prostaglandin.

22. Enzymes are organic catalysts that (a) alter the direction in which a chemical reaction proceeds, (b) determine the nature of the products of a reaction, (c) increase the speed of a chemical reaction, (d) are essential raw materials for a chemical reaction that are converted into some of its products.

Short Answer Essay Questions

23. Define or describe energy, and explain the relationship between potential and kinetic energy.

24. Some energy is lost in every energy conversion. Explain the meaning of this statement. (Direct your response to answering the question, Is it really lost? If not, what then?)

25. Provide the atomic symbol for each of the following elements: (a) calcium, (b) carbon, (c) hydrogen, (d) iron, (e) nitrogen, (f) oxygen, (g) potassium, (h) sodium.

26. Consider the following information about the structure of three atoms:

$$^{12}_{6}C \quad ^{13}_{6}C \quad ^{14}_{6}C$$

(a) How are they similar to one another? (b) How do they differ from one another? (c) What are the members of such a group of atoms called? (d) Using the planetary model, draw the atomic configuration of $^{12}_{6}C$ showing the relative position and numbers of its subatomic particles.

27. How many moles of aspirin, $C_9H_8O_4$, are there in a bottle containing 450 g by weight? (*Note:* The approximate atomic weights of its atoms are C = 12, H = 1, and O = 16.)

28. Given the following types of atoms, decide which type of bonding, ionic or covalent, is most likely to occur: (a) two oxygen atoms; (b) four hydrogen atoms and one carbon atom; (c) a potassium atom ($^{39}_{19}K$) and a fluorine atom ($^{19}_{9}F$).

29. What are hydrogen bonds and how are they important in the body?

30. The following equation, which represents the oxidative breakdown of glucose by body cells, is a reversible reaction.

glucose + oxygen $\rightarrow$ carbon dioxide + water + ATP

(a) How can you indicate that the reaction is reversible? (b) How can you indicate that the reaction is in chemical equilibrium? (c) Define chemical equilibrium.

31. (a) Differentiate clearly between primary, secondary, and tertiary protein structure. (b) What level of structure do structural proteins achieve? (c) How do functional proteins relate to the structural levels?

32. Dehydration and hydrolysis reactions are essentially opposite reactions. How do these two types of reactions relate to the synthesis and degradation (breakdown) of biological molecules?

33. Describe the mechanism of enzyme activity. In your discussion, explain how enzymes decrease activation energy requirements.

Clinical Application Questions

34. Some antibiotics act by binding to certain essential enzymes in the target bacteria. (a) How might these antibiotics influence the chemical reactions controlled by the enzymes? (b) What is the anticipated effect on the bacteria? On the person taking the antibiotic prescription?

35. Mrs. Roberts, in a diabetic coma, has just been admitted to Noble Hospital. Determination of her blood pH indicates that she is in severe acidosis, and measures are quickly instituted to bring her blood pH back within normal limits. (a) Define pH and note the normal pH of blood. (b) Why is severe acidosis problematic?

CHAPTER 3

Cells: The Living Units

Chapter Outline and Student Objectives

Overview of the Cellular Basis of Life (pp. 61–62)

1. Define cell.
2. List the three major regions of a generalized cell and indicate the general functional role(s) of each region.

The Plasma Membrane: Structure (pp. 62–65)

3. Describe the chemical composition of the plasma membrane.
4. Compare the structure and function of tight junctions, desmosomes, and gap junctions.

The Plasma Membrane: Functions (pp. 66–74)

5. Relate plasma membrane structure to active and passive transport mechanisms. Differentiate clearly between these transport processes relative to energy source, substances transported, direction, and mechanism.
6. Define membrane potential and explain how the resting membrane potential is maintained.

The Cytoplasm (pp. 74–81)

7. Describe the composition of the cytosol; define inclusions and list several types.
8. Discuss the structure and function of mitochondria.
9. Discuss the structure and function of ribosomes, the endoplasmic reticulum, and the Golgi apparatus; note functional interrelationships among these organelles.
10. Compare the functions of lysosomes and peroxisomes.
11. Name and describe the structure and function of cytoskeletal elements.
12. Describe the roles of centrioles in mitosis and in formation of cilia and flagella.

The Nucleus (pp. 82–84)

13. Describe the chemical composition, structure, and function of the nuclear membrane, nucleolus, and chromatin.

Cell Growth and Reproduction (pp. 84–93)

14. List the phases of the cell life cycle and describe the events of each phase.
15. Describe the process of DNA replication. Explain the importance of this process.
16. Define gene and explain the function of genes. Also explain the meaning of "genetic code."
17. Name the two phases of protein synthesis and describe the roles of DNA, mRNA, tRNA, and rRNA in each phase. Contrast triplets, codons, and anticodons.

Extracellular Materials (p. 93)

18. Name and describe the composition of extracellular materials.

Developmental Aspects of Cells (pp. 93–96)

19. Discuss some of the theories of cell aging.

Preview of Selected Key Terms

Cell Living structural and functional unit of all organisms.

Nucleus (noo′-klē-is) (*nucle* = pit, kernel) The control center of a cell; contains genetic material in the form of diffuse chromatin threads or condensed chromosomes (*chrom* = colored; *soma* = body).

Cytoplasm (sī′-tō-plaz-zim) (*cyto* = cell; *plasm* = shaped, molded) The cellular material surrounding the nucleus and enclosed by the plasma membrane.

Organelles (or″-guh-nelz′) (*elle* = little) Small cellular structures (ribosomes, mitochondria, and others) that perform specific metabolic functions for the cell as a whole.

Hydrophilic (hī″-drō-fih′-lik) (*hydro* = water; *phil* = like, love) and /**hydrophobic** (hī″-drō-fō′-bik) (*phob* = dislike, fear) Terms that refer to molecules, or portions of molecules, that interact with water and charged particles (hydrophilic) or that interact only with nonpolar molecules (hydrophobic).

Concentration gradient The difference in the concentration of some substance between two different areas.

Passive transport Membrane transport processes that do not require cellular energy (ATP); e.g., diffusion, which is driven by kinetic energy.

Tonicity (tuh-nih'-sih-tē) (*ton* = strength) A measure of the ability of a solution to cause a change in cell shape or tone by promoting osmotic flows of water.

Active transport Membrane transport processes for which ATP is provided; e.g., solute pumping and endocytosis.

Mitosis (mī-tō'-sis) (*mit* = thread; *osis* = process) Process during which the chromosomes are redistributed to two daughter nuclei; nuclear division. Typically followed by cytoplasmic division (cytokinesis) (*kines* = movement).

Genetic code Information encoded in nucleotide base sequence.

All organisms are cellular in nature, be they one-celled independent "generalists" like amoebas or complex multicellular organisms such as humans, dogs, or fir trees. Just as bricks and timbers are the structural units of a house, cells are the structural units of all living things. The human body has trillions of these tiny building blocks.

This chapter focuses on structural similarities among different kinds of cells and discusses some of the functional processes common to all of our cells. Specialized cells and their unique functions are considered in detail in later chapters.

Overview of the Cellular Basis of Life

The English scientist Robert Hooke first observed plant cells with a crude microscope in the late 1600s, but it was not until the mid-1800s that two German scientists, Matthias Schleiden and Theodor Schwann, were bold enough to assert that all living things are composed of cells. The German pathologist Rudolf Virchow extended this idea by suggesting that cells arise only from other cells. Virchow's proclamation was a true landmark in biological history because the theory of spontaneous generation, which stated that organisms arise spontaneously from garbage or other nonliving matter, was rampant at the time. Since the late 1800s, cell research has been exceptionally fruitful, and our present understanding of the cell provides us with four very solid concepts:

1. A cell is the basic structural and functional unit of living organisms; when you define the properties of a cell, you are in fact defining the properties of life.
2. The activity of an organism is dependent on both the individual and collective activity of its cells.
3. According to the *principle of complementarity*, the biochemical activities of cells are determined and made possible by the specific subcellular structures of cells.
4. The continuity of life has a cellular basis.

All of these concepts will be expanded upon, but for now, let us just consider the idea that the cell is the minimum amount of matter that is alive and as such is the fundamental unit from which the biological hierarchy is fashioned and on which life processes depend. Thus, whatever its form, however it behaves, the cell is the microscopic package that contains all the parts and processes necessary to survive in an ever-changing world. Indeed, loss of cellular homeostasis underlies virtually every disease that can possibly confront us.

Perhaps the most striking property of a cell is its complex organization. Chemically, cells are composed chiefly of carbon, hydrogen, nitrogen, oxygen, and trace amounts of several other elements. These substances exist in the air around us and the ground beneath our feet, but within the cell, they take on the unique characteristics of life. Life, then, relates to the way living matter is organized and carries out metabolic processes—considerations far beyond chemical composition.

The microscope has revealed an amazing diversity in both cell size and shape. Cell diameter ranges from about 2 micrometers (1/12,000 of an inch) in smallest cells to 10 centimeters (4 inches) or more in the largest cell (the yolk of an ostrich egg). The "typical" human cell has a diameter of about 10 micrometers (μm); the largest (the fertilized egg) is nearly 100 μm across and is still just a speck, nearly impossible to see with the naked eye.

The range in cell length is even greater—from a few micrometers to a meter or more. Some skeletal muscle cells are 30 centimeters (cm) long, and the nerve cells that cause your foot muscles to contract run from the end of your spinal cord to your foot—well over a meter!

There is also tremendous variability in cell shape. Some cells are spherical (fat cells), some disk-shaped (red blood cells), some branching (nerve cells), and some cubelike (kidney tubule cells), and this by no means exhausts all the possibilities. A cell's shape reflects its function. For example, the flat, tilelike epithelial cells that line the inside of your cheek fit closely together, forming a living barrier that protects underlying tissues from bacterial invasion.

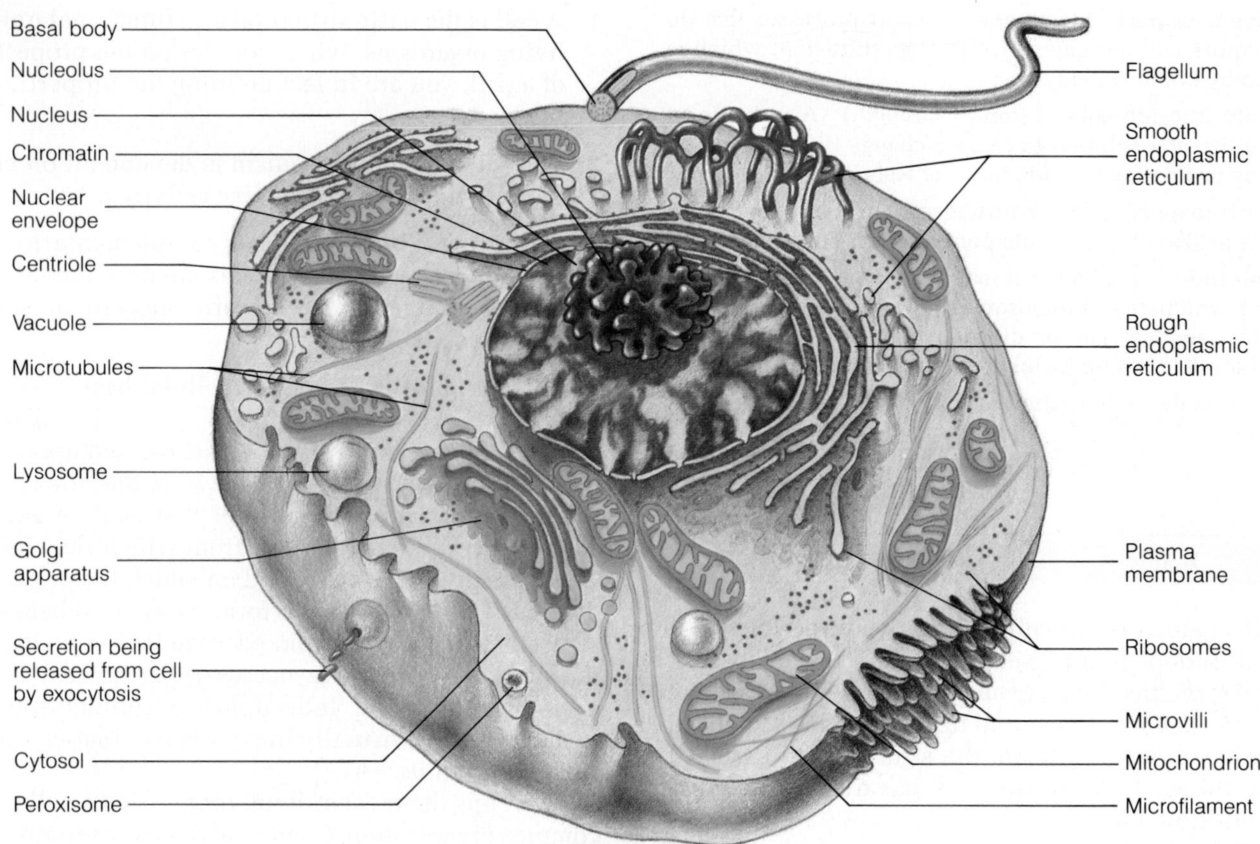

Figure 3.1 Structure of the generalized cell. No cell is exactly like this one, but this composite illustrates features common to many human cells.

While no one cell type is exactly like all others, cells *do* have many common structural and functional features that can be described in terms of a **generalized,** or **composite, cell** (Figure 3.1). All cells have three major parts: a nucleus, cytoplasm, and a plasma membrane. The *nucleus* (noo′-klē-is), or control center of the cell, is easily seen with a light microscope and is usually centrally located. The nucleus is surrounded by *cytoplasm* (sī′-tō-pla-zim), which is packed with *organelles,* small structures that perform specific functions in the cell. The cytoplasm is enclosed by the *plasma membrane,* which forms the external cell boundary. Functions of the various cell structures are summarized in Table 3.2 (p. 85) and described in detail later in the chapter.

The Plasma Membrane: Structure

The flexible **plasma membrane** defines the extent of the cell and acts as a fragile barrier. (The term *cell membrane* is often used instead, but since nearly all cellular organelles are membranous, we will specifically designate the cell's surface or outer limiting membrane as the plasma membrane.) Although the plasma membrane is important in maintaining the integrity of the cell, it is much more than a passive envelope. As you will see, its unique structure allows it to play a dynamic role in many cellular activities.

The Fluid Mosaic Model

The *fluid mosaic model* of membrane structure (Figure 3.2) depicts the plasma membrane as an exceedingly thin (7–8 nm) but stable structure composed chiefly of a double layer, or bilayer, of phospholipid molecules with protein molecules dispersed in it. The lipid bilayer forms the basic "fabric" of the membrane and is relatively impermeable to most water-soluble molecules. The proteins are responsible for most specialized functions of the plasma membrane. Some mark the cell for recognition by the immune system; some serve as receptors for hormones and other chemical messengers; some are enzymes; and others transport nutrients and other substances across the membrane.

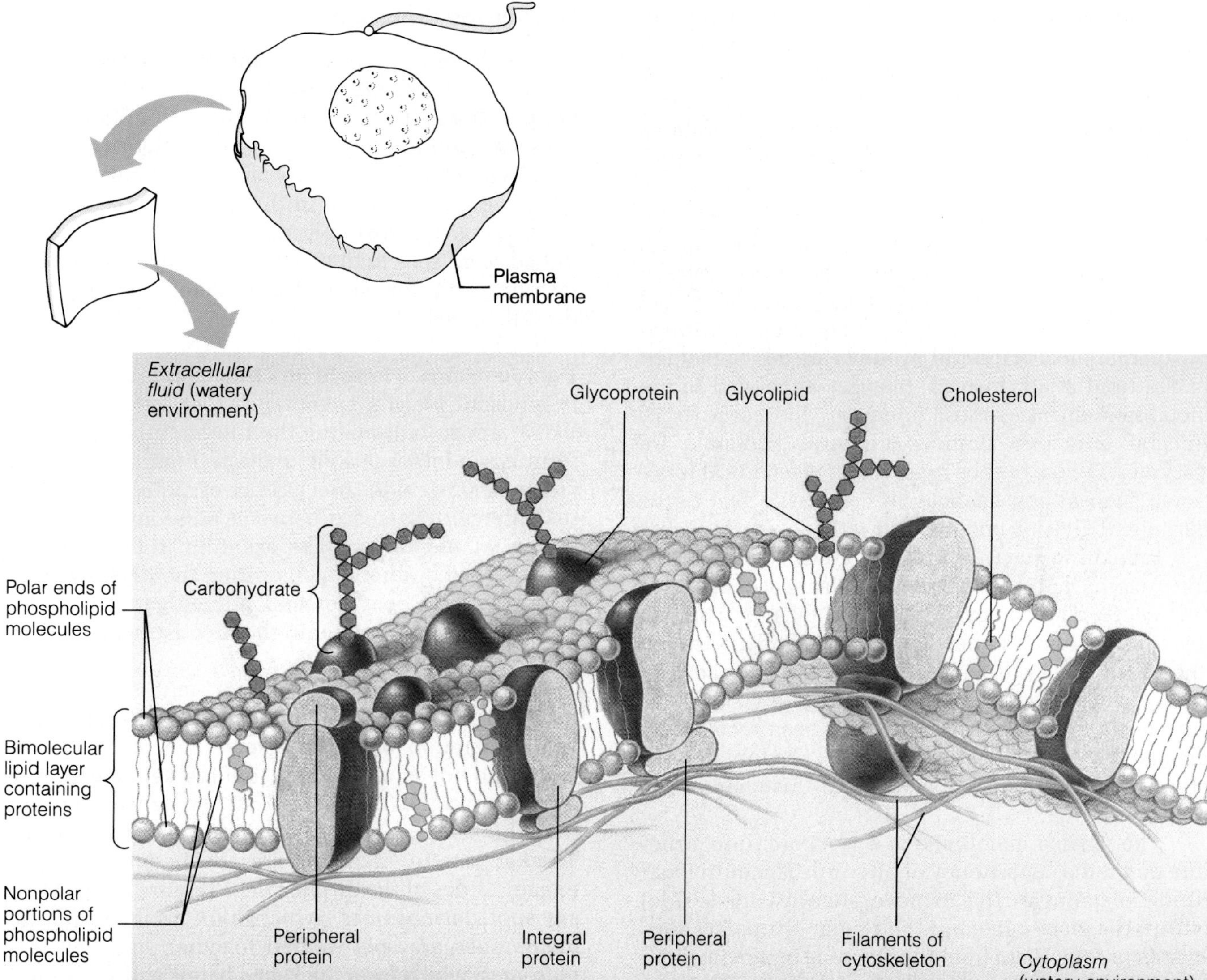

Figure 3.2 Structure of the plasma membrane according to the fluid mosaic model.

As explained in Chapter 2, phospholipids have a polar phosphorus-containing end, or "head," and a nonpolar "tail" composed of hydrocarbon fatty acid chains. The polar "head" interacts with water and is said to be *hydrophilic* (water-loving). The nonpolar "tail" interacts only with other nonpolar substances, avoiding water and charged particles; this end is *hydrophobic* (water-fearing). Because of these phospholipid properties, all biological membranes share a common basic structure: They are "sandwiches" composed of two layers of phospholipid molecules in which the polar heads are exposed to water inside and outside the cell, and the nonpolar tails face each other and are buried in the internal portion of the membrane. This self-orienting property of phospholipids allows the lipid portion of a biological membrane to form by self-assembly and to reseal (repair) itself quickly when torn.

The two lipid layers differ somewhat in the specific kinds of phospholipids they contain, and about 10% of the externally facing lipid molecules have attached sugar groups and are called *glycolipids* (glī″-kō-lih′-pidz) (see Figure 3.2). The function of the glycolipids is not yet fully understood. The membrane also contains substantial amounts of cholesterol. Cholesterol is believed to stabilize the lipid membrane by wedging its platelike hydrocarbon rings between the phospholipid tails and immobilizing parts of them; since this also prevents the phospholipids from aggregating, it helps to keep the membrane more fluid.

Most membrane proteins also have both hydrophobic and hydrophilic regions, allowing them to interact with nonpolar parts of lipid molecules within the membrane and with water inside and outside the cell. Notice also that there are two distinct protein populations: integral and peripheral (see Figure 3.2). **Integral proteins** are firmly inserted in the lipid bilayer. Some face the aqueous environment on one membrane face only, but most are *transmembrane proteins* that span the entire width of the membrane and protrude on both sides. Transmembrane proteins are mainly involved in transport functions. Some aggregate to form channels, or pores, through which small, water-soluble molecules or ions can pass, thereby bypassing the lipid part of the membrane. Others act as carriers that bind to a substance and move it through the membrane. **Peripheral proteins** are not embedded in the lipid at all. Instead, they are appended to the membrane surface, most often to exposed parts of integral proteins. Some peripheral proteins are enzymes. Others may be involved in mechanical functions, such as the changes in cell shape that occur during cell division and the contraction of muscle cells.

Branching sugar groups are attached to most of the proteins that abut the extracellular space. The term *cell coat* or *glycocalyx* (glī″-kō-kā′-liks) is used to describe the fuzzy, somewhat sticky carbohydrate-rich area at the cell surface; you can think of your cells as being "sugar-coated." In addition, the glycocalyx is enriched by glycoproteins that have been secreted by the cell and then cling to its surface. The functions of glycoproteins are described later in this chapter (on p. 74).

The plasma membrane is a dynamic fluid structure about the consistency of olive oil. The individual lipid molecules are free to move laterally (side to side) within the membrane, but their polar-nonpolar interactions prevent them from flip-flopping or moving from one lipid layer to the other. Some of the proteins, however, appear to be "tethered" to intracellular structures (the cytoskeleton) and are much more restricted in their movement.

Specializations of the Plasma Membrane

Microvilli

Microvilli (mī″-krō-vih′-lē) are minute, fingerlike extensions of the plasma membrane that project from a free, or exposed, cell surface (see Figure 3.1). They increase the plasma membrane surface area tremendously and are most often found on the surface of absorptive cells such as kidney tubule and intestinal cells. Microvilli have a core of actin filaments. Actin is a contractile protein, but in microvilli, it appears to function as a mechanical "stiffener."

Membrane Junctions

Although certain cell types—blood cells, sperm cells, and some phagocytic cells—are "footloose" within the body, many cells, particularly those of epithelial tissues, are bound together into rather tight communities. Typically, three factors act to bind cells closely: adhesive glycoproteins in the cell coat; the wavy, or undulating, courses of plasma membranes, which permit adjacent cells to fit together in a tongue-and-groove fashion; and the specialized membrane junctions described next.

Tight Junctions. In **tight junctions,** protein molecules in adjacent plasma membranes fuse together tightly like a zipper, obliterating the intercellular space and forming an impermeable junction (Figure 3.3a). Tight junctions prevent the free passage of molecules through the intercellular space between adjacent cells of an epithelial membrane. For example, tight junctions between the epithelial cells lining the digestive tract keep digestive enzymes and microorganisms in the intestine from seeping into the bloodstream.

Desmosomes. **Desmosomes** (dez′-muh-sōmz) act as mechanical couplings or adhesion junctions that prevent separation of tissue layers. The plasma membranes do not actually touch, but are held together by fine glycoprotein filaments stretched between button-like thickenings of the inner plasma membranes. Thicker keratin filaments are attached to the cytoplasmic sides of the desmosomes (Figure 3.3b). There are *spot desmosomes,* which look as if the plasma membranes are spot-welded together, and *belt desmosomes,* which form thickened bands entirely around the cells. Desmosomes are abundant in tissues subjected to great mechanical stress, such as skin, heart muscle, and the neck of the uterus.

Gap Junctions. The basic function of **gap junctions** is to provide for direct passage of chemical substances between adjacent cells. At gap junctions, the adjacent plasma membranes are very close, and the cells are connected by hollow cylinders composed of transmembrane proteins, called *connexons* (kuh-nek′-sonz). Ions, sugars, and other small molecules pass through these channels from one cell to the next (Figure 3.3c). Gap junctions between embryonic cells are thought to be vital for distributing nutrients before the circulatory system is established. In adult tissues, gap junctions are found in electrically excitable tissues, such as the heart and smooth muscle, where passage of ions from cell to cell helps synchronize electrical activity and contraction. They also occur between some nerve cells.

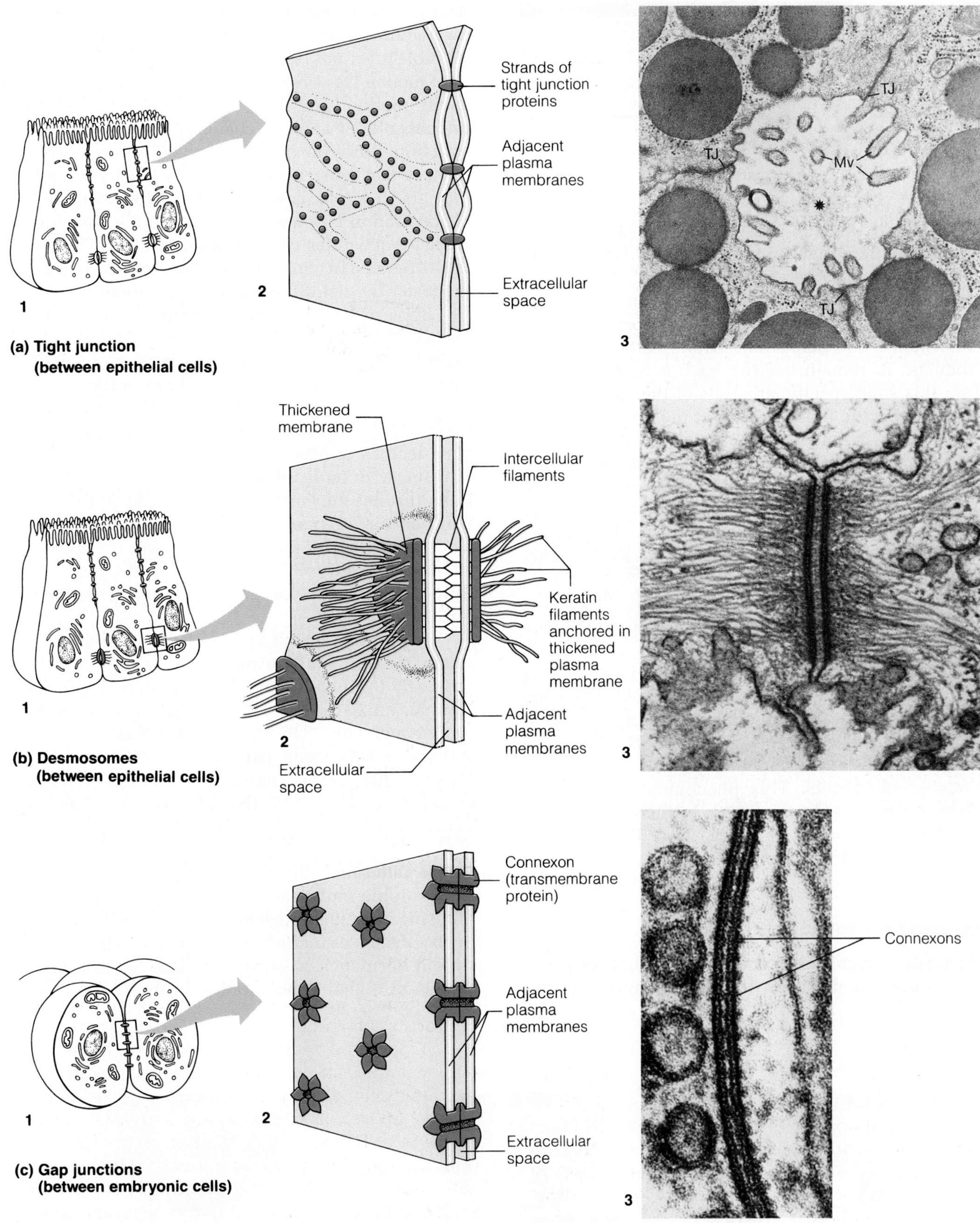

Figure 3.3 Cell junctions. In each case, view 1 provides an orienting simplified diagram of the location of the junction. View 2 is an enlarged diagrammatic view of the same junction; view 3 is an electron micrograph (EM) of the junction. (**a**) Tight junctions (TJ) between epithelial cells (magnification of EM = approx. 25,000 ×). Mv = microvilli. (**b**) Desmosomes connecting epithelial cells (magnification of EM = 58,000 ×). (**c**) Gap junctions connecting embryonic cells (magnification of EM = 200,000 ×).

The Plasma Membrane: Functions

Membrane Transport

Our cells are continuously bathed by a fluid called *interstitial* (in′-ter-stih′-shul) *fluid* that is derived from the blood. Interstitial fluid can be thought of as a rich, nutritious, and rather unusual "soup." It contains thousands of ingredients, including amino acids, sugars, fatty acids, vitamins, regulatory substances such as hormones and neurotransmitters, salts, and waste products. To remain healthy, each cell must extract from this soup specific amounts of the substances it needs at specific times and reject the rest.

The plasma membrane is a *selectively,* or *differentially, permeable* barrier, meaning that it allows some substances to pass while excluding others. When substances penetrate the membrane without any energy input from the cell, their movement is called *passive transport.* When the cell must provide metabolic energy (ATP) to drive the movement of a substance, the process is called *active transport.* The various transport processes that occur in cells are summarized in Table 3.1 on p. 71.

Selective permeability is a characteristic of living cells with intact plasma membranes. When the cell (or its plasma membrane) is severely damaged, the membrane becomes permeable to virtually everything, and substances flow into and out of cells in an unrestricted manner. This phenomenon is evident when someone has been severely burned. Fluids, proteins, and precious ions "weep" from the dead and damaged cells of the burned areas. ■

Passive Processes

Most passive transport processes depend on the physical process of **diffusion** (dih-fyoo′-zhun), the tendency of a molecule or ion to scatter itself evenly throughout its environment (Figure 3.4). Recall that all molecules possess kinetic energy and are in constant motion (see Chapter 2, p. 25). As molecules move about randomly at high speeds, they collide and ricochet off one another, changing direction with each collision. Since the overall effect of this erratic movement is that molecules move away from areas where they are in greater concentration to areas where their concentration is lower, we say that molecules diffuse along (or *down*) the *concentration gradient.* The greater the difference in concentration between the two areas, the faster the net diffusion of the particles.

Because the driving force for diffusion is the kinetic energy of the molecules themselves, the rate of diffusion is also influenced by the size of molecules (the smaller, the faster) and by temperature (the warmer, the faster). In a closed container, diffusion will eventually produce a uniform mixture of the different kinds of molecules. In other words, the system reaches equilibrium, with molecules moving equally in all directions (no net movement). Examples of diffusion are familiar to everyone. When you peel onions, you get teary because the cut onion releases volatile substances that diffuse through the air, dissolve in the fluid film covering your eyes, and form irritating sulfuric acid.

The plasma membrane represents a physical barrier to free diffusion. However, a molecule will diffuse passively through the plasma membrane if it is lipid soluble, uncharged and very small, or assisted by a carrier molecule. The unassisted diffusion of lipid-soluble or very small particles is called *simple diffusion,* whereas the unassisted diffusion of water is called *osmosis* (os-mō′-sis). The assisted process is known as *facilitated diffusion.*

Simple Diffusion. Substances that are nonpolar and lipid-soluble move easily into and out of the cell by diffusing directly through the lipid bilayer (Figure 3.5a). Such substances include oxygen, carbon dioxide, fats, steroid hormones, urea, and alcohol. Since oxygen is always in higher concentration in the blood than in

Figure 3.4 Diffusion. Molecules in solution move continuously and collide constantly with other molecules. As a result, molecules tend to move away from areas of their highest concentration and become evenly distributed, as illustrated by the diffusion of sugar molecules in a cup of coffee.

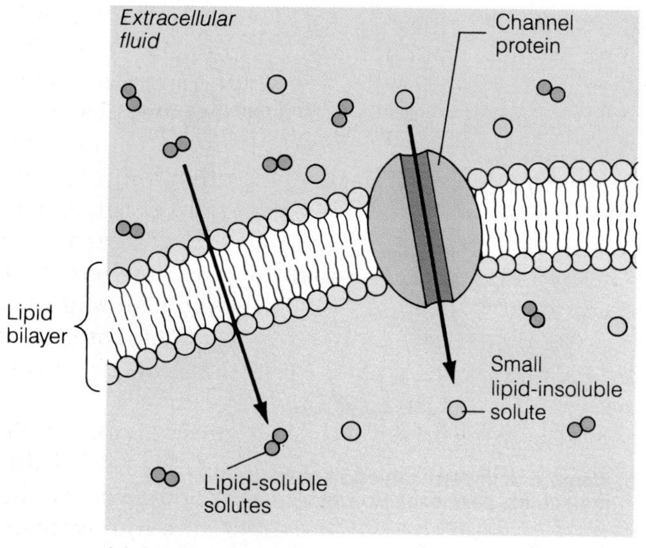

Figure 3.5 Diffusion through the plasma membrane. (**a**) Simple diffusion. As depicted on the left, fat-soluble molecules diffuse directly through the lipid bilayer of the plasma membrane, in which they can dissolve. On the right, small polar or charged particles (water molecules or small ions) are shown diffusing through the membrane channels constructed by channel proteins. (**b**) Facilitated diffusion moves large, lipid-insoluble molecules (e.g., glucose) across the membrane. The substance to be transported binds to a lipid-soluble membrane carrier protein.

tissue cells, oxygen continuously moves into the cells, whereas carbon dioxide (in higher concentration within the cells) diffuses from cells into the blood. Most water-soluble particles are unable to diffuse (unassisted) through the lipid bilayer because they are repelled by the nonpolar hydrocarbon chains.

Polar and charged particles that are insoluble in the lipid bilayer can diffuse through the membrane if they are small enough to pass through the water-filled pores constructed by the channel proteins (Figure 3.5a). Pore size varies, but it is estimated to be no more than 0.8 nm in diameter, and the channels tend to be selective. For example, sodium channels allow only sodium ions to pass. Additionally, while some pores are always open, others can be opened or closed in response to various chemical or electrical signals.

Osmosis. The diffusion of a solvent, such as water, through a selectively permeable membrane like the plasma membrane is called **osmosis.** Since water is a highly polar molecule, it is restricted from passing through the lipid bilayer, but it is small enough to move easily through the membrane pores of most plasma membranes. Osmosis occurs whenever there is a difference in water concentration on the two sides of the membrane. We will use some examples from nonliving systems to illustrate the process of osmosis and then go on to specifics of osmosis across living membranes.

If distilled water is present on both sides of a selectively permeable membrane, no net osmosis occurs, even though water molecules continue to move in both directions through the membrane. If, however, the solute concentration on the two sides of the membrane differs, water concentration differs as well, because as solute concentration increases, water concentration decreases. The extent to which water's concentration is decreased by solutes depends only on the number of solute particles in a solution, not the type of solute, because one molecule or one ion of solute (theoretically) displaces one water molecule. The total concentration of all solute particles in a solution is referred to as the solution's *osmolarity* (oz″-mō-layr′-ih-tē). When equal volumes of solutions of different osmolarity are separated by a membrane that is permeable to *all* molecules in the system, net diffusion of both solute and water occurs, each moving down its own concentration gradient (Figure 3.6a). Eventually, equilibrium will be reached, and the water and solute concentrations will be the same in both compartments. If we consider the identical system, but make the membrane impermeable to the solute molecules, we see quite a different result (Figure 3.6b). Water quickly diffuses from compartment 1 into compartment 2 and will continue to do so until its concentration (and that of the solution) is equal on both

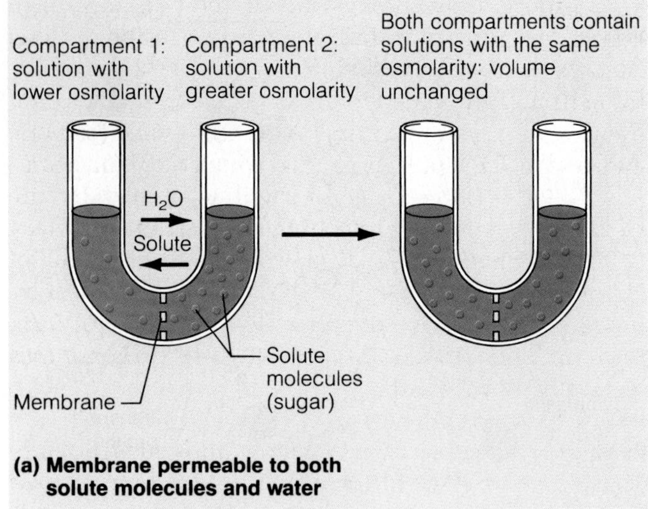
Compartment 1: solution with lower osmolarity
Compartment 2: solution with greater osmolarity
Both compartments contain solutions with the same osmolarity: volume unchanged
H_2O
Solute
Solute molecules (sugar)
Membrane
Both compartments contain solutions of identical osmolarity, but volume of compartment 2 is greater because only water is free to move
Compartment 1
Compartment 2
H_2O
Membrane

(a) Membrane permeable to both solute molecules and water

(b) Membrane impermeable to solute molecules, permeable to water

Figure 3.6 Influence of membrane permeability on diffusion and osmosis. (**a**) In this system, the membrane is permeable to both water and solute (sugar) molecules. Water moves from the solution with lower osmolarity (compartment 1) to the solution with greater osmolarity (compartment 2). The solute moves along its own concentration gradient in the opposite direction. When the system comes to equilibrium (right), the solutions have the same osmolarity and volume. (**b**) This system is identical to that in (a) except that the membrane is impermeable to the solute. Water moves by osmosis from compartment 1 to compartment 2, until its concentration and that of the solutions are identical. Since the solute is prevented from moving, the volume of the solution in compartment 2 increases.

sides of the membrane. Notice, however, that in this case, equilibrium results from the movement of water alone (the solutes are prevented from moving) and leads to dramatic changes in the volumes of the two compartments.

The last example is similar to osmotic events that occur across plasma membranes of living cells, with one major difference. In our examples, the volumes of the compartments are infinitely expandable and the effect of pressure exerted by the added weight of the higher fluid column is not considered. In living cells, this is not the case. As water diffuses into a cell, the point is finally reached where the **hydrostatic pressure** (the back pressure exerted by water against the membrane) within the cell becomes high enough to resist further water entry. At this point, we say that the intracellular solution's hydrostatic pressure is equal to and balanced by its **osmotic pressure**—its tendency to "pull" or attract water entry. In general, the higher the amount of nondiffusible (or nonpenetrable) solutes within the cell, the higher the osmotic pressure and the greater the hydrostatic pressure that must be developed to resist further net water entry.

This leads us to another important concept: tonicity. As noted, many molecules, particularly intracellular proteins and selected ions, are prevented from diffusing through the plasma membrane. Consequently, any changes in their concentration produces a water concentration gradient across the membrane and results in net loss or gain of water by the cell. The ability of a solution to change the tone or shape of cells by altering their internal water volume is called **tonicity** (tun-nih'-sih-tē). Solutions with concentrations of nonpenetrating solutes equal to that found in cells (0.9% saline or 5% glucose) are *isotonic* (literally, the same tonicity). When exposed to such solutions, cells retain their normal shape and neither lose nor gain water (Figure 3.7a). As you might expect, extracellular fluids of the body and most intravenous solutions (solutions infused into the body via a vein) are isotonic. Solutions that contain a higher concentration of nonpenetrating solutes than the cell are *hypertonic;* when immersed in such solutions, cells lose water by osmosis, which causes them to shrink, or *crenate* (krē'-nāt) (Figure 3.7b). Solutions that are more dilute (have lower concentrations of nonpenetrating solutes) than cells are called *hypotonic*. Cells placed in hypotonic solutions plump up rapidly as water rushes into them (Figure 3.7c). Distilled water represents the most extreme example of hypotonicity. Since it contains *no* nonpenetrating solutes, water continues to enter the cell until it finally bursts.

Hypertonic solutions are sometimes infused into the bloodstream of edematous patients (those swollen because of water retained in their tissues) to draw the excess water out of the extracellular (interstitial) space and move it into the bloodstream so that it can be eliminated by the kidneys. Hypotonic solutions are used (with care) to rehydrate the tissues of extremely dehydrated patients. ■

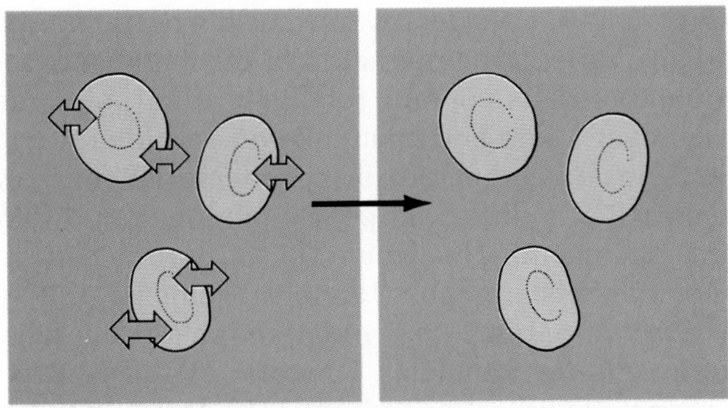

(a) Cells in isotonic solution

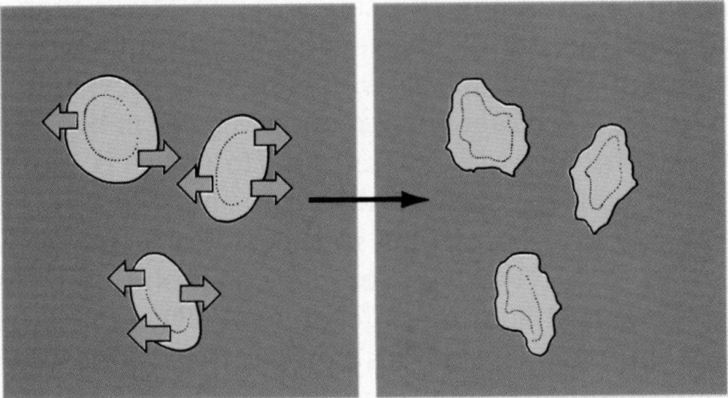

(b) Cells in hypertonic solution

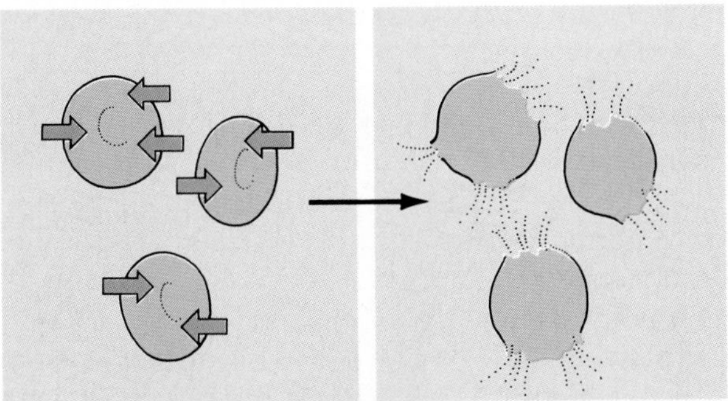

(c) Cells in hypotonic solution

Figure 3.7 The effect of solutions of varying tonicities on living red blood cells. (**a**) In isotonic solutions (same solute/water concentrations as inside cells), cells retain their normal size and shape. (**b**) When in a hypertonic solution (containing more solutes than are present inside the cells), the cells lose water and shrink (crenate). (**c**) In a hypotonic solution (containing fewer solutes than are present inside the cell), the cells take in water by osmosis until they become bloated and may burst (lysis).

Notice that osmolarity and tonicity are not the same. The key factor in tonicity is the nonpenetrating solutes, whereas osmolarity considers the total solute spectrum. Osmolarity is expressed as osmoles per liter (osmol/L); 1 osmol is equal to 1 mole of nonionizing molecules. A 0.3 osmol/L solution of NaCl is isotonic because sodium ions are usually prevented from diffusing freely through the plasma membrane. But if the cell is immersed in a 0.3 osmol/L solution of a penetrating solute, the solute will enter the cell and water will follow. Thus, the cell will swell and burst, just as if it had been placed in pure water.

Osmotic water flows are extremely important in determining the distribution of water in the various fluid-containing compartments of the body (cells, interstitial fluid, blood, etc.). In general, osmosis continues until pressures acting at the membrane (osmotic and hydrostatic pressures) are equalized. For example, water is forced out of capillary blood by the hydrostatic pressure of the blood against the capillary wall. On the other hand, the presence in the blood of solutes that are too large to cross the capillary membrane exerts osmotic pressure that draws water back into the bloodstream. As a result, very little net loss of plasma fluid occurs.

Facilitated Diffusion. Certain molecules, most notably glucose, are both lipid-insoluble and too large to pass through plasma membrane pores. However, they move through the membrane very rapidly by a passive transport process called **facilitated diffusion,** in which they combine with lipid-soluble protein *carrier* molecules in the plasma membrane and are released into the cytoplasm. Just how this translocation is achieved is still a puzzle. But biologists are fairly sure that the carrier does not "flip-flop" or physically move across the membrane like a ferryboat. The most popular model, shown in Figure 3.5b, suggests that changes in the shape of the carrier occur that allow it to first envelop and then release the transported substance.

The diffusion processes previously described are not very selective; in those processes, whether a molecule can pass through the membrane depends chiefly on its size or solubility in lipid, not on its unique structure. Facilitated diffusion, on the other hand, is highly selective; the carrier for glucose combines specifically with glucose, much in the way an enzyme binds to its specific substrate. Although it is presumed that the cell does not expend energy to drive facilitated diffusion, it does provide assistance in the form of protein carrier molecules. (However, it is interesting to note that upon entry into the cytoplasm, a phosphate group is attached to glucose during a coupled reaction with ATP. Thus, glucose transport may in fact prove to be an example of active transport.) As in any diffusion process, glucose is transported down its concentration gradient. Glucose is normally in higher concentrations in the blood than in the cells, where it is rapidly used for ATP synthesis; so the direction of glucose transport is typically unidirectional—into the cells. Carrier-mediated transport is limited by the number of receptors present. For example, if all the glucose carriers are "engaged," they are said to be *saturated,* and glucose transport is occurring at its maximum rate.

When you consider how vitally important oxygen, water, and glucose are to cellular homeostasis, you can see that their passive transport by diffusion represents a tremendous saving of cellular energy.

Indeed, if these substances (and carbon dioxide) had to be transported actively, cell expenditures of ATP would increase exponentially!

Filtration. **Filtration** is the process by which water and solutes are forced through a body membrane or vessel wall by the hydrostatic pressure of blood. Like diffusion, filtration is a passive transport process and involves a gradient. In filtration, however, the gradient is a *pressure gradient* that actually pushes solute-containing fluid (filtrate) from the higher-pressure area to the lower-pressure area. We have already mentioned that hydrostatic pressure forces fluid from the capillaries, and this fluid contains solutes that are vital to the tissues. Filtration also provides the fluid ultimately excreted by the kidneys as urine. Filtration is not a selective process; only blood cells and protein molecules too large to pass through pores in the capillary membrane are held back.

Active Processes

Substances moved across the plasma membrane by active means are usually unable to pass in the necessary direction by any of the passive diffusion processes. They may be too large to pass through the pores, unable to dissolve in the bilipid membrane core, or required to move against rather than with a concentration gradient. There are two major mechanisms of active membrane transport: active transport and bulk transport.

Active Transport. **Active transport,** or **solute pumping,** is similar to facilitated transport in that both processes require carrier proteins that combine *specifically* and *reversibly* with the transported substances. However, facilitated diffusion always honors concentration gradients because its driving force is kinetic energy. In contrast, the enzyme-like protein carriers, or *solute pumps,* that mediate active transport move solutes, most importantly amino acids and ions (such as Na^+, K^+, and Ca^{2+}) "uphill," or against their concentration gradients. To do this work, cells must harness the energy of ATP supplied by cellular metabolism. Although the exact mechanism of active transport is not fully understood, it is believed that the energized protein carrier changes its conformation in a manner that moves the bound solute across the membrane.

Active transport allows cells to take up nutrients that are unable to pass by other means; for example, amino acids, crucial to cell survival, are insoluble in the lipid bilayer. Furthermore, since cells actively accumulate, or "hoard," amino acids, these nutrients must be transported into cells against a concentration gradient. The ability of cells to maintain an internal pool of ions that differs from that of the interstitial fluid also reflects the operation of the membrane solute pumps. Cells contain relatively high potassium ion concentrations and relatively low sodium ion concentrations as compared to the interstitial fluid. As noted earlier, sodium and potassium ions can diffuse across the membrane when the appropriate membrane channels are open. Rapid movements of these ions (Na^+ into the cell, K^+ out of the cell) occur when a nerve ending stimulates a muscle cell to contract. Once inside the cells, sodium ions have little tendency to diffuse in the reverse direction, because their concentration inside the cell is much lower than in the interstitial fluid. The reverse condition exists for potassium ions; once outside the cell, they tend to remain there, rather than reentering the cell, where they are in high concentration. However, if muscle contraction is to continue, these ions must be returned to their original positions so that the process can be repeated. An ATP-driven sodium-potassium pump, an enzyme called *sodium-potassium ATPase* (ā″-tē-pē′-ās), simultaneously moves both of these ions across the plasma membrane (Figure 3.8). The operation of the sodium-potassium pump is described in more detail shortly.

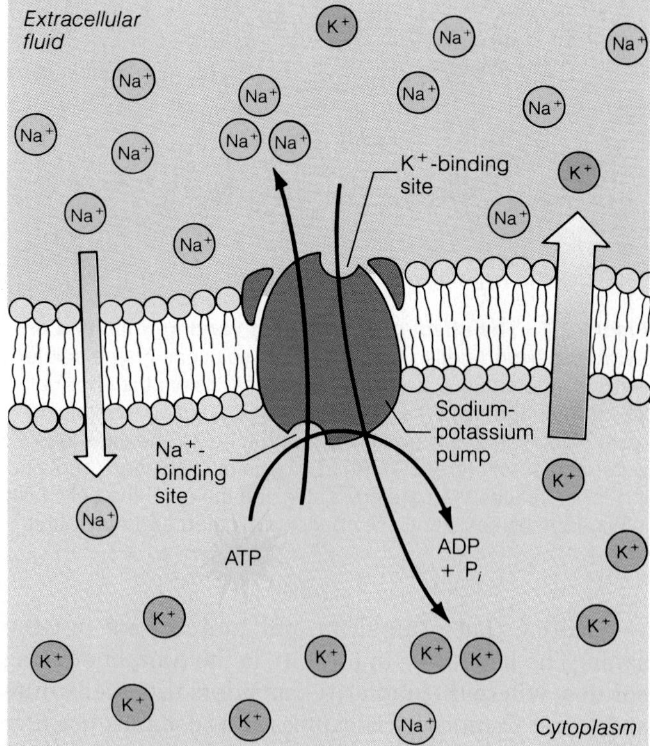

Figure 3.8 Operation of the sodium-potassium pump. Hydrolysis of a molecule of ATP provides energy for a "pump" protein to move three sodium ions out of the cell and two potassium ions into the cell. Both ions are moved against their concentration gradients, indicated by colored arrows moving through the membrane (yellow arrow = Na^+ gradient; green arrow = K^+ gradient).

Table 3.1 Membrane Transport Processes

Process	Energy source	Description	Examples
Passive processes			
Simple diffusion	Kinetic energy	Net movement of particles (ions, molecules, etc.) from an area of their higher concentration to an area of their lower concentration, that is, along their concentration gradient	Movement of fats, oxygen, carbon dioxide through the lipid portion of the membrane, and ions through protein channels under certain conditions
Osmosis	Kinetic energy	Simple diffusion of water through a selectively permeable membrane	Movement of water into and out of cells via membrane pores.
Facilitated diffusion	Kinetic energy	Same as simple diffusion, but the diffusing substance is attached to a lipid-soluble membrane carrier protein	Movement of glucose into cells
Filtration	Hydrostatic pressure	Movement of water and solutes through a semipermeable membrane from a region of higher hydrostatic pressure to a region of lower hydrostatic pressure, that is, along a pressure gradient	Movement of water, nutrients, and gases through a capillary wall; formation of kidney filtrate
Active processes			
Active transport (solute pumping)	ATP (cellular energy)	Movement of a substance through a membrane against a concentration (or electrochemical) gradient; requires a membrane carrier protein	Movement of amino acids and most ions across the membrane
Bulk transport			
Exocytosis	ATP	Secretion or ejection of substances from a cell; the substance is enclosed in a membranous vesicle, which fuses with the plasma membrane and ruptures, releasing the substance to the exterior	Secretion of neurotransmitters, hormones, mucus, etc.; ejection of cell wastes
Phagocytosis (endocytosis)	ATP	"Cell eating": A large external particle (proteins, bacteria, dead cell debris) is surrounded by a "seizing foot" and becomes enclosed in a plasma membrane sac	In the human body, occurs primarily in protective phagocytes (some white blood cells, macrophages)
Pinocytosis (endocytosis)	ATP	"Cell drinking": Plasma membrane sinks beneath an external fluid droplet containing small solutes; membrane edges fuse, forming a fluid-filled vesicle	Occurs in most cells; important for taking in solutes by absorptive cells of the kidney and intestine
Receptor-mediated endocytosis	ATP	Selective endocytosis process; external substance binds to membrane receptors, and coated pits are formed	Means of intake of some hormones, cholesterol, iron, and other molecules

Bulk Transport. Large particles and macromolecules are transported through plasma membranes by **bulk transport.** Like solute pumping, bulk transport is energized by ATP. The two kinds of bulk transport are exocytosis and endocytosis.

Exocytosis (ek″-sō-sī-to′-sis) is the mechanism by which substances are moved from the cell interior into the extracellular space. It accounts for hormone secretion, neurotransmitter release, mucus secretion, and, in some cases, the ejection of wastes. In exocytosis, the substance or cell product to be released is first enclosed within a membranous sac. The sac migrates

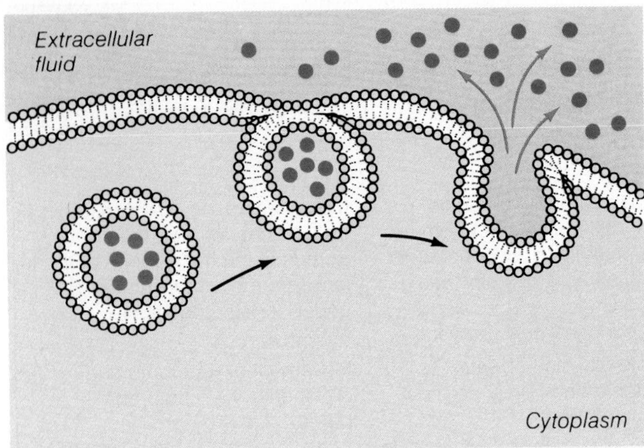

Figure 3.9 Exocytosis. The membrane-bounded vesicle containing the substance to be secreted migrates to the plasma membrane, and the two membranes fuse. The fused site opens and releases the contents of the secretory vesicle into the intercellular space.

to the plasma membrane and fuses with it. The fused area then ruptures, spilling the sac contents out of the cell (Figure 3.9).

Endocytosis (en″-dō-sī-tō′-sis) provides a means for large particles or macromolecules to enter cells. The substance to be taken into the cell is progressively enclosed by a portion of the plasma membrane. Once the membranous sac is formed, it pinches off from the plasma membrane and moves into the cytoplasm, where its contents are digested. Three types of endocytosis are recognized: phagocytosis, pinocytosis, and receptor-mediated endocytosis.

In **phagocytosis** (fa″-gō-sī-tō′-sis), literally, "cell eating," parts of the plasma membrane and cytoplasm expand and flow around some relatively large or solid material, such as a clump of bacteria or cell debris, and engulf it (Figure 3.10a). The membranous sac thus formed is called a *phagosome* (fa′-guh-sōm). In most cases, the phagosome becomes fused with a *lysosome* (lī′-suh-sōm), a specialized cell structure containing digestive enzymes, and its contents are digested.

Phagocytic cells in the human body include macrophages and some white blood cells. These phagocytic "professionals" help to police and protect the body by ingesting and disposing of bacteria, other foreign substances, and dead tissue cells. Most phagocytes have the ability to move about by *amoeboid* (uh-mē′-boyd) *motion;* that is, the flowing of their cytoplasm into temporary pseudopods (soo′-duh-podz) (*pseudo* = false; *pod* = foot) allows them to creep along.

If we can say that cells eat, we can also say that they drink, and **pinocytosis** (pē″-nō-sī-tō′-sis) is "cell drinking" (Figure 3.10b). In pinocytosis, a minute droplet of interstitial (or external) fluid containing dissolved solutes is surrounded by a bit of invaginated

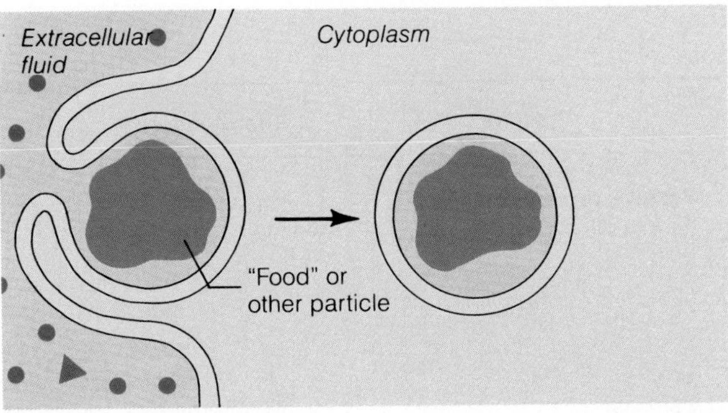

(a) Phagocytosis

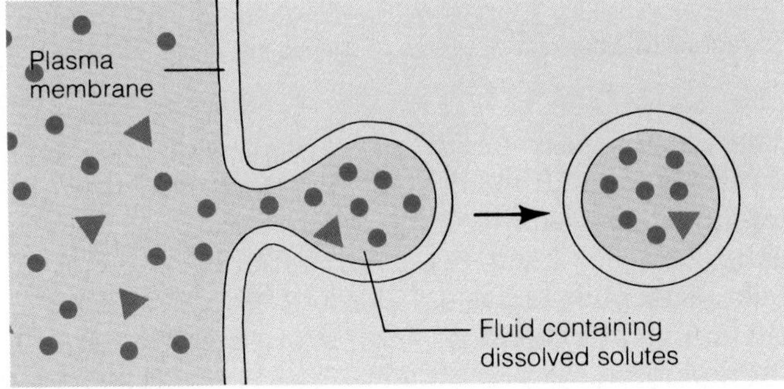

(b) Pinocytosis

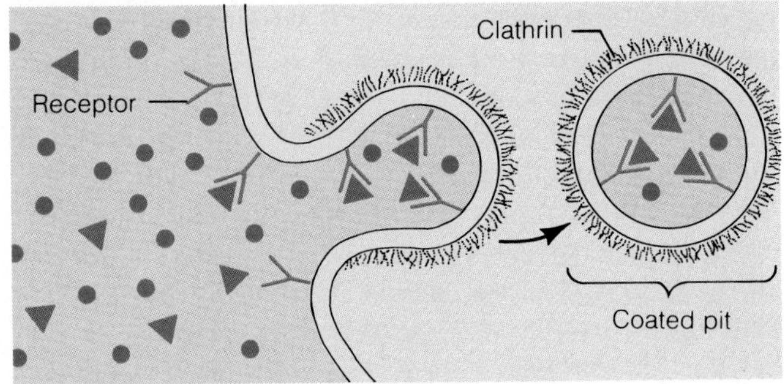

(c) Receptor-mediated endocytosis

Figure 3.10 Three types of endocytosis.

plasma membrane and then taken into the cell within a tiny membranous *pinocytotic* (pē′-nō-sī-tah-tik) *vesicle.* Unlike phagocytosis, pinocytosis is a routine activity of most cells. It is particularly important in cells that function in absorption, such as intestinal cells.

In both phagocytosis and pinocytosis, bits of the plasma membrane are removed when the membranous sacs are internalized. However, since these membranes are recycled back to the plasma membrane (probably by the process of exocytosis), the surface area of the plasma membrane remains remarkably constant.

Unlike phagocytosis and pinocytosis, which are often nonspecific ingestion processes, **receptor-mediated endocytosis** is exquisitely selective (Figure 3.10c). The receptors are plasma membrane proteins that bind only with certain molecules. Both the receptors and attached material are internalized in a small vesicle called a *coated pit*, a term that refers to the bristlelike *clathrin* (klā′-thrin) protein coating on the cytoplasmic face of the vesicle.

Substances taken up by receptor-mediated endocytosis include the hormone insulin, low-density lipoproteins (such as cholesterol attached to a transport protein), and iron. When the coated pit combines with a lysosome, the hormone (or cholesterol or iron) is released, and the membranes containing the attached receptors are recycled back to the plasma membrane.

Membrane Potential

The plasma membrane is more permeable to some molecules than to others. We explained how this differential permeability can lead to dramatic osmotically induced changes in cell tone, but that is not its only consequence. An equally important result is the generation of a **membrane potential,** or voltage, across the membrane. A *voltage* is electrical potential energy resulting from the separation of oppositely charged particles. In regard to cells, the oppositely charged particles are ions, and the barrier that keeps them apart is the plasma membrane.

In their resting state, all body cells exhibit a *resting membrane potential* that ranges from -50 to -200 millivolts (mV), depending on cell type. The sign before the voltage indicates that the inside of the cell is negative compared to its outside. This voltage (or charge separation) exists only at the membrane, however, because if we were to add all the negative and positive charges in the cell cytoplasm, we would find that the cell interior is electrically neutral. Likewise, the positive and negative charges in the extracellular fluid balance exactly. So how does the resting membrane potential come about, and how is it maintained?

Although many different kinds of ions are found both inside cells and in the extracellular fluid, the resting membrane potential is determined mainly by concentration gradients of sodium and potassium ions and by the differential permeability of the plasma membrane to those ions. As illustrated in Figure 3.11, body cells have a predominance of K^+ inside and are surrounded by extracellular fluid containing relatively more Na^+. At rest, the plasma membrane is slightly permeable to K^+, but nearly impermeable to Na^+. As you can see, potassium diffuses out of the cell along its concentration gradient. Sodium is strongly attracted to the cell interior by its concentration gradient, but because sodium permeability is only 1/25 that of potassium, sodium influx is much less and is inadequate to balance potassium outflow. The result of this unequal Na^+-K^+ diffusion across the membrane is a slight excess of negative ions within the cell, which establishes the resting membrane potential.

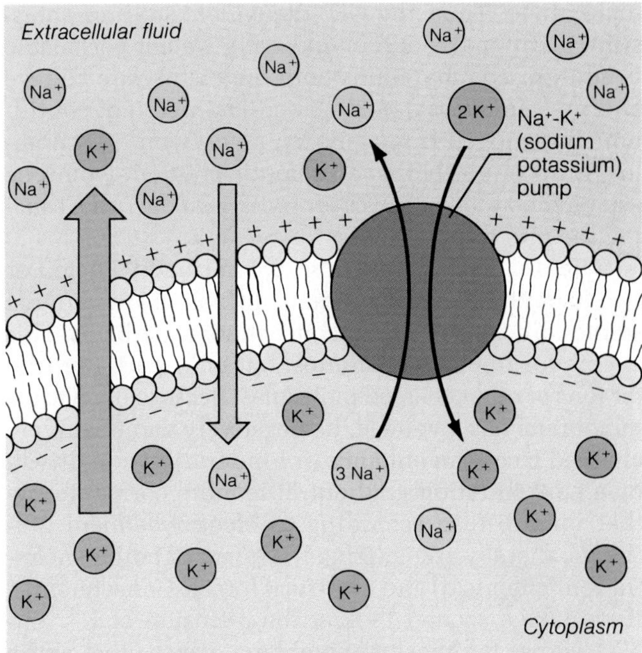

Figure 3.11 Summary of forces leading to the generation and maintenance of membrane potentials. The ionic imbalances that produce the membrane potential reflect passive diffusion of ions (sodium diffuses into the cell more slowly than potassium diffuses out of the cell because of differential membrane permeability to these two ions) and the active transport of sodium and potassium ions (in a ratio of 3:2) by the Na^+-K^+ pump.

Although it is tempting to believe that massive flows of ions take place to generate the resting potential, this is not the case. Actually, the number of ions producing the membrane potential is so small that it does not change ion concentrations in any significant way.

In the polarized state, sodium and potassium concentrations are not at equilibrium, and if only passive forces were at work, sodium and potassium concentrations would eventually become equal inside and outside the cell. Instead, what exists is a *steady state* in which the ionic distribution of the polarized state is maintained. This steady state results from both passive and active processes—namely, diffusion and active transport. As we have seen, potassium and sodium ions leak passively according to their driving forces, but the active transport rate is equal to, and depends on, the rate of diffusion of sodium ions into the cell. If more sodium enters, more is pumped out. (This is like being in a leaky boat. The more water that comes in, the faster you bail!) The sodium-potassium pump couples sodium and potassium transport, and each "turn" of the pump ejects $3Na^+$ out of the cell and carries $2K^+$ back in (see Figure 3.11). Because the membrane is always slightly more permeable to K^+, the ionic imbalance and membrane potential are maintained. Thus, the ATP-dependent sodium-potassium pump makes it look like cells are not permeable to sodium and maintains both the membrane voltage and the osmotic balance. (If sodium were not continuously removed from cells, so much would accumulate intracellularly that an osmotic gradient would be established that would draw water into the cells, causing them to burst.)

Before leaving this topic, we must add a detail or two to our earlier discussion of diffusion. We said that solutes diffuse down their concentration gradients. This is true for uncharged solutes, but only partially true for ions or other charged molecules. Because the plasma membrane has a voltage, its negatively and positively charged faces can enhance or hinder diffusion driven by a concentration gradient. It is more correct to say that ions diffuse according to *electrochemical gradients,* thereby recognizing the effect of both concentration (chemical) and electrical forces. Consequently, if we take a second look at the diffusion of K^+ and Na^+ across the plasma membrane, we see that while diffusion of potassium is facilitated by the membrane's greater permeability to it and by its concentration gradient, its diffusion is resisted somewhat by the positivity of the cell exterior. On the other hand, Na^+ is drawn into the cell by a steep electrochemical gradient; the limiting factor is the membrane's relative impermeability to it.

Functions of the Glycocalyx

The externally facing glycoproteins of the cell membrane (see Figure 3.2) act as highly specific biological markers that aid cellular interactions. Although the full range of activities of these sugar-coated proteins still escapes us, we do know that they are involved in the following activities:

1. Determining the ABO and other blood groups. The difference between the A and B blood groups results from a variation in a single sugar group at the end of the carbohydrate chain of a specific glycoprotein on the red blood cell surface. (In type O blood, there is *no* sugar in that position.) These simple sugars differ by only a few atoms, but that minute difference may be a matter of life or death, because transfusing the wrong blood type can have fatal results.

2. Providing binding sites for certain toxins. Several toxins, such as cholera and tetanus toxin, recognize some of the carbohydrates present on cell surfaces and bind selectively to those sugars.

3. Recognition of the egg by sperm. Cell surface sugars on an ovum signal sperm that "this is the place" to deposit their genes. (In humans, the sperm penetrates a precursor of the ovum, but this will be dealt with later in Chapter 29.)

4. Determining cellular life span. In red blood cells, the thickness of the cell coat is dramatically reduced as the cells age. There is some experimental evidence suggesting that a reduction in the sugar coating of red blood cells (and perhaps of other cell types) serves as a signal to liver and other phagocytes that the cells are old and should be destroyed.

5. Serving in the immune response. Cell surface glycoproteins of cells involved in the immune response are receptors that bind with great specificity to bacteria, viruses, and some cancer cells. The binding activates the immune response, which mounts an attack against the foreign invader.

6. Helping to guide and direct embryonic development. Pronounced changes in the glycocalyx occur during embryonic development as tissue growth and specialization occur. The sticky sugar groups also promote cell adhesion at every stage of development, thus helping to establish body form.

Definite changes in the glycocalyx also occur in a cell that is becoming transformed into a cancerous cell. In fact, a cancer cell's glycocalyx may change almost continuously, allowing it to keep ahead of immune system recognition mechanisms and avoid destruction. ■

The Cytoplasm

The **cytoplasm** is the cellular material inside the plasma membrane and outside the nucleus. It is the major functional area, the site where most cellular activities are accomplished. Although early microscopists thought that the cytoplasm was a structureless gel, the electron microscope has revealed that it consists of three major elements: the cytosol, organelles, and inclusions (Table 3.2). The **cytosol** is the viscous, semitransparent fluid environment of the cytoplasm within which the other elements are suspended. Dissolved in the cytosol, which is largely water, are soluble proteins, salts, sugars, and a variety of other solutes. Hence, the cytosol is a complex mixture with both colloidal and true solution properties.

The **organelles,** described in detail shortly, are the metabolic machinery of the cell. Each type of organelle is "engineered" to carry out a specific function for the cell as a whole; some synthesize proteins, others package those proteins, and so on. **Inclusions** are not functioning units, but instead are a number of chemical substances that may or may not be present, depending on the specific cell type. Examples of inclusions are stored nutrient substances, such as the glycogen granules abundant in liver and muscle cells;

the pigment (melanin) granules typically seen in certain cells of the skin and hairs; and the zymogen (enzyme-containing) granules synthesized in pancreas cells and then transported into the small intestine. Other secretory products (mucus), water vacuoles, and crystals of various types are also inclusions.

Cytoplasmic Organelles

The organelles, literally, "little organs," are specialized cellular compartments, each performing its own job to maintain the life of the cell. Most organelles are bounded by a selectively permeable membrane similar in composition to the plasma membrane, which allows them to maintain an internal environment quite different from that of the surrounding cytosol. This compartmentalization is crucial to the functioning of the cell; without it, thousands of enzymes would be randomly mixed and biochemical activity would be chaotic. Let us consider what goes on in each of the workshops of our cellular factory.

Mitochondria

Mitochondria (mī″-tō-kon′-drē-uh), about 1 μm in diameter and 5–10 μm long, are usually depicted as tiny kidney-bean- or sausage-shaped organelles (Figure 3.12), but in living cells they squirm, elongate, and change shape almost continuously. Mitochondria are called the powerhouses of the cell because they provide most of a cell's ATP supply. Mitochondria are found in most of our cells, but their density in a particular cell reflects that cell's energy requirements; mitochondria are generally clustered "where the action is." Busy cells (like muscle and liver cells) have hundreds of mitochondria, whereas cells that are relatively inactive (such as lymphocytes) have just a few.

Each mitochondrion is surrounded by two lipid bilayer membranes, each with the general structure of the plasma membrane. The outer membrane is smooth and featureless, but the inner membrane has projections called *cristae* (kris′-tē) that protrude into the gel-like substance within the mitochondrion, called the *matrix*. Enzymes dissolved in the matrix, together with those forming part of the crista membranes, cooperate to break down glucose and other foodstuffs to water and carbon dioxide. Some of the energy released as glucose is broken down, captured, and used to attach phosphate groups to ADP molecules to form ATP. This multistep mitochondrial process, generally referred to as *aerobic* (ayr-ō′-bik) *cellular respiration* because it requires oxygen, is described in detail in Chapter 25.

It is interesting to note that mitochondria contain DNA and RNA and are self-replicating organelles. When cellular requirements for ATP increase, the mitochondria simply pinch in half (a process called fission) to increase their number, then grow to their former size.

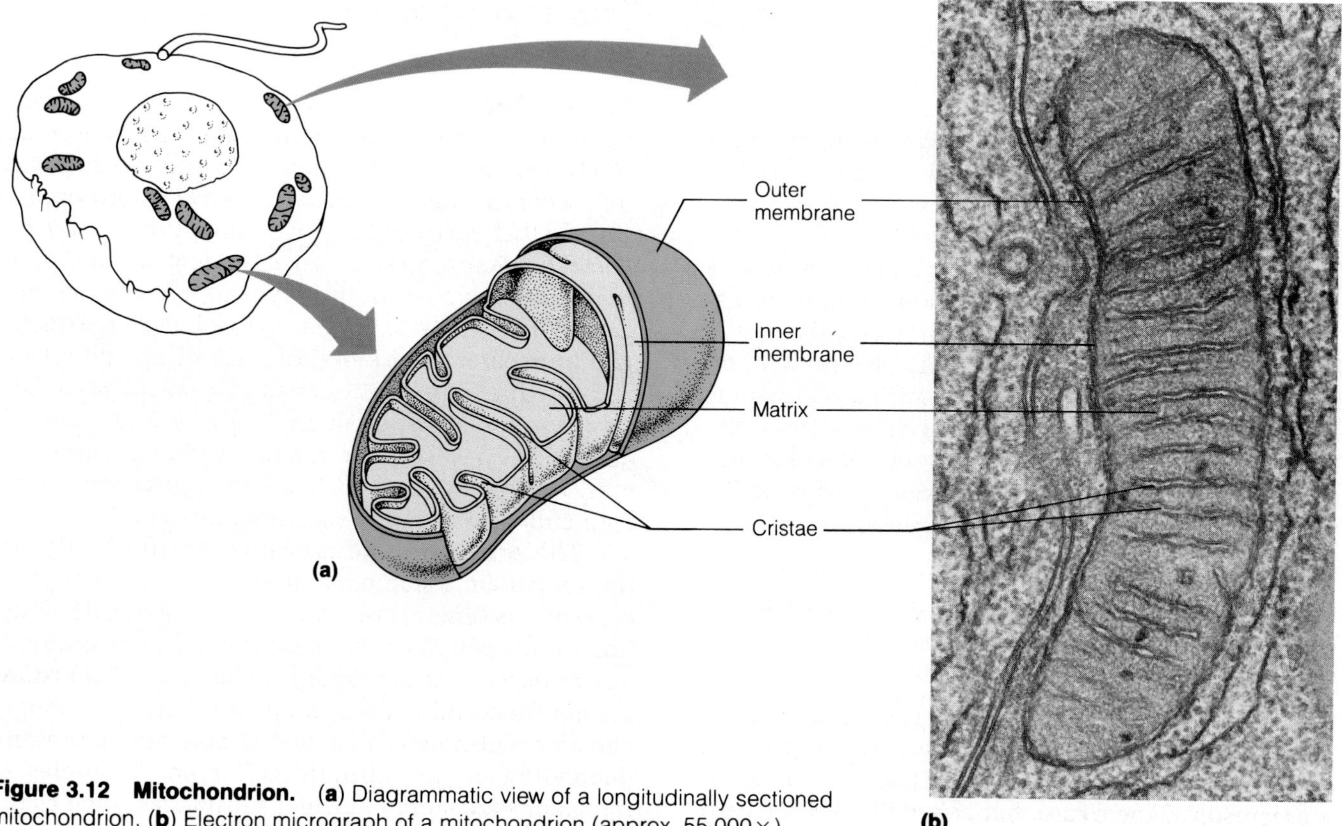

Figure 3.12 Mitochondrion. (**a**) Diagrammatic view of a longitudinally sectioned mitochondrion. (**b**) Electron micrograph of a mitochondrion (approx. 55,000×).

Ribosomes

Ribosomes (rī′-buh-sōmz) are minute (25 nm diameter), dark-staining granules composed of proteins and one variety of RNA (specifically called *ribosomal RNA*). Each ribosome is constructed of two roughly globular subunits that fit together when the small unit sits atop the larger one. Ribosomes are sites of protein synthesis, a function we will discuss in detail later in this chapter.

Some ribosomes float freely in the cytoplasm; others are attached to membranes, forming a complex called the rough endoplasmic reticulum. Although there are exceptions, these ribosomal populations appear to divide the chore of protein synthesis. *Free ribosomes* make proteins for cellular use, whereas *membrane-bound ribosomes* are chiefly involved in synthesizing protein products destined for cellular membranes or for export from the cell.

Endoplasmic Reticulum

The **endoplasmic reticulum** (en″-dō-plaz′-mik rih-tik′-yoo-lum) **(ER)** is an extensive system of interconnected parallel membranes that coils and twists through the cytoplasm, enclosing fluid-filled cavities, or *cis-*

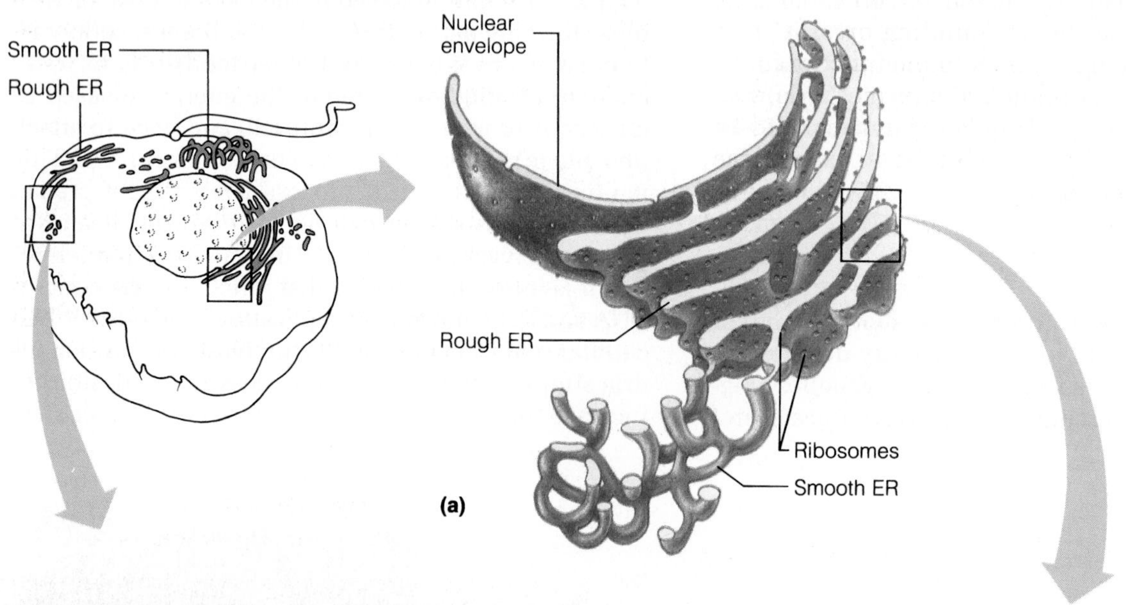

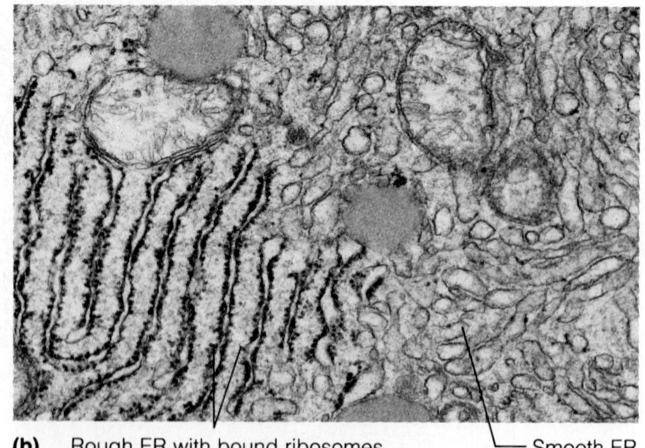

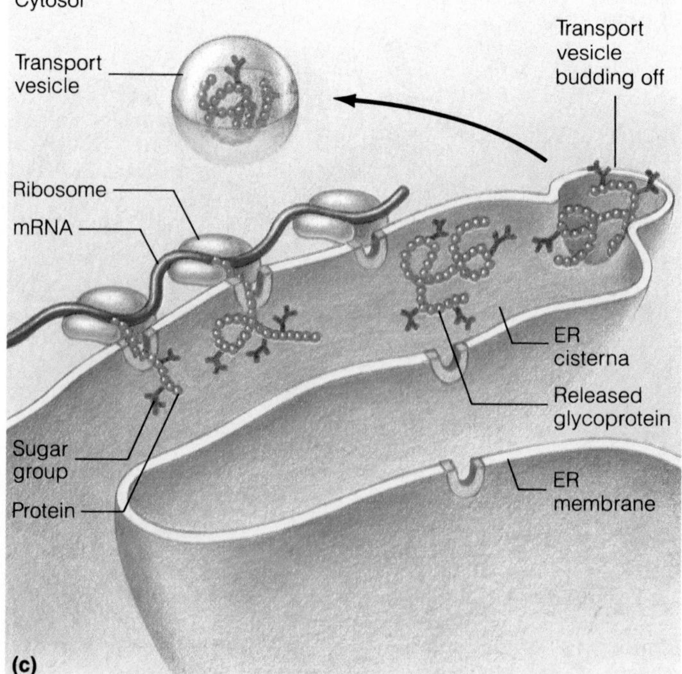

Figure 3.13 The endoplasmic reticulum. (**a**) Three-dimensional diagram of the rough ER in a liver cell: its connections to smooth ER areas are also illustrated. (**b**) Electron micrograph of smooth and rough endoplasmic reticulum (approx. 30,000×). (**c**) Enlarged view of a portion of the rough ER membrane bearing ribosomes and the ER cisterna where sugar groups are attached. Some of the glycoproteins then become encapsulated by bits of ER membrane, bud off, and migrate to the Golgi apparatus.

ternae (sis-ter′-nē). It is continuous with both the nuclear membrane and the plasma membrane and accounts for about half of the cell's membranes. There are two distinct varieties of ER: rough ER and smooth ER. One variety or the other may predominate in a given cell type, depending on the specific functions of the cell.

The external surface of the **rough ER** is studded with ribosomes (Figure 3.13). As proteins are assembled on the ribosomes, they thread their way into the ER cisternae, where sugar groups are attached to the still-growing polypeptide chains, and they begin to fold or coil into their higher levels of structural form (see Figure 3.13c). Some of these proteins are enzymes that will be retained to function within the cisternae. Other proteins are secreted from the cell or become part of the membrane proteins of other organelles or the plasma membrane; these become enclosed in membranous sacs that pinch off from the ER. These sacs, or transport vesicles, make their way to the cell's *Golgi apparatus* (gol′-jē a″-puh-ra′-tis) for further processing of the proteins (see Figure 3.15). Rough ER is particularly abundant in cells specialized to make and release proteins that function at other locations, such as most secretory cells, antibody-producing plasma cells, and liver cells, which produce most blood proteins.

The **smooth ER** (see Figures 3.1 and 3.13) is a continuation of the rough ER and contains enzymes delivered by the rough ER, but it plays no role in protein synthesis. Instead, its basic products are lipid molecules. Specific functions of the smooth ER include (1) synthesis of phospholipids and cholesterol (between them, the smooth and rough ERs construct all of a cell's membrane components); (2) synthesis of steroid-based hormones such as sex hormones (testosterone-synthesizing cells of the testes are full of smooth ER); (3) synthesis of fats and lipoproteins (in liver cells); (4) absorption, synthesis, and transport of fats (in intestinal cells); (5) synthesis and storage of glycogen (in liver and skeletal muscle); and (6) detoxification of drugs (in the liver and kidneys). Additionally, skeletal and cardiac muscle cells have an elaborate smooth ER called the sarcoplasmic reticulum that plays an important role in calcium ion storage and release during muscle contraction.

Golgi Apparatus

The **Golgi apparatus** appears as flattened membranous sacs, stacked like dinner plates, associated with swarms of tiny membranous vesicles (Figure 3.14). It is generally situated close to the nucleus. The Golgi apparatus is probably the principal traffic director for

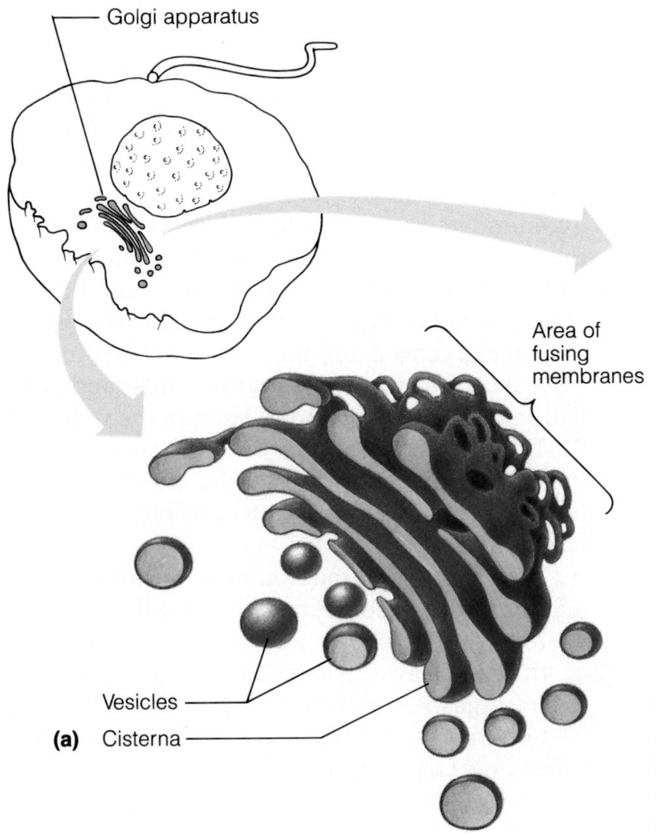

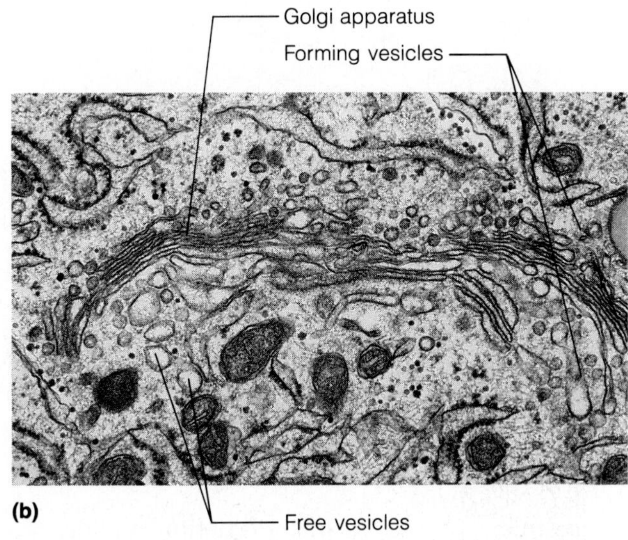

Figure 3.14 Golgi apparatus. (**a**) Three-dimensional diagrammatic view of the Golgi apparatus. (**b**) Electron micrograph of the Golgi apparatus (approx. 85,000×). Notice the secretory vesicles in the process of pinching off from the Golgi membranous stacks.

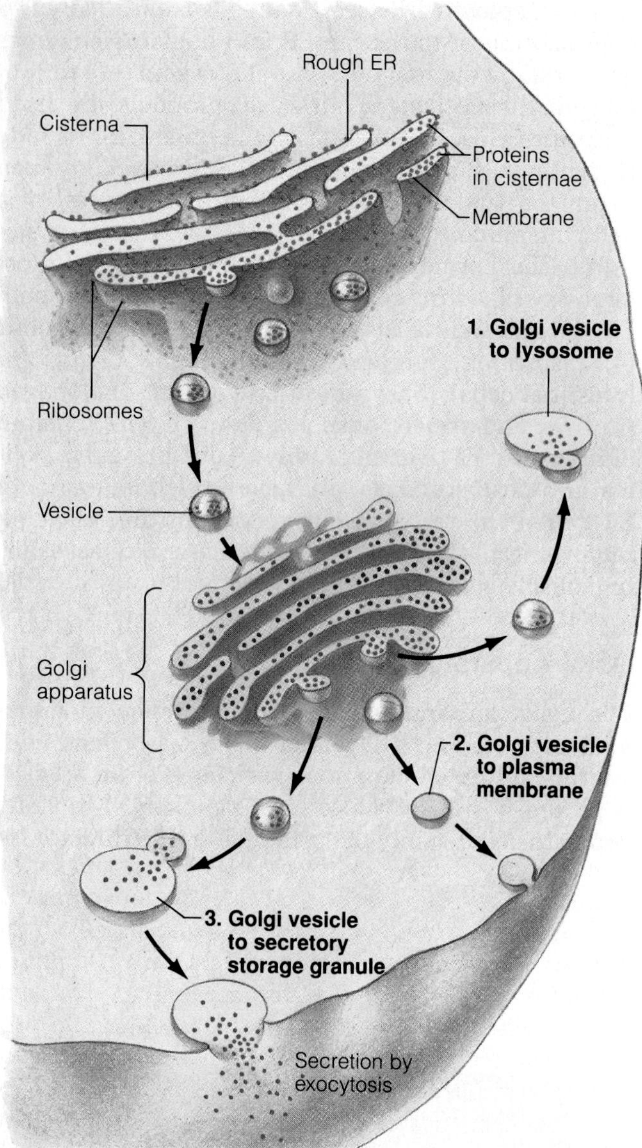

Figure 3.15 Role of the Golgi apparatus in packaging proteins for cellular use and for secretion. The sequence of events from protein synthesis on the rough ER to the final distribution of those proteins. Protein-containing vesicles pinch off the rough ER and migrate to fuse with the face of the Golgi apparatus. Within the Golgi compartments, the proteins are modified; they are then accumulated in different Golgi vesicle types, depending on their ultimate destination.

cellular proteins: Its major function is to modify, concentrate, and package proteins in specific ways, depending on their ultimate destination. The transport vesicles that bud off from the rough ER migrate to and fuse with the membranes of the Golgi apparatus (Figure 3.15). Inside the apparatus, glycoproteins are modified: Some sugar groups are trimmed while others are added, and in some cases, phosphate groups are added. The various proteins are "tagged" for delivery to a specific address, sorted, and accumulated in at least three distinct types of vesicles attached to the Golgi membranes.

Vesicles containing proteins destined for export pinch off as *secretory vesicles*, which migrate to the plasma membrane and discharge their contents from the cell by exocytosis. Specialized secretory cells such as the enzyme-producing cells of the pancreas have a very prominent Golgi apparatus. In addition to its packaging-for-release function, the Golgi apparatus pinches off vesicles containing proteins destined for the plasma membrane and packages hydrolytic (digestive) enzymes into membranous sacs called lysosomes that remain in the cell.

Lysosomes

Lysosomes are spherical membranous bags containing hydrolytic enzymes (Figure 3.16). As already noted, their enzymes are synthesized on the ribosomes of the rough ER and packaged by the Golgi apparatus. Lysosomes provide sites where digestion can proceed safely within a cell. As you might guess, lysosomes are large and abundant in phagocytes. The lysosomal enzymes are capable of digesting all varieties of biological molecules. They work best in an acid pH (approximately pH 5) and thus are called *acid hydrolases* (hī′-drō-lā-siz). The lysosomal membrane is adapted to serve lysosomal functions in two important ways: (1) It contains hydrogen ion (proton) "pumps" that sequester hydrogen ions from the surrounding cytosol to maintain the organelle's low internal pH, and (2) it retains the dangerous acid hydrolases, while permitting the final products of digestion to escape so that they can be used by the cell or excreted. The usual functions of lysosomes include (1) digestion of particles ingested by endocytosis, particularly important in rendering ingested bacteria, viruses, and toxins harmless to the cell; (2) digestion of worn-out or nonfunctional organelles; (3) metabolic functions, such as the breakdown of stored glycogen and the release of thyroid hormone from its storage form in thyroid cells; and (4) degradation of nonuseful tissues, such as the webs between the fingers and toes of a developing fetus and the uterine lining during menstruation. Breakdown of bone to release calcium ions into the blood also reflects lysosomal activity.

The lysosomal membrane is ordinarily quite stable, but it becomes fragile when the cell is injured or deprived of oxygen and when excessive amounts of vitamin A are present. Lysosomal rupture results in self-digestion of the cell, a process called *autolysis* (a″-tol′-eh-sis). For this reason, lysosomes are sometimes referred to as suicide sacs. ■

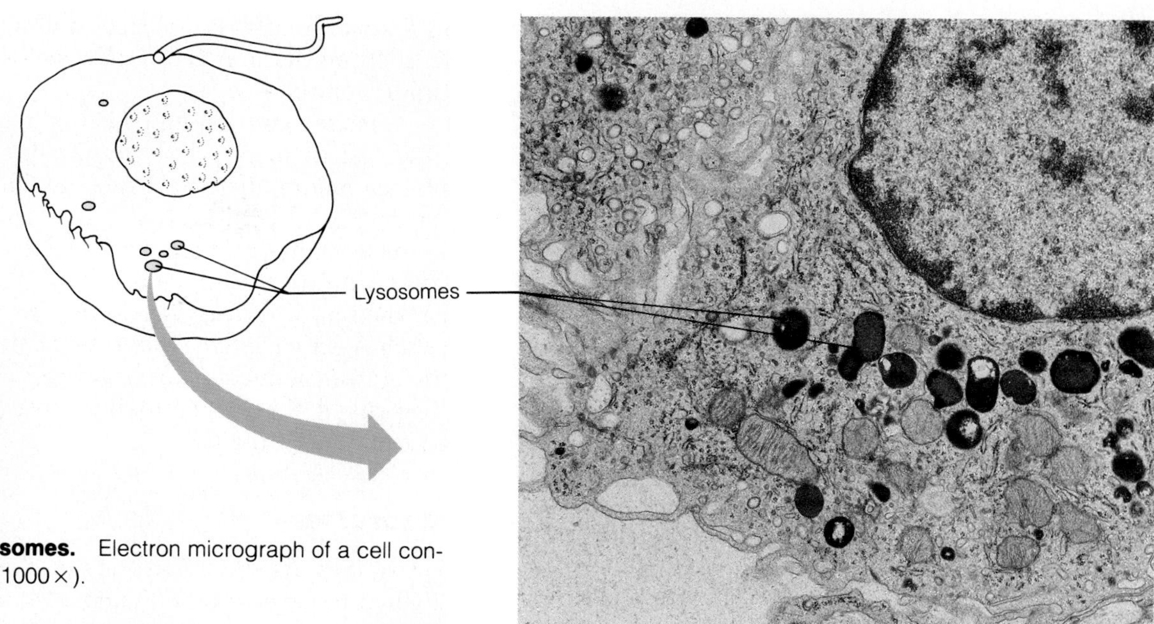

Figure 3.16 Lysosomes. Electron micrograph of a cell containing lysosomes (1000×).

Peroxisomes

Peroxisomes (per-ok′-sih-sōmz), also called **microbodies,** are membranous sacs containing powerful enzymes called *peroxidases* (per-ok″-sih-dā′-siz). Peroxidases detoxify a number of harmful or toxic substances, including alcohol and formaldehyde. However, their most important function is to "disarm" *oxygen-free radicals* such as superoxide radical by converting them to hydrogen peroxide (H_2O_2); the enzyme *catalase* (ka′-tuh-lās) then reduces hydrogen peroxide to water. Free radicals are highly reactive chemicals with unpaired electrons that can scramble the structure of proteins, lipids, and nucleic acids. Although free radicals and hydrogen peroxide are normal by-products of cellular metabolism, if allowed to accumulate, they can have devastating effects on cells. Peroxisomes are numerous in liver and kidney cells, which are active in detoxification. Although peroxisomes look like small lysosomes (see Figure 3.1), they are thought to originate directly from the budding of the rough ER.

Cytoskeletal Elements

An elaborate network of protein structures called microfilaments, microtubules, and intermediate filaments ramifies throughout the cytoplasm (Figure 3.17a and b). This network, or **cytoskeleton,** acts as an internal scaffolding that helps support the soft cell substance and also provides the "muscles" for various types of cellular movements.

Microfilaments (mī″-krō-fih′-luh-ments) are thin strands of the contractile protein *actin.* Each cell appears to have its own arrangement of microfilaments; thus, no two cells are alike. Most microfilaments are involved in cell motility or movement of cell parts. For example, actin microfilaments form the core of microvilli and in some microvilli promote a pumping motion; provide for amoeboid movement; and, together with myosin, another contractile protein, form the cleavage ring that pinches a cell in two during cell division. Microfilaments are most highly developed in muscle cells, where actin and myosin filaments are arranged in parallel bundles. The cell shortening that leads to muscle contraction reflects the sliding of these filaments past one another. (Muscle contraction is discussed in Chapter 9.) Some microfilaments appear to cooperate with the other cytoskeletal elements to brace the cell and maintain its shape.

Intermediate filaments are tough, insoluble protein fibers, with diameters intermediate in size between the microfilaments and microtubules. They are the most stable of the cytoskeletal elements and have high tensile strength. They act as internal guy wires to resist mechanical stress on the cell, and they play a role in forming desmosomes (the adhesive junctions described on p. 64). The protein composition of intermediate filaments varies in different cell types, which has led to numerous names for these cytoskeletal elements. For example, they are called neurofilaments in nerve cells and keratin filaments (or tonofilaments) in epithelial cells.

Microtubules (mī″-krō-too′-byoo-ulz), the largest of the cytoskeletal elements, are long, hollow tubes

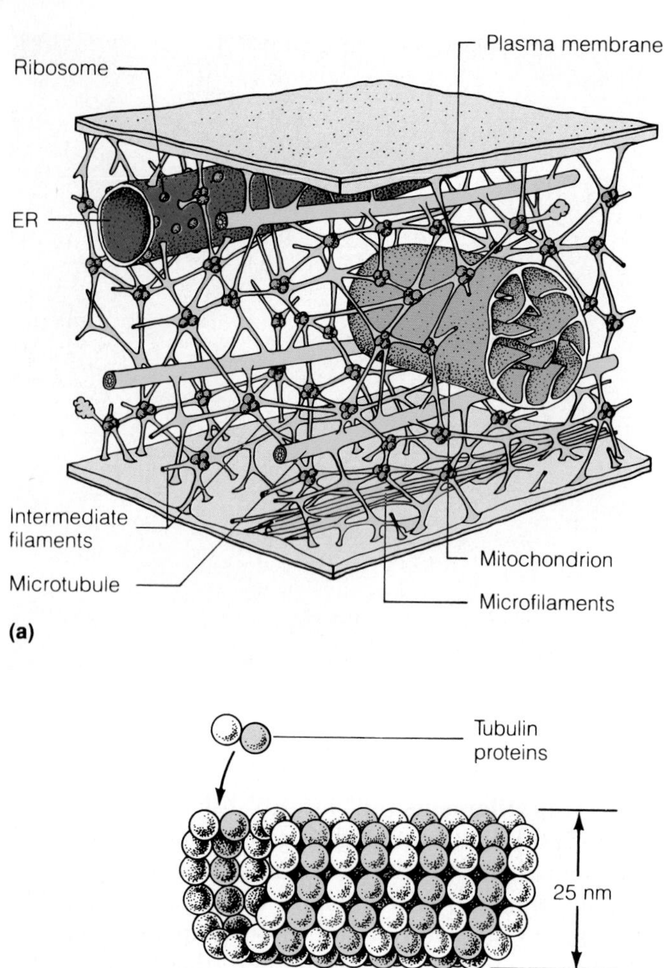

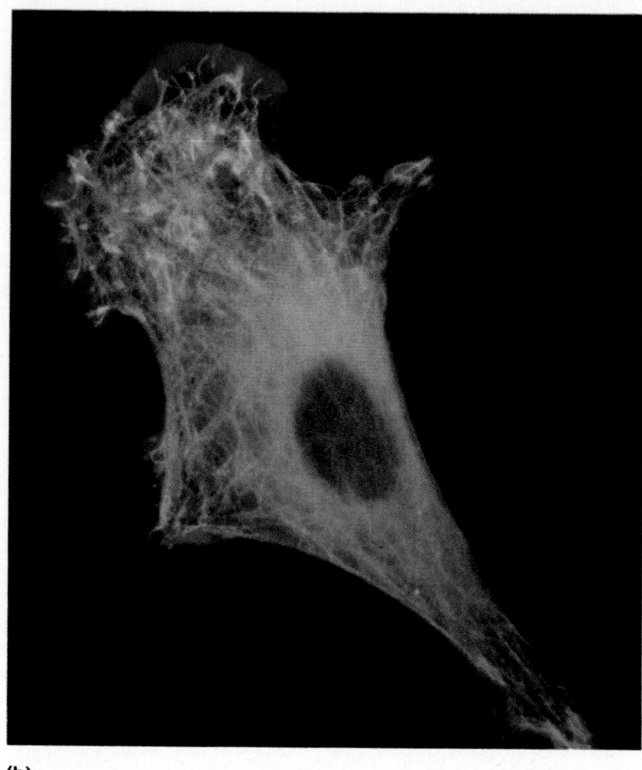

Figure 3.17 Cytoplasmic skeleton. (**a**) Three-dimensional reconstruction of the cytoskeleton. (**b**) Micrograph of the cytoskeleton (approx. 1200 ×); microtubules are in green and microfilaments are in blue. Intermediate filaments form most of the rest of the network. (**c**) Structure of a microtubule showing its globular protein subunits (tubulins).

formed of globular proteins called *tubulins* (Figure 3.17c). They are very difficult to study because they self-assemble spontaneously and then disassemble in an instant into their molecular subunits, which dissolve in the cytosol. Because of their dynamic and changing nature, microtubules have a variety of cellular roles; most importantly, they appear to be the overall "organizers" of the cytoskeleton. When present in parallel bundles, they help to maintain cell shape and rigidity. Their disassembly or rearrangement leads to changes in cell shape, allows the cytoplasm to flow, and plays a role in exocytosis. Microtubules also help to suspend or anchor organelles at specific locations within the cell and function in intracellular transport. For example, the flow of substances down the long nerve cell extensions is believed to be guided by microtubule "tracks." Microtubules also form the walls of more complex organelles called *centrioles* (sen′-trē-ōlz) which organize the construction of a microtubular structure (*spindle*) to which the chromosomes (krō′-muh-somz) attach during cell division.

Centrosome and Centrioles

Cytoskeletal elements would be of limited value as structural supports or promoters of cell movement unless linked to other parts of the cell. Many microtubules appear to be anchored at one end in a region near the nucleus called the *centrosome* (sen′-truh-sōm) or *cell center.* The centrosome contains paired organelles called **centrioles,** small cylindrical bodies oriented at right angles to each other (Figure 3.18). Each centriole consists of an array of nine triplets of microtubules, arranged to form a hollow tube (see also Figure 3.20c).

Centrioles are best known for their role in cell division. They also form the basis of two types of motile cellular projections: cilia and flagella.

Cilia and Flagella

Cilia (sih′-lē-uh) are hairlike, motile cellular extensions that occur, typically in large numbers, on the free (exposed) surfaces of certain cells (Figure 3.19).

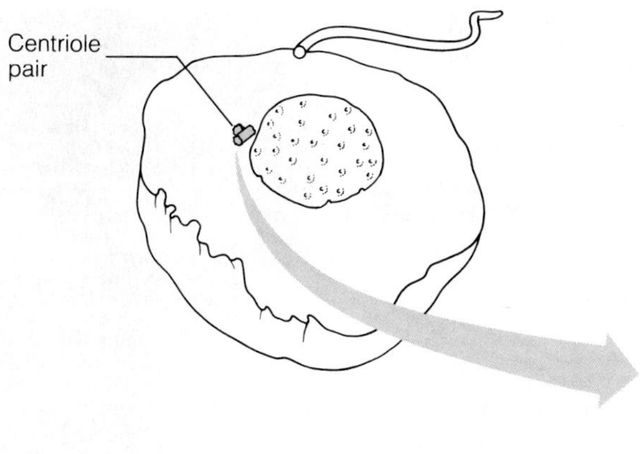

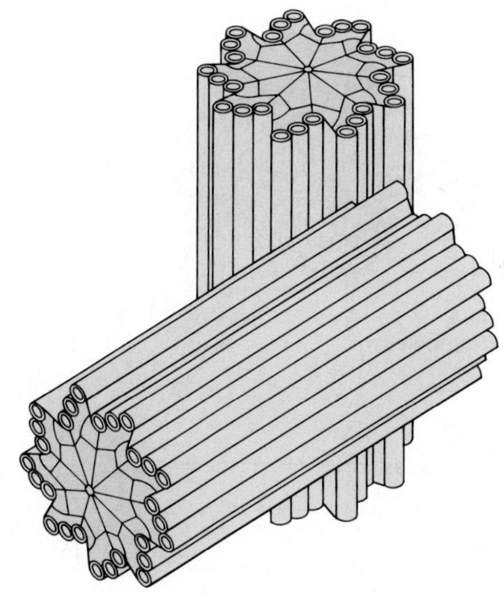

Figure 3.18 Centrioles. Three-dimensional view of a centriole pair oriented at right angles, as they are usually seen in the cell. Notice that each is composed of nine microtubule triplets.

The cilia move in unison like a well-disciplined rowing crew, creating a current that propels substances in one direction across the cell surface. For example, the ciliated cells that line the respiratory tract propel mucus laden with dust particles and bacteria upward away from the lungs.

When a cell is about to form cilia, the centrioles multiply and line up beneath the plasma membrane at the free cell surface. Microtubules then begin to "sprout" from each centriolar region and exert pressure on the plasma membrane, forming the ciliary projections. The sliding movement of the microtubules upon one another is responsible for the bending movements of each cilium, but how the activity of the cilia is coordinated is not understood. When the projections formed by centrioles are substantially longer, they are called **flagella** (fluh-jeh′-luh). The single example of a flagellated cell in the human body is a sperm, which has one propulsive flagellum, commonly called a "tail."

Centrioles forming the bases of cilia and flagella are commonly referred to as **basal** (bā′-sul) **bodies** (Figure 3.20a and c), since they were thought to be different from the structures seen in the centrosome. We now know that centrioles and basal bodies are identical structures. However, the pattern of microtubules in the cilium or flagellum itself differs slightly from that of a centriole; instead of nine microtubule triplets, there are nine microtubule pairs, with an additional microtubule pair in the center (Figure 3.20a and b).

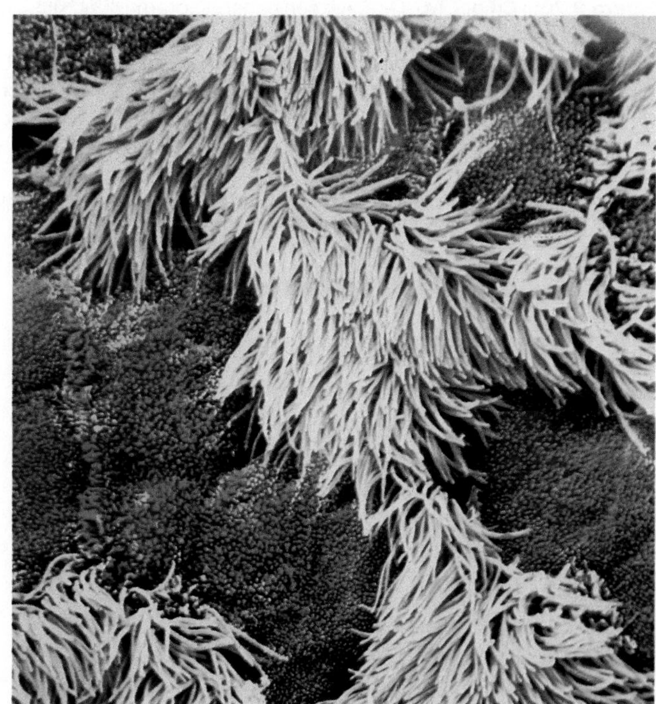

Figure 3.19 Cilia. Scanning electron micrograph of cilia in the trachea (221,000 ×). The cilia appear as yellow, grasslike projections. Mucus-secreting goblet cells (orange) with short microvilli are interspersed between the ciliated cells.

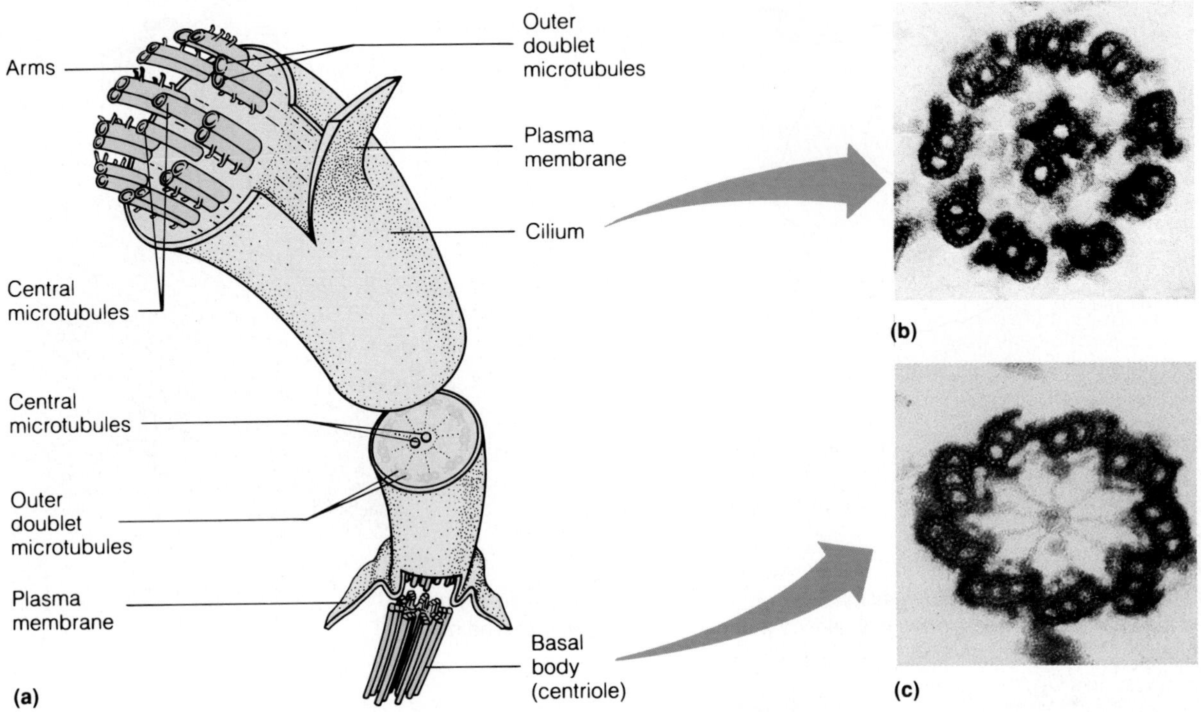

Figure 3.20 Cilia. (**a**) Three-dimensional diagram of a cross section through a cilium, showing its nine pairs of peripheral microtubules and one central microtubule pair. (**b**) Electron micrograph of a cilium, cross-sectional view (approx. 150,000×). (**c**) Electron micrograph of a basal body (centriole), cross-sectional view (approx. 150,000×).

The Nucleus

Anything that works, works best when it is controlled. For cells, the control center is the gene-containing **nucleus.** The nucleus can be compared to a computer, design department, construction boss, and board of directors—all rolled into one unique organelle. While most cells have only one nucleus, some, including skeletal muscle cells, bone destruction cells, and some liver cells, are *multinucleate* (mul′-tī-noo′-klē-ut); that is, they have many nuclei. The presence of more than one nucleus usually signifies that a larger-than-usual cytoplasmic mass must be regulated. With one exception, all of our body cells are *nucleated*. The exception is mature red blood cells, whose nuclei are ejected before the cells enter the bloodstream. These *anucleate* (ā″-noo′-klē-ut) (*a* = without) cells are incapable of reproducing themselves and can live in the bloodstream for only three to four months before they begin to deteriorate. You could say that anucleate cells are "programmed to die." Without a nucleus, a cell cannot make any more proteins, and when its enzymes and cell structures start to break down (as all biological molecules eventually do), they cannot be replaced.

The nucleus, averaging 5 μm in diameter, is the largest organelle in the cell. While most often spherical or oval, its shape usually conforms to the shape of the cell; for example, if a cell is elongated, the nucleus may be extended as well. The nucleus has three distinct regions or structures: the nuclear membrane, nucleoli, and chromatin (Figure 3.21).

The nuclear membrane encloses a colloidal fluid called *nucleoplasm* (noo′-klē-ō-plaz″-im), within which nucleoli and chromatin are suspended. Like the cytosol, the nucleoplasm contains salts, nutrients, and other needed chemicals.

Nuclear Membrane

The nucleus is bounded by the **nuclear membrane,** or **nuclear envelope,** which is a *double membrane* barrier (each membrane is a phospholipid bilayer) similar to the mitochondrial membrane. Between the two membranes is a fluid-filled space called the *perinuclear* (payr″-ih-noo′-klē-er) *cisterna*. The outer nuclear

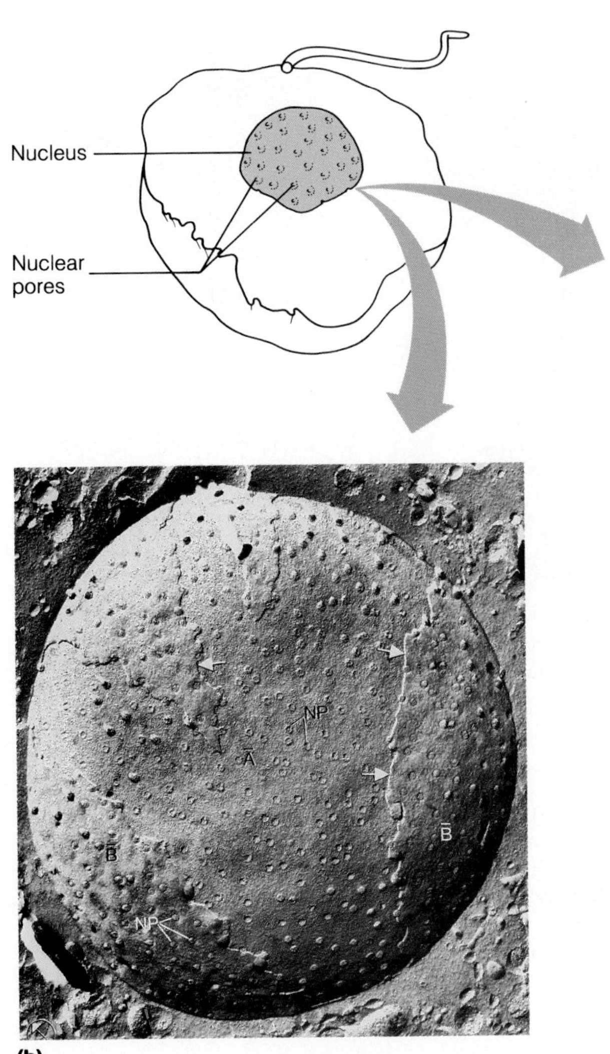

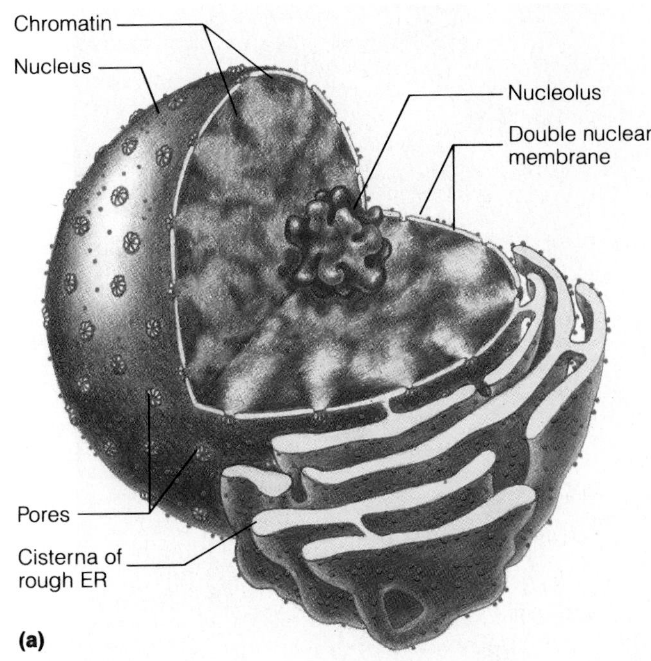

Figure 3.21 The nucleus. (**a**) Three-dimensional diagrammatic view of the nucleus, showing the continuity of its double membrane with the ER. (**b**) Freeze-fracture micrograph (11,600×) of the surface of the nucleus, showing the nuclear pores (NP). The outer nuclear membrane is labeled B̄; the inner nuclear membrane, labeled Ā, has been revealed by the fracturing process.

membrane is continuous with the ER of the cytoplasm and may be studded with ribosomes.

At various points, the two layers of the nuclear membrane fuse, and *nuclear pores*, each about 30–100 nm in diameter, penetrate through the fused regions. Like other cellular membranes, the nuclear membrane is selectively permeable, but the passage of substances is much freer than elsewhere because of its relatively large pores. Protein molecules imported from the cytoplasm and RNA molecules exported from the nucleus pass easily through the pores.

Nucleoli

Nucleoli (noo-klē′-uh-lī) are dark-staining spherical bodies found within the nucleus (see Figure 3.21). They are made up of ribosomal RNA (rRNA) and proteins, and they are not membrane bounded. Typically, there are one or two nucleoli per cell, but there may be more. Nucleoli are "ribosome-producing machines"; consequently, they are usually very large in growing cells that are actively making large amounts of tissue proteins. Nucleoli are associated with chromatin regions containing the DNA that issues genetic instructions for synthesizing ribosomal RNA. Such DNA segments are called *nucleolar organizer regions.* As molecules of rRNA are synthesized within a nucleolus, they are combined there with proteins forming the two kinds of ribosomal subunits. (The proteins are manufactured on ribosomes in the cytoplasm and "imported" into the nucleus.) The subunits leave the nucleus through the nuclear pores and enter the cytoplasm, where they are assembled into functional ribosomes.

Chromatin

Seen through a light microscope, **chromatin** (krō′-muh-tin) appears as a fine, lightly stained network, but special techniques reveal it as a system of bumpy threads

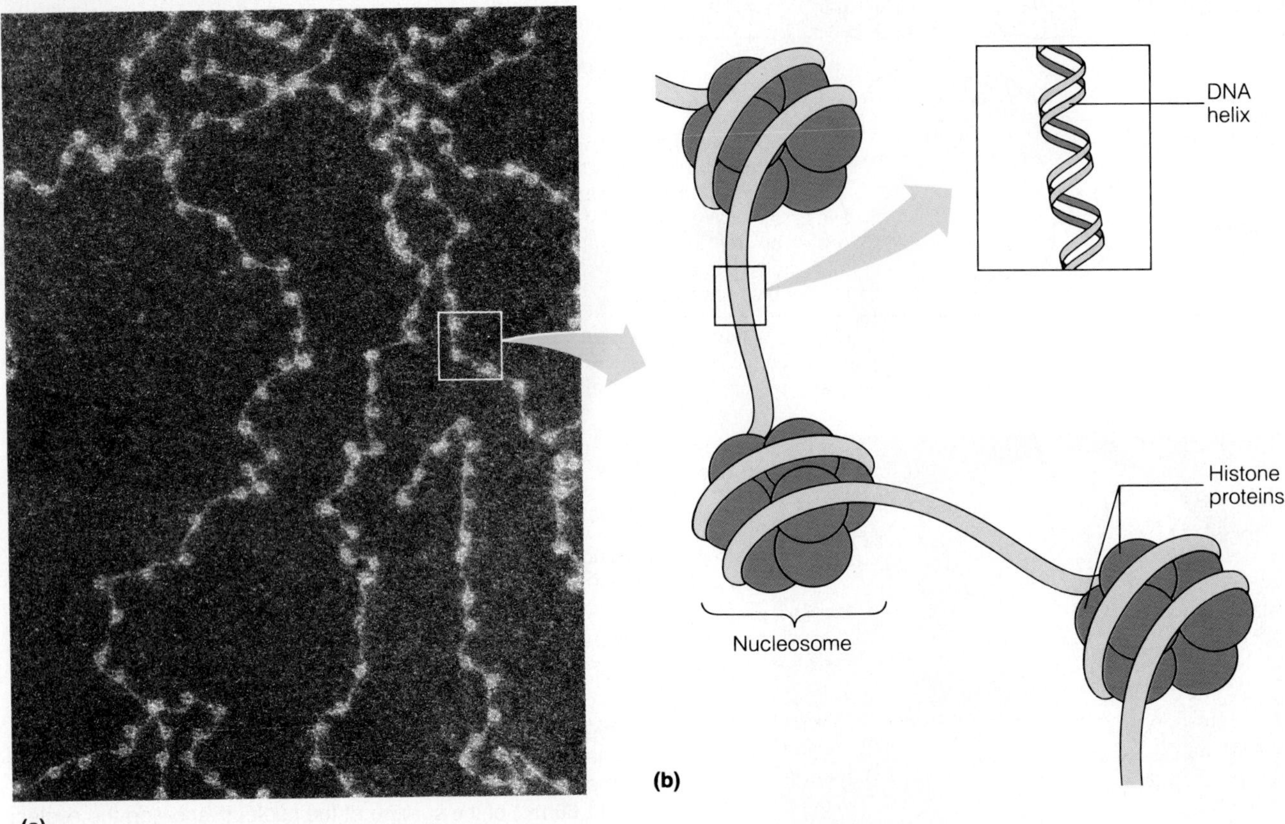

Figure 3.22 Chromatin. (**a**) Electron micrograph of chromatin fibers, which have the appearance of beads on a string (216,000×). (**b**) A proposed model for the structure of a chromatin fiber. The DNA molecule winds around spherical particles called nucleosomes; each nucleosome is composed of eight histone proteins.

weaving their way through the nucleoplasm (Figure 3.22a). Chromatin is composed of approximately equal amounts of DNA, which comprises the *chromosomes*, and globular *histone* (his′-tōn) *proteins*. *Nucleosomes* (noo′-klē-uh-sōmz), the fundamental units of chromatin, are spherical clusters of eight histone proteins connected like beads on a string by a DNA molecule that winds around each of them (Figure 3.22b). In addition to providing a physical site for the attachment of the chromosomal DNA, the histone proteins are believed to mediate many aspects of DNA function. Changes in the shape of the histones in a nondividing cell expose different DNA segments, or genes, so that they can "dictate" the specifications for protein synthesis. When a cell is preparing to divide, the chromatin threads coil and condense enormously to form short, barlike bodies. The functions of DNA and the mechanism of cell division are described later in the next section.

Cell Growth and Reproduction

The Cell Life Cycle

The **cell life cycle** is the series of changes a cell goes through from the time it is formed until it reproduces itself. The cycle can be divided into two major periods: interphase, in which the cell grows and carries on most of its activities, and cell division, or the mitotic phase, during which it reproduces itself (Figure 3.23).

Interphase

Interphase encompasses the total period from cell formation to cell division. Early cytologists, unaware of the constant molecular activity in cells and impressed by the obvious movements of cell division, called interphase the resting phase of the cell cycle. (The term *interphase* reflects this idea that it is merely a stage between cell divisions.) However, this is grossly

Table 3.2 Parts of the Cell: Structure and Function

Cell part	Structure	Functions
Plasma membrane (Figure 3.2)	Membrane composed of a double layer of lipids (phospholipids, cholesterol, etc.) within which proteins are embedded; proteins may extend entirely through the lipid bilayer or protrude on only one face; externally facing proteins and some lipids have attached sugar groups	Serves as an external cell barrier; acts in transport of substances into or out of the cell; externally facing proteins act as receptors (for hormones, neurotransmitters, etc.) and in cell-to-cell recognition
Cytoplasm	Cellular region between the nuclear and plasma membranes; consists of fluid *cytosol,* containing dissolved solutes, *inclusions* (stored nutrients, secretory products, pigment granules), and *organelles,* the metabolic machinery of the cytoplasm	
Cytoplasmic organelles		
• Mitochondria (Figure 3.12)	Rodlike, double-membrane structures; inner membrane folded into projections called cristae	Site of ATP synthesis; powerhouse of the cell
• Ribosomes (Figure 3.13)	Dense particles consisting of two subunits, each composed of ribosomal RNA and proteins; free or attached to rough ER	The sites of protein synthesis
• Rough endoplasmic reticulum (rough ER) (Figure 3.13)	Membrane system enclosing a cavity, the cisterna, and coiling through the cytoplasm; externally studded with ribosomes	Sugar groups are attached to proteins within the cisternae; proteins are bound in vesicles for transport to the Golgi apparatus and other sites
• Smooth endoplasmic reticulum (smooth ER) (Figure 3.13)	Membranous system of sacs and tubules; free of ribosomes	Site of lipid and steroid synthesis
• Golgi apparatus (Figure 3.14)	A stack of smooth membrane sacs close to the nucleus	Packages, modifies, and segregates proteins for secretion from the cell, inclusion in lysosomes, and incorporation into the plasma membrane
• Lysosomes (Figure 3.16)	Membranous sacs containing acid hydrolases	Sites of intracellular digestion
• Peroxisomes	Membranous sacs of oxidase enzymes	The enzymes detoxify a number of toxic substances; the most important enzyme, catalase, breaks down hydrogen peroxide
• Microfilaments (Figure 3.17)	Fine filaments of the contractile protein actin	Involved in muscle contraction and other types of intracellular movement; help form the cell's cytoskeleton
• Intermediate filaments (Figure 3.17)	Protein fibers; composition varies	The stable cytoskeletal elements; resist mechanical forces acting on the cell
• Microtubules (Figures 3.17, 3.18, 3.20)	Cylindrical structures composed of tubulin proteins	Support the cell and give it shape; involved in intracellular and cellular movements; form centrioles
• Centrioles (Figure 3.18)	Paired cylindrical bodies, each composed of nine triplets of microtubules	Organize a microtubule network during mitosis to form the spindle and asters; form the bases of cilia and flagella
• Cilia (Figures 3.19, 3.20)	Short, cell surface projections; each cilium composed of nine pairs of microtubules surrounding a central pair	Move in unison, creating a unidirectional current that propels substances across cell surfaces
• Flagellum	Like cilium, but longer; only example in humans is the sperm tail	Propels the cell
Nucleus (Figure 3.21)	Largest organelle; surrounded by the nuclear membrane; contains fluid nucleoplasm, nucleoli, and chromatin	Control center of the cell; responsible for transmitting genetic information and providing the instructions for protein synthesis
• Nuclear membrane (Figure 3.21)	Double bilipid membrane containing proteins; pierced by pores; outer membrane continuous with the cytoplasmic ER	Separates the nucleoplasm from the cytoplasm and regulates passage of substances to and from the nucleus
• Nucleoli (Figure 3.21)	Dense spherical (non-membrane-bounded) bodies; composed of ribosomal RNA and proteins	Site of ribosome subunit manufacture
• Chromatin (Figures 3.21, 3.22)	Granular, threadlike material composed of DNA and histone proteins	DNA constitutes the genes

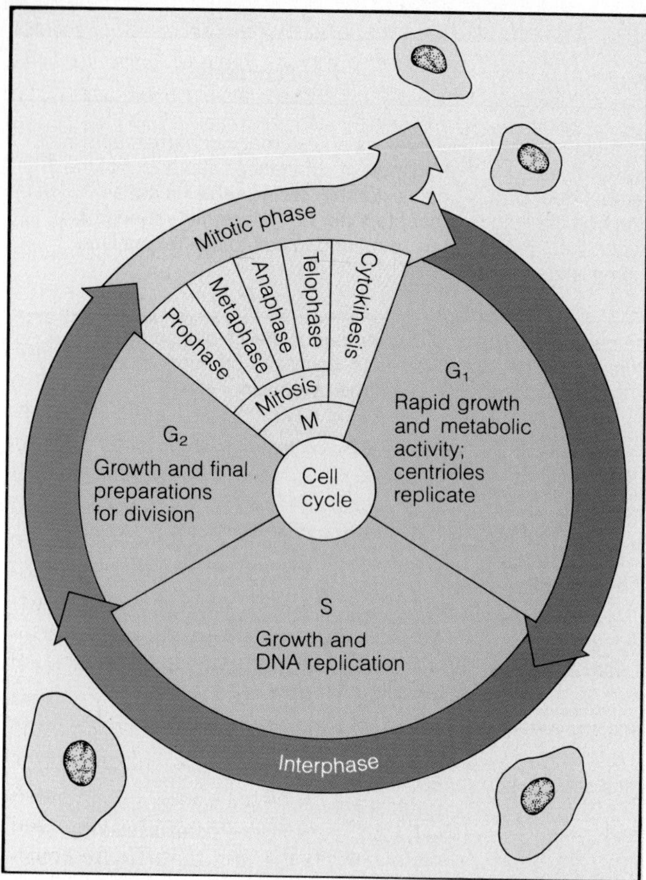

Figure 3.23 The cell cycle. During G_1, cells grow rapidly and carry out their routine functions; the centrioles replicate. The S phase begins when DNA synthesis starts, and it ends when the DNA has replicated. In the short G_2 phase, materials needed for cell division are synthesized and growth continues. During the M phase (cell division), mitosis and cytokinesis occur. The length of the cell cycle varies with different cell types, the G_1 phase is the longest and most variable phase.

misleading, because during interphase, a cell is carrying out all its routine activities and is "resting" only from dividing. Perhaps a more accurate name for this phase would be *metabolic phase* or *growth phase*.

In addition to carrying on its life-sustaining reactions, an interphase cell engages in preparations for the next cell division. Interphase is divided into G_1, S, and G_2 subphases. During G_1, the first portion of interphase, cells are metabolically active, synthesize proteins rapidly, and grow vigorously. This is the most variable phase in terms of length. Cells that have rapid division rates have a G_1 phase of hours; those that divide more slowly can be in this phase for days or even years.

For most of G_1, no cell activities occur that are directly related to cell division, but at the end of this phase, the centrioles replicate in preparation for their role in cell division. During the next phase, the S (synthetic) phase, DNA replicates itself, ensuring that the two future cells will receive identical copies of the genetic material, and new histones are made and assembled into chromatin. (We will describe DNA replication next.) The final phase of interphase, G_2, is very brief. During this period, enzymes and other proteins needed for the division process are synthesized and moved to their proper sites. Throughout the S and G_2 phases, the cell continues to grow and carries on with business as usual.

DNA Replication Before a cell can divide, its DNA must replicate itself exactly, so that identical copies of the cell's genes can be passed on to each of its offspring, thereby ensuring that all body cells have the same heritage. While the precise trigger for DNA synthesis is unknown, once it begins, it is an all-or-none phenomenon. The replication process begins simultaneously on several chromatin threads and continues until all of the DNA has been replicated.

The process begins with the uncoiling of the DNA helices. Then the hydrogen bonds between the paired nitrogen bases are broken by enzymes, so that a DNA molecule gradually separates into its two complementary nucleotide chains (Figure 3.24). The sites where this is occurring are generally Y shaped and are called *replication forks*. Each freed nucleotide strand then serves as a *template*, or set of instructions, for building a new complementary nucleotide strand from DNA nucleotides in the nucleoplasm. The positioning and joining of nucleotides into the new strand is catalyzed by enzymes and energized by ATP.

Recall that nucleotide base pairing is always complementary: adenine (A) always bonds to thymine (T), and guanine (G) always bonds to cytosine (C) (see p. 56). Such specific base pairing ensures the precision of DNA replication: The order of the nucleotides on the template strand determines the order on the new strand being built. For example, a TACTGC sequence on a template strand would bond to new nucleotides with the order ATGACG; and the comparable region on the other template strand, which has the base sequence ATGACG, bonds to new nucleotides with the order TACTGC. The end result is that two DNA molecules are formed from and are identical to the original DNA helix, and each consists of one old and one newly assembled nucleotide strand. As soon as replication ends, histones associate with the DNA, completing the formation of two new chromatin strands. The chromatin strands (or chromatids, as they are called after they condense) are united by a centromere and remain attached until the cell has entered the anaphase stage of mitotic cell division. They are then distributed to the daughter cells, as next described, ensuring that each has identical genetic information.

Cell Division

Cell division is essential for body growth and tissue repair. Short-lived cells that continually wear away,

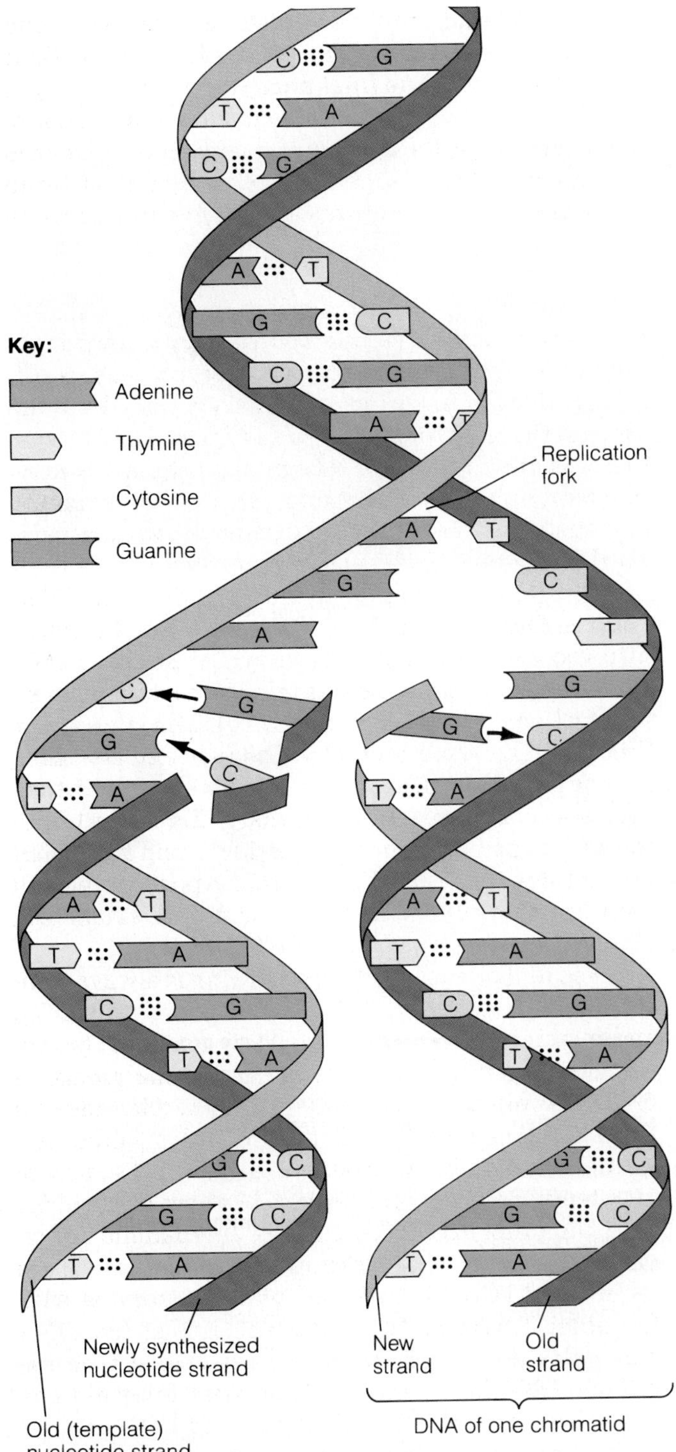

Figure 3.24 Replication of DNA. The DNA helix uncoils, and the hydrogen bonds between its base pairs are broken. Then each nucleotide strand of the DNA acts as a template for the construction of a complementary nucleotide strand, as illustrated in the bottom portion of the diagram. After replication is completed, two DNA molecules exist, identical to the original DNA molecule and to each other. Each DNA molecule formed consists of one old (template) strand and one newly assembled strand and constitutes a chromatid of a chromosome.

such as cells of the skin and intestinal lining, reproduce themselves almost continuously. Others, such as liver cells, divide only until the organ they compose reaches a particular size, but retain the ability to reproduce themselves if the organ is damaged. Cells of nervous tissue and skeletal and heart muscle totally lose their ability to divide when they are fully mature, and repair is accomplished by inelastic scar tissue (a fibrous type of connective tissue).

The signals that prod cells to divide are poorly understood, but we know that *surface-volume relationships* are important. The amount of nutrients a growing cell requires is directly related to its volume. As a cell increases in volume, its surface area does not increase proportionately. Volume increases with the cube of cell radius, whereas surface area increases with the square of the radius; thus, a 64-fold increase in cell volume would be accompanied by only a 16-fold increase in surface area. Consequently, the surface area of the plasma membrane becomes inadequate for nutrient and waste exchange when a cell reaches a certain critical size. This problem is solved by cell division because the smaller daughter cells once again have a favorable surface-volume relationship. These surface-volume relationships explain why most cells are microscopic in size. Other mechanisms that help to determine when cells divide include the availability of space (normal cells stop proliferating when they begin touching) and the release of inhibitory or stimulatory chemical signals by other cells in the area. Cancer cells escape many of the normal controls of cell division and divide wildly, making them dangerous to their host (see the box on pp. 94–95).

In most body cells, cell division or the M phase of the cell life cycle (see Figure 3.23), involves two distinct events: *mitosis* (mī-tō′-sis), or division of the nucleus, and *cytokinesis* (sī″-tō-kuh-nē′-sis), or division of the cytoplasm. A somewhat different process of nuclear division called *meiosis* (mī-ō′-sis) produces sex cells (ova and sperm) and ensures that they have only half the number of genes found in other body cells. (When the two sex cells combine in fertilization, the full complement of genetic material is restored.) The details of meiosis are discussed in Chapter 28. Here we will concentrate on mitotic cell division.

Mitosis **Mitosis** refers to the series of events that parcel out the replicated DNA of the mother cell to two daughter cells. (Use of the terms *mother cell* and *daughter cells* has a long history and carries no sexual implications.) Mitosis is usually discussed in terms of four phases: **prophase, metaphase, anaphase,** and **telophase.** However, mitosis is actually a continuous process, with one phase merging smoothly into the next. Its duration varies according to cell type, but it typically takes about 2 hours from start to finish. Mitosis is described in detail in Figure 3.25.

Figure 3.25 The stages of mitosis. (Micrographs approx. 600×.)

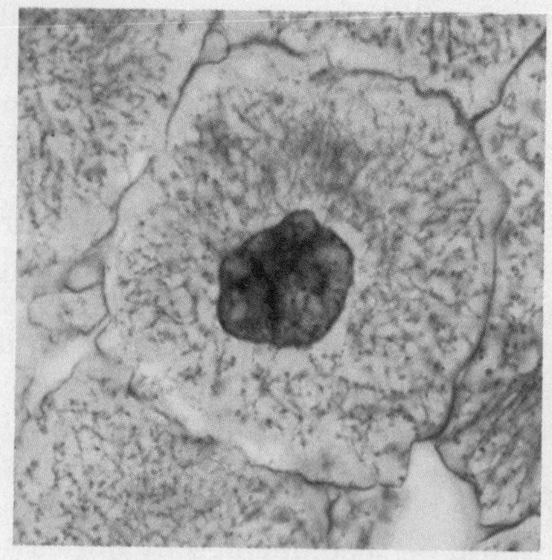

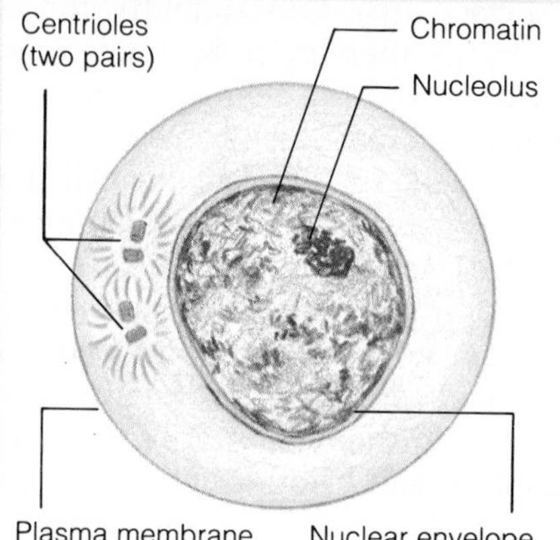

Interphase

Interphase is the period of a cell's life when it is carrying out its normal metabolic activities and growing. During various periods of this phase, the centrioles are replicated (G_1), DNA is replicated (S), and the final preparations for mitosis are completed (G_2). During interphase, the chromosomal material is seen in the form of diffuse chromatin, and the nuclear membrane and nucleolus are intact and visible.

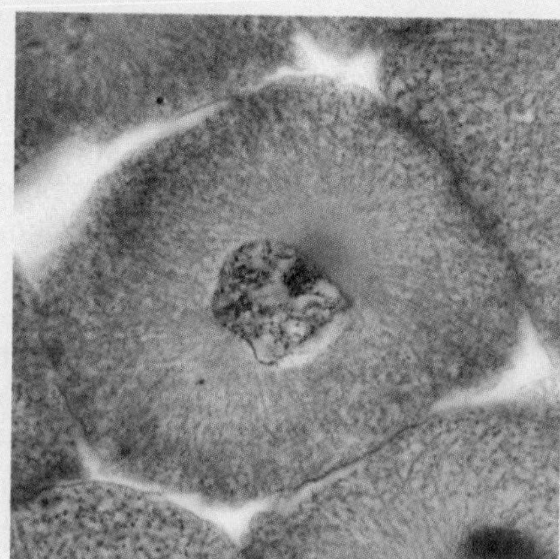

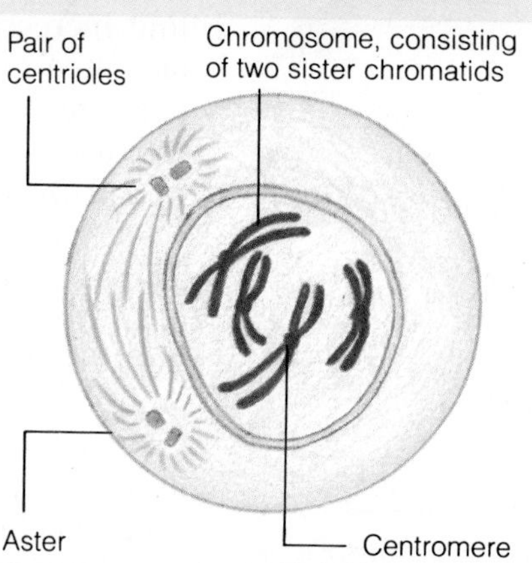

Early prophase

Prophase, the first and longest phase of mitosis, begins when the chromatin threads start to condense and shorten, forming barlike *chromosomes* that are visible with a light microscope. Since DNA replication has occurred during interphase, each chromosome is actually made up of two identical chromatin threads, now called *chromatids*. The chromatids of each chromosome are held together by a small, buttonlike body called a *centromere*. Chromatids can be thought of as half-chromosomes; after they separate, each will be considered a new chromosome.

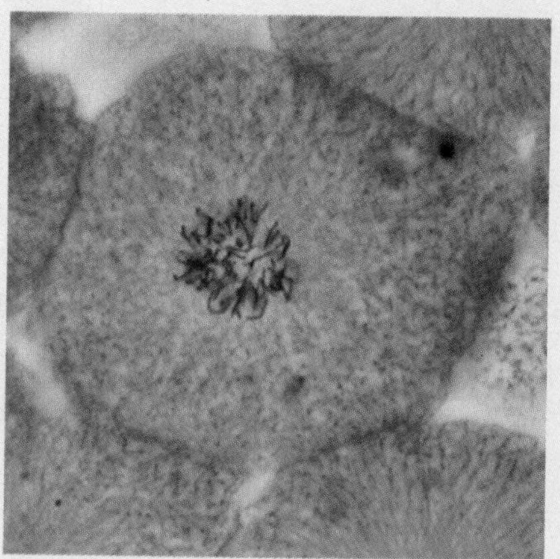

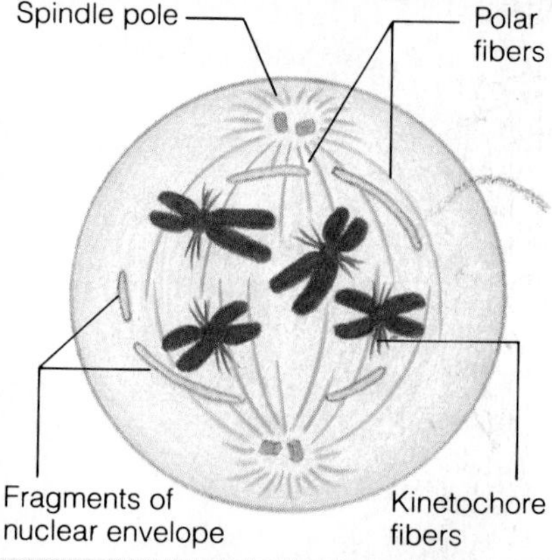

Late prophase

As the chromosomes appear, the nucleoli begin to break down, the cytoplasmic cytoskeleton disassembles and disappears, and the centriole pairs separate from one another. The centrioles act as focal points for "reeling out" an assembly of microtubules called the **mitotic spindle**. As the microtubules lengthen, the centrioles are pushed farther and farther apart and toward opposite ends (poles) of the cell until the spindle finally extends from pole to pole. While the centrioles are still moving, the nuclear membrane breaks up into fragments, allowing the spindle to occupy the center of the cell and to interact with the chromosomes. The centrioles also direct the formation of *asters*, systems of microtubules that radiate outward from the ends of the spindle. The asters are distinct from the *polar fibers* (microtubules) forming the spindle; they act to stabilize the spindle by anchoring it to the plasma membrane. Meanwhile, the centromeres of the chromosomes "sprout" microtubules called *kinetochore* (kih-neh′-tih-kor) *fibers* that interact with and contribute to the spindle fibers. The interaction of the kinetochore fibers with the spindle fibers sets the chromosomes into vigorous motion. The spindle is essential not only as an attachment site for chromosomes, but also for distributing the chromosomes to opposite ends of the cell.

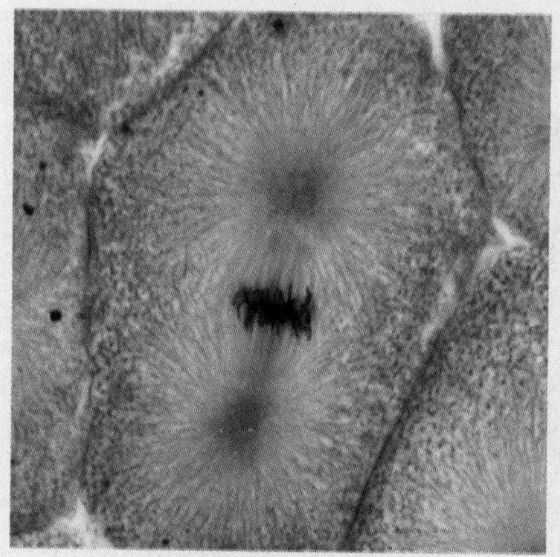

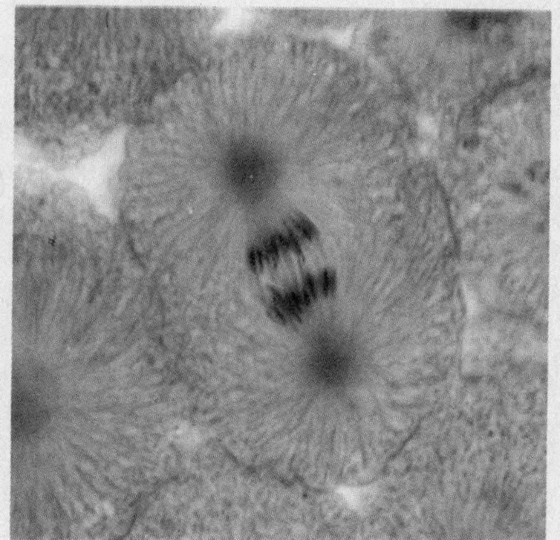

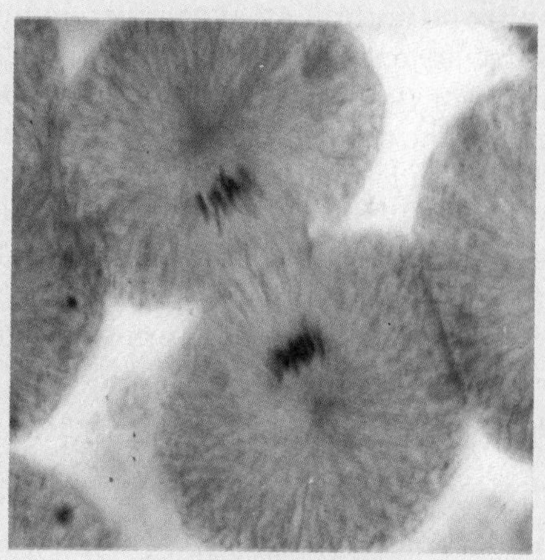

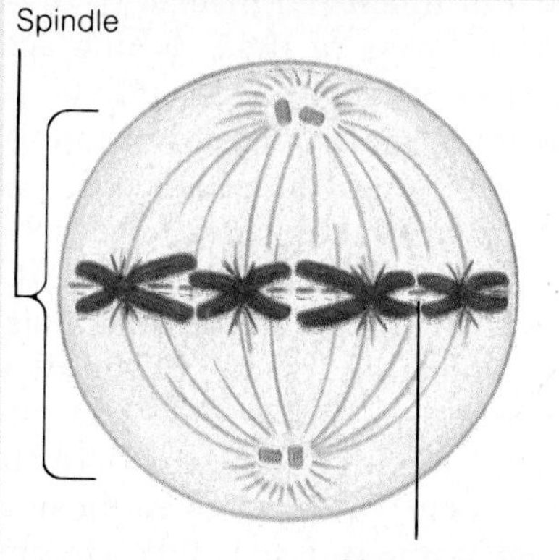

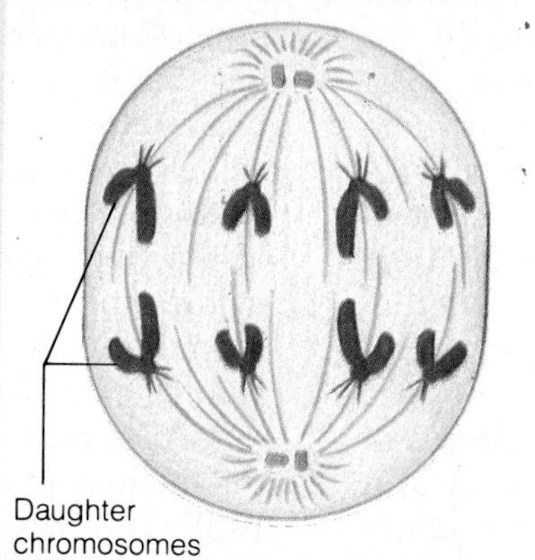

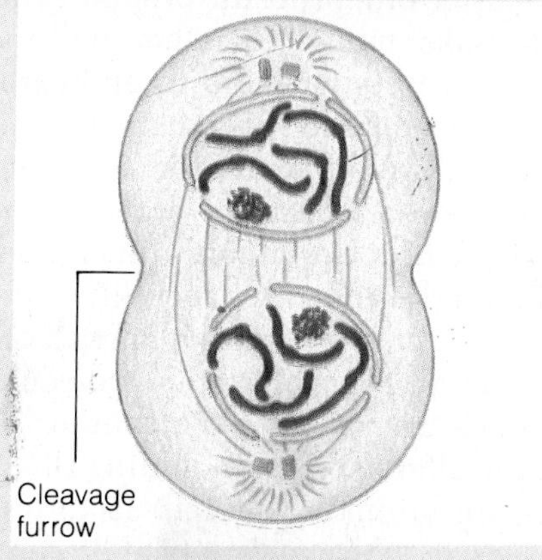

Metaphase

Metaphase, the second phase of mitosis, is quite short. The chromosomes cluster at the middle of the cell, with their centromeres precisely aligned at the exact center, or *equator*, of the spindle. By the end of metaphase, a straight line of chromosomes can be seen along the spindle equator.

Anaphase

Anaphase, the third phase of mitosis, begins abruptly as the centromeres of the chromosomes split, and each chromatid now becomes a chromosome in its own right. The microtubules attached to the half-centromere of each chromosome shorten and gradually pull the chromosome toward the pole, like a recoiling rubber band. The sister chromatids of each original chromosome are pulled toward opposite ends of the cell. Anaphase is easy to recognize because the moving chromosomes look V-shaped: The centromeres, which are attached to the spindle fibers, lead the way, and the chromosomal "arms" dangle behind them. Anaphase is the shortest mitotic phase; it typically lasts only a few minutes.

This process of parceling out chromosomes is aided by the fact that the chromosomes being moved about are short, compact bodies. Diffuse chromatin threads would tend to tangle and trail, increasing the possibility of breakage and imprecise distribution of the genetic material.

Telophase and cytokinesis

Telophase (tē′-luh-fāz) begins as soon as chromosomal movement stops. This final phase is like prophase in reverse. The identical sets of chromosomes at the opposite ends of the cell uncoil and resume their threadlike chromatin form; a new nuclear membrane, derived from the endoplasmic reticulum, re-forms around each chromatin mass; nucleoli reappear within the nuclei; and the spindle breaks down and disappears. Mitosis is now ended. The cell, for just a brief period, is a binucleate (two-nucleus) cell, and each new nucleus is identical to the original mother nucleus.

As a rule, as mitosis draws to a close, *cytokinesis* occurs, completing the division of the cell into two daughter cells.

Cytokinesis Division of the cytoplasm, or **cytokinesis** (sī″-tō-kih-nē′-sis), begins during late anaphase or early telophase (see Figure 3.25). The plasma membrane over the cell center (the spindle equator) is drawn inward to form a *cleavage furrow* by the activity of microfilaments. The cleavage furrow deepens until the original cytoplasmic mass is pinched into two parts; thus, at the end of cytokinesis, there are two daughter cells. Each is smaller and has less cytoplasm than the mother cell, but is *genetically* identical to it. The daughter cells then enter the interphase portion of their life cycle and grow and carry out normal cell activities until it is their turn to divide.

Protein Synthesis

In addition to its self-replication role, DNA serves as the master blueprint for protein synthesis. Although cells make numerous other molecules needed for homeostasis, such as lipids and carbohydrates, most of their metabolic machinery is concerned in some way with protein synthesis. This should not be surprising, since structural proteins constitute most of the dry cell material, and functional proteins direct and underlie all cellular activities. DNA does not dictate the structure of carbohydrates or fats; it specifies only the structure of protein molecules, including the enzymes that catalyze the synthesis of all classes of biological molecules, including DNA itself. Basically, cells are miniature protein factories that constantly synthesize the huge variety of proteins that determine the chemical and physical nature of cells—and therefore of the body as a whole.

Proteins, you will recall, are composed of polypeptide chains, which in turn are made up of amino acids (see pp. 49–50). For purposes of this discussion, we can define a **gene** as a segment of a DNA molecule that carries the instructions for one polypeptide chain. However, there are selected genes that specify the structure of certain varieties of RNA as their final product, rather than a polypeptide chain.

The four nucleotide bases (A, G, T, and C) are the "letters" of the genetic dictionary, and the information of DNA is found in the sequence of these bases. Each sequence of three bases, called a *triplet*, can be thought of as a "word" that specifies a particular amino acid (Figure 3.26). For example, the triplet AAA calls for the amino acid phenylalanine, while CCT calls for glycine. The sequence of the triplets in each gene forms a "sentence" that tells exactly how a particular polypeptide is to be made; it tells the number, kinds, and order of amino acids needed to build that protein. Variations in the arrangement of A, T, C, and G allow our cells to make all the different kinds of proteins needed. It has been estimated that a typical gene has between 300 and 3000 base pairs in sequence. Since the ratio between DNA bases in the gene and amino acids in the polypeptide is 3:1, polypeptides typically contain from 100 to 1000 amino acids.

The Role of RNA

By itself, DNA is rather like a strip of magnetic recording tape: The information it contains cannot be expressed without a decoding mechanism. Furthermore, the manufacturing sites for polypeptides are the ribosomes in the cytoplasm, but in an interphase cell, DNA never leaves the nucleus. Hence, DNA requires not only a decoder, but a messenger as well. The decoding and messenger functions are carried out by the second type of nucleic acid: RNA.

As you learned in Chapter 2, RNA differs from DNA in being single stranded and in having ribose instead of deoxyribose and uracil (U) instead of thymine (T). There are three forms of RNA, which act together to carry out DNA's instructions for polypeptide synthesis: (1) **transfer RNA (tRNA),** whose molecules are small and roughly cloverleaf shaped; (2) **ribosomal RNA (rRNA),** whose molecules form part of the ribosomes; and (3) **messenger RNA (mRNA),** whose molecules are relatively long nucleotide strands resembling "half-DNA" molecules, or one of the two strands of a DNA molecule.

All three types of RNA are formed on the DNA in the nucleus in much the same way as DNA replicates itself: The DNA chain separates and one of its strands serves as a template for synthesis of a complementary RNA strand. Once formed, the RNA molecule is released from the DNA template strand and migrates into the cytoplasm. Its job done, DNA simply recoils into its helical form. Most of the nuclear DNA codes for the synthesis of short-lived mRNA, so named because it carries the "message" containing the instructions for building a polypeptide from the gene to the ribosomes. Thus, the end product of most genes is a polypeptide. Relatively small portions of the DNA code for the synthesis of rRNA (the nucleolar organizer regions mentioned previously) and tRNA, both of which are long-lived and stable types of RNA. Since rRNA and tRNA do not carry codes for synthesizing other molecules, they are the final products of the genes that code for them. Ribosomal RNA and tRNA act together to "translate" the message carried by mRNA.

The rules by which the base sequence of a DNA gene are translated into protein structure (amino acid sequence) are referred to as the **genetic code.** Essentially, polypeptide synthesis involves two major steps: (1) *transcription*, in which information in DNA is encoded in mRNA, and (2) *translation*, in which the information carried by mRNA is decoded and used to assemble polypeptides.

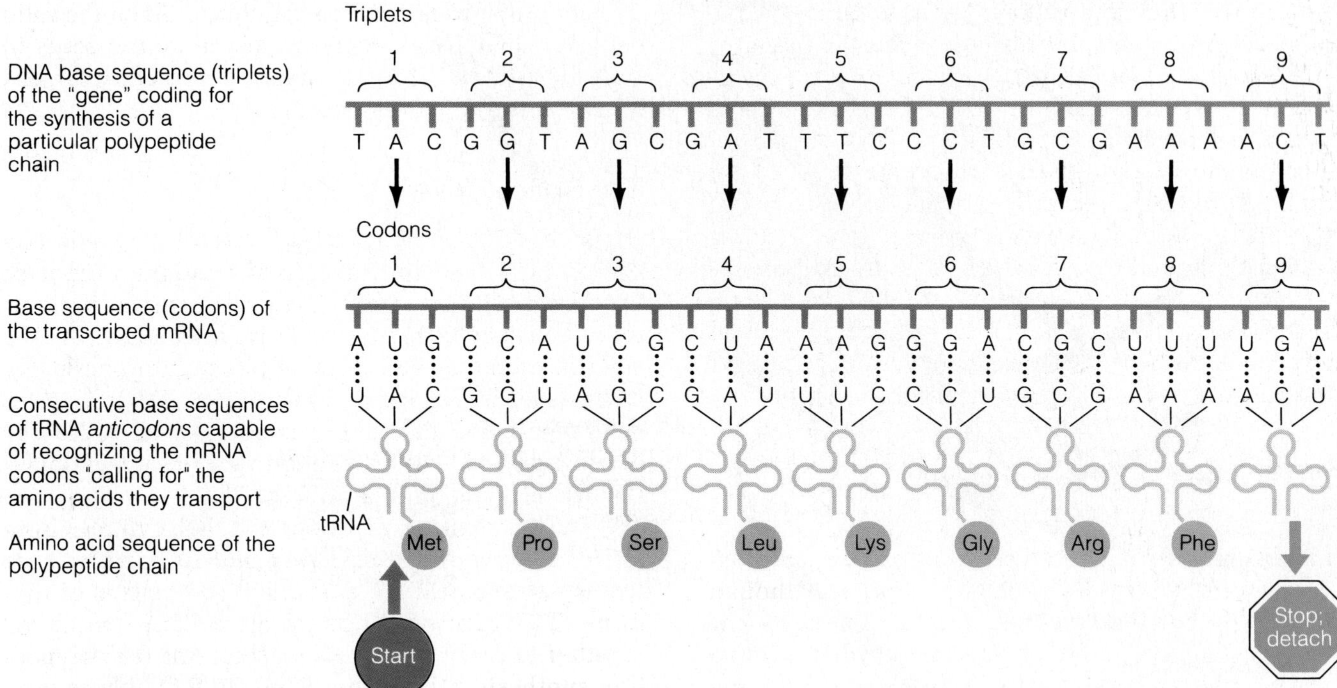

Figure 3.26 Information transfer from DNA to RNA. Information is transferred from the DNA of the gene to the complementary messenger RNA molecule, whose codons are then "read" by transfer RNA anticodons. Notice that the "reading" of the mRNA by tRNA anticodons reestablishes the base (triplet) sequence of the DNA genetic code, except that T is replaced with U.

Transcription

The word *transcription* often refers to one of the jobs done by a secretary—converting notes taken in shorthand into a longhand or typewritten copy. In other words, the same information is converted from one form (or format) to another. In cells, **transcription** involves the transfer of information from a DNA gene's base sequence to the complementary base sequence of an mRNA molecule (Figure 3.26). The form is different, but the same information is being conveyed. Once the mRNA molecule is made, it detaches and leaves the nucleus. Only DNA and mRNA are involved in the transcription process.

Let us follow the transcription process when a particular polypeptide is to be made. The DNA segment coding for that protein is uncoiled, and one DNA nucleotide strand serves as the template for the construction of a complementary mRNA molecule (see Figure 3.27, step 1). For example, if a particular DNA triplet sequence is AGC, the mRNA sequence synthesized at that site will be UCG. While each three-base sequence on DNA is called a triplet, the corresponding three-base sequence on mRNA is called a *codon,* reflecting the fact that it is the coded information of mRNA that is actually used in protein synthesis.

Since there are four kinds of RNA (or DNA) nucleotides, there are 4^3, or 64, possible codons. Three of these 64 codons are "stop signs" that specify the termination of a polypeptide; all the rest code for amino acids. Since there are only about 20 amino acids, some are specified by more than one codon. A few examples of mRNA codons and the amino acids they specify are provided in Figure 3.26.

Translation

A translator takes a message in one language and restates it in another. In the **translation** step of protein synthesis, the language of nucleic acids (base sequence) is translated into the language of proteins (amino acid sequence). The process of translation occurs in the cytoplasm and involves all three varieties of RNA (see Figure 3.27, steps 2–5).

When the mRNA molecule carrying instructions for a particular protein reaches the cytoplasm, it binds to a small ribosomal subunit by base pairing to rRNA. Then transfer RNA comes into the picture. As its name suggests, tRNA has the job of transferring amino acids to the ribosome. There are approximately 20 different types of tRNA, each capable of binding with a specific amino acid. The attachment process is controlled by

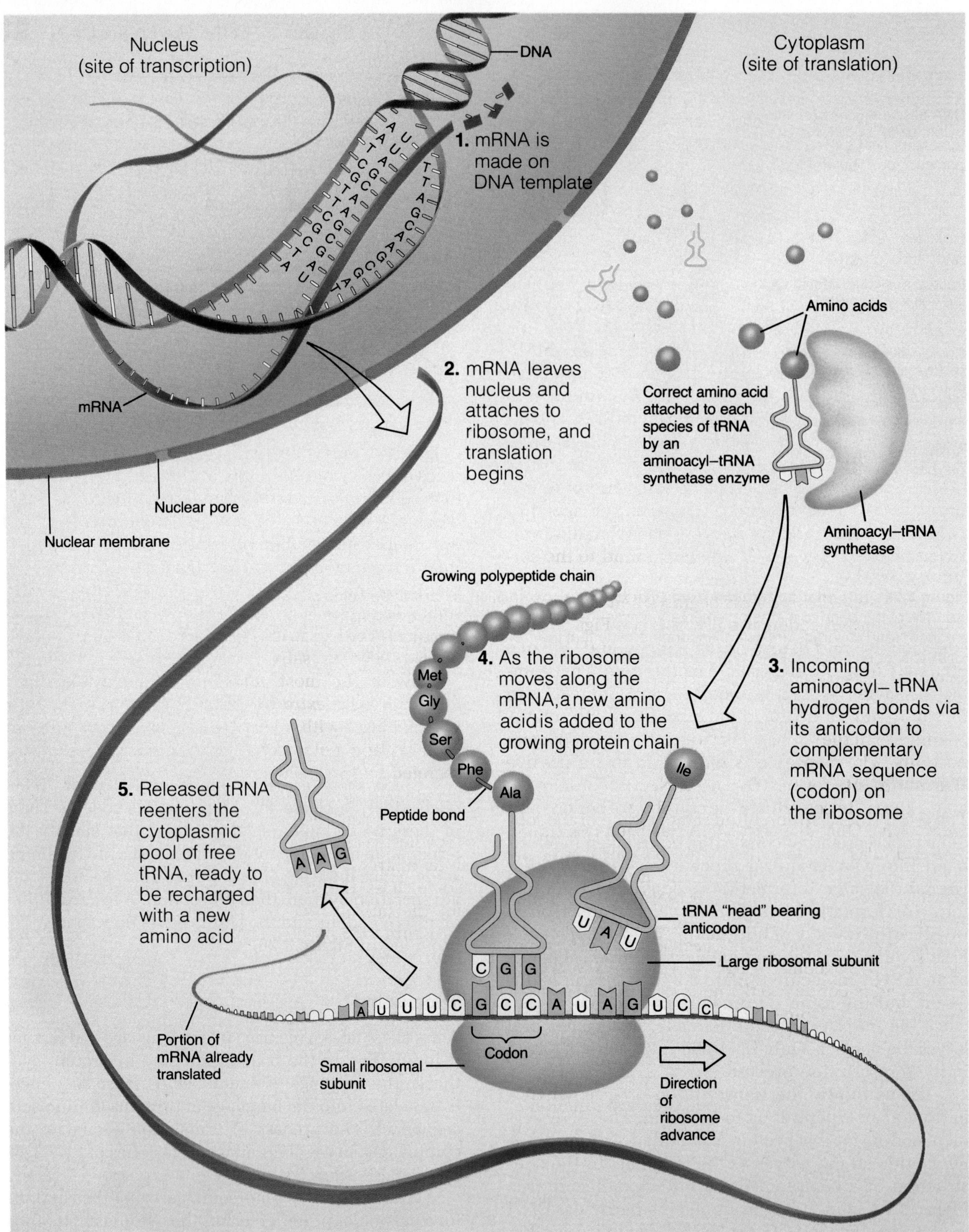

Figure 3.27 Protein synthesis. (1) *Transcription.* The DNA segment, or gene, specifying one polypeptide uncoils, and one strand acts as a template for the synthesis of a complementary mRNA molecule. (**2–5**) *Translation.* Messenger RNA from the nucleus attaches to a small ribosomal subunit in the cytoplasm (2). Transfer RNA transports amino acids to the mRNA strand and recognizes the mRNA codon calling for its amino acid by its ability to base pair with the codon (via its anticodon). The ribosome then assembles and translation begins (3). The ribosome moves along the mRNA strand as each codon is read sequentially (4). As each amino acid is bound to the next by a peptide bond, its tRNA is released (5). The polypeptide chain is released when the termination (stop) codon is read.

a synthetase enzyme in each case and is activated by ATP. Once the tRNA molecule has picked up its amino acid from the cytosol, it migrates to the ribosome, where it maneuvers the amino acid into the proper position, as specified by the codons on the mRNA strand. This is a bit more complex than it seems; not only must tRNA bring an amino acid to the protein synthesis site, but it must also "recognize" the codon calling for that amino acid.

The structure of the tiny tRNA molecule is well suited for its dual function. The amino acid is bound to one end of tRNA, at a region called the "tail." At the other end, at the "head," is its *anticodon* (an″-tī-kō′-don), a specific three-base sequence complementary to the mRNA codon calling for the amino acid carried by that particular tRNA. Since tRNA anticodons are able to form hydrogen bonds with complementary codons, the tiny tRNAs serve as the links between the languages of nucleic acids and proteins. Thus, if the mRNA codon is UUU, which specifies phenylalanine, the tRNA carrying phenylalanine will have the anticodon AAA, which can bind to the correct codon.

The translation process begins when the first, or "initiator," mRNA codon, usually AUG (see Figure 3.26), is recognized and bound by the anticodon (UAC) of a tRNA carrying the amino acid methionine. This stimulates a large ribosomal subunit to bind and results in the formation of a functional ribosome with mRNA positioned in the "groove" between the two ribosomal subunits. The ribosome is more than just a passive attachment site for mRNA and tRNA; its function in translation is to coordinate the coupling of codons and anticodons. Once the first tRNA has been positioned on a codon as just described, the ribosome moves the mRNA strand along, bringing the next codon into position to be "read" by another tRNA. As successive amino acids are brought to their proper positions, peptide bonds are formed between them and the polypeptide chain becomes longer and longer. When each amino acid is bonded to the next, its tRNA is released and moves away from the ribosome to pick up another amino acid. As mRNA is progressively read and its initial portion moved off the ribosome, that portion may become attached successively to several other ribosomes, all reading the same message simultaneously. Such a multiple ribosome-mRNA complex is called a *polyribosome,* and it provides an efficient system for producing many copies of the same protein. The mRNA strand continues to be read sequentially until its last codon, the *terminal codon,* or *stop codon,* has entered the ribosomal groove; this is the "period" at the end of the mRNA sentence that triggers the release of the completed polypeptide chain from the ribosome. The ribosome then dissociates into its subunits, if not engaged by another mRNA molecule.

The genetic information of a cell is translated into the production of proteins via a sequence of information transfer that is completely directed by complementary base pairing. Notice that if the anticodons of all the tRNAs reading the mRNA codons could be aligned at the same time, the DNA base sequence would be reestablished, except that T of DNA is represented by U in tRNA (see Figure 3.26). Thus, the transfer of information goes from DNA base sequence (triplets) to the complementary base sequence of mRNA (codons) and back to the DNA base sequence (anticodons).

Extracellular Materials

Many substances contributing to body mass are found outside the cells. These substances are referred to collectively as **extracellular materials.** One class of extracellular materials is *body fluids*. Interstitial fluid, blood plasma, cerebrospinal fluid, and humors of the eye are all representative body fluids. These extracellular fluids are important as transport and dissolving media. *Cellular secretions* are also extracellular materials. Secretions include substances that aid in digestion (intestinal, gastric, and pancreatic fluids) and that act as lubricants (saliva, mucus, and serous fluids).

By far the most abundant of the extracellular materials is the *extracellular matrix*. Most body cells are in contact with a jellylike substance composed of proteins and polysaccharides. These molecules are secreted by the cells and self-assemble into an organized mesh in the extracellular space, where they serve as a universal "cell glue" that helps to hold body cells together. In connective tissues, however, the extracellular matrix is particularly abundant—so much so, in some cases, that it (rather than living cells) accounts for the bulk of that tissue type. Depending on the structure to be formed, connective tissue extracellular matrix may be fluid, as in blood; soft and crystalline, as in the cornea of the eye; tough and ropelike, as in tendons; or rock-hard, as in bones and teeth. In the next chapter, we will pay a good deal of attention to the structure and function of the matrix of specialized connective tissues.

Developmental Aspects of Cells

We all begin life as a single cell, the fertilized egg, from which arises the daughter cells that form our muticellular embryonic body. Very early in development, the cells begin to specialize, some forming liver cells, some nerve cells, and others becoming the transparent lens of the eye. Since all our cells carry the same genes, how can one cell be so different from

A CLOSER LOOK Cancer—The Intimate Enemy

The word **cancer** strikes dread into nearly everyone. Why does cancer strike some and not others? Are its "seeds" part of our genetic dowry?

Research has revealed a surprisingly close relationship between cancer and the fundamental life processes. Although perceived as an example of rapid disorganized cell growth, this disease is now known to be a logical, coordinated process in which certain cells acquire special abilities. Indeed, cancer is one of life's biological marvels and has an extraordinary life history.

Before cancer can start, a precise sequence of tiny alterations must occur that change a normal cell into a killer. Improbable as this is, it happens all the time. Certain chemicals and radiation can be *carcinogenic* (cancer causing) and serve as triggers. The chemical carcinogens themselves do not cause cancer; they must be modified within a cell before they can promote cancer, a process called *activation*. Most carcinogens never gain entry into our cells because of the alertness of scavenger cells, but for those that do, the peroxisomal enzymes usually render potentially harmful chemicals harmless. However, some people have unusual peroxidases that actually change carcinogens in such a way that they can more easily enter the cell's nucleus and bind irreversibly to DNA. But this activation of a carcinogen by the "renegade enzymes" is just the first step toward cancer.

Even when an activated carcinogen binds to DNA, cancer is unlikely because of our natural built-in DNA repair system, the SOS system. The molecules of this system detect faulty DNA and perform a type of molecular surgery, clipping out the altered DNA sequence and prompting synthesis of a new normal DNA patch, thus blocking the cancer process. If the SOS mechanism is faulty or cell division occurs before the DNA is repaired, the carcinogen-modified portion may be copied abnormally. The daughter cells then inherit a *mutated* gene. Still, unless the mutation occurs at precisely the right place on the right gene, it will play no role in cancer. At worst, it may cripple or kill that single cell. But when the mutation concerns a critical DNA region, *initiation*, the first step toward the cancer process is completed.

Initiation sensitizes a cell to the action of *promoters*, which encourage the cell to divide more rapidly than other cells of its type. Many chemicals act as promoters, and each seems to exert its strongest effects on particular cell types. Cigarette tars are strong promoters for lung cells, and saccharine appears to target bladder cells. Mechanical factors such as abrasion and wounds may also act as promoters. Normal cells stop dividing when the wound is healed; but some promoted cells keep dividing. Various dietary substances, such as vitamin A, seem to counteract the effects of promoters and to increase one's resistance to their cancer-promoting effects.

Promoted cells still behave normally, but their growing numbers increase the likelihood that one or more of them will be "hit" by carcinogens. Consequently, promoted cells tend to accumulate mutations until one occurs that causes a cell to keep dividing even when promoters are absent. The key mutation is the one that removes or disables the cell's normal controls over cell division.

Promoted cells form an abnormal mass of proliferating cells called a **neoplasm,** literally a "new growth." Neoplasms are classified as benign or malignant. A *benign neoplasm*, commonly called a *tumor*, is strictly a local affair. Its cells remain compacted, and it is often encapsulated. Benign neoplasms tend to grow in a leisurely way and seldom kill their hosts if removed before vital organs are compressed. By contrast, *malignant (cancerous) neoplasms* are nonencapsulated "killers" and grow more relentlessly. Their cells invade their surroundings rather than simply pushing them aside, as reflected in the word *cancer*, from the Latin meaning "crab." The most devastating property of malignant cells is their ability to break away from the parent mass (the primary tumor) and enter the blood or lymphatic system to travel to and invade other body organs, where they form secondary cancer masses. This ability is called *metastasis*. As you can see, the main characteristics that differentiate malignant from benign neoplasms are the invasive and metastatic properties of cancer cells. The next question is, What causes the transformation of a benign neoplasm into a cancerous one?

The newest twist in the cancer story is the discovery of **oncogenes,** or cancer-causing genes, found in rapidly spreading types of cancers. This is perplexing and became even more so when it was discovered that *proto-oncogenes* are found in the normal genetic makeup of all human cells. The only logical conclusion is that these proto-oncogenes play important roles in the normal lives of cells. Proto-oncogenes have fragile sites that break when exposed to chemicals or radiation. Thus, oncogenes are not alien guests (originally thought to be delivered by viruses), but essential components of the cell's genetic apparatus that betray the cell only when their structure or control is disturbed.

Although little is known about how oncogenes *transform* benign tumor cells into malignant ones, the key appears to be activation of additional genes that allow the cells to become invasive and to metastasize. For metastasis to occur, substances must be produced that can break down the extracellular matrix and digest holes in blood vessel walls. This ability is lacking in normal adult cells but is present in some embryonic cells, suggesting that at least one of the mutations switches on long-dormant genes.

In the blood and lymph, the immune system has one last chance to prevent the cancer cells from finding new homes in the body. When the seemingly immortal cancer cells evade immune surveillance and multiply greatly, the body's food supply and ultimately its tissues are consumed. Since the "seeds" of cancer may well be our own genes, cancer is an intimate enemy indeed.

Key:

Events that defend the body against cancer

Events that lead or may lead to cancer

Formation of a benign tumor

In this example, carcinogens are inhaled into the lungs in cigarette smoke.

First line of defense: Scavenger cells detoxify carcinogens (●). Any carcinogens that gain entry to body cells are usually destroyed by cell enzymes.

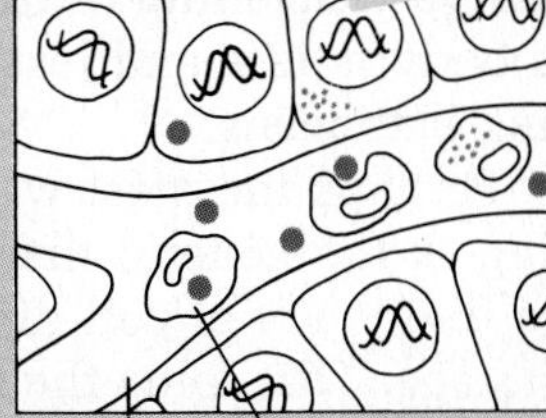

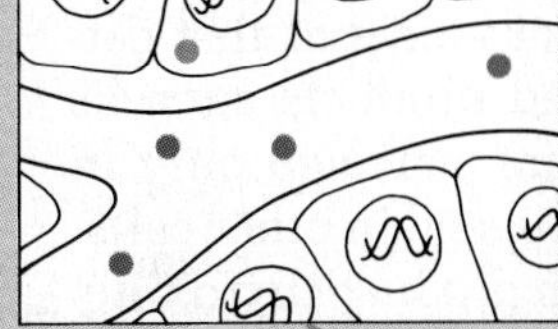

Activation: Rarely, cell enzymes change a carcinogen (●) into a more active form (●). Activated carcinogen enters cell nucleus and binds to DNA.

Second line of defense: SOS system. Faulty DNA is detected and removed; DNA repair occurs before the cell divides. The cell and its daughter cells are normal.

Initiation: Carcinogen-bound DNA is copied incorrectly, resulting in a mutation in daughter cells. If a critical DNA region is altered, initiation is completed.

Promotion: Initiated cells are more sensitive to the effects of promoters (▲), cigarette tars in this example. Promoters stimulate the cell to proliferate rapidly. Promoted cells tend to accumulate mutations. If mutations occur which remove controls of cell division, a cell mass (neoplasm) begins to form.

Formation of a benign neoplasm (tumor): Cells remain compacted; tumor is usually encapsulated. However, the tumor may be transformed into a cancer.

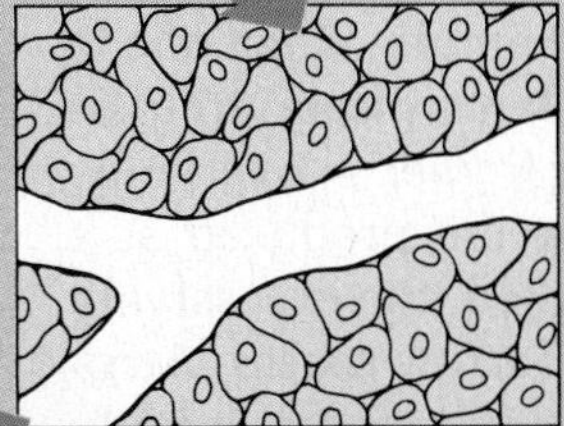

Formation of a malignant neoplasm (cancer): Cells become invasive and spread crab-like into surrounding tissue.

Third line of defense: Immune cells attack and kill cancer cells.

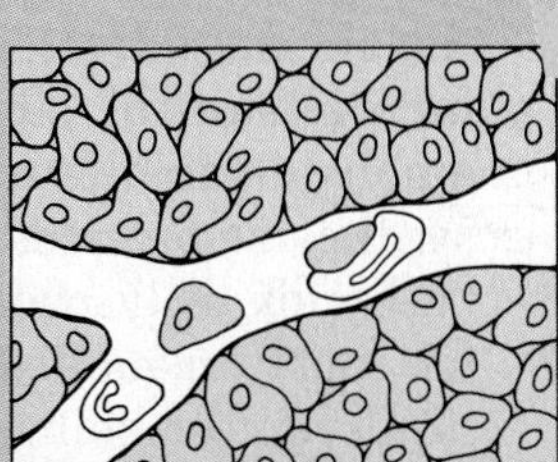

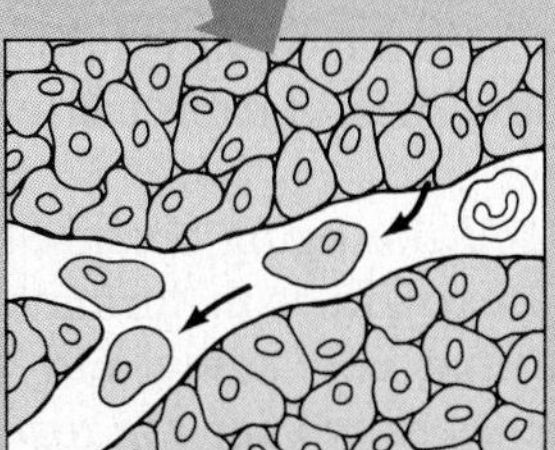

Metastasis: Malignant cells enter blood (and/or lymph) and are carried to new body sites.

another? This question has fascinated developmental biologists for years and is the subject of much current research. Apparently, cells in various regions of the embryo are exposed to different chemical signals that channel them into specific pathways of development. When the embryo consists of just a few cells, slight differences in oxygen and carbon dioxide concentrations between the more superficial and the deeper cells may be the major environmental difference. But as development continues, cells begin to produce various chemicals that can influence development of neighboring cells by switching some of their genes "on" or "off." Some genes are active in all cells; for example, all cells have to carry out protein synthesis and make ATP. However, genes for enzymes that catalyze the synthesis of specialized products such as hormones or neurotransmitters are activated only in certain cell populations and inactivated in other cells. Only cells of the thyroid gland can produce thyroxine, and only nerve cells can secrete neurotransmitters. Hence, the secret of cell specialization lies in the kinds of proteins made and reflects differential gene activation in the different cell types.

During early development, cell death and destruction is a normal event. Apparently, nature takes few chances. More cells than are needed are produced; then, those in excess are eliminated. This is particularly true in the nervous system. By birth, most organs are well formed and functional, but the body continues to grow and enlarge by forming new cells throughout childhood and adolescence. Once adult size has been reached, cell division is important mainly for replacement of short-lived cells and for wound repair.

During young adulthood, cell numbers remain fairly constant. However, local cellular responses leading either to accelerated cell division and rapid tissue growth or to slowed cell division and tissue wasting are fairly common. For example, when one is anemic, bone marrow undergoes **hyperplasia** (hī″-per-plā′-zhuh), or accelerated growth (*hyper* = over; *plas* = grow), so that red blood cells are produced at a faster rate. If the anemia is remedied, the excessively rapid production of cells in the marrow ceases. **Atrophy** (a′-truh-fē), a decrease in size of an organ or body tissue, can result from loss of normal stimulation. For example, muscles that lose their nerve supply begin to atrophy and waste away, and lack of exercise leads to thinned, brittle bones.

There is no doubt that cellular aging occurs, but no one has yet been able to explain it in terms of a single cause. Some believe that cell aging and the accelerated cell loss that occurs as we age are results of little chemical insults that have cumulative effects throughout life. For example, environmental toxins such as alcohol, carbon monoxide, and bacterial toxins may damage cell membranes, poison enzyme systems, or prevent DNA replication. Temporary deprivation of oxygen, which occurs increasingly with age as our blood vessels become clogged with fatty materials, leads to accelerated rates of cell death throughout the body. X-rays and other types of radiation, as well as some chemicals, can generate huge numbers of free radicals, which can overwhelm the peroxisomal enzymes and cause irreversible injury to DNA.

Another theory of cell aging attributes it to progressive disorders in the immune system. According to this theory, cell damage results from (1) autoimmune responses, or events in which the immune system turns against one's own tissues, and (2) a progressive weakening of the immune response, so that the body is less and less able to get rid of cell-damaging pathogens.

Perhaps the most widely accepted theory of cell aging is the genetic theory, which implicates DNA. The frequency of cell division decreases with a person's age. This, plus the fact that normal cells grown in culture are only capable of a definite number of cell divisions, has led to the suggestion that cessation of mitosis is a genetically programmed event and reflects a problem with the genetic material. The fragile structure of DNA is easily damaged by physical and chemical agents, but in young, healthy cells, DNA has a remarkable ability to repair itself. The repair process, called **SOS response,** is mediated by enzymes that are activated by DNA damage. (The repair process is described briefly in the box on pp. 94–95.) Progressive failure of the SOS response may be at the root of cellular aging.

A variation of the genetic theory suggests that we run out of DNA in old age. Our genetic material is redundant; that is, we have multiple copies of the same genes. As our DNA breaks down, we use the extra copies, but eventually there are no more copies to use. The "DNA crash time" differs in various tissues, which may account for differences in onset of age-related changes in various body tissues.

* * *

We have described the cell, the structural and functional unit that fashions and maintains our bodies, from the vantage point of the generalized cell. One of the wonders of the cell is the disparity between its minute size and its extraordinary activity, which reflects the amazing diversity of its organelles. The evidence for division of labor and functional specialization among organelles is inescapable: Only ribosomes can synthesize proteins, and intracellular digestion is the bailiwick of lysosomes. Membranes compartmentalize most organelles, allowing them to work without hindering or being hindered by other cell activities going on at the same time, and the plasma membrane regulates molecular traffic across the cell's boundary. Now that you know what cells have in common, you are ready to explore how they differ in the various body tissues.

Related Clinical Terms

Anaplasia (a″-nuh-plā′-zē-uh) (*an* = without, not; *plas* = to grow) Abnormalities in cell structure; e.g., cancer cells typically lose the appearance of the parent cells and come to resemble undifferentiated or embryonic cells.

Dialysis (dī-al′-ih-sis) (*dialys* = separate, break apart) The process of separating smaller solute molecules from larger ones in a solution by means of diffusion through a selectively permeable membrane; used by the artificial kidney to cleanse the blood of metabolic wastes, such as urea, and excess ions.

Dysplasia (dis-plā′-zē-uh) (*dys* = abnormal) A change in cell size, shape, or arrangement due to chronic irritation or inflammation (infections, etc.).

Hypertrophy (hī-per′-truh-fē) Growth of an organ or tissue due to an increase in the size of its cells; hypertrophy is a normal response of skeletal muscle cells when they are challenged to lift excessive weight; differs from hyperplasia, which is an increase in size due to increase in cell number.

Mutation A change in DNA base sequence leading to incorporation of "incorrect" amino acids in particular positions in the resulting protein; the affected protein may remain unimpaired or may function abnormally, or not at all, leading to disease.

Necrosis (neh-krō′-sis) (*necros* = death; *osis* = process) Death of a cell or group of cells due to injury or disease.

Chapter Summary

OVERVIEW OF THE CELLULAR BASIS OF LIFE (pp. 61–62)

1. All living organisms are composed of cells—the basic structural and functional units of life.

2. The principle of complementarity states that the biochemical activity of cells reflects the operation of organelles.

3. Cells vary widely in both shape and size.

4. The generalized cell is a concept that typifies all cells. The generalized cell has three major regions—the nucleus, cytoplasm, and plasma membrane.

THE PLASMA MEMBRANE: STRUCTURE (pp. 62–65)

1. The plasma membrane encloses cell contents, mediates exchanges with the extracellular environment, and plays a role in cellular communication.

The Fluid Mosaic Model (pp. 62–64)

2. The fluid mosaic model depicts the plasma membrane as a fluid bilayer of lipids (phospholipids, cholesterol, and glycolipids) within which proteins are inserted.

3. The lipid substances have both hydrophilic and hydrophobic regions that organize their aggregation and self-repair. The lipids form the structural part of the plasma membrane.

4. Most proteins are integral transmembrane proteins that extend entirely through the membrane. Some, appended to the integral proteins, are peripheral proteins.

5. The proteins are responsible for most specialized membrane functions: i.e., some are enzymes, some receptors for hormones or neurotransmitters, and others mediate transport functions of the membrane. Externally facing glycoproteins contribute to the glycocalyx.

Specializations of the Plasma Membrane (p. 64–65)

6. Microvilli are extensions of the plasma membrane that typically increase its surface area for absorption.

7. Membrane junctions join cells together and may aid or inhibit movement of molecules between or past cells. Tight junctions are impermeable junctions; desmosomes mechanically couple cells into a functional community; gap junctions allow communication between joined cells.

THE PLASMA MEMBRANE: FUNCTIONS (pp. 66–74)

Membrane Transport (pp. 66–73)

1. The plasma membrane acts as a selectively permeable barrier. Substances are moved across the plasma membrane by passive processes, which depend on the kinetic energy of molecules or on pressure gradients, and by active processes, which depend on the provision of cellular energy (ATP).

2. Diffusion is the movement of molecules (driven by kinetic energy) from an area of high concentration to an area where they are lower in concentration, down a concentration gradient. Fat-soluble solutes can diffuse directly through the membrane by dissolving in the lipid. Charged small molecules or ions move by diffusion through the membrane if they are small enough to pass through the protein channels. Some protein channels are selective.

3. Osmosis is the diffusion of a solvent, such as water, through a selectively permeable membrane. Water diffuses through the membrane pores from a solution of lesser osmolarity to a solution of greater osmolarity.

4. The presence of impermeable solutes leads to changes in cell tone because water moves by osmosis until its movement is resisted by hydrostatic pressure. Net osmosis ceases when a solution's osmotic pressure is balanced by its hydrostatic pressure.

5. Solutions that cause a net loss of water from cells are called hypertonic; those causing net water gain are hypotonic; those causing neither gain nor loss of cellular water are isotonic.

6. Facilitated diffusion is the passive movement of certain solutes across the membrane by their combination with a membrane carrier protein. As with other diffusion processes, it is driven by kinetic energy, but the carriers are selective.

7. Filtration occurs when a filtrate is forced across a membrane by hydrostatic pressure. It is nonselective and limited only by pore size. The pressure gradient is the driving force.

8. Active transport (solute pumping) depends on a carrier protein and ATP. In most cases, substances transported (amino acids and ions) move against concentration or electrical gradients.

9. Endocytotic processes also require that ATP be provided. If the substance is particulate, the process is called phagocytosis; if the substance is dissolved molecules, the process is pinocytosis. Receptor-mediated endocytosis is selective; engulfed particles (minerals, cholesterol, and some hormones) attach to receptors on the membrane before endocytosis occurs.

Membrane Potential (pp. 73–74)

10. All cells in the resting stage exhibit a voltage across their membrane, called the resting membrane potential.

11. The membrane potential is generated by concentration gradients of and differential permeability of the plasma membrane to sodium and potassium ions. Sodium is in high extracellular–low intracellular concentration, and the membrane is poorly permeable to it. Potassium is in high concentration in the cell and low concentration in the extracellular fluid. The membrane is more permeable to potassium than to sodium.

12. The greater outward diffusion of potassium (than inward diffusion of sodium) leads to a charge separation at the membrane (inside negative). This charge separation is maintained by the operation of the sodium-potassium pump.

13. Because of the membrane potential, both concentration and electrical gradients determine the ease of an ion's diffusion.

Functions of the Glycocalyx (p. 74)

14. The glycocalyx determines blood type, provides entry sites for certain toxins, allows recognition of the egg by sperm, is involved in the immune response, and helps to guide embryonic development. It also acts as an adhesive agent between cells.

THE CYTOPLASM (pp. 74–81)

1. The cytoplasm, the cellular region between the nuclear and plasma membranes, consists of the cytosol (fluid cytoplasmic environment), inclusions (nonliving nutrient stores, secretory products, pigment granules, crystals, etc.), and cytoplasmic organelles.

Cytoplasmic Organelles (pp. 75–81)

2. The cytoplasm is the major functional cellular area of the cell. These functions are mediated by the highly specialized cytoplasmic organelles.

3. Mitochondria are the sites of ATP formation. Their internal enzymes carry out the oxidative reactions of cellular respiration.

4. Ribosomes, composed of two subunits containing ribosomal RNA and proteins, are the sites of protein synthesis. They may be free or attached to membranes.

5. The rough endoplasmic reticulum is a ribosome-studded membrane system. Its cisternae act as sites for protein modification and transport the proteins to other cell sites.

6. The smooth endoplasmic reticulum synthesizes lipid and steroid molecules. It also acts in fat and glycogen synthesis and in drug detoxification. In muscle cells, it is an important calcium ion depot.

7. The Golgi apparatus is a membranous system close to the nucleus that packages proteins for export, packages enzymes into lysosomes for cellular use, and modifies proteins destined to become part of cellular membranes.

8. Lysosomes are membranous sacs of acid hydrolases packaged by the Golgi apparatus. Sites of intracellular digestion, they also degrade no longer useful tissues and release ionic calcium from bone.

9. Peroxisomes are membranous sacs containing peroxidase enzymes that protect the cell from the destructive effects of oxygen-free radicals and hydrogen peroxide.

10. Cytoskeletal elements include microfilaments, intermediate filaments, and microtubules. Microfilaments, formed of contractile proteins, are important in cell motility or movement of cell parts. Microtubules are the overall organizers of the cytoskeleton and are important in cell shape, suspension of organelles, and intracellular transport. Intermediate filaments help cells resist mechanical stress.

11. The centrioles form the mitotic spindle and are the basis of cilia and flagella.

THE NUCLEUS (pp. 82–84)

1. The nucleus is the control center of the cell. Most cells have a single nucleus. Without a nucleus, a cell cannot divide or synthesize more proteins; thus, it is destined to die.

2. The nucleus is surrounded by the nuclear membrane, a double bilipid membrane penetrated by fairly large pores.

3. Chromatin is a complex network of slender threads containing histone proteins and DNA. The chromatin units are called nucleosomes. When a cell begins to divide, the chromatin coils and condenses.

4. Nucleoli are nuclear sites of ribosomal synthesis.

CELL GROWTH AND REPRODUCTION (pp. 84–93)

The Cell Life Cycle (pp. 84–90)

1. The cell life cycle is the series of changes that a cell goes through from the time it is formed until it divides.

2. Interphase is the nondividing phase of the cell life cycle. Interphase consists of G_1, S, and G_2 subphases. During G_1, the cell grows vigorously and the centrioles are replicated; during the S phase, DNA is replicated; and during G_2, the final preparations for division are made.

3. DNA replication occurs before cell division. The DNA helix uncoils and each DNA nucleotide strand acts as a template for forming a complementary strand from DNA nucleotides present in the nucleus. Base pairing provides the guide for the proper positioning of the nucleotides.

4. The products of replication of a single DNA molecule are two DNA molecules identical to the parent molecule and each formed of one "old" and one "new" strand.

5. Cell division, essential for body growth and repair, occurs during the M phase. Cell division is stimulated by certain chemicals and increasing cell size. Lack of space and inhibitory chemicals deter cell division. Cell division consists of two distinct phases: mitosis and cytokinesis.

6. Mitosis, consisting of prophase, metaphase, anaphase, and telophase, results in the parceling out of the replicated chromosomes to two daughter nuclei, each identical in genetic makeup to the mother nucleus. Cytokinesis, which usually follows mitosis, is the division of the cytoplasmic mass into two parts.

Protein Synthesis (pp. 90–93)

7. A gene is defined as a DNA segment that provides the instructions for the synthesis of one polypeptide chain. Since the major structural materials of the body are proteins, and since all enzymes are proteins, this amply covers the synthesis of all biological molecules.

8. The base sequence of DNA provides the information for the structure of proteins. Each three-base sequence (triplet) calls for a particular amino acid to be built into a polypeptide chain.

9. The three varieties of RNA are synthesized on the DNA. One DNA nucleotide strand serves as the template in each case. RNA nucleotides are joined following base-pairing rules.

10. Ribosomal RNA forms part of the protein synthesis sites; messenger RNA carries instructions for making a polypeptide chain from the DNA to the ribosomes; transfer RNA ferries amino acids to the ribosomes and recognizes codons on the mRNA strand specifying its amino acid.

11. Protein synthesis involves transcription, synthesis of a complementary mRNA, and translation, "reading" of the mRNA by tRNA and peptide bonding of the amino acids into the polypeptide chain. The ribosome coordinates the process.

EXTRACELLULAR MATERIALS (p. 93)

1. Substances found outside the cells are called extracellular materials. These include body fluids, cellular secretions, and extracellular matrix. Extracellular matrix is particularly abundant in connective tissues.

DEVELOPMENTAL ASPECTS OF CELLS (pp. 93–96)

1. The first cell of an organism is the fertilized egg. Early in development, cell specialization begins and is thought to reflect differential gene activation.

2. During adulthood, cell numbers remain fairly constant; cell division occurs primarily to replace lost cells.

3. Cellular aging may reflect chemical insults, progressive disorders of immunity, a genetically programmed decline in the rate of cell division with age, and/or a failure in the SOS system.

Review Questions

Multiple Choice/Matching

1. The smallest unit capable of life by itself is (a) the organ, (b) the organelle, (c) the tissue, (d) the cell, (e) the nucleus.

2. The most abundant types of lipid found in the plasma membranes are (choose two) (a) cholesterol, (b) neutral fats, (c) phospholipids, (d) fat-soluble vitamins.

3. Membrane junctions that allow nutrients or ions to flow from cell to cell are (a) desmosomes, (b) gap junctions, (c) tight junctions, (d) all of these.

4. A person drinks a six-pack of beer and has to make several trips to the bathroom. This increase in urination reflects an increase in what process occurring in the kidneys? (a) diffusion, (b) osmosis, (c) dialysis, (d) filtration.

5. The term used to describe the type of solution in which cells will lose water to their environment is (a) isotonic, (b) hypertonic, (c) hypotonic, (d) catatonic.

6. Osmosis always involves (a) a differentially permeable membrane, (b) a difference in solvent concentration, (c) diffusion, (d) active transport, (e) a, b, and c.

7. A physiologist observes that the concentration of sodium inside a cell is decidedly lower than that outside of it. Yet sodium does diffuse easily across the plasma membrane of such cells when they are dead but *not* when they are alive. Which of the following terms applies best to this cellular function that is lacking in dead cells? (a) osmosis, (b) diffusion, (c) active transport (solute pumping), (d) dialysis.

8. The solute-pumping variety of active transport is accomplished by (a) pinocytosis, (b) phagocytosis, (c) electrical forces in the cell membrane, (d) conformational-positional changes in carrier molecules in the plasma membrane.

9. The endocytotic process in which particulate matter is engulfed and brought into the cell is called (a) phagocytosis, (b) pinocytosis, (c) exocytosis.

10. The nuclear substance composed of histone proteins and DNA is (a) chromatin, (b) the nucleolus, (c) nuclear sap, or nucleoplasm, (d) nuclear pores.

11. The information sequence that determines the nature of a protein is called (a) the nucleotide, (b) the gene, (c) the triplet, (d) the codon.

12. Mutations may be caused by (a) X-rays, (b) certain chemicals, (c) radiation from ionizing radioisotopes, (d) all of these.

13. The phase of mitosis during which centrioles reach the poles and chromosomes attach to the spindle is (a) anaphase, (b) metaphase, (c) prophase, (d) telophase.

14. Final preparations for cell division are made during the life cycle subphase called (a) G_1, (b) G_2, (c) M, (d) S.

15. The RNA synthesized on one of the DNA strands is (a) mRNA, (b) tRNA, (c) rRNA, (d) all of these.

16. The RNA species that carries the coded message, specifying the sequence of amino acids in the protein to be made, from the nucleus to the cytoplasm is (a) mRNA, (b) tRNA, (c) rRNA, (d) all of these.

17. If DNA has a sequence of AAA, then a segment of mRNA synthesized on it will have a sequence of (a) TTT, (b) UUU, (c) GGG, (d) CCC.

18. A nerve cell and a lymphocyte are presumed to differ in their (a) specialized structure, (b) suppressed genes and embryonic history, (c) genetic information, (d) a and b, (e) a and c.

Short Answer Essay Section

19. (a) Name the organelle that is the major site of ATP synthesis. (b) Name three organelles involved in protein synthesis and/or modification. (c) Name two organelles that contain enzymes, and describe their relative functions.

20. Explain why mitosis can be thought of as cellular immortality.

21. If a cell loses or ejects its nucleus, what is its fate and why?

22. The external face of some of the proteins in the plasma membrane have carbohydrate groups attached to them. What role do such "sugar-coated" proteins play in the life of the cell?

23. Cells are living units. However, there are three classes of nonliving substances found outside the cells. What are these classes, and what functional roles do they play?

24. Comment on the role of the sodium-potassium pump in maintaining a cell's resting membrane potential.

Clinical Application Questions

25. A "red-hot" bacterial infection of the intestinal tract irritates the intestinal cells and interferes with normal digestion. Such a condition is often accompanied by diarrhea, which causes loss of body water. On the basis of what you have learned about osmotic water flows, explain why diarrhea may occur.

26. Two examples of chemotherapeutic drugs (used to treat cancer) and their cellular actions are listed below. Explain why each drug could be fatal to a cell.

- Vincristine: Damages the mitotic spindle.
- Adriamycin: Binds to DNA and blocks mRNA synthesis.

CHAPTER 4

Tissue: The Living Fabric

Chapter Outline and Student Objectives

Epithelial Tissue (pp. 101–112)

1. List several characteristics that typify epithelial tissue.
2. Describe the criteria used to classify epithelia structurally.
3. Name and describe the various types of epithelia, indicate their chief function(s) and location(s).
4. Contrast a sheet of covering or lining epithelium with an epithelial membrane. Name and describe the three varieties of epithelial membranes.
5. Compare endocrine and exocrine glands relative to their general structure, product(s), and mode of secretion.
6. Describe how multicellular exocrine glands are classified structurally and functionally.

Connective Tissue (pp. 112–123)

7. Describe common characteristics of connective tissue.
8. List the structural elements of connective tissue and describe each element.
9. Explain the bases for classification of connective tissue.
10. Describe the types of connective tissue found in the body, and note their characteristic functions.

Muscle Tissue (pp. 123–125)

11. Compare and contrast the structures and body locations of the three types of muscle tissue.

Nervous Tissue (p. 125)

12. Note the general characteristics and functions of nervous tissue.

Tissue Repair (pp. 125–127)

13. Describe the process of tissue repair involved in the normal healing of a superficial wound.

Developmental Aspects of Tissues (pp. 127–129)

14. Name the three primary germ layers, and indicate the primary tissue classes arising from each.
15. Briefly mention tissue changes that occur with age. Cite possible causes of such changes.

Preview of Selected Key Terms

Tissue (*tissu* = woven) A group of similar cells (and their intercellular substance) specialized to perform a specific function; primary tissue types of the body are epithelial, connective, muscle, and nervous tissue.

Epithelial (eh″-pih-the′-lē-ul) (*epi* = upon, over; *thel* = delicate) Pertaining to a primary tissue that covers the body surface, lines its internal cavities, and forms glands.

Gland One or more cells specialized to secrete a particular product called a secretion.

Mesenchyme (meh′-zin-kīm) Common embryonic tissue from which all connective tissues arise.

Matrix Specialized extracellular substance secreted by connective tissue cells that determines the specialized function of each connective tissue type; typically includes ground substance (fluid to hard) and fibers (collagen, elastic, and/or reticular).

Fibroblast/fibrocyte (*blast* = bud, forming; *cyte* = cell) The fibroblast is a young, actively mitotic cell that forms the fibers of connective tissue; in its mature state, it becomes a fibrocyte, which maintains the matrix.

Chondroblast/chondrocyte The actively mitotic and mature cell forms, respectively, of cartilage.

Osteoblast/osteocyte The actively mitotic and mature cell forms, respectively, of bone.

Macrophage (ma′-krō-fāj″) (*macro* = large; *phago* = eat) A protective cell type common in connective tissue, lymphatic tissue, and certain body organs that phagocytizes tissue cells, bacteria, and other foreign debris.

Germ layers Three cellular layers (ectoderm, mesoderm, and endoderm) that represent the initial specialization of cells in the embryonic body and from which all body tissues arise.

Cells that exist as isolated unicellular (one-cell) organisms, such as amoebas, are rugged individualists. They alone obtain and digest their food, carry out gas exchange, and perform all the other activities necessary to keep themselves alive and healthy. But in the multicellular human body, cells do not operate independently or in isolation. Instead, they form tight, interdependent cell communities that function cooperatively. Individual cells are specialized, with each type performing specific functions that help maintain homeostasis and benefit the body as a whole. The specialization of our cells is obvious: Muscle cells look and function quite differently from skin cells, which in turn are easily distinguished from cells of the brain.

Cell specialization allows the various parts of the body to function in very sophisticated ways, but this division of labor also has certain hazards. When a particular group of cells is indispensable, its loss can severely disable or even destroy the body.

Groups of closely associated cells that are similar in structure and perform a common function are called **tissues** (*tissu* = woven). There are four primary tissue types that interweave to form the "fabric" of the body: epithelial, connective, muscle, and nervous tissue, and each has several subdivisions or varieties. If we had to assign a single term to each primary tissue type that would best describe its overall role, the terms would most likely be *protection* (epithelial), *support* (connective), *movement* (muscle), and *control* (nervous). However, these terms reflect only a small part of the total tissue functions.

As explained in Chapter 1, tissues are organized into organs such as the kidneys and the heart. Most organs contain several tissue types, and the arrangement of the tissues determines the organ's structure and what it is able to do. The study of tissues, or histology, complements the study of gross anatomy. Together they provide the structural basis for understanding organ physiology.

The close correlation between tissue structure and function makes the study of tissues intriguing, but the value of any learning is to increase our ability to perceive relationships and see how things "fit" or work together. Learning the characteristic patterns of each tissue type will allow you to predict the function of an organ once its structure is known, and vice versa.

Epithelial Tissue

Epithelial (eh″-pih-thē′-lē-ul) **tissue**, or **epithelium**, occurs in the body as (1) *covering and lining epithelium* and (2) *glandular epithelium*. Covering and lining epithelium is found on free surfaces of the body such as the outer layer of the skin, dipping into and lining the open cavities of the digestive and respiratory systems, lining blood vessels and the heart, and covering the walls and organs of the closed ventral body cavity. Since epithelium forms the boundaries that mark us off from the outside world, nearly all substances received or given off by the body must pass through epithelium. Glandular epithelium fashions the glands of the body.

Epithelium is highly specialized to accomplish many functions, including protection, absorption, filtration, and secretion. Each of these functions will be described in detail as we consider various types of epithelium, but briefly, the epithelium of the skin protects underlying tissues from mechanical and chemical injury and bacterial invasion; that lining the digestive tract is specialized to absorb substances; and that found in the kidneys performs nearly the whole functional "menu"—absorption, secretion, and filtration. Secretion is the chief specialty of glands.

Special Characteristics of Epithelium

Epithelial tissues have many characteristics that distinguish them from other tissue types.

1. **Cellularity.** Epithelial tissue is composed almost entirely of cells. In muscle and connective tissues, cells are often widely separated by extracellular matrix. This is not true of epithelium, where cells are close together.

2. **Specialized contacts.** Epithelial cells fit close together to form continuous sheets. Adjacent cells are bound by lateral contacts, including tight junctions and desmosomes (see Chapter 3), which reduce or eliminate the extracellular space between them.

3. **Polarity.** Epithelium always has one free (apical) surface—the portion of the epithelium exposed to the body exterior or the cavity of an internal organ. Some exposed plasma membrane surfaces are smooth and slick; others exhibit cell surface modifications such as *microvilli* or *cilia*. Microvilli, fingerlike extensions of the plasma membrane, increase tremendously the exposed surface area and are common in epithelia that absorb or secrete substances (intestinal lining and kidney tubules). Cilia projecting from the epithelial lining of the trachea and certain other internal tracts propel substances along the epithelial surface.

4. **Avascularity** (ā″-vas″-kyoo-layr′-ih-tē). Epithelium may be well supplied by nerve fibers but is avascular (has no blood vessels within it). Epithelial cells receive their nourishment by diffusion of substances from blood vessels in the underlying connective tissue layer.

5. **Basement membrane.** Epithelium rests on a thin supporting *basal lamina* (bā′-zul la′-mih-nuh),

which separates it from the underlying connective tissue. The basal lamina is a nonliving, adhesive material formed largely of glycoproteins secreted by the epithelial cells. The connective tissue cells, just deep to the basal lamina, secrete a similar extracellular material containing fine collagenous or reticular fibers; this layer is called the *reticular* (rih-tih′-kyoo-ler) *lamina.* The basal lamina of the epithelium and the reticular lamina of the connective tissue together form the **basement membrane.** The basement membrane reinforces the epithelial sheet, helping it to resist stretching and tearing forces, and defines the space that may be occupied by the epithelial cells.

An important characteristic of cancerous epithelial cells is their failure to respect this boundary, which they penetrate to invade the tissues beneath. ■

6. **Regeneration.** Epithelium has a high regenerative capacity. Some epithelia are exposed to friction, and their surface cells tend to abrade off; others are damaged by hostile substances (bacteria, acids, smoke) in the external environment and die. As long as epithelial cells receive adequate nutrition, they are able to replace lost cells rapidly by cell division.

Classification of Epithelia

The many types of epithelia are identified structurally according to two criteria: the shape of the cells and the number of cell layers present (Figure 4.1).

All epithelial cells are irregularly polyhedral (many-sided) in cross section, but differ in cell height. On the basis of height, there are three common shapes of epithelial cells. **Squamous** (skwā′-mus) **cells** are flattened and scalelike (*squam* = scale); **cuboidal** (kyoo-boy′-dul) **cells** are approximately as tall as they are wide; and **columnar** (kuh-lum′-nur) **cells** are tall and column-shaped. In each case, the shape of the nucleus conforms to that of the cell. The nucleus of a squamous cell is thin and flattened; that of a cuboidal cell is spherical; and a columnar cell nucleus is elongated from top to bottom and is usually located close to the cell base. Nuclear shape is an important structural characteristic to keep in mind when you attempt to distinguish epithelial types.

On the basis of cell arrangement (layers), there are two major varieties of epithelium: simple and stratified. **Simple epithelia** are composed of a single layer of cells. Simple epithelia are typically found where selective absorption and filtration occur and where the thinness of the barrier helps to speed the process. **Stratified epithelia** consist of multiple cell layers stacked one on top of the other. They are typically found in high abrasion areas, where protection is important, such as the skin surface and the lining of the mouth.

The terms denoting shape and arrangement of epithelial cells are combined to describe epithelia fully, as shown in Figure 4.2. There are four major classes of simple epithelia: simple squamous, simple cuboidal, simple columnar, and a highly modified simple epithelium called pseudostratified (*pseudo* = false).

There are also four major classes of stratified epithelia: stratified squamous, stratified cuboidal, stratified columnar, and a modified stratified squamous epithelium called transitional epithelium. In terms of body distribution and abundance, only stratified squamous and transitional epithelia are significant. Stratified epithelia are named according to the shape of the cells at the free surface, not according to deeper cell types. For example, the surface cells of stratified squamous epithelium are squamous cells, but its basal cells are cuboidal or columnar.

As you read the descriptions of the individual epithelial classes and compare these descriptions with the illustrations in Figure 4.2, keep in mind that tissues are three-dimensional, but that their structure is studied using stained tissue sections mounted on slides and viewed through the microscope. Therefore, a cross-sectional view will differ from a longitudinal view of the same tissue. Depending on the precise plane of the cut made when tissue slides are prepared, the nucleus of a particular cell may or may not be visible, and (frustratingly) the boundaries between epithelial cells are often indistinct.

Simple Epithelia

The simple epithelia are most concerned with absorption, secretion, and filtration. Because they are usually very thin, protection is not one of their "specialties."

Simple Squamous Epithelium. The cells of a **simple squamous epithelium** are flattened laterally, and their cytoplasm is sparse (Figure 4.2a). When viewed from the surface, the close-fitting cells resemble a tiled floor; when cut perpendicular to their free surface, the cells resemble fried eggs seen from the side, with their cytoplasm wisping out from the slightly bulging nucleus. Thin and often permeable, this epithelium is found where filtration or the exchange of substances by diffusion is a high priority. Capillary walls, through which exchanges occur between the blood and tissue cells, are composed exclusively of a simple squamous epithelium. In the kidneys, simple squamous epithelium forms part of the filtration membrane; in the lungs, it forms the walls of the air sacs across which gas exchange occurs. *Mesothelium* (meh″-zō-thē′-lē-um) is the name given to the simple squamous epithelium lining the ventral body cavity and covering its organs.

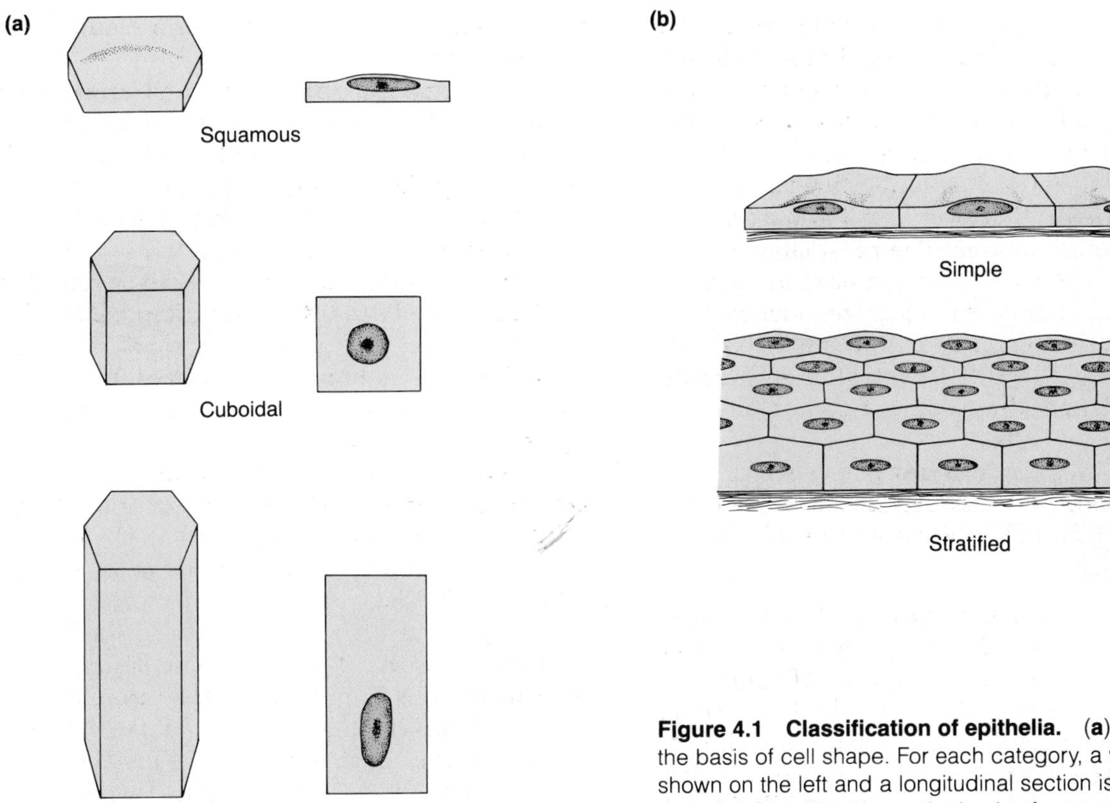

Figure 4.1 Classification of epithelia. (**a**) Classification on the basis of cell shape. For each category, a whole cell is shown on the left and a longitudinal section is shown on the right. (**b**) Classification on the basis of arrangement.

Simple Cuboidal Epithelium. **Simple cuboidal epithelium** consists of a single layer of cubical cells (Figure 4.2b). When viewed microscopically, the spherical nuclei stain darkly, causing the layer to look like a string of beads. Important functions of simple cuboidal epithelium are secretion and absorption. In glands, it forms both the secretory portions and the ducts that deliver secretions to their destinations. The simple cuboidal epithelium in the kidney tubules has dense microvilli, betraying its active role in absorption.

Simple Columnar Epithelium. **Simple columnar epithelium** is seen as a single layer of tall, closely packed cells, aligned like soldiers in a row (Figure 4.2c). Columnar cells are most associated with absorption and secretion. Cells actively involved in secretion have an elaborate Golgi apparatus and usually abundant rough endoplasmic reticulum. This epithelial type lines the digestive tract from the stomach to the rectum. The digestive tract lining has two distinct modifications that reflect its dual function: (1) dense microvilli on the surface of absorptive cells and (2) **goblet cells** that secrete a protective lubricating mucus. The goblet cells contain "cups," or goblets, of mucus that occupy most of the apical cell volume (see Figure 4.4).

Some simple columnar epithelia display cilia on their free surfaces. This more unusual variety, called **simple columnar ciliated epithelium**, lines the oviducts and limited areas of the respiratory tract.

Pseudostratified Columnar Epithelium. The cells of **pseudostratified** (soo″-dō-stra′-tih-fīd) **columnar epithelium** are varied (Figure 4.2d). All of its cells rest on the basement membrane, but some are shorter than others and, as seen in the figure, may not reach the surface of the cell layer. Their nuclei vary in shape and are seen at different levels above the basement membrane, giving the false impression that several cell layers are present. This epithelium may contain goblet cells and is often ciliated, in which case, the epithelium is more precisely called **pseudostratified columnar ciliated epithelium.** Both cilia and goblet cells are found in the pseudostratified epithelium that lines most of the respiratory tract. The mucus produced traps inhaled dust and other debris, and the cilia act to propel it away from the lungs.

Stratified Epithelia

Stratified epithelia consist of two or more cell layers. Considerably more durable than the simple epithelia, their major (but not their only) function is protection.

Stratified Squamous Epithelium. **Stratified squamous epithelium** is the most widespread stratified epithelium (Figure 4.2e). Composed of several layers, it is thick and well suited for its protective role in the body. Its free surface cells are squamous; cells of the deeper

Figure 4.2 Epithelial tissues. (**a**) to (**d**), Simple epithelia.

(a) Simple squamous epithelium

Description: Single layer of flattened cells with disk-shaped central nuclei and sparse cytoplasm; the simplest of the epithelia.

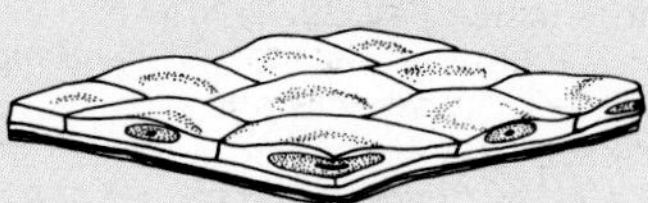

Location: Air sacs of lungs; kidney glomeruli; lining of heart, blood vessels, and lymphatic vessels; lining of ventral body cavity (serosae).

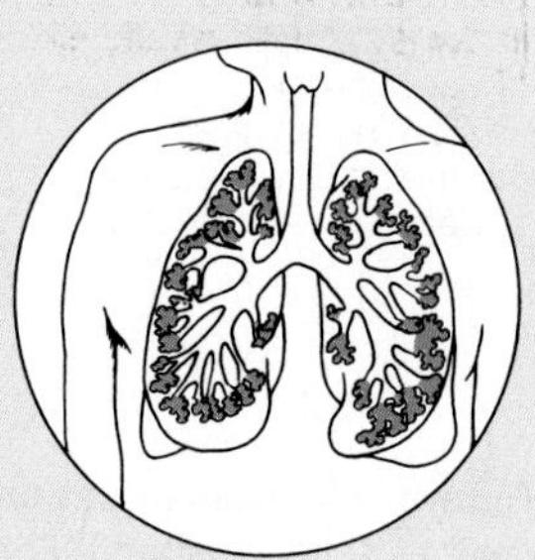

Function: Allows passage of materials by diffusion and filtration in sites where protection is not important; secretes lubricating substances in serosae.

Photomicrograph: Simple squamous epithelium forming walls of alveoli (air sacs) of the lung (280 ×).

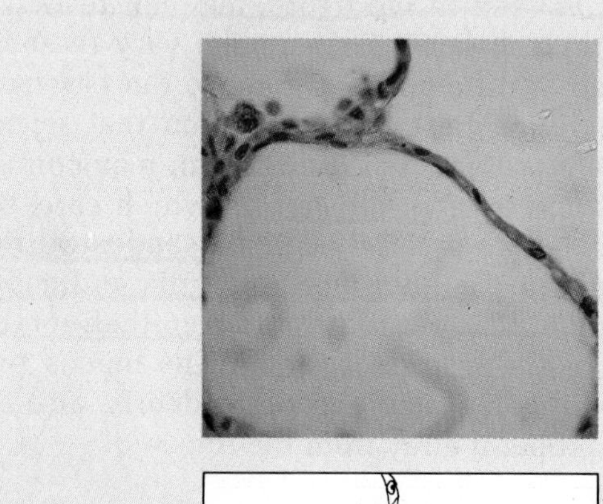

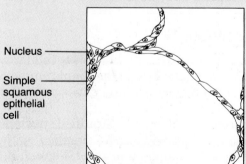

(b) Simple cuboidal epithelium

Description: Single layer of cubelike cells with large, spherical central nuclei; may have microvilli.

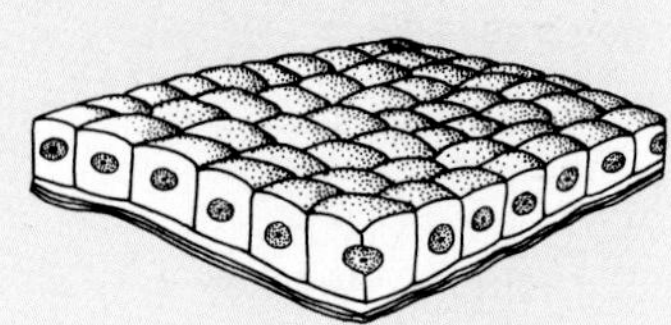

Location: Kidney tubules; ducts and secretory portions of small glands; ovary surface.

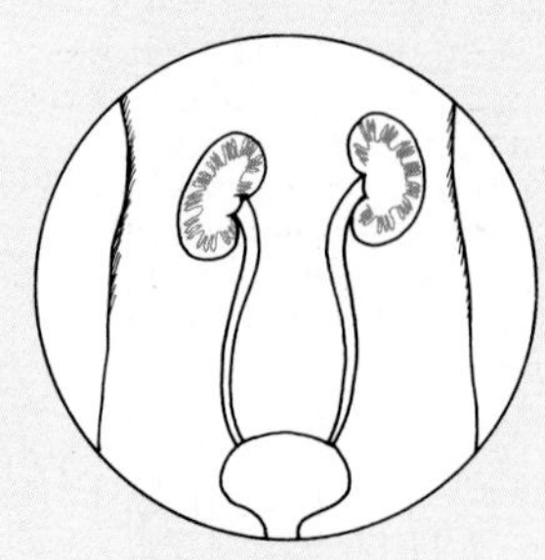

Function: Secretion and absorption.

Photomicrograph: Simple cuboidal eipthelium in kidney tubules (110 ×); the strand of cells on the left has become detached from their basement membrane.

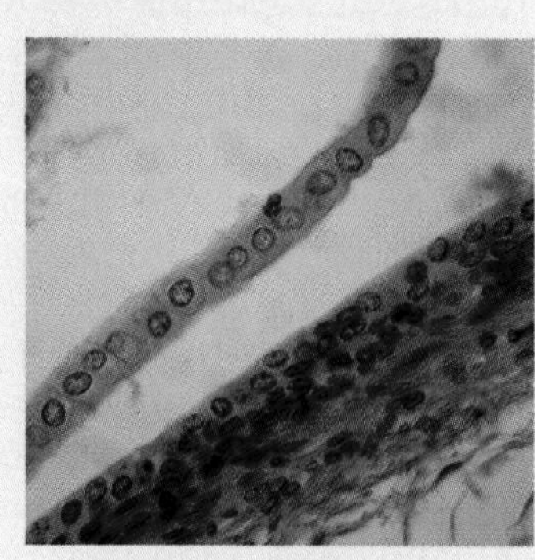

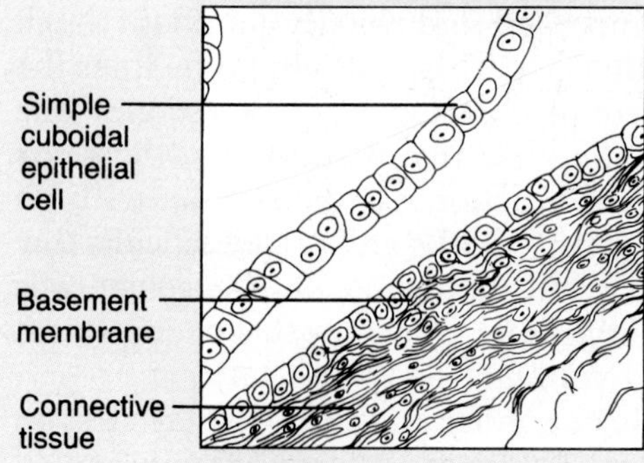

(c) Simple columnar epithelium

Description: Single layer of tall cells with *oval* nuclei; some cells bear microvilli; layer may contain mucus-secreting glands (goblet cells).

Location: Nonciliated type lines most of the digestive tract (stomach to anal canal), gallbladder and excretory ducts of some glands; ciliated variety lines small bronchi, uterine tubes, and some regions of the uterus.

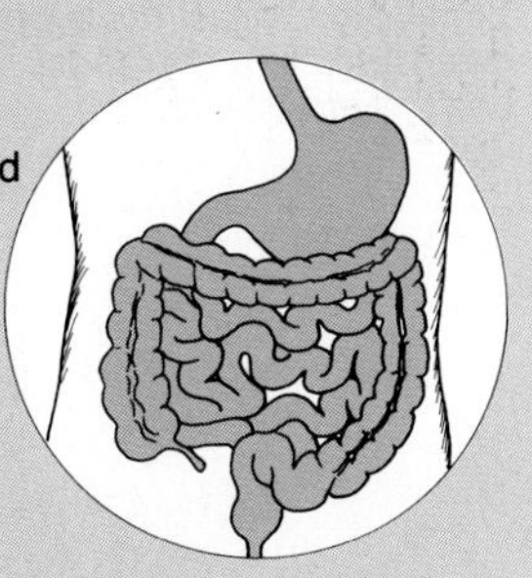

Function: Absorption; secretion of mucus, enzymes, and other substances; ciliated type propels mucus (or reproductive cells) by ciliary action.

Photomicrograph: Simple columnar epithelium of the stomach mucosa (280×).

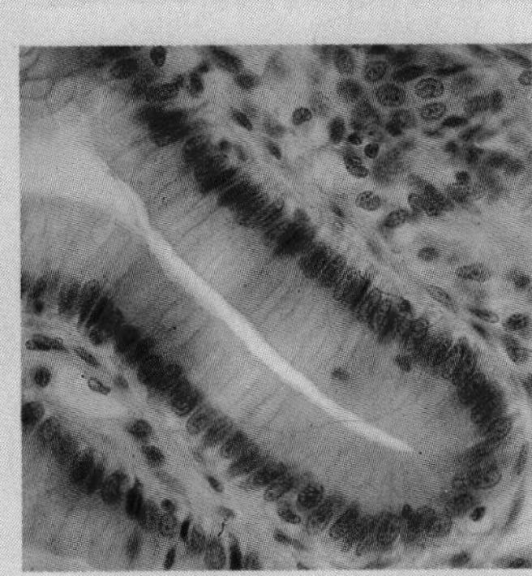

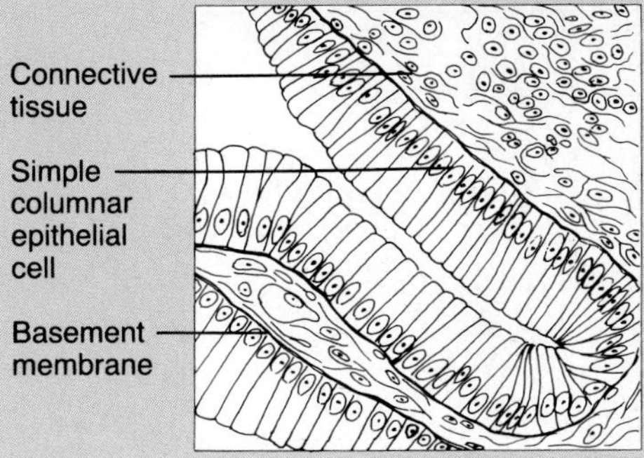

(d) Pseudostratified epithelium

Description: Single layer of cells of differing heights, some not reaching the free surface; nuclei seen at different levels; may contain goblet cells and bear cilia.

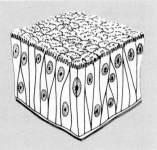

Location: Nonciliated type in ducts of large glands, parts of male urethra; ciliated variety lines the trachea, most of the upper respiratory tract.

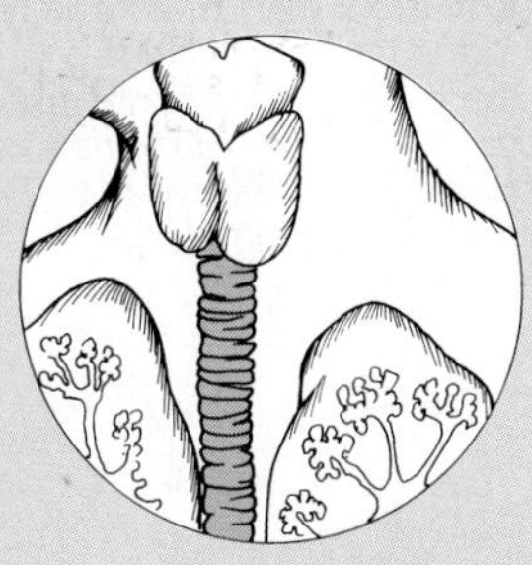

Function: Secretion, particularly of mucus; propulsion of mucus by ciliary action.

Photomicrograph: Pseudostratified ciliated columnar epithelium lining the human trachea (430×).

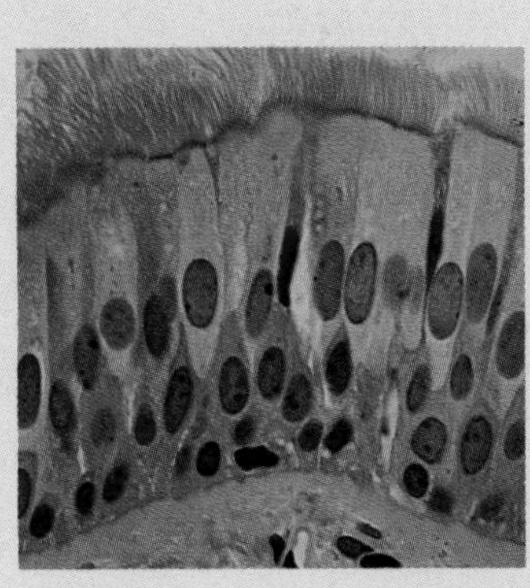

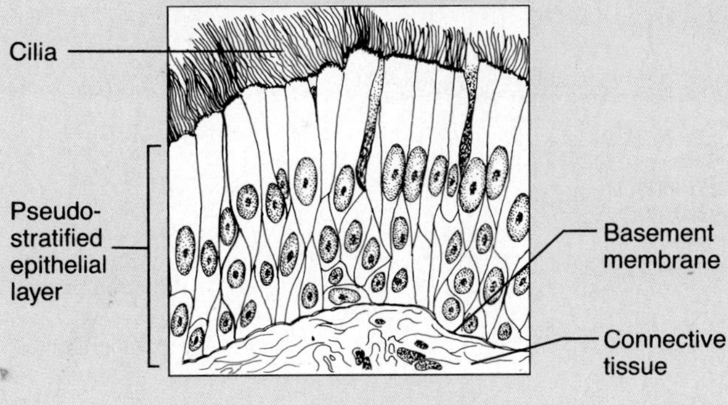

Figure 4.2 (continued) (**e**) to (**h**), Stratified epithelia.

(e) Stratified squamous epithelium

Description: Thick membrane composed of several cell layers; basal cells are cuboidal or columnar and metabolically active; surface cells are flattened (squamous); in the keratinized type, the surface cells are full of keratin and dead; basal cells are active in mitosis and produce the cells of the more superficial layers.

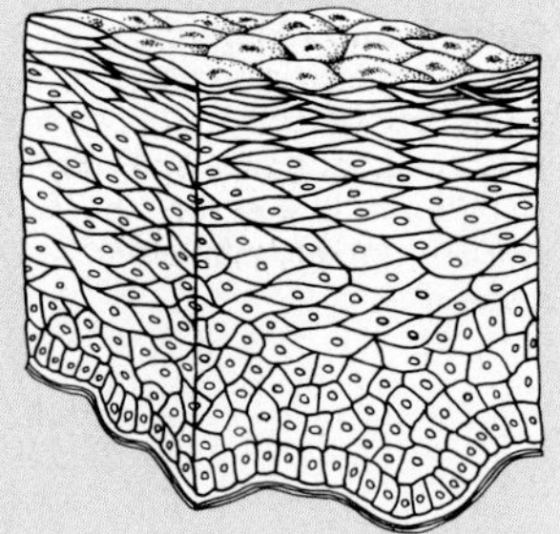

Location: Nonkeratinized type forms the moist linings of the esophagus, mouth, and vagina; keratinized variety forms the epidermis of the skin, a dry membrane.

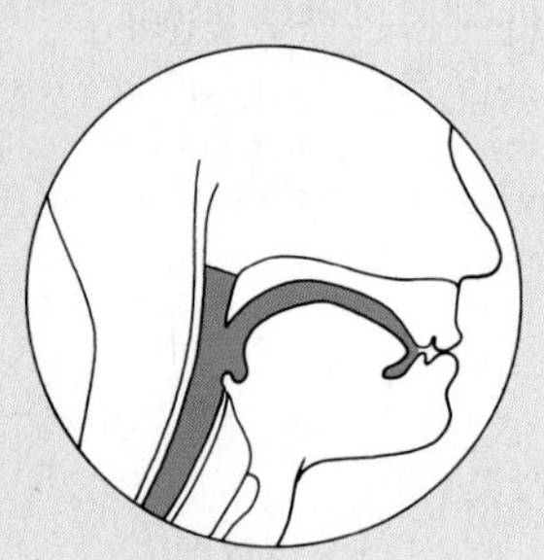

Function: Protects underlying tissues in areas subjected to abrasion.

Photomicrograph: Stratified squamous epithelium lining of the esophagus (173 ×).

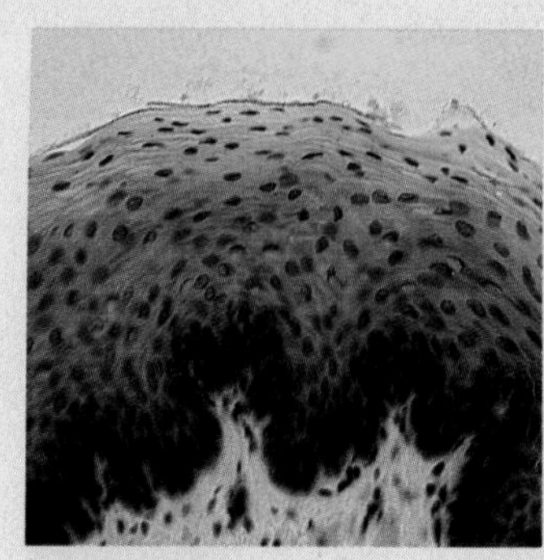

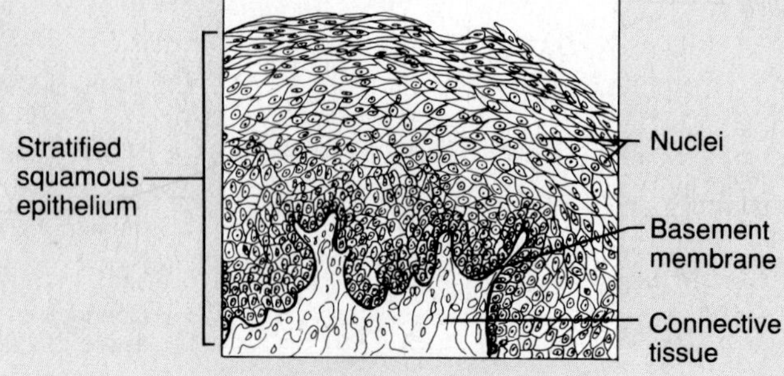

(f) Stratified cuboidal epithelium

Description: Generally two layers of cube-like cells.

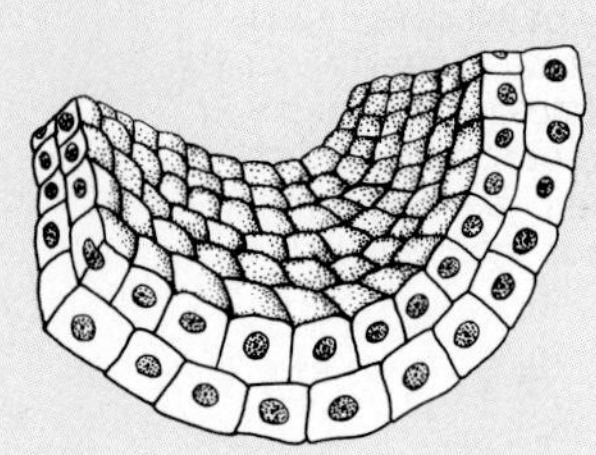

Location: Largest ducts of sweat glands, mammary glands, and salivary glands.

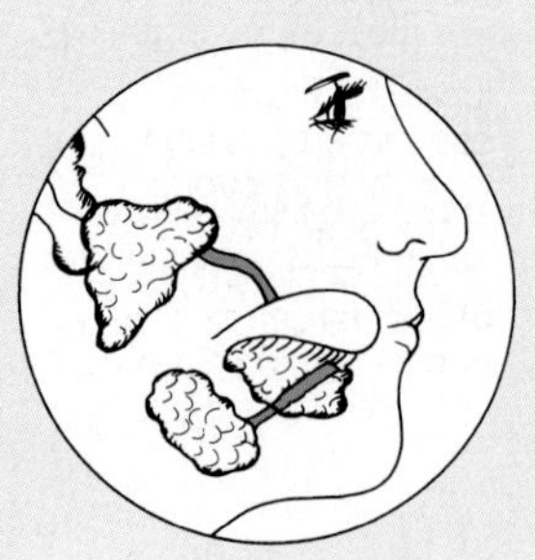

Function: Protection.

Photomicrograph: Stratified cuboidal epithelium forming a salivary gland duct (1200 ×).

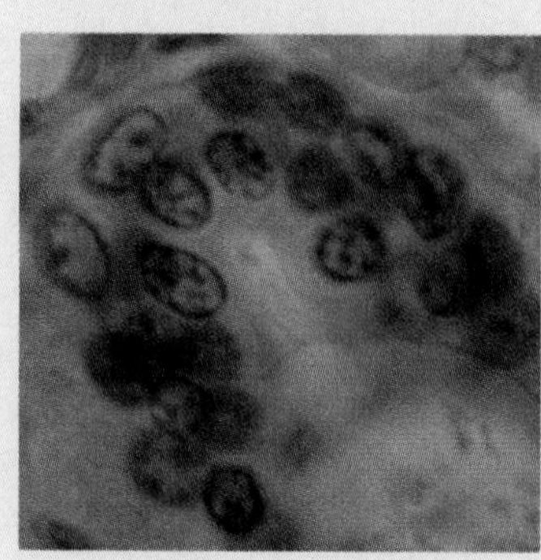

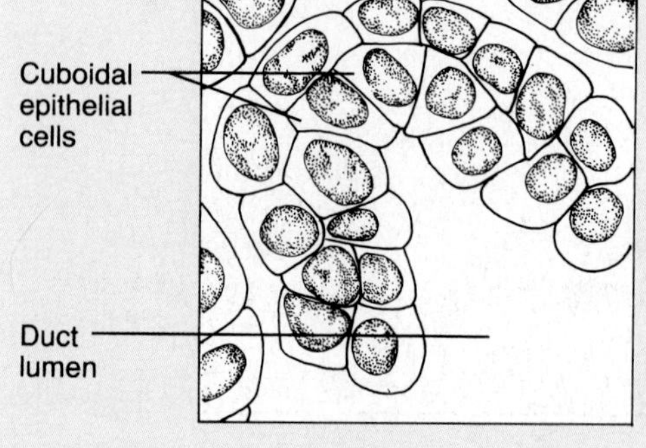

(g) Stratified columnar epithelium

Description: Several cell layers; basal cells usually cuboidal; superficial cells elongated and columnar; may contain goblet cells or bear cilia.

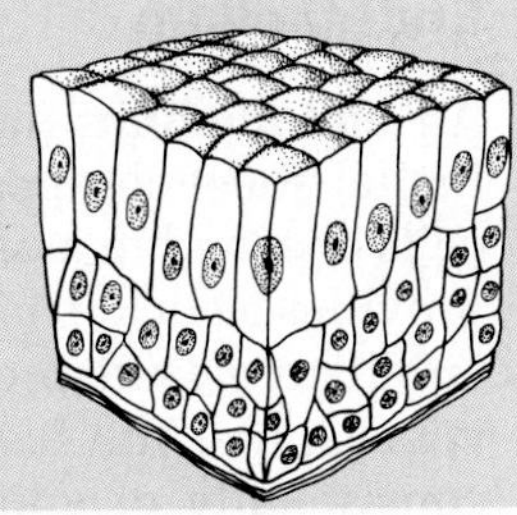

Location: Rare in the body; small amounts at epithelial junctions (stomach-esophagus junction, anorectal junction), in the pharynx, larynx, conjunctiva of eye, larger excretory ducts (e.g., urethra).

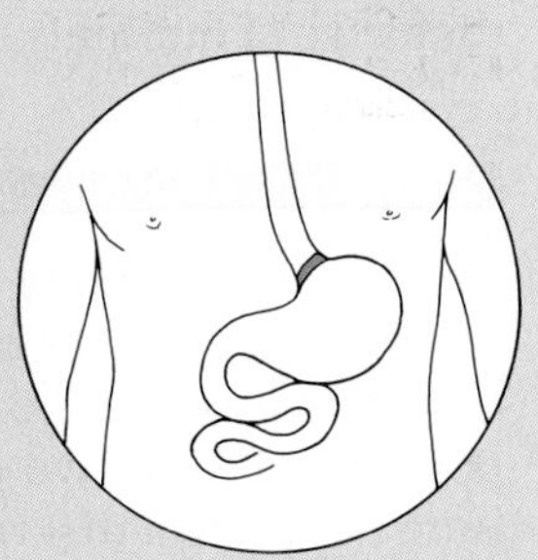

Function: Protection; secretion.

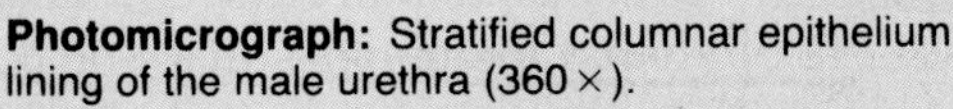

Photomicrograph: Stratified columnar epithelium lining of the male urethra (360 ×).

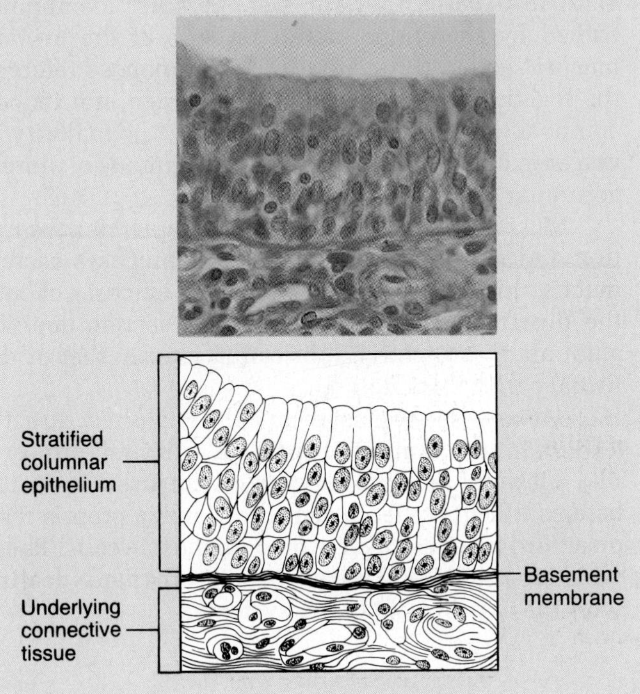

(h) Transitional epithelium

Description: Resembles both stratified squamous and stratified cuboidal; basal cells cuboidal or columnar; surfaces cells dome-shaped or squamous-like, depending on degree of organ stretch.

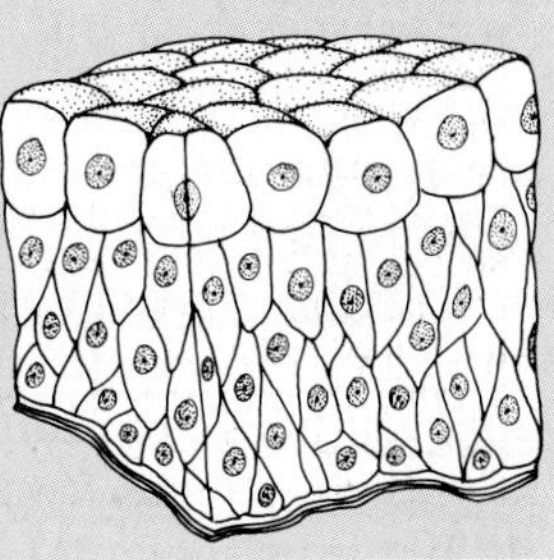

Location: Lines the ureters, bladder, and part of the urethra.

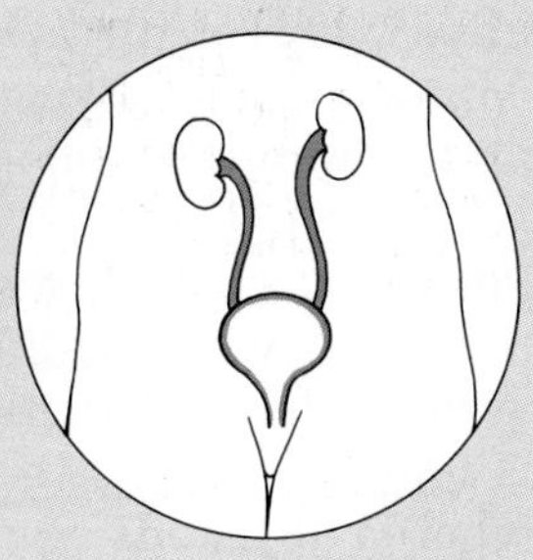

Function: Stretches readily and permits distension of urinary organ by contained urine.

Photomicrograph: Transitional epithelium lining of the bladder, relaxed state (170 ×); note the bulbous, or rounded, appearance of the cells at the surface; these cells flatten and become elongated when the bladder is filled with urine.

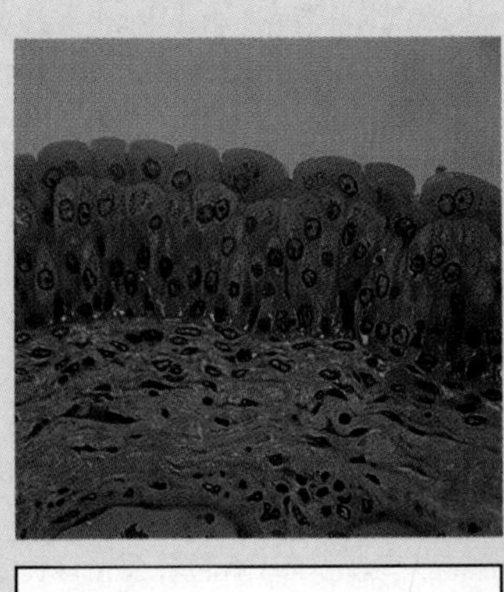

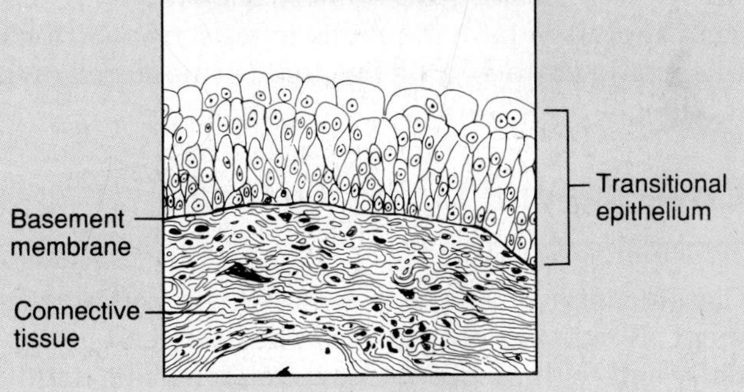

layers are cuboidal or, less commonly, columnar. This epithelium is found in areas subjected to wear and tear, and its surface cells are constantly being rubbed away and replaced by mitotic division of the cells of its basal layer. Since epithelium depends on diffusion of nutrients from a deeper connective tissue layer, the epithelial cells farthest from the basement membrane are less viable, and those at the free apical (outer) surface are flattened and atrophied.

Stratified squamous epithelium covers the tongue and lines the mouth, pharynx, esophagus, anal canal, and vagina. A modified form of this tissue, *keratinized (keh′-ruh-tin-nizd″) stratified squamous epithelium*, forms the outer layer, or *epidermis*, of the skin. The surface of the epidermis consists of dead cells filled with keratin, a waterproofing protein. The epidermis, which provides a tough yet resilient covering for the body surface, is discussed in Chapter 5.

Stratified Cuboidal Epithelium. Generally formed of only two cell layers, **stratified cuboidal epithelium** has a very limited distribution in the body (Figure 4.2f). It is found primarily in the ducts of sweat glands and other large glands.

Stratified Columnar Epithelium. True **stratified columnar epithelium** is rare. Its specific locations are listed in Figure 4.2g. Its free surface cells are columnar; those in deeper layers are small and vary in shape.

Transitional Epithelium. **Transitional epithelium** forms the lining of urinary organs, which are subjected to considerable stretching and varying internal pressure (Figure 4.2h). Cells of its basal layer are cuboidal or columnar; those at the free surface vary in appearance, depending on the degree of distension of the organ. When the organ is not stretched, the membrane is many-layered and the superficial cells are rounded and domelike. When the organ is distended with urine, fewer cell layers are obvious, and the surface cells become flattened and squamouslike. This ability of transitional cells to slide past one another and change their shape accommodates the increasing surface area of a stretching ureter wall as a greater volume of urine flows through the organ; in the bladder, it allows more urine to be stored. Additionally, transitional epithelium appears to have the ability to resist osmotic forces that would act to dilute hypertonic urine stored in the bladder.

Covering and Lining Epithelia

Classification of epithelia by cell type and arrangement allows each epithelium to be described individually and with precision. However, it reveals nothing about the tissue's body location. In this section, we will describe the covering and lining epithelia using terms that indicate their special locations in the body and/or denote general functional qualities. According to this scheme, there are endothelium and the more complex epithelial membranes.

Endothelium

An **endothelium** (en-dō-thē′-li-um) is a simple epithelial sheet composed of a single layer of squamous cells attached to a basement membrane. Endothelium provides a slick, friction-reducing lining in all hollow circulatory system organs—lymphatic vessels, blood vessels, and the heart (Figure 4.3a). Capillary walls consist only of endothelium, which, because of its permeability and extreme thinness, encourages the exchange of nutrients and wastes across capillary walls.

Epithelial Membranes

There is tremendous variety in the way the term *epithelial membrane* is used. Here, **epithelial membrane** is defined as a continuous multicellular sheet composed of at least two primary tissue types: an epithelium bound to a discrete underlying connective tissue layer. Hence, epithelial membranes can be considered to be simple organs. Most of the covering and lining epithelia take part in forming one of three common types of epithelial membranes: mucous, cutaneous, or serous.

Mucous Membranes. **Mucous membranes,** or **mucosae** (myoo-kō′-sē), are epithelial membranes that line body cavities that are open to the exterior, such as those of the digestive, respiratory, and urogenital tracts (Figure 4.3b). In all cases, they are "wet," or moist, membranes bathed by secretions or, in the case of the urinary mucosa, urine. Notice that the term *mucosa* refers to the location of the epithelial membrane, not its cell composition, which varies. However, the majority of mucosae contain either stratified squamous or simple columnar epithelia.

Mucous membranes are often adapted for absorption and secretion. Although many mucosae secrete mucus, this is not a requirement. The mucosae of both the digestive and respiratory tracts secrete copious amounts of protective lubricating mucus; that of the urinary tract does not.

All mucosae consist of an epithelial sheet directly underlain by a **lamina propria** (la′-mih-nuh prō′-prē-uh), a layer of loose connective tissue just deep to the basement membrane. In some, the lamina propria rests on a third (deeper) layer of smooth muscle cells. These variations will be covered in later chapters dealing with the appropriate organ systems.

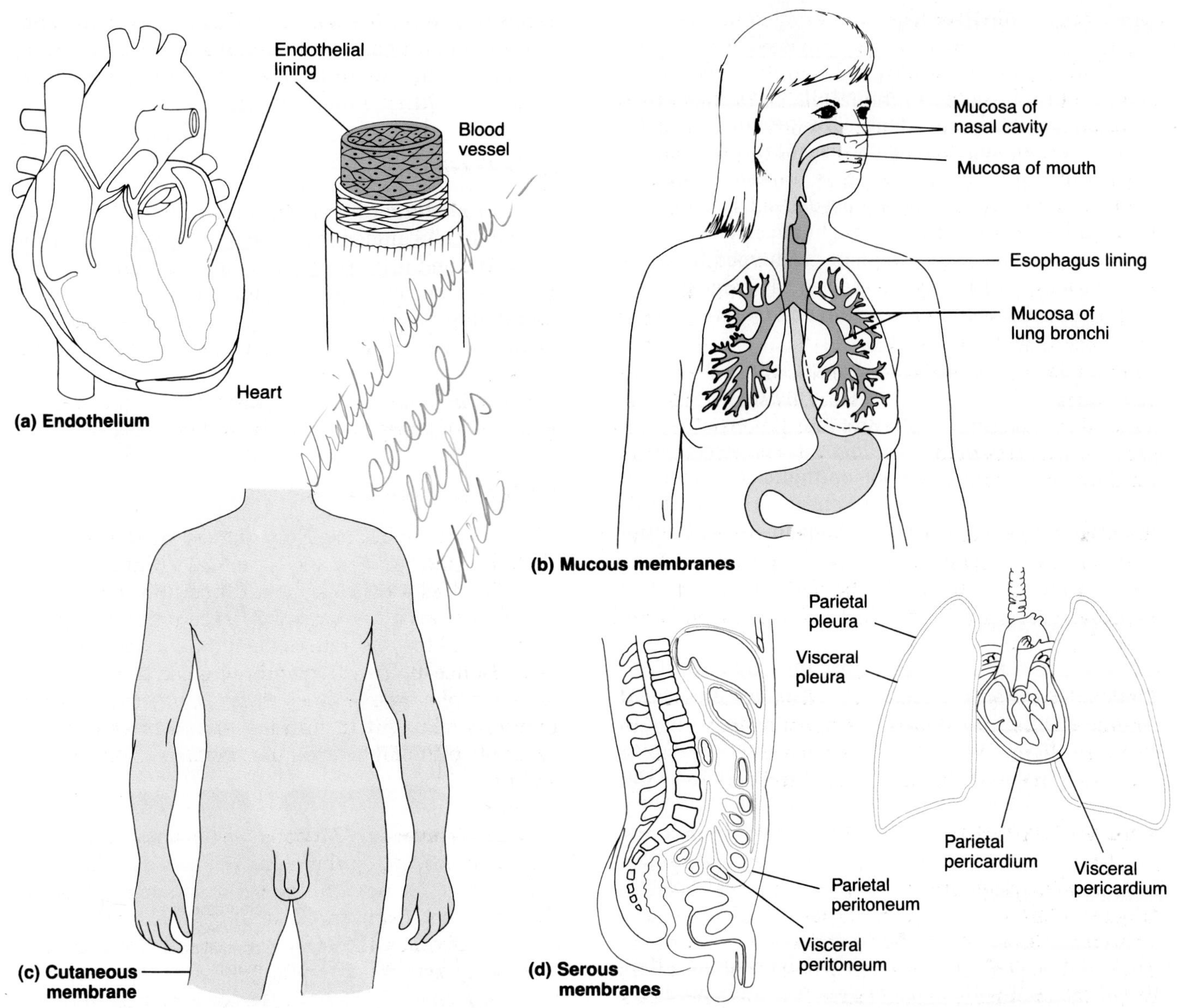

Figure 4.3 Classes of covering and lining epithelial membranes. (**a**) Endothelium lines the lumen of heart, blood vessels, and lymphatic vessels. (**b**) Mucous membranes line body cavities that are open to the exterior. (**c**) Cutaneous membrane, or skin. (**d**) Serous membranes line ventral body cavities that are closed to the exterior.

Cutaneous Membrane. The **cutaneous** (kyoo-tā′-nē-us) **membrane** is your skin (Figure 4.3c). It is an organ consisting of a keratinized stratified squamous epithelium (epidermis) firmly attached to a thick connective tissue layer (dermis). Unlike other epithelial membranes, the cutaneous membrane is exposed to the air and is a dry membrane. Chapter 5 is devoted to this unique organ.

Serous Membranes. Serous membranes, or **serosae** (suh-rō′-sē), are the moist membranes found in closed ventral body cavities (Figure 4.3d). Each serosa consists of a *parietal layer* that lines the cavity wall and then reflects back as the *visceral layer* to cover the outer surface(s) of organs within the cavity. Each of these layers consists of simple squamous epithelium (mesothelium) resting on a tiny amount of loose connective tissue. The mesothelial cells secrete thin, clear *serous fluid* that lubricates the facing surfaces of the parietal and visceral layers, so that they slide across each other easily. This reduction of friction prevents organs from sticking to the cavity walls and to each other.

The serosae are named according to site and specific organ associations. For example, the serosa lining the thoracic wall and covering the lungs is the *pleura*; that enclosing the heart is the *pericardium*; and those of the abdominopelvic cavity and viscera are the *peritoneums*.

Glandular Epithelia

A **gland** consists of one or more cells that produce and secrete a particular product. This product, called a *secretion*, is an aqueous (water-based) fluid, typically containing proteins. Secretion is an active process whereby glandular cells obtain needed substances from the blood and transform them chemically into their secretory product, which is then discharged from the cell. Notice that the term *secretion* can refer to both the *process* of secretion formation and release and the *product* of glandular activity.

Glands are classified as *endocrine* (en′-duh-krin) or *exocrine* (ek′-suh-krin), depending on their route of secretion and the general function of their products, and as *unicellular* or *multicellular* on the basis of their structure. Most multicellular epithelial glands form by invagination of an epithelial sheet and, at least initially, have ducts connecting them to the epithelial sheet.

Endocrine Glands

Endocrine glands eventually lose their ducts and are often called *ductless glands*. They produce regulatory chemicals called *hormones*, which they secrete directly into the extracellular space. The hormones then enter the blood or lymphatic fluid. Since not all endocrine glands are epithelial derivatives, consideration of their structure and function is deferred to Chapter 17.

Exocrine Glands

Exocrine glands are far more numerous than endocrine glands, and many of their products are familiar ones. The multicellular glands secrete their products through a duct onto body surfaces or into body cavities. Exocrine glands are a diverse lot. They include sweat and oil glands, salivary glands, the liver (which secretes bile), the pancreas (which synthesizes digestive enzymes), mammary glands, mucous glands, and many others.

Unicellular Exocrine Glands. **Unicellular exocrine glands** are single cells interposed in an epithelium between cells with other functions. They have no ducts. In humans, all such glands produce *mucin* (myoo′-sin), a complex glycoprotein that dissolves in water. Once dissolved, mucin forms a slimy coating (mucus) that both protects and lubricates surfaces. Unicellular glands include the goblet cells of the intestinal and respiratory mucosae (Figure 4.4), as well as mucin-producing cells found in other body regions. Although unicellular glands probably outnumber multicellular glands, unicellular glands are the less common of the two gland types.

Multicellular Exocrine Glands. **Multicellular exocrine glands** have three common structural elements: an epithelium-derived *duct*, a *secretory unit*, and, in all but the simplest glands, *supportive connective tissue* that surrounds the secretory unit and supplies it with blood vessels and nerve fibers. Often, the connective tissue forms a fibrous capsule that extends into the gland proper and divides the gland into lobes.

Multicellular glands can be divided into two major categories on the basis of their duct structures. **Simple glands** have a single unbranched duct, whereas **compound glands** have a branching or divided duct. The glands can be further described according to the struc-

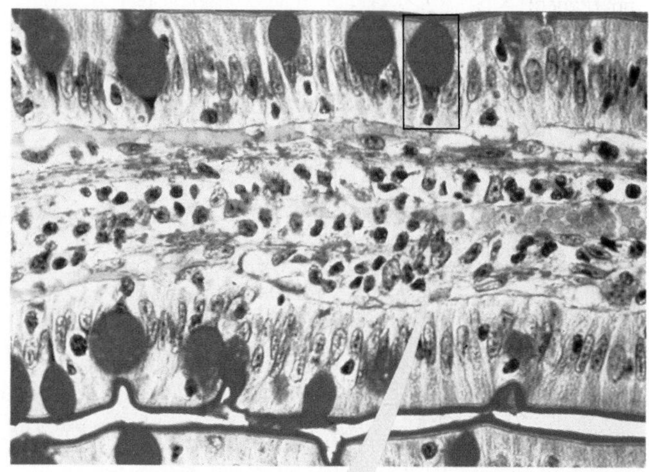

(a)

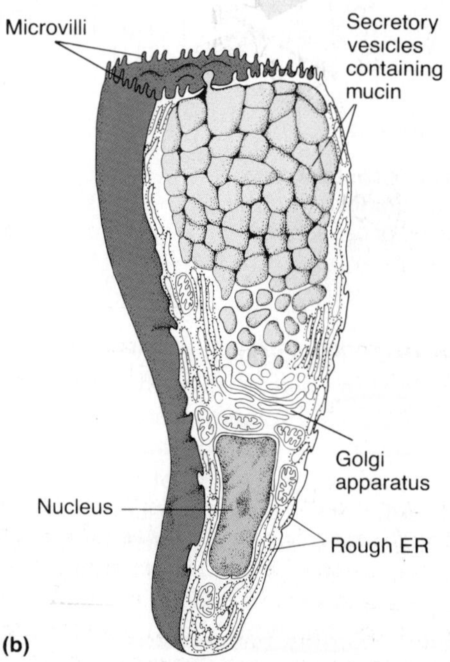

(b)

Figure 4.4 Goblet cells, examples of unicellular exocrine glands. **(a)** Photomicrograph of simple columnar (intestinal) epithelium with goblet cells (approx. 300×). **(b)** Diagram of the fine structure of a goblet cell. Notice the large Golgi apparatus that is involved in packaging the mucin secretion in vesicles.

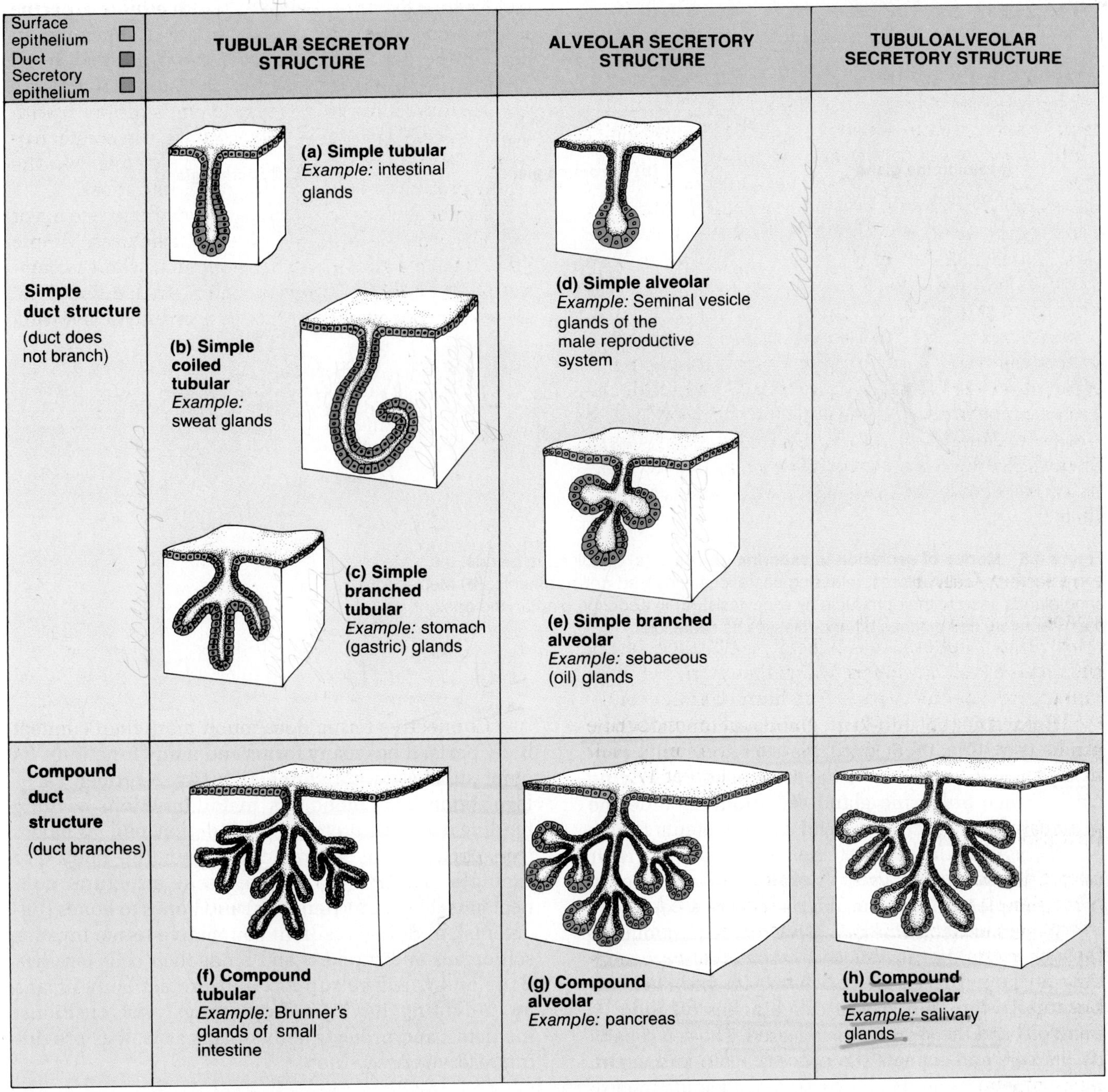

Surface epithelium / Duct / Secretory epithelium	TUBULAR SECRETORY STRUCTURE	ALVEOLAR SECRETORY STRUCTURE	TUBULOALVEOLAR SECRETORY STRUCTURE
Simple duct structure (duct does not branch)	**(a) Simple tubular** *Example:* intestinal glands **(b) Simple coiled tubular** *Example:* sweat glands **(c) Simple branched tubular** *Example:* stomach (gastric) glands	**(d) Simple alveolar** *Example:* Seminal vesicle glands of the male reproductive system **(e) Simple branched alveolar** *Example:* sebaceous (oil) glands	
Compound duct structure (duct branches)	**(f) Compound tubular** *Example:* Brunner's glands of small intestine	**(g) Compound alveolar** *Example:* pancreas	**(h) Compound tubuloalveolar** *Example:* salivary glands

Figure 4.5 Type of multicellular exocrine glands. Multicellular glands are classified according to duct type (simple or compound) and the structure of their secretory units (tubular, alveolar, or tubuloalveolar).

ture of their secretory parts as (1) **tubular,** with the secretory cells forming a tube; (2) **alveolar** (al-vē′-uh-ler), or **acinar** (a′-sih-ner), with the secretory cells forming small, flasklike sacs (*alveolus* = small hollow cavity); and (3) **tubuloalveolar,** with the secretory parts having both tubular and alveolar portions. Terms denoting duct and secretory part structure are combined to describe the gland fully (Figure 4.5). Be sure you understand the type called *compound tubuloalveolar*, as it is very common and important.

Since multicellular glands secrete their products in different ways, they can also be classified functionally, according to their secretory behavior. Most exocrine glands are **merocrine** (mayr′-ō-krin) **glands,** which secrete their products by exocytosis shortly after the products are produced. The secretory cells are not altered in any way. The pancreas, most sweat glands, and salivary glands belong to this class (Figure 4.6b).

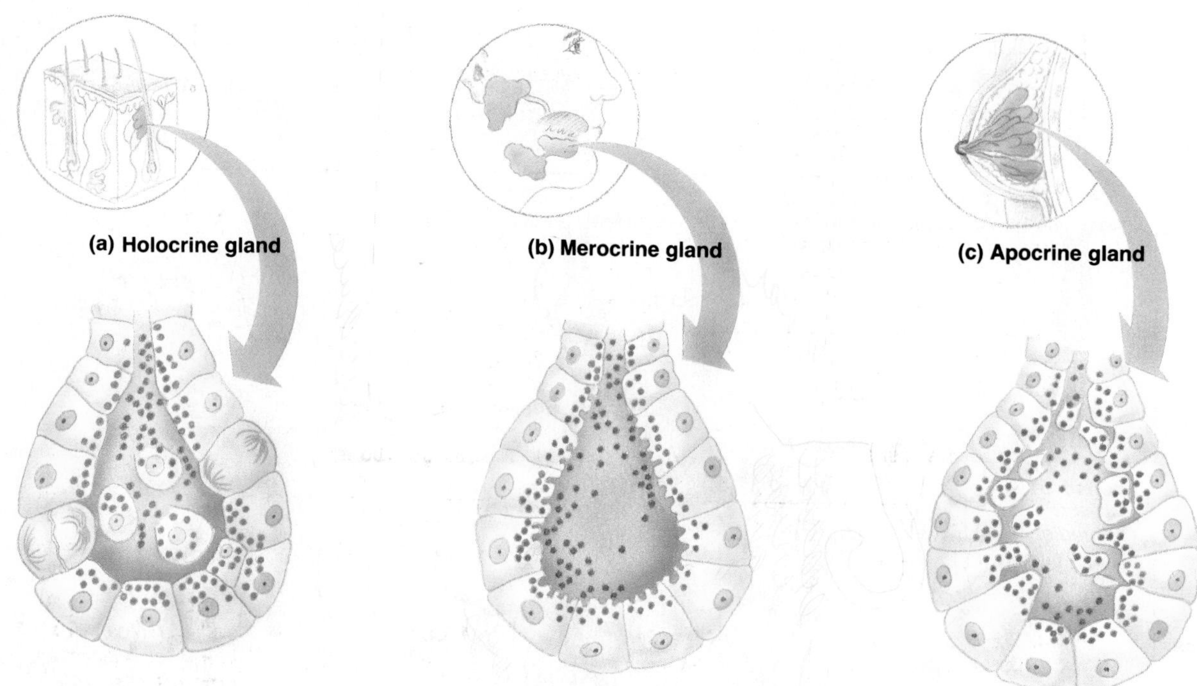

Figure 4.6 Modes of secretion in exocrine glands. (**a**) In holocrine glands, the entire secretory cell ruptures, releasing secretions and dead cell fragments. (**b**) Merocrine glands secrete their products by exocytosis. (**c**) In apocrine glands, the apex of each secretory cell pinches off and releases its secretions.

Holocrine (hō′-luh-krin) **glands** accumulate their products within them until the secretory cells rupture. (They are replaced by the division of underlying cells.) Since holocrine gland secretions include the synthesized product plus dead cell fragments (*holos* = all), you could say that their cells "die for their cause." Sebaceous (oil) glands of the skin are the only true example of holocrine glands (Figure 4.6a).

Apocrine (a′-puh-krin) **glands** also accumulate their products, but in this case, accumulation occurs only at the cell apex (just beneath its free surface). Eventually, the apex of the cell pinches off (*apo* = from, off) and the secretion is released. The cell repairs its damage and repeats the process again and again. The mammary glands and some sweat glands release their secretions by this mechanism (Figure 4.6c).

Connective Tissue

Connective tissue is found everywhere in the body. It is the most abundant and widely distributed of the primary tissues, but its amount in particular organs varies greatly. For example, bone and skin are made up primarily of connective tissue, whereas the brain contains very little.

Connective tissue does much more than connect body parts; it has many forms and many functions. Its chief subclasses are connective tissue proper, cartilage, bone, and blood. Its major functions include *binding, support, protection, insulation,* and, as blood, *transportation* of substances within the body. For example, cordlike connective tissue structures connect muscle to bone (tendons) and bones to bones (ligaments), and fine, resilient connective tissue invades soft organs and supports and binds their cells together. Bone and cartilage support and protect body organs by providing hard "underpinnings"; fat cushions, insulates, and protects body organs as well as providing reserve energy fuel.

Common Characteristics of Connective Tissue

Despite their multiple and varied functions in the body, connective tissues have certain common properties that set them apart from other primary tissues:

1. **Common origin.** All connective tissues arise from *mesenchyme,* an embryonic tissue derived from the mesoderm germ layer, and hence have a common kinship (Figure 4.7).

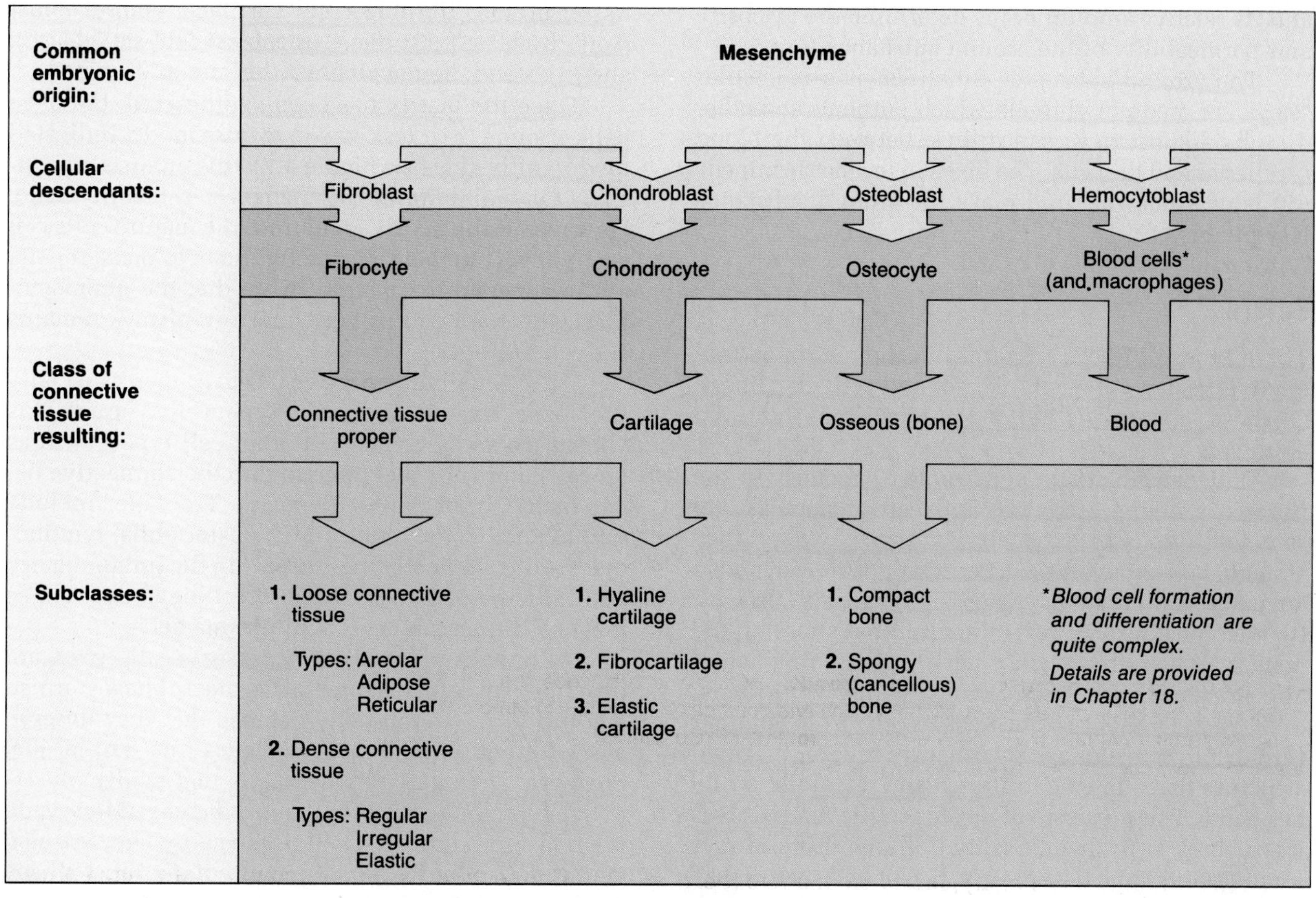

Figure 4.7 Major classes of connective tissue. All of these classes arise from the same common embryonic cell type (mesenchyme).

2. **Degrees of vascularity.** Unlike epithelium, which is avascular, and muscle and nervous tissue, which have a rich vascular supply, connective tissues run the entire gamut of vascularity. Cartilage is avascular; dense connective tissue is poorly vascularized; and the other types have a rich supply of blood vessels.

3. **Matrix.** Whereas all other primary tissues are composed mainly of cells, connective tissues are composed largely of nonliving **extracellular matrix,** which separates, often widely, the living cells of the tissue. Because of this matrix, connective tissue is able to bear weight, withstand great tension, and endure abuses, such as physical trauma and abrasion, that no other tissue could withstand.

Structural Elements of Connective Tissue

In any type of connective tissue, three elements must be considered: ground substance, fibers, and cells. The ground substance and fibers make up the extracellular matrix. (However, you should be aware that some authors use the term *matrix* to indicate the ground substance only.) The properties of the cells and the composition and arrangement of extracellular matrix elements vary tremendously, giving rise to an amazing diversity of connective tissues, each uniquely adapted to perform its specific function in the body. For example, the matrix can be delicate and fragile to form a soft "packing" around an organ, or it can form "ropes" (tendons and ligaments) of incredible strength.

Ground Substance

Ground substance is an amorphous (unstructured) material that fills the space between the cells and contains the fibers. It is composed of interstitial fluid, glycoproteins, and *glycosaminoglycans* (glī″-kō-suh-mē″-nō-glī-kanz″) (GAGs), a diverse group of large, negatively charged polysaccharides. The long, strandlike GAGs coil, intertwine, and trap water, forming a substance that varies from a fluid to a semistiff hydrated gel. One type of GAG, *hyaluronic* (hī″-yul-yoo-rah′-nik) *acid*, is found in virtually all connective tissues, and its relative amount helps determine the viscosity and permeability of the ground substance.

The ground substance functions as a molecular "sieve," or medium, through which nutrients and other dissolved substances can diffuse between the blood capillaries and the cells. The fibers in the matrix impede diffusion somewhat and make the ground substance less pliable.

Fibers

Three types of fibers are found in the matrix of connective tissue: collagen, elastic, and reticular fibers. Of these, collagen is by far the most important and abundant.

Collagen fibers are constructed primarily of the fibrous protein *collagen*. Collagen molecules are secreted into the extracellular space, where they assemble spontaneously into fibers. Collagen fibers are extremely tough and provide high tensile strength (that is, the ability to resist longitudinal stress) to the matrix. When fresh, they have a glistening white appearance; they are therefore also called *white fibers*.

Elastic fibers are formed largely from another fibrous protein, *elastin*. Elastin has a randomly coiled structure that allows it to stretch and recoil like a rubber band. The presence of elastin in the matrix gives it a rubbery, or resilient, quality. Collagen fibers, always found in the same tissue, stretch a bit and then "lock" in full extension, which limits the extent of stretch and prevents the tissue from tearing. Elastic fibers then snap the connective tissue back to its normal length when the tension lets up. Elastic fibers are found where greater elasticity is needed, for example, in the skin, lungs, and blood vessel walls. Since fresh elastic fibers appear yellow, they are sometimes called *yellow fibers*.

Reticular fibers are believed to be fine collagenous fibers (with a slightly different chemistry) and are continuous with collagen fibers. They branch extensively, forming a netlike *reticulum* in the matrix. They construct a fine mesh around small blood vessels, support the soft tissue of organs, and are particularly abundant at the junction between connective tissue and other tissue types, for example, in the basement membrane of epithelial tissues.

Cells

Each major class of connective tissue has a fundamental cell type that exists in immature and mature forms (see Figure 4.7). The undifferentiated cells, indicated by the suffix *blast* (literally, "bud," or "sprout," but meaning "forming"), are actively mitotic cells that secrete both the ground substance and the fibers characteristic of their particular matrix. The primary blast cell types by connective tissue class are (1) connective tissue proper: **fibroblast;** (2) cartilage: **chondroblast** (kon′-drō-blast″); (3) bone: **osteoblast** (ah′-stē-ō-blast″); and (4) blood: **hemocytoblast** (hē″-mō-sī′-tō-blast).

Once the matrix has been synthesized, the blast cells assume their less active, mature mode, indicated by the suffix *cyte* (see Figure 4.7); this mode is responsible for maintaining the matrix in a healthy state. However, if the matrix is injured, the mature cells can easily revert to their more active state to make repairs and regenerate the matrix. (Note that the hemocytoblast, the stem cell of bone marrow, always remains actively mitotic.)

Additionally, connective tissue proper, especially the loose connective tissue type called areolar, is "home" to an assortment of other cell types, such as fat cells and cells that migrate into the connective tissue matrix from the bloodstream. The latter include white blood cells (neutrophils, eosinophils, lymphocytes) and other cell types that act in the inflammatory and immune responses that protect the body, such as mast cells, macrophages, and plasma cells.

Although all of these accessory cell types are described in later chapters, the macrophages are so significant to overall body defense that they deserve a brief mention here. **Macrophages** (ma′-krō-fā″-juz) are large, irregularly shaped cells that avidly phagocytize both foreign matter that has managed to invade the body and dying or dead tissue cells. They are also central actors in the immune system, as you will see in Chapter 22. In connective tissues, they may be fixed (attached to the connective tissue fibers) or they may migrate freely through the matrix. However, macrophages are not limited to connective tissue. In fact, their body distribution is so broad and their numbers so vast that they are often referred to collectively as the **macrophage system.**

Macrophages are peppered throughout loose connective tissue, bone marrow, lymphatic tissue, the spleen, and the mesentery that suspends the abdominal viscera. Those in certain sites are given specific names; they are called *histiocytes* (his′-tē-ō-sīts) in loose connective tissue, *Kupffer* (koop′-fer) *cells* in the liver, and *microglial* (mī″-krō′-glē-ul) *cells* in the brain. Although all these cells are phagocytes, some have selective appetites. For example, the macrophages of the spleen function primarily to engulf aging red blood cells; but they will not turn down other "delicacies" that come their way.

Types of Connective Tissue

As noted, all classes of connective tissue consist of living cells surrounded by a matrix. Their major differences reflect cell type, fiber type, and proportion of the matrix contributed by fibers. Collectively, these three factors determine not only major connective tissue classes, but also their subclasses and types. The connective tissue classes described in this section are illustrated in Figure 4.8. Additionally, since the mature connective tissues arise from a common embryonic tissue, it seems appropriate to describe this here as well.

Embryonic Connective Tissue: Mesenchyme

Mesenchyme (meh'-zin-kīm), or **mesenchymal tissue,** is embryonic connective tissue and represents the first definitive tissue formed from the mesoderm germ layer. It arises during the early weeks of development and eventually differentiates (specializes) into all other connective tissues. Mesenchyme is composed of star-shaped mesenchymal cells and a fluid ground substance containing fine fibrils (Figure 4.8a).

Mucous connective tissue is a temporary tissue, derived from mesenchyme and similar to it, that appears in the fetus in very limited amounts. *Wharton's jelly,* which supports the umbilical cord, is the best representative of this scant embryonic tissue.

Connective Tissue Proper

Connective tissue proper has two subclasses: the **loose connective tissues** (areolar, adipose, and reticular) and **dense connective tissues** (dense regular, dense irregular, and elastic). Except for bone, cartilage, and blood, all mature connective tissues belong to this class.

Areolar Connective Tissue. **Areolar** (uh-rē'-uh-ler) **connective tissue** has a semifluid ground substance formed primarily of hyaluronic acid in which all three fiber types are loosely dispersed (Figure 4.8b). **Fibroblasts,** flat, branching cells that appear spindle-shaped in profile, are the predominant cell type of this tissue. Numerous macrophages are also seen, but other cell types are scattered throughout.

Fat cells appear singly or in small clusters. *Mast cells* are identified easily by the large, darkly stained cytoplasmic granules that often obscure their nucleus. Mast cell granules contain (1) *heparin* (heh'-puh-rin), an anticoagulant that is released into the capillaries and helps prevent blood clotting, and (2) *histamine* (his'-tuh-mēn), which is released during inflammatory reactions and makes the capillaries leaky. (The inflammatory process is discussed in Chapter 22.)

Perhaps the most obvious structural feature of this tissue is the loose arrangement of its fibers, which account for only small portions of matrix. The rest of the matrix, occupied by fluid ground substance, appears to be empty space when viewed through the microscope; in fact, the Latin term *areola* means "a small open space." Because of its loose and fluid nature, areolar connective tissue provides a reservoir of water and salts for surrounding body tissues. If extracellular fluids accumulate in excess, the affected areas swell and become puffy, a condition called *edema.*

Areolar connective tissue is soft and pliable and serves as a kind of universal packing material between other tissues. The most widely distributed connective tissue in the body, it separates muscles, allowing them to move freely over one another; wraps small blood vessels and nerves; surrounds glands; and forms the subcutaneous tissue, which attaches the skin to underlying structures. It is present in all mucous membranes as the lamina propria.

Adipose (Fat) Tissue. **Adipose** (a'-dih-pōs) **tissue** is basically an areolar connective tissue in which the *adipocytes* (a'-dih-pō-sīts), commonly called fat cells, have accumulated in large numbers. A glistening oil droplet (almost pure neutral fat), which occupies most of a fat cell's volume, compresses the nucleus and displaces it to one side; only a thin rim of surrounding cytoplasm is seen. Since the oil-containing region looks empty, and the thin cytoplasm containing the bulging nucleus looks like a ring with a seal, fat cells have been called "signet ring" cells (Figure 4.8c). Mature adipocytes are among the largest cells in the body and are fully specialized cells that are incapable of cell division. As they take up or release fat, they become more plump or more wrinkled looking, respectively.

Compared to other connective tissues, adipose tissue is very cellular; adipose cells account for approximately 90% of the tissue mass and are packed closely together, giving a chicken wire appearance to the tissue. Very little matrix is seen, except for that separating the adipose cells into lobules (cell clusters) and permitting the passage of blood vessels and nerves to the cells. Adipose tissue is richly vascularized, indicating its high metabolic activity, and it has many functions; most importantly, it acts as a storehouse of nutrients. Without stored fat, we could not live for more than a few days without eating.

Adipose tissue may develop almost anywhere areolar tissue is plentiful, but it usually accumulates in subcutaneous tissue, where it acts as a shock absorber and as insulation. Since fat is a poor conductor of heat, it helps prevent heat loss from the body. Other sites of fat accumulation include genetically determined fat depots such as the abdomen and hips, the bone marrow, around the kidneys, and behind the eyeballs.

Some nutritionists believe that obesity in later life results from overfeeding during infancy and childhood. Since unused nutrients are converted to fat for storage, excessive food intake may encourage differentiation of excessive numbers of fat cells, which are capable of storing large amounts of fat throughout

Figure 4.8 Connective tissues.

Embryonic connective tissue

(a) Mesenchyme

Description: Embryonic connective tissue; gel-like ground substance containing fine fibers; star-shaped mesenchymal cells.

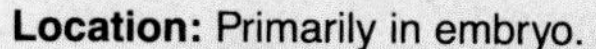

Location: Primarily in embryo.

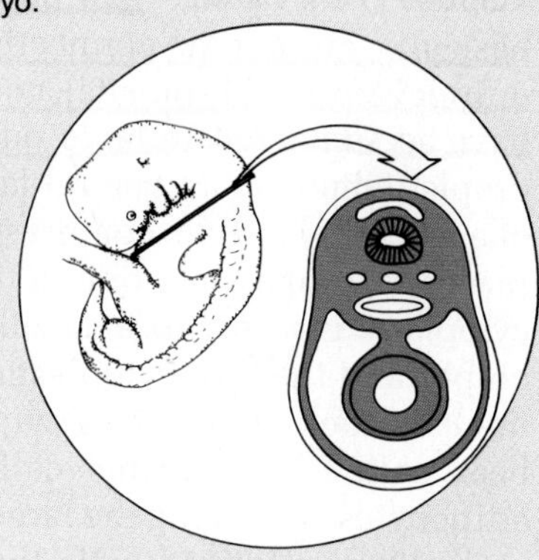

Function: Gives rise to all other connective tissue types.

Photomicrograph: Mesenchymal tissue, an embryonic connective tissue (475×); the clear-appeari background is the fluid ground substance of the matrix; notice the fine, sparse fibers.

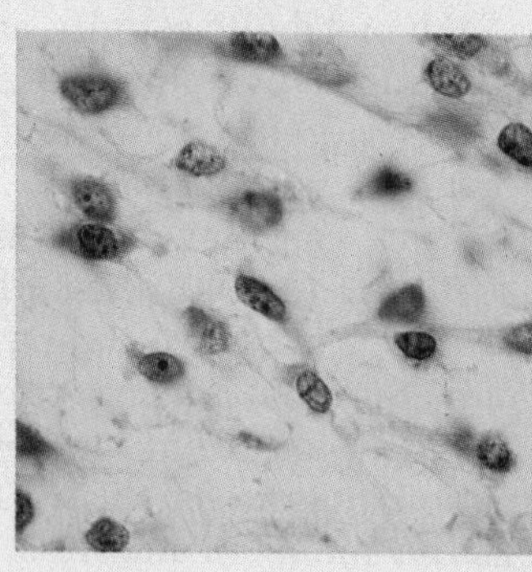

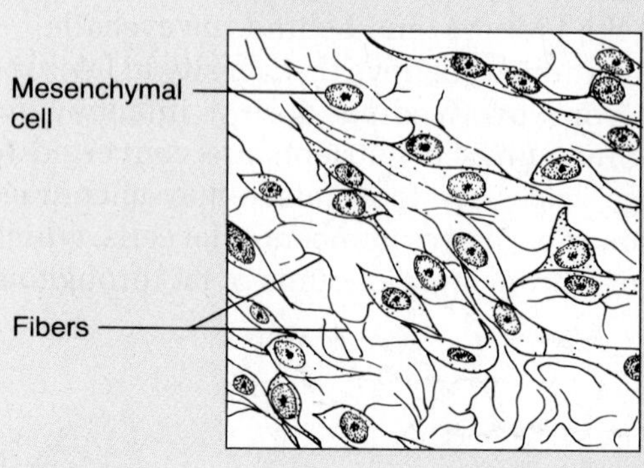

Connective tissue proper: Loose connective tissue (b to d)

(b) Areolar connective tissue

Description: Gel-like matrix with all three fiber types; cells: fibroblasts, macrophages, mast cells, and some white blood cells.

Location: Widely distributed under epithelia of body, e.g., forms lamina propria of mucous membranes; packages organs; surrounds capillaries.

Function: Wraps and cushions organs; its macrophages phagocytize bacteria; plays important role in inflammation; holds and conveys tissue fluid.

Photomicrograph: Areolar connective tissue, a soft packaging tissue of the body (170×).

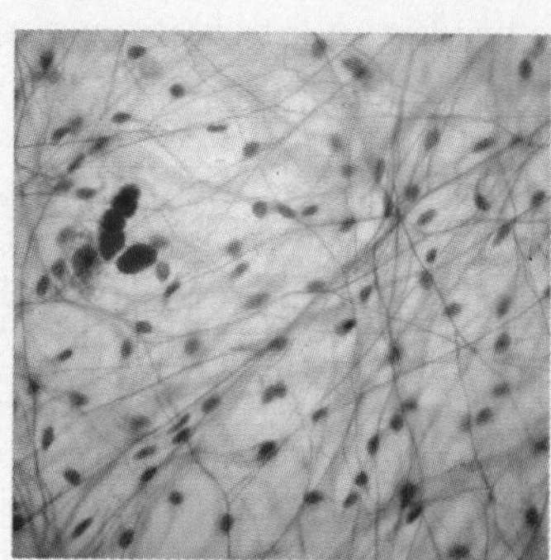

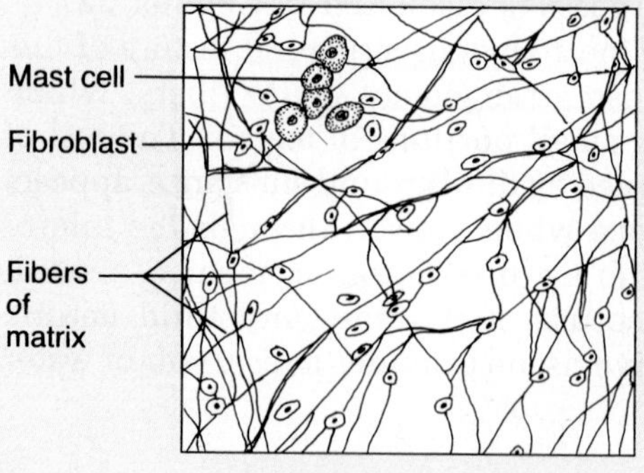

(c) Adipose tissue

Description: Matrix as in areolar, but very sparse; closely packed adipocytes, or fat cells, have nucleus pushed to the side by large fat droplet.

Location: Under skin; around kidneys and eyeballs; in bones and within abdomen; in breasts.

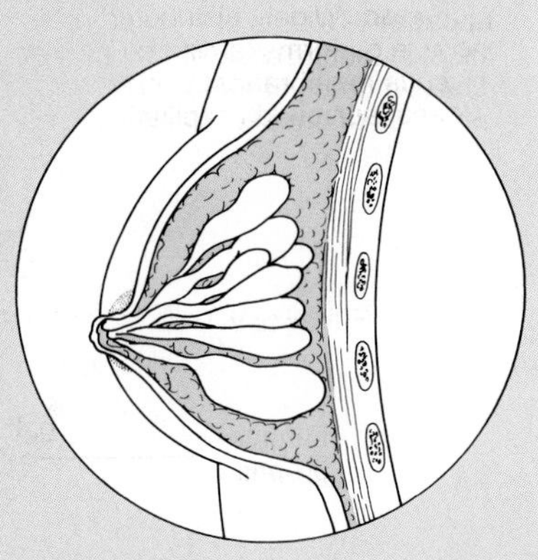

Function: Provides reserve food fuel; insulates against heat loss; supports and protects organs.

Photomicrograph: Adipose tissue from the subcutaneous layer under the skin (500×).

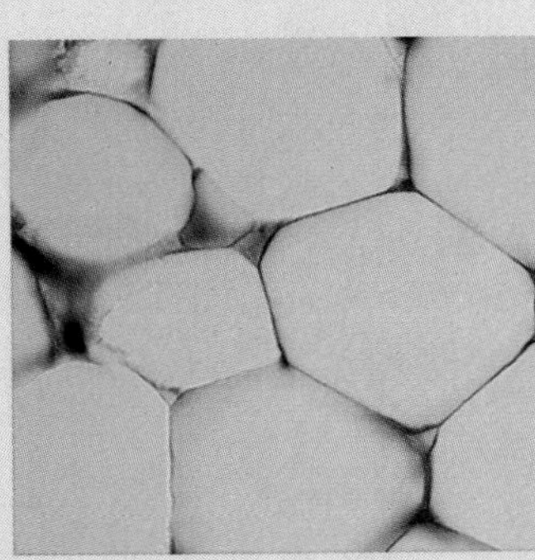

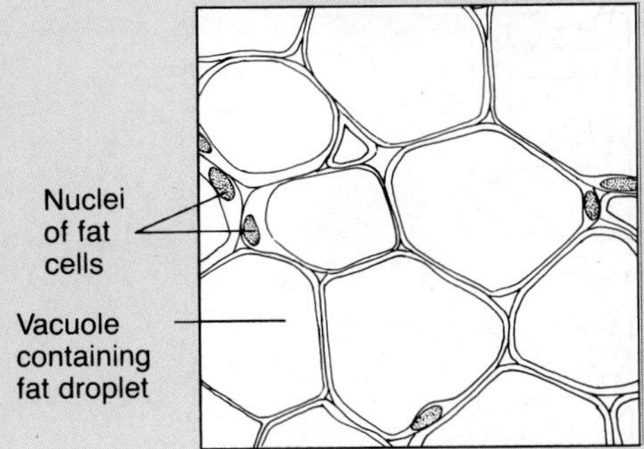

(d) Reticular connective tissue

Description: Network of reticular fibers in a typical loose ground substance; reticular cells predominate.

Location: Liver, lymph nodes, bone marrow, and spleen.

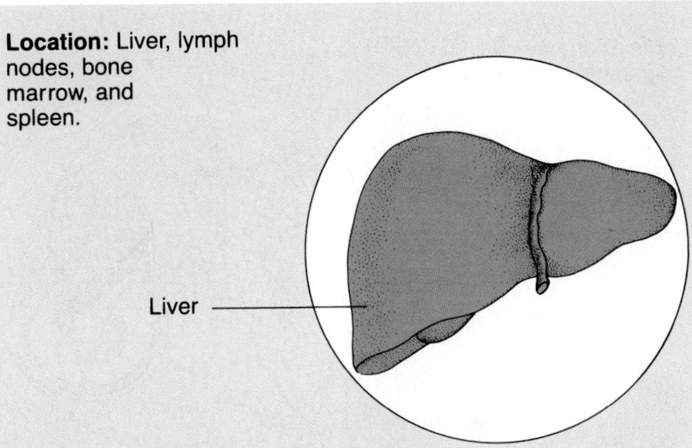

Function: Fibers form a soft internal skeleton that supports other cell types; some reticulocytes are phagocytic and play a role in body protection.

Photomicrograph: Dark-staining network of reticular connective tissue fibers forming the internal skeleton of the spleen (625×).

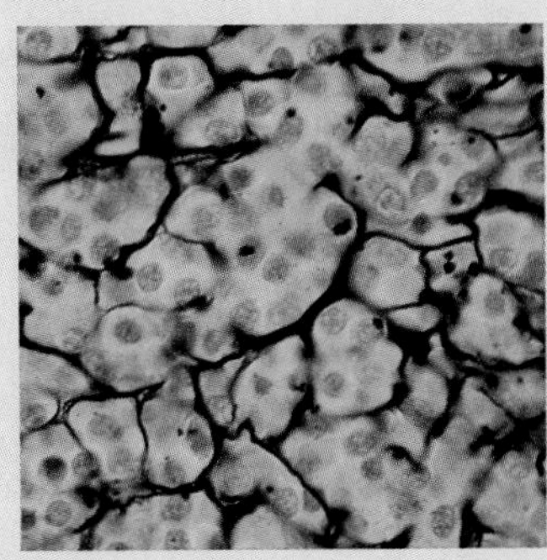

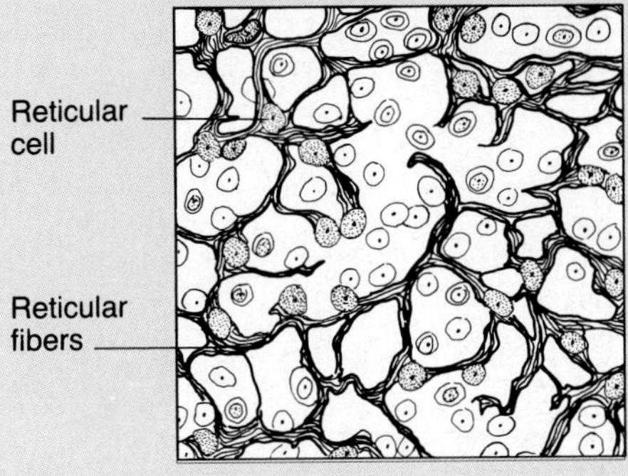

Figure 4.8 (continued)

Connective tissue proper: Dense connective tissue (e to g)

(e) Dense regular connective tissue

Description: Primarily parallel collagen fibers; a few elastin fibers; major cell type is the fibroblast.

Location: Tendons, most ligaments, aponeuroses.

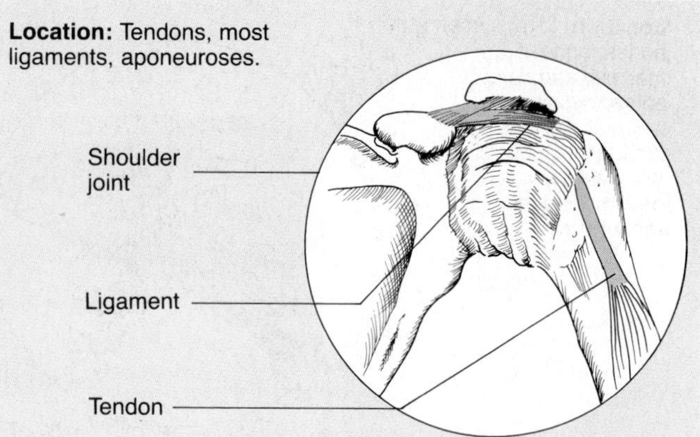

Function: Attaches muscles to bones or to muscles; attaches bones to bones; withstands great tensile stress when pulling force is applied in one direction.

Photomicrograph: Dense regular connective tissue from a tendon (200×).

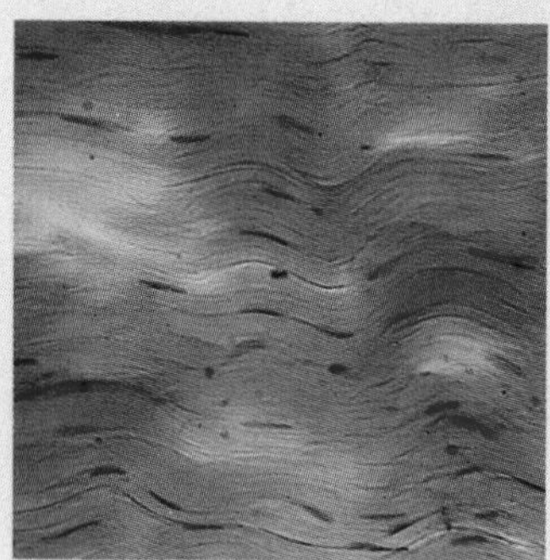

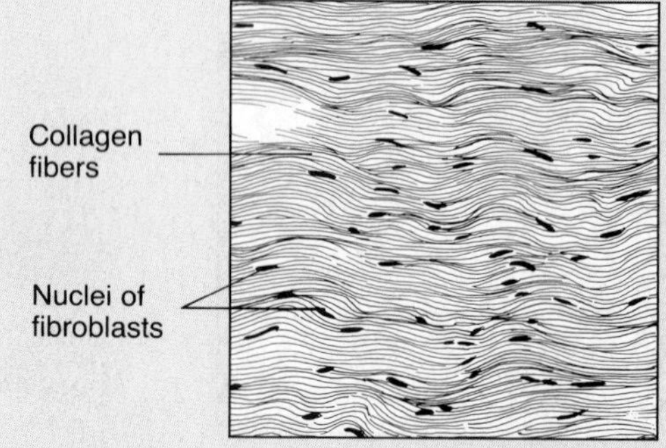

(f) Dense irregular connective tissue

Description: Primarily irregularly arranged collagen fibers; some elastic fibers; major cell type is the fibroblast.

Location: Dermis of the skin; submucosa of digestive tract; fibrous capsules of organs and of joints; fascia.

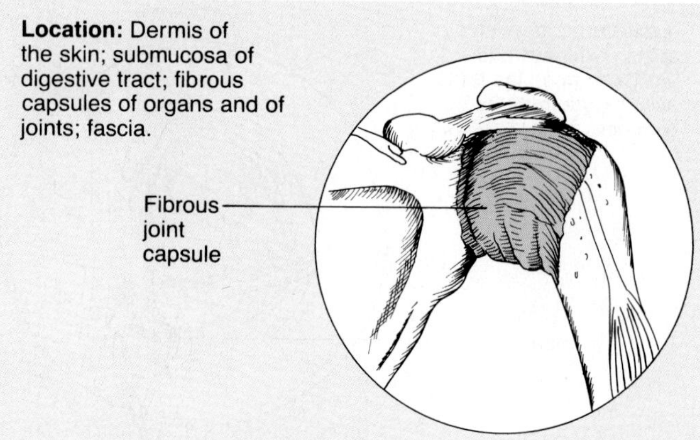

Function: Able to withstand tension exerted in many directions; provides structural strength.

Photomicrograph: Dense irregular connective tissue from the dermis of the skin (475×).

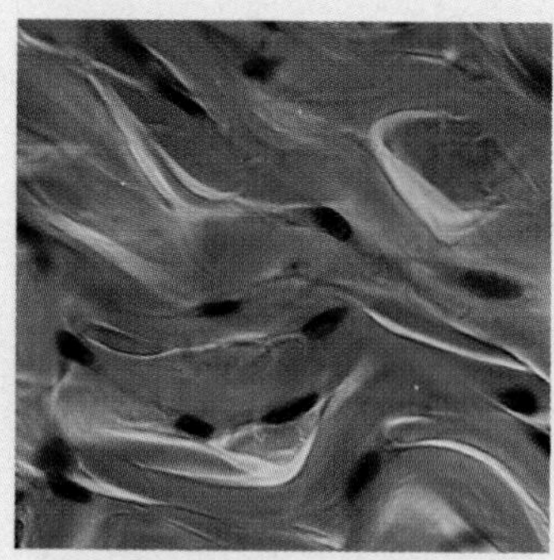

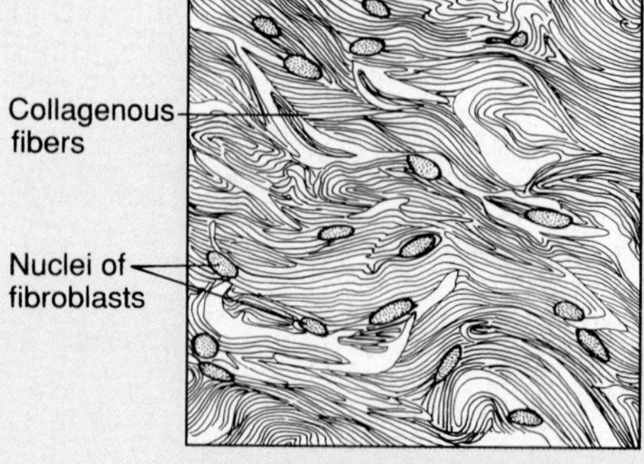

(g) Elastic connective tissue

Description: Same as for the other dense connective tissues, but predominant fiber type is elastin.

Location: Walls of the aorta, some parts of trachea and bronchi; forms the vocal cords and the ligamenta flava connecting the vertebrae.

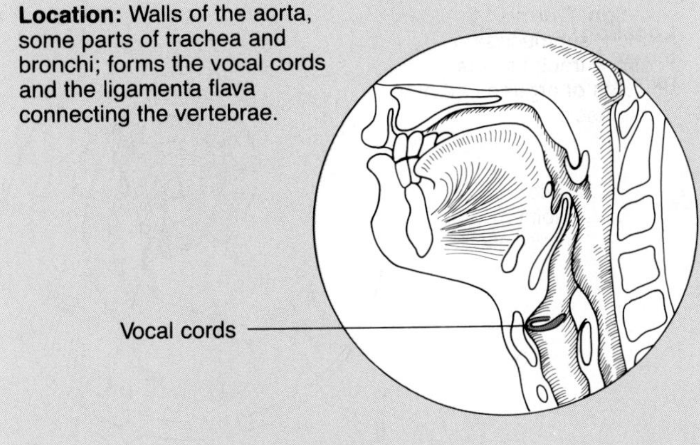

Function: Provides durability with stretch.

Photomicrograph: Elastic connective tissue in a wall of the aorta (190×); notice the wavy appearance of the elastin fibers.

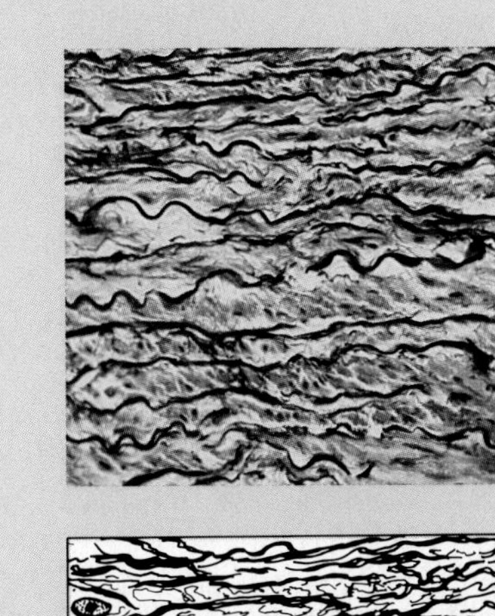

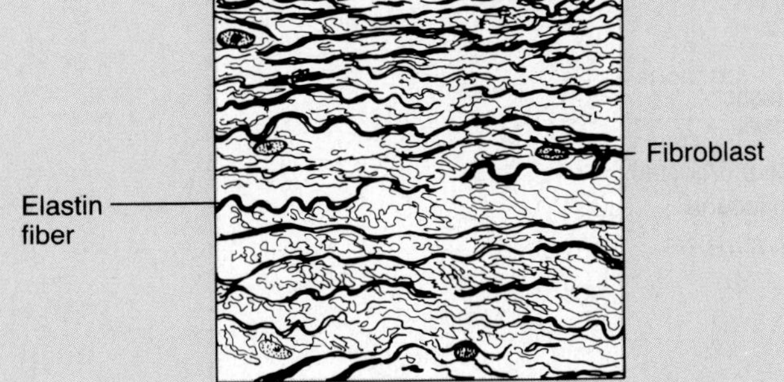

Cartilage: (h to j)

(h) Hyaline cartilage

Description: Amorphous but firm matrix; collagen fibers form an imperceptible network; chondroblasts produce the matrix and when mature (chondrocytes) lie in lacunae.

Location: Forms most of the embryonic skeleton; covers the ends of long bones in joint cavities; forms costal cartilages of the ribs; cartilages of the nose, trachea, and larynx.

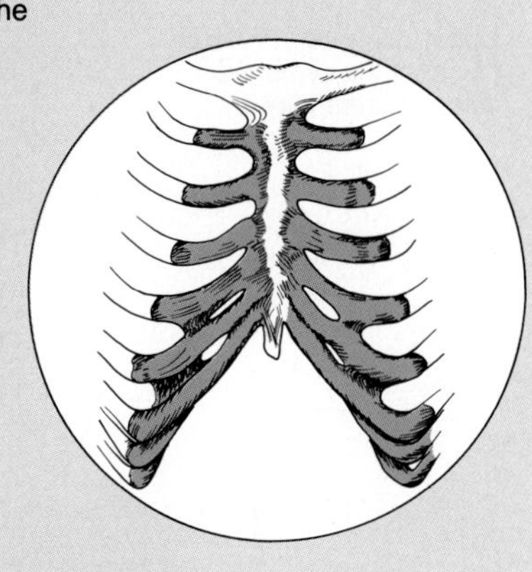

Function: Supports and reinforces; stands up to wear and tear; has resilient cushioning properties; resists compressive stress.

Photomicrograph: Hyaline cartilage from the trachea (475×).

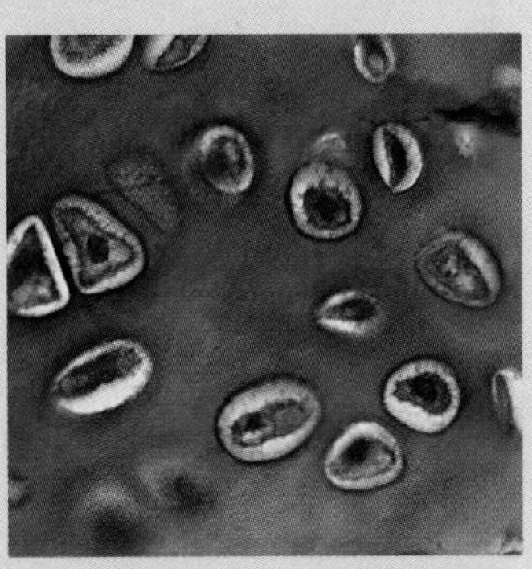

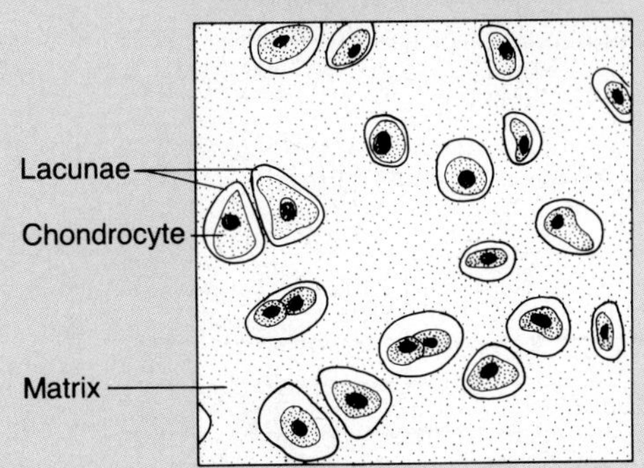

Figure 4.8 (continued)

(i) Fibrocartilage

Description: Matrix similar but less firm than in hyaline cartilage; thick collagen fibers predominate; ligament-like.

Location: Intervertebral discs; pubic symphysis; discs of knee joint.

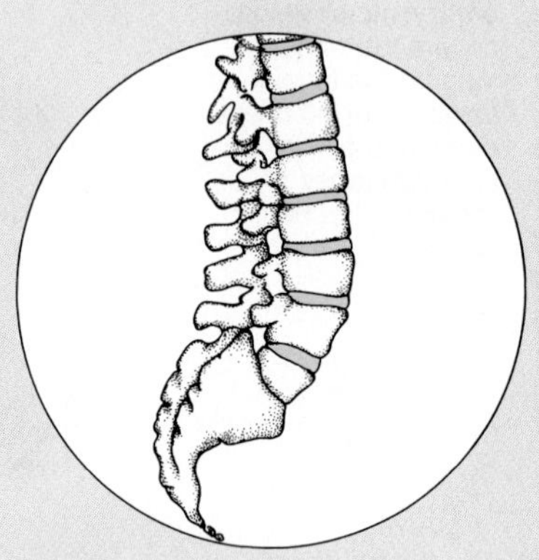

Function: Tensile strength with the ability to absorb compressive shock.

Photomicrograph: Fibrocartilage of an intervertebral disc (500×).

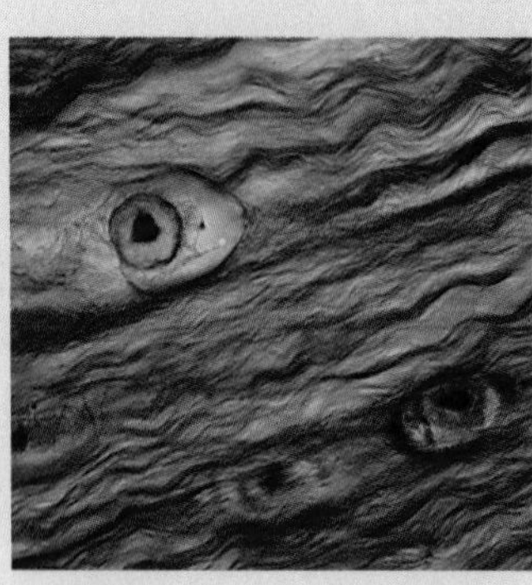

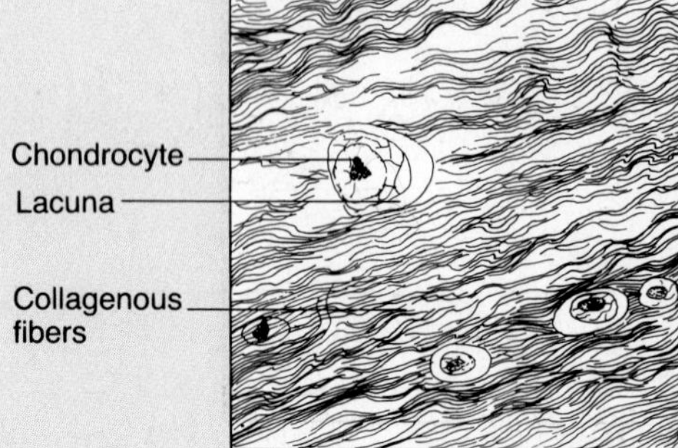

(j) Elastic cartilage

Description: Similar to hyaline cartilage, but more elastic fibers in matrix.

Location: Supports the external ear (pinna); epiglottis.

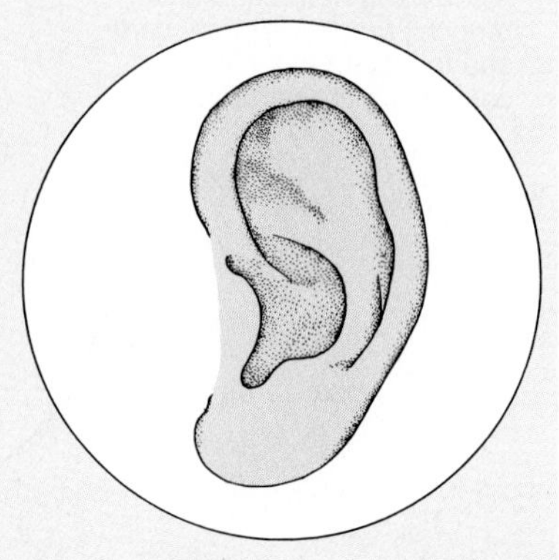

Function: Maintains the shape of a structure while allowing great flexibility.

Photomicrograph: Elastic cartilage from the human ear pinna; forms the flexible skeleton of the ear.

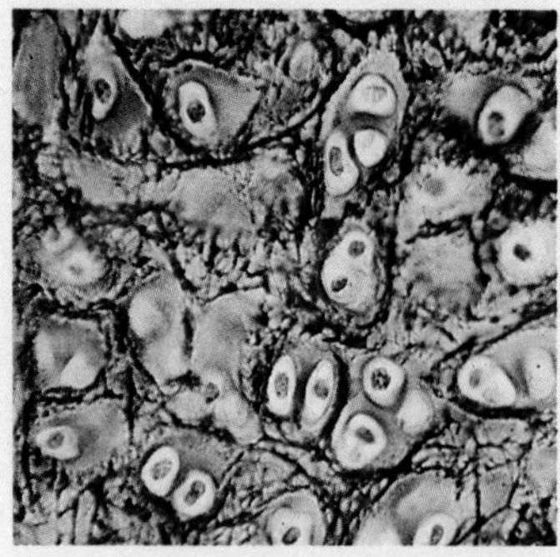

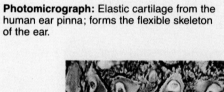

Others: (k and l)

(k) Bone (osseous tissue)

Description: Hard, calcified matrix containing many collagen fibers; osteocytes lie in lacunae. Very well vascularized.

Location: Bones

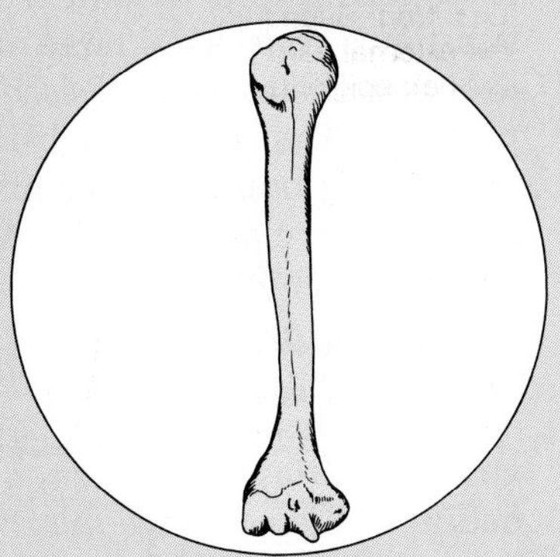

Function: Bone supports and protects (by enclosing); provides levers for the muscles to act on; stores calcium and other minerals and fat; marrow inside bones is the site for blood cell formation (hematopoiesis).

Photomicrograph: Cross-sectional view of bone (100×).

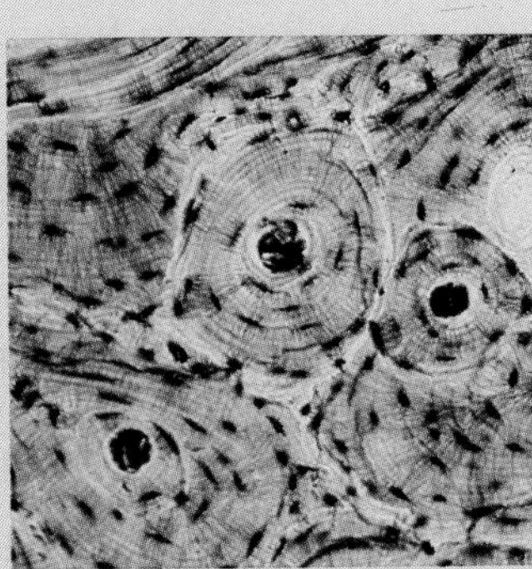

(l) Blood

Description: Red and white blood cells in a fluid matrix (plasma).

Location: Contained within blood vessels.

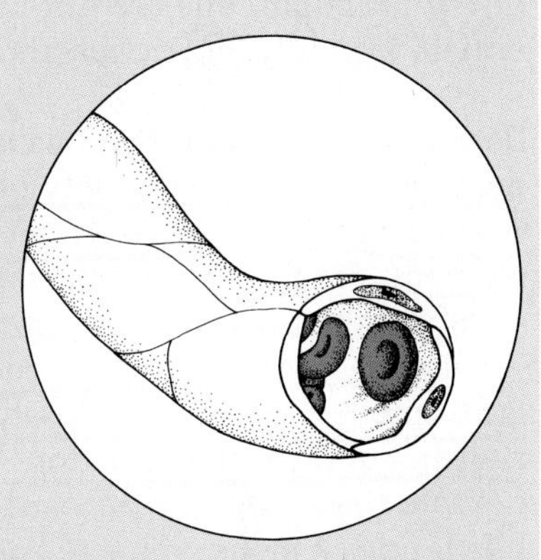

Function: Transport of respiratory gases, nutrients, wastes, and other substances.

Photomicrograph: Smear of human blood (1000×); two white blood cells (neutrophil in upper left and lymphocyte in lower right) are seen surrounded by red blood cells.

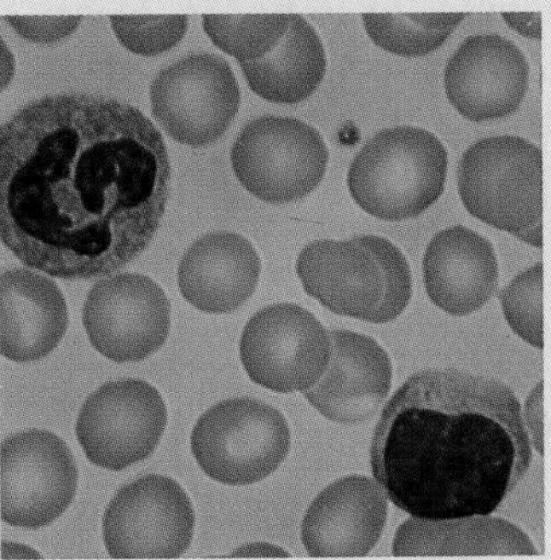

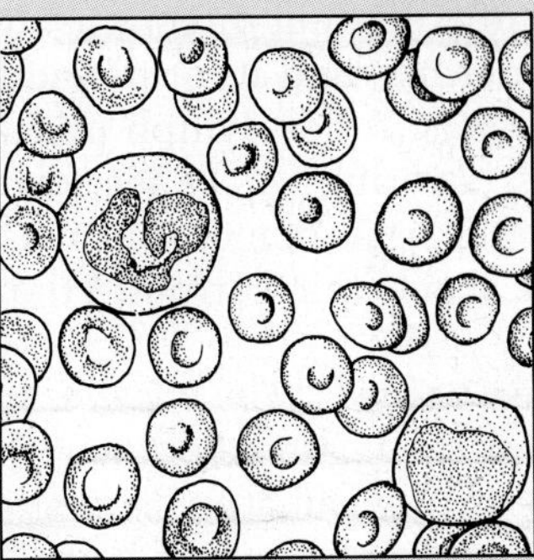

life. Fat cells may even release chemicals into the blood that make you hungry. Obese people have millions of these little "gluttons" screaming for food. Notice, however, that these theories are still controversial! ■

Reticular Connective Tissue. **Reticular connective tissue** consists of a delicate network of interwoven reticular fibers associated with primitive *reticular cells,* which resemble mesenchymal cells (Figure 4.8d). Although reticular fibers are widely distributed in the body, reticular tissue is limited to certain sites. It forms the *stroma,* or internal supporting framework, of lymph nodes, the spleen, bone marrow, and the liver. Some of the reticular cells are fibroblast-like; others differentiate into phagocytic macrophages.

Dense Regular Connective Tissue. **Dense regular connective tissue** is one variety of the dense connective tissues, all of which have fibers as their predominant element. For this reason, the dense connective tissues are often referred to as dense *fibrous* connective tissues.

Dense regular connective tissue contains regularly arranged bundles of closely packed collagen fibers running in the same direction. This results in a white, flexible tissue with great resistance to pulling forces. It is found in areas where tension is always exerted in a single direction. Crowded between the collagen fibers are rows of fibroblasts that continue to form fibers and scant ground substance. As seen in Figure 4.8e, collagen fibers are slightly wavy. This allows the tissue to stretch somewhat, but once the fibers are straightened out, there is no further "give" to this tissue.

With its enormous tensile strength, dense regular connective tissue forms the *tendons,* cords that attach muscles to bones, and flat, sheetlike tendons called *aponeuroses* (a″-pō-noo-rō′-sēs), which attach muscles to other muscles or to bones. Dense regular connective tissue also forms the *ligaments* that bind bones together at joints. Ligaments contain more elastic fibers than do tendons and thus are slightly more stretchy.

Dense Irregular Connective Tissue. **Dense irregular connective tissue** has the same structural elements as the regular variety, but the collagen fibers are interwoven and arranged irregularly, that is, they run in more than one plane (Figure 4.8f). This type of tissue usually forms sheets in body areas where tension is exerted from many different directions. It is found in the skin as the *dermis,* and it forms the fibrous capsules of some organs (testes, lymph nodes, and liver) and the fibrous coverings of bones, cartilages, and nerves. It is also the basis of most **fasciae** (fash′-e-ah), glistening white sheets that surround the muscles.

Elastic Connective Tissue. The vocal cords and some ligaments, such as the *ligamenta flava* (lih-guh-men′-tuh flā′-vuh) connecting adjacent vertebrae, are composed almost entirely of *elastin* fibers. These structures combine strength with elasticity. They yield easily to a pulling force (or pressure) and then recoil to their original length as soon as the tension is released. This dense, fibrous tissue is called *elastic connective tissue* to distinguish it from the dense varieties in which collagen fibers predominate (Figure 4.8g).

Cartilage

Cartilage (kar′-tih-lij) has qualities intermediate between dense connective tissue and bone; it is tough and yet flexible, providing a resilient rigidity to the structures it supports (see the box on p. 128). Cartilage is avascular and devoid of nerve fibers. Its ground substance consists of large amounts of the GAG chondroitin sulfate, as well as hyaluronic acid bound to proteins. The ground substance is heavily invested with firmly bound collagen fibers and, in some cases, reticular or elastic fibers. As a result, the matrix is usually quite firm.

Chondroblasts produce the matrix and are the predominant cell type in cartilage. Their mature forms, **chondrocytes,** are found singly or in small groups within cavities called *lacunae* (luh-koo′-nē). The rigid nature of the cartilage matrix prevents the cells from becoming widely separated. The surfaces of most cartilage structures are surrounded by a well-vascularized dense irregular connective tissue membrane called a *perichondrium* (payr″-ih-kon′-drē-um) (*peri* = around; *chondro* = cartilage), from which the nutrients diffuse through the matrix to the chondrocytes. This mode of nutrient delivery limits cartilage thickness, which is why we have no large cartilage structures.

Cartilages heal slowly when injured—a phenomenon excruciatingly familiar to those experiencing sports injuries. During later life, cartilages tend to calcify or even ossify (become bony). In such cases, the chondrocytes are poorly nourished and die. ■

There are three varieties of cartilage: hyaline cartilage, fibrocartilage, and elastic cartilage.

Hyaline Cartilage. **Hyaline** (hī′-uh-lin) **cartilage,** or *gristle,* is very resistant to wear and tear. Although it contains large amounts of collagen fibers, they are not apparent and the matrix appears amorphous and glassy white (Figure 4.8h).

The most widely distributed cartilage type in the body, hyaline cartilage provides firm support with some pliability. It covers the ends of long bones as *articular cartilage,* providing springy pads that absorb compression stresses at joints. Hyaline cartilage also supports the tip of the nose, connects the ribs to the sternum, and forms most of the larynx and supporting cartilages of the trachea and bronchial tubes. Most of the embryonic skeleton is formed of hyaline cartilage before bone is formed. Hyaline cartilage persists during childhood as the *epiphyseal* (eh-pih″-fih-sē′-ul) *plates,* actively growing regions near the end of long bones that provide for continued growth in length.

Fibrocartilage. The coarse collagenic fibers of **fibrocartilage** are arranged in thin, roughly parallel bundles that give the matrix a grainy fibrous appearance. The chrondrocytes are seen squeezed between the collagen bundles (Figure 4.8i). Fibrocartilage looks quite similar to dense irregular connective tissue. Because it is compressible and resists tension well, fibrocartilage is found where strong support and the ability to withstand heavy pressure are required. For example, the intervertebral discs, which provide resilient cushions between the bony vertebrae, and the spongy cartilages of the knee are fibrocartilage structures.

Elastic Cartilage. Histologically, **elastic cartilage** resembles hyaline cartilage (Figure 4.8j). However, elastic cartilage contains more elastin fibers than other cartilage varieties, which gives this tissue a yellow color in the fresh state. It is found where strength and exceptional ability to stretch are needed. Elastic cartilage forms the "skeletons" of the auditory tubes, the external ear, and the epiglottis. The epiglottis is the flap that covers the opening to the respiratory passageway when we swallow, preventing food or fluids from entering the lungs.

Bone (Osseous Tissue)

Because of its hardness, **bone,** or **osseous** (ah′-sē-us) **tissue,** has an exceptional ability to support and protect softer tissues. Bones of the skeleton also provide cavities for fat storage and synthesis of blood cells. The matrix of bone is similar to that of cartilage, but it is harder and more rigid because bone matrix has far more collagen fibers and deposits of inorganic calcium salts (bone salts).

Osteoblasts produce the organic portion of the matrix; then, bone salts are deposited on and between the fibers. Mature bone cells, or **osteocytes,** reside in the lacunae within the matrix they have made (Figure 4.8k). Unlike cartilage, the next firmest connective tissue, bone is very well supplied by blood vessels, which invade the bone tissue. We will consider the structure and metabolism of bone further in Chapter 6.

Blood

Blood or vascular tissue is considered a connective tissue because it has living cells, called *formed elements* or *blood cells,* surrounded by a fluid matrix called *plasma* (Figure 4.8l). The "fibers" of blood are soluble protein molecules that only become visible during blood clotting. Still, we must recognize that blood is quite atypical as connective tissues go. Blood acts as the transport vehicle for the cardiovascular system, carrying nutrients, wastes, respiratory gases, and many other substances throughout the body. Blood is considered in detail in Chapter 18.

Muscle Tissue

Muscle tissues are highly cellular, well-vascularized tissues that are responsible for most types of body movement. Among the most important characteristics of muscle cells are their elongated shape, which enhances their shortening (contraction) function, and their possession of specialized *myofibrils* (mī″-ō-fī′-brulz), composed of the contractile proteins *actin* and *myosin* (mī′-ō-sin). There are three types of muscle tissue: skeletal, cardiac, and smooth.

Skeletal muscle is packaged by connective tissue sheets into organs called skeletal muscles that are attached to the bones of the skeleton; these muscles form the flesh of the body. As the muscles contract, they pull on bones or skin, causing gross body movements or facial expressions. Skeletal muscle cells are long, cylindrical cells that contain many nuclei. Their obvious banded, or *striated,* appearance reflects the alignment of their myofibrils (Figure 4.9a).

Cardiac muscle makes up the walls of the heart; it is found nowhere else in the body. Its contractions propel blood through the blood vessels to all parts of the body. Like skeletal muscle cells, cardiac muscle cells are striated. However, they differ structurally in that (1) they are uninucleate cells and (2) they are branching cells that fit together tightly at unique junctions called *intercalated* (in-ter′-kuh-lā″-tid) *discs* (Figure 4.9b).

Smooth muscle is so named because no externally visible striations can be seen. Individual smooth muscle cells are spindle-shaped and contain one centrally located nucleus (Figure 4.9c). Smooth muscle occurs in the walls of hollow organs (digestive and urinary tract organs, uterus, and blood vessels). It generally acts to propel substances through the organ by alternately contracting and relaxing.

Since skeletal muscle contraction is under our conscious control, skeletal muscle is often called *voluntary muscle,* while the other two types are called *involuntary muscle.* Skeletal and smooth muscle are described in detail in Chapter 9; cardiac muscle is discussed in Chapter 19.

Figure 4.9 Muscle tissues.

(a) Skeletal muscle

Description: Long, cylindrical, multinucleate cells; obvious striations.

Location: In skeletal muscles attached to bones or occasionally to skin.

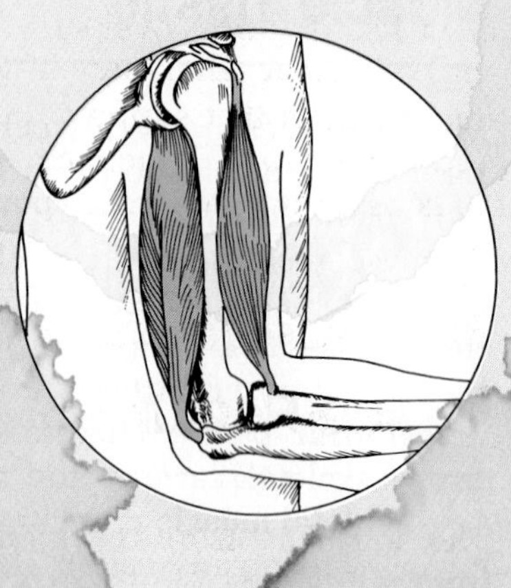

Function: Voluntary movement; locomotion; manipulation of the environment; facial expression. Voluntary control.

Photomicrograph: Skeletal muscle (approx. 300x). Notice the obvious banding pattern and the fact that these large cells are multinucleate.

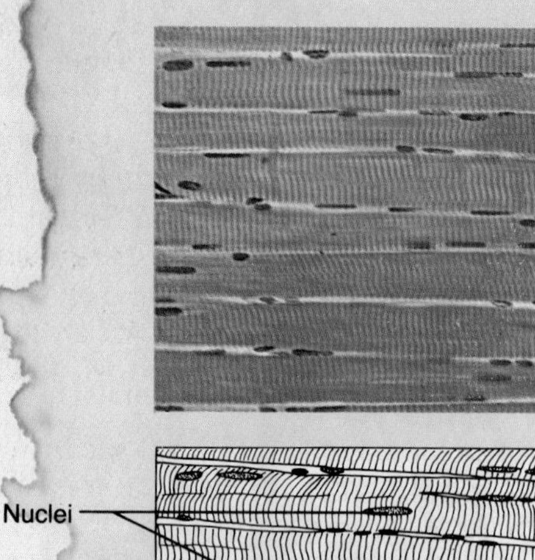

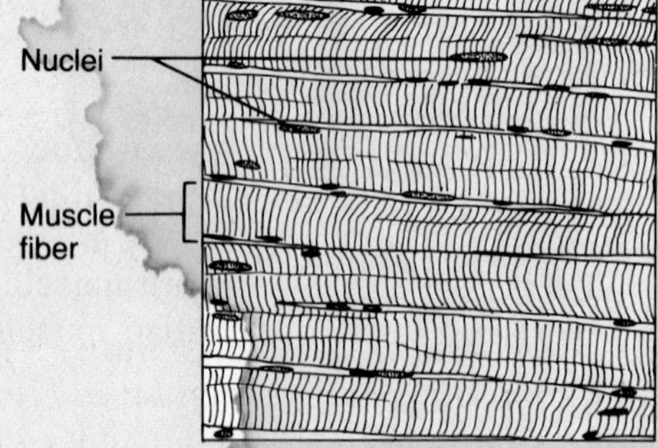

(b) Cardiac muscle

Description: Branching, striated, generally uninucleate cells that interdigitate at specialized junctions (intercalated discs).

Location: The walls of the heart.

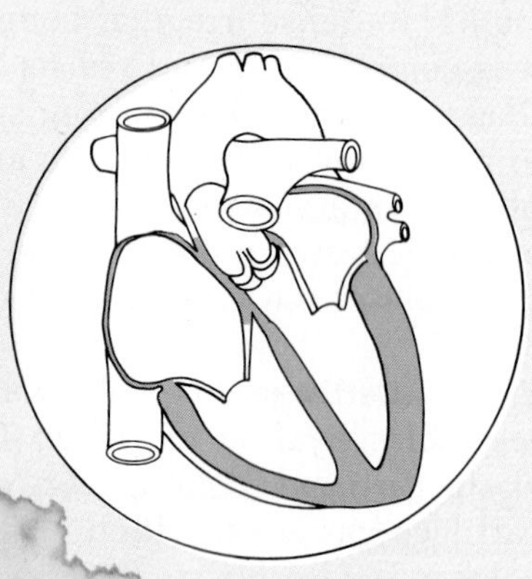

Function: As it contracts, it propels blood into the circulation; involuntary control.

Photomicrograph: Cardiac muscle (300×); notice the striations, branching of fibers, and the intercalated discs.

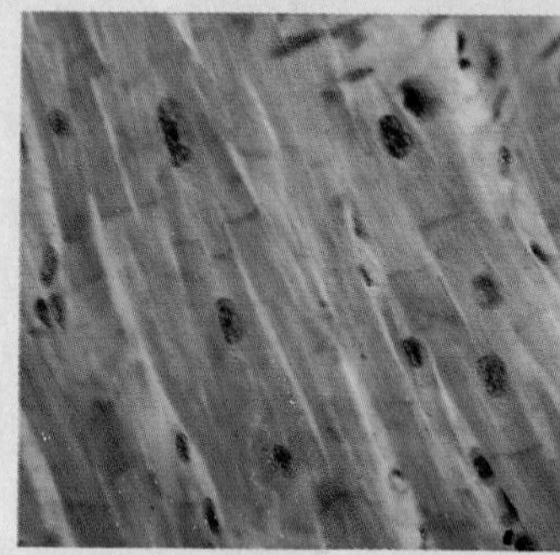

Inter-
calated
disc

Nucleus

(c) Smooth muscle

Description: Spindle-shaped cells with central nuclei; cells arranged closely to form sheets; no striations.

Location: Mostly in the walls of hollow organs.

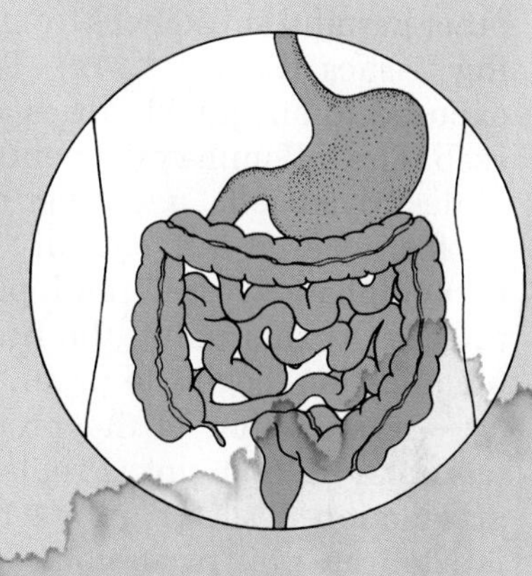

Function: Propels substances or objects (foodstuffs, urine, a baby) along internal passageways.

Photomicrograph: Sheet of smooth muscle (approx. 800x).

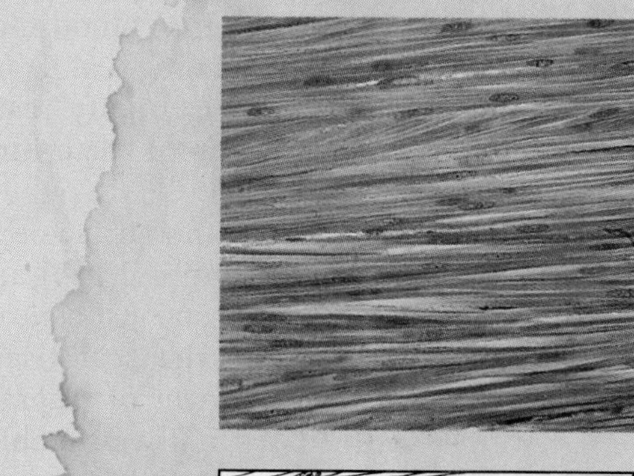

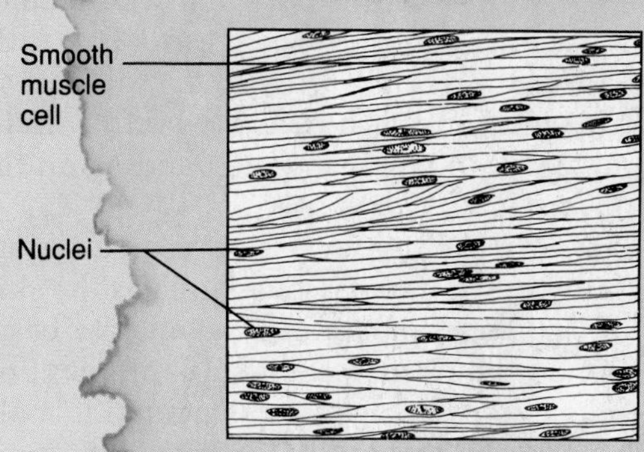

Nervous Tissue

Nervous tissue makes up the nervous system: the brain, spinal cord, and nerves that conduct impulses to and from the various body organs. It is composed of two major cell types. **Neurons** are highly specialized cells that generate and conduct nerve impulses (Figure 4.10). Typically, they are branching or stellate cells; their cytoplasmic extensions allow them to conduct the electrical impulses over substantial distances within the body. **Neuroglial** (ner-ah′-glē-ul) **cells** are nonconducting cells that support, insulate, and protect the delicate neurons. A more complete discussion of nervous tissue appears in Chapter 11.

Tissue Repair

The body has many techniques for protecting itself from uninvited "guests" or injury. Intact mechanical barriers such as the skin and mucosae, the ciliary activity of epithelial cells lining the respiratory tract, and a strong acid (chemical barrier) produced by stomach glands represent three defenses exerted at the local tissue level. When tissue injury does occur, it stimulates the body's inflammatory and immune responses. The inflammatory response is a relatively nonspecific reaction that occurs whenever and wherever tissues are injured. Essentially, inflammation represents an offensive action on the part of the body to eliminate the injurious agent, prevent further injury, and restore the tissue to a healthy condition. The immune response, on the other hand, is extremely specific. Cells of the immune system, programmed to identify foreign substances such as microorganisms, toxins, and cancer cells, mount a vigorous attack against recognized invaders, either by interacting directly with them or by releasing antibodies. The inflammatory and immune responses are considered in detail in Chapter 22.

Tissue repair occurs in two major ways: by regeneration and by fibrosis. Which of these occurs depends on (1) the type of tissue damaged and (2) the severity of the injury. **Regeneration** is replacement of destroyed tissue by proliferation of the same kind of cells, whereas **fibrosis** involves proliferation of fibrous connective tissue, that is, scar tissue formation. In most tissues, repair involves both activities.

Figure 4.11 illustrates the process of tissue repair of a skin wound, showing both regeneration and fibrosis. The process begins while the inflammatory reac-

Figure 4.10 Nervous tissue.

Description: Branching cells; cytoplasmic extensions that may be quite long extend from the nucleus-containing cell body.

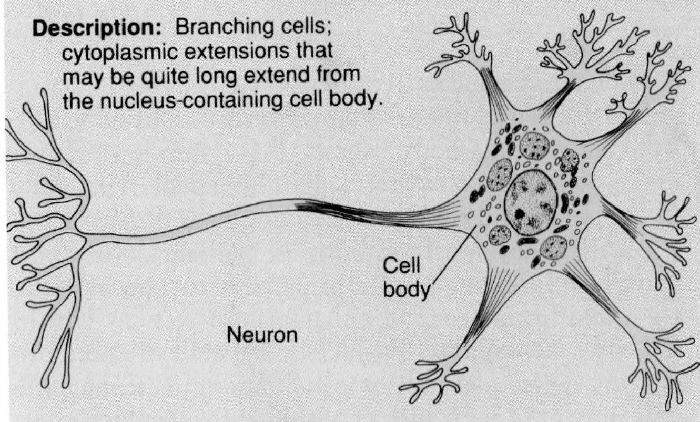

Location: Brain, spinal cord, and nerves.

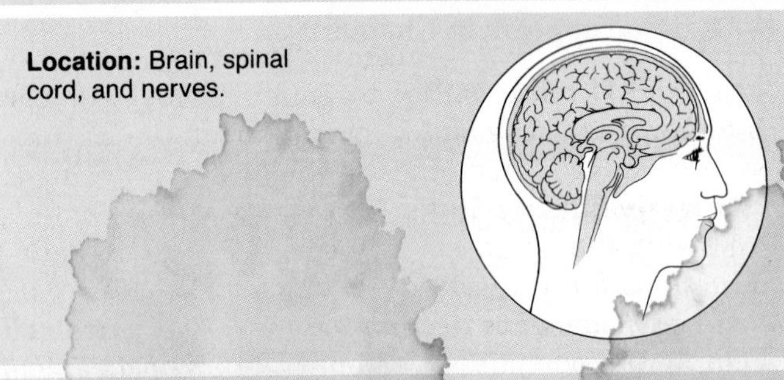

Function: Transmit electrical signals from sensory receptors and to effectors (muscles and glands) which control their activity.

Photomicrograph: Neuron (170×).

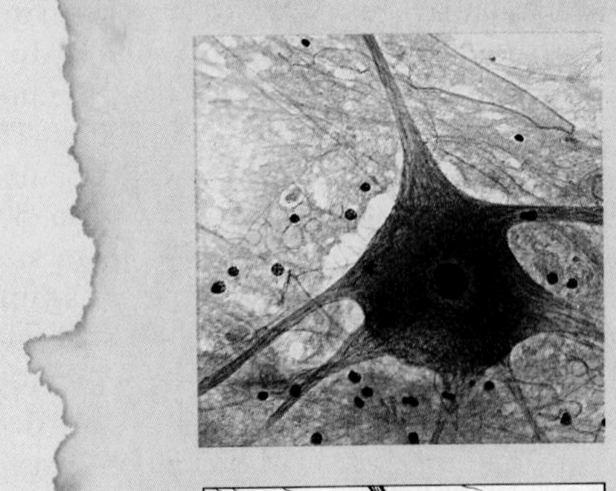

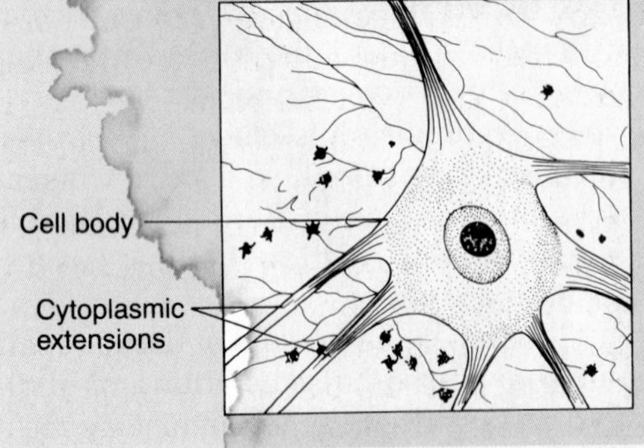

tion is still going on. Since the inflammatory reaction sets the stage for repair, it is necessary to note briefly what has happened up to this point. Tissue injury sets the following events into motion. First, the capillaries dilate and become very permeable, which allows white blood cells and plasma fluid rich in clotting proteins, antibodies, and other substances to seep into the injured area. Then the leaked clotting proteins construct a clot, which stops the loss of blood, holds the edges of the wound together, and effectively "walls off," or isolates, the injured area, preventing bacteria, toxins, or other harmful substances from spreading to surrounding tissues (Figure 4.11a). The portion of the clot exposed to air quickly dries and hardens, forming a scab. The inflammatory events leave excess fluid, bits of destroyed cells, and other debris in the area. Most of this material is eventually removed from the area by lymphatic vessels and phagocytized by macrophages. At this point, the first step of tissue repair, organization, begins to occur.

Organization is the process during which the temporary blood clot is replaced by the ingrowth of granulation tissue (Figure 4.11b). Granulation tissue is a delicate pink tissue composed of several elements. Thin, extremely permeable capillaries bud from undamaged capillaries and enter the damaged area, laying down a new capillary bed; they protrude nublike from the surface of the granulation tissue, giving it a granular appearance. These capillaries are fragile and bleed freely, as when a scab is picked away from a skin wound. Also present in granulation tissue are scattered macrophages and large, immature fibroblasts that synthesize new collagen fibers to permanently bridge the gap. As organization continues, macrophages digest and remove the original blood clot. The granulation tissue, destined to become scar tissue (a permanent fibrous tissue patch), is highly resistant to infection, owing to its secretion of bacterium-inhibiting substances.

While organization is going on, the surface epithelium begins to regenerate (see Figure 4.11b) and makes its way across the granulation tissue just beneath the scab, which soon detaches. As the scar tissue beneath matures and contracts, the layer of regenerating epithelium thickens until it finally resembles that of the adjacent skin. The final result is a fully regenerated epithelial surface (Figure 4.11c), and an underlying area of fibrosis (the scar), which is invisible or visible as a thin white line, depending on the severity of the wound.

The ability of the different tissue types to regenerate varies widely. Epithelial tissues such as the skin epidermis and mucous membranes regenerate beautifully. So, too, do the fibrous connective tissues and bone. But skeletal muscle regenerates poorly, if at all, and cardiac muscle and nervous tissue within the brain and spinal cord can be replaced only by scar tissue formation.

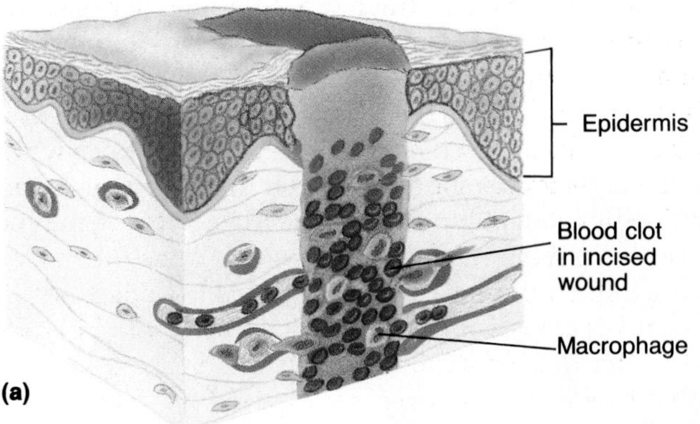

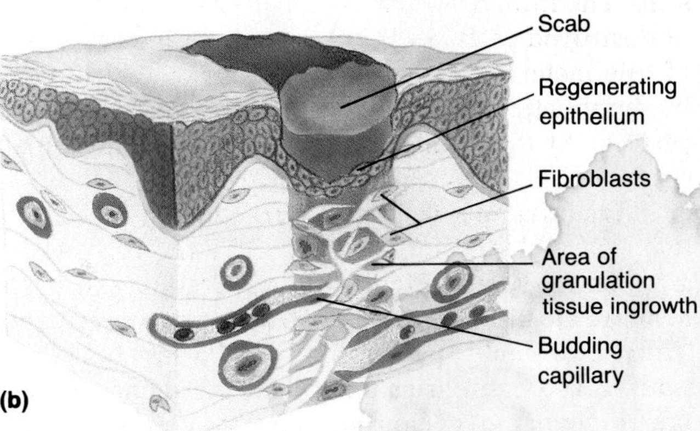

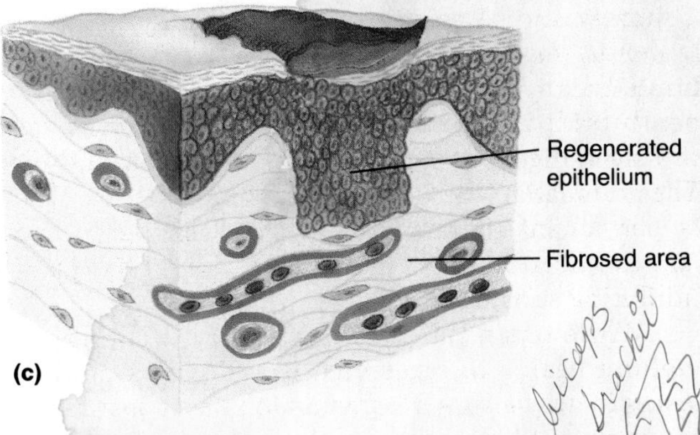

Figure 4.11 Tissue repair of a nonextensive skin wound: regeneration and fibrosis. (**a**) Severed blood vessels bleed. Inflammatory chemicals are released, and local blood vessels dilate and become more permeable, allowing white blood cells, fluid, clotting proteins, and other plasma proteins to invade the injured site. Clotting proteins initiate clotting; surface dries and forms a scab. (**b**) Granulation tissue is formed. Capillary buds invade the clot, restoring a vascular supply. Fibroblasts invade the region and secrete soluble collagen, which constructs collagen fibers that bridge the gap. Macrophages phagocytize dead and dying cell debris. Surface epithelial cells proliferate and migrate over the granulation tissue. (**c**) Approximately one week later, the fibrosed area (scar) has contracted, and regeneration of the epithelium is in progress. The scar tissue may or may not be visible beneath the epidermis.

In exceptionally severe wounds or in nonregenerating tissues, collagen formation (fibrosis) continues and totally replaces the lost tissue. Over a period of months, the mass of fibrous tissue contracts and becomes more and more compact. The scar formed appears as a pale, often shiny area. It is composed mostly of collagen fibers and contains few cells or capillaries. Scar tissue is very strong, but it lacks the flexibility and elasticity of most normal tissues. Additionally, it cannot perform the normal functions of the tissue it has replaced.

If scar tissue forms in the wall of the bladder, heart, or other muscular organ, it may severely hamper organ function. The normal shrinking of scar tissue eventually reduces the size and internal volume of the organ and may cause abnormal flow patterns through hollow organs. Scar tissue not only hampers muscle's ability to contract, but also may interfere with its normal excitation by the nervous system. In the heart, these problems may result in progressive heart failure. In some cases, particularly following abdominal surgery, scar tissue bands called *adhesions* form, which may connect adjacent organs together. Adhesions in the abdominal cavity can prevent the normal shifting about of loops of the intestine, dangerously obstructing the flow of substances through it. Other possible consequences of adhesion formation include restriction of heart movements and joint immobility. ■

Developmental Aspects of Tissues

One of the first events of embryonic development is the formation of the three **primary germ layers**, which lie one atop the next like a three-layered cellular pancake. From superficial to deep, these layers are the **ectoderm**, **mesoderm** (meh′-zō-derm), and **endoderm** (Figure 4.12). These primary germ layers then begin to specialize to form the four primary tissues from which all body organs are derived. Epithelial tissues are formed from all three germ layers: the mucosae from endoderm; endothelium and mesothelium from mesoderm; and the epidermis from ectoderm. Muscle and connective tissue develop from mesoderm, and nervous tissue arises from ectoderm.

By the end of the second month of development, the primary tissues have developed and most organs are laid down, at least in rudimentary form. In general, tissue cells remain mitotic and provide for the rapid growth that occurs throughout the embryonic and fetal periods. Most tissue cells, except neurons,

A CLOSER LOOK Cartilage—Strong Water

Cartilage forms the framework for building most bones, provides growth plates for continued long bone growth in children, and fashions spongy cushions between the bones. Throughout life, broken bones are repaired by cartilage before new bone tissue is in place—much like replaying a tape of embryonic events.

Although embryonic cartilage-forming tissue is initially sprayed with capillaries, these disappear by the fourth day of development, so that the increasingly crowded cartilage cells have no direct supply of oxygen and nutrients. Low cell density, high oxygen content, and high levels of vitamin B_3 have actually been shown to *inhibit* cartilage formation. What a strange tissue this is. Crowding and oxygen and nutrient deprivation are necessary if its cells are to become cartilage cells!

The cartilages of professional tennis players, like Ivan Lendl, should be well nourished because of the tremendous physical demands placed on body joints by this sport.

Cartilage performs its crucial functions even though it is very unlike other body tissues. It has no nerves, blood vessels, or lymphatics. The properties of cartilage are determined not by its cells but by the elaborate network of giant molecules the cells deposit around themselves. Its matrix includes proteoglycans (prō″-tē-ō-glī′-kanz″), some of the largest modified proteins made by any living cells (proteins attached to large polysaccharide molecules, or GAGs), abundant collagen fibers, and huge volumes of water. Indeed, the structuring of water is responsible for this tissue's functional characteristics. The collagen fibers in cartilage form a framework much like the steel girders supporting a bridge. Within this mesh are the central organizing molecules of the matrix, the hyaluronic acid molecules. Up to 100 glycoproteins are attached to and extend from each hyaluronic acid molecule; around these cores are proteoglycans. The sugars and negative charges of these gigantic proteoglycan molecules avidly attract water molecules, causing them to form interacting "water shells" around them. Thus, proteoglycans structure massive amounts of water—many times their own weight—and water becomes the chief component of cartilage.

In the developing embryo, this contained water "saves space" for the development of the long bones of the body, and incredible as it seems, the resiliency, or springiness, of cartilage results directly from its water-storing property. When pressure is applied to cartilage, water is forced away from the negatively charged areas, but then as the charged areas are pressed closer and closer together, they tend to repel one another and resist further compression. When the pressure is released, water molecules rush back to their original positions. Pressure also plays a vital role in the nourishment of joint cartilage, which receives nutrients from the flow of liquids that occurs when it is alternately compressed and released. This explains the observation that long periods of inactivity can weaken joint cartilage.

Like all body cells, chondrocytes age and seem to respond to some programmed biological clock. Proteoglycans produced by aging chondrocytes are distinctly different from those of young cells: They may account for some forms of osteoarthritis in elderly people, in which the cartilage thins and becomes less resilient. If the pressure on the joint exceeds the ability of the aging cartilage to respond, the cartilage can fragment, which can lead to the swelling and pain of osteoarthritis. The aging chondrocyte theory raises the possibility that arthritis could be prevented or reversed by transplanting young chondrocytes to joint surfaces.

The unique properties of cartilage raise some other interesting research prospects. Tumors require a rich blood supply and are known to release a factor that causes blood vessels to proliferate at the site. Viable cartilage, on the other hand, contains a substance that specifically excludes blood vessels. Could this substance be applied to a growing tumor to throttle its blood supply and kill it? We do not yet know, but the race to isolate the blood vessel–inhibiting factor is now on.

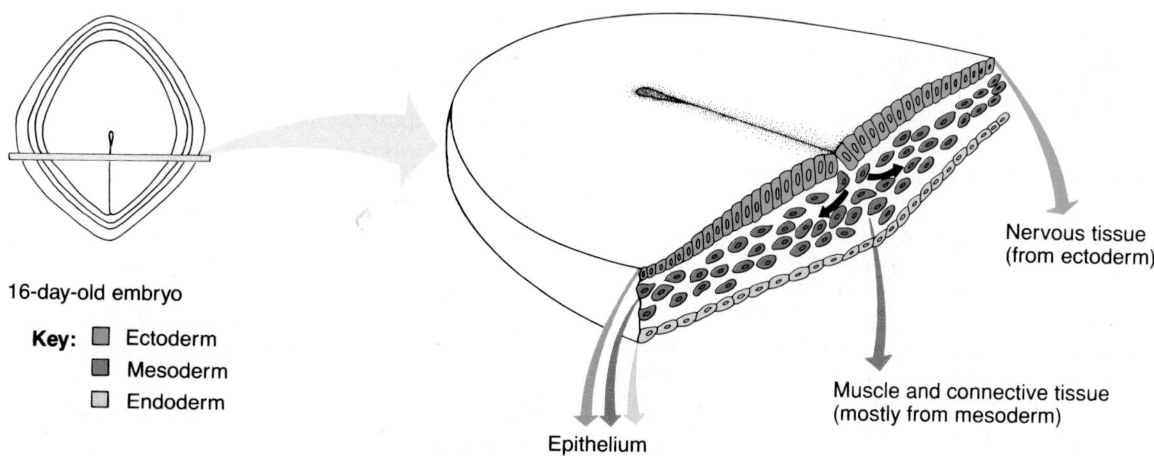

Figure 4.12 Embryonic germ layers and the primary tissue types they produce. The three embryonic layers collectively form the very early embryonic body.

continue to undergo mitosis until adult body size is achieved, after which only epithelial and connective tissues are routinely mitotic. Given good nutrition, good circulation, and relatively infrequent encounters with inflammatory agents, our tissues normally maintain optimal health and function efficiently through youth and middle age. The importance of nutrition for tissue health cannot be overemphasized. For example, vitamin A is needed for normal epithelial regeneration; vitamin C is essential for synthesis of the intercellular cement and collagen (thus for connective tissue integrity in general); and since proteins are the structural material of the body, adequate intake of high-quality protein is absolutely essential if the tissues are to retain their structural integrity.

With increasing age, epithelial sheets become thin and are more easily breached; the amount of collagen in the body declines, making tissue repair less efficient; and bone, muscle, and nervous tissues begin to atrophy. These events are due in part to decreased circulatory efficiency, which reduces nutrient delivery to the tissues, but in some cases, diet is a contributing factor. As income declines or as chewing becomes more difficult, older people tend to eat soft foods, which may be low in protein and vitamins. As a result, tissue health suffers.

* * *

As we have seen, body cells combine to form four discrete tissue types: epithelial, connective, muscle, and nervous tissues. The cells making up each of these tissues share certain features but are by no means identical. They "belong" together because they have basic functional similarities. The connective tissues assume many guises, but perhaps the most versatile cells are those of epithelium: They protect our outer and inner surfaces, permit us to obtain oxygen, allow our kidneys to excrete wastes, and absorb vital nutrients into the blood. The important concept to carry away with you is that tissues, despite their unique abilities, cooperate to keep the body safe, healthy, and whole.

Related Clinical Terms

Fat embolism Freely floating fat globule in the bloodstream; may result from extensive injury to fatty subcutaneous tissue or a bone fracture in which the fat in the bone marrow cavity is released; the danger is that the fatty masses may obstruct circulation to vital organs (heart, lungs, brain).

Healing by first intention The simplest type of healing; occurs when the edges of the wound are brought together by sutures, staples, or other means used to close surgical incisions; only small amounts of granulation tissue need be formed.

Healing by second intention The wound edges remain separated, and the gap is bridged by granulation tissue; the manner in which unattended wounds heal (slower than in healing by first intention); more granulation tissue is formed, and a greater degree of epithelial proliferation occurs than in wounds in which the edges are brought together; larger scars result.

Keloid (kē′-loyd) Abnormal proliferation of connective tissue during healing of skin wounds; results in large, unsightly mass of scar tissue at the skin surface.

Proud flesh Excessive granulation tissue formation such that it protrudes above the surrounding surfaces; common in large skin wounds; may have to be removed surgically to allow normal epithelial regeneration to occur.

Pus A collection of tissue fluid, bacteria, dead and dying tissue cells, white blood cells, and macrophages in an inflamed area.

Scurvy A nutritional deficiency caused by lack of adequate vitamin C needed to synthesize collagen; signs and symptoms include blood vessel disruption, delay in wound healing, weakness of fibrous scar tissue, and loosening of teeth.

Suction lipectomy (*lip* = fat: *ectomy* = removal) A surgical procedure in which subcutaneous fat is suctioned from fatty areas of the body, such as the thighs, abdomen, buttocks, and breasts; largely a cosmetic procedure that does not result in permanent fat reduction.

Chapter Summary

Cells of the multicellular body aggregate to form tissues, collections of similar cells that are specialized to perform a specific function for the body. The four primary tissue types are epithelial, connective, muscle, and nervous tissues.

EPITHELIAL TISSUE (pp. 101–112)

1. Epithelial tissue is the covering, lining, and glandular tissue of the body. Its functions include protection, absorption, secretion, and filtration.

Special Characteristics of Epithelium (pp. 101–102)

2. Epithelial cells exhibit a huge degree of cellularity, specialized lateral contacts, polarity, avascularity, a basement membrane, and high regenerative capacity.

Classification of Epithelia (pp. 102–108)

3. Epithelium is classified by cell shape as squamous, cuboidal, or columnar and by arrangement as simple (one layer) or stratified (more than one). The terms denoting cell shape and arrangement are combined to describe the epithelium fully.

4. Simple squamous epithelium consists of a simple layer of squamous cells. Highly adapted for filtration and exchange of substances. Forms walls of air sacs of the lungs, lines blood vessels, and contributes to serosae.

5. Simple cuboidal epithelium is commonly active in secretion and absorption. Found in glands and in kidney tubules.

6. Simple columnar epithelium, specialized for secretion and absorption, consists of a single layer of tall columnar cells that often exhibit microvilli and goblet cells. Lines most of the digestive tract.

7. Pseudostratified columnar epithelium is a simple columnar epithelium that appears stratified. Its ciliated variety, rich in goblet cells, lines most of the upper respiratory passages.

8. Stratified squamous epithelium is multilayered; cells at the free edge are squamous. It is adapted to resist abrasion. It lines the esophagus; its keratinized variety forms the skin epidermis.

9. Stratified cuboidal epithelium is found chiefly in ducts of large glands.

10. Stratified columnar epithelium is rare in the body; commonly appears at junctions between other epithelial types.

11. Transitional epithelium is a modified stratified squamous epithelium. Adapted for responding to stretch; lines hollow urinary system organs.

Covering and Lining Epithelia (pp. 108–109)

12. Endothelium is a simple sheet of squamous epithelium. It forms the lining of all hollow circulatory system organs.

13. Epithelial membranes are simple organs, consisting of an epithelium bound to an underlying connective tissue layer. Include the mucosae, serosae, and the cutaneous membrane.

Glandular Epithelia (pp. 110–112)

14. A gland is one or more cells specialized to secrete a particular product.

15. On the basis of route of secretion and type of product, glands` are classified as exocrine or endocrine. Glands are classed structurally as unicellular or multicellular.

16. Multicellular exocrine glands are classified according to duct structure as simple or compound, and according to the structure of their secretory parts as tubular, alveolar, and tubuloalveolar.

17. Multicellular exocrine glands are classified functionally as merocrine, holocrine, or apocrine.

CONNECTIVE TISSUE (pp. 112–123)

1. Connective tissue is the most abundant and widely distributed tissue of the body. Its functions include support, protection, binding, insulation, and transportation (blood).

Common Characteristics of Connective Tissue (pp. 112–113)

2. Connective tissues originate from embryonic mesenchyme and exhibit matrix. Depending on type, a connective tissue may be well vascularized (most), poorly vascularized (dense connective tissues), or avascular (cartilage).

Structural Elements of Connective Tissue (pp. 113–114)

3. The structural elements of all connective tissues are matrix and cells.

4. Matrix consists of ground substance and fibers. It may be fluid, gel-like, or firm.

5. Each connective tissue type has a primary cell type that can exist as a mitotic, matrix-secreting cell (blast) or as a mature cell (cyte) responsible for maintaining the matrix. The chief cell type of connective tissue proper is the fibroblast; that of cartilage is the chondroblast; that of bone is the osteoblast; and that of blood-forming tissue is the hemocytoblast.

Types of Connective Tissue (pp. 114–123)

6. Embryonic connective tissue is called mesenchyme.

7. Connective tissue proper consists of loose and dense varieties. The loose connective tissues are:

- areolar: semifluid ground substance; all three fiber types loosely interwoven; contains a variety of cells; forms a soft packing around body organs and lamina propria.
- adipose: consists largely of adipocytes; scant matrix; insulates and protects body organs; provides reserve energy fuel.
- reticular: finely woven reticular fibers in soft ground substance; the stroma of lymphoid organs and bone marrow.

8. Dense connective tissue proper includes:

- dense regular: dense parallel bundles of collagen fibers; few cells, little ground substance; high tensile strength; forms tendons, ligaments, aponeuroses.
- dense irregular: like regular variety, but fibers are arranged in different planes; resists tension exerted from many different directions; forms the dermis of the skin and fascia.
- elastic: largely elastin fibers; has both regular and irregular forms; found in elastic ligaments and blood vessels.

9. Cartilage exists as:

- hyaline: firm ground substance containing collagen fibers; high resistance to wear; found in fetal skeleton, at articulating surfaces of bones, and trachea; most abundant type.

- fibrocartilage: coarse parallel collagen fibers; provides support with compressibility; forms intervertebral discs and knee cartilages.
- elastic cartilage: elastic fibers predominate; provides flexible support of the external ear and epiglottis.

10. Bone (osseous tissue) consists of a firm, collagen-containing matrix embedded with calcium salts, which confer rigidity; forms the body skeleton.

11. Blood consists of red and white blood cells in a fluid matrix (plasma).

MUSCLE TISSUE (pp. 123–125)

1. Muscle tissue consists of elongated cells specialized to contract and cause movement.

2. Based on structure and function, there are:

- skeletal muscle: attached to and moves the bony skeleton.
- cardiac muscle: forms the walls of the heart; pumps blood.
- smooth muscle: in the walls of hollow organs; propels substances through the organs.

NERVOUS TISSUE (p. 125)

1. Nervous tissue forms organs of the nervous system. It is composed of neurons and supportive neuroglial cells.

2. Neurons are branching cells that receive and transmit electrical impulses; involved in body regulation.

TISSUE REPAIR (pp. 125–127)

1. Inflammation is the body's response to injury. Tissue repair begins during the inflammatory process. It may lead to regeneration, fibrosis, or both.

2. Tissue repair begins with organization, during which the blood clot is replaced by granulation tissue. If the wound is small and the damaged tissue is actively mitotic, the tissue will regenerate and cover the fibrous tissue. When a wound is extensive or the damaged tissue amitotic, it is repaired only by fibrous connective (scar) tissue.

DEVELOPMENTAL ASPECTS OF TISSUES (pp. 127–129)

1. All body tissues develop from one or more of the three primary germ layers that appear in the embryo. Epithelium arises from all three germ layers (ectoderm, mesoderm, endoderm); muscle and connective tissue from mesoderm; and nervous tissue from ectoderm.

2. The decrease in mass and viability seen in most body tissues during old age often reflects circulatory deficits or poor nutrition.

Review Questions

Multiple Choice/Matching

1. Use the key to classify each of the following described tissue types into one of the four major tissue categories.

Key: (a) connective tissue (b) epithelium
(c) muscle (d) nervous tissue

_____ Tissue type composed largely of nonliving intercellular matrix; important in protection and support

_____ The tissue immediately responsible for body movement

_____ The tissue that allows us to be aware of the external environment and to react to it; specialized for communication

_____ The tissue that lines body cavities and covers surfaces

2. An epithelium that has several layers, with a layer of flattened cells most superficial, is called (choose all that apply): (a) cilated, (b) columnar, (c) stratified, (d) simple, (e) squamous.

3. Match the epithelial types named in column B with the appropriate description(s) in column A.

	Column A	Column B
_____	Lines most of the digestive tract	(a) pseudostratified ciliated columnar
_____	Lines the esophagus	(b) simple columnar
_____	Lines much of the respiratory tract	(c) simple cuboidal
		(d) simple squamous
_____	Forms the walls of the air sacs of the lungs	(e) stratified columnar
		(f) stratified squamous
_____	Found in urinary tract organs	(g) transitional

4. The gland type that secretes products such as milk, saliva, bile, or sweat through a duct is (a) an endocrine gland, (b) an exocrine gland.

5. The membranous lining of the body cavities open to the exterior is: (a) endothelium, (b) cutaneous membrane, (c) mucous membrane, (d) serous membrane

Short Answer Review Questions

6. Define tissue.

7. Name four important functions of epithelial tissue and provide at least one example of a tissue that exemplifies each function.

8. Describe the criteria used to classify covering and lining epithelia.

9. Explain the functional classification of multicellular exocrine glands and supply an example for each class.

10. Name four important functions of connective tissue and provide examples from the body that illustrate each function.

11. Name the primary cell type in connective tissue proper; in cartilage; in bone.

12. Name the two major components of matrix and, if applicable, subclasses of each component.

13. Matrix is extracellular. How does the matrix get to its characteristic position?

14. Name the specific connective tissue type found in the following body locations: (a) forming the soft packing around organs, (b) supporting the ear pinna, (c) forming the vocal cords, (d) in the embryo, (e) forming the intervertebral discs, (f) covering the ends of bones at joint surfaces.

15. Define macrophage system.

16. Compare and contrast skeletal, cardiac, and smooth muscle tissue relative to structure, body location, and specific function.

17. Differentiate clearly between the roles of neurons and neuroglial cells.

18. Describe the process of tissue repair, making sure you indicate factors that influence this process.

19. Name the three embryonic germ layers, and indicate which primary tissue classes derive from each layer.

Clinical Application Questions

20. John has sustained a severe injury during football practice and is told that he has a torn knee cartilage. Can John expect a quick, uneventful recovery? Explain your response.

21. Dr. Hutton, an anatomy professor, refers to "fixed" cells that remain stationary and "freely mobile" cells that enter and leave areolar connective tissue. To which cells or cell types found in areolar tissue can these terms be applied?

Chapter 5 describes the body covering (skin) and its contribution to body integrity as a whole. The remaining chapters in this unit, Chapters 6–10, present the skeletal and muscular systems and their interaction in promoting body support, protection, mobility, and form.

Light micrograph of bone tissue in the thigh.

UNIT 2

COVERING, SUPPORT, AND MOVEMENT OF THE BODY

CHAPTER 5

The Integumentary System

Chapter Outline and Student Objectives

The Skin (pp. 135–139)

1. Name the specific tissue types composing the epidermis and dermis. List the major layers of each and relate the function of each layer.
2. List and describe the factors that normally contribute to skin color. Briefly describe how changes in skin color may be used as clinical signs of certain disease states.

Derivatives of the Skin (pp. 139–144)

3. List the parts of a hair follicle and explain the function of each part. Also describe the functional relationship of arrector pili muscles to the hair follicle.
4. Explain the basis of hair color. Describe the distribution, growth, and replacement of hairs and the changing nature of hair during the life span.
5. Compare the structure, distribution, and most common locations of sweat and oil glands. Also compare the composition and functions of their secretions.
6. Compare and contrast eccrine and apocrine glands.

Functions of the Integumentary System (pp. 144–145)

7. Name at least five different functions of the skin, and specify how these functions are accomplished by the various skin components.

Homeostatic Imbalances of Skin (pp. 145–147)

8. Explain why a serious burn represents a loss of homeostasis and a threat to life. Describe a technique that may be used to determine the extent of a burn and differentiate between first-, second-, and third-degree burns.
9. Summarize the characteristics of basal cell carcinoma, squamous cell carcinoma, and malignant melanoma.

Developmental Aspects of the Integumentary System (pp. 149)

10. Briefly describe the changes that occur in the skin from birth to old age. Attempt to explain the causes of such changes.

Preview of Selected Key Terms

Integumentary system (in-teh″-gyoo-men′-tah-rē) (*integere* = to cover) The skin (and its derivatives), which provides the external protective covering of the body.

Epidermis (eh″-peh-der′-mis) (*epi* = upon) Superficial layer of the skin, composed of keratinized stratified squamous epithelium.

Dermis (*derm* = skin) The layer of the skin deep to the epidermis, composed of dense irregular connective tissue.

Keratin (kayr′-uh-tin) (*kera* = horn) Water-insoluble protein found in the epidermis, hair, and nails that makes those structures hard and water-repellent.

Melanin (meh′-luh-nin) (*melan* = black) The dark pigment formed by cells called melanocytes; imparts color to the skin and hair.

Papilla (puh-pih′-luh) (*papill* = nipple) Small, nipplelike projection; e.g., dermal papillae are projections of dermal tissue into the epidermis.

Hair follicle (*folli* = bag) The epithelial-connective tissue sheath within which a hair is formed by epithelial cells.

Sudoriferous gland (soo″-duh-rih′-fer-us) (*sudori* = sweat) An epidermal gland that produces sweat.

Sebaceous gland (suh-bā-shus) (*seb* = grease) An epidermal gland that produces an oily secretion called sebum.

Would you be enticed by an advertisement for a coat that was waterproof, stretchable, washable, and permanent-press, that automatically repaired small cuts, rips, and burns with invisible mending, and that was guaranteed to last a lifetime with reasonable care? Sounds too good to be true, but you already have such a coat—your skin. The skin and its derivatives (sweat and oil glands, hairs, and nails) make up a very complex set of organs that serves a number of functions, mostly protective. Together, these organs are called the **integumentary** (in-teh″-gyoo-men′-tah-rē) **system.**

The Skin

The skin ordinarily receives very little respect from its inhabitants, but architecturally, it is a marvel. It covers the entire body, has a surface area of 15–20 square feet, and weighs approximately 9 pounds. It has been estimated that every square inch of the skin contains 15 feet of blood vessels, 4 yards of nerves, 650 sweat glands, 100 oil glands, 1500 sensory receptors, and over 3 million cells that are constantly dying and being replaced (Figure 5.1). The skin is also called the **integument**, which simply means "covering," but its functions go well beyond serving as a large, opaque bag for the body contents. Without our skin, we would quickly fall prey to bacteria and perish from water and heat loss.

The skin, which varies in thickness from 0.5 to 4.0 mm or more in different parts of the body, is composed of two distinct regions, the **epidermis** (eh″-peh-der′-mis) and the **dermis.** These two areas, which are firmly attached to one another, are separated by an undulating, or wavy, borderline (see Figure 5.1). The epidermis is the surface coat, composed of many layers of epithelial cells, and is the outermost protective shield of the body. The underlying dermis, making up the bulk of the skin, is connective tissue. Only the dermis is vascularized; nutrients reach the epidermis by diffusing through tissue fluid from the dermal blood vessels.

The subcutaneous tissue (the tissue beneath the skin) is known as the **hypodermis** or **superficial fascia** and is made up of loose connective tissue that varies from areolar to adipose in nature. The hypodermis anchors the skin to the underlying organs and allows the skin to move relatively freely over those structures. Because of its fatty composition, the hypodermis also acts as a shock absorber and insu-

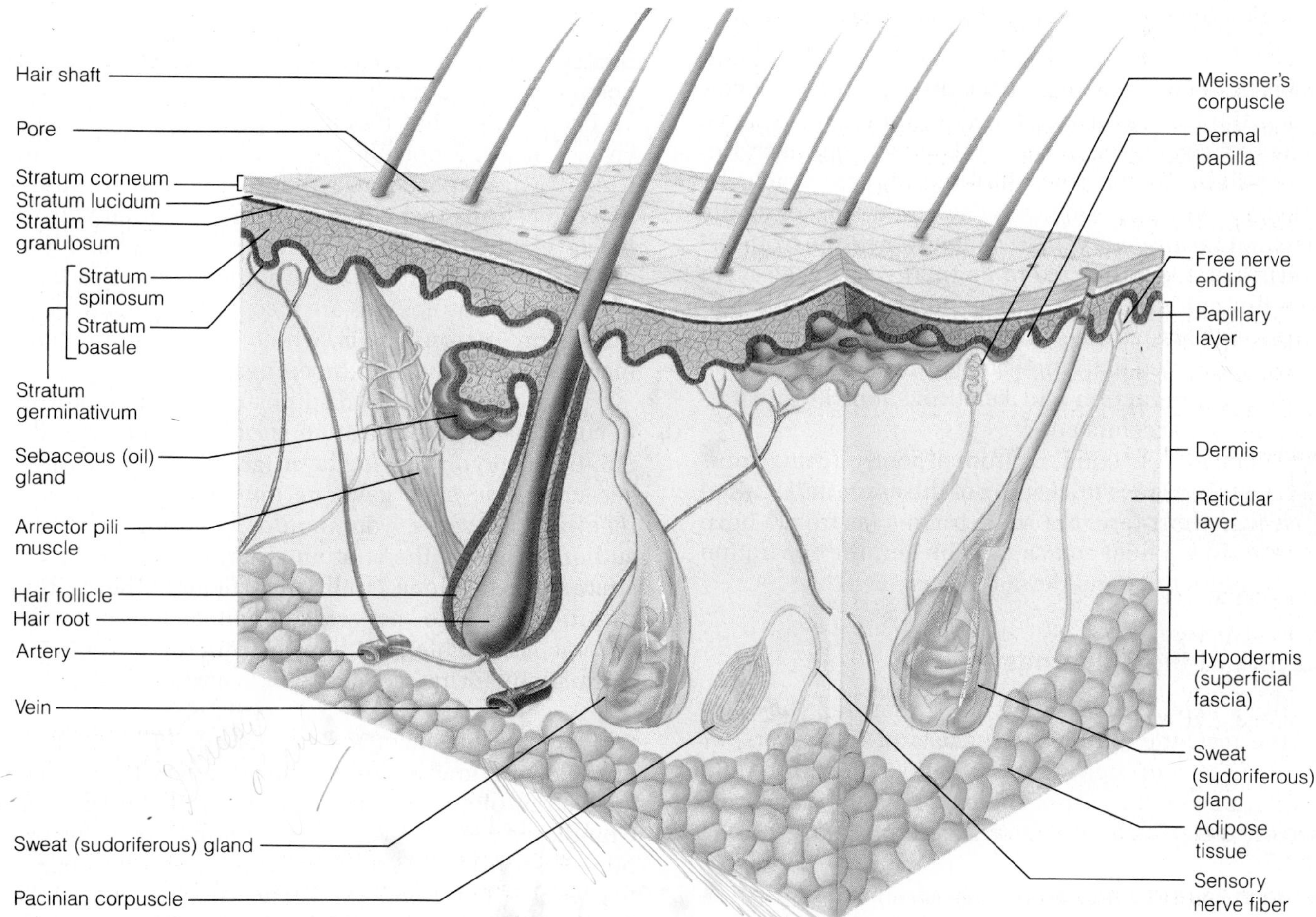

Figure 5.1 Skin structure. Three-dimensional view of the skin and underlying subcutaneous tissue. The epidermal and dermal layers have been pulled apart at the right front corner to reveal the dermal papillae.

lates the deeper body tissues from heat loss. Strictly speaking, the hypodermis is not considered part of the skin.

Epidermis

Structurally, the epidermis is a keratinized stratified squamous epithelium consisting of four distinct cell types: keratinocytes, melanocytes, Merkel's cells, and Langerhans' cells. Since most epidermal cells are keratinocytes, we will discuss them first; we will discuss the other cell types when we examine the epidermal layers.

Keratinocytes

The chief role of **keratinocytes** (kayr″-uh-tin′-uh-sīts″) is to produce *keratin* (kayr′-uh-tin), the fibrous protein that makes the epidermis a protective layer. Tightly connected to one another by desmosomes, the keratinocytes arise in the deepest layer of the epidermis from cells that undergo almost continuous mitosis. As the keratinocytes are pushed toward the skin surface by the production of new cells beneath them, they begin to produce the keratin that eventually dominates their cell contents. By the time the keratinocytes reach the free surface of the skin, they have become scalelike structures that are little more than keratin-filled plasma membranes. Millions of these dead cells are worn off by abrasion each day, and we have a totally new epidermis every 35 to 45 days, the time from the "birth" of a keratinocyte to its final wearing away. In healthy epidermis, the production of new cells balances cell loss at the skin surface. In body areas regularly subjected to friction such as the hands and feet, both cell production and keratinization (keratin formation) are accelerated.

Persistent friction, as from a poorly fitting shoe, causes a gross thickening of the epidermis, called a *callus.* Short-term but acute trauma (as from a burn or wielding a hoe) can cause a **blister,** the separation of the epidermal and dermal layers. ■

Layers of the Epidermis

In *thick skin,* which covers the palms and the soles of the feet, the epidermis consists of five layers, or *strata* (stra′-tuh) (see Figures 5.1 and 5.2). In *thin skin,* which covers the rest of the body, the various layers are thinner, and only four strata are present.

Stratum Basale (Basal Layer). The **stratum basale** (stra′-tum buh-sah′-lē), the deepest epidermal layer, is separated from the dermis only by a thin basement membrane. For the most part, it consists of a single row of cells, mostly columnar keratinocytes, directly abutting the dermis below. The many mitotic nuclei seen in this layer reflect the rapid division of these cells. Occasional **Merkel's cells** appear amid the keratinocytes. Merkel's cells are intimately associated with sensory nerve endings and are thought to be involved in touch reception.

About 25% of the cells in the stratum basale are **melanocytes** (meh-la′-nō-sīts), specialized epithelial cells that synthesize the pigment **melanin** (meh′-luh-nin). Melanocytes have numerous branching processes that contact all the keratinocytes in the basal layer. As melanin is formed, it passes through these projections to the keratinocytes, which phagocytize the ends of the processes and ingest the pigment. The melanin granules accumulate on the superficial, or "sunny," side of a keratinocyte nucleus, forming a pigment shield that protects the keratinocytes (and underlying nerve endings in the dermis) from the destructive effects of ultraviolet radiation in sunlight. Melanocytes are most numerous in the genital area, less so in the face, and very sparse on the body trunk. Since all humans have the same relative number of these cells, individual and racial differences in skin coloring are probably due to differences in melanocyte activity.

Stratum Spinosum (Spiny Layer). The **stratum spinosum** (spī-nō′-sum) is several cell layers thick. Mitosis occurs there but less frequently than in the basal layer. The keratinocytes are somewhat flattened and irregular in shape; because they display minute spine-like projections that connect them with other cells of the layer, they are often called *prickle cells.* These cells actively synthesize tonofilaments. Scattered among the keratinocytes are macrophage-like **Langerhans'** (lon′-ger-onz) **cells,** which arise from the bone marrow and migrate to the epidermis.

The strata basale and spinosum are often referred to collectively as the **stratum germinativum** (jer-mih″-nuh-tih′-vum) or **growing layer** (*germinate* = to grow), because these cells generate epidermal growth and renewal. (However, you should be aware that some authors consider the stratum germinativum to be the stratum basale alone.) Cells superficial to the stratum germinativum do not receive adequate nutrients, become less viable, and finally begin to die. This is a completely normal sequence of events.

Stratum Granulosum (Granular Layer). The **stratum granulosum** (gra″-nyoo-lō′-sum) is a very thin region consisting of two or three layers of flattened cells. Granules of a substance called *keratohyalin* (kayr″-uh-tuh-hī′-uh-lin) within these cells give the layer its name. It is in this third epidermal layer that keratinization begins. Langerhans' cells are also found in this layer.

Stratum Lucidum (Clear Layer). The **stratum lucidum** (loo′-sih-dum) appears through the light microscope as a thin translucent band just above the stratum gra-

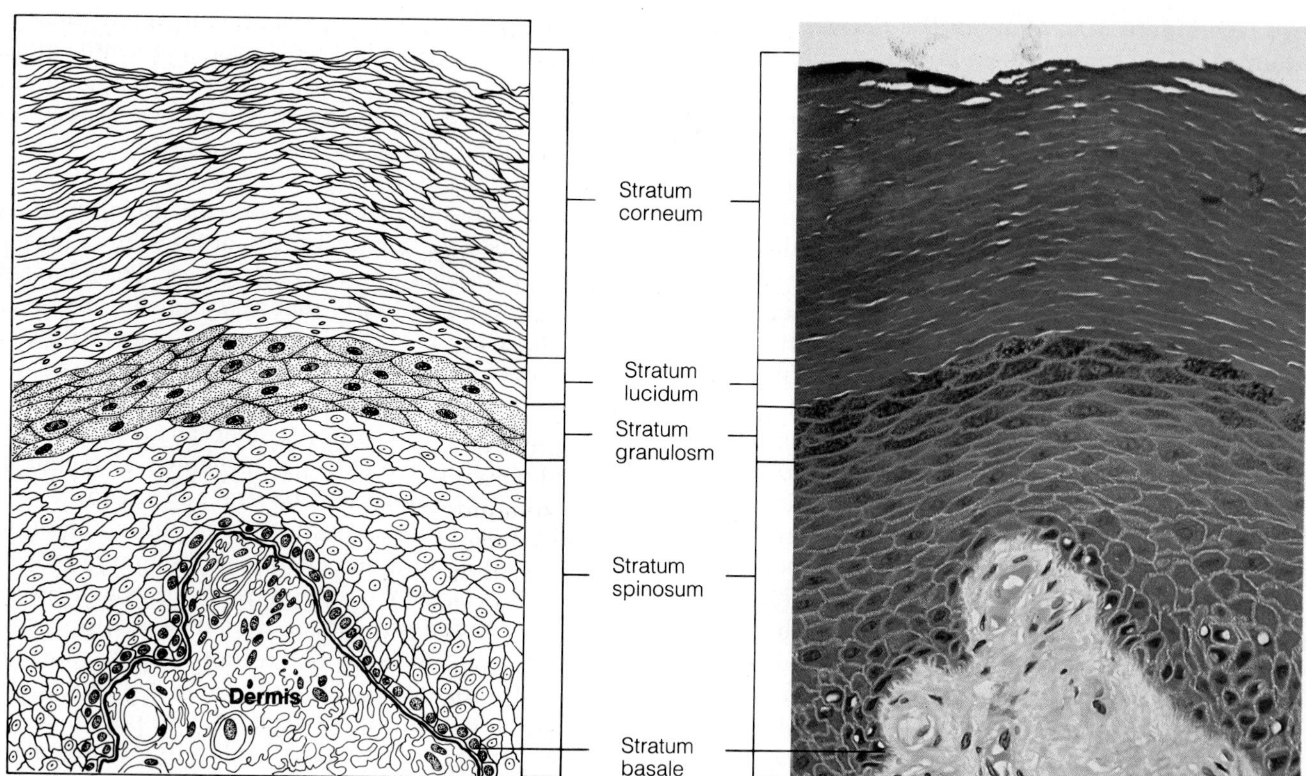

Figure 5.2 Layers of the epidermis. The photomicrograph (right) is magnified 300 x.

nulosum. It consists of a few rows of flattened anucleate keratinocytes with indistinct boundaries. Their sparse cytoplasm contains semifluid keratohyalin, apparently the product of the granules present in the granular layer. Here, or in the stratum corneum above, keratohyalin becomes intimately associated with parallel arranged tonofilaments (intermediate filaments) present in the cells. It is this keratohyalin–tonofilament combination that forms the keratin fibrils. The stratum lucidum is present only in areas where the epidermis is quite thick.

Stratum Corneum (Horny Layer). The outermost layer, the **stratum corneum** (kor′-nē-um), is a broad zone 20 to 30 cell layers thick. It accounts for about three-quarters of the epidermal thickness. The shinglelike dead cell remnants, completely filled with keratin fibers, are referred to as *cornified* or *horny cells* (*cornu* = horn). The common saying "Beauty is only skin deep" is especially interesting in light of the fact that nearly everything we see when we look at someone is dead! Keratin is a tough, water-repellent protein. Its abundance in the stratum corneum allows that layer to provide a durable "overcoat" for the body, which protects deeper cells from the hostile external environment (air) and from water loss and renders the body relatively insensitive to biological, chemical, and physical assaults.

Dermis

The dermis is the second major skin region. It is a strong but flexible connective tissue layer composed of a gel-like matrix heavily embedded with collagen, elastin, and reticular fibers. The cell types found in the dermis are typical of those found in any connective tissue proper: fibroblasts, macrophages, and occasional mast cells and white blood cells. The dermis is your "hide" and corresponds exactly to animal hides used to make highly prized leather products.

The dermis is richly supplied with nerve fibers (many of which are equipped with sensory receptors), blood vessels, and lymphatic vessels. The major portions of hair follicles, as well as oil and sweat glands, reside in the dermis but are derived from epidermal tissue, as will be described later. Like the epidermis, the dermis varies in thickness among individuals and body regions; however, the terms *thick skin* and *thin skin* refer only to differences in epidermal thickness. The dermis has two major layers: the papillary and reticular layers (see Figure 5.1).

The thin superficial **papillary** (pa′-pih-layr-ē) **layer,** is a loose (areolar-like) connective tissue in which the fibers form a loosely woven mat that is heavily invested with blood vessels. Its superior surface is thrown into projections called *dermal papillae* (pah-pih′-lē) that protrude into, or indent, the epidermis above (see Figure 5.1). Many of the dermal papillae contain capillary loops; others house free nerve endings (pain recepters) and touch receptors called *Meissner's corpuscles* (mīs′-nerz kor′-puh-sulz). On the ventral aspect of the hands and feet, the papillae are arranged in definite patterns that are reflected in looped and whorled ridges on the epidermal surface. These ridges increase friction and enhance the gripping ability of the fingers and feet. Papillary patterns are genetically determined and unique to each individual. Well provided with sweat pores, the ridges of the fingertips leave unique, identifying *fingerprints* on almost anything they touch.

The thick **reticular layer** is a typical dense irregular connective tissue. It contains bundles of interlocking collagen fibers that run in various planes parallel to the skin surface, as well as elastin and reticular fibers. Separations, or less dense regions, between these bundles form *lines of cleavage,* or *tension lines,* in the skin, which tend to run longitudinally in the limbs and in circular patterns around the neck and trunk. Cleavage lines are important to both surgeons and their patients. When an incision is made parallel to these lines, the skin tends to gape less and heals more readily than when the incision is made across cleavage lines.

In addition to the papillary ridges and tension lines, a third type of skin marking, *flexure lines,* reflects dermal modifications. Flexure lines (essentially dermal folds) result from the fact that the dermis is secured to deeper structures by the hypodermis. Since, in some regions, the skin cannot slide easily to accommodate joint movement, folding of the dermis occurs. Flexure lines are especially obvious on the wrists, palms, soles of the feet, fingers, and toes.

The connective tissue fibers of the dermis give skin its strength and resiliency. In addition, collagen tends to bind water, thus helping to maintain the hydration of the skin. Elastin fibers provide for the stretch-recoil properties of skin.

Extreme stretching of the skin, such as occurs during pregnancy, can tear the dermis. Dermal tearing is indicated by silvery white scars called *striae* (strī′-ē), commonly referred to as "stretch marks." ■

Skin Color

Three pigments contribute to skin color: melanin, carotene, and hemoglobin. **Melanin** (meh′-luh-nin), a polymer made up of tyrosine amino acids, ranges in color from yellow to orange to brown. As noted earlier, it is made by melanocytes and transferred to keratinocytes. Racial differences in skin coloring reflect the relative kind and amount of melanin made. The melanocytes of black- and brown-skinned people produce much more and darker melanin than those of fair-skinned individuals. *Freckles* and *pigmented moles* reflect local accumulations of melanin. Melanocytes are stimulated to greater activity when we expose our skin to sunlight. Prolonged sun exposure causes a substantial melanin buildup, which helps protect viable skin cells from ultraviolet radiation and in fair-skinned people causes visible darkening of the skin (a tan).

Despite melanin's protective effects, excessive sun exposure eventually damages the skin. It causes a clumping of the elastin fibers (solar elastosis), leading to "leathery skin," temporarily depresses the immune system, and can alter the DNA of skin cells and in this way lead to skin cancer. The fact that black people seldom have skin cancer attests to melanin's amazing effectiveness as a natural sunscreen.

There are other consequences of ultraviolet radiation as well. Numerous chemicals induce photosensitivity; that is, they heighten the skin's sensitivity to ultraviolet radiation, setting up sun worshippers for a skin rash the likes of which they would rather not see. Such substances include some antibiotic and antihistamine drugs, many chemicals in perfumes and detergents, and a chemical found in limes and celery. Small, itchy, blisterlike lesions erupt all over the body; then the peeling begins, in sheets! ■

Carotene (kayr′-uh-tēn) is a yellow to orange pigment found primarily in the stratum corneum and in adipose tissue of the hypodermis. Its color is most obvious in the palms and soles, where the stratum corneum is thickest. Carotene is particularly abundant in the skin of Oriental peoples; together with melanin, it accounts for the yellowish tinge of their skin.

The pinkish hue of Caucasian skin reflects the crimson color of oxygenated **hemoglobin** (hē′-muh-glō-bin) in the red blood cells circulating through the dermal capillaries. Since Caucasian skin contains only small amounts of melanin, the epidermis is quite transparent and allows hemoglobin's rosy color to show through.

When hemoglobin is poorly oxygenated, both the blood and the skin of Caucasians appear blue, a condition called *cyanosis* (sī″-uh-nō′-sis). Skin often becomes cyanotic during heart failure and severe respiratory disorders. In black individuals, the skin does not appear cyanotic because of the masking effects of melanin, but cyanosis is apparent in the mucous membranes and nail beds.

Skin color is also influenced by emotional stimuli, and many alterations in skin color signal certain disease states:

- *redness, or erythema* (ayr″-uh-thē′-muh): A reddened skin may indicate embarrassment (blushing), fever, hypertension, inflammation, or allergy.

- *pallor, or blanching:* Under certain types of emotional stress (fear, anger, and others), some people become pale. Pale skin may also signify anemia or low blood pressure.
- *jaundice (jon'-dis), or yellow cast:* The presence of an abnormal yellow skin tone usually signifies a liver disorder, in which bile pigments are absorbed into the blood and circulated throughout the body.
- *bronzing:* A bronze, almost metallic appearance of the skin is a sign of Addison's disease, hypofunction of the adrenal cortex.
- *black-and-blue marks,* or *bruises*: Black-and-blue marks reveal sites where blood has escaped from the circulation and clotted in the tissue spaces. Such clotted blood masses are called *hematomas.* An unusual susceptibility to bruising may signify a deficiency of vitamin C in the diet or bleeder's disease. ■

Derivatives of the Skin

Skin derivatives include hairs and hair follicles, nails, sweat glands, and sebaceous glands. Each of these is derived from the epidermis and has a unique role in maintaining body homeostasis.

Hairs and Hair Follicles

Although hair served primitive people well by helping to keep them warm, our body hair has become less luxuriant and has lost much of its utility. Hair still has some minor protective functions, such as guarding the head against physical trauma, heat loss, and sunlight; shielding the eyes (via eyelashes); and helping to keep dust and other foreign particles out of the upper respiratory tract (via nose hairs).

Structure of a Hair

A **hair** is a flexible structure, produced by a hair follicle, that consists largely of fused, keratinized cells. Its chief regions are the *shaft,* which projects from the skin (Figure 5.3), and the *root,* the portion embedded in the skin (Figure 5.4a). The shape of the hair shaft determines whether hair is straight or curly. If the shaft is flat and ribbonlike in cross section, the hair is kinky; if it is oval, the hair is wavy; if it is perfectly round, the hair is straight.

A hair has three concentric layers of keratinized cells (Figure 5.4b). Its central core, the *medulla* (meh-dul'-luh), consists of large cells that are partially separated by air spaces. The *cortex,* a bulky layer surrounding the medulla, consists of several layers of flattened cells. The outermost *cuticle* is formed from a single layer of cells that overlap one another from below like shingles on a roof (Figure 5.3). The cuticle is most heavily keratinized; it provides strength and helps keep the inner layers tightly compacted. Because it is subjected to the most abrasion, the cuticle tends to wear away at the tip of the hair shaft, allowing the keratin fibrils in the cortex and medulla to frizz out, a phenomenon called "split ends."

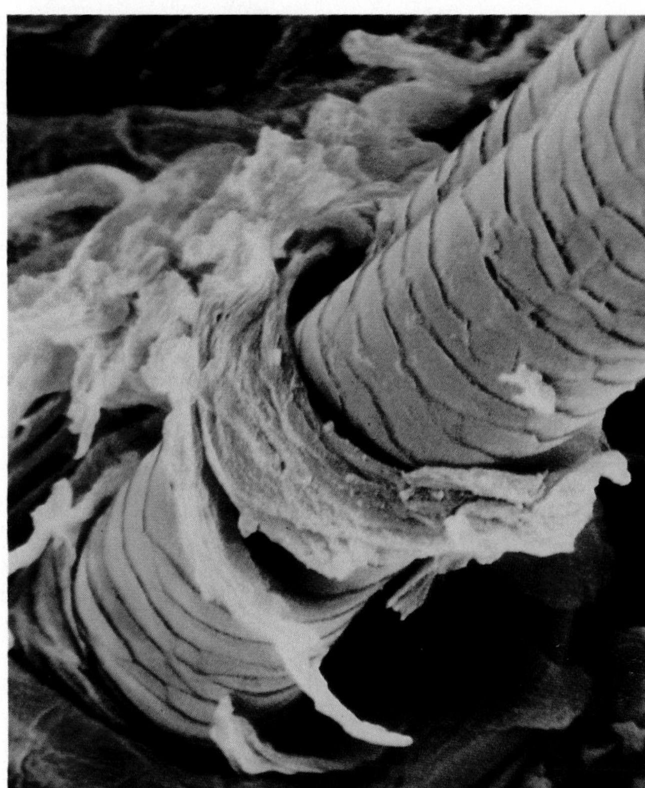

Figure 5.3 Scanning electron micrograph showing a hair shaft emerging from a follicle at the epidermal surface. Notice how the scalelike cells of the cuticle overlap one another (1,500 x).

Hair pigment is made by melanocytes at the base of the hair follicle and transferred to the cortical and medullary cells. Different proportions of three types of melanin (black, brown, and yellow) combine to produce all varieties of hair color from blond to pitch black. Gray or white hair results from a decrease in melanin production (mediated by delayed-action genes) and the replacement of melanin by air bubbles in the hair shaft.

Structure of a Hair Follicle

Hair follicles extend down into the dermis, and in the scalp, they may even extend into the hypodermis. The deep end of the follicle is expanded, forming a *hair*

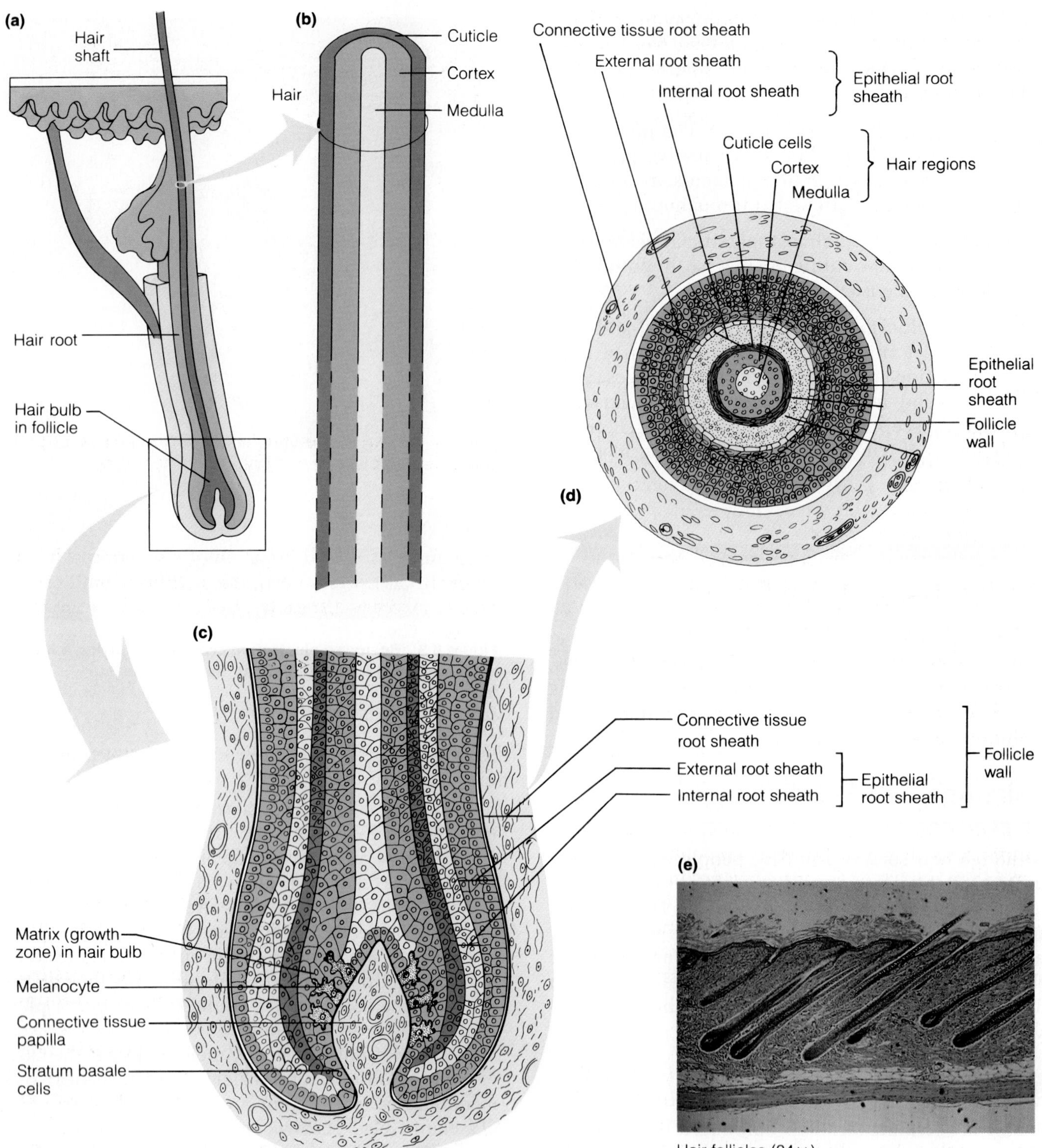

Figure 5.4 Structure of a hair and hair follicle. (**a**) Longitudinal section of a hair within its follicle. (**b**) Enlarged longitudinal section of a hair. (**c**) Enlarged longitudinal view of the expanded hair bulb of the follicle, which encloses the *matrix*, actively dividing epithelial cells that produce the hair. (**d**) Cross section of a hair and hair follicle. (**e**) Photomicrograph of scalp tissue showing numerous hair follicles.

bulb (Figure 5.4c). A *papilla*, a nipplelike bit of dermal connective tissue containing a knot of capillaries, protrudes into the hair bulb and provides the nutrient supply for the growing hair. Except for its specific location, this papilla is similar to the dermal papillae underlying other epidermal regions.

The wall of a hair follicle is actually a compound structure composed of an outer *connective tissue root sheath*, derived from the dermis, and an inner *epithelial root sheath*, formed by an invagination of the epidermis (Figure 5.4c and d). The epithelial root sheath, which has external and internal parts, thins as it approaches the hair bulb, so that only a single layer of stratum basale cells remains to wind around the papilla and produce the *matrix*, or actively dividing germinal area of the hair bulb that produces the hair.

As new hair cells are produced by the matrix, the older part of the hair is pushed upward, and its fused cells become increasingly keratinized and die. Thus, the hair shaft is made almost entirely of keratin.

Structures associated with hair follicles include nerve endings surrounding the bulb, sebaceous glands, and smooth muscle structures called arrector pili muscles. As you can see in Figures 5.1 and 5.4e, most hair follicles approach the skin surface at a slight angle, or obliquely. The **arrector pili** (uh-rek′-ter pih′-lē) muscles, which are bundles of smooth muscle cells, are attached in such a way that their contraction pulls the hair follicle into an upright position, dimpling the skin surface and producing goose bumps. The arrector pili muscles are regulated by the nervous system, and either cold external temperatures or fright can lead to their activation. This is an important heat retention and protection mechanism in other animals. For example, this response helps to keep furry animals warm in the winter by adding a layer of insulating air in their fur; and a scared animal with its hair on end looks larger and more formidable to its enemy. However, this response is not very useful to humans because most of our body hairs are short and sparse.

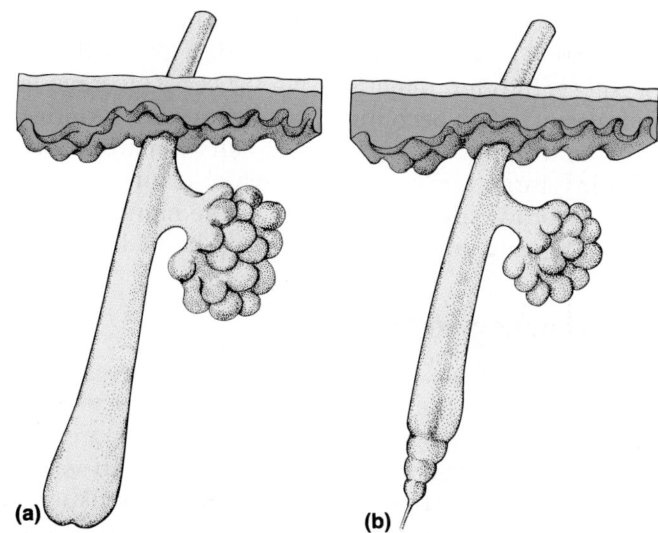

Figure 5.5 Overall shape of (a) an active and (b) a resting hair follicle. During the resting phase, the hair is shed.

Distribution and Growth of Hair

Millions of hairs are scattered over nearly all of the body. There are about 100,000 of them in the scalp and another 30,000 in a man's beard. Only the lips, nipples, parts of the external genitalia, and thick-skin areas such as the palms of the hands and soles of the feet totally lack hair. Hairs come in various sizes and shapes, but as a rule, they can be classified as vellus (*vell* = wool, fleece) or terminal. The body hair of children and women is of the pale, fine **vellus** (veh′-lus) **hair** variety. The coarser, often longer hair of the eyebrows and scalp is **terminal hair;** terminal hair may also be darker. At puberty, sex hormones (primarily androgens, the male sex hormones of which testosterone is the most important) cause terminal hairs to grow in the axillary and pubic regions of both sexes and on the face and chest (and, typically, the arms and legs) of males.

Hair growth and density are influenced by many factors, but most importantly by nutrition and hormones. As a rule, poor nutrition means poor hair growth. On the other hand, any condition that increases the local dermal blood flow (such as chronic physical irritation or inflammation) may enhance hair growth in that area. Many old-time bricklayers who carried their hod on one shoulder all the time developed one hairy shoulder. As noted, testosterone also encourages hair growth, and when male hormones are present in large amounts, hair growth is luxuriant.

In women, small amounts of androgens are normally produced by both the ovaries and the adrenal glands, but excessive hairiness, or *hirsutism* (hersoo′-tih-zim), as well as other signs of masculinization, may be caused by an adrenal gland tumor that produces abnormally large amounts of male hormones. Since few women want a beard or hairy chest, such tumors are surgically removed as soon as possible. ■

The rate of hair elongation varies from one body region to another, as well as with sex and age, but it ranges from 1.5 to 2.2 millimeters (mm) per week. Given ideal conditions, hair grows fastest from the teen years to the 40s and then declines; this slowed growth accounts for the thinning of hair in old age, since hairs are not replaced as fast as they are shed. Each follicle goes through *growth cycles* (Figure 5.5): An active growth phase is followed by a resting phase, when the matrix is inactive and the follicle atrophies somewhat. After the resting phase, the matrix proliferates again and forms a new hair to replace the old one that has fallen out or will be pushed out by the new hair. The life span of hairs varies. The follicles of the scalp often remain active for years (average is four years) before becoming inactive for a few months. But the follicles of the eyebrow hairs remain active for only three to four months, which explains why your eyebrows are never as long as the hairs on your head. We shed an average of 90 scalp hairs daily.

Hair Thinning and Baldness

The declining rate of hair growth that begins in the fourth decade of life reflects a natural age-related atrophy of hair follicles and results in some degree of baldness, or *alopecia* (a″-luh-pē′-shuh), in both sexes. Much less dramatic in women, the process usually begins at the anterior hairline and progresses posteriorly. Coarse terminal hairs are replaced by vellus hairs, and the hair becomes increasingly wispy.

The most common type of frank baldness, *male pattern baldness,* is a genetically determined, sex-influenced condition. It is thought to be caused by a delayed-action gene that alters the normal ratio of different testosterone receptors in follicles, causing the follicular growth cycles to become so short that many hairs never even emerge from their follicles, and those that do are fine vellus hairs that look like peach fuzz in the "bald" area. Until recently, the only "cure" for male pattern baldness has been drugs that inhibit testosterone production, but they also cause loss of sex drive—a solution few men would select for vanity's sake. Quite by accident, it was discovered that minoxidil, a drug used to reduce high blood pressure, has an interesting side effect in some bald men; it stimulates hair regrowth. Although its results are variable, minoxidil is now available in ointment form for application to the scalp.

Hair thinning can be induced by a number of other factors that lengthen follicular resting periods and upset the normal balance between the rate of hair loss and the rate of replacement. Outstanding examples are stressors, such as acutely high fever, surgery, or severe emotional trauma, and certain drugs (excessive vitamin A, some antidepressants, and most chemotherapy drugs). Protein-deficient diets lead to hair thinning because new hair growth stops when protein needed for keratin synthesis is not available. In all of these cases, hair regrows if the cause of thinning is removed or corrected; however, hair loss due to prolonged physical trauma, excessive radiation, or genetic factors is permanent. ■

Nails

A **nail** is a scalelike modification of the epidermis. It corresponds to the hoof or claw of other animals and forms a clear protective covering on the dorsum of the distal part of a finger or toe. Each nail has a *free edge,* a *body* (visible attached portion), and a *root* embedded in the skin (Figure 5.6).

Nails normally appear pink because of the rich bed of capillaries in the underlying dermis. At the proximal end of the nail body, the stratum germinativum is thickened; the white crescent that appears at this site is called the *lunula* (loo′-nyoo-luh). The proximal and lateral borders of the nail are overlapped by skin folds, called *nail folds.* The proximal nail fold is thickened and projects over the nail body; this region, the *eponychium* (eh-pō-nih′-kē-um), is commonly called the *cuticle.*

The stratum germinativum of the epidermis extends beneath the nail as the *nail bed.* The thickened portion beneath the lunula, called the *nail matrix,* is responsible for nail growth. As the nail cells are produced by the matrix, they become heavily keratinized, and the nail body slides distally over the nail bed.

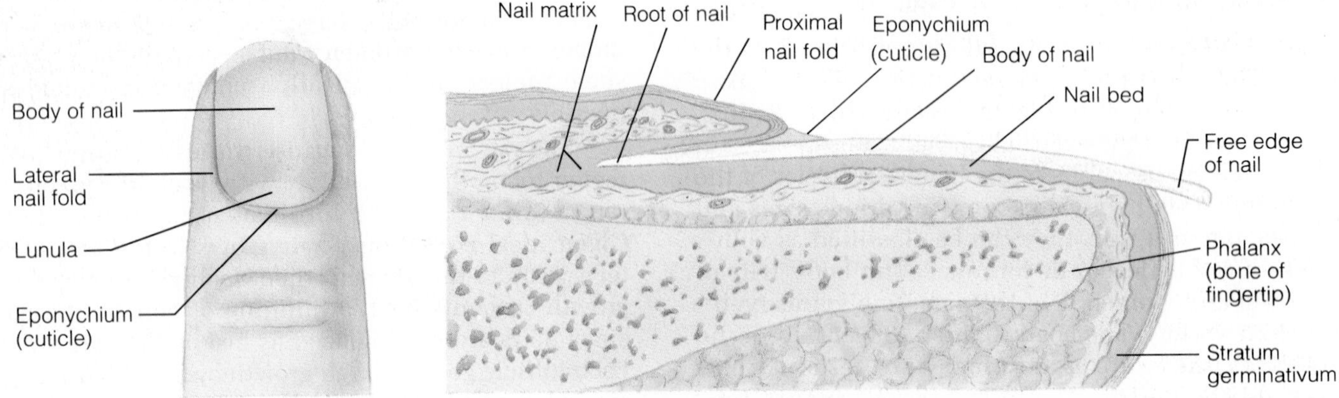

Figure 5.6 Structure of a nail. Longitudinal section of the distal part of a finger, showing nail parts and the nail matrix that forms the nail.

Sweat (Sudoriferous) Glands

Humans have more than 2.5 million **sweat,** or **sudoriferous** (soo-duh-rih′-fer-us), **glands** distributed over the entire body surface except the lips, nipples, and parts of the external genitalia. There are two types of sweat glands: eccrine and apocrine (see Figure 5.1).

Eccrine (ek′-rin) **sweat glands** are by far the more numerous and are particularly abundant on the palms of the hands, soles of the feet, and forehead. Each is a simple, coiled, tubular gland of the merocrine variety (that is, their secretion is released by exocytosis, and the structure of the secretory cells is not changed in any way by their secretory activity). The secretory part lies coiled in the dermis; the duct extends upward to open in a funnel-shaped *pore* (*por* = channel) at the skin surface (see Figure 5.1).

Eccrine gland secretion, the substance we normally call *sweat,* is largely (99%) water, with some salts (mostly sodium chloride), antibodies, traces of metabolic wastes (urea, uric acid, ammonia), lactic acid, and vitamin C. The exact composition depends on heredity and diet. Small amounts of drugs currently being taken may also be excreted by this route. Normally, the pH of sweat ranges from 4 to 6.

Sweating is regulated by the sympathetic division of the autonomic nervous system, over which we have little control; its major role is to assist in thermoregulation (maintaining normal body temperature). Heat-induced sweating begins on the forehead and then spreads downward over the rest of the body. Emotionally induced sweating—the so-called cold sweat brought on by fright, embarrassment, or nervousness—begins on the palms, soles, and axillae (armpits) and then spreads to other body areas.

Apocrine (a′-puh-krin) **sweat glands** are largely confined to the axillary and anogenital areas. They are larger than eccrine glands, and their ducts terminate in hair follicles (see Figure 5.1). Apocrine secretion contains the same basic components as true sweat, but it also contains fatty acids and proteins; consequently, it is quite viscous and sometimes a milky or yellowish color. The secretion is odorless, but when its organic molecules are decomposed by bacteria on the skin, it takes on a musky, unpleasant odor.

Apocrine glands begin to function at puberty under the influence of androgens. Although their secretion is produced almost continuously, apocrine glands have little role to play in thermoregulation. Their precise function is not yet known, but they are activated by sympathetic nerve fibers during pain and stress. Since their activity is also increased by sexual foreplay, and they enlarge and recede with the phases of a woman's menstrual cycle, they may be analogous to the sexual scent glands of other animals.

Ceruminous (ser-oo′-mih-nus) **glands** are modified apocrine glands found in the lining of the external ear canal. They secrete a rather sticky substance called *cerumen,* or earwax, that is thought to repel insects and block entry of foreign material.

Sebaceous (Oil) Glands

The **sebaceous** (suh-bā′-shus) **glands,** or **oil glands** (see Figure 5.1), are simple alveolar glands found all over the body except on the palms and soles. They are small on the body trunk and limbs, but quite large on the face, neck, and upper chest. These glands secrete an oily secretion called *sebum* (sē′-bum). The central glandular cells accumulate triglycerides, cholesterol, and other lipids until they become so engorged that they begin to degenerate and fragment, thus, functionally these glands are *holocrine glands.* The accumulated lipids and cell fragments constitute the sebum product that is usually ducted into a hair follicle (Figure 5.7), or occasionally to a pore on the skin surface. Sebum softens and lubricates the hair and skin, prevents hair from becoming brittle, and impedes water loss from the skin when the external humidity is low. Perhaps even more important is its bactericidal (bacterium-killing) action.

The secretory activity of sebaceous glands is stimulated by hormones, especially androgens. These

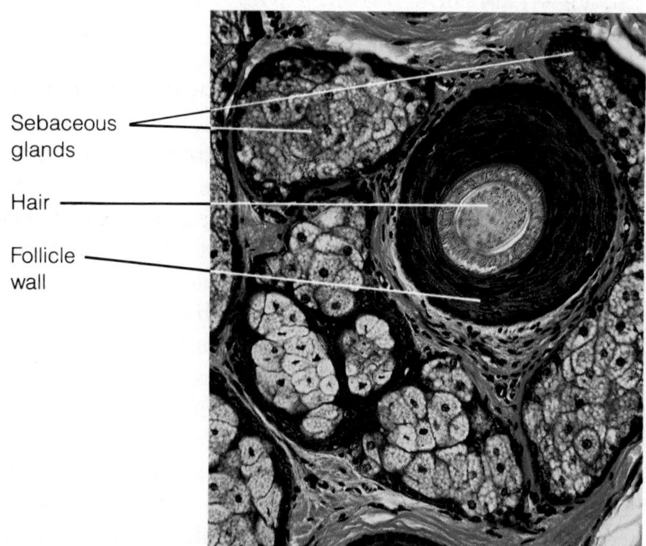

Figure 5.7 Sebaceous glands. In this cross-section, sebaceous glands surround a hair follicle into which their sebum product is secreted (150 x).

glands are relatively inactive during childhood, but are activated during puberty in both sexes, when testosterone production begins to rise.

If a sebaceous gland duct becomes blocked by accumulated sebum, a *whitehead* appears on the skin surface; if the material oxidizes and dries, it darkens, forming a *blackhead*. *Acne* is an active inflammation of the sebaceous glands accompanied by "pimples" (pustules or cysts) on the skin. It is usually caused by bacterial infection, particularly staphylococcus, and can be mild or extremely severe, leading to permanent scarring. *Seborrhea* (seh″-buh-rē′-uh), known as "cradle cap" in infants, is caused by overactivity of the sebaceous glands. It begins on the scalp as pink, raised lesions that gradually become yellow to brown and begin to slough off oily scales. Careful washing to remove the excessive oil often helps. ■

Functions of the Integumentary System

The skin and its derivatives perform a variety of functions that prevent external factors such as bacteria, abrasion, heat, cold, and chemicals from upsetting body homeostasis.

Protection

The skin puts up at least three types of barriers: chemical, physical, and biological.

The *chemical barriers* include skin secretions and melanin. Although the skin's surface is always teeming with bacteria, the acidity of skin secretions, or the so-called *acid mantle*, retards their multiplication. In addition, many bacteria are killed outright by bactericidal substances in sebum. As discussed earlier, melanin provides a chemical pigment shield to prevent ultraviolet damage to the viable skin cells.

Physical, or *mechanical, barriers* are provided by the continuity of skin and the hardness of its keratinized cells. As a physical barrier, the skin is a remarkable compromise. If the epidermis were thicker, it would certainly be more impenetrable, but we would pay the price in loss of suppleness and agility. Epidermal continuity works hand in hand with the acid mantle to ward off bacterial invasion. It also effectively blocks the diffusion of water and water-soluble substances, preventing both their loss from and entry into the body through the skin. Substances that do penetrate the skin in limited amounts include (1) *lipid-soluble substances*, such as oxygen, carbon dioxide, fat-soluble vitamins (A, D, E, and K), and steroids; (2) *oleoresins* (ō″-lē-ō-reh′-zinz) *of certain plants*, such as poison ivy and poison oak; (3) *organic solvents*, such as acetone, dry-cleaning fluid, and paint thinner, which dissolve the cell lipids; and (4) *salts of heavy metals*, such as lead, mercury, and nickel.

Both organic solvents and heavy metals are devastating to the body and can be lethal. Absorption of organic solvents through the skin into the bloodstream can cause the kidneys to shut down and can also cause brain damage; absorption of lead leads to anemia and neurological defects. These substances should never be handled with bare hands. ■

Biological barriers include the Langerhans' cells of the epidermis and also phagocytic cells (macrophages) in the dermis. The Langerhans' cells are active elements of the immune system. For the immune response to be activated, the foreign substances, or antigens, must be presented to specialized white blood cells called lymphocytes. In the epidermis, it is the macrophage-like Langerhans' cells that recognize antigens, ingest them, and present them to the lymphocytes for eventual destruction. Dermal macrophages constitute a second line of defense to dispose of viruses and bacteria that have managed to penetrate the epidermis. They, too, act as antigen "presenters".

Even a mild sunburn disrupts the normal immune response, because ultraviolet radiation abolishes the ability of the presenter cells to perform their function. This may help to explain why many people infected by the *herpes simplex* (her′-pēz sim′-plex), cold sore, virus are more likely to have a cold sore eruption after sunbathing. ■

Excretion

Limited amounts of nitrogen-containing wastes (ammonia, urea, and uric acid) are eliminated from the body in sweat, although most such wastes are excreted in urine. Sweat is an important avenue for sodium chloride loss when sweating is profuse.

Body Temperature Regulation

Even when we exercise vigorously, and even when the environmental temperature is high or low, body temperature remains within homeostatic limits. Like car engines, we need to get rid of the heat generated by our internal reactions. As long as the external temperature is lower than body temperature, the skin surface loses heat to the air and to cooler objects in its environment, just as a car radiator loses heat to the air and other nearby engine parts.

Under normal resting conditions, and as long as the environmental temperature is below 88°– 90° F, the sweat glands continuously secrete unnoticeable amounts of sweat (about 500 ml of sweat per day). When body temperature begins to rise, the dermal blood vessels become more dilated, and the sweat glands are stimulated into vigorous secretory activity. Sweat output increases dramatically and can account for the loss of up to 12 L of body water in one day. The evaporation of sweat from the skin surface dissipates body heat very efficiently and is an important mechanism for preventing overheating of the body.

When the external environment is cold, dermal blood vessels are constricted, which causes the warm blood to temporarily bypass the skin and allows skin temperature to drop to that of the external environment. Once this has happened, further passive heat loss from the body is slowed. This conserves body heat and helps maintain homeostatic body temperature. Body temperature regulation is discussed in greater detail in Chapter 25.

Cutaneous Sensation

The skin provides a "seat" for the cutaneous sensory receptors, which are actually part of the nervous system. The cutaneous receptors provide us with a great deal of information about our external environment—its temperature, the texture and pressure exerted by objects, and the presence of tissue-damaging factors. When activated, these receptors transmit nerve impulses to the brain for evaluation of whether or not a response is needed. Discussion of these cutaneous receptors is deferred to Chapter 11, but a few of them (Meissner's corpuscles, Pacinian corpuscles, and free nerve endings) are illustrated in Figure 5.1.

Vitamin D Synthesis

When modified cholesterol molecules in epidermal cells are irradiated by ultraviolet radiation, they are converted to vitamin D. The vitamin D is then absorbed into the dermal capillaries and transported to other body areas to play various roles in calcium metabolism. For example, calcium cannot be absorbed from the digestive tract in the absence of vitamin D.

Blood Reservoir

The skin vascular supply is quite extensive and can hold large volumes of blood. When other body organs, such as vigorously working muscles, need a greater blood supply, the dermal blood vessels are constricted by the nervous system. This shunts more blood into the general circulation, making it available to the muscles and other body organs.

Homeostatic Imbalances of Skin

It is difficult to scoff at anything that goes wrong with the skin; when it rebels, it is quite a visible revolution. Loss of homeostasis in body cells and organs can reveal itself on the skin in ways that are sometimes almost too incredible to believe. Because of the skin's complexity and extensiveness, it can develop more than 1000 different conditions and ailments. The most common skin disorders result from bacterial, viral, or yeast infections; a number of these are summarized in Related Clinical Terms on p. 149. Less common, but far more damaging to body well-being, are burns and skin cancer, considered next.

Burns

Burns represent a devastating threat to the body primarily because of their effects on the skin. A burn is tissue damage inflicted by intense heat, electricity, radiation, or certain chemicals, all of which denature cell proteins and cause cell death in the affected areas. Burns are the major cause of death for those under 40 years of age and the third leading cause of death in all age groups.

The immediate threat to life resulting from severe burns is a catastrophic loss of body fluids that contain proteins and electrolytes. As fluid seeps from the burned surfaces, dehydration and electrolyte imbalance result. These, in turn, lead to renal shutdown and circulatory shock (inadequate blood circulation resulting from reduced blood volume). To save the patient, the lost fluids must be replaced immediately. The volume of fluid lost can be estimated indirectly by computing the percentage of body surface burned (extent of the burns) using the *rule of nines*. This method divides the body into a number of areas, each accounting for 9% (or a multiple of 9%) of total body area (Figure 5.8).

Metabolism is grossly altered in burn patients. In addition to fluid and electrolyte replacement, burn patients need thousands of extra food calories daily to replace lost proteins and allow tissue repair. There is no possible way that anyone can eat enough food to provide these calories, so burn patients are given supplementary nutrients through gastric tubes or intravenous (IV) lines. Later, infection becomes the most important threat and is the leading cause of death in burn victims. Burned skin is sterile for about 24 hours, but thereafter, bacteria, fungi, and other pathogens can easily invade areas where the mechanical barrier of the skin has been destroyed, and they multiply rapidly in the nutrient-rich environment of dead tissues and protein-containing fluid. Adding to this problem is the fact that the patient's immune system becomes deficient within one to two days after severe burn injury.

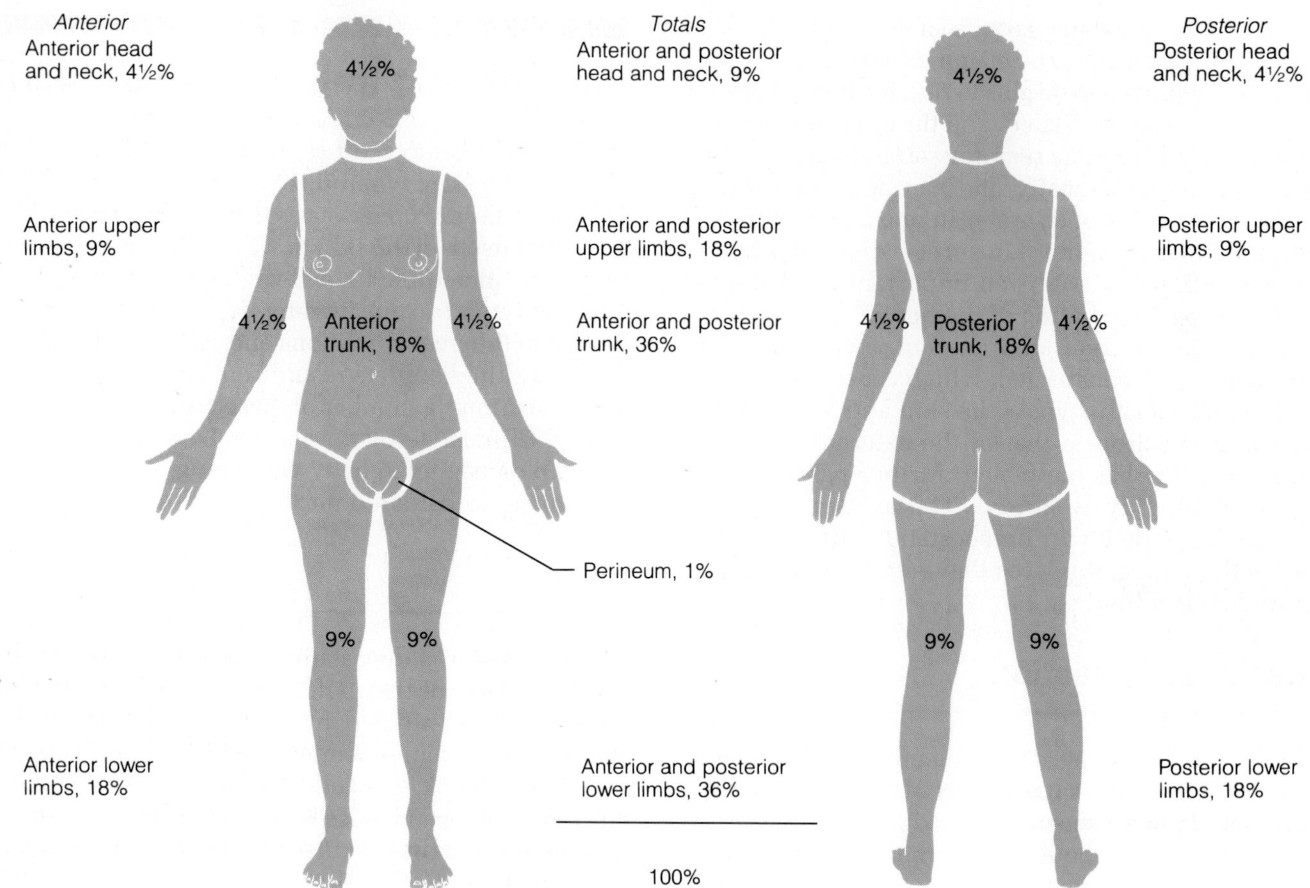

Figure 5.8 Estimating the extent of burns by using the rule of nines.

Burns are classified according to their severity (depth) as first-, second-, or third-degree burns. In *first-degree burns,* only the epidermis is damaged. Signs and symptoms include localized redness, swelling, and pain. Generally, first-degree burns heal in two to three days without any special attention. Sunburn is usually a first-degree burn. *Second-degree burns* involve injury to the epidermis and the upper region of the dermis. Signs and symptoms mimic those of first-degree burns, but blisters also appear. Since sufficient numbers of epithelial cells are still present, skin regeneration will occur with little or no scarring within three to four weeks if care is taken to prevent infection. First- and second-degree burns are referred to as *partial-thickness burns.*

Third-degree burns consume the entire thickness of the skin; these burns are also called *full-thickness burns.* The burned area appears blanched (gray-white) or blackened and there is little or no edema. Since the nerve endings in the area have been destroyed, the burned area is not painful. Although skin regeneration might *eventually* occur by the proliferation of epithelial cells at the edges of a third-degree burn, it is usually impossible to wait this long because of fluid loss and infection; thus, skin grafting is usually necessary. To prevent infection and fluid loss while the burned area is being prepared for grafting—the *eschar* (es′-kar), or burned skin, must first be debrided (removed)—the area is then deluged with antibiotics and covered temporarily with a synthetic membrane, animal (pig) skin, cadaver skin, or "living bandage" made from amniotic sac membrane (the thin, transparent membrane that surrounds a fetus). However, even when grafting is done quickly, extensive scar tissue often forms in the burned areas.

An exciting new technique that uses artificial skin in combination with epidermal cells cultured from the patient's own body is eliminating many of the traditional problems of skin grafting, such as widespread scarring and rejection in the case of cadaver or animal tissue grafts. Once the burned area has been debrided, a unique synthetic skin (Figure 5.9) made of a protective plastic "epidermis" bound to a spongy "dermal" layer composed of collagen and ground cartilage (obtained from cowhide and shark skeletons, respectively) is applied to the burned area. In time, the arti-

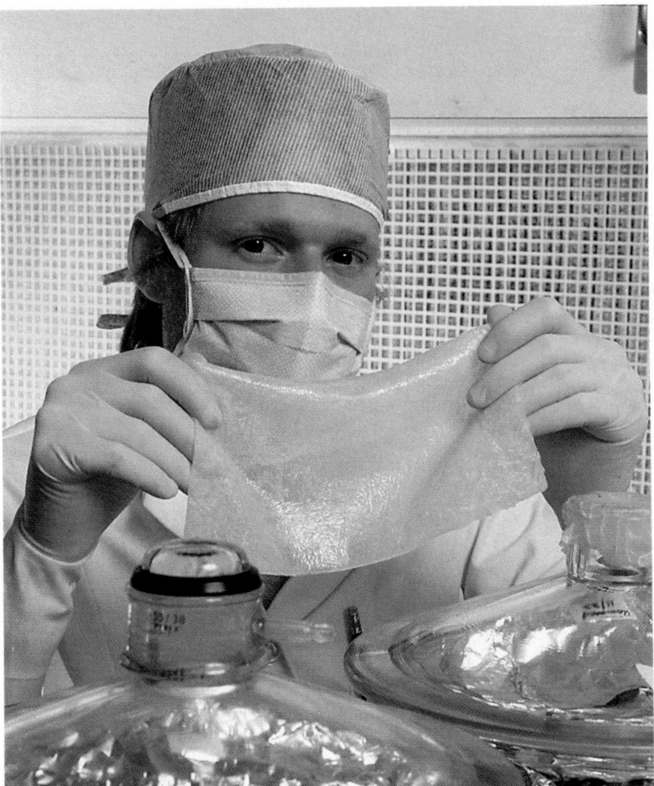

Figure 5.9 Synthetic skin.

ficial dermis is invaded by a rich vascular supply and by fibroblasts that produce collagen fibers. These newly formed collagen fibers replace those of the synthetic dermis, which naturally degrade. While this dermal reconstruction is going on, tiny bits of epidermal tissue are harvested from unburned parts of the patient's body, minced, and treated with enzymes to separate the cells. The isolated cells are then coaxed to proliferate in test tubes, forming virtually thousands of new cells. Once dermal reconstruction is complete (this usually takes two to three months), sheets of smooth, pink, bottle-grown epidermis are grafted on the dermal surface to seed new epidermal growth; this eventually forms a completely new epidermal barrier.

In general, burns are considered *critical* if any of the following conditions exists: (1) 30% of the body has second-degree burns, (2) over 10% of the body has third-degree burns, or (3) there are third-degree burns of the face, hands, or feet.

Facial burns introduce the possibility of burned respiratory passageways, which can swell and cause suffocation. Joint injuries are troublesome because scar tissue formation can severely limit joint mobility.

Skin Cancer

Numerous types of tumors arise in the skin. Most are benign and do not spread (metastasize) to other body areas. (A wart is one such example.) However, some skin tumors are malignant, or cancerous; that is, they tend to invade other body areas. The cause of most skin cancers is not known, but the most important risk factor is overexposure to the ultraviolet radiation in sunlight. In limited numbers of cases, frequent irritation of the skin by infections, chemicals, or physical trauma seems to be a predisposing factor.

Basal Cell Carcinoma

Basal cell carcinoma (kar″-sih-nō′-muh) is the least malignant and the most common of the skin cancers. Cells of the stratum basale, altered so that they cannot form keratin, no longer honor the boundary between epidermis and dermis. They proliferate, invading the dermis and hypodermis and causing tissue erosions or ulcers. The cancer lesions occur most often on sun-exposed areas of the face and appear as shiny, raised areas that later develop a central ulcer with a "pearly" beaded edge. Basal cell carcinoma is relatively slow-growing, and metastasis seldom occurs before it is noticed. Full cure by surgical excision is the rule in 99% of cases.

Squamous Cell Carcinoma

Squamous cell carcinoma arises from the prickle cells of the stratum spinosum. The lesion appears as a scaly reddened papule (small, rounded elevation) that tends to grow rapidly and metastasize to adjacent lymph nodes if not removed. This type of epidermal cancer is also believed to be sun-induced. If it is caught early and removed surgically or by radiation therapy, the chance of complete cure is good.

Malignant Melanoma

Malignant melanoma (meh″-luh-nō′-muh) is a cancer of melanocytes. It is fairly rare and only accounts for about 1% of cancer cases, but it is deadly. Melanoma can begin wherever there is pigment; most such cancers appear spontaneously, but some develop from pigmented moles. It usually appears as a spreading brown to black patch that metastasizes rapidly to surrounding lymph and blood vessels. Even when it is diagnosed early, the chance of cure is poor. The usual therapy for malignant melanoma is wide surgical excision accompanied by chemotherapy. ■

Integumentary system
External body covering

Lymphatic system

◀ Picks up excessive leaked fluid; prevents edema

▶ Protects organs

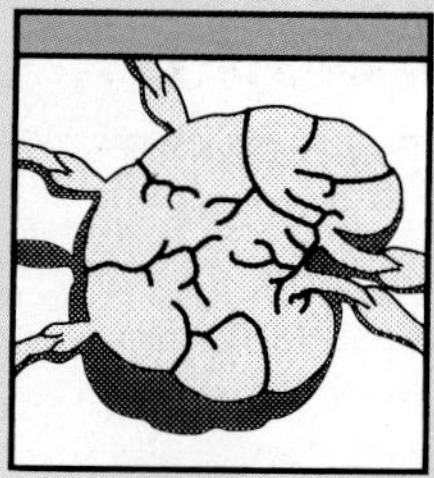

Skeletal system

Provides support for body organs, including the skin ▶

Protects; synthesizes vitamin D needed for normal calcium absorption and metabolism ◀

Immune system

◀ Protects skin cells

▶ Prevents pathogen invasion; Langerhans cells act in activating the immune response

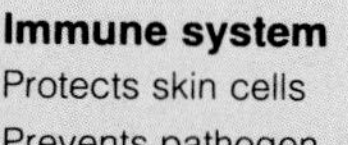

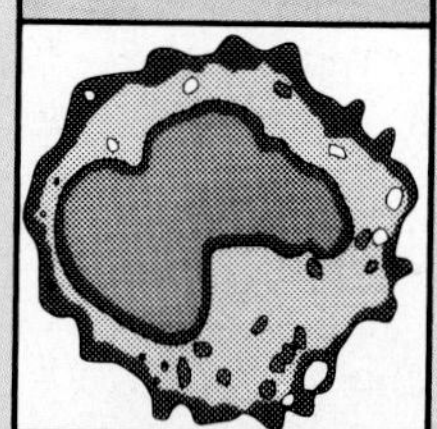

Muscular system

Active muscles generate large amounts of heat which increase blood flow to the skin and may promote activation of sweat glands ▶

Protects ◀

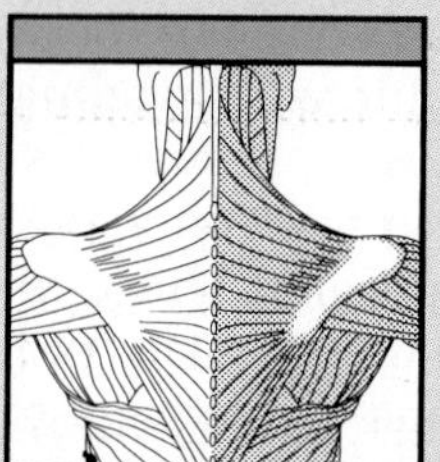

Respiratory system

◀ Furnishes oxygen to skin cells and removes carbon dioxide via gas exchange with blood

▶ Protects organs

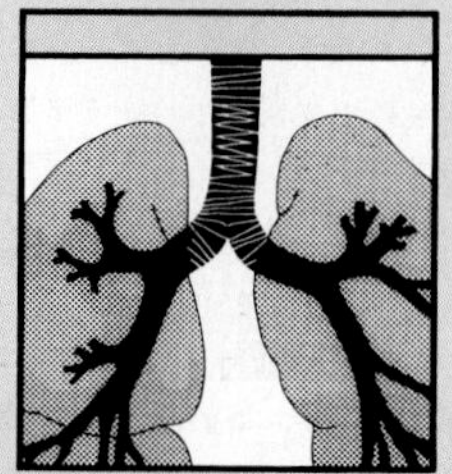

Nervous system

Regulates blood vessel diameter; activates sweat glands contributing to thermoregulation; interprets cutaneous sensation; activates arrector pili muscles ▶

Protects; seat of cutaneous receptors ◀

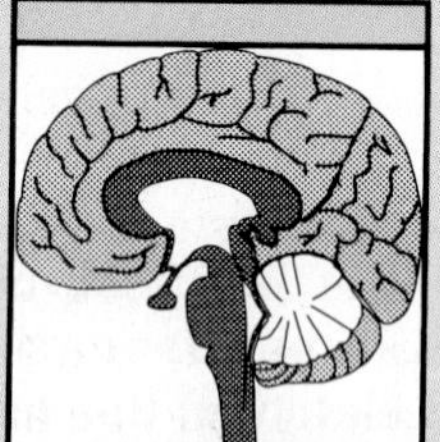

Digestive system

◀ Provides needed nutrients

▶ Protects organs; provides vitamin D needed for calcium absorption

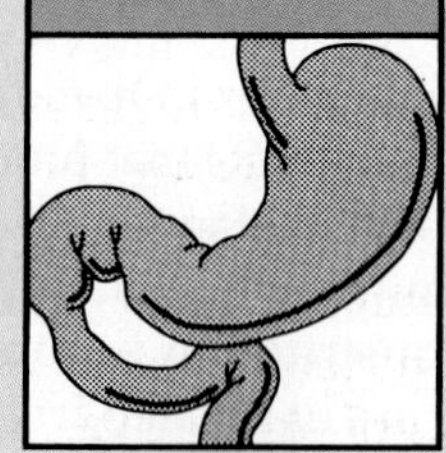

Endocrine system

Androgens activate sebaceous glands; involved in regulation of hair growth ▶

Protects ◀

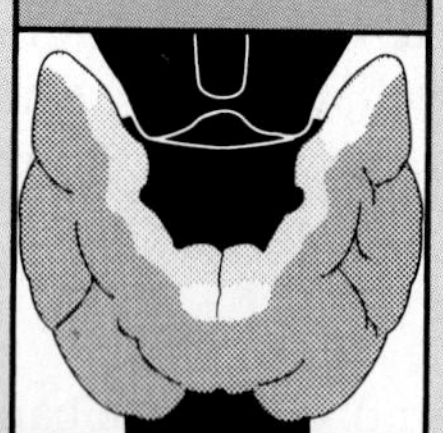

Urinary system

◀ Removes metabolic wastes and maintains electrolyte and acid-base balance of blood

▶ Protects organs; excretes salts and some nitrogenous wastes

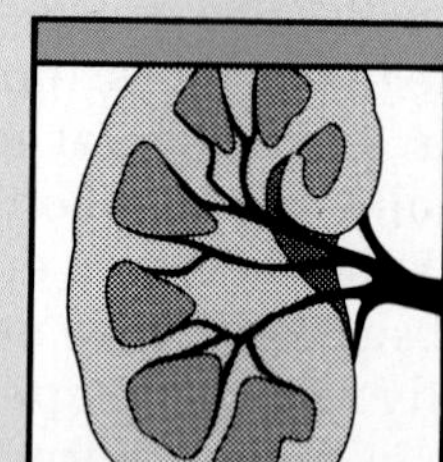

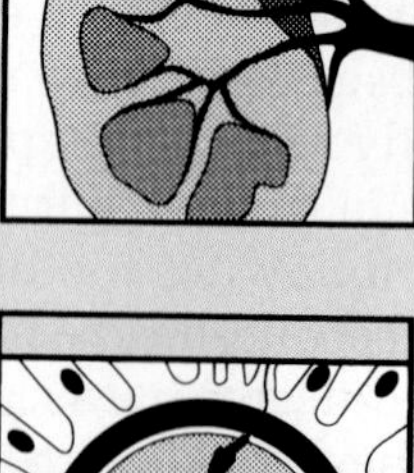

Cardiovascular system

Transports oxygen and nutrients to skin, and wastes from skin; provides substances needed by glands to make their secretions ▶

Protects; prevents fluid loss from body; blood reservoir ◀

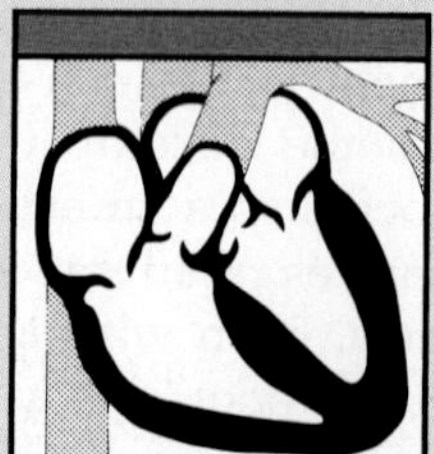

Reproductive system

▶ Protects sex organs; cutaneous receptors respond to erotic stimuli

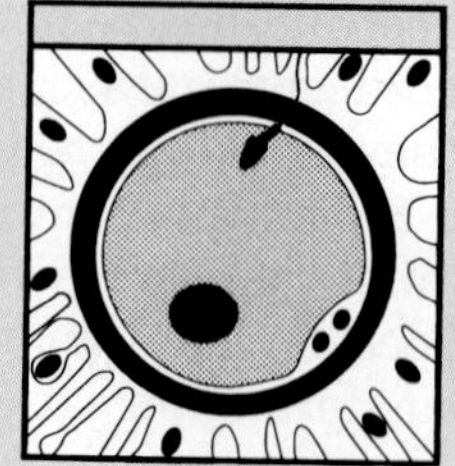

Figure 5.10 Homeostatic interrelationships between the integumentary system and other body systems. The arrowheads pointing toward the center of the diagram show the effects of the various body systems on the integumentary system. The arrowheads pointing away from the center indicate effects of the integumentary system on other body systems.

Developmental Aspects of the Integumentary System

The epidermis and dermis develop from the ectoderm and mesoderm germ layers, respectively. By the fourth month of development, the skin is fairly well formed, dermal papillae are obvious, and rudimentary epidermal derivatives are present. During the fifth and sixth months, the fetus is covered with a downy coat of delicate hairs called the *lanugo (luh-noo′-gō) coat.* This hairy cloak is shed at seven months gestational age, and vellus hairs make their appearance.

When a baby is born, its skin is covered with *vernix caseosa* (ver′-niks ka-sē-ō′-suh), a white, cheesy-looking substance produced by the sebaceous glands that protects the fetus's skin while inside the water-filled sac (the amnion). The newborn's skin is very thin and often shows small white spots called *milia* (mih′-lē-uh), accumulations in the sebaceous glands on the forehead and nose. These normally disappear by the third week after birth. During infancy and childhood, the skin thickens, and more subcutaneous fat is deposited.

With the onset of adolescence, the skin and hair become oilier as sebaceous glands are activated, and acne becomes more common. Generally, acne subsides in early adulthood, and skin reaches its optimal appearance when we reach our 20s and 30s. Then visible changes in the skin associated with the effects of continued environmental assaults (abrasion, wind, sun, chemicals) begin to appear, and scaling and various kinds of skin inflammation, or *dermatitis* (der-muh-tī′-tis), become more common.

As old age approaches, the rate of epidermal cell replacement slows, the skin thins, and its susceptibility to bruises and other types of injury increases. All of the lubricating substances produced by the skin glands that make young skin so soft start to become deficient; as a result, the skin becomes dry and itchy. Elastin fibers begin to clump and degenerate, and collagen fibers become fewer and stiffer. These alterations in dermal fibers can be hastened by prolonged exposure to the wind and sun. The hypodermal fat layer starts to diminish, leading to the intolerance to cold so common in elderly people. The decreasing elasticity of the skin, along with the loss of subcutaneous tissue, inevitably leads to wrinkling. Melanocytes also decrease in number and activity, resulting in less protection from UV radiation and a higher incidence of skin cancer in this age group. As a general rule, redheads and fair-haired individuals, who have less melanin to begin with, show age-related changes more rapidly than do those with darker skin and hair. For the same reason, black people look youthful much longer than Caucasians.

Although there is no known way to avoid the aging of the skin, one of the best ways to slow the process is to shield your skin from the sun by means of sunscreens and protective clothing. Good nutrition, plenty of fluids, and cleanliness may also help to delay the process.

By the age of 50 years, the number of hair follicles has dropped by one-third and continues to decline, resulting in hair thinning. Hair loses its luster in old age, and the delayed-action genes responsible for graying of hair and male pattern baldness become activated.

* * *

The skin is only about as thick as a paper towel—not too impressive as organ systems go. And yet, when it is severely damaged, nearly every body system reacts. Metabolism accelerates or may be impaired, changes in the immune system occur, bones may soften, the cardiovascular system may falter—the list goes on and on. On the other hand, when the skin is intact and performing its myriad functions, the body as a whole benefits. The important homeostatic interrelationships between the integumentary system and other organ systems of the body are summarized in Figure 5.10.

Related Clinical Terms

Albinism (al′-bih-nih-zim) (*alb* = white) Inherited condition reflecting an inability of melanocytes to synthesize melanin; an albino's skin is pink; the hair is very pale or white.

Athlete's foot Itchy red, peeling condition of the skin between the toes resulting from fungus infection.

Boils and carbuncles (kar′-bunk-ulz) Inflammation of hair follicles and sebaceous glands; common on dorsal neck; carbuncles are composite boils; common cause is bacterial infection (often *Staphylococcus aureus*).

Cold sores (fever blisters) Small fluid-filled blisters that itch and smart; usually occur around the lips and in oral mucosa of mouth; caused by a herpes simplex infection; the virus localizes in a cutaneous nerve, where it remains dormant until activated by emotional upset, fever, or UV radiation.

Contact dermatitis Itching, redness, and swelling, progressing to blister formation; caused by exposure of the skin to chemicals (e.g., poison ivy) that provoke an allergic response in sensitive individuals.

Dermatology The branch of medicine that studies and treats disorders of the skin.

Impetigo (im-puh-tē′-gō) (*impet* = an attack) Pink, water-filled, raised lesions (common around the mouth and nose) that develop a yellow crust and eventually rupture; caused by staphylococcus infection; contagious; common in school-age children.

Mongolian spots Blue-black spots, often appearing in the skin of the sacral region; results from the escape of melanin pigment into the dermis, where it is phagocytized by dermal macrophages; the light-scattering effect of the epidermis causes these dermal melanin-containing regions to appear blue.

Psoriasis (ser-ī′-uh-sis) A chronic condition, characterized by reddened epidermal lesions covered with dry silvery scales: when severe, may be disfiguring and debilitating; cause is unknown; may be hereditary in some cases; attacks often triggered by trauma, infection, hormonal changes, and stress.

Vitiligo (vih-tih′-lih-gō) (*viti* = a vine, winding) An abnormality of skin pigmentation characterized by a loss and uneven dispersal of melanin, i.e., unpigmented skin regions (light spots) surrounded by normally pigmented areas; can be cosmetically disfiguring, particularly in those with darker skin.

Chapter Summary

THE SKIN (pp. 135–139)

1. The skin, or integument, is composed of two discrete tissue layers, an outer epidermis and a deeper dermis, resting on subcutaneous tissue, the hypodermis.

Epidermis (pp. 136–137)

2. The epidermis is an avascular, keratinized stratified squamous epithelial sheet. From deep to superficial, its strata, or layers, are the basale, spinosum, granulosum, lucidum, and corneum. The stratum lucidum is absent in thin skin.

3. The viable and actively mitotic stratum germinativum (basale and spinosum) is the source of new cells for epidermal growth and renewal. The more superficial layers are increasingly keratinized and less viable. The stratum corneum consists of fully keratinized, dead cells that continuously slough off.

4. Scattered among the keratinocytes in the deepest epidermal layers are melanocytes, Merkel's cells, and Langerhans' cells.

Dermis (pp. 137–138)

5. The dermis, composed mainly of dense, irregular connective tissue, is well supplied with blood vessels, lymphatic vessels, and nerves. Cutaneous receptors, glands, and hair follicles reside within the dermis.

6. The superior papillary layer exhibits dermal papillae that protrude into the epidermis above. The patterns of the dermal papillae produce fingerprints.

7. In the inferior thicker reticular layer, the connective tissue fibers are much more densely interwoven. Less dense regions between the bundles produce cleavage, or tension, lines in the skin. Points of dermal attachment to the hypodermis often cause dermal folds, or flexure lines.

Skin Color (pp. 138–139)

8. Skin color reflects the amount of pigments (melanin and/or carotene) in the skin and the oxygenation level of hemoglobin.

9. Melanin production is stimulated by exposure to ultraviolet radiation in sunlight. Melanin, produced by melanocytes and phagocytized by keratinocytes, protects the keratinocyte nuclei from the damaging effects of UV radiation.

10. Skin color is affected by emotional state, and alterations in normal skin color (jaundice, bronzing, erythema, and others) may indicate certain disease states.

DERIVATIVES OF THE SKIN (pp. 139–144)

1. Skin derivatives, which arise from the epidermis, include hairs and hair follicles, nails, and glands (sweat and sebaceous).

Hairs and Hair Follicles (pp. 139–142)

2. A hair, produced by a hair follicle, consists of heavily keratinized cells. Each hair consists of a central medulla, a cortex, and an outer cuticle and has root and shaft portions. Hair color reflects the amount and variety of melanin present.

3. A hair follicle consists of an inner epidermal root sheath, which forms the matrix (region of the hair bulb that produces the hair), and an outer connective tissue sheath derived from the dermis. A hair follicle is richly vascularized and well supplied with nerve fibers. Contraction of arrector pili muscles pull the follicle into an upright position.

4. Hairs formed initially are fine vellus hairs; at puberty, under the influence of androgens, coarser, darker terminal hairs appear.

5. The rate of hair elongation varies in different body regions and with sex and age. Differences in life span of hairs account for differences in length on different body regions. Hair thinning reflects factors that cause long follicular resting periods, natural age-related atrophy of hair follicles, and a delayed-action gene.

Nails (p. 142)

6. A nail is a scalelike modification of the epidermis that covers the dorsum of a finger (or toe) tip. The actively growing region is the nail matrix.

Sweat (Sudoriferous) Glands (p. 143)

7. Eccrine sweat glands, with a few exceptions, are distributed over the entire body surface. They are most involved with thermoregulation. They are simple coiled tubular glands that secrete a salt solution containing small amounts of other solutes. Their ducts usually empty to the skin surface via pores.

8. Apocrine sweat glands are found primarily in the axillary and anogenital areas. Their secretion is similar to eccrine secretion, but it also contains proteins and fatty acids. Apocrine glands begin to function at puberty under the influence of androgens; their role is still not fully known.

Sebaceous (Oil) Glands (pp. 143–144)

9. Sebaceous glands occur all over the body surface except for the palms and soles. They are simple alveolar glands; their oily holocrine secretion is called sebum. Sebaceous gland ducts usually empty into hair follicles.

10. Sebum lubricates the skin and hair, prevents water loss from the skin, and acts as a bactericidal agent. Sebaceous glands are activated (at puberty) and controlled by androgens.

FUNCTIONS OF THE INTEGUMENTARY SYSTEM (pp. 144–145)

1. **Protection.** The skin protects by chemical barriers (the antibacterial nature of sebum and the acid mantle, and melanin), physical barriers (the hardened keratinized surface) and biological barriers (phagocytes).

2. **Excretion.** Sweat contains small amounts of nitrogenous wastes and plays a minor role in excretion.

3. **Body Temperature Regulation.** The skin vasculature and sweat glands, regulated by the nervous system, play an important role in maintaining body temperature homeostasis.

4. **Cutaneous Sensation.** Cutaneous sensory receptors in the dermis respond to temperature, touch, pressure, and pain stimuli.

5. Vitamin D Synthesis. Vitamin D is synthesized from cholesterol by epidermal cells.

6. Blood Reservoir. The extensive vascular supply of the dermis allows the skin to act as a blood reservoir.

HOMEOSTATIC IMBALANCES OF SKIN (pp. 145–148)

1. The most common skin disorders result from infections.

Burns pp. (145–147)

2. In severe burns, the initial threat is loss of protein- and electrolyte-rich body fluids, which may lead to circulatory collapse; the second threat is overwhelming bacterial infection.

3. The extent of a burn may be evaluated by using the rule of nines. The severity of burns is indicated by the terms first degree, second degree, and third degree. Third-degree burns require grafting for successful recovery.

Skin Cancer (p. 147)

4. The most common cause of skin cancer is exposure to ultraviolet radiation.

5. Cure of basal cell carcinoma and squamous cell carcinoma is complete if they are removed before metastasis. Malignant melanoma, a cancer of melanocytes, is rare, but almost always fatal.

DEVELOPMENTAL ASPECTS OF THE INTEGUMENTARY SYSTEM (p. 149)

1. The epidermis develops from the embryonic ectoderm; the dermis develops from the mesoderm.

2. The fetus exhibits a downy lanugo coat. Fetal sebaceous glands produce vernix caseosa, which helps to protect the fetus's skin from its watery environment.

3. A newborn's skin is thin, but during childhood the skin thickens and more subcutaneous fat is deposited. At puberty, sebaceous glands are activated and terminal hairs appear.

4. In old age, the rate of epidermal cell replacement declines and the skin and hair thin. Skin glands become less active. Loss of collagen and elastin fibers and subcutaneous fat lead to wrinkling; delayed-action genes cause graying and balding.

Review Questions

Multiple Choice/Matching

1. Which of the epidermal cell types is most numerous? (a) keratinocytes, (b) melanocytes, (c) Langerhans' cells, (d) Merkel's cells.

2. Which is probably a macrophage? (a) a keratinocyte, (b) a melanocyte, (c) a Langerhans' cell, (d) a Merkel's cell.

3. Match the names of epidermal cell layers provided in the key with the appropriate description(s) given below.

Key: (a) stratum basale (b) stratum corneum (d) stratum granulosum (d) stratum lucidum (e) stratum spinosum

____ Cells often called prickle cells
____ Cells are literally flat plates of keratin
____ Together form the stratum germinativum and are mitotic, viable cells

4. The barrier nature of the epidermis is largely due to the presence of (a) melanin, (b) carotene, (c) collagen, (d) keratin.

5. Skin color is determined by (a) the amount of blood, (b) pigments, (c) oxygenation level of the blood, (d) all of these.

6. The sensations of touch and pressure are picked up by receptors located in (a) the stratum germinativum, (b) the dermis, (c) the hypodermis, (d) the stratum corneum.

7. Which is not a true statement about the papillary layer of the dermis? (a) It produces the pattern for fingerprints, (b) it is most responsible for the toughness of the skin, (c) it contains nerve endings that respond to stimuli, (d) it is highly vascular.

8. Skin surface markings that reflect points of tight dermal attachment to underlying tissues are called (a) tension lines, (b) papillary ridges, (c) flexure lines, (d) epidermal papillae.

9. Which of the following is not an epidermal derivative? (a) hair, (b) sweat gland, (c) sensory receptor, (d) sebaceous gland.

10. You can cut hair without feeling pain because (a) there are no nerves associated with hair, (b) the shaft of the hair consists of dead cells, (c) hair follicles develop from epidermal cells and the epidermis lacks a nerve supply, (d) hair follicles have no source of nourishment and therefore cannot respond.

11. An arrector pili muscle (a) is associated with each sweat gland, (b) can cause the hair to stand up straight, (c) enables each hair to be stretched when wet, (d) provides new cells for continued growth of its associated hair.

12. The product of this type of sweat gland includes protein and lipid substances that become odiferous as a result of bacterial action: (a) apocrine gland, (b) eccrine gland, (c) sebaceous gland, (d) pancreatic gland.

13. Sebum (a) lubricates the surface of the skin and hair, (b) consists of dead cells and fatty substances, (c) may function in an infant's recognition of the nipple, (d) all of these.

14. The "rule of nines" is helpful clinically in (a) diagnosing skin cancer, (b) estimating the extent of a burn, (c) estimating how serious a cancer is, (d) preventing acne.

Short Answer Review Questions

15. Is a bald man really hairless? Explain.

16. Both newborn infants and aged individuals have a deficit of subcutaneous tissue. How does this affect their sensitivity to cold environmental temperature?

17. You go to the beach to swim on an extremely hot, sunshiny July afternoon. Describe two different ways in which your integumentary system acts to preserve homeostasis during your outing on this particular day.

18. Distinguish clearly between first-, second-, and third-degree burns.

19. Describe the process of hair formation, and list several factors that may influence (a) growth cycles and (b) hair texture.

20. What is cyanosis and what does it reflect?

21. Why does skin wrinkle and what factors accelerate the wrinkling process?

Clinical Application Questions

22. Victims of third-degree burns demonstrate the vital functions performed by the skin. What are the two most important problems encountered clinically with such patients? Explain each in terms of the absence of skin.

23. A mother, disturbed about the bluish spots just above her infant daughter's buttocks, is told not to worry because they are just Mongolian spots. She is unconvinced and looks the term up in a medical dictionary. What does she learn?

24. A 30-year-old mental patient has an abnormal growth of hair on the dorsum of her right index finger. The orderly comments that she gnaws on that finger continuously. What is the relationship between this patient's gnawing activity and her "hairy" finger?

CHAPTER 6

Bones and Bone Tissue

Chapter Outline and Student Objectives

Functions of the Bones (p. 153)

1. List and describe five important functions of bones.

Classification of Bones (pp. 153–154)

2. Compare and contrast the structure of the four classes of bones and provide examples of each class.

Bone Structure (pp. 154–158)

3. Describe the gross anatomy of a typical long bone and short bone. Indicate the locations and functions of red and yellow marrow, articular cartilage, and periosteum.
4. Distinguish clearly between the several types of bone markings and note their relative functions.
5. Describe bone histology (compact and spongy bone).
6. Discuss the chemical composition of bone. Note the relative advantages conferred by both its organic and its inorganic components.

Bone Development (Osteogenesis) (pp. 158–161)

7. Compare and contrast the two types of bone formation: endochondrial and intramembranous ossification.
8. Describe the process of long bone growth that occurs at the epiphyseal discs.

Bone Homeostasis: Remodeling and Repair (pp. 161–165)

9. Compare the locations and functions of the osteoblasts, osteocytes, and osteoclasts in bone remodeling.
10. Explain how physical stress and hormonal controls regulate bone remodeling.
11. Describe the steps of fracture repair.

Homeostatic Imbalances of Bone (p. 167)

12. Contrast the disorders of bone remodeling seen in osteoporosis, osteomalacia, and Paget's disease.

Developmental Aspects of Bones: Timing of Events (p. 168)

13. Describe the timing and cause of changes in bone architecture and bone mass throughout life.

Preview of Selected Key Terms

Diaphysis (dī-a'-fih-sis) (*dia* = through; *physis* = growth) The elongated shaft of a long bone.

Epiphysis (ih-pih'-fih-sis) (*epi* = upon; *physis* = growth) The ends of a long bone, attached to the shaft.

Epiphyseal plate Plate of hyaline cartilage at the junction of the diaphysis and epiphysis that provides for growth in length of a long bone.

Periosteum (peh-rē-ah'-stē-um) (*peri* = around; *os* = bone) Dense connective tissue covering a bone.

Osteon (*osteo* = a bone) The structural and functional unit of compact bone; also called a Haversian system.

Osteogenesis (ah"-stē-ō-jeh'-neh-sis) (*genesis* = beginning) The process of bone formation.

Osteoid (os'-tē-oyd) Unmineralized bone matrix.

Bone remodeling The process involving bone formation and destruction in response to hormonal and mechanical factors.

Fracture A break in a bone.

At one time or another, all of us have heard expressions like "bone tired," "dry as a bone," and "bag of bones"—pretty unflattering and inaccurate images of one of our most phenomenal tissues. Our brains, not our bones, convey feelings of fatigue, and bones are far from dry. As for "bag of bones," they are indeed more prominent in some of us, but without them to form our internal skeleton, we would creep along the ground like slugs, lacking any defined shape or form.

Individual bones that fashion the skeleton and joints that allow skeletal mobility are the topics of Chapters 7 and 8. Here we will focus on the general structure and function of bone tissue and on the dynamics of its formation and remodeling throughout life.

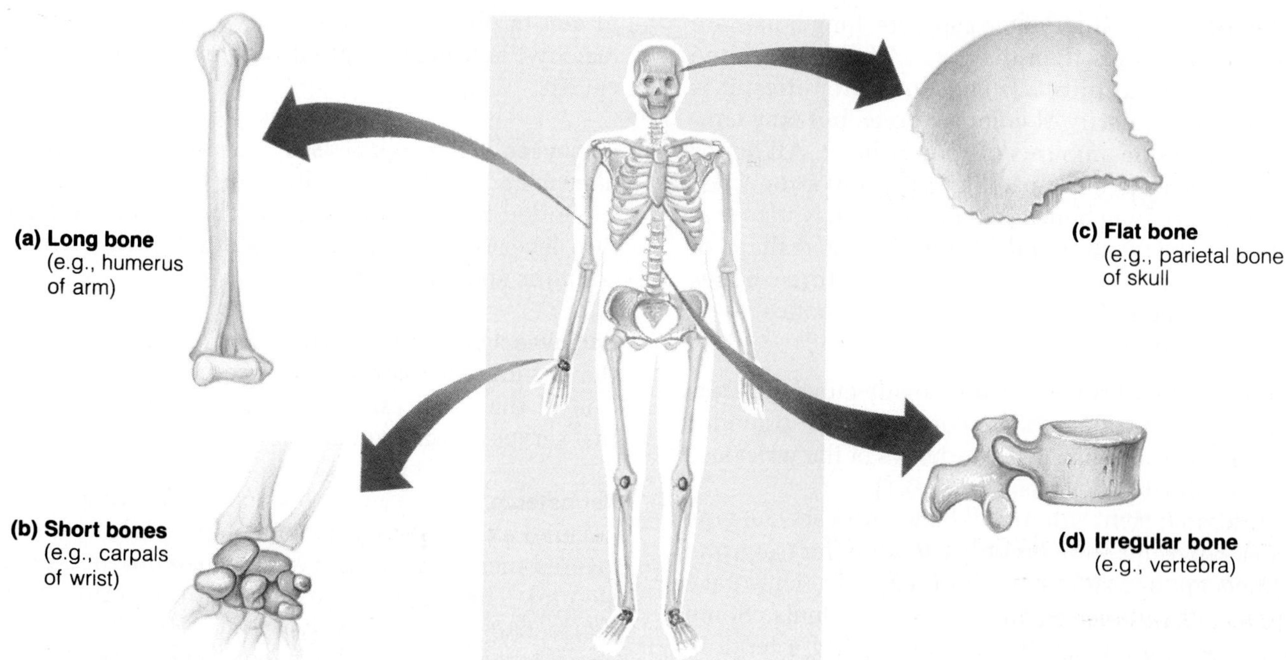

Figure 6.1 Classification of bones on the basis of shape.

Functions of the Bones

Besides contributing to body shape and form, our bones perform several important functions for the body:

1. Support. Bones provide a hard framework that supports and anchors all soft organs of the body. The bones of the legs act as pillars to support the body trunk when we stand, and the rib cage supports the thorax wall.

2. Protection. The fused bones of the skull provide a snug enclosure for the brain. The vertebrae surround the spinal cord, and the rib cage helps protect the vital organs of the thorax.

3. Movement. Skeletal muscles, attached to bones by tendons, use the bones as levers to move the body and its parts. As a result, we can walk, grasp, and breathe. The arrangement of bones and the design of joints determine the types of movement possible.

4. Storage. Fat is stored in the internal cavities of bones. Bone matrix itself serves as a storehouse, or reservoir, for minerals, the most important being calcium and phosphorus, although potassium, sodium, sulfur, magnesium, and copper are also stored. The stored minerals can be mobilized and released into the bloodstream in ionic form for distribution to all parts of the body as needed; indeed, "deposits" and "withdrawals" of minerals to and from the bony matrix go on almost continuously.

5. Blood cell formation. The bulk of blood cell formation, or *hematopoiesis* (hē″-mah-to-poy-ē′-sis), occurs within the marrow cavities of certain bones.

Classification of Bones

Bones come in many sizes and shapes. For example, the tiny pisiform bone of the wrist is the size and shape of a pea, whereas the femur, or thighbone, is nearly 2 feet long and has a large ball-shaped head. The unique shape of each bone fulfills a particular need. The femur, for example, must withstand great weight and pressure, and its cylindrical design provides maximum strength with minimum weight.

Bones are classified according to shape as long, short, flat, and irregular bones (Figure 6.1). Bones of different shapes contain different proportions of the two basic types of osseous tissue: compact and spongy bone. **Compact bone** is dense and looks smooth and homogeneous. **Spongy,** or **cancellous, bone** is composed of small needlelike or flat pieces of bone called *trabeculae* (truh-beh′-kyoo-lē), literally, "little beams," and has a good deal of open space. We will describe these bone textures in more detail when we discuss the microscopic structure of bone.

1. Long Bones. As their name suggests, long bones are considerably longer than they are wide. A long bone consists of a shaft plus two ends, or extremities; it is constructed primarily of compact bone, but may contain considerable amounts of spongy bone. All bones of the limbs, except the patella and those of the wrist and ankle, are long bones (see Figure 6.1). Notice that this bone classification reflects the elongated shape of the bones, not their overall size. The three bones structuring each of your fingers are long bones, even though they are very small.

2. Short Bones. Short bones are roughly cubelike. They contain mostly spongy bone; compact bone provides just their thin surface layer. The bones of the wrist and ankle are short bones (see Figure 6.1).

Sesamoid (seh'-suh-moyd) *bones* are a special type of short bone embedded within a tendon (for example, the kneecap, or patella) or a joint capsule. They vary in size and number in different individuals. Some clearly act to alter the direction of pull of a tendon; the function of others is not known.

3. Flat Bones. Flat bones are thin, flattened, and usually curved. They have two roughly parallel compact bone surfaces, with a layer of spongy bone between them. The sternum (breastbone), ribs, and most skull bones are flat bones (see Figure 6.1). Some flat bones are so thin that they consist only of a wafer-like layer of compact bone.

4. Irregular Bones. Bones that fit none of the preceding classes are called irregular bones. Irregular bones include some skull bones, vertebrae, and the hipbones (see Figure 6.1). All of these bones consist mainly of spongy bone enclosed by thin layers of compact bone.

Bone Structure

We will consider bone anatomy at three structural levels: the gross, microscopic, and chemical levels.

Gross Anatomy

Structure of a Typical Long Bone

With few exceptions, all long bones have the same general structure (Figure 6.2).

Diaphysis. The tubular **diaphysis** (dī-a'-fih-sis), or *shaft*, constitutes the long axis of the bone. It is constructed of a relatively thick *collar* of compact bone that surrounds a **medullary** (meh'-duh-layr"-ē) **cavity.** In adults, the medullary cavity contains fat (yellow marrow) and is also called the **yellow bone marrow cavity.**

Epiphyses. The **epiphyses** (ih-pih'-fih-sēz) are the bone ends, or extremities. In many cases, they are more expanded than the diaphysis. The epiphyses have a thin layer of compact bone externally; their interior contains spongy bone.

Epiphyseal Line. The **epiphyseal line** is a remnant of the epiphyseal plate, a cartilage present at the junction of the diaphysis and epiphysis in young bones that serves as a growth area for long bone lengthening.

Periosteum. The outer surface of the diaphysis is covered and protected by a glistening white, double-layered membrane called the **periosteum** (peh-rē-ah'-stē-um). The outer *fibrous layer* of the periosteum is dense irregular connective tissue; the inner *osteogenic layer*, abutting the bone surface, consists primarily of bone-forming cells, or **osteoblasts** (ah'-stē-ō-blasts). The periosteum is richly supplied with nerve fibers, lymphatic vessels, and blood vessels, which enter the bone via *nutrient canals*, or *nutrient foramina* (for-a'-mih-nuh). It is secured to the underlying bone by *Sharpey's fibers*, tufts of collagen fibers that extend from the fibrous layer into the bone matrix. The periosteum also provides an insertion or anchoring point for tendons and ligaments, and at these points, the Sharpey's fibers are exceptionally dense.

Endosteum The medullary cavity, composed of the cavities of spongy bone and the canals passing through the bones, is lined with a thin membrane called the **endosteum** (en-dah'-stē-um) (*endo* = within). The endosteum contains both osteoblasts and bone-destroying cells, or **osteoclasts.** Osteoclasts are large multinucleate cells of uncertain origin. The present theory is that they arise from immature hematopoietic cells, possibly the same cells that differentiate into macrophages. Bone reabsorption by osteoclasts is discussed later in the chapter.

Articular Cartilage. Where long bones articulate with one another at epiphyseal surfaces, the bony surfaces are covered with **articular** (hyaline) **cartilage** rather than periosteum. This glassy-textured cartilage cushions the bone ends and absorbs stress during joint movement.

Structure of a Typical Flat Bone

Flat bones (and many of the flattened irregular bones) are easy to describe. They consist of two relatively thin plates of compact bone separated by a middle

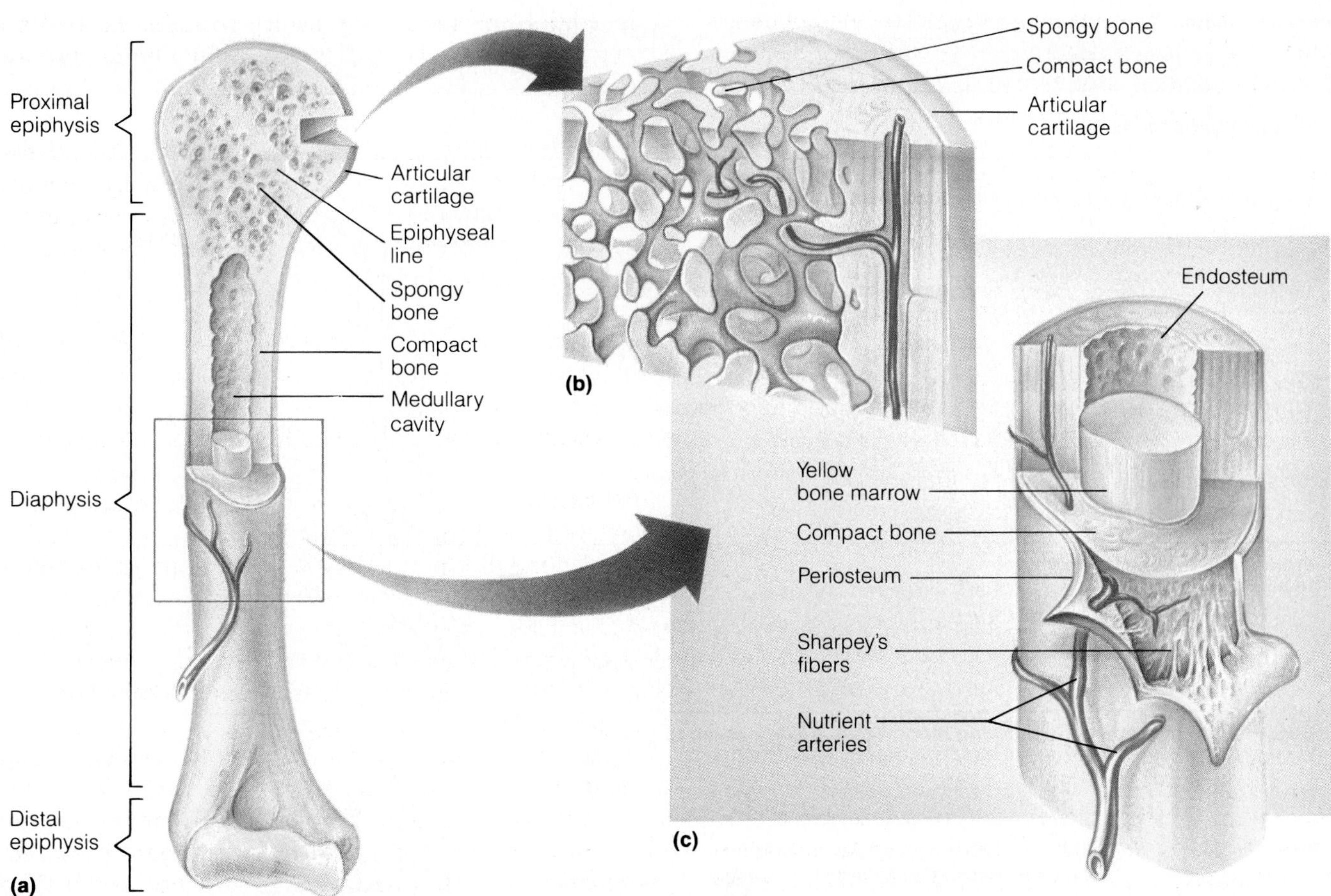

Figure 6.2 The structure of a long bone (tibia). (**a**) Anterior view with longitudinal section cut away at the proximal end. (**b**) Pie-shaped, three-dimensional view of spongy bone and compact bone of the epiphysis. (**c**) Cross section of shaft (diaphysis). Note that the external surface of the diaphysis is covered by a periosteum, but the articular surface of the epiphysis is covered with hyaline cartilage.

region of spongy bone called the **diploe** (dih′-pluh-ē″). Like the shaft of long bones, flat bone surfaces are covered with a periosteum, and endosteum lines internal cavities.

Location of Hematopoietic Tissue

The tissue promoting hematopoiesis is **red marrow,** typically found within the cavities of spongy bone of long bones and in the diploe of flat bones; these cavities are often referred to as **red marrow cavities.** In newborn infants, both the medullary cavity and all cavities of spongy bone contain red bone marrow. In most adult long bones, the fat-containing medullary cavity extends well into the epiphysis, and little red marrow is present in the spongy bone cavities; hence, blood cell production in adult long bones routinely occurs only in the head of the femur and humerus. Much more important and more active in hematopoiesis is the red marrow found in the diploe of flat bones (such as the sternum) and in some irregular bones (such as the hipbone). However, yellow marrow of the medullary cavity can revert to red marrow if a person becomes very anemic and needs enhanced red blood cell production.

Bone Markings

The external surfaces of bones are rarely smooth and featureless. Instead, they display bulges, depressions, and holes, which serve as sites of muscle, ligament, and tendon attachment, as points of articulation, or as conduits for blood vessels and nerves. These *markings* are named in different ways. Projections that grow outward from the bone surface include heads, trochanters, spines, and others, and each has certain distinguishing features and functions. Depressions and openings are termed fossae, sinuses, foramina, and

Table 6.1 Bone Markings

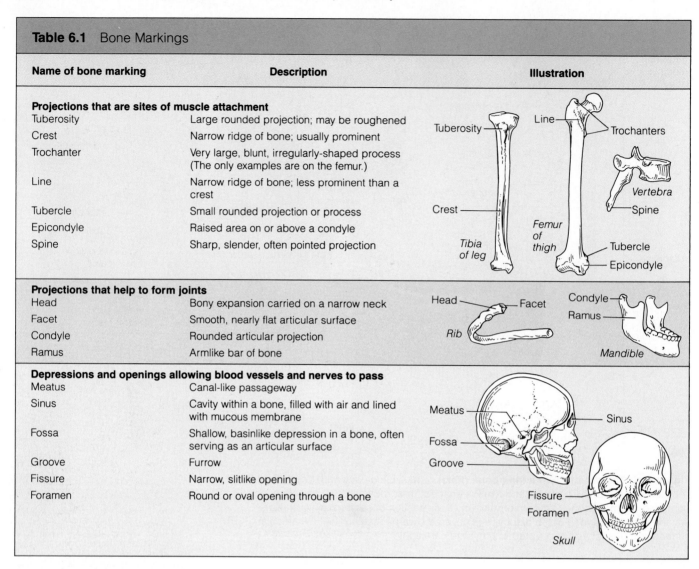

Name of bone marking	Description
Projections that are sites of muscle attachment	
Tuberosity	Large rounded projection; may be roughened
Crest	Narrow ridge of bone; usually prominent
Trochanter	Very large, blunt, irregularly-shaped process (The only examples are on the femur.)
Line	Narrow ridge of bone; less prominent than a crest
Tubercle	Small rounded projection or process
Epicondyle	Raised area on or above a condyle
Spine	Sharp, slender, often pointed projection
Projections that help to form joints	
Head	Bony expansion carried on a narrow neck
Facet	Smooth, nearly flat articular surface
Condyle	Rounded articular projection
Ramus	Armlike bar of bone
Depressions and openings allowing blood vessels and nerves to pass	
Meatus	Canal-like passageway
Sinus	Cavity within a bone, filled with air and lined with mucous membrane
Fossa	Shallow, basinlike depression in a bone, often serving as an articular surface
Groove	Furrow
Fissure	Narrow, slitlike opening
Foramen	Round or oval opening through a bone

grooves. The most important types of bone markings are described in Table 6.1. You should familiarize yourself with these terms because you will encounter them again as identifying marks of the individual bones studied in Chapter 7.

Microscopic Structure of Bone

Compact Bone

Observed with the unaided eye, compact bone appears very dense. However, a microscope reveals that it is riddled with canals and passageways serving as conduits for nerves, blood vessels, and lymphatic vessels (Figure 6.3). The structural unit of compact bone is called the **osteon** or **Haversian system.** Each osteon consists mostly of hard bone matrix arranged in concentric rings, or *lamellae* (luh-meh′-lē), around a central canal, the *Haversian canal*, oriented along the long axis of the bone. Based on this structural characteristic, compact bone is sometimes called **lamellar bone.** *Volkmann's canals* run at right angles to the long axis of the bone, connecting the vascular and nerve supply of the periosteum to those of the Haversian canals and the medullary cavity.

Spider-shaped **osteocytes** (mature bone cells) lie in small concavities, or *lacunae* (luh-koo′-nē), between the lamellae. Hairlike canals called *canaliculi* (ka″-nuh-lih′-kyoo-lī) connect the lacunae to each other and to the central Haversian canal. The manner in which these canaliculi are formed is interesting. In forming bone, the osteoblasts secreting the bone matrix maintain contact with one another by tentacle-like cytoplasmic projections. Then, as the matrix hardens and the maturing cells become trapped within it, an entire system of tiny canals (filled with tissue fluid and containing the osteocyte extensions) is formed. The canaliculi tie all the osteocytes in an osteon together, permitting easy diffusion of nutrients and wastes from and to the blood vessels in the Haversian canal.

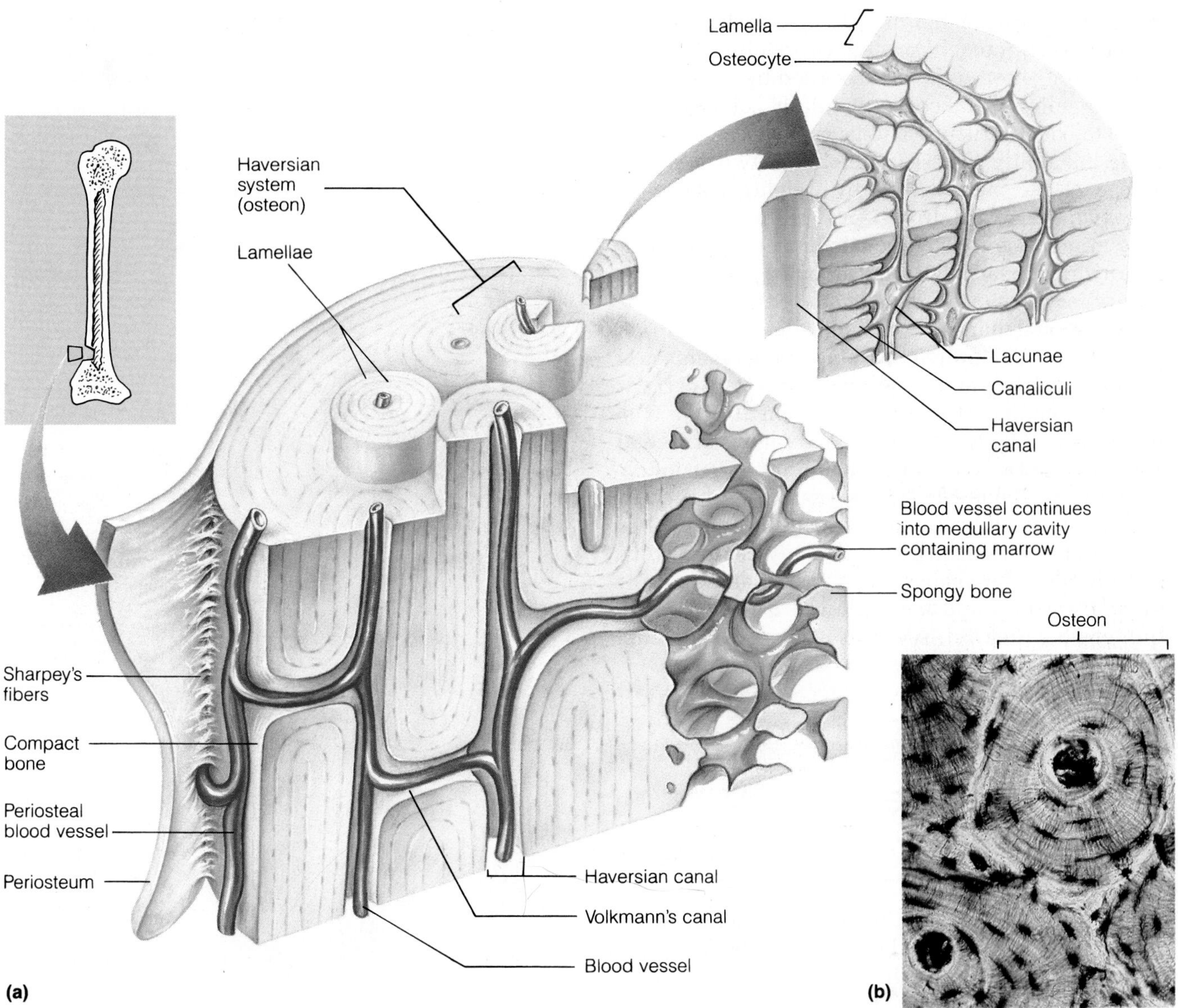

Figure 6.3 Microscopic structure of compact bone. (**a**) Diagrammatic view of a pie-shaped segment of compact bone, illustrating its structural units (osteons). The inset shows a more highly magnified view of a portion of one osteon. Note the position of osteocytes in lacunae (cavities in the matrix). (**b**) Photomicrograph of a cross-sectional view of one osteon and portions of others (90 x).

Although bone matrix is hard, its canaliculi allow bone cells to be very well nourished.

Matrix areas between intact osteons contain incomplete lamellae called *interstitial* (in-ter-stih'-shul) *lamellae*. These fill the gaps between forming osteons or represent remnants of osteons that have been cut through by bone remodeling (discussed later).

Spongy Bone

In contrast to compact bone, spongy bone, consisting of trabeculae, looks like a poorly organized tissue (see Figure 6.2b). However, the arrangement of the trabeculae is far from haphazard: It reveals where stress is exerted on the bone and helps the bone resist the stress as much as possible. Only a few cell layers thick, trabeculae contain irregularly arranged lamellae and osteocytes interconnected by canaliculi. No osteons are present. Nutrients reach the osteocytes of spongy bone by diffusion through the canaliculi from the marrow spaces between the bony spicules.

Chemical Composition of Bone

Bone is made of both organic and inorganic components. The *organic components* include the cells—osteoblasts, osteocytes, and osteoclasts—and approximately one-third of the matrix. The organic matrix elements are proteoglycans, glycoproteins, and collagen fibers, all of which are secreted by osteoblasts. These organic substances, particularly collagen, contribute to bone's structure and its great tensile strength, that is, its ability to resist stretch and twisting. The balance of the matrix (65% by weight) consists of *hydroxyapatites* (hī-drok″-sē-a′-puh-tīts), or *inorganic mineral salts,* largely calcium phosphate, calcium hydroxide, and calcium carbonate. Calcium salts account for the most notable characteristic of bone—its exceptional hardness, which allows it to resist compression. The proper combination of organic and inorganic matrix elements allows bones to be exceedingly durable and strong without being brittle. It is always surprising to learn that bone can resist 25,000 lb/in^2 of compression and 15,000 lb/in^2 of tension.

Because of bone salts, bones persist long after death and provide an enduring "monument." In fact, bone is so durable that skeletal remains many centuries old have revealed the shapes, sizes, races, and sexes of ancestral peoples and have allowed us to know the kinds of work they did, as well as many of the ailments (arthritis and others) they suffered.

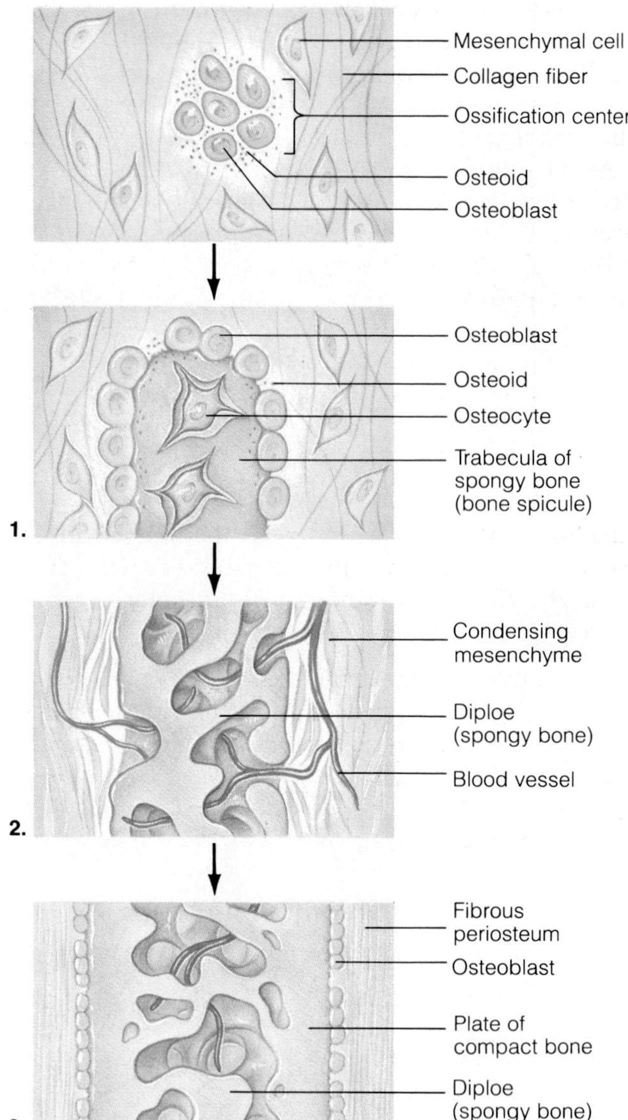

Figure 6.4 The stages in intramembranous ossification. The unnumbered top diagram illustrates the composition of the fibrous membrane as intramembranous ossification begins. Events depicted in diagrams 1 through 3 correspond with the description on p. 159. Note that diagrams 2 and 3 represent much lower magnification than the diagrams preceding them.

Bone Development (Osteogenesis)

Osteogenesis (ah″-stē-ō-jeh′-nuh-sis) and **ossification** are synonyms indicating the process of bone formation. This process occurs in two somewhat different forms. In embryos it leads to the *formation of the bony skeleton* from existing fibrous or cartilaginous structures. *Bone growth,* another form of ossification, goes on until early adulthood as we continue to increase in size. In fact, bones are capable of growing in thickness throughout life. However, ossification in adults serves mainly for *remodeling* and repair of bones and involves modification of existing bone structures.

Formation of the Bony Skeleton

At six weeks, the skeleton of a human embryo is constructed entirely from fibrous membranes and hyaline cartilage. Bone begins to develop at about this time and eventually replaces most of the fibrous or cartilage structures. When a bone forms from a fibrous membrane, the process is called *intramembranous ossification,* and the bone is called a *membrane bone.* Bone formation from hyaline cartilage structures is called *endochondral ossification* (*endo* = within; *chondro* = cartilage), and the bone is called a *cartilage bone.*

Intramembranous Ossification

Intramembranous ossification results in the formation of flat bones and some irregular bones, including bones of the skull and the clavicles. Fibrous connective tissue membranes formed by mesenchymal cells serve as the initial supporting structures on which ossification occurs. Essentially, the process involves the following steps (Figure 6.4).

1. Formation of spongy bone within the fibrous membrane. Mesenchymal cells in the fibrous membrane differentiate into osteoblasts and begin to secrete the organic bone matrix, called *osteoid* (os′-tē-oyd), around the collagen fibers. This process may begin at several sites, but typically one major *ossification center* appears in the center of the membrane.

As the deposits of osteoid accumulate, trabeculae typical of spongy bone appear. The osteoblasts also release an enzyme called *alkaline phosphatase* (al′-kuh-līn fos′-fuh-tās) that encourages deposit of calcium salts within the osteoid. Within a few days, the osteoid is mineralized and converted to a true bone matrix. Once trapped in lacunae, the osteoblasts differentiate into mature bone cells or osteocytes, and the vascular tissue within the spongy bone differentiates into hematopoietic tissue (red marrow), completing the formation of the diploe.

2. Formation of the periosteum. As the diploe forms, a layer of vascular mesenchyme condenses on the outside of the fibrous membrane and becomes transformed into a two-layered periosteum.

3. Formation of compact bone plates. After the periosteum forms, mesenchymal cells in its inner, osteogenic layer become osteoblasts and secrete osteoid against the membrane surface, forming trabeculae. Thus, initially, the bone collar, like the bone formed within the membrane, is spongy bone and is referred to as *woven bone*. Later, the plates of woven bone are replaced with compact bone as the areas between the trabeculae are filled with concentric lamellae (osteons). The compact bone is mineralized and eventually totally encloses the diploe.

Endochondral Ossification

The long and short bones—that is, most bones—form by the process of **endochondral** (en-dō-kon′-drul) **ossification,** which uses hyaline cartilage "bones" as models, or patterns, for bone construction. The process begins in the third month of development and is more complex than intramembranous ossification because the hyaline cartilage must be broken down as ossification proceeds. The formation of a long bone typically begins at a *primary ossification center* at the center of the hyaline cartilage shaft. First, the perichondrium (the fibrous connective tissue membrane covering the hyaline cartilage "bone") becomes infiltrated with blood vessels, converting it to a vascularized periosteum. As a result of the change in nutrition, the underlying chondroblasts respecialize into osteoblasts. The stage is now set for ossification to begin, as illustrated in Figure 6.5.

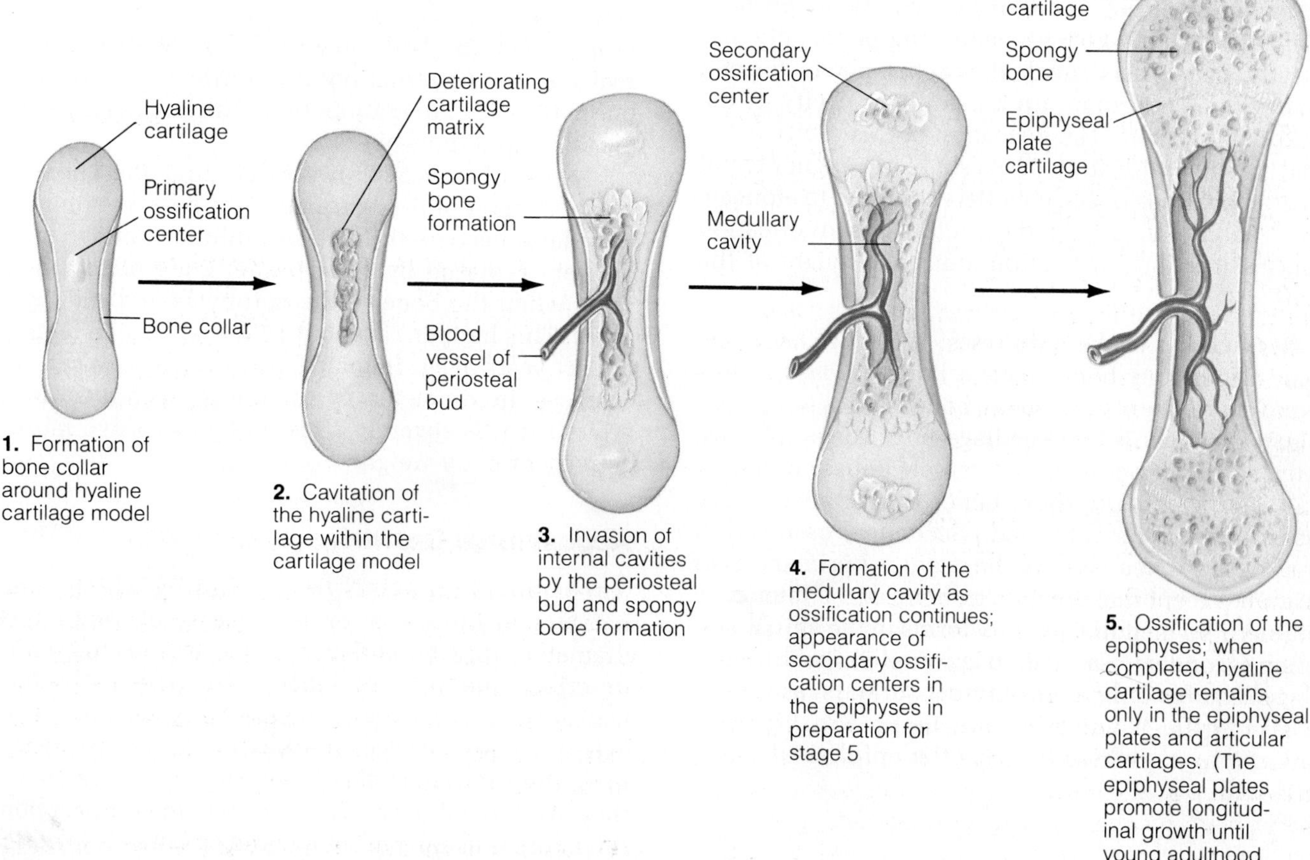

Figure 6.5 Stages in endochondral ossification occurring in a long bone.

1. Formation of a bone collar around the hyaline cartilage model. Osteoblasts of the newly "converted" periosteum begin to secrete osteoid against the hyaline cartilage shaft, forming woven bone. The mineralization of the matrix, the entrapment of osteoblasts in lacunae and their conversion to osteocytes, and the eventual transformation of woven bone into compact bone occur in the same manner as in intramembranous ossification.

2. Cavitation of the hyaline cartilage shaft. While the compact bone collar forms externally, the chondrocytes within the shaft begin to hypertrophy (enlarge) and secrete alkaline phosphatase, promoting mineralization of the matrix. As the avascular cartilage matrix becomes calcified and impermeable to diffusing nutrients, the chondrocytes die and the cartilage matrix they maintained begins to deteriorate. Although this opens up cavities, the deteriorating hyaline cartilage shaft is stabilized externally by the bone collar.

3. Invasion of internal cavities by the periosteal bud and spongy bone formation. The forming cavities are quickly invaded by a collection of elements (blood vessels, lymphatics, nerve fibers, red marrow elements, osteoblasts, and osteoclasts) called the *periosteal bud*. The entering osteoblasts secrete osteoid around the remaining fragments of hyaline cartilage, forming bone-covered trabeculae.

4. Formation of the medullary cavity. As the primary ossification center enlarges, spreading proximally and distally, osteoclasts break down the newly formed spongy bone and open up a medullary cavity in the center of the shaft. The formation of this cavity is the final step in ossification of the shaft. Throughout fetal life, the hyaline cartilage models continue to elongate by division of viable cartilage cells; thus, ossification "chases" cartilage formation along the length of the shaft.

5. Ossification of the epiphyses. When we are born, most of our long bones have a bony diaphysis, surrounding remnants of spongy bone, a widening medullary cavity, and two cartilage ends, or epiphyses. Later, secondary ossification centers appear in one or both epiphyses. (In short bones, only the primary ossification center is formed.) Secondary ossification reproduces almost exactly the events of primary ossification, except that the spongy bone in the interior is retained: No medullary cavity forms in the epiphyses. When secondary ossification is complete, hyaline cartilage remains only at epiphyseal surfaces, as the *articular cartilage*, and at the junction of the diaphysis and epiphysis, where it forms the *epiphyseal plate*, or growth plate.

Bone Growth

During youth, long bones continue to grow in length through continued activity of the epiphyseal plates, and all bones grow in thickness by a process called *appositional* (a-puh-zih'-shuh-nul) *growth*. Most bones stop growing during adolescence or in early adulthood. However, some facial bones, such as those of the nose and lower jaw, continue to grow almost imperceptibly throughout life.

Growth in Length of Long Bones

The process of **longitudinal bone growth** mimics many of the events of endochondral ossification. The hyaline cartilage of the epiphyseal plate grows by mitosis of the cells on its *distal face* (the portion of the plate farthest from the medullary cavity), as shown in Figures 6.6 and 6.7. At the same time, the chondrocytes on the proximal face of the plate (next to the medullary cavity) age and become hypertrophied. Their matrix calcifies and the cells die, resulting in the deterioration of their matrix. Osteoblasts in the medullary cavity then ossify the cartilage spicules, forming spongy bone that is eventually digested by osteoclasts.

Longitudinal growth is accompanied by almost continuous remodeling of the epiphyseal ends to maintain the proper proportions between the diaphysis and epiphyses (see Figure 6.7). Bone remodeling, involving both new bone formation and bone reabsorption (destruction) is described in more detail on p. 162–164, in conjunction with the changes that occur in adult bones.

As adolescence draws to an end, the chondroblasts of the epiphyseal plates divide less often and the plates become thinner and thinner until they are entirely replaced by bone tissue. Long bone growth ends when the bone of the epiphysis and diaphysis fuses. This happens at about 18 years of age in females and 21 years of age in males. However, a bone can still increase in diameter or thickness by appositional growth if it is severely stressed by excessive muscle activity or body weight.

Appositional Growth

Appositional growth is the process by which bones increase in thickness, or, in the case of long bones, diameter. *Appositional* means "placed next to," which describes the process fairly well. The osteoblasts beneath the periosteum secrete bony matrix on the external bone surface; grooves on the bone surface form the central canals of new Haversian systems as they are "boned over." This increase in compact bone thickness is usually accompanied by bone destruction by osteoclasts on the endosteal surface of the compact

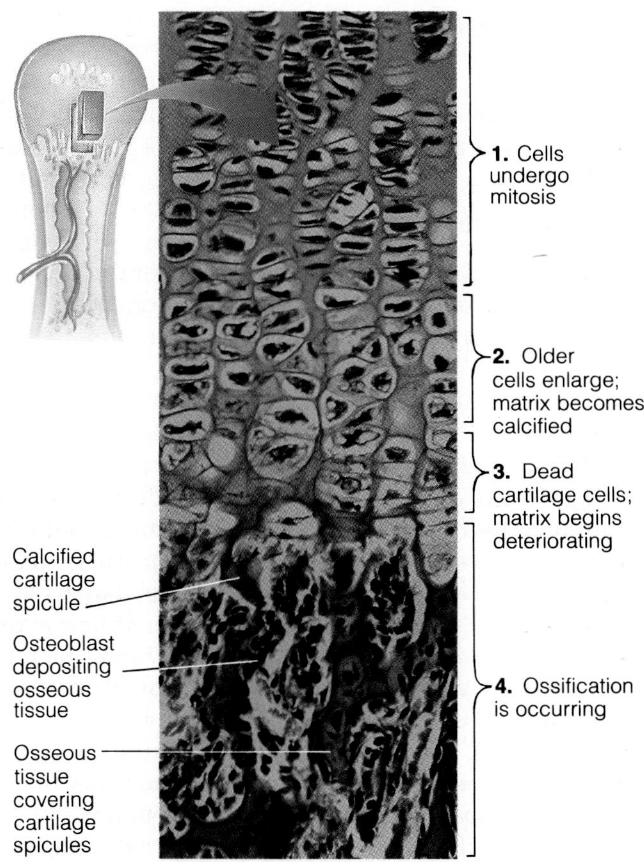

Figure 6.6 Long bone growth. The region of the epiphyseal plate closest to the epiphysis (distal face) contains resting cartilage cells. As shown in the photomicrograph (250 x), the cells of the epiphyseal plate proximal to the resting cartilage area are arranged in four functionally differing layers. (**1**) Consecutive layers of mitotic cells. (**2**) Cartilage cells undergoing hypertrophy, followed by calcification of their matrix. (**3**) An area of dead cartilage cells; matrix begins to deteriorate. (**4**) On the proximal face of the epiphyseal plate next to the medullary cavity, ossification is occurring as osteoblasts deposit bone matrix around the cartilage remnants, forming bony trabeculae.

bone (see Figure 6.7); however, there is normally less breaking down than building up. This unequal process produces a thicker and stronger bone but prevents it from becoming too heavy.

Hormonal Regulation of Bone Growth During Youth

The growth of bones that occurs throughout childhood and ends in early adulthood is exquisitely controlled by a symphony of hormones acting collectively. During infancy and childhood, the single most important stimulus of epiphyseal plate activity is *growth hormone* released by the anterior pituitary gland. Growth hormone actually acts indirectly, prodding the liver to release a growth factor called *somatomedin,* which in turn stimulates growth of the epiphyseal plate cartilage. Thyroid hormones (T_3 and T_4) modulate the activity of growth hormone, ensuring

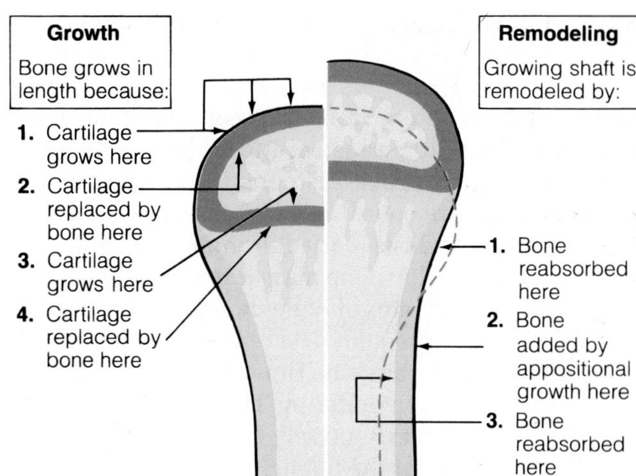

Figure 6.7 Long bone growth and remodeling during youth. The events indicated at the left depict the process of endochondral ossification that occurs at the articular cartilages and epiphyseal plates as the bone grows in length. The events indicated at the right reveal the process of bone remodeling that occurs during long bone growth to maintain proper bone proportions.

that the skeleton has proper proportions as growth occurs. At puberty, male and female sex hormones (testosterone and estrogens, respectively) are released in increasing amounts and promote the growth spurt typical of the adolescent, as well as the masculinization or feminization of specific parts of the skeleton. Excesses or deficits of any of these hormones can result in obvious skeletal malproportion. For example, hypersecretion of growth hormone in children results in excessive height (gigantism), and deficits of growth hormone or thyroid hormone result in characteristic types of dwarfism. Hormone activity is considered in detail in Chapter 17.

Bone Homeostasis: Remodeling and Repair

Bones appear to be the most lifeless of body organs, and once they are formed, they seem set for life. But appearances can be deceiving. Bone is a very dynamic and active tissue. As much as half a gram of calcium may enter or leave the adult skeleton each day! Unlike predominantly cellular tissues, in which renewal takes place mainly at the cellular (or even molecular) level, renewal, or revitalization, of bone occurs at the tissue level. Large volumes of bone are removed and replaced, and bone architecture continually changes. And when we break bones—the commonest disorder of bone homeostasis—they undergo a remarkable process of self-repair that we will describe shortly.

Bone Remodeling

In the adult skeleton, bone deposit and bone reabsorption occur at all periosteal and endosteal surfaces. Together, the two processes constitute **bone remodeling,** and they are coordinated by "packets" of osteoblasts and osteoclasts called *remodeling units.* In healthy young adults, bone mass remains constant, an indication that the rates of bone deposit and reabsorption are essentially equal. The remodeling process is not uniform and, at any one time, most bone surfaces are *not* involved in remodeling. However, some bones (or bone areas) are very actively remodeled. For example, the distal part of the femur, or thighbone, is fully replaced every five to six months, whereas its shaft undergoes alteration much more slowly.

Bone deposit occurs at sites of bone injury or where added bone strength is required. Sites of new matrix deposit are revealed by the presence of an *osteoid seam,* an unmineralized band of bone matrix that is typically 10–12 μm wide. Between the osteoid seam and the older mineralized bone, there is an abrupt transition called the *calcification front.* Because the osteoid seam is always of constant width and the change from unmineralized to mineralized matrix is sudden, some authorities theorize that the osteoid must "mature" before it can be calcified. This maturation is under the control of the osteoblasts and appears to involve several biochemical changes that occur over a period of 10 to 12 days. Once the proper conditions are present, calcium salt crystallization occurs simultaneously throughout the "matured" matrix.

Bone reabsorption is accomplished by osteoclasts, which secrete lysosomal and perhaps other catabolic enzymes onto the free bone surface. There is also some evidence that osteoclasts may phagocytize the matrix. Osteoclasts have a clear *ruffled border,* a region rich in actin filaments that anchor the cells to the surface being digested (Figure 6.8). Various metabolic acids, such as lactic and carbonic acids, accumulate in the area during the erosion process. As the organic matrix is degraded and the mineral salts are dissolved, the released calcium enters the interstitial fluid and then the blood in ionic form.

Control of Remodeling

Bone remodeling is regulated by two control loops. One is a negative feedback hormonal mechanism; the other involves mechanical and gravitational forces acting on the skeleton. The hormonal mechanism involves the interaction of *parathyroid hormone (PTH),* produced by the parathyroid glands, and *calcitonin* (kal″-sih-tō′-nin), produced by parafollicular cells (C cells) of the thyroid gland (Figure 6.9). PTH, released when blood levels of ionic calcium decline, enhances calcium release from bone matrix by stimulating osteoclast activity and bone reabsorption. The osteoclasts are no "respecters" of matrix age. When activated, they break down both old and fairly new matrix. Only osteoid, devoid of calcium salts, escapes digestion. As blood concentrations of calcium increase, the stimulus for PTH release ends. Calcitonin, secreted when blood calcium levels rise, inhibits bone reabsorption and causes calcium salts to be deposited in bone matrix, effectively reducing blood calcium levels. As blood calcium levels fall, calcitonin release wanes as well.

The hormonal control loop acts to maintain blood calcium homeostasis, rather than the skeleton's strength or well-being. In fact, if blood calcium levels are low for an extended time, the bones will become so demineralized that they develop large, "punched-out"-looking holes. Thus, the bones serve as a storehouse from which ionic calcium is drawn to meet the body's needs.

Ionic calcium is necessary for an amazing number of physiological processes, including transmission of nerve impulses and release of neurotransmitters, con-

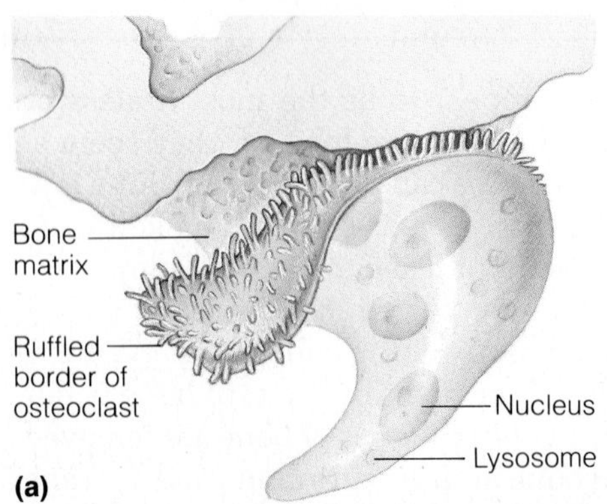

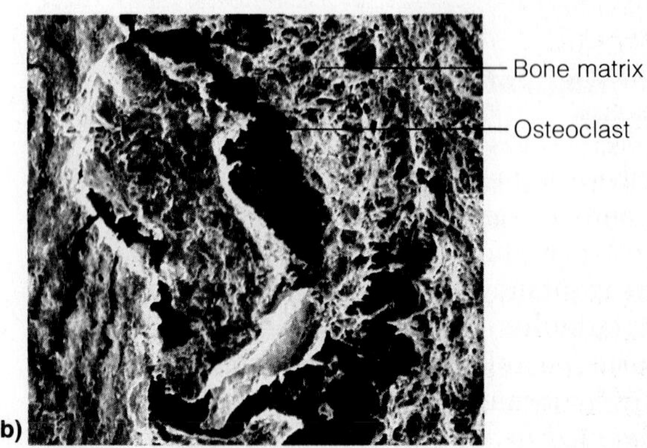

Figure 6.8 Osteoclasts are bone-degrading cells. (**a**) Osteoclasts are large, multinucleate cells that exhibit a ruffled border on the surface being reabsorbed. (**b**) Scanning electron micrograph of an osteoclast involved in matrix breakdown (500×).

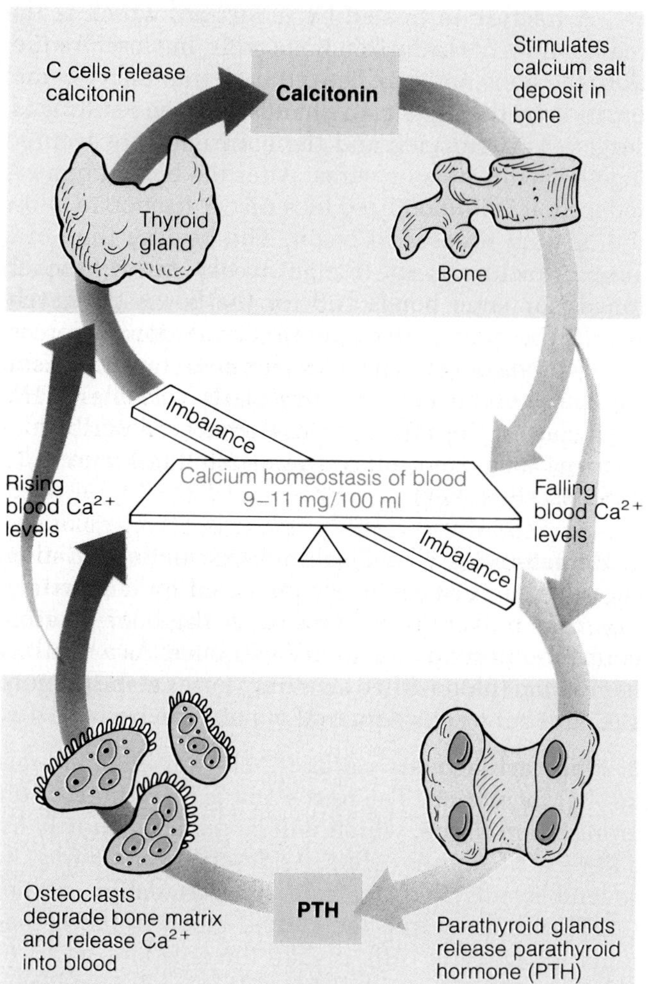

Figure 6.9 Hormonal controls of ionic calcium levels in the blood. PTH and calcitonin operate in negative feedback control systems that influence each other.

traction of muscles, blood coagulation, secretion by glands, and cell division. It is also an essential component of all plasma membranes. The human body contains about 1200–1400 g of calcium, over 99% of which is present as bone minerals. Most of the balance is intracellular, that is, within body cells. Less than 1.5 g is present in blood, and ionic blood calcium levels are normally maintained within the very narrow range of 9–11 mg per 100 ml of blood by the hormonal control loop. Calcium is absorbed from the intestine under the control of vitamin D metabolites; the daily requirement for young adults is 600–1000 mg.

Minute changes from the homeostatic range for ionic blood calcium can lead to severe neuromuscular problems ranging from hyperexcitability to inability to function. In addition, sustained *hypercalcemia* (hī″-per-kal-sē′-mē-uh), high blood levels of ionic calcium, can lead to the deposit of calcium salts in the blood vessels, kidneys, and other soft organs, which may hamper the functioning of these organs. ■

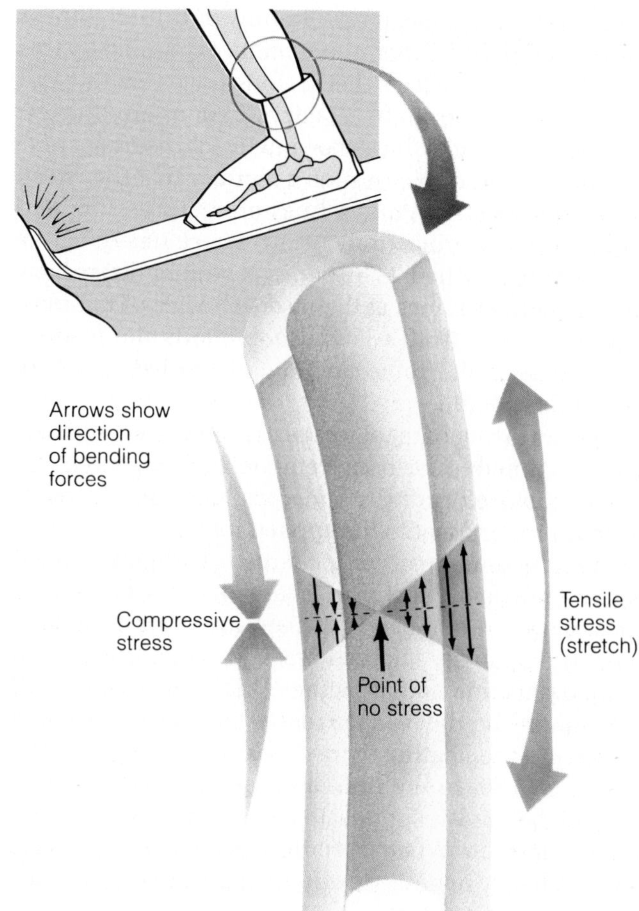

Figure 6.10 Effect of bending on a long bone. The event shown here illustrates the consequence when a skier is stopped abruptly by some solid object or by deep grooves in the snow. As the feet are "stopped in their tracks," the body continues its forward momentum, resulting in compression of the anterior aspects and stretching of the posterior aspects of bones of the leg. Since forces due to tension (stretch) and compression dissipate toward the bone center, the long bone interior may be hollowed out or may contain less-dense spongy bone without impairing the bone's ability to accommodate bending stress.

The second mechanism for regulating bone remodeling involves the response of bone to mechanical stress (muscle pull) and gravity. Unlike the hormonal mechanism, this set of controls serves the needs of the skeleton itself, keeping the bones strong where stressors are acting. *Wolff's law*—accepted by some, but not all, scientists—holds that a bone grows or remodels in response to the forces or stresses placed on it. Perhaps the very first concept to understand is that long bone structure itself quite specifically reflects the common stresses placed on such bones. For example, most long bones are stressed, or loaded, by bending, a combination of *compression* and *tension* (stretch) forces acting on opposite sides of the shaft (Figure 6.10). But since both forces are minimal toward the center of the bone, it can "hollow out" for lightness without jeopardy. Other phenomena explained by this theory include the facts that (1) long bones are thickest midway along the shaft; (2) curved bones are thickest where they are most likely to buckle; (3) the trabeculae of bone are laid down in such a manner that they form buttresses, or struts, along lines of compression; and (4) large, bony projections occur where heavy, active muscles are attached. (The bones of weight lifters have enormous thickenings at the insertion sites of the most used muscles.) Wolff's law also explains the featureless bones of the fetus and the atrophied bones of bedridden people.

How do mechanical forces communicate with the cells responsible for bone remodeling? Research has shown that deforming a bone produces an electrical current proportional to the applied force; regions under tension become positively charged, and compressed regions become negatively charged. This has led to the suggestion that newly formed matrix is deposited around negatively charged areas, causing the growth of compact bone in that region. This principle, as well as some of the devices currently used to speed bone repair and the healing of fractures, is discussed in the box on p. 166. In the final analysis, the mechanisms by which bone responds to mechanical stimuli are still uncertain. What is certain is that heavy usage leads to heavy bones (or bony areas), and nonuse leads to rapid bone wasting.

The skeleton is continuously subjected to both hormonal influences and mechanical forces. At the risk of constructing too large a building on too small a foundation, one can speculate that PTH (along with calcitonin during youth) is the major determinant of *whether* and *when* remodeling will occur in response to changing blood calcium levels, while mechanical and gravitational forces determine *where* remodeling occurs. For example, if bone must be reabsorbed to increase blood calcium levels, PTH is released and targets the osteoclasts. However, mechanical forces (perhaps through electrical signals) determine which osteoclasts are most sensitive to PTH stimulation, so that bone in the least stressed areas (which is temporarily dispensable) is broken down.

Repair of Fractures

Despite their remarkable strength, bones are susceptible to **fractures,** or breaks, all through life. During youth, most fractures result from exceptional trauma that twists or smashes the bones. Sports activities such as football and skiing jeopardize the bones, and automobile accidents certainly take their toll. In old age, bones thin and weaken, and fractures occur with increased frequency. There are many common types of fractures, as exemplified in Table 6.2.

A fracture is treated by *reduction,* which is the realignment of the broken bone ends. In *closed reduction,* the bone ends are coaxed back into their normal position by the physician's hands; in *open reductions,* surgery is performed and the bone ends are secured together with pins or wires. After the broken bone is reduced, it is immobilized by a cast or traction to allow the healing process to begin. The healing time of a simple fracture is six to eight weeks, but it is much longer for large bones and for the bones of elderly people (because of their poorer circulation).

Four major cell types take part in the repair process: capillary endothelial cells, fibroblasts, osteoblasts, and osteoclasts. The repair process can be described in terms of four major phases, illustrated in Figure 6.11 and described next.

1. Hematoma (hē″-muh-tō′-muh) **formation.** In most cases, bone breakage is accompanied by hemorrhaging from broken blood vessels in the bone, periosteum, and perhaps surrounding tissues. As a result, a *hematoma* (blood-filled swelling) forms at the fracture site, and bone cells deprived of nutrition begin to die.

2. Fibrocartilaginous callus (fī′-brō″-kar-tih-la′-jih-nus ka′-lis) **formation.** The next event is formation of soft *granulation tissue,* which will be replaced shortly by a fibrocartilaginous callus. As described in Chapter 4, several events help to form the granulation tissue: Capillaries grow into the clotted blood of the hematoma; phagocytic cells invade the area and begin to clean up the debris; and fibroblasts and osteoblasts migrate into the fracture site from the periosteal and endosteal membranes. The fibroblasts produce collagen fibers that span the break and connect the broken bone ends, and some differentiate into chondroblasts that secrete cartilage matrix. Within this fibrocartilage, osteoblasts begin forming spongy bone, but those farthest from the capillary supply secrete an externally bulging cartilage-like matrix that later becomes calcified. This entire mass of repair tissue, called the *fibrocartilaginous callus,* splints the broken bone. The portion that directly connects the broken bone ends is called the **internal callus;** the part protruding from the outer bone surface is referred to as the **external callus.**

3. Bony callus formation. Osteoblasts and osteoclasts continue to migrate inward and multiply rapidly in the fibrocartilaginous callus, which, as a result of their labors, is gradually converted to a *bony callus* of spongy bone.

4. Remodeling. Over a period of months, the bony callus is remodeled. The excess material on the outside of the bone shaft and within the medullary cavity is removed, and compact bone is laid down to reconstruct the shaft. The final structure of the remodeled area represents its response to mechanical stimuli acting on the repaired bone.

Table 6.2 Common Types of Fractures

Fracture type	Illustration	Description	Comments
Simple		Bone breaks cleanly, but does not penetrate the skin	Sometimes called a closed fracture
Compound		Broken ends of the bone protrude through soft tissues and the skin	More serious than a simple fracture; may result in a severe bone infection (osteomyelitis), requiring massive doses of antbiotics
Comminuted		Bone fragments into many pieces	Particularly common in the aged, whose bones are more brittle
Compression		Bone is crushed	Common in porous bones (i.e., osteoporotic bones)
Depressed		Broken bone portion is pressed inward	Typical of skull fracture
Impacted		Broken bone ends are forced into each other	Commonly occurs when one falls and attempts to break the fall with outstretched arms; also common in hip fractures
Spiral		Ragged break occurs when excessive twisting forces are applied to a bone	Common sports fracture
Greenstick		Bone breaks incompletely, much in the way a green twig breaks	Common in children, whose bones have relatively more inorganic matrix and are more flexible than those of adults

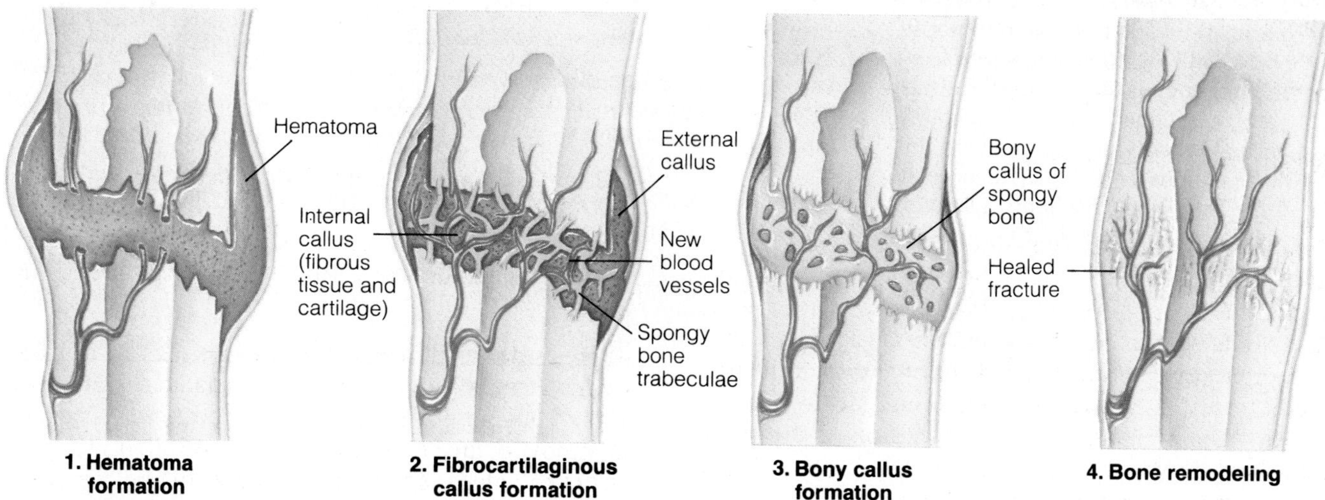

Figure 6.11 Phases of fracture healing.

A CLOSER LOOK

Them Bones, Them Bones Goin' to Walk Around—Clinical Advances in Bone Repair

Although bones have remarkable self-regenerative powers, some conditions frustrate even their most heroic efforts. Outstanding examples include extensive shattering (as in automobile accidents), inadequate circulation in old bones, severe bone atrophy, and certain birth defects. However, recent medical advances are allowing us to deal with some problems that bones cannot handle themselves.

Large fractures and fractures of old bones tend to heal very slowly and often poorly. **Electrical stimulation of fracture sites** has dramatically increased the rapidity and completeness of bone healing in these cases, but just how electricity accomplishes this feat was not known until very recently. It now appears that the electrical fields prevent parathyroid hormone (PTH) from acting on the osteoclasts at the fracture site, thus allowing the formation of bony tissue to increase and accumulate.

A delicate, experimental procedure called the **free vascular fibular graft technique** uses pieces of the fibula (the thin, sticklike bone of the leg) to replace missing or severely damaged bone areas. The fibula is a nonessential bone that does not bear weight in humans; it serves primarily to stabilize the ankle. In the past, extensive bone grafts usually failed because an adequate blood supply could not reach their interior, necessitating eventual amputation. But this new technique, which transplants normal blood vessels along with the bone section, is being used successfully to replace large bone segments. Free vascular fibular grafts have been used to construct a radius (forearm bone) for a child born without one and to replace long bones destroyed by osteomalacia or accidental trauma. Subsequent remodeling of the bones leads to an almost perfect replica of the normally present bone.

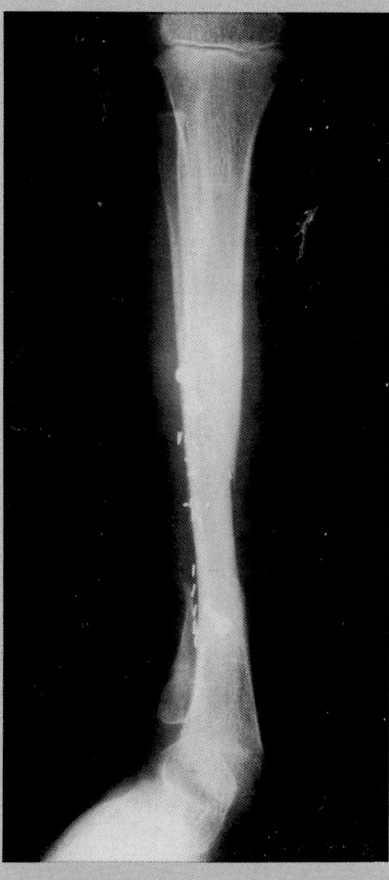

Anteroposterior x-ray of a healed fibular graft transplant.

Research on bone replacement materials has produced **artificial bone** made of a biodegradable ceramic substance known as TCP. This product is soft enough to be carved into the desired shape, but it is not very strong. Its biggest application has been to replace parts of non-weight-bearing bones, such as skull bones. Since the material is porous, osteoblasts enter it and lay down bone matrix, and in most instances, the TCP has been completely degraded and replaced by new bone within two years.

Another exciting technique, using **crushed bone** from human cadavers, also induces the body to form new bone of its own. This means that surgeons can mold bones where none previously existed and avoid the often painful and time-consuming procedure of bone transplant or grafting. The bone powder is mixed with water to form a paste that can be molded to the shape of the desired bone. When implanted, each bone speck becomes surrounded by connective tissue cells, which form cartilage; later, the cartilage is replaced by bone, incorporating the product into the new bony matrix. The major use of this technique thus far has been to treat children born with skeletal defects such as a misshapen skull, missing nose, or cleft palate. However, its widest application will probably be to treat periodontal disease, a condition of bone loss around the teeth, which often results in tooth loss. Advantages of crushed bone include ease of use, ability to pack it into small, difficult-to-reach areas, and the fact that it can be stored (and thus available for immediate use in victims of bone trauma).

Homeostatic Imbalances of Bone

Interferences with the remodeling process, that is, imbalances between bone formation and bone reabsorption, underlie nearly every disease that influences the adult skeleton. The most common skeletal disorders are those associated with decreased skeletal mass—when bone reabsorption exceeds bone formation. It is presumed that most such disorders are caused by imbalances in hormones or other chemicals in the blood.

Osteoporosis

Osteoporosis (os″-tē-ō-por-ō′-sis) refers to a group of diseases in which bone reabsorption on endosteal surfaces outpaces bone deposit beneath the periosteum. Although the bone mass is reduced, the chemical composition of the matrix remains normal. Osteoporotic bones become more porous and lighter. On an X-ray, the compact bone looks thinner and less dense than normal, and spongy bone has fewer trabeculae. The loss of bone mass is not itself a problem, but it frequently leads to specific types of fractures and deformities. For example, even though the entire skeleton is involved in the osteoporotic process, the spine is most vulnerable, and compression fractures of the vertebrae are frequent. The thighbone, particularly its neck, is also increasingly susceptible to fracture.

Osteoporosis occurs most often in the aged, particularly post-menopausal women; in fact, it is the most important cause of fractures in women over the age of 50 years. Estrogen helps to maintain the health and normal density of a woman's skeleton. However, after menopause, estrogen secretion wanes, and estrogen deficiency is strongly implicated in osteoporosis in older women. Other factors that may contribute to age-related osteoporosis include insufficient exercise to stress the bones, inadequate intake of calcium and protein, abnormalities of vitamin D metabolism, malabsorption problems, and hormone-related conditions (such as the normal decrease in calcitonin secretion that occurs with age, the intake of corticosteroid drugs, and diabetes mellitus). In addition, osteoporosis can occur at any age as a result of immobility. At present, no one abnormality or deficiency can explain all cases of osteoporosis. The condition is usually treated by supplemental calcium and vitamin D, increased exercise, and, in selected cases, estrogen replacement.

Osteomalacia

The term **osteomalacia** (os″-tē-ō-muh-lā′-shē-uh), literally, "soft bones," is applied to a number of disorders in which the bones are inadequately mineralized. Osteoid is produced, but since calcium deposit does not occur, bones soften and become structurally weak. Hence, bones may fracture or bend and deform under weight bearing. This is most obvious in the legs and pelvis. The characteristic symptom is pain when weight is put on the affected bones. **Rickets,** the analogous disease in children, is accompanied by many of the same signs and symptoms, such as bowed legs and a deformed pelvis; additionally, malformities of the head and rib cage are common. However, since bones are still growing rapidly, the condition is much more severe. Because the epiphyseal plates cannot be calcified, they continue to widen, and the ends of long bones become visibly enlarged.

The usual cause of osteomalacia is vitamin D deficiency, and vitamin D supplementation (often accompanied by calcium supplements) is the most common treatment for osteomalacia. The major dietary sources of vitamin D are vitamin D-supplemented dairy products. In addition, vitamin D is formed in skin exposed to sunlight.

Paget's Disease

Paget's (pa′-jits) **disease** is characterized by excessive and abnormal bone reabsorption and formation. Newly formed bone, called *Pagetic bone,* has an abnormally high ratio of woven bone to compact bone; this, along with reduction in bone mineralization, leads to bone softening and weakness. Late in the disease, osteoclast activity wanes, but osteoblasts continue to function, often forming irregular bone thickenings or filling the marrow cavity with Pagetic bone.

Although Paget's disease may affect many parts of the skeleton, it is usually a localized and intermittent condition. Any bone can be affected, but the spine, pelvis, femur, and skull are most often involved and become increasingly deformed. The condition, equally common in both sexes, is rarely seen before the age of 40 years. Its cause is unknown, but it is believed that a latent virus may initiate the disease. Its treatment is complex and beyond the scope of this book. ■

Developmental Aspects of Bones: Timing of Events

It can be said that the bones are on a timetable from the time they are formed until death. The mesodermal germ layer gives rise to embryonic mesenchymal cells, which in turn produce the fibrous membranes and cartilages that form the embryonic skeleton. These structures are then nearly entirely replaced by bone according to an amazingly predictable schedule. Although each bone has its own developmental schedule, ossification of long bones generally begins by 8 weeks, and by 12 weeks most hyaline cartilage "long bones" exhibit primary ossification centers (Figure 6.12). So precise is the ossification timetable that fetal age can be determined easily from X-rays of the fetal skeleton.

At birth, most long bones of the appendicular skeleton are well ossified except for their epiphyses. Shortly after birth, secondary ossification centers develop in a predictable sequence from infancy to the preschool years. The epiphyseal plates persist and provide for long bone growth during childhood and the sex-hormone–mediated growth spurt at adolescence. All bones are relatively featureless at birth, but as the child uses its muscles more and more, the bone markings develop and become increasingly prominent. By the age of 25 years, nearly all bones are completely ossified and skeletal growth ceases. In children and adolescents, the rate of bone formation exceeds the rate of bone reabsorption; in young adults, these processes are in balance; in old age, reabsorption predominates. Beginning in the fourth decade of life, the mass of both compact and spongy bone decreases with age; the only exception appears to be in bones of the skull. Among young adults, skeletal mass is generally greater in males than in females, and in blacks than in whites. As we age, the rate of bone loss is faster in whites than in blacks (who have greater bone density to begin with), and in females than in males.

In addition to the mass changes that occur with age, qualitative changes also occur. An increasing number of osteons remain incompletely formed, and mineralization is less complete. There is also an increase in the amount of nonviable compact bone, reflecting a diminished blood supply to the bones in old age.

It cannot be emphasized too strongly that bones have to be mechanically stressed to remain healthy. When we remain physically active and muscles and gravity pull on the skeleton, the bones become heavier and stronger; if we are totally inactive, they begin to atrophy. Although we cannot escape the osteoporotic process completely, we can defer it with proper activity and diet, as amply illustrated by many older people who enjoy active, rewarding lives. ■

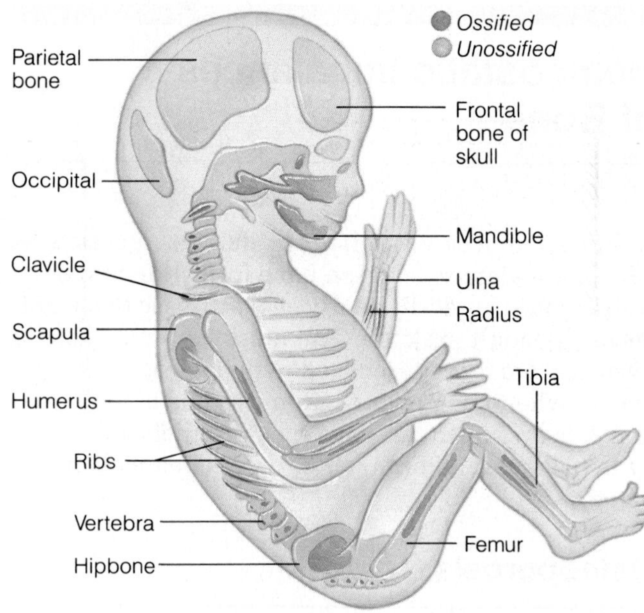

Figure 6.12 Primary ossification centers in the skeleton of a ten-week-old fetus.

* * *

Bones—their architecture, composition, and dynamic nature—have been examined in some detail in this chapter. We have also discussed the role of bones in maintaining overall body homeostasis, as summarized in Figure 6.13. Now we are ready to look at how they are assembled into the skeleton and contribute to its functions, both collectively and individually.

Skeletal system

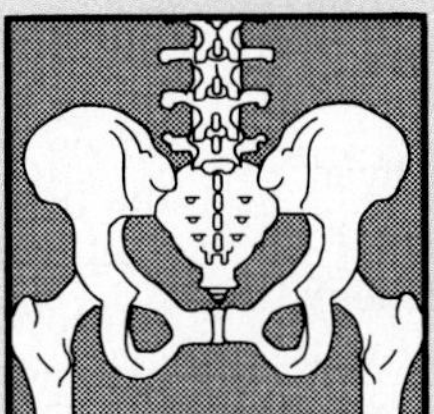

Integumentary system

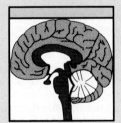

- Provides vitamin D needed for proper calcium absorption and use
- Provides support for body organs including the skin

Muscular system

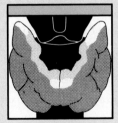

- Muscle pull on bones increases bone strength and viability; helps determine bone shape
- Provides levers plus calcium for muscle activity

Nervous system

- Innervates bone and joint capsules providing for pain and joint sense
- Protects brain and spinal cord; depot for calcium ions needed for neural function

Endocrine system

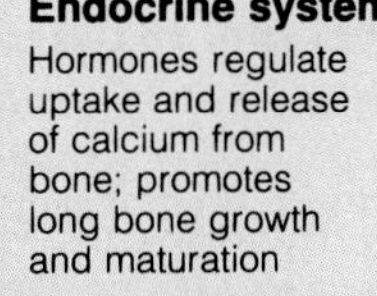

- Hormones regulate uptake and release of calcium from bone; promotes long bone growth and maturation
- Provides some bony protection

Cardiovascular system

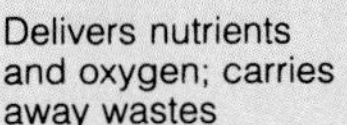

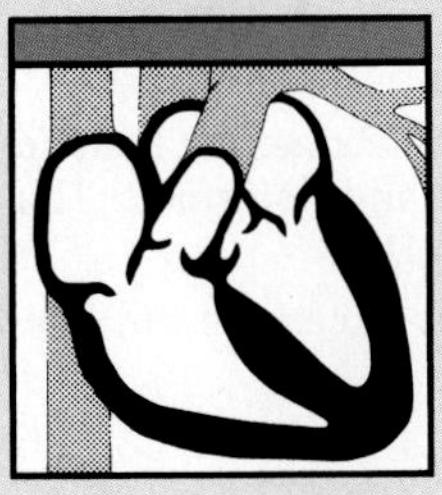

- Delivers nutrients and oxygen; carries away wastes
- Bone marrow cavaties provide site for blood cell formation

Lymphatic system

- Drains leaked tissue fluids
- Provides some protection

Immune system

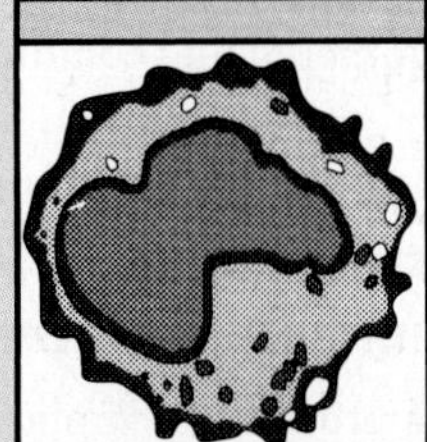

- Protects against pathogens
- Bone marrow is site of origin for lymphocytes involved in immune response

Respiratory system

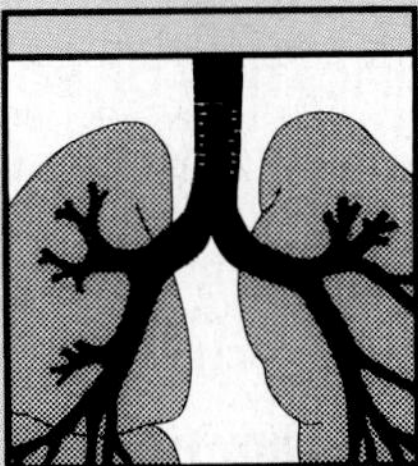

- Provides oxygen; disposes of carbon dioxide
- Protects lungs by enclosure (rib cage)

Digestive system

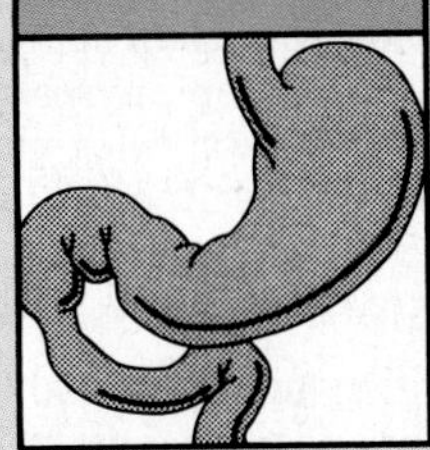

- Provides nutrients needed for bone health and growth
- Provides some bony protection to intestines, pelvic organs, and liver

Urinary system

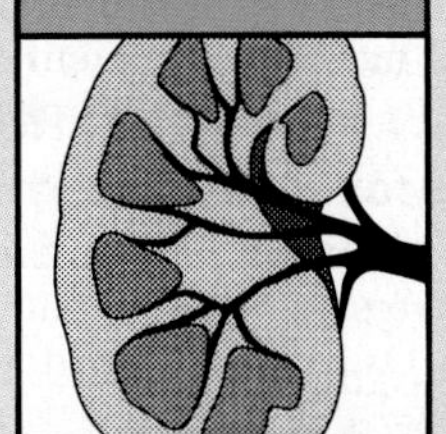

- Activates vitamin D; disposes of nitrogenous wastes
- Protects pelvic organs (bladder, etc.)

Reproductive system

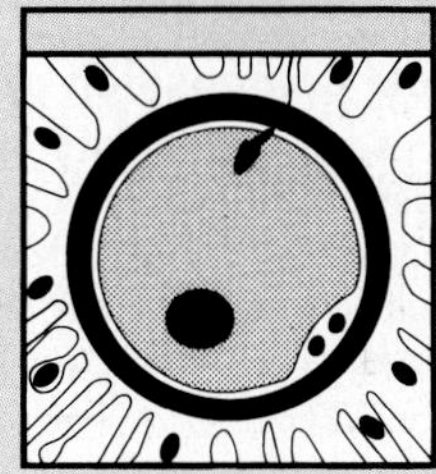

- Gonads produce hormones that influence form of skeleton and epiphyseal closure
- Protects some reproductive organs by enclosure

Figure 6.13 Homeostatic interrelationships between the skeletal system and other body organ systems.

Related Clinical Terms

Achondroplasia (ā-kon-drō-plā′-zhuh) (*a* = without; *chondro* = cartilage; *plasi* = mold, shape) A congenital condition involving an inability to form endochondral bones; results in dwarfism.

Bony spur Abnormal projection from a bone due to bone overgrowth; common in aging bones.

Osteitis (os-tē-ī′-tis) (*itis* = inflammation) Inflammation of bony tissue.

Metastatic calcification Deposit of bone salts in tissues that do not normally become calcified; a consequence of sustained hypercalcemia resulting from hyperparathyroidism or other bone-demineralizing conditions.

Chapter Summary

FUNCTIONS OF THE BONES (p. 153)

1. Bones protect and support body organs; serve as levers for muscles to pull on; store calcium, fats, and other substances; and are the site of blood cell production.

CLASSIFICATION OF BONES (p. 153–154)

1. Bones are classified as long, short, flat, or irregular on the basis of their shape and the proportion of compact or spongy bone they contain.

BONE STRUCTURE (pp. 154–158)

Gross Anatomy (pp. 154–156)

1. A long bone is composed of a diaphysis and two epiphyses. The medullary cavity of the diaphysis contains yellow marrow; the epiphyses contain spongy bone. The epiphyseal line is the remnant of the epiphyseal plate. Periosteum covers the diaphysis; endosteum lines medullary cavities and the spaces of spongy bone. Hyaline cartilage covers articular surfaces.

2. Flat bones consist of two thin plates of compact bone enclosing a diploe (spongy bone layer).

3. In adults, hematopoietic tissue is found within the diploe of flat bones and occasionally within the epiphyses of long bones. In infants, red marrow is also found in the medullary cavity.

4. Bone markings are important anatomical landmarks that reveal sites of muscle attachment, points of articulation, and sites of blood vessels and nerve passage.

Microscopic Structure of Bone (pp. 156–157)

5. The structural unit of compact bone is the osteon, consisting of a Haversian canal surrounded by concentric lamellae of bone matrix. Osteocytes, embedded in lacunae, are connected to each other and the Haversian canal by canaliculi.

6. Spongy bone has slender trabeculae containing irregularly arranged lamellae that enclose (red-marrow-filled) cavities.

Chemical Composition of Bone (pp. 157–158)

7. Bone is composed of living cells (osteoblasts, osteocytes, and osteoclasts) and matrix. The matrix includes organic substances that are secreted by osteoblasts and give bone tensile strength. Its inorganic components are hydroxyapatites (calcium salts), which make bone hard.

BONE DEVELOPMENT (OSTEOGENESIS) (pp. 158–161)

Formation of the Bony Skeleton pp. (158–160)

1. Intramembranous ossification forms flat and some irregular bones. Bone matrix is deposited between collagen fibers within the fibrous membrane to form spongy bone. Eventually, compact bone plates enclose the diploe.

2. Most bones are formed by endochondral ossification of a hyaline cartilage model. Osteoblasts beneath the periosteum secrete bone matrix on the hyaline cartilage model, forming the bone collar. Deterioration of the cartilaginous matrix opens up cavities, allowing entry of the periosteal bud. Bone matrix is deposited around the cartilage remnants, but is later broken down.

Bone Growth (pp. 160–161)

3. Increase in long bone length is accomplished by growth of the epiphyseal plate cartilage and its subsequent replacement by bone.

4. Appositional growth increases bone diameter/thickness.

BONE HOMEOSTASIS: REMODELING AND REPAIR (pp. 161–165)

Bone Remodeling (pp. 162–164)

1. New bone is continually deposited and reabsorbed in response to hormonal and mechanical stimuli. Together these processes are called bone remodeling.

2. An osteoid seam appears at areas of new bone formation; calcium salt is deposited 10 to 12 days later.

3. Multinucleate osteoclasts secrete catabolic enzymes on bone surfaces to be reabsorbed.

4. Control of bone remodeling involves hormonal and mechanical influences. The hormonal mechanism serves blood calcium homeostasis. When blood calcium levels decline, PTH is released and stimulates osteoclasts to digest bone matrix, releasing ionic calcium. Calcitonin appears important only in children. When blood calcium levels rise, calcitonin is released, stimulating osteoblasts to remove calcium from the blood. Mechanical stress and gravity acting on the skeleton help maintain skeletal strength. Bones thicken or develop heavier prominences in sites where they are stressed.

Repair of Fractures (pp. 164–165)

5. Fractures are treated by open or closed reduction. The healing process involves formation of a hematoma, a fibrocartilaginous callus, a bony callus, and bone remodeling in succession.

HOMEOSTATIC IMBALANCES OF BONE (p. 167)

1. Imbalances between bone formation and reabsorption underlie all skeletal disorders.

2. **Osteoporosis** is any condition in which bone breakdown outpaces bone formation, causing bones to weaken. Postmenopausal women are particularly susceptible.

3. **Osteomalacia** and rickets occur when bones are inadequately mineralized. The bones become soft and deformed. The most frequent cause is inadequate vitamin D.

4. **Paget's disease** is characterized by excessive but abnormal bone remodeling.

DEVELOPMENTAL ASPECTS OF BONES: TIMING OF EVENTS (p. 168)

1. Osteogenesis occurs in a predictable and precisely timed manner.

2. Longitudinal long bone growth continues until the end of adolescence. Skeletal mass increases dramatically during puberty and adolescence, when formation exceeds reabsorption.

3. Bone mass is constant in young adulthood, but beginning in the 40s, bone reabsorption exceeds formation.

Review Questions

Multiple Choice/Matching

1. Which is a function of the skeletal system? (a) support, (b) hematopoeitic site, (c) storage, (d) provision of levers for muscle activity, (e) all of these.

2. A bone that has essentially the same width, length, and height is most likely (a) a long bone, (b) a short bone, (c) a flat bone, (d) an irregular bone.

3. The shaft of a long bone is properly called the (a) epiphysis, (b) periosteum, (c) diaphysis, (d) compact bone.

4. Sites of hematopoiesis include all but (a) red marrow cavities of spongy bone, (b) the diploe of flat bones, (c) medullary cavities in bones of infants, (d) medullary cavities in bones of a healthy adult.

5. The osteon exhibits (a) a central Haversian canal carrying a blood vessel, (b) concentric lamellae of matrix, (c) osteocytes in lacunae, (d) canaliculi that connect lacunae to the Haversian canal, (e) all of these.

6. The organic portion of matrix is important in providing all but (a) tensile strength, (b) hardness, (c) ability to resist stretch, (d) flexibility.

7. The flat bones of the skull develop from (a) areolar tissue, (b) hyaline cartilage, (c) a fibrous connective tissue membrane, (d) compact bone.

8. The following events apply to the endochondral ossification process as it occurs in the primary ossification center. Put these events in their proper order by assigning each a number (1–6).

_____ Cavity formation occurs within the hyaline cartilage

_____ Collar of bone is laid down around the hyaline cartilage model just beneath the periosteum

_____ Periosteal bud invades the marrow cavity

_____ Perichondrium becomes vascularized to a greater degree and becomes a periosteum

_____ Osteoblasts lay down bone around the cartilage spicules in the bone's interior

_____ Osteoclasts remove the cancellous bone from the shaft interior, leaving a marrow cavity that then houses fat

9. The remodeling of bone is a function of which of these cells? (a) chondrocytes and osteocytes, (b) osteoblasts and osteoclasts, (c) chondroblasts and osteoclasts, (d) osteoblasts and osteocytes.

10. Bone growth during childhood and in adults is regulated and directed by (a) growth hormones, (b) thyroxine, (c) sex hormones, (d) mechanical stress, (e) all of these.

11. Wolff's law is concerned with (a) calcium homeostasis of the blood, (b) the thickness and shape of a bone being determined by mechanical and gravitational stresses placed on it, (c) the electrical charge on bone surfaces.

12. Formation of the bony callus in fracture repair is followed by (a) hematoma formation, (b) fibrocartilaginous callus formation, (c) bone remodeling to convert woven bone to compact bone, (d) formation of granulation tissue.

13. A fracture in which the bone ends penetrate soft tissue is (a) greenstick, (b) compound, (c) simple, (d) comminuted, (e) compression.

14. The disorder in which bones are porous and thin but bone composition is normal is (a) osteomalacia, (b) osteoporosis, (c) Paget's disease.

Short Answer Essay Questions

15. What information can be gained by examining bone markings?

16. Compare compact and spongy bone in macroscopic appearance, microscopic structure, and relative location.

17. As we grow, our long bones increase in diameter, but the thickness of the bony collar of the shaft remains relatively constant. Explain.

18. Describe the process of new bone formation in an adult bone. Use the terms *osteoid seam* and *calcification front* in your discussion.

19. Compare and contrast controls of bone remodeling exerted by hormones and mechanical/gravitational forces relative to the actual purpose of each control system and changes in bone architecture that might occur.

20. (a) During what period of life does skeletal mass increase dramatically? Begin to decline? (b) Why are fractures most common in elderly individuals? (c) Why are greenstick fractures most common in children?

Clinical Application Questions

21. In his radiology laboratory, Jim was handed an X-ray of the femur of a ten-year-old boy and asked to indicate the epiphyseal plate. What is the epiphyseal plate and what part does it play in long bone growth?

22. Following a motorcycle accident, a 22-year-old man was rushed to the emergency room. X-rays revealed a spiral fracture of his right tibia. When the X-rays were repeated two months later, he was told that good bony callus formation was visible. What is bony callus?

23. A mother brought her four-year-old daughter to the doctor, complaining that her daughter didn't "look right." The examination revealed that the child had severe rickets: Her forehead was enlarged, her rib cage was knobby, and her lower limbs were bent and deformed. The mother was advised to increase dietary amounts of vitamin D and milk and to "shoo" the girl outside to play in the sun. Explain these instructions. Why are the epiphyseal plates thickened in a child with rickets? Explain the bent leg bones seen in this child.

CHAPTER 7

The Skeleton

Chapter Outline and Student Objectives

1. Name the major parts of the axial and appendicular skeletons and describe their most important functions.

PART 1: THE AXIAL SKELETON

The Skull (pp. 173–187)

2. Name, describe, and identify the bones of the skull. Identify their important markings.
3. Distinguish between the major functions of the cranium and the facial skeleton.
4. Identify bones containing paranasal sinuses. Discuss the importance of the paranasal sinuses.

The Vertebral Column (pp. 187–193)

5. Describe the general structure and components of the vertebral column.
6. Indicate a common function of the spinal curvatures and the intervertebral discs.
7. Discuss the structure of a typical vertebra and then relate the special characteristics of cervical, thoracic, and lumbar vertebrae.

The Bony Thorax (pp. 194–195)

8. Name and describe the bones of the thorax.
9. Differentiate between true and false ribs.

PART 2: THE APPENDICULAR SKELETON

The Pectoral (Shoulder) Girdle (pp. 196–198)

10. Identify the bones forming the pectoral girdle and relate their structure and arrangement to the function of this girdle.
11. Identify important shoulder girdle bone markings.

The Upper Limb (pp. 198–202)

12. Identify or name the bones of the upper limb and their important markings.

The Pelvic (Hip) Girdle (pp. 202–207)

13. Name the bones contributing to the os coxa and relate the strength of the pelvic girdle to its function in the body.
14. Describe differences in the anatomy of the male and female pelves and relate these to functional differences.

The Lower Limb (pp. 207–211)

15. Identify the bones of the lower limb and their important markings.
16. Name the three supporting arches of the foot and explain their importance.

Developmental Aspects of the Skeleton (pp. 211–212)

17. Describe changes in facial contour and body height and proportion that occur during childhood and adolescence and relate them to the growth of specific regions of the skeleton.
18. Compare and contrast the skeleton of an aged person with that of a young adult. Discuss how age-related skeletal changes may interfere with body function as a whole.

Preview of Selected Key Terms

Axial skeleton (*ax* = axis) Portion of the skeleton that forms the central (longitudinal) axis of the body; includes the bones of the skull, the vertebral column, and the bony thorax.

Appendicular skeleton (a″-pen-dih′-kyoo-ler) The bones of the limbs and limb girdles that are attached to the axial skeleton.

Skull Bones of the head; encloses and protects the brain.

Suture (soo′-cher) An immovable joint; with one exception, all bones of the skull are united by sutures.

Vertebral column The spine; formed of a number of individual bones called vertebrae and two composite bones, the sacrum and coccyx.

Bony thorax Bones that form the framework of the thorax, including the sternum, ribs, and thoracic vertebrae.

Pectoral girdle Bones that attach the upper limbs to the axial skeleton; includes the clavicle and scapula.

Pelvic girdle Consists of the paired coxal bones that attach the lower limbs to the axial skeleton.

The word **skeleton** comes from the Greek word meaning "dried up body" or "mummy," a rather unflattering description. Nonetheless, the internal human framework is a triumph of design and engineering that puts any skyscraper to shame. It is strong, yet light, and is almost perfectly adapted for the locomotor and manipulative functions it must perform.

Our skeleton's present form was shaped more than 3 million years ago, when ancestors of human beings first stood erect on hind legs. As opposed to animals that walk on all fours, our upright stance increases the ability of our skeletal muscles to resist the pull of gravity. Though the human infant's backbone is like an arch, it soon changes to the swayback structure required for the upright posture. The skeleton has its flaws, to be sure: that S-shaped spine burdens us with lower back pain, an almost universal plague of middle and old age. However, this structural imperfection is offset by the skeleton's capacity to respond to most demands we place on it.

The skeleton, composed of bones, cartilage, joints, and ligaments, accounts for about 20% of body mass, weighing about 30 pounds in a 160-pound man. Bones make up the bulk of the skeleton. Cartilage occurs only in isolated areas, such as the bridge of the nose, parts of the ribs, and the joints. Ligaments connect bones and reinforce joints, allowing required movements while restricting motions in other directions. Joints, discussed separately in Chapter 8, provide for the remarkable mobility of the skeleton without impairing its immense strength.

For descriptive purposes, the 206 bones of the human skeleton are divided into the axial and appendicular skeletons (see Figure 7.1). The **axial skeleton** forms the long axis of the body and includes the bones of the skull, vertebral column, and rib cage. The **appendicular** (a″-pen-dih′-kyoo-ler) **skeleton** consists of the bones of the upper and lower limbs and the girdles (shoulder bones and hip bones) that attach the limbs to the axial skeleton.

PART 1: THE AXIAL SKELETON

The axial skeleton is structured from 80 bones segregated into three major regions: the *skull*, *vertebral column*, and *bony thorax* (Figure 7.1). Besides forming the upright axis of the body trunk, the axial skeleton helps protect the brain and spinal cord and the organs housed within the thorax.

The Skull

The **skull** is the body's most complex bony structure. It is formed by two sets of bones, the *cranial bones* and the *facial bones*, 22 in all. The tiny bones of the middle ear cavity are sometimes counted as skull bones, too, but since they are involved in hearing, we will consider them with the special senses in Chapter 16. Besides providing a site for attachment of head muscles, the cranial bones, or **cranium** (krā′-nē-um), enclose and protect the fragile brain and the organs of hearing and equilibrium. The facial bones (1) form the framework of the face, (2) hold the eyes in an anterior position, (3) provide cavities for the organs of taste and smell and openings for the passage of air and food, (4) secure the teeth, and (5) anchor the facial muscles of expression, which we use to show our feelings. As you will see, the individual skull bones are well suited to their assignments.

Most skull bones are flat bones. Except for the mandible, which is connected to the rest of the skull by a freely movable joint, all bones of the adult skull are firmly united by immobile interlocking joints called **sutures** (soo′-cherz). The suture lines have a tortuous or serrated appearance, particularly obvious on the external bone surfaces. The major skull sutures, the coronal, sagittal, squamous, and lambdoidal sutures, connect cranial bones (see Figure 7.3a). Most other skull sutures connect facial bones and are named according to the specific bones that they connect.

Overview of Skull Geography

It seems worthwhile to spend a few moments on skull "geography" before describing the individual skull bones. With the lower jaw removed, the skull resembles a lopsided, hollow, bony sphere. The facial bones form its anterior aspect, and the cranium forms the rest of the skull (see Figure 7.3a). The cranium can be divided into the cranial *vault*, also called the *calvaria* (kal-vayr′-ē-uh) or *skullcap*, forming the top, lateral, and posterior aspects of the skull, and the cranial *floor*, or *base*, forming the skull bottom. Internally, prominent bony ridges divide the base into three distinct "steps" or fossae—the anterior, middle, and posterior cranial fossae (see Figure 7.4b). The brain sits snugly in these cranial fossae, completely enclosed by the cranial vault. In addition to the large, well-hidden cranial cavity, the skull has many other cavities, including those of the middle and inner ears (carved into the lateral side of the cranium), and anteriorly, the nasal cavities and the orbits. The orbits house the eyeballs. Several bones of the skull have sinuses, mucosa-lined cavities, within them. The sinuses in bones sur-

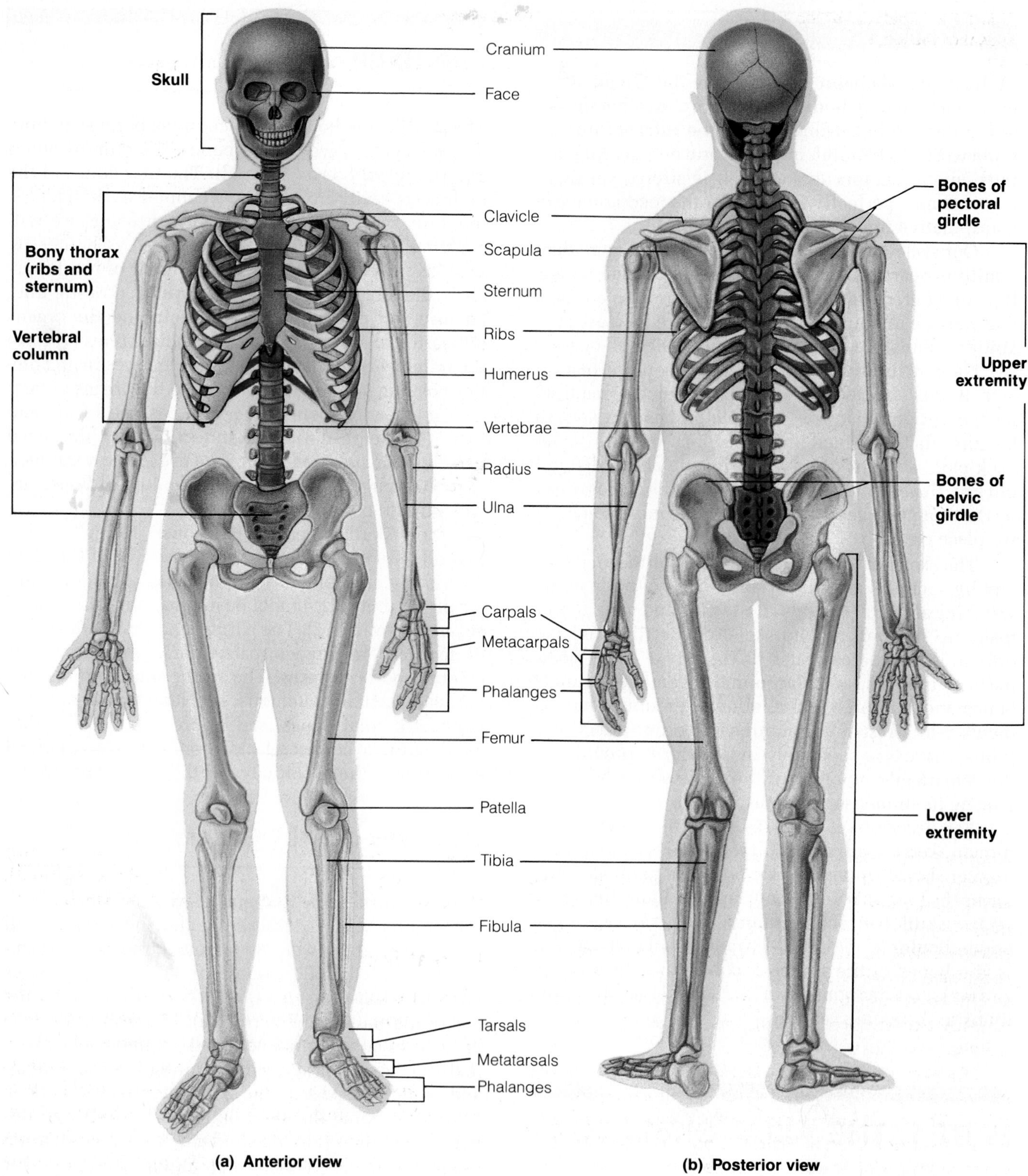

Figure 7.1 The human skeleton. The bones of the axial skeleton are shaded green; the bones of the appendicular skeleton are shaded gold.

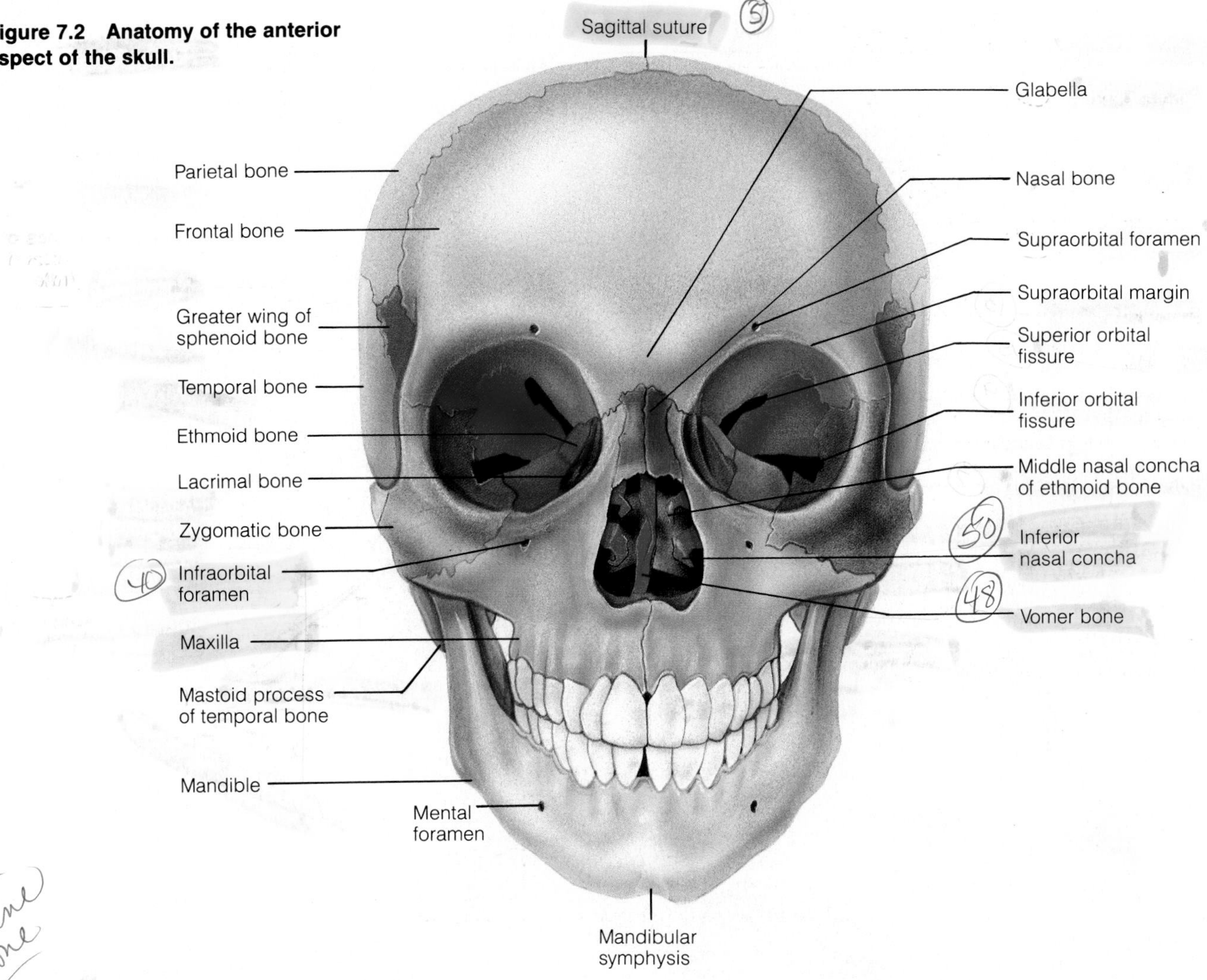

Figure 7.2 Anatomy of the anterior aspect of the skull.

rounding the nasal cavity are referred to as the *paranasal sinuses*. The sinuses lighten the skull and have additional roles that we will describe shortly. Moreover, the skull has about 85 named openings (foramina, canals, fissures, etc.). The most important are those that provide openings for the spinal cord, the major blood vessels serving the brain (internal carotid arteries and internal jugular veins), and the cranial nerves.

As you read about the bones of the skull, locate each bone on the different skull views in Figures 7.2 to 7.4. The skull bones and their important markings are also summarized in Table 7.1 on pp. 186–187. The color-coding dot beside a bone's name in the table corresponds to the color of that bone in the figures.

Cranium

The eight cranial bones are the paired parietal and temporal bones and the unpaired frontal, occipital, sphenoid, and ethmoid bones. Together, these construct the brain's protective bony "helmet." Because its superior aspect is curved, the cranium is self-bracing. This allows the bones to be thin, and, like an eggshell, the cranium is remarkably strong for its weight.

Frontal Bone

The **frontal bone** (Figures 7.2, 7.3, and 7.4b) forms the anterior portion of the cranium, the forehead and roofs of the orbits, and the anterior part of the cranial floor. It articulates posteriorly with the paired parietal bones via the prominent *coronal suture*. The frontal bone is thickened and protrudes slightly at the **supraorbital margins** (superior margins of the orbits) underlying the eyebrows. From here, the frontal bone extends posteriorly, forming the superior wall of the orbits and the **anterior cranial fossa** (see Figure 7.4b), which supports the frontal lobes of the brain within the cranial cavity. Each supraorbital margin is pierced by a **supraorbital foramen,** or **notch,** which allows the supraorbital artery and nerve to pass to the forehead region.

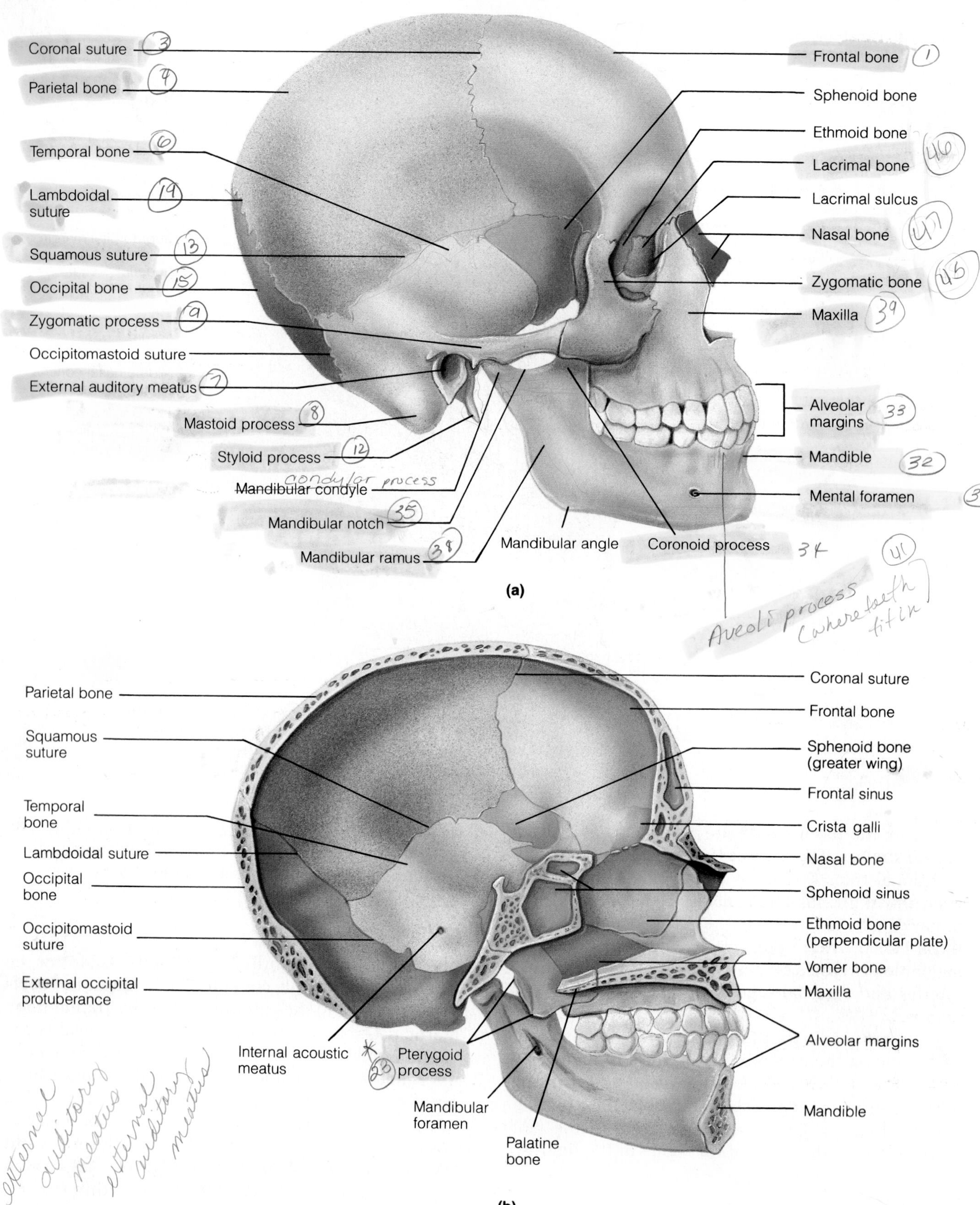

Figure 7.3 Anatomy of the lateral aspects of the skull. (**a**) External anatomy of the right lateral aspect of the skull. (**b**) Sagittal view of the skull showing the internal anatomy of the left side of the skull.

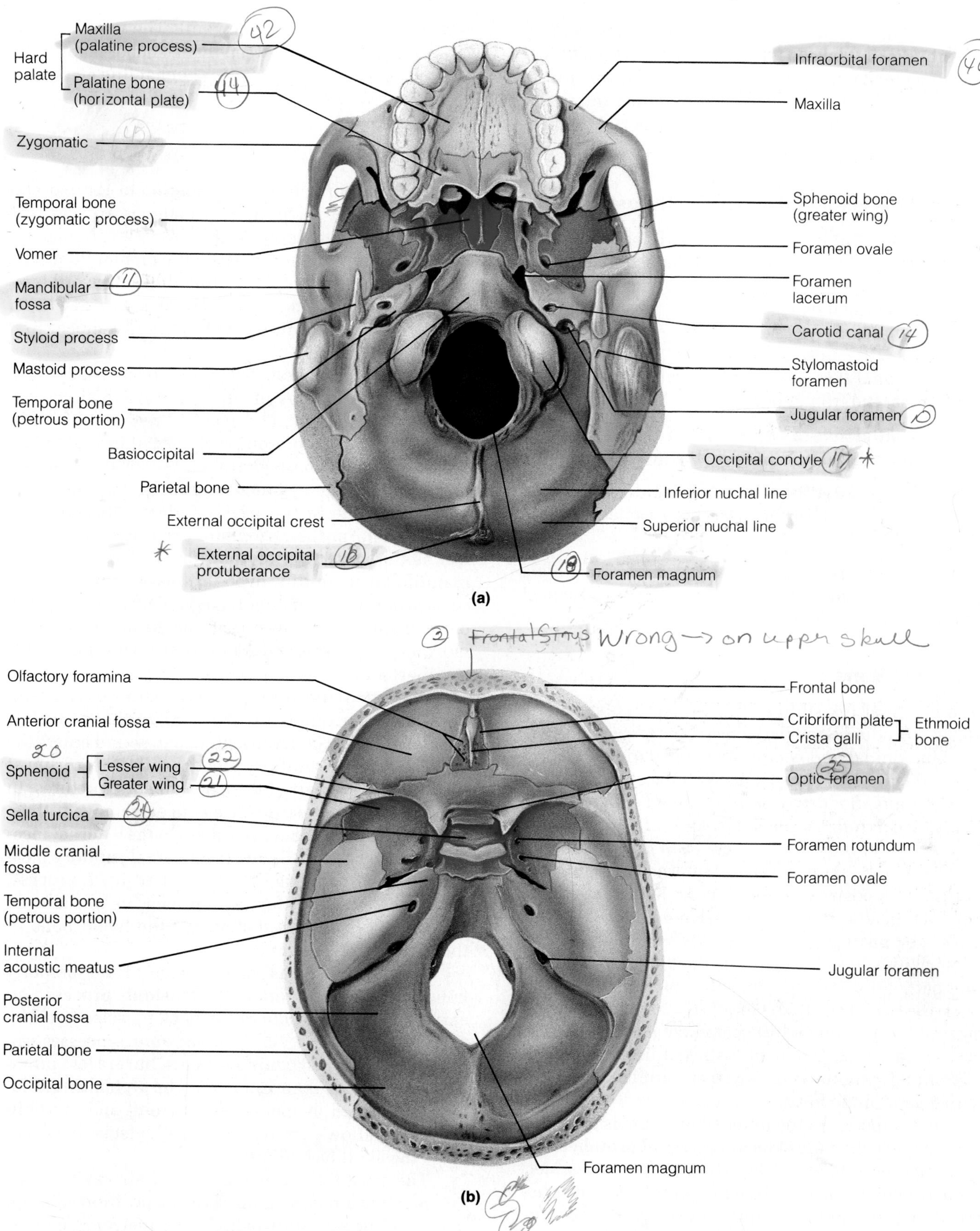

Figure 7.4 Anatomy of the inferior portion of the skull. (**a**) Inferior superficial view of the skull (mandible has been removed). (**b**) Superior view of the floor of the cranial cavity (calvaria has been removed).

The smooth portion of the frontal bone between the orbits is the **glabella** (gluh-beh′-luh); here the frontal bone meets the nasal bones at the *frontonasal suture.* The areas lateral to the glabella are riddled internally with sinuses, called the **frontal sinuses** (see Figures 7.3b and 7.11).

Parietal Bones and the Major Sutures

The two **parietal bones** form the bulk of the cranial vault, that is, most of the top and superolateral aspects of the skull. The four major sutures of the skull occur where the parietal bones articulate (form a joint) with other cranial bones:

1. The **coronal** (kor-ō′-nul) **suture**, where the parietal bones meet the frontal bone anteriorly (Figure 7.3).

2. The **sagittal suture**, where the two parietal bones meet at the cranial midline (see Figure 7.2).

3. The **lambdoidal** (lam-doy′-dul) **suture**, where the parietal bones meet the occipital bone posteriorly (see Figure 7.3).

4. The **squamous suture** (one on each side), where a parietal and a temporal bone meet on the lateral aspect of the skull (see Figure 7.3).

Occipital Bone

The **occipital** (ok-sih′-pih-tul) **bone** articulates anteriorly with the paired parietal and temporal bones via the lambdoidal and *occipitomastoid sutures,* respectively (Figure 7.3). It also joins with the sphenoid bone in the cranial floor via a narrowed band of bone called the *basioccipital* (bā″-zē-ok-sih′-pih-tul) (Figure 7.4a). The occipital bone forms most of the posterior wall and base of the skull externally. Internally, it forms the walls of the **posterior cranial fossa**. In the base of the occipital bone is the **foramen magnum,** literally, "large hole," the passageway between the vertebral and cranial cavities. It is through this opening that the inferior part of the brain (the medulla oblongata) connects with the spinal cord. The foramen magnum is flanked laterally by two rockerlike **occipital condyles** (see Figure 7.4a), which articulate with the first vertebra of the spinal column in a way that permits a nodding movement of the head.

Just superior to the foramen magnum is a median protrusion called the **external occipital protuberance** (see Figures 7.3b and 7.4a). You can feel this projection just below the most bulging part of your posterior skull. A number of inconspicuous ridges, the *external occipital crest* and the *superior* and *inferior nuchal* (noo′-kul) *lines* mark the occipital bone near the foramen magnum (see Figure 7.4a). The external occipital crest secures the *ligamentum nuchae* (lih-guh-men′-tum noo′-kē), a large ligament that connects the vertebrae of the neck to the skull. The nuchal lines, and the bony regions between them, anchor many neck and back muscles. The superior nuchal line marks the upper limit of the neck.

Temporal Bones

The two **temporal bones** lie inferior to the parietal bones and meet them at the squamous sutures (see Figure 7.3). The temporal bones form the inferolateral aspects of the skull and parts of the cranial floor. The terms *temple* and *temporal,* from the Latin word *temporum,* meaning "time," came about because gray hairs, a common sign of time's passing, usually appear first in the temple areas.

Each temporal bone has a complicated shape and is described in terms of its four major regions, the squamous, tympanic, and mastoid regions shown in Figure 7.5 and the petrous region seen in Figure 7.4. The flaring **squamous region** abuts the squamous suture and has a barlike **zygomatic process** that meets the zygomatic bone of the face anteriorly. Together, the two bones form the **zygomatic arch**, which you can feel as the projection of your cheek. The small oval **mandibular** (man-dih′-byoo-ler) **fossa** on the inferior surface of the zygomatic process receives the condyle of the mandible, or lower jawbone, forming the freely movable *temporomandibular* (tem′-per-ō-man-dih′-byoo-ler) *joint.*

The **tympanic** (tym-pa′-nik) **region** of the temporal bone surrounds the **external auditory meatus,** or **external ear canal**, through which sound enters the ear. The external auditory meatus and the eardrum at its deep end are part of the *external ear* structures. In a dried skull, the eardrum has been removed; thus, part of the middle ear cavity deep to the external meatus can also be seen. Below the external auditory meatus is the sharp, needlelike **styloid** (stī′-loyd) **process,** an attachment point for several muscles of the neck and for the ligaments that secure the hyoid bone in the neck region (see Figure 7.12).

The **mastoid** (mas′-toyd) **region** of the temporal bone exhibits a prominent **mastoid process,** an anchoring site for some of the neck muscles (see Figures 7.3a, 7.4a, and 7.5). The mastoid processes can be felt as lumps just behind the ears, but are less prominent in females than in males. The **stylomastoid foramen,** located between the styloid and mastoid processes, allows a cranial nerve (the facial nerve) to leave the skull (see Figure 7.4a).

The mastoid process is full of air cavities, the **mastoid sinuses,** or **air cells**. Its position adjacent to the middle ear cavity (a high-risk area for infections spreading from the throat) makes it extremely susceptible to infection itself, and a mastoid sinus infection, or *mastoiditis,* is difficult to treat. Since the mastoid air cells are separated from the brain by only a very

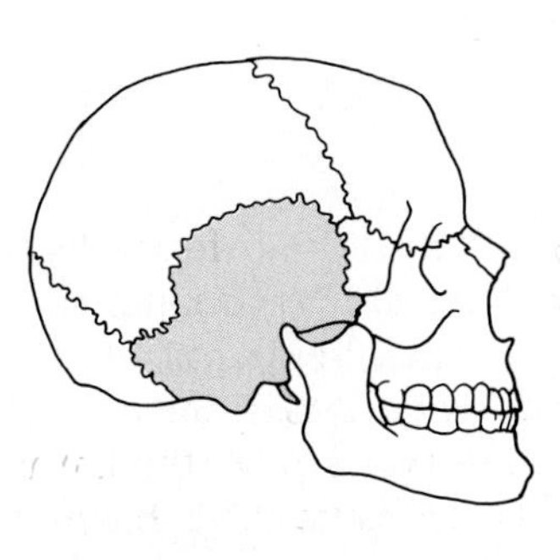

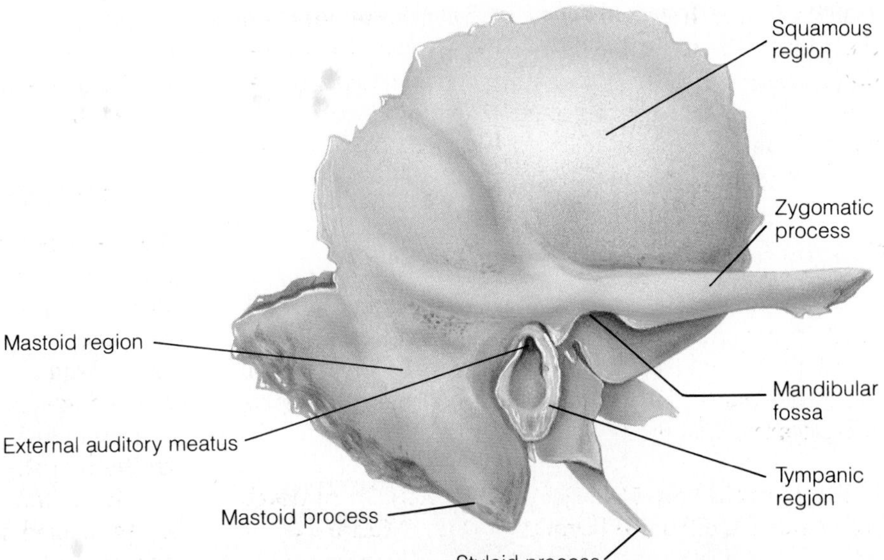

Figure 7.5 The temporal bone. Right superficial view.

thin bony plate, mastoid infections may spread to the brain itself. In people susceptible to repeated bouts of mastoiditis, surgical removal of the mastoid process may be the best way to prevent life-threatening brain inflammations. ■

The inferiormost **petrous** (peh′-trus) **region** of the temporal bone (see Figure 7.4) appears as a bony wedge between the occipital bone posteriorly and the sphenoid bone anteriorly. Together, the sphenoid bone and the petrous portions of the temporal bones construct the **middle cranial fossa** (see Figure 7.4b), which supports the temporal lobes of the brain on each side. Housed within the petrous regions are the *middle* and *inner ear cavities,* which contain the sensory receptors for hearing and balance.

Several foramina penetrate the bone of the petrous region (see Figure 7.4a). The large **jugular foramen** at the junction of the occipital and petrous temporal bones allows passage of the internal jugular vein and three cranial nerves. The **carotid** (kuh-rah′-tid) **canal**, just anterior to the jugular foramen, transmits the internal carotid artery into the cranial cavity. The two internal carotid arteries provide the arterial blood supply of over 80% of the cerebral hemispheres of the brain; their closeness to the inner ear cavities explains why during excitement or exertion, we sometimes hear our rapid pulse as a thundering sound in the head. The **foramen lacerum** (la′-ser-um), a jagged opening (*lacerum* = torn or lacerated) between the petrous temporal bone and the sphenoid bone, is a passageway for a number of small nerves. This foramen is relatively unimportant and is almost completely closed by cartilage in a living person. However, it is conspicuous in a dried skull, and students usually ask its name. The **internal acoustic meatus,** positioned superolateral to the jugular foramen (see Figures 7.3b and 7.4b), transmits the internal auditory artery and cranial nerves VII and VIII.

Figure 7.6 The sphenoid bone. (**a**) Superior view. (**b**) Posterior view.

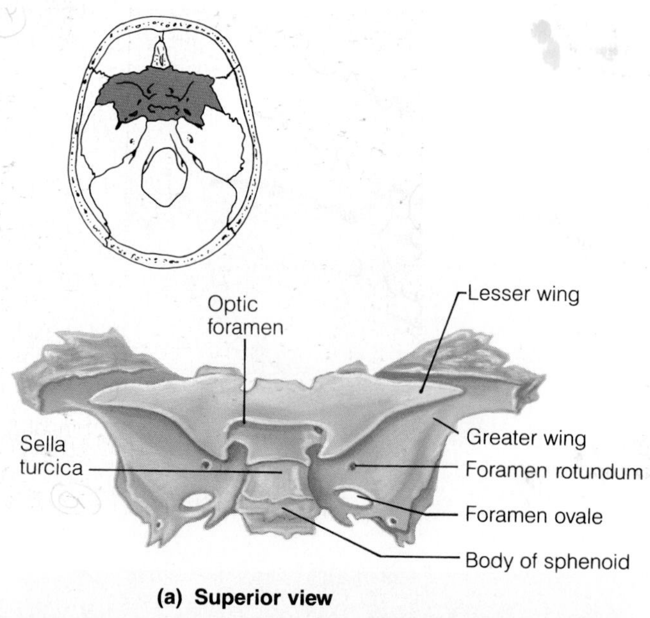

(a) Superior view

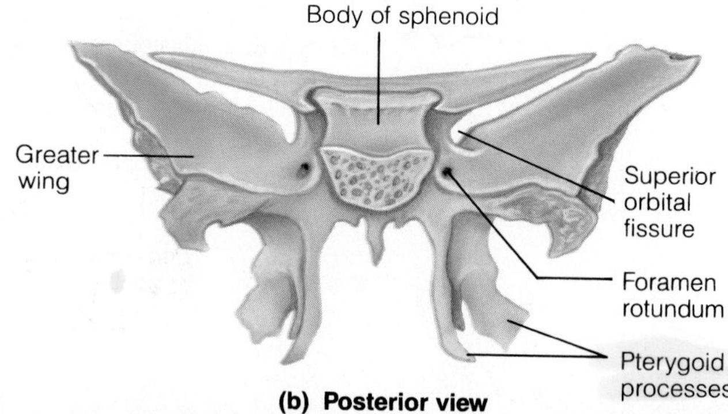

(b) Posterior view

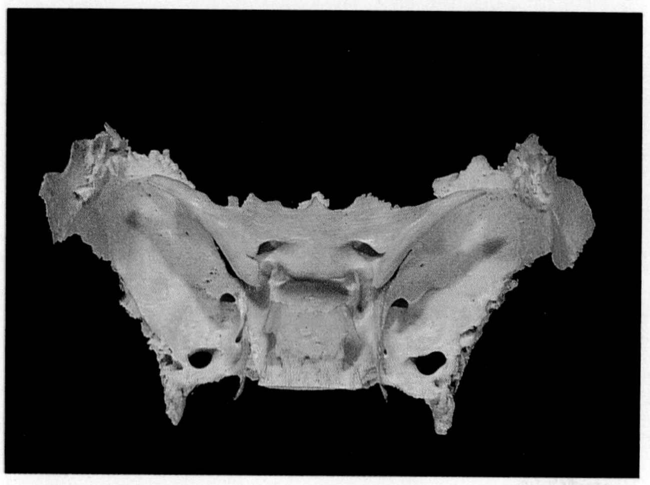

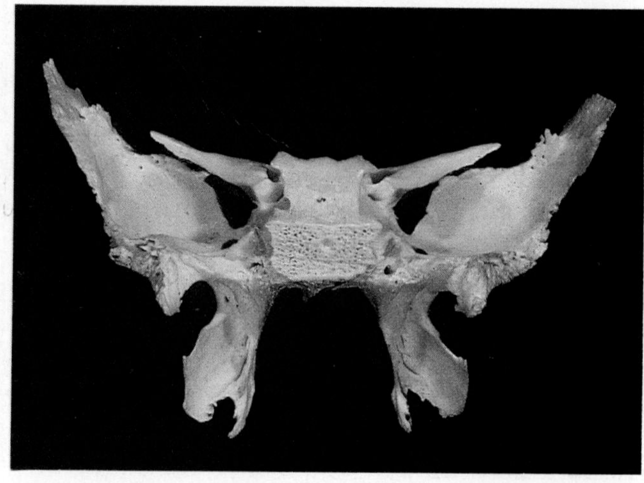

Sphenoid Bone

The butterfly-shaped **sphenoid** (sfē′-noyd) **bone** spans the width of the cranial floor (see Figure 7.4b). The sphenoid is considered the keystone of the cranium because it articulates as the central wedge with all other cranial bones. It consists of a central body and three pairs of processes, or "wings," that extend laterally toward the sides of the skull: the greater wings, lesser wings, and pterygoid processes (Figure 7.6). Within the **body** of the sphenoid are the paired **sphenoid sinuses** (see Figures 7.3b and 7.11). The superior surface of the body bears a saddlelike depression, the **sella turcica** (seh′-luh ter′-sih-kuh), commonly called the "Turk's saddle," in which the pituitary gland is firmly seated. The **greater wings** project laterally from the body, forming parts of (1) the middle cranial fossa (see Figure 7.4b), (2) the laterodorsal walls of the orbits (see Figure 7.2), and (3) the external wall of the skull, where they are seen as small flag-shaped, bony areas medial to the zygomatic arch (see Figure 7.3a). The **lesser wings** form part of the anterior fossa of the cranial floor (see Figure 7.4b) and a portion of the medial walls of the orbits. The platelike **pterygoid** (tayr′-ih-goyd) **processes** project inferiorly from the body (see Figure 7.6b). They form part of the lateral walls of the nasopharynx.

A number of important openings in the sphenoid bone are visible in Figures 7.4b and 7.6. The **optic foramina** are anterior to the sella turcica at the junctions of the sphenoid body with the lesser wings; they allow the optic nerves and the ophthalmic arteries to pass. On each side of the sphenoid body is a crescent-shaped series of three openings. The **superior orbital fissure,** a long slit between the greater and lesser wings, allows cranial nerves that control eye movements to enter the orbit. These fissures are most obvious in an anterior view of the skull (see Figure 7.2). The fora-

Figure 7.7 The ethmoid bone. Anterior view.

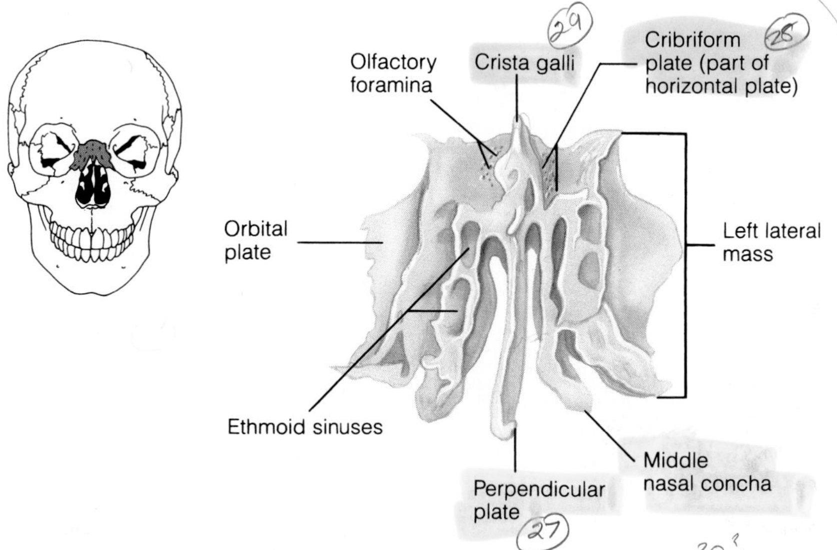

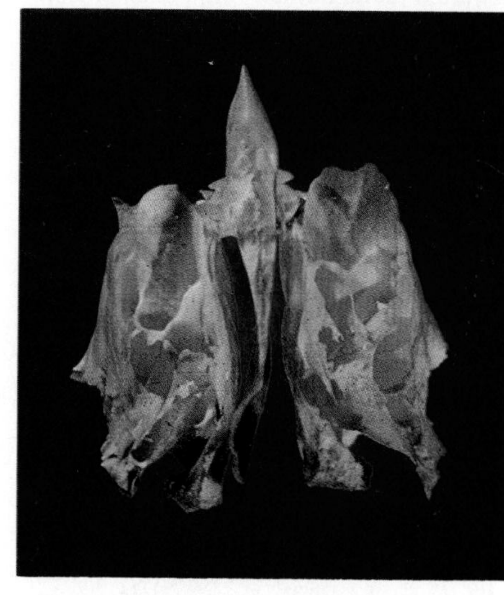

men rotundum and foramen ovale provide passageways for branches of cranial nerve V to reach the face. The **foramen rotundum** appears at the base of the greater wing and is usually oval, despite its name meaning "round opening." The **foramen ovale** (ō-vah′-lē), a large, oval foramen posterior to the foramen rotundum, is the only one of the three openings that can be seen in an inferior view of the skull (see Figure 7.4a).

Ethmoid Bone

Like the temporal and sphenoid bones, the **ethmoid bone** has a very complex shape (Figure 7.7). Situated between the sphenoid and the nasal bones of the face, it forms most of the bony area between the nasal cavity and the orbit.

The superior surface of the ethmoid, the **horizontal plate** (see also Figure 7.4b), has a triangular projection called the **crista galli** (kris′tuh gah′-lē), literally "cock's comb." The outermost covering of the brain attaches to the crista galli and helps to secure the brain in the cranial cavity. The **cribriform** (krih′-brih-form) **plates**, which form the bulk of the horizontal plate, are punctured by tiny holes (*olfactory foramina*) that allow the olfactory nerves to pass from the smell receptors in the nasal cavities to the brain. The horizontal plate also helps form the roof of the nasal cavities and the floor of the anterior cranial fossa. The downward-projecting **perpendicular plate** of the ethmoid bone forms the superior part of the nasal septum (see Figure 7.3b). Flanking the perpendicular plate on each side is a **lateral mass** riddled with the **ethmoid sinuses** (see Figures 7.7 and 7.11) for which the bone is named (*ethmos* = sieve). Extending medially from the lateral masses are the delicately coiled **superior** and **middle nasal conchae** (kon′-kē) (*concha* = shell), or **turbinates,** which protrude into the nasal cavity. The lateral aspects (*orbital plates*) of the lateral masses form part of the medial walls of the orbits.

Facial Bones

The facial skeleton is made up of 14 bones (see Figures 7.2 and 7.3a), of which only the mandible and the vomer are unpaired. The maxillae, zygomatics, nasals, lacrimals, palatines, and inferior conchae are paired bones. As a rule, the facial skeleton of men is more elongated than that of women; therefore, women's faces are rounder and somewhat less angular.

Mandible

The V-shaped **mandible** (man′-dih-bul), or lower jaw (Figures 7.2, 7.3 and 7.8b), is the largest, strongest bone of the face. It has a *body,* which forms the chin, and two upright *rami*. Each ramus meets the body at a **mandibular angle**. At the superior margin of each ramus are two processes separated by the *mandibular notch*. The anterior **coronoid** (kor′-uh-noyd) **process** is an insertion point for the large temporalis muscle that elevates the lower jaw during chewing. The posterior **mandibular condyle** articulates with the mandibular fossa of the temporal bone, forming the temporomandibular joint on the same side. The superior border of the mandible, the **alveolar** (al-vē′-uh-ler) **margin**, contains sockets called *alveoli* in which the teeth are found. In the midline of the mandibular body is a slight depression, the *mandibular symphysis* (sim′-fih-sis), which indicates the point of fusion of the two mandibular bones (see Figure 7.2).

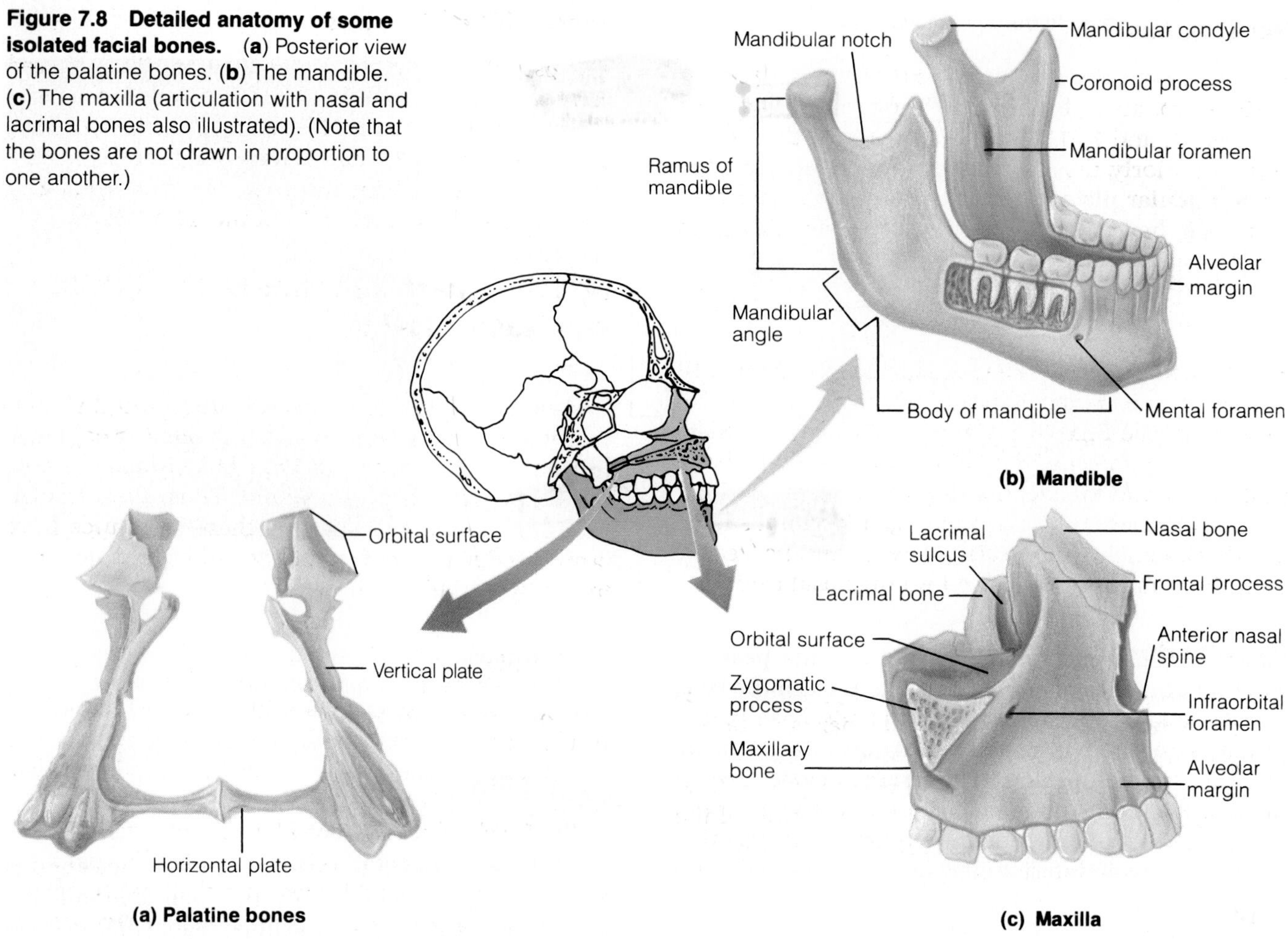

Figure 7.8 Detailed anatomy of some isolated facial bones. (**a**) Posterior view of the palatine bones. (**b**) The mandible. (**c**) The maxilla (articulation with nasal and lacrimal bones also illustrated). (Note that the bones are not drawn in proportion to one another.)

Large **mandibular foramina**, one on the medial surface of each ramus, permit the nerves responsible for tooth sensation to pass to the teeth in the lower jaw. Dentists inject novocaine into these foramina to prevent pain while working on the lower teeth. The **mental foramina**, openings on the lateral aspects of the mandibular body, allow blood vessels and nerves to pass to the chin (*ment* = chin) and lower lip.

Maxillary Bones

The **maxillary bones,** or **maxillae** (mak-sih′-lē) (see Figures 7.2, 7.3, and 7.8c), are fused medially. They form the upper jaw and the central portion of the facial skeleton. All facial bones except the mandible articulate with the maxillae; hence, the maxillae are considered the keystone bones of the facial skeleton.

The maxillae carry the upper teeth in their **alveolar margins**. The **palatine** (pa′-luh-tīn) **processes** of the maxillae project posteriorly from these margins and fuse medially, forming the anterior two-thirds of the hard palate, or bony roof, of the oral cavity (see Figures 7.3b and 7.4a). The **frontal processes** extend upward to the frontal bone, forming part of the lateral aspects of the bridge of the nose (see Figure 7.2 and 7.8c). The regions that flank the nasal cavity laterally contain the **maxillary sinuses** (see Figure 7.11), the largest of the paranasal sinuses. They extend from the orbits to the upper teeth. The maxillae articulate with the zygomatic bones laterally via their **zygomatic processes.**

The **inferior orbital fissure** is located deep within the orbit (see Figure 7.2) at the junction of the maxilla with the greater wing of the sphenoid. It permits a branch of cranial nerve V, the zygomatic nerve, and blood vessels to pass to the maxillary and zygomatic regions of the face. Just below the eye socket on each side is an *infraorbital foramen* that allows the infraorbital nerve and artery to reach the face.

Zygomatic Bones

The irregularly shaped **zygomatic,** or **malar** (mā″-ler), **bones** (see Figures 7.2 and 7.3a) are commonly called the cheekbones. They articulate with the zygomatic processes of the temporal bones posteriorly and with the zygomatic processes of the maxillae anteriorly, forming the prominences of the cheeks and part of the inferolateral margins of the orbits.

Nasal Bones

The thin, basically rectangular **nasal** (nā′-zul) **bones** are fused medially, forming the bridge of the nose (see Figures 7.2 and 7.3). They articulate with the frontal bone superiorly, the maxillary bones laterally, and the perpendicular plate of the ethmoid bone posteriorly. Inferiorly they attach to the cartilage plates that form the lower skeleton of the nose.

Lacrimal Bones

The delicate fingernail-shaped **lacrimal** (la′-krih-mul) **bones** are minor contributors to the medial walls of each orbit (see Figures 7.2 and 7.3a). They articulate with the frontal bone superiorly, the ethmoid bone posteriorly, and the maxillae anteriorly. Each lacrimal bone is pierced by a large opening, a *lacrimal sulcus* (sul′-kus), which permits tears (*lacrima* means "tears") to drain from the eye surface into the nasal cavity.

Palatine Bones

The **horizontal plates** of the L-shaped **palatine bones** complete the posterior portion of the hard palate (see Figures 7.3b, 7.4a, and 7.8a). Projecting superiorly from the horizontal plate of each palatine bone is a vertical plate that forms part of the posterolateral wall of the nasal cavity and a very small part of the orbit wall.

Vomer

The slender, plow-shaped **vomer** (vō′-mer) is located within the nasal cavity, where it forms part of the nasal septum (see Figures 7.2 and 7.10b). It will be discussed later in connection with the nasal cavity.

Inferior Nasal Conchae

The paired **inferior nasal conchae** are thin, curved bones seen projecting from the lateral walls of the nasal cavity just below the middle conchae of the ethmoid bone (see Figures 7.2 and 7.10a). They are the largest of the three pairs of conchae, and like the others, they form part of the lateral walls of the nasal cavity.

Special Characteristics of the Orbits and Nasal Cavity

Two restricted skull regions, the orbits and the nasal cavity, are formed from an amazing number of bones. In fact, it is sometimes difficult to comprehend how the many parts form the whole. Thus, even though the individual bones forming these structures have been described, a brief summary will be provided here to pull the parts together.

The Orbits

The **orbits** are bony cavities within which the eyes are firmly encased and cushioned by fatty tissue. The muscles of eye movement and the tear-producing lacrimal glands are also housed within the orbits. The walls of each orbit are formed by parts of seven bones—the frontal, sphenoid, zygomatic, maxilla, palatine, lacrimal, and ethmoid bones, and their relationships are shown in Figure 7.9. Seen in the orbits are the external openings of the superior and inferior orbital fissures and the optic foramina, described earlier.

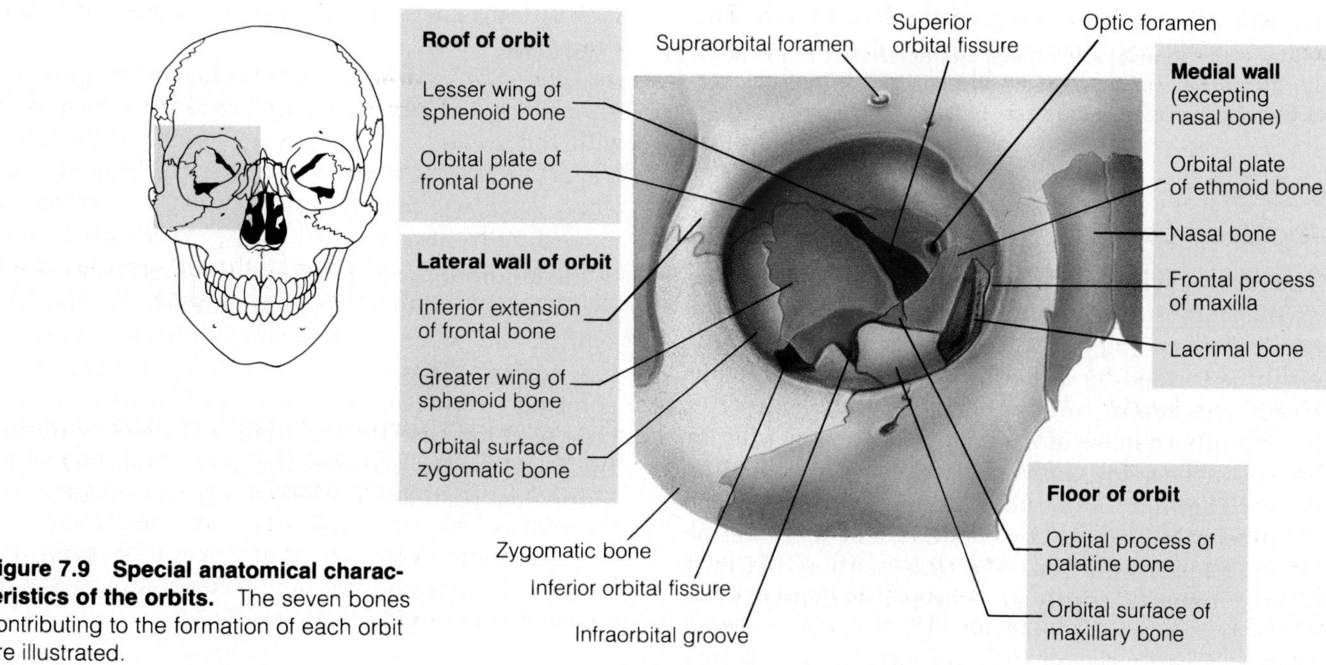

Figure 7.9 Special anatomical characteristics of the orbits. The seven bones contributing to the formation of each orbit are illustrated.

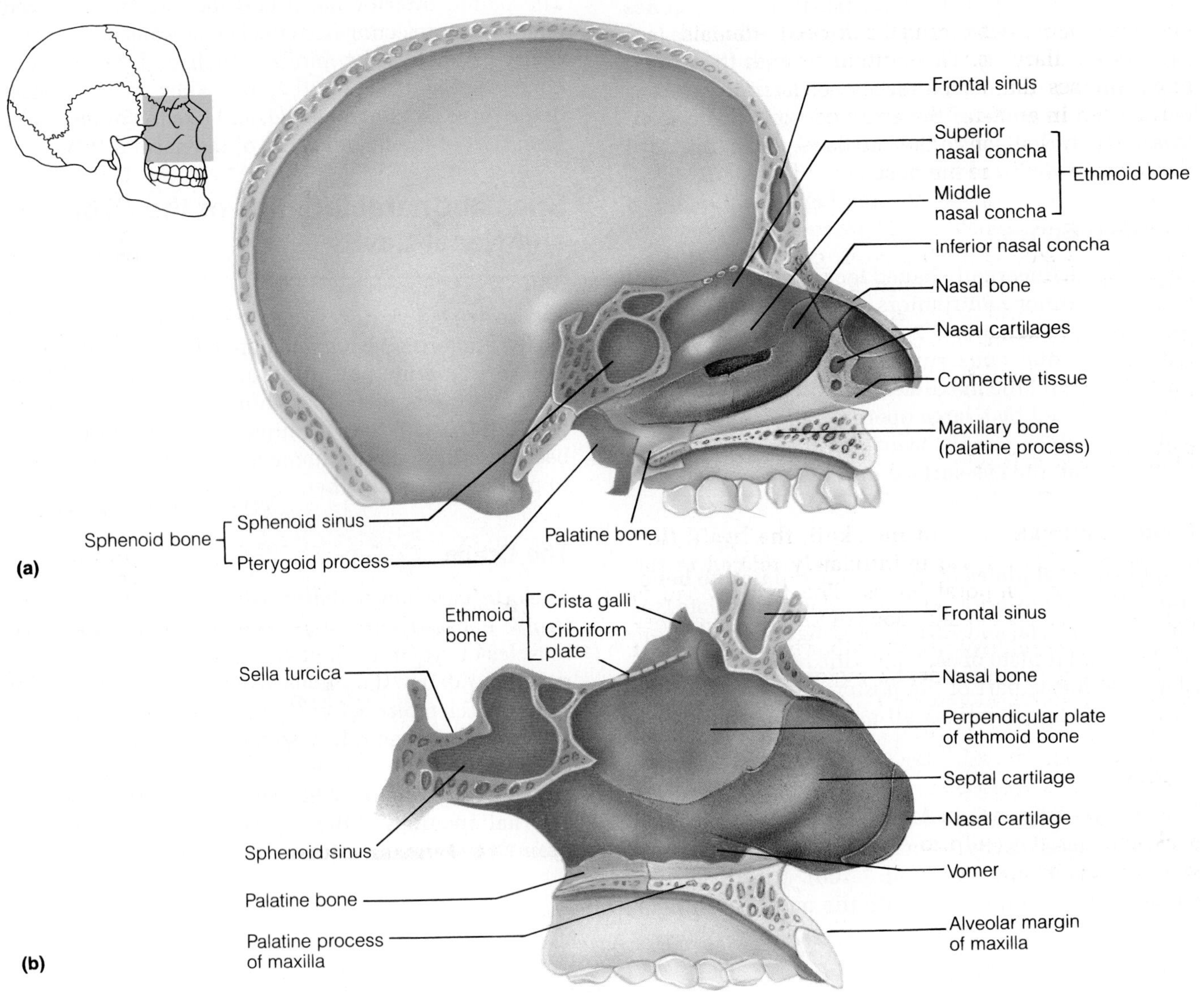

Figure 7.10 Special anatomical characteristics of the nasal cavity. (**a**) Bones forming the left lateral wall of the nasal cavity. (**b**) The contribution of the ethmoid, vomer, and cartilage to the nasal septum.

The Nasal Cavity

The **nasal cavity** is constructed of bone and hyaline cartilage; most of its bony structure is formed by the ethmoid bone (Figure 7.10a). The *roof* of the nasal cavity is formed by the cribriform plates of the ethmoid. The *lateral walls* are shaped by the superior and middle conchae of the ethmoid bone, the inferior nasal conchae, and the vertical plates of the palatine bones. The *floor* of the nasal cavity is formed by the palatine processes of the maxillae and the palatine bones. The bony portion of the nasal septum is formed by the vomer inferiorly and the perpendicular plate of the ethmoid bone superiorly (Figure 7.10b). Septal cartilage completes the septum anteriorly. The nasal septum and conchae are covered with a mucus-secreting mucosa. This mucosa moistens and warms the entering air and helps cleanse it of debris. The conchae also increase the turbulence of air flow through the nasal cavity and encourage the trapping of airborne dust particles and bacteria in the sticky mucus.

Paranasal Sinuses

Five skull bones—the frontal, sphenoid, ethmoid, and paired maxillary bones—contain mucosa-lined, air-filled sinuses that cause these bones to look rather moth-eaten in an X-ray. As noted previously, these sinuses are called **paranasal sinuses**, since they are clustered around the nasal cavity (Figure 7.11). The paranasal sinuses lighten the skull and are thought to enhance the resonance of the voice. Small openings connect the sinuses to the nasal cavity. These passageways act as "two-way streets"; air enters the sinuses from the nasal cavity and mucus formed by the sinus mucosae drains into the nasal cavity. The mucosa of the sinus also helps to warm and humidify inspired air.

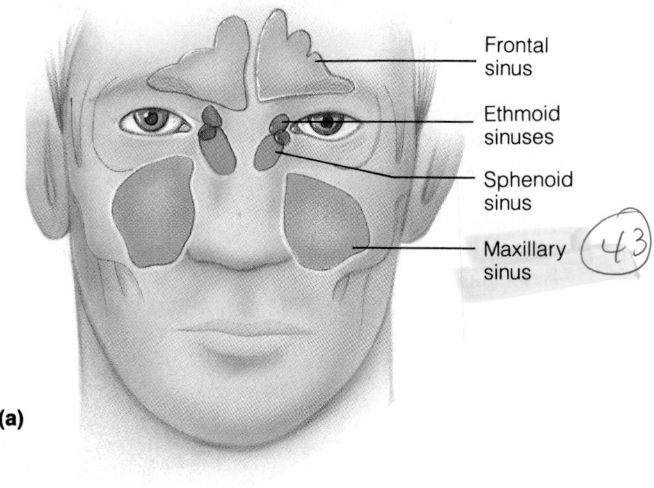

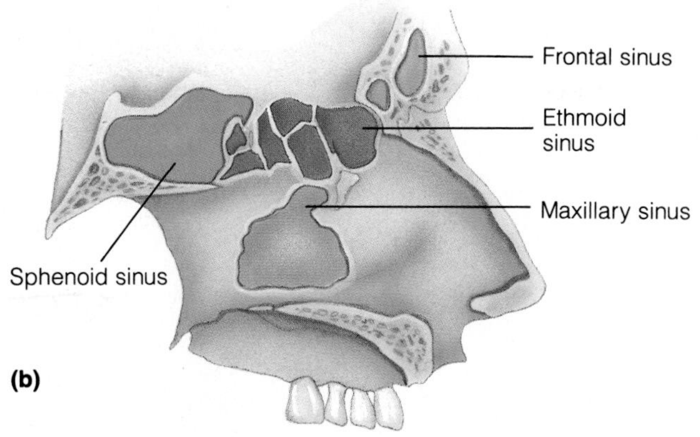

Figure 7.11 Paranasal sinuses.

The Hyoid Bone

Though not really part of the skull, the **hyoid** (hī′-oyd) **bone** (Figure 7.12) is intimately related to the mandible and temporal bones. The hyoid bone is unique in that it is the only bone of the body that does not articulate directly with any other bone. Instead, it is suspended in the midneck region about 2 cm (1 inch) above the larynx, where it is anchored by ligaments to the styloid processes of the temporal bones. Horseshoe-shaped, with a *body* and two pairs of *horns*, or *cornua*, the hyoid bone serves as a movable base for the tongue and an attachment point for several neck muscles that help to elevate the larynx during swallowing and speech.

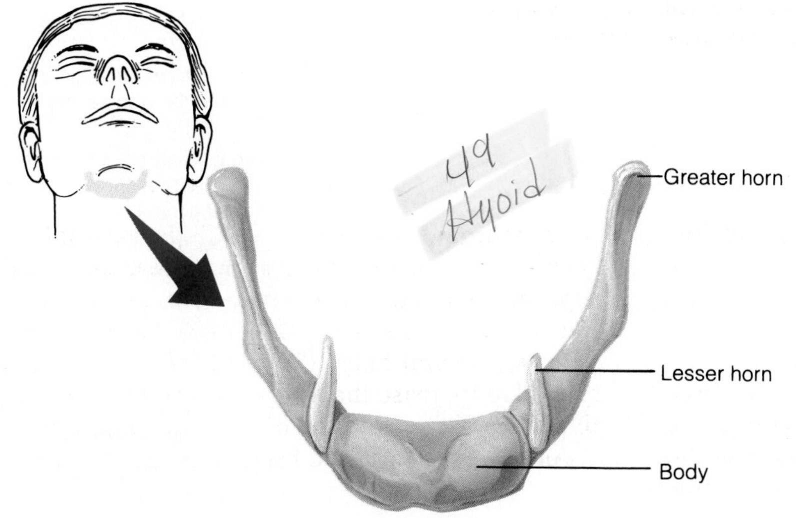

Figure 7.12 Anatomical location and structure of the hyoid bone. The hyoid bone is suspended in the midanterior neck by ligaments attached to the lesser horns (cornua) and the styloid processes of the temporal bones.

Table 7.1 Bones of the Skull

Bone*	Comments	Important markings
CRANIAL BONES		
Frontal (1) (Figures 7.2, 7.3, and 7.4b)	Forms forehead; superior part of orbits and anterior cranial fossa; contains sinuses	Supraorbital foramina: allow the supraorbital arteries and nerves to pass
Parietal (2) (Figures 7.2 and 7.3)	Forms most of the superior and lateral aspects of the skull	
Occipital (1) (Figures 7.3 and 7.4)	Forms posterior aspect and most of the base of the skull	Foramen magnum: allows the spinal cord to issue from the brain stem to enter the vertebral canal Occipital condyles: articulate with the atlas External occipital protuberance and nuchal lines: sites of muscle attachment External occipital crest: attachment site of ligamentum nuchae
Temporal (2) (Figures 7.3, 7.4, and 7.5)	Forms inferolateral aspects of the skull and contributes to the middle cranial fossa	Zygomatic process: helps to form the zygomatic arch, which forms the prominence of the cheek Mandibular fossa: articular point of the mandibular condyle External auditory meatus: canal leading from the external ear to the eardrum Styloid process: attachment site for several neck muscles Mastoid process: attachment site for several neck muscles Stylomastoid foramen: allows cranial nerve VII (facial nerve) to pass Jugular foramen: allows passage of the jugular vein and cranial nerves IX, X, and XI Internal acoustic meatus: allows passage of cranial nerves VII and VIII Carotid canal: allows passage of the internal carotid artery
Sphenoid (1) (Figures 7.2, 7.3, 7.4b, and 7.6)	Keystone of the cranium; contributes to the middle cranial fossa and orbits	Sella turcica: seat of the pituitary gland Optic foramina: allow passage of cranial nerve II and the ophthalmic arteries Superior orbital fissures: allow passage of cranial nerves III, IV, VI, and part of V (ophthalmic division) Foramen rotundum (2): allows passage of the maxillary division of cranial nerve V Foraman ovale (2): allows passage of the mandibular division of cranial nerve V
Ethmoid (1) (Figures 7.2, 7.3, 7.4b, 7.7, and 7.10)	Helps to form the anterior cranial fossa; forms most of the nasal septum and the lateral walls and roof of the nasal cavity; contributes to the medial wall of the orbit	Crista galli: attachment point for the falx cerebri, a dural membrane fold Cribriform plates: allow passage of nerve fibers of the olfactory nerves Superior and middle conchae: form part of lateral walls of nasal cavity; increase turbulence of air flow

Table 7.1 (continued)

Bone*	Comments	Important markings
Ear ossicles (malleus, incus, and stapes) (2 each)	Found in middle ear cavity; involved in sound transmission; see Chapter 16	
FACIAL BONES		
Mandible (1) (Figures 7.2, 7.3, and 7.8b)	The lower jaw	Coronoid processes; insertion points for the temporalis muscles Mandibular condyles: articulate with the temporal bones in freely movable joints Alveoli: sockets for the teeth Mandibular foramina: permit the alveolar nerves to pass Mental foramina: allow blood vessels and nerves to pass to the chin and lower lip
Maxilla (2) (Figures 7.2, 7.3, 7.4, and 7.8c)	Keystone bones of the face; form the upper jaw and parts of the hard palate, orbits, and nasal cavity walls	Alveoli: sockets for the teeth Palatine processes: form the anterior hard palate Inferior orbital fissures: permit maxillary branch of cranial nerve V, the zygomatic nerve, and blood vessels to pass
Zygomatic (2) (Figures 7.2, 7.3a, and 7.4a)	Form the cheek and part of the orbit	
Nasal (2) (Figures 7.2 and 7.3)	Construct the bridge of the nose	
Lacrimal (2) (Figures 7.2 and 7.3a)	Form part of the medial orbit wall	Lacrimal sulcus: allows tears to drain into the nasal cavity
Palatine (2) (Figures 7.3b, 7.4a, and 7.8a)	Form posterior part of the hard palate and a small part of nasal cavity and orbit walls	
Vomer (1) (Figures 7.2 and 7.10)	Part of the nasal septum	
Inferior nasal concha (2) (Figures 7.2 and 7.10a)	Form part of the lateral walls of the nasal cavity	

*The color-coded dot beside each bone name corresponds to the bone's color in Figures 7.2 to 7.10. The number in parentheses ( ) following each bone name denotes the number of such bones in the body.

The Vertebral Column

General Characteristics

Some people think of the **vertebral column** as a rigid supporting rod, but this picture is inaccurate. The vertebral column, also called the **spine**, is formed from 26 irregular bones connected in such a way that a flexible, curved structure results (Figure 7.13). Serving as the axial support of the body trunk, the spine extends from the skull, which it supports, to its anchoring point in the pelvis, where it transmits the weight of the trunk to the lower limbs. Running through its central cavity is the delicate spinal cord, which it surrounds and protects. Additionally, the vertebral column provides attachment points for the ribs and for the muscles of the back. In the fetus and infant, the vertebral column consists of 33 separate bones, or **vertebrae** (ver′-tuh-brē). Nine of these eventually fuse to form two composite bones, the sacrum and the tiny coccyx. The remaining 24 bones persist as individual vertebrae separated by connective tissue discs.

Like a tall, tremulous TV transmitting tower, the vertebral column cannot possibly stand upright by itself; it must be held in place by an elaborate system of supports. In the case of the TV tower, the supports are steel guy wires; for the vertebral column, straplike ligaments and the trunk muscles assume this role. The reinforcing muscles are discussed in Chapter 10. The major connecting ligaments are the *anterior* and *posterior longitudinal ligaments* (Figure 7.14), which run as continuous bands down the front and back surface

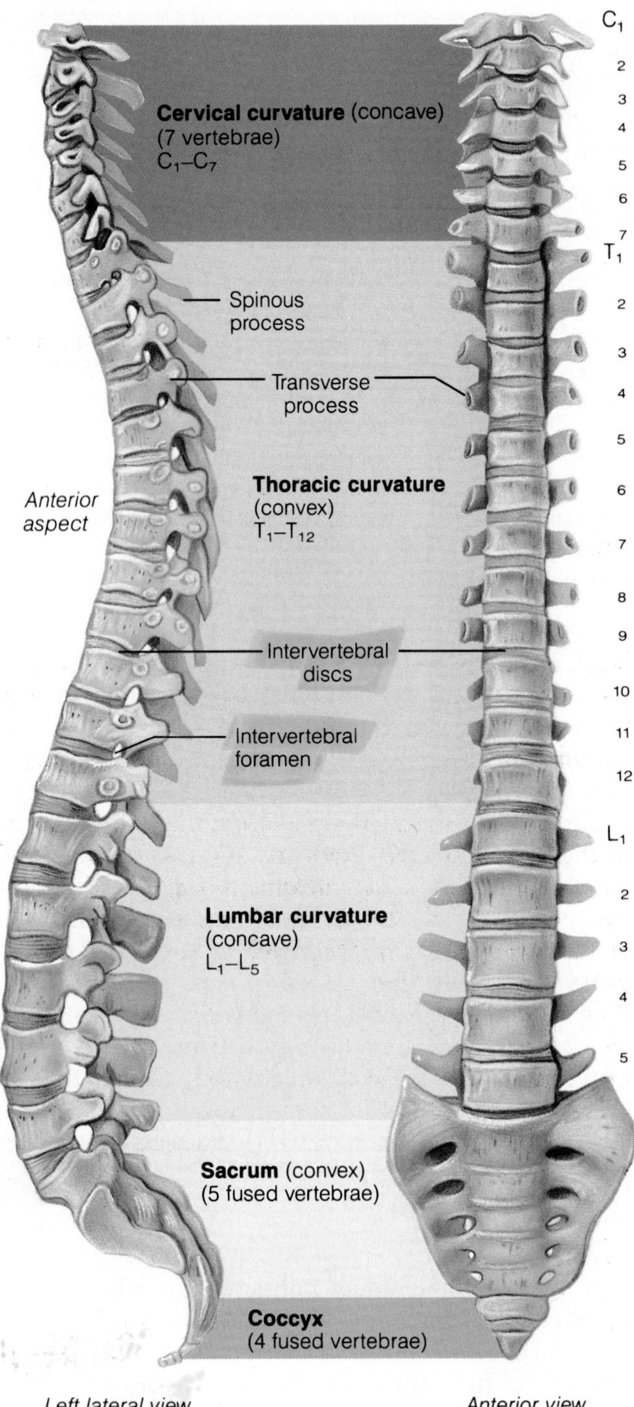

Figure 7.13 The vertebral column. Notice the curvatures in the lateral view. (The terms *convex* and *concave* refer to the curvature of the posterior aspect of the vertebral column.)

of the spine from the neck to the sacrum. The anterior ligament is wide and strongly attached to both the bony vertebrae and the discs. The posterior ligament is narrow and relatively weak; it is attached only to the discs. Additionally, short ligaments (ligamenta flava and others) connect each vertebra to those immediately above and below.

Discs

Each **intervertebral disc** is a cushionlike pad composed of an inner semifluid **nucleus pulposus** (pul-pō′-sus), which gives the disc its elasticity and compressibility, and a strong outer ring of fibrocartilage, the **anulus fibrosus** (a′-nyoo-lis fī-brō′-sis), which contains the nucleus pulposus and limits its expansion (see Figure 7.14b). The anulus fibrosus also holds together successive vertebrae. The discs act as shock absorbers during walking, jumping, and running, and they allow the spine to flex and extend and (to a lesser extent) bend from side to side. At points of compression, the discs flatten and bulge out a bit from the intervertebral spaces. The discs are thickest in the lumbar and cervical regions, which enhances the flexibility of these regions.

Severe or sudden physical trauma to the spine—for example, from bending forward while lifting a heavy object—may result in herniation of one or more discs. A *herniated disc* (commonly called a slipped disc) usually involves rupture of the anulus fibrosus followed by protrusion of the spongy nucleus pulposus (see Figure 7.14b). If the protrusion presses on the spinal cord or on spinal nerves exiting from the cord, numbness, excruciating pain, or even destruction of these nervous system structures may result. Herniated discs are generally treated with bed rest, traction, and pain killers. If this conservative therapy is unsuccessful, the protruding disc or part of a vertebra may have to be surgically removed. ■

Divisions and Curvatures

The vertebral column, about 70 cm (28 inches) long in an average adult, has five major divisions (see Figure 7.13). The 7 vertebrae of the neck are the *cervical* (ser′-vih-kul) *vertebrae*, the next 12 are the *thoracic* (thor-a′-sik) *vertebrae*, and the 5 supporting the lower back are the *lumbar* (lum′-bar) *vertebrae*. Inferior to the lumbar vertebrae is the *sacrum* (sā′-krum), which articulates with the pelvis. The terminus of the vertebral column is the tiny *coccyx* (kok′-siks). The number of cervical vertebrae is constant in all humans. Variations in the number of vertebrae in other regions occur in about 5% of otherwise normal people.

When you view the vertebral column from the side, you can see the four curvatures that give it its S, or springlike, shape. The **cervical** and **lumbar curvatures** are convex anteriorly; the **thoracic** and **sacral curvatures** are convex posteriorly. These curvatures increase the strength, resilience, and flexibility of the spine.

There are several types of abnormal spinal curvatures. Some are congenital (present at birth); others result from disease, poor posture, or unequal muscle pull on the spine. *Scoliosis* (scō-lē-ō′-sis) is an abnormal lateral curvature that occurs most often

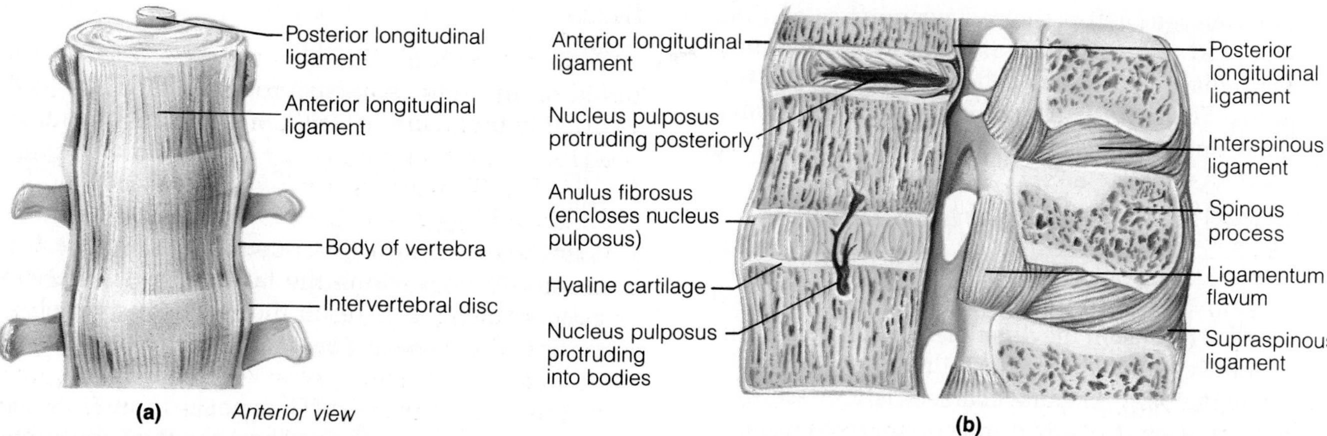

Figure 7.14 Ligaments and fibrocartilage discs uniting the vertebrae. **(a)** Anterior view of portions of three articulated vertebrae, showing the difference in width of the anterior and posterior longitudinal ligaments. **(b)** Median section of vertebrae, illustrating the ligaments and the composition of the discs. Two protrusions of the nuclei pulposus (commonly called "slipped discs") are also illustrated.

in the thoracic region. It is quite common during late childhood (particularly in girls for some unknown reason). The most severe cases result from abnormal vertebral structure or diseases that cause muscle paralysis. If muscles on one side of the body are nonfunctional, those of the opposite side will exert an unopposed pull on the spine and force it out of alignment. Severe cases must be treated (with body braces or surgically) before growth ends if permanent deformity and breathing difficulties are to be avoided.

Kyphosis (kī-fō'-sis), or hunchback, is a dorsally exaggerated thoracic curvature. It is particularly common in aged individuals because of osteoporosis, but may also reflect tuberculosis of the spine, rickets, or osteomalacia. *Lordosis,* or swayback, is an accentuated lumbar curvature; it, too, can result from spinal tuberculosis or rickets. Temporary lordosis is common in those carrying a "large load up front," such as obese people and pregnant women, who automatically accentuate their lumbar curvature in an attempt to preserve their center of gravity. ■

General Structure of Vertebrae

All vertebrae have a common structural pattern (Figure 7.15). Each vertebra consists of a discoid weight-bearing **body,** or **centrum,** anteriorly and a **vertebral arch** posteriorly. Together, the body and vertebral arch enclose an opening called the **vertebral foramen;** successive vertebral foramina of the articulated vertebrae form the **vertebral canal,** through which the spinal cord passes.

The vertebral arch is a composite structure formed by two pedicles and two laminae. The **pedicles,** short bony cylinders projecting posteriorly from the vertebral body, form the sides of the vertebral arch. The **laminae** (la'-mih-nē) are flattened plates that fuse in the median plane completing the arch posteriorly. Seven processes project from the vertebral arch. The **spinous process** is a single midline posterior projection arising at the junction of the two laminae. The two **transverse processes** issue from the lateral aspects of the vertebral arch. Both the spinous process and the transverse processes are attachment sites for muscles that move the vertebral column and for the ligaments that stabilize it. The **superior** and **inferior articular processes** are paired processes protruding upward and downward, respectively, from the pedicle-lamina junction on each side. The surfaces of the articular processes are covered with hyaline cartilage. The inferior articular processes of each vertebra form joints with the superior articular processes of the vertebra immediately below.

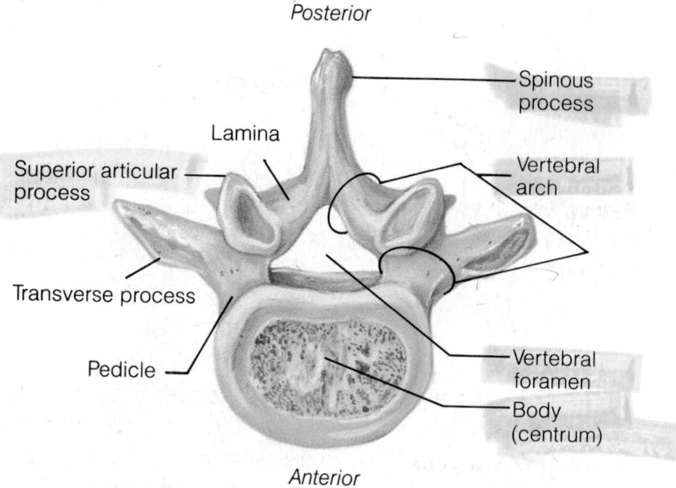

Figure 7.15 Structure of a typical vertebra. Superior view (inferior articulating surfaces not shown).

The right and left pedicles have notches on their superior and inferior borders, providing openings between adjacent pedicles called **intervertebral foramina** (see Figure 7.13). Nerves issuing from the spinal cord pass through these foramina.

Regional Vertebral Characteristics

In addition to the common structural features just described, vertebrae in the different regions of the spine have modifications that reflect their specific functions. The regional vertebral characteristics described in this section are illustrated and summarized in Table 7.2 on p. 192.

Cervical Vertebrae

The **cervical vertebrae**, identified as C_1–C_7, are the smallest, lightest vertebrae. The first two (C_1 and C_2) are atypical and we will skip them for the moment. The "typical" cervical vertebrae (C_3–C_7) have the following distinguishing features:

1. The body is broader from side to side than in the anteroposterior dimension.
2. Except in C_7, the spinous process is short, projects directly back, and is *bifid* (bī′-fid), split at its tip.
3. The vertebral foramen is large and generally triangular.
4. The transverse processes contain **transverse foramina** through which the large vertebral arteries ascend the neck to reach the brain. The vertebral veins and nerves also pass through these foramina.

The spinous process of C_7 is not bifid and is much larger and more prominent than those of the other cervical vertebrae (see Figure 7.17a). Because its spinous process is visible through the skin, C_7 can be used as a landmark for counting the vertebrae and is called the *vertebra prominens* ("prominent vertebra").

The first two cervical vertebrae, the atlas and the axis, are highly modified, reflecting their special functions, and have no intervertebral disc between them. The **atlas** (C_1) has no body and no spinous process (Figure 7.16a and b). Essentially, it is a ring of bone consisting of *anterior* and *posterior arches* and *lateral masses* on each side. Each lateral mass has articular

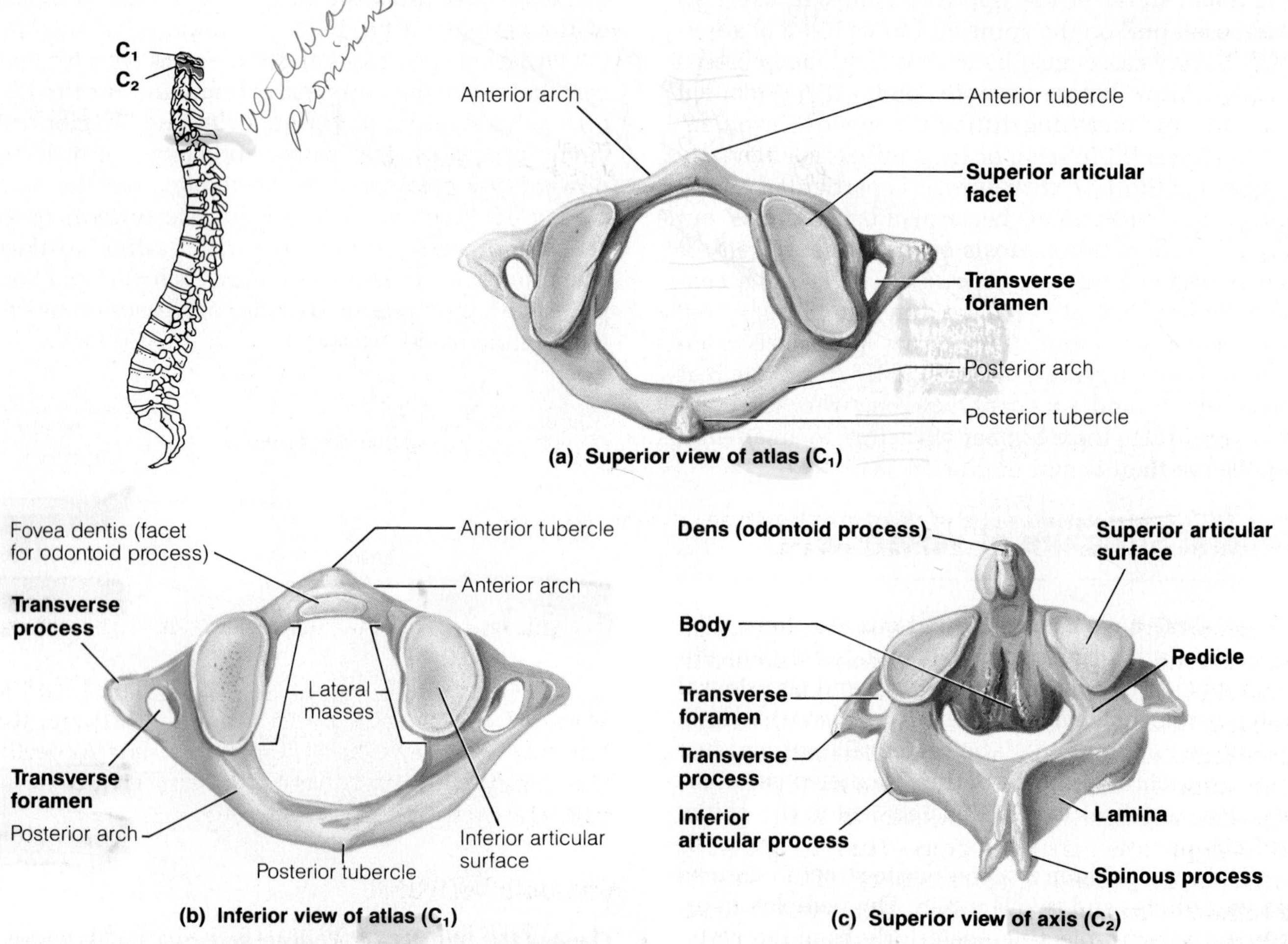

(a) Superior view of atlas (C_1)

(b) Inferior view of atlas (C_1)

(c) Superior view of axis (C_2)

Figure 7.16 The first and second cervical vertebrae.

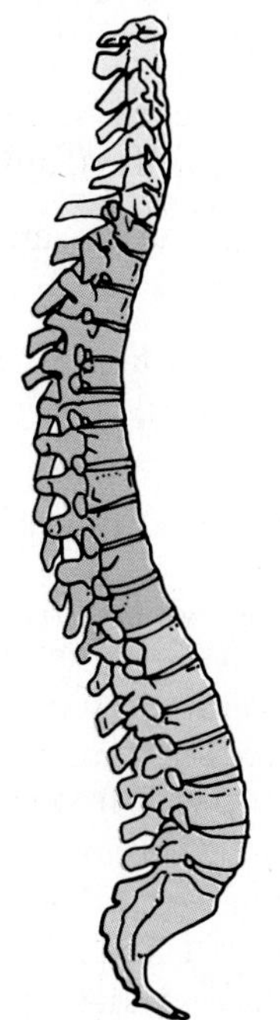

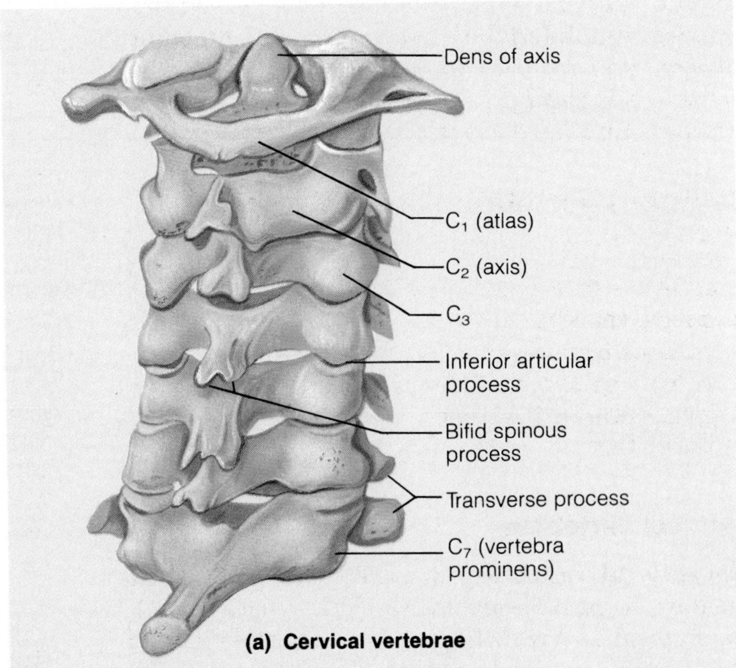

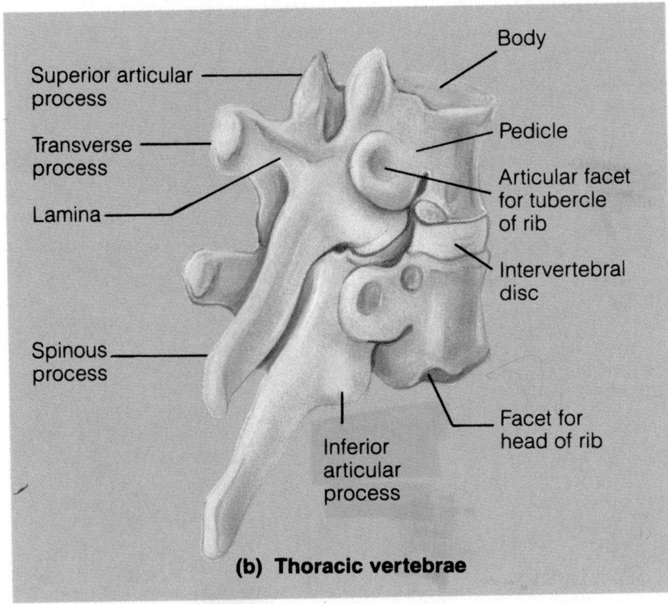

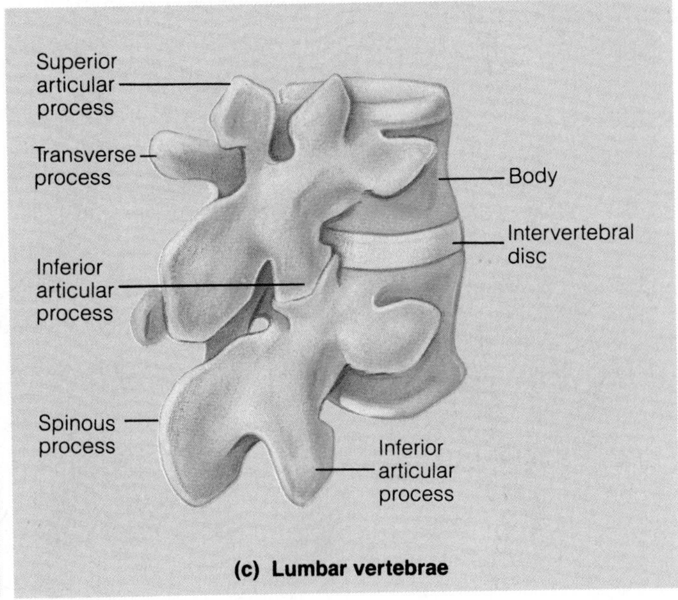

Figure 7.17 Posterolateral views of articulated vertebrae. Notice the bulbous tip on the spinous process of C_7, the vertebra prominens.

surfaces on both its superior and inferior surfaces. The superior articular facets receive the occipital condyles of the skull; thus, they "carry" the skull, just as Atlas supported the heavens according to Greek mythology. These joints allow you to nod "yes." The inferior articular surfaces form joints with the **axis** (C_2) below. Projecting superiorly from the body of the axis is a toothlike process called the **dens** (denz) or **odontoid** (ō-don′-toyd) **process** (Figure 7.16c). (The dens is actually the "missing" body of the atlas, which fuses with the axis during embryonic development.) The dens acts as a pivot for the rotation of the atlas; hence, this joint allows you to rotate your head from side to side to indicate "no."

In rare cases of severe head trauma in which the skull is driven inferiorly toward the spine, the dens may be forced into the brain stem, causing death. This is the most serious potential consequence of whiplash injuries in automobile accidents. ■

Thoracic Vertebrae

The 12 **thoracic vertebrae** (T_1–T_{12}), all of which articulate with the ribs, are closest to being "typical" vertebra (see Table 7.2 and Figure 7.17b). However, the

Table 7.2 Regional Characteristics of Cervical, Thoracic, and Lumbar Vertebrae

Characteristic	Cervical (3–7)	Thoracic	Lumbar
Body	Small, wide side to side	Larger than cervical; heart shaped; bears two costal facets	Massive; kidney shaped
Spinous process	Short; projects directly backward; bifid	Long; sharp; projects downward	Short; blunt; projects directly backward
Vertebral foramen	Triangular	Circular	Triangular
Transverse processes	Contain foramina	Bear facets for ribs (except T_{11}–T_{12})	No special features
Superior and inferior articulating processes	Superior: facets directed posterolaterally Inferior: facets directed anteromedially	As for cervical vertebrae	Superior: facets directed medially Inferior: facets directed laterally Provides immobility
Superior view			
Right lateral view			

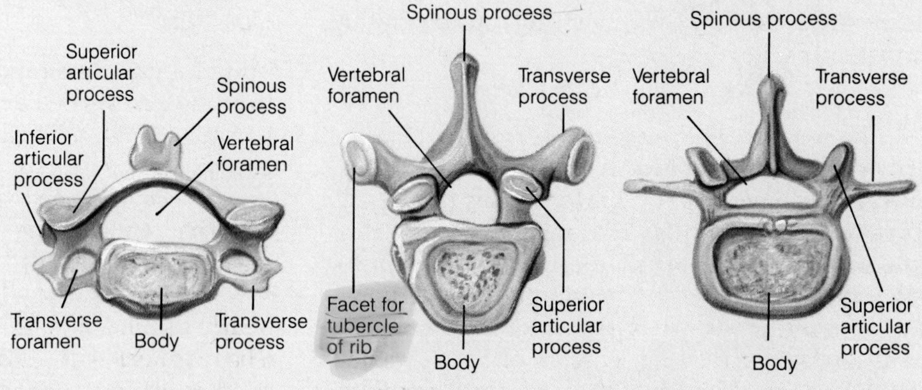

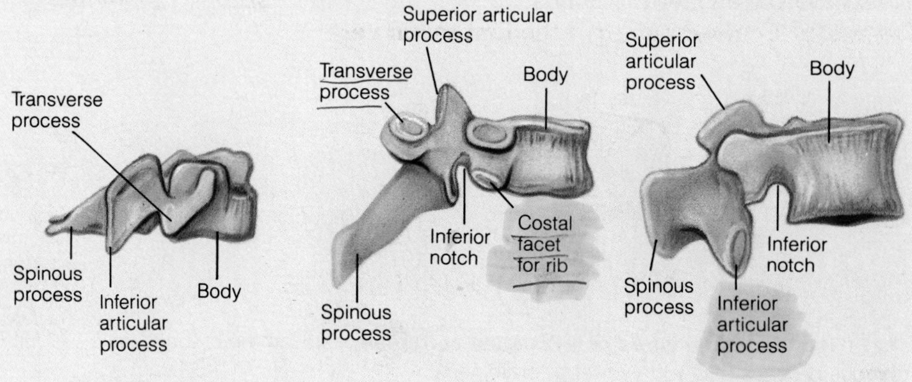

first looks much like C_7, and the last four show a progression toward lumbar vertebral structure. Generally speaking, they increase in size from the first to the last. Unique characteristics of the thoracic vertebrae include the following:

1. The body is roughly heart shaped and typically bears two *costal facets* on each side, which receive the heads of the ribs.
2. The intervertebral foramen is circular.
3. The spinous process is long and hooks sharply downward.
4. With the exception of T_{11} and T_{12}, the transverse processes have facets that articulate with the tubercles of the ribs.

Lumbar Vertebrae

The lumbar region of the vertebral column, commonly called the small of the back, receives the most stress. The enhanced weight-bearing function of the

lumbar vertebrae (L_1–L_5) is reflected in their sturdier structure. Their bodies are massive and kidney-shaped (see Table 7.2 and Figure 7.17c). Other characteristics typical of these vertebrae include the following:

1. The pedicles and laminae are shorter and thicker than those of other vertebrae.
2. The spinous processes are short, flat, and hatchet-shaped and are easily seen when a person bends forward. They project directly backward, an adaptation for the attachment of the large back muscles.
3. The vertebral foramen is triangular.
4. Atypically the facets of the superior articular processes face posteromedially instead of posterolaterally, and those of inferior articular processes face anterolaterally instead of anteromedially. These modifications lock the lumbar vertebrae together and provide stability by preventing rotation of the lumbar spine.

Sacrum

The triangular **sacrum** (Figure 7.18), which shapes the posterior wall of the pelvis, is formed by five fused vertebrae (S_1–S_5) in adults. It strengthens and stabilizes the pelvis. It articulates superiorly (via its *superior articular surfaces*) with L_5 and inferiorly with the coccyx. Laterally, the sacrum's two winglike **alae** (fused remnants of the transverse processes of S_1–S_5) articulate with the two hip bones to form the *sacroiliac* (sa″-krō-ih′-lē-ak) *joints* of the pelvis.

The anterosuperior margin of the first sacral vertebra bulges forward into the pelvic cavity; this marking is known as the **sacral promontory** (prah′-mun-tor″-ē). The body's center of gravity lies about 1 cm behind this marking, and as you will see, the sacral promontory is an important anatomical landmark for obstetricians. Four ridges (lines of vertebral fusion) cross its concave anterior aspect, and *sacral foramina* penetrate the sacrum at the lateral ends of these ridges. The sacral foramina transmit blood vessels and nerves.

Dorsally, the median sacral surface is roughened by the **median sacral crest** (the fused spinous processes of the sacral vertebrae). The vertebral canal continues inside the sacrum as the **sacral canal**. Since the laminae of the fifth (and sometimes the fourth) sacral vertebrae fail to fuse medially, an enlarged external opening called the **sacral hiatus** (hī-ā′-tis) is obvious at the end of the sacral canal.

Coccyx

The **coccyx,** or vestigial tailbone, consists of four (or in some cases three or five) vertebrae fused together to form a small triangular bone (see Figure 7.18). The coccyx articulates superiorly with the sacrum. (The name *coccyx* comes from the Greek word meaning "cuckoo" and was so named because of its fancied resemblance to a bird's beak.) Except for the slight support the coccyx affords the pelvic organs, it is a nearly useless bone in the human body. Occasionally, a baby is born with an unusually long coccyx. In many such cases, this extraneous bony "tail" is discretely snipped off by the physician.

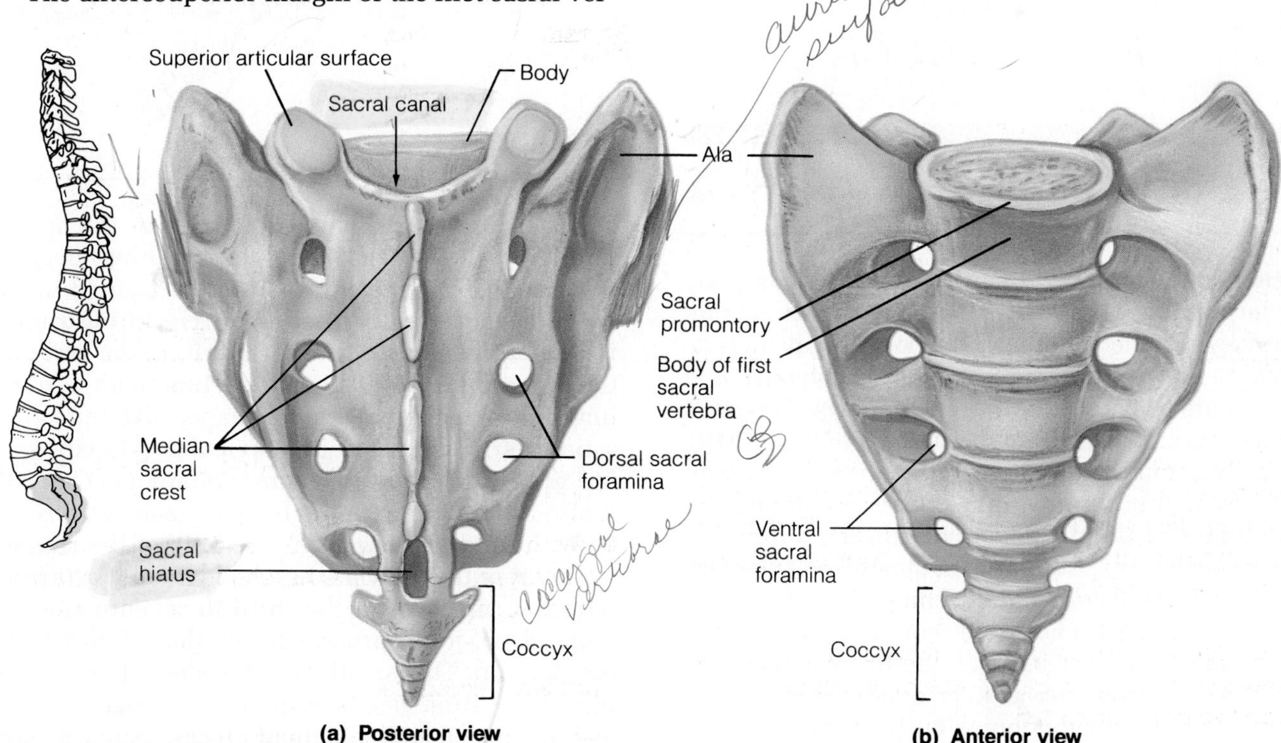

Figure 7.18 The sacrum and coccyx.

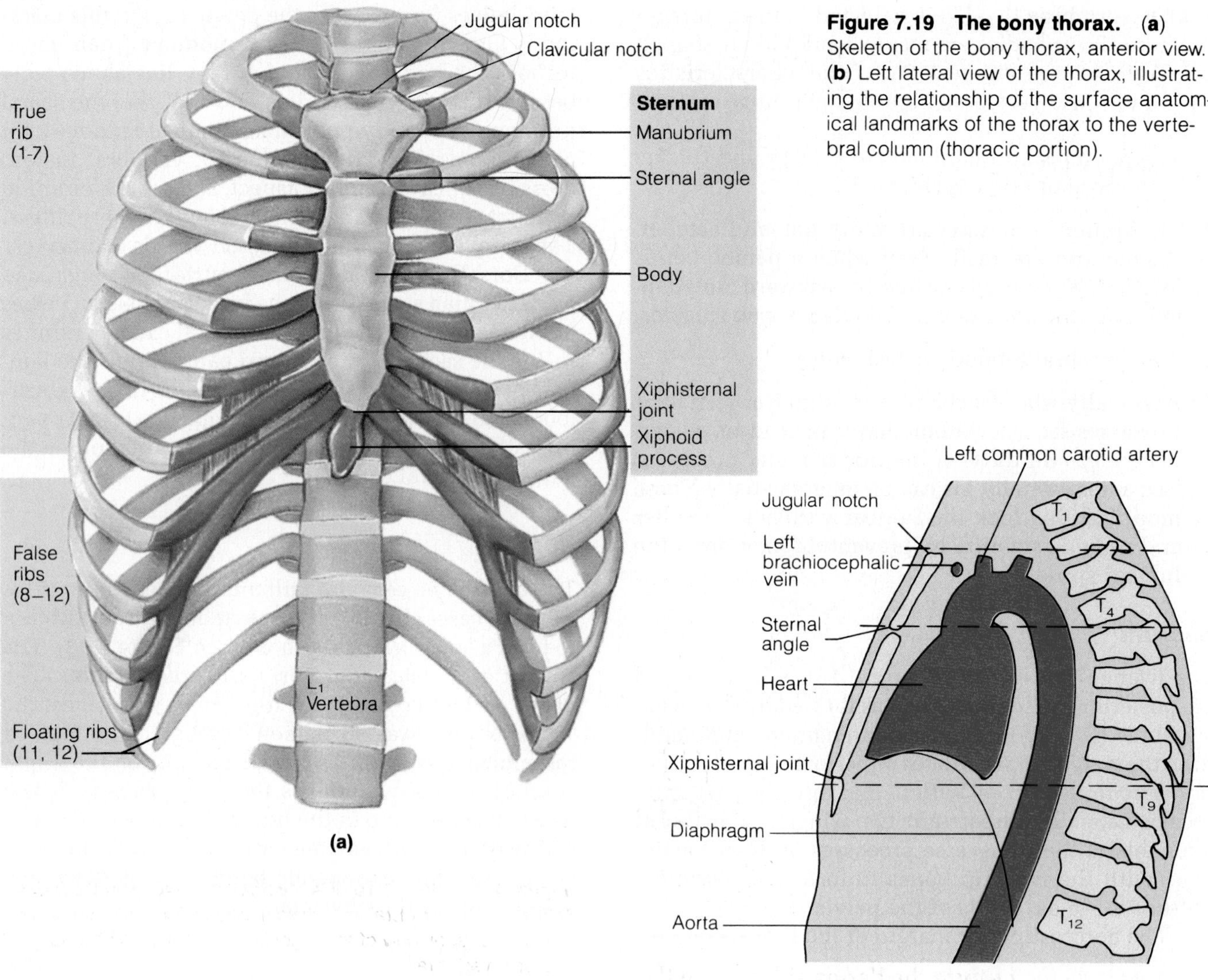

Figure 7.19 The bony thorax. (**a**) Skeleton of the bony thorax, anterior view. (**b**) Left lateral view of the thorax, illustrating the relationship of the surface anatomical landmarks of the thorax to the vertebral column (thoracic portion).

The Bony Thorax

Anatomically, the thorax is the chest, and its bony "underpinnings" are called the **bony thorax** or **thoracic cage**. Elements forming the bony thorax include the thoracic vertebrae dorsally, the ribs laterally, and the sternum and costal cartilages anteriorly. The costal cartilages secure the ribs to the sternum (Figure 7.19a). Basically a cone with its broad dimension inferiorly, the bony thorax forms a protective cage around the vital organs of the thoracic cavity (heart, lungs, and great blood vessels), supports the shoulder girdles and upper limbs, and provides attachment points for the muscles of the back, chest, and shoulders. In addition, the *intercostal spaces* between the ribs are occupied by the external and internal intercostal muscles, which elevate and depress the thorax during breathing.

Sternum

The **sternum** (breastbone) is located in the anterior midline of the thorax. Vaguely resembling a dagger, it is a typical flat bone approximately 15 cm (6 inches) long and consisting of three regions resulting from the fusion of three bones: the manubrium, the body, and the xiphoid process. The **manubrium** (muh-noo′-brē-um), the superior portion, is shaped like the knot in a tie; it articulates via its **clavicular** (kla-vih′-kyoo-ler) **notches** with the clavicles (collarbones) laterally; it also articulates with the first two pairs of ribs. The **body**, or midportion, forms the bulk of the sternum; the sides of the body are notched where it articulates with the cartilages of the third to seventh ribs. The **xiphoid** (zī′-foyd) **process** forms the inferior end of the sternum. This small, variably shaped process is a plate of hyaline cartilage in youth, but it is usually ossified in adults. The xiphoid process articulates only with the sternal body and serves as an attachment point for the diaphragm and some abdominal muscles.

In some people, the xiphoid process tends to project dorsally. This may present a problem because chest trauma can push such a xiphoid into the heart or liver (both immediately deep to the xiphoid process), causing massive hemorrhage. ■

The sternum has three important anatomical landmarks: the jugular notch, the sternal angle, and the xiphisternal joint (see Figure 7.19). The easily palpated **jugular (suprasternal) notch** is the central indentation in the upper border of the manubrium. It is generally in line with the disc between the second and third thoracic vertebrae and the point where the left common carotid artery issues from the aorta (see Figure 7.19b). The manubrium is joined to the sternal body by a cartilaginous joint that acts like a hinge, allowing the sternal body to swing anteriorly when we inspire. The sternum is slightly angled at this joint, and this **sternal angle** can be felt as a horizontal ridge across the front of the sternum. The sternal angle lies at the same height as the disc between the fourth and fifth thoracic vertebrae and at the level of the second pair of ribs. It provides a handy reference point for finding the second rib and thus for counting the ribs during a physical examination. The **xiphisternal** (zih″-fih-ster′-nul) **joint,** the fusion point of the sternal body and xiphoid process, lies opposite the ninth thoracic vertebra.

Ribs

Twelve pairs of **ribs** form the flaring sides of the thoracic cage. All attach posteriorly to the thoracic vertebrae and then curve downward and forward toward the anterior body surface. The upper seven rib pairs are attached directly to the sternum by individual costal cartilages (bars of hyaline cartilage) and are called the **true** or **vertebrosternal** (ver″-tih-brō-ster′-nul) **ribs**. The remaining five pairs of ribs are called **false ribs** because they either attach indirectly to the sternum or lack a sternal attachment entirely. Rib pairs 8–10 attach to the sternum indirectly by joining to each other via the costal cartilages immediately above; these ribs are also called **vertebrochondral** (ver″-tih-brō-kon′-drul) **ribs**. Rib pairs 11 and 12 are called **vertebral ribs** or **floating ribs** because they have no anterior attachments; their costal cartilages lie embedded in the muscles of the lateral body wall. The ribs gradually increase in length from the first to the seventh pairs; from pair 8 to pair 12 they decrease in length.

A typical rib is a bowed flat bone (Figure 7.20). The bulk of a rib is simply called the *shaft*. Its superior border is smooth, but its inferior border is sharp and thin and contains a *costal groove* on its inner face that lodges the intercostal nerves and blood vessels. In addition to the shaft, each rib has a head, neck, and tubercle. The wedge-shaped *head*, the posteriormost end, has two facets: One articulates with the body of the same-numbered thoracic vertebra, the other with the body of the vertebra immediately above. The *neck* is the constricted portion of the rib beyond the head. Lateral to this, the knoblike *tubercle* articulates with the transverse process of the same-numbered thoracic vertebra. Beyond the tubercle, the shaft angles sharply forward and then extends to attach to its costal cartilage anteriorly. The costal cartilages provide secure but flexible sternum-rib attachments.

The first pair of ribs is quite atypical; they are flattened superiorly to inferiorly and are quite broad. There are also other exceptions to the typical rib pattern. Ribs 1 and 10–12 articulate only with one vertebral body, and the eleventh and twelfth ribs do not articulate with transverse processes of the corresponding vertebrae. Except for the first rib, which lies deep to the clavicle, the ribs are easily felt in people of normal weight.

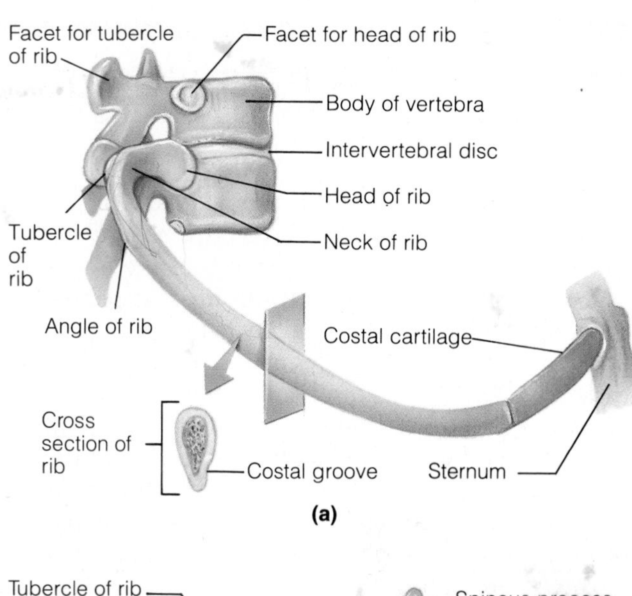

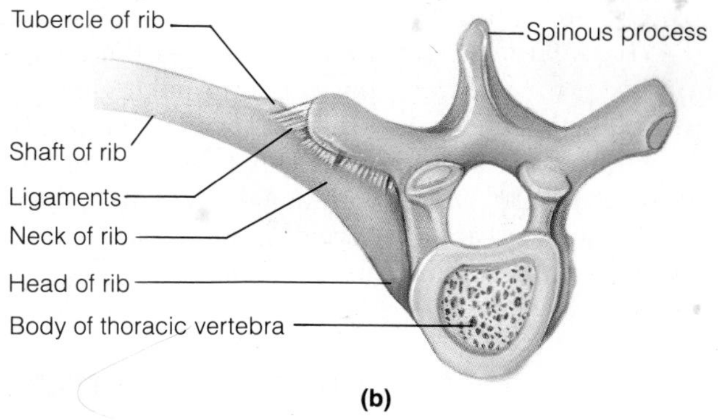

Figure 7.20 Structure of a "typical" true rib and its articulations. (**a**) Vertebral and sternal articulations of a typical true rib. (**b**) Superior view of the articulation between a rib and a thoracic vertebra.

PART 2: THE APPENDICULAR SKELETON

The bones of the **appendicular skeleton** hang suspended from two yokelike, bony girdles anchored to the axial skeleton. Thus, the bones of the appendicular skeleton are (as the name implies) "appended" to the longitudinal axis of the body (see Figure 7.1). The *pectoral* (pek'-ter-ul) *girdles* attach the upper limbs to the body trunk; the more firmly anchored *pelvic girdle* secures the lower limbs. Although the bones of the upper and lower limbs are quite different in their functions and mobility, they have the same fundamental structural plan, with each limb composed of three major segments connected together by freely movable joints.

Whereas the axial skeleton provides the central support for the body and protects internal organs, the bones of the appendicular skeleton are adapted to carry out the movements typical of our freewheeling and manipulative life-style. Each time we take a step, throw a ball, or pop a caramel into our mouth, we are making good use of our appendicular skeleton.

The Pectoral (Shoulder) Girdle

The **pectoral,** or **shoulder, girdle** consists of two bones, the anterior *clavicle* (kla'-vih-kul) and the posterior *scapula* (ska'-pyoo-luh) (Figure 7.21 and Table 7.3 on p. 202). The two pectoral girdles and their associated muscles form your shoulders. Although the term *girdle* usually signifies a beltlike structure encircling the body, a single pectoral girdle, or even the pair, does not quite satisfy this description. Anteriorly, the medial end of each clavicle joins the sternum; the distal ends of the clavicles meet the scapulae laterally. However, the scapulae fail to complete the ring posteriorly and are attached to the thorax and vertebral column only by the muscles that clothe their surfaces.

The pectoral girdles attach the upper limbs to the axial skeleton and provide insertion points for many muscles that move the upper limbs. These girdles are very light and allow the upper limbs a degree of flexibility and mobility not seen anywhere else in the body. This is due to the following factors:

1. The only attachment point of the pectoral girdles to the axial skeleton is at the sternoclavicular joints anteriorly.
2. The relative looseness of the scapular attachments allows the scapulae to move rather freely across the thorax. They can be elevated, depressed, and moved side to side by attached muscles.
3. The socket of the shoulder joint (the glenoid cavity) is small, shallow, and poorly reinforced by ligaments. Although this arrangement is good for flexibility, it is bad for stability, and shoulder dislocations are fairly common.

Clavicles

The **clavicles**, or collarbones, are slender, doubly curved long bones that can be felt along their entire course as they extend horizontally across the upper thorax (Figure 7.21b and c). Each clavicle is rounded on its medial *sternal end,* which attaches to the sternal manubrium, and flattened on its lateral *acromial* (uh-krō'-mē-ul) *end,* where it articulates with the acromion of the scapula. The medial two-thirds of the clavicle is convex anteriorly; its lateral third is concave anteriorly. Its superior surface is smooth, but the inferior surface is ridged and grooved. Besides providing attachment points for many thoracic and shoulder muscles, the clavicles act as anterior braces: They hold out the scapulae and their appended limbs away from the upper, narrower portion of the thorax. This bracing function becomes immediately obvious when a clavicle is fractured: The entire shoulder region collapses medially. The clavicles also transmit forces acting on the upper limbs to the axial skeleton; but the clavicles resist compression poorly and are likely to fracture, for example, when a person uses their outstretched arms to break a fall. However, the clavicles are exceptionally sensitive to muscle pull and become noticeably larger and stronger in those who perform manual labor involving the shoulder and arm muscles.

Scapulae

The **scapulae,** or *shoulder blades,* are complicated, triangular flat bones located on the dorsal thorax between ribs 2 and 7 (Figure 7.21d–f). Each scapula has three borders. The *superior border* is the shortest, sharpest border. The *vertebral,* or *medial, border* parallels the vertebral column; the *axillary,* or *lateral, border* abuts the armpit and has a small, shallow fossa, the **glenoid** (gleh'-noyd) **cavity**. The glenoid cavity articulates with the humerus of the arm, forming the relatively unstable shoulder joint. The superior scapular border meets the medial border at the *superior angle* and the lateral border at the *lateral angle*. The medial and lateral borders join at the *inferior angle*.

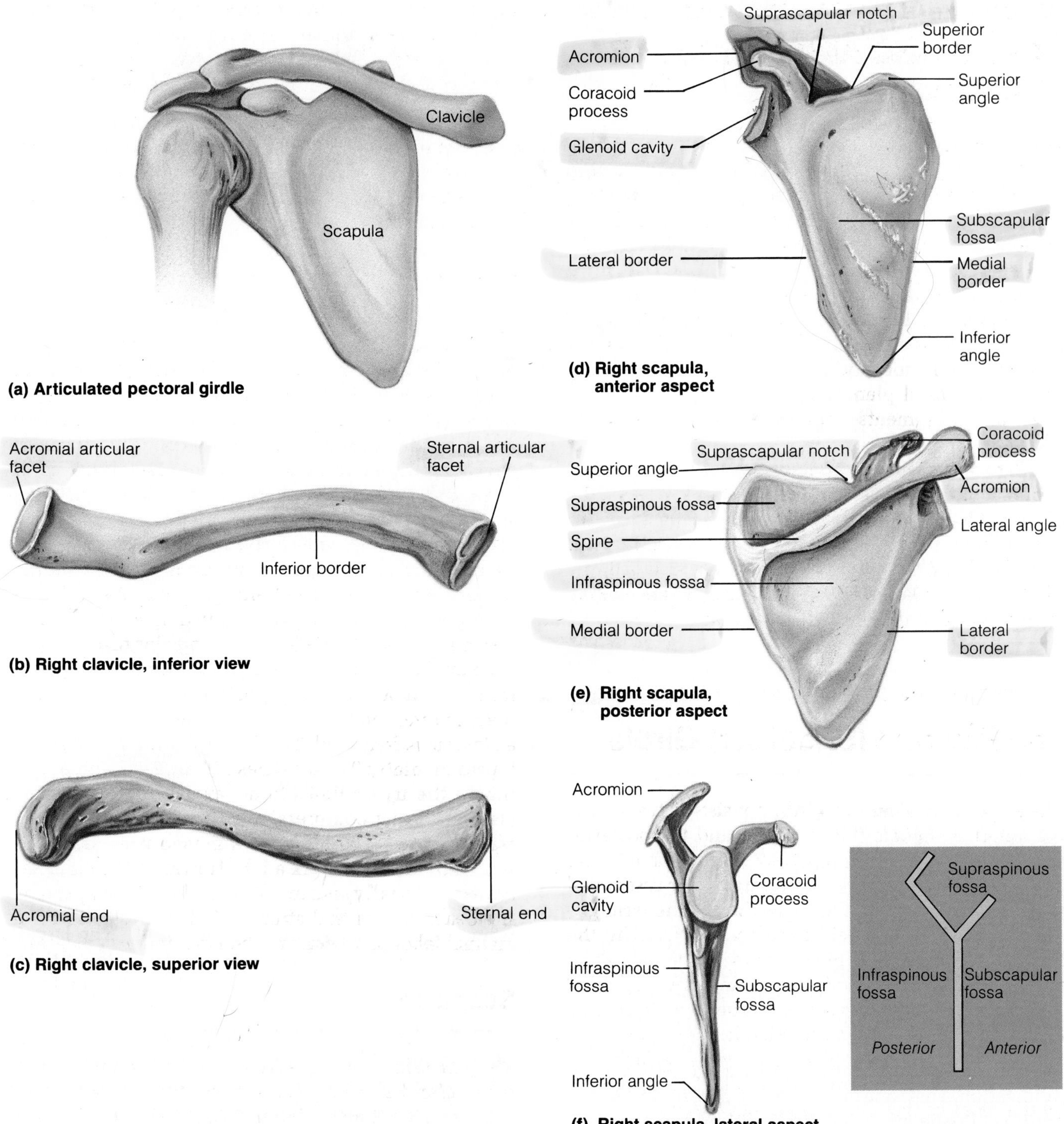

Figure 7.21 Bones of the pectoral girdle.

The inferior angle moves extensively when the arm is raised and is an important anatomical landmark for studying scapular movements.

The anterior, or costal, surface of the scapula is concave and relatively featureless. Its posterior surface bears a sharp, prominent **spine**, which ends laterally in an enlarged, roughened anterior projection called the **acromion** (uh-krō′-mē-un), which can be seen as the "point of the shoulder." The acromion articulates with the acromial end of the clavicle, forming the *acromioclavicular joint.* Although each acromioclavicular joint is only about the size of a big

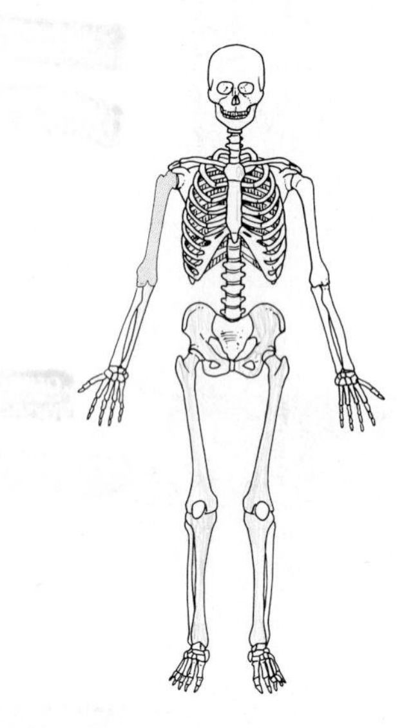

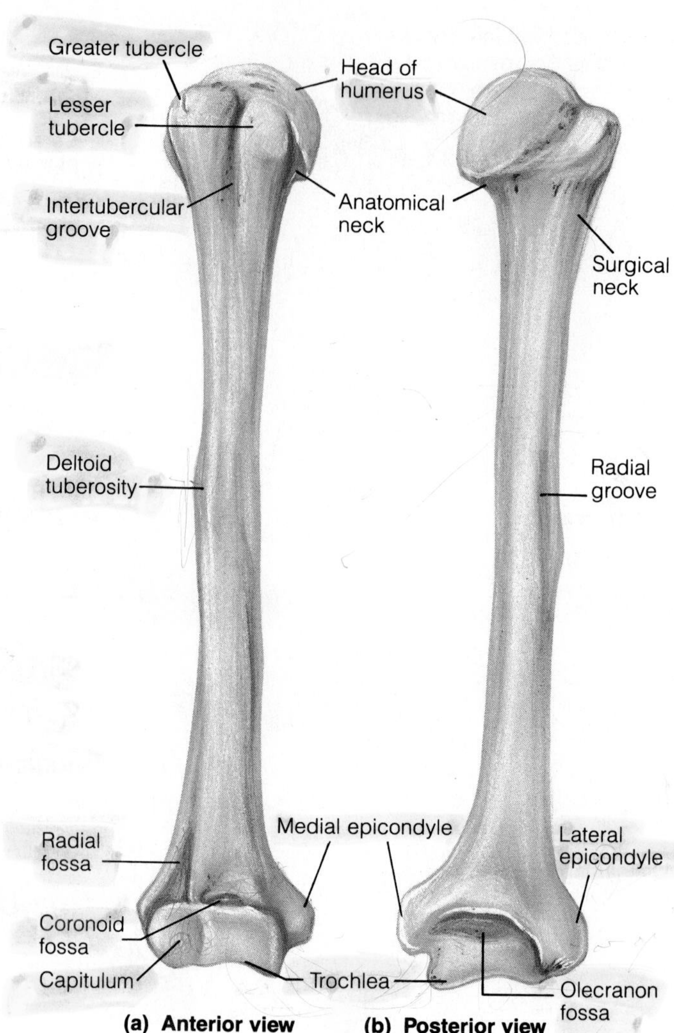

Figure 7.22 The humerus of the arm. (**a**) Anterior view of the right humerus. (**b**) Posterior view of the right humerus.

toe joint, it plays a crucial role in attaching the upper limb to the body trunk. Projecting anteriorly from the superior scapular border is the **coracoid** (kor′-uh-koyd) **process**; *corac* means "beaklike," but the coracoid process looks more like a bent little finger. The coracoid process helps to anchor the biceps muscle of the arm and is bounded by the **suprascapular notch** (a nerve passage) medially and by the glenoid fossa laterally. Several shallow depressions, or fossae, appear on both sides of the scapula and are named according to location. For example, the *infraspinous* and *supraspinous fossae* are respectively below and above the spine, and the *subscapular fossa* is the concavity of the anterior scapular surface.

The Upper Limb

Thirty separate bones form the skeletal framework of each upper limb (see Figures 7.22 to 7.25, and Table 7.3 on p. 202). Each of these bones may be described regionally as a bone of the arm, forearm, or hand.

Arm

The **humerus** (hyoo′-mer-us), the sole bone of the upper arm, is a typical long bone (Figure 7.22). The largest, longest bone of the upper limb, it articulates with the scapula at the shoulder and with the radius and ulna (forearm bones) at the elbow.

At the proximal end of the humerus is its smooth, hemispherical *head*, which fits into the glenoid cavity of the scapula in a manner that allows the arm to hang freely at one's side. Immediately below the head is a slight constriction, the **anatomical neck**; just inferior to the anatomical neck are the lateral **greater tubercle** and medial **lesser tubercle**, separated by the **intertubercular,** or **bicipital** (bī-sih′-pih-tul), **groove**. The tubercles serve as muscle attachment points; the intertubercular groove guides a tendon of the biceps muscle of the arm to its attachment point at the rim of the glenoid cavity. Just distal to the tubercles, where the epiphysis joins the shaft, is the **surgical neck**, so named because it is the most frequently fractured part of the humerus. The shaft of the humerus is cylindrical

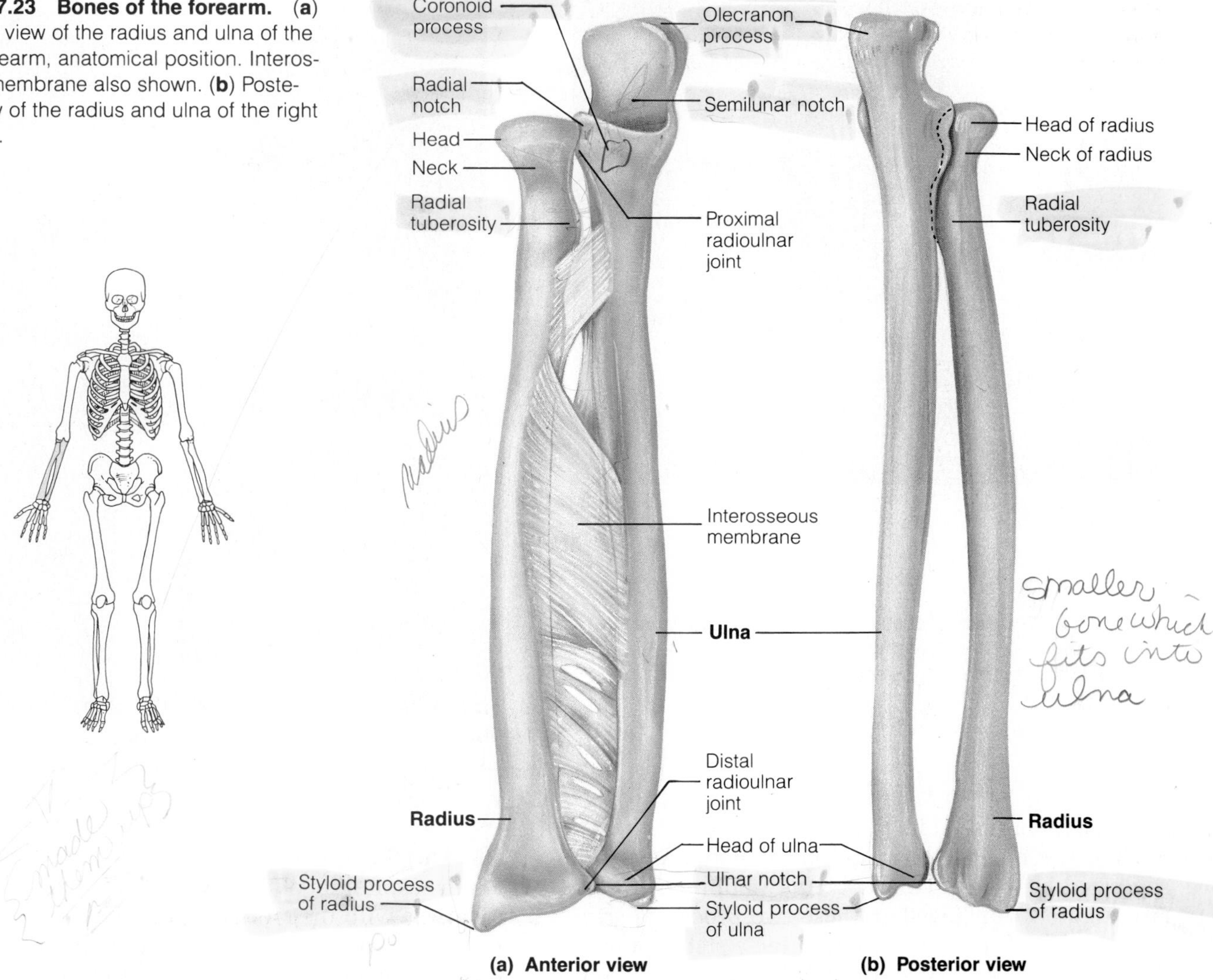

Figure 7.23 Bones of the forearm. (**a**) Anterior view of the radius and ulna of the right forearm, anatomical position. Interosseous membrane also shown. (**b**) Posterior view of the radius and ulna of the right forearm.

proximally, but distally it becomes triangular in cross section, and ridges appear externally. A roughened area of the shaft, the **deltoid** (del′-toyd) **tuberosity,** is the attachment site of the fleshy deltoid muscle of the shoulder. A **radial groove** runs obliquely down the posterior aspect of the shaft, marking the course of the radial nerve.

At the distal end of the humerus are two condyles, a medial pulley-shaped **trochlea** (trok′-lē-uh) and a lateral ball-like **capitulum** (kuh-pih′-chuh-lum), which articulate with the ulna and radius, respectively. The condyle pair is flanked by the *medial* and *lateral epicondyles.* The ulnar nerve runs behind the medial epicondyle and is responsible for the painful, tingling sensation you experience when you hit your "funny bone." Above the trochlea on the anterior surface is the **coronoid** (kor′-uh-noyd) **fossa**; on the posterior surface is the deeper **olecranon** (ō-leh′-kruh-non) **fossa**. These two depressions allow the corresponding prominences of the ulna to move freely when the elbow is flexed and extended. A small *radial fossa,* lateral to the coronoid fossa, receives the head of the radius when the elbow is flexed.

Forearm

Two parallel long bones, the *radius* and the *ulna,* form the skeleton of the forearm, or *antibrachium* (an″-tīh-brā′-kē-um) (Figure 7.23), and are easily palpated along their entire length. Their proximal ends articulate with the humerus; their distal ends form joints with bones of the wrist. The radius and ulna articulate with each other both proximally and distally at small *radioulnar* (rā″-dē-ō-ul′-nar) *joints,* and they are connected along their entire length by a flexible *interosseous* (in″-ter-ah′-sē-us) *membrane.* In the anatomical position, the radius lies laterally (on the thumb side) and the ulna medially. However, when you rotate your hand so that the palm faces posteriorly (a movement called pronation), the distal end of the radius crosses over the ulna and the two bones form an X (see Figure 8.7a).

Ulna

The **ulna** is slightly longer than the radius and has the major responsibility for forming the elbow joint

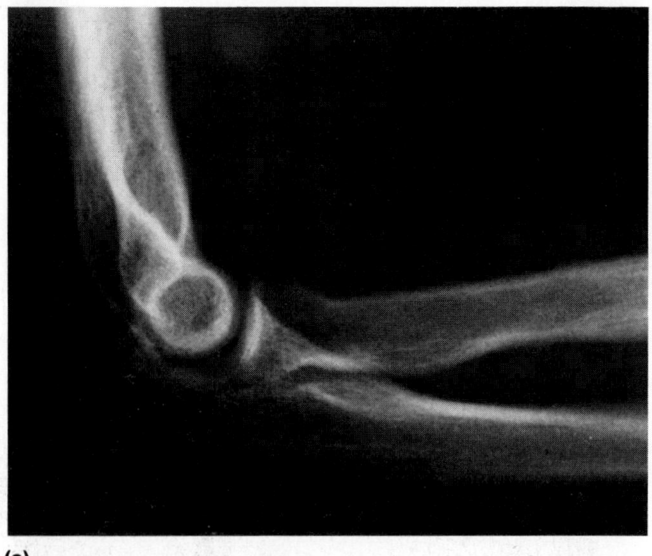

(a)

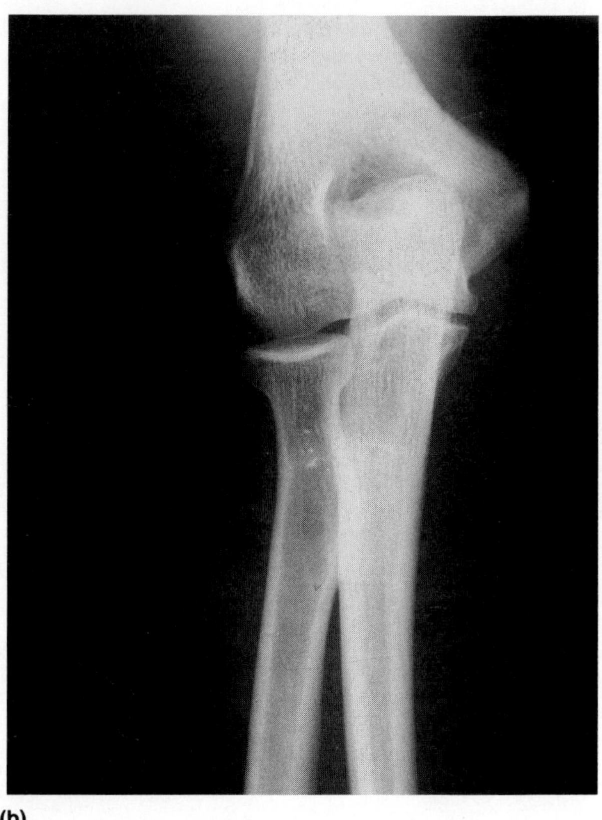

(b)

Figure 7.24 X-rays of the left elbow joint formed by articulation of the humerus and ulna. (**a**) Elbow flexed, lateral view. (**b**) Elbow extended, posterior view.

with the humerus. At its proximal end are two prominent processes, the **olecranon** and **coronoid processes**, separated by a deep concavity, the **semilunar notch** (see Figure 7.23). Together, these two processes grip the trochlea of the humerus, forming a stable pliers-like hinge joint that allows the forearm to be bent upon the arm (flexed) or extended. When the arm is fully extended (Figure 7.24b), the olecranon process "locks" into the olecranon fossa, preventing hyperextension of the forearm (that is, preventing its movement posteriorly beyond the elbow joint). The posterior olecranon process forms the angle of the elbow when the forearm is flexed and is the bony part that rests on the table when you lean on your elbows. On the lateral side of the coronoid process is a small articular depression, the **radial notch,** which articulates with the head of the radius.

The ulnar shaft is triangular and ridged, narrowing as it runs distally to its smaller, knoblike *head.* Medial to the head is a **styloid process,** to which some ligaments of the wrist muscles attach. Laterally, the head articulates with the radius, forming the distal radioulnar joint. The ulnar head is separated from the bones of the wrist by a fibrocartilage disc and plays little or no role in hand movements.

Radius

The **radius,** too, is triangular in cross section and marked by lengthwise ridges. The *head* of the radius is shaped like the end of a spool of thread (see Figure 7.23), and the upper concavity of the head articulates with the capitulum of the humerus. Medially, the head articulates with the radial fossa of the ulna. Just inferior to the head is a rough projection, the **radial tuberosity**, which anchors the biceps muscle of the arm. Distally, the radius is expanded. It has a medial **ulnar notch,** which articulates with the ulna, and a lateral **styloid process**. Between these two markings, the radius is concave where it articulates with carpal bones of the wrist. The radius is the major forearm bone contributing to the wrist joint; when the radius moves, the hand moves with it.

Hand

The skeletal framework of the hand includes the bones of the carpus, or wrist; the bones of the metacarpus, or palm; and the phalanges, or bones of the digits (Figure 7.25).

Carpus (Wrist)

A "wrist" watch is actually worn on the distal forearm, over the lower ends of the radius and ulna. The anatomical wrist, or **carpus,** is the proximal portion of the structure we generally call our "hand." The carpus consists of a group of eight marble-size short bones,

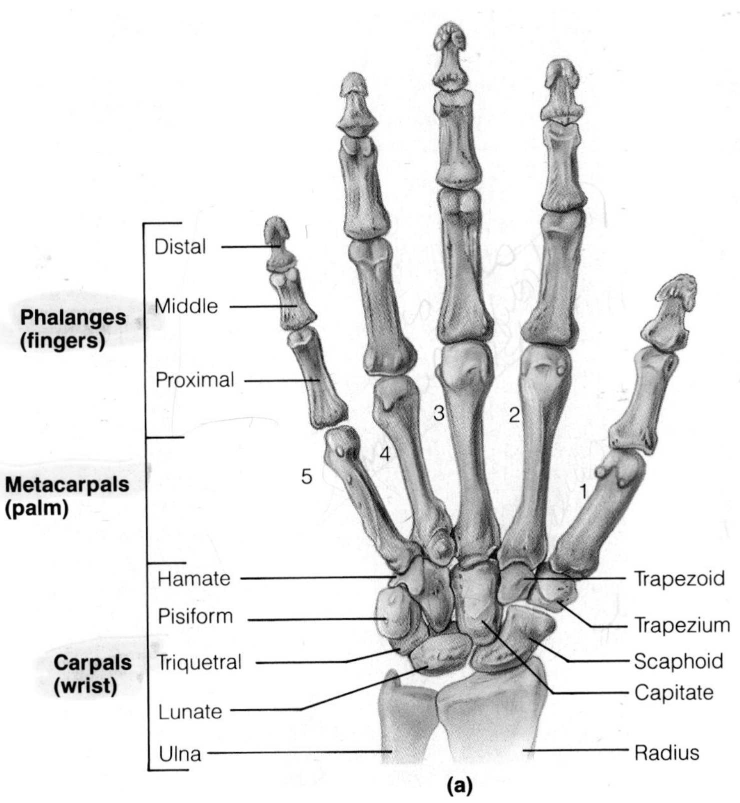

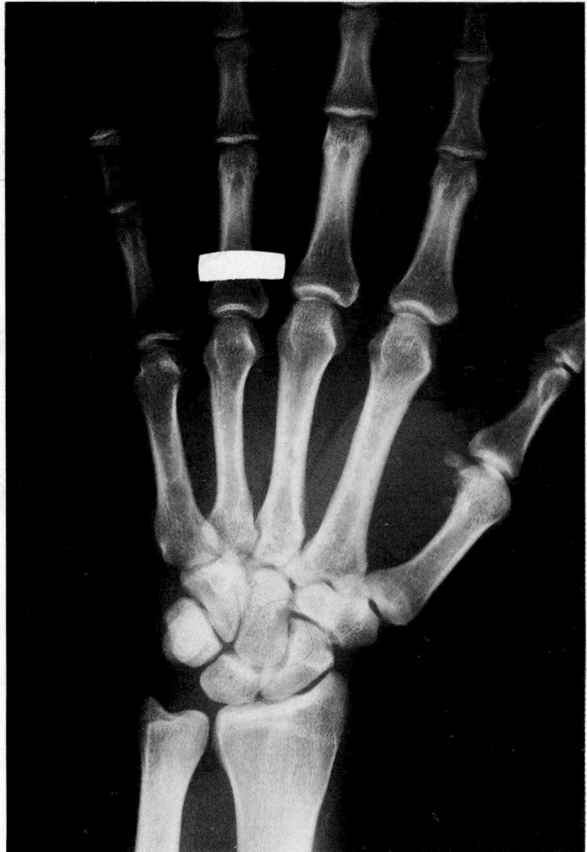

Figure 7.25 Bones of the hand. (**a**) Ventral view of the right hand, illustrating the anatomical relationships of the carpals, metacarpals, and phalanges. (**b**) X-ray of the right hand. Notice the white bar on phalanx 1 of the ring finger, showing the position at which a ring would be worn.

or **carpals** (kar′-pulz), closely united by ligaments. Although the mobility of these bones is restricted to gliding movements, the carpus as a whole is extremely flexible. The carpals are arranged in two irregular rows of four bones each (Figure 7.25). In the proximal row (lateral to medial) are the **scaphoid** (ska′-foyd), **lunate** (loo′-nāt), **triquetral** (trī-kweh′-trul), and **pisiform** (pī′-sih-form). The scaphoid and lunate articulate with the distal end of the radius to form the wrist joint. The carpals of the distal row (lateral to medial) are the **trapezium** (truh-pē′-zē-um), **trapezoid** (tra′-puh-zoyd), **capitate**, and **hamate** (ha′-māt). Each carpal, except for the pisiform, has several facets for articulating with neighboring carpals.

Metacarpus (Palm)

Five **metacarpal bones,** which radiate from the wrist like spokes, form the palm of the hand (see Figure 7.25). These small long bones are not named, but instead are numbered 1 to 5 from thumb to little finger. The bases of the metacarpals articulate with the carpals proximally and each other laterally; their bulbous heads articulate with the proximal phalanges of the fingers. When you clench your fist, the heads of the metacarpals become prominent as your knuckles. Metacarpal 1, associated with the thumb, is the shortest and most mobile. It does not lie in the same plane as the other metacarpals, but occupies a more anterior position. The joint between metacarpal 1 and the carpals is a unique saddle joint that allows the thumb to oppose the fingers. The special mobility of the thumb is what makes the human hand such an efficient tool for grasping and manipulating.

Phalanges (Fingers)

The **digits**, or fingers, numbered 1 to 5 beginning with the thumb, or *pollex* (pah′-leks), are made up of miniature long bones called phalanges. In most people, the third finger is the longest. Each hand contains 14 **phalanges** (fuh-lan′-jez), and with the exception of the thumb, each finger has three: the distal, middle, and proximal phalanges. The thumb has no middle phalanx. The proximal phalanges articulate with the heads of the metacarpals.

Table 7.3 Bones of the Appendicular Skeleton

Body region	Bones*	Illustration	Location	Markings
PART I: Bones of the Pectoral Girdle and Upper Limb				
Pectoral girdle (Figure 7.21)	Clavicle (2)		Superoanterior thorax; articulates medially with sternum and laterally with scapula	Acromial end; sternal end
	Scapula (2)		Posterior thorax; forms part of the shoulder; articulates with humerus and clavicle	Glenoid cavity; spine; acromion; coracoid process
Upper limb Arm (Figure 7.22)	Humerus (2)		Sole bone of arm; between scapula and elbow	Head; greater and lesser tuberosities; intertubercular groove; deltoid tuberosity; trochlea; capitulum; coronoid and olecranon fossae
Forearm (Figure 7.23)	Ulna (2)		Medial bone of forearm between elbow and wrist; forms elbow joint	Coronoid process; olecranon process; semilunar notch; styloid process
	Radius (2)		Lateral bone of forearm; carries wrist	Radial tuberosity; styloid process
Hand (Figure 7.25)	8 Carpals (16)		Form a bony crescent at the wrist; arranged in two rows of four bones	
	5 Metacarpals (10)		Form the palm; one in line with each digit	
	14 Phalanges (28)	Anterior view–right side of body	Form the fingers; three in digits 2–5; two in digit 1 (the thumb)	

The Pelvic (Hip) Girdle

The **pelvic girdle**, or **hip girdle**, attaches the lower limbs to the axial skeleton and supports the visceral organs of the pelvic cavity (Figure 7.26 and Table 7.3). While the shoulder girdle moves somewhat freely across the thorax and allows the upper limb a high degree of mobility, the pelvic girdle is secured to the axial skeleton by some of the strongest ligaments in the body, and its sockets, which articulate with the thigh bones, are deep and cuplike and heavily reinforced by ligaments. Thus, even though both the shoulder and hip joints are ball-and-socket joints, very few of us can wheel our legs and arms about with the same degree of freedom.

The pelvic girdle is formed by a pair of **coxal** (*coxa* = hip) **bones** (Figure 7.26), or the **ossa coxae** (ah′-suh cok′-sē), commonly called the *hip bones* or *innominate* ("no name") *bones.* Each coxal bone unites with its partner anteriorly and with the sacrum pos-

Table 7.3 (continued)

Body region	Bones*	Illustration	Location	Markings
PART II: Bones of the Pelvic Girdle and Lower Limb				
Pelvic girdle (Figure 7.26)	Coxa (2)		Each coxal bone is formed by the fusion of an ilium, ischium, and pubic bones; each coxal bone meets its partner anteriorly at the pubic symphysis and forms a sacroiliac joint with the sacrum posteriorly; girdle consisting of both coxal bones is basinlike	Iliac crest; anterior and posterior iliac spines; auricular surface; greater and lesser sciatic notches; obturator foramen; ischial tuberosity and spine; acetabulum
Lower limb				
Thigh (Figure 7.28)	Femur (2)		Sole bone of thigh; between hip joint and knee; largest bone of the body	Head; greater and lesser trochanters; neck; lateral and medial condyles; gluteal tuberosity; linea aspera
Leg (Figure 7.29)	Tibia (2)		Larger and more medial bone of leg; between knee and foot	Medial and lateral condyles; tibial tuberosity; anterior crest; medial malleolus
	Fibula (2)		Lateral bone of leg	Head; lateral malleolus
Foot (Figure 7.30)	7 Tarsals (14)		Seven bones forming each ankle; the talus articulates with the leg bones; ankle joints reinforced by the malleoli; calcaneus is the largest tarsal; forms the heel	
	5 Metatarsals (10)		Five bones forming each instep	
	14 Phalanges (28)	Anterior view of pelvic girdle and left lower limb	Form the toes; three in digits 2–5, two in digit 1 (the great toe)	

*The number in parentheses ( ) following the bone name denotes the total number appearing in the body.

teriorly; the deep, basinlike structure formed by the ossa coxae, together with the sacrum and coccyx, is called the **pelvis**. Each coxal bone is large and irregularly shaped. During childhood, it consists of three separate bones: the ilium, ischium, and pubis. In adults, these bones are firmly fused and indistinguishable, but their names are retained to refer to different regions of the composite coxal bone. At the point of fusion of the ilium, ischium, and pubis, a deep hemispherical socket called the **acetabulum** (a″-sih-ta′-byoo-lum) is formed on the lateral surface of the os coxa (see Figure 7.26b). The acetabulum articulates with the head of the femur, or thigh bone.

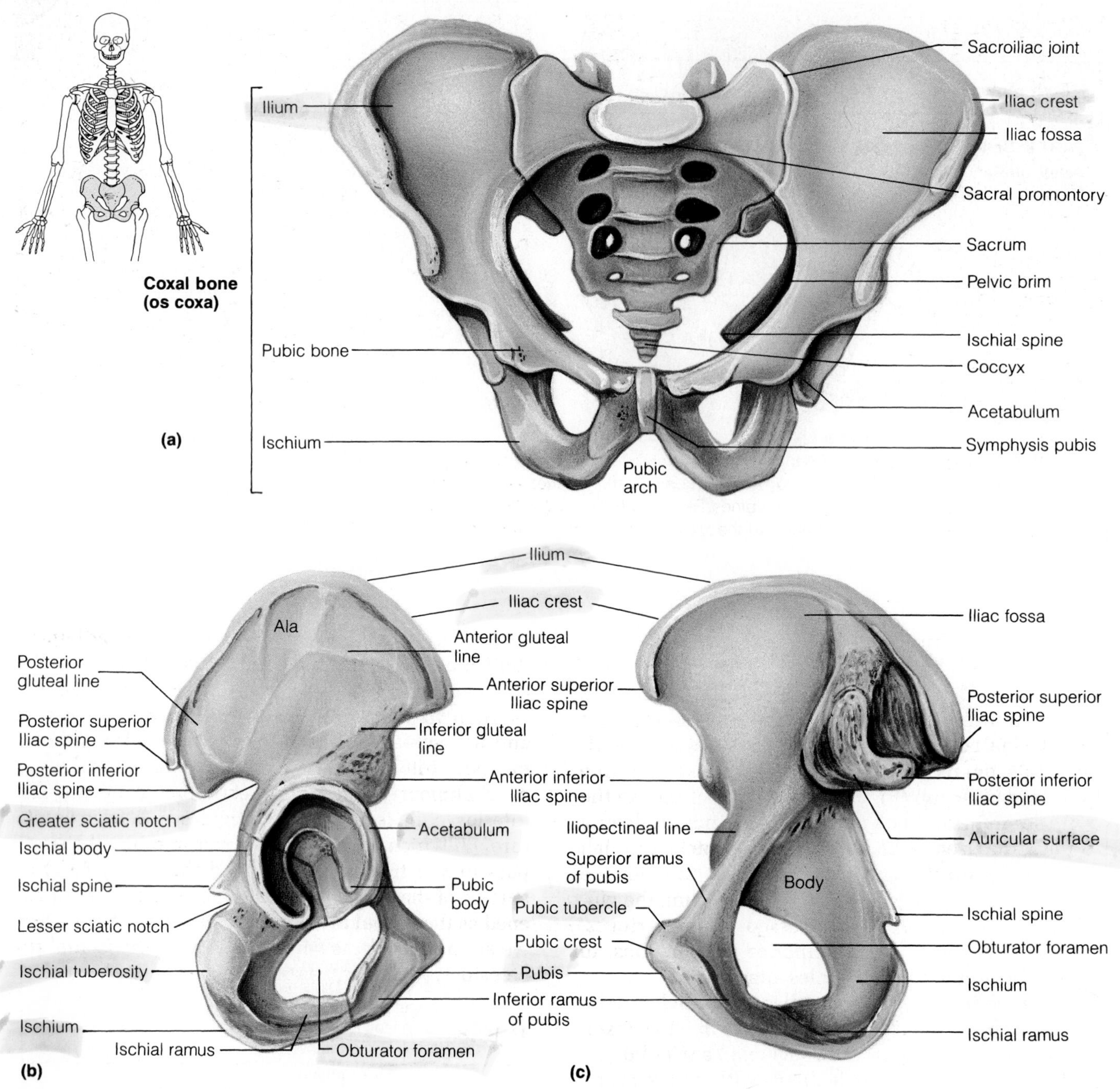

Figure 7.26 Bones of the pelvic girdle. (**a**) Articulated pelvis showing the two coxal bones and the sacrum. (**b**) Lateral view of right coxal bone showing the point of fusion of the ilium, ischium, and pubic bones at the acetabulum. (**c**) Right os coxa, medial view.

Ilium

The **ilium** is a large flaring bone that forms the major portion of a coxal bone. It consists of a *body* and a superior winglike portion called the *ala* (ā′-luh). When you rest your hands on your hips, you are resting them on the thickened superior margins of the alae, the **iliac crests**. Each iliac crest terminates anterosuperiorly in the blunt **anterior superior iliac spine** and posterosuperiorly in the sharp **posterior superior iliac spine**. Located below these are the two less prominent *anterior* and *posterior inferior iliac spines*. All of these

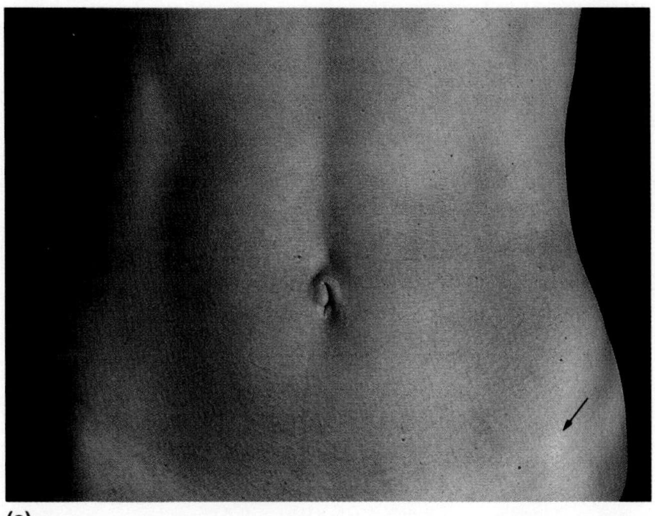

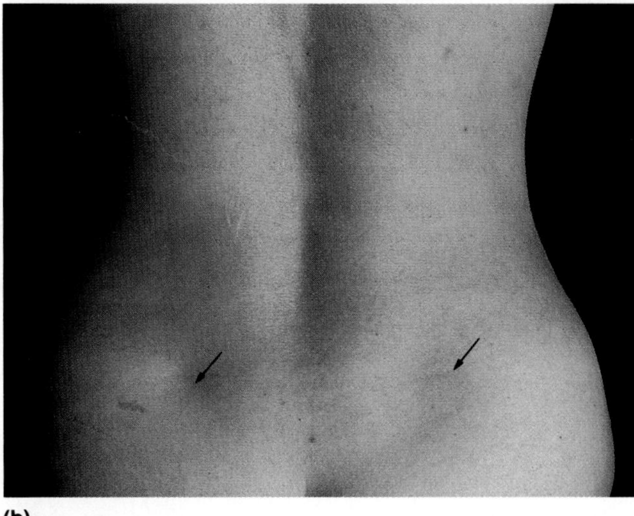

(a) (b)

Figure 7.27 Surface anatomy of the pelvis. (**a**) Anterior view clearly showing the protrusion created by the left anterior superior iliac spine. (**b**) Posterior view illustrating the position of the posterior superior iliac spines. The attachment of the deep fascia to the posterior spines creates skin dimples at the level of the second sacral vertebra.

spines are attachment points for the muscles of the trunk, hip, and thigh. The anterior superior iliac spine is an especially important anatomical landmark; it is easily felt through the skin and may be visible (Figure 7.27a). The posterior superior iliac spine is difficult to palpate, but its position is revealed by a skin dimple in the sacral region (Figure 7.27b). Just inferior to the posterior inferior iliac spine, the ilium indents deeply to form the **greater sciatic** (sī-a′-tik) **notch**, through which the cordlike sciatic nerve passes to enter the thigh. The posterolateral surface of the ilium, the *gluteal* (gloo′-tē-ul) *surface,* is crossed by three ridges, the *posterior, anterior,* and *inferior gluteal lines,* to which the gluteal (hip) muscles attach.

The medial surface of the iliac ala is slightly concave, and this concavity is referred to as the **iliac fossa**. Its posteriormost region, the roughened **auricular** (or-ih′-kyoo-ler) **surface,** articulates with the sacrum, forming the **sacroiliac joint**. The weight of the body trunk is transmitted from the spine to the pelvis through the sacroiliac joints. Running inferiorly and anteriorly from the auricular surface is a ridge called the **iliopectineal** (ih″-lē-ō-pek-tih′-nē-ul) or **arcuate** (ar′-kyoo-āt) **line**. Together with the sacral promontory, the iliopectinal line defines the **pelvic brim**. The pelvic brim is the superior margin of the *true pelvis,* which we will discuss shortly. Anteriorly, the body of the ilium articulates with the ischium and pubic bones.

Ischium

The **ischium** (is′-kē-um) forms the posteroinferior part of the hipbone (see Figure 7.26). Roughly L- or arc-shaped, it possesses an upper thicker *body* adjoining the ilium and a thinner *ramus* that joins the pubic bone anteriorly. Three important markings on the ischium are the ischial spine, the lesser sciatic notch, and the ischial tuberosity. The **ischial spine** projects medially into the pelvic cavity and serves as a point of attachment of a major (*sacrospinous*) ligament. Just inferior to the ischial spine is the **lesser sciatic notch,** through which a number of nerves and blood vessels pass to and from the pelvis and thigh. The inferior surface of the ischial body is rough and grossly thickened as the **ischial tuberosity** (see Figure 7.26b). When we sit, our weight is borne entirely by the ischial tuberosities, which are the strongest parts of the hipbones.

Pubis

The **pubis** (pyoo′-bis) or **pubic bone,** forms the anterior portion of the os coxa (see Figure 7.26). Essentially, it is V-shaped with two *rami* issuing from a flattened medial *body.* In its anatomical position, it lies nearly horizontally and the bladder rests upon it. The body of the pubis lies medially and its anterior border is thickened to form a *pubic crest.* At the lateral end of the pubic crest is the **pubic tubercle,** the main pelvic attachment for the inguinal ligament. The joining of the rami of the pubic bone with the body and ramus of the ischium posterolaterally forms an opening in the os coxa, the **obturator** (ob″-ter-ā′-ter) **foramen,** through which blood vessels and nerves pass. Although the obturator foramen is large, it is nearly closed by a fibrous membrane in life.

Table 7.4 Comparison of the Male and Female Pelvis

Characteristic	Female	Male
General structure and functional modifications	Tilted forward; modified for childbearing; defines the birth canal; cavity of the true pelvis is broad, shallow, and has a greater capacity	Tilted backward; adapted for support of a strong body; cavity of the true pelvis is narrow and deep
Bone thickness	Less; bones lighter, thinner, and smoother	Greater; bones heavier and thicker, and markings are more prominent
Acetabula	Smaller; farther apart	Larger; closer
Pubic angle/arch	Broader (over 90°); more rounded	Angle is more acute (less than 90°)
Anterior view	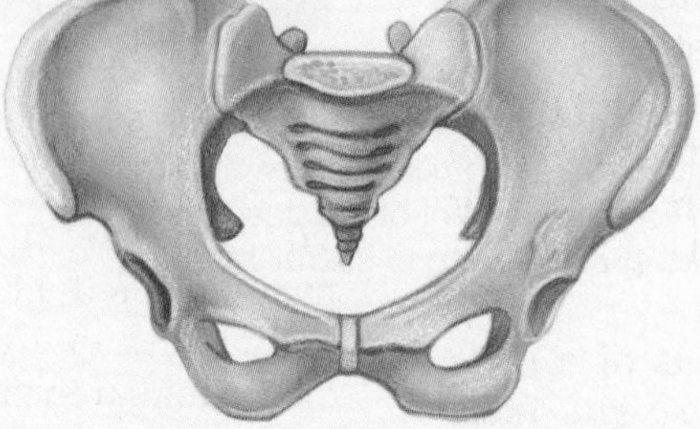	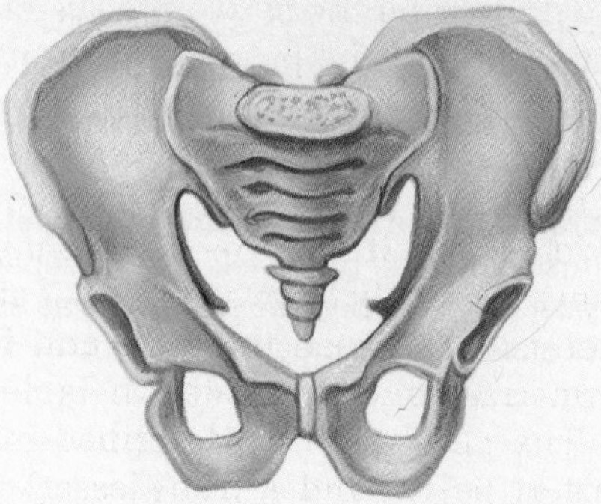
Sacrum	Wider; shorter; flatter; sacral curvature is accentuated	Narrow; longer; sacral promontory more ventral
Coccyx	More movable; straighter	Less movable; curves ventrally
Left lateral view	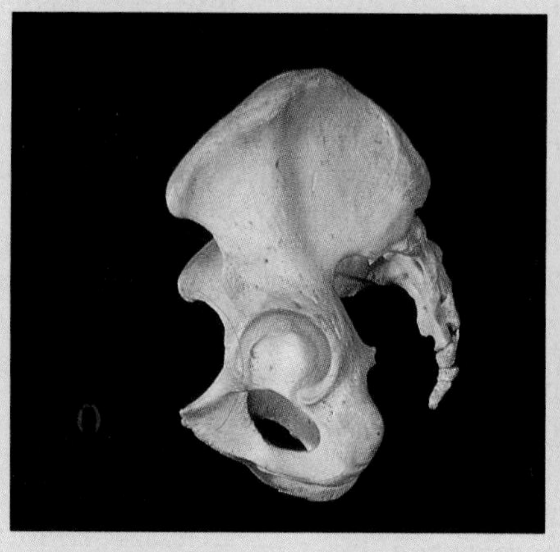	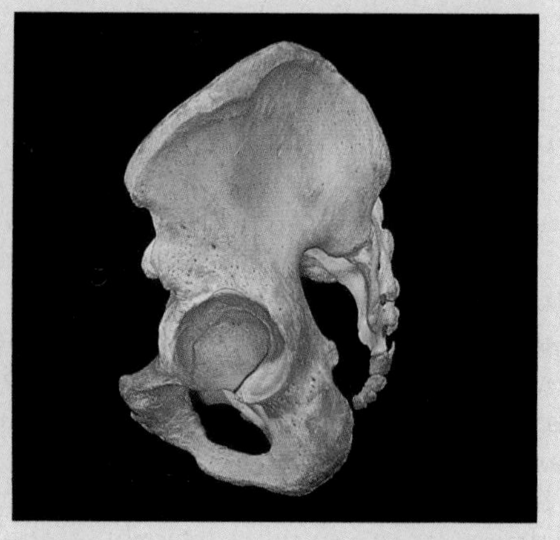
Pelvic inlet	Wider; oval from side to side	Narrow; basically heart shaped
Pelvic outlet	Wider; ischial tuberosities shorter, farther apart and everted	Narrower; ischial tuberosities longer, sharper, and point more medially
Posteroinferior view	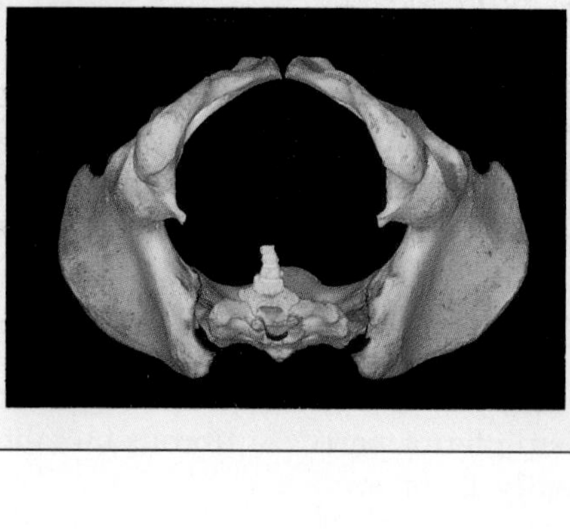	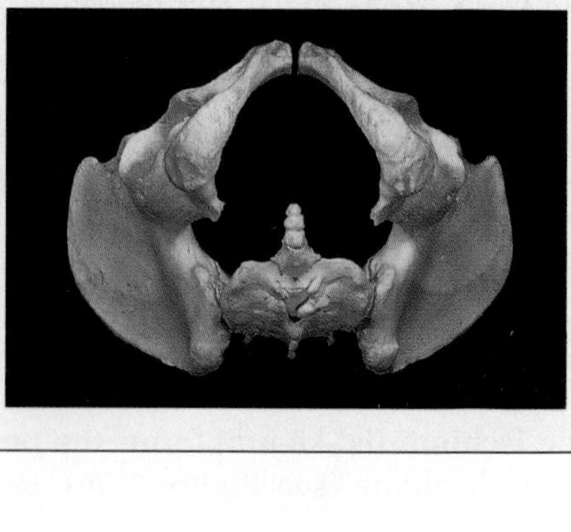

The bodies of the two pubic bones are joined by a fibrocartilage disc, forming the midline **symphysis pubis** joint. Inferior to this joint, the inferior pubic rami angle laterally, forming an inverted V-shaped arch called the **pubic arch** (Table 7.4). The sharpness of the pubic arch helps to differentiate the male and female pelvis.

Pelvic Structure and Childbearing

So striking are the differences between the male and female pelvis that a trained anatomist can immediately determine the sex of the skeleton from a casual examination of the pelvis. The female pelvis reflects modifications for childbearing: It tends to be wider, shallower, lighter, and rounder than that of a male. Not only must the female pelvis accommodate a growing fetus, but it must be large enough to allow the infant's relatively large head to exit at birth. The major differences between the male and female pelvis are summarized and illustrated in Table 7.4.

The pelvis can be described in terms of a false (greater) pelvis and a true (lesser) pelvis. The **false pelvis** is that portion superior to the pelvic brim; it is bounded by the alae of the ilia laterally and the lumbar vertebrae posteriorly. The false pelvis is really part of the abdomen and helps support the abdominal viscera. It does not restrict childbirth in any way. The **true pelvis** is the region inferior to the pelvic brim that is almost entirely surrounded by bone and that forms a deep "bowl" containing the pelvic organs. Its dimensions, particularly those of its inlet and outlet, are critical to the uncomplicated delivery of a baby, and are carefully measured by an obstetrician.

The **pelvic inlet** is the **pelvic brim**, and its widest dimension is along the frontal plane (see Table 7.4). Generally, the infant's head enters the inlet with the forehead facing one ilium and the occipital region facing the other. A sacral promontory that is particularly large can impair the infant's entry into the true pelvis. The **pelvic outlet** is the inferior margin of the true pelvis. It is bounded anteriorly by the pubic arch, laterally by the ischia, and posteriorly by the sacrum and coccyx. Since both the coccyx and the ischial spines protrude into the outlet opening, a sharply angled coccyx or large sharp spines can cause problems in childbirth. The largest dimension of the outlet is the anteroposterior diameter. Generally, as the baby's head passes through the inlet, it rotates so that the forehead faces posteriorly and the occiput anteriorly. Thus, the position of the head during birth follows the widest dimensions of the true pelvis.

The Lower Limb

The lower limbs carry the entire weight of the erect body and are subjected to exceptional forces when we jump or run. Thus, it is not surprising that the bones of the lower limbs are thicker and stronger than comparable bones of the upper limbs. Whereas the lighter bones of the upper limbs are adapted for flexibility and mobility, the more massive bones of the lower limbs are specialized for stability and weight bearing. The three functional segments of each lower limb are the thigh, the leg, and the foot (see Table 7.3 on page 203).

Thigh

The **femur** (fē′-mer), the single bone of the thigh (Figure 7.28), is the largest, longest, strongest bone in the body. Its durable anatomy reflects the fact that the stress on the femur during vigorous jumping can reach 2 tons per square inch! The femur is clothed by bulky muscles that prevent us from palpating its course down the length of the thigh. Its length is roughly one-quarter of a person's height. Proximally, the femur articulates with the hip bone and then courses medially toward its articulation point with the leg bones at the knee. This arrangement allows the knee joints to be closer to the body's center of gravity and provides for better balance. The medial course of the two femurs is more pronounced in women because of their wider pelvis.

The ball-like *head* of the femur has a small central pit called the **fovea capitis** (fō′-vē-uh ka′-pih-tis). A short ligament, the *ligamentum teres* (tayr′-ēz), runs from this pit to the acetabulum, where it helps secure the femur. The head is carried on a short *neck* that angles superiorly and medially from the shaft. This arrangement reflects the fact that the femur articulates with the side (rather than the inferior aspect) of the pelvis. The neck is the weakest part of the femur. At the junction of the head and neck are the lateral **greater trochanter** (trō-kan′-ter) and medial **lesser trochanter**, which serve as sites of thigh and buttock muscle attachment. The two trochanters are connected by the **intertrochanteric line** anteriorly and the prominent **intertrochanteric crest** posteriorly. Just below the greater trochanter on the posterior femur surface is the **gluteal tuberosity**, which blends into a ridge, the **linea aspera** (lih′-nē-uh as′-per-uh), distally; both markings are sites of muscle attachment. Except for the linea aspera, the femur shaft is smooth and rounded.

Figure 7.28 The femur of the right thigh.

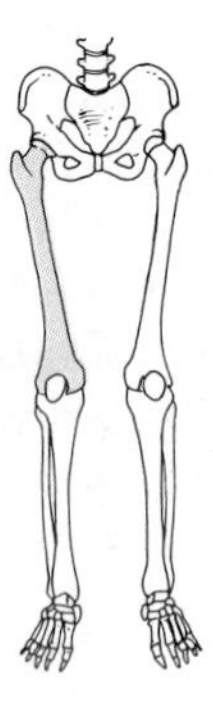

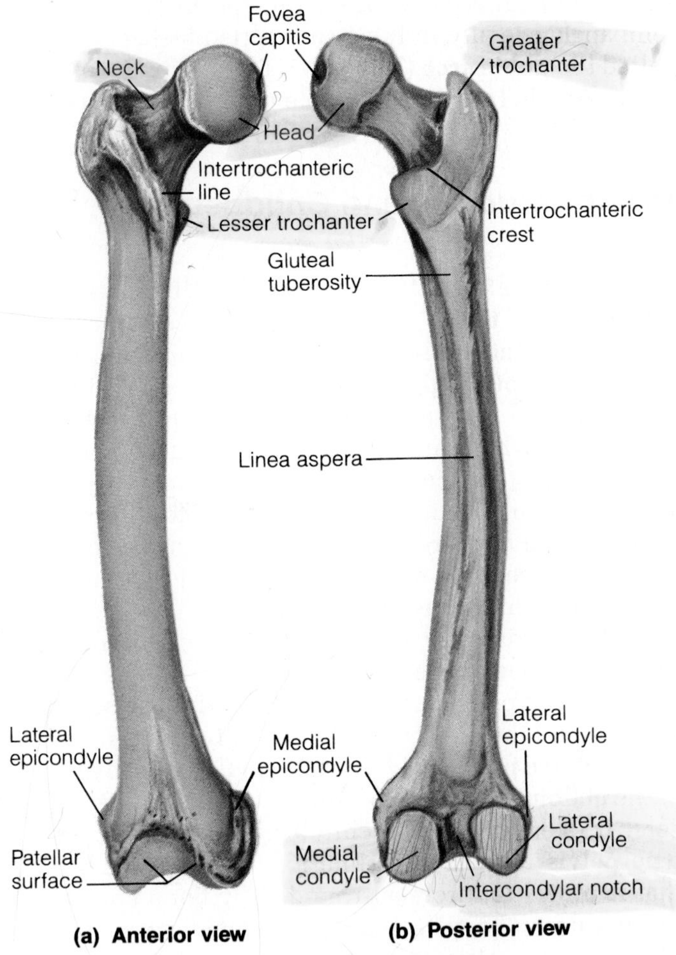

(a) Anterior view **(b) Posterior view**

Distally, the femur shaft broadens and terminates in the **lateral** and **medial condyles**, which articulate with the tibia of the leg. The smooth **patellar surface**, between the condyles on the anterior femur surface, articulates with the **patella** (puh-teh′-luh), or **kneecap** (see Figure 7.1). The patella is a small, flat, triangular sesamoid bone enclosed in the tendon that secures the anterior thigh muscles to the tibia. It guards the knee joint anteriorly and improves the leverage of the thigh muscles acting at the joint. Between the condyles on the posterior aspect of the femur is the deep, U-shaped **intercondylar notch.** The *medial* and *lateral epicondyles* (sites of muscle attachment) flank the condyles superiorly.

Leg

Two parallel bones, the tibia and fibula, form the skeleton of the leg (Figure 7.29). These two bones are connected by an interosseous membrane and articulate with each other both proximally and distally. However, unlike the joints between the radius and ulna of the forearm, the *tibiofibular* (tih″-bē-ō-fih′-byoo-ler) *joints* of the leg allow essentially no movement. Thus, the bones of the leg are less flexible but stronger and more stable that those of the forearm. The large medial tibia articulates proximally with the femur to form the hinge joint of the knee and distally with the talus bone of the ankle. The role of the fibula in joint formation is limited to stabilization of the ankle joint.

Tibia

The **tibia** (tih′-bē-uh), or *shinbone*, receives the weight of the body from the femur and transmits it to the foot. Except for the femur, it is the largest, strongest bone in the body. At its broad proximal end are the concave **medial** and **lateral condyles**; these are separated by an irregular projection, the **intercondylar eminence**, which varies in size from person to person and may

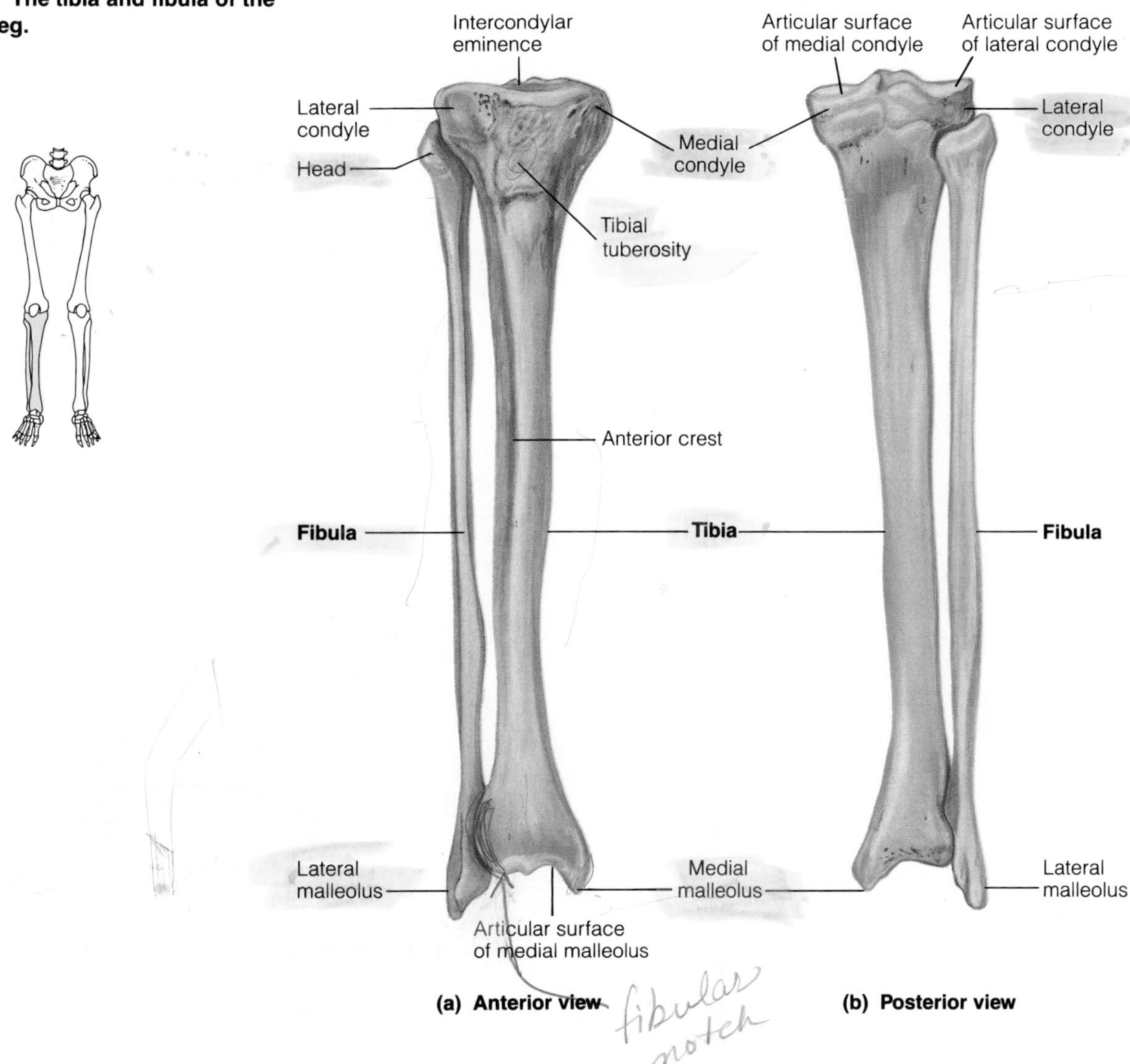

Figure 7.29 The tibia and fibula of the right lower leg.

even be absent. The tibial condyles articulate with the corresponding condyles of the femur. The lateral tibial condyle bears a facet that indicates the site of the proximal tibiofibular joint. Just inferior to the condyles on the anterior surface is the large, rough **tibial tuberosity**, to which the patellar ligament attaches.

The tibial shaft is triangular in cross section. Its anteriormost border, the sharp **anterior crest**, and its medial surface are unprotected by muscles and can be felt just beneath the skin along their entire length. The anguish of a "bumped" shin is an experience familiar to nearly everyone. The distal end of the tibia is blunt where it articulates with the talus of the ankle, but medial to that is an inferior projection, the **medial malleolus** (muh-lē′-uh-lus), which forms the inner, or medial, bulge of the ankle. The *fibular notch* is found on the lateral surface of the tibia, opposite the medial malleolus.

Fibula

The non-weight-bearing **fibula** (fih′-byoo-luh) is a stick-like bone with slightly expanded ends. It articulates proximally and distally with the lateral aspects of the tibia. Its upper end is its *head*; its lower end is called the **lateral malleolus**. The lateral malleolus forms the prominent lateral ankle bulge and articulates with the talus. The fibular shaft is heavily ridged and appears to have been twisted a quarter turn.

Foot

The skeleton of the foot includes the bones of the tarsus, or ankle, the bones of the metatarsus, or instep, and the phalanges, or toe bones (Figure 7.30). The foot has two important functions: It supports our body weight and it serves as a lever to propel the body for-

Figure 7.30 Bones of the right foot.

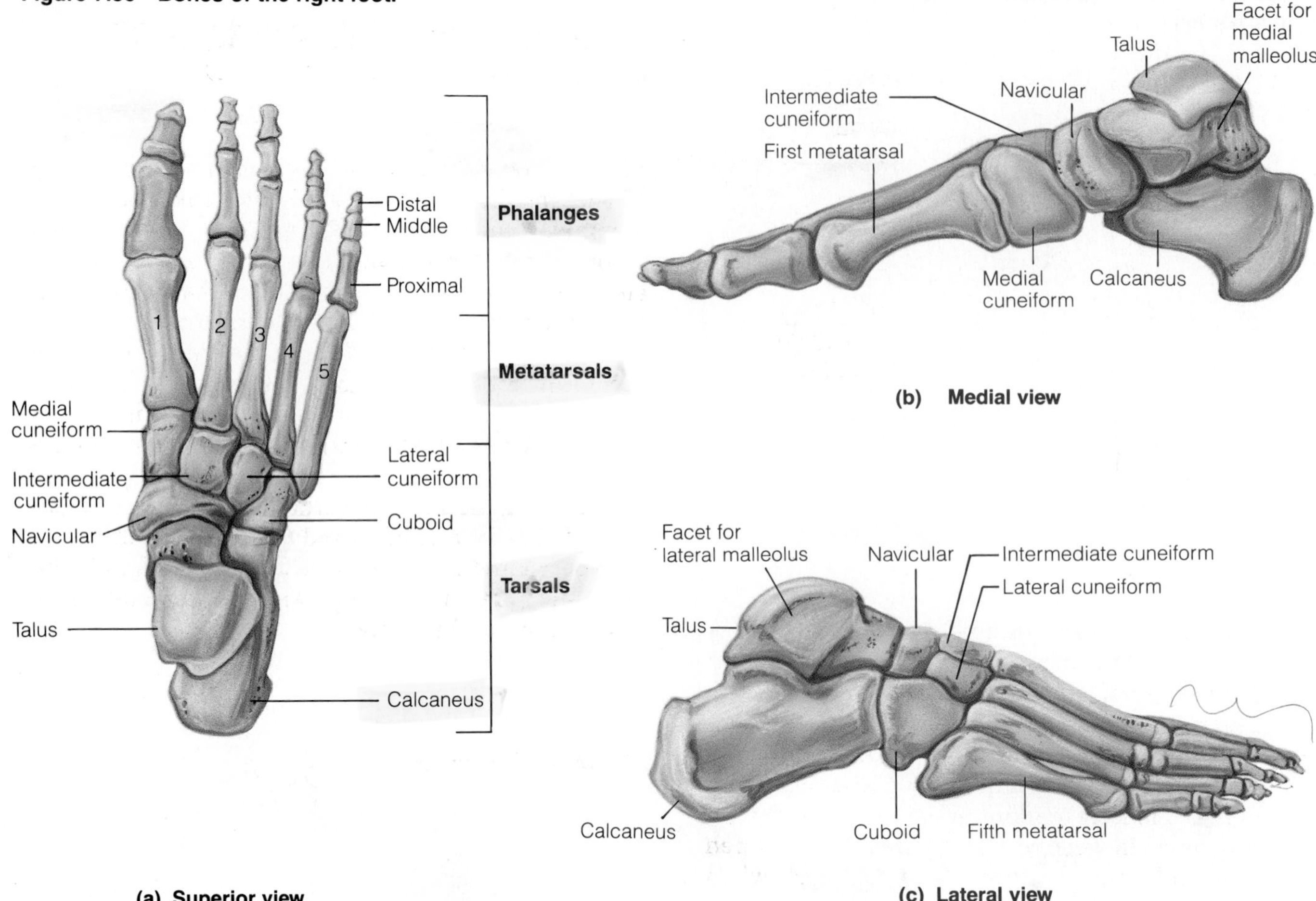

ward when we walk and run. A single bone could serve both purposes but would adapt poorly to uneven surfaces. This problem is avoided by the segmented nature of the foot, which allows it to be more pliable.

Tarsus (Ankle)

The seven **tarsal** (tar′-sul) **bones** of the ankle, or **tarsus**, correspond to the carpals of the wrist. Body weight is carried primarily by the two largest, most posterior tarsals: the **talus** (tā′-lus), which articulates with the tibia and fibula superiorly, and the strong **calcaneus** (kal-kā′-nē-us), which forms the heel of the foot and carries the talus on its superior surface. The heavy, cordlike *Achilles tendon* of the calf muscles attaches to the posterior surface of the calcaneus. The remaining tarsals are the lateral **cuboid**, the medial **navicular** (nuh-vih′-kyoo-ler), and most anteriorly the **medial**, **intermediate**, and **lateral cuneiform bones**.

Metatarsus (Instep)

The **metatarsus,** or instep, consists of five small long bones called **metatarsal bones**. These are numbered 1 to 5 beginning with the medial (great toe) side of the foot. The first metatarsal is large and strong and plays an important role in supporting the weight of the body. The arrangement of the metatarsals is more parallel than that of the metacarpals of the hands. Distally, where the metatarsals articulate with the proximal phalanges of the toes, the enlarged heads of the metatarsals form the "ball" of the foot. Proximally, metatarsal 1 articulates with the medial cuneiform; metatarsals 2 and 3 articulate with the intermediate and lateral cuneiform bones, respectively; and metatarsals 4 and 5 both articulate with the cuboid bone.

Phalanges (Toes)

The 14 phalanges of the toes are a good deal smaller than those of the fingers and thus much less nimble. But their general structure and arrangement are the same. There are three phalanges in each digit except for the great toe, which has only two, proximal and distal.

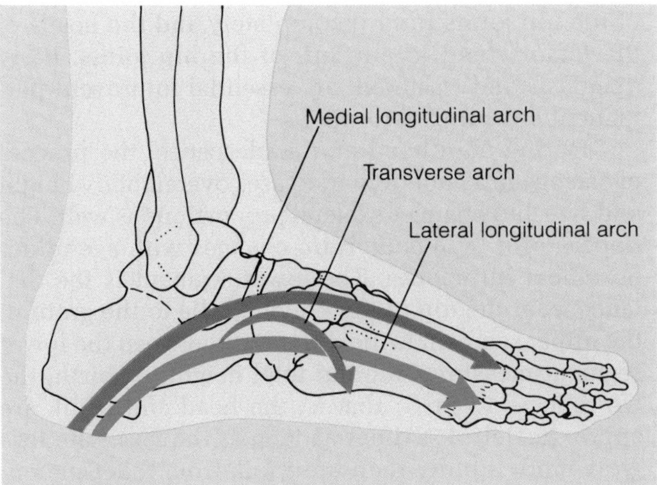

Figure 7.31 Arches of the foot.

Arches of the Foot

A segmented structure can hold up weight only if it is arched. The foot has three arches: two *longitudinal arches* (the *medial* and *lateral*) and one *transverse arch* (Figure 7.31). These arches are maintained by the shape of the foot bones, by strong ligaments, and by the pull of some tendons during muscle activity. The ligaments and muscle tendons allow a certain amount of springiness. In general, the arches "give" when weight is applied to the foot and spring back when the weight on them is removed.

If you examine your wet footprints, you will see that the medial margin from the heel to the head of the first metatarsal leaves no print because the medial longitudinal arch curves well above the ground. The talus bone serves as the keystone of the medial arch, which originates at the calcaneus, rises to the talus, and then descends to the three medial metatarsals. The lateral longitudinal arch is very low and elevates the lateral part of the foot just enough to redistribute some of the weight to the calcaneus and the head of the fifth metatarsal (to the ends of the arch). The cuboid is the keystone bone of the lateral arch. The longitudinal arches serve as pillars for the transverse arch, which runs obliquely from one side of the foot to the other. The transverse arch is formed by the bases of the metatarsals anteriorly and the cuboid and cuneiform bones posteriorly. Together, the arches of the foot form a half-dome that distributes about half our standing and walking weight to the heel bones and half to the heads of the metatarsals.

Standing immobilized for extended periods places excessive strain on the bones and ligaments of the feet (because the muscles are inactive) and can result in fallen arches, or "flat feet," particularly if one is overweight. Running on hard surfaces without proper arch supports can also cause arches to fall because the supporting structures are progressively weakened. ■

Developmental Aspects of the Skeleton

The skull bones of newborn infants are connected by as yet unossified remnants of fibrous membranes called *fontanels* (fon″-tuh-nelz′) (Figure 7.32). The fontanels accommodate brain growth in the fetus and infant, and they allow the infant's head to be compressed slightly during birth. The baby's pulse can be felt in these "soft spots"; hence their name (*fontanel* = little fountain). The large, diamond-shaped frontal fontanel is palpable for 1½ to 2 years after birth; the others have usually been replaced by bone by the end of the first year.

At birth, the skull bones are very thin and single-layered. The frontal bone and the mandible are composed of paired bones that fuse during childhood. The tympanic part of the temporal bone is merely a C-shaped ring in the newborn.

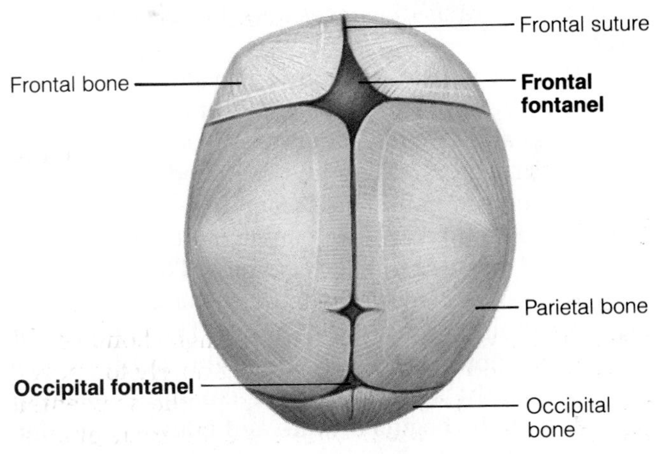

(a) Superior view

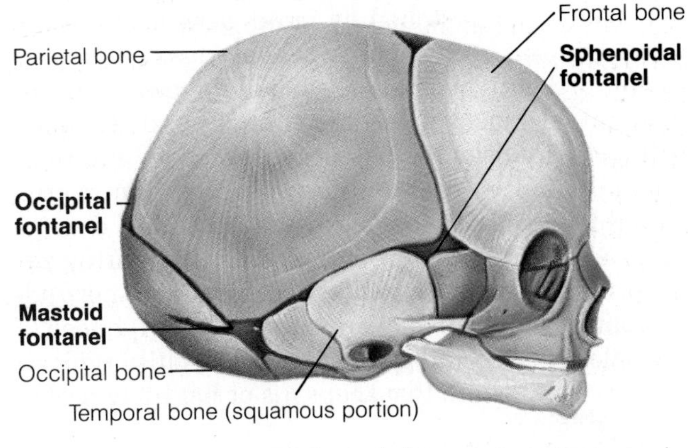

(b) Lateral view

Figure 7.32 The fetal skull.

Several congenital abnormalities may distort the axial skeleton, but perhaps the most common is *cleft palate*, a condition in which the palatine processes of the maxillae or the palatine bones (or both) fail to fuse medially. The persistent opening between the oral and nasal cavities interferes with sucking and can lead to aspiration (inhalation) of food into the lungs and *aspiration pneumonia*.

The skeleton changes throughout life, but the changes in childhood are most dramatic. At birth, the baby's cranium is huge relative to its face. The maxillae and mandible are foreshortened, and the contours of the face are flat. The rapid growth of the cranium before and after birth is related to the growth of the brain. By nine months, the skull is already half of its adult size (volume). By two years, it is three-quarters of its adult size, and by eight to nine years, the skull has almost reached adult proportions. However, between the ages of 6 and 11, the head appears to enlarge substantially as the face literally grows out from the skull. The jaws increase in size and mass and the cheekbones and nose become more prominent. These facial changes are correlated with the expansion of respiratory passages and development of the teeth.

Only two of the four spinal curvatures are present at birth: the thoracic and sacral curvatures. Both of these so-called **primary curvatures** are convex posteriorly, and an infant's spine is arched, like that of a four-legged animal. The **secondary curvatures**—cervical and lumbar—are convex anteriorly and are associated with a child's development. They result from reshaping of the intervertebral discs rather than from modifications of the bony vertebrae. The cervical curvature develops when the baby begins to lift its head independently, and the lumbar curvature develops when the baby begins to walk and allows the weight of the trunk to be brought over the body's center of gravity.

Vertebral problems (scoliosis or lordosis) may appear during the early school years, when the muscles are being stretched by rapid growth of the long limb bones. During the preschool years, lordosis is often present, but this is usually rectified as the abdominal muscles become stronger and the pelvis tilts forward. The thorax becomes broader and flatter laterally, but a true "military posture" (head erect, shoulders back, abdomen in, and chest out) does not appear until further growth of the trunk during adolescence. Vertebral curvatures and posture are not under voluntary control, but are primarily influenced by muscle strength and general health. A child typically adopts the posture that keeps his or her body parts in proper balance.

Like the axial skeleton, the appendicular skeleton can suffer from a number of congenital abnormalities. One that is fairly common and quite severe is *dysplasia* (dis-plā′-zhuh) *of the hip*, in which the acetabula of the hip joints form incompletely and the heads of the femurs tend to slip out of the hip joints. Early diagnosis and treatment are essential to prevent permanent crippling deformities.

During childhood and adolescence, the process of osteogenesis not only increases overall body height and size, but changes skeletal proportions as well. The *upper-lower (UL) body ratio* changes with age under hormonal influences. The lower segment is the distance from the top of the pelvic girdle to the ground; the upper segment is the difference between the lower segment measurement and total height. At birth, the UL ratio is 1.7 to 1; that is, the head and trunk are approximately 1½ times as long as the legs. The legs grow more rapidly than the trunk from this time on, and by the age of ten, the UL ratio is approximately 1 to 1 and changes little thereafter. During puberty, the pelvis of young girls broadens in preparation for childbearing. Adult height is reached by the end of adolescence. Given good nutrition and proper exercise, the skeleton changes very little from young adulthood to old age.

With age, the water content of the intervertebral discs declines (as it does in other tissues throughout the body). As the discs become thinner and less elastic, the risk of disc herniation increases. By 55 years, a loss of ½ to ¾ of an inch in stature is common. Further shortening of the trunk can be produced by osteoporosis of the spine or by kyphosis (called "dowager's hump" in the elderly). What was done during youth is undone in old age as the vertebral column gradually resumes its initial arc shape.

The thorax becomes more rigid with increasing age, largely because the costal cartilages ossify. This loss of rib cage elasticity tends to cause shallow breathing, which leads to less efficient gas exchange.

All bones, you will recall, lose mass with age. Although cranial bones lose less mass than most, changes in facial contours of the aged are common. As the bony tissue of the jaws declines, the jaws appear foreshortened once again and the bony structure of the face becomes more childlike. As bones become more porous, they become more susceptible to fracture, especially the vertebrae and the neck of the femur. Fracture of the femoral neck, usually called a broken hip, is a common and serious event in people with severe osteoporosis.

* * *

Our skeleton is a marvelous substructure, to be sure, but it is much more than that. It is a protector and supporter of other body systems, and without it (and the joints considered in Chapter 8), our muscles would be almost useless. The homeostatic relationships between the skeletal system and other body systems are illustrated in Figure 6.13.

Related Clinical Terms

Clubfoot A relatively common congenital defect in which the soles of the feet face medially and the toes point inferiorly; may be genetically induced or reflect abnormal position of the foot during fetal development.

Colles's fracture Fracture of the radius, typically about 1 cm proximal to the wrist; usually results from forceful trauma to the hands, such as falling on outstretched hands; since the ulna plays no part in forming the wrist joint, the radius receives the brunt of the force.

Laminectomy Surgical removal of a vertebra lamina; most often done to relieve the symptoms of a ruptured disc.

Lumbago (lum-bā′-gō) Acute lower back pain; its cause varies, but it may be due to a herniated disc; because of the muscle spasms associated with low back pain, the lumbar region of the spine becomes rigid and movement is painful.

Ostealgia (os-tē-al′-jē-uh) Pain in a bone.

Prolapsed disc A herniated disc.

Spina bifida (spī′-nuh bī′-fih-duh) Congenital defect of the vertebral column resulting from failure of the vertebral laminae to fuse medially; ranges in severity from inconsequential to severe conditions that impair neural functioning and encourage nervous system infections.

Spinal fusion Surgical procedure involving insertion of bone chips (or crushed bone) to immobilize and stabilize a specific region of the vertebral column, particularly in cases of vertebral fracture and herniated discs.

Chapter Summary

1. The axial skeleton forms the longitudinal axis of the body. Its principle subdivisions are the skull, vertebral column, and thorax. It provides support and protection (by enclosure).

2. The appendicular skeleton consists of the bones of the pectoral and pelvic girdles and the limbs. It allows mobility for manipulation and locomotion.

PART 1: THE AXIAL SKELETON

THE SKULL (pp. 173–187)

1. The skull is formed by 22 bones. The cranium forms the vault and base of the skull, which protects the brain. The facial skeleton provides openings for the respiratory and digestive passages, and attachment points for facial muscles.

2. With the exception of the temporomandibular joints, all bones of the adult skull are joined by immovable sutures.

3. **Cranium.** The eight bones of the cranium include the paired parietal and temporal bones and the single frontal, occipital, ethmoid, and sphenoid bones (see Table 7.1, pp. 186–187).

4. **Facial Bones.** The 14 bones of the face include the paired maxillae, zygomatics, nasals, lacrimals, palatines, and inferior conchae and the single mandible and vomer bones (see Table 7.1).

5. **Orbits and Nasal Cavity.** Both the orbits and the nasal cavities are complicated bony regions formed of several bones.

6. **Paranasal Sinuses.** Paranasal sinuses occur in the frontal, ethmoid, sphenoid, and maxillary bones.

7. **The Hyoid Bone.** The hyoid bone, supported in the neck by ligaments, serves as an attachment point for the tongue, pharynx muscles, and larynx.

THE VERTEBRAL COLUMN (pp. 187–193)

1. **General Characteristics.** The vertebral column includes 24 individual vertebrae (7 cervical, 12 thoracic, and 5 lumbar), and the sacrum and coccyx.

2. The fibrocartilage intervertebral discs act as shock absorbers and provide flexibility to the vertebral column.

3. The primary curvatures of the vertebral column are the thoracic and sacral; the secondary curvatures are the cervical and lumbar.

4. **General Structure of Vertebrae.** With the exception of C_1 and C_2, all vertebrae have a body, two transverse processes, two superior and two inferior articular processes, a spinous process, and a vertebral arch.

5. **Regional Vertebral Characteristics.** Special features distinguish the regional vertebrae (see Table 7.2, p. 192).

THE BONY THORAX (pp. 194–195)

1. The bones of the thorax include the 12 rib pairs, the sternum, and the thoracic vertebrae. The thoracic cage protects the organs of the thoracic cavity.

2. **Sternum.** The sternum consists of the fused manubrium, the body, and the xiphoid process.

3. **Ribs.** The first seven rib pairs are called true ribs; the rest are called false ribs.

PART 2: THE APPENDICULAR SKELETON

THE PECTORAL (SHOULDER) GIRDLE (pp. 196–198)

1. Each pectoral girdle consists of one clavicle and one scapula. The pectoral girdles attach the upper limbs to the axial skeleton.

2. **Clavicles.** The clavicles hold the scapulae laterally away from the thorax. The sternoclavicular joints are the only attachment points of the pectoral girdle to the axial skeleton.

3. **Scapulae.** The scapulae articulate with the clavicles and with the humerus bones of the arms.

THE UPPER LIMB (pp. 198–202)

1. Each upper limb consists of 30 bones and is specialized for mobility.

2. **Arm/Forearm/Hand.** The skeleton of the arm is composed solely of the humerus; the skeleton of the forearm is composed of the radius and ulna; and the skeleton of the hand consists of the carpals, metacarpals, and phalanges.

THE PELVIC (HIP) GIRDLE (pp. 202–207)

1. The pelvic girdle, a heavy structure specialized for weight bearing, is composed of two coxal bones that secure the lower limbs to the axial skeleton. Together with the sacrum, the ossa coxae form the basinlike pelvis.

2. Each coxal bone consists of three fused bones: ilium, ischium, and pubis; the acetabulum occurs at the point of fusion.

3. **Ilium/Ischium/Pubis.** The ilium is the superior flaring portion of the os coxa. Each ilium forms a secure joint with the sacrum posteriorly. The ischium is a curved bar of bone; we sit on the ischial tuberosities. The V-shaped pubic bones articulate anteriorly at the symphysis pubis.

4. **Pelvic Structure and Childbearing.** The male pelvis is deep and narrow with larger, heavier bones than those of the female. The female pelvis, which forms the birth canal, is shallow and wide.

THE LOWER LIMB (pp. 207–211)

1. Each lower limb consists of the thigh, leg, and foot and is specialized for weight bearing and locomotion.

2. **Thigh.** The femur is the sole bone of the thigh. Its ball-shaped head articulates with the acetabulum.

3. **Leg.** The bones of the leg are the tibia, which participates in forming both the knee and ankle joints, and the slender fibula.

4. **Foot.** The bones of the foot include the tarsals, metatarsals, and phalanges. The most important tarsals are the calcaneus (heel bone) and the talus, which articulates with the tibia superiorly.

5. The foot is supported by three arches that distribute body weight to the heel and ball of the foot.

DEVELOPMENTAL ASPECTS OF THE SKELETON (pp. 211–212)

1. Fontanels, which allow brain growth and ease birth passage, are present in the skull at birth. Growth of the cranium after birth is related to brain growth; increase in size of the facial skeleton follows tooth development and enlargement of respiratory passageways.

2. The vertical column is C-shaped at birth (thoracic and sacral curvatures are present); the secondary curvatures form when the baby begins to lift its head and walk.

3. Long bones continue to grow in length until late adolescence. The UL ratio changes from 1.7:1 to 1:1 by the age of ten years.

4. Changes in the female pelvis (preparatory for childbirth) occur during puberty.

5. Once adult height is reached, the skeleton changes little until late middle age. With old age, the intervertebral discs thin; this, along with osteoporosis, leads to a gradual loss in height. Loss of bone mass predisposes elderly individuals to fractures.

Review Questions

Multiple Choice/Matching

1. Using the letters from column B, match the bone descriptions in column A. (Note that some require more than a single choice.)

Column A	Column B
_____ (1) connected by the frontal suture	(a) ethmoid
_____ (2) keystone bone of the cranium	(b) frontal
_____ (3) keystone bone of the face	(c) mandible
_____ (4) form the hard palate	(d) maxillary
_____ (5) allows the spinal cord to pass	(e) occipital
_____ (6) forms the chin	(f) palatine
_____ (7) contain paranasal sinuses	(g) parietal
_____ (8) contains mastoid sinuses	(h) sphenoid
	(i) temporal

2. Match the key terms with the bone descriptions that follow.
Key: (a) clavicle (b) ilium (c) ischium (d) pubis
(e) sacrum (f) scapula (g) sternum

_____ (1) bone of the axial skeleton to which the pectoral girdle attaches

_____ (2) markings include the glenoid fossa and acromion process

_____ (3) features include the ala, crest, and greater sciatic notch

_____ (4) doubly curved; acts as a shoulder strut

_____ (5) pelvic girdle bone that articulates with the axial skeleton

_____ (6) the "sit-down" bone

_____ (7) anteriormost bone of the pelvic girdle

3. Use key choices to identify the bone descriptions that follow.
Key: (a) carpals (b) femur (c) fibula (d) humerus
(e) radius (f) tarsals (g) tibia (h) ulna

_____ (1) articulates with the acetabulum and the tibia

_____ (2) forms the lateral aspect of the ankle

_____ (3) bone that "carries" the hand

_____ (4) the wrist bones

_____ (5) has a plierslike head

_____ (6) articulates with the capitulum of the humerus

Short Answer Essay Questions

4. Name the cranial and facial bones and compare and contrast the functions of the cranial and facial skeletons.

5. How do the relative proportions of the cranium and face of a fetus compare with those of an adult skull?

6. Name and diagram the normal vertebral curvatures. Which are primary and which are secondary curvatures?

7. List at least two specific anatomical characteristics each for typical cervical, thoracic, and lumbar vertebrae that would allow *anyone* to identify each type correctly.

8. (a) What is the function of intervertebral discs? (b) Distinguish between the anulus fibrosis and nucleus pulposus regions of a disc. (c) Which provides durability and strength? (d) Which provides resilience?

9. Name the major components of the bony thorax.

10. (a) What is a true rib? A false rib? (b) Is a floating rib a true rib or a false rib? (c) Why are floating ribs easily broken?

11. The major function of the shoulder girdle is flexibility. What is the major function of the pelvic girdle? Relate these functional differences to anatomical differences seen in these girdles.

12. List three important differences between the male and female pelvis.

13. Describe the function of the arches of the foot.

14. Briefly describe the anatomical characteristics and impairment of function seen in cleft palate and hip dysplasia.

15. Compare a young adult skeleton to that of an extremely aged person relative to bone mass in general and the bony structure of the skull, thorax, and vertebral column.

Clinical Application Questions

16. Mark was hit in the face with a football during practice. An X-ray revealed multiple fractures of the bones around his left orbit. Name the bones that form the margins of an orbit.

17. An exhausted biology student was attending a lecture; after 30 minutes or so, he lost interest and began to doze. As the lecture ended, the hubbub aroused him and he let go with a tremendous yawn. To his great distress, he couldn't close his mouth—his lower jaw was "stuck" open. What do you think had happened?

18. Ralph had polio as a boy and was partially paralyzed in one lower limb for over a year. Although no longer paralyzed, he now has a severe lateral curvature of the lumbar spine. Explain what has happened and identify his condition.

19. Mary's grandmother slipped on a scatter rug and fell heavily to the floor. Her left leg was laterally rotated and noticeably shorter than the right, and when she attempted to get up, she grimaced with pain. Mary surmised that her grandmother might have "fractured her hip," which later proved to be true. What bone was probably fractured and at what site? Why is a "fractured hip" a common type of fracture in the elderly?

CHAPTER

8 JOINTS

Chapter Outline and Student Objectives

1. Define joint or articulation.

Classification of Joints (p. 216)

2. Describe the criteria used to classify joints structurally and functionally.

Fibrous Joints (pp. 216–217)

3. Describe the general structure of fibrous joints. Name and give an example of each of the three common types of fibrous joints.

Cartilaginous Joints (pp. 217–218)

4. Describe the general structure of cartilaginous joints. Name and give an example of each of the two common types of cartilaginous joints.

Synovial Joints (pp. 218–232)

5. Describe the anatomical characteristics common to all synovial joints.
6. List three factors that help make synovial joints stable.
7. Compare the structures and functions of bursae and tendon sheaths.
8. Name and describe (or perform) the common types of body movements.
9. Name five types of synovial joints based on the type of movement allowed. Provide examples of each type.
10. Describe the elbow, knee, hip, and shoulder joints. Consider in each case the articulating bones, anatomical characteristics of the joint, types of movement allowed, and relative joint stability.

Homeostatic Imbalance of Joints (pp. 232–235)

11. Name the most common joint injuries and discuss the symptoms and problems associated with each.
12. Compare and contrast osteoarthritis, rheumatoid arthritis, and gouty arthritis with respect to the population affected, possible causes, structural joint changes, disease outcome, and therapy.

Developmental Aspects of the Joints (p. 235)

13. Discuss briefly the factors that promote or disturb joint homeostasis.

Preview of Selected Key Terms

Articulation A joint; point where two bones meet.

Synarthrosis (sin″-ar-thrō′-sis) (*syn* = together; *arthro* = joint) Immovable joint.

Amphiarthrosis (am″-fē-ar-thrō′-sis) (*amphi* = on both sides) A slightly movable joint.

Diarthrosis (dī″-ar-thrō′-sis) (*dia* = through) A freely movable joint; also called a synovial joint.

Synostosis (sin″-os-tō′-sis) (*syn* = together, with; *ost* = bone; *osis* = condition) A completely ossified joint; a fused joint.

Syndesmosis (sin″-dez-mō′-sis) (*syndesmos* = ligament) A joint in which the bones are united by a ligament or a sheet of fibrous tissue.

Synchondrosis (sin″-kon-drō′-sis) (*chondro* = cartilage) A joint in which the bones are united by hyaline cartilage.

Symphysis (sim′-fih-sis) (*symphysis* = a growing together) A joint in which the bones are connected by fibrocartilage.

Articular capsule Double-layered capsule composed of an outer fibrous capsule lined by synovial membrane; encloses the joint cavity of a synovial joint.

Bursa (*bursa* = purse). A fibrous sac lined with synovial membrane and containing synovial fluid; occurs between bones and muscle tendons (or other structures), where it acts to decrease friction during movement.

Our **joints,** or **articulations,** have two fundamental functions: They secure our bones together and they allow our skeleton of rigid bone to be mobile. The graceful movements of a ballet dancer and the rough-and-tumble grapplings of football players attest to the great variety of motion allowed by joints, the sites where two or more bones meet. With fewer joints, we would move like robots. Nevertheless, the bone-binding function of joints is just as important as their role in providing mobility. The rigid joints of the skull, for example, make it a secure enclosure for our vital "gray matter." With the exception of the hyoid bone of the neck, every bone in the body forms a joint with at least one other bone.

Classification of Joints

Joints may be classified structurally or functionally. *Structural classification* is based on the material binding the bones together and whether or not a joint cavity is present. Structurally, there are *fibrous, cartilaginous,* and *synovial joints.*

Functional classification focuses on the amount of movement allowed at the joint. On this basis, there are **synarthroses** (sin″ar-thrō′-sēz), which are immovable joints; **amphiarthroses** (am″-fē-ar-thrō′-sēz), slightly movable joints; and **diarthroses** (dī″-ar-thrō′-sēz), or freely movable joints. Freely movable joints predominate in the limbs, whereas immovable and slightly movable joints are largely restricted to the axial skeleton, where firm bony attachments and protection of enclosed organs are a priority.

In general, fibrous joints are immovable, and all synovial joints are freely movable. However, cartilaginous joints offer both rigid and slightly movable examples. Since the structural categories are more clear-cut, we will use the structural classification for the present discussion, indicating functional properties where appropriate. The characteristics of selected body joints are summarized in Table 8.1 (pp. 220–221).

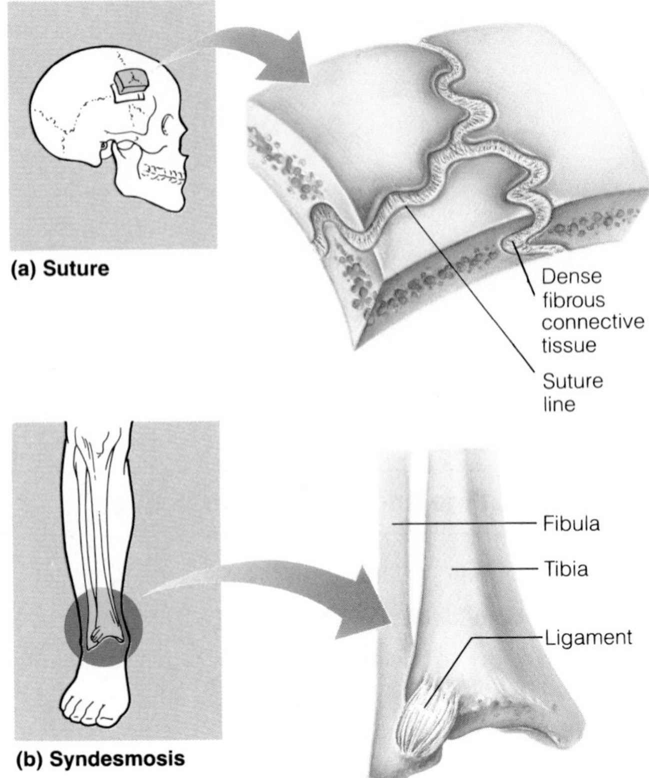

Figure 8.1 Fibrous joints. (**a**) Sutures are fibrous joints found only in the skull. The interconnecting fibers of dense connective tissue are very short, and the bone edges interlock so that the joint is immovable (synarthrotic). (**b**) In the syndesmosis at the distal tibiofibular joint, the fibrous tissue (ligament) connecting the bones is longer than that in sutures; "give," but no true movement, is permitted.

Fibrous Joints

In **fibrous joints,** the bones are joined by fibrous tissue; no joint cavity is present. The amount of movement allowed depends on the length of the fibers uniting the bones. Although a few fibrous joints are slightly movable, most are synarthrotic and permit essentially no movement. The three types of fibrous joints, based on type and length of intervening fibers, are sutures, syndesmoses, and gomphoses.

Sutures

Sutures occur only between bones of the skull (Figure 8.1a). The wavy articulating bone edges interdigitate by either overlapping or interlocking, and the junction is completely filled by very short connective tissue fibers that penetrate deeply into the substance of the articulating bones. The result is nearly rigid splices that bind the bones tightly together. During adulthood, the fibrous tissue becomes completely ossified and the skull bones fuse into a single unit; the suture joints are then more precisely called **synostoses** (sin″-os-tō′-sēz), literally, "bony joints." Since any movement of the skull bones would severely damage the brain, the immovable nature of their joints is an exquisite adaptation for their protective function.

Syndesmoses

Syndesmoses (sin″-dez-mō′-sēz) are fibrous joints in which the bones are connected by a cord or sheet of fibrous tissue, called a *ligament* or an *interosseous membrane*, respectively. Although the connecting fibers are always longer than those in sutures, they do vary quite a bit in length. Since the amount of movement possible increases with increasing length of the connecting fibers, slight to considerable movement can be achieved. For example, the ligament connecting the

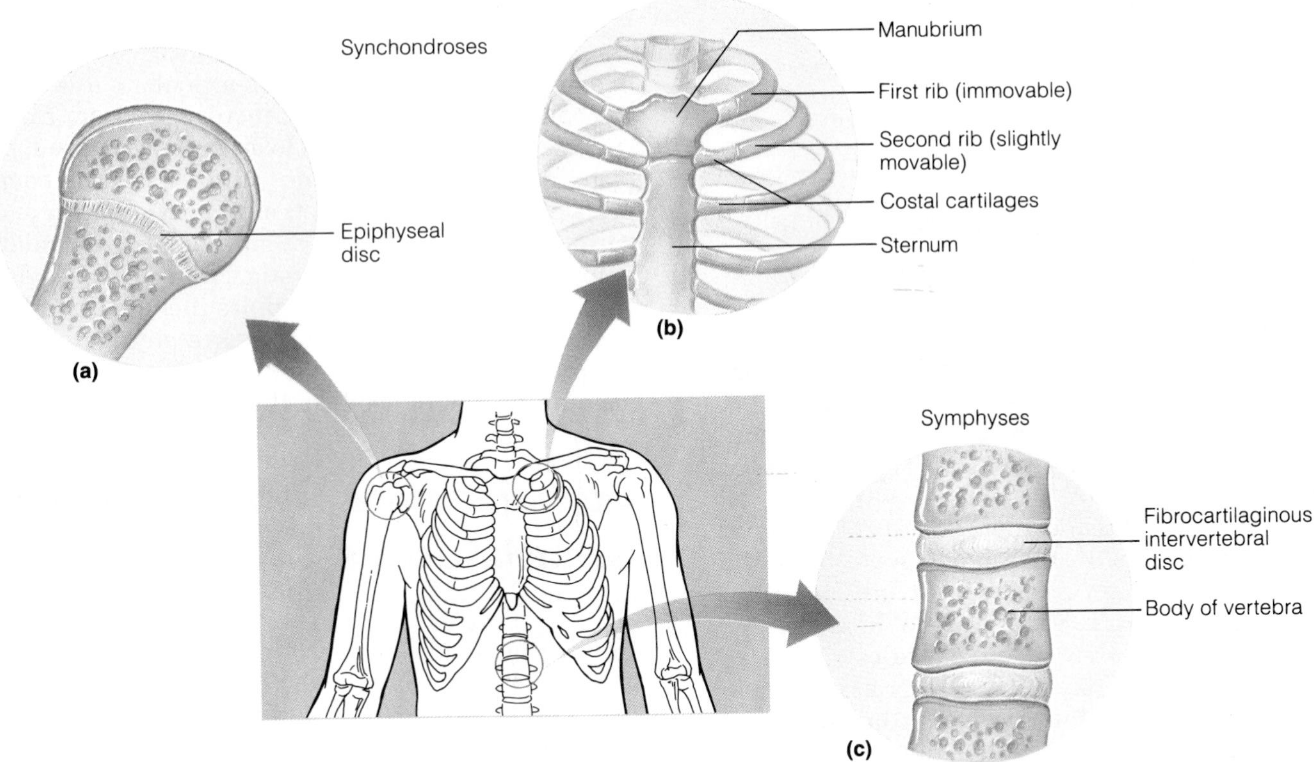

Figure 8.2 Cartilaginous joints. (**a**) The epiphyseal disc seen in a growing long bone is a temporary synchondrosis; the shaft and epiphysis are connected by hyaline cartilage that is later completely ossified. (**b**) The costal cartilages connecting the ribs to the sternum and the joint between rib 1 and the manubrium of the sternum form permanent synchondroses. (**c**) The intervertebral joints, in which the vertebrae are connected by fibrocartilaginous discs, are symphyses.

distal ends of the tibia and fibula is quite short (Figure 8.1b), and this tibiofibular joint is only slightly more resilient than a suture, a characteristic best described as "give." True movement is still prevented, so the joint is classed functionally as an immovable joint, or synarthrosis. On the other hand, the interosseous membrane connecting the radius and ulna along their length (see Figure 7.23a) is sufficiently broad and flexible to permit the ample movements of supination and pronation.

Gomphoses

A **gomphosis** (gom-fō′-sis) is a fibrous joint represented solely by the articulation of a tooth with its bony alveolar socket. The term *gomphosis* comes from the Greek *gompho,* meaning "nail" or "bolt," and refers to the manner in which the teeth are embedded in their sockets (as if hammered in). The fibrous connection in this case is the thin *periodontal ligament* (see Figure 24.12.)

Cartilaginous Joints

In **cartilaginous** (kar″-tih-la′-jih-nus) **joints,** the articulating bones are united by cartilage. Like fibrous joints, they lack a joint cavity. The two types of cartilaginous joints are synchondroses and symphyses.

Synchondroses

A bar or plate of *hyaline cartilage* unites the bones at a **synchondrosis** (sin″-kon-drō′-sis). Depending on the specific joint, synchondroses may be functionally synarthrotic or amphiarthrotic.

An epiphyseal plate connecting the diaphysis with the epiphysis in a child's long bone is structurally a synchondrosis and functionally a synarthrosis (Figure 8.2a). Epiphyseal plates are temporary joints, and when full growth is achieved, they are totally ossified, becoming synostoses.

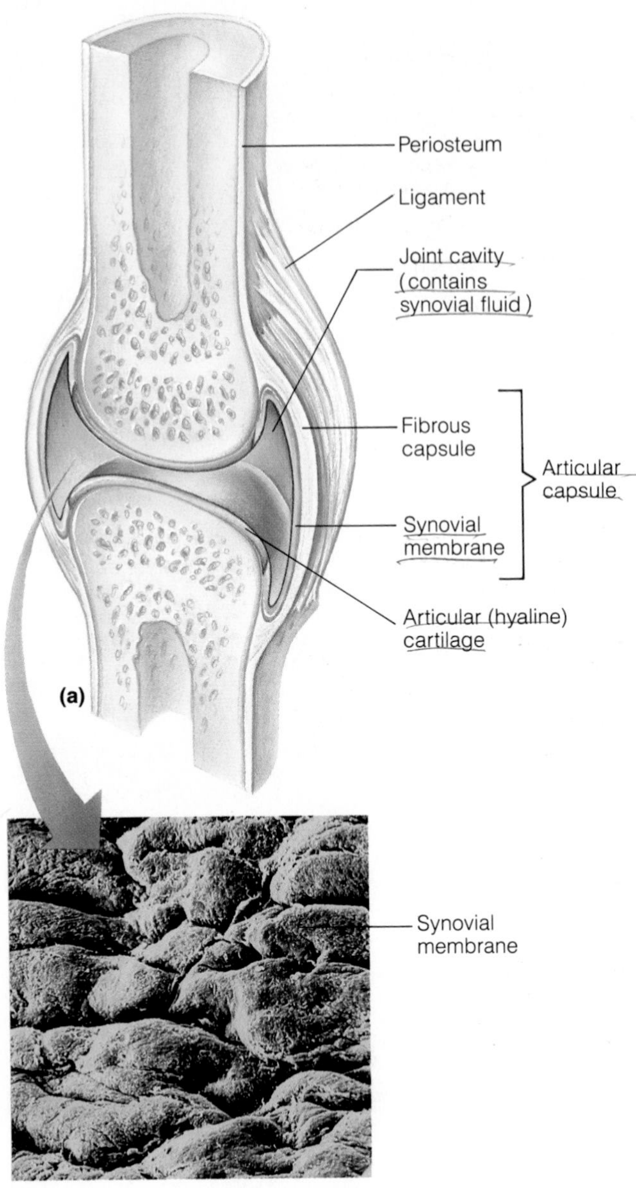

Figure 8.3 A synovial joint. (**a**) The articulating bone ends are covered with articular cartilage and are enclosed within an articular capsule. The external portion of the articular capsule, the fibrous capsule, is continuous with the periostea of the bones. Internally, the fibrous capsule is lined with a smooth synovial membrane that secretes synovial fluid. Typically, the joint is reinforced by ligaments (not shown). (**b**) Scanning electron micrograph of the synovial membrane from a knee joint.

The joint between the first rib and the manubrium of the sternum is a hyaline cartilage joint that eventually ossifies. It is functionally a synarthrosis (Figure 8.2b). In contrast, the costal cartilages uniting ribs 2–10 to the sternum form synchondroses that allow a slight degree of rib movement (see Figure 8.2b). These joints are amphiarthrotic.

Symphyses

In **symphyses** (sim′-fih-sēz), the articular surfaces of the bones are covered with articular (hyaline) cartilage, which in turn is fused to an intervening pad, or plate, of *fibrocartilage*. Since fibrocartilage is compressible and resilient, it acts as a shock absorber and permits a limited amount of movement at the joint. Symphyses are amphiarthrotic joints designed for strength with flexibility. Examples include the intervertebral joints (Figure 8.2c), the symphysis pubis of the pelvis (see Table 8.1), and the joint between the manubrium and the body of the sternum.

Synovial Joints

Synovial joints are those in which the articulating bones are separated by a fluid-containing joint cavity (Figure 8.3). This arrangement permits substantial freedom of movement, and all synovial joints are freely movable diarthrotic joints. All joints of the limbs—indeed, most joints of the body—fall into this class.

General Structure

Typically, synovial joints have five distinguishing features:

1. **Articular cartilage.** The articular surfaces of the bones are covered with glassy-smooth articular (hyaline) cartilage.

2. A **joint cavity.** The unique feature of synovial joints, the joint cavity is more a potential space than a real one because it is filled with synovial fluid.

3. An **articular capsule.** The joint cavity is enclosed by a double-layered articular capsule. The external layer is a tough, flexible **fibrous capsule** that is continuous with the periostea of the articulating bones. A **synovial membrane,** composed of loose connective tissue, lines the fibrous capsule internally. It extends from the margin of one articular cartilage to the margin of the other and covers all internal joint surfaces not covered by hyaline cartilage.

4. **Synovial fluid.** A small amount of slippery synovial fluid occupies all free spaces within the joint capsule. Its viscous, egg-white consistency (*ovum* = egg) is due to hyaluronic acid secreted by the cells of the synovial membrane. The hyaluronic acid is thinned by interstitial fluid derived from blood plasma, and synovial fluid becomes less viscous as it warms during joint activity. Synovial fluid bears most of the weight

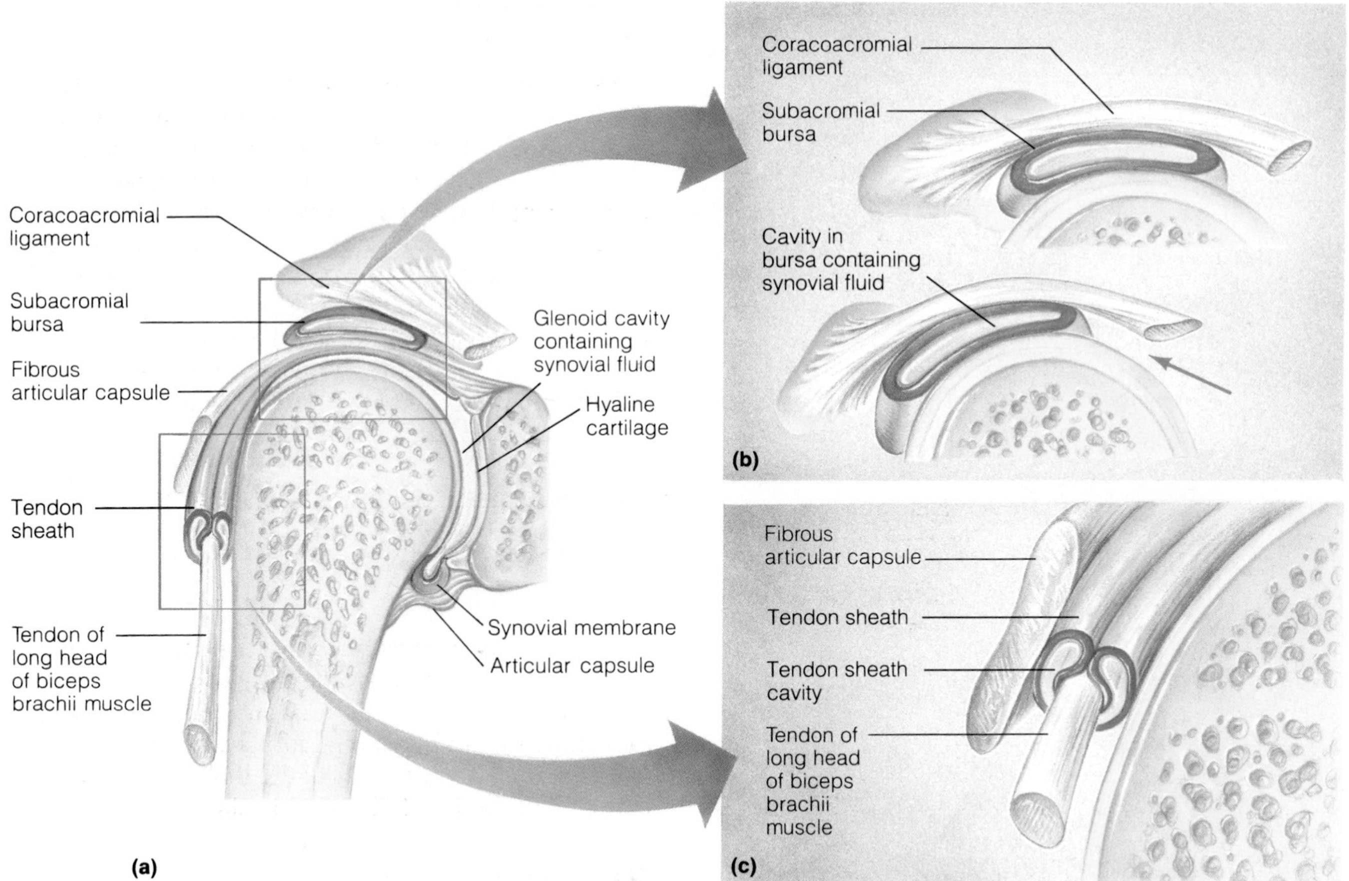

Figure 8.4 Friction-reducing structures: Bursae and tendon sheaths. (**a**) Longitudinal section through the synovial joint of the shoulder showing the saclike bursa and the tendon sheath around a muscle tendon. (**b**) The manner in which a bursa eliminates friction when a tendon (or other structure) is liable to rub against a bone. The synovial fluid within the bursa acts as a lubricant that allows its walls to slide easily across each other. (**c**) Three-dimensional view of a tendon sheath.

at the joint surfaces and keeps the articular cartilages from touching each other, preventing cartilage erosion. A fine lubricant, it minimizes friction in the moving joint. Synovial fluid also nourishes the articular cartilage and contains phagocytic cells that rid the joint cavity of microbes or cellular debris resulting from wear and tear during joint activity.

5. **Reinforcing ligaments.** Synovial joints are reinforced and strengthened by a number of ligaments. Sometimes, the ligaments are *intrinsic,* or *capsular,* that is, part of the fibrous capsule. In other cases, they are found outside the capsule *(extracapsular ligaments)* or deep to it *(intracapsular ligaments).* Since intracapsular ligaments are covered with synovial membrane, they do not actually penetrate the joint cavity.

Certain synovial joints have other structural features. Some, such as the hip and knee joints, have cushioning *fatty pads* between the fibrous capsule and the synovial membrane or bone. Others have discs or wedges of fibrocartilage separating the articular surfaces of the bones. These so-called **articular discs,** or **menisci** (muh-nih′-skī), extend inward from the articular capsule and subdivide the synovial cavity into two separate cavities. The articular discs modify the shape of the articular surfaces to improve their fit, increasing the stability of the joint. Articular discs are found in the knee, jaw, and sternoclavicular joints.

Bursae and Tendon Sheaths

Bursae and tendon sheaths are not strictly part of synovial joints, but they are frequently found closely associated with them (Figure 8.4). They act as "ball bearings" to prevent friction on adjacent structures during joint activity. **Bursae** (bur′-sē), from the Latin word meaning "purse," are flattened fibrous sacs lined with synovial membrane and containing a thin film of synovial fluid. They are common in sites where

Table 8.1 Structural and Functional Characteristics of Body Joints

Illustration	Joint	Articulating bones	Structural type*	Functional type; movements allowed
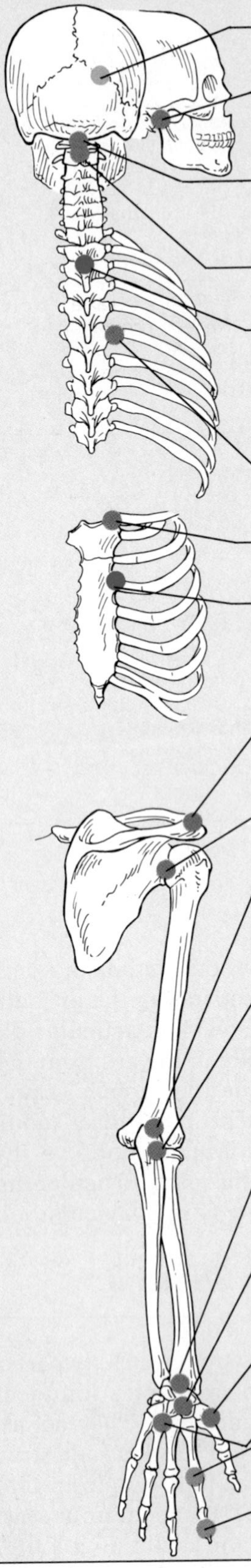	Skull	Cranial and facial bones	Fibrous; suture	Synarthrotic; no movement
	Temporomandibular	Temporal bone of skull and mandible	Synovial (contains articular disc)	Diarthrotic; biaxial; elevation, depression, protraction, retraction
	Atlanto-occipital	Occipital bone of skull and atlas	Synovial; condyloid	Diarthrotic; biaxial; flexion, extension, abduction, adduction, circumduction
	Atlantoaxial	Atlas (C_1) and axis (C_2)	Synovial; pivot	Diarthrotic; uniaxial; rotation of the head.
	Intervertebral	Between adjacent vertebral bodies	Cartilaginous; symphysis	Amphiarthrotic; slight movement
		Between articular processes	Synovial; plane	Diarthrotic; gliding
	Vertebrocostal	Vertebrae and ribs	Synovial; plane	Diarthrotic; gliding
	Sternoclavicular	Sternum and clavicle	Synovial; double plane	Diarthrotic; biaxial gliding
	Sternocostal	Sternum and ribs 2–7	Cartilaginous; synchondrosis	Amphiarthrotic; slight movement.
	Acromioclavicular	Acromion process of scapula and clavicle	Synovial; plane	Diarthrotic; gliding; elevation, depression, protraction, retraction
	Shoulder (glenohumeral)	Scapula and humerus	Synovial; ball and socket	Diarthrotic; multiaxial; flexion, extension, abduction, adduction, circumduction, rotation
	Elbow (humeroulnar)	Humerus and ulna	Synovial; hinge	Diarthrotic; uniaxial; flexion, extension
	Radioulnar (proximal)	Radius and ulna	Synovial; pivot	Diarthrotic; rotation around the long axis of forearm to allow pronation and supination distally.
	Wrist (radiocarpal)	Radius and proximal carpals	Synovial; condyloid	Diarthrotic; biaxial; flexion, extension, abduction, adduction, circumduction
	Intercarpal	Adjacent carpals	Synovial; plane	Diarthrotic; gliding
	Carpometacarpal of digit 1 (thumb)	Carpal and metacarpal 1	Synovial; saddle	Diarthrotic; biaxial; flexion, extension, abduction, adduction, circumduction
	of digits 2–5	Carpal(s) and metacarpal(s)	Synovial; plane	Diarthrotic; gliding
	Knuckle (metacarpophalangeal)	Metacarpal and proximal phalanx	Synovial; condyloid	Diarthrotic; biaxial; flexion, extension, abduction, adduction
	Finger (interphalangeal)	Adjacent phalanges	Synovial; hinge	Diarthrotic; uniaxial; flexion; extension

Table 8.1 (continued)

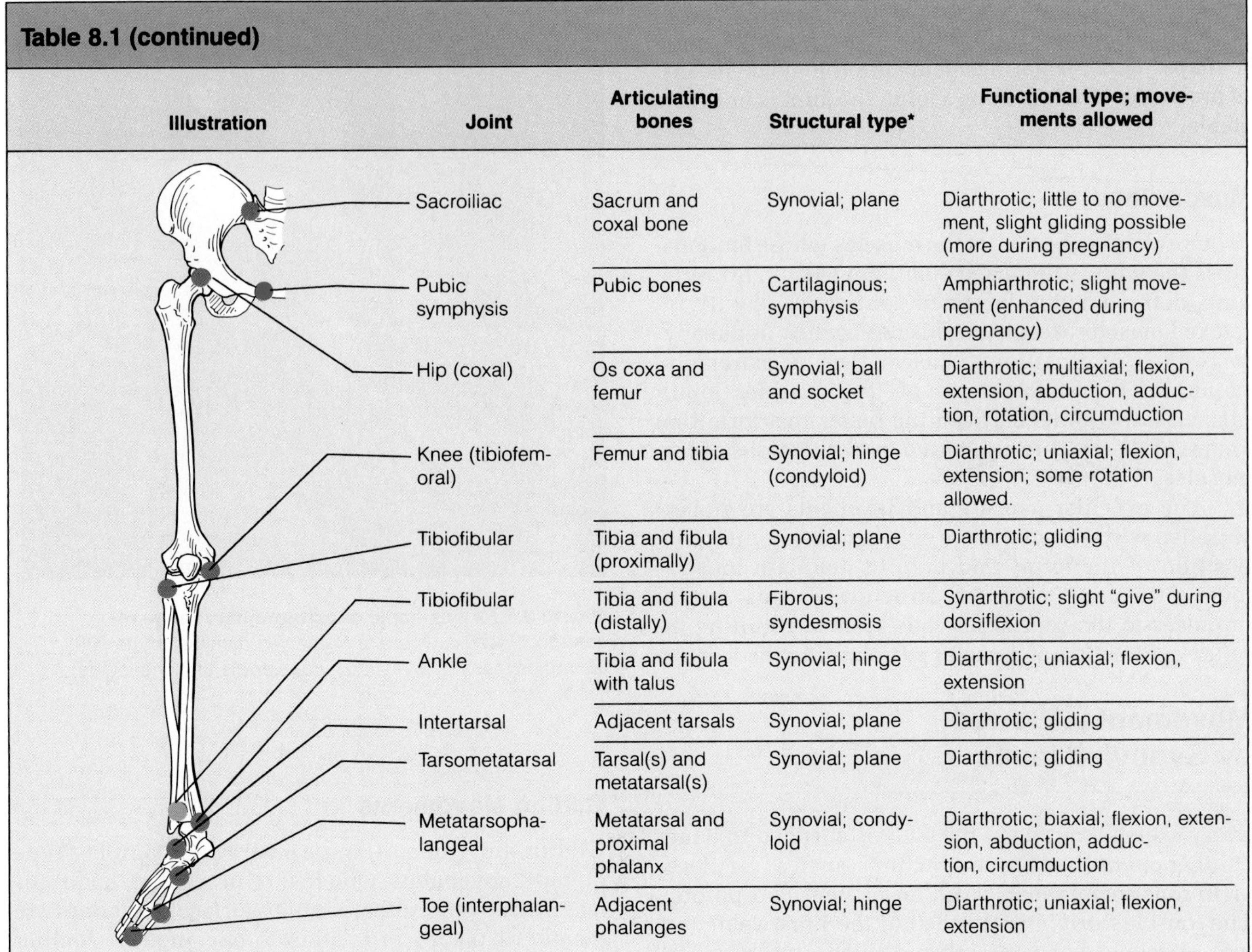

Illustration	Joint	Articulating bones	Structural type*	Functional type; movements allowed
	Sacroiliac	Sacrum and coxal bone	Synovial; plane	Diarthrotic; little to no movement, slight gliding possible (more during pregnancy)
	Pubic symphysis	Pubic bones	Cartilaginous; symphysis	Amphiarthrotic; slight movement (enhanced during pregnancy)
	Hip (coxal)	Os coxa and femur	Synovial; ball and socket	Diarthrotic; multiaxial; flexion, extension, abduction, adduction, rotation, circumduction
	Knee (tibiofemoral)	Femur and tibia	Synovial; hinge (condyloid)	Diarthrotic; uniaxial; flexion, extension; some rotation allowed.
	Tibiofibular	Tibia and fibula (proximally)	Synovial; plane	Diarthrotic; gliding
	Tibiofibular	Tibia and fibula (distally)	Fibrous; syndesmosis	Synarthrotic; slight "give" during dorsiflexion
	Ankle	Tibia and fibula with talus	Synovial; hinge	Diarthrotic; uniaxial; flexion, extension
	Intertarsal	Adjacent tarsals	Synovial; plane	Diarthrotic; gliding
	Tarsometatarsal	Tarsal(s) and metatarsal(s)	Synovial; plane	Diarthrotic; gliding
	Metatarsophalangeal	Metatarsal and proximal phalanx	Synovial; condyloid	Diarthrotic; biaxial; flexion, extension, abduction, adduction, circumduction
	Toe (interphalangeal)	Adjacent phalanges	Synovial; hinge	Diarthrotic; uniaxial; flexion, extension

*Fibrous joints indicated by orange circles; cartilaginous joints indicated by blue circles; synovial joints indicated by purple circles.

ligaments, muscles, skin, or muscle tendons overlie and rub against bone. Most bursae are present from birth, but additional ones, called *false bursae,* may develop at any site where there is excessive motion, and they function in the same manner as true bursae. **Tendon sheaths** are essentially elongated bursae that have become wrapped completely around a tendon subjected to friction, like a hot dog bun around a hot dog.

Factors Influencing the Stability of Synovial Joints

The stability of a synovial joint depends chiefly on three factors: the shape, size, and arrangement of the articular surfaces; the number and positioning of ligaments; and muscle tone.

Articular Surfaces

Surprisingly, articular surfaces play a minimal role in stabilizing most joints. Most joints, like the acromioclavicular joint and the knee joint, have shallow sockets and flat or noncomplementary articulating surfaces that contribute little to joint stability. But when articular surfaces are large and fit snugly together, or when the socket is deep, stability is vastly improved. The ball and deep socket of the hip joint provide the best example of a joint made extremely stable by the shape of its articular surfaces.

Ligaments

The ligaments of synovial joints help to direct bone movement and prevent excessive or undesirable motion. As a rule, the more ligaments a joint has, the stronger it is. However, when other stabilizing factors are absent or inadequate, undue tension or strain is placed on the ligaments and they stretch. Once stretched, ligaments stay stretched, like taffy, and a ligament can stretch only about 6% of its length before it snaps. Thus, when ligaments are the major means of bracing or strengthening a joint, the joint is not very stable.

Muscle Tone

For most joints, the tone of the muscles whose tendons cross the joint is the major stabilizing factor. (Muscle tone, defined as low levels of contractile activity in relaxed muscles, keeps the muscles healthy and ready to react to stimulation.) Muscle tone is extremely important in reinforcement of the shoulder joint. Likewise, the joints between the bones that form the arches of the foot are supported by the tendons of leg muscles.

The articular capsule and ligaments are richly supplied with sensory nerve endings that monitor the position of the joints and help to maintain muscle tone. Stretching of these structures sends nerve impulses to the central nervous system, resulting in reflex contraction of muscles surrounding the joint.

Movements Allowed by Synovial Joints

Every skeletal muscle of the body is attached to bone or other connective tissue structures, such as cartilage or fibrous membranes, at no fewer than two points. The muscle's *origin* is attached to the immovable (or less movable) bone; its other end, the *insertion*, is attached to the movable bone. Body movement occurs when muscles contract across joints and their insertion moves toward their origin. The movements can be described in directional terms relative to the lines, or *axes*, around which the body part moves and the planes of space along which the movement occurs, that is, along the transverse, frontal, or sagittal plane. (These planes were described in Chapter 1.)

Range of motion allowed by synovial joints varies from *nonaxial movement* (slipping movements only, since there is no axis around which movement can occur) to *uniaxial movement* (movement in one plane) to *biaxial movement* (movement in two planes) to *multiaxial movement* (movement in or around all three planes and axes). Range of motion varies greatly in different people. In some, such as trained gymnasts or acrobats, range of joint movement may be extraordinary (Figure 8.5). The ranges of motion at the major joints are given in Table 8.1. There are three general types of movements: gliding of one bone surface across another; movements that change the angle between two bones; and rotation about a longitudinal axis. The most common types of body movements allowed by synovial joints are described next and illustrated in Figure 8.6.

Figure 8.5 An example of extraordinary range of motion. Aurelia Dobre, a Rumanian gymnast in the 1988 Olympics, has hip joints nearly as flexible as her shoulder joints.

Gliding Movements

Gliding movements (Figure 8.6a) are the simplest type of joint movements. One flat, or nearly flat, bone surface slips over another similar surface; the bones are merely displaced in relation to one another. Neither angular nor rotatory movement is allowed. Gliding movements occur at the intercarpal, intertarsal, and sternoclavicular joints.

Angular Movements

Angular movements (Figure 8.6b–g) increase or decrease the angle between two bones. Angular movements may occur along any plane of the body and include flexion, extension, abduction, adduction, and circumduction.

Flexion. **Flexion** (flek′-shun) is a bending movement that decreases the angle of the joint and brings the two articulating bones closer together. The movement usually occurs along the sagittal plane. Examples include bending the head forward on the chest (Figure 8.6b) and bending a finger, the elbow, the body trunk, or the knee from a straight to an angled position (Figure 8.6c and d). Flexion of the shoulder and ankle require a bit more explanation. The shoulder is flexed when the arm is moved to a position anterior to the shoulder (Figure 8.6d). Flexion of the ankle so that the superior aspect of the foot approaches the shin is specifically called **dorsiflexion** (Figure 8.6e).

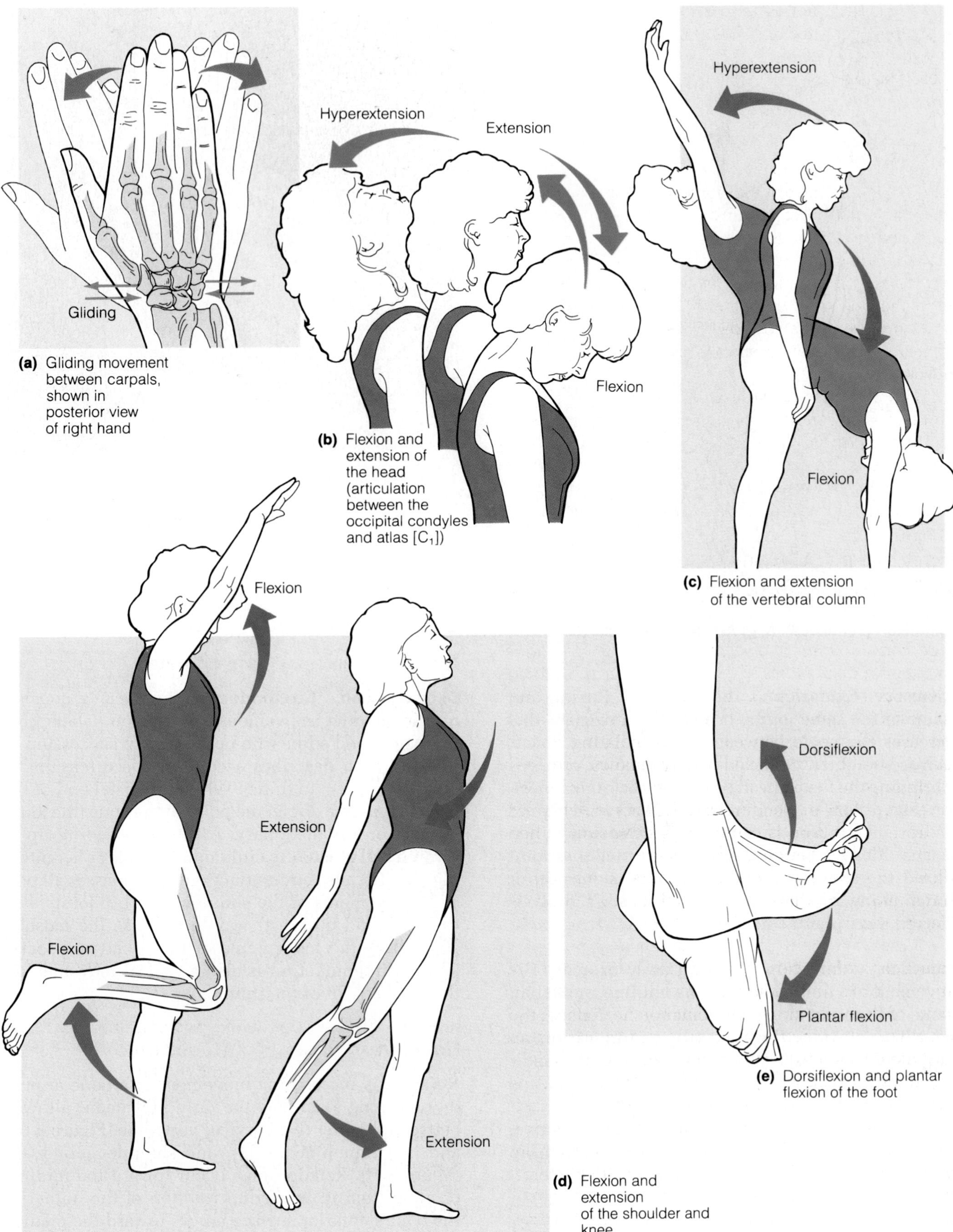

(a) Gliding movement between carpals, shown in posterior view of right hand

(b) Flexion and extension of the head (articulation between the occipital condyles and atlas [C_1])

(c) Flexion and extension of the vertebral column

(d) Flexion and extension of the shoulder and knee

(e) Dorsiflexion and plantar flexion of the foot

Figure 8.6 Movements allowed by synovial joints. (a) Gliding movements. **(b–g)** Angular movements. **(h, i)** Rotation. (See p. 224 for **f–i.**)

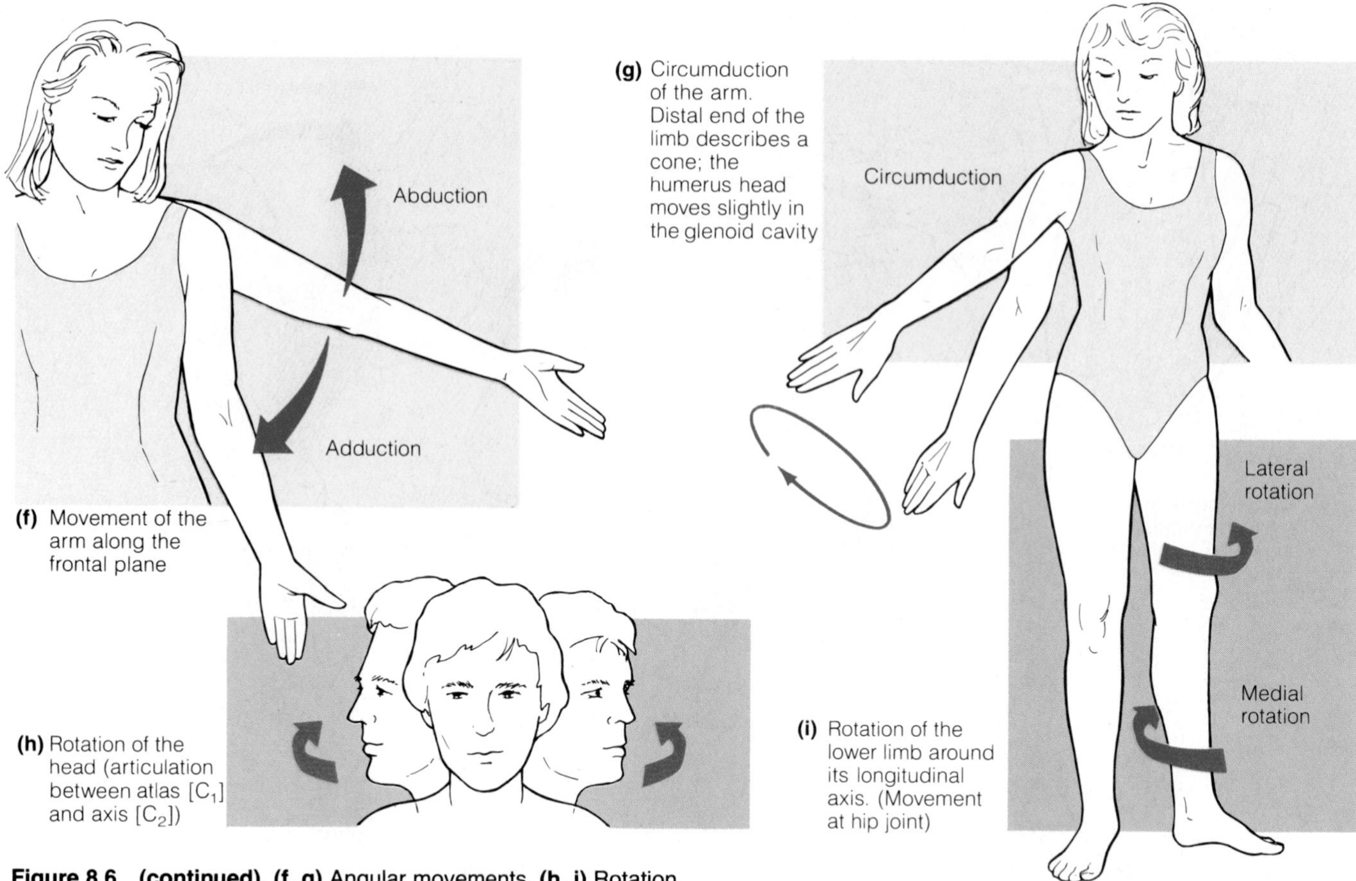
Abduction
Adduction
(f) Movement of the arm along the frontal plane
(g) Circumduction of the arm. Distal end of the limb describes a cone; the humerus head moves slightly in the glenoid cavity
Circumduction
Lateral rotation
Medial rotation
(h) Rotation of the head (articulation between atlas [C1] and axis [C2])
(i) Rotation of the lower limb around its longitudinal axis. (Movement at hip joint)

Figure 8.6 (continued). (f, g) Angular movements, **(h, i)** Rotation.

Extension. **Extension** is the reverse of flexion and occurs at the same joints. It involves movement that increases the angle between the articulating bones, such as straightening flexed fingers, elbows, or knees or bringing the flexed head back to its anatomical position (Figure 8.6b–e). Bending the head backward beyond the upright position is called **hyperextension.** When the shoulder is extended, the arm is carried to a point behind the shoulder joint. In the foot, extension or straightening of the ankle (pointing one's toes) is referred to as **plantar flexion.**

Abduction. **Abduction** ("moving away or apart") is movement of a limb away from the midline, or median plane, of the body, along the frontal plane. Raising the arm (Figure 8.6f) or thigh laterally and upward is an example of abduction. When the term is used to indicate the movement of the fingers or toes, it means spreading them apart. The reference point in this case is the midline of the hand or foot (digit 3). Notice, however, that lateral bending of the trunk away from the body midline in the frontal plane is called (lateral) flexion, not abduction.

Adduction. **Adduction** ("moving toward") is the opposite of abduction, so it is the movement of a limb toward the body midline (Figure 8.5f), or in the case of the digits, toward the midline of the hand or foot.

Circumduction. **Circumduction** (Figure 8.6g) consists of the angular movements of flexion, abduction, extension, and adduction performed in succession so that the limb describes a cone in space (*circum* = around; *duco* = to draw). While the distal end of the limb moves in a circle, the point of the cone (the joint) is more-or-less stationary. A pitcher winding up to throw a ball is actually circumducting his or her pitching arm. Because circumduction encompasses all possible movements of the shoulder and hip joints, it is the best (and quickest) way to exercise the muscles controlling the movements of those ball-and-socket joints. Circumduction is also a typical movement of the saddle joint of the thumb.

Rotation

Rotation is the turning movement of a bone around its own long axis. It is the only movement allowed between the first two cervical vertebrae (Figure 8.6h) and is common at the hip and shoulder joints (see Figure 8.5i). Rotation may occur toward the midline or away from it; in *medial rotation* of the humerus, the bone's anterior surface moves toward the median plane of the body.

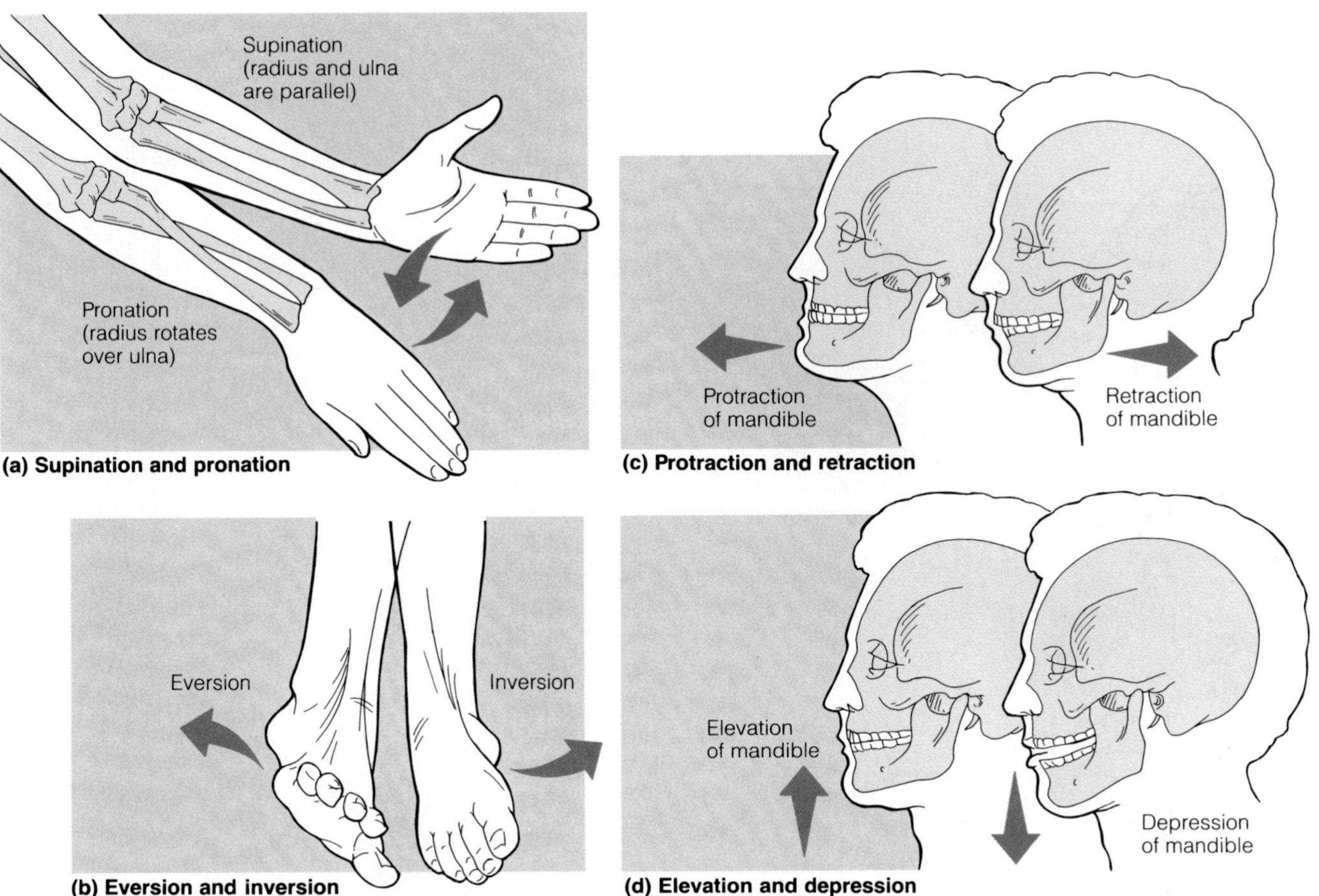

Figure 8.7 Special body movements.

Special Movements

Certain movements occur only at specific joints or areas of the body (Figure 8.7). Since these movements involve more than a single joint or cannot be accurately regarded as any of the movements already described, they are classed as special movements.

Supination and Pronation. The terms **supination** (soo″-pih-nā′-shun) and **pronation** (prō-nā′-shun) refer only to the movements of the radius around the ulna (Figure 8.7a). Movement of the forearm so that the palm faces anteriorly or superiorly is supination. When the body is in the anatomical position, the hand is supinated and the radius and ulna are parallel. In pronation, the palm is moved to a posterior- or inferior-facing position. During this action, the distal end of the radius moves across the ulna toward the body midline. This is the normal position for the forearm when we are standing in a relaxed manner.

Inversion and Eversion. The terms **inversion** and **eversion** refer to special movements of the foot (Figure 8.7b). In inversion, the sole of the foot is turned medially; in eversion, the sole faces laterally.

Protraction and Retraction. Nonangular forward and backward movements in a transverse plane are called **protraction** and **retraction,** respectively (Figure 8.7c). Protraction is anterior movement of a body part, such as the forward projection of the mandible when you jut out your jaw. Retraction is the posterior movement that returns a protracted bone or body part to its original position. "Squaring" your shoulders in a military stance is an example of retraction.

Elevation and Depression. **Elevation** means lifting or moving superiorly along a frontal plane (Figure 8.7d). For example, the scapulae are elevated when you shrug your shoulders. When the elevated part is moved downward to its original position, the movement is called **depression.** During chewing, the mandible is alternately elevated and depressed.

Types of Synovial Joints

Although all synovial joints have common structural features, they do not have a common structural plan. Based on their structure and movements allowed, synovial joints can be classified further into the six

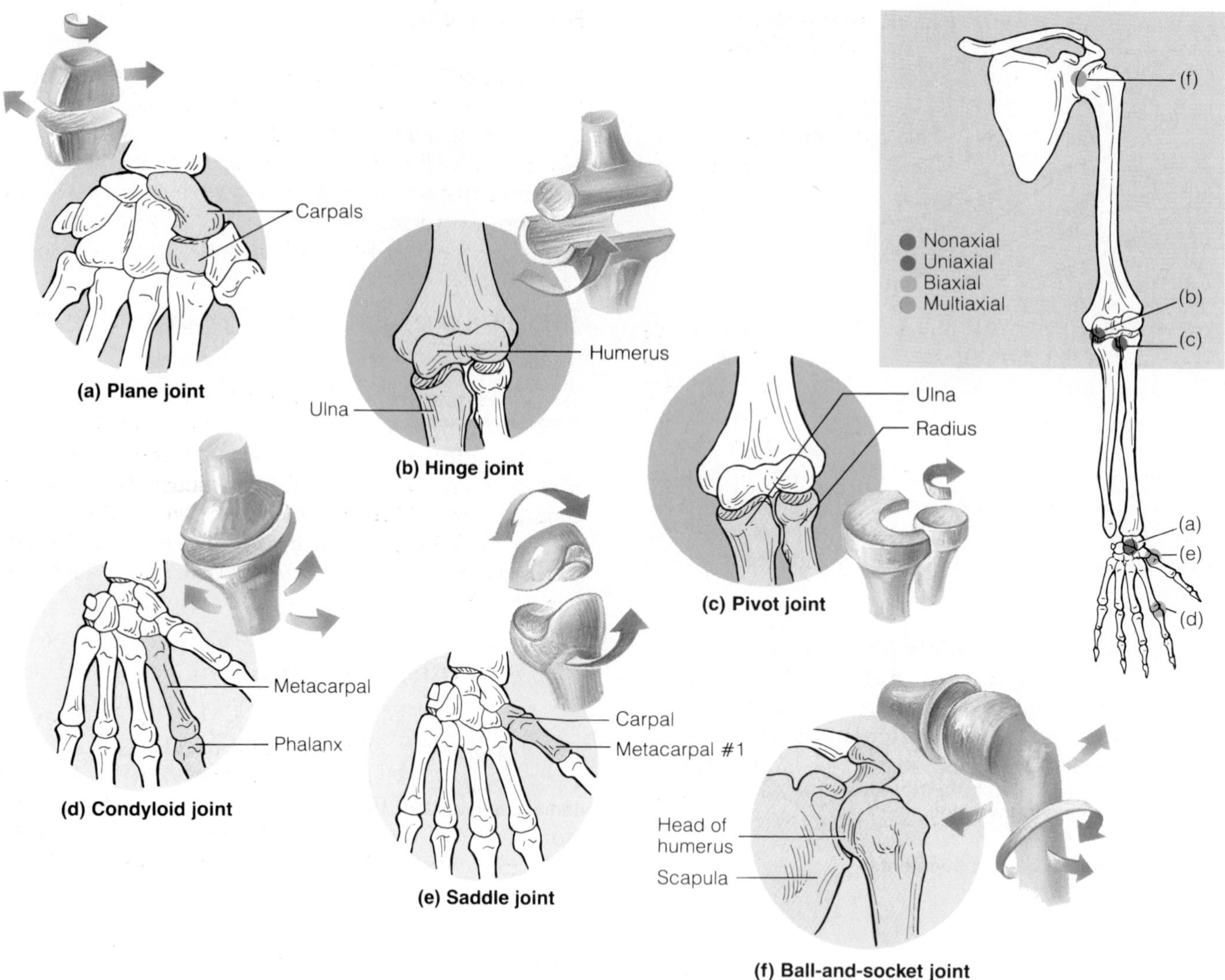

Figure 8.8 Types of synovial joints. (**a**) Plane joint (e.g., intercarpal and intertarsal joints). (**b**) Hinge joint (e.g., elbow joints and interphalangeal joints). (**c**) Pivot joint (e.g., proximal radioulnar joint). (**d**) Condyloid joint (e.g., metacarpophalangeal joints). (**e**) Saddle joints (e.g., carpometacarpal joint of the thumb). (**f**) Ball-and-socket joint (e.g., shoulder joint).

major categories illustrated in Figure 8.8. Plane joints are nonaxial; hinge and pivot joints are uniaxial; condyloid and saddle joints are biaxial; and ball-and-socket joints are multiaxial.

Nonaxial Joints

Plane Joints. **Plane joints** (Figure 8.8a) are the only examples of nonaxial joints. The articular surfaces are flat or only slightly curved; they allow slipping or gliding movements in all directions, including the twisting of one bone on another. However, because such joints are bound by tight ligaments, movement at gliding joints is generally very limited; for example, the joint may move only side to side, or only back and forth. Examples of plane joints include the intercarpal and intertarsal joints, the sternoclavicular joint, and the joints between vertebral articular processes.

Uniaxial Joints

Hinge Joints. In **hinge joints** (Figure 8.8b), a convex or cylindrical projection of one bone fits into a concave surface or cylindrical groove of another. Motion is along a single plane and is similar to that of a mechanical hinge. Hinge joints permit flexion and extension only, typified by the bending and straightening of the elbow, interphalangeal, and knee joints.

Pivot Joints. In a **pivot joint** (Figure 8.8c), a rounded or conical surface of one bone protrudes into a "sleeve" or ring composed of bone (and possibly ligaments) of another. The only movement allowed is uniaxial rotation of one bone around or against the other. Examples include the joint between the atlas and axis, which allows you to move your head from side to side to indicate "no," and the proximal radioulnar joint, where the head of the radius rotates within a ringlike ligament secured to the ulna.

Biaxial Joints

Condyloid Joints. In **condyloid** (kon′-dih-loyd) ("knucklelike") **joints,** the oval articular surface of one bone fits into a complementary elongated concavity in another (Figure 8.8d). Condyloid joints permit all angular motions, that is flexion and extension, abduction and adduction, and circumduction. The radiocarpal joints, the metacarpophalangeal (knuckle) joints, and the atlantooccipital joint (between the occipital condyles of the skull and the first cervical vertebra) are condyloid joints.

Saddle Joints. Saddle joints (Figure 8.8e) are structurally much like condyloid joints, but the freedom of movement allowed is greater. Each articular surface has both concave and convex areas, that is, is shaped like a saddle. The articular surfaces then fit together, concave to convex surfaces. The only saddle joints of the body are the carpometacarpal joints of the thumbs.

Multiaxial Joints

Ball-and-Socket Joints. Ball-and-socket joints (Figure 8.8f) are the most freely moving synovial joints. Universal movement is allowed (that is, in all axes and planes, including rotation). In such joints, the spherical or hemispherical head of one bone articulates with the concave socket of another. The shoulder and hip joints are the only ball-and-socket joints in the body.

Structure of Selected Synovial Joints

In this section, we will investigate the structure of four joints (shoulder, elbow, hip, and knee) in detail. All of these joints have a synovial membrane and articular cartilage, and we will not discuss these characteristics again. Rather, we will emphasize the unique structural features of each of these joints and relate them to their functional abilities and, in certain cases, functional weaknesses.

Shoulder (Glenohumeral) Joint

Shoulder joint stability has been sacrificed to provide the most freely moving joint of the body. The shoulder joint is a ball-and-socket joint formed by the head of the humerus and the small, shallow, pear-shaped glenoid cavity of the scapula (Figure 8.9). Although the glenoid cavity is slightly deepened by a rim of fibro-

Acromion process
Coracoacromial ligament
Subacromial bursa
Coracohumeral ligament
Greater tubercle of humerus
Transverse humeral ligament
Tendon sheath
Tendon of long head of biceps brachii muscle
Coracoid process
Articular capsule reinforced by glenohumeral ligaments
Subscapular bursa
Tendon of the subscapularis muscle
Scapula
(a)

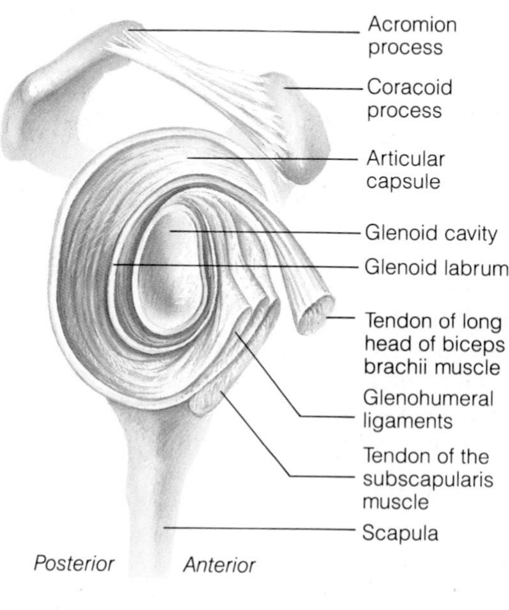

Figure 8.9 Shoulder joint relationships. (**a**) Anterior view of the shoulder joint (superficial aspect) illustrating some of the ligamentous reinforcements, associated muscles, and bursae. (**b**) The right shoulder joint, cut open and viewed from the lateral aspect; the humerus has been removed.

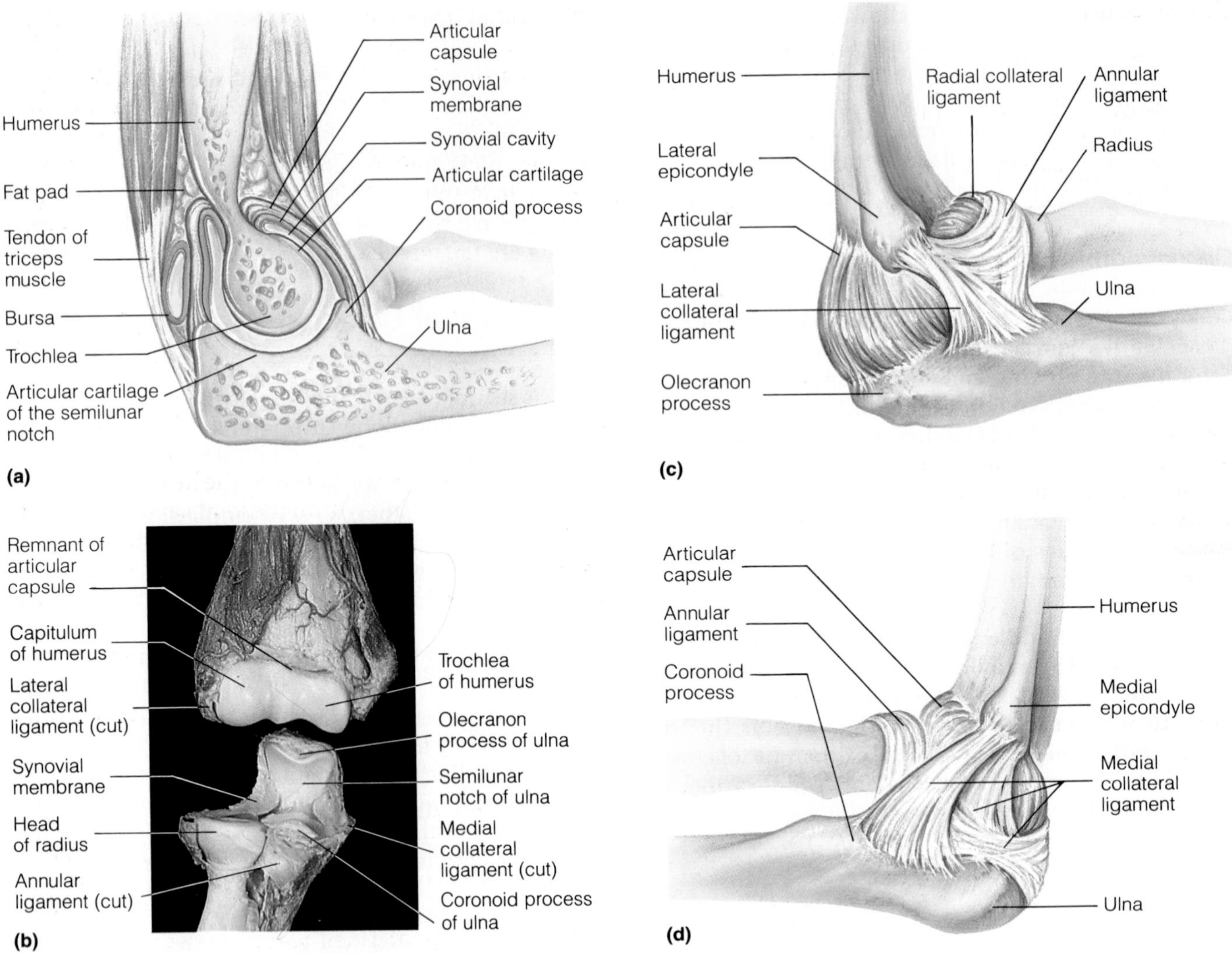

Figure 8.10 Elbow joint relationships. (**a**) Midsagittal section of right elbow joint. (**b**) Photograph of opened, disarticulated elbow joint, anterior view. (**c**) Lateral view of right elbow joint. (**d**) Medial view of right elbow joint.

cartilage called the *glenoid labrum* (gleh′-noyd lā′-brum), the articular surfaces contribute little to joint stability. The thin, lax articular capsule that encloses the joint cavity extends from the margin of the glenoid cavity to the anatomical neck of the humerus. The joint receives some protection superiorly from the coracoid process and acromion of the scapula.

The ligaments reinforcing the shoulder joint are located primarily on its anterior aspect. These include the **coracohumeral** (kor′-uh-kō-hyoo′-mer-ul) **ligament,** which extends from the coracoid process to the greater tubercle of the humerus and reinforces the capsule superiorly and anteriorly; the **glenohumeral** (gleh″-nō-hyoo′-mer-ul) **ligaments,** three weak, fibrous bands that strengthen the front of the capsule; and the **transverse humeral ligament,** a narrow band spanning the gap between the humeral tubercles. Several bursae are associated with the shoulder joint.

Although the ligaments help reinforce the shoulder joint, the muscle tendons surrounding the shoulder are far more important. The "superstabilizer" is the tendon of the long head of the biceps brachii muscle of the arm. This tendon runs from the superior aspect of the glenoid labrum, enters and travels within the joint cavity, and then exits the cavity to enter the intertubercular groove of the humerus. This arrangement secures the head of the humerus tightly against the glenoid cavity. Other tendons, collectively called the "rotator cuff," encircle the joint and fuse with the articular capsule. These include the tendons of the subscapularis, supraspinatus, infraspinatus, and teres minor muscles. Because of their arrangement, these muscle tendons can be severely stretched when the arm is vigorously circumducted; they are commonly injured by baseball pitchers. Since the shoulder's reinforcements are weakest inferiorly, shoulder dislocations most often occur in the inferior direction.

Elbow Joint

Our upper limbs are flexible extensions that permit us to reach out and manipulate things in our environment. Besides the shoulder joint, the most prominent of the upper limb joints is the elbow. The careful fitting of the humeral and ulnar bone ends forming this joint provides a stable and smoothly operating hinge that allows flexion and extension only (Figure 8.10). (Construction cranes exhibit the same type of flexibility.) Within the joint, the humeral trochlea and capitulum articulate respectively with the semilunar notch of the ulna and the head of the radius, but it is the gripping of the trochlea by the semilunar notch that constitutes the "hinge" and provides the main stabilizing factor of this joint. A relatively loose-fitting fibrous capsule extends from the coronoid and olecranon fossae of the humerus inferiorly to the coronoid and olecranon processes of the ulna and to the **annular** (a′-nyoo-ler) **ligament** enclosing the head of the radius.

Two important ligaments strengthen the sides of the articular capsule. These are the **medial collateral ligament,** three strong bands reinforcing the capsule on its medial aspect, and the **lateral collateral ligament,** a triangular ligament coursing through the lateral part of the capsule. Tendons of several muscles (the biceps, triceps, brachialis, and others) cross the elbow joint and provide security.

Flexion at the elbow is limited by the meeting of the soft tissues of the forearm and arm. Extension is checked by the tension of the medial ligament and by tendons of the forearm flexor muscles. The radius is a passive "onlooker" in the angular elbow movements, but its head rotates within the annular ligament during supination and pronation of the forearm.

Hip (Coxal) Joint

The hip joint, like the shoulder joint, is a ball-and-socket joint; it has good range of motion, but not nearly as wide as the shoulder's range. All angular movements, as well as rotation, occur, but are limited by the joint ligaments. The joint is formed by the articulation of the spherical head of the femur with the deeply cupped acetabulum of the coxal bone (Figure 8.11). The depth of the acetabulum is enhanced by a horseshoe-shaped fibrocartilage rim called the **acetabular** (a″-sih-ta′-byoo-ler) **labrum.** In contrast to the articulation at the shoulder joint, these articular surfaces fit snugly together, and hip joint dislocations are rare. The thick articular capsule extends from the rim of the acetabulum to the anatomical neck of the femur and completely encloses the joint.

Several strong ligaments reinforce the hip joint. These include the **iliofemoral** (ih″-lē-ō-feh′-mer-ul) **ligament,** a strong V-shaped ligament extending from the anterior inferior iliac spine to the intertrochanteric line of the femur; the **pubofemoral** (pyoo″-bō-feh′-mer-ul) **ligament,** a triangular fibrous thickening of the articular capsule that extends from the pubic ramus to the intertrochanteric line, providing anteromedial reinforcement; and the **ischiofemoral** (is″-kē-ō-feh′-mer-ul) **ligament,** a spiral, posteriorly located ligament extending superiorly and laterally from the ischium to the greater trochanter of the femur. These ligaments are arranged in such a way that they "screw" the femur head into the acetabulum when a person stands up straight, providing more stability.

The *ligamentum teres,* also called the *capitate ligament,* is a flat intracapsular ligament that runs from the femur head to the lower lip of the acetabulum. Since the ligamentum teres is slack during most hip movements, it is not important in stabilizing the joint. However, it does help to prevent the head of the femur from gliding upward within the joint cavity. The small artery that travels within the ligamentum teres provides most of the arterial blood supply to the head of the femur. Any damage to this artery leads to severe arthritis.

Muscle tendons that cross the joint and the bulky hip and thigh muscles that surround it contribute to its stability and strength, but much less than do the articular surfaces and ligaments.

Knee (Tibiofemoral) Joint

The knee joint is the largest joint in the body and is exceptionally complex (Figure 8.12). It allows extension, flexion, and some rotation. It consists of three joints in one: an intermediate one between the patella and the lower end of the femur, and lateral and medial joints between the femoral condyles above and the C-shaped semilunar cartilages, or menisci, of the tibia below. Besides deepening the shallow tibial articular surfaces, the menisci help prevent side-to-side rocking of the femur on the tibia and absorb shock transmitted to the knee joint. (However, the menisci are attached only at their outer margins and are frequently torn free.) The tibiofemoral joint is functionally a hinge joint because its movements are strenuously restricted by ligaments. However, it is a condyloid joint structurally, and some rotation is possible. The shape of the articular surfaces contributes little or nothing to joint stability. The femoropatellar joint is a plane joint, and the patella glides across the distal end of the femur during knee movements.

The knee joint is unique in that its joint cavity is only partially enclosed by a capsule. On the sides and posterior aspects of the knee, the relatively thin articular capsule extends from the margins of the articular surfaces of the distal femur to those of the tibial head. It is absent anteriorly, which allows the synovial membrane to protrude upward beneath the tendon of the quadriceps (bulky anterior thigh muscle group) forming the *suprapatellar bursa.* The capsule is heav-

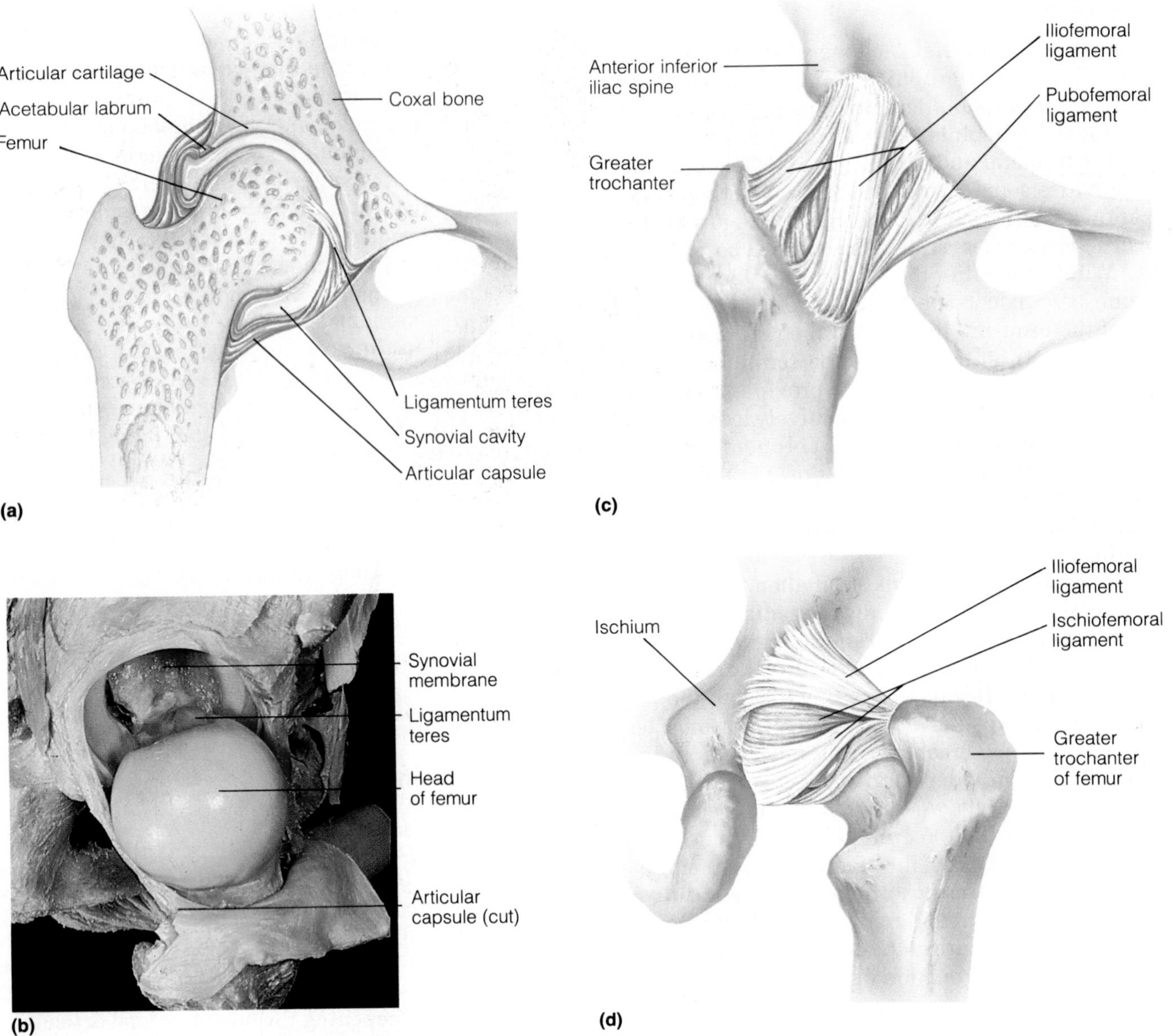

Figure 8.11 Hip joint relationships. (**a**) Frontal section through the right hip joint. (**b**) Photograph of the interior of the hip joint, lateral view. (**c**) Anterior superficial view of the right hip joint. (**d**) Posterior superficial view of the right hip joint.

ily reinforced by ligaments and surrounding muscle tendons. Four bursae are situated inferiorly to the joint, and six are behind it. Some of these bursae are illustrated in Figure 8.12.

Both extra- and intracapsular ligaments are important in stabilizing and strengthening the tenuous knee joint. The *extracapsular ligaments* include the following:

1. The **patellar ligament,** which extends from the patella to the tibial tuberosity inferiorly and is actually a continuation of the tendon of the quadriceps muscle. It is separated from the synovial membrane of the joint by a fatty pad and a small bursa.

2. The **lateral and medial collateral ligaments,** which are critical in preventing lateral or medial angular movements of the knee joint. The lateral collateral ligament extends from the lateral epicondyle of the femur to the head of the fibula. The broad medial collateral ligament extends from the medial epicondyle of the femur to the medial condyle and surface of the tibial shaft below and is fused to the medial semilunar cartilage.

3. The **oblique popliteal** (pop-lih-tē′-ul) **ligament,** actually an expansion of the tendon of the semimembranosus muscle crossing the posterior aspect of the knee joint.

4. The **arcuate popliteal ligament,** which extends from the lateral condyle of the femur to the head of the fibula posteriorly.

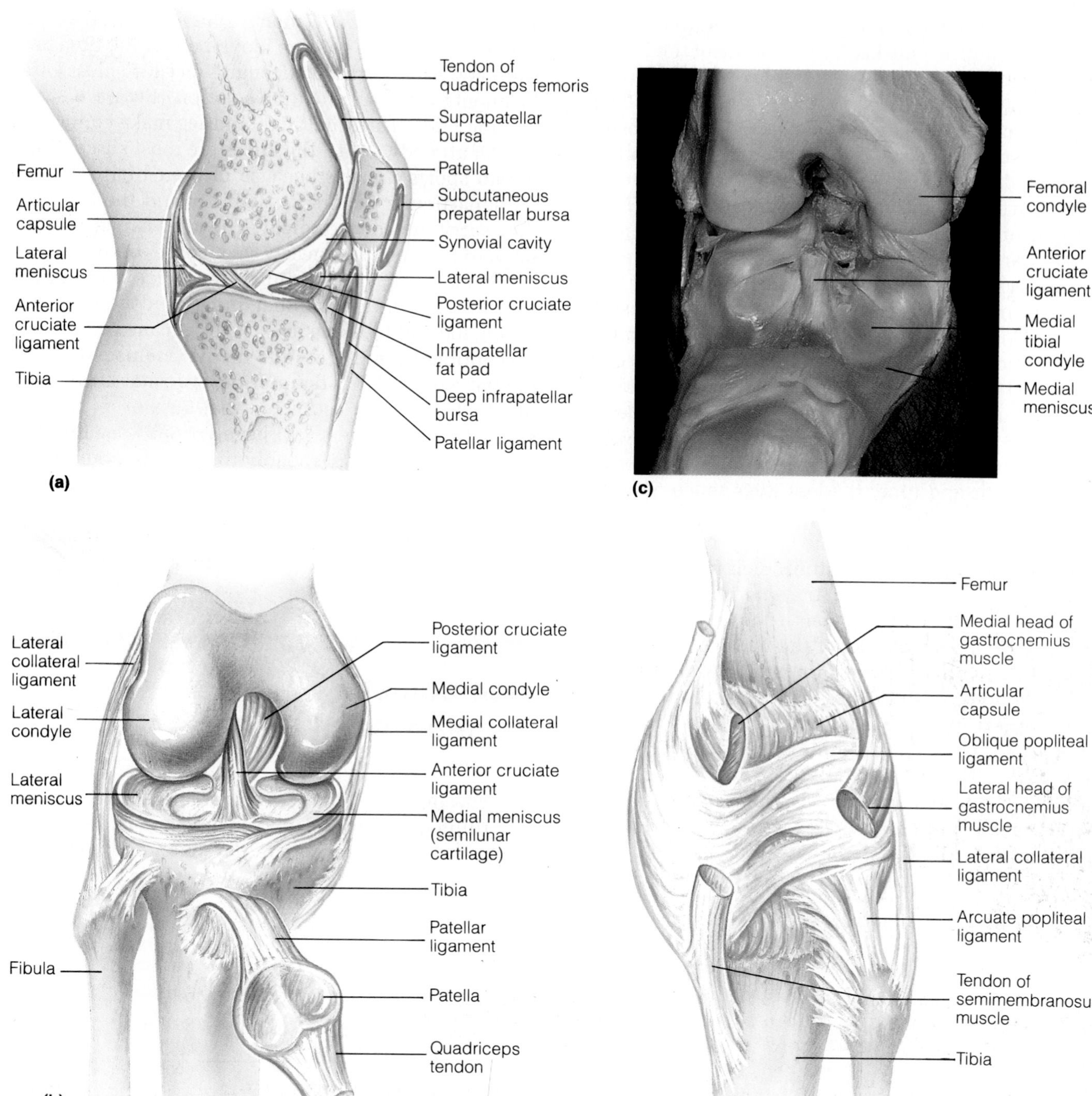

Figure 8.12 Knee joint relationships. (**a**) Midsagittal section of right knee joint. (**b**) Anterior view of slightly flexed right knee joint. Joint has been flexed to show cruciate ligaments. Articular capsule has been removed; the quadriceps tendon has been cut and reflected distally. (**c**) Photograph of an opened knee joint corresponds to view shown in **(b)**. **(d)** Posterior superficial view of the ligaments clothing the knee joint.

The *intracapsular ligaments* are called *cruciate* (kroo′-shē-ut) *ligaments* because they cross each other, forming an X (*cruci* = cross) within the notch between the femoral condyles. They help prevent displacement of the articular surfaces and secure the articulating bones together (see Figure 8.12b). Note that the two cruciate ligaments are named for their tibial attachment site. The **anterior cruciate ligament** passes posteriorly, laterally, and upward from the anterior intercondylar area of the tibia to attach to the femur on the posteromedial surface of its lateral condyle. This ligament prevents backward sliding of the femur and overextension of the knee. It is lax when the knee is flexed and taut when the knee is extended. The **posterior cruciate ligament** is attached to the posterior intercondylar area of the tibia and passes anteriorly, medially, and upward to attach to the anterolateral surface of the medial femoral condyle. This strong ligament prevents forward sliding of the femur or backward displacement of the tibia; thus, it helps prevent overflexion of the knee joint.

The knee capsule is strongly reinforced by muscle tendons. Most important are the expanded tendons of the quadriceps muscles of the anterior thigh and the tendon of the semimembranosus muscle posteriorly (the oblique popliteal ligament already described). Since the muscles associated with the joint are extremely important in stabilizing the knee, the greater their strength and tone, the less the chance of knee injury.

During knee extension, the femoral condyles roll like ball bearings on the tibia until their further movement is limited by tension in the anterior cruciate ligament (Figure 8.13a). Further extension is accomplished by medial rotation of the femur on the tibia, which effectively locks the joint into a rigid structure. When the joint is fully extended (or slightly hyperextended), all major ligaments of the knee joint are twisted and taut and the semilunar cartilages are compressed. For flexion to occur, the ligaments must be untwisted and become slack. This is accomplished by the popliteus muscle, which rotates the femur laterally on the tibia. When the knee is flexed at a right angle, considerable movement is possible—medial and lateral rotation and passive movement of the tibia forward and backward on the femur—because most joint ligaments are slack in this position (Figure 8.13b). Flexion ends when the thigh and leg make contact.

Of all body joints, the knees are most susceptible to sports injuries because of their high reliance on nonarticular factors for stability and the fact that they carry the body's weight. The knee can absorb an upward force equal to nearly seven times body weight; however, it is apt to be injured by direct blows or by the high-pressure rotational movements that occur during football blocking and tackling maneuvers. Most susceptible to injury are the medial menisci and the medial collateral and anterior cruciate knee ligaments. It has been estimated that 50% of all professional football players will have serious knee injuries during their careers. Probably the most famous knees in football history are those of former New York Jets quarterback Joe Namath. During his professional football career, Namath had two ligaments and nearly all of the cartilage removed from both knees! ■

Homeostatic Imbalances of Joints

Few of us pay attention to our joints unless something goes wrong with them; yet considering the stress that our joints endure day in and day out, it is

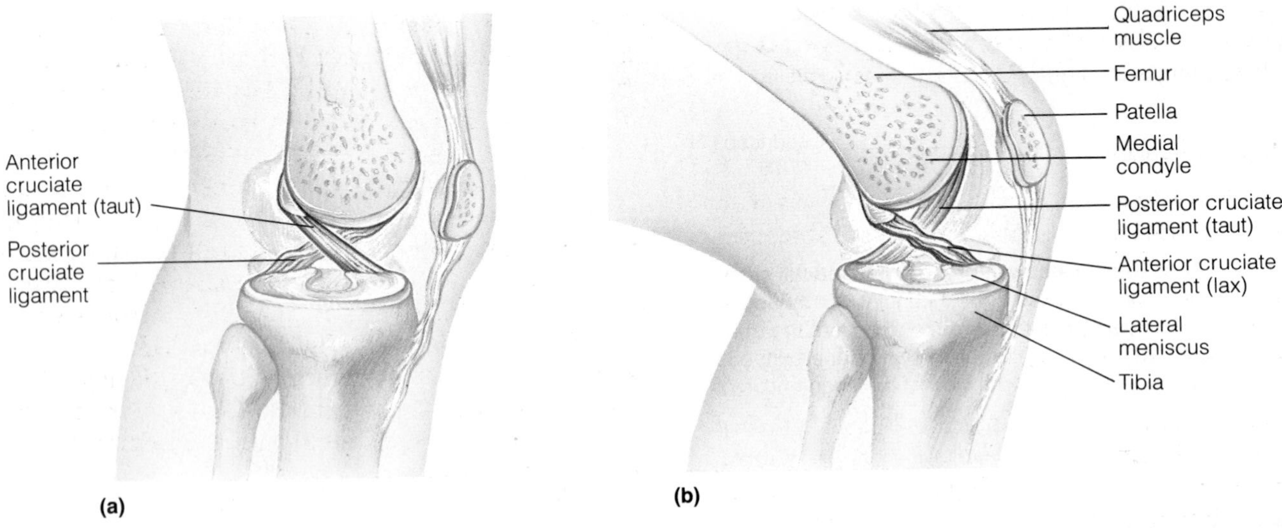

Figure 8.13 Knee joint movement. The cruciate ligaments of the knee function to prevent undesirable movements at the knee joint. (**a**) When the knee is extended, the anterior cruciate ligament becomes taut and prevents hyperextension of the joint. (**b**) When the knee is flexed, the posterior cruciate ligament prevents posterior slipping movements of the tibia.

a wonder that joint disorders are so infrequent. Joint pain and malfunction can be caused by a number of factors, but the most common joint problems result from injuries and from inflammatory or degenerative conditions.

Common Joint Injuries

For most of us, sprains and dislocations are the most common trauma-induced joint injuries, but cartilage injuries are equally important in athletes.

Sprains

In a **sprain,** the ligaments reinforcing a joint are stretched or torn. The lumbar region of the spine, the ankle, and the knee are common sprain sites. A partially torn ligament will repair itself, but since ligaments are so poorly vascularized, sprains heal slowly and tend to be excruciatingly painful and immobilizing. Completely ruptured ligaments must be repaired surgically and quickly, because the inflammatory response in the joint will break down neighboring tissues and turn the injured ligament to "mush." Repair can be a difficult task indeed: A ligament consists of hundreds of fibrous strands, and sewing one back together has been compared to trying to sew two hairbrushes together.

When important ligaments are too severely damaged to be repaired, they must be removed and substitute ligaments implanted. For example, a piece of tendon from a muscle, or collagen extracted from cattle hide and woven into bands, can be stapled to the margins of the articulating bones. In another technique, carbon fibers are implanted in the torn ligament to form a supporting mesh. The mesh is then invaded by fibroblasts, which lay down new collagen fibers that reconstruct the ligament.

Cartilage Injuries

Most cartilage injuries involve tearing of the knee menisci. This can result from a fall or from subjecting the cartilage to high pressure and twisting motions simultaneously. For example, tennis players who lunge to return a ball sometimes twist their knee medially, grinding the meniscus cartilage between the tibia and femur.

Cartilage is avascular and it rarely can obtain sufficient nourishment to repair itself; thus, it usually stays torn. Because cartilage fragments can interfere with joint function by causing locking or binding of the joint, most sports physicians recommend that damaged portions of cartilage be removed. Today, this can be done by a procedure called *arthroscopy* (ar-thrāh'-skuh-pē), which enables patients to be out of the hospital the same day. The arthroscope, a small instrument bearing a tiny lens and light source and attached to a television camera, enables the surgeon to view the interior of a joint through a small incision. Removal of cartilage fragments or ligament repair is performed through the same tiny slit, minimizing tissue damage and scarring. Removal of some or all of the cartilage does not severely impair knee joint mobility, but the joint is definitely less stable.

Dislocations

A **dislocation (luxation)** occurs when bones are forced out of their normal positions in a joint cavity. It is usually accompanied by sprains, considerable inflammation, and joint immobilization. Dislocations may occur during a serious fall and are common contact sports injuries. The joints of the shoulders, fingers, and thumbs are common dislocation sites. Like fractures, dislocations must be *reduced;* that is, the bone ends must be returned to their proper positions by a physician. Attempts by an untrained person to "snap the bone back into its socket" are usually more harmful than helpful.

Inflammatory and Degenerative Conditions

Inflammatory conditions affecting joint areas include inflammations of bursae and tendon sheaths and various forms of arthritis. Arthritis is inflammation or degeneration of the joints, and some types of arthritis are severely disabling.

Bursitis and Tendonitis

Bursitis is an inflammation of a bursa and is usually caused by excessive stress or friction. Falling on one's knee may result in a painful bursitis of the patellar bursa, known as *housemaid's knee* or *water on the knee.* A long, strenuous tennis match that traumatizes the bursa close to the olecranon process may produce *tennis elbow.* Bursitis may also result from bacterial infection. Symptoms of bursitis include pain accentuated by joint movement, redness, and local swelling. In severe cases, antiinflammatory corticosteroid drugs may be injected. If fluid accumulation is excessive, removing some of the fluid by needle aspiration may relieve the pressure. **Tendonitis** is an inflammation of tendon sheaths. Its causes, symptoms, and treatment mirror those of bursitis.

Arthritis

The term *arthritis* describes over 100 different types of inflammatory or degenerative diseases that damage the joints. In all its forms, arthritis is the most widespread crippling disease in the United States. Approximately 35 million Americans (one out of seven) suffer its ravages. All forms of arthritis have, to a greater or lesser degree, the same initial symptoms: pain, stiffness, and swelling of the joint. Depending on the specific form, pathological changes within the joint may affect the synovial membrane, articular cartilages, articulating bones, or all of these.

Acute forms of arthritis usually result from bacterial invasion and are treated with antibiotic drugs. The synovial membrane thickens and fluid production decreases, leading to increased friction and pain. Chronic forms of arthritis include osteoarthritis, rheumatoid arthritis, and gouty arthritis, which differ substantially from one another in their later symptoms and consequences.

Osteoarthritis. **Osteoarthritis (OA),** the most common arthritis, is a chronic degenerative condition that is often referred to as "wear-and-tear arthritis." Although seen occasionally in younger people, it is most prevalent in the aged and is probably related to the normal aging process. Local ischemia (inadequate blood supply), which diminishes nutrient delivery to the cartilage, is believed to be one of the culprits instigating cartilage deterioration; normal age-related changes in the cartilage matrix may be another. Although inflammatory symptoms may be present, osteoarthritis is generally considered a noninflammatory type of arthritis.

OA affects the articular cartilages. Over the years, cumulative effects of weight bearing and abrasion at joint surfaces lead to softening, roughening, and eventual erosion of the cartilage. As the disease progresses, the exposed bone thickens, forming bony spurs that cause the bone ends to enlarge. Since this encroaches on the joint space, joint movement may be restricted. Patients complain of stiffness on arising that lessens with activity. The affected joints may make a crunching noise when moved; this sound, called *crepitus* (kreh′-pih-tus), results from contact of the roughened articular surfaces. The joints most commonly affected include the phalangeal joints, the cervical and lumbar intervertebral joints, and the large weight-bearing joints of the lower limbs (knees and hips).

The course of osteoarthritis is usually slow and irreversible. In most cases, its symptoms are controllable with a mild analgesic such as aspirin, moderate activity to maintain joint mobility, and rest when the joint becomes very painful. Osteoarthritis is rarely crippling, but it can be, particularly when the hip or knee joints are involved.

Rheumatoid Arthritis. **Rheumatoid** (roo′-muh-toyd) **arthritis (RA)** is a chronic inflammatory disorder. In addition to characteristic joint symptoms, manifestations include anemia, osteoporosis, muscle atrophy, and cardiovascular problems. Its onset is insidious and usually between the ages of 30 and 40, but it may occur at any age. It affects three times as many women as men. Although not as common as osteoporosis, rheumatoid arthritis is far from a rare disease; more than 6.5 million Americans seek medical help for this condition each year. In the early stages of RA, fatigue, joint tenderness, and joint stiffness are common. Many joints, particularly the small joints of the fingers, wrists, ankles, and feet, are afflicted at the same time and in a symmetrical manner. For example, if the right elbow is affected, most likely the left elbow will be affected also. The course of RA is variable and marked by remissions and exacerbations (*rheumat* = susceptible to change or flux).

RA is an autoimmune disease—a disorder in which the body's immune system attempts to destroy its own tissues. The reaction involves an astonishing number of the body's immune elements, T lymphocytes, the rheumatoid factor (an abnormal antibody), and antirheumatoid antibodies (antibodies formed against the rheumatoid factor) to name a few. The initial trigger for this reaction is unknown, but the streptococcus bacterium and viruses have been suspect. Stress and genetic factors seem to be involved in some cases.

The initial pathology is inflammation of the synovial membrane (synovitis) of the affected joints, but ultimately all of the joint tissues may become involved. Without treatment, the synovial membrane gradually thickens, an abnormally thin synovial fluid accumulates, causing joint swelling, and inflammatory cells (lymphocytes, macrophages, and others) migrate into the joint cavity from the blood (Figure 8.14). The proliferation of synovial membrane cells and inflammatory cells gives rise to *pannus*, an abnormal tissue that clings to the articular cartilages. Eventually, the cartilage (and sometimes the underlying bony tissue) is eroded away by enzymes released by the inflammatory cells. As the cartilage is destroyed, fibroblasts elaborate fibrous tissue that connects the bone ends. This fibrous tissue eventually ossifies, and the bone ends become firmly fused (ankylosis) and often deformed. Not all cases of RA progress to the severely crippling *ankylosis* stage, but all cases do involve restriction of joint movement and extreme pain.

Current therapy for RA involves many different drug classes. Drug therapy begins with the least toxic drug, aspirin, which in large doses is an effective anti-inflammatory agent. Sadly, none of the drugs currently used has proved to be the wonder drug that RA sufferers hope for, and all of these drugs have some toxic effects, some very severe. Exercising the affected joints is recommended to maintain as much joint mobility as possible. Cold packs are used to relieve the swelling and pain, and heat helps to relieve morning stiffness. Joint prostheses, if available, are the last resort for severely crippled RA patients.

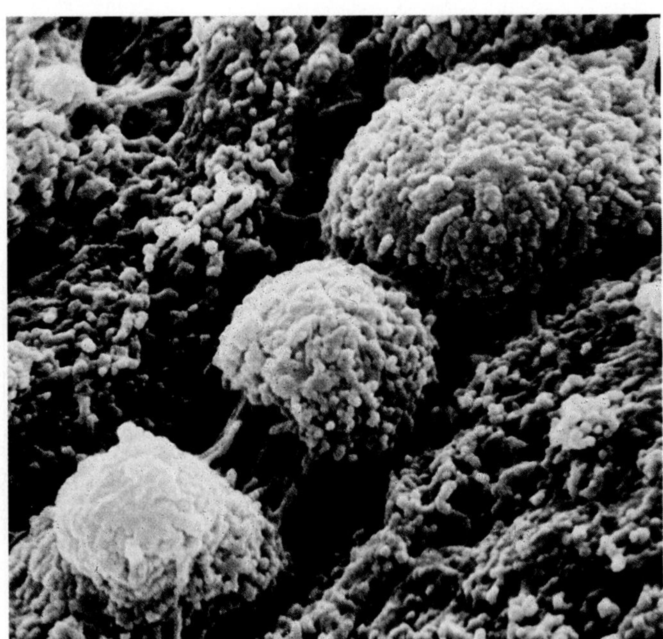

Figure 8.14 Scanning electron micrograph of the synovial membrane from a joint affected by rheumatoid arthritis. Notice the inflammatory cells, exhibiting numerous microvilli, that have accumulated on the membrane surface (X 3500). (Compare this view with the normal synovial membrane in Figure 8.3b.)

Gouty Arthritis. Uric acid, a normal waste product of nucleic acid metabolism, is ordinarily eliminated in urine without any problems. However, when blood levels of uric acid rise excessively, a condition called *hyperuricemia*, uric acid may be deposited as urate crystals in the soft tissues of joints. This leads to an agonizingly painful attack of **gouty** (gow′-tē) **arthritis,** or gout. Gouty arthritis typically affects a single joint and is most common in males. The cause of the hyperuricemia that precipitates gout is still a mystery. Some gout sufferers have an excessive rate of uric acid production; in others, uric acid is flushed out of the body in urine more slowly than normal; and some patients exhibit both problems. Because gout seems to run in families, genetic factors are definitely implicated.

Untreated gout can be very destructive; the articulating bone ends fuse and the joint becomes immobilized. Fortunately, several drugs (colchicine, allopurinol, and others) are currently used with high success to terminate or prevent gout attacks. Since some individuals suffer only one attack of gouty arthritis in their lifetime, drug therapy is usually withheld until at least two such attacks have occurred. Patients are advised to lose weight if obese, rest the affected joint, use cold packs to reduce swelling, avoid foods such as liver, kidneys, and sardines, which are high in purine-containing nucleic acids, and avoid alcohol and excessive vitamin C. (Alcohol and vitamin C inhibit renal excretion of uric acid.) ■

Developmental Aspects of the Joints

Joints develop from mesenchyme, as do other connective tissue elements. As bones form during the first two months of development, the joints develop in parallel. By eight weeks, the synovial joints resemble adult joints in form and arrangement: Joint cavities are present, synovial membranes are developed, and synovial fluid is being secreted.

Traumatic injuries aside, relatively few interferences with joint function occur until late middle age. Eventually, however, advancing years take their toll on the joints. The intervertebral discs become more susceptible to damage and rupture, and osteoarthritis rears its ugly head. Virtually everyone has osteoarthritis of one or more joints by the time they are in their 70s. The middle years also see an increased incidence of rheumatoid arthritis, and the crippling effects of this disease account for most cases of joint prosthesis implants in elderly individuals (see the box on p. 236).

Just as bones must be stressed to maintain their strength, joints must be exercised prudently to maintain their mobility. The key word is "prudently," because excessive or abusive use of the joints guarantees early onset of osteoarthritis. While exercise does not prevent arthritis, it helps to maintain joint integrity by increasing the strength of muscles that help support and stabilize the joints. The buoyancy of water relieves much of the stress on weight-bearing joints, and people who swim or exercise in a pool often retain good joint function as long as they live. As with so many medical problems, it is easier to prevent joint problems than to cure or correct them.

* * *

The importance of joints is obvious: The ability of the skeleton to protect other organs and to move smoothly reflects their presence. Now that we have scrutinized joint structure and are familiar with the movements that they allow, we are ready to consider how the muscles attached to the skeleton cause body movements by acting across its joints.

A CLOSER LOOK Joints: From Knights in Shining Armor to Bionic Man

There is a stark contrast between the time it took to develop joints for armor and the modern development of joint prostheses. The greatest challenge faced by visionary men of the Middle Ages and Renaissance was to design armor joints that allowed mobility while still protecting the human joints beneath. The ball-and-socket joints that they found so difficult to protect were the first to be engineered by modern-day "visionaries" for implantation into the body.

The armor worn by the Crusaders in the twelfth century was entirely made of chain mail—little rings of metal linked to one another. It was not until the sixteenth century that a complete suit of plate armor was developed that provided for both the protection and the mobility of the shoulder and hip joints. Since this was the period in which firearms became available (against which armor is essentially useless), the newly birthed suit of plate armor quickly became a relic.

Although armor took centuries to reach its peak of protective excellence, the history of joint prostheses (artificial joints) is less than 50 years old. It dates back to the 1940s and 1950s, when World War II and the Korean War yielded large numbers of wounded who needed artificial limbs and joints. Today, about a quarter of a million arthritis patients receive total joint replacements each year, mostly because of the destructive effects of osteoarthritis or rheumatoid arthritis.

The human body attempts to reject any implanted foreign object or causes its corrosion. Thus, several problems had to be surmounted to produce strong, mobile, lasting joints. What was needed was a substance that was strong, nontoxic to the body, and resistant to the corrosive effects of organic acids present in blood. In 1963, Sir John Charnley, an English orthopedic surgeon, performed the first total hip replacement and revolutionized the therapy of arthritic hips. His device consisted of a Vitallium metal ball on a stem and a cup-shaped polyethylene plastic socket anchored to the pelvis by methyl methacrylate cement. This cement proved to be exceptionally strong and relatively problem-free. It was not until ten years later that total knee joint replacements that operated smoothly and allowed all natural movements of the knee became available. Earlier devices locked abruptly when the knee was extended, causing the patient to fall or the joint to be wrenched free.

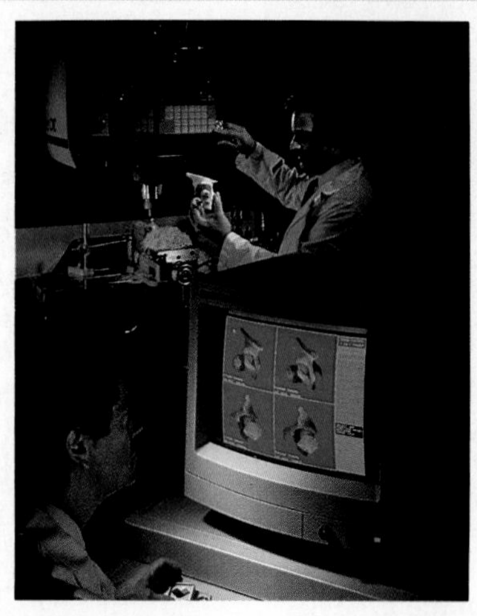

Joint replacement therapy employing both metal and plastic is now available for a number of other joints, including those of the fingers, elbow, and shoulder. Current techniques have produced total hip and knee replacements that last about ten years in elderly patients who do not excessively stress the joint. Most such operations are done to reduce pain and restore about 80% of original joint function.

Replacement joints are not yet strong or durable enough for young, active people, but dramatic changes are occurring in the way joints are designed, manufactured, and installed. CAD/CAM (computer-aided design and computer-aided manufacturing) techniques are being used to design and create individualized artificial implants in several research centers (see photo). Information about available implantable devices is fed into the computer along with the patient's X-rays and details of the patient's problems. The computer uses a database of hundreds of normal joints and generates possible designs and modifications that can be reviewed in less than a minute. Once the design is selected, the computer produces a program that directs the machines that cut and modify the required implant or totally fabricate a new joint prosthesis.

The influence of CAD/CAM on the joint replacement business is likely to be enormous. Previously, designing and producing a customized joint took 12 weeks or more. The wait has been reduced to two weeks by this technique, and those "in the know" predict that a two-day wait will soon be the norm. Some prostheses, such as an artificial knee, can cost as much as $4500 to produce manually, but the CAD/CAM system can produce it for under $3000, and soon the cost may be even lower.

Joint replacement therapy is coming of age, but perhaps equally exciting is research into the capability of joint tissues to regenerate. For example, we now know that if a joint with localized areas of articular cartilage erosion is slowly flexed and extended over a period of weeks, the joint surfaces sometimes regenerate themselves. This offers hope for younger patients, since it can stave off the need for a joint prosthesis for several years.

And so, through the centuries, the focus has shifted from jointed armor to artificial joints that can be put inside the body to restore lost joint function. Modern technology has in fact accomplished what the armor designers of the Middle Ages never even dreamed of.

Related Clinical Terms

Ankylosing spondylitis (an′-kih-lō″-sing spon″-dih-lī′-tis) (*ankyl* = crooked, bent; *spondyl* = vertebra) An unusual variant of rheumatoid arthritis that chiefly affects males; it usually begins in the sacroiliac joints and progresses superiorly along the spine; the vertebrae become squared and connected by fibrous tissue, causing the spine to become rigid ("poker back").

Arthrology (ar″-thrah′-luh-jē) (*arthros* = joint; *logos* = study) The study of joints.

Loose bodies Small pieces of bone or cartilage within a joint capsule; usually reflect joint trauma or the wearing away of part of the articular cartilage, which exposes the surface of bone beneath, causing it to die and separate; symptoms are painful catching or locking of the joint; unless surgically removed, loose bodies promote osteoarthritis.

Synovitis (sih-nō-vī′-tis) Inflammation of the synovial membrane of a joint; in healthy joints, only small amounts of synovial fluid are present; but when the joint is inflamed (resulting from a blow or an infection), copious amounts may be produced, leading to swelling and limitation of joint movement.

Chapter Summary

1. Joints, or articulations, are sites where bones meet. Their functions are to hold bones together and to allow various degrees of skeletal movement.

CLASSIFICATION OF JOINTS (p. 216)

1. Joints are classified structurally as fibrous, cartilaginous, or synovial. They are classed functionally as synarthrotic, amphiarthrotic, or diarthrotic.

FIBROUS JOINTS (pp. 216–217)

1. Fibrous joints occur where bones are connected by fibrous tissue; no joint cavity is present. Nearly all fibrous joints are synarthrotic.

2. **Sutures/Syndesmoses/Gomphoses.** The major types of fibrous joints are sutures, syndesmoses, and gomphoses.

CARTILAGINOUS JOINTS (pp. 217–218)

1. In cartilaginous joints, the bones are united by cartilage; no joint cavity is present.

2. **Synchondroses/Symphyses.** Cartilaginous joints include synchondroses and symphyses. Most synchondroses and all symphyses are amphiarthrotic.

SYNOVIAL JOINTS (pp. 218–232)

1. Most body joints are synovial joints, all of which are diarthrotic.

General Structure (pp. 218–219)

2. All synovial joints have a joint cavity enclosed by a fibrous capsule lined with synovial membrane and reinforced by ligaments, articulating bone ends covered with articular cartilage, and synovial fluid in the joint cavity. Some (e.g., the knee) contain fibrocartilage discs that absorb shock.

Bursae and Tendon Sheaths (pp. 219–221)

3. Bursae are fibrous sacs lined with synovial membrane and containing synovial fluid. Tendon sheaths are similar to bursae but are cylindrical structures that surround muscle tendons. Both allow adjacent structures to move smoothly over one another.

Factors Influencing the Stability of Synovial Joints (pp. 221–222)

4. Articular surfaces providing the most stability have large surfaces and deep sockets and fit snugly together.

5. Ligaments prevent undesirable movements and help to direct joint movement.

6. The tone of muscles whose tendons cross the joint is an important stabilizing factor in many joints.

Movements Allowed by Synovial Joints (pp. 222–225)

7. When a skeletal muscle contracts, the insertion (movable attachment) moves toward the origin (immovable attachment). Three common types of movements can occur when muscles contract across joints: (a) gliding movements, (b) angular movements (which include flexion, extension, abduction, adduction, and circumduction), and (c) rotation.

8. Special movements include supination and pronation, inversion and eversion, protraction and retraction, and elevation and depression.

Types of Synovial Joints (pp. 225–227)

9. Synovial joints differ in their range of motion. Motion may be nonaxial (gliding), uniaxial (in one plane), biaxial (in two planes), or multiaxial (in three planes).

10. The six major categories of synovial joints are plane joints (movement nonaxial), hinge joints (movement uniaxial), pivot joints (movement uniaxial, rotation permitted), condyloid joints (movement biaxial with angular movements in two planes), saddle joints (biaxial, like condyloid joints, but with freer movement), and ball-and-socket joints (multiaxial and rotational movement).

Structure of Selected Synovial Joints (pp. 227–232)

11. The shoulder joint is a ball-and-socket joint formed by the glenoid cavity of the scapula and the humeral head. The most freely movable joint of the body, it allows all angular and rotational movements. Its articular surfaces are shallow. Its capsule is lax and reinforced by ligaments anteriorly and superiorly only. The tendons of the biceps brachii and rotator cuff muscles help to stabilize it.

12. The elbow is a hinge joint formed by the articulation of the ulna and humerus and allowing flexion and extension. Its articular surfaces are highly complementary and are the most important factor contributing to joint stability.

13. The hip joint is a ball-and-socket joint formed by the acetabulum of the coxal bone and the femoral head. It is highly adapted for weight bearing. Its articular surfaces are deep and secure. Its capsule is heavy and strongly reinforced by ligaments.

14. The knee joint is the largest joint in the body. It is a hinge joint formed by the articulation of the tibial and femoral condyles (and anteriorly by the patella and patellar surface of the femur). Extension, flexion, and (some) rotation are allowed. Its articular surfaces are shallow and condyloidlike. C-shaped menisci deepen the articular surfaces. The joint cavity is enclosed by a capsule only on the sides and posterior aspects. Several extracapsular ligaments and the intracapsular anterior and posterior cruciate ligaments help prevent displacement of the joint surfaces. Muscle tone of the quadriceps and semimembranosus muscles is important in knee stability.

HOMEOSTATIC IMBALANCES OF JOINTS (pp. 232–235)

Common Joint Injuries (p. 233)

1. Sprains involve stretching or tearing of joint ligaments. Since ligaments are poorly vascularized, healing is slow.

2. Cartilage injuries are common in contact sports and may result from excessive twisting or high pressure. Because it is avascular, cartilage is unable to repair itself.

3. Dislocations involve displacement of the articular surfaces of bones. They must be reduced.

Inflammatory and Degenerative Conditions (pp. 233–235)

4. Bursitis and tendonitis are inflammations of a bursa and a tendon sheath respectively.

5. Arthritis is joint inflammation or degeneration accompanied by stiffness, pain, and swelling. Acute forms generally result from bacterial infection. Chronic forms include osteoarthritis, rheumatoid arthritis, and gouty arthritis.

6. Osteoarthritis is a degenerative condition most common in the aged. Weight-bearing joints are most commonly affected.

7. Rheumatoid arthritis, the most crippling arthritis, is an autoimmune disease involving severe inflammation of the joints.

8. Gouty arthritis, or gout, is joint inflammation caused by the deposit of urate salts in soft joint tissues.

DEVELOPMENTAL ASPECTS OF THE JOINTS (p. 235)

1. Joints form from mesenchyme and in tandem with bone development in the embryo.

2. Excluding traumatic injury, joints usually function well until late middle age at which time symptoms of osteoarthritis begin to appear in most people.

Review Questions

Multiple Choice/Matching

1. Match the terms given in the key to the appropriate descriptions. (More than one choice applies in some cases.)

Key: (a) fibrous joints (b) cartilaginous joints (c) synovial joints

_____ (1) exhibit a joint cavity
_____ (2) types are sutures and syndesmoses
_____ (3) bones connected by collagen fibers
_____ (4) types include synchondroses and symphyses
_____ (5) all are diarthrotic
_____ (6) nearly all are amphiarthrotic
_____ (7) bones connected by a disk of hyaline cartilage or fibro-cartilage
_____ (8) nearly all are synathrotic

2. Freely movable joints are (a) synarthroses, (b) diarthroses, (c) amphiarthroses.

3. Anatomical characteristics of a synovial joint include (a) articular cartilage, (b) a joint cavity, (c) an articular capsule, (d) all of these.

4. Factors that influence the stability of a synovial joint include (a) shape of articular surfaces, (b) presence of strong reinforcing ligaments, (c) presence of muscle tendons, (d) all of these.

5. The following description—"Articular surfaces deep and secure; capsule heavily reinforced by ligaments and muscle tendons; intracapsular ligamentum teres; extremely stable joint"—best describes (a) the elbow joint, (b) the hip joint, (c) the knee joint, (d) the shoulder joint.

6. Ankylosis means (a) twisting of the ankle, (b) tearing of ligaments, (c) displacement of a bone, (d) immobility of a joint due to fusion of its articular surfaces.

7. An autoimmune disorder in which joints are affected bilaterally and which involves pannus formation and gradual joint immobilization is (a) bursitis, (b) gout, (c) osteoarthritis, (d) rheumatoid arthritis.

Short Answer Essay Questions

8. Define joint.

9. Discuss the relative value (to body homeostasis) of immovable, slightly movable, and freely movable joints.

10. Compare the structure, function, and common body locations of bursae and tendon sheaths.

11. Joint movements may be nonaxial, uniaxial, biaxial, or multiaxial. Define what each of these terms means.

12. Compare and contrast the paired movements of flexion and extension with adduction and abduction.

13. How does rotation differ from circumduction?

14. Name two types of uniaxial, biaxial, and multiaxial joints.

15. What is the specific role of the menisci of the knee? of the anterior and posterior cruciate ligaments?

16. The knee has been called "a beauty and a beast." Provide several reasons that might explain the negative beast part of this description.

17. Why are sprains and cartilage injuries particularly problematic?

Clinical Applications

18. Harry was jogging down the road when he tripped in a pothole. As he fell, his left ankle violently twisted to the side. When he picked himself up, he found he was unable to put any weight on that ankle. The diagnosis was severe dislocation and sprains of the left ankle. The orthopedic surgeon stated that he would perform a closed reduction of the dislocation and attempt ligament repair by using arthroscopy. (a) Is the ankle joint normally a stable joint? (b) What does its stability depend on? (c) What is a closed reduction? (d) Why is ligament repair necessary? (e) What does arthroscopy entail? (f) How will the use of this procedure minimize Harry's recuperation time (and suffering)?

19. A 45-year-old woman appeared at her physician's office complaining of unbearable pain in the distal interphalangeal joint of her right great toe. The joint was very reddened, swollen, and warm to the touch. When asked if she had ever had such an episode in the past, she claimed to have had a similar attack two years earlier that had disappeared as suddenly as it had come. Her diagnosis was arthritis. (a) What type? (b) What is the precipitating cause of joint inflammation in this particular type of arthritis?

CHAPTER 9

Muscles and Muscle Tissue

Chapter Outline and Student Objectives

Overview of Muscle Tissues (pp. 240—241)

1. Compare and contrast the basic types of muscle tissue.
2. List three important functions of muscle tissue.

Skeletal Muscle (pp. 241—265)

3. Describe the gross structure of a skeletal muscle with respect to location, names of its connective tissue coverings, and patterns of fascicle arrangement.
4. Describe the microscopic structure and functional roles of the myofibrils, sarcoplasmic reticulum, and T tubules of skeletal muscle cells.
5. Explain how muscle cells are stimulated to contract, and explain the sliding filament mechanism of skeletal muscle contraction.
6. Define muscle twitch and describe the events occurring during its three phases.
7. Explain how the smooth, graded contractions of a skeletal muscle are produced.
8. Differentiate between isometric and isotonic contractions.
9. Describe three ways in which ATP is regenerated during skeletal muscle contraction.
10. Define oxygen debt and muscle fatigue. List possible causes of muscle fatigue.
11. List and describe factors that influence the force, velocity, and duration of skeletal muscle contraction.
12. Name and describe three types of skeletal muscle fibers. Explain the relative value of each fiber type in the body.
13. Compare and contrast the effects of aerobic and resistance exercise on the skeletal muscles and other body systems.

Smooth Muscle (pp. 265—268)

14. Compare the gross and microscopic anatomy of smooth muscle cells to that of skeletal muscle fibers.
15. Compare and contrast the contractile mechanisms and the means of activation of skeletal and smooth muscles in the body.
16. Distinguish between single-unit and multiunit smooth muscle structurally and functionally.

Developmental Aspects of Muscles (pp. 268—269)

17. Describe briefly the embryonic development of muscle tissues and the changes that occur in skeletal muscles with age.

Preview of Selected Key Terms

Muscle fiber A single muscle cell.

Fascicle (fa′-sih-kul) (*fasci* = bundle) A bundle of muscle cells bound together by connective tissue.

Myofibril (*myo* = muscle) A rodlike bundle of contractile filaments found in muscle cells; composed of individual sarcomeres.

Sarcomere (sar′-kō-mēr) (*sarco* = flesh; *mer* = part) The smallest contractile unit of muscle; contains *myofilaments* composed mainly of contractile proteins, actin and myosin.

T tubules Extensions of the muscle cell plasma membrane (sarcolemma) that protrude deeply into the muscle cell.

Sarcoplasmic reticulum (sar″-kō-plaz′-mik rih-tik′-yoo-lum) The specialized endoplasmic reticulum of muscle cells.

Motor unit A motor neuron and all the muscle cells it stimulates.

Latent period (*laten* = hidden) Period of time between stimulation and the onset of muscle contraction.

Tetanus (tet′-nus) (*tetan* = rigid, tense) A smooth, sustained muscle contraction resulting from high-frequency stimulation.

Isotonic contraction (*iso* = same; *ton* = tone, tension) A contraction in which muscle tension remains constant and the muscle shortens.

Isometric contraction (*metr* = measure) A contraction in which the muscle does not shorten (the load is too heavy) but its internal tension increases.

For most of us, the word *muscle* evokes an image of a professional boxer with bulging muscle masses, but muscle is also the dominant tissue in the heart and the bladder wall tissue that propels urine out of the body. In all of its forms, muscle is one of the most remarkable of all body tissues, and it makes up nearly half the body's mass. Its most distinguishing functional characteristic is its ability to transform chemical energy (in the form of ATP) into directed mechanical energy. In so doing, muscle becomes capable of exerting force.

Muscles can be viewed as the "machines" of the body. Mobility of the body as a whole reflects the activity of skeletal muscles. These are distinct from the muscles of internal organs, most of which force fluid and other substances through internal body channels.

Overview of Muscle Tissues

Muscle Types

The three types of muscle tissue are skeletal, cardiac, and smooth. In all three types, the individual muscle cells are also referred to as **muscle fibers.** Muscle tissues differ in the structure of their cells, their body location, their function, and the means by which they are activated to contract.

Skeletal muscle tissue is packaged into *skeletal muscles,* discrete organs that are attached to and cover the bony skeleton. Skeletal muscle fibers are typically the longest of the muscle cell types, have obvious bands called *striations,* and can be controlled voluntarily. Although it is often activated by reflexes without our willed "command," skeletal muscle is called *voluntary muscle* because it is the only type subject to conscious control. When you think of skeletal muscle tissue, the key words to keep in mind are *skeletal, striated,* and *voluntary.* Skeletal muscle can contract rapidly and vigorously, but the fibers tire easily and must rest after short periods of activity. These muscles can also exert tremendous power, a fact revealed by reports of people lifting cars to save their loved ones. Skeletal muscle is also remarkably adaptable. For example, your hand muscles can exert a force of a fraction of an ounce to pick up a dropped paper clip or a force of about 70 pounds to pick up this book, and the same muscles can exert a hand squeeze of more than 100 pounds!

Cardiac muscle tissue is found only in the heart, where it constitutes most of the mass of the heart walls. Cardiac muscle, like skeletal muscle, is striated, but it is not voluntary: Most of us have no conscious control over the beating of our heart. Key words to remember for this type of muscle are *cardiac, striated,* and *involuntary.* Cardiac muscle usually contracts at a fairly steady rate set by the heart's pacemaker, but neural controls allow the heart to "shift into high gear" for brief periods, as when you race across the tennis court to make that overhead smash, for instance.

Smooth muscle tissue is found in the walls of visceral organs, such as the stomach, urinary bladder, and bronchi of the lungs. It has no striations, and, like cardiac muscle, it is not subject to voluntary control. We can describe smooth muscle tissue most precisely as *visceral, nonstriated,* and *involuntary.* Contractions of smooth muscle fibers are slow and sustained. If skeletal muscle is like a speedy windup car that quickly runs down, then smooth muscle is like a steady, heavy-duty engine that lumbers along tirelessly.

Skeletal and smooth muscle will be discussed in this chapter; cardiac muscle is discussed in Chapter 19 (The Heart). Table 9.3 on pp. 270–271 summarizes the most important characteristics of each type of muscle tissue.

Functions

Muscle performs three important functions for the body: It provides for movement, maintains posture, and generates heat.

Movement

With the exception of the amoeboid movement of certain white blood cells, the movement of cilia, and the flagellar movement of sperm cells, all movements of the human body and its parts are a result of muscle contraction. Skeletal muscles are responsible for all locomotion and manipulation, and they enable you to respond quickly to changes in the external environment. For example, their speed and power enable you to jump out of the way of a runaway car. Vision relies partly on skeletal muscles that direct your eyeballs, and your ability to express joy or fury often reflects the silent language of contracting facial muscles.

Coursing of blood through your body reflects the work of the rhythmically beating cardiac muscle of your heart and the smooth muscle in the walls of your blood vessels, which helps maintain blood pressure. Smooth muscle in the organs of the digestive, urinary, and reproductive tracts propels, or squeezes, substances (foodstuffs, urine, a baby) through organs and along the tract.

Posture Maintenance

We are rarely aware of the workings of the skeletal muscles that maintain body posture. Yet, these muscles function almost continuously, making one tiny adjustment after another that enables us to maintain an erect or seated posture despite the never-ending downward pull of gravity.

Heat Generation

The third function of muscle, the production of body heat, is a by-product of muscle metabolism and contractile activity. As the energy of ATP is liberated to power muscle contraction, nearly three-quarters of that energy escapes as heat, which is vitally important in maintaining normal body temperature. Since skeletal muscle accounts for at least 40% of body mass, it is the muscle type most responsible for heat generation.

Functional Characteristics of Muscle

Muscle tissue is endowed with some special functional properties that enable it to perform its duties. These properties include excitability, contractility, extensibility, and elasticity.

Excitability is the ability to receive and respond to a stimulus. (A stimulus is an environmental change; it may arise within the body or externally.) In the case of muscle, the stimulus is usually a chemical—for example, a neurotransmitter released by a nerve cell, a hormone, or a local change in pH—and the response is the generation and transmission of an electrical current called an action potential along the muscle cell plasma membrane.

Contractility is the ability to shorten (forcibly) when an adequate stimulus is received. It is this property that sets muscle apart from all other tissue types.

Extensibility is the ability to be stretched or extended. Muscle fibers shorten when contracting and lengthen (are extended) when relaxed.

Elasticity is the ability of a muscle fiber to resume its resting length after it has contracted or has been stretched.

In the following section, we will examine the structure and functioning of skeletal muscle in detail; then we will consider smooth muscle more briefly, largely by comparing it with skeletal muscle. Because cardiac muscle is described in Chapter 19, its further treatment in this chapter is limited to the summary of its characteristics provided in Table 9.3 on pp. 270–271.

Skeletal Muscle

The levels of skeletal muscle organization, gross to microscopic, are summarized in Table 9.1.

Gross Anatomy of a Skeletal Muscle

Depending on its size, a skeletal muscle may be made up of hundreds to thousands of muscle fibers. Each skeletal muscle is a discrete organ; although muscle fibers predominate, substantial amounts of connective tissue, blood vessels, and nerve fibers are also present. A skeletal muscle's shape, the arrangement of its fibers, and the means of its attachment within the body can be examined easily without the help of a microscope.

Connective Tissue Wrappings

Within an intact muscle, the individual muscle fibers are wrapped by several different layers of connective tissue (Figure 9.1). Each muscle fiber is surrounded by a fine reticular connective tissue sheath called the **endomysium** (en″-dō-mī′-sē-um). Several ensheathed muscle fibers are then gathered together side by side into bundles of muscle cells called **fascicles** (fa′-sih-kulz) (*fasci* = bundle, band), and each fascicle is bound by a collagenic sheath referred to as a **perimysium** (payr″-ih-mī′-sē-um). The fascicles are in turn bound together by an even coarser "overcoat" of dense fibrous connective tissue called the **epimysium** (eh″-pih-mī′-sē-um), or **deep fascia** (fa′-shē-uh). The epimysium surrounds the entire muscle. All of these connective tissue sheaths are continuous; extensions of the epimysium form the perimysia, and delicate invaginations of the perimysia form the endomysia.

Like all body cells, skeletal muscle fibers are soft and fragile. The connective tissue coverings support each cell and reinforce the muscle as a whole so that force can be exerted during contraction. These sheaths also provide access and exit routes for the blood vessels and nerve fibers that serve the muscle.

Nerve and Blood Supply

The normal activity of a skeletal muscle is absolutely dependent on its nerve supply and its rich blood supply. Unlike the other muscle tissues, which can contract in the total absence of nerve stimulation, each skeletal muscle fiber is supplied with a nerve ending that controls its activity.

Contracting muscle cells use tremendous amounts of energy, which requires a more or less continuous delivery of oxygen, glucose, and other nutrients by way of arteries. In addition, muscle cells give off large amounts of metabolic wastes that must be removed through veins if contraction is to remain efficient. In general, each muscle is served by an artery and one or more veins.

Table 9.1 Structure and Organizational Levels of Skeletal Muscle

Structure and organizational level	Description	Connective tissue wrappings
Muscle (organ) 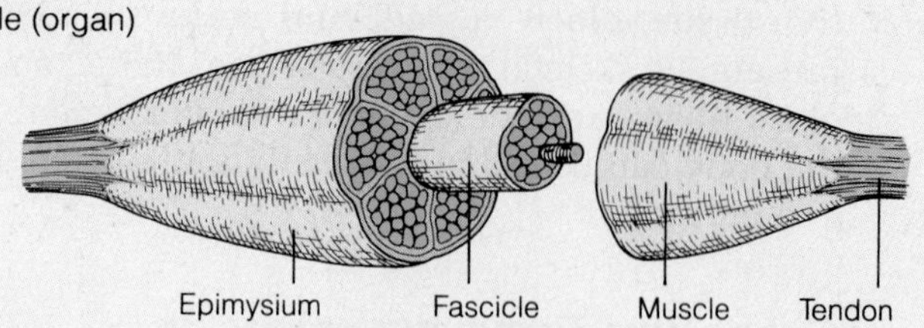 Epimysium · Fascicle · Muscle · Tendon	Consists of hundreds to thousands of muscle cells, plus connective tissue wrappings, blood vessels, and nerve fibers	Covered externally by the epimysium
Fascicle (a portion of the muscle) 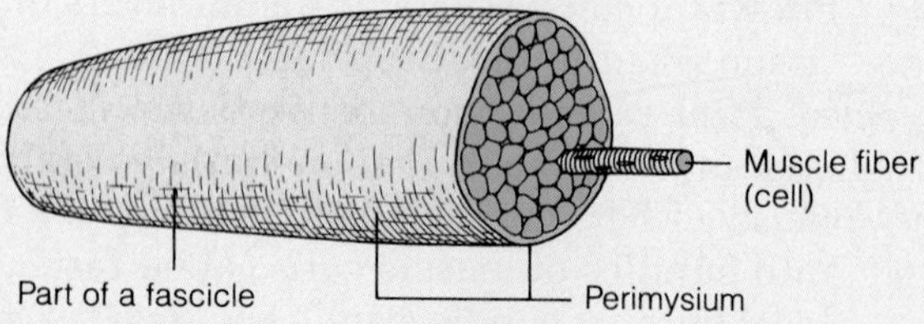 Muscle fiber (cell) · Part of a fascicle · Perimysium	Discrete bundle of muscle cells, segregated from the rest of the muscle by a connective tissue sheath	Surrounded by a perimysium
Muscle fiber (cell) 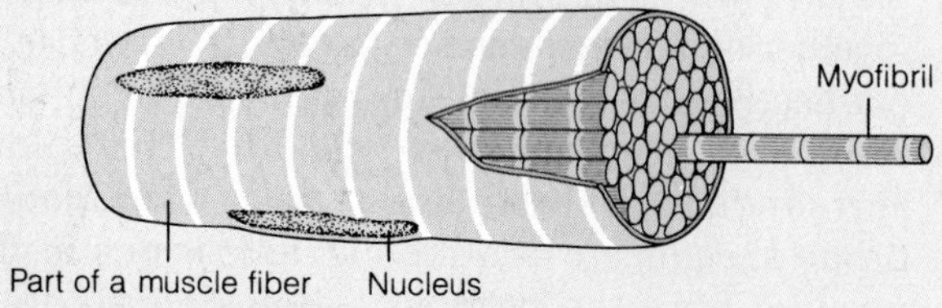 Myofibril · Part of a muscle fiber · Nucleus	Elongated multinucleate cell; has a banded (striated) appearance	Surrounded by the endomysium
Myofibril or fibril (complex organelle composed of bundles of myofilaments) 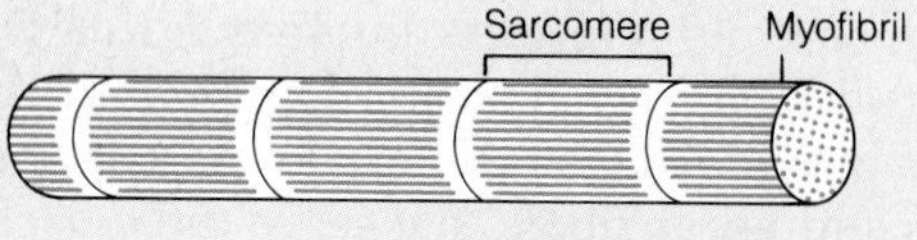 Sarcomere · Myofibril	Rodlike contractile element; myofibrils occupy most of the muscle cell volume; appear banded, and bands of adjacent myofibrils are aligned; composed of sarcomeres, end to end	
Sarcomere (a segment of a myofibril) 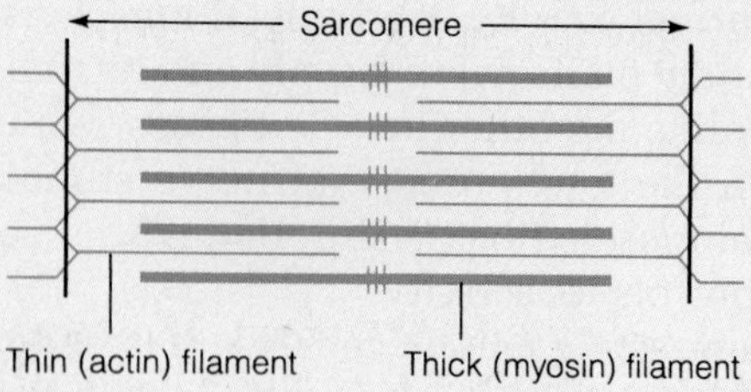 Sarcomere · Thin (actin) filament · Thick (myosin) filament	The contractile unit, composed of myofilaments made up of contractile proteins	
Myofilament or filament (extended macromolecular structure) 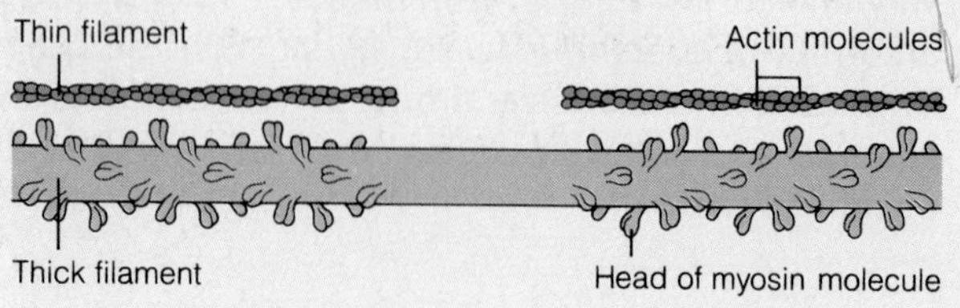 Thin filament · Actin molecules · Thick filament · Head of myosin molecule	Myofilaments are of two types—thick and thin—composed of contractile proteins; the thick filaments contain bundled myosin molecules; the thin filaments contain actin molecules (plus other proteins); the sliding of the thin filaments past the thick filaments produces muscle shortening	

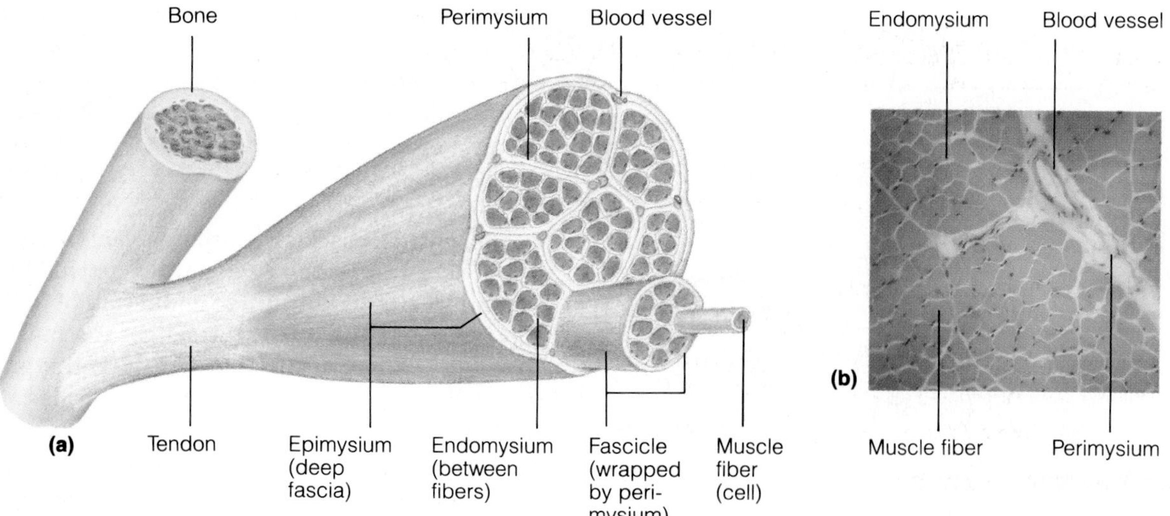

Figure 9.1 Connective tissue wrappings of skeletal muscle. (**a**) In a skeletal muscle, each muscle fiber is wrapped with a delicate connective tissue sheath, the endomysium. Bundles, or fascicles, of muscle fibers are bound by a collagenic sheath called a perimysium. The entire muscle is strengthened and wrapped by a coarse epimysium, or deep fascia sheath. (**b**) Photomicrograph of a cross section of a skeletal muscle (64 ×).

The blood vessels and nerve fibers supplying each muscle fiber enter the muscle and branch profusely through its connective tissue septa. The capillaries (Figure 9.2) and final nerve endings invade the delicate endomysium surrounding each muscle fiber.

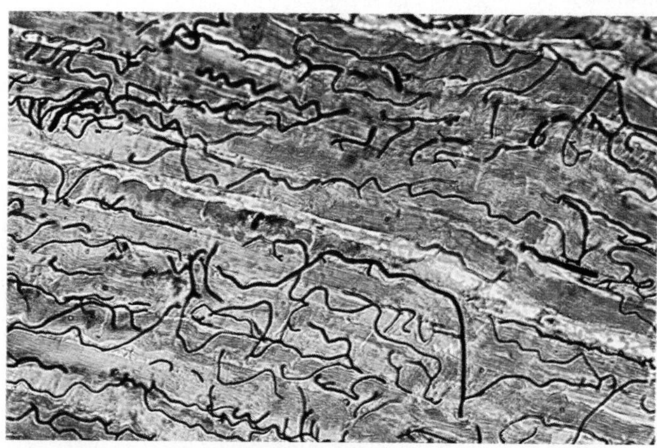

Figure 9.2 Photomicrograph of the capillary network of a portion of a skeletal muscle. The arterial supply has been injected with purple gelatin to demonstrate the capillary bed. Notice the numerous cross connections between the capillaries. The muscle fibers are stained blue (150 ×).

Attachments

Muscles are attached to bones (or other structures) either directly or indirectly. In *direct attachments*, the epimysium of the muscle is fused to the periosteum of a bone or perichondrium of a cartilage. In *indirect attachments*, the muscle fascia extends beyond the muscle as a ropelike tendon or a flat, broad **aponeurosis** (a″-pō-ner-rō′-sis). The tendon or aponeurosis anchors the muscle to the connective tissue covering of a skeletal element (bone or cartilage) or to the fascia of other muscles.

Of the two, indirect attachments are much more common in the body. This no doubt reflects the size and durability of such attachments. Tendons, composed almost entirely of tough collagen fibers, can withstand friction at points of contact with bone much better than can soft muscle tissue.

Arrangement of Fascicles

Although all skeletal muscles consist of fascicles, fascicle arrangements vary. This results in muscles with different shapes and functional capabilities. The most common patterns of fascicle arrangement are parallel, pennate, convergent, and circular (Figure 9.3).

In a *parallel* arrangement, the long axes of the fascicles run with the longitudinal axis of the muscle. Such muscles are either *straplike*, such as the sartorius muscle of the thigh, or *fusiform* (fyoo′-zih-form), with an expanded belly, like the biceps brachii muscle of the arm.

In a *pennate* (peh′-nāt) pattern, the muscle fibers are short and the fascicles are attached obliquely

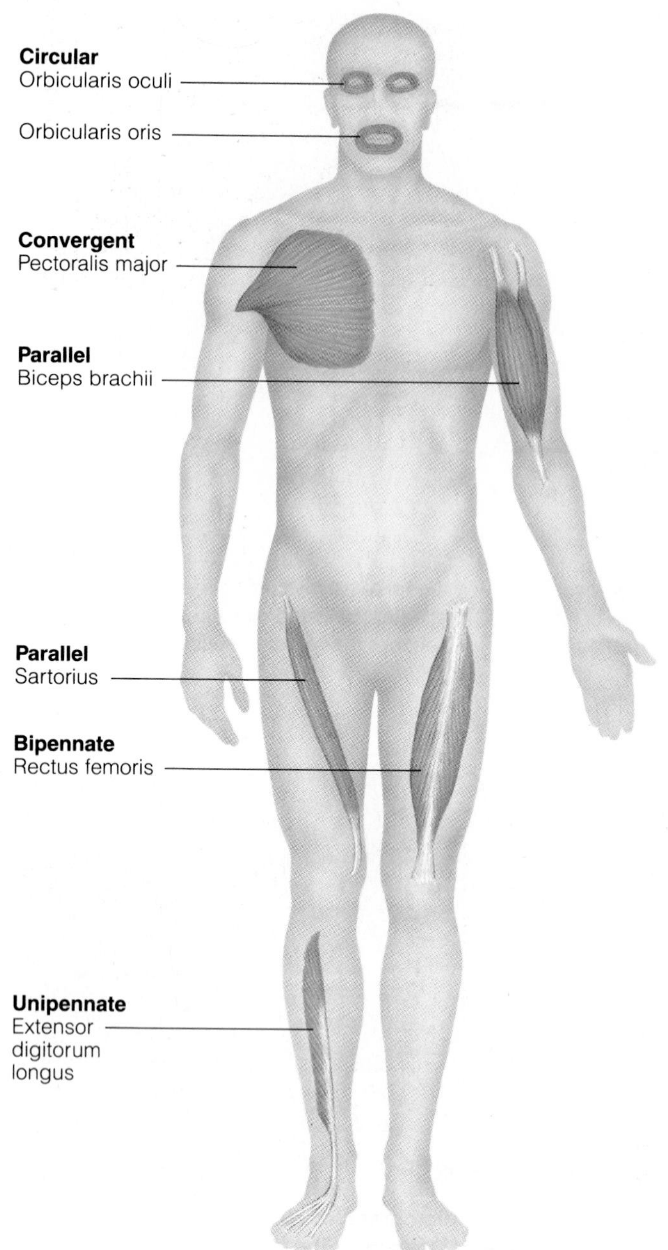

Figure 9.3 Relationship of fascicle arrangement to muscle structure.

(*penna* = feather, wing) to a central tendon running the length of the muscle. If, as seen in the extensor digitorum muscle of the leg, the fascicles insert into only one side of the tendon, the muscle is *unipennate*. If the fascicles insert into the tendon from opposite sides, so that the muscle's "grain" resembles a feather, the arrangement is *bipennate*. The rectus femoris of the thigh is bipennate.

A *convergent* muscle has a broad origin, and its fascicles converge toward a single tendon. Such a muscle is essentially triangular. The pectoralis major muscle of the anterior thorax has a convergent pattern.

The fascicular pattern is *circular* if the fascicles are arranged to form a circle of muscle fibers. Muscles with this arrangement are always found surrounding an external body opening, where they control its opening or closing. Examples are the orbicularis oris muscle surrounding the mouth and the orbicularis oculi muscles surrounding the eyes.

The pattern of fascicle arrangement influences both the muscle's range of motion and its power. Since skeletal muscle fibers shorten to about half of their resting length when contracted, the longer the muscle fibers in the direction of the muscle's axis, the greater the muscle's range of motion will be. Muscles with parallel fascicle arrangement provide the greatest degree of shortening, but they are not usually very powerful. Muscle power is more dependent on the number of muscle cells in the muscle; the greater the number, the greater the power. The stocky bipennate muscles shorten very little, but they tend to be very powerful.

Microscopic Anatomy of Skeletal Muscle

Each skeletal muscle fiber is a cylindrical cell with multiple oval nuclei arranged just beneath its membrane surface (see Figure 9.4a and Table 9.1). Skeletal muscle cells are large compared to most other human cells, ranging in length from 1 to 40 μm; however, some may be as long as 30 cm (1 foot). Their diameter typically ranges from 10 to 100 μm.

The **sarcolemma** (sar-kō-leh′-muh) of the muscle fiber is its plasma membrane. (Note that whenever you see the prefixes *myo, mys,* or *sarco,* the reference is to muscle). The **sarcoplasm** of a muscle fiber is similar to the cytoplasm of other cells, but it contains unusually large amounts of stored glycogen and a unique oxygen-binding protein called **myoglobin** that is not found in other cell types. Myoglobin, a red pigment that stores oxygen within the muscle cells, is similar to hemoglobin, the pigment that transports oxygen in the blood. The usual organelles are present, as well as some that are highly modified in muscle cells. These are the myofibrils and sarcoplasmic reticulum. The T tubules are unique modifications of a skeletal muscle cell's sarcolemma.

Myofibrils

When viewed at high magnification, each muscle cell is seen to contain a large number of rodlike structures called **myofibrils** that run in parallel fashion and extend the entire length of the cell (see Figure 9.4b). The myofibrils are so densely packed that mitochondria and other organelles appear to be squeezed between them. There are hundreds to thousands of myofibrils in a single cell, depending on its size, and they account

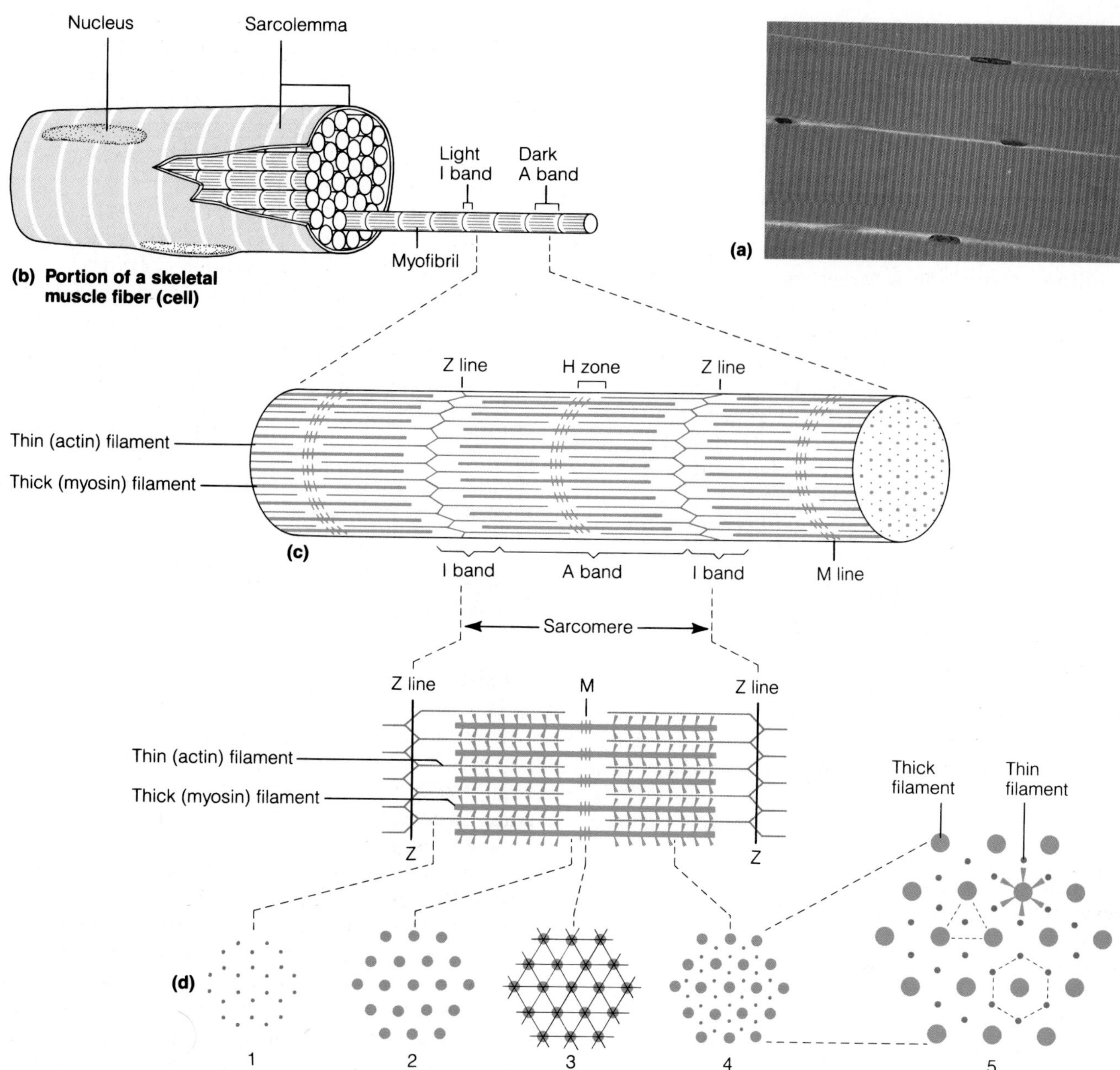

Figure 9.4 Microscopic anatomy of a muscle fiber. (**a**) Photomicrograph of portions of two isolated skeletal muscle cells (250 ×). Notice the obvious cross striations (alternating light and dark bands). (**b**) One myofibril is shown extending from the cut end of a muscle fiber. Both the cell (fiber) and the myofibril have a banded appearance. (**c**) A small portion of one myofibril is enlarged to show the myofilaments responsible for the banding pattern. Each sarcomere, or contractile unit, extends from one Z line to the next. (**d**) The latticelike arrangement of the actin and myosin filaments becomes obvious when the myofibril is cross-sectioned. (1) In a cut made through the I band, only actin filaments are seen; (2) in a cut made through the H zone, only myosin filaments are seen; (3) in a cut made through the M line, myosin filaments and connecting fibers are obvious; (4) if a cut is made through a portion of the A band containing both actin and myosin filaments and this region is enlarged, we see (5) the hexagonal arrangement of actin filaments around each myosin filament and the triangular arrangement of myosin filaments around each actin filament.

for about 80% of cellular volume. The myofibrils are the contractile elements of skeletal muscle cells, and each myofibril consists of a chain of even smaller contractile units called sarcomeres.

Striations, Sarcomeres, and Myofilaments. Along the length of each myofibril, a repeating series of alternating dark and light bands is evident. The dark bands are called **A bands** because they are *anisotropic* (an-ī″-suh-trō′-pik); that is, they can polarize visible light. The light bands, called **I bands,** are *isotropic* (ī″-suh-trō′-pik), or nonpolarizing. In an intact muscle cell, the myofibril bands are nearly perfectly aligned with one another and are continuous across the width of the cell. This causes the cell as a whole to appear striated.

As illustrated in Figure 9.4c, each A band is interrupted in its midsection by a lighter stripe called the *H zone* (*H* stands for *helle*, which means "bright"). The H zones are visible only in relaxed muscle fibers for reasons that will soon become obvious. To make things a bit more complicated, each H zone is bisected by a dark line called the *M line*. The I bands also have a midline interruption, a darker area called the **Z line.** A **sarcomere** (sar′-kō-mēr), literally, "muscle part or segment," is the region of a myofibril between two successive Z lines (see Figure 9.4c) and is the smallest contractile unit of a muscle cell. Thus, the *functional units* of skeletal muscle are actually very minute portions of the myofibrils, and each myofibril can be considered a chain of sarcomere units laid end to end.

If we examine the banding pattern of the myofibril at the molecular level, we see that it arises from an orderly arrangement of two types of protein filaments (*myofilaments*) within the sarcomeres. The central **thick filaments** extend the entire length of the A band. The more lateral **thin filaments** extend across the I band and partway into the A band. The Z line is actually a disklike protein sheet that serves as the point of attachment of the thin filaments and also connects each myofibril to the next throughout the width of the muscle cell. The H zone of the A band appears less dense when viewed microscopically because the thick filaments are not overlapped by the thinner ones in this region. The M line in the center of the H zone is slightly darker because of the presence of fine strands that connect adjacent thick filaments together in this area.

A longitudinal view of the myofilaments, such as that depicted in Figure 9.4c, is somewhat misleading because it gives the impression that each thick filament interdigitates along a plane with only four thin filaments. We can get a more accurate picture of the three-dimensional relationships of the filaments by observing cross-sectional views of a myofibril taken at different regions of a sarcomere (see Figure 9.4d). In areas where thick and thin filaments overlap, each thick filament is surrounded by a hexagonal arrangement of six thin filaments, and each thin filament is bordered by a triangular arrangement of three thick filaments.

Ultrastructure and Molecular Composition of the Myofilaments. Thick filaments are composed primarily of the protein **myosin** (Figure 9.5a). Each myosin molecule has a distinctive structure; it has a rodlike tail, or axis, terminating in two globular heads, sometimes called *cross bridges* because they interact with specific sites on the thin filaments. As will be explained shortly, it is these cross bridges that generate the tension developed by a muscle cell when it contracts.

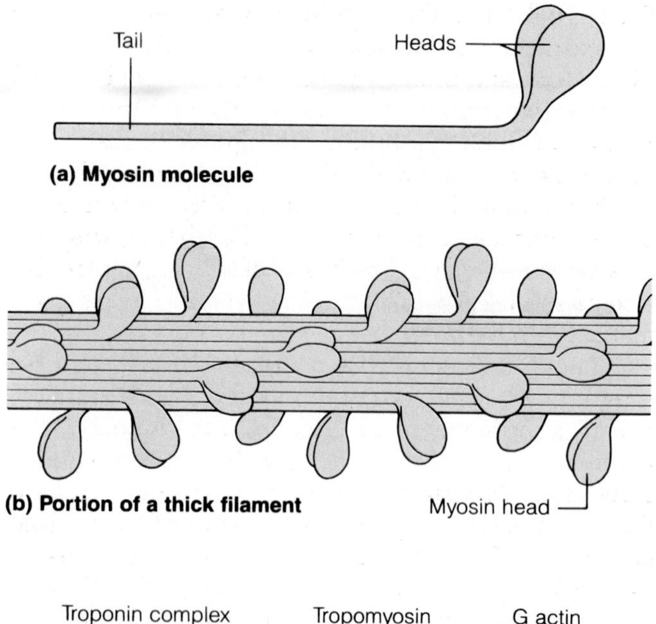

(a) Myosin molecule

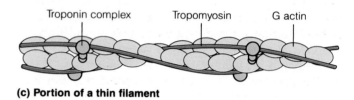

(b) Portion of a thick filament

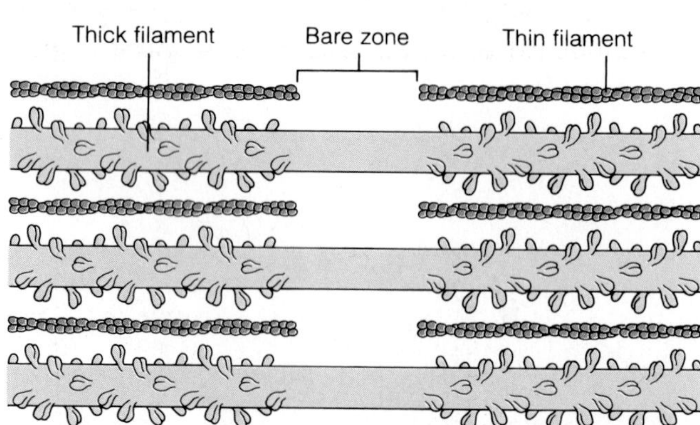

(c) Portion of a thin filament

(d) Longitudinal section of filaments within one sarcomere of a myofibril

Figure 9.5 Myofilament composition in skeletal muscle. (**a**) An individual myosin molecule has a stalklike tail, from which two "heads" protrude. (**b**) Each thick filament consists of many myosin molecules, whose heads protrude at opposite ends of the filament, as shown in (d). (**c**) A thin filament contains two strands of F actin twisted together. Each strand is made up of G actin subunits. Tropomyosin molecules coil around the F actin, helping to reinforce the filament. A troponin complex is attached to each tropomyosin molecule. (**d**) Arrangement of the filaments in a sarcomere (longitudinal view). In the center of the sarcomere, the thick filaments are devoid of myosin heads, but at points of thick and thin filament overlap, the heads extend toward the actin with which they interact during contraction.

Each thick filament within a sarcomere contains approximately 200 myosin molecules. As shown in Figure 9.5b and d, the myosin molecules are bundled together in such a way that their tails form the central portion of the filament and their globular heads face outward and in opposite directions at each end. As a result, the central portion of a thick filament is smooth, but its ends are studded with myosin heads arranged in a staggered array. The heads of myosin molecules also contain ATPases that have the ability to act enzymatically to split ATP.

The thin filaments are composed chiefly of **actin** (Figure 9.5c). The polypeptide subunits of actin are called *globular actin* or *G actin,* and these G actin monomers are polymerized into long linear strands called *fibrous actin* or *F actin*. The backbone of each thin filament is formed by two strands of F actin that coil around each other, forming a helical structure that looks like a twisted double strand of pearls. Several regulatory proteins are also present. **Tropomyosin** (trō″-puh-mī′-uh-sin), a rod-shaped protein, spirals about the F actin core and helps to stiffen it. Successive tropomyosin molecules are arranged end to end along the F actin chains. The other major protein in the thin filament, **troponin** (trō′-puh-nin), is actually a complex of three polypeptides. One of these polypeptides (TnI) binds to actin; another (TnT) binds to tropomyosin and helps to position it on actin; the third (TnC) binds calcium ions. Both troponin and tropomyosin play a role in controlling the myosin-actin interactions involved in contraction.

The Sarcoplasmic Reticulum and T Tubules

The **sarcoplasmic reticulum** (sar″-kō-plaz′-mik rih-tik′-yoo-lum) **(SR)** inside each muscle cell is an elaborate form of smooth endoplasmic reticulum. Its interconnecting tubules lie in the narrow spaces between the myofibrils, run parallel to them, and surround each of them, much as the sleeve of a loosely crocheted sweater surrounds your arm (Figure 9.6). At the level of the H zones and the A-I junctions, the tubules fuse to provide lateral channel connections. The saclike channels abutting the A-I junctions are called *terminal cisternae*.

At each Z line, the sarcolemma of the muscle cell penetrates deeply into the cell interior to form a hollow elongated tube called the **transverse tubule** or **T tubule,** the lumen of which is continuous with the extracellular space. As each T tubule protrudes deep into the cell interior, it runs between the terminal cisternae of the SR so that *triads*, successive groupings of the three membranous structures (terminal cisterna, T tubule, terminal cisterna), are formed. The thousands of T tubules of each muscle cell are collectively called the *T system*.

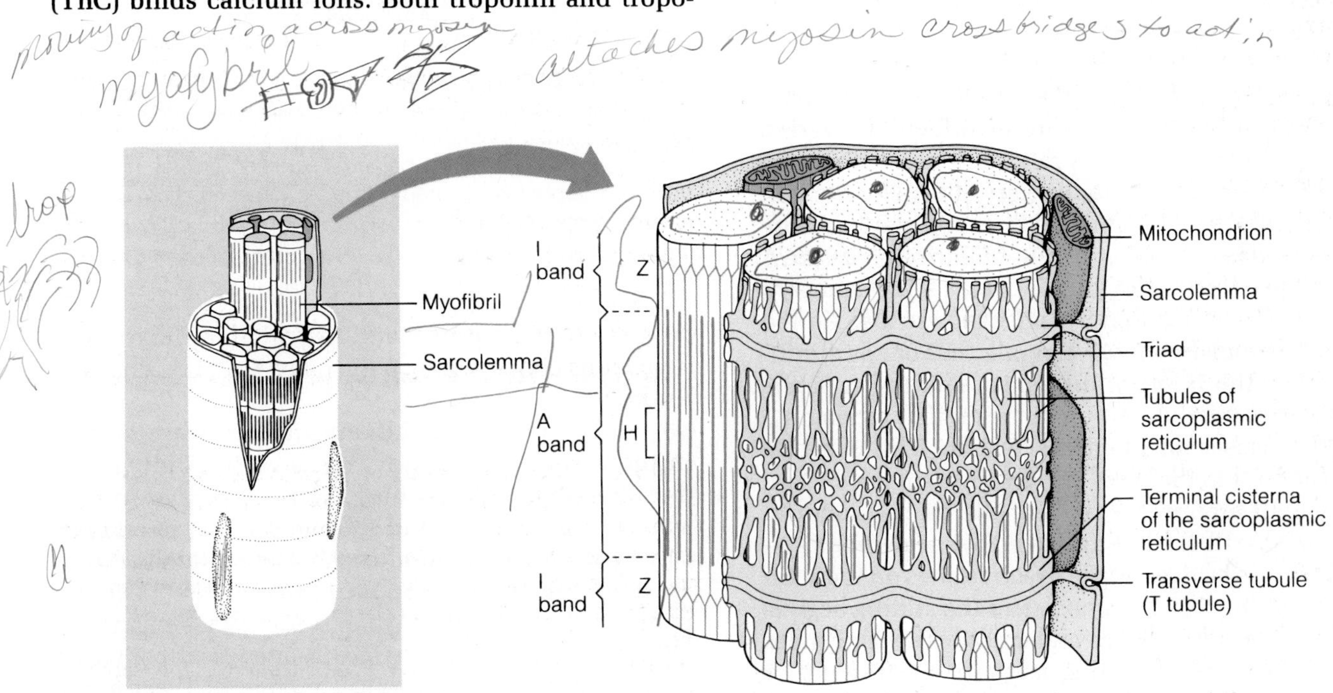

Figure 9.6 Relationship of the sarcoplasmic reticulum and T tubules to the myofibrils of skeletal muscle. The tubules of the sarcoplasmic reticulum encircle each myofibril like a "holey" sleeve. At certain points, the tubules fuse laterally to form communicating channels; this occurs primarily at the level of the H zone and in the regions abutting the A-I junctions where the saclike elements are called terminal cisternae. The T tubules, inward invaginations of the sarcolemma, run deep into the cell between the terminal cisternae at the A-I junctions. Sites of close contact of these three elements (terminal cisterna, T tubule, and terminal cisterna) are called triads.

The tubules of the T system can be thought of as having a "rapid telegraphy" role: Since they are continuations of the sarcolemma, which receives the nerve stimulus, they can conduct the stimulus deep into the cell to virtually every sarcomere. Additionally, the T tubules provide inlets through which extracellular fluid (containing glucose, oxygen, and various ions) can be brought into close contact with deeper parts of the muscle cell. The major role of the SR is to regulate intracellular levels of ionic calcium: It sequesters calcium and releases it "on demand" when the muscle fiber is stimulated to contract. The importance of this function will become apparent shortly.

Contraction of a Skeletal Muscle Fiber

Mechanism of Contraction

Contraction reflects the activity of individual sarcomeres. When a muscle cell contracts, its sarcomeres shorten and the distance between successive Z lines is reduced. As the length of their sarcomeres decreases, the myofibrils shorten as well, resulting in shortening of the cell as a whole.

Close examination of the contractile event reveals that the actin- and myosin-containing filaments do not change in length as the sarcomeres shorten. So how can muscle cell shortening be explained? According to the **sliding filament theory of contraction,** first proposed in 1954 by Hugh Huxley, the contraction mechanism involves a sliding of the thin filaments past the thick ones so that the extent of myofilament overlap increases (Figure 9.7). In a relaxed muscle fiber, the thick and thin filaments overlap only slightly; but during contraction, the thin filaments penetrate more and more deeply into the central portion of the A band. Notice that as the thin filaments slide centrally, the Z lines to which they are attached are also pulled toward the thick filaments. Overall, the I bands diminish in size, the H zones disappear, and the A bands move closer together but do not diminish in length.

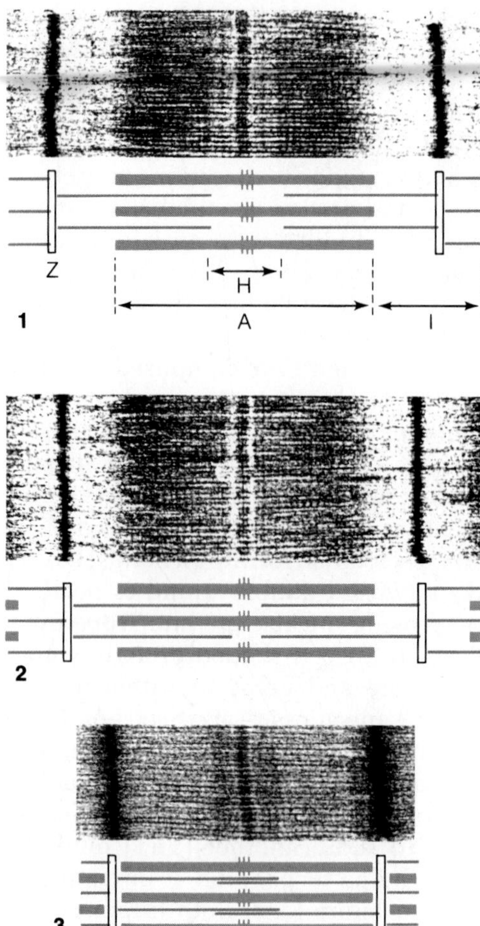

Figure 9.7 Sliding filament model of contraction. At full contraction, the Z lines abut the myosin filaments and the actin filaments overlap each other. The numbers indicate the sequence of events, with 1 being relaxed and 3 being fully contracted. The photomicrographs (top view in each case) show enlargements of 20,000 ×.

What causes the filaments to slide? This question brings us back to the cross bridges (myosin heads) that protrude all around the ends of the thick filaments. When muscle fibers are stimulated by the nervous system, the cross bridges attach to active sites on the actin subunits of the thin filaments, and the sliding begins. Each cross bridge attaches and detaches several times during a contraction, acting much like a tiny oar or ratchet to generate tension and pull the thin filaments toward the center of the sarcomere. As this event occurs simultaneously in sarcomeres throughout the cell, the muscle cell shortens. The attachment of the myosin cross bridges to actin requires calcium ions, and the nerve impulse leading to contraction causes an increase in calcium ions within the muscle cell. We will discuss the nerve impulse and the calcium fluxes later; here let us concentrate on the contraction mechanism itself.

Although many details are unclear, it appears that in the absence of calcium, the myosin-binding sites on actin are physically blocked by tropomyosin molecules (Figure 9.8a), and the muscle cell is relaxed. However, when calcium ions become available, they are avidly bound by troponin (Figure 9.8b). The troponin-calcium complex then undergoes a change in its conformation that physically moves tropomyosin into the center of the helical groove of F actin and away from the myosin-binding sites (Figure 9.8c, 9.8d). Thus, the tropomyosin "blockade" is removed when calcium is present.

Once binding sites on actin are exposed, the following events occur in rapid succession (Figure 9.9).

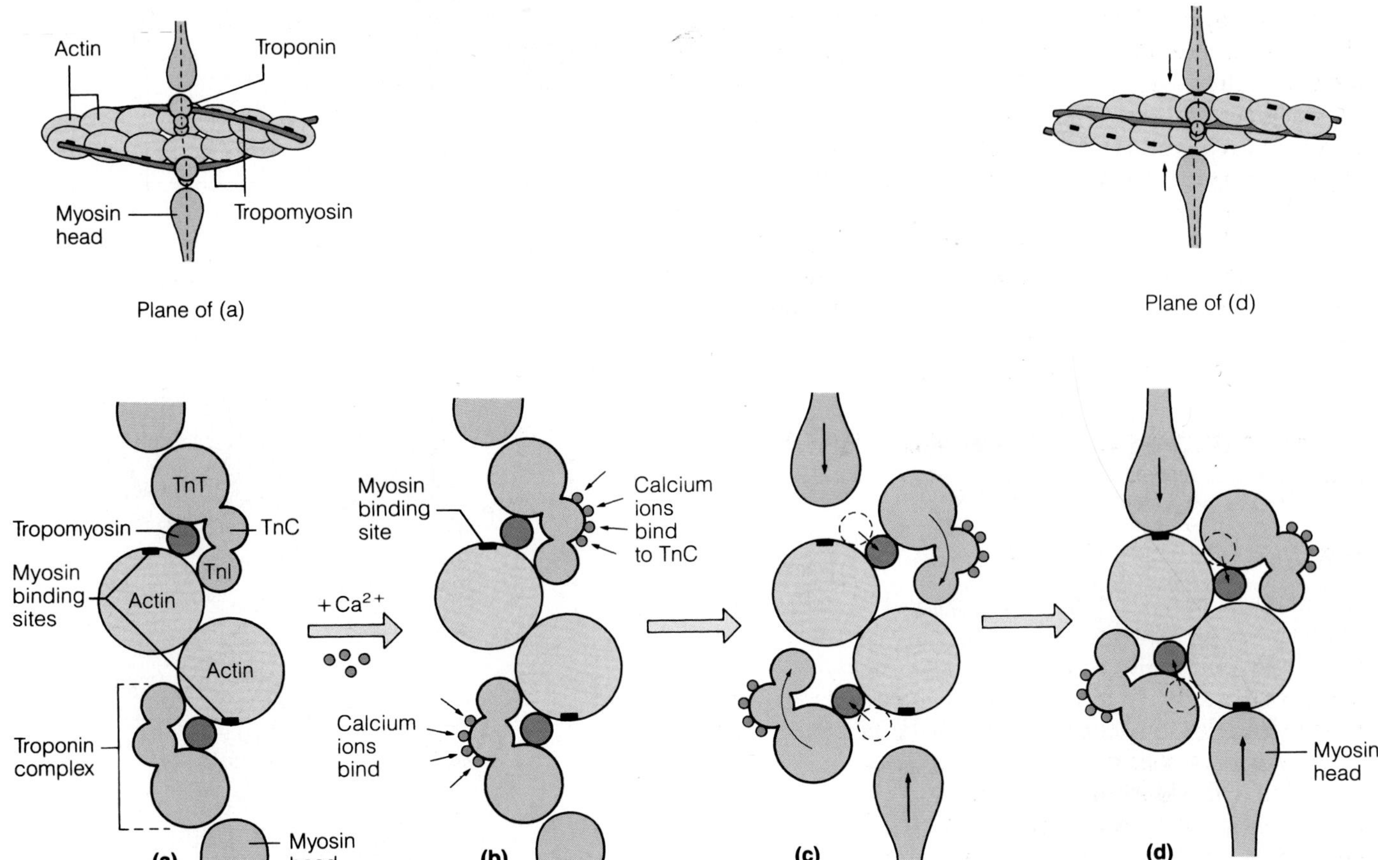

Figure 9.8 Role of ionic calcium in the contraction mechanism. Views (a)–(d) are cross-sectional views of the thin (actin) filament. (**a**) At low intracellular Ca^{2+} concentrations, tropomyosin blocks the binding sites on actin and prevents myosin cross bridge attachment, which enforces the relaxed muscle state. (**b**) At higher intracellular Ca^{2+} concentrations, calcium binds to troponin. (**c**) Calcium-bound troponin undergoes a conformational change that moves the tropomyosin away from the myosin-binding sites. (**d**) This allows the myosin heads to bind, permitting contraction (sliding of the thin filaments by the myosin cross bridges) to occur.

1. Cross bridge attachment. The activated myosin heads are strongly attracted to the exposed binding sites on actin and cross bridge binding occurs.

2. Power stroke. As a myosin head binds, it changes from its high-energy configuration to its bent, low-energy shape, which causes the head to pull on the thin filament, sliding it toward the center of the sarcomere. At the same time, ADP and inorganic phosphate (P_i) generated during the *prior* contraction cycle are released from the myosin head.

3. Cross bridge detachment. As a new ATP molecule binds to the myosin head, the myosin cross bridge is released from actin.

4. "Cocking" of the myosin head. Hydrolysis of ATP to ADP and P_i by ATPase provides the energy needed to return the myosin head to its high-energy, or "cocked," position, which gives it the potential energy needed for the next attachment–power stroke sequence. (The ADP and P_i remain attached to the myosin head during this phase.) At this point, we are back where we started, with the myosin head in its high-energy configuration, ready to attach to another actin site farther along the thin filament, and the cycle is repeated. You could say that the myosin cross bridges "walk along" the adjacent actin filaments when a muscle shortens. A single power stroke of all the cross bridges in a muscle results in a shortening of only about 1%. Since contracting muscles routinely shorten 35% to 50% of their total resting length, it is obvious that each myosin cross bridge attaches and detaches many times during the course of a single contraction. It has been estimated that only half of the myosin heads of a thick filament are actively exerting a pulling force at the same instant; the balance are randomly seeking their next binding site. Sliding of thin filaments continues as long as the calcium signal is present. Removal of

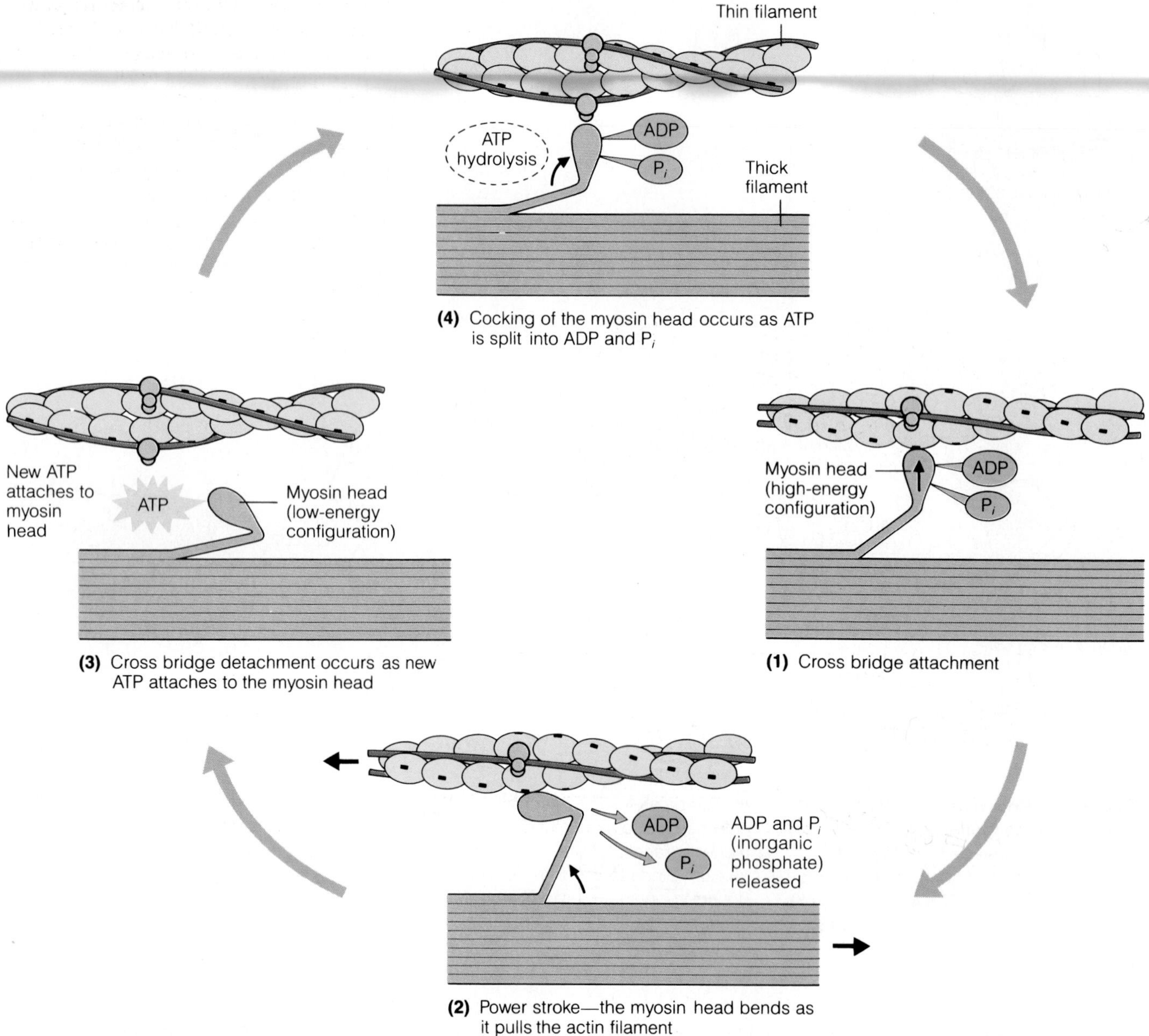

Figure 9.9 Sequence of events involved in the sliding of the actin filaments during contraction. A small segment of adjacent thick and thin filaments is used to illustrate the interactions that occur between the two types of myofilaments. These events occur only in the presence of ionic calcium (Ca^{2+}).

calcium ions from the sarcoplasm by the SR restores the tropomyosin inhibition, contraction ends, and the muscle fiber relaxes.

The phenomenon of *rigor mortis* (death rigor) illustrates the fact that cross bridge detachment is ATP-driven. Muscles begin to stiffen 3 to 4 hours after death. Peak rigidity occurs at 12 hours and then gradually dissipates over the next 48 to 60 hours. Because injured and dying cells are unable to exclude calcium ions (which are in higher concentration in the extracellular fluid), calcium influx into muscle cells promotes myosin cross bridge binding. However, in the absence of ATP synthesis, which occurs shortly after breathing stops, cross bridge detachment is impossible. Actin and myosin become irreversibly cross-linked, producing the stiffening of dead muscle. The gradual disappearance of rigor mortis reflects the breakdown of biological molecules, including actin and myosin, several hours after death. ■

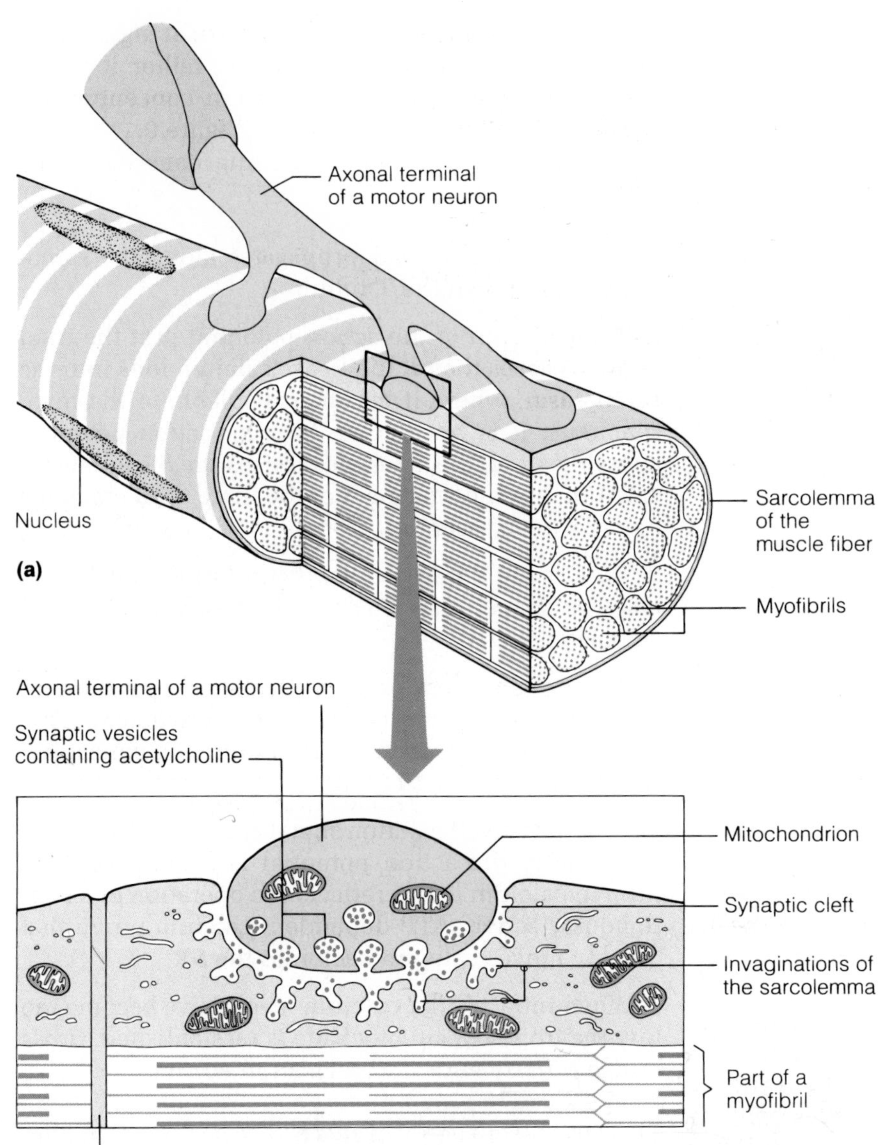

Figure 9.10 The neuromuscular junction. (**a**) Axonal terminal of a motor neuron forming a neuromuscular junction with a muscle fiber. (**b**) The axonal terminal contains vesicles filled with the neurotransmitter acetylcholine (ACh), which is released when the action potential reaches the axonal terminal. Acetylcholine diffuses across the synaptic cleft and attaches to ACh receptors on the sarcolemma, initiating depolarization of the sarcolemma. The sarcolemma is highly invaginated adjacent to the synaptic cleft, allowing it to have many acetylcholine receptors in this region.

Regulation of Contraction

For a skeletal muscle fiber to contract, a nerve stimulus must result in the propagation of an electrical current, or *action potential*, along the sarcolemma. This electrical event results in the transient rise in intracellular calcium ion levels, which is the final trigger for contraction. The series of events linking the electrical signal to contraction is called excitation-contraction coupling.

The Nerve Stimulus and the Action Potential. Skeletal muscle cells are stimulated by *motor neurons* of the somatic (voluntary) division of the nervous system. These motor neurons have nucleus-containing cell bodies that reside in the brain or spinal cord and long extensions called *axons* that travel, bundled within nerves, to the muscle cells that they serve. The axon of each motor neuron divides into a number of terminals as it enters the muscle, and each of these terminals forms a **neuromuscular junction** with a single muscle fiber (Figure 9.10). Usually, each muscle fiber has only one neuromuscular junction. Although the plasma membranes of the axonal ending and the muscle fiber are exceedingly close, they do not actually touch, but remain separated by a fluid-filled extracellular space called the *synaptic cleft*. Within the axonal ending are *synaptic vesicles*, small membranous sacs containing a neurotransmitter called *acetylcholine* (uh-sē″-tul-kō′-lēn) or *ACh*.

When a nerve impulse reaches the end of an axon, some of the synaptic vesicles fuse with the axonal membrane and release ACh into the synaptic cleft by exocytosis. ACh diffuses across the cleft and attaches to ACh receptors on the sarcolemma. The electrical events triggered in a muscle cell membrane when ACh binds are similar to those that take place in excited nerve cell membranes. Since we will consider these events in detail in Chapter 11, just a brief summary is provided here.

Like the plasma membranes of all cells, a resting sarcolemma is *polarized;* that is, there is a voltage across the membrane and the inside is negative. (The resting membrane potential is described in Chapter 3, pp. 73–74.) Attachment of ACh molecules to appropriate receptors on the sarcolemma causes a change in membrane permeability that allows Na^+, driven by its electrochemical gradient, to diffuse across the membrane into the cell. The net result of sodium's entry is that the muscle cell interior becomes slightly *less negative,* an event called *depolarization.* Initially, depolarization is a local event (occurring only at the receptor site), but if the nerve stimulus is strong enough, a muscle cell action potential is generated that passes in all directions from the neuromuscular junction across the sarcolemma. The action potential is a predictable sequence of electrical changes that occurs along the sarcolemma. First, the membrane is depolarized (a consequence of Na^+ channel opening and Na^+ entry into the cell); then it is *repolarized.* The repolarization wave, which quickly follows the depolarization wave along the sarcolemma, is a consequence of K^+ channel opening and K^+ efflux from the cell. Repolarization restores the sarcolemma to its initial polarized state. During repolarization, a muscle cell is said to be in its *refractory period,* because until repolarization has occurred, the cell is insensitive to further stimulation. Once initiated, the action potential is unstoppable and ultimately results in the full contraction of the muscle cell. This phenomenon, referred to as the *all-or-none response,* means that muscle fibers contract to the full extent of their ability or not at all. Although the action potential itself is very brief (1–2 milliseconds [ms]), the contraction phase of a muscle fiber may persist for 100 ms or more and far outlasts the electrical event that triggers it.

After ACh is released from the motor neuron and binds to the ACh receptors, it is swiftly destroyed by acetylcholinesterase (AChE), an enzyme located on the sarcolemma in the neuromuscular junction. This prevents continued muscle fiber contraction in the absence of additional nervous system stimulation.

The period between action potential generation and the beginning of mechanical activity, or muscle cell shortening, is several milliseconds long and is referred to as the *latent period.* It is during the latent period that the events of excitation-contraction coupling occur.

Excitation-Contraction Coupling. **Excitation-contraction coupling** is the sequence of events by which electrical excitation of the sarcolemma leads to the contraction of a muscle fiber. The electrical signal does not act directly on the myofilaments; rather, it causes the rise in intracellular calcium ion concentrations that allows the filaments to slide (Figure 9.11).

Excitation-contraction coupling consists of the following steps:

1. The action potential propagates along the sarcolemma and down the T tubules.

2. Transmission of the action potential past the triad regions causes the SR to release calcium ions into the sarcoplasm, where it becomes available to the myofilaments. It is not known how an action potential coursing along the T tubules causes the SR to release calcium, but the closeness of the three triad elements appears to be essential.

3. As already described, troponin binds to calcium, changes shape, and removes the blocking action of tropomyosin.

4. The myosin cross bridges attach and pull the thin filaments toward the center of the sarcomere. This occurs when the intracellular calcium ion concentration is in excess of 10^{-6} *M*.

5. The short-lived calcium signal ends, usually within 30 ms after the action potential is terminated. The drop in calcium levels reflects the operation of a continuously active, ATP-dependent calcium pump that quickly moves calcium back into the SR.

6. When intracellular calcium levels have become too low, the tropomyosin blockade is reestablished. Cross bridge activity ends and relaxation occurs.

This entire sequence of events is repeated when another nerve impulse arrives at the neuromuscular junction. When the impulses are delivered very rapidly, calcium ions continue to be made available and intracellular calcium levels increase greatly due to successive "rounds" of its release from the SR. Under such circumstances, the muscle cells do not completely relax between successive stimuli and contraction will be stronger and more sustained (within limits) until nervous stimulation ceases.

Except for the brief period following the action potential, calcium ion concentrations in the sarcoplasm are kept almost undetectably low. There is a reason for this: ATP provides the energy source for cellular processes, and as we have seen, its hydrolysis yields inorganic phosphates (P_i). If the intracellular concentration of ionic calcium were always high, the calcium and phosphate ions would combine to form hydroxyapatite crystals, the stony-hard salts found in bone matrix. Such calcified cells would die. Also, since calcium's physiological roles are so vital, its cytoplasmic concentration is exquisitely regulated by intracellular proteins, such as *calsequestrin* (found

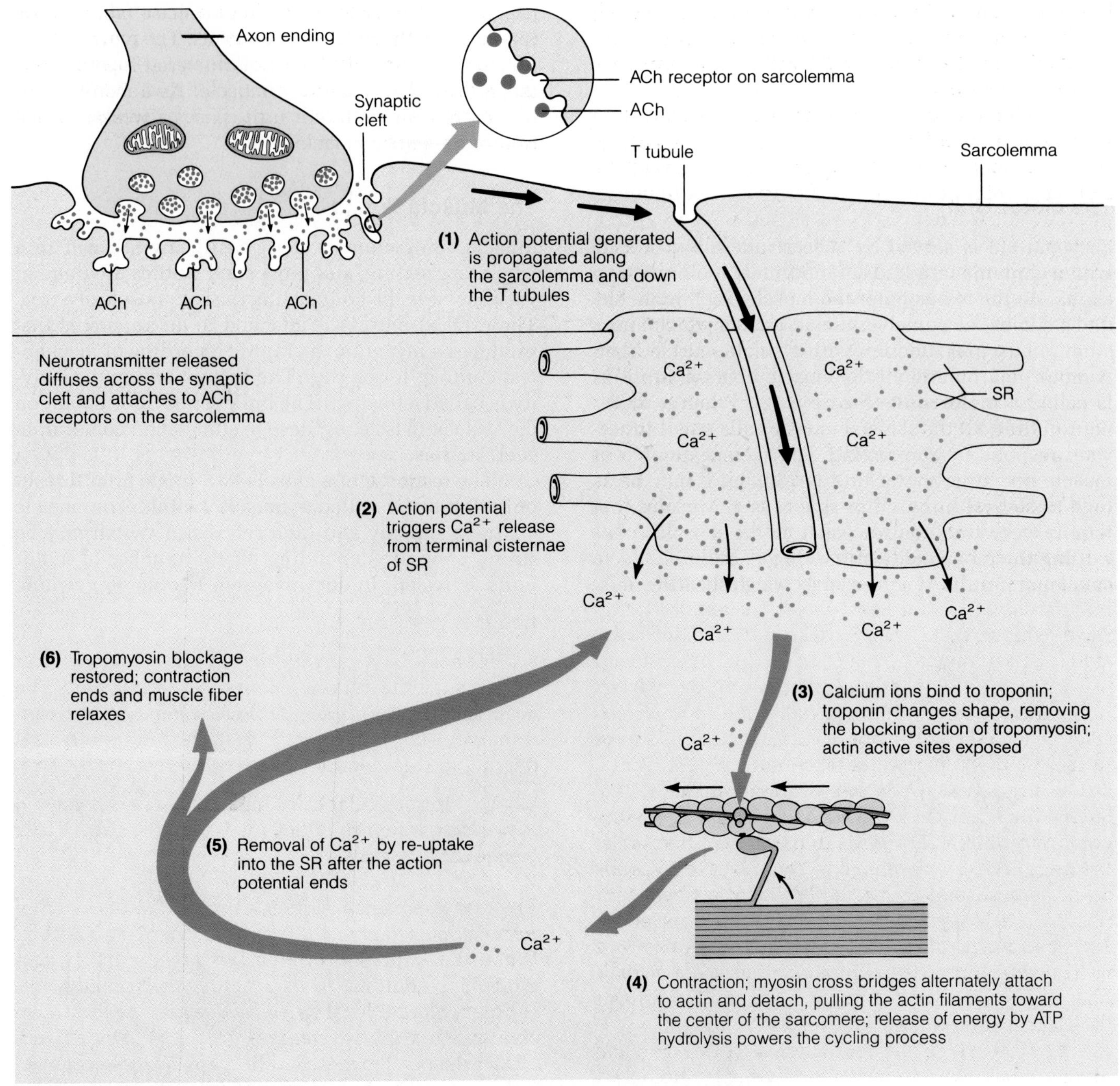
Axon ending
Synaptic cleft
ACh receptor on sarcolemma
ACh
T tubule
Sarcolemma
ACh
ACh
ACh
(1) Action potential generated is propagated along the sarcolemma and down the T tubules
Neurotransmitter released diffuses across the synaptic cleft and attaches to ACh receptors on the sarcolemma
Ca2+
SR
(2) Action potential triggers Ca2+ release from terminal cisternae of SR
(6) Tropomyosin blockage restored; contraction ends and muscle fiber relaxes
(3) Calcium ions bind to troponin; troponin changes shape, removing the blocking action of tropomyosin; actin active sites exposed
(5) Removal of Ca2+ by re-uptake into the SR after the action potential ends
(4) Contraction; myosin cross bridges alternately attach to actin and detach, pulling the actin filaments toward the center of the sarcomere; release of energy by ATP hydrolysis powers the cycling process

Figure 9.11 Sequence of events in excitation–contraction coupling. Events (1) through (5) indicate the sequence of events in the coupling process. As shown in the flow of events to the left, contraction continues until the calcium signal ends.

within the SR) and *calmodulin,* which can alternately bind calcium (removing it from solution) and release it to provide a metabolic signal.

Contraction of a Skeletal Muscle

In its completely relaxed state, a muscle is unimpressive. It is soft and not at all what you would expect of a "prime mover" of the body. However, within a few milliseconds, it can become transformed into a hard elastic structure with dynamic characteristics that intrigue not only biologists, but engineers and physicists as well.

Now that you are familiar with the cellular events of muscle contraction, we can consider muscle contraction on the gross level. Although single muscle cells respond to stimulation in an "all-or-none" fashion, a skeletal muscle, consisting of huge numbers of cells, may contract with varying degrees of force, and for different periods of time. To understand why this is so, we must look at the nerve-muscle functional unit called a motor unit and investigate a muscle's response to stimuli of varying frequencies and intensities.

The Motor Unit

Each muscle is served by at least one motor nerve, which contains hundreds of individual motor neuron axons. As the axons enter the muscle, each branches into a number of axonal terminals, each of which forms a neuromuscular junction with a single muscle fiber. A motor neuron and all the muscle fibers it supplies is called a **motor unit** (Figure 9.12). When a motor neuron fires, all the skeletal muscle cells that it innervates respond by contracting. The average number of muscle fibers per motor unit is 150, but it may be as high as several hundred or as few as 4. Muscles that require very fine control (such as the muscles controlling the movements of the fingers and eyes) have small motor units, whereas large, weight-bearing muscles, whose movements are less precise (such as the hip muscles), have large motor units. The muscle fibers in a single motor unit are not clustered together, but are spread throughout the muscle. As a result, stimulation of a single motor unit causes a weak contraction of the entire muscle.

The Muscle Twitch

Muscle contraction is most easily investigated in a laboratory setting, and most such studies are done *in vitro* (outside the body) using the calf muscle of a frog. The excised muscle is attached to an apparatus that produces a *myogram*, a graphic recording of mechanical contractile activity. (The line recording the activity is called a tracing.) The bulk of our information on the skeletal muscle response to stimulation comes from such studies.

The response of a muscle to a single brief threshold stimulus is called a **muscle twitch.** The muscle contracts quickly and then relaxes. A twitch may be strong or weak, depending on the number of motor units activated. In any myogram tracing of a twitch,

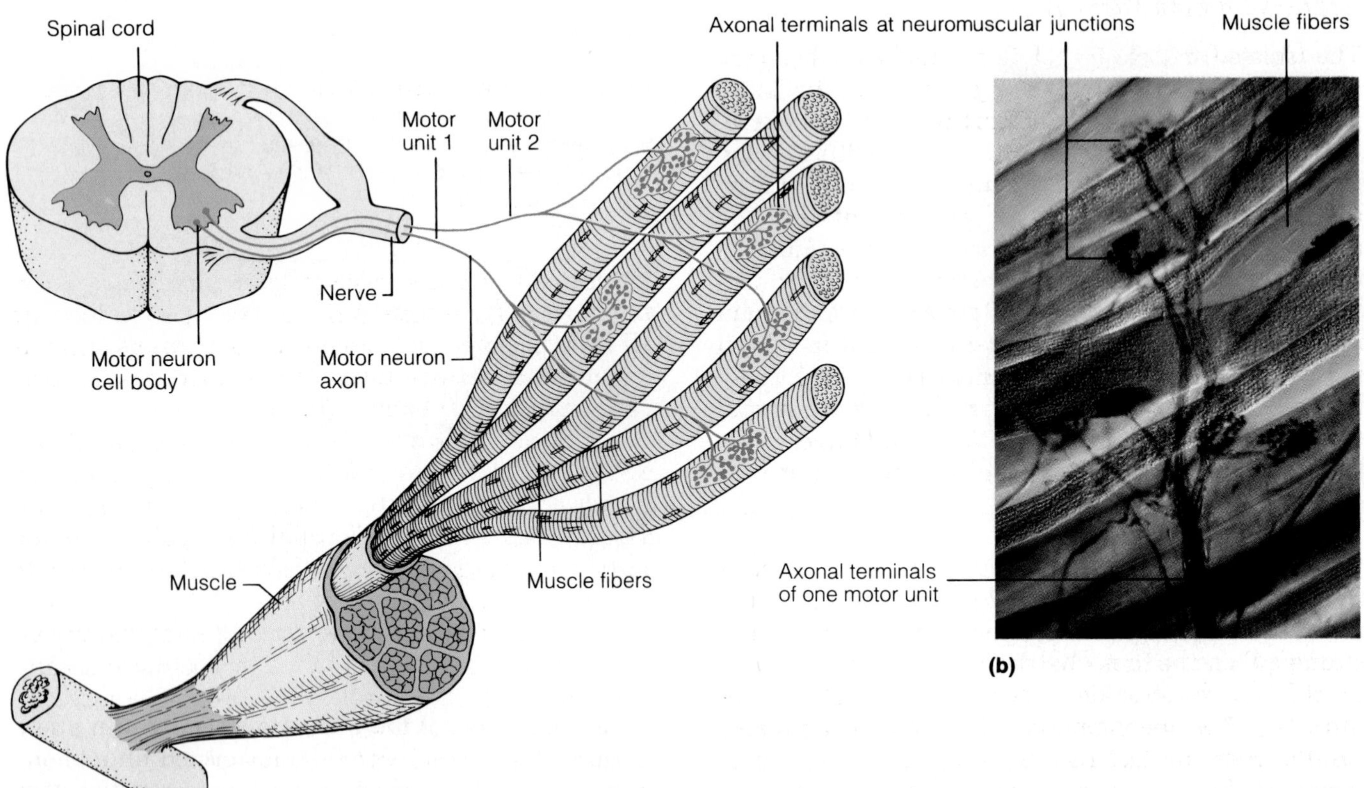

Figure 9.12 Motor units. Each motor unit consists of a motor neuron and all of the muscle fibers it innervates. (**a**) Schematic view of portions of two motor units. The cell bodies of the motor neurons reside in the spinal cord, and their axons extend to the muscle. Within the muscle, each axon divides into a number of axonal terminals, which are distributed to muscle fibers scattered throughout the muscle. (**b**) Photomicrograph of a portion of a motor unit (115×). Notice the diverging axonal terminals and the neuromuscular junctions with the muscle fibers.

three distinct contractile phases are obvious (Figure 9.13a). The **latent period** is the time following stimulation when excitation-contraction coupling is occurring in the muscle; during this period (lasting a few milliseconds), no response is seen on the myogram. The **period of contraction** is the time from the onset of shortening to the peak of tension development, during which the myogram tracing rises to a peak. This period lasts 10–100 ms. If the tension becomes great enough to overcome the resistance of an attached load, the muscle shortens. The period of contraction is followed by the **period of relaxation,** during which contractile force is no longer being generated; muscle tension gradually decreases to zero, and the tracing returns to the baseline. If the muscle has shortened during contraction, it now resumes its initial length.

As you can see in Figure 9.13b, the twitch contractions of some muscles are rapid and brief, as with the extraocular muscles controlling eye movements. In contrast, the fleshy calf muscles, the gastrocnemius and soleus, contract more slowly and usually remain contracted for much longer periods. These differences between muscles reflect the metabolic properties of the myofibrils and enzymes present in the different muscles.

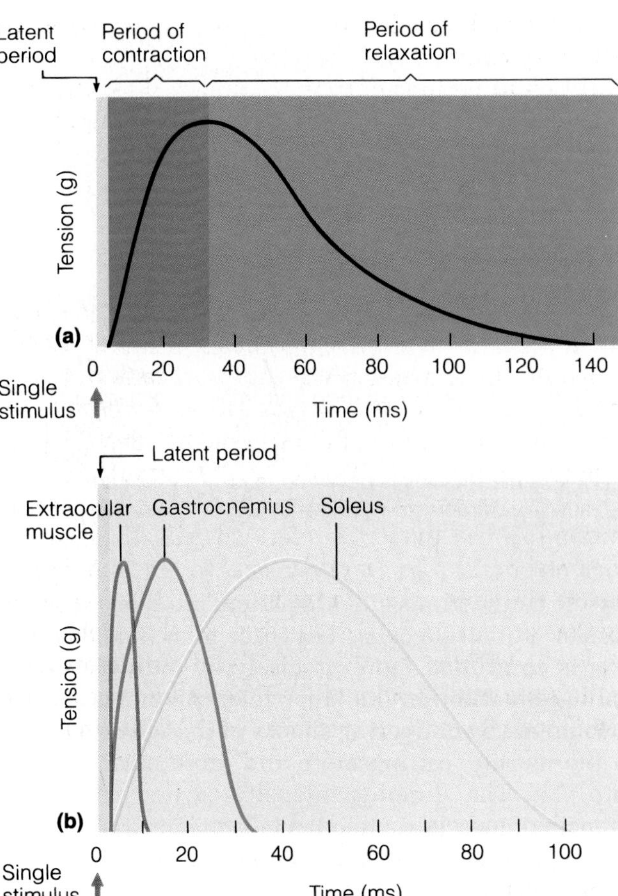

Figure 9.13 The muscle twitch. (**a**) Myogram tracing of a twitch contraction, showing its three phases: the latent period, the period of contraction, and the period of relaxation. (**b**) Comparison of the twitch responses of the extraocular, gastrocnemius, and soleus muscles.

Graded Muscle Responses

The isolated muscle twitch is primarily a laboratory phenomenon. Muscular activity *in vivo* (in the body) is rarely expressed as such short-lived jerky contractions unless neuromuscular abnormalities exist. Instead, our muscle contractions are relatively long and smooth, varying in strength as different demands are placed on them. Variations in the degree of muscle contraction (an obvious requirement for proper control of skeletal movement) are referred to as **graded responses.** In general, there are two ways in which muscle contraction can be graded: (1) by increasing the rapidity of stimulation to produce wave summation, and (2) by recruitment of larger and larger numbers of motor units to produce multiple motor unit summation.

Wave Summation and Tetanus. If two identical electrical shocks (or nerve impulses) are delivered to a muscle in rapid succession, the second twitch will be stronger than the first. On the myogram it will appear to "ride" on the shoulders of the first contraction (Figure 9.14). This phenomenon, called **wave summation,** results from the fact that the second contraction is induced before the muscle has completely relaxed after the first. Since the muscle is already partially contracted, tension produced during the second contraction produces more shortening than the first; in other words, the contractions are summed. If the stimulus (voltage) is held constant and the muscle is stimulated at an increasingly faster rate, the relaxation time between the twitches becomes shorter and shorter and the degree of summation greater and greater. Finally, all evidence of muscle relaxation disappears and the contractions become fused into a smooth, sustained contraction called **tetanus** (tet′-nus).

Tetanus (not to be confused with the bacterial disease of the same name) reflects the usual manner of muscle contraction in the body. That is, motor neurons usually deliver *volleys* of impulses (impulses in rapid succession) rather than the single impulses that result in twitch contractions.

Vigorous muscle activity cannot continue indefinitely. Prolonged tetanus inevitably leads to a condition in which the muscle is unable to contract and its tension drops to zero, a condition called **muscle fatigue.** One pivotal factor in muscle fatigue is the muscle's inability to generate sufficient ATP to power the contractile process. This and other factors causing muscle fatigue are discussed later in the chapter.

Multiple Motor Unit Summation. Although wave summation contributes to the force of muscle response, its primary function is to produce smooth, continuous

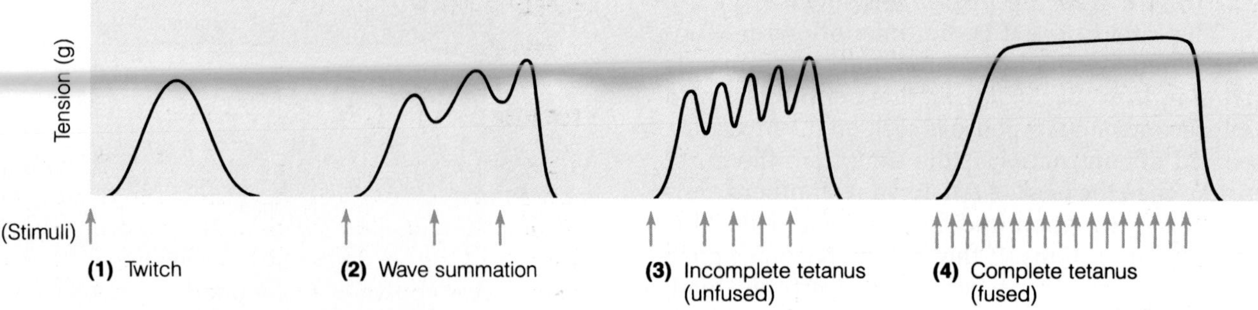

Figure 9.14 Wave summation and tetanus. A whole muscle's response to different stimulus frequencies is illustrated. In (1), a single stimulus is delivered, and the muscle contracts and relaxes (twitch contraction). In (2), stimuli are delivered more frequently, so that the muscle does not have adequate time to completely relax, and contraction force increases (wave summation). In (3), more complete twitch fusion (incomplete tetanus) occurs as stimuli are delivered at a more rapid rate. Complete tetanus, a smooth, continuous contraction without any evidence of relaxation, is shown in (4).

muscle contractions via rapid stimulation of a specific number of muscle cells. The force of muscle contraction is controlled more precisely by **multiple motor unit summation.** In the laboratory, this phenomenon is achieved by delivering shocks of increasing voltage to the muscle, calling more and more muscle fibers into play. The stimulus intensity beyond which the strength of muscle contraction fails to increase is called the *maximal stimulus* and represents the point at which all the motor units have been recruited. In the body, the same result is accomplished by neural activation of an increasingly larger number of motor units serving the muscle.

In weak and precise muscle contractions, relatively few motor units are stimulated. Conversely, when many motor units are activated, the muscle contracts forcibly. This explains how the same hand that pats your cheek can deliver a stinging slap. In any muscle, the smallest motor units (those with the fewest muscle fibers) are controlled by the most excitable motor neurons. These motor units tend to be activated first; the larger motor units, controlled by less excitable neurons, are activated only if a stronger contraction is necessary.

Even when individual motor units of a muscle are not being stimulated at a high frequency, the contraction exerted by a muscle may still be smooth rather than jerky, because while one group of motor units is actively contracting, others are relaxing. As a result of this *asynchronous motor unit summation*, even weak muscle contractions (promoted by infrequent stimuli) usually remain smooth (Figure 9.15).

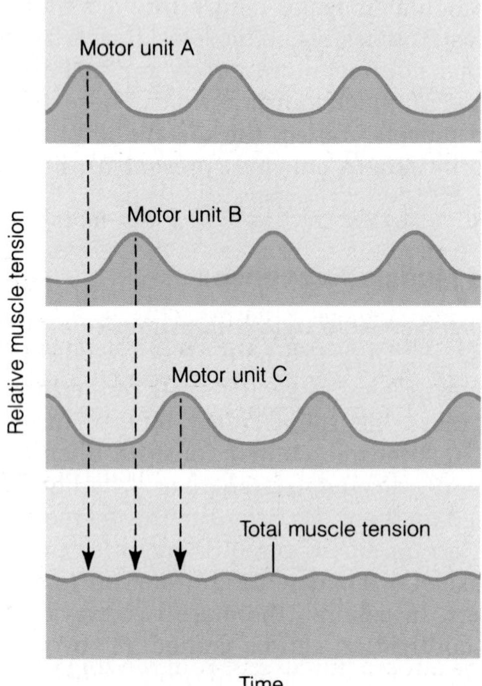

Figure 9.15 Asynchronous motor unit summation. Asynchronous, or staggered, activation of different motor units results in a nearly constant state of tension in the entire muscle, even though the stimulation of the motor units is not frequent.

Treppe: The Staircase Effect

When a muscle first begins to contract, its contractions may be only half as strong as those that occur shortly thereafter; thus, the recording shows a staircase pattern called **treppe** (trep′-eh). It is believed that treppe reflects the sudden increased availability of Ca^{2+}. Additionally, as the muscle begins to work and heat is liberated, the muscle's enzyme systems become more efficient. Together, these factors result in a slightly stronger contraction with each successive stimulus during the beginning phase of muscle activity. This is the basis of the warm-up period required of athletes.

Isometric and Isotonic Contractions

Until now, we have been discussing contraction in terms of shortening behavior, but muscles do not always shorten during contraction. It is imperative to understand that the term *contraction* refers to the active process of generating force within a muscle by cross bridge activity. The force exerted by a contracting muscle on some object is referred to as *muscle tension* and the weight, or reciprocal force (resistance to movement), exerted by the object on the muscle is called the *load*. Since muscle tension and load are opposing forces, to move a load the muscle tension must be greater than the load.

The most familiar muscle contractions, and those we have discussed so far, are called **isotonic contractions** (*iso* = same; *tonic* = tension). During isotonic contractions, the muscle shortens and moves the load. The tension in the muscle and the load remain constant and equal throughout most of the period of shortening, but shortening of the muscle does not occur until the tension exceeds the load. Thus, the heavier the load, the longer the latent period and the slower the velocity of contraction. When, however, a muscle develops tension but does not shorten, the contraction is said to be **isometric** (*iso* = same; *metric* = measure). Muscle tension continues to increase throughout an isometric contraction. Isometric contractions occur when a muscle attempts to move a load that is greater than the tension (force) that the muscle is able to develop. For example, when you attempt to lift a piano single-handedly, your arm muscles are contracting isometrically (without causing movement).

The electrochemical events occurring within a muscle are identical in both types of contractions, and activated cross bridges exert force on the thin filaments. However, the result is different. In isotonic contractions, the thin filaments are sliding, whereas in isometric contractions, the cross bridge forces are generated but are unsuccessful in moving the thin filaments. (You could say that they are "spinning their wheels" on the same actin binding site.)

Purely isotonic or isometric contractions are largely a laboratory phenomenon. In the body, most movements involve both types of contractile activity. Few muscles operate in isolation (most movements require the coordinated activity of several muscles), and the load on a particular muscle is likely to change as the muscle shortens. However, muscle contractions that cause obvious movements, such as the movements of the legs when walking or kicking, are usually called isotonic; and those that act primarily to maintain upright posture and to hold joints in stationary positions while movements occur at other joints are referred to as isometric.

Muscle Metabolism

Providing Energy for Contraction

As a muscle contracts, the energy of ATP is directly coupled to contractile events (cross bridge movement and detachment) and to the activity of the calcium pump. So long as ATP synthesis balances ATP use, muscles can continue to respond to low-frequency stimuli for long periods of time. Surprisingly, however, muscles store very limited reserves of ATP. When contraction begins, ATP reserves are soon exhausted, and ATP must be regenerated continuously if contraction is to continue. There are three pathways by which ATP is regenerated during muscle activity: (1) by interaction of ADP with creatine phosphate, (2) by aerobic respiration, and (3) by anaerobic respiration. Aerobic and anaerobic respiration are the two metabolic pathways by which energy fuels are utilized to produce ATP in all body cells. These pathways, just touched upon here, will be described in detail in Chapter 25.

Coupled Reaction with Creatine Phosphate. As we begin to exercise vigorously, stored ATP in working muscles is depleted in about 6 seconds. At this point, a rapid supplementary system for regenerating ATP goes into effect until the metabolic pathways can adjust to the sudden increased demands for ATP. The reaction that occurs couples ADP (a product of ATP hydrolysis) with a unique high-energy compound stored in muscles, called **creatine** (krē′-uh-tēn) **phosphate (CP).** The overall result is the almost instantaneous transfer of energy and a phosphate group to ADP to form ATP:

$$\text{Creatine phosphate} + \text{ADP} \rightarrow \text{creatine} + \text{ATP}$$

Substantial amounts of creatine phosphate are stored in muscle cells, and the coupled reaction, catalyzed by the enzyme *creatine kinase*, is so effective that the concentration of ATP in muscle cells changes very little during the initial periods of contraction. However, even though there is approximately three times as much stored CP as ATP, reserves of CP are also quickly exhausted.

Together, stored ATP and CP can provide for maximum muscle power for about 10–15 seconds—about long enough to energize a 100-meter dash. The coupled reaction is readily reversible, and CP reserves are replenished during periods of inactivity, when the muscle fibers are producing by other means more ATP than they need. High levels of ATP favor its reaction with creatine to re-form creatine phosphate (Figure 9.16a).

Aerobic Respiration. Even as stored ATP and CP are being used, more ATP is being generated by aerobic

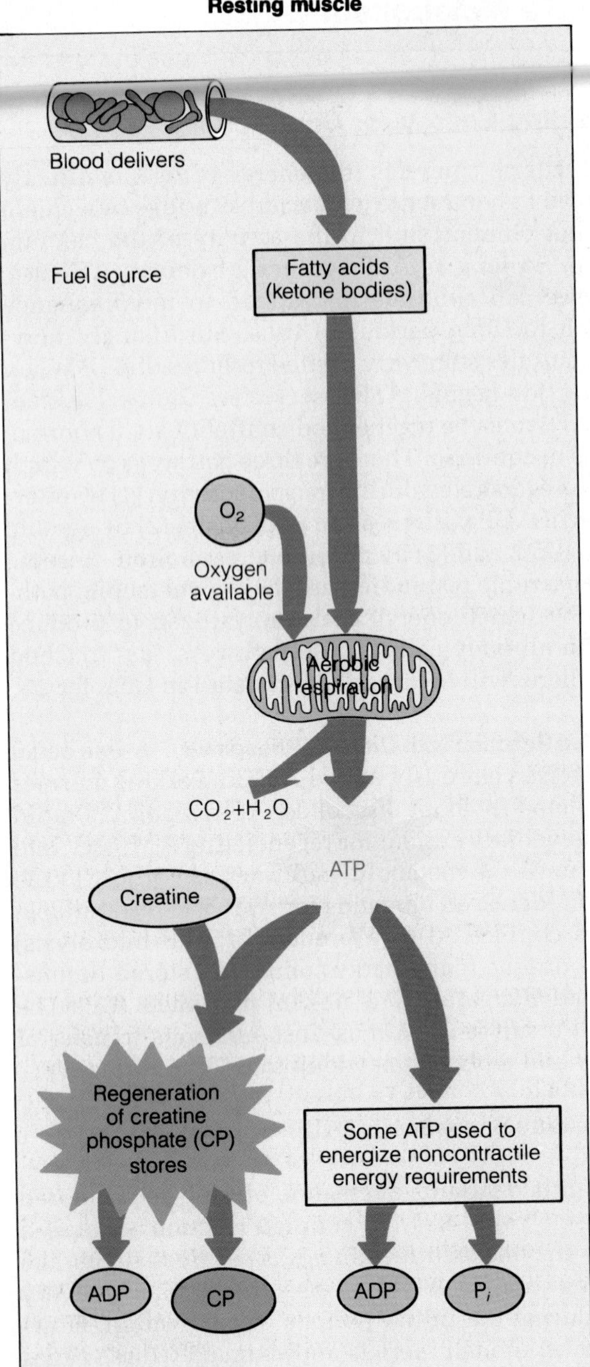

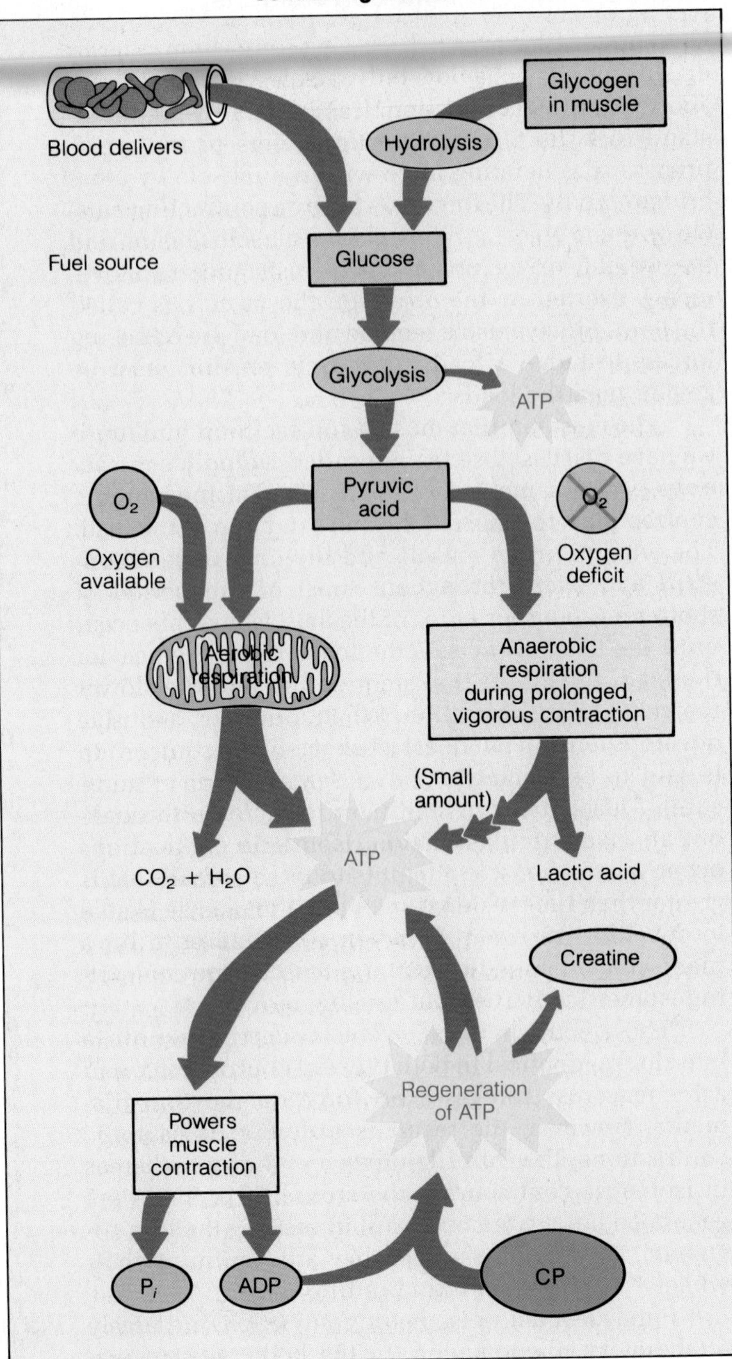

Figure 9.16 Skeletal muscle energy metabolism. (**a**) In resting muscle, muscle fibers rely heavily on ketone bodies (breakdown products of fatty acids) to fuel the aerobic pathways that lead to ATP production. ATP is used to regenerate the secondary high-energy compound creatine phosphate (CP) and to energize cellular activities other than those involved in contraction. (**b**) As muscle cells begin to contract, ATP is used to power cross bridge activity. As stored ATP is exhausted, CP is used to rapidly regenerate ATP during the interval when aerobic pathways are adjusting to the sudden increased demands for ATP. Once this adjustment has been made, the primary source of ATP is provided by the aerobic respiration of glucose obtained from the blood and glycogen stored in muscle. When oxygen delivery and usage fail to keep pace with the requirements of the aerobic pathways, glucose is metabolized only anaerobically, leading to the formation and release of lactic acid. The anaerobic pathway produces much less ATP than the aerobic pathway.

(oxygen-requiring) processes. Resting and slowly contracting muscles obtain the bulk of their ATP supply via aerobic respiration of fatty acids, but when muscles are actively contracting, glucose (obtained from the bloodstream and from the breakdown of stored glycogen) becomes the primary fuel source (see Figure 9.16).

Aerobic respiration occurs in the mitochondria, requires oxygen, and involves a number of chemical reactions in which the bonds of fuel molecules are broken and the energy released is used to make ATP. These reactions are known collectively as *oxidative phosphorylation* (fos″-for-uh-lā′-shun). During aerobic respiration, glucose is broken down entirely, yielding water, carbon dioxide, and large amounts of ATP:

Glucose + oxygen → carbon dioxide + water + ATP

The carbon dioxide released diffuses out of the muscle tissue into the blood and is removed from the body by the lungs.

Anaerobic Respiration and Lactic Acid Formation. The initial pathway of glucose respiration is called *glycolysis* (glī-kah′-luh-sis). During glycolysis, glucose is broken down to two pyruvic acid molecules and some of the energy released is captured to form small amounts of ATP. Since this pathway does not use oxygen, it is an *anaerobic* (an″-ayr-ō′-bik) pathway (see Figure 9.16b). Ordinarily, pyruvic acid then enters the aerobic pathways within the mitochondria and reacts with oxygen to produce still more ATP, as previously described. In such cases, since oxygen is ultimately used, the respiration process as a whole is described as being *aerobic*. As long as it gets enough oxygen and glucose, a muscle cell will form ATP by aerobic reactions. But when muscles have been contracting vigorously for an extended period, the circulatory system begins to "lose ground" in oxygen and glucose delivery, and the aerobic pathways cannot function fast enough to keep pace with muscle demands. Under these conditions, most of the pyruvic acid produced during glycolysis is converted into *lactic acid*. Thus, during oxygen deficit, lactic acid, rather than carbon dioxide and water, is the end product of cellular respiration of glucose. Since the only pathway used under such conditions is glycolysis, which does not use oxygen, the entire process is referred to as **anaerobic respiration.** Most of the lactic acid diffuses out of the muscles into the bloodstream. When oxygen is again available, the lactic acid is reconverted to pyruvic acid and oxidized via the aerobic pathways to carbon dioxide and water (or converted to glycogen).

The aerobic pathway harvests nearly 20 times more ATP from each glucose molecule than the anaerobic pathway. However, the totally anaerobic pathway produces ATP about 2½ times faster than do the oxidative pathways in the mitochondria. Thus, when large amounts of ATP are needed for moderate periods (30–40 seconds) of strenuous muscle activity, the anaerobic pathway can provide most of the ATP needed. Together, the stored high-energy compounds (ATP and CP) and the glycolysis–lactic acid system can support high-exertion muscle activity for nearly a minute.

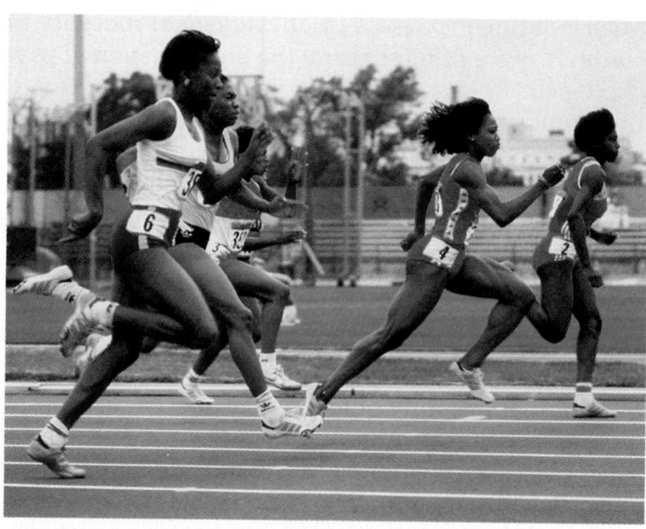

Figure 9.17 Stored ATP and ATP regeneration. Sprinters rely almost entirely on energy supplied by stored ATP and ATP regeneration by a coupled reaction of ADP with creatine phosphate to power their short-lived (10–15 seconds) but powerful muscle activity.

Energy Systems Used During Sports Activities. By evaluating the effort and duration of various athletic activities, exercise physiologists have been able to estimate the relative importance of each energy-producing system to athletic performance. Generally speaking, activities that require a surge of power but last only a few seconds, such as weight lifting, diving, and sprinting (Figure 9.17) rely entirely on ATP and creatine phosphate stores. The somewhat more sustained, but less vigorous activities of tennis, soccer, and a 100-meter swim appear to be fueled almost entirely by the anaerobic mechanism producing lactic acid. Prolonged activities, such as marathon runs and jogging, in which endurance rather than power is the goal, depend mainly on aerobic mechanisms. ■

Muscle Fatigue

Glycogen stored in muscle cells provides some independence from blood-delivered glucose; but with continued exertion, even those reserves are exhausted. When ATP production fails to keep pace with ATP use, muscle fatigue sets in and muscle activity ceases, even though the muscle may still be receiving stimuli.

Muscle fatigue is a state of *physiological inability to contract*, quite different from the phenomenon of psychological fatigue, in which we voluntarily discontinue muscular activity when we feel tired. In psychological fatigue, the flesh (muscles) is still willing, but the spirit is not! It is the "will to win" in the face of psychological fatigue that sets athletes apart. Notice that muscle fatigue results from a relative deficit of ATP, not its total absence. When no ATP is available, *contractures*, or states of continuous contraction, result because of the inability of the cross bridges to detach (not unlike the scenario of rigor mortis). Writer's cramp is a familiar example of temporary contractures.

Other factors contributing to muscle fatigue include excessive accumulation of lactic acid and ionic imbalances. Lactic acid, which causes the pH in muscles to drop (and the muscles to ache), causes extreme fatigue, which limits the usefulness of the anaerobic mechanism for ATP production. During the transmission of action potentials, potassium is lost from the muscle cells, and excess sodium enters. So long as ATP is available to energize the Na^+–K^+ pump, these slight ionic imbalances are corrected. However, in the absence of ATP, the pump is inactive, and ionic imbalances finally cause the muscle cells to become nonresponsive to stimulation.

Oxygen Debt

Even when fatigue does not occur, intense muscle work causes dramatic changes in a muscle's chemical environment. For a muscle to be restored to its resting state, its oxygen stores must be replenished, accumulated lactic acid must be reconverted to pyruvic acid, glycogen stores must be replaced, and ATP and creatine phosphate reserves must be resynthesized. Additionally, the liver must reconvert the lactic acid produced by muscle activity to glucose or glycogen. All of these processes require oxygen. During anaerobic muscle contraction, these oxygen-requiring activities have been occurring more slowly and have been (at least partially) deferred until oxygen is once again available. Thus, an **oxygen debt** is incurred, which must be repaid before resting conditions are restored. Oxygen debt is defined as the extra amount of oxygen that must be taken in by the body to provide for these restorative processes, and it represents the difference between the amount of oxygen that would be needed for totally aerobic respiration during muscle activity and the amount that is actually used. All nonaerobic sources of ATP used during muscle activity contribute to this debt.

A simple example will help illustrate oxygen debt. If you were to run the 100-yard dash in 12 seconds, your body would require approximately 6 L of oxygen to allow totally aerobic respiration. However, the amount of oxygen that could actually be delivered to and used by your muscles during that 12-second interval would be about 1.2 L, far short of the required amount. Thus, you would have incurred an oxygen debt of about 4.8 L, which would be repaid by rapid deep breathing for some time after exertion was ended. This heavy breathing is triggered primarily by high levels of lactic acid in the blood, which stimulates the respiratory center of the brain. Oxygen consumption during exertion depends on several factors, including age, size, previous athletic training, and health. In general, the more exercise to which a person is accustomed, the higher his or her oxygen consumption during exercise will be and the lower the oxygen debt. For example, the rate of oxygen use of most athletes is at least 10% greater than that of a normally sedentary person, and that of a trained marathon runner may be as much as 45% greater. ■

Heat Production During Muscle Activity

If muscles contracted with 100% efficiency, no heat would be produced and all of the energy supplied by ATP cleavage would be converted into the mechanical sliding of the filaments. But muscles are only about as efficient as the best human-made engine, and only 20% to 25% of the energy liberated is actually converted to useful work. The rest is given off as heat, which has to be dealt with if body homeostasis is to be maintained. When you exercise vigorously, you start to feel uncomfortably hot as your blood is warmed by the liberated heat. Ordinarily, heat buildup is prevented from reaching dangerous levels by several homeostatic mechanisms, including radiation of heat from the skin surface and sweating. Shivering represents the opposite end of homeostatic balance, in which muscle contractions are being used to produce more heat. Mechanisms of body temperature regulation will be described in Chapter 25.

Force, Velocity, and Duration of Muscle Contraction

Factors affecting the force, velocity, and duration of muscle contraction are summarized in Figure 9.18. Most of these factors are discussed in this section, but one of them, lever systems, is described in Chapter 10.

Force of Contraction

The strength, or force, of muscle contraction is affected by (1) the number of muscle fibers contracting, (2) the relative size of the muscle, (3) series-elastic elements, and (4) the degree of muscle stretch (see Figure 9.18).

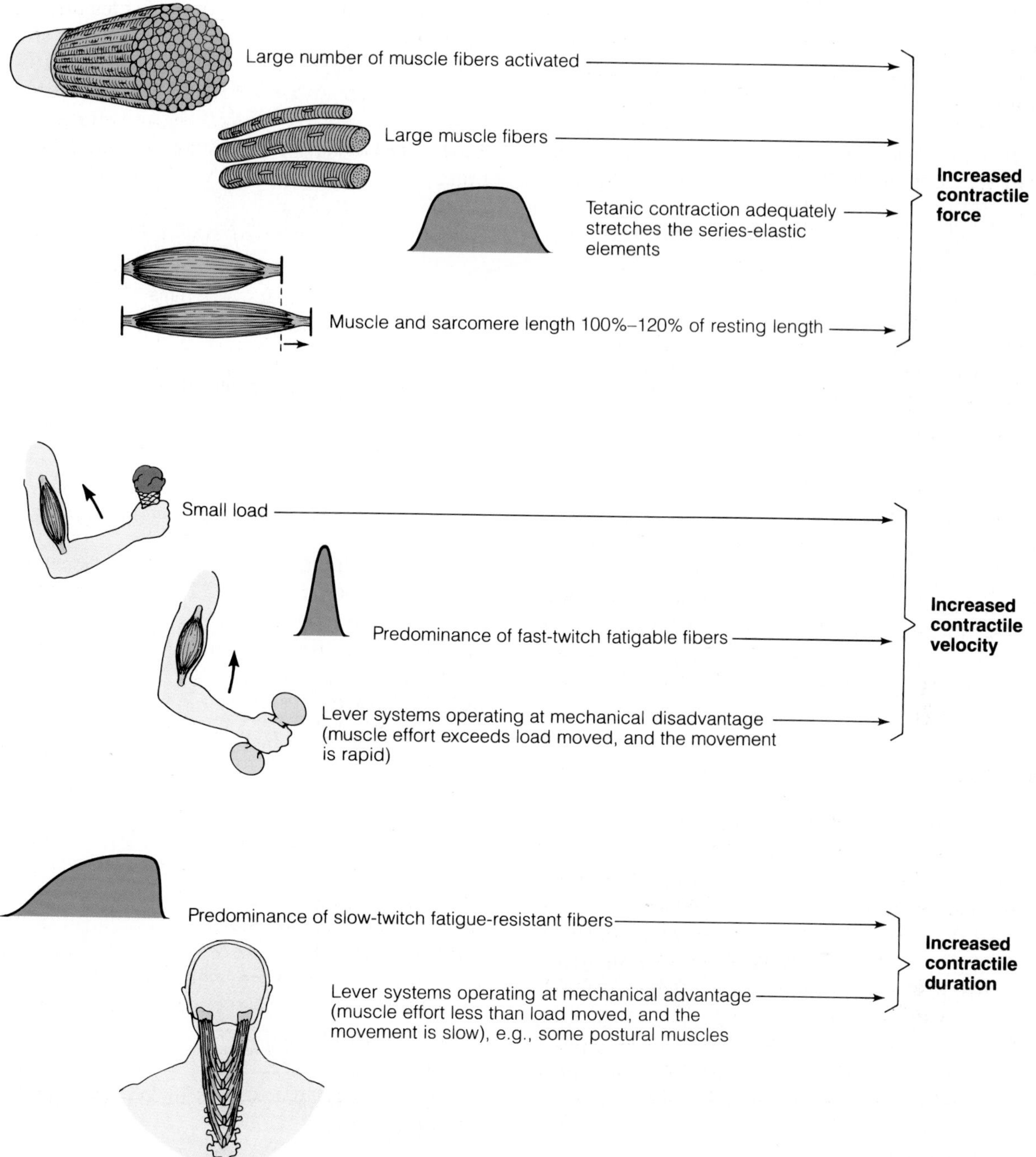

Figure 9.18 Factors influencing force, velocity, and duration of skeletal muscle contraction.

Number of Muscle Fibers Stimulated. As we have already discussed, the more motor units that are recruited, the greater the force of muscle contraction.

Relative Size of the Muscle. The greater the cross-sectional area of a muscle, the greater the tension it is capable of developing and the greater its strength. Regular exercise leads to increased muscle strength by causing muscle cells to hypertrophy (hī-per′-truh-fē) or increase in size.

Series-Elastic Elements Muscle shortening in and of itself does not promote movement of body parts: The muscle must be attached to other movable structures if it is to do useful work. As you will remember, skeletal muscles have connective tissue coverings that attach the muscles to bones. Tension generated by cross bridge activity must be transmitted to the muscle cell surfaces and then through the connective tissue wrappings to the load (the muscle insertion) before work can be done. All of these noncontractile structures are collectively called the **series-elastic elements,** since

Figure 9.19 Relationship of stimulus frequency to external tension exerted on the load. (**a**) During a single-twitch contraction, the internal tension developed by the contractile elements peaks and begins to decline well before the series-elastic elements can be stretched to an equal tension. As a result, the external tension exerted on the load is always less than the internal tension. (**b**) When the muscle is stimulated rapidly so that tetanus occurs, the internal tension is maintained long enough for the series-elastic components to be stretched to a similar tension, so the external tension exerted approaches and finally approximates the internal tension.

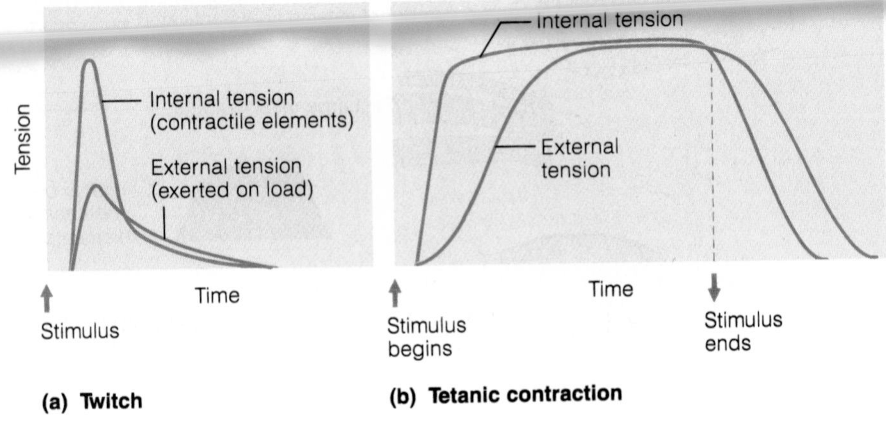

they are able to stretch and recoil. The force generated by the contractile elements (myofibrils), called the *internal tension*, stretches the series-elastic elements; these then transfer their tension, called *the external tension*, to the load.

The critical point about all this is that it takes time to take up slack and stretch the series-elastic elements. Even as the series-elastic elements are being stretched, the stretching forces (internal tension) have already begun to decline. Therefore, in twitch contractions, the maximum external tension exerted on the load is always less than the internal tension generated by the sliding filaments (Figure 9.19a). However, when a muscle is stimulated repeatedly in rapid succession, more time is available to stretch the series-elastic elements, and the external tension approaches that generated by the cross bridges. During tetanus, the external tension becomes similar to the internal tension, and the total force exerted on the load becomes equal (or nearly so) to the total force generated internally (Figure 9.19b). Thus, the more rapidly a muscle is stimulated, the greater its exerted force becomes.

Degree of Muscle Stretch. There is an ideal resting length for muscle fibers at which they are able to generate maximum force (Figure 9.20b). The ideal **length-tension relationship** within a sarcomere occurs when a muscle is slightly stretched and the actin and myosin filaments barely overlap, because this permits sliding along nearly the entire length of the actin filaments. If a muscle cell is stretched so that the myofibrils do not overlap at all (Figure 9.20c), the cross bridges have nothing to attach to and tension cannot be generated. At the opposite extreme is the situation in which the sarcomeres are compressed so that the Z lines abut the myosin myofilaments and the actin filaments touch and interfere with one another (Figure 9.20a). Under these conditions, no further shortening can occur.

Identical relationships exist in a whole muscle. In *in vitro* studies, if a muscle is severely stretched

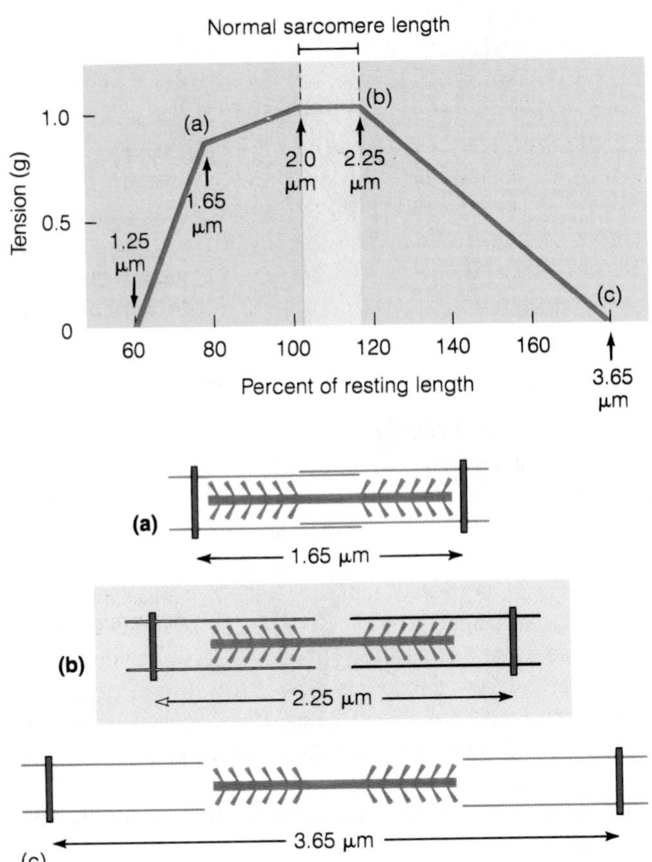

Figure 9.20 Length–tension relationships in skeletal muscles. The average length of a sarcomere is 2.0–2.25 μm and maximum force generation is possible when the muscle is 100% to 120% of its resting length. Increases and decreases beyond the optimal range result in decreased force and finally an inability to generate tension. Depicted here is the relative sarcomere length in a muscle that is (a) strongly contracted, (b) at normal resting length, and (c) excessively stretched.

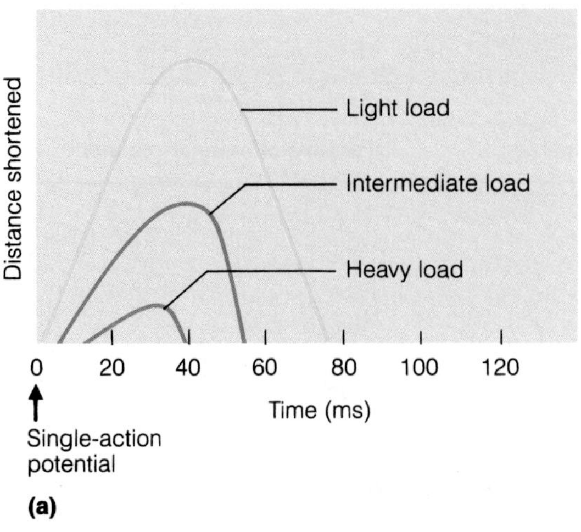

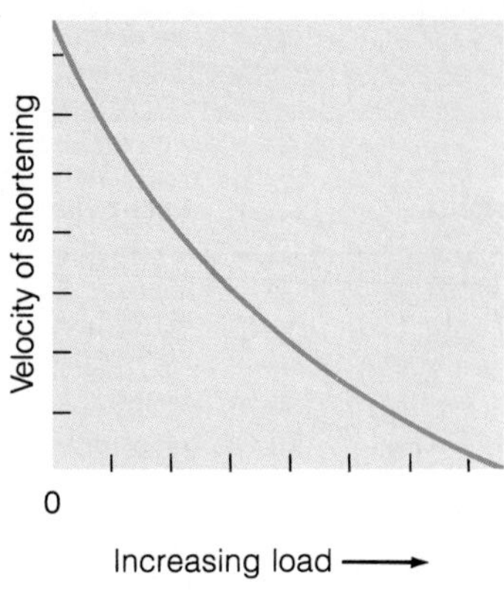

Figure 9.21 Influence of load on velocity and duration of contraction. (**a**) Relationship of load to degree of contraction (shortening) and duration of contraction. (**b**) Relationship of load to velocity of shortening. As the load increases, the speed of contraction decreases.

(say to 175% of its optimal length), it cannot develop any tension, and once a muscle has contracted to 60% of its resting length, further shortening is limited. In the body, skeletal muscles are maintained very near their optimum length by the manner in which they are attached to bones. Additionally, a skeletal muscle harnessed to bone in the body can undergo only about a 30% increase or decrease in length. Thus, degree of stretch contributes only slightly to *in vivo* skeletal muscle strength, although it is very important in determining the force of contraction of heart muscle.

Velocity and Duration of Contraction

Muscles vary in how fast they can contract and in how long they can remain contracted before fatigue sets in. Both load and muscle fiber type influence the velocity and duration of muscle response (see Figure 9.18).

Load. Since muscles are attached to bones, they are always pitted against some resistance, or have some load when they contract. Not surprisingly, they contract fastest when there is no additional load on them. The greater the load becomes, the longer the latent period, the slower the velocity of contraction, and the shorter the duration of contraction (Figure 9.21). If the load exceeds the ability of the muscle to move it, the velocity of shortening is zero and the contraction is isometric.

Muscle Fiber Type. All muscle fibers are not alike. Three discrete types of muscle fibers can be identified on the basis of differences in the (1) efficiency of their myosin-ATPases, (2) the amount of myoglobin present, and (3) the predominant pathway used for ATP synthesis (Table 9.2).

Slow-twitch fatigue-resistant fibers are typically small *red* cells that contain *slow-acting myosin ATPases* and contract slowly; hence the name *slow-twitch* fibers. Their red color reflects a plentiful supply of myoglobin, which stores oxygen and increases the rate of oxygen diffusion throughout the muscle cell. The slow-twitch fibers have abundant mitochondria and a rich capillary supply, and the enzymes catalyzing the aerobic pathways for ATP synthesis are very active. All of these features, along with the abundant myoglobin supply, reveal the oxygen dependence of these fibers. Because these fibers can meet virtually all their energy needs via aerobic pathways (so long as adequate oxygen is available), they are extremely *fatigue resistant* and highly specialized for endurance, that is, for the delivery of prolonged, strong contractions.

Fast-twitch fatigable fibers are generally large pale cells (*white fibers*) with a diameter about twice that of the slow-twitch fibers. They contain *fast-acting myosin ATPases* and contract rapidly. They have few mitochondria, but large glycogen reserves, and they depend on the anaerobic lactic acid pathway to generate ATP during contraction. Since the anaerobic pathway yields relatively small amounts of ATP, glycogen reserves are short-lived, and lactic acid accumulates quickly in these cells, they fatigue quickly (hence, *fatigable* fibers). However, their large diameter, indicative of large numbers of contractile filaments, allows them to generate extremely powerful contractions before they "poop out." Thus, the fast-twitch fatigable fibers are best suited for delivering rapid, intense movements for brief periods.

Fast-twitch fatigue-resistant fibers are *red* cells intermediate in size between the two other fiber types. They contain *fast-acting myosin ATPases* like the fast-twitch fatigable fibers, but they are more like the slow-twitch fibers in their oxygen dependence, high myoglobin content, and rich capillary supply. Since they depend largely on aerobic mechanisms, they are fatigue-resistant, but less so than the slow-twitch fibers.

Most body muscles contain a mixture of fiber types, which allows them to exhibit a range of contractile speeds and fatigue resistance. For example, certain leg

Table 9.2 Structural and Functional Characteristics of the Three Types of Skeletal Muscle Fibers

Feature	Slow-twitch fatigue-resistant	Fast-twitch fatigable	Fast-twitch fatigue-resistant
Metabolic characteristics:			
Twitch rate	Slow	Fast	Fast
Myosin ATPase activity	Slow	Fast	Fast
Primary pathway for ATP synthesis	Aerobic	Anaerobic	Aerobic
Myoglobin content	High	Low	High
Glycogen stores	Low	High	Intermediate
Rate of fatigue	Slow	Fast	Intermediate
Structural characteristics:			
Color	Red	White (pale)	Red (pink)
Fiber diameter	Small	Large	Intermediate
Mitochondria	Many	Few	Many
Capillaries	Many	Few	Many

muscles promote running at some times and support upright posture at others; more motor units containing fast-twitch fibers are stimulated for running, and more units containing slow-twitch fatigue-resistant fibers are stimulated for standing. As might be expected, all muscle fibers in a particular motor unit are of the same type.

Muscles with specialized functions may contain a predominance of one fiber type. For example, disproportionately large amounts of fast-twitch fatigable fibers are found in muscles of the arms and hands, which are specialized for speed. The external eye muscles that position the eyes also have mostly fast fibers, permitting rapid, precisely controlled movement. On the other hand, the posterior back muscles contain larger amounts of slow-twitch fatigue-resistant fibers well suited for the almost continuous contractions needed to sustain upright posture.

Although everyone's muscles contain "mixtures" of the three fiber types, some people have relatively more of the fast-twitch variety. These differences are genetically controlled and no doubt determine athletic capabilities to a large extent. For example, muscles of marathon runners have been shown to have a high percentage of slow-twitch fibers (about 80%), while those of sprinters contain a higher percentage (about 60%) of the fast-twitch fibers. Weight lifters appear to have approximately equal amounts of fast- and slow-twitch fibers. ■

Effect of Exercise on Muscles

The amount of work done by a muscle is reflected in changes in the muscle itself. When used actively or strenuously, muscles may increase in size or strength or become more efficient and fatigue-resistant. On the other hand, muscle inactivity, whatever the cause, always leads to muscle weakness and wasting.

Adaptations to Exercise

Aerobic, or *endurance, exercise* such as swimming, jogging, fast walking, and biking result in several recognizable changes in skeletal muscles. Capillaries surrounding the muscle fibers, as well as mitochondria within the cells, increase in number, and the fibers synthesize more myoglobin. Such changes occur in all fiber types, but they are most dramatic in the fast-twitch fatigue-resistant fibers that depend primarily on aerobic pathways. These changes result in more efficient muscle metabolism and in greater endurance, strength, and resistance to fatigue.

However, aerobic exercise benefits far more than the skeletal muscles. It makes overall body metabolism and neuromuscular coordination more efficient, improves gastrointestinal mobility (and elimination), and enhances the health and strength of the skeleton. Aerobic exercise also leads to changes in cardiovascular and respiratory system functioning that improve the delivery of oxygen and nutrients to all body tissues. The heart hypertrophies and develops a greater stroke volume (more blood is pumped out with each beat), fatty deposits are cleared from the blood vessel walls, and ventilation of the lungs and gas exchange become more efficient. These beneficial effects may be permanent or temporary, depending on the duration and intensity of the exercise.

The moderately weak, but sustained, type of muscle activity required for endurance exercise does not promote significant skeletal muscle hypertrophy (increased size), even though the exercise may go on for hours. Muscle hypertrophy, illustrated by the bulging biceps and chest muscles of a professional weight lifter, results mainly from high-intensity exercise such as weight lifting or isometric exercise, in which the muscles are pitted against high resistance or immovable forces. Such *resistance exercises* need not be prolonged; indeed, a few minutes every other day, during which 18 contractions (three sets of six contractions each) of the muscle or muscle group are performed is sufficient. The key is forcing the muscles to exert more than 75% of their maximum possible force. The increased muscle bulk that results appears to reflect increases in the size of individual muscle fibers (particularly the fast-twitch fatigable variety) rather than an increase in the number of muscle fibers. (However, some investigators claim that some of the increased muscle size results from longitudinal splitting of the hypertrophied fibers and subsequent growth of these "split" cells.) Vigorously stressed muscle fibers contain more mitochondria, form more myofilaments and myofibrils, and lay down larger glycogen reserves. The amount of connective tissue between the cells also increases. These changes in both the contractile and series-elastic elements promote significant increases in muscle strength and size.

Because endurance and resistance exercises produce different patterns of muscular response, it is important to know what your exercise goals are. Lifting weights will not improve your endurance for a triathalon. By the same token, jogging will do little to improve muscle definition for competition in the Mr. or Ms. Muscle contest, nor will it enhance your strength for moving furniture. ■

Disuse Atrophy

Muscles must be physically active if they are to remain healthy. Complete immobilization of muscle, resulting from enforced bed rest or loss of neural stimulation, results in muscle atrophy (degeneration and loss of mass), which begins almost as soon as the muscles are immobilized. Under such conditions, muscle strength can decrease at the rate of 5% per day!

Even in the inactive state, muscles receive weak intermittent stimuli from the nervous system, which is essential to maintain a firm, relatively normal muscle. When a muscle is totally deprived of neural stimulation, the result is devastating; the paralyzed muscle may ultimately atrophy to about one-quarter of its initial size. As atrophy occurs, most of the muscle tissue is replaced by fibrous connective tissue, which makes muscle rehabilitation impossible. Atrophy of a denervated muscle may be delayed by periodic electrical stimulation of the muscle while waiting to see whether the damaged nerve fibers will regenerate. ■

Smooth Muscle

With the exception of the heart, which is made up of cardiac muscle, *visceral muscle* is composed almost exclusively of smooth muscle. Although the chemical and mechanical events of contraction are essentially the same in all muscle tissues, smooth muscle is distinctive in several important ways (see Table 9.3 on pp. 270–271).

Microscopic Structure and Arrangement of Smooth Muscle Fibers

Smooth muscle fibers are small, spindle-shaped cells, each with one centrally located nucleus (see Table 9.3). Typically, they have a diameter of 2–10 μm and a length of 50–200 μm. Skeletal muscle fibers, in contrast, are about 20 times as wide and thousands of times as long.

The sarcoplasmic reticulum of smooth muscle fibers is poorly developed, and T tubules are notably absent. No striations are visible, as the name *smooth muscle* indicates. Smooth muscle fibers do contain thick and thin filaments, but these differ somewhat from those of skeletal muscle. The proportion and organization of the myofilaments are also different:

1. The ratio of thick to thin actin filaments is 1:16, as opposed to the 1:2 ratio of skeletal muscle.

2. Tropomyosin is associated with the thin filaments, but no troponin appears to be present.

3. There are no sarcomeres, but the thick and thin filaments are collected into bundles that correspond to myofibrils.

4. Smooth muscle fibers contain noncontractile *intermediate filaments* that attach to dark-staining *dense bodies* that are distributed throughout the cell and occasionally anchor to the sarcolemma. The dense bodies also serve as attachment points for the thin filaments, and they are considered counterparts of Z lines in skeletal muscle. The intermediate filament–dense body network forms a strong, cablelike intracellular cytoskeleton that harnesses the pull generated by the sliding of the myofilaments during contraction (Figure 9.22).

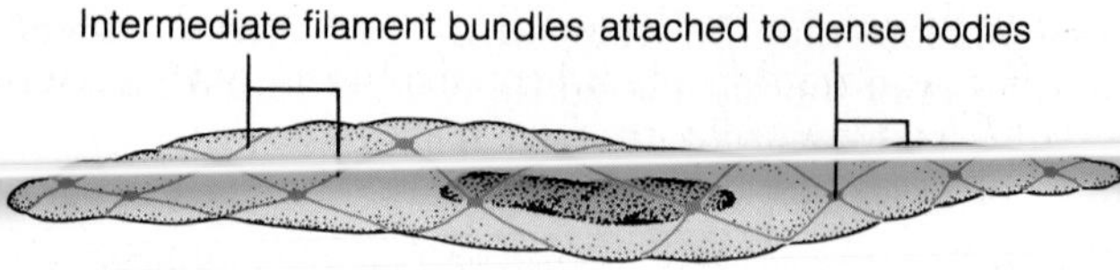

(a) Relaxed smooth muscle cell

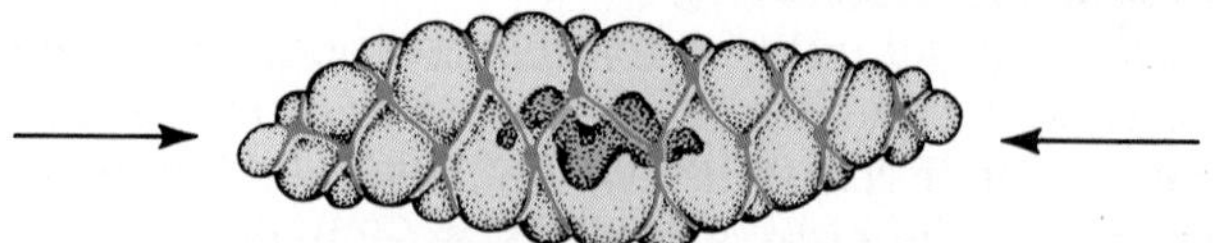

(b) Contracted smooth muscle cell

Figure 9.22 Intermediate filaments and dense bodies of smooth muscle fibers. Intermediate filaments and dense bodies of smooth muscle fibers harness the pull generated during myosin cross bridge activity. The dense bodies are attached to the sarcolemma, the intermediate filaments, and the actin-containing filaments. (**a**) A relaxed smooth muscle cell. (**b**) A contracted smooth muscle cell.

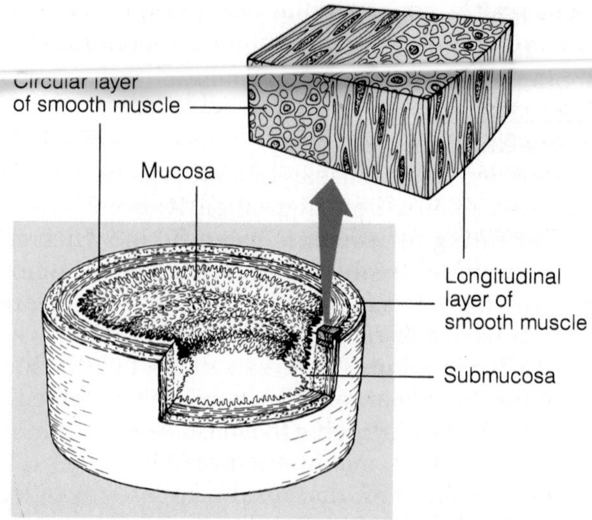

Figure 9.23 Arrangement of smooth muscle in the walls of hollow organs. As seen in this simplified cross-sectional view of the intestine, two smooth muscle layers are present (one circular and the other longitudinal) and run at right angles to each other.

Smooth muscle cells are surrounded by a small amount of fine connective tissue (endomysium) and bound together by strands of collagen and elastin. They lack the elaborate connective tissue coverings seen in skeletal muscle. In rare instances, smooth muscle fibers occur singly or in small fascicles, but usually they are organized into sheets of closely apposed fibers. Such smooth muscle sheets occur in the walls of all but the smallest blood vessels and in the walls of hollow organs of the respiratory, digestive, urinary, and reproductive tracts. In most cases, at least two smooth muscle sheets are present and oriented at right angles to each other (Figure 9.23). One of these, the *longitudinal layer*, runs with the long axis of the organ; the other, the *circular layer*, runs around the circumference of the organ. The cyclic contraction and relaxation of these opposing layers allows the lumen (cavity) of the organ to alternately constrict and dilate so that substances are mixed and "squeezed" along the internal pathway. This phenomenon is called *peristalsis* (payr″-ih-stal′-sis). Contraction of smooth muscle in the rectum, urinary bladder, and uterus helps those organs to expel their contents.

Smooth muscle lacks the highly structured neuromuscular junctions of skeletal muscle. Instead, the innervating nerve fibers approach the smooth muscle fibers and, via many bulbous endings called varicosities, release the neurotransmitter into a wide synaptic cleft in the general area of the smooth muscle cells. Such junctions are called *diffuse junctions*.

Contraction of Smooth Muscle

Mechanism and Characteristics of Contraction

In many cases, adjacent smooth muscle cells exhibit slow, synchronized contractions; the whole sheet responds to a stimulus in unison. This phenomenon reflects electrical coupling of smooth muscle cells by gap junctions, specialized cell connections described in Chapter 4. Whereas skeletal muscle cells are electrically isolated from one another, with each stimulated to contract by its own neuromuscular junction, gap junctions allow smooth muscles to transmit action potentials from cell to cell. Some smooth muscle fibers appear to be *pacemaker cells*, and once excited, they act as "drummers" to set the contractile pace for the entire sheet of smooth muscle. Additionally, some of these pacemakers can depolarize spontaneously in the absence of external stimuli; that is, their membranes are self-excitatory. However, both the rate and intensity of smooth muscle contraction may be modified by neural and chemical stimuli.

The mechanism of contraction in smooth muscle parallels that described for skeletal muscle in the following ways: (1) actin and myosin interact by the sliding filament mechanism; (2) the final trigger for contraction is a rise in intracellular calcium ion level; and (3) the sliding process is energized by ATP.

During excitation-contraction coupling, ionic calcium is released by the tubules of the SR, but it also moves into the cell from the extracellular space. In smooth muscle, however, ionic calcium appears to bind to the *thick filaments* (probably to calmodulin molecules associated with myosin) rather than to the thin filaments as in skeletal muscle. Additionally, myosin ATPase activity in smooth muscle is one-tenth that in skeletal muscle, even under optimal conditions. This, along with the lower numbers of myosin filaments in smooth muscle, results in much slower contraction and a greater economy of energy usage. Typically, smooth muscle begins to contract 50–100 ms after being excited and remains contracted for 1–3 sec (approximately 30 times longer than a single skeletal muscle contraction). Additionally, smooth muscle can maintain the same tension of contraction as skeletal muscle at less than 1% of the energy cost.

The ability of smooth muscle to maintain tension for a prolonged period while using relatively small amounts of ATP is extremely important to overall body homeostasis. The small arterioles and other visceral organs are routinely called on to maintain a moderate degree of contraction (without fatiguing) day in and day out. This tonic contraction of smooth muscle is referred to as *smooth muscle tone*. Since the energy requirements of smooth muscle are low, smooth muscle can generate adequate ATP to support its contractile activity even in the total absence of oxygen, and most ATP synthesis occurs via anaerobic pathways.

Regulation of Contraction

Events leading to the activation of smooth muscle by a neural stimulus are identical to those described for skeletal muscle. An action potential is generated by the binding of neurotransmitter molecules to membrane receptors and is coupled to the release of calcium ions into the sarcoplasm. However, not all neural signals received by smooth muscle result in its activation, and not all smooth muscle activation is the result of neural signals.

All somatic nerve endings release acetylcholine, which always excites skeletal muscle. However, nerve fibers of different autonomic nerves serving smooth muscle release different neurotransmitters, each of which may excite or inhibit a particular group of smooth muscle cells. The effect of a given neurotransmitter on a given type of smooth muscle depends on the type of receptor molecules—stimulatory or inhibitory—on the smooth muscle cell sarcolemma. For example, when acetylcholine binds to ACh receptors on smooth muscle cells in the bronchioles (small air passageways of the lungs), the smooth muscle contracts strongly, narrowing the bronchioles. When norepinephine, released by a different type of autonomic fiber, binds to norepinephine receptors on the *same* smooth muscle cells, the effect is inhibitory, and the smooth muscle relaxes, dilating the air passageways. On the other hand, when norepinephine binds to smooth muscle in the walls of most blood vessels, the effect is stimulating and causes the smooth muscle cells to contract and constrict the vessel.

The extent to which smooth muscles are activated by the nervous system varies extensively among different body organs. Some smooth muscle layers receive no neural stimulation at all and instead depolarize spontaneously or in response to chemical stimuli. Others respond both to neural and chemical stimuli. Chemical factors that can promote smooth muscle contraction or relaxation without an action potential (by enhancing or inhibiting calcium ion entry into the sarcoplasm) include the presence of certain hormones, lack of oxygen, excess carbon dioxide, and low pH. The direct response of smooth muscle to these chemical stimuli probably is most responsible for smooth muscle tone and can cause changes in smooth muscle activity according to local tissue needs. For example, the presence of the hormone gastrin stimulates increased contractile activity of stomach smooth muscle so that it can more efficiently churn foodstuffs. We will consider how the smooth muscle of specific organs is activated as we discuss each organ in subsequent chapters.

Special Features of Smooth Muscle Contraction

Located within the walls of hollow organs, smooth muscle is intimately involved with the functioning of those organs and has a number of unique characteristics. Some of these—smooth muscle tone; slow, prolonged contractile activity; and low energy requirements—have already been considered. But smooth muscle also responds differently to stretch, is capable of shortening more than other muscle types, and has secretory functions.

Response to Stretch. When skeletal or cardiac muscles are stretched, they respond with more vigorous contractions. Stretching of smooth muscle also elicits a contraction, which automatically moves substances along an internal tract. However, the increased tension persists only briefly, and within a few minutes, the tension returns to its original level. This response, called the **stress-relaxation response,** allows a hollow organ to become filled or to expand slowly (within certain limits) to accommodate an increased internal volume without promoting expulsive contractions. This is an important attribute, since organs such as the stomach and bladder must be able to store their contents temporarily. If this did not occur, then the stretching of your stomach and intestines as you ate would lead to vigorous contractions that would rush the foodstuffs through your digestive tract, providing insufficient time for digestion and absorption of the nutrients. And although your kidneys form urine continuously, the stress-relaxation response allows the urinary bladder to store the urine temporarily until you are able to empty your bladder.

Length and Tension Changes. Smooth muscle not only stretches a good deal more than skeletal muscle, but also generates more tension than skeletal muscles that have been stretched to a comparable extent. As you may remember (see Figure 9.20c), precise, highly organized sarcomeres limit the extent to which a skeletal muscle can be stretched before it becomes incapable of generating force. In contrast, smooth muscle cells appear to contract in a helical, or corkscrewlike, manner. Their lack of sarcomeres and the irregular, highly overlapping arrangement of their filaments allow for considerable filament overlap and force generation, even when the cells are substantially stretched. The total length change that skeletal muscles can undergo and still function efficiently is 60% (from 30% shorter to 30% longer than resting length). In contrast, smooth muscle can contract from twice its normal length to half its normal (resting) length—a total length change of 150%. This allows organ cavities to respond to tremendous changes in volume without becoming flabby when they are emptied.

Hyperplasia. In contrast to both skeletal and cardiac muscle, certain smooth muscle fibers are able to divide to increase their numbers, that is, to undergo *hyperplasia* (hī″-per-plā′-zē-uh). Excellent examples are seen in the response of the uterus to estrogen. At puberty, girls' plasma estrogen levels begin to rise. Estrogen binding to uterine smooth muscle receptors stimulates the synthesis of more smooth muscle, causing the uterus to grow to adult size. Then, during pregnancy, high blood levels of estrogen stimulate uterine hyperplasia to accommodate the increasing size of the fetus.

Secretory Functions. Smooth muscle cells also synthesize and secrete the connective tissue proteins collagen and elastin. Thus, the smooth muscle cells themselves, not fibroblasts, secrete the soft connective tissue that surrounds them.

Types of Smooth Muscle

Smooth muscle in different body organs varies substantially in (1) fiber arrangement and organization, (2) responsiveness to various stimuli, and (3) innervation. However, for simplicity's sake, smooth muscle is usually categorized into two major types: single-unit and multiunit smooth muscle.

Single-Unit Smooth Muscle

Single-unit smooth muscle is by far the more common type and is the type commonly called *visceral muscle.* Its cells tend to contract as a unit and rhythmically, are electrically coupled to one another by gap junctions, and often exhibit spontaneous action potentials. All the smooth muscle characteristics described so far pertain to single-unit smooth muscle. Thus, the cells of single-unit smooth muscle are arranged in sheets, exhibit the stretch-relaxation response, and so on.

Multiunit Smooth Muscle

The smooth muscles in the large airways to the lungs and in large arteries, the arrector pili muscles attached to hair follicles of the skin, and the internal eye muscles that adjust your pupil size and allow you to focus visually are all examples of **multiunit smooth muscle.**

In contrast to single-unit muscle, this type consists of muscle fibers that are structurally independent of each other and closely regulated by the nervous system. Gap junctions are rare, and spontaneous or synchronous depolarizations infrequent. Multiunit smooth muscle (like skeletal muscle) is richly supplied with nerve endings, each of which forms a motor unit with a number of muscle fibers, and it responds to neural stimulation with graded contractions. However, while skeletal muscle is served by the somatic (voluntary) division of the nervous system, multiunit smooth muscle (like single-unit smooth muscle) is innervated by the autonomic (involuntary) division and is susceptible to hormonal controls.

Developmental Aspects of Muscles

With rare exceptions, all muscle tissues develop from embryonic mesoderm cells called *myoblasts.* Myoblasts destined to become smooth or cardiac muscle cells migrate to and surround the rudimentary linings of the visceral organs with which they are associated. Skeletal muscle develops from mesodermal "blocks" (somites) that flank the immature nerve cord and from small clusters of mesodermal cells that form the embryonic limb buds.

The multinucleate skeletal muscle fibers form by the fusion of a number of small mononucleated myoblasts. Once fusion has occurred, the cells form actin and myosin filaments and become capable of contraction. Ordinarily, this has happened by the seventh week of development, when the embryo is only 1 inch long. Spinal nerves grow into the muscle masses, bringing them under the control of the somatic nervous system from this point on. It is at this time that the relative number of fast-twitch and slow-twitch fiber types is determined.

In contrast, myoblasts producing cardiac and smooth muscle fibers do not fuse during embryonic development. The details of development of these muscle tissues are less known, but both types develop gap junctions at a very early embryonic stage. Cardiac muscle is functioning as a blood pump by the end of the third week of development (which is often before a woman even realizes that she is pregnant).

Specialized skeletal and cardiac muscle fibers lose their ability to divide, but do retain the ability to hypertrophy given appropriate stimulation. By birth, their specialization generally has been completed, and injured muscles (and the heart) are repaired by scar tissue formation after this time. However, smooth muscle retains the ability to reproduce itself to a very limited extent throughout life.

At birth, the skeletal muscles are mostly straplike, and a baby's movements are all gross reflex types of movements. As a rule, muscular development reflects the level of neuromuscular coordination, which develops in a head-to-toe and proximal-to-distal direction. A baby can lift its head before it can walk. Gross movements precede fine ones, and infancy sees the rapid progress from random arm waving to such precise movements as the pincer grasp (the ability to pick up a pin between the index finger and the opposed thumb). All through childhood, our control of our skeletal muscles becomes more and more sophisticated. By midadolescence, we have reached the peak of natural neural control of muscles, or neuromuscular development, and can either accept that level of development or improve upon it by athletic or other types of training.

A frequently asked question is whether the difference in strength between women and men has a biological basis. It does. There are individual variations, but on the average, women's skeletal muscles make up approximately 36% of their body weight, whereas men's account for about 42%. Men's greater muscular development is due primarily to the effects of testosterone on skeletal muscle, not to the effects of exercise. Since men are usually larger than women, the actual difference in strength is even greater than the percent difference in muscle mass, but muscle strength per unit mass is the same. Strenuous muscle exercise causes substantially more muscle hypertrophy in males than in females, again because of the influence of male hormones.

Because of its rich blood supply, skeletal muscle is amazingly resistant to infection throughout life, and given good nutrition and moderate exercise, relatively few problems afflict skeletal muscles. However, muscular dystrophy is one such "problem," and because of its devastating effects and incurable nature, it deserves more than a passing mention.

The term *muscular dystrophy* refers to a group of inherited muscle-destroying diseases that affect specific muscle groups. The muscles increase in size owing to fat and connective tissue deposit, but the muscle fibers degenerate and atrophy.

The most common and serious form is *Duchenne muscular dystrophy,* which is inherited as a sex-linked recessive disease. (Females carry and transmit the abnormal gene, which is expressed in males.) This tragic disease is usually diagnosed in young males between the age of two and ten years. Active, normal-appearing children become clumsy and begin to fall frequently as their muscles weaken. The disease progresses relentlessly from the extremities upward, finally affecting the head and chest muscles. At present, there is no cure, and its treatment is largely palliative (for easing of symptoms). Victims rarely live beyond early adulthood and usually die of respiratory failure.

As we age, the amount of connective tissue in our skeletal muscles increases, the number of muscle fibers decreases, and the muscles become stringier, or more sinewy. Since skeletal muscles represent so much of body mass, body weight begins to decline in the elderly person. Loss in muscle mass leads to decreased muscle strength, usually about a 50% decrease by the age of 80 years. Regular exercise helps offset the effects of aging on the muscular system, but muscles can also suffer indirectly. The aging of the cardiovascular system affects nearly every organ in the body, and muscles are no exception. As arteriosclerosis takes its toll and begins to occlude the distal arteries, a circulatory condition called *intermittent claudication* occurs in some individuals. Restriction of blood supply to the legs leads to excruciating pains in the legs during walking, and the affected person must often stop and rest to get relief.

Smooth muscles are remarkably trouble free. Most problems that impair their functioning result from external factors such as irritants. In the gastrointestinal tract, irritation might result from ingestion of excess alcohol, very spicy foods, or bacterial infection. Under such conditions, smooth muscle mobility increases in an attempt to rid the body of the irritating agents, and diarrhea or vomiting occurs. In other visceral organs, spasmatic contractions may occur. Specific congenital and developmental problems of the heart are considered in Chapter 19.

No body system acts in isolation. Certain systems age faster than others, but age-related deficits in any body system affect all others to a greater or lesser extent. When trying to maintain the health and well-being of our muscles or of any body system, we should take a holistic (total) view. Moderate exercise, good nutrition, and healthy habits should be lifelong priorities.

Table 9.3 Comparison of Skeletal, Cardiac, and Smooth Muscle

Characteristic	Skeletal	Cardiac	Smooth
Body location	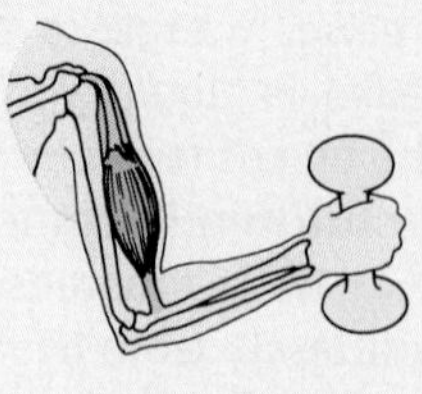 Attached to bones or (some facial muscles) to skin	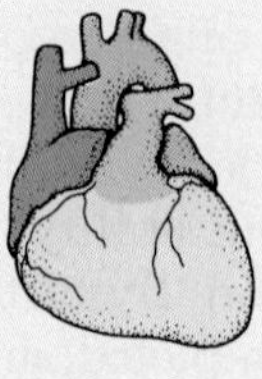 Walls of the heart	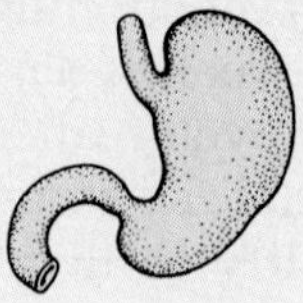 Single-unit muscle in walls of hollow visceral organs (other than the heart); multiunit muscle in intrinsic eye muscles
Cell shape and appearance	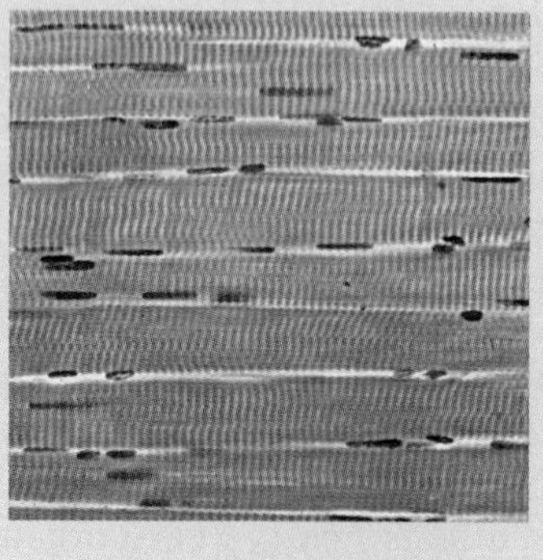 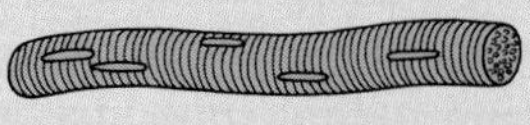 Single, very long, cylindrical, multinucleate cells with very obvious striations	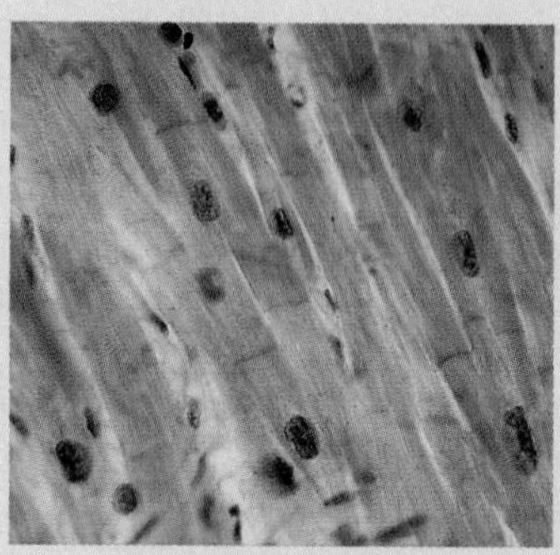 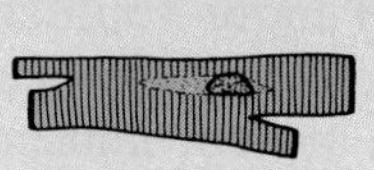 Branching chains of cells; uni- or binucleate; striations	 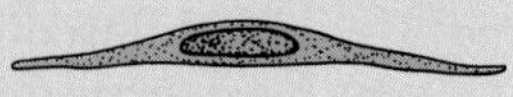 Single, fusiform, uninucleate; no striations
Connective tissue components	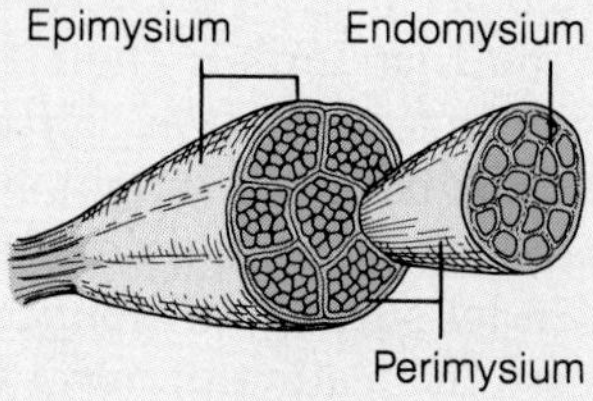 Epimysium, perimysium, and endomysium	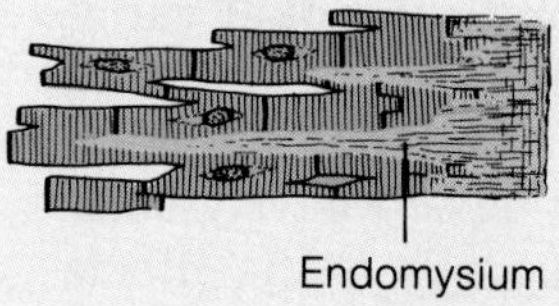 Endomysium attached to fibrous skeleton of heart	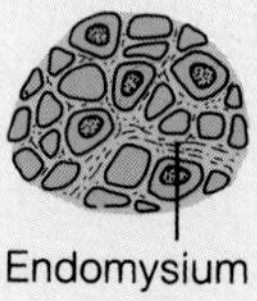 Endomysium
Presence of myofibrils composed of sarcomeres	Yes	Yes	No, but actin and myosin filaments are present throughout
Presence of T tubules and site of invagination	Yes; at A-I junctions 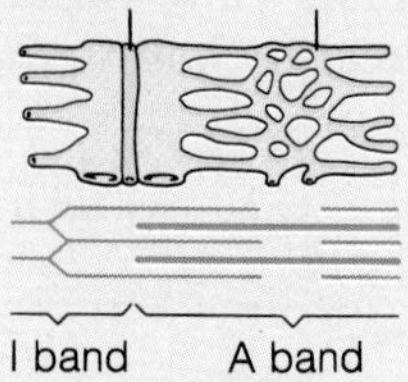	Yes; at Z lines 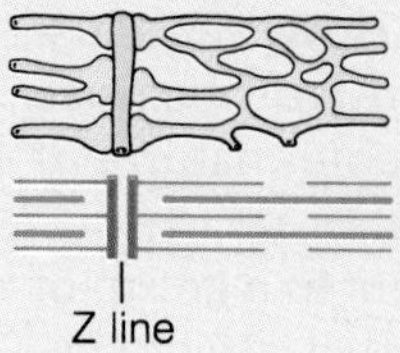	No

Table 9.3 (continued)

Characteristic	Skeletal	Cardiac	Smooth
Elaborate sarcoplasmic reticulum	Yes	Yes	No; SR rudimentary
Presence of gap junctions	No	Yes; at intercalated discs	Yes; in single-unit muscle
Fibers exhibit individual neuromuscular junctions	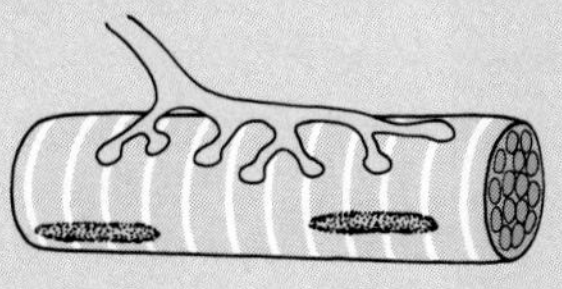 Yes	No	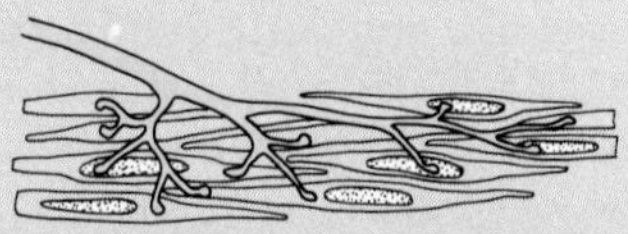 Not in single-unit muscle; yes in multiunit muscle
Regulation of contraction	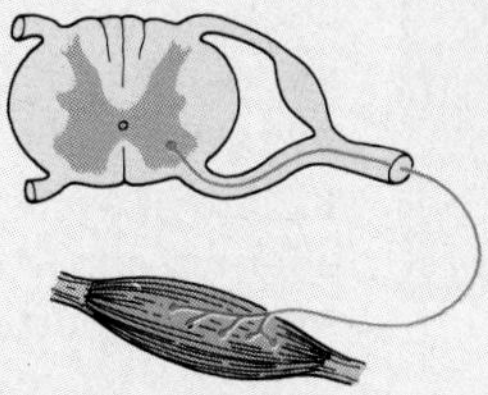 Voluntary via axonal endings of the somatic nervous system	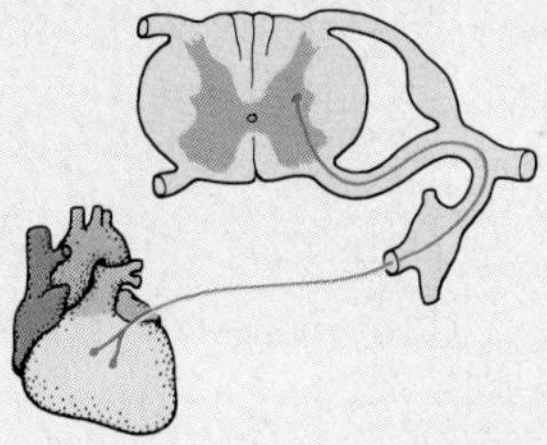 Involuntary; intrinsic system regulation; also autonomic nervous system controls; hormones	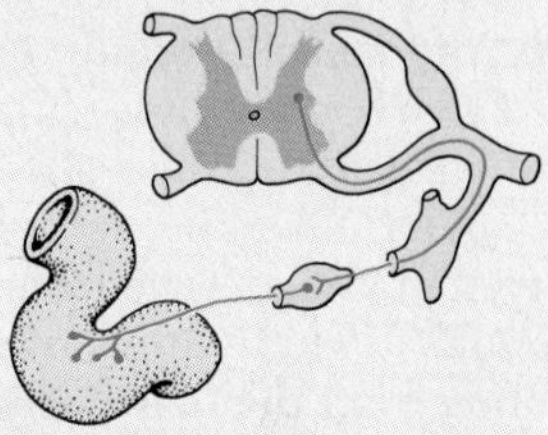 Involuntary; autonomic nerves, hormones, local chemicals, stretch
Source of ionic calcium for calcium pulse	Sarcoplasmic reticulum (SR)	SR and from extracellular space	SR and from extracellular space
Site of calcium regulation	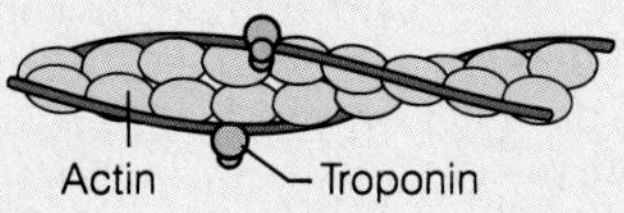 Actin Troponin Troponin on actin-containing thin filaments	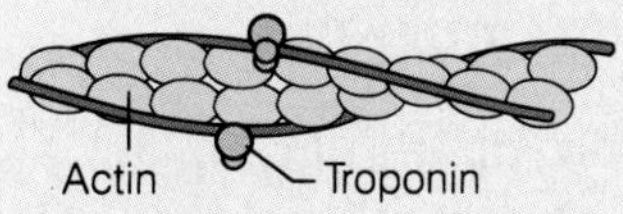 Actin Troponin Troponin on actin-containing thin filaments	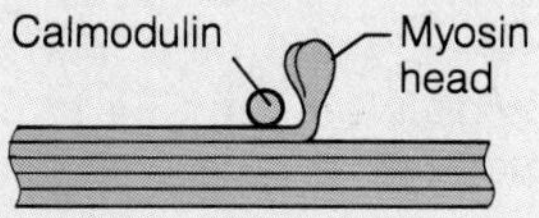 Calmodulin Myosin head Calmodulin on myosin-containing thick filaments
Presence of pacemaker(s)	No	Yes	Yes (in single-unit muscle only)
Effect of nervous system stimulation	Excitation	Excitation or inhibition	Excitation or inhibition
Speed of contraction	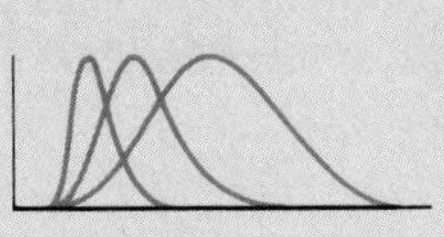 Slow to fast	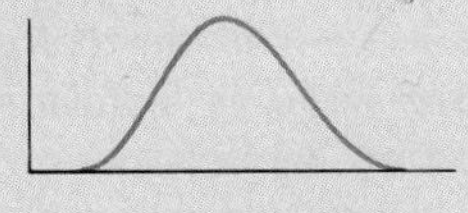 Slow	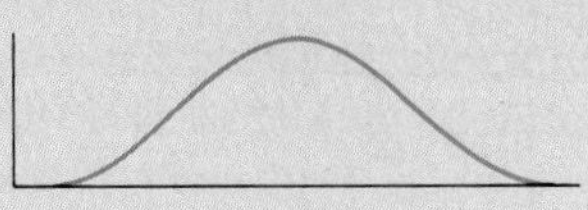 Very slow
Rhythmic contraction	No	Yes	Yes in single-unit muscle
Response to stretch	Vigorous contraction	Vigorous contraction	Stress–relaxation response
Respiration	Aerobic and anaerobic	Aerobic	Mainly anaerobic

A CLOSER LOOK Muscle Sounds: Speak Up—I Can't Hear You!

Although most of us have never heard our working muscles, they do make rumbling sounds, at a frequency very near the lower limits of human hearing. If you are raising your eyebrow and don't quite believe this, you can perform a simple and historical experiment that will allow you to hear your forearm muscles at work. Insert each thumb *gently* into an ear and then make tight fists with your hands. The tighter the fist, the louder the sounds will become.

The first recorded account of hearing muscle sounds in this way was in a book written in 1663 by the Italian Jesuit Francesco Grimaldi. Grimaldi, like the second century Greek physician Galen, attributed the sounds to fluid "animal spirits" emanating from the brain. After Grimaldi's death, muscle sounds were largely forgotten until the early 1800s, when the English chemist-physician William Wollaston took up the chant and went one better. Wollaston drove his carriage over the uniform cobblestones of London streets at different speeds until the noise he heard exactly matched the rumble he heard through his thumbs. This occurred when the speed was 8 mph, and the carriage wheels were hitting 23 cobblestones per second. He concluded that the frequency of muscle sounds was 23 hertz (cycles per second), a figure exceedingly close to that reported by modern instruments. (Good show, Wollaston!)

Recent studies of muscle sounds have used electronic stethoscopes that transform muscle sounds to electrical signals and a computerized treatment of the data that cancels out the effects of background "noise." Such techniques have shown the dominant frequency of muscle sounds to be 25 hertz. As a result of these studies, we also know the following:

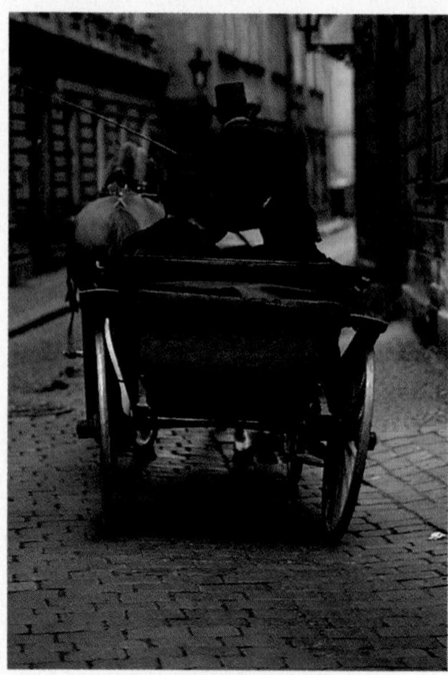

1. The loudness of a muscle sound is proportional to the load. For example, the gastrocnemius (a calf muscle that raises the heel) produces almost no sound when a person is lying down. The sound becomes fairly loud when one is standing and very loud when the whole weight of the body is balanced on the toes.

2. Measurement of muscle sounds can indicate which portions of a muscle are active in a certain physical maneuver. A ballerina's calf muscles generate different and substantially louder sounds than those of an untrained person, indicating that a dancer's stance "on point" is distinctive.

3. Muscle sounds can be readily heard under water. This may help to explain why a shark is attracted to a thrashing fish (or person).

4. It is possible that muscle sounds can serve as a means of communication between embryos. For example, one researcher detected low-frequency sounds coming from quail eggs four days before hatching and suggested that the sounds may explain why eggs in contact with one another usually hatch earlier than isolated eggs.

5. Most muscle sound is generated by fast-twitch fibers, and the frequency of the sounds can be linked to the events of energy use in fast fibers. These fibers typically respond to stimulation in 1/25th of a second (40 milliseconds), which is roughly the time for the utilization and regeneration of ATP. Of course, 1/25th of a second is also the period required for one cycle of the 25 hertz muscle tone.

6. If a muscle contracts only very slightly, *clicks* are heard, but as the force of contraction increases, the clicks occur closer and closer together until they fuse into the familiar rumbling sound. The clicks have been shown to result from the activation of the muscle's motor units. In the orbicularis oculi, the circular muscles surrounding the eyes, each motor unit is so distinct that its operation can be felt and heard independently of the others. If you place your finger tip lightly on your eyelid, you will feel an occasional jerk or throb, which results from the contraction of a single motor unit and produces a click-like sound.

Muscle sounds are now being studied for their usefulness in athletics, science, and medicine. Because the heart too is a muscle, heart muscle sounds might some day become an extremely valuable indicator of the health of that organ and help to predict those susceptible to myocardial infarct or heart attack.

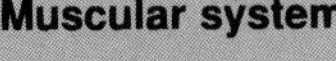

Muscular system

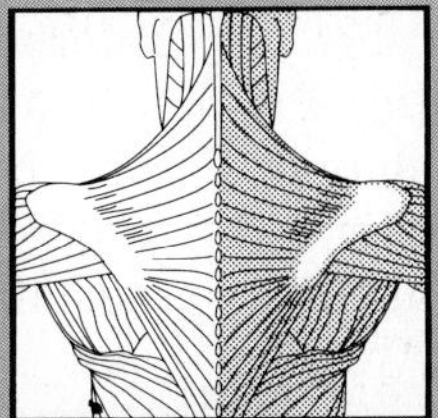

Integumentary system

- ▶ Protects by external enclosure
- ◀ Exercise enhances circulation to skin and improves skin health

Skeletal system

- ▶ Provides levers for muscle activity
- ◀ Activity of skeletal muscles maintains bone health and strength

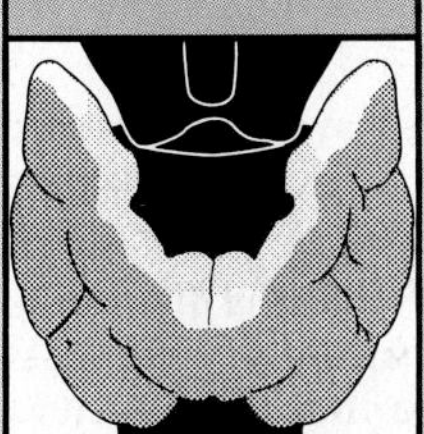

Nervous system

- ▶ Stimulates and regulates muscle activity
- ◀ Activity of facial muscles allow emotions to be expressed

Endocrine system

- ▶ Growth hormone and androgens influence skeletal muscle strength and mass; other hormones help regulate cardiac and smooth muscle activity

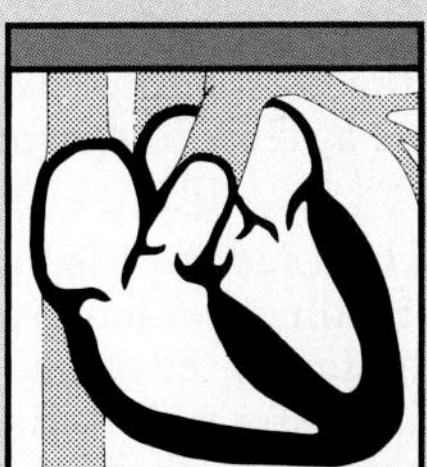

Cardiovascular system

- ▶ Delivers oxygen and nutrients to muscles
- ◀ Skeletal muscle activity increases efficiency of cardiovascular functioning; helps prevent atherosclerosis and causes cardiac hypertrophy

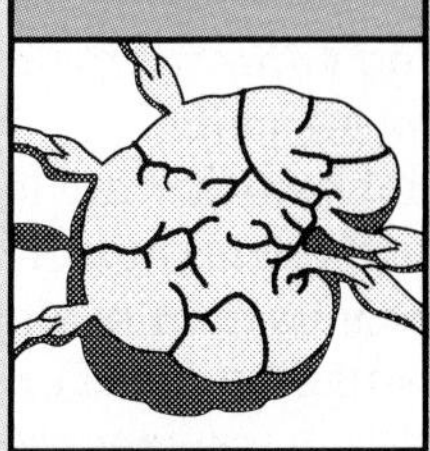

Lymphatic system

- ◀ Drains leaked tissue fluids
- ▶ Contraction of skeletal muscles may help promote flow of lymph

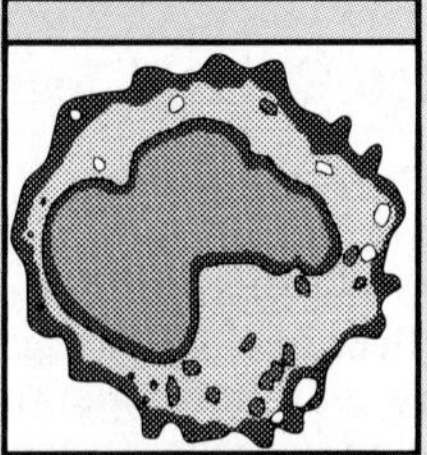

Immune system

- ◀ Protects muscles from disease

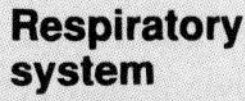

Respiratory system

- ◀ Provides oxygen; disposes of carbon dioxide
- ▶ Muscular exercise increases respiratory capacity

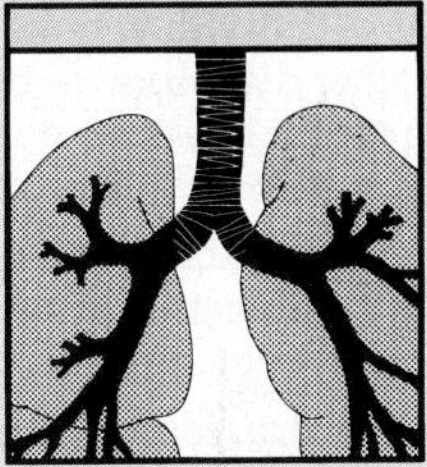

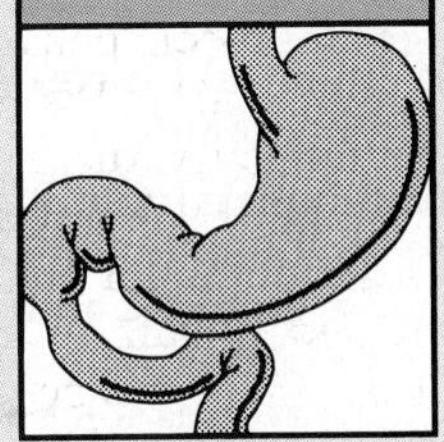

Digestive system

- ◀ Provides nutrients needed for muscle health; liver metabolizes lactic acid
- ▶ Physical activity increases gastrointestinal mobility

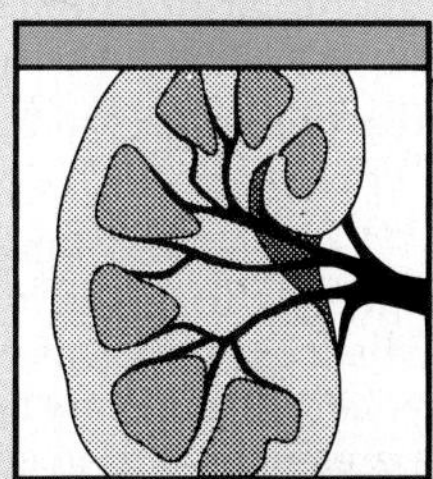

Urinary system

- ◀ Disposes of nitrogenous wastes
- ▶ Physical activity promotes normal voiding behavior; skeletal muscle forms the voluntary sphincter of the urethra

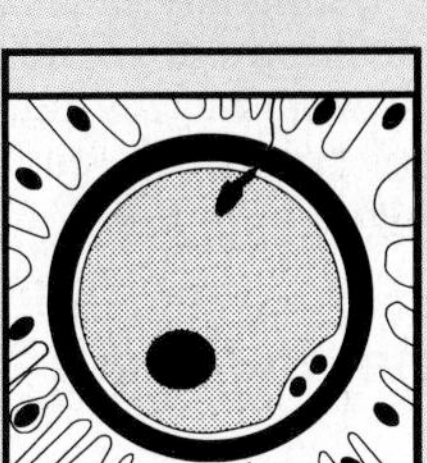

Reproductive system

- ◀ Testicular androgen promotes increased skeletal muscle size
- ▶ Helps support pelvic organs (e.g., uterus); assists erection of penis and clitoris

Figure 9.24 Homeostatic interrelationships between the muscular system and other body systems.

* * *

The capacity for movement is an intrinsic property of all cells, but with the exception of muscle, these movements are largely restricted to intracellular events. Skeletal muscles, the major focus of this chapter, allow us to interact with our external environment in an amazing number of ways, but they also contribute to our internal homeostasis, as summarized in Figure 9.24. We have covered muscle structure from the gross to molecular levels in this chapter and have considered its physiology in some detail. Chapter 10 continues from this point to explain how muscles interact with bones and with each other, and then describes the individual skeletal muscles that contribute to the muscular system of the body.

Related Clinical Terms

Cramp Prolonged spasm, or tetanic contraction, of a muscle, causing it to become taut and painful; common in calf, thigh, and hip muscles, and usually occurs at night or after exercise; may reflect low blood sugar levels, electrolyte depletion (particularly sodium or calcium), dehydration, or irritability of the spinal cord neurons; stretching of the cramped muscle may help.

Fasciculations (fuh-sih″-kyoo-lā′-shunz) Spontaneous contractions of individual motor units in a muscle, causing a visible twitching or dimpling of the overlying skin; may start and stop for no apparent reason, but may reflect extreme muscle irritation or degenerative nervous system disease.

Fibromyositis (*fibro* = fiber; *myo* = muscle; *itis* = inflammation) A combination of a strain and inflammation of a muscle's connective tissue coverings.

Hernia Protrusion of an organ through its body cavity wall; may be congenital (owing to failure of muscle fusion during development), but most often is caused by heavy lifting or obesity and subsequent muscle weakening.

Myalgia (mī-al′-juh) (*algia* = pain) Muscle pain resulting from any muscle disorder.

Myasthenia gravis (mī″-us-thē′-nē-uh gra′-vis) (*asthen* = weakness; *gravi* = heavy) A progressively crippling disease characterized by drooping of the upper eyelids, difficulty in swallowing and talking, and generalized muscular weakness and fatigability; involves some abnormality (possibly a shortage) of acetylcholine receptors at the neuromuscular junction, and it is believed to be an autoimmune disease (a disorder in which the immune system turns against body tissues).

Myopathy (mī-ah′-puh-thē) (*path* = disease, suffering) Any disease of muscle.

Spasm A sudden involuntary muscle twitch ranging in severity from merely irritating to very painful; may be due to chemical imbalances; in spasms of the eyelid or facial muscles, called tics, psychological factors have been implicated; massaging the affected area may help to end the spasm.

Strain Commonly called a "pulled muscle," a strain is excessive stretching and possibly tearing of a muscle due to muscle overuse or abuse; the injured muscle becomes painfully inflammed (myositis), and adjacent joints are usually immobilized.

Chapter Summary

OVERVIEW OF MUSCLE TISSUES (pp. 240–241)

Muscle Types (p. 240)

1. Skeletal muscle is attached to the skeleton, is striated, and can be controlled voluntarily.

2. Cardiac muscle forms the heart, is striated, and is controlled involuntarily.

3. Smooth muscle, located chiefly in the walls of hollow organs, is controlled involuntarily. Its fibers are not striated.

Functions (pp. 240–241)

4. Muscles move internal and external body parts, maintain posture, and generate heat.

5. Special functional characteristics of muscle include excitability, contractility, extensibility, and elasticity.

SKELETAL MUSCLE (pp. 241–265)

Gross Anatomy of a Skeletal Muscle (pp. 241–244)

1. Skeletal muscle fibers (cells) are protected and strengthened by connective tissue coverings, including the endomysium, perimysium, and epimysium (deep fascia).

2. Skeletal muscle attachments may be direct or indirect via tendons or aponeuroses. Indirect attachments withstand friction better.

3. Common patterns of fascicle arrangement are parallel, pennate (uni-, bi-, and multi-), convergent, and circular.

Microscopic Anatomy of Skeletal Muscle (pp. 244–248)

4. Skeletal muscle fibers are long, striated, and multinucleate.

5. Myofibrils are contractile elements that occupy most of the cell volume. Their banded appearance results from a regular alteration of dark (A) and light (I) bands. Myofibrils are chains of sarcomeres; each sarcomere contains thick (myosin) and thin (actin) myofilaments arranged in a regular array. The heads of myosin molecules form cross bridges that interact with the thin filaments.

6. The sarcoplasmic reticulum (SR) is a system of membranous tubules surrounding each myofibril. Its function is to release and then sequester calcium ions.

7. T tubules are invaginations of the sarcolemma that run between the terminal cisternae of the SR and allow the electrical stimulus and extracellular fluid to come into close contact with deep cell regions.

Contraction of a Skeletal Muscle Fiber (pp. 248–253)

8. According to the sliding filament theory, the thin filaments are pulled toward the sarcomere centers by cross bridge (myosin head) activity of the thick filaments.

9. Sliding of the filaments is triggered by a sudden rise in intracellular calcium ion levels. Troponin binding of calcium moves tropomyosin away from the myosin-binding sites on actin, allowing cross bridge binding. Myosin ATPases split ATP, which energizes the power strokes and is necessary for cross bridge detachment.

10. Neural regulation of skeletal muscle cell contraction involves (a) generation and transmission of an action potential along the sarcolemma and (b) excitation-contraction coupling.

11. The action potential is set up when acetylcholine released by a nerve ending binds to sarcolemma receptors, causing changes in membrane permeability that allow ion flows that depolarize and then repolarize the membrane. Once initiated, the action potential is self-propagating and unstoppable.

12. In excitation-contraction coupling, the action potential is propagated down the T tubules, causing calcium to be released from the SR into the cell interior. Calcium initiates cross bridge activity and sliding of the filaments. Cross bridge activity ends when calcium is pumped back into the SR.

Contraction of a Skeletal Muscle (pp. 253–257)

13. A motor unit is one motor neuron and all of the muscle cells it innervates. The neuron's axon has several terminals, each of which forms a neuromuscular junction with one muscle cell.

14. A skeletal muscle's response to a single brief threshold stimulus is a twitch. A twitch has three phases: the latent period, the period of contraction, and the period of relaxation.

15. Graded responses of muscles to rapid stimuli are wave summation and tetanus. A graded response to increasingly strong stimuli is multiple motor unit summation.

16. Isotonic contractions occur when the muscle shortens and the load is moved. Isometric contractions occur when no muscle shortening occurs.

Muscle Metabolism (pp. 257–260)

17. The energy for muscle contraction is ATP, obtained from a coupled reaction of creatine phosphate with ADP and from aerobic and anaerobic respiration of glucose. When ATP use exceeds its production, muscle fatigue occurs.

18. When ATP is produced by nonaerobic pathways, lactic acid accumulates and an oxygen debt occurs. To return the muscles to their resting state, ATP must be produced aerobically and used to regenerate creatine phosphate and glycogen reserves and to oxidize accumulated lactic acid.

19. Only 20%–25% of energy released during ATP hydrolysis is used to power contractile activity. The rest is liberated as heat.

Force, Velocity, and Duration of Muscle Contraction (pp. 260–264)

20. The force of muscle contraction is affected by the number and size of contracting muscle cells (the more and the larger the cells, the greater the force), the series-elastic elements, and the degree of muscle stretch.

21. In twitch contractions, the external tension exerted on the load is less than the internal tension. When a muscle is tetanized, the external tension equals the internal tension.

22. When the thick and thin filaments are slightly overlapping, the muscle can generate maximum force. With excessive increase or decrease in muscle length, force declines.

23. Factors determining the velocity and duration of muscle contraction include the load (the greater the load, the slower the contraction) and muscle fiber types.

24. There are three types of muscle fibers: (1) fast-twitch fatigable fibers, (2) slow-twitch fatigue-resistant fibers, and (3) fast-twitch fatigue-resistant fibers. Most muscles contain a mixture of fiber types.

Effect of Exercise on Muscles (pp. 264–265)

25. Regular aerobic exercise results in increased efficiency, endurance, strength, and resistance to fatigue of skeletal muscles and more efficient cardiovascular, respiratory, and neuromuscular functioning.

26. Resistance exercises cause skeletal muscle hypertrophy and large gains in skeletal muscle strength.

27. Complete immobilization of muscles leads to muscle weakness and severe atrophy.

SMOOTH MUSCLE (pp. 265–268)

Microscopic Structure and Arrangement of Smooth Muscle Fibers (pp. 265–266)

1. Smooth muscle fibers are small, spindle-shaped, and uninucleate; they display no striations.

2. The SR is poorly developed; T tubules are absent. Actin and myosin filaments are present, but sarcomeres are not. Intermediate filaments and dense bodies form an intracellular network that harnesses the pull generated during cross bridge activity.

3. Smooth muscle cells are most often arranged in sheets. They lack elaborate connective tissue coverings.

Contraction of Smooth Muscle (pp. 266–268)

4. Smooth muscle fibers may be electrically coupled by gap junctions; the pace of contraction may be set by pacemaker cells.

5. Smooth muscle contraction is energized by ATP and is activated by a calcium pulse. However, calcium appears to bind on the thick filaments rather than on the thin filaments.

6. Smooth muscle contracts for extended periods at low-energy cost and without fatigue.

7. Neurotransmitters of the autonomic nervous system may inhibit smooth muscle fibers or stimulate them. Smooth muscle contraction may also be initiated by pacemaker cells, hormones, or other local chemical factors that influence intracellular calcium levels, and by mechanical stretch.

8. Special features of smooth muscle contraction include the stress-relaxation response, the ability to generate large amounts of force when extensively stretched, hyperplasia under certain conditions, and synthesis and secretion of collagen and elastin.

Types of Smooth Muscle (pp. 268)

9. Single-unit smooth muscle has electrically coupled fibers that contract synchronously and often spontaneously.

10. Multiunit smooth muscle has independent, well-innervated fibers that lack gap junctions and pacemaker cells. Stimulation is via autonomic nerves (or hormones). Multiunit muscle contractions are rarely synchronous.

DEVELOPMENTAL ASPECTS OF MUSCLES (pp. 268–269)

1. Muscle tissue develops from embryonic mesoderm cells called myoblasts. Skeletal muscle fibers are formed by the fusion of several myoblasts. Smooth and cardiac fibers develop from single myoblasts and display gap junctions.

2. Specialized skeletal and cardiac muscle cells lose their ability to divide but retain the ability to hypertrophy. Smooth muscle has a limited ability to regenerate and to undergo hyperplasia.

3. Skeletal muscle development reflects maturation of the nervous system and occurs in cephalocaudal and proximal-to-distal directions. Peak development of natural neuromuscular control is achieved in midadolescence.

4. Women's muscles account for about 36% of their total body weight and men's for about 42%, a difference due chiefly to the effects of male hormones on skeletal muscle growth.

5. Skeletal muscle is richly vascularized and quite resistant to infection, but in old age, skeletal muscles become fibrous, decline in strength, and atrophy.

Review Questions

Multiple Choice/Matching

1. The connective tissue covering that encloses the sarcolemma of an individual muscle cell is called the (a) epimysium, (b) perimysium, (c) endomysium, (d) periosteum.

2. A fascicle is (a) a muscle, (b) a bundle of muscle cells enclosed by a connective tissue sheath, (c) a bundle of myofibrils, (d) a group of myofilaments.

3. A muscle in which the fibers are arranged at an angle to a central longitudinal tendon has (a) a circular arrangement, (b) a longitudinal arrangement, (c) a pennate arrangement, (d) a convergent arrangement.

4. Thick and thin myofilaments have different compositions. For each descriptive phase, indicate whether the filament is (a) thick or (b) thin.

_____ (1) contains actin _____ (3) contains myosin

_____ (2) contains ATPases _____ (4) contains troponin

5. The function of the T tubules in muscle contraction is (a) to make and store glycogen, (b) to release Ca^{2+} into the cell interior and then pick it up again, (c) to provide a rapid telegraphy system to transmit the action potential deep into the muscle cells, (d) to form proteins.

6. The sites where the motor nerve impulse is transmitted from the nerve endings to the skeletal muscle cell membranes are the (a) neuromuscular junctions, (b) sarcomeres, (c) myofilaments, (d) Z lines.

7. Contraction elicited by a single brief stimulus is called (a) a twitch, (b) wave summation, (c) multiple motor unit summation, (d) tetanus.

8. A smooth, sustained contraction resulting from very rapid stimulation of the muscle, in which no evidence of relaxation is seen, is called (a) a twitch, (b) wave summation, (c) multiple motor unit summation, (d) tetanus.

9. Characteristics of isometric contractions include all but (a) shortening, (b) increased muscle tension throughout, (c) absence of shortening, (d) use in resistance training.

10. During muscle contraction, ATP is provided by (a) a coupled reaction of creatine phosphate with ADP, (b) aerobic respiration of glucose, and (c) anaerobic respiration of glucose.

_____ (1) Which provides ATP fastest?

_____ (2) Which does (do) not require that oxygen be available?

_____ (3) Which (aerobic or anaerobic pathway) provides the highest yield of ATP per glucose molecule?

_____ (4) Which results in the formation of lactic acid?

_____ (5) Which has carbon dioxide and water by-products?

_____ (6) Which is most important in endurance sports?

11. The neurotransmitter released by somatic motor neurons is (a) acetylcholine, (b) acetylcholinesterase, (c) norepinephrine.

12. The ions that enter the muscle cell during action potential generation are (a) calcium ions, (b) chloride ions, (c) sodium ions, (d) potassium ions.

13. Use the choices in the key to identify the factors requested below.

Key: **a.** load **d.** number of fibers activated
b. muscle fiber type **e.** degree of muscle stretch
c. series-elastic elements

_____ (1) all factors that influence contractile force

_____ (2) all factors that influence the velocity and duration of muscle contraction

14. Myoglobin has a special function in muscle tissue. It (a) breaks down glycogen, (b) is a contractile protein, (c) holds a reserve supply of oxygen in the muscle.

15. Aerobic exercise is desirable because it results in all of the following except: (a) increased cardiovascular system efficiency, (b) increase in the number of mitochondria in the muscle cells, (c) increase in the size and strength of existing muscle cells, (d) increased neuromuscular system coordination.

16. The smooth muscle type found in the walls of digestive and urinary system organs and which exhibits gap junctions and pacemaker cells is (a) multiunit, (b) single-unit.

Short Answer Essay Questions

17. Name and describe the four special functional characteristics of muscle that are the basis for muscle response.

18. Distinguish between (a) direct and indirect muscle attachments and (b) a tendon and an aponeurosis.

19. (a) Describe the structure of a sarcomere and indicate the relationship of the sarcomere to the myofilament. (b) Explain the sliding filament theory of contraction using appropriately labeled diagrams of a relaxed and a contracted sarcomere.

20. What is the importance of acetylcholinesterase in muscle cell contraction?

21. Explain how a slight (but smooth) contraction differs from a vigorous contraction of the same muscle using the principal understandings of multiple motor unit summation.

22. Explain what is meant by the term excitation-contraction coupling.

23. Define motor unit.

24. Describe the three distinct types of skeletal muscle fibers.

25. True or false. Most muscles contain a predominance of one skeletal muscle fiber type. Explain the reasoning behind your choice.

26. Describe the cause(s) of muscle fatigue and define this term clearly.

27. Define oxygen debt.

28. Smooth muscle has some unique properties, e.g., low energy usage, ability to maintain contraction over long periods, and the stress-relaxation response. Tie these properties to the function of smooth muscle in the body.

Clinical Application Questions

29. A 30-year-old male decided that his physique left much to be desired. In an attempt to rectify this shortcoming, he joined a local health club and began to "pump iron" three times weekly. After three months of training, during which he was able to lift increasingly heavier weights, he noticed that his arm and chest muscles were substantially larger. Explain the structural and functional basis of these changes.

30. When a suicide victim was found, the coroner was unable to remove the drug vial clutched tightly in his hand. Explain the reasons for this. If the victim had been discovered three days later, would the coroner have had the same difficulty? Explain.

CHAPTER 10

The Muscular System

Chapter Outline and Student Objectives

Lever Systems: Bone–Muscle Relationships (pp. 277–279)

1. Name the three types of lever systems and indicate the arrangement of elements (effort, fulcrum, and load) in each. Also note the advantages of each type of lever system.

Interactions of Skeletal Muscles in the Body (pp. 279–280)

2. Explain the function of prime movers, antagonists, synergists, and fixators, and describe how each promotes normal muscular function.

Naming of Skeletal Muscles (pp. 280–282)

3. List and define the criteria used in naming muscles. Provide an example to illustrate the use of each criterion.

Major Skeletal Muscles of the Body (pp. 282–327)

4. Name and identify (on an appropriate diagram or torso model) each of the muscles described in Tables 10.1 to 10.15. State the origin and insertion for each, and describe the action of each.

Preview of Selected Key Terms

Muscular system The organ system consisting of the skeletal muscles of the body and their connective tissue attachments.

Prime mover Muscle that bears the major responsibility for effecting a particular movement; agonist.

Antagonist (an-ta'-guh-nist) Muscle that reverses, or opposes, the action of another muscle.

Synergist (sih'-ner-jist) Muscle that aids the action of a prime mover by effecting the same movement or by stabilizing joints across which the prime mover acts to prevent undesirable movements.

Fixator (fix'-ā-ter) Muscle that immobilizes one or more bones, allowing other muscles to act from a stable base.

Muscle tissue includes all contractile tissues—skeletal, cardiac, and smooth muscle alike. However, when the **muscular system** is studied, **skeletal muscles,** organs composed of skeletal muscle fibers and their associated connective tissue wrappings and attachments, take center stage. As we have seen, the body activated by its skeletal muscles enjoys an incredibly wide range of movements. The gentle blinking of an eye, standing on tiptoe, and wielding a sledgehammer are just a sampling of the different activities promoted by the muscular system. It is these activities and the muscular "machines" that bring them about that are the focus of this chapter. However, before describing the individual muscles in detail, we will explain the principles of leverage, describe the manner in which muscles "play" with or against each other to promote, restrict, or modify movements, and note the criteria used for naming muscles.

Lever Systems: Bone–Muscle Relationships

The operation of most skeletal muscles involves the use of leverage. A **lever** is a rigid bar that moves on a fixed point, or *fulcrum,* when a force is applied. The applied force, or *effort,* is used to move a resistance, or *load.* In your body, your joints are the fulcrums, and the bones of your skeleton act as levers. The effort, provided by muscle contraction, is applied where a muscle attaches to a bone. The load that is moved is the bone itself, along with overlying tissues and anything else you are trying to move with that particular lever.

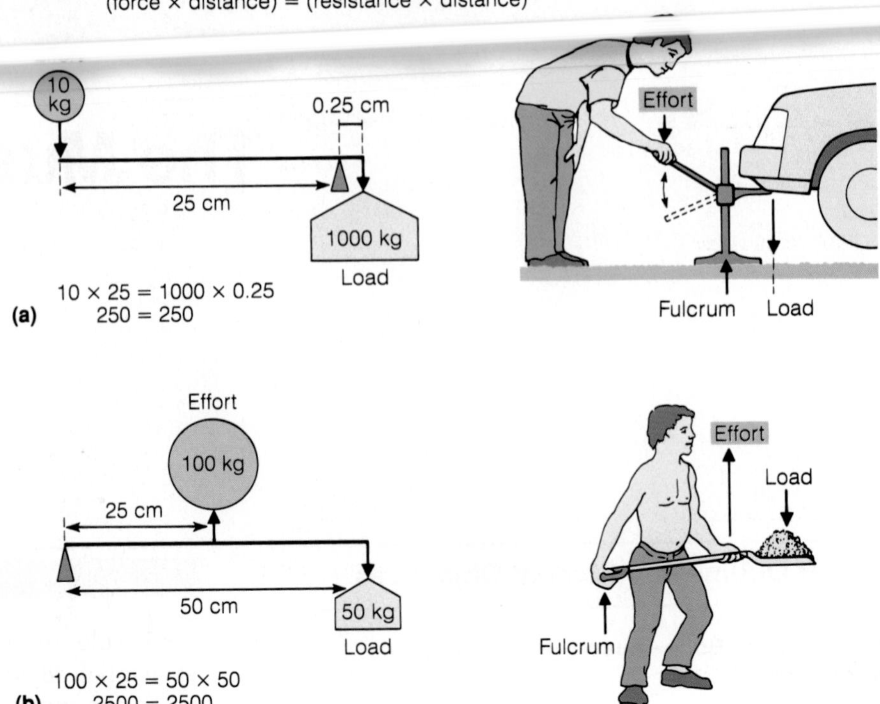

Figure 10.1 Lever systems operating at a mechanical advantage and mechanical disadvantage. The equation at th[illegible] relationships among the forces and distances in any lever system. (**a**) 10 kg of force (the effort) is used to lift a 1000 kg car (the load). This lever system, which employs a jack, operates at a mechanical advantage: The load lifted is greater than the applied muscular effort. (**b**) The lever system exemplified by the use of a shovel to lift dirt operates at a mechanical disadvantage. A muscular force (effort) of 100 kg is used to lift 50 kg of dirt (the load). Levers operating at a mechanical disadvantage are common in the body because they permit a muscle to be inserted close to the load and provide rapid contractions that have a wide range of motion.

A lever allows a given effort to move a heavier load, or to move a load farther, than it otherwise could. If, as shown in Figure 10.1a, the load is close to the fulcrum and the effort is applied far from the fulcrum, a small effort exerted over a relatively large distance can be used to move a large load over a small distance. We say that such a lever operates at a *mechanical advantage*. For example, as shown to the right in Figure 10.1a, a man can lift a car with such a lever, in this case, a jack. The car only moves up a fraction of an inch with each downward movement of the jack handle, but relatively little muscle effort is needed. If, on the other hand, the load is far from the fulcrum and the effort is applied near the fulcrum, the effort applied must be greater than the load to be moved (Figure 10.1b). Such a lever is said to operate at a *mechanical disadvantage*. Nevertheless, it can be advantageous to us, for it allows the load to be moved rapidly through a large distance. Wielding a shovel provides such an example. As you can see, small differences in the site of a muscle's insertion (relative to the fulcrum or joint) can translate into large differences in the amount of force a muscle must generate to move a given load or resistance.

Depending on the relative position of the three elements—the effort, fulcrum, and load—a lever belongs to one of three classes. In **first-class levers,** the effort is applied at one end of the lever and the load to be moved is at the other, with the fulcrum somewhere between them (Figure 10.2a). Seesaws, scissors, and the use of a crowbar provide familar examples of first-class levers. An example of first-class leverage in the body occurs when you lift your head off your chest. Some first-class levers in the body operate at a mechanical advantage (effort less than the load to be moved), but others, such as the operation of the triceps muscle in extending the forearm against resistance, operate at a mechanical disadvantage (effort greater than the load to be moved).

In a **second-class lever,** the effort is applied at one end of the lever and the fulcrum is located at the other, with the load at some intermediate point between them (Figure 10.2b). A wheelbarrow demonstrates this type of lever system. Second-class leverage is uncommon in the body; the best example is the act of standing on your toes. Joints forming the ball of the foot act together as the fulcrum, the load is the entire body weight, and the calf muscles inserted into the calcaneus of the heel exert the effort, pulling the heel upward. Second-class levers in the body work at a mechanical advantage because the muscle insertion is always farther from the fulcrum than is the load to be moved. Second-class levers are levers of strength, but speed and range of motion are sacrificed for that strength.

In **third-class levers,** the effort is applied at a point between the load and the fulcrum (Figure 10.2c). These levers operate with great speed and always at a mechanical disadvantage. A tweezers or forceps provides this type of leverage. Most skeletal muscles of the body interact in third-class lever systems, as illustrated by the activity of the biceps muscle of the arm. The fulcrum is the elbow joint, the force is exerted on

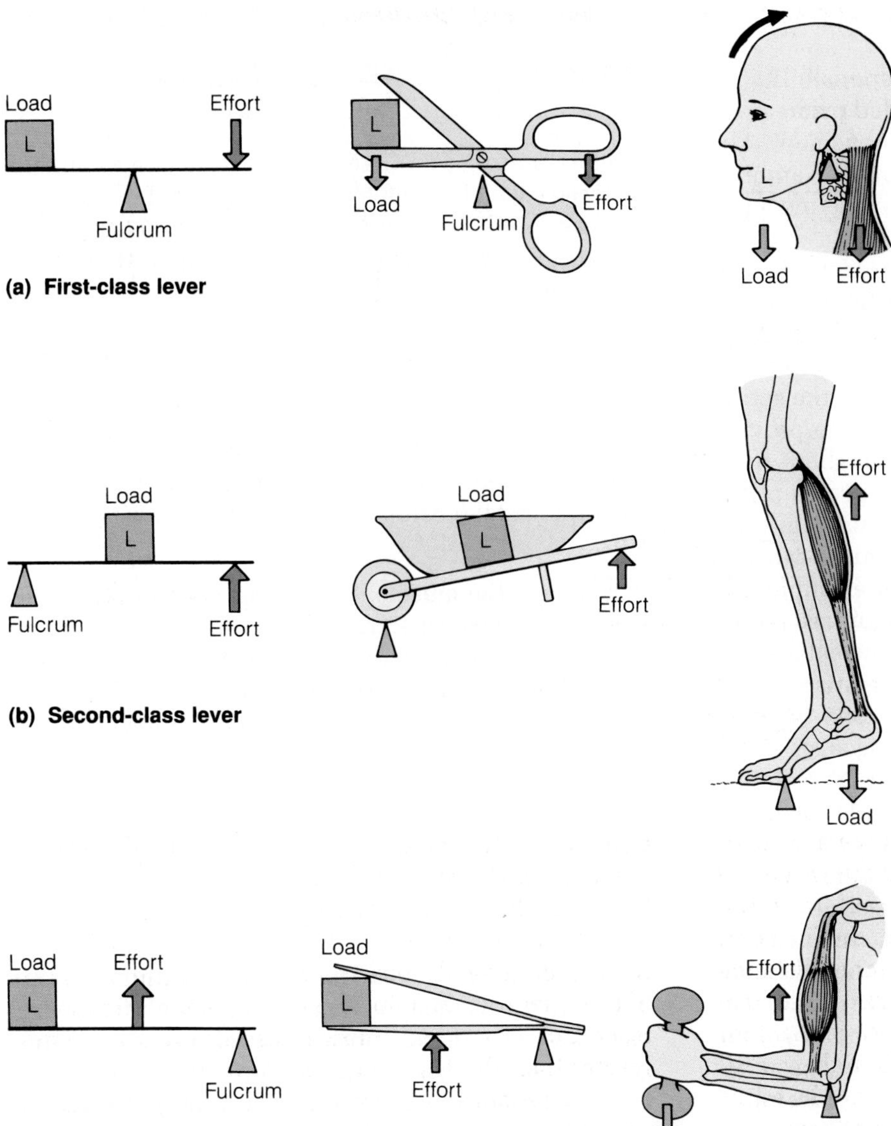

Figure 10.2 Lever systems. (**a**) In a first-class lever, the arrangement of the elements is load-fulcrum-effort. A scissors exhibits this lever class. In the body, a first-class lever system is activated when you raise your head off your chest. The posterior neck muscles provide the effort, the occipitoatlanto joint is the fulcrum, and the weight to be lifted is the facial skeleton. (**b**) In second-class levers, the arrangement is fulcrum-load-effort, as exemplified by a wheelbarrow. In the body, second-class leverage is exerted when you stand on tip-toe. The joints of the ball of the foot are the fulcrum, the weight of the body is the load, and the effort is exerted by the calf muscles pulling upward on the heel (calcaneus) of the foot. (**c**) In a third-class lever, the arrangement is load-effort-fulcrum. A pair of tweezers or forceps uses this type of leverage. The operation of the biceps brachii muscle in flexing the forearm exemplifies third-class leverage. The load is the hand and distal end of the forearm, the effort is exerted on the proximal radius of the forearm, and the fulcrum is the elbow joint.

the proximal radius, and the load to be lifted is the distal part of the forearm (and anything carried in the hand or over the forearm). Third-class lever systems permit a muscle to be inserted very close to the joint (fulcrum) across which movement occurs, providing for rapid and extensive movement with relatively little shortening of the muscle.

In conclusion, differences in the positioning of the three elements modify muscle activity with respect to (1) speed of contraction, (2) range of movement, and (3) the weight of the load that can be lifted. In lever systems that operate at a mechanical disadvantage, force is lost but speed is gained, and this is often a distinct advantage. Systems that operate at a mechanical advantage tend to be slower, more stable, and used where strength is a priority.

Interactions of Skeletal Muscles in the Body

The arrangement of body muscles permits groups of muscles to work either together or in opposition to achieve a wide variety of movements. As you eat, for example, you alternately raise your fork to your lips and lower it to your plate, and both sets of actions are accomplished by your arm and hand muscles. But muscles can only *pull;* they never *push.* Muscle contraction causes shortening, not lengthening, of the muscle and, as a muscle shortens, its insertion (attachment on the movable bone) moves toward its origin (its fixed or immovable point of attachment). Thus, whatever one muscle (or muscle group) can do, there must be another muscle or group of muscles that can "undo" the action.

Muscles that assume the major responsibility for producing a specific movement are called **prime movers**, or **agonists** (a′-guh-nists), of that movement. The biceps brachii muscle, which fleshes out the anterior arm (and inserts on the radius), is a prime mover of elbow flexion.

Muscles that oppose, or reverse, a particular movement are called **antagonists** (an-ta′-guh-nists). When a prime mover is active, the fibers of the antagonist muscles are often stretched and in the relaxed state. Antagonists can also help to regulate the action of a prime mover by partially contracting to provide some resistance, helping to prevent overshoot or to slow or stop the action. As you might expect, a prime mover and its antagonist are located on opposite sides of the joint across which they act. Antagonists can also be prime movers in their own right. For example, the biceps brachii muscle of the arm is antagonized by the triceps brachii, the prime mover of elbow extension.

In addition to agonists and antagonists, most movements also involve the action of one or more **synergists** (sih′-ner-jists). Synergists aid agonists either by (1) promoting the same movement or (2) reducing undesirable or unnecessary extra movements that might result as the prime mover contracts. This latter function deserves a bit more explanation. When a muscle crosses two or more joints, its contraction causes movement at all of the spanned joints unless other muscles act as joint stabilizers. For example, the finger flexor muscles cross both the wrist and the phalangeal joints, but you can make a fist without bending your wrist because synergistic muscles stabilize the wrist joint. Additionally, as some flexors act, undesirable rotatory movements may occur; synergists can prevent this movement, allowing all of the prime mover's force to be exerted in the desired direction.

When synergists immobilize a bone, or a muscle's origin, they are more specifically called **fixators** (fix′-ā-ters). You may recall from Chapter 7 that the scapula is held to the axial skeleton by muscles only and is quite freely movable. Many muscles that move the arm originate on the scapula, and if they are to be effective, movements of the scapula must be minimized. The fixator muscles that run from the axial skeleton to the scapula can immobilize the scapula so that only the desired movements occur at the shoulder joint. Additionally, muscles that help to maintain upright posture are fixators.

In summary, although prime movers seem to "get all the credit" for causing certain movements, the actions of antagonistic and synergistic muscles are also important in effecting smooth, coordinated, and precise movements.

Naming of Skeletal Muscles

Skeletal muscles are named according to a number of criteria, each of which focuses on particular structural or functional characteristics of a muscle. Paying close attention to these cues can simplify the task of learning muscle names and actions.

1. Location of the muscle. Some muscle names indicate the bone or body region with which the muscle is associated. For example, the temporalis (tem″-por-a′-lis) muscle overlies the temporal bone, the brachialis (brā″-kē-a′-lis) muscle is found in the arm, and the intercostal (*costal* = rib) muscles run between the ribs.

2. Shape of the muscle. Since muscles often have distinctive shapes, this criterion is sometimes used for naming. For example, the deltoid (del′-toyd) muscle is roughly triangular (*deltoid* = triangle), and the right and left trapezius (truh-pē′-zē-is) muscles together form a trapezoid.

3. Relative size of the muscle. Terms such as *maximus* (largest), *minimus* (smallest), *longus* (long), and *brevis* (short) are often used in muscle names—as in gluteus maximus and gluteus minimus (the large and small gluteus muscles, respectively).

4. Direction of muscle fibers. The naming of some muscles reveals the direction in which their fibers (and fascicles) run in reference to some imaginary line, usually the midline of the body or the longitudinal axis of a limb bone or the vertebral column. In muscles with the term *rectus* (straight) in their name, the fibers run parallel to that imaginary line, whereas the terms *transversus* and *oblique* indicate that the muscle fibers run respectively at right angles and obliquely to that line. Muscles named according to fiber direction include the rectus femoris (straight muscle of the thigh, or femur) and transversus abdominis (transverse muscle of the abdomen).

5. Number of origins. When the term *biceps, triceps,* or *quadriceps* forms part of a muscle's name, you can assume that the muscle has two, three, or four origins, respectively. For example, the biceps brachii (brā′-kē-ī) muscle of the arm has two origins, or *heads,* and the quadriceps femoris muscle of the thigh has four.

6. Location of the muscle's origin and/or insertion. Some muscles are named according to their attachment points. When this criterion is used and both origin and insertion points are noted, the origin is always named first. For instance, the sternocleidomastoid (ster″-nō-klī″-dō-mas′-toyd) muscle of the neck

A CLOSER LOOK Athletes Looking Good—Doing Better With Anabolic Steroids?

American society loves a winner and its top athletes reap large social and monetary rewards. Thus, it is not surprising that some will grasp at any thing that will increase their performance—including anabolic steroids. Anabolic steroids, variants of testosterone engineered by pharmaceutical companies, were introduced in the 1950s to treat victims of certain muscle-wasting diseases and anemia, and to prevent muscle atrophy in patients immobilized after surgery. Testosterone is responsible for the increase in muscle and bone mass and other physical changes that occur during puberty and convert boys into men. Convinced that megadoses of the steroids could produce enhanced masculinizing effects in grown men, many athletes were using the steroids by the early 1960s and the practice is still going on today.

It has been difficult to determine the incidence of anabolic steroid use among athletes because the use of drugs has been banned by most international competitions and users (and prescribing physicians or drug dealers) are naturally reluctant to talk about it. Nonetheless, there is little question that many professional body builders and athletes competing in events that require great muscle strength (e.g., shot put, discus throwing, and weight lifting) are heavy users. Sports figures such as football players have also admitted to using steroids as an adjunct to training, diet, and psychologic preparation for games. Advantages of anabolic steroids cited by athletes include increased muscle mass and strength, increased oxygen carrying capability owing to enhanced red blood cell volume, and an increase in agressive behavior (i.e., the urge to "steamroller" the other guy.)

Typically body builders who use steroids combine high doses (up to 200 mg/day) with heavy resistance training. Intermittent use begins several months before an event and both oral and injected steroid doses (a method called stacking) are increased gradually as the competition nears.

But do the drugs do all that is claimed for them? Research studies have reported increases in isometric strength and body weight increases in steroid users. While these are results weight lifters dream about, there is a hot dispute over whether this also translates into athletic performance requiring the fine muscle coordination and endurance needed by runners, etc. Present evidence indicates that it does not, and nonusers handily outperform users in such athletic endeavors.

Do the seemingly minimal advantages conferred by steroid use outweight the risks? This too is doubtful. Physicians say they cause bloated faces (Cushingoid sign of steroid excess); shriveled testes and infertility; damage the liver and promote liver cancer; and cause changes in blood cholesterol levels (which may predispose longterm users to coronary heart disease). Additionally, it now appears that the psychiatric hazards of anabolic steroids may be equally threatening and recent studies have indicated that one-third of anabolic steroid users have serious mental problems. Manic behavior in which the users undergo Jekyll-Hyde personality swings and become extremely violent is common, as are depression and delusions.

The question of why athletes use these drugs is easy to answer. Some admit to a willingness to do almost anything to win, short of killing themselves. Are they unwittingly doing this as well?

has a dual origin on the sternum (*sterno*) and clavicle (*cleido*), and it inserts on the *mastoid* process of the temporal bone.

7. Action of the muscle. When muscles are named for their action, descriptive words such as *flexor, extensor, adductor,* or *abductor* appear in the muscle's name. For example, the adductor longus, located on the medial thigh, brings about thigh adduction; all extensor muscles of the wrist effect wrist extension; and the supinator (soo'-pih-nā-ter) muscle supinates the forearm.

Often, several criteria are combined in the naming of a muscle. For instance, the name *extensor carpi radialis longus* tells us the muscle's action (extensor), what joint it acts on (*carpi* = wrist), and that it lies close to the radius of the forearm (radialis); it also hints at its relative size (longus) relative to other wrist extensor muscles. It would help if all muscle names were this descriptive; sadly, they are not.

Major Skeletal Muscles of the Body

The grand plan of the muscular system is all the more impressive because of the sheer number of skeletal muscles in the body—there are over 600 of them! Obviously, trying to commit to memory all the names, locations, and actions of these muscles would be a monumental task. Take heart; only the principal muscles (approximately 125 pairs of them) will be considered here. Although this number is far fewer than 600, the job of learning about all these muscles will still require a concerted effort on your part to memorize the necessary information. Yet, memorization will only be useful if you can apply what you have learned in a practical, or clinical, way. That is, memorization should be done with a *functional anatomy focus.* Once you are satisfied that you have learned the name of a muscle and can identify it on a cadaver, model, or diagram, you must then flesh out your learning by asking yourself, "What does it do?"

In the tables that follow, the muscles of the body have been grouped according to their body location and function. Each table is keyed to a particular figure (or group of figures) illustrating the muscles it describes. The legend at the beginning of each table provides an orienting overview of the types of movements effected by the muscles listed and gives pointers on the way those muscles interact with one another. The table itself describes each muscle's shape, location relative to other muscles, origin and insertion, primary actions, and innervation. (Since some instructors want students to defer learning the muscle innervations until the nervous system has been studied, you might be wise to determine what is expected of you in this regard.)

As you consider each individual muscle, be aware of the information its name provides. Then, after you have read through its entire description, identify the muscle on the figure that corresponds to the table and, in the case of superficial muscles, also on Figure 10.3 or 10.4. This will help you to link the descriptive material in the table to a visual image of the muscle's relative location in a specific body region and in the body as a whole. As you scrutinize the muscle's position, try to relate its attachments and location to the actions allowed by the joint(s) it acts across. This will

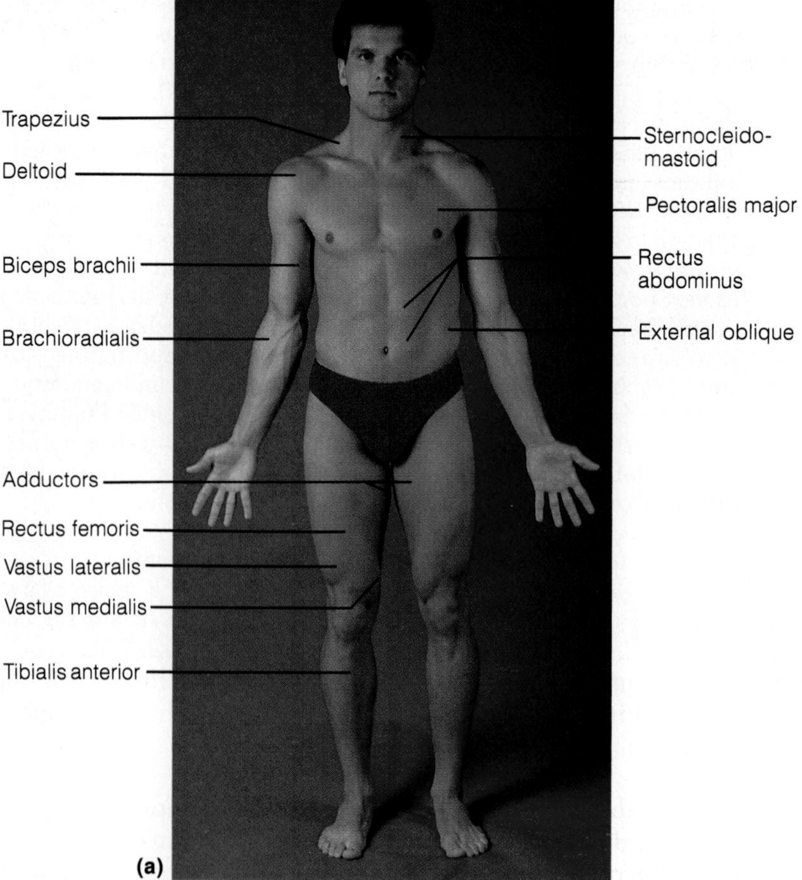

Figure 10.3 Anterior view of superficial muscles of the body. (**a**) Photograph of surface anatomy. (**b**) Diagrammatic view. The abdominal surface has been partially dissected on the right side to show somewhat deeper muscles.

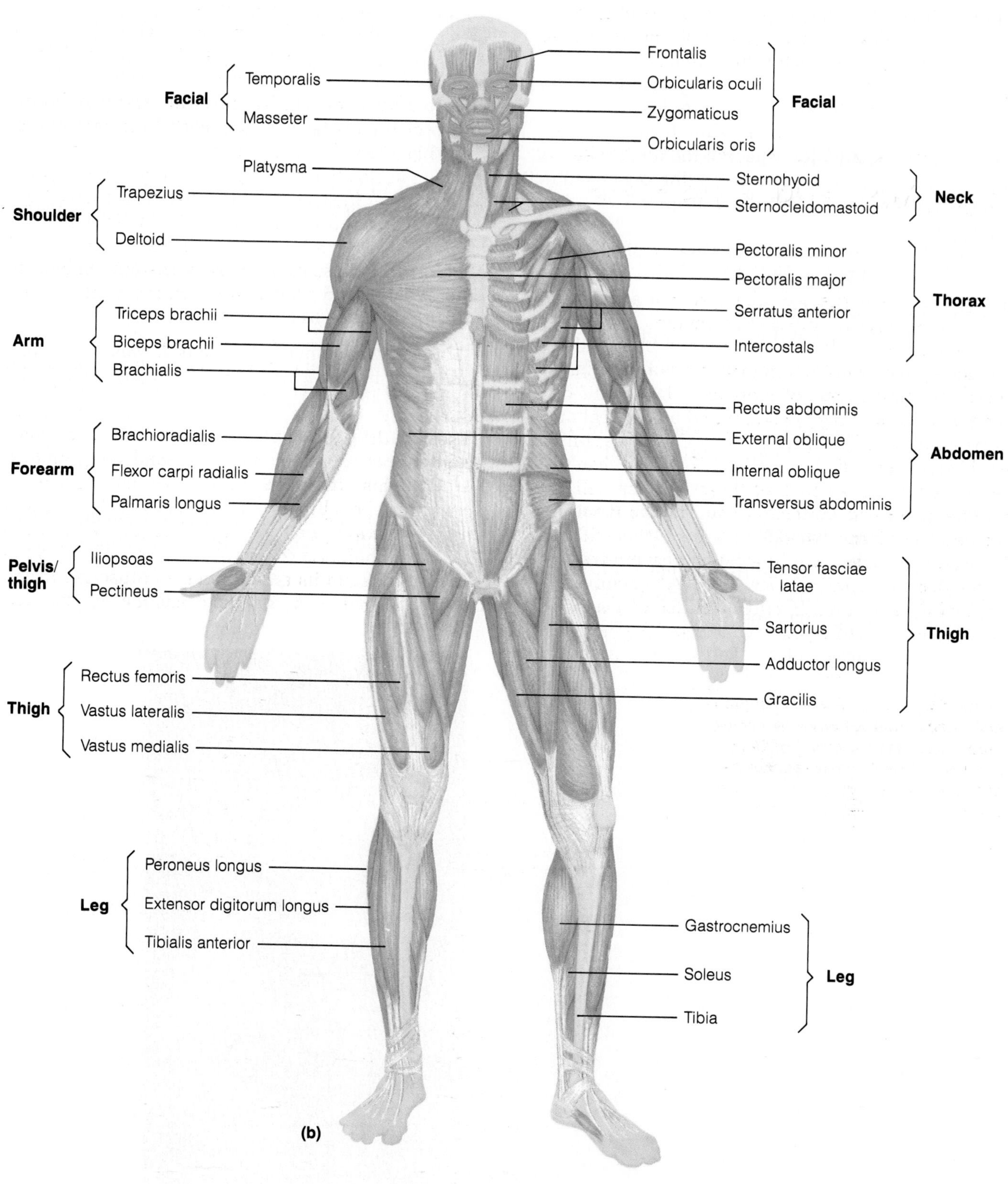
Facial
Temporalis
Masseter
Frontalis
Orbicularis oculi
Zygomaticus
Orbicularis oris
Facial
Platysma
Shoulder
Trapezius
Deltoid
Sternohyoid
Sternocleidomastoid
Neck
Pectoralis minor
Pectoralis major
Serratus anterior
Intercostals
Thorax
Arm
Triceps brachii
Biceps brachii
Brachialis
Forearm
Brachioradialis
Flexor carpi radialis
Palmaris longus
Rectus abdominis
External oblique
Internal oblique
Transversus abdominis
Abdomen
Pelvis/thigh
Iliopsoas
Pectineus
Tensor fasciae latae
Sartorius
Adductor longus
Gracilis
Thigh
Thigh
Rectus femoris
Vastus lateralis
Vastus medialis
Leg
Peroneus longus
Extensor digitorum longus
Tibialis anterior
Gastrocnemius
Soleus
Tibia
Leg
(b)

help you to focus in on functional details that often escape student awareness. For example, both the elbow and knee joints are hinge joints that allow flexion and extension. However, the knee flexes to the dorsum of the body (the calf moves toward the posterior thigh), whereas elbow flexion carries the forearm toward the anterior aspect of the arm. Therefore, leg flexors are located on the posterior thigh, while forearm flexors are found on the anterior aspect of the humerus.

The following list summarizes the organization and sequence of the tables in this chapter:

- **Table 10.1** Muscles of the Head, Part I: Facial Expression (Figure 10.5); pp. 286–287
- **Table 10.2** Muscles of the Head, Part II: Mastication and Tongue Movement (Figure 10.6); pp. 288–289
- **Table 10.3** Muscles of the Anterior Neck and Throat: Swallowing (Figure 10.7); pp. 290–291
- **Table 10.4** Muscles of the Neck and Vertebral Column: Head and Trunk Movements (Figure 10.8); pp. 292–295
- **Table 10.5** Deep Muscles of the Thorax: Breathing (Figure 10.9); pp. 296–297
- **Table 10.6** Muscles of the Abdominal Wall: Trunk Movements and Compression of Abdominal Viscera (Figure 10.10); pp. 298–299
- **Table 10.7** Muscles of the Pelvic Floor and Perineum: Support of Abdominopelvic Organs (Figure 10.11); pp. 300–301
- **Table 10.8** Superficial Muscles of the Anterior and Posterior Thorax: Movements of the Scapula (Figure 10.12); pp. 302–303
- **Table 10.9** Muscles Crossing the Shoulder Joint: Movements of the Arm (Humerus) (Figure 10.13); pp. 304–306
- **Table 10.10** Muscles Crossing the Elbow Joint: Flexion and Extension of the Forearm (Figure 10.13); pp. 306–307
- **Table 10.11** Muscles of the Forearm: Movements of the Wrist, Hand, and Fingers (Figures 10.14 and 10.15); pp. 308–311
- **Table 10.12** Summary of Actions of Muscles Acting on the Arm, Forearm, and Hand (Figure 10.16); pp. 312–313
- **Table 10.13** Muscles Crossing the Hip and Knee Joints: Movements of the Thigh and Leg (Figures 10.17 and 10.18); pp. 314–319
- **Table 10.14** Muscles of the Leg: Movements of the Ankle and Toes (Figures 10.19 to 10.21); pp. 320–325
- **Table 10.15** Summary of Actions of Muscles Acting on the Thigh, Leg, and Foot (Figure 10.22); pp. 326–327

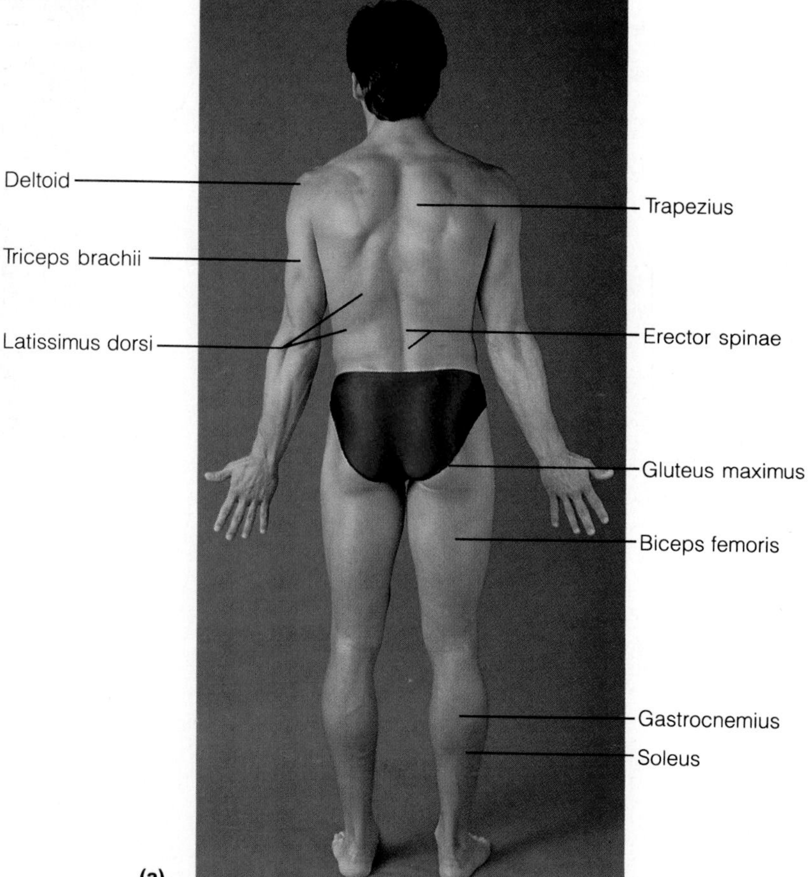

Figure 10.4 Posterior view of superficial muscles of the body. (**a**) Photograph of surface anatomy. (**b**) Diagrammatic view.

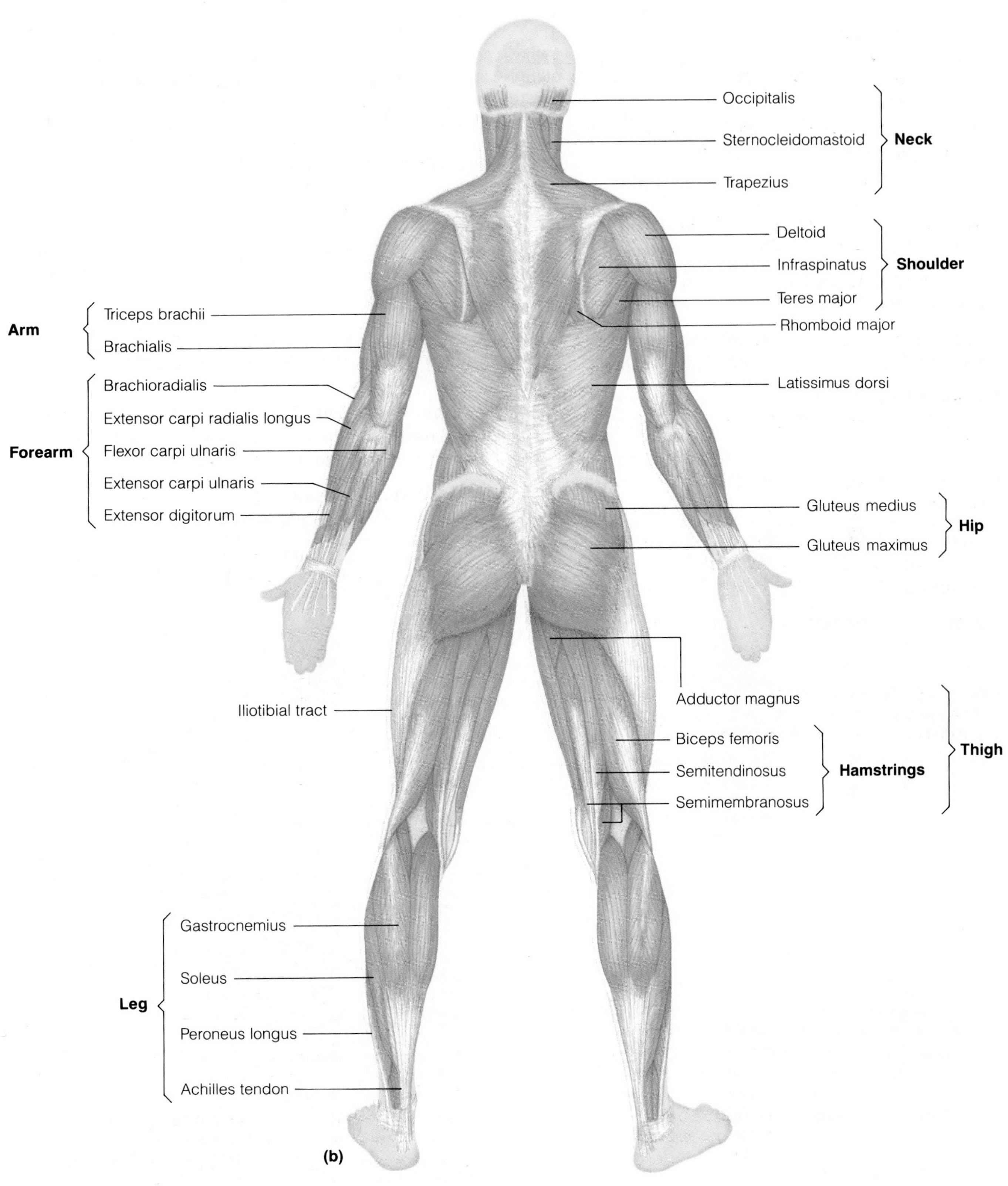
Occipitalis
Sternocleidomastoid
Trapezius
Neck
Deltoid
Infraspinatus
Teres major
Shoulder
Rhomboid major
Latissimus dorsi
Arm
Triceps brachii
Brachialis
Forearm
Brachioradialis
Extensor carpi radialis longus
Flexor carpi ulnaris
Extensor carpi ulnaris
Extensor digitorum
Gluteus medius
Gluteus maximus
Hip
Iliotibial tract
Adductor magnus
Biceps femoris
Semitendinosus
Semimembranosus
Hamstrings
Thigh
Leg
Gastrocnemius
Soleus
Peroneus longus
Achilles tendon

(b)

Table 10.1 Muscles of the Head, Part I: Facial Expression (Figure 10.5)

The superficial muscles of the head involved in promoting facial expression include those of the scalp and face. They are highly variable in shape and strength, and adjacent muscles tend to be fused. The bulk of the scalp musculature is provided by the bipartite *epicranius* muscle; the lateral scalp muscles are vestigial in humans. Muscles clothing the facial skeleton allow the eyebrows to be lifted and the nostrils to be flared, allow opening and closing of the eyes and mouth, and provide one of the most effective tools for influencing others—the smile. The tremendous importance of facial muscles in nonverbal communication becomes glaringly clear when they are paralyzed, as in some stroke victims. All muscles listed in this table are innervated by *cranial nerve VII (the facial nerve)*. The external muscles of the eye, which are contained within the orbit and act to direct the eyeball, and the levator palpebrae superioris muscles that raise the eyelids are described in Chapter 16.

Muscle	Description	Origin (O) and insertion (I)	Action	Nerve supply
MUSCLES OF THE SCALP				
Epicranius (Occipitofrontalis) (eh″-pih-krā′-nē-us; ok-sih″-pih-tō-frun-ta′-lis) (*epi* = over; *cran* = skull)	Bipartite muscle consisting of the frontalis and occipitalis muscles connected by a cranial aponeurosis, the galea aponeurotica; the alternate action of these two muscles pulls scalp forward and backward			
• **Frontalis** (frun-ta′-lis) (*front* = forehead)	Covers forehead and dome of skull; no bony attachments	O—galea aponeurotica I—skin of eyebrows and root of nose	With aponeurosis fixed, raises the eyebrows (as in surprise); antagonist of orbicularis oculi	Facial nerve
• **Occipitalis** (ok-sih″-pih-ta′-lis) (*occipito* = base of skull)	Overlies base of occiput; by pulling on the galea, fixes origin of frontalis	O—occipital bone I—galea aponeurotica	Fixes aponeurosis and pulls scalp posteriorly	Facial nerve
MUSCLES OF THE FACE				
Corrugator supercilii (kor′-uh-gā″-ter soo-per-sih′-lē-ī) (*corrugo* = wrinkle; *supercilium* = eyebrow)	Small muscle; activity associated with that of orbicularis oculi	O—arch of frontal bone above nasal bone I—skin of eyebrow	Draws eyebrows together; wrinkles skin of forehead vertically (as in frowning)	Facial nerve
Orbicularis oculi (or-bih″-kyoo-layr′-is o′-kyoo-lī) (*orb* = circular; *ocul* = eye)	Thin, flat sphincter muscle of eyelid; surrounds rim of the orbit; paralysis results in drooping of lower eyelid and spilling of tears	O—frontal and maxillary bones and ligaments around orbit I—tissue of eyelid	Protects eyes from intense light and injury; various parts can be activated individually; produces blinking, squinting, and draws eyebrows downward; strong closure of eyelids causes creasing of skin at lateral aspect of eye (these folds, called "crow's feet," become permanent with aging)	Facial nerve
Zygomaticus (zī-gō-ma′-tih-kus), major and minor (*zygomatic* = cheekbone)	Muscle pair extending diagonally from corner of mouth to cheekbone	O—zygomatic bone I—skin and muscle at corner of mouth	Raises lateral corners of mouth upward (smiling muscle)	Facial nerve
Risorius (rih-zor′-ē-us) (*risor* = laughter)	Slender muscle running beneath and laterally to zygomaticus	O—lateral fascia associated with masseter muscle I—skin at angle of mouth	Draws corner of lip laterally; tenses lips; synergist of zygomaticus	Facial nerve
Levator labii superioris (leh-vā′-ter la′-bē-ī soo-pēr″-ē-or′-is) (*leva* = raise; *labi* = lip; *superior* = above, over)	Thin muscle between orbicularis oris and inferior eye margin	O—zygomatic bone and infraorbital margin of maxilla I—skin and muscle of upper lip	Opens lips; raises and furrows the upper lip; flares nostril (as in disgust)	Facial nerve
Depressor labii inferioris (dē-preh′-ser la′-bē-ī in-fēr″-ē-or′-is) (*depressor* = depresses; *infer* = below)	Small muscle running from lower lip to jawbone	O—body of mandible lateral to its midline I—skin and muscle of lower lip	Draws lower lip downward (as in a pout)	Facial nerve

Muscle	Description	Origin (O) and Insertion (I)	Action	Nerve supply
Depressor anguli oris (an′-gyoo-lē or′-is) (*angul* = angle, corner)	Small muscle lateral to depressor labii inferioris	O—body of mandible below incisors I—skin and muscle at angle of mouth below insertion of zygomaticus	Zygomaticus antagonist; draws corners of mouth downward and laterally (as in a "tragedy mask" grimace)	Facial nerve
Orbicularis oris (*or* = mouth)	Complicated, multilayered muscle of the lips with fibers that run in many different directions	O—arises indirectly from maxilla and mandible; fibers blended with fibers of other facial muscles associated with the lips I—encircles mouth; inserts into muscle and skin at angles of mouth	Closes lips; purses and protrudes lips (kissing muscle)	Facial nerve
Mentalis (men-ta′-lis) (*ment* = chin)	One of the muscle pair forming a V-shaped muscle mass on chin	O—incisor fossa of mandible I—skin of chin	Protrudes lower lip; wrinkles chin	Facial nerve
Buccinator (byoo′-sih-nā″-ter) (*bucc* = cheek or "trumpeter")	Thin, horizontal cheek muscle; principal muscle of cheek; deep to masseter (see also Figure 10.6.)	O—molar region of maxilla and mandible I—orbicularis oris; fibers form deep portion of lips	Draws corner of mouth laterally; compresses cheek (as in whistling, blowing, and sucking); holds food between teeth during chewing; well developed in nursing infants	Facial nerve
Platysma (pluh-tiz′-muh) (*platy* = broad, flat)	Unpaired, thin, sheetlike superficial neck muscle; not strictly a head muscle, but plays a role in facial expression	O—fascia of chest (over pectoral muscles and deltoid) I—lower margin of mandible, and skin and muscle at corner of mouth	Helps depress mandible; pulls lower lip back and down, i.e., produces downward sag of mouth; takes up slack in skin when head is bent forward	Facial nerve

Figure 10.5 Lateral view of muscles of the scalp, face, and neck.

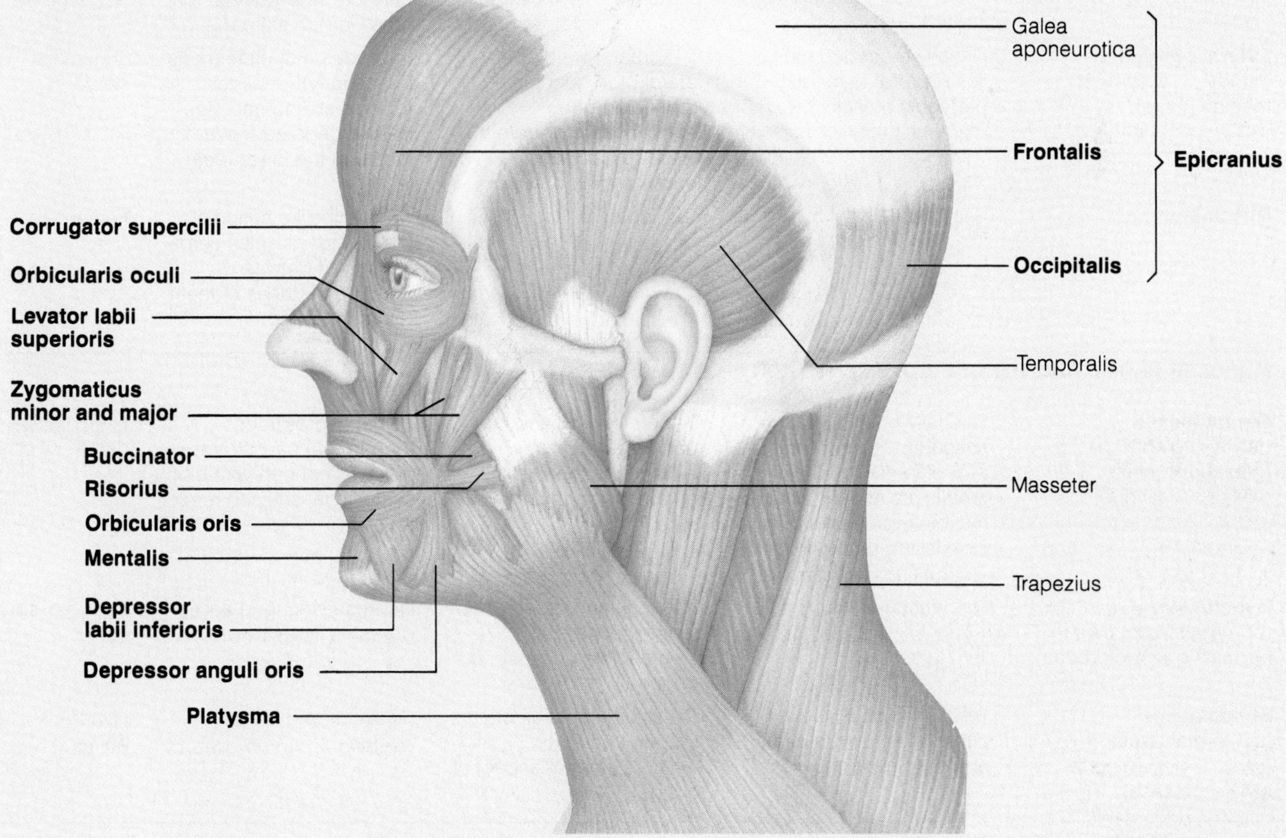

Table 10.2 Muscles of the Head, Part II: Mastication and Tongue Movement (Figure 10.6)

Four pairs of muscles are involved in mastication (chewing and biting) activities, and all are innervated by the mandibular division of *cranial nerve V* (the *trigeminal nerve*). The prime movers of jaw closure (and biting) are the powerful *masseter* and *temporalis* muscles, which can be palpated easily when the teeth are clenched. Grinding movements are provided for by the *pterygoid* muscles. The *buccinator* muscles (see Table 10.1) also play a role in chewing. Normally, gravity is sufficient to depress the mandible, but if there is resistance to jaw opening, neck muscles (digastric and mylohyoid muscles; see Table 10.3) are activated.

The tongue is composed of muscle fibers peculiar to itself that curl, squeeze, and fold the tongue during speaking and chewing. These *intrinsic tongue muscles* are arranged in several planes (superior and inferior longitudinal, transverse, and vertical), which contributes to the exceptional nimbleness of the tongue. Only the *extrinsic tongue muscles,* which serve to anchor and move the tongue, are considered in this table. The extrinsic tongue muscles are all innervated by *cranial nerve XII* (the *hypoglossal nerve*).

Muscle	Description	Origin (O) and Insertion (I)	Action	Nerve supply
MUSCLES OF MASTICATION				
Masseter (ma′-sih-ter) (*maseter* = chewer)	Powerful muscle that covers lateral aspect of the jawbone	O—zygomatic process and arch I—angle and ramus of mandible	Prime mover of jaw closure; elevates mandible	Trigeminal nerve
Temporalis (tem″-por-a′-lis) (*tempora* = time)	Fan-shaped muscle over temporal bone	O—temporal fossa I—coronoid process of mandible via a tendon that passes beneath zygomatic arch	Closes jaw; elevates and retracts mandible; maintains position of the mandible at rest	Trigeminal nerve
Medial pterygoid (mē′-dē-ul tayr′-ih-goyd) (*medial* = toward median plane; *pterygoid* = winglike)	Deep two-headed muscle that runs along internal surface of mandible and is largely concealed by that bone	O—medial surface of lateral pterygoid plate of sphenoid bone, maxilla, and palatine bone I—medial surface of mandible near its angle	Synergist of temporalis and masseter muscles in elevation of the mandible; acts with the lateral pterygoid muscle to promote side-to-side (chewing) movements	Trigeminal nerve
Lateral pterygoid (*lateral* = away from median plane)	Deep two-headed muscle; lies superior to medial pterygoid muscle	O—greater wing and lateral pterygoid plate of sphenoid bone I—condyle of mandible and capsule of temporomandibular joint	Protrudes mandible (pulls it anteriorly); provides forward sliding and side-to-side grinding movements of the lower teeth	Trigeminal nerve
Buccinator	See Table 10.1	See Table 10.1	Trampolinelike action of buccinator muscles helps keep food between grinding surfaces of teeth during chewing	Facial nerve
MUSCLES PROMOTING TONGUE MOVEMENTS (EXTRINSIC MUSCLES)				
Genioglossus (jē″-nē-ō-glah′-sus) (*geneion* = chin; *glossus* = tongue)	Vertically positioned fan-shaped muscle; forms bulk of inferior part of tongue; its attachment to mandible prevents tongue from falling backward and obstructing respiration	O—internal surface of mandible near symphysis I—inferior aspect of the tongue and body of hyoid bone	Primarily protrudes tongue, but can depress or act in concert with other extrinsic muscles to retract tongue	Hypoglossal nerve
Hyoglossus (hī′-ō-glah″-sus) (*hyo* = pertaining to hyoid bone)	Flat, quadrilateral muscle	O—body and greater horn of hyoid bone I—inferiolateral aspect of tongue	Depresses tongue and draws its sides downward	Hypoglossal nerve
Styloglossus (stī′-lō-glah″-sus) (*stylo* = pertaining to styloid process)	Slender muscle running superiorly to and at right angles to hyoglossus	O—styloid process of temporal bone I—lateral inferior aspect of tongue	Retracts (and elevates) tongue	Hypoglossal nerve

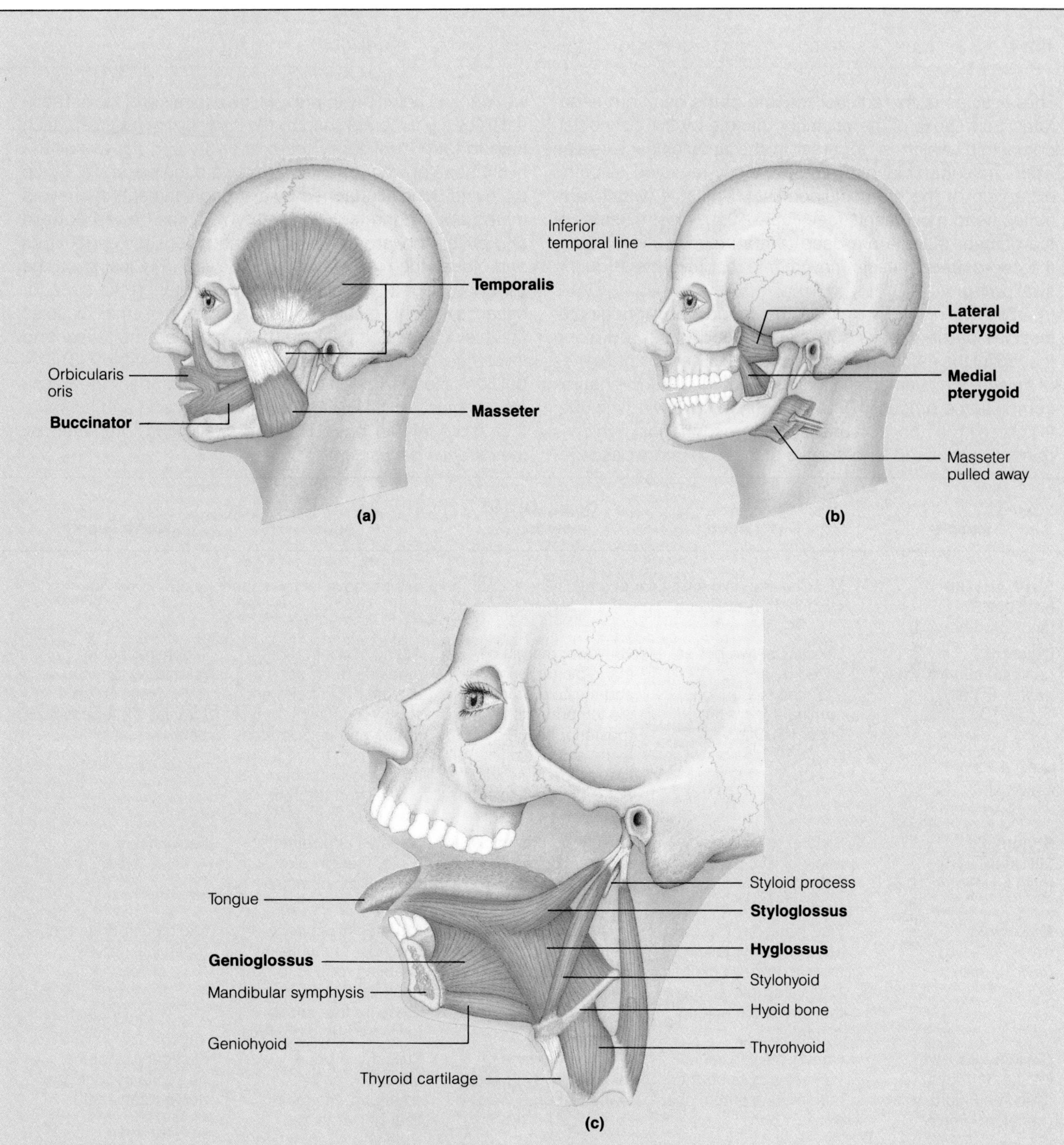

Figure 10.6 Muscles promoting mastication and tongue movements. (**a**) Lateral view of the temporalis, masseter, and buccinator muscles. (**b**) Lateral view of the deep chewing muscles, the medial and lateral pterygoid muscles. (**c**) Extrinsic muscles of the tongue. Some suprahyoid muscles of the throat are also illustrated.

Table 10.3 Muscles of the Anterior Neck and Throat: Swallowing (Figure 10.7)

The muscles of the anterior triangle of the neck are separated from those of the posterior triangle by the sternocleidomastoid, which is apparent in the anterior neck region when the superficial platysma muscle is removed. With the exception of the sternocleidomastoid muscle, which promotes head movements (see Table 10.4), most muscles of the anterior triangle are deep (throat) muscles involved in the coordinated muscle movements that take place during swallowing.

Swallowing begins when the tongue and buccinator muscles of the cheeks squeeze the food backward along the roof of the mouth toward the pharynx. Then a rapid series of muscular movements in the posterior mouth and pharynx complete the process. These events of swallowing include: (1) Opening of the food passageway (esophagus), which is normally collapsed, and closure of the respiratory passageway (larynx) anteriorly to prevent the aspiration of food. These activities are accomplished by the *suprahyoid muscles*, which help to form the floor of the oral cavity and act to pull the hyoid bone upward and forward toward the mandible. Since the hyoid bone is attached by a strong ligament (thyrohyoid membrane) to the larynx, the larynx is also raised upward and pulled forward, a maneuver that opens the esophagus and closes the respiratory passageway. (2) Closing off the nasal passages to prevent food from entering the superior nasal cavity. This results from the activation of small muscles that elevate the soft palate. (These muscles, the *tensor* and *levator veli palatini*, are not described in the table.) (3) Propulsion of food through the pharynx by the pharyngeal *constrictor muscles*. (4) Return of the hyoid bone and larynx to their more inferior positions after swallowing, promoted by the *infrahyoid muscles*.

Muscle	Description	Origin (O) and insertion (I)	Action	Nerve supply
SUPRAHYOID MUSCLES (soo″-pruh-hī′-oyd)	Muscles that help form floor of oral cavity, anchor tongue, and move larynx superiorly during swallowing; lie superior to hyoid bone			
Digastric (dī″-gas′-trik) (*di* = two; *gaster* = belly)	Consists of two bellies united by an intermediate tendon, forming a V shape under the chin	O—lower margin of mandible (anterior belly) and mastoid process of the temporal bone (posterior belly) I—by a connective tissue loop to hyoid bone	Acting in concert, the digastric muscles elevate hyoid bone and steady it during swallowing and speech; acting from behind, they open mouth (prime mover) and depress mandible	Trigeminal nerve (cranial V) for anterior belly; facial nerve (cranial VII) for posterior belly
Stylohyoid (stī′-lō-hī″-oyd) (also see Figure 10.6)	Slender muscle below angle of jaw; parallels posterior belly of digastric muscle	O—styloid process of temporal bone I—hyoid bone	Elevates and retracts hyoid, thereby elongating floor of mouth during swallowing	Facial nerve
Mylohyoid (mī′-lō-hī″-oyd) (*myle* = mill)	Flat, triangular muscle just deep to digastric muscle; this muscle pair forms a sling that forms the floor of the anterior mouth	O—medial surface of mandible I—hyoid bone	Elevates hyoid bone and floor of mouth, enabling tongue to exert backward and upward pressure that forces food bolus into pharynx	Mandibular branch of trigeminal nerve
Geniohyoid (jē′-nē-ō-hī″-oyd) (also see Figure 10.6) (*geneion* = chin)	Narrow muscle in contact with its partner medially; runs from chin to hyoid bone	O—inner surface of mandibular symphysis I—hyoid bone	Pulls hyoid bone superiorly and anteriorly, shortening floor of mouth and widening pharynx for receiving food during swallowing	First cervical spinal nerve via hypoglossal nerve (cranial XII)
INFRAHYOID MUSCLES	Muscles causing depression of hyoid bone and larynx during swallowing and speaking; because of ribbonlike appearance, often called *strap muscles*			
Sternohyoid (ster′-nō-hī″-oyd) (*sterno* = sternum)	Most medial muscle of the neck; thin; superficial except inferiorly, where covered by sternomastoid	O—manubrium and medial end of clavicle I—lower margin of hyoid bone	Depresses larynx and hyoid bone if mandible is fixed; may also flex skull	C_1–C_3 through ansa cervicalis (slender nerve root in cervical plexus)
Sternothyroid (ster′-nō-thī′-royd) (*thyro* = thyroid gland)	Lateral and deep to sternohyoid	O—posterior surface of manubrium of sternum I—thyroid cartilage	Pulls thyroid cartilage (plus larynx and hyoid bone) inferiorly	As for sternohyoid

Muscle	Description	Origin (O) and insertion (I)	Action	Nerve supply
Omohyoid (ah′-mō-hī″-oyd) (*omo* = shoulder)	Straplike muscle with two bellies united by an intermediate tendon; lateral to sternohyoid	O—superior surface of scapula I—hyoid bone, lower border	Depresses and retracts hyoid bone	As for sternohyoid
Thyrohyoid (thī′-rō-hī″-oyd) (also see Figure 10.6)	Appears as a superior continuation of sternothyroid muscle	O—thyroid cartilage I—hyoid bone	Depresses hyoid bone and elevates thyroid cartilage	First cervical nerve via hypoglossal
Pharyngeal constrictor muscles—superior, middle, and inferior (fayr-in-jē′-ul)	Composite of three paired muscles whose fibers run circularly in pharynx wall; arranged so that the superior muscle is innermost and inferior one is outermost; substantial overlap	O—attached anteriorly to mandible and medial pterygoid plate (superior), hyoid bone (middle), and laryngeal cartilages (inferior), and run around to back of pharynx I—posterior median raphe of pharynx	Working as a group, all constrict pharynx during swallowing, which propels a food bolus to esophagus	Pharyngeal plexus [branches of vagus (X) and glossopharyngeal (IX) nerves]

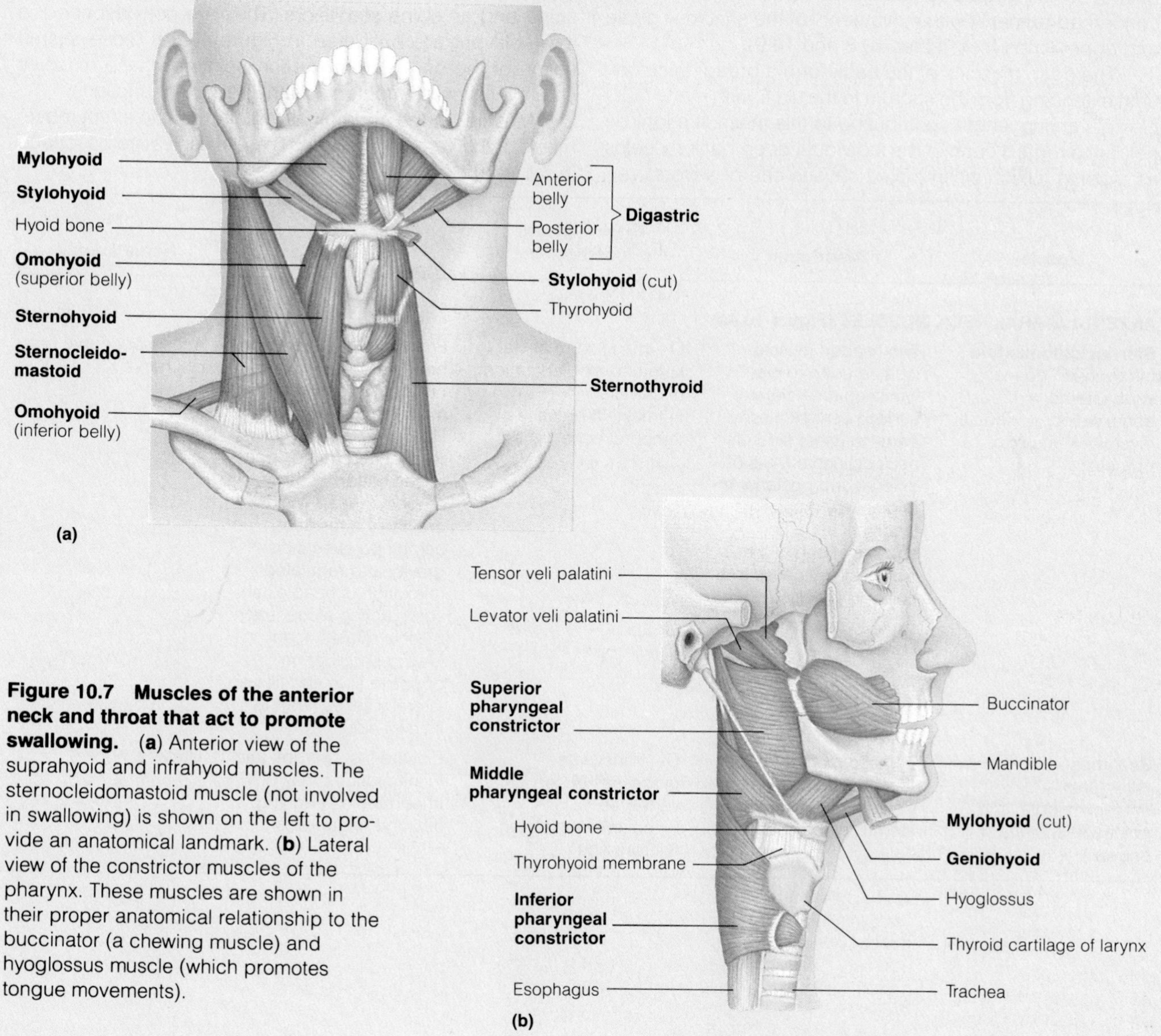

Figure 10.7 Muscles of the anterior neck and throat that act to promote swallowing. (**a**) Anterior view of the suprahyoid and infrahyoid muscles. The sternocleidomastoid muscle (not involved in swallowing) is shown on the left to provide an anatomical landmark. (**b**) Lateral view of the constrictor muscles of the pharynx. These muscles are shown in their proper anatomical relationship to the buccinator (a chewing muscle) and hyoglossus muscle (which promotes tongue movements).

Table 10.4 Muscles of the Neck and Vertebral Column—Head and Trunk Movements (Fi

Movements of the head are promoted by muscles originating from the axial skeleton. The major head flexors are the sternocleidomastoid muscles, but the suprahyoid and infrahyoid muscles described in Table 10.3 act as synergists in this action. Lateral head movements are effected by the sternocleidomastoids and a number of deeper neck muscles, including the *scalenes,* and by several straplike muscles of the vertebral column at the back of the neck. Head extension is aided by the superficial trapezius muscles of the back, but the *splenius* muscles underlying the trapezius muscles are much more important in head extension.

Trunk movements are effected by the *deep* or *intrinsic back muscles* associated with the bony vertebral column, muscles that also have an important role in maintaining the normal curvatures of the spine. Muscles of the intermediate layer overlying the deep back muscles are involved in breathing movements (see Table 10.5), while the superficial back muscles (trapezius, latissimus dorsi, and others) are concerned primarily with movements of the shoulder girdle and upper limbs (see Tables 10.8 and 10.9).

The deep muscles of the back form a broad, thick column extending from the sacrum to the skull, with many muscles of varying length contributing to this mass. It might be helpful to regard each of the individual deep back muscles as a string, which when pulled causes one or several vertebrae to be extended or to rotate on the vertebrae below. The largest and most extensive of the deep back muscle groups is the *erector spinae* (literally, spine erectors). Since the origins and insertions of the different muscle groups overlap extensively, entire regions of the vertebral column can be moved simultaneously and smoothly. Acting in concert, the deep back muscles can extend (or hyperextend) the spine, while contraction of muscles on only one side can cause lateral bending of the back, neck, or head. Lateral flexion is automatically accompanied by some degree of rotation of the vertebral column. During vertebral movements, the articular facets of the vertebrae glide or slip on each other.

In addition to the larger, more prominent muscles of the back, there are a number of small muscles that extend from one vertebra to the next. These muscles (the rotatores, multifidus, interspinalis, and intertransversarius muscles) function primarily as synergists in extension and rotation of the spine and as spine stabilizers. They are not described in the table but are illustrated in Figure 10.8d. Close examination of the origin and insertion points of these muscles should allow you to determine their specific functions.

Muscles of the abdominal wall also help to effect movements of the vertebral column. These muscles are described in Table 10.6.

Muscle	Description	Origin (O) and insertion (I)	Action	Nerve supply
ANTEROLATERAL NECK MUSCLES (Figure 10.8a)				
Sternocleidomastoid (ster″-nō-klī″-dō-mas′-toyd) (*sterno* = breastbone; *cleido* = clavicle; *mastoid* = mastoid process)	Two-headed muscle located deep to platysma on anterolateral surface of neck; fleshy parts on either side of neck delineate limits of anterior and posterior triangles; key muscular landmark in neck; spasms of one of these muscles may cause torticollis (wryneck)	O—manubrium of sternum and medial portion of clavicle I—mastoid process of temporal bone	Prime mover of active head flexion; simultaneous contraction of both muscles causes neck flexion, generally against resistance as when one raises head when lying on back; (head flexion is ordinarily a result of combined effects of gravity and regulated relaxation of head extensors); acting alone, each muscle rotates head toward shoulder on opposite side and tilts or laterally flexes head to its own side	Accessory nerve (cranial nerve XI)
Scalenes (skā′-lēnz)—anterior, middle, and posterior (*scalene* = uneven)	Located more laterally than anteriorly on neck; deep to platysma and sternocleidomastoid	O—transverse processes of cervical vertebrae I—posterolaterally on first two ribs	Elevate first two ribs (aid in inspiration); may be important in coughing; flex and rotate neck	Cervical nerves

Muscle	Description	Origin (O) and insertion (I)	Action	Nerve supply
INTRINSIC MUSCLES OF THE BACK (Figure 10.8b–d)				
Splenius (spleh′-nē-us)—capitis and cervicis portions (ka′-pih-tus; ser′-vih-kis) (*splenion* = bandage; *caput* = head; *cervi* = neck) (Figure 10.8b)	Broad bipartite superficial muscle (capitis and cervicis parts) extending from upper thoracic vertebrae to skull; capitis portion known as "bandage muscle" because it covers and holds down deeper neck muscles	O—ligamentum nuchae,* spinous processes of vertebrae C_7–T_6 I—mastoid process of temporal bone and occipital bone (capitis); transverse processes of C_2–C_4 vertebrae (cervicis)	Act as a group to extend or hyperextend head; when splenius muscles on one side are activated, head is rotated and bent laterally toward same side	Cervical spinal nerves (dorsal rami)

Figure 10.8 Muscles of the neck and vertebral column causing movements of the head and trunk. (**a**) Muscles of the anterolateral neck. The superficial platysma muscle and the deeper neck muscles have been removed to show the origins and insertions of the sternocleidomastoid and scalene muscles clearly. (**b**) Deep muscles of the posterior neck. Superficial muscles have been removed.

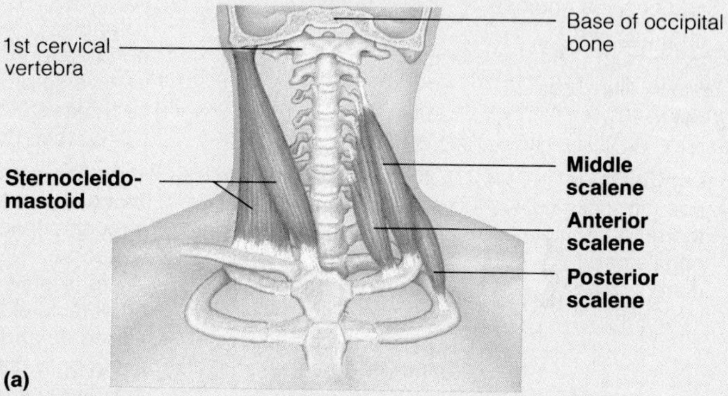

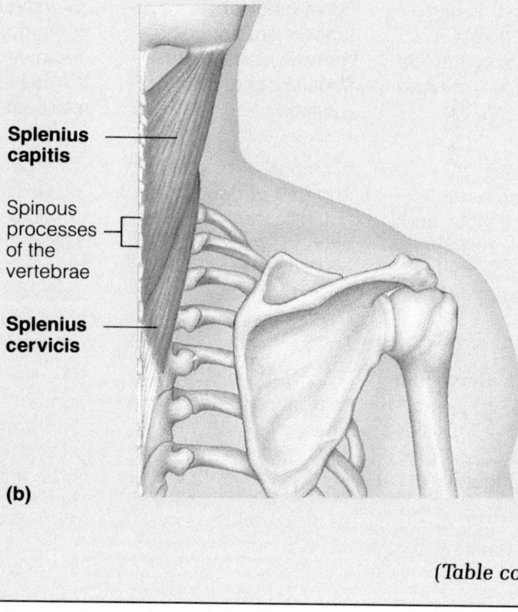

(Table continues)

*The ligamentum nuchae (lih″-guh-men′-tum noo′-kē)is a strong, elastic ligament extending from the occipital bone of the skull along the tips of the spinous processes of the cervical vertebrae. It binds the cervical vertebrae together and inhibits excessive head and neck flexion, thus preventing damage to the spinal cord in the vertebral canal.

Table 10.4 (continued)

Muscle	Description	Origin (O) and insertion (I)	Action	Nerve supply
Erector spinae (ē-rek′-ter spih′-nē) (Figure 10.8c, left side)	Prime mover of back extension; erector spinae muscles, each consisting of three columns—the iliocostalis, longissimus, and spinalis muscles—form intermediate layer of intrinsic back muscles; erector spinae provide resistance that helps control action of bending forward at the waist and act as powerful extensors to promote return to erect position; during full flexion (i.e., when touching fingertips to floor), erector spinae are relaxed and strain is borne entirely by ligaments of back; on reversal of the movement, these muscles are initially inactive, and extension is initiated by hamstring muscles of thighs and gluteus maximus muscles of buttocks. As a result of this peculiarity, lifting a load or moving suddenly from a bent-over position is potentially dangerous (in terms of possible injury) to muscles and ligaments of back and intervertebral discs; erector spinae muscles readily go into painful spasms following injury to back structures			
• **Iliocostalis** (ih″-lē-ō-kos-ta′-lis)—lumborum, thoracis, and cervicis portions (lum-bor′-um; thor-a′-kis) (*ilio* = ilium; *cost* = rib)	Most lateral muscle group of erector spinae muscles; extend from pelvis to neck	O—iliac crests (lumborum); inferior 6 ribs (thoracis); ribs 3 to 6 (cervicis) I—angles of ribs (lumborum and thoracis); transverse processes of cervical vertebrae C_6–C_4 (cervicis)	Extend vertebral column, maintain erect posture; acting on one side, bend vertebral column to same side	Spinal nerves (dorsal rami)
• **Longissimus** (lon-gih′-sih-mus)—thoracis, cervicis, and capitis parts (*longissimus* = longest)	Intermediate tripartite muscle group of erector spinae; extend by many muscle slips from lumbar region to skull; mainly pass between transverse processes of the vertebrae	O—transverse processes of lumbar through cervical vertebrae I—transverse processes of thoracic or cervical vertebrae and to ribs superior to origin as indicated by name; capitis inserts into mastoid process of temporal bone	Thoracis and cervicis act together to extend vertebral column and acting on one side, bend it laterally; capitis extends head and turns the face toward same side	Spinal nerves (dorsal rami)
• **Spinalis** (spih-na′-lis)—thoracis and cervicis parts (*spin* = vertebral column, spine)	Most medial muscle column of erector spinae; cervicis usually rudimentary and poorly defined	O—spines of upper lumbar and lower thoracic vertebrae I—spines of upper thoracic and cervical vertebrae	Extends vertebral column	Spinal nerves (dorsal rami)
Semispinalis (seh′-mē-spih-na′-lis)—thoracis, cervicis, and capitis regions (*semi* = half; *thorac* = thorax) (Figure 10.8c, right side)	Composite muscle forming part of deep layer of intrinsic back muscles; extends from thoracic region to head	O—transverse processes of C_7–T_{12} I—occipital bone (capitis) and spinous processes of cervical (cervicis) and thoracic vertebrae T_1 to T_4 (thoracis)	Extends vertebral column and head and rotates them to opposite side; acts synergistically with sternocleidomastoid muscles of opposite side	Spinal nerves (dorsal rami)
Quadratus lumborum (kwah-drā′-tis lum-bor′-um) (*quad* = four-sided; *lumb* = lumbar region) (See also Figure 10.17a)	Fleshy muscle forming part of posterior abdominal wall	O—iliac crest and iliolumbar fascia I—transverse processes of upper lumbar vertebrae and lower margin of 12th rib	Flexes vertebral column laterally when acting separately; when pair acts jointly, lumbar spine is extended and 12th rib is fixed; maintains upright posture	T_{12} and upper lumbar spinal nerves (ventral rami)

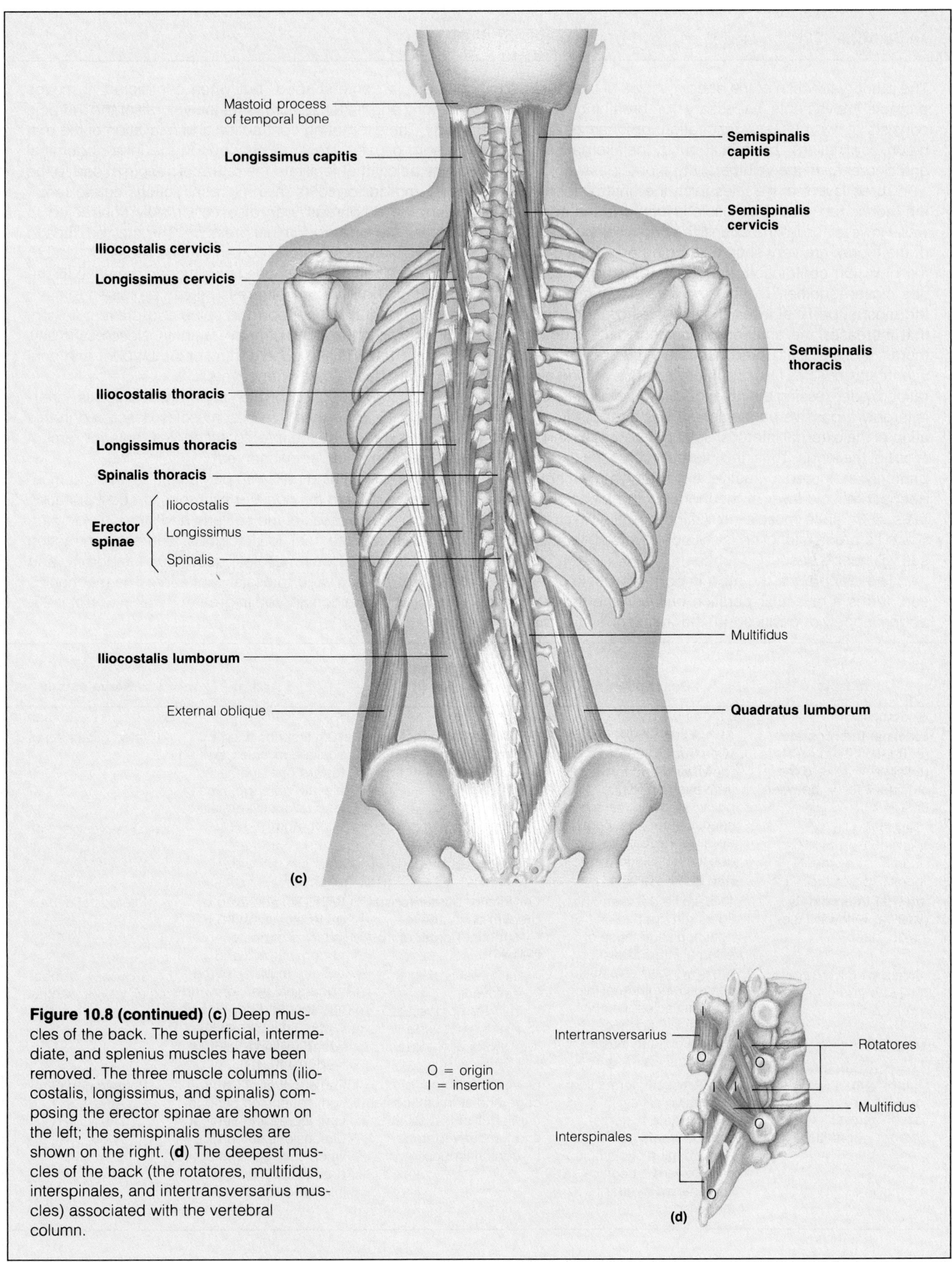

Figure 10.8 (continued) (**c**) Deep muscles of the back. The superficial, intermediate, and splenius muscles have been removed. The three muscle columns (iliocostalis, longissimus, and spinalis) composing the erector spinae are shown on the left; the semispinalis muscles are shown on the right. (**d**) The deepest muscles of the back (the rotatores, multifidus, interspinales, and intertransversarius muscles) associated with the vertebral column.

Table 10.5 Deep Muscles of the Thorax: Breathing (Figure 10.9)

The primary function of the deep muscles of the thorax is to promote movements necessary for breathing. Breathing consists of two phases—inspiration, or inhaling, and expiration, or exhaling—brought about by the alternate increase and decrease in the volume of the thoracic cavity.

Three layers of muscles form the anterolateral wall of the thorax, just as three muscle layers form the abdominal wall. However, unlike the extensive abdominal muscles, those of the thorax are very short, extending only from one rib to the next. On contraction, they draw the somewhat flexible ribs closer together. The *external intercostal* muscles form the most superficial layer; they lift the rib cage, an action that increases the anterior to posterior dimensions of the thorax, and are considered inspiratory muscles. The *internal intercostal* muscles form the intermediate layer and aid expiration by depressing the rib cage. However, quiet expiration is largely a passive phenomenon; i.e., it results from relaxation of the external intercostals and diaphragm and elastic recoil of the lungs. Thus, the internal intercostal muscles act primarily in forced or active expiratory movements. The deepest muscle layer of the thorax is the *transversus thoracis,* a tripartite muscle consisting of many muscle slips. Since its precise function is still controversial, the transversus will not be described further.

The *diaphragm,* the most important muscle of respiration, forms a muscular partition between the thoracic and abdominopelvic cavities. In the relaxed state, the diaphragm is dome-shaped, but when contracted, it moves inferiorly and flattens, increasing the volume of the thoracic cavity. The alternating contraction and relaxation of the diaphragm causes pressure changes in the intra-abdominal cavity below that facilitate the return of venous blood to the heart. In addition to its rhythmic contractions during respiration, the diaphragm can also be strongly contracted to increase the intra-abdominal pressure. This may be done to help evacuate pelvic organ contents (urine, feces, or a baby) or for weight lifting. When one takes a deep breath to fix the diaphragm, the abdominal pressure can be raised to such an extent that it will support the spine and prevent flexion, which helps in the lifting of heavy weights. Needless to say, it is important to have good control of the bladder and anal sphincters during such maneuvers.

With the exception of the diaphragm, which is innervated by the *phrenic nerves,* all muscles described in this table are served by the *intercostal nerves* (anterior rami of the first 11 thoracic spinal nerves).

Forced breathing calls into play a number of other muscles that insert into the ribs. For example, forced inspiration may involve activation of the scalene and sternocleidomastoid muscles of the neck to lift the ribs, whereas contraction of the abdominal wall muscles, quadratus lumborum, and others may be enlisted to assist expiration. The mechanics of respiration are considered in greater detail in Chapter 23.

Muscle	Description	Origin (O) and insertion (I)	Action	Nerve supply
External intercostals (in″-ter-kos′-tulz) (*external* = toward the outside; *inter* = between; *cost* = rib)	11 pairs lie between ribs; fibers run obliquely (downward and forward) from each rib to rib below; in lower intercostal spaces, fibers are continuous with external oblique muscle forming part of abdominal wall	O—inferior border of rib above I—superior border of rib below	With first ribs fixed by scalene muscles, pull ribs toward one another to elevate rib cage; aids in inspiration; synergists of diaphragm	Intercostal nerves
Internal intercostals (*internal* = toward the inside, deep)	11 pairs lie between ribs; fibers run deep to and at right angles to those of external intercostals (i.e., run downward and posteriorly); lower internal intercostal muscles are continuous with fibers of internal oblique muscle of abdominal wall	O—inferior border (costal groove) of rib above I—superior border of rib below	With 12th ribs fixed by quadratus lumborum, muscles of posterior abdominal wall and oblique muscles of the abdominal wall, draw ribs together and depress rib cage; aid in expiration; antagonistic to external intercostals	Intercostal nerves
Diaphragm (dī′-uh-fram) (*dia* = across; *phragm* = partition)	Broad muscle; forms floor of thoracic cavity; in relaxed state is dome-shaped; fibers converge from margins of thoracic cage toward a boomerang-shaped central tendon	O—inferior border of rib cage and sternum, costal cartilages of last six ribs and lumbar vertebrae I—central tendon	Prime mover of inspiration; flattens on contraction, increasing vertical dimensions of thorax; when strongly contracted, dramatically increases intra-abdominal pressure	Phrenic nerves

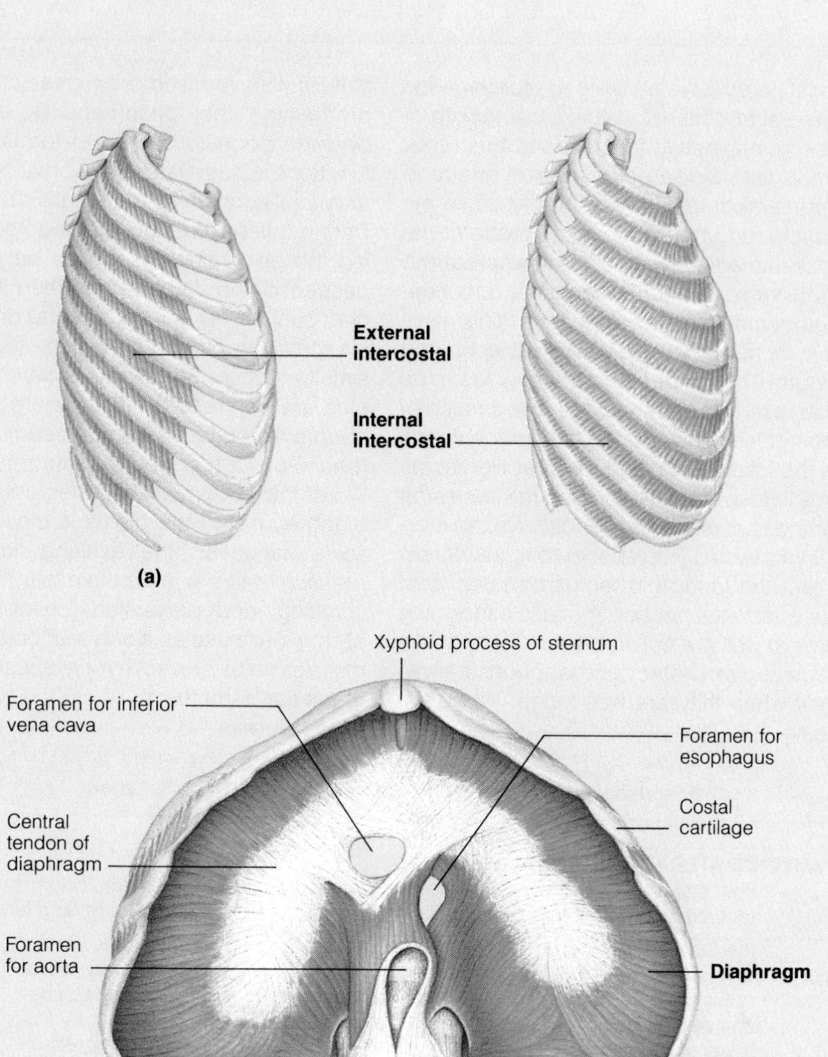

Figure 10.9 Muscles of quiet respiration. (**a**) Deep muscles of the thorax. The external intercostals (inspiratory muscles) are shown on the left and the internal intercostals (expiratory muscles) on the right. These two muscle layers run obliquely and at right angles to each other. (**b**) Inferior view of the diaphragm, the prime mover of inspiration. Notice that its muscle fibers converge toward a central tendon, an arrangement that causes the diaphragm to flatten and move inferiorly as it contracts. The diaphragm and its tendon are pierced by the great vessels (aorta and inferior vena cava) and the esophagus.

Table 10.6 Muscles of the Abdominal Wall: T[illegible]pression of Abdominal Viscera (Fi[illegible])

Walls of the abdominal cavity have no bony reinforcements. Instead, the anterolateral abdominal wall is a composite of four paired muscles, their investing fasciae, and their aponeuroses. Three broad flat muscle pairs, layered one atop the next, form the lateral abdominal walls: The fibers of the *external oblique* muscle run at right angles to those of the *internal oblique,* which immediately underlies it, whereas the grain of the deep *transversus abdominis* muscle runs horizontally across the abdomen at an angle to both. This alternation of grains is similar to the construction of plywood and provides great strength. The oblique and transversus muscles blend into broad insertion aponeuroses along much of their margins. The anterior extensions of these aponeuroses are continuous with the fascia enclosing the straplike *rectus abdominis* muscles and fuse medially, forming the *linea alba* ("white line"), a tendinous seam that extends from the sternum to the pubic symphysis. The aponeurotic reinforcements help to prevent the long, thin rectus muscles from "bowstringing." (The quadratus lumborum muscles forming the posterior abdominal wall are considered in Table 10.4.)

The abdominal muscles protect and support the viscera most effectively when they are well toned. When not sufficiently exercised or when severely stretched (as during pregnancy), they become weak, allowing the abdomen to become pendulous (i.e., to form a "potbelly"). Additional functions include lateral flexion and rotation of the trunk and anterior flexion of the trunk against resistance (as in sit-ups). During quiet expiration, the abdominal muscles relax, allowing the abdominal viscera to be pushed inferiorly by the descent of the diaphragm. When all these abdominal muscles contract in unison, several different activities may be effected, depending on which other muscles are activated simultaneously. For example, when all the abdominal muscles are contracted, the ribs are pulled inferiorly and the abdominal contents compressed. This pushes the visceral organs upward on the diaphragm aiding forced expiration. When the abdominal muscles contract in concert with the diaphragm and the glottis is closed (an action called Valsalva maneuver), the resulting increase in intra-abdominal pressure helps to promote micturition (voiding), defecation, vomiting, and parturition (childbirth). Contraction of the abdominal muscles along with contraction of the deep back muscles helps prevent hyperextension of the spine and splints the entire body trunk.

Muscle	Description	Origin (O) and insertion (I)	Action	Nerve supply
MUSCLES OF THE ANTEROLATERAL ABDOMINAL WALL				
	Four paired flat muscles; very important in supporting and protecting abdominal viscera and play an important role in promoting movement of vertebral column (extension, flexion, and lateral binding)			
Rectus abdominis (rek′-tus ab-dah′-mih-nis) (*rectus* = straight; *abdom* = abdomen)	Medial superficial muscle pair; extend from pubis to rib cage; ensheathed by aponeuroses of lateral muscles; segmented by three reinforcing tendinous intersections	O—pubic crest and symphysis I—xiphoid process and costal cartilages of ribs 5–7	Flex and rotate lumbar region of vertebral column; fix and depress ribs, stabilize pelvis during walking, increase intra-abdominal pressure	Thoracic nerves T_6 (or T_7)–T_{12}
External oblique (ō-blēk′) (*external* = toward outside; *oblique* = running at an angle)	Largest and most superficial of the three lateral muscles; fibers run downward and medially (same direction outstretched fingers take when hands put into pants pockets); aponeurosis turns under inferiorly forming inguinal ligament	O—by fleshy strips from outer surfaces of lower eight ribs I—most fibers into linea alba; some into pubic tubercle and iliac crest; majority of fibers insert anteriorly via a broad aponeurosis	When pair contracts simultaneously, aid rectus abdominis muscles in flexing vertebral column and in compressing abdominal wall and increasing intra-abdominal pressure; acting individually, aid muscles of back in trunk rotation and lateral flexion	Thoracic nerves T_7–T_{12}
Internal oblique (*internal* = toward the inside; deep)	Fibers fan upward and forward and run at right angles to those of external oblique, which it underlies	O—lumbodorsal fascia, iliac crest, and inguinal ligament I—linea alba, pubic crest, last three ribs	As for external oblique	T_7–T_{12} and L_1
Transversus abdominis (tranz-ver′-sus) (*transverse* = running straight across or transversely)	Deepest (innermost) muscle of abdominal wall; fibers run horizontally	O—inguinal ligament, lumbodorsal fascia, cartilages of last six ribs I—linea alba, pubic crest	Compresses abdominal contents	T_7–T_{12} and lumbar nerves

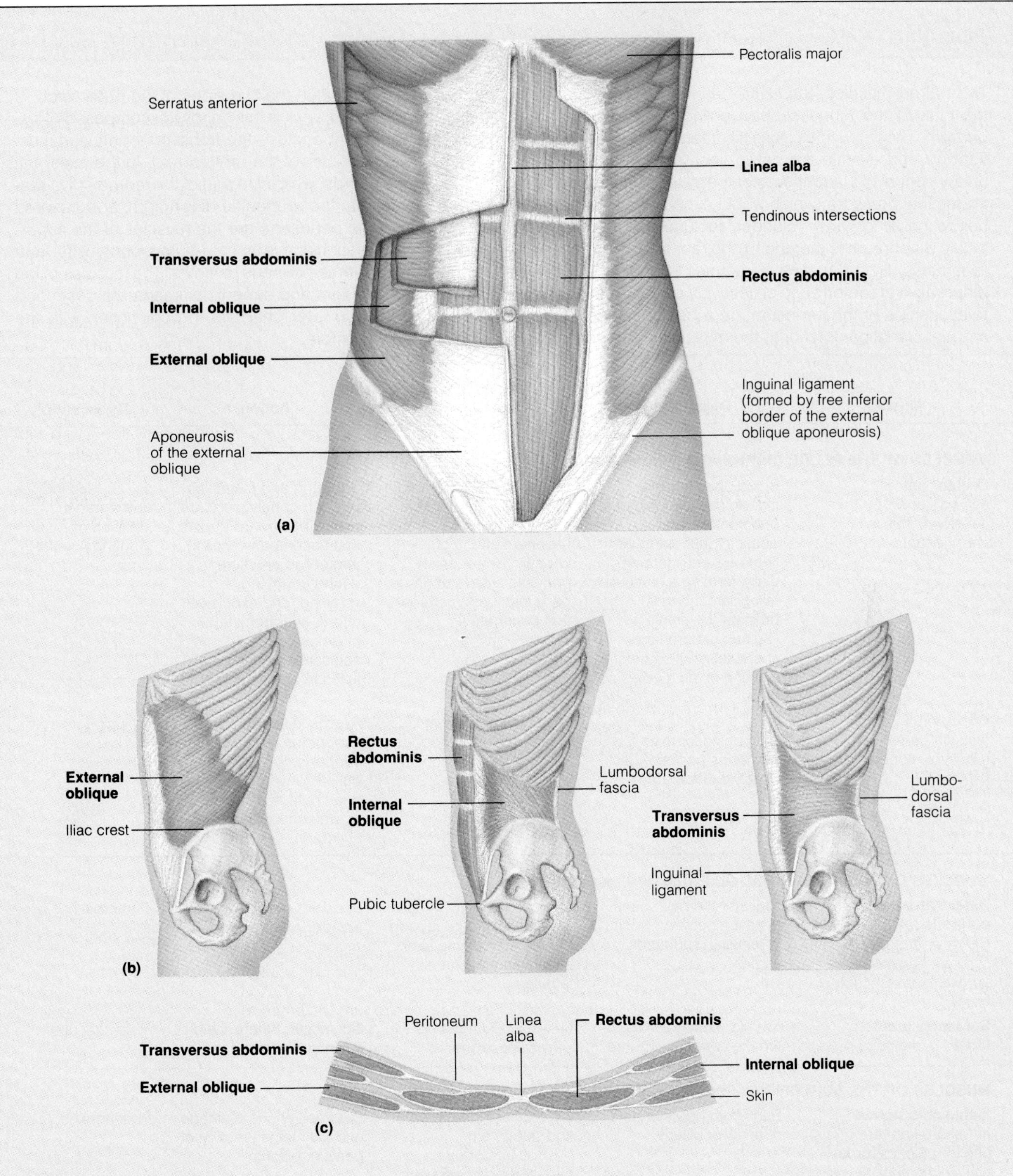

Figure 10.10 Muscles of the abdominal wall. (**a**) Anterior view of the muscles forming the anterolateral abdominal wall. The superficial muscles have been partially cut away on the left side of the diagram to reveal the deeper internal oblique and transversus abdominis muscles. (**b**) Lateral view of the trunk, illustrating the fiber direction and attachments of the external oblique, internal oblique, and transversus abdominis muscles. (**c**) Transverse section through the anterolateral abdominal wall (midregion), showing how the aponeuroses of the lateral abdominal muscles contribute to the rectus abdominis sheath.

Table 10.7 Muscles of the Pelvic Floor and Perineum: Support of Abdominopelvic Organs (Figure 10.11)

Two paired muscles, the *levator ani* and *coccygeus*, form the bulk of the funnel-shaped pelvic floor, or *pelvic diaphragm*. These muscles close the inferior outlet of the pelvis, support and elevate the pelvic floor, are important in voluntary control of bladder emptying and resist increased intra-abdominal pressure (which would expel the contents of the hollow pelvic organs—bladder, rectum, and vagina). The pelvic diaphragm is pierced by the rectum and urethra and, in females, by the vagina. Structures inferior to the pelvic diaphragm are referred to collectively as the perineum. The relationships of the perineum are a bit complex and worth more explanation. Inferior to the muscles of the pelvic floor, and stretching between the two sides of the pubic arch, is the *urogenital diaphragm*, a thin sandwich composed of two layers of fascia that enclose the transverse perineus muscles and the sphincter of the urethra and, in the male, the bulbourethral glands, which are part of the reproductive system. Superficial to the urogenital diaphragm, and covered by the skin of the perineum, are the muscles of the superficial space, the ischiocavernosus, bulbospongiosus, and superficial transverse perineus muscles. The muscles of the urogenital diaphragm and superficial space are described in greater detail in later chapters in conjunction with the functions they promote.

Muscle	Description	Origin (O) and insertion (I)	Action	Nerve supply
MUSCLES OF THE PELVIC DIAPHRAGM (Figure 10.11a)				
Levator ani (leh-va′-ter ā′nē) (*levator* = raises; *ani* = anus)	Broad, thin, tripartite muscle (pubococcygeus, puborectalis, and iliococcygeus parts); its fibers extend inferomedially, forming a muscular "sling" around male prostate (or female vagina), urethra, and anorectal junction before meeting in the median plane	O—extensive linear origin inside pelvis from pubis to ischial spine. I—inner surface of coccyx, levator ani of opposite side, and (in part) into the structures that penetrate it	Supports and maintains position of pelvic viscera; resists downward thrusts that accompany rises in intrapelvic pressure during coughing, vomiting, and expulsive efforts of abdominal muscles; forms sphincters at anorectal junction and vagina	S_3, S_4, and inferior rectal nerve (branch of pudendal nerve)
Coccygeus (kok-sih′-jē-us) (*coccy* = coccyx)	Small triangular muscle lying posterior to levator ani; forms posterior part of pelvic diaphragm	O—spine of ischium I—sacrum and coccyx	Assists levator ani in supporting pelvic viscera; supports coccyx and pulls it forward after it has been reflected posteriorly by defecation and childbirth	S_4 and S_5
MUSCLES OF THE UROGENITAL DIAPHRAGM (Figure 10.11b)				
Deep transverse perineus (payr-ih-nē′-is) (*deep* = far from surface; *transverse* = across; *perine* = near anus)	Together the pair spans distance between ischia; in females, surrounds vagina	O—ischial rami I—midline central tendon of perineum; some fibers into vaginal wall in females	Supports pelvic organs; steadies central tendon	Pudendal nerve
Sphincter urethrae (*sphin* = squeeze)	Circular muscle surrounding membranous urethra	O—ischiopubic rami I—midline raphe	Constricts urethra; helps support pelvic organs	Pudendal nerve
MUSCLES OF THE SUPERFICIAL SPACE (Figure 10.11c)				
Ischiocavernosus (ih′-shē-ō-ka″-ver-no′-sis) (*ischi* = hip; *caverna* = hollow chamber)	Runs from pelvis to base of penis or clitoris	O—ischial tuberosities and pubic rami I—crus of corpus cavernosa of male penis or female clitoris	Retards venous drainage and maintains erection of penis or clitoris	Pudendal nerve
Bulbospongiosus (bul″-bō-spun″-jē-o′-sis) (*bulbon* = groin; *spongio* = sponge)	Encloses base of penis (bulb) and lies deep to labia in females	O—midline raphe of penis and central tendon of perineum I—Anteriorly into corpus cavernosa of penis or clitoris; posteriorly into central tendon	Empties urethra; assists in erection of penis in males and of clitoris in females	Pudendal nerve
Superficial transverse perineus (*superficial* = closer to surface)	Paired muscle bands posterior to urethral (and in females, vaginal) opening	O—ischial tuberosity I—central tendon of perineum	Stabilize and strengthen midline tendon of perineum	Pudendal nerve

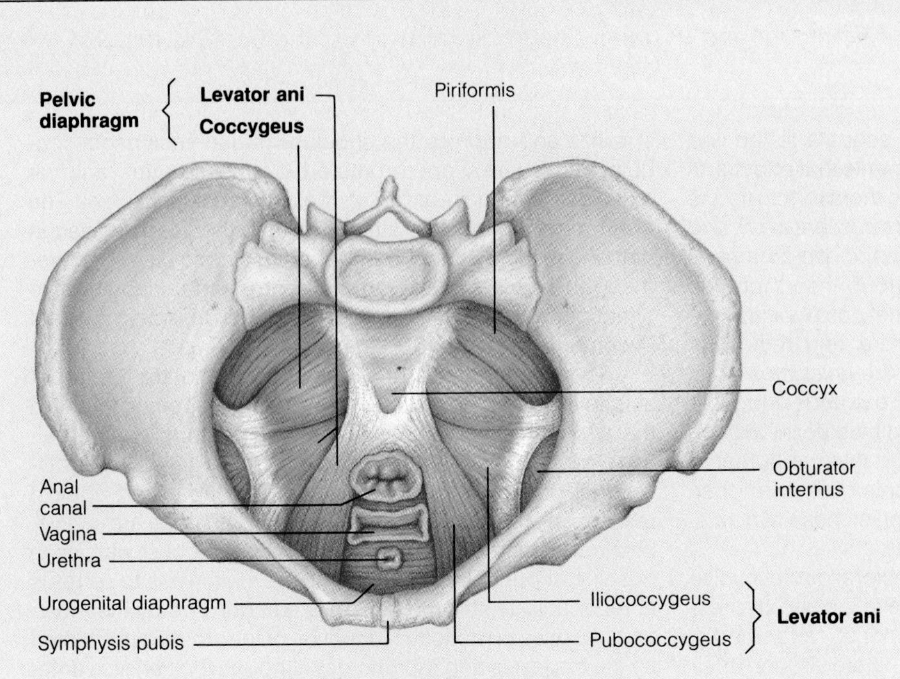

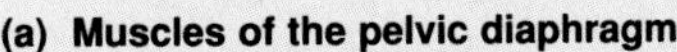
(a) Muscles of the pelvic diaphragm

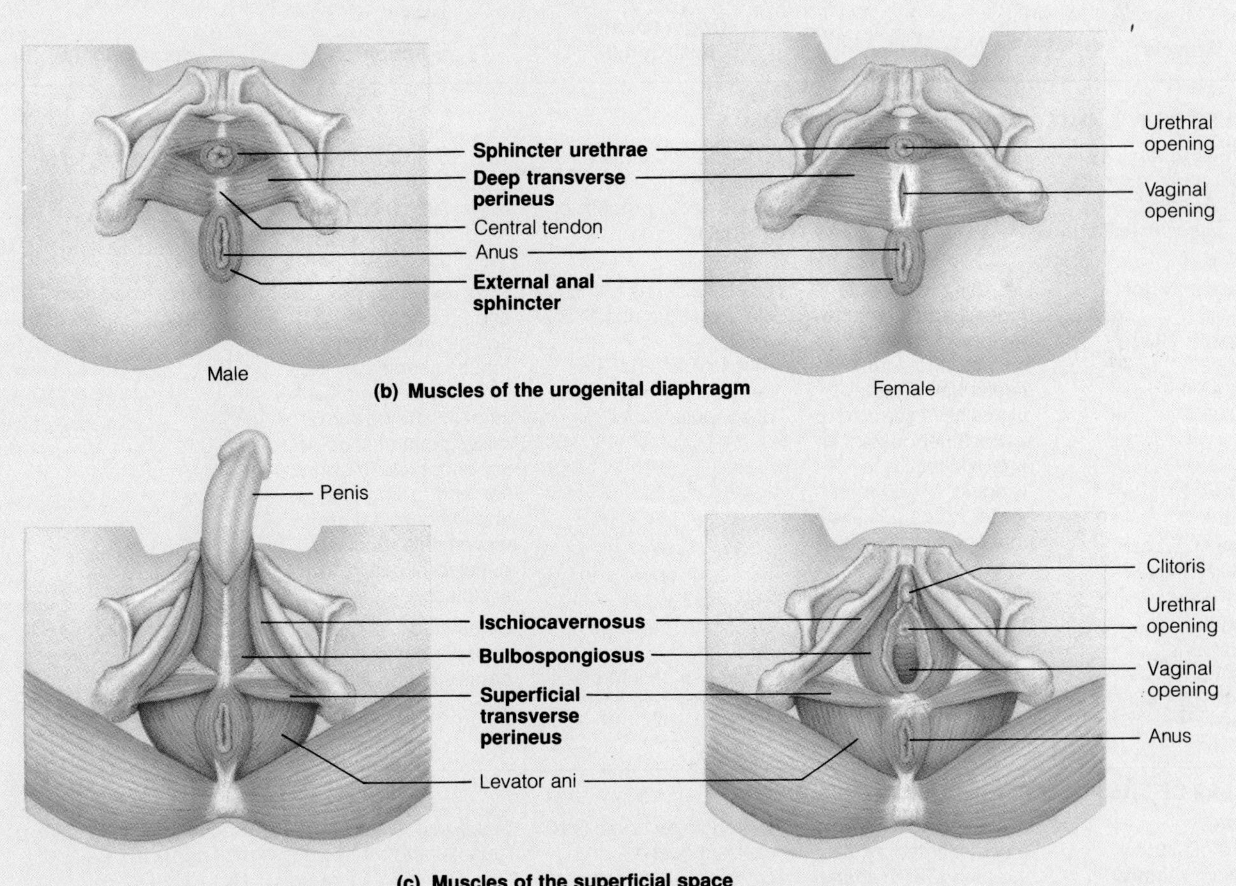

(b) Muscles of the urogenital diaphragm

(c) Muscles of the superficial space

Figure 10.11 Muscles of the pelvic floor and perineum. (**a**) Muscles of the pelvic floor (levator ani and coccygeus) seen from above in a female pelvis. (**b**) Muscles of the urogenital diaphragm (sphincter urethrae and deep transverse perineus) of the perineum, which make up the second (and more superficial) layer of muscles. (**c**) Muscles of the superficial space of the perineum (ischiocavernosus, bulbospongiosus, and superficial transverse perineus), which lie just deep to the skin of the perineum.

Table 10.8 Superficial Muscles of the Anterior and Posterior Thorax: Movem[illegible] 10.12)

Most superficial thorax muscles fix the scapula to the wall of the thorax or promote scapular movements that effect arm movement. The muscles of the anterior thorax include the *pectoralis major, pectoralis minor, serratus anterior,* and *subclavius*. Except for the pectoralis major, which inserts into the humerus, all muscles of this group insert into the pectoral girdle. Extrinsic muscles of the posterior thorax, which secure the upper limb to the trunk and restrict or mobilize scapular movements, include the *latissimus dorsi* and *trapezius* muscles superficially and the underlying *levator scapulae* and *rhomboids*. The latissimus dorsi, like the pectoralis major muscles anteriorly, insert into the humerus and are more concerned with movements of the arm than of the scapula. We will defer consideration of these two muscle pairs to Table 10.9 (arm movements).

The important movements of the pectoral girdle involve displacements of the scapula, i.e., its elevation and depression, rotation, lateral (forward) movements, and medial (backward) movements. The clavicles rotate around their own axes during scapular movements, and provide both stability and precision to scapular movements.

Except for the serratus anterior, the anterior muscles stabilize and depress the shoulder girdle. Thus, most scapular movements are promoted by the serratus anterior muscles anteriorly and by the posterior muscles. The arrangement of muscle attachments to the scapula is such that one muscle cannot bring about a simple (linear) movement on its own; to elevate or depress the scapula or to effect other scapular movements, several muscles must act in combination.

The prime movers of shoulder elevation are the trapezius and levator scapulae. When acting together to shrug the shoulder, their opposite rotational effects counterbalance each other. Scapular depression is largely due to gravitational pull, but when the scapula is depressed against resistance, the trapezius and serratus anterior are active. Forward movements of the scapula on the chest wall, as in pushing or punching movements, are mainly due to serratus anterior activity; this movement is antagonized by the rhomboid muscles posteriorly. Adduction of the scapula is effected by the trapezius and rhomboids. Although the serratus anterior and trapezius muscles are antagonists in forward/backward movements of the scapulae, they act in unison to coordinate the *rotational* scapular movements.

Muscle	Description	Origin (O) and insertion (I)	Action	Nerve supply
MUSCLES OF THE ANTERIOR THORAX (Figure 10.12a)				
Pectoralis minor (pek″-tor-a′-lis mī′-ner) (*pectus* = chest, breast; *minor* = lesser)	Flat, thin muscle directly beneath and obscured by pectoralis major	O—anterior surfaces of ribs 3–5 I—coracoid process of scapula	With ribs fixed, draws scapula forward and downward; with scapula fixed, draws rib cage superiorly	Medial pectoral nerve (C_8 and T_1)
Serratus anterior (ser-ā′-tis) (*serratus* = saw)	Lies deep to scapula beneath and inferior to pectoral muscles on lateral rib cage; forms medial wall of axilla; origins have serrated, or sawtooth, appearance; paralysis results in "winging" of vertebral border of scapula away from chest wall, making arm elevation impossible	O—by a series of muscle slips from ribs 1–8 (or 9) I—entire anterior surface of vertebral border of scapula	Prime mover to protract and hold scapula against chest wall; rotates scapula so that its inferior angle moves laterally and upward; raises point of shoulder; important role in abduction and raising of arm and in horizontal arm movements (pushing, punching); called "boxer's muscle"	Long thoracic nerve (C_5–C_7)
Subclavius (sub-klā′vē-is) (*sub* = under, beneath; *clav* = clavicle)	Small cylindrical muscle extending from rib 1 to lateral clavicle	O—costal cartilage of rib 1 I—groove on inferolateral surface of clavicle	Helps stabilize and depress pectoral girdle; paralysis produces no obvious effects	Nerve to subclavius (C_5 and C_6)
MUSCLES OF THE POSTERIOR THORAX (Figure 10.12b)				
Trapezius (truh-pē′-zē-is) (*trapezion* = irregular four-sided figure)	Most superficial muscle of posterior thorax; flat, and triangular in shape; upper fibers run downward to scapula; middle fibers run horizontally to scapula; lower fibers run superiorly to scapula	O—occipital bone, ligamentum nuchae, and spines of C_7 and all thoracic vertebrae I—a continuous insertion along acromion and spine of scapula and lateral third of clavicle	Stabilizes, raises, retracts, and rotates scapula; middle fibers retract (adduct) scapula; superior fibers elevate scapula or can help extend head; inferior fibers depress scapula (and shoulder)	Accessory nerve (cranial nerve XI); C_3 and C_4

Muscle	Description	Origin (O) and insertion (I)	Action	Nerve supply
Levator scapulae (ska′-pyoo-lē) (*levator* = raises)	Located at back and side of neck, deep to trapezius; thick, strap-like muscle	O—transverse processes of C_1–C_4 I—superior vertebral border of scapula	Elevates/adducts scapula in concert with superior fibers of trapezius; tilts glenoid cavity downward; when scapula is fixed, flexes neck to same side	Dorsal scapular nerve C_3–C_5
Rhomboids (rom′-boydz)—major and minor (*rhomboid* = diamond-shaped)	Two rectangular muscles lying deep to trapezius and inferior to levator scapulae; rhomboid minor is the more superior muscle; both muscles extend from vertebral column to scapula	O—spinous processes of C_7 and T_1 (minor) and spinous processes of T_2–T_5 (major) I—medial border of scapula	Act together (and with middle trapezius fibers) to retract scapula, thus "squaring shoulders"; rotate scapulae so that glenoid cavity rotates downward (as when arm is lowered against resistance; e.g., paddling a canoe); stabilize scapula	Dorsal scapular nerve (C_5)

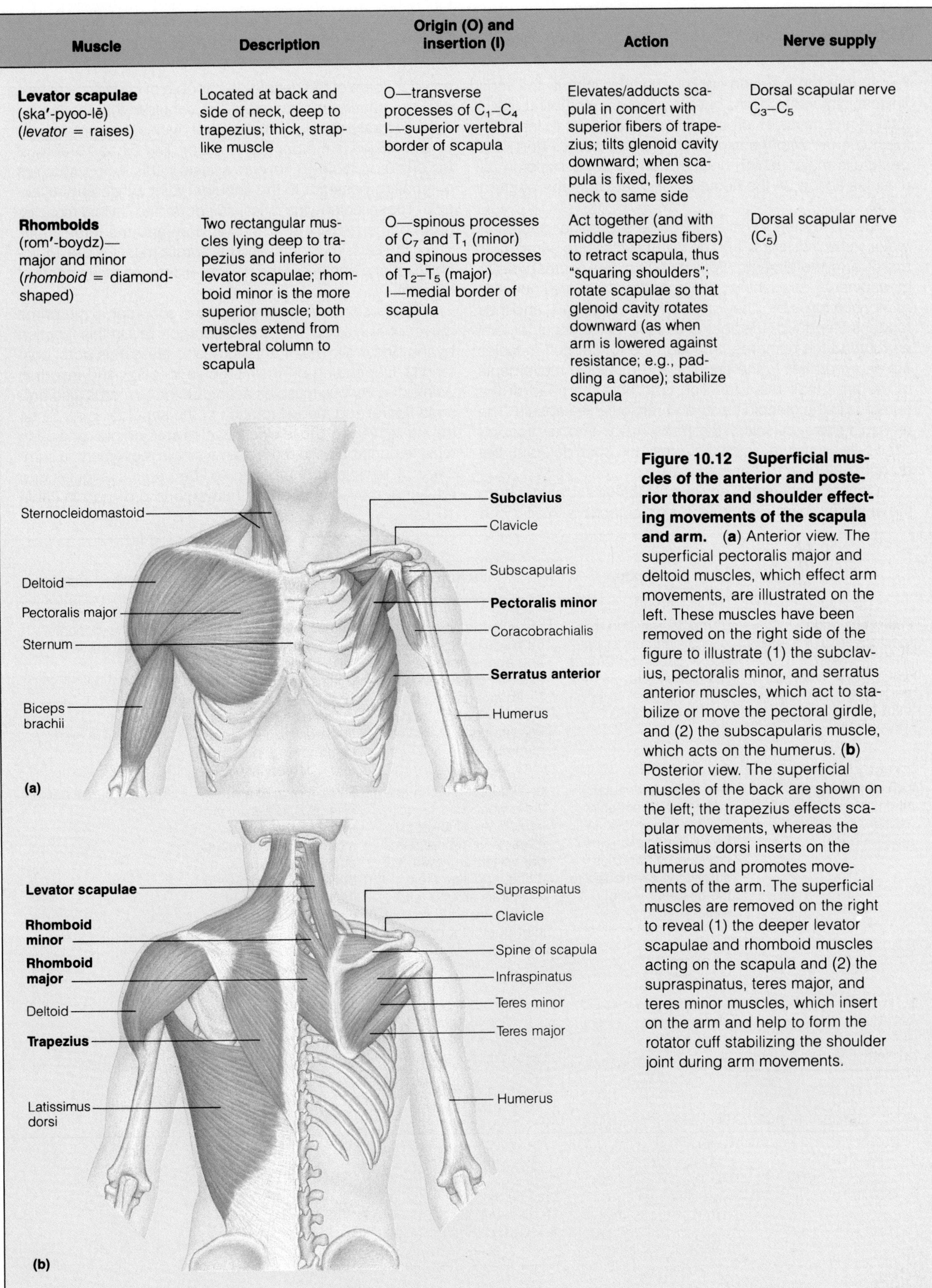

Figure 10.12 Superficial muscles of the anterior and posterior thorax and shoulder effecting movements of the scapula and arm. (**a**) Anterior view. The superficial pectoralis major and deltoid muscles, which effect arm movements, are illustrated on the left. These muscles have been removed on the right side of the figure to illustrate (1) the subclavius, pectoralis minor, and serratus anterior muscles, which act to stabilize or move the pectoral girdle, and (2) the subscapularis muscle, which acts on the humerus. (**b**) Posterior view. The superficial muscles of the back are shown on the left; the trapezius effects scapular movements, whereas the latissimus dorsi inserts on the humerus and promotes movements of the arm. The superficial muscles are removed on the right to reveal (1) the deeper levator scapulae and rhomboid muscles acting on the scapula and (2) the supraspinatus, teres major, and teres minor muscles, which insert on the arm and help to form the rotator cuff stabilizing the shoulder joint during arm movements.

Table 10.9 Muscles Crossing the Shoulder Joint: Movements of the Arm (Humerus) (Figure 10.13)

flexible joint in the body, but pays the price of instability. A total of nine muscles cross each shoulder joint to insert on the humerus. With the exception of the latissimus dorsi and pectoralis major, which originate on the axial skeleton, all muscles acting on the humerus originate from the pectoral girdle.

Of these nine muscles, only the superficial *pectoralis major, latissimus dorsi,* and *deltoid* muscles, are prime movers of arm movements. Four muscles, the *supraspinatus, infraspinatus, teres minor,* and *subscapularis*, are known as *rotator cuff muscles*. They originate on the scapula, and their tendons blend with the fibrous capsule of the shoulder joint en route to the humerus. Although the rotator cuff muscles act as synergists in the angular and rotational movements of the arm, their main function is to hold the head of the humerus in the glenoid cavity and reinforce its capsule. The remaining two muscles, the *teres major* and *coracobrachialis*, are small muscles that cross the shoulder joint, but do not contribute to it anatomically.

Generally speaking, any muscle that arises anteriorly to the shoulder joint (pectoralis major, coracobrachialis, and

ion, i.e., angular movement at the shoulder that moves the arm anteriorly, usually in the sagittal plane. The prime mover of arm flexion is the pectoralis major. The biceps brachii of the arm (see Table 10.10) also assists in this action. Muscles originating posterior to the shoulder joint cause arm extension. These include the latissimus dorsi and deltoid muscles (both prime movers of arm extension) and the teres major. Thus, the latissimus dorsi and pectoralis muscles are *antagonists* of one another in the flexion–extension movements of the arm.

The fleshy deltoid muscle of the shoulder is the prime mover of arm abduction and is antagonized in this function by the pectoralis major anteriorly and latissimus dorsi posteriorly. Depending on their precise location and insertion points, the various muscles acting on the humerus also promote lateral and medial rotation of the shoulder joint. Since the interactions of these nine muscles are complex and each muscle contributes to more than a single movement, a summary of muscles contributing to the various angular and rotational movements of the humerus is provided in Table 10.12.

Muscle	Description	Origin (O) and insertion (I)	Action	Nerve supply
Pectoralis major (pek″-tor-a′-lis mā′-jer) (*pect* = breast, chest; *major* = larger)	Large, fan-shaped muscle covering upper portion of chest; forms anterior axillary fold	O—clavicle, sternum, cartilage of ribs 1–6, and aponeurosis of external oblique muscle I—fibers converge to insert by a short tendon into greater tubercle of humerus	Prime mover of arm flexion; rotates arm medially; adducts arm against resistance; with scapula (and arm) fixed, pulls rib cage upward, thus can help in climbing, throwing, pushing, and in forced inspiration	Lateral and medial pectoral nerves
Latissimus dorsi (la-tih′-sih-mis dor′-sī) (*latissimus* = widest; *dorsi* = back)	Broad, flat, triangular muscle of lower back (lumbar region); extensive superficial origins; covered by trapezius superiorly; contributes to the posterior wall of axilla	O—indirect attachment via lumbodorsal fascia into spines of lower six thoracic vertebrae, lumbar vertebrae, lower 3 to 4 ribs, and iliac crest I—spirals around teres major to insert in floor of intertubercular groove of humerus	Prime mover of arm extension; powerful arm adductor; medially rotates arm at shoulder; depresses scapula; because of its power in these movements, it plays an important role in bringing the arm down in a power stroke, as in striking a blow, hammering, swimming, and rowing	Thoracodorsal nerve

Muscle	Description	Origin (O) and insertion (I)	Action	Nerve supply
Deltoid (del′-toyd) (*delta* = triangular)	Thick, multipennate muscle forming rounded shoulder muscle mass; responsible for roundness of shoulder; a site commonly used for intramuscular injection, particularly in males, where it tends to be quite fleshy	O—embraces insertion of the trapezius; lateral third of clavicle; acromion and spine of scapula I—deltoid tuberosity of humerus	Prime mover of arm abduction when all its fibers contract simultaneously; antagonist of pectoralis major and latissimus dorsi, which adduct the arm; if only anterior fibers are active, can act powerfully in flexion and medial rotation of humerus, therefore synergist of pectoralis major; if only posterior fibers are active, effects extension and lateral rotation of arm; active during rhythmic arm swinging movements during walking	Axillary nerve (C_5 and C_6)
Subscapularis (sub-sca″-pyoo-layr′-is) (*sub* = under; *scapular* = scapula)	Forms part of posterior wall of axilla; tendon of insertion passes in front of shoulder joint; a rotator cuff muscle	O—subscapular fossa of scapula I—lesser tubercle of humerus	Chief medial rotator of humerus; assisted by pectoralis major; helps to hold head of humerus in glenoid cavity, thereby stabilizing shoulder joint	Subscapular nerves (C_5–C_7)
Supraspinatus (soo″-pruh-spih-nah′-tus) (*supra* = above, over; *spin* = spine)	Named for its location on posterior aspect of scapula; deep to trapezius; a rotator cuff muscle	O—supraspinous fossa of scapula I—greater tubercle of humerus	Stabilizes shoulder joint; helps to prevent downward dislocation of humerus, as when carrying a heavy suitcase; assists in abduction	Suprascapular nerve
Infraspinatus (in″-fruh-spih-nah′-tus) (*infra* = below)	Partially covered by deltoid and trapezius; named for its scapular location; a rotator cuff muscle	O—infraspinous fossa of scapula I—greater tubercle of humerus posterior to insertion of supraspinatus	Helps to hold head of humerus in glenoid cavity; rotates humerus laterally	Suprascapular nerve
Teres minor (tayr′-ēz) (*teres* = round; *minor* = lesser)	Small, elongated muscle; lies inferior to infraspinatus and may be inseparable from that muscle; a rotator cuff muscle	O—lateral border of dorsal scapular surface I—greater tubercle of humerus inferior to infraspinatus insertion	Same action(s) as infraspinatus muscle	Axillary nerve
Teres major	Thick, rounded muscle; located inferior to teres minor; helps to form posterior wall of axilla (along with latissimus dorsi and subscapularis)	O—posterior surface of scapula at inferior angle I—crest of lesser tubercle on anterior humerus; insertion tendon fused with that of latissimus dorsi	Extends, medially rotates, and adducts humerus; synergist of latissimus dorsi	Lower subscapular nerve
Coracobrachialis (kor″-uh-kō-brā′-kē-a″-lis) (*coraco* = coracoid; *brachi* = arm)	Small, cylindrical muscle	O—coracoid process of scapula I—medial surface of humerus shaft	Flexion and adduction of the humerus; synergist of pectoralis major	Musculocutaneous nerve

(Table continues)

Table 10.9 (continued)

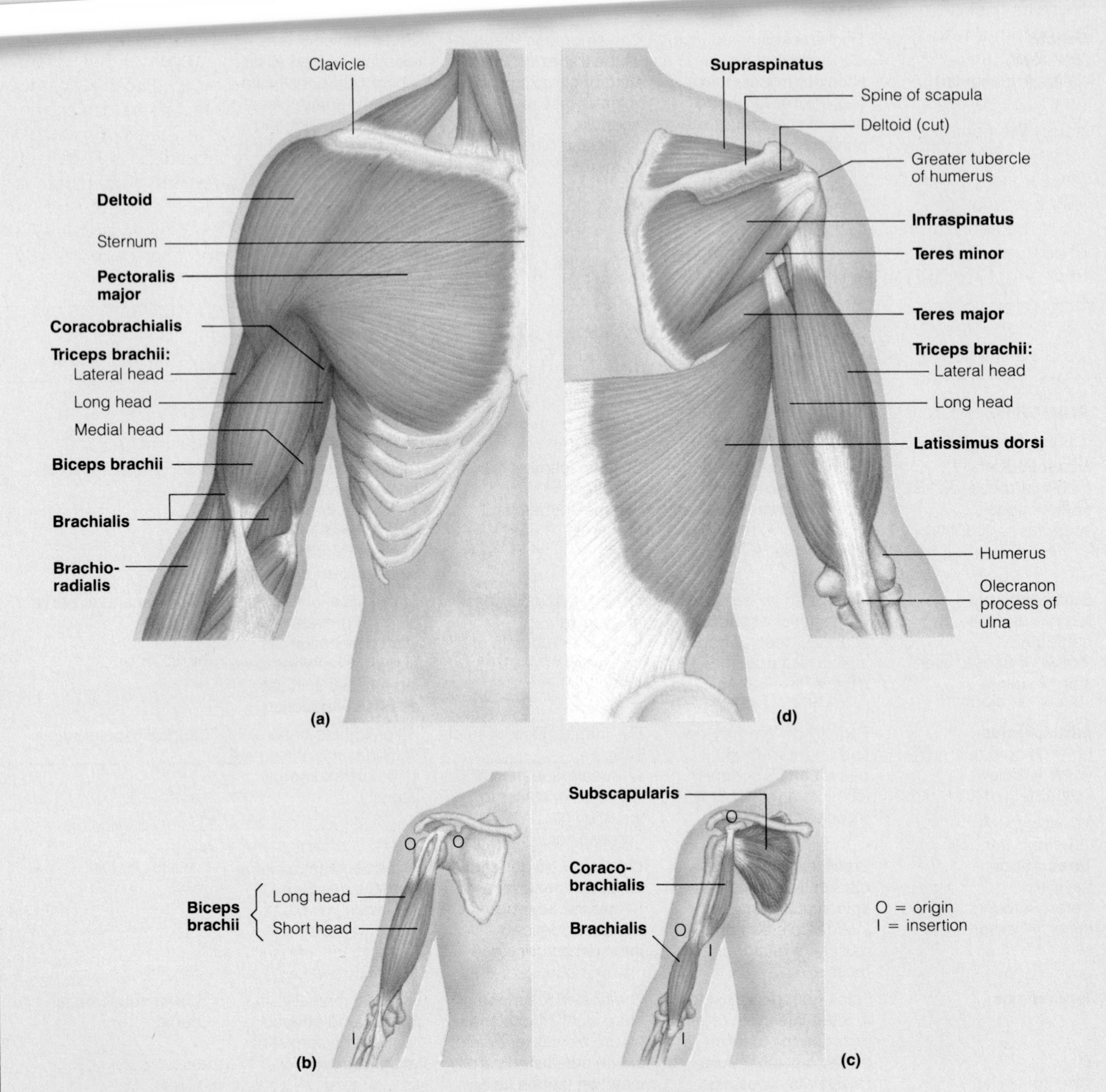

Figure 10.13 Muscles crossing the shoulder and elbow joint, respectively, causing movements of the arm and forearm. (**a**) Superficial muscles of the anterior thorax, shoulder, and arm, anterior view. (**b**) The biceps brachii muscle of the anterior arm, shown in isolation. (**c**) The brachialis muscle arising from the humerus and the coracobrachialis and subscapularis muscles arising from the scapula, shown in isolation. (Note, however, that the coracobrachialis effects movements of the arm, not the forearm, and the subscapularis stabilizes the shoulder joint.) (**d**) The extent of the triceps brachii muscle of the posterior arm, shown in relation to the deep scapular muscles. The deltoid muscle of the shoulder has been removed.

Table 10.10 Muscles Crossing the Elbow Joint: Flexion and Extension of the Forearm (Figure 10.13)

Muscles fleshing out the arm cross the elbow joint to insert on the forearm bones. Since the elbow is a hinge joint, movements promoted by these muscles are limited almost entirely to flexion and extension of the forearm. Posterior arm muscles are extensors. The bulky *triceps brachii* muscle forms nearly the entire musculature of that region and is the prime mover of elbow extension. It is assisted (minimally) by the tiny *anconeus* muscle, which just about spans the elbow joint posteriorly.

All anterior arm muscles cause elbow flexion. In order of decreasing strength, these are the *brachialis, biceps brachii,* and *brachioradialis.* The brachialis and biceps are inserted (respectively) into the ulna and radius and contract simultaneously during flexion; they are considered the chief forearm flexors. The biceps brachii, a muscle that bulges when the forearm is flexed, is familiar to almost everyone; the brachialis, which lies deep to the biceps, is less known, but is as important in flexing the elbow. The biceps muscle also supinates the forearm. The brachioradialis is a relatively weak forearm flexor. Since it arises from the distal humerus and inserts into the distal forearm, its force is exerted far from the fulcrum. It plays little or no role in initiating flexion and only becomes active when the elbow has been partially flexed by the prime movers.

Muscle	Description	Origin (O) and insertion (I)	Action	Nerve supply
POSTERIOR MUSCLES				
Triceps brachii (trī′-seps brā′-kē-ī) (*triceps* = three heads; *brachi* = arm)	Large fleshy muscle; the only muscle of posterior compartment of arm; three-headed origin; long and lateral heads lie superficial to medial head	O—long head: infraglenoid tubercle of scapula; lateral head: posterior humerus; medial head: posterior humerus distal to radial groove I—by common tendon into olecranon process of ulna	Powerful forearm extensor (prime mover, particularly medial head); antagonist of forearm flexors; long head tendon may help stabilize shoulder joint and assist in arm adduction	Radial nerve
Anconeus (an-kō′-nē-is) (*ancon* = elbow) (see Figure 10.15)	Short triangular muscle; closely associated (blended) with distal end of triceps on posterior humerus	O—lateral epicondyle of humerus I—lateral aspect of olecranon process of ulna	Abducts ulna during forearm pronation; synergist of triceps brachii in elbow extension	Radial nerve
ANTERIOR MUSCLES				
Biceps brachii (bī-seps) (*biceps* = two heads)	Two-headed fusiform muscle; bellies unite as insertion point is approached; tendon of long head helps stabilize shoulder joint	O—short head: coracoid process; long head: tubercle above glenoid cavity and lip of glenoid cavity; tendon of long head runs within capsule and descends into intertubercular groove of humerus I—by common tendon into radial tuberosity	Flexes elbow joint and supinates forearm; these actions usually occur at same time (e.g., when you open a bottle of wine, it turns the corkscrew and pulls the cork); weak flexor of arm at shoulder	Musculocutaneous nerve
Brachialis (brā″-kē-a′-lis)	Strong muscle that is immediately deep to biceps brachii on distal humerus	O—front of distal half of humerus; embraces insertion of deltoid muscle I—coronoid process of ulna and capsule of elbow joint	A major forearm flexor (lifts ulna as biceps lifts the radius)	Musculocutaneous nerve
Brachioradialis (brā″-kē-ō-rā″-dē-a′-lis) (*radi* = radius, ray) (also see Figure 10.14)	Superficial muscle of lateral forearm; forms lateral boundary of cubital fossa; extends from distal humerus to distal forearm	O—lateral supracondylar ridge at distal end of humerus I—base of styloid process of radius	Synergist in forearm flexion; acts to best advantage when forearm is partially flexed already	Radial nerve (an important exception: the radial nerve typically serves extensor muscles)

Table 10.11 Muscles of the Forearm: Movements of Wrist, Hand, and Fingers (Figures 10.14 and 10.15)

[illegible] divided into two approximately equal groups: those causing wrist movements and those moving the fingers and thumb. In most cases, their fleshy portions contribute to the roundness of the proximal forearm and then they taper to long insertion tendons. Their insertions are securely anchored by strong ligaments (the flexor and extensor retinacula) and surrounded by synovial tendon sheaths that facilitate their movements.

Although many of the forearm muscles actually arise from the humerus (and thus cross both the elbow and wrist joints), their actions on the elbow are so slight that they are normally disregarded. Flexion and extension are the movements typically effected at both the wrist and finger joints. In addition, the wrist can be abducted and adducted.

The forearm muscles are subdivided into two main compartments (the anterior and posterior) by fascia sheets, and each compartment has superficial and deep muscle layers. The muscles of each compartment differ not only in their location, but also in function. Those of the anterior fascial group arise from a common tendon on the humerus and are innervated largely by the median nerve. Most anterior forearm muscles are wrist or finger *flexors*. However, two muscles of this group, the superficial *pronator teres* and the deep *pronator quadratus*, pronate the forearm, one of the most important arm movements.

Muscles of the posterior compartment are chiefly wrist and finger *extensors;* the only exception is the *supinator* muscle, which assists the biceps brachii muscle of the arm in supination of the forearm. Like those of the anterior compartment, most muscles of the posterior compartment arise from a common origin tendon on the humerus: however, the location on the humerus differs. All posterior forearm muscles are supplied by the radial nerve. Most movements of the hand are promoted by the forearm muscles; these movements are assisted and made more precise by a number of small *intrinsic* muscles of the hand. These include the four *lumbrical* muscles that lie between the metacarpals (visible on the palmar surface of the hand) and the eight *interosseous* muscles that lie deep to the lumbricals. Together, the lumbricals and interossei flex the metacarpophalangeal joints (knuckles) and help extend the fingers. Additionally, there are four *thenar* muscles, short muscles that act exclusively to circumduct and oppose the thumb. The more superficial lumbrical and thenar muscles are illustrated in Figure 10.14c. The deeper interossei are shown in Figure 10.15b.

The locations and actions of the forearm muscles are the focus of the table. Since the forearm muscles are numerous and their interactions varied, a summary of their actions is provided in Table 10.12.

Muscle	Description	Origin (O) and insertion (I)	Action	Nerve supply
PART I: ANTERIOR MUSCLES (Figure 10.14)	These eight muscles of the anterior fascial compartment are listed from the lateral to the medial aspect. Most arise from a common flexor tendon attached to the medial epicondyle of the humerus and have additional origins as well. Most of the tendons of these flexors are held in place at the wrist by a thickening of deep fascia called the *flexor retinaculum.*			
SUPERFICIAL MUSCLES				
Pronator teres (prō′-nā″-ter tayr-ēz) (*pronation* = turning palm posteriorly, or down; *teres* = round)	Two-headed muscle; seen in superficial view between proximal margins of brachioradialis and flexor carpi radialis; forms medial boundary of cubital fossa	O—medial epicondyle of humerus; coronoid process of ulna I—by common tendon into lateral radius, midshaft	Pronates forearm; weak flexor of elbow	Median nerve
Flexor carpi radialis (flek′-ser kar′-pē rā″-dē-a′-lis) (*flex* = decrease angle between two bones; *carpi* = wrist; *radi* = radius)	Runs diagonally across forearm; midway, its fleshy belly is replaced by a flat tendon that becomes cordlike at wrist	O—medial epicondyle of humerus I—base of second and third metacarpals; insertion tendon easily seen and provides guide to position of radial artery (used for pulse taking) at wrist	Powerful flexor of wrist; abducts hand; synergist of elbow flexion	Median nerve
Palmaris longus (pal-mayr′-is lon′-gus) (*palma* = palm; *longus* = long)	Small fleshy muscle with a long insertion tendon; often absent; may be used as guide to find median nerve that lies lateral to it at wrist	O—medial epicondyle of humerus I—palmar aponeurosis	Weak wrist flexor; may tense fascia of palm (palmar aponeurosis) during hand movements	Median nerve

Muscle	Description	Origin (O) and insertion (I)	Action	Nerve supply
Flexor carpi ulnaris (ul-nayr′-is) (*ulnar* = ulna)	Most medial muscle of this group; two-headed; ulnar nerve lies lateral to its tendon	O—medial epicondyle of humerus, olecranon process, and posterior surface of ulna I—pisiform bone and base of fifth metacarpal	Powerful flexor of wrist; also adducts hand in concert with extensor carpi ulnaris (posterior muscle); stabilizes wrist during finger extension	Ulnar nerve
Flexor digitorum superficialis (dih″-jih-tor′-um soo″-per-fih″-shē-a′-lis) (*digit* = finger, toe; *superficial* = close to surface)	Two-headed muscle; more deeply placed (therefore, actually forms an intermediate layer); overlain by muscles above but visible at distal end of forearm	O—medial epicondyle of humerus, medial surface, and coronoid process of ulna; shaft of radius I—by four tendons into middle phalanges of fingers 2–5	Flexes wrist and middle phalanges of fingers 2–5; the important finger flexor when speed and flexion against resistance are required	Median nerve
DEEP MUSCLES				
Flexor pollicis longus (pah′-lih-kis) (*pollix* = thumb)	Partly covered by flexor digitorum superficialis; parallels flexor digitorum profundus laterally	O—anterior surface of radius and interosseous membrane I—distal phalanx of thumb	Flexes distal phalanx of thumb; weak flexor of wrist	Median nerve
Flexor digitorum profundus (prō-fun′-dus) (*profund* = deep)	Extensive origin; overlain entirely by flexor digitorum superficialis	O—anteromedial surface of ulna and interosseous membrane I—by four tendons into distal phalanges of fingers 2–5	Slow-acting finger flexor; assists in flexing wrist; the only muscle that can flex distal interphalangeal joints	Medial half by ulnar nerve; lateral half by median nerve

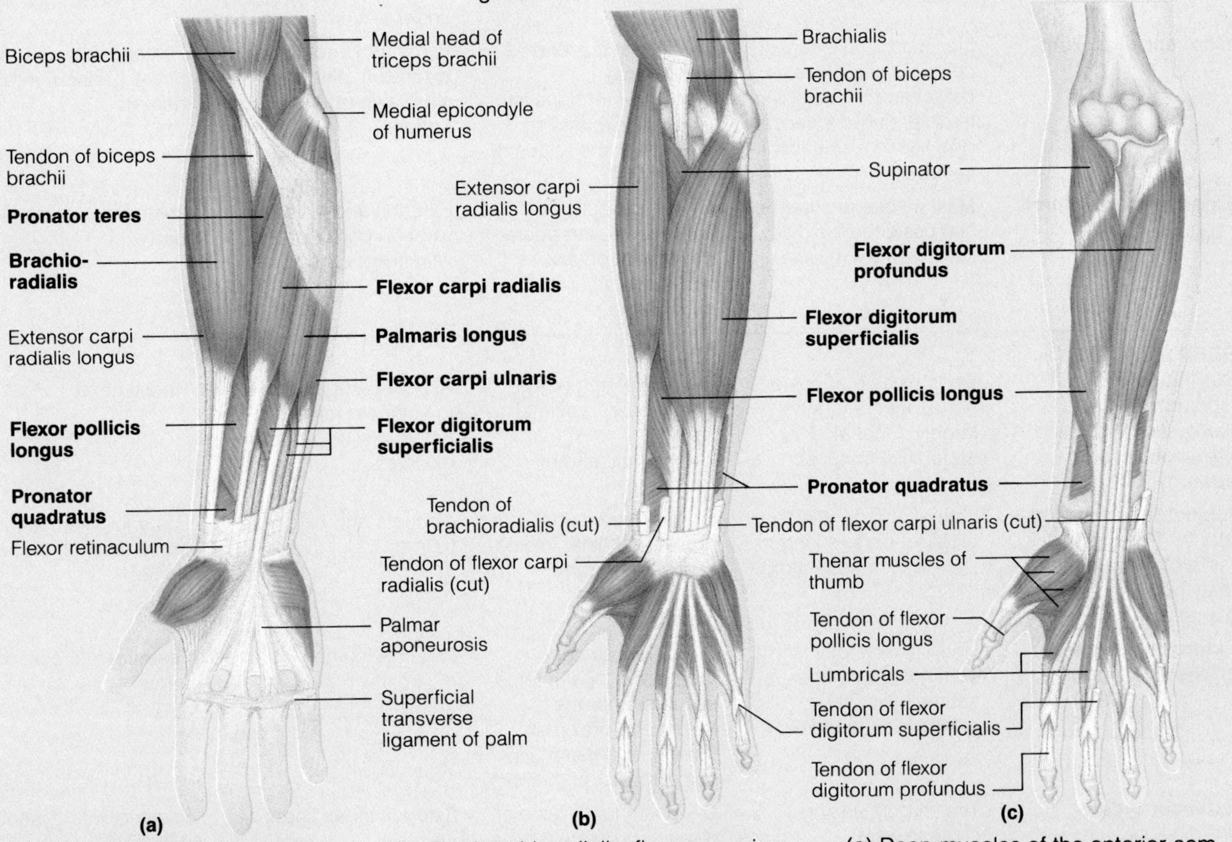

Figure 10.14 Muscles of the anterior fascial compartment of the forearm acting on the wrist and fingers. (**a**) Superficial view of the muscles of the right forearm and hand. (**b**) The brachioradialis, flexors carpi radialis and ulnaris, and palmaris longus muscles have been removed to reveal the position of the somewhat deeper flexor digitorum superficialis. (**c**) Deep muscles of the anterior compartment. Superficial muscles have been removed. The lumbricals and thenar muscles of the hand (intrinsic hand muscles) are also illustrated.

Table 10.11 (continued)

Muscle	Description	Origin (O) and insertion (I)	Action	Nerve supply
Pronator quadratus (kwah-drā′-tis) (*quad* = square, four-sided)	Deepest muscle of distal forearm; passes downward and laterally; only muscle that arises solely from ulna and inserts solely into radius	O—distal portion of anterior ulnar shaft I—distal surface of anterior radius	Pronates forearm; acts with pronator teres; also helps hold ulna and radius together	Median nerve
PART II: POSTERIOR MUSCLES (Figure 10.15)	These 11 muscles of the posterior fascial compartment are listed from the lateral to the medial aspect. They are all innervated by the radial nerve. More than half of the posterior compartment muscles arise from a common extensor origin tendon attached to the posterior surface of the lateral epicondyle of the humerus and adjacent fascia. The extensor tendons are held in place at the posterior aspect of the wrist by the *extensor retinaculum*, which prevents "bowstringing" of these tendons when the wrist is hyperextended.			
SUPERFICIAL MUSCLES				
Extensor carpi radialis longus (ek-sten′-ser) (*extend* = increase angle between two bones)	Parallels brachioradialis on lateral forearm, and may blend with it	O—lateral supracondylar ridge of humerus I—base of second metacarpal	Extends and abducts wrist	Radial nerve
Extensor carpi radialis brevis (breh′-vis) (*brevis* = short)	Somewhat shorter than extensor carpi radialis longus and lies deep to it	O—lateral epicondyle of humerus I—base of third metacarpal	Extends and abducts wrist; acts synergistically with extensor carpi radialis longus to steady wrist during finger flexion	Deep branch of radial nerve
Extensor digitorum	Lies medial to extensor carpi radialis brevis; a detached portion of this muscle, called *extensor digiti minimi*, extends little finger	O—lateral epicondyle of humerus I—by four tendons into extensor expansions and distal phalanges of fingers 2–5	Prime mover of finger extension; extends wrist; can abduct (flare) fingers	Posterior interosseous nerve (branch of radial nerve)
Extensor carpi ulnaris	Most medial of superficial posterior muscles; long, slender muscle	O—lateral epicondyle of humerus and posterior border of ulna I—base of fifth metacarpal	Extends and adducts wrist (in conjunction with flexor carpi ulnaris)	Posterior interosseous nerve
DEEP MUSCLES				
Supinator (soo′-pih-nā-ter) (*supination* = turning palm anteriorly or upward)	Deep muscle at posterior aspect of elbow; largely concealed by superficial muscles	O—lateral epicondyle of humerus; proximal ulna I—proximal end of radius	Assists biceps brachii to supinate forearm; antagonist of pronator muscles	Radial nerve
Abductor pollicis longus (ab′-duk-ter) (*abduct* = movement away from median plane)	Lateral and parallel to extensor pollicis longus; just distal to supinator	O—posterior surface of radius and ulna; interosseous membrane I—base of first metacarpal	Abducts and extends thumb	Posterior interosseous nerve (branch of radial nerve)
Extensor pollicis brevis and **longus**	Deep muscle pair with a common origin and action; overlain by extensor carpi ulnaris	O—dorsal shaft of radius and ulna; interosseous membrane I—base of proximal (brevis) and distal (longus) phalanx of thumb	Extends thumb	Posterior interosseous nerve
Extensor indicis (in′-dih-kis) (*indicis* = index finger)	Tiny muscle arising close to wrist	O—posterior surface of distal ulna; interosseous membrane I—extensor expansion of index finger; joins tendon of extensor digitorum	Extends index finger	Posterior interosseous nerve

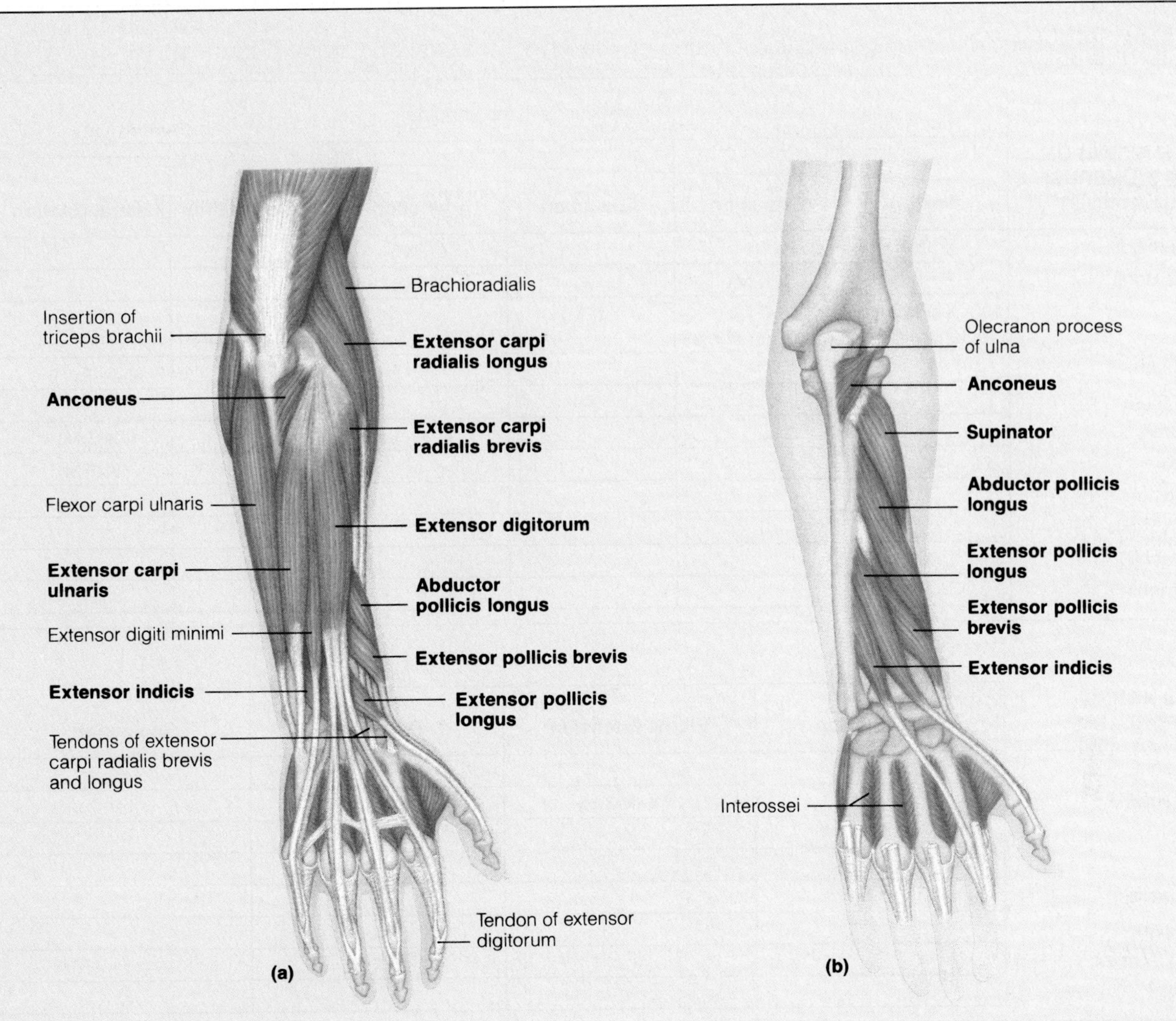

Figure 10.15 Muscles of the posterior fascial compartment of the forearm acting on the wrist and fingers. The extensor retinaculum that secures the extensor tendons at the wrist is not illustrated. (**a**) Superficial muscles of the right forearm, posterior view. (**b**) Deep posterior muscles of the right forearm; superficial muscles have been removed. The interossei, the deepest layer of intrinsic hand muscles, are also illustrated.

Table 10.12 Summary of Actions of Muscles Acting on the Arm, Forearm, and Hand (Figure 10.16)

PART I: MUSCLES ACTING ON THE ARM (HUMERUS) (PM = prime mover)	Actions at the shoulder					
	Flexion	**Extension**	**Abduction**	**Adduction**	**Medial rotation**	**Lateral rotation**
Pectoralis major	X (PM)			X (PM)	X	
Latissimus dorsi		X (PM)		X (PM)	X	
Deltoid	X (PM) (anterior fibers)	X (PM) (posterior fibers)	X (PM)		X (anterior fibers)	X (posterior fibers)
Subscapularis					X (PM)	
Supraspinatus			X			X
Infraspinatus				X		X (PM)
Teres minor						X (PM)
Teres major		X		X	X	
Coracobrachialis	X			X		
Biceps brachii	X					
Triceps brachii				X		

PART II: MUSCLES ACTING ON THE FOREARM	Actions			
	Elbow flexion	**Elbow extension**	**Pronation**	**Supination**
Biceps brachii	X (PM)			X
Triceps brachii		X (PM)		
Anconeus		X		
Brachialis	X (PM)			
Brachioradialis	X			
Pronator teres			X	
Pronator quadratus			X	
Supinator				X

PART II: MUSCLES ACTING ON THE WRIST AND FINGERS	Actions on the wrist				Actions on the fingers	
	Flexion	**Extension**	**Abduction**	**Adduction**	**Flexion**	**Extension**
Anterior Compartment:						
Flexor carpi radialis	X (PM)		X			
Palmaris longus	X (weak)					
Flexor carpi ulnaris	X (PM)			X		
Flexor digitorum superficialis	X (PM)				X	
Flexor pollicis longus	X (weak)				X (thumb)	
Flexor digitorum profundus	X				X	
Posterior Compartment:						
Extensor carpi radialis longus and brevis		X	X			
Extensor digitorum		X (PM)				X (and abducts)
Extensor carpi ulnaris		X		X		
Abductor pollicis longus			X		(Abducts thumb)	
Extensor pollicis longus and brevis						X (thumb)
Extensor indicis						X (little finger)

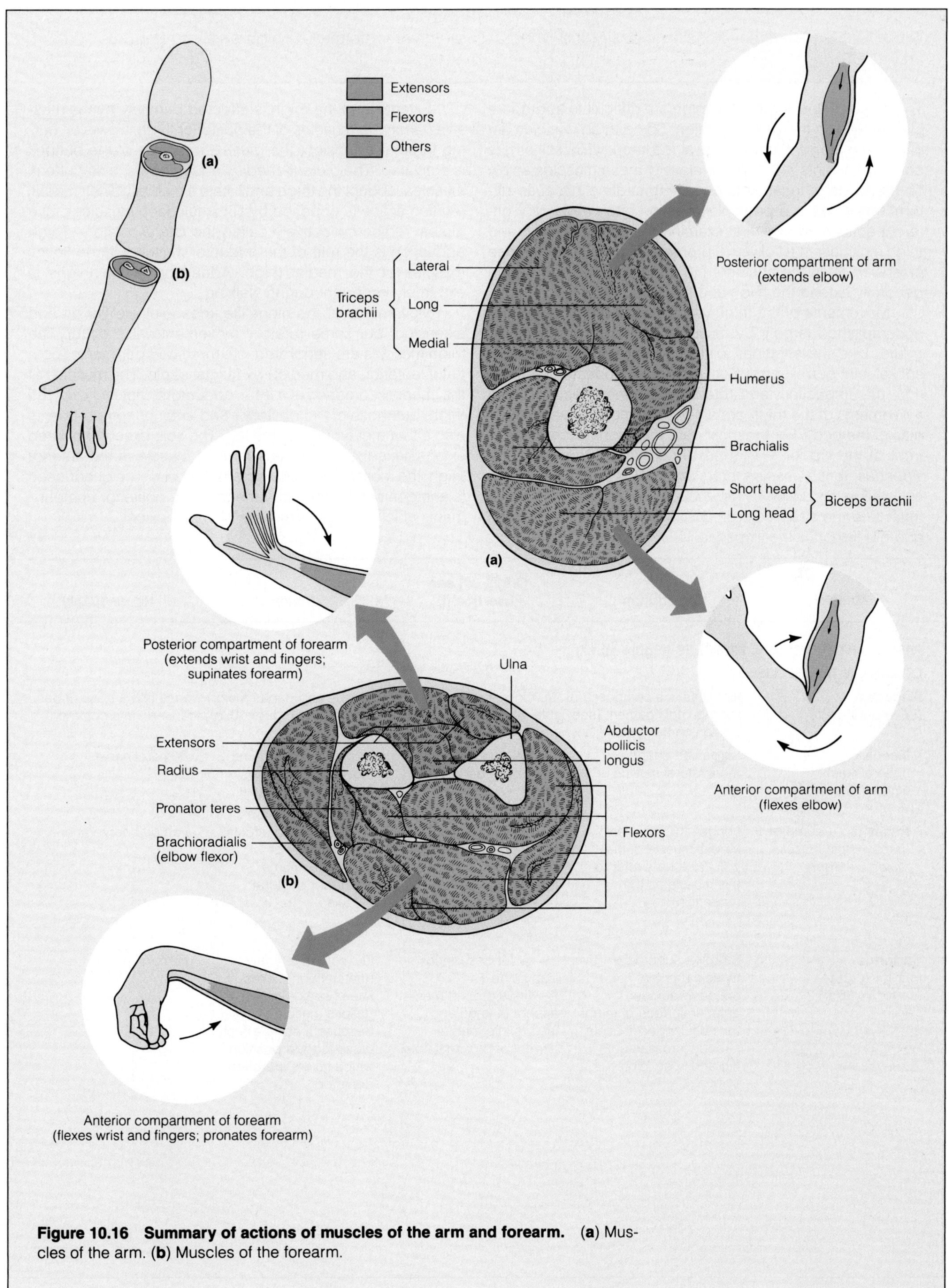

Figure 10.16 Summary of actions of muscles of the arm and forearm. (**a**) Muscles of the arm. (**b**) Muscles of the forearm.

Table 10.13 Muscles Crossing the Hip and Knee Joints: Movements of the Thigh and Leg (Figures 10.17 and 10.18)

The muscles fleshing out the thigh are difficult to segregate into groups on the basis of action. Some thigh muscles act only at the hip joint, others only at the knee, while still others act at both joints. Attempts to classify these muscles solely on the basis of location is equally frustrating because different muscles in a particular location often have very different actions. We will first examine the anterior pelvic and thigh muscles, most of which promote thigh flexion and knee extension, then the posterior hip and thigh muscles, which generally extend the thigh and flex the knee.

Movements of the thigh (occurring at the hip joint) are accomplished largely by muscles anchored to the pelvic girdle. Like the shoulder joint, the hip joint is a ball-and-socket joint permitting flexion, extension, abduction, adduction, circumduction, and rotation. Muscles effecting these movements of the thigh are among the most powerful muscles of the body. For the most part, the thigh flexors pass in front of the hip joint. The most important of these are the *iliopsoas, tensor fasciae latae,* and *rectus femoris;* they are assisted in this action by the *adductor* muscles of the medial thigh and the straplike *sartorius.* The prime mover of thigh and hip flexion is the iliopsoas.

Extension of the thigh is effected primarily by the massive *hamstring* muscles of the posterior thigh. However, during forceful extension, the *gluteus maximus* of the buttock is activated. The *gluteus medius* and *minimus,* also buttock muscles, abduct the thigh and rotate it medially. Their medial rotation action is opposed by six small deep muscles of the gluteal region collectively called the *lateral rotators.* Thigh adduction is the role of the adductor muscles, three fleshy muscles of the medial thigh. Adduction of the thighs is extremely important during walking.

Movements at the hingelike knee joint are flexion and extension, but some rotatory movements also occur. The thigh muscles are separated into three compartments (posterior, anterior, and medial) by fascial septa. The muscles of the medial compartment (the adductors) act only on the thigh, but those of the posterior and anterior compartments also effect movements of the leg. The sole executor of knee extension is the *quadriceps femoris* muscle of the anterior thigh, the most powerful muscle in the body. The quadriceps is antagonized by the hamstrings of the posterior compartment, which are prime movers of knee flexion.

Muscle	Description	Origin (O) and insertion (I)	Action	Nerve supply
PART I: ANTEROMEDIAL MUSCLES (Figure 10.17)				
ORIGIN ON THE PELVIS				
Iliopsoas (ih″-lē-ō-sō′-us)	Iliopsoas is a composite of two closely related muscles (iliacus and psoas major) whose fibers pass under the inguinal ligament [see Table 10.6 (external oblique muscle) and Figure 10.10] to insert via a common tendon on the femur			
• **Iliacus** (ih-lē-a′-kus) (*iliac* = ilium)	Large, fan-shaped, more lateral muscle	O—iliac fossa I—lesser trochanter of femur via iliopsoas tendon	Iliopsoas is the prime mover of hip flexion; flexes thigh on trunk when pelvis is fixed	Femoral nerve
• **Psoas** (sō′-us) **major** (*psoa* = loin muscle; *major* = larger)	Longer, thicker, more medial muscle of the pair. (Butchers refer to this muscle as the tenderloin)	O—by fleshy slips from transverse processes, bodies, and discs of lumbar vertebrae I—lesser trochanter of femur via iliopsoas tendon	As above; also effects lateral flexion of vertebral column; important postural muscle	Femoral nerve and L_1
Sartorius (sar-tor′-ē-us) (*sartor* = tailor)	Straplike superficial muscle running obliquely across anterior surface of thigh to knee; longest muscle in body; crosses both hip and knee joints	O—anterior superior iliac spine I—winds around medial aspect of knee and inserts into medial aspect of proximal tibia	Flexes and laterally rotates thigh; flexes knee (weak); known as "tailor's muscle" because it helps effect cross-legged position in which tailors are often depicted	Femoral nerve

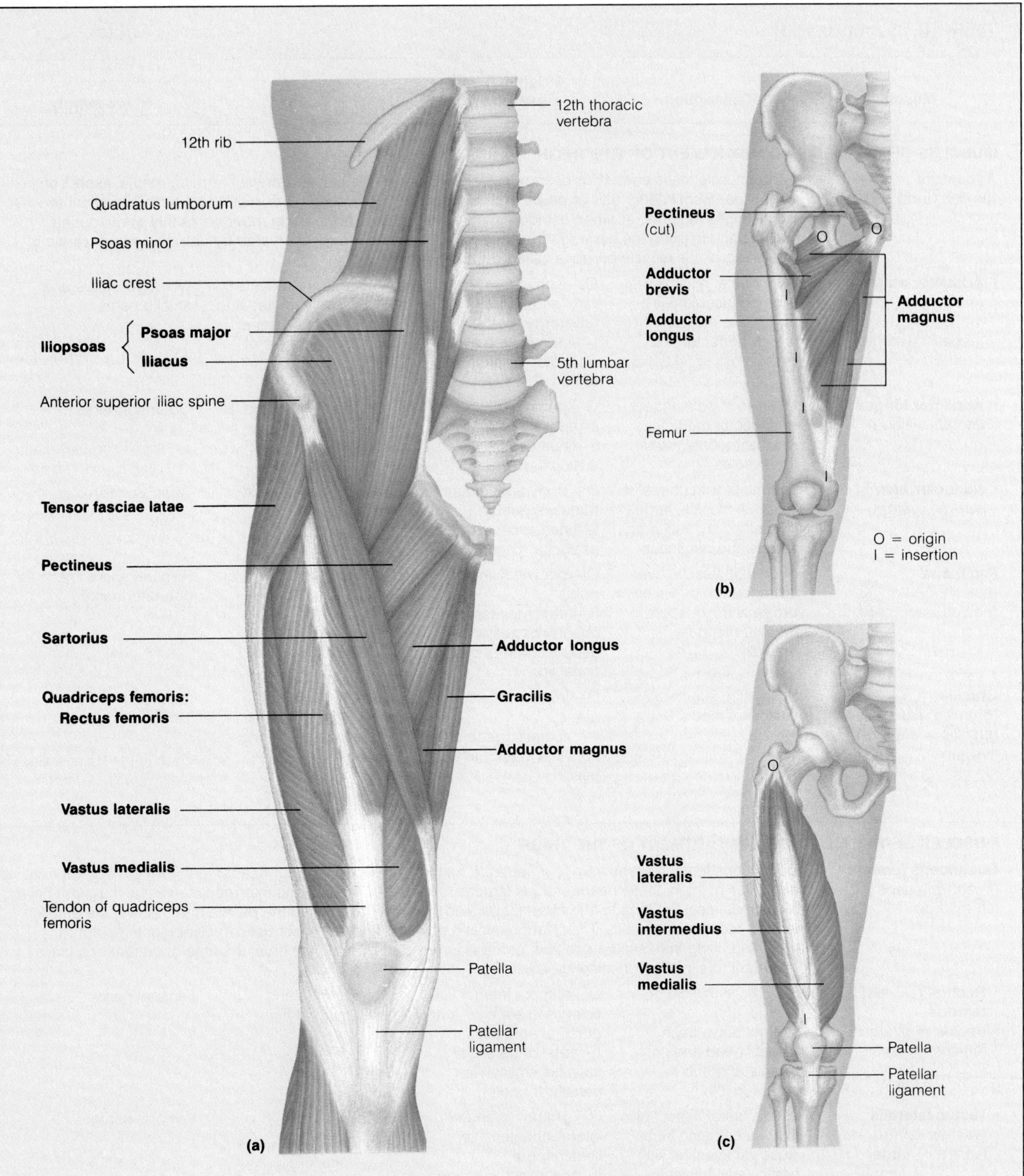

Figure 10.17 Anteromedial muscles promoting movements of the thigh and leg. (**a**) Anterior view of the deep muscles of the pelvis and superficial muscles of the right thigh. (**b**) Adductor muscles of the medial compartment of the thigh. Other muscles have been removed so that the origins and insertions of the adductor muscles can be seen. (**c**) The vastus muscles of the quadriceps group. The rectus femoris muscle of the quadriceps group and surrounding muscles have been removed to reveal the attachments and extent of the vastus muscles.

(Table continues)

Table 10.13 (continued)

Muscle	Description	Origin (O) and insertion (I)	Action	Nerve supply
MUSCLES OF THE MEDIAL COMPARTMENT OF THE THIGH				
Adductors (uh-duk′-terz)	Large muscle mass consisting of three muscles (magnus, longus, and brevis) forming medial aspect of thigh; arise from inferior part of pelvis and insert at various levels on femur; all used in movements that press thighs together, as when astride a horse; important in pelvic tilting movements that occur during walking and in fixing the hip when the knee is flexed; entire group innervated by obturator nerve. Strain or stretching of this muscle group is called a "pulled groin."			
• **Adductor magnus** (mag′-nus) (*adduct* = move toward midline; *magnus* = large)	A triangular muscle with a broad insertion; is a composite muscle that is part adductor and part hamstring in action	O—ischial and pubic rami and ischial tuberosity I—linea aspera and adductor tubercle of femur	Anterior part adducts and laterally rotates and flexes thigh; posterior part is a synergist of hamstrings in thigh extension	Obturator nerve and sciatic nerve
• **Adductor longus** (*longus* = long)	Overlies middle aspect of adductor magnus; most anterior of adductor muscles	O—pubis near pubic symphysis I—linea aspera above adductor magnus	Adducts, flexes, and laterally rotates thigh	Obturator nerve
• **Adductor brevis** (*brevis* = short)	In contact with obturator externus muscle; largely concealed by adductor longus and pectineus	O—body and inferior ramus of pubis I—linea aspera above adductor longus	Adducts and laterally rotates thigh	Obturator nerve
Pectineus (pek-tih′-nē-is) (*pecten* = comb)	Short, flat muscle; overlies adductor brevis on proximal thigh; abuts adductor longus medially	O—pectineal line of pubis I—inferior to lesser trochanter of posterior aspect of femur on pectineal line	Adducts, flexes, and laterally rotates thigh	Femoral and sometimes obturator nerve
Gracilis (gruh-sih′-lis) (*gracilis* = slender)	Long, thin, superficial muscle of medial thigh	O—inferior ramus of pubis I—medial surface of tibia just inferior to tibial head	Adducts thigh, flexes, and medially rotates leg, especially during walking	Obturator nerve
MUSCLES OF THE ANTERIOR COMPARTMENT OF THE THIGH				
Quadriceps femoris (kwah′-drih-seps feh-mor′-is)	Quadriceps femoris arises from four separate heads (*quadriceps* = four heads) that form the flesh of front and sides of thigh; these heads (rectus femoris, and lateral, medial, and intermediate vastus muscles) have a common insertion tendon, the quadriceps tendon, which inserts into the patella and then via the patellar tendon into tibial tuberosity. The quadriceps is a powerful knee extensor used in climbing, jumping, running, and rising from seated position; group is innervated by femoral nerve; the tone of quadriceps plays important role in strengthening knee joint			
• **Rectus** (rek′-tus) **femoris** (*rectus* = straight; *femoris* = femur)	Superficial muscle of anterior thigh; runs straight down thigh; longest head and only muscle of group to cross hip joint	O—anterior inferior iliac spine and superior margin of acetabulum I—base of patella and tibial tuberosity via patellar tendon	Extends knee and flexes thigh at hip	Femoral nerve
• **Vastus lateralis** (vas′-tis la-ter-a′-lis) (*vastus* = large; *lateralis* = lateral)	Forms lateral aspect of thigh; a common intramuscular injection site, particularly in infants (who have poorly developed buttock and arm muscles)	O—greater trochanter, intertrochanteric line, linea aspera I—as for rectus femoris	Extends knee	Femoral nerve
• **Vastus medialis** (mē-dē-a′-lis) (*medialis* = medial)	Forms inferomedial aspect of thigh	O—linea aspera, intertrochanteric line I—as for rectus femoris	Extends knee; lower fibers stabilize patella	Femoral nerve
• **Vastus intermedius** (in″-ter-mē′-dē-us) (*intermedius* = intermediate)	Obscured by rectus femoris; lies between vastus lateralis and vastus medialis on anterior thigh	O—anterior and lateral surfaces of proximal femur shaft I—as for rectus femoris	Extends knee	Femoral nerve

Muscle	Description	Origin (O) and insertion (I)	Action	Nerve supply
Tensor fasciae latae (ten′-ser fa′-shē-ē lā′-tē) (*tensor* = to make tense; *fascia* = band; *lata* = wide)	Enclosed between fascia layers of anterolateral aspect of thigh; functionally associated with medial rotators and flexors of thigh	O—anterior aspect of iliac crest and anterior superior iliac spine I—iliotibial tract*	Flexes and abducts thigh (thus a synergist of the iliopsoas and gluteus medius and minimus muscles); rotates thigh medially; steadies trunk on thigh by making iliotibial tract taut	Superior gluteal nerve
PART II: POSTERIOR MUSCLES (Figure 10.18)				
GLUTEAL MUSCLES—ORIGIN ON PELVIS				
Gluteus maximus (gloo′-tē-us mak′-sih-mus) (*glutos* = buttock; *maximus* = largest)	Largest and most superficial of gluteus muscles; forms bulk of buttock mass; fibers are thick and coarse; important site of intramuscular injection (dorsal gluteal site); overlies large sciatic nerve; covers ischial tuberosity only when standing; when sitting, moves superiorly leaving ischial tuberosity exposed in the subcutaneous position	O—ilium and dorsal sacrum and coccyx I—gluteal tuberosity of femur; iliotibial tract	Major extensor of thigh; complex, powerful, and most effective when thigh is flexed and force is necessary, as in rising from a forward flexed position and in thrusting the thigh posteriorly in climbing stairs and running; generally inactive during walking; laterally rotates thigh; antagonist of iliopsoas muscle	Inferior gluteal nerve
Gluteus medius (mē′-dē-us) (*medius* = middle)	Thick muscle largely covered by gluteus maximus; important site for intramuscular injections (ventral gluteal site); considered safer than dorsal gluteal site because there is less chance of injuring sciatic nerve	O—upper lateral surface of ilium I—by short tendon into lateral aspect of greater trochanter of femur	Abducts and medially rotates thigh; steadies pelvis; its action is extremely important in walking; e.g., muscle of limb planted on ground tilts or holds pelvis in abduction so that pelvis on side of swinging limb does not sag; the foot of swinging limb can thus clear the ground	Superior gluteal nerve
Gluteus minimus (mih′-nih-mus) (*minimus* = smallest)	Smallest and deepest of gluteal muscles	O—external surface of ilium I—anterior border of greater trochanter of femur	As for gluteus medius	Superior gluteal nerve
LATERAL ROTATORS				
Piriformis (pih″-rih-for′-mis) (*piri* = pear; *forma* = shape)	Pyramidal muscle located on posterior aspect of hip joint; inferior to gluteus minimus; issues from pelvis via greater sciatic notch	O—anterolateral surface of sacrum (opposite greater sciatic notch) I—superior border of greater trochanter of femur	Rotates thigh laterally; since inserted above head of femur, can also assist in abduction of thigh when hip is flexed; stabilizes hip joint	S_1 and S_2
Obturator externus (ob′-ter-ā″-ter ek-ster′-nus) (*obturator* = obturator foramen; *externus* = outside)	Flat, triangular muscle deep in upper medial aspect of thigh	O—outer surface of obturator membrane, external surface of pubis and ischium, and margins of obturator foramen I—by a tendon into trochanteric fossa of posterior femur	Rotates thigh laterally and stabilizes hip joint	Obturator nerve

(Table continues)

Table 10.13 (continued)

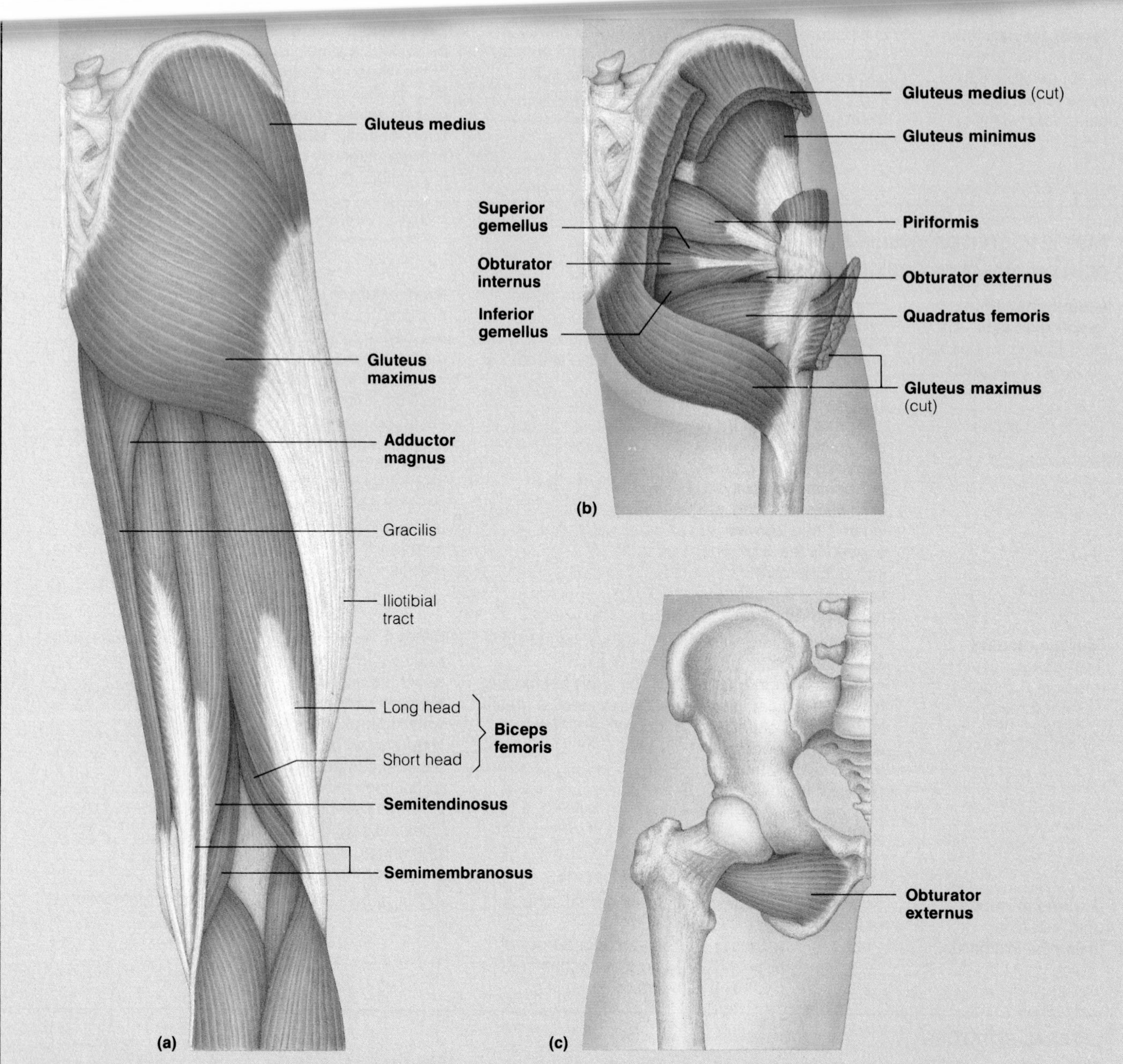

Figure 10.18 Posterior muscles of the right hip and thigh. (**a**) Superficial view showing the gluteus muscles of the buttock and hamstring muscles of the thigh. (**b**) Deep muscles of the gluteal region, which act primarily to rotate the thigh laterally. The superficial gluteus maximus and medius have been removed. (**c**) Anterior view of the isolated obturator externus muscle, showing its course as it travels from its origin on the anterior pelvis to the posterior aspect of the femur.

Muscle	Description	Origin (O) and insertion (I)	Action	Nerve supply
Obturator internus (in-ter′-nus) (*internus* = inside)	Surrounds obturator foramen within pelvis; leaves pelvis via lesser sciatic notch and turns acutely forward to insert in femur	O—inner surface of obturator membrane, greater sciatic notch, and margins of obturator foramen I—greater trochanter in front of piriformis	As for obturator externus	L_5 and S_1
Gemellus (geh-meh′-lis)—superior and inferior (*gemin* = twin, double; *superior* = above; *inferior* = below)	Two small muscles with common insertions and actions; considered extrapelvic portions of obturator internus	O—ischial spine (superior); ischial tuberosity (inferior) I—greater trochanter of femur	Rotate thigh laterally and stabilize hip joint	L_4, L_5, and S_1
Quadratus femoris (*quad* = four-sided square)	Short, thick muscle; most inferior of lateral rotator muscles; extends laterally from pelvis	O—ischial tuberosity I—greater trochanter of femur	Rotates thigh laterally and stabilizes hip joint	L_5 and S_1
MUSCLES OF THE POSTERIOR COMPARTMENT OF THE THIGH				
Hamstrings	The hamstrings are fleshy muscles of the posterior thigh (biceps femoris, semitendinosus, and semimembranosus); they cross both the hip and knee joints and are prime movers of thigh extension and knee flexion; group has a common origin site and is innervated by sciatic nerve; ability of hamstrings to act on one of the two joints spanned depends on which joint is fixed; i.e., if knee is fixed (extended), they promote hip extension; if hip is extended, they promote knee flexion; however, when hamstrings are stretched, they tend to restrict full accomplishment of antagonistic movements; e.g., if knees are fully extended, it is difficult to flex the hip fully (and touch your toes), and when the thigh is fully flexed as in kicking a football, it is almost impossible to extend the knee fully at the same time (without considerable practice); name of this muscle group comes from old butchers' practice of using their tendons to hang hams for smoking; "pulled hamstrings" are common sports injuries in those who run very hard, e.g., football halfbacks			
• **Biceps femoris** (*biceps* = two heads)	Most lateral muscle of the group; arises from two heads	O—ischial tuberosity (long head); linea aspera and distal femur (short head) I—common tendon passes downward and laterally (forming lateral border of popliteal fossa) to insert into head of fibula and lateral condyle of tibia	Extends thigh and flexes knee; laterally rotates leg, especially when knee is flexed	Sciatic nerve
• **Semitendinosus** (seh″-mē-ten″-dih-nō′-sus) (*semi* = half; *tendinosus* = tendon)	Lies medial to biceps femoris; although its name suggests that this muscle is largely tendinous, it is quite fleshy; its slender tendon begins about ⅔ the way down thigh	O—ischial tuberosity in common with long head of biceps femoris I—medial aspect of upper tibial shaft	Extends thigh at hip; flexes knee; with semimembranosus, medially rotates leg	Sciatic nerve
• **Semimembranosus** (seh″-mē-mem″-bruh-nō′-sus) (*membranosus* = membrane)	Deep to semitendinosus	O—ischial tuberosity I—medial condyle of tibia	Extends thigh and flexes knee; medially rotates leg	Sciatic nerve

*The iliotibial tract is a thickened lateral portion of the *fascia lata* (the fascia that ensheaths all the muscles of the thigh). It extends as a tendinous band from the iliac crest to the knee.

Table 10.14 Muscles of the Leg: Movements of the Ankle and Toes (Figures 10.19 to 10.21)

The deep fascia of the leg is continuous with the fascia lata that ensheaths the thigh. It is quite dense and binds the leg muscles tightly, helping to prevent excessive swelling of muscles during exercise and facilitating venous return. The leg fascia is attached to the anterior border of the tibia, and its inward extensions segregate the leg muscles into anterior, lateral, and posterior compartments, each with its own nerve and blood supply. Distal fascial extensions form the retinaculae (flexor, extensor, and peroneal), which secure the muscle tendons in place as they cross the ankle to run to the foot.

Depending on their location and placement, the various muscles of the leg promote movements at the ankle joint (dorsiflexion and plantar flexion), the tarsal joints (inversion and eversion of the foot), and/or the toes (flexion, extension, abduction, and adduction). Muscles in the anterior compartment of the leg (*tibialis anterior, extensor digitorum longus, extensor hallucis longus*, and *peroneus tertius)* are primarily ankle dorsiflexors. Although dorsiflexion is not a powerful movement, it is extremely important in preventing the dragging of the toes during walking. Lateral compartment muscles *(peroneus longus* and *brevis)* effect plantar flexion and foot eversion, whereas muscles of the posterior compartment *(gastrocnemius, soleus, tibialis posterior, flexor digitorum longus*, and *flexor hallucis longus)* are primarily ankle plantar flexors. Plantar flexion is the most powerful movement of the ankle (and foot). It is necessary for standing on tip-toe and for providing the necessary forward thrust when walking and running. The popliteus muscle, which crosses the knee, is important in "unlocking" the extended knee in preparation for flexion.

Intrinsic muscles of the sole of the foot (lumbricals, interossei, and many others) help in flexion, extension, abduction, and adduction of the toes. Collectively, they are very important (along with the tendons of leg muscles) in supporting the arches of the foot. However, since these muscles are numerous, their arrangement complex, and their individual actions relatively unimportant, we will not consider them further.

Muscle	Description	Origin (O) and insertion (I)	Action	Nerve supply
PART I: MUSCLES OF THE ANTERIOR COMPARTMENT (Figures 10.19 and 10.20)				
	All muscles of the anterior compartment are dorsiflexors of the ankle and have a common innervation, the deep peroneal nerve. Paralysis of the anterior muscle group causes *foot drop*, which requires that the leg be lifted unusually high during walking to prevent tripping over one's toes. "Shinsplints" is a painful inflammatory condition of the muscles of the anterior compartment.			
Tibialis anterior (tih″-bē-a′-lis) (*tibial* = tibia; *anterior* = toward the front)	Superficial muscle of anterior leg; laterally parallels sharp anterior margin of tibia	O—lateral condyle and upper ⅔ of tibia; interosseous membrane I—by tendon into inferior surface of first cuneiform and first metatarsal bone	Prime mover of dorsiflexion; inverts foot; assists in supporting medial longitudinal arch of foot	Deep peroneal nerve
Extensor digitorum longus (*extensor* = increases angle at a joint; *digit* = finger or toe; *longus* = long)	On anterolateral surface of leg; lateral to tibialis anterior muscle	O—lateral condyle of tibia; proximal ¾ of fibula; interosseous membrane I—second and third phalanges of toes 2–5	Dorsiflexes and everts foot; prime mover of toe extension (acts mainly at metatarsophalangeal joints)	Deep peroneal nerve
Peroneus tertius (payr″-ō-nē′-is ter′-shus) (*perone* = fibula; *tertius* = third)	Small muscle; usually continuous and fused with distal part of extensor digitorum longus; not always present	O—distal anterior surface of fibula and interosseous membrane I—tendon passes anterior to lateral malleolus and inserts on dorsum of fifth metatarsal.	Dorsiflexes and everts foot	Deep peroneal nerve
Extensor hallucis (hah-lo′-kis) **longus** (*hallux* = great toe)	Deep to extensor digitorum longus and tibialis anterior; narrow origin	O—anteromedial fibula shaft and interosseous membrane I—tendon inserts on distal phalanx of great toe	Extends great toe; dorsiflexes foot and aids in foot inversion	Deep peroneal nerve

(Table continues)

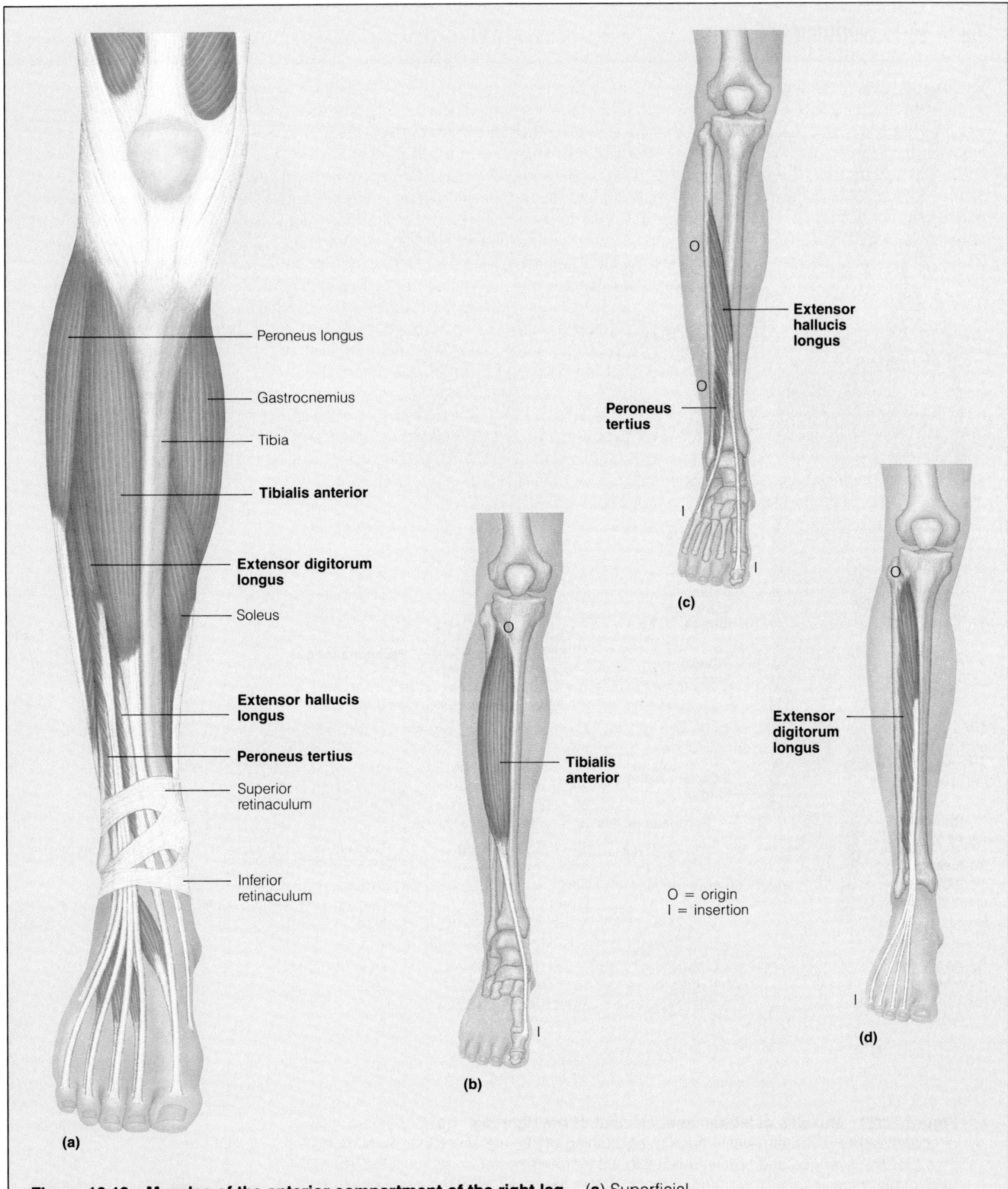

Figure 10.19 Muscles of the anterior compartment of the right leg. (**a**) Superficial view of anterior leg muscles. (**b–d**) Some of the same muscles shown in isolation to allow visualization of their origins and insertions.

(Table continues)

Table 10.14 (continued)

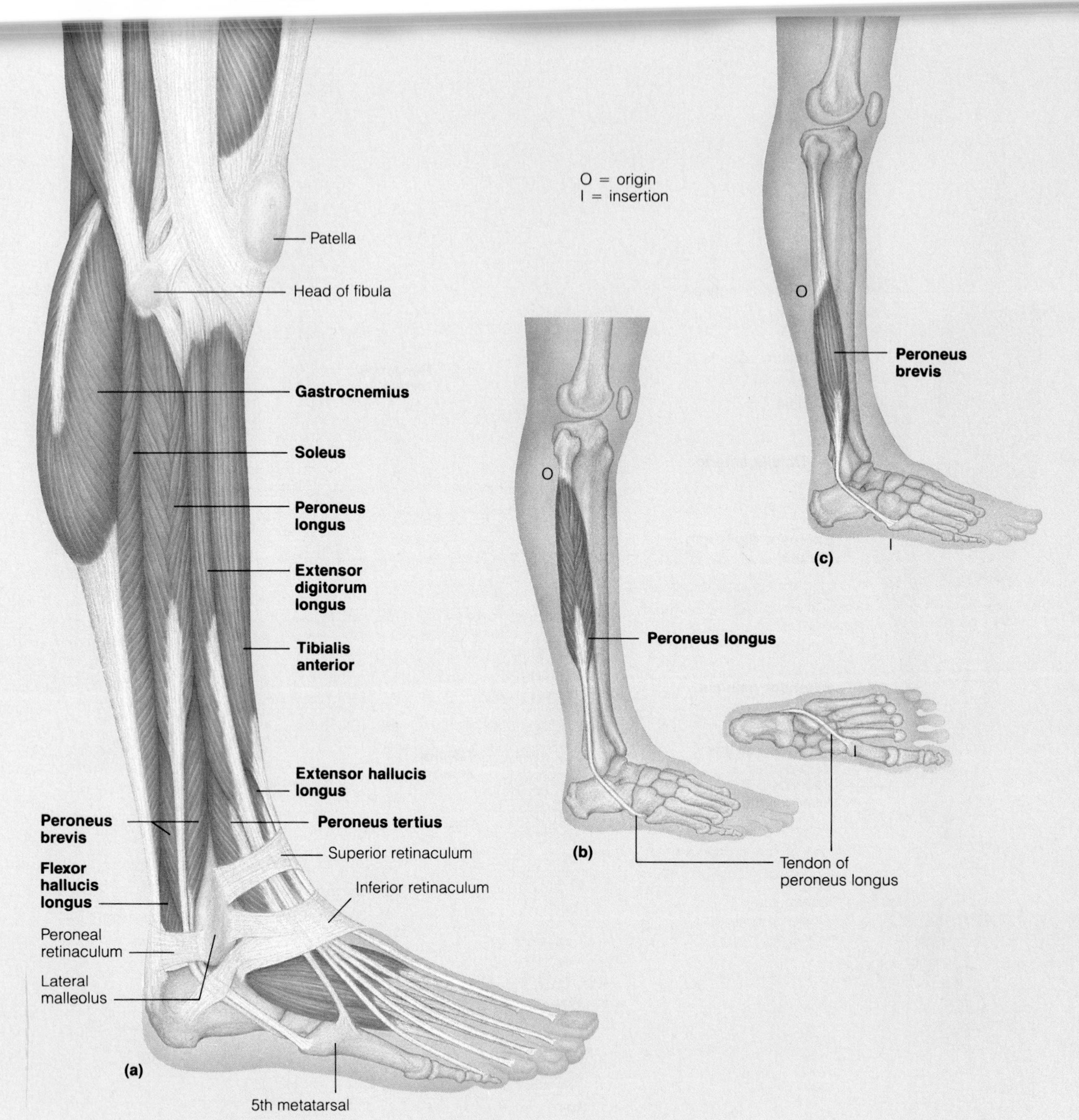

Figure 10.20 Muscles of lateral compartment of the right leg. (**a**) Superficial view of lateral aspect of the leg, illustrating the positioning of the lateral compartment muscles (peroneus longus and brevis) relative to anterior and posterior leg muscles. (**b**) View of peroneus longus in isolation; inset illustrates the insertion of the peroneus longus on the plantar surface of the foot. (**c**) Isolated view of the peroneus brevis muscle.

Muscle	Description	Origin (O) and insertion (I)	Action	Nerve supply
PART II: MUSCLES OF THE LATERAL COMPARTMENT (Figures 10.20 and 10.21)				
	These muscles have a common innervation, the superficial peroneal nerve. Besides plantar flexion and foot eversion, these muscles stabilize the lateral ankle and lateral longitudinal arch of the foot.			
Peroneus longus (See also Figure 10.19)	Superficial lateral muscle; overlies fibula	O—head and upper portion of fibula I—by long tendon that curves under foot to first metatarsal	Plantar flexes and everts foot; helps keep foot flat on ground	Superficial peroneal nerve
Peroneus brevis (*brevis* = short)	Smaller muscle; deep to peroneus longus; enclosed in a common sheath	O—distal fibula shaft I—by tendon running behind lateral malleolus to insert on proximal end of fifth metatarsal	Plantar flexes and everts foot	Superficial peroneal nerve
PART III: MUSCLES OF THE POSTERIOR COMPARTMENT (Figure 10.21)				
	The muscles of the posterior compartment have a common innervation, the tibial nerve. They act in concert to plantar flex the ankle.			
SUPERFICIAL MUSCLES				
Triceps surae (trī′-seps sur′-ē) (See also Figure 10.20)	Refers to muscle pair (gastrocnemius and soleus) that shapes the posterior calf and inserts via a common tendon into the calcaneus of the heel; this Achilles, or calcaneal, tendon is the largest tendon in the body; prime movers of ankle plantar flexion			
• **Gastrocnemius** (gas″-truk-nē′-mē-is) (*gaster* = belly; *kneme* = leg)	Superficial muscle of pair; two prominent bellies that form proximal curve of calf	O—by two heads from medial and lateral condyles of femur I—calcaneus via Achilles tendon	Plantar flexes foot when knee is extended; since it also crosses knee joint, it can flex knee when foot is dorsiflexed	Tibial nerve
• **Soleus** (sō′-lē-us) (*soleus* = sole of foot)	Deep to gastrocnemius on posterior surface of calf	O—extensive cone-shaped origin from superior tibia, fibula, and interosseous membrane I—calcaneus via Achilles tendon	Plantar flexes ankle; important locomotor and postural muscle during walking, running, and dancing	Tibial nerve
Plantaris (plan-tar′-is) (*planta* = sole of foot)	Generally a small feeble muscle, but varies in size and extent; may be absent	O—posterior femur above lateral condyle I—via a long, thin tendon into calcaneus or calcaneal tendon	Assists in knee flexion and plantar flexion of foot	Tibial nerve
DEEP MUSCLES				
Popliteus (pop-lih′-tē-us) (*poplit* = back of knee)	Thin, triangular muscle; passes downward and medially to tibial surface	O—lateral condyle of femur I—proximal tibia	Flexes and rotates leg medially to unlock knee from full extension when flexion begins	Tibial nerve
Flexor digitorum longus (*flexor* = decreases angle at a joint)	Long, narrow muscle; runs medial to and partially overlies tibialis posterior	O—posterior tibia I—tendon runs behind medial malleolus and splits into four parts to insert into distal phalanges of toes 2–5	Plantar flexes and inverts foot; flexes toes; helps foot "grip" ground	Tibial nerve
Flexor hallucis longus (See also Figure 10.20)	Bipennate muscle; lies lateral to inferior aspect of tibialis posterior	O—middle part of shaft of fibula; interosseous membrane I—tendon runs under foot to distal phalanx of great toe	Plantar flexes and inverts foot; flexes great toe at all joints; "push off" muscle during walking	Tibial nerve
Tibialis posterior (*posterior* = toward the back)	Thick, flat muscle deep to soleus; placed between posterior flexors	O—extensive origin from superior tibia and fibula and interosseous membrane I—tendon passes behind medial malleolus and under arch of foot; inserts into several tarsals and metatarsals 2–4	Prime mover of foot inversion; plantar flexes ankle; stabilizes medial longitudinal arch of foot (as during ice skating)	Tibial nerve

(Table continues)

Table 10.14 (continued)

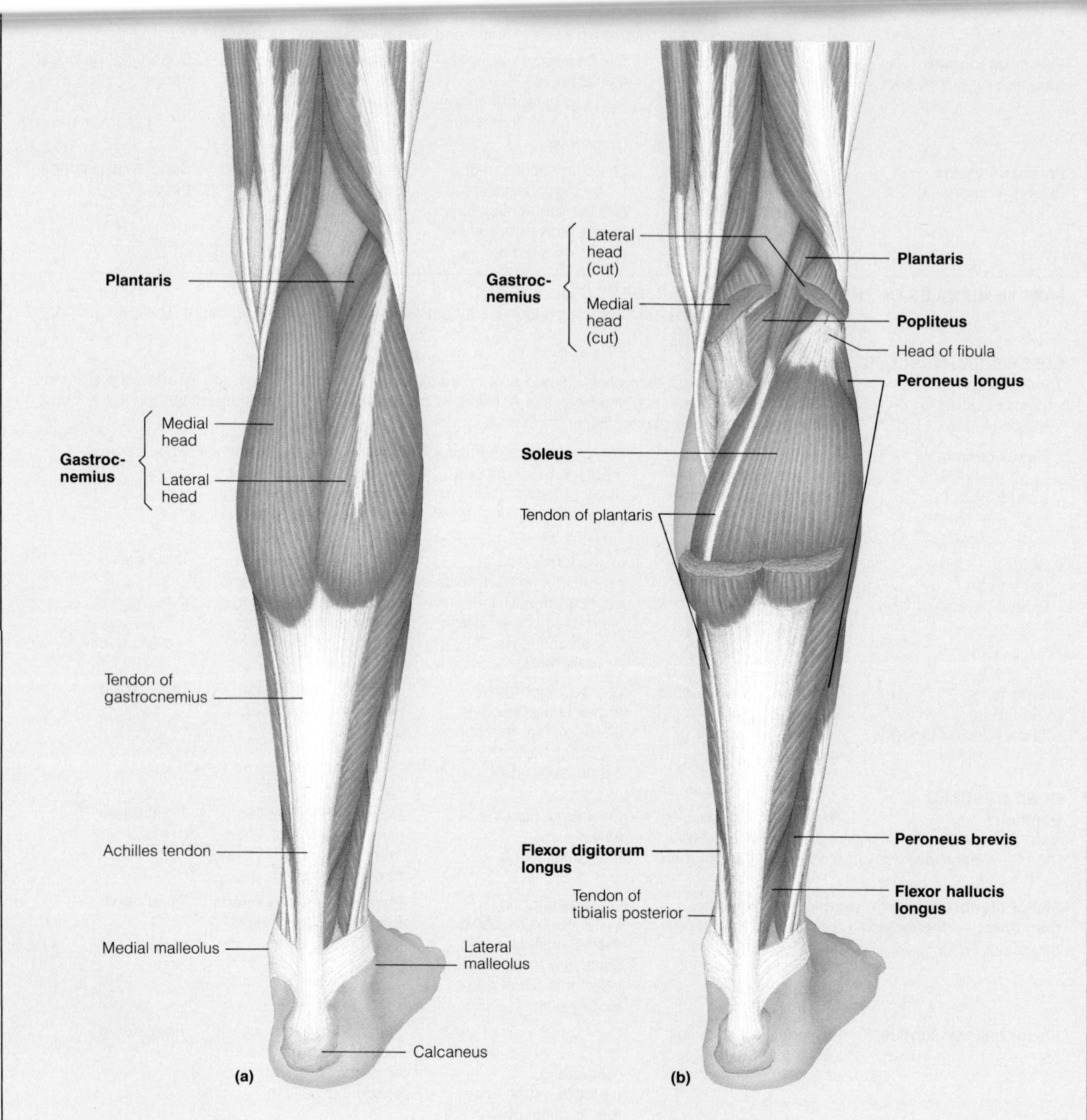

Figure 10.21 Muscles of the posterior compartment of the right leg. (**a**) Superficial view of the posterior leg. (**b**) The fleshy gastrocnemius has been removed to show the soleus immediately deep to it. (**c**) The triceps surae has been removed to show the deep muscles of the posterior compartment. (**d–f**) Individual deep muscles are shown in isolation so that their origins and insertions may be visualized.

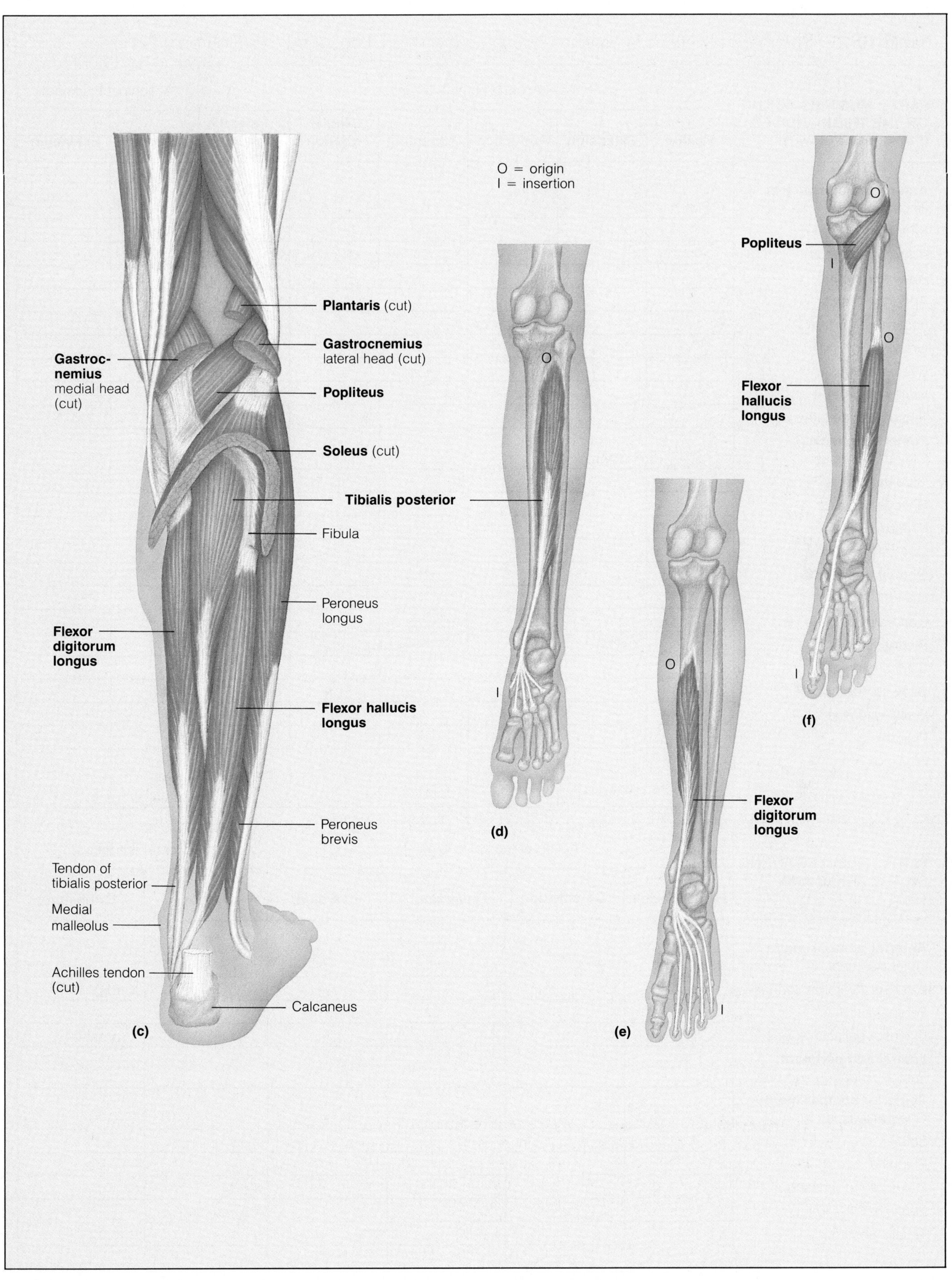
O = origin
I = insertion
Plantaris (cut)
Gastrocnemius lateral head (cut)
Gastrocnemius medial head (cut)
Popliteus
Soleus (cut)
Tibialis posterior
Fibula
Peroneus longus
Flexor digitorum longus
Flexor hallucis longus
Peroneus brevis
Tendon of tibialis posterior
Medial malleolus
Achilles tendon (cut)
Calcaneus
(c)
O
I
(d)
Popliteus
I
O
O
Flexor hallucis longus
I
(f)
O
Flexor digitorum longus
I
(e)

Table 10.15 Summary of Actions of Muscles Acting on the Thigh, Leg, and Foot (Figure 10.22)

PART I: MUSCLES ACTING ON THE THIGH AND LEG (PM = prime mover)	Actions at the hip joint						Actions at the knee	
	Flexion	**Extension**	**Abduction**	**Adduction**	**Medial rotation**	**Lateral rotation**	**Flexion**	**Extension**
Anteromedial muscles:								
Iliopsoas	X (PM)							
Sartorius	X					X	X	
Adductor magnus		X		X		X		
Adductor longus	X			X		X		
Adductor brevis	X			X		X		
Pectineus	X			X		X		
Gracilis				X			X	
Rectus femoris	X							X
Vastus muscles								X
Tensor fasciae latae	X				X			
Posterior muscles:								
Gluteus maximus		X (PM)				X		
Gluteus medius			X		X			
Gluteus minimus			X		X			
Piriformis			X			X		
Obturator internus						X		
Obturator externus						X		
Gemelli						X		
Quadratus femoris						X		
Biceps femoris		X (PM)					X (PM)	
Semitendinosus		X					X (PM)	
Semimembranosus		X					X (PM)	
Gastrocnemius							X	
Plantaris							X	
Popliteus							X (and rotates medially)	

PART II: MUSCLES ACTING ON THE ANKLE AND TOES	Action at the ankle joint				Action at the toes	
	Plantar flexion	**Dorsiflexion**	**Inversion**	**Eversion**	**Flexion**	**Extension**
Anterior compartment:						
Tibialis anterior		X (PM)	X			
Extensor digitorum longus		X		X		X (PM)
Peroneus tertius		X		X		
Extensor hallucis longus		X	X			X (great toe)
Lateral compartment:						
Peroneus longus and brevis	X			X		
Posterior compartment:						
Gastrocnemius	X					
Soleus	X					
Plantaris	X					
Flexor digitorum longus	X		X		X (PM)	
Flexor hallucis longus	X		X		X (great toe)	
Tibialis posterior	X		X (PM)			

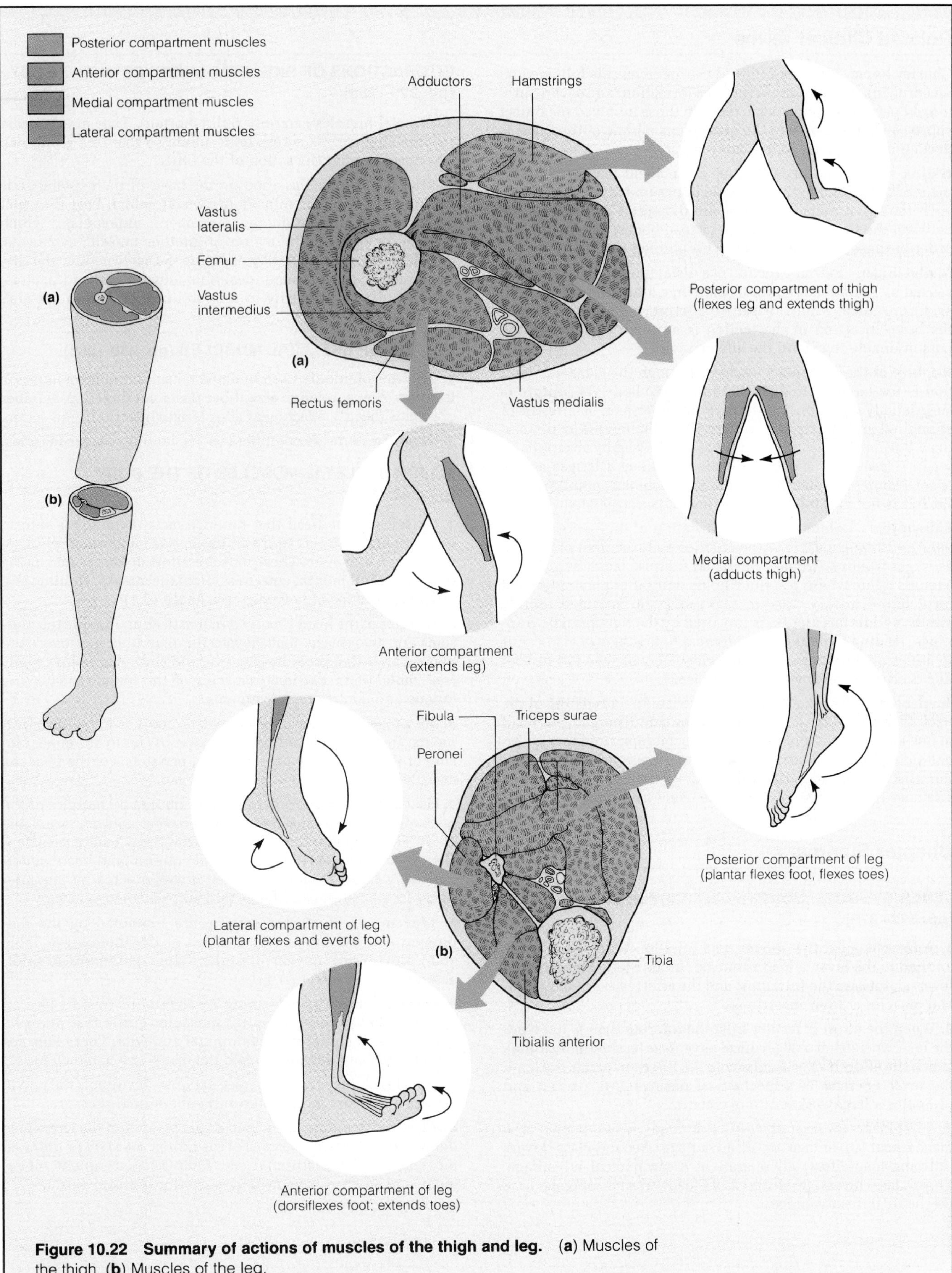

Figure 10.22 Summary of actions of muscles of the thigh and leg. (**a**) Muscles of the thigh. (**b**) Muscles of the leg.

Related Clinical Terms

bleeding into the tissues (hematoma formation) and severe, prolonged pain; a common occurrence in those involved in contact sports; a charley horse of the quadriceps muscle of the thigh is particularly frequent in football players.

Hallux valgus (ha′-lux val′-gus) Permanent lateral displacement of the great toe; the sesamoid bones under the head of the great toe's first metatarsal bone are displaced so that they lie between the first and second metatarsal bones; a common disorder in those who wear shoes with pointed toes.

Mallet finger Extreme flexion of a distal interphalangeal joint, caused by a sudden forceful tension on a long extensor tendon, as, for instance, when a baseball is improperly caught; part of the bony insertion of the tendon is stripped off, leaving the person unable to extend the affected joint.

Rupture of the calcaneus tendon Although the calcaneus tendon is the largest, strongest tendon in the body, its rupture is surprisingly common, particularly in older men as a result of stumbling and in young sprinters when the tendon is traumatized during the takeoff; the rupture is followed by abrupt pain; a gap is seen just above the heel, and the calf bulges as the triceps surae are released from their attachment point; plantar flexion is not possible, but dorsiflexion is exaggerated.

Shinsplints Common term for irritation of the periostea and anterior tibialis muscle of the anterior compartment of the leg, as might follow extreme or unusual exercise (such as going to a tennis clinic where you run 6 hours daily after playing at most for 2 hours weekly prior to that time); the inflamed muscle swells, and its circulation is impaired by the tight fascial wrappings, leading to pain and tenderness to touch; may also occur in athletes who fail to warm up adequately or who fail to cool down properly after vigorous exercise.

Torticollis (tor″-tih-kah′-lis) (*tort* = twisted) A twisting of the neck in which there is a fixed rotation and tilting of the head to one side, due to injury of the sternocleidomastoid muscle on one side; also called wryneck; sometimes presents at birth when the muscle fibers are torn during difficult delivery; exercise that stretches the affected muscle is the usual treatment.

Chapter Summary

LEVER SYSTEMS: BONE–MUSCLE RELATIONSHIPS (pp. 277–279)

1. A lever is a bar that moves on a fulcrum. When an effort is applied to the lever, a load is moved. In the body, bones are the levers, joints are the fulcrums, and the effort is exerted by skeletal muscles at their insertions.

2. When the effort is farther from the fulcrum than is the load, the lever operates at a mechanical advantage (is slow and strong). When the effort is exerted closer to the fulcrum than is the load, the lever operates at a mechanical disadvantage (is fast and promotes a large degree of movement).

3. First-class levers (effort-fulcrum-load) may operate at a mechanical advantage or disadvantage. Second-class levers (fulcrum-load-effort) all operate at a mechanical advantage. Third-class levers (fulcrum-effort-load) always operate at a mechanical disadvantage.

(pp. 279–280)

1. Skeletal muscles can only pull (shorten). They are arranged in opposing groups across body joints so that one group can reverse or modify the action of the other.

2. Muscles may be classified on the basis of their interactions in the body as prime movers (agonists), which bear the chief responsibility for producing movement; antagonists, which reverse, or oppose, the action of another muscle; synergists, which aid a prime mover by effecting the same action, stabilizing joints, or preventing undesirable movements; and fixators, which function primarily to immobilize a bone or a muscle's origin.

NAMING OF SKELETAL MUSCLES (pp. 280–282)

1. Criteria commonly used to name muscles include a muscle's location, shape, relative size, fiber (fascicle) direction, number of origins (heads), attachment sites (origin/insertion), and action.

2. Several criteria are combined in the naming of some muscles.

MAJOR SKELETAL MUSCLES OF THE BODY (pp. 282–327)

1. Muscles of the head that produce facial expression tend to be small and to insert into soft tissue (skin and other muscles) rather than into bone. These muscles allow opening and closing of the eyes and mouth, compression of the cheeks, smiling, and other types of facial language (see Table 10.1).

2. Muscles of the head involved in mastication include the masseter and temporalis that elevate the mandible and two deep muscle pairs that promote grinding and sliding jaw movements (see Table 10.2). Extrinsic muscles of the tongue anchor the tongue and control its movements.

3. Deep muscles of the anterior neck promote swallowing movements, including elevation/depression of the hyoid bone, closure of the respiratory passages, and peristalsis of the pharynx (see Table 10.3).

4. Head and trunk movements are promoted by muscles of the neck and by deep muscles of the vertebral column (see Table 10.4). The deep muscles of the posterior trunk can cause extension of large regions of the vertebral column (and head) simultaneously. Head flexion and rotation are effected by the anteriorly located sternocleidomastoid and scalene muscles.

5. Movements of quiet breathing are promoted by the diaphragm and the intercostal muscles of the thorax (see Table 10.5). Downward movement of the diaphragm increases intra-abdominal pressure.

6. The four muscle pairs forming the abdominal wall are layered like plywood to form a natural muscular girdle that protects, supports, and compresses abdominal contents. These muscles can also flex and laterally rotate the trunk (see Table 10.6).

7. Muscles of the pelvic floor (see Table 10.7) support the pelvic viscera and resist increases in intra-abdominal pressure.

8. With the exception of the pectoralis major and the latissimus dorsi, the superficial muscles of the thorax act to fix or promote movements of the scapula (see Table 10.8). Scapular movements are effected primarily by posterior thoracic muscles.

9. Nine muscles cross the shoulder joint to effect movements of the humerus (see Table 10.9). Of these, seven originate on the scapula and two arise from the axial skeleton. Four muscles contribute to the "rotator cuff" helping to stabilize the multiaxial shoulder joint. Generally, muscles located anteriorly flex, rotate, and adduct the arm. Those located posteriorly extend, rotate, and adduct the arm. The deltoid muscle of the shoulder is the prime mover of shoulder abduction.

10. Muscles causing movements of the forearm form the flesh of the arm (see Table 10.10). Anterior arm muscles are elbow flexors; posterior muscles are elbow extensors.

11. Movements of the wrist, hand, and fingers are effected mainly by muscles originating on the forearm (see Table 10.11). Except for the two pronator muscles, the anterior forearm muscles are wrist and/or finger flexors; those of the posterior compartment are wrist and finger extensors. Intrinsic hand muscles aid in finger (and thumb) movements.

12. Muscles crossing the hip and knee joints effect thigh and leg movements (see Table 10.13). Anteromedial muscles include thigh flexors and/or adductors and knee extensors. Muscles of the posterior gluteal region extend and rotate the thigh. The posterior thigh muscles both extend the hip and flex the knee.

13. Muscles of the leg act on the ankle and toes (see Table 10.14). Anterior compartment muscles are largely ankle dorsiflexors. Lateral compartment muscles are plantar flexors and foot everters. Those of the posterior leg are plantar flexors. Intrinsic foot muscles support foot arches and help effect toe movements.

Review Questions

Multiple Choice/Matching

1. A muscle that assists an agonist by causing a like movement or by stabilizing a joint over which an agonist acts is (a) an antagonist, (b) a prime mover, (c) a synergist, (d) a fixator.

2. Match the muscle names in column B to the facial muscles described in column A.

Column A	Column B
______(1) squints the eyes	(a) corrugator supercilii
______(2) furrows the forehead horizontally	(b) depressor anguli oris
______(3) smiling muscle	(c) frontalis
______(4) puckers the lips	(d) occipitalis
______(5) pulls the scalp posteriorly	(e) orbicularis oculi
	(f) orbicularis oris
	(g) zygomaticus

3. The prime mover(s) of inspiration is (are) the (a) diaphragm, (b) internal intercostals, (c) external intercostals, (d) abdominal wall muscles.

4. The arm muscle that both flexes the elbow and supinates the forearm is the (a) brachialis, (b) brachioradialis, (c) biceps brachii, (d) triceps brachii.

5. The muscles of mastication that protrude the mandible and produce side-to-side grinding movements are the (a) buccinators, (b) masseters, (c) temporalis, (d) pterygoids.

6. Muscles that depress the hyoid bone and larynx include all but the (a) sternohyoid, (b) omohyoid, (c) geniohyoid, (d) sternothyroid.

7. Intrinsic muscles of the back that promote extension and lateral flexion of the spine (or head) include all but (a) splenius muscles, (b) semispinalis muscles, (c) scalene muscles, (d) erector spinae.

8. Several muscles act to move and/or stabilize the scapula. Which of the following are small rectangular muscles that square the shoulders as they act together to retract the scapula? (a) levator scapulae, (b) rhomboids, (c) serratus anterior, (d) trapezius.

9. The quadriceps include all but (a) vastus lateralis, (b) vastus intermedius, (c) vastus medialis, (d) biceps femoris, (e) rectus femoris.

10. A prime mover of hip flexion is the (a) rectus femoris, (b) iliopsoas, (c) vasti muscles, (d) gluteus maximus.

11. The prime mover of hip extension *against* resistance is the (a) gluteus maximus, (b) gluteus medius, (c) biceps femoris, (d) semimembranosus.

12. Muscles that cause plantar flexion include all but the (a) gastrocnemius, (b) soleus, (c) tibialis anterior, (d) tibialis posterior, (e) peroneus muscles.

Short Answer Review Questions.

13. Name four criteria used in naming muscles, and provide an example that illustrates each criterion.

14. Differentiate clearly between the arrangement of elements (load, fulcrum, and effort) in first-, second-, and third-class levers.

15. What does it mean when we say that a lever operates at a mechanical disadvantage, and what benefits does such a lever system provide?

16. What muscles act to propel a food bolus down the length of the pharynx to the esophagus?

17. Name and describe the action of muscles used to shake your head no. To nod yes.

18. (a) Name the four muscle pairs that act in unison to compress the abdominal contents. (b) How does their arrangement (fiber direction) contribute to the strength of the abdominal wall? (c) Which of these muscles can effect lateral rotation of the spine? (d) Which can act alone to flex the spine?

19. List all possible movements that can occur at the shoulder joint and name the prime mover(s) of each movement. Then name their antagonists.

20. (a) Name two forearm muscles that are powerful extensors and abductors of the wrist. (b) Name the sole forearm muscle that can flex the distal interphalangeal joints.

21. Name the muscles usually grouped together as the lateral rotators of the hip.

22. Name three thigh muscles that help you keep your seat astride a horse.

23. (a) Name three muscles or muscle groups used as sites for intramuscular injections. (b) Which of these is used most often in infants, and why?

Clinical Application Questions

24. Susan, a student nurse, was giving Mr. Graves a back rub. What two broad superficial muscles of the back were receiving most of her attention?

25. When Mrs. O'Brien returned to her doctor for a follow-up visit after childbirth, she complained that she was having problems controlling her urine flow (was incontinent) when she sneezed. The physician asked his nurse to give Mrs. O'Brien instructions on how to perform exercises to strengthen the muscles of the pelvic floor. To what muscles was he referring?

26. An obese, out-of-shape 45-year-old man was advised by his physician to lose weight and to exercise on a regular basis. He followed his diet faithfully and began to jog daily. One day, while on his morning jog, he heard a snapping sound that was immediately followed by a severe pain in his right lower calf. When examined, a gap was seen between his swollen upper calf region and his heel, and the patient was unable to plantar flex that ankle. What do you think happened? Why was the upper part of his calf swollen?

The seven chapters of Unit 3 provide a comprehensive treatment of the two principal regulatory systems of the body: the nervous and endocrine systems. The chief focus, as you will see, is on the nervous system because of its tremendous versatility and complexity. However, each system is given its due attention relative to the structures involved, mechanisms of control, and types of activities integrated and regulated. Of all the units covered in this book, this should be the most challenging and exciting.

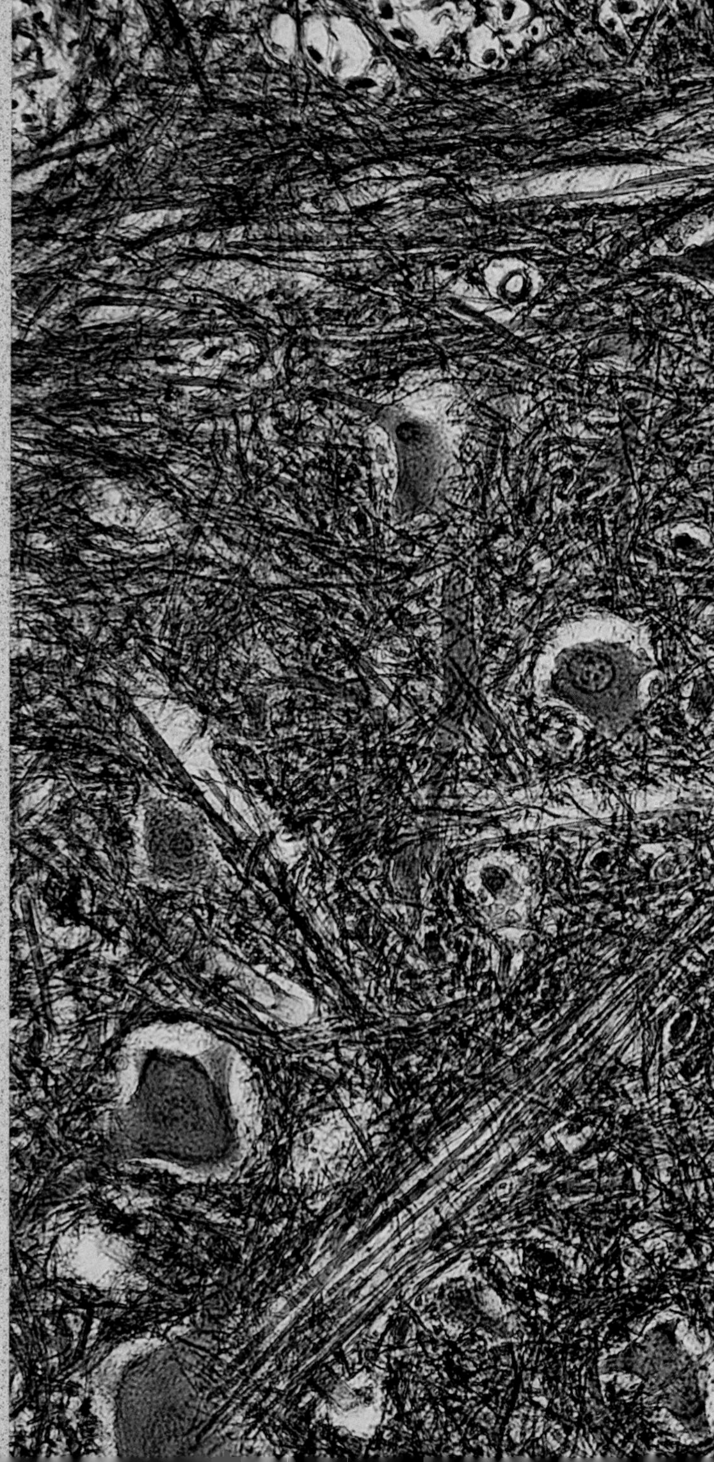

Light Micrograph of a cross-section through the spinal cord

UNIT 3

REGULATION AND INTEGRATION OF THE BODY

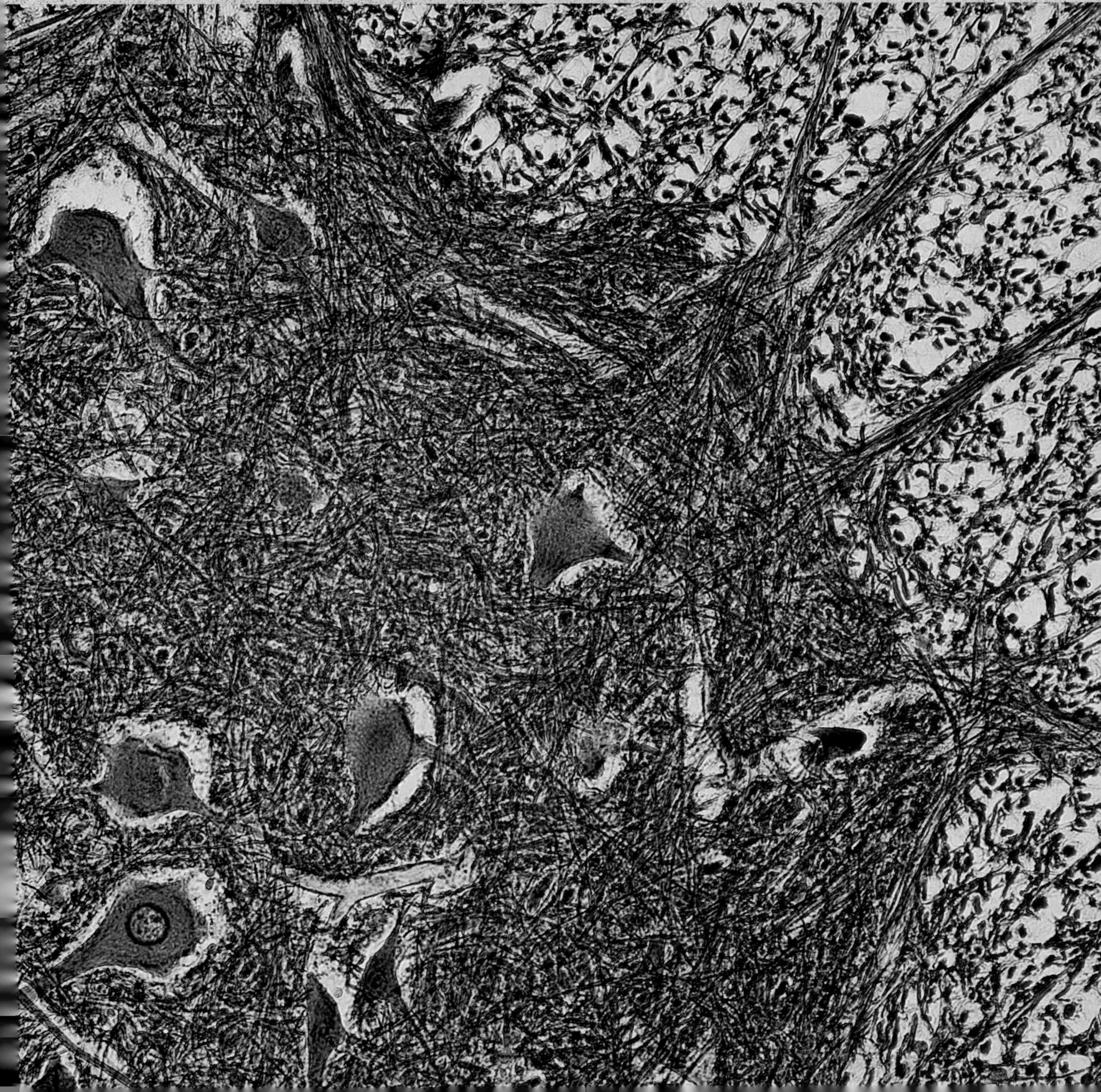

CHAPTER 11

Fundamentals of the Nervous System and Nervous Tissue

Chapter Outline and Student Objectives

1. List the basic functions of the nervous system.

Organization of the Nervous System (pp. 333–334)

2. Explain the structural and functional divisions of the nervous system.

Histology of Nervous Tissue (pp. 334–347)

3. List the types of supporting cells and cite their functions.
4. Describe the important anatomical structures of a neuron and relate each structure to a physiological role.
5. Explain the importance of the myelin sheath and describe how it is formed in the central and peripheral nervous systems.
6. Classify neurons, both structurally and functionally.
7. Define nerve and describe the general structure of a nerve.
8. Describe the process of nerve fiber regeneration.
9. Classify sensory receptors according to body location, stimulus detected, and structure.

Neurophysiology (pp. 347–367)

10. Define resting membrane potential and describe its electrochemical basis.
11. Compare and contrast graded and action potentials.
12. Explain how action potentials are generated and propagated along neurons.
13. Define absolute and relative refractory periods.
14. Define saltatory conduction and contrast it to conduction along unmyelinated fibers.
15. Define synapse. Distinguish between electrical and chemical synapses structurally and in their mechanism of information transmission.
16. Distinguish between excitatory and inhibitory postsynaptic potentials.
17. Describe how synaptic events are integrated and modified.
18. Define neurotransmitter and name several classes of neurotransmitters.
19. Describe receptor potentials, and define adaptation.

Basic Concepts of Neural Integration (pp. 367–370)

20. Describe common patterns of neuron organization and neuronal processing.
21. Define reflex and list the minimum number of components in a reflex arc.

Developmental Aspects of Neurons (p. 370)

22. Describe the role of nerve cell adhesion molecules (N-CAMs) in neuronal differentiation.

Preview of Selected Key Terms

Central nervous system (CNS) The brain and spinal cord.

Peripheral nervous system (PNS) All nervous system structures outside of the CNS; i.e., nerves, ganglia, and sensory receptors.

Neuroglia (ner-ah′-glē-uh) (*neuro* = nerve; *glia* = glue) Nonexcitable cells of neural tissue that support, protect, and insulate the neurons.

Neuron Cell of the nervous system specialized to generate and transmit nerve impulses.

Dendrite (*dendr* = tree) Branching neuron process that serves as a receptive, or input, region.

Axon (*axo* = axis, axle) Neuron process that conducts impulses.

Myelin sheath (mī′-uh-lin) Fatty insulating sheath that surrounds all but the smallest nerve fibers.

Sensory receptor Dendritic end organs, or parts of other cell types, specialized to respond to a stimulus.

Graded potential A local change in membrane potential that declines with distance and is not conducted along the nerve fiber.

Action potential A large transient depolarization event, including polarity reversal, that is conducted along the nerve fiber; also called the nerve impulse.

Saltatory conduction Transmission of an action potential along a myelinated fiber in which the nerve impulse appears to leap from node to node.

Synapse (sih′-naps) (*synaps* = a union) Functional junction or point of close contact between two neurons or between a neuron and an effector cell.

Neurotransmitter Chemical released by neurons that may, upon binding to receptors of neurons or effector cells, stimulate or inhibit those cells.

Sensory transduction Conversion of stimulus energy into a nerve impulse.

You are driving down the freeway, and a horn blares to your right. You swerve to your left. Charlie leaves a note on the kitchen table: "See you later—have the stuff ready at 6." You know that the "stuff" is chili with taco chips. You are dozing, and your infant son makes a soft cry. Instantly, you awaken. What do all these events have in common? They are all everyday examples of the functioning of your nervous system, which has your body cells humming with activity nearly all of the time.

The nervous system is the master controlling and communicating system of the body; every thought, action, and emotion reflects its activity. Along with the endocrine system, it is responsible for regulating and maintaining body homeostasis; of the two systems, it is by far the more rapid acting and complex. Cells of the nervous system communicate by means of electrical signals, which are rapid and specific, usually causing almost immediate responses. In contrast, the endocrine system typically brings about its effects in a more leisurely way through the activity of hormones released into the blood.

The nervous system has three overlapping functions: (1) It uses its millions of sensory receptors to *monitor changes* occurring both inside and outside the body; the gathered information is called *sensory input;* (2) it processes and interprets the sensory input and makes decisions about what should be done at each moment—a process called *integration;* and (3) it *effects a response* by activating muscles or glands; the response is called *motor output.* An example will illustrate how these functions work together. When you are driving and see a red light just ahead (sensory input), your nervous system integrates this information (red light means "stop"), and your foot goes for the brake (motor output).

This chapter begins with a brief overview of the organization of the nervous system. It then focuses on the functional anatomy of nervous tissue, especially that of the nerve cells, or *neurons,* which are the key to the subtle, efficient system of neural communication.

Organization of the Nervous System

We have only a single, highly integrated nervous system; however, for the sake of convenience, the nervous system is divided into two principal parts (Figure 11.1). The **central nervous system (CNS)** consists of the brain and spinal cord, which occupy the dorsal body cavity and act as the integrating and command centers of the nervous system. They interpret incoming sensory information and dictate responses based on past experience, reflexes, and current conditions. The **peripheral nervous system (PNS),** the part of the nervous system *outside* the CNS, consists mainly of the nerves that extend from the brain and spinal cord. *Spinal nerves* carry impulses to and from the spinal cord; *cranial* (krā′-nē-ul) *nerves* carry impulses to and from the brain. These peripheral nerves serve as communication lines linking all parts of the body to the central nervous system.

The peripheral nervous system has two functional subdivisions (see Figure 11.1). The **sensory,** or **afferent** (a′-fer-ent), **division** consists of nerve fibers that convey impulses *to* the central nervous system from sensory receptors located in various parts of the body. Sensory fibers conveying impulses from the skin, skeletal muscles, and joints are called *somatic afferents* (*soma* = body), whereas those transmitting impulses from the visceral organs are called *visceral* afferents. The sensory division keeps the CNS constantly informed of events going on both within and outside the body. The **motor,** or **efferent** (eh′-fer-ent), **division** transmits impulses *from* the CNS to effector organs, the muscles and glands. These impulses activate muscles and glands; that is, they *effect* (bring about) a motor response.

The motor division in turn has two subdivisions (see Figure 11.1):

1. The **somatic nervous system** is composed of motor nerve fibers that conduct impulses from the CNS to skeletal muscles. It is often referred to as the **voluntary nervous system,** since it allows us to consciously control our skeletal muscles.

2. The **autonomic nervous system (ANS)** consists of motor nerve fibers that regulate the activity of smooth muscles, cardiac muscles, and glands. *Autonomic* literally means "a law unto itself," and because we generally cannot control activities such as the pumping of our hearts or the movement of food through our digestive tracts, the ANS is also referred to as the **involuntary nervous system.** As indicated in Figure 11.1 and described in Chapter 14, the ANS has two functional subdivisions, the *sympathetic* and the *par-*

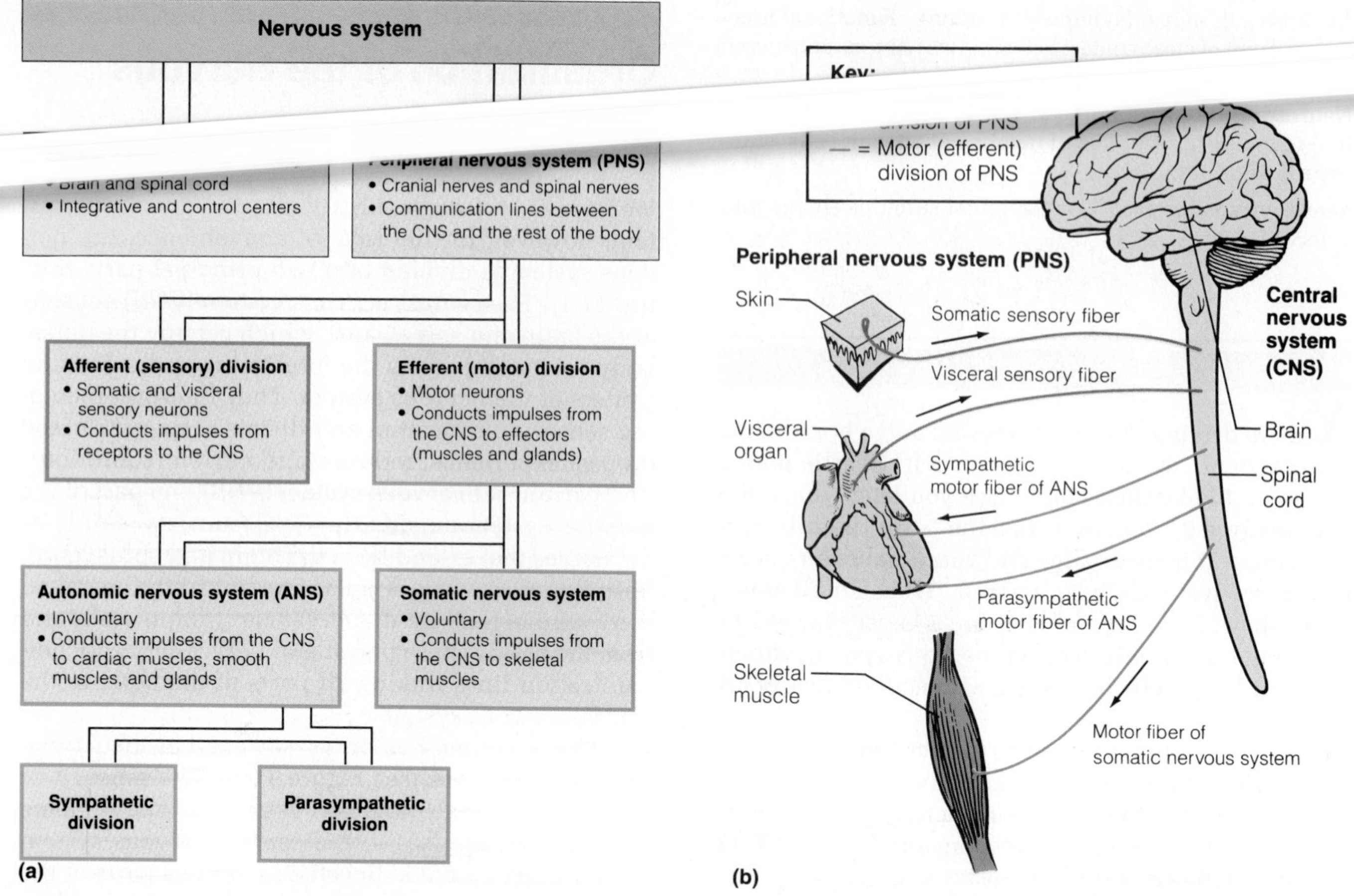

Figure 11.1 Levels of organization in the nervous system. (**a**) Organizational chart. (**b**) The *somata* (limbs and body wall) are served by motor fibers of the somatic nervous system and by somatic sensory fibers. *Visceral organs* (for the most part located in the ventral body cavity) are served by motor fibers of the autonomic nervous system (ANS) and by visceral sensory fibers. Arrows indicate the direction of nerve impulses.

asympathetic, which typically bring about opposite effects on the activity of the same visceral organs. What one subdivision stimulates, the other inhibits.

Histology of Nervous Tissue

Nervous tissue is highly cellular. In the CNS, for example, less than 20% is extracellular space, which means that the cells there are very densely packed and tightly intertwined. Although exceedingly complex, nervous tissue is made up of just two principal types of cells. *Supporting cells* are nonexcitable cells that surround and wrap the more delicate *neurons* that transmit nerve impulses. With one exception (the microglia derived from mesoderm), both types of cells are derived from the same embryonic tissue (neuroectoderm) and form the structures of both the central and peripheral nervous systems.

Supporting Cells

The **supporting cells** form the scaffolding of nervous tissue and generally assist, segregate, and insulate neurons. Additionally, as described shortly, each of the six types of supporting cells has special functions. Schwann cells and satellite cells are restricted to the peripheral nervous system. The other types of supporting cells are found in the central nervous system and are collectively called **neuroglia** (ner-ah′-glē-uh), literally, "nerve glue," or simply **glial** (glē′-ul) **cells.** (However, you should be aware that some authorities classify all supporting cells as neuroglia.) Supporting cells are approximately nine times more numerous than neurons, and unlike neurons, which are amitotic, they retain their ability to reproduce themselves throughout life. For this reason, most brain tumors are *gliomas* (tumors formed by uncontrolled proliferation of glial cells).

The star-shaped **astrocytes** are particularly abundant in the CNS and account for nearly half of neural tissue volume. They have numerous radiating projec-

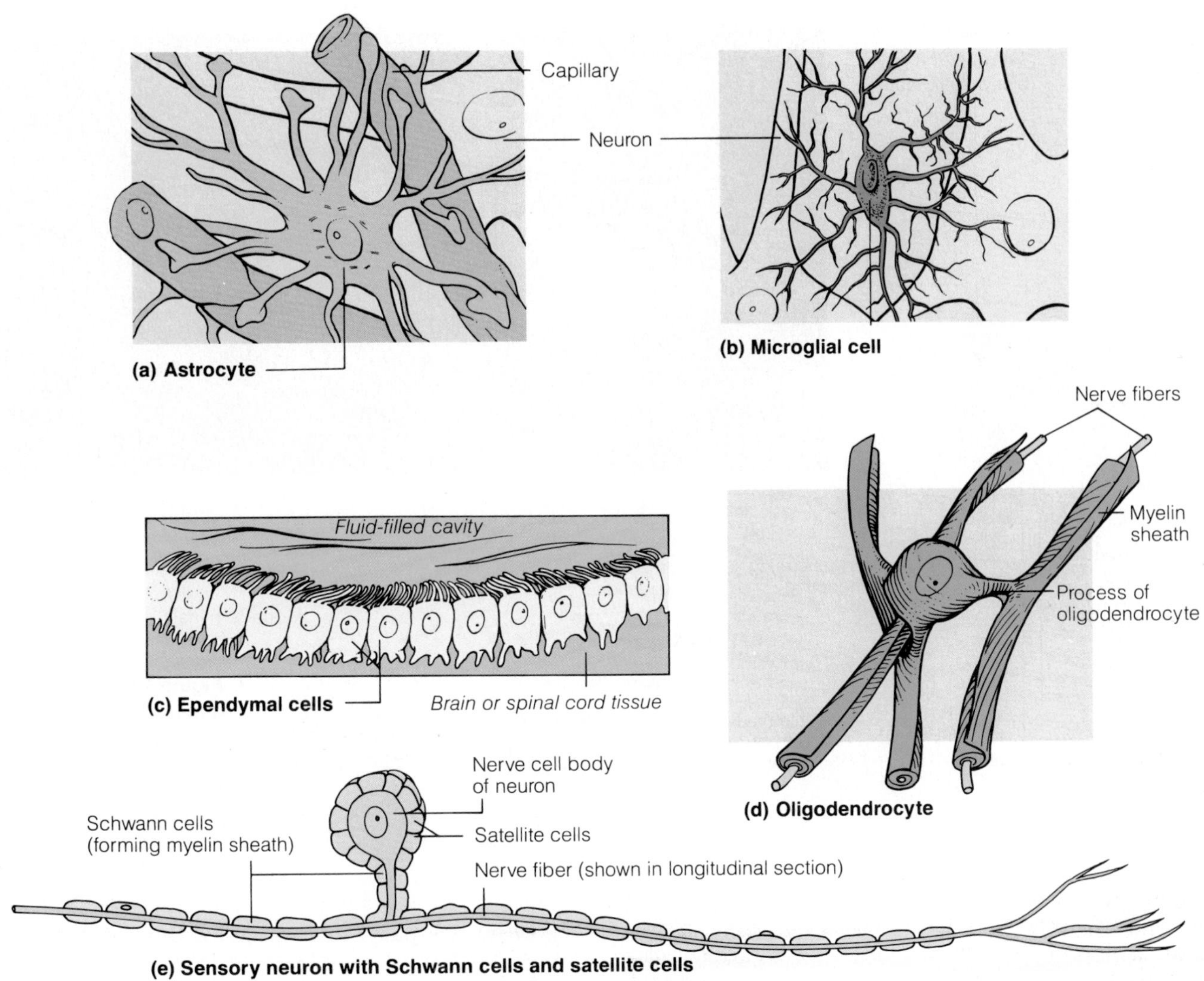

Figure 11.2 Supporting cells of nervous tissue. (**a–d**) Types of glial cells (supporting cells) found in the central nervous system. Notice in (d) that it is the processes of the oligodendrocytes that form the myelin sheaths around nerve fibers in the CNS. (**e**) The relationships of Schwann cells (myelinating cells) and satellite cells to a sensory neuron (nerve cell) in the peripheral nervous system.

tions with bulbous ends that cling to neurons and capillaries, bracing the neurons and anchoring them to their nutrient supply lines, the blood capillaries (Figure 11.2a). Because the astrocytes form a living barrier between capillaries and neurons, it is assumed that they can make exchanges between the two. Another astrocyte role is to control the ionic environment around neurons. As described later, the charge and ion types outside nerve fibers must be just right for nerve impulses to be conducted.

Microglia (mī-krog′lē-uh) are ovoid cells with relatively long processes (Figure 11.2b). They help protect the CNS by engulfing invading microorganisms and dead neural tissues and are a special type of macrophage. There is also evidence that microglia can differentiate into and replace astrocytes or oligodendrocytes as those cells die.

Ependymal (uh-pen′-dih-mul) **cells** line the central cavities of the brain and the spinal cord. They play an active role in forming cerebrospinal fluid, which fills those cavities and surrounds and cushions the brain and spinal cord. The beating of their cilia helps to circulate the cerebrospinal fluid (Figure 11.2c).

Small branching glia called **oligodendrocytes** (ah-lih″-gō-den′-drō-sīts) align along thick neuron fibers in the central nervous system. They wrap their cytoplasmic extensions tightly around the nerve fibers, producing insulating coverings called myelin sheaths (Figure 11.2d).

Schwann cells, which form myelin sheaths around the larger nerve fibers in the peripheral nervous system (Figure 11.2e), are functionally similar to oligodendrocytes. (The formation of myelin sheaths is

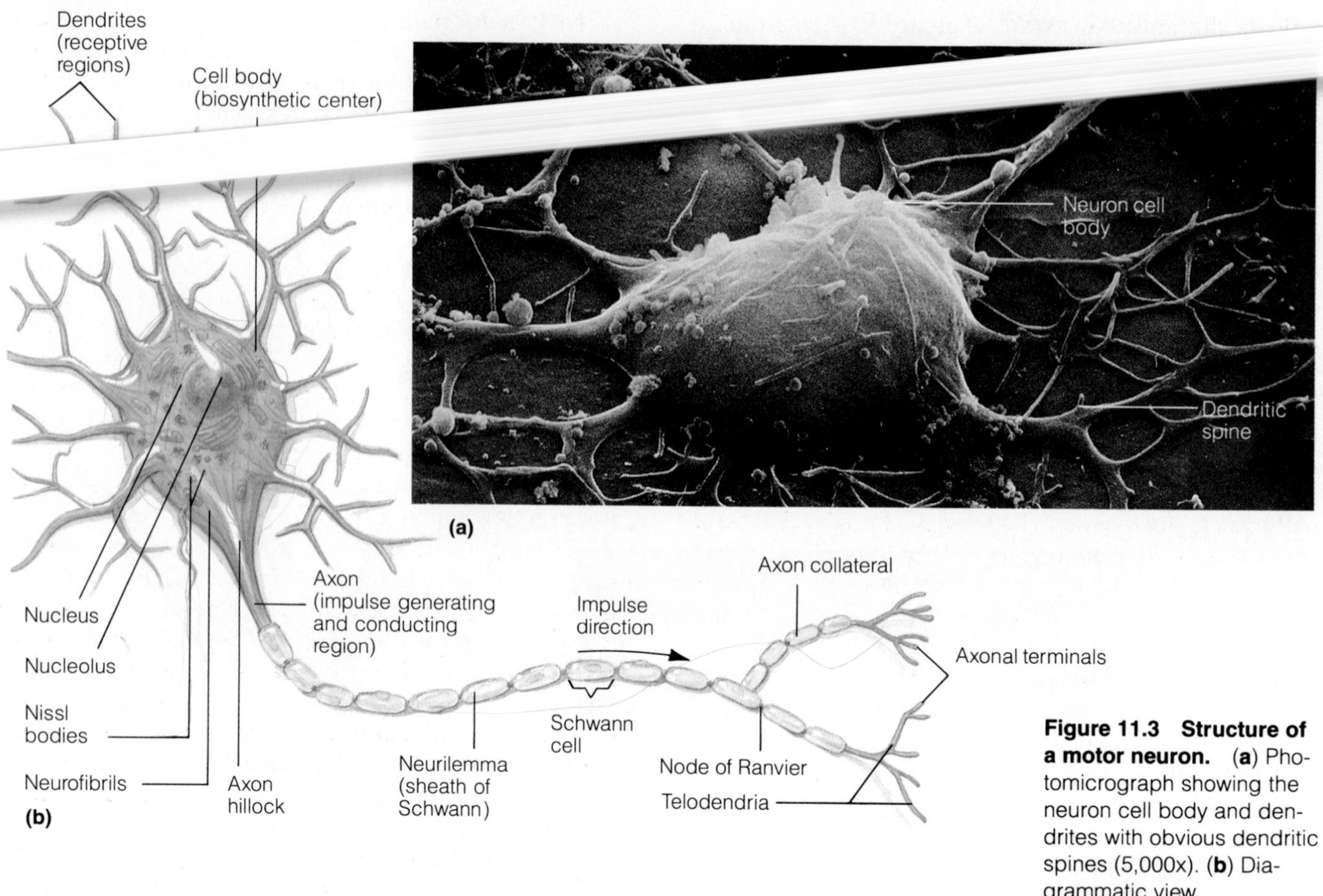

Figure 11.3 Structure of a motor neuron. (**a**) Photomicrograph showing the neuron cell body and dendrites with obvious dendritic spines (5,000x). (**b**) Diagrammatic view.

described later in this chapter.) Schwann cells also act as phagocytes to rid a damaged nerve of deteriorating cell debris and are vital to the process of peripheral nerve fiber regeneration. **Satellite cells,** often found closely associated with Schwann cells (see Figure 11.2e), are thought to play some role in controlling the chemical environment of neurons with which they are associated.

Neurons

Neurons, also called **nerve cells,** are highly specialized cells that conduct messages in the form of nerve impulses from one part of the body to another. Neurons are the structural units of the nervous system and have three special characteristics:

1. They have extreme longevity. Given good nutrition, they can live and function optimally for a lifetime (over 100 years).

2. They are amitotic. Specialization of cells often leads to a loss of certain other cellular functions. As neurons assume their roles as communicating links of the nervous system, they lose their ability to undergo mitosis. We pay a dear price for this neuron feature; because they are unable to reproduce themselves, they cannot be replaced if destroyed.

3. They have an exceptionally high metabolic rate and require continuous and abundant supplies of oxygen and glucose. Neurons cannot survive for more than a few minutes without oxygen.

Neurons are typically large, complex cells. Although they show some variety in structure, they all have a cell body and one or more slender processes extending from it (Figure 11.3). The plasma membrane of neurons is the site of electrical signaling, and it plays a crucial role in cell-to-cell interactions that occur during development.

Neuron Cell Body

The **cell body,** also called the **perikaryon** (*peri* = around, *kary* = nucleus) or **soma,** ranges in diameter from 5 to 140 μm. It contains a large, spherical nucleus with a conspicuous nucleolus and abundant granular cytoplasm. The cell body is the *biosynthetic center* of a neuron. It contains the usual organelles with the exception of centrioles. (The lack of centrioles, which play an important role in forming the mitotic spindle, reflects the amitotic nature of neurons.) Its protein-making machinery, free ribosomes and rough endoplasmic reticulum (ER), is probably the most active and best developed of any cell in the body. This rough ER, referred to as **Nissl** (nih′-sul) **bodies** or **chromatophilic substance,** stains darkly with basic dyes and is quite obvious microscopically. Many **neurofibrils,** bundles of microtubules and microfilaments important in intracellular transport, are seen throughout the cell body. The cell body of some neurons also contains pigment inclusions. For example, some contain a black melanin or red iron-containing pigments. *Lipofuscin* (lih″-pō-fyoo′-shun), a yellow-brown pigment, is believed to be a harmless by-product of lysosomal activity. Because lipofuscin is most abundant in neurons of elderly individuals, it may be related to aging. The cell body is the focal point for the outgrowth of neuron processes during embryonic development. In many neurons, the plasma membrane of the cell body acts as a receptive surface to receive information from other neurons.

Neuron cell bodies are usually located within the CNS, where they are protected by the bones of the cranium and vertebral column. A few clusters of cell bodies, called **ganglia** (gan′-glē-uh), are found in the PNS.

Neuron Processes

Cytoplasmic extensions called **processes,** or nerve *fibers,* extend from the cell body. Organs of the CNS contain both cell bodies and their processes. The PNS, for the most part, consists chiefly of neuron processes. Bundles of neuron processes are called **tracts** in the CNS and **nerves** in the PNS.

Neuron processes are of two types, dendrites and axons, which differ from each other both in structure and in the functional properties of their outer membranes. The convention is to describe neuron processes using a motor neuron as an example of a typical neuron. We shall follow this practice, but keep in mind that many CNS neurons and sensory neurons differ considerably from the "typical" pattern presented here.

Dendrites (*dendr* = tree) of motor neurons are short, thick, diffusely branched extensions. Typically, motor neurons have hundreds of dendrites clustering close to the cell body. Virtually all organelles present in the cell body are also found in dendrites. Dendrites function as *receptive sites;* they provide an enlarged surface area for reception of signals from other neurons. In many brain areas, the finer dendrites are highly specialized for information collection; they bristle with thorny appendages called *dendritic spines* (see Figure 11.3a), which represent points of close contact (synapses) with other neurons. Dendrites conduct electrical signals toward the cell body.

Each motor neuron has a single **axon** (*axo* = axis, axle). The axon arises from a cone-shaped region of the cell body called the **axon hillock** and then tapers to form a slender process that remains uniform in diameter for the rest of its length (Figure 11.3). Except for Nissl bodies and Golgi apparatus, the *axoplasm* (axonal cytoplasm) contains the same organelles found in the cell body.

In some CNS neurons, the axon is very short, but in others, it is long and accounts for nearly the entire length of the neuron. For example, axons of motor neurons controlling the skeletal muscles of your great toe extend from the lumbar region of your spine to your foot, a distance of a meter or more (3–4 feet), making them the longest cells in the body. Axon diameter also varies considerably, and axons with the largest diameters conduct impulses the most rapidly.

Although each neuron has only one axon, axons may give off infrequent branches along their length. These branches, called *axon collaterals,* extend from the axon at more or less right angles. Whether an axon remains undivided or has collaterals, it usually branches profusely at its end (terminus): 10,000 or more of these *telodendria,* or axonal branches, per neuron is not unusual. The bulbous distal endings of the telodendria are variously called *axonal terminals, synaptic knobs,* or *boutons* (see Figure 11.3).

Functionally, axons are *impulse generators and conductors* that transmit nerve impulses away from the cell body. In motor neurons, the nerve impulse is usually generated at the axon hillock and conducted along the axon away from the cell body. When the impulse reaches the axonal terminals, it causes the release of chemicals called *neurotransmitters* into the extracellular space. Neurotransmitters excite or inhibit neurons (or effector cells) with which the axon is in close contact. Since each neuron both receives signals from and sends signals to scores of other neurons, it carries on "conversations" with many different neurons at the same time.

Axonal Transport and Axoplasmic Flow. All large molecules are synthesized in the cell body, and since axons are often very long processes, it would seem that the distribution of molecules and organelles from one end of a neuron to the other could be a problem. However, two unique transport mechanisms—axonal transport and axoplasmic flow—move needed materials along the axon and return materials from these fibers to the neuron's cell body for recycling.

Axonal transport is selective and fast (300–400 mm per day) and conducts substances in both directions along the axon. Regardless of direction, this type of transport is intermittent and is the principal mechanism for transporting macromolecules. Substances moved toward the axonal terminals include receptor proteins, synaptic vesicles containing neurotransmitters manufactured in the cell body, and membrane components to be used for renewing the axonal plasma membrane, or *axolemma* (ak″-sō-leh′-muh). Materials moved in the opposite direction are m[illegible] being r[illegible] [illegible] it is known that the process requires ATP, the mechanism of axonal transport is not completely understood. Some researchers suggest that cytoskeletal elements are involved; for example, microtubules may provide static tracks along which the substances are pulled by the contractile activity of the microfilaments.

Certain viruses and bacterial toxins that damage neural tissues use axonal transport as a vehicle to reach the cell body. This transport mechanism has been demonstrated for polio, rabies, and herpes simplex viruses and for tetanus toxin. ■

Axoplasmic flow is a slow (0.5–4.0 mm per day), unidirectional transport process that carries cytoskeletal elements and soluble proteins (mostly enzymes) down the axon's length. Axoplasmic flow is not ATP dependent but is too fast to be accounted for by diffusion. Overall, it resembles the slow movement of a gel-type toothpaste from a squeezed tube, and it may be a microstreaming process that depends on sol–gel changes of the cytoplasm.

Myelin Sheath and Neurilemma. Many nerve fibers, particularly those that are long or large in diameter, are covered with a whitish, fatty (phospholipid), segmented sheath called the **myelin** (mī′-uh-lin) **sheath.** Myelin protects and electrically insulates fibers from one another, and it increases the speed of transmission of nerve impulses. *Myelinated fibers* (axons bearing a myelin sheath) conduct nerve impulses rapidly, whereas *unmyelinated fibers* tend to conduct impulses quite slowly. (The order of difference may be more than 150 times—from 150 m/s to less than 1 m/s.)

Myelin sheaths in the peripheral nervous system are formed by Schwann cells. The Schwann cells first become indented to receive the axon and then wrap themselves around it in a jelly roll fashion (Figure 11.4). Initially, the wrappings are loose, but the Schwann cell cytoplasm is gradually squeezed from between the membrane layers. When the wrapping process is complete, many concentric layers of Schwann cell plasma membrane enclose the axon. This tight coil of wrapped membranes is the myelin sheath, and its thickness depends on the number of spirals. The nucleus and most of the cytoplasm of the Schwann cell ends up just beneath the outermost part of its plasma membrane, external to the myelin sheath; this portion of the Schwann cell, which surrounds the myelin sheath, is called the **neurilemma** or **sheath of Schwann.** Adjacent Schwann cells along an axon do not touch one another, so there are gaps in the sheath, called **nodes of Ranvier** (ran′-vēr), at regular intervals along the myelinated axon. It is at these nodes that axon collaterals can emerge from th[illegible] [illegible] periph-[illegible] but the coiling process does not occur. In such instances, a single Schwann cell can partially enclose 15 or more axons, and each of these fibers occupies a separate tubular recess in the surface of the Schwann cell (see Figure 11.4f). Nerve fibers associated with Schwann cells in this manner are said to be *unmyelinated.*

Both myelinated and unmyelinated axons are also found in the central nervous system. However, it is oligodendrocytes that form CNS myelin sheaths (see Figure 11.2d). In contrast to Schwann cells, each of which can form only one segment (internode) of a myelin sheath, oligodendrocytes have multiple flat extensions that can coil around as many as 60 different axons at the same time. Nodes of Ranvier are present, though more widely spaced than in the PNS. However, CNS myelin sheaths lack a neurilemma. This reflects the fact that their cell extensions are doing the coiling rather than entire cells, as is the case when myelin sheaths are formed by Schwann cells. As the oligodendrocyte processes coil around axons, the squeezed-out cytoplasm is forced back toward the centrally located nucleus. When the oligodendrocytes "embrace" but do not wrap CNS axons, the fibers are unmyelinated.

Regions of the brain and spinal cord containing dense collections of myelinated fibers are referred to as *white matter* and are primarily fiber tracts. *Gray matter* contains mostly nerve cell bodies and unmyelinated fibers.

The importance of myelin to nerve transmission is painfully clear to those with demyelinating diseases. In people with *multiple sclerosis,* for example, the myelin sheaths gradually disappear, impulse conduction ceases, and the affected individuals lose the ability to control their muscles. ■

Classification of Neurons

Neurons may be classified according to structure or function. We will describe both classifications here, but we will use the functional classification in most subsequent discussions.

Structural Classification. Neurons are grouped structurally according to the number of processes that extend from their cell body. Three major neuron groups make up this classification: multipolar, bipolar, and unipolar neurons.

Multipolar neurons characteristically have numerous branching dendrites and one axon (Figure 11.5a). They are the most common neuron type in humans and the major neuron type in the central nervous system.

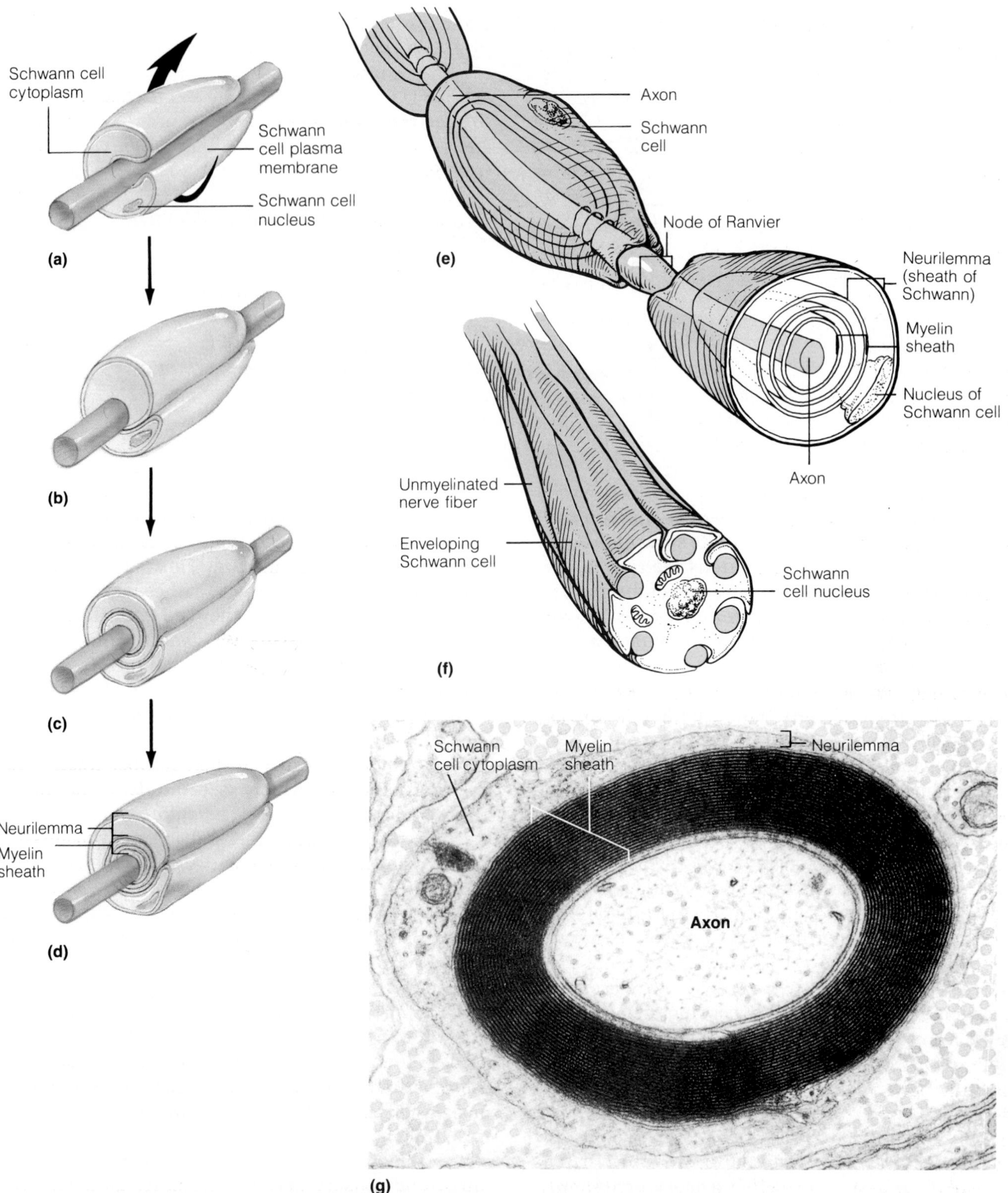

Figure 11.4 Relationship of Schwann cells to axons in the peripheral nervous system. (**a–d**) Myelination of a nerve fiber (axon). As illustrated in the three-dimensional views, a Schwann cell envelops an axon in a trough. It then begins to rotate around the axon, enveloping it loosely in successive layers of its plasma membrane. Eventually, the Schwann cell cytoplasm is forced from between the membranes and comes to lie peripherally just beneath the exposed portion of the Schwann cell membrane. The tight membrane wrappings surrounding the axon form the myelin sheath; the area of Schwann cell cytoplasm and its exposed membrane is referred to as the neurilemma, or sheath of Schwann. (**e**) Three-dimensional "see-through" view of a myelinated axon, showing portions of adjacent Schwann cells and the node of Ranvier (an area of exposed axolemma) between them. (**f**) Unmyelinated fibers. Schwann cells may associate loosely with several axons (usually of small diameter), which they partially invest. In such cases, Schwann cell coiling around the axons does not occur. (**g**) Photomicrograph of a myelinated axon, cross-sectional view (20,200X).

Figure 11.5 The three neuron types based on structural classification (number of processes projecting from the nerve cell body).

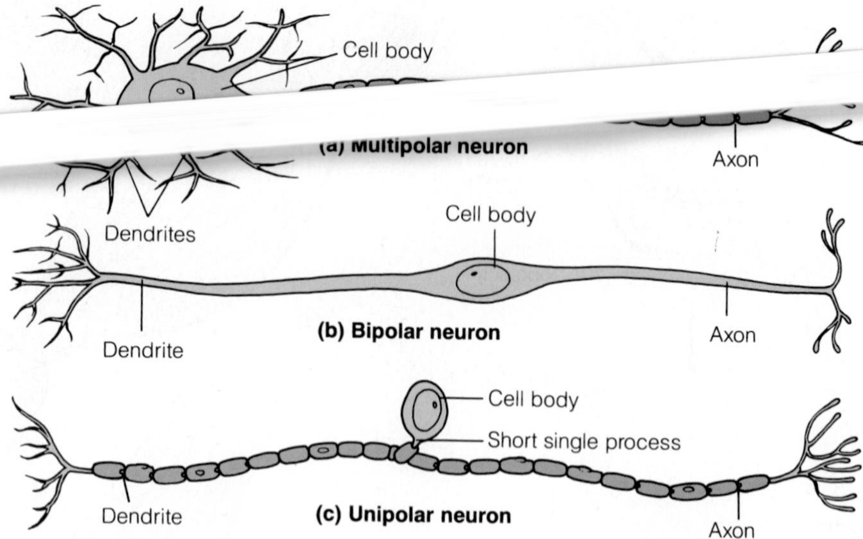

Bipolar neurons have two processes—an axon and a dendrite—that extend from opposite sides of the cell body (Figure 11.5b). Bipolar neurons are rare in the adult body; they are found only in some of the special sense organs, where they act as receptor cells. Examples include some neurons in the retina of the eye and in the olfactory mucosa.

Unipolar neurons have a single process that emerges from the cell body (Figure 11.5c). However, it is very short and divides into a proximal and distal fiber—axon and dendrite, respectively. Unipolar neurons are sometimes called **pseudounipolar neurons** (*pseudo* = false) because they originate as bipolar neurons. Then, during early embryonic development, the two processes converge and partially fuse to form the single process that issues from the cell body. Unipolar neurons are found chiefly in the peripheral nervous system, where they function as sensory neurons.

Functional Classification. The functional classification scheme groups neurons according to the direction the nerve impulse is traveling relative to the central nervous system. Based on this criterion, there are sensory neurons, motor neurons, and association neurons (Figure 11.6).

Neurons that transmit impulses from sensory receptors in the skin or internal organs toward the central nervous system are called **sensory,** or **afferent, neurons.** Except for the bipolar neurons found in some of the special sense organs, sensory neurons of the body are unipolar, and their cell bodies are located in sensory ganglia outside the CNS. Functionally, only the most distal parts of these unipolar neurons act as impulse receptor sites. Where the dendritic branches join, they form a single fiber (dendrite), which may be very long. For example, fibers carrying sensory impulses from the skin of your big toe travel for more than a meter before they reach their cell bodies in a ganglion close to the spinal cord. These long dendrites look and behave like axons in nearly every way; when they are large in diameter, they are typically heavily myelinated, and they can both generate and transmit nerve impulses. Consequently, they are frequently referred to as axons, which is technically incorrect because they carry impulses toward rather than away from the nerve cell body.

Although dendritic endings of some sensory neurons are naked, in which case the dendrites themselves function as sensory receptors, many exhibit receptors that include other cell types. The various types of sensory receptor end organs are described shortly.

Neurons that carry impulses away from the CNS to the effector organs (muscles and glands) of the body periphery are called **motor,** or **efferent, neurons.** Motor neurons are multipolar, and except for some neurons of the autonomic nervous system, their cell bodies are located in the CNS. All motor neurons form junctions with their effector cells, but the *neuromuscular junctions* between somatic motor neurons and skeletal muscle cells are particularly elaborate (see Chapter 9, p. 251).

Association neurons, or **interneurons,** lie between motor and sensory neurons in neural pathways and shuttle signals, often through complex CNS pathways where integration occurs. Association neurons are typically multipolar and are confined entirely within the CNS. They make up over 99% of the neurons of the body, including most of those in the central nervous system. There is considerable diversity among association neurons both in size and in fiber-branching patterns. Several types of association neurons are shown in Figure 11.7 so you can appreciate their variety.

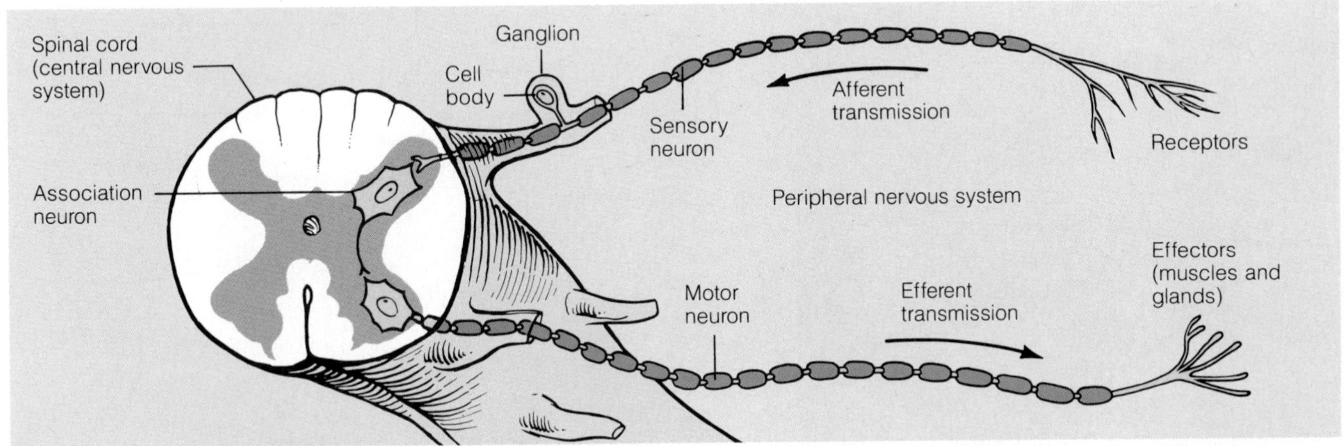

Figure 11.6 Neurons classified by function. Sensory (afferent) neurons conduct impulses from sensory receptors (in skin, viscera, muscles, and tendons) to the central nervous system; most are unipolar neurons with their nerve cell bodies in ganglia in the PNS. Motor (efferent) neurons transmit impulses from the CNS (brain or spinal cord) to effectors such as muscles and glands in the body periphery. Association neurons complete the communication line between sensory and motor neurons; they are typically multipolar, and their nerve cell bodies reside in the CNS.

Dendrites
Cell body
Axon
(a) Pyramidal cell

Dendrites
Cell body
Axon
(b) Purkinje cell

(c) Stellate cell

(d) Granule cell

(e) Basket cell

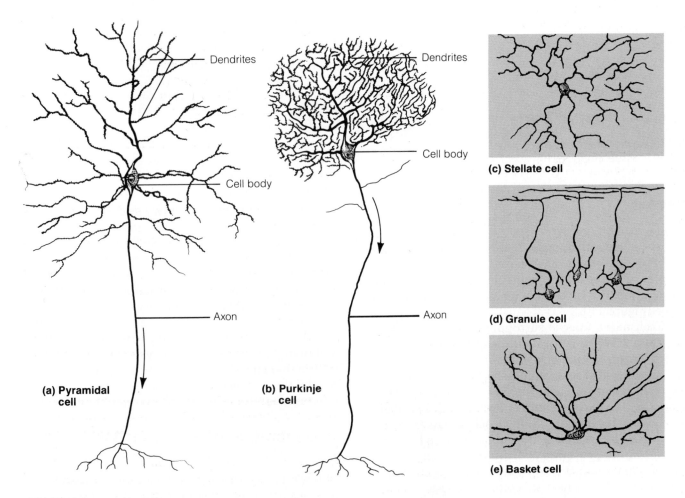

Figure 11.7 Structural variations in selected association neurons of the brain. Notice that (**a**) pyramidal cells and (**b**) Purkinje cells have extremely long axons and distinctive dendrite-branching patterns, whereas (**c**) stellate, (**d**) granule, and (**e**) basket cells have short (or even absent) axons and profuse dendrites.

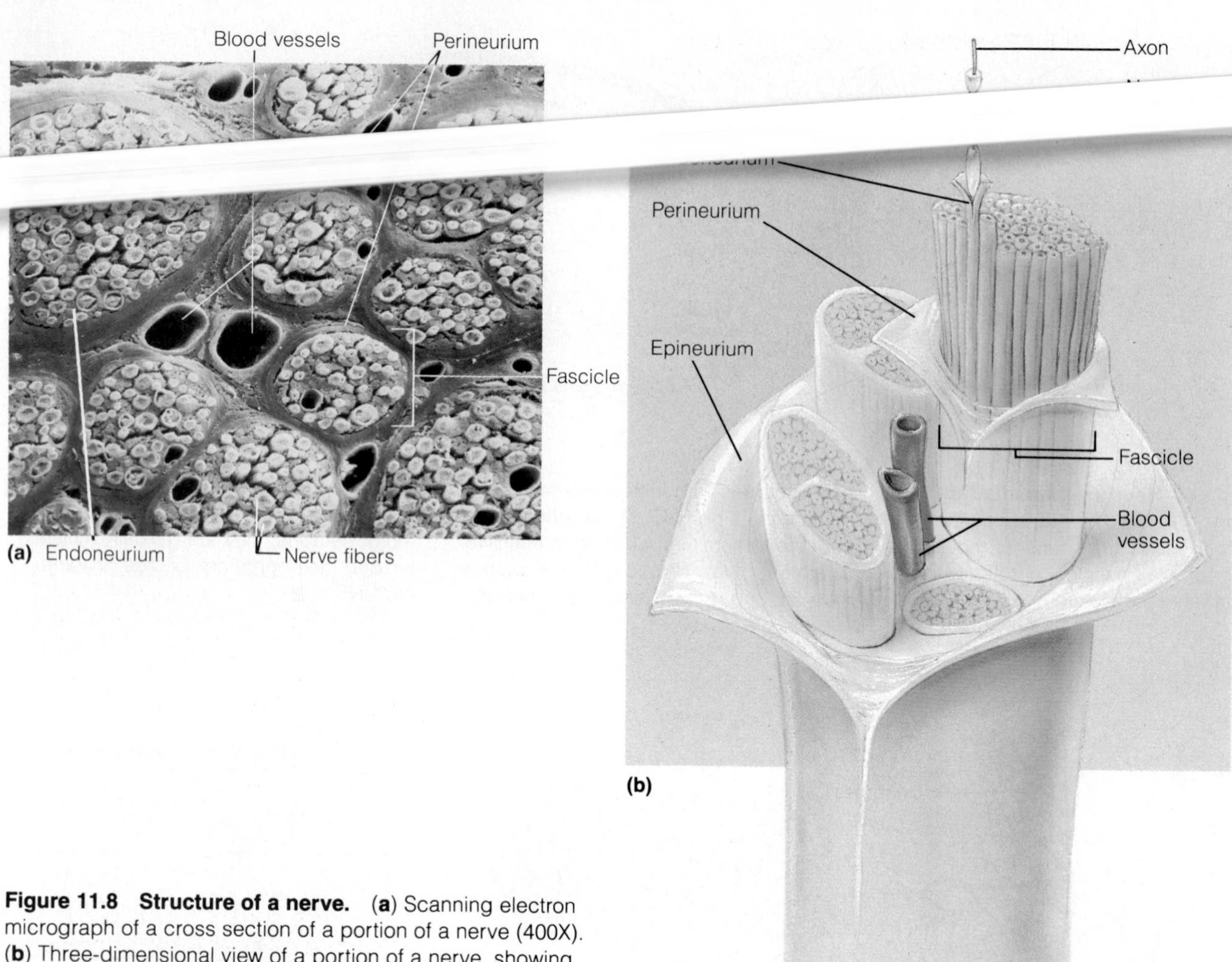

Figure 11.8 Structure of a nerve. (**a**) Scanning electron micrograph of a cross section of a portion of a nerve (400X). (**b**) Three-dimensional view of a portion of a nerve, showing conective tissue wrappings.

Nerves

A nerve is a cordlike organ that is part of the peripheral nervous system. Each nerve consists of parallel bundles of nerve fibers (some myelinated and some not) enclosed by successive wrappings of connective tissue (Figure 11.8).

Each nerve fiber is surrounded by a loose, delicate connective sheath called the **endoneurium** (en″-dō-ner′-ē-um), which also encloses the fiber's associated myelin and/or neurilemma sheath. Groups of fibers are bound together by a coarser connective tissue wrapping, the **perineurium,** to form bundles of fibers called **fascicles.** Finally, all the fascicles are enclosed by a tough fibrous sheath, the **epineurium,** to form the nerve. Neuron processes constitute only a small fraction of a nerve's bulk; the balance consists chiefly of myelin and the protective connective tissue wrappings. Blood vessels and lymphatic vessels are also found within the substance of a nerve.

Nerves vary in size and are classified according to the direction in which they transmit impulses. Nerves containing both sensory and motor fibers and transmitting impulses both to and from the central nervous system are called **mixed nerves;** nerves that carry impulses toward the CNS only are **sensory (afferent) nerves;** and nerves carrying impulses only away from the CNS are **motor (efferent) nerves.** Most nerves are mixed; purely sensory or motor nerves are extremely rare.

Since mixed nerves often carry both somatic and autonomic (visceral) nervous system fibers, the fibers within them may be classified further according to the region they innervate as *somatic afferent, somatic efferent, visceral afferent,* and *visceral efferent.*

Regeneration of Nervous Tissue

Damage to nervous tissue is serious because mature neurons are incapable of cell division. If the damage is severe or close to the cell body, the entire neuron may die, and in some cases, other neurons that are normally stimulated by its axon die as well. However, in certain cases, cut or compressed axons in peripheral nerves can regenerate successfully.

Almost immediately after a peripheral axon has been severed or crushed, the separated ends seal themselves off and then swell as organelles and other substances being transported along the axon begin to accumulate in the sealed ends (Figure 11.9a). Within a few minutes, the axon and its myelin sheath distal to the site of injury begin to disintegrate since they cannot receive nutrients and other needed materials from the cell body. This process, called *Wallerian degeneration*, spreads distally from the injury site. By the third to fourth day, the distal part of the axon is completely fragmented (Figure 11.9b). Schwann cells and macrophages that migrate into the trauma zone from surrounding connective tissue begin to phagocytize disintegrating myelin and axonal debris. Generally, the entire axon distal to the injury will be degraded within about one week. Once the debris has been disposed of, the Schwann cells align themselves within the endoneurium, forming tunnels to guide regenerating axonal "sprouts" to their original contacts (Figure 11.9c and d). Schwann cells also release growth factors that encourage axonal growth.

Characteristic changes also occur in the neuronal cell body after destruction of the axon. Within two days, the cell body swells (often doubling in size), its nucleolus becomes more prominent, and its chromatophilic substance breaks apart and disperses to the cell periphery. All these events signal that the neuron is preparing to synthesize proteins to support regeneration of its axon.

Axonal regeneration occurs at the rate of 1–5 mm a day. If proper connections are remade, the nerve cell body regains its former appearance; if not, the cell may die. The greater the distance between the severed nerve endings, the less the chance of recovery because adjacent tissues block growth by protruding into the larger gaps, and axonal sprouts tend to escape into surrounding areas and form a tissue mass called a *neuroma*. Neurosurgeons align cut nerve endings surgically to enhance the chance of successful regeneration. Pinpoint accuracy in nerve fiber–effector cell alignment is obviously impossible, and much of the learning (or relearning) after nerve injury involves training the nervous system to respond appropriately so that stimulus and response are coordinated. Post-trauma regrowth is never exactly the same as what existed before the injury.

In contrast to peripheral nerve fibers, those within the CNS never regenerate. This difference in regenerative capacity seems to have less to do with the neurons themselves than with the "company they keep"—their supporting cells. After CNS injury, microglial

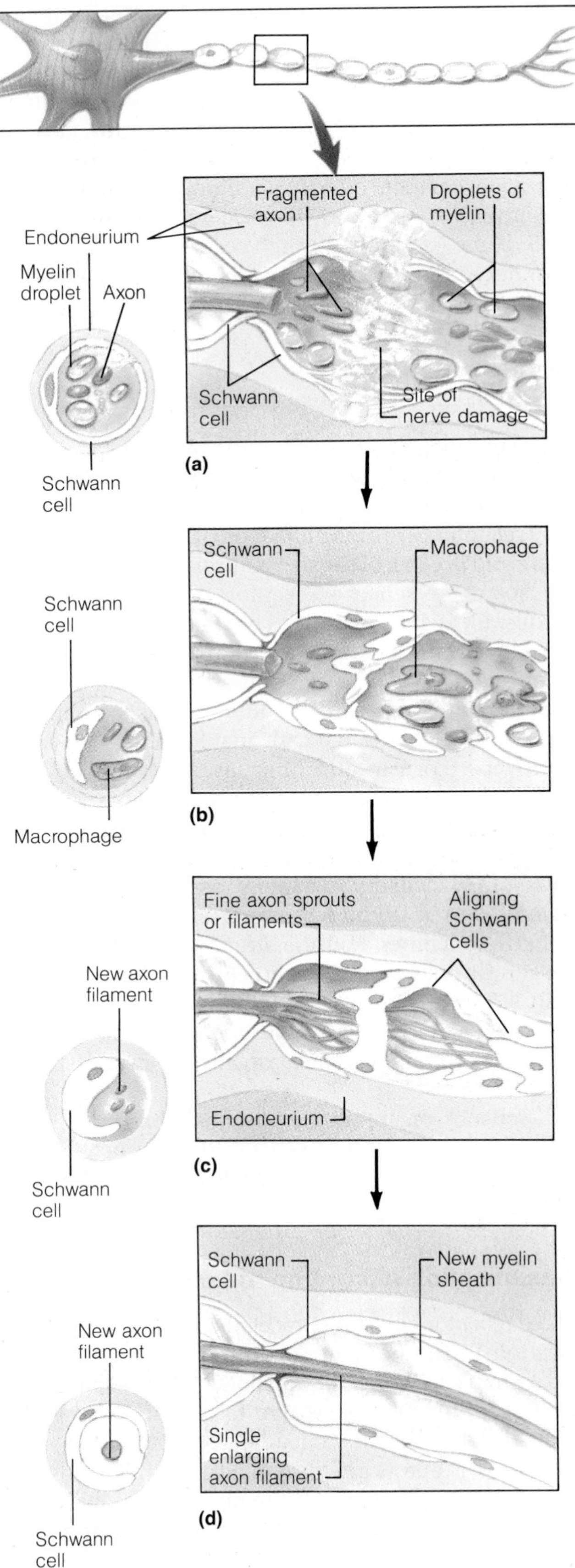

Figure 11.9 Events leading to regeneration of nerve fibers in the peripheral nerves.

cells (and probably astrocytes) rid the area of deteriorating cell debris. However, the oligodendrocytes surrounding the injured fibers die, an event called *secondary demyelination*, and [illegible] by a dense *glial scar* formed by astrocytes, which effectively blocks regeneration of sprouting axons. However, research has shown that CNS axons *will* grow through segments of cut peripheral nerve containing Schwann cells and establish functional connections.

Sensory Receptors

Sensory receptors are structures that are specialized to respond to changes in their environment; such environmental changes are called *stimuli*. Most sensory receptors are modified dendritic endings, or *dendritic end organs*, of sensory neurons. In certain cases, the so-called sensory receptors are actually *sense organs*, localized collections of cells (usually of many types) working together to accomplish a particular receptive process. For example, the eye is a complex sense organ composed not only of sensory neurons, but also of nonneuronal cells that form its supporting wall, lens, cornea, and other associated structures. Although these large, complex sense organs associated with the special senses of vision, hearing, smell, and taste are most familiar to us, the simpler, more widespread sensory receptors associated with the *general senses* are no less important. In this section, we will concentrate on the structure and function of the simple receptors, the tiny sentinels that act to keep the CNS apprised of what is happening at the body surface and in its skeletal muscles and joints. We will discuss the complex sense organs in more detail in Chapter 16.

Sensory receptors may be described from three vantage points: (1) their general location in the body, (2) the type of stimulus they detect, and (3) their structure.

Classification According to Location

Three receptor classes are recognized according to their location in the body or according to the location of the stimulus to which they respond. **Exteroceptors** (ek″-ster-ō-sep′-terz) are sensitive to stimuli arising outside the body (*extero* = outside). As might be expected, most exteroceptors are located near or at the body surface. Exteroceptors include touch, pressure, pain, and temperature receptors located in the skin and most receptors of the special sense organs.

Interoceptors (in″-ter-ō-sep′-terz), also called **visceroceptors,** respond to stimuli arising from within the body (*intero* = inside), such as from viscera, blood vessels, and the like. The different [illegible] tissues, and pressure. Sometimes, their activation causes us to feel pain, discomfort, hunger, or thirst; however, we are usually unaware of the workings of our visceral receptors. **Proprioceptors** (pro″-prē-ō-sep′-terz) can be considered a subclass of interoceptors because they respond to internal stimuli; however, their location is quite restricted. Proprioceptors are located in skeletal muscles, tendons, and joints. (Some authorities also include the equilibrium receptors of the inner ear in this class.) *Propria* is from the Latin, meaning "one's own," and the proprioceptors constantly advise the brain of our own movements. This information allows us to be aware of the physical state and position of our skeletal muscles and bones without having to look at them for clues.

Classification According to the Stimulus Type Detected

The receptor classes named according to the stimuli that activate them are easy to remember, because the class name usually indicates the stimulus. **Mechanoreceptors** generate nerve impulses when they, or adjacent tissues, are deformed by mechanical forces. Examples are pressure and touch receptors of the skin, proprioceptors, inner ear receptors that respond to sound, and stretch receptors of the lungs, blood vessels, and bladder.

Thermoreceptors are sensitive to changes in temperature, and **photoreceptors,** such as those of the retina of the eye, respond to light energy. **Chemoreceptors** respond to chemicals in solution. They include taste buds, which allow us to taste foods; olfactory receptors of the nose, which detect odor molecules; and interoceptors that respond to changes in levels of oxygen, carbon dioxide, and various ions in the blood.

Nociceptors (nō″-sih-sep′-terz), from the Latin *nocere*, meaning "to injure," respond selectively to potentially damaging stimuli. Their activation results in our perception of pain. Since nearly every type of stimulus, when vigorously applied, has the capacity to do harm and cause pain, virtually all receptor types function as nociceptors at one time or another. For example, searing heat, extreme cold, excessive pressure, and chemicals released at sites of inflammation are all interpreted as painful.

Classification According to Structure

The general sensory receptors can be grouped anatomically into (1) free (naked) nerve endings and (2) corpuscular (encapsulated) nerve endings. These receptors are illustrated in Figure 11.10 and are classified

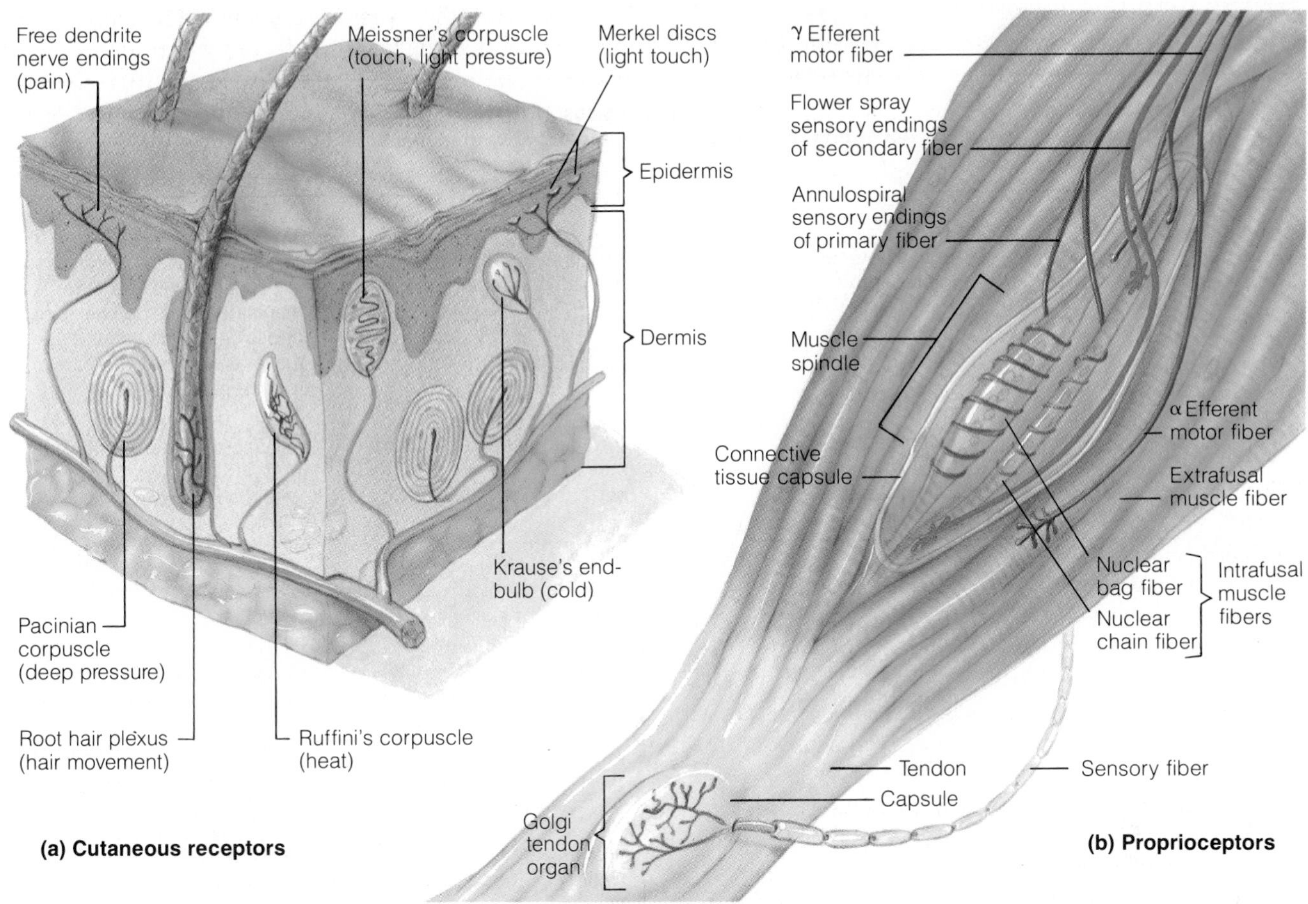

Figure 11.10 Sensory receptors. (**a**) Cutaneous receptors: free dendritic nerve endings, Merkel disc, root hair plexus, Meissner's corpuscle, Pacinian corpuscle, Krause's end-bulb, and Ruffini's corpuscle. (**b**) Proprioceptors: muscle spindle and Golgi tendon organ.

in Table 11.1 according to both the structural and functional schemes. Except for proprioceptors, all these receptors are found in the skin and are commonly referred to as **cutaneous receptors.**

Free Sensory Nerve Endings. **Free,** or **naked, dendritic endings** of sensory neurons invade virtually all body tissues, but they are particularly abundant in epithelia and connective tissues. The sensory fibers all have a small diameter, and their distal dendrites usually have small knoblike swellings that respond chiefly to pain (usually resulting from chemical stimuli), but some of these endings respond to touch, heat, and/or cold as well. Thus, although the bare nerve endings are considered "the" pain receptors (nociceptors) that warn of tissue injury, they also function as mechanoreceptors and thermoreceptors.

Certain free dendritic endings form disclike structures called **Merkel discs** that attach to the deeper layers of the skin epidermis and function as light touch receptors. **Root hair plexuses,** dendritic endings that entwine basketlike in hair follicles, detect movements of the hair.

Corpuscular Receptors. All **corpuscular** (kor-pus′-kyoo-ler) **receptors,** or **encapsulated nerve endings,** exhibit one or more terminal nerve fibers enclosed in a connective tissue capsule. However, they vary greatly in shape, size, and distribution in the body.

Meissner's corpuscles (mīs′-nerz kor′-puh-sulz) are small (about 100 μm in length), egg-shaped mechanoreceptors in which a few spiraling dendrites are surrounded by a thin capsule. They respond to low-frequency vibrations and are sensitive to light pressure. Meissner's corpuscles are found just beneath the skin epidermis in the dermal papillae and are especially numerous in sensitive skin areas such as the lips and fingertips.

Pacinian corpuscles, scattered deep in the subcutaneous tissue underlying the skin and in the viscera, are mechanoreceptors stimulated by deep pressure and vibration. They are the largest of the corpuscular receptors; some are over 2 mm long and visible to the naked eye. In longitudinal section, a Pacinian corpuscle resembles a cut onion. Its single dendrite is surrounded by up to 60 layers of flattened

Table 11.1 General Sensory Receptors Classified by Structure and Function

Anatomical class		according to stimulus type detected	Body location
Unencapsulated			
Free dendritic nerve endings of sensory neurons	Both exteroceptors and interoceptors	Mechanoreceptors (pressure); possible thermoreceptors	Most body tissues; most dense in connective tissues (ligaments, tendons, dermis, joint capsules, periostea) and epithelia (epidermis, cornea, mucosae, and glands)
Modified free dendritic endings:			
Merkel discs	Exteroceptors	Mechanoreceptors (light pressure)	At base of epidermis of skin
Root hair plexuses	Exteroceptors	Mechanoreceptors (deflection)	In and surrounding hair follicles
Encapsulated			
Meissner's corpuscles	Exteroceptors	Mechanoreceptors (light pressure, discriminative touch, vibration of low frequency)	Dermal papillae of the skin, particularly lips, nipples, external genitalia, fingertips, eyelids
Pacinian corpuscles	Exteroceptors and interoceptors	Mechanoreceptors (deep pressure, stretch, vibration of high frequency)	Subcutaneous tissue of the skin; periostea, mesentery, tendons, ligaments, joint capsules; most abundant on fingers, soles of feet, external genitalia, nipples
Krause's end-bulbs	Exteroceptors	Thermoreceptors (cold?); possible mechanoreceptors	Superficial dermis, joint capsules, and other connective tissues
Ruffini's corpuscles	Exteroceptors	Thermoreceptors (heat?); possible mechanoreceptors	Deep in dermis, conjunctiva, lips, oral cavity mucosa
Muscle spindles	Proprioceptors	Mechanoreceptors (muscle stretch)	Skeletal muscles, particularly those of the extremities
Golgi tendon organs	Proprioceptors	Mechanoreceptors (tendon stretch)	Tendons
Joint kinesthetic receptors	Proprioceptors	Mechanoreceptors	Joint capsules of synovial joints

Schwann cells, which in turn are enclosed by a connective tissue capsule.

Krause's end-bulbs* are similar in size to Meissner's corpuscles and are abundant in the dermis, conjunctiva of the eye, lips, and oral cavity. Their somewhat bulbous capsule surrounds dendrites with clublike endings. Krause's end-bulbs are thought to be thermoreceptors sensitive to cold and to be activated by temperatures below 20°C (68°F).

Ruffini's corpuscles,* located in the skin dermis and various other body sites, contain a spray of dendritic endings enclosed by a flattened capsule. They are heat receptors most sensitive to temperatures in the range of 25°–45°C (77°–113°F), and when they are stimulated, the brain interprets their impulses as a painful burning sensation. Ruffini's corpuscles are located more deeply than Krause's end-bulbs and are less numerous, explaining our greater sensitivity to cold. Because Ruffini's corpuscles are present in joint capsules, some authorities believe that they also function as mechanoreceptors.

Muscle spindles, also called **neuromuscular spindles,** are fusiform proprioceptors interspersed throughout the skeletal muscles. Each muscle spindle consists of three to ten modified skeletal muscle fibers, called **intrafusal** (in″-truh-fyoo′-zul) **fibers,** enclosed within a connective tissue capsule (see Figure 11.10b). Since the central region of an intrafusal fiber is devoid of myofilaments (actin and myosin), it is noncontractile. Instead, these regions serve as receptive surfaces

*The description of Krause's end-bulbs and Ruffini's corpuscles as thermoreceptors has a long history, but there is controversy over the anatomical nature of thermoreceptors. Some researchers are convinced that cold and heat receptors are bare nerve endings and that Krause's end-bulbs and Ruffini's corpuscles are actually mechanoreceptors and just variations of Meissner's corpuscles.

for the spindle and are wrapped by afferent endings of two types (see Figure 11.10b). The contractile ends of the muscle spindles are innervated by *gamma* (γ) *efferent fibers* that arise from small motor neurons in the ventral horn of the spinal cord. These motor fibers are distinct from the *alpha* (α) *efferent* motor fibers that cause contraction of the large effector muscle fibers (*extrafusal fibers*) of the skeletal muscle.

When a muscle is stretched, as when you fall off balance, the intrafusal fibers are extended. This activates the associated sensory neurons and causes impulses to be transmitted to the spinal cord (and ultimately to the brain). As a result, the motor neurons are stimulated and cause the stretched muscle to contract and shorten back to its resting length, thus resisting the stretch. This stretch reflex, which is very important for maintaining muscle tone, is described in more detail in Chapter 13.

Golgi tendon organs are located in tendons, close to the point of skeletal muscle insertion. They consist of small bundles of tendon fibers enclosed in a layered capsule, with dendrites coiling between and around the fibers (see Figure 11.10b). Golgi tendon organs are stimulated when the associated muscle contracts and stretches the tendon, and since they monitor muscle tension, their activity is functionally associated with that of the muscle spindles. When Golgi tendon organs are activated, the contracting muscle is inhibited, which causes it to relax, ending the stimulation of the Golgi tendon organs.

Joint kinesthetic (kih″-nes-theh′-tik) **receptors** are distributed in the articular capsules that enclose synovial joints. Although their structure varies, most are similar to Meissner's corpuscles. These receptors provide information on joint position and motion (*kines* = movement), a sensation of which we are highly conscious. (Close your eyes and flex and extend your elbow joint—you can really *feel* this movement.)

Neurophysiology

Neurons are highly irritable (responsive to stimuli). When a neuron is adequately stimulated, an electrical impulse is conducted along the length of its axon. This response is always the same, regardless of the source or type of stimulus. This electrical phenomenon underlies virtually all functional activities of the nervous system.

In this section, we will consider how neurons become excited or inhibited and how they communicate with other cells in the body. However, we first need to explore some basic principles of electricity.

Basic Principles of Electricity

The human body as a whole is electrically neutral; it has the same number of positive and negative charges. However, there are sites where one type of charge predominates, making such regions positively or negatively charged. Because opposite charges attract each other, energy must be used (work must be done) to separate them. On the other hand, when positive and negative charges come together, energy is liberated that can be used to perform work. The stored energy of a battery, for example, is released when the two points are connected (the circuit is completed), allowing electrons to flow from the negatively charged area to the positively charged area. Thus, separated electrical charges of opposite sign have potential energy. The measure of this potential energy is called **voltage** and is measured in either *volts* or *millivolts* (1 mV = 0.001 V). Voltage is always measured between two points and is called the **potential difference** or simply the **potential** between these two points. The greater the difference in charge between two points, the higher the voltage.

The flow of electrical charge from one point to another is called a **current,** and it can be used to do work—for example, to power a flashlight. The amount of charge that moves between the two points depends on two factors: the voltage and the resistance. **Resistance** is the hindrance to charge flow provided by intervening substances through which the current must pass. Substances with high electrical resistance are called *insulators;* those with low resistance are called *conductors.*

The relationship between voltage (V), current (I), and resistance (R) is given by Ohm's law: $I = V/R$. As you can see, the current (I) is directly proportional to the voltage: The greater the voltage (potential difference), the greater the current. And no net current flows between points that have the same potential. Ohm's law also tells us that current is inversely related to the resistance: The greater the resistance, the less the current.

In the body, electrical events typically occur in an aqueous medium, and electrical currents reflect the flow of ions rather than free electrons across cellular membranes. (There are no free electrons "running around" in living systems.) As described in Chapter 3, there is a slight difference in the numbers of positive and negative ions on the two sides of cellular plasma membranes (there is a charge separation), so there is a voltage across those membranes. The resistance to current flow is provided by the plasma membranes themselves.

Plasma membranes contain a variety of ion channels constructed by membrane proteins. Some of these membrane channels are *passive,* or *leakage, channels*

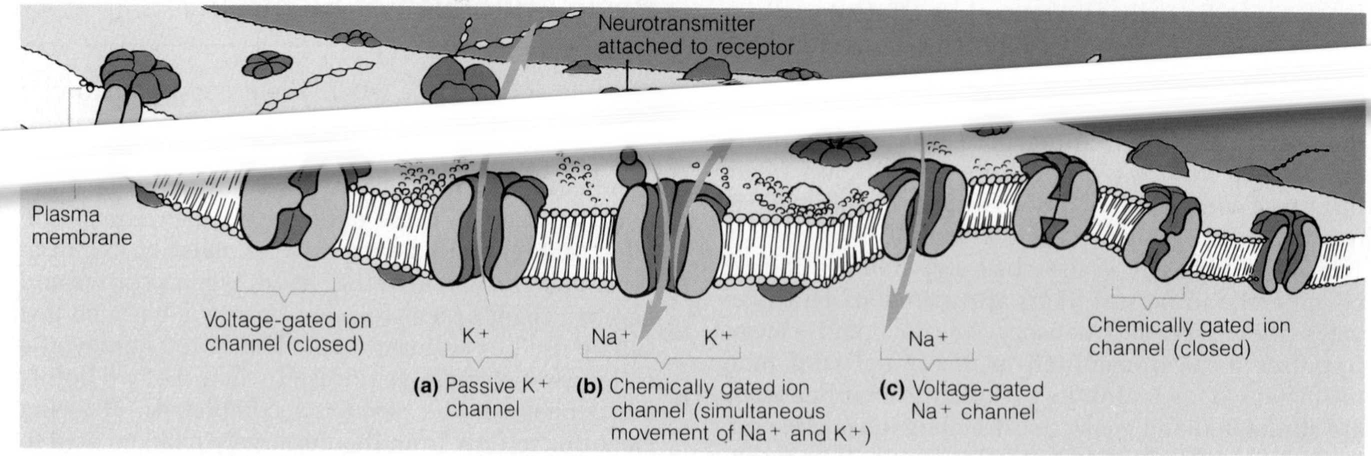

Figure 11.11 Types of plasma membrane ion channels. Plasma membranes contain both passive (leakage) ion channels and active (gated) channels. Depending on their composition and location, gated channels may be opened in response to voltage changes, attachment of chemicals (neurotransmitters), or various physical stimuli. **(a)** Passive channels for potassium ions (K^+); always open. **(b)** A chemically gated channel that allows passage of both Na^+ and K^+; channel opened in response to neurotransmitter binding. **(c)** Voltage-gated Na^+ channel; channel opens or closes in response to voltage changes across the membrane.

that are always open; others are *active*, or *gated, channels* (Figure 11.11). Gated channels are characterized by a molecular "gate," usually one or more protein molecules, that can undergo shape changes to open or close the channel in response to various signals. Chemically gated channels open when the appropriate neurotransmitter binds. Voltage-gated channels open and close in response to changes in the membrane potential, or voltage. We will speak more of this later. Each type of channel is selective; for example, a potassium ion channel generally allows only potassium ions to pass.

When gated ion channels are open, ions diffuse quickly across the membrane following their electrochemical gradients, creating electrical currents and voltage changes across the membrane according to $V = I \times R$. Ions move along *chemical gradients* when they diffuse passively from an area of higher concentration to an area of lower concentration, and along *electrical gradients* when they move toward an area of opposite electrical charge. Together, the electrical and chemical gradients constitute the *electrochemical gradient*. It is these ion flows along electrochemical gradients that underlie all electrical phenomena in neurons.

The Resting Membrane Potential: The Polarized State

The potential difference between two points is measured by using two electrodes connected to a voltmeter (Figure 11.12). When one electrode is inserted into the neuron and the other is allowed to rest on its outside surface, a voltage across the membrane of approximately −70 mV is recorded. The minus sign indicates that the cytoplasmic side (inside) of a neuron's membrane is negatively charged with respect to the outside. This potential difference at rest is called the **resting membrane potential** (V_m), and the membrane is said to be **polarized.** All cells of the body exhibit a resting membrane potential, but its value varies (from −40 mV to −90mV) in different cell types.

The resting membrane potential exists only across the membrane; that is, the bulk solutions inside and outside the cell are electrically neutral. The resting membrane potential is generated by differences in the ionic composition of the intracellular and extracellular fluids, as shown in Figure 11.13. The cell cytoplasm contains a lower concentration of sodium (Na^+) and a higher concentration of potassium (K^+) than the extracellular fluid surrounding it. Although there are many other substances (glucose, urea, and other ions) in both fluids, sodium and potassium play the most important roles in generating the membrane potential. In the extracellular fluid, the positive charges of sodium and other cations are balanced chiefly by chloride ions (Cl^-). In the cytoplasm, negatively charged (anionic) proteins (A^-) help to balance the positive charges of intracellular cations (primarily K^+).

These ionic differences are the consequence of (1) the differential permeability of the plasma membrane to sodium and potassium ions and (2) the operation of the sodium–potassium pump, which actively transports Na^+ out of the cell and K^+ into the cell (see Figure 11.13). The membrane at rest is impermeable to the large anionic cytoplasmic proteins, slightly permeable to sodium, approximately 75 times more

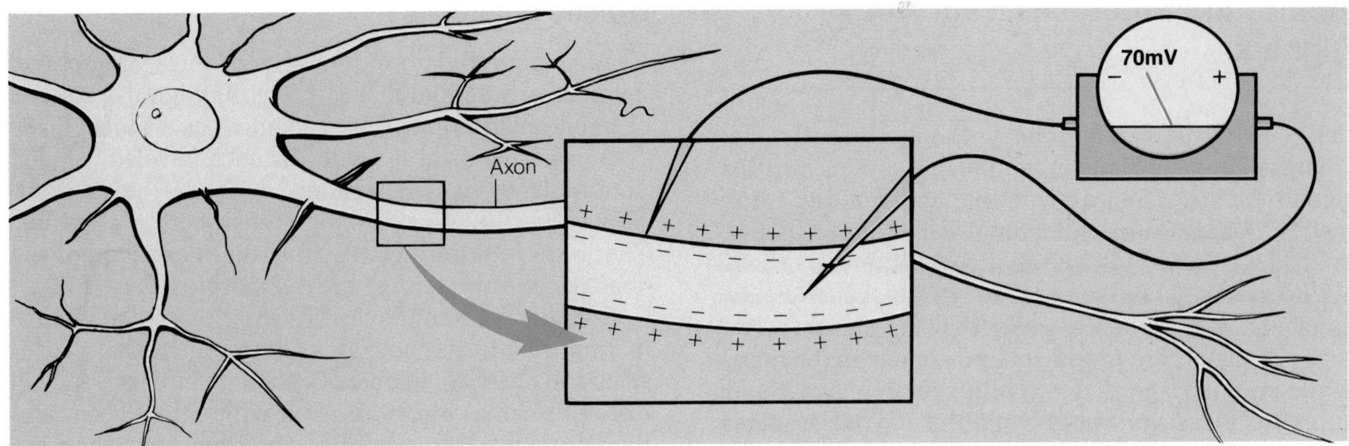

Figure 11.12 Measuring potential difference between two points in neurons. When one electrode of a voltmeter is placed on the external membrane and the other electrode is inserted just inside the membrane, a voltage (membrane potential) of approximately −70 mV (inside negative) is recorded.

permeable to potassium than to sodium, and quite freely permeable to chloride ions. These resting permeabilities reflect the properties of the passive (leakage) ionic channels present in the membrane. Consequently, potassium ions diffuse out of the cell along their *concentration gradient* much more easily and quickly than sodium ions can enter the cell along theirs. The net result is that a slightly greater number of positive ions diffuse out of the cell than in, leaving the cell interior with a slight excess of negative charge; this chemical event results in an *electrical gradient* that produces the resting membrane potential. Because some K^+ is always leaking out of the cell and Na^+ is always leaking in, it would appear that the concentration gradients would eventually "run down," resulting in equal concentrations of Na^+ and K^+ inside and outside the cell. This does not happen because of the ATP-driven sodium–potassium pump, which ejects $3Na^+$ from the cell while simultaneously transporting $2K^+$ back into the cell. Thus, the sodium–potassium pump stabilizes the resting membrane potential by maintaining the diffusion gradients for sodium and potassium.

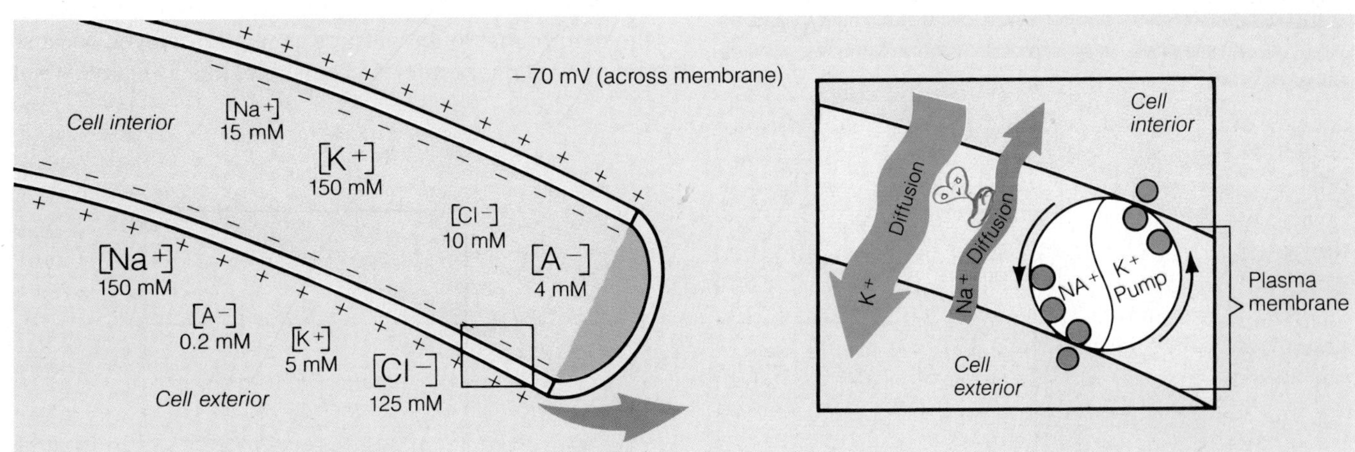

Figure 11.13 Passive and active forces that establish and maintain the resting membrane potential. Approximate ion concentrations of sodium (Na^+), potassium (K^+), chloride (Cl^-), and protein anions (A^-) in the intracellular and extracellular fluids are indicated in millimoles per liter (mM). Although many other ions exist (e.g., Ca^{2+}, PO_4^{3-}, and HCO_3^-), those shown here are most significant in terms of neural function. Diffusion of K^+ to the cell exterior is strongly promoted by its concentration (chemical) gradient. Na^+ is strongly attracted to the cell interior by its concentration gradient, but it is less able to cross the cell membrane. The relative differences in the amounts of Na^+ and K^+ crossing the membranes are indicated by diffusion arrows of different thickness. The resulting net outward diffusion of positive charge leads to a state of relative negativity on the inner membrane face. This membrane potential (−70 mV, inside negative) is maintained by the sodium-potassium pump, which transports $3Na^+$ out of the cell for each $2K^+$ transported back into the cell. Notice that the electrical gradient thus maintained resists K^+ efflux, but enhances the drive for Na^+ influx.

Membrane Potentials That Act as Signals

C[illegible] ...potential as communication signals for receiving, integrating, and sending information. A change in membrane potential can be produced by (1) anything that changes membrane permeability to any type of ion or (2) anything that alters ion concentrations on the two sides of the membrane. Two types of signals are produced by a change in the membrane potential: *graded potentials,* which signal over short distances, and *action potentials,* which are long-distance signals.

Because the terms *depolarization* and *hyperpolarization* are used frequently in the following sections to describe the direction of changes in membrane potential *relative to the resting membrane potential,* it is important to understand these terms clearly. **Depolarization** is a reduction in membrane potential: The inside of the membrane becomes *less negative,* and the potential approaches zero (Figure 11.14a). For instance, a change from a resting potential of −70 mV to a potential of −50 mV is a depolarization. By convention, depolarization also includes those events where the membrane potential reverses and moves above zero to become positive. **Hyperpolarization** occurs when the membrane potential or voltage increases, becoming *more negative* than the resting potential. For example, a change from −70 mV to −90 mV (increased negativity inside) is hyperpolarization (Figure 11.14b). As described shortly, depolarization increases the probability of producing nerve impulses, whereas hyperpolarization decreases this probability.

Graded Potentials

Graded potentials are short-lived, local changes in membrane potential th[illegible] ...of current that decrease with distance traveled. Graded potentials are called "graded" because their magnitude varies directly with the intensity of the stimulus. The more intense the stimulus, the more the voltage changes and the farther the current flows.

Graded potentials are triggered by some change (a stimulus) in the neuron's environment that causes gated ion channels to open. Graded potentials are given different names, depending on where they occur and the functions they perform. For example, when the receptor of a sensory neuron is excited by some form of energy (heat, light, or others), the resulting graded potential is called a *receptor potential.* When the stimulus is a neurotransmitter released by another neuron, the graded potential is called a *postsynaptic potential,* because the neurotransmitter is released into a fluid-filled gap called a synapse and influences the neuron beyond (post) the synapse.

Fluids inside and outside cells are fairly good conductors, and current, carried by ions, flows through them whenever voltage changes occur. Let us assume that a small area of the membrane has been depolarized by the opening of gated ion channels (Figure 11.15). A local response will occur on both sides of the membrane: Positive ions will flow toward adjacent more-negative areas (this is designated as the direction of current flow), and negative ions will migrate in the opposite direction, toward more-positive areas. As the current flows to adjacent membrane areas, it changes the membrane potential there as well. However, because the plasma membrane is permeable, like a "leaky" water

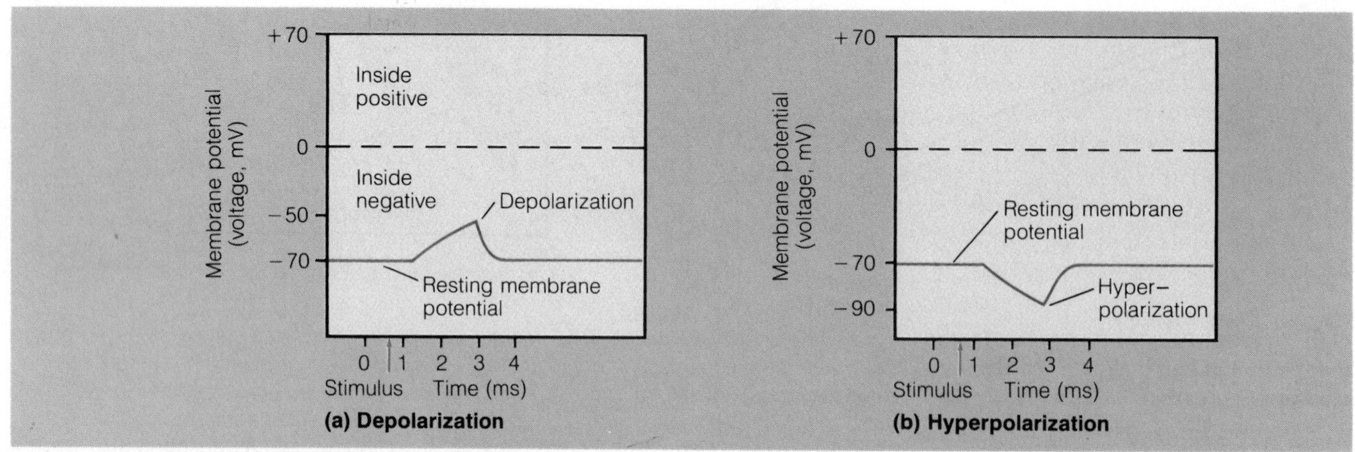

Figure 11.14 Depolarization and hyperpolarization of the resting membrane potential. The resting membrane potential is approximately −70 mV in neurons. Changes in this potential result in depolarization or hyperpolarization of the membrane. (**a**) In depolarization, the membrane potential moves toward 0 mV, the inside becoming less negative (more positive). (**b**) In hyperpolarization, the membrane potential increases, the inside becoming more negative.

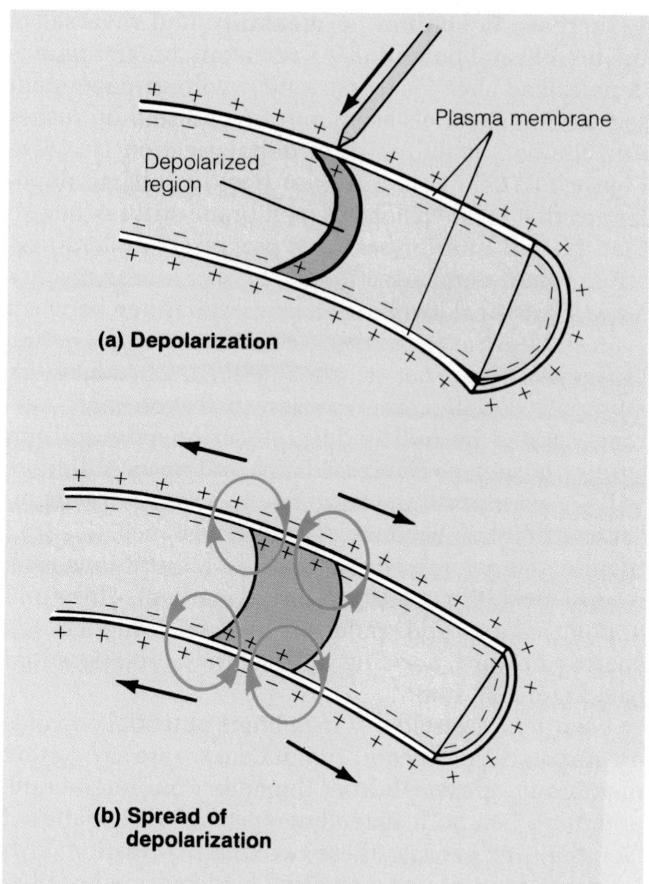

Figure 11.15 The mechanism of a graded potential. (**a**) A small patch of the membrane has become depolarized, resulting in a change in polarity at that point. (**b**) As positive ions flow toward negative areas (and negative ions flow toward adjacent more positive areas), local currents are created that depolarize adjacent membrane areas and allow spreading of the wave of depolarization. The electrical currents conducted in this manner are dissipated with distance.

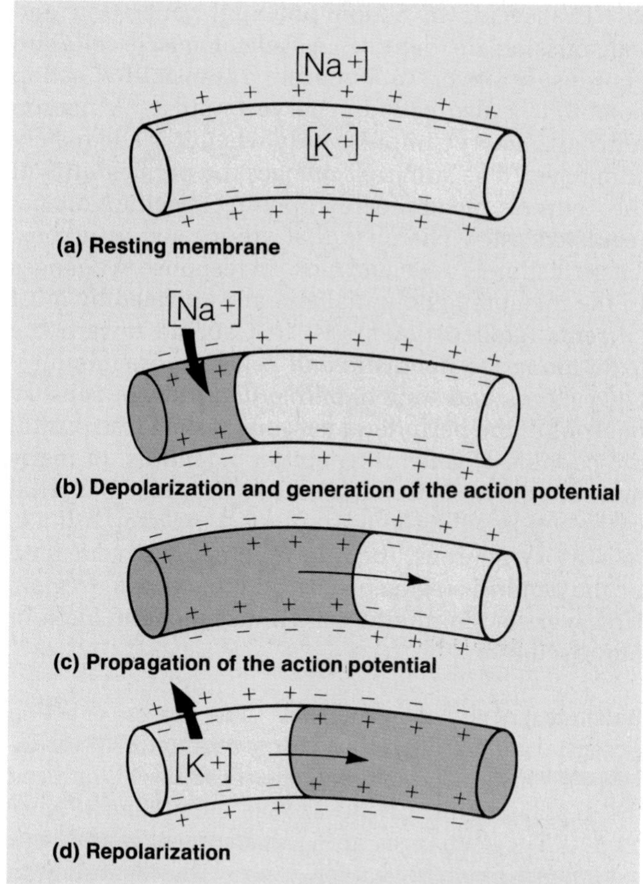

Figure 11.16 Summary of events in the generation and propagation of an action potential in skeletal muscle cells and neurons. (**a**) Electrical conditions of a resting (polarized) membrane. The predominant extracellular ion is sodium (Na^+), whereas the predominant ion inside the cells is potassium (K^+). The membrane is relatively impermeable to both ions. (**b**) Depolarization and generation of the action potential. A stimulus changes the permeability of a patch of membrane, and sodium ions diffuse rapidly into the cell. The resting potential is decreased. If the stimulus is strong enough, an action potential is initiated. (**c**) Propagation of the action potential. The positive charge inside the initial patch of membrane changes the permeability of an adjacent patch, and the events described in (b) are repeated. Thus, the action potential propagates itself rapidly along the entire length of the membrane. (**d**) Repolarization. Potassium ions diffuse out of the cell as the membrane permeability changes once again. This restores the negative charge on the inside of the membrane and the positive charge on its external surface. Repolarization occurs in the same direction as depolarization. The ionic concentrations of the resting state are restored later by the sodium-potassium pump.

hose, most of the charge is quickly lost through the membrane and the current dies out within a few millimeters of its origin. Because the current dissipates quickly and dies out with increasing distance, graded potentials can act as signals only over very short distances. Nonetheless, they are essential in initiating action potentials, the long-distance signals.

Action Potentials

The principal way neurons communicate is by generating and propagating **action potentials,** and for the most part, only cells with excitable membranes—neurons and muscle cells—can generate action potentials. An action potential is a brief, but large, depolarization with a total amplitude (change in voltage) of about 100 mV (from −70 mV to +30 mV). When we considered skeletal muscle physiology in Chapter 9, we indicated briefly that muscle fibers are stimulated to contract when an action potential is transmitted along their sarcolemma. A summary of that discussion is provided in Figure 11.16. Now, let us examine how this happens in more detail.

The events of action potential generation and transmission are identical in skeletal muscle cells and neurons; however, in a neuron, a transmitted action potential is also called a nerve [illegible] The stimulus changes the permeability of the neuron's membrane by opening specific voltage-regulated gated channels that are located on axons. These channels open and close in response to changes in the membrane potential and are activated by local currents (graded potentials) that spread toward the axon along the dendritic and cell body membranes. *Only axons and long axonlike dendrites of sensory neurons of the peripheral nervous system* (see p. 340) *are capable of generating action potentials.* In many neurons, the transition from the local graded potential to the action potential takes place at the axon hillock. In sensory neurons, the action potential is generated by the dendritic region just beyond the receptor region. However, for simplicity, we will use the term *axon* in our discussion.

Generation of an Action Potential. The process of action potential generation reflects three sequential but overlapping changes in membrane permeability, all induced by depolarization of the axonal membrane (Figure 11.17). These changes are (1) a transient increase in sodium ion permeability due to the opening of sodium gates, (2) the restoration of relative sodium ion impermeability by the closing of the sodium gates, and (3) a short-lived increase in potassium ion permeability due to the opening of potassium gates.

1. Increase in sodium permeability and reversal of the membrane potential. As the axonal membrane is depolarized by local [illegible] point of depolarization (see also Figure 11.16b). This influx of positive charge depolarizes that local "patch" of membrane still further, so that the cell interior becomes progressively less negative. When depolarization of the membrane reaches a certain critical level called *threshold* (often between −55 and −50 mV), the depolarization process becomes self-generating. That is, after being initiated by the stimulus, depolarization is driven by the ionic currents created by sodium ion influx. As more sodium enters, the voltage changes further and opens still more voltage-gated sodium channels. As a result, the membrane potential becomes less and less negative and then overshoots about +30 mV as sodium ions rush inward along their electrochemical gradient. This rapid depolarization and polarity reversal produce the sharply upward, or rising, spike of the action potential (see Figure 11.17a).

Earlier, we stated that membrane potential depends on membrane permeability, but here we are saying that membrane permeability depends on membrane potential. Can both statements be true? As a matter of fact, they are, because these two distinct relationships interlock to establish a *positive feedback* cycle called the *Hodgkin cycle* (Figure 11.18) that is responsible for the rising (depolarizing) phase of action potentials.

2. Decrease in sodium permeability. The explosively rising phase of the action potential (period of sodium permeability) persists for only about 1 ms and is self-

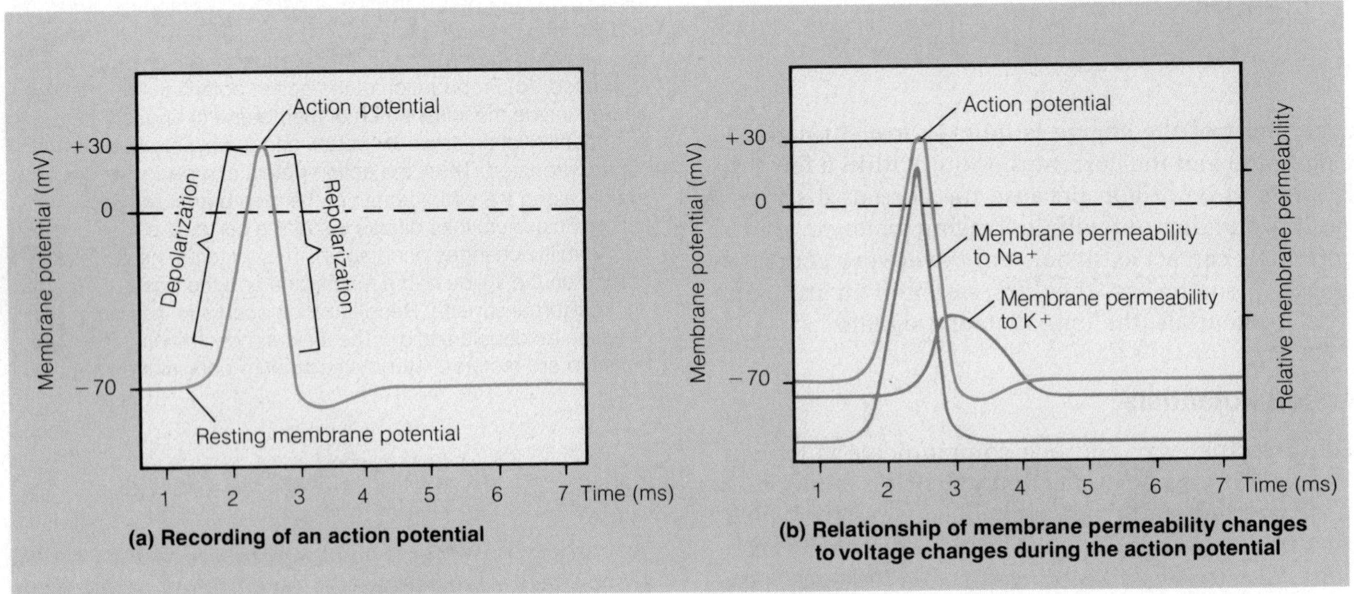

Figure 11.17 Voltage and membrane permeability changes occurring during action potential generation.

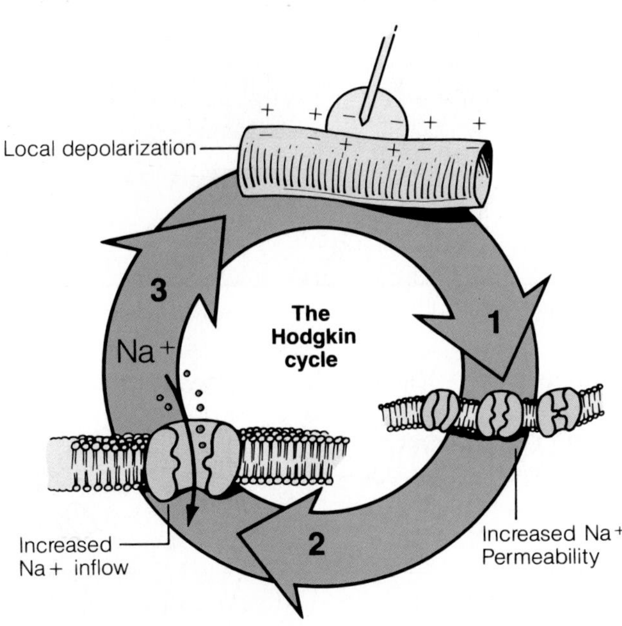

Figure 11.18 The Hodgkin cycle, which produces the rising phase of the action potential. The critical feature of this positive feedback cycle is that Na^+ permeability is *voltage dependent.* (**1**) Local depolarization increases Na^+ permeability; (**2**) the increased Na^+ permeability allows influx of Na^+ down its electrochemical gradient, which (**3**) further depolarizes the membrane and opens more voltage-dependent Na^+ channels. Once threshold is reached, the cycle is self-sustaining and driven by voltage changes created by the ionic current (passage of Na^+ into the cell).

limiting. As the membrane potential passes 0 mV and becomes increasingly more positive, the positive intracellular charge resists further sodium entry. In addition, the sodium gates close after a few milliseconds of depolarization; consequently, the membrane becomes increasingly impermeable to sodium, and the net influx of sodium declines and finally stops completely (see Figure 11.17b).

3. Increase in potassium permeability and repolarization. As sodium entry declines, voltage-regulated potassium gates open and potassium rushes out of the cell, following its electrochemical gradient (see Figure 11.17b). As potassium leaves, the cell interior becomes progressively less positive, and the membrane potential moves back toward the resting level, an event called **repolarization** (see Figures 11.17a and 11.16d). Both the abrupt decline in sodium permeability and the increased permeability to potassium contribute to the repolarization process.

Although repolarization restores resting electrical conditions, it does *not* restore the original ionic distributions of the resting state. Following repolarization, this is accomplished by activation of the sodium-potassium pump. While it might appear that opening of the voltage-gated channels allows tremendous numbers of sodium and potassium ions to change places during action potential generation, nothing could be further from the truth. Only small amounts of sodium and potassium are exchanged, and since an axonal membrane has thousands of sodium-potassium pumps, these small ionic changes are quickly corrected.

Propagation of an Action Potential. If the generated action potential is to serve as the neuron's signaling device, it must be *propagated* (sent or transmitted) along the axon's entire length. This process is illustrated in Figure 11.19.

As we have seen, the action potential is generated by the inward movement of sodium ions, and the local patch of depolarized axonal membrane undergoes polarity reversal: The inside becomes positive, and the outside becomes negative. The sodium ions that have diffused into the axon then move laterally from the area of polarity reversal toward the neuron area that is still polarized, move outward through the membrane, and then migrate back toward the area of greatest negative charge (the area of polarity reversal) to complete the circuit. As a result, local current flows are established.

These local current flows depolarize adjacent membrane areas in the forward direction (away from the origin of the nerve impulse), which opens voltage-gated channels and triggers an action potential there. Since the area in the opposite direction has just generated an action potential, the sodium gates are closed, so no new action potential is generated there. Thus, the impulse always propagates away from its point of origin. If an isolated axon is stimulated by an electrode, the nerve impulse will move away from the stimulus in *both* directions; but in the body, action potentials are always initiated at one end of the axon and conducted away from that point toward the axon's terminals. Once initiated, an action potential is a *self-propagating* event that continues along the axon at a constant velocity—something like a "domino effect."

Following depolarization, each segment of axonal membrane undergoes repolarization, which restores the resting membrane potential in that region. Since these electrical changes also set up local current flows, the repolarization wave chases the wave of depolarization down the length of the axon. The propagation process just described occurs on unmyelinated axons (and on muscle fiber sarcolemmas). The unique propagation process that occurs in myelinated axons, called saltatory conduction, will be described shortly. Although the phrase *conduction of a nerve impulse*

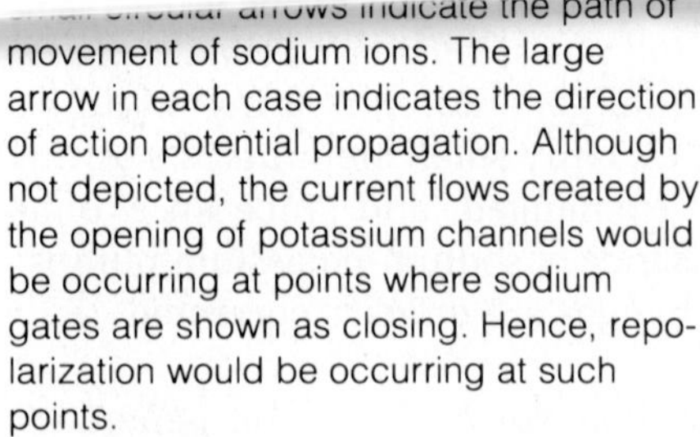

Figure 11.19 Propagation of an action potential. The distance of propagation of the action potential along the axon is shown at time 0 ms, 1 ms, and 2 ms. The condition of the sodi[illegible] circular arrows indicate the path of movement of sodium ions. The large arrow in each case indicates the direction of action potential propagation. Although not depicted, the current flows created by the opening of potassium channels would be occurring at points where sodium gates are shown as closing. Hence, repolarization would be occurring at such points.

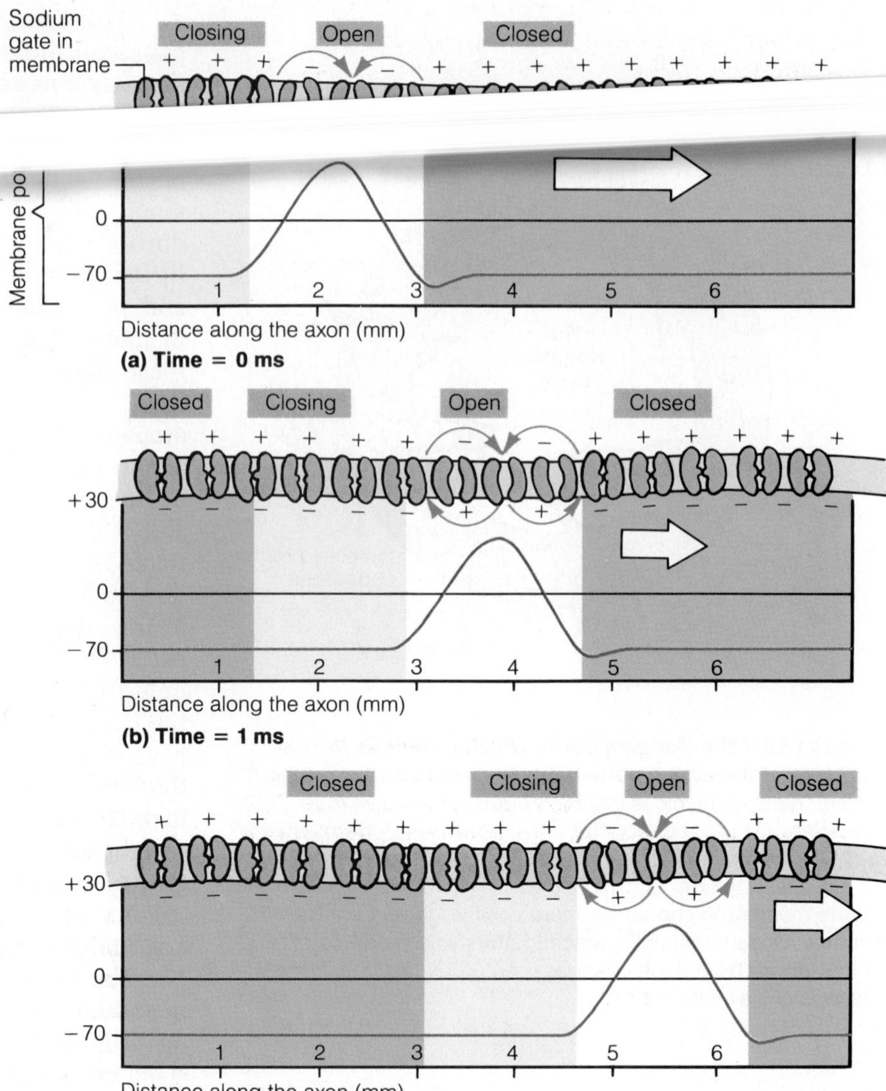

is commonly used, nerve impulses are not really conducted in the same way that an insulated wire conducts current. In fact, neurons are fairly poor conductors of electrical current, and local current flows decline with distance because the charges leak through the membrane. The expression *propagation of a nerve impulse* is more accurate, because the action potential is regenerated anew at each membrane patch, and every subsequent action potential is identical to that generated initially.

Threshold and the All-or-None Phenomenon. Not all local depolarization events produce action potentials. The depolarization must reach threshold values if an axon is to "fire." What determines the **threshold point?** One explanation is that the threshold value is that membrane potential at which the outward current carried by K^+ is exactly equal to the inward current created by Na^+. Threshold is typically reached when the membrane has been depolarized by 15–20 mV from the resting value. It seems to represent an unstable equilibrium state at which one of two things can happen. If an extra sodium ion enters, further depolarization occurs, opening more sodium gates and allowing more sodium ions entry. If, on the other hand, another potassium ion leaves, the membrane potential is driven away from threshold, sodium gates are closed, and potassium ions continue to diffuse outward until the potential returns to its resting value.

Recall that the local depolarizations are graded potentials and that their magnitude increases with increasing stimulus intensity. Weak stimuli (*subthreshold stimuli*) produce subthreshold depolarizations that are not translated into nerve impulses. On the other hand, strong stimuli (*threshold stimuli*) produce depolarizing currents that push the membrane potential toward and beyond the threshold value and increase sodium permeability to such an extent that sodium entry "swamps" (exceeds) the outward movement of potassium ions, allowing the Hodgkin cycle to become established and generating an action potential.

The action potential is an **all-or-none phenomenon;** it either happens completely or it doesn't happen at all. It might help to compare the generation of the action potential to lighting a match under a small dry twig. The heating of part of the twig can be compared to the change in membrane permeability that initially allows more sodium ions to enter the cell. When that part of the twig becomes hot enough (when enough sodium ions have entered the cell), the critical flash point (threshold) is reached, and the flame will consume the entire twig, even if you then blow out the match (the action potential will be generated and propagated regardless of whether the stimulus continues). But if the match is extinguished just before the twig has reached the critical temperature, ignition will not take place. Likewise, if too few sodium ions enter the cell to achieve threshold, no action potential will occur.

Coding for Stimulus Intensity. Once generated, all action potentials are independent of stimulus strength, and all action potentials are alike. So how can the CNS determine whether a particular stimulus is intense or weak—information it needs to effect an appropriate response? The answer is really quite simple: Strong stimuli cause nerve impulses to be generated more often in a given time interval than do weak stimuli. Thus, stimulus intensity is coded for by the number of impulses generated per second—that is, by the *frequency of impulse transmission*—rather than by increases in the strength (amplitude) of the action potential.

Absolute and Relative Refractory Periods. When a patch of neuron membrane is generating an action potential and its sodium gates are open, the neuron is incapable of responding to another stimulus, no matter how strong. This period, called the **absolute refractory period,** ensures that each action potential is a separate, all-or-none event. This makes sense if you keep in mind that once an action potential is generated, the sodium gates are kept open by sodium ion currents, not by the stimulus.

The interval following the absolute refractory period, when the sodium gates are closed, the potassium gates open, and repolarization is occurring, is the **relative refractory period.** During this time, the axon's threshold for impulse generation is substantially elevated. A threshold stimulus is unable to trigger an action potential during the relative refractory period, but an exceptionally strong stimulus can reopen the sodium gates and allow another impulse to be generated. Thus, by intruding into the relative refractory period, strong stimuli cause action potentials to be generated more frequently.

Conduction Velocities of Axons. Propagation of impulses by an axon is influenced by its membrane resistance and by the threshold of the neuron. Conduction velocities of neurons vary by a factor of more than 100. Fibers that transmit impulses most rapidly (100 m/s or more) are found in neural pathways where speed is essential, such as those that mediate some postural reflexes. More slowly conducting axons most often serve internal organs (the gut, glands, blood vessels), where slower responses are not usually a handicap. The rate of impulse propagation depends largely on two structural factors: axon diameter and degree of myelination.

1. Influence of axon diameter. As a general rule, the larger the axon's diameter, the faster it conducts impulses. This is because larger axons have a greater cross-sectional and membrane surface area, which together provide a lower resistance, larger pathway for current flow permitting adjacent membrane areas to be depolarized more quickly.

2. Influence of a myelin sheath. On unmyelinated axons, action potentials are generated at sites immediately adjacent to each other and conduction is relatively slow. The presence of a myelin sheath dramatically increases the rate of impulse propagation, because myelin acts as an insulator to prevent almost all leakage of charge. Current can pass through the membrane of a myelinated axon only at the nodes of Ranvier, where the myelin sheath is interrupted and the axon is bare, and practically all the active sodium channels are concentrated at the nodes. Consequently, when an action potential is generated in a myelinated fiber, the local depolarizing current does not dissipate through the adjacent (nonexcitable) membrane regions but instead is forced to move as much as 1 mm to the next node, where it triggers another action potential. This type of conduction is called **saltatory conduction** (*saltare* = to leap) because the electrical signal appears to jump from node to node along the axon (Figure 11.20). Saltatory conduction is much faster than the conduction produced by the continuous spread of local depolarizing currents along unmyelinated membranes.

Nerve fibers may be classified as group A, B, or C fibers based on their diameter, degree of myelination, and speed of conduction. Group A fibers, mostly somatic sensory and motor fibers serving the skin, skeletal muscles, and joints, have the largest diameter and thick myelin sheaths, and they conduct impulses at speeds ranging from 15 to 130 m/s. Autonomic nervous system motor fibers serving the visceral organs, visceral sensory fibers, and the smaller somatic sen-

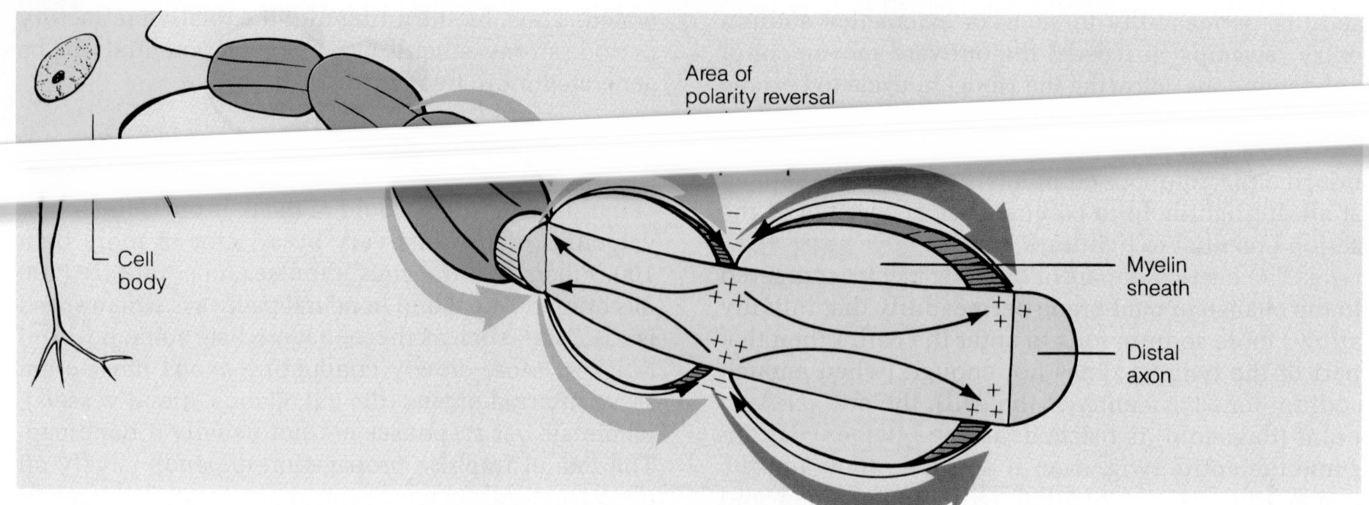

Figure 11.20 Saltatory conduction in a myelinated axon. In myelinated fibers, the local current flows (small black arrows in longitudinally sectioned axon) give rise to a propagated action potential (large light to dark pink arrows) that appears to jump from node to node.

sory fibers transmitting afferent impulses from the skin (such as pain and small touch fibers) belong in the B and C fiber groups. Group B fibers, lightly myelinated fibers of intermediate diameter, transmit impulses at the rate of 3–15 m/s. Group C fibers have the smallest diameter and are unmyelinated; hence, these fibers are incapable of saltatory conduction and conduct impulses very slowly, 1 m/s or less.

A number of chemical and physical factors can impair the conduction of impulses. Although their mechanisms of action differ, alcohol, sedatives, and anesthetics all block nerve impulses by reducing membrane permeability to sodium ions. As we have seen, no sodium entry—no action potential.

Cold and continuous pressure interrupt blood circulation (and hence the delivery of oxygen and nutrients) to the neuronal processes, thus impairing their ability to conduct impulses. For example, your fingers get numb when you hold an ice cube for more than a few seconds. Likewise, when you sit on your foot, it "goes to sleep." When you remove the pressure, the impulses begin to be transmitted once again, leading to an unpleasant prickly feeling. ■

The Synapse

The operation of the nervous system depends on the flow of information through elaborate circuits consisting of chains of neurons functionally connected by synapses. A **synapse** (sih′-naps), from the Greek *syn*, meaning "to clasp or join," is a unique junction that mediates the transfer of information from one neuron to the next or from a neuron to an effector cell.

Most synapses occur between the axonal endings of one neuron and the dendrites or the cell bodies (soma) of other neurons and are called *axodendritic* or *axosomatic synapses*, respectively. Less common (and far less understood) are the synapses formed between two axons (*axoaxonic*), between two dendrites (*dendrodendritic*), or between a dendrite and a cell body (*dendrosomatic*). Some of these synaptic patterns are illustrated in Figure 11.21.

The neuron conducting impulses toward the synapse is called the **presynaptic neuron.** Presynaptic neurons are the information senders. The neuron that transmits the electrical activity away from the synapse is called the **postsynaptic neuron.** Postsynaptic neurons are the information recipients. As you might anticipate, most neurons (including all interneurons) function as both presynaptic and postsynaptic neurons, receiving information from some neurons and dispatching it to others. A typical neuron has anywhere from 1000 to 10,000 axonal terminals making synapses and is stimulated by an equal number of other neurons. In the body periphery, the postsynaptic cell may be either another neuron or an *effector cell* (a muscle cell or gland cell). Synapses between neurons and muscle cells are specifically called *neuromuscular junctions* (described in Chapter 9); those between neurons and gland cells are *neuroglandular junctions*.

There are two varieties of synapses: electrical and chemical. The major structural and functional properties of these synapses are described next.

Electrical Synapses

Electrical synapses are *bridged junctions* that correspond to the gap junctions found between certain other body cells (see Figure 3.3c). They contain protein

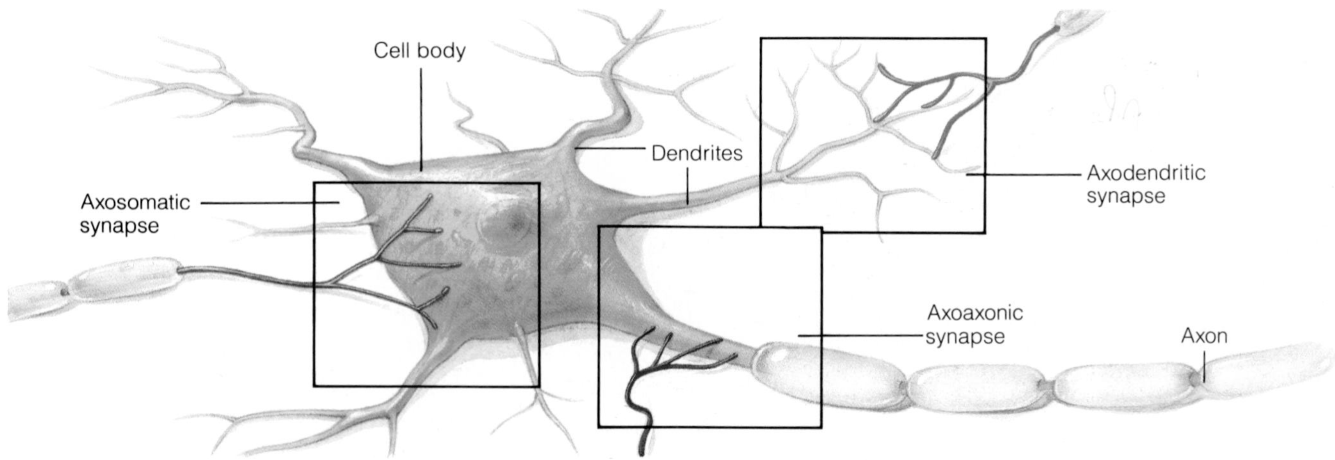

Figure 11.21 Examples of types of chemical synapses. Axodendritic, axosomatic, and axoaxonic synapses. Axonal terminals of the presynaptic neurons are indicated in different colors at the different synapse types.

channels that interconnect the cytoplasm of adjacent neurons, providing low-resistance electrical pathways through which ions can flow directly from one neuron to the next. Neurons joined in this way are said to be *electrically coupled*, and transmission across these synapses is very rapid. Depending on the nature of the synapse, communication is unidirectional or bidirectional.

A key feature of electrical synapses is that they provide a simple means of synchronizing the activity of all interconnected neurons. In adults, electrical synapses are found in regions of the brain responsible for certain stereotyped movements, such as the normal jerky movements of the eyes. They are far more abundant in embryonic nervous tissues, where they are thought to allow for the exchange of guiding clues during early neuronal development so that neurons can connect properly with one another. Most electrical synapses are replaced by chemical synapses during development. However, electrical synapses are abundant in certain nonnervous tissues, such as cardiac and smooth muscle, where they allow sequential and rhythmic excitations.

Chemical Synapses

In contrast to electrical synapses, which are specialized to allow the flow of ions between neurons, **chemical synapses** are specialized for release and reception of chemical **neurotransmitters.** Neurotransmitters (described in detail shortly) function to open or close ion channels that influence membrane permeability, and consequently, membrane potential.

A typical chemical synapse is made up of two parts: a knoblike *axonal terminal* of the presynaptic (transmitting) neuron and a *receptor region* located on the membrane of a dendrite or the cell body of the postsynaptic neuron (Figure 11.22). Although close, the presynaptic and postsynaptic membranes are always separated by the **synaptic cleft,** a fluid-filled space approximately 20–30 nm (about one-millionth of an inch) wide. Because the current from the presynaptic membrane is dissipated in the fluid-filled cleft, chemical synapses effectively prevent a nerve impulse from being directly transmitted from one neuron to another. Although transmission of nerve impulses along an axon is a purely electrical phenomenon, the transmission of signals across chemical synapses is a chemical event that depends on the release, diffusion, and receptor binding of neurotransmitter molecules, and results in unidirectional communication between neurons.

Within an axonal terminal are many tiny, membrane-bounded sacs called **synaptic vesicles,** each containing thousands of neurotransmitter molecules. Some neurotransmitters are synthesized locally in the terminal and enclosed there within vesicles. Others are made in the cell body and dispatched, enclosed within membranous vesicles, to the axonal terminals.

Information Transfer Across Chemical Synapses. When a nerve impulse reaches the axon terminal, a chain of events is set into motion that triggers neurotransmitter release. The neurotransmitter crosses the synaptic cleft and upon binding to receptors on the postsynaptic membrane causes changes in the postsynaptic membrane permeability. The successive events appear to be as follows (Figure 11.22):

1. When the nerve impulse reaches the axon terminal, voltage-regulated calcium ion gates in the axonal membrane open briefly, and calcium ions flood into the terminal from the extracellular fluid.

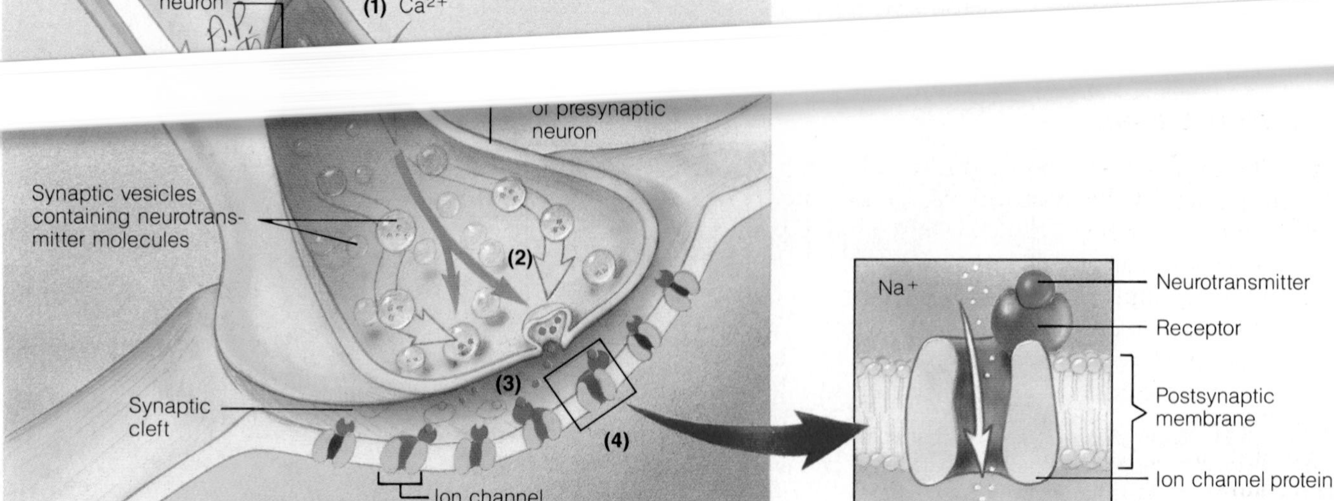

Figure 11.22 Events occurring at a chemical synapse in response to depolarization of the axonal terminal. (**1**) Arrival of the nerve impulse results in the influx of calcium (Ca^{2+}) into the axonal terminal. (**2**) Calcium ions promote the fusion of synaptic vesicles with the presynaptic membrane and the exocytosis of neurotransmitter. (**3**) The neurotransmitter diffuses across the synaptic cleft and attaches to receptors on the postsynaptic membrane. (**4**) Binding of neurotransmitter causes ion channels to open, which results in voltage changes in the postsynaptic membrane. (Effects are short-lived because the neurotransmitter is quickly destroyed by enzymes or taken back up into the presynaptic terminal.)

2. Increased intracellular levels of ionic calcium (Ca^{2+}) promote fusion of synaptic vesicles with the axonal membrane and the emptying of neurotransmitter by exocytosis into the synaptic cleft. The Ca^{2+} is then quickly removed, either taken up into the mitochondria or ejected to the outside by an active calcium membrane pump.

3. The neurotransmitter diffuses across the synaptic cleft and binds to specific protein receptors that are densely clustered on the postsynaptic membrane.

4. As the receptor proteins bind neurotransmitter molecules, their three-dimensional shape changes. This leads to an ion channel opening, and the resulting current flows cause local changes in the membrane potential. Depending on the types of neurotransmitters released and the receptor proteins to which they bind, the result may be either excitation or inhibition of the postsynaptic neuron. The changes prompted by neurotransmitter-receptor binding are extremely brief because the neurotransmitter is quickly removed from the postsynaptic membrane by enzymatic degradation or reuptake into the presynaptic axon. This limits the effect of a single synaptic event to a few milliseconds.

For each nerve impulse reaching the presynaptic terminal, many vesicles (perhaps 300) are emptied into the synaptic cleft. The higher the frequency of impulses reaching the terminals (that is, the more intense the stimulus), the greater the total number of synaptic vesicles that fuse and spill their contents, and the greater the effect on the postsynaptic cell.

Synaptic Delay. Although some neurons can transmit impulses at rates that approach 250 mph (100 mm/ms), neural transmission across a chemical synapse is comparatively slow and reflects the time required for neurotransmitter release, diffusion across the synapse, and binding to receptors. Typically, this **synaptic delay** in transmission lasts 0.3–5.0 ms, and it is the *rate-limiting* step of neural transmission. Synaptic delay helps to explain why transmission along short neural pathways involving only two or three neurons occurs rapidly, while that typical of higher mental functioning and involving many neurons and synapses occurs much more slowly. (However, in practical terms these differences are not noticeable.)

Postsynaptic Potentials and Synaptic Integration

Two kinds of chemical synapses, excitatory and inhibitory, are distinguished according to how they affect the postsynaptic neuron. In both types, binding of neurotransmitter results in local changes in the postsynaptic membrane potential (Figure 11.23). *Excitatory postsynaptic potentials (EPSPs)* occur at excitatory synapses and *inhibitory postsynaptic potentials (IPSPs)* are generated at inhibitory synapses. Action potentials are compared with the two types of postsynaptic potentials in Table 11.2 on p. 360.

Excitatory Synapses and EPSPs

At excitatory synapses, neurotransmitter binding causes depolarization of the postsynaptic membrane. However, events occurring here differ in several ways from those happening on axonal membranes. The ion channels on postsynaptic membranes (membranes of dendrites and neuronal cell bodies) are activated by binding of an appropriate neurotransmitter; that is, they are *chemically gated,* rather than voltage-gated, channels. Further, neurotransmitter binding opens a single type of channel that allows both sodium and potassium ions to diffuse *simultaneously* through the membrane in opposite directions, rather than first opening sodium gates and later potassium gates. Although this may appear to be self-defeating when depolarization is the goal, remember that the electrochemical gradient for sodium is much steeper than that for potassium. Hence, the influx of Na^+ is greater than the efflux of K^+, and net depolarization occurs.

If enough neurotransmitter is bound, depolarization of the postsynaptic membrane can successfully reach 0 mV. However, postsynaptic membranes do not generate or transmit action potentials; only axons have this capability. The dramatic polarity reversal seen in axons never occurs in membranes containing *only* chemically gated channels because the opposite movements of K^+ and Na^+ prevent accumulation of excessive positive charge inside the cell. Hence, instead of action potentials, local depolarization events called **excitatory postsynaptic potentials (EPSPs)** occur at excitatory postsynaptic membranes (see Figure 11.23a). EPSPs are graded potentials that vary in amplitude (strength) with the amount of neurotransmitter bound to the receptors. Each EPSP lasts only a few milliseconds and then the membrane returns to its resting potential. Although the current flows created by these individual depolarization events decline with distance, they do spread all the way to the axon hillock. If the currents reaching the axon hillock are strong enough to depolarize the axon to threshold, the axonal voltage-gated channels will open and an action potential will be generated. The only function of EPSPs is to help trigger an action potential distally at the axon hillock of the postsynaptic neuron.

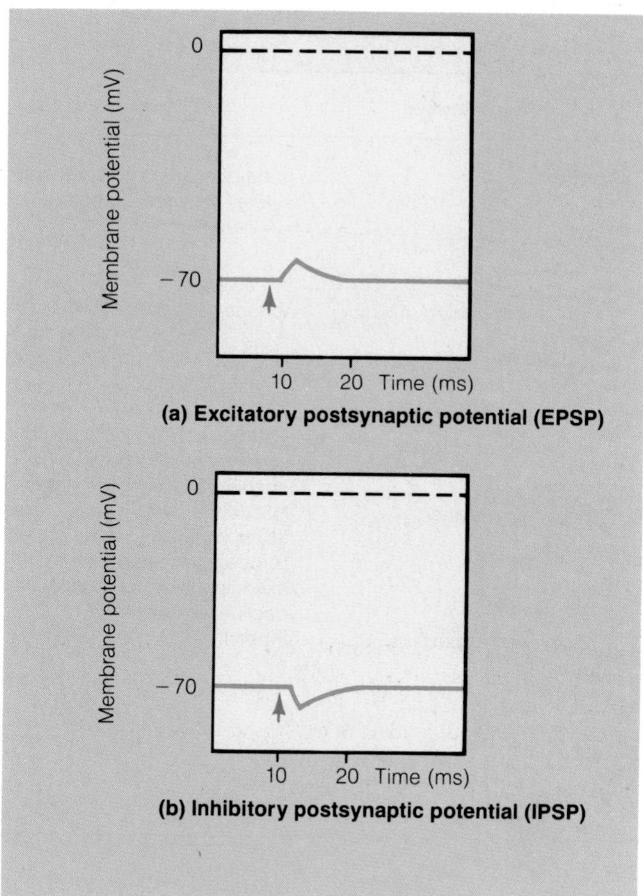

Figure 11.23 Postsynaptic potentials. (**a**) An excitatory postsynaptic potential (EPSP) is a local depolarization of the postsynaptic membrane. It is mediated by neurotransmitter binding, which causes the opening of channels that allow simultaneous passage of sodium and potassium through the postsynaptic membrane. (**b**) An inhibitory postsynaptic potential results in the hyperpolarization of the postsynaptic neuron. It is mediated by neurotransmitter binding, which opens potassium or chloride gates or both. Vertical arrows represent stimulation.

Inhibitory Synapses and IPSPs

The binding of neurotransmitter at inhibitory synapses reduces a postsynaptic neuron's ability to generate an action potential. Most inhibitory neurotransmitters induce hyperpolarization of the postsynaptic membrane by making the membrane more permeable to potassium ions, chloride ions, or both. Sodium ion permeability is not affected. If potassium gates are opened, potassium ions move out of the cell; if chloride gates are opened, chloride ions move in. In either case, the charge on the inner face of the membrane becomes relatively more negative, so that the membrane potential is driven from the resting level of −70 mV toward −90 mV, farther from the axon's threshold. This means that larger depolarizing currents are required to induce an action potential. Such changes in potential are called **inhibitory postsynaptic potentials (IPSPs)** (see Figure 11.23b).

Table 11.2 Comparison of Action Potentials with Postsynaptic Potentials

Characteristic	Action potential	[illegible]	[illegible]
[illegible]	[illegible] stitutes the nerve impulse	[illegible]-distance signaling; depolarization that spreads to axon hillock; moves membrane potential *toward* threshold for generation of action potential	Short-distance signaling; hyperpolarization that spreads to axon hillock; moves potential *away from* threshold for generation of action potential
Stimulus for opening of ionic gates	Voltage (depolarization)	Chemical (neurotransmitter)	Chemical (neurotransmitter)
Initial effect of stimulus	First opens sodium gates, then potassium gates	Open channels that allow simultaneous sodium and potassium fluxes	Opens potassium and/or chloride channels
Repolarization	Voltage regulated; closing of sodium gates followed by opening of potassium gates	Dissipation of membrane changes with time and distance	
Conduction distance	Not conducted by local current flows; continually regenerated (propagated) along entire axon; intensity does not decline with distance	1–2 mm; local electrical events; intensity declines with distance	
Hodgkin cycle (positive feedback)	Present	Absent	Absent
Peak membrane potential	+40 to +50 mV	0 mV	Becomes hyperpolarized; moves toward −90 mV
Summation	None; an all-or-none phenomenon	Present; produces graded depolarization	Present; produces graded hyperpolarization
Refractory period	Present	Absent	Absent

Integration and Modification of Synaptic Events

Summation by the Postsynaptic Neuron. The activity of a single excitatory axonal terminal cannot induce an action potential in the postsynaptic neuron. But if thousands of excitatory axonal terminals are firing on the same postsynaptic membrane, or if a smaller number of terminals are delivering impulses very rapidly, the probability of reaching threshold depolarization increases greatly. Thus, EPSPs add together, or **summate**, to collectively influence the activity of a postsynaptic neuron. In fact, nerve impulses would never be initiated if this were not so.

Two types of summation can occur: temporal and spatial. We will first explain these events in terms of EPSPs. **Temporal summation** occurs when one or more presynaptic neurons transmit impulses in rapid-fire order (Figure 11.24b) and waves of neurotransmitter release occur in quick succession. The first impulse produces a slight EPSP, and before it dissipates, successive impulses trigger more EPSPs. These summate, producing a greater depolarization of the postsynaptic membrane than would result from a single EPSP.

Spatial summation occurs when the postsynaptic neuron is being stimulated by a large number of terminals from the same or (more commonly) different neurons at the same time. As a result, huge numbers of its receptors can bind neurotransmitter and simultaneously initiate EPSPs, which sum up and dramatically enhance the depolarization (Figure 11.24c).

Although we have focused on the summation of EPSPs here, the effects of IPSPs can also be summated, both spatially and temporally. In this case, a greater degree of inhibition of the postsynaptic neuron is produced.

Most neurons receive both stimulatory and inhibitory inputs from thousands of other neurons. How is all of this conflicting information sorted out? It appears that each neuron's axon hillock keeps a "running account" of all the signals it receives (Figure 11.24d). Not only do EPSPs summate and IPSPs summate, but EPSPs summate with IPSPs. If the stimulatory effects of EPSPs dominate the membrane potential enough to reach threshold, the neuron will fire. If, on the other hand, the summing process yields only subthreshold depolarization or hyperpolarization, the neuron will fail to generate an action potential. However, partially depolarized neurons are *facilitated*—that is, more easily excited by successive depolarization events—because they are already nearer to threshold. Thus, axon hillock membranes function as *neural integrators*, and their potential at any time reflects the sum of all incoming neural information. Since EPSPs and IPSPs are graded potentials that diminish in strength the farther they spread, the most effective synapses are those closest to the axon hillock. Synapses on distal

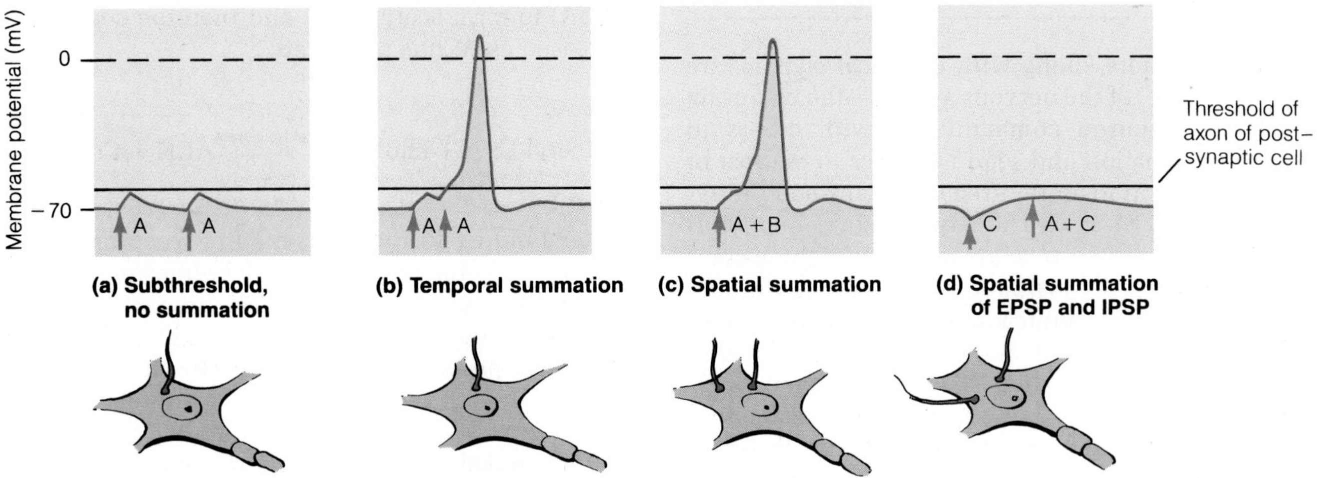

Figure 11.24 Neural integration of EPSPs and IPSPs at the axonal membrane of the postsynaptic cell. Synapses A and B are excitatory; synapse C is inhibitory. Each individual impulse is itself subthreshold. (**a**) Synapse A is stimulated and then is stimulated again shortly thereafter. The two EPSPs do not overlap in time, so no interaction (summation) occurs; threshold is not reached in the axon of the postsynaptic neuron. (**b**) Synapse A is stimulated a second time before the initial EPSP has died away; *temporal summation* occurs, and the axon's threshold is reached, causing an action potential to be generated. (**c**) Synapses A and B are stimulated simultaneously (*spatial summation*), resulting in a threshold depolarization. (**d**) Synapse C is stimulated, resulting in a short-lived IPSP (hyperpolarization). When A and C are simultaneously stimulated, the changes in potential cancel each other out.

dendrites have far less influence on firing the axon than do synapses on the cell body.

Presynaptic Inhibition and Neuromodulation. Postsynaptic activity can also be influenced by events occurring at the presynaptic membrane, such as presynaptic inhibition and neuromodulation. **Presynaptic inhibition** occurs when the release of excitatory neurotransmitter by one neuron is inhibited by the activity of another neuron via an axoaxonic synapse (Figure 11.25). More than one mechanism is involved, but the result is: less neurotransmitter released and bound and smaller EPSPs. In contrast to postsynaptic inhibition by IPSPs, which decreases the excitability of the postsynaptic neuron, presynaptic inhibition is more like a functional synaptic "pruning"; it acts to reduce excitatory stimulation of the postsynaptic neuron.

Neuromodulation occurs when chemicals other than neurotransmitters enhance or dampen neuronal activity. Some *neuromodulators* influence the synthesis, release, degradation, or reuptake of neurotransmitter by a presynaptic neuron. Others influence the sensitivity of the postsynaptic membrane to the neurotransmitter. Many neuromodulators are hormones that act at sites relatively far from their release site.

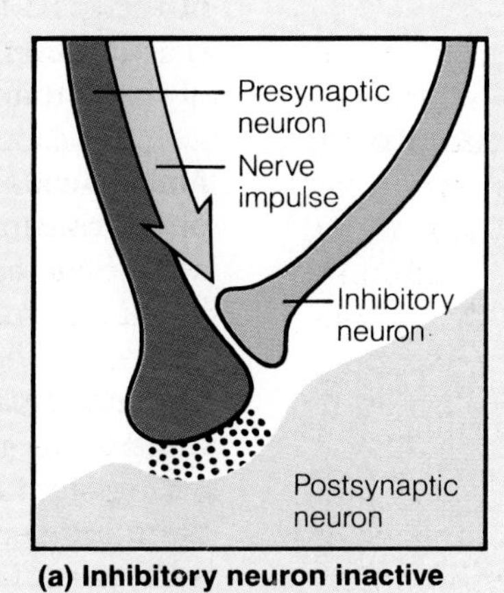

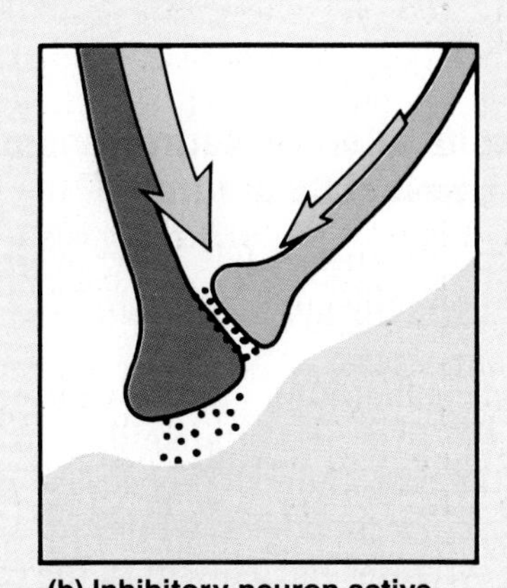

Figure 11.25 Presynaptic inhibition. Presynaptic inhibition is mediated by inhibitory neurons that form axoaxonal synapses with presynaptic terminals at excitatory synapses. (**a**) When the inhibitory neuron is inactive, it exerts no effect on the release of neurotransmitter by the presynaptic neuron. (**b**) When the inhibitory axon is firing, less neurotransmitter is released by the presynaptic axonal terminal.

Neurotransmitters

Neurotransmitters, along with electrical signals, are the "languages" of the nervous system—the means by

the body. Neurotransmitters regulate many body activities and states. Sleep and hunger, memory and mobility, anger and joy all reflect the workings of these communicator molecules. Just as speech defects may hinder interpersonal communication, interferences with neurotransmitter activity may short-circuit the brain's "conversations" or internal talk (see the box on p. 366).

At present, over 100 different chemicals are either known neurotransmitters or considered neurotransmitter candidates. To be recognized as a neurotransmitter, a chemical must meet several criteria. The most important of these are: (1) It must be present in the presynaptic terminal; (2) it must produce ion fluxes (and EPSPs or IPSPs) when applied experimentally to the postsynaptic membrane; and (3) there must be some natural means of removing it from the synaptic junction (enzymatic inactivation or reuptake into the presynaptic axon terminal).

Although most neurons produce and release only one kind of neurotransmitter, some neurons make two, three, or more and may release any one or all of them. It is difficult to understand how multiple neurotransmitters (particularly if they are released simultaneously) interact without producing a jumble of nonsense messages. This question has not yet been resolved, but it seems that the coexistence of more than one neurotransmitter in a single neuron makes it possible for that cell to exert several influences rather than a single discrete effect. We will consider only the effects of single neurotransmitters.

Neurotransmitters are classified in two ways: chemically and functionally. Table 11.3 on p. 365 provides a comprehensive overview of neurotransmitters, some of which are described here. This table will provide a handy reference for you to look back to when the neurotransmitters are mentioned in subsequent chapters.

Classification of Neurotransmitters According to Chemical Structure

Neurotransmitters fall into several chemical classes based on molecular structure.

Acetylcholine (ACh). **Acetylcholine** (a″-sen-til-kō′-lēn) was the first neurotransmitter to be identified. It is still the best understood because it is released at neuromuscular junctions, which are much easier to study than synapses buried within the CNS. ACh is synthesized in axonal terminals from acetic acid and choline in a reaction catalyzed by the enzyme *choline acetyltransferase.* Acetic acid is first bound to coenzyme A (CoA) to form acetyl-CoA, and then the coenzyme is released as ACh is produced:

After binding to the postsynaptic receptors, ACh is degraded by the enzyme *acetylcholinesterase* (a″-sih-tēl-kō-lih-neh′-ster-ās), or *AChE,* present in the synaptic cleft and on postsynaptic membranes. The choline thus produced is recaptured by the presynaptic terminals and reused to synthesize more ACh.

Acetylcholine is released by all neurons that stimulate skeletal muscles and by some neurons of the autonomic nervous system. ACh-releasing neurons are less prevalent in the central nervous system.

Biogenic Amines. The **biogenic** (bī-ō-jeh′-nik) **amines** include the *catecholamines* (ka″-tih-kō-luh-mēnz), exemplified by dopamine, norepinephrine, and epinephrine, and the *indolamines,* which include serotonin and histamine. As illustrated in Figure 11.26, *dopamine* and *norepinephrine* (nor″-eh-pih-neh′-frin) (NE) are synthesized from the amino acid tyrosine in a common pathway consisting of several steps. Apparently, neurons contain only the enzymes controlling the steps needed to produce their own neurotransmitter. Thus, the sequence stops at dopamine in dopamine-releasing neurons, but continues on to NE in NE-releasing neurons. The same pathway is used by the epinephrine-releasing cells of the adrenal medulla. However, since epinephrine is released to the blood, it is considered a hormone rather than a neurotransmitter. *Serotonin* (sayr″-uh-tō′-nin) is also synthesized from tyrosine (in a somewhat different set of enzymatic reactions). *Histamine* is synthesized from the amino acid histidine.

Neurotransmitters of the biogenic amine class are broadly distributed in the brain, where they appear to play a role in emotional behavior and in the regulation of the biological clock. Additionally, catecholamines (particularly NE) are released by some motor neurons of the autonomic nervous system.

Amino Acids. It has been very difficult to prove a neurotransmitter role when the suspect is an **amino acid.** Whereas ACh and the biogenic amines can be localized only in neurons, amino acids occur in all cells of the body and are important in many biochemical reactions. Those amino acids for which a neurotransmitter role is certain include *gamma (γ)-aminobutyric acid (GABA), glycine,* and *glutamate,* but there are sure to be others. Amino acid neurotransmitters have so far been found only in the CNS.

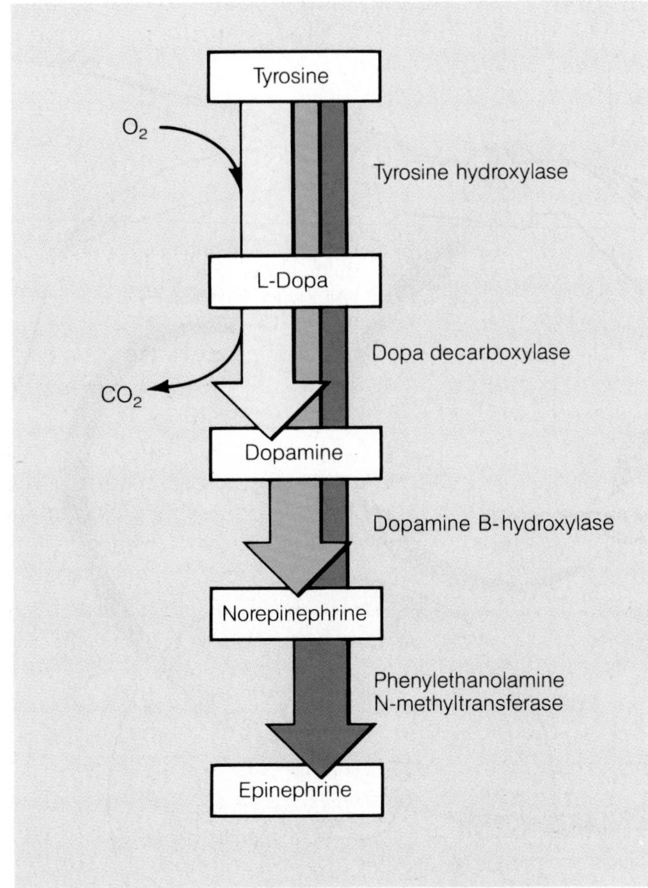

Figure 11.26 Common pathway for synthesis of dopamine, norepinephrine, and epinephrine. How far synthesis proceeds along the pathway depends on the enzymes present in the cell. The production of dopamine and norepinephrine takes place in the axonal terminals of neurons releasing these neurotransmitters. Epinephrine, a hormone, is released (along with norepinephrine) by the cells of the adrenal medulla.

Peptides. The **neuropeptides,** essentially strings of amino acids, include a broad spectrum of molecules with diverse effects. For example, the *beta* (β)-*endorphins* and *enkephalins* (en-keh′-fuh-linz) act as natural opiates or euphorics, reducing our perception of pain under certain stressful conditions. Enkephalin activity increases dramatically in pregnant women in labor. Endorphin release is enhanced when an athlete gets his or her second wind and is probably responsible for the "runner's high." These pain-killing neurotransmitters remained undiscovered until investigators began to study why morphine and other opiates reduce anxiety and pain. As it turned out, these drugs attach to the same receptors that bind natural opiates, producing similar but stronger effects.

Some of the peptides, such as somatostatin and cholecystokinin, are also produced by nonneural body tissues and are widespread in the gastrointestinal tract. Such peptides are commonly referred to as *gut–brain peptides.*

Classification of Neurotransmitters According to Function

This text cannot begin to describe the incredible diversity of functions that neurotransmitters mediate. Therefore, we will limit our discussion to two broad ways of classifying neurotransmitters according to function; more details will be added as appropriate in subsequent chapters.

Excitatory and Inhibitory Neurotransmitters. We can summarize this classification scheme by saying that some neurotransmitters are excitatory, some are inhibitory, and others exert both types of effects, depending on the specific receptor types with which they are bound. For example, the amino acids GABA and glycine are usually inhibitory, while glutamate is typically excitatory. On the other hand, ACh and NE each bind to at least two receptor types that cause opposite effects. For example, acetylcholine is excitatory at neuromuscular junctions with skeletal muscle and inhibitory when released on cardiac muscle. These effects of ACh and NE will be described more fully in Chapter 14.

Ionotropic and Metabotropic Neurotransmitters. Neurotransmitters that open ion channels, as described earlier, are classed as **ionotropic** (ī-ah″-nuh-trō′-pik). These neurotransmitters mediate very rapid responses in the postsynaptic cells by promoting changes in membrane potential. ACh and the amino acid neurotransmitters are ionotropic.

Metabotropic (muh-ta″-buh-trō′-pik) neurotransmitters promote much broader and longer-lasting effects by acting through intracellular second-messenger molecules, and in this way their mechanism of action is similar to that of many hormones. When the second messenger is *cyclic AMP* (Figure 11.27), binding of the neurotransmitter to the receptor activates *adenylate cyclase* (a′-deh-nil-āt sī′-klās), an enzyme that catalyzes the conversion of ATP molecules to cyclic AMP. Cyclic AMP, in turn, can trigger a variety of intracellular effects by activating various enzymes. Some of these enzymes modify membrane proteins, thereby altering membrane permeability and indirectly producing ionotropic effects. Others interact with nuclear proteins that activate genes and induce synthesis of new proteins in the target cell. The biogenic amines and the peptides are metabotropic neurotransmitters.

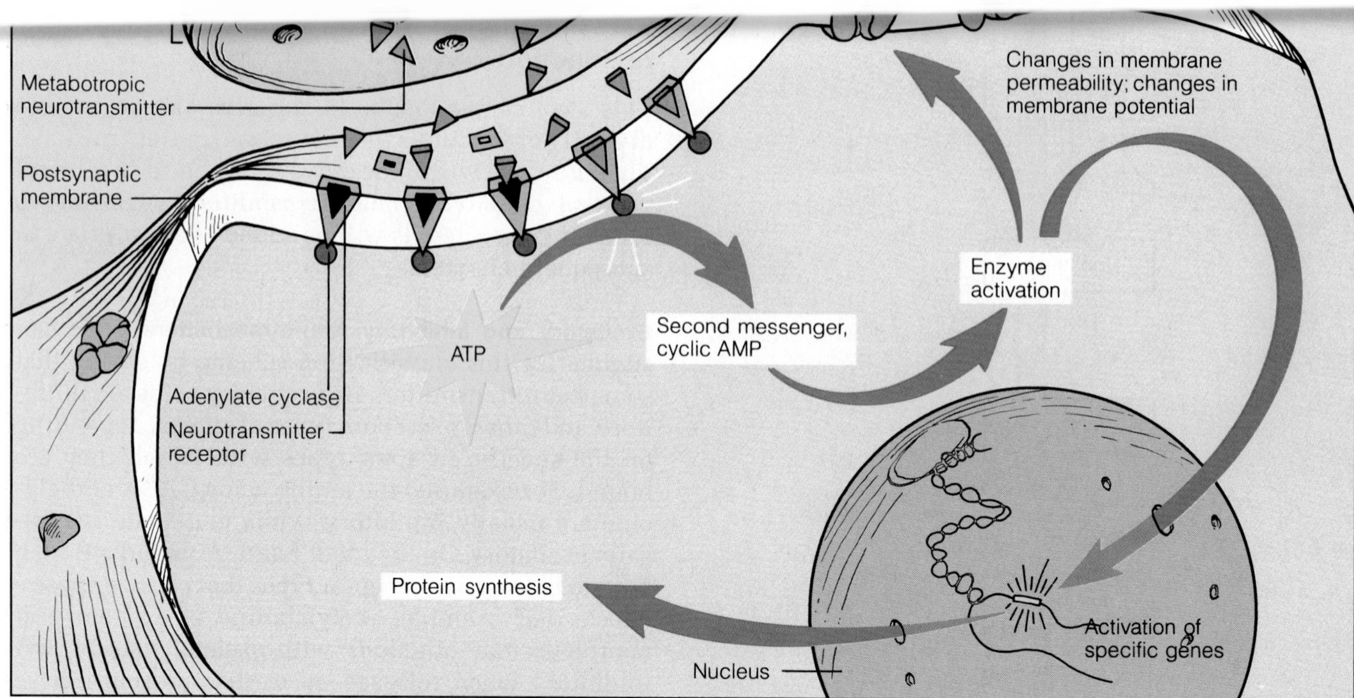

Figure 11.27 Mechanism of a metabotropic neurotransmitter. In the example shown, effects are mediated by cyclic AMP, which acts as a second messenger.

Receptor Potentials

Information about our external and internal worlds presents itself in different energy forms—sound, pressure, chemicals, and so on. Only the sensory receptors are specialized to respond to these energetic stimuli; the rest of the nervous system can respond only to action potentials (which cause neurotransmitter release) or, over short distances, to local graded potentials. If the sensory receptors are to communicate with other neurons, they must translate the information of their stimulus into nerve impulses, the universal neural language.

As the energy of the stimulus is absorbed by the receptor, it is converted, or *transduced*, into an electrical event. That is, the stimulus causes changes in the membrane permeability of the receptor region that result in a local graded potential, essentially the same as an EPSP generated at a postsynaptic membrane in response to binding of neurotransmitters. In both cases, one type of ion channel is opened that allows ion fluxes (usually of sodium and potassium) across the membrane, and summation of local potentials occurs. The graded potential occurring at a receptor membrane is called a **receptor** or **generator potential.**

If the receptor potential is of or above threshold strength when it reaches the voltage-gated sodium ion channels (these are usually just proximal to the receptor membrane, often at the first node of Ranvier), the sodium channels will be opened and an action potential will be generated and propagated to the CNS. Action potentials will continue to be generated as long as a threshold stimulus is applied, and, as explained earlier, the strength of the stimulus is encoded in the frequency of impulse transmission. For example, a blow to the hand would cause faster action potential generation than a soft stroke. Thus, stronger stimuli result in more nerve impulses per second reaching the CNS, which enables the integrative centers to determine the intensity of stimuli acting on the body.

Although receptor potentials are graded potentials that can summate and vary with stimulus intensity, an unusual phenomenon called **adaptation** may occur in certain sensory receptors when they have been responding to an unchanging stimulus. Adaptation uncouples the relation between stimulus strength and frequency of impulse generation; though its mechanism is not fully understood, researchers believe that the responsiveness of the receptor membranes declines with time. As a result, the receptor potentials decrease in frequency, or stop completely. Some

Table 11.3 Neurotransmitters

Neurotransmitter	Functional classes	Sites where secreted	Comments
Acetylcholine	Excitatory to skeletal muscles; excitatory or inhibitory to visceral effectors, depending on receptors bound Ionotropic effects	CNS: basal nuclei and some neurons of motor cortex of brain PNS: all neuromuscular junctions with skeletal muscle; some autonomic motor endings (all preganglionic and parasympathetic postganglionic fibers)	Release increased by insecticides; decreased ACh levels in certain brain areas in Alzheimer's disease
Biogenic Amines			
Norepinephrine	Excitatory or inhibitory, depending on receptor type bound Metabotropic effects	CNS: brain stem, particularly reticular activating system; limbic system; some areas of cerebral cortex PNS: some autonomic motor neurons (postganglionic sympathetic neuron endings)	Release enhanced by amphetamines; removal from synapse blocked by cocaine
Dopamine	Generally excitatory; may be inhibitory in sympathetic ganglia Metabotropic effects	CNS: Substantia nigra of midbrain; hypothalamus; is the principal neurotransmitter of extrapyramidal system PNS: some sympathetic ganglia	Release enhanced by L-dopa and amphetamines; reuptake blocked by cocaine; deficient in Parkinson's disease; may be involved in pathogenesis of schizophrenia
Serotonin	Generally inhibitory Metabotropic effects	CNS: brain stem; hypothalamus; limbic system; cerebellum; pineal gland; spinal cord	Activity blocked by LSD; may play a role in sleep.
Histamine	Metabotropic effects	CNS: Hypothalamus	Also released by mast cells during inflammation and acts as powerful vasodilator
Amino Acids			
GABA (γ-aminobutyric acid)	Generally inhibitory Ionotropic effects	CNS: Purkinje cells of cerebellum; spinal cord	Important in presynaptic inhibition at axoaxonal synapses
Glycine	Generally inhibitory Ionotropic effects	CNS: spinal cord; retina	Inhibited by strychnine
Glutamate	Generally excitatory Ionotropic effects	CNS: Spinal cord; widespread in brain	
Peptides			
Endorphins, enkephalins	Generally inhibitory Metabotropic effects	CNS: widely distributed in brain; hypothalamus; limbic system; pituitary; spinal cord	Effects mimicked by morphine, heroin, and methadone
Substance P	Excitatory Metabotropic effects	CNS: basal nuclei; midbrain; hypothalamus; cerebral cortex PNS: certain sensory neurons of dorsal root ganglia of spinal cord (pain afferents)	Mediates pain transmission
Somatostatin	Generally inhibitory Metabotropic effects	CNS: hypothalamus; retina and other parts of brain Pancreas	Inhibits release of growth hormone; A gut-brain peptide
Cholecystokinin	Possible neurotransmitter	Cerebral cortex Small intestine	A gut-brain peptide

receptors, such as those responding to light pressure, adapt rapidly. Such receptors are referred to as *phasic* (fā-zik) receptors. This explains why we are not usually aware (after a short period) of the feel of our clothing against our skin. However, not all receptors adapt quickly, and some do not adapt at all. Pain receptors and proprioceptors respond continuously to threshold stimuli. Since pain typically warns of potential or actual tissue damage and proprioceptive sense is essential to balance and coordinated skeletal muscle activity, the nonadaptability of these receptors is important. Slowly adapting, or *tonic*, receptors include many of the interoceptors that respond to changing levels of chemicals in the blood.

A CLOSER LOOK Neurotransmitters: Chemical Handcuffs and Boosters

molecules that chemically connect neurons, affect both body and brain. Sleep, thought, rage, movement, and even your smile reflect their "doings." Most factors that specifically affect synaptic transmission do so by enhancing or inhibiting neurotransmitter release or destruction or by blocking the binding of neurotransmitters to receptors. As you might expect, all such factors are chemicals of some type, and in their presence, physical and emotional well-being is always altered. The number of these chemical "handcuffs" or "boosters" of neurotransmitter activity is vast. What is provided here is just a representative sampling.

This brightly colored arrow-poison frog, *Dendrobates pumilio*, lives in rain forests of Central America. Its skin glands secrete highly neurotoxic alkaloids.

Drugs

Extremely small amounts of neurotransmitters underlie all of our moods, from ecstasy to the depths of depression, and the balance is a very delicate one. Norepinephrine and dopamine, both biogenic amines, are our primary "feeling good" neurotransmitters. When brain levels of norepinephrine are too low, we become depressed. This came to light when it was discovered that people taking *reserpine*, an antihypertensive drug used to treat high blood pressure, suffered severe depression, some to the point of becoming suicidal. Investigations revealed that reduced levels of norepinephrine in these patients' brains accounted for the depression.

The revelation that mood has a biochemical basis paved the way for the development of many new classes of *psychotropic* drugs that could be used to modify behavior. The *antipsychotic drugs* (Thorazine and others), used to reduce aggression or anxiety in severely schizophrenic patients, block dopamine receptors and thus reduce dopamine's excitatory effects. However, prolonged use of these drugs exerts a harsh penalty: They reduce dopamine levels in parts of the brain controlling our skeletal muscles, and motor problems result.

The *minor tranquilizers*, such as Valium and Librium, are thought to bind to GABA receptors, enhancing the inhibitory effects of GABA on neural pathways involved with anxiety. The *tricyclic antidepressants* (Elavil and others) prevent reuptake of norepinephrine and serotonin from the synaptic cleft and prolong and enhance their mood-elevating effects.

Not all drugs used clinically target mood. Most *antihypertensive* drugs used today act on peripheral nervous system structures to reduce blood pressure. For example, Inderal blocks norepinephrine's stimulation of smooth muscle of the blood vessels by binding to NE receptors. *Muscle paralytic agents*, such as curare, which blocks ACh receptors on skeletal muscle, are used to induce respiratory paralysis during surgery.

Many substances considered drugs are not prescribed. LSD, a *hallucinogen*, binds to serotonin receptors, preventing the normal serotonin-induced inhibition of certain neural pathways. *Narcotics*, such as heroin and morphine, produce euphoria by attaching to natural

mines and *cocaine* affect, in different ways, the influence of the "pleasure" neurotransmitters norepinephrine, dopamine, and serotonin (see the box in Chapter 14 on p. 461).

Toxins

Strychnine, a poison, blocks glycine receptors in the spinal cord. Since glycine is an inhibitor of motor neurons, blocking its activity results in uncontrolled muscle spasms, convulsions, and respiratory arrest. The botulism toxin (sometimes formed by botulism bacteria in improperly processed canned goods) interferes with the release of acetylcholine, ultimately resulting in respiratory paralysis and cessation of breathing. Many animal species also produce protective chemicals considered to be neurotoxins, which interfere with nervous system activity (see photo). Nerve gas and organophosphate insecticides (malathion) act by interfering with AChE activity. The result is tetanic muscle spasms and neuromuscular "frying."

Antibodies

Autoantibodies formed against one's own receptors can also cause severe muscular problems, as indicated by the disease called *myasthenia gravis*. In myasthenia gravis, skeletal muscle ACh receptors are destroyed by one's own antibodies. Consequently, the skeletal muscles cannot be activated.

Foods

Capsaicin, a chemical irritant found in jalapeño peppers and Hungarian red peppers, causes a massive "dump" of substance P from pain-mediating afferent neuron endings of the tongue and mouth. The result is an excruciatingly (but, to some, delightfully!) painful "hot" sensation in the mouth when these peppers are chewed.

Basic Concepts of Neural Integration

Until now, we have been concentrating on the activities of individual neurons or, at most, a pair of nerve cells that make synaptic contacts with each other. However, neurons function in groups, and each group contributes to still broader neural functions. Thus, the organization of the nervous system is hierarchical, or ladderlike.

Any time you have a large number of *anything*—people included—there must be *integration*; that is, the parts must be fused into a smoothly operating whole. In this section, we will begin to ascend the ladder of **neural integration,** dealing only with the first rung, *neuronal pools* and their basic patterns of communicating information to other parts of the nervous system. Chapter 15 picks up the thread of neural integration once again to examine how sensory inputs are interfaced with motor activity, and looks at the highest levels of neural integration—how we think, remember, and communicate with language.

Organization of Neurons: Neuronal Pools

The millions of neurons in the central nervous system are organized into **neuronal pools,** functional groups of neurons that process and integrate incoming information received from other sources (receptors or different neuronal pools) and then forward, or transmit, the processed information to other destinations.

The composition of a simple type of neuronal pool is shown in Figure 11.28. In this example, one incoming presynaptic fiber branches profusely as it enters the pool and then synapses with several different neurons in the pool. When the incoming fiber is excited, it will excite some postsynaptic neurons and facilitate others. Neurons likely to be most excited and to generate impulses are those closely associated with the incoming fiber, because they receive the bulk of the synaptic contacts. Those neurons are said to be in the *discharge zone* of the neuronal pool. Neurons farther away from the center are not usually excited to threshold by the incoming fiber, but they are facilitated and can easily be brought to threshold by stimuli from another source. Thus, the periphery of the pool is the *facilitated zone*. Keep in mind, however, that our figure is a gross oversimplication. Most neuronal pools consist of thousands of neurons and include inhibitory as well as excitatory neurons.

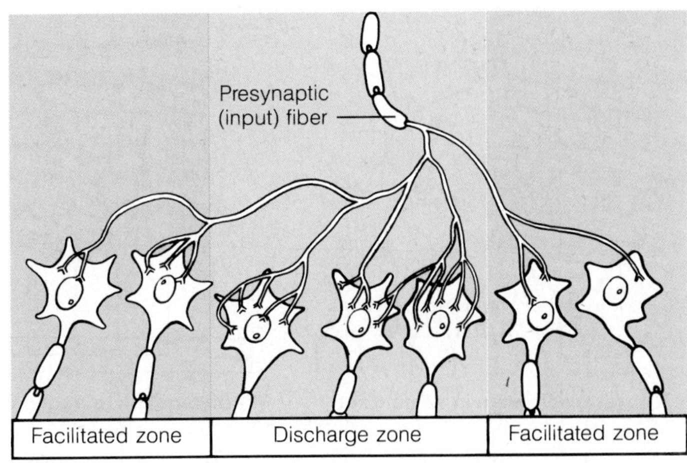

Figure 11.28 Simple neuronal pool. This simplified representation of a neuronal pool contains seven neurons and shows the relative position of postsynaptic neurons in the discharge and facilitated zones. Notice that the presynaptic fiber makes more synapses per neuron with neurons in the discharge zone.

Types of Circuits

Individual neurons within a neuronal pool act both as information recipients and as information transmitters, and synaptic contacts may cause either excitation or inhibition. The patterns of synaptic connections in neuronal pools are called *circuits*, and they determine the functional capabilities of each type of pool. Although some circuits are very complicated, we can gain some understanding of their properties by studying four basic types; diverging, converging, reverberating, and parallel after-discharge circuits. These circuit patterns are shown in simplified form in Figure 11.29.

In **diverging circuits,** one incoming fiber may trigger responses in ever-increasing numbers of neurons farther and farther along in the circuit; thus, diverging circuits are often *amplifying* circuits. Divergence can occur along a single pathway or along several (see Figure 11.29a and b, respectively). These circuits are common in both sensory and motor systems. For example, input from a single sensory receptor may be relayed up the spinal cord and on to several different brain regions at the same time. Likewise, impulses traveling downward from a single neuron of the brain can activate a hundred or more motor neurons in the spinal cord and, consequently, thousands of skeletal muscle fibers.

The pattern of **converging circuits** is opposite that of diverging circuits, but they too are common in both both sensory and motor pathways. In these circuits, the pool neurons receive inputs from several presynaptic neurons, and the circuit as a whole has a funneling, or *concentrating*, effect. Incoming stimuli may

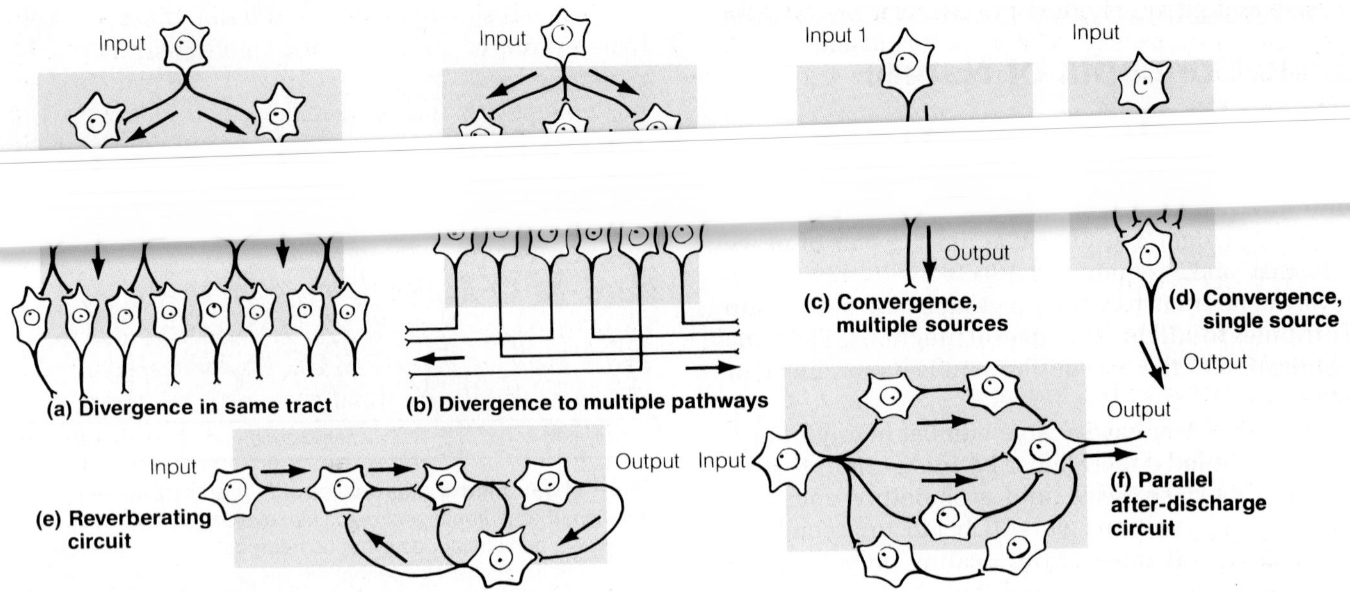

Figure 11.29 Types of circuits in neuronal pools.

converge from many different areas (see Figure 11.29c) or from the same source (see Figure 11.29d), which results in strong stimulation or inhibition. The former condition underlies the observation that different types of sensory stimuli can cause the same ultimate effect or reaction. For instance, seeing the smiling face of their infant, smelling the baby's freshly powdered skin, or hearing the baby gurgle can all trigger a flood of loving feelings in parents.

In **reverberating,** or **oscillating, circuits** (see Figure 11.29e), the incoming signal travels through a chain of neurons, each of which makes collateral synapses with neurons in the previous part of the pathway. The impulses reverberate (are sent through the circuit again and again), giving a continuous output signal until one neuron in the circuit is inhibited and fails to fire. Reverberating circuits may be involved in control of rhythmic activities, such as the sleep-wake cycle, breathing, and certain motor activities (such as arm-swinging when walking). Some researchers believe that such circuits underlie short-term memory. Depending on the arrangement and number of neurons in the circuit, reverberating circuits may continue to oscillate for seconds, hours, or (in the case of the circuit controlling respiratory rhythm) even a lifetime.

In **parallel after-discharge circuits,** the incoming fiber stimulates several neurons arranged in parallel arrays that eventually stimulate a common output cell (see Figure 11.29f). Since each array has a different number of synapses, impulses reach the output cell at different times, creating a burst of impulses called an *after discharge* that lasts 15 ms or more after the initial input has ended. Unlike the reverberating circuit, this type of circuit has no positive feedback, and once all the neurons have fired, circuit activity ends. Parallel after-discharge circuits may be involved in complex, exacting types of mental processing, such as performing mathematics or other types of problem solving.

Patterns of Neural Processing

Processing of inputs in the various circuits is both *serial* and *parallel*. In serial processing, the input travels along a single pathway to a specific destination. In parallel processing, the input travels along several different pathways to be integrated in different CNS regions. Each mode of information handling has unique advantages in the overall scheme of neural functioning.

Serial Processing and the Reflex Arc

In **serial processing,** the whole system works in a predictable all-or-nothing manner. One neuron stimulates the next in sequence, which stimulates the next, and so on, eventually causing a specific, anticipated response. The most clear-cut examples of serial processing are provided by spinal reflexes, but straight-through sensory pathways from the receptors to the brain provide additional examples. Since reflexes are the functional units of the nervous system, and they will be referred to often in later chapters, it is important that they be considered and understood early on.

Reflexes are rapid, automatic responses to stimuli, in which a particular stimulus always causes the same motor response. Reflexes are classified as *somatic*, those that result in the contraction of skeletal muscles, or *autonomic*, those activating cardiac and smooth muscle and glands. Reflexes occur over neural pathways called **reflex arcs.** There are five essential components of all reflex arcs (Figure 11.30a):

1. The *receptor,* which is the site of the stimulus action—the bare or modified dendritic endings of a sensory neuron, for example.

2. The *sensory neuron,* which transmits the afferent impulses to the CNS.

3. The *integration center,* which in the simplest reflex arcs may be a single synapse between the sensory neuron and a motor neuron. In more complex reflexes, it involves multiple synapses with association neurons. The integration center is always within the CNS, exemplified by the spinal cord in Figure 11.30.

4. The *motor neuron,* which conducts efferent impulses from the integration center to an effector organ.

5. The *effector,* the muscle fiber or gland cell that responds to the efferent impulses in a characteristic way (by contraction or secretion).

The left side of Figure 11.30b illustrates a simple two-neuron, or *monosynaptic* (mo″-nō-sih-nap′-tik), reflex mediated by the spinal cord. As you can see, the integration center here is the synapse between the sensory and motor neurons. A simple *polysynaptic reflex arc* involving a single association neuron is shown on the right side of the same figure.

Parallel Processing

In **parallel processing,** inputs are segregated into many different pathways, and information delivered by each pathway is dealt with simultaneously by different parts of the neural circuitry. For example, smelling a pickle (the input) may cause you to remember picking cucumbers on a farm; or it may remind you that you don't like pickles or that you must buy some at the market; or perhaps it will call to mind *all* these thoughts. For each person, parallel processing may trigger some pathways that are unique. The same stimulus—smell, in our example—promotes many responses beyond simple awareness of the pickle smell.

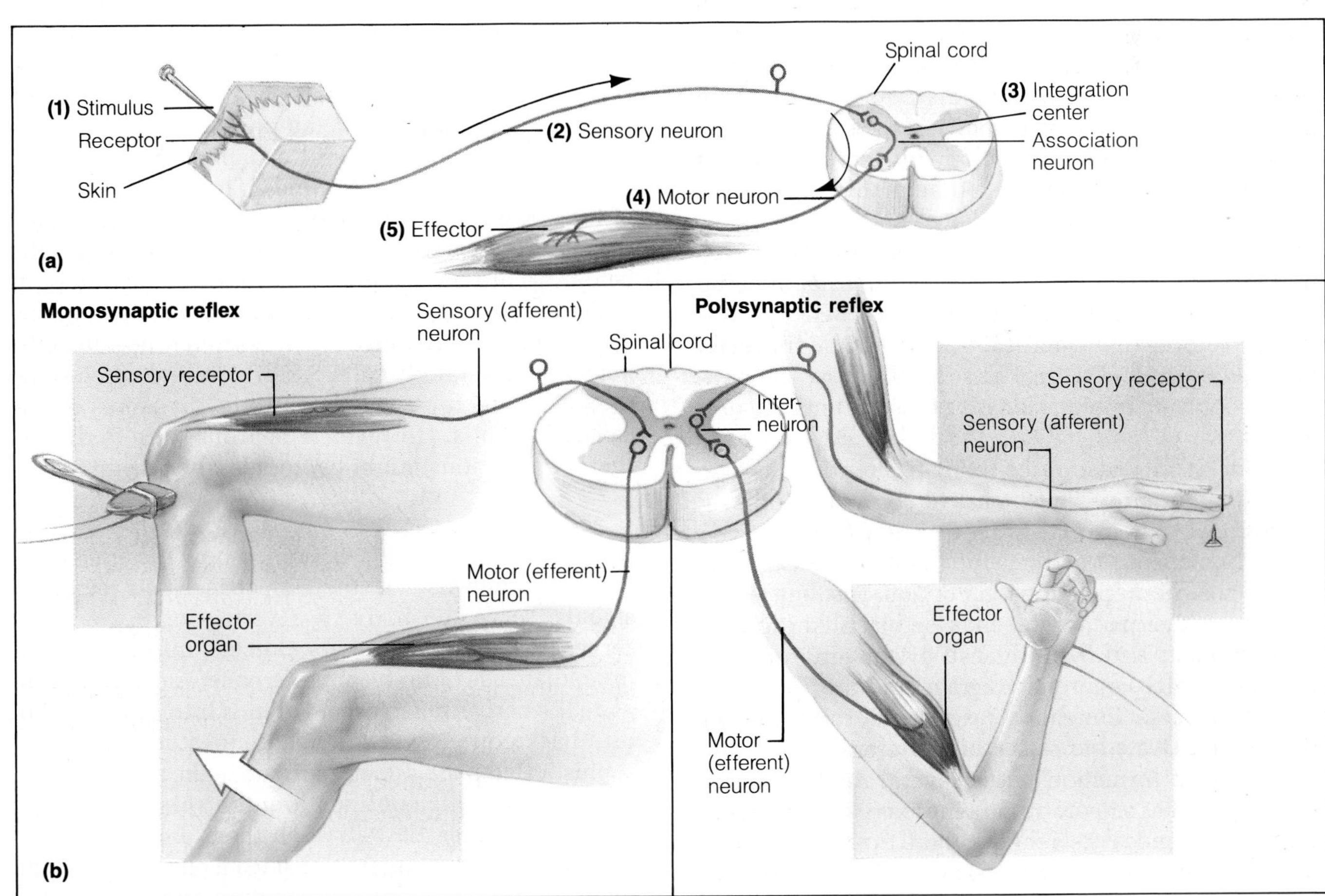

Figure 11.30 Simple reflex arcs. (**a**) Components of all human reflex arcs: receptor, sensory neuron, integration center (one or more synapses in the CNS), motor neuron, and effector. (**b**) Monosynaptic reflex arc (left) and polysynaptic reflex arc (right). The integration center is the spinal cord.

Parallel processing is not redundant, because the circuits do different things with the information, and each "channel" is decoded in relation to all the others to produce a total picture. As will be explained in more detail in Chapters 13 and 15, even simple reflex [illegible] at the cord level, whereas parallel processing of the same input is going on at the same time at higher brain levels, allowing the sensory event to be consciously perceived and responded to if necessary.

Developmental Aspects of Neurons

The nervous system is covered in several chapters, so we will limit our attention here to the development of neurons, beginning with the questions, How do nerve cells originate? and How do they mature or grow up? The second question involves some crucial understandings about how neurons form the correct connections with each other and with effector organs so that appropriate behavior can occur.

As described in greater detail in Chapter 12, the nervous system originates from a dorsal *neural tube* (see Figure 12.2), formed from surface ectoderm. By the fourth week after conception, the neural tube cells are *determined* in their development; that is, they can only become neural tissue. The potential neurons, or *neuroblasts*, then begin a three-phase process of differentiation. First, they proliferate to establish the appropriate number of cells needed for nervous system development. Once this phase ends, the cells become amitotic. During the second phase, the neuroblasts migrate to characteristic positions in the neural tube.

The third phase is cellular differentiation. The appropriate neuron types must be formed, synthesis of appropriate neurotransmitters must occur, and the correct synapses must be made. Little is known about the biochemical specialization of neurons, but we do know that the process depends on establishing synaptic contacts and is influenced by the environment through which neurons migrate and by chemical growth factors. Somehow, specific neurotransmitter genes are activated to cause this differentiation.

Synapse formation poses several difficult questions. How does a neuron's growing axon "know" where to go—and, once it gets there, where to stop and make the appropriate connection? The tortuous growth of an axon toward an appropriate target appears to be guided by multiple signals: pathways laid down by similar, older "pathfinder" neurons; orienting glial fibers; and attracting substances released by the target. For example, skeletal muscles release chemicals that attract motor axons to them. Further, a "sticky" molecule called *nerve cell adhesion molecule (N-CAM)* has been identified on skeletal muscle cells and on the surfaces of glial cells. N-CAM seems to provide [illegible] patterns that when it is blocked by antibodies, developing neural tissue falls into a tangled, spaghetti-like mass, and neural function is hopelessly impaired.

The growth of an axon and its ability to interact with and sense its environment resides in a special structure at its growing tip, the *growth cone* (which also bears N-CAM molecules). The membrane of the growth cone has motile extensions that attach to adjacent structures and pull it along. The growth cone also engulfs molecules from its environment that are transported to the nerve cell body and inform it of the correct pathway for growth and synapse formation. Once the axon has reached its general target area, it must select the right site on the right target cell to form a synapse. The recognition of one nerve cell by another is quite specific and requires signaling by both presynaptic and postsynaptic cells. The exact mechanism of this interchange is far from clear, but trial and error occurs, and gap junctions appear and disappear before permanent chemical synapses are formed.

Neurons that fail to make appropriate synaptic contacts act as if they have been deprived of some essential nutrient and generally die. Besides cell death resulting from unsuccessful synapse formation, *programmed cell death* also appears to be a normal part of the developmental process. Of the neurons formed during the embryonic period, perhaps two-thirds die before we are born. Those that remain constitute our total neural endowment for life. With the notable exception of adult cardiac and skeletal muscle cells, nearly all other body cells can divide, and when some die, they are replaced by mitotic divisions of the remaining cells. However, if neurons were to divide, their connections might be hopelessly disrupted.

* * *

In this chapter, we have examined how the amazingly complex neurons, via their electrical and chemical signals, serve the body in a variety of ways: Some serve as "lookouts," others process information for immediate use or for future reference, and others stimulate the body's muscles and glands into activity. With this background, we are ready to study the most sophisticated mass of neural tissue in the entire body—the brain (and its continuation, the spinal cord), the focus of Chapter 12. Although there are still several nervous system chapters to navigate our way through, Figure 11.31 provides an orienting summary of the ways in which the nervous system and other body systems interact to provide for overall body homeostasis.

Nervous system

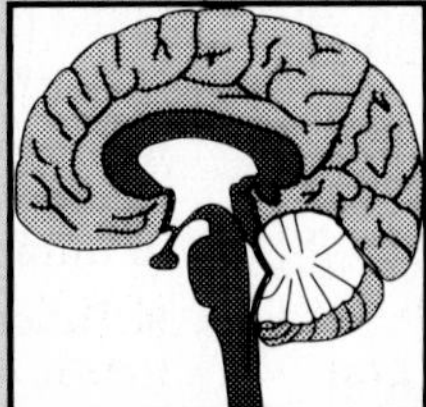

Integumentary system

Serves as heat loss surface

Sympathetic division of the ANS regulates sweat glands and blood vessels of skin (therefore heat loss/ retention)

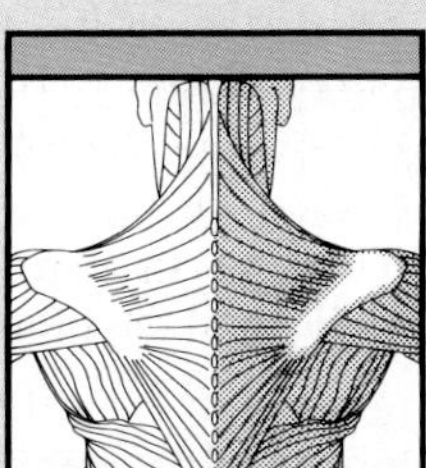

Skeletal system

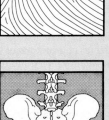

Bones serve as depot for calcium needed for neural function; protects CNS structures

Innervates bones

Muscular system

Skeletal muscles are the effectors of the somatic division

Activates skeletal muscles; maintains muscle health

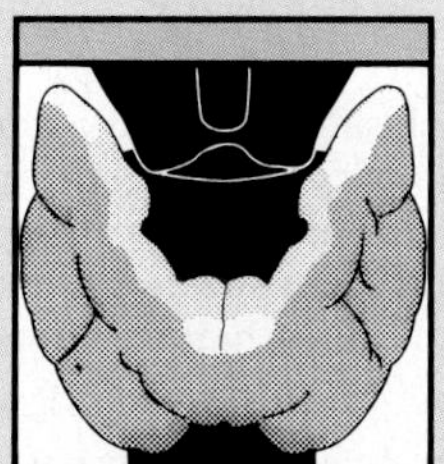

Endocrine system

Hormones influence neuronal metabolism

Sympathetic division of the ANS activates the adrenal medulla; hypothalamus helps regulate the activity of the anterior pituitary gland and produces two hormones

Cardiovascular system

Provides blood containing oxygen and nutrients to the nervous system; carries away wastes

ANS helps regulate heart rate and blood pressure

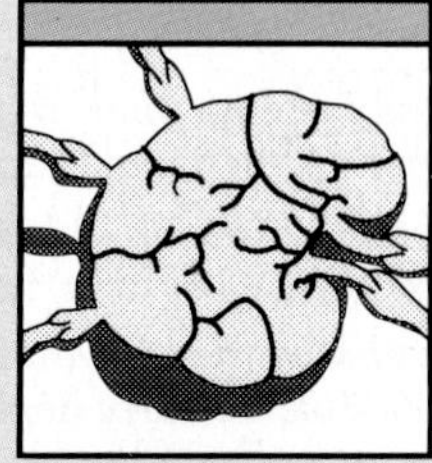

Lymphatic system

Carries away leaked tissue fluids from tissues surrounding nervous system structures

Innervates lymphoid organs

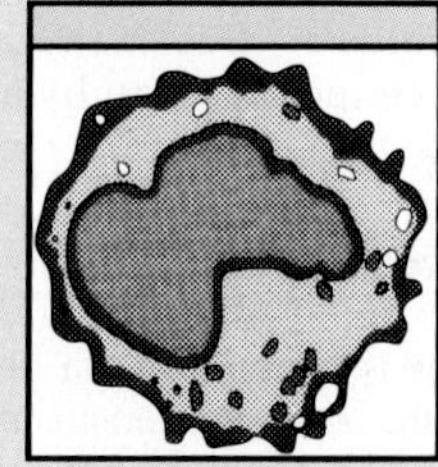

Immune system

Protects all body organs from pathogens (CNS has additional mechanisms as well)

Plays a role in regulating immune function

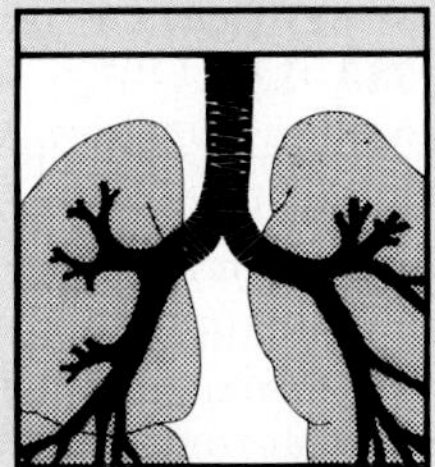

Respiratory system

Provides life sustaining oxygen; disposes of carbon dioxide

Regulates respiratory rhythm and depth

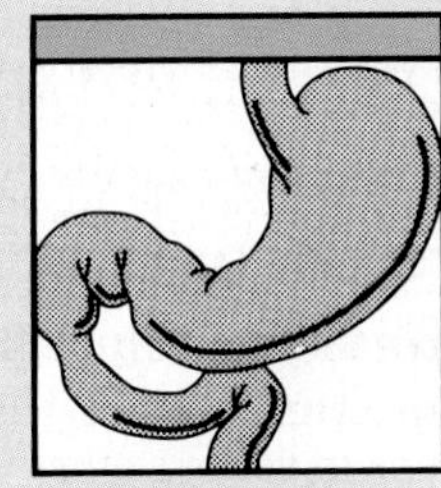

Digestive system

Provides nutrients needed for neuronal health

ANS (particularly the parasympathetic division) regulates digestive mobility and glandular activity

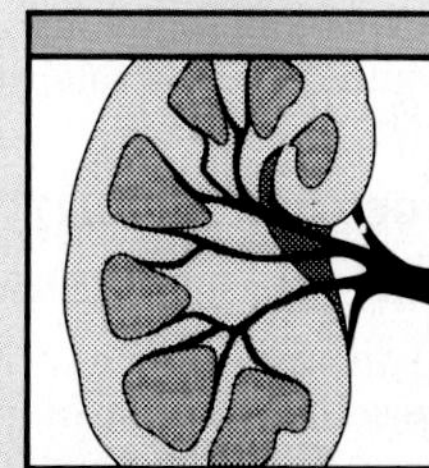

Urinary system

Helps to dispose of metabolic wastes and maintains proper electrolyte composition and pH of blood for neural functioning

ANS regulates bladder emptying, and renal blood pressure

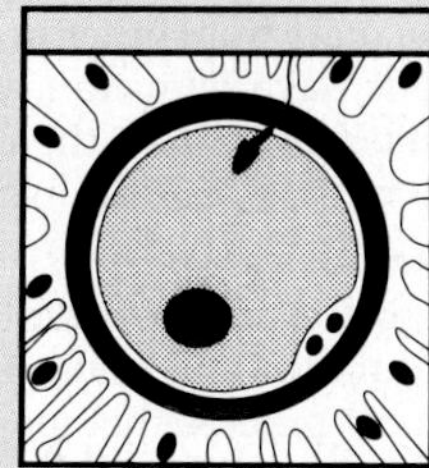

Reproductive system

Testosterone causes masculinization of the brain and underlies sex drive and agressive behavior

ANS regulates sexual erection and ejaculation in males; erection of the clitoris in females

Figure 11.31 Homeostatic interrelationships between the nervous system and other body systems.

Related Clinical Terms

[illegible] specialist in the study of the nervous system, its functions, and its disorders.

Neuropathy (ner-ah'-puh-thē) Any disease of the nervous tissue, but particularly degenerative disease of nerves.

Neuropharmacology (ner"-ō-far"-muh-kah'-luh-jē) Scientific study of the effects of drugs on the nervous system.

Neurotoxin Substance that is poisonous or destructive to nervous tissue; e.g., botulism and tetanus toxins.

Chapter Summary

1. The nervous system bears a major responsibility for maintaining body homeostasis. Its chief functions are to monitor, integrate, and respond to information in the environment.

ORGANIZATION OF THE NERVOUS SYSTEM (pp. 333–334)

1. The nervous system is divided anatomically into the central nervous system (brain and spinal cord) and the peripheral nervous system (cranial and spinal nerves and ganglia).

2. The major functional divisions of the nervous system are the sensory (afferent) division, which conveys impulses to the CNS, and the motor (efferent) division, which conveys impulses from the CNS.

3. The efferent division is subdivided into the somatic (voluntary) system, which serves skeletal muscles, and the autonomic (involuntary) system, which innervates smooth and cardiac muscle and glands.

HISTOLOGY OF NERVOUS TISSUE (pp. 334–347)

Supporting Cells (pp. 334–336)

1. Supporting cells segregate and insulate neurons and assist neurons in various other ways.

2. Neuroglial cells, supporting cells found in the CNS, include astrocytes, microglia, ependymal cells, and oligodendrocytes. Schwann cells and satellite cells are supporting cells found in the PNS.

Neurons (pp. 336–341)

3. Neurons have a cell body and cytoplasmic processes called axons and dendrites.

4. The cell body is the biosynthetic center of the neuron. Except for those found in ganglia, cell bodies are found in the CNS.

5. Some neurons have many dendrites, receptive sites that conduct signals from other neurons toward the nerve cell body; all neurons have one axon, which generates and conducts nerve impulses away from the nerve cell body. Terminal endings of axons release neurotransmitter.

6. Axonal transport is a fast, bidirectional, ATP-dependent process that conducts particulate material; it probably involves microtubules and microfilaments. Axoplasmic flow is a slow transport process that moves soluble proteins and cytoskeletal elements away from the soma. Its mechanism is unknown.

7. Large nerve fibers are myelinated. The myelin sheath is formed in the PNS by Schwann cells and in the CNS by oligodendrocytes. The sheath has gaps called nodes of Ranvier. Unmyeli- [illegible] does not occur.

8. Anatomically, neurons are classified according to the number of processes issuing from the cell body as multipolar, bipolar, or unipolar.

9. Functionally, neurons are classified according to the direction of nerve impulse conduction. Sensory neurons conduct impulses toward the CNS, motor neurons conduct away from the CNS, and association neurons lie between sensory and motor neurons in the neural pathways.

Nerves (p. 342)

10. A nerve is a bundle of neuron fibers in the PNS. Each fiber is enclosed by an endoneurium, fascicles of fibers are wrapped by a perineurium, and the whole nerve is bundled by the epineurium.

11. Nerves are classified according to the direction of impulse conduction of their fibers as sensory, motor, or mixed; most nerves are mixed.

Regeneration of Nervous Tissue (pp. 342–344)

12. Injured PNS fibers may regenerate if Schwann cells proliferate, phagocytize the debris, and form a channel to guide axon sprouts to their original contacts. Fibers in the CNS do not regenerate because the oligodendrocytes fail to aid the process.

Sensory Receptors (pp. 344–347)

13. Sensory receptors are specialized to respond to environmental changes (stimuli). Most sensory receptors consist of modified dendrites or dendritic end organs of sensory neurons. A sense organ is a complex structure, consisting of sensory receptors and other cells, that serves a specific receptive process.

14. The general sensory receptors include the simple receptors for pain, touch, pressure, and temperature found in the skin, as well as those found in skeletal muscles and tendons and in the visceral organs. Complex sense organs serve the special senses (vision, hearing, equilibrium, smell, and taste).

15. Receptors are classified according to location as exteroceptors, interoceptors, and proprioceptors and according to stimulus detected as mechanoreceptors, thermoreceptors, photoreceptors, chemoreceptors, and nociceptors.

16. The general sensory receptors are classified structurally as free or corpuscular nerve endings. The corpuscular (encapsulated) nerve endings include Meissner's corpuscles, Pacinian corpuscles, Krause's end-bulbs, Ruffini's corpuscles, muscle spindles, Golgi tendon organs, and joint kinesthetic receptors.

NEUROPHYSIOLOGY (pp. 347–367)

Basic Principles of Electricity (pp. 347–348)

1. The measure of the potential energy of separated electrical charges is called voltage (V) or potential. Current (I) is the flow of electrical charge from one point to another. Resistance (R) is hindrance to current flow. The relationship among these is given by Ohm's law: $I = V/R$.

2. In the body, electrical charges are provided by ions; cellular plasma membranes provide resistance to ion flow. The membranes contain passive (open) and active (gated) channels.

The Resting Membrane Potential: The Polarized State (pp. 348–349)

3. A resting neuron exhibits a voltage called the resting membrane potential, which is -70 mV (inside negative) owing to differences in sodium and potassium ion concentrations inside and outside the cell.

4. The ionic differences result from greater permeability of the membrane to potassium than to sodium and from the operation of the sodium-potassium pump, which ejects $3Na^+$ from the cell for each $2K^+$ transported in.

Membrane Potentials That Act as Signals (pp. 350–356)

5. Depolarization is a reduction in membrane potential (inside becomes less negative); hyperpolarization is an increase in membrane potential (inside becomes more negative).

6. Graded potentials are small, brief, local changes in membrane potential that act as short-distance signals. The current produced dissipates with distance.

7. An action potential, or nerve impulse, is a large, but brief, depolarization signal that underlies long-distance neural communication. It is an all-or-none phenomenon.

8. Generation of an action potential involves three phases: (1) Increase in sodium permeability and reversal of the membrane potential to approximately +30 mV (inside positive). Local depolarization opens voltage-regulated sodium gates; at threshold, depolarization becomes self-generating (driven by sodium ion influx). (2) Decrease in sodium permeability. (3) Increase in potassium permeability and repolarization.

9. In nerve impulse propagation, each action potential provides the depolarizing stimulus for triggering an action potential in the next membrane patch. Regions that have just generated action potentials are refractory; hence, the nerve impulse is propagated in one direction only.

10. If threshold is reached, an action potential is generated; if not, the depolarization remains local.

11. Action potentials are independent of stimulus strength: strong stimuli cause action potentials to be generated more frequently but not with greater amplitude.

12. During the absolute refractory period, a neuron is incapable of responding to another stimulus because it is already generating an action potential. The relative refractory period is the time when the neuron's threshold is elevated because repolarization is ongoing.

13. In unmyelinated fibers, action potentials are produced in a wave all along the axon; in myelinated fibers, depolarization occurs only at nodes of Ranvier and is propagated more rapidly by saltatory conduction.

14. Nerve fibers are classified according to their diameter, degree of myelination, and speed of conduction as group A, B, or C fibers.

The Synapse (pp. 356–358)

15. A synapse is a functional junction between neurons. The information-transmitting neuron is the presynaptic neuron; the neuron beyond the synapse is the postsynaptic neuron.

16. Electrical synapses allow ions to flow directly from one neuron to another; the cells are electrically coupled.

17. Chemical synapses are sites of neurotransmitter release and binding. When the impulse reaches the presynaptic axonal terminals, Ca^{2+} enters the cell and mediates neurotransmitter release. Neurotransmitters diffuse across the synaptic cleft and attach to postsynaptic membrane receptors, opening ion channels.

Postsynaptic Potentials and Synaptic Integration (pp. 358–361)

18. Binding of neurotransmitter at excitatory chemical synapses results in local graded depolarizations called EPSPs, caused by the opening of channels that allow simultaneous passage of Na^+ and K^+.

19. Neurotransmitter binding at inhibitory chemical synapses results in hyperpolarizations called IPSPs, caused by the opening of K^+ and/or Cl^- gates. IPSPs drive the membrane potential farther from threshold.

20. EPSPs and IPSPs summate temporally and spatially. The membrane of the axon hillock acts as a neuronal integrator.

21. Presynaptic inhibition is mediated by axoaxonal synapses that reduce the amount of neurotransmitter released by the inhibited neuron. Neuromodulation occurs when chemicals (often other than neurotransmitters) alter neuronal or neurotransmitter activity.

Neurotransmitters (pp. 362–363)

22. The major classes of neurotransmitters based on chemical structure are acetylcholine, biogenic amines, amino acids, and peptides.

23. Functionally, neurotransmitters are classified as (1) inhibitory or excitatory (or both) and (2) ionotropic or metabotropic. Ionotropic neurotransmitters cause channel opening. The effects of metabotropic neurotransmitters are mediated through second messengers and are complex.

Receptor Potentials (pp. 364–365)

24. Sensory receptors transduce (convert) stimulus energy into action potentials. As the receptor membrane absorbs energy, a single type of ion channel is opened, which allows ionic movements across the receptor membrane and results in a receptor potential that resembles an EPSP. If the stimulus is of threshold strength, an action potential is generated and propagated.

25. Adaptation is the reduced responsiveness of a sensory receptor to unchanging stimulation. Adaptation does not occur in pain receptors or proprioreceptors.

BASIC CONCEPTS OF NEURAL INTEGRATION (pp. 367–370)

Organization of Neurons: Neuronal Pools (p. 367)

1. CNS neurons are organized into several types of neuronal pools.

Types of Circuits (pp. 367–368)

2. Distinguishing patterns of synaptic connections are called circuits. The four basic circuit types are diverging, converging, reverberating, and parallel after-discharge.

Patterns of Neural Processing (pp. 368–370)

3. In serial processing, one neuron stimulates the next in sequence, producing specific, predictable responses, as in spinal reflexes. A reflex is a rapid, involuntary motor response to a stimulus. Reflexes are classified as autonomic or somatic.

4. Reflexes are mediated over neural pathways called reflex arcs. The minimum number of elements in a reflex arc is five: receptor, sensory neuron, integration center, motor neuron, and effector.

5. In parallel processing, which underlies complex mental functions, impulses are sent along several pathways to different integration centers.

DEVELOPMENTAL ASPECTS OF NEURONS (p. 370)

1. Neuron development involves proliferation, migration, and cellular differentiation. Cellular differentiation entails neuron specialization, specific neurotransmitter synthesis, and synapse formation.

2. Axonal outgrowth and synapse formation are guided by other neurons, glial fibers, and chemicals (such as N-CAM).

3. Neurons that do not make appropriate synapses die, and approximately two-thirds of neurons formed in the embryo undergo programmed cell death before birth.

Review Questions

Multiple Choice/Matching

2. Match the names of the supporting cells found in column B with the appropriate descriptions in column A.

Column A	Column B
______ (1) myelinates nerve fibers in the CNS	(a) astrocyte
______ (2) lines brain cavities	(b) ependymal cell
______ (3) myelinates nerve fibers in the PNS	(c) microglia
______ (4) CNS phagocytes	(d) oligodendrocyte
______ (5) may regulate the ionic composition of the extracellular fluid	(e) satellite cell
	(f) Schwann cell

3. The connective tissue sheath that surrounds a fascicle of nerve fibers is the (a) epineurium, (b) endoneurium, (c) perineurium, (d) neurilemma.

4. Assume that an EPSP is being generated on the dendritic membrane. Which will occur? (a) specific Na^+ gates will open, (b) specific K^+ gates will open, (c) a single type of channel will of channel will open, permitting simultaneous flow of Na^+ and K^+, (d) Na^+ gates will open first and then close as K^+ gates open.

5. The velocity of nerve impulse conduction is greatest in (a) heavily myelinated, large-diameter fibers, (b) myelinated, small-diameter fibers, (c) unmyelinated, small-diameter fibers, (d) unmyelinated, large-diameter fibers.

6. Chemical synapses are characterized by all of the following except (a) the release of neurotransmitter by the presynaptic membranes, (b) the postsynaptic membranes bear receptors that bind neurotransmitter, (c) ions flow through protein channels from the presynaptic to the postsynaptic neuron, (d) a fluid-filled gap separating the neurons.

7. Biogenic amine neurotransmitters include all but (a) norepinephrine, (b) acetylcholine, (c) dopamine, (d) serotonin.

8. The neuropeptides that act as natural opiates are (a) substance P, (b) somatostatin, (c) cholecystokinin, (d) enkephalins.

9. Different types of receptor activity are described below. Characterize each by choosing the appropriate letter and number from keys A and B.

Key A: (a) exteroceptor
(b) interoceptor
(c) proprioceptor

Key B: (1) chemoreceptor
(2) mechanoreceptor
(3) nociceptor
(4) photoreceptor
(5) thermoreceptor

______, ______ You are enjoying an ice cream cone.
______, ______ You have just scalded yourself with hot coffee.
______, ______ The retinas of your eyes are stimulated.
______, ______ You bump (lightly) into someone.
______, ______ You have been lifting weights and experience a certain sensation in your upper limbs.

10. The large onion-shaped receptors that are found in the dermis and subcutaneous tissue and that respond to deep pressure are (a) Merkel discs, (b) Pacinian receptors, (c) free nerve endings, (d) Krause's end-bulbs.

11. Proprioceptors include all of the following except (a) muscle spindles, (b) Golgi tendon organs, (c) Merkel discs, (d) joint kinesthetic receptors.

12. Identify the neuronal circuits described by choosing the correct response from the key.

Key: (a) converging, (b) diverging, (c) parallel after-discharge, (d) reverberating

______ (3) Many neurons influence a few neurons.
______ (4) May be involved in exacting types of mental activity.

Short Answer Essay Questions

13. Explain both the anatomical and functional divisions of the nervous system. Include the subdivisions of each.

14. (a) Describe the composition and function of the cell body. (b) How are axons and dendrites alike? In what ways (structurally and functionally) do they differ?

15. (a) Compare and contrast axoplasmic flow and axonal transport relative to the substances transported, direction, rate, and mechanism if known.

16. (a) What is myelin? (b) How does the myelination process differ in the CNS and PNS?

17. Explain why damage to peripheral nerve fibers is often reversible, whereas damage to CNS fibers rarely is.

18. (a) Contrast unipolar, bipolar, and multipolar neurons structurally. (b) Indicate where each is most likely to be found.

19. What is the polarized membrane state? How is it maintained? (Note the relative roles of both passive and active mechanisms.)

20. Describe the events that must occur to generate an action potential. Indicate how the ionic gates are controlled, and explain why the action potential is an all-or-none phenomenon.

21. Since all action potentials generated by a given nerve fiber have the same magnitude, how does the CNS "know" whether a stimulus is strong or weak?

22. (a) Explain the difference between an EPSP and an IPSP. (b) What specifically determines whether an EPSP or IPSP will be generated at the postsynaptic membrane?

23. Since at any moment, a neuron is likely to have thousands of neurons releasing neurotransmitters at its surface, how is neuronal activity (to fire or not to fire) determined?

24. The effects of neurotransmitter binding are very brief. Explain.

25. During a neurobiology lecture, a professor repeatedly refers to type A and type B fibers, absolute refractory period, and nodes of Ranvier. Define these terms.

26. How is a receptor potential like an EPSP? How is it different?

27. Define reflex and list the minimum number of components in a reflex arc.

28. Briefly describe the three stages of neuron development.

29. What factors appear to guide the outgrowth of an axon and its ability to make the "correct" synaptic contacts?

Clinical Application Questions

30. General and local anesthetics block action potential generation, thereby rendering the nervous system quiescent while surgery is performed. What specific process do anesthetics impair, and how does this interfere with nerve transmission?

31. When admitted to the emergency room, John was holding his right hand, which had a deep puncture hole in its palm. He explained that he had fallen on a nail while exploring a barn. John was given an antitetanus shot to prevent neural complications. Tetanus bacteria fester in deep, dark wounds, but how do they travel to neural tissue?

CHAPTER 12

The Central Nervous System

Chapter Outline and Student Objectives

The Brain (pp. 376–404)

1. Describe the process of brain development.
2. Name the major brain regions.
3. Define the term *ventricle* and indicate the location of the four ventricles of the brain.
4. Indicate the major lobes, fissures, and functional areas of the cerebral cortex.
5. Explain lateralization of hemispheric function.
6. Differentiate between commissures, association fibers, and projection fibers.
7. State the general function of the cerebral nuclei.
8. Describe the location of the diencephalon, and name its subdivisions.
9. Identify the three major regions of the brain stem, and note the general function of each area.
10. Describe the structure and function of the cerebellum.
11. Localize the reticular formation and the limbic system and explain the role of each functional system.
12. Describe how meninges, cerebrospinal fluid, and the blood-brain barrier protect the central nervous system.
13. Describe the formation of cerebrospinal fluid, and follow its circulatory pathway.
14. Distinguish between a concussion and a contusion.
15. Describe the cause (if known) and major signs and symptoms of cerebrovascular accidents, Alzheimer's disease, and multiple sclerosis.

The Spinal Cord (pp. 404–413)

16. Describe the embryonic development of the spinal cord.
17. Describe the gross and microscopic structure of the spinal cord.
18. List the major spinal cord tracts, and describe each in terms of its origin, termination, and function.
19. Distinguish between flaccid and spastic paralysis, and between paralysis and paresthesia.

Diagnostic Procedures for Assessing CNS Dysfunction (p. 414)

20. List and explain several techniques used to diagnose brain disorders.

Developmental Aspects of the Central Nervous System (pp. 414–415)

21. Describe several maternal factors that can impair development of the nervous system in an embryo.
22. Explain how true senility and reversible senility differ.

Preview of Selected Key Terms

Central nervous system (CNS) The brain and spinal cord.

Brain ventricle (*ventr* = hollow cavity) Fluid-filled cavity of the brain.

Cerebrum (suh-rē′-brum) The cerebral hemispheres.

Brain stem Collectively the midbrain, pons, and medulla of the brain.

Cerebral cortex (suh-rē′-brul) The outer gray matter region of the cerebral hemispheres.

Cerebellum Brain region most involved in producing smooth, coordinated skeletal muscle activity.

Basal (cerebral) nuclei Gray matter areas located deep within the white matter of the cerebral hemispheres.

Limbic system (*limbus* = ring) Functional brain system involved in emotional response.

Reticular formation Functional system that spans the brain stem; involved in regulating sensory input to the cerebral cortex, cortical arousal, and control of motor behavior.

Choroid plexus (kor′-oyd plek′-sus) (*plex* = interweaving) A capillary knot that protrudes into a brain ventricle; involved in forming cerebrospinal fluid.

Meninges (meh-nin′-jēz) (*mening* = membrane) Protective coverings of the CNS; from the most external to the most internal, the dura mater, arachnoid, and pia mater.

Cerebrospinal fluid (suh-rē″-brō-spī′nul) Plasmalike fluid that fills the cavities of the CNS and surrounds the CNS externally; protects the brain and spinal cord.

H[illegible] a dizzying number of incoming and outgoing calls. Nowadays, many envision the CNS as a kind of computer. These analogies help explain the workings of the spinal cord, which acts as a relay and reflex station, and some of the input–output (sensory–motor) responses of the brain. But the computer that can duplicate the complexity and flexibility of the human brain has yet to be developed. Whether we view it as an evolved biological organ, an impressive computer, or simply a miracle, the human brain is certainly one of the most amazing things known and is the basis of each person's unique behavior.

Cephalization (seh″-fuh-lih-zā′-shun) has occurred during the course of brain evolution. That is, there has been an elaboration of the rostral, or anterior, portion of the central nervous system, accompanied by an increased number of neurons in the head. This phenomenon has reached its highest level in the human brain.

This chapter focuses on the structure of the central nervous system (CNS) and touches on the functions associated with its specific anatomical regions. Its more complex integrative functions, such as sensory and motor integration, sleep-wake cycles, and higher mental functions such as consciousness, memory, and language processing are deferred to Chapter 15.

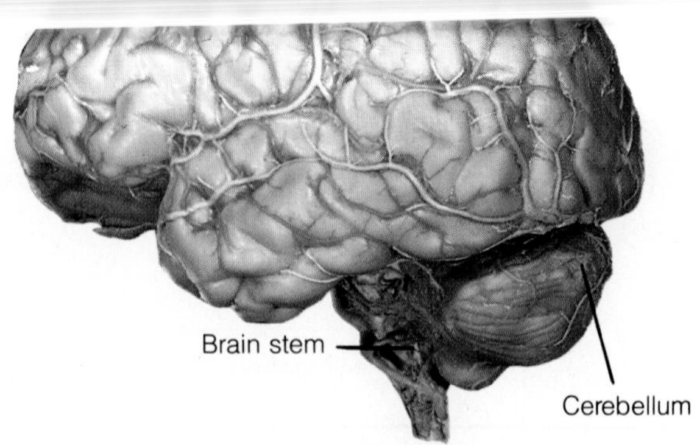

Figure 12.1 Lateral aspect of a human brain.

The Brain

The unimpressive appearance of the human brain (see Figure 12.1) gives few hints of its remarkable abilities. It is about two good fistfuls of pinkish gray tissue, wrinkled like a walnut, and somewhat the consistency of cold oatmeal. The average adult male's brain weighs about 1600 g (3.5 pounds); that of a woman averages 1450 g; but in terms of brain weight per body weight, males and females have equivalent brain sizes. However, size does not seem to be the determinant of brainpower; the complexity of the neural "wiring" appears to be much more important. Einstein's brain was only average in size, and he was an acknowledged genius, while Neanderthal man's brain (which produced, at most, a limited technology) is estimated to have been 15% larger than that of humans living today.

Embryonic Development of the Brain

We will break from our usual pattern here and discuss embryonic brain development first, because the terminology used for the structural divisions of the adult brain is easier to understand when you comprehend brain embryology. The earliest phase of brain development is shown in Figure 12.2. Starting in the third week of pregnancy, the ectoderm thickens along the dorsal midline axis of the embryo to form the **neural plate,** which eventually gives rise to all neural tissues. The neural plate then invaginates, forming a groove flanked by **neural folds.** As the groove deepens, the superior edges of the neural folds fuse to form the **neural tube,** which soon detaches from the surface ectoderm and assumes a deeper position. The neural tube is formed by the fourth week of pregnancy and differentiates rapidly into the organs of the CNS, the brain forming anteriorly (rostrally) and the spinal cord developing from the posterior (caudal) portion of the neural tube. Small masses of neural fold cells migrate laterally from between the surface ectoderm and the neural tube to form the tissue called the **neural crest,** which gives rise to sensory neurons and some autonomic neurons destined to reside in ganglia.

As soon as the neural tube is formed, its anterior end begins to expand more rapidly than the remaining portion, and constrictions appear that delineate three **primary brain vesicles** (Figure 12.3): the **prosencephalon** (pro″-zin-seh′-fuh-lon), or **forebrain;** the **mesencephalon** (meh″-zin-seh′-fuh-lon), or **midbrain;** and the **rhombencephalon** (rom″-bin-seh′-fuh-lon), or **hindbrain.** (Notice that the word stem *encephalo* always refers to the brain.) The remainder of the neural tube becomes the spinal cord; we will discuss its development later in the chapter.

Figure 12.2 Development of the neural tube from embryonic ectoderm. Dorsal superficial views of the embryo are shown at the left; transverse sections of the embryo are shown at the right. (**a**) Formation of the neural plate from surface ectoderm. (**b–d**) Progressive development of the neural plate into the neural groove (flanked by neural folds) and then the neural tube. The neural tube forms CNS structures; the neural crest cells form some PNS structures.

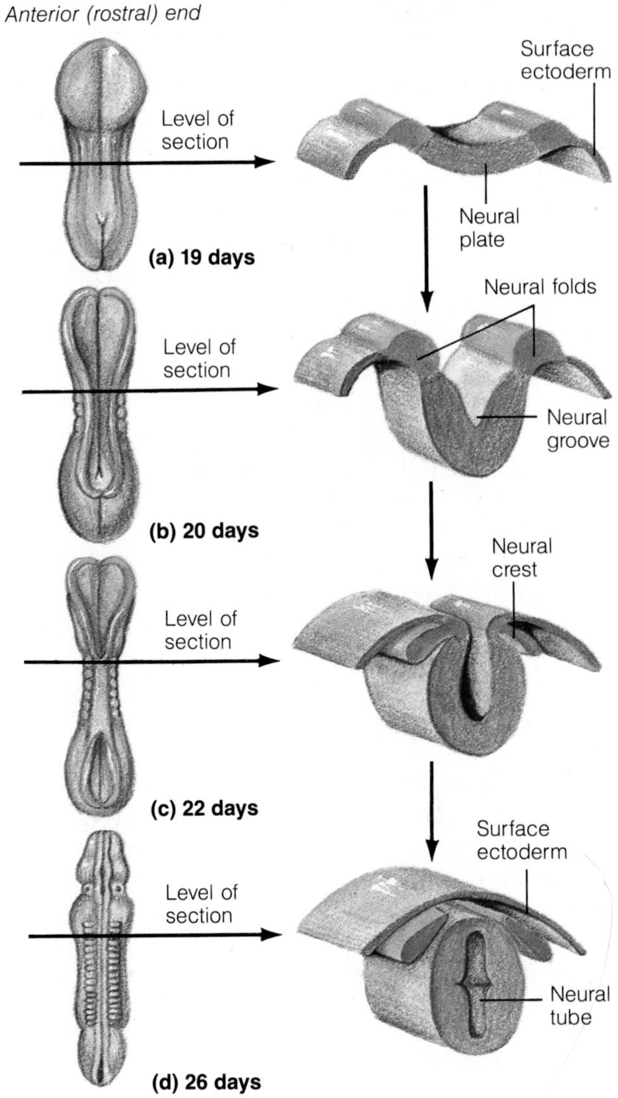

Anterior (rostral) end
Level of section
Surface ectoderm
Neural plate
(a) 19 days
Neural folds
Level of section
Neural groove
(b) 20 days
Neural crest
Level of section
(c) 22 days
Surface ectoderm
Level of section
Neural tube
(d) 26 days

Figure 12.3 Embryonic development of the human brain. (**a**) The neural tube becomes subdivided into (**b**) the primary brain vesicles, which subsequently form (**c**) the secondary brain vesicles, which differentiate into (**d**) the adult brain structures. (**e**) The adult structures derived from the neural canal.

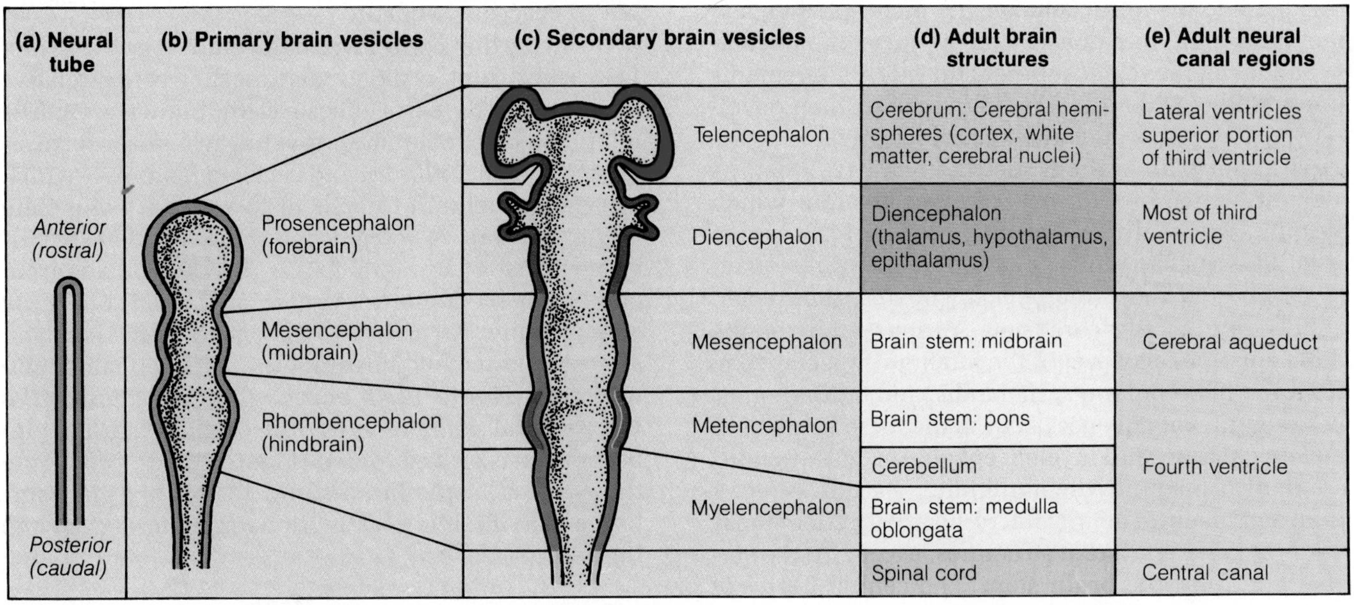

(a) Neural tube	(b) Primary brain vesicles	(c) Secondary brain vesicles	(d) Adult brain structures	(e) Adult neural canal regions
Anterior (rostral)	Prosencephalon (forebrain)	Telencephalon	Cerebrum: Cerebral hemispheres (cortex, white matter, cerebral nuclei)	Lateral ventricles superior portion of third ventricle
		Diencephalon	Diencephalon (thalamus, hypothalamus, epithalamus)	Most of third ventricle
	Mesencephalon (midbrain)	Mesencephalon	Brain stem: midbrain	Cerebral aqueduct
	Rhombencephalon (hindbrain)	Metencephalon	Brain stem: pons	Fourth ventricle
			Cerebellum	
		Myelencephalon	Brain stem: medulla oblongata	
Posterior (caudal)			Spinal cord	Central canal

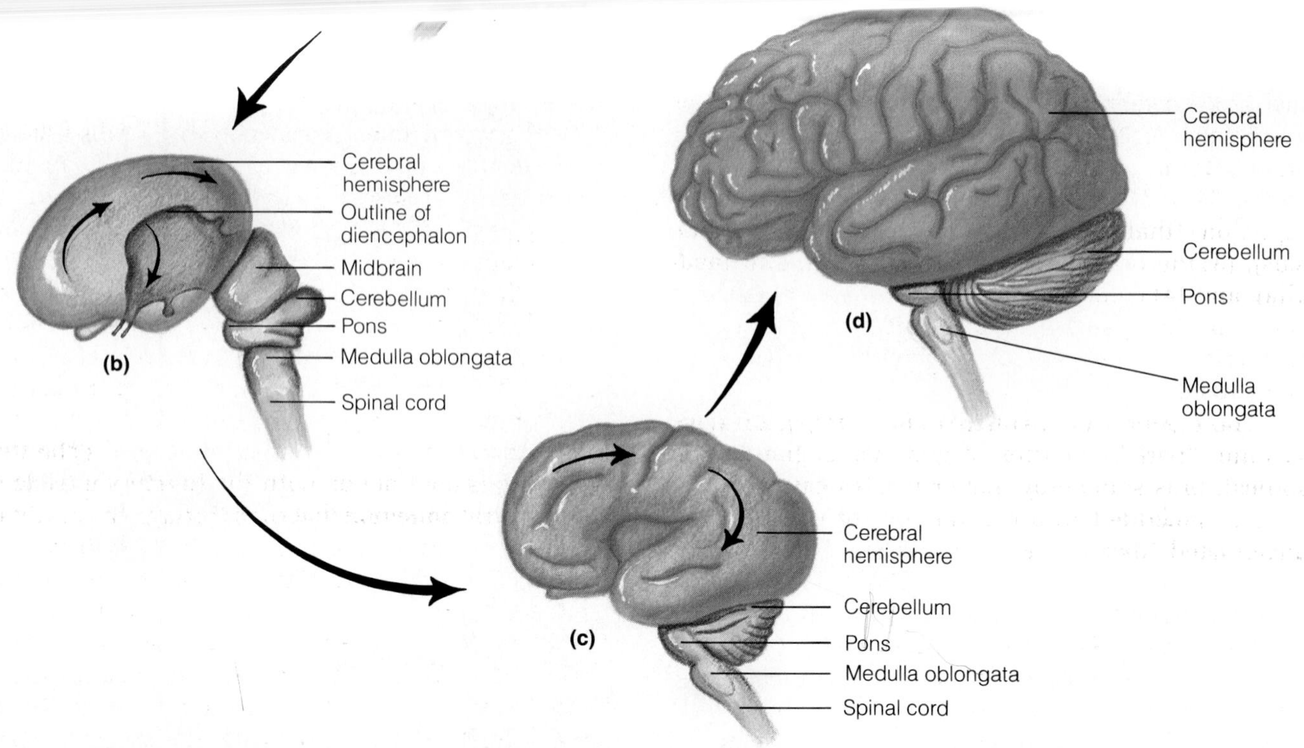

Figure 12.4 Effect of space restriction on brain development. (**a**) The appearance of the two major flexures by the fifth week of development causes the telencephalon and diencephalon to be angled toward the brain stem. Development of the cerebral hemispheres at (**b**) 13 weeks, (**c**) 26 weeks, and (**d**) birth. Initially, the cerebral surface is smooth; convolutions become more obvious as development continues. The superolateral growth of the cerebral hemispheres ultimately completely encloses the diencephalon and the superiormost aspect of the brain stem.

By the fifth week, five brain regions, the **secondary brain vesicles,** are evident. The forebrain has divided into the **telencephalon** and **diencephalon,** and the hindbrain has constricted to form the **metencephalon** and **myelencephalon;** the midbrain remains undivided. Each of the five brain regions then develops rapidly to produce the major recognizable structures of the adult brain. Notice in Figure 12.3 that the greatest change occurs in the telencephalon, which sprouts two swellings that project anteriorly, somewhat like Mickey Mouse's ears. These paired expansions become the *cerebral hemispheres,* referred to collectively as the **cerebrum.** Various areas of the diencephalon, also part of the forebrain, specialize to form the *hypothalamus, thalamus,* and *epithalamus.* Less significant changes occur in the mesencephalon, metencephalon, and myelencephalon as these regions are transformed into the *midbrain, pons and cerebellum,* and *medulla oblongata,* respectively. All of these midbrain and hindbrain structures, except for the cerebellum, form the **brain stem.** The central cavity of the neural tube remains continuous and becomes enlarged in four areas to form the *ventricles* of the brain, described shortly.

During this period when the brain is growing rapidly, significant changes also occur in the relative positions of its parts. Because the brain's growth is restricted by a membranous skull, two major flexures develop—the *midbrain* and *cervical flexures*—which cause the forebrain to be bent toward the brain stem (Figure 12.4a). A second consequence of restricted space is that the cerebral hemispheres, hindered from continuing their anterior course, are forced superiorly and laterally. As a result, they grow back over and almost completely envelop the diencephalon and midbrain (Figure 12.4b and c). Continued growth of the cerebral hemispheres causes their surfaces to become creased and folded (Figure 12.4d), producing their typical *convolutions* and increasing their surface area, which allows more neurons to occupy the limited space.

Regions of the Brain

There are various approaches to describing brain structure. Some authors discuss brain anatomy in terms of the five secondary vesicles (see Figure 12.3c); this is the embryonic scheme. In clinical institutions, one more often hears reference to the adult brain regions, such as the cerebral hemispheres, cerebellum, and pons (see Figure 12.3d). In this text, we will consider the brain in terms of the following subdivisions shown in Figure 12.5: (1) cerebral hemispheres, (2) the diencephalon (thalamus, hypothalamus, and epithalamus), (3) the brain stem (midbrain, pons, and medulla), and (4) cerebellum. Most neuroanatomists and neurophysiologists favor this scheme, but some gross anatomists include the diencephalon as part of the brain stem.

The cerebral hemispheres and cerebellum have an outer "bark" or cortex of gray matter (nerve cell bodies), plus some gray matter masses called *nuclei* deeply embedded in a central core of white matter (myelinated fiber tracts). This pattern changes with descent through the brain stem. At the caudal end of the brain stem, the basic pattern is that of the spinal cord: gray matter inside and white matter outside.

To help you picture the spatial relationships of the brain regions, we will first consider the fluid-filled ventricles that lie deep within the brain. Then the location and structure of each of the brain regions will be described in turn to flesh out the brief summary provided in Table 12.1 on p. 398.

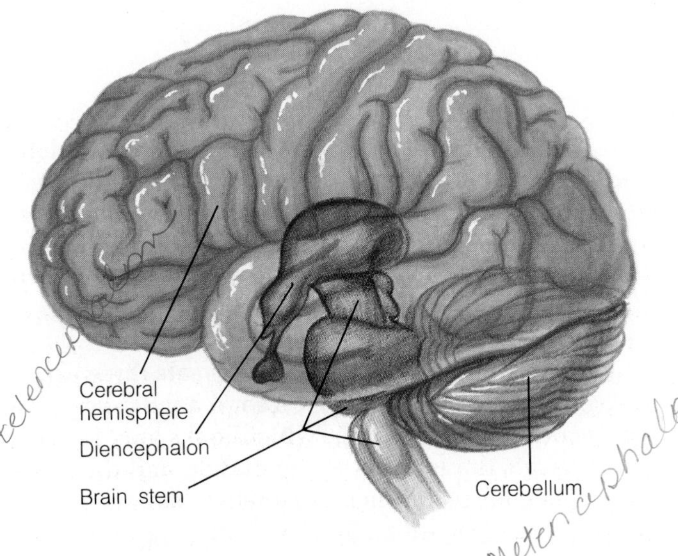

Figure 12.5 Regions of the brain. The brain is a bilateral and essentially symmetrical group of structures that can be considered in terms of four main parts: the cerebral hemispheres, the diencephalon, the brain stem, and the cerebellum.

Ventricles of the Brain

As noted earlier, the **ventricles** of the brain arise from expansions of the lumen of the embryonic neural tube. They are interconnected with each other and with the central canal of the spinal cord (Figure 12.6). The ventricular chambers are lined by *ependymal cells*, a type of neuroglia (see Figure 11.2c).

The **lateral ventricles** (occasionally called the first and second ventricles) are large C-shaped chambers that reflect the pattern of cerebral growth. There is one lateral ventricle located deep within each cerebral hemisphere. For part of their length, the lateral ventricles lie side by side and are separated medially only by a thin translucent membrane called the **septum pellucidum** (puh-loo′-sih-dum). Each lateral ventricle communicates with the narrow **third ventricle** in the diencephalon via a small opening called the **interventricular foramen** (*foramen of Monro*). The third ventricle is continuous with the **fourth ventricle** via the **cerebral aqueduct** that runs through the midbrain. The fourth ventricle, which lies anterior to the cerebellum, is continuous with the central canal of the spinal cord inferiorly. Three openings mark the walls of the fourth ventricle: the paired **lateral apertures** in its lateral walls and the **median aperture** in its roof. These apertures provide continuity between the ventricles and the *subarachnoid* (sub″-uh-rak′-noyd) *space*, a fluid-filled space surrounding the brain. The importance of the cerebrospinal (suh-re″-brō-spī′-nul) fluid, which fills both the internal brain cavities and the subarachnoid space, is discussed later in the chapter.

The Cerebral Hemispheres

The **cerebral hemispheres** form the superiormost part of the brain (Figure 12.7). Together they account for more than 60% of total brain weight and without a doubt are the most conspicuous parts of an intact brain. Picture how a mushroom cap covers the top of its stalk, and you have a fairly good analogy of how the paired cerebral hemispheres cover and obscure the diencephalon and the top of the brain stem (see Figure 12.5).

Nearly the entire surface of the cerebral hemispheres is marked by elevated ridges of tissue called **gyri** (jī′-rī) (singular, gyrus) separated by shallow grooves called **sulci** (sul′kī) (singular, sulcus) or by deeper grooves called **fissures.** The more prominent gyri and sulci are similar in all people and are important anatomical landmarks. The median **longitudinal fissure** separates the cerebral hemispheres from one

Figure 12.6 Three-dimensional views of the ventricles of the brain. (**a**) Left lateral view. (**b**) Anterior view. Note that

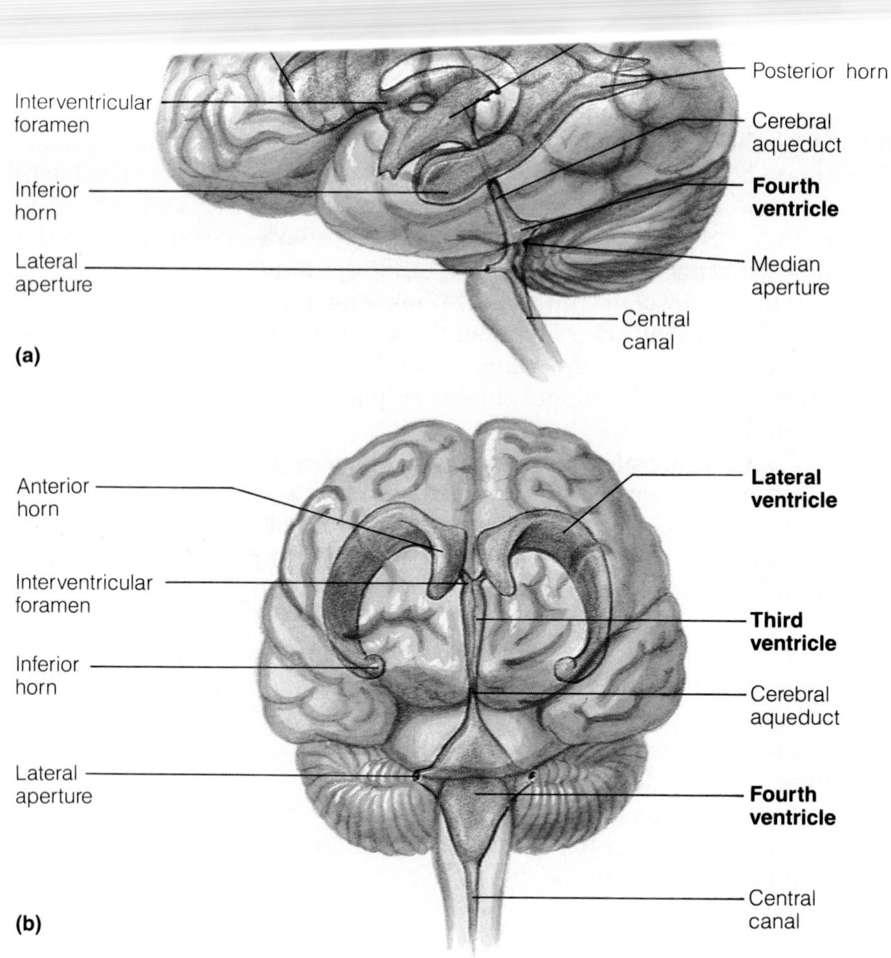

another (see Figure 12.7b); another large fissure, the **transverse fissure,** separates the cerebral hemispheres from the cerebellum below (see Figure 12.7a).

Some of the deeper sulci divide each hemisphere into five lobes, most of which are named for the cranial bones that overlie them (see Figure 12.7a). The **central sulcus** runs perpendicular to the longitudinal fissure and separates the **frontal lobe** from the **parietal lobe.** Two functionally important gyri parallel the central sulcus across each hemisphere. These gyri, named for their position relative to the central sulcus, are the *precentral gyrus,* immediately anterior to the sulcus, and the *postcentral gyrus,* just behind it. The boundary between the parietal lobe and the more posterior **occipital lobe** is established by several landmarks, but perhaps the most identifiable of these is the **parietooccipital** (puh-rī″-ih-tō-ok-sih′-pih-tul) **sulcus,** located on the internal, or medial, surface of the cerebral hemisphere.

The deep **lateral sulcus** outlines the **temporal lobe** and separates it from the inferior aspects of the parietal and frontal lobes. A fifth lobe of the cerebral hemisphere, the **insula (island of Reil),** is buried within the lateral sulcus, forming part of its floor, and is covered by portions of the temporal, parietal, and frontal lobes (Figure 12.8).

A frontal section of the brain reveals three easily recognizable regions in each cerebral hemisphere: the outer *cortex* of gray matter, which looks gray in fresh brain tissue; the internal *white matter;* and the *basal nuclei,* islands of gray matter situated deep within the white matter (Figure 12.8).

Cerebral Cortex

The **cerebral cortex** is the "executive suite" of the nervous system. It enables us to perceive, communicate, remember, understand, appreciate, and initiate voluntary movements. Composed mainly of nerve cell

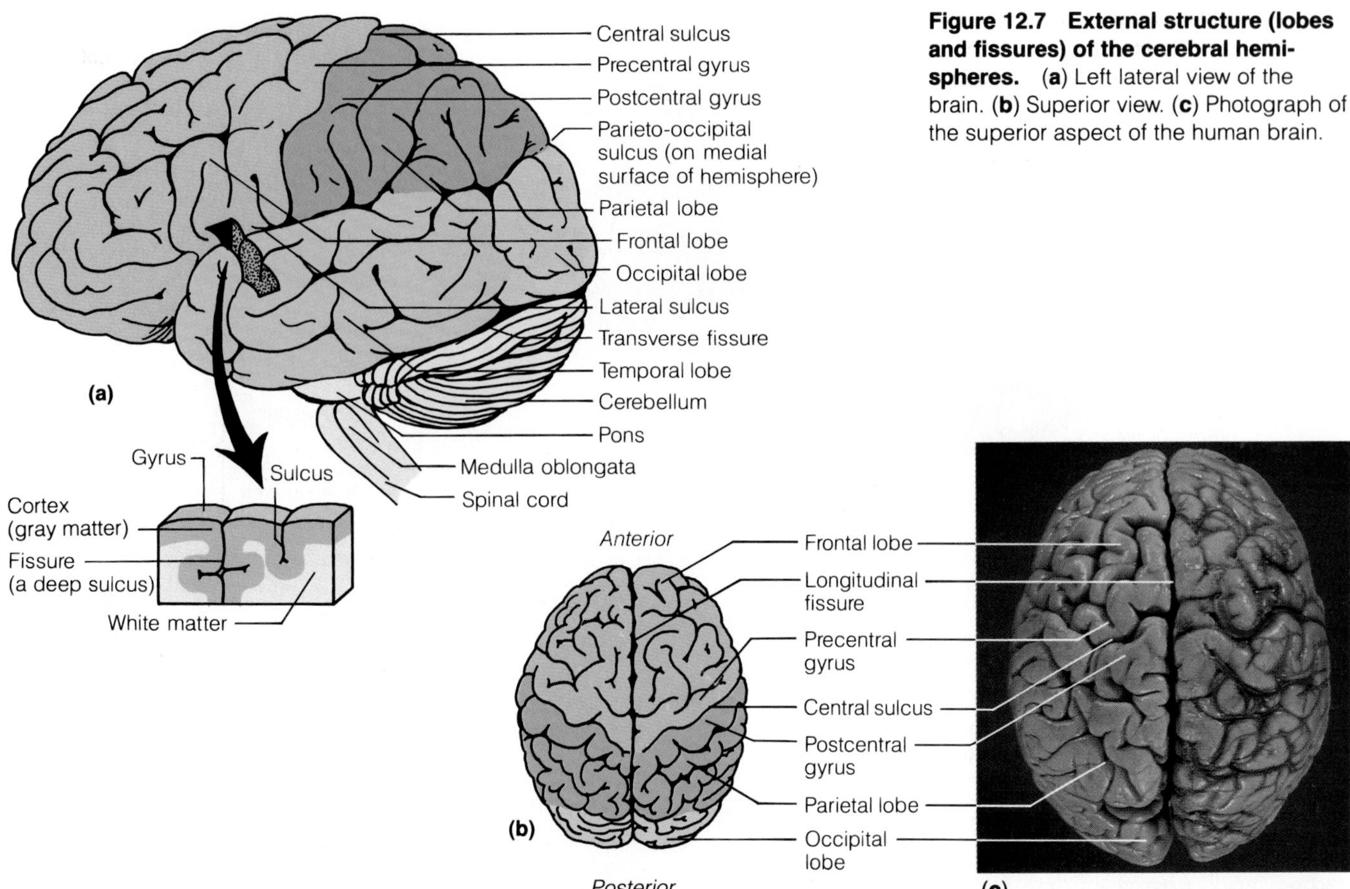

Figure 12.7 External structure (lobes and fissures) of the cerebral hemispheres. (**a**) Left lateral view of the brain. (**b**) Superior view. (**c**) Photograph of the superior aspect of the human brain.

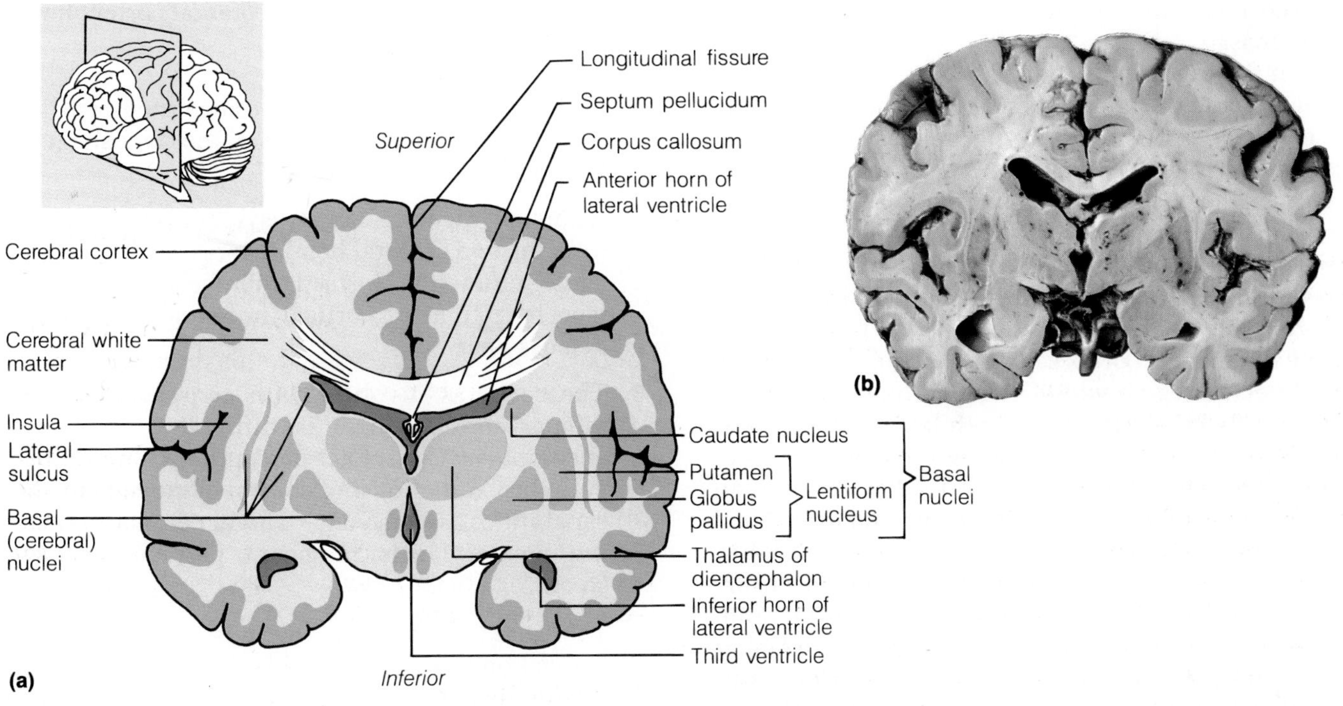

Superior

Longitudinal fissure

Septum pellucidum

Corpus callosum

Anterior horn of lateral ventricle

Cerebral cortex

Cerebral white matter

Insula

Lateral sulcus

Basal (cerebral) nuclei

Caudate nucleus

Putamen

Globus pallidus

Lentiform nucleus

Basal nuclei

Thalamus of diencephalon

Inferior horn of lateral ventricle

Third ventricle

(a)

Inferior

(b)

Figure 12.8 Major regions of the cerebral hemispheres. The human brain has been sectioned coronally to reveal the relative positions of the (cerebral) cortex, white matter, and basal (cerebral) nuclei within the cerebral hemispheres. Since the cerebral hemispheres enclose the structures of the diencephalon, that region of the cerebrum is also shown. Compare the labeled diagram in (**a**) to the photograph in (**b**).

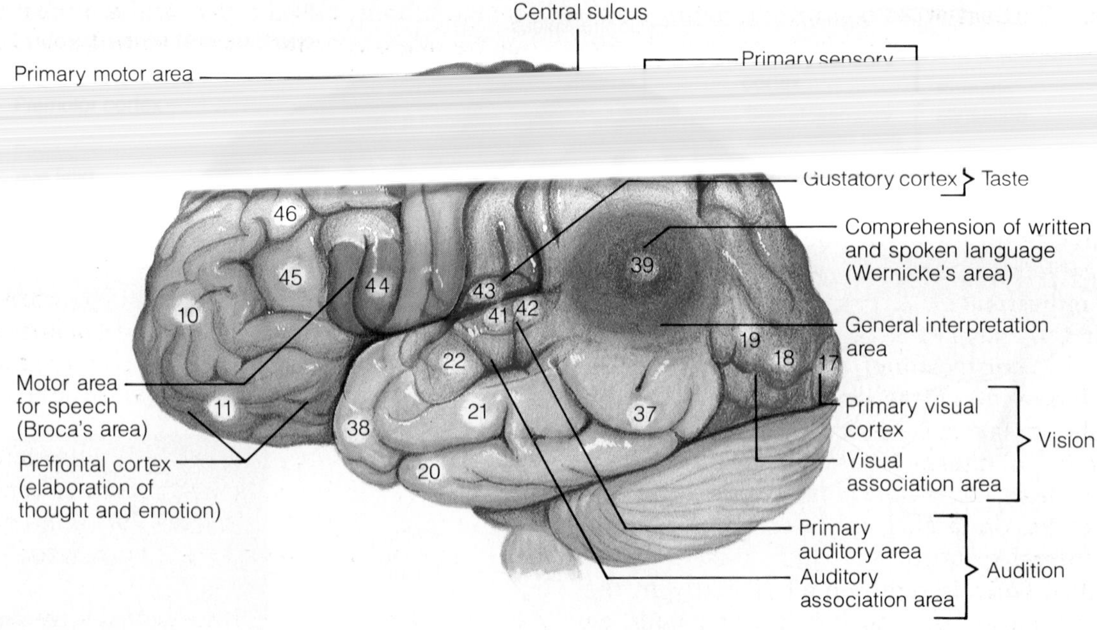

Figure 12.9 Functional areas of the left cerebral cortex. The olfactory area, which is deep to the temporal lobe on the medial hemispheric surface, is not identified. Numbers indicate brain regions plotted by the Brodman system.

bodies and unmyelinated fibers (plus associated glia and blood vessels), the cerebral cortex is only 2–4 mm (less than 1/16 inch) thick. However, its many convolutions effectively triple the cortical surface area, and it accounts for roughly 40% of total brain mass.

In the nineteenth century, there were two dominant schools of thought concerning the localization of cerebral cortical functions. The *regional specialization theory,* put forth by Paul Broca and many others, assumed that histologically distinct areas of the cortex carried out particular functions. The *aggregate field view,* based on reactions observed when parts of the brain were cut out, offered quite another hypothesis. It proposed that mental functions were not localized at all, but that the cortex acted as a whole. Thus, injury to a specific region would affect all higher functions equally. Today we know that both theories have some merit. It has been proved that specific motor and sensory areas are indeed localized in discrete *cortical areas,* or *domains.* However, many higher mental functions, such as memory and language, appear to have overlapping domains and are more diffusely located. By examining the histology of specific brain regions, neuroscientists have identified a large number of cortical areas, called *Brodmann areas,* that have functional significance, and they have produced a map of the cerebral cortex that is an elaborate numbered mosaic of these areas. Although some of the most important Brodmann areas are numbered in Figure 12.9, we will examine the functional regions of the cerebral cortex only in the broadest sense. But first, some generalizations about the cerebral cortex should be presented:

1. Each hemisphere is chiefly concerned with the sensory and motor functions of the opposite (contralateral) side of the body.
2. Although they are largely symmetrical in structure, the two cerebral hemispheres are not entirely equal in function. Instead, there is a lateralization (specialization) of cortical functions, as we will discuss later.
3. The cortex is involved with *consciousness.* Although consciousness is difficult to define, it encompasses conscious perception of sensations, voluntary initiation and control of movement, and capabilities associated with higher mental processing (memory, logic, judgment, perseverance, and so on).
4. The final, and perhaps the most important, generalization to keep in mind is that our approach is a gross oversimplification, because *no* functional area of the cortex acts alone, and consciousness involves the entire cortex in one way or another.

Motor Areas. Cortical areas controlling motor functions are located in the frontal lobes and include the primary motor cortex, the premotor cortex, and Broca's area.

1. Primary (somatic) motor cortex. The primary, or somatic, motor cortex is located in the precentral gyrus of the frontal lobe of each hemisphere. Neurons called pyramidal cells in these gyri allow us to consciously control the movements of our skeletal muscles. Their axons form the major voluntary motor tracts, the *pyramidal,* or *corticospinal* (kor″-tih-kō-spī′-nul), *tracts,* which provide a direct connection between the cortex and the spinal cord. All other descending motor tracts—that is, all motor tracts issuing from brain regions other than the primary motor cortex (or the premotor cortex, described next)—are called *extrapyramidal* (literally, "outside the pyramidal") *tracts.*

The entire body is represented spatially in the somatic motor cortex of each hemisphere. Such an orderly mapping of the body in CNS structures is called *somatotopy* (sō-ma′-tōh-tō-pē). As illustrated in Figure 12.10, the body is represented upside down—the head at the bottom of the precentral gyrus near the temporal lobe, and the toes at the medial end. Most of the neurons in these gyri control muscles in body areas having the most precise motor control—that is, the face, lips, tongue, and hands. The distorted, or caricature, nature of the *homunculus* (hō-mun′-kyoo-lus), or "little man," drawn on the gyrus in Figure 12.10 indicates the cortical location of neurons controlling muscles in different parts of the body and represents the variations in muscle innervation in the different body areas. Areas with the greatest motor innervation, such as the tongue and hands, are shown disproportionally large. The left primary motor gyrus chiefly controls muscles on the right side of the body, and vice versa.

Damage to localized areas of the primary motor cortex paralyzes the body muscles controlled by those areas. If the lesion is in the right hemisphere, the left side of the body will be paralyzed. ■

2. Premotor cortex. Just anterior to the precentral gyrus in the frontal lobe is the premotor cortex (see Figure 12.9). This region controls learned motor skills of a repetitious or patterned nature, such as playing a musical instrument and typing. Neurons in the pre-

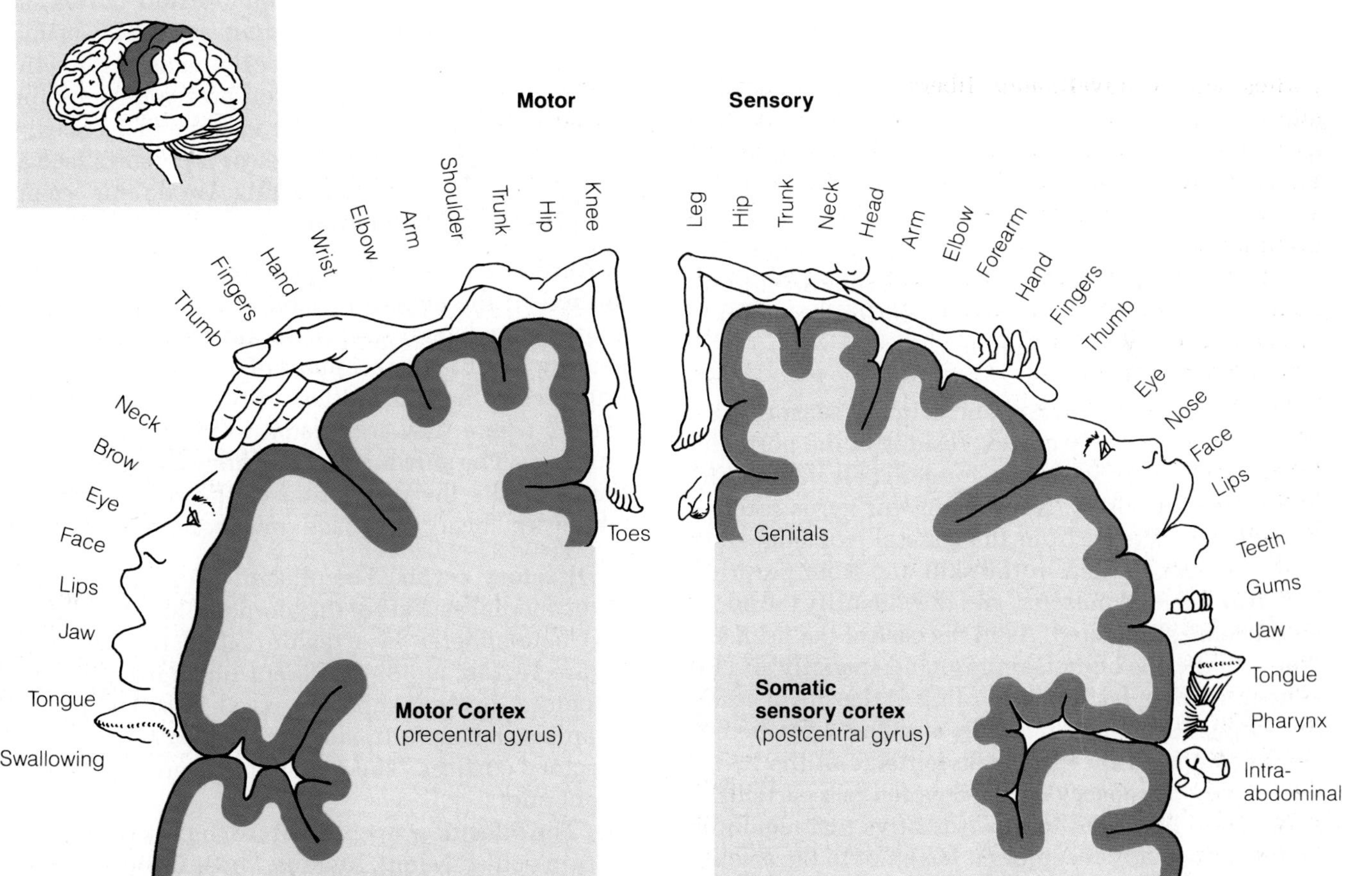

Figure 12.10 Sensory and motor areas of the cerebral cortex. The relative amount of cortical tissue devoted to each function is indicated by the amount of the gyrus occupied by the body area diagrams. The primary motor cortex is represented on the left, and the somatic sensory cortex on the right.

motor cortex coordinate the movements of several muscles simultaneously and/or sequentially by sending activating impulses both to the primary motor cor- [illegible] premotor area, or part of it, results in a loss of the motor skill(s) programmed in that particular region, but muscle strength and the ability to perform the discrete individual movements are not hindered. For example, if the premotor area controlling the flight of your fingers over the typewriter keyboard were damaged, you would not be able to type with your usual speed, but you could still make the same movements with your fingers. Reprogramming the skill into another set of premotor neurons would require practice and repetition, just as the initial learning process did. ■

3. Broca's (brō′-kuz) **area. Broca's area,** also called the **motor speech area,** is situated at the base of the premotor area of the frontal lobe in one cerebral hemisphere only, usually the left. The neurons of Broca's area produce a detailed program for all the countless combinations of muscular movements needed to articulate every word you have ever spoken. They direct the motor cortex to activate, in proper sequence, the muscles of the lips, jaw, mouth, throat, and tongue used in speech. Say the word *surreptitiously* out loud while trying to be aware of all the muscles you are using. As you can see, speech represents an incredible task for Broca's area.

Sensory Areas. Unlike the motor areas, which are confined to the frontal lobe cortex, various areas concerned with sensation occur in the parietal, temporal, and occipital lobes (see Figure 12.9).

1. Primary sensory cortex. The primary sensory cortex, or **somatosensory cortex,** resides in the postcentral gyrus of the parietal lobe, immediately behind the primary motor cortex. Neurons in this gyrus receive information, relayed from the general (somatic) sensory receptors located in the skin and from proprioceptors in skeletal muscles, and they identify the body region being stimulated. As in the case of the primary motor cortex, the body is represented spatially and in an upside-down fashion according to the site of stimulus input (see Figure 12.10), and the right hemisphere receives input from the left side of the body. The amount of sensory cortex devoted to a particular body region is related to how sensitive that region is (that is, how many receptors it has), not to the size of the body region. In humans, the face (especially the lips), tongue, and fingertips are the most sensitive body areas, and hence, these are the largest parts of the sensory homunculus.

2. Somatosensory association area. The somatosensory association area, or **somesthetic cortex** (sō″-mis- [illegible] this area is to integrate and analyze different somatic sensory inputs, such as temperature, touch, pressure, and pain. For example, the primary sensory cortex tells you that certain impulses are coming from your right hand, and then the somatosensory association area allows you to know without looking that you have a spool of thread in your right hand, rather than a ball. This accomplishment (perception) requires integration of concepts of size, texture, and relationship of parts of the object with memories of similar sensory experiences.

Damage to the somatosensory association area handicaps one's ability to analyze the different characteristics of a sensory experience, whereas damage to the primary sensory cortex causes a loss in ability to determine where the stimulus is acting on the body. ■

3. Visual cortex. The occipital lobe of each cerebral hemisphere houses the **primary visual cortex,** surrounded by a **visual association area.** The primary area receives information that originates in the retinas of the eyes, and the association area interprets these visual stimuli in light of past visual experiences, enabling us to recognize a flower or a person's face and to appreciate what we are seeing. We do our "seeing" with these cortical neurons.

4. Auditory cortex. Each **primary auditory cortex** is located in the superior margin of the temporal lobe abutting the lateral sulcus. Sound energy exciting the hearing (cochlear) receptors of the inner ear causes impulses to be transmitted to the primary auditory cortex, where they are related to pitch, rhythm, and loudness. The surrounding **auditory association area** then permits the perception of the sound stimulus, which we "hear" as speech, music, noise, and so on.

5. Olfactory cortex. The olfactory cortices are found in the medial aspects of the temporal lobes of the cerebral hemispheres in a region called the *uncus* (see Figure 12.13b, p. 389). Afferent fibers from the smell receptors in the superior nasal cavities transmit impulses that are ultimately relayed to neurons in the olfactory cortices. The outcome is perception of different odors.

The olfactory cortex is part of the *rhinencephalon* (rī″-nin-seh′-fuh-lon), literally, "nose brain," which in more primitive vertebrates is entirely concerned with olfactory reception and perception. The rhinencephalon includes parts of the cerebral cortex located on or in the medial aspects of the temporal lobes, as well as the protruding olfactory tracts and bulbs that extend to the nose region. During the course of evolution, most of the old rhinencephalon has taken on new functions concerned chiefly with emotions and memory. This "newer" emotional brain, called the *limbic system*, is considered on p. 396. The only portions of the rhinencephalon still devoted to smell in humans are the olfactory bulbs and tracts (described in Chapter 16) and the greatly reduced olfactory cortex.

6. Gustatory cortex. The **gustatory** (gus′-tuh-tor-ē) **cortex** (see Figure 12.9), a region involved in the perception of taste stimuli, is located in the parietal lobe deep to the temporal lobe. Logically, it occurs at the tip of the tongue of the somatosensory homunculus.

7. Wernicke's area. A specialized sensory integration area called **Wernicke's** (wer′-nih-kēz) **area** (see Figure 12.9) is found in the posterior aspect of the temporal lobe of one cerebral hemisphere (the left, in most people). Sometimes referred to as the **speech area,** Wernicke's area receives impulses from the visual and auditory association cortices that adjoin it and is involved in the *comprehension* of written and spoken language. A fiber tract runs from Wernicke's area to Broca's area anteriorly, allowing the language comprehension area to communicate with the language articulation area.

Association Areas. As described, the primary sensory cortex and each of the special sensory areas have nearby association areas with which they communicate. The association areas, in turn, communicate with the motor cortex and with other sensory association areas to analyze, recognize, and act on sensory inputs. Also considered an association area is Wernicke's area. Each of these areas has multiple inputs and outputs quite independent of the primary sensory and motor areas, indicating that their function is complex indeed. In general, any cortical area not specifically pinpointed as a primary motor or sensory area is considered to have some sort of association or integrative function. The most important of the remaining association areas are described next.

1. Prefrontal cortex. The prefrontal cortex, which occupies the anterior portions of the frontal lobes (see Figure 12.9), is most involved with elaboration of thought, intelligence, motivation, and personality. It associates experiences necessary for the production of abstract ideas, judgment, persistence, planning, concern for others, and conscience. These qualities all develop slowly in children, so it appears that the prefrontal cortex matures slowly and is heavily dependent on positive and negative feedback from one's social environment. It is the elaboration of this region that sets human beings apart from other animals.

Tumors or other lesions of the prefrontal cortex often cause wide mood swings and loss of attentiveness, initiative, and judgment. The affected individual may become oblivious to social restraints, perhaps becoming careless about personal appearance, or rashly fighting a 7-foot opponent rather than running.

A surgical technique called *prefrontal lobotomy,* which severs the connections to the prefrontal cortex, has been used to treat severe mental illness. The early results of this procedure appeared to be favorable because patients became less anxious. However, it later became apparent that the cure might be worse than the disease, because many people so treated developed epilepsy and abnormal personality changes, such as a lack of inhibition or a loss of initiative. ■

2. General interpretation area. The general interpretation area, or **gnostic** (nos′-tik) ("knowing") **area,** is an ill-defined region at the superoposterior limits of the temporal lobe where the temporal, parietal, and occipital lobes meet. It is found in one hemisphere only, usually the left. This region receives impulses from all the sensory association areas as well as from lower brain centers, and it appears to be a storage site for complex memory patterns associated with sensation. It integrates all incoming signals into a single thought or understanding and activates other parts of the cortex to cause an appropriate response. Suppose, for example, you drop a bottle of hydrochloric acid and it splashes on you. You see the bottle drop and shatter; you hear the crash; you feel the burning on your skin; and you smell the acid fumes. However, these individual perceptions do not really dominate your consciousness. What *does* is the message "danger"—by which time your leg muscles have already been activated and are propelling you to the shower.

Injury to the gnostic area causes one to become imbecilic—even if all the other sensory association areas are unharmed—because one's ability to interpret the entire situation is lost. ■

3. Affective language areas. As we have seen, Broca's and Wernicke's areas exist only in one cerebral hemisphere. However, the opposite hemisphere has *affective language areas,* that is, comparable regions involved in the nonverbal, emotional components of language. Thus, the area comparable to Broca's area controls our expression of the emotions related to the vocalization—for example, the lilt or intonation of our voice or the gestures we make when talking with others. Likewise, while Wernicke's area allows us to comprehend the cognitive aspects of the written or spoken word, its mirror image in the opposite cortex allows us to comprehend the emotional content of language. For example, a soft, melodious response to your question conveys quite a different meaning than a reply voiced in a sharp, curt manner.

Disorders of these affective regions are collectively called *aprosodia* (literally, "lacking expression"). An individual telling you (quite honestly) that [illegible]

Lateralization of Cortical Functioning. Although we use both cerebral hemispheres for almost every activity, and the hemispheres appear nearly identical, there is a division of labor, and each hemisphere has unique abilities not shared by its partner. One cerebral hemisphere or the other "dominates" each task, but the term *cerebral dominance* designates the hemisphere that is dominant for language. In most people (about 90%), it is the left hemisphere that controls language abilities, as well as mathematical abilities and the ability to reason logically and analytically. This so-called dominant hemisphere is working when we form thoughts into words, balance a checkbook, and memorize a list of names. The other (usually the right) hemisphere is more involved in motor activities, visual-spatial skills, intuition, emotion, and appreciation of art and music. It is the poetic, creative, and the "Ah-ha!" (insightful) side of our nature, and it is far better at recognizing faces. Most individuals with left cerebral dominance are right-handed.

In the remaining 10% of people, the dominant (language) hemisphere is the right cerebral hemisphere, and the spatial hemisphere is the left. Typically, these right-dominant people are left-handed and more often male. However, in a few left-handed individuals, neither hemisphere dominates; that is, the cerebral cortex functions bilaterally for almost all functions. In some cases, the mutuality (or duality) of brain control results in more than the usual dexterity and strength in the nondominant hand and in ambidexterity. In other cases, however, this phenomenon results in cerebral confusion ("Is it your turn, or mine?") and learning disabilities. The reading disorder *dyslexia,* in which otherwise intelligent people reverse the order of letters or syllables in words (and of words in sentences), has been attributed to such a lack of cerebral dominance.

Another aspect of cerebral hemispheric interaction is that each side of the brain exerts controls over the other. The dominant, more intellectual hemisphere prevents rash emotional displays by the nondominant hemisphere. On the other hand, the emotional side of the brain urges us to take time out from routine and to fantasize or do something spontaneous for a change. This is an important concept, because it lays to rest the common misconception that we have "two brains" and that one dominates the other. It is crucial to understand that the two cerebral hemispheres have perfect and almost instant communication with one another via connecting fiber tracts, as well as complete integration of their functions. Fur- [illegible]

[illegible] are well known, it is often possible to determine the area of brain damage after a stroke by observing the symptoms of the patient. For example, if the patient has left-sided paralysis, the right primary motor cortex of the precentral gyrus is most likely involved. However, if the person has speech problems, it is not usually a problem with his or her intellect, but rather with Wernicke's or Broca's area. ■

Cerebral White Matter

The **cerebral white matter** underlying the gray matter of the cortex accounts for about 60% of cerebral hemispheric volume (see Figure 12.8). It consists largely of myelinated fibers bundled into large tracts that allow communication between cerebral areas or between the cerebral cortex and lower CNS centers. These fibers and the tracts they form are classified as *commissural, association,* or *projection* according to the direction in which they run.

Commissures (kō′-mih-shūrz), composed of commissural fibers, connect corresponding regions of the two hemispheres, enabling the cerebral hemispheres to function as a coordinated whole. The most important and largest commissure is the **corpus callosum** (kal-lō′-sum) (Figure 12.11); a less important example is the **anterior commissure.**

Association fibers transmit impulses between gyri within a single hemisphere. Short association fibers connect adjacent gyri; long association fibers are bundled into tracts and connect different cortical lobes.

Projection fibers include afferent fibers entering the cerebral hemispheres from lower brain or cord centers, and efferent fibers leaving the cortex to travel to lower areas. Projection fibers tie the cortex to the rest of the nervous system and to the receptors and effectors of the body.

At the upper limit of the brain stem, the projection fibers on each side form a compact band known as the **internal capsule** (see Figures 12.11b and 12.12a), which passes between the thalamus and some of the basal nuclei. Beyond that point, they radiate to the cerebral cortex. These radiating portions of the projection tracts are known as the **corona radiata** (radiating crown) because of this distinctive arrangement.

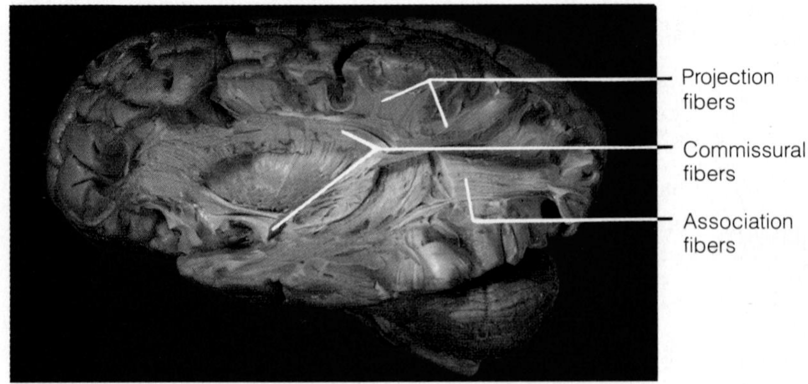

Figure 12.11 Types of fiber tracts composing the white matter of the cerebral medulla. (**a**) Medial view of the right cerebral hemisphere; photograph above, diagrammatic view below. Several association tracts (tracts connecting different parts of the same hemisphere) are depicted. The corpus callosum is a commissure that allows communication between the two cerebral hemispheres. (**b**) Frontal section of the brain showing some commissural fibers of the cerebrum and projection fibers that run between the cerebrum and lower centers (brain stem and spinal cord). The ascending and descending projection tracts are bundled together into a tight band called the internal capsule as they pass between the thalamus and the basal nuclei.

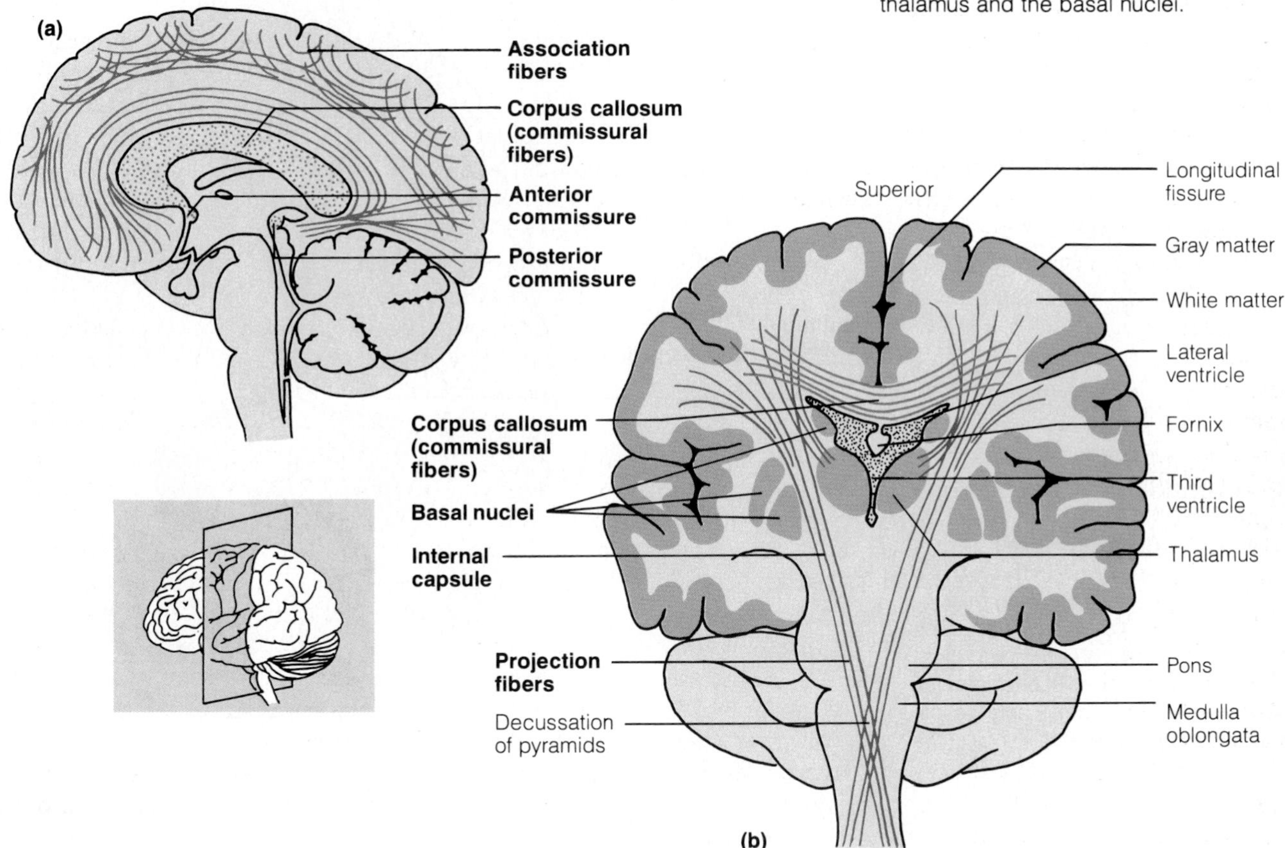

Basal (Cerebral) Nuclei

Deep within the cerebral white matter of each hemisphere are bilateral collections of subcortical motor nuclei called the **basal (cerebral) nuclei** or basal ganglia.* The basal nuclei are part of the extrapyramidal system of motor control.

Although there is controversy about the precise structures forming the cerebral nuclei, most agree that the **caudate** (kah′-dāt) **nucleus, putamen** (pyoo-tā′-min), and **globus pallidus** (glō′-bis pal′-lih-dis) constitute their main mass (see Figure 12.12). Together, the putamen and globus pallidus (*pallidus* = pale) form a lens-shaped nuclear mass commonly called the **lentiform nucleus;** this aggregate flanks the internal capsule laterally. The comma-shaped caudate nucleus arches superiorly over the diencephalon and lies medial to the internal capsule. Collectively, the lentiform and caudate nuclei are called the **corpus striatum** (stri-ā′-tum) because the fibers of the internal capsule that

*Since a nucleus is a collection of nerve cell bodies within the CNS, the terms *basal nuclei* and *cerebral nuclei* are correct. The more frequently used term *basal ganglia* is a misnomer, because ganglia are PNS structures.

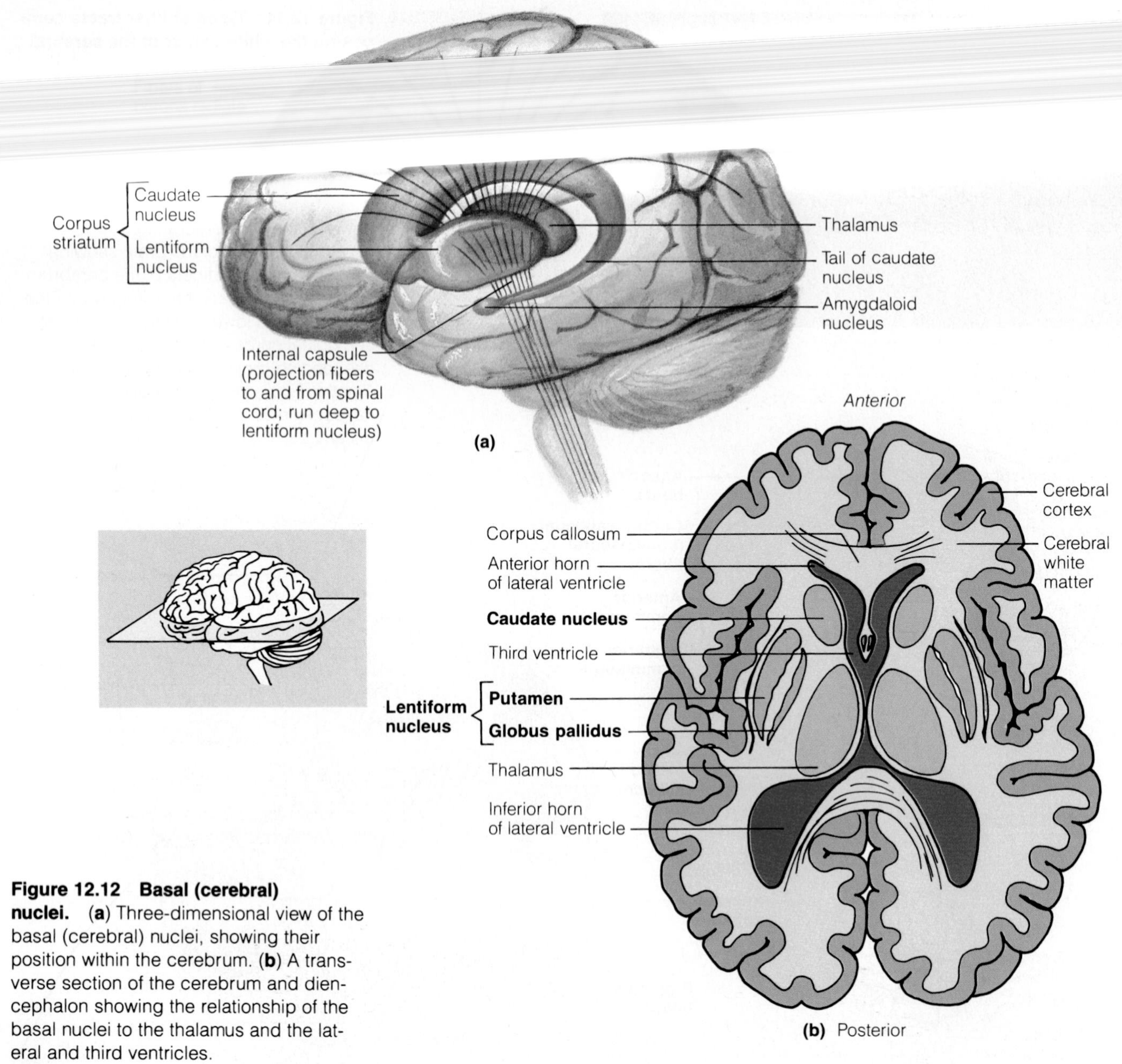

Figure 12.12 Basal (cerebral) nuclei. (**a**) Three-dimensional view of the basal (cerebral) nuclei, showing their position within the cerebrum. (**b**) A transverse section of the cerebrum and diencephalon showing the relationship of the basal nuclei to the thalamus and the lateral and third ventricles.

course past these nuclei give them a striped appearance (see Figure 12.12a).

The almond-shaped **amygdaloid** (uh-mig′-duh-loyd) **nucleus,** or **amygdala** (uh-mig′-duh-luh), sits on the tail of the caudate nucleus. Traditionally, it has been grouped with the basal nuclei, but functionally it belongs to the limbic system (see p. 396).

The basal nuclei are closely "wired" to the neurons of the cerebral cortex and receive extensive inputs from the cortical motor, sensory, and association areas, as well as from the thalamus and each other. Via relays through the thalamus, the basal nuclei indirectly help to initiate and control muscle movements directed by the primary motor cortex. However, the basal nuclei have no direct access to motor pathways.

The precise role of the basal nuclei has been elusive because of their inaccessible location and because their functions overlap to some extent with those of the cerebellum. However, they appear to be particularly important in initiating movements, especially those that are relatively slow and sustained, or stereotyped. For example, the caudate nucleus and putamen contribute to involuntary control of arm-swinging during walking. When the basal nuclei are impaired, the result is postural disturbances, muscle tremors at rest, or uncontrolled muscle contractions. The regulatory activities and most common disorders of the basal nuclei are described in more detail in Chapter 15.

The Diencephalon

The **diencephalon** forms the central core of the forebrain and is surrounded by the cerebral hemispheres. It consists largely of three bilaterally symmetric structures—the thalamus, hypothalamus, and epithalamus—which collectively enclose and form the boundaries of the third ventricle (see Figures 12.8 and 12.13).

The Thalamus

The egg-shaped **thalamus,** which makes up 80% of the diencephalon, forms the superolateral walls of the third ventricle. It is composed of bilateral masses of gray matter held together by a medial commissure called the **intermediate mass** or **massa intermedia.** The term *thalamus* is from the Greek, meaning "inner room," which well describes this hidden brain region.

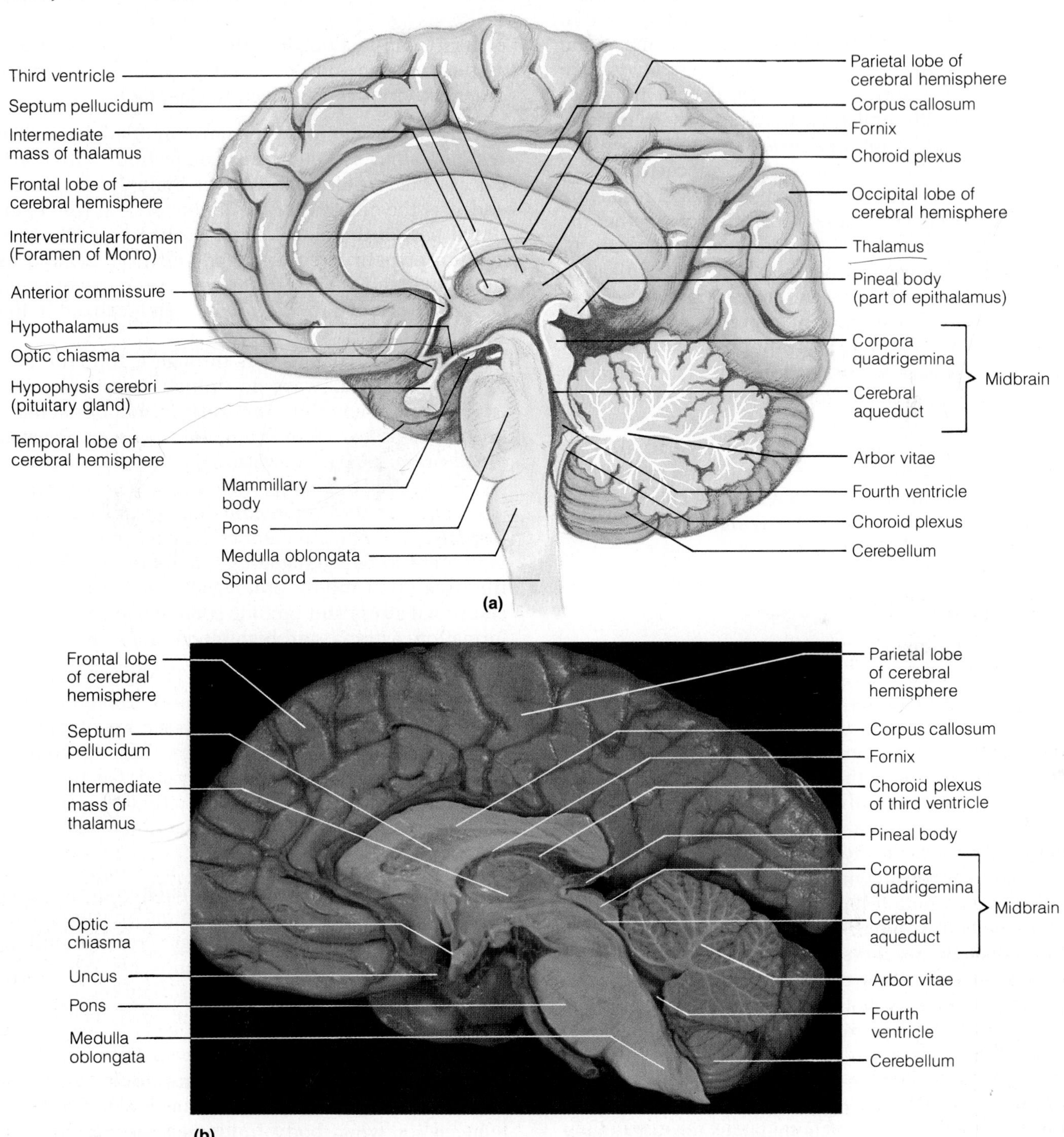

Figure 12.13 Diencephalon and brain stem structures as seen in a midsagittal section of the brain.

The thalamus contains many different nuclei, each functionally specialized. One of its major roles is to serve as a relay station for sensory information [illegible] one of its nuclei. For example, the *ventral posterior lateral nucleus* is a major synapse point for fibers carrying impulses from the general sensory receptors (touch, pressure, etc.), and the *lateral* and *medial geniculate* (juh-nih′-kyoo-lāt) *bodies* are important visual and auditory relay centers, respectively. Within the thalamus, a sorting-out process occurs. Impulses having to do with similar functions are grouped together and relayed to the appropriate area of the sensory cortex as well as to other cortical and subcortical association centers. As the afferent impulses reach the thalamus, we have a crude recognition of whether the sensation we are about to experience is pleasant or unpleasant. However, specific stimulus localization and discrimination occur in the cerebral cortex.

Not only are sensory inputs relayed through the thalamus, but virtually *all* inputs ascending to the cerebral cortex are funneled through the thalamic nuclei, including some from the cerebellum and subcortical nuclei (basal, nuclei and others). The thalamus plays a key role in mediating cortical arousal, memory, and somatic motor activities by controlling which inputs are allowed to enter the cortex. Thus, via its various connections, the thalamus is in contact with the entire cerebral cortex.

The Hypothalamus

The **hypothalamus** (hī″-pō-tha′-luh-mis), named for its position below the thalamus, caps the top of the brain stem and forms the inferolateral walls of the third ventricle, merging into the midbrain inferiorly (see Figure 12.13). It extends from the optic chiasma (point of crossover of the optic nerves) to the posterior margin of the mammillary bodies. The **mammillary** (ma′-mih-layr-ē) **bodies,** paired pealike nuclei that bulge anteriorly from the hypothalamus, are involved in olfaction and swallowing reflexes. Between the optic chiasma and mammillary bodies is the **infundibulum** (in″-fun-dih′-byoo-lum), a stalk of the hypothalamic tissue that connects the **hypophysis cerebri** (hī-pah′-fih-sis ser′-eh-brī), or pituitary gland, to the base of the hypothalamus. Notice that while the term *pituitary gland* may be easier to remember, analyzing *hypophysis cerebri* in terms of its word roots (*physis* = little sprout; *hypo* = beneath; *cerebri* = the cerebrum) makes its location much more understandable. Like the thalamus, the hypothalamus contains many functionally important nuclei.

Despite its small size, the hypothalamus is the main visceral control center of the body and is vitally [illegible]

1. Autonomic center. The hypothalamus regulates autonomic (involuntary) nervous system activity by controlling the activity of autonomic centers in the brain stem and spinal cord. In this role, the hypothalamus can influence blood pressure, rate and force of heart contraction, motility of the digestive tract, respiratory rate and depth, eye pupil size, and many other activities as described in Chapter 14.

2. Center for emotional response and behavior. The hypothalamus has numerous connections with cortical association areas and lower brain stem centers, and it is the "motor" of the limbic system (the emotional part of the brain). Nuclei involved in the perception of pain, pleasure, fear, and rage, as well as those involved in biological rhythms and drives (such as the sex drive) have definitely been localized in the hypothalamus.

The hypothalamus acts through autonomic nervous system pathways to initiate most physical expressions of emotion. For example, emotional manifestations of fear include a pounding heart, elevated blood pressure, pallor, sweating, and a dry mouth.

Since the hypothalamus is the neural clearing house for both autonomic function and emotional response, the phenomenon of *psychosomatic illness* is thought to be mediated by this brain region. Such illnesses often follow prolonged or acute mental or emotional stress and include some instances of ulcer formation, asthma, and high blood pressure. ■

3. Body temperature regulation. The body's thermostat is found within certain nuclei of the hypothalamus. If the blood flowing through the hypothalamus is above normal body temperature, then the cooling mechanisms of the body are activated. On the other hand, cooler than normal blood circulating through the hypothalamus triggers heat retention mechanisms (see Chapter 25).

4. Regulation of food intake. Stimulation of a lateral hypothalamic nucleus initiates eating behavior. This region, called the *feeding* or *hunger center,* monitors blood levels of glucose and other nutrients. When adequate food has been ingested, the *satiety center* in the medial hypothalamic region is activated; as a result, the desire to eat declines.

5. Regulation of water balance and thirst. When the volume of blood (and other extracellular fluids) is reduced, or when body fluids become too concentrated, specific hypothalamic neurons called *osmoreceptors* are activated. They, in turn, excite hypothalamic nuclei that trigger the release of antidiuretic hormone (ADH) from the posterior pituitary. This hormone targets the kidneys, causing them to retain water. Under the same conditions, hypothalamic neurons in the *thirst center* are also stimulated, causing us to become thirsty and seek water or other fluids.

6. Regulation of sleep-wake cycles. The *sleep* and *wakefulness centers* of the hypothalamus act together with a number of other brain centers to produce alternating cycles of sleeping and wakefulness (see Chapter 15, pp. 476–478).

7. Control of endocrine system functioning. The hypothalamus acts as a kind of helmsman of the endocrine system in two important ways. First, by producing *releasing factors,* it *controls* the release of hormones produced by the anterior pituitary gland. Second, two of its nuclei—specifically, the *supraoptic* and *paraventricular* nuclei, produce hormones—ADH and oxytocin. The endocrine roles of the hypothalamus are described in more detail in Chapter 17.

Hypothalamic disturbances cause a number of disorders in body homeostasis, such as severe body wasting or obesity, sleep disturbances, dehydration, and a broad range of emotional imbalances. For example, infants deprived of a warm, nurturing relationship may develop sleep disorders and fail to thrive. ■

The Epithalamus

The **epithalamus** is the most dorsal portion of the diencephalon and forms the roof of the third ventricle. Extending from its posterior border and visible externally is the **pineal** (pih'-nē-ul) **gland** or **body** (see Figure 12.13). The role of the pineal gland is still being investigated, but it seems to be involved with body rhythms; that is, it is a biological clock that acts through hypothalamic pathways. (The pineal gland is discussed further in Chapter 17.) A cerebrospinal fluid-forming structure called the **choroid plexus** (kor'-oyd plek'-sus) is also part of the epithalamus.

The Brain Stem

From most superior to most inferior, the brain stem regions are the midbrain, pons, and medulla oblongata (see Figures 12.13 and 12.14). Brain stem centers produce the rigidly programmed, automatic behaviors necessary for our survival. Since the brain stem is positioned between the cerebrum and the spinal cord, it also provides transmission pathways for fibers carrying information between higher and lower neural centers. Additionally, the brain stem nuclei are associated with 10 of the 12 pairs of cranial nerves (see Chapter 13).

The Midbrain

Located between the diencephalon superiorly and the pons inferiorly (see Figure 12.14), the **midbrain** is a short region, only 2.5 cm (1 inch) long. On its ventral aspect are two bulging **cerebral peduncles** (peh-dung'-kulz), literally, "little feet of the cerebrum," within which the great descending voluntary motor tracts—

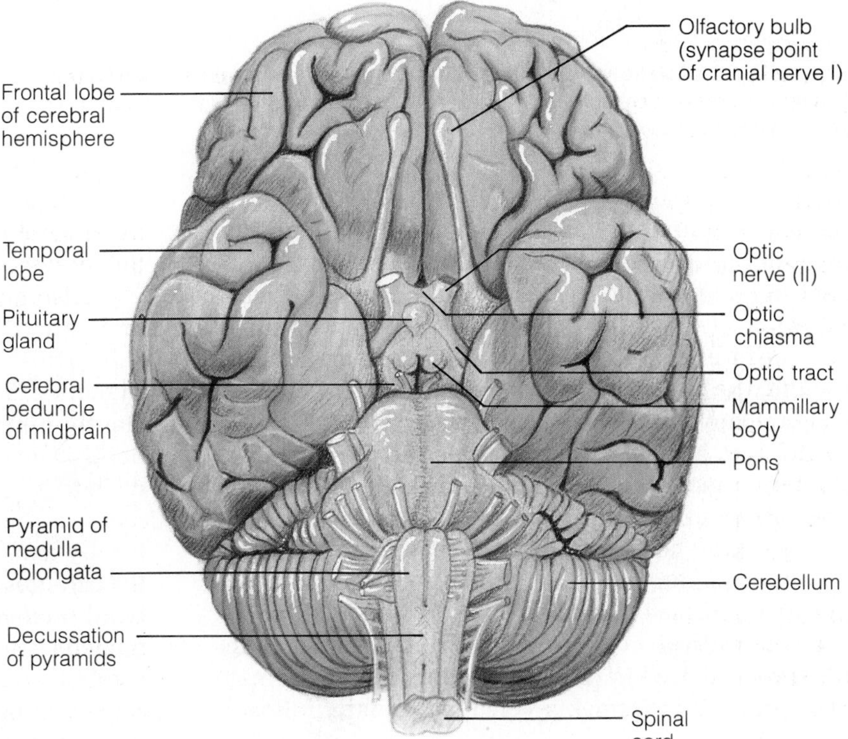

Figure 12.14 Ventral aspect of the human brain, showing the three regions of the brain stem. Only a small portion of the midbrain can be seen; the rest is surrounded by other brain regions.

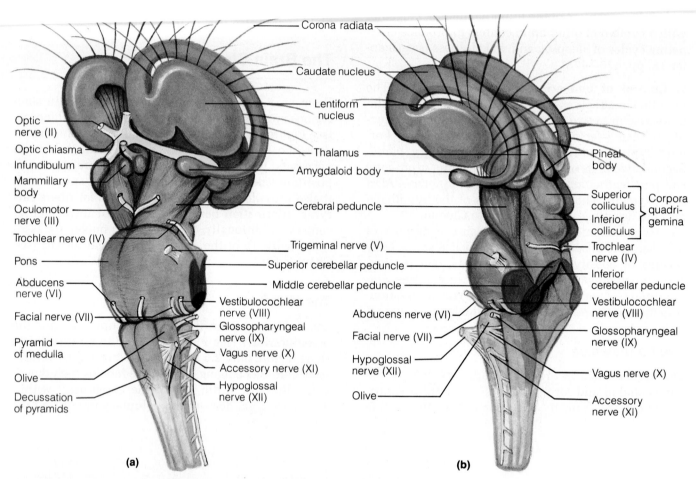

Figure 12.15 Relationship of the brain stem to the diencephalon and basal nuclei. Cerebral hemispheres have been removed. (**a**) Anterolateral view. (**b**) Posterolateral view.

the corticospinal (pyramidal) tracts—are found. The *superior cerebellar peduncles*, also fiber tracts, connect the midbrain to the cerebellum dorsally (Figure 12.15).

As in all parts of the CNS, there is gray matter around the hollow central region (in this case, the **cerebral aqueduct**, which connects the third and fourth ventricles). Nuclei are also scattered in the surrounding white matter. The largest of these nuclei are the **corpora quadrigemina** (kor′-por-uh kwah″-drih-jeh′-mih-nuh), or "four-twin bodies" (two pairs of twins), which raise four dome-like protrusions on the dorsal midbrain surface (see Figures 12.13 and 12.15). The superior pair of nuclei, the **superior colliculi** (kuh-lih′-kyoo-lī), are visual reflex centers that coordinate head and eye movements when we visually follow a moving object. The **inferior colliculi,** immediately below, contain auditory reflex centers.

Also embedded in the white matter of the midbrain are the substantia nigra and red nucleus, paired subcortical motor nuclei that contribute to the extrapyramidal motor pathways (Figure 12.16a). The **substantia nigra** (sub-stan′-shuh nī′-gruh) is a bandlike, darkly pigmented (*nigr* = black) gray mass located deep to the cerebral peduncle; it is the largest nuclear mass in the midbrain. The substantia nigra is structurally and functionally linked to the basal nuclei of the cerebral hemispheres and is considered part of the basal nuclear complex by many authorities. The **red nucleus** is found between the substantia nigra and the cerebral aqueduct. Its natural reddish hue is due to its rich vascular supply and the presence of iron pigment in the cell bodies of its neurons. The red nuclei are

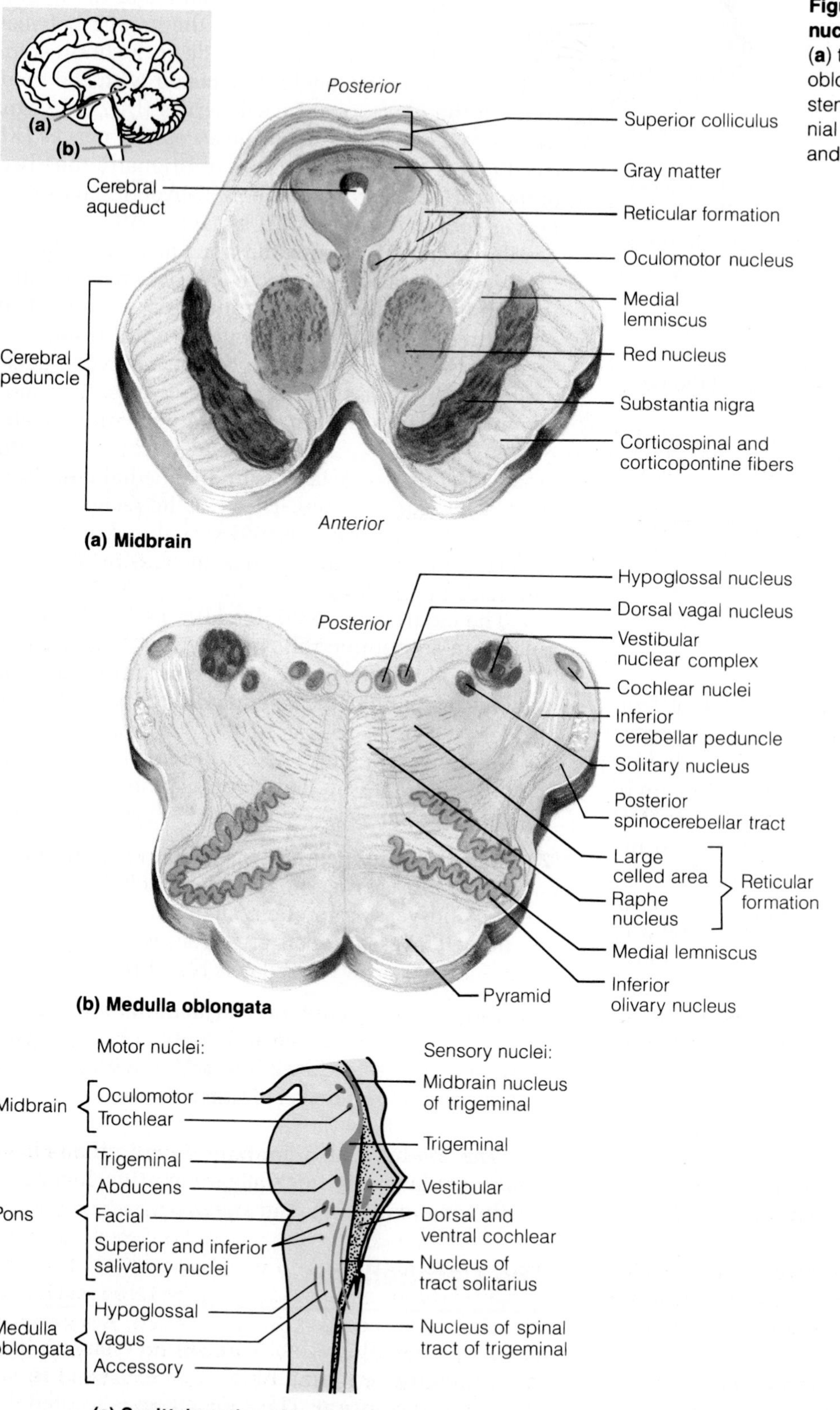

Figure 12.16 Important brain stem nuclei. Horizontal sections through (**a**) the midbrain and (**b**) the medulla oblongata. (**c**) Sagittal section of the brain stem showing the positions of several cranial nerve nuclei in the midbrain, pons, and medulla.

important relay nuclei in the descending motor pathways. The midbrain also contains part of the reticular formation (described later) and nuclei associated with two pairs of cranial nerves, the *oculomotor nerves* (III) and *trochlear* (trok′-lē-er) *nerves* (IV) (see Figures 12.15 and 12.16).

The Pons

The **pons** is the bulging brain stem region wedged [illegible] the pons is chiefly composed of conduction tracts. These fiber pathways course in two directions through the pons. The deep projection fibers run longitudinally and complete the superior-inferior pathway between higher brain centers and the spinal cord. The more superficial fibers are oriented transversely and dorsally as the *middle cerebellar peduncles* and connect the pons bilaterally with the cerebellum posteriorly.

Several cranial nerve pairs issue from pons nuclei (see Figures 12.15 and 12.16c). These include the *trigeminal* (trī-jeh′-mih-nul) *nerves* (V), the *abducens* (ab-doo′-sinz) *nerves* (VI), and the *facial nerves* (VII). Other important pons nuclei are respiratory centers; together with medullary respiratory centers, they help to maintain the normal rhythm of breathing.

The Medulla Oblongata

The conical **medulla oblongata** (meh-dul′-ah ob-lon-gah′-tuh), or simply **medulla,** is the most inferior part of the brain stem. About 2.5 cm long, it blends into the spinal cord at the level of the foramen magnum of the skull (see Figures 12.13 and 12.14). The central canal of the spinal cord continues upward into the lower part of the medulla, and there it broadens out to form the cavity of the fourth ventricle. Thus, both the medulla and the pons help to form the ventral wall of the fourth ventricle. (The dorsal wall of the ventricle is formed by the base of the cerebellum and a choroid plexus.)

The medulla has several externally visible landmarks (see Figures 12.14 and 12.15). Particularly obvious on its ventral aspect are two enlarged fiber tracts called the **pyramids,** which house the large corticospinal tracts descending from the motor cortex. Just above the medulla–spinal cord junction, most of these fibers cross over to the opposite side of the medulla before continuing their descent into the spinal cord. This externally visible crossover point is called the **decussation** (deh″-kuh-sā′-shun) **of the pyramids.** As noted earlier, the consequence of this crossover is that each cerebral hemisphere chiefly controls voluntary muscles on the opposite side of the body.

Also obvious externally are the inferior cerebellar peduncles, the olives, and several cranial nerves. The *inferior cerebellar peduncles* are fiber tracts allowing communication between the medulla and the cerebellum posteriorly. Situated lateral to the pyramids, the **olives** are oval swellings produced by the underlying *olivary nuclei* (see Figure 12.16b). The olivary nuclei relay sensory information on the state of stretch of our muscles and joints. The rootlets of the *hypoglossal nerves* (XII) emerge from the [illegible] and 12.16c). Additionally, the fibers of the *vestibulocochlear* (veh-stih″-byoo-lō-kok′-lē-er) *nerves* (cranial nerves VIII) synapse with numerous vestibular nuclei in both the pons and medulla (see Figures 12.16b and c). Collectively, these nuclei, called the **vestibular nuclear complex,** are involved in mediating responses that maintain equilibrium.

Also housed within the medulla are several nuclei associated with ascending sensory tracts. The most prominent of these are the inferodorsally located *nucleus gracilis* (gra-sih′-lis) and *nucleus cuneatus* (koo″-nē-a′-tis), associated with the medial lemniscus tract (see Figure 12.16b). These nuclei receive impulses from the sensory receptors of the skin and the proprioceptors of skeletal muscles and in turn send impulses to the thalamus.

The small size of the medulla belies its crucial role as an autonomic reflex center involved in maintaining body homeostasis. The most important visceral nuclei found in the medulla include the following:

1. *The cardiac center.* The cardiac center adjusts the force and rate of heart contraction to meet the body's needs.
2. *The vasomotor center.* This center regulates blood pressure by acting on smooth muscle in the walls of blood vessels to effect changes in blood vessel diameter. Vessel constriction causes blood pressure to increase; dilation reduces blood pressure.
3. *The respiratory centers.* The medullary respiratory centers control the rate and depth of breathing and (in a negative feedback interaction with centers of the pons) maintain respiratory rhythm.

The medulla also contains many other nuclei, which regulate activities such as vomiting, hiccuping, swallowing, coughing, and sneezing.

The Cerebellum

The large, cauliflowerlike **cerebellum** (seh″-rih-beh′-lum), literally, our "small brain," is exceeded in size only by the cerebrum. The cerebellum is located dorsal to the pons and medulla (and to the intervening fourth ventricle) and protrudes under the occipital lobes of the cerebral hemispheres, from which it is separated by the transverse fissure (see Figure 12.7). It rests in the posterior cranial fossa of the skull.

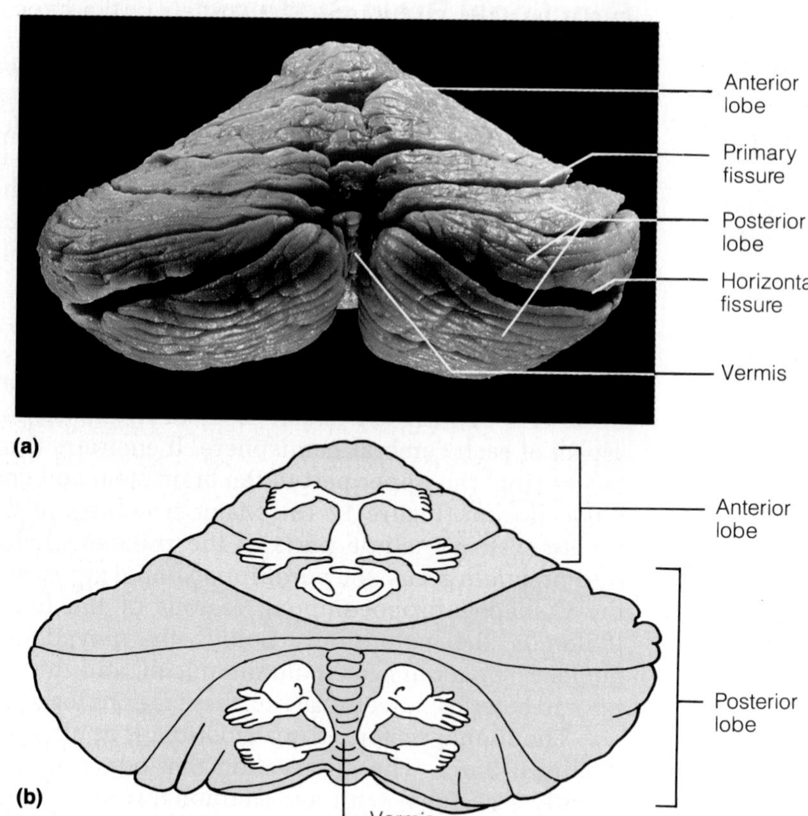

Figure 12.17 Cerebellum. (**a**) Photograph of the superior aspect of the cerebellum. (**b**) Positioning of congruent motor and sensory maps of the body in the cerebellum. The midline cerebellar region corresponding to the vermis coordinates gross movements of the body girdles and trunk. The intermediate portions of each hemisphere coordinate fine movements of the limbs. The lateral hemisphere regions are involved in the planning of movements.

The cerebellum continuously processes inputs received from the cerebral motor cortex, various brain stem nuclei, and sensory receptors to provide the precise timing and appropriate patterns of skeletal muscle contraction needed for smooth, coordinated movements. Cerebellar activity occurs subconsciously; that is, we have no awareness of its functioning. We will examine cerebellar function in more detail shortly, but first let us consider cerebellar anatomy.

Anatomy of the Cerebellum

The cerebellum is bilaterally symmetrical; its two lateral portions, the **cerebellar hemispheres,** are connected medially by the wormlike **vermis** (Figure 12.17). Its surface is heavily convoluted, but since the fissures are all transversely oriented, the cerebellar surface exhibits fine, parallel, pleatlike gyri known as *folia.* Deep fissures subdivide each hemisphere into three lobes: the **anterior, posterior,** and **flocculonodular** (flah″-kyoo-lō-nah-joo-ler) lobes. The flocculonodular lobe, which is situated deep to the vermis and posterior lobe, cannot be seen in a surface view.

Like the cerebrum, the cerebellum has a thin outer cortex of gray matter, internal white matter, and small, deeply situated, paired masses of gray matter, the most familiar of which are the *dentate* nuclei. These nuclei mediate most of the output of the cerebellum (which, however, is initiated by cells of the cerebellar cortex). The pattern of white matter in the cerebellum is quite distinctive. Seen in sagittal section (see Figure 12.13b), it resembles a branching tree, a pattern fancifully called the *arbor vitae* (ar′-ber vī′-tē), literally, "tree of life."

The anterior and posterior lobes of the cerebellum have completely overlapping sensory and motor maps of the entire body (see Figure 12.17b). In both areas, the medial portions (corresponding to the vermis) receive information from the axial (trunk) portion of the body and influence the motor activities of the trunk and girdle muscles by relaying information to the cerebral motor cortex. The intermediate parts of each hemisphere are more concerned with the distal parts of the limbs. In primates, including human beings, the lateral parts of each hemisphere are greatly elaborated. They receive inputs from the association areas of the cerebral cortex and are thought to play a role in planning rather than executing movements; hence, these regions are integrative areas. The small flocculonodular lobes, which receive inputs from the equilibrium apparatus of the inner ears, are concerned with maintaining balance and controlling certain eye movements.

As noted earlier, three paired fiber tracts—the cerebellar peduncles—connect the cerebellum to the brain stem (see Figure 12.15). The **superior cerebellar peduncles,** connect the cerebellum and midbrain. Fibers in these peduncles originate from neurons in the deep cerebellar nuclei and course to the red nucleus, the thalamus, and/or the reticular nuclei of the mid-brain. These brain regions, in turn, [illegible]

[illegible] one-way communication between the pons and the cerebellar neurons (allowing the cerebellum to be advised of voluntary motor activities initiated by the motor cortex). The **inferior cerebellar peduncles** connect the cerebellum with the medulla. These peduncles contain afferent tracts conveying sensory information to the cerebellum from (1) muscle proprioceptors throughout the body and (2) the vestibular nuclei of the brain stem, which are concerned with equilibrium and balance.

Cerebellar Processing

The functional scheme of cerebellar processing seems to be as follows:

1. The cerebral motor cortex initiates voluntary muscle contractions, and through collateral fibers of the pyramidal tracts, simultaneously notifies the cerebellum of its activity.
2. At the same time, the cerebellum is receiving information from proprioceptors throughout the body (regarding the state of tension in the muscles and tendons, and joint position) and from the visual and equilibrium pathways. This information enables the cerebellum to determine where the body is and where it is going.
3. The cerebellar cortex assesses this information and calculates the best way to coordinate the force, direction, and extent of muscle contraction to prevent overshoot, maintain posture, and ensure smooth, coordinated movements.
4. Then, via the superior peduncles, the cerebellum dispatches its "blueprint" for coordination to the cerebral motor cortex as well as to various extrapyramidal motor centers that help regulate skeletal muscle movements.

The cerebellum can be compared to the control system of an automatic pilot. It continually compares the higher brain's intention with the body's actual performance and sends out messages to initiate the appropriate corrective measures. In this way, it helps promote smooth voluntary movements that are precise and economical in terms of muscular effort. Cerebellar injury results in loss of muscle tone and clumsy, disorganized movements. Cerebellar disturbances are described in more detail in Chapter 15.

Functional Brain Systems

Functional [illegible] *reticular system*, which traverses the brain stem, are excellent examples of functional networks (see Table 12.1).

The Limbic System

The limbic system consists of a complex group of fiber tracts and subcortical nuclei located on the medial aspect of each cerebral hemisphere. It encircles (*limbus* = ring) the upper part of the brain stem and corpus callosum (Figure 12.18). Major structures of the limbic system include parts of the rhinencephalon (the *cingulate* gyrus, the *parahippocampal* gyrus, and the C-shaped *hippocampus*), regions of the *hypothalamus*, the *mammillary bodies*, the *amygdaloid nucleus*, some *anterior thalamic nuclei*, and the *fornix* and other fiber tracts that link these regions together.

The limbic system is our *emotional*, or *affective* (feelings) *brain*. The observation that odors are frequently associated with our emotional reactions and memories reflects the origin of much of this system in the primitive "smell brain" (rhinencephalon). Our reactions to odors are rarely neutral (a skunk smells *bad* and repulses us), and odors are intimately linked to our memory traces of emotionally laden experiences.

Extensive connections between the limbic system and lower and higher brain centers allow the system to integrate and respond to a wide variety of environmental stimuli. The hypothalamus is the gatekeeper of these responses, and our emotional states are frequently associated with visceral responses that involve both autonomic and hormonal alterations. The most extreme consequence of severe emotional upheaval (paralyzing fear, extreme joy, or overwhelming grief, for example) is cardiac arrest. Since the limbic system also interacts with higher cerebral cortical areas (frontal lobes and other association areas), there is an intimate relationship between our feelings (mediated by the affective brain) and our thoughts (mediated by the cognitive brain). As a result, we are consciously aware of the emotional richness of our lives. Communications between the cerebral cortex and the limbic system also permit us to refrain from acting out our emotions (rage, hatred, and so forth) when reason warns that a particular response would be unwise.

Because the anatomical connections of the limbic system are so complex, it is difficult to pinpoint discrete pathologies of this system. However, specific lesions in the amygdaloid nuclei and temporal lobe areas result in certain personality changes, such as docile behavior, emotional instability, restlessness, and increased interest in fighting, food, or sex. ■

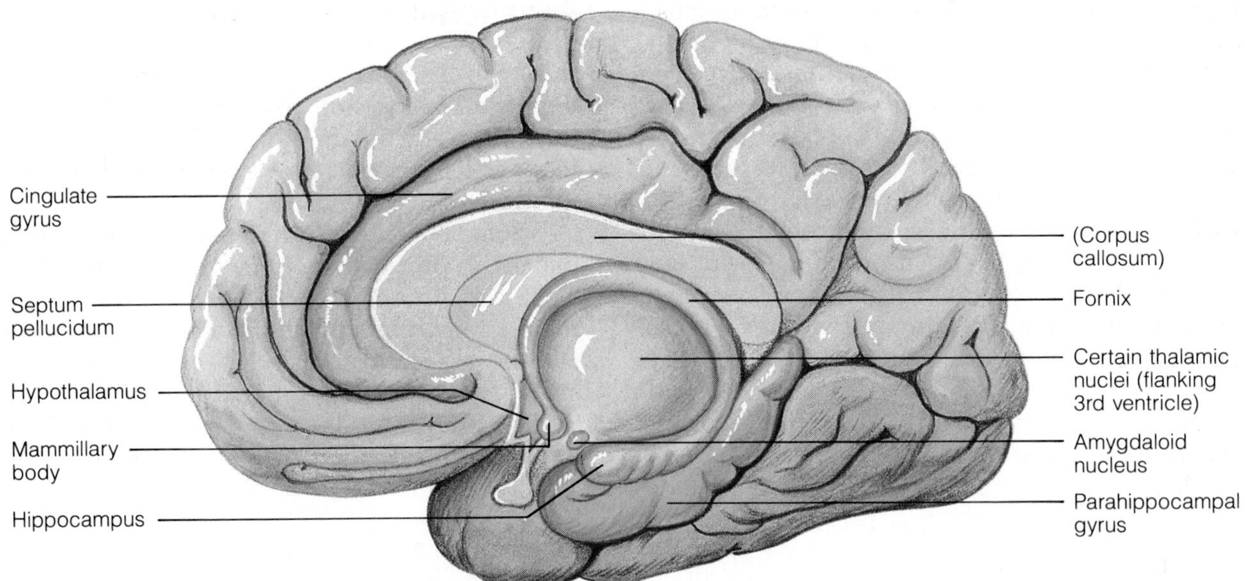

Figure 12.18 Medial view of the brain, showing some of the structures that constitute the limbic system, the emotional-visceral brain. The brain stem is not illustrated.

The Reticular Formation

The **reticular formation** (Figure 12.19) extends through the central core of the medulla oblongata, pons, and midbrain. It is an intricate system composed of fiber pathways, individual neurons, and small scattered nuclei that are separate from the major nuclei of the brain stem. The clustered reticular neurons can be localized into three broad groups along the length of the brain stem: (1) the midline *raphe* (rā′-fē) *nuclei* (*raphe* = seam or crease), which are flanked laterally by (2) the *large-celled region* (see Figure 12.16b) and then (3) the *small-celled region*. The *locus ceruleus*, another important reticular nucleus, is located in the pons and midbrain. You will be hearing more about some of these nuclei when we discuss sleep in Chapter 15.

The outstanding feature of the reticular neurons is their far-flung axonal connections: Individual reticular neurons may connect with cells in the cerebral cortex, hypothalamus, thalamus, cerebellum, and spinal cord. Their widespread connections make reticular neurons ideal for governing the activity of the nervous system as a whole. For example, certain reticular neurons, unless inhibited by other brain areas, send a continuous stream of impulses to the cerebral cortex that is believed to maintain the cortex in an aroused, conscious state. This particular arm of the reticular formation is known as the **reticular activating system (RAS).** Impulses from all the great ascending sensory tracts pass through the RAS, keeping its neurons active. The RAS neurons, in turn, transmit impulses to the cerebral cortex that keep it alert and "awake." The RAS

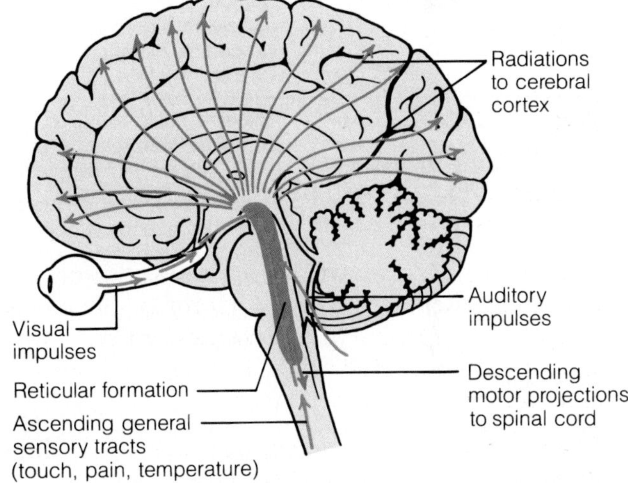

Figure 12.19 The reticular formation. The reticular formation extends the length of the brain stem. A portion of this formation, the reticular activating system (RAS), maintains alert wakefulness of the cerebral cortex. Ascending arrows on the diagram indicate input of sensory systems to the RAS and then output from the reticular formation to the cerebral cortex. Other reticular nuclei are involved in the coordination of muscle activity. Their output is indicated by the blue arrow descending the brain stem.

also appears to act rather like a filter for this flood of sensory inputs. Repetitive, familiar, or weak signals are damped (filtered out), but unusual, significant, or [illegible] the RAS and the cerebral cortex disregard perhaps 99% of all sensory stimuli as unimportant. If this did not occur, the sensory overload would drive us crazy. The drug LSD removes these sensory dampers, promoting a similar type of sensory overload. (Take just a moment to become aware of all the stimuli in your environment. Notice all the colors, shapes, odors, sounds, and so on. How many of these sensory stimuli are you usually aware of?) The activity of the RAS is inhibited by sleep centers located in [illegible]

Some nuclei of the reticular formation are motor nuclei that project to motor neurons in the spinal cord. They act in concert with the cerebellar centers to maintain muscle tone and produce coordinated skeletal muscle contractions. Still other reticular nuclei, such as the vasomotor and cardiac regulatory centers of the medulla, are involved in autonomic regulation of the cardiovascular system.

Table 12.1 Functions of Major Brain Regions

Region	Function
Cerebral hemispheres (pp. 379–388)	Cortical gray matter localizes and interprets sensory inputs, controls gross and skilled skeletal muscle activity, and functions in intellectual and emotional processing; basal (cerebral) nuclei are subcortical motor centers important in initiation of skeletal movements and skills memory (see Chapter 15)
Diencephalon (pp. 389–391)	Thalamic nuclei are relay stations in conduction of (1) sensory impulses (except smell) to cerebral cortex for interpretation, and (2) impulses to and from cerebral motor cortex and lower (subcortical) motor centers, including cerebellum; thalamus is also involved in memory processing (see Chapter 15)
	Hypothalamus is chief integration center of autonomic (involuntary) nervous system; it functions in regulation of body temperature, food intake, water balance, thirst, and biological rhythms and drives; regulates hormonal output of anterior pituitary gland and is an endocrine organ in its own right (produces ADH and oxytocin); part of limbic system
Limbic system (p. 396)	Emotional brain; a functional system involving cerebral and diencephalon structures that mediates emotional response
Brain stem	
Midbrain (pp. 391–393)	Conduction pathway between higher and lower brain centers; its superior and inferior colliculi are visual and auditory reflex centers; substantia nigra and red nuclei are subcortical motor centers; contains nuclei for cranial nerves III and IV
Pons (p. 394)	Conduction pathway between higher and lower brain centers; pontine nuclei relay information from the cerebrum to the cerebellum; its respiratory nuclei cooperate with medullary respiratory centers to control respiratory rate and depth; houses nuclei of cranial nerves V–VII
Medulla oblongata (p. 394)	Conduction pathway between higher brain centers and spinal cord; site of decussation of the pyramidal tracts; houses nuclei of cranial nerves VIII–XII, olivary nuclei, nuclei cuneatus and gracilis (synapse points of ascending sensory pathways transmitting sensory impulses from skin and proprioceptors), and visceral nuclei controlling heart rate, blood vessel diameter, respiratory rate, vomiting, coughing, etc.
Reticular formation (pp. 397–398)	A functional brain stem system that maintains cerebral cortical alertness (reticular activating system) and filters out repetitive stimuli; its motor nuclei help regulate skeletal muscle activity
Cerebellum (pp. 394–396)	Processes information received from cerebral motor cortex and from proprioceptors and visual and equilibrium pathways, and provides "instructions" to cerebral motor cortex and subcortical motor centers that result in proper balance and posture and smooth, coordinated skeletal muscle movements

Protection of the Brain

Nervous tissue is soft and delicate, and the irreplaceable neurons are injured by even slight pressure. However, the brain is protected by a bony enclosure (the skull), membranes (the meninges), and a watery cushion (cerebrospinal fluid). Protection from harmful substances in the blood is provided by the so-called blood-brain barrier. We have already considered the calvarium, the brain's bony encasement, in Chapter 7. Here we will turn our attention to the other protective devices.

Meninges

The **meninges** (meh-nin′-jēz) (singular, meninx) are three connective tissue membranes, intimately associated with one another, that (1) cover and protect the central nervous system organs, (2) enclose venous sinuses, (3) contain cerebrospinal fluid, and (4) form partitions within the skull. From the external to the internal aspect, the meninges are the dura mater, arachnoid, and pia mater (Figure 12.20).

The Dura Mater. The leathery **dura mater** (doo′-ruh mā′-ter), meaning "tough (or hard) mother," is a double-layered membrane where it surrounds the brain. Its outer layer, the *periosteal layer,* is an inelastic membrane attached to the inner surface of the skull; that is, it is the periosteum. Note that there is no dural periosteal layer surrounding the spinal cord. The deeper *meningeal layer* forms the outermost brain covering and continues caudally in the vertebral canal as the dural sheath of the spinal cord. The brain's dural layers are fused together except in certain areas, where they enclose the *dural sinuses* that collect venous blood from the brain and direct it into the internal jugular veins of the neck.

In four places, the meningeal dura mater extends inward to form membranous septa that securely anchor the brain to the skull. These septa, which limit exces-

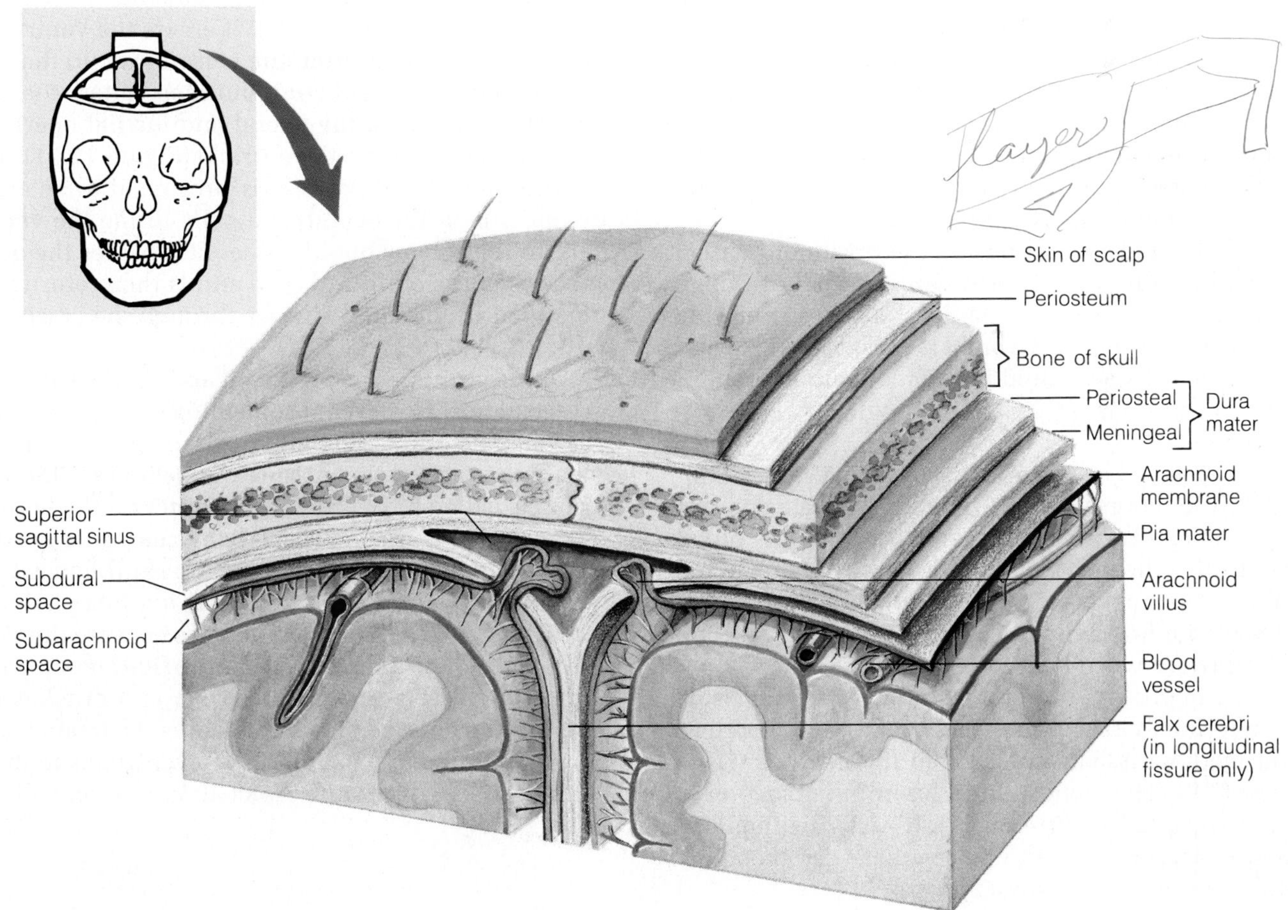

Figure 12.20 Meninges of the brain. Three-dimensional frontal section showing the relationship of the dura mater, arachnoid, and pia mater. The meningeal dura forms the falx cerebri fold, which extends into the longitudinal fissure and attaches the brain to the ethmoid bone of the skull. A dural sinus, the superior sagittal sinus, is enclosed by the dural membranes superiorly. Arachnoid villi, which return cerebrospinal fluid to the dural sinus, are also shown.

sive movement of the brain within the cranium, include the following:

- The falx cerebri (f

- The **falx cerebelli** (sayr″-eh-beh′-lī), a small vertical sheet in the sagittal plane that forms a midline partition that runs along the vermis of the cerebellum.
- The **tentorium** (ten-tor′-ē-um) **cerebelli,** a crescent-shaped dural fold that extends into the transverse fissure (see Figure 12.21). It forms a partition between the cerebral hemispheres (which it helps to support) and the cerebellum.

The Arachnoid. The middle meninx, the **arachnoid** (uh-rak′-noyd), forms a loose brain covering. The only fissure it dips into is the longitudinal fissure between the cerebral hemispheres. It is separated from the dura mater by a very narrow **subdural space.** Beneath the arachnoid membrane is the rather wide **subarachnoid space,** which is spanned by threadlike arachnoid extensions that secure this membrane to the innermost pia mater. (*Arachnida* means "spider," and this membrane was named for its weblike extensions.) The subarachnoid space is filled with cerebrospinal fluid and is the site of all major cerebral arteries and veins. Since the arachnoid is so fine and elastic, these blood vessels are rather poorly protected.

Specialized projections of the arachnoid called **arachnoid villi** (vih′-lī) protrude externally through the dura mater and into the dural sinuses overlying the superior aspect of the brain (see Figures 12.20 and 12.21). Cerebrospinal fluid is absorbed into the venous blood of the sinuses through these villi.

The Pia Mater. The **pia mater** (pē′-uh mā′-ter), meaning "gentle mother," is composed of delicate connective tissue and is richly invested with minute blood vessels. It is the only meninx that clings tightly to the brain, following every convolution. Small arteries entering the brain tissue carry ragged sheaths of pia mater inward with them for short distances.

Meningitis, an inflammation of the meninges, is a serious threat to the brain because a bacterial or viral meningitis may spread into the nervous tissue of the CNS. This condition of brain inflammation is called *encephalitis* (in-seh″-fuh-lī′-tis). Meningitis is usually diagnosed by taking a sample of cerebrospinal fluid from the subarachnoid space. ■

Cerebrospinal Fluid

Cerebrospinal fluid (CSF) f

from blows and other trauma. Additionally, although the brain has a rich blood supply, cerebrospinal fluid helps to nourish the brain and to absorb and remove waste products of neuronal metabolic activity.

CSF is similar in composition to blood plasma, from which it arises. However, it contains less protein and its ion concentration is different. For example, CSF contains more sodium, chloride, magnesium, and hydrogen ions than blood plasma, and fewer calcium and potassium ions. CSF composition, particularly its pH, is important in the control of cerebral blood flow and breathing, as described in later chapters. CSF also transports hormones along the ventricular channels.

The **choroid plexuses,** clusters of capillaries surrounded by ependymal cells that hang from the roof of each ventricle, continuously form CSF from blood plasma by a combination of filtration and active transport processes. Approximately 800 ml of CSF are formed daily; once produced, it enters the ventricles. Some CSF circulates from the ventricles into the central canal of the spinal cord, but most enters the subarachnoid space via the lateral and medial apertures in the walls of the fourth ventricle (Figure 12.21). The constant motion of the CSF is presumed to be aided by the cilia of the ependymal cells lining the ventricles. In the subarachnoid space, CSF bathes the outer surfaces of the brain and cord and is then returned to the blood in the dural sinuses through the arachnoid villi.

Ordinarily, CSF is produced and drained at a constant rate. However, if something (such as a tumor) obstructs its circulation and/or drainage, it begins to accumulate and exert pressure on the brain. This condition is called *hydrocephalus* ("water on the brain"). Hydrocephalus in a newborn baby causes its head to enlarge; this is possible because the skull bones have not yet fused. In adults, however, hydrocephalus is more likely to result in brain damage because the skull is rigid and hard, and accumulating fluid compresses the blood vessels serving the brain and crushes the soft nervous tissue. Hydrocephalus is treated surgically by inserting a shunt in the ventricles to drain off the excess fluid and direct it into a vein in the neck. ■

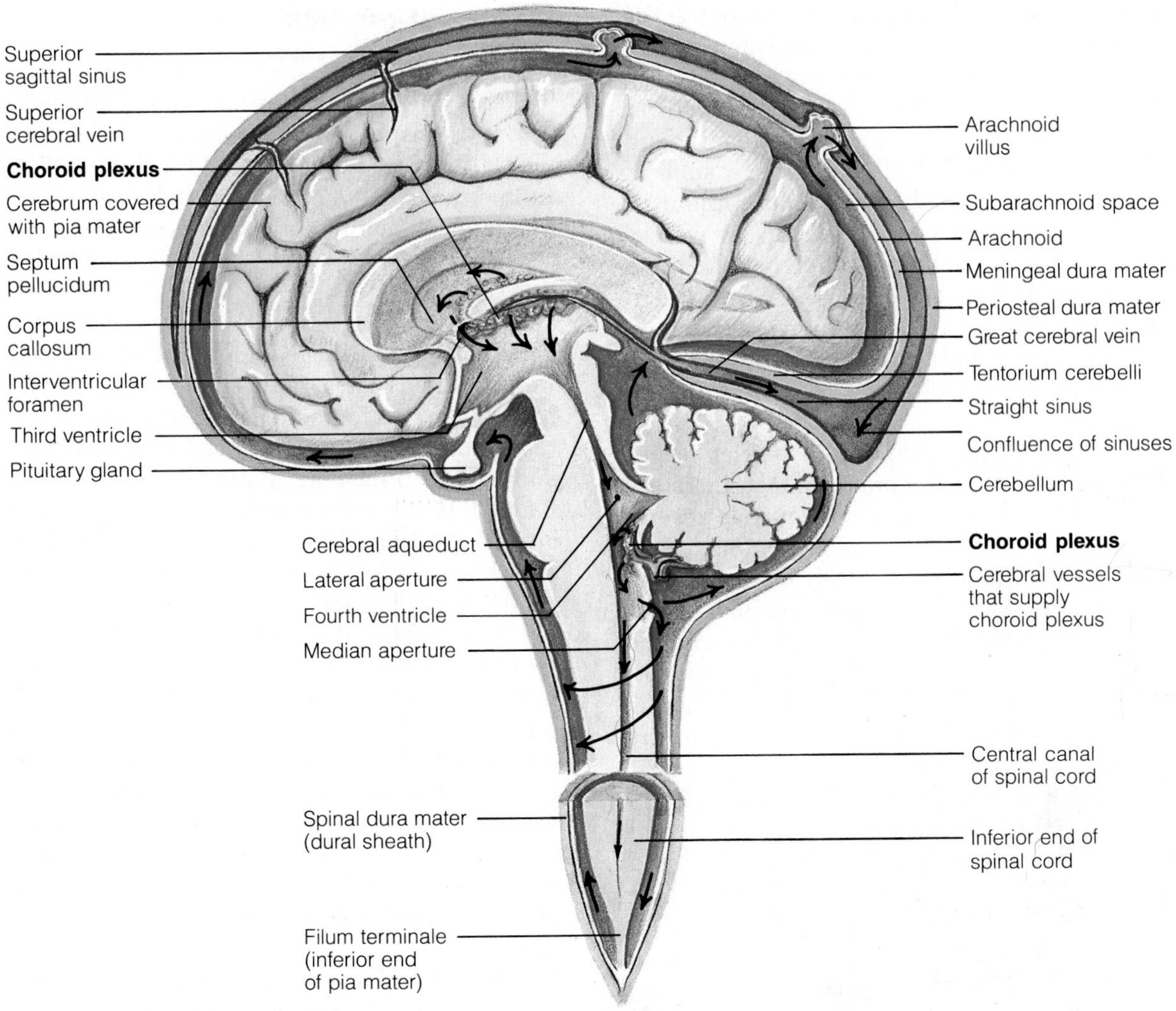

Figure 12.21 Location and circulatory pattern of cerebrospinal fluid. Direction of circulation is indicated by the arrows. (The relative position of the right lateral ventricle is indicated by the pale blue area deep to the corpus callosum and septum pellucidum.)

Blood-Brain Barrier

The **blood-brain barrier** is a protective mechanism that helps ensure that the brain's environment remains stable. No other body tissue is so absolutely dependent on a constant internal milieu as is the brain. In other body regions, the extracellular concentrations of hormones, amino acids, and ions continuously undergo small fluctuations, particularly after eating or exercise. If the brain were exposed to such chemical variations, uncontrolled neural activity might result, because some hormones and amino acids serve as neurotransmitters and certain ions (particularly potassium) modify the threshold for neuronal firing.

Blood-borne substances within the brain's capillaries are separated from the extracellular space and neurons by the following structures:

1. The continuous endothelium of the capillary wall.
2. A relatively thick basement membrane surrounding the external face of the capillary.
3. The bulbous "feet" of the astrocytes that cling to the capillaries.

Since the capillary endothelial cells are joined all around by *tight junctions* (see Chapter 20, p. 625), they are the least permeable capillaries in the entire body; this relative impermeability of brain capillaries constitutes most (if not all) of the blood-brain barrier. Although it has long been assumed that the astrocytes contribute to the blood-brain barrier, it now appears that their major role is to provide developmental signals for the transformation of brain capillaries to their relatively impermeable state.

The only substances that pass freely through the blood-brain barrier are water; nutrients (such as glucose, essential amino acids, and some electrolytes), [illegible] proteins, certain toxins, and most antibiotics and other drugs, are prevented from entering brain tissue. Small nonessential amino acids and potassium ions not only are prevented from entering the brain, but also are actively pumped from the brain across the capillary endothelium.

The structure of the blood-brain barrier is not identical in all regions of the brain. In fact, there are areas where the blood-brain barrier is entirely absent, and in such areas, the capillary endothelium is quite porous, allowing proteins and small organic molecules easy access to the neural tissue. One such region is in the hypothalamus. Since the hypothalamus is involved in regulating water balance and many metabolic activities, lack of a blood-brain barrier here is essential to allow the hypothalamus to sample the chemical composition of the blood.

The blood-brain barrier is incompletely developed in newborn and premature infants, and potentially toxic substances can readily enter the CNS and cause problems not seen in adults. For example, bilirubin, a normal breakdown product of the hemoglobin pigment found in red blood cells, is toxic to neurons. In adults, bilirubin is excluded from brain tissue, and the liver converts it to a form that can be excreted in urine. In newborn and premature infants, however, the liver is often immature, and bilirubin may accumulate in levels that cause it to be deposited in fatty tissues of the body, causing them to become *jaundiced*, or yellow. The brain also becomes jaundiced, and if enough bilirubin is deposited, signs of neurological damage appear, a condition called *kernicterus* (ker-nik′-ter-is). Kernicterus is evidenced by diminished or absent infant reflexes (sucking and others), lethargy, reduced motor tone, and a high-pitched cry. If untreated, this condition leads to permanent mental retardation.

Injury to the brain, whatever the cause—direct physical trauma, infections, chemical toxins, and so on—may cause a localized breakdown of the blood-brain barrier. Most likely, this reflects actual destruction of capillary endothelial cells or their tight junctions. This can be fortunate in the case of CNS infections, since antibiotics are now allowed access to the brain. ■

Homeostatic Imbalances of the Brain

[illegible]

Traumatic Brain Injuries

Head injuries are the leading cause of accidental death in the United States. Consider, for example, what happens when you forget to fasten your seat belt and then rear-end another car. Your head is moving and then is suddenly stopped as it hits the windshield. Brain damage is caused not only by localized injury at the site of the blow (the coup injury), but also by the ricocheting effect as the brain hits the opposite end of the skull (contrecoup injury).

A *concussion* occurs when brain injury is slight and the symptoms are mild and transient. The victim may be dizzy, "see stars," or lose consciousness briefly, but no permanent neurological damage is sustained. The result of marked tissue destruction is a brain *contusion*, which produces a broad variety of signs and symptoms. In cortical contusions, the individual may remain conscious, but severe brain stem contusions always result in unconsciousness (coma), ranging in duration from hours to a lifetime because of injury to the reticular activating system.

After head blows, fatal consequences may result from subdural or subarachnoid hemorrhage (bleeding from ruptured vessels into those spaces). Individuals who are initially lucid following head trauma and then begin to deteriorate neurologically are, in all probability, hemorrhaging intracranially. Accumulating blood within the skull increases intracranial pressure and compresses brain tissue. When the brain stem is forced inferiorly through the foramen magnum by the increasing pressure, control of blood pressure, heart rate, and respiration are lost. Intracranial hemorrhages are treated by removal of the hematoma (localized blood mass) and surgical repair of the ruptured blood vessels.

Another consequence of traumatic head injury is *cerebral edema*, or swelling of the brain. This results both from exudate formation during the inflammatory response and from water uptake by brain tissue (particularly by astrocytes). At present, anti-inflammatory steroids such as *prednisone* and *cortisol* are routinely administered to head injury patients in an attempt to prevent cerebral edema from aggravating brain injury.

Degenerative Brain Diseases

The most important degenerative diseases of the central nervous system are cerebrovascular accidents, Alzheimer's disease, and multiple sclerosis.

Cerebrovascular Accidents. *Cerebrovascular* (suh-rē″-brō-vas′-kyoo-ler) *accidents (CVA)*, commonly called *strokes*, are the single most common nervous system disorder and the third leading cause of death in the United States. Strokes occur when blood circulation to a brain area is blocked and vital brain tissue dies. (Deprivation of blood supply to any tissue is called *ischemia* [is-kē′-mē-uh] and results in deficient oxygen and nutrient delivery to cells.)

A particularly common cause of CVA is blockage of a cerebral artery by a fixed or floating blood clot. Other causes include compression of brain tissue by hemorrhage or edema and progressive narrowing of brain vessels by arteriosclerosis. Whatever the cause, the type and extent of neurological deficits that follow reflect the area and extent of damage. Those who survive a CVA typically exhibit paralysis of one side of the body and sensory deficits. If the language areas are damaged, difficulties in understanding or vocalizing speech develop.

Fewer than 35% of those surviving an initial CVA are alive three years later, because those suffering CVAs from blood clots (the most common cause) are likely to have recurrent clotting problems. Even so, the picture is not hopeless. Some patients recover at least a portion of their lost faculties, because after neurons die and are removed, adjacent undamaged neurons spread into the affected CNS area and take over some lost functions. Indeed, most of the recovery seen after brain injury is due to this phenomenon. However, neuronal spread is limited to only about 5 mm (0.2 inch); thus, large lesions always result in less than complete regeneration.

Not all strokes are "completed." Temporary episodes of reversible cerebral ischemia, called *transient ischemic attacks (TIAs)*, last from 5–50 minutes and are characterized by symptoms such as numbness, temporary paralysis, and impaired speech. While these deficits are not permanent, they do constitute "red flags" that warn of impending, more serious CVAs.

Alzheimer's Disease. **Alzheimer's** (alz′-hī″-merz) **disease** is a progressive degenerative disease of the brain that ultimately results in dementia (mental deterioration). CVA patients represent nearly half of nursing home patients; Alzheimer's patients represent most of the remaining half. Although this disorder is usually seen in elderly people (in fact, the condition was originally called *senile dementia*), it may begin in middle age. Victims of Alzheimer's disease exhibit widespread cognitive deficits, including memory loss (particularly for recent events), short attention span, and loss of orientation to their environment. Formerly good-natured individuals become irritable, moody, and confused. Ultimately, hallucinations occur.

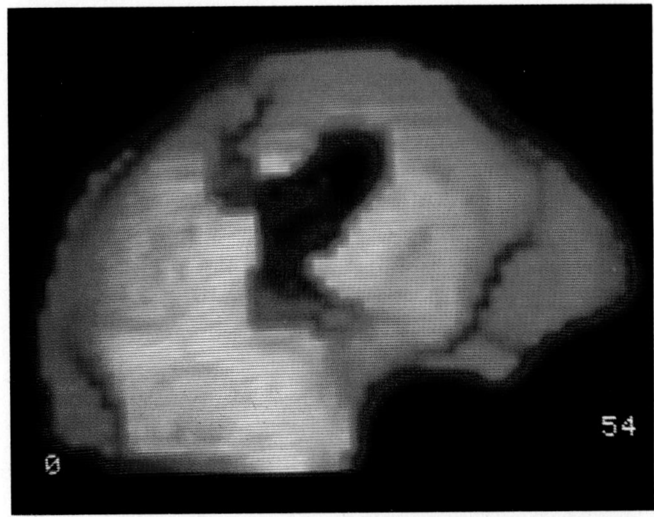

Figure 12.22 PET (positron emission tomography) scan of the brain of a person with Alzheimer's disease. This scan shows a predominance of blues and greens, which indicates a decline in glucose use in the cerebral cortex and hence reduced cortical activity.

Alzheimer's disease is associated with obvious structural changes in the brain, particularly in the cerebral cortex and hippocampus, areas involved with cognitive functions and memory. Examples include the presence of abnormal protein deposits that form so-called *senile plaques*, *neurofibrillar tangles* (twisted fibrils) within neurons, and localized brain atrophy. These degenerative changes develop over a period of several years, during which time the family members watch the person they love "disappear." It is a long and painful process. Additionally, PET scans show significant declines in glucose use in the cortex of Alzheimer's patients (Figure 12.22), information that can be used to identify the incipient disease before the patient exhibits its actual symptoms.

Although there are deficits of acetylcholine neurotransmitter in brain areas affected by Alzheimer's disease, thus so far acetylcholine replacement therapy has been unsuccessful in reversing or stopping the course of the disease. However, infusing a natural chemical called *nerve growth factor* appears to partially reverse brain shriveling and to improve memory. This is an exciting discovery that may hold promise for making unhealthy neurons healthy again.

Multiple Sclerosis. **Multiple sclerosis** (skler-ō′-sis), a demyelinating disease, is believed to be an autoimmune disorder triggered by viral infection. Its onset is [illegible]

Experimental treatments being tested include transplants of myelin-producing oligodendrocytes into the brain and the use of drugs that appear to enhance neural transmission in the absence of myelin.

As the disease progresses, lesions of the myelin sheaths become hardened and scarlike, forming *scleroses*, which severely impair normal transmission of nerve impulses. This short circuiting results in symptoms that vary with the extent and locality of the demyelinating lesions, but few neural functions escape unaffected. Neurological deficits include impaired vision and speech, muscular weakness, tremors, paralysis, diminished sensation, confusion, loss of memory, and mood changes ranging from apathy and depression to euphoria. Eventually, voluntary motor control is completely lost, and the patient becomes bedridden. Inability to empty the urinary bladder leads to frequent urinary tract infections, and overwhelming infection is the usual cause of death. ■

THE SPINAL CORD

Structure and Protection of the Spinal Cord

About 42 cm (17 inches) long and 1.8 cm (¾ of an inch) thick, the glistening-white *spinal cord* provides a two-way conduction pathway to and from the brain (Figure 12.23). In addition, it is a major reflex center: Spinal reflexes are initiated and completed at the spinal cord level. We will discuss reflex functions of the cord in Chapter 13. The following section focuses on the anatomy of the cord and on the location and naming of its ascending and descending tracts.

Enclosed within the vertebral column, the spinal cord extends from the foramen magnum of the skull (where it is continuous with the medulla of the brain stem) to the level of the first lumbar vertebra, just inferior to the ribs. In young children, the spinal cord is relatively longer, usually terminating at the level of the third lumbar vertebra.

Like the brain, the spinal cord is protected by bone, cerebrospinal fluid, and meninges. The single-layered dura mater of the spinal cord, called the *spinal dural sheath* (see Figures 12.21 and 12.24), is not attached to the bony walls of the vertebral column. Between the bony vertebrae and the dural sheath is a [illegible]

well beyond the end of the spinal cord in the vertebral canal, approximately to the second sacral vertebra (S_2). Since the spinal cord ends at L_1, there is usually no danger of damaging the spinal cord beyond L_3; thus, the subarachnoid space within the meningeal sac inferior to that point provides a nearly ideal spot for removing cerebrospinal fluid for testing. This procedure is called a *lumbar puncture* or *tap*.

Inferiorly, the spinal cord terminates in a tapering structure called the **conus medullaris** (kō′-nus meh″-dyoo-layr′-is). A fibrous extension of the pia mater, the **filum terminale** (fī′-lum ter″-mih-nah′-lē) ("terminal filament") extends downward from the conus medullaris to the posterior surface of the coccyx, where it attaches (Figure 12.23).

In humans, 31 pairs of spinal nerves arise from the cord by paired roots and exit from the vertebral column via the intervertebral foramina to travel to the body regions they serve. The spinal cord is about the width of a thumb for most of its length, but it has obvious enlargements in the cervical and lumbosacral regions, where the nerves serving the upper and lower limbs issue from the cord. These enlargements are the *cervical* and *lumbosacral enlargements*, respectively (see Figure 12.23a). Because the cord does not reach the end of the vertebral column, the lumbar and sacral spinal nerve roots angle sharply downward and travel inferiorly through the vertebral canal for some distance before exiting through their intervertebral foramina. The collection of nerve roots at the inferior end of the vertebral canal is named the **cauda equina** (kah′-duh ē-kwī′nuh) because of its resemblance to a horse's tail. This rather strange arrangement reflects the different rates of growth of the vertebral column and spinal cord. In embryos, the spinal cord occupies the entire length of the vertebral canal. Subsequently, however, the vertebral column grows inferiorly more rapidly than the spinal cord, forcing the lower spinal nerve roots to "chase" their exit points inferiorly through the vertebral canal.

The spinal cord is somewhat flattened from front to back, and two grooves mark its surface: the **anterior median fissure** and the more shallow **posterior median sulcus** (see Figure 12.25b and c). These grooves extend the length of the cord and partially divide it into right and left portions. As mentioned earlier, the gray matter of the cord is located on the inside, the white matter outside.

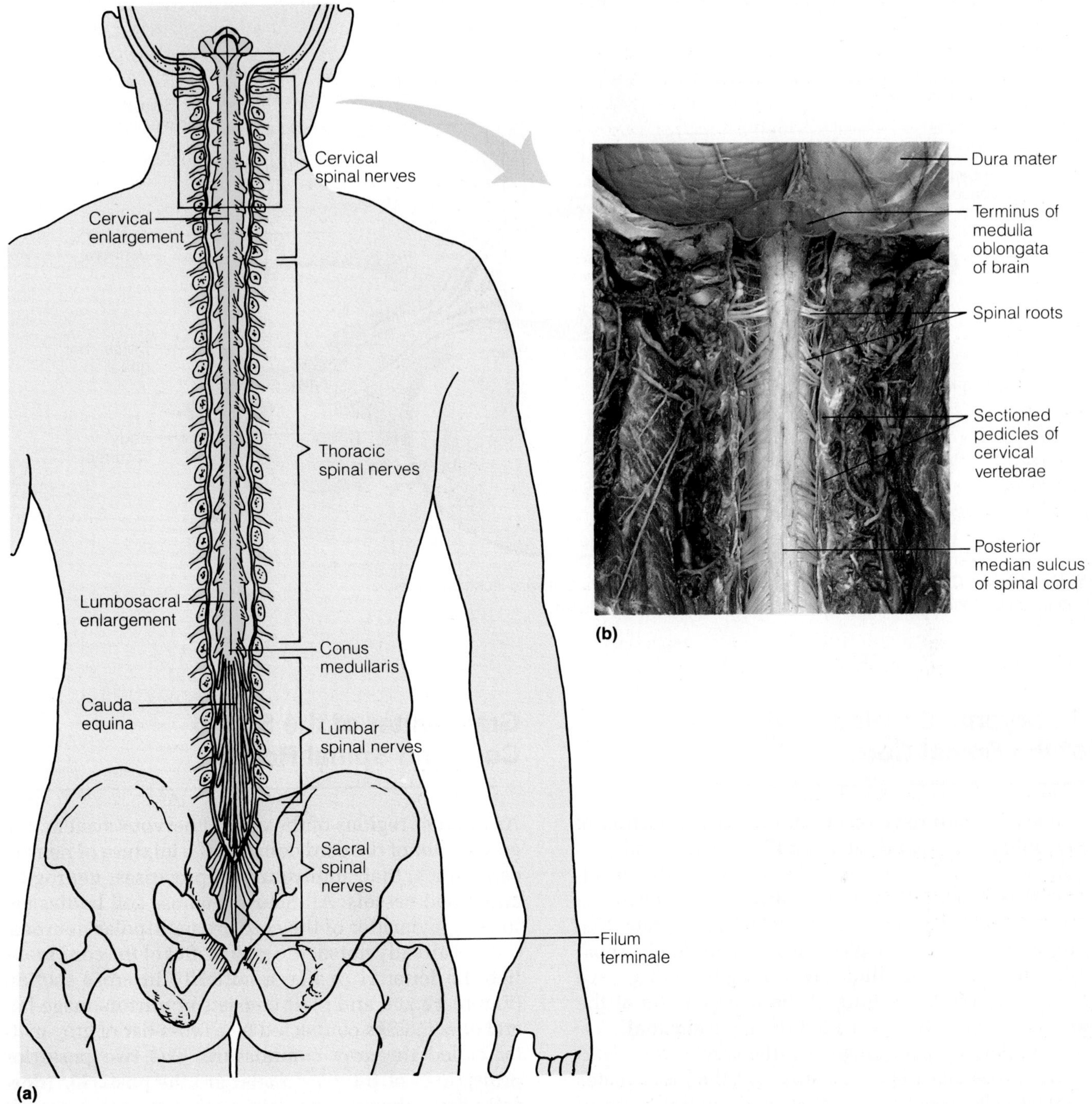

Figure 12.23 Structure of the spinal cord. (**a**) The vertebral arches have been removed to show the dorsal aspect of the spinal cord (and its nerve roots). The dura mater is cut and reflected laterally. The various regions of the spinal cord are indicated in relation to the vertebral column as cervical, thoracic, lumbar, and sacral. (**b**) Photograph of the cervical region of the spinal cord. The meningeal coverings have been removed to show the spinal roots that give rise to the spinal nerves.

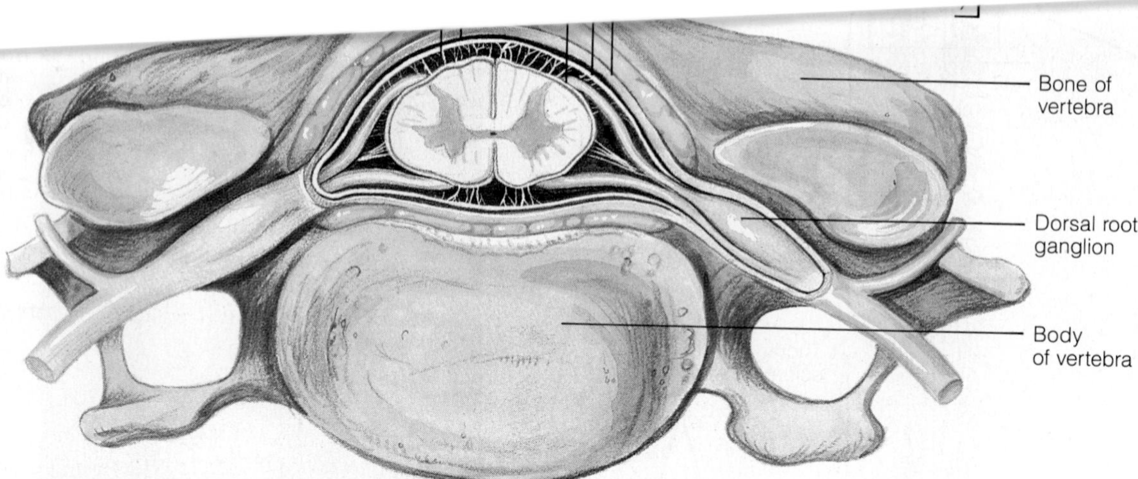

Figure 12.24 Cross section through the spinal cord. This diagram illustrates the relationship of the spinal cord to the surrounding vertebral column.

Embryonic Development of the Spinal Cord

The spinal cord develops from the caudal portion of the embryonic neural tube (see Figures 12.2 and 12.3 on p. 377). By the sixth week, each side of the developing cord exhibits two recognizable gray masses: a dorsal **alar** (a′-ler) **plate** and a ventral **basal plate** (Figure 12.25a). The two plates are incompletely separated by the **sulcus limitans,** a longitudinal groove formed by the thickening of the lateral walls of the neural tube cavity, now called the **central canal.**

As development progresses, the gray matter plates expand (the alar plates dorsally and the basal plates ventrally) to produce the H-shaped central mass of gray matter of the adult spinal cord (Figure 12.25b). Neural crest cells that come to lie alongside the cord form the *dorsal root ganglia* containing sensory neuron cell bodies, which send their axons into the dorsal aspect of the cord. The external white matter of the cord is formed by ascending and descending fibers.

Gray Matter of the Spinal Cord and Spinal Roots

As in other regions of the central nervous system, the gray matter of the cord consists of a mixture of neuron cell bodies, their unmyelinated processes, neuroglia, and blood vessels. All neurons whose cell bodies are in the gray matter of the cord are multipolar neurons.

As already noted, the gray matter of the cord looks like the letter H or like a butterfly in cross section (Figure 12.25b and c). It consists of mirror-image lateral gray masses connected by a cross-bar of gray matter called the **gray commissure.** The two posterior projections of the gray matter are the **posterior (dorsal) horns;** the anterior pair are the **anterior (ventral) horns.** An additional pair of gray matter columns, the small **lateral horns,** is seen in the thoracic and upper lumbar segments of the cord.

The posterior horns contain association neurons. The anterior horns contain nerve cell bodies of somatic motor neurons, which send their axons out via the **ventral root** of the spinal cord (see Figure 12.25c) to ultimately reach the skeletal muscles (effector organs). The amount of ventral gray matter present at a given level of the spinal cord reflects the amount of skeletal

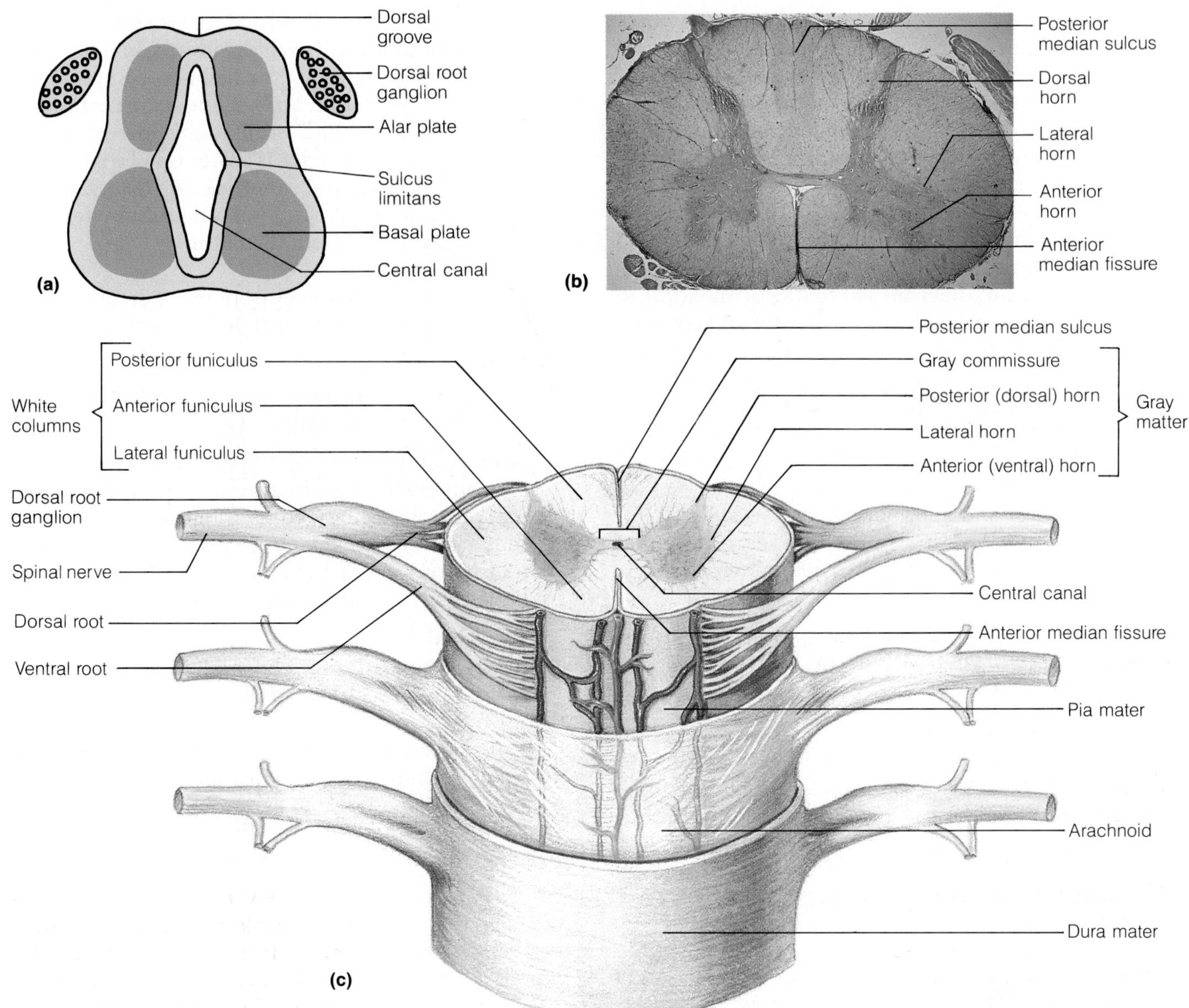

Figure 12.25 The spinal cord: embryonic development and adult structure. The spinal cord develops from the caudal end of the neural tube. (**a**) The spinal cord at six weeks of development. The alar and basal plates (aggregations of gray matter) have formed, and dorsal root ganglia have arisen from neural crest cells. (**b**) Photomicrograph of the adult spinal cord, cross-sectional view (10×). (**c**) Three-dimensional view of the adult spinal cord and its meningeal coverings.

muscle innervated at that particular level. Thus, the anterior horns are largest in the limb-innervating cervical and lumbosacral regions of the cord and are responsible for the enlargements seen in those regions.

The symptoms of *poliomyelitis* (pō″-lē-ō-mī″-uh-lī′-tis) (*polio* = gray matter; *myelitis* = inflammation of the spinal cord) result from the destruction of anterior horn motor neurons by the poliovirus. Initial symptoms include fever, headache, muscle pain and weakness, and loss of certain somatic reflexes. However, as the disease progresses, the patient develops paralysis and atrophy of the muscles served. The victim may die from paralysis of the respiratory muscles or from cardiac arrest if neurons in the medulla oblongata are destroyed. In most cases, the poliovirus enters the body in feces-contaminated water (such as might occur in a public swimming pool), and the incidence of the disease has traditionally been highest in children and during the summer months. Fortunately, the Salk and Sabin polio vaccines have nearly eliminated this disease. ■

The lateral horn neurons are autonomic (sympathetic division) motor neurons that serve visceral organs. Their axons leave the cord via the ventral root

... form the **dorsal roots** of the spinal cord (see Figure 12.25c). The nerve cell bodies of the associated sensory neurons are found in an enlarged region of the dorsal root called the **dorsal root ganglion** or **spinal ganglion.** After entering the cord, their axons may take a number of routes. For example, some enter the posterior white matter of the cord directly and travel to synapse at higher cord or brain levels, while others synapse with neurons in the spinal cord gray matter at the level they enter.

The dorsal and ventral roots are very short and fuse laterally to form the **spinal nerves.** The spinal nerves, which are part of the peripheral nervous system, are considered in Chapter 13.

White Matter of the Spinal Cord

The white matter of the cord is composed of myelinated and unmyelinated nerve fibers running in three directions: (1) ascending to higher centers, (2) descending to the cord from the brain (or within the cord to lower levels), and (3) running from one side of the cord to the other. The vertical (ascending and descending) tracts predominate.

The white matter on each side of the cord is divided into three **funiculi** (fuh-nih'-kyoo-lī) (singular, funi-

... ascending and descending pathways to and from the brain:

1. Most of them cross over from one side of the CNS to the other (decussate) at some point along their journey.
2. Most consist of a chain of two or three neurons that contribute to successive tracts.
3. Most exhibit *somatotopy,* a precise spatial relationship among the tract fibers that reflects the orderly mapping of the body. For example, in an ascending sensory tract, somatotopy refers to the fact that fibers transmitting inputs from the upper parts of the body lie laterally and in a specific relationship to those conveying information from lower body regions.
4. All tracts are paired (right and left) with a member of the pair on each side of the spinal cord (or brain).

For the most part, the names of the spinal tracts reveal both their origin and destination, but there are exceptions. The principal ascending and descending tracts of the spinal cord are illustrated schematically in Figure 12.26.

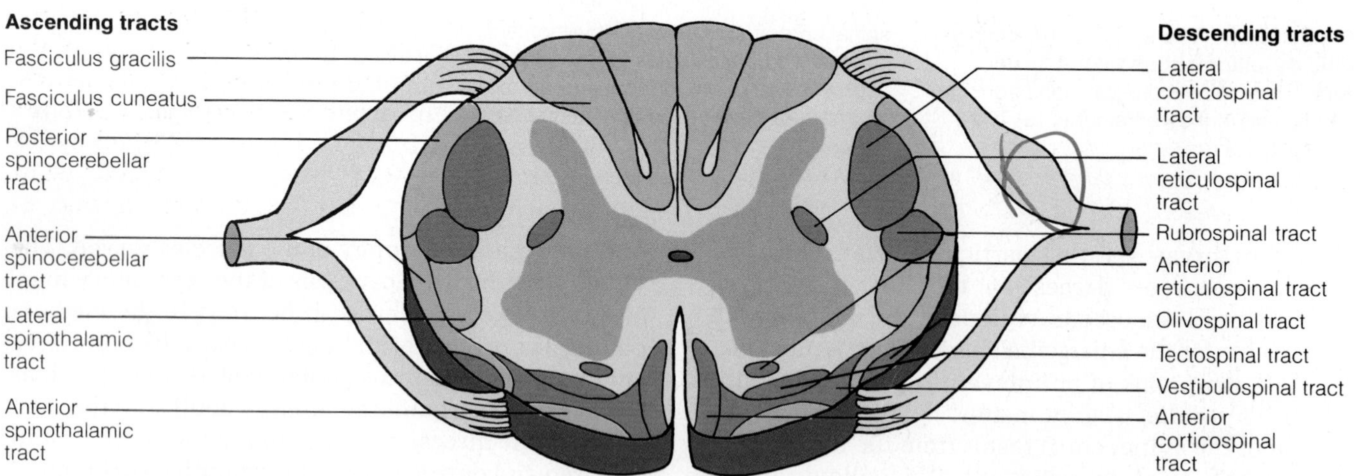

Figure 12.26 Major ascending (sensory) and descending (motor) tracts of the spinal cord, cross-sectional view. Ascending tracts are shown in pink and labeled only on the left side of the figure. Descending tracts are shown in blue and labeled on the right.

Ascending (Sensory) Pathways and Tracts

The ascending pathways conduct afferent impulses that enter the spinal cord upward via chains of two or three successive neurons (*first-*, *second-*, and *third-order neurons*) to various parts of the brain. Most of this information results from stimulation of touch, pressure, temperature, and pain receptors in the skin and from stimulation of proprioceptors, which report on the degree of stretch in the muscles, tendons, and joints. In general, this information is conveyed along six main pathways on each side of the spinal cord. Four of these pathways, the **fasciculi** (fuh-sih′-kyoo-lī) **cuneatus** and **gracilis** and the **lateral** and **anterior spinothalamic tracts,** transmit impulses to the sensory cortex for conscious interpretation; the last pair, the **anterior** and **posterior spinocerebellar tracts,** convey information from proprioceptors (muscle or tendon stretch) to the cerebellum, which uses this information to coordinate skeletal muscle activity. Theseascending pathways and the fiber tracts contributing to them are described in Table 12.2; four of these tracts are illustrated in Figure 12.27. Notice that only the fasciculi cuneatus and gracilis are formed by the axons of sensory (first-order) neurons. The other ascending tracts described are formed by axons of second-order neurons (association neurons) of the spinal cord with which the transmitting sensory neurons synapse.

Recall that our ability to identify and appreciate the kind of sensation being transmitted—that is, whether it is touch, temperature, pain, and so on—depends on the specific location of the target neurons in the cerebral sensory cortex, with which the ascending pathway fibers synapse, and not on the nature of the message, which is always action potentials. Each sensory nerve fiber is analogous to a "labeled line" in a telephone system and is identified by the brain as associated with a particular sensory modality. Receiving information via a certain line always tells the brain "who" is calling, whether it is a taste bud or a pressure receptor. Furthermore, when a sensory neuron is excited, the brain interprets its activity as a specific sensory modality, regardless of how the receptor is activated. If we press a Pacinian receptor of the index fingertip, jolt it with an electrical shock, or electrically stimulate the area of the somatosensory cortex that recognizes it, the result will be the same: We will perceive deep touch or pressure and will interpret it as coming from the index fingertip. This phenomenon, by which the brain refers sensations to their usual point of stimulation, is called *projection*.

Descending (Motor) Pathways and Tracts

The descending tracts that deliver efferent impulses from the brain to the spinal cord are grouped as **pyramidal** or **extrapyramidal tracts.** These tracts will be briefly overviewed here, but most of the detailed information is provided in Table 12.3 on p. 413.

The pyramidal or **corticospinal tracts** are the major motor pathways concerned with voluntary movement (Figure 12.28a on p. 412). The neurons of these tracts, called **upper motor neurons** (but actually interneurons), descend without synapsing from the pyramidal cells of the primary motor area all the way to the spinal cord. There they synapse both with interneurons and with **lower motor neurons** in the anterior horns, principally with those neurons controlling limb muscles. The anterior horn motor neurons activate the skeletal muscles with which they are associated.

The remaining descending tracts are lumped together as the extrapyramidal tracts, or system, because their fibers lie outside the medullary pyramids (pyramidal tracts). Extrapyramidal tracts include the **rubrospinal** (roo″-brō-spī′-nul), **vestibulospinal, reticulospinal,** and **tectospinal tracts,** which originate in different subcortical motor nuclei of the brain stem.

Although the cerebellum orchestrates the activity in most extrapyramidal pathways, no motor efferents descend directly from the cerebellum to the spinal cord. Rather, as already described, the cerebellum influences motor activity by acting through relays on the motor cortex. The cerebellum also interacts with the basal nuclei and with lower brain stem motor centers such as the red nucleus of the midbrain and various reticular nuclei. These nuclei, in turn, influence lower motor neurons via the extrapyramidal tracts. The extrapyramidal system is complex, and the pathways are multisynaptic. They are most involved in regulating (1) the axial muscles involved in posture, (2) the muscles controlling coarse movements of the proximal portions of the limbs, and (3) head, neck, and eye movements that act to follow objects in the visual field. Many of the activities controlled by the extrapyramidal system are heavily dependent on reflex activity. One of these tracts, the rubrospinal tract, is illustrated in Figure 12.28b.

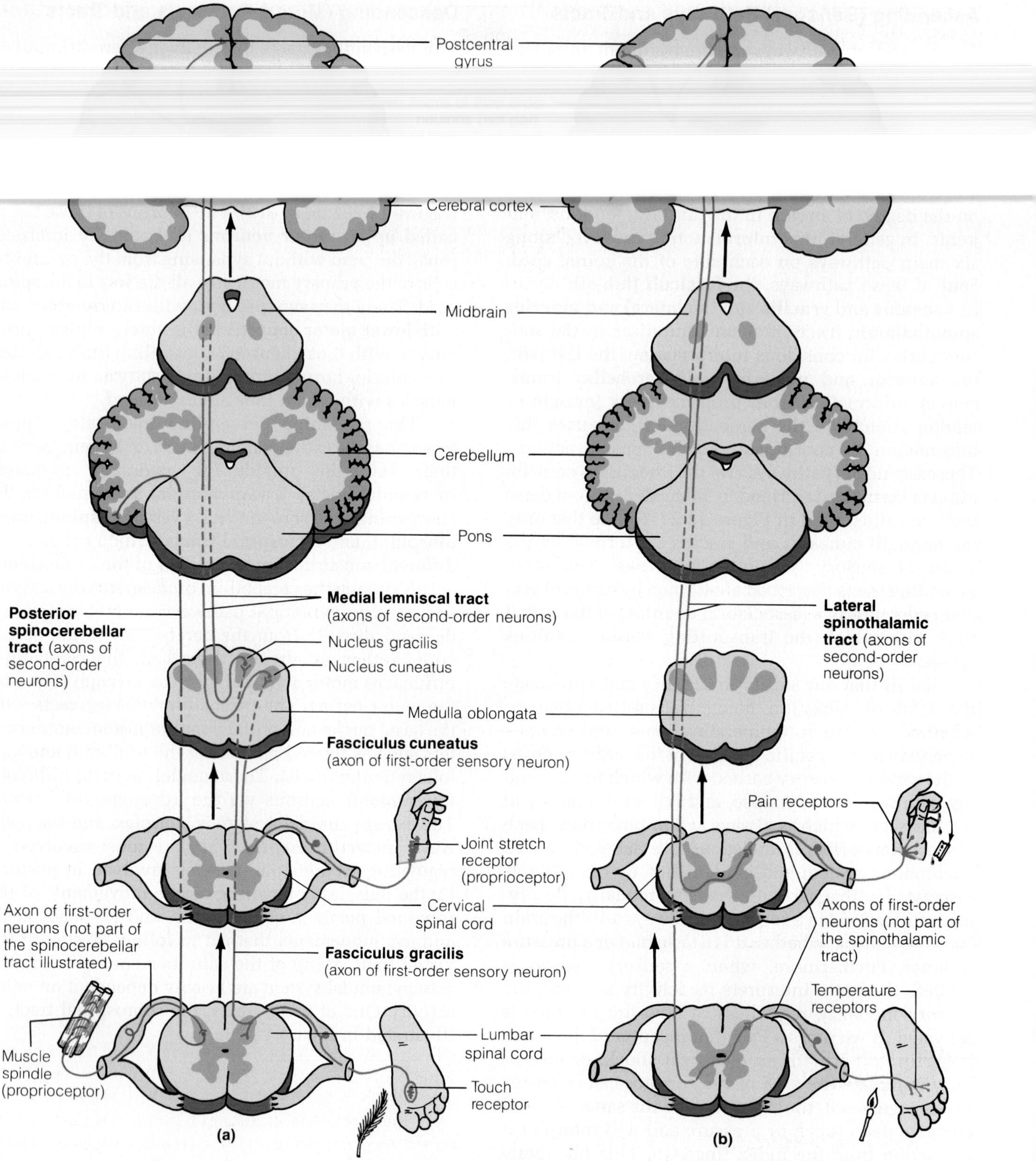

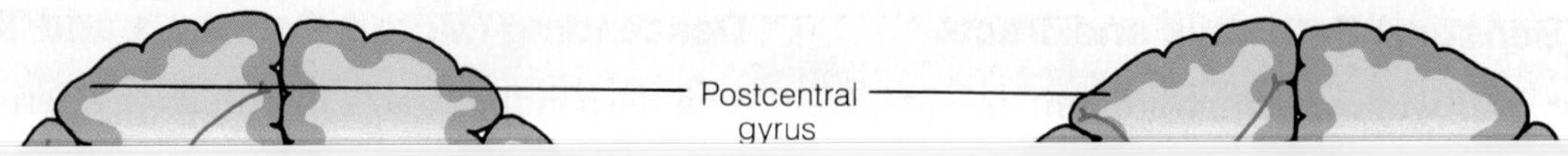

Figure 12.27 Pathways of selected ascending spinal cord tracts.
(**a**) Sensory pathways for touch, pressure, and proprioception (conscious and unconscious) within the fasciculus gracilis and fasciculus cuneatus tracts (each tract extends only to the medulla oblongata) and the posterior spinocerebellar tract (tract extends only to the cerebellum). (**b**) Sensory pathways for pain and temperature within the lateral spinothalamic tract (tract extends only to the thalamus). Notice that the entire ascending pathway is shown in each case.

Table 12.2 Major Ascending (Sensory) Pathways and Spinal Cord Tracts

Spinal cord tract	Location (funiculus)	Origin	Termination	Function
Fasciculus cuneatus and fasciculus gracilis	Posterior	Central axons of sensory (first-order) neurons enter dorsal root of the spinal cord and branch; branches enter posterior white column on same side without synapsing	By synapse with second-order neurons in nucleus cuneatus and nucleus gracilis in medulla; fibers of medullary neurons cross over and ascend in medial lemniscal tracts to thalamus, where they synapse with third-order neurons; thalamic neurons then transmit impulses to somatosensory cortex	Both tracts transmit sensory impulses from general sensory receptors of skin and proprioceptors of joints, which are interpreted as discriminative touch, pressure, and "body sense" (limb and joint position) in opposite somatosensory cortex; cuneatus transmits afferent impulses from upper limbs, upper trunk, and neck; it is not present in spinal cord below level of T_6; gracilis carries impulses from lower limbs and inferior body trunk
Lateral spinothalamic	Lateral	Association (second-order) neurons of posterior horn; fibers cross to opposite side before ascending	By synapse with third-order neurons in thalamus; impulses then conveyed to somatosensory cortex by thalamic neurons	Transmits impulses concerned with pain and temperature to opposite side of brain; eventually interpreted in somatosensory cortex
Anterior spinothalamic	Anterior	Association neurons in posterior horns; fibers cross to opposite side before ascending	By synapse with third-order neurons in thalamus; impulses eventually conveyed to somatosensory cortex by thalamic neurons	Transmits impulses concerned with crude touch and pressure to opposite side of brain for interpretation by somatosensory cortex
Posterior spinocerebellar	Lateral (posterior part)	Association (second-order) neurons in posterior horn on same side of cord; fibers ascend without crossing	By synapse in cerebellum	Transmits impulses from proprioceptors (except joint sense) on one side of body to same side of cerebellum; subconscious proprioception
Anterior spinocerebellar	Lateral (anterior part)	Association (second-order) neurons of posterior horn; contains both crossed and uncrossed fibers	By synapse in cerebellum	Transmits impulses from both sides of body to cerebellum; subconscious proprioception

Homeostatic Imbalances of the Spinal Cord

Any localized damage to the spinal cord or spinal roots is associated with some form of functional loss, either *paralysis* (loss of motor function) or *paresthesias* (payr″-es-thē′-zhuz) (sensory losses). Damage to the ventral root or anterior horn cells results in a *flaccid* (fla′-sid) *paralysis* of the skeletal muscles served. Because the lower motor neurons are damaged, nerve impulses do not reach these muscles, which consequently cannot move either voluntarily or involuntarily. When no longer stimulated, the muscles begin to atrophy and waste away. This type of paralysis is quite different from that produced when only the upper motor neurons of the primary motor cortex are damaged. In those cases, as long as the lower neural structures are undamaged, the muscles will continue to be stimulated by spinal reflex activity. Thus, the muscles remain healthy, but their movements are no longer subject to voluntary control. This type of paralysis is called *spastic paralysis*.

Transection of the spinal cord at any level results in total motor and sensory loss in body regions inferior to the site of damage. If the transection occurs

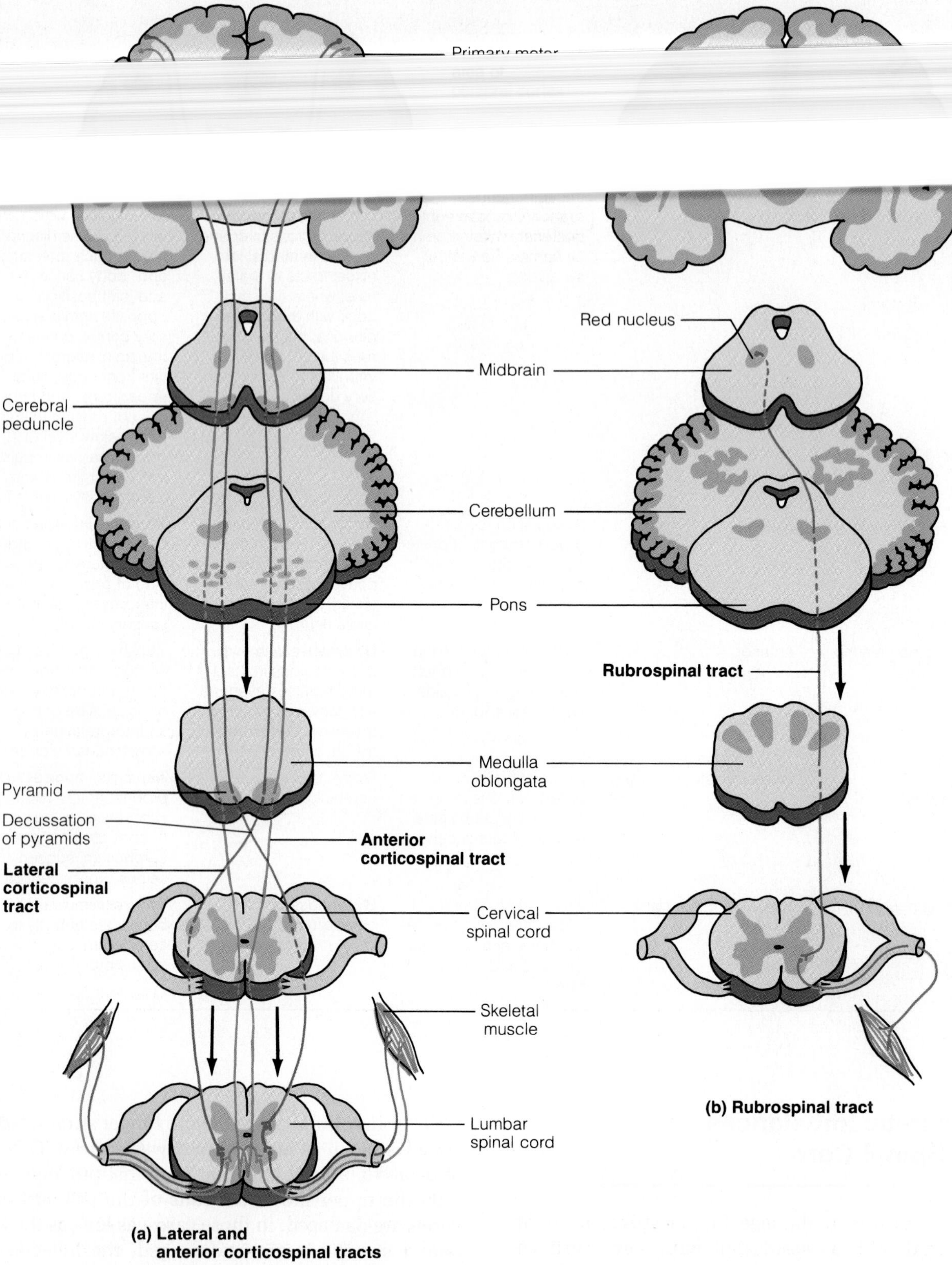

Figure 12.28 Pathways of selected descending spinal cord tracts.
(**a**) Pathways of the pyramidal tracts (lateral and anterior corticospinal tracts) carrying motor impulses to skeletal muscles. (**b**) Extrapyramidal motor pathway: The rubrospinal tract helps regulate muscle tone of muscles on opposite side of body.

Table 12.3 Major Descending (Motor) Pathways and Spinal Cord Tracts

Spinal cord tract	Location (funiculus)	Origin	Termination	Function
Pyramidal				
Lateral corticospinal	Lateral	Pyramidal neurons of motor cortex of the cerebrum; descend ipsilaterally (on same side) and then decussate in pyramids of medulla	Interneurons and anterior horn motor neurons; anterior horn motor neurons conduct impulses to skeletal muscles via spinal nerves	Transmits motor impulses from cerebrum to spinal cord motor neurons (which activate skeletal muscles on opposite side of body); voluntary motor tract
Anterior corticospinal	Anterior	Pyramidal neurons of motor cortex; fibers descend ipsilaterally	Interneurons and anterior horn motor neurons; anterior horn motor neurons conduct impulses to skeletal muscles via spinal nerves	Transmits motor impulses from cerebrum to spinal cord motor neurons (which activate skeletal muscles on same side of body); voluntary motor tract
Extrapyramidal				
Tectospinal	Anterior	Superior colliculus of midbrain of brain stem (fibers cross to opposite side of cord)	Anterior horn	Transmits motor impulses from one side of brain stem that eventually control skeletal muscles of opposite side of body in response to visual, auditory, and cutaneous stimuli
Vestibulospinal	Anterior	Vestibular nuclei in medulla of brain stem (fibers descend without crossing)	Anterior horn	Transmits motor impulses that regulate muscle tone (and posture) of muscles on same side of body in response to head movements and adjustments for balance
Rubrospinal	Lateral	Red nucleus of midbrain of brain stem (fibers cross to opposite side of cord)	Anterior horn	Transmits motor impulses concerned with muscle tone and posture to skeletal muscles on opposite side of body
Reticulospinal (anterior, medial, and lateral)	Anterior and lateral	Reticular formation of brain stem	Anterior horn	Transmits impulses concerned with muscle tone and activation of sweat glands

between T_{11} and L_1, both lower limbs are affected and the condition is called *paraplegia* (payr″-uh-plē′-zhuh). If the injury occurs in the cervical region, all four limbs are affected and *quadraplegia* occurs. *Hemiplegia*, paralysis of one side of the body, usually reflects brain, rather than spinal cord, injury (see p. 403).

Anyone with traumatic spinal cord injury must be watched for symptoms of *spinal shock*, a transient period of functional loss that follows injury to the cord. This phenomenon, commonly seen in *whiplash* (flexion or extension) injuries of the neck, results in an immediate depression of all reflex activity caudal to the lesion site. Bowel and bladder reflexes stop, blood pressure falls, and all muscles (somatic and visceral alike) below the injured site are paralyzed and insensitive. Since the individual does not perspire in the paralyzed body region, a fever may ensue. If the damage is not permanent, function usually returns within a few hours following injury. Delay in return of neural function for more than 48 hours is an ominous sign that forecasts permanent paralysis in most cases. Over 10,000 Americans are paralyzed yearly in automotive or sports accidents. ■

Diagnostic Procedures for Assessing CNS Dysfunction

function. The doctor taps your patellar or Achilles tendon with a reflex hammer and your leg muscles contract, resulting in the knee- or ankle-jerk response. These responses show that the spinal cord and upper brain centers are functioning normally. However, when reflex tests are abnormal or when brain cancer, intracranial hemorrhage, multiple sclerosis, or hydrocephalus are suspected, other more sophisticated neurological tests may be ordered to try to localize and identify the problem.

Pneumoencephalography (noo″-mō-en-seh″-fuh-lah′-gruh-fē) provides a fairly clear X-ray picture of the brain ventricles and has been the procedure of choice for diagnosing hydrocephalus. A small amount of cerebrospinal fluid is withdrawn by lumbar puncture. Then, air (or another gas) is injected into the subarachnoid space and allowed to float upward and into the ventricles of the brain, allowing them to be visualized. Although the procedure is remarkably simple, it can cause a blinding headache.

A *cerebral angiogram* (an′-jē-ō-gram) is used to assess the condition of the cerebral arteries serving the brain (or the carotid arteries of the neck, which feed most of those vessels). A radiopaque dye is injected into an artery, and time is allowed for the dye to become dispersed to the brain. This phase is somewhat uncomfortable for the patient, because as the dye flows through the blood vessels, a sensation of intense heat is experienced. Then, an X-ray is taken of the arteries of interest; the dye allows visualization of arteries narrowed by arteriosclerosis. This procedure is commonly ordered for individuals who have suffered a stroke or who have a history of transient ischemic attacks.

The new imaging techniques described in Chapter 2 (pp. 32–33) have revolutionized the diagnosis of brain lesions. *CAT scans* allow most tumors, intracranial lesions, and areas of dead brain tissue (infarcts) to be identified quickly. The CAT scanner is also becoming an important tool to enhance the precision and safety of brain surgery. Tumors in the posterior cranial fossa and multiple sclerosis plaques that are poorly revealed by CAT scans are being "flushed out" by *MRI scans*, which use powerful magnets to reveal the brain's problems. While CAT scans and MRI excel at revealing detailed maps of brain anatomy, *PET scans* employ high-energy gamma (γ) rays to monitor the brain's biochemical activity. PET scans are also being used to diagnose Alzheimer's disease (see Figure 12.22) and to analyze functional brain illness.

Developmental Aspects of the Central Nervous System

out prenatal development. Radiation, various drugs (alcohol, opiates, and others), and infections can have harmful effects on a developing infant's nervous system, particularly during the initial formative stages. For example, rubella (German measles) often leads to deafness and other types of CNS damage in the newborn. Also, since nervous tissue has the highest metabolic rate in the body, lack of oxygen for even a few minutes causes death of neurons. Because smoking decreases the amount of oxygen in the blood, a smoking mother may be sentencing her infant to some degree of brain damage.

In difficult deliveries, a temporary lack of oxygen may lead to *cerebral palsy*, but any of the factors listed in the preceding paragraph may also be a cause. Cerebral palsy is a neuromuscular disability in which the voluntary muscles are poorly controlled or paralyzed as a result of brain damage. In addition to spasticity, speech difficulties, and other motor impairments, about half of cerebral palsy victims have seizures, half are mentally retarded, and about a third have some degree of deafness. Visual impairments are also common. Although cerebral palsy does not get worse over time, its deficits are nonreversible. Cerebral palsy is the largest single cause of crippling in children, affecting six out of every thousand births.

The central nervous system is plagued by a number of other congenital malformations triggered by genetic or environmental factors during early embryonic development. The most serious are congenital hydrocephalus (see p. 400), anencephaly, and spina bifida.

In *anencephaly* (a″-nin-seh′-fuh-lē) , literally, without brain, the cerebrum fails to develop, presumably because the neural folds in the most rostral region fail to fuse. The child is totally vegetative, unable to see, hear, or process sensory inputs. Muscles are flaccid, and no voluntary movement is possible. Brain stem functions are unimpaired, but mental life as we know it does not exist. Mercifully, death occurs soon after birth.

Spina bifida (spī′-nuh bih′-fih-duh) results from incomplete formation of the vertebrae and involves various malformations of the spine that typically involve the lumbosacral region. There are several varieties, but the common abnormality is that vertebral laminae are absent. If severe, neural deficits occur as well. *Spina bifida occulta*, the least serious type, involves only one or a few vertebrae and causes no neural problems. Other than a small dimple or tuft of hair over the site of nonfusion, it has no external manifestations. In the more severe form, *spina bifida cystica,* a saclike cyst protrudes dorsally from the child's spine. The cyst may contain meninges and cerebrospinal fluid, in which case it is called a *meningocele* (meh-ning′-gō-sēl), and it may also contain portions of the spinal cord and spinal nerve roots and be called a *myelomeningocele* (mī′-uh-lō-meh-ning″-gō-sēl). The larger the cyst and the more neural structures it contains, the greater the degree of neurological impairment. In the worst case, where the inferior spinal cord is rendered functionless, bowel incontinence, bladder muscle paralysis (which predisposes the infant to urinary retention, urinary tract infection, and renal failure), and lower limb paralysis occur. Problems with infection are continual because the cyst tends to rupture or leak. Spina bifida cystica is accompanied by hydrocephalus in 90% of all cases.

One of the last CNS areas to mature is the hypothalamus. Since the hypothalamus contains body temperature regulatory centers, premature babies usually have problems in controlling their loss of body heat and must be kept in temperature-controlled environments. Interesting studies using PET scans of infants' brains have revealed that the thalamus is active in a five-day-old baby, but that the visual cortex is not—biochemical evidence supporting the observation that infants of this age respond to touch but have very poor vision. By 11 weeks, more of the cortex is active, allowing the baby to reach for a rattle. By eight months, the cortex is highly active, and the child can think about what he or she sees. Growth and maturation of the nervous system continue throughout childhood and largely reflect progressive myelination. A good indication of the degree of myelination of a particular neural pathway is the level of neuromuscular control in that body area. As described in Chapter 9, neuromuscular coordination progresses in a superior to inferior direction and in a proximal to distal direction, and we know that myelination also occurs in this sequence.

The brain reaches its maximum weight in the young adult. Over the next 60 years or so, neurons are damaged and die, and since they cannot reproduce themselves, our store of neurons continually decreases. The consequence of this erosive loss of neurons is a steady decline in brain weight and volume. However, the number of neurons lost over the decades is normally only a small percentage of the total, and the remaining neurons can change their synaptic connections, providing for continued learning throughout our lives.

Although age brings some cognitive declines in spatial ability, speed of perception, decision making, reaction time, and memory, these losses are not significant in the *healthy individual* until after the seventh decade. Then a fairly rapid decline occurs in some, but not all, of these abilities. *Crystallized intelligence,* which involves the ability to build on experience, mathematical skills, and verbal fluency, does not decline with age, and many people continue to enjoy intellectual lives and to work at mentally demanding tasks their entire life. In fact, fewer than 5% of people over 65 years old demonstrate true senility.

Sadly, there are many cases of "reversible senility" caused by low blood pressure, constipation, poor nutrition, depression, dehydration, and hormone imbalances that go undiagnosed. Therefore, the best way to maintain one's mental abilities in old age may be to seek regular medical checkups on one's physical condition throughout life.

Although eventual shrinking of the brain is normal, it seems that some individuals (professional boxers and chronic alcoholics) accelerate the process long before aging plays its part. Whether a boxer wins the match or not, the likelihood of brain damage and atrophy increases with each fight as the brain bounces and rebounds within the skull with every blow. The expression "punch drunk" reflects the symptoms of slurred speech, tremors, abnormal gait, and senile dementia (age-related mental illness) seen in many members of this group.

Everyone recognizes that alcohol has a profound effect on the mind as well as on the body. However, these effects may not be temporary. CAT scans of chronic alcoholics reveal reduced brain size and density appearing at a fairly early age. Like boxers, chronic alcoholics tend to exhibit signs of senile dementia unrelated to the aging process.

* * *

The human cerebral hemispheres—our "thinking caps"—are awesome in their complexity. But no less amazing are the brain regions that oversee all our subconscious, autonomic body functions—the diencephalon and brain stem—particularly when you consider their relatively insignificant size. The spinal cord, which acts as a reflex center and a communication link between the brain and body periphery, is equally important to body homeostasis.

A good deal of new terminology has been introduced in this chapter, and as you will see, much of it will come up again in one or more of the remaining chapters dealing with nervous system functions. Chapter 13, your next challenge, considers the structures of the peripheral nervous system that work hand in hand with the CNS to keep it informed and to deliver its orders to the effectors of the body.

A CLOSER LOOK "He-Brain" versus "She-Brain"

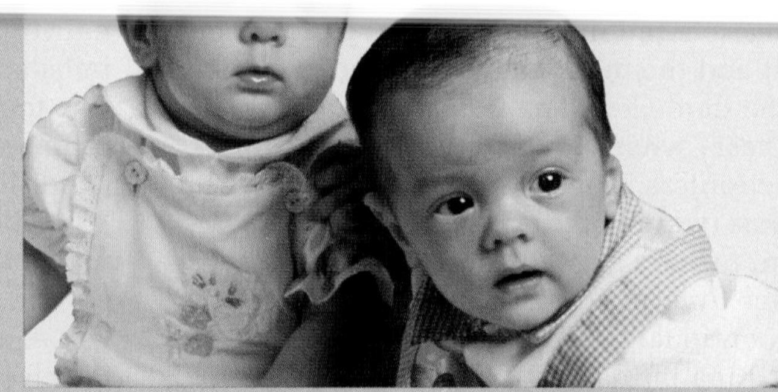

human race. Yet, when all sexism is finally stripped away and differences in nurture are disregarded, there *is* something different about males and females, and the difference is grounded in their biology.

As boys enter puberty, they tend to exhibit more aggressive behavior, and there is little doubt that the dramatic rise in male sex hormones surging through their blood is responsible for this change. The question now is how testosterone promotes or enables aggressive or violent behavior. For any hormone to influence behavior, it must first affect the brain. Radioisotope techniques have demonstrated that both male and female sex hormones (testosterone and estrogen, respectively) concentrate selectively in certain brain regions that play an important role in courtship, sex and mating behaviors, and aggression—behaviors in which the sexes differ most. There is also evidence that sexual differences in behavior exist long before the onset of puberty and its raging hormones. For example, newborn boys show more motor strength and muscle tone, while girls show greater sensitivity to touch, taste, and light and exhibit more reflex smiles. Comprehensive studies of school children have demonstrated that girls do better on verbal skills tests, whereas boys are more proficient in spatial-visual tasks. Further, although it was once presumed that damage to the left hemisphere would produce greater deficits in verbal tasks and that right hemispheric injury would impair spatial abilities, this has been borne out only for males. As a result of these studies, it has been hypothesized that females have more hemispheric overlap in verbal and spatial functions than do males, who demonstrate earlier lateralization of cortical functioning.

Does greater lateralization for a given function imply superior performance for that function? Apparently not. Females show less lateralization of language functions, yet they tend to be superior to males in language skills. Also, greater lateralization carries a steep price tag, because if a highly lateralized functional area is damaged, the function is terminated. For example, aphasia follows damage to the left hemisphere three times more often in males than in females.

How can these differences in behavior and skills be explained? The accounting may lie deep within the brain. In 1973, it was demonstrated for the first time that male and female brains differ structurally in many ways. The best example concerns a specific region (now called the *sexually dimorphic nucleus*) in the anterior part of the hypothalamus that has a distinctive synaptic pattern in each sex. Castration of male monkeys shortly after birth produced the female hypothalamic pattern, while injection of testosterone into females triggered the development of the male pattern. This discovery rocked the neuroscience community because it was the first evidence that there were structural brain differences in the sexes and that sex hormones circulating before, at, or after birth could *change* the brain. Based on these and later observations, scientists concluded that the basic plan of the mammalian brain is female and stays that way unless "told to do otherwise" by masculinizing (male) hormones. The production of testosterone by male fetuses is the key to anatomical development as a male. It now appears that fetal testosterone is also responsible for development of the male brain pattern.

Since these discoveries, a number of other sex-related differences have been found in the central nervous system including the following:

1. Females tend to have a longer left temporal lobe.

2. There are sex-related differences in the human corpus callosum. In females, its posterior portion is bulbous and wide, while in males the corpus callosum is generally cylindrical and fairly uniform in diameter. This may indicate that there are more communicating fibers between the hemispheres in females, which could support the hypothesis that the female brain is less lateralized.

3. Sex differences have been found in the spinal cord. Certain clusters of neurons that serve the external genitals are much larger in males than in females and they contain receptors for testosterone, but not estrogen. Testosterone exposure before birth produces the male pattern of these particular spinal cord neurons.

What should we make of these findings? Do these observed differences in the two sexes have little or nothing to do with behavioral subtleties, or do they underlie many (or most) manifestations of maleness or femaleness? It is too soon to know and studies of sexual dimorphism of the brain are still "in diapers."

Related Clinical Terms

Cordotomy (kor-dah'-tuh-mē) A procedure during which a nerve tract within the spinal cord is severed surgically; usually done to relieve unremitting pain.

Encephalopathy (en-sef"-ah-lop'-ah-thē) (*enceph* = brain; *path* = disease) Any disease or disorder of the brain.

Functional brain disorders Psychological disorders for which no structural cause can be found; include neuroses and psychoses (see below).

Microencephaly (mī"-krō-en-seh'-fuh-lē) (*micro* = small) Congenital condition involving the formation of a small brain, as evidenced by reduced skull size; most microencephalic children are severely retarded.

Monoplegia (*mono* = one; *plegia* = blow, strike) Paralysis of one limb.

Myelitis (mī"-uh-lī'-tis) (*myel* = spinal cord; *itis* = inflammation) Inflammation of the spinal cord.

Myelogram (*gram* = recording) X-ray of the spinal cord after injection of a contrast medium.

Neuroses (ner-ō'-sēz) The less debilitating class of functional brain disorders; examples include severe anxiety (panic attacks), phobias (irrational fears), and compulsive-obsessive behaviors (e.g., washing one's hands every few minutes); the affected individual, however, retains contact with reality.

Psychoses (sī-kō'-sēz) Class of functional brain disorders in which the affected individuals detach themselves from reality and exhibit bizarre behaviors; the *legal* word for psychotic behavior is *insanity;* the psychoses include *schizophrenia* (skit"-sō-frē'-nē-uh), *depression,* and *manic-depressive disease.*

Chapter Summary

THE BRAIN (pp. 376–404)

1. The brain provides for voluntary movements, interpretation and integration of sensation, consciousness, and cognitive function.

Embryonic Development of the Brain (pp. 376–378)

2. The brain develops from the rostral portion of the embryonic neural tube.

3. Early brain development yields the three primary brain vesicles: the prosencephalon (cerebral hemispheres and diencephalon), mesencephalon (midbrain), and rhombencephalon (pons, medulla, and cerebellum).

4. Cephalization results in the envelopment of the diencephalon and superior brain stem by the cerebral hemispheres.

Regions of the Brain (p. 379)

5. In a widely used system, the brain is divided into the cerebral hemispheres, diencephalon, brain stem, and cerebellum.

6. The cerebral hemispheres and cerebellum have gray matter nuclei surrounded by white matter and an outer cortex of gray matter. The brain stem lacks a cortex.

Ventricles of the Brain (p. 379)

7. The brain contains four ventricles filled with cerebrospinal fluid. The lateral ventricles are in the cerebral hemispheres; the third ventricle is in the diencephalon; the fourth ventricle is in the brain stem and connects with the central canal of the spinal cord.

The Cerebral Hemispheres (pp. 379–388)

8. The two cerebral hemispheres exhibit gyri, sulci, and fissures. The longitudinal fissure partially separates the hemispheres; the other fissures subdivide each hemisphere into lobes.

9. Each cerebral hemisphere consists of the cerebral cortex, the cerebral white matter, and the basal (cerebral) nuclei.

10. Each cerebral hemisphere receives sensory impulses from, and dispatches motor impulses to, the opposite side of the body. The body is represented in an upside-down fashion on the sensory and motor cortices.

11. Functional areas of the cerebral cortex include: (1) motor areas: primary motor and premotor areas of the frontal lobe and Broca's (motor speech) area in the frontal lobe of one hemisphere (usually the left); (2) sensory areas: primary sensory cortex, somatosensory association area, gustatory area, and general integration area in the parietal lobe; visual areas in the occipital lobe; and olfactory and auditory areas in the temporal lobe; (3) association areas: Wernicke's area in one hemisphere (temporal lobe) only, usually the left; prefrontal cortex in the frontal lobe; general interpretation area at junction of temporal, parietal, and occipital lobe on one side (usually the left); and affective language areas in one hemisphere (usually the right).

12. The cerebral hemispheres show lateralization of cortical function. In most people, the left hemisphere is dominant (i.e., specialized for language and mathematical skills), and the right hemisphere is more concerned with motor and visual-spatial skills and creative endeavors.

13. Fiber tracts of the cerebral white matter include commissures, association fibers, and projection fibers.

14. The paired cerebral nuclei include the lentiform nucleus (globus pallidus and putamen) and caudate nucleus. The cerebral nuclei are subcortical motor nuclei that help control muscular movements.

The Diencephalon (pp. 389–391)

15. The diencephalon consists of the thalamus, hypothalamus, and epithalamus, and encloses the third ventricle.

16. The thalamus is the major relay station for (1) afferent impulses ascending to the sensory cortex and (2) inputs of subcortical nuclei and the cerebellum to the cerebral motor cortex.

17. The hypothalamus is the most important autonomic nervous system control center and a pivotal part of the limbic system. It maintains water balance and regulates thirst, eating behavior, gastrointestinal activity, body temperature, and the activity of the anterior pituitary gland.

18. The epithalamus consists of the pineal gland and the choroid plexus of the third ventricle.

The Brain Stem (pp. 391–394)

19. The brain stem includes the midbrain, pons, and medulla oblongata.

20. The midbrain contains the corpora quadrigemina (visual and auditory reflex centers), the red nucleus (extrapyramidal motor center), and nuclei of cranial nerves III and IV. The cerebral peduncles on its ventral face house the pyramidal fiber tracts. The midbrain surrounds the cerebral aqueduct.

21. The pons is mainly a conduction area. Its nuclei contribute to regulation of respiration and cranial nerves V–VII.

22. The pyramids (descending corticospinal tracts) form the ventral face of the medulla oblongata; these fibers cross over (decussation of the pyramids) before entering the spinal cord. Important nuclei regulate respiratory rhythm, heart rate, and blood pressure and serve cranial nerves VIII–XII. The olivary nucleus and the cough, sneezing, swallowing, and vomiting centers are in the medulla.

Functional Brain Systems (pp. 396–398)

25. The limbic system consists of numerous structures that encircle the diencephalon. It is the "emotional-visceral brain."

26. The reticular formation includes nuclei and fiber tracts spanning the length of the brain stem. It is involved in maintaining the alert state of the cerebral cortex (RAS) and is part of the extrapyramidal motor system.

Protection of the Brain (pp. 399–402)

27. The brain is protected by bone, meninges, cerebrospinal fluid, and the blood-brain barrier.

28. The meninges consist of the outermost dura mater, the arachnoid, and the innermost pia mater. They enclose the brain and spinal cord and their blood vessels. Inward projections of the inner layer of the dura mater form septa that secure the brain to the skull.

29. The cerebrospinal fluid (CSF), formed by the choroid plexuses from blood plasma, circulates through the ventricles and into the subarachnoid space. It is returned to the dural venous sinuses by the arachnoid villi. CSF supports and cushions the brain and cord and helps to nourish them and remove metabolic wastes.

30. The blood-brain barrier reflects the relative impermeability of the epithelium of capillaries of the brain. It allows water, respiratory gases, essential nutrients, and a few fat-soluble molecules to enter the neural tissue, but prevents entry of other, potentially harmful substances.

Homeostatic Imbalances of the Brain (pp. 402–404)

31. Head trauma may cause brain injuries called concussions (reversible damage) or contusions (nonreversible damage). When the brain stem is affected, unconsciousness (temporary or permanent) occurs. Trauma-induced brain injuries may be aggravated by intracranial hemorrhage or cerebral edema, both of which compress brain tissue.

32. Cerebrovascular accidents (strokes) result when blood circulation to brain neurons is blocked and brain tissue dies. The result may be hemiplegia, sensory deficits, or speech impairment.

33. Alzheimer's disease is a degenerative brain disease in which abnormal protein deposits and neurofibrillar tangles appear. It results in slow, progressive loss of memory and motor control and increasing dementia.

34. Multiple sclerosis is a demyelinating disease that reflects an autoimmune response. Demyelination impairs nerve impulse transmission and eventually leads to complete disability.

THE SPINAL CORD (pp. 404–413)

Structure and Protection of the Spinal Cord (pp. 404–405)

1. The spinal cord, a two-way impulse conduction pathway and a reflex center, resides within the vertebral column and is protected by meninges and cerebrospinal fluid. It extends from the foramen magnum to the end of the first lumbar vertebra.

2. Thirty-one pairs of spinal nerve roots issue from the cord. The cord is enlarged in the cervical and lumbosacral regions, where spinal nerves serving the limbs arise.

4. The central gray matter of the cord is H shaped. Anterior horns contain somatic motor neurons. Lateral horns contain autonomic motor neurons. Posterior horns contain interneurons.

5. Axons of neurons of the lateral and anterior horns emerge in common from the cord via the ventral roots. Axons of sensory neurons (with cell bodies located in the dorsal root ganglion) enter the posterior aspect of the cord and form the dorsal roots. The ventral and dorsal roots combine to form the spinal nerves.

White Matter of the Spinal Cord (pp. 408–410)

6. Each side of the white matter of the cord has posterior, lateral, and anterior funiculi, and each funiculus contains a number of ascending and descending tracts. All tracts are paired and most decussate.

7. Ascending (sensory) tracts include the fasciculi gracilis and cuneatus (touch and joint proprioception), spinothalamic tracts (pain, touch, temperature), and spinocerebellar tracts (muscle/tendon proprioception).

8. Descending tracts include the pyramidal tracts (anterior and lateral corticospinal tracts that originate from the primary motor cortex) and the extrapyramidal tracts originating from extrapyramidal motor nuclei.

Homeostatic Imbalances of the Spinal Cord (pp. 411–413)

9. Injury to the anterior horn neurons or the ventral roots results in flaccid paralysis. (Injury to the upper motor neurons in the brain results in spastic paralysis.) If the dorsal roots or sensory tracts are damaged, paresthesias occur.

DIAGNOSTIC PROCEDURES FOR ASSESSING CNS DYSFUNCTION (p. 414)

1. Diagnostic procedures used to assess neurological condition and function range from routine reflex testing to sophisticated techniques such as pneumoencephalography, cerebral angiography, CAT scans, MRI scars, and PET scans.

DEVELOPMENTAL ASPECTS OF THE CENTRAL NERVOUS SYSTEM (pp. 414–415)

1. Maternal and environmental factors may impair embryonic brain development, and oxygen deprivation destroys brain cells. Severe congenital brain diseases include cerebral palsy, anencephaly, hydrocephalus, and spina bifida.

2. Premature babies have trouble regulating body temperature because the hypothalamus is one of the last brain areas to mature prenatally.

3. Development of motor control indicates the progressive myelination and maturation of a child's nervous system.

4. Brain growth ends in young adulthood. Neurons die throughout life and are not replaced; thus, brain weight and mass decline with age.

5. Healthy aged people maintain nearly optimal intellectual function. Disease—particularly cardiovascular disease—is the major cause of declining mental function with age.

Review Questions

Multiple Choice/Matching

1. The somatic motor cortex, Broca's area, and the premotor area are located in (a) the frontal lobe, (b) the parietal lobe, (c) the temporal lobe, (d) the occipital lobe.

2. Choose the proper response from the key for the statements describing various brain areas.

Key (a) cerebellum **(b)** corpora quadrigemina **(c)** corpora striata **(d)** corpus callosum **(e)** hypothalamus **(f)** medulla **(g)** midbrain **(h)** pons **(i)** thalamus

_____ **(1)** basal nuclei involved in fine control of motor activities; lesions may lead to Parkinson's disease
_____ **(2)** region where there is a gross crossover of fibers of descending pyramidal tracts
_____ **(3)** control of temperature, autonomic nervous system reflexes, hunger, and water balance
_____ **(4)** houses the substantia nigra and cerebral aqueduct
_____ **(5)** relay stations for visual and auditory stimuli input; found in midbrain
_____ **(6)** houses vital centers for control of the heart, respiration, and blood pressure
_____ **(7)** brain area through which (nearly) all the sensory input must travel to get to the cerebral cortex; acts as a sensory relay
_____ **(8)** brain area most concerned with equilibrium, body posture, and coordination of motor activity

3. The innermost layer of the meninges, delicate and closely apposed to the brain tissue, is the (a) dura mater, (b) corpus callosum, (c) arachnoid, (d) pia mater.

4. Cerebrospinal fluid is formed by (a) arachnoid villi, (b) the dura mater, (c) choroid plexuses, (d) all of these.

5. A patient has suffered a cerebral hemorrhage that has caused dysfunction of the precentral gyrus of his right cerebral cortex. As a result, (a) He cannot voluntarily move his left arm or leg, (b) he feels no sensation on the left side of his body, (c) he feels no sensation on the right side of his body.

6. Ascending pathways in the spinal cord convey (a) motor impulses, (b) sensory impulses, (c) voluntary impulses, (d) all of these.

7. Destruction of the anterior horn cells of the spinal cord results in loss of (a) integrating impulses, (b) sensory impulses, (c) voluntary motor impulses, (d) all of these.

8. Fiber tracts that allow neurons within the same hemisphere to communicate are (a) association tracts, (b) commissures, (c) projection tracts.

Short Answer Essay Questions

9. Make a diagram showing the three primary (embryonic) brain vesicles. Name each and then use clinical terminology to name the resulting adult brain regions.

10. (a) What is the advantage of having a cerebrum that is highly convoluted? (b) What term is used to indicate its grooves? Its outward folds? (c) What groove divides the cerebrum into two hemispheres? (d) What divides the parietal from the frontal lobe? The parietal from the temporal lobe?

11. (a) Make a rough drawing of the lateral aspect of the left cerebral hemisphere. (b) Locate the following areas and provide the major function of each: primary motor cortex, premotor cortex, somatosensory association area, primary sensory area, visual and auditory areas, prefrontal cortex, Wernicke's and Broca's areas.

12. (a) What does lateralization of cortical functioning mean? (b) Why is the term *cerebral dominance* a misnomer?

13. (a) What is the function of the basal nuclei? (b) Which basal nuclei form the lentiform nucleus? (c) Which arches over the diencephalon?

14. (a) Explain how the cerebellum is physically connected to the brain stem. (b) List ways in which the cerebellum is very similar to the cerebrum.

15. Describe the role of the cerebellum in maintaining smooth, coordinated skeletal muscle activity.

16. (a) Where is the limbic system? (b) What structures make up this system? (c) How is the limbic system important in behavior?

17. (a) Localize the reticular formation in the brain. (b) What does RAS mean, and what is its function?

18. List four ways in which the CNS is protected.

19. (a) How is cerebrospinal fluid formed and drained? Follow its pathway within and around the brain. (b) What happens if CSF is not drained properly? Why is this consequence more harmful in adults?

20. What constitutes the blood-brain barrier?

21. (a) Differentiate between a concussion and a contusion. (b) Why do severe brain stem contusions result in unconsciousness?

22. Describe the spinal cord, depicting its extent, its distribution and composition of gray and white matter, and its spinal roots.

23. Which ascending spinal cord tracts carry sensory impulses concerned with touch and pressure? Concerned with proprioception only? Concerned with pain and temperature?

24. (a) Name the descending pathways that are concerned with voluntary skeletal movements and that form the pyramidal tracts. (b) Name three tracts that form part of the extrapyramidal motor system.

25. Differentiate clearly between spastic and flaccid paralysis.

26. Explain how paraplegia, hemiplegia, and quadriplegia differ.

27. (a) Define cerebrovascular accident or CVA (b) Describe its possible causes and consequences.

28. (a) What factors account for brain growth after birth? (b) List some structural brain changes observed with aging.

Clinical Application Questions

29. Mrs. Jones has had a progressive decline in her mental capabilities in the last five or six years. At first her family attributed her occasional memory lapses, confusion, and agitation to grief over her husband's death six years earlier. When examined, Mrs. Jones was aware of her cognitive problems and was shown to have an IQ score of approximately 30 points less than would be predicted by her work history. A CAT scan showed diffuse cerebral atrophy. The physician prescribed a mild tranquilizer for Mrs. Jones and told her family that there was little else he could recommend. What is Mrs. Jones's problem?

30. Robert, a brilliant computer analyst, suffered a blow to his anterior skull, delivered by a falling rock while mountain climbing. It was immediately obvious to his co-workers that his behavior had undergone a dramatic change. Although he had previously been a smart dresser, he was now unkempt. One morning, he was observed defecating into the wastebasket. His supervisor ordered Robert to report to the company's doctor immediately. What region of Robert's brain was affected by the cranial blow?

CHAPTER 13

Chapter Outline and Student Objectives

1. Define peripheral nervous system and list its components.
2. Distinguish clearly between sensory, motor, and mixed nerves.

Cranial Nerves (pp. 421–428)

3. Name the 12 pairs of cranial nerves and describe the body region and structures innervated by each.

Spinal Nerves (pp. 428–438)

4. Describe the formation of a spinal nerve, and distinguish between spinal roots and rami. Describe the general distribution of the ventral and dorsal rami.
5. Define plexus. Name the major plexuses, their origin sites, and the major nerves arising from each, and describe the distribution and function of the peripheral nerves.

Reflex Activity (pp. 438–444)

6. Distinguish between autonomic and somatic reflexes.
7. Compare and contrast stretch, flexor, and crossed extensor reflexes.

Developmental Aspects of the Peripheral Nervous System (p. 444)

8. Describe the developmental relationship between the segmented arrangement of peripheral nerves, skeletal muscles, and skin dermatomes.

Preview of Selected Key Terms

Peripheral nervous system Portion of the nervous system consisting of nerves and ganglia that lie outside of the brain and spinal cord.

Somatic nervous system (*soma* = body) Division of the peripheral nervous system that provides the motor innervation of skeletal muscles.

Cranial nerves The 12 nerve pairs that arise from the brain.

Spinal nerves The 31 nerve pairs that arise from the spinal cord.

Plexus (*plexus* = braid) A network of converging and diverging nerve fibers.

Spinal reflex A somatic reflex mediated through the spinal cord.

The human brain, for all its sophistication, would be useless without its links to the outside world. Our very sanity depends on a continual flow of information from the outside. When blindfolded volunteers were suspended in a tank of warm water (a situation that limited sensory inputs), they began to hallucinate: One saw charging herds of pink and purple elephants; another heard a singing chorus; others had taste hallucinations. The structures of the **peripheral nervous system (PNS)** provide our links to the real world. Its ghostly white nerves thread through virtually every part of the body, enabling the CNS to receive information and to carry out its decisions.

The peripheral nervous system includes all nervous tissue found outside the brain and spinal cord, that is, the peripheral nerves, their associated ganglia, and sensory receptors. The peripheral nervous system is divided into the **sensory (afferent)** and **motor (efferent) divisions.** Peripheral nerve fibers are described as afferent if they conduct impulses toward the CNS, efferent if they transmit impulses away from it. However, nearly all periphral nerves are mixed; that is, they contain both types of fibers. The motor division of the PNS includes both somatic and autonomic nerve fibers. Somatic motor fibers innervate skeletal muscles, whereas autonomic motor fibers innervate smooth and cardiac muscle and glands (that is, primarily visceral organs). PNS **ganglia** contain nerve cell bodies. Ganglia associated with afferent nerve fibers contain the nerve cell bodies of sensory neurons, whereas ganglia associated with efferent fibers contain nerve cell bodies of motor neurons of the autonomic nervous system.

For convenience, peripheral nerves are classified as *cranial nerves* or *spinal nerves,* depending on whether they arise from the brain or spinal cord, respectively. This chapter considers the cranial and spinal nerves and then goes on to describe some important spinal reflex arcs. Although autonomic efferents of cranial nerves are mentioned, this chapter focuses on somatic functions. A detailed consideration of the autonomic nervous system is deferred to Chapter 14.

Cranial Nerves

Twelve pairs of **cranial nerves** are associated with the brain and pass through various foramina of the skull (Figure 13.1). The first two pairs originate from the forebrain; the rest originate from the brain stem. Other than the vagus nerves, which extend well into the ventral body cavity, the cranial nerves serve only head and neck structures.

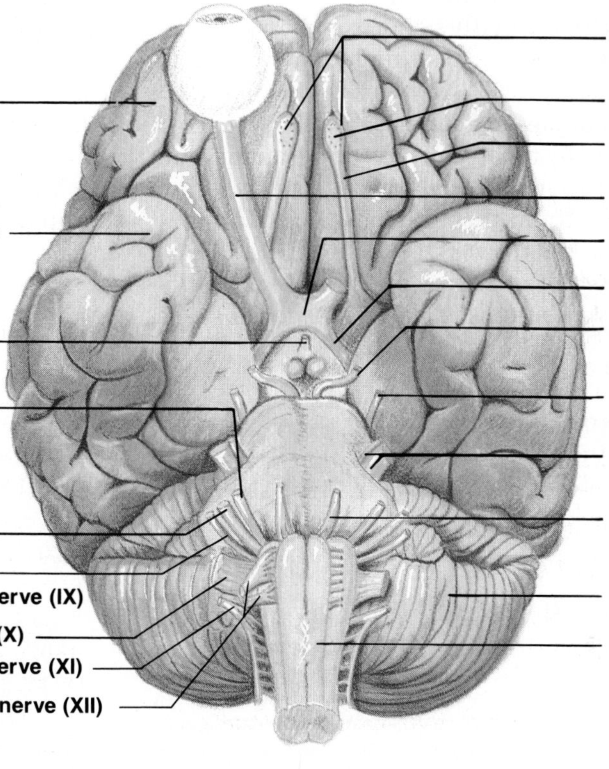

Figure 13.1 Summary of location and function of the cranial nerves. (**a**) Ventral view of the human brain, showing the cranial nerves. (**b**) Summary of cranial nerves by function. All cranial nerves that have a motor function also carry afferent fibers from proprioceptors in the muscles served; only sensory functions other than proprioception are indicated. Notice that three cranial nerves (I, II, and VIII) have sensory function only and no motor function. Notice also that four nerves (III, VII, IX, and X) carry parasympathetic fibers that serve visceral muscles and glands.

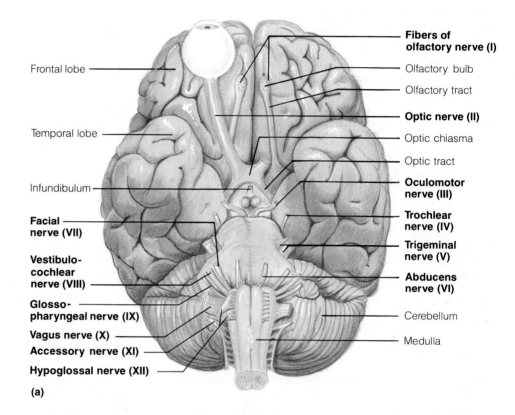

(a)

Cranial nerve	Name	Sensory function	Motor function	Parasympathetic fibers
I	Olfactory	Yes (smell)	No	No
II	Optic	Yes (vision)	No	No
III	Oculomotor	No	Yes	Yes
IV	Trochlear	No	Yes	No
V	Trigeminal	Yes (general sensation)	Yes	No
VI	Abducens	No	Yes	No
VII	Facial	Yes (taste)	Yes	Yes
VIII	Vestibulocochlear	Yes (hearing and balance)	No	No
IX	Glossopharyngeal	Yes (taste)	Yes	Yes
X	Vagus	Yes (taste)	Yes	Yes
XI	Accessory	No	Yes	No
XII	Hypoglossal	No	Yes	No

(b)

The cranial nerves, listed in order, are the *olfactory, optic, oculomotor, trochlear, trigeminal, abducens, facial, vestibulocochlear* (formerly called the auditory), *glossopharyngeal, vagus, accessory* (formerly called the spinal accessory), and *hypoglossal* nerves. In most cases, their names reveal the most

mer names of the cranial nerves in order: "*On old Olympus's towering top, a Finn and German viewed some hops.*" It is presented here because of its venerable history. However, since two of the cranial nerve names have been changed, I suggest you make up your own saying or use the following: "*Oh, oh, oh, to touch and feel very good velvet, ah.*" The cranial nerves are also numbered (traditionally using Roman numerals) from anterior to posterior according to site of origin in the brain.

In the last chapter, we described how all spinal nerves are formed by the fusion of ventral (motor) and dorsal (sensory) roots. Cranial nerves, on the other hand, vary markedly in their composition. Most cranial nerves are mixed nerves (see Figure 13.1b); however, three pairs (the olfactory, optic, and vestibulocochlear) are intimately associated with special sense organs and are generally considered purely sensory nerves. The cell bodies of the sensory neurons of the olfactory and optic nerves are located within the special sense organs with which they are associated. In all other cases, the cell bodies of sensory neurons contributing to cranial nerves are located in ganglia out-

several, and still others have none.

Several of the mixed cranial nerves contain both somatic and autonomic motor fibers and hence serve both skeletal muscles and visceral organs. Except for some autonomic nervous system motor neurons located in ganglia (see Chapter 14), the cell bodies of motor neurons contributing to the cranial nerves are located in gray matter regions (nuclei) of the brain.

The cranial nerves are described by name, number, origin, course, and function in Table 13.1. Notice that the pathways of the purely sensory nerves (I, II, and VIII) are described from the receptors to the brain, while those of the other nerves are described in the opposite direction (from the brain distally).

Table 13.1 Cranial Nerves

I The olfactory (ol-fak'tō-re) **nerves**

Origin and course: Olfactory nerve fibers arise from olfactory receptor cells located in olfactory epithelium of nasal cavity and pass through cribriform plate of ethmoid bone to synapse in olfactory bulb; fibers of olfactory bulb neurons extend posteriorly as olfactory tract, which runs beneath frontal lobe to enter cerebral hemispheres and terminates in primary olfactory cortex; see also Figure 16.3

Function: Purely sensory; carry afferent impulses for sense of smell

Fracture of ethmoid bone or lesions of olfactory fibers may result in partial or total loss of smell, a condition known as *anosmia* (a-noz'-me-uh). ■

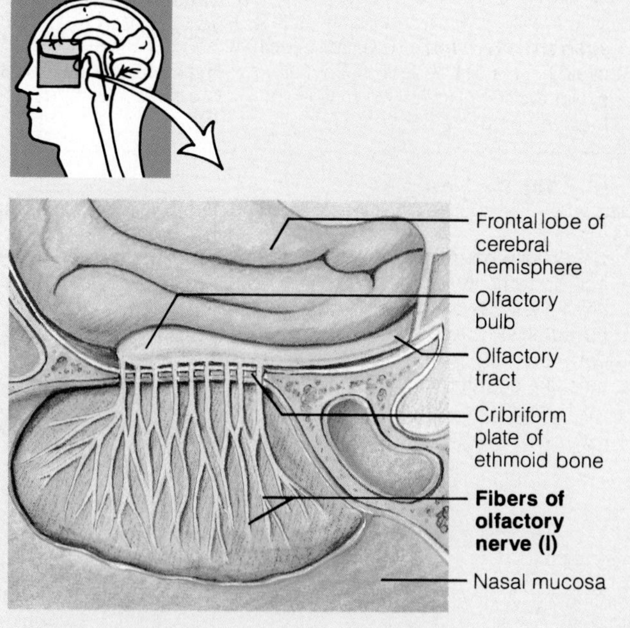

II The optic nerves

Origin and course: Fibers arise from retina of eye to form optic nerve, which passes through optic foramen of orbit; the optic nerves converge to form the optic chiasma (kī-az′-muh), where partial crossover of fibers occurs, continue on as optic tracts, enter thalamus, and synapse there; thalamic fibers run (as the optic radiation) to occipital (visual) cortex, where visual interpretation occurs; see also Figure 16.23
Function: Purely sensory; carry afferent impulses for vision
Damage to optic nerve results in blindness in eye served by nerve; damage to visual pathway distal to optic chiasma results in partial visual losses; visual defects are called *anopsias* (an-op′-sē-ahs). ■

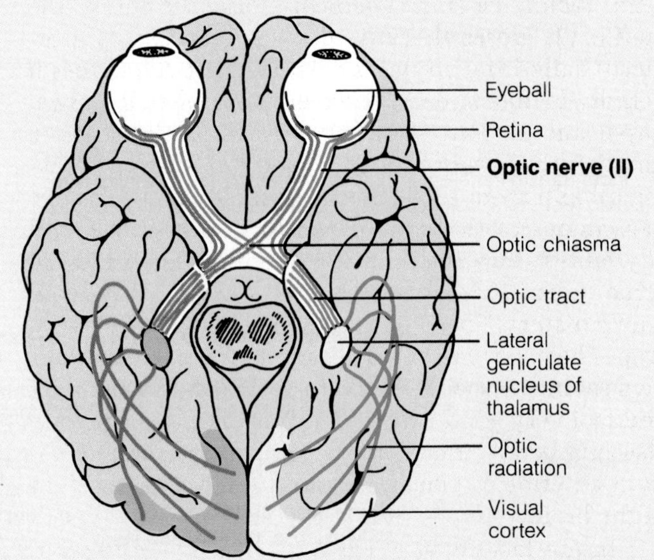

III The oculomotor (ok″-yū-lō-mō′-ter) **nerves**

Origin and course: Fibers extend from ventral midbrain (near its junction with pons) and pass through bony orbit, via superior orbital fissure, to eye
Function: Mixed nerves; although oculomotor nerves contain a few proprioceptive* afferents, they are chiefly motor nerves (as implied by their name, meaning "motor to the eye"); each nerve includes:

- Somatic motor fibers to four of the six extrinsic eye muscles (inferior oblique and superior, inferior, and medial rectus muscles) that help direct eyeball, and to levator palpebrae superioris muscle, which raises upper eyelid
- Parasympathetic (autonomic) motor fibers to constrictor muscles of iris, which cause pupil to constrict, and to ciliary muscle, controlling lens shape for visual focusing
- Sensory (proprioceptor) afferents, which run from same four extrinsic eye muscles to midbrain

In oculomotor nerve paralysis, eye cannot be moved up, down, or inward, and at rest, eye rotates laterally (*external strabismus* [struh-biz′-mis]) because the actions of the two extrinsic eye muscles not served by cranial nerve III are unopposed; upper eyelid droops (*ptosis*), and the person has double vision and trouble focusing on close objects. ■

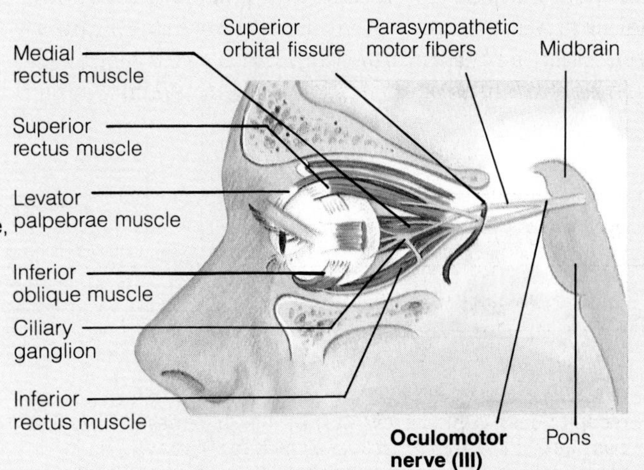

IV The trochlear (trok′-lē-ar) **nerves**

Origin and course: Fibers emerge from dorsal midbrain and course ventrally around midbrain to enter orbits through *superior orbital fissures* along with oculomotor nerves
Function: Mixed nerves; supply somatic motor fibers to, and carry proprioceptor* fibers from, one of the extrinsic eye muscles, the superior oblique muscle
Trauma to, or paralysis of, a trochlear nerve results in double vision and reduced ability to rotate eye inferolaterally ■

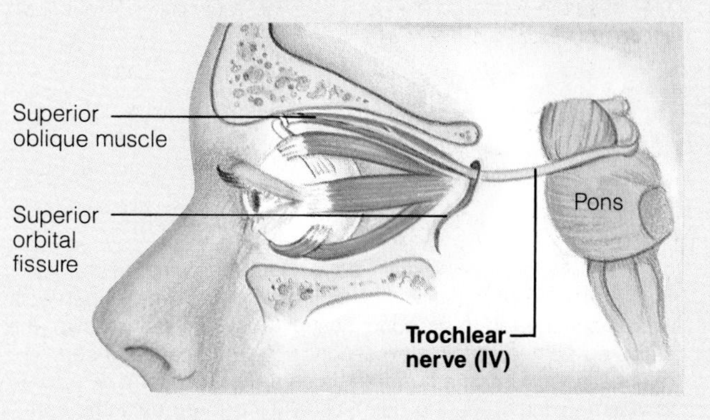

*Proprioceptors (prō′-prē-ō-sep″-terz) are sensory receptors found within muscles and their tendons, which monitor stretch and tension within those structures. These receptors are described later in this chapter and in Chapter 11.

Table 13.1 (continued)

sions are located in large *trigeminal* (also called *semilunar* or *gasserian*) *ganglion*; the mandibular division also contains some

numb

	Ophthalmic division (V_1)	Maxillary division (V_2)	Mandibular division (V_3)
Origin and course:	Fibers run from face to pons via superior orbital fissure	Fibers run from face to pons via foramen rotundum	Fibers pass through skull via foramen ovale
Function:	Conveys sensory impulses from skin of anterior scalp, upper eyelid, and nose, and from nasal cavity mucosa, cornea, and lacrimal gland	Conveys sensory impulses from nasal cavity mucosa, palate, upper teeth, skin of cheek, upper lip, lower eyelid	Conveys sensory impulses from tongue (except taste buds), lower teeth, skin of chin, temporal region of scalp; supplies motor fibers to, and carries proprioceptor fibers from, muscles of mastication

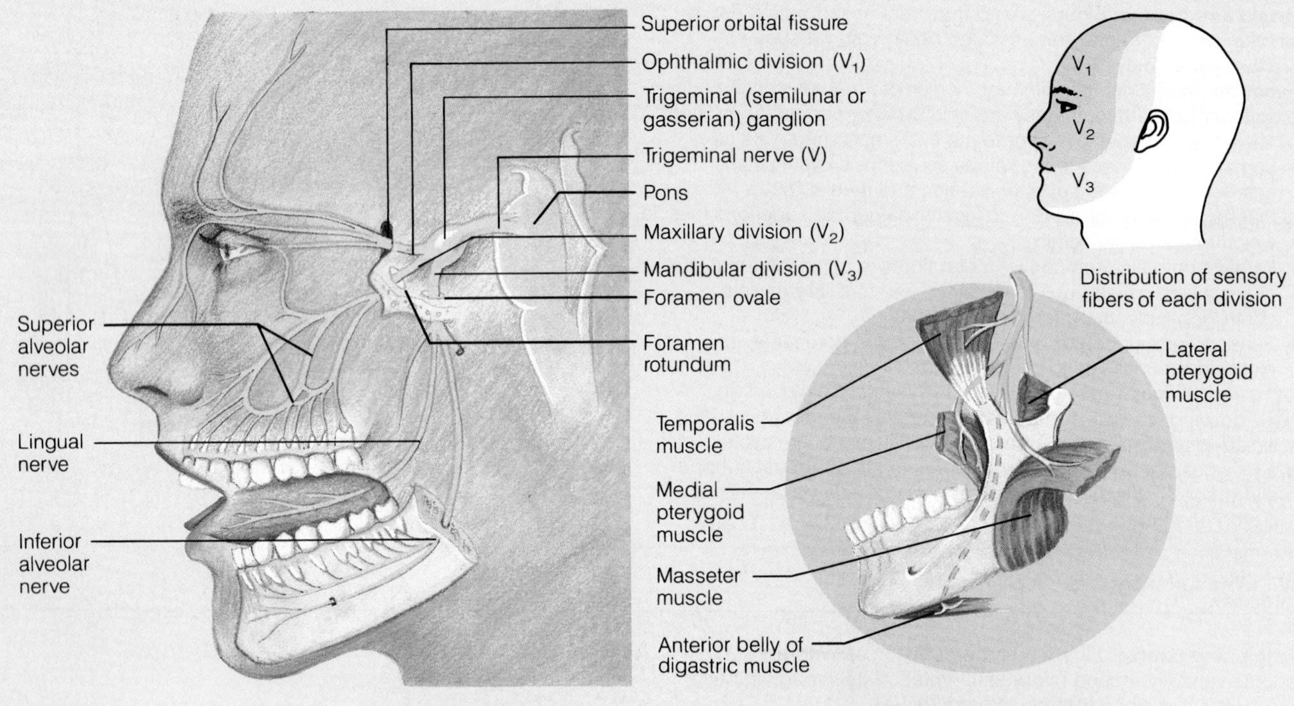

Distribution of sensory fibers of each division

Inset shows motor branches of the mandibular division (V_3)

Tic douloureux (tik doo″-loo-roo′), or *trigeminal neuralgia* (ner-al′-juh), caused by inflammation of trigeminal nerve, is widely considered to produce most excruciating pain known; the stabbing pain lasts for a few seconds to a minute, but it can be relentless, occurring a hundred times a day; (the term *tic* refers to victim's wincing during pain); usually triggered by some sensory stimulus, such as brushing teeth or even a puff of air hitting the face, but may reflect pressure on trigeminal nerve root; analgesics only partially effective; in severe cases, nerve is cut proximal to trigeminal ganglion; this relieves the agony, but also results in loss of sensation on that side of face ■

VI The abducens (ab-doo′-sinz) nerves

Origin and course: Fibers leave inferior pons and enter orbit via superior orbital fissure to run to eye

Function: Mixed nerve, primarily motor; supplies somatic motor fibers to lateral rectus muscle, an extrinsic muscle of the eye; convey proprioceptor impulses from same muscle to brain.

In abducens nerve paralysis, eye cannot be moved laterally; at rest, affected eyeball rotates medially (*internal strabismus*) ■

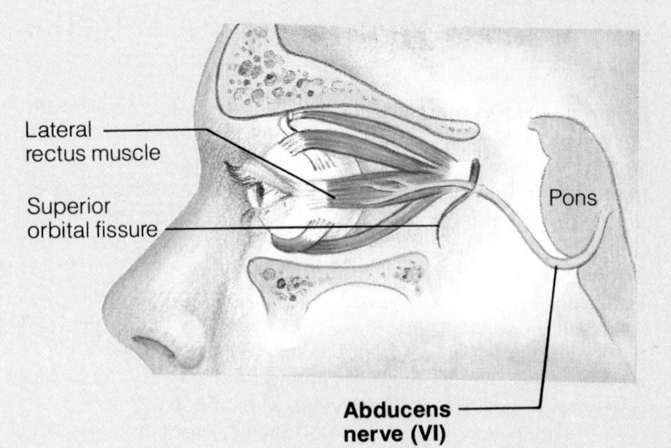

VII The facial nerves

Origin and course: Fibers issue from pons, just lateral to abducens nerves (see Figure 13.1), enter temporal bone via *internal auditory meatus,* and run within bone (and through inner ear cavity) before emerging through *stylomastoid foramen*; nerve then courses to lateral aspect of face

Function: Mixed nerves that are the chief motor nerves of face; have five major branches: temporal, zygomatic, buccal, mandibular, and cervical (see **a**)

- Convey motor impulses to skeletal muscles of face (muscles of facial expression), except for chewing muscles served by trigeminal nerves, and transmit proprioceptor impulses from same muscles to pons (see **c**)
- Transmit parasympathetic (autonomic) motor impulses to lacrimal (tear) glands, nasal and palatine glands, and submandibular and sublingual salivary glands. Some of the cell bodies of these parasympathetic motor neurons are in *sphenopalatine* (sfe″-nō-pa′-luh-tīn) and *submandibular ganglia* (see **b**)
- Convey sensory impulses from taste buds of anterior two-thirds of tongue; cell bodies of these sensory neurons are in *geniculate ganglion* (see **b**)

Bell's palsy, characterized by paralysis of facial muscles on affected side and partial loss of taste sensation, may develop rapidly (often overnight); cause is usually unknown, but inflammation of facial nerve is often suspect; lower eyelid droops, corner of mouth sags (making it difficult to eat or speak normally), eye constantly produces tears and cannot be completely closed; condition may disappear spontaneously without treatment ■

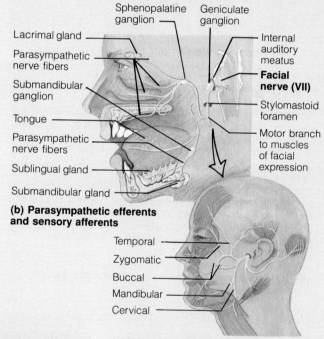

Sphenopalatine ganglion
Geniculate ganglion
Lacrimal gland
Internal auditory meatus
Parasympathetic nerve fibers
Facial nerve (VII)
Submandibular ganglion
Stylomastoid foramen
Tongue
Motor branch to muscles of facial expression
Parasympathetic nerve fibers
Sublingual gland
Submandibular gland

(b) Parasympathetic efferents and sensory afferents

Temporal
Zygomatic
Buccal
Mandibular
Cervical

(c) Motor branches to muscles of facial expression and scalp muscles

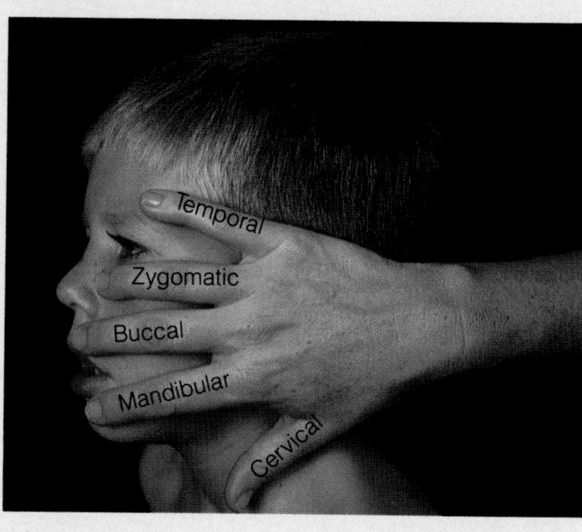

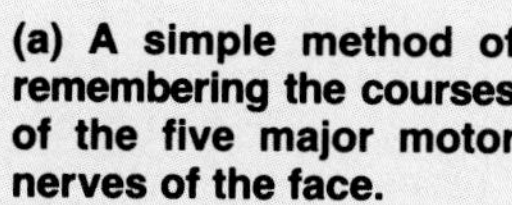

(a) A simple method of remembering the courses of the five major motor nerves of the face.

Table 13.1 (continued)

form *vestibular division;* the two divisions merge to form vestibulocochlear nerve; see also Figure 16.26
Function: Purely sensory; vestibular branch transmits afferent impulses for sense of equilibrium, and sensory nerve cell bodies are located in *vestibular ganglia;* cochlear branch transmits afferent impulses for sense of hearing, and sensory nerve cell bodies are located in *spiral ganglia* within cochlea

Lesions of cochlear nerve or cochlear receptors result in *central* or *nerve deafness,* whereas damage to vestibular division produces dizziness, rapid involuntary eye movements, loss of balance, nausea and vomiting. ■

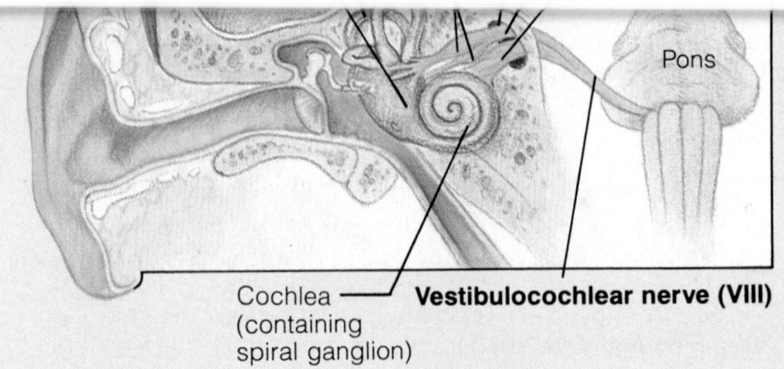

IX The glossopharyngeal (glah″-sō-fayr-in′-jē-ul) **nerves**

Origin and course: Fibers emerge from medulla and leave skull via *jugular foramen* to run to throat
Function: Mixed nerves that innervate part of tongue and pharynx; provide motor fibers to, and carry proprioceptor fibers from, superior pharyngeal muscles involved in swallowing and gag reflex; provide parasympathetic motor fibers to parotid salivary gland; (some of the nerve cell bodies of these parasympathetic motor neurons are located in *otic ganglion*)

Sensory fibers conduct taste and general sensory (touch, pressure, pain) impulses from pharynx and posterior tongue, and from pressure receptors of carotid artery (which help to regulate blood pressure by providing feedback information); sensory neuron cell bodies are located in *superior* and *inferior ganglia*

Injury or inflammation of glossopharyngeal nerves impairs swallowing and taste, particularly for sour and bitter substances. ■

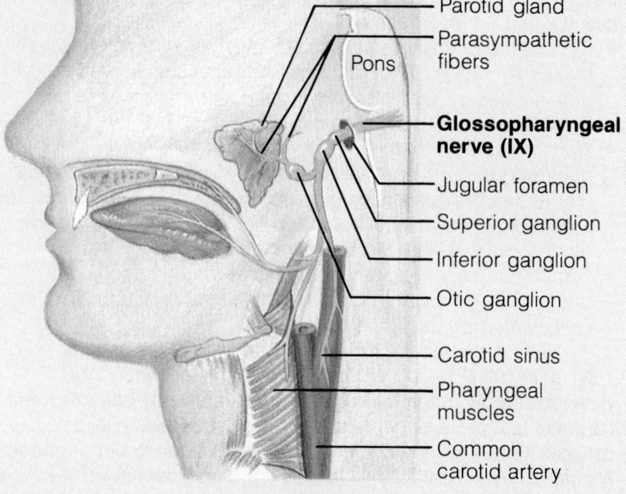

X The vagus (vā′-gus) nerves

Origin and course: The only cranial nerves to extend beyond head and neck region; fibers emerge from medulla, pass through skull via jugular foramen, and descend through neck region into thorax and abdomen; see also Figure 14.2
Function: Mixed nerves; nearly all motor fibers are parasympathetic efferents, except those serving skeletal muscles of pharynx and larynx (involved in swallowing); parasympathetic motor fibers supply heart, lungs, and abdominal viscera and are involved in regulation of heart rate, breathing, and digestive system activity; transmit sensory impulses from thoracic and abdominal viscera and taste buds of posterior tongue and pharynx; carry proprioceptor fibers from muscles of larynx and pharynx

Since nearly all muscles of the larynx ("voice box") are innervated by laryngeal branches of the vagus, vagal nerve paralysis can lead to hoarseness or loss of voice; other symptoms are difficult swallowing and impaired digestive system mobility. Total destruction of vagus nerve is incompatible with life, because these parasympathetic nerves are crucial in maintaining normal state of visceral organ activity; without their influence, the activity of the sympathetic nerves, which mobilize and accelerate vital body processes (and shut down digestion), would be unopposed. ■

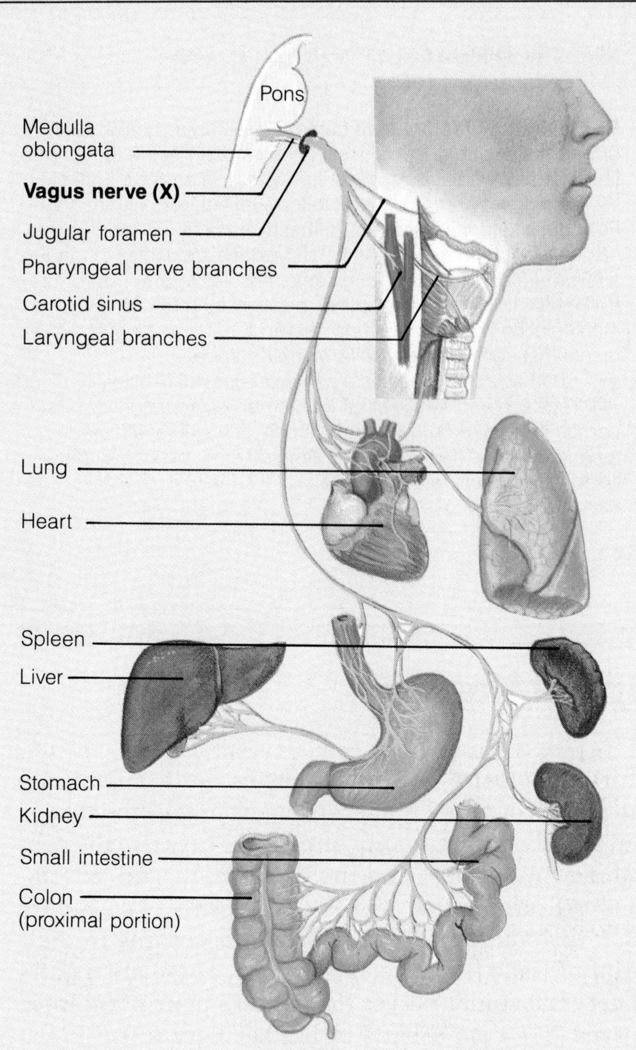

XI The accessory nerves

Origin and course: Unique in that they are formed by union of a *cranial root* and a *spinal root;* cranial root emerges from lateral aspect of medulla of brain stem; spinal root arises from superior region (C_1–C_5) of spinal cord. Spinal portion passes upward along spinal cord, enters skull via foramen magnum, and temporarily joins cranial root; the resulting accessory nerve exits from skull through *jugular foramen;* then cranial and spinal fibers diverge to travel to neck and shoulder regions, respectively
Function: Mixed nerves, but primarily motor in function; cranial division joins with fibers of vagus nerve (X) to supply motor fibers to larynx, pharynx, and soft palate; spinal root supplies motor fibers to trapezius and sternocleidomastoid muscles, which together move head and neck, and conveys proprioceptor impulses from same muscles

Injury to the spinal root of one accessory nerve causes head to turn toward injury site as result of sternocleidomastoid muscle paralysis; shrugging of that shoulder (role of trapezius muscle) becomes difficult. ■

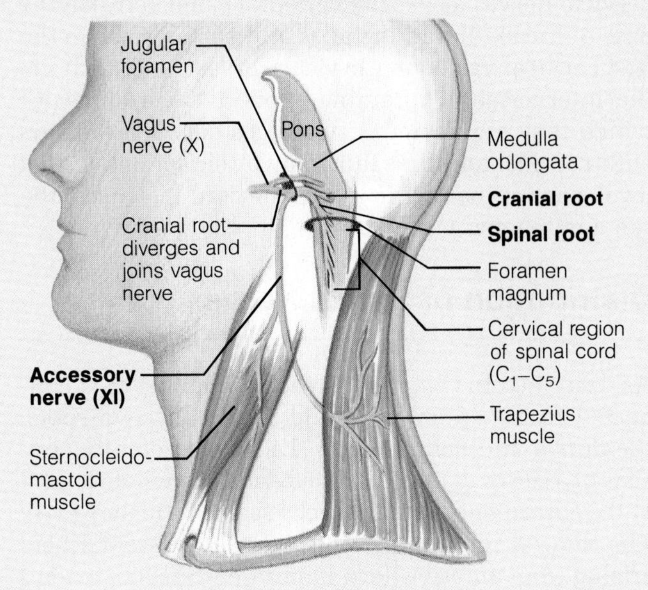

Table 13.1 (continued)

and proprioceptor fibers from same muscles to brain stem; hypoglossal nerve control allows not only food mixing and manipulation by tongue during chewing, but also tongue movements that contribute to swallowing and speech.

Damage of hypoglossal nerves causes difficulties in speech and swallowing; if both nerves are impaired, the person cannot protrude tongue; if only one side is affected, tongue deviates (leans) toward affected side; eventually paralyzed side begins to atrophy. ■

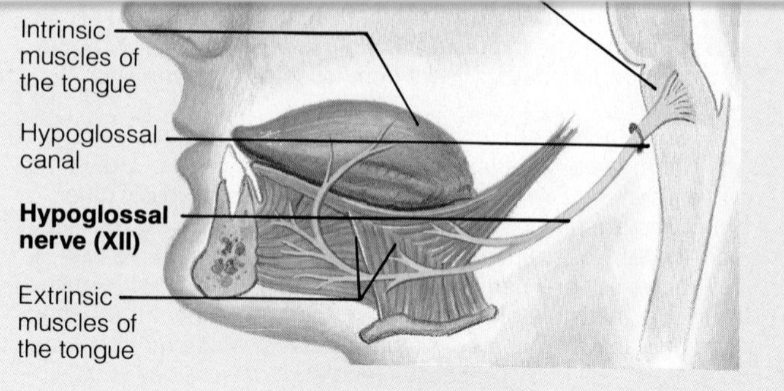

Spinal Nerves

Thirty-one pairs of **spinal nerves,** each containing thousands of nerve fibers, spring from the spinal cord and supply the communicating links between the central nervous system and the neck, trunk, and extremities. All are mixed nerves. As illustrated in Figure 13.2, the spinal nerves are named according to their point of issue from the spinal cord. There are 8 pairs of cervical spinal nerves (C_1–C_8), 12 pairs of thoracic nerves (T_1–T_{12}), 5 pairs of lumbar nerves (L_1–L_5), 5 pairs of sacral nerves (S_1–S_5), and 1 pair of tiny coccygeal nerves (designated C_0).

Notice that there are eight pairs of cervical nerves but only seven cervical vertebrae. The first pair of cervical nerves leaves the vertebral canal between the base of the skull and the atlas and emerges above the first cervical vertebra. Cervical nerves C_2–C_7 exit via the intervertebral foramina above the vertebrae for which they are named. C_8 emerges between the seventh cervical and first thoracic vertebrae. Below that level, each spinal nerve leaves the cord inferior to the same-numbered vertebra.

Distribution of Spinal Nerves

As described in Chapter 12, each spinal nerve is connected to the spinal cord and formed by two **roots,** the *dorsal* and *ventral* roots. Each root actually consists of a series of rootlets that extend the whole length of the corresponding spinal cord segment (Figure 13.3). The ventral, or motor, roots contain efferent fibers arising from anterior horn motor neurons that extend

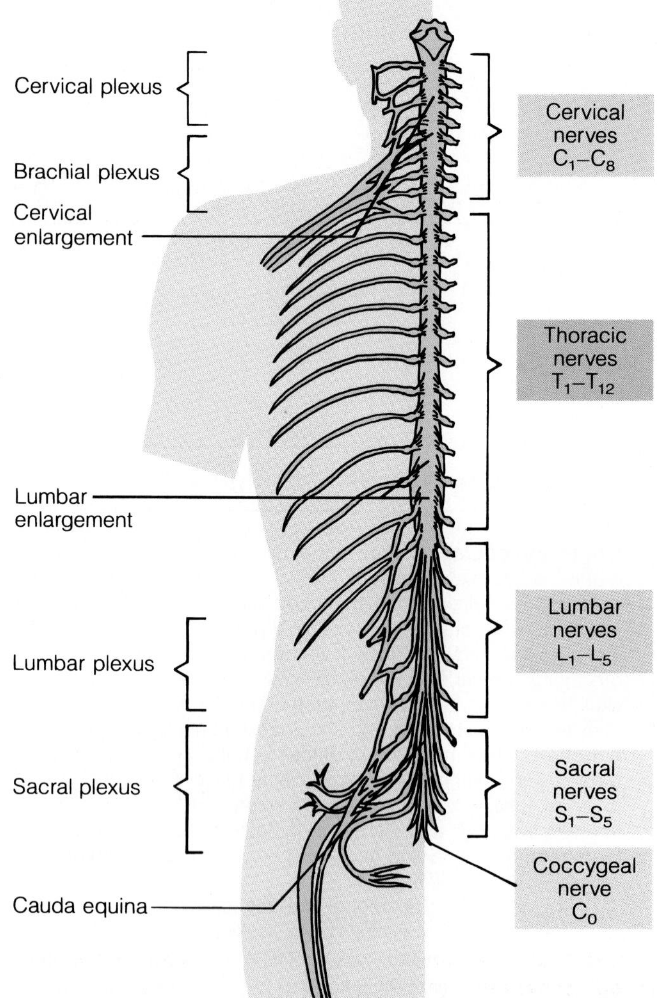

Figure 13.2 Distribution of spinal nerves, posterior view. Note that the spinal nerves are named according to their points of issue.

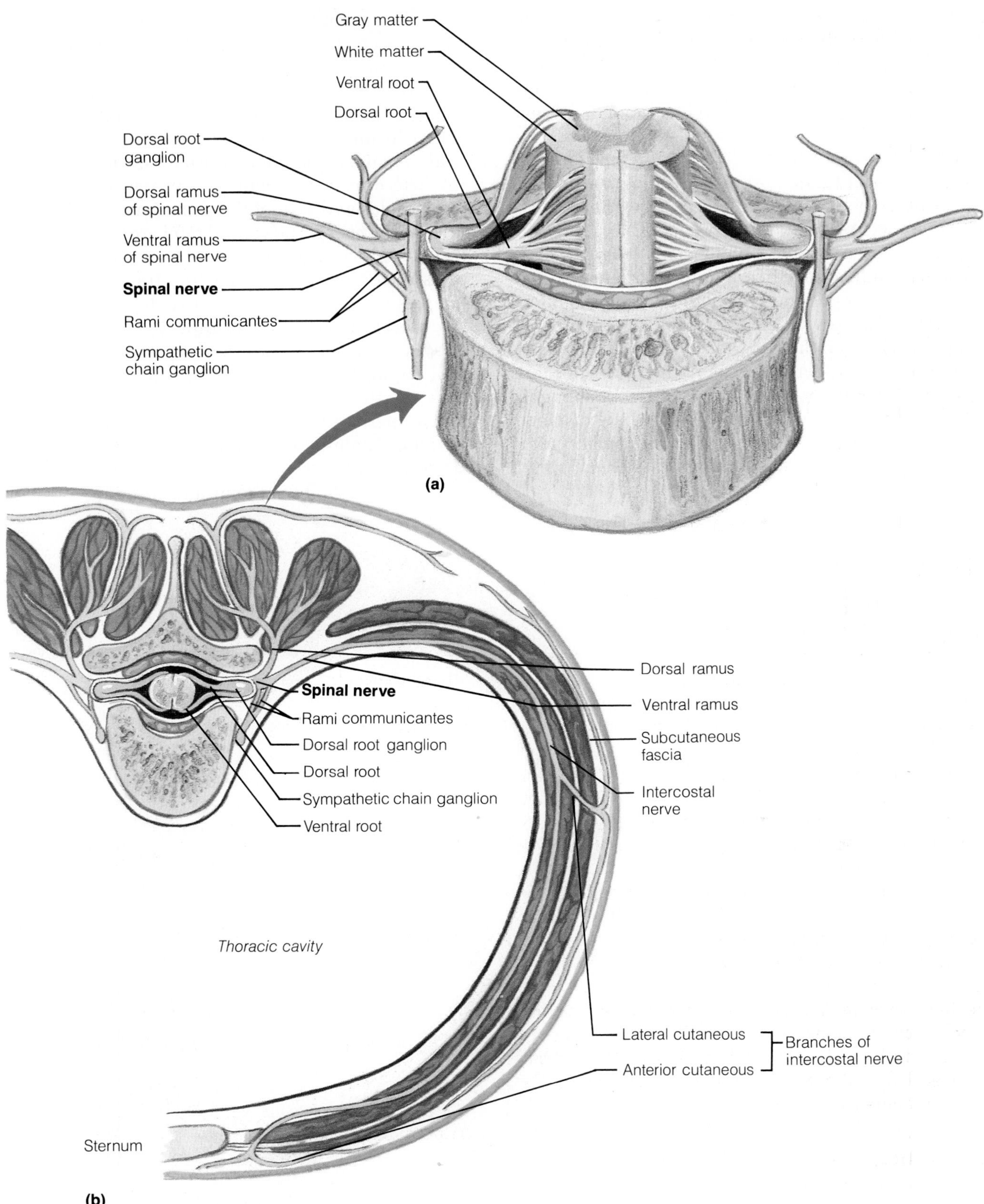

Figure 13.3 Formation and branches of spinal nerves. (**a**) A diagrammatic view of a spinal cord segment, illustrating the formation of one pair of spinal nerves by the union of the ventral and dorsal roots of the spinal cord. (**b**) Cross-sectional view of the left side of the body at the level of the thorax, showing the distribution of the dorsal and ventral rami of a spinal nerve. Notice also the rami communicantes branches of the spinal nerve. (The small meningeal branch, which reenters the vertebral canal to serve the vertebrae and associated structures, is not illustrated.)

to and innervate the skeletal muscles. (The autonomic nervous system efferents contained in the ventral roots are described in Chapter 14.) Dorsal, or sensory, roots contain afferent fibers arising from sensory neurons in the dorsal root ganglia that conduct impulses from peripherally located receptors to the spinal cord.

the spinal nerve, which is quite short (only 1–2 cm), the motor and sensory fibers mingle together, so the spinal nerve proper contains both efferent and afferent fibers. The length of the spinal roots increases progressively from the superior to the inferior aspect of the cord. In the cervical region, the roots are short and run horizontally, but the roots of the lumbar and sacral nerves extend inferiorly for some distance through the lower vertebral canal as the cauda equina before exiting the vertebral column (see Figure 13.2).

Almost immediately after emerging from its foramen, each spinal nerve divides into a large **ventral ramus**, a smaller **dorsal ramus**, and a tiny meningeal (meh-nin'je-ul) branch that reenters the vertebral canal to innervate the meninges, vertebrae, and asso-

they supply a larger region—the anterior and lateral parts of the body trunk and the limbs. The ventral rami of T_2–T_{12} course anteriorly deep to the ribs as the **intercostal nerves** that supply the intercostal muscles and skin of the anterior thorax and abdominal wall. Ventral rami of nearly all other spinal nerves form interlacing networks of nerves called *plexuses* that primarily serve the limbs (see Figure 13.2). *Rami communicantes* (communicating branches), branches of the spinal nerves in the thoracic region, are shown in Figure 13.3b. These rami and the unique ganglia

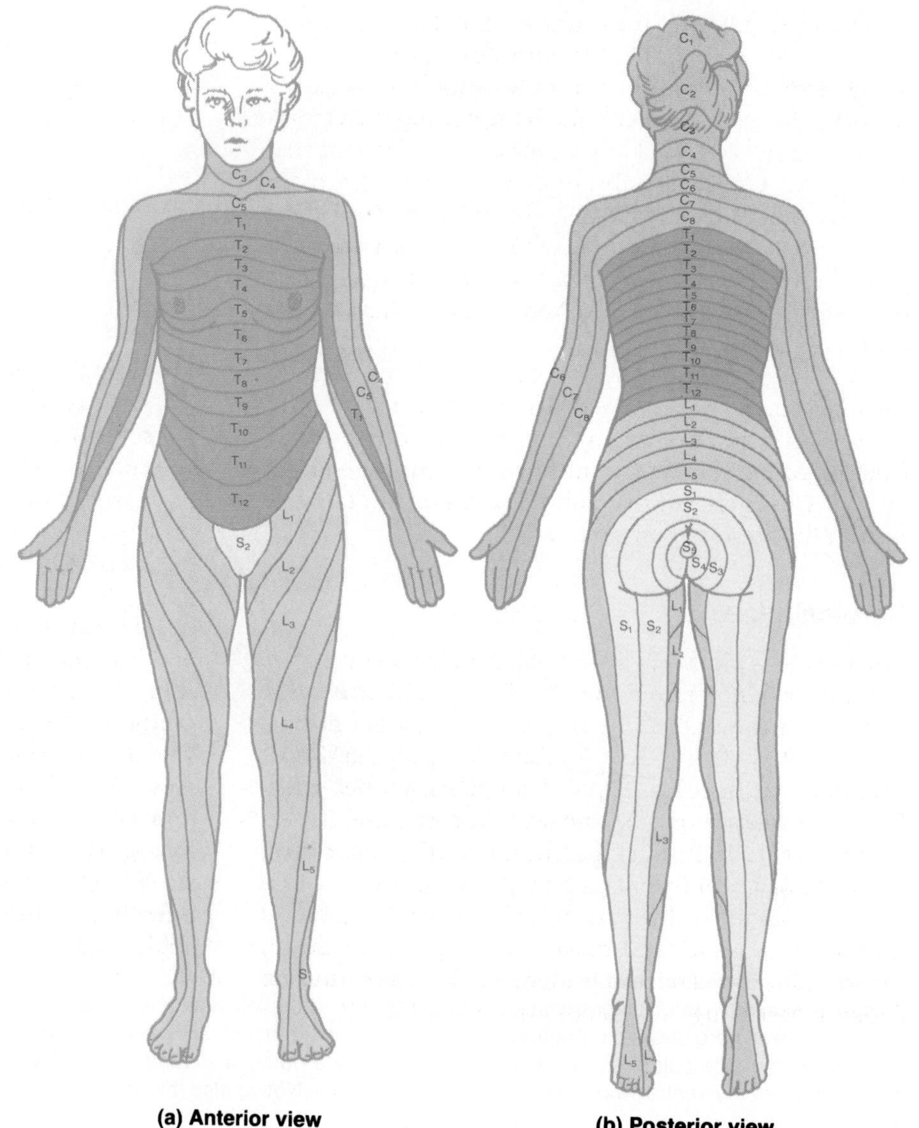

Figure 13.4 Dermatomes (skin segments) of the body are related to the sensory innervation regions of the spinal nerves.

with which they are associated are part of the sympathetic division of the autonomic nervous system and are described in Chapter 14.

The area of skin innervated by cutaneous branches of a single spinal nerve is called a **dermatome** (der'-muh-tōm). Adjacent dermatomes on the body trunk are fairly uniform in width, and the spinal nerve rami supply horizontal body trunk segments in line with their emergence points (Figure 13.4). The arrangement of the dermatomes in the limbs is less obvious. Ventral rami of cervical nerves supply the entire upper limbs except for the anteromedial surfaces, which are supplied by T_1. The ventral rami of the lumbar nerves supply the anterior surfaces of the buttocks, thighs, and legs; the ventral rami of sacral nerves serve most of the posterior surfaces of the lower limbs. However, there is considerable overlap (about 50%) of the dermatome regions; thus, the picture is much more complex than indicated by a typical dermatome map.

Plexuses and Peripheral Nerves

A feature of all spinal nerves except T_2–T_{12} is that their ventral rami branch to join one another lateral to the vertebral column, forming complicated nerve **plexuses** (see Figure 13.2). Such branching nerve networks occur in the cervical, brachial, lumbar, and sacral regions. Only ventral (not dorsal) rami form plexuses. Within the plexuses, fibers of the ventral rami criss-cross each other and become redistributed so that (1) each resulting branch of the plexus contains fibers from several different nerve roots and (2) fibers from each nerve root are carried to the body periphery via several different pathways or branches. Thus, each of the muscles in the limb receives its nerve supply from more than one spinal nerve root. An advantage of this fiber regrouping is that damage to one spinal segment or root does not lead to complete motor or sensory loss in the limb region served.

Cervical plexus

The **cervical plexus** is buried deep in the neck under the sternocleidomastoid muscle. This plexus is formed by the ventral rami of the upper four cervical nerves (Figure 13.5). The plexus forms an irregular series of interconnecting loops, from which the branches arise. Its single most important nerve is the **phrenic** (freh'-nik) **nerve,** (which receives fibers from C_5 in addition to its major input from C_3 and C_4). The phrenic nerve courses inferiorly through the thorax and transmits motor impulses to (and most sensory impulses from) the diaphragm. The diaphragm is the chief muscle causing breathing movements (see Chapter 23).

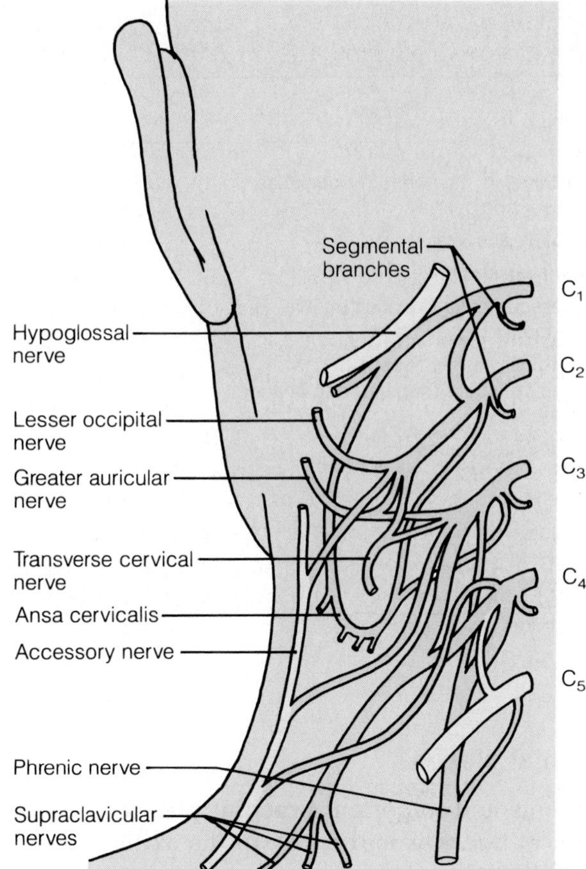

Figure 13.5 The cervical plexus. (See Table 13.2.)

Irritation of the phrenic nerve causes spasms of the diaphragm, or hiccups. If both phrenic nerves are severed, or if the C_3–C_5 region of the spinal cord is crushed or destroyed, the diaphragm becomes paralyzed and respiratory arrest occurs. Victims can be kept alive only by mechanical respirators that force air into their lungs and do their breathing for them. ■

The many superficial (cutaneous) branches of the cervical plexus transmit sensory impulses from the skin of the neck, the ear area, and the shoulder. The motor branches (other than the phrenic nerves) innervate the anterior muscles of the neck (infrahyoid muscles). Additionally, some of the fibers of the cervical plexus combine with those of the accessory and hypoglossal nerves and help to provide the motor supply to muscles served by those cranial nerves. The branches of the cervical plexus are summarized in Table 13.2.

Table 13.2 Branches of the Cervical Plexus (See Figure 13.5)

		chest
Motor branches (deep)		
Ansa cervicalis (superior and inferior roots)	C_1–C_4	Geniohyoid and infrahyoid muscles of neck
Segmental branches	C_1–C_5	Deep muscles of neck (omohyoid, geniohyoid, thyrohyoid, sternohyoid, and sternothyroid) and portions of scalenes, levator scapulae, trapezius, and sternocleidomastoid muscles
Phrenic	C_3–C_5	Diaphragm (sole motor nerve supply)

Brachial plexus

The large and important **brachial plexus** is situated partly in the neck and partly in the axilla and gives rise to virtually all the nerves that innervate the upper limb. It can be palpated in a living person in the angle between the clavicle and sternocleidomastoid muscle; it lies deep to both structures as it descends toward the arm.

This plexus is formed by the intermixing of the ventral rami of the lower four cervical nerves (C_5–C_8) and most of the T_1 root (Figure 13.6a). Additionally, it often receives fibers from C_4 or T_2 or both.

The brachial plexus is very complex (some consider it to be the anatomy student's nightmare). As it moves away from the vertebral column, its ventral rami unite to form **upper, middle,** and **lower trunks,** each of which almost immediately divides into an **anterior** and a **posterior division.** The divisions provide a general indication of which fibers will ultimately serve the front of the limb and which its posterior aspect. The divisions pass deep to the clavicle and over the first rib and then enter the axilla, where they reunite into three large fiber bundles called the **lateral, medial,** and **posterior cords.** All along the pathway of the plexus, nerves are given off that provide the motor and sensory supply of the muscles and skin of the shoulder, and of the superior part of the anterolateral thorax.

The brachial plexus terminates within the axilla, where its three cords wind along the axillary artery and then issue terminal branches that serve the arm. The most important of the many terminal branches are the axillary, musculocutaneous, median, ulnar, and radial nerves. These nerves are described here and illustrated in Figure 13.6c. Students required to learn the sequence of joining and dividing branches of the brachial plexus will find this information in an easy-to-follow flowchart in Figure 13.6b. If you are required to learn only the initial ventral roots (rami), the cords, and their terminal branches, Table 13.3 provides all the information you need. The table also lists a few of the nerves issuing from the more proximal parts of the plexus.

The **axillary nerve** innervates the deltoid and teres minor muscles and the skin and joint capsule of the shoulder. The **musculocutaneous nerve,** the major terminal branch of the lateral cord, courses within the anterior arm, supplying motor fibers to the arm muscles that flex the forearm along its course. Beyond the elbow it provides for cutaneous sensation of the lateral forearm surface.

The **median nerve** effects pronation of the forearm, flexion of the wrist, fingers, and thumb, and thumb opposition. It runs through the arm to the anterolateral forearm, where it gives off branches to the skin, to all the muscles in the anterior forearm (except the flexor carpi ulnaris and the medial part of the flexor digitorum profundus), and to five muscles in the hand. Median nerve injury makes it difficult to oppose the thumb and little finger and thus to pick up small objects. ■

A CLOSER LOOK **Does Polio "Haunt" Overachievers?**

It is now some 30-plus years after the great polio epidemics of the late 1940s and 1950s. But it appears that many of its "recovered" survivors are not yet done with the viral disease that destroys spinal and brain motor neurons, rendering associated nerves useless. Vital people, today in their late 30s to mid-40s, who fought their way back to health and active, vigorous lives have begun to notice a new, disturbing set of symptoms. Examples include extreme lethargy; sensitivity to cold; sharp, burning pains in their muscles and joints; and weakness and gradual loss of mass in the muscles that for so many years after their initial illness served them well.

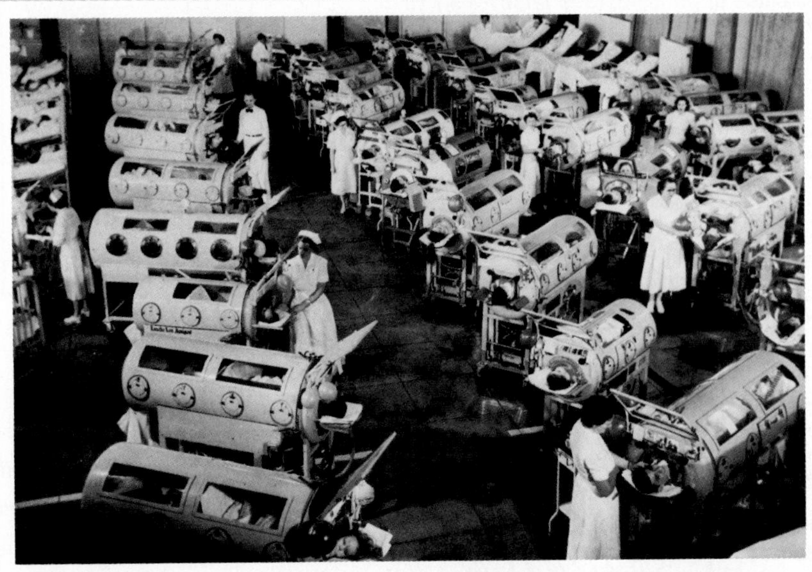

Photograph of a multitude of iron lungs in a March of Dimes Respiratory Center in the 1950s. The iron lungs provided mechanical breathing assistance for polio patients whose respiratory muscles were paralyzed.

Initially, these baffling symptoms were dismissed as the flu or as psychosomatic illness, or they were misdiagnosed as multiple sclerosis, arthritis, lupus, or Lou Gehrig's disease, a devastating neurological disease that kills its victims swiftly. But as the numbers of similar complaints has increased, more attention has been given to this group of symptoms, which is now named *postpoliomyelitis muscular atrophy (PPMA)* or *postpolio syndrome*.

According to a 1983 study conducted by the Mayo Clinic, 25% of polio survivors are affected. It now appears that this estimate is extremely low. Moreover, certain survivors seem to be particularly vulnerable: Those who contracted the disease after the age of ten, those who were severely affected, those who needed ventilatory support, and those who had all four limbs affected. The cause of PPMA is not known. The earlier belief that the polio virus was reactivated has now been discarded. A more likely explanation is that polio survivors, like all of us, continue to lose neurons throughout life. While unimpaired nervous systems can enlist nearby neurons to compensate for the losses, polio survivors who have already lost many neurons have drawn on that "pool" and may have few neurons left to take over. What is particularly ironic is that PPMA victims seem to be overachievers; that is, they are the ones who worked hardest to overcome their disease. The grueling hours of exercise that they assumed would make them strong may be the factor that ultimately caused their overworked motor neurons to "burn out."

Postpolio syndrome progresses slowly, and it is not life threatening. Currently, its victims are advised to conserve their energy by resting more and by using devices such as canes, walkers, and braces to support their wasting muscles. A major concern is determining how much muscle rest is required, so that the muscle atrophy caused by PPMA is not accentuated by disuse atrophy.

Many PPMA sufferers are emotionally devastated—surprised by the return of an old enemy, angry because no remedy presently exists, and understandably bitter because all of their hard work has resulted in a new affliction. Their plight has finally begun to receive the attention it deserves from the medical and research community, and grants for research and postpolio assessment clinics have been established across the nation. But this is cold comfort to those experiencing the limbo of PPMA.

Figure 13.6 The brachial plexus. (**a**) Roots, trunks, and cords of the brachial plexus. (**b**) Flowchart showing the consecutive branches formed in the brachial plexus from

Major terminal branches (peripheral nerves)	Cords	Divisions	Trunks	Roots (ventral rami)
Musculocutaneous, Median	Lateral	Anterior, Posterior	Upper	C_5, C_6
Median, Ulnar	Medial	Anterior, Posterior	Middle	C_7
Radial, Axillary	Posterior	Anterior, Posterior	Lower	C_8, T_1

(b)

(c)

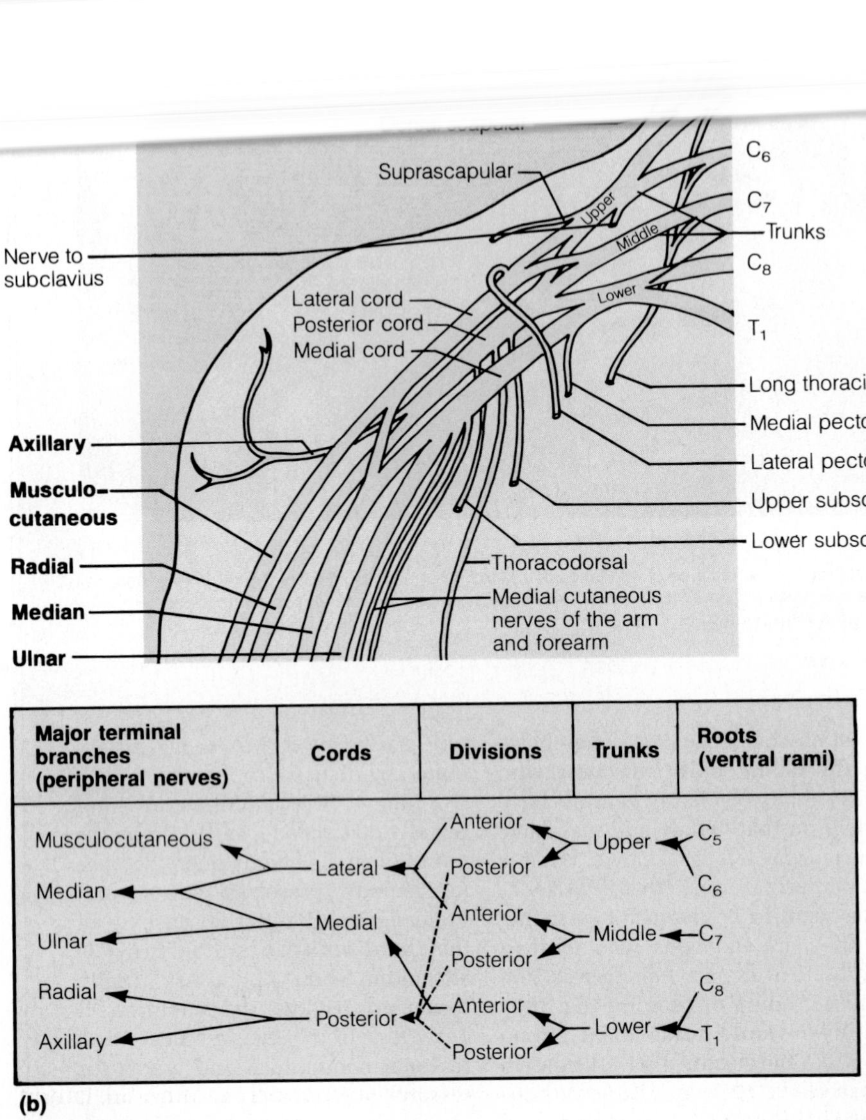

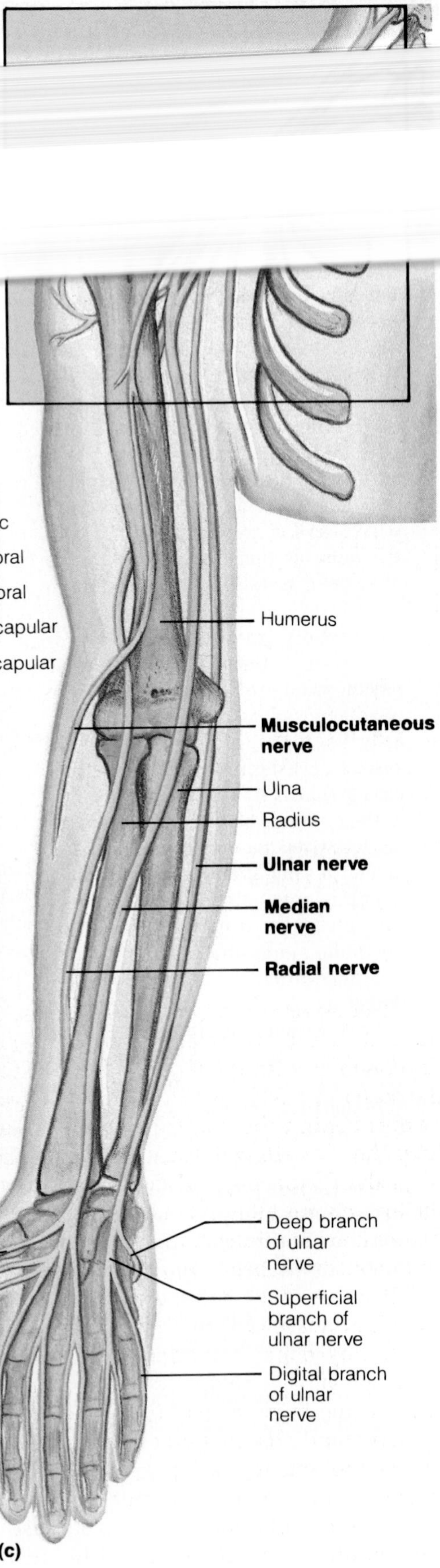

Table 13.3 Branches of the Brachial Plexus (See Figure 13.6)

Nerves	Cord and spinal roots (ventral rami)	Structures served
Musculocutaneous	Lateral cord (C_5–C_7)	Muscular branches: flexor muscles in anterior arm (biceps brachii, brachialis, coracobrachialis) Cutaneous branches: skin on anterolateral forearm; (extremely variable)
Median	By two roots from medial cord (C_8, T_1) and lateral cord (C_5–C_7)	Muscular branches: palmaris longus, flexor carpi radialis, flexor digitorum superficialis, flexor pollicis longus, lateral half of flexor digitorum profundus, and pronator muscles in anterior forearm; intrinsic muscles of lateral palm and first two fingers. Cutaneous branches: skin of lateral two-thirds of hand, palm side
Ulnar	Medial cord (C_8 and T_1)	Muscular branches: flexor muscles in anterior forearm (flexor carpi ulnaris and medial half of flexor digitorum profundus); most intrinsic muscles of hand Cutaneous branches: skin of medial third of hand, both anterior and posterior aspects
Radial	Posterior cord (C_5–C_8, T_1)	Muscular branches: posterior muscles of arm, forearm, and hand (triceps brachii, anconeus, supinator, brachioradialis, extensors carpi radialis longus and brevis, extensor carpi ulnaris, and several muscles that move the fingers) Cutaneous branches: skin of posterolateral surface of entire limb
Axillary	Posterior cord (C_5, C_6)	Muscular branches: deltoid and teres minor muscles Cutaneous branches: skin of shoulder
Dorsal scapular	Branches of C_5 rami	Rhomboid muscles and levator scapulae
Long thoracic	Branches of C_5–C_7 rami	Serratus anterior muscle
Subscapular	Upper trunk; branches of C_5 and C_6 rami	Teres major and subscapular muscles
Suprascapular	Upper trunk (C_5, C_6)	Shoulder joint; supra- and infraspinatus muscles
Pectoral (lateral and medial)	Branches of all three trunks (C_5–T_1)	Pectoralis major and minor muscles

The **ulnar nerve** produces wrist and finger flexion and adduction and abduction of the medial fingers (in common with the median nerve). It moves down the posterior medial aspect of the arm toward the elbow, swings behind the medial epicondyle, and then follows the ulna along the medial forearm, where it supplies the flexor carpi ulnaris and part of the flexor digitorum profundus. It continues into the hand, where it supplies most intrinsic hand muscles and provides cutaneous branches to the medial aspect of the hand.

Because of its exposed position at the elbow, the ulnar nerve is very vulnerable to injury. Striking the "funny bone"—the spot where the ulnar nerve rests against the medial epicondyle—causes tingling of the little finger. Severe or chronic damage (caused by perpetually leaning on one's elbows, for example) can lead to sensory loss, paralysis, and atrophy of the muscles served by the ulnar nerve. Affected individuals have trouble making a fist because they cannot flex the little and ring fingers at the distal interphalangeal joints. The typical appearance of the hand is called *clawhand.* ■

The **radial nerve,** the largest branch of the brachial plexus, is a continuation of the posterior cord. It produces elbow extension, supination of the forearm, extension of the wrist and fingers, and abduction of the thumb. This nerve wraps around the humerus (in the spiral groove), curves anteriorly around the lateral epicondyle at the elbow, and then follows the lateral edge of the radius to the hand. It supplies the posterior skin of the limb along its entire course. Its motor branches innervate the extensor muscles of the limb—the triceps brachii of the arm and the dorsolateral forearm muscles.

Trauma to the radial nerve results in *wristdrop,* an inability to extend the hand at the wrist. Many fractures of the humerus will follow the spiral groove, causing radial nerve damage. Improper use of a crutch or "Saturday night paralysis," in which an intoxicated person falls asleep with an arm draped over a chair or sofa edge, cause radial nerve compression and ischemia. ■

Lumbar plexus

The **lumbar plexus** arises from the first four lumbar spinal nerves and lies within the psoas major muscle (Figure 13.7). Its proximal branches innervate parts of the abdominal wall muscles and the psoas major,

of large branches. The motor branches innervate the anterior thigh muscles, which are the principal thigh flexors and knee extensors. The cutaneous branches serve the skin of the anterior thigh and the medial surface of the leg from the knee to the foot.

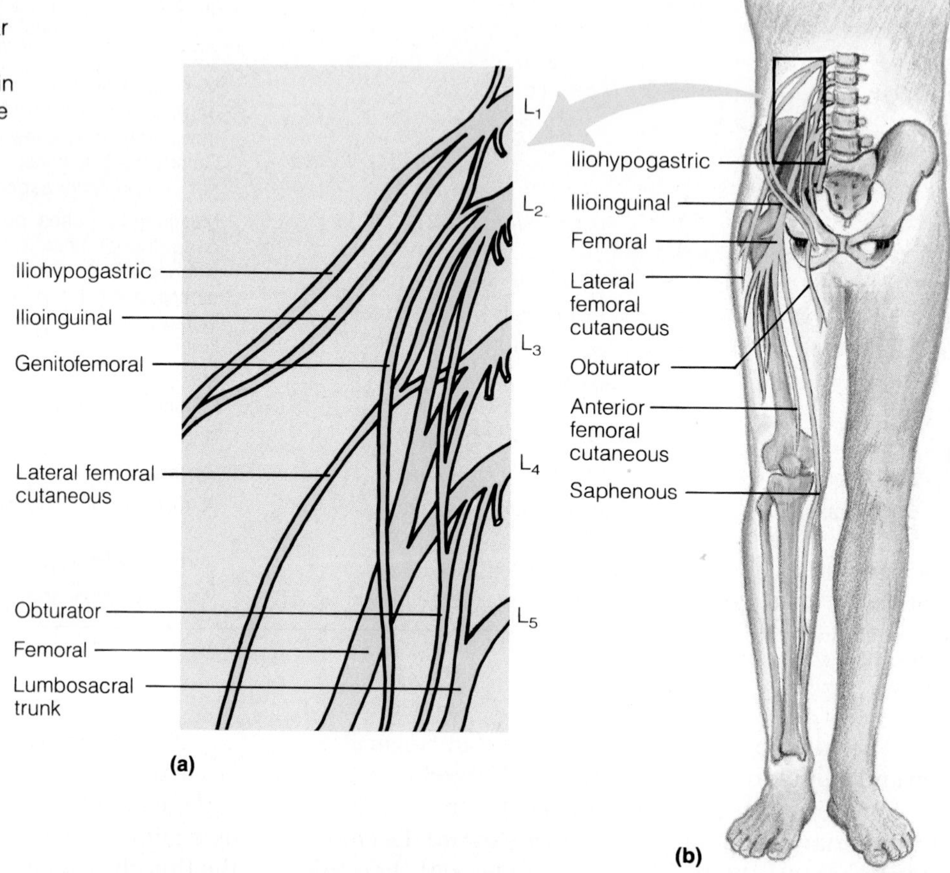

Figure 13.7 The lumbar plexus. (**a**) Roots and major branches of the lumbar plexus. (**b**) Distribution of the major peripheral nerves of the lumbar plexus in the lower limb (anterior view). (See Table 13.4.)

Table 13.4 Branches of the Lumbar Plexus (See Figure 13.7)

Nerves	Spinal roots	Structures served
Femoral	L_2–L_4	Skin of anterior and medial thigh via *anterior femoral cutaneous* branch; skin of leg, foot, hip, and knee joints via *saphenous* branch; motor to anterior muscles (quadriceps and sartorius) of thigh; psoas major, pectineus, iliacus, and parts of lower abdominal muscles
Obturator	L_2–L_4	Motor to adductor magnus (part), longus and brevis muscles, gracilis muscle of medial thigh, obturator externus; sensory for skin of medial thigh and hip joint
Lateral femoral cutaneous	L_2, L_3	Skin of lateral thigh; some sensory branches to peritoneum
Iliohypogastric	L_1	Skin of lower abdomen, lower back, and hip; muscles of anterolateral abdominal wall (obliques and transversus) and pubic region
Ilioinguinal	L_1	Skin of external genitalia and proximal medial aspect of the thigh; inferior abdominal muscles
Genitofemoral	L_1, L_2	Skin of scrotum in males, of labia majora in females, and of anterior thigh and inferior to middle portion of inguinal region; cremaster muscle in males

Sacral plexus

The **sacral plexus** arises from spinal nerves L_4–S_4 and lies immediately caudal to the lumbar plexus (Figure 13.8). Since the sacral and lumbar plexuses overlap substantially, and since many fibers of the lumbar plexus contribute to the sacral plexus via the **lumbosacral trunk,** the two plexuses are often referred to as the **lumbosacral plexus.** The sacral plexus has 12 named branches. Seven of these serve the buttock and lower limb; the others innervate pelvic structures. The most important branches are described here. Table 13.5 summarizes all but the smallest branches.

The major branch of the sacral plexus is the huge **sciatic** (sī-a′-tik) **nerve,** the largest and longest nerve in the body. Actually, two nerves (the tibial and common peroneal) wrapped in a common sheath, the sciatic nerve leaves the pelvis via the greater sciatic notch. It descends deep to the gluteus maximus muscle of the buttock and enters deeply into the posterior thigh. There it gives off motor branches to the hamstring muscles (all thigh extensors and knee flexors) and to part of the adductor magnus. Immediately above the knee, its two divisions diverge.

The **tibial nerve** courses through the popliteal fossa and supplies the posterior compartment muscles and the skin of the posterior calf and also the sole. Important branches of the tibial nerve are the **sural,** which serves the skin of the posterolateral leg, and the **plantar nerves,** which serve most of the foot. The **common peroneal nerve** innervates the knee joint and provides cutaneous branches to the lateral calf and dorsum of the foot and motor branches to all the muscles of the anterolateral leg.

Together, the *superior* and *inferior gluteal nerves* innervate the buttock (gluteal) and tensor fasciae latae muscles. Other branches of the sacral plexus supply the thigh rotators and muscles of the pelvic floor and perineum (the area encompassing the external genitalia and the rectum).

Injury to the proximal sciatic nerve, as might follow improper administration of an injection into the buttock, a fall, or a herniated disc, results in a number of lower limb impairments, depending on the precise nerve roots injured. *Sciatica* (sī-a′-tih-kuh), characterized by stabbing pain radiating over the course of the sciatic nerve, is common. When the sciatic nerve is completely transected, the leg becomes nearly useless. Knee flexion is difficult, and all foot and ankle

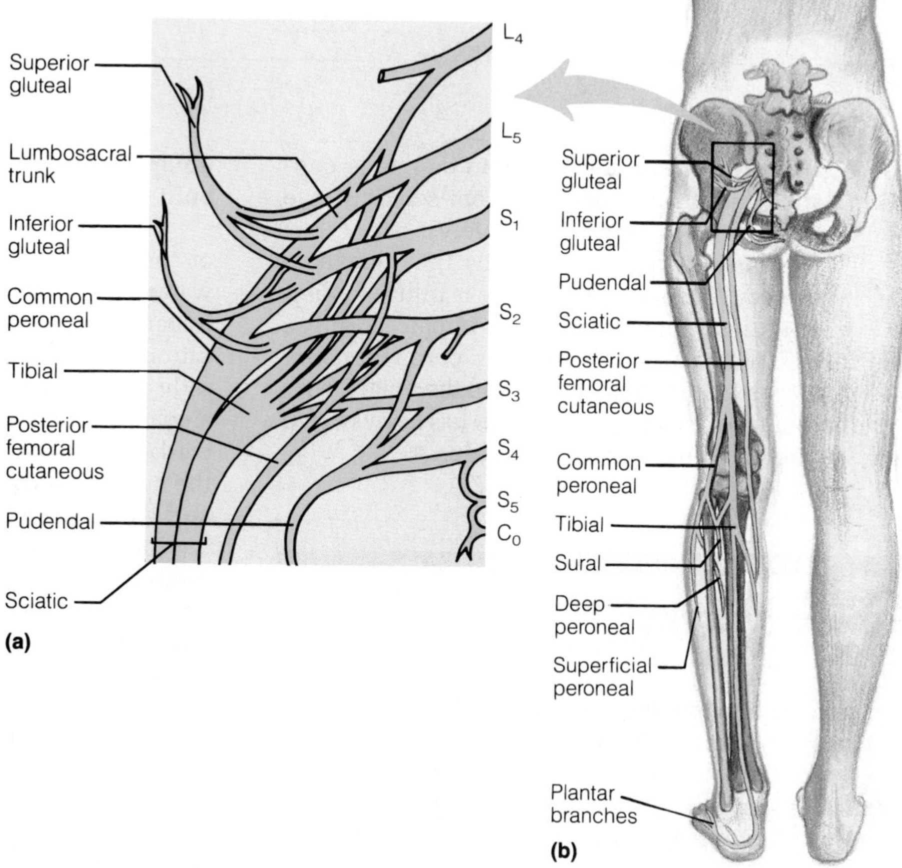

Figure 13.8 The sacral plexus. (**a**) The roots and major branches of the sacral plexus. (**b**) Distribution of the major peripheral nerves in the lower limb (posterior view). (See Table 13.5.)

Table 13.5 Branches of the Sacral Plexus (See Figure 13.8)

		... of leg and foot Motor branches: to all muscles of back of thigh, leg, and foot (biceps femoris, semimembranosus, semitendinosus, triceps surae, tibialis posterior, popliteus, flexor digitorum longus, flexor hallucis longus, and intrinsic muscles of foot)
• Common peroneal (superficial and deep branches)	L_4–S_2	Cutaneous branches: to skin of anterior surface of leg and dorsum of foot Motor branches: to peroneal muscles of lateral compartment of leg, tibialis anterior, and extensor muscles of toes (extensor hallucis longus, extensors digitorum longus and brevis)
Superior gluteal	L_4, L_5, S_1	Motor branches: to gluteus medius and minimus and tensor fasciae latae
Inferior gluteal	L_5–S_2	Motor branches: to gluteus maximus
Posterior femoral cutaneous	S_1–S_3	Skin of buttock, posterior thigh, and popliteal region; length variable; may also innervate part of skin of calf and heel
Pudendal	S_2–S_4	Supplies most of skin and muscles of perineum (region encompassing external genitalia and anus and including clitoris, labia, and vaginal mucosa in females, and scrotum and penis in males); external anal sphincter

movements are lost; the foot tends to dangle, a condition called *foot drop*. Recovery from sciatic nerve injury is usually slow and incomplete.

If the lesion occurs below the knee, the thigh muscles are spared. When the tibial nerve is injured, plantar flexion is lost, foot inversion is impaired, and a shuffling gait develops. The common peroneal nerve is much more susceptible to injury than the tibial nerve; even the weight of a blanket on the toes has been known to cause foot drop in bedridden people. ■

Reflex Activity

Many of the body's control systems belong to the general category of stimulus-response sequences known as reflexes. In the most restricted definition of the term, a **reflex** is a rapid, predictable motor response to a stimulus; it is unlearned, unpremeditated, and involuntary. Such basic reflexes may be considered to be built into our neural anatomy.

In many cases, we are aware of the final response of the reflex activity. If you splash a pot of boiling water on your arm, you are likely to drop the pot instantly and involuntarily before you have any sensation of pain. This response is triggered by a spinal reflex without any help from the brain. But the pain signals picked up by the interneurons of the spinal cord are quickly transmitted to the brain, so that within the next few seconds you do become aware of the pain, and you also know what happened to cause it.

In other cases, reflex activities go on without any awareness on our part. This is typical of many reflexively controlled visceral activities. Thus, although many students believe that only the conscious part of the brain causes muscle contraction, most motor activities are actually regulated by the subconscious lower regions of the central nervous system, specifically, the basal nuclei, brain stem, and spinal cord.

In addition to these basic, inborn types of reflexes, there are many *learned*, or *acquired, reflex responses* that result from practice or repetition. Take, for instance, the complex sequence of reactions that occurs when an experienced driver drives a car. These manipulations are largely automatic, but only because substantial time and conscious effort were expended to acquire the driving skill. However, most reflex actions, no matter how basic they may appear to be, are subject to modification by learning and conscious effort. If a three-year-old child was standing by your side when you scalded your arm, you most likely would set the pot down (rather than just letting go) because of your conscious recognition of the danger to the child. Thus, there is no clear-cut distinction between basic and learned reflexes.

As you learned in Chapter 11 (p. 368), reflexes are classified functionally as somatic and autonomic, and all reflex arcs have five elements: a *receptor*, a *sensory neuron*, an *integration center*, a *motor neuron*, and an *effector*. Here we will examine some of the more common types of somatic reflexes mediated by the spinal cord.

Spinal Reflexes

Somatic reflexes mediated by the spinal cord are called **spinal reflexes.** Many spinal reflexes occur without the involvement of higher brain centers. These reflexes work equally well in decerebrate animals (those in which the brain has been destroyed) as long as the spinal cord is functional. Other spinal reflexes require the activity of the brain for their successful completion. Additionally, though many spinal reflexes do not require the participation of higher centers, the brain is frequently "advised" of spinal cord reflex activity and can alter it by facilitating or inhibiting reflexes. Moreover, it appears that continuous facilitating signals from the brain are required for normal cord activity because, as noted in Chapter 12, if the spinal cord is suddenly transected, all functions controlled by the cord are immediately depressed, a phenomenon called *spinal shock*.

Testing of somatic reflexes is important clinically to assess the condition of the nervous system. Exaggerated, distorted, or absent reflexes indicate degeneration or pathology of specific regions of the nervous system, often before other signs are apparent.

Stretch and Deep Tendon Reflexes

If skeletal muscles are to perform normally, the brain must be continually informed of the current state of the muscles, and the muscles must exhibit healthy tone (resistance to active or passive stretch). The first requirement depends on transmission of information from *muscle spindles* and *Golgi tendon organs* (proprioceptors located in the skeletal muscles and their associated tendons) to the cerebellum and cerebral cortex. Normal muscle tone, the second requirement, depends on **stretch reflexes** initiated by the muscle spindles. These processes are important to normal skeletal muscle function, posture, and locomotion and deserve more than a passing mention.

Muscle spindles monitor changes in the length of a muscle by responding to both the rate and magnitude of changes in muscle length. When the muscle spindles are activated by muscle stretch (this is usually due to contraction of the antagonistic muscles), their associated sensory neurons transmit impulses at a higher frequency to the spinal cord. There the sensory neurons synapse directly (monosynaptically) with large anterior horn motor neurons called *alpha* (α) *motor neurons*, which in turn send efferent impulses to the muscle fibers in the stretched muscle (Figure 13.9b). The reflexive muscle contraction that follows resists further stretching. The afferent neurons also synapse with interneurons that inhibit motor neurons controlling the antagonistic muscles; the resulting inhibition is called *reciprocal inhibition*. Consequently, the stretch stimulus causes the antagonists to relax so that they cannot resist the shortening of the "stretched" muscle caused by the main reflex arc.

While this spinal reflex is occurring, impulses are also being relayed via the dorsal white columns to higher brain centers to provide information on muscle length and the velocity of muscle shortening. In this way, muscle tone is maintained reflexively and is adjusted to the requirements of posture and movement. Should either the afferent or efferent fibers be cut, the muscle immediately loses its tone and becomes flaccid. The stretch reflex is most important in the large antigravity muscles, such as the quadriceps, which sustain upright posture.

Although we will not get ensnared in details, a few comments must be made about the unique nature of the muscle spindles. The muscle spindles themselves receive efferent (motor) innervation via small motor fibers called *gamma* (γ) *efferent fibers* (see Figure 13.9a) that issue from anterior motor neurons distinct from the α motor neurons that stimulate mos
(*extrafusal*) fibers of the skeletal muscle. The γ effe
ents regulate the response of the *intrafusal mus*
fibers of the muscle spindle. Since descending fi
of motor pathways synapse with both α and γ
neurons, motor impulses are sent simultaneo
the large skeletal muscle fibers and the intrafu
of the muscle spindles. Stimulation of the
fibers maintains the muscle spindle's tensi
sitivity during muscle contraction, so tha
continually advised of the state of the
absence of such a system, information o
and the rate of changes in muscle len
to be transmitted in contracting mu

The motor supply to the muscl
the CNS to regulate the stretch
stimulating or inhibiting the γ

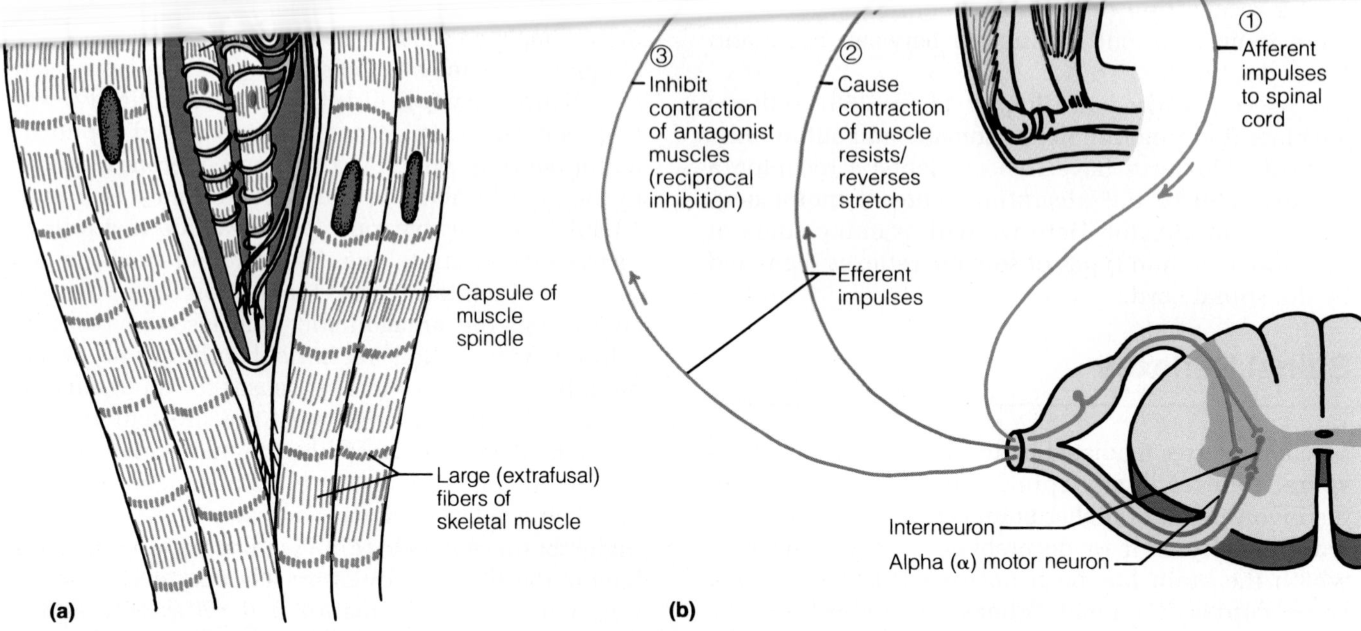

Figure 13.9 Structures and events of the stretch reflex. (**a**) Structure of a muscle spindle. Notice the afferent fibers to, and the efferent fibers from, the muscle spindle. (**b**) Events of the stretch reflex by which muscle stretch is damped. The events are shown in circular fashion. (*1*) Stretching of the muscle activates a muscle spindle. (*2*) Impulses transmitted by afferent fibers from muscle spindle to α motor neurons in the spinal cord result in activation of the stretched muscle causing it to contract. (*3*) Impulses transmitted by afferent fibers from muscle spindle to interneurons in the spinal cord result in reciprocal inhibition of the antagonist muscle.

…ns are stimulated, the spindle is stretched and …nsitive; when they are inhibited, the muscle …mes flaccid (like a loose rubber band) and …The ability to modify the stretch reflex …y situations. For example, if you …h a baseball, it is essential that …ssed so that your muscles …notion.

…mple of the stretch …**nee-jerk, reflex,** …ith a reflex …tendon …ne mus- … contrac- …ition of its

…n any skeletal …king its tendon or …pping the Achilles tendon (ankle jerk) assesses the L_5–S_2 segments of the cord, and striking the triceps brachii tendon at the elbow assesses the C_6–C_8 region. All stretch reflexes are monosynaptic and *ipsilateral* (*ipsi* = same; *lateral* = side); that is, they involve a single synapse and motor activity on the same side of the body.

A positive knee jerk (or a positive result for any stretch reflex test) provides two important pieces of information. First, it provides evidence that the sensory and motor connections between that muscle and the spinal cord are intact. Second, the vigor of the motor response provides a good indication of the degree of excitability of the spinal cord. When the spinal motor neurons are highly facilitated by impulses descending from higher centers, just touching the muscle tendon will elicit a vigorous reflex response. On the other hand, if the lower motor neurons are besieged by inhibitory signals, even pounding on the tendon may fail to elicit the reflex response.

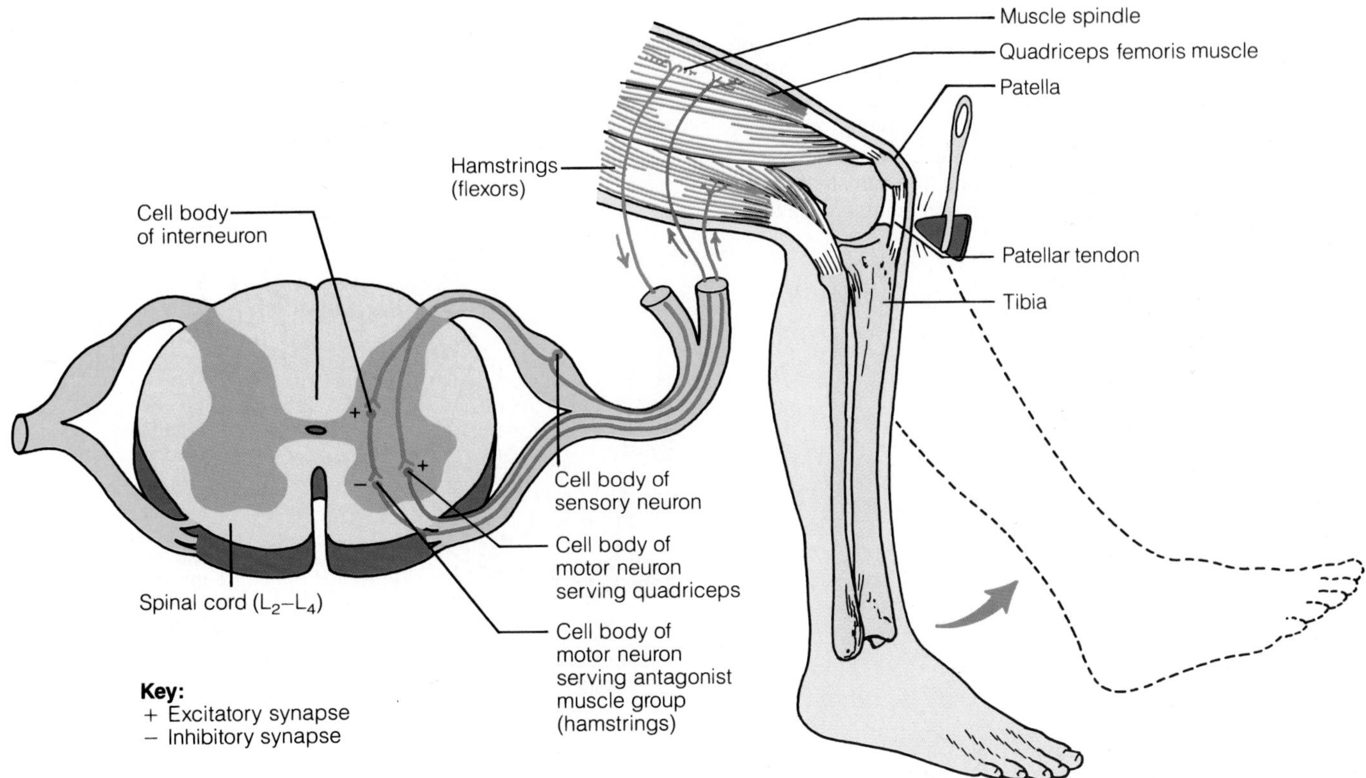

Figure 13.10 The patellar reflex. Tapping the patellar tendon excites muscle spindles within the quadriceps muscle. Afferent impulses travel to the spinal cord, where synapses occur with motor neurons and interneurons. The motor neurons send activating impulses to the same muscle, resulting in contraction of the quadriceps muscle, which causes extension of the knee and a forward movement of the foot, counteracting the initial stretch. The interneurons make inhibitory synapses with anterior horn neurons that serve the antagonistic muscle group (hamstrings), preventing it from resisting the contraction.

Stretch reflexes tend to be hypoactive or absent in cases of peripheral nerve damage or ventral horn injury involving the tested area; they are absent in those with chronic diabetes mellitus and neurosyphilis and during coma. On the other hand, these reflexes are hyperactive when lesions of the corticospinal tract reduce the inhibitory effect of the brain on the spinal cord. ■

While the net effect of stretch reflexes mediated by muscle spindles is to cause muscle contraction in response to increased muscle length (stretch), **deep tendon reflexes** cause just the opposite effect: muscle relaxation and lengthening in response to the muscle's contraction (Figure 13.11). When muscle tension increases moderately during muscle contraction or passive stretching, Golgi tendon receptors in the muscle's tendon are activated, and afferent impulses are transmitted to the spinal cord and from there to the cerebellum, where this information is used to help adjust muscle tension to precisely the amount needed to carry out the required movement. Simultaneously, in the spinal cord circuits, the motor neurons supplying the contracting muscle are inhibited and the antagonist muscles are activated, a phenomenon called *reciprocal activation*. As a result, the contracting muscle quickly relaxes as its antagonist is activated.

Although it was originally believed that the Golgi tendon organs primarily served a protective function, preventing muscle tension from becoming so great that muscle or tendon tearing was a possibility, it is now generally accepted that they help to ensure smooth onset and termination of muscle contraction and are particularly important in motor activities that involve rapid switching between flexion and extension, such as running. Notice that although the Golgi tendon organs are also stimulated during a clinically induced stretch reflex, such as the flexor reflex, the brevity of the stimulus and the fact that the stretched muscle is already relaxed prevent it from being inhibited by the deep tendon reflex.

The Flexor Reflex

The **flexor,** or **withdrawal, reflex** is initiated by a painful stimulus (actual or perceived) and causes automatic withdrawal of the threatened body part from the stimulus. The response that occurs when you prick

Figure 13.11 The deep tendon reflex. When the quadriceps muscle of the thigh contracts, the tension in the patellar tendon increases, activating Golgi tendon organs in the tendon. Affer-

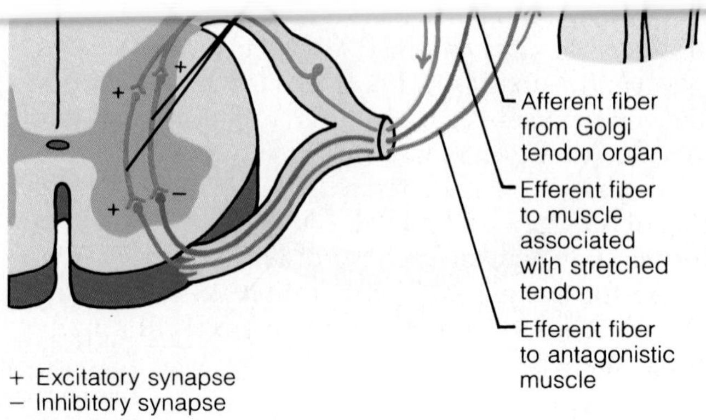

your finger or step barefoot on a piece of broken glass is a good example (Figure 13.12). So, too, is the trunk flexion that occurs when someone pretends to "throw a punch" to your abdomen. Like stretch reflexes, flexor reflexes are ipsilateral reflexes, but unlike stretch reflexes, they involve more than a single synapse—they are polysynaptic reflexes. Since flexor reflexes are protective reflexes important to our survival, it is not surprising that they are *prepotent*; that is, they override the spinal pathways and prevent any other reflexes from using them at the same time.

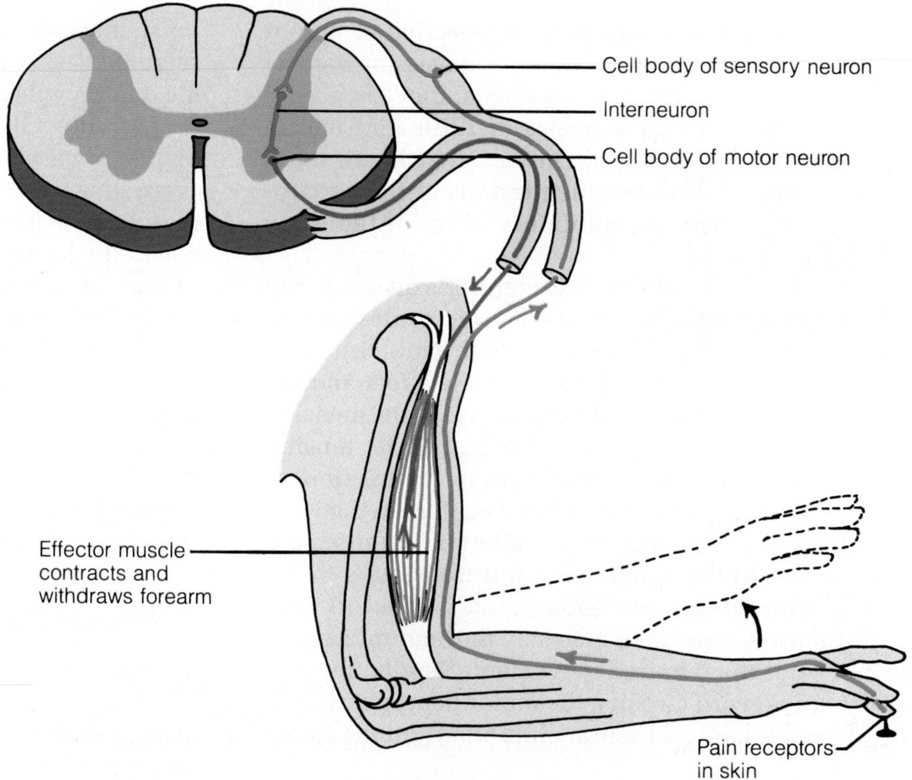

Figure 13.12 The flexor, or withdrawal, reflex. Contact with a noxious (painful) stimulus triggers the flexor reflex. The example shown here illustrates the pricking of a finger on a tack. Afferent impulses travel to the spinal cord and then (via a synapse with an interneuron and a motor neuron) efferent impulses are transmitted back to the flexor muscles, causing withdrawal of the hurt limb.

Crossed Extensor Reflex

The **crossed extensor reflex** is a complex spinal reflex consisting of an ipsilateral withdrawal reflex and a contralateral (opposite side) extensor reflex. Incoming afferent fibers synapse with interneurons that control the flexor withdrawal response on the same side of the body and with other interneurons that control the extensor muscles on the opposite side (Figure 13.13). This reflex is quite obvious when a stranger suddenly grabs your arm. The flexor reflex is the immediate withdrawal of your clutched arm; the extensor reflex causes you to simultaneously push the intruder away with your opposite arm. Another, and perhaps more important, example occurs when you step barefoot on broken glass. The ipsilateral response causes rapid lifting of the cut foot, while the contralateral response activates the extensor muscles of the opposite leg to support the weight suddenly shifted to it. Crossed extensor reflexes are particularly important in maintaining balance.

Superficial Reflexes

Superficial reflexes are elicited by gentle cutaneous stimulation, such as that produced by stroking the skin with a tongue depressor or a blunt key. The superficial reflexes are clinically important reflexes that depend both on functional upper motor pathways and on cord-level reflex arcs. The best known of these are the plantar and abdominal reflexes.

The **plantar reflex** is a complex response that specifically tests the integrity of the spinal cord from L_4 to S_2 and indirectly determines if the corticospinal tracts are functioning properly. It is elicited by drawing a blunt object downward along the lateral aspect of the plantar surface (sole) of the foot (Figure 13.14a). The normal response is downward flexion (curling) of the toes. However, if the primary motor cortex or corticospinal tract is damaged, the plantar reflex is replaced by an abnormal reflex called *Babinski's sign* (Figure 13.14b), in which the great toe dorsiflexes and the smaller toes fan laterally. Infants exhibit Babinski's

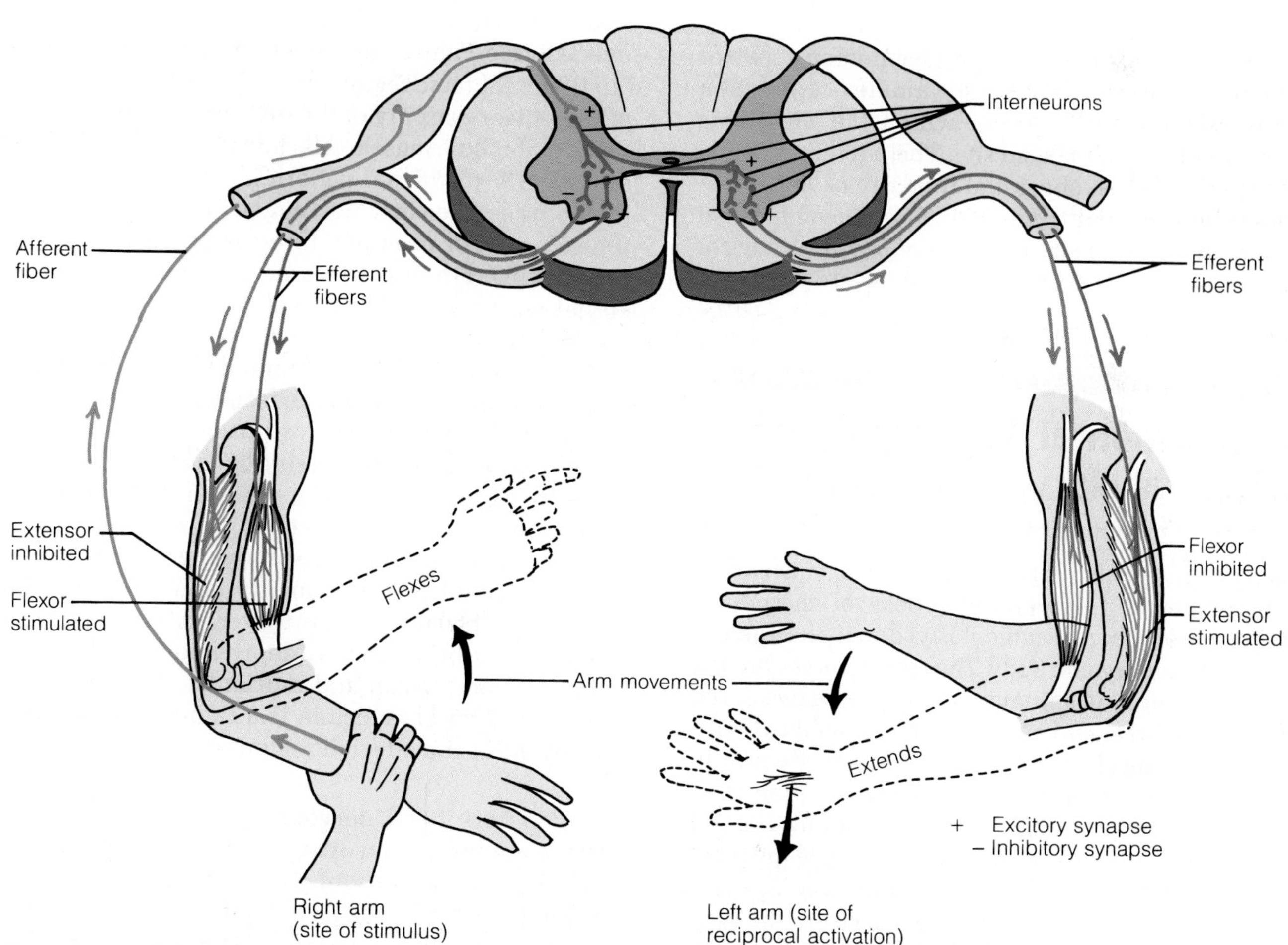

Figure 13.13 The crossed extensor reflex. The crossed extensor reflex causes reflex withdrawal of the body part on the stimulated side (flexor reflex) and extension of the muscles on the opposite side of the body (extensor reflex). Thus, reciprocal activation of muscles on the opposite side of the body is an important part of the reflex.

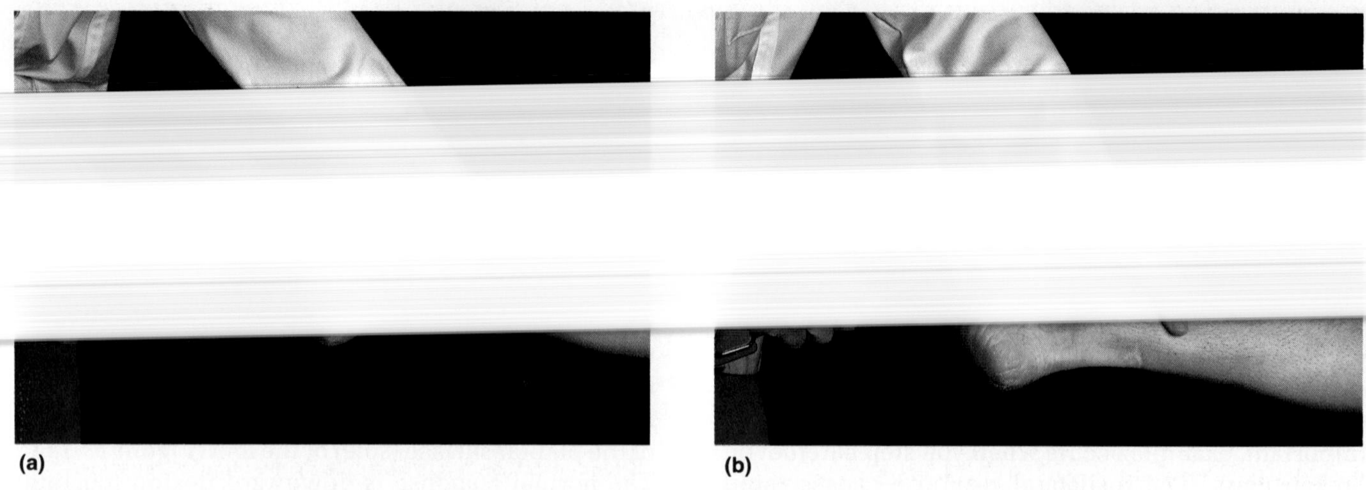

(a) (b)

Figure 13.14 The plantar reflex. (**a**) The normal reflex. (**b**) Babinski's sign.

sign because their nervous systems are incompletely myelinated. Myelination is usually completed by the first year, and Babinski's sign is replaced by the plantar reflex. Despite its clinical significance, the physiological mechanism of Babinski's sign is not understood.

Stroking the skin of the lateral abdomen above or below the umbilicus may induce a reflex contraction of the abdominal muscles in which the umbilicus moves toward the stimulated site. These reflexes, called **abdominal reflexes,** check the integrity of the spinal cord and dorsal rami from T_8 to T_{12}. The abdominal reflexes vary in intensity in different people. They are absent when corticospinal tract lesions are present.

Developmental Aspects of the Peripheral Nervous System

As described in Chapter 9, most skeletal muscles of the body derive from paired blocks of mesoderm (somites) distributed segmentally down the medial aspect of the embryonic body. The spinal nerves branch from the developing spinal cord and exit between the forming vertebrae, and each becomes associated with the adjacent muscle mass. The spinal nerves supply both sensory and motor fibers to the developing muscles and also help to direct their maturation. Cranial nerves issuing from the brain stem innervate the muscles of the head in a comparable manner.

The distribution of cutaneous nerves to the skin follows a similar pattern. Most of the scalp and skin of the face is innervated by the trigeminal nerves. Except for C_1, all spinal nerves supply cutaneous branches to specific (adjacent) dermatomes that eventually become dermal segments. Thus, the distribution and growth of the spinal nerves correlate with the segmented body plan, which is established by the fourth week of embryonic development.

Growth of the limbs and unequal growth of other body areas result in an adult pattern of dermatomes with unequal sizes and shapes and varying degrees of overlap (see Figure 13.4, p. 430). There is little overlap in the thoracic region, but substantial overlapping of the sensory supply of the lower body trunk occurs. Since embryonic muscle cells migrate extensively, much of the early segmental pattern is lost. Understanding the pattern of sensory nerve distribution is critical to physicians. For example, in areas of substantial dermatome overlap, it is necessary to block two or three spinal nerves to perform local surgery.

Reflexes occur a bit slower during old age. Sensory receptors atrophy to some degree, and there is some lessening of tone in the facial and neck muscles. However, since it has been estimated that 80% of total reaction time reflects processing by the CNS, this points more to a general loss in neurons and an increasing sluggishness of central processing circuits than to any specific age-related changes in the peripheral nerves. The peripheral nerves remain viable and normally functional throughout life unless subjected to traumatic injury or ischemia (deprivation of blood carrying needed oxygen and nutrients). The most common symptom of ischemia is a tingling sensation or numbness in the affected region.

* * *

The peripheral nervous system, composed primarily of nerves, is an essential part of any functional nervous system. Without it, the central nervous system would entirely lack its rich information bank about events of the external and internal environments. Now that we have connected the CNS to both of these environments, we are ready to consider the autonomic nervous system, a special subdivision of the peripheral nervous system and the topic of Chapter 14.

Related Clinical Terms

Neuralgia (ner-al′-juh) (*neuro* = a nerve) Sharp spasmlike pain along the course of one or more nerves; usually caused by inflammation or traumatic injury to the nerve(s).

Neuritis (nū-rī′-tis) Inflammation of a nerve; there are many forms with different effects (e.g., increased or decreased nerve sensitivity, paralysis of part served, and pain).

Neurologist (nū-ral′-ah-jist) A specialist in the scientific study of the nervous system, its functions, and disorders.

Nerve conduction studies Diagnostic tests that assess the integrity of nerves as reflected by their conduction velocities; the nerve is stimulated separately at two points, and the time required for the stimulus to reach the muscle it serves is recorded; used to assess suspected peripheral neuropathies.

Paresthesia (payr-is-thē′-zhuh) An abnormal (burning, tingling) sensation resulting from a disorder of a sensory nerve.

Chapter Summary

1. The peripheral nervous system consists of nerves conducting impulses to and from the CNS, their associated ganglia, and sensory receptors.

2. Most peripheral nerves include both afferent (sensory) fibers and efferent (motor) fibers. The efferent fibers may be somatic or autonomic.

CRANIAL NERVES (pp. 421–428)

1. Twelve pairs of cranial nerves originate from the brain and issue through the skull to innervate the head and neck. Only the vagus nerves extend into the thoracic and abdominal cavities.

2. Cranial nerves are numbered from anterior to posterior in order of emergence from the brain. Their names reflect structures served and/or function.

3. The cranial nerves include:

- The olfactory nerves (I): purely sensory; concerned with the sense of smell.
- The optic nerves (II): purely sensory; transmit visual impulses from the retina to the thalamus.
- The oculomotor nerves (III): mixed nerves; primarily motor; emerge from the midbrain and serve four extrinsic eye muscles, the levator palpebrae superioris of the eyelid, and the intrinsic ciliary muscle and constrictor fibers of the iris; also carry proprioceptive impulses from the skeletal muscles served.
- The trochlear nerves (IV): mixed nerves; emerge from the dorsal midbrain and carry motor and proprioceptor impulses to and from superior oblique muscles of the eyeballs.
- The trigeminal nerves (V): mixed nerves; emerge from the lateral pons; the major sensory nerves of the face; each has three sensory divisions—ophthalmic, maxillary, and mandibular; the mandibular branch also contains motor fibers that innervate the chewing muscles.
- The abducens nerves (VI): mixed nerves; emerge from the pons and serve the motor and proprioceptive functions of the lateral rectus muscles of the eyeballs.
- The facial nerves (VII)--mixed nerves; emerge from the pons; major motor nerves of the face; also carry sensory impulses from the taste buds of anterior two-thirds of the tongue.
- The vestibulocochlear nerves (VIII): purely sensory; transmit impulses from the hearing and equilibrium receptors of the inner ears.
- The glossopharyngeal nerves (IX): mixed nerves; issue from the medulla; transmit sensory impulses from the taste buds of the posterior tongue, from the pharynx, and from chemoreceptors of the carotid sinuses; innervate the pharyngeal muscles and parotid glands.
- The vagus nerves (X): mixed nerves; arise from the medulla; almost all motor fibers are autonomic parasympathetic fibers; motor efferents to, and sensory fibers from, the pharynx, larynx, and visceral organs of the thoracic and abdominal cavities.
- The accessory nerves (XI): mixed nerves; consist of a cranial root arising from the medulla and a spinal root arising from the cervical spinal cord; cranial root supplies motor fibers to the pharynx and larynx; spinal root supplies somatic efferents to the trapezius and sternocleidomastoid muscles of the neck and carries proprioceptor afferents from the same muscles.
- The hypoglossal nerves (XII): mixed nerves; issue from the medulla; carry somatic motor efferents to, and proprioceptive fibers from, the intrinsic and extrinsic tongue muscles.

SPINAL NERVES (pp. 428–438)

1. The 31 pairs of spinal nerves (all mixed nerves) are numbered successively according to the region of the spinal cord from which they issue.

Distribution of Spinal Nerves (pp. 428–431)

2. Spinal nerves are formed by the union of dorsal and ventral roots of the spinal cord and are short, confined to the intervertebral foramina.

3. Branches of each spinal nerve include dorsal and ventral rami, a meningeal branch, and rami communicantes (ANS branches).

4. Dorsal rami serve the muscles and skin of the posterior body trunk. Ventral rami, except T_2–T_{12}, form plexuses that serve the limbs. T_2–T_{12} ventral rami serve the thorax wall and abdominal surface.

5. All spinal nerves except C_1 innervate specific segments of the skin called dermatomes.

Plexuses and Peripheral Nerves (pp. 431–438)

6. The cervical plexus (C_1–C_4) innervates the muscles and skin of the neck and shoulder. Its phrenic nerve serves the diaphragm.

7. The brachial plexus serves the shoulder, some thorax muscles, and the upper limb. It arises primarily from C_5–T_1. From proximal to distal, the brachial plexus has roots, trunks, divisions, and cords. The principal nerves arising from the cords are the axillary, musculocutaneous, median, radial, and ulnar nerves.

8. The lumbar plexus (L_1–L_4) provides the motor supply to the anteromedial thigh muscles and the cutaneous supply to the anterior thigh and part of the leg. Its chief nerves are the femoral and obturator.

9. The sacral plexus (L_4–S_4) supplies the muscles and skin of the posterior aspect of the lower limb. Its principal nerve is the large sciatic nerve (composed of the tibial and common peroneal nerves).

REFLEX ACTIVITY (pp. 438–444)

1. A reflex is a rapid, involuntary motor response to a stimulus. The reflex arc has five elements: receptor, sensory neuron, integration center, motor neuron, and effector.

Spinal Reflexes (pp. 439–444)

2. Testing of somatic spinal reflexes provides information on the integrity of the reflex pathway and the degree of excitability of the spinal cord.

3. Somatic spinal reflexes include stretch, deep tendon, flexor, crossed extensor, and superficial reflexes.

4. A stretch reflex is initiated by stretching of muscle spindles and causes contraction of the stimulated muscle and inhibition of its antagonist. It is a monosynaptic, ipsilateral reflex. Stretch

[illegible]

8. Superficial reflexes are elicited by cutaneous stimulation. They require functional cord reflex arcs and corticospinal pathways. Examples include the plantar and abdominal reflexes.

DEVELOPMENTAL ASPECTS OF THE PERIPHERAL NERVOUS SYSTEM (p. 444)

1. Each spinal nerve provides the sensory and motor supply of an adjacent muscle mass (destined to become skeletal muscles) and the cutaneous supply of a dermatome (skin segment).

2. Reflexes decline in speed with age; this probably reflects neuronal loss.

Review Questions

Multiple Choice/Matching

1. Match the names of the cranial nerves in column B to the appropriate description in column A.

Column A	Column B
______ (1) causes pupillary constriction	(a) abducens
______ (2) is the major sensory nerve of the face	(b) accessory
______ (3) serves the sternocleidomastoid and trapezius muscles	(c) facial
______ (4) are purely sensory (three nerves)	(d) glossopharyngeal
______ (5) serves the tongue muscles	(e) hypoglossal
______ (6) allows you to chew your food	(f) oculomotor
______ (7) is impaired in tic douloureux	(g) olfactory
______ (8) helps to regulate heart activity	(h) optic
______ (9) serves your ability to hear and to maintain your balance	(i) trigeminal
______ (10) contain parasympathetic motor fibers (four nerves)	(j) trochlear
	(k) vagus
	(l) vestibulocochlear

2. For each of the following muscles or body regions, identify the plexus and the peripheral nerve (or branch of one) involved. Use choices from keys A and B.

Key A: Plexuses
(a) brachial
(b) cervical
(c) lumbar
(d) sacral

Key B: Nerves
(1) common peroneal

______, ______ (1) the diaphragm
______, ______ (2) muscles of the posterior thigh and leg
______, ______ (3) anterior thigh muscles
______, ______ (4) medial thigh muscles

[illegible]

wrist and digits
______, ______ (8) skin and extensor muscle of the posterior arm and forearm
______, ______ (9) peroneal muscles, tibialis anterior, and toe extensors

3. An ipsilateral reflex that produces a rapid withdrawal of the body part from a painful stimulus is (a) a crossed extensor reflex, (b) a flexor reflex, (c) a Golgi tendon reflex, (d) a muscle stretch reflex.

Short Answer Essay Questions

4. What is the functional relationship of the peripheral nervous system to the central nervous system?

5. List the structural components of the peripheral nervous system, and describe the function of each component.

6. (a) Describe the formation and composition of a spinal nerve. (b) Name the branches of a spinal nerve (other than the rami communicantes), and explain what body region each supplies.

7. (a) Define plexus. (b) Indicate the spinal roots of origin of the four major nerve plexuses, and name the general body regions served by each.

8. Differentiate between ipsilateral and contralateral reflexes.

9. What is the homeostatic value of flexor reflexes?

10. Compare and contrast the flexor and crossed extensor reflexes.

11. What clinical information can be gained by conducting somatic reflex tests?

12. What is the structural and functional relationship between spinal nerves, skeletal muscles, and dermatomes?

Clinical Application Questions

13. Harry fell off a ladder and was found to have a fracture of the right anterior cranial fossa of his skull and a watery, blood-tinged discharge from the right nostril. Several days later, Harry complained that he could no longer smell. What was damaged in his fall?

14. Mr. Frank, a former stroke victim who had made a remarkable recovery, suddenly began to have problems reading. He complained that he was seeing double and also had problems navigating steps. He was unable to move his left eye downward and laterally. What cranial nerve was found to be the site of lesion? (Right or left?)

15. Joseph, a man in his early 70s, was having problems chewing his food. He was asked to stick out his tongue. It deviated to the right, and its right side was quite wasted. What cranial nerve was impaired?

CHAPTER 14

The Autonomic Nervous System

Chapter Outline and Student Objectives

Overview of the Autonomic Nervous system (pp. 448–450)

1. Compare the somatic and autonomic nervous systems relative to effectors, efferent pathways, and neurotransmitters released.
2. Compare and contrast the general functions of the sympathetic and parasympathetic divisions.

Anatomy of the Autonomic Nervous System (pp. 450–455)

3. Describe the site of CNS origin, locations of ganglia, and general fiber pathways of both the sympathetic and parasympathetic divisions.

Physiology of the Autonomic Nervous System (pp. 455–460)

4. Define cholinergic and adrenergic fibers, and list the different types of cholinergic and adrenergic receptors.
5. Briefly describe the clinical importance of drugs that mimic or inhibit adrenergic or cholinergic effects.
6. Note the effects of the sympathetic and parasympathetic divisions on the following organs: heart, blood vessels, gastrointestinal tract, lungs, adrenal medulla, and external genitalia.
7. Describe the levels of control of autonomic nervous system functioning.

Homeostatic Imbalances of the Autonomic Nervous System (pp. 460–462)

8. Explain the relationship between some types of hypertension, Raynaud's disease, and the mass reflex reaction to disorders of autonomic functioning.

Developmental Aspects of the Autonomic Nervous System (p. 462)

9. Describe the effects of aging on the autonomic nervous system.

Preview of Selected Key Terms

Autonomic nervous system (*auto* = self; *nom* = govern) The efferent division of the peripheral nervous system that innervates cardiac and smooth muscles and glands; also called the involuntary or visceral motor system.

Ganglion (gan'-glē-on) (*ganglion* = knot on a string, swelling) Collection of nerve cell bodies outside the CNS.

Preganglionic neuron Autonomic motor neuron that has its cell body in the CNS and projects its axon to a peripheral ganglion.

Postganglionic neuron Autonomic motor neuron that has its cell body in a peripheral ganglion and projects its axon to an effector.

Sympathetic division (*sym* = together; *pathos* = feeling) Division of the autonomic nervous system that activates the body to cope with some stressor (danger, excitement, etc.); the fight, fright, and flight subdivision.

Parasympathetic (*para* = beside, alongside) The division of the autonomic nervous system that oversees digestion, elimination, and glandular function; the resting and digesting subdivision.

Sympathetic tone State of partial vasoconstriction of the blood vessels maintained by sympathetic fibers.

Some means of internal regulation is a fundamental part of any government. In the United States, a system of checks and balances exists between the executive branch and Congress. Balance is continually sought. The body's attempt to maintain balance is also unending, but the body is far more sensitive to changing events than any human invention or cultural institution. Although all body systems contribute, the relative stability of our internal environment depends largely on the orchestrations of the **autonomic nervous system (ANS).** At every moment, signals flood from the visceral organs into the central nervous system, and autonomic nerves make adjustments as necessary to ensure optimal support for body activities [illegible] As the term *autonomic* implies, this motor subdivision of the peripheral nervous system has a certain amount of functional independence. The autonomic nervous system is also called the *involuntary nervous system,* which reflects its subconscious control, or the *general visceral motor system,* which indicates the location of most of its effectors.

Overview of the Autonomic Nervous System

Comparison of the Somatic and Autonomic Nervous Systems

Our previous discussions of motor nerves have focused largely on the activity of the somatic nervous system. So, before plunging into a description of autonomic nervous system anatomy, we will take the time to point out important differences between the somatic and autonomic divisions, as well as some areas of functional overlap. Although both systems have motor fibers, the somatic and autonomic nervous systems differ in their effectors, in their efferent pathways, and to some degree in target organ responses to their neurotransmitters. These differences are summarized in Figure 14.1. The figure also points out some of the differences between the sympathetic and parasympathetic divisions of the ANS, which we will consider shortly.

Effectors

The somatic nervous system stimulates skeletal muscles, whereas the ANS innervates cardiac muscle, smooth muscle, and glands. Differences in the physiological characteristics of the effector organs themselves account for most of the remaining differences between somatic and autonomic effects on their target organs.

Efferent Pathways and Ganglia

In the somatic division, cell bodies of the mot[illegible] [illegible] ron, the **preganglionic neuron,** resides in the brain or spinal cord. Its axon (the *preganglionic axon*) synapses with the second motor neuron, the **postganglionic neuron,** in a ganglion outside the central nervous system. The *postganglionic axon* then extends to the effector organ. (If you take the time to think about the meanings of all these terms while referring to Figure 14.1, understanding the rest of the chapter will be much easier. Note that the postganglionic neuron is somewhat misnamed, since its nerve cell body is in the ganglion, not beyond ["post"] it.) Preganglionic axons are usually lightly myelinated, thin fibers; the postganglionic fibers are even thinner and are unmyelinated. Consequently, conduction through the autonomic efferent chain is slower than conduction in the somatic system. Many pre- and postganglionic fibers are incorporated into spinal or cranial nerves for most of their course.

Notice that the somatic motor division lacks ganglia. The dorsal root ganglia are part of the sensory, not the motor, division of the peripheral nervous system. The composition of autonomic ganglia is much more complex than that of the dorsal root ganglia. Autonomic ganglia consist of (1) an outer connective tissue capsule that encloses cell bodies of postganglionic motor neurons and sometimes interneurons, (2) collections of synapses between the pre- and postganglionic fibers, and (3) afferent and efferent fibers that pass through the ganglion without synapsing. Technically, autonomic ganglia are synapse sites of the pre- and postsynaptic neurons; however, the presence of interneurons within some of these ganglia suggests that more than just information transmission occurs there. The interneurons appear to exert local feedback effects, so the autonomic ganglia may have certain integrative functions as well.

Neurotransmitter Effects

All somatic motor neurons release acetylcholine at their junctions with skeletal muscle fibers. The effect is always excitatory, and if stimulation reaches threshold, the skeletal muscle fiber contracts.

Neurotransmitters released onto visceral effector organs by postganglionic autonomic fibers include norepinephrine (by most sympathetic fibers) and acetylcholine (by parasympathetic fibers). Depending on

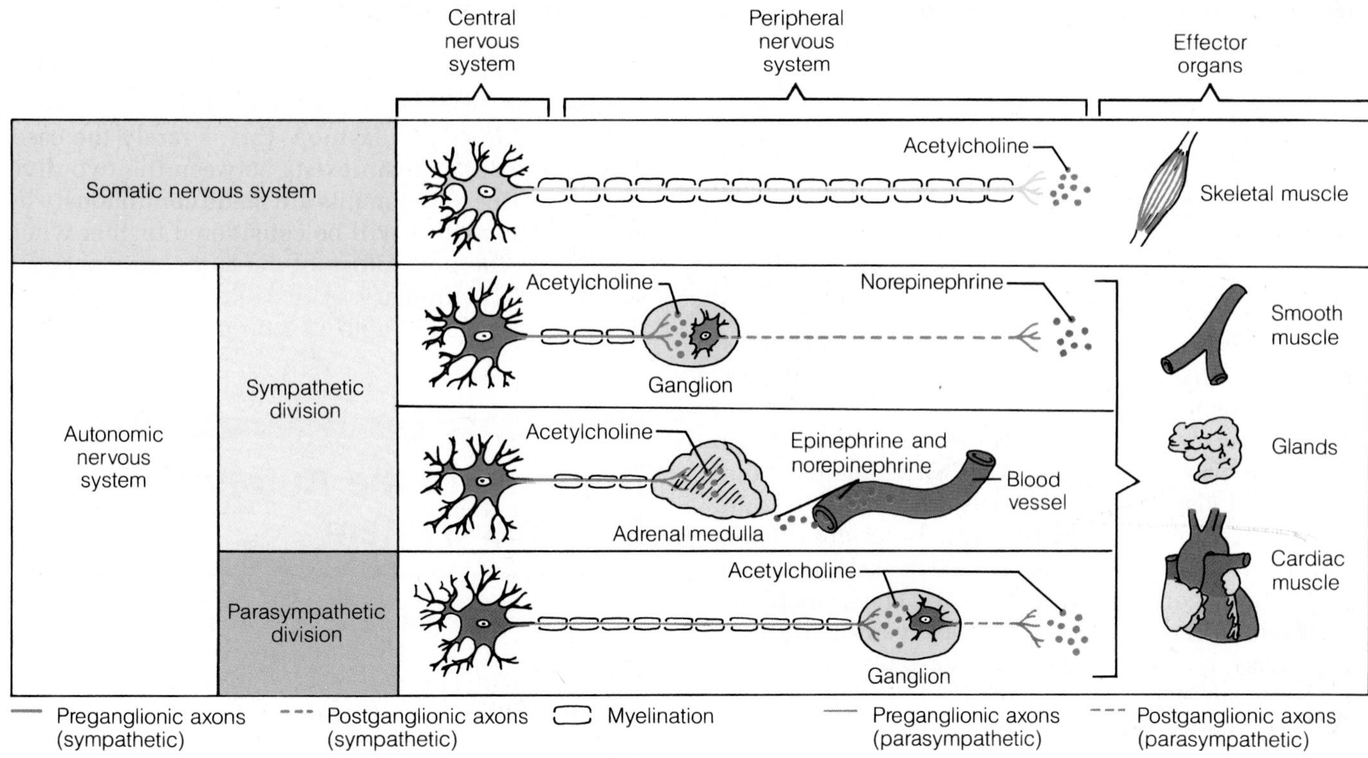

Figure 14.1 Comparison of the somatic and autonomic nervous systems. *Somatic Division:* Axons of somatic motor neurons extend from the CNS to the effectors (skeletal muscle cells). Somatic motor neurons release acetylcholine. The effect is always stimulatory. *Autonomic Division:* Axons of most preganglionic neurons run from the CNS to synapse in a peripheral autonomic ganglion with a postganglionic neuron. A few sympathetic preganglionic axons synapse with cells of the adrenal medulla. Axons of postganglionic neurons run from the ganglion to the effectors (cardiac and smooth muscle fibers and glands). All autonomic preganglionic fibers release acetylcholine; all parasympathetic postganglionic fibers release acetylcholine; most sympathetic postganglionic fibers release norepinephrine. Stimulated adrenal medullary cells release norepinephrine and epinephrine into the blood. Autonomic effects are stimulatory or inhibitory, depending on the postganglionic neurotransmitter released and the receptor types on the effectors.

the receptors present on the target organ, the effect of these neurotransmitters may be either excitation or inhibition.

Overlap of Somatic and Autonomic Function

Higher brain centers regulate both somatic and visceral motor activities, and nearly all spinal nerves (and many of the cranial nerves) contain both somatic and autonomic fibers. Moreover, most body responses to changing internal and external conditions involve both skeletal muscle activity and enhanced responses of certain visceral organs. Coordination of these activities is essential if homeostasis is to be maintained. For example, when muscles are working hard, they need more oxygen and glucose, and the autonomic control mechanisms speed up the heart and breathing to meet these needs. Although the autonomic nervous system is only one part of our integrated nervous system, in accordance with convention we will consider it an individual entity and describe its role in isolation in the sections that follow.

The Divisions of the Autonomic Nervous System

The two arms of the autonomic nervous system, the sympathetic and parasympathetic divisions, generally serve the same visceral organs but cause essentially opposite effects, counterbalancing each other's activities to keep body systems running smoothly. The sympathetic part mobilizes the body during extreme situations (such as fear, exercise, or rage), whereas the parasympathetic division allows us to unwind and works to conserve body energy. To emphasize the relative roles of the two autonomic divisions in maintaining body homeostasis, we will focus briefly on situations in which each division is exerting primary control.

Role of the Sympathetic Division

The **sympathetic division** is often referred to as the fight-or-flight system. Its activity is evident when we are excited or find ourselves in emergency or threatening situations, such as being frightened by street

electrical resistance of the skin (galvanic skin resistance)—events that are most frequently recorded during lie detector examinations.

During exercise, the sympathetic division also promotes a number of other adjustments. Cutaneous and visceral blood vessels are constricted, and vessels of the heart and skeletal muscles are dilated. This results in the shunting of blood to skeletal muscle and to the vigorously working heart. The bronchioles in the lungs are dilated to increase ventilation (and ultimately oxygen delivery to the body cells), and nutrients are mobilized from the liver to increase blood glucose levels. At the same time, temporarily nonessential activities, such as gastrointestinal and urinary tract motility, are damped (although paradoxical fear-induced defecation and urination are known to happen). If you are running from a mugger, it is far more important that your muscles be provided with everything they need to get you out of danger's way than that you finish digesting lunch. ■

The sympathetic division generates a head of steam that enables the body to cope rapidly and vigorously with situations that threaten homeostasis. Its function is to provide the optimal conditions for an appropriate response to some threat, whether that response is to run, to see more clearly, or to think more critically.

Role of the Parasympathetic Division

The **parasympathetic division** is most active when the body is at rest. This division, sometimes called the resting and digesting system, is chiefly concerned with activating the digestive viscera and with conserving body energy. (This explains why it is a good idea to relax after a heavy meal so that digestion is not inhibited or interfered with by sympathetic activity.) Its activity is best illustrated in a person who relaxes after a meal and reads the newspaper. Blood pressure and heart and respiratory rates are being regulated at low normal levels, the gastrointestinal tract is active in digestive activities, and the skin is warm (indicating that there is no need to divert blood to skeletal muscles or vital organs). The eye pupils are constricted to protect the retinas from excessive damaging light, and the lenses of the eyes are accommodated for close vision. We might consider the parasympathetic division the housekeeping system of the body.

While it is easiest to think of the operation of the sympathetic and parasympathetic divisions as working in an all-or-none fashion, this is rarely the case.

division appears in Table 14.3 on p. 459.

Anatomy of the Autonomic Nervous System

The sympathetic and parasympathetic divisions have unique origin sites. They are also distinguished by the different lengths of their pre- and postganglionic fibers and by the locations of their ganglia. These and other anatomical considerations are summarized in Table 14.1 on p. 455 and illustrated in Figure 14.2.

Sympathetic (Thoracolumbar) Division

All preganglionic fibers of the sympathetic division arise from cell bodies of preganglionic neurons located in the lateral horns (the so-called *visceral motor zones*) of spinal cord segments T_1 through L_2 (see Figure 14.2, left). For this reason, the sympathetic division is also referred to as the **thoracolumbar** (thō″-rah-kō-lum′-bar) **division.** After leaving the cord via the ventral root, these preganglionic fibers separate from the somatic motor fibers by passing through a **white ramus communicans** (plural, rami communicantes [kuh-myoo-nih-kan′-tēz]) to enter an adjoining **paravertebral ganglion** in the **sympathetic trunk** or **chain** (Figure 14.3). Looking much like strands of glistening white beads, the sympathetic trunks flank each side of the vertebral column (see Figure 14.4, p. 453). Thus, the naming of these ganglia really describes their location: *para* = beside, or close by; *vertebral* = the vertebral column.

Although the sympathetic trunks extend from the neck to the pelvis, their sympathetic fibers arise only from the thoracic and lumbar spinal cord segments, as shown in Figure 14.2. The ganglia may vary in size, position, and number, but typically, there are 22 ganglia in each sympathetic chain: 3 cervical, 11 thoracic, 4 lumbar, and 4 sacral.

Once a preganglionic axon reaches a paravertebral ganglion, one of three things can happen to it:

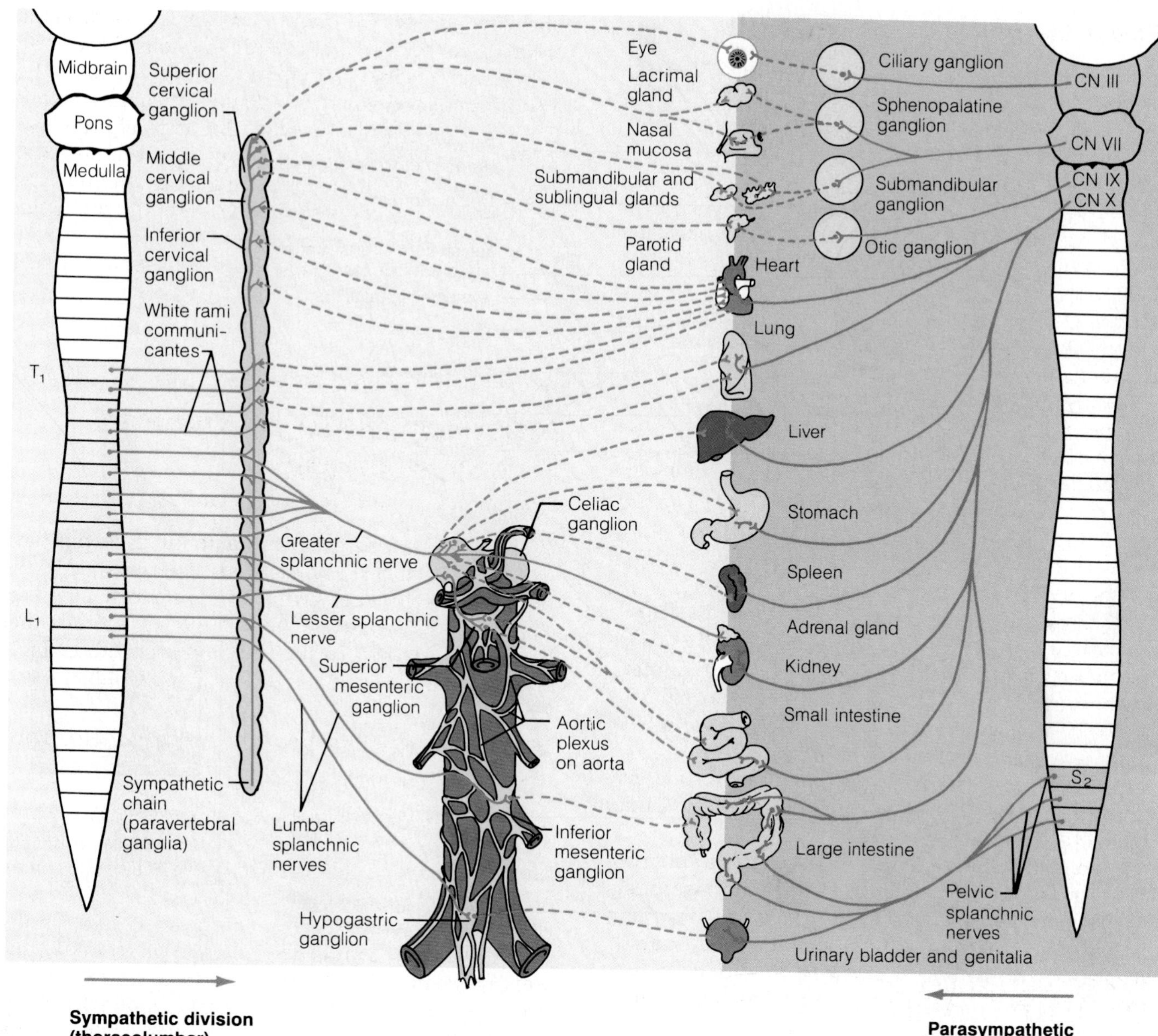

Figure 14.2 The autonomic nervous system. Solid lines indicate preganglionic nerve fibers; broken dotted lines indicate postganglionic nerve fibers. Terminal ganglia of the vagus nerve fibers are not shown; most of these ganglia are located in or on the target organ. (Although there are four sympathetic lumbar splanchnic nerves, only two are illustrated for simplicity. Note: CN = cranial nerve.)

1. It can synapse with a postsynaptic neuron within that ganglion (see Figure 14.3a).
2. It can ascend or descend within the chain to synapse in another ganglion (see Figure 14.3b). (It is these fibers running from one ganglion to another that connect the ganglia into the sympathetic trunk.)
3. It can pass through the ganglion without synapsing (see Figure 14.3c).

Preganglionic fibers exhibiting the last pattern help to form the *splanchnic* (splank′-nik) *nerves,* which synapse with **prevertebral,** or **collateral, ganglia** located anterior to the vertebral column. Unlike paravertebral (chain) ganglia, the prevertebral ganglia are not segmentally arranged. Regardless of where the synapse occurs, all sympathetic ganglia are close to the spinal cord, and the sympathetic postganglionic fibers, which run from the ganglion to the organs they supply, are typically much longer than the preganglionic fibers.

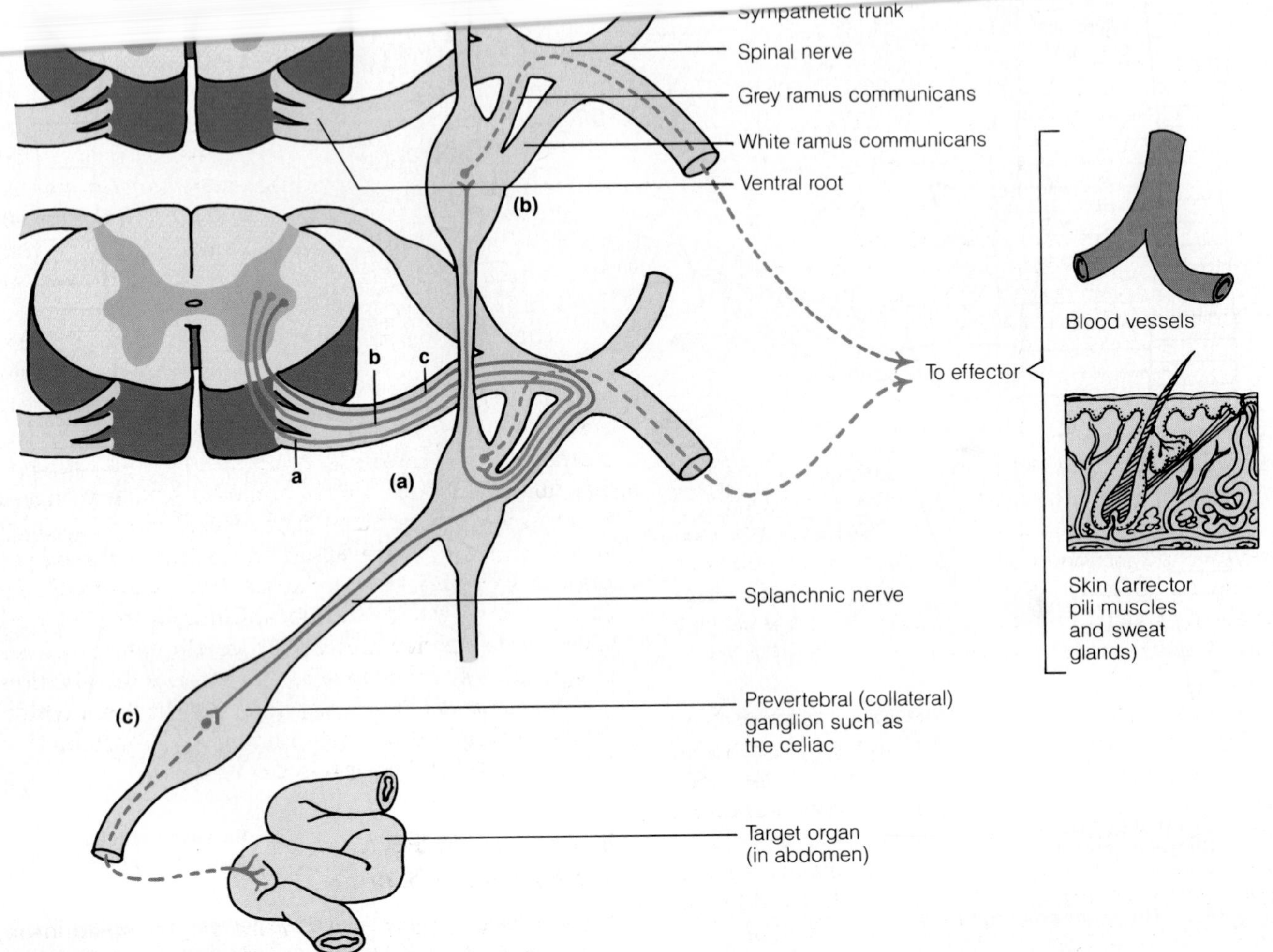

Figure 14.3 Sympathetic pathways. (**a**) Synapse in a paravertebral (sympathetic chain) ganglion at the same level. (**b**) Synapse in a paravertebral ganglion at a different level. (**c**) Synapse in a prevertebral (collateral) ganglion anterior to the vertebral column.

Pathways with Synapses in a Paravertebral (Sympathetic Chain) Ganglion

When synapses are made in the paravertebral ganglia, the axons of the postganglionic neurons enter the spinal nerves by way of communicating branches called **gray rami communicantes** (see Figure 14.3a). After entering the spinal nerve proper, they are distributed in the branches of the spinal nerves to the smooth muscle in the walls of blood vessels and to the sweat glands and arrector pili muscles of the skin.

Notice that the naming of the communicating branches as white or gray reflects whether or not their fibers are myelinated. The preganglionic fibers composing the white rami communicantes are myelinated, whereas postganglionic axons forming the gray rami communicantes are unmyelinated and appear gray.

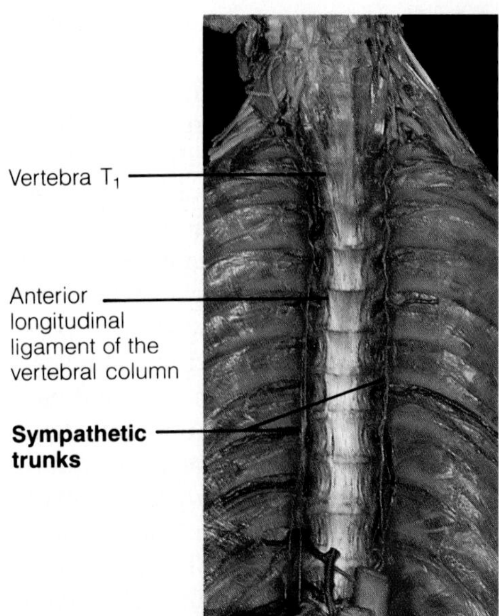

Figure 14.4 The sympathetic trunks. The organs of the anterior thorax have been removed to allow visualization of the paravertebral ganglia of the sympathetic trunks.

The white rami, through which the preganglionic axons enter the sympathetic chain, are found only in the T_1–L_2 cord segments. However, gray rami connect each paravertebral ganglion to at least one, and often to two, nearby spinal nerves. Thus, postganglionic sympathetic fibers are found in essentially all spinal nerves, which allows sympathetic output to reach all parts of the body.

Postganglionic fibers issuing from the cervical ganglia (see Figure 14.2) also furnish the sympathetic outflow to several additional structures of the head and neck region. The large **superior cervical ganglion** contributes sympathetic fibers to a number of cranial nerves and to the upper three or four cervical spinal nerves. Besides being distributed to the skin and blood vessels, these fibers are distributed to the dilator muscles of the irises of the eyes, to the salivary glands, and to organs of the respiratory tract. The superior cervical ganglion also gives off direct branches to the carotid body and sinus, to the *pharyngeal* (fah-rin′-jē-al) *plexus* (and thus to the larynx and pharynx), and provides several cardiac nerves to the heart.

Postganglionic fibers emerging from the **middle** and **inferior cervical ganglia** enter cervical nerves C_4–C_8. Most of these fibers innervate the heart. (The inferior cervical ganglion is also called the *stellate* (steh′-lāt) *ganglion*.) Additionally, some postganglionic fibers emerging from the first five or so thoracic chain ganglia pass directly to the heart, aorta, lungs, and esophagus.

Pathways with Synapses in a Prevertebral (Collateral) Ganglion

It is the preganglionic fibers of T_5–L_2 that synapse in the prevertebral ganglia. These fibers enter and leave the sympathetic chains without synapsing, and they form several nerves, called **splanchnic nerves.** The sphanchnic nerves contribute to a number of interweaving nerve plexuses known collectively as the **abdominal aortic** (ā-or′-tik) **plexus,** which clings to the surface of the abdominal aorta (see Figure 14.2). This complex plexus contains several ganglia, some large and some small, that together serve the entire abdominopelvic viscera (*splanchni* = viscera). From the superior to the inferior aspect, the most important of these ganglia are the *celiac, superior mesenteric, inferior mesenteric,* and *hypogastric ganglia,* named for the arteries with which they are most closely associated. Postganglionic fibers issuing from these ganglia generally travel to their target organs in the company of arteries serving these organs.

The large, paired celiac ganglia are the synapse point of the fibers of the **greater splanchnic nerves,** and postganglionic fibers issuing from these ganglia are distributed to most of the abdominal viscera: stomach, small intestine, liver, spleen, and kidneys. The fiber distribution of the other splanchnic nerves is less easily traced and highly variable. Generally speaking, fibers of the **lesser splanchnic nerves** synapse in the celiac, superior mesenteric, and small aorticorenal ganglia. The **lumbar splanchnic nerves** send fibers to the inferior mesenteric ganglion, from which postganglionic fibers extend to serve the distal colon and rectum, and to the hypogastric ganglia, from which fibers serving the urinary bladder and reproductive organs located in the pelvic cavity issue.

Pathways with Synapses in the Adrenal Medulla

A few fibers traveling within the greater splanchnic nerves pass through the celiac ganglion without synapsing and terminate by synapsing with the hormone-producing medullary cells of the adrenal gland (see Figures 14.1 and 14.2). When stimulated by the preganglionic fibers, the medullary cells secrete norepinephrine and epinephrine (also called noradrenaline and adrenaline, respectively) into the blood. Embryologically, sympathetic ganglia and the adrenal medulla arise from the same tissue. For this reason, the adrenal medulla is sometimes viewed as a "misplaced" sympathetic ganglion, and its hormone-releasing cells, although lacking nerve processes, are considered equivalent to postganglionic sympathetic neurons.

Parasympathetic (Craniosacral) Division

synapse with postganglionic neurons located in **terminal ganglia** within or very close to the target organ. Very short postganglionic axons issue from the terminal ganglia and complete the transmission pathway by synapsing with effector cells in their immediate area. Notice that this pattern is opposite to that seen in the sympathetic division, where the preganglionic fibers are short and the postganglionic fibers are long.

Cranial Outflow

The cell bodies of cranial preganglionic neurons are located in the brain stem visceromotor nuclei associated with cranial nerves III, VII, IX, and X. Axons of these neurons travel in their respective cranial nerves to the smooth muscle and glands of the head and to the viscera of the thorax and abdomen. Four pairs of terminal ganglia are associated with parasympathetic fibers of cranial nerves III, VII, and IX (see Figure 14.2):

1. **Ciliary ganglia** within the orbits (see Table 13.1); synapse sites of *oculomotor (III) nerve* fibers that innervate the smooth muscles within the eyes, causing the pupils to become constricted and the lenses of the eyes to bulge—actions that allow focusing on close objects in the field of vision (accommodation).
2. **Sphenopalatine** (sfē″-nō-pa′-luh-tēn) **ganglia** posterior to the maxillae; synapse sites of the *facial (VII) nerve* fibers that serve the nasal mucosa and the lacrimal glands.
3. **Submandibular ganglia** deep to the mandibular angles; synapse sites of facial nerve fibers that innervate the submandibular and sublingual salivary glands.
4. **Otic ganglia** just inferior to the foramen ovale of the skull; synapse sites of the *glossopharyngeal (IX) nerve* fibers that serve the parotid salivary glands.

Although the parasympathetic fibers of cranial nerves III, VII, and IX initially course along within their respective cranial nerves, their postganglionic fibers "jump over" to travel with fibers of the trigeminal nerves (V) to reach the face. This "hitchhiking" no doubt reflects the fact that the trigeminal nerves have the broadest distribution to the face, but it is

abdominal cavities. The vagal nerve fibers terminate by synapsing in terminal ganglia that are usually located within the target organ. Many of these terminal ganglia are not individually named but as a group are called *intramurine* (in″-truh-myer′-ēn) ("within the walls") *ganglia*. As the vagus nerves pass into the thorax, they contribute to the following plexuses located within or close to the organs served:

1. *Cardiac plexuses*, supplying postganglionic fibers to the heart.
2. *Pulmonary plexuses*, serving the lungs and bronchi.
3. *Aortic plexuses*, supplying the thoracic aorta (and continuing in the abdominal cavity to supply the aorta there).
4. *Esophageal* (eh-sah″-fuh-jē′-al) *plexuses*, supplying the esophagus.

When the main trunks of the vagus nerves reach the esophagus, they intermingle and form the **anterior** and **posterior vagal trunks,** each containing fibers from both vagus nerves. These vagal trunks then enter the abdominal cavity and give off fibers to a number of plexuses (celiac and superior and inferior mesenteric) that serve the abdominal viscera. (Notice that the sympathetic and parasympathetic fibers are intermingled in these abdominal plexuses, but the parasympathetic fibers do not synapse there.) Abdominal organs receiving vagal innervation include the liver, stomach, small intestine, proximal two-thirds of the large intestine, kidneys, and pancreas. The distal portion of the large intestine and the pelvic organs are served by the sacral outflow, as described next.

Sacral Outflow

Parasympathetic preganglionic fibers from the spinal cord arise from neurons located in the lateral gray matter of sacral segments S_2–S_4. Since the parasympathetic motor neurons in the sacral cord are less numerous than the sympathetic motor neurons in the thoracolumbar regions, no obvious lateral horn appears in the sacral region of the cord.

The sacral preganglionic axons issue from the cord via the ventral roots of the spinal nerves and then branch off in the **pelvic splanchnic nerves** (see Figure

14.2). The pelvic splanchnic nerves synapse with postganglionic neurons located in ganglia within the walls of the following organs: distal one-third of the large intestine, the urinary bladder, ureters, and reproductive organs. Because parasympathetic fibers do not become incorporated into the spinal nerves, no rami communicantes are associated with the parasympathetic system.

Visceral Sensory Neurons

Because the autonomic nervous system is a visceral motor system, the presence of visceral sensory fibers (mostly pain afferents) in autonomic nerves is often overlooked. Nearly all the sympathetic and parasympathetic fibers described thus far are accompanied by afferent fibers conducting sensory impulses from glands or muscular structures served by autonomic motor fibers. Thus, peripheral processes of visceral sensory neurons are found in cranial nerves VII, IX, and X, the splanchnic nerves, and the sympathetic trunk, as well as in spinal nerves. Like sensory neurons serving somatic structures (skeletal muscles and skin), the cell bodies of visceral sensory neurons traveling along with autonomic fibers arising from the cord are located within dorsal root ganglia, and their central processes reach the spinal cord via the dorsal roots. Cell bodies of sensory afferents accompanying parasympathetic fibers in cranial nerves are located in the sensory ganglia associated with those nerves.

Physiology of the Autonomic Nervous System

Neurotransmitters and Receptors

Acetylcholine (a″-seh-til-kō′-lēn) (*ACh*) and *norepinephrine* (nor″-ep-ih-nef′-rin) (*NE*) are the major neurotransmitters released by autonomic nervous system neurons. Acetylcholine, the same neurotransmitter secreted by somatic motor neurons, is released by (1) all preganglionic axons of both ANS divisions and (2) all parasympathetic postganglionic axons at synapses with their effectors (see Figure 14.1 and Table 14.1). ACh-releasing fibers are called **cholinergic** (kō-lih-ner′-jik) **fibers.**

Table 14.1 Anatomical and Physiological Differences Between the Sympathetic and Parasympathetic Divisions

Characteristic	Sympathetic	Parasympathetic
Origin	Thoracolumbar outflow: lateral horn of gray matter of spinal cord segments T_1–L_2	Craniosacral outflow: brain stem nuclei of cranial nerves III, VII, IX, and X; spinal cord segments S_2–S_4
Location of ganglia	Ganglia within a few cm of CNS: alongside vertebral column (paravertebral ganglia) and anterior to vertebral column (prevertebral ganglia)	Terminal ganglia in or close to visceral organ served
Relative length of pre- and postganglionic fibers	Short preganglionic; long postganglionic	Long preganglionic; short postganglionic
Rami communicantes	Gray and white rami communicantes; white contain myelinated preganglionic fibers; gray contain unmyelinated postganglionic fibers	None
Degree of branching of preganglionic fibers	Extensive	Minimal
Functional goal	Prepares body to cope with emergencies and intense muscular activity	Conserves and stores energy
Neurotransmitters	All preganglionic fibers release ACh; most postganglionic fibers release norepinephrine (adrenergic fibers); some postganglionic fibers (e.g., those serving sweat glands and blood vessels of skeletal muscles) release ACh; neurotransmitter activity augmented by release of adrenal medullary hormones (norepinephrine and epinephrine)	All fibers release ACh (cholinergic fibers)

Table 14.2 Cholinergic and Adrenergic Receptors

[illegible]	[illegible]	[illegible]	[illegible]
[illegible]	[illegible]	[illegible] target organs	Excitation in most cases; inhibition of cardiac muscle
		Selected sympathetic targets:	
		• Sweat glands	Activation
		• Blood vessels in skeletal muscles	Inhibition (causes vasodilation)
Norepinephrine (and epinephrine released by adrenal medulla)	**Adrenergic** β_1	Heart, kidneys, adipose tissue	Increases heart rate and strength; stimulates secretion of renin; stimulates lipolysis
	β_2	Lungs and most other sympathetic target organs; abundant on blood vessels serving skeletal muscles	Inhibition: dilation of blood vessels and bronchioles; relaxes smooth muscle walls of digestive and urinary visceral organs
	α_1	Most importantly blood vessels, but virtually all sympathetic target organs except heart	Activation: constricts blood vessels and visceral organ sphincters
	α_2	Membrane of adrenergic axon terminals; blood platelets	Mediates inhibition of NE release from adrenergic terminals; promotes blood clotting

In contrast, most sympathetic postganglionic axons release NE and are classified as **adrenergic** (ad″-ren-er′-jik) **fibers.** The only exceptions are the sympathetic postganglionic fibers innervating the sweat glands of the skin, the blood vessels within skeletal muscles, and the external genitalia, all of which secrete ACh. A third neurotransmitter, *dopamine* (dō-puh-mēn), is released by some interneurons found in sympathetic ganglia, but its effects are exerted locally and will not be considered here.

The naming of NE-releasing fibers as adrenergic reflects a double nomenclature for neurohormones (substances that function both as neurotransmitters and as hormones). The older terms, *adrenaline* and *noradrenaline,* are based on the fact that large amounts of adrenaline and smaller amounts of noradrenaline are produced by the adrenal medulla. The newer terms, *epinephrine* and *norepinephrine,* have never completely replaced the old, and NE-releasing fibers continue to be called adrenergic fibers (perhaps because the term is easier to pronounce than "epinephrinergic fibers").

Unfortunately, there is no sweeping generalization that can explain the effects of ACh and NE on their effectors as straightforward excitation or inhibition. The response of visceral effectors to these neurotransmitters depends not on the neurotransmitters themselves but on the receptors to which they attach. There are two or more kinds of receptors for each of the autonomic neurotransmitters, and these allow the same chemicals to exert different effects (activation or inhibition) at different body targets (Table 14.2).

Cholinergic Receptors

The two types of receptors that bind ACh have different characteristics and are named for drugs that bind to them and mimic acetylcholine's effects. The first of these receptors to be identified were the **nicotinic** (nih″-kuh-tin′-nik) **receptors.** When nicotine binds to these receptors, it produces the same effects as ACh. A mushroom poison, *muscarine* (mus′-kuh-rin), activates a different set of ACh receptors, which were named **muscarinic receptors.** Muscarine has no effect on nicotinic receptors, and vice versa. All acetylcholine receptors are either nicotinic or muscarinic.

Nicotinic receptors are found on (1) the neuromuscular junctions of skeletal muscle cells, (2) all postganglionic neurons, both sympathetic and parasympathetic, and (3) the hormone-producing cells of the adrenal medulla. Binding of ACh to nicotinic receptors is always stimulatory and results in excitation of the neuron or effector cell. Muscarinic receptors occur on all effector cells stimulated by postganglionic cholinergic fibers—that is, all parasympathetic target organs and a few sympathetic targets (sweat glands and blood vessels of skeletal muscles). The effect of ACh binding to muscarinic receptors is inhibitory or stimulatory, depending on the target organ. For example, binding of ACh to cardiac muscle receptors slows heart activity, whereas ACh binding to receptors on smooth muscle of the gastrointestinal tract causes increased motility.

Adrenergic receptors

By the same token, there are two major classes of adrenergic (NE-binding) receptors: **alpha** (γ) and **beta** (β). Organs that respond to NE (or epinephrine) display one or both types of receptor. In general, NE or epinephrine binding to α receptors is stimulatory, while their binding to β receptors is inhibitory. But there are notable exceptions. For example, binding of NE to the β receptors of cardiac muscle stimulates the heart into more vigorous activity. These differences reflect the fact that both α and β receptors have two receptor subclasses (α_1 and α_2, β_1 and β_2), and each receptor type tends to predominate in certain target organs. The chief sites of these receptor subclasses are indicated in Table 14.2, but the picture is far more complex and well beyond the scope of an introductory anatomy and physiology course.

The Effects of Drugs

Knowing the locations of the cholinergic and adrenergic receptor subtypes is very important clinically because it allows specific drugs to be prescribed to obtain the desired blocking or stimulatory effects on selected target organs. For example, the anticholinesterase drug neostigmine inhibits the enzyme acetylcholinesterase, thus preventing the enzymatic breakdown of ACh and allowing it to accumulate in the synapses. This drug is used to treat myasthenia gravis, a condition in which skeletal muscle activity is impaired for lack of ACh stimulation. The drugs called *tricyclic antidepressants* work in treating depression because they prolong the activity of NE on the postsynaptic membrane. As described in the box on p. 461, NE is one of our "feeling good" neurotransmitters. Much of the research going on in pharmaceutical companies is directed toward finding drugs that affect only one subclass of receptor without upsetting the whole adrenergic or cholinergic system. An important breakthrough was finding adrenergic blockers that attach specifically to the β_1 receptors of cardiac muscle. Also called beta-blockers, propranolol (Inderal) and other drugs of its class are used when the goal is to reduce heart rate in cardiac patients without interfering with other sympathetic effects.

Interactions of the Autonomic Divisions

Although most visceral organs are innervated by both sympathetic and parasympathetic fibers—that is, they receive *dual innervation*—the two divisions do not always exert equal control over body organs. Normally, both ANS divisions are partially active, with the sympathetic division providing the "accelerator" and the parasympathetic division providing the "brakes." This dynamic antagonism results in very precise controls of visceral activity. However, one division or the other usually exerts the predominant effects in given circumstances. In a few cases, the autonomic divisions actually cooperate with one another to achieve a specific ANS result. The major effects of the two divisions are summarized in Table 14.3 on p. 459.

Antagonistic Interactions

Antagonistic effects were described earlier in the chapter, and these effects are most clearly seen on the activity of the heart, respiratory system, and gastrointestinal organs. In a fight-or-flight situation, the sympathetic division increases both your respiratory rate and your heart rate, while inhibiting digestion and elimination. When the emergency situation has passed, the parasympathetic division restores heart and breathing rates to resting levels and then can attend to processes that refuel the body cells and discard wastes.

Sympathetic and Parasympathetic Tone

Although we have described the parasympathetic division as the "at rest" division, the sympathetic division is the major actor in controlling blood pressure, even at rest. With few exceptions, the vascular system is entirely innervated by sympathetic fibers that keep the blood vessels in a continual state of partial constriction called *sympathetic* or *vasomotor tone*. When faster blood delivery is needed, these fibers deliver impulses more rapidly, causing blood vessels to constrict and blood pressure to rise. Conversely, when blood pressure is to be decreased, sympathetic stimulation declines and the vessels are allowed to dilate. Drugs that interfere with the activity of these vasomotor fibers (such as phentolamine) are often used to treat hypertension. During circulatory shock (inadequate blood delivery to body tissues), or when more blood must be provided to meet the increased needs of working skeletal muscles, the blood vessels serving the skin and abdominal viscera are strongly constricted. This shunts blood into the major circulatory channels so that circulation to vital organs can be maintained or blood delivery to skeletal muscles enhanced.

[illegible] also determines the normal activity levels of the digestive and urinary tracts. However, the sympathetic division can override these parasympathetic effects during times of stress. Drugs that block parasympathetic responses increase heart rate and cause fecal and urinary retention. With the exception of the adrenal glands and sweat glands of the skin, most glands are activated by parasympathetic fibers.

Cooperative Effects

The best example of cooperative ANS effects is seen in autonomic controls of the external genitalia. Parasympathetic stimulation causes vasodilation of blood vessels in the external genitalia and is responsible for erection of the male penis or female clitoris during sexual excitement. (This may help explain why sexual performance is sometimes impaired when people are anxious and the sympathetic division is dominant.) Sympathetic stimulation then causes ejaculation of semen by the male or reflex peristalsis of a female's vagina.

Unique Roles of the Sympathetic Division

The adrenal medulla, the sweat glands and arrector pili muscles of the skin, the kidneys, and most blood vessels receive only sympathetic fibers. Consequently, the sympathetic division regulates many functions not subject to parasympathetic influence. We have already described how the sympathetic control of blood vessels allows blood pressure regulation and the shunting of blood in the vascular system. We will now mention several other uniquely sympathetic functions.

Thermoregulatory Responses to Heat. The sympathetic division mediates reflexes that help to regulate body temperature. For example, application of heat to the skin results in reflex dilation of blood vessels in that area. When systemic body temperature is elevated, sympathetic nerves effect widespread dilation of the skin vasculature, allowing the skin to become flushed with warm blood, and activate the sweat glands to help cool the body. Conversely, when body temperature is falling, skin blood vessels are strongly constricted, and blood is restricted or shunted to deeper, more vital organs.

Release of Renin from the Kidneys. Sympathetic impulses stimulate the kidneys to release renin, an enzyme that promotes an increase in blood pressure. The mechanism by which this occurs, called the renin-angiotensin mechanism, is described in Chapter 26.

[illegible] increases in the metabolic rate of body cells; (2) increases in blood glucose levels; (3) increased hydrolysis of fats to be used as fuels; and (4) increased mental alertness via stimulation of the reticular activating system (RAS) of the brain stem. It is estimated that circulating adrenal medullary hormones produce 25% to 50% of all the sympathetic effects acting on the body at a given time.

Localized Versus Diffuse Effects

In the parasympathetic division, one preganglionic neuron synapses with one (or at most a few) postganglionic neurons. Additionally, all parasympathetic fibers release ACh, which is quickly destroyed by enzymatic (acetylcholinesterase) hydrolysis. Consequently, the parasympathetic division exerts short-lived, localized control over its effectors. In contrast, preganglionic sympathetic axons typically branch profusely on entering the sympathetic trunk and synapse with many postganglionic neurons at several levels. Additionally, the postganglionic neurons within the ganglia receive inputs from many different preganglionic neurons. Thus, when the sympathetic division is activated, it responds in a diffuse and highly interconnected way.

Effects of sympathetic activation are also much longer-lasting than those of parasympathetic activation. This reflects three phenomena: (1) NE is inactivated more slowly than ACh because it must be taken back up into the presynaptic ending before it is hydrolyzed (or stored); (2) NE, being a metabotropic neurotransmitter that exerts its effects via a second messenger system (see p. 363), acts more slowly than ACh, which is ionotropic; and (3) when the sympathetic division is mobilized, small amounts (15%) of norepinephrine and larger amounts (85%) of epinephrine are secreted into the blood by the adrenal medullary cells. These hormones cause essentially the same effects as NE released by sympathetic neurons and continue to exert these effects for several minutes before being destroyed by the liver. Thus, although sympathetic nerve impulses themselves may act only briefly, the hormonal effects they provoke linger. The widespread and prolonged effect of sympathetic activation helps explain why we need time to "come down" after an extremely stressful situation.

Table 14.3 Effects of the Sympathetic and Parasympathetic Divisions on Various Organs

Target organ/system	Sympathetic effects	Parasympathetic effects
Eye (iris)	Stimulates dilator muscles, dilates eye pupils	Stimulates constrictor muscles; constricts eye pupils
Eye (ciliary muscle)	No effect	Stimulates muscles, which results in bulging of the lens for accommodation and close vision
Glands (nasal, lacrimal, salivary, gastric, pancreas)	Inhibits secretory activity; causes vasoconstriction of blood vessels supplying the glands	Stimulates secretory activity
Sweat glands	Stimulates copious sweating (cholinergic fibers)	No effect
Adrenal medulla	Stimulates medulla cells to secrete epinephrine and norepinephrine	No effect
Arrector pili muscles attached to hair follicles	Stimulates to contract (erects hairs and produces "goosebumps")	No effect
Heart muscle	Increases rate and force of heartbeat	Decreases rate; slows and steadies heart
Heart: coronary blood vessels	Causes vasodilation	Constricts coronary vessels
Bladder/urethra	Causes relaxation of smooth muscle of bladder wall; constricts urethral sphincter; inhibits voiding	Causes contraction of smooth muscle of bladder wall; relaxes urethral sphincter; promotes voiding
Lungs	Dilates bronchioles and mildly constricts blood vessels	Constricts bronchioles
Digestive tract organs	Decreases activity of glands and muscles of digestive system and constricts sphincters (e.g., anal sphincter)	Increases motility (peristalsis) and amount of secretion by digestive organs; relaxes sphincters to allow movement of foodstuffs along tract
Liver	Epinephrine stimulates liver to release glucose to blood	No effect
Gallbladder	Inhibits (gallbladder is relaxed)	Excites (gallbladder contracts to expel bile)
Kidney	Causes vasoconstriction; decreases urine output; promotes renin formation	No effect
Penis	Causes ejaculation	Causes erection (vasodilation)
Vagina/clitoris	Causes reverse peristalsis (contraction) of vagina	Causes erection (vasodilation) of clitoris
Blood vessels	Constricts most vessels and increases blood pressure; constricts vessels of abdominal viscera and skin to divert blood to systemic circulation when necessary; dilates vessels of the skeletal muscles (cholinergic fibers) during exercise	Little or no effect
Blood coagulation	Increases coagulation	No effect
Cellular metabolism	Increases metabolic rate	No effect
Adipose tissue	Stimulates lipolysis (fat breakdown)	No effect
Mental activity	Increases alertness	No effect

Control of Autonomic Functioning

Although the ANS is not usually considered to be under voluntary control, its activity is nevertheless regulated. Several levels of controls exist—in the spinal cord, brain stem, hypothalamus, and cerebral cortex.

Brain Stem and Spinal Cord Controls

The reticular formation nuclei in the brain stem appear to exert the most direct control over autonomic functioning. For example, many autonomic responses can be elicited by stimulating centers in the medulla that reflexively regulate heart rate (cardiac centers), blood vessel diameter (vasomotor center), respiration, and many types of gastrointestinal reflexes. Most sensory impulses involved in eliciting these autonomic reflexes are delivered to the brain stem via vagus nerve afferents. The pons also contains respiratory centers that

tor—except that a visceral reflex arc has a *two-neuron* motor chain. All of these autonomic reflexes are described in later chapters in relation to the organ systems they serve.

Hypothalamic Controls

The main integration center of the autonomic nervous system is the hypothalamus. Medial and anterior hypothalamic regions appear to direct parasympathetic functions, whereas lateral and posterior areas contain sympathetic centers. Both types of centers exert their effects via relays through brain stem (reticular formation) centers. Not only does the hypothalamus contain centers that coordinate heart activity, blood pressure, body temperature, water balance, and endocrine activity, but also it contains centers that help regulate various emotional states (rage, pleasure) and biological drives (thirst, hunger, sex). Afferent impulses travel through the hypothalamus on their way to the cerebral cortex, and the cerebral cortical centers involved in emotional processing also communicate with the hypothalamus. Because of its central position in the limbic system, the hypothalamus serves as the keystone of the emotional/visceral (autonomic) brain, and it is through the hypothalamic centers that emotions influence autonomic nervous system functioning and behavior. The relationship of certain drug abuse behaviors to hypothalamic pleasure centers is the topic of the box on p. 461.

Cortical Controls

It was originally believed that the autonomic nervous system was not subject to voluntary controls. However, studies of meditation and biofeedback have shown that cortical control of visceral activities is possible—a capability that is untapped by most people.

Influence of Meditation on Autonomic Function. Studies of meditating yogis have revealed that they are in a physiological state that is almost the exact opposite of sympathetic-induced hyperactivity. Heart and breathing rates are low, oxygen use and metabolism drop, and EEG recordings show a predominance of alpha waves

ral activities because we have little or no conscious awareness of what is happening within our bodies. During such training, subjects are connected to monitoring devices that provide such awareness, which is called **biofeedback.** The devices detect and amplify changes in physiological processes such as heart rate, blood pressure, and skeletal muscle tone, and these data are then "fed back" to the subject in the form of flashing lights or audible tones. Subjects are asked to try to alter or control some "involuntary" function by concentrating on calming, pleasant thoughts. Since the monitor allows them to identify physiological changes in the desired direction, they gradually develop the ability to recognize the feelings associated with these changes and to produce them at will.

Biofeedback techniques have been quite successful in helping individuals plagued by migraine headaches. They are also being used to manage stress by cardiac patients, many of whom have learned to lower their blood pressure and heart rate, reducing their risk of disabling or fatal heart attack. However, it is doubtful that biofeedback will ever become a widely used clinical tool. Biofeedback training is a time-consuming and often frustrating process, and the training equipment is prohibitively expensive and difficult to use.

Homeostatic Imbalances of the Autonomic Nervous System

Since the autonomic nervous system is involved in nearly every important process that goes on within the body, it is not surprising that abnormalities of autonomic functioning can have far-reaching effects. Such abnormalities can impair blood delivery and elimination processes and even threaten life itself.

Most autonomic disorders reflect exaggerated or deficient controls of smooth muscle activity. Of these, the most devastating involve blood vessels and include conditions such as Raynaud's disease, hypertension, and the mass reflex reaction.

A CLOSER LOOK **Pleasure Me, Pleasure Me!**

Deep inside the hypothalamus is a small bundle of neural tissue, called the *pleasure center*, that drives much of human behavior. We humans are programmed to eat, drink, and procreate, and it is still the pleasure center that reinforces these behaviors. However, because of it, we are also perilously vulnerable. Few would deny that much of what we do and value is driven by the "pleasure principle," and herein lies the genesis of addiction.

Our ability to feel good involves brain neurotransmitters such as norepinephrine, dopamine, and serotonin. For example, romantic love has been described as a "brain bath" of norepinephrine and dopamine, which affect the activation threshold of the pleasure center and account for all of the decidedly pleasurable sensations associated with that emotion. Both norepinephrine and dopamine chemically resemble amphetamines, and people who use "speed" and other amphetamines are artificially stimulating their brains to provide their pleasure flush. However, the pleasure is short lived, because when the brain is flooded with neurotransmitter-like chemicals from the outside, it begins to produce less and less of its natural chemicals (why bother?). The wisdom of the body mediates against unnecessary effort.

Cocaine, another pleasure center titillater, has been around for a long time in its granular form, which is inhaled, or "snorted." Enter "crack"—the cheaper, much more potent, smokable form of cocaine. For ten dollars, a novice user can get "wasted for the night" and flooded with a rush of intense pleasure. But crack is treacherous; intensely addictive, it represents perhaps the worst street-drug threat to date. It produces not only a higher high than the inhaled form of cocaine, but also a deeper crash that leaves the user desperate for more.

Lines of cocaine (granular form).

Researchers are beginning to zero in on how cocaine produces its effects. It now appears that by interfering with brain neurotransmitters, this drug first stimulates the pleasure center and then "squeezes it dry." Cocaine produces the rush by blocking the reabsorption of dopamine, norepinephrine, and serotonin. (Of these, dopamine is the one most linked to the feeling of euphoria.) Since the neurotransmitters remain in the synapse, they stimulate the receptor cells again and again, allowing the body to feel their effects over a prolonged period. This sensation is accompanied by increases in heart rate, blood pressure, and sexual appetite. But over time—and sometimes after just one use—the effect changes. As the reabsorption of dopamine continues to be blocked, that accumulated in the synapses is washed away and degraded, and the brain's neurotransmitter supply becomes depleted and inadequate to maintain normal mood. The sending cells cannot produce neural dopamine fast enough to make up for its loss, and the pleasure circuits go dry. Meanwhile, the receptor cells become hypersensitive, and new receptors begin to sprout up in a desperate effort to pick up dopamine signals. The user becomes depressed, anxious, and, in a very real sense, unable to experience pleasure without cocaine. A vicious cycle of addiction is established. Cocaine is needed to experience pleasure, but using it further depletes the neurotransmitter supply. Heavy, prolonged cocaine use may deplete the supply so totally that pleasure is impossible to achieve and deep depression sets in. Cocaine addicts tend to lose weight, have trouble sleeping, develop cardiac abnormalities, and become increasingly susceptible to infections.

Cocaine addiction carries with it a desperate craving that is notoriously difficult to manage or treat. Various drugs have been used with mixed success to treat cocaine addicts. For example, bromocriptine provides a temporary substitute for dopamine by stimulating the dopamine receptors, but it does not reproduce the cocaine "high." Somewhat more successful has been the use of imipramine in conjunction with amino acid therapy. Imipramine eases the craving and blocks the euphoria while the brain uses the amino acids to manufacture dopamine and other neurotransmitters.

Crack use in the United States is a national problem. Unlike heroin use, which has been most prevalent among the poor and has had a sordid reputation, cocaine use has been seen as more acceptable, and it has permeated all economic strata. Even small-town middle America is getting hooked, and cocaine-related cardiovascular deaths (due to acceleration of the heart and enhanced blood pressure) are increasing daily.

The pleasure center, with its cries of "pleasure me! pleasure me!," is a powerful Pied Piper. But crack addiction is only the latest and most disastrous manifestation of its powers. We want to feel young and powerful forever, and we want instant solutions to our problems. But drugs are only temporary solutions. The brain, with its complex biochemistry, always wins and circumvents attempts to keep it in a euphoric haze. Perhaps this means that pleasure must be transient and can be experienced only against a background of its absence.

Raynaud's disease is characterized by intermittent attacks, during which the skin of the extremities becomes pale, then cyanotic and painful. Commonly

[illegible]

Hypertension, or high blood pressure, may result from arteriosclerosis (the filling of blood vessels with fatty, calcified material), renal disease, or an overactive sympathetic nervous system response promoted by continual high levels of stress. Whatever the cause, hypertension is always serious because (1) it increases the work load on the heart and may precipitate hypertensive heart disease and (2) it increases the wear and tear on the artery walls. When the suspected cause is overwhelming stress responses, the condition is treated with adrenergic-blocking drugs.

The *mass reflex reaction* is a life-threatening condition involving both somatic and autonomic nerves that occurs in a majority of individuals with quadriplegia and in others with spinal cord injuries above the level of T_6. The cord injury is followed by spinal shock (see p. 413) accompanied by paralysis of skeletal muscles and loss of autonomic reflexes below the level of injury. Eventually, reflex activity returns, but usually in an exaggerated state because of the lack of inhibitory input from higher centers. Then episodes of mass reflex begin to occur—a phenomenon involving large regions of the spinal cord. The usual trigger for the mass reflex is a painful stimulus to the skin or overfilling of a visceral organ. The body goes into flexor spasms, the colon and bladder empty reflexively, and profuse sweating begins. The arterial blood pressure rises to life-threatening levels (200 mm Hg or more), indicating massive sympathetic nervous system activation, which may precipitate a stroke or cerebrovascular accident. The precise mechanism of the mass reflex is unknown, but it has been envisioned as a type of epilepsy of the spinal cord. ■

Developmental Aspects of the Autonomic Nervous System

Preganglionic neurons of the autonomic nervous system derive from the embryonic neural tube, whereas postganglionic neurons and all autonomic ganglia derive from the *neural crest tissue* that has separated from the neural tube by the first month of development (see Figure 12.2). Some neural crest cells form the sympathetic chains; others

[illegible]

entiated into the medulla of the adrenal gland. Neural crest cells also contribute to the ciliary and other autonomic ganglia of the head and to the terminal parasympathetic ganglia of the viscera. The neural crest cells appear to reach their ultimate destinations by migrating along growing axons.

Congenital abnormalities reflecting autonomic nervous system disorders are rare, but they do occur. Perhaps the most notable example is *Hirschsprung's disease (megacolon)*, in which the parasympathetic plexus supplying the distal part of the large intestine (colon) fails to develop normally. As a result, the affected portion of the colon is immobile, and feces accumulate proximal to the inactive bowel segment, causing the area to become enormously distended. The condition is corrected surgically by removing the inactive bowel segment.

During youth, impairments of autonomic nervous function result primarily from traumatic injuries to the spinal cord or autonomic nerves. In old age the efficiency of the autonomic nervous system begins to decline. Elderly people often complain of constipation (due to reduced gastrointestinal tract motility), and tend to have dry eyes and frequent eye infections (a consequence of diminished tear formation). Additionally, the sympathetic vasoconstrictor centers are slower to respond to changes in body position, which can lead to fainting episodes. While most of these problems are distressing, they are not usually life threatening, and most can be managed by life-style changes or artificial aids. For example, changing position slowly gives the sympathetic nervous system more time to bring about appropriate adjustments in blood pressure; drinking ample fluids helps to alleviate constipation; and eye drops (artificial tears) are available for the dry eye problem.

* * *

In this chapter, we have described the structure and function of the autonomic nervous system, a unique arm of the motor division of the peripheral nervous system. Because virtually every organ system still to be considered depends on autonomic neural controls to function normally, you will be hearing more about autonomic nervous system in the chapters that follow.

Related Clinical Terms

Atonic bladder (ā″-tah′-nik) (*a* = without; *ton* = tone, tension) A condition in which the urinary bladder becomes flaccid and overfills, allowing urine to dribble through the sphincters; results from the temporary loss of autonomic innervation following spinal cord injury.

Horner's syndrome A condition due to destruction of the superior sympathetic trunk on one side of the body; the affected person exhibits drooping of the upper eyelid, constricted pupil, and does not sweat on the affected side.

Orthostatic hypotension (*ortho* = straight, upright; *stat* = standing, placed) Also called postural hypotension; a form of low blood pressure that occurs when assuming a standing position; particularly common in the elderly because of the sluggishness of the sympathetic nervous system in old age.

Vagotomy (vā-gah′-tuh-mē) Cutting or severing of a vagus nerve; often done to help relieve symptoms of gastrointestinal ulcers.

Chapter Summary

1. The autonomic nervous system is the motor division of the PNS that controls visceral activities, with the goal of maintaining internal homeostasis.

OVERVIEW OF THE AUTONOMIC NERVOUS SYSTEM (pp. 448–450)

Comparison of the Somatic and Autonomic Nervous Systems (pp. 448–449)

1. The somatic (voluntary) nervous system provides motor fibers to skeletal muscles. The autonomic (involuntary or visceral motor) nervous system provides motor fibers to smooth and cardiac muscles and glands.

2. In the somatic division, a single motor neuron forms the efferent pathway from the CNS to the effectors. The efferent pathway of the autonomic division consists of a two-neuron chain: the preganglionic neuron in the CNS and the postganglionic neuron in a ganglion.

3. Acetylcholine, the neurotransmitter of somatic motor neurons, is stimulatory to skeletal muscle fibers. Neurotransmitters released by autonomic motor neurons (acetylcholine and norepinephrine) may cause excitation or inhibition.

The Divisions of the Autonomic Nervous System (pp. 449–450)

4. The autonomic nervous system consists of two divisions, the sympathetic and parasympathetic, which normally exert antagonistic effects on many of the same target organs.

5. The sympathetic division activates the body under conditions of emergency and is called the fight-or-flight system.

6. Sympathetic responses include dilated pupils, increased heart and respiratory rates, increased blood pressure, dilation of the bronchioles of the lungs, increased blood glucose levels, and sweating. During exercise, sympathetic vasoconstriction shunts blood from the skin and digestive viscera to the skeletal muscles.

7. The parasympathetic division (the resting–digesting system) conserves body energy and maintains body activities at basal levels.

8. Parasympathetic effects include pupillary constriction, glandular secretion, increased digestive tract mobility, and muscle actions leading to elimination of feces and urine.

ANATOMY OF THE AUTONOMIC NERVOUS SYSTEM (pp. 450–455)

Sympathetic (Thoracolumbar) Division (pp. 450–453)

1. Preganglionic sympathetic neurons arise from the lateral horn of the spinal cord from the level of T_1 to L_2.

2. Preganglionic axons leave the cord via white rami communicantes and enter the paravertebral ganglia in the sympathetic trunk. An axon may synapse in a paravertebral ganglion at the same or at a different level, or it may issue from the sympathetic chain without synapsing. Preganglionic fibers are short; postganglionic fibers are long.

3. When the synapse occurs in a paravertebral ganglion, the postganglionic fiber may enter the spinal nerve via the gray ramus communicans to travel to the body periphery. Postganglionic fibers issuing from the cervical paravertebral ganglia also serve visceral organs and blood vessels of the head, neck, and thorax.

4. When synapses do not occur in the paravertebral ganglia, the preganglionic fibers form splanchnic nerves (greater, lesser, and lumbar). Most splanchnic nerve fibers synapse in prevertebral ganglia, and the postganglionic fibers serve the abdominal viscera. A few splanchnic nerve fibers synapse with cells of the adrenal medulla.

Parasympathetic (Craniosacral) Division (pp. 454–455)

5. Parasympathetic preganglionic neurons arise from the brain stem and from the sacral (S_2–S_4) region of the cord.

6. Preganglionic fibers synapse with postganglionic neurons in terminal ganglia located in or close to their effector organs. Preganglionic fibers are long; postganglionic fibers are short.

7. Cranial fibers arise in the brain stem nuclei of cranial nerves III, VII, IX, and X and synapse in ganglia of the head, thorax, and abdomen. The vagus nerve serves virtually all organs of the thoracic and abdominal cavities.

8. Sacral fibers (S_2–S_4) issue from the lateral region of the cord and form pelvic splanchnic nerves that innervate the pelvic viscera. The preganglionic axons do not travel within rami communicantes or spinal nerves.

Visceral Sensory Neurons (p. 455)

9. Virtually all autonomic nerves and ganglia contain visceral sensory fibers.

PHYSIOLOGY OF THE AUTONOMIC NERVOUS SYSTEM (pp. 455–460)

Neurotransmitters and Receptors (pp. 455–457)

1. Two major neurotransmitters, acetylcholine (ACh) and norepinephrine (NE), are released by autonomic motor neurons. On the basis of the neurotransmitter released, the fibers are classified as cholinergic or adrenergic.

2. ACh is released by all preganglionic fibers and all parasympathetic postganglionic fibers. NE is released by all sympathetic postganglionic fibers except those serving the sweat glands of the skin, the blood vessels within skeletal muscles, and the external genitalia (those fibers secrete ACh).

3. Neurotransmitter effects depend on the receptors to which the transmitter binds. Cholinergic (ACh) receptors are classified as nicotinic and muscarinic. Adrenergic (NE) receptors are classified as α_1 and α_2 and β_1 and β_2.

The Effects of Drugs (p. 457)

4. Drugs that mimic, enhance, or inhibit the action of ANS neurotransmitters are used to treat conditions caused by excessive, inadequate, or inappropriate ANS

and exhibit vasomotor tone. Parasympathetic activity dominates the heart and muscles of the gastrointestinal tract (which normally exhibit parasympathetic tone) and glands.

7. The two ANS divisions exert cooperative effects on the external genitalia.

8. Roles unique to the sympathetic division are blood pressure regulation, shunting of blood in the vascular system, thermoregulatory responses, stimulation of release of renin by the kidneys, and metabolic effects.

9. Activation of the sympathetic division causes widespread, long-lasting mobilization of the visceral system. Parasympathetic effects are highly localized and short lived.

Control of Autonomic Functioning (pp. 459–460)

10. Autonomic function is controlled at several levels: (1) Reflex activity is mediated by the spinal cord and brain stem (particularly medullary) centers; (2) hypothalamic integration centers interact with both higher and lower centers to orchestrate autonomic, somatic, and endocrine responses; and (3) cortical centers influence autonomic functioning via connections with the limbic system; conscious controls of autonomic function are possible, as illustrated by meditation and biofeedback techniques.

HOMEOSTATIC IMBALANCES OF THE AUTONOMIC NERVOUS SYSTEM (pp. 460–462)

1. Most autonomic disorders reflect problems with smooth muscle control. Abnormalities in vascular control, such as occurs in Raynaud's disease, hypertension, and the mass reflex reaction, are most devastating.

DEVELOPMENTAL ASPECTS OF THE AUTONOMIC NERVOUS SYSTEM (p. 462)

1. Preganglionic neurons develop from the neural tube; postganglionic neurons develop from the embryonic neural crest.

2. Hirschsprung's disease involves functional blockage of the large intestine due to failure of parasympathetic innervation.

3. The efficiency of the autonomic nervous system declines in old age, as reflected by decreased glandular secretory activity, decreased gastrointestinal motility, and slowed sympathetic vasomotor responses to changes in position.

Review Questions

Multiple Choice/Matching

1. All of the following characterize the ANS except (a) two-neuron efferent chain, (b) presence of nerve cell bodies in the CNS, (c) presence of nerve cell bodies in the ganglia, (d) innervation of skeletal muscles.

2. Relate each of the following terms or phrases to either the sympathetic (S) or parasympathetic (P) division of the autonomic nervous system:

______ **(9)** increases gastric motility and secretion of lacrimal, salivary, and digestive glands

______ **(10)** innervates blood vessels

______ **(11)** active when you are swinging in a hammock

______ **(12)** active when you are competing in the Boston Marathon

Short Answer Essay Questions

3. Briefly explain why the following terms are sometimes used to refer to the autonomic nervous system: involuntary nervous system and emotional-visceral system.

4. Describe the anatomical relationship of the white and gray rami to the spinal nerve, and note the kind of fibers found in each ramus type.

5. Indicate the results of sympathetic activation of the following structures: sweat glands, eye pupils, adrenal medulla, heart, lungs, liver, blood vessels of skeletal muscles, blood vessels of digestive viscera, salivary glands.

6. Which of the effects listed (in question 5) would be reversed by parasympathetic activity?

7. Which ANS fibers release acetycholine? Which release norepinephrine?

8. Describe the meaning and importance of sympathetic tone and parasympathetic tone.

9. List the receptor subtypes for ACh and NE, and indicate the major sites where each type is found.

10. What area of the brain is most involved in mediating autonomic reflexes?

11. Describe the importance of the hypothalamus in controlling the autonomic nervous system.

12. Describe the basis and uses of biofeedback training.

13. What manifestations of decreased autonomic nervous system efficiency are seen in elderly individuals?

Clinical Applications

14. A 32-year-old woman complains that she has been experiencing aching pains in the medial two fingers of both hands and that, during such episodes, the fingers become blanched and then blue. Her history is taken, and it is noted that she is a heavy smoker. The physician advises her that she must stop smoking and that he will not prescribe any medication until she has discontinued smoking for a month. What is this woman's problem, and why was she told to stop smoking?

15. Two-year-old Jimmy has a swollen abdomen and is continuously constipated. On examination, a mass was felt over the descending colon, and the X-ray showed the colon to be greatly distended in that region. What do you think is wrong with Jimmy, and how does this relate to the innervation of the colon?

CHAPTER **15**

Neural Integration

Chapter Outline and Student Objectives

Sensory Integration: From Reception to Perception (pp. 466–471)

1. Describe the important role of receptors in sensory processing.
2. Compare and contrast specific and nonspecific ascending somatosensory pathways.
3. Describe the main features of perceptual processing of sensory inputs.

Motor Integration: From Intention to Effect (pp. 471–474)

4. Describe the levels of the motor control hierarchy.
5. Define central pattern generator and command neuron.
6. Compare the roles of the pyramidal and extrapyramidal systems in controlling motor activity.
7. Explain the function of the cerebellum and basal nuclei in somatic sensory and motor integration.
8. Describe symptoms of cerebellar and basal nuclear disease.

Higher Mental Functions (pp. 474–483)

9. Define EEG and distinguish between alpha, beta, theta, and delta waves.
10. Explain the importance of slow-wave and REM sleep, and note how their patterns change through life.
11. Explain what is meant by the term *holistic processing* in relation to human consciousness.
12. Compare and contrast the stages and categories of memory.
13. Indicate the major brain structures believed to be involved in fact and skill memories and describe their relative roles.
14. List the brain areas involved in language understanding and articulation, and indicate the role of each.

Preview of Selected Key Terms

Sensation Awareness of internal and external events.

Perception Assigning a meaning to sensation.

Receptor potential A graded potential that occurs at a sensory receptor membrane.

Sensory transduction Conversion of stimulus energy into a nerve impulse.

Command neuron An interneuron located in a brain stem extrapyramidal nucleus that helps regulate the spinal cord motor circuits.

Synergy (sih′-ner-jē) (*synerg* = work together, cooperate) Coordinated activity of agonist and antagonist muscles that results in smooth, well-controlled movements.

Brain waves Patterns of electrical activity of the neurons of the brain, recordable with an electroencephalograph.

Although we seldom look at nervous system functioning in its entirety, virtually all of its activities—sensory, integrative, and motor alike—go on simultaneously. Take, for example, a leisurely drive with a friend. Sensory inputs, such as what you see in your visual field and the pressure of the pedals on your feet, keep your CNS constantly apprised of what is going on both inside and outside your body. Your muscles respond to CNS commands to brake or accelerate, and all the while, you carry on an animated conversation with your passenger. Let us analyze the neural events that underlie similar examples of our daily activities.

Because sensory experiences are so immediate, they somehow seem more understandable than many other types of neural functions, so they are good starting points for our studies of neural integration. After considering sensory integration, we will examine how some of the motor activities we take so much for granted (like walking and maintaining our upright posture) are accomplished. Then these topics will be tied together at the highest levels of

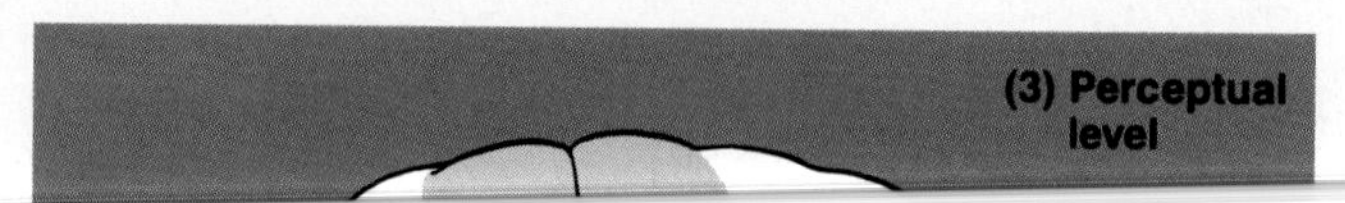

Reception to Perception

Our survival depends not only on **sensation** (awareness of changes in the internal and external environments), but also on **perception** (conscious interpretation of those stimuli), which in turn determines how we will respond to the various stimuli. For example, when we hear (sense) the honk of an automobile horn, we might perceive the honk to mean that the car we are driving has strayed across the white line and is dangerously close to causing an accident.

In this section, we will follow sensory inputs from the sensory receptors to the somatosensory cortex and examine the role of the neural structures at each level of the pathway.

General Organization of the Somatosensory System

The **somatosensory system,** or that portion of the sensory system dealing with reception in the body wall and limbs, receives inputs from exteroceptors, proprioceptors, and interoceptors. Consequently, it is concerned with transmitting information about several different sensory modalities, each mediated by a particular set of receptors.

As described in Chapter 12, ascending sensory pathways between the receptor and the cerebral cortex exhibit a chain of three neurons: the cell body of the first-order neuron in the dorsal root ganglion; that of the second-order neuron in the dorsal horn of the spinal cord or in the medulla oblongata; and that of the third-order neuron in the thalamus (see Figure 12.27). Although this is the basic scheme for upward flow of information, collateral synapses are made all along the pathway.

There are three primary levels of neural integration involved in the somatosensory (or any sensory) system. These are the *receptor level*, the *circuit level*, and the *perceptual level*, corresponding to the sensory receptors, the ascending pathways, and the neuronal circuits in the cerebral cortex (Figure 15.1).

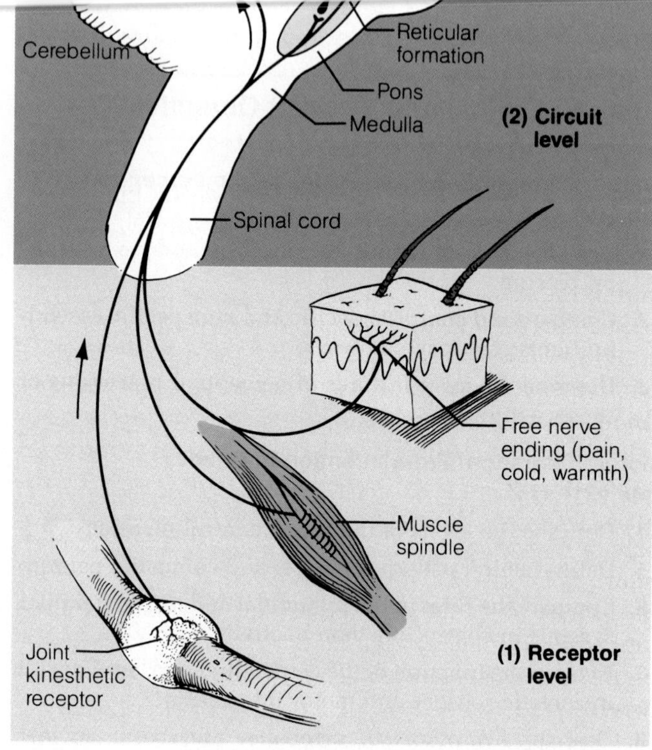

Figure 15.1 General organization of the somatosensory system. The three basic levels of neural integration are the receptor level (level 1), the circuit level (level 2), and the perceptual level (level 3). The circuit level involves all CNS centers with the exception of the somatosensory cerebral cortex.

Processing at the Receptor Level

Information about our external and internal world presents itself in different energy forms: sound, mechanical, chemical, and so on. The sensory receptors are specialized to respond to these energetic stimuli; the rest of the nervous system responds chiefly to action potentials or, over short distances, to local graded potentials. If sensory neurons are to communicate with other neurons, they must translate the information of their stimulus into nerve impulses, the universal neural language.

As stimulus energy is absorbed by the receptor, it is converted, or *transduced*, into an electrical event. That is, the stimulus causes changes in the permeability of the receptor membrane that result in a graded potential, called a **receptor potential**, that is essentially the same as an EPSP generated at a postsynaptic membrane in response to binding of neurotransmitters (see Chapter 11).

If the receptor potential is of threshold strength, an action potential will be generated and propagated to the CNS by the sensory fibers, and nerve impulse transmission will continue as long as a threshold stimulus is applied. Since, as described in Chapter 11, the strength of the stimulus is encoded in the frequency of impulse transmission, stronger stimuli cause more impulses per second to reach the CNS.

Processing at the Circuit Level

Sensory fibers, carrying impulses from cutaneous receptors of the skin and from proprioceptors, branch diffusely as they enter the spinal cord. Some branches end in the cord as they become involved in local spinal cord reflexes, and these fibers directly initiate motor activities. The remaining sensory afferents synapse with second-order neurons in the dorsal horn or continue upward in the white columns to synapse in medullary nuclei. Small-diameter pain fibers synapse with superficially located dorsal horn (substantia gelatinosa) neurons; the large myelinated fibers from pressure and touch receptors make collateral synapses with neurons deep within the dorsal horns. As described in the box on pp. 468–469, interactions between the touch and pain afferents and small dorsal horn interneurons allow pain inputs to be damped at the circuit level.

Somatic sensory information reaches the somatosensory cortex via two major ascending routes: the nonspecific and the specific pathways, found respectively within the anterolateral and dorsal white columns of the spinal cord. Also located in the lateral white columns are the *ascending spinocerebellar* (spī″-nō-sayr″-eh-bel′-ar) *tracts*, which serve muscle sense. Since these tracts terminate in the cerebellum (see Figure 12.27a), they do not contribute to conscious sensory perception. Instead, the proprioceptor inputs they carry from the muscles and tendons serve motor function rather than sensation per se.

Nonspecific Ascending Pathways. The evolutionarily older *nonspecific pathways* are so named because they receive inputs from many different types of sensory receptors and make multiple synapses in the brain stem. These pathways form the lateral and anterior *spinothalamic tracts*, introduced in Chapter 12 (see Table 12.2 and Figure 12.27b). Most of these fibers transmit pain and temperature impulses, but some convey light touch, pressure, or joint information. The spinothalamic tract fibers ultimately synapse with third-order neurons in the *nonspecific nuclei* of the thalamus; but en route, they give off numerous branches to the reticular formation. The reticular neurons, in turn, form synapses with most parts of the brain—various brain stem motor nuclei, the reticular activating system, and most areas of the cerebral cortex.

The nonspecific somatosensory system is involved in emotional aspects of perception (pleasure, aversion) and in pain perception (see the box on p. 468), and participates in cortical arousal as well as in certain higher level motor reflexes and orienting responses (such as turning the head toward a novel stimulus).

Specific Ascending Pathways. The *specific pathways* are chiefly concerned with the precise, straight-through transmission of inputs from a single type (or a few related types) of sensory receptor. These pathways are formed by the *fasciculus cuneatus, fasciculus gracilis*, and *medial lemniscal* (lem-nis′-kul) *tracts*. The latter arise in the medulla and terminate in the *ventrobasal nuclei* of the thalamus (see Table 12.2 and Figure 12.27a). From the thalamus, impulses are forwarded to specific areas of the somatosensory cortex. The trigeminal nerve (cranial nerve V) also feeds into these pathways. The specific pathways are mainly concerned with sending information about discriminative touch, pressure, vibration, and conscious proprioception (limb and joint position) to the cortex. These pathways also give off some collaterals to the reticular nuclei and so contribute to the arousal mechanism.

Tabes dorsalis (ta′-bēz dor-sa′-lis) is a slowly progressive condition caused by deterioration of the posterior white matter tracts (gracilis and cuneatus) and associated dorsal roots. It is a late sign of the neurological damage caused by the syphilis bacterium. Because joint proprioceptor tracts are destroyed, affected individuals have poor muscle coordination and an unstable gait, and they must constantly watch the ground when walking to avoid falling. Bacterial invasion of the sensory (dorsal) roots results in pain, which ends when dorsal root destruction is complete. ■

The specific and nonspecific divisions of the somatosensory system are activated simultaneously, and their interactions with each other and the cerebral cortex are numerous. Remember, however, that brain stem centers, particularly the reticular nuclei, can damp sensory inputs destined for the cerebral cortex, so not all ascending sensory impulses are perceived.

Processing at the Perceptual Level

Perception, the final stage of sensory processing, involves awareness of stimuli and discrimination of their characteristics. As sensory information ap-

A CLOSER LOOK Pain: Sound the Alarm, But Pain Me Not!

Clinically, pain cannot be measured except by some roundabout and inaccurate techniques, such as tapping the painful spot and observing the patient's degree of wincing. This seems a bit strange, particularly since the other most common sign of disease or injury—fever—is readily measured with a thermometer.

Pain reception

The principal pain receptors are the several million bare sensory nerve endings that weave through all the tissues and organs of the body except the brain and respond to noxious stimuli—a catchall term for anything damaging to tissues. But evidence is mounting that bradykinin is the universal pain stimulus. Bradykinin is the most potent pain-producing chemical known, and it is generated whenever and wherever body tissue is injured. At such sites, enzymes are activated that release bradykinin from larger precursor molecules present in blood and many other tissues. Bradykinin, in turn, triggers the production of inflammatory chemicals, such as histamine and prostaglandins, that initiate healing. Some scientists think that bradykinin also binds to pain receptor endings, causing them to fire. These scientists are working to create a variant of bradykinin that will prevent the body's natural bradykinin from binding to the pain receptor endings, extinguishing pain at its source.

Pain transmission and perception

Clinically, pain is classified as *somatic* or *visceral*. Somatic pain, arising from

somatic pain is more likely to be burning or aching pain; it results from stimulation of pain receptors in the deep skin layers, muscles, or joints. Deep somatic pain is both more diffuse and longer lasting than superficial somatic pain, and it always indicates tissue destruction. Impulses from the deep pain receptors are transmitted slowly (0.4–1 m/s, or up to 3.5 ft/s) along the small, unmyelinated C fibers. When you burn your hand severely with boiling water, you experience both superficial and deep somatic pain. The initial excruciatingly sharp pain is followed by burning, gnawing pain that lingers on for hours.

Visceral pain results from noxious stimulation of receptors in the organs of the thorax and abdominal cavity. It is similar or identical to deep somatic pain—usually a dull ache, burning, or a feeling of pressure. The most important stimuli for visceral pain are extreme stretching of tissue, ischemia, irritating chemicals, and muscle spasms. Visceral pain inputs travel along the same pathways as somatic pain, and projection by the brain may cause visceral pain to be perceived as somatic in origin. This phenomenon is called *referred pain*. For example, a heart attack may produce a sensation of pain that radiates along the medial aspect of the left arm. Tissue damage in the heart gives rise to pain impulses that enter the spinal cord in the superior thoracic (T_1–T_4) region, which also receives afferent impulses from the left side of the chest and arm. Hence, the brain interprets most such signals as coming from the arm. Cutaneous areas to which visceral pain is commonly referred are shown in illustration 1.

that the axons of most of the second-order neurons cross the cord and enter the anterolateral spinothalamic tracts, which ascend to the thalamus. From the thalamus, impulses are relayed to the somatosensory cortex for interpretation. Some second-order fibers "take the express"—that is, they ascend directly to the ventrobasal nuclei of the thalamus. These inputs allow the sensory cortex to make a swift and precise analysis of what hurts and how much. Other fibers of the spinothalamic tracts "take the local," making abundant synapses in the brain stem, hypothalamus, and limbic system structures before reaching the intralaminar nuclei of the thalamus. This second set of pathways is important in mediating arousal and our emotional reactions to pain. It also exerts more general and longer-lasting influences on the CNS.

We all have the same *threshold* for pain; that is, we perceive it at the same stimulus intensity. For example, heat is perceived as painful at 44°–46°C, the narrow range at which the thermal stimulus begins to cause tissue damage. However, reactions to pain, or *pain tolerance*, vary widely and are heavily influenced by cultural and psychological factors. When we say that someone is "very sensitive to pain," we are referring to that person's pain tolerance rather than to the person's pain threshold. Pain tolerance seems to increase with age, but this, too, may reflect the sociocultural notion that we have to expect pain when we get old. Pain is also modified by emotions and mental state. In a disaster, many who struggle to help others and who feel no pain may actually be severely injured.

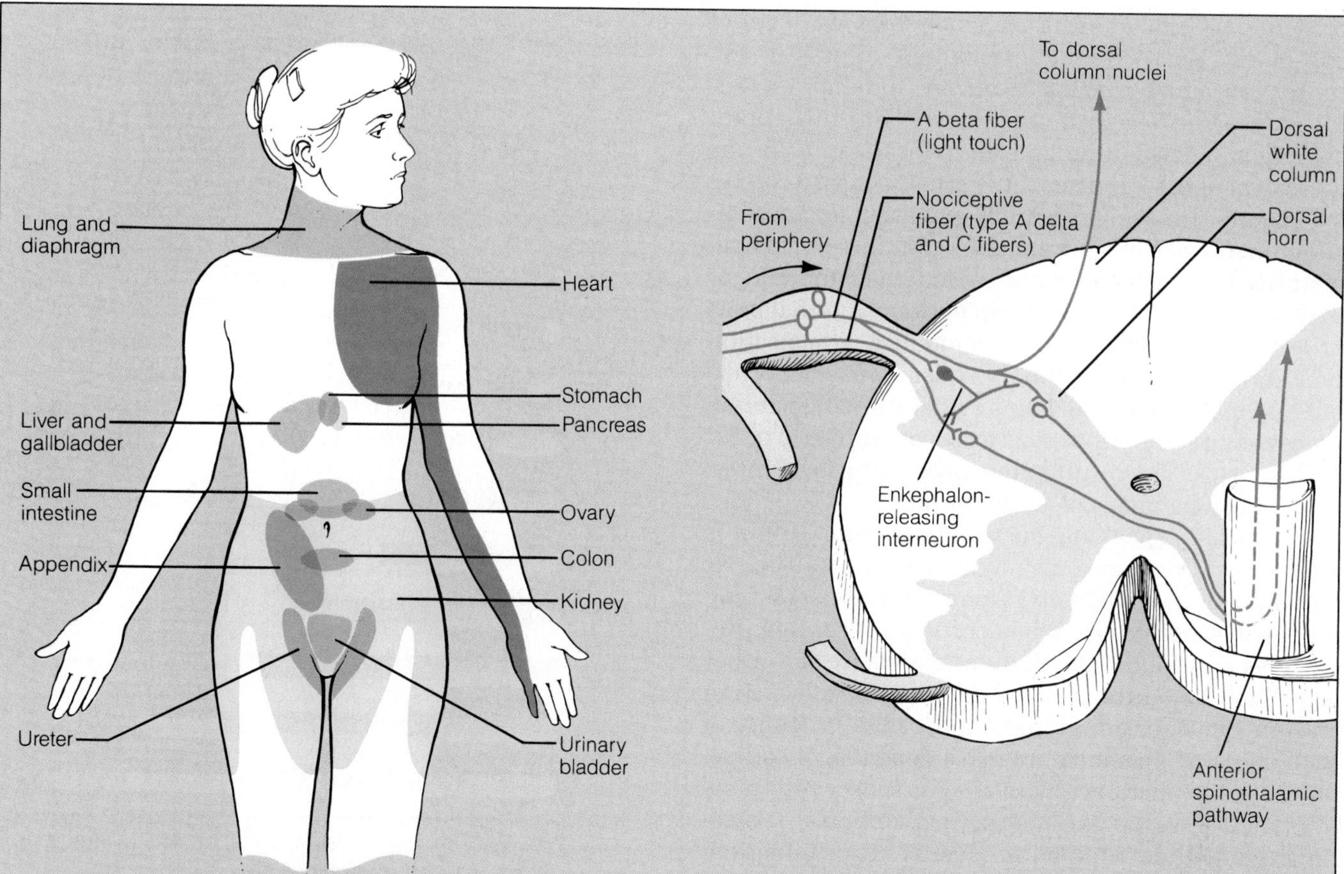

(1) Anterior cutaneous areas to which pain from certain visceral organs is referred.

(2) When type A delta and type C pain fibers transmit through to their transmission neurons in the spinothalamic pathway, pain impulses are transmitted to the cerebral cortex. When the gating mechanism is activated, the pain pathway is blocked. The gating mechanism appears to be mediated by small enkephalin-releasing interneurons in the dorsal horn that make inhibitory synapses with both the large touch afferents (A beta fibers) and the pain fibers. Activation of these interneurons by the touch afferents inhibits both touch and pain.

Pain modulation

The extraordinary plasticity of human pain suggests that there must be natural neural mechanisms that can modulate pain transmission and perception. It was discovered some time ago that our natural opiates (beta-endorphins and enkephalins) are released within the brain when we are in pain and act to reduce its perception. Hypnosis, natural childbirth techniques, and stimulus-induced analgesia are all believed to tap into these natural-opiate pathways. More recently, it was discovered that the opiates also modulate pain transmission at the spinal cord level under certain conditions. According to the gate control theory of pain put forth by Melzcak and Wall in the 1960s:

1. Pain results from a complex pattern of interacting slow and fast pain fibers and descending fibers from the brain.

2. The "gate" appears to be in the dorsal horn (substantia gelatinosa), where nerve impulses from peripheral receptors (both types of pain fibers and the large touch [A beta] fibers) enter the cord.

3. If impulses along the slow (type C) pain fibers outnumber inputs along the fast A beta fibers (touch afferents), the gate is opened and pain impulses are transmitted and perceived.

4. Stimulation of more A beta fibers closes the gate, inhibiting transmission of pain impulses and reducing pain perception. This is referred to as the *counterirritant theory.*

These events reflect the fact that collaterals of the A beta afferents form excitatory presynapses with small *enkephalin-releasing interneurons* located in the dorsal horns. These interneurons, in turn, have inhibitory presynaptic terminals that synapse with both the touch and the pain fibers (see Illustration 2). When the touch fibers are active, the interneurons release enkephalins that prevent the transmission of both pain and touch signals to the brain along the spinothalamic tracts. These events help explain why a massage, or even just rubbing a bumped elbow, can lessen the intensity of pain. You could say that artificial pain-reducing techniques—drugs, electrical stimulation, acupuncture—work because the body allows it!

proaches the thalamus, it begins to enter the level of consciousness. Within the thalamus, origins of the sensory input are roughly localized, and their modal-

in the cortex. The result of this processing is an internal, conscious image of the stimulus.

Generally speaking, the result of sensory input is a behavioral response. In higher vertebrates such as human beings, however, a response is not necessarily reflexive or obligatory. When cortical processing produces a conscious image of the stimulus, we can either act or not act on the basis of the image's information and on our conscious prediction of the outcome of each choice. The choice we make depends, of course, on our past experience with similar sensory inputs.

The main aspects of sensory perception are detection, magnitude estimation, spatial discrimination, feature abstraction, quality discrimination, and pattern recognition.

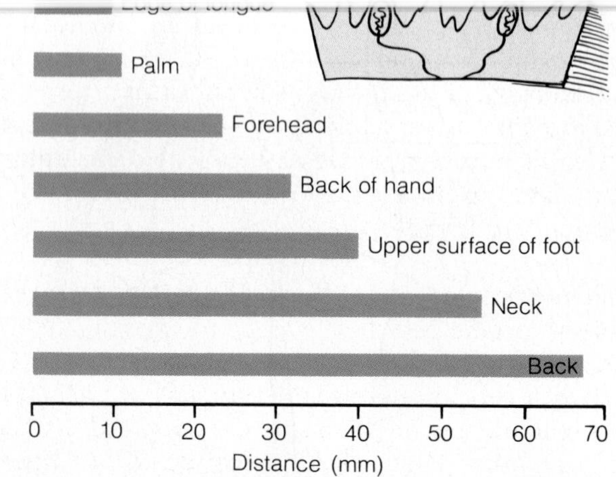

Figure 15.2 Two-point thresholds in the adult. The length of each bar represents the smallest distance between the tips of the caliper for which the two points were perceived as separate.

Perceptual Detection. Detecting that a stimulus has occurred is the simplest level of perception. As a general rule, several receptor impulses must be summated for **perceptual detection** to occur.

Magnitude Estimation. Magnitude estimation is the ability of the brain to detect how *much* of the stimulus is acting on the body. Because of frequency coding, perception increases as stimulus intensity increases.

Spatial Discrimination. Spatial discrimination allows us to identify the site or pattern of stimulation. A common tool for studying this quality in the laboratory is the *two-point discrimination test*. The test determines how close together two points on the skin can be and still be perceived as two points, when stimulated, rather than one. This test provides a crude map of the density of tactile receptors in the various regions of the skin. As illustrated in Figure 15.2, the distance between perceived points varies from less than 1–5 mm on highly sensitive body areas to more than 50 mm on less sensitive areas.

Feature Abstraction. Somatosensory stimulation usually involves an interplay of several stimulus properties. This implies that a unit of perception, or module, is tuned to a coordinated set of several stimulus properties, called a *feature*. The mechanism by which a neuron or circuit is tuned to one feature in preference to others is called **feature abstraction.** For instance, when we run our fingers over a slab of marble, we may become attentive first to its coolness, then to its hardness, then to its smoothness—each a feature that contributes to our perception of the marble surface.

It is important to understand that there are no texture receptors. The skin has only touch, pressure, pain, and temperature receptors. But when their inputs are integrated in parallel fashion, we can appreciate the hardness, coolness, and smoothness of the marble. It is feature abstraction that gives us our ability to identify a substance or object that has specific texture or shape. Velvet is warm, compressible, and smooth but not completely continuous, and we recognize it with one soft touch.

Quality Discrimination. Each sensory modality has several **qualities,** or submodalities; for example, the submodalities of taste are the qualities sweet, salt, bitter, and sour. **Quality discrimination** is the ability to distinguish the submodalities of a particular sensation as distinctly different. It is one of the major achievements of our sensory system.

Quality discrimination may be analytic or synthetic. Some senses allow only analytic discrimination, some allow only synthetic discrimination, and others allow both. In *analytic discrimination*, each quality retains its individual nature. If we mix sugar and salt, analytic discrimination allows us to taste each quality; they do not merge into a new sensation. On the other hand, chocolate is also a mixture of qualities (sweet, bitter, and perhaps some salt), but our perception of it is synthesized from the primary qualities and is distinct from them. *Synthetic discrimination* is also very important in vision. Our color photoreceptors respond chiefly to red, blue, and green wavelengths of light; yet, through synthetic processes, we are able to see yellow, purple, and orange when different numbers of each type of color receptor are stimulated. Both vision and olfaction use only synthetic discrimination.

Pattern Recognition. **Pattern recognition** refers to our ability to take in the scene around us and instantly recognize a familiar pattern, an unfamiliar one, or one that has special significance for us. For example, a figure made just of dots may be recognized as a familiar face, and when we listen to music, we hear the melody, not just a string of notes. Indeed, most of our sensory experiences consist of complex patterns that we tend to perceive as wholes, but how this is accomplished is teasingly out of reach.

Motor Integration: From Intention to Effect

In the motor system, we have effectors (muscle fibers) instead of sensory receptors, descending efferent circuits instead of ascending afferent circuits, and motor behavior instead of perception. However, like sensory systems, the basic mechanisms of motor systems operate at three levels. These three levels form a hierarchy of control of motor activity.

Levels of Motor Control

The idea of multiple and successive levels of motor control, or a **motor hierarchy,** was first proposed by the English neurologist John Jackson in 1873. According to his thesis, the most elementary level of motor control is in the reflex activity of the spinal cord and brain stem, the next is in the cerebellum, and the highest level is in the motor cortex of the cerebrum, with each level acting through lower levels, adding to but never replacing them.

But modern research has revealed that the actual pattern of information flow in the motor control hierarchy is somewhat different from that proposed by Jackson. The cerebral cortex is at the highest level of the conscious motor pathways, but is not the ultimate planner and coordinator of complex motor activities. The cerebellum and basal nuclei play this role and are therefore placed at the top of the motor control hierarchy. Further, the present understanding of motor control exerted by lower levels is that whereas some motor activities are mediated by *reflex arcs* (simple stimulus-motor responses), more complex motor behavior, walking and swimming for example, appears to depend on fixed-action patterns. *Fixed-action patterns* are stereotyped sequential motor actions generated internally or triggered by appropriate environmental stimuli. Once triggered, the entire sequence is released in an all-or-none fashion. Currently, we define three levels of motor control: the segmental level, the projection level, and the programs/instructions level. These levels, their interactions, and the structures they involve are depicted in Figure 15.3.

The Segmental Level

The lowest level of the motor hierarchy, the **segmental level,** consists of the **segmental circuits of the spinal cord**. A segmental circuit activates anterior horn neurons of a single cord segment, causing them to stimulate a specific group of muscle fibers. Those circuits that control locomotion are called **central pattern generators (CPGs).**

The manner in which upright posture and locomotion are controlled by the CNS has fascinated scientists for over a century, and most of our information about the control of motor activity comes from locomotion studies. Early investigators found that isolated portions of nerve cords, freed of all afferent and efferent connections and placed in a physiological solution, generated rhythmic bursts of motor impulses that would excite extensors and flexors rhythmically and in the proper sequence to cause normal patterns of locomotor activity in an intact animal. This was surprising. A nerve cord, sitting in a dish with no sensory input, was continuing to "regulate locomotion." These and later studies revealed that the spinal cord has an intrinsic ability to excite muscles of locomotion in the appropriate pattern and led to the concept of the CPG. The mechanism of the CPG is not yet clear, but it probably involves networks of spinal cord neurons arranged in reverberating circuits. Although all such information has come out of animal studies, it is assumed that human locomotion is controlled in much the same

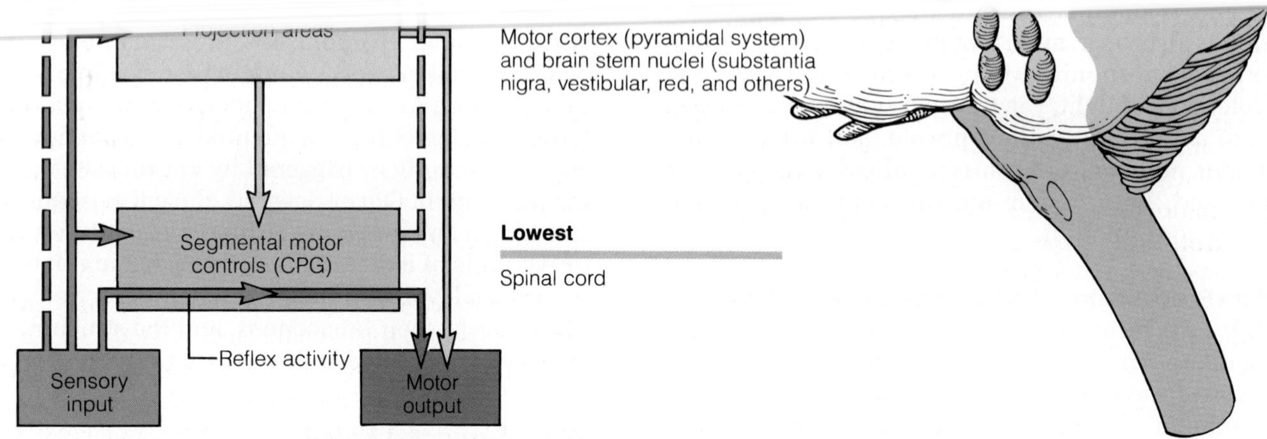

Figure 15.3 Hierarchy of motor control.

way and that the pattern for walking is imprinted in our spinal cords. It is inherited, not learned, and is present even in infants.

What controls a CPG (or other segmental circuit) so that it produces purposeful movements? Clearly, we are not locomoting continuously, and when we are, it may be an amble, a jog, or a flat-out sprint. The most accepted theory to explain how the segmental apparatus is turned on or off is that it is triggered by a "switch" located in the higher neural centers. It appears that this switch is made up of brain stem interneurons called *command neurons* that form part of the projection level.

The Projection Level

The spinal cord is under the immediate control of the **projection level** of motor control, consisting of cortical motor areas (the pyramidal system) and brain stem motor areas (the extrapyramidal system). The axons of these neurons *project* to the spinal cord, where they help control reflex and fixed-action pattern activities and produce discrete voluntary movements. As previously noted, the projection level houses the command neurons that control the segmental apparatus.

Command Neurons. A **command neuron** is an interneuron that can activate a unique fragment of coordinated behavior by exciting or inhibiting a specific group of spinal cord motor neurons. Its function as a higher-order neuron is to start, stop, or modify the basic rhythm of the CPG or other segmental circuits regulating one of the following categories of fixed action: postures, episodic (one-time) movements, repetitive movements, or progressive sequences of movements. In vertebrates, command neurons have been localized in the reticular and red nuclei and in the vestibular nuclear complex of the brain stem. All of these nuclei are part of the extrapyramidal system.

How are command neurons themselves activated? In some cases (for instance, control of breathing rhythm), they are always "on"; their activity is simply accelerated or decelerated to accommodate our different activity levels. In other cases, one motor behavior must be chosen over another. Sensory inputs, though not the trigger for central pattern generation, do act on the command neurons to influence the decision-making process.

The Pyramidal System. The projection-level neurons of the cerebral cortex are the **pyramidal neurons,**

located in the precentral gyri of the frontal lobes (see Chapter 12). These neurons send impulses through the brain stem via the large **pyramidal (corticospinal) tracts** (see Figure 12.28a). Most pyramidal tract fibers synapse in the spinal cord with interneurons or directly with anterior horn motor neurons. Activation of the anterior horn neurons produces voluntary muscle contractions. As the pyramidal tracts descend through the brain stem, they send collaterals to brain stem motor nuclei, basal nuclei, and the cerebellum. Part of the pyramidal tracts called the *corticobulbar tracts* innervate the cranial nerve nuclei of the brain stem.

The Extrapyramidal System. The most important brain stem nuclei of the extrapyramidal arm of the projection level are the reticular, vestibular, and red nuclei and the superior colliculi. Collectively, these nuclei set the main patterns of day-by-day motor behavior. The **reticular nuclei,** which transmit via the descending **reticulospinal tracts,** are particularly important in controlling the segmental (spinal cord) apparatus during standing—that is, in supporting the body against gravity. The **vestibular nuclei** receive inputs from the equilibrium apparatus of the ears and from the cerebellum. Impulses conducted downward along the **vestibulospinal tracts** from the vestibular nuclei help to maintain balance by varying muscle tone of postural muscles. The **red nuclei** send facilitating impulses via the **rubrospinal tracts** to motor neurons controlling flexor muscles, whereas the **superior colliculi** and the **tectospinal tracts** are involved in mediating head movements in response to visual stimuli. It is assumed that the interactions between these brain stem nuclei control the CPGs of the cord during locomotion and other rhythmic activities, such as arm-swinging and scratching.

Notice that the cerebral motor cortex can bypass the segmental apparatus of the spinal cord to directly activate the anterior horn motor neurons (see Figure 15.3). Additionally, there is feedback from every level of the motor system. Projection motor pathways not only convey information to lower motor neurons, but also send a copy of that information (*internal feedback*) back to higher command levels, continually informing them of what is happening. The pyramidal and extrapyramidal systems provide, to some extent, separate and parallel pathways for controlling the spinal cord, but these systems are interrelated at all levels.

The Programs/Instructions Level

Two other vast systems of brain neurons, located in the cerebellum and basal nuclei, are needed to regulate motor activity—to start or stop movements in a precise manner, to coordinate movements with posture, to block unwanted movements, and to monitor muscle tone. These systems, collectively called **precommand areas,** control the outputs of the cortex and brain stem motor centers and stand at the highest level of the motor hierarchy, the **programs/instructions level.** At this level, the sensory and motor systems finally merge, and sensory and motor integration occurs.

The key center for sensorimotor integration and control is the **cerebellum,** which stores the information for motor routines. Remember from Chapter 12 that the cerebellum is the ultimate target of ascending proprioceptor, tactile, equilibrium, and visual inputs. It also receives information from the motor cortex, via collaterals from the descending pyramidal tracts, and from various brain stem nuclei. Since the cerebellum lacks direct connections to the spinal cord, it acts through the projection areas of the brain stem on motor pathways and via the thalamus on the motor cortex. In general, the midline vermis region of the cerebellum exerts control over axial and girdle muscles via the vestibular and reticular nuclei of the brain stem. The intermediate cerebellar regions influence the muscles controlling the distal parts of the limbs by acting through the cerebral cortex and the red nuclei of the brain stem.

The **basal nuclei** are also involved in regulating motor activities initiated by cortical neurons, and like the cerebellum, they are at the crossroads of many afferent and efferent pathways. They receive fibers from both the primary and association motor areas of the cerebral cortex and send their output back to the cerebral cortex via thalamic connections. The basal nuclei influence the flow of motor information in the extrapyramidal pathways primarily by sending impulses to the substantia nigra of the midbrain. The substantia nigra, in turn, influences other brain stem nuclei and the spinal cord via the superior colliculi and the reticular formation. Cells in both the basal nuclei and the cerebellum discharge in advance of willed movements.

When we look at motor controls, we have to rid ourselves of the idea that higher functions are the exclusive province of the cerebral cortex. The cortex is the instrument of volition and the highest level of the brain in both the anatomical and the evolutionary sense, but the cerebellum and the basal nuclei actually stand at the top of the motor control hierarchy. Once the frontal motor association areas of your cerebral cortex have indicated their intent to perform a movement such as drumming your fingers, planning the movement (which may involve thousands of synapses in different parts of your brain) occurs in the precommand areas; the primary motor cortex is quiet during this phase. Most involved in this unconscious planning are the lateralmost portions of the cerebellar hemispheres (and the dentate nuclei) and the caudate and putamen nuclei of the basal nuclei. When you actually move your fingers, both the precommand areas and the primary motor cortex are active. The precommand area is the director of the enterprise; it provides the "orders." The motor cortex executes those orders by sending the activating commands to the appropriate muscle groups.

At the risk of oversimplifying, it appears that the cortex says, "I want to do this," and then lets the precommand areas take over. The basal nuclei appear to function as a liaison between the cerebral cortex and the cerebellum in the planning and initiation of motor

work has already been laid.

Homeostatic Imbalances of Motor Integration

The cerebellum is an amazing brain region. Although it contains complete motor and sensory maps, its injury produces neither muscle weakness nor disorders of perception. Cerebellar disorders affect the same side of the body and fall into three major groups, categorized by the precise location of the lesion: disorders of synergy and muscle tone, disturbances of equilibrium, and speech disorders.

Synergy is the coordination of agonist and antagonist muscles to produce a smooth, well-controlled movement. Normally, the cerebellum takes the credit for synergy, but when synergy is disrupted, *ataxia* (ah-tak'-sē-uh) results and cannot be compensated by conscious effort. Ataxia sufferers exhibit slow, tentative, inaccurate movements. They are unable to touch their finger to their nose with their eyes closed—a feat that normal individuals accomplish easily. Typically, they have a wide stance and an unsteady "drunken sailor" gait, which predisposes them to falling.

When muscles have a healthy tone, manipulated limbs offer some resistance to movement. But when the cerebellum is impaired, muscle tone declines. This leads to *lack of check*, or overshoot (a rapidly moving limb is unable to stop quickly and sharply), and to tremors at the beginning and end of movements.

Cerebellar damage may also produce equilibrium problems and speech difficulties. The so-called *scanning speech* that results is slurred and somewhat slow and singsong in nature.

Diseases involving the basal nuclei reflect some problem in the controls of the basal nuclei, which are either released from inhibitory influences or more strongly stimulated. Characteristic symptoms, collectively called *dyskinesia* (dis'-kih-nē'-zhuh), include disorders of muscle tone and posture, as well as involuntary movements. The abnormal movements range from *tremor*, to slow, writhing movements of the fingers and hands, to violent, flailing movements of the limbs. Although there are many basal nuclear diseases, perhaps the most familiar are Parkinson's disease and Huntington's disease.

Largely a disorder of the elderly, Parkinson's disease results from inadequate dopamine synthesis by the substantia nigra, a brain stem nucleus that projects

ease. Afflicted individuals have a rhythmic tremor at rest (exhibited by head nodding and a "pill-rolling" movement of their fingers), a forward-bent walking posture, lack of facial expression, and they are slow in initiating (but not in executing) movement. They begin walking with a shuffling gait, but once started, they pass to a running pace that is difficult to stop; this phenomenon is called "cogwheel walking." The drug *L-dopa*, a precursor of dopamine, is often helpful in alleviating some of these symptoms temporarily. Much more promising for long-term results are the recent intrabrain transplants of adrenal medulla tissue (which produces dopamine as well as epinephrine and norepinephrine) and substantia nigra tissue obtained from fetuses. (However, the use of fetal tissue is highly controversial and riddled with ethical and legal roadblocks.)

Huntington's disease is a hereditary disorder that leads to degeneration of the basal nuclei and later of the cerebral cortex. Its initial symptoms—abrupt, jerky, and almost continuous movements called *chorea* (Greek for "dance")—begin to appear in early adulthood. Although the movements appear to be voluntary, they are not. These manifestations are essentially the opposite of the motor impairments of Parkinson's disease, and Huntington's disease is usually treated with drugs that block, rather than enhance, dopamine's effects. Huntington's disease is progressive and usually fatal within 15 years of onset of symptoms. Late in the disease, marked mental deterioration (dementia) occurs. ■

Higher Mental Functions

During the last two decades, the exploration of outer space has captured the public's attention like no other scientific endeavor. Yet, during these same years, an equally exciting exploration of our "inner space" has

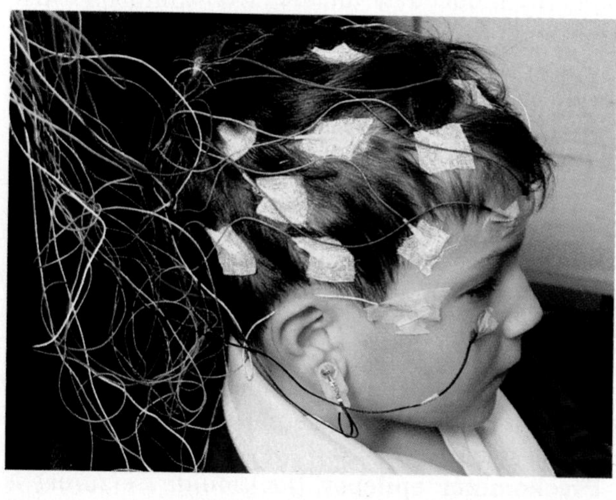

(a)

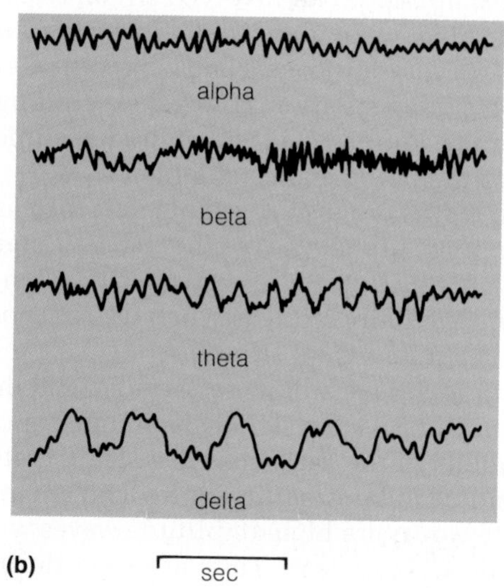

(b)

Figure 15.4 Electroencephalography and brain waves. (**a**) To obtain a recording of brain wave activity (an EEG), electrodes are positioned on the patient's scalp and attached to a recording device called an electroencephalograph. (**b**) Typical EEGs. Alpha waves are typical of the awake but relaxed state; beta waves occur in the awake, alert state; theta waves are common in children but not in normal adults; and delta waves occur during deep sleep.

nearly escaped public notice. Psychology and biology have begun to converge in the investigation of the higher mental functions, or what we commonly call the mind, including consciousness, memory, reasoning, and language. The value of these investigations is both theoretical (to determine biological mechanisms of higher mental functioning) and practical (to identify drugs that might cure or ease certain types of mental illness). But workers in the field of cognition are still struggling to understand how the mind's presently incomprehensible qualities might spring from living tissue and electrical impulses. Souls and synapses are hard to reconcile!

In Chapter 11, we described the simple neuronal pools that provide the basis for nervous system activity, and in this chapter, we have so far considered the more complex mechanisms of sensorimotor integration. Now we will take another giant step, to the awe-inspiring highest levels of mental functioning. Since brain waves reflect the electrical activity on which such functioning is based, we will consider them first, along with the related topic of sleep. We will then examine, very briefly, the topics of consciousness, memory, and language. However, be aware that as we ascend the ladder of complexity from a single neuron to the vast constellations of neurons involved in higher mental functions, we move into the realm of uncertainty and are forced to hypothesize or speculate on what must (or should) happen.

Brain Wave Patterns and the EEG

Normal brain function involves continuous electrical activity of the neurons. An **electroencephalogram** (ih-lek′-trō-en-seh′-fuh-lō-gram), or **EEG,** is a record of some aspects of this activity, although it is not known how much each kind of neuronal activity (action potentials, synaptic potentials, etc.) contributes to it. An EEG can be made by placing electrodes at various locations on the scalp and then connecting the electrodes to an apparatus that measures electrical potential differences between various cortical areas (Figure 15.4a). The resulting patterns of neuronal electrical activity are called **brain waves.**

Because people differ genetically, and because everything we have ever experienced has left its imprint in our brain, each of us has a brain wave pattern that is as unique as our fingerprints. However, for simplicity, we can describe these complex waves on the basis of the four frequency classes illustrated in Figure 15.4b.

Alpha waves are low-amplitude, slow, synchronous waves with an average frequency of 8–13 Hz (hertz, or cycles per second). In most cases, they indicate a brain that is "idling"—a calm, relaxed state of wakefulness.

In light of the fact that alpha waves are typically recorded when a person is sitting quietly with eyes closed, the observation that alpha waves are recorded in the left half of an "elite" sprinter's brain just prior to a race is particularly interesting. Alpha waves also appear to be involved in the euphoria, or "runner's high," experienced both during and after running by

and in adults in the early stages of sleep, theta waves are considered abnormal in adults who are awake.

Delta waves are high-amplitude waves with a frequency of 4 Hz or less. They are seen during deep sleep and when the reticular activating system is damped, such as during anesthesia.

Brain waves have a normal frequency range of 1–30 Hz, a dominant rhythm of 10 Hz, and an average amplitude of 20–100 μV. The amplitude reflects the number of neurons firing together in synchrony, not the degree of electrical activity of individual neurons. When the brain is active and neurons are involved in many different activities, complex, low-amplitude brain waves are seen. But when the brain is inactive, as during sleep, large numbers of neurons fire synchronously, producing similar, high-amplitude brain waves.

Brain waves change with age, sensory stimuli, brain disease, and the chemical state of the body. EEG recordings have long been used for diagnosis and localization of many types of brain lesions, such as tumors, infarcts, infections, abscesses, and epileptic lesions. Interference with cerebral cortical functions is suggested by brain waves that are too fast or too slow, and unconsciousness occurs at both extremes. Brain-depressing drugs and coma result in brain wave patterns that are abnormally slow, whereas fright, various types of drug intoxication, and epilepsy are associated with excessively fast brain waves. Since spontaneous brain waves are always present, even during unconsciousness and coma, their absence—called a "flat EEG"—is clinical evidence of brain death.

Abnormal Electrical Activity of the Brain: Epilepsy

Almost without warning, the human brain can fail. The victim may stare into space or lose consciousness and fall stiffly to the ground, the body wracked by uncontrollable jerking. These seizures, called **epileptic seizures,** reflect abnormal electrical discharges of groups of brain neurons, and while their uncontrolled activity is occurring, no other messages can get through. Epilepsy is not associated with, nor does it cause, intellectual impairments. Some cases of epilepsy are induced by genetic factors, but it can also result from brain injuries caused by

sciousness do not. The hypothalamus and thalamus exhibit depressed brain wave patterns (around 3 Hz) during the seizure.

Psychomotor epilepsy (*Jacksonian seizures*) is accompanied by extremely rapid temporal lobe brain waves. The victim becomes disoriented, loses contact with reality, and exhibits uncontrolled motor activity of isolated muscle groups, such as hand clapping or lip smacking; sometimes aimless wandering occurs. The strange manifestations of psychomotor epilepsy have caused many such cases to be misdiagnosed as mental illness.

Grand mal, the most severe form of epilepsy, is accompanied by rapid brain waves of 30 Hz or more. The person loses consciousness and displays intense convulsions. Bones are often broken during such seizures, showing the incredible strength of the muscle contractions that occur. Loss of bowel and bladder control and severe biting of the tongue are common. The seizure lasts for a few minutes and then the muscles relax and the person awakens, but the victim remains disoriented for several minutes thereafter. Many grand mal sufferers experience a sensory hallucination, such as a taste, smell, or flashes of light, just before the seizure begins. This phenomenon, called an *aura*, is helpful because it gives the person time to lie down and avoid falling heavily to the floor. Grand mal can usually be controlled by anticonvulsive drugs, which act in different ways to limit seizure activity. ■

Sleep and Sleep-awake Cycles

Although most people spend about a third of their lives asleep, we know little about sleep's biological basis. But we do know that the alternating cycles of sleep and wakefulness involve the whole brain and that they reflect a natural *circadian*, or 24-hour, *rhythm.* Sleep is defined as a state of changed consciousness, or partial unconsciousness, from which a person can be aroused by stimulation. This distinguishes sleep from coma, a state of unconsciousness from which a person cannot be aroused by even the most vigorous stimuli. Although cortical activity is depressed during sleep, brain stem functions, such as control of respi-

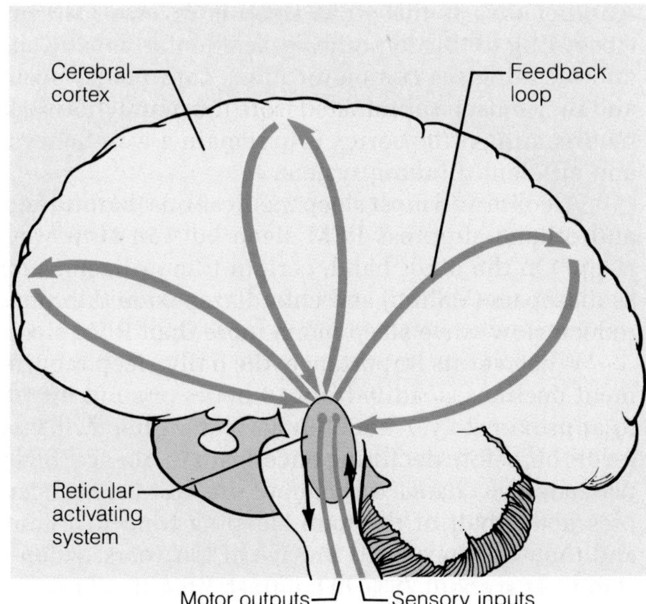

Figure 15.5 Feedback loop between the reticular activating system (RAS) and the cerebral cortex. Inputs ascending from the RAS maintain the alertness of the cerebral cortex. The cortex, in turn, feeds back to influence the activity of the reticular system.

ration, heart rate, and blood pressure, continue. Even environmental monitoring continues to some extent, as illustrated by the fact that strong stimuli ("things that go bump in the night") immediately arouse us. In fact, people who sleepwalk can avoid objects and navigate stairs perfectly well while truly asleep.

In the awake state, alertness of the cerebral cortex (the "conscious brain") is mediated by the reticular activating system, or RAS (Figure 15.5). When the activity of the RAS declines, cerebral cortical activity declines as well; thus, lesions of the RAS nuclei result in unconsciousness. However, sleep is much more than simply turning off the arousal (RAS) mechanism. It is an actively induced process in which the brain is guided into sleep. RAS centers are involved not only in maintaining the awake state, but also in mediating some stages of sleep, especially dreaming sleep. The hypothalamus is probably responsible for the actual timing of the sleep cycle.

Stages of Sleep

The stages of sleep are defined in terms of their EEG patterns. The two major types of sleep, which alternate through most of the sleep cycle, are **nonrapid eye movement (NREM) sleep** and **rapid eye movement (REM) sleep.** During the first 30–45 minutes of the sleep cycle, we pass through four stages of sleep, culminating in **slow-wave sleep.** As we pass through these stages and slip into deeper and deeper sleep, the frequency of the EEG pattern declines, but the wave amplitude increases. REM sleep, recognized by an EEG pattern more typical of the awake state, does not appear until after slow-wave sleep has been achieved.

NREM Sleep. All stages of sleep other than REM are collectively referred to as nonrapid eye movement (NREM) sleep. The four stages of NREM sleep are as follows:

Stage 1. Our eyes are closed and we begin to relax. Thoughts flit in and out, and we have a drifting sensation. Vital signs (body temperature, respiration, pulse, and blood pressure) are normal, and the EEG pattern shows alpha waves. If stimulated, we immediately become aroused and usually deny being asleep.

Stage 2. The EEG begins to show a more irregular pattern. *Sleep spindles*—sudden, short (1–2 second) wave bursts—appear, and arousal becomes more difficult.

Stage 3. Sleep deepens, and theta and delta waves appear. Vital signs have begun to decline, and the skeletal muscles are very relaxed. Dreaming is common. Usually, stage 3 is reached some 20 minutes after the onset of stage 1.

Stage 4. We are now in slow-wave sleep, and the EEG pattern is dominated by delta waves. All vital signs reach their lowest normal levels, and gastrointestinal motility increases. The skeletal muscles are relaxed, but normal sleepers turn approximately every 20 minutes. Arousal is difficult, and if we are aroused, we are usually confused. Bedwetting and sleepwalking occur during this phase.

REM Sleep. About 90 minutes after sleep begins, an abrupt change occurs in the EEG pattern. It becomes very irregular and appears to backtrack quickly through the stages until alpha waves, characteristic of stage 1 and indicating the onset of REM, appear. This brain wave change is coupled with increases in body temperature, heart rate, respiratory rate, and blood pressure and a decrease in gastrointestinal motility. REM sleep is also called **paradoxical sleep** because its EEG pattern is more typical of the awake state. Oxygen use by the brain is tremendous during REM—greater than during the awake state.

Although the eyes move rapidly under the lids during REM, most of the body's muscles are temporarily paralyzed (actively inhibited), which prevents us from acting out our dreams. Most dreaming occurs during REM sleep, and some suggest that the eye movements are associated with following the visual imagery of our dreams. (Note, however, that most nightmares and night terrors occur during stages 3 and 4 of NREM sleep.) In adolescent and adult males, REM episodes are frequently associated with erections of the penis. The threshold for arousal is highest during REM sleep; on the other hand, the sleeper is more

[illegible]

more and more sleeping time, and we tend to awaken more frequently in the early morning hours. When the neurons of the dorsal raphe nuclei of the midbrain reticular formation are firing at maximal rates, we awaken for the day. According to some researchers, this event is rigidly tied to a rise in core body temperature.

In addition to changes in brain wave patterns during sleep, there are changes in levels of neurotransmitters in certain brain regions. Norepinephrine levels decline and serotonin levels rise during deep sleep. This is not surprising, because norepinephrine is involved in maintaining an alert state of consciousness, and serotonin (released by reticular neurons of the raphe nuclei of the medulla) has long been implicated as the "sleep neurotransmitter," important in slow-wave sleep. During REM, norepinephrine and acetylcholine levels rise in nuclei that are active during such periods. For example, the activity of ACh-secreting cells of the giant cell nucleus, part of the reticular formation in the pons, is believed to cause the arousal pattern of REM sleep, and norepinephrine released by the locus ceruleus of the pons is thought to induce the transient paralysis of REM sleep.

Importance of Sleep

Although we do not really understand the significance of sleep, slow-wave and REM sleep seem to be important to health in different ways. Shakespeare called sleep "nature's soft nurse," referring to its restorative effects on body functioning. Slow-wave sleep is presumed to be the restorative stage—the time when most neural mechanisms can wind down to basal levels—and when we are deprived of sleep, we spend more time than usual in slow-wave sleep during the next sleep episode.

Volunteers persistently deprived of REM sleep become emotionally unstable and exhibit various personality disorders. It may be that REM sleep gives the brain an opportunity to analyze the day's events and to work through emotional problems in dream imagery. Sigmund Freud saw dreams as symbolic expressions of frustrated desires and attempts to fulfill wishes. Another idea is that REM sleep is reverse learning. According to this hypothesis, accidental, repetitious, and meaningless communications

[illegible]

hours in early adulthood. It then levels off before declining once again in old age. Sleep patterns also change throughout life. REM sleep occupies about half of the total sleeping time in infants and then declines until the age of ten years, when it stabilizes at about 25%. In contrast, time spent in stage 4 sleep declines steadily from birth, and stage 4 sleep often disappears completely in those over 60 years of age. The elderly, forced into a perpetual state of light sleep, awaken more frequently during the night; ultimately, most add an afternoon nap to try to make up the difference.

Homeostatic Imbalances of Sleep

Two clinically important disorders of sleep are narcolepsy and insomnia. *Narcolepsy* is a sleep disorder in which people lapse involuntarily into sleep during waking hours. These sleep episodes last about 15 minutes and can occur without warning at any time. The sleep attack often seems to be triggered by a pleasurable event—a good joke, a game of poker, a sports event. This condition can be extremely hazardous when the person must drive, operate machinery, or take a bath. Narcoleptics seem to have an inability to control the circuits responsible for REM; they begin sleep with REM, and in all daytime narcoleptic episodes they show the typical EEG of REM sleep. However, during the evening sleep cycle, they spend much less time in REM than those with normal sleep patterns.

Insomnia is a chronic inability to obtain the *amount* or *quality* of sleep needed to function adequately during the daytime. Since the sleep requirement varies from 4 to 9 hours a day among normal people, there is no way to determine the "right" amount. Self-professed insomniacs tend to exaggerate the extent of their sleeplessness, and they have a notorious tendency toward self-medication and barbiturate abuse.

True insomnia often reflects normal age-related changes. In frequent travelers, it may be caused by *jet lag*. But perhaps the most common cause of insomnia is psychological disturbance. We have difficulty falling asleep when we are anxious or upset, and depression is often accompanied by early awakening ■

Consciousness

Clinically, **consciousness** is defined on a continuum that grades levels of behavior in response to stimuli as (1) *alertness*, (2) *drowsiness* or *lethargy* (which proceeds to sleep), (3) *stupor*, and (4) *coma*. Alertness is the highest state of consciousness and cortical activity, whereas coma is the most depressed state. But when we get away from the clinical definition, consciousness is very difficult to define. There is little argument that a sleeping person lacks something that he or she has when awake, and we call this "something" consciousness. It is equally obvious that human consciousness, with its rich tapestry of perceptions and concepts, is much more than the opposite of sleep. It is the complexity of human consciousness that sets us apart from all other animals.

Scientists have been thinking about thinking for some time. But human consciousness is still a mystery, and most of what you are likely to read about it is speculation—a synthesis provided in an attempt to explain the impressive capabilities of the conscious brain. For example, cognitive scientists have proposed that consciousness is a manifestation of *holistic information processing*. Holistic processing encompasses many lines of reasoning, but its major suppositions are as follows:

1. *Consciousness involves simultaneous activity of large areas of the cerebral cortex*. Localized damage to any specific region of the cerebral cortex does *not* destroy consciousness.

2. *Consciousness is superimposed on other types of neural activity*. At any time, specific neurons and neuronal pools are involved both in specific localized activities (such as motor control or sensory perception) and in mediating conscious cognitive behaviors.

3. *Consciousness is totally interconnected*. Information for "thought" can be claimed from many locations in the cerebrum simultaneously. For example, retrieval of a specific memory can be triggered by one of several routes—a smell, a place, an association with a particular person, and so on. The cortical cross-file is enormous.

With the exception of sleep, unconsciousness is always a signal of some impairment of brain function. A transient loss of consciousness is called **fainting** or **syncope** (sing′-kō-pē). Most often, it reflects inadequate cerebral blood flow resulting from low blood pressure, as might follow hemorrhage or sudden emotional stress. Fainting is usually preceded by a feeling of lightheadedness as the blood pressure falls to critically low levels.

When a person is totally unresponsive to sensory stimuli for an extended period—eyes closed and no recognizable speech—the condition is called **coma.** Coma is not deep sleep. During sleep, the brain is active and oxygen consumption resembles that of the waking state. In contrast, oxygen use is always below resting levels in coma patients.

Blows to the head may induce coma by causing widespread cerebral trauma, due to the damage done by the blow itself or resulting from hemorrhage or edema, or by causing similar trauma to the brain stem structures, particularly the reticular formation. Coma may also be produced by tumors or infections that invade brain stem structures. Metabolic disturbances such as hypoglycemia (abnormally low blood sugar levels), drug overdose (from opiates, barbiturates, aspirin, or alcohol), or liver and/or kidney failure depress and interfere with brain function over a wide area and can result in coma. Cerebral infarctions, unless massive and accompanied by extreme swelling of the brain, do not often cause coma, since substantial amounts of cerebral tissue may be damaged without impairing consciousness.

When the brain has suffered irreparable damage, irreversible coma occurs, even though life-support measures may have restored vitality to other body organs. The result is a dead brain (*brain death*) in an otherwise living body. Physicians must determine whether a patient in an irreversible coma is legally alive or dead, because life support can be removed only after death. ■

Memory

Memory is the storage and retrieval of previous experience, or, more simply, the ability to recall our thoughts. Memories are essential for learning and incorporating our experiences into behavior and are part and parcel of our consciousness. Stored somewhere in your 3 pounds of wrinkled brain are ZIP codes, the face of your grandfather, and the taste of yesterday's pizza. In essence, your memory bank reflects your lifetime.

Most of what is known about memory and learning can be summarized in three principles: (1) Memory storage occurs in stages and is continually changing; (2) the hippocampus and surrounding structures play unique roles in memory processing; and (3) memory traces—chemical or structural changes encoding memory—are widely distributed in the brain.

Stages of Memory

Memory storage, or fixation, involves at least two distinct stages: short-term memory and long-term memory (Figure 15.6). **Short-term memory (STM)** is a fleeting memory of the events that continually parade before you. Lasting for seconds to a few hours, it is the preliminary step to long-term memory, as well as the power that lets you look up a telephone number, dial it, and then never think of it again. The capacity

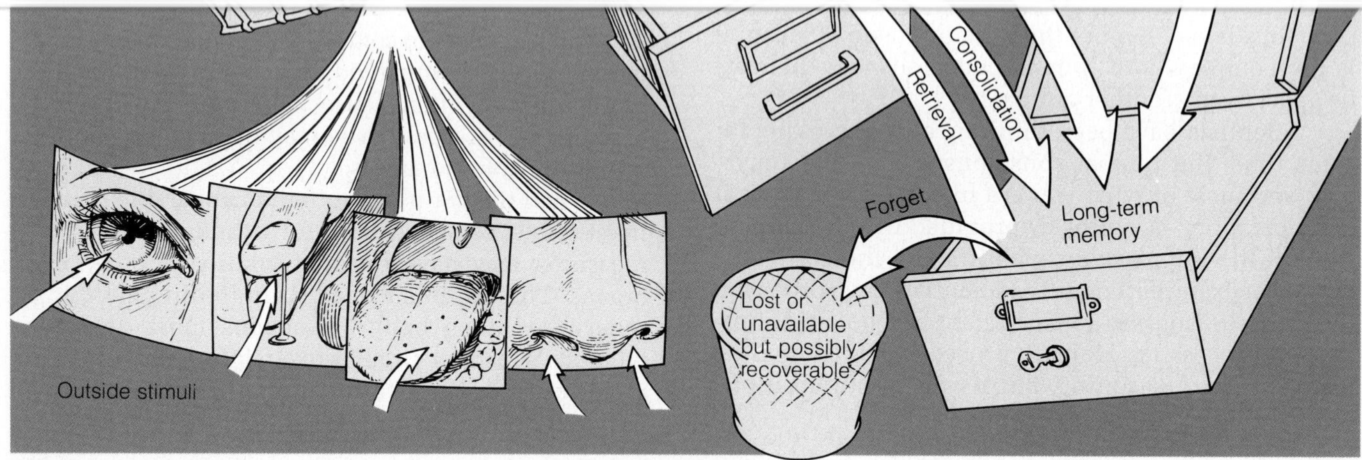

Figure 15.6 Memory processing. Outside stimuli act on the mind via the sensory receptors. This information is processed by the cerebral cortex (indicated in the diagram as the temporary storage or buffer area), which selects what is to be noticed and what is to be sent to short-term memory. Repetition (rehearsal) helps the transfer of information from short-term memory to long-term memory. For long-term memory to become permanent, the new facts must be linked with information already in memory; this process is called consolidation. A certain amount of information not attended to goes directly into long-term memory (automatic memory).

of STM appears to be limited to seven or eight chunks of information, such as the digits of a telephone number or the sequence of words in an elaborate sentence.

In contrast to STM, **long-term memory (LTM)** seems to have a limitless capacity. Although our short-term memory devices cannot recall numbers much longer than a telephone number, we can remember scores of telephone numbers by committing them to LTM. However, our ability to store and to retrieve information declines with aging. Thus, long-term memories too can be forgotten, and our memory bank continually changes with time.

We do not remember or even consciously notice much of what is going on around us. As sensory inputs flood into our cerebral cortex, they are processed, and some of this information is selected for transfer to STM (Figure 15.6). STM serves as a sort of temporary holding bin for data that we may or may not want to retain. The transfer of information from STM to LTM is affected by many factors, including:

1. *Emotional state.* We learn best when we are alert, motivated, and/or aroused. For example, when we witness shocking events, transferal is almost immediate. Norepinephrine is known to be involved in memory processing, and more norepinephrine is released during excited or stressful states, which may help to explain this phenomenon.

2. *Rehearsal.* An opportunity to repeat or rehearse the material enhances memory.

3. *Association of new information with information already stored in LTM.* A staunch football fan can tell you who did what in all of the important plays of a particular football game, which someone unfamiliar with the sport would find difficult to understand and therefore to remember.

4. *Automatic memory.* Not all impressions that become part of LTM are consciously formed. For example, when concentrating on a lecturer's speech, automatic memory of the pattern of his tie may also be recorded.

Memory transfer to LTM takes time to become permanent. The process of *memory consolidation* apparently involves the fitting of new facts into the various categories of preexisting knowledge stored in the cerebral cortex.

Categories of Memory

It appears that the brain makes a fundamental distinction between factual knowledge and skills, and that we process and store these different kinds of information in different ways. **Fact memory** is the ability to learn explicit information, such as names, faces, words, and dates. It is related to our conscious thoughts and our ability to manipulate symbols and language. Fact memories are learned and are often forgotten quickly, but when committed to LTM, they are usually filed along with the context in which they were learned. For instance, when you think of your new friend Joe, you probably picture him at the basketball game where you met him.

Skill memory is concerned with less conscious learning. It usually involves motor activities and is acquired only through practice, as when we learn to ride a bike or play a bass fiddle. Skill memories do not preserve the circumstances of learning; in fact, they are best remembered in the doing. A batter does not consciously recall the moves needed to properly time and swing the bat, nor do you have to think through how to tie your shoes. Once learned, skill memories are hard to unlearn.

Brain Structures Involved in Memory

Much of what is known about how we learn or remember has come from two sources: experimental work done on macaque monkeys and studies of amnesia in humans. Such studies have revealed that different brain structures are involved in the two categories of memory (Figure 15.7).

If the human brain uses holistic processing, then fact memory processing must be holistic as well. Memories are not stored just in the temporal lobes, as was once thought. Instead, specific pieces of each memory must be stored near regions of the brain that need them so that the new inputs can be quickly associated with the old. Accordingly, visual memories would be stored in the occipital cortex, memories of music in the temporal cortex, and so on. Thus, your memory of a dear aunt—the scent of her cologne, the softness of her hands—is found in bits and pieces scattered all over your cortex. But how are the memory connections made?

Extensive research has indicated that the crucial structures for incorporating and storing sensory perceptions into fact memory are the *hippocampus* and the *amygdala* (both part of the limbic system), the *diencephalon* (thalamus and hypothalamus), the *ventromedial prefrontal cortex* (a cortical region tucked beneath the front of the brain), and the *basal forebrain* (a cluster of acetylcholine-secreting neurons that lies anterior to the hypothalamus) (see Figure 15.7a). The proposed scheme of information flow is as follows. When a sensory perception is formed in the cerebral sensory cortex, the cortical neurons dispatch impulses along two parallel circuits to the hippocampus and the amygdala, both of which encompass connections with the diencephalon, basal forebrain, and prefrontal cortex. The basal forebrain, via its many connections with the sensory cortex, then closes the memory loop by sending impulses back to the sensory cortical areas, initially forming the perception. This feedback presumably causes changes that transform the perception into a more durable memory of the sensory event that has just occurred. This scheme suggests that the subcortical structures make the initial connections between previous memories stored in the sensory association areas and the new perception by connecting the cortical regions where the LTM is stored in a kind of "conference call" until consolidation of the new information can occur. Later recall of the new memory occurs when the same cortical neurons are stimulated.

The hippocampus seems to oversee circuitry that is especially important for learning and remembering spatial relationships, while the amygdala seems to be the crossroads or gatekeeper of the memory system. The amygdala has widespread connections with all cortical sensory areas, as well as with the thalamus and the emotional response centers of the hypothalamus. It is speculated that the amygdala bears the responsibility for associating memories formed through different senses and linking them to emotional states generated in the hypothalamus.

Damage to either the hippocampus or amygdala results in only slight memory loss, but bilateral destruction of both structures causes global (widespread) amnesia. Consolidated memories are not lost, but new sensory inputs cannot be associated with old, and the person lives in the here and now from that point on. This phenomenon is called *anterograde* (an′-ter-ō-grād) *amnesia*, in contrast to *retrograde amnesia*, which is the loss of memories formed in the distant past. You could carry on an animated conversation with a person with anterograde amnesia, excuse yourself, return 5 minutes later, and that person would not remember you. The same situation occurs when the connections of the hippocampus or amygdala to the diencephalon or prefrontal cortex are cut. Thus, it appears that none of the structures named makes a totally independent contribution to memory; the integrity of the entire circuit is essential. ■

Individuals suffering from anterograde amnesia can still learn skills. Consequently, a second learning circuit, independent of the pathways used in fact memory, has been suggested (see Figure 15.7b). It

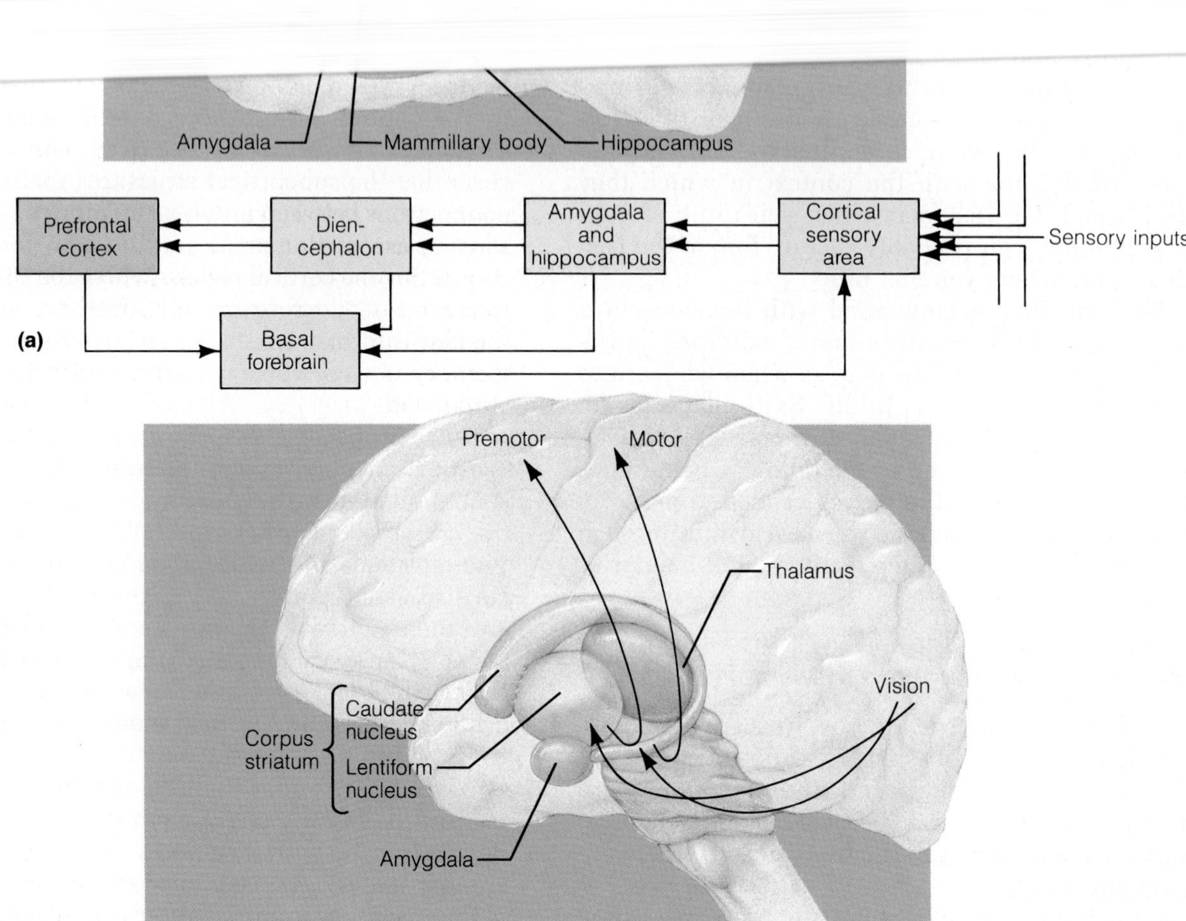

Figure 15.7 Proposed memory circuits. (**a**) The essential structures of the fact memory circuits include the hippocampus and amygdala (limbic system structures), the diencephalon, the basal forebrain, and the prefrontal cortex. The flowchart below the illustration indicates how these structures may interact in memory formation. Sensory inputs from the cortex flow through parallel circuits, one involving the hippocampus and the other involving the amygdala. Both circuits encompass parts of the diencephalon, basal forebrain, and prefrontal cortex. The basal forebrain feeds back to the sensory cortex, closing the memory loop. (**b**) Major structures involved in skills memory. The corpus striatum appears to mediate the automatic connections between a stimulus and a motor response. (A visual stimulus is used in this example.)

appears that the cerebral cortex, activated by sensory inputs, signals the *corpus striatum* (lentiform and caudate nuclei of the basal nuclei) of its intent to mobilize skill memory. The corpus striatum, which is abundantly connected to extrapyramidal motor nuclei of the brain stem, then communicates with one or more of those nuclei as well as with the motor regions of the cerebral cortex (via thalamic projections) to promote the desired movement. Thus, the corpus striatum forms the link between a perceived stimulus and a motor response. Because conditioning of voluntary muscle contractions is involved in many skills, it is assumed that the cerebellum is also involved in the skill memory pathways. Some authorities prefer to call these pathways the *habit system*, rather than memory in the true sense.

Mechanisms of Memory

In the early 1900s, Karl Lashley, a pioneer neuropsychologist, set out to find the *engram*, the hypothetical unit of memory or permanent memory trace. Now, near the end of the century, the mysterious engram still eludes us. Since STM is so fleeting, it is not believed to involve permanent changes in neural circuits. Proposed mechanisms of STM include (1) reverberating circuits that maintain the thought for a limited time and (2) facilitation mediated by intracellular regulatory molecules, such as cyclic AMP, which cause enhanced neurotransmitter release.

If an engram exists, it will probably be found in long-term memory depots. Among the many suggestions arising from experimental studies are that during learning, (1) neuronal RNA content is altered, (2) dendritic spines change in shape, (3) unique extracellular proteins are deposited at synapses involved in LTM, and (4) inhibitory controls of certain neurons are removed. However, human memory is notoriously difficult to study, and animals, largely limited to skill memories, do not provide a good substitute for studying how the human brain reconstructs fact memories, which may involve many different cortical areas. To date, little or no progress has been made in either localizing or determining the specific biochemistry of memory traces in humans.

Language

Most distinctly human brain functions involve **language,** either directly or indirectly. Conscious thought, memory, and language are inextricably interwoven; one function cannot exist without the others. These capabilities are a direct consequence of (1) the elaboration of the cerebral cortex, which provides the machinery for manipulating symbolic information, and (2) cerebral dominance, which, as far as we know, occurs only in humans.

Spoken language is remarkably complex. To utter a thought, we must choose words that fit the intended meaning, dredge up the grammar to organize the words, and activate the muscles of articulation (larynx, tongue, lips, etc.). All of this occurs in a split second, so it is unlikely that the process is entirely consciously controlled. On the other hand, what people say is usually so original (in form, if not in content) that purely reflexive activity cannot explain it either. Human speech has been extremely difficult to investigate. However, the major brain areas involved in speech—Wernicke's and Broca's areas—have been identified, and their roles are partially understood.

Wernicke's area, the perceptual speech area on the upper border of one temporal lobe, is not a "word storage box" per se, but an area where rules of grammar are used to form meaningful messages. Some suggest that stored words are linked to associated words in the cortex and that when a particular word is selected, a spreading wave of activation readies associated areas for utterance. For example, the number 2 might be associated with 1, 3, bicycle, biped, numeral, bilateral, twins, and countless other symbols. Wernicke's area is also the region that allows us to determine the meaning of new or unfamiliar words from natural contexts and thus to acquire new information. Wernicke's area is connected by conduction pathways to Broca's (motor speech) area anteriorly in the same hemisphere. Recall that Broca's area is where the complex patterns for controlling the muscles of articulation are stored.

Most studies of language have come from studies of aphasic people with derangements in their motor or perceptual speech areas. *Aphasia* (ah-fā′-zhuh) means lacking speech. People with *motor aphasia*, resulting from lesions of Broca's area, speak very little or not at all because they cannot activate the muscles that produce speech, even though they may fully comprehend spoken or written language. Motor aphasia is common in stroke victims exhibiting right-sided paralysis. Like nonverbal motor skills, speech patterns can be reprogrammed. But it is a difficult task, because we must begin with "mama" all over again.

Lesions in Wernicke's area, also prevalent in stroke victims, result in *sensory aphasia,* in which a person can speak but cannot understand spoken or written language. Speech is effortless, yet it conveys little or no information and involves incorrect word usage and deranged grammar. Since these speakers are usually unaware of their errors (because of diminished comprehension of their own speech), they are unable to correct them. Damage to the conduction pathway between Wernicke's and Broca's areas disrupts communication between them, resulting in *conduction aphasia*. People with this deficit omit parts of words and use words incorrectly. They cannot repeat simple phrases, even though they understand what they hear and read and their motor speech area itself is unimpaired. It is a maddening condition.

* * *

In this chapter, we have brought together most of the neural machinery studied in earlier chapters and have examined how sensory inputs are integrated with motor activity in the central nervous system. We have also taken a look at the most complex aspects of neural integration, which allow us to think, remember, and converse with others. Now that the nervous system has been viewed as an integrated whole, we are ready to consider the special senses in Chapter 16.

Related Clinical Terms

Analgesia (a″-nul-jē′-zhuh) (an

...involving muscles of the limbs; myoclonal jerks occurring in normal individuals as they are falling asleep are thought to reflect a fleeting reactivation of the RAS; may also be caused by diseases of the reticular formation and/or cerebellum.

Phantom pain A phenomenon in which amputees have chronic pain in their nonexistent (amputated) limb; although not well understood, apparently the cut sensory nerves that previously transmitted inputs from the limb continue to respond to the trauma of amputation, and the brain interprets (projects) these painful stimuli as coming from the missing limb.

Chapter Summary

SENSORY INTEGRATION: FROM RECEPTION TO PERCEPTION (pp. 466–471)

1. Sensation is awareness of internal and external stimuli; perception is conscious interpretation of those stimuli.

General Organization of the Somatosensory System (pp. 466–471)

2. The three levels of sensory integration are the receptor, circuit, and perceptual levels. These levels are functions of the sensory receptors, the ascending pathways, and the cerebral cortex, respectively.

3. Sensory receptors transduce (convert) stimulus energy into action potentials. Stimulus strength is frequency coded.

4. Some sensory fibers entering the spinal cord act in local reflex arcs. Some synapse with the dorsal horn neurons (nonspecific ascending pathways), and others continue upward to synapse in medullary nuclei (specific ascending pathways). Second-order neurons of both specific and nonspecific ascending pathways terminate in the thalamus.

5. The specific sensory pathway consists of the fasciculus cuneatus, the fasciculus gracilis, and the medial lemniscal tracts, which are concerned with straight-through, precise transmission of one or a few related sensory modalities. The nonspecific pathway, consisting of the spinothalamic tracts, is a multimodal pathway that permits brain stem processing of ascending impulse.

6. The origins and modalities of sensory impulses are crudely identified in the thalamus and then projected to the somatosensory cortex and other sensory association areas.

7. Perception—the internal, conscious image of the stimulus that serves as the basis for response—is the result of cortical processing. Regions of the somatosensory cortex devoted to a particular body area reflect the number of receptors in that area.

8. The major aspects of sensory perception are perceptual detection, magnitude estimation, spatial discrimination, feature abstraction, quality discrimination, and pattern recognition.

...and fixed-action patterns. Segmental circuits controlling locomotion are called central pattern generators (CPGs).

4. The projection level consists of descending fibers that project to and control the segmental level. These fibers issue from the brain stem motor areas (extrapyramidal system) and cortical motor areas (pyramidal system). Command neurons (interneurons) in the brain stem appear to turn CPGs on and off, or to modulate them.

5. The programs/instructions level consists of the cerebellum and basal nuclei. These constitute the precommand areas that subconsciously integrate mechanisms mediated by the projection level.

Homeostatic Imbalances of Motor Integration (p. 474)

6. Cerebellar deficits cause ipsilateral symptoms, which include disorders in synergy and muscle tone, equilibrium problems, and speech problems. Ataxia is the most common symptom.

7. Basal nuclear deficits result in abnormal muscle tone and involuntary movements (tremors, athetosis, and chorea). Parkinson's disease and Huntington's disease are examples.

HIGHER MENTAL FUNCTIONS (pp. 474–483)

Brain Wave Patterns and the EEG (pp. 475–476)

1. Patterns of electrical activity of the brain are called brain waves; a record of this activity is an electroencephalogram (EEG). Brain wave patterns, identified by their frequencies, include alpha, beta, theta, and delta waves.

2. Epilepsy results from abnormal electrical activity of brain neurons. The three types of seizure are petit mal, psychomotor epilepsy, and grand mal.

Sleep and Sleep-awake Cycles (pp. 476–478)

3. Sleep is a state of altered consciousness from which one can be aroused by stimulation.

4. The two major types of sleep are nonrapid eye movement (NREM) sleep (stages 1–4) and rapid eye movement (REM) sleep.

5. During stages 1–4 of NREM sleep, brain waves become more irregular and increase in amplitude until delta wave sleep (stage 4) is achieved. REM sleep is indicated by a return to a stage 1 EEG. During REM, the eyes move rapidly under the lids. NREM and REM sleep alternate throughout the night.

6. Slow-wave sleep (stage 4) appears to be restorative. REM sleep is important for emotional stability.

7. REM occupies half of an infant's sleep time and then declines to about 25% of sleep time by the age of ten years. Time spent in slow-wave sleep declines steadily throughout life.

8. Narcolepsy is involuntary lapses into sleep that occur without warning during waking periods. Insomnia is a chronic inability to obtain the amount or quality of sleep needed to function adequately.

Consciousness (p. 479)

9. Consciousness is described clinically on a continuum from alertness to drowsiness to stupor and finally to coma.

10. Human consciousness is thought to reflect a mode of information processing called holistic, which is (1) not localizable, (2) superimposed on other types of neural activity, and (3) totally interconnected.

11. Fainting (syncope) is a temporary loss of consciousness that usually reflects inadequate blood delivery to the brain. Coma is loss of consciousness in which the victim is unresponsive to stimuli.

Memory (pp. 479–483)

12. Memory is the ability to recall one's thoughts. It is essential for learning and is part of consciousness.

13. Memory storage has two stages: short-term memory (STM) and long-term memory (LTM). Transfer of information from STM to LTM takes minutes to hours, but more time is required for LTM consolidation.

14. Fact memory is the ability to learn and consciously remember information. Skill memory is the learning of motor skills, which are then performed without conscious thought.

15. Fact memory appears to involve the hippocampus, amygdala, diencephalon, basal forebrain, and prefrontal cortex. Skill memory pathways are thought to be mediated by the corpus striatum.

16. The nature of memory traces in the human brain has yet to be explained. Widespread cortical damage does not impair memory.

Language (p. 483)

17. Language underlies human brain functions and is inseparable from memory and consciousness. We remember and think in words and other symbols.

18. Language symbols appear to be stored holistically in the cerebral cortex. Wernicke's area is the grammar area. Patterns of activation of muscles used in articulating speech are stored in Broca's area.

19. Impairments of Wernicke's area, Broca's area, or conduction pathways between them lead to aphasias.

REVIEW QUESTIONS

Multiple Choice/Matching

1. All of the following descriptions refer to *specific* ascending pathways except (a) include the fasciculus gracilis and fasciculus cuneatus, which terminate in the medulla, (b) include a chain of three neurons, (c) connections are diffuse and polymodal, (d) are concerned with precise transmission of one or a few related sensory modalities.

2. The aspect of sensory perception by which the cerebral cortex identifies the site or pattern of stimulation is (a) perceptual detection, (b) feature abstraction, (c) pattern recognition, (d) spatial discrimination.

3. The neural machinery of the spinal cord is at the (a) programs level, (b) projection level, (c) segmental level.

4. Which of the following is part of the extrapyramidal system? (a) vestibular nuclei, (b) red nucleus, (c) superior colliculi, (d) reticular nuclei, (e) all of these.

5. Brain waves typical of the alert, wide-awake state are (a) alpha, (b) beta, (c) delta, (d) theta.

6. Identify the stage of sleep described by using choices from the key. **Key** (**a**) stage 1, (**b**) stage 2, (**c**) stage 3, (**d**) stage 4, (**e**) REM

______ (1) the stage when vital signs (blood pressure, heart rate, and body temperature) reach their lowest levels

______ (2) indicated by movement of the eyes under the lids; dreaming occurs

______ (3) when sleepwalking is likely to occur.

______ (4) when the sleeper is very easily awakened; EEG shows alpha waves

Short Answer Essay Questions

7. Differentiate clearly between sensation and perception.

8. How do synthetic and analytic quality discrimination differ?

9. Central pattern generators (CPGs) are found at the segmental level of motor control. (a) What is the job of the CPGs? (b) What controls them, and where is this control localized?

10. Make a diagram of the hierarchy of motor control. Position the CPGs, the command neurons, and the cerebellum/basal nuclei in this scheme.

11. How do the types of motor activity controlled by the pyramidal and extrapyramidal systems differ?

12. Why are the cerebellum and basal nuclei called *precommand areas*?

13. Define EEG.

14. How do the patterns of sleep, the extent of sleeping time, and the amount of time spent in REM and NREM sleep change through life?

15. (a) Define narcolepsy and insomnia. (b) Why is narcolepsy thought to reflect a problem with the sleep control centers? (c) Why is narcolepsy a dangerous condition?

16. (a) Define epilepsy. (b) Which form of this disorder is least distressing? Which is most distressing? Explain.

17. What is an aura?

18. What is holistic information processing, and what are its essential characteristics?

19. Compare and contrast short-term memory (STM) and long-term memory (LTM) relative to storage capacity and duration of the memory.

20. (a) Name several factors that can enhance the transfer of information from STM to LTM. (b) Define memory consolidation.

21. Compare and contrast fact and skill memory relative to the types of things remembered and the importance of conscious retrieval.

22. Explain why consciousness, memory, and language are inseparable.

Clinical Application Questions

23. Mrs. Long has been complaining of a nagging pain in her leg that is keeping her awake. In response to her complaint, the nursing aide gives her a back rub, and shortly thereafter, Mrs. Long drifts comfortably into sleep. Explain how a back rub (or brisk rubbing of your "bumped" skin) might ease pain in terms of the gate theory of pain control.

24. Mr. Jake was admitted to the hospital with excruciating pain in his left shoulder and arm. He was found to have suffered a heart attack. Explain the phenomenon of referred pain as exhibited by Mr. Jake.

25. The medical chart of a 68-year-old man includes the following notes: "Slight tremor of right hand at rest; stony facial expression; difficulty in initiating movements." (a) Based on your present knowledge, what is the diagnosis? (b) What brain areas are most likely involved in this man's disorder, and what is the deficiency? (c) How is this condition currently treated?

CHAPTER **16**

The Special Senses

Chapter Outline and Student Objectives

The Chemical Senses: Taste and Smell (pp. 487–492)

1. Describe the location, structure, and afferent pathways of taste and olfactory receptors, and explain how these receptors are activated.

The Eye and Vision (pp. 492–511)

2. Describe the structure and function of accessory eye structures, the eye tunics, the lens, and the humors of the eye.
3. Trace the pathway of light through the eye to the retina, and explain how light is focused for distant and close vision.
4. Describe the events involved in the stimulation of photoreceptors by light, and compare and contrast the roles of rods and cones in vision.
5. Note the cause and consequences of astigmatism, cataract, glaucoma, hyperopia, myopia, and color blindness.
6. Compare and contrast light and dark adaptation.
7. Briefly describe the process of visual processing, and trace the visual pathway to the optic cortex.

The Ear: Hearing and Balance (pp. 512–525)

8. Describe the structure and general function of the outer, middle, and inner ears.
9. Explain how the balance organs of the semicircular canals and the vestibule help to maintain dynamic and static equilibrium.
10. Describe the sound conduction pathway to the fluids of the inner ear, and follow the auditory pathway from the organ of Corti to the temporal cortex.
11. Explain how one is able to differentiate pitch and loudness of sounds and to localize the source of sounds.
12. Note the causes and symptoms of motion sickness, Ménière's syndrome, otitis media, and deafness.

Developmental Aspects of the Special Senses (pp. 525–526)

13. Describe changes that occur in the special sense organs with aging.

Preview of Selected Key Terms

Special senses The senses of taste, smell, vision, hearing, and equilibrium.

Taste buds Sensory receptor organs that house gustatory cells, which respond to dissolved food chemicals.

Olfactory epithelium (*olfact* = to smell) Sensory receptor region in the nasal cavity containing olfactory neurons that respond to volatile chemicals that enter the nasal passages in air.

Extrinsic eye muscles (*extrins* = from the outside) The six skeletal muscles that are attached to and move each eyeball.

Sclera (sklé'-ruh) (*scler* = hard) The outer fibrous tunic of the eyeball.

Choroid (kō'-royd) (*choroid* = membranelike) The vascular middle tunic of the eyeball.

Retina (reh'-tih-nuh) (*retin* = a net) The neural tunic of the eyeball; contains the photoreceptors, the rods and cones.

Accommodation Focusing for near vision.

Labyrinth (*labyrinth* = maze) The bony cavities and membranes of the inner ear.

Cochlea (kōk'-lē-ah) (*cochlea* = snail) The snail-shaped chamber of the bony labyrinth that houses the receptor for hearing (the organ of Corti).

Static equilibrium The sense of head position in space with respect to gravity.

Dynamic equilibrium The sense that reports on angular or rotatory movements of the head in space.

Macula (ma'-kyoo-luh) (*macula* = spot) Sensory receptor organ within the vestibule of the inner ear; static equilibrium receptor.

Crista ampullaris (*crista* = crest; *ampulla* = flask) Sensory receptor organ within a semicircular canal of the inner ear; dynamic equilibrium receptor.

People are responsive creatures. Hold a sizzling steak before us and our mouths water. Whistle loudly into our ears, and we recoil. These irritants and many others are the stimuli that continually greet us and are interpreted by our nervous system.

We are usually told that we have five senses: touch, taste, smell, sight, and hearing. Actually, touch is made up of a complex of general senses that we considered in Chapter 11. The other four traditional senses—*smell, taste, sight,* and *hearing*—are called special senses. Receptors for a fifth special sense, *equilibrium,* are housed in the ear, along with the organ of hearing. In contrast to the widely distributed general receptors, the **special sense receptors** are either large, complex sensory organs (eyes and ears) or small, localized clusters of receptors (taste buds and olfactory epithelium).

This chapter considers the functional anatomy of each of the special sense organs. But keep in mind that our perceptions of sensory inputs are overlapping. What we finally experience—our "feel" of the world—is a blending of stimulus effects.

The Chemical Senses: Taste and Smell

The receptors for taste and olfaction are classified as **chemoreceptors** because they respond to chemicals in an aqueous solution. The taste receptors are excited by food chemicals dissolved in saliva, the smell receptors by airborne chemicals that dissolve in fluids produced by the nasal membranes. The receptors for taste and smell complement each other and respond to many of the same stimuli.

Taste Buds and the Sense of Taste

The word *taste* comes from the Latin *taxare,* meaning "to touch, estimate, or judge." When we taste things, we are in fact testing or judging our environment in an intimate way, and the sense of taste is considered by many to be the most pleasurable of the special senses.

Localization and Structure of Taste Buds

The **taste buds,** the sensory receptor organs for taste, are located primarily in the oral cavity. Of our 10,000 or so taste buds, most are on the tongue. A few are scattered on the soft palate, inner surface of the cheeks, pharynx, and epiglottis of the larynx.

Characteristically, taste buds are found in projections of the tongue mucosa called **papillae** (pah-pil′-ē), which give the tongue surface a slightly abrasive feel. These papillae are of three major types: filiform, fungiform, and circumvallate. Only the latter two types house taste buds. The mushroom-shaped **fungiform** (fun′-jih-form) **papillae** are scattered over the entire tongue surface, but are most abundant at its tip and along its sides (Figure 16.1a). The round **circumvallate** (ser″-kum-vā′-lat) **papillae** are the largest and least numerous of the papillae; 7 to 12 of these papillae form a V at the back of the tongue (see Figure 16.1a). Taste buds are located in the side walls of the circumvallate papillae (Figure 16.1b) and on the top of the fungiform papillae.

Each taste bud is a globular structure consisting of 40 to 60 epithelial cells of three major types: supporting cells, taste cells, and basal cells (Figure 16.1c). **Supporting cells** form the bulk of the taste bud and surround receptor cells called **taste,** or **gustatory, cells.** Microvilli called **gustatory hairs** protrude from the tips of the taste cells through a **taste pore** to the surface of the taste bud, where they are bathed by saliva. Apparently, the gustatory hairs are the sensitive portions (receptor membranes) of the taste cells. Coiling intimately around the taste cells are sensory dendrites that represent the initial part of the gustatory pathway to the brain. Taste cells are among the most dynamic cells in the body. They are shed every week to ten days and are replaced by new sensory cells derived from the division of the **basal cells.**

Basic Taste Sensations

Normally, our taste sensations are complicated mixtures of qualities. However, when taste is tested with pure chemical compounds, taste sensations can all be grouped into one of four basic qualities: *sweet, sour, salty,* and *bitter.* The sweet taste is elicited by many organic substances including sugars, saccharin, and some amino acids. Sour taste is produced by acids, specifically their hydrogen ions (H^+), whereas a salty taste is produced by metal ions (inorganic salts) in solution. Table salt (sodium chloride) tastes the "saltiest." The bitter taste is elicited by alkaloids (such as quinine, nicotine, caffeine, and strychnine) as well as a number of nonalkaloid substances, such as aspirin.

Sensitivity to one or another of these basic qualities varies in different regions of the tongue (Figure 16.2). Generally speaking, the tip of the tongue is most sensitive to sweet tastes, the sides to sour, and the back of the tongue (near its root) to bitter. The ability to respond to salty tastes appears to be more evenly distributed, with some emphasis on the anterolateral tongue edges. However, taste sensation is not quite as clear-cut as this discussion may imply. Most taste buds respond to two, three, or all four taste qualities, and

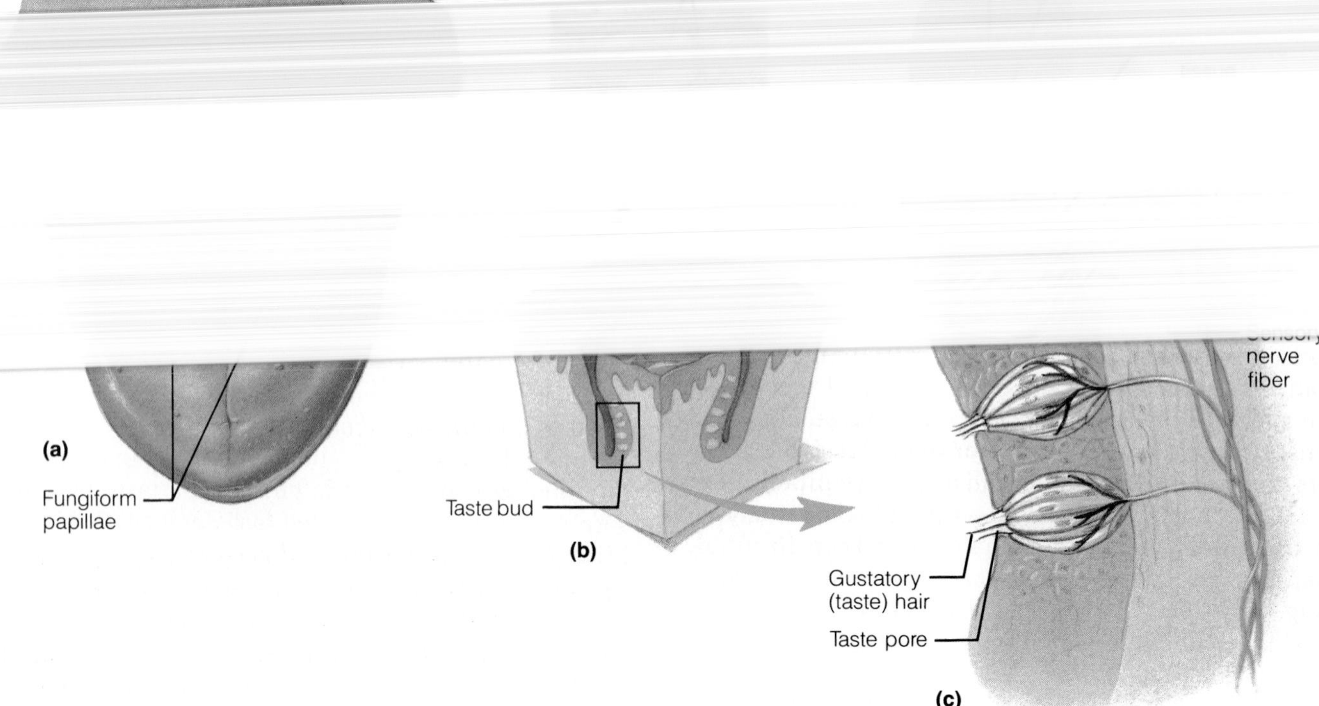

Figure 16.1 Location and structure of taste buds. (**a**) Taste buds on the tongue are associated with papillae, peglike projections of the tongue mucosa. (**b**) A longitudinally sectioned circumvallate papilla shows the positioning of the taste buds in its lateral walls. (**c**) An enlarged view of four taste buds, illustrating the gustatory (taste), supporting, and basal cells.

many substances produce a mixture of the basic taste sensations. Furthermore, some substances change in flavor as they move through the mouth. Saccharin is initially tasted as sweet, but acquires a bitter aftertaste at the back of the tongue.

Taste likes and dislikes have homeostatic value. A liking for sugar and salt will satisfy the body's need for carbohydrates and minerals (as well as some amino acids). Many sour, naturally acidic foods (such as oranges, lemons, and tomatoes) are rich sources of vitamin C, an essential vitamin. Since many natural poisons and spoiled foods are bitter, our dislike for bitterness is protective. (The position of the bitter receptors on the most posterior part of the tongue seems strange, because usually by the time we actually taste a bitter substance, we have swallowed some of it.) Although many people learn to enjoy bitter foods such as olives and quinine water, most animals scrupulously avoid bitter substances.

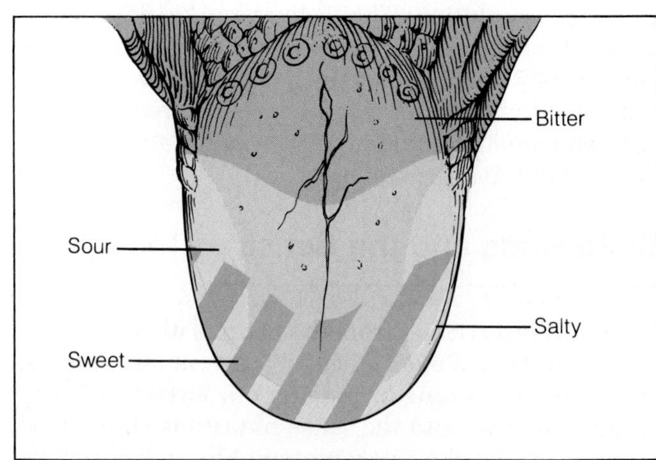

Figure 16.2 Relative patterns of taste sensitivity on the tongue dorsum. (Note: since the area most sensitive to sweet overlaps the sour and salty areas, the sweet area is indicated by stripes.)

Activation of Taste Receptors

For a chemical to be tasted, it must be dissolved in saliva and diffuse into the taste pore to contact the gustatory hairs. It appears that the chemical binds to receptors on the gustatory hairs, resulting in the development of a receptor potential in the taste cell (receptor potentials are described in Chapter 11). Taste cells are incapable of generating action potentials, but their receptor potentials summate to trigger action potentials in the associated sensory neurons.

The higher the concentration of a given chemical, the more intense its perceived taste. However, the different types of gustatory cells have different thresholds for activation. The bitter receptors (in line with their protective nature) detect substances present in minute amounts. The receptor cells for sour are less sensitive, and the sweet and salt receptor cells are the least sensitive. Taste receptors adapt very rapidly, with partial adaptation in 3–5 seconds and complete adaptation in 1–5 minutes.

The Gustatory Pathway

As described in Chapter 13, afferent fibers carrying taste information from the tongue are found primarily in two cranial nerve pairs. The *facial nerve* (cranial nerve VII) transmits impulses from taste receptors in the anterior two-thirds of the tongue, whereas the *glossopharyngeal nerve* (cranial nerve IX) services the posterior third. Taste impulses from the few taste buds in the epiglottis and pharynx are conducted primarily by the vagus nerve (cranial nerve X). These afferent fibers synapse in one of the medullary nuclei, and from there impulses are transmitted to the thalamus and ultimately to the somatosensory cortex in the parietal lobes (see Figure 16.4).

Damage to the glossopharyngeal nerves reduces one's ability to detect bitter substances, whereas blockage of the facial nerves interferes with one's ability to taste sweet, sour, and salty substances. ■

An important role of taste is to trigger reflexes involved in digestion. As taste impulses pass through the medullary nuclei, they initiate reflexes that increase secretion of saliva into the mouth and secretion of gastric juice into the stomach. Saliva contains mucus that moistens food and digestive enzymes that begin the digestion of starch. Acidic foods are particularly strong stimulants of the salivary reflex. Additionally, when we eat noxious or foul-tasting substances, gagging or even reflexive vomiting may be initiated.

Influence of Other Senses on Taste Perception

What we commonly call taste depends heavily on the stimulation of olfactory receptors. Indeed, taste is 80% smell: The subtleties of flavor that allow us to distinguish a fine port wine from grape juice and beef Wellington from hamburger actually result from aromas passing through the nose as we eat or drink. When the olfactory receptors in the nasal cavity are blocked by nasal congestion, food is bland. Without the sense of smell, our morning coffee would lack its richness and simply taste bitter.

The mouth also contains thermoreceptors, mechanoreceptors, and nociceptors, and the temperature and texture of foods can enhance or detract from their taste. Some people will not eat foods that have a pasty texture (avocados) or are gritty (pears), and almost everyone considers a cold, greasy hamburger unfit to eat. "Hot" foods such as chili peppers actually excite pain receptors in the mouth.

The Olfactory Epithelium and the Sense of Smell

Although our sense of smell is far less acute than that of many other animals, the human nose is still much keener in picking up small differences in odors than any machine. Some people capitalize on this ability by becoming tea and coffee blenders or wine tasters.

Localization and Structure of the Olfactory Receptors

Olfaction, like taste, detects chemicals in solution. The organ of smell is a yellow-tinged patch of nasal mucosa called the **olfactory epithelium,** located in the roof of each nasal cavity (Figure 16.3a). Since air entering the nasal cavities must make a hairpin turn to stimulate olfactory receptors before entering the respiratory passageway below, the human olfactory epithelium is in a very poor position for doing its job. (This is why sniffing, which flushes more air superiorly across the olfactory epithelium, intensifies the sense of smell.) About ½ square inch in area and covering the whole of the upper concha, each olfactory epithelium contains some 5 million **olfactory receptor cells,** surrounded and cushioned by columnar **supporting cells,** which make up the bulk of the thin epithelial membrane (Figure 16.3b). The supporting cells contain large amounts of lipofuscin granules (importance unknown), which give the olfactory epithelium its yellow hue.

The olfactory receptor cells are bipolar neurons, each with a thin dendrite that terminates in a knob bearing several long cilia called **olfactory hairs.** These olfactory hairs, which substantially increase the receptive surface area, protrude outward from the nasal epithelium and are covered by a coat of thin mucus produced by the olfactory glands in the lamina propria of the nasal mucosa. The secretion keeps the olfactory membrane moist and serves as a solvent for the odor molecules. Unlike the cilia of other parts of

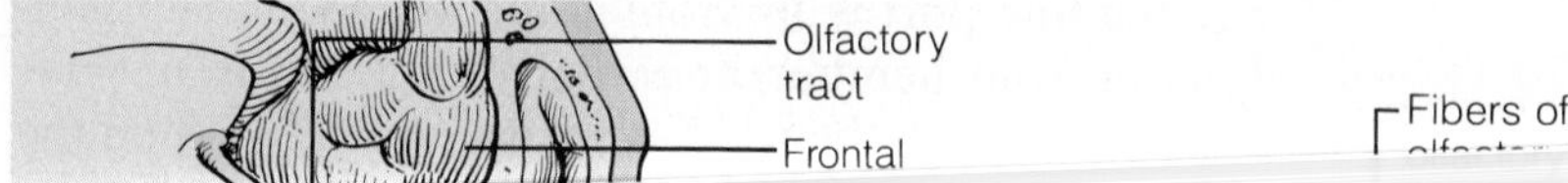

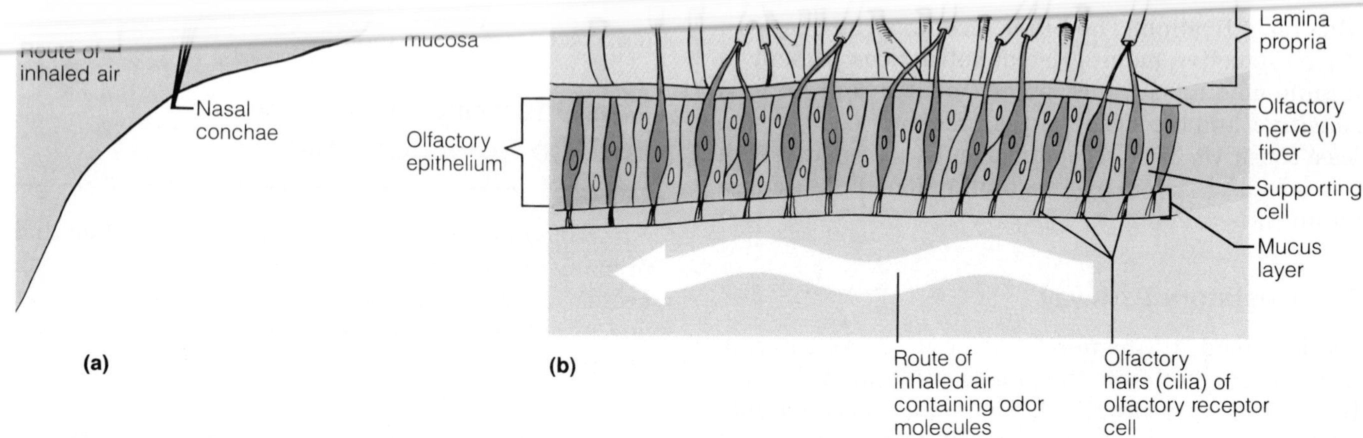

Figure 16.3 Olfactory receptors. (**a**) Site of olfactory epithelium in the superior aspect of the nasal cavity. (**b**) Enlarged view illustrating the cellular composition of the olfactory epithelium and the course of the fibers of the olfactory nerve (I) through the ethmoid bone to synapse in the overlying olfactory bulb. The glomeruli and mitral cells (output cells) within the olfactory bulb are also shown.

the respiratory tract, which beat rapidly and in a coordinated manner, the olfactory cilia exhibit a gentle, asynchronous motion that stirs, rather than propels, the mucus. The unmyelinated axons of the olfactory receptor cells constitute the fibers of the olfactory nerve (cranial nerve I). They project superiorly through the openings in the cribriform plate of the ethmoid bone, where they synapse within the overlying olfactory bulbs.

Olfactory receptor cells are unique in that they are the only neurons known to undergo turnover throughout adult life. Their typical life span is 60 days, after which they are replaced through the differentiation of **basal cells** in the olfactory epithelium.

Specificity of the Olfactory Receptors

Research on smell is difficult because any given odor (say, tobacco smoke) may be made up of hundreds of different chemicals. Taste has been neatly packaged into four taste qualities, but science has yet to discover any similar means for classifying smell. Our keen sense of smell can distinguish tens of thousands of chemicals, but research suggests that we have just 15 to 30 kinds of receptors, which are stimulated in different combinations.

The olfactory neurons are exquisitely sensitive—in some cases, just a few molecules can cause their activation. However, the majority of olfactory receptors appear to have relatively low specificity; that is, they respond to a variety of chemicals. An idea that is now gaining acceptance is that odor recognition depends on the simultaneous activation of the entire olfactory epithelium to varying degrees. As with taste, some of what we call smell is really pain. The nasal cavities contain pain receptors that respond to irritating substances such as ammonia and chili peppers. Impulses from these pain receptors are delivered to the central nervous system by afferent fibers of the trigeminal nerves.

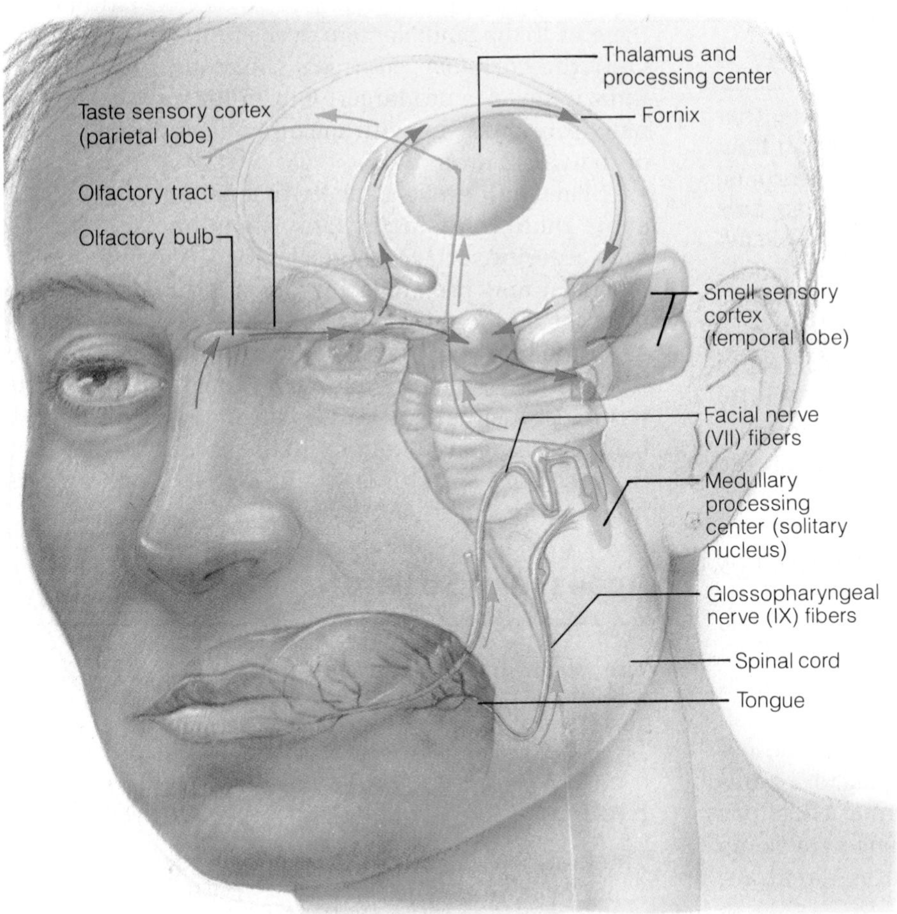

Figure 16.4 Idealized view of the gustatory and olfactory pathways to the cerebral cortex of the brain.

Activation of the Olfactory Receptors

In order for us to smell a particular chemical, it must be *volatile*—that is, it must be in the gaseous state as it enters the nasal cavity. Additionally, it must be water-soluble so that it can dissolve in the fluid coating of the olfactory epithelium. It is presumed that the dissolved chemicals stimulate the olfactory receptors by binding to protein receptors in the olfactory hair membranes. This leads to a receptor potential and ultimately (assuming threshold stimulation) to an action potential that is conducted to the first relay station in the olfactory bulb.

The sense of smell adapts quickly. However, olfactory adaptation is not due to a fading of the receptor response, but to inhibitory influences in the central olfactory pathways (olfactory bulbs), as described next.

The Olfactory Pathway

Axons of the olfactory receptor cells constitute the olfactory nerves and synapse in the overlying olfactory bulbs. The relationships in the olfactory bulbs are somewhat complex. Fibers of the olfactory nerves synapse with **mitral** (mī′-trul) **cells** in complex synaptic structures called *glomeruli* (glō-mer′-ū-lī) within the olfactory bulbs (see Figure 16.3b). The mitral cells appear to play an important integration role in olfaction, and depending on local conditions, they may be activated or inhibited. The olfactory bulbs also house **granule cells,** which seem to inhibit mitral cells (via dendrodendritic synapses), so that only highly excitatory olfactory impulses can be transmitted. This inhibition is believed to contribute to the mechanism of olfactory adaptation and helps explain how someone working in a sewage treatment plant can still enjoy his lunch! The granule cells also receive efferent impulses from the brain, which modify the reaction of the olfactory bulbs to smell under certain conditions. For example, the aromas of food are perceived differently depending on whether we are ravenously hungry or have just finished eating.

When the mitral cells are activated, impulses flow from the olfactory bulbs via the olfactory tracts (composed of mitral cell axons) to the olfactory cortex, where smell interpretation occurs (Figure 16.4). Olfactory tract fibers also project to the amygdaloid nuclei, hypothalamus, and other regions of the limbic system, where the emotional aspects of smells are analyzed and responded to.

Homeostatic Imbalances of the Chemical Senses

deficiency, and the cure is almost instant once the right form of zinc supplement is prescribed. Zinc, normally found in saliva, is a known growth factor for taste buds, and it is strongly suspected that it performs the same function for the olfactory epithelium.

Brain disorders can distort the sense of smell. Some people have *uncinate* (uns′-ih-nāt) *fits,* during which they experience some particular (and usually repugnant) odor, such as gasoline, iodine, or rotting meat. Although some such cases are undeniably psychological, many result from irritation of the olfactory pathway due to brain surgery or head trauma. Olfactory auras—olfactory hallucinations experienced by some epileptics just before they go into seizures—are transient uncinate fits. ■

The Eye and Vision

Of all the sensory receptors in the body, 70% are in the eyes. The photoreceptors sense and encode the patterns made by light in our surroundings. The massive optic tracts that carry the coded information from the eyes to the brain contain over a million nerve fibers. Only th

actually involved in photoreception. But before turning our attention to the eye, let us consider the accessory structures that protect it or enhance its functioning.

Accessory Structures of the Eye

The **accessory structures** of the eye include the eyebrows, eyelids, conjunctiva, lacrimal apparatus, and extrinsic eye muscles.

Eyebrows

The **eyebrows** are short, coarse hairs that overlie the superior orbital ridges of the skull (Figures 16.5 and 16.6a). They help to shade the eyes from sunlight and prevent perspiration trickling downward from the forehead from reaching the eyes. Deep to the skin of the eyebrows are parts of the orbicularis oculi and corrugator muscles. Contraction of the orbicularis muscle moves the eyebrows inferiorly, whereas contraction of the corrugator promotes medial movements of the eyebrows.

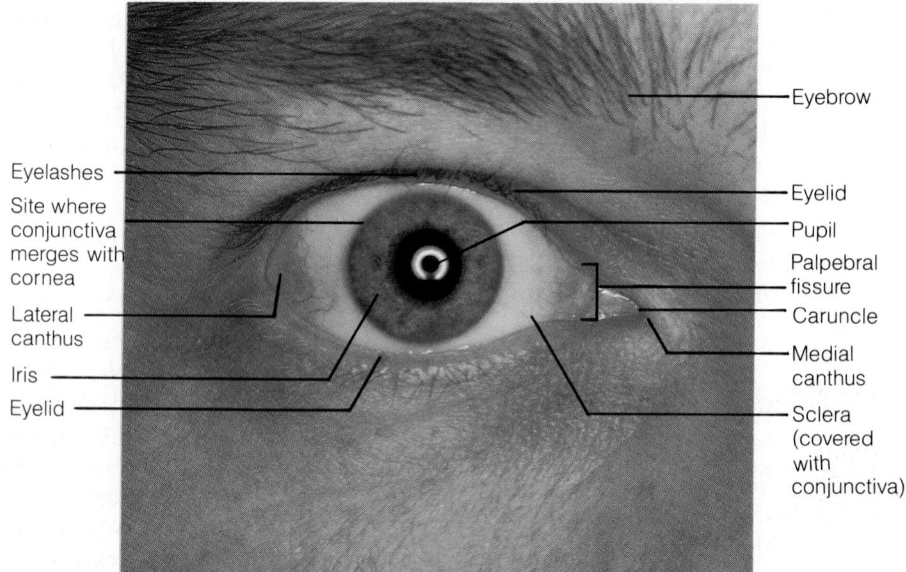

Figure 16.5 Surface anatomy of the eye and associated accessory structures. The eyelids are open more widely than normal to show the entire iris.

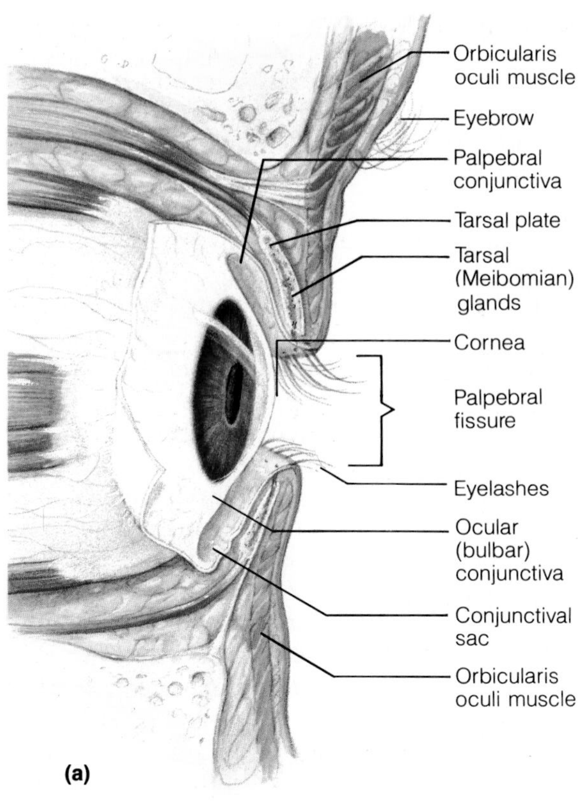

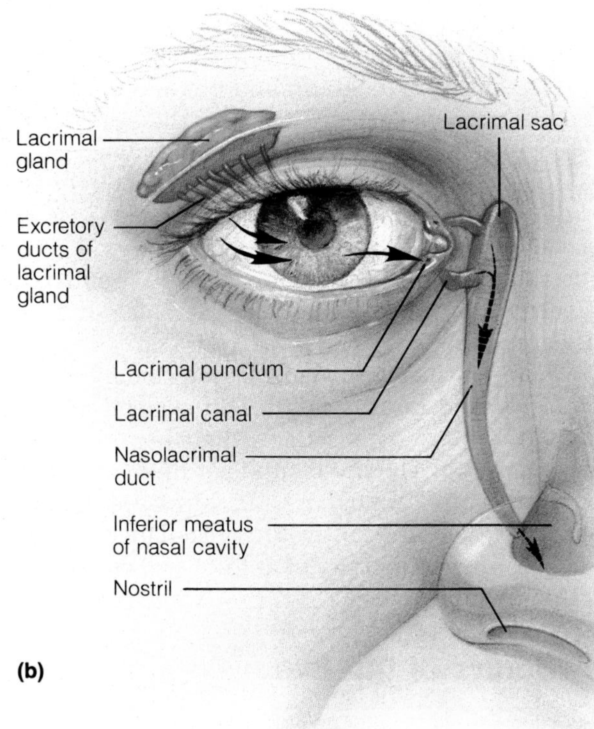

Figure 16.6 Accessory structures of the eye. (**a**) Sagittal section of the accessory structures associated with the anterior portion of the eye. (**b**) The lacrimal apparatus. Arrows indicate the direction of flow of lacrimal fluid after its release by the lacrimal gland.

Eyelids

Anteriorly, the eyes are protected by the mobile **eyelids** or **palpebrae** (pal′-peh-brē), which meet at the medial and lateral angles of the eye—the **medial** and **lateral canthi,** respectively (see Figure 16.5). The medial canthus sports a fleshy elevation (a remnant of the third eyelid found in many lower animals) called the **caruncle** (kayr′-un-kel). The caruncle contains sebaceous and sudoriferous glands and produces the whitish, oily secretion that sometimes collects at the medial canthus.

The eyelids are thin, skin-covered folds supported internally by flat connective tissue sheets called **tarsal plates** (see Figure 16.6a). The tarsal plates also anchor the orbicularis oculi and levator palpebrae superioris muscles that run within the eyelid. The orbicularis muscle encircles the eye and functions as a sphincter; when it contracts, the eye closes. Of the two eyelids, the upper one is much more mobile, largely because of the levator palpebrae superioris muscle, which raises that eyelid to open the eye. The eyelid muscles are activated reflexively to cause blinking every 3–7 seconds and to protect the eye when it is threatened by foreign objects. Reflex blinking of the eyelids helps prevent desiccation (drying) of the eyes: Each time we blink, accessory structure secretions (oil, mucus, and saline solution) are spread across the eyeball surface.

Projecting from the border of each eyelid are the **eyelashes.** The follicles of the eyelash hairs are richly innervated by nerve endings (hair root plexuses), and when anything touches the eyelashes (even a puff of air), reflex blinking is triggered.

Several types of glands are associated with the exposed inner edges of the eyelids. The **tarsal,** or **Meibomian** (mī-bō′-mē-un), **glands** are embedded in the tarsal plates (see Figure 16.6a), and their ducts open at the eyelid edge just posterior to the eyelashes. These modified sebaceous glands produce an oily secretion that lubricates the eyelid and the eye and prevents the eyelids from sticking together. Associated with the eyelash follicles are large sebaceous glands called **ciliary glands** or **glands of Zeis,** and modified sweat glands lie between the hair follicles.

Infection of a tarsal gland results in an unsightly cyst called a *chalazion* (kuh-lā′-zē-un), whereas inflammation of a ciliary gland is called a *sty.* ■

The Conjunctiva

The **conjunctiva** (kon″-junk-tī′-vuh) is a delicate mucous membrane that lines the inner surfaces of the eyelids as the *palpebral conjunctiva* and reflects (folds back) over the anterior surface of the eyeball as the *ocular,* or *bulbar, conjunctiva* (see Figure 16.6a). The ocular conjunctiva is very thin, and blood vessels are clearly visible beneath it. Recesses formed at the junctions of the palpebral and ocular conjunctivas are called the **inferior** and **superior conjunctival sacs.** Eye medications are often administered into these recesses. The major function of the conjunctiva is to produce a lubricating mucus that prevents the eyes from drying out.

apparatus consists of the lacrimal gland and a number of ducts that drain the lacrimal secretions into the nasal cavity (Figure 16.6b). The **lacrimal gland** is located within the orbit above the lateral end of the eye. It continually releases a dilute saline solution called *lacrimal secretion*—or, more commonly, **tears**—into the superior conjunctival sac through several small excretory ducts. Blinking spreads the tears downward and across the eyeball to the medial canthus, where they enter the paired **lacrimal canals** via two tiny orifices called *lacrimal puncta* (singular, punctum), visible as tiny red dots on the medial margin of each eyelid. From the lacrimal ducts, the tears drain into the **lacrimal sac** and then into the **nasolacrimal duct,** which empties into the nasal cavity just below the inferior nasal concha.

Lacrimal secretion contains mucus, antibodies, and **lysozyme,** an antibacterial enzyme. Thus, it cleanses and protects the eye surface as it moistens and lubricates it. Since the activity of the lacrimal glands decreases with age, the eyes tend to become less moist and more vulnerable to infection and irritation during the later years of life. When lacrimal secretion increases substantially, tears spill over the eyelids and fill the nasal cavities, causing congestion. This happens when the eyes are irritated by foreign objects or noxious chemicals and when we are emotionally upset (in which case, the parasympathetic nervous system triggers crying). In the case of eye irritation, the enhanced tearing serves to wash away or dilute the irritating substance. The importance of emotionally induced tears is poorly understood. Recently, it has been shown that lacrimal secretion contains enkephalins (natural opiates); this, coupled with the fact that only humans shed emotional tears, has led to the suspicion that crying is important in reducing stress. Although anyone who has had a good cry would probably agree, this has been difficult to demonstrate scientifically.

Because the nasal cavity mucosa is continuous with that of the lacrimal duct system, a cold or nasal inflammation often causes the lacrimal mucosa to become inflamed and swell. This impairs tear drainage of tears from the eye surface, causing "watery" eyes. ■

Extrinsic Eye Muscles

The movement of each eyeball

points and the movements that they promote are clearly indicated by their names: **superior, inferior, lateral,** and **medial rectus muscles.** The movements produced by the two *oblique muscles* are less easily deduced because of their rather strange modes of attachment to the eyeball. Their action is particularly important in moving the eye in the vertical plane when the eyeball is already turned medially. The **superior oblique muscle** originates in common with the rectus muscles, runs along the medial wall of the orbit, and then makes a right-angle turn and passes through a pulleylike cartilage loop, called the *trochlea* (trok′-lē-uh), before inserting on the superior aspect of the eyeball. Its contraction rotates the eye downward and somewhat laterally. The **inferior oblique muscle** originates from the medial orbit surface and runs laterally and obliquely to insert on the inferior eye surface. It rotates the eye up and laterally. Except for the lateral rectus and superior oblique muscles, which are innervated respectively by the *abducens* and *trochlear nerves,* all extrinsic eye muscles are served by the *oculomotor nerves.* The action and nerve supply of these muscles are summarized in Figure 16.7c. (The courses of the associated nerves are illustrated in Table 13.1.)

The extrinsic eye muscles are among the most precisely and rapidly controlled skeletal muscles in the entire body. This reflects their low axon-to-muscle-fiber ratio: The motor units of these muscles contain only 8 to 12 muscle cells. Most eye movements involved in aiming the eyes for visual focusing require the coordinated activity of at least two of the muscles acting in concert.

Structure of the Eye

The eye itself, commonly called the *eyeball,* is a hollow sphere (Figure 16.8). Its wall is composed of three coats, or tunics, and its interior is filled with fluids called *humors* that help to maintain its shape. The lens, the main focusing apparatus of the eye, is supported vertically within the eye cavity, dividing it into anterior and posterior cavities.

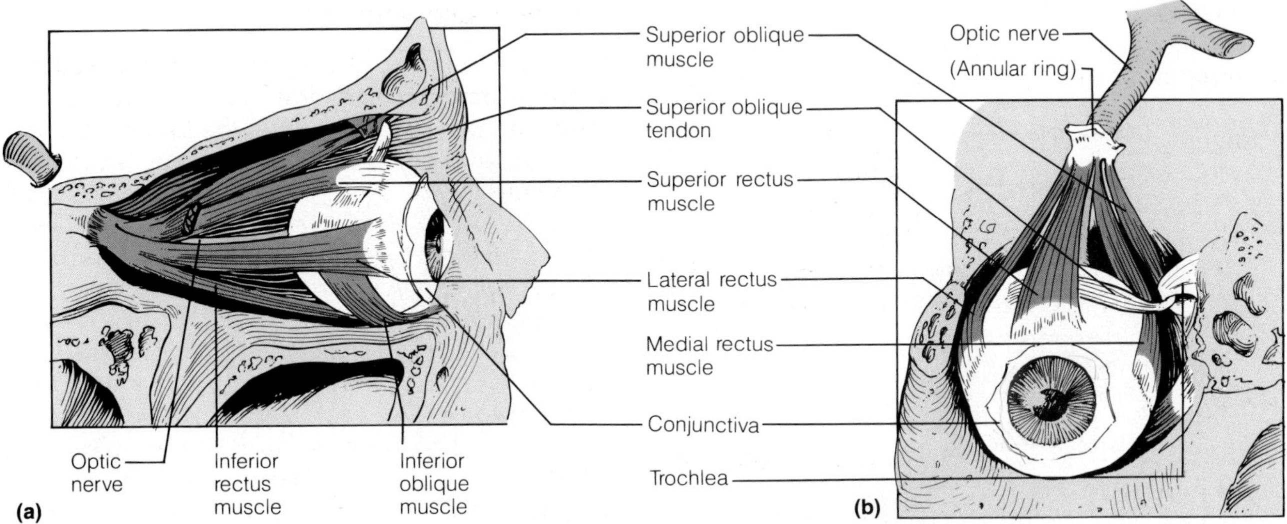

Name	Controlling cranial nerve	Action
Lateral rectus	VI (abducens)	Moves eye laterally
Medial rectus	III (oculomotor)	Moves eye medially
Superior rectus	III (oculomotor)	Elevates eye or rolls it superiorly
Inferior rectus	III (oculomotor)	Depresses eye or rolls it inferiorly
Inferior oblique	III (oculomotor)	Elevates eye and turns it laterally
Superior oblique	IV (trochlear)	Depresses eye and turns it laterally

Figure 16.7 Extrinsic muscles of the eye. (**a**) Lateral view of the right eye. (**b**) Anterosuperior view of the right eye. (**c**) Summary of cranial nerve innervations and muscle actions.

Tunics Forming the Wall of the Eye

The Fibrous Tunic. The outermost coat of the eye, the **fibrous tunic,** is composed of dense avascular connective tissue. It has two obviously different regions: the sclera and the cornea. The **sclera** (skle′-ruh), forming the posterior portion and the bulk of the fibrous tunic, is glistening white and opaque. Seen anteriorly as the "white of the eye," the tough, tendonlike sclera protects and shapes the eyeball and provides a sturdy site for anchoring the extrinsic eye muscles. Posteriorly, where the sclera is pierced by the optic nerve, it is continuous with the dura mater of the brain. The anterior portion of the fibrous tunic is modified to form the transparent **cornea,** which bulges anteriorly from its junction with the sclera. The regular arrangement of the connective tissue fibers makes the cornea crystal clear, allowing it to form a window that allows light to enter the eye. As explained shortly, the cornea is also part of the light-bending apparatus of the eye.

The cornea is covered by epithelial membranes on both faces (see Figure 16.12, p. 500). The external membrane merges with the ocular conjunctiva at the sclera-cornea junction and provides the cornea with extra protection from abrasion. A simple squamous epithelium lines the inner face of the cornea and has a very special property: Its cells have a very active sodium pump that continually ejects sodium ions from the cornea into the fluids of the eyeball. Water follows (along its osmotic gradients), and as a result, the collagen fibers of the cornea remain tightly and regularly packed, undisturbed by accumulations of tissue fluid that could force them apart. Hence, the nearly perfect clarity of the cornea is maintained.

The cornea is well supplied with nerve endings, most of which are pain fibers, and when the cornea is touched, reflex blinking and increased lacrimal fluid secretion occur. Even so, the cornea is the most exposed part of the eye, and it is very vulnerable to damage. Luckily, its ability for regeneration and repair is extraordinary; furthermore, it can be replaced. The cornea is the only tissue in the body that can be transplanted from one person to another without the worry of rejection. Since it has no blood vessels, it is beyond the reach of the immune system.

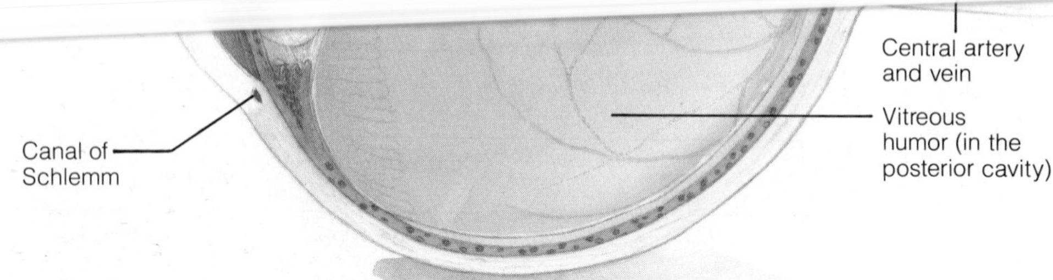

Figure 16.8 Internal structure of the eye (sagittal section). The vitreous humor is illustrated only in the bottom half of the eyeball.

The Vascular Tunic (Uvea). The **vascular tunic,** or middle coat of the eyeball, is also called the **uvea.** The uvea is pigmented and has three distinct regions: the choroid, the ciliary body, and the iris (see Figure 16.8).

The **choroid** (kō′-royd) is a highly vascular, deeply pigmented membrane that occupies the posterior part of the eyeball. Its blood vessels provide nutrition to the eye tunics. Its brown pigment, produced by melanocytes, helps absorb light, preventing it from scattering within the eye or leaving the eye. The choroid is incomplete posteriorly where the optic nerve leaves the eye; anteriorly, it is continuous with the **ciliary body,** a thickened ring of tissue that encircles the lens. The ciliary body is composed chiefly of interlacing smooth muscle bundles called **ciliary muscles,** which are important in controlling lens shape. Its surface is thrown into folds called **ciliary processes** that secrete the fluid that fills the anterior cavity of the eyeball. A **suspensory ligament** extending from the ciliary body encircles the lens in its upright position.

The **iris,** the visible colored part of the eye, is the most anterior portion of the uvea. Shaped like a flattened doughnut, it lies between the cornea and the lens and is continuous with the ciliary body posteriorly. Its round central opening, the **pupil,** allows light to enter the eye. The iris is made up of circularly and radially arranged smooth muscle fibers and acts as a reflexively activated diaphragm to vary pupil size. In close vision and bright light, the circular muscles contract and the pupil constricts. In distant vision and dim light, the radial muscles contract and the pupil dilates, allowing more light to enter the eye. As described in Chapter 14, pupillary dilation and constriction are controlled by sympathetic and parasympathetic fibers, respectively. The rapid constriction of the pupils when the eyes are exposed to intense (potentially damaging) light is called the **pupillary light reflex.**

Interestingly, reflex changes in pupil size may also reflect our interests or emotional reactions to what is viewed. Pupillary dilation frequently occurs when the subject matter is appealing and during problem solving. (Computing your taxes should make your pupils get bigger and bigger.) On the other hand, boredom or subject matter that is personally repulsive causes pupillary constriction.

Although irises come in different colors, they contain only brown pigment. When they contain a lot of pigment, the eyes appear brown or black. If the amount of pigment is small, the shorter wavelengths of light are scattered from the unpigmented parts of the iris, and the eyes appear blue, green, or gray. This is similar to the scattering of light that makes the sky appear blue. Most newborn babies' eyes are slate gray or blue because their iris pigment is not yet developed.

The Sensory Tunic (Retina). The innermost tunic is the delicate, two-layered **sensory tunic,** or **retina** (reh′-tih-nuh). The outer **pigmented layer** abuts the choroid. Its pigment cells, like those of the choroid, absorb light; they also act as phagocytes and store vitamin A needed by the photoreceptor cells. The inner **neural (nervous) layer** is transparent and contains the mil-

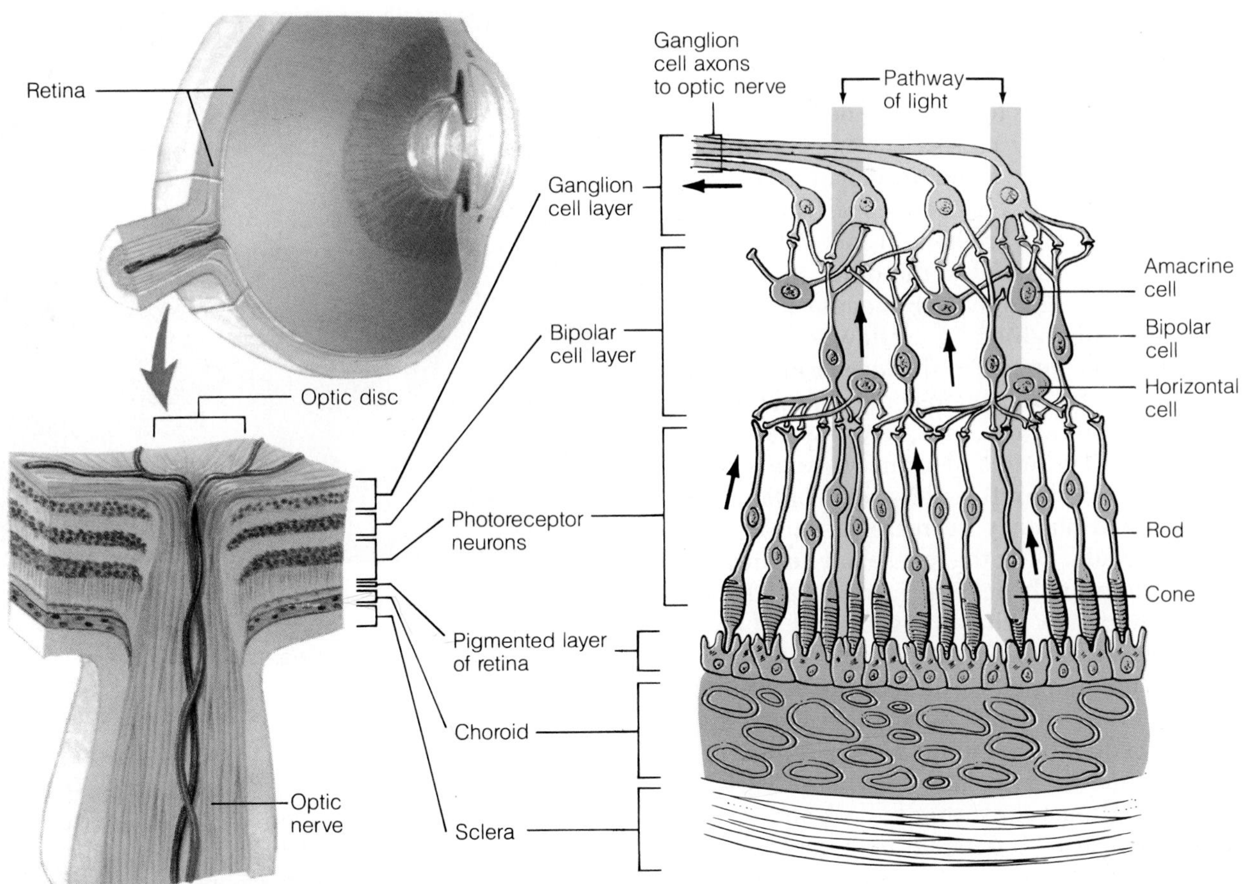

Figure 16.9 Schematic view of the three major neuronal populations (photoreceptors, bipolar cells, and ganglion cells) composing the sensory layer of the retina. The axons of the ganglion cells form the optic nerve, which leaves the back of the eyeball at the optic disc. Notice that light passes through the retina to excite the photoreceptor cells. The flow of electrical signals occurs in the opposite direction: from the photoreceptors to the bipolar cells and finally to the ganglion cells.

lions of photoreceptors that transduce light energy. Although these two layers are very close together, they are not fused.

Separation of the pigmented and nervous layers, called *retinal detachment*, can cause permanent blindness. It usually happens when the retina is torn during a traumatic blow to the head, allowing the jelly-like vitreous humor from the back of the eye to seep through the tear and between the two retinal layers. The symptom that victims most often describe is "a curtain being drawn across the eye," but some people see sootlike spots or light flashes. If the condition is diagnosed early, it is often possible to reattach the retina surgically or with a laser before photoreceptor damage becomes permanent. ■

Functionally, the retina has two regions. Its posterior receptive part, which contains the photoreceptors and other neurons involved in the processing of light stimuli, extends anteriorly to the ciliary body, where it ends by forming a wavy ring called the *ora serrata* (see Figure 16.8). The nonreceptive portion, containing only the pigmented layer, extends beyond the ora serrata to cover the ciliary body and the inner face of the iris.

The nervous layer is composed of three main types of neurons: **photoreceptors, bipolar cells,** and **ganglion cells** (Figure 16.9). Local currents produced in response to light spread from the photoreceptors (abutting the pigmented layer) to the bipolar neurons and then to the innermost ganglion cells, where action potentials are generated. The ganglion cell axons make a right-angle turn at the inner face of the retina, then leave the posterior aspect of the eye as the thick optic nerve. The other types of neurons in the retina (horizontal cells and amacrine cells) will be described shortly. The **optic disc,** where the optic nerve exits the eye, is a weak spot in the **fundus** (posterior wall) of the eye because it is not reinforced by the sclera. The optic disc is also called the **blind spot,** because it lacks photoreceptors, and light focused on it cannot be seen.

The major blood vessels serving the eyes are the *central artery* (a branch of the ophthalmic artery) and the *central vein,* which enter and leave the eye through the center of the optic nerve. These give rise to a rich vascular network, clearly seen when the eyeball interior is examined with an ophthalmoscope (Figure 16.10). The fundus of the eye is the only place in the body where living capillaries can be observed directly.

The photoreceptors found in the neural reti[illegible]

[illegible] declines gradually. The retina periphery contains only rods, which continuously decrease in density from there to the macula. Only the foveae have a sufficient cone density to provide detailed color vision, so anything we wish to view critically is focused on the foveae. Because each fovea is only about the size of the head of a pin, not more than a thousandth of the entire visual field is in hard focus (foveal focus) at a given moment. Consequently, for us to visually comprehend a scene that is rapidly changing (as when we drive in traffic), our eyes must flick rapidly back and forth to provide the foveae with detailed images of different parts of the visual field.

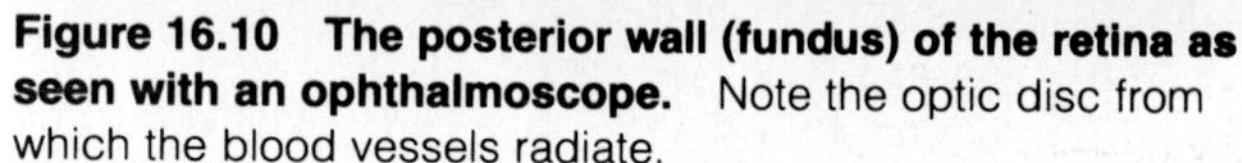

Figure 16.10 The posterior wall (fundus) of the retina as seen with an ophthalmoscope. Note the optic disc from which the blood vessels radiate.

Structure of the Photoreceptors

As shown in Figure 16.11, photoreceptors of both types have an **inner segment** to which an **outer segment** (the receptor region) is connected by a stalk containing a cilium. These outer segments are partially immersed in the pigmented retinal layer.

The rods are slender neurons with a long cylindrical outer segment; their inner segment is connected to the cell body, or nuclear region, by a slender dendrite. By contrast the fatter cones have a short conical outer segment and their inner segment and cell body are adjacent to one another and continuous.

The light-absorbing visual pigments are packaged in membrane-bounded discs within the outer segments. This coupling of the photoreceptor pigments to cellular membranes greatly increases the surface area available for light trapping. In rods, the discs are discontinuous and stacked like pennies within the plasma membrane, but in cones, the discs become increasingly smaller toward the end of the cell, and their membranes are continuous with the plasma membrane. Thus, the interiors of the cone discs are continuous with the extracellular space.

The photoreceptor cells are highly vulnerable to damage. If the retina becomes detached, the photoreceptor cells immediately begin to degenerate. They are also readily destroyed by intense heat—or light, the very energy they are designed to detect. How is it, then, that we do not all gradually go blind from a progressive accumulation of injuries to the visual cells? Like all neurons, the photoreceptors are nonreplicating, but they do have a unique system for renewal of their light-trapping outer segment. In rods, new discs are assembled at the base of the outer segment from materials synthesized in the cell body. As new discs are made, they push the others distally. The discs at the tip of the outer segment continually fragment off and are degraded by the phagocytic cells of the pigmented layer. The outer segments of cones are also renewed, but in this case, the discs are not moved or destroyed. Instead, the photoreceptor pigments appear to be transported slowly along the existing outer segment discs.

Internal Chambers and Fluids

As noted earlier, the lens and its suspensory ligament "halo" divide the eye into two cavities, the anterior cavity in front of the lens and the larger posterior cavity behind it (see Figure 16.12). The posterior cavity is filled with a clear gel called **vitreous humor** (*vitre* = glassy), consisting of a network of fine collagenic fibrils embedded in a viscous connective tissue matrix. Vitreous humor (1) transmits light, (2) supports the posterior surface of the lens and presses the neural retina firmly against the pigmented layer, and (3) contributes to intraocular pressure, helping to counteract the pulling force of the extrinsic eye muscles.

The anterior cavity (Figure 16.12), partially subdivided by the iris into the **anterior chamber** (between

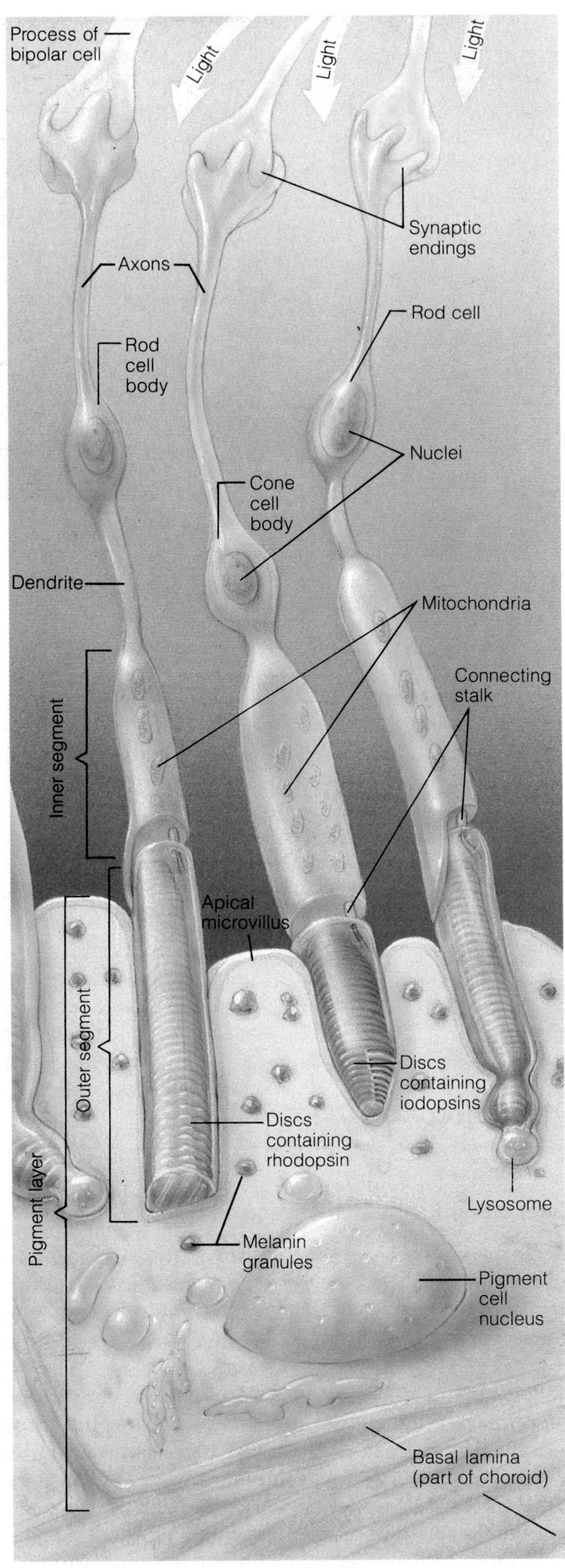

Figure 16.11 Diagrammatic view of the photoreceptors (rods and cones) of the retina. The relationship of the outer segments of the photoreceptors to the pigmented layer of the retina is also illustrated. Notice that the tip of the outer segment of the rod cell on the right is sloughing off.

the cornea and the iris) and the **posterior chamber** (between the iris and the lens), is filled with **aqueous humor,** a clear fluid similar in composition to blood plasma. Unlike the vitreous humor, which is formed during embryonic life and lasts for a lifetime, aqueous humor is continually formed and drained and is in constant motion. It is secreted by the epithelium of the ciliary processes into the posterior chamber, flows through the pupil into the anterior chamber, and is drained into the venous blood by the **canal of Schlemm,** a network of channels that encircles the eye in the angle at the sclera-cornea junction. Normally, the rates of production and drainage of aqueous humor are equal, and a constant intraocular pressure is maintained, which helps to support the eyeball interiorly. The aqueous humor supplies nutrients and oxygen to the lens and cornea, and it carries away their metabolic wastes.

If the drainage of aqueous humor is blocked, pressure within the eye increases and may cause compression of the retina and optic nerve—a condition called *glaucoma* (glah-kō′-muh). The resulting destruction of the neural structures eventually causes blindness unless the condition is detected early. Unfortunately, many forms of glaucoma steal sight so slowly and painlessly that people do not realize they have a problem until the damage is done. Late signs include seeing halos around lights, headaches, and blurred vision. The glaucoma examination is simple (a puff of air is directed at the sclera, and the amount of deformation is measured) and should be done yearly in those over 40 years of age. Early glaucoma is treated with eye drops (miotics) that increase the rate of aqueous humor drainage. ■

Lens

The **lens** is a biconvex, transparent, flexible structure that can change shape to allow precise focusing of light on the retina. It is enclosed in a thin, elastic capsule and suspended just behind the iris by the suspensory ligament (see Figure 16.12). Like the cornea, the lens is avascular; blood vessels interfere with transparency.

The lens has two regions: the lens epithelium and the lens fibers (see Figure 16.12). The **lens epithelium,** confined to the anterior lens surface, is composed of cuboidal cells that eventually differentiate into anucleate **lens fibers** forming the bulk of the lens. The tightly packed lens fibers contain precisely folded proteins called *crystallins,* which make them transparent. Since new lens fi

unfolding of the crystalline proteins. Fortunately, the offending lens can be surgically removed and an artificial lens implanted to save the patient's sight. ■

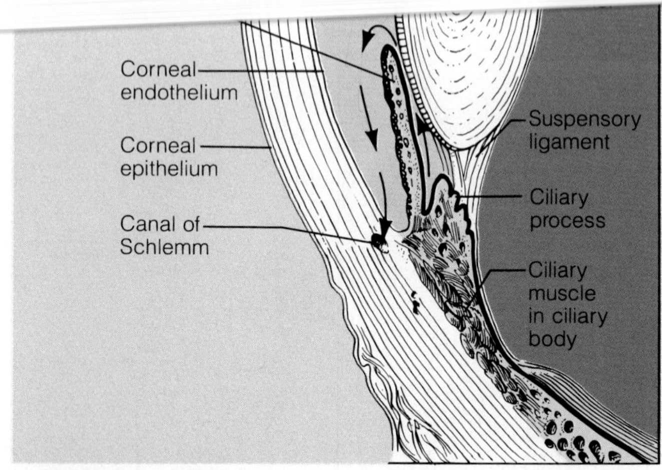

Figure 16.12 Structures seen in a transverse section through the anterior portion of the eye. The anterior cavity in front of the lens is incompletely divided into the anterior chamber (anterior to the iris) and the posterior chamber (behind the iris), which are continuous through the pupil. Aqueous humor, which fills the cavity, is formed by the ciliary processes and reabsorbed into the venous blood by the canal of Schlemm. The arrows indicate the circulation pathway.

Embryonic Development of the Eye

Embryologically, the eye is an outpocketing of the brain (Figure 16.13). By the fourth week of development, two lateral outgrowths, the **optic vesicles,** can be seen protruding from the diencephalon. Their tips indent to form the double-layered **optic cups;** the proximal parts of the outgrowths, called the **optic stalks,** become part of the optic nerves.

Once a growing optic vesicle reaches the overlying surface ectoderm, it induces the ectoderm to thicken and form a **lens placode** (pla′-kōd). By the fifth week, the lens placode has invaginated to form a **lens vesicle.** Shortly thereafter, the lens vesicle pinches off and comes to lie in the cavity of the optic cup, where it differentiates into the lens.

The lining (internal) epithelium of the optic cups differentiates into the neural retina, and the outer optic cup layer forms the pigmented retinal layer. The *optic fissure,* a groove on the underside of each optic cup and stalk, serves as a direct pathway for blood vessels to reach and supply the interior of the developing eye. When the fissure closes, the now tubelike optic stalk furnishes a tunnel through which the optic nerve fibers, originating in the neural retina, can pass to the brain, and the blood vessels come to lie in the center of the developing optic nerve. The balance of the eye tissues and the vitreous humor are formed by mesenchymal cells derived from the mesoderm. The interior of the eyeball has a rich blood supply during development, but virtually all of these blood vessels (with the exception of those serving the choroid) degenerate before birth.

Physiology of Vision

Overview: Light and Optics

To really comprehend the function of the eye as a photoreceptor organ, we need to have some understanding of the properties of light. We will therefore devote a few words to this important topic.

Wavelength and Color. *Electromagnetic radiation* encompasses all energy wavelengths, from the long radio waves (whose wavelengths can be measured in meters) to the very short gamma (γ) waves and X-rays with wavelengths of 0.001 μm and less. The only wavelengths in this spectrum to which our eyes respond

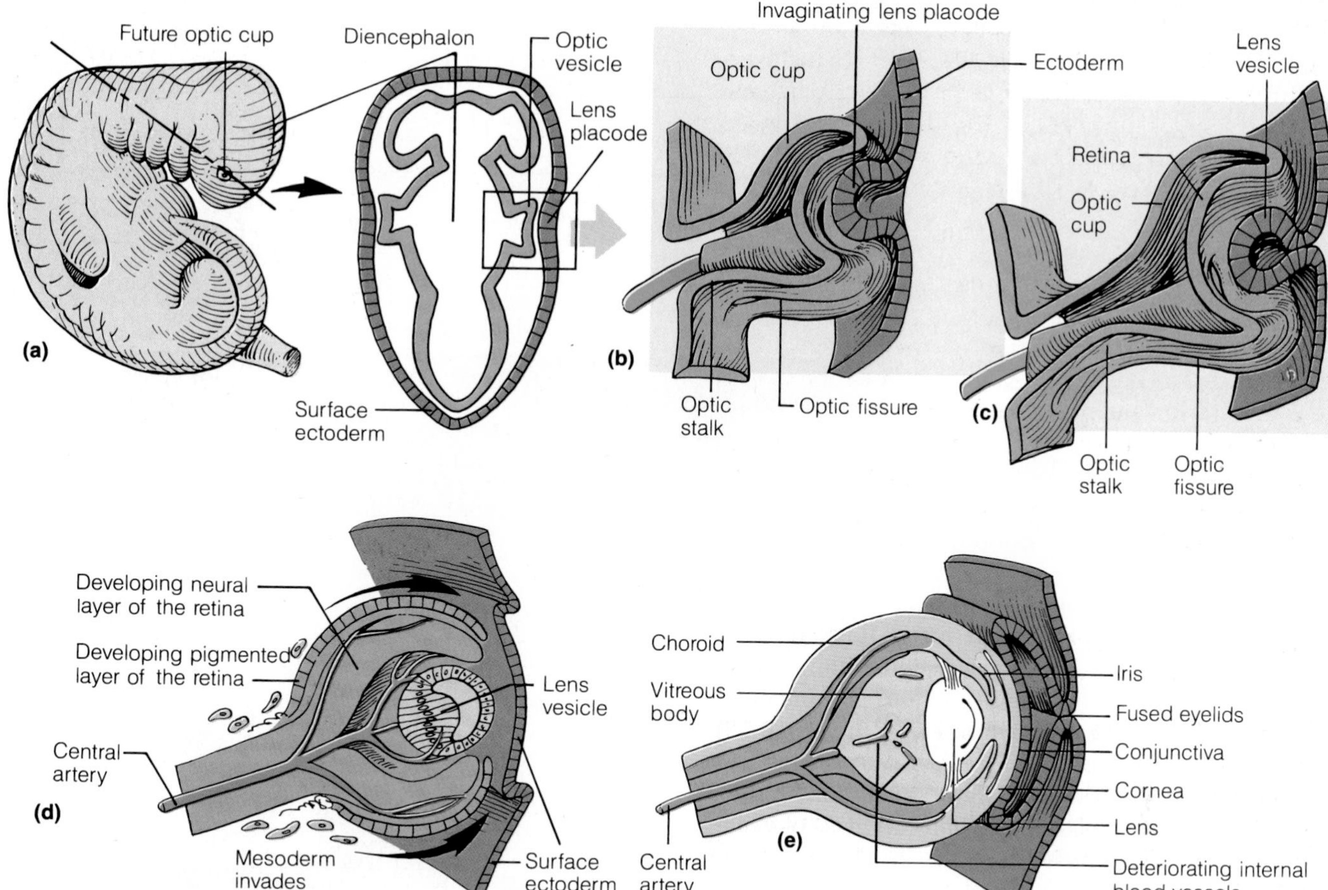

Figure 16.13 Embryonic development of the eye. (**a**) Section through the diencephalon and optic vesicle at the time of lens placode formation. (Broken line on diagram to the left shows the plane of the section.) (**b**) Contact of the optic vesicle with the lens placode causes the lens placode to invaginate. (**c**) The lens placode forms the lens vesicle, which becomes enclosed by the optic cup and then detaches. (**d**) Blood vessels reach the eye interior via the optic fissure and then become enclosed in the optic stalk (which becomes the optic nerve). The optic cup forms the nervous and pigmented layers of the retina, and invading mesoderm forms the other eye tunics and the vitreous humor. (**e**) The lens vesicle differentiates to form the lens; the surrounding accessory structures develop from the surface ectoderm.

are those in the portion designated as **visible light,** which has a wavelength range of approximately 400–700 nm (Figure 16.14). (1 nm = 10^{-9}m, or one-billionth of a meter.) Visible light has been shown to travel in the form of waves, and its wavelengths can be measured very accurately. On the other hand, many experiments have shown that light is composed of small particles or packets of energy called **photons** or **quanta.** Attempts to reconcile these two views have led to the present concept of light energy as packets of energy (photons) traveling in a wavelike fashion at very rapid speeds (186,000 miles per second). Thus, light can be thought of as a vibration of pure energy rather than a material substance.

When visible light is passed through a prism, each of its component waves is bent to a different degree, so that the beam of light is dispersed and a **visible spectrum,** or band of colors, is seen (see Figure 16.14). (In the same manner, the rainbow seen during a summer shower represents the collective prismatic effects of all the tiny water droplets suspended in air.) The visible spectrum progresses from red to violet; the red wavelengths are the longest and have the lowest energy, whereas the violet wavelengths are the shortest and most energetic. Objects have color because they absorb some wavelengths and reflect others. Things that look white are reflecting all wavelengths of light, whereas black objects are absorbing them all. A red apple reflects mostly red light; grass reflects more of the green.

Refraction and Lenses. Light travels in straight lines and is easily blocked by any nontransparent object. Like sound, light can be reflected, or bounced, off a surface. **Reflection** of light by objects in our environment accounts for most of the light reaching our eyes.

When light is traveling in a given medium, its speed is constant. But when it passes from one transparent medium into another with a different density, its speed is altered. Light speeds up as it passes into

Figure 16.14 The electromagnetic spectrum. The electromagnetic spectrum extends from the very short gamma waves to the long radio waves. The wave

a less dense medium and slows as it passes into a denser medium. As a consequence of these changes in speed, a light ray is bent or refracted when it meets the surface of a different medium at an oblique angle rather than perpendicular to its surface (Figure 16.15). The greater the incident angle, the greater the amount of bending. To demonstrate light **refraction,** hold a pencil in a slanted position in a half-full glass of water: The pencil will appear to bend at the air–water interface.

A lens is a piece of transparent material, curved on one or both surfaces. Since light hits the curve at an angle, it is refracted. If the lens surface is convex, that is, thickest in the center, like a camera lens, the light is bent so that it converges, or comes together, at a single point called the **focal point** (Figure 16.16). In general, the thicker (more convex) the lens, the more the light is bent and the shorter the focal distance (distance between the lens and focal point). The image formed by a convex lens, called a **real image,** is upside down and reversed from left to right. On the other hand, concave lenses, which are thicker at the edges than at the center, like those of magnifying glasses, diverge the light (bend it outward) and prevent it from focusing.

Focusing of Light on the Retina

The eye is frequently compared to a camera (Figure 16.17). A camera contains a variable aperture to control light entry and a convex lens to focus the light on a light-sensitive surface (the film), and it is aimed by the operator to capture the desired scene. The eye also contains a variable aperture (the pupil) and refractive media (the lens, cornea, and others). Like a camera lens, the lens of the eye is the dynamic focusing apparatus. However, unlike a camera lens, which is moved back and forth to gain a precise focus, the lens of the eye adjusts focal length by changing shape—flatten-

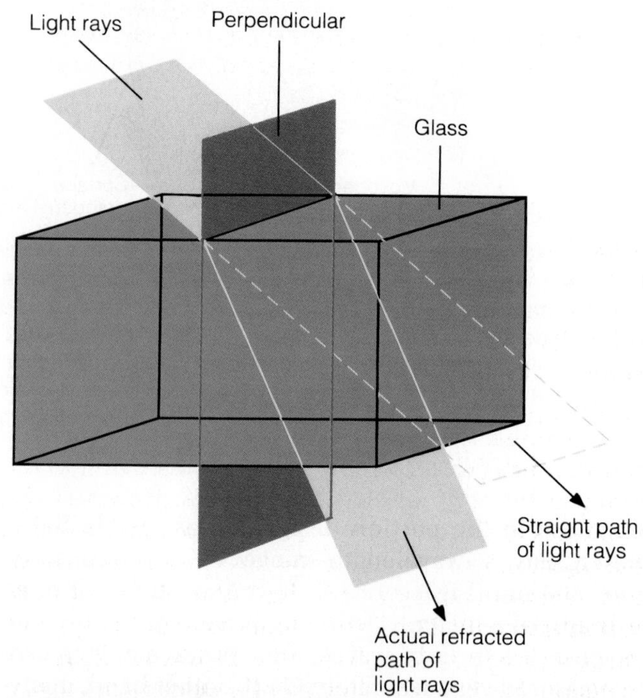

Figure 16.15 Refraction of light as it passes from one transparent medium to another. Light rays passing obliquely from air into the more dense medium of glass. Since their speed is slowed, the light rays are bent toward the perpendicular. (In the reverse case—that is, passing obliquely from a more dense to a less dense medium—the rays would be bent away from the perpendicular.)

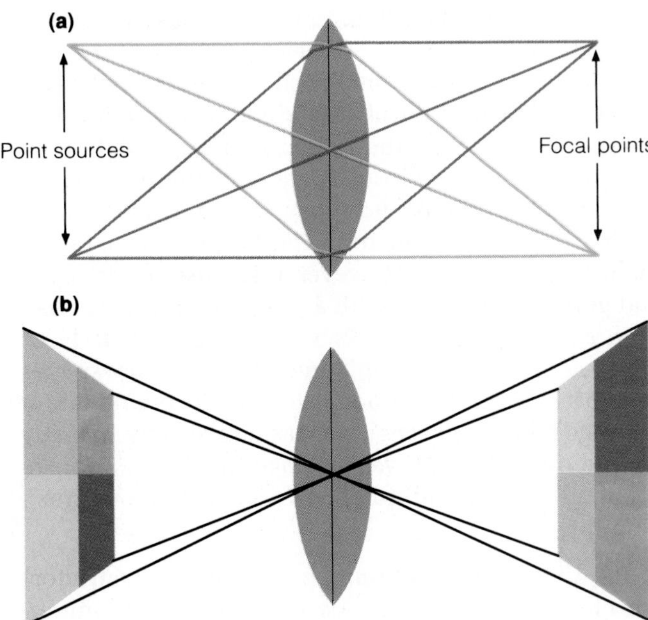

Figure 16.16 Bending of light points by a convex lens. (**a**) How two light points are focused on the opposite side of the lens. (**b**) The formation of an image by a convex lens. Notice that the image is upside down and reversed from right to left.

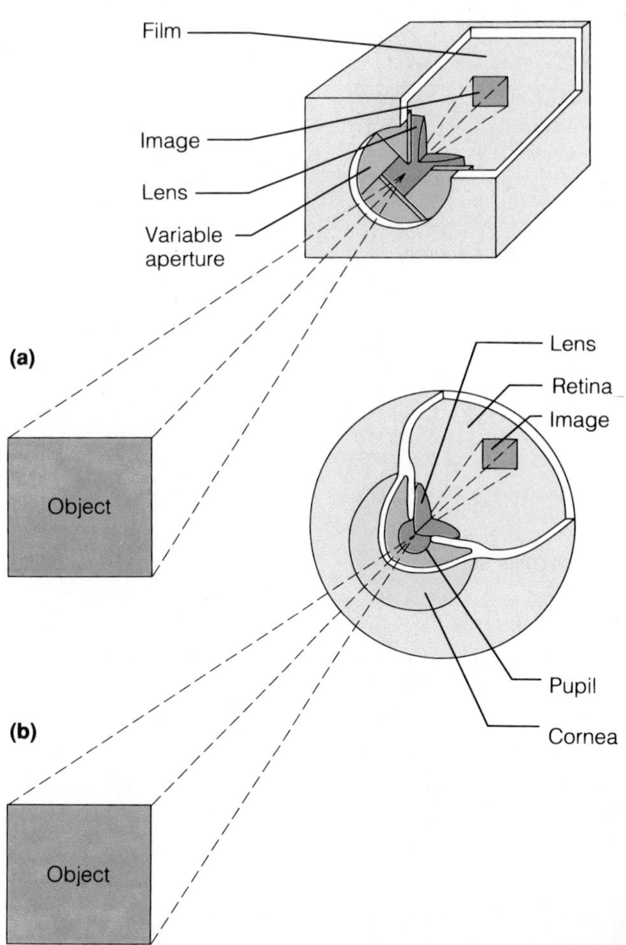

Figure 16.17 Comparison of the eye to a simple pinhole camera. (**a**) Optics of a camera. (**b**) Optics of the eye.

ing or bulging. The retina is comparable to the light-sensitive film, and the eye is aimed by the extrinsic eye muscles.

Since a camera utilizes a convex lens, the image that falls on the film is a real image. The final "hard-copy" picture is produced when the film is developed by the photo lab. Likewise, a real image is projected on the retina, and information about the energies and trajectories of the light hitting it is transmitted to the brain in the form of electrical signals. However, the workings of the conscious brain allow us to perceive the object in its true position (rather than upside down and reversed) and to "develop" the picture almost instantaneously. Thus, the comparison of the eye to a camera is fairly accurate for the mechanical aspects of vision, but it breaks down when we consider visual processing, which is far more complex than film processing.

As light passes from air into the eye, it moves successively through the cornea, aqueous humor, lens, and vitreous humor, and then passes through the entire thickness of the neural layer of the retina to excite the photoreceptors that abut the pigmented layer (see Figure 16.9). During its passage, light is bent three times: on entry into the cornea and on entering and leaving the lens. The cornea is responsible for most of the light refraction in the eye, but since its thickness is constant, its refractive power is unchanging. On the other hand, the lens is highly elastic, and its curvature can be actively changed to allow fine focusing of the image. The humors are of minimal importance in light refraction.

Focusing for Distance Vision. Our eyes are best adapted for distance vision. To look at distant objects, we need only aim our eyeballs so that they are both fixated on the same spot. The **far point of vision** is that distance beyond which no accommodation, or change in lens shape, is needed for focusing. For the normal, or **emmetropic** (eh″-muh-trō′-pik), eye, the far point is 6 m (20 feet).

Since distant objects appear smaller, light from an object at or beyond the far point of vision approaches the eyes as nearly parallel rays and is focused precisely on the retina by the fixed refractory apparatus (cornea and humors) and the at-rest lens (Figure 16.18a). During distance vision, the ciliary muscles are completely relaxed, and the lens is as flat as it gets. Consequently, its refractory power is at its lowest point.

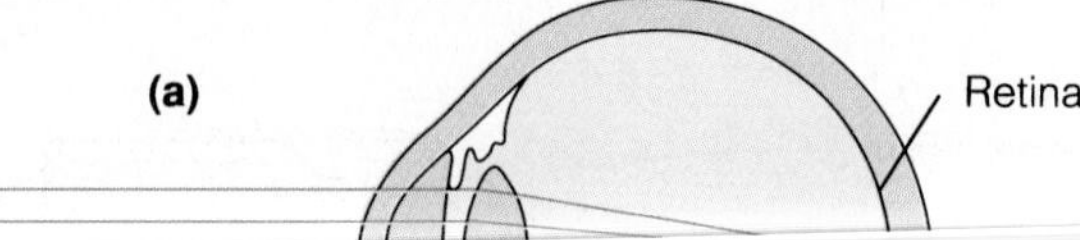

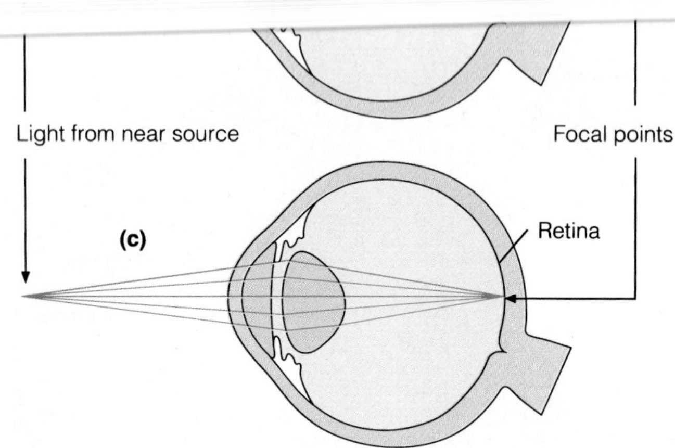

Figure 16.18 Influence of distances of the object viewed and lens convexity on focal length. (**a**) Light from distant objects approaches the eyes as parallel rays, whereas (**b**) light from objects closer than 6 m diverges as it approaches the eyes. The lenses in (a) and (b) are of equal strength. Notice the difference in the focal points. (**c**) When the lens's curvature is increased, the focal length is decreased.

Focusing for Close Vision. Light from objects less than 6 m away diverges as it approaches the eyes and comes to a focal point farther from the lens, as shown in Figure 16.18b. To restore focus, three processes—accommodation of the lenses, constriction of the pupils, and convergence of the eyeballs—must be initiated simultaneously, and the signal that induces this trio of reflex responses appears to be a blurring of the retinal image.

1. Accommodation. The process by which the refractive power of the lens is increased is called accommodation. To achieve this, the ciliary muscles contract, pulling the ciliary body forward and inward and allowing the fibers of the suspensory ligament to go slack. Because the tension on the lens capsule is reduced, the lens bulges, providing the shorter focal length neeeded to focus the image of a close object on the retina (Figure 16.18c). Contraction of the ciliary muscles is controlled mainly by the parasympathetic fibers of the oculomotor nerves.

The closest point on which we can

commodating, a condition known as *presbyopia* (prez″-bē-ō′-pē-ah).

2. Pupillary constriction. The circular (constrictor) muscles of the iris enhance the effect of accommodation by reducing the size of the pupil toward 2 mm. This **accommodation pupillary reflex** prevents the most divergent light rays from entering the eye and passing through the lens edge; such rays would not be focused properly and would cause blurred vision. The constriction of the pupils for close vision makes the eye act like a pinhole camera and greatly increases the clarity and depth of focus.

3. Convergence of the eyeballs. Synchronized movements of the eyeballs (promoted by the extrinisic eye muscles) allow us to fix our gaze on something. The goal is always to keep the object being viewed focused on the retinal fovea of each eye. When we look at distant objects, both eyes are directed either straight ahead or to one side to the same degree, but when we fixate on a close object, our eyes converge. **Convergence** is the medial rotation of the eyeballs by the medial rectus muscles so that each is directed toward the object being viewed. The closer the object viewed, the greater the degree of convergence required; when you focus on the tip of your nose, you "go cross-eyed."

Reading or other close work requires almost continuous accommodation and effort on the part of the intrinsic eye muscles (iris and ciliary muscles) and the medial recti. This is why prolonged periods of reading tire the eyes and can result in *eyestrain*. If you are reading for an extended time, it is helpful to look up from time to time and stare into the distance. This relaxes all the eye muscles temporarily.

Homeostatic imbalances of refraction. Visual problems related to refraction can result from an overconverging or underconverging lens or from structural abnormalities of the eyeball.

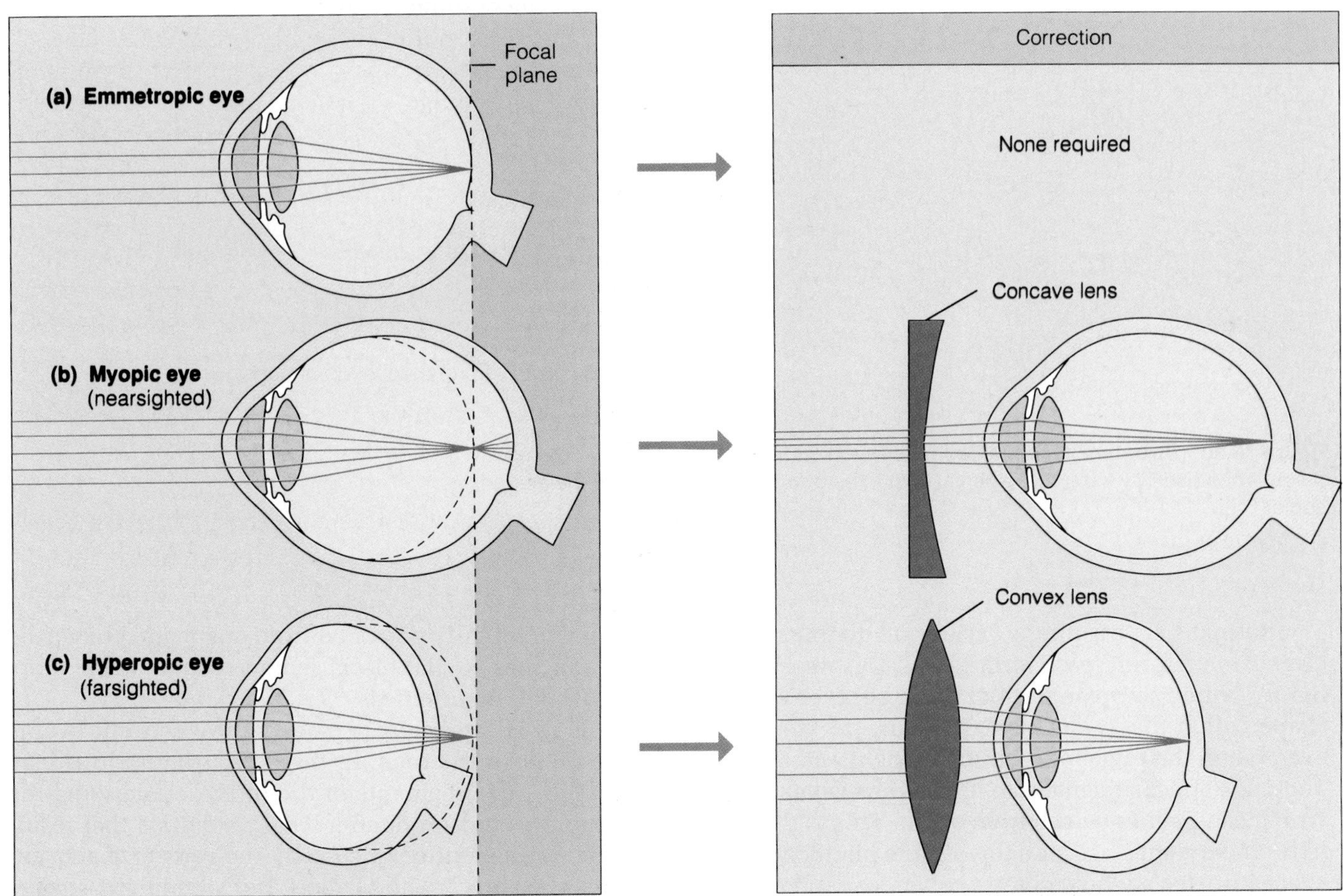

Figure 16.19 Problems of refraction. (**a**) In the emmetropic (normal) eye, light is focused properly on the retina. (**b**) In a myopic eye, light is brought to a focal point before reaching the retina and then diverges again. (**c**) In the hyperopic eye, light is brought to a focal point behind (past) the retina.

Myopia (mī-ō′-pē-uh) occurs when distant objects are focused not on, but in front of, the retina (Figure 16.19b). Myopic people can see close objects without problems (because of accommodation), but distant objects are blurred. The common name for myopia is *nearsightedness.* (Notice that this terminology names the aspect of vision that is *unimpaired.*) Myopia results from a lens that is too strong or an eyeball that is too long. Correction requires a concave lens that diverges the light before it enters the eye.

If a close image is normally focused behind the retina, the individual is said to have *hyperopia* (hī-per-ah′-pē-uh), or *farsightedness* (Figure 16.19c). Hyperopic individuals see distant objects perfectly well, but need corrective glasses with convex lenses to converge the light more strongly for close vision. Hyperopia usually results from a lazy lens (one with poor refractory power) or an eyeball that is too short.

Unequal curvatures in different parts of the lens (or cornea) lead to a blurred vision problem called *astigmatism.* Special cylindrically ground lenses are used to correct this problem. ■

Stimulation of the Photoreceptors

Once light has been focused on the retina, the photoreceptors come into play. We have already seen how the rods and cones are built; now we will consider how they are excited by light.

The Chemistry of Visual Pigments. How does light affect the photoreceptors so that it is ultimately translated into electrical signals? The key is a light-absorbing molecule called **retinal** (or **retinene**) that combines with proteins called **opsins** to form four types of visual pigments. Depending on the type of opsin to which it is bound, retinal preferentially absorbs different wavelengths of the visible spectrum. Retinal is chemically related to vitamin A and is synthesized from it. The liver contains the body's stores of vitamin A and releases it as needed by the photoreceptors to make their visual pigments. The cells of the pigmented layer of the retina absorb vitamin A from the blood and serve as a vitamin A depot for the rods and cones.

Figure 16.20 Structure of retinal isomers involved in photoreception. The effect of light absorption by the visual pigment is to transform the 11-*cis* retinal to the all-*trans* isomer.

Retinal can assume a variety of distinct three-dimensional forms, each form called an isomer. When tightly bound to opsin, retinal has a bent, or kinked, shape called the **11-*cis* isomer** (Figure 16.20). However, when the pigment is struck by light and absorbs photons, retinal straightens to its **all-*trans* isomer** form, which causes it to detach from opsin. This is the *only* light-dependent stage, and this simple photochemical event initiates a whole chain of chemical and electrical reactions in rods and cones that ultimately causes electrical impulses to be transmitted along the optic nerve.

Excitation of Rods. The visual pigment of rods is a deep purple pigment called **rhodopsin** (rō-dop′-sin), which is arranged in a single layer in each of the thousands of discs in the rods' outer segments. Rhodopsin is formed and accumulated in the dark in the sequence of reactions shown in Figure 16.21. As illustrated, vitamin A is oxidized to the 11-*cis* retinal form and then combined with **scotopsin** (skō-top′-sin), the opsin found in rods, to form rhodopsin.

When rhodopsin absorbs light, retinal changes shape from the 11-*cis* to the all-*trans* form, and the retinal-scotopsin combination then breaks down, allowing retinal and scotopsin to separate from each other. This breakdown is known as the *bleaching* of the pigment. Rhodopsin is regenerated when all-*trans* retinal is converted back to its 11-*cis* form and joined to opsin once again. Actually, the story is a good deal more complex than this, and the breakdown of rhodopsin that triggers the transduction process involves a rapid cascade of intermediate steps (shown on the right side of Figure 16.21) that converts rhodopsin to an unstable intermediate called metarhodopsin II in less than 2 ms. The remaining steps in which all-*trans* retinal is actually detached from opsin and converted by enzymes to its 11-*cis* isomer occur much more slowly, taking minutes.

Now we are ready to discuss the electronic events that occur in the rods. In the dark, sodium ions leak into the outer segments of the rods, depolarizing the membrane and producing local potentials that result in neurotransmitter release by the rods at their synapses with bipolar nerve cells. But when light triggers rhodopsin breakdown, the sodium permeability of the outer segment membranes decreases abruptly. The mechanism of this change is still unclear, but inhibition of cyclic GMP, an intracellular second messenger, is believed to be involved. Whatever the mechanism, the net effect of light is to "turn off" sodium entry and to hyperpolarize the rod cells, which thus develop a *hyperpolarizing receptor potential* that inhibits their release of neurotransmitter. This is bewildering, to say the least. Here we have receptors built to detect light that are depolarized in the dark—producing the so-called *dark current*—and hyperpolarized when exposed to light! Nevertheless, as we will explain in the section called *Visual Processing*, turning off electrical signals (hyperpolarization) can convey information just as effectively as turning them on.

The photoreceptors do not generate action potentials. In fact, none of the retinal neurons generate large action potentials except the ganglion cells. This is consistent with the function of ganglion cells as the output neurons of the retina. In all other cases, only local currents (graded potentials) are produced. This is not surprising if you remember that the primary function of action potentials is to carry information rapidly over long distances. Since the retinal cells are very close together, graded potentials serve quite adequately as signals to regulate neurotransmitter release.

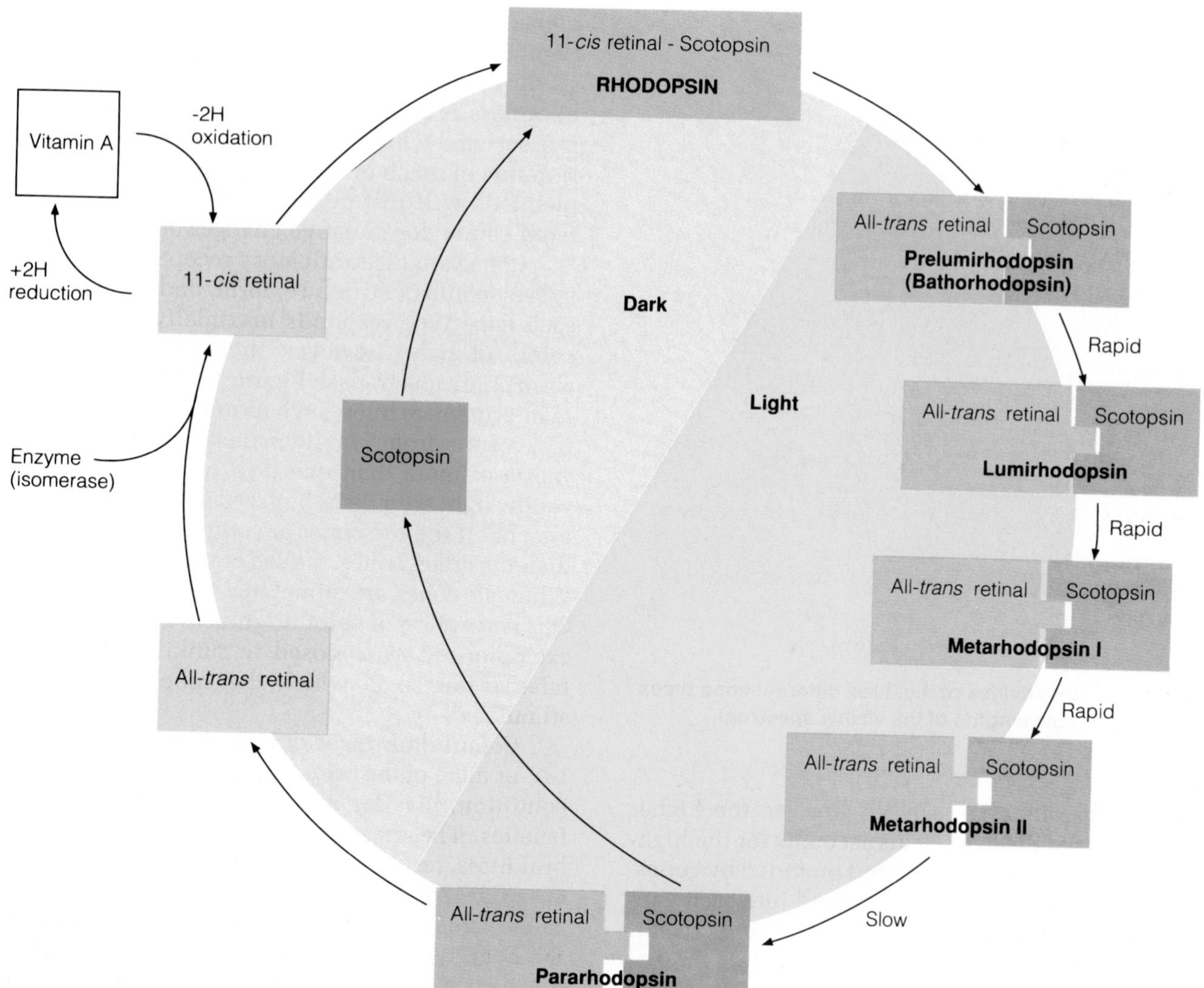

Figure 16.21 Sequence of chemical events in the formation and breakdown of rhodopsin. When struck by light, rhodopsin breaks down and is converted to its precursors via a number of intermediate steps. Absorption of light causes a rapid conversion of 11-*cis* retinal to all-*trans* retinal as the unstable intermediate prelumirhodopsin is formed; then lumirhodopsin, metarhodopsin I, and metarhodopsin II are generated. This cascade occurs rapidly and is responsible for transduction of the light stimulus. When the next intermediate, pararhodopsin, is formed, all-*trans* retinal and scotopsin become separated. The 11-*cis* retinal is regenerated by enzymatic action from all-*trans* retinal or from vitamin A and recombined with scotopsin to form rhodopsin.

Excitation of Cones. The visual pigments of cones, like those of rods, are a combination of retinal and opsins. The opsins of cones are collectively referred to as **photopsins.** However, there are three distinct types of cones, and each type is most sensitive to a different wavelength range, which reflects the specific properties of the photopsin to which the retinal is bound. The three types of photopsins differ both from the scotopsin of the rods and from one another. Each of the different cone opsins provides a unique environment for the retinal by surrounding it with negative electrical charges, and retinal's sensitivity to different wavelengths of light depends on the number and location of these negative charges. The naming of the cones reflects the colors (that is, wavelengths) of light that the cones absorb best and that are most efficient at causing the retinal to change shape and detach from its opsin. The blue cones respond maximally to wavelengths of 455 nm, the green cones to wavelengths of 530 nm, and the red cones to wavelengths of 625 nm (Figure 16.22).

The chemistry of cone function is not completely worked out, but the breakdown and regeneration of visual pigments is believed to be essentially the same as for rhodopsin. However, the threshold for cone activation is much higher than for rods, and cones respond only to high-intensity (bright) light. Another important difference is that in the fovea, one cone (or at most a few) feeds into each ganglion cell; thus, each cone

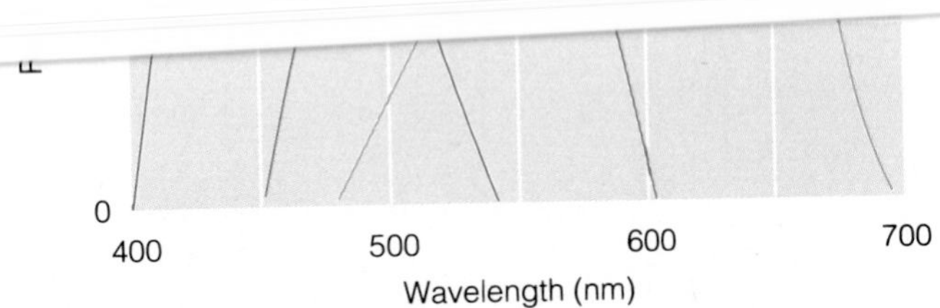

Figure 16.22 Sensitivities of the three different cone types to the different wavelengths of the visible spectrum.

essentially has its own "labeled line" to the higher visual centers of the brain. This accounts for the high-acuity vision (sharpness of vision) provided by cones. In contrast, large numbers of rods feed into each ganglion cell, and their effects are summated and considered collectively.

Role of the Rods and Cones in Vision. Rods enable us to see best under conditions of low illumination and are responsible for night vision and peripheral vision. Rhodopsin is amazingly sensitive; even starlight causes some of the molecules to become bleached. As long as the light is of low intensity, the rods quickly regenerate rhodopsin, and the retina continues to respond to light stimuli. However, in high-intensity light, rhodopsin is bleached nearly as rapidly as it is made, and the rods become nonfunctional. At this point, cones begin to take over.

Although rhodopsin absorbs light throughout the visible spectrum, the rods do not confer color discrimination, and rod-mediated visual inputs are perceived as gray tones. Since rods are absent from the foveae, and since cones do not respond to low-intensity light, we see dimly lit objects best when we do not look at them directly. However, without foveal focusing, or *hard focus*, our vision is nondiscriminatory. We see only vague outlines of objects and recognize them best when they move. If you doubt this, go out into your backyard on a moonlit evening and see how much you can actually discriminate.

Anything that interferes with rod function hinders our ability to see at night, a condition called *night blindness* or *nyctalopia* (nik″-tuh-lō′-pē-uh). Night blindness dangerously impairs the ability to drive safely at night. The most

differential activation of more than one type of cone. For example, yellow light stimulates both red and green cone receptors, but if the red cones are stimulated more strongly than the green cones, we see orange instead of yellow. When all cones are stimulated equally, we see white. Our perception of color brightness and saturation (for example, red as opposed to pink) also reflects the relative degree to which the cone populations are stimulated.

Color blindness is due to a congenital lack of one or more of the cone types. Inherited as a sex-linked condition, it is far more common in males than in females. The most common type is red-green color blindness, resulting from a deficit or absolute absence of either red or green cones. Red and green are seen as the same color—either red or green, depending on the cone type present. Many color-blind people are unaware of their condition because they have learned to rely on other cues—such as differences in intensities of the same color—to distinguish something green from something red, such as traffic signals. ■

Light and Dark Adaptation. There is an automatic adjustment of retinal sensitivity to the amount of light present, which can only be partially explained by the light-induced breakdown of photoreceptor pigments. The other part of the story appears to involve retinal neurons other than the photoreceptors. **Light adaptation** occurs when we move from darkness into bright light, as when leaving a movie matinee. We are momentarily dazzled—all we see is white light—because the sensitivity of the retina is still "set" for dim light. Both rods and cones become strongly stimulated, and significant amounts of the photopigments are broken down almost instantaneously, producing a flood of signals that accounts for the glare. Under such conditions, compensations occur: (1) The sensitivity of the retina decreases dramatically (actually, less than a 1% breakdown of rhodopsin results in a huge loss in rod sensitivity), and (2) retinal neurons undergo rapid adaptation that inhibits rod functioning when light conditions favor cone function. These events raise the threshold of the retina, and within about 60 seconds, the cones are sufficiently excited by the bright light to take over. Visual acuity and color vision continue to improve over the next 5–10 minutes. Thus, during light adaptation, retinal sensitivity is lost, but visual acuity is gained.

Dark adaptation is essentially the reverse of light adaptation, and it occurs when we go from a well-lit area into a dark one. Initially, we see nothing but velvety blackness because (1) our cones nearly cease functioning in low-intensity light, and (2) our rod pigments have been bleached out by the bright light, and the rods are still initially inhibited. But once we are in the dark, rhodopsin begins to accumulate, and sensitivity of the retina increases. Dark adaptation is much slower than light adaptation and can go on for hours. However, usually enough rhodopsin has been accumulated within 20–45 minutes to allow adequate dim-light vision.

During these adaptation phenomena, reflexive changes also occur in pupil size. Bright light shining in one or both eyes causes both pupils to constrict (elicits the *consensual* and *pupillary light reflexes*). These pupillary reflexes are mediated by the pretectal nucleus of the midbrain and by parasympathetic fibers. In dim light, the pupils are dilated, allowing more light to enter the eye interior.

The Visual Pathway to the Brain

As described in Chapter 13, the axons of the retinal ganglion cells issue from the back of the eyeballs in the **optic nerves,** which merge to form the **optic chiasma** (Figure 16.23). The fibers from the medial aspect of each eye cross over at the optic chiasma, then continue on to the thalamus via the **optic tracts.** Thus, each optic tract (1) contains fibers from the lateral (temporal) aspect of the eye on the same side and fibers from the medial (nasal) aspect of the opposite eye, and (2) carries all the information from the same half of the visual field. Thus, because the image formed on the retina is a real image, that is, reversed left to right, the left optic tract carries a complete representation of the right half of the visual field. The ganglion cell axons synapse with neurons in the **lateral geniculate body** of the thalamus, which maintains this fiber separation. Axons of the thalamic neurons then extend to the occipital lobes of the cerebrum (*visual,* or *striate, cortex*) via the **optic radiation,** which completes the visual pathway to the brain.

Many nerve fibers are projected from the retina directly to the midbrain. These fibers terminate in the **superior colliculi,** visual reflex centers controlling the extrinsic muscles of the eyes, and in the **pretectal nuclei,** which, as already noted, mediate the pupillary light reflexes.

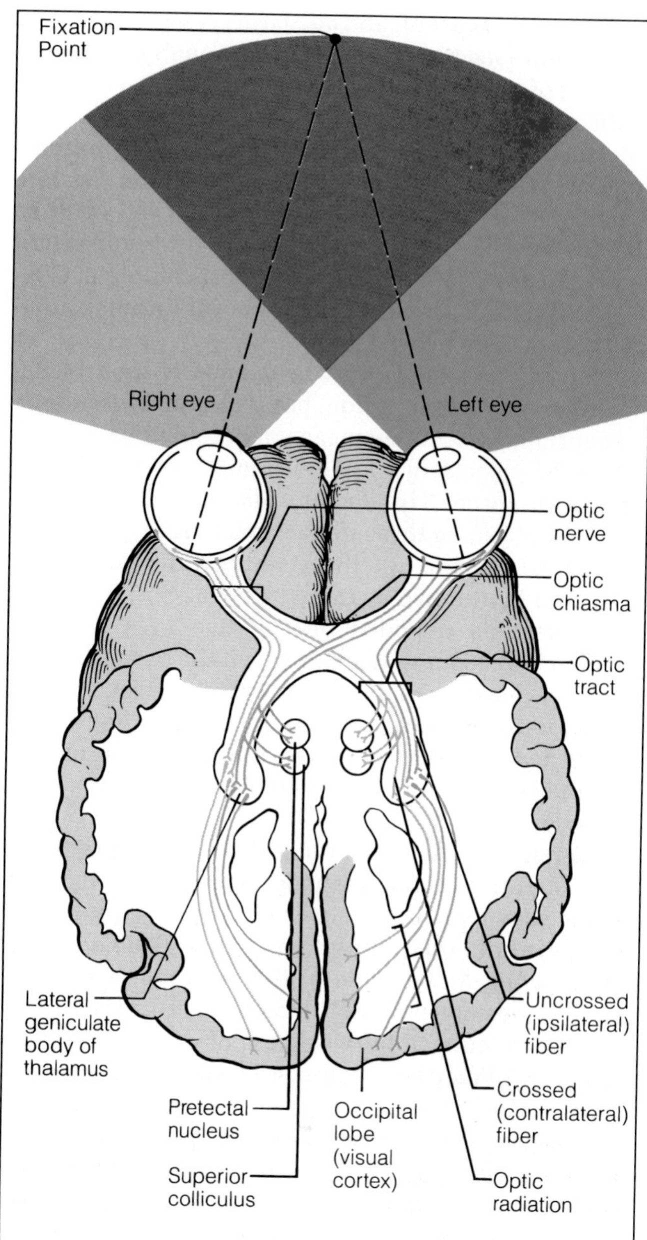

Figure 16.23 Visual fields of the eyes and visual pathway to the brain. Notice that the visual fields overlap considerably (area of binocular vision). Also notice the retinal sites at which a real image would be focused when both eyes are fixated on a close, pointlike object. (For each eye, light from an object in the nasal portion of the visual field is inverted by the lens and projected to the temporal side of the retina.) The full lateral extension of the visual fields is not illustrated here.

Binocular Vision and Stereopsis

Binocular vision (literally, "two-eyed vision") is found in humans and most other primates, as well as in predatory birds and cats. Although both eyes are set anteriorly and look in approximately the same direction, their visual fields, each about 170 degrees, overlap to a considerable extent, and each eye sees a slightly different view (see Figure 16.23). Additionally, the crossing over of approximately half of the optic nerve fibers at the optic chiasma provides

... also called **stereopsis** (steh-rē-op′-sis), or **three-dimensional vision,** is a result of cortical "fusion" of the slightly different images delivered by the two eyes.

Stereopsis depends on the two eyes working together and accurately focusing on the object. If only one eye is used, then depth perception is lost, and the person must learn to judge an object's position based on learned cues (for example, the nearer the object, the larger it appears, and parallel lines converge with distance).

A person who cannot focus the same area of the visual field on corresponding regions of the two retinas sees two images instead of one. This condition, called *diplopia* (dih-plō-pē-uh), or *double vision,* occurs when movements of the external muscles of the two eyes are not coordinated. It can result from paralysis or congenital weakness of certain extrinsic muscles, or it may be a temporary consequence of acute alcohol intoxication.

Congenital weakness of the external eye muscles, called *strabismus* (struh-biz′-mis), can also lead to walleye, in which one eye rotates medially or laterally. In some cases, the normal and deviant eyes alternate in focusing on objects. In other cases, only the controllable eye is used, and the visual cortex begins to disregard inputs from the deviant eye, which becomes functionally blind. ■

Visual Processing

How does information received by the rods and cones become vision? In the past few years, a tremendous amount of research has been devoted to this question. Some of the basic information that has come out of these studies is presented here.

Retinal Processing. The retinal ganglion cells generate action potentials at a fairly steady rate (20–30 Hz), even in the dark. Illuminating the entire retina evenly has no effect on the basal rate of discharge, but the activity of individual ganglion cells changes dramatically when a tiny spot of light falls just on certain portions of their receptive field. The receptive field of any ganglion cell consists of

... ganglion cells with off-center fields are inhibited by light hitting rods in the field center and are stimulated by light hitting rods in the periphery. Equal illumination of the "on" and "off" regions causes little change in the basal rate of ganglion cell firing. However, the ganglion cells do respond to differing illumination of parts of their fields, which leads to changes in their rate of impulse generation (see Figure 16.24).

The mechanisms of retinal processing as currently visualized can be summarized as follows:

1. The action of light on photoreceptors hyperpolarizes them.

2. Bipolar neurons in the "on" regions depolarize when the rods feeding into them are illuminated (and hyperpolarized); bipolar neurons in the "off" regions hyperpolarize when the rods feeding into them are stimulated.

3. Bipolar neurons in the center of a receptive field feed straight through to the ganglion cell; this is the on-line pathway. Thus, depolarizing bipolar neurons excite the ganglion cell, whereas hyperpolarizing bipolar cells inhibit the ganglion cell.

4. The activity of bipolar neurons in the field periphery, the off-line pathway, is modified (subjected to *lateral inhibition*) by the horizontal cells of the retina (see Figure 16.9). The horizontal cells, connected to both the photoreceptors and bipolar neurons, cause a "sign reversal" in the field periphery. If the bipolar neurons in the straight-through pathway are excited, the horizontal cells inhibit the bipolar neurons in the periphery. On the other hand, if the bipolar neurons in the field center are hyperpolarized (off), the horizontal cells excite bipolar neurons in the field periphery. Thus, the horizontal cells are local integrators that cause the ganglion cells to receive information from the field periphery that is *antagonistic* to the information they are receiving from the field center. This "selective destruction" of some of the rod inputs allows the retina to convert the pointlike light inputs into

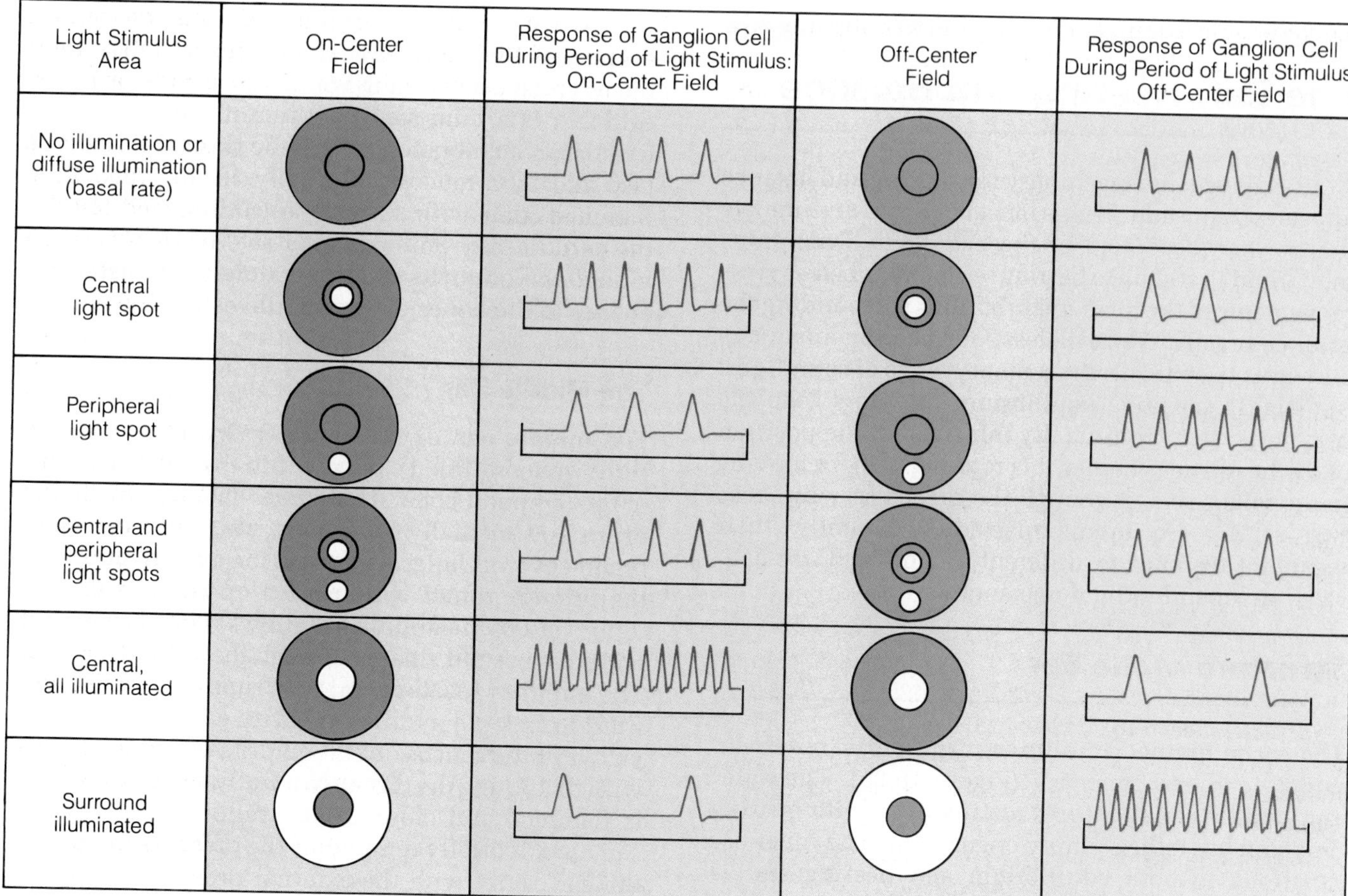

Light Stimulus Area
On-Center Field
Response of Ganglion Cell During Period of Light Stimulus: On-Center Field
Off-Center Field
Response of Ganglion Cell During Period of Light Stimulus: Off-Center Field
No illumination or diffuse illumination (basal rate)
Central light spot
Peripheral light spot
Central and peripheral light spots
Central, all illuminated
Surround illuminated

Figure 16.24 Responses of on-center and off-center retinal ganglion cells to different types of illumination.

perceptually more meaningful contrast and contour information and accentuates bright/dark contrasts, or edges.

5. The amacrine cells, also local integrator cells, exert transient inhibitory effects on the ganglion cells. Their activity is complex and will not be considered further here.

Thalamic Processing. The lateral geniculate bodies of the thalamus are most concerned with segregating the retinal axons in preparation for depth perception and with sharpening the contrast information received from the retina. The inputs from the two eyes are projected to discrete layers of each lateral geniculate nucleus, and it is in the thalamus that the signals from the two eyes first come together. However, not all parts of the retina are represented equally: Many more synapses are made with axons coursing from the center of the retina than with those from the retina periphery. Thus, it appears that these thalamic nuclei are most concerned with the high-acuity aspects of vision, and their projections to the visual cortex emphasize visual inputs from the region of high cone density.

Cortical Processing. There are two major populations of cortical neurons involved in processing retinal inputs: the simple and complex cortical neurons. The receptive field of a **simple cortical neuron** consists of a group of retinal ganglion cell fields, all with the same orientation and center type. The simple cortical cells respond to straight edges of light and dark contrast. Since we can see the spoke of a wheel, there must be simple cortical cells that respond to every possible radius on the retinal surface. The **complex cortical neurons** have receptive fields encompassing those of several simple cortical neurons, and these neurons continue to respond as long as any of their "feeder" simple cortical cells are activated. This allows the complex cells to monitor the motion of objects in the visual field.

The Ear: Hearing and Balance

two senses are interconnected structurally, their receptors respond to different stimuli and are activated independently of one another.

Structure of the Ear

The ear is divided into three major areas: outer ear, middle ear, and inner ear (Figure 16.25). The outer and middle ear structures are involved with hearing only and are rather simply engineered. The inner ear functions in both equilibrium and hearing and is extremely complex.

The Outer (External) Ear

The **outer (external) ear** consists of the auricle and the external auditory canal. The **auricle,** or **pinna,** is what most people call the ear—the shell-shaped projection surrounding the opening of the external auditory canal. The auricle is elastic cartilage covered with thin skin containing sebaceous and sweat glands and, occasionally, hairs. Its rim, the *helix,* is somewhat thicker, and its fleshy, dangling *lobe* lacks supporting cartilage. The function of the auricle is to direct sound waves into the external auditory canal. Some animals (coyotes, for example) can move their auricles toward a sound source, but in humans, the muscles controlling the auricle's movement are vestigial and inefficient.

The **external auditory canal** is a short, narrow chamber (about 2.5 cm long by 0.6 cm wide) that extends from the auricle to the tympanic membrane. Near the auricle, its framework is elastic cartilage; the remainder of the canal is carved into the temporal bone. The entire canal is lined with skin bearing hairs, sebaceous glands, and modified apocrine sweat glands called **ceruminous** (ser-oo′-mih-nus) **glands.** The ceruminous glands secrete yellow–brown waxy **cerumen** (ser-oo′-min), or earwax, which provides a sticky trap for foreign bodies.

Sound waves entering the external auditory canal eventually hit the **tympanic membrane,** or **eardrum,** the boundary between the outer and middle ears. The

of the temporal bone. It is flanked laterally by the eardrum and medially by a bony wall with two membrane-covered holes, the superior **oval window** and the inferior **round window.** An opening in the posterior wall of the middle ear allows it to communicate with the mastoid sinuses through the **mastoid antrum.** The **auditory (eustachian) tube** runs obliquely downward to link the middle ear cavity with the nasopharynx, and the mucosa of the middle ear is continuous with that lining the throat. Normally, the auditory tube is flattened and closed, but swallowing or yawning can open it briefly to equalize the pressure in the middle ear cavity with the external pressure. This is an important function because the eardrum does not vibrate freely unless the pressure on both of its surfaces is the same. When the pressures are unequal, the eardrum bulges inward or outward, causing hearing difficulty (voices may sound far away) and sometimes earaches. The ear-popping sensation of the pressures equalizing is familiar to anyone who has flown in an airplane.

Inflammation of the middle ear lining, *otitis media* (mē-dē-uh), is a fairly common result of a sore throat, especially in children, whose auditory tubes run more horizontally. In otitis media, the eardrum bulges and often becomes inflamed and red. When large amounts of fluid or pus accumulate in the cavity, an emergency *myringotomy* (lancing of the eardrum) may be required to relieve the pressure. ■

The tympanic cavity is spanned by the three smallest bones in the body: the **ossicles** (see Figure 16.25). These bones, named for their shape, are the **malleus** (ma′-lē-us), or **hammer;** the **incus** (in′-kis), or **anvil;** and the **stapes** (stā′-pēz), or **stirrup.** The "handle" of the malleus is secured to the eardrum, and the base of the stapes fits into the oval window. The incus articulates with the malleus laterally and the stapes medially at synovial joints. The ossicles transmit the vibratory motion of the eardrum to the oval window, which in turn sets the fluids of the inner ear into motion, eventually exciting the hearing receptors.

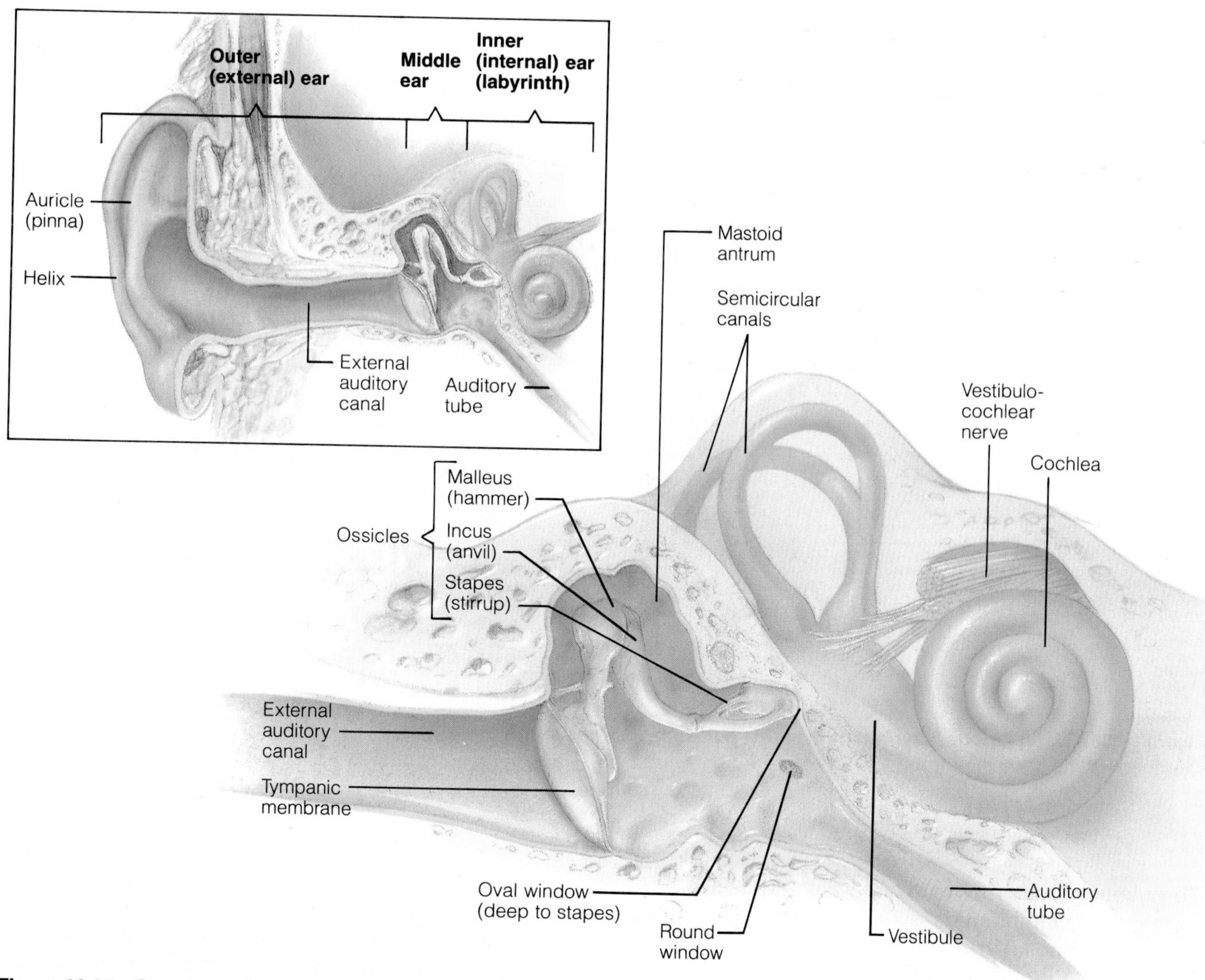

Figure 16.25 Structure of the ear. Notice that the semicircular canals, vestibule, and cochlea of the inner ear form the bony labyrinth.

Two tiny skeletal muscles are found within the tympanic cavity. The **tensor tympani** (ten′-ser tim′-puh-nē) arises from the wall of the auditory tube and inserts into the handle of the malleus. The **stapedius** (stuh-pē′-dē-us) originates from the posterior wall of the cavity and inserts into the stapes. When the ears are assaulted by exceptionally loud sounds, these muscles are reflexively activated and pull the ossicles to which they are attached slightly away from their membrane contacts. This *tympanic*, or *sound attenuation, reflex* reduces the conduction of sound to the inner ear, but since it has a lag period of about 40 ms, it does not protect the receptors from *sudden* loud noises.

The Inner (Internal) Ear

The **inner (internal) ear** is also called the **labyrinth** because of its mazelike, highly complicated shape (see Figure 16.25). It lies deep within the temporal bone behind the eye socket and provides a secure site for all of the delicate receptor machinery housed there. The inner ear consists of two major divisions: the bony labyrinth and the membranous labyrinth. The **bony,** or **osseous, labyrinth** is a system of tortuous channels worming through the bone; its three structurally and functionally unique regions are the vestibule, the cochlea, and the semicircular canals. The views of the bony labyrinth typically seen in most textbooks, including this one, are somewhat misleading because we are talking about a cavity here. The representation seen in Figure 16.25 can be compared to a cast of the shape of a bony labyrinth that has been filled with plaster of Paris and then had the bony walls removed.

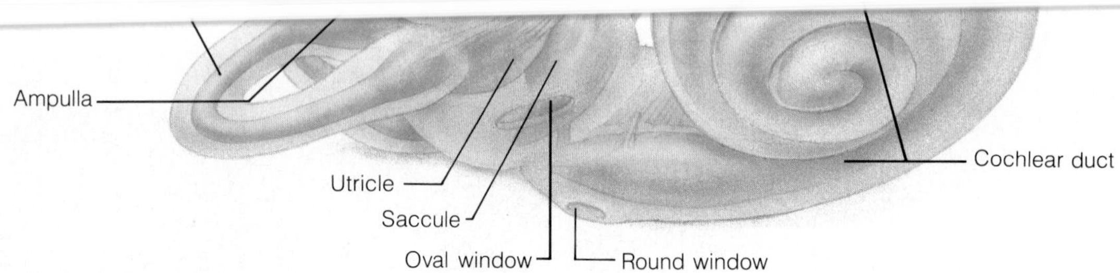

Figure 16.26 Membranous labyrinth of the inner ear, shown in relation to the chambers of the bony labyrinth.

The **membranous labyrinth** is a collection of interconnecting membranous sacs and ducts contained within the bony labyrinth and (more or less) following its contours (Figure 16.26).

The bony labyrinth is lined with endosteum and filled with **perilymph,** a fluid similar to cerebrospinal fluid. The membranous labyrinth is surrounded by and floats in the perilymph; its interior contains **endolymph,** which is chemically similar to intracellular fluid. These two fluids conduct the sound vibrations involved in hearing and respond to the mechanical forces occurring during changes in body position and acceleration. They have no relationship to the lymph circulating in the lymphatic vessels of the body.

The Vestibule. The **vestibule** is the central part of the bony labyrinth. It lies posterior to the cochlea and anterior to the semicircular canals and flanks the middle ear medially. In its lateral wall are the oval and round windows. Suspended within its perilymph and united by a small duct are two membranous labyrinth sacs, the **saccule** and **utricle** (yoo′-trih-kul) (see Figure 16.26). The smaller saccule is continuous with the membranous labyrinth extending anteriorly into the cochlea (the *cochlear duct*), whereas the utricle is continuous with the ducts extending into the semicircular canals posteriorly. The saccule and utricle house equilibrium receptors called *maculae* that respond to the pull of gravity and report on changes of head position.

The Semicircular Canals. The cavities of the bony *semicircular canals* project from the posterior aspect of the vestibule; each is oriented in one of the three planes of space. Accordingly, there is an *anterior (superior), posterior,* and *lateral (horizontal)* semicircular canal in each inner ear. The anterior and posterior canals are oriented at right angles to each other in the vertical plane, whereas the lateral canal is set horizontally and at a right angle to the other two (see Figure 16.26). Snaking through each semicircular canal is a membranous **semicircular duct,** which communicates with the utricle anteriorly. Each of these ducts has an enlarged swelling at one end called an *ampulla,* which houses an equilibrium receptor called a *crista ampullaris.* These receptors respond to angular movements of the head.

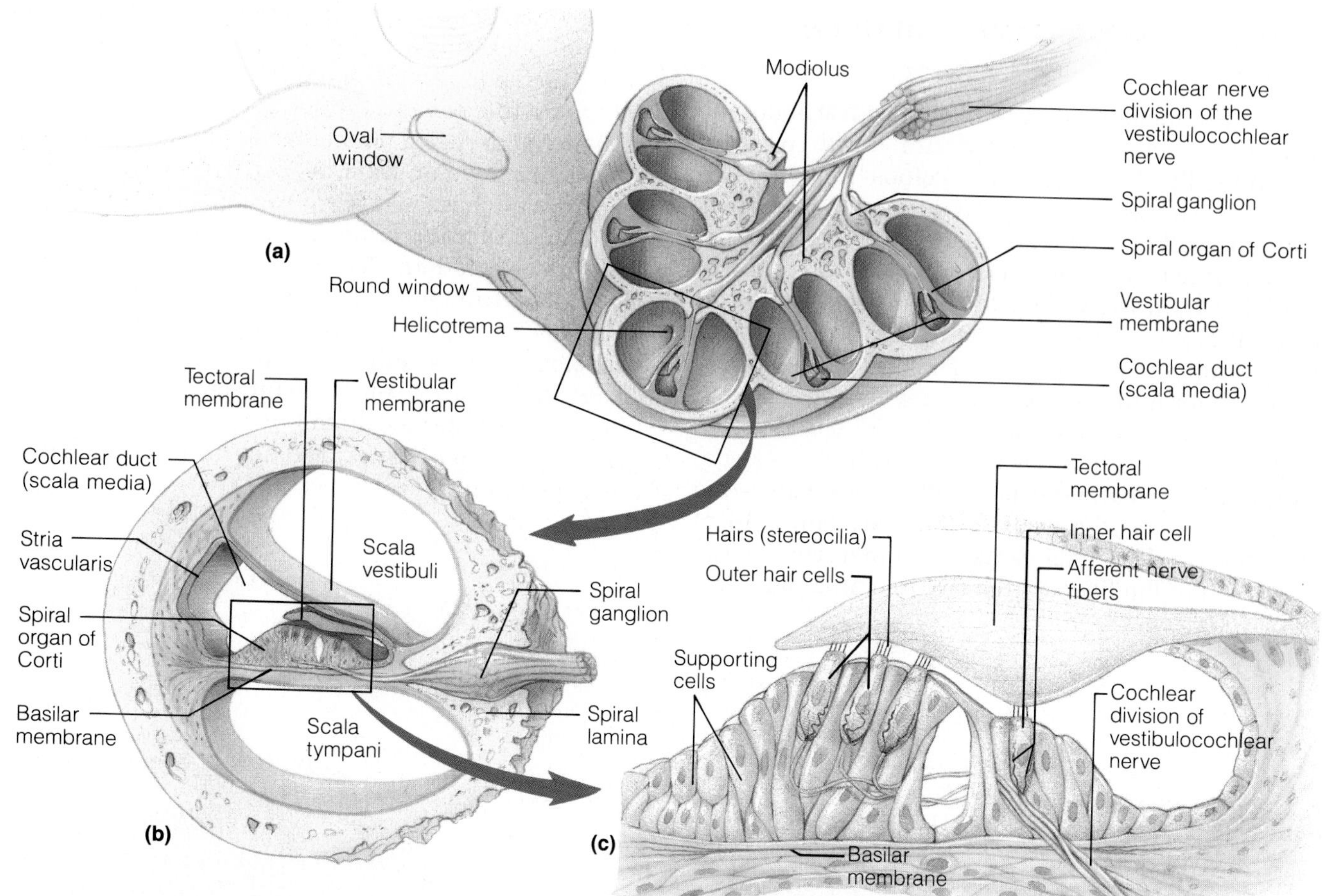

Figure 16.27 Anatomy of the cochlea. (**a**) A section through the coiled cochlea. (**b**) Magnified cross-sectional view of one turn of the cochlea, showing the relationship of the three scalae. The scalae vestibuli and tympani contain perilymph; the cochlear duct (scala media) contains endolymph. (**c**) Detailed structure of the spiral organ of Corti.

The Cochlea. The **cochlea** (kōk′-lē-ah), from the Greek meaning "snail," is a spiral, conical, bony chamber about half the size of a split pea. It extends from the anterior part of the vestibule and then coils for approximately 2½ turns around a bony pillar called the **modiolus** (mō-dī′-uh-lus) (Figure 16.27a). Running through its center like a wedge-shaped worm is the membranous **cochlear duct,** which ends blindly at the cochlear apex. The cochlear duct houses the **spiral organ of Corti,** the receptor organ for hearing (Figure 16.27b). The cochlear duct and the **spiral lamina,** a shelflike extension of the modiolus, together divide the cavity of the bony cochlea into three separate chambers: a superior **scala vestibuli** (ska′-luh veh-stih′-byoo-lī), which is continuous with the vestibule and abuts the oval window; a middle **scala media** (the cochlear duct); and the inferior **scala tympani,** which terminates at the round window. The scala vestibuli and the scala tympani contain perilymph; the scala media is filled with endolymph. The perilymph-containing chambers are continuous with each other at the apex, a region called the **helicotrema** (hē″-lih-kō-trē′-muh).

The "roof" of the cochlear duct, separating the scala media from the scala vestibuli, is the **vestibular membrane** (see Figure 16.27b). Its lateral wall is the **stria vascularis,** composed of a richly vascularized mucosa that is thought to secrete the endolymph. The cochlear duct "floor" is composed of the spiral lamina and the flexible, platelike **basilar membrane,** which supports the organ of Corti. (We will describe the organ of Corti when we discuss the mechanism of hearing.) The basilar membrane is narrow and thick near the oval window and gradually becomes wider and thinner as it approaches the cochlear apex; as you will see, its structure plays a critical role in sound reception. The cochlear nerve, a division of the vestibulocochlear (eighth cranial) nerve, runs from the receptor organ through the modiolus on its way to the brain.

Embryonic Development of the Ear

The development of the ear and eye begin at about the

mesoderm forms the walls of the bony labyrinth of the inner ear. The inner ear is then innervated by the outgrowth of the fibers of the eighth cranial nerve.

While the inner ear structures are developing, lateral outpocketings, called *pharyngeal pouches*, form from the endoderm of the pharynx. The bony cavity

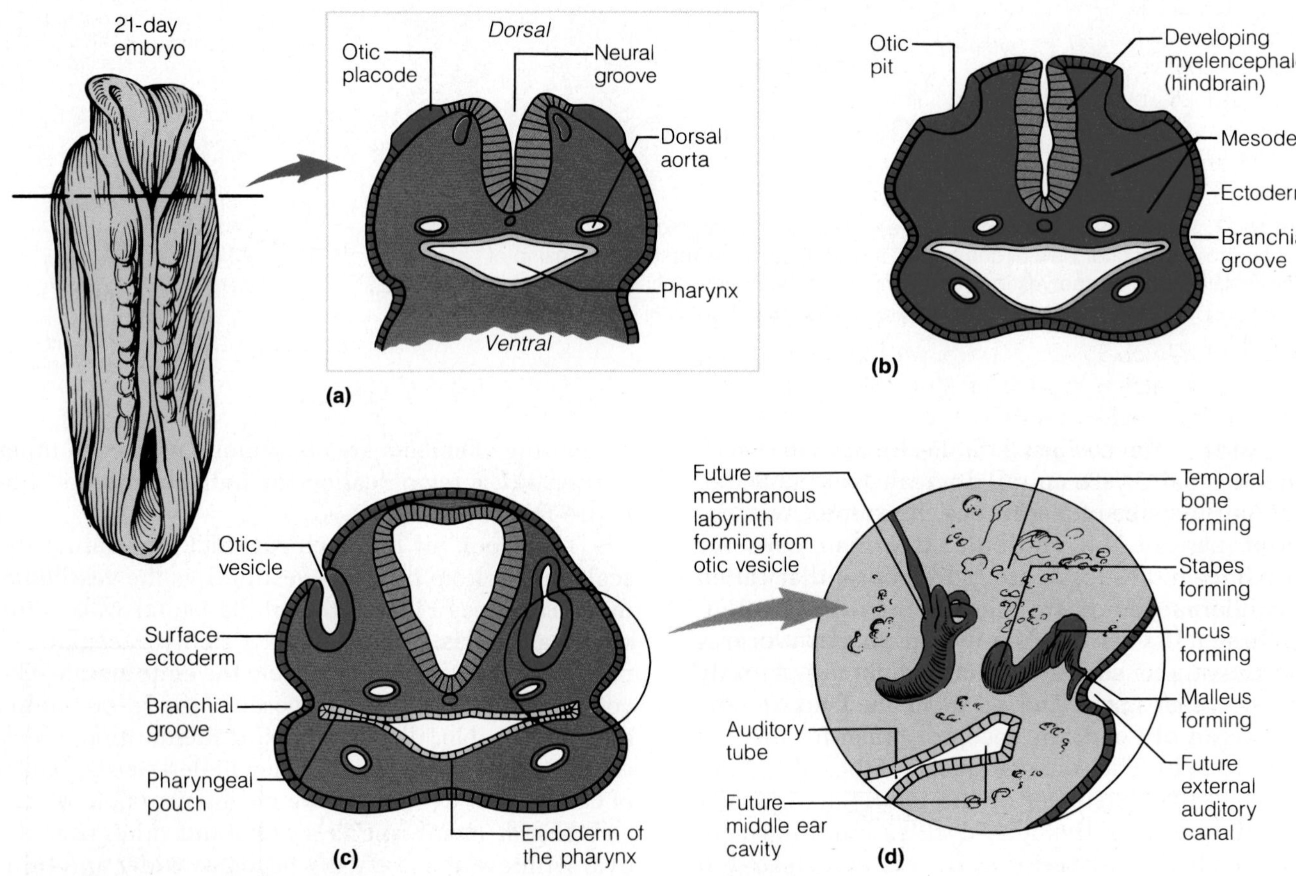

Figure 16.28 Embryonic development of the ear. The surface view of the embryo shows the plane of sections. (**a**) By about the twenty-first day of development, the otic placodes have formed from thickenings of the surface ectoderm, and the pharynx has been formed from endoderm. (**b**)The otic placodes invaginate, forming the otic pits, and the branchial grooves begin to invaginate from the surface ectoderm. (**c**) At approximately 28 days, the otic vesicles are forming from the otic pits, and the branchial grooves have deepened. (**d**) Between the fifth and eighth weeks of development, the inner ear structures form from the otic vesicles, the endoderm of the pharyngeal pouches gives rise to the auditory tubes and the middle ear cavity, and the branchial grooves develop into the external auditory canals. Mesoderm forms the surrounding bony structures.

Mechanisms of Equilibrium and Orientation

If a cat is released upside down from a height, it will land on all fours. If an infant is tilted backward, its eyes will roll downward so that its gaze remains fixed, a reflex movement called the *doll's eyes reflex.* Both of these reactions, and countless others, are compensations for a disturbance in balance, reflexes that depend on the sensory receptors within the vestibule and semicircular canals. The term **vestibular apparatus** refers collectively to these equilibrium receptors of the inner ear.

The equilibrium sense is not easy to describe because it does not "see," "hear," or "feel," but responds (and frequently without our awareness) to various head movements. Furthermore, this sense depends not only on inputs from the ear, but also on vision and on information from stretch receptors of the muscles and tendons. If the vestibular apparatus is damaged, the system can adapt and learn to function without it. Thus, it is difficult to credit any one particular set of receptors with our ability to maintain our orientation and balance in space. Nevertheless, we do know that the equilibrium receptors of the inner ear can be divided into two functional arms, with the receptors in the vestibule and semicircular canals monitoring **static** and **dynamic equilibrium,** respectively.

Static Equilibrium: Function of the Maculae

The sensory receptors for static equilibrium are the **maculae** (ma'-kyoo-lē), one in the wall of the saccule and one in the utricle wall. These receptors report on the position of the head with respect to gravity when the body is not moving, but they also respond to *linear* acceleration forces. That is, they respond to straight-line changes in speed and direction, but not to rotational movement.

Each macula is a flat epithelial patch containing scattered **supporting cells** and 20 to 50 receptor cells called **hair cells** (Figure 16.29a). Displacement of the head in any linear direction activates the receptors in either the utricles or the saccules. Named for the hairy

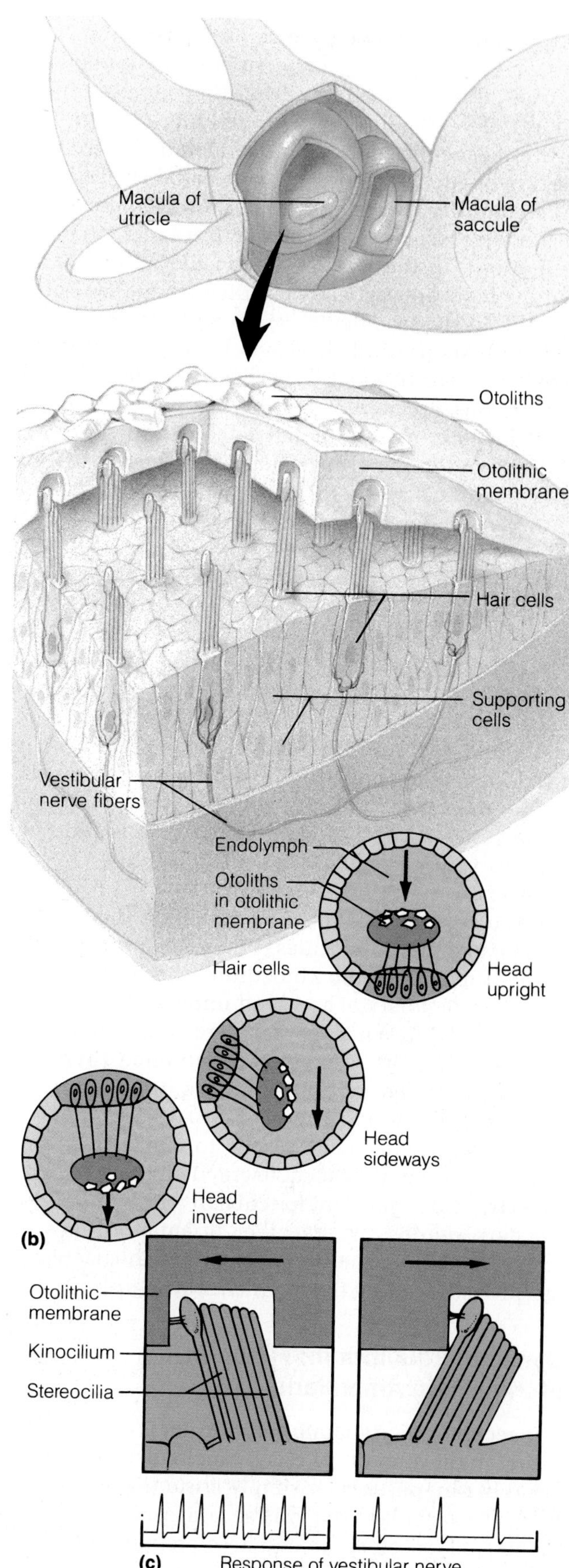

Figure 16.29 Structure and function of a macula. (**a**) The "hairs" of the receptor cells of a macula project into the gelatinous otolithic membrane. Vestibular nerve fibers surround the base of the hair cells. (**b**)The effect of gravitational pull on a utricular macula when the head is in the upright position, when the head is held horizontally, and when the head is inverted. (**c**) When movement of the otolithic membrane (direction indicated by the arrow) bends the hair cells in the direction of the kinocilium, depolarization and action potential generation occur. When the hairs are bent in the direction away from the kinocilium, the hair cells become hyperpolarized, and the nerve fibers are unresponsive.

look of their free surface, the hair cells have numerous *stereocilia* (microvilli) and a single long *kinocilium* (a true cilium) protruding from their apices. These "hairs" are embedded in the [illegible]

[illegible] the [illegible] excitement is transmitted to the *vestibular nerve* endings coiling around their bases. The cell bodies of the sensory neurons are located in the nearby *superior* and *inferior vestibular ganglia* (see Figure 16.26).

When your head starts or stops moving in a linear direction, inertia causes the otolithic membrane to slide backward or forward like a greased plate over the hair cells, bending the hairs (see Figure 16.29b). For example, when you run, the otolithic membranes of the utricle maculae lag behind, bending the hairs backward; when you suddenly stop, the otolithic membrane slides abruptly forward (just as you slide forward in your car when you brake), bending the hair cells forward. Likewise, when you nod your head or fall, the otoliths roll inferiorly, bending the hairs of the maculae in the saccules. When the hairs are bent in the direction of the kinocilium, the hair cells are activated (depolarize) and send impulses to the brain (Figure 16.29c). When the hairs are bent in the opposite direction, the receptors are inhibited (hyperpolarized). In either case, the brain is informed of the position of the head in space.

Thus, the maculae help us to maintain normal head position with respect to gravitational pull. They are extremely important to scuba divers swimming in the dark depths (where other orientation cues are absent), enabling the divers to orient themselves to the surface—telling them which way is up!

Dynamic Equilibrium: Function of the Crista Ampullaris

The receptor for dynamic equilibrium is a minute elevation in the ampulla of each semicircular canal called the **crista ampullaris,** or simply *crista* (Figure 16.30b). Like the maculae, the cristae are excited by head movement (acceleration and deceleration), but in this case, the major stimuli are rotatory (angular) movements. When you twirl on the dance floor or suffer through a rough boat ride, these receptors are working overtime. Since the semicircular canals are located in all three planes of space, all rotatory movements of the head disturb one or another pair of cristae (one in each ear). Each crista is composed of [illegible]

[illegible] of the body continues at a constant rate, the endolymph eventually comes to rest—it moves along at the same speed as the body—and stimulation of the hair cells ends. Consequently, if we are blindfolded, we cannot tell whether we are moving at a constant speed or not moving at all after the first few seconds of continuous rotatory movement. However, when we suddenly stop moving, the endolymph keeps on going, in effect reversing its direction within the canal. This sudden reversal in the direction of hair bending results in hyperpolarization of the receptor cells and a reduced rate of impulse transmission, which tells the brain that we have slowed or stopped.

The key point to remember when considering both the dynamic and the static equilibrium receptors is that the rigid bony labyrinth moves with the body, while the fluids (and gels) within the membranous labyrinth are free to move at various rates, depending on the forces (gravity, acceleration, etc.) acting on them.

Impulses transmitted from the semicircular canals are tied to reflex movements of the eyes and skeletal muscles necessary to maintain equilibrium and to keep the eyes fixed on an object in the visual field. (We will describe some of these pathways in the next section.) *Vestibular nystagmus* is a complex of rather strange eye movements that occurs during and immediately after rotational motion. As you rotate, your eyes slowly drift in the opposite direction of rotation, as though they were fixed on some object in the environment. This reaction is related to the backflow of endolymph in the semicircular canals. Then, because of compensating mechanisms of the central nervous system, the eyes jump rapidly toward the direction of rotation to establish a new fixation point. This alternation of eye movements continues until the endolymph comes to rest. When you stop rotating, your eyes initially continue to move in the direction of the previous spin, and then they jerk rapidly in the opposite direction. This sudden change is caused by the change in the direction that the cristae are bending after you stop. Nystagmus is often accompanied by a feeling of loss of balance called *vertigo*.

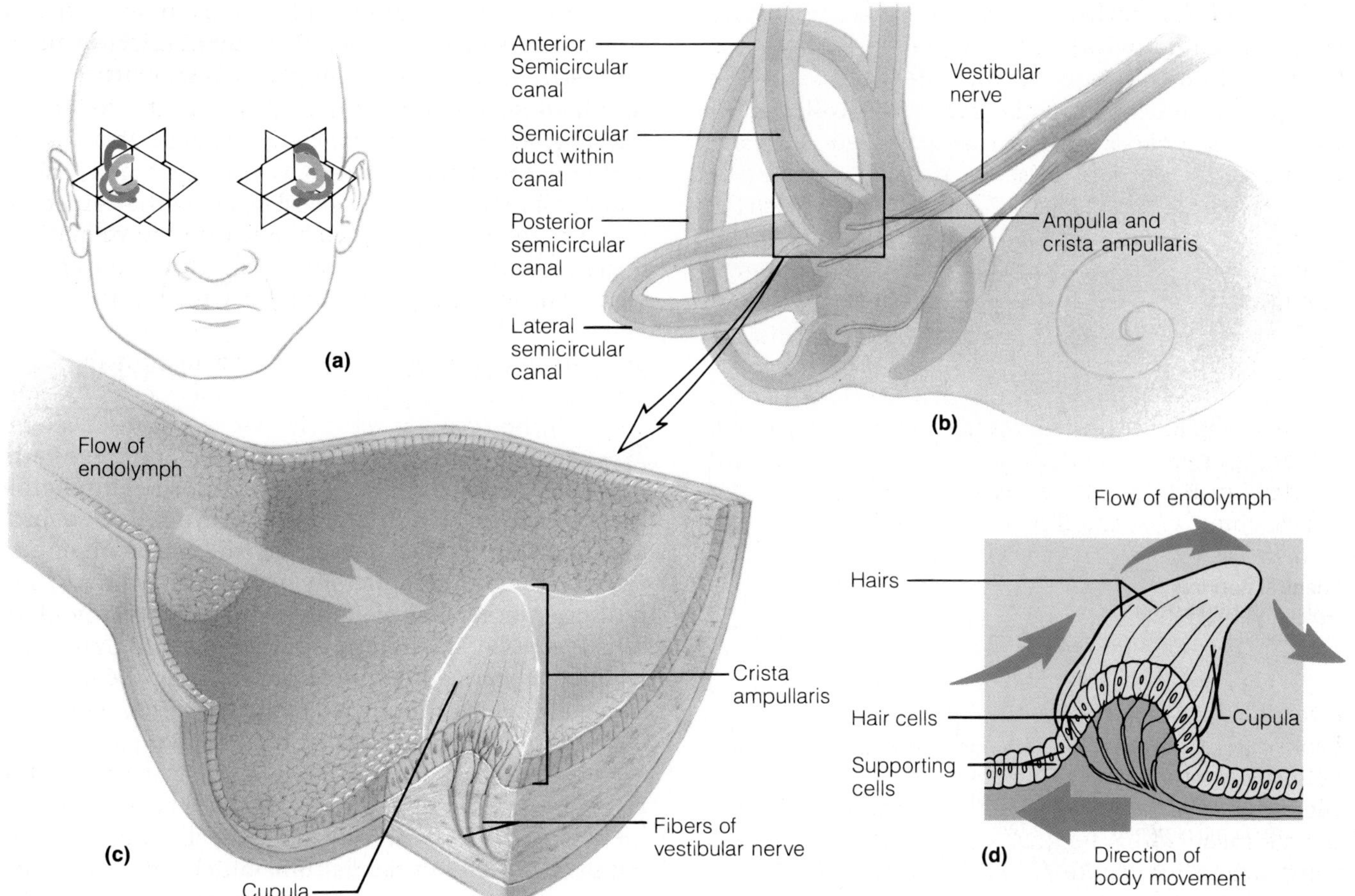

Figure 16.30 Localization and structure of a crista ampullaris. (**a**) Localization of the semicircular canals within the temporal bones of the skull. (**b**) View of the three semicircular canals of one ear, showing the positioning of the receptor organs in the ampullae. (**c**) Anatomy of a crista ampullaris. (**d**) Movement of the endolymph during acceleratory rotation displaces the cupula, stimulating the hair cells. The direction of endolymph and cupula movement is always contrary to the motion of the body because of the inertia of the endolymph.

The Equilibrium Pathway to the Brain

The nerve pathways that connect the vestibular apparatus with the brain are complex and poorly traced. The transmission sequence begins when the hair cells in the static or dynamic equilibrium receptors (or both) are activated. As depicted in Figure 16.31, impulses travel initially to one of two destinations: the *vestibular nuclear complex* in the brain stem or the *cerebellum* (flocculonodular node). In addition to impulses from the vestibular apparatus, the vestibular nuclei receive inputs from the visual and somatic receptors, particularly from proprioceptors in the neck muscles that report on the angle or inclination of the head. This information is integrated by the vestibular nuclei, which in turn send signals to brain stem centers that control the extrinsic eye muscles (cranial nerve nuclei III, IV, and VI) and reflex movements of the neck, limb, or body muscles (via the vestibulospinal tracts). Reflex movements of the eyes and body produced by these pathways allow us to remain focused on the visual field and to constantly adjust body position to maintain balance. The cerebellum, too, integrates inputs from the eyes and somatic receptors (as well as from the cerebrum). It coordinates the activity of skeletal muscles and regulates muscle tone so that head position, posture, and balance are maintained, often in the face of rapidly changing inputs.

The vestibular nuclei are in constant communication with both the cerebellum and the reticular nuclei of the brain stem, and a few vestibular fibers reach the cerebral cortex, allowing us to become aware of changes in body acceleration or balance. Inputs from the eyes and somatic receptors are also transmitted to the reticular nuclei, but their precise function in the equilibrium scheme is unclear.

Notice that the vestibular apparatus does not act as a tachometer or governor that automatically compensates for forces acting on the body. Rather, its job is to send warning signals to the central nervous system, which initiates the appropriate compensations

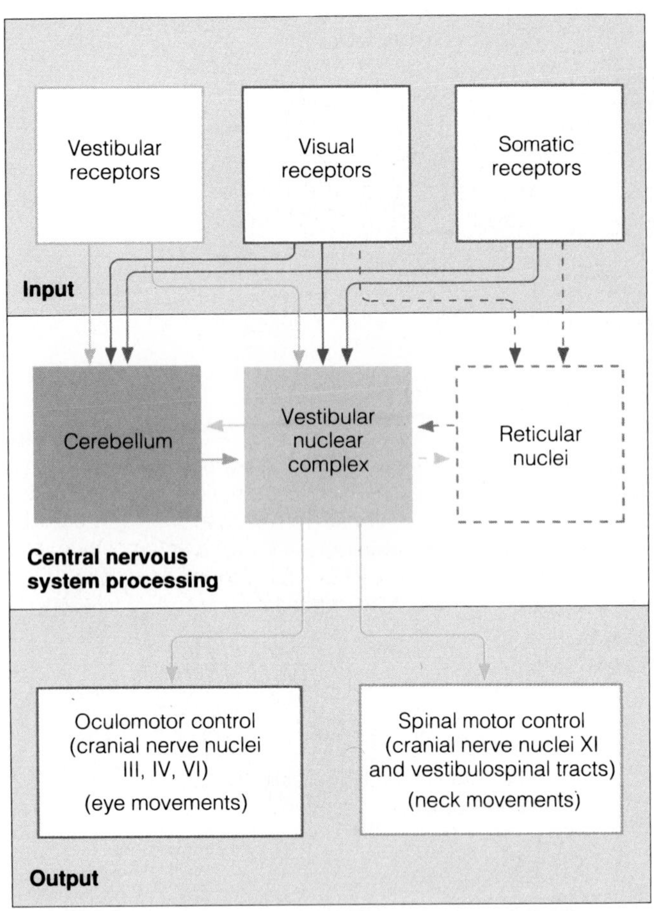

Figure 16.31 Pathways of the balance and orientation system. The three modes of input—vestibular, visual, and somatic (skin, muscle, and joint) receptors—transmit impulses to two major processing areas: the vestibular nuclei of the brain stem and the cerebellum. (The pathways to and the precise function of the reticular nuclei are unclear.) After vestibular nuclear processing, impulses are sent to one of two outputs—areas controlling eye movements (oculomotor control) or those controlling the skeletal muscles of the neck (spinal motor control).

(righting) to keep your body balanced, your weight evenly distributed, and your eyes focused on what you were looking at when the disturbance occurred. Since balance calls for fast timing and intricate muscle control, responses to equilibrium are totally reflexive. For the most part, we only become aware of vestibular apparatus activity when its functioning is less than optimal.

Equilibrium problems are usually obvious. Nausea, dizziness, and problems in maintaining balance are common symptoms. Also, there may be nystagmus in the absence of rotational stimuli.

Motion sickness, a common equilibrium disorder, has been difficult to explain. The most popular explanation is that it is due to sensory input mismatch. For example, if you are inside a ship during a storm, visual inputs indicate that your body is fixed with reference to a stationary environment (your cabin). But as the ship is tossed about by the rough seas, impulses from the vestibular apparatus "disagree" with the visual information, somehow resulting in motion sickness. Warning signals, which precede nausea and vomiting, include excessive salivation, pallor, rapid deep breathing, and profuse sweating. Over-the-counter drugs such as meclizine HCl (Bonine) and dimenhydrinate (Dramamine), antimotion drugs that depress vestibular inputs, help many sufferers. ■

Sound and Mechanisms of Hearing

The mechanics of human hearing can be summed up in a single sprawling sentence: Sounds set up vibrations in air that beat against the eardrum that pushes a chain of tiny bones that press fluid in the inner ear against membranes that set up shearing forces that pull on the tiny hair cells that stimulate nearby neurons that give rise to impulses that travel to the brain, which interprets them—and you hear. Before we unravel this intriguing sequence, let us pause to describe sound, the stimulus for hearing.

Properties of Sound

Unlike light, which can be transmitted through a vacuum (for instance, outer space), sound depends on an elastic medium for its transmission. Sound also travels much more slowly than light. Its speed in dry air is only about 0.2 miles per second (331 m/s), as opposed to the 186,000 miles per second of light. A lightning flash is almost instantly visible, but the sound it creates (thunder) reaches our ears much more slowly. The velocity of sound transmission is constant in a given medium; it is greatest in solids and lowest in gases, including air.

Sound is a pressure disturbance originating from a vibrating object and propagated by the molecules of the medium. Consider a sound arising from a vibrating tuning fork (Figure 16.32a). If the tuning fork is struck on the left, its prongs will move first to the right, creating an area of high pressure by compressing the air molecules in that region. As the compressed air molecules bump into others in front of them, they push the other molecules outward, creating a new area of compression. Then, as the prongs rebound to the left, the air on the left will be compressed, and the region on the right will be a rarefied, or low-pressure, area (since most of its air molecules have been pushed farther to the right). As the fork vibrates alternately from right to left, it produces a series of compressions and rarefactions, collectively called a sound wave, which continues to move outward in all directions (Figure 16.32b). However, the individual air molecules just vibrate back and forth for short distances as

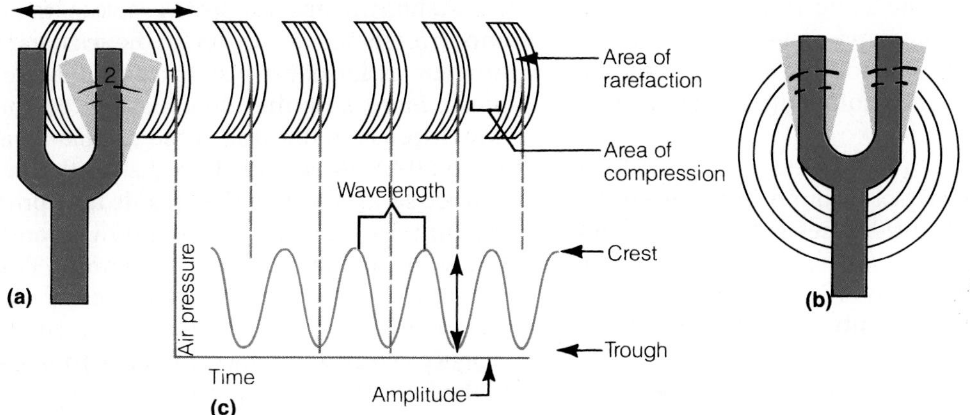

Figure 16.32 Sound: source and propagation. The source of sound is a vibrating object. (**a**) A tuning fork is struck on the left with a mallet. Its tines move to the right, compressing the air molecules in that region. (The tine movement is shown on the right side only and is exaggerated.) Then the tines move in the opposite direction, compressing the air there and leaving a rarefied area to the right side of the tuning fork. (**b**) In this way, sound waves, consisting of alternate areas of compression and rarefaction, are set up that radiate outward in all directions from the sound source. (**c**) The sound wave may be depicted as a sine wave. The peaks of the wave represent areas of high pressure (compressed areas), and the troughs represent areas of low pressure (rarefied areas). For any tone, the distance between two corresponding points on the wave (two crests or two troughs) is called the wavelength. The height (amplitude) of the crests is related to the energy, or intensity, of the sound wave.

they bump other molecules and rebound. Since the outward-moving molecules give up kinetic energy to the molecules they bump, there is always a transfer of energy in the direction that the sound wave is traveling. Thus, with time and distance, the energy of the wave declines, and the sound dies a "natural death."

We can depict a sound wave graphically as an S-shaped curve, or *sine wave,* in which the compressed areas are the crests and the rarefied areas are the troughs (Figure 16.32c). Sound can be described in terms of two physical properties inferred from this sine wave graph: frequency and amplitude.

Frequency. The sine wave of a pure tone is *periodic;* that is, its crests and troughs repeat at definite distances. The distance between two consecutive crests (or troughs) is called the **wavelength** of the sound and is constant for a particular tone. **Frequency** (expressed in hertz) is defined as the number of waves that pass a given point in a given time. The shorter the wavelength, the higher the frequency of the sound (Figure 16.33a).

The frequency range of human hearing is from 20 to 20,000 Hz, but our ears are most sensitive to frequencies between 1000 and 4000 Hz, and in that range we can distinguish frequencies differing by only 2–3 Hz. We perceive different sound frequencies as differences in **pitch:** The higher the frequency, the higher the pitch. A tuning fork produces a pure (but bland and colorless) sound with a single frequency, but most sounds are mixtures of several frequencies. It is this characteristic of sound, called *quality,* that enables us

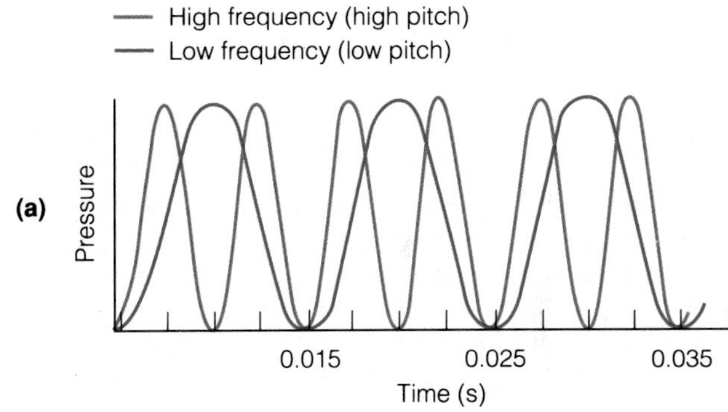

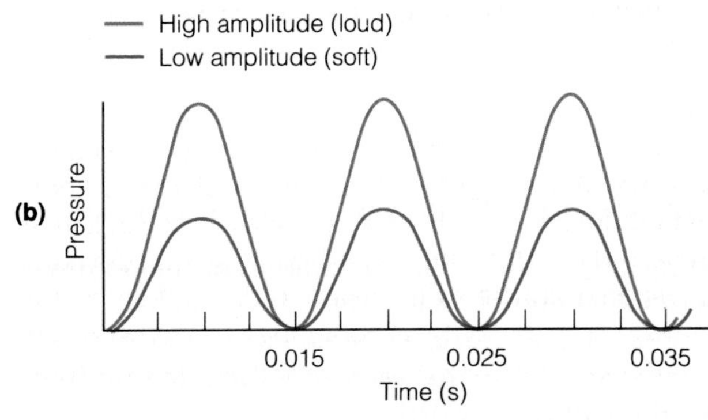

Figure 16.33 Frequency and amplitude of sound waves. (**a**) The wave shown in red has a shorter wavelength, and thus a greater frequency, than that shown in blue. The frequency of sound is perceived as pitch. (**b**) The wave shown in red has a greater amplitude (intensity) than that shown in blue and is perceived as being louder.

to distinguish between the same musical note—say, high C—sung by a soprano and played on a piano or clarinet. It is the quality of sound that provides the richness and complexity of sounds (and music) that we hear.

Amplitude. The **intensity** of a sound is related to its energy, or the pressure differences between its compressed and rarefied areas. When we represent the sound wave graphically, as in Figure 16.33b, its intensity corresponds to the **amplitude,** or height, of the sine wave crests.

Whereas intensity is an objective, precisely measurable property of sound, *loudness* refers to our subjective interpretation of sound intensity. Because we can hear such an enormous range of intensities, from the proverbial pin drop to a steam whistle 100 trillion times as loud, sound intensity (and loudness) is measured in logarithmic units called **decibels** (deh′-sih-bulz) **(dB).** On a clinical audiometer, the decibel scale is arbitrarily set to begin at 0 dB, which is the threshold of hearing (barely audible sound) for normal ears. Each 10 dB increase represents a tenfold increase in sound intensity. A sound of 10 dB has ten times more energy than one of 0 dB, and a 30 dB sound has 1000 times ($10 \times 10 \times 10$) more energy than one of 0 dB. However, the same 10 dB increase represents only a twofold increase in loudness. In other words, most people would report that a 20 dB sound seems about twice as loud as a 10 dB sound. The healthy adult ear can pick up differences in sound intensity as small as 0.1 dB, and the normal range of hearing (from barely audible to the loudest sound we can process without excruciating pain) extends over a range of 120 dB.

Severe hearing loss occurs with frequent or prolonged exposure to sounds with intensities greater than 90 dB. This figure becomes more meaningful when you realize that the average home has background noise in the 50 dB range, a noisy restaurant has 80 dB levels, and amplified rock music is far above the 90 dB danger zone.

Transmission of Sound to the Inner Ear

Hearing results from the stimulation of the auditory area of the temporal lobe cortex. However, before this can occur, sound waves must be propagated through air, bone, and fluids to reach and stimulate receptor cells in the organ of Corti.

A sound wave entering the external auditory canal strikes the tympanic membrane and sets it vibrating at the same frequency. The motion of the tympanic membrane is conveyed to the oval window by the ossicles, which both transfer and amplify the energy. If sound hit the oval window directly, most of its energy would be reflected and lost because of the high impedance (resistance to motion) of the cochlear fluid. However, the ossicle lever system acts much like a hydraulic press, increasing the force exerted on the oval window. Moreover, the area of the tympanic membrane is 17 to 20 times greater than that of the oval window, and these factors together result in a pressure (force per unit area) on the oval window approximately 22 times that on the tympanic membrane. This increased pressure overcomes the impedance of cochlear fluid and sets it into wave motion. This situation can be roughly compared to the difference in pressure relayed to the floor by the broad rubber heels of a man's shoes and a woman's tiny spike heels. The man's weight—say, 150 pounds—is spread over several square inches, and his heels will not make dents in a pliable vinyl floor. But spike heels will concentrate the same 150-pound force in an area of about 1 square inch and will dent the floor, causing it to take on a "pock-marked" look.

Resonance of the Basilar Membrane

When the stapes oscillates back and forth against the oval window, it sets the perilymph in the scala vestibuli into a similar back-and-forth motion. During these movements, the basilar membrane swings up and down, causing the nearby (basal) part of the cochlear duct to oscillate in time, and a pressure wave travels through the perilymph from the basal end toward the helicotrema, much as a piece of rope held horizontally can be set into wave motion by movements initiated at one end. Sounds of very low frequency arriving at the oval window create pressure waves that take the complete route through the cochlea—up the scala vestibuli, around the helicotrema, and back toward the round window through the scala tympani (Figure 16.34a). Such sounds do not activate the organ of Corti and are thus below the threshold of hearing. But sounds of higher frequency (and shorter wavelengths) create pressure waves that fail to reach the helicotrema and instead are transmitted through the flexible vestibular membrane, the endolymph of the cochlear duct, and then through the basilar membrane into the perilymph of the scala tympani. Since fluids are incompressible, each time the oval window is pressed medially by the stapes, the round window bulges laterally into the middle ear cavity and acts as a pressure valve.

As a pressure wave descends through the basilar membrane, it sets the entire membrane into vibrations, but maximal displacement of the membrane occurs where the fibers of the basilar membranes are "tuned" to a particular sound frequency (Figure 16.34c). The fibers of the basilar membrane span its width like the strings of a guitar. The fibers near the oval window

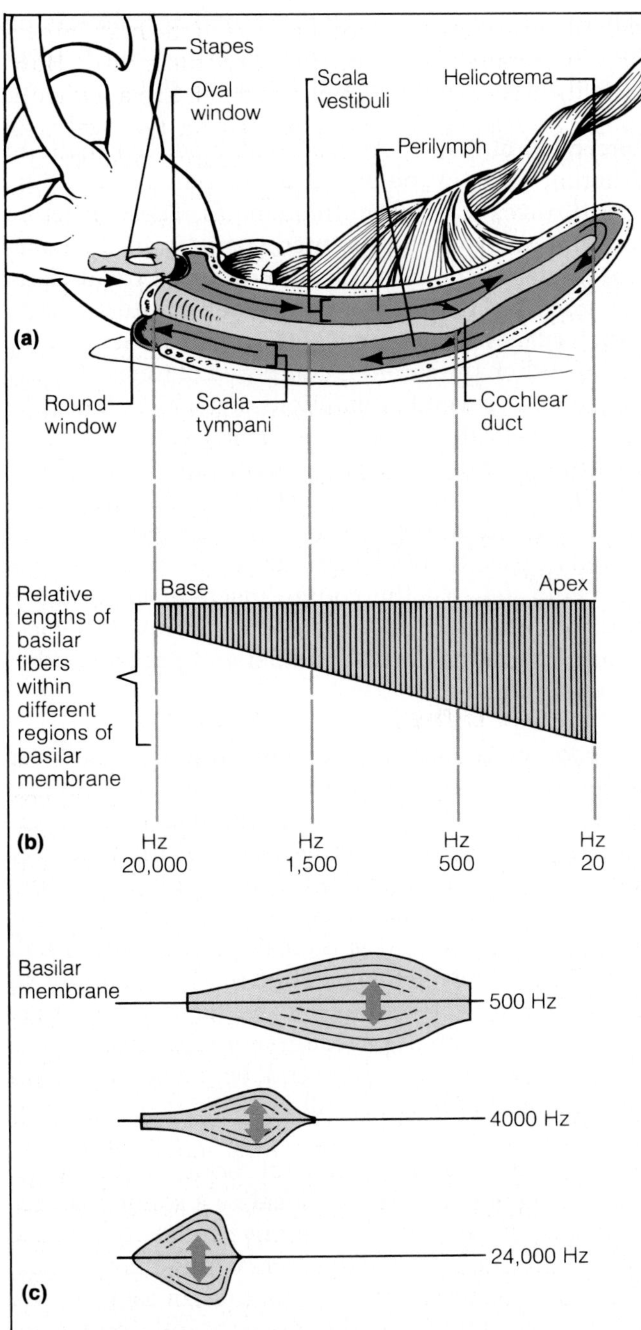

Figure 16.34 Resonance of the basilar membrane and activation of the cochlear hair cells. (**a**) The cochlea is depicted as if it uncoiled to make the events of sound transmission occurring there easier to follow. Sound waves of low frequency that are below the level of hearing travel around the helicotrema without exciting hair cells. But sounds of higher frequency result in pressure waves that penetrate through the cochlear duct and basilar membrane to reach the scala tympani. This sets up vibrations in the basilar membrane, causing it to vibrate maximally in certain areas in response to certain frequencies of sound. (**b**) Relative lengths of basilar fibers in different regions of the basilar membrane. (**c**) High-frequency waves cause maximal basilar membrane displacement and resonance (and excitation of hair cells) near the oval window, whereas low-frequency sounds excite hair cells near the cochlear apex. The red arrows indicate the relative degrees of upward and downward displacement.

(cochlear base) are short and rigid, and they resonate (vibrate at the same frequency) in response to high-frequency pressure waves (Figure 16.34b). The longer, more floppy basilar membrane fibers near the cochlear apex resonate in response to lower-frequency pressure waves. Thus, sound signals are mechanically processed, before reaching the receptors, by the resonance of the basilar membrane.

Excitation of Hair Cells in the Organ of Corti

The organ of Corti, which rests atop the basilar membrane, is remarkably like the receptor organs for equilibrium. It is composed of supporting cells and several long rows of *cochlear hair cells,* which have afferent fibers of the *cochlear nerve* coiled about their bases. The hairs (stereocilia) of the hair cells are enmeshed in the overlying, gel-like **tectorial membrane** (see Figure 16.27c). The up-and-down vibrations of the basilar membrane "tweak," or pull the hairs, causing ion channels in the hair cell membranes to open and leading to the generation of receptor potentials. Neurotransmitters released by the activated hair cells excite the cochlear nerve fibers, initiating action potentials that are transmitted to the brain for auditory interpretation. As you might guess, activation of the hair cells occurs at points of vigorous basilar membrane vibration. Hair cells nearest the oval window are activated by the highest-pitch sounds, and those located at the cochlear apex are stimulated by the lowest-frequency tones.

The Auditory Pathway to the Brain

The ascending auditory pathways to the brain are a bit bewildering because there are several brain stem auditory nuclei. Here we will stress only the major way stations. Impulses generated in the cochlea pass through the **spiral ganglion** (the sensory ganglion of the cochlear nerve) along the afferent fibers of the vestibulocochlear nerve to reach the *dorsal* and *ventral cochlear nuclei* of the medulla (Figure 16.35). From there, impulses are sent on to the *superior olivary nucleus,* then via the lateral lemniscal tract to the auditory cortex in the superior temporal gyrus via relays in the *inferior colliculus* (auditory reflex center in the midbrain) or the *medial geniculate body* of the thalamus. Since some of the fibers from each ear decussate (cross), each auditory cortex receives impulses from both ears.

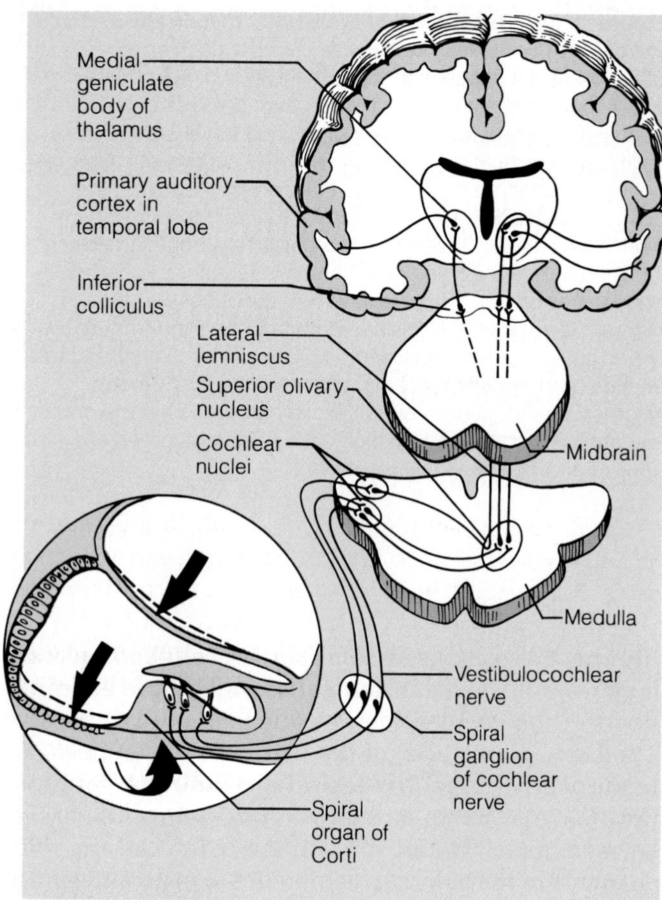

Figure 16.35 Simplified diagram of the auditory pathway from the organ of Corti to the auditory cortex in the temporal lobe of the brain. For clarity, only the pathway from the right ear is illustrated.

Auditory Processing

If you are at a Broadway musical, the sound of the instruments, the voices of the actors, the rustling of clothing, and the closing of doors are all intermingled in your sound awareness. Yet, the auditory cortex can distinguish the separate parts of this auditory jumble. In contrast to color perception, which uses synthetic analysis (see p. 471), auditory processing is analytic, and whenever the differences between sound wavelengths is sufficient for discrimination, you hear two separate and distinct tones. In fact, the analytic powers of the auditory cortex are so great that we are able to pick single instruments out of a whole orchestra and to detect changes in tone frequency as small as 0.3%.

Cortical processing of sound stimuli seems to be quite complex. For example, certain cortical cells depolarize at the beginning of a particular tone, others depolarize when the tone ends, some depolarize continuously, others appear to have high thresholds (low sensitivity), and so on. Here we will concentrate on the more straightforward aspects of cortical perception of pitch, loudness, and sound location.

Perception of Pitch. As just explained, hair cells in different parts of the organ of Corti are activated by sound waves of different frequencies. These different receptor areas are represented spatially in the cochlear nuclei of the medulla, and different frequencies are processed in different parts of the auditory cortex, which suggests a mapping of the organ of Corti on the cortex. When the sound is composed of tones of many frequencies, several populations of cochlear hair cells and cortical cells are activated nearly simultaneously, resulting in the perception of multiple tones.

Over a century ago, investigators of hearing concluded that impulses from specific hair cells are interpreted as specific pitches. However, most researchers now concede that the coding mechanism for pitch perception is much more complex and involves local processing in the cochlear nuclei and probably other areas as well.

Detection of Loudness. Our perception of loudness suggests that certain cochlear cells have higher thresholds than others for responding to a tone of the same frequency. For example, some of the receptors for a tone with a frequency of 540 Hz might be stimulated by a sound wave of very low intensity. As the intensity of the sound is increased, more of the hair cells at the same relative position on the basilar membrane would be recruited by the membrane's more vigorous vibration. As a result, more impulses would reach the auditory cortex and be recognized as a louder sound of the same pitch.

Localization of Sound. Several brain stem nuclei process signals that help us localize a sound's source by means of two cues: the *relative intensity* and the *relative timing* of sound waves reaching the two ears. If the sound source is directly in front, in back, or over the midline of the head, the intensity and timing cues will be the same for both ears. However, when sound is coming from one side, the receptors of the nearer ear will be activated slightly earlier and more vigorously (because of the greater intensity of the sound waves impinging on that ear).

Homeostatic Imbalances of Hearing

Deafness. Any hearing loss, no matter how slight, is **deafness** of some sort. Hearing loss may range from the inability to hear sound of a certain pitch or intensity to a complete inability to detect sound. Deafness is classified as conduction or sensineural deafness, according to its cause.

Conduction deafness results when something interferes with the conduction of sound vibrations to the fluids of the inner ear. The possibilities are almost limitless. For example, impacted earwax or a cold blocking the auditory canal impairs hearing by interfering with vibration of the eardrum. A ruptured eardrum prevents conduction of sound vibrations from the eardrum to the ossicles. But perhaps the most common causes of conduction deafness are middle ear inflammations (otitis media) and *otosclerosis* (ō″-tō-skler-ō′-sis) of the ossicles. Otosclerosis ("hardening of the ear"), a common age-related problem, occurs when connective tissue overgrowth fuses the stapes foot plate to the oval window or fuses the ossicles to one another. The fused ossicles cannot vibrate, so sound conduction to the receptors of that ear occurs through vibrations of the bones of the skull, which is far less satisfactory. Otosclerosis is routinely treated by surgery to remove the excess tissue or to replace one or more of the ossicles or the oval window.

Sensineural deafness results from damage to neural structures at any point from the cochlear hair cells to and including the auditory cortical cells. This type of deafness may be partial or complete and typically results from the gradual loss of the hearing receptor cells throughout life. These cells can also be destroyed at an earlier age by a single explosively loud noise or prolonged exposure to high-amplitude sounds, such as rock bands or airport noise. Degenerative lesions of the cochlear nerve, cerebral infarcts, and tumors affecting the auditory cortex are other causes. For cases of age- or noise-related cochlear damage, cochlear prostheses can be implanted into a drilled recess in the temporal bone. These miniature transducer devices convert sound wave energy into electrical stimuli that are delivered directly to the cochlear nerve fibers. The cochlear implants provide less than normal hearing; for example, they make human voices sound tinny and robotlike. But the profoundly deaf will agree that partial hearing is better than none.

Tinnitus. Tinnitus (tin-ni′-tus) is a ringing or clicking sound in the ears in the absence of auditory stimuli. It is more a symptom of pathology than a disease. For example, tinnitus is one of the first symptoms of cochlear nerve degeneration. It may also result from inflammation of the middle or inner ears and is a side effect of some medications, such as aspirin.

Ménière's Syndrome. Classic *Ménière's* (men″-eh-airs′) *syndrome* is a labyrinth disorder that affects both the semicircular canals and the cochlea. The afflicted person has transient but repeated attacks of vertigo, nausea, and vomiting. Balance is so disturbed that standing erect is nearly impossible. A "howling" tinnitus is common, so hearing is impaired (and ultimately lost) as well. The cause of the syndrome is uncertain, but it may result from distortion of the membranous labyrinth by excessive endolymph accumulation. Less severe cases can usually be managed by antimotion drugs. For more debilitating attacks, salt restriction and diuretics are used to decrease overall extracellular fluid volumes and, consequently, endolymph fluid volume. A last resort is surgical removal of the malfunctioning labyrinth, which is usually deferred until hearing loss is nearly complete. ■

Developmental Aspects of the Special Senses

All the special senses are functional, to a greater or lesser degree, at birth. Taste buds are the earliest sense organs to become functional, responding to chemicals in the amniotic fluid surrounding the fetus by the sixth month of fetal development. The olfactory system also appears to function before birth, since injection of an odorous substance into the amniotic fluid induces sucking behavior. Smell and taste are sharp at birth, and infants relish food that adults consider bland or tasteless. Some researchers claim that smell is just as important as touch in guiding newborn infants to their mother's breast. However, very young children seem indifferent to odors and can play happily with their own feces. As they get older, their emotional responses to specific odors and foods increase.

There are few interferences with the chemical senses throughout childhood and young adulthood. Women generally have a more acute sense of smell than men, and nonsmokers have a sharper sense of smell than smokers. Beginning in the fourth decade of life, our ability to taste and smell declines. This reflects the gradual loss of receptors, which continue to be replaced, but much more slowly than in younger people. Almost half of people over the age of 80 years cannot smell at all, and their sense of taste is poor, which may explain their loss of appetite and inattention to formerly disagreeable odors.

In the darkness of the uterus, the fetus cannot see. Nonetheless, even before the light-sensitive portions of the photoreceptors develop, the central nervous system connections have been made and are functional. During infancy, many of the synapses made during development disappear, and the typical cortical fields that allow binocular vision are established.

Congenital problems of the eyes are relatively uncommon, but their incidence is dramatically increased by certain maternal infections, particularly rubella (German measles) occurring during the critical first three months of pregnancy. Common sequelae of rubella include congenital blindness and cataracts.

As a rule, the sense of vision is the only special sense that is not fully functional at birth. The eyeballs continue to enlarge until the age of eight or nine years, but the lens continues to grow throughout life. At birth, the eyeballs are foreshortened, and all babies are hyperopic. The newborn infant sees only in gray tones, eye movements are uncoordinated, and often only one eye at a time is used. Because the lacrimal glands are not completely developed until about two weeks after birth, babies are tearless for this period, even though they may cry lustily. By five months, infants can follow moving objects with their eyes, but visual acuity is still poor (20/200). By the age of five years, depth perception is present, color vision is well developed, and visual acuity has improved to about 20/30, providing a readiness to begin reading. Hyperopia has usually been replaced by emmetropia, which continues until presbyopia begins to set in around age 40 owing to increasing loss of lens elasticity.

With age, the lens loses its crystal clarity and becomes discolored. Since the dilator muscles of the iris become less efficient, the pupils are always somewhat constricted. These two conditions together decrease by half the amount of light reaching the retina, and visual acuity is dramatically lower by the 70s. In addition, elderly persons are susceptible to certain conditions that cause blindness, such as glaucoma, cataracts, arteriosclerosis, and diabetes mellitus.

Newborn infants can hear after their first cry, but early responses to sound are mostly reflexive—for example, crying and clenching the eyelids in response to a startling noise. By the third or fourth month, infants can localize sounds and will turn to the voices of family members. Critical listening begins to occur in toddlers as they increase their vocabulary, and good language skills are very closely tied to the ability to hear well. Oddly, even deaf babies babble, but they stop after a few months and then go silent.

Congenital abnormalities of the ears are fairly common. Examples include partly or completely missing pinnae and closed or absent external ear canals. Maternal rubella during the first trimester commonly results in sensineural deafness. Except for ear inflammations resulting from bacterial infections or allergies, few problems affect the ears during childhood and adult life. By the 60s, however, deterioration and atrophy of the organ of Corti become noticeable. We are born with approximately 20,000 hair cells in each ear, but they are not replaced when damaged or destroyed by loud noises, disease, or drugs. It is estimated that if we live to the age of 140 years, we will have lost all of our hearing receptors.

The ability to hear high-pitched sounds leaves us first. This condition, called *presbycusis* (prez″-beh-kyoo′-sis), is a type of sensineural deafness. Although presbycusis is considered a disability of old age, it is becoming much more common in younger people as our world grows noisier. Since noise is a stressor, one of its physiological consequences is vasoconstriction, and when delivery of blood to the ear is reduced, the ear becomes even more sensitive to the damaging effects of noise.

* * *

Our abilities to see, hear, taste, and smell—and some of our responses to the effects of gravity—are largely the work of our brain. However, as we have discovered in this, the last of the nervous system chapters, the large and often elaborate sensory receptor organs that serve the special senses are works of art in and of themselves.

The final chapter of this unit describes how the body's functions are controlled by chemicals called hormones in a manner quite different from what we have described for neural control.

Related Clinical Terms

Blepharitis (bleh″-fer-ī-tis) (*blephar* = eyelash; *itis* = inflammation) Inflammation of the margins of the eyelids.

Enucleation (ē-nyū″-klē-ā′-tion) Surgical removal of an eyeball.

Exophthalmos (ek″-sof″-thal′-mus) (*exo* = out; *phthalmo* = the eye) Anteriorly bulging eyeballs; this condition is seen in some cases of hyperthyroidism.

Labyrinthitis (la″-bih-rin-thī′-tis) Inflammation of the labyrinth.

Papilledema (pa″-pil-eh-dē′-muh) (*papill* = nipple) Protrusion of the optic disc into the eyeball, which can be observed by ophthalmoscopic examination of the eye interior; caused by conditions that increase intracranial pressure.

Synesthesia (sih″-nes-thē′-zhuh) (*syn* = together; *esthesi* = sensation) Literally, mingling of the senses; a rare condition in which sensory perceptions are mixed up; e.g., some with this condition can "see" sounds as colors or other visual perceptions or can "touch" tastes (feel certain shapes when tasting certain foods); cause is unknown, but the limbic region of the brain appears to be involved.

Scotoma (skō-tō′-muh) (*scoto* = darkness) A visual blind spot other than the normal (optic disc) blind spot; often reflects the presence of a brain tumor pressing on fibers of an afferent visual pathway.

Chapter Summary

THE CHEMICAL SENSES: TASTE AND SMELL (pp. 487–492)

Taste Buds and the Sense of Taste (pp. 487–489)

1. The taste buds are scattered in the oral cavity and pharynx but are most abundant on the papillae of the tongue.

2. Gustatory cells, the receptor cells of the taste buds, have gustatory hairs (microvilli) that serve as the receptor regions. The gustatory cells are excited by the binding of chemicals to their receptors.

3. The four basic taste qualities—sweet, sour, salty, and bitter—are sensed best at different regions on the tongue.

4. The taste sense is served by cranial nerves VII, IX, and X, which send impulses to the medulla. From there, impulses are transmitted to the thalamus and the somatosensory cortex.

The Olfactory Epithelium and the Sense of Smell (pp. 489–491)

5. The olfactory epithelium is located in the roof of each nasal cavity. The receptor cells are ciliated bipolar neurons. Their axons are the fibers of the olfactory nerve (cranial nerve I).

6. Individual olfactory neurons show a range of responsiveness to different chemicals.

7. Olfactory neurons are excited by volatile chemicals that bind to receptors in the olfactory hairs (cilia).

8. Action potentials of the olfactory nerve fibers are transmitted to the olfactory bulb and then via the olfactory tract to the olfactory cortex. Fibers carrying impulses from the olfactory receptors also project to the limbic system.

Homeostatic Imbalances of the Chemical Senses (p. 492)

9. Most chemical sense dysfunctions are olfactory disorders. Common causes are nasal damage or obstruction and zinc deficiency.

THE EYE AND VISION (pp. 492–511)

1. The eye is enclosed within the bony orbit and cushioned by fat.

Accessory Structures of the Eye (pp. 492–494)

2. The eyebrows help to shade and protect the eyes.

3. The eyelids protect and lubricate the eyes by reflex blinking. Within the eyelids are the orbicularis oculi and levator palpebrae muscles, modified sebaceous glands, and sweat glands.

4. The conjunctiva is a mucosa that lines the eyelids and covers the anterior eyeball surface. Its mucus lubricates the eyeball surface.

5. The lacrimal apparatus consists of the lacrimal gland (which produces a saline solution containing mucus, lysozyme, and antibodies), the lacrimal canals, the lacrimal sac, and the nasolacrimal duct.

6. The extrinsic eye muscles (superior, inferior, lateral, and medial rectus and superior and inferior oblique) move the eyeballs.

Structure of the Eye (pp. 494–500)

7. The wall of the eye is made up of three layers, or tunics. The outermost fibrous tunic consists of the sclera and the cornea. The sclera protects the eye and gives it shape; the cornea allows light to enter the eye.

8. The middle pigmented vascular tunic (uvea) consists of the choroid, the ciliary body, and the iris. The choroid provides nutrients to the eye and prevents light scattering within the eye. The ciliary muscles of the ciliary body control lens shape; the iris controls the size of the pupil.

9. The sensory tunic, or retina, consists of an outer pigmented layer and an inner nervous layer. The nervous layer contains photoreceptors (rods and cones), bipolar cells, and ganglion cells. Ganglion cell axons form the optic nerve, which exits via the optic disc ("blind spot").

10. The outer segments of the photoreceptors contain the light-absorbing pigment within membrane-bounded discs.

11. The biconvex lens is suspended within the eye by the suspensory ligaments attached to the ciliary body. It is the major refractory (focusing) structure of the eye.

12. The posterior cavity, behind the lens, contains vitreous humor, which helps support the eyeball and keep the retina in place. The anterior cavity, anterior to the lens, is filled with aqueous humor, formed by the ciliary processes and drained into the canal of Schlemm. Aqueous humor is a major factor in maintaining intraocular pressure.

Embryonic Development of the Eye (p. 500)

13. The eye starts as an optic vesicle, an outpocketing of the diencephalon that invaginates to form the optic cup, which becomes the retina. Overlying ectoderm folds to form the lens vesicle, which comes to lie in the optic cup, where it gives rise to the lens. The remaining eye tissues and the accessory structures are formed by mesoderm.

Physiology of Vision (pp. 500–511)

14. Light is made up of those wavelengths of the electromagnetic spectrum that excite the photoreceptors.

15. Light is refracted (bent) when passing from one transparent medium to another of different density, or when it strikes a curved surface. Concave lenses disperse light; convex lenses converge light and bring its rays to a focal point.

16. As light passes through the eye, it is bent by the cornea and the lens and focused on the retina. The cornea accounts for most of the refraction, but the lens allows focusing for different distances.

17. Focusing for distance vision requires no special movements of the eye structures. Focusing for close-up vision requires accommodation (bulging of the lens), pupillary constriction, and convergence of the eyeballs. All three reflexes are controlled by parasympathetic fibers of cranial nerve III.

18. Refractory problems include myopia, hyperopia, and astigmatism.

19. The light-absorbing molecule retinal is combined with various opsins in the photoreceptor pigments. When struck by light, retinal changes shape (11-*cis* to all-*trans*) in a series of events leading to transduction and the release of opsin. The pigment is then regenerated.

20. Rod visual pigment, rhodopsin, is a combination of retinal and scotopsin. The light-triggered changes in retinal cause hyperpolarization of the rods. Photoreceptors and bipolar cells generate local potentials only; action potentials are generated by the ganglion cells.

21. The three types of cones all contain retinal, but each has a different type of opsin. Each cone type responds maximally to one color of light: red, blue, or green. The chemistry of cone function is similar to that of rods.

22. Rods respond to low-intensity light and provide night and peripheral vision. Cones are bright-light, high-discrimination receptors that provide for color vision. Anything that must be viewed precisely is focused on the fovea centralis.

23. During light adaptation, the photopigments are bleached, the rods are inactivated, and then, as cones begin to respond to high-intensity light, high-acuity vision ensues. In dark adaptation, the cones cease functioning, and visual acuity decreases; rod function begins when sufficient rhodopsin has accumulated.

24. The visual pathway to the brain begins with the optic nerve fibers (ganglion cell axons) from the retina. At the optic chiasma, fibers from the medial half of each retinal field cross over and continue on in the optic tracts to the thalamus. Thalamic neurons project to the optic cortex via the optic radiation. Fibers also project from the retina to the midbrain pretectal nuclei and the superior colliculi.

25. In binocular vision, each eye receives a slightly different view of the visual field. These views are fused by the optic cortices to provide for depth perception (stereopsis).

26. Retinal processing involves the selective destruction of rod inputs that emphasizes bright/dark contrasts and edges. (The horizontal cells of the retina act in lateral inhibition of rod inputs to the ganglion cells.) Thalamic processing subserves high-acuity vision and depth perception. Cortical processing involves simple cortical cells, which receive inputs from the retinal ganglion cells, and complex cortical neurons, which receive inputs from several simple cortical cells.

THE EAR: HEARING AND BALANCE (pp. 512–525)

Structure of the Ear (pp. 512–515)

1. The auricle and external auditory canal compose the outer ear. The tympanic membrane, the boundary between the outer ear and the middle ear, transmits sound waves to the middle ear.

2. The middle ear is a small chamber within the temporal bone, connected by the auditory tube to the nasopharynx. The ossicles span the middle ear cavity and transmit sound vibrations from the eardrum to the oval window.

3. The inner ear consists of the bony labyrinth, within which the membranous labyrinth is suspended. The bony labyrinth chambers contain perilymph; the membranous labyrinth ducts contain endolymph.

4. The vestibule contains the saccule and utricle. The semicircular canals extend posteriorly from the vestibule in three planes. They contain the semicircular ducts.

5. The cochlea houses the cochlear duct (scala media), which contains the organ of Corti (hearing receptor). Within the cochlear duct, the hair (receptor) cells rest on the basilar membrane, and their hairs project into the gelatinous tectorial membrane.

Embryonic Development of the Ear (p. 516)

6. The membranous labyrinth develops from the otic placode, an ectodermal thickening lateral to the hindbrain. Mesoderm forms the surrounding bony structures.

7. Pharyngeal pouch endoderm, in conjunction with mesoderm, forms the middle ear structures. The outer ear is formed by ectoderm.

Mechanisms of Equilibrium and Orientation (pp. 517–520)

8. The equilibrium receptors of the inner ear are called the vestibular apparatus.

9. The receptors for static equilibrium are the maculae of the saccule and utricle. A macula consists of hair cells with stereocilia and a kinocilium embedded in an overlying otolithic membrane. Linear movements cause the otolithic membrane to move, pulling on the hair cells and initiating action potentials in the vestibular nerve fibers.

10. The dynamic equilibrium receptor, the crista ampullaris within each semicircular duct, responds to angular or rotatory movements in one plane. It consists of a tuft of hair cells whose microvilli are embedded in the gelatinous cupula. Rotatory movements cause the endolymph to flow in the opposite direction, bending the cupula and exciting the hair cells.

11. Impulses from the vestibular apparatus are transmitted along vestibular nerve fibers to the vestibular complex of the brain stem and the cerebellum. These centers initiate responses that fix the eyes on objects and activate muscles to maintain balance.

Sound and Mechanisms of Hearing (pp. 520–525)

12. Sound originates from a vibrating object and travels in waves consisting of alternating areas of compression and rarefaction of the medium.

13. The distance from crest to crest on a sine wave is the sound's wavelength; the shorter the wavelength, the higher the frequency, which is measured in hertz. Frequency is perceived as pitch.

14. The amplitude of sound is the height of the peaks of the sine wave, which reflect the sound's intensity. Sound intensity is measured in decibels. Intensity is perceived as loudness.

15. Sound passing through the external auditory canal sets the eardrum into vibration at the same frequency. The ossicles amplify and deliver the vibrations to the oval window.

16. Pressure waves in the cochlear fluids set specific fibers of the basilar membrane into resonance. At points of maximal basilar membrane vibration, the hair cells of the organ of Corti are excited. High-frequency sounds excite hair cells near the oval window; low-frequency sounds excite hair cells near the cochlear apex.

17. Impulses generated along the cochlear nerve travel to the cochlear nuclei of the medulla and from there through several brain stem nuclei to the auditory cortex. Each auditory cortex receives impulses from both ears.

18. Auditory processing is analytic; each tone is perceived separately. Perception of pitch is related to the position of the excited hair cells along the basilar membrane. Intensity perception suggests that as sound intensity increases, more hair cells respond. Cues for sound localization include the intensity and timing of sound arriving at each ear.

19. Conduction deafness results from interference with conduction of sound vibrations to the fluids of the inner ear. Sensineural deafness reflects damage or destruction of neural structures.

20. Tinnitus is an early sign of sensineural deafness and may also result from the use of certain drugs.

21. Ménière's syndrome is a disorder of the membranous labyrinth. Symptoms include tinnitus, deafness, and vertigo. Excessive endolymph accumulation is the suspected cause.

DEVELOPMENTAL ASPECTS OF THE SPECIAL SENSES (pp. 525–526)

1. The chemical senses are sharpest at birth and gradually decline with age as the replacement of the receptor cells becomes more sluggish.

2. Congenital eye problems are uncommon, but maternal rubella or gonorrhea can cause blindness.

3. The eye is foreshortened at birth and reaches adult size at the age of eight or nine years. Depth perception and color vision develop during early childhood.

4. With age, the lens loses its elasticity and clarity, there is a decline in the ability of the iris to dilate, and visual acuity decreases. The elderly are at risk for eye problems resulting from disease.

5. Congenital ear problems are fairly common. Maternal rubella can cause deafness.

6. Response to sound in infants is reflexive. By five months, an infant can locate sound. Critical listening develops in toddlers.

7. Gradual deterioration of the organ of Corti occurs throughout life as noise, disease, and drugs destroy the cochlear hair cells. Obvious age-related loss of hearing (presbycusis) occurs in the 60s and 70s.

Review Questions

Multiple Choice/Matching

1. Olfactory tract damage would probably affect your ability to (a) taste, (b) hear, (c) feel pain, (d) smell.

2. Sensory impulses transmitted over the facial, glossopharyngeal, and vagus nerves are involved in the sensation of (a) taste, (b) touch, (c) equilibrium, (d) smell.

3. Taste buds are found on the (a) anterior part of the tongue, (b) posterior part of the tongue, (c) palate, (d) all of these.

4. Gustatory cells are stimulated by (a) the movement of otoliths, (b) stretch, (c) substances in solution, (d) photons of light.

5. The accessory glands that produce an oily secretion are the (a) conjunctiva, (b) lacrimal glands, (c) tarsal glands.

6. The portion of the fibrous tunic that is white and opaque is the (a) choroid, (b) cornea, (c) retina, (d) sclera.

7. Which sequence best describes a normal route for the flow of tears from the eyes into the nasal cavity? (a) lacrimal canals, nasolacrimal ducts, nasal cavity, (b) lacrimal ducts, lacrimal canals, nasolacrimal ducts, (c) nasolacrimal ducts, lacrimal canals, lacrimal sacs.

8. Four refractory media of the eye, listed in the sequence in which they refract light, are (a) vitreous humor, lens, aqueous humor, cornea, (b) cornea, aqueous humor, lens, vitreous humor, (c) cornea, vitreous humor, lens, aqueous humor, (d) lens, aqueous humor, cornea, vitreous humor.

9. Damage to the medial recti muscles would probably affect (a) accommodation, (b) refraction, (c) convergence, (d) pupil constriction.

10. The phenomenon of light adaptation is best explained by the fact that (a) rhodopsin does not function in dim light, (b) rhodopsin breakdown occurs slowly, (c) rods exposed to intense light need time to generate rhodopsin, (d) cones are stimulated to function by bright light.

11. Blockage of the canal of Schlemm might result in (a) a sty, (b) glaucoma, (c) conjunctivitis, (d) a cataract.

12. Nearsightedness is more properly called (a) myopia, (b) hyperopia, (c) presbyopia, (d) emmetropia.

13. Of the neurons in the retina, the axons of which of these form the optic nerve? (a) bipolar neurons, (b) ganglion cells, (c) cone cells, (d) horizontal cells.

14. Which sequence of reactions occurs when a person looks at a distant object? (a) pupils constrict, suspensory ligaments relax, lenses become less convex, (b) pupils dilate, suspensory ligaments become taut, lenses become less convex, (c) pupils dilate, suspensory ligaments become taut, lenses become more convex, (d) pupils constrict, suspensory ligaments relax, lenses become more convex.

15. During embryonic development, the lens of the eye forms (a) as part of the choroid coat, (b) from the surface ectoderm overlying the optic cup, (c) as part of the sclera, (d) from mesodermal tissue.

16. The blind spot of the eye is (a) where more rods than cones are found, (b) where the macula lutea is located, (c) where only cones occur, (d) where the optic nerve leaves the eye.

17. Conduction of sound from the middle ear to the inner ear occurs via vibration of the (a) malleus against the tympanic membrane, (b) stapes in the oval window, (c) incus in the round window, (d) stapes against the tympanic membrane.

18. The transmission of sound vibrations through the inner ear occurs chiefly through (a) nerve fibers, (b) air, (c) fluid, (d) bone.

19. Which of the following statements does not correctly describe the organ of Corti? (a) Sounds of high frequency stimulate hair cells at the basal end of the basilar membrane, (b) the "hairs" of the receptor cells are embedded in the tectorial membrane, (c) the basilar membrane acts as a resonator, (d) sounds of high frequency stimulate cells at the apex of the basilar membrane.

20. Pitch is to frequency of sound as loudness is to (a) quality, (b) intensity, (c) overtones, (d) all of these.

21. The structure that allows pressure in the middle ear to be equalized with atmospheric pressure is the (a) cochlear duct, (b) auditory tube, (c) tympanic membrane, (d) pinna.

22. Which of the following is important in maintaining the balance of the body? (a) visual cues, (b) semicircular canals, (c) the saccule, (d) proprioceptors, (e) all of these.

23. Static equilibrium receptors that report the position of the head in space relative to the pull of gravity are (a) organ of Corti, (b) maculae, (c) crista ampullaris.

24. Which of the following is not a possible cause of conduction deafness? (a) impacted cerumen, (b) middle ear infection, (c) cochlear nerve degeneration, (d) otosclerosis.

Short Answer Essay Questions

25. Name the four primary taste qualities, and indicate the regions of the tongue that are most sensitive to each quality.

26. Where are the olfactory receptors, and why is that site poorly suited for their job?

27. Why do you often have to blow your nose after crying?

28. What is the major difference in the function of rods and cones?

29. What and where is the fovea centralis, and why is it important?

30. Describe the response of rhodopsin to light stimuli. What is the outcome of this cascade of events?

31. Since there are only three types of cones, how can you explain the fact that we see many more colors?

32. Describe the effect of aging on the special sense organs.

Clinical Application Questions

33. Mr. Gaspe appeared at the eye clinic complaining of a chip of wood in his eye. Examination of the eye surface revealed no foreign body, but the conjunctiva was obviously inflamed. What name is given to this inflammatory condition, and where would you look for a foreign body that has been floating around on the eye surface for a while?

34. Mrs. Orlando has been noticing flashes of light and tiny specks in her right visual field. When she begins to see a "veil" floating before her right eye, she makes an appointment to see the eye doctor. What is your diagnosis? Is the condition serious? Explain.

35. An engineering student has been working in a disco to earn money to pay for his education. After about eight months, he notices that he is having problems hearing high-pitched tones. What is the cause-and-effect relationship here?

CHAPTER 17

The Endocrine System

Chapter Outline and Student Objectives

1. Indicate important differences between endocrine and neural controls of body functioning.

The Endocrine System and Hormone Function: An Overview (pp. 531–532)

2. List the endocrine organs, and briefly describe their location in the body.

Hormones (pp. 532–537)

3. Describe how hormones are classified chemically.
4. Describe the two major mechanisms by which hormones bring about their effects on their target tissues, and explain how hormone release is regulated.

Endocrine Organs of the Body (pp. 537–561)

5. Describe the structural and functional relationships between the hypothalamus and the pituitary gland.
6. List and describe the chief effects of adenohypophyseal hormones.
7. Discuss the role of the neurohypophysis, and note the effects of the two hormones it releases.
8. Describe the important effects of the two groups of hormones produced by the thyroid gland. Follow the process of thyroxine formation and release.
9. Note the general functions of parathyroid hormone.
10. List the hormones produced by the cortical and medullary regions of the adrenal gland, and cite their physiological effects.
11. Compare and contrast the effects of the two major pancreatic hormones.
12. Describe the functional roles of the hormone products of the testes and ovaries.
13. Briefly describe the importance of thymic hormones in the operation of the immune system.

Developmental Aspects of the Endocrine System (pp. 561–562)

14. Describe the effect of aging on endocrine system functioning.

Preview of Selected Key Terms

Endocrine system (*endo* = within, internal; *crine* = to secrete) The body system that includes internal organs that secrete hormones.

Hormone (*hormon* = to excite) Steroidal or amino acid–based molecules released to the blood that act as chemical messengers to regulate specific body functions.

Target cell A cell that is capable of responding to a hormone because it bears receptors to which the hormone can bind.

Second messenger Intracellular molecule generated by the binding of a chemical (hormone or neurotransmitter) to a plasma membrane receptor; mediates intracellular responses to the chemical messenger.

Cyclic AMP Important intracellular second messenger that mediates hormonal effects; formed from ATP by the action of adenylate cyclase, an enzyme associated with the plasma membrane.

Tropic hormone (trō′-pik) (*tropi* = turn, change) A hormone that regulates the function of another endocrine organ.

When insulin molecules, carried passively along in the blood, attach tenaciously to protein receptors of nearby cells, the response is dramatic: Glucose molecules begin to disappear into the cells, and cellular activity accelerates. Such is the power of the second great controlling system of the body, the **endocrine system.** However, the endocrine system does not act alone. It interacts with the nervous system to coordinate and integrate the activity of body cells, although the means and speed of control used by these two great regulatory systems are very different. The nervous system regulates the activity of muscles and glands by means of electrochemical impulses delivered by neurons; those organs respond within milliseconds. The endocrine system, on the other hand, influences the metabolic activities of cells by means of **hor-**

mones: chemical messengers released into the blood to be transported in a more leisurely fashion throughout the body. Tissue or organ responses to hormones typically occur after a lag period of seconds or even days, but once initiated, the responses tend to be much more prolonged than those initiated by the nervous system. Hormonal targets ultimately include most cells of the body.

The Endocrine System and Hormone Function: An Overview

Compared to other organs of the body, the organs of the endocrine system are small and unimpressive. Indeed, to collect 1 kg of hormone-producing tissue, you would need to collect *all* of the endocrine tissue from eight or nine adults! In addition, the anatomical continuity typical of most organ systems does not exist between the endocrine organs. Instead, endocrine organs are scattered about in widely separated areas of the body (Figure 17.1).

As explained in Chapter 4, there are two kinds of glands: endocrine and exocrine. Endocrine glands, also called *ductless glands,* release hormones into the blood or lymph, and they typically have a rich vascular and lymphatic drainage. In contrast, exocrine glands have ducts through which their nonhormonal products are routed to a membrane surface. The endocrine glands of the body include the pituitary, thyroid, parathyroid, adrenal, pineal, and thymus glands. In addition, several organs of the body contain discrete areas of endocrine tissue within their substance and produce hormones as well as exocrine products. These organs, which include the pancreas and gonads (the ovaries and testes), are also major endocrine organs. In this text, the hypothalamus is also recognized as a major endocrine organ. There is no question that the hypothalamus is an integral part of the nervous system. On the other hand, it produces and releases hormones, so we can consider the hypothalamus a *neuroendocrine* (ner′-ō-en′-dō-krin) *organ.*

Besides the major endocrine organs, various other tissues and organs produce hormones. For example, pockets of hormone-producing cells are found in the walls of the small intestine, stomach, kidneys, and heart—organs whose chief functions have little to do with hormone production. The placenta, a temporary organ formed in the uterus of a pregnant woman, produces hormones generally thought of as ovarian hormones (estrogen and progesterone). Additionally, certain tumor cells, such as those of some cancers of the lung or pancreas, synthesize hormones identical to those made in normal endocrine glands, but in an excessive and uncontrolled fashion.

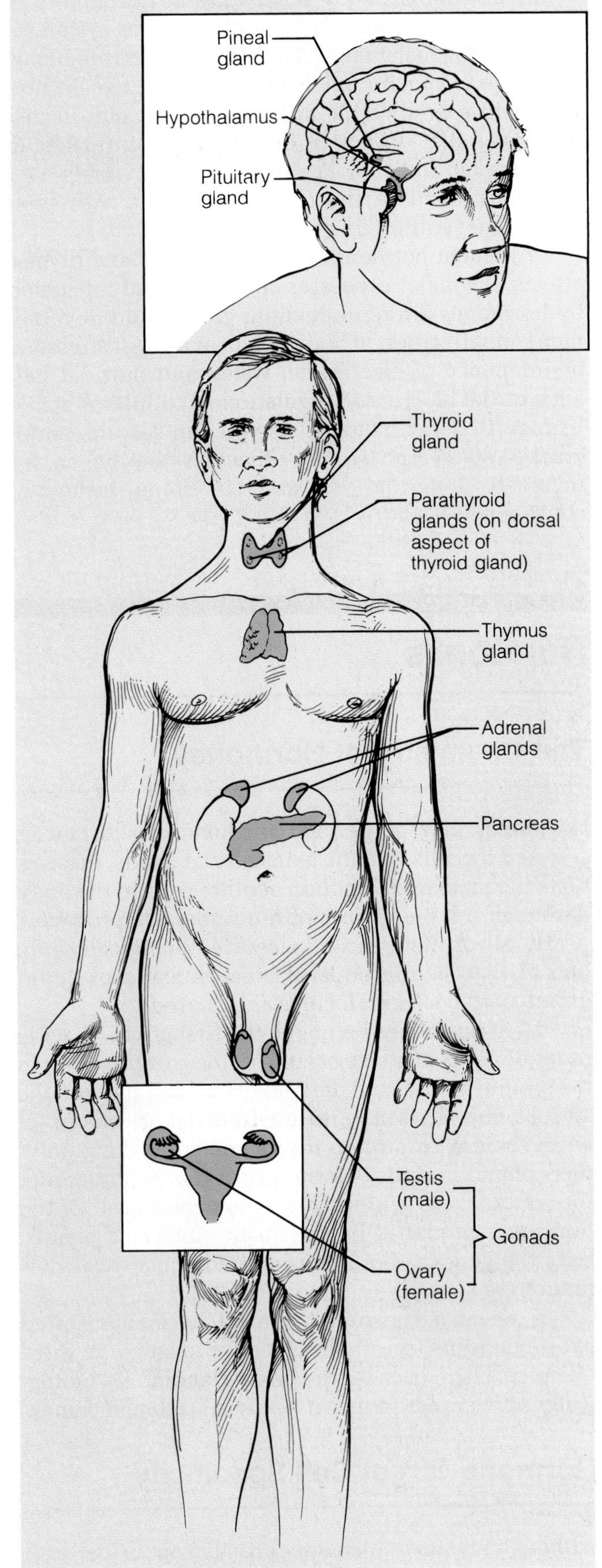

Figure 17.1 Location of the major endocrine organs of the body.

Even though many other sites of hormone production exist, we consider the *endocrine system* to consist of the major endocrine organs, which are highly receptive to specific internal signals and release hormones in a predictable fashion. We will not discuss hormones that do not fit this description further in this chapter, but they are summarized in Table 17.1 and considered in the chapters dealing with their respective synthesizing organs.

Although hormones have widespread and diverse effects, the major processes controlled and integrated by hormones are reproduction; growth and development; mobilization of body defenses against stressors; maintenance of electrolyte, water, and nutrient balance of the blood; and regulation of cellular metabolism and energy balance. As you can see, the endocrine system orchestrates processes that go on for relatively long periods and, in some instances, continuously.

Hormones

The Chemistry of Hormones

Hormones may be defined as chemical substances, secreted by cells into the extracellular fluids, that regulate the metabolic function of other cells in the body. Although a large variety of hormones are produced, nearly all of them can be classified chemically into one of two large groups of biochemical molecules: *amino acid-based molecules* and *steroids*.

Most hormones belong to the first group. A wide range of molecular size occurs in this grouping—from the simple amino acid derivatives, which include the amines and thyroxine, to peptides (short chains of amino acids), to protein macromolecules (long polymers of amino acids). Hormones of the second group, the steroids, are synthesized from cholesterol. Of the hormones produced by the major endocrine organs, only the gonadal hormones and adrenocortical hormones are steroids.

If we also consider the local hormones called **prostaglandins** (see Table 17.1), we must add a third chemical class, because the prostaglandins are biologically active lipids found in nearly all cell membranes.

Hormone-Target Cell Specificity

Although all major hormones circulate to virtually all tissues, a given hormone influences the activity of only certain tissue cells, referred to as its **target cells.** The ability of a target cell to respond to a hormone depends on the presence of specific protein *receptors* on its plasma membrane or in its interior, to which that hormone can bind in a complementary manner. For example, receptors for adrenocorticotropic hormone (ACTH) are normally found only on certain cells of the adrenal cortex. By contrast, thyroxine is the principal hormonal stimulant of cellular metabolism, and nearly all body cells have thyroxine receptors. An analogy can be drawn between the endocrine gland–target cell response and a radio's transmitter-receiver system. The endocrine gland can be viewed as the "transmitter," sending signals to the rest of the body, and the receptors on target cells are the "receivers." Like a radio tuned to a single station, the receptors respond only to a particular signal, even though many other signals may be present at the same time. While a radio receiver responds to the radio signal by producing sound, a hormone receptor responds to hormone binding by prompting the cell to perform, or "turn on," some genetically determined function.

Although hormone-receptor binding is the crucial first step, the extent of target cell activation by hormone-receptor interaction depends equally on three factors: blood levels of the hormone, the relative numbers of receptors for that hormone on or in the target cells, and the affinity (strength) of the union between the hormone and the receptor. Changes in all three factors occur rapidly in response to various stimuli and changes within the body. As a rule, a large number of high-affinity receptors produces a pronounced hormonal effect, whereas a smaller number of low-affinity receptors results in reduced target cell response or outright endocrine dysfunction at the same blood hormone levels. Furthermore, receptors are dynamic structures. In some instances, the target cells form more receptors in response to increasingly higher levels of the specific hormones to which they respond, a phenomenon called *up-regulation*. But in other cases, prolonged exposure to high hormone concentrations may desensitize the target cells, so that they begin to respond less vigorously to hormonal stimulation; this *down-regulation* is thought to involve a loss in the number of receptors and effectively prevents the target cells from overreacting to persistently high hormone levels. In addition, hormones may influence the number and affinity not only of their own receptors, but also of receptors responsive to other hormones. For example, progesterone induces a loss of estrogen receptors in the uterus, thus antagonizing estrogen's actions, whereas estrogen promotes an increased production of progesterone receptors in the same cells, enhancing their ability to respond to progesterone.

Mechanisms of Hormone Action

Hormones bring about their characteristic effects on target cells by altering cell activity; that is, they increase

Table 17.1 Hormones Produced by Organs Other Than the Major Endocrine Organs

Hormone	Chemical composition	Source/trigger	Target organ/effects
Prostaglandins (PGs)	20-carbon fatty acid derivatives synthesized from arachidonic acid; several groups, designated by letters A–I (PGA–PGI) and subgroups within major groupings indicated by numbers (e.g., PGE_2)	Associated with plasma membrane of virtually all body cells; various triggers (local irritation, hormones, etc.)	Multiple targets; act locally; myriad effects include mediation of hormone response and stimulation of smooth muscles of arterioles (increase blood pressure) or uterus (increase expulsive contractions); increase HCl and pepsin secretion by stomach; cause platelet aggregation (enhance blood clotting); cause constriction of bronchioles; increase inflammation and pain; induce fever
Gastrin	Peptide	Stomach; secreted in response to food	Stomach; stimulates glands to release hydrochloric acid (HCl)
Enterogastrin	Peptide	Duodenum of small intestine; secreted in response to food, especially fats	Stomach; inhibits HCl secretion and gastrointestinal tract mobility
Secretin	Peptide	Duodenum; secreted in response to food	Pancreas: stimulates release of bicarbonate-rich juice; liver: increases release of bile; stomach: inhibits secretory activity
Cholecystokinin	Peptide	Duodenum; secreted in response to food	Pancreas: stimulates release of enzyme-rich juice; gall bladder: stimulates expulsion of stored bile; sphincter of Oddi: causes sphincter to relax, allowing bile and pancreatic juice to enter duodenum
Erythropoietin	Glycoprotein	Kidney*; secreted in response to hypoxia	Bone marrow; stimulates production of red blood cells
Active vitamin D_3	Steroid	Kidney activates vitamin D made by epidermal cells of skin; activated and released in response to parathyroid hormone	Intestine; stimulates active transport of dietary calcium across intestinal cell membranes
Atrial natriuretic hormone	Peptide	Atrium of heart; secreted in response to stretching of atria	Kidney: inhibits sodium ion reabsorption and renin release; adrenal cortex: inhibits secretion of aldosterone

*The kidneys release an enzyme that modifies a circulating blood protein to produce erythropoietin.

or decrease the rates of normal cellular processes. The precise hormonal response is dictated by the target cell type. For example, epinephrine binding to smooth muscle cells in the walls of blood vessels stimulates them to contract; epinephrine binding to cells other than muscle cells may have a different effect, but it will not cause those noncontractile cells to shorten.

A hormonal stimulus typically produces one or more of the following changes:

1. Changes in plasma membrane permeability or electrical state.
2. Synthesis of proteins or certain regulatory molecules (such as enzymes) within the cell.
3. Enzyme activation or deactivation.
4. Induction of secretory activity.

In general, there are two major mechanisms by which hormone binding is harnessed to the specific intracellular machinery needed for hormone action. One involves the formation of one or more intracellular second messengers, which mediate the target cell's response to the hormone; the other involves direct gene (DNA) activation by the hormone itself.

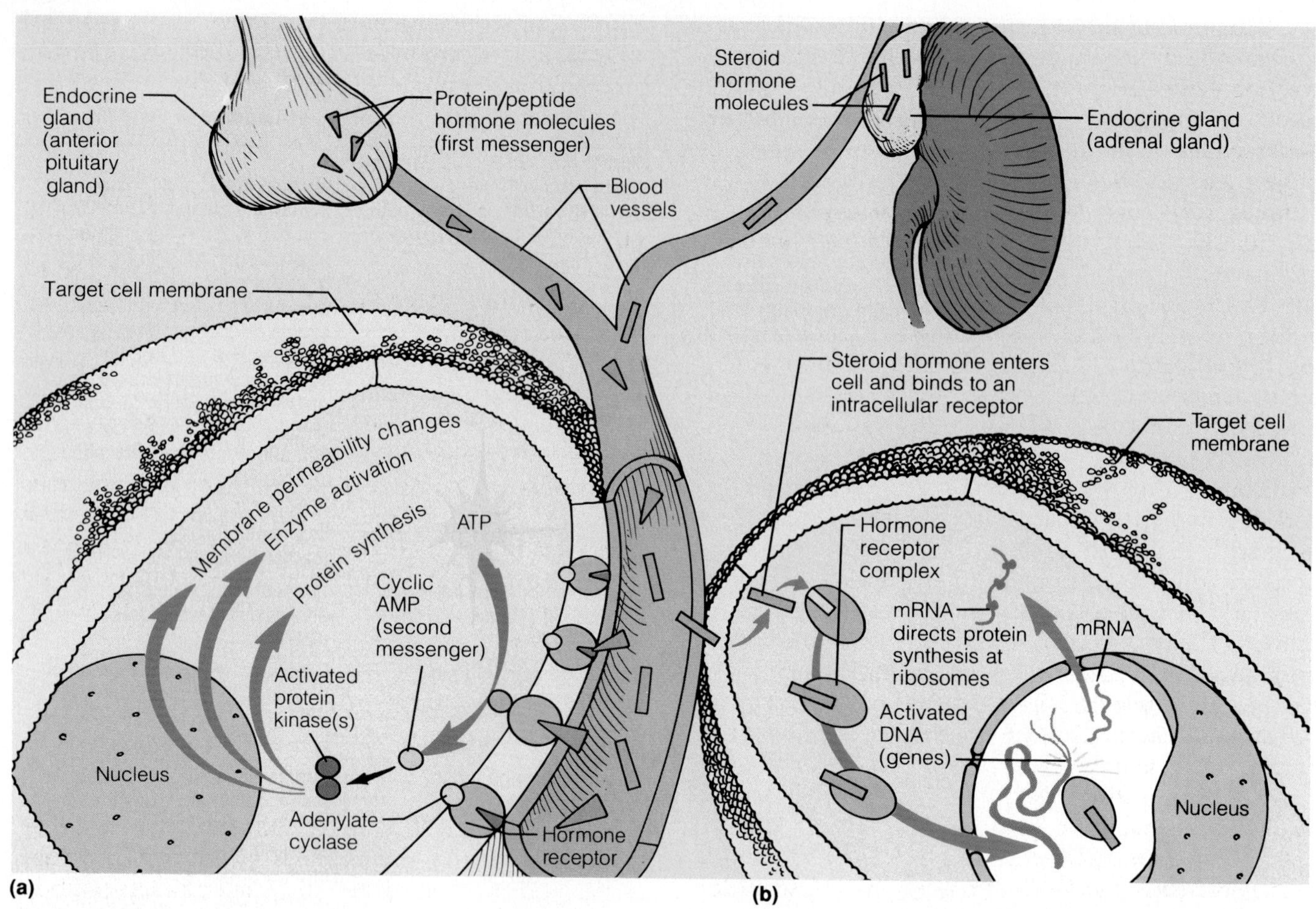

Figure 17.2 Mechanisms of hormone action. (**a**) Second-messenger mechanism. Amino acid–based hormones act through intracellular second messengers—cyclic AMP in this example. Binding to membrane receptors activates adenylate cyclase to convert ATP to cyclic AMP, which acts intracellularly as a second messenger to activate one or more protein kinase enzymes that mediate the response to the hormone. (**b**) Direct gene activation, the mechanism of action of most steroid hormones. Steroids enter the target cell and bind to intracellular receptors. The receptor-hormone complex enters the nucleus and causes specific genes to be transcribed into messenger RNA. The RNA migrates to the cytoplasm, where it directs synthesis of specific proteins.

Second-Messenger Systems

Virtually all of the protein or amino acid–based hormones exert their effects through intracellular **second messengers** generated by hormone binding to plasma membrane receptors. Of the second messengers, cyclic AMP, which also mediates the effects of certain neurotransmitters, is by far the best understood, so it will receive most of our attention. When hormones bind to receptors coupled to the membrane-bound enzyme **adenylate cyclase** (Figure 17.2a), adenylate cyclase is activated and catalyzes the conversion of intracellular ATP to **cyclic AMP** (cyclic adenosine 3′,5′-monophosphate). This particular type of hormone–target cell interaction is called the *cyclic AMP mechanism.* In this scheme, the hormone, called the first messenger, delivers the signal to adenylate cyclase to become activated. Cyclic AMP, generated by adenylate cyclase activity and acting as a second messenger, can then diffuse throughout the cell to initiate a cascade of chemical reactions in which one or more enzymes, called *protein kinases,* are activated. These intracellular events mediate the cell's response to the hormone. A given cell may have several types of protein kinases, each of which has distinct substrates. The protein kinases phosphorylate (add a phosphate group to) different proteins, many of which are other enzymes. Since phosphorylation activates some of these proteins and inhibits others, a variety of reactions may be occurring in the same target cell at the same time. For example, a fat cell responds to epinephrine binding by breaking down glycogen and stored fat, reactions mediated by different enzymes.

This type of intracellular enzymatic cascade has a tremendous amplification effect because a single enzyme can catalyze literally hundreds of reactions. Hence, as the reaction cascades through one enzyme intermediate after another, the number of product molecules increases dramatically at each step. Theoretically, receptor binding of a single hormone molecule could generate millions of the final product molecules!

The sequence of biochemical reactions set into motion by cyclic AMP depends on the target cell type, the specific protein kinases it contains, and the hormone acting as first messenger. For example, in thyroid cells, cyclic AMP generated in response to the binding of thyroid-stimulating hormone promotes the synthesis of the thyroid hormone thyroxine; in bone and muscle cells, cyclic AMP generated in response to growth hormone binding activates anabolic reactions in which amino acids are built into tissue proteins. Since cyclic AMP is rapidly degraded by an intracellular enzyme (*cyclic AMP phosphodiesterase*), its action persists only briefly. While this may, at first glance, appear to be a problem, it is quite the reverse. Because of the amplification effect just explained, most hormones need to be present for only short periods of time to promote the desired results. Thus, in this system, only one mechanism—activation—is necessary. Continued hormone production causes continued cellular activity; no extracellular controls are necessary to stop the activity.

Although cyclic AMP is known to be the activating second messenger in some tissues for at least 12 of the amino acid–based hormones, some of these same hormones are known to act through a different second-messenger system in other tissues. One such mechanism, the *phosphatidyl inositol (PIP) mechanism,* activates a membrane-bound enzyme that splits a membrane phospholipid called PIP_2 (phosphatidyl inositol 4,5-biphosphate) into *diacylglycerol* and *inositol triphosphate,* both of which ultimately act as second messengers to activate specific protein kinases. Antidiuretic hormone, for example, influences kidney tubule cells via cyclic AMP, but when antidiuretic hormone binds to liver cells, the phosphatidyl inositol mechanism is activated. Other hormones act on their target cells through different (and in some cases, unknown) mechanisms or messengers. Insulin, for example, decreases rather than increases cyclic AMP levels. Previously, it was believed that insulin's effects were mediated by cyclic GMP (cyclic guanosine 3′,5′-monophosphate). Although this no longer seems to be the case, cyclic GMP is a suspected second messenger for some hormones. In certain instances, any of the second messengers mentioned—and the hormone receptor itself—can cause changes in intracellular calcium ion levels. The calcium ions then act as a third messenger to effect enzyme activation and hormonal responses by binding to intracellular proteins or to an intracellular receptor called calmodulin.

Direct Gene Activation

Being lipid soluble, steroid hormones (and, strangely, thyroxine, an iodinated amine) can diffuse easily into their target cells: Once inside, they bind to a receptor located within either the cytoplasm or nucleus. The activated hormone-receptor complex then interacts with the nuclear chromatin, where the hormone binds to a DNA-associated protein specific for it. This interaction "turns on" DNA transcription of messenger RNA that in turn directs the synthesis of specific protein molecules. These protein products include enzymes that promote the metabolic activities induced by that particular hormone and, in some cases, the synthesis of structural proteins or proteins that are exported by the target cell. The steroidal gene activation mechanism is depicted in Figure 17.2b.

Onset and Duration of Hormone Activity

Hormones are exceptionally potent chemicals, and they exert profound effects on their target organs at very low blood concentrations. The concentration of a hormone in blood at any time reflects not only its rate of release, but also the speed with which it is inactivated or removed from the body. Some hormones are rapidly degraded by enzymes within their target cells, but most are removed from the blood by kidney and liver enzyme systems, and their breakdown products are quickly excreted from the body in urine or, to a lesser extent, in feces. As a result, the persistence of a hormone in the blood, referred to as its *half-life,* is usually brief—from a fraction of a minute to 30 minutes. The time required for onset of hormone effects varies greatly. Some hormones promote target organ responses almost immediately; others, particularly the steroid hormones, may require hours to days before their effects are seen. Additionally, some hormones, such as testosterone secreted by the testes and thyroxine released by the thyroid gland, are secreted in a relatively inactive form and must be activated in the target cells. The duration of hormone action is limited, ranging from 20 minutes to several hours, depending on the hormone. Effects may disappear rapidly as blood levels drop, or they may persist for hours after very low levels have been reached. Because of these many variations, hormonal blood levels must be precisely and individually controlled to meet the continuously changing needs of the body.

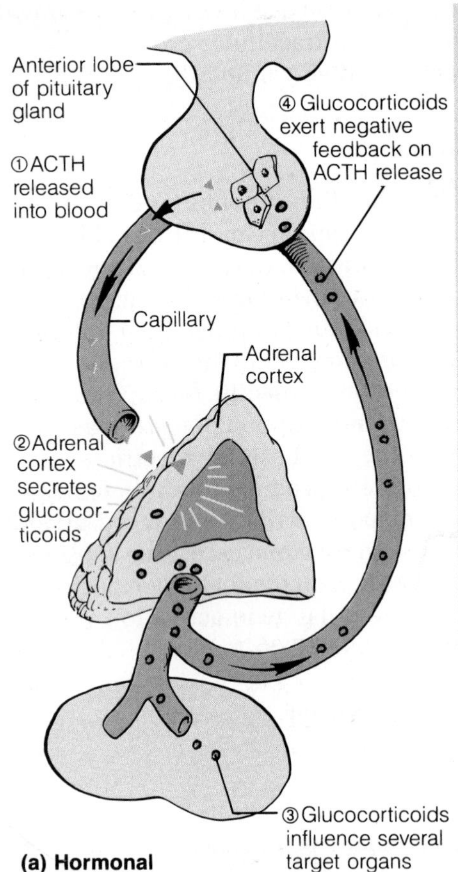

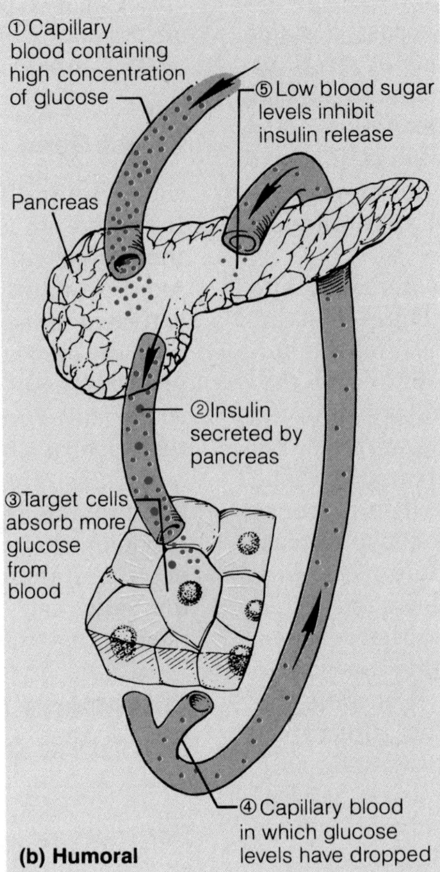

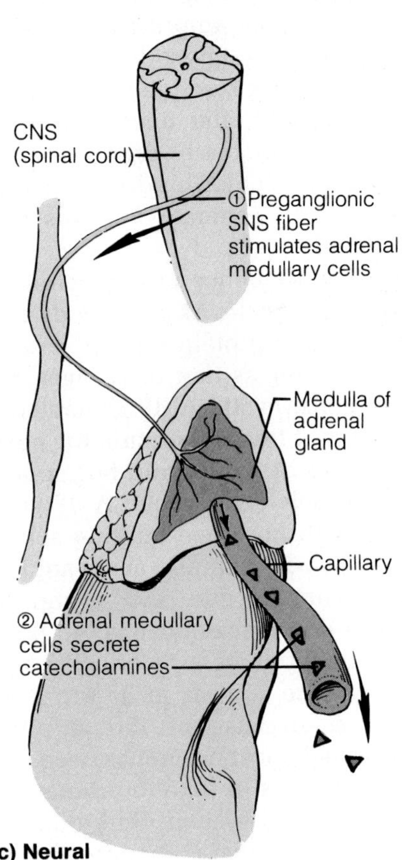

Figure 17.3 Endocrine gland stimuli. (**a**) Hormonal stimulus of endocrine gland activity. In the example illustrated, ACTH (adrenocorticotropic hormone) stimulates the release of glucocorticoid hormones by the adrenal cortex. As blood levels of glucocorticoids rise, they inhibit ACTH release. (**b**) Humoral stimulus of endocrine gland activity. High blood glucose levels trigger insulin release from the pancreas. Insulin causes a rapid uptake of glucose by its target cells, causing blood sugar levels to drop and the stimulus for insulin release to end. (**c**) Neural stimulus of endocrine gland activity. The stimulation of adrenal medullary cells by sympathetic nervous system (SNS) fibers triggers the release of catecholamines (epinephrine and norepinephrine) to the blood.

Control of Hormone Release

The synthesis and release of most hormones is regulated by some type of *negative feedback system* (see Chapter 1, p. 13). In such a system, hormone secretion is triggered by some internal or external stimulus, after which rising hormone levels (even while causing target organ effects) inhibit further hormone release. As a result, blood levels of many hormones vary only within a very narrow range.

Endocrine Gland Stimuli

Endocrine gland stimuli fall into three major categories: **hormonal, humoral** (hyoo-mor″-ul), and **neural stimuli.**

Hormonal Stimuli. The release of most anterior pituitary hormones is regulated by releasing and inhibiting hormones produced by the hypothalamus. Many anterior pituitary hormones in turn stimulate other endocrine organs to release their hormones into the blood. As the hormones produced by the final target glands increase in the blood, they inhibit the release of anterior pituitary hormones and thus their own release. Hormone release promoted by this mechanism tends to be rhythmic, with hormone blood levels rising and falling in a specific pattern. Figure 17.3a illustrates the release of glucocorticoid hormones by the adrenal cortex in response to ACTH, an anterior pituitary hormone. As shown, rising blood levels of glucocorticoids feed back to inhibit ACTH release.

Humoral Stimuli. Changing blood levels of certain ions and nutrients may also stimulate hormone release. Such stimuli are referred to as humoral stimuli to distinguish them from hormonal stimuli, which are also blood-borne chemicals. The term *humoral* harkens back to the ancient use of the term *humor* to refer to various body fluids (blood, bile, and others). For example, the release of insulin by pancreatic cells is prompted by increasing blood sugar levels. Since insulin promotes glucose entry into tissue cells, blood sugar levels soon decline, ending the initiative for insulin release (Figure 17.3b). Other hormones released in response to humoral stimuli include calcitonin, released by the thyroid gland; parathyroid hormone, produced by the parathyroid glands; and aldosterone, one of the adrenal cortex hormones.

Neural Stimuli. In isolated cases, nerve fibers stimulate hormone release. The classic example is sympathetic nervous system stimulation of the adrenal medulla to release catecholamines (norepinephrine and epinephrine) during periods of stress (Figure 17.3c). In addition, oxytocin and antidiuretic hormone are released from the posterior pituitary in response to nerve impulses from the hypothalamic neurons.

While these three mechanisms typify most systems that control hormone release, they are by no means all-inclusive, and some endocrine organs respond to multiple stimuli.

Nervous System Modulation

Both "turn on" factors (hormonal, humoral, and neural stimuli) and "turn off" factors (feedback inhibition and others) may be modulated by the activity of the nervous system. Without this added safeguard, endocrine system activity would be strictly mechanical, much like a household thermostat. A thermostat can maintain the house temperature at or around its set value, but it cannot sense that your visiting grandmother from Florida feels cold at that temperature and reset itself accordingly. *You* must make that adjustment. Similarly, the nervous system can, in certain cases, override normal endocrine controls as necessary to maintain homeostasis. For example, blood sugar levels are normally maintained within a range of 80–120 mg glucose per 100 ml of blood by the action of insulin and various other hormones. However, when the body is under severe stress, the hypothalamus and sympathetic nervous system centers are strongly activated and cause blood sugar levels to rise much higher. This ensures that the cells will have sufficient fuel for the more vigorous activity required of them during such periods.

Remember that the hypothalamus is an autonomic center concerned with homeostatic control of such body functions as water balance and temperature as well as an integrating center for biological rhythms and emotions. This helps explain how a single external stimulus, such as acute blood loss or severe physical trauma, can be followed by body-wide neuroendocrine adjustments.

Endocrine Organs of the Body

The Pituitary Gland (Hypophysis)

Securely seated in the sella turcica of the sphenoid bone is the pea-size **pituitary gland,** or **hypophysis** (literally, "to grow under"). Although the pituitary gland is structurally and functionally a distinct organ, it is connected to the hypothalamus by a stalk of tissue called the *infundibulum* (in″-fun-dih′-byoo-lum) (Figure 17.4). The human pituitary gland has two major lobes. The **posterior lobe,** or **neurohypophysis** (nyoo″-rō-hī-pof′-ih-sis), is composed largely of neuroglia, and it releases neurohormones that it receives ready-made from the hypothalamus. Thus, the neurohypophysis is a hormone-storage area and not a true endocrine gland in the precise sense. The **anterior lobe,** or **adenohypophysis** (a″-dih-nō-hī-pof′-ih-sis), is composed of glandular tissue, and it manufactures and releases a number of hormones (Table 17.2 on pp. 542–543).

Pituitary-Hypothalamic Relationships

The contrasting histology of the two pituitary lobes reflects the dual origin of this tiny gland. The neurohypophysis derives from a down growth of hypothalamic (nervous) tissue, and it maintains its connection with the hypothalamus via a nerve bundle called the *hypothalamic-hypophyseal tract,* which runs through the infundibulum. Neurosecretory cells in the hypothalamus synthesize two *neurohormones,* oxytocin and antidiuretic hormone, which are transported down the length of the axons (by axonal transport) to the neurohypophysis. There they are stored temporarily and released on demand into the capillary blood in response to appropriate neural stimuli from hypothalamic neurons.

In contrast, the anterior lobe originates from a superior outpocketing of the oral mucosa (Rathke's pouch) and is formed from epithelial tissue. After touching the posterior lobe, the adenohypophysis loses its connection with the oral mucosa and adheres to the neurohypophysis to form the composite organ. There is no direct neural connection between the adenohypophysis and the hypothalamus, but rather a vascular connection, called the *hypophyseal portal system.* Via this portal system, *releasing* and *inhibiting hormones* secreted by neurons in the ventral hypothalamus circulate directly to the adenohypophysis, where they regulate the secretory activity of its hormone-producing cells. All of these hypothalamic regulatory hormones are amino acid–based, but they vary widely in size from amines to polypeptides.

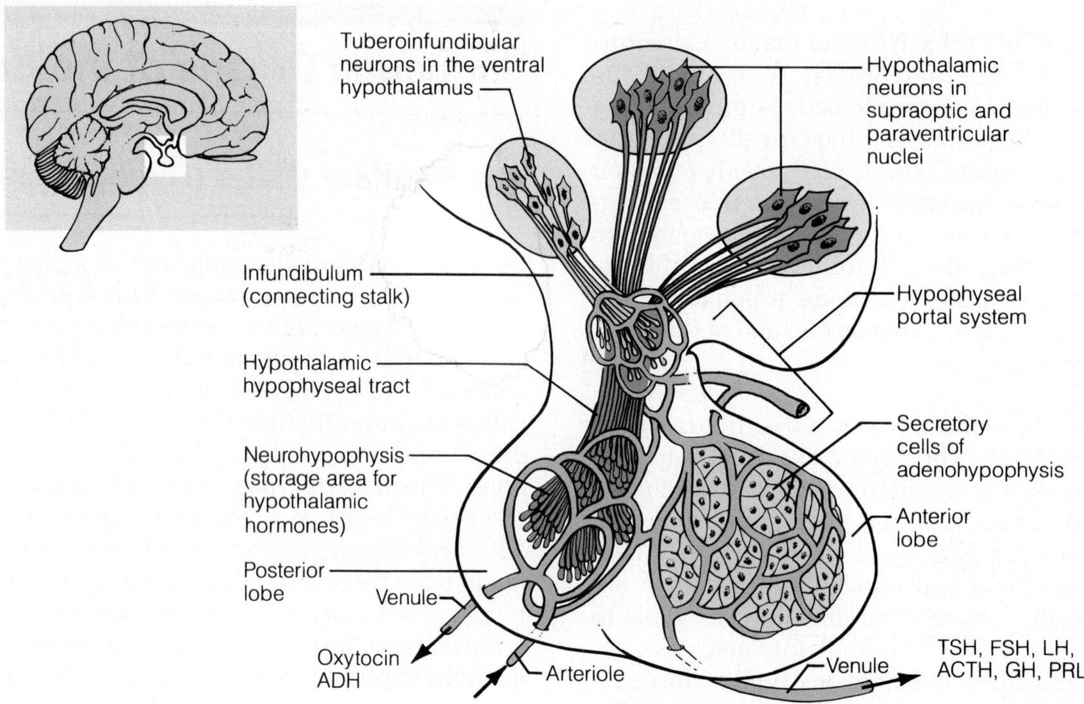

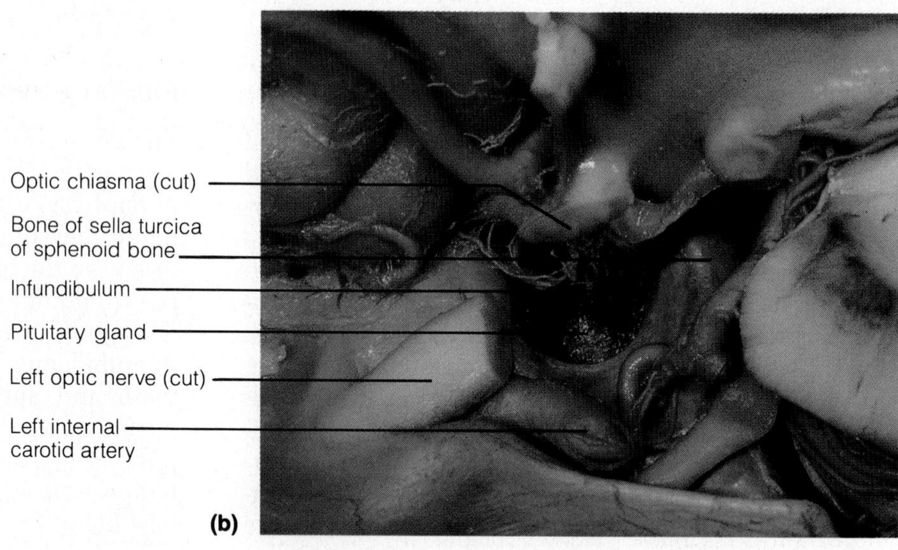

Figure 17.4 Structural and functional relationships of the pituitary and the hypothalamus. (**a**) Hypothalamic neurons in the supraoptic and paraventricular nuclei synthesize oxytocin and ADH, which are transported along their axons (hypothalamic-hypophyseal tract) to the neurohypophysis for storage. Tuberoinfundibular neurons in the ventral hypothalamus have very short axons that release releasing and inhibiting factors to the capillaries of the hypophyseal portal system, which runs to the adenohypophysis. These factors influence the adenohypophyseal secretory cells to release (or not release) their hormones. (**b**) Photograph of part of the pituitary gland in its anatomical location (sella turcica of the sphenoid bone) in the body.

One of the inhibiting hormones, *somatostatin* (sō-ma″-tō-sta′-tin), or *growth hormone–inhibiting hormone (GHIH)*, is distributed widely in cells throughout the nervous system as well as in many nonneural tissues, including the intestine and pancreas. Somatostatin has far-reaching effects that deserve special mention. In addition to inhibiting growth hormone secretion, it blocks the release of several other adenohypophyseal hormones (see Table 17.2). Additionally, it inhibits the release of virtually all gastrointestinal and pancreatic secretions—both endocrine and exocrine.

Adenohypophyseal Hormones

The adenohypophysis has traditionally been called the "master endocrine gland" because of its numerous hormonal products, many of which regulate the activity of other endocrine glands. However, the adenohypophysis has been dethroned by the hypothalamus, which is now known to control anterior pituitary activity. Six distinct adenohypophyseal hormones with specific physiological actions in humans are known. In addition, a large molecule with the tongue-twisting name *pro-opiomelanocortin* (prō″-ō″-pē-ō-muh-la′-nō-kor″-tin) *(POMC)* has been isolated from the anterior pituitary. POMC is a *prohormone;* that is, it is a large

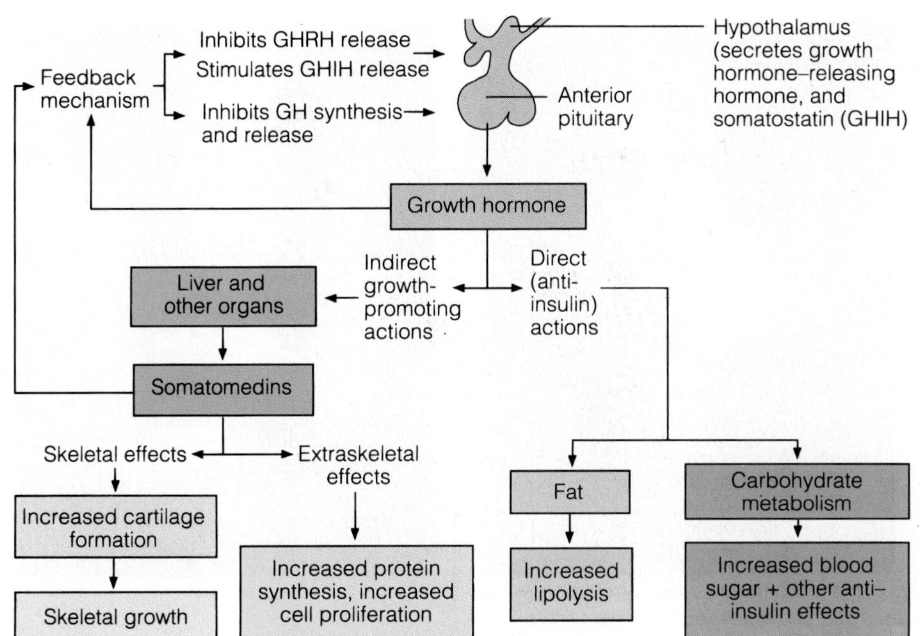

Figure 17.5 Proposed scheme for classifying the metabolic actions of growth hormone (GH). The direct actions of GH on target cells stimulate fat breakdown (lipolysis) and release from adipose tissues and hinder glucose uptake from the blood by tissue cells, maintaining blood sugar levels fairly high. (Because these actions antagonize those of the pancreatic hormone insulin, they are referred to as anti-insulin actions.) Indirect, largely anabolic effects of GH are mediated through somatomedins. Elevated levels of GH and somatomedins feed back to promote GHIH release (and depress GHRH release) by the hypothalamus and to inhibit GH release by the anterior pituitary gland.

precursor molecule from which other active molecules are split by enzymatic cleavage. POMC is the source of adrenocorticotropic hormone, some of our natural opiates (an enkephalin and beta-endorphin, described in Chapter 11), and melanocyte-stimulating hormone, which is presumed to be a hypothalamic neurotransmitter.

When the adenohypophysis receives an appropriate chemical stimulus from the hypothalamus, one or more of its hormones is released by certain of its cells. Although many different releasing and inhibiting hormones pass from the hypothalamus to the anterior lobe, the specific anterior pituitary target cells distinguish the messages directed to them and respond in kind—synthesizing and secreting the proper hormones in response to specific releasing hormones, and closing down hormone release in response to inhibiting hormones. The releasing hormones are far more important as regulatory factors because little hormone is stored by the secretory cells of the anterior lobe.

Four of the six anterior pituitary lobe hormones—thyroid-stimulating hormone (TSH), adrenocorticotropic hormone (ACTH), follicle-stimulating hormone (FSH), and luteinizing hormone (LH)—are **tropic** (trō′-pik) **hormones.** Tropic hormones regulate the hormonal functioning of other endocrine glands. The remaining two adenohypophyseal hormones—growth hormone (GH) and prolactin (PRL)—exert their major effects on nonendocrine targets. All adenohypophyseal hormones affect their target cells via second-messenger systems. (The adenohypophyseal hormones, their effects, and their relationships to hypothalamic regulatory factors are summarized briefly in Table 17.2.)

Growth Hormone (GH). Growth hormone (GH), or **somatotropin** (*soma* = body), is a protein hormone that is produced by the *somatotropic cells.* Although GH stimulates most body cells to increase in size and divide, its major targets are the bones and skeletal muscles. Stimulation of epiphyseal plate activity leads to long bone growth; effects on skeletal muscles promote increased muscle mass.

Essentially an anabolic hormone, GH promotes protein synthesis, and it encourages the use of fats for cellular fuel, thus conserving glucose (Figure 17.5). The second-messenger system through which GH influences its target cells is still controversial, but the growth-promoting effects of GH are known to be mediated indirectly by *somatomedins* (sō-ma″-tō-mē′-dins), growth-promoting proteins produced by the liver and perhaps by the kidneys and muscles as well. Specifically, GH (1) stimulates cellular uptake of amino acids from the blood and their incorporation into proteins; (2) stimulates uptake of sulfur (needed for the synthesis of chondroitin sulfate) into cartilage matrix; (3) mobilizes fats from fat depots for transport to cells, increasing blood levels of fatty acids; and (4) decreases the rate of glucose uptake and metabolism, helping to maintain homeostatic blood glucose levels.

Secretion of GH is regulated chiefly by two hypothalamic hormones with antagonistic effects. *Growth hormone–releasing hormone (GHRH)* stimulates GH release, while feedback of GH and somatomedin inhibit it, presumably by stimulating release of *growth hormone–inhibiting hormone (GHIH),* or *somatostatin.*

Typically, GH secretion has a diurnal cycle, with the highest levels occurring during evening sleep, but the total amount secreted daily declines with age. Moreover, as indicated in Table 17.2, GH release is

Figure 17.6 An individual with acromegaly. This condition results from hypersecretion of growth hormone in the adult. Note the enlarged jaw, nose, and hands. Left to right, the same person is shown at age 16, age 33, and age 52.

highly variable and responsive to a number of secondary triggers. Apparently, relative concentrations of blood-borne nutrients and various stressors, such as exercise and emotional deprivation, act directly on the hypothalamus to influence its release of the regulatory hormones.

Both hypersecretion and hyposecretion of GH may result in structural abnormalities. Hypersecretion in children results in *gigantism,* a condition in which growth is exceptionally rapid and the person becomes abnormally tall (often reaching a height of 8 feet) but has relatively normal body proportions. If excessive amounts of GH are secreted after adult height and epiphyseal plate closure are achieved, *acromegaly* (a″-krō-meg′-ah-lē) results. Acromegaly is characterized by enlargement and thickening of bony areas still responsive to GH, namely, bones of the hands, feet, and face (Figure 17.6). Thickening of soft tissues may lead to coarse or malformed facial features and an enlarged tongue. Hypersecretion of GH usually results from an adenohypophyseal tumor, which is insensitive to normal hypothalamic controls and churns out excessive amounts of GH. The usual treatment is surgical removal of the tumor; however, anatomical changes that have already occurred are not reversible.

Hyposecretion of GH in adults usually causes no problems, but in rare instances, the deficit is so severe that body tissues begin to atrophy and clinical signs of premature aging appear. Growth hormone deficiency in children results in slowed long bone growth and the production of *pituitary dwarfs* with a maximum height of 4 feet. Such people usually have fairly normal body proportions. However, failure of GH secretion is often accompanied by deficiencies of other adenohypophyseal hormones, and if thyroid-stimulating hormone and gonadotropins are lacking, the individual will be malproportioned and will fail to mature sexually as well. When pituitary dwarfism is diagnosed before puberty, growth hormone replacement therapy can promote nearly normal somatic growth. Fortunately, human growth hormone is now being produced commercially by genetic engineering techniques. ■

Thyroid-Stimulating Hormone (TSH). **Thyroid-stimulating hormone (TSH),** or **thyrotropin,** is a glycoprotein tropic hormone that stimulates normal development and secretory activity of the thyroid gland. TSH release from the *thyrotropic cells* of the anterior pituitary is triggered by the hypothalamic peptide *thyrotropin-releasing hormone (TRH)* and inhibited by feedback inhibition exerted by rising blood levels of thyroid hormones acting on both the pituitary and the hypothalamus. The hypothalamus, in response, releases somatostatin, which reinforces the blockade of anterior pituitary release of TSH. As described later in the discussion of thyroid hormones, certain external factors, acting through the hypothalamic controls, may also influence the release of TSH.

Adrenocorticotropic Hormone (ACTH). **Adrenocorticotropic** (uh-drē″-nō-kor-tih-kō-trō′-pik) **hormone (ACTH),** or **corticotropin,** is secreted by the *corticotropic cells* of the adenohypophysis. ACTH stimulates the adrenal cortex to release corticosteroid hormones, most importantly the glucocorticoid hormones that help the body to resist stressors. ACTH release, elicited by hypothalamic *corticotropin-releasing hormone (CRH),* has a basically diurnal rhythm, with highest levels occurring in the morning, shortly after one arises. Rising levels of glucocorticoids feed back and block secretion of CRH and consequently ACTH release. Internal and external factors that alter the normal ACTH rhythm by triggering CRH release include fever, hypoglycemia, and stressors of all types. Because CRH is both the ACTH regulator and a neurotransmitter that is active in the central nervous system, some believe it is the stress response integrator.

Gonadotropins. **Follicle-stimulating hormone (FSH)** and **luteinizing** (loo′-tē-in-īz″-ing) **hormone (LH),** referred to collectively as **gonadotropins,** regulate the function of the gonads (ovaries and testes). In both sexes, FSH stimulates gamete (sperm or egg) production, while LH promotes production of gonadal hormones. In females, LH works cooperatively with FSH to cause maturation of an egg-containing ovarian follicle; it then independently triggers ovulation (the expulsion of the egg from the follicle) and promotes synthesis and release of ovarian hormones (estrogens and progesterone). In males, LH stimulates the interstitial cells of the testes to produce the male hormone testosterone. For this reason, LH is sometimes referred to as **interstitial** (in″-ter-stih′-shul) **cell-stimulating hormone (ICSH)** in males.

Gonadotropins are virtually absent from the blood of prepubertal boys and girls. However, during puberty, the *gonadotropic cells* are activated and gonadotropin levels begin to rise, causing the gonads to mature to the adult state. In both sexes, gonadotropin release by the adenohypophysis is prompted by *gonadotropin-releasing hormone (GnRH)* produced by the hypothalamus. Gonadal hormones, produced in response to the gonadotropins, feed back to suppress FSH and LH release.

Prolactin (PRL). **Prolactin (PRL)** (*pro* = for; *lact* = milk) is a protein hormone structurally similar to growth hormone. Produced by the *lactotropic cells,* it stimulates the ovaries of some animals and is considered a gonadotropin by some researchers. However, its only known effect in humans is to stimulate milk production by the breasts. Like GH, prolactin release is controlled by hypothalamic releasing and inhibiting hormones. *Prolactin-releasing hormone (PRH)* causes prolactin synthesis and release, whereas *prolactin-inhibiting hormone (PIH)* prevents prolactin secretion. (PIH is believed to be the hypothalamic neurotransmitter *dopamine.*) In males and nonlactating women, the influence of PIH predominates, but in women, prolactin levels rise and fall in concert with estrogen blood levels. High estrogen stimulates PIH release, while low estrogen promotes the release of PRH and thus prolactin. The brief rise in prolactin levels just before the menstrual period partially accounts for the breast swelling and tenderness some women experience at that time, but since this PRL stimulation is so brief, the breasts do not become functional in milk production. In pregnant women, PRL blood levels rise dramatically toward the end of pregnancy, and milk production by the breasts becomes possible. After birth, the infant's suckling stimulates PRH release, encouraging continued milk production and availability.

Hypersecretion of prolactin is more common than its hyposecretion (which is not a problem in anyone except women who choose to nurse). In fact, hyperprolactinemia is the most frequent abnormality of adenohypophyseal tumors. Clinical signs include inappropriate lactation (galactorrhea), lack of menses in women, and impotence in males. ■

The Neurohypophysis and Hypothalamic Hormones

The neurohypophysis, constructed largely of supportive neuroglial cells, serves as a storehouse for antidiuretic hormone and oxytocin, synthesized and released by hypothalamic neurons of the supraoptic and paraventricular nuclei. The stored hormones are later released in response to nerve impulses from the same hypothalamic neurons.

Antidiuretic hormone and oxytocin, each composed of nine amino acids, are almost identical, differing only in two amino acids, and yet they have very different physiological effects on their target organs. Antidiuretic hormone influences body water balance, whereas oxytocin stimulates the contraction of smooth muscle, particularly that of the uterus (see Table 17.2). Both hormones exert their effects on their target cells by acting through second-messenger systems.

Oxytocin. **Oxytocin** (ok″-sih-tō′-sin) is a strong stimulant of uterine contraction; to a lesser extent, it initiates constriction of vascular smooth muscle. Oxytocin synthesis and release are significant only during childbirth and in nursing women. The number of oxytocin receptors in the uterus peaks near the end of pregnancy, and the uterine smooth muscle becomes more and more sensitive to the stimulatory effects of oxytocin. The stretching of the uterus and cervix as birth nears causes afferent impulses to be transmitted to the hypothalamus; the hypothalamus responds by synthesizing and releasing oxytocin and triggering the release of stored oxytocin from the neurohypophysis. As oxytocin levels rise in the blood, it promotes the expulsive contractions of labor. Oxytocin also acts as the hormonal trigger for milk ejection (the "letdown" reflex) in women whose breasts are actively producing

Table 17.2 Pituitary Hormones: Regulation and Effects

Hormone	Chemical structure	Regulation of release	Target/effects	Effects of hyposecretion and hypersecretion
Anterior Pituitary Hormones				
Growth hormone (GH)	Protein	Stimulated by GHRH* release, which is triggered by low blood levels of GH as well as by a number of secondary triggers including hypoglycemia, increases in blood levels of amino acids, low levels of fatty acids, exercise and other types of stressors; inhibited by feedback inhibition exerted by GH and somatomedin, and by hyperglycemia, hyperlipidemia, and emotional deprivation, all of which elicit GHIH* release	Body cells, primarily bone and muscle; anabolic hormone; stimulates somatic growth; mobilizes fats; spares glucose	*Hyposecretion:* Pituitary dwarfism in children *Hypersecretion:* Gigantism in children; acromegaly in adults
Thyroid-stimulating hormone (TSH)	Glycoprotein	Stimulated by TRH* and indirectly by pregnancy and cold temperature; inhibited by feedback inhibition exerted by thyroid hormones on anterior pituitary and hypothalamus and by GHIH* (somatostatin)	Thyroid gland; stimulates thyroid gland to release thyroid hormone	*Hyposecretion:* Cretinism in children; myxedema in adults *Hypersecretion:* Graves's disease; exophthalmos
Adrenocorticotropic hormone (ACTH)	Polypeptide (39 amino acids)	Stimulated by CRH*; stimuli that increase CRH release include fever, hypoglycemia, and other stressors; inhibited by feedback inhibition exerted by glucocorticoids	Adrenal cortex; promotes release of glucocorticoids and androgens (mineralocorticoids to a lesser extent)	*Hyposecretion:* Rare *Hypersecretion:* Cushing's disease
Follicle-stimulating hormone (FSH)	Glycoprotein	Stimulated by GnRH*; inhibited by feedback inhibition exerted by estrogen in females and testosterone and inhibin in males	Ovaries and testes; in females, stimulates ovarian follicle maturation and estrogen production; in males, stimulates sperm production	*Hyposecretion:* Failure of sexual maturation *Hypersecretion:* No important effects
Luteinizing hormone (LH)	Glycoprotein	Stimulated by GnRH*; inhibited by feedback inhibition exerted by estrogen and progesterone in females and testosterone in males	Ovaries and testes; in females, triggers ovulation and stimulates ovarian production of progesterone; in males, promotes testosterone production	As for FSH
Prolactin (PRL)	Protein	Stimulated by PRH*; PRH release enhanced by estrogens, birth control pills, opiates, and breast-feeding; inhibited by PIH*	Breast secretory tissue; promotes lactation	*Hyposecretion:* Poor milk production in nursing women *Hypersecretion:* Galactorrhea; cessation of menses in females; impotence in males

Table 17.2 (continued)

Hormone	Chemical structure	Regulation of release	Target/effects	Effects of hyposecretion and hypersecretion
Posterior Pituitary Hormones (made by hypothalamic neurons and stored in posterior pituitary)				
Oxytocin	Peptide	Stimulated by impulses from hypothalamic neurons in response to cervical/uterine stretching and suckling of infant at breast; inhibited by lack of appropriate neural stimuli	Uterus; stimulates uterine contractions; initiates labor; initiates milk ejection from breast	Unknown
Antidiuretic hormone (ADH)	Peptide	Stimulated by impulses from hypothalamic neurons in response to increased osmolarity of blood or decreased blood volume; also stimulated by pain, some drugs, low blood pressure; inhibited by adequate hydration of the body and by alcohol	Kidneys; stimulates kidney tubule cells to reabsorb water	*Hyposecretion:* Diabetes insipidus *Hypersecretion:* Unknown

*Indicates hypothalamic releasing and inhibiting hormones: GHRH = growth hormone–releasing hormone; GHIH = growth hormone–inhibiting hormone; PRH = prolactin-releasing hormone; PIH = prolactin-inhibiting hormone; TRH = thyrotropin-releasing hormone; GnRH = gonadotropin-releasing hormone; CRH = corticotropin-releasing hormone.

milk in response to prolactin. Suckling causes a reflex-initiated release of oxytocin, which targets the specialized myoepithelial cells surrounding the milk-producing glands. As these cells contract, milk is forced from the breast into the mouth of the suckling infant. Both of these mechanisms are positive feedback mechanisms and are described in more detail in Chapter 29.

Both natural and synthetic oxytocic drugs (Pitocin and others) are used to induce labor or to hasten labor that is progressing normally but at a slow pace. Less frequently, oxytocics are used to stop postpartum bleeding (by causing constriction of the ruptured blood vessels at the placental site) and to stimulate the milk ejection reflex.

Antidiuretic Hormone (ADH). *Diuresis* (dī″-yer-ē′-sis) is urine production; thus, an *antidiuretic* (an″-tī-dī″-yer-eh′-tik) is a chemical substance that inhibits or prevents urine formation. **Antidiuretic hormone (ADH)** targets the kidney tubules, which respond by reabsorbing more water from the forming urine and returning it to the bloodstream. As a result, less urine is produced and blood volume increases.

ADH prevents wide swings in water balance, which might result in dehydration or excessive fluid volumes. Highly specialized hypothalamic neurons, called *osmoreceptors,* continually monitor the solute concentration (and thus the water concentration) of the blood. When solutes become too concentrated (as might follow excessive perspiration or inadequate fluid intake), the osmoreceptors transmit excitatory impulses to hypothalamic neurons responsible for ADH synthesis and release. ADH, liberated into the blood by the neurohypophysis, binds to kidney tubule cells, causing them to reabsorb more water, and blood volume rises to homeostatic levels. As solute concentration declines, the osmoreceptors stop depolarizing, effectively ending ADH release. Other stimuli triggering ADH release include pain, low blood pressure, and certain drugs (such as nicotine, morphine, and barbiturates).

Ingestion of alcoholic beverages inhibits ADH secretion and results in copious urine output. The dry mouth and intense thirst experienced "the morning after" reflects this dehydrating effect of alcohol. As might be expected, ADH release is also inhibited by drinking excessive amounts of water. Certain drugs, classed together as *diuretics,* antagonize the effects of ADH and cause water to be flushed from the body. These drugs are used to manage the edema (water retention in tissues) typical of congestive heart failure.

Under certain conditions, such as severe blood loss, exceptionally large amounts of ADH are released. At very high blood concentrations, ADH has a decided pressor effect; that is, it causes vasoconstriction, primarily of the visceral blood vessels. This, in turn, causes an elevation in systemic blood pressure. The alternative name for this hormone (*vasopressin*) reflects this particular effect.

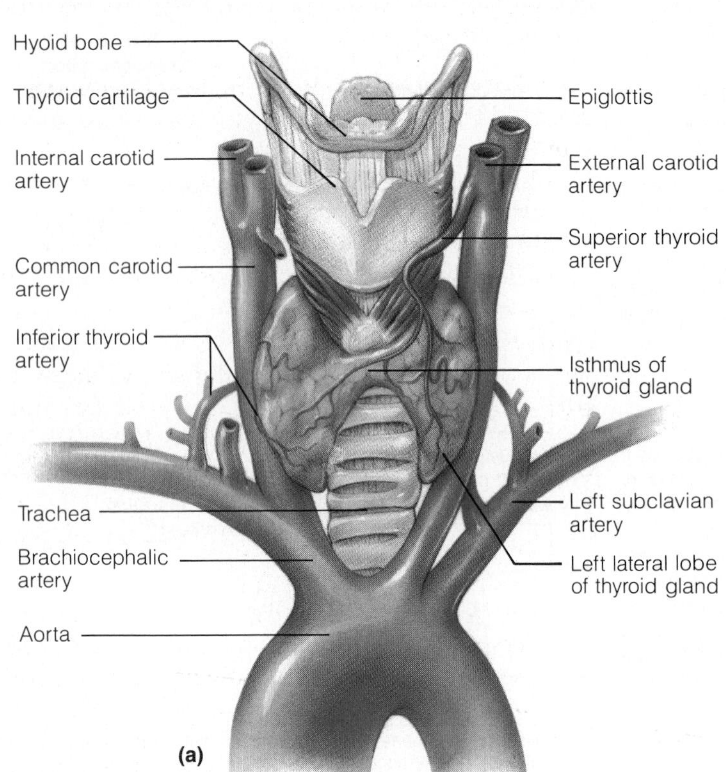

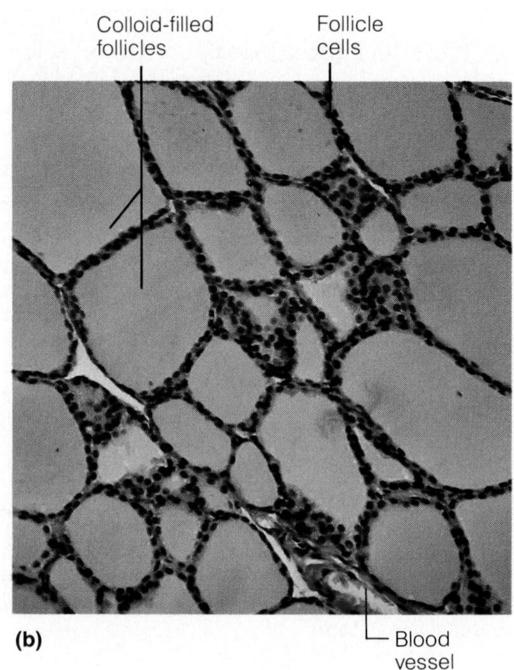

Figure 17.7 Gross and microscopic anatomy of the thyroid gland. (**a**) Location and arterial blood supply of the thyroid gland, anterior view. (**b**) Photomicrograph of the thyroid gland (150X).

The only important imbalance of ADH secretion is *diabetes insipidus,* a syndrome marked by the output of large urine volumes (polyuria) and intense thirst. The name of this condition (*diabetes* = overflow; *insipidus* = tasteless) distinguishes it from diabetes mellitus (*mel* = honey), in which an insulin deficiency causes large amounts of blood sugar to be lost in the urine. At one time, urine was tasted to determine the cause of the polyuria.

The usual cause of diabetes insipidus is hypothalamic trauma, as from a blow to the head; less commonly, damage to the posterior pituitary releasing site is at fault. In either case, ADH release is deficient. The condition, though inconvenient, is not serious when the thirst center is operating properly and the person drinks enough water to prevent dehydration. However, it can be life threatening in unconscious or comatose patients, so accident victims with head trauma must be carefully monitored. ■

The Thyroid Gland

Location and Structure

Formed from a thickening of the endoderm that forms the mucosa of the pharynx, the butterfly-shaped thyroid gland is located in the anterior throat, overlying the inferior border of the larynx (Figure 17.7a). Its two lateral lobes are connected by a medial tissue mass called the *isthmus*; an additional medial tissue extension, the pyramidal lobe, may or may not be present as an upward (slightly lateral) extension of the isthmus. The thyroid gland is one of the largest endocrine glands in the body and has a prodigious blood supply, making thyroid surgery an extremely painstaking endeavor.

Internally, the thyroid gland is composed of hollow spherical structures called *follicles* (Figure 17.7b). The walls of each follicle are formed by cuboidal epithelial cells called follicle cells, which produce the glycoprotein **thyroglobulin** (thī-rō-glob′-yoo-lin). The central cavity, or lumen, of the follicle stores an amber-colored, sticky colloidal material consisting of thyroglobulin molecules with attached iodine atoms. Two hormones collectively called thyroid hormone are derived from this iodinated thyroglobulin. The follicles are separated from one another by connective tissue containing *parafollicular cells,* which produce the hormone *thyrocalcitonin* (described shortly).

Thyroid Hormone

Thyroid hormone, often referred to as the body's major metabolic hormone, is actually two active iodine-containing hormones, **thyroxine** (thī-rok′-sin), or $\mathbf{T_4}$, and

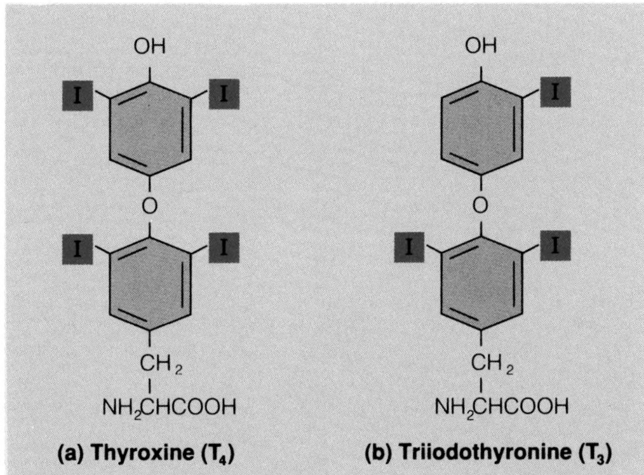

Figure 17.8 Molecular structures of thyroxine (T_4) and triiodothyronine (T_3), the iodine-containing hormones of the thyroid gland. Each hormone consists of two tyrosine molecules linked together; T_4 has four attached iodine atoms, whereas T_3 has three.

triiodothyronine (trī″-ī″-ō-dō-thī′-rō-nēn), or **T_3**. Thyroxine is the major hormone secreted by the thyroid follicles; most triiodothyronine is formed at the target tissues by conversion of T_4 to T_3. These two hormones are very much alike (Figure 17.8): Each is constructed from two tyrosine amino acids linked together, but thyroxine has four bound iodine atoms, whereas triiodothyronine has three (thus, T_4 and T_3).

Thyroid hormone accelerates the rate of cellular metabolism throughout the body. Except for the adult brain, spleen, testes, uterus, and the thyroid gland itself, it affects virtually every cell in the body. Additionally, it is an important regulator of tissue growth and development; it is especially critical for normal skeletal and nervous system development and maturation and reproductive capabilities. Generally speaking, thyroid hormone stimulates enzymes concerned with glucose oxidation. In this way, it acts to increase basal metabolic rate (and oxygen consumption) and body heat production; this is known as the hormone's *calorigenic effect* (*calorigenic* = heat producing). Thyroid hormone also provokes an increase in the number of adrenergic receptors in blood vessels, so it plays an important role in maintaining blood pressure (see Chapter 14). Decline in thyroid hormone levels in the blood produces the opposite effects. A summary of the broad effects of thyroid hormone appears in Table 17.3 on p. 547.

Synthesis of thyroid hormone involves three interrelated processes that are set into motion when thyroid-stimulating hormone (TSH) binds to follicular cell receptors:

1. *Formation of thyroglobulin.* Thyroglobulin is synthesized on the ribosomes of the follicular cells and then transported to the Golgi apparatus, where sugar residues are attached and the molecules are packed into vesicles. These transport vesicles then move to the apex of the follicle cell where their contents are discharged into the lumen and become part of the storage colloid.

2. *Iodine attachment to thyroglobulin (Figure 17.9, steps 1–4).* To produce the functional iodinated hormones, the follicle cells must actively accumulate iodides (anions of iodine) from the blood by means of a very efficient ATP-dependent iodide pump. Once inside the cell, the iodides are converted (oxidized) to iodine, which is then extruded into the follicle lumen and attaches to the tyrosine amino acids forming part of the thyroglobulin colloid. The attachment of two iodides to a tyrosine produces diiodotyrosine (DIT); the attachment of one iodide produces monoiodotyrosine (MIT). Enzymes within the colloid link DIT and MIT together. Two linked DITs result in T_4, while a coupling of MIT and DIT produces T_3; at this point, however, the hormones are still part of the thyroglobulin colloid.

3. *Cleavage of the hormones for release (see Figure 17.9, steps 5–6).* Hormone secretion requires that the iodinated thyroglobulin be taken back into the follicle cells by endocytosis and combined with lysosomes. The hormones are cleaved out of the colloid by lysosomal enzymes and then diffuse from the follicle cells into the bloodstream. Some of the T_4 is converted to T_3 before secretion.

The thyroid gland is unique among the endocrine glands in its ability to store and slowly release its hormones. In the normal thyroid gland, the amount of stored colloid is sufficient to provide normal levels of hormone release for over three months.

Most released thyroid hormone immediately binds to plasma proteins, of which the most important is *thyroxine-binding globulin (TBG)* produced by the liver. Both T_4 and T_3 bind to target tissue receptors, but T_3 binds much more avidly and is about ten times more active. Most peripheral tissues have the enzymes needed to convert T_4 to the more active T_3, a process that entails the enzymatic removal of one iodine group. There is no doubt that cyclic AMP is involved as a second messenger in some target cell responses to thyroid hormone. However, this hormone, particularly T_3, is known to enter some target cells and bind to intracellular receptors within the mitochondria (stimulating oxygen uptake) and within the nucleus (initiating DNA transcription). Hence, several mechanisms of thyroid hormone activity exist, and the one used depends on the specific receptor types on or in the various target cells.

Falling thyroxine blood levels trigger TSH, and ultimately thyroxine, release. Rising levels of thyroxine feed back to inhibit the hypothalamic-adenohypophyseal axis, temporarily shutting off the stimulus for releasing TSH. Conditions that increase body energy

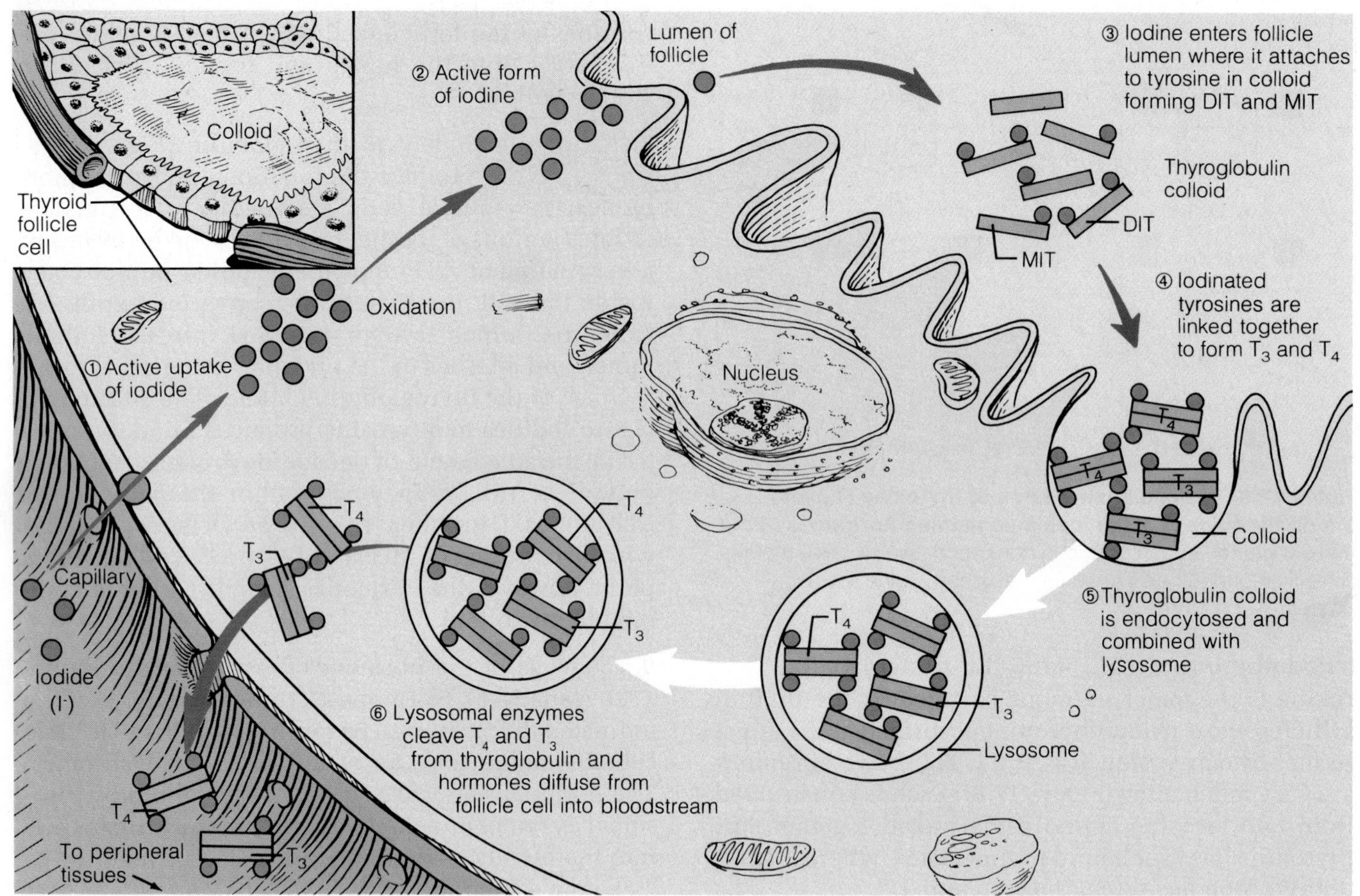

Figure 17.9 Biosynthesis of thyroid hormones. The sequence of steps in thyroxine (T_4) and triiodothyronine (T_3) synthesis. Binding of TSH to the follicle cell receptors stimulates synthesis of the thyroglobulin colloid (not illustrated), the steps illustrated here, and the subsequent release of T_4 and T_3 from the cell. A complete description of these events appears in the text. (Note: only iodinated tyrosine amino acids of the thyroglobulin colloid are illustrated. The rest of the colloid is indicated as the unstructured yellow substance.)

requirements, such as pregnancy and prolonged cold, cause the hypothalamus to trigger TSH release. In such instances, TRH overcomes the negative feedback controls. Enhanced body metabolism and heat generation result as the thyroid gland is prodded to release larger amounts of thyroid hormones. Factors that inhibit TSH release include rising levels of glucocorticoids and sex hormones (estrogen or testosterone).

Both overactivity and underactivity of the thyroid gland can cause severe metabolic disturbances (see Table 17.3). Hypothyroid disorders may result from some defect within the thyroid gland itself or secondarily from deficits of TSH or TRH release. They also occur when the thyroid gland is removed surgically (*thyroidectomy*) or when inadequate iodine is present in the diet.

In adults, the full-blown hypothyroid syndrome is called *myxedema* (mik″-seh-dē′-muh). Symptoms include a low basal metabolic rate, feeling chilled, constipation, thick, dry skin and puffy eyes, edema, lethargy, and mental sluggishness (but not mental retardation). If the condition is caused by a lack of iodine, the thyroid gland enlarges, producing an *endemic* (en-deh′-mik), or *colloidal, goiter.* The follicular cells produce the colloid but are unable to iodinate it; thus, the follicles become congested with colloid but do not make functional hormones. Without feedback inhibition by thyroid hormones, TSH levels remain high and continue to stimulate the thyroid, which enlarges in its futile attempt to meet the demand. If untreated, the thyroid cells eventually "burn out" from frantic activity, and the gland atrophies. In the meantime, hyperthyroidism may cause serious heart problems. Before the marketing of iodized salt, parts of the midwestern United States were called the "goiter belt." Because these areas had iodine-poor soil and no access to iodine-rich shellfish, goiters were very common there. Depending on the cause, myxedema can be reversed by iodine supplements or hormone replacement therapy.

Severe hypothyroidism in infants is called *cretinism* (krē′-tih-nih-zim). The child has a short, disproportionate body and a thick tongue and neck and is mentally retarded. Cretinism may arise from some genetic deficiency of the fetal thyroid gland or from maternal factors, such as the lack of dietary iodine.

Table 17.3 Major Effects of Thyroid Hormone (T_4 and T_3) in the Body

Process or system affected	Normal physiological effects	Effects of hyposecretion	Effects of hypersecretion
Basal metabolic rate (BMR)/ temperature regulation	Promotes normal oxygen consumption and BMR; calorigenesis; enhances effects of catecholamines (and sympathetic nervous system)	BMR below normal; decreased body temperature; cold intolerance; decreased appetite; weight gain; decreased sensitivity to catecholamines	BMR above normal; increased body temperature; heat intolerance; increased appetite; weight loss; increased sensitivity to catecholamines; may lead to hypertension (high blood pressure)
Carbohydrate/lipid/protein metabolism	Promotes glucose catabolism; mobilizes fats; essential for protein synthesis; enhances liver secretion of cholesterol	Decreased glucose metabolism; elevated cholesterol/triglyceride levels in blood; decreased protein synthesis; edema	Enhanced catabolism of glucose and fats; weight loss; increased protein catabolism; loss of muscle mass
Nervous system	Promotes normal development of nervous system in fetus and infant; necessary for normal adult nervous system function	In infant, slowed/deficient brain development, retardation; in adult, mental dulling, depression, paresthesias, memory impairment, listlessness, hypoactive reflexes	Irritability, restlessness, insomnia, overresponsiveness to environmental stimuli, exophthalmos, personality changes
Cardiovascular system	Promotes normal functioning of the heart	Decreased efficiency of pumping action of the heart; low heart rate and blood pressure	Rapid heart rate and possible palpitations; high blood pressure; if prolonged, leads to heart failure
Muscular system	Promotes normal muscular development, tone, and function	Sluggish muscle action; muscle cramps; myalgia	Muscle atrophy and weakness
Skeletal system	Promotes normal growth and maturation of the skeleton	In child, growth retardation, skeletal stunting/malproportion, retention of child's body proportions; in adult, joint pain	In child, excessive skeletal growth initially, followed by early epiphyseal closure and short stature; in adult, demineralization of skeleton
Gastrointestinal system	Promotes normal GI motility and tone; increases secretion of digestive juices	Depressed GI motility, tone, and secretory activity; constipation	Excessive GI motility; diarrhea; loss of appetite
Reproductive system	Promotes normal female reproductive ability and normal lactation	Depressed ovarian function; sterility; depressed lactation	In females, depressed ovarian function; in males, impotence
Integumentary system	Promotes normal hydration and secretory activity of skin	Skin pale, thickened, and dry; facial edema; hair coarse and thin; nails hard and thick	Skin flushed, thin, and moist; hair fine and soft; nails soft and thin

Cretinism is preventable by thyroid hormone replacement therapy, but once developmental abnormalities and mental retardation appear, they are not reversible.

The most common (and puzzling) hyperthyroid pathology is *Graves's disease*. Since patients' serum often contains abnormal thyroid gland–stimulating antibodies (TSAbs), Graves's disease is believed to be an autoimmune disease, in which one's immune system attacks one's own body tissues. Typical symptoms include an elevated metabolic rate, excessive perspiration, rapid, irregular heartbeat, nervousness, and weight loss despite adequate food intake. Exophthalmos, a protruding of the eyeballs, may occur if the tissue behind the eyes becomes edematous and then fibrous. Treatment is surgical removal of the thyroid or ingestion of radioisotope-tagged iodine, which localizes in the gland and destroys the most active thyroid cells. ■

Thyrocalcitonin

The most important effect of **thyrocalcitonin,** or simply **calcitonin,** produced by the parafollicular cells of the thyroid gland, is to depress blood calcium levels. (Calcitonin is a direct antagonist of parathyroid hormone, produced by the parathyroid glands.) Upon release, calcitonin acts on the skeleton, where it (1) inhibits bone reabsorption and release of ionic calcium from the bony matrix and (2) stimulates calcium uptake and incorporation into bone matrix by osteoblasts (bone-forming cells). Thus, calcitonin has a bone-sparing effect. Calcitonin also increases the excretion of calcium and phosphate ions by the kidneys.

An excessive blood level of ionic calcium (approximately 20% above normal) acts as a humoral stimulus for calcitonin release, whereas declining blood calcium levels inhibit parafollicular cell secretory activity. Calcitonin regulation of blood calcium levels is short-lived but extremely rapid.

The Parathyroid Glands

The tiny, yellow-brown **parathyroid glands** are embedded (nearly hidden from view) in the posterior aspect of the thyroid gland (Figure 17.10). There are usually four of these glands, but the precise number varies. As many as eight glands have been reported, and they may be located in other regions of the neck or even in the thorax. Each is about 6 mm long, 4 mm wide, and 2 mm deep.

Discovery of the parathyroid glands was quite accidental. Years ago, surgeons were confounded by the observation that while some patients recovered uneventfully after partial (or even total) thyroid gland removal, others exhibited uncontrolled muscle spasms and severe pain and rapidly deteriorated to their death. It was only after several such tragic deaths that the parathyroid glands were discovered and their hormonal function, quite different from that of the thyroid gland hormones, became known.

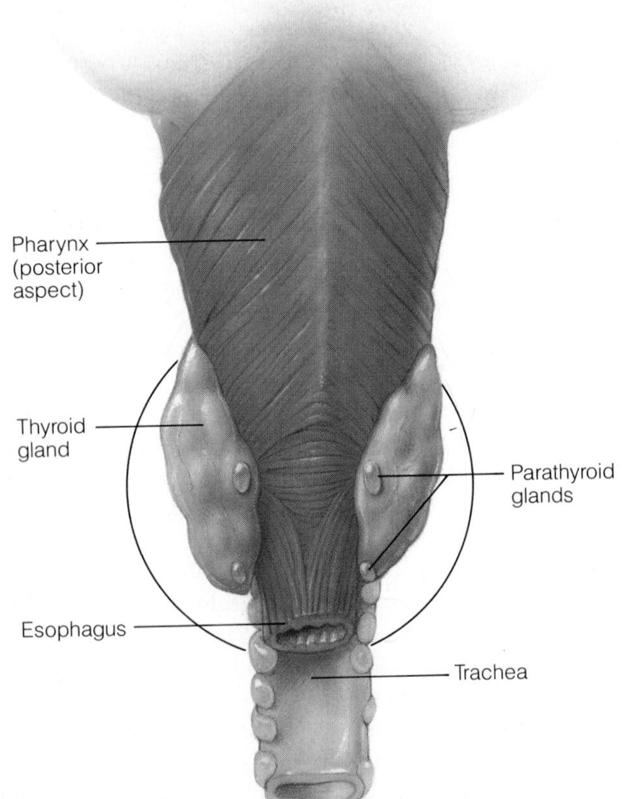

Figure 17.10 Location of the parathyroid glands on the posterior aspect of the thyroid gland. These tiny glands may be much more inconspicuous than depicted.

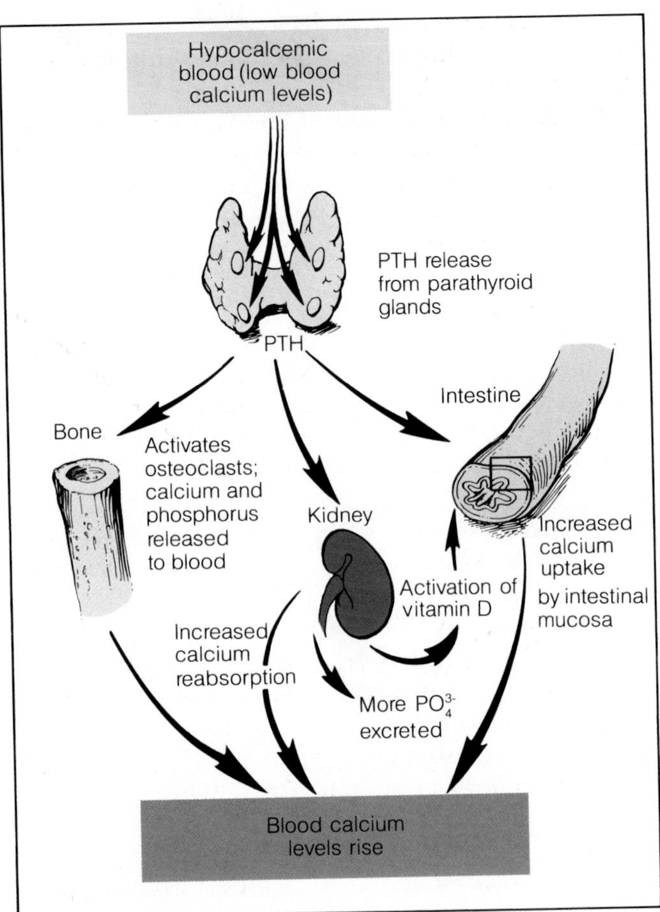

Figure 17.11 Summary of parathyroid hormone's effects on bone, the kidneys, and the intestine. (Note that PTH's effect on the intestine is indirect via activated vitamin D.)

Parathyroid hormone (PTH), or **parathormone,** the protein hormone of these glands, is the single most important hormone controlling the calcium balance of the blood. Its release is triggered by falling blood calcium levels and inhibited by hypercalcemia. The major effect of PTH is to increase ionic calcium levels by stimulating three target organs: the skeleton (which contains considerable amounts of calcium salt in its matrix), the kidneys, and the intestine (Figure 17.11).

PTH release results simultaneously in (1) activation of osteoclasts (bone-reabsorbing cells) to digest some of the bony matrix and release ionic calcium and phosphate to the blood; (2) enhanced reabsorption of calcium ions (and decreased retention of phosphate) by the kidney tubules; and (3) increased absorption of calcium by the intestinal mucosal cells. Calcium absorption by the intestine is enhanced indirectly, via parathormone's effect on vitamin D activation. Vitamin D is required for intestinal absorption of calcium from ingested food, but the form in which the vitamin is ingested or produced by the skin is relatively inactive. For vitamin D to exert its physiological effects, it must first be converted by the kidneys to its active form, 1,25-dihydroxycholecalciferol, a transformation stimulated by PTH.

Since plasma calcium ion homeostasis is essential for so many functions, including conduction of nerve impulses, muscle contraction, and blood clotting, precise control of ionic calcium levels is critical.

Hyperparathyroidism is quite rare and usually results from a parathyroid gland tumor. When it occurs, calcium is leached from the bones, which soften and deform as the mineral salts are replaced by fibrous connective tissue. The resulting abnormally elevated blood calcium level (hypercalcemia) has many outcomes, but the two most notable are (1) depression of nervous system activity, which leads to abnormal reflexes and weakness of the skeletal muscles, and (2) formation of kidney stones as excess calcium salts precipitate in the kidney tubules. Calcium deposits may also form in soft tissues throughout the body and severely impair vital organ functioning. In severe cases, widespread bone matrix destruction occurs; the bones have a moth-eaten appearance on X-rays and tend to fracture spontaneously.

Hypoparathyroidism, a result of PTH deficiency, most often follows parathyroid gland trauma or removal during thyroid surgery. The resulting hypocalcemia increases the excitability of neurons, causing them to depolarize in an uncontrolled manner, and accounts for such classic symptoms as loss of sensation, muscle twitches (uncontrolled spasms), and convulsions. Untreated, the symptoms progress to spasms of the larynx, respiratory paralysis, and death. ■

The Adrenal Glands

The paired adrenal glands, each approximately the size of an almond, are perched atop the kidneys, where they are enclosed in a fibrous capsule and a cushion of fat (see Figure 17.1). For this reason, they are often referred to as the *suprarenal glands* (*supra* = above; *renal* = kidneys).

Each adrenal gland is structurally and functionally two endocrine glands in one. The inner *adrenal medulla,* more like a 'knot' of nervous tissue than a gland, is derived from neuroectoderm and is functionally part of the sympathetic nervous system. The outer *adrenal cortex,* encapsulating the medullary region and forming the bulk of each gland, is derived from embryonic mesoderm. Each of these regions produces hormones (Table 17.4, p. 555).

The Adrenal Cortex

Well over two dozen steroid hormones, collectively called **corticosteroids,** are synthesized from cholesterol (or, less frequently, from acetate) by the adrenal cortex in a multistep pathway involving varying intermediates, depending on the hormone being formed. The structures of several corticosteroid hormones are illustrated in Figure 17.12. The large, lipid-laden cortical cells are arranged in three concentric regions (Figure 17.13). The clusters of cells forming the super-

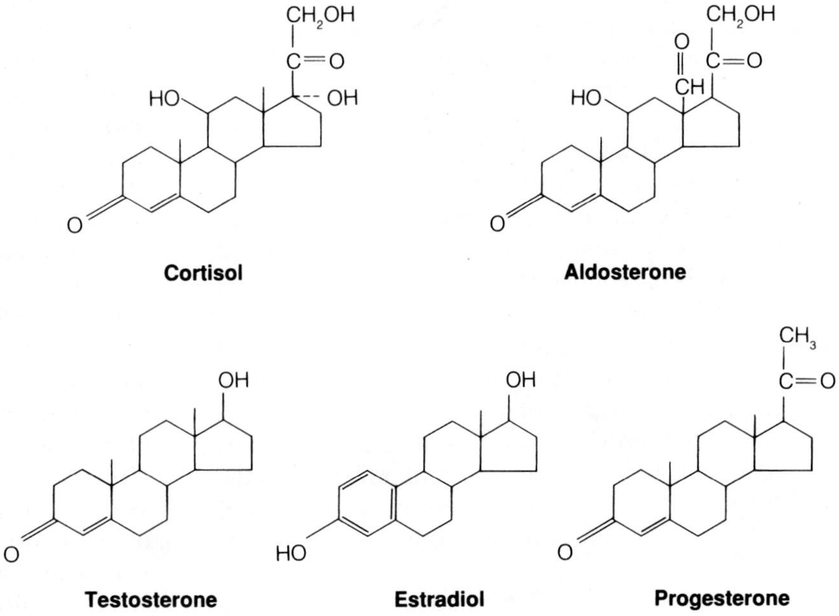

Figure 17.12 Structures of some steroid hormones. Cortisol (a glucocorticoid) and aldosterone (a mineralocorticoid) are the major hormones made in the adrenal cortex. Testosterone (an androgen), estradiol (an estrogen), and progesterone (a progestin) are synthesized by both the adrenal cortex and the gonads.

CH_2OH C=O OH HO O — **Cortisol**

CH_2OH O C=O CH HO O — **Aldosterone**

OH O — **Testosterone**

OH HO — **Estradiol**

CH_3 C=O O — **Progesterone**

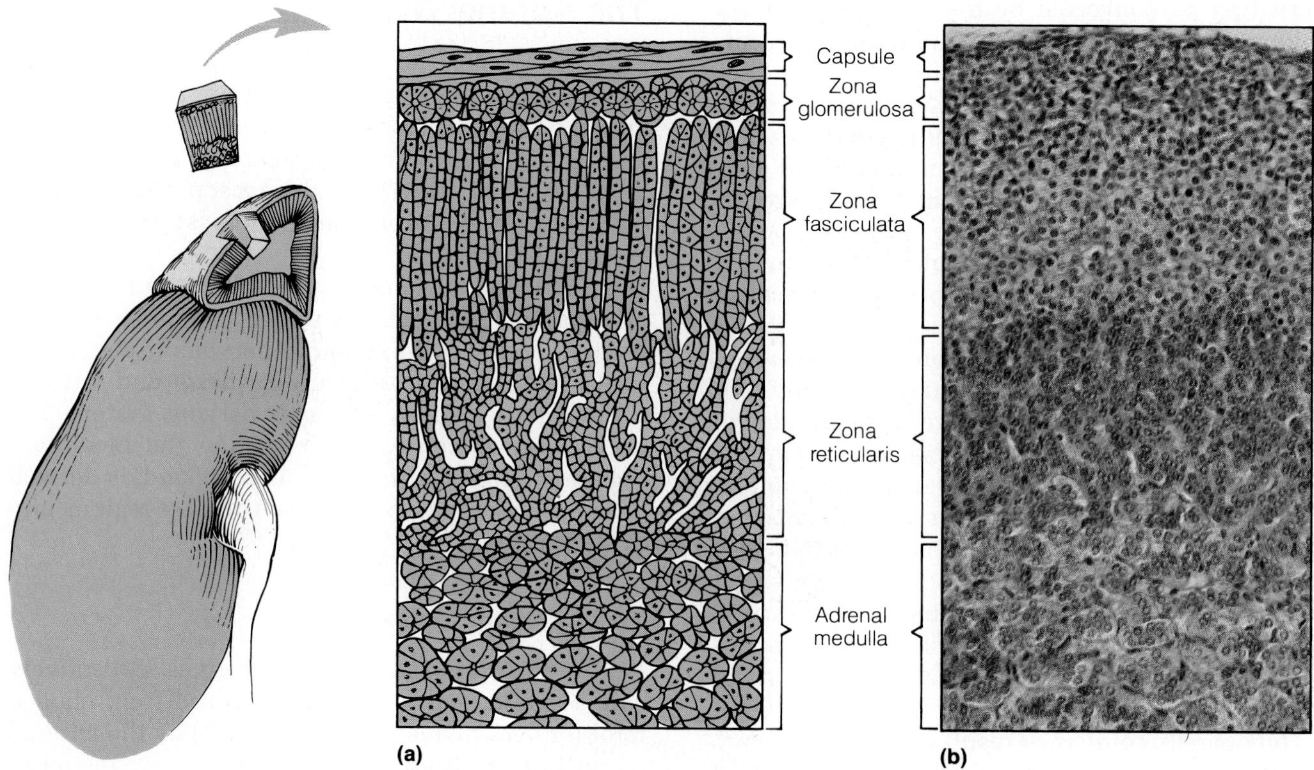

Figure 17.13 Microscopic structure of the adrenal gland. (**a**) Diagram and (**b**) photomicrograph of the zona glomerulosa, zona fasciculata, and zona reticularis (cortical regions) (50X). (Part of the adrenal medulla is also shown.)

ficial *zona glomerulosa* (zō′-nuh glah-mer″-yoo-lō′-suh) produce mineralocorticoids, hormones that help control the balance of minerals and water in the blood. The middle *zona fasciculata* (fuh-sik″-yoo-lah′-tuh) cells, arranged in more or less linear cords, secrete the metabolic hormones called glucocorticoids. The cells of the innermost *zona reticularis* (reh′-tik″-yoo-layr′-is), abutting the adrenal medulla, have a netlike arrangement; these cells produce glucocorticoids and small amounts of adrenal sex hormones, or gonadocorticoids. Although, as just described, there is a division of labor relative to the major corticosteroids produced by each zone, the entire spectrum of corticosteroids is produced by all three cortical layers.

Mineralocorticoids. The most important function of the mineralocorticoids is regulation of the electrolyte (mineral salt) concentrations in extracellular fluids, particularly concentrations of sodium and potassium ions. Although there are several mineralocorticoids, **aldosterone** (al-dah′-ster-ōn) is the most potent and accounts for over 95% of the mineralocorticoids produced.

Maintaining sodium ion balance is the primary goal of aldosterone activity. The single most abundant positive ion (cation) in the extracellular fluid is sodium, and while sodium is vital to homeostasis, excessive sodium intake and retention may promote high blood pressure (hypertension) in susceptible individuals.

Aldosterone decreases the excretion of sodium from the body in general. Its primary target is the distal parts of the kidney tubules, where it stimulates reabsorption of sodium ions from the forming urine and their return into the bloodstream. Aldosterone also enhances sodium ion reabsorption from perspiration, saliva, and gastric juice. The mechanism of aldosterone activity appears to involve the synthesis of an enzyme required for sodium transport.

The regulation of a number of other ions—including potassium, hydrogen, bicarbonate, and chloride—is coupled to sodium reabsorption; and where sodium goes, water follows—an event that leads to changes in blood volume and blood pressure. Hence, it is not surprising that sodium ion regulation is crucial to overall body homeostasis. Briefly, aldosterone's effects on the renal tubules lead to sodium and water retention accompanied by potassium ion elimination and in some instances alterations in the acid-base balance of the blood. Since aldosterone's regulatory effects are very brief (lasting approximately 20 minutes), plasma electrolyte balance can be very precisely controlled and continuously modified and manipulated. These

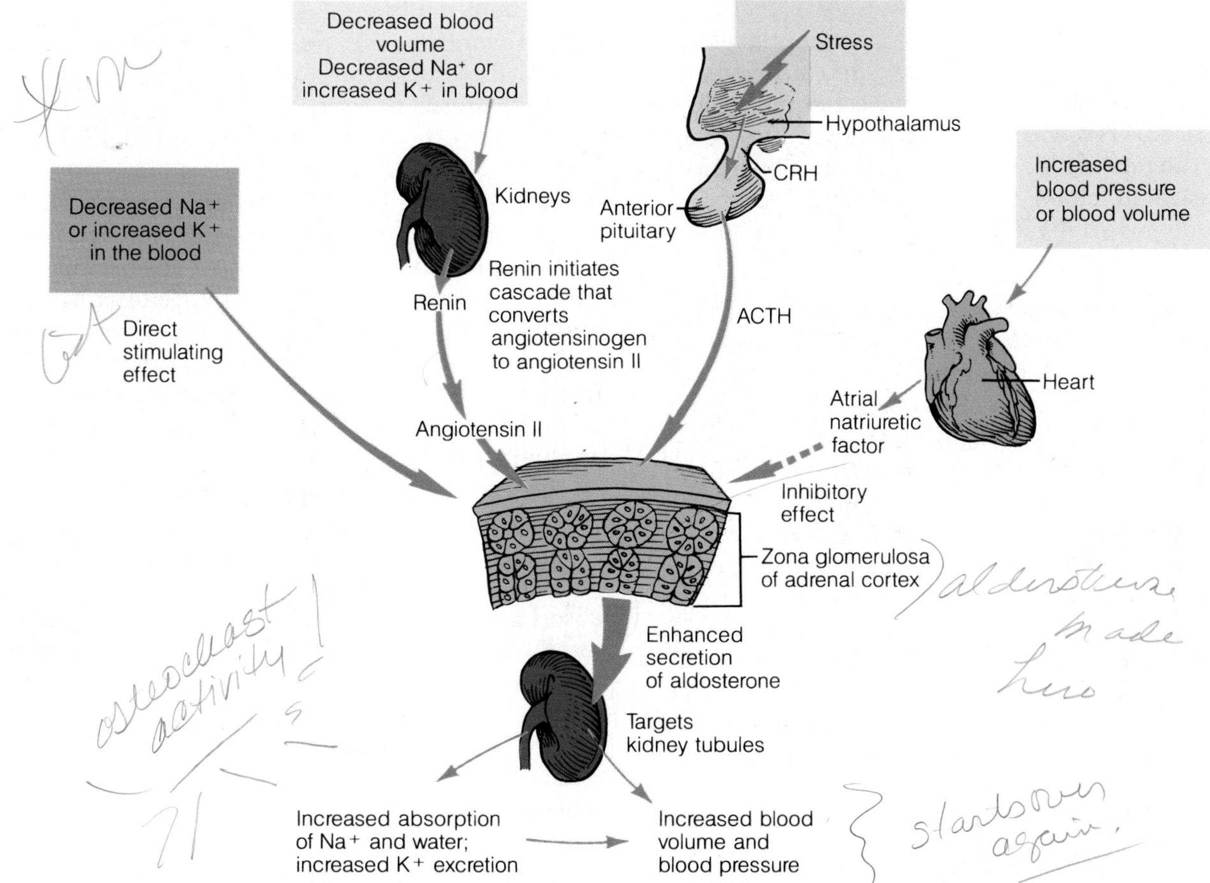

Figure 17.14 Major mechanisms controlling aldosterone release from the adrenal cortex.

matters are described in greater detail in Chapter 27, which considers the regulation of water, electrolyte, and acid-base balance of the blood.

Aldosterone secretion is stimulated by a number of interlocking factors: elevated blood levels of potassium ions, low blood levels of sodium, and decreases in blood volume and blood pressure. The reverse conditions inhibit aldosterone secretion. Four mechanisms are involved in the complex regulation of aldosterone secretion: the renin-angiotensin system, plasma concentrations of sodium and potassium ions, controls exerted by ACTH, and plasma concentrations of atrial natriuretic factor (Figure 17.14).

1. The renin-angiotensin (rē′-nin an″-jē-ō-ten′-sin) **system.** The renin-angiotensin system, the major regulator of aldosterone release, influences both the electrolyte-water balance of the blood and the blood pressure. Specialized cells of the *juxtaglomerular apparatus* in the kidneys become excited when blood pressure declines or plasma osmolarity (solute concentration) drops. They respond by releasing the enzyme *renin* into the blood. Renin cleaves off part of the plasma protein *angiotensinogen* (an″-jē-ō-ten-sin′-ō-jin), triggering an enzymatic cascade leading to the formation of *angiotensin II,* a powerful stimulator of aldosterone release by the glomerulosa cells of the adrenal cortex. However, the renin-angiotensin system does much more than trigger aldosterone release, and all of its widespread effects are ultimately involved in raising the systemic blood pressure. These additional effects are described in detail in Chapters 26 and 27.

2. Plasma concentration of sodium and potassium ions. Fluctuations in blood levels of sodium and potassium ions can directly influence the zona glomerulosa cells. Increased potassium and decreased sodium are stimulatory; the opposite conditions are inhibitory.

3. ACTH. Under normal circumstances, ACTH released by the anterior pituitary has an inconsequential effect on aldosterone release. However, when a person is severely stressed, the hypothalamus secretes more CRH. The rise in ACTH blood levels that follows effectively steps up the rate of aldosterone secretion. The resulting increase in blood volume and blood pressure ensures adequate delivery of nutrients and respiratory gases during the stressful period.

4. Atrial natriuretic factor. Atrial natriuretic factor (ANF), a hormone secreted by the heart when blood pressure rises, acts to fine-tune blood pressure and the sodium-water balance of the body by modifying the effects of the renin-angiotensin system. ANF's major effects on that system are inhibitory: It blocks renin and aldosterone secretion and inhibits other angiotensin-induced mechanisms that enhance water and sodium reabsorption. Consequently, the overall influence of ANF is to decrease blood pressure.

Hypersecretion of aldosterone, a condition called *aldosteronism,* most often results from adrenal neoplasms. Two major sets of problems result: (1) hypertension and edema caused by excessive sodium and water retention and (2) accelerated excretion of potassium ions. If potassium loss is extreme, neurons become nonresponsive and muscle weakness (and eventually paralysis) occurs. Hyposecretory disease of the adrenal cortex generally involves a deficient output of both mineralocorticoids and glucocorticoids. This syndrome, called *Addison's disease,* is described shortly in the discussion of glucocorticoids. ■

Glucocorticoids. The **glucocorticoids** influence the metabolism of most body cells and help provide resistance to stressors. They are absolutely essential to life. Under normal circumstances, the glucocorticoids allow the body to adapt to external changes and intermittent food intake by keeping blood sugar levels fairly constant, and they maintain blood volume by preventing the shift of water into tissue cells. However, severe stress due to hemorrhage, infections, or physical or emotional trauma evokes a dramatically higher output of glucocorticoids, which help the body to negotiate the crisis. The glucocorticoid hormones include **cortisol (hydrocortisone), cortisone,** and **corticosterone,** but only cortisol is secreted in significant amounts by the human adrenal cortex. The basic mechanism of glucocorticoid activity on its target cells is not fully understood, but it is known to alter the DNA transcription process.

Control of glucocorticoid secretion is a typical negative feedback system. Cortisol release is promoted by ACTH, triggered in turn by the hypothalamic releasing hormone CRH; rising cortisol levels act on both the hypothalamus and the anterior pituitary, preventing CRH release and shutting off ACTH and cortisol secretion. Cortisol secretory bursts, driven by patterns of eating and activity, occur in a definite pattern throughout the day and night. Peak cortisol blood levels occur shortly after arising in the morning, and the lowest levels are seen in the evening just before and shortly after sleep ensues. Although the cortisol rhythm is related to sleep or day/night light cycles, it is interrupted by acute stress of any variety as the sympathetic nervous system overrides the inhibitory effects of elevated cortisol levels and triggers CRH release. The resulting increase in ACTH blood levels causes an outpouring of cortisol from the adrenal cortex.

Stress results in a dramatic rise in blood levels of glucose, fatty acids, and amino acids, all provoked by the metabolic effects of cortisol. Cortisol's prime metabolic effect is *gluconeogenesis,* the formation of glucose from noncarbohydrate molecules (fats and proteins). In addition, cortisol causes fatty acids to be mobilized from adipose tissue, encouraging increased use of fatty acids and decreased use of glucose for energy needs. Under cortisol's influence, stored proteins are broken down and their amino acids are released into the blood, providing building blocks for repair or for making enzymes used in metabolic processes. Cortisol enhances epinephrine's vasoconstrictive effects. The increased blood pressure and circulatory efficiency that result help ensure that these nutrients are distributed to the cells in short order.

Excessively high levels of glucocorticoids (1) depress cartilage and bone formation, (2) inhibit the inflammatory response by stabilizing lysosomal membranes and preventing vasodilation, (3) depress the activity of the immune system, and (4) promote changes in cardiovascular, neural, and gastrointestinal function.

It is important to keep in mind that ideal amounts of glucocorticoids promote *normal* function. Significant anti-inflammatory and anti-immune effects are rarely seen when glucocorticoids are present at physiological levels; these effects are associated primarily with cortisol excess. Recognition of this group of effects has led to the widespread use of glucocorticoid drugs to control symptoms of many chronic inflammatory disorders, such as rheumatoid arthritis or allergic responses. However, the use of these potent drugs at higher than normal (pharmacological) doses is a double-edged sword. It may relieve some of the patient's symptoms, but it also causes the undesirable effects of excessive levels of these hormones.

The pathology of cortisone excess, *Cushing's disease,* or *syndrome,* may be caused by an ACTH-releasing tumor of the pituitary or by a glucocorticoid-releasing tumor of the adrenal cortex. However, it most often results from the administration of large doses of glucocorticoid drugs. The syndrome is characterized by persistent hyperglycemia (steroid diabetes), dramatic losses in muscle and bone protein, and water and salt retention, leading to hypertension and edema. The so-called cushingoid signs include a swollen "moon" face, redistribution of fat to the abdomen and the posterior neck (causing a "buffalo hump"), a tendency to bruising, and poor wound healing. Because of enhanced anti-inflammatory effects, infections may be masked and become overwhelmingly severe before producing recognizable symptoms. As the disease progresses, muscles weaken and spontaneous fractures force the person to become bedridden. The only treatment is removal of the cause—be it surgical removal of the offending tumor or discontinuation of the drug.

Addison's disease, the major hyposecretory disorder of the adrenal cortex, usually involves deficits in both glucocorticoids and mineralocorticoids. People with Addison's disease tend to lose weight; their plasma glucose and sodium levels drop, and potassium levels rise. Severe dehydration and hypotension are common. Corticosteroid replacement therapy at physiological doses is the usual treatment. ■

Gonadocorticoids (Sex Hormones). The **gonadocorticoids** are *androgens,* or male sex hormones, for the most part, but the adrenal cortex also makes small amounts of female hormones (estrogens and progesterone). Substantial amounts of these steroid hormones are secreted in the fetus and again during early puberty, but then production declines rapidly. The amount of sex hormones produced by the adrenal cortex is insignificant compared to the amounts made by the gonads during late puberty and adulthood. The exact role of the adrenal sex hormones is still a question, but the androgens are thought to be responsible for the sex drive in adult women and the adrenal cortex may provide small amounts of estrogen after menopause. Control of gonadocorticoid secretion is not completely understood. Release appears to be stimulated by ACTH, but the gonadorcorticoids do not appear to exert feedback inhibition on ACTH release.

Since androgens predominate, hypersecretion of gonadocorticoids usually causes *masculinization,* or *virilization,* the so-called *androgenital syndrome.* In adult males, these effects are often obscured, since virilization by testicular testosterone has already been completed; but in prepubertal males and in females, the results are often dramatic. In the young man, maturation of the reproductive organs and appearance of the secondary sex characteristics occur rapidly, and the sex drive emerges with a vengeance. Females develop a beard and a masculine pattern of body hair distribution, and the clitoris grows to resemble a small penis. ■

The Adrenal Medulla

The *chromaffin* (krō'-muh-fin) *cells,* the hormone-producing cells of the adrenal medulla, are arranged in irregular clusters that crowd around blood-filled capillaries and sinusoids. The adrenal medulla is related to the nervous system in many ways. As noted in Chapter 14, it is formed from neural crest tissue, and the medullary cells are directly stimulated by sympathetic preganglionic fibers.

The two powerful adrenal medulla hormones are **epinephrine** (eh"-pih-neh'-frin), or **adrenaline,** and **norepinephrine,** or **noradrenaline;** referred to collectively as *catecholamines* (ka"-tih-kōl'-uh-mēnz). As shown in Figure 17.15, these medullary hormones are very similar in structure. They are formed from the amino acid tyrosine by a common enzymatic pathway in which the last step is the formation of epinephrine from norepinephrine (see Figure 11.26).

Since norepinephrine is also a neurotransmitter released by the postganglionic endings of sympathetic neurons, perhaps a more precise name for these hormones is **sympathomimetic** (sim"-pah-thō-mih-meh'-tik) **amines,** which translates to "amines that mimic the effects of the sympathetic nervous system." When the body is activated to its fight-or-flight status by some short-term stressor or emergency, the sympathetic nervous system is mobilized by hypothalamic centers. As a result, blood sugar levels rise, blood vessels constrict and the heart beats faster (together raising the blood pressure), and blood is diverted from temporarily nonessential organs to the brain, heart, and skeletal muscles. At the same time, sympathetic nerve endings stimulate the adrenal medulla to release the

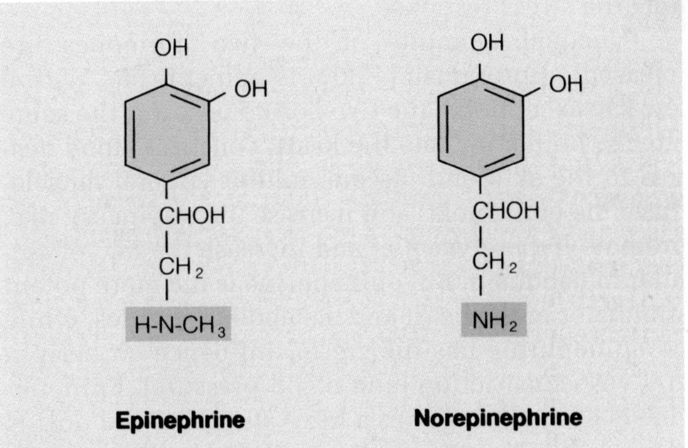

Figure 17.15 Structure of the adrenal medullary hormones, epinephrine and norepinephrine. Norepinephrine is both a hormone in its own right and a precursor of epinephrine. (Norepinephrine is also considered a neurotransmitter when released by sympathetic nerve endings.) The regions of structural difference between epinephrine and norepinephrine are shown in different colors.

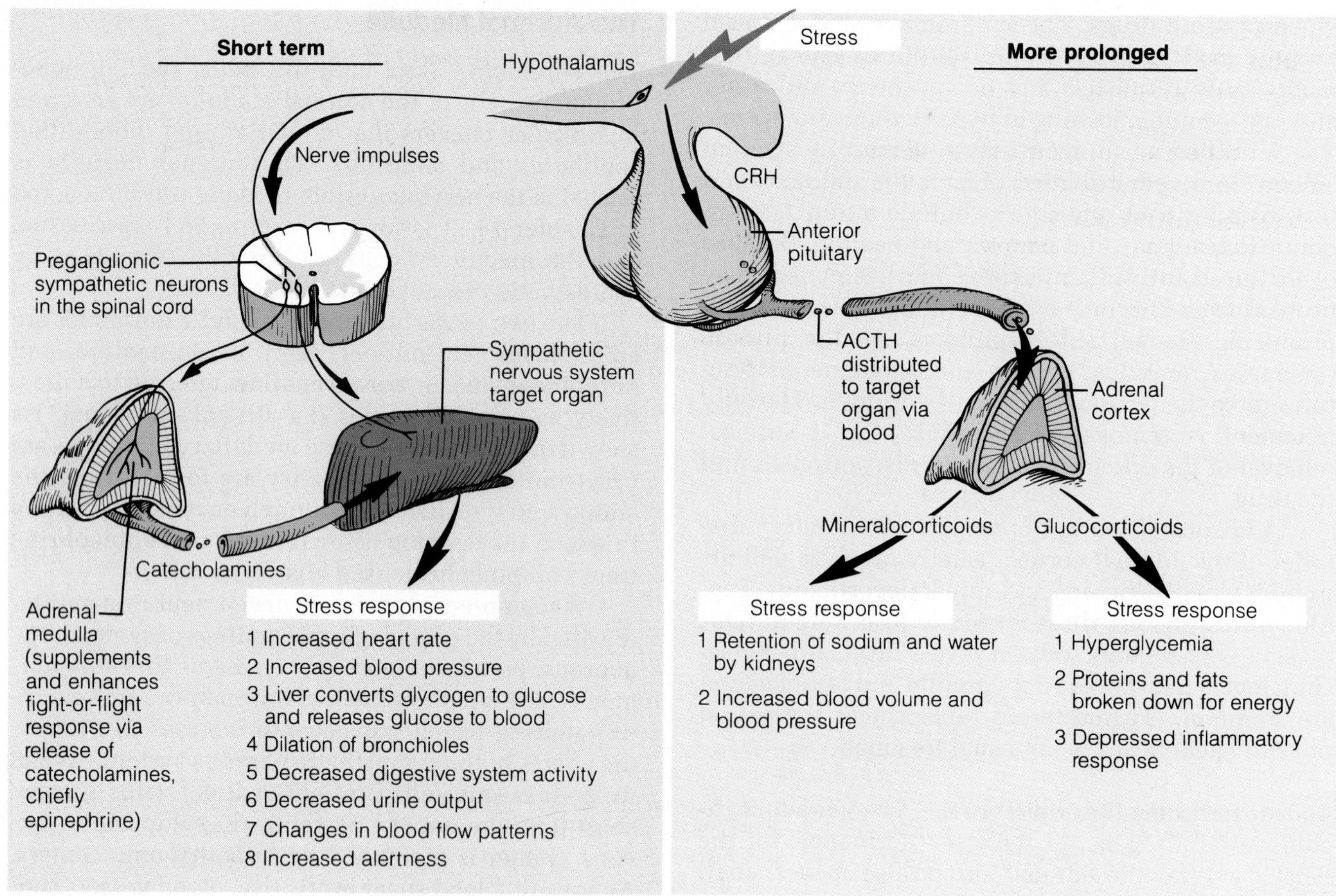

Figure 17.16 Role of the hypothalamus, adrenal medulla, and adrenal cortex in the stress response. Note that ACTH is only a weak stimulator of mineralocorticoid release under normal conditions.

sympathomimetic amines, which reinforce and prolong the stress response.

Unequal amounts of the two hormones are released; approximately 80% is epinephrine. With a few key exceptions, the two hormones exert the same effects: Both stimulate the heart, constrict blood vessels in the skin and viscera, inhibit visceral muscle, dilate the bronchioles and increase the respiratory rate, promote hyperglycemia, and increase the rate of cellular metabolism. But epinephrine is the more potent stimulator of the heart and metabolic activities, while norepinephrine has the greater influence on peripheral vasoconstriction (and blood pressure). Epinephrine is used clinically as a heart stimulant and during acute asthmatic attacks to cause bronchodilation.

Unlike the adrenocortical hormones, which promote long-lasting body responses to stressors, the catecholamines cause fairly brief responses. The interrelationships of the adrenal hormones and the hypothalamus are depicted in Figure 17.16.

Since hormones of the adrenal medulla merely intensify the activities set into motion by the sympathetic nervous system neurons, their deficiency is not a problem. Unlike the glucocorticoids, the catecholamines are not essential for life. However, hypersecretion of catecholamines, sometimes arising from a rare chromaffin cell tumor called a *pheochromocytoma* (fē-ō-krō″-mō-sī-tō′-muh), is quite another story. Symptoms of massive and uncontrolled sympathetic nervous system activity appear, including hyperglycemia, increased metabolic rate, rapid heart beat and palpitations, hypertension, intense nervousness, and sweating. ■

Table 17.4 Adrenal Gland Hormones: Regulation and Effects

Hormone	Regulation of release	Target/effects	Effects of hyposecretion	Effects of hypersecretion
Adrenocortical Hormones				
Mineralocorticoids (chiefly aldosterone)	Stimulated by renin-angiotensin mechanism (activated by decreasing blood volume or blood pressure), elevated K^+ or low Na^+ blood levels, and ACTH (minor influence); inhibited by increased blood volume and pressure, increased Na^+ and decreased K^+ blood levels	Kidneys; increased blood levels of Na^+ and decreased blood levels of K^+; since water reabsorption accompanies sodium retention, blood volume and blood pressure rise	Addison's disease	Aldosteronism
Glucocorticoids (chiefly cortisol)	Stimulated by ACTH; inhibited by feedback inhibition exerted by cortisol	Body cells; promote gluconeogenesis and hyperglycemia; mobilize fats for energy metabolism; stimulate protein catabolism; assist body to resist stressors; depress inflammatory and immune responses	Addison's disease	Cushing's disease
Gonadocorticoids (chiefly androgens)	Stimulated by ACTH; mechanism of inhibition incompletely understood, but feedback inhibition not seen	Insignificant effects in adults; may be responsible for female libido and source of estrogen after menopause	No effects known	Virilization of females (androgenital syndrome)
Adrenal Medullary Hormones				
Epinephrine and norepinephrine	Stimulated by preganglionic fibers of the sympathetic nervous system	Sympathetic nervous system target organs; sympathomimetics; effects mimic sympathetic nervous system activation; increase heart rate and metabolic rate; increase blood pressure by promoting vasoconstriction	Unimportant	Prolonged flight-or-fight response; hypertension

The Pancreas

Located partially behind the stomach in the abdomen, the soft, triangular **pancreas** is a mixed gland composed of both endocrine and exocrine gland cells (Figure 17.17). Like the thyroid and parathyroids, it develops as an outpocketing of the embryonic endoderm, which forms the mucosa of the gastrointestinal and respiratory tracts. Acinar cells, forming the bulk of the gland, produce an enzyme-rich juice that is ducted into the small intestine during food digestion. This exocrine product is discussed in Chapter 24.

Scattered among the acinar cells are up to 2 million **islets of Langerhans,** minute clusters of cells that produce pancreatic hormones. The islets contain two major populations of hormone-producing cells, the glucagon-synthesizing *alpha* (α) *cells* and the more numerous insulin-producing *beta* (β) *cells.* Islet cells synthesize other peptides in small amounts, including somatostatin (secreted by the *delta* [δ] *cells*), which is produced in larger quantities by the hypothalamus. Insulin and glucagon are intimately but independently involved in various phases of metabolism and regulation of blood glucose levels. Their effects are opposite: Insulin is a *hypoglycemic hormone,* whereas glucagon is a *hyperglycemic hormone* (Figure 17.18).

Glucagon

Glucagon (gloo′-kuh-gon), a 29-amino-acid polypeptide, is an extremely potent hyperglycemic agent: One molecule can cause the release of 100 million mole-

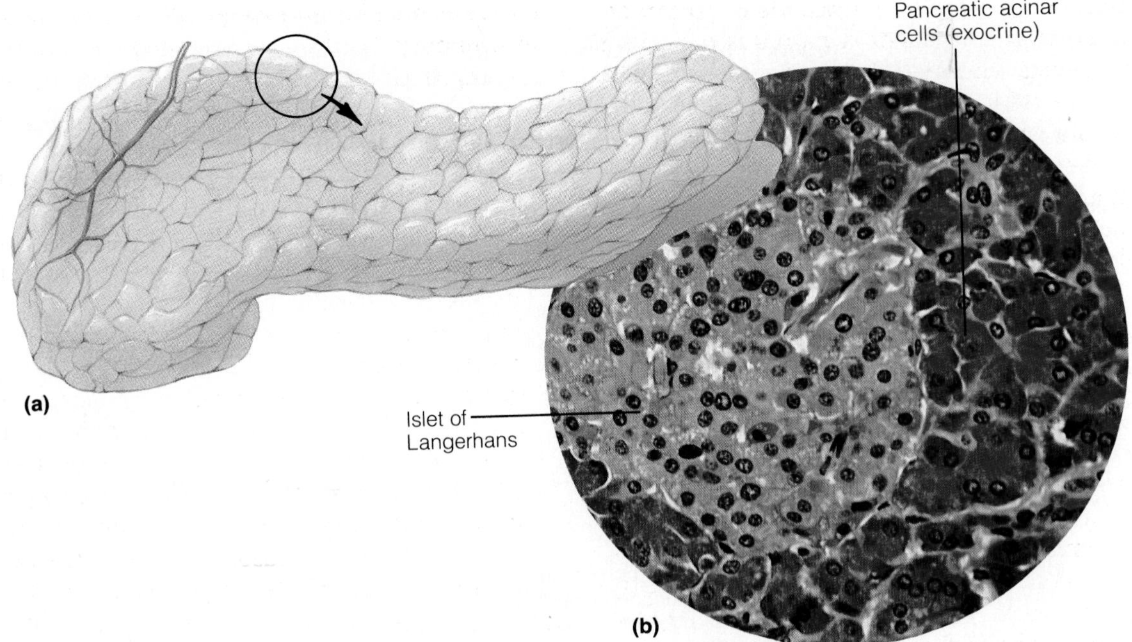

Figure 17.17 The pancreas, a mixed gland. (**a**) Surface anatomy of the pancreas. (**b**) Photomicrograph of pancreas tissue (300X) showing a more lightly staining islet of Langerhans surrounded by the acinar cells, which produce the exocrine product (pancreatic juice). The β cells of the islets produce insulin; the α cells produce glucagon.

cules of glucose into the blood! The major target of glucagon is the liver, where it acts through cyclic AMP to promote (1) glycogenolysis, or conversion of glycogen to glucose, and (2) gluconeogenesis, the formation of glucose from fatty acid and amino acid molecules. The liver then releases glucose to the bloodstream, raising the blood sugar level. A secondary outcome is a drop in the amino acid levels of the blood, since liver cells actively take up amino acids from the blood to carry out their gluconeogenic function.

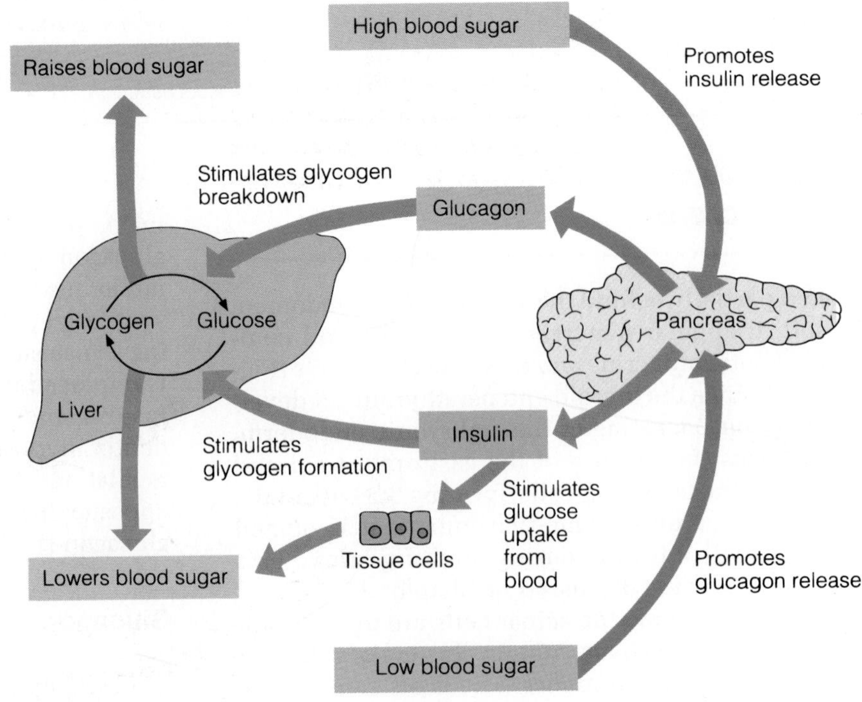

Figure 17.18 Regulation of blood sugar levels by insulin and glucagon. When blood sugar levels are high, the pancreas releases insulin. Insulin stimulates sugar uptake by cells and glycogen formation in the liver, which lowers blood sugar levels; that is, insulin exerts hypoglycemic effects. Glucagon, released when blood sugar levels are low, stimulates glycogen breakdown and thereby raises blood sugar; that is, glucagon is a hyperglycemic hormone.

Secretion of glucagon by the α cells is prompted by humoral stimuli. The major stimulus is declining blood sugar levels; however, increasing amino acid levels (as might follow a protein-rich meal) are also stimulatory. Glucagon release is suppressed by rising blood sugar levels and somatostatin. Because glucagon is such an important hyperglycemic agent, it has been speculated that people with persistently low blood sugar levels are deficient in glucagon.

Insulin

Insulin is a small (51-amino-acid) protein consisting of two amino acid chains linked by disulfide bonds (Figure 17.19). Insulin's effects are most obvious when we have just eaten. Although insulin's most prominent effect is to lower blood sugar levels, it has a substantial influence on protein and fat metabolism as well. Circulating insulin lowers blood sugar levels by enhancing membrane transport of glucose (and other simple sugars) into muscle cells, fat and other connective tissue cells, and white blood cells. (It does not accelerate glucose entry into liver, kidney, and brain tissue, all of which have easy access to blood glucose regardless of insulin levels.) Insulin also inhibits the breakdown of glycogen to glucose and the conversion of amino acids or fatty acids to glucose; thus, it counters any metabolic activity that would increase plasma levels of glucose. Insulin's mechanism of action is still being investigated, but whatever the mechanism, it does not act through cyclic AMP; in fact, as noted earlier, insulin decreases cyclic AMP levels.

After glucose enters the target cells, insulin binding triggers enzymatic activities that (1) catalyze the oxidation of glucose for ATP production, (2) join glucose together to form glycogen, and (3) convert glucose to fat (particularly in adipose tissue). As a rule, energy needs are met first, then glycogen deposit occurs, and finally, if excess glucose is still available, fat deposit occurs. Insulin also induces amino acid uptake and protein synthesis in muscle tissue; thus, insulin activity is essential for normal growth and repair processes. In summary, insulin sweeps glucose out of the blood, causing it to be utilized for energy or converted to other forms (glycogen or fats), and it promotes protein synthesis and fat storage.

The β cells are stimulated to secrete insulin chiefly by elevated blood sugar levels, but increased plasma levels of amino acids and fatty acids also trigger insulin release. As the cells avidly take up sugar and other nutrients, the plasma levels of these substances drop, and insulin secretion is suppressed. Other hormones may also influence insulin release. For example, any hyperglycemic hormone (such as glucagon, epinephrine, growth hormone, thyroxine, or glucocorticoids) that is called into action as blood sugar levels drop will indirectly stimulate insulin release by promoting glucose entry into the bloodstream. Somatostatin depresses insulin release. Thus, blood sugar levels ultimately reflect a balance of both humoral and hormonal influences. Insulin and (indirectly) somatostatin are the hypoglycemic factors countering and counterbalancing the many hyperglycemic hormones.

Diabetes mellitus is the pathology arising from hyposecretion or hypoactivity of insulin. When insulin activity is absent or deficient, blood sugar levels remain high after a meal because glucose is unable to gain entry to most tissue cells. Ordinarily, when blood sugar levels rise, hyperglycemic hormones are not released, but when hyperglycemia becomes excessive, the person begins to feel nauseated, which activates the sympathetic nervous system and precipitates the fight-or-flight response. This results, inappropriately, in all the reactions that normally occur in the hypoglycemic (fasting) state to make glucose available—that is, glycogenolysis, lipolysis, and gluconeogenesis. Thus, the already high blood sugar levels rise even higher, and excesses of glucose begin to be lost from the body in the urine (glycosuria).

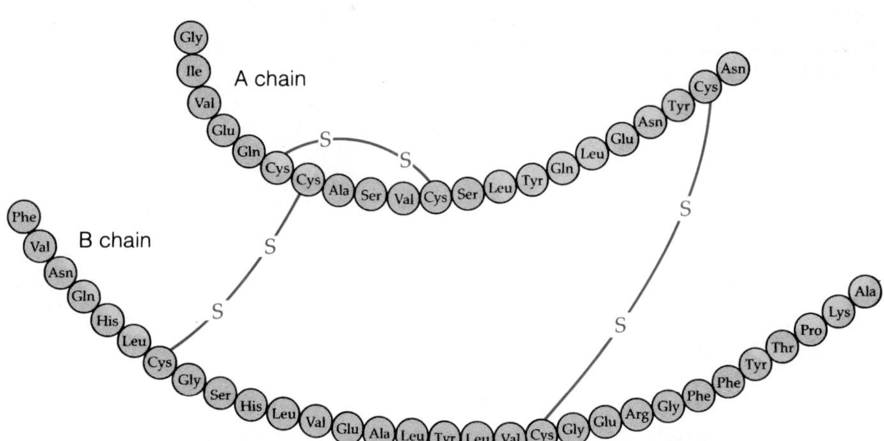

Figure 17.19 Structure of insulin. Each "bead" in this model is an amino acid.

When sugars cannot be used as cellular fuel, more fats are mobilized. In severe cases of diabetes mellitus, blood levels of fatty acids and their metabolites (acetoacetic acid, acetone, and others) rise dramatically. The fatty acid metabolites, collectively called *ketones* (kē′-tōnz) or *ketone bodies,* are strong organic acids. When they accumulate faster than they can be used or excreted, the blood pH drops, resulting in a condition called *ketosis* or *ketoacidosis.* Severe ketosis is a life-threatening situation. The nervous system responds by initiating rapid deep breathing to blow off carbon dioxide from the blood and increase blood pH. (The physiological basis of this mechanism is explained in Chapter 23.) If untreated, ketosis disrupts virtually all metabolic processes, including heart activity and oxygen transport. Severe depression of the nervous system leads to coma and, finally, death.

The three cardinal signs of diabetes mellitus are polyuria, polydipsia, and polyphagia. The excessive glucose (and, in ketosis, fatty acids) in the kidney filtrate substantially prevents water reabsorption by the kidney tubules, resulting in *polyuria,* a huge urine output that leads to decreased blood volume and dehydration. Serious electrolyte losses also occur because of the need to rid the body of excess ketones. Generally, ketone bodies are negatively charged, and they carry positive ions out with them. As a result, sodium and potassium ions are also depleted from the body. Because of the electrolyte imbalance, the person gets abdominal pains and may vomit, and the stress reaction spirals even higher. Dehydration stimulates hypothalamic thirst centers, causing *polydipsia,* or excessive thirst. The final cardinal sign, *polyphagia,* refers to excessive hunger and food consumption, a sign that the person is "starving in the land of plenty"; that is, although large amounts of glucose are available, it cannot be used, and the body starts to utilize its fat and protein stores for energy metabolism. Figure 17.20 provides an abbreviated flowchart of these consequences of insulin deficiency.

Hereditary factors play an important role in many cases of diabetes mellitus. Islet cell destruction by viral infections has also been implicated in some forms. Two major forms of diabetes mellitus have been distinguished: type I and type II.

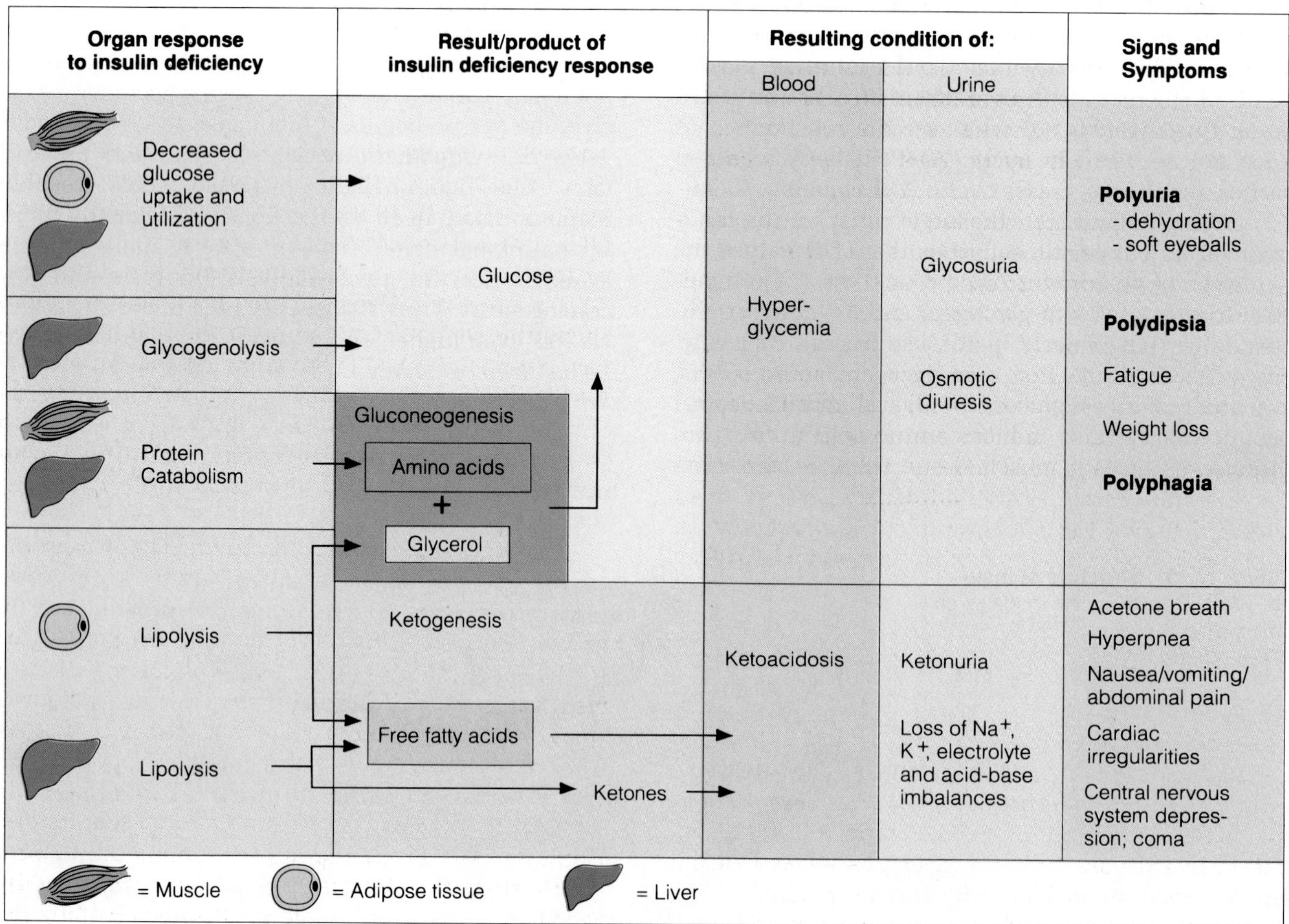

Organ response to insulin deficiency
Result/product of insulin deficiency response
Resulting condition of:
Blood
Urine
Signs and Symptoms
Decreased glucose uptake and utilization
Glucose
Glycogenolysis
Gluconeogenesis
Protein Catabolism
Amino acids
+
Glycerol
Hyper-glycemia
Glycosuria
Osmotic diuresis
Polyuria
- dehydration
- soft eyeballs
Polydipsia
Fatigue
Weight loss
Polyphagia
Lipolysis
Ketogenesis
Free fatty acids
Lipolysis
Ketones
Ketoacidosis
Ketonuria
Loss of Na+, K+, electrolyte and acid-base imbalances
Acetone breath
Hyperpnea
Nausea/vomiting/ abdominal pain
Cardiac irregularities
Central nervous system depression; coma
= Muscle
= Adipose tissue
= Liver

Figure 17.20 Symptomatic results of insulin deficit (diabetes mellitus).

Type I diabetes mellitus, or insulin-dependent diabetes mellitus (IDDM), was formerly known as *juvenile-onset diabetes.* Type I diabetes develops suddenly, usually before the age of 15 years, and often reflects pathology of the β islet cells resulting from viral infection or an autoimmune response. Type I diabetics totally lack insulin activity, and their disease is extremely difficult to control. Insulin injections must be given several times daily to manage ketosis and, to a lesser extent, hyperglycemia. Because of the early onset of their disease, type I diabetics typically exhibit long-term vascular and neural problems. Complications resulting from vascular problems include atherosclerosis, strokes, heart attacks, gangrene, and blindness; consequences of neuropathies include loss of sensation, impaired bladder function, and impotence.

Type II diabetes, or non-insulin-dependent diabetes mellitus (NIDDM) was formerly called *mature-onset diabetes* because it occurs mostly after the age of 40 years and is increasingly common with age. Heredity or a familial predisposition is particularly striking in this diabetic group; if an identical twin has type II diabetes mellitus, the probability that the other twin will have the disease is 100%. Although most type II diabetics produce insulin, the amount is inadequate or there is some abnormality of the insulin receptors. Type II diabetics are almost always overweight and account for over 90% of the known cases of diabetes mellitus. Ketosis is not a major problem for this group, and in many cases the symptoms can be managed solely by diet and exercise. Weight control is very important, because obesity alone causes the insulin receptors to become less sensitive to insulin.

Hyperinsulinism, or excessive secretion of insulin, results in low blood sugar levels, or hypoglycemia. Hypoglycemia triggers the release of hyperglycemic hormones, which cause anxiety, nervousness, tremors, and a feeling of weakness. Insufficient glucose delivery to the brain leads to disorientation, progressing to convulsions, unconsciousness, and even death. In rare cases, hyperinsulinism results from an islet cell tumor. More commonly, it is caused by an overdose of insulin and is easily treated by ingesting some sugar. ■

The Gonads

The male and female **gonads** (see Figure 17.1) produce gonadal sex hormones, identical in every way to those produced by adrenal cortical cells. The major distinction is the source and relative amounts produced. The paired ovaries are small, oval organs located in the female's abdominopelvic cavity. Besides producing ova, or eggs, the ovaries produce two major steroidal hormones: **estrogen** and **progesterone** (prō-jes′-ter-ōn). Estrogen alone is responsible for maturation of the reproductive organs and the appearance of the secondary sex characteristics of females at puberty. Acting in concert with progesterone, estrogen promotes breast development and cyclic changes in the uterine mucosa (the menstrual cycle). During pregnancy, levels of both hormones rise in the mother's blood and help maintain the pregnancy; however, the hormone source at this time is the placenta, not the ovaries.

The testes, located in an inferior extra-abdominal skin pouch called the scrotum, produce sperm and male sex hormones, primarily **testosterone** (tes-tah′-ster-ōn). Testosterone initiates the maturation of the reproductive organs of the pubertal boy and the appearance of the male secondary sex characteristics and sex drive. In addition, testosterone is necessary to promote normal sperm production, as well as to maintain the reproductive organs in their mature functional state in adult males.

The release of gonadal hormones is regulated by gonadotropins, as described earlier. Hormonal interaction between the gonadotropins and their target organs is complex and delicately timed. We will discuss the roles of the gonadal, placental, and gonadotropic hormones in detail in Chapters 28 and 29, where we will consider the reproductive system organs and pregnancy.

The Pineal Gland

The tiny, pine cone-shaped **pineal gland,** hanging from the roof of the third ventricle within the diencephalon (see Figure 17.1), is composed of supportive neuroglial cells and secretory cells called *pineal cells.* The endocrine function of the pineal gland is still somewhat of a mystery. Although many peptides and amines (including serotonin, norepinephrine, and histamine) have been isolated from the minute gland, only **melatonin** (meh″-luh-tō′-nin) is known to be a major secretory product. Melatonin concentrations in the blood wax and wane in a diurnal cycle; peak levels occur during the night, and the lowest levels occur during the daylight hours around noon.

The pineal gland receives neural input from the visual pathways concerning the intensity and duration of daylight. In some animals, mating behavior and gonadal size vary with changes in the relative lengths of light and dark periods, and melatonin acts as a mediator in these effects. In humans, melatonin appears to have an antigonadotropic effect; that is, it acts on the hypothalamus to inhibit the release of gonadotropin-releasing hormone (GnRH). This apparently inhibits precocious (too early) sexual maturation. Clinical findings that support this notion include the observations that (1) blind females mature sexually later than sighted individuals, (2) some children exhibiting puberty changes at an abnormally early age have been shown to have pineal tumors, and (3) mela-

A CLOSER LOOK Sunlight and the Clock Within

It has been known for a long time that many body rhythms move in step with one another. Body temperature, pulse, and the sleep-wake cycles seem to follow the same "beat" over approximate 24-hour cycles, while other processes follow a different "drummer." What can throw these rhythms out of whack? Illness, drugs, jet travel, and changing to the night shift are all candidates. So is sunlight.

Light exerts its internal biochemical effects through the eye. Light hitting the retina generates nerve impulses along the optic nerve, tract, and radiation to the visual cortex (where "seeing" occurs) and via the retinohypothalamic tracts to the suprachiasmatic nucleus (SCN), the so-called biological clock of the hypothalamus. The SCN regulates multiple-drive rhythms, including the rhythmic melatonin output of the pineal gland and the output of various anterior pituitary hormones. Generally speaking, release of melatonin is inhibited by light and enhanced during darkness, and historically, the pineal gland has been called the "third eye."

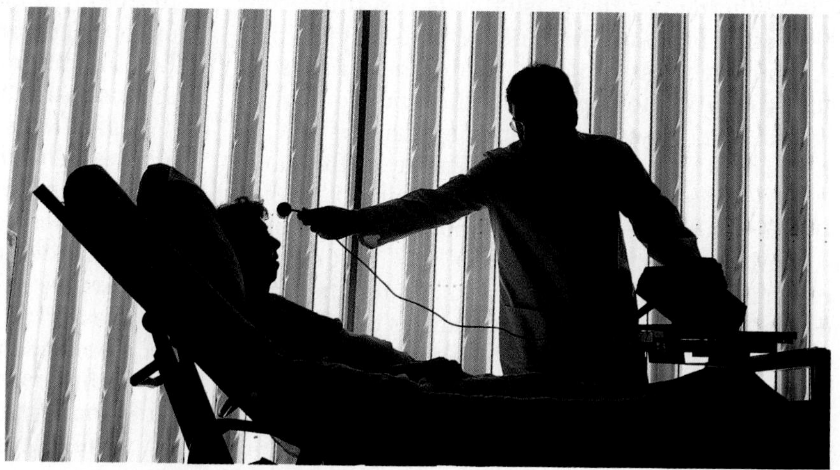

A clinician is shown measuring the light intensity preliminary to giving a treatment with light of a specific brightness to reset the patient's biological clock.

Light produces melatonin-mediated effects on reproductive, eating, and sleeping patterns of other animals, but until very recently, humans were believed to have evolved free of such effects. Thanks to the research of many scientists, however, we now know that people are influenced by three major variables of light: its intensity, its spectrum (color mixture), and its timing (day/night and seasonal changes).

A number of human processes are known to be influenced by light:

1. *Mood.* Many people have seasonal mood rhythms, particularly those of us who live far from the equator, where the day/night cycle changes dramatically during the course of a year. We seem to feel better during the summer and become cranky and depressed in the long, gray days of winter. Is this just our imagination? Apparently not. Researchers have found a relatively rare emotional disorder called seasonal affective disorder, or SAD, in which these mood swings are grossly exaggerated. As the days grow shorter each fall, people with SAD become irritable, anxious, sleepy, and socially withdrawn. Their appetite becomes insatiable; they crave carbohydrates and gain weight. Phototherapy, the use of very bright lights for 6 hours daily, reversed these symptoms in nearly 90% of patients studied in two to four days (considerably faster than any antidepressant drug could do it). When patients stopped receiving therapy or were given melatonin, their symptoms returned as quickly as they had lifted, indicating that melatonin may be the key to seasonal mood changes.

2. *Night work schedules and jet lag.* People who work at night (the graveyard shift) exhibit reversed melatonin secretion patterns, with no hormone released during the night (when they are exposed to light) and high levels secreted during daytime sleeping hours. Waking such people and exposing them to bright light causes their melatonin levels to drop. The same sort of melatonin inversion (but much more precipitous) occurs in those who fly from coast to coast.

3. *Behavior.* Researchers have found that large doses of melatonin provide a sedating effect similar to that produced by diazepam (Valium), but free of the undesirable side effects of that tranquilizer.

4. *Immunity.* Ultraviolet light activates white blood cells called suppressor T cells, which partially block the immune response. UV therapy has been found to stop rejection of tissue transplants from unrelated donors in animals. This technique offers the hope that diabetics who have become immune to their own pancreas islet tissue may be treated with pancreas transplants in the future.

People have worshipped sunlight since the earliest times. Scientists are just now beginning to understand the reasons for this, and as they do, they are increasingly distressed about windowless offices, restricted and artificial illumination of work areas, and the growing numbers of institutionalized elderly who rarely feel the sun's warm rays. Artificial lights do not provide the full spectrum of sunlight: Incandescent bulbs used in homes provide primarily the red wavelengths, and fluorescent bulbs of institutions provide yellow-green; neither provides the invisible UV or infrared wavelengths that are also components of sunlight. Animals exposed for prolonged periods to artificial lighting exhibit reproductive abnormalities and an enhanced susceptibility to cancer. Could it be that some of us are unknowingly expressing the same effects?

tonin secretion decreases somewhat at puberty. Changing melatonin levels may also be a means by which the day/night cycles influence physiological processes that show rhythmic variations, such as body temperature, sleep, appetite, and hypothalamic activity in general (see the box on p. 560).

The Thymus

Located deep to the sternum in the thorax is the lobulated **thymus gland.** Large and conspicuous in infants and children, the thymus diminishes in size throughout adulthood. By old age, it is composed largely of adipose and fibrous connective tissues.

The major hormonal product of the thymic epithelial cells is a family of peptide hormones, including **thymopoietin** and **thymosin** (thī'-mō-sin), which appear to be essential for the normal development of the immune response. The role of thymic hormones is described in Chapter 22 in conjunction with a discussion of the immune system.

Developmental Aspects of the Endocrine System

Hormone-producing glands arise from all three embryonic tissues. Most arise from the endoderm, but there are ectodermal and mesodermal representatives as well. Mesoderm-derived endocrine glands produce steroid hormones; all others produce amines or amino acid and protein hormones.

Barring outright pathology of the endocrine glands (hypersecretory and hyposecretory disorders), most endocrine organs seem to operate smoothly throughout life until old age. Growing old may bring about changes in hormone secretion rates, in endocrine organ response to stimuli, or in the rate of hormone breakdown and excretion. In addition, the sensitivity of target cell receptors may be altered with age. Research on usual endocrine functioning in the elderly is difficult, however, because functioning is frequently altered by the chronic illnesses common in that age group.

Structural changes in the anterior pituitary occur with age; the amount of connective tissue and aging pigment in the gland increases, vascularization decreases, and the number of hormone-secreting cells declines. This may or may not affect hormone production; for example, blood levels and the release rhythm of ACTH remain constant, whereas levels of TSH and gonadotropins increase.

The adrenal gland also shows structural changes with age, but normal controls of cortisol appear to persist as long as a person is healthy. Plasma levels of aldosterone are reduced by half in old age; however, this may reflect a decline in renin release by the kidneys, which become less responsive to renin-evoking stimuli. No age-related differences have been found in the release of catecholamines by the adrenal medulla.

The gonads, particularly the ovaries, undergo significant weight changes with age. In late middle age, the ovaries decrease in size and weight, and they become insensitive to gonadotropins. As female hormone production declines dramatically, the ability to bear children ends, and problems associated with estrogen deficiency, such as arteriosclerosis and osteoporosis, begin to occur. Testosterone production by the testes also wanes with age, but this effect usually is not seen until a man is in his 60s.

Glucose tolerance (the ability to dispose of a glucose load effectively) begins to deteriorate as early as the fourth decade of life. Blood glucose levels rise higher and return to resting levels more slowly in the elderly than in young adults. On the basis of this test for glucose tolerance, a high proportion of elderly have been found to have *chemical*, or *borderline, diabetes.* Since near-normal amounts of insulin continue to be secreted by the islet cells, it is believed that the decreasing glucose tolerance reflects a declining receptor sensitivity to insulin (pre–type II diabetes).

The synthesis and release of thyroid hormones diminishes somewhat with age. Typically, the follicles are loaded with colloid in the elderly, and fibrosis of the gland occurs. The basal metabolic rate lessens as one gets older, but the mild hypothyroidism is only one cause of this decline; the increase in body fat relative to muscle is equally important, since muscle tissue is much more active metabolically than fat.

The parathyroid glands show little change with age, and PTH levels remain at fairly normal values. However, since estrogen protects women against the demineralizing effects of PTH, and osteoporosis is rapidly accelerated in postmenopausal women, the postmenopausal estrogen decline may sensitize women to the effects of PTH.

* * *

In this chapter, we have covered the general mechanisms of hormone action and have provided an overview of the major endocrine organs, their chief targets, and their most important physiological effects, as summarized in Figure 17.21. However, every one of the hormones discussed here comes up in at least one other chapter in this text, where its actions are described as part of the functional framework of a particular organ system. For example, the effects of PTH and calcitonin on bone mineralization are described in Chapter 6 along with the discussion of bone remodeling, and gonadal hormones take center stage as the regulators of reproductive system maturation and function.

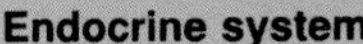

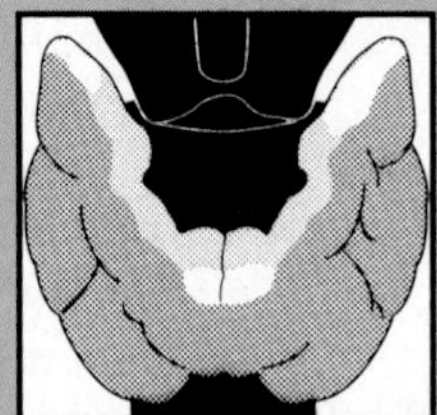

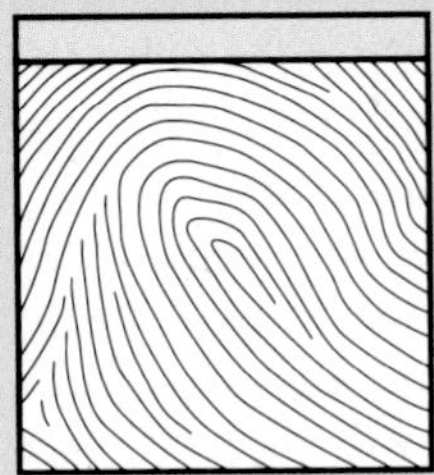

Integumentary system
Androgens cause activation of sebaceous glands; estrogen increases skin hydration

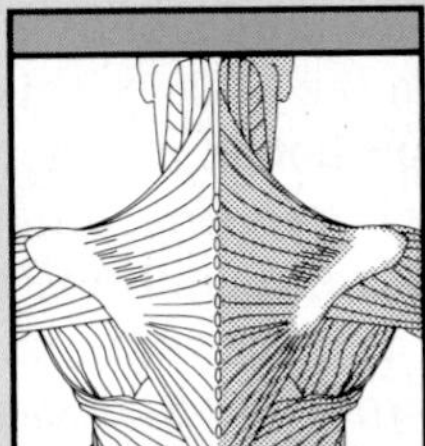

Skeletal system
Provides some protection to endocrine organs

PTH and calcitonin regulate calcium blood levels; growth hormone, T_3, T_4, and sex hormones necessary for normal skeletal development

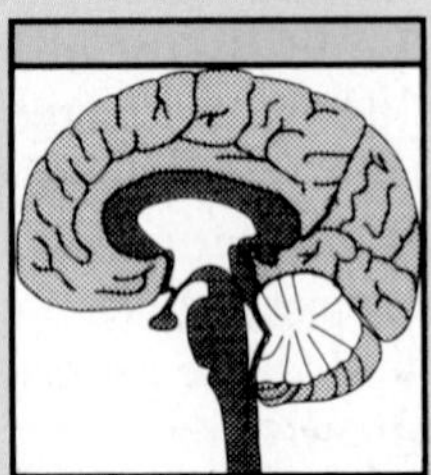

Muscular system
Growth hormone essential for normal muscular development

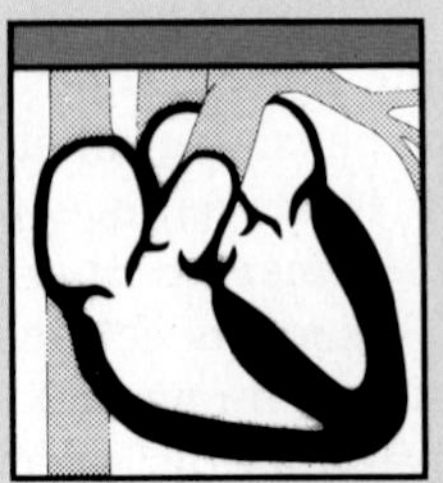

Nervous system
Hypothalamus controls anterior pituitary function and produces two hormones

Many hormones (growth hormone, thyroxine, sex hormones) influence normal maturation and function of the nervous system

Cardiovascular system
Blood is the transport medium of hormones; heart produces atrial natriuretic hormone

Several hormones influence blood volume, blood pressure and heart contractility

Lymphatic system
Lymph provides a route for transport of hormones

Lymphocytes "programmed" by thymic hormones seed the lymph nodes

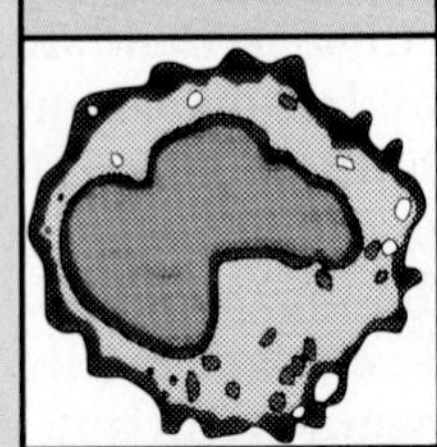

Immune system
Thymosin of the thymus gland involved in "programming" of the immune response; glucocorticoids depress the immune response

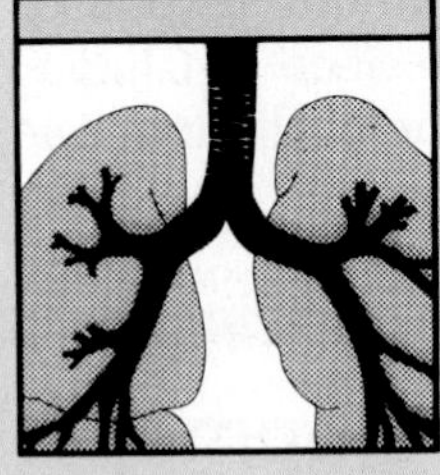

Respiratory system
Provides oxygen; disposes of carbon dioxide

Epinephrine influences ventilation (dilates bronchioles)

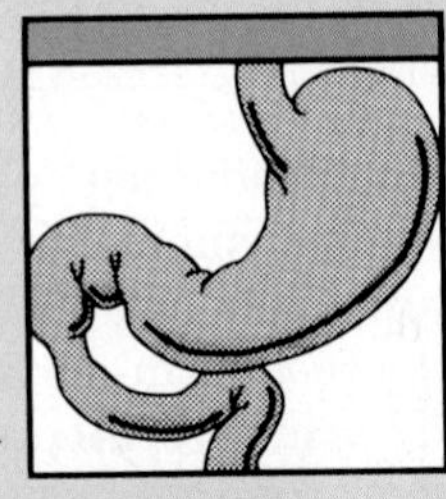

Digestive system
Provides nutrients to endocrine organs

Local GI hormones influence GI function; activated vitamin D necessary for absorption of calcium from diet

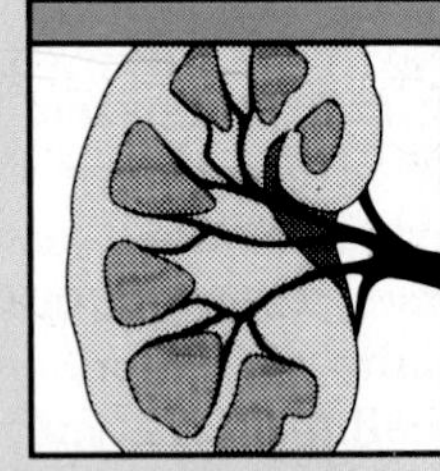

Urinary system
Kidneys activate vitamin D (considered a hormone)

Aldosterone and ADH influence renal function; renin released by kidneys influences blood pressure, and water and salt balance

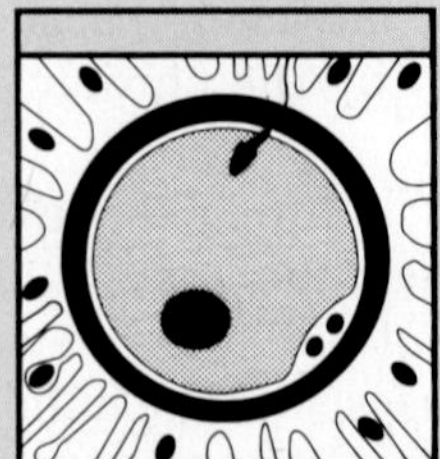

Reproductive system
Gonadal hormones feed back to influence endocrine system function

Hypothalamic, anterior pituitary and gonadal hormones direct reproductive system development and function

Figure 17.21 Homeostatic interrelationships between the endocrine system and other body systems.

Related Clinical Terms

Hirsutism (her′-sūt-izm) (*hirsut* = hairy, rough) Excessive hair growth; usually refers to this phenomenon in women and reflects excessive androgen production.

Hypophysectomy (hī-pah″-fih-sek′-tuh-mē) Surgical removal of the pituitary gland.

Idiopathy (*idio* = one's own, peculiar) Pathology of undetermined or unknown cause.

Prolactinoma (prō-lak″-tih-nō′-muh) (*oma* = tumor) The most common type (30% to 40% or more) of pituitary gland tumor; evidenced by hypersecretion of prolactin and menstrual disturbances in women.

Psychosocial dwarfism Dwarfism (and failure to thrive) resulting from stress and emotional disorders that suppress hypothalamic release of growth hormone–releasing hormone and thus anterior pituitary secretion of growth hormone.

Chapter Summary

1. The nervous and endocrine systems are the major controlling systems of the body. The nervous system exerts rapid controls via nerve impulses; the endocrine system's effects are mediated by hormones and are more prolonged.

THE ENDOCRINE SYSTEM AND HORMONE FUNCTION: AN OVERVIEW (pp. 531–532)

1. Endocrine organs are ductless glands that release hormones directly into the blood or lymph. They are small and widely separated in the body.

2. The major endocrine organs are the pituitary, thyroid, parathyroid, adrenal, pineal, and thymus glands, as well as the pancreas and gonads. The hypothalamus is a neuroendocrine organ. Many body organs not normally considered endocrine organs contain isolated cell clusters that secrete hormones; examples include the stomach, small intestine, kidneys, and heart.

3. Hormonally regulated processes include reproduction; growth and development; mobilization of body defenses to stressors; maintenance of electrolyte, water, and nutrient balance; and regulation of cellular metabolism.

HORMONES (pp. 532–537)

The Chemistry of Hormones (p. 532)

1. Most hormones are steroids or amino acid derivatives.

Hormone–Target Cell Specificity (p. 532)

2. The ability of a target cell to respond to a hormone depends on the presence of receptors within the cell or on its plasma membrane, to which the hormone can bind.

3. Hormone receptors are dynamic structures. Changes in number and sensitivity of hormone receptors may occur in response to high levels of stimulating hormones.

Mechanisms of Hormone Action (pp. 532–535)

4. Hormones alter cell activity by stimulating or inhibiting characteristic cellular processes.

5. Cell responses to hormone stimulation may involve changes in membrane permeability, enzyme synthesis, activation or inhibition, secretory activity, and others.

6. Second-messenger mechanisms employing intracellular messengers are a common means by which protein, peptide, and amine hormones interact with their target cells. In the cyclic AMP system, the hormone binds to a plasma membrane receptor coupled to adenylate cyclase, which catalyzes the synthesis of cyclic AMP from ATP. Cyclic AMP initiates reactions in which protein kinases and other enzymes are activated, leading to the cellular response. The phosphatidyl inositol mechanism is another important second-messenger system. Other presumed second messengers are cyclic GMP and calcium.

7. Steroid hormones (and thyroxine) effect target cell responses by activating DNA, initiating messenger RNA formation leading to protein synthesis.

Onset and Duration of Hormone Activity (p. 535)

8. Blood levels of hormones reflect a balance between secretion and degradation/excretion. The liver and kidneys are the major organs that degrade hormones; breakdown products are excreted in urine and feces.

9. Hormone half-life and duration of activity are limited and vary from hormone to hormone.

Control of Hormone Release (pp. 536–537)

10. Endocrine organs are activated to release their hormones by hormonal, humoral, or neural stimuli. Negative feedback is important in regulating hormone levels in the blood.

11. The nervous system, acting through hypothalamic controls, can in certain cases override or modulate hormonal effects.

ENDOCRINE ORGANS OF THE BODY (pp. 537–561)

The Pituitary Gland (Hypophysis) (pp. 537–544)

1. The pituitary gland hangs from the base of the brain by a stalk and is enclosed by bone. It consists of a hormone-producing glandular portion (anterior pituitary) and a neural portion (posterior pituitary), which is an extension of the hypothalamus.

2. The hypothalamus (a) regulates the hormonal output of the anterior pituitary via releasing and inhibiting hormones and (b) synthesizes two hormones that it exports to the posterior pituitary for storage and later release.

3. Four of the six adenohypophyseal hormones are tropic hormones that regulate the function of other endocrine organs. Most anterior pituitary hormones exhibit a diurnal rhythm of release, which is subject to modification by stimuli influencing the hypothalamus.

4. Growth hormone (GH) is an anabolic hormone that stimulates growth of all body tissues but especially skeletal muscle and bone. It may act directly or via somatomedins made by the liver. It mobilizes fats, stimulates protein synthesis, and inhibits glucose uptake and metabolism. Secretion is regulated by growth hormone–releasing hormone (GHRH) and growth hormone–inhibiting hormone (GHIH), or somatostatin. Hypersecretion causes gigantism in children and acromegaly in adults; hyposecretion causes pituitary dwarfism in children.

5. Thyroid-stimulating hormone (TSH) promotes normal development and activity of the thyroid gland. Thyrotropin-releasing hormone (TRH) stimulates its release; negative feedback of thyroid hormone inhibits it.

6. Adrenocorticotropic hormone (ACTH) stimulates the adrenal cortex to release corticosteroids. ACTH release is triggered by corticotropin-releasing hormone (CRH) and inhibited by feedback inhibition of rising glucocorticoid levels.

7. The gonadotropins, follicle-stimulating hormone (FSH) and luteinizing hormone (LH), regulate the functions of the gonads in both sexes. FSH stimulates sex cell production; LH stimulates gonadal hormone production. Gonadotropin levels rise in response to gonadotropin-releasing hormone (GnRH). Negative feedback of gonadal hormones inhibits gonadotropin release.

8. Prolactin (PRL) promotes milk production in humans. Its secretion is prompted by prolactin-releasing hormone (PRH) and inhibited by prolactin-inhibiting hormone (PIH).

9. The neurohypophysis stores and releases two hypothalamic hormones, oxytocin and antidiuretic hormone (ADH).

10. Oxytocin stimulates powerful uterine contractions, which trigger labor and delivery of an infant, and milk ejection in nursing women. Its release is mediated reflexively by the hypothalamus and represents a positive feedback mechanism.

11. Antidiuretic hormone stimulates the kidney tubules to reabsorb and conserve water; as urine output declines, blood volume and blood pressure rise. ADH is released in response to high solute concentrations in the blood and inhibited by low solute concentrations in the blood. Hyposecretion results in diabetes insipidus.

The Thyroid Gland (pp. 544–548)

12. The thyroid gland is located in the anterior throat. Thyroid follicles store thyroglobulin, a colloid from which thyroid hormone is derived.

13. Thyroid hormone includes thyroxine (T_4) and triiodothyronine (T_3), which increase the rate of cellular metabolism. Consequently, oxygen use and heat production rise.

14. Secretion of thyroid hormone, prompted by TSH, requires the reuptake of the stored colloid by the follicle cells and the splitting of the hormones from the colloid for release. Rising levels of thyroid hormone feed back to inhibit the pituitary and hypothalamus.

15. Most T_4 is converted to T_3 (the more active form) in the target tissues. These hormones appear to attach to multiple receptors and to act via several mechanisms.

16. Hypersecretion of thyroid hormone results most importantly in Graves's disease; hyposecretion causes cretinism in infants and myxedema in adults.

17. Thyrocalcitonin (calcitonin), produced by the parafollicular cells of the thyroid gland in response to rising blood calcium levels, depresses blood calcium levels by inhibiting bone matrix reabsorption and enhancing calcium uptake into bone.

The Parathyroid Glands (pp. 548–549)

18. The parathyroid glands, located on the dorsal aspect of the thyroid gland, secrete parathyroid hormone (PTH), which causes an increase in blood calcium levels by targeting bone, the intestine, and the kidneys. PTH is the antagonist of thyrocalcitonin.

19. PTH release is triggered by falling blood calcium levels and is inhibited by rising blood calcium levels.

20. Hyperparathyroidism results in hypercalcemia and all its effects and extreme bone wasting. Hypoparathyroidism leads to hypocalcemia, evidenced by tetany and respiratory paralysis.

The Adrenal Glands (pp. 549–554)

21. The paired adrenal glands sit atop the kidneys. Each adrenal gland has two functional portions, the cortex and the medulla.

22. Three groups of steroid hormones are produced by the cortex from cholesterol.

23. Mineralocorticoids (primarily aldosterone) regulate sodium ion reabsorption by the kidneys and thus indirectly regulate levels of other electrolytes that are coupled to sodium transport. Release of aldosterone is stimulated by the renin-angiotensin mechanism, rising potassium ion or falling sodium ion levels in the blood, and ACTH. Atrial natriuretic factor (ANF) inhibits aldosterone release.

24. Glucocorticoids (primarily cortisol) are important metabolic hormones that enable the body to resist stressors by increasing blood glucose, fatty acid, and amino acid levels and enhancing blood pressure. High levels of glucocorticoids depress the immune system and the inflammatory response. ACTH is the major stimulus for glucocorticoid release.

25. Gonadocorticoids (mainly androgens) are produced in small amounts throughout life.

26. Hypoactivity of the adrenal cortex results in Addison's disease. Hypersecretion can result in aldosteronism, Cushing's disease, and/or masculinization.

27. The adrenal medulla produces catecholamines (epinephrine and norepinephrine) in response to sympathetic nervous system stimulation. Its catecholamines enhance and prolong the fight-or-flight response to short-term stressors. Hypersecretion leads to symptoms typical of sympathetic nervous system overactivity.

The Pancreas (pp. 555–559)

28. The pancreas, located in the abdomen close to the stomach, is both an exocrine and an endocrine gland. The endocrine portion (islets of Langerhans) releases insulin and glucagon to the blood.

29. Glucagon, released by alpha (α) cells when blood levels of glucose are low, stimulates the liver to release glucose to the blood.

30. Insulin is released by beta (β) cells when blood levels of glucose (and amino acids) are rising. It increases the rate of glucose uptake and metabolism by most body cells. Hyposecretion of insulin results in diabetes mellitus; cardinal signs are polyuria, polydipsia, and polyphagia.

The Gonads (p. 559)

31. The ovaries of the female, located in the pelvic cavity, release two hormones. Estrogen secretion by the ovarian follicles begins at puberty under the influence of FSH. Estrogen stimulates the maturation of the female reproductive system and the development of the secondary sex characteristics. Progesterone is released in response to high blood levels of LH. It works with estrogen in establishing the menstrual cycle.

32. The testes of the male begin to produce testosterone at puberty in response to LH. Testosterone promotes the maturation of the male reproductive organs, development of secondary sex characteristics, and the production of sperm by the testes.

The Pineal Gland (pp. 559–561)

33. The pineal gland is located in the third ventricle of the brain. Its primary hormone is melatonin, which appears to have an antigonadotropic effect in humans.

The Thymus (p. 561)

34. The thymus gland, located in the upper thorax, declines in function with age. Its hormones, thymosin and thymopoietin, are important to the normal development of the immune response.

DEVELOPMENTAL ASPECTS OF THE ENDOCRINE SYSTEM (pp. 561–562)

1. Endocrine glands are derived from all three germ layers. Those derived from mesoderm produce steroidal hormones; the others produce the amino acid derivatives and protein and peptide hormones.

2. The natural decrease in function of the female's ovaries during late middle age results in menopause.

3. The efficiency of all endocrine glands seems to gradually decrease as aging occurs. This leads to a generalized increase in the incidence of diabetes mellitus and a lower metabolic rate.

Review Questions

Multiple Choice/Matching

1. The major stimulus for release of parathyroid hormone is (a) hormonal, (b) humoral, (c) neural.

2. Choose from the following key to identify the hormones described.
T and T

Key (a) aldosterone (b) antidiuretic hormone
(c) growth hormone (d) luteinizing hormone
(e) oxytocin (f) prolactin
(g) T_4 and T_3

_____ (1) important anabolic hormone; many of its effects mediated by somatomedin
_____ (2) involved in water balance; causes the kidneys to conserve water
_____ (3) stimulates milk production
_____ (4) tropic hormone that stimulates the gonads to secrete sex hormones
_____ (5) increases uterine contractions during birth
_____ (6) major metabolic hormone(s) of the body
_____ (7) causes reabsorption of sodium ions by the kidneys

3. The anterior pituitary does not secrete (a) antidiuretic hormone, (b) growth hormone, (c) gonadotropins, (d) TSH.

4. A hormone not involved with sugar metabolism is (a) glucagon; (b) cortisone, (c) aldosterone, (d) insulin.

5. Parathyroid hormone (a) increases bone formation and lowers blood calcium levels, (b) increases calcium excretion from the body, (c) decreases calcium absorption from the gut, (d) demineralizes bone and raises blood calcium levels.

6. A hypodermic injection of epinephrine would (a) increase heart rate, increase blood pressure, dilate the bronchi of the lungs, and increase peristalsis, (b) decrease heart rate, decrease blood pressure, constrict the bronchi, and increase peristalsis, (c) decrease heart rate, increase blood pressure, increase peristalsis, and decrease peristalsis, (d) increase heart rate, increase blood pressure, dilate the bronchi, and decrease peristalsis.

7. Testosterone is to the male as what hormone is to the female? (a) luteinizing hormone, (b) progesterone, (c) estrogen, (d) prolactin.

8. If anterior pituitary secretion is deficient in a growing child, the child will (a) develop acromegaly, (b) become a dwarf, but have fairly normal body proportions, (c) mature sexually at an earlier than normal age, (d) be in constant danger of becoming dehydrated.

9. If there is adequate carbohydrate intake, secretion of insulin results in (a) lower blood sugar levels, (b) increased cell utilization of glucose, (c) storage of glycogen, (d) all of these.

10. Hormones (a) are produced by exocrine glands, (b) are carried to all parts of the body in blood, (c) remain at constant concentration in the blood, (d) affect only non-hormone-producing organs.

11. Some hormones act by (a) increasing the synthesis of enzymes, (b) converting an inactive enzyme into an active enzyme, (c) affecting only specific target organs, (d) all of these.

12. Absence of thyroxine would result in (a) increased heart rate and increased force of heart contraction, (b) depression of the CNS and lethargy, (c) exophthalmos, (d) high metabolic rate.

Short Answer Essay Questions

13. Define hormone.

14. (a) Describe the body location for each of the following endocrine organs: anterior pituitary, pineal gland, pancreas, ovaries, testes, and adrenal glands. (b) Name the hormones produced by each organ.

15. Name two endocrine glands (or regions) that are important in the stress response, and explain why they are important.

16. The anterior pituitary is often referred to as the master endocrine organ, but it, too, has a "master." What controls the release of anterior pituitary hormones?

17. The posterior pituitary is not really an endocrine gland. Why not? What is it?

18. A colloidal, or endemic, goiter is not really the result of malfunction of the thyroid gland. What does cause it?

19. List some problems that elderly people might have as a result of decreasing hormone production.

Clinical Application Questions

20. Mary Morgan has just been brought into the emergency room of City General Hospital. She is perspiring profusely and is breathing rapidly and irregularly. Her breath smells like acetone (sweet and fruity), and her blood sugar tests out at 650 mg/100 ml blood. She is in acidosis. What hormone drug should be administered, and why?

21. Johnny, a five-year-old boy, has been growing by leaps and bounds. His height is 100% above normal for his age group. A CAT scan reveals a pituitary tumor. (a) What hormone is being secreted in excess? (b) What name is given to the condition that Johnny will exhibit if corrective measures are not taken?

In the ten chapters of this unit, we cover a relatively large part of the body's functional landscape. As we examine the workings of body systems that maintain bodywide homeostasis on a day-to-day basis, we look at mechanisms that help to ensure that cells have sufficient oxygen and nutrients to perpetuate their energy harvest and provide the proper internal environment for them to do so.

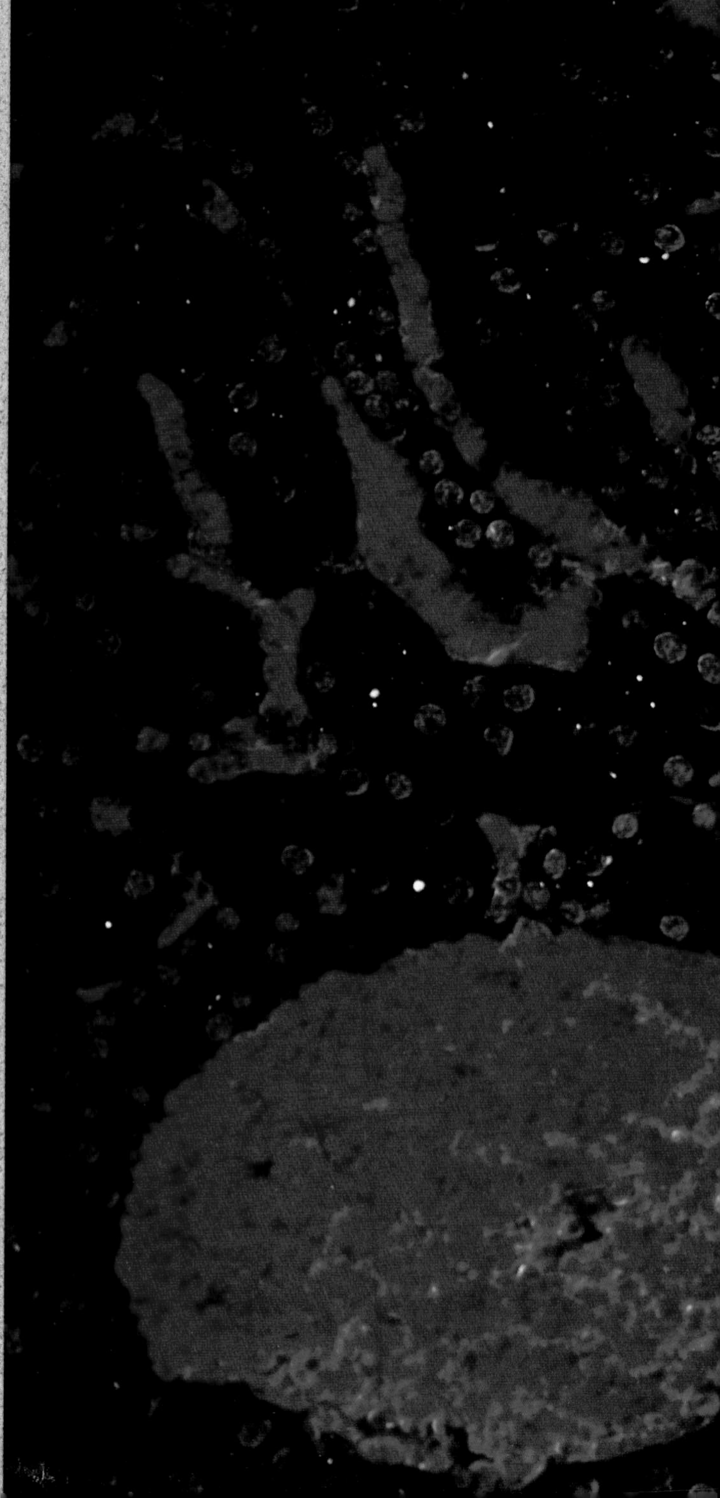

Light micrograph of a blood vessel in the kidney.

UNIT 4

MAINTENANCE OF THE BODY

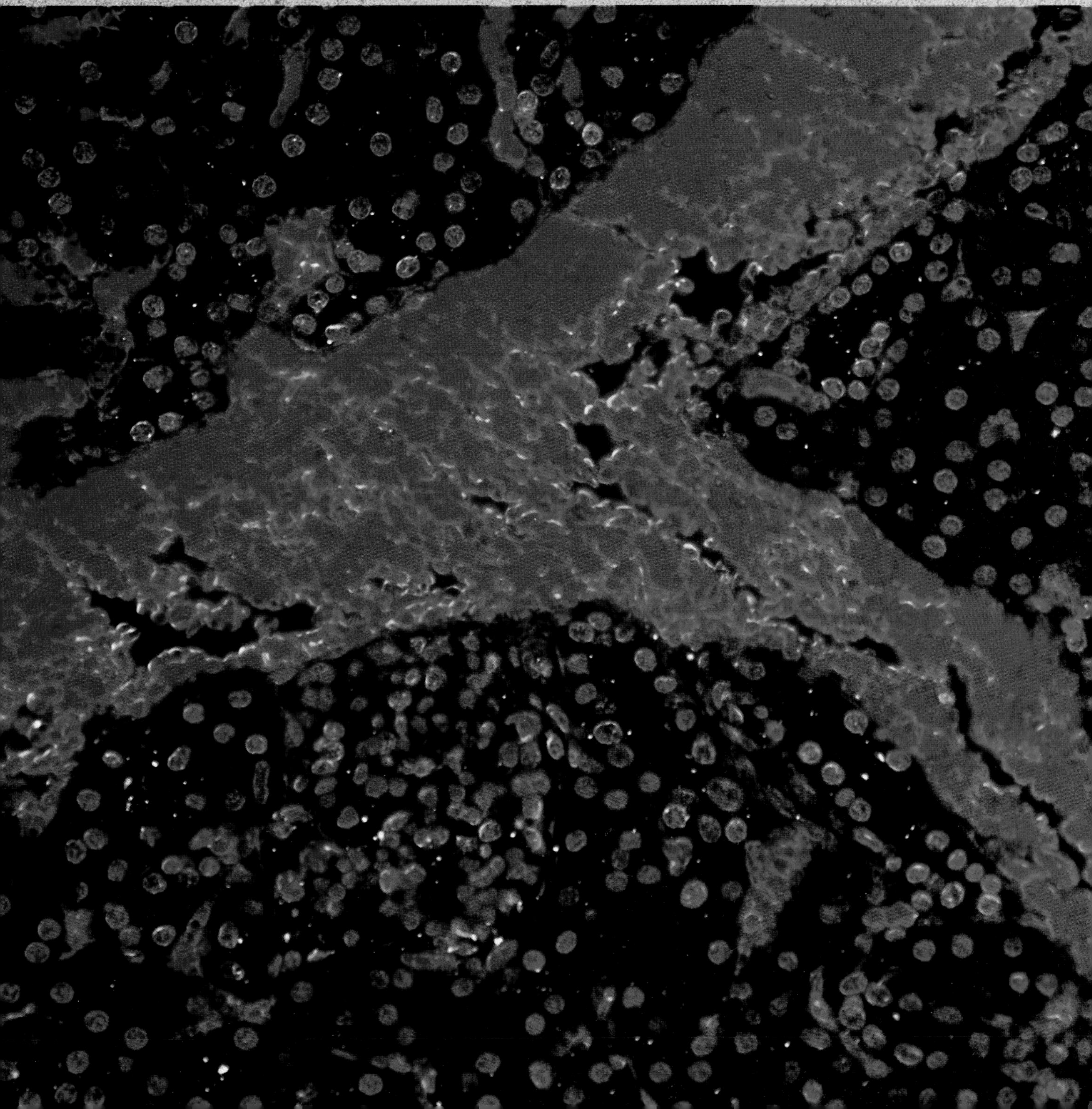

CHAPTER 18

Blood

Chapter Outline and Student Objectives

Overview: Composition and Function of Blood (pp. 569–570)

1. Describe the composition and physical characteristics of whole blood. Explain why it is classified as a connective tissue.
2. List six functions of blood.

Formed Elements (pp. 570–581)

3. Describe the structural characteristics, function, and production of erythrocytes.
4. List the classes, structural characteristics, and functions of leukocytes. Also describe leukocyte genesis.
5. Describe the structure and function of thrombocytes.
6. Give examples of disorders caused by abnormalities of each of the formed elements. Explain the mechanism of each disorder.

Plasma (pp. 581–582)

7. Discuss the composition and functions of plasma.

Hemostasis (pp. 582–587)

8. Describe the processes involved in hemostasis. Note the factors that limit clot formation and prevent undesirable clotting.
9. Give examples of hemostatic disorders. Note the cause of each condition.

Transfusion and Blood Replacement (pp. 587–590)

10. Describe the ABO and Rh blood groups. Explain the basis of transfusion reactions.
11. List several examples of blood expanders. Describe their function and the circumstances in which they are usually used.

Diagnostic Blood Tests (pp. 590–591)

12. Explain the importance of blood testing as a diagnostic tool.

Developmental Aspects of Blood (p. 591)

13. Describe changes in the sites of blood production and in the type of hemoglobin produced after birth.

Preview of Selected Key Terms

Formed elements The cellular portion of blood.

Plasma The nonliving fluid component of blood within which formed elements and various solutes are suspended and circulated.

Hemocytoblast (*hemo* = blood; *cyt* = cell; *blast* = bud) The bone marrow stem cell that gives rise to all formed elements.

Erythrocytes (eh-rih′-thrō-sīts) (*erythr* = red) Red blood cells; when mature, erythrocytes are literally sacs of hemoglobin, the iron-containing pigment that transports oxygen in the blood.

Leukocytes (*leuk* = white) White blood cells. Formed elements involved in body protection that take part in inflammatory and immune responses.

Thrombocytes (*thrombo* = clot) Platelets; cell fragments that participate in blood coagulation.

Hematopoiesis (hem″-ah-tō-poy-ē′-sis) (*poiesis* = make, produce) Blood cell formation.

Erythropoietin (eh-rih″-thrō-poy′-eh-tin) Hormone that stimulates production of red blood cells.

Anemia (*an* = without; *emia* = blood) Reduced oxygen-carrying ability of blood resulting from too few erythrocytes or abnormal hemoglobin.

Hemostasis (*stasis* = halting) Stoppage of bleeding or blood flow.

Blood is the river of life that surges within us, transporting nearly everything that must be carried from one place to another within the body. For centuries, long before modern medicine, blood was viewed as

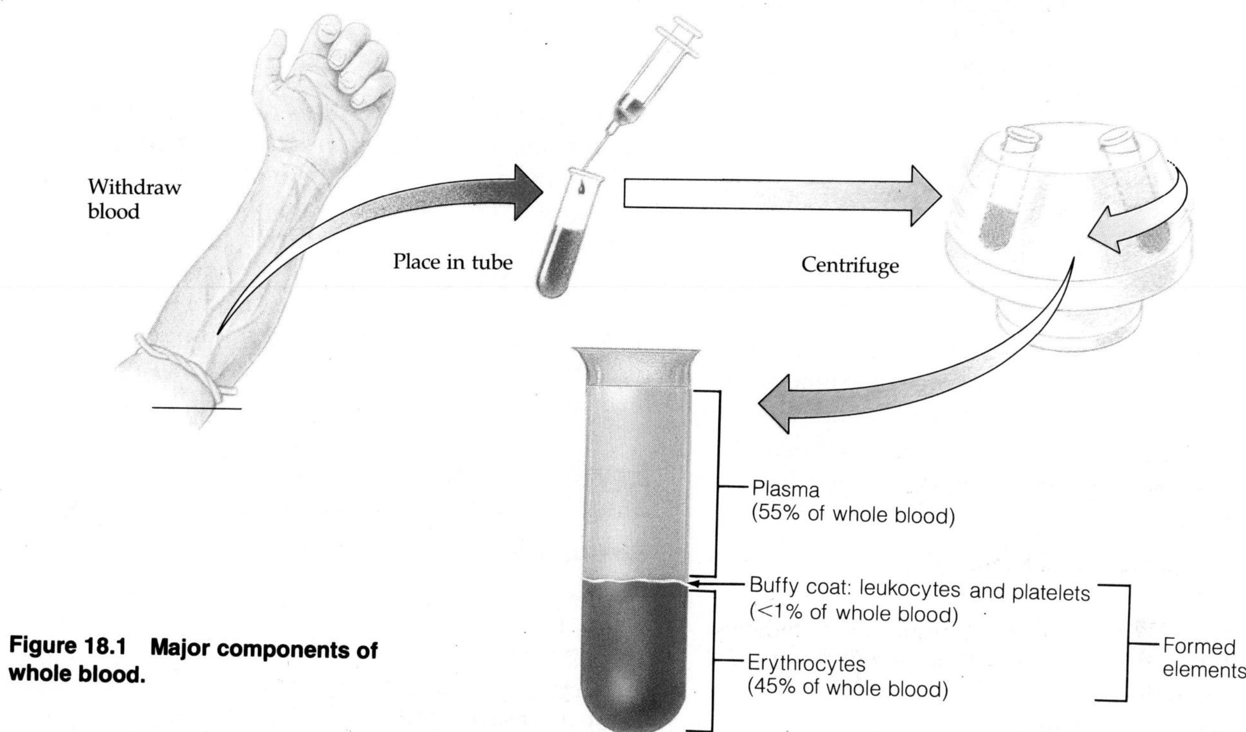

Figure 18.1 Major components of whole blood.

magical—an elixir of immortality or wisdom. But even then, people recognized that blood was vital: When blood left the body, life departed as well. In this chapter, we describe the composition and functions of this life-sustaining fluid.

To get started, we need a brief overview of blood circulation. Blood leaves the heart via the *arteries*, which branch repeatedly until they become tiny *capillaries*. By moving across the capillary walls, oxygen and nutrients leave the blood and enter the body tissues, and carbon dioxide and wastes move from the tissues to the bloodstream. As oxygen-deficient blood leaves the capillary beds, it flows into *veins*, which return it to the heart. Blood flows from the heart to the lungs, where it picks up oxygen, and then returns to the heart to be pumped throughout the body once again. Now let us look more closely at the nature of blood.

Overview: Composition and Functions of Blood

Components

Among all of the body's tissues, blood is unique: It is the only fluid tissue. Although blood appears to be a thick, homogeneous liquid, it has both solid and liquid components. Essentially, blood is a complex connective tissue in which living blood cells, the **formed elements**, are suspended in a nonliving fluid matrix called **plasma.** The collagen and elastic fibers typically seen in other connective tissues are absent from blood, but dissolved fibrous proteins become visible as fibrin strands when blood clotting occurs.

If a sample of blood is spun in a centrifuge, the heavier formed elements are packed down by centrifugal force and the less dense plasma rises to the top (Figure 18.1). Most of the reddish mass at the bottom of the tube consists of *erythrocytes* (eh-rih′-thrō-sīts), the red blood cells that function in oxygen transport. A thin, whitish layer called the *buffy coat* is present at the erythrocyte–plasma junction. This layer contains *leukocytes* (white blood cells), which act in various ways to protect the body, and *platelets*, cell fragments that function in the blood-clotting process. Erythrocytes normally constitute about 45% of the total volume of a blood sample, a percentage known as the *hematocrit*. White blood cells and platelets contribute less than 1% of blood volume. Plasma makes up most of the remaining 55% of whole blood.

Physical Characteristics and Volume

Blood is a sticky, opaque fluid with a characteristic salty taste. As children, we discover its saltiness the first time we stick a cut finger into our mouth. Depending on the amount of oxygen it is carrying, the color of blood varies from scarlet (oxygen-rich) to a dark red (oxygen-poor). Blood is heavier than water and about five times more viscous, largely because of its formed elements. Blood is slightly alkaline, with a pH between 7.35 and 7.45.

Blood accounts for approximately 8% of body weight. Its volume in healthy adult males (5–6 L) is somewhat greater than in females (4–5 L).

Functions

Blood performs a number of functions, all concerned in one way or another with substance distribution or body protection. These functions overlap and interact to maintain the constancy of our internal environment.

The distribution functions of blood include:

1. Delivery of oxygen from the lungs and of nutrients from the digestive tract to all body cells.

2. Transport of metabolic waste products from cells to elimination sites (to the lungs for elimination of carbon dioxide, and to the kidneys for elimination of nitrogenous wastes in urine).

3. Transport of hormones from the endocrine organs to their target organs.

4. Maintenance of body temperature through absorption and distribution of body heat.

The protection functions of blood include:

1. Maintenance of normal pH in body tissues. Many blood proteins and other blood-borne solutes act as buffers to prevent excessive or abrupt changes in blood pH, which could jeopardize normal cellular activities. Additionally, blood acts as the reservoir for the body's "alkaline reserve" of bicarbonate atoms.

2. Maintenance of an adequate fluid volume in the circulatory system. Sodium chloride and other salts act with blood proteins, such as albumin, to prevent excessive fluid loss from the bloodstream into the tissue spaces. As a result, the fluid volume in the blood vessels remains ample enough to support efficient blood circulation to all parts of the body.

3. Prevention of blood loss. When a blood vessel is damaged, platelets and plasma proteins initiate clot formation, halting blood loss.

4. Prevention of infection. Drifting along in the blood are antibodies, complement proteins, and white blood cells, all of which help defend the body against foreign invaders such as bacteria, viruses, toxins, and tumor cells.

Formed Elements

If you observe a smear of human blood under the light microscope, you will see smooth, disk-shaped red blood cells, a variety of gaudily stained white blood cells, and, most likely, some scattered platelets that look like debris. As shown in a photomicrograph of a typical blood smear (Figure 18.2), erythrocytes vastly outnumber the other types of formed elements. Table 18.1 (p. 578) provides a summary of the important structural and functional characteristics of the various formed elements.

Erythrocytes

Structural Characteristics

Erythrocytes, or **red blood cells (RBCs),** which function primarily to ferry oxygen in blood to all cells of the body, are superb examples of the complementarity of cellular structure and function. Mature erythrocytes lack a nucleus and have few organelles. In fact, they are little more than bags of hemoglobin molecules. Hemoglobin, the RBC protein that binds reversibly with oxygen, constitutes about 33% of cell weight. Other proteins are present, but they function mainly to facilitate gas exchange, maintain the plasma membrane, or promote changes in RBC shape. Moreover, because erythrocytes lack mitochondria and generate ATP by anaerobic mechanisms, they do not consume any of the oxygen they are transporting, making them very efficient oxygen transporters indeed.

Erythrocytes are small cells, about 8 μm in diameter. Shaped like biconcave disks—flattened disks with depressed centers (Figure 18.3)—their thin centers appear lighter in color than their edges. Consequently, erythrocytes look like miniature doughnuts when viewed with a microscope. Their small size and peculiar shape provide a large surface area relative to their volume. Because no point within the cytoplasm is far from the surface, the disk structure is ideally suited for gas exchange. However, erythrocytes are far from rigid; their shape is dynamic, changing continually as they are swept passively through the body's blood vessels. This flexibility results from an abundance of *spectrin,* a fibrous protein located on the cytoplasmic side of the erythrocyte's plasma membrane. Spectrin allows red blood cells to change shape when necessary—to twist, turn, and become cup-shaped as they travel through capillaries with diameters smaller than themselves and then to resume their biconcave shape. Spectrin also allows red blood cells to aggregate in stacks in slowly flowing blood, a phenomenon called *rouleaux* (roo-lō') *formation* (Figure 18.3c).

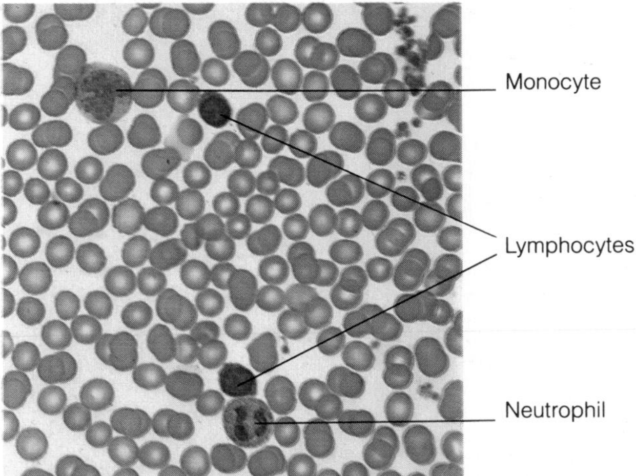

Figure 18.2 Photomicrograph of a blood smear. Most of the cells in the field of view are the disk-shaped red blood cells. Three varieties of white blood cells are also present and are identified individually. (500×.)

Erythrocytes outnumber leukocytes by 800 to 1 and are the major factor contributing to blood viscosity. Women typically have a red blood cell count between 4.3 and 5.2 million cells per cubic millimeter of blood, while the male erythrocyte count is ordinarily between 5.1 and 5.8 million cells per cubic millimeter. When the number of red blood cells increases beyond the normal range, blood viscosity increases and blood may flow more slowly. Similarly, as the number of red blood cells drops below the lower end of the range, the blood thins and flows more rapidly.

Functions

The major function of erythrocytes is to pick up oxygen in the lungs and distribute it to the body cells. Their contained hemoglobin binds easily and reversibly with oxygen, and most oxygen carried in blood is bound to hemoglobin. Normal values for hemoglobin in grams per 100 milliliters of blood (g/100 ml) are 14 to 20 in infants, 14 to 18 in adult males, and 12 to 16 in adult females.

Hemoglobin is made up of the protein **globin** bound to the red pigment **heme.** Every hemoglobin molecule contains four ringlike *heme groups,* each group bearing an atom of iron set like a jewel in its center (Figure 18.4). Globin itself is rather complex, consisting of four polypeptide chains: two alpha (α) and two beta (β). Each of its chains is bound to one of the heme groups. Each iron atom can combine reversibly with one molecule of oxygen; thus, a hemoglobin molecule can transport four molecules of oxygen. A single red blood cell contains about 250 million hemoglobin molecules, so each of these tiny cells can carry about 1 billion molecules of oxygen!

Hemoglobin is enclosed in erythrocytes rather than existing free in plasma, which presumably prevents it from breaking into fragments that would leak out of the bloodstream through the rather porous capillary membranes. Also, its enclosure within red blood cells prevents it from contributing to blood viscosity and osmotic pressure. As oxygen-deficient blood moves through the lungs, oxygen diffuses from the air sacs

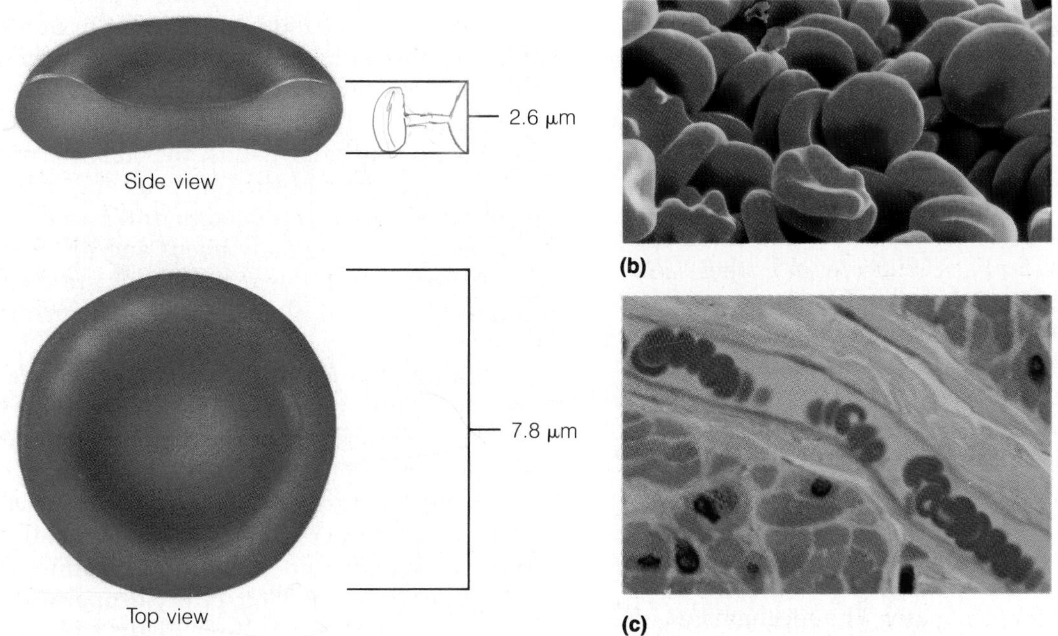

Figure 18.3 Structure of erythrocytes. (**a**) An erythrocyte in a cut side view and in a superficial surface view. Note the distinctive biconcave shape. (**b**) Scanning electron micrograph of human erythrocytes (2000X). (**c**) Photomicrograph of erythrocytes in rouleaux formation within a capillary (800×).

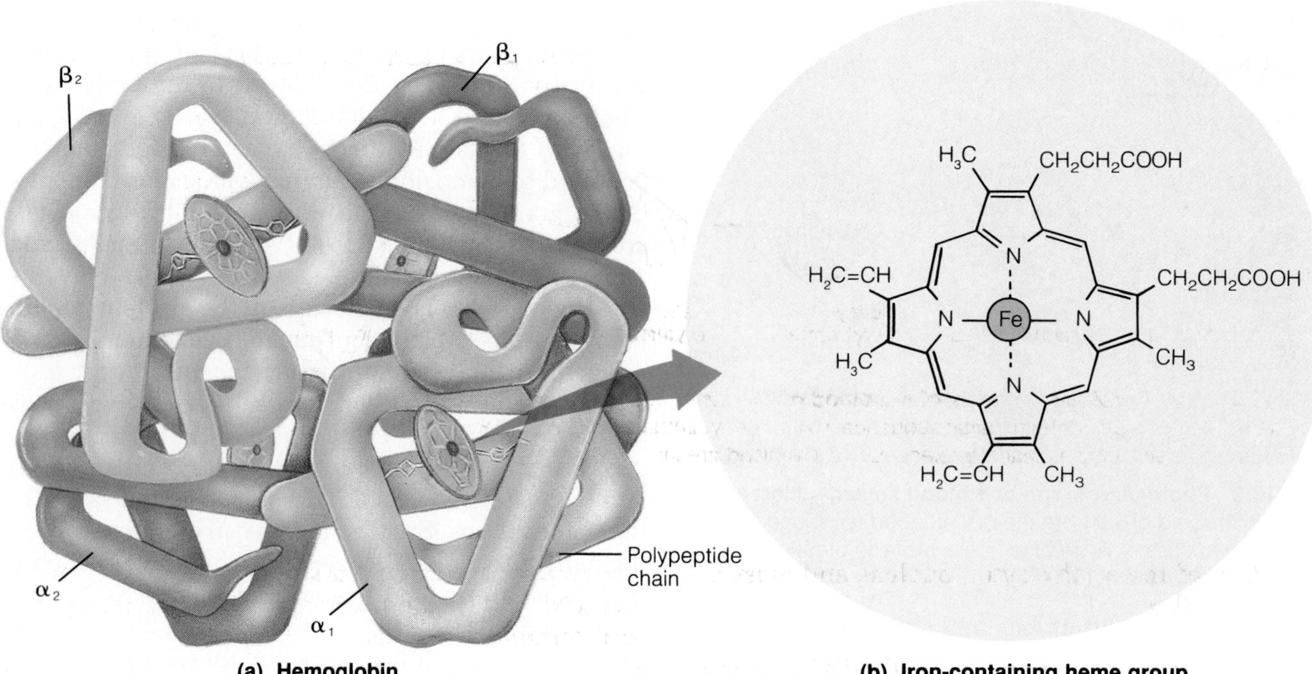

Figure 18.4 Structure of hemoglobin. (**a**) The intact hemoglobin molecule is composed of the protein globin bound to the iron-containing heme pigments. Each globin molecule has four polypeptide chains: two alpha (α) chains and two beta (β) chains. Each chain is complexed with a heme group, shown as a green beaded structure. (**b**) Structure of a single heme group.

of the lungs into the blood and then into the erythrocytes, where it binds to hemoglobin. When oxygen binds to iron, the hemoglobin assumes a new three-dimensional shape; now called *oxyhemoglobin,* it becomes bright red. In the tissues, the process is reversed. Oxygen dissociates from iron, causing hemoglobin to resume its former shape, and the resulting *deoxyhemoglobin,* or *reduced hemoglobin,* becomes dark red. The released oxygen diffuses from the blood into the tissue fluid and then into the tissue cells.

About 20% of the carbon dioxide transported in the blood combines with hemoglobin, but it binds to amino acids of globin rather than with the heme group. This formation of *carbamino* (kar″-buh-mē′-nō) *hemoglobin* occurs more readily when hemoglobin is in the reduced state (dissociated from oxygen). Whereas oxygen loading occurs in the lungs and the direction of transport is from lungs to tissue cells, carbon dioxide loading occurs in the tissues and the direction of transport is from tissues to lungs, where carbon dioxide is eliminated from the body. The mechanisms of loading and unloading these respiratory gases are described in Chapter 23.

Production of Erythrocytes

Blood formation, or **hematopoiesis** (hem″-ah-tō-poy-ē′-sis), occurs in the red bone marrow, or myeloid tissue, where the immature blood cells are supported on a soft network of fat cells, fibroblasts, and reticular fibers. In adults, this tissue is found chiefly in the flat bones of the skull and pelvis, the ribs, sternum, and proximal epiphyses of the humerus and femur. Each type of blood cell is produced in different numbers in response to changing body needs and different regulatory factors. After they mature, they are discharged into the thin-walled blood vessels that surround the area.

Although the various formed elements have differing functions, there are important similarities in their life histories. All of the formed elements arise from the same type of *stem cell,* the **hemocytoblast,** which resides in the red bone marrow. However, their maturation pathways differ; and once a cell is committed to a specific blood pathway, it cannot change.

Erythrocyte production, or **erythropoiesis** (eh-rih″-thrō-poy-ē′-sis), involves three distinct phases:

1. Production by the immature erythrocytes of substantial numbers of ribosomes.
2. Synthesis of hemoglobin at these ribosomes and its accumulation in the cell's cytoplasm.

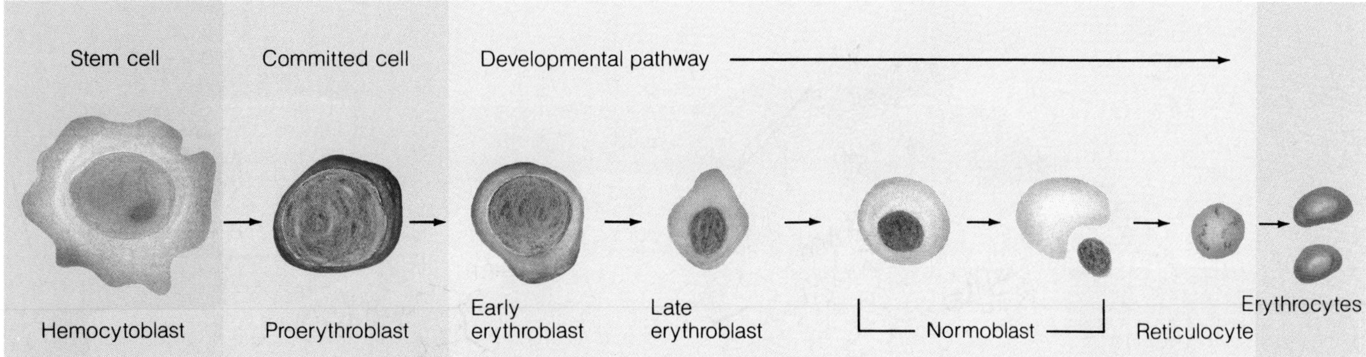

Figure 18.5 Erythropoiesis: genesis of red blood cells. Erythropoiesis proceeds through a proliferation and differentiation sequence involving erythroblasts and the normoblast to form the reticulocytes that are released into the bloodstream.

3. Ejection of the erythrocyte's nucleus and most of its organelles.

During the first two phases, the cells divide many times.

Erythropoiesis begins when a hemocytoblast is transformed into a *proerythroblast* (Figure 18.5). Proerythroblasts, in turn, give rise to the *early (basophilic) erythroblasts* that serve as ribosome-producing factories. Hemoglobin synthesis and accumulation occur as the early erythroblast is transformed into a *late (polychromatophilic) erythroblast* and then a *normoblast*. When a normoblast has accumulated a hemoglobin concentration of about 34%, its nuclear functions end. Its nucleus degenerates and is ejected—an event that causes the cell to collapse inward and assume the biconcave shape. The result is the *reticulocyte*. Reticulocytes (essentially young erythrocytes) are so named because they still contain some rough endoplasmic reticulum. The entire process from hemocytoblast to reticulocyte takes three to five days. The reticulocytes enter the peripheral circulation to begin their task of oxygen transport and usually become fully mature within two days of release.

Regulation and Requirements for Erythropoiesis

The number of circulating erythrocytes in a given individual is remarkably constant and reflects a balance between red blood cell production and destruction. This balance is extremely important because having too few erythrocytes leads to tissue hypoxia (oxygen deprivation), whereas having too many makes the blood undesirably viscous. To ensure that the number of erythrocytes in the blood remains within the homeostatic range, new cells are produced at the incredibly rapid rate of more than 2 million per second in healthy people. This process is controlled hormonally and depends on adequate supplies of iron and certain B vitamins.

Hormonal Controls. The direct stimulus for erythrocyte formation is provided by **erythropoietin,** a glycoprotein hormone. Normally, a small amount of erythropoietin circulates in the blood at all times and sustains red blood cell production at a basal rate. The kidneys play the major role in erythropoietin production. When kidney cells become hypoxic, they release an enzyme called *renal erythropoietin factor (REF),* which cleaves (splits apart) a blood protein to form erythropoietin (Figure 18.6). The drop from normal blood oxygen levels that triggers erythropoietin formation can result from

1. Declining numbers of red blood cells caused by hemorrhage or excess RBC destruction.
2. Reduced availability of oxygen to the blood, as might occur at high altitudes or during pneumonia.
3. Increased tissue demands for oxygen (common in those who engage in aerobic exercise).

Conversely, an overabundance of erythrocytes or an excessive amount of oxygen in the bloodstream depresses REF release and erythropoietin production. The easily missed point to keep in mind is that the concentration of erythrocytes in the blood does *not* control the rate of erythropoiesis; control is based on their ability to transport adequate oxygen to meet tissue demands.

Blood-borne erythropoietin stimulates red marrow cells that *have already become committed* or channelized in their development to become erythrocytes, causing them to mature at a more rapid rate. A marked increase in the rate of reticulocyte release occurs from one to two days after erythropoietin levels rise in the blood.

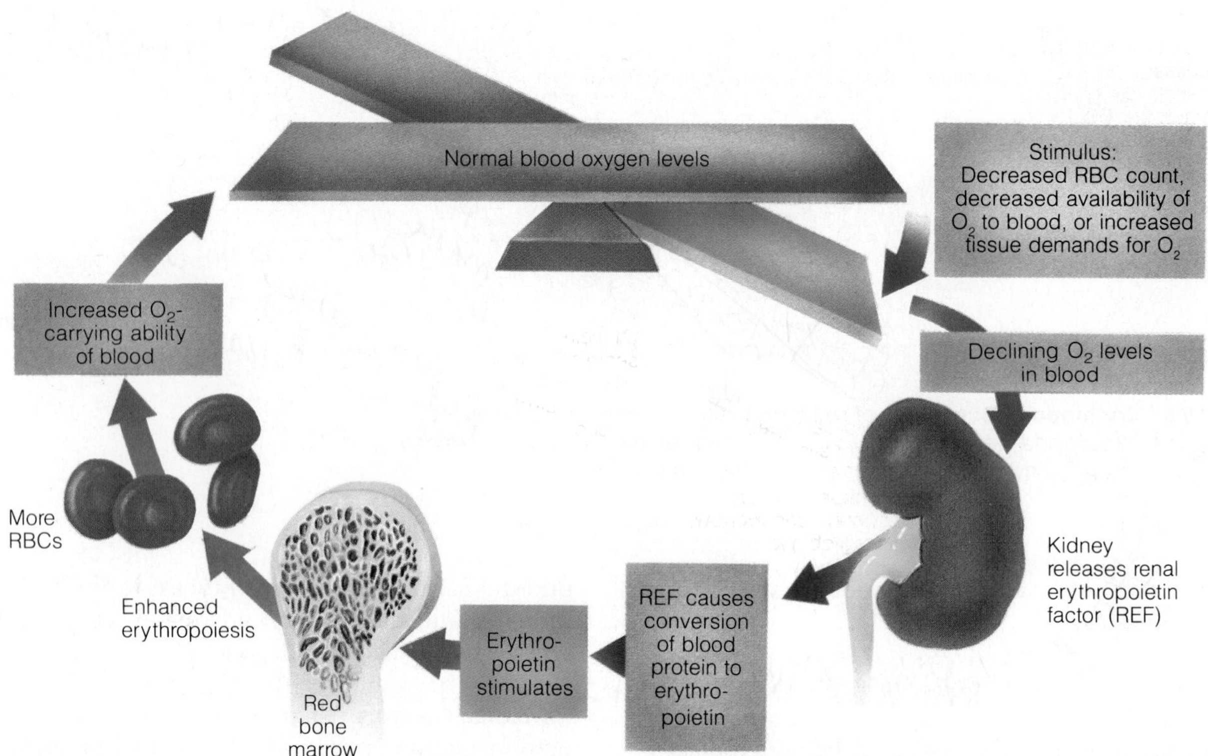

Figure 18.6 Erythropoietin mechanism for regulating the rate of erythropoiesis. Notice that increased erythropoietin release, which stimulates erythropoiesis in bone marrow, occurs when oxygen levels in the blood become inadequate to support normal cellular activity, whatever the cause.

Notice that hypoxia, the initial stimulus for the regulatory sequence, does not activate the erythropoietic site directly. Instead it stimulates the kidney, which, in turn, by releasing REF, provides the hormonal stimulus that activates the bone marrow. Patients whose kidneys have failed and who undergo renal dialysis routinely have red blood cell counts less than half that of normal individuals.

The male sex hormone *testosterone* also enhances the erythropoietin-generating function of the kidneys. Because female sex hormones do not have similar stimulatory effects, testosterone may be at least partially responsible for the higher red blood cell counts and hemoglobin levels in males. Additionally, some chemicals released by leukocytes are believed to stimulate bursts of red blood cell production.

Dietary Requirements: Iron and B-Complex Vitamins. The raw materials required for erythropoiesis include the usual nutrients and structural materials—proteins, lipids, and carbohydrates. In addition, iron and B-complex vitamins are essential for hemoglobin synthesis (Figure 18.7). Iron is made available through diet, and iron absorption into the bloodstream is precisely controlled by intestinal cells. When body stores of iron are ample, iron uptake is slow; but when reserves begin to drop, the rate of iron absorption is accelerated.

Approximately 65% of the body's iron supply (about 4000 mg) is in hemoglobin. The remainder is stored in the liver, spleen, and (to a much lesser extent) bone marrow. Because free iron is toxic to body cells, iron is stored within cells as protein–iron complexes such as *ferritin* (fayr′-ih-tin) and *hemosiderin* (hē″-mō-sid′-er-in). In the blood, iron is loosely bound to a transport protein called *transferrin*, and developing erythrocytes take up iron as needed to form functional hemoglobin molecules.

Small amounts of iron are lost each day in feces, urine, and perspiration. In women, the menstrual flow accounts for additional losses. The average daily loss of iron is 1.7 mg in women and 0.9 mg in men.

Two B-complex vitamins—vitamin B_{12} and folic acid—must be available if normal erythropoiesis is to occur. Although they affect different steps of the process, minute amounts of both vitamins are essential for DNA synthesis. Even slight deficits jeopardize rapidly dividing cell populations, such as developing erythrocytes. Under such conditions, the cells do not divide, but they do increase in size and become large, abnormal cells.

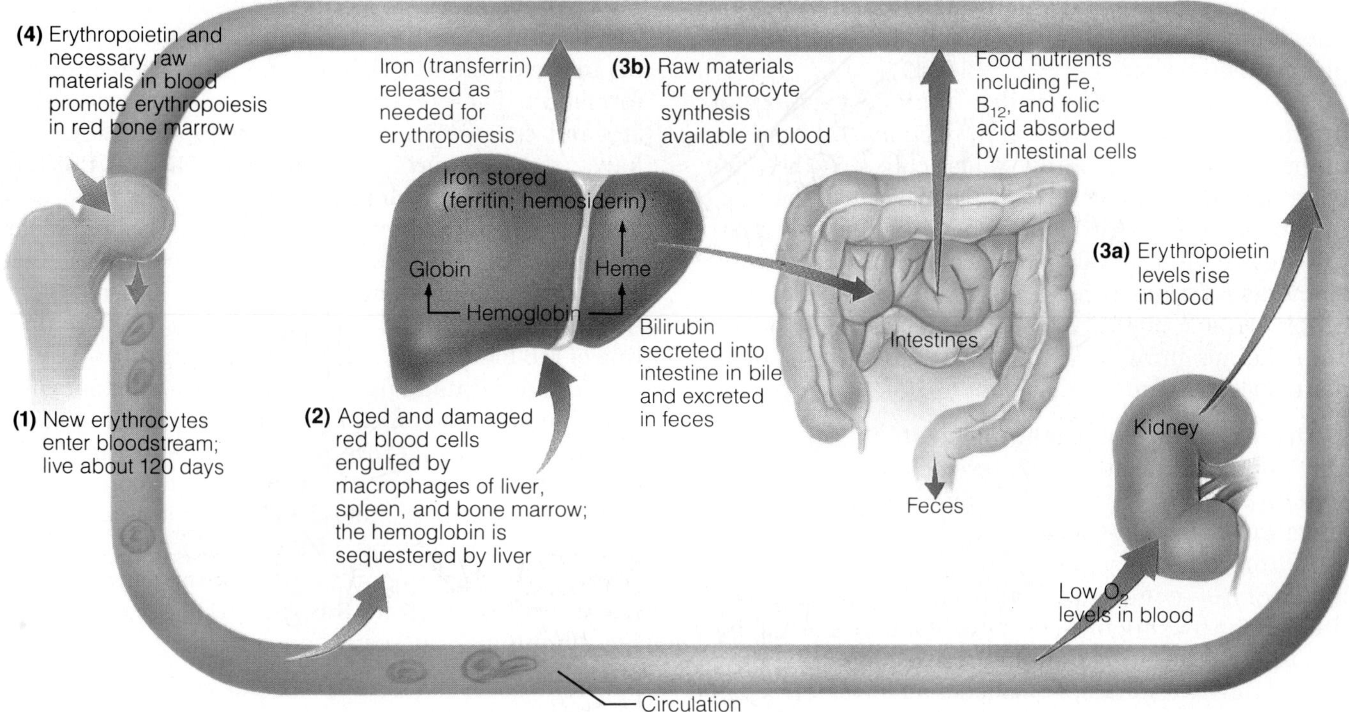
(4) Erythropoietin and necessary raw materials in blood promote erythropoiesis in red bone marrow
Iron (transferrin) released as needed for erythropoiesis
(3b) Raw materials for erythrocyte synthesis available in blood
Food nutrients including Fe, B_{12}, and folic acid absorbed by intestinal cells
Iron stored (ferritin; hemosiderin)
Globin
Heme
Hemoglobin
Bilirubin secreted into intestine in bile and excreted in feces
Intestines
(3a) Erythropoietin levels rise in blood
Kidney
(1) New erythrocytes enter bloodstream; live about 120 days
(2) Aged and damaged red blood cells engulfed by macrophages of liver, spleen, and bone marrow; the hemoglobin is sequestered by liver
Feces
Low O_2 levels in blood
Circulation

Figure 18.7 Life cycle of red blood cells.

Fate and Destruction of Erythrocytes

The anucleate condition of erythrocytes carries with it a number of important limitations. Red blood cells are unable to synthesize proteins, to grow, or to divide. Erythrocytes become "old" as the enzymes needed to maintain ATP synthesis gradually degrade. Because they cannot replace these enzymes, energy-dependent processes such as the calcium pump stop operating. As intracellular calcium levels rise and calcium becomes cross-linked with intracellular proteins, the erythrocytes lose their flexibility and become increasingly rigid and fragile. Red blood cells have a useful life span of 100 to 120 days, after which they become trapped and damaged in the smaller circulatory channels, particularly those of the spleen. For this reason, the spleen is sometimes called the "red blood cell graveyard." Dying erythrocytes are engulfed and destroyed by phagocytes, and their hemoglobin is degraded to *bilirubin* (bih″-lih-roo′-bin), a yellow pigment secreted in bile by the liver (see Figure 18.7). The released iron is salvaged and recycled; it is complexed with protein and stored for later reuse. In healthy people the rate of erythrocyte destruction is balanced by new red blood cell production via the erythropoietin hormonal mechanism (see Figure 18.6).

Erythrocyte Disorders

Most erythrocyte disorders can be classified as anemias or polycythemias. The many varieties and causes of these conditions are described next.

Anemias. **Anemia** is a condition of reduced oxygen-carrying ability of the blood, whatever the cause. It is a symptom of some disorder rather than a disease in and of itself. Its hallmark is oxygen blood levels that are inadequate to support normal metabolism. Consequently, anemic individuals are perpetually fatigued and often pale, short of breath, and chilly. Common causes of anemia include:

1. Insufficient numbers of red blood cells. Conditions that reduce the red blood cell count include blood loss, destruction of red blood cells, and bone marrow failure.

Hemorrhagic (hem″-ō-ra′-jik) *anemias*, result from blood loss. Acute hemorrhagic anemia, as might follow a severe wound, is treated by blood replacement. Slight, persistent blood loss, as might result from hemorrhoids or an undiagnosed bleeding ulcer, causes chronic hemorrhagic anemia. Once the primary problem is resolved, normal erythropoietic mechanisms replace the deficient blood cells.

In *hemolytic* (hē″-mō-lih′-tik) *anemias*, erythrocytes are ruptured, or lysed, prematurely. This may result from hemoglobin abnormalities, transfusion of mismatched blood, certain bacterial and parasitic infections, or congenital defects in the erythrocyte plasma membrane.

Aplastic anemia results from destruction or inhibition of the red marrow. In bone marrow cancer, the marrow is replaced by cancer cells. Toxins, ionizing radiation, and certain drugs may cause the marrow to be replaced by connective tissue. Because marrow destruction impairs the formation of *all* formed elements, anemia is just one of its signs; bleeding problems and inability to fight infection also occur. Transfusions provide a stopgap treatment until bone marrow transplants can be done.

2. Decreases in hemoglobin content. When normal hemoglobin molecules are present, but erythrocytes contain fewer than the usual number, a nutritional anemia is always suspected.

Iron-deficiency anemia results from inadequate intake of iron-containing foods, impaired iron absorption, or, most commonly, chronic blood loss from hidden internal bleeding or from profuse menstrual flow. The erythrocytes produced, called *microcytes,* are small and pale. The obvious treatment is iron supplements. If chronic hemorrhage is the cause, blood transfusions may also be needed.

Pernicious anemia reflects a deficiency of vitamin B_{12}, which is needed by developing red blood cells. Because the vitamin is amply provided by meats, poultry, and fish, diet is rarely the problem except for strict vegetarians. A substance called *intrinsic factor,* produced by the stomach mucosa, must be present for vitamin B_{12} to be absorbed by intestinal cells. In most cases of pernicious anemia, intrinsic factor is deficient, rather than the vitamin. In pernicious anemia the developing erythrocytes grow but do not divide, resulting in large, pale cells called *macrocytes.* Treatment involves regularly scheduled intramuscular injections of vitamin B_{12}.

3. Abnormal hemoglobin. Abnormal hemoglobin formation usually has a genetic basis. Two such examples, thalassemia and sickle-cell anemia, are serious, incurable, and often fatal diseases. In both diseases the globin portion of the hemoglobin molecule is abnormal and the erythrocytes produced are fragile and rupture prematurely. Standard treatment for both of these anemias is blood transfusion.

Thalassemias (tha″-luh-sē′-mē-uhz) typically are seen in people of Mediterranean ancestry, such as Greeks and Italians. The erythrocytes are thin and delicate, and the red blood cell count is generally less than 2 million cells per cubic millimeter.

In *sickle-cell anemia,* the abnormal hemoglobin formed, *hemoglobin S,* becomes spiky and sharp and causes the red blood cells to become crescent-shaped when they unload oxygen molecules or when the oxygen content of the blood is lower than normal, as during vigorous exercise, anxiety, or other types of stress. The deformed erythrocytes, which rupture easily, dam up in small blood vesssels and promote blood clot formation. These events interfere with oxygen delivery and cause extreme pain. It is amazing that this havoc results from a change in just one of the 287 amino acids in the globin molecule!

Sickle-cell anemia occurs chiefly in black descendants of peoples who lived in the malaria belt of Africa and in some parts of Asia and southern Europe. Apparently, the same gene that causes sickling of red blood cells acts to prevent rupture of erythrocytes confronted with the malaria-causing parasite. Individuals with the sickle-cell gene are more resistant to malaria than are noncarriers and, therefore, have a better chance of survival in regions where malaria is prevalent. Sickle-cell anemia is exhibited only by those carrying two copies of the defective gene. Those carrying only one sickling gene are said to have sickle-cell trait; they generally do not display the symptoms, but can transmit the gene to their offspring.

Polycythemia. Polycythemia (pah″-lē-sī-thē′-mē-uh) is an excessive or abnormal increase in the number of erythrocytes. This increases the viscosity of the blood, causing it to sludge, or flow sluggishly. *Polycythemia vera,* most often a result of bone marrow cancer, is a serious condition characterized by dizziness and an exceptionally high red blood cell count (8 to 11 million cells per cubic millimeter). The hematocrit may be as high as 80% and blood volume may double, causing the circulatory vascular system to become engorged with blood and severely impairing circulation.

Secondary polycythemias develop because of decreased availability of oxygen or increased erythropoietin production. The secondary polycythemia that appears in individuals living at high altitudes is a normal physiological response that compensates for decreased atmospheric pressure and reduced oxygen content of the air in such areas. Red blood cell counts of 6 to 8 million per cubic millimeter are common in such people. ■

Blood doping, practiced by some athletes competing in aerobic events, is essentially artificially induced polycythemia. Some of the athlete's red blood cells are drawn off and then reinjected a few days before the event. Because the erythropoietin mechanism is triggered shortly after the blood is removed, the erythrocytes are quickly replaced. Then, when the stored blood is reinfused, a temporary polycythemia results. The reasoning behind this practice is that since red blood cells carry oxygen, the additional infusion should translate into greater endurance and faster time. Although blood doping appears to work, the practice is unethical and has been banned from the Olympic Games. ■

Leukocytes

General Structural and Functional Characteristics

Although **leukocytes,** or **white blood cells (WBCs),** are far less numerous than red blood cells, they are crucial to body defense against disease. On average, there are 4000–11,000 WBCs per cubic millimeter, and leukocytes account for about 1% of total blood volume. White blood cells contain nuclei and the usual organelles.

Leukocytes form a protective mobile army that helps protect the body from damage by bacteria, viruses, parasites, toxins, and tumor cells. As such, they have some very special functional characteristics. Red blood cells are confined to the bloodstream, and they carry out their functions in the blood. But white blood cells are able to slip into and out of the blood vessels—a process called *diapedesis* (dī″-uh-puh-dē′-sis)—and they use the circulatory system chiefly as a transportation system to ferry them to areas of the body (largely connective tissue spaces or lymphoid tissues) where their protective services are required to mount inflammatory or immune responses (as described in Chapter 22). Leukocytes move through the tissue spaces by *amoeboid motion* (by forming flowing cytoplasmic extensions that help move them along). They are also able to locate areas of tissue damage and infection by responding to chemicals released by damaged cells or other leukocytes. By following the chemical trail, a phenomenon called *positive chemotaxis,* they are able to pinpoint areas of tissue damage and gather in large numbers to destroy foreign substances or dead cells.

Whenever white blood cells are mobilized for action, the body speeds up their production and twice the normal number may appear in the blood within a few hours. A *white blood cell count* of over 11,000 cells per cubic millimeter is referred to as *leukocytosis* and generally indicates that a bacterial or viral infection is brewing in the body.

Leukocytes are grouped into two major categories on the basis of structural and chemical characteristics: *Granulocytes* contain specialized granules; *agranulocytes* lack obvious granules. General information about the various leukocytes is provided next. More detailed information concerning their diameter, concentration in blood, and percentage of the white blood cell population appears in Figure 18.8 and Table 18.1.

Granulocytes

Granulocytes (gran′-yoo-lō-sīts) characteristically have lobed nuclei (rounded nuclear masses connected by thin strands of nuclear material), and their cytoplasmic granules stain quite specifically with Wright's stain. The granulocytes include neutrophils, basophils, and eosinophils.

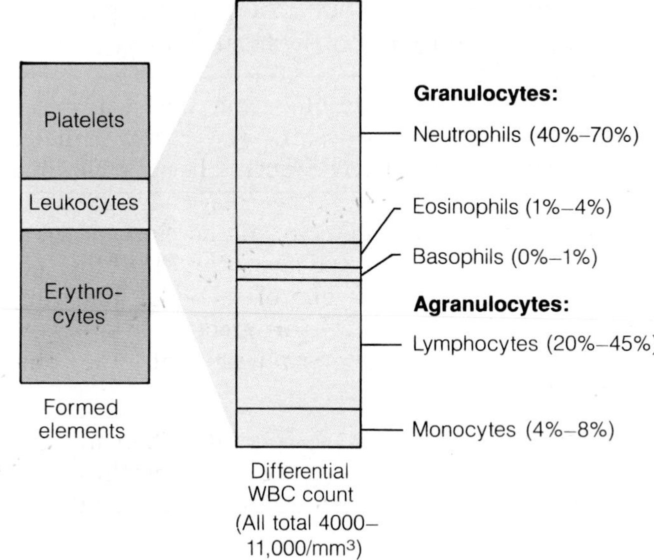

Figure 18.8 Types of leukocytes.

Agranulocytes

Neutrophils. **Neutrophils,** the most numerous of the white blood cells, typically account for half or more of the WBC population. Neutrophils are about twice as large as erythrocytes.

The neutrophil cytoplasm stains pale pink and contains very fine granules that are difficult to see (see Table 18.1 and Figure 18.2). Their naming as neutrophils (literally, "neutral-loving") indicates that their granules take up neither basic (blue) nor acidic (red) dyes. Some of these granules contain lysozyme and other antibacterial substances; others, especially the larger, darker granules, are lysosomes. Neutrophil nuclei consist of three to five lobes; because of this they are often called *polymorphonuclear* ("many shapes of the nucleus") *leukocytes (PMNs)* or simply *polys.*

Neutrophils are chemically attracted to sites of inflammation and are active phagocytes. They are particularly partial to bacteria and some fungi, which they can ingest and destroy. Bacterial killing is promoted by a process called a *respiratory burst,* in which oxygen is actively metabolized to produce potent germ-killing substances. Neutrophil numbers increase explosively during acute infections such as appendicitis.

Basophils. **Basophils** are the least numerous white blood cells, usually constituting less than 1% of the leukocyte population. Basophils are just slightly larger than erythrocytes and have a distinctive cell structure (see Table 18.1). Their pale cytoplasm contains a few coarse granules, which have an affinity for the basic dyes and stain purplish-black. The deep purple nucleus is generally U- or S-shaped with two or three conspicuous constrictions.

Table 18.1 Summary of Formed Elements of the Blood

Cell type	Illustration	Description*	Number of cells/ mm^3 (μL) of blood	Duration of development (D) and lifespan (LS)	Function
Erythrocytes (red blood cells; RBCs)		Biconcave, anucleate disk; salmon-colored; diameter 6–8 μm	4–6 million	D: 5–7 days LS: 100–120 days	Transport oxygen and carbon dioxide
Leukocytes (white blood cells, WBCs)		Spherical, nucleated cells	4000–11,000		
Granulocytes					
• Neutrophil		Nucleus multilobed; inconspicuous cytoplasmic granules; diameter 12–14 μm	3000–7000	D: 6–9 days LS: 6 hours to a few days	Phagocytize bacteria
• Eosinophil		Nucleus bilobed; red cytoplasmic granules; diameter 10–14 μm	100–400	D: 6–9 days LS: 8–12 days	Destroy antigen–antibody complexes; inactivate some inflammatory chemicals
• Basophil		Nucleus lobed; large blue-purple cytoplasmic granules; diameter 8–10 μm	20–50	D: 3–7 days LS: ? (a few hours to a few days)	Release heparin and histamine
Agranulocytes					
• Lymphocyte		Nucleus spherical or indented; pale blue cytoplasm; diameter 5–17 μm	1500–3000	D: 1–2 days LS: hours to years	Mount immune response by direct cell attack or via antibodies
• Monocyte		Nucleus U- or kidney-shaped; gray-blue cytoplasm; diameter 14–24 μm	100–700	D: 2–3 days LS: months	Phagocytosis; develop into macrophages in tissues
Thrombocytes (platelets)		Discoid cytoplasmic fragments containing granules; stain deep purple; diameter 2–4 μm	250,000 – 500,000	D: 4–5 days LS: 5–10 days	Instrumental in blood clotting

*Appearance when stained with Wright's stain.

Tissue basophils are called *mast cells,* although there is some controversy over this point. Some investigators believe that basophils and mast cells are identical cells in different locations; others believe that each type is unique. The cells are nearly indistinguishable microscopically, and both bind to a particular antibody (immunoglobin E) that causes the granules to release heparin and histamine. *Histamine* is a vasodilator that makes blood vessels "leaky" (a phenomenon usually seen at sites of inflammation); *heparin* prevents blood clotting. Both chemicals enhance the migration of white blood cells to the inflamed site.

Eosinophils. Eosinophils (ē″-ō-sin′-ō-filz) account for 1% to 4% of the white blood cell population and are approximately twice the size of RBCs. Their blue-red nucleus usually has two lobes that have the appearance of a figure 8 (see Table 18.1). Large, coarse granules that stain from brick red to crimson with acid dyes pack the cytoplasm; these granules are thought to be elaborate lysosomes.

The exact function of eosinophils is incompletely understood. What is known is that they tend to reside in the intestinal and pulmonary mucosae and in the dermis of the skin, and they increase in number when the body is invaded by parasitic worms or protozoans and during allergy attacks. They are phagocytic, but appear to ingest foreign proteins and immune complexes rather than bacteria. Eosinophils also release chemicals that counteract the effect of certain inflammatory chemicals released during allergic reactions.

Agranulocytes

Agranulocytes lack visible cytoplasmic granules. After being formed in the bone marrow, most agranulocytes migrate to the lymphatic tissues and reproduce further there. Their nuclei are spherical, oval, or kidney-shaped. The agranulocytes include lymphocytes and monocytes.

Lymphocytes. **Lymphocytes** are the second most numerous leukocytes. Although large numbers of lymphocytes exist in the body, only a small proportion of them is found in the bloodstream. In fact, lymphocytes are so called because most are firmly enmeshed in lymphoid tissues, where they play a crucial role in immunity. T lymphocytes function in the immune response by acting directly against virus-infected cells and tumor cells. B lymphocytes give rise to *plasma cells,* which produce antibodies (immunoglobulins) that are released to the blood. (The specifics of B and T lymphocyte function are described in Chapter 22.)

Lymphocyte diameter ranges from 5 to 17 μm, but they are often classified according to size as small (5 to 8 μm), medium (10 to 12 μm), and large (14 to 17 μm). The small lymphocytes predominate in blood and represent the end cell in lymphocyte differentiation. Large lymphocytes are found mainly in lymphoid organs and rarely circulate in the blood. When stained, a typical lymphocyte has a large, dark purple nucleus that occupies most of the cell volume. The nucleus is usually spherical, but may be slightly indented, and is surrounded by a thin rim of pale blue cytoplasm (see Table 18.1 and Figure 18.2).

Monocytes. **Monocytes,** with an average diameter of 18 μm, are the largest of the leukocytes. They have abundant gray-blue cytoplasm and a darkly staining blue-purple nucleus, which is usually U- or kidney-shaped (see Table 18.1 and Figure 18.2). Once in the tissues, monocytes differentiate into highly mobile macrophages with prodigious appetites. The macrophages increase in number and are actively phagocytic in chronic infections such as tuberculosis and they are crucial in the body's defense against viruses and certain intracellular bacterial parasites. As explained in Chapter 22, they are also important in activating lymphocytes to mount the immune response.

Production and Life Span of Leukocytes

Like erythropoiesis, **leukopoiesis** or the production of white blood cells is hormonally stimulated. These hematopoietic hormones, known collectively as **colony-stimulating factors,** or **CSFs,** not only prompt the white blood cell precursors to divide and mature, but also enhance the protective potency of mature leukocytes. Macrophages and T lymphocytes are the most important sources of CSF's, but they are also produced by selected other cell types, such as fibroblasts and endothelial cells. CSF's are named for the leukocyte population they stimulate: Thus far, *macrophage-monocyte CSF (M-CSF), granulocyte-CSF (G-CSF), granulocyte-macrophage CSF (GM-CSF),* and *multi-CSF* or *interleukin 3 (IL-3),* which stimulates colony formation of a mixture of white blood cells, have been identified.

Apparently the genes specifying the CSF's are activated by specific chemical signals in the environment. For example, G-CSF production is turned on in monocytes and macrophages when they are exposed to bacterial endotoxins, and both IL-3 and GM-CSF are released by T lymphocytes which have been activated by binding to a particular antigen. The network of chemical interactions that marshalls up an army of leukocytes to fight off some attack on the body is complex and very much tied into the immune response.

Figure 18.9 shows the pathways of leukocyte differentiation. Notice that beyond the hemocytoblast there is an early major branching of the differentiation pathway that divides the precursor cells of lymphocytes, the *lymphoid stem cells,* from the *myeloid stem cells* that give rise to all other types of leukocytes (as well as the thrombocytes and erythrocytes). As granulocytes begin their specialization, both the stem cell and its nucleus decrease in size. Then the nucleus becomes more dense, and functional proteins accumulate in the cytoplasm. In this case, the myeloid stem cells become *myeloblasts* (mī′-uh-lō-blasts), which in turn develop into *promyelocytes* (prō″-mī′-uh-lō-sīts). Further development occurs along three separate pathways as the granules specific to each granulocyte cell type are formed. Just before granulocytes are discharged from the marrow into the circulation, their nuclei constrict, beginning the process of nuclear-lobe formation. The bone marrow stores mature granulocytes (but not erythrocytes) and usually contains 10 to 20 times more granulocytes than are found in the blood. The normal ratio of granulocytes to erythro-

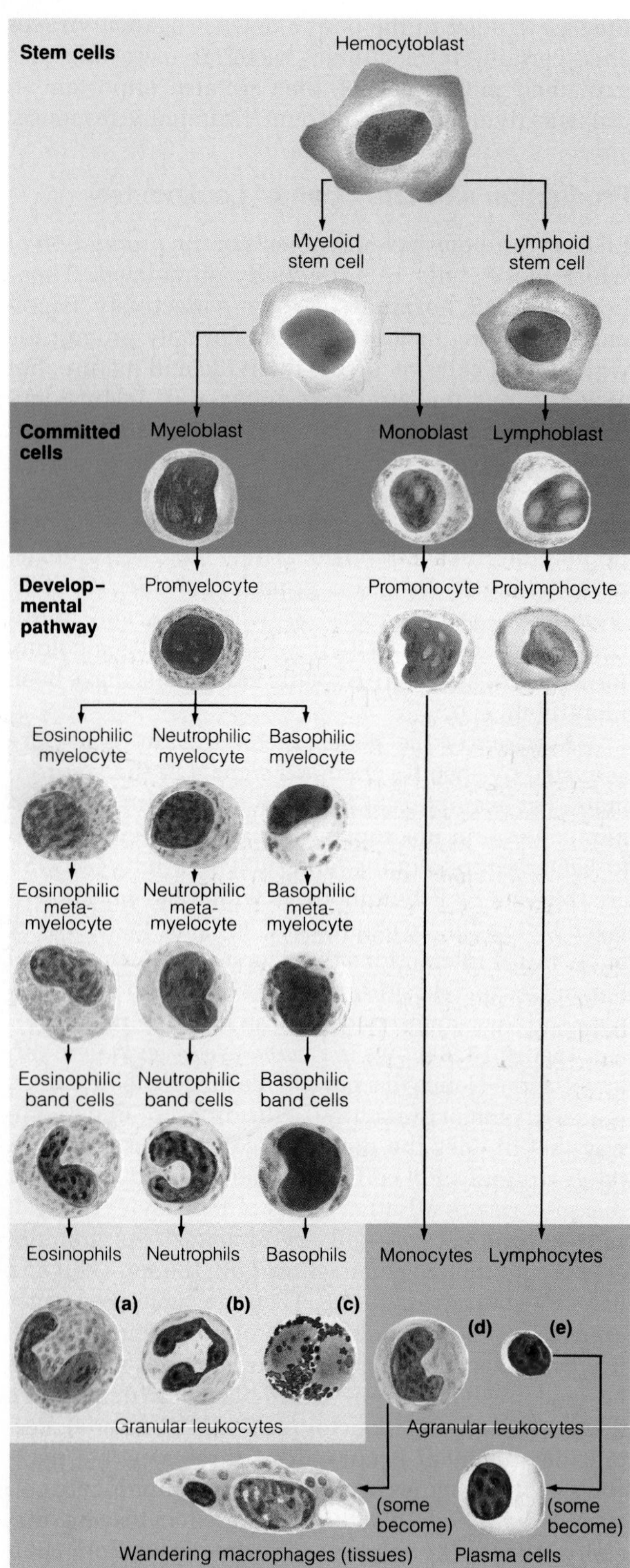

Figure 18.9 Formation of leukocytes. Leukocytes, like all other formed elements, arise from ancestral stem cells called hemocytoblasts. (**a–c**) Granular leukocytes (eosinophils, neutrophils, and basophils) develop via a sequence involving myeloblasts. The developmental pathway is common until the granules typical of each granulocyte begin to be formed. (**d–e**) Agranular leukocytes (monocytes and lymphocytes) develop from monoblasts and lymphoblasts, respectively. However, notice that the monocytes, like granular leukocytes, are progeny of the myeloid stem cell. Only lymphocytes arise via the lymphoid stem cell line. The lymphocytes released from the bone marrow are immature; their further differentiation occurs in the lymphoid organs.

cytes produced is about 3:1, which reflects the much shorter life span (0.5 to 9.0 days) of the granulocytes. Most granulocytes are believed to die combating invading microorganisms.

Agranulocytes arise from hemocytoblasts, but then their lineage diverges. Monocytes have the same lineage as granulocytes since they diverge from common multipotent myeloid stem cells; they then progress through monoblast and promonocyte stages (see Figure 18.9). By contrast, lymphocytes derive from the lymphoid stem cell and progress through the lymphoblast and prolymphocyte stages. The promonocytes and prolymphocytes leave the bone marrow and travel to the lymphoid tissues, where they function in the immune response (and in phagocytosis, in the case of the monocyte macrophages). Monocytes may live for several months, whereas the life span of lymphocytes varies from a few days to several years.

Leukocyte Disorders

Leukocytosis, the controlled rise in numbers of leukocytes that typically follows bacterial or viral attacks on the body, is a normal homeostatic response. In contrast, excessive production of abnormal leukocytes, as occurs in leukemia and infectious mononucleosis, is distinctly pathological. At the opposite pole, *leukopenia* (loo″-kō-pē′-nē-uh) is an abnormally low white blood cell count. It is commonly induced by drugs, particularly glucocorticoids.

Leukemias. The term *leukemia,* literally "white blood," refers to a group of cancerous conditions of white blood cells. As a rule, the abnormal leukocytes are members of a single clone (descendants of a single cell) that (1) fail to respond to normal regulatory mechanisms, (2) tend to remain unspecialized, (3) have an enhanced ability to divide, and (4) suppress or impair normal bone marrow function. The leukemias are named according to the abnormal cell type primarily involved. For example, myelocytic leukemia involves descendants of the myeloblasts, whereas lymphocytic leukemia involves the lymphocytes. Leukemia may be acute or chronic. The more serious acute form primarily affects children; chronic leukemia is seen more often in elderly people. Without therapy, all leukemias are fatal; only the time course differs.

In acute leukemias, the bone marrow becomes almost totally occupied by leukocytes. Because eryth-

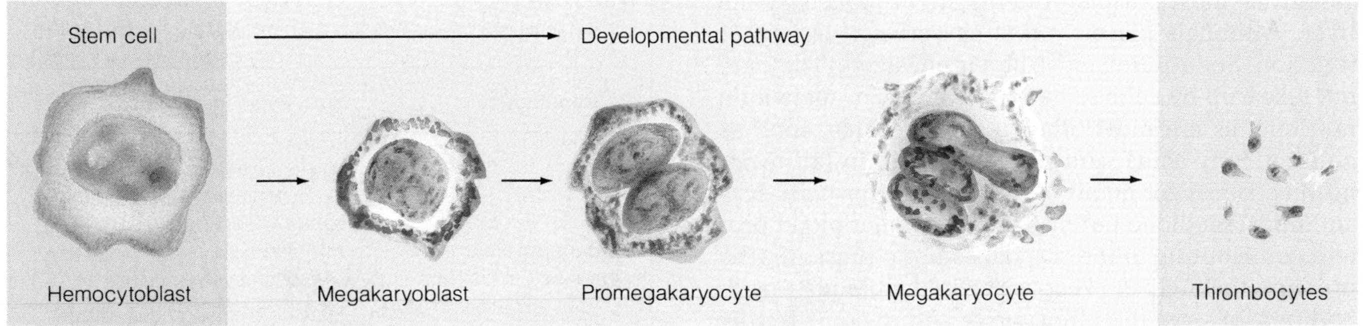

Figure 18.10 Genesis of thrombocytes (platelets). The stem cell (hemocytoblast) gives rise to cells that undergo several mitotic divisions unaccompanied by cytoplasmic division to produce the megakaryocytes. The cytoplasm of the megakaryocyte becomes compartmentalized by membranes, and the plasma membrane then fragments, liberating the thrombocytes.

rocyte- and platelet-generating cells are crowded out, severe anemia, enlargement of the spleen, and bleeding problems result. Other symptoms include fever, weight loss, and bone pain. Although tremendous numbers of leukocytes are produced, they are immature or abnormal and are incapable of defending the body in the usual way. The most common causes of death are internal hemorrhage and devastating infections.

Current treatment involves irradiation and administration of antileukemic drugs to destroy the rapidly dividing cells. These therapies do not cure the disease, but have been successful in inducing remissions (symptom-free periods) lasting from months to several years. Bone marrow transplants are used in selected patients when compatible donors are available.

Infectious Mononucleosis. Once called the kissing disease, *infectious mononucleosis* is a highly contagious viral disease that is most often seen in children and young adults. It is caused by the Epstein–Barr virus. The hallmark of infectious mononucleosis is excessive numbers of monocytes and lymphocytes, many of which are atypical. The affected individual complains of being tired and achy, and has a chronic sore throat and a low-grade fever. There is no cure, but the condition typically runs its course in a few weeks and recovery is usually complete. ■

Thrombocytes

Thrombocytes, or **platelets,** are not cells in the strict sense. They are fragments of large, multinucleate cells called **megakaryocytes** (meh-guh-kayr′-ē-ō-sīts). Megakaryocytes are unique progeny of the hemocytoblast and the myeloid stem cell that give rise to all of the formed elements except the lymphocytes, but their maturation is quite unusual (Figure 18.10). Mitotic cell division occurs several (2 to 15) times, but cytokinesis does not. The final result, the megakaryocyte, is a multinucleate cell with a very large cytoplasmic mass. When the megakaryocyte stage is reached, a network of membranes invaginates into the cytoplasm from the plasma membrane and divides the cytoplasm into 50 or more compartments. Finally, the cell ruptures, releasing the platelet fragments. The plasma membranes associated with each fragment quickly seal around the cytoplasm to form grainy, roughly disk-shaped platelets (see Table 18.1), each with a diameter of 2 to 4 μm. Each cubic millimeter of blood contains between 250,000 and 500,000 of these tiny platelets.

Platelets are essential for the clotting process that occurs in plasma when blood vessels are ruptured or their lining is injured. By clumping together at the damaged site, platelets form a temporary plug that helps seal the break. (This mechanism will be explained shortly.) Because platelets are anucleate, they age quickly in the bloodstream and degenerate in about ten days if they are not involved in clotting. Platelet formation is regulated by a hormone called *thrombopoietin*.

Plasma

When formed elements are removed from the blood, the remaining volume (about 55%) is a straw-colored, sticky fluid called **plasma** (see Figure 18.1). Plasma, which is approximately 90% water, contains over 100 different dissolved solutes, including nutrients, gases, hormones, various wastes and products of cell activity, ions, and proteins (albumin, clotting proteins, and globulins). Table 18.2 provides a summary of the major plasma components.

Plasma proteins, accounting for about 8% by weight of plasma volume, are the most abundant plasma solutes. Except for blood-borne hormones and gamma globulins, most plasma proteins are produced by the liver. Although plasma proteins serve a variety of functions, one point must be emphasized: They are *not* taken up by cells to be used as fuels or metabolic nutrients as are most other plasma solutes, such as glucose, fatty acids, and oxygen. *Albumin* (al′-byoo-min) accounts for about 60% of plasma protein. It is an important blood buffer and is the major blood protein contributing to the plasma osmotic pressure (the pressure that helps to keep water in the bloodstream). Sodium ions are the other major solute contributing to blood osmotic pressure.

The composition of plasma varies continuously as cells remove or add substances to the blood. However, assuming a healthy diet, plasma composition is kept relatively constant by various homeostatic mechanisms. For example, when blood protein levels drop to undesirable levels, the liver is stimulated to make more proteins; and when the blood starts to become too acidic (acidosis), both the respiratory system and the kidneys are called into action to restore the normal, slightly alkaline pH range of plasma. There are literally dozens of adjustments that various body organs make, day in and day out, to maintain the many plasma solutes at life-sustaining levels. In addition to transporting various substances around the body, plasma helps to distribute heat (a by-product of cellular metabolism) throughout the body.

Table 18.2 Composition of Plasma

Constituent	Description/importance
Water	90% of plasma volume; dissolving and suspending medium for solutes of blood; absorbs heat
Solutes	
Proteins	8% (by weight) of plasma volume
• Albumin	60% of plasma proteins; produced by liver; exerts osmotic pressure to maintain water balance between blood and tissues
• Fibrinogen	4% of plasma proteins; produced by liver; acts in blood clotting
• Globulins	36% of plasma proteins
alpha, beta	Produced by liver; transport proteins that bind to lipids and fat-soluble vitamins
gamma	Released by B lymphocytes and plasma cells; equivalent to antibodies released during immune response
• Others	Metabolic enzymes, antibacterial proteins, hormones
Nonprotein and nitrogenous substances	Metabolic by-products of cellular metabolism, such as lactic acid, urea, uric acid, creatinine, and ammonium salts
Nutrients (organic)	Materials absorbed from digestive tract and transported for use throughout body; include glucose and other simple carbohydrates, amino acids (digestion products of proteins), fatty acids, glycerol and triglycerides (fat products), cholesterol, and vitamins
Electrolytes	Cations include sodium, potassium, calcium, magnesium; anions include chloride, phosphate, sulfate, and bicarbonate; help to maintain plasma osmotic pressure and normal blood pH
Respiratory gases	Oxygen and carbon dioxide; some dissolved oxygen (most bound to hemoglobin inside RBCs); carbon dioxide transported bound to hemoglobin in RBCs and as bicarbonate ion dissolved in plasma

Hemostasis

Normally, blood flows smoothly past the intact lining (endothelium) of the blood vessel walls; but if a blood vessel wall breaks, a whole series of reactions is set in motion to accomplish **hemostasis,** or stoppage of blood flow. This response, which is fast, localized, and precisely controlled, involves many substances that are normally present in plasma, as well as some that are released by platelets and injured tissues.

Hemostasis involves three phases that occur in rapid sequence: (1) vascular spasms, (2) platelet plug formation, and (3) coagulation, or blood clotting. Blood loss at the site is permanently prevented when fibrous tissue grows into the clot and seals the hole in the blood vessel.

Vascular Spasms

The immediate response to blood vessel injury is constriction of the blood vessel (vasoconstriction). Direct injury to vascular smooth muscle, compression of the vessel by escaping blood, chemical factors released by platelets, and reflexes initiated by the stimulation of local pain receptors are all thought to play a role in initiating the spasm response. The value of this response is obvious: A strongly constricted artery can significantly reduce blood loss for 20–30 minutes, during which time platelet plug formation and blood clotting can occur.

The spasm mechanism is most efficient when the vessel is crushed or damaged by a blunt instrument. When the vessel is cut cleanly by a sharp object, bleeding is both more profuse and more prolonged because less tissue is damaged and fewer of the stimuli triggering vasoconstriction occur.

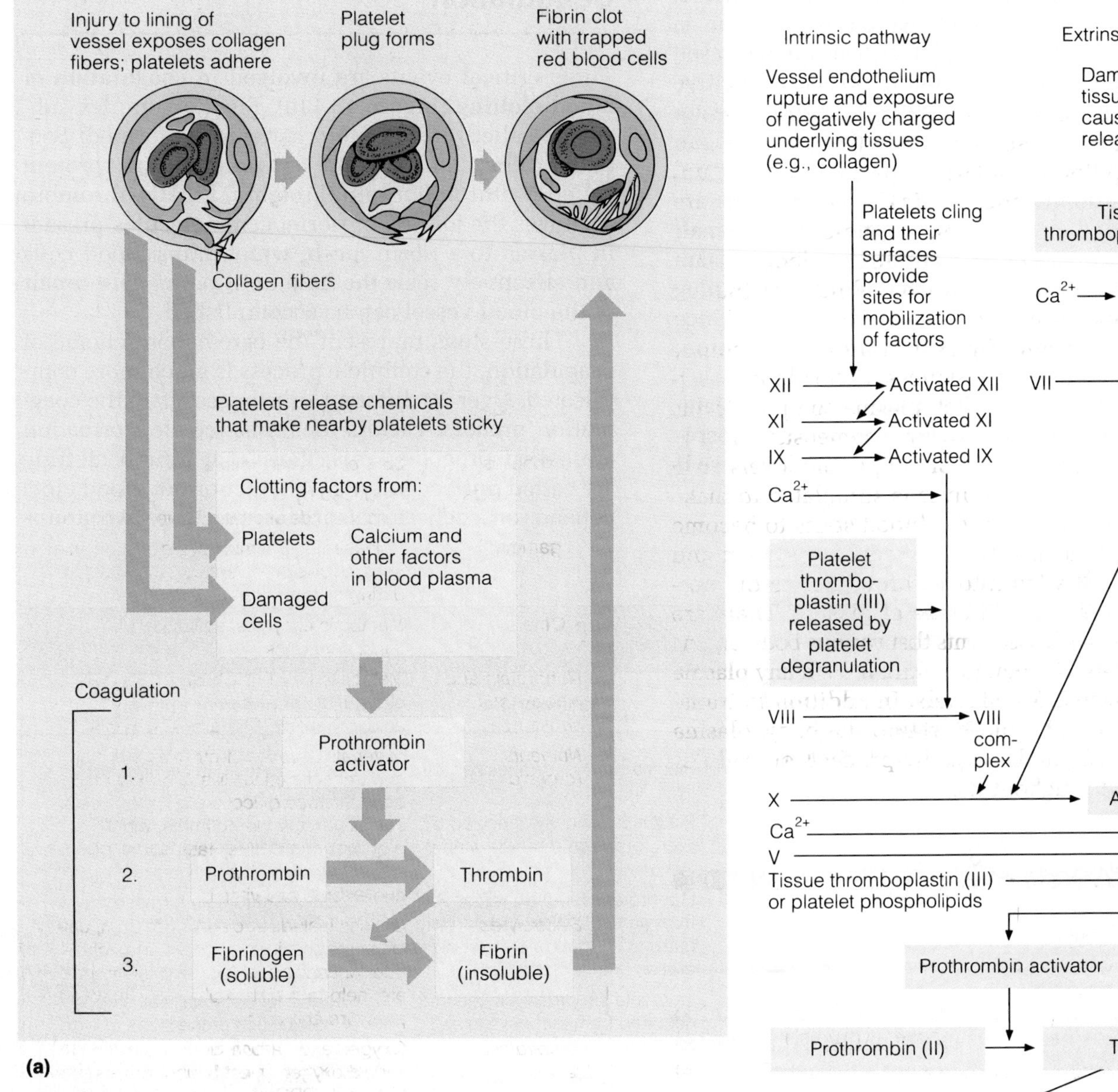

Figure 18.11 Events of platelet plug formation and blood clotting. (**a**) Simplified schematic of events. Steps numbered 1–3 represent the major events of coagulation. (**b**) Detailed flowchart indicating the intermediates and events involved in platelet plug formation and the intrinsic and extrinsic mechanisms of blood clotting (coagulation).

Formation of a Platelet Plug

Platelets play a central role in hemostasis by forming a plug that temporarily seals the break in the vessel wall. They also help to orchestrate subsequent events that lead to blood clot formation. These events are shown in simplified form in Figure 18.11a.

As a rule, platelets do not cling to each other or to the blood vessel endothelium because they are repelled by the positive charge on the endothelial surface. However, when the endothelium is ruptured and the underlying negatively charged collagen fibers are exposed, platelets undergo some remarkable changes. They swell and begin to form numerous spiky processes. They also become sticky and adhere tenaciously to the negatively charged area. Once attached, degranulation occurs; that is, the platelet granules begin to break down and release several chemicals. Some, like *serotonin* and *thromboxane* (throm-bok′-sān) A_2, a short-lived prostaglandin derivative, enhance the vascular spasm. In addition thromboxane A_2 and *adenosine diphosphate (ADP)*, also released from platelet granules, are potent aggregating agents that attract more platelets to the area and cause them to release their contents. Phospholipids are also released during the degranulation process. Thus, a positive feedback cycle of activation and attraction of greater and greater numbers of platelets to the area is begun and, within one minute, a platelet plug is built up, which further reduces blood loss. Platelet plugs are loosely knit, but are quite effective in sealing small tears in a blood vessel. Since small vascular nicks are a frequent consequence of normal activity, platelet plug formation is very important on a day-to-day basis. Once the platelet plug is formed, the next stage, coagulation, is triggered.

Aspirin inhibits prostaglandin release by platelets and platelet plug formation. Excessive aspirin intake leads to prolonged bleeding time because it inhibits clot formation. ■

Coagulation

Three critical events are involved in **coagulation** or **blood clotting** (Figure 18.11a). First, a complex substance called *prothrombin activator* is formed. Second, prothrombin activator converts a plasma protein called *prothrombin* into *thrombin*. Finally, thrombin catalyzes the joining of *fibrinogen* molecules present in plasma to a *fibrin mesh*, which traps blood cells and effectively seals the hole until permanent repair of the blood vessel can be accomplished.

These steps represent the bare-bones scheme of coagulation; the complete process is much more complicated. Over 30 different substances affect the coagulation process. Factors that enhance clot formation are called *procoagulants;* those that inhibit clotting are called *anticoagulants*. Whether or not blood clots depends on a delicate balance between these two groups

Table 18.3 Blood Coagulation Factors (Procoagulants)

Factor number	Factor name	Nature/origin	Function or pathway
I	Fibrinogen	Plasma protein; synthesized by liver	Converted to fibrin, weblike substance of clot
II	Prothrombin	Plasma protein; synthesized by liver; formation requires vitamin K	Converted to thrombin, which enzymatically converts fibrinogen to fibrin
III	Thromboplastin	Lipoprotein complex; released from damaged tissues (extrinsic pathway) or during platelet disintegration (intrinsic pathway)	Catalyzes thrombin formation
IV	Calcium ions	Inorganic ion present in plasma; acquired from diet or released from bone	Needed for essentially all stages of coagulation process
V	Proaccelerin, labile factor, or platelet accelerator	Plasma protein; synthesized in liver; also released by platelets	Both extrinsic and intrinsic mechanisms
VI	Number no longer used; substance now believed to be same as factor V		
VII	Serum prothrombin conversion accelerator (SPCA) or stable factor	Plasma protein; synthesized in liver in process that requires vitamin K	Extrinsic pathway
VIII	Antihemophilic factor	Globulin synthesized in liver; deficiency causes hemophilia A	Intrinsic mechanism
IX	Plasma thromboplastin component (PTC) or Christmas factor	Plasma protein; synthesized in liver; deficiency results in hemophilia B; synthesis requires vitamin K	Intrinsic mechanism
X	Stuart factor or Stuart–Prower factor	Plasma protein; synthesized in liver; synthesis requires vitamin K	Both extrinsic and intrinsic pathways
XI	Plasma thromboplastin antecedent (PTA)	Plasma protein; synthesized in liver; deficiency results in hemophilia C	Intrinsic mechanism
XII	Hageman factor	Plasma protein; proteolytic enzyme; source unknown	Intrinsic mechanism; activates plasmin; known to be activated by contact with glass and may initiate clotting in vitro
XIII	Fibrin stabilizing factor (FSF)	Protein present in plasma and platelets; plasma source unknown	Cross-links fibrin and renders it insoluble

of factors. Normally, anticoagulants dominate and clotting is prevented; but when a vessel is ruptured, procoagulant activity in that area increases dramatically and clot formation begins. Although the procoagulants are numbered I to XIII (Table 18.3), the numerical order does not reflect the reaction sequence, and thromboplastin (III) and Ca^{2+}(IV) are usually indicated by their names, rather than by numerals. Most of these factors are plasma proteins made by the liver.

Clotting may be initiated by either the *intrinsic* or the *extrinsic pathway*, and in the body both pathways are usually triggered by the same vessel-damaging events. However, clotting of blood *in vitro* (outside the body, such as in a test tube) is initiated only by the intrinsic mechanism.

A pivotal molecule in both mechanisms is **thromboplastin.** In the intrinsic pathway, thromboplastin is generated in the bloodstream by platelet degranulation. In such cases, the intermediates of the pathway are activated on the platelet surfaces. If, however, thromboplastin is released by injured cells in the vessel wall or in surrounding tissues (that is, its source is *extrinsic* to the bloodstream), the shorter extrinsic mechanism is triggered. Each pathway requires ionic calcium and involves the activation of a series of procoagulants, each serving to activate the next procoagulant in the sequence. The intermediate steps of each pathway cascade toward a common intermediate, factor X. From this point on, the clotting mechanisms have a common pathway.

The intermediate steps of these complex mechanisms leading to factor X activation are indicated in Figure 18.11b. Once factor X has been activated, four sequential events occur:

1. Factor X complexes with tissue thromboplastin or platelet phospholipids, factor V, and calcium ions to form **prothrombin activator.** This step is usually the slowest step of the blood clotting process, but once accomplished, the clot forms within 10 to 15 seconds.

2. Prothrombin activator acts enzymatically on the plasma protein **prothrombin** to release the active enzyme **thrombin.**

3. Thrombin causes **fibrinogen** (another plasma protein made by the liver) to undergo a remarkable transformation. The altered fibrinogen molecules spontaneously align themselves into long, hairlike, insoluble **fibrin** strands, which glue the platelets together and intertwine to make a web that forms the structural basis of the clot. In the presence of fibrin, plasma becomes gel-like and traps formed elements that pass through it (Figure 18.12). The fibrin mesh forms only in the immediate vicinity of the clumping platelets because that is the site of the thromboplastin release.

Figure 18.12 Scanning electron micrograph of erythrocytes trapped in a fibrin mesh. The more-or-less spherical gray object (top center of photo) is a platelet (15,000×).

4. In the presence of calcium ions, thrombin also activates factor XIII, the cross-linking enzyme that binds the fibrin strands tightly together and strengthens and stabilizes the clot.

Clot formation is normally complete within 3 to 6 minutes after blood vessel damage. Because the extrinsic pathway involves fewer intermediate steps, it is more rapid than the intrinsic pathway, and in cases of severe tissue trauma, the extrinsic mechanism can promote clot formation within 15 seconds.

Clot Retraction and Fibrinolysis

Within 30 to 60 minutes the clot is stabilized further by a platelet-induced process called **clot retraction** or **syneresis** (sih-ner′-eh-sis). Platelets contain a contractile protein complex called *actomyosin*, or *thrombosthenin*, and they contract in much the same manner as muscle cells. As the platelets contract, they create a pull on the surrounding fibrin strands, squeezing *serum* (plasma minus the clotting proteins) from the mass. As a result, the clot is compacted and the ruptured edges of the blood vessel are drawn more closely together. Even as clot retraction is occurring, vessel healing is also taking place. Fibroblasts form a connective tissue patch in the injured area, and endothelial cells multiply to restore the endothelial lining to an intact condition.

A process called **fibrinolysis** disposes of clots when healing has occurred and the clots are no longer needed. Because small clots are formed continually in vessels throughout the body, this cleanup detail is crucial. Without it, blood vessels would gradually become occluded by unnecessary clots.

The critical natural "clot buster" is a protein-digesting enzyme called *plasmin*, which is produced when the blood protein *plasminogen* is activated. Large amounts of plasminogen are incorporated into a forming clot, where it remains inactive until appropriate signals reach it. The presence of a clot in and around the blood vessel causes the endothelium (and, to a lesser extent, adjacent tissue cells) to secrete the activator. Activated factor XII and thrombin released during clotting also serve as plasminogen activators. As a result, most plasmin activity is confined to the clot, and any plasmin that strays into the plasma is quickly destroyed by circulating enzymes. Fibrinolysis begins within two days and continues over several days until the clot is finally dissolved.

Factors Limiting Clot Growth or Formation

Factors Limiting Normal Clot Growth

Once the clotting cascade has begun, it continues until a clot is formed. Normally, two important homeostatic mechanisms prevent clots from becoming unnecessarily enlarged: (1) rapid removal of coagulation factors, and (2) inhibition of activated clotting factors. For clotting to occur in the first place, the concentration of activated procoagulants must reach certain critical levels. Any tendency toward clot formation in rapidly moving blood is usually unsuccessful because the activated clotting factors are diluted, washed away, and prevented from accumulating. For the same reasons, further growth of a forming clot is hindered when it contacts normally flowing blood.

Other mechanisms block the final step in which fibrinogen is polymerized into fibrin by restricting thrombin to the clot or inactivating it if it manages to escape into the general circulation. As a clot forms, almost all of the thrombin produced is adsorbed onto the fibrin threads. Thus, fibrin effectively acts as an anticoagulant to prevent enlargement of the clot and prevents thrombin from acting elsewhere. Thrombin not adsorbed to fibrin is bound by *antithrombin III*, an alpha-globulin protein present in plasma. Antithrombin III inhibits thrombin's enzymatic activity and within 20 to 30 minutes completely inactivates it. Any thrombin reaching the liver is inactivated there.

Heparin, the natural anticoagulant contained in the granules of basophils and mast cells, is ordinarily secreted in small amounts into the plasma. It acts as a thrombin inhibitor by enhancing the activity of antithrombin III. Heparin, like most other clotting inhibitors, also inhibits the intrinsic pathway.

Recent biochemical experiments conducted on smokers have revealed that their blood contains two unique factors that inhibit blood clotting. One factor slows and disturbs the linking together of fibrinogen to form fibrin. The second factor inhibits cross linking of fibrin by preventing activation of factor XIII. Since fibrin cross linking protects the newly formed clot from being destroyed by protein-digesting enzymes always present in the blood, this interference can severely impair clot formation, and may be an important factor in certain disease processes that afflict smokers. ■

Factors Preventing Undesirable Clotting

Factors that normally act to ward off unnecessary clotting include both structural and molecular characteristics of the endothelial lining of blood vessels. As long as the endothelium is intact, platelets are prevented from clinging and piling up. Also, the endothelial surface is positively charged, which actively repels the platelets as they pass by in the bloodstream.

Disorders of Hemostasis

The two major disorders of hemostasis are at opposite poles. *Thromboembolytic disorders* result from conditions that cause undesirable clot formation. *Bleeding disorders* arise from abnormalities that prevent proper clot formation.

Thromboembolytic Conditions

Despite the safeguards that the body possesses to protect itself from abnormal clotting, undesirable intravascular clotting sometimes occurs. A clot that develops and persists in an unbroken blood vessel is called a *thrombus*. If the thrombus is large enough, it may block circulation to the cells beyond the occlusion and lead to the death of those tissues. For example, if the blockage occurs in the coronary circulation of the heart (coronary thrombosis), the consequences may be the death of heart muscle and a fatal heart attack. If the thrombus breaks away from the vessel wall and floats freely in the bloodstream, it becomes an *embolus* (plural, *emboli*). An embolus is usually no problem until it encounters a blood vessel that is too narrow for it to pass through. For example, emboli that become trapped in the lungs (pulmonary emboli) can dangerously impair the ability of the body to obtain oxygen, while a cerebral embolus may cause a stroke. Some new drugs that dissolve blood clots and innovative medical techniques that remove clots are described in the box on pp. 634–635 in Chapter 20.

Persistent roughening of the vessel endothelium, as might result from arteriosclerosis, severe burns, or inflammation, cause thromboembolytic disease by allowing the platelets to gain a foothold or to cling. A slowdown in the flow of blood, or blood stasis, is another risk factor, particularly in immobilized patients. In this case, clotting factors are *not* washed away as usual and accumulate so that clot formation finally becomes possible. Anticoagulants such as heparin and dicumarol are usually prescribed for thrombus-prone patients.

Bleeding Disorders

Anything that interferes with the clotting mechanism can cause abnormal bleeding, but the most common causes are platelet deficiency (thrombocytopenia) and deficits of some of the procoagulants such as might result from impaired liver function or certain genetic conditions.

Thrombocytopenia. *Thrombocytopenia*, a condition in which the number of circulating platelets is reduced, causes spontaneous bleeding from small blood vessels all over the body. Even normal movement leads to widespread hemorrhage, evidenced by many small purplish blotches, called *petechiae* (peh-tē′-kē-ī), on the skin. Thrombocytopenia can arise from any condition that suppresses or destroys the myeloid tissue, such as bone marrow malignancy, exposure to ionizing radiation, or certain drugs. A platelet count of under 50,000 platelets per cubic millimeter of blood is usually diagnostic for this condition. Temporary relief from bleeding is provided by whole blood transfusion.

Impaired Liver Function. When the liver is unable to synthesize its usual supply of procoagulants, abnormal, and often severe, bleeding episodes occur. The resulting problem can range from an easily resolved vitamin K deficiency to nearly total impairment of liver function (as in hepatitis or cirrhosis). Vitamin K is required by the liver cells for production of the clotting factors and, because vitamin K is produced by bacteria that reside in the intestines, dietary deficiencies are rarely a problem. However, vitamin K deficiency can occur if fat absorption is impaired, because vitamin K is a fat-soluble vitamin that is absorbed into the blood along with fats. In liver disease the nonfunctional liver cells fail to produce not only the procoagulants but also bile, which is required for fat and vitamin K absorption.

Hemophilias. The term *hemophilia* is loosely applied to several different hereditary bleeding disorders that exhibit similar signs and symptoms. Hemophilia A, or classical hemophilia, results from a deficiency of factor VIII (antihemophilic factor). This is the most common type of hemophilia and accounts for 83% of cases. Hemophilia B results from a deficiency of factor IX (plasma thromboplastin component). Both types are sex-linked conditions occurring primarily in males.

Symptoms of hemophilia begin early in life; even minor tissue trauma causes prolonged bleeding that can be life threatening. Commonly, the person's joints become seriously disabled and painful because of repeated bleeding into the joint cavities after exercise or trauma.

Regardless of type, hemophilia is managed clinically by transfusions of fresh plasma or injections of the appropriate purified clotting factor. Both therapies provide relief for several days but are expensive and inconvenient. Because hemophiliacs are absolutely dependent on blood transfusions or factor injections, they have become recipients of the hepatitis virus and more recently the AIDS virus, a blood-transmitted virus that severely depresses the immune system. (AIDS is described in Chapter 22.) However, this dilemma may soon be resolved because a genetically engineered source of factor VIII is undergoing clinical trials and is expected to be commercially available soon. ■

Transfusion and Blood Replacement

Transfusion of Whole Blood

The human cardiovascular system is designed to minimize the effects of blood loss by (1) reducing the volume of the blood vessels, which helps to maintain normal circulation, and (2) stepping up the production of red blood cells. However, there is a limit to the amount that the body can compensate for blood loss. Losses of 15% to 30% cause pallor and weakness. A loss of more than 30% of blood volume results in severe shock, which can be fatal. Considering these consequences, it is no small wonder that the development of safe blood transfusion techniques has been one of the most important achievements of modern medicine.

Whole blood transfusions are routinely given to compensate for substantial blood loss and to treat severe anemia or thrombocytopenia. The usual blood bank procedure involves collecting blood from a donor and then mixing it with an anticoagulant, such as certain citrate and oxalate salts, that prevent clotting by binding with calcium ions. The blood can then be stored for several weeks at 4°C until needed. When freshly collected blood is to be transfused almost immediately, heparin (which acts in the same way as the salts) is used as the anticoagulant.

Human Blood Groups

People have different blood types and transfusion of incompatible blood can be fatal. Red blood cell plasma membranes, like those of all body cells, bear highly specific glycoproteins (antigens), which identify each person as unique from all others. One person's erythrocyte proteins will be recognized as foreign if transfused into another person with a different red blood cell type, and the transfused cells will be agglutinated (clumped) and destroyed. Since these RBC antigens promote agglutination, they are more specifically called *agglutinogens* (uh-gloo′-tih-nō-jinz).

There are at least 30 common varieties of naturally occurring red blood cell antigens in humans. Besides these, there are perhaps 100 others that occur in individual families ("private antigens") rather than in the general population. The presence or absence of each antigen allows each person's blood cells to be classified into several different blood groups. Some of the common antigens, such as those determining the ABO and Rh blood groups, cause vigorous transfusion reactions in which the foreign erythrocytes are destroyed when they are improperly transfused. Thus, blood typing for these antigens is always done before blood is transfused. Other antigens (such as the M, N. P, S, Kell, and Lewis factors) are mainly of legal or academic importance, that is, for determining or studying paternity, inheritance, race, and so forth. Because these factors cause weak or no transfusion reactions, blood is not usually typed for them unless the person is expected to need several transfusions, in which case, the many weak transfusion reactions could have cumulative effects. Only the ABO and Rh blood groups are described here.

ABO Blood Groups. As shown in Table 18.4, the **ABO blood groups** are based on the presence or absence of two agglutinogens, type A and type B. Depending on which of these a person inherits, the ABO blood group will be one of the following: A, B, AB, or O. The O blood group, which has neither agglutinogen, is the most common ABO blood group in Caucasian Americans; AB, with both antigens, is the least prevalent. The presence of either the A or the B agglutinogen results in group A or B, respectively.

Unique to the ABO blood groups is the presence in the plasma of *preformed antibodies* called *agglutinins*. The agglutinins act against RBCs carrying ABO antigens that are *not* present on a person's own red blood cells. A newborn has none of these antibodies, but they begin to appear in the plasma within two months. They reach peak levels between 8 and 10 years of age and then slowly decline throughout the rest of one's life. As indicated in Table 18.4, a baby with neither the A or the B antigen (group O) forms both anti-A and anti-B agglutinins. Those with group A blood form anti-B antibodies, while those with group B form anti-A antibodies. Neither antibody is produced by AB infants.

Just why these antibodies develop without exposure to the foreign antigen is a mystery. One explanation is that small amounts of type A or B antigens enter the body in food or bacteria and stimulate formation of antibodies against antigens different from one's own.

Rh Blood Groups. There are at least eight different types of Rh agglutinogens, each of which is called an *Rh factor.* Only three of these, the C, D, and E antigens, are fairly common. The Rh blood typing system is so named because one Rh antigen (agglutinogen D) was originally identified in *Rh*esus monkeys; later the same antigen was discovered in human beings. Most Americans (about 85%) are Rh^+ (Rh positive), meaning that their RBCs carry the Rh antigen.

Unlike the ABO system, anti-Rh antibodies are not spontaneously formed in the blood of Rh^- (Rh negative) individuals. However, if an Rh^- person receives Rh^+ blood, the immune system becomes sensitized and, shortly after the transfusion, begins producing anti-Rh^+ antibodies against the foreign antigen. Hemolysis does not occur after the first such transfusion because it takes time for the body to react and start making antibodies. But the second time, and every time thereafter, a typical transfusion reaction occurs in which the recipient's antibodies attack and rupture the donor RBCs. In most cases, the intensity of these reactions is much less severe than in anti-A or anti-B transfusion reactions.

An important problem related to the Rh factor occurs in pregnant Rh^- women who are carrying Rh^+ babies. The first such pregnancy usually results in the delivery of a healthy baby. But because the mother is sensitized by her baby's Rh^+ antigens that have passed through the placenta into her bloodstream, she will form anti-Rh^+ antibodies unless treated with RhoGAM shortly after she has given birth. (RhoGAM is a serum containing anti-Rh^+ agglutinins. Because it agglutinates the Rh factor, it blocks the mother's immune response and prevents her sensitization.) If the mother is not treated and becomes pregnant again with an Rh^+ baby, her antibodies will cross through the placenta and destroy the baby's RBCs, producing a condition known as hemolytic disease of the newborn. The baby becomes anemic and hypoxic. Brain damage and even death may result unless transfusions are done *before* birth to provide more erythrocytes for oxygen transport. Additionally, one or two exchange

Table 18.4 ABO Blood Groups

Blood group	Frequency (Caucasian population of the U.S.)*	RBC agglutinogens	Plasma agglutinins	Blood that can be received
O	47%	None	Anti-A and Anti-B	O only (universal donor)
A	41%	A	Anti-B	A and O
B	9%	B	Anti-A	B and O
AB	3%	A and B	None	A, B, AB, and O (universal recipient)

*Relative frequency of ABO blood groups differs in different ethnic groups/populations. For example, O, A, B, and AB frequency in North American Indians is 92%, 8%, 1%, 0%, and that in Americans of Chinese ancestry is 30%, 25%, 35%, and 10%, respectively.

transfusions are done after birth. The baby's Rh^+ blood is removed, and Rh^- blood is infused. Within approximately six weeks, the transfused Rh^- erythrocytes have been broken down and replaced with the baby's own Rh^+ cells. ■

Transfusion Reactions: Agglutination and Hemolysis

When mismatched blood is transfused, the donor's red blood cells are avidly attacked by the recipient's plasma agglutinins. (Although the donor's plasma antibodies are also agglutinating the host's RBCs, they are so diluted in the recipient's circulation that this does not usually present a serious problem.) The initial event, agglutination of the foreign red blood cells, leads to clogging of small blood vessels throughout the body. During the next few hours, the clumped red blood cells begin to rupture or are destroyed by phagocytes, and their hemoglobin is released into the bloodstream. (When the transfusion reaction is exceptionally severe, the RBCs are lysed almost immediately.) These events lead to two easily recognized problems: (1) The oxygen-carrying capability of the transfused blood cells is disrupted, and (2) the clumping of red blood cells in small vessels hinders blood flow to tissues beyond those points. Not so apparent, but more devastating, is the consequence of hemoglobin release into the bloodstream. Circulating hemoglobin passes freely into the kidney tubules and, when in high concentration, it precipitates, blocking the kidney tubules and causing renal shutdown. If shut down is complete (acute renal failure), the person may die.

Transfusion reactions can also cause fever, chills, nausea, vomiting, and general toxicity; but in the absence of renal shutdown, these reactions are rarely lethal. Treatment of transfusion reactions is directed toward preventing kidney damage by infusing alkaline fluids to dilute and dissolve the hemoglobin and wash it out of the body. Diuretics, which increase urine output, are also given. ■

As indicated in Table 18.4, transfusion of group O blood rarely triggers a transfusion reaction because neither the A nor the B antigen is present on the RBCs; thus, this blood group is called the *universal donor.* Since group AB plasma is devoid of antibodies to both A and B antigens, group AB people, *universal recipients,* can receive blood transfusions from any of the ABO groups. Details of other successful transfusion possibilities are given in the table.

Blood Typing

The importance of determining the blood group of both the donor and the recipient *before* blood is transfused is glaringly obvious. The general procedure for determining ABO blood type is briefly outlined in Figure 18.13. Because it is critical that blood groups be compatible, cross matching is also done. *Cross matching* involves testing for agglutination of donor RBCs by the recipient's serum, and of the recipient's RBCs by the donor serum. Typing for the Rh factors is done in the same manner as ABO blood typing.

Plasma and Blood Volume Expanders

Frequently, when the patient's blood volume is so low that death from shock may be imminent, there just isn't time to type blood or appropriate whole blood may be unavailable. Such emergency circumstances demand that blood *volume* be replaced immediately to restore circulation.

Plasma can be administered to anyone without concern about a transfusion reaction because the antibodies it contains become harmlessly diluted in the recipient's blood. With the exception of red blood cells, plasma provides a complete and natural blood replacement. When plasma is not available, various colloidal solutions, or plasma expanders, such as *purified human serum albumin, plasminate,* and *dextran* can be infused. All of these have osmotic properties that directly increase the fluid volume of blood. Still another option to increase circulating blood volume quickly is to infuse isotonic salt solutions. *Normal saline* or a *multiple electrolyte solution* that mimicks the electrolyte composition of plasma (for example, *Ringer's solution*) is a common choice.

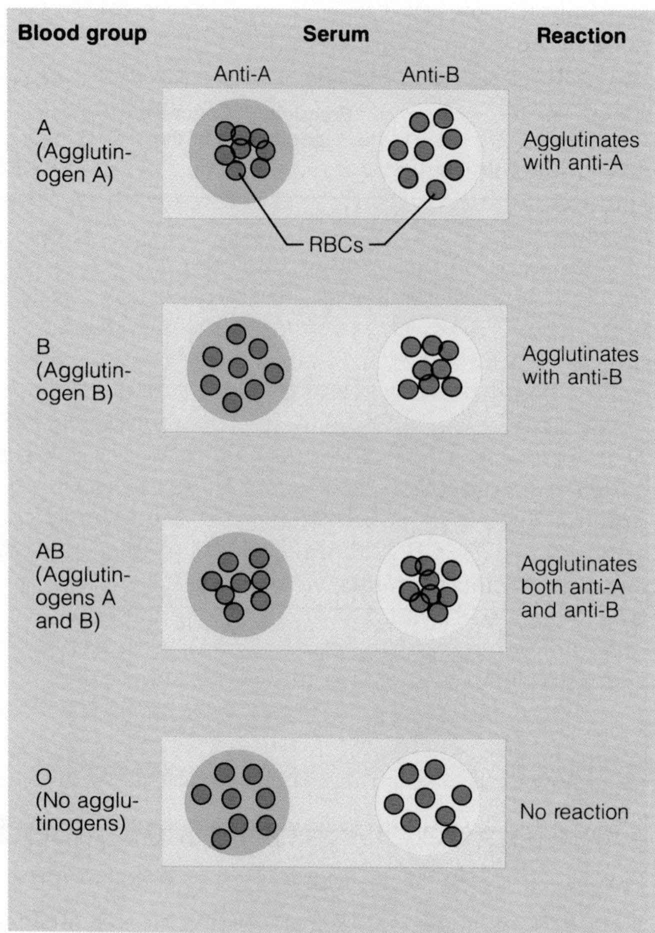

Figure 18.13 Blood typing of ABO blood types. When serum containing anti-A or anti-B agglutinins is added to a blood sample diluted with saline, agglutination will occur between the agglutinin and the corresponding agglutinogen (A or B). As illustrated, agglutination occurs with anti-A serum in blood group A, with anti-B serum in blood group B, with both sera in blood group AB, and with neither serum in blood group O.

Diagnostic Blood Tests

A laboratory examination of blood yields information that can be used to evaluate a person's current state of health. A few examples are given here to illustrate this point. In some anemias, the blood is pale and yields a low hematocrit. A high fat level (*lipidemia*) gives the blood a yellowish hue and forecasts possible problems in those with heart disease. How well a diabetic is controlling diet and blood sugar levels is routinely determined by blood glucose tests. Infections are signaled by leukocytosis and larger than normal buffy coats in the hematocrit. In some types of leukemias, the leukocyte-containing layer may be larger than the erythrocyte layer.

Microscopic studies of blood can reveal variations in the size and shape of erythrocytes that predict iron deficiency or pernicious anemia. Furthermore, a determination of the relative proportion of individual leukocyte types, a *differential white blood cell count,* is a valuable diagnostic tool; for example, a high eosinophil count may indicate an allergic response somewhere in the body.

A number of tests provide information on the status of the hemostasis system. For example, the amount of prothrombin present in blood is assessed by determining the *prothrombin time*, and a *platelet count* is done when thrombocytopenia is suspected.

When one considers all of the indispensable attributes of blood, it is little wonder that a battery of tests called the *complete blood count (CBC)* is routinely ordered during physical examinations and hospital admissions. The CBC includes counts of the different types of formed elements, a hematocrit, tests for clotting factors, and a number of other indicators that provide a comprehensive picture of one's general health status in relation to normal blood values.

Developmental Aspects of Blood

Before birth, there are many sites of blood cell formation—the fetal yolk sac, liver, and spleen, among others—but by the seventh month of development, the red marrow has become the primary hematopoietic area and remains so (barring serious illness) throughout life. In adults, repression of myeloid tissue or widespread destruction of red blood cells can cause inactive yellow bone marrow (essentially, fatty tissue) to be converted to active red marrow. In severe cases, the liver, spleen, and lymph nodes may also resume their fetal blood-forming roles.

Blood cells develop from collections of mesenchymal cells (*blood islands*), derived from the mesoderm germ layer. During fetal life, a unique hemoglobin, hemoglobin F, is formed that has a higher affinity for oxygen than does adult hemoglobin (hemoglobin A). It contains two alpha and two gamma (γ) polypeptide chains per globin molecule, instead of the paired alpha and beta chains typical of hemoglobin A. After birth, fetal erythrocytes carrying hemoglobin F are rapidly destroyed by the liver, and the baby's erythrocytes begin to produce hemoglobin A.

The most common blood-related problems that appear during aging are chronic types of leukemias, anemias, and thromboembolytic disease. However, these and most other age-related blood disorders are usually precipitated by disorders of the heart, blood vessels, or immune system. For example, the increased incidence of leukemias in old age is believed to result from the waning efficiency of the immune system, whereas abnormal thrombus and embolus formation reflects the insidious progress of arteriosclerosis, which roughens the blood vessel walls. Anemias that appear in the elderly are often secondary to nutritional deficiencies, drug therapy, or some disorder such as cancer or leukemia. However, the elderly are particularly at risk for pernicious anemia because the stomach mucosa (which produces intrinsic factor) atrophies with age.

* * *

Because blood serves as the vehicle by which the cardiovascular system transports substances throughout the body, it could be considered the servant of the cardiovascular system. On the other hand, without blood, the normal functions of the heart and blood vessels would be impossible. So perhaps the organs of the cardiovascular system, described in Chapters 19 and 20, are subservient to blood. The point of this circular thinking is that blood and the cardiovascular system organs are vitally intertwined in their common functions—to see that nutrients, oxygen, and other vital substances reach all tissue cells of the body and to relieve the cells of their wastes.

Related Clinical Terms

Reticulocyte (reh-tih'kyoo-lō-sīt) **count** Count of reticulocytes in a blood sample. Provides a rough estimate of the rate of erythropoiesis.

Hemochromatosis (hē″-mō-krō″-muh-tō'sis) A disorder of iron overload.

Plasmapheresis (plaz″-muh-feh'-rē-sis) Chemical filtering technique used to cleanse plasma of specific blood components such as proteins or toxins. Most important application appears to be for removal of antibodies or immune complexes from the blood of individuals with autoimmune disorders (multiple sclerosis, myasthenia gravis, and others).

Embolus (*em* = in; *bolus* = mass) Any abnormal mass carried freely in the bloodstream. May be a blood clot, bubbles of air, fat masses, clumps of tissue cells, or aggregates of debris.

Disseminated intravascular clotting (DIC) Infrequent, and often fatal, condition in which the clotting mechanism is activated throughout the circulation. Clots are produced in such tremendous numbers that small peripheral blood vessels become plugged. Ultimately, hemorrhage occurs because clotting factors have been used up in the initial widespread clotting reaction. Occurs primarily in patients in circulatory shock or with retention of a dead fetus.

Chapter Summary

OVERVIEW: COMPOSITION AND FUNCTIONS OF BLOOD (pp. 569–570)

Components (p. 569)

1. Blood is composed of formed elements and plasma.

Physical Characteristics and Volume (pp. 569–570)

2. Blood is a viscous, slightly alkaline fluid representing about 8% of total body weight. Blood volume of a normal adult is about 5 L.

Functions (p. 570)

3. Distribution functions are delivery of oxygen and nutrients to body tissues, removal of metabolic wastes, transport of hormones, and maintenance of body temperature.

4. Protective functions are maintenance of a constant blood pH and adequate fluid volume, hemostasis, and prevention of infection.

FORMED ELEMENTS (pp. 570–581)

1. Formed elements, accounting for 45% of whole blood, are erythrocytes, leukocytes, and thrombocytes. All formed elements arise from hemocytoblasts in red bone marrow.

Erythrocytes (pp. 570–576)

2. Erythrocytes (red blood cells) are small, biconcave cells containing large amounts of hemoglobin. They have no nucleus and few organelles. Spectrin allows the cells to change shape as they pass through tiny capillaries.

3. Oxygen transport is the major function of erythrocytes. In the lungs, oxygen binds to iron atoms in hemoglobin molecules, producing oxyhemoglobin. In body tissues, oxygen dissociates from iron, producing deoxyhemoglobin.

4. Red blood cells begin as hemocytoblasts and, through erythropoiesis, proceed from the proerythroblast (committed cell) stage to the erythroblast (early and late), normoblast, and reticulocyte stages. During this process, hemoglobin accumulates and the organelles and nucleus are extruded. Differentiation of reticulocytes is completed in the bloodstream.

5. Erythropoietin and testosterone enhance erythropoiesis.

6. Iron, vitamin B_{12}, and folic acid are essential for production of hemoglobin.

7. Red blood cells have a life span of approximately 120 days. Old and damaged erythrocytes are removed from the circulation by phagocytes of liver and spleen. Released iron from hemoglobin is stored as ferritin or hemosiderin to be reused.

8. Erythrocyte disorders include anemias and polycythemia.

Leukocytes (pp. 577–581)

9. Leukocytes are white blood cells. All are nucleated, and all have crucial roles in defending against disease. Two main categories exist: granulocytes and agranulocytes.

10. Granulocytes include neutrophils, basophils, and eosinophils. Basophils contain heparin and histamine, which enhance migration of leukocytes to sites of inflammation. Neutrophils are active phagocytes. The function of eosinophils is unclear, although their numbers increase during allergic reactions.

11. Agranulocytes are lymphocytes, which have crucial roles in immunity, and monocytes, which differentiate into macrophages.

12. Leukopoiesis is directed by colony-stimulating factors released mainly by lymphocytes.

13. Leukocyte disorders include leukemias and infectious mononucleosis.

Thrombocytes (p. 581)

14. Thrombocytes, or platelets, are fragments of large, multinucleate megakaryocytes formed in red marrow. When a blood vessel is damaged, platelets form a plug to help prevent blood loss.

PLASMA (pp. 581–582)

1. Plasma is a straw-colored, viscous fluid composed of 90% water. The remaining 10% is solutes, such as nutrients, respiratory gases, salts, hormones, and proteins. Plasma makes up 55% of whole blood.

2. Plasma proteins, most made by the liver, include albumin, globulins, and clotting proteins. Albumin is an important blood buffer and contributes to the osmotic pressure of blood.

HEMOSTASIS (pp. 582–587)

1. Hemostasis is prevention of blood loss. The three major phases of hemostasis are **vascular spasms, platelet plug formation,** and **blood coagulation.**

Coagulation (pp. 584–585)

2. Coagulation of blood may be initiated by either the intrinsic or the extrinsic pathway. Thromboplastin is common to both pathways. Thromboplastin is released by platelets in the intrinsic pathway; thromboplastin generated by tissue cell injury triggers the extrinsic pathway. A series of activated procoagulants are formed in the intermediate steps of each cascade.

Clot Retraction and Fibrinolysis (pp. 585–586)

3. After a clot is formed, clot retraction occurs. Serum is squeezed out and the ruptured vessel edges are drawn together. The vessel is repaired by connective tissue and endothelial cell proliferation and migration. When healing is complete, clot digestion (fibrinolysis) occurs.

Factors Limiting Clot Growth or Formation (p. 586)

4. Abnormal expansion of clots is prevented by removal of coagulation factors in contact with rapidly flowing blood and by inhibition of activated blood factors.

Disorders of Hemostasis (pp. 586–587)

5. Thromboembolytic disorders involve undesirable clot formation, which can occlude vessels.

6. Thrombocytopenia, a deficit of platelets, causes spontaneous bleeding from small blood vessels. Hemophilia is caused by a genetic deficiency of certain coagulation factors. Liver disease can also cause bleeding disorders because many coagulation proteins are formed by the liver.

TRANSFUSION AND BLOOD REPLACEMENT (pp. 587–590)

Transfusion of Whole Blood (pp. 587–590)

1. Whole blood transfusions are given to replace severe blood loss and to treat anemia or thrombocytopenia.

2. Blood group is based on the presence of agglutinogens (antigens) present on red blood cell membranes.

3. When mismatched blood is transfused, the recipient's agglutinins (plasma antibodies) clump the foreign RBCs; the clumped cells are then lysed. Blood vessels may be blocked by clumped RBCs; released hemoglobin may precipitate in the kidney tubules, causing renal shutdown.

4. Before whole blood can be transfused, it must be typed and cross matched so that transfusion reactions are avoided. The most important blood groups for which blood must be typed are the ABO and Rh groups.

Plasma and Blood Volume Expanders (p. 590)

5. Transfusions of plasma alone or plasma expanders are given when rapid replacement of blood volume is necessary.

DIAGNOSTIC BLOOD TESTS (pp. 590–591)

1. Diagnostic blood tests can provide large amounts of information about the current status of the blood and of the body as a whole.

DEVELOPMENTAL ASPECTS OF BLOOD (p. 591)

1. Fetal hematopoietic sites include the yolk sac, liver, and spleen. By the seventh month of development, the red bone marrow is the primary blood-forming site.

2. Blood cells develop from blood islands derived from mesoderm. Fetal blood contains hemoglobin F. After birth, hemoglobin A is formed.

3. The major blood-related problems associated with aging are leukemia, anemia, and thromboembolytic disease.

Review Questions

Multiple Choice/Matching

1. The blood volume in an adult averages approximately (a) 1 L, (b) 3 L, (c) 5 L, (d) 7 L.

2. The hormonal stimulus that prompts red blood cell formation is (a) serotonin, (b) heparin, (c) erythropoietin, (d) thrombopoietin.

3. All of the following are true of RBCs except (a) biconcave disk shape, (b) life span of approximately 120 days, (c) contain hemoglobin, (d) contain nuclei.

4. Most numerous WBC is the (a) eosinophil, (b) neutrophil, (c) monocyte, (d) lymphocyte.

5. Blood proteins play an important part in (a) blood clotting, (b) immunity, (c) maintenance of blood volume, (d) all of the above.

6. The white blood cell that releases histamine and heparin is the (a) basophil, (b) neutrophil, (c) monocyte, (d) eosinophil.

7. The blood cell that is said to be immunologically competent is the (a) lymphocyte, (b) megakaryocyte, (c) neutrophil, (d) basophil.

8. The normal erythrocyte count (per cubic millimeter) for adults is (a) 3 to 4 million, (b) 4.5 to 5 million, (c) 8 million, (d) 500,000.

9. The normal pH of the blood is about (a) 8.4, (b) 7.8, (c) 7.4, (d) 4.7.

10. Suppose your blood was found to be AB positive. This means that (a) agglutinogens A and B are present on your red blood cells, (b) there are no anti-A or anti-B agglutinins in your plasma, (c) your blood is Rh^+, (d) all of the above.

Short Answer Essay Questions

11. (a) Define formed elements and list their three major categories. (b) Which is least numerous? (c) Which comprises the buffy coat in a hematocrit tube?

12. Discuss hemoglobin relative to its chemical structure, its function, and the color changes it undergoes during loading and unloading of oxygen.

13. If you had a high hematocrit, would you expect your hemoglobin determination to be low or high? Why?

14. What nutrients are needed for erythropoiesis?

15. (a) Describe the process of erythropoiesis. (b) What name is given to the immature cell type released to the circulation? (c) How does it differ from a mature erythrocyte?

16. Besides the ability to move by amoeboid action, what other physiological attributes contribute to the function of white blood cells in the body?

17. (a) If you had a severe infection, would you expect your WBC count to be closest to 5000, 10,000, or 15,000 per cubic millimeter? (b) What is this condition called?

18. (a) Describe the appearance of platelets and state their major function. (b) Why should platelets not be called "cells"?

19. (a) Define hemostasis. (b) List the three major steps of coagulation. Explain what initiates each step and what the step accomplishes. (c) In what general way do the intrinsic and extrinsic mechanisms of clotting differ? (d) What ion is essential to virtually all stages of coagulation?

20. (a) Define fibrinolysis. (b) What is the importance of this process?

21. (a) How is clot overgrowth usually prevented? (b) List two conditions that may lead to unnecessary (and undesirable) clot formation.

22. How can liver dysfunction cause bleeding disorders?

23. (a) What is a transfusion reaction and why does it happen? (b) What are its possible consequences?

24. How can poor nutrition lead to anemia?

25. What blood-related problems are most common in the aged?

Clinical Application Questions

26. A young woman with severe vaginal bleeding is admitted to the emergency room. She is three months pregnant, and the physician is concerned about the volume of blood she is losing. (a) What type of transfusion will probably be given to this patient? (b) What blood tests will be performed before starting the transfusion?

27. A middle-aged college professor from Boston is in the Swiss Alps studying astronomy during his sabbatical leave. He has been there for two days and plans to stay for the entire year. However, he notices that he is short of breath when he walks up steps and that he tires easily with any physical activity. His symptoms gradually disappear, and after two months, he feels fine. Upon returning to the United States, he has a complete physical exam and is told that his erythrocyte count is higher than normal. (a) Attempt to explain this finding. (b) Will his RBC count remain at this higher-than-normal level? Why or why not?

28. A young child is diagnosed as having acute lymphocytic leukemia. Her parents cannot understand why infection is a major problem for Janie when her WBC count is so high. Could you provide an explanation for Janie's parents?

CHAPTER 19

The Cardiovascular System: The Heart

Chapter Outline and Student Objectives

Heart Anatomy (pp. 595–604)

1. Describe the size and shape of the heart, and indicate its location in the thorax.
2. Name the coverings of the heart.
3. Describe the structure and function of each of the three layers of the heart wall.
4. Describe the structure and functions of the four heart chambers. Name each chamber and provide the name and general route of its associated great vessel(s).
5. Trace the pathway of blood through the heart.
6. Name the heart valves and describe their location, function, and mechanism of operation.
7. Name the major branches of the coronary arteries and describe their distribution.

Heart Physiology (pp. 604–617)

8. Describe the structural and functional properties of cardiac muscle, and explain how it differs from skeletal muscle.
9. Briefly describe the events of cardiac muscle cell contraction.
10. Name the components of the conduction system of the heart, and trace the conduction pathway.
11. Draw a diagram of a normal electrocardiogram tracing; name the individual waves and intervals. Name some of the abnormalities that can be detected on an ECG tracing.
12. Describe the timing and events of the cardiac cycle.
13. Describe normal heart sounds, and explain how heart murmurs differ from normal sounds.
14. Name and explain the various factors involved in regulation of stroke volume and heart rate.
15. Explain the role of the autonomic nervous system in regulating cardiac output.

Developmental Aspects of the Heart (pp. 617–618)

16. Describe the formation of the fetal heart, and indicate how the fetal heart differs from the adult heart.
17. Provide examples of age-related changes in heart function.

Preview of Selected Key Terms

Pericardium (payr″-rih-kar′-dē-um) (*peri* = around; *cardi* = the heart) Double-layered serosa enclosing the heart and forming its superficial layer.

Myocardium (mī″-ō-kar′-dē-um) (*myo* = muscle) Layer of the heart wall composed of cardiac muscle.

Atria (ā′-trē-uh) (plural of *atrium* = vestibule) Paired, superiorly located heart chambers that receive blood returning to the heart from the circulations.

Ventricles (*ventr* = underside) Paired, inferiorly located heart chambers that function as the major blood pumps.

Intercalated discs (in-ter′-kuh-lā-tid) (*intercala* = insert) Gap junctions connecting muscle cells of the myocardium.

Pacemaker Region of specialized cardiac tissue (the SA node) that has the fastest rate of spontaneous depolarization, and hence sets the rate of contraction for the heart as a whole.

Electrocardiogram (*gram* = recording) Graphic record of the electrical activity of the heart.

Systole (sis′-tō-lē) (*systol* = contraction) Period when either the ventricles or the atria are contracting.

Diastole (dī-as′-tuh-lē) (*diastol* = dilation) Period when either the ventricles or the atria are relaxing.

Cardiac cycle Sequence of events encompassing one complete contraction and relaxation of the atria and ventricles of the heart.

Cardiac output Amount of blood pumped out of a ventricle in one minute.

Stroke volume Amount of blood pumped out of a ventricle during one contraction.

The ceaselessly beating heart within the chest has intrigued people for centuries. The ancient Greeks believed that the heart was the seat of intelligence; others have thought that it was the seat of the emotions. Popular songs and expressions like "heart throb" and "heart's delight" attest to the belief—and scientific fact—that heart rate is affected by the emotions. When your heart pounds or occasionally skips a beat, you become acutely aware of how much you depend on this single dynamic organ for your very life.

Despite its vital importance, the heart is not an organ working in isolation; it is part of the cardiovascular system, which also includes the blood vessels of the body. All day and all night, tissue cells take in nutrients and oxygen and excrete wastes. Because cells can make such exchanges only with their immediate environment, some means of changing and renewing that environment is necessary to ensure a continual supply of nutrients and to prevent pollution from the buildup of ejected wastes. The cardiovascular system provides the transport system "hardware" that keeps blood in continuous circulation to fulfill this critical homeostatic need.

When stripped of its romantic cloak, the heart is no more than the transport system pump; the delivery routes are the hollow blood vessels. Using blood as the transport medium, the heart continually propels oxygen, nutrients, wastes, and many other substances into the interconnecting blood vessels that move to and past the body cells.

Blood and blood vessels are considered separately in Chapters 18 and 20, respectively. In this chapter we focus on the structure and function of the heart.

Heart Anatomy

Size and Location

The relative size and weight of the heart belie its incredible strength and fortitude. Approximately the size of a person's fist, the hollow, cone-shaped heart is enough like the popular valentine image to satisfy the sentimentalists among us (Figure 19.1). Typically, the heart weighs less than a pound (between 250 and 350 grams).

Midsternal line
2nd rib
Diaphragm
Point of maximal intensity (PMI)
(a)
Right lung
Heart
(b)
Anterior
Superior vena cava
Left lung
Aorta
Parietal pleura (cut)
Pulmonary trunk
Parietal pericardium (cut)
Apex of heart
Diaphragm
(c)

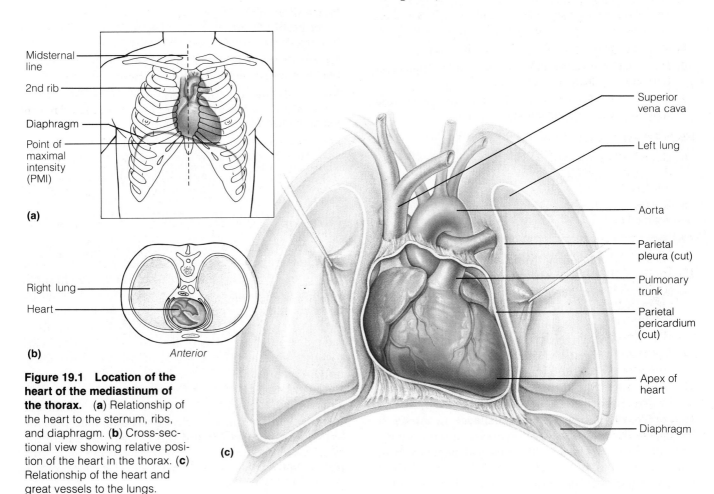

Figure 19.1 Location of the heart of the mediastinum of the thorax. (**a**) Relationship of the heart to the sternum, ribs, and diaphragm. (**b**) Cross-sectional view showing relative position of the heart in the thorax. (**c**) Relationship of the heart and great vessels to the lungs.

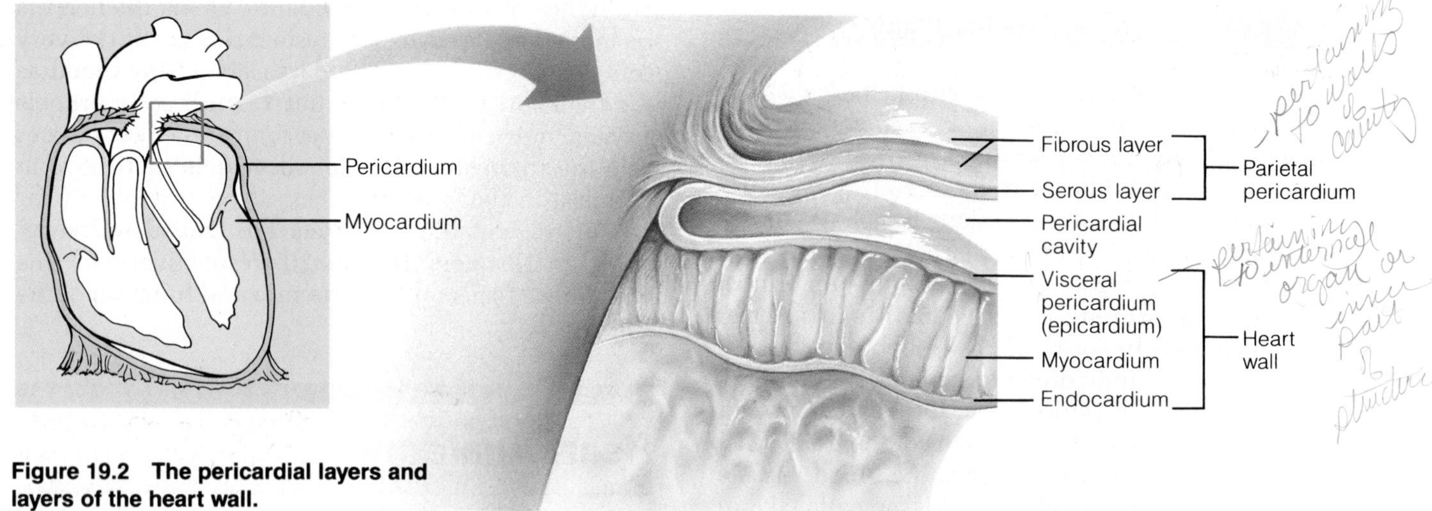

Figure 19.2 The pericardial layers and layers of the heart wall.

The heart, snugly enclosed within the mediastinum (medial cavity of the thorax, extends for 12 to 14 cm (about 5 inches) from the second rib to the fifth intercostal space distally (Figure 19.1a). The heart lies anterior to the vertebral column and posterior to the sternum; it is flanked laterally, and partially obscured, by the overlapping lungs. Because the heart tips slightly to the left, it assumes an oblique position in the thorax. Approximately two-thirds of its mass can be seen to the left of the midsternal line; the balance projects to the right. Its broad superior portion is about 9 cm wide and is directed toward the right shoulder. Its more pointed *apex* is directed inferiorly toward the left hip and rests on the diaphragm. If you press your fingers between the fifth and sixth ribs (just below the left nipple, where the heart apex comes into contact with the chest wall), you can easily feel the beating of your heart; hence, this site is referred to as the *point of maximal intensity (PMI).*

Coverings of the Heart

The heart is enclosed within a double sac of serous membrane called the pericardium (Figure 19.2). The loosely fitting outer **parietal pericardium** (puh-rī′-eh-tul payr″-rih-kar′-dē-um) has two layers: a fibrous and a serous layer. The *fibrous layer* is composed of tough, white fibrous connective tissue that protects the heart and anchors it to surrounding structures such as the diaphragm, sternum, and the great vessels issuing from the heart base. The thin, smooth *serous layer,* composed of squamous epithelial cells underlain by sparse connective tissue, lines the fibrous layer. At the heart base, the serous layer turns downward and continues over the heart surface as the **visceral pericardium,** which is also called the **epicardium.** The epicardium is an integral part of the heart wall; it is often infiltrated with fat, particularly in older people.

Between the epicardium and the serous layer of the parietal pericardium is a potential space, the *pericardial cavity,* which contains a small amount of thin serous fluid produced by the pericardial cells. The serous membranes, lubricated by the fluid, glide smoothly against one another during heart activity and allow the mobile heart to operate in a relatively friction-free environment.

Inflammation of the pericardium, *pericarditis,* interferes with this well-oiled machinery by hindering the production of serous fluid. Pericarditis may lead to painful adhesions in which the pericardia stick together and impede heart activity. ■

Heart Wall

The heart wall is composed of three layers: the outermost epicardium (described above), the middle myocardium, and the innermost endocardium (Figure 19.2). All three layers are richly supplied with blood and lymphatic vessels.

The **myocardium,** composed mainly of cardiac muscle, forms the bulk of the heart and is the layer that actually contracts. Within this layer, the branching cardiac muscle cells are tethered to each other by crisscrossing connective tissue fibers and arranged in spiral or circular bundles (Figure 19.3). These interlacing bundles effectively link all parts of the heart together, and the connective tissue fibers form a dense network, called the *fibrous skeleton* or the *skeleton of the heart,* that reinforces the myocardium internally. This network is thicker in some areas than others and forms ropelike rings of fibrous tissue that provide additional support around the valves and where the great vessels issue from the heart (see also Figure 19.7 on p. 601).

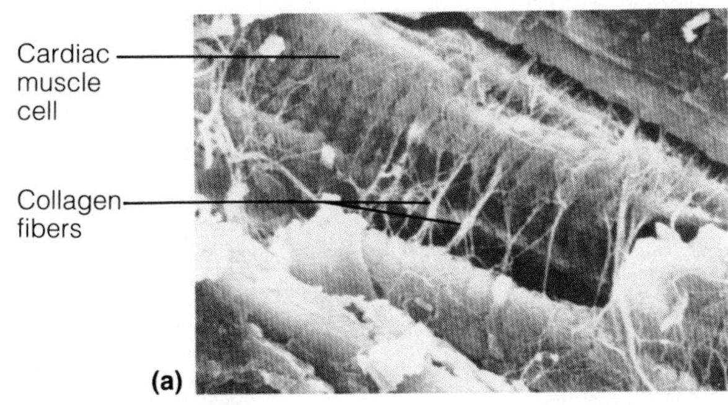

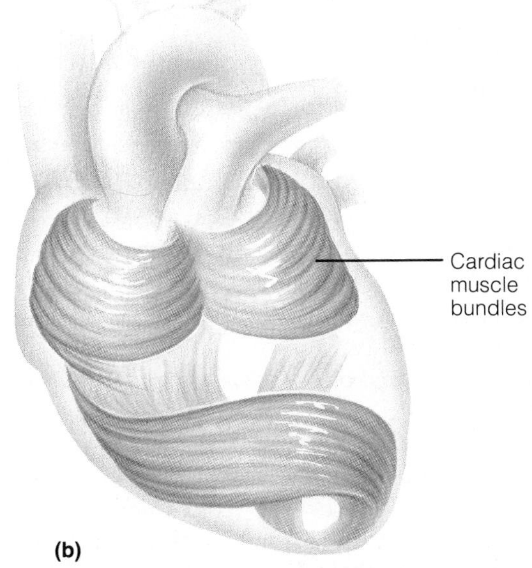

Figure 19.3 Packaging of cardiac muscle in the heart. (**a**) Scanning electron micrograph showing the thin struts of collagen (and elastic) fibers linking the lateral surfaces of cardiac muscle fibers together (1700X). (**b**) Longitudinal view of the heart showing the spiral and circular arrangement of cardiac muscle bundles.

The **endocardium** is a glistening white sheet of endothelium resting on a thin connective tissue layer. Located on the inner myocardial surface, it lines the heart chambers and covers the connective tissue skeleton of the valves. The endocardium is continuous with the endothelial linings of the blood vessels leaving and entering the heart.

Chambers and Associated Great Vessels

The heart has four chambers, or cavities: two superior **atria,** and two inferior **ventricles** (Figure 19.4). The partition that divides the heart longitudinally is called either the **interatrial septum** or the **interventricular septum,** depending on which chambers it separates. The right ventricle forms most of the anterior surface of the heart; the left ventricle forms the heart apex. Two grooves visible on the anterior heart surface indicate the limits of its chambers and carry the blood vessels that supply the myocardium. The *atrioventricular grooves* are at the junction of the atrium and ventricle on each side; the *anterior interventricular sulcus* (sul′-kus) marks the anterior position of the interventricular septum separating the right and left ventricles. The *posterior interventricular sulcus* provides a similar landmark on the posterior aspect of the heart.

Functionally, the atria are receiving chambers for blood returning to the heart from the circulation. Because they need contract only minimally to push blood "next door" into the ventricles, the atria are relatively small, thin-walled chambers. As a rule, they contribute little to the propulsive pumping activity of the heart. Protruding externally from each atrial chamber is a flaplike, wrinkled appendage called an *auricle* (*auris* = ear), which increases the atrial volume somewhat. Except for their posterior surfaces, which are quite smooth, the internal atrial walls have ridges of muscle tissue. Apparently, some scientist thought the walls looked as if they had been raked by the tines of a comb, so he called these muscle bundles *pectinate* (*pectin* = comb) *muscles.* The interatrial septum bears a shallow depression, the *fossa ovalis* (fah′-suh ō-va′-lis), which marks the spot where an opening, the *foramen ovale* (for-a′-min ō-va′lē), existed in the fetal heart (see Figure 19.19, p. 617).

Blood enters the right atrium via three veins: (1) The *superior vena cava* returns blood from body regions superior to the heart; (2) the *inferior vena cava* returns blood from body areas below the heart; and (3) the *coronary sinus* collects blood draining from the myocardium itself. Four *pulmonary veins* enter the left atrium. The pulmonary veins transport blood from the lungs back to the heart. These vessels are best seen in a posterior view of the heart (Figure 19.4d).

The ventricles are the discharging chambers or actual pumps of the heart; and when they contract, blood is propelled out of the heart into the circulation. Most discussions concerning the work of the heart refer to ventricular activity. This difference in function between the atria and the ventricles is reflected in the much more massive ventricular walls.

The right ventricle pumps blood into the *pulmonary trunk,* which routes the blood to the lungs, where gas exchange occurs. The left ventricle ejects blood into the *aorta* (ā-or′-tuh), the largest artery in the body, whose branches ultimately deliver blood to all body organs.

Marking the inner walls of the ventricular chambers are collections of irregular muscle folds called *trabeculae carnae* (truh-beh′-kyoo-lē kar′-nē). Some of these muscle bundles, the *papillary muscles,* are projecting and stalklike; as described shortly, the papillary muscles play a role in valvular function.

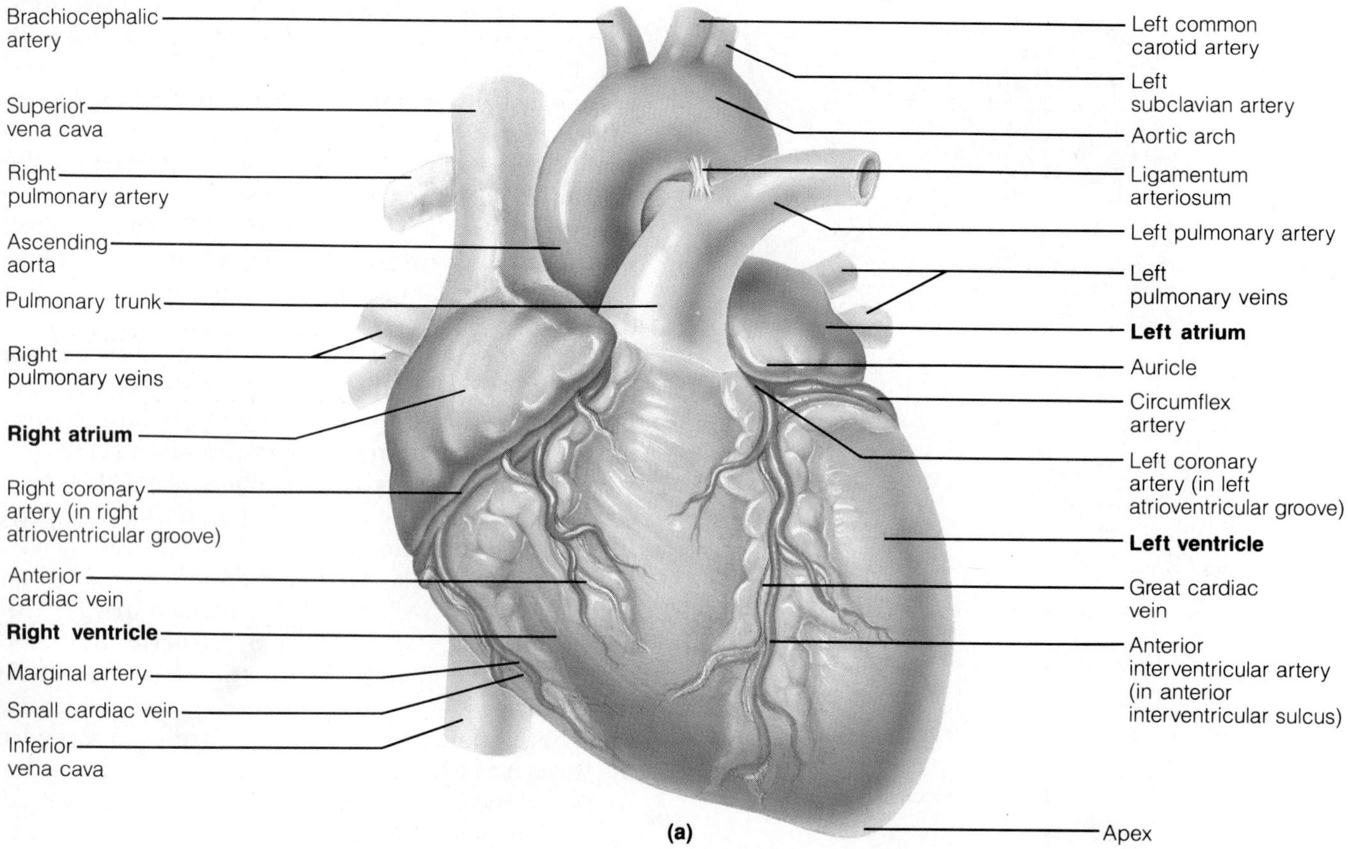

Brachiocephalic artery
Superior vena cava
Right pulmonary artery
Ascending aorta
Pulmonary trunk
Right pulmonary veins
Right atrium
Right coronary artery (in right atrioventricular groove)
Anterior cardiac vein
Right ventricle
Marginal artery
Small cardiac vein
Inferior vena cava
Left common carotid artery
Left subclavian artery
Aortic arch
Ligamentum arteriosum
Left pulmonary artery
Left pulmonary veins
Left atrium
Auricle
Circumflex artery
Left coronary artery (in left atrioventricular groove)
Left ventricle
Great cardiac vein
Anterior interventricular artery (in anterior interventricular sulcus)
Apex
(a)

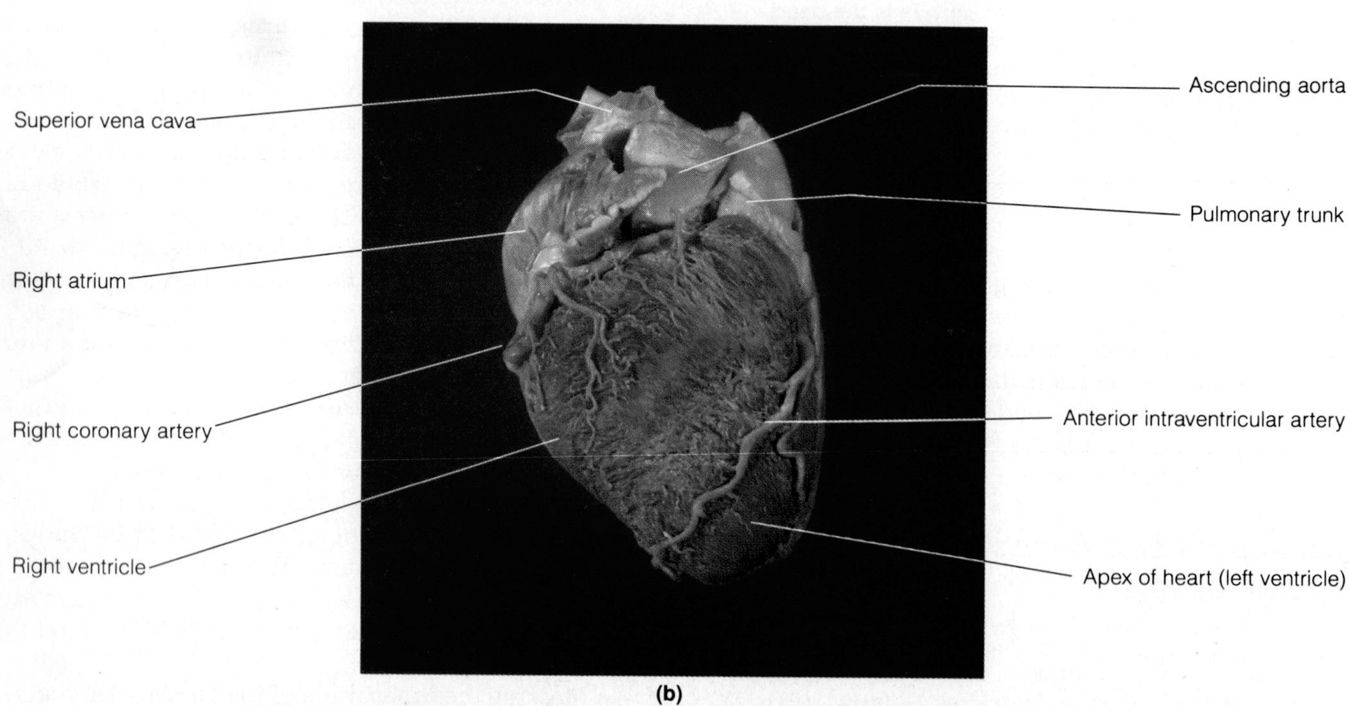

Superior vena cava
Right atrium
Right coronary artery
Right ventricle
Ascending aorta
Pulmonary trunk
Anterior intraventricular artery
Apex of heart (left ventricle)
(b)

Figure 19.4 Gross anatomy of the heart. (**a**) Anterior view. (**b**) Photograph of anterior aspect of heart (pericardium removed). (**c**) Frontal section showing interior chambers and valves. (**d**) Posterior view.

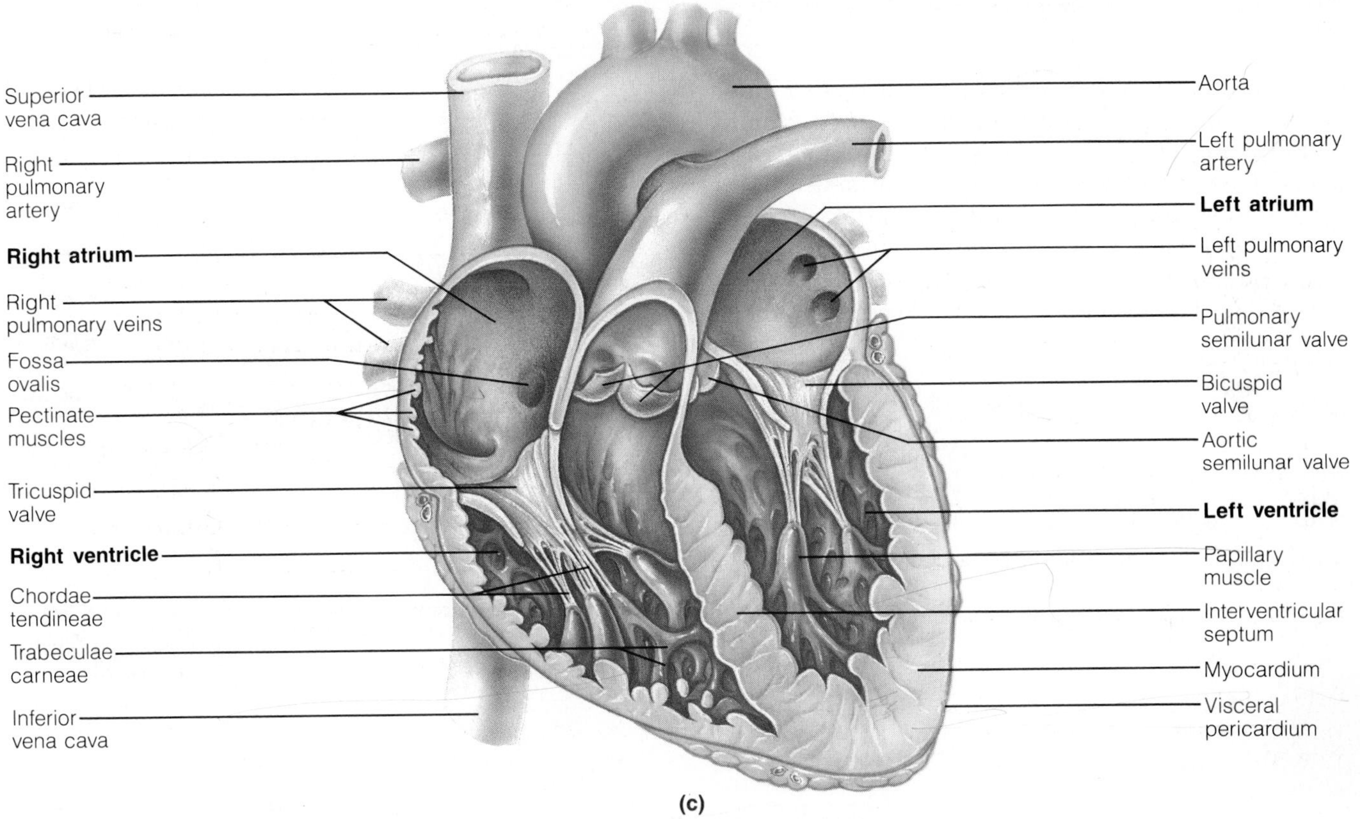

Superior vena cava
Right pulmonary artery
Right atrium
Right pulmonary veins
Fossa ovalis
Pectinate muscles
Tricuspid valve
Right ventricle
Chordae tendineae
Trabeculae carneae
Inferior vena cava
Aorta
Left pulmonary artery
Left atrium
Left pulmonary veins
Pulmonary semilunar valve
Bicuspid valve
Aortic semilunar valve
Left ventricle
Papillary muscle
Interventricular septum
Myocardium
Visceral pericardium

(c)

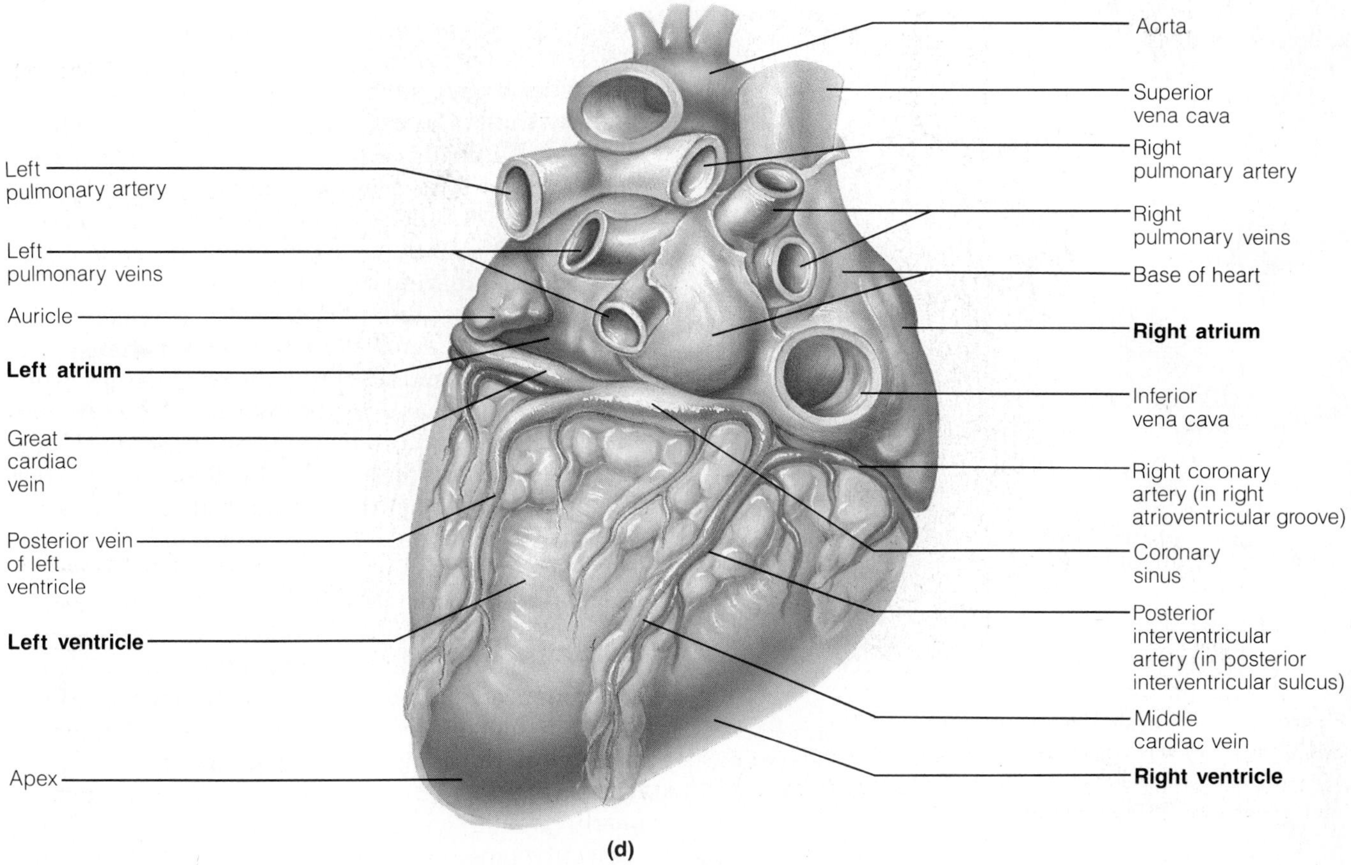

Left pulmonary artery
Left pulmonary veins
Auricle
Left atrium
Great cardiac vein
Posterior vein of left ventricle
Left ventricle
Apex
Aorta
Superior vena cava
Right pulmonary artery
Right pulmonary veins
Base of heart
Right atrium
Inferior vena cava
Right coronary artery (in right atrioventricular groove)
Coronary sinus
Posterior interventricular artery (in posterior interventricular sulcus)
Middle cardiac vein
Right ventricle

(d)

Pathway of Blood Through the Heart

Until the sixteenth century, it was believed that blood moved through the heart in a side-to-side fashion by seeping through pores in the septum. We now know that the heart passages open not from side to side (atrium to atrium) but vertically (atrium to ventricle) and that the heart is not one pump, but two side-by-side pumps, each serving a separate blood circuit (Figure 19.5). The *pulmonary circuit* collects blood returning to the heart from the body and sends it through the lungs for oxygenation; this circuit strictly serves a gas exchange function. The *systemic circuit* supplies the entire body with oxygenated blood.

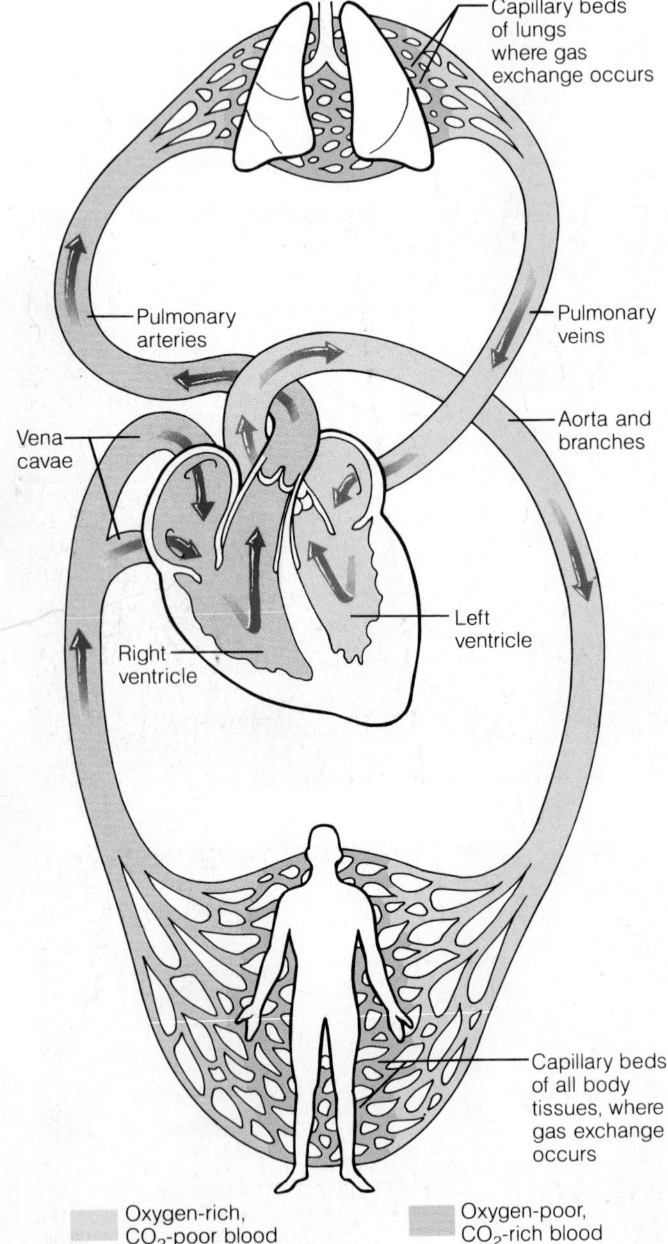

Figure 19.5 The systemic and pulmonary circuits. The left side of the heart is the systemic pump; the right side is the pulmonary circuit pump. (Although there are two pulmonary arteries, one each to the right and left lung, only one is shown for simplicity.)

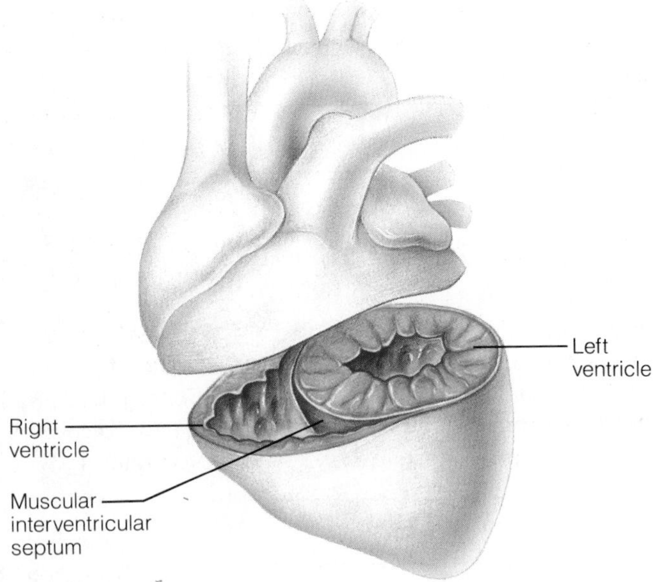

Figure 19.6 Anatomical differences in the right and left ventricles. The left ventricular chamber is basically circular; the right is crescent-shaped and wraps around the left ventricle.

The right side of the heart is the *pulmonary circuit pump.* Blood returning from the body, which is oxygen-poor, carbon dioxide-rich (relatively speaking), enters the right atrium and passes into the right ventricle, which pumps it to the lungs via the pulmonary trunk. Gas exchange takes place in the lungs: The blood unloads carbon dioxide and picks up oxygen. The freshly oxygenated blood is carried by the pulmonary veins back to the left side of the heart.

The left side of the heart is the *systemic circuit pump.* Blood leaving the lungs is returned to the left atrium and passes into the left ventricle, which pumps it into the aorta. From there the blood is transported via smaller systemic arteries to the body tissues, where exchange of gases and nutrients occurs across the capillary walls. The oxygen-depleted blood returns through the systemic veins to the right side of the heart, where it enters the right atrium through the superior and inferior venae cavae.

The ventricles serving the two circuits have very unequal work loads: The pulmonary circuit is a short, low-pressure circulation, whereas the systemic circuit takes a long pathway through the entire body and encounters about five times as much friction, or resistance to blood flow. This functional difference is revealed in the comparative anatomy of the two ventricles (Figure 19.6). The left ventricular chamber is nearly circular, while the right one is flattened into a crescent shape that partially encloses the left ventricle, much the way a hand might wrap around a clenched fist. In addition, the walls of the left ventricle are at least twice as thick as those of its counterpart on the right. Consequently, the left ventricle can generate much more pressure than the right and is a much more powerful pump.

Heart Valves

Blood flows through the heart in one direction: from the atria to the ventricles and out the great arteries leaving the base of the heart. This one-way traffic is enforced by the presence of four heart valves: the paired atrioventricular and semilunar valves (Figures 19.4c and 19.7).

The two **atrioventricular (AV) valves,** located at the junction of the atrial and ventricular chambers on each side, prevent backflow into the atria when the ventricles are contracting. The right AV valve, the **tricuspid** (trī-kus′-pid) **valve,** has three flexible valve flaps, or cusps. The left AV valve, with two flaps, is called the **bicuspid valve** or the **mitral** (mī′-trul) **valve** (because of its resemblance to a bishop's miter or hat). Attached to each of the AV valve flaps are tiny white collagen cords called **chordae tendineae** (literally, heart strings) that anchor the cusps to the papillary muscles protruding from the ventricular walls.

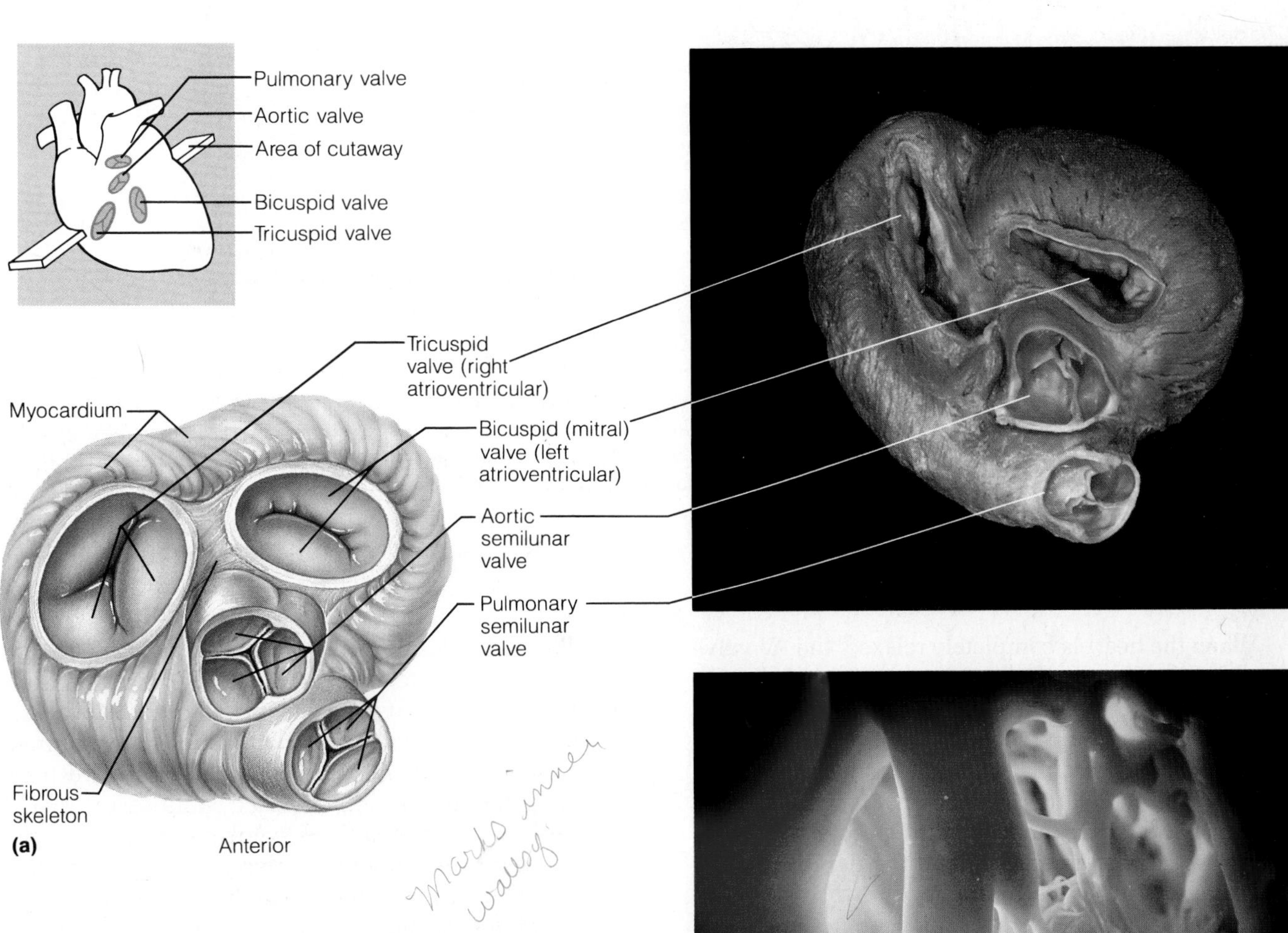

Figure 19.7 Heart valves. (**a**) Superior view of the two sets of heart valves (atria removed). The paired atrioventricular (AV) valves are located between the atria and the ventricles; the two semilunar valves are located at the junction of the ventricles and the arteries that issue from them. (**b**) Photograph of the heart valves, superior view. (**c**) Photograph of the right AV valve. This bottom-to-top view begins in the right ventricle and faces toward the right atrium.

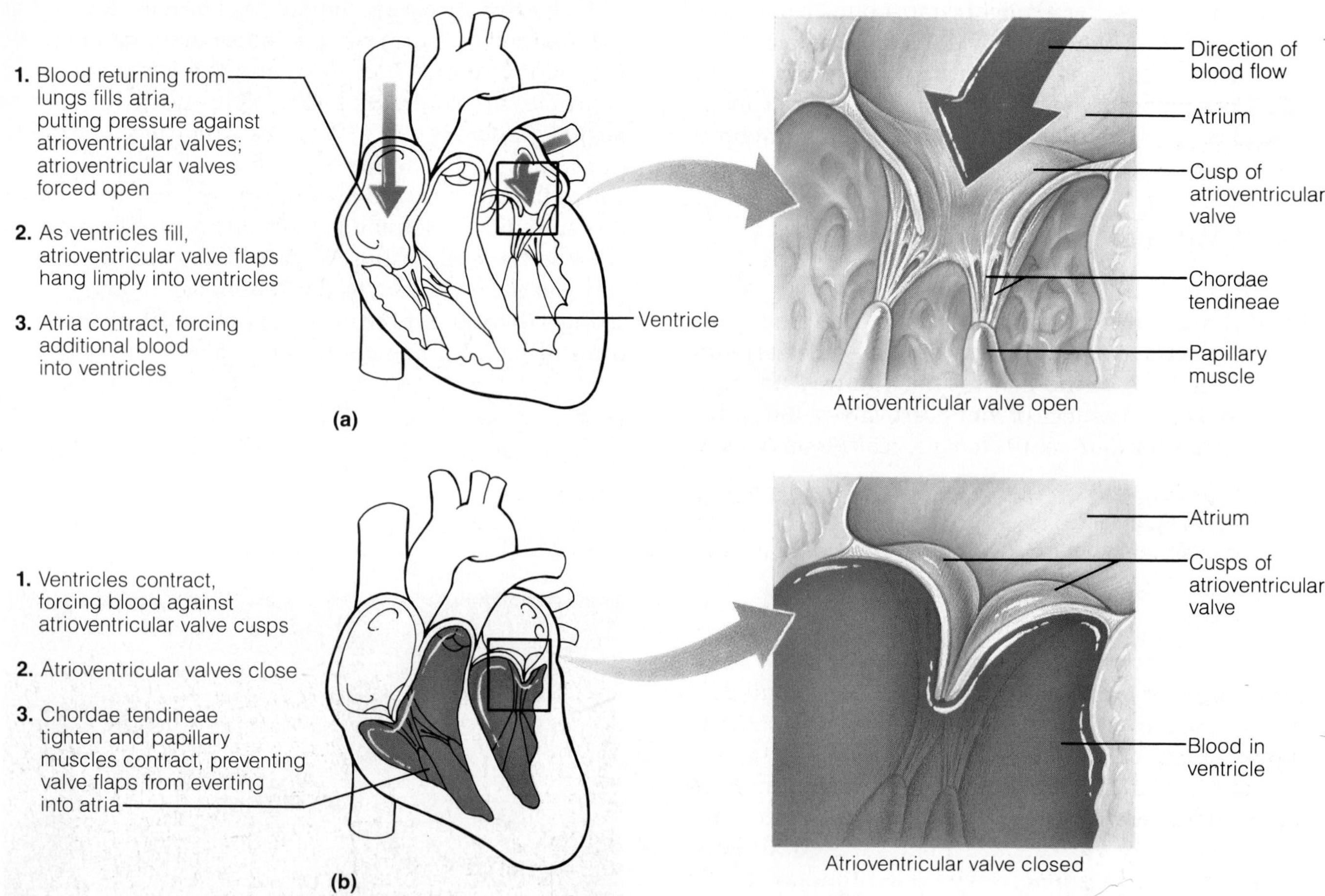

1. Blood returning from lungs fills atria, putting pressure against atrioventricular valves; atrioventricular valves forced open
2. As ventricles fill, atrioventricular valve flaps hang limply into ventricles
3. Atria contract, forcing additional blood into ventricles
Ventricle
(a)
Direction of blood flow
Atrium
Cusp of atrioventricular valve
Chordae tendineae
Papillary muscle
Atrioventricular valve open
1. Ventricles contract, forcing blood against atrioventricular valve cusps
2. Atrioventricular valves close
3. Chordae tendineae tighten and papillary muscles contract, preventing valve flaps from everting into atria
(b)
Atrium
Cusps of atrioventricular valve
Blood in ventricle
Atrioventricular valve closed

Figure 19.8 Operation of the atrioventricular valves of the heart. (**a**) The valves open when the blood pressure exerted on their atrial side is greater than that exerted on their ventricular side. (**b**) The valves are forced closed when the ventricles contract, moving their contained blood superiorly. The action of the chordae tendineae and papillary muscles maintains the valve flaps in the closed position.

When the heart is completely relaxed, the AV valve flaps hang limply into the ventricular chambers below; blood flows into the atria and then through the open AV valves into the ventricles. When the ventricles begin to contract, compressing the blood in their chambers, the blood is forced superiorly against the valve flaps, causing their edges to meet and close the valve (Figure 19.8). The chordae tendineae and the papillary muscles serve as guy wires to anchor the valve flaps in their closed position. If the cusps were not anchored in this manner, they would be blown upward into the atria, in the same way that an umbrella is blown inside out by a gusty wind.

The two **semilunar valves** guard the bases of the large arteries issuing from the ventricles and prevent backflow into the ventricles. The **aortic semilunar valve** is at the junction of the aorta and left ventricle; the **pulmonary semilunar valve** guards the opening between the right ventricle and the pulmonary trunk.

Each of the semilunar valves is fashioned from three pocketlike endocardial cusps; however, their mechanism of action differs from that of the AV valves. When the ventricles are at the peak of their contraction, the semilunar valves are forced open and their cusps flatten against the arterial walls as the blood rushes past them. When the ventricles relax, and the blood (no longer propelled forward by the pressure of ventricular contraction) begins to flow backward toward the heart, it fills the cusps and effectively closes the valves (Figure 19.9).

Heart valves are basically simple devices, and the heart—like any mechanical pump—can function with "leaky" valves as long as the impairment is not too great. However, severe valve deformities can seriously hamper cardiac function. For example, an *incompetent valve* forces the heart to pump and repump the same blood because the valve does not close properly and blood backflows. In valvular *stenosis*, the valve flaps become stiff (often because of

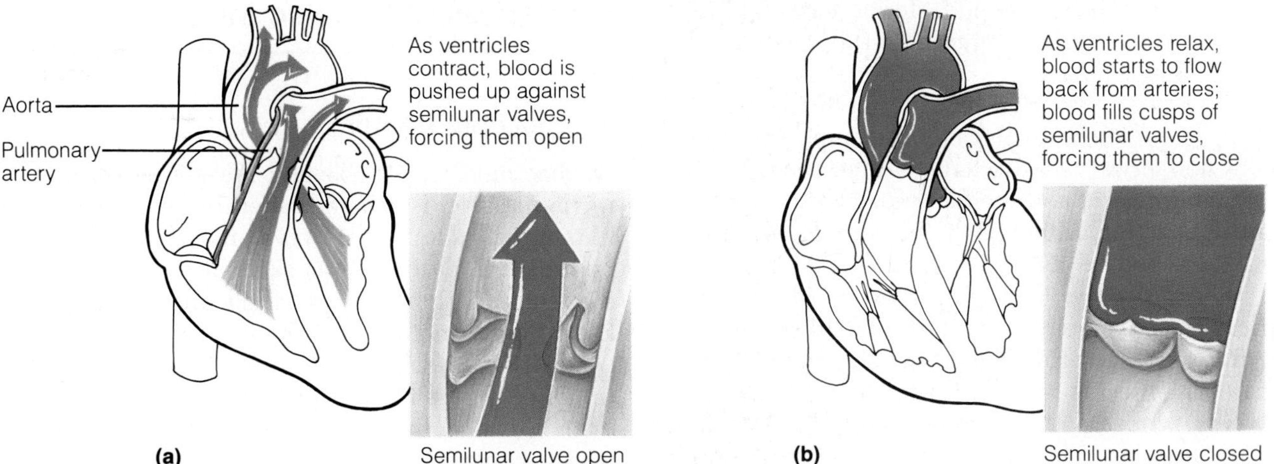
Aorta
Pulmonary artery
As ventricles contract, blood is pushed up against semilunar valves, forcing them open
Semilunar valve open
(a)
As ventricles relax, blood starts to flow back from arteries; blood fills cusps of semilunar valves, forcing them to close
Semilunar valve closed
(b)

Figure 19.9 Operation of the semilunar valves. (**a**) During ventricular contraction, the valves are open and their flaps are flattened against the artery walls. (**b**) When the ventricles relax, the backflowing blood fills the valve cusps and closes the valves.

scar tissue formation following endocarditis) and constrict the opening; this stiffness compels the heart to contract more forcibly than normal. In both instances, the work load of the heart increases and its functioning becomes less efficient. Ultimately, the heart may become severely weakened. Under such conditions, the faulty valve is replaced with a synthetic valve or with a valve taken from a pig heart. ■

Cardiac Circulation

Although the heart chambers are continuously bathed with blood, this contained blood does not nourish the heart tissues. The myocardium is too thick to make diffusion a practical means of providing nutrition. The functional blood supply of the heart is provided by the right and left **coronary arteries,** which branch from the root of the aorta where it emerges from the left ventricle (Figure 19.10). The coronary arteries run in the atrioventricular grooves and encircle the heart like a crown (see Figure 19.4). The left coronary artery runs toward the left side of the heart and then divides into its major branches: the **anterior interventricular artery,** which follows the anterior ventricular sulcus and supplies blood to the interventricular septum and anterior walls of both ventricles; and the **circumflex artery,** which serves the laterodorsal walls of the left atrium and left ventricle.

The right coronary artery courses to the right side of the heart, where it also divides into two branches: The **marginal artery** serves the myocardium of the lateral part of the right side of the heart; the more important **posterior interventricular artery** runs to the heart apex and supplies the posterior ventricular

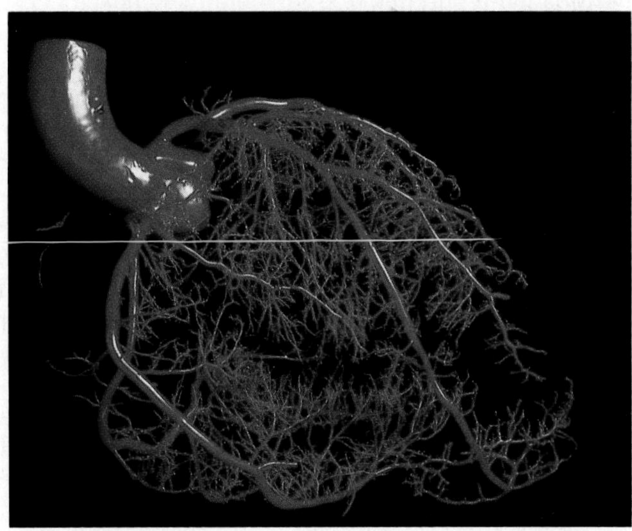

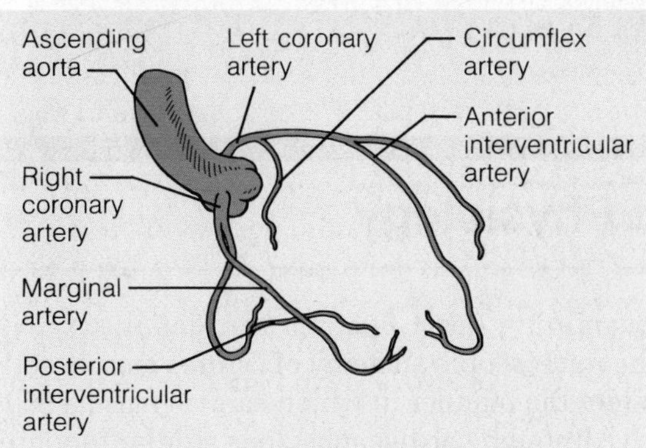
Ascending aorta
Left coronary artery
Circumflex artery
Anterior interventricular artery
Right coronary artery
Marginal artery
Posterior interventricular artery

Figure 19.10 A resin cast of the coronary arterial system supplying the heart, anterior view. Each coronary artery gives off two main branches: the circumflex and anterior interventricular arteries from the left coronary artery; and the marginal and posterior interventricular artery from the right. The courses of these arteries in the cast are outlined in the diagram below the photo.

walls. The branches of the posterior and anterior interventricular arteries merge near the apex of the heart.

The coronary arteries provide an intermittent, pulsating blood flow to the myocardium. They are actively delivering blood when the heart is relaxed and are compressed and ineffective when the ventricles are contracting. Although the heart represents only about 1/200 of body weight, it requires about 1/20 of the body's blood supply. As might be expected, the left ventricle receives the most plentiful blood supply.

After passing through the capillary beds of the myocardium, the venous blood is collected by the **cardiac veins,** whose paths roughly follow those of the coronary arteries. The cardiac veins, in turn, join together to form an enlarged vessel called the **coronary sinus,** which empties the blood into the right atrium. The coronary sinus is obvious on the posterior aspect of the heart (Figure 19.4d).

Any blockage of the coronary arterial circulation can be serious; in some cases, it is fatal. *Angina pectoris* (an-jī-nuh pek'-tor-is), literally, a choked chest, is chest pain caused by fleeting, deficient blood delivery that may result from stress-induced spasms of the coronary arteries or increased physical demands on the heart. The myocardial cells are weakened by the temporary lack of oxygen, but they do not die. Far more serious is prolonged coronary blockage resulting from an occluding blood clot or severe atherosclerosis. Under such conditions, the ischemic cardiac cells die, forming an *infarct.* The resulting *myocardial infarction* is commonly referred to as a heart attack or a coronary. Because adult cardiac muscle is amitotic, any areas of cell death are repaired with noncontractile scar tissue. Whether or not a person survives a myocardial infarction depends on the extent of cell death and on the location of the damage. Damage to the left ventricle is most serious. Some of the current approaches to treating damaged hearts are described in the box on p. 614. ■

Heart Physiology

Before examining heart operation in detail, we will review the microscopic anatomy of cardiac muscle cells and consider the manner in which cardiac muscle cells function. Although cardiac muscle is similar in many ways to skeletal muscle, it displays special anatomical features that reflect its unique blood-pumping role. Because skeletal muscle anatomy and physiology have already been covered in some detail in Chapter 9, it seems appropriate to compare and contrast these two muscle types.

Properties of Cardiac Muscle

Microscopic Anatomy

Cardiac muscle, like skeletal muscle, is striated, and its contraction is accomplished by the same sliding-filament mechanism. However, in contrast to the long, cylindrical, multinucleate skeletal muscle fibers, cardiac cells are short, fat, branched, and interconnected. Each fiber contains one or at most two large, pale, centrally located nuclei (Figure 19.11a). The intercellular spaces are filled with a loose connective tissue matrix, called the *endomysium* (en"-dō-mī'-sē-um), which contains numerous capillaries. This delicate matrix is connected, in turn, to the previously mentioned more dense fibrous connective tissue skeleton of collagen and elastin fibers that link the cardiac cells together in coiling array and reinforces the basketlike walls of the heart (see Figures 19.3 and 19.7a).

Unlike skeletal muscle fibers, which are independent of one another both in structure and function, adjacent cardiac cells interconnect at dark-staining junctions called **intercalated discs** (Figure 19.11). These discs contain anchoring desmosomes and gap junctions (specialized cell junctions discussed in Chapter 4). The desmosomes prevent separation of adjacent cardiac cells during contraction. Gap junctions allow ions to pass from cell to cell, allowing direct transmission of the depolarizing impulse across the entire heart. Because all cardiac fibers are electrically coupled by the gap junctions, the myocardium behaves as a single unit or *functional syncytium.*

Large mitochondria account for 30% to 40% of the volume of cardiac muscle fibers, and most of the remaining volume is occupied by myofibrils composed of fairly typical sarcomeres. The sarcomeres exhibit A bands and I bands, which reflect the arrangement of the thick (myosin) and thin (actin) myofilaments composing them. (Figure 19.11c is provided to refresh your memory of these relationships.) However, in contrast to what is seen in skeletal muscle, the myofibrils of cardiac muscle cells vary in diameter and tend to fuse together. This results in a banding pattern less dramatic than that of skeletal muscle. An elaborate T system is also present, but the T tubules are wider and fewer than those of skeletal muscle. They enter the fibers at the regions of the intercalated discs, which correspond to the Z lines of skeletal muscle. (Remember that T tubules are invaginations of the sarcolemma, or muscle cell plasma membrane.) The cardiac sarcoplasmic reticulum is smaller and less well developed than that of skeletal muscle.

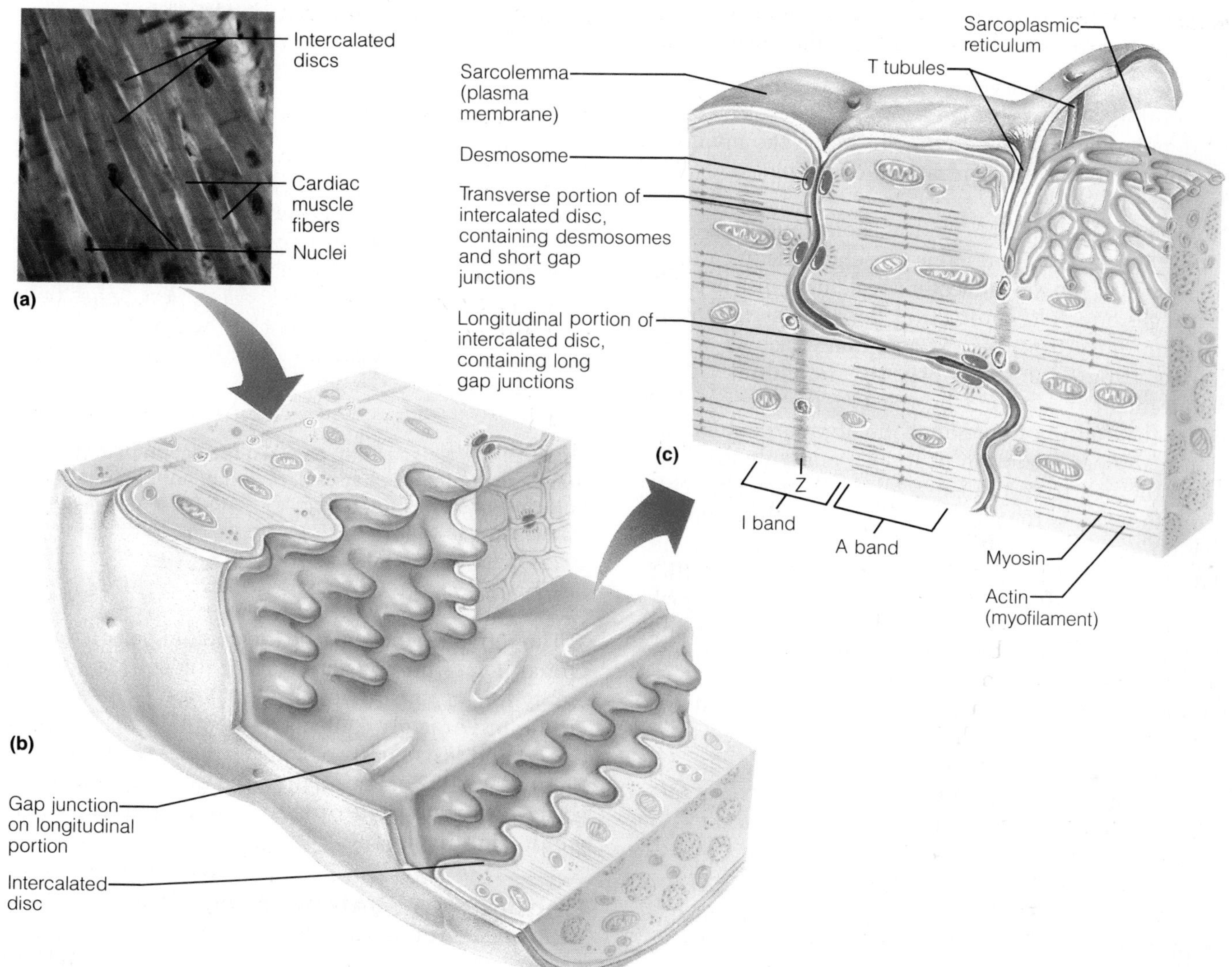

Figure 19.11 Microscopic anatomy of cardiac muscle. (**a**) Photomicrograph of cardiac muscle (300X). Notice that the cardiac muscle fibers are short, branched, and striated. Also note the dark-staining intercalated discs or junctions between the adjacent cardiac fibers. (**b**) Three-dimensional diagrammatic view of the relationship of cardiac fibers at the intercalated discs. (**c**) Diagrammatic view of a section of cardiac muscle cells illustrating the banding pattern enforced by the myofilaments. Portions of the sarcoplasmic reticulum and T system are also depicted, as are the positioning of the desmosomes and gap junctions at the intercalated disc connecting two adjoining cardiac muscle cells.

Energy Requirements

Cardiac muscle has more mitochondria (per unit volume) and depends more on a continual supply of oxygen for its energy metabolism than does skeletal muscle. As described in Chapter 9, skeletal muscles can contract for prolonged periods, even during oxygen deficits, by carrying out anaerobic respiration and incurring an oxygen debt. In contrast, the heart relies almost exclusively on aerobic respiration; hence, it cannot incur much of an oxygen debt and still operate effectively.

Both types of muscle tissue use multiple fuel molecules, including glucose, fatty acids, lactic acid, and others; but cardiac muscle appears to use fatty acids most effectively for forming ATP, even when other energy fuels are equally available. Skeletal muscle, on the other hand, seems to have a preference for glucose when it is most active.

Mechanism and Events of Contraction

The sequence of events of cardiac muscle contraction is similar to that of skeletal muscle:

1. Influx of sodium ions from the extracellular fluid into the cells initiates the action potential by opening voltage-regulated sodium channels.

2. Transmission of the depolarization wave down the T tubules causes calcium entry into the sarcoplasm.

3. Ionic calcium serves as the signal (via troponin binding) for cross bridge activation and couples the depolarization wave to the sliding of the myofilaments (excitation–contraction coupling).

However, as noted, the sacroplasmic reticulum (the ionic calcium storage area) is less elaborate in cardiac muscle cells, and the intracellular calcium pulse needed to trigger effective contraction depends as much on calcium entry from the extracellular space as on its release from the sarcoplasmic reticulum. This requirement is satisfied by the fact that, in cardiac cells, the sodium-dependent membrane depolarization *also* opens calcium channels, allowing Ca^{2+} to enter the cell from the extracellular space. This calcium flux across the membrane prolongs the depolarization potential briefly, resulting in a *plateau* in the action potential tracing (Figure 19.12). Anything that interferes with the binding or movement of calcium ions through the calcium channels of the sarcolemma and T tubules diminishes heart contractile activity. During repolarization, calcium ions are pumped back into the sarcoplasmic reticulum and the extracellular space.

There are also some other fundamental differences between skeletal and cardiac muscle:

1. In skeletal muscle, the all-or-none law applies to contractile activity at the cellular level; impulses do not spread from cell to cell. In cardiac muscle, the law applies at the organ level; that is, either the heart contracts as a unit or it doesn't contract at all. This is assured by the transmission of the depolarization wave across the heart from cell to cell through the gap junctions, which act to tie all cardiac muscle cells together into a single contractile unit.

2. While each skeletal muscle cell must be independently stimulated to contract by a nerve ending, some cardiac muscle cells are self-excitable and can initiate their own depolarization. This property, called *automaticity*, results from the fact that plasma membranes of these cardiac cells are more "leaky" than those of skeletal muscle. Extracellular sodium (and calcium) ions can enter cardiac muscle cells more easily. This results in a gradual depolarization of the membrane; and when threshold levels are reached, spontaneous action-potential generation and propagation (followed by contraction) occur. (There is also some evidence that these cardiac cells have lower permeability to potassium ions.) Further, cardiac muscle cells exhibit *rhythmicity*; that is, they contract in a definite rhythm. However, in most cases, the automaticity and rhythmicity of the healthy heart reflect the beat set by its pacemaker (described below) rather than the beat of individual heart cells.

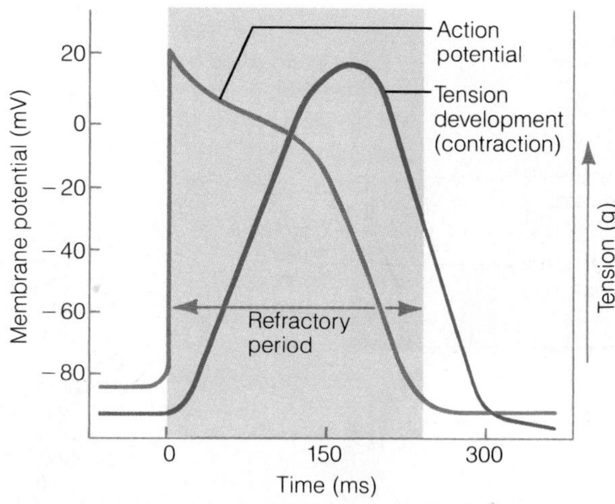

Figure 19.12 Relationship between the action potential (membrane potential changes), period of contraction (tension development), and refractory period in a single ventricular cardiac cell.

3. The refractory period (the inexcitable period during which potassium ions are leaving the cell) is considerably longer in cardiac muscle than in skeletal muscle (approximately 250 ms versus 1–2 ms, respectively) and lasts nearly as long as the contraction (see Figure 19.12). This normally prevents the heart from going into prolonged or tetanic contractions, which would stop its pumping action.

Conduction System of the Heart

The ability of cardiac muscle to depolarize and contract is intrinsic; that is, it is a property of the heart muscle itself and does not depend on extrinsic nerve impulses. In fact, even if all nerve connections to the heart are severed, the heart continues to beat rhythmically, as amply demonstrated by transplanted hearts.

This very independent, but coordinated, activity of the heart results from two factors: the presence of gap junctions, and the activity of the heart's "in-house" conduction system. The intrinsic **cardiac conduction system,** or **nodal system,** consists of specialized nervelike, but noncontractile, cardiac cells that initiate and distribute impulses throughout the heart, so that the myocardium depolarizes and contracts in an orderly, sequential manner from atria to ventricles. Thus, the heart beats (nearly) as one cell.

The components of the conduction system are the sinoatrial (SA) node, the atrioventricular (AV) node, the atrioventricular (AV) bundle (bundle of His), the right and left bundle branches, and the Purkinje fibers (Figure 19.13). The crescent-shaped **sinoatrial** (sī″-nō-ā′-trē-ul) **node** is the **pacemaker,** a minute cell mass with a mammoth job. Located in the right atrial wall,

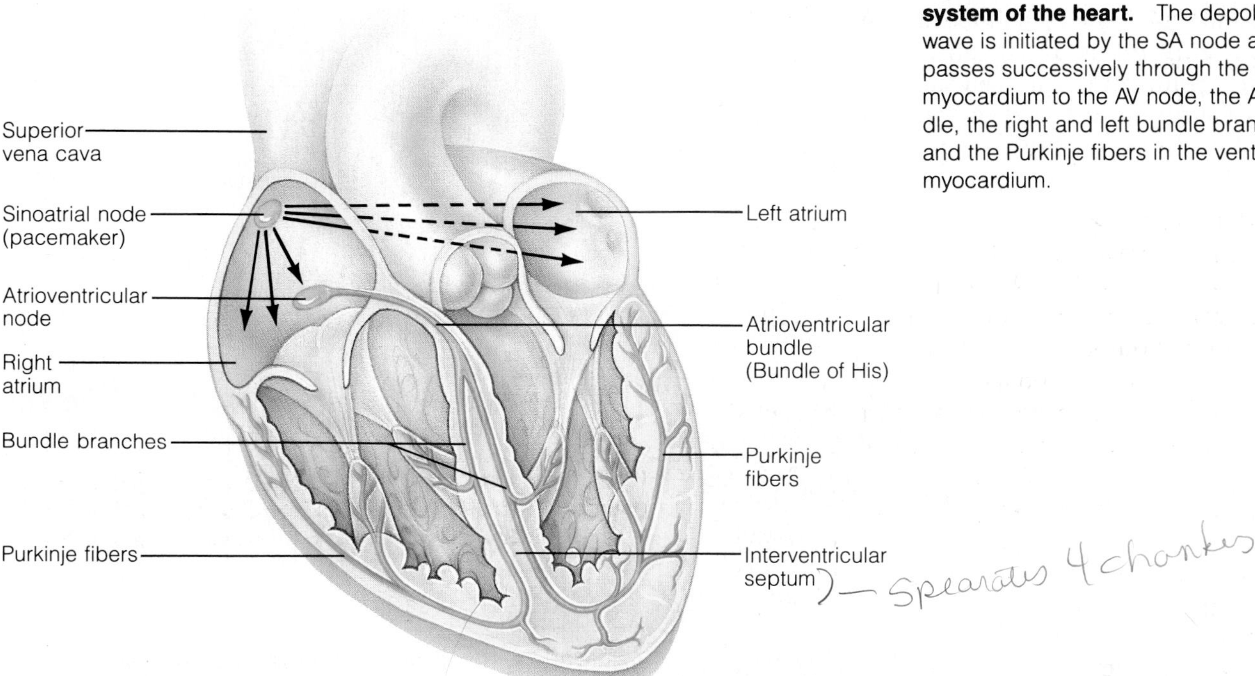

Figure 19.13 The intrinsic conduction system of the heart. The depolarization wave is initiated by the SA node and then passes successively through the atrial myocardium to the AV node, the AV bundle, the right and left bundle branches, and the Purkinje fibers in the ventricular myocardium.

just inferior to the entrance of the inferior vena cava, the SA node depolarizes spontaneously at the rate of 70 to 80 times every minute. Because no other region of the conduction system or the myocardium has a faster depolarization rate, the SA node initiates each depolarization wave that travels across the heart and sets the pace for the heart as a whole. Its characteristic rhythm is called the *sinus rhythm.*

From the SA node, the depolarization wave spreads via gap junctions throughout the atrial syncytium and reaches the **atrioventricular node** located in the inferior portion of the interatrial septum (via internodal pathways) in approximately 0.04 s. The impulse is delayed momentarily (for about 0.1 s) at the AV node, allowing the atria to complete their contraction. It then passes rapidly through the **atrioventricular bundle** and **bundle branches** in the interventricular septum and then the **Purkinje fibers** that penetrate into the ventricular myocardium. The AV bundle and bundle branches ensure excitation of the septal cells, but the bulk of ventricular depolarization depends on the Purkinje fibers and, ultimately, on cell-to-cell transmission of the impulse by the ventricular muscle cells themselves. Because the left ventricle is much larger than the right, the Purkinje network is more elaborate in that side of the heart. The total time elapsing between initiation of the impulse by the SA node and depolarization of the last of the ventricular muscle cells is approximately 0.22 s in a healthy human heart.

Ventricular contraction almost immediately follows the ventricular depolarization wave. A wringing contraction of the ventricles begins at the heart apex and moves toward the atria. This effectively ejects some of the contained blood superiorly into the large arteries leaving the base of the heart.

Although cardiac cells that exhibit automaticity and rhythmicity are found in all heart chambers, their rates differ. Nonnodal atrial myocardial cells typically depolarize 60 times per minute, whereas ventricular cells depolarize spontaneously at the much slower rate of 20–40 times per minute. The cardiac conduction system not only enforces a faster contraction rate on the heart as a whole, but it coordinates and synchronizes heart activity as well. Without it, the impulse would travel much more slowly through the myocardium—at the rate of 0.3–0.5 m/s as opposed to several meters per second in most parts of the conduction system. This would allow some of the muscle fibers to contract well ahead of others, resulting in reduced pumping effectiveness.

External nerve stimulation is not required for heart contraction. Nonetheless, considerable nervous control of heart activity is exerted by fibers of the autonomic nervous system, which act as "brakes" and "accelerators" of the heart. For example, if the body has been aroused to fight-or-flight status by sympathetic nervous system activity, both the rate and the force of heartbeat are increased. Conversely, parasympathetic stimulation causes a decrease in heart rate. Thus, autonomic nervous system controls modify the activity of the intrinsic conduction system. (These neural controls are explained in more detail later in the chapter.)

Defects in the intrinsic conduction system can cause irregular heart rhythms, or *arrhythmias* (ah-rith′-mē-uz), resulting in uncoordinated atrial and ventricular contractions or even fibrillation. *Fibrillation* is a condition of rapid and irregular or out-of-phase contractions in which the heart has been compared with a writhing bag of worms. Fibrillating ventricles are useless as pumps; and unless the heart can be defibrillated quickly, circulation stops and brain death occurs. Defibrillation is accomplished by exposing the heart to a strong electric shock, which interrupts the chaotic twitching of the heart by depolarizing the entire myocardium. The hope is that ("with the slate wiped clean") the SA node will again begin to function normally and sinus rhythm will be reestablished.

A defective pacemaker may have several consequences. An *ectopic* (ek-tah′-pik) *focus* or abnormal atrial pacemaker may appear and become the heart's "new" pacemaker, or the AV node may become the pacemaker. The pace set by the AV node (*nodal rhythm*) is slower than sinus rhythm, but is still adequate to maintain circulation. Occasionally, ectopic pacemakers appear even when the conduction system is operating normally. A small region of the heart becomes hyperexcitable, sometimes as a result of too much caffeine (several cups of coffee) or nicotine (excessive smoking), and generates impulses even more quickly than the SA node. This leads to a *premature contraction* or *extrasystole* (eks″-truh-sis′-tuh-lē) before the SA node paces the next contraction.

Because the atria and ventricles are almost entirely separated from one another by the electrically inert tissue of the fibrous skeleton, the only route for impulse transmission from the atria to the ventricles is through the AV node. Thus, any damage to the AV node, referred to as a *heart block*, may seriously interfere with the ability of the ventricles to receive the pacing depolarization wave. In total heart block (no atrioventricular transmission), the ventricles beat at their intrinsic rate, which is too slow to maintain adequate circulation. In such cases, a fixed-rate artificial pacemaker, set to discharge at a constant rate, is usually implanted. Those suffering from a partial heart block, in which a fraction of the atrial impulses reach the ventricles, commonly receive demand-type pacemakers, which deliver impulses only when the heart is not transmitting on its own. ■

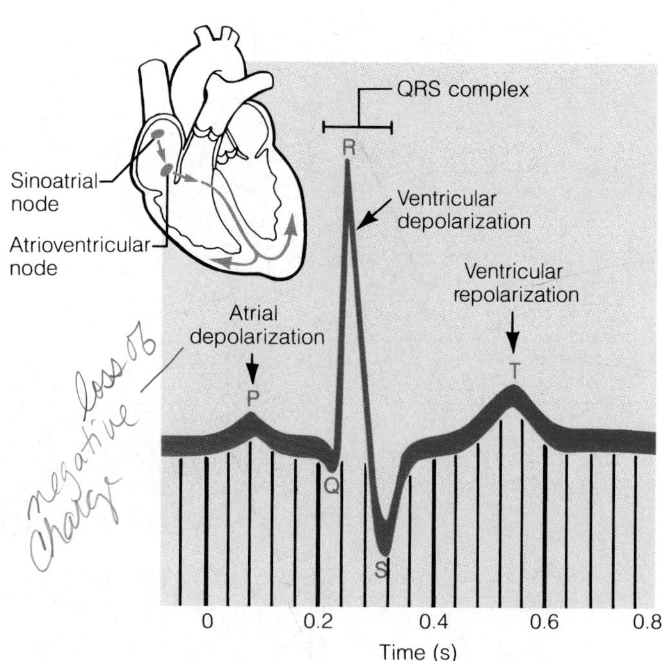

Figure 19.14 An electrocardiogram tracing illustrating the three normally recognizable deflection waves.

Electrocardiography

Because body fluids are good conductors, the electrical currents generated and transmitted through the heart also spread throughout the body and can be picked up, amplified, and recorded with an instrument called an *electrocardiograph*. The graphic recording of electrical changes during heart activity is called an **electrocardiogram (ECG).**

A typical ECG consists of a series of three distinguishable waves called *deflection waves* (Figure 19.14). The first wave, the small **P wave,** lasting about 0.08 s, results from impulse conduction from the SA node through the atria. Approximately 0.1 s after the P wave begins, the atria contract.

The large **QRS complex** results from ventricular depolarization and precedes ventricular contraction. Its complicated shape reveals the different size of the two ventricles and the time required for each to depolarize. Average duration of the QRS complex is 0.08 s.

The **T wave** is caused by ventricular repolarization and typically lasts about 0.16 s. Because atrial repolarization takes place during the period of ventricular excitation, its occurrence is normally obscured by the large QRS complex being recorded at the same time.

The *P–R interval* represents the time (about 0.16 s) from the beginning of atrial excitation to the beginning of ventricular excitation. It includes atrial depolarization (and contraction) as well as the passage of the depolarization wave through the AV node and the rest of the conduction system. The *Q–T interval*, lasting about 0.36 s, is the period from the beginning of ventricular depolarization through their repolarization and includes the time of ventricular contraction.

In a healthy heart, the size, duration, and timing of the deflection waves tend to be consistent. Thus, any changes in the pattern or timing of the ECG may

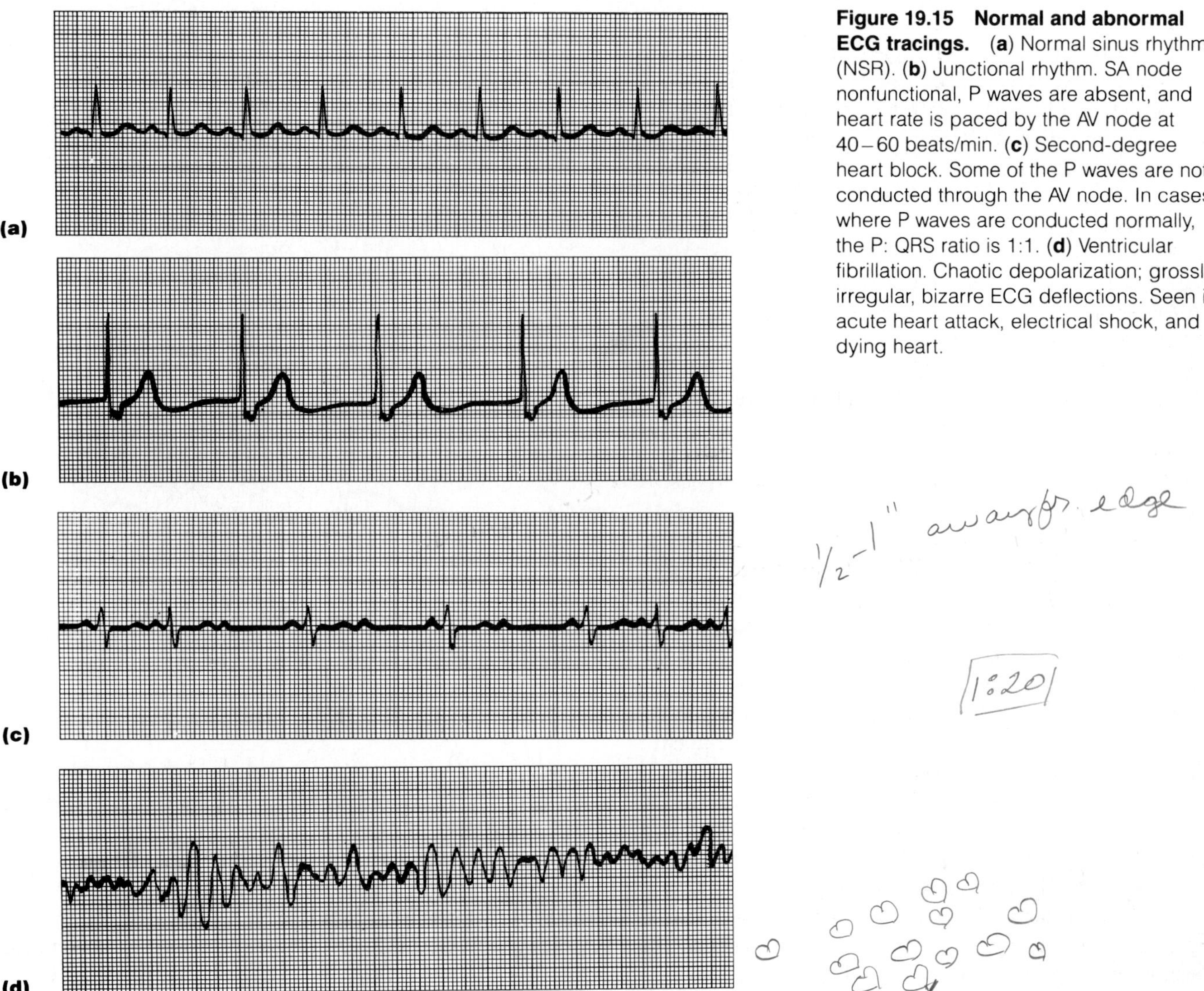

Figure 19.15 Normal and abnormal ECG tracings. (**a**) Normal sinus rhythm (NSR). (**b**) Junctional rhythm. SA node nonfunctional, P waves are absent, and heart rate is paced by the AV node at 40–60 beats/min. (**c**) Second-degree heart block. Some of the P waves are not conducted through the AV node. In cases where P waves are conducted normally, the P: QRS ratio is 1:1. (**d**) Ventricular fibrillation. Chaotic depolarization; grossly irregular, bizarre ECG deflections. Seen in acute heart attack, electrical shock, and dying heart.

reveal a "silent" region of myocardial infarct or problems with the heart's conduction system. For example, an inverted or lost P wave indicates that the AV node has become the pacemaker. A few examples of abnormalities detectable with an ECG are illustrated in Figure 19.15.

Cardiac Cycle

The heart undergoes some fairly dramatic writhing movements as it alternately contracts, forcing blood out of its chambers, and then relaxes, allowing its chambers to refill with blood. The terms **systole** (sis′-tō-lē) and **diastole** (dī-as′-tō-lē) refer respectively to these contraction and relaxation periods of heart activity. The **cardiac cycle** includes *all* events associated with the flow of blood through the heart during one complete heartbeat, that is, atrial systole and diastole followed by ventricular systole and diastole. However, the cycle is usually described in terms of ventricular events, and this is the approach that we use here.

The cardiac cycle is marked by a succession of pressure and blood volume changes within the heart. Although pressure changes occurring in the right side of the heart are only about one-fifth as great as those occurring in the left, each ventricle pumps the same volume of blood per beat and the overall relationships are the same for both chambers (Figure 19.16).

Because blood circulates endlessly, we will arbitrarily choose a starting point to follow its route through the heart during a single cardiac cycle. We begin our explanation with the heart in total relaxation: Both the atria and ventricles are quiet, and it is mid-to-late diastole.

1. Period of ventricular filling: mid-to-late diastole. Pressure within the heart is low, and blood returning from the circulation is flowing passively into and through the atria into the ventricles below. The AV valves are open, and the semilunar valves (aortic and pulmonary) are shut. This phase is indicated by interval 1 in Figure 19.16. Approximately 70% of ventricular filling occurs during this period, and the AV valve flaps begin to drift upward toward their closed posi-

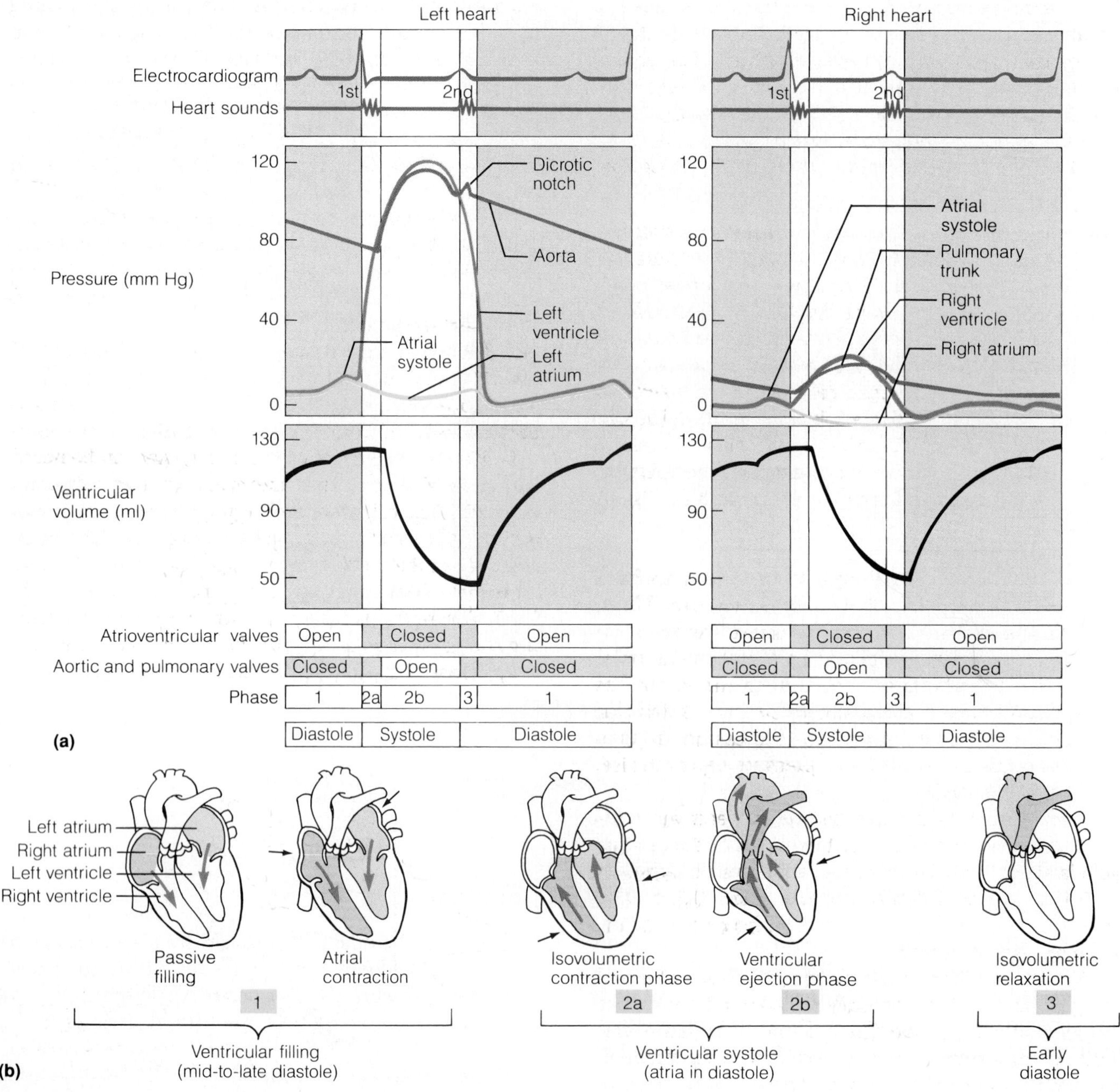

Figure 19.16 Summary of events occurring in the heart during the cardiac cycle. (**a**) Events in the left side of the heart are shown on the left; events occurring in the right side are summarized graphically on the right. An ECG tracing is superimposed on the graph (top) so that pressure and volume changes can be related to electrical events occurring at any point. Time occurrence of heart sounds is also indicated. (**b**) Events of phase 1 through 3 of the cardiac cycle are depicted in diagrammatic views of the heart.

tion. (The remaining 30% is delivered to the ventricles when the atria contract toward the end of this phase.) Now the stage is set for atrial systole. Following depolarization, the atria contract, compressing the blood contained in their chambers. This causes a sudden slight rise in atrial pressure, which propels residual blood out of the atrial chambers into the ventricles. Then the atria relax, and atrial diastole persists through the rest of the cycle.

2. Ventricular systole. As the atria go into diastole, the ventricles begin their contraction phase. Their walls close in on the blood in their chambers, and ventricular pressure rises rapidly and sharply, closing the AV valves. For a split second, the ventricles are completely closed chambers and blood volume in the chambers remains constant; this **isovolumetric** (ī″-sō-vol-yoo-meh′-trik) **contraction phase** is indicated as phase 2a in Figure 19.16. As ventricular pressures continue to rise and finally exceed those in the large arteries issuing from them, the semilunar valves are forced open and blood is expelled from the ventricles into the aorta and pulmonary trunk. During this **ventricular ejection phase,** indicated as phase 2b in Figure 19.16, the pressure in the aorta normally reaches about 120 mm Hg.

3. Isovolumetric relaxation: early diastole. During this brief phase, the ventricles begin to relax. Because the blood remaining in their chambers is no longer compressed, ventricular pressure drops rapidly and blood in the aorta and pulmonary trunk begins to backflow toward the heart, closing the semilunar valves. Closure of the aortic semilunar valve causes a brief rise in aortic pressure, indicated in Figure 19.16 as the *dicrotic notch*. Once again the ventricles are totally closed chambers. This isovolumetric ventricular relaxation phase is indicated by interval 3 in Figure 19.16.

All during ventricular systole, the atria have been in diastole. During that time, they have been filling with blood and the intra-atrial pressure has been rising. When the pressure exerted by blood on the atrial side of the AV valves exceeds that in the ventricles below, the AV valves are forced open and ventricular filling, phase 1, begins again. Atrial pressure drops to its lowest point and ventricular pressure begins to rise, completing the cycle.

Assuming that the average heart beats approximately 75 times each minute, the length of the cardiac cycle is about 0.8 s. Of this time, atrial systole accounts for 0.1 s and ventricular systole takes 0.3 s. The remaining 0.4 s is the period of total heart relaxation, or the *quiescent period*.

Notice two important points: (1) Flow of blood through the heart is controlled entirely by pressure changes, and (2) blood flows along a pressure gradient, always from higher to lower pressure through any available opening. The pressure changes, in turn, reflect the alternating contraction and relaxation of the myocardium and cause the opening and closing of the heart valves, which keep blood flowing in the forward direction.

Heart Sounds

During such cardiac cycle, two distinguishable sounds can be heard if the thorax is auscultated with a stethoscope. These **heart sounds,** often described as lub-dup, result from the closing of heart valves. The timing of heart sounds in the cardiac cycle is shown in Figure 19.16.

The basic rhythm of the heart sounds is lub-dup, pause, lub-dup, pause, and so on. The pause indicates the quiescent period. The first heart sound, reflecting AV valve closure, tends to be louder, longer, and more resonant compared to the second: the short, sharp sound of the semilunar valves snapping shut. Because the mitral valve closes slightly before the tricuspid valve, and the aortic semilunar valve generally snaps shut just before the pulmonary valve, it is possible to distinguish the individual valve sounds by auscultating specific regions of the thorax (Figure 19.17). Notice that these four points also define the four corners of the normal heart. Knowing normal heart size and location is essential for recognizing an enlarged (and often diseased) heart.

Abnormal or unusual heart sounds are called *murmurs*. Blood flows silently as long as the flow is smooth and uninterrupted. If, however, it strikes obstructions, its flow becomes turbulent and generates sounds, such as heart murmurs, that can be heard with a stethoscope. Heart murmurs are relatively common in young children (and some elderly people) with perfectly healthy hearts, probably because their heart walls are relatively thin and vibrate with rushing blood. However, most often murmurs indicate valve problems. For example, if a valve is incompetent, a swishing sound will be heard *after* that valve has (supposedly) closed, as the blood backflows or regurgitates

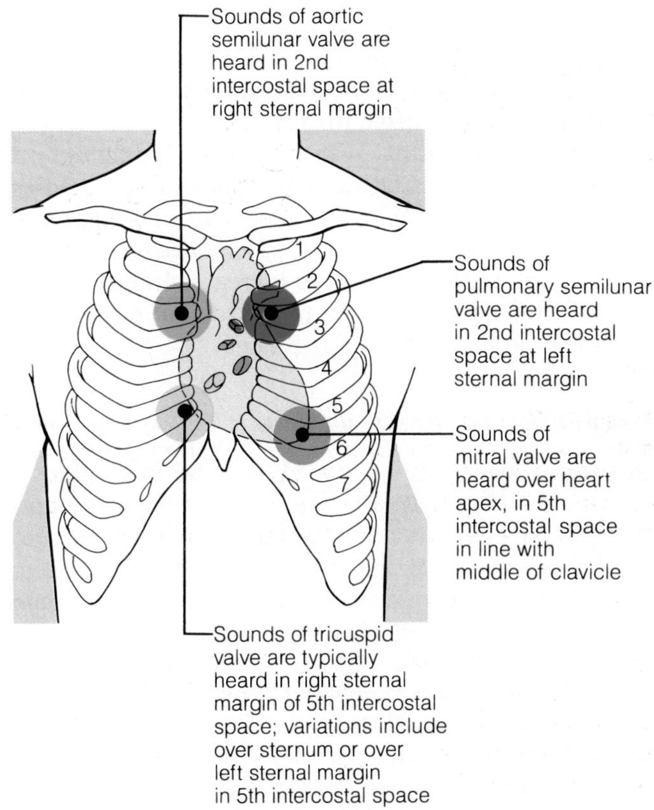

Sounds of aortic semilunar valve are heard in 2nd intercostal space at right sternal margin
Sounds of pulmonary semilunar valve are heard in 2nd intercostal space at left sternal margin
Sounds of mitral valve are heard over heart apex, in 5th intercostal space in line with middle of clavicle
Sounds of tricuspid valve are typically heard in right sternal margin of 5th intercostal space; variations include over sternum or over left sternal margin in 5th intercostal space
1
2
3
4
5
6
7

Figure 19.17 Areas of the thoracic surface where the sounds of the heart can be best detected.

through the partially open valve. A stenotic valve, in which the valvular opening is narrowed, restricts blood flow *through* the valve. In this case, a high-pitched sound can be detected during systole just before valve closure, as blood flows through the constricted opening. ■

Cardiac Output

Cardiac output (CO) is the amount of blood pumped out by *each* ventricle in 1 minute. It is the product of heart rate (HR) and stroke volume (SV). **Stroke volume** is the volume of blood pumped out by a ventricle with each contraction. In general, stroke volume is correlated with force of ventricular contraction.

Using the normal resting values for heart rate (75 beats/min) and stroke volume (70 ml/beat), the average adult cardiac output can be easily computed:

$$\text{CO (ml/min)} = \text{HR (75 beats/min)} \times \text{SV (70 ml/beat)}$$

$$\text{CO} = 5250\ \text{ml/min}$$

Because normal adult blood volume is about 5000 ml, the entire blood supply passes through the body once each minute. Cardiac output varies with the demands of the body: It rises when the stroke volume is increased or the heart beats faster or both; it drops when either or both of these factors decrease.

Cardiac output can increase markedly in response to special demands, such as making a dash to catch the bus. This ability of the heart to push its CO above normal levels is referred to as **cardiac reserve.** In nonathletic people, cardiac reserve is typically about four times their normal CO; but in trained athletes, cardiac output may reach 35 L/min (a seven-times increase) during vigorous competition. How does the heart accomplish such tremendous increases in output? To answer this question, we'll take a look at how stroke volume and heart rate are regulated.

Regulation of Stroke Volume

A healthy heart pumps out about 60% of the blood that enters its chambers (see Figure 19.16a). Essentially, SV represents the difference between *end diastolic volume (EDV)*, the amount of blood that collects in a ventricle during diastole, and *end systolic volume (ESV)*, the volume of blood remaining in a ventricle *after* it has contracted. The EDV is normally about 120 ml; the ESV is approximately 50 ml. To figure normal stroke volume, these values are simply plugged into the equation:

$$\text{SV (ml/beat)} = \text{EDV (120 ml)} - \text{ESV (50 ml)}$$

$$\text{SV} = 70\ \text{ml/beat}$$

Hence, each ventricle pumps out about 70 ml (close to 2 ounces) of blood with each heartbeat.

So what is important here—how do we make sense out of this alphabet soup (SV, ESV, EDV)? According to the *Frank–Starling law of the heart*, the critical factor controlling stroke volume is the *degree of stretch of the cardiac muscle cells just before they contract.* The more they are stretched, within physiological limits, the greater the force of contraction will be. Hence, cardiac muscle exhibits a length–tension relationship much like that described for skeletal muscle in Chapter 9. Stretching the muscle fibers (and sarcomeres) increases the number of active cross bridge attachments that can be made between the myosin and actin filaments and the degree of shortening. While skeletal muscles are maintained near or at the optimal length for developing maximum tension by their attachments and the action of antagonists, the important factor stretching cardiac muscle is the amount of blood returning to the heart and distending its ventricles (venous return). The function of this mechanism is to ensure equal outputs of the ventricles and proper distribution of blood volume between the systemic and pulmonary circuits. Should one side of the heart suddenly begin to pump more blood than the other, the increased venous return to the opposite ventricle would force it to pump out an equal volume, thus preventing backup or accumulation of blood in the circulation.

Anything that increases the volume or speed of venous return, such as a slow heart rate or exercise, also increases stroke volume and force of contraction. A slow heartbeat allows more time for ventricular filling. Exercise speeds venous return because of an increased heart rate and the squeezing action of the skeletal muscles on the veins returning blood to the heart. Conversely, low venous return, such as might result from severe blood loss or an extremely rapid heart rate, decreases stroke volume, causing the heart to beat less forcefully.

Although the Frank–Starling law is the model currently accepted to explain the mechanism by which the heart is stretched, it is being challenged by researchers who view the heart as a dynamic suction pump. They propose that deformation of the heart's connective tissue skeleton during contraction results in a natural recoil or rebound of the ventricle walls during diastole, which causes the ventricles to expand. They believe that this sudden expansion creates a negative pressure that literally sucks blood into the heart from the circulation. In either case, the central understanding is that enlargement of the ventricular chambers drives stroke volume and the force of heart contraction. However, as described in the next section, stimulation of the heart by sympathetic nerves also plays a role in determining the force of heartbeat.

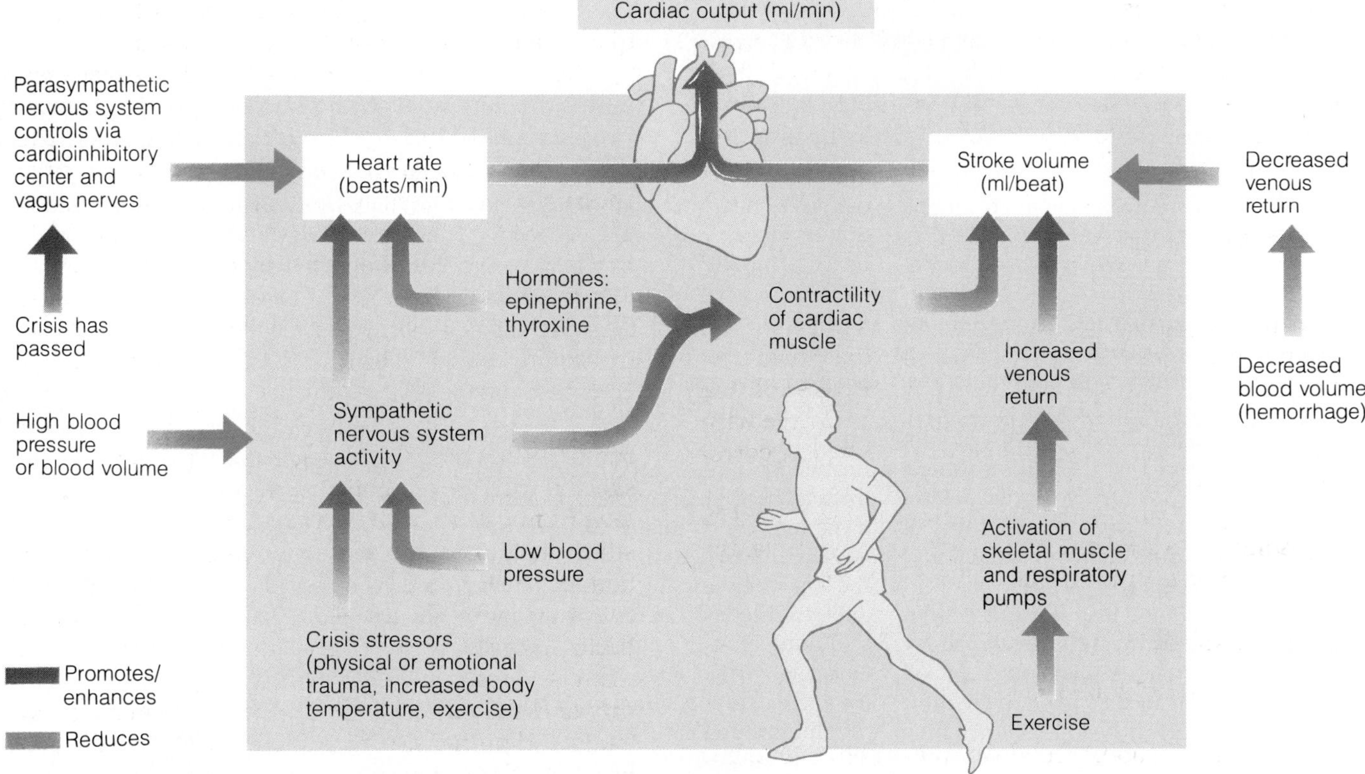

Figure 19.18 Influence of neural controls (autonomic nervous system) and selected other factors on cardiac output. Note that the direction of the flow chart is from the bottom up.

Regulation of Heart Rate

Given a healthy cardiovascular system, stroke volume tends to be relatively constant. However, when blood volume drops precipitously or when the heart has been seriously weakened, stroke volume declines and cardiac output is maintained by increasing the rate of heartbeat. Temporary stressors can also influence heart rate—and consequently cardiac output—by acting through neurally, chemically, and physically induced homeostatic mechanisms. The influence of these various factors on cardiac output is depicted in Figure 19.18 and described briefly next.

Autonomic Nervous System Regulation. The most important extrinsic influences on heart rate are exerted by the autonomic nervous system via cardiac centers located in the medulla oblongata. Activation of the **cardioacceleratory center,** the sympathetic nervous system (SNS) area, results from emotional or physical stressors, such as fright, anxiety, excitement, or exercise. Sympathetic nerve fibers, arising from the T_1–T_5 level of the spinal cord, release norepinephrine, which stimulates the SA and AV nodes, as well as the myocardium itself. The heart responds by beating faster and more vigorously. Notice that sympathetic stimulation enhances *both* the force of myocardial contraction, resulting in a more complete ejection of blood, and the rate of heartbeat; consequently, stroke volume does not decline, as would occur if only heart rate were increased. Because the increased contractile strength resulting from sympathetic stimulation is independent of the EDV, some prefer to refer to this response of the myocardium as increased *contractility,* rather than increased contractile strength (SV) which *is* modified by the EDV.

Activation of the **cardioinhibitory center,** the parasympathetic area, opposes sympathetic effects and effectively reduces heart rate and contractility when the stressful situation has passed. However, the parasympathetic division may be persistently activated in certain emotional conditions, such as grief and severe depression. These cardiac responses are mediated by the release of acetylcholine by vagus nerve fibers that synapse with the SA and AV nodes.

Under resting conditions, both autonomic divisions continuously send impulses to the SA node of the heart, but the predominant influence is inhibitory. Thus, the heart is said to exhibit *vagal tone,* and heart rate is generally slower than it would be if the vagal nerves were not stimulating it. Cutting the vagal nerves results in an almost immediate increase in heart rate of 30 to 40 beats per min, which reflects the inherent depolarization rate of the pacemaking SA node.

A CLOSER LOOK Heart Boosters, Retreads, and Replacements

The sharp, squeezing pain of angina is a red flag that warns of more dire consequences. When a heart attack hits you, it seems as though a lightning bolt has struck. Few events in life are more frightening. A sharp pain crashes through your chest (and perhaps your arm and neck), and unlike an angina attack, the pain doesn't let up. Between 1 and 2 million Americans will suffer a myocardial infarct this year, and perhaps one-third of them will die almost immediately as a result of it.

Strange as it may seem, those who suffer angina or survive a heart attack should consider themselves lucky: They have been given a painful warning that, if heeded, allows time to make dietary changes, take some of the new heart-saving drugs, or have coronary bypass surgery to increase the blood supply to their oxygen-deprived heart. Indeed, coronary bypass surgery is the most frequently performed operation in the United States. Not all people with repeated myocardial ischemia are so fortunate. For as many as 250,000 Americans yearly, their fatal but painless heart attack is the first and last symptom of heart disease. These victims of silent ischemia have no apparent symptoms of heart disease and are at much greater risk of suffering a heart attack, which, even if not fatal, permanently damages the myocardium. What causes silent heart disease is still a mystery; some suggest that these people have abnormal pain mechanisms. Whatever the cause, a severely damaged heart typically begins its downward spiral toward congestive heart failure and death.

The key to prevention of cardiac death is to identify those at risk, early if possible, by using sophisticated medical imaging techniques, such as CAT scans and MRI techniques (see pp. 32–33), as well as other diagnostic techniques and monitors. Traditional cardiac drugs, such as nitroglycerine (which dilates the coronary vessels) and digitalis (which slows the heart, enhancing venous return and sparing the heart's energy), are still popular; but a number of newer, "miracle" drugs, such as beta blockers and calcium channel blockers, are revolutionizing cardiac medicine.

The beta blockers are so named because they interfere with sympathetic nervous system stimulation of the heart. When sympathetic neurotransmitters (or epinephrine) attach to beta receptors, the heart increases its rate and force of contraction. Because beta blockers inhibit this activation and successfully prevent increases in blood pressure, they generally reduce the strain on the heart. Propranolol was the first beta blocker approved by the FDA.

The calcium channel blockers are calcium antagonists that have been useful in preventing the coronary artery spasms that often trigger heart attack. Because calcium is necessary for the vascular smooth muscle contraction that produces the spasm in the first place, the drugs prevent vasoconstriction and dilate the arteries, making it easier for the weakened heart to manage its work load. Calcium channel blockers also inhibit transport of calcium into cardiac muscle cells, but these effects are believed to be counterbalanced by the vasodilating effects of the drugs.

What are the options for someone whose heart is so devastated by disease that these measures are too little, too late? Until recently, heart transplant surgery has been the only hope for a dying heart. But this procedure is riddled with problems. First, an acceptable tissue match must be found. Then, after the painstaking, complicated surgery, the heart recipient must be dosed with drugs that suppress his immune system—enough to thwart rejection of the donor heart, but not enough to be toxic to the recipient. Although heart transplants have been done for the past two decades, the survival rate still leaves much to be desired. For example, the one-year survival rate at the world's leading heart transplant centers is about 37%, but very few heart recipients live to see their tenth postsurgical year.

Newer to the scene are artificial hearts intended to be permanent heart replacements. These mechanical devices have a long history, dating back to the awkward-looking but ingenious perfusion pump—a glass cylinder 2 feet high sprouting a forest of tubes—devised by the famous pilot Charles Lindbergh some 50 years ago. Since then, many mechanical devices have become available, including

When either division of the autonomic nervous system is stimulated more strongly by sensory inputs relayed from various parts of the cardiovascular system, the alternate division is temporarily inhibited. Most such sensory receptors are *baroreceptors*, or *pressoreceptors*, which respond to changes in systemic blood pressure. Because increased cardiac output results in increased systemic blood pressure and vice versa, the regulation of blood pressure often involves reflex controls of heart rate. The neural mechanisms that regulate blood pressure are described in more detail in Chapter 20, but are summarized briefly in Table 19.1 on p. 616.

Chemical Regulation. Chemicals normally present in the blood and other body fluids may influence heart rate, particularly if they become excessive or deficient. Some of these chemical factors are described briefly next.

1. Hormones. Epinephrine, a hormone liberated by the adrenal medulla during periods of SNS activation,

some powered by motorized pendulums or electromagnetic energy. None, however, has managed to overcome the very real problems of undesirable clotting and cell destruction that follow their implantation. In December of 1982, our nation was startled by the news that the first permanent artificial heart had been implanted in a human being, a retired dentist, Dr. Barney Clark. The device, called the Jarvik-7 after its developer, Dr. Robert Jarvik, is an air-driven, aluminum and plastic apparatus that, like the natural heart, has two "ventricles." Barney Clark survived for only ten days with his artificial heart, but since 1982, several more patients have received a Jarvik-7 heart intended as permanent transplants. Hospitalized and tethered to the huge air compressor that drives the artificial heart, such patients have succumbed to strokes, overwhelming infections, and kidney failure.

Because of these disappointing results, the federal government has revised its stance on permanent artificial heart transplants. Each transplant case is reviewed before another is permitted, and new regulations have been instituted that allow the use of an artificial heart only as a lifesaving stopgap measure while the patient awaits a permanent human heart transplant. At its present state of development, the artificial heart does not appear to ensure a healthy person with a long life expectancy. But research efforts are intensive, and a fully implantable device with its own miniature power pack is now undergoing animal tests.

The newest and most exciting technique is the autotransplant, in which the patient's own skeletal muscle is used to form a viable patch in the heart wall or to augment its pumping ability. The procedure, first done in February 1985 on a woman with a large tumor in her heart, involves surgical removal of the damaged portion of the heart wall, suturing the cut edges together, and reinforcing the wall with borrowed latissimus dorsi muscle from the back of the body (see illustration). The graft of the latissimus dorsi muscle is freed from its attachments and wrapped around the heart area to be reinforced and stitched into place, but the rest of the muscle is left attached to the posterior body wall and to its blood and nerve supply. A pacemaker is then attached to the skeletal muscle patch to stimulate it intermittently, gradually working the pacemaker rate up to synchrony with the heartbeat. Because the muscle tissue is "home grown," there is, obviously, no problem of finding a suitable donor heart for transplant. Autotransplants also avoid the cumbersome external power source (compressor) needed by artificial hearts. The major hurdle in autotransplants has been in coaxing the skeletal muscle, which is built to work in short spurts, to perform in the same manner as cardiac muscle—around the clock. But researchers are working on conditioning skeletal muscle with electrical shocks, thereby increasing its percentage of slow-twitch, fatigue-resistant fibers. Also being investigated in animal studies are autotransplant "pouches," made of specifically arranged skeletal muscle bundles and connected to the circulation. It is hoped that this technique may someday be the most promising option for cardiac patients with debilitated hearts.

Although medical science is trying mightily to reproduce the human heart and will no doubt succeed, the best course is still to practice healthy habits and have routine medical checkups to prevent the heart from reaching the point of no return.

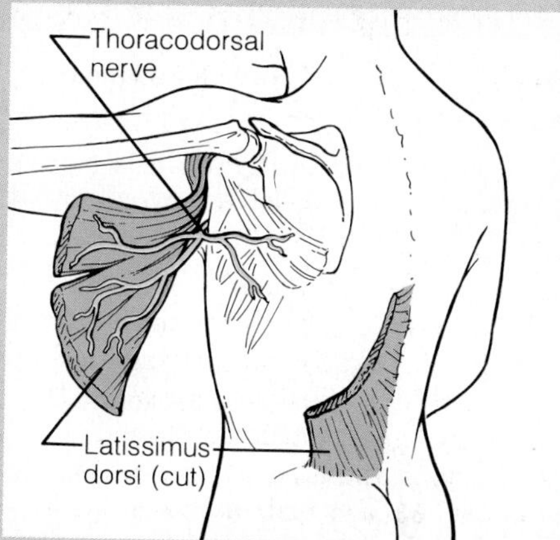

(a)

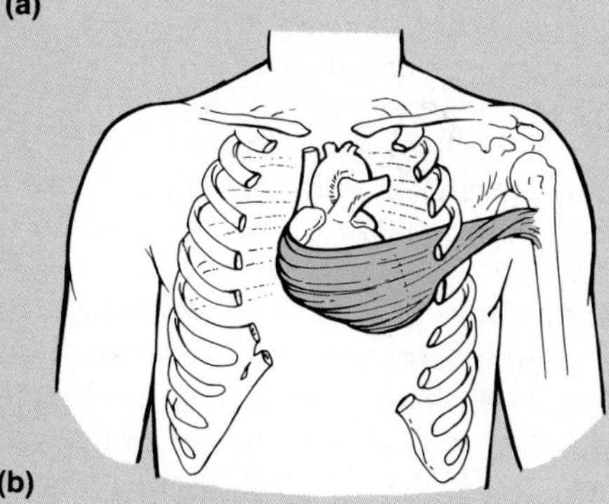
(b)

Schematic representation of the autograft heart wrap procedure.

produces the same cardiac effects as does norepinephrine released by the sympathetic nerves; that is, it enhances heart rate and contractility. Thyroxine, a thyroid gland hormone, causes a slower but more sustained increase in heart rate when it is released in large quantities. It also enhances the effect of epinephrine and norepinephrine on the heart. Severely hyperthyroid individuals may develop a weakened heart when these effects are prolonged.

2. Ions. Physiologic relationships between intracellular and extracellular ions must be maintained for normal heart function. Electrolyte imbalances pose real dangers to the heart.

Reduced blood levels of ionic calcium (hypocalcemia) depress the heart. On the other hand, hypercalcemia tightly couples the excitation–contraction mechanism, causing heart irritability to increase dramatically. In extreme cases, contractions may be so prolonged that the heart stays contracted (calcium rigor), which is fatal.

Table 19.1 Reflex Regulation of Heart Rate by Mechanisms Concerned with Blood Pressure Regulation

Reflex	Location of receptor	Receptor stimulus	Result
Carotid sinus	Site of common carotid artery bifurcation into the internal and external carotid arteries (in neck)	Rising systemic arterial blood pressure stretches the carotid sinus, activating baroreceptors	Stimulation of the cardioinhibitory center causes impulses to be sent, via vagus nerves, to the heart, resulting in decreased heart rate and force and a decline in arterial blood pressure
Aortic	Aortic arch	Rising systemic arterial blood pressure stretches the aortic arch, activating baroreceptors	Same as for the carotid sinus
Bainbridge	Venae cavae and right atrium	Declining venous pressure stimulates baroreceptors	Stimulation of the cardioacceleratory center results in increased rate of sympathetic impulses to the heart that increase the rate and force of heartbeat and blood pressure, thus resulting in faster circulation, which relieves blood congestion in veins

Excesses of ionic sodium and potassium are equally dangerous. Too much sodium (hypernatremia) inhibits transport of ionic calcium into the cardiac cells, thus blocking heart contraction. Excessive potassium (hyperkalemia) interferes with the depolarization mechanism and may lead to heart block and cardiac arrest. Hypokalemia is also life threatening, in that the heart beats feebly and abnormal rhythms appear. ■

Physical Factors. A number of physical factors, including age, gender, exercise, and body temperature, influence heart rate. Resting heart rate is fastest in the fetus (between 140 and 160 beats per min) and gradually decreases throughout life. Average heart rate is faster in females (72 to 80 beats per min) than in males (64 to 72 beats per min).

Exercise promotes increased heart rate by acting through the sympathetic nervous system, which also increases systemic blood pressure and routes more blood to the working muscles. However, in physically fit individuals, resting heart rate tends to be substantially lower than in people out of condition; in trained athletes, it may be as slow as 40 to 60 beats per min. This apparent paradox is explained below.

Heat increases heart rate by increasing the metabolic rate of the cardiac cells. This explains the rapid, pounding heartbeat you feel when you have a high fever and accounts, in part, for the effect of exercise on heart rate (remember, working muscles generate heat). Cold has the opposite effect: It directly decreases heart rate.

Although the heart exhibits normal variations in rate with changes in activity, marked and persistent rate changes usually signal cardiovascular disease. *Tachycardia* (ta″-kih-kar′-dē-uh) is an abnormally fast heart rate, over 100 beats per minute, which may result from elevated body temperature, stress, certain drugs, or heart disease. Because tachycardia occasionally promotes fibrillation, persistent tachycardia is considered pathological.

Bradycardia (brā″-dih-kar′-dē-uh) is a heart rate slower than 60 beats/min. It may be the result of low body temperature, certain drugs, or parasympathetic nervous system activation. It is an anticipated consequence of endurance-type athletic training. As physical and cardiovascular conditioning increases, the heart hypertrophies and its stroke volume increases; thus, resting heart rate can be lower and still provide the same cardiac output. However, persistent bradycardia in poorly conditioned people may result in grossly inadequate blood circulation to body tissues, and bradycardia is a frequent warning of brain edema after head trauma. ■

Homeostatic Imbalance of Cardiac Output

The pumping action of the heart ordinarily maintains a balance between cardiac output and venous return. An important understanding about the dynamics of the cardiovascular system lies in this statement because, were this not so, a dangerous damming up of blood (blood congestion) would occur in the return vessels of the body.

When the pumping efficiency of the heart is depressed so that circulation is inadequate to meet tissue needs, the heart is said to be in *congestive heart*

failure (CHF). Congestive heart failure is usually a progressively worsening condition that reflects weakening of the myocardium by coronary atherosclerosis, persistent high blood pressure, or multiple myocardial infarcts. The manner in which these various conditions damage the myocardium differs, but they are all equally devastating:

1. Coronary atherosclerosis, clogging of the coronary vessels with fatty buildup, impairs blood and oxygen delivery to the cardiac cells. The heart becomes increasingly hypoxic and begins to contract ineffectively.

2. Normally, the pressure in the aorta during diastole is 80 mm Hg, and the left ventricle must exert only slightly over that amount of force to eject blood from its chamber. When aortic blood pressure during diastole rises to 90 mm Hg or more, the myocardium must exert more power to force the aortic valve open to pump out the same amount of blood. Short-term this is no problem; but if this situation becomes chronic, the ESV rises and the myocardium hypertrophies to meet the challenge. Eventually, the stress takes its toll and the myocardium becomes progressively weaker.

3. A succession of myocardial infarcts leads to depressed pumping efficiency because the dead heart cells are replaced by noncontractile fibrous tissue.

Because the heart is a double pump, each side can initially fail independently of the other. If the left side fails, *pulmonary congestion* occurs. The right side of the heart continues to propel blood to the lungs, but the left side does not adequately eject the returning blood into the systemic circulation. Thus, the blood vessels within the lungs become engorged with blood, the pressure within them increases, and fluid leaks from the circulation into the lung tissue, causing pulmonary edema. If untreated, the person suffocates.

If the right side fails, *peripheral congestion* occurs. Blood stagnates within the organs, and pooled fluids in the tissue spaces impair the ability of body cells to obtain adequate nutrients and oxygen and to rid themselves of wastes. Edema is most noticeable in the distal parts of the body: The feet, ankles, and fingers become swollen and puffy.

Failure of one side of the heart puts a greater strain on the opposite side, and ultimately the whole heart fails. A seriously weakened, or *decompensated*, heart is irreparable; treatment is directed primarily toward conserving heart energy and removing the excess leaked fluid with diuretics (drugs that increase excretion of water by the kidneys). However, heart transplants and the use of skeletal muscle to replace damaged heart muscle have provided additional hope to selected cardiac patients (see the box on p. 614). ■

Developmental Aspects of the Heart

The human heart, derived from mesoderm, begins its existence as two simple endothelial tubes that quickly fuse to form a single chamber that is busily pumping blood by the fourth week of gestation (Figure 19.19). During the next three weeks, it undergoes dramatic contortions and major structural changes that convert it into a four-chambered organ capable of acting as a double pump—all without missing a beat! After this time, few changes other than growth occur until birth.

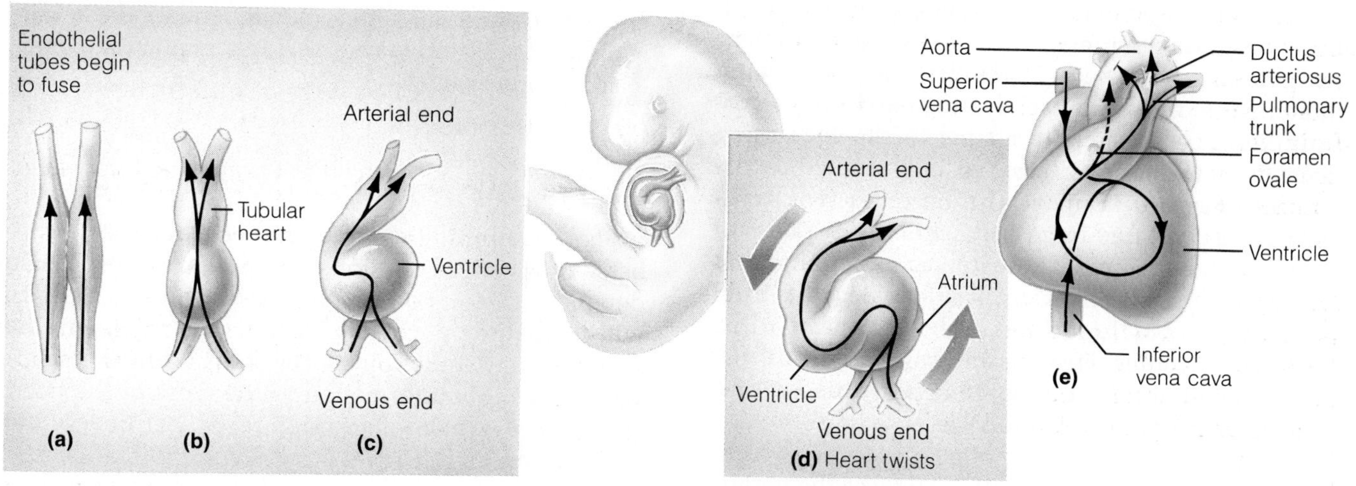

Figure 19.19 Development of the human heart. Arrows show the direction of blood flow.

The interatrial septum of the fetal heart is incomplete. The *foramen ovale* (literally, "oval door,") connects the two atria and allows blood entering the right heart to bypass the pulmonary circuit and the collapsed, nonfunctional fetal lungs. Another lung bypass, the *ductus arteriosus*, exists between the pulmonary trunk and the aorta. (These and other fetal circulatory bypasses are described in more detail in Chapter 29.) At or shortly after birth, these shunts close, completing the separation between the right and left sides of the heart. In the adult heart, the position of the foramen ovale is revealed by the fossa ovalis, and the *ligamentum arteriosum* is the fibrous remnant of the ductus arteriosus (see Figure 19.4c and a).

Congenital heart defects account for nearly half of all infant deaths arising from congenital anomalies. The most common defects are *patent ductus arteriosus*, in which the connection between the aorta and pulmonary trunk remains open, and septal defects. A rare but exceptionally serious inborn heart defect is *tetralogy of Fallot*, which involves four defects: a stenosed pulmonary valve, an aorta that arises from both ventricles, an interventricular septal opening, and an enlarged right ventricle. Children born with tetralogy of Fallot are cyanotic within minutes of birth; that is, they are "blue babies" because their blood is almost entirely bypassing the lungs. Surgical repair must be done as soon as possible if the baby is to survive. A patent ductus arteriosus or foramen ovale allows some mixing of oxygenated and unoxygenated blood, so these problems are much less serious; and the children become cyanotic only after physical exertion. Surgical repair or drug therapy is still necessary if the child is to grow normally. Most congenital heart defects can be traced to environmental influences such as maternal infection or drug intake during the first few weeks of pregnancy when the fetal heart is forming.

In the absence of congenital heart problems, the resilient heart usually functions admirably throughout a long lifetime for most people. Homeostatic mechanisms are normally so effective and efficient that people rarely notice when the heart is working harder. In people who exercise regularly and vigorously, the heart gradually adapts to the increasing demands by increasing in size. Not only does it become a more powerful pump, it also becomes more efficient: Stroke volume increases, while heart rate and blood pressure decrease. Researchers have found that aerobic exercise also helps to clear fatty deposits from blood vessel walls throughout the body, thus retarding the process of atherosclerosis and coronary heart disease. Barring some chronic illnesses, this cardiac response persists into ripe old age.

The key word on the benefits of exercise is *regularity*; regular exercise gradually enhances the endurance and strength of the myocardium. Intermittent vigorous exercise, as enjoyed by weekend athletes, may push an unconditioned heart beyond its ability to respond to unexpected demands and may bring on a myocardial infarct.

Because of the incredible amount of work the heart does during the course of a lifetime, certain anatomical changes are inevitable. Age-related changes that occur in the heart include:

1. **Sclerosis and thickening of the valve flaps.** This occurs particularly where the stress of blood flow is greatest. Because this is the left ventricle, the mitral valve tends to become increasingly incompetent with age. Thus, heart murmurs are more frequently detected in elderly people.

2. **Decline in cardiac reserve.** Although the passing years seem to cause few changes in resting heart rate, the aged heart reacts more poorly to both sudden and prolonged stressors. Sympathetic controls of the heart become less efficient, and heart rate gradually becomes more variable. This decline in maximal heart rate is much less of a problem in physically active people.

3. **Fibrosis of cardiac muscle.** In cases where the nodes of the conduction system become fibrosed, initiation and transmission of the impulse may be hindered, leading to arrhythmias and other conduction problems.

4. **Atherosclerosis.** Although the insidious progress of atherosclerosis begins in childhood, it is accelerated by inactivity, smoking, and stress. The most serious consequences to the heart are hypertensive heart disease and coronary artery occlusion. These conditions, in turn, increase the risk of heart attack and stroke. Although the aging process itself leads to changes in blood vessel walls that result in atherosclerosis, many investigators feel that diet, not aging, is the single most important contributor to cardiovascular disease. There seems to be some agreement that the risk is lowered if people consume less animal fat, cholesterol, and salt. Other recommendations include avoiding stress and taking part in a regular, moderate exercise program.

* * *

As we have seen, the heart is an exquisitely engineered double pump that operates with precision to propel blood into the large arteries leaving its chambers. However, continuous circulation of blood also depends critically on the pressure dynamics existing within the blood vessels. Chapter 20 considers the structure and function of the body's blood vessels and relates this information to the work of the heart to provide a complete picture of cardiovascular functioning.

Related Clinical Terms

Asystole (ā-sis'-tuh-lē) Situation in which the heart fails to contract.

Cardiac catheterization Diagnostic procedure involving passage of a fine catheter (tubing) through a blood vessel into the heart; blood samples are withdrawn, and pressures within the heart are also determined; findings help to detect valve problems, heart deformities, and other heart malfunctions.

Cardiac tamponade (tam'-puh-nād) Compression of the heart by fluids accumulating within the pericardial cavity.

Cor pulmonale (kor pul-mun-nah'-lē) (*cor* = heart; *pulmo* = lung) A life-threatening condition of right-sided heart failure resulting from elevated blood pressure in the pulmonary circuit (pulmonary hypertension); acute cases may develop suddenly as a result of a pulmonary embolism; chronic cor pulmonale is usually associated with chronic lung disorders such as emphysema.

Heart palpitation A heartbeat that is unusually strong, fast, or irregular so that the person becomes aware of it; may be caused by certain drugs, emotional pressures ("nervous heart"), or heart disorders.

Mitral valve prolapse One of the most common valve disorders; one or more of the valve flaps of the mitral valve become incompetent and bulge into the left atrium during ventricular systole, allowing blood regurgitation; mitral valve prolapse is a common indication for valve replacement surgery.

Myocarditis (mī"-ō-kar-dī'-tis) (*myo* = muscle; *card* = heart; *itis* = inflammation) Inflammation of the cardiac muscle layer (myocardium) of the heart; sometimes follows an untreated streptococcal infection in children; may be extremely serious because it can weaken the heart and impair its ability to act as an effective pump.

Chapter Summary

HEART ANATOMY (pp. 595–604)

Size and Location (pp. 595–596)

1. The human heart, about the size of a clenched fist, is located obliquely within the mediastinum of the thorax.

Coverings of the Heart (p. 596)

2. The heart is enclosed within a double sac made up of the outer parietal pericardium (fibrous and serous layers) and the visceral pericardium (epicardium). The pericardial cavity between contains lubricating serous fluid.

Heart Wall (p. 596–597)

3. Layers of the heart wall, from the interior out, are the endocardium, the myocardium (reinforced by a fibrous skeleton), and the epicardium.

Chambers and Associated Great Vessels (pp. 597–599)

4. The heart has two superior atria and two inferior ventricles. Functionally, the heart is a double pump.

5. Entering the right atrium are the superior vena cava, the inferior vena cava, and the coronary sinus. Four pulmonary veins enter the left atrium.

6. The right ventricle discharges blood into the pulmonary trunk; the left ventricle pumps blood into the aorta.

Pathway of Blood Through the Heart (pp. 600–601)

7. The right heart is the pulmonary circuit pump, which serves gas exchange. Oxygen-poor systemic blood enters the right atrium, passes into the right ventricle, through the pulmonary trunk to the lungs, and back to the left atrium via the pulmonary veins.

8. The left heart is the systemic circuit pump. Oxygen-laden blood entering the left atrium from the lungs flows into the left ventricle and then into the aorta, which provides the functional supply of all body organs. Systemic veins return the oxygen-depleted blood to the right atrium.

Heart Valves (pp. 601–603)

9. The atrioventricular valves (tricuspid and bicuspid) prevent backflow into the atria when the ventricles are contracting; the pulmonary and aortic semilunar valves prevent backflow into the ventricles when they are relaxing.

Cardiac Circulation (pp. 603–604)

10. The right and left coronary arteries branch from the aorta to supply the heart itself. Venous blood, collected by the cardiac veins, is emptied into the coronary sinus.

11. Blood delivery to the myocardium occurs during heart relaxation.

HEART PHYSIOLOGY (pp. 604–617)

Properties of Cardiac Muscle (pp. 604–606)

1. Cardiac muscle cells are branching, striated, generally uninucleate cells. They contain myofibrils consisting of typical sarcomeres.

2. Adjacent cardiac cells are connected by intercalated discs containing desmosomes and gap junctions. The myocardium behaves as a functional syncytium because of electrical coupling provided by gap junctions.

3. Cardiac muscle has abundant mitochondria and depends primarily on aerobic respiration to form ATP.

4. The means of action-potential generation in cardiac muscle mimics that of skeletal muscle. Depolarization of the membrane causes opening of sodium channels and sodium entry. The action potential is coupled to sliding of the myofilaments by ionic calcium released by the SR and entering from the extracellular space. Compared to skeletal muscle, cardiac muscle has a prolonged refractory period that prevents tetanization.

5. Certain cardiac muscle cells have automaticity and rhythmicity and can independently initiate action potentials.

Conduction System of the Heart (pp. 606–608)

6. The conduction, or nodal, system of the heart consists of the SA and AV nodes, the AV bundle and bundle branches, and the Purkinje fibers. This system coordinates the depolarization of the heart and ensures that the heart beats as a unit. The SA node has the fastest rate of spontaneous depolarization and acts as the heart's pacemaker; it sets the sinus rhythm.

7. Defects in the intrinsic conduction system can cause arrhythmias, fibrillation, and heart block.

Electrocardiography (pp. 608–609)

8. An electrocardiogram is a graphic representation of the cardiac conduction cycle. The P wave reflects atrial depolarization. The QRS complex indicates ventricular depolarization; the T wave represents ventricular repolarization.

Cardiac Cycle (pp. 609–611)

9. Cardiac cycle refers to events occurring during one heartbeat. During mid-to-late diastole, the ventricles fill and the atria contract. Ventricular systole consists of the isovolumetric contraction phase and the ventricular ejection phase. During early diastole, the ventricles are relaxed and are closed chambers until

increasing atrial pressure forces the AV valves open and the cycle begins again. At a normal heart rate of 75 beats per min, a cardiac cycle lasts 0.8 s.

10. Pressure changes promote blood flow and valve opening and closing.

Heart Sounds (pp. 611–612)

11. Normal heart sounds arise chiefly from the closing of heart valves. Abnormal heart sounds, called murmurs, usually reflect valve problems.

Cardiac Output (pp. 612–617)

12. Cardiac output, typically 5 L per min, is the amount of blood pumped out by each ventricle in 1 minute. Stroke volume is the amount of blood pumped out by a ventricle with each contraction. Cardiac output = heart rate × stroke volume.

13. Stroke volume depends to a large extent on the degree of stretch of cardiac muscle by venous return. Approximately 70 ml, it is the difference between end diastolic volume (EDV) and end systolic volume (ESV). Anything that influences heart rate or blood volume influences venous return, hence stroke volume.

14. Activation of the sympathetic nervous system increases heart rate and contractility; parasympathetic activation decreases heart rate and contractility. Ordinarily, the heart exhibits vagal tone.

15. Chemical regulation of the heart is affected by hormones (epinephrine and thyroxine) and ions (sodium, potassium, and calcium). Imbalances in ions severely impair heart activity.

16. Physical factors influencing heart rate are age, sex, exercise, and body temperature.

17. Congestive heart failure occurs when the pumping ability of the heart is inadequate to provide normal circulation to meet body needs. Right heart failure leads to systemic edema; left heart failure results in pulmonary edema.

DEVELOPMENTAL ASPECTS OF THE HEART (pp. 617–618)

1. The heart begins as a simple (mesodermal) tube that is pumping blood by the fourth week of gestation. The fetal heart has two lung bypasses: the foramen ovale, and the ductus arteriosus.

2. Congenital heart defects account for more than half of infant deaths. The most severe of these is tetralogy of Fallot.

3. Age-related changes include sclerosis and thickening of the valve flaps, declines in cardiac reserve, fibrosis of cardiac muscle, and atherosclerosis.

4. Risk factors for cardiac disease include dietary factors, excessive stress, cigarette smoking, and lack of exercise.

Review Questions

Multiple Choice/Matching

1. When the semilunar valves are open, which of the following are occurring? (a) 2,3,5,6, (b) 1,2,3,7, (c) 1,3,5,6, (d) 2,4,5,7.

(1) coronary arteries fill	(5) blood enters aorta
(2) AV valves are closed	(6) blood enters pulmonary arteries
(3) ventricles are in systole	
(4) ventricles are in diastole	(7) atria contract

2. The portion of the intrinsic conduction system located in the interventricular septum is the (a) AV node, (b) SA node, (c) bundle of His, (d) Purkinje fibers.

3. An ECG provides information about (a) cardiac output, (b) movement of the excitation wave across the heart, (c) coronary circulation, (d) valve impairment.

4. In heart failure, venous return is slowed. Edema results because (a) blood volume is increased, (b) venous pressure is increased, (c) osmotic pressure is increased, (d) none of the above.

5. The fact that the left ventricular wall is thicker than the right reveals that it (a) pumps a greater volume of blood, (b) pumps blood against greater resistance, (c) expands the thoracic cage, (d) pumps blood through a smaller valve.

6. The chordae tendineae (a) close the atrioventricular valves, (b) prevent the AV valve flaps from everting, (c) contract the papillary muscles, (d) open the semilunar valves.

7. In the heart, (1) action potentials are conducted from cell to cell across the myocardium via gap junctions, (2) the SA node sets the pace for the heart as a whole, (3) spontaneous depolarization of cardiac cells can occur in the absence of nerve stimulation, (4) cardiac muscle can continue to contract for long periods of time in the absence of oxygen.
(a) all of the above, (b) 1, 3, 4, (c) 1, 2, 3, (d) 2, 3.

8. The activity of the heart depends on intrinsic properties of cardiac muscle and on neural factors. Thus, (a) vagus nerve stimulation of the heart reduces heart rate, (b) sympathetic nerve stimulation of the heart decreases time available for ventricular filling, (c) sympathetic stimulation of the heart increases its force of contraction, (d) all of the above.

9. Freshly oxygenated blood is received by the (a) right atrium, (b) left atrium, (c) right ventricle, (d) left ventricle.

Short Answer Essay Questions

10. Describe the location and position of the heart in the body.

11. Distinguish between the parietal and visceral pericardia relative to histological structure and location.

12. Trace one drop of blood from the time it enters the right atrium until it enters the left atrium. What is this circuit called?

13. (a) Describe how coronary blood flow is regulated by heart contraction and relaxation. (b) Name the major branches of the coronary arteries, and note the heart regions served by each.

14. The refractory period of cardiac muscle is much longer than that of skeletal muscle. Why is this a desirable functional property?

15. (a) Name the elements of the intrinsic conduction system of the heart in order, beginning with the pacemaker. (b) What is the important function of this conduction system?

16. Draw a normal ECG pattern. Label and explain the significance of its deflection waves.

17. Define cardiac cycle, and follow the events of one cycle.

18. What is cardiac output, and how is it calculated?

19. Discuss how the Frank–Starling law of the heart helps to explain the influence of venous return on stroke volume.

20. (a) Describe the common function of the foramen ovale and the ductus arteriosus in a fetus. (b) What problems result if these shunts remain patent (open) after birth?

Clinical Application Questions

21. You have been called upon to demonstrate the technique for listening to valve sounds. (a) Explain where you would position your stethoscope to auscultate (1) the aortic valve of a patient with severe aortic semilunar valve incompetence and (2) a stenosed mitral valve. (b) During which period(s) would you hear these abnormal valve sounds most clearly? (During atrial diastole, ventricular systole, ventricular diastole, or atrial systole?) (c) What cues would you use to differentiate between an incompetent and a stenosed valve?

22. A middle-aged woman is admitted to the coronary care unit with a diagnosis of left ventricular failure resulting from a myocardial infarction. Her history indicates that she was aroused in the middle of the night by severe chest pain. Her skin is pale and cold, and moist sounds are heard over the lower regions of both lungs. Explain how failure of the left ventricle can cause these signs and symptoms.

CHAPTER 20

The Cardiovascular System: Blood Vessels

Chapter Outline and Student Objectives

Overview of Blood Vessel Structure and Function (pp. 622–628)

1. Describe the three layers that typically form the wall of a blood vessel, and state the function of each.
2. Define vasoconstriction and vasodilation.
3. Compare and contrast the structure and function of the three types of arteries.
4. Describe the structure and function of veins, and explain how veins differ from arteries.
5. Describe the structure and function of a capillary bed.

Physiology of Circulation (pp. 628–640)

6. Define blood flow, blood pressure, and resistance, and explain the relationships between these factors.
7. List and explain the factors that influence blood pressure, and describe how blood pressure is regulated.
8. Define hypertension. Note both its symptoms and consequences.
9. Explain how blood flow is regulated in the body in general and in its specific organs.
10. Define circulatory shock. Note several possible causes.
11. Outline the factors involved in capillary dynamics, and explain the significance of each.

Circulatory Pathways (pp. 640–658)

12. Trace the pathway of blood through the pulmonary circulation, and state the importance of this special circulation.
13. Describe the general functions of the systemic circulation. Name and give the location of the major arteries and veins in the systemic circulation.
14. Describe the structure and special function of the hepatic portal circulation.

Developmental Aspects of the Blood Vessels (pp. 658–659)

15. Explain how blood vessels develop in the fetus.
16. Provide examples of changes that often occur in blood vessels as a person ages.

Preview of Selected Key Terms

Arteries Blood vessels that conduct blood away from the heart and into the circulation.

Capillaries The smallest of the blood vessels and the sites of exchange between the blood and tissue cells.

Veins Blood vessels that return blood toward the heart from the circulation.

Blood flow The amount of blood flowing through a vessel or organ at a particular time.

Blood pressure The force exerted by blood against a unit area of the blood vessel walls; differences in blood pressure between different areas of the circulation provide the driving force for blood circulation.

Peripheral resistance A measure of the amount of friction encountered by blood as it flows through the blood vessels.

Vasomotor fibers Sympathetic nerve fibers that regulate the contraction of smooth muscle in the walls of blood vessels, thereby regulating blood vessel diameter.

Autoregulation The automatic adjustment of blood flow to a particular body area in response to its current requirements.

The **blood vessels** of the body form a closed delivery system that begins and ends at the heart. The idea that blood circulates in the body dates back only about 300 years to the inspired experiments of William Harvey, a seventeenth-century English physician. Prior to that time, it was thought, as proposed by the ancient Greek physician Galen, that blood moved through the body like an ocean tide, first moving out from the heart (to get rid of its impurities in the lungs) and then ebbing back into the heart via the same vessels.

Although blood vessels are often compared to a system of plumbing pipes within which blood is circulated, this analogy serves only as a starting point. Unlike stationary pipes, blood vessels are dynamic structures that pulsate, constrict and relax, and even proliferate, as demanded by the changing needs of the body. In this chapter, we will examine the structure and function of these important circulatory pathways through the body.

Overview of Blood Vessel Structure and Function

The three major types of blood vessels are *arteries*, *capillaries*, and *veins*. As the heart alternately contracts and relaxes, blood is forced into the large arteries leaving its chambers. It then moves into successively smaller arteries, finally reaching their smallest branches, the *arterioles* (ar-tēr′-ē-ōlz), which feed into the capillary beds of all body organs and tissues. Blood draining from the capillaries is collected by *venules* (ven′-yoolz), small veins that merge to form larger veins that ultimately empty into the great veins converging on the heart. Altogether, the blood vessels carry blood on a journey that stretches for about 60,000 miles through the internal body landscape!

Arteries and veins simply act as conduits for blood. Only the hairlike capillaries come into intimate contact with tissue cells and directly serve cellular needs, because exchanges between the blood and tissue cells occur primarily through the gossamer-thin capillary walls.

The Structure of Blood Vessel Walls

The walls of all blood vessels, except capillaries, are composed of three distinct layers, or *tunics*, surrounding a central tubular opening, the vessel *lumen* (Figure 20.1).

The innermost layer that lines the vessel lumen is called the **tunica intima** (too′-nih-kuh in′-tih-muh), or **tunica interna** (in-ter′-nuh). The tunica intima consists of two layers: a thin layer of *endothelium* (simple squamous epithelial cells) underlain by a sparse connective tissue *basement membrane*. Essentially, the endothelium is a continuation of the endocardial lining of the heart, and it is the only tunic present in all blood vessels. The endothelial cells fit closely together, forming a slick surface that helps decrease friction as blood moves through the vessel lumen.

The middle layer, the **tunica media** (mē′-dē-uh), consists mostly of circularly arranged smooth muscle cells and elastic connective tissue fibers. The activity of the smooth muscle is regulated by *vasomotor fibers* of the sympathetic division of the autonomic nervous system. Depending on the precise needs of the body at any moment, the vasomotor fibers can cause **vasoconstriction** (reduction in lumen diameter due to smooth muscle contraction) or **vasodilation** (widening of the lumen due to smooth muscle relaxation). Since small changes in blood vessel diameter greatly influence blood flow and blood pressure, the activities of the tunica media are critical in regulating circulatory dynamics. Generally, the tunica media is the bulkiest layer in arteries, which bear the greatest responsibility for maintaining blood pressure and continuous blood circulation. These functional considerations are described shortly.

The outermost layer of a blood vessel wall, the **tunica adventitia** (ad-ven-tih′-shuh), or **tunica externa** (ex-ter′-nuh), is composed largely of loosely woven collagen fibers that protect the blood vessel and anchor it to surrounding structures. The tunica adventitia is infiltrated with nerve fibers and lymphatic vessels and, in the larger arteries and veins, a system of tiny blood vessels. These vessels, called the *vasa vasorum* (va′-suh va-sor′-um)—literally, "vessels of the vessels"—supply nutrient-rich blood to the tissues of the blood vessel wall.

The three vessel types vary in length, diameter, and the relative thickness and tissue makeup of their walls. These differences are described next.

Arteries

Arteries are vessels that transport blood away from the heart. Some people have the misconception that arteries always carry oxygenated blood, whereas veins always carry oxygen-poor blood. Although this is true in most instances, it is not true for the vessels of the pulmonary circulation (see Figure 20.13) or for the special umbilical vessels of a fetus (see Figure 29.13a).

In terms of relative size and function, arteries can be divided into three groups—elastic arteries, muscular arteries, and arterioles.

Elastic Arteries

Elastic arteries are the large, thick-walled arteries close to the heart, such as the aorta and its major branches. These arteries are both the largest in diameter and the most elastic. Their large-diameter lumen allows them

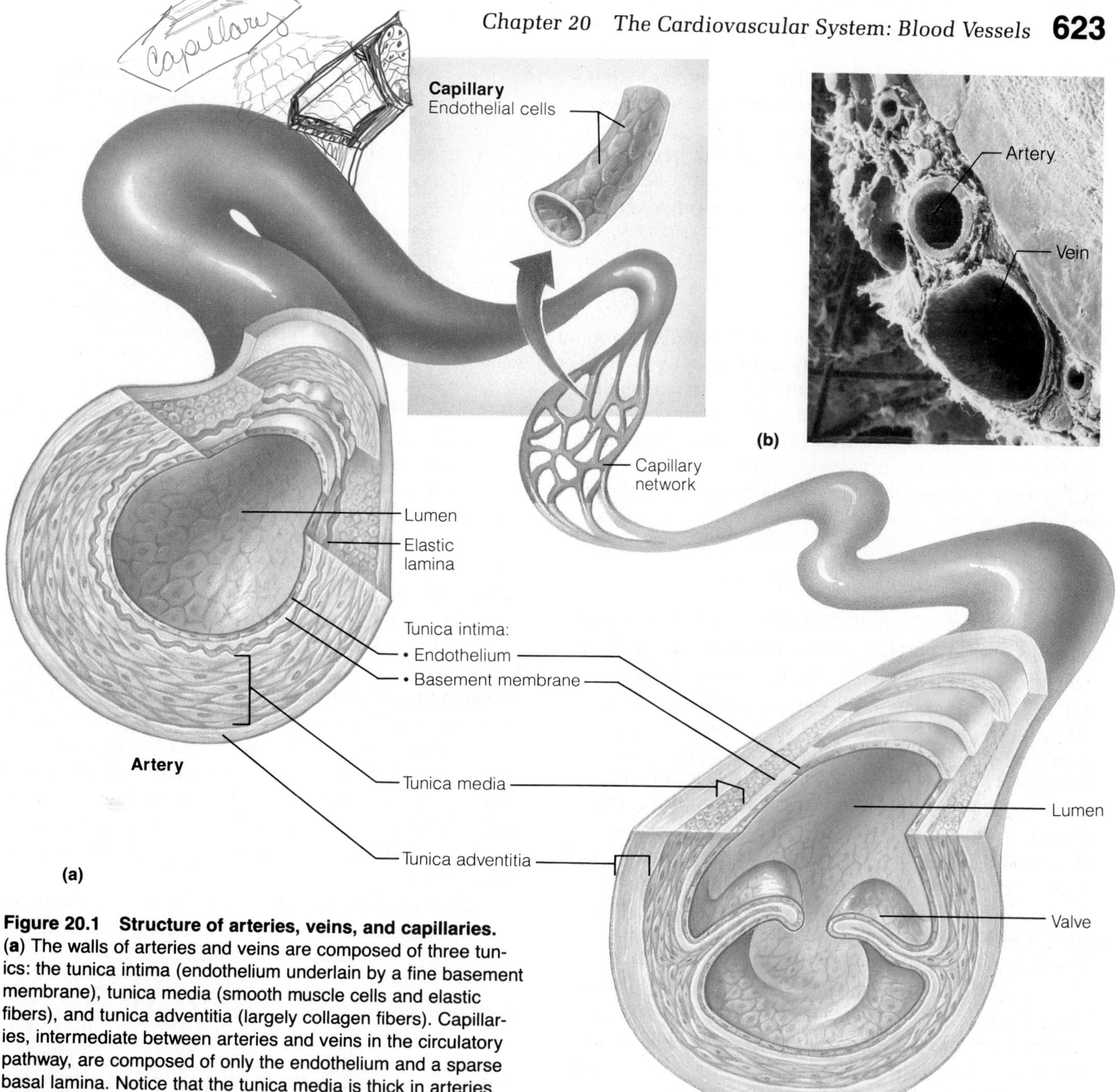

Figure 20.1 Structure of arteries, veins, and capillaries. (**a**) The walls of arteries and veins are composed of three tunics: the tunica intima (endothelium underlain by a fine basement membrane), tunica media (smooth muscle cells and elastic fibers), and tunica adventitia (largely collagen fibers). Capillaries, intermediate between arteries and veins in the circulatory pathway, are composed of only the endothelium and a sparse basal lamina. Notice that the tunica media is thick in arteries and thin in veins, while the tunica adventitia is thin in arteries and relatively thicker in veins. (**b**) Scanning electron micrograph of an artery and vein in cross section (120X).

to serve as low-resistance conduits to convey blood from the heart to medium-sized arteries; for this reason, they are sometimes referred to as *conducting arteries*. In addition to a generous amount of elastic tissue in the tunica media, these arteries have a stretchy sheet of elastic fibers called the *elastic lamina* (la′-mih-nuh) on each face of the media, which enables the arteries to withstand large pressure fluctuations by expanding when the heart contracts, forcing blood into them, and recoiling as blood flows forward into the circulation during heart relaxation. Although elastic arteries also contain substantial amounts of smooth muscle, they are relatively inactive in vasoconstriction. Thus, functionally, the elastic arteries can be visualized as simple elastic tubes.

Because the elastic arteries can expand and recoil passively to accommodate changes in blood volume, the blood is kept under continuous pressure. Consequently, blood flows fairly continuously rather than starting and stopping with the rhythm of the heartbeat. If the blood vessels become hard and unyielding, as in arteriosclerosis, then blood flows more intermittently, similar to the way water flows through a hard rubber garden hose attached to a faucet. When the faucet (the pump) is on, the water pressure in the hose is high, and water gushes out with a good deal of force. But when the faucet is shut off, the water flow abruptly becomes a trickle and then stops, because the hose walls cannot recoil to keep the water under pressure.

Muscular Arteries

Muscular arteries are medium- and smaller-sized arteries farther along in the circulatory pathway that carry blood to specific body organs. Their tunica media contains proportionately more smooth muscle and less elastic tissue than that of elastic arteries; therefore, they are more active in vasoconstriction and are less distensible. Because they distribute blood throughout the body, they are also called *distributing arteries*.

Arterioles

Arterioles have a lumen diameter smaller than 0.5 mm and are the smallest of the arterial vessels. The larger arterioles exhibit all three tunics, and their tunica media is chiefly smooth muscle with a few scattered elastic fibers. The walls of the smallest arterioles, which directly feed into the capillary beds, are little more than smooth muscle cells that coil like a spiral staircase around the tunica intima lining (Figure 20.2).

As described shortly, vasoconstriction and vasodilation of arterioles in response to changing stimuli from vasomotor nerve fibers and local chemical influences are the most important factors determining blood flow into the capillary beds on a minute-to-minute basis. When the arterioles constrict, the tissues served are largely bypassed; when the arterioles dilate, blood flow into the local capillaries increases dramatically.

Arterial Pulse

The alternating expansion and recoil of elastic arteries during each cardiac cycle creates a pressure wave—a **pulse**—that is transmitted through the arterial tree with each heartbeat. You can feel a pulse in any artery lying close to the body surface by compressing the artery against firm tissue, and this provides an easy way of counting heart rate. Because it is so accessible, the point where the radial artery surfaces at the wrist (the radial pulse) is routinely used to take a pulse measurement, but there are several other clinically important arterial pulse points (Figure 20.3). Because these same points are compressed to stop blood flow into distal tissues during hemorrhage, they are also called *pressure points*. For example, if you seriously lacerate your hand, you can stop the bleeding somewhat by compressing the radial artery or the brachial artery.

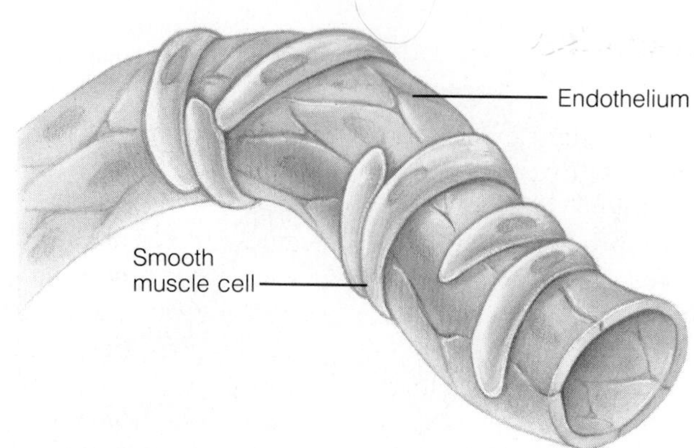

Figure 20.2 Structure of a small arteriole.

Since the arterial pulse rate reflects heart rate, it is influenced by activity, postural changes, and emotions. For example, the pulse of a normal man may be around 66 beats per minute when he is lying down, increase to 70 when he sits up, and rise to 80 when he suddenly stands. During vigorous exercise or emotional upset, pulse rates between 140 and 180 are not unusual.

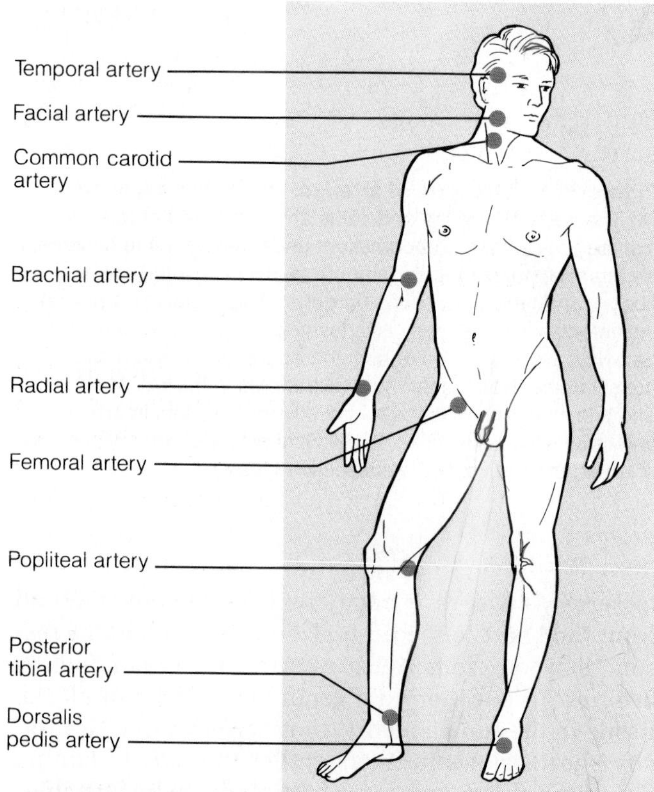

Figure 20.3 Body sites where the pulse is most easily palpated. (The specific arteries indicated are discussed on pp. 644–651.)

Capillaries

The microscopic **capillaries** form dense networks that branch throughout nearly all body tissues. Their exceedingly thin walls consist of just the tunica intima (see Figure 20.1a), and in some cases, one endothelial cell forms the entire circumference of the capillary wall. The average capillary length is 1 mm; average lumen diameter is 0.01 mm, just large enough for red blood cells to slip through in single file.

If blood vessels are compared to a system of expressways and roads, then the capillaries are the back alleys and driveways providing direct access to every cell in the body. Although most tissues have a rich capillary supply, there are exceptions. Tendons and ligaments are poorly vascularized; cartilage and the transparent cornea and lens of the eye have no capillary supply at all.

Devoid of muscle or connective tissue, capillaries are ideally suited for the exchange of materials between the blood and interstitial fluid. The mechanisms of these exchanges are described later in this chapter; here we will focus on capillary structure. Structurally, capillaries are classified as either continuous or fenestrated. *Continuous capillaries,* abundant in the skin and muscles, are common; they are continuous in the sense that their endothelial cells provide an uninterrupted lining. Adjacent cells are joined laterally by tight junctions that are, however, usually incomplete, leaving intercellular *clefts* just large enough to allow limited passage of fluids and small solutes. Typically, the endothelial cell cytoplasm contains numerous small pinocytotic vesicles believed to ferry fluids across the capillary wall by bulk transport. The major exception to the structure just described is seen in the brain. There the tight junctions of the continuous capillaries are complete and constitute the structural basis of the blood-brain barrier, described in Chapter 12.

In *fenestrated* (feh′-neh-strā-tid) *capillaries,* the endothelial cells are joined by gap junctions and are riddled with oval pores or *fenestrations* (Figure 20.4), which are usually covered by a very thin membrane or diaphragm. The pores of this capillary variety increase their permeability to fluids and small solutes. Fenestrated capillaries are found where active capillary absorption occurs, such as in the mucosa of the small intestine and in endocrine organs, where hormones gain rapid entry into the blood. Fenestrated capillaries with perpetually open pores occur in the glomerular capillaries of the kidneys, where rapid filtration of blood plasma is essential.

In most body regions, a capillary bed consists of two types of vessels: (1) *thoroughfare channels,* short vessels that directly connect the arteriole and venule at opposite ends of the bed, and (2) *true capillaries* (Figure 20.5). The true capillaries typically number 10 to 100 per capillary bed, depending on the organ or tissues served. They usually branch off the proximal end of the thoroughfare channel and return to its distal end, but occasionally, they spring from the feeder arteriole and empty directly into the venule. Typically, a cuff of smooth muscle, called a *precapillary sphincter,* surrounds the origin of each branching group of true capillaries and acts as a valve to regulate the flow of blood into those capillaries. However, there is variation: In some organs, such as the intestine, precapillary sphincters encircle the origin of each true capillary; and in the heart, no precapillary sphincters are found. Blood flowing through a terminal arteriole may take one of two routes: through the true capillaries or through the thoroughfare channel. Blood flowing through true capillaries takes part in exchanges with tissue cells, whereas blood flowing through the thoroughfare shunts bypasses the tissue cells and thus forgoes the exchange process.

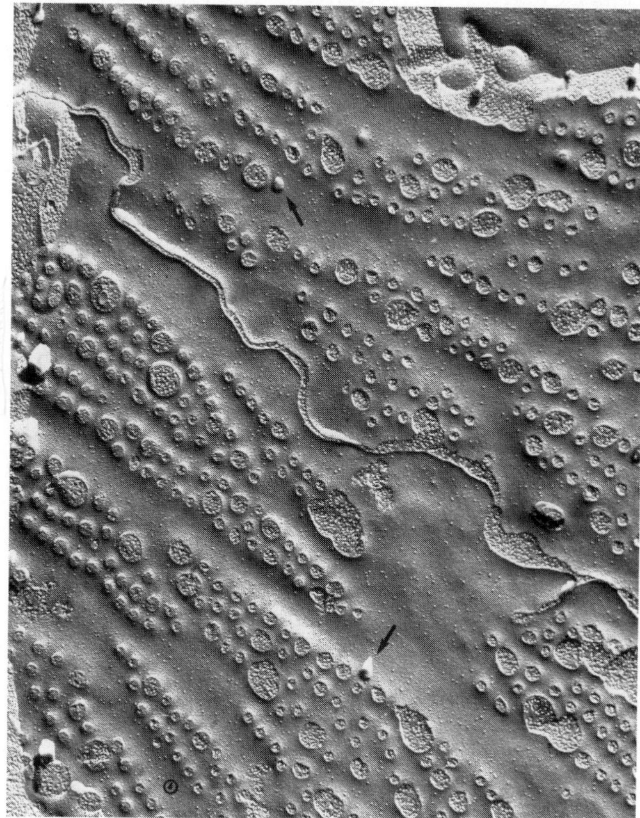

Figure 20.4 Freeze-fracture preparation of the outer face of two fenestrated endothelial cells. The fenestrae appear as granular-looking circular areas (119,000X).

The relative amount of blood flowing into the true capillaries is regulated by vasomotor nerve fibers and local chemical conditions. A capillary bed may be flooded with blood or almost completely bypassed, depending on conditions in the body or in that spe-

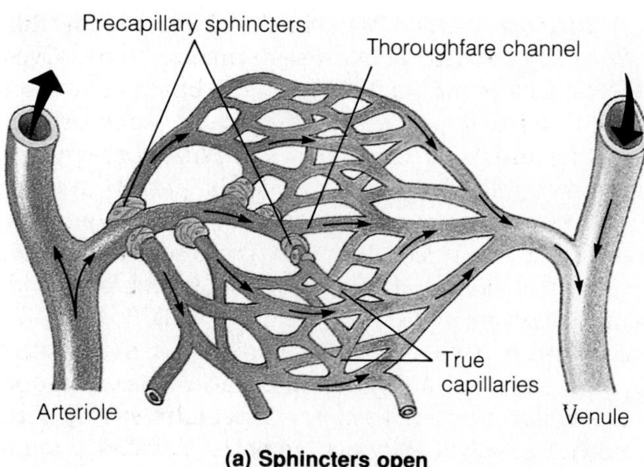

(a) Sphincters open

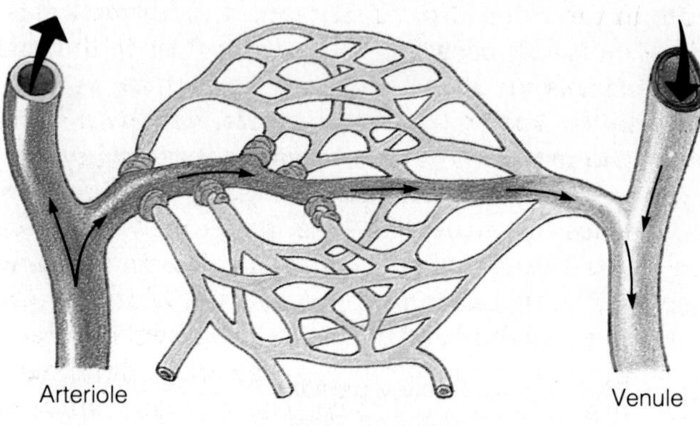

(b) Sphincters closed

Figure 20.5 Anatomy of a capillary bed. Thoroughfare channels act as shunts to bypass the true capillaries when precapillary sphincters controlling blood entry into the true capillaries are constricted.

cific organ. For example, suppose that you have just eaten and are sitting relaxed, listening to your favorite musical group. Digestion is going on, and blood is circulating freely through the true capillaries of your gastrointestinal organs to receive the breakdown products of digestion. On the other hand, most of these same capillary pathways are closed between meals, the blood bypassing most of the true capillaries in the digestive viscera. Further, when the skeletal muscles are engaged in vigorous activity, blood is rerouted from the digestive organs (food or no food) to the muscle capillary beds, where it is more immediately needed to meet the needs of the working muscle cells. This helps to explain why vigorous exercise immediately after a meal can cause indigestion or abdominal cramps.

Highly modified capillaries called **sinusoids** (sī′-nyoo-soydz″) connect the arterioles and venules in the liver, lymphatic tissues, and some endocrine organs. Sinusoids have large, irregularly shaped lumens, and they are often lined, at least in part, by phagocytic rather than endothelial cells. Blood flows sluggishly through the tortuous sinousoid channels, allowing it to be processed in various ways. For example, the liver receives venous blood, draining from the digestive organs, which contains small amounts of bacteria. The slow blood flow allows time for the phagocytic cells to perform their cleanup job of engulfing and destroying the bacteria.

Veins

Venules, which represent the initial part of the venous return to the heart, are formed when capillaries unite. The smallest venules consist entirely of the tunica intima and tunica adventitia; in larger venules, spiraling smooth muscle fibers form a sparse tunica media layer. Venules join to form **veins,** which usually have three distinct tunics, but their walls are always much thinner and their lumens much larger than those of corresponding arteries (see Figure 20.1). The tunica media, in particular, tends to be thin, and there is relatively little smooth muscle, even in the largest veins. The tunica adventitia is the heaviest wall layer and is often several times thicker than the tunica media. In the largest veins—the venae cavae, which return blood directly to the heart—the tunica adventitia is further thickened by longitudinal bands of smooth muscle.

With their large lumens and thin walls, veins can accommodate a fairly large blood volume. Since up to 65% of the body's total blood supply is found in the veins at any time, veins are called *capacitance vessels*, or *blood reservoirs*. Even so, the veins are normally only partially filled with blood; consequently, they tend to be partially collapsed. Because the blood pressure within them is low, veins can be much thinner walled than arteries without danger of bursting. However, the low pressure in the veins demands some adaptations if venous return to the heart is to equal cardiac output. The large-diameter lumens of veins (which offer relatively little resistance to blood flow) are one structural adaptation; another is the presence of valves that prevent blood from flowing backward. Valves are especially prevalent in the veins of the limbs, which must return blood to the heart against gravity. Venous valves are formed from folds of the tunica intima, and they resemble the semilunar valves of the heart in both structure and function.

To demonstrate the effectiveness of your venous valves in preventing the backflow of blood, conduct the following simple experiment. Allow one hand to hang by your side until the blood vessels on its dorsal aspect become distended with blood. Then place two fingertips against one of the distended veins, and pressing firmly, move the superior finger proximally along the vein and then release that finger. The vein will remain flattened and collapsed, despite the pull

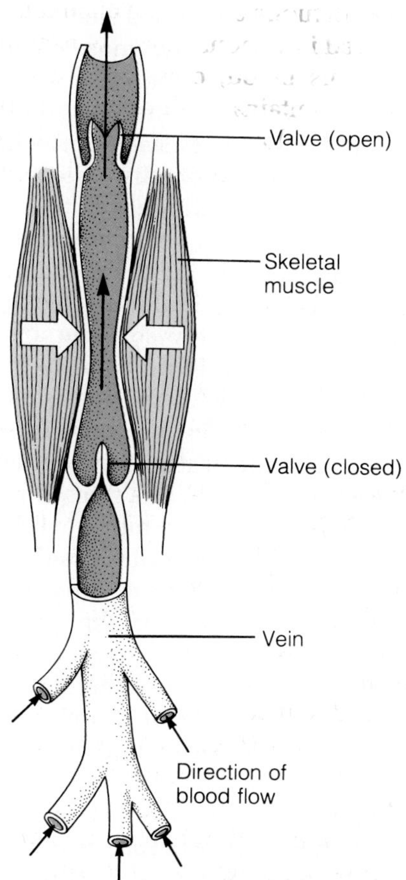

Figure 20.6 Relationship of skeletal muscle activity to venous valve function. As illustrated, when skeletal muscles contract and press against the flexible veins, the valves are forced open, and blood is propelled toward the heart. The valves are closed by the backflowing blood behind (distal to) the point of contraction when the skeletal muscles relax.

of gravity. Finally, remove the distal fingertip and watch as the vein rapidly refills with blood.

Functional adaptations, the respiratory and skeletal muscle "pumps," also aid venous return. Pressure changes occurring in the ventral body cavity during breathing create the *respiratory pump* that sucks blood upward toward the heart. As we inhale, abdominal pressure increases, squeezing the local veins; since venous valves prevent backflow, blood is forced toward the heart. At the same time, the pressure within the chest decreases, allowing thoracic veins to expand and speeding blood entry into the right atrium of the heart. Skeletal muscle activity is even more important: As the skeletal muscles surrounding the veins contract and relax, they "milk" blood toward the heart, and once blood has passed each successive valve, it cannot flow back (Figure 20.6). People who earn their livings in "standing professions," such as hairdressers and dentists, often have swollen ankles. This is a result of blood pooling in their feet and legs during prolonged periods of skeletal muscle inactivity.

Varicose veins are veins that have become tortuous and dilated because of incompetent valves. Several factors may contribute, including heredity and conditions that hinder venous return, such as prolonged standing in one position, obesity, or pregnancy. Both the potbelly of an overweight person and the enlarged uterus of a pregnant woman exert downward pressure on the vessels of the groin, restricting the return of blood to the heart. Consequently, blood tends to accumulate in the lower limbs, and with time, the valves weaken and the venous walls stretch and become floppy. Superficial veins, which receive less support from surrounding tissues, are especially susceptible. Another cause of varicose veins is elevated venous pressure. For example, straining to have bowel movements in chronic constipation creates high blood pressure in the veins of the anal canal; venous wall deterioration finally occurs, causing varicosities in the anal canal, or *hemorrhoids* (hem′-mer″-oydz). ■

Highly specialized veins called **venous sinuses** are flattened veins with extremely thin walls composed only of endothelium. They are supported by the tissues that surround them, rather than by tunica media and tunica adventitia layers. The intracranial venous sinuses, called the *dural sinuses,* which receive cerebrospinal fluid and blood draining from the brain, are underlaid and reinforced by the tough dura mater that covers the brain surface. The *coronary sinus* of the heart, which drains the myocardium, is another example.

Vascular Anastomoses

When vascular channels unite or interconnect, **vascular anastomoses** (uh-nas″-tō-mō′-sēz) (*anastomos* = coming together) are formed. Most organs receive blood from more than one arterial branch, and the arteries supplying the same territory often merge with one another, forming *arterial anastomoses.* Arterial anastomoses permit free communication between the vessels involved and provide alternate pathways for blood to reach a given organ. If one branch becomes blocked by a clot or atheroma, the alternate—or collateral—channel helps ensure that the organ receives adequate nutrition. Arterial anastomoses are abundant around joints, where active movement may hinder blood flow through one channel, and in the abdominal organs. Arteries that do not anastomose, such as those supplying the toes and fingers, are called *end arteries.* If their blood flow is interrupted, the cells supplied by that branch die and decompose, eventually causing gangrene. The tissue first takes on a bruised appearance and then becomes cold and black.

The thoroughfare channels of capillary beds that connect arterioles and venules are examples of *arteriovenous anastomoses.* Veins anastomose much more freely than arteries, and *venous anastomoses* are common; that is, many veins interconnect freely with others along their entire length. As a result, occlusion of venous channels is rarely a cause of tissue death and deterioration (necrosis).

Physiology of Circulation

To sustain life, blood must be kept circulating. It is easy to accept this statement, but to understand how circulation is accomplished takes a bit more effort. By now, you are aware that the heart is the pump, the arteries are conduits, the arterioles are resistance vessels, the capillaries are exchange sites, and the veins are blood reservoirs. Now we need to define three physiologically important terms—blood flow, blood pressure, and resistance—and examine just how these factors are related to the physiology of blood circulation.

Blood Flow, Blood Pressure, and Resistance

Blood flow is the actual volume of blood flowing through a vessel, an organ, or the entire circulation in a given period of time. When we are considering the entire vascular system, blood flow is equivalent to cardiac output, and under resting conditions, it is relatively constant. At a given time, however, blood flow to individual body organs may vary widely and is intimately related to their immediate needs.

Blood pressure is the force per unit area exerted on the wall of a blood vessel by its contained blood. Differences in blood pressure within the vascular system provide the immediate driving force that keeps blood moving through the body. Blood pressure is expressed in terms of millimeters of mercury (mm Hg). For example, a blood pressure of 120 mm Hg would be equal to the pressure exerted by a column of mercury 120 mm high. Unless stated otherwise, the term *blood pressure* is generally understood to mean systemic arterial blood pressure in the largest arteries near the heart.

Resistance is opposition to flow and is a measure of the amount of friction the blood encounters as it passes through the vessels. Since most friction is encountered in the peripheral circulation, away from the heart, we generally use the term *peripheral resistance*. There are three important sources of resistance: blood viscosity, vessel length, and vessel diameter.

Viscosity (vis-kah′-sih-tē) is the internal resistance to flow of a fluid and may be thought of as the thickness or "stickiness" of a fluid. The greater the viscosity, the less easily molecules slide past one another and the more difficult it is to get and keep the fluid moving. Blood is substantially more viscous than water because it contains formed elements and plasma proteins; consequently, it flows more slowly under the same conditions. The relationship between vessel length and resistance is very straightforward: the longer the vessel, the greater the resistance encountered. Since both blood viscosity and blood vessel length are normally unchanging, the influence of these factors can be considered constant in healthy people.

On the other hand, blood vessel diameter changes frequently and is a very important factor in altering peripheral resistance. How so? The answer lies in principles of fluid flow. Fluid close to the walls of a tube or channel is slowed by friction as it passes along the wall, whereas fluid in the center of the channel flows more freely and faster. You can verify this by watching the flow of water in a river. The water close to the bank hardly seems to move, while that in the middle of the river flows quite rapidly. Further, the smaller the tube (or blood vessel), the greater the friction, because relatively more of the fluid contacts the tube walls. Because arterioles are small-diameter vessels to begin with and can enlarge or constrict in response to neural and chemical controls, they are the major determinants of peripheral resistance on a minute-to-minute basis.

Now that we have defined these terms, let us summarize briefly the relationships between them. Blood flow is directly proportional to the difference in blood pressure (ΔP) between two points in the circulation; that is, if the ΔP increases, then blood flow increases, and vice versa. In addition, blood flow is inversely proportional to the resistance (R); that is, if R increases, then blood flow decreases. These relationships can be stated by the formula

$$\text{Blood flow} = \frac{\Delta P}{R}$$

Of the two factors influencing blood flow, resistance is far more important than ΔP, because if the diameters of the arterioles serving a particular tissue increase (thus decreasing the resistance), then blood flow to that tissue also increases, even though the systemic pressure is unchanged or may actually be declining.

Systemic Blood Pressure

Any fluid driven by a pump through a circuit of closed channels operates under pressure, and the nearer the fluid is to the pump, the greater the pressure it is under. The dynamics of blood flow in blood vessels is no exception, and blood flows through the blood vessels along a pressure gradient, always moving from higher- to lower-pressure areas. Fundamentally, the pumping action of the heart generates blood flow. Pressure results when flow is opposed by resistance.

As the left ventricle contracts and expels blood into the aorta, it imparts kinetic energy to the blood, which in turn stretches the elastic walls of the aorta and causes aortic pressure to reach its peak. Indeed, if the aorta were opened during this period, blood would spurt upward for 5 or 6 feet! This pressure peak, called the **systolic** (sis-tō′-lik) **arterial blood pressure,** averages about 120 mm Hg in healthy adults. Blood moves forward into the arterial bed because the pressure in the aorta is higher than the pressure in the more distal vessels. During diastole, closure of the aortic semilunar valve prevents blood from flowing back into the heart, and the walls of the aorta (and other elastic arteries) recoil, maintaining continuous pressure on the reducing blood volume. Consequently, blood continues to flow into the smaller vessels. During this time, aortic pressure drops to its lowest level (approximately 70–80 mm Hg in healthy adults), called the **diastolic** (dī-us-tō′-lik) **pressure.** Thus, you can picture the elastic arteries as auxiliary pumps that operate throughout diastole, when the heart relaxes, to deliver to the vascular system the volume and energy of blood stored within them during systole.

Since aortic pressure oscillates up and down with each heartbeat, the important pressure figure to consider is the *average*, or *mean, arterial pressure,* because it is this pressure that drives the blood to the tissues throughout the cardiac cycle. Because diastole lasts longer than systole, the mean pressure is not simply the value halfway between systolic and diastolic pressures; instead, it is approximately equal to the diastolic pressure plus one-third of the *pulse pressure* (the difference between the systolic and diastolic pressures). Thus, a person with a systolic blood pressure of 120 mm Hg and a diastolic pressure of 80 mm Hg would have a mean arterial pressure of approximately 93 mm Hg [80 + ⅓(40)]. The pulse pressure is the actual working pressure of the blood and is significant because it is the pressure difference that drives blood through the circulatory system. Without such pressure differences, blood would cease to flow.

As illustrated in Figure 20.7, systemic blood pressure is highest in the aorta and declines throughout the length of the pathway to finally reach 0 mm Hg in the right atrium. The steepest change (drop) in blood pressure occurs in the arterioles, which offer the greatest resistance to blood flow. However, so long as a pressure gradient exists, blood continues to flow from higher- to lower-pressure areas until it completes the circuit back to the heart. Unlike arterial pressure, which pulsates with each contraction of the left ventricle, venous blood pressure is steady and changes very little during the cardiac cycle. The entire pressure gradient in the veins, from venules to the termini of the venae cavae, is only about 10 mm Hg as opposed to an average pressure gradient of 50 mm Hg in the arteries. This very low pressure in the venous system reflects the cumulative effects of peripheral resistance, which dissipates most of the energy of blood pressure (as heat) during each circuit.

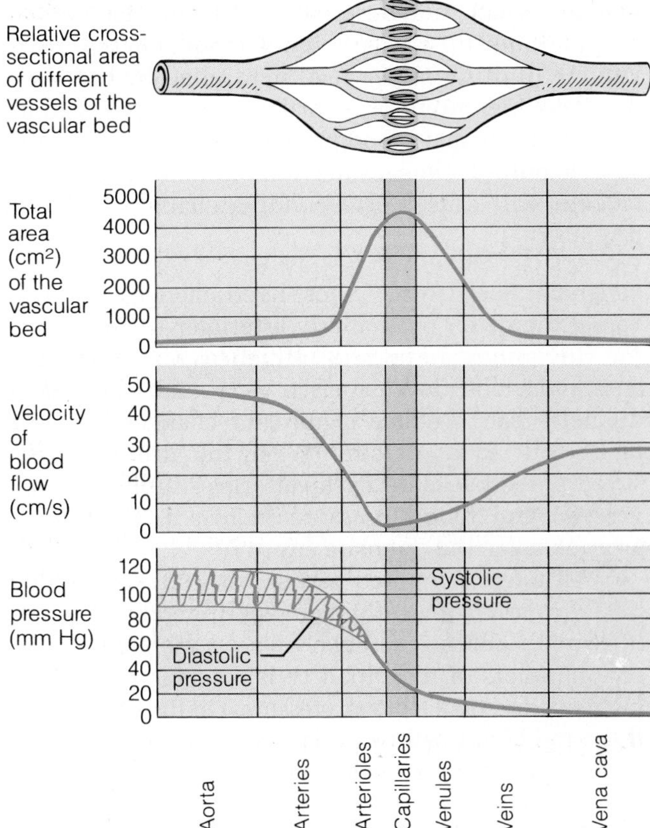

Figure 20.7 Relationship between blood pressure, blood flow velocity, and total cross-sectional area in various blood vessels of the systemic circulation.

These pressure differences between arteries and veins become very clear when these vessels are cut. If a vein is cut, the blood flows evenly from the wound; a lacerated artery produces rapid spurts of blood. ■

Factors Affecting Blood Pressure

The main factors influencing blood pressure are *cardiac output, peripheral resistance,* and *blood volume.* If we rearrange slightly the formula pertaining to blood flow presented on p. 628, we can see the relationship of cardiac output (blood flow of the entire circulation) and peripheral resistance to blood pressure:

$$\text{Blood pressure} = \text{cardiac output} \times \text{peripheral resistance}$$

Let us examine each of these factors in a bit more detail.

Cardiac Output. As described in Chapter 19, cardiac output (in milliliters per minute) is equal to the stroke volume (milliliters per beat) multiplied by the heart rate (beats per minute); normal cardiac output is about 5.5 L/min. Because blood pressure varies directly with cardiac output, changes in stroke volume or heart rate or both will cause corresponding changes in blood pressure.

Peripheral Resistance. Short-lived changes in blood vessel diameter, particularly arteriolar diameter, are the single most important influences on blood pressure and blood flow patterns. Very small changes in diameter can produce significant changes in resistance and blood pressure. When the arterioles constrict, blood backs up in the arteries supplying those vessels, and blood pressure increases. Dilation of arterioles has the opposite effect.

Although, as noted earlier, viscosity is a rather constant factor influencing peripheral resistance, infrequent conditions such as polycythemia (excessive numbers of red blood cells) can increase blood viscosity and hence blood pressure. On the other hand, if the red blood cell count is low, as in some types of anemia, blood becomes less viscous, and both peripheral resistance and blood pressure decline. ■

Blood Volume. Blood pressure varies directly with the amount of blood in the vascular system. Although blood volume varies with age and sex, it is usually maintained at approximately 5 L in adults by the homeostatic functions of the kidneys. But blood volume becomes an important consideration when it is reduced by hemorrhage or extreme dehydration. Indeed, a sudden drop in blood pressure often signals internal bleeding and the need to restore blood volume. Conversely, anything that increases blood volume, such as excessive salt intake (which enhances water retention), increases arterial blood pressure because of the greater fluid load in the circulation.

Regulation of Blood Pressure

Maintaining a steady flow of blood from the head to the toes is vital for proper organ function. But making sure that the energetic person who jumps out of bed in the morning does not keel over from inadequate blood flow to the brain requires the finely tuned cooperation of the heart, blood vessels, and kidneys—all under close supervision of the brain.

Blood pressure is regulated by neural, chemical, and renal controls that act continuously to modify and adjust cardiac output, peripheral resistance, and/or blood volume. The various controls of cardiac output were considered in Chapter 19. In this chapter, we will focus on factors that regulate blood pressure by altering peripheral resistance and blood volume, but a flowchart showing the influence of all important factors is provided in Figure 20.8.

Nervous System Controls. Neural controls of blood vessels are directed primarily at (1) maintaining adequate systemic blood pressure and (2) altering blood distribution to achieve specific functions. For example, during exercise, blood is shunted temporarily from the skin and digestive organs to the skeletal muscles, heat loss from the body is enhanced when blood vessels in the skin are dilated, and under conditions of low blood volume, all vessels except those supplying the heart and brain are constricted to allow as much blood as possible to flow to those vital organs. Essentially, the nervous system controls blood pressure and blood distribution by altering the diameter of arterioles.

Most neural controls operate via reflex arcs chiefly involving the following components: pressoreceptors and the associated afferent fibers, the vasomotor center of the medulla, vasomotor (efferent) fibers, and vascular smooth muscle. Occasionally, inputs from chemoreceptors and higher brain centers also influence the neural control mechanism. Each of these factors is described next:

- **Vasomotor** (vā-zō-mō'-ter) **fibers.** Vasomotor nerve fibers are sympathetic nervous system efferents that innervate the smooth muscle layer of blood vessels, most importantly the arterioles. Most vasomotor fibers release norepinephrine, a potent vasoconstrictor. In skeletal muscle, however, some of the vasomotor fibers release acetylcholine, causing vasodilation. These vasodilator fibers, though important to local controls of muscle blood flow during exercise, are not important to overall regulation of systemic blood pressure.
- **Vasomotor center.** The vasomotor center, a cluster of sympathetic neurons in the medulla, is the integrating center for blood pressure control. This center transmits impulses in a fairly steady stream along the vasomotor fibers; as a result, arterioles are nearly always in a state of partial constriction, called *vasomotor tone.* Any increase in the rate of vasomotor impulse delivery intensifies vasoconstriction and leads to a rise in systemic blood pressure; a decreased rate allows the vascular muscle to relax somewhat, causing vasodilation and a drop in blood pressure. The activity of the vasomotor center is modified by inputs coming from pressoreceptors, chemoreceptors, and higher brain centers, as well as by certain hormones and other blood-borne chemicals.
- **Pressoreceptors.** Pressoreceptors that detect changes in arterial pressure are located not only in the carotid and aortic sinuses, but also in nearly every large artery of the neck and thorax. When arterial blood pressure rises and stretches these receptors, they send off a faster stream of impulses to the vasomotor center. This inhibits the vasomotor center, reducing impulse transmission along the vasomotor fibers, and results in vasodilation and a decline in blood

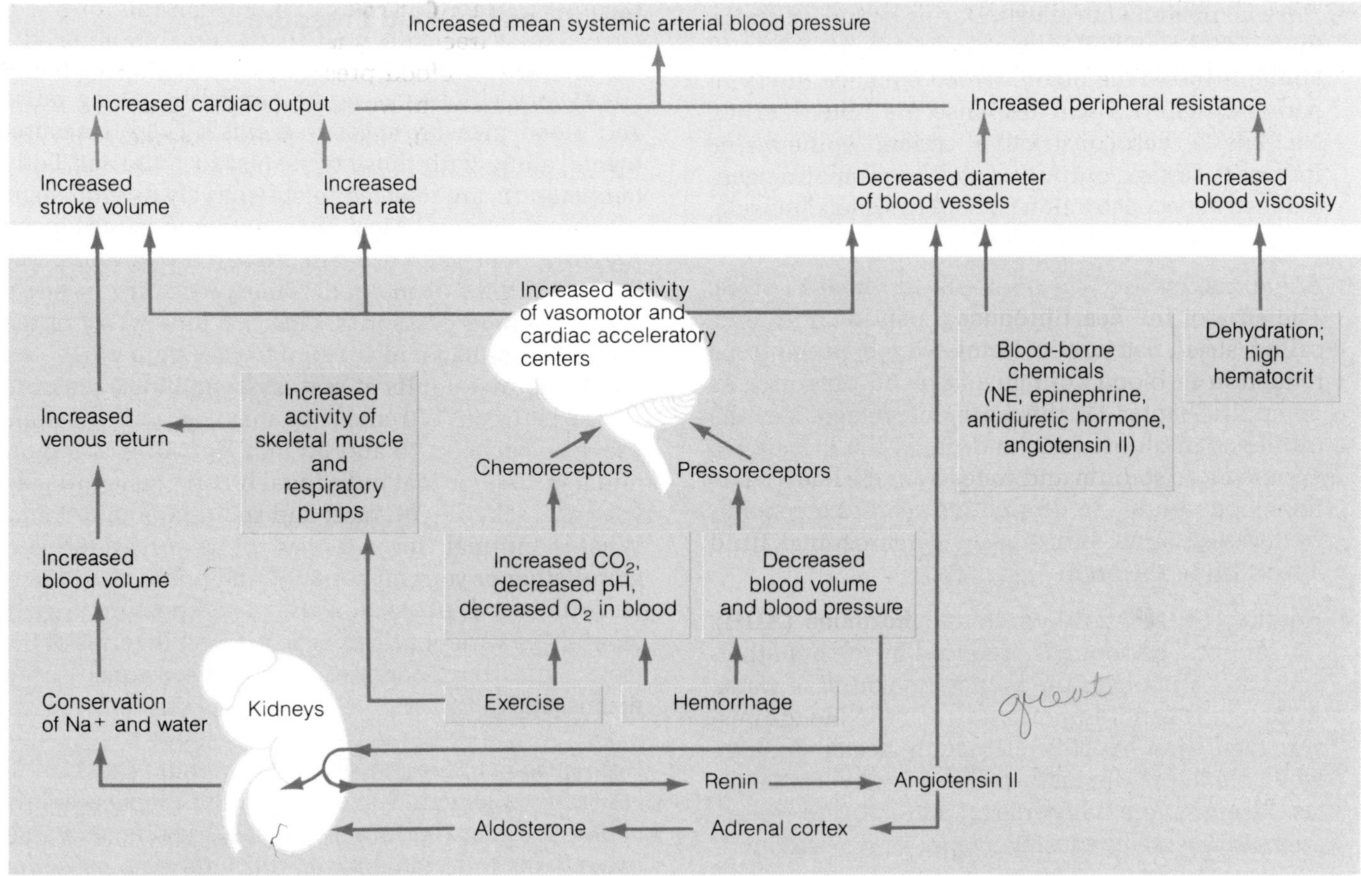

Figure 20.8 Factors that lead to increased systemic arterial blood pressure.

pressure. Afferent impulses from the pressoreceptors also reach the cardiac inhibitory center in the medulla, leading to a reduction in heart rate and contractile force (see Chapter 19). Conversely, a decline in mean arterial pressure initiates reflex vasoconstriction and increased cardiac output, which cause blood pressure to rise. Thus, peripheral resistance and cardiac output are regulated in tandem according to pressoreceptor inputs.

The central function of the pressoreceptors is to protect the circulation against short-term changes in blood pressure, such as those occurring with changes in posture. Pressoreceptors are relatively ineffective in providing the same protection against sustained pressure changes, as evidenced by the fact that some people do have chronic hypertension. In such cases, the pressoreceptors apparently adapt to the higher pressures and become reset to monitor pressure changes at a higher set point.

- **Chemoreceptors.** When the oxygen content of the blood drops sharply, or when hydrogen ion levels rise, chemoreceptors in the aortic arch and large arteries of the neck region transmit impulses to the vasomotor center, and reflex vasoconstriction occurs. The rise in blood pressure that follows helps to speed blood return to the heart and lungs. The most prominent of these receptors are the carotid and aortic bodies. Because they are more important in regulating respiratory rate and depth than they are in blood pressure regulation, their function is considered more fully in Chapter 23.

- **Higher brain centers.** Reflexes that regulate blood pressure are integrated at the brain stem (medulla) level. Although the cerebral cortex and hypothalamus are not required for routine blood pressure maintenance, these higher brain centers can modify arterial blood pressure via relays to the medullary centers. For example, the fight-or-flight response mediated by the hypothalamus has profound effects on blood pressure as well as on other body systems. (Even the simple act of speaking can make your blood pressure jump if the person you are talking to causes you to become anxious.) The hypothalamus also mediates the redistribution of blood flow and other cardiovascular responses that occur during exercise and changes in body temperature.

Chemical Controls. As we have seen, changing levels of oxygen and carbon dioxide help regulate blood pressure via the chemoreceptor reflexes. But numerous other blood-borne chemicals influence blood pressure by acting directly on vascular smooth muscle or on the vasomotor center. These include:

- **Adrenal medulla hormones.** During periods of stress, the adrenal gland releases norepinephrine (NE) and epinephrine to the blood, which enhance the sympathetic fight-or-flight response. As noted earlier, NE has a vasoconstrictive action; epinephrine increases cardiac output, as well as promoting generalized vasoconstriction (except in skeletal muscle, where it causes vasodilation).
- **Atrial natriuretic** (nā″-trē-ur-eh′-tik) **factor (ANF).** The atria of the heart produce a peptide hormone, called atrial natriuretic factor, which promotes a reduction in blood volume and blood pressure. As noted in Chapter 17, this hormone antagonizes the influence of aldosterone and prods the kidneys to excrete more sodium and water from the body, causing blood volume to drop. It also causes a generalized vasodilation and reduces cerebrospinal fluid formation in the brain.
- **Antidiuretic** (an″-tī-dī″-yer-eh′-tik) **hormone (ADH).** Antidiuretic hormone is produced by the hypothalamus and stimulates the kidneys to conserve water. Although it is not important in routine blood pressure regulation, ADH is released in greater amounts when blood pressure falls to dangerously low levels (as during severe hemorrhage) and helps to restore arterial pressure by causing intense vasoconstriction.
- **Alcohol.** Ingestion of alcohol causes a drop in blood pressure by inhibiting ADH release and by depressing the vasomotor center and promoting vasodilation, especially in the skin. This accounts for the flushed appearance of someone who has drunk a generous amount of an alcoholic beverage.

Renal Regulation. The kidneys act both directly and indirectly to regulate arterial pressure and provide the major long-term mechanism of blood pressure control. The direct mechanism reflects their ability to alter blood volume. When blood volume or blood pressure rise, the kidneys respond by allowing more water to flush from the body in urine; blood pressure falls as blood volume declines. Conversely, when blood pressure or blood volume is low, water is conserved and returned to the bloodstream.

The indirect renal mechanism involves the renin-angiotensin mechanism. When arterial blood pressure declines, special cells in the kidneys release the enzyme *renin* (rē′-nin) into the blood. Renin triggers a series of enzymatic reactions that results in the formation of *angiotensin* (an-jē-ō-ten′-sin) *II*, a potent vasoconstrictor chemical. Angiotensin II also stimulates the adrenal cortex to release *aldosterone* (al-dah′-ster-ōn), a hormone that enhances renal reabsorption of sodium. As sodium moves into the bloodstream, water follows; thus, both blood volume and blood pressure rise. These renal mechanisms are described in detail in Chapter 27.

Variations in Blood Pressure

A fairly good indication of the status of a person's circulatory efficiency can be gained by taking pulse and blood pressure measurements. (These measurements, along with those of respiratory rate and body temperature, are referred to collectively as *vital signs* in clinical settings.) Most often, systemic arterial blood pressure is measured indirectly by the *auscultatory* (os-kul′-tuh-tor-ē) *method*. This procedure, used to measure blood pressure in the brachial artery of the arm, is described and illustrated in Figure 20.9.

In normal adults at rest, systolic blood pressure varies between 110 and 140 mm Hg, and diastolic pressure between 75 and 80 mm Hg, but it is important to recognize that individual blood pressure varies with age, sex, weight, race, and socioeconomic status. What is "normal" for you may not be normal for your grandfather or your neighbor. Blood pressure also varies with mood, physical activity, and posture. Nearly all of these variations can be explained in terms of the factors affecting blood pressure that have already been discussed.

Hypotension. *Hypotension*, or low blood pressure, is generally considered to be a systolic blood pressure below 100 mm Hg. In many cases, hypotension simply reflects individual variations and is no cause for concern. In fact, low blood pressure is often associated with long life and an old age free of illness.

Elderly people sometimes experience temporary low blood pressure and dizziness when they rise suddenly from a reclining or sitting position—a condition called *orthostatic hypotension*. Because the aging sympathetic nervous system does not react as quickly as it once did in response to postural changes, blood pools briefly in the lower extremities, reducing blood pressure and consequently blood delivery to the brain. Making postural changes more slowly to give the nervous system time to make the necessary adjustments usually prevents this problem.

Chronic hypotension may hint at poor nutrition, because those poorly nourished are often anemic and have inadequate levels of blood proteins. Because blood viscosity is low, blood pressure is also lower than normal. Chronic hypotension may also reflect Addison's disease (inadequate adrenal cortex function), hypothyroidism, or severe tissue wasting, as might occur with cancer. Acute hypotension is one of the most important warnings of circulatory shock (see p. 639). ■

Hypertension. Transient elevations in systolic blood pressure occur as normal adaptations during fever, physical exertion, and emotional upset, such as during anger and fear. Persistent *hypertension*, or high blood pressure, is common in obese people because the total length of their blood vessels is relatively greater than that in thinner individuals. For each pound of

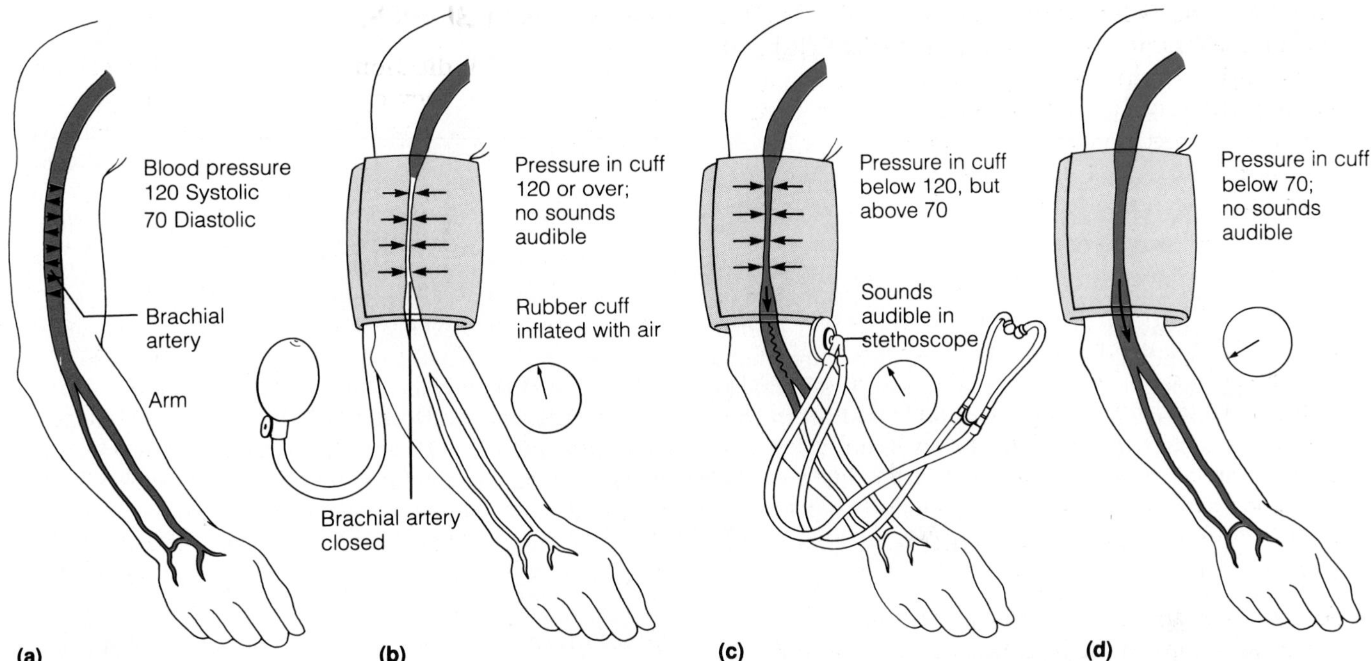

Figure 20.9 Measurement of blood pressure. (**a**) The course of the brachial artery of the arm. Assume a blood pressure of 120/70 in a young, healthy individual. (**b**) The cuff of the sphygmomanometer is wrapped snugly around the arm just superior to the elbow and inflated until the cuff pressure exceeds the systolic blood pressure. At this point, blood flow into the arm is stopped and a brachial pulse cannot be felt or heard. (**c**) The pressure of the cuff is gradually reduced while the examiner listens (auscultates) carefully for sounds in the brachial artery with a stethoscope. The pressure read as the first soft tapping sounds are heard (the first point at which a small amount of blood is spurting through the constricted artery) is recorded as the systolic pressure. (**d**) As the pressure is reduced still further, the sounds become louder and more distinct, but when the artery is no longer constricted and blood flows freely, the sounds can no longer be heard. The pressure at which the sounds disappear is recorded as the diastolic pressure.

fat, miles of additional blood vessels are required to service the extra tissue.

Chronic hypertension is a common and dangerous disease that warns of increased peripheral resistance. Approximately 30% of people over the age of 50 years are hypertensive. Although hypertension is usually asymptomatic for the first 10 to 20 years, it slowly but surely strains the heart and damages the arteries; for this reason, hypertension is often called the silent killer. Prolonged hypertension is the most significant factor in cardiovascular disease and is the major cause of heart failure, renal failure, and stroke. Because the heart is forced to pump against increased resistance, it must work harder, and in time, the myocardium enlarges. When finally strained beyond its capacity to respond, the heart weakens and its walls become flabby. Hypertension also ravages blood vessels, causing small tears in the endothelium that accelerate the progress of atherosclerosis and ultimately cause arteriosclerosis, or hardening of the arteries (see the box on pp. 634–635). As the vessels become increasingly blocked (Figure 20.10), blood flow to the tissues becomes inadequate, and vascular complications of the brain, retinas of the eyes, heart, and kidneys begin to appear.

Although hypertension and arteriosclerosis are often linked, it is difficult to blame hypertension on any distinct anatomical pathology. Hypertension is defined physiologically as a condition of sustained elevated arterial pressure of 140/90 or higher, and the higher the blood pressure, the greater the risk for serious cardiovascular problems. As a general rule, elevated diastolic pressures are more significant medically, because they always represent progressive occlusion and/or hardening of the arterial tree.

Approximately 90% of hypertensive people have *primary*, or *essential*, *hypertension*, which cannot be accounted for by any specific organic cause. However, factors such as diet, obesity, heredity, race, and stress are believed to be involved. Clinical signs of the disease usually appear after the age of 40 years, and its course varies in different population groups. For instance, more women than men and more blacks than whites are hypertensive. A tendency toward hypertension runs in families; the child of a hypertensive parent is twice as likely to develop high blood pressure as is a child of normotensive parents. Dietary factors presumed to contribute to hypertension include

A CLOSER LOOK Arteriosclerosis? Get Out the Cardiovascular Drāno

When pipes get clogged, it is usually because something gets stuck in them—a greasy mass of bone bits or a hair ball in the sink drain. But in arteries narrowed by **arteriosclerosis,** the damming-up process occurs from the inside out: The walls of the vessels thicken and protrude into the vessel lumen. Once this happens, it does not take much to close the vessel completely. A roaming blood clot or arterial spasms can do it.

Although all blood vessels are susceptible to this serious degenerative condition of blood vessel walls, for some unknown reason, the aorta and the coronary arteries serving the heart are most often affected. The disease progresses through many stages before the arterial walls actually become hard and approach the rigid tube system described in the text, but some of the earlier stages are just as lethal as those that come later.

What triggers this scourge of blood vessels? According to the *response to injury* hypothesis, the initial event is damage to the tunica intima caused by blood-borne chemicals such as carbon monoxide (present in cigarette smoke or auto exhaust); by viruses; or by physical factors such as a blow or persistent hypertension. Once a break has occurred, blood platelets begin to cling to the injured site and initiate clotting to prevent blood loss, but they also release other chemicals that increase the permeability of the endothelium to fats and cholesterol, allowing them entry into the deeper tunica media and producing the so-called *fatty streak stage.* Another platelet-liberated chemical, platelet-derived growth factor, prompts smooth muscle cells to divide rapidly, producing greasy gray to yellow lesions called *atheromas* or *atherosclerotic plaques.* When these small, fatty mounds of muscle begin to protrude into the vessel lumen, the condition is called *atherosclerosis* (see Figure 20.10).

Arteriosclerosis is the end stage of the disease. As enlarging plaques hinder diffusion of nutrients from the blood to the deeper tissues of the artery wall, smooth muscle cells in the tunica media die, and the elastic fibers deteriorate. These elements are gradually replaced by nonelastic scar tissue, and calcium salts are deposited in the lesions. Collectively, these events cause the arterial wall to become frayed and, often, ulcerated; the increased rigidity of the vessels leads to hypertension. Together, these events increase the risk of myocardial infarcts, strokes, and aneurysms. (Besides their suspected role in lowering blood levels of cholesterol and triglycerides, omega-3 fatty acids found in the flesh of cold water fish have been shown to reduce the production of platelet-derived growth factor by 50%–85%. This finding may help to explain why those groups that eat large amounts of fish (Greenland Eskimos, etc.) have a low incidence of coronary artery disease.)

So what can be done when the damage is done and the heart is at risk because of arteriosclerotic coronary vessels? Traditionally, the choice has been coronary bypass surgery, in which vessels removed from the legs are implanted in the heart to restore adequate myocardial circulation. More recently, intravascular devices, threaded through blood vessels to obstructed sites, have become part of the ammunition of cardiovascular medicine. *Balloon angioplasty* uses a catheter with a balloon tightly

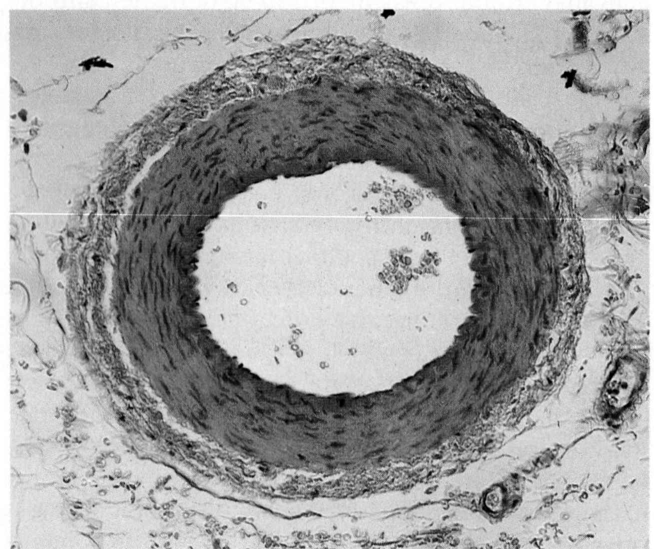
(a)

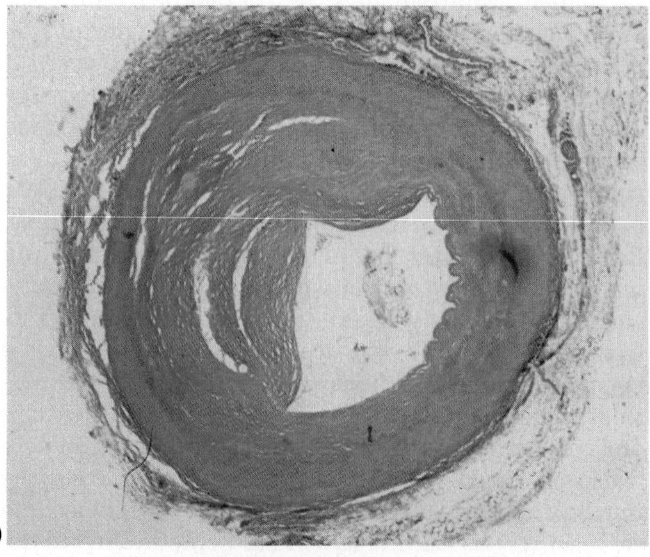
(b)

Figure 20.10 Comparison of a normal artery and an atherosclerotic artery. (**a**) Cross section of a normal artery (170X). (**b**) Cross section of an artery that is partially occluded by atherosclerotic plaque (30X).

packed into its tip (see photos). When the catheter reaches the obstruction, the balloon is inflated, and the fatty mass is compressed against the vessel wall. However, this procedure is useful when only a few very localized obstructed areas are present. Another catheter device uses a laser beam to vaporize the arterial clogs. Although these intravascular devices are faster, cheaper, and much less risky than coronary bypass surgery, they carry with them the same major shortcoming: They do nothing to stop the underlying disease, and after a few years, new blockages occur.

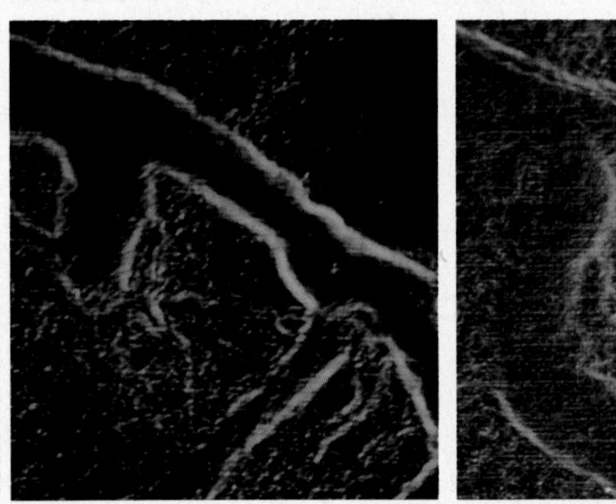

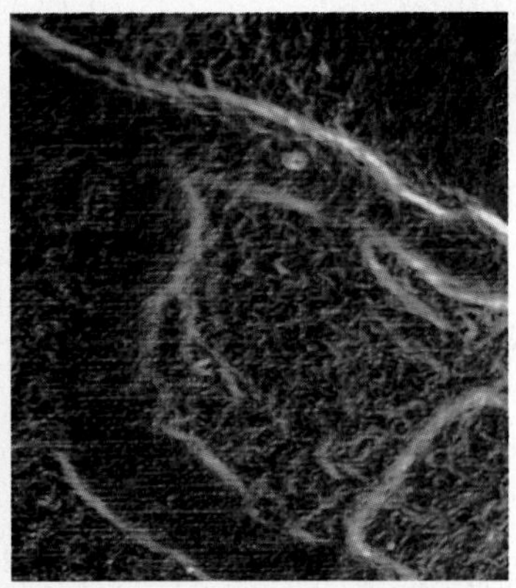

Angiographs of an occluded artery (left) and the same artery after being cleared by balloon angioplasty (right). These views were prepared by a computer imaging technique, digital subtraction angiography (DSA).

It was hoped that the cholesterol-lowering drugs such as cholestyramine and lovastatin would act as cardiovascular Drāno and simply wash the fatty plaques off the walls. However, these drugs are only used in those where dietary restrictions have failed to reduce blood cholesterol levels, and their side effects—nausea, bloating, and constipation—are so miserable that many people simply stop taking them. Another type of Drāno used when the problem blockage is a clot trapped by the diseased vessel walls are the *thrombolytic (clot-dissolving) agents*. These drugs include the powerful enzyme *streptokinase* and *tissue plasminogen activator (t-PA)*, a naturally occurring substance produced by genetic engineering techniques. Injecting the enzyme or t-PA directly into the heart via a catheter restores blood flow quickly and puts an early end to many heart attacks in progress.

There is little doubt that life-style factors—emotional stress and smoking (both capable of promoting vascular spasms and unwanted clot formation), obesity, high-fat and high-cholesterol diets, and lack of exercise—contribute to both arteriosclerosis and hypertension. If these are the risk factors (and indeed they are), then why not just advise patients at risk to change their life-style? This is not as easy as it seems. Old habits die hard, and Americans like their burgers and butter. Further, stress management techniques and behavior modification are just beginning to come into the mainstream of cardiovascular medicine. The goal: to reverse heart disease by reversing arteriosclerosis. Can it be done? One thing is sure: If it is successful, many more people with diseased arteries may be more willing to trade lifelong habits for a healthy old age!

high sodium, saturated fat, and cholesterol intake and deficiencies in certain metal ions (potassium, calcium, and magnesium). Factors that most accurately predict risk are the severity of blood pressure elevation, blood cholesterol levels, cigarette smoking, presence of diabetes mellitus, and stress levels. Primary hypertension cannot be cured, but most cases can be controlled by dietary measures (salt and cholesterol restriction and weight reduction) and antihypertensive drugs. Drugs most commonly used are diuretics, sympathetic nerve blockers, and calcium channel blockers.

Secondary hypertension, which accounts for 10% of cases, is related to identifiable disorders, such as excessive renin secretion by the kidneys, arteriosclerosis, and endocrine gland disorders such as hyperthyroidism and Cushing's disease. Treatment for secondary hypertension is directed toward correcting the causative problem. ■

Blood Flow

Blood flow is responsible for (1) delivery of oxygen and nutrients to, and removal of wastes from, tissue cells, (2) gas exchange in the lungs, (3) absorption of nutrients from the digestive tract, and (4) processing of blood by the kidneys. Furthermore, the rate of blood flow to each tissue and organ is almost exactly the right amount to provide for proper function—no more, no less. When the body is at rest, the brain receives about 13% of total blood flow, the heart 4%, kidneys 20%, and the abdominal organs 24%. The skeletal muscles, which make up almost half of body mass, normally receive about 20%; during exercise, however, nearly all of the increased cardiac output flushes into the skeletal muscles (Figure 20.11).

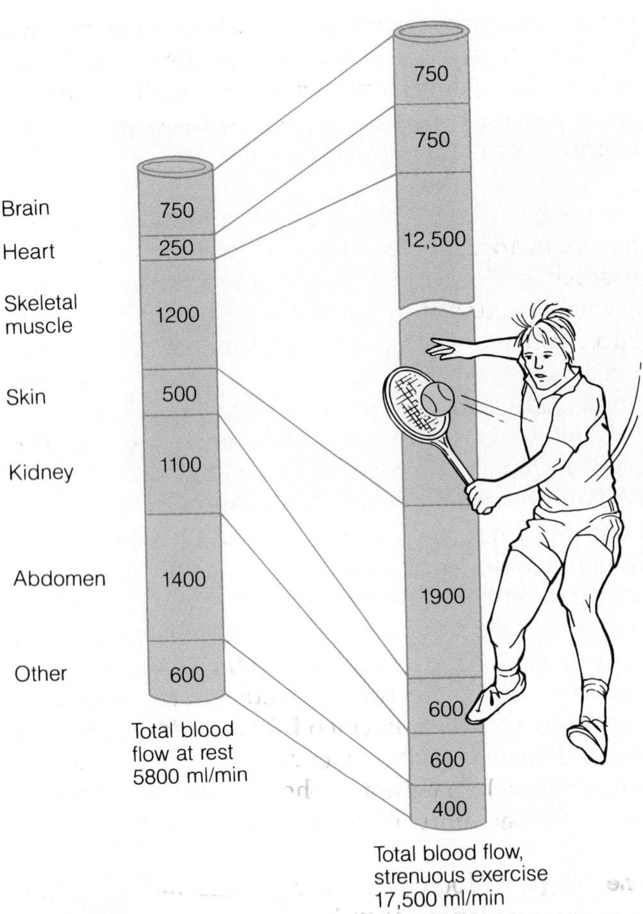

Figure 20.11 Distribution of blood flow to selected body organs at rest and during strenuous exercise.

Velocity of Blood Flow

As shown in Figure 20.7, the velocity of blood flow changes as it travels through the systemic circulation. It is fastest in the aorta and other large arteries, slowest in the capillaries, and then speeds up again somewhat in the veins leading back to the heart.

Velocity is inversely related to the *cross-sectional area* of the blood vessels to be filled. Blood flows fastest where the total cross-sectional area is least. As the arterial system continues to branch, the cross-sectional area of the vascular bed increases, and the velocity of blood flow declines proportionately. Even though the individual branches have smaller lumens, their combined cross-sectional area is much greater than that of the aorta. For example, the cross-sectional area of the aorta is 2.5 cm^2, and the average velocity of blood flow in the aorta is 40–50 cm/s. Since the total cross-sectional area of the capillaries is 2500 cm^2, it is not surprising that capillary blood flow is very slow (approximately 0.03 cm/s). Slow capillary blood flow is beneficial, because it allows adequate time for exchanges to be made between the blood and tissue cells.

As capillaries combine to form venules and then veins, total cross-sectional area declines and velocity increases. The cross-sectional area of the venae cavae is 8 cm^2, and the velocity of blood flow varies from 10 to 30 cm/s in those vessels, depending on how active the skeletal muscles are.

Autoregulation: Local Regulation of Blood Flow

Autoregulation is the automatic adjustment of blood flow to each tissue in proportion to its requirements at any point in time. This process is regulated by local chemical conditions and is largely independent of systemic factors. Since the mean arterial pressure is identical throughout the body, changes in blood flow through individual organs are controlled intrinsically by modifications in the diameter of local arterioles feeding the capillary beds. Thus, organs self-regulate their own blood flows by varying the resistance of their arterioles.

In most tissues, declining levels of nutrients, particularly oxygen, are the strongest stimuli for autoregulation; however, in the brain, a localized increase in carbon dioxide (accompanied by a decrease in pH) is an even more potent trigger, because neuronal activity changes with alterations in carbon dioxide levels. Certain substances released by metabolically active tissues (such as potassium and hydrogen ions, adenosine, and lactic acid), prostaglandins, and inflammatory chemicals (histamine and kinins) can also serve as autoregulation stimuli.

Whatever the precise stimulus, the net result is immediate vasodilation of the arterioles serving the capillary beds of the "needy" tissues, and blood flow to the local area is temporarily increased. This is accompanied by relaxation of the precapillary sphincters (see Figure 20.5), allowing blood to surge through the true capillaries and become available to the tissue cells. The local autoregulation of blood flow in the brain, heart, and kidneys is extraordinarily efficient, and adequate perfusion is maintained even in the face of fluctuating mean arterial pressure.

If the nutrient requirements of a tissue are more than the short-term autoregulatory mechanism can easily supply, a long-term autoregulation mechanism may evolve over a period of weeks or months to enhance the local blood flow still more. The number of blood vessels in the region increases, and existing vessels enlarge. This phenomenon is particularly common in the heart when a coronary vessel is partially occluded; it occurs throughout the body in people who live in high-altitude areas, where the air contains less oxygen.

Blood Flow in Special Areas

Each organ has special requirements and functions that are revealed in its pattern of autoregulation.

Skeletal Muscles. Blood flow in skeletal muscle is extremely changeable and varies with the degree of muscle activity. During resting conditions, skeletal muscles receive about 1 L of blood per minute, and only about 25% of their capillaries are open. As muscles become active, blood flow increases (hyperemia) in direct proportion to the enhanced metabolic activity of the muscles, a phenomenon called *active* or *exercise hyperemia* (hī″-per-ē′-mē-uh). During vigorous physical exertion, blood flow can increase tenfold or more in response to acetylcholine release by sympathetic vasodilator fibers, and virtually all capillaries open to accommodate the increased flow (Figure 20.11). Without question, strenuous exercise is one of the most demanding conditions the cardiovascular system faces.

Muscular autoregulation occurs almost entirely in response to the decreased oxygen concentrations (accompanied by local increases in carbon dioxide, lactic acid, and other metabolites) that result from the enhanced metabolism of working muscles. However, systemic adjustments, mediated by the sympathetic vasomotor center, must also occur to ensure that increased blood volume reaches the muscles. Strong vasoconstriction of the vessels of blood reservoirs such as those of the digestive viscera diverts blood away from these regions temporarily, ensuring that an increased blood supply reaches the muscles. Ultimately, the major factor determining how long muscles can continue vigorous activity is the ability of the cardiovascular system to deliver adequate oxygen and nutrients.

The Brain. Blood flow to the brain averages about 750 ml/min and is maintained at relatively constant levels. (Contrary to what you might imagine, it does not take more "brain energy" to think!) Cerebral blood flow is regulated by one of the most precise autoregulatory systems in the entire body and is tailored to local neuronal need. Thus, when you make a fist with your right hand, the neurons in the left cerebral motor cortex controlling that movement are receiving a more abundant blood supply than are the adjoining neurons. Brain tissue is exceptionally sensitive to declining pH, and increased blood carbon dioxide levels (resulting in acidic conditions in brain tissue) promote marked vasodilation. Although it seems strange, oxygen deficits are a much less potent stimulus for autoregulation. However, excess carbon dioxide causes vasoconstriction and severely depresses brain activity.

In addition to the autoregulation that occurs in response to metabolites, the brain also has a myogenic mechanism that protects it from possibly damaging changes in blood pressure. *Myogenic* (my″-ō-jeh′-nik) *responses* are direct responses of vascular smooth muscle to changes in blood pressure. When the mean arterial pressure declines, cerebral vessels dilate, thereby ensuring adequate brain perfusion. On the other hand, when arterial pressure rises, the cerebral vessels constrict, thereby protecting the small, more fragile vessels farther along in the pathway from being ruptured by the excessive pressure.

The Skin. Blood flow through the skin supplies nutrients to the cells and aids in body temperature regulation. The skin vessels also serve as a blood reservoir. The first function is served by autoregulation in response to the need for oxygen; the second and third require neural intervention. We will concentrate on the skin's temperature-related function here.

Below the skin surface are extensive venous plexuses, in which the blood flow can be changed from 50 ml/min to as much as 2500 ml/min, depending on body temperature. When heat loss is required, the hypothalamic "thermostat" signals for reduced vasomotor stimulation of the skin vessels; warm blood flushes into the plexuses, and heat radiates from the skin surface. When body temperature drops and heat must be conserved, the superficial skin vessels are strongly constricted, and blood almost entirely bypasses the cutaneous venous plexuses (via arteriovenous anastomoses), maintaining homeostatic temperature of the deeper, more vital body organs.

The Lungs. Blood flow through the lungs, or pulmonary circulation, is unique in many ways. Pulmonary arteries and arterioles are similar to veins and venules in structure; that is, they have thin walls and large lumens. As a result, there is little resistance to blood flow, and less pressure is needed to move blood through the pulmonary arterial system. Consequently, arterial pressure in the pulmonary circulation is much lower than in the systemic circulation (22/8 versus 120/80).

In addition, the autoregulatory mechanism in the pulmonary circulation is exactly the opposite of that seen in most tissues: Low oxygen levels cause vasoconstriction, and high levels promote vasodilation. While this may seem odd, it is perfectly consistent with the role of the pulmonary circulation in gas exchange. When the air sacs of the lungs are flooded with oxygen-rich air, the pulmonary capillaries become flushed with blood and ready to receive their oxygen load. If the air sacs are collapsed or blocked with mucus, the oxygen concentration in those areas will be low, and pulmonary capillary blood will largely bypass those nonfunctional areas.

The Heart. Movement of blood through the heart, or coronary circulation, is influenced by aortic pressure and the pumping activity of the ventricles. When the ventricles contract, the coronary vessels are compressed, and blood flow through the myocardium stops.

As the heart relaxes, the high aortic pressure forces blood through the coronary circulation. Under normal circumstances, the myoglobin content of the myocardium contains sufficient stored oxygen to satisfy the oxygen requirements of the cardiac cells during systole. However, an abnormally rapid heartbeat may seriously hinder their ability to receive adequate oxygen and nutrients during diastole.

Under resting conditions, blood flow through the heart is about 250 ml/min. During strenuous exercise, the coronary vessels dilate (in response to a local accumulation of carbon dioxide), and blood flow can increase three to four times (see Figure 20.11). This enhanced blood flow during more vigorous heart activity is important because cardiac cells use about 65% of the oxygen carried to them in blood under resting conditions. (Most other tissue cells use about 25% of the oxygen delivered during each circulatory round.) Thus, increased blood flow is the only way in which sufficient additional oxygen can be made available to a more vigorously working heart.

Capillary Dynamics

Blood flow through capillary networks is slow and intermittent, rather than steady. This phenomenon, called *vasomotion,* reflects the "on-off" opening and closing of precapillary sphincters in response to local autoregulatory controls.

Exchanges of Respiratory Gases and Nutrients. Oxygen, carbon dioxide, most nutrients (amino acids, glucose, lipids), and metabolic wastes pass between the blood and interstitial fluid by *diffusion.* In diffusion, movement always occurs along a concentration gradient, each substance moving from an area of higher concentration to an area of lower concentration. Oxygen and nutrients pass from the blood, where their concentration is fairly high, through the interstitial fluid to the tissue cells. Carbon dioxide and metabolic wastes leave the cells, where their content is high, and diffuse into the capillary blood. In general, water-soluble solutes, such as amino acids and sugars, pass through fluid-filled capillary clefts, while lipid-soluble molecules, such as respiratory gases and fats, diffuse directly through the lipid bilayer of the endothelial cell plasma membranes. As noted earlier, capillaries differ in their "leakiness," or permeability. For example, liver capillaries have large fenestrations that allow even proteins to pass freely, whereas brain capillaries are impermeable to most substances.

Fluid Movements. Fluid flows also occur at capillary beds. Fluid (water plus dissolved solutes to which the capillary wall is permeable) is forced out of the capillaries through the clefts at the arterial end of the bed, but most of it returns to the bloodstream at the venous end. According to the traditional view of capillary dynamics, the opposing forces of hydrostatic and osmotic pressure determine the direction and amount of fluid that flows across capillary walls.

Hydrostatic pressure is the force exerted by a fluid pressing against a wall. In the case of capillaries, hydrostatic pressure is the same as capillary blood pressure—the pressure of blood against the capillary walls. Capillary hydrostatic pressure is also called *filtration pressure* because it forces fluids through the porous capillary walls. Because blood pressure drops as blood flows through the length of a capillary bed, capillary hydrostatic pressure is higher at the arterial end of the bed (25 mm Hg) than at the venous end (10 mm Hg).

In theory, blood pressure—which forces fluid out of the capillaries—is opposed by hydrostatic pressure (back pressure) exerted by the interstitial fluid acting outside the capillaries. To determine the net hydrostatic force acting on the capillaries at any point, we need to find the difference between the hydrostatic force that is pushing fluid *out* of the capillary (blood pressure) and the hydrostatic force that is pushing fluid *into* the capillary (interstitial fluid pressure). However, there is usually very little fluid in the interstitial space, because it is constantly withdrawn by the lymphatic vessels (see Chapter 21). As a result, the hydrostatic pressure in the interstitial space is actually negative (−6 mm Hg). This negative hydrostatic pressure enhances rather than opposes the movement of fluid out of the capillaries. The *net effective hydrostatic pressures* at the arterial and venous ends of the capillary bed are summarized in Figure 20.12a.

Osmotic pressure is created by the presence in a fluid of large nondiffusible molecules, such as plasma proteins, that are prevented from moving through the capillary membrane. Such substances draw water toward them; that is, they encourage osmosis (see Chapter 3) whenever the water concentration in their vicinity is relatively lower than it is on the opposite side of the capillary membrane. Plasma proteins in the capillary blood (primarily albumin molecules) develop an osmotic pressure of approximately 28 mm Hg. Since interstitial fluid contains few proteins, its osmotic pressure is substantially lower—only about 5 mm Hg. Unlike hydrostatic pressure, osmotic pressure does not vary from one end of the capillary bed to the other; thus, the *net osmotic pressure* that pulls fluid back into the capillary blood is 23 mm Hg (Figure 20.12b).

At any point along a capillary, fluids will leave the capillary if net hydrostatic pressure is greater than net osmotic pressure, and fluids will enter the capillary if net osmotic pressure exceeds net hydrostatic pressure. As shown in Figure 20.12c, hydrostatic forces dominate at the arterial end, while osmotic forces

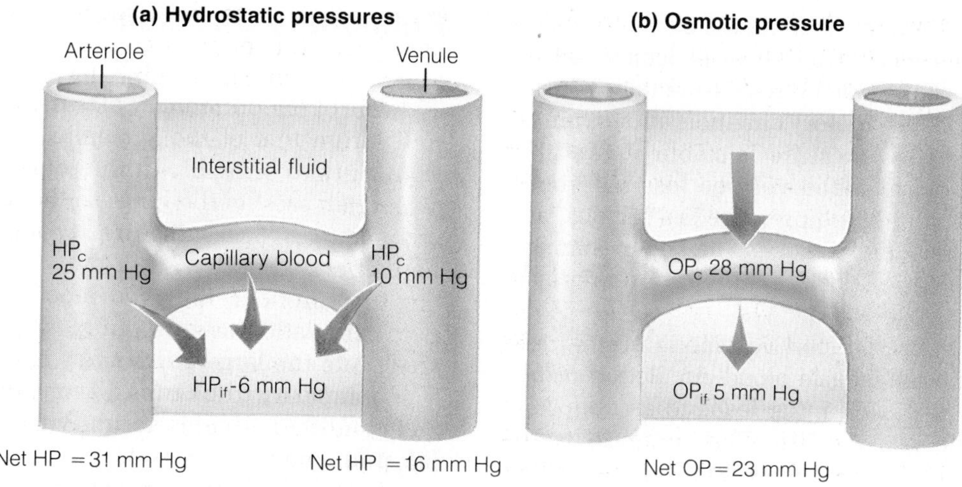

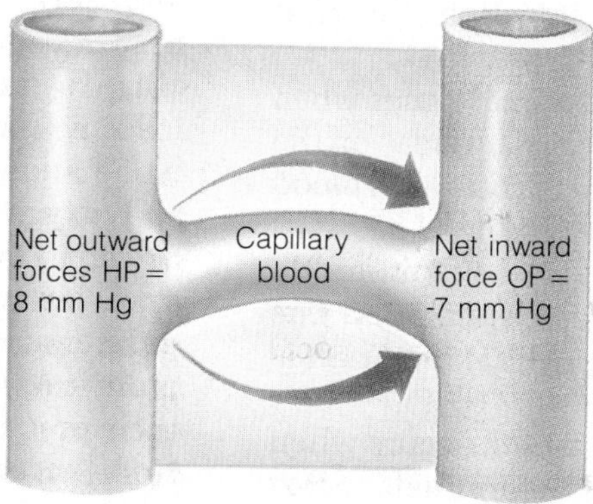

Figure 20.12 Forces acting at the capillary bed determine the direction of fluid flows. (HP_c = hydrostatic pressure in the capillary; HP_{if} = hydrostatic pressure in the interstitial fluid; OP_c = osmotic pressure in the capillary; OP_{if} = osmotic pressure in the interstitial fluid.)

dominate at the venous end. Thus, fluid flows out of the circulation at the arterial ends of capillary beds and back into the bloodstream at the venous ends. Note, however, that there is a slight imbalance. More fluid enters the tissue spaces than is returned to the blood, resulting in a net loss of fluid from the circulation of about 1.7 ml/min. This fluid and any leaked proteins are picked up by the lymphatic vessels and returned to the vascular system, which accounts for the relatively low levels of both fluid and proteins in the interstitial space. Were this not so, this seemingly insignificant fluid loss would empty your blood vessels of plasma in about 24 hours!

More recently, an alternative theory has been offered to explain fluid flows at capillary beds. It has been suggested that when the precapillary sphincters are opened and the capillaries are filled with blood, the increased capillary pressure leads to a leaking of fluid all along the capillary length. Then, when the sphincters close and capillary pressure declines, the osmotic force predominates, promoting net fluid absorption back into the blood. In either case, fluid flows occur, and there is incomplete pickup of lost capillary fluid.

Circulatory shock

Circulatory shock is a condition in which blood vessels are inadequately filled and blood cannot circulate normally—whatever the cause. This results in inadequate blood flow to meet tissue needs, and if the condition persists, cell death and organ damage ensue.

The simplest form of shock, *hypovolemic* (hī″-pō-vō-leh′-mik) *shock* (*hypo* = low, inadequate; *volemia* = blood volume), results from large-scale loss of blood volume, such as might follow acute hemorrhage, severe vomiting or diarrhea, or extensive burns. If blood volume drops rapidly, heart rate increases; thus, a weak, "thready" pulse is often the first sign of hypovolemic shock. Intense vasoconstriction also occurs, which allows blood in various blood reservoirs to be rapidly added to the major circulatory channels and enhances venous return. Blood pressure is stable at first, but eventually it drops if blood volume loss continues; thus, a sharp drop in blood pressure is a serious, and late, sign of hypovolemic shock. The key to management of hypovolemic shock is fluid volume replacement as quickly as possible.

In *vascular shock,* blood volume is normal and remains constant. Inadequate blood circulation results from extreme vasodilation that leads to an abnormal expansion of the vascular bed. The immense reduction in peripheral resistance that follows is revealed by rapidly falling blood pressure. The two most common causes of vascular shock are loss of vasomotor tone because of failure of autonomic nervous system regulation and septicemia. Septicemia is a widespread, severe systemic bacterial infection; bacterial toxins are notorious vasodilators.

A transient type of vascular shock may occur when you sunbathe for a prolonged time. The heat of the sun on your skin causes vasodilation of cutaneous blood vessels, and if you stand up abruptly, blood remains in the dilated vessels rather than returning promptly to the heart. Consequently, blood pressure falls. The dizziness felt at this point is a signal that the brain is not receiving enough oxygen.

Cardiogenic shock, or pump failure, occurs when the pumping action of the heart is so inefficient that it cannot sustain adequate circulation. The usual cause of this condition is myocardial damage, as might follow numerous myocardial infarcts. ■

Circulatory Pathways

When referring to the body's complex network of blood vessels, the term *vascular system* is often used. However, the heart is actually a double pump that serves two distinct circulations, each with its own set of arteries, capillaries, and veins. The pulmonary circulation is the short loop that runs from the heart to the lungs and back to the heart. The systemic circulation routes blood through a long loop to all parts of the body before returning it to the heart.

Besides the two major circuits, a number of special systemic subdivisions aid the functioning of particular organs. These local circulations, such as the coronary, renal, and cerebral circulations, are discussed shortly. The special vessels and shunts of the fetal circulation are described in Chapter 29.

Pulmonary Circulation

The role of the pulmonary circulation differs substantially from that of the systemic circulation. It functions only to bring blood into close contact with the alveoli (air sacs) of the lungs so that gaseous exchanges can occur; it does not directly serve the metabolic needs of body tissues.

Oxygen-poor, dark red blood enters the pulmonary circulation as it is pumped from the right ventricle into the large **pulmonary trunk** (Figure 20.13). The pulmonary trunk runs diagonally upward for about 8 cm and then divides abruptly to form the **right** and **left pulmonary arteries.** In the lungs, the pulmonary arteries subdivide into the **lobar** (lō′-bar) **arteries** (three in the right lung and two in the left lung), each of which serves one lung lobe. The lobar arteries accompany the main bronchi into the lungs and then branch profusely, forming arterioles and, finally, the dense networks of **pulmonary capillaries** that surround and cling to the delicate air sacs. It is here that oxygen loading and carbon dioxide unloading between the blood and alveolar air occur. As gas exchanges occur and oxygen content of the blood rises, the blood becomes bright red. The pulmonary capillary beds drain into venules, which in turn join to form the two **pulmonary veins** exiting from each lung. The four pulmonary veins complete the circuit by unloading their precious cargo into the left atrium of the heart. Note that any vessel with the term *pulmonary* or *lobar* in its name is part of the pulmonary circulation. All others are part of the systemic circulation.

It is important to remember that pulmonary arteries carry oxygen-poor, carbon dioxide–rich blood, and pulmonary veins carry oxygen-rich blood. This is exactly opposite the situation in the systemic circulation where arteries carry oxygen-rich blood and veins carry carbon dioxide–rich, relatively oxygen-poor blood.

Systemic Circulation

The systemic circulation provides the *functional blood supply* to all body tissues; that is, it delivers oxygen, nutrients, and other needed substances while carrying away carbon dioxide and other metabolic wastes. Freshly oxygenated blood returning from the pulmonary circuit is pumped out of the left ventricle into the aorta. From the aorta, blood can take various routes, since essentially all systemic arteries branch from this single great vessel. The aorta arches upward from the heart and then curves and runs downward along the body midline to its terminus in the pelvis, where it splits to form the two large arteries serving the lower extremities. The various branches of the aorta continue to subdivide to produce the arterioles and, finally,

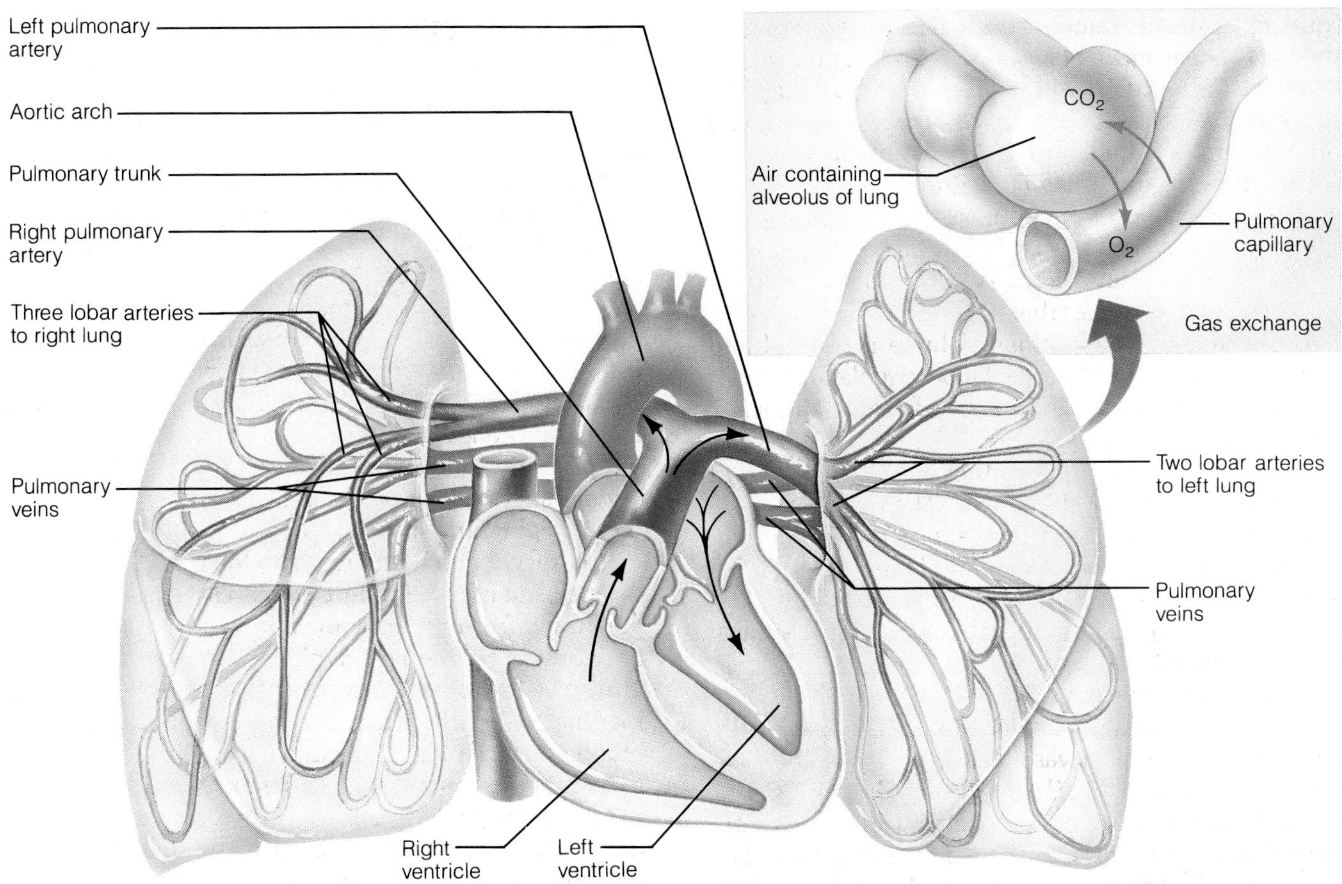

Figure 20.13 Pulmonary circulation. The pulmonary trunk divides into right and left pulmonary arteries, which subdivide to form the lobar arteries (three on the right and two on the left) serving the lobes of the lungs. After passing through the pulmonary capillaries, where gas exchange occurs, blood drains into the pulmonary veins (two per lung), which return the blood to the left atrium of the heart. The arterial system is shown in blue to indicate that the blood carried is oxygen-poor; the venous drainage is shown in red to indicate that the blood transported is oxygen-rich.

the seemingly countless capillaries that ramify through the body organs. As the blood yields its oxygen to the tissue cells and takes on carbon dioxide and other cellular wastes, it changes from bright to dull red.

Blood draining from the capillary beds starts the journey back to the heart as it flows into venules and then into larger and larger veins. Ultimately, all systemic veins empty into one of the two great veins of the body, the superior vena cava or the inferior vena cava, which collect blood draining from body areas above or below the heart, respectively, and empty it into the right atrium.

Two points concerning the two major circulations must be emphasized: (1) Blood passes from systemic veins to systemic arteries only after first moving through the pulmonary circuit, and (2) although the entire cardiac output of the right ventricle passes through the pulmonary circulation, only a small fraction of the output of the left ventricle flows through any single organ. The systemic circulation can be viewed as multiple circulatory channels functioning in parallel to distribute blood to all body organs at virtually the same time.

Because the systemic circulation contains such a large number of vessels, presenting its gross anatomy in a comprehensible way is much more difficult than describing the pulmonary circulation. As you study the various systemic arteries and veins and locate them in the illustrations, be aware of cues that make your memorization task easier. In many cases, the name of a vessel reflects the body region traversed (axillary, brachial, femoral, etc.), the organ served (renal, hepatic, gonadal, etc.), or the bone followed (vertebral, radial, tibial, etc.). Also, since many of the smaller blood vessels are described, you might want to ask your instructor which vessels are essential to know for the course you are taking.

Systemic Arteries

The arteries of the systemic circulation are illustrated in Figure 20.14 and summarized in Table 20.1.

Figure 20.14 Major arteries of the systemic circulation, anterior view.

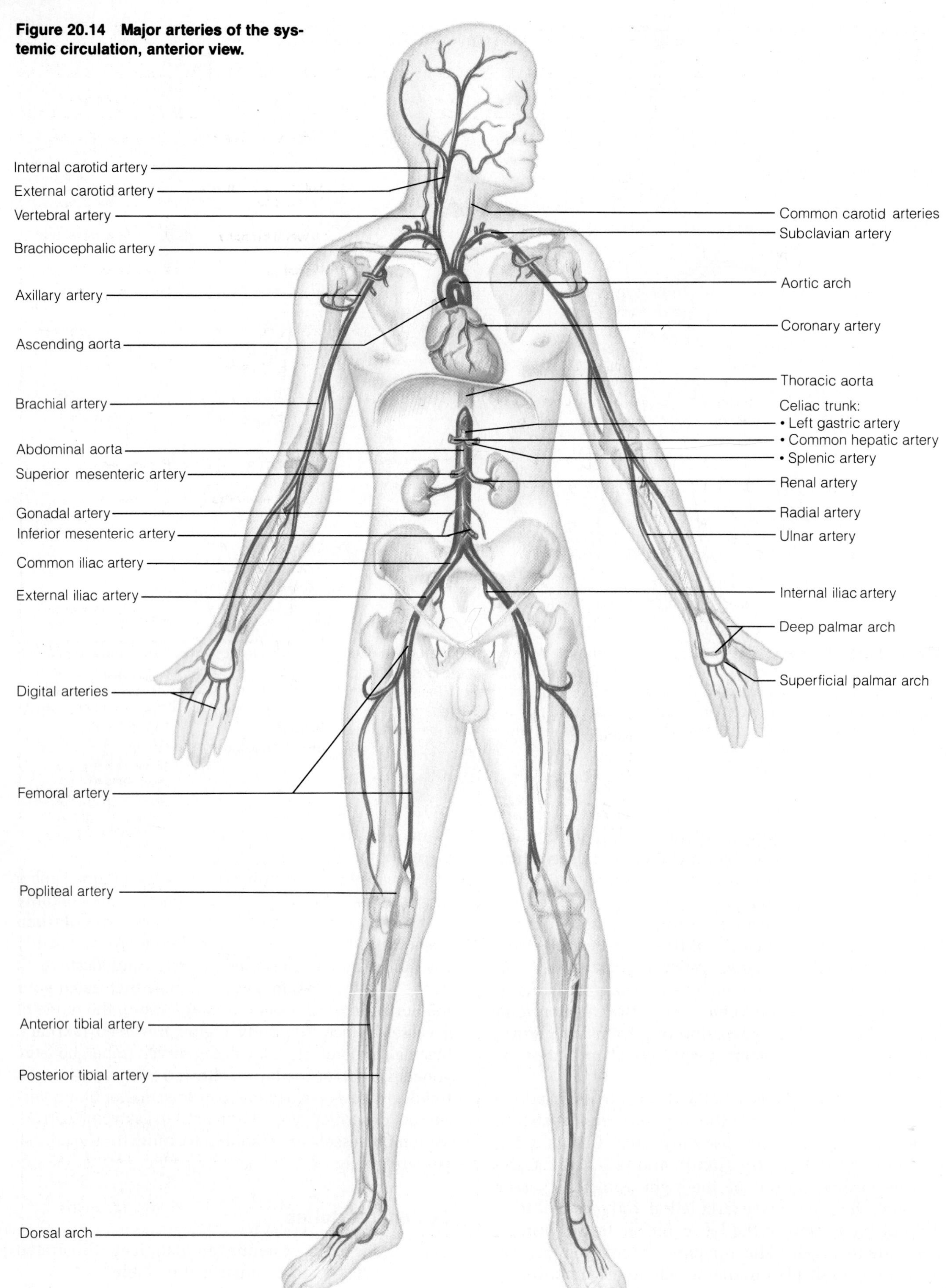
Internal carotid artery
External carotid artery
Vertebral artery
Brachiocephalic artery
Axillary artery
Ascending aorta
Brachial artery
Abdominal aorta
Superior mesenteric artery
Gonadal artery
Inferior mesenteric artery
Common iliac artery
External iliac artery
Digital arteries
Femoral artery
Popliteal artery
Anterior tibial artery
Posterior tibial artery
Dorsal arch
Common carotid arteries
Subclavian artery
Aortic arch
Coronary artery
Thoracic aorta
Celiac trunk:
• Left gastric artery
• Common hepatic artery
• Splenic artery
Renal artery
Radial artery
Ulnar artery
Internal iliac artery
Deep palmar arch
Superficial palmar arch

Table 20.1 The Aorta and Its Branches

Aortic region	Major branches	Area served	Aortic region	Major branches	Area served
Ascending aorta	**Coronary arteries** (right and left)	Myocardium of the heart		**Left common carotid** and its branches	Same relative regions as right common carotid artery
Aortic arch	**Brachiocephalic** (innominate)	Branches provide arterial blood supply of head, neck, and upper extremity (right side)		**Left subclavian** and its branches	Same relative regions as right subclavian artery
	Right common carotid		Thoracic aorta	**Visceral branches**	Together serve viscera of thorax
	Right external carotid			Pericardial	Pericardium
	• Superficial thyroid	Anterior neck and thyroid gland		Bronchial	Lungs and bronchi
	• Lingual	Tongue		Esophageal	Esophagus
	• Facial	Anterior facial structures		Mediastinal	Structures of posterior mediastinum
	• Occipital	Posterior scalp		**Parietal branches**	Thorax wall and diaphragm
	• Maxillary	Deep facial structures, chewing muscles, teeth, nasal mucosae, dura mater		Posterior intercostals	Intercostal and pectoral muscles, pleurae, and overlying skin
	• Superficial temporal	Lateral and superior scalp		Superior phrenics	Posterior and superior surface of diaphragm
	Right internal carotid		Abdominal aorta	**Visceral branches**	Together supply abdominal and pelvic viscera
	• Ophthalmic	Orbit and eye structures, anterior scalp, nasal structures		Celiac trunk	
				Common hepatic	Liver
				Splenic	Spleen
	• Anterior cerebral	Anterior medial surface of cerebral hemispheres		*Left gastric*	Stomach and inferior esophagus
	• Middle cerebral	Lateral and superior portion of cerebral hemispheres		Inferior phrenics	Inferior diaphragmatic surface
	Right subclavian			Superior mesenteric	Small intestine, ascending and part of transverse colon
	Vertebral	Deep structures of posterior neck		Suprarenals	Adrenal glands
	• Basilar	Cerebellum, pons, inner ear, and via its terminal branches, the posterior cerebral arteries, the occipital and part of the temporal lobes of cerebral hemispheres		Renals	Kidneys
				Gonadals	Testes in males, ovaries in females
				Inferior mesenteric	Distal colon (part of transverse, sigmoid, and descending portions)
				Parietal branches	
	Internal thoracic	Anterior thorax wall (deep)		Lumbars (several pairs)	Posterior abdominal wall
	Thyrocervical trunk	Inferior neck structures and superior intercostal muscles		Middle sacral	Sacrum, coccyx, and rectum
	Costocervical trunk	Inferior neck structures and superior intercostal muscles		Right and left common iliacs	
	Right axillary	Axilla		*Internal iliacs*	Lower abdominal wall, pelvic organs
	• Lateral thoracic	Wall of lateral thorax		*External iliacs*	
	• Supscapular	Scapula, shoulder, and pectoral region		• Femoral	Thighs (and posterior knee where femoral arteries become popliteal arteries)
	• Thoracoacromial	Scapula, shoulder, and pectoral region		• Anterior tibials	Anterior leg muscles, ankles, and dorsum of feet
	• Thoracodorsal	Dorsal thorax; latissimus dorsi			
	• Brachial	Arm		• Posterior tibials	Flexor muscles of leg, plantar surfaces of feet, toes
	Radial	Forearm and hand			
	Ulnar	Forearm and hand		Peroneals	Lateral leg muscles

Aorta. The **aorta** (ā-or′-tuh), the largest artery in the body, is indeed the king of vessels. In adults, the aorta is approximately the size of a garden hose where it issues from the left ventricle of the heart. Its internal diameter is 2.5 cm, and its wall is about 2 mm thick; it decreases in size only slightly as it runs to its terminus. The aortic semilunar valve guards the base of the aorta and prevents backflow of blood during diastole. Opposite each of the semilunar valve cusps is an *aortic sinus,* a slight enlargement of the aortic wall containing pressoreceptors important in reflex regulation of blood pressure (see pp. 630–631).

Different portions of the aorta are named according to their relative shape or location. Major branches issuing from each aortic area are shown in Figure 20.14 and described briefly here.

- **Ascending aorta.** The first portion of the aorta, called the ascending aorta, runs superiorly in front of the pulmonary trunk. It persists for only about 5 cm (2 inches) before curving to the left to become the aortic arch. The only branches of the ascending aorta are the **right** and **left coronary arteries.** These arteries issue from the aortic sinuses, run to encircle the heart in its grooves, and then branch throughout the myocardium, which they supply. The distribution of the coronary arteries is described in Chapter 19, p. 603.
- **Aortic arch.** The aortic arch extends from the sternal angle to the level of the fourth thoracic vertebra, where it lies dorsally at the left side of the vertebral column deep to the sternal manubrium. Three major arteries branch off the aortic arch. The first and largest branch is the **brachiocephalic** (brā″-kē-ō-seh-fa′-lik), or **innominate** (ih-nah′-mih-nit), **artery,** which passes superiorly under the right clavicle and branches into the **right common carotid** (kuh-rah′-tid) **artery** and the **right subclavian artery.** The second and third aortic arch branches are the **left common carotid artery** and the **left subclavian artery,** respectively. These three vessels provide the arterial supply of the head, neck, upper extremities, and part of the thorax wall. (See Arteries of the Head and Neck, this page, and Arteries of the Thorax, p. 646.)
- **Thoracic aorta.** The thoracic portion of the descending aorta runs from the fourth to the twelfth thoracic vertebra. Along its course, it sends off numerous small arteries to the organs of the thoracic cavity and to portions of the thorax wall (see Table 20.1).
- **Abdominal aorta.** As the aorta pierces the diaphragm and enters the abdominal cavity, it becomes the abdominal aorta. This portion of the aorta overlies the vertebral column. As it moves inferiorly, it sends branches to the inferior surface of the diaphragm, the posterior abdominal walls, and virtually all the organs of the abdominal viscera (see Table 20.1 and Arteries of the Abdomen, p. 648). The abdominal aorta terminates at the level of the fourth lumbar vertebra by dividing into the **right** and **left common iliac arteries,** which supply the pelvis and the lower extremities.

Arteries of the Head and Neck. Four paired arteries provide the functional blood supply of the head and neck. These are the common carotid arteries, the vertebral arteries, and the thyrocervical and costocervical trunks.

- **Common carotid arteries.** Most head and neck tissues receive their blood supply from the paired common carotid arteries. The right common carotid artery arises from the brachiocephalic artery; the left member is the second branch of the aortic arch. The common carotid arteries run upward through the lateral neck, and at the superior border of the larynx, each divides into its two major branches, the *external* and *internal carotid arteries* (Figure 20.15). At the division point, each internal carotid artery has a slight dilation, the *carotid sinus.* The carotid sinuses (like the aortic sinuses) contain pressoreceptors that react to changes in arterial blood pressure and assist in reflex blood pressure control. The *carotid bodies,* chemoreceptors that are part of the neural mechanism involved in the control of respiratory rate (and to a lesser extent blood pressure), are located close by.

  The **external carotid arteries** supply most of the extracranial tissues of the head and neck (Figure 20.15a). As each runs superiorly, it sends branches to the thyroid gland and larynx **(superior thyroid artery),** the tongue **(lingual artery),** the skin and muscles of the anterior face **(facial artery),** and the posterior scalp **(occipital artery).** Each external carotid artery terminates by splitting into **maxillary** and **superficial temporal arteries,** which supply the lower jaw and chewing muscles, the teeth, the nasal and pharyngeal mucosae, most of the scalp, lateral aspects of the face, and the dura mater.

  The larger **internal carotid arteries** provide the major blood supply of the cerebral hemispheres of the brain. They assume a deep course and enter the skull through the carotid canals of the temporal bones. Once inside the cranium, each gives off one main branch, the ophthalmic artery, and then divides into the anterior and middle cerebral arteries. The **ophthalmic** (of-thal′-mik) **arteries** supply the structures within and around the eye sockets, the anterior aspect of the scalp, and the nasal cavity. Each **anterior cerebral artery** supplies the medial surface of

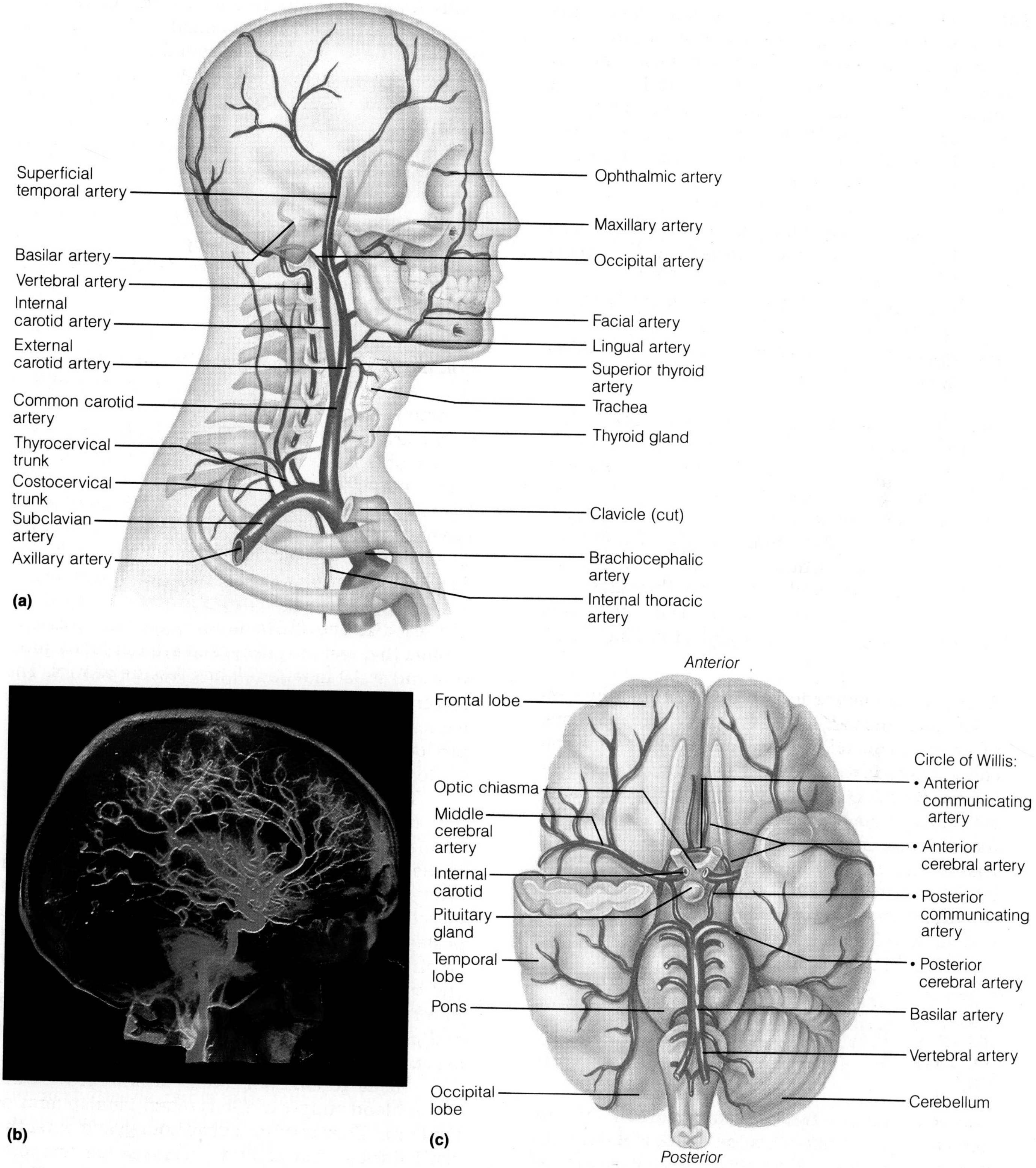

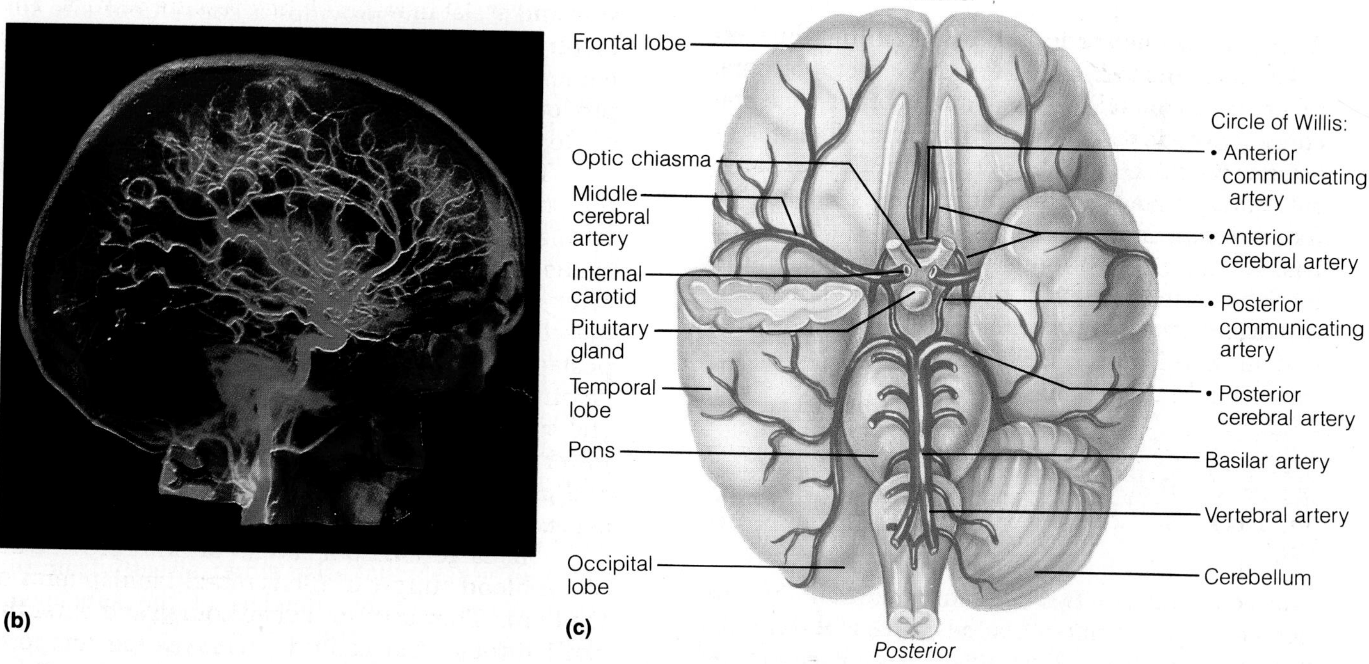

Figure 20.15 Arteries of the head, neck, and brain. (**a**) Arteries of the head and neck, right aspect. (**b**) Arteriograph of the arterial supply of the brain. (**c**) Major arteries serving the brain and the circle of Willis. In this inferior view of the brain, the right side of the cerebellum and part of the right temporal lobe have been removed to show the distribution of the posterior cerebral artery.

the cerebral hemisphere on its side and then anastomoses with its partner on the opposite side via a short arterial shunt called the **anterior communicating artery** (Figure 20.15c). The **middle cerebral arteries** run in the lateral fissures of their respective cerebral hemispheres and give off branches to the lateral parts of the temporal and parietal lobes.

- **Vertebral arteries.** The blood supply of the posterior head and neck is provided by the right and left vertebral arteries. The vertebral arteries spring from the subclavian arteries at the root of the neck and ascend through the foramina in the transverse processes of the cervical vertebrae to enter the skull through the foramen magnum. En route, they send branches to the deep structures of the neck. Within the cranium, the vertebral arteries join to form the **basilar** (ba′-sih-ler) **artery,** which, as it ascends the anterior aspect of the brain stem, gives off several branches that supply the cerebellum, pons, and structures of the inner ear (see Figure 20.15c). At the base of the cerebrum, the basilar artery divides to produce the **right** and **left posterior cerebral arteries,** which supply the occipital lobes and the inferior portions of the temporal lobes of the cerebral hemispheres.

  Short arterial shunts called the **posterior communicating arteries** connect the posterior cerebral arteries to the middle cerebral arteries anteriorly. The two posterior communicating arteries and the single anterior communicating artery complete the formation of an arterial anastomosis called the **circle of Willis.** The circle of Willis encircles the pituitary gland and optic chiasma and unites the brain's anterior and posterior blood supplies provided by the internal carotid and vertebral arteries. This anastomosis provides alternate routes for blood to reach the brain tissue in the case of an arterial occlusion of either a carotid or vertebral artery.

- **Thyrocervical and costocervical trunks.** The balance of the head and neck arterial system is provided by two smaller branches of the paired subclavian arteries, the thyrocervical and costocervical trunks. These short vessels arise just lateral to the vertebral arteries on each side (see Figures 20.15a and 20.16). Together, they supply branches to structures of the inferior neck (the neck muscles, thyroid and parathyroid glands, larynx, pharynx, trachea, and esophagus), muscles attaching the scapula to the vertebral column, and the superior intercostal muscles.

Arteries of the Upper Extremities. The upper extremities are supplied entirely by arteries arising from the subclavian arteries (see Figure 20.16). After giving off branches to the neck, each subclavian artery courses laterally between the clavicle and first rib to enter the axilla, where it becomes known as the axillary artery.

- **Axillary artery.** Through its various branches, each axillary artery supplies the structures of the axilla and chest wall, including the shoulder joint, scapula, part of the deltoid muscle, and muscles of the lateral thorax (pectoralis, scapular, and latissimus dorsi).

- **Brachial artery.** As the axillary artery emerges from the axilla, it becomes the brachial artery (see Figure 20.16), which runs down the medial aspect of the humerus and supplies the anterior flexor muscles of the arm. One major branch, the **deep brachial artery,** serves the posterior triceps brachii muscle. As the brachial artery nears the elbow, it gives off several small branches that contribute to an anastomosis connecting it to the major arteries of the forearm. As the brachial artery crosses the anterior aspect of the elbow, it provides an easily palpated pulse point (brachial pulse) that is used routinely for obtaining blood pressure measurements (see Figures 20.3 and 20.9). Immediately beyond the elbow, the brachial artery splits to form the radial and ulnar arteries, which follow the course of similarly named bones down the length of the anterior forearm.

- **Radial artery.** The radial artery supplies the posterior superficial extensor muscles of the forearm, the lateral wrist, and the thumb and index finger. At the root of the thumb, the radial artery is very close to the surface and provides a convenient site for taking the radial pulse (see Figure 20.3).

- **Ulnar artery.** The ulnar artery supplies the medial aspect of the forearm and fingers 3–5, as well as the medial aspect of the index finger. As the ulnar artery passes toward the wrist, it gives off a short branch, the **common interosseous** (in″-ter-ah′-sē-us) **artery,** which runs between the radius and ulna to serve the deep flexors and extensors of the forearm. In the palm of the hand, branches of the radial and ulnar arteries anastomose to form two palmar arches, the **superficial** and **deep palmar arches,** formed mainly by the ulnar and radial arteries, respectively. The **digital arteries** that supply the fingers arise from these palmar arches.

Arteries of the Thorax. The thorax wall is supplied by an array of vessels that arise either directly from the thoracic aorta or from branches of the subclavian arteries (see Figure 20.16).

- **Posterior and anterior intercostal arteries.** Nine or ten pairs of *posterior intercostal arteries* issue from the thoracic aorta and course around the rib cage to anastomose anteriorly with the *anterior intercostal arteries*. The upper two to three pairs of posterior intercostal arteries are derived from the **costocervical trunk;** the anterior intercostal arteries are branches of the **internal thoracic (mammary) arter-**

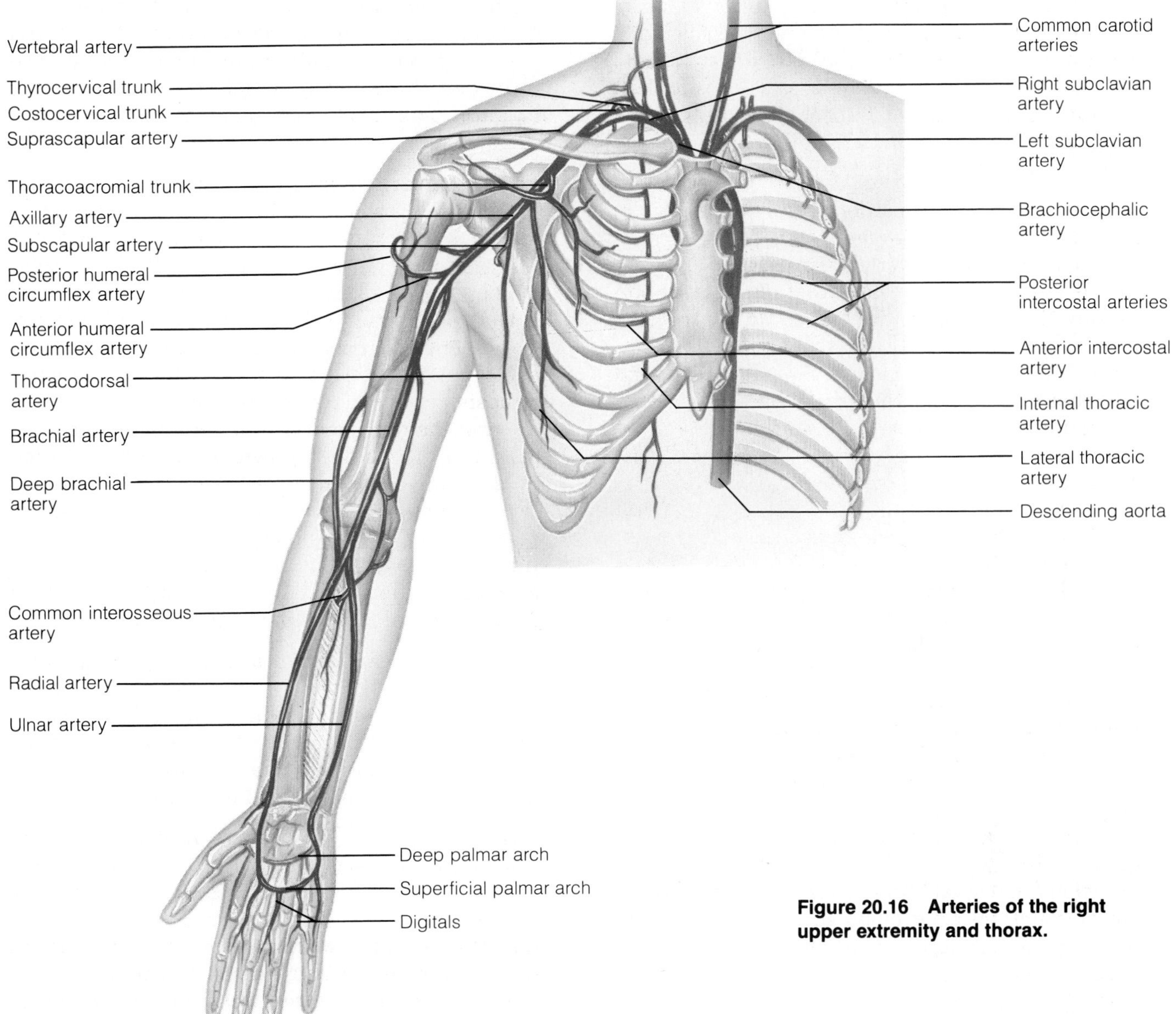

Figure 20.16 Arteries of the right upper extremity and thorax.

ies. (The internal thoracic arteries arise from the subclavian arteries.) Together, the posterior and anterior intercostal arteries supply the intercostal muscles, the deep muscles of the back, the vertebrae, the spinal cord, the muscles of the anterior thorax, and the mammary glands. Twiglike branches of the internal thoracic arteries also contribute to the blood supply of the mediastinal structures, the diaphragm, and the rectus abdominis muscles of the abdomen.

- **Branches of the axillary arteries.** Several branches of the axillary arteries serve the thorax region as well. These include the **lateral thoracic arteries,** which serve the lateral chest wall; the **subscapular arteries** and **thoracoacromial** (thor′-uh-kō-uh-krō-mē-ul) **trunks,** which supply the scapula, shoulder region, and pectoralis muscles; and the **thoracodorsal arteries,** which serve the dorsal thorax and latissimus dorsi muscles (see Figure 20.16).

Most visceral organs of the thorax receive their functional blood supply directly from small branches issuing from the thoracic aorta. Because these vessels are so small and tend to vary in number (with the exception of the bronchial arteries), they are not illustrated in Figure 20.16.

- **Pericardial arteries.** Several tiny branches supply the posterior pericardium.
- **Bronchial arteries.** Two left and one right bronchial artery serve the lungs, bronchi, and pleurae.

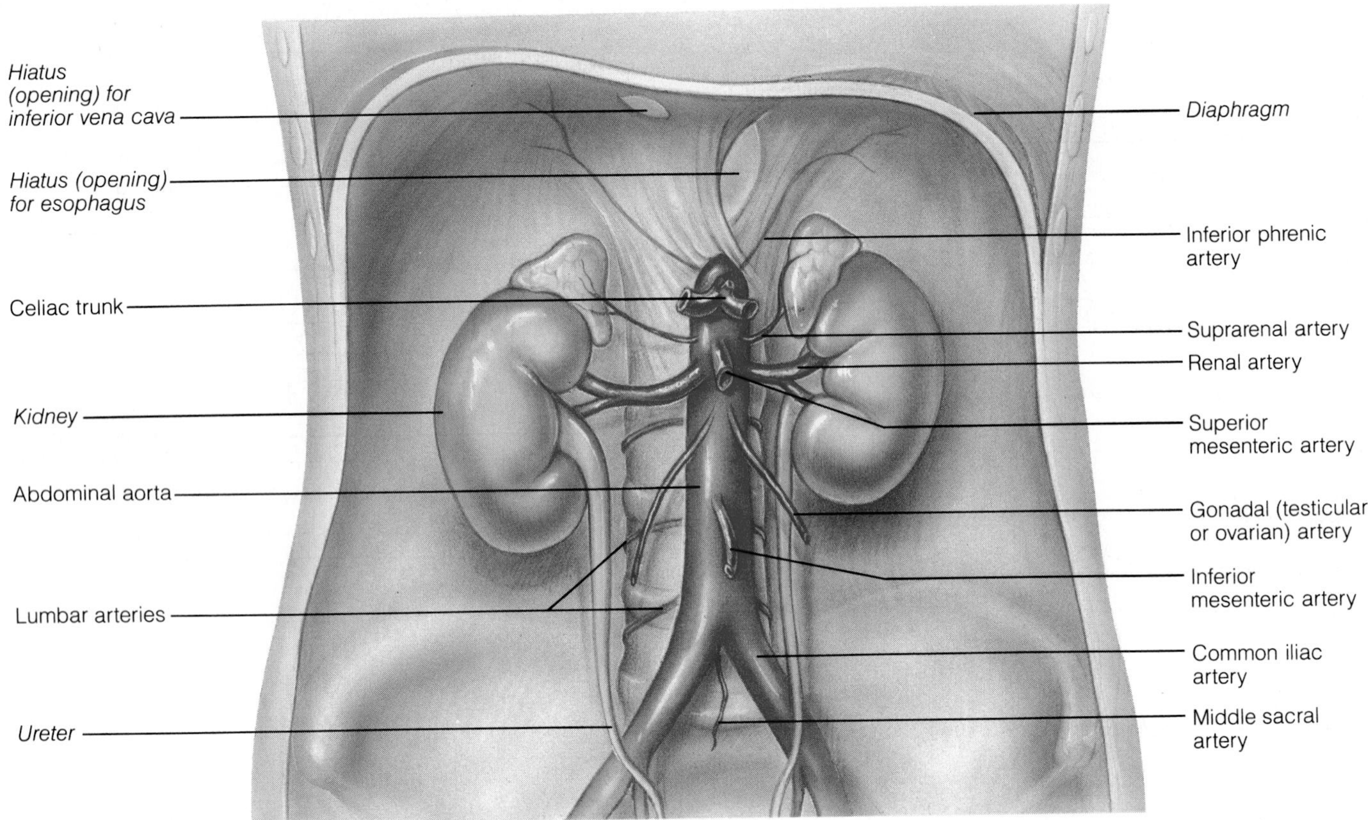

Figure 20.17 Major branches of the abdominal aorta.

- **Esophageal arteries.** Four to five esophageal arteries supply the esophagus.
- **Mediastinal arteries.** Many small mediastinal arteries serve the contents of the posterior mediastinum.
- **Superior phrenic arteries.** Several paired superior phrenic arteries serve the posterior part of the superior diaphragm surface.

Arteries of the Abdomen. The arterial supply to the abdominal organs arises from the abdominal aorta. Under resting conditions, about half of the entire arterial flow is found in these vessels. These branches are given here in order of their issue.

- **Celiac trunk.** This first and very large branch of the abdominal aorta divides almost immediately into three branches: the common hepatic, splenic, and left gastric arteries (Figures 20.17 and 20.18). The **common hepatic** (heh-pa′-tik) **artery** runs superiorly, giving off branches to the stomach, duodenum, and pancreas before it splits into the **right** and **left hepatic arteries** that serve the liver. As the **splenic** (spleh′-nik) **artery** passes deep to the stomach, it sends branches to the pancreas and stomach and then terminates in branches to the spleen. The **left gastric artery** supplies part of the stomach and the inferior esophagus. The right and left gastroepiploic arteries (formed by branches of the hepatic and splenic arteries, respectively) serve the left (major) curvature of the stomach.
- **Inferior phrenic arteries.** The inferior phrenics emerge from the aorta at about the same level as the celiac trunk. They serve the inferior diaphragm surface.
- **Superior mesenteric** (meh-zen-teh′-rik) **artery.** This large, unpaired artery arises from the abdominal aorta immediately below the celiac trunk (see Figures 20.17 and 20.18). It runs deep to the pancreas and then enters the mesentery, where it forms numerous anastomosing branches that serve virtually all of the small intestine and a large portion of the large intestine—the appendix, cecum, ascending colon (via the *ileocolic artery*), and part of the transverse colon (via the *right* and *middle colic arteries*) (Figure 20.19).
- **Suprarenal** (sū″-pruh-rē′-nul) **arteries.** The right and left suprarenal arteries flank the origin of the superior mesenteric artery as they emerge from the abdominal aorta (see Figure 20.17). They supply blood to the adrenal (suprarenal) glands overlying the kidneys.

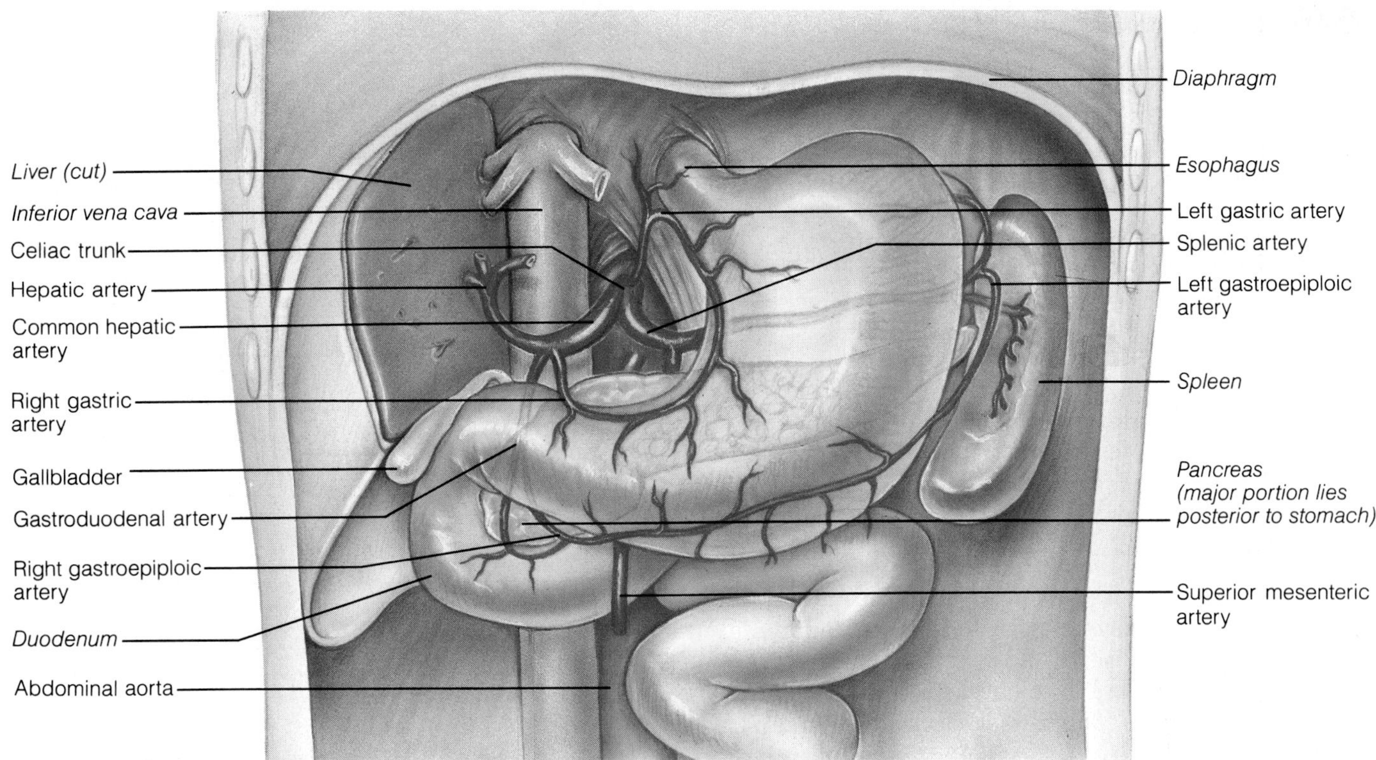

Figure 20.18 The celiac trunk and its major branches.

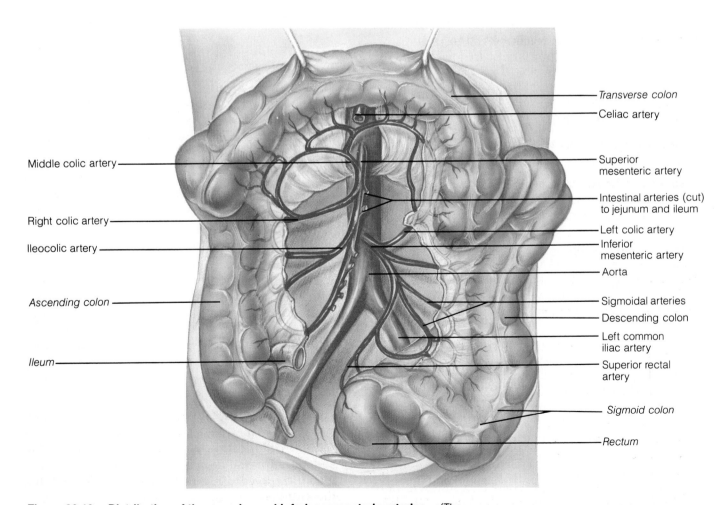

Middle colic artery
Right colic artery
Ileocolic artery
Ascending colon
Ileum
Transverse colon
Celiac artery
Superior mesenteric artery
Intestinal arteries (cut) to jejunum and ileum
Left colic artery
Inferior mesenteric artery
Aorta
Sigmoidal arteries
Descending colon
Left common iliac artery
Superior rectal artery
Sigmoid colon
Rectum

Figure 20.19 Distribution of the superior and inferior mesenteric arteries. (The transverse colon has been reflected superiorly to provide a better view of these arteries.)

- **Renal arteries.** The short, but wide, right and left renal arteries issue from the lateral surfaces of the aorta slightly below the superior mesenteric artery (see Figure 20.17). Each serves the kidney on its side.

- **Gonadal** (gō-na′-dul) **(ovarian or testicular) arteries.** The paired gonadal arteries are more specifically called the *testicular arteries* in males and the *ovarian arteries* in females to indicate the type of gonad served (see Figure 20.17). In females, these vessels extend into the pelvis to serve the ovaries and part of the uterine tubes. The much longer testicular arteries descend through the pelvis and inguinal canal to enter the scrotal sac, where they serve the testes.

- **Inferior mesenteric artery.** The inferior mesenteric artery is the final major branch of the abdominal aorta (see Figure 20.17). Like the superior mesenteric artery, it is unpaired and arises from the anterior aortic surface. This vessel serves the distal part of the large intestine—the transverse, descending, and sigmoid colons and the rectum—via its *left colic, sigmoidal,* and *superior rectal* branches (see Figure 20.19). Looping anastomoses between the superior and inferior mesenteric arteries help ensure that blood will continue to reach the various regions of the digestive viscera in cases of trauma to one of these abdominal arteries.

- **Lumbar arteries.** Four pairs of lumbar arteries arise from the posterior surface of the aorta in the lumbar region (see Figure 20.17). These segmental arteries supply the posterior abdominal wall.

- **Middle sacral artery.** The middle sacral artery issues from the posterior medial surface of the abdominal aorta at its terminus (see Figure 20.17). This tiny artery carries blood to the sacrum, coccyx, and rectum.

- **Common iliac arteries.** At the level of the pelvic brim, the abdominal aorta splits into the right and left common iliac arteries, which supply blood to the lower abdominal wall, pelvic organs, and lower limbs (see Figures 20.17 and 20.20).

Arteries of the Pelvis and Lower Limbs. At the level of the sacroiliac joints, the **common iliac arteries** divide into two major branches, the internal and external iliac arteries (Figure 20.20).

- **Internal iliac arteries.** The paired internal iliac arteries run into the pelvis and distribute blood to the pelvic walls and viscera (bladder, rectum, uterus and vagina in the female, prostate gland and ductus deferens in the male), the external genitalia, and the medial aspects of the thighs.

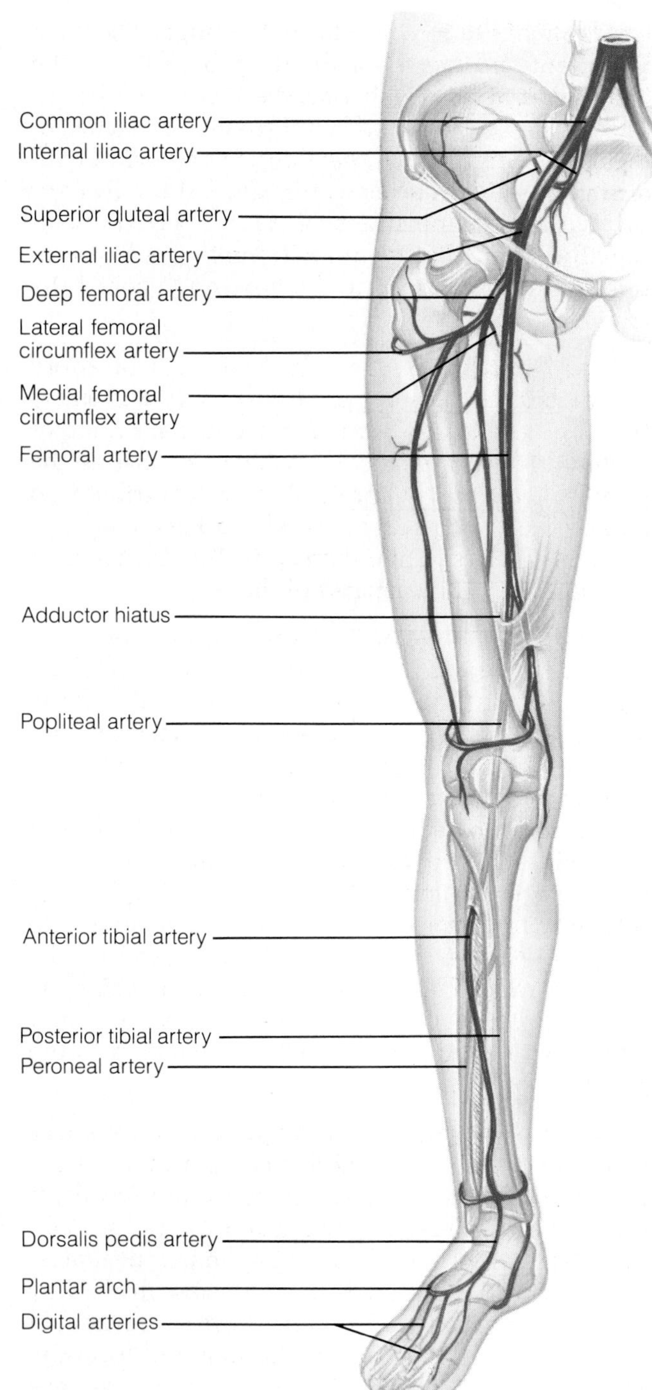

Figure 20.20 Arteries of the right pelvis and lower extremity.

- **External iliac arteries.** The right and left external iliac arteries supply the lower limbs. As they course through the pelvis, they give off branches to the muscles and skin of the inferior abdominal wall. They then pass under the inguinal ligaments to enter the thigh, where they become the femoral arteries.

- **Femoral artery.** As each femoral artery passes down the anteromedial aspect of the thigh, it gives off several branches to the muscles of the thigh. The most important and largest of the deep branches is the **deep femoral artery** that serves the posterolateral (flexor) muscles of the thigh. Proximal branches of the deep femoral artery, the **lateral** and **medial femoral circumflex arteries,** circle around the head of the femur and supply muscles in that region. Additionally, the medial femoral circumflex artery supplies most of the blood to the head and neck of the femur.

- **Popliteal** (pop-lih′-tē-ul) **artery.** As the femoral artery approaches the knee, it passes posteriorly and enters the popliteal fossa through a canal, the *adductor hiatus* (hī-ā′-tis), where it becomes known as the popliteal artery. The popliteal artery contributes to an arterial anastomosis around the knee and supplies the knee region. It then splits into the anterior and posterior tibial arteries of the leg.

- **Anterior tibial artery.** The smaller anterior tibial artery runs through the anterior compartment of the leg, supplying the extensor muscles en route. At the ankle, it becomes the **dorsalis pedis artery,** which supplies the ankle and dorsum of the foot and helps to form the plantar arch of the foot.

The dorsalis pedis artery is quite superficial and provides a clinically important pulse point, the pedal pulse (see Figure 20.3). If the pedal pulses are easily felt, it is fairly certain that the blood supply to the leg is good. When these pulses are absent, the nutritional state of the leg is in jeopardy. Checking the pedal pulses is routine after leg or groin surgery and for those known to have impaired circulation to the legs. ■

- **Posterior tibial artery.** The large posterior tibial artery courses through the posterior compartment of the leg and supplies the flexor muscles. Proximally, it gives off a large branch, the **peroneal** (payr″-uh-nē′-ul) **artery,** which supplies the lateral peroneal muscles of the leg. At the ankle, the posterior tibial artery divides into **lateral** and **medial plantar arteries** that serve the plantar surface of the foot and form the **plantar arch.** The **digital arteries** serving the toes arise from the plantar arch.

Systemic Veins

Although there are many similarities between the systemic arteries and veins, there are also important differences:

1. Whereas the heart pumps all of its blood into a single systemic artery—the aorta—blood returning to the heart is delivered by two terminal systemic veins, the superior and inferior venae cavae. The single exception to this is the blood draining from the myocardium of the heart, which is collected by the cardiac veins and reenters the right atrium via the coronary sinus (see p. 604).

2. All arteries run deep and are well protected by body tissues along most of their course, but both deep and superficial veins exist. Deep veins parallel the course of the systemic arteries, and with a few exceptions, the naming of these veins is identical to that of their companion arteries. Superficial veins run just beneath the skin and are readily seen, especially in the limbs, face, and neck. Since there are no superficial arteries, the names of the superficial veins do not correspond to the names of any of the arteries.

3. Unlike the fairly clear arterial pathways, venous pathways tend to have numerous interconnections, and many veins are represented by not one but two similarly named vessels. As a result, venous pathways are more difficult to follow.

4. In most body regions, there is a similar and predictable arterial supply and venous drainage. However, the pattern of venous drainage in at least two important body areas is quite unique. First, venous blood draining from the brain enters large *dural sinuses* rather than typical veins. Second, blood draining from the digestive organs enters a special subcirculation, called the *hepatic portal circulation,* and perfuses through the liver before it reenters the general systemic circulation.

In our survey of systemic veins, the major branches of the venae cavae will be noted first, followed by a description of the venous pattern of the various body regions. Since veins run toward the heart, the most distal veins will be named first and those closest to the heart last to emphasize the difference in direction of blood flow in veins and arteries. Deep veins will be named, but since they generally drain the same areas served by their companion arteries, they will not be described in detail. The systemic veins are illustrated in Figure 20.21 and summarized in Table 20.2.

Venae Cavae and Their Major Tributaries.

- **Superior vena cava.** The superior vena cava receives blood draining from all areas superior to the diaphragm, *except the lungs.* (Remember, blood leaving the lungs via the pulmonary veins enters the left atrium of the heart.) The superior vena cava is formed by the union of the **right** and **left brachiocephalic veins** and empties into the superior aspect of the right atrium (see Figure 20.21). Notice that there are two brachiocephalic veins, but only one brachiocephalic artery. Each brachiocephalic vein is formed in turn by the joining of the **internal jugular** and **subclavian veins** on its side.

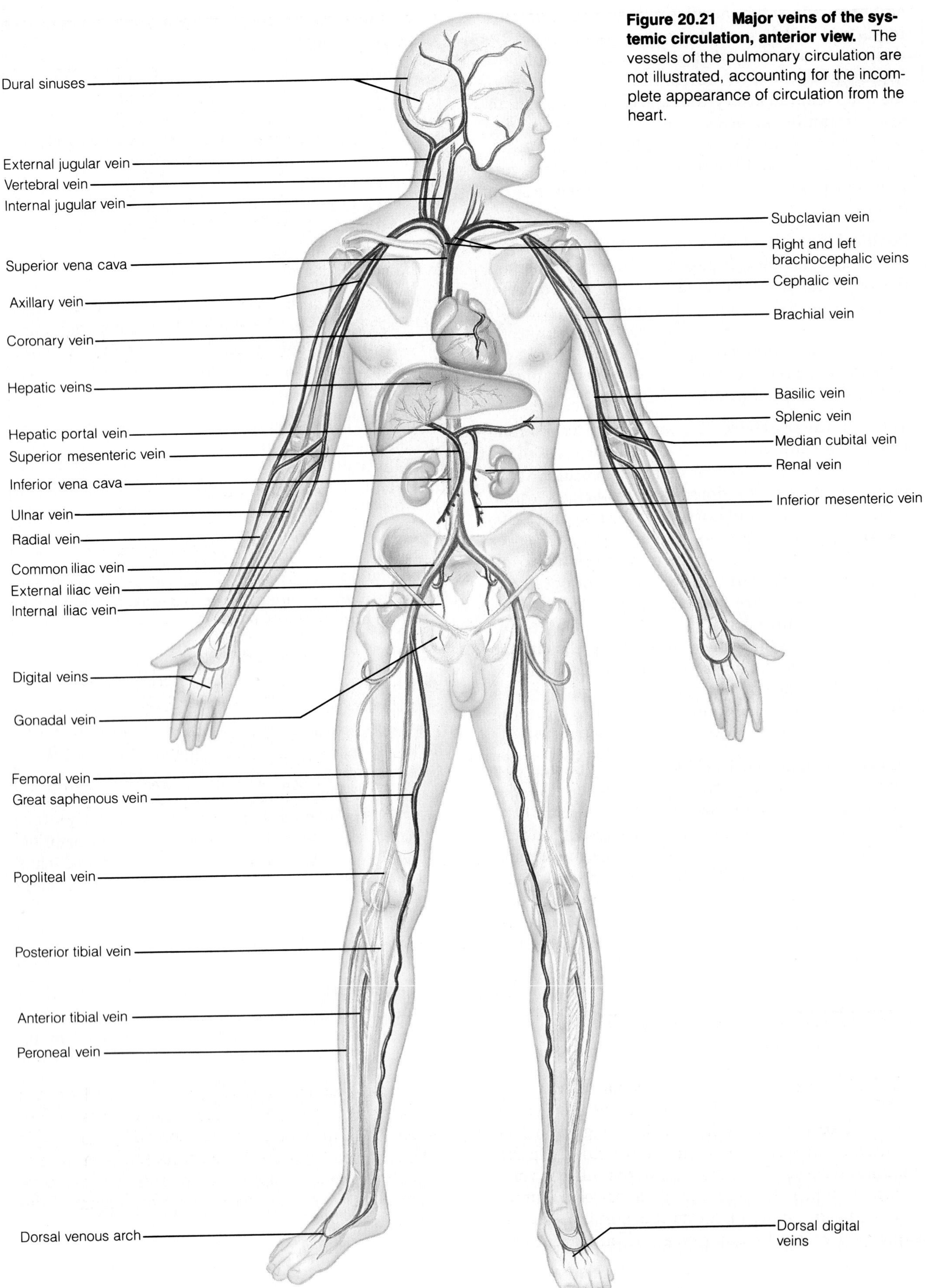

Figure 20.21 Major veins of the systemic circulation, anterior view. The vessels of the pulmonary circulation are not illustrated, accounting for the incomplete appearance of circulation from the heart.

Table 20.2 Major Systemic Veins Emptying into the Venae Cavae*

Vena cava	Major tributaries	Area drained
Superior vena cava	**Brachiocephalic**	Formed by union of internal jugular and subclavian on each side; the pair join to form the superior vena cava
	Internal jugular	Dural venous sinuses of the brain
	Vertebral	Posterior extracranial tissues
	Subclavian	
	External jugular	Drains superficial tissues of head and neck
	Axillary	Continuation of subclavian; branches together drain the shoulder, axilla, arm, and forearm (Note that the median cubital vein connects the cephalic and basilic veins.)
	• Brachial	
	Radial	
	Ulnar	
	Basilic†	
	• Cephalic†	
	Median cubital†	
	Median antebrachial	
	Azygos	Vessels of the azygos system together drain the thorax
	Hemiazygos	
	Accessory azygos	
Inferior vena cava	**Hepatic**	The hepatic portal system drains the digestive viscera via the superior and inferior mesenteric veins and splenic vein. That blood is shunted to the liver via the hepatic portal vein; the liver is drained by hepatic veins
	Liver sinusoids	
	Hepatic portal	
	Right suprarenal	Right adrenal gland
	Renal	Drain the kidneys; left renal vein also receives blood from left gonadal and left suprarenal veins
	Right gonadal	Right ovary or testis
	Lumbars (several pairs)	Posterior abdominal wall
	Common iliac	Together the common iliac veins receive the drainage of the pelvis and lower extremities
	Internal iliac	Pelvis; medial thigh
	External iliac	Lower limb
	Femoral	
	• Great saphenous†	
	• Popliteal	
	Radial	
	Ulnar	
	Small saphenous†	

*Unless otherwise specified, each vein has right and left members.

†Indicates a superficial vein.

- **Inferior vena cava.** The inferior vena cava is substantially longer than the superior vena cava. It returns blood to the heart from all body regions below the diaphragm and corresponds most closely to the abdominal aorta. The distal end of the inferior vena cava is formed by the junction of the paired **common iliac veins** in the pelvis. From this point, it courses superiorly along the anterior aortic surface. Along its way, it receives venous blood draining from the abdominal walls, gonads, and kidneys. Just before it penetrates the diaphragm, it is joined by the hepatic veins, which transport blood from the liver. Immediately above the diaphragm, the inferior vena cava enters the inferior aspect of the right atrium of the heart.

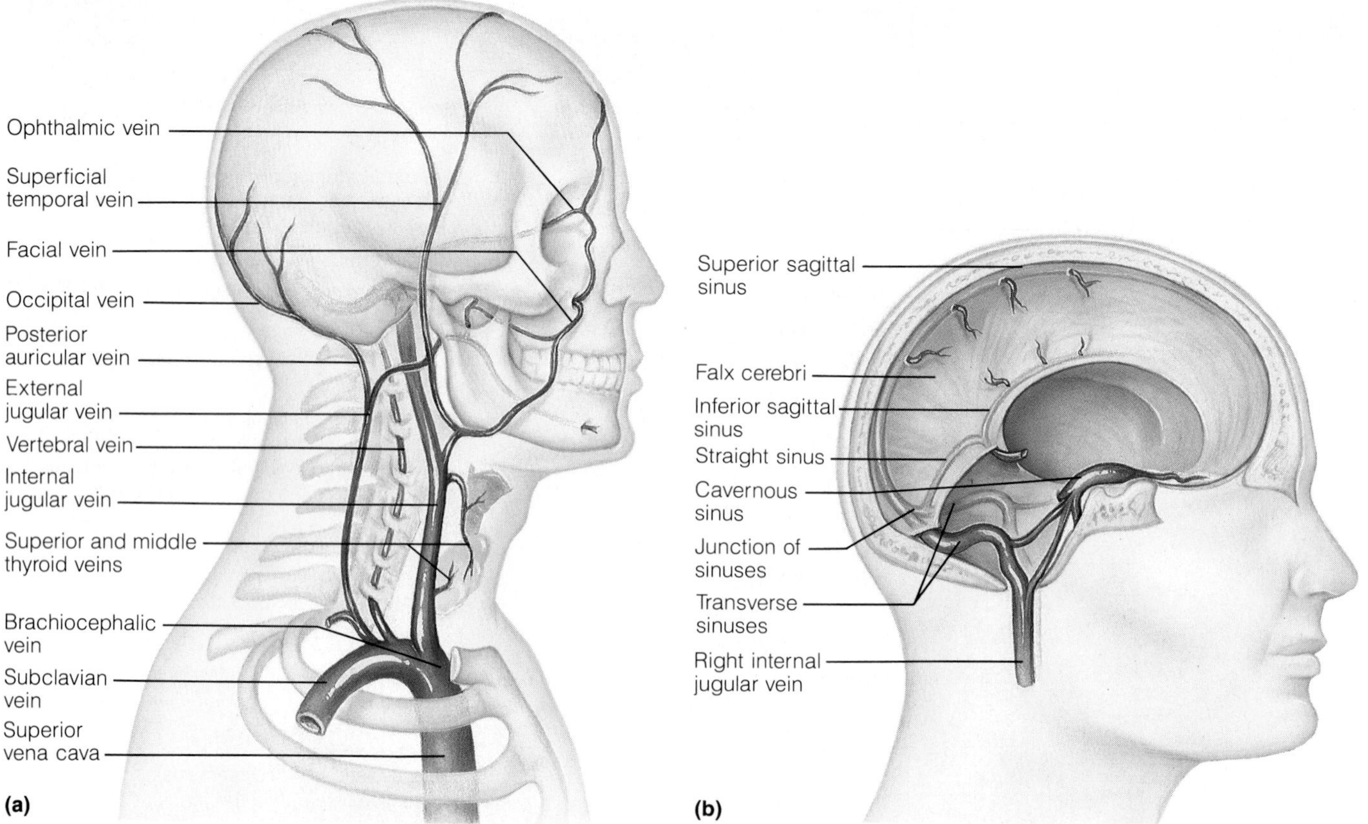

Figure 20.22 Venous drainage of the head, neck, and brain. (**a**) Veins of the head and neck, right superficial aspect. (**b**) Dural sinuses of the brain, right aspect.

Veins of the Head and Neck. Most blood draining from the head and neck is collected by three pairs of veins: the internal jugular veins, external jugular veins, and vertebral veins (Figure 20.22a). Although most of the extracranial veins have the same names as the extracranial arteries (for example, facial, ophthalmic, occipital, and superficial temporal), their courses and interconnections differ substantially.

- **External jugular veins.** The right and left external jugular veins drain superficial neck and head (scalp and face) structures served by the external carotid arteries. However, their tributaries anastomose frequently, and some of the superficial drainage enters the internal jugular veins as well. As the external jugular veins descend through the lateral neck, they pass obliquely over the sternocleidomastoid muscles and then empty into the subclavian veins.

- **Internal jugular veins.** The paired internal jugular veins, which receive the bulk of blood draining from the brain, are considerably larger than the external jugulars. Most of the venous drainage from the brain initially enters the **dural sinuses,** enlarged chambers located between the dura mater layers. The most important of these venous sinuses are the **superior** and **inferior sagittal sinuses** found within the falx cerebri, which dips down between the cerebral hemispheres, but several smaller sinuses exist (Figure 20.22b). The internal jugular veins arise from these venous sinuses, exit from the skull via the *jugular foramina,* and then descend through the neck alongside the internal carotid arteries. As they move inferiorly, they receive blood from some of the deep veins of the face and neck—branches of the facial and superficial temporal veins (see Figure 20.22a). At the base of the neck, each internal jugular vein joins the subclavian vein on its own side to form a brachiocephalic vein. As already noted, the two brachiocephalic veins unite to form the superior vena cava.

- **Vertebral veins.** In contrast to the vertebral arteries, which serve substantial portions of the brain, the vertebral veins primarily drain extracranial tissues. Some blood draining from the posterior **occipital veins** and from the deep structures of the neck (vertebrae and muscles) enters the vertebral veins as they run inferiorly through the transverse foramina of the cervical vertebrae. The vertebral veins join the brachiocephalic veins in the neck.

Veins of the Upper Extremities.

- **Deep veins.** The deep veins of the upper limbs follow the paths of their companion arteries and have the same names; however, most are paired vessels. The deep and superficial **palmar venous arches** of the hand empty into the **radial** and **ulnar veins** of

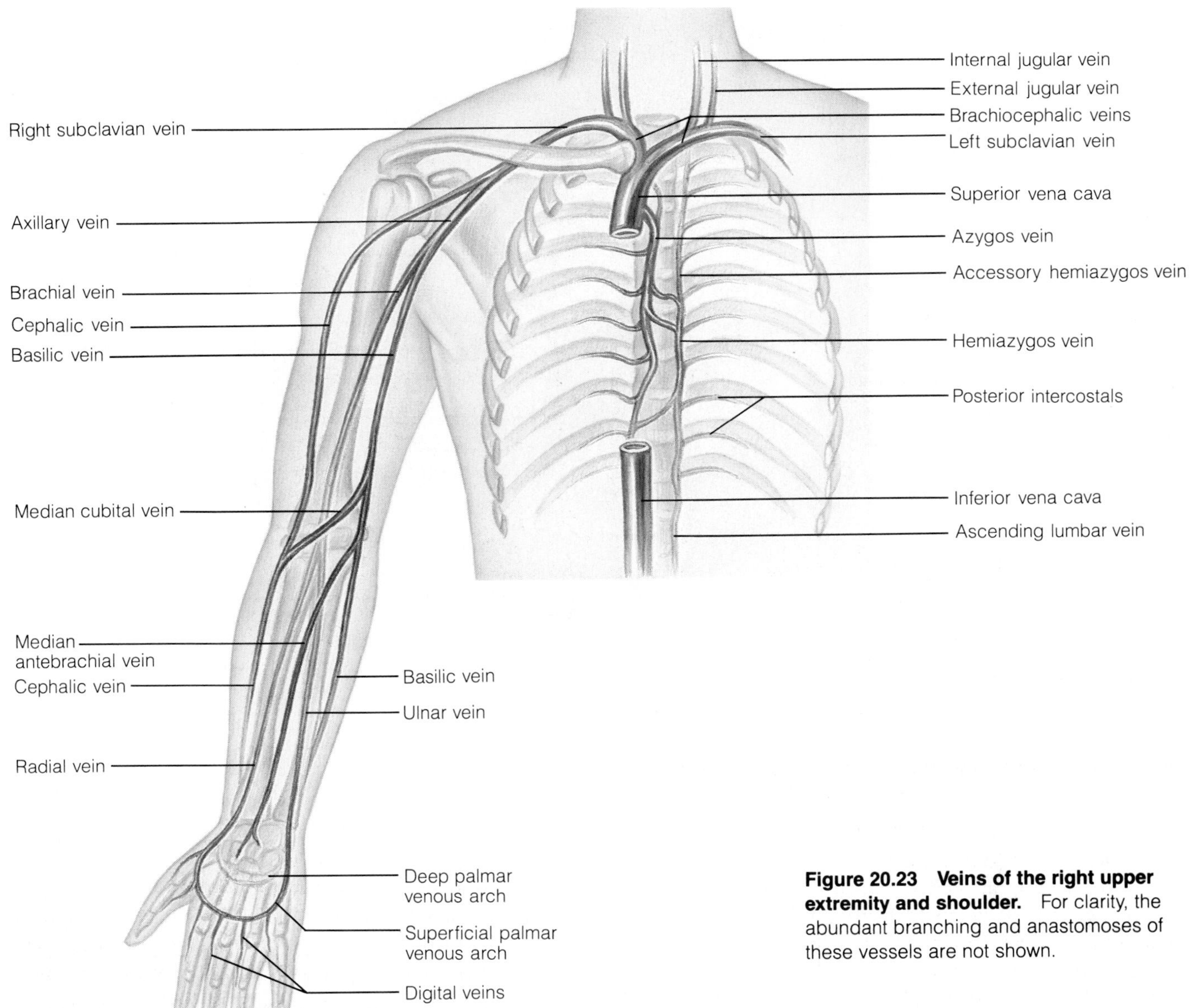

Figure 20.23 Veins of the right upper extremity and shoulder. For clarity, the abundant branching and anastomoses of these vessels are not shown.

the forearm, which then unite to form the **brachial vein** of the arm. As the brachial vein enters the axilla, it becomes the **axillary vein,** which in turn becomes the **subclavian vein** (Figure 20.23).

- **Superficial veins.** The superficial veins of the upper extremities are larger than the deep veins and are easily seen just beneath the skin. The superficial venous system begins with the **dorsal venous arch,** a plexus of superficial veins in the dorsum of the hand. In the distal forearm, this plexus drains into three major superficial veins—the cephalic, basilic, and median antebrachial veins—that anastomose frequently as they course upward (see Figure 20.23). The **cephalic vein** coils around the radius as it travels superiorly and then continues up the lateral superficial aspect of the arm to the shoulder, where it passes deep between the deltoid and pectoralis muscles to empty into the axillary vein. The **basilic vein** courses along the medial posterior aspect of the forearm, crosses the elbow, and then takes a deep course to reach the brachial vein, which it joins. At the anterior aspect of the elbow, the **median cubital vein** connects the basilic and cephalic veins. The median cubital vein is quite obvious and is commonly used to obtain venous blood samples or for administering intravenous medications or blood transfusions. The **median antebrachial vein** is located between the radial and ulnar veins in the forearm. Its termination point at the elbow is highly variable.

Veins of the Thorax. Blood draining from the mammary glands and the first two to three intercostal spaces enters the **brachiocephalic veins.** However, the vast majority of thoracic tissues and the thorax wall are drained by a rather complex network of veins collectively called the **azygos** (ā-zī′-gus) **system,** which flanks the vertebral column and ultimately empties directly into the superior vena cava (see Figure 20.23). The azygos system consists of the following vessels:

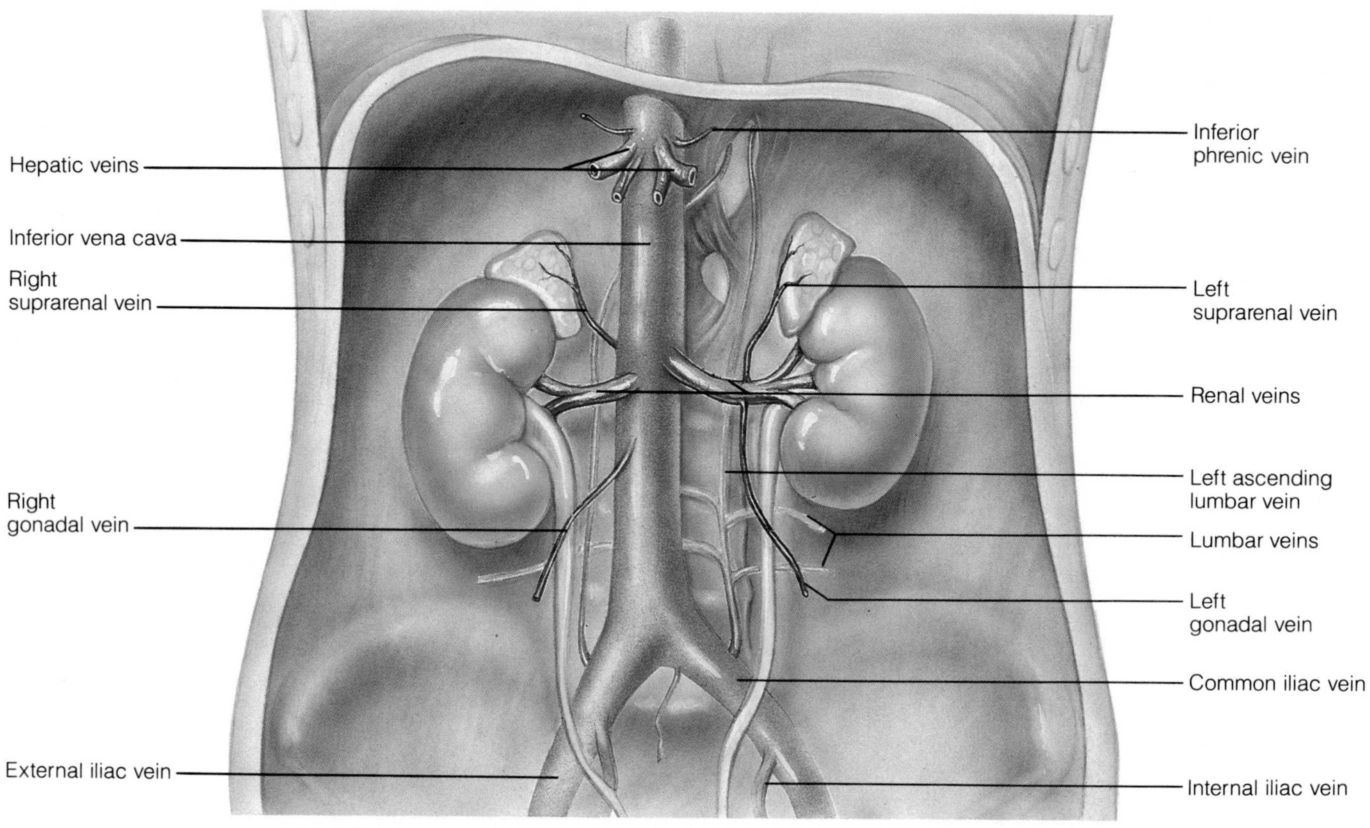

Figure 20.24 Tributaries of the inferior vena cava. Venous drainage of abdominal organs not drained by the hepatic portal vein.

- **Azygos vein.** The azygos vein, found against the right side of the vertebral column, originates from the **ascending lumbar veins** that drain most of the right abdominal cavity wall and from the **right intercostal veins** that drain the chest muscles. As it runs superiorly, it receives venous blood from the *right bronchial vein* and from tributaries from the esophagus, mediastinum, and pericardium, as well as from other veins of the azygos system. At the level of the fourth thoracic vertebra, it empties into the dorsal aspect of the superior vena cava.
- **Hemiazygos** (heh′-mē-ā-zī′-gus) **vein.** The hemiazygos vein ascends on the left side of the vertebral column. Its origin, from the **left ascending lumbar vein** and the lower **intercostal veins,** mirrors that of the inferior portion of the azygos vein on the right. Approximately midthorax, the hemiazygos vein passes in front of the vertebral column and empties into the azygos vein.
- **Accessory hemiazygos vein.** The accessory hemiazygos completes the venous drainage of the left (middle) thorax and then crosses to the right to empty into the azygos vein.

The branching nature of the azygos system provides a collateral circulation for draining the abdominal wall and other areas served by the inferior vena cava. The ascending lumbar veins drain the same areas of the abdominal trunk as do the lumbar branches of the inferior vena cava, and there are numerous anastomoses between the azygos system and the inferior vena cava.

Veins of the Abdomen. Blood draining from the abdominopelvic viscera and abdominal walls is returned to the heart by the **inferior vena cava** (Figure 20.24). Most of its venous tributaries have names that correspond to the arteries serving the abdominal organs.

- **Lumbar veins.** Several pairs of lumbar veins drain the posterior abdominal wall. They empty both directly into the inferior vena cava and into the ascending lumbar veins, which contribute to the azygos system of the thorax.
- **Gonadal (testicular or ovarian) veins.** The right gonadal vein drains the ovary or testis on the right side of the body. The left member drains into the left renal vein superiorly.
- **Renal veins.** The right and left renal veins drain the kidneys.

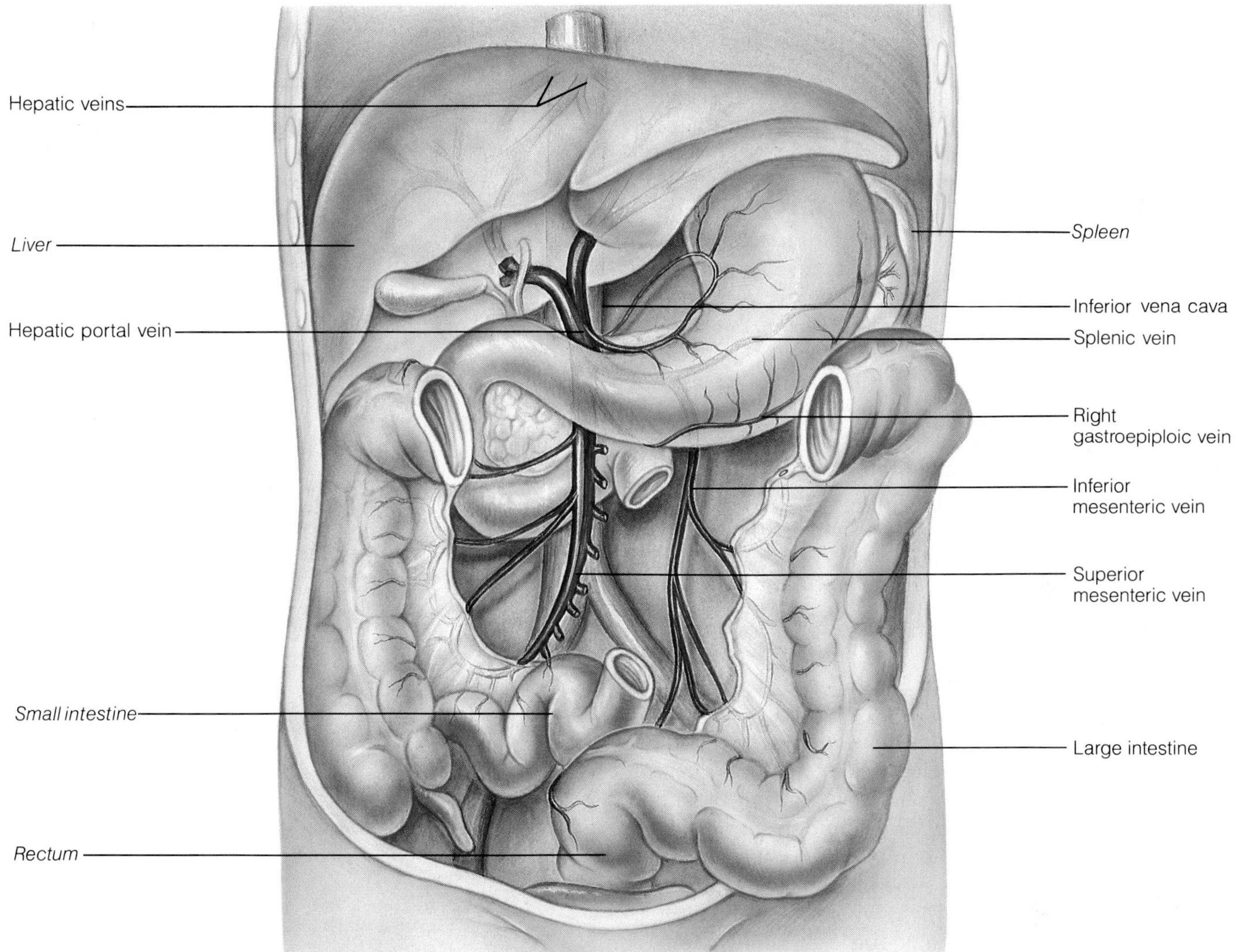

Figure 20.25 The hepatic portal circulation.

- **Suprarenal veins.** The right suprarenal vein drains the adrenal gland on the right and empties into the inferior vena cava. The left suprarenal vein drains into the left renal vein.
- **Hepatic veins.** The right and left hepatic veins carry venous blood from the liver to the inferior vena cava.
- **Hepatic portal system.** Veins draining the digestive viscera empty their blood into a common vessel, the **hepatic portal vein,** which transports this venous blood into the liver before it is allowed to enter the major systemic circulation via the hepatic veins (Figure 20.25). Such a venous system—veins to capillaries (or sinusoids) to veins—is called a *portal system*. Portal systems always serve very specific regional tissue needs. The hepatic portal system delivers blood laden with nutrients from the digestive organs to the liver. As the blood percolates slowly through the liver sinusoids, the hepatic parenchymal cells remove nutrients required for their various metabolic functions, and phagocytic cells lining the sinusoids deftly rid the blood of bacteria and other foreign matter that has penetrated the digestive mucosa. The role of the liver in nutrient processing and metabolism is described in detail in Chapter 25.

Numerous tributaries from the stomach and pancreas contribute to the hepatic portal system (see Figure 20.25), but the major vessels are as follows:

- **Superior mesenteric vein.** This vein drains the small intestine, part of the large intestine (ascending and transverse regions), and stomach.
- **Splenic vein.** This vessel collects blood from the spleen, parts of the stomach, and pancreas and then joins with the superior mesenteric vein to form the hepatic portal vein.
- **Inferior mesenteric vein.** This vein drains the distal portions of the large intestine and rectum and joins with the splenic vein just before that vessel unites with the superior mesenteric vein.

Veins of the Pelvis and Lower Extremities.

- **Deep veins.** As is the case of the upper limbs, most deep veins of the lower extremities have the same names as the arteries they accompany. After being formed by the union of the small **medial** and **lateral plantar veins,** the **posterior tibial vein** ascends deep within the calf muscle (Figure 20.26b). The **anterior tibial vein** is the superior continuation of the **dorsalis pedis vein** of the foot. It runs to the knee, where it unites with the posterior tibial vein to form the **popliteal vein,** which crosses the back of the knee. As the popliteal vein emerges from the knee, it becomes the **femoral vein,** which drains the deep structures of the thigh. The femoral vein becomes the **external iliac vein** as it enters the pelvis. In the pelvis, the external iliac vein unites with the **internal iliac vein** to form the **common iliac vein.** The distribution of the internal iliac veins parallels that of the internal iliac arteries: They drain the gluteus muscles, medial thigh, bladder, rectum, and reproductive organs of the pelvis.
- **Superficial veins.** Two large superficial veins, the **great** and **small saphenous** (suh-fē′-nus), issue from the **dorsal venous arch** of the foot (Figure 20.26b). These veins anastomose frequently with each other and with the deep veins along their course. Because the saphenous veins are poorly supported by surrounding tissues, they are very common sites of varicosities. The great saphenous vein is the longest vein in the body. It travels superiorly along the medial aspect of the leg to the thigh, where it empties into the femoral vein just distal to the inguinal ligament. The great saphenous vein is frequently excised and used as a coronary bypass vessel. In such cases, the vein is reversed, so that its valves do not obstruct circulation. The small saphenous vein runs along the lateral aspect of the foot and then through the superficial calf muscles, which it drains. At the knee, it empties into the popliteal vein.

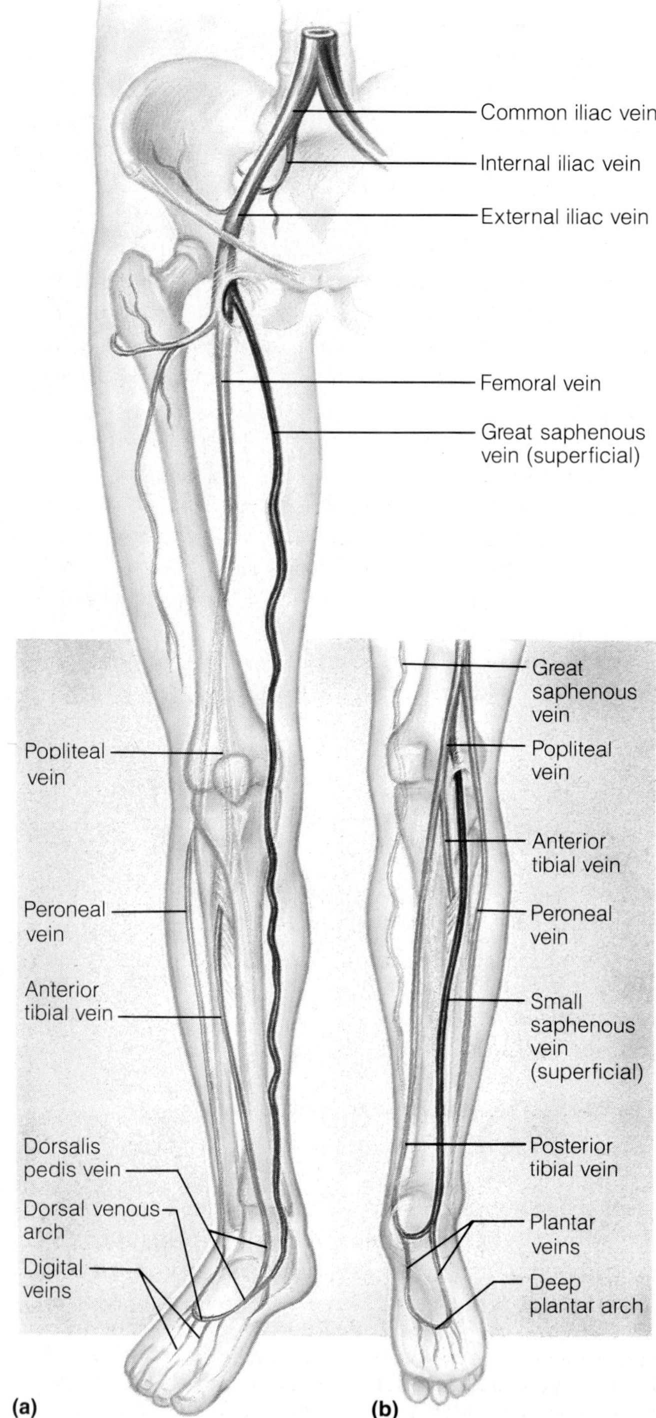

Figure 20.26 Veins of the right lower extremity. (**a**) Anterior view of the lower extremity. (**b**) Posterior view of the leg and foot.

Developmental Aspects of the Blood Vessels

The endothelial lining of blood vessels is formed by mesoderm-derived cells called *angioblasts*. Early in development, these primitive cells collect in little masses called *blood islands* throughout the microscopic embryo. They then begin to reach toward one another and the forming heart to lay down the rudimentary circulatory system. While this is occurring, adjacent mesenchymal cells surround the endothelial tubes, forming the muscular and fibrous coats of the vessel walls. The arteries become recognizable as their muscular coat thickens, and as noted in Chapter 19, the heart is pumping blood through the vascular system by the fourth week of development.

In addition to the fetal shunts that bypass the nonfunctional lungs (the *foramen ovale* and *ductus arteriosus*), other vascular modifications are found in the fetus. A special vessel, the ductus venosus, largely bypasses the liver, and the umbilical vein and arteries circulate blood back and forth between the fetal circulation and the placenta (see Chapter 29, pp. 970–972). Once the fetal circulatory pattern has been laid down, few vascular changes occur until birth, when the umbilical vessels and shunts are occluded.

In contrast to congenital heart diseases, congenital vascular problems are rare, and blood vessels are remarkably trouble free during youth. But as we age, signs of vascular disease begin to appear, giving a ring of truth to the old saying, "You are only as old as your arteries." In some, venous problems appear: The valves weaken and purple, snakelike varicose veins become obvious. In others, more insidious signs of inefficient circulation appear: tingling in the fingers and toes and cramping of muscles.

Although the degenerative process of atherosclerosis begins in youth, its consequences are rarely apparent until middle to old age, when it may take the form of a myocardial infarct or stroke. Until puberty, the blood vessels of boys and girls look alike, but from puberty to about the age of 45 years, women have strikingly less atherosclerosis than men, probably because of the protective effects of estrogen on a woman's blood vessels.

Blood pressure changes with age. In a newborn baby, arterial pressure is normally about 90/55. Blood pressure rises steadily during childhood to finally reach the adult value (120/80). In old age, "normal" blood pressure averages 150/90, which is a hypertensive pressure in younger people. After the age of 40 years, the incidence of hypertension increases dramatically. Unlike arteriosclerosis, which is increasingly important in old age, hypertension claims many more youthful victims and is the single most important precipitating cause of sudden cardiovascular death in men in their early 40s and 50s.

There is little doubt that most vascular disease is disease of advanced culture. "Blessed" with high-protein, lipid-rich diets, empty-calorie snacks, all manner of devices to save energy, and high-stress jobs, many of us are struck down prematurely. Cardiovascular disease can be prevented somewhat by diet modifications, regular aerobic exercise, and the elimination of cigarette smoking. Poor diet and exercise habits and smoking are probably more detrimental to your blood vessels than aging itself could ever be!

* * *

Now that we have described the structure and function of blood vessels, our survey of the cardiovascular system is finally complete. The pump, the plumbing, and the circulating fluid form a dynamic organ system that ceaselessly services every other organ system of the body, as summarized by Figure 20.27 on p. 660. However, our study of the so-called *circulatory system* is still unfinished. We have yet to examine the lymphatic system, which acts cooperatively with the cardiovascular system to ensure continuous circulation and to provide sites from which lymphocytes can police the body. These are the topics of Chapter 21.

Related Clinical Terms

Aneurysm (an′-yoo-rizm) (*aneurysm* = a widening) A balloonlike outpocketing of an artery wall that places the artery at risk for rupture; may reflect a congenital weakness of the artery wall, but more often reflects gradual weakening of the artery by chronic hypertension or arteriosclerosis; the most common sites of aneurysm formation are the abdominal aorta and arteries feeding the brain and kidneys.

Angiogram (an′-jē-ō-gram) (*angio* = a vessel; *gram* = writing) Diagnostic technique involving the infusion of a radiopaque substance into the circulation for X-ray examination of specific blood vessels.

Diuretic (*diure* = urinate) A chemical substance that promotes urine formation, thus reducing blood volume; diuretic drugs are frequently prescribed to manage hypertension.

Microangiopathic lesion (*micro* = small) An abnormal thickening of a capillary basement membrane due to the deposit of glycoproteins; the result is thickened—but leaky—capillary walls; one of the vascular hallmarks of long-standing diabetes mellitus.

Phlebitis (fleh-bī′-tis) (*phleb* = vein; *itis* = inflammation) Inflammation of a vein accompanied by painful throbbing and redness of the skin over the inflamed vessel; most often caused by bacterial infection or local physical trauma.

Phlebotomy (fleh-bot′-ō-mē) (*tomy* = cut) A venous incision made for the purpose of withdrawing blood or bloodletting.

Raynaud's disease A vascular (vasomotor) disorder in which the extremities—particularly the fingers and toes—become ischemic, cyanotic, and painful; often occurs during exposure to cold and is accentuated by anything that increases vasoconstriction, such as nicotine; treated with vasodilator drugs.

Sounds of Korotkoff Sounds made by blood as it flows through a constricted artery; the sounds heard through the stethoscope during blood pressure auscultation.

Thrombophlebitis Condition of undesirable intravascular clotting initiated by a roughening of a venous lining; often follows severe episodes of phlebitis; an ever-present danger is that the clot may become detached and form an embolus.

Integumentary system
Skin vasculature is an important blood reservoir and provides a site for heat loss from body ▶
Delivers oxygen and nutrients; carries away wastes ◀

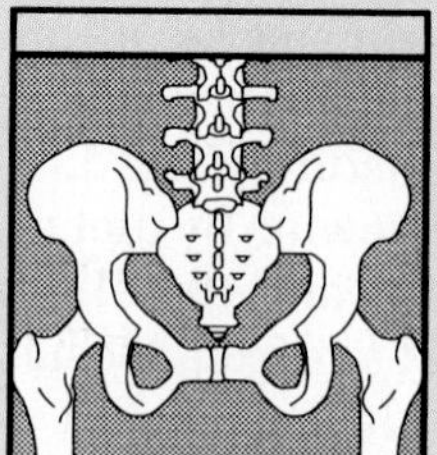

Skeletal system
Site of hematopoiesis; protects cardiovascular organs by enclosure; calcium depot ▶
Delivers oxygen and nutrients; carries away wastes ◀

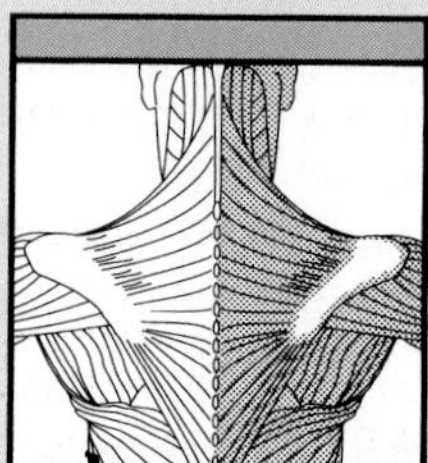

Muscular system
Aerobic exercise enhances cardiovascular efficiency and helps prevent arteriosclerosis; muscle "pump" aids venous return ▶
Delivers oxygen and nutrients; carries away wastes ◀

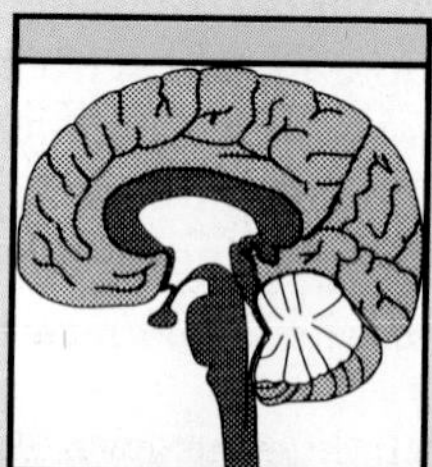

Nervous system
ANS regulates cardiac rate and force; sympathetic division maintains blood pressure and controls blood distribution according to need ▶
Delivers oxygen and nutrients; carries away wastes ◀

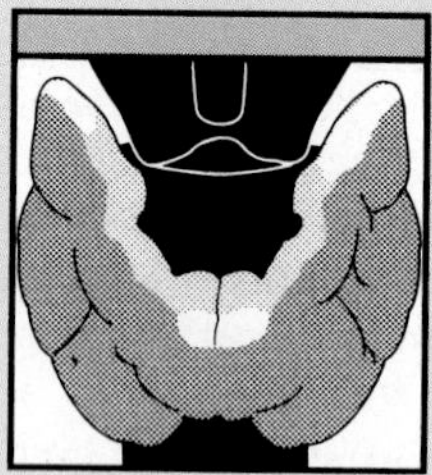

Endocrine system
Various hormones influence blood pressure (epinephrine, ANF, thyroxine, ADH); estrogen maintains vascular health ▶
Delivers oxygen and nutrients; carries away wastes; blood serves as a transport vehicle for hormones ◀

Cardiovascular system

Lymphatic system
◀ Picks up leaked fluid and plasma proteins and returns them to the cardiovascular system
▶ Delivers oxygen and nutrients; carries away wastes

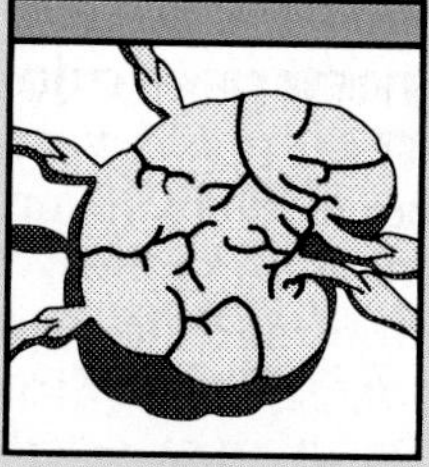

Immune system
◀ Protects cardiovascular organs from specific pathogens
▶ Delivers oxygen and nutrients to lymphatic organs, which house immune cells; provides transport medium for lymphocytes

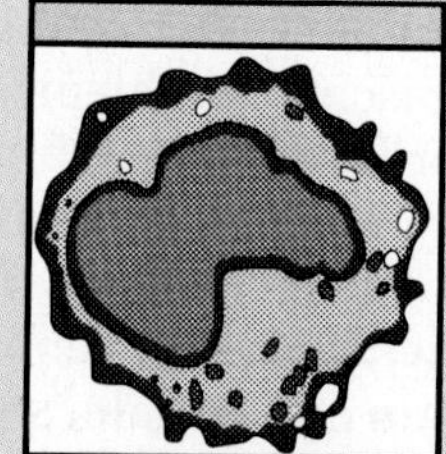

Respiratory system
◀ Carries out gas exchange: loads oxygen and unloads carbon dioxide from the blood; respiratory "pump" aids venous return
▶ Delivers oxygen and nutrients; carries away wastes

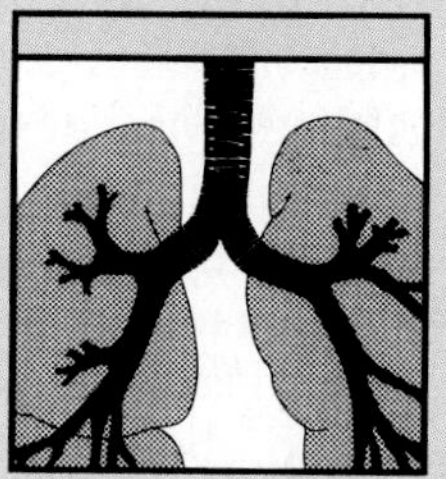

Digestive system
◀ Provides nutrients to the blood including iron and B vitamins essential for RBC (& hemoglobin) formation
▶ Delivers oxygen and nutrients; carries away wastes

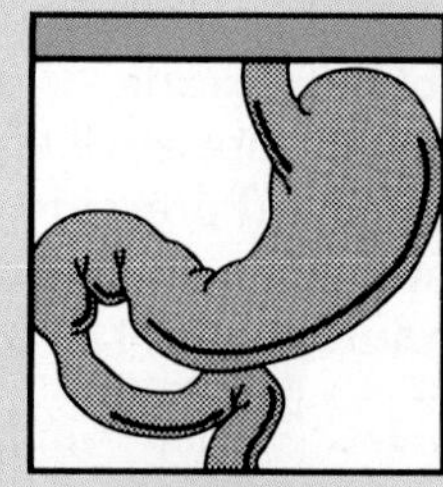

Urinary system
◀ Helps regulate blood volume and pressure by altering urine volume and releasing renin
▶ Delivers oxygen and nutrients; carries away wastes; blood pressure maintains kidney function

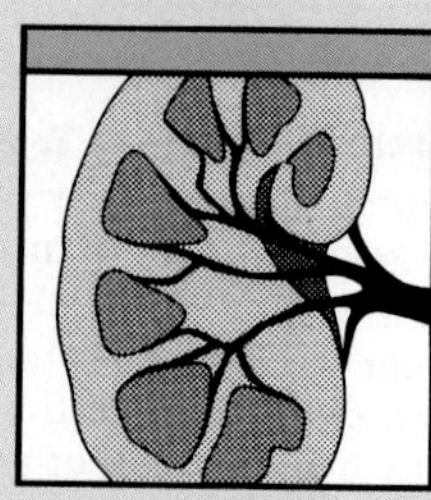

Reproductive system
◀ Estrogen maintains vascular health
▶ Delivers oxygen and nutrients; carries away wastes

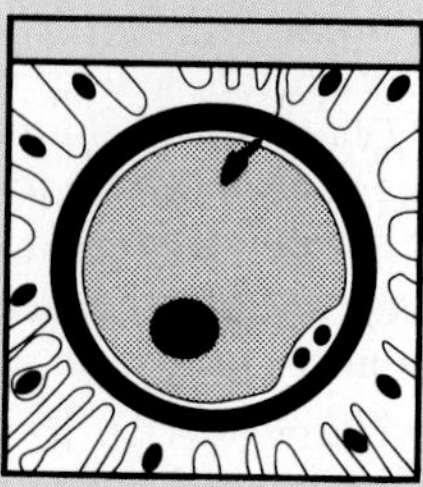

Figure 20.27 Homeostatic interrelationships between the cardiovascular system and other body systems.

Chapter Summary

OVERVIEW OF BLOOD VESSEL STRUCTURE AND FUNCTION (pp. 622–628)

1. Blood is transported throughout the body via a continuous system of blood vessels. Arteries transport blood away from the heart; veins carry blood back to the heart. Capillaries carry blood to tissue cells and are exchange sites.

The Structure of Blood Vessel Walls (p. 622)

2. All blood vessels except capillaries have three layers: the tunica intima, the tunica media, and the tunica adventitia. Capillary walls are composed of the tunica intima only.

Arteries (pp. 622–624)

3. Elastic (conducting) arteries are the large arteries close to the heart that expand and recoil to accommodate changing blood volume. Muscular (distributing) arteries carry blood to specific organs; they are less stretchy and more active in vasoconstriction. Arterioles regulate blood flow into capillary beds.

4. The pulse is the alternating expansion and recoil of arterial walls with each heartbeat. Pulse points are also pressure points.

5. Arteriosclerosis is a degenerative vascular disease. Initiated by endothelial lesions, it progresses through fatty streak, atherosclerotic, and arteriosclerotic stages.

Capillaries (pp. 625–626)

6. Capillaries are microscopic vessels with very thin walls. Most exhibit clefts, which aid in the exchange between the blood and interstitial fluid. Capillaries at sites of active absorption have fenestrations that enhance their permeability.

7. Thoroughfare channels connect the terminal arteriole and venule at opposite ends of a capillary bed. Most true capillaries arise from and rejoin the thoroughfare channels. The amount of blood flowing into the true capillaries is regulated by precapillary sphincters.

Veins (pp. 626–627)

8. Veins have comparatively larger lumens than arteries, and a system of valves prevents backflow of blood. The respiratory and skeletal muscle pumps are important in ensuring venous return of blood to the heart.

9. Most veins are normally only partially filled with blood and tend to be collapsed; thus, they can serve as blood reservoirs.

Vascular Anastomoses (pp. 627–628)

10. The joining together of arteries serving a common organ is called an arterial anastomosis. Such vascular patterns provide alternate channels for blood to reach the same organ. Vascular anastomoses also form between veins and between arterioles and venules.

PHYSIOLOGY OF CIRCULATION (pp. 628–640)

Blood Flow, Blood Pressure, and Resistance (p. 628)

1. Blood flow is the amount of blood flowing through a vessel or the entire circulation in a given period of time. Blood pressure is the force per unit area exerted on a vessel wall by the contained blood. Resistance is opposition to blood flow; blood viscosity and blood vessel length and diameter contribute to resistance.

2. Blood flow is directly proportional to blood pressure and inversely proportional to resistance.

Systemic Blood Pressure (pp. 628–635)

3. Normal blood pressure in adults is 120/80 (systolic/diastolic). It is highest in the aorta and lowest in the venae cavae. Venous pressure is low because of the cumulative effects of resistance.

4. Blood pressure is influenced by cardiac output, peripheral resistance, and blood volume. Vessel diameter is the most important of these factors, and small changes in vessel (chiefly arteriolar) diameter significantly affect blood pressure.

5. Blood pressure varies directly with both cardiac output and blood volume; it varies inversely with vessel diameter.

6. Blood pressure is regulated by autonomic neural reflexes (involving pressoreceptors, chemoreceptors, the vasomotor center, and vasomotor fibers acting on vascular smooth muscle), inputs from higher CNS centers, chemicals such as hormones, and renal mechanisms.

7. Blood pressure is routinely measured by the auscultatory method. Hypotension is rarely a problem. Hypertension is the major cause of myocardial infarct, stroke, and renal disease.

Blood Flow (pp. 635–640)

8. Blood flows fastest where the cross-sectional area of the vascular bed is least (aorta), and slowest where the cross-sectional area is greatest (capillaries). The slow flow in capillaries allows time for nutrient-waste exchanges.

9. Autoregulation is the intrinsic adjustment of blood flow to individual organs based on their immediate requirements. It is largely controlled by local chemical factors that cause vasodilation of the arterioles serving the area and open the precapillary sphincters.

10. In most instances, autoregulation is controlled by oxygen deficits and accumulation of local metabolites. However, autoregulation in the brain is controlled primarily by a drop in pH and by myogenic mechanisms; and vasodilation of pulmonary circuit vessels occurs in response to high levels of oxygen.

11. Nutrients, gases, and other solutes smaller than plasma proteins cross the capillary wall by diffusion. Water-soluble substances move through the clefts; fat-soluble substances pass through the lipid portion of the endothelial cell membrane.

12. Fluid flows occurring at capillary beds reflect the relative effect of outward (net hydrostatic pressure) forces minus the effect of inward (net osmotic pressure) forces. In general, fluid flows out of the capillary bed at the arterial end and reenters the capillary blood at the venule end.

13. The small net loss of fluid and protein into the interstitial space is collected by lymphatic vessels and returned to the cardiovascular system.

14. Circulatory shock occurs when blood perfusion of body vessels is inadequate. Shock may reflect low blood volume (hypovolemic shock), abnormal vasodilation of the vasculature (vascular shock), or pump failure (cardiogenic shock).

CIRCULATORY PATHWAYS (pp. 640–658)

Pulmonary Circulation (p. 640)

1. The pulmonary circulation transports oxygen-poor, carbon dioxide–laden blood to the lungs for oxygenation and carbon dioxide unloading. Blood returning to the right atrium of the heart is pumped via the pulmonary trunk, pulmonary arteries, and lobar arteries to the pulmonary capillaries by the right ventricle. Blood issuing from the lungs is returned to the left atrium by the pulmonary veins.

Systemic Circulation (pp. 640–658)

2. The systemic circulation transports oxygenated blood from the left ventricle to all body tissues via the aorta and its branches. Venous blood returning from the systemic circuit is finally returned to the right atrium via the superior and inferior venae cavae.

3. Tables 20.1 and 20.2 and Figures 20.14 to 20.26 illustrate the arteries and veins of the systemic circulation.

DEVELOPMENTAL ASPECTS OF THE BLOOD VESSELS (pp. 658–659)

1. The fetal vasculature develops from embryonic blood islands and mesenchyme and is functioning in blood delivery by the fourth week.

2. Fetal circulation differs from circulation after birth with respect to pulmonary and hepatic shunts and special umbilical vessels. These are normally occluded shortly after birth.

3. Blood pressure is low in infants and rises to reach adult values.

4. Age-related vascular problems include varicose veins, hypertension, and arteriosclerosis. Hypertension is the most important cause of sudden cardiovascular death in middle-aged men. Arteriosclerosis is the most important cause of cardiovascular disease in the aged.

Review Questions

Multiple Choice/Matching

1. Which statement does not accurately describe veins? (a) They have less elastic tissue and smooth muscle than arteries, (b) they contain more fibrous tissue than arteries, (c) most veins in the extremities have valves, (d) they always carry deoxygenated blood.

2. Which of the following tissues is mainly responsible for vasoconstriction? (a) elastic tissue, (b) smooth muscle, (c) collagenic tissue, (d) adipose tissue.

3. Peripheral resistance in the cardiovascular system (a) is inversely related to the diameter of the arterioles, (b) tends to increase if blood viscosity increases, (c) is directly proportional to the length of the vascular bed, (d) all of these.

4. Which of the following can lead to decreased venous return of blood to the heart? (a) an increase in blood volume, (b) an increase in venous pressure, (c) damage to the venous valves, (d) increased muscular activity.

5. Which of the following can contribute to increased arterial blood pressure? (a) increasing stroke volume, (b) increasing heart rate, (c) arteriosclerosis, (d) increasing blood volume, (e) all of these.

6. Exchange of nutrients and wastes in the capillary beds is regulated by local chemical controls. Which of the following would *not* result in the dilation of the feeder arterioles and opening of the precapillary sphincters in any capillary bed? (a) a decrease in oxygen content of the blood, (b) an increase in carbon dioxide content of the blood, (c) a local increase in histamine, (d) a local increase in pH.

7. The structure of a capillary wall differs from that of a vein or an artery because (a) it has two tunics instead of three, (b) there is less smooth muscle, (c) it is single layered—only the tunica intima, (d) none of these.

8. The pressoreceptors in the carotid and aortic bodies are sensitive to (a) a decrease in carbon dioxide, (b) changes in arterial pressure, (c) a decrease in oxygen, (d) all of these.

9. The myocardium receives its blood supply directly from (a) the aorta, (b) the coronary arteries, (c) the coronary sinus, (d) the pulmonary arteries.

10. Blood flow in the circulation is steady despite the rhythmic pumping action of the heart because of (a) the elasticity of the large arteries, (b) the small diameter of the capillaries, (c) the thin walls of the veins, (d) the venous valves.

11. Tracing the blood from the heart to the right hand, we find that blood leaves the heart and passes through the ascending aorta, the right subclavian artery, the axillary and brachial arteries, and through either the radial or ulnar artery to arrive at the hand. Which artery is missing from this sequence? (a) coronary, (b) innominate, (c) cephalic, (d) right common carotid.

12. Which of the following do not drain directly into the inferior vena cava? (a) lumbar veins, (b) hepatic veins, (c) inferior mesenteric vein, (d) renal veins.

Short Answer Essay Questions

13. How is the anatomy of capillaries and capillary beds well suited to their function?

14. Distinguish between elastic arteries, muscular arteries, and arterioles relative to location, histology, and functional adaptations.

15. Write an equation showing the relationship between peripheral resistance, blood flow, and blood pressure.

16. (a) Define blood pressure. Differentiate between systolic and diastolic blood pressure. (b) What is the normal blood pressure value for a young adult?

17. Describe the neural mechanisms responsible for controlling blood pressure.

18. Explain the reasons for the observed changes in blood flow velocity in the different regions of the circulation.

19. How does the control of blood flow to the skin for the purpose of regulating body temperature differ from the control of nutrient blood flow to skin cells?

20. Describe neural and chemical (both systemic and local) effects exerted on the blood vessels when one is fleeing from a mugger. (Be careful, this is more involved than it appears at first glance.)

21. How are nutrients, wastes, and respiratory gases transported to and from the blood and tissue spaces?

22. (a) What blood vessels contribute to the formation of the hepatic portal circulation? (b) What is the function of this circulation? (c) Why is a portal circulation a "strange" circulation?

Clinical Application Questions

23. Mrs. Johnson is brought to the emergency room after being involved in an auto accident. She is hemorrhaging and has a rapid, thready pulse, but her blood pressure is still within normal limits. Describe the compensatory mechanisms that are being promoted to maintain her blood pressure in the face of blood loss.

24. Mr. Connors has just had a right femoral arterial graft and is resting quietly. Orders are given to check his popliteal and pedal (dorsalis pedis) pulses and the color and temperature of his right leg four times daily. Why were these orders given?

25. A 60-year-old man is unable to walk more than 100 yards without experiencing severe pain in his left leg; the pain is relieved by resting for 5–10 minutes. He is told that the arteries of his leg are becoming occluded with fatty material and is advised to have the sympathetic nerves serving that body region severed. Explain how such surgery might help to relieve this man's problem.

CHAPTER 21

The Lymphatic System

Chapter Outline and Student Objectives

Lymphatic Vessels (pp. 663–666)

1. Describe the structure and distribution of lymphatic vessels, and note their important functions.
2. Describe the source of lymph and the mechanism(s) of lymph transport.

Lymph Nodes (pp. 666–667)

3. Describe the general location, histological structure (including major cell populations), and functions of lymph nodes.

Other Lymphoid Organs (pp. 667–669)

4. Name and describe the other lymphoid organs of the body. Compare and contrast them with lymph nodes, structurally and functionally.

Developmental Aspects of the Lymphatic System (p. 670)

5. Outline the development of the lymphatic system.

Preview of Selected Key Terms

Lymphatic system (lim-fa-tik) (*lympha* = clear water) System consisting of lymphatic vessels, lymph nodes, and other lymphoid organs and tissues; drains excess tissue fluid from the extracellular space and provides a site for immune surveillance.

Lymph Protein-containing fluid transported by lymphatic vessels.

Lymphatics General term used to designate lymphatic vessels.

Thoracic duct Large duct that receives lymph drained from the entire lower body, the left upper extremity, and the left side of the head and thorax.

Lymphocyte An agranular white blood cell that arises from the bone marrow and becomes functionally mature in the lymphoid organs of the body.

When we mentally tick off the names of the body's organ systems, some—like the nervous, cardiovascular, and digestive systems—trip instantly into our thoughts. We are not likely to remember with such ease the lymphatic system, which consists of a meandering network of lymphatic vessels, lymph nodes, and various other lymphoid organs and tissues scattered throughout the body. Yet without this quietly working system, our cardiovascular system would cease working and our immune system would be hopelessly impaired because the lymphatic vessels transport fluids that have escaped from the blood vascular system back to the blood. Other lymphatic organs house phagocytic cells and lymphocytes, which play essential roles in body defense.

Lymphatic Vessels

As blood circulates through the body, exchanges of nutrients, wastes, and gases occur between the blood and the interstitial fluid. As explained in Chapter 20, the hydrostatic and osmotic pressures operating at capillary beds force fluid out of the blood at the arterial ends of the capillaries and cause most of it to be reabsorbed at the venous ends. The fluid that remains behind in the tissue spaces becomes part of the interstitial fluid. Because as much as 3 L of plasma seeps into the interstitial space daily, these leaked fluids, as well as any plasma proteins that have escaped from the bloodstream, must be carried back to the blood if the vascular system is to have sufficient blood volume to operate properly. This problem of circulatory dynamics is resolved fairly well by the **lymphatic vessels**, or **lymphatics**, a special system of drainage vessels, which absorb the excess protein-containing interstitial fluid and return it to the bloodstream.

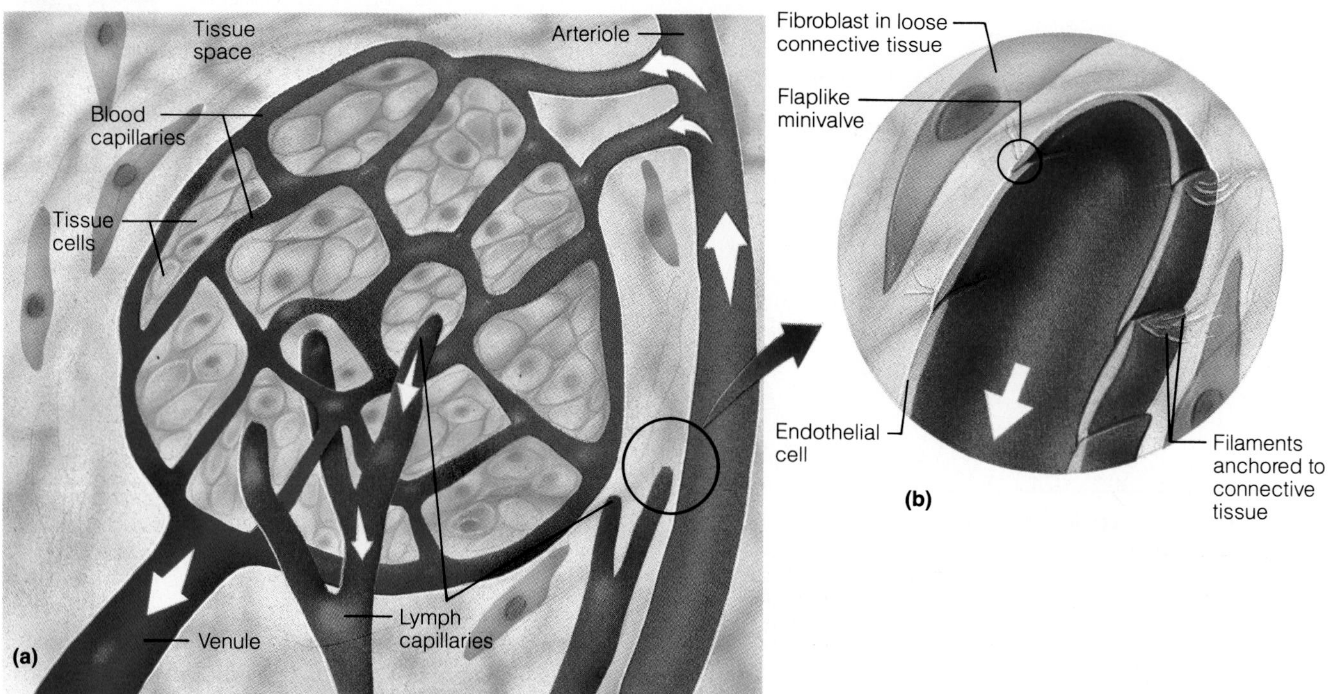

Figure 21.1 Distribution and special structural features of lymphatic capillaries. (**a**) Structural relationship between a capillary bed (blood vascular system) and lymphatic capillaries. (**b**) Lymphatic capillaries begin as blind-ended tubes. The endothelial cells forming their walls tend to overlap one another, forming flaplike minivalves.

Distribution and Structure of Lymphatic Vessels

The lymphatic vessels form a one-way system in which their contained fluid, **lymph,** flows only toward the heart. This transport system begins in microscopic blind-ended *lymphatic capillaries* (Figure 21.1a), which weave between the tissue cells and blood capillaries of nearly all tissues and organs of the body. The few exceptions include the central nervous system, bone, cartilage, bone marrow, the epidermis, and teeth.

Although similar to blood capillaries, lymphatic capillaries are structurally unique in two important ways:

1. The endothelial cells forming the walls of lymphatic capillaries are not tightly joined; instead, their edges loosely overlap one another, forming flaplike minivalves (Figure 21.1b).
2. Bundles of fine filaments anchor the endothelial cells to surrounding structures so that any increase in interstitial fluid volume separates these cell flaps, exposing gaps in the wall, rather than causing the lymphatic capillary to collapse.

These structural specializations create a system analogous to one-way swinging doors in the lymphatic capillary wall. The flaps gape open when the fluid pressure is greater in the interstitial space, allowing fluid to enter the lymphatic capillary. However, when the pressure is greater inside the lymphatic vessels, the endothelial cell flaps adhere closely together, preventing the plasma-like fluid, now called lymph, from leaking back into the interstitial space and forcing it along the vessel.

Proteins in the interstitial space are normally prevented from entering the blood capillaries, but lymphatic capillaries are exceptionally permeable to proteins. In addition, when tissues are inflamed, lymphatic capillaries develop openings that permit the uptake of even larger particles such as cell debris and pathogens. In such instances, virtually all large molecules and particulate matter can enter the lymphatic capillaries with ease.

Highly specialized lymphatic capillaries called *lacteals* (lak′-te-als) are present in the fingerlike villi of the intestinal mucosa. The lacteals absorb the digested fats from the intestine, which causes the lymph draining from the digestive viscera to become milky-white. This creamy lymph, called *chyle,* is also delivered to the blood via the lymphatic stream.

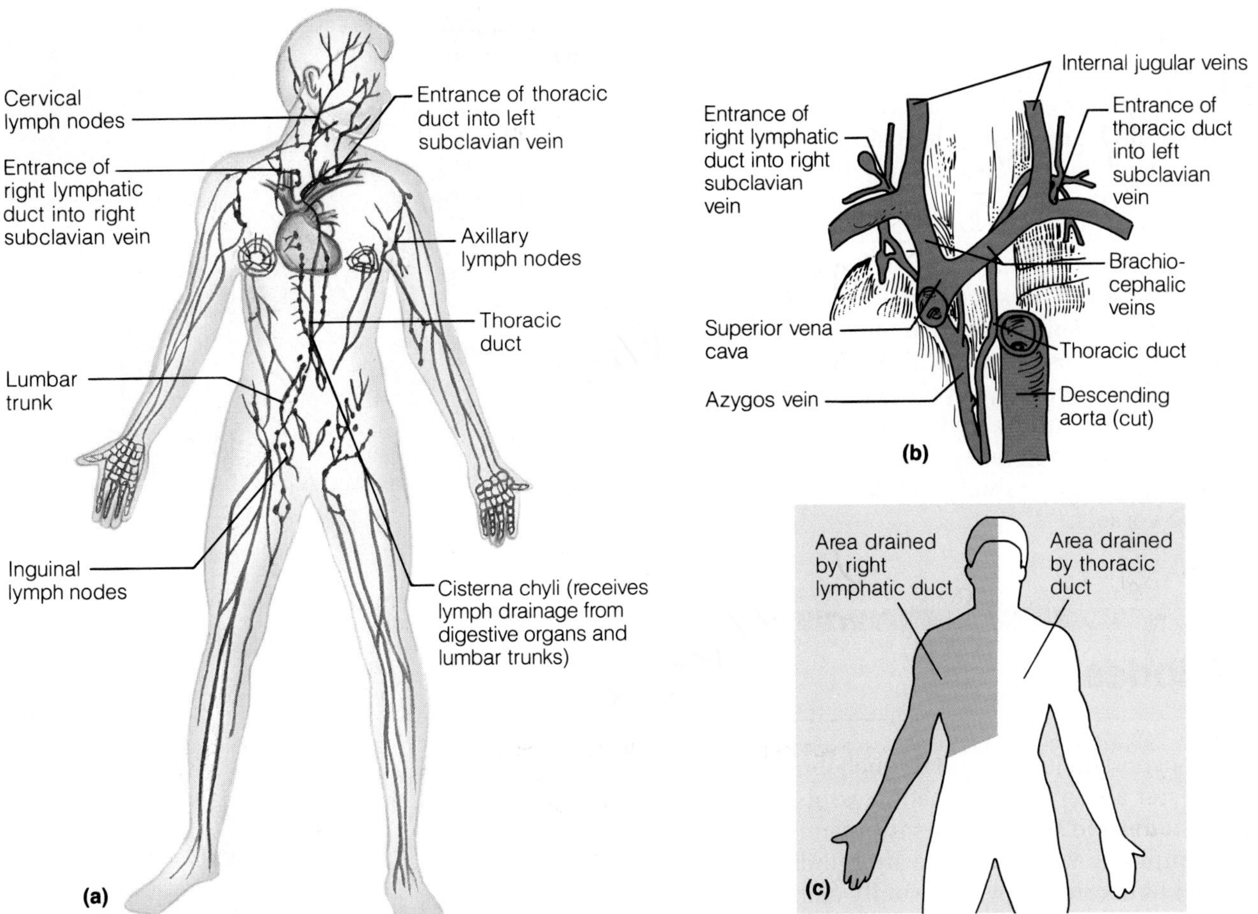

Figure 21.2 The lymphatic system. (**a**) Distribution of lymphatic vessels and lymph nodes. (**b**) Enlarged view of the major veins in the superior thorax showing the entry points of the thoracic and right lymphatic ducts. (**c**) The darkened area represents body area drained by the right lymphatic duct; the rest of the body is drained by the thoracic duct.

Once lymph enters the lymphatic capillaries, it flows through successively larger and thicker-walled lymphatic vessels (Figure 21.2a), the largest of which are called *lymphatic trunks*. Though these lymphatic vessels closely resemble veins, the lymphatics tend to be thinner-walled and wider in diameter and to have more internal valves than do veins. In general, the lymphatics in the skin form diffuse networks that travel along with veins; those of the trunk and digestive viscera parallel and form anastomoses around the deep arteries.

Lymph is eventually delivered to one of two large ducts in the thoracic region. The **right lymphatic duct** drains lymph from the right upper arm and the right side of the head and thorax (Figure 21.2c). The much larger **thoracic duct,** which receives lymph from the rest of the body, arises anterior to the second lumbar vertebra as an enlarged sac, the *cisterna chyli* (see Figure 21.2a). The cisterna chyli collects lymph from the two large lumbar trunks that drain the lower limbs and from the intestinal trunk that drains the digestive organs. As the thoracic duct runs superiorly along the anterior surface of the vertebral column, it accepts lymphatic drainage from the left side of the thorax and left upper limb and head region. Each terminal duct empties the lymph into the venous circulation at the junction of the internal jugular vein and subclavian vein on its own side of the body (Figure 21.2b).

Lymph Transport

Unlike the blood vascular system, the lymphatic system is a pumpless system and, under normal conditions, lymphatic vessels are very low pressure conduits. The same mechanisms that promote venous return in blood vessels act here as well: the milking action of active skeletal muscles, pressure changes within the thorax during breathing, and valves to prevent backflow. Because lymphatics are usually packaged together in connective tissue sheaths with blood vessels, the pulsating expansions of the nearby arteries also promote lymph flow. It has recently been suggested that, in addition to these mechanisms, smooth muscle in the lymphatic vessel walls contracts rhythmically, actually helping to pump the lymph along. Even so, lymph transport is sporadic and much slower than that occurring in veins. About 3 L of lymph enters the bloodstream every 24 hours, a volume almost exactly equal to the amount of fluid lost to the tissue spaces from the bloodstream in the same time period. When physical activity increases, the transport mechanism becomes more efficient and lymph flow is much more rapid, balancing the greater rate of fluid outflow from the vascular system in such situations.

Anything that prevents the normal return of lymph to the blood, such as blockage of the lymphatics by tumors or surgical removal of regional lymphatics during cancer surgery, results in severe localized edema. However, when lymphatics are removed surgically, lymphatic drainage is eventually reestablished by regrowth of the vessels. ■

Lymph Nodes

As lymph is transported, it is filtered through the lymph nodes that cluster along the lymphatic vessels of the body. **Lymph nodes** are small organs intimately associated with lymphatic vessels. There are hundreds of lymph nodes; but because they are usually embedded in connective tissue, they are not ordinarily seen. However, those associated with the gastrointestinal tract are large and quite obvious (Figure 21.3). Particularly large clusters of lymph nodes occur near the body surface in the inguinal, axillary, and cervical regions of the body—places where the lymphatic vessels converge to form large trunks (see Figure 21.2a).

Within the lymph nodes are macrophages, which engulf and destroy bacteria, cancer cells, and any other particulate matter in the lymphatic stream. Thus, the lymph nodes act as filters to cleanse lymph before it is allowed to reenter the blood. Collections of **lymphocytes** are also strategically located in lymph nodes, where they play a role in mounting an immune response to pathogenic microorganisms or other types of foreign matter carried in the lymphatic stream.

Structure of a Lymph Node

Lymph nodes vary in both shape and size, but most of them are kidney-shaped and less than 2.5 cm (1 inch) in length. Each node is surrounded by a dense fibrous *capsule* from which connective tissue strands called *trabeculae* (truh-beh'-kyoo-lē) extend inward to divide the node into a number of compartments (Figure 21.3b and c). The basic internal framework or

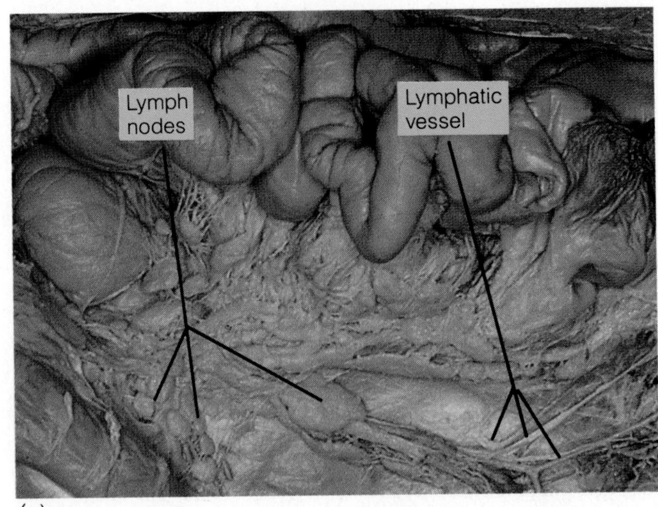

(a)

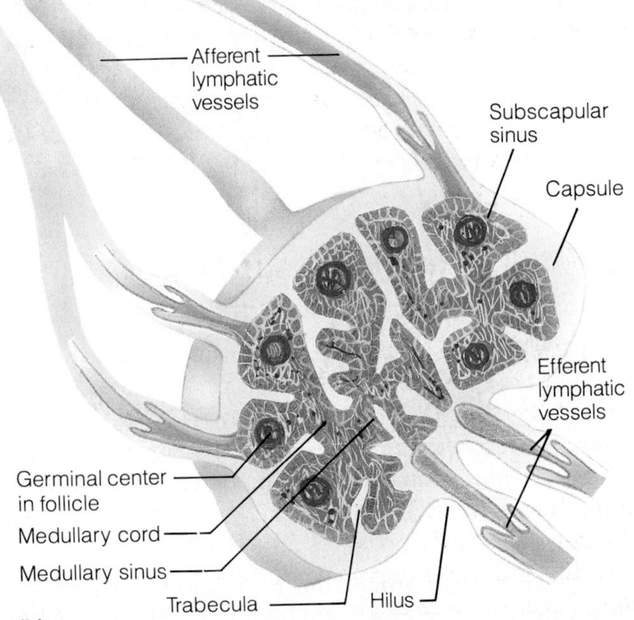

(b)

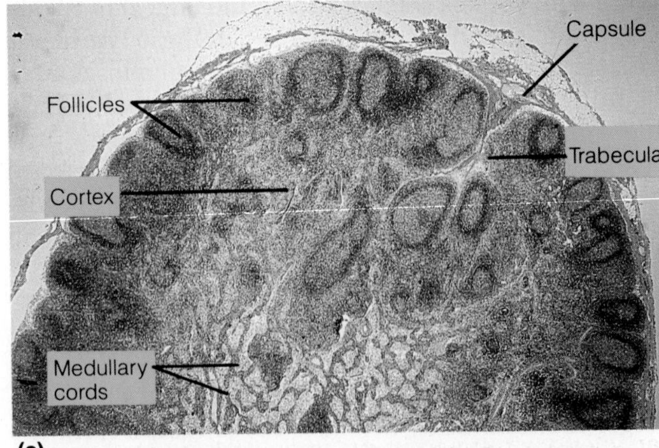

(c)

Figure 21.3 Structure of a lymph node. (**a**) Lymph nodes and associated lymphatic vessels associated with the gastrointestinal tract organs and mesentery. (**b**) Longitudinal view of the internal structure of a lymph node and associated lymphatics. Notice that several afferent lymphatics converge on its convex side, whereas fewer efferent lymphatics exit at its hilus. (**c**) Photomicrograph of part of a lymph node (20X).

stroma (strō′-muh) is provided by a soft, open network of reticular fibers that physically support the ever-changing population of lymphocytes. (As described in Chapter 18, lymphocytes initially arise from the bone marrow but migrate to the lymphatic organs, where they proliferate further.)

Two histologically distinct regions in a lymph node are the *cortex* and the *medulla*. The outer cortex contains densely packed spherical collections of lymphocytes called *follicles*, which frequently have lighter-staining centers called *germinal centers*. B lymphocytes predominate in the germinal centers. These centers enlarge when the B lymphocytes are in rapid cell division, generating daughter cells called *plasma cells*, which release antibodies that act against a particular foreign substance. The rest of the cortical cells are primarily T lymphocytes in transit; they circulate continuously between the blood, lymph nodes, and lymphatic stream, performing their surveillance role. Cordlike extensions of the cortex, called *medullary cords*, invade the medulla area where macrophages are located. (The precise roles of the cellular inhabitants of the lymph nodes in immunity are discussed in Chapter 22.)

Circulation in the Lymph Nodes

Lymph enters the convex side of a lymph node through a number of *afferent lymphatic vessels*. It then moves into a large, baglike sinus, the *cortical* or *subcapsular sinus*, from which it flows into a number of smaller interconnected sinuses that cut through the cortex and enter the medulla. The lymph meanders through these sinuses and finally exits from the node at its *hilus* (hī′-lus), the indented region, via *efferent lymphatic vessels*. Because there are fewer efferent vessels draining the node than afferent vessels feeding it, the flow of lymph through the node stagnates somewhat, allowing time for the lymphocytes and macrophages to perform their protective functions. Additionally, the reticular fibers spanning the sinusoids act as baffles to whirl the lymph, increasing the chance that contained foreign particles will come in contact with the macrophages that rest on the fiber network. In general, lymph must pass through several nodes before its cleansing process is completed.

Lymph nodes help rid the body of infectious agents and cancer cells, but sometimes they are overwhelmed by the very agents they are trying to destroy. For example, when large numbers of bacteria or virus particles are trapped in the nodes, the nodes become inflamed and very tender to the touch. The lymph nodes can also become secondary cancer sites, particularly in cancers that use lymphatic vessels to spread throughout the body. The fact that cancer-infiltrated lymph nodes are swollen but not painful helps to distinguish cancerous lymph nodes from those infected by microorganisms. ■

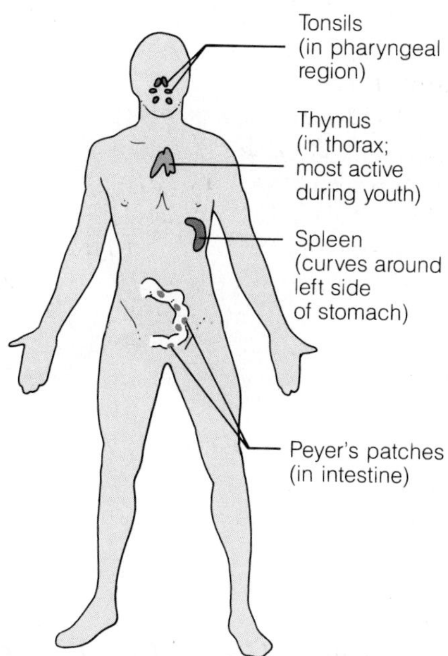

Figure 21.4. Lymphoid organs. Location of the tonsils, spleen, thymus gland, and Peyer's patches.

Other Lymphoid Organs

Lymph nodes are just one example of many types of **lymphoid organs** or aggregates of lymphatic tissue in the body. Others are the spleen, thymus gland, tonsils, and Peyer's patches of the intestine (Figure 21.4), as well as bits of lymphatic tissue scattered in the epithelial and connective tissues. The common feature of all these organs is their tissue makeup or composition: All are distinguished by a predominance of reticular connective tissue and free cells (primarily lymphocytes). Depending on the specific organ, these elements may be diffusely arranged or compacted into more dense lymph nodules. Although all lymphoid organs have roles in protecting the body, only the lymph nodes filter lymph. The other lymphoid organs and tissues typically have efferent lymphatics draining them, but have no associated afferent lymphatic vessels. The distinctive features of these other lymphoid organs are described next.

Spleen

The soft, blood-rich **spleen** is the largest lymphoid organ. Located in the left side of the abdominal cavity

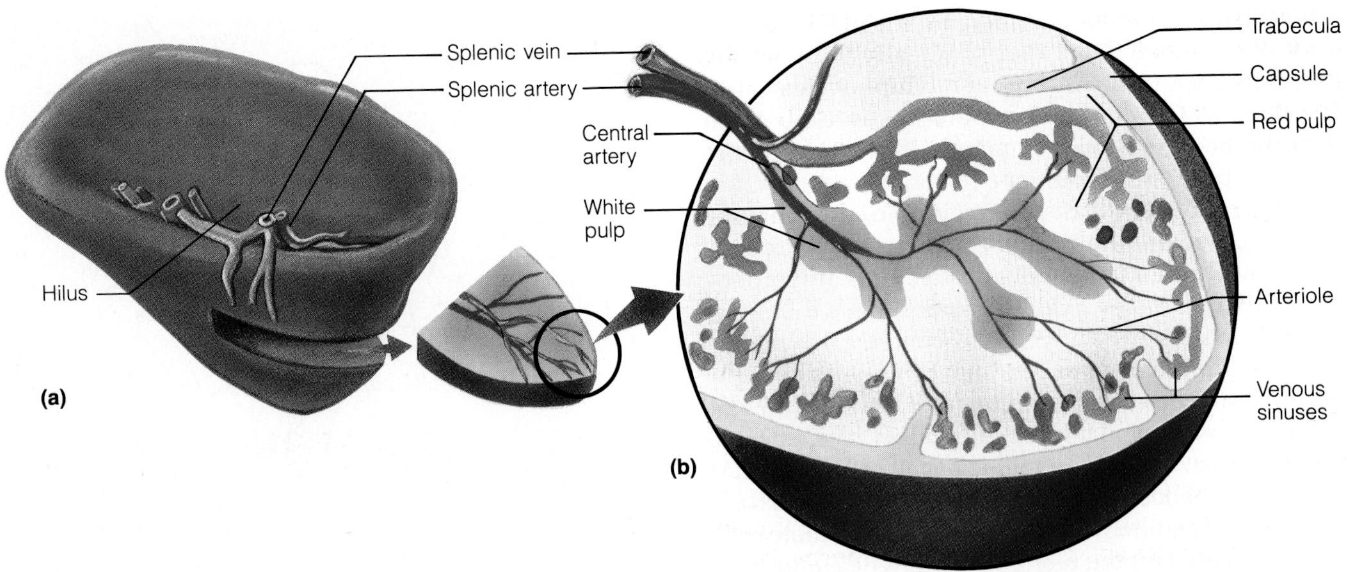

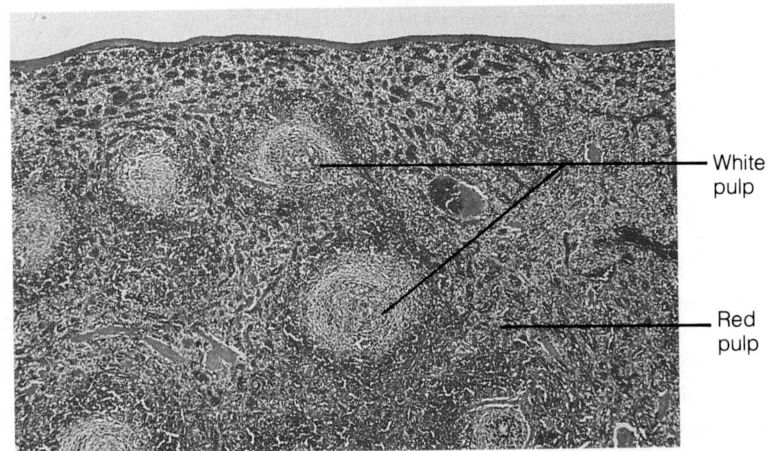

Figure 21.5 Structure of the spleen. (**a**) Gross structure. (**b**) Diagrammatic representation of the histological structure of one lobule of the spleen. (**c**) Photomicrograph of spleen tissue showing white and red pulp regions (60X). The white pulp, consisting mainly of lymphocytes, is surrounded by red pulp regions containing abundant erythrocytes.

just beneath the diaphragm, it extends to curl around the anterior aspect of the stomach (Figure 21.4). It is served by the splenic artery and vein.

The spleen's main functions are to remove aged or defective blood cells and platelets from the blood and to store or release some of the breakdown products of red blood cells to the blood for processing by the liver. For example, iron is salvaged and stored within spleen macrophages for later use by the bone marrow in making hemoglobin; the remainder of the hemoglobin molecule is first broken down to bilirubin and then secreted by the liver in bile. Additional functions of the spleen include:

- Acting as a blood filter. As blood flows through the sinuses, the spleen's macrophages remove debris, bacteria, and other foreign matter.
- Serving as a site for erythrocyte production in the developing embryo. Although this capability ceases after birth, it can be reactivated under conditions of severe red blood cell depletion, such as hemolytic anemia.
- Storing blood platelets. In some animals, the spleen also serves as a blood reservoir, but this function is trivial in humans.
- Providing a site for lymphocyte proliferation and immune surveillance and response.

Despite its imposing size and important roles, surgical removal of the spleen seems to have few lasting detrimental effects. In such circumstances, its functions are taken over by the liver and bone marrow.

The spleen, like the lymph nodes, is surrounded by a fibrous capsule, has trabeculae that divide the organ into lobules, and contains both lymphocytes and macrophages. However, consistent with its blood-processing functions, it also contains huge numbers of erythrocytes. Venous sinuses and other regions that contain red blood cells are referred to as *red pulp*, whereas areas composed mostly of lymphocytes and macrophages suspended on reticular fibers are called *white pulp*. The white pulp clusters around small branches of the splenic artery within the organ and forms what appear to be islands in a sea of red pulp.

The naming of the pulp regions reflects their appearance in fresh spleen tissue rather than their staining properties. Indeed, as can be seen in the photomicrograph in Figure 21.5c, the white pulp takes up a reddish stain and appears even darker than the red pulp.

Thymus Gland

The bilobed **thymus** (thī′-mus) **gland** has important functions primarily during the early years of life. In infants, it is found in the lower neck region and extends into the thorax, where it lies deep to the sternum (see Figure 21.4). By secreting hormones, the thymus causes T lymphocytes to become immunocompetent; that is, it enables them to function against specific pathogens in the immune response (see Chapter 22). The thymus varies in size with age. Prominent in newborns, it continues to increase in size during childhood, when it is most active. During adolescence its growth stops, and it starts to atrophy very gradually. By old age it has been replaced entirely by fibrous and fatty tissue and is usually impossible to distinguish from surrounding connective tissue.

The structure of the thymus is generally similar to that of the spleen and lymph nodes; that is, it contains numerous lobules, each containing cortex and medullary elements (Figure 21.6a). In the cortex, the lymphocytes are densely packed. The medulla of the thymus contains fewer lymphocytes and appears lighter; however, it also contains some rather bizarre cystlike structures called *Hassal's* or *thymic corpuscles* (Figure 21.6b). Hassal's corpuscles are believed to be areas of degenerating epithelial cells, but their significance is unknown.

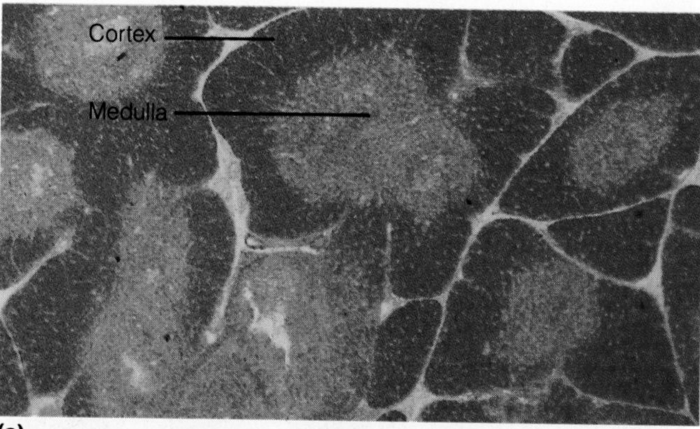

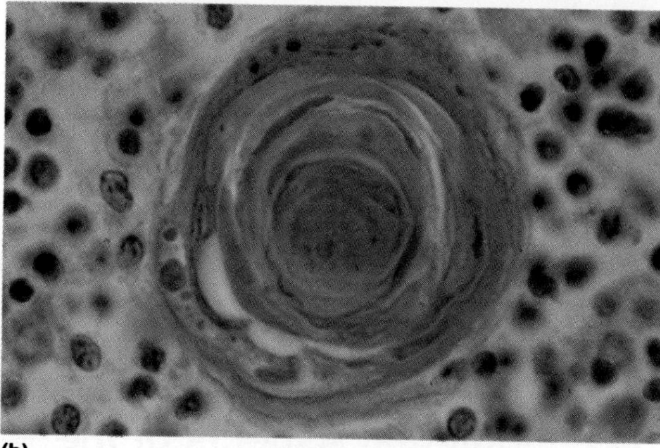

Figure 21.6 Histological structure of the thymus. (**a**) Photomicrograph of a portion of the thymus showing its lobules with cortical and medullary regions (25x). (**b**) Photomicrograph of a Hassal's corpuscle of the thymus (950X); it appears as the whorled structure in the center of the photomicrograph.

Tonsils

The **tonsils** form a ring of lymphatic tissue around the entrance to the pharynx, where they are found in the mucosa (Figure 21.4). Essentially, the tonsils are named according to location. The *palatine tonsils* are located on either side at the posterior end of the oral cavity. These are the largest of the tonsils and the most often infected. The *lingual tonsils* lie at the base of the tongue, and the *pharyngeal tonsils* (*adenoids*) are in the posterior wall of the nasopharynx. The tonsils gather and remove many of the pathogens entering the throat region.

The lymphoid tissue of the tonsils contains nodules with obvious germinal centers surrounded by diffusely scattered lymphocytes. The tonsil masses are not fully encapsulated, and the epithelium overlying them occasionally invaginates deep into their interior, forming blind-ended structures called *crypts* in which bacteria and particulate matter tend to become trapped (Figure 21.7). Lymphocytes from within the tonsils also migrate into the oral cavity through these crypts, but the functional importance of this migratory behavior is not known.

Peyer's Patches

Peyer's patches, isolated clusters of lymph nodules structurally similar to the tonsils, are found in the wall of the distal portion (ileum) of the small intestine (Figures 21.4 and 21.8). Macrophages of Peyer's patches are in an ideal position to capture and destroy bacteria (which are always present in tremendous numbers in the intestine), thereby preventing them from breaching the intestinal wall. Peyer's patches and the tonsils are part of the collection of small lymphoid tissues referred to as **gut-associated lymphatic tissue (GALT).** Collectively, GALT acts as a sentinel to protect the upper respiratory and digestive tracts from the never-ending onslaughts of foreign matter that enters those cavities.

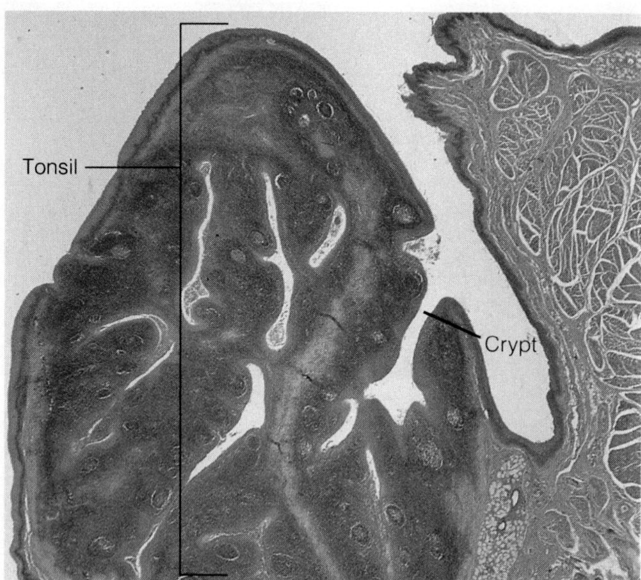

Figure 21.7 Histological structure of the palatine tonsil. The exterior surface of the tonsil is covered by squamous epithelium, which invaginates deep into the tonsil. The spherical regions seen are the nodules (10X).

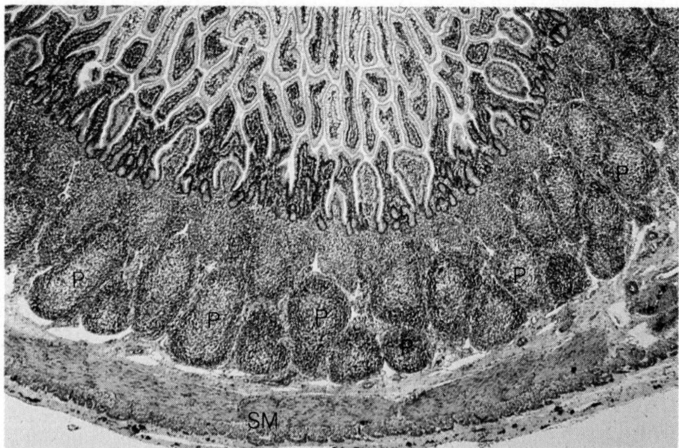

Figure 21.8 Histological structure of Peyer's patches (indicated by P) located in the wall of the ileum (20X). SM = submucosa.

Developmental Aspects of the Lymphatic System

By the fifth week of embryonic development, the beginnings of the lymphatic vessels are apparent in the budding of irregular **lymph sacs** from developing veins. The first of these lymph sacs, the *jugular lymph sacs*, arise at the junctions of the internal jugular and subclavian veins in the superior thorax and spread throughout the thorax, upper extremities, and head, forming a branching system of lymphatic vessels. The main connections of the jugular lymph sacs to the venous system are retained and become the right lymphatic duct and, on the left, the superior part of the thoracic duct. Development continues caudally: The elaborate system of abdominal lymphatics buds largely from the primitive inferior vena cava, and the lymphatics of the pelvic region and lower extremities form from the iliac veins.

Except for the thymus, which is an endodermal derivative, the lymphoid organs, like the lymphatic vessels, develop from mesoderm. Mesenchymal cells migrate to characteristic body sites where they develop into reticular connective tissue clusters. The thymus is the first lymphoid organ to appear. It forms as an outgrowth of the lining of the primitive pharynx that proceeds to grow caudally and, soon thereafter, it becomes infiltrated with immature lymphocytes derived from the hematopoietic tissues elsewhere in the embryonic body. Except for the spleen, the other lymphoid organs are poorly developed before birth. However, shortly after birth, they become heavily populated by lymphocytes, and their development parallels the maturation of the immune system (described in Chapter 22). It has been suggested that the embryonic thymus produces hormones that control the development of the lymphoid organs during infancy.

* * *

As we have seen, the lymphatic system serves two chief masters: the cardiovascular system and the immune system. Although the functions of the lymphatic vessels and lymphoid organs overlap, each helps to maintain body homeostasis in unique ways, as summarized in Figure 21.9. The lymphatic vessels return protein-containing interstitial fluid to the bloodstream, helping to maintain blood volume. The macrophages of lymph nodes remove and destroy foreign matter in the lymphatic stream, whereas those within the spleen rid the blood of foreign substances and aging red blood cells; in each case, these vital body fluids are continuously purified. The lymphoid organs and tissues also provide sites from which immunocompetent cells (lymphocytes) can monitor body fluids and mount an attack against specific antigens by releasing antibodies or by direct cellular interaction. In Chapter 22, we examine the inflammatory and immune responses that allow us to resist the constant barrage of pathogens.

Related Clinical Terms

Elephantiasis (eh-leh-fen-tī′-uh-sis) Typically a tropical disease in which the lymphatics (particularly those of the male's lower limbs and scrotum) become clogged with parasitic worms, and swelling reaches enormous proportions.

Hodgkin's disease A malignancy of the lymph nodes characterized by swollen, nonpainful lymph nodes, fatigue, and often persistent fever and night sweats. Treated with radiation therapy; high cure rate.

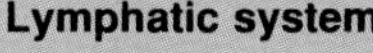

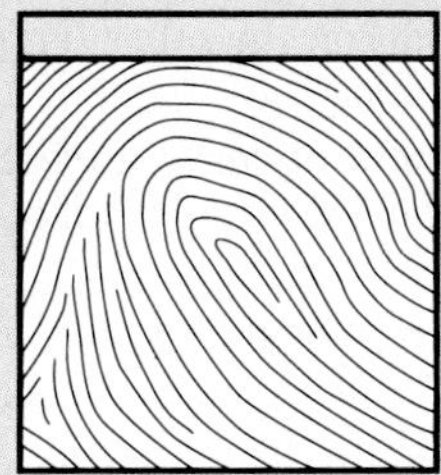

Integumentary system

Provides protective shield for body as a whole ▶

Picks up leaked plasma fluid and proteins from dermis ◀

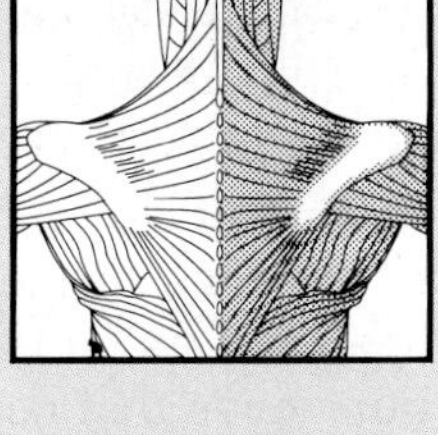

Skeletal system

Hematopoietic tissue in bone marrow provides lymphocytes (and macrophages), which populate lymphoid organs ▶

Picks up leaked plasma fluid and proteins from periostea ◀

Muscular system

Skeletal muscle "pump" aids flow of lymph ▶

Picks up leaked plasma fluids and proteins from skeletal muscle tissue ◀

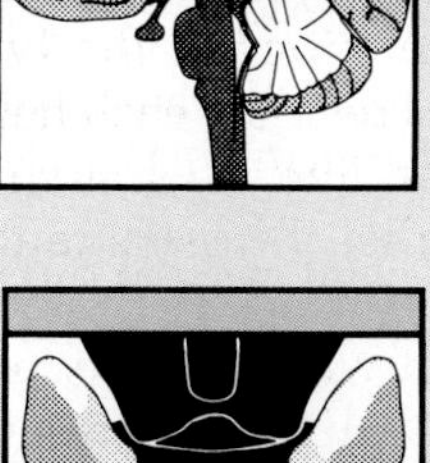

Nervous system

Provides motor and sensory fibers to larger lymphatic vessels ▶

Picks up leaked plasma fluid and proteins in peripheral nervous system structures ◀

Endocrine system

Thymus produces hormones that promote development of lymphatic organs and T cells ▶

Lymph distributes some hormones ◀

Cardiovascular system

Leaked plasma = lymph; lymphatics develop from veins ▶

Spleen destroys deteriorating RBCs, stores iron, removes debris from blood; lymphatics return leaked plasma and proteins to bloodstream ◀

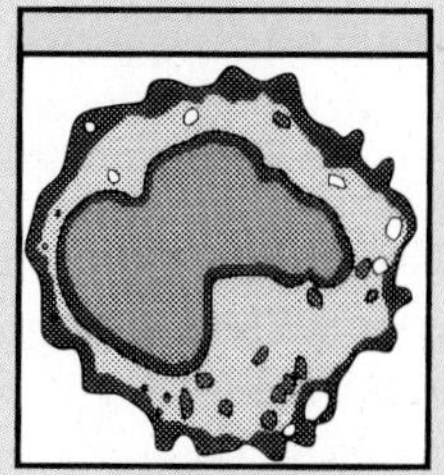

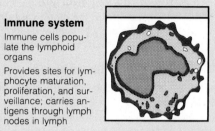

Immune system

◀ Immune cells populate the lymphoid organs

▶ Provides sites for lymphocyte maturation, proliferation, and surveillance; carries antigens through lymph nodes in lymph

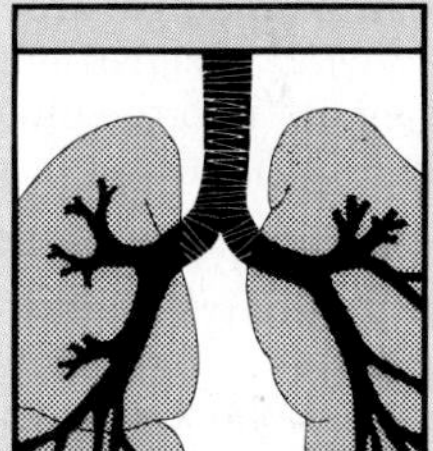

Respiratory system

◀ Action of respiratory "pump" aids lymph flow

▶ Picks up leaked plasma fluid and proteins

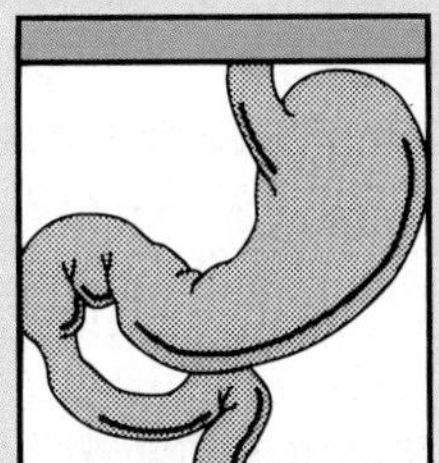

Digestive system

◀ Absorbs nutrients needed by lymphatic organs and tissues

▶ Picks up some products of fat digestion; delivers them to blood

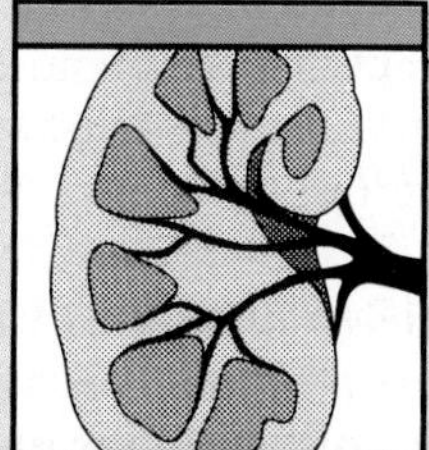

Urinary system

◀ Excretes metabolic wastes, excess ions, etc. present in blood

▶ Picks up leaked fluid, plasma, and proteins from urinary system organs

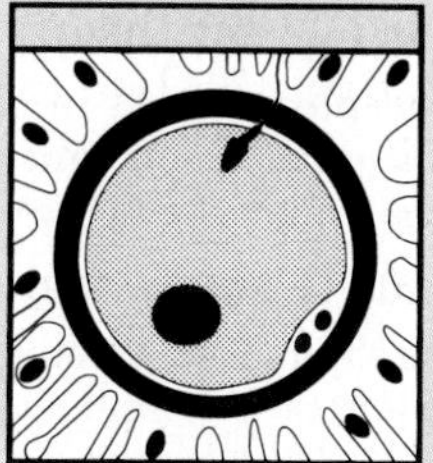

Reproductive system

▶ Picks up leaked fluid and plasma proteins from reproductive organs

Figure 21.9 Homeostatic interrelationships between the lymphatic system and the other body systems.

Lymphadenopathy (lim-fad-eh-nop-ah-thē) (*adeno* = a gland; *pathy* = disease) Disease of the lymph nodes.

Lymphedema Swelling due to accumulation of lymphatic fluid in the interstitial space.

Tonsillitis (ton-sih-lī′-tis) (*itis* = inflammation) Congestion of the tonsils, typically with bacteria, which causes them to become red, swollen, and sore.

Chapter Summary

1. Lymphatic vessels, lymph nodes, and other lymphoid organs and tissues make up the lymphatic system. This system returns fluids that have leaked from the blood vascular system back to the blood, protects the body by removing foreign material from the lymph stream, and provides a site for immune surveillance.

LYMPHATIC VESSELS (pp. 663–666)

Distribution and Structure of Lymphatic Vessels (pp. 664–665)

1. Lymphatic vessels form a one-way network in which fluid flows only toward the heart. The right lymphatic duct drains lymph from the right arm and right side of the upper body; the thoracic duct receives lymph from the rest of the body. These ducts empty into the blood vascular system at the junction of the internal jugular and subclavian veins in the neck.

Lymph Transport (pp. 665–666)

2. The flow of lymphatic fluid is slow; it is maintained by skeletal muscle contraction, pressure changes in the thorax, and (possibly) contractions of the lymphatic vessels. Backflow is prevented by valves.

3. Lymphatic capillaries are exceptionally permeable, admitting proteins and particulate matter from the interstitial space.

LYMPH NODES (pp. 666–667)

Structure of a Lymph Node (pp. 666–667)

1. Lymph nodes, clustered along lymphatic vessels, filter lymph. Each lymph node has a fibrous capsule, a cortex, and a medulla. The cortex contains lymphocytes, which act in immune responses; the medulla contains macrophages, which engulf and destroy viruses, bacteria, and other foreign debris.

Circulation in the Lymph Nodes (p. 667)

2. Lymph enters the lymph nodes via afferent lymphatic vessels and exits via efferent vessels. There are fewer efferent vessels, therefore, lymph flow stagnates within the lymph node allowing time for its cleansing.

OTHER LYMPHOID ORGANS (pp. 667–669)

1. Unlike lymph nodes, the spleen, thymus gland, tonsils, and Peyer's patches do not filter lymph. However, most lymphoid organs contain both macrophages and lymphocytes.

Spleen (pp. 667–669)

2. The spleen's most important function is to destroy aged or defective red blood cells. It also stores platelets and acts as a hematopoietic site in the embryo and in cases of severe anemia.

Thymus Gland (p. 669)

3. The thymus gland is most functional during youth. Its hormones cause T lymphocytes to become immunocompetent.

Tonsils and Peyer's Patches (p. 669)

4. The tonsils of the pharynx and the Peyer's patches of the intestinal wall are small organs known as GALT (gut-associated lymphatic tissue). They prevent pathogens in the respiratory and digestive tract from breaching the mucous membrane lining.

DEVELOPMENTAL ASPECTS OF THE LYMPHATIC SYSTEM (p. 670)

1. Lymphatics develop as outpocketings of developing veins. The thymus gland develops from endoderm; the other lymphoid organs derive from mesenchymal cells.

2. The thymus gland is the first lymphoid organ to appear in the embryo. It appears to play a role in the development of other lymphoid organs.

3. Lymphoid organs are populated by lymphocytes, which arise from hematopoietic tissue.

Review Questions

Multiple Choice/Matching

1. Lymphatic vessels (a) serve as sites for immune surveillance, (b) filter lymph, (c) transport leaked plasma proteins and fluids to the cardiovascular system, (d) are represented by vessels that resemble arteries, capillaries, and veins.

2. The saclike initial portion of the thoracic duct is the (a) lacteal, (b) right lymphatic duct, (c) cisterna chyli, (d) lymph sac.

3. Entry of lymph into the lymphatic capillaries is promoted by which of the following? (More than one answer may be appropriate.) (a) one-way minivalves formed by overlapping endothelial cells, (b) the respiratory pump, (c) the skeletal muscle pump, (d) greater fluid pressure in the interstitial space.

4. The structural framework of lymphoid organs is (a) areolar connective tissue, (b) hematopoietic tissue, (c) reticular tissue, (d) adipose tissue.

5. Lymph nodes are densely clustered in all of the following body areas *except* (a) the brain, (b) the axillae, (c) the groin, (d) the cervical region.

6. The germinal centers in lymph nodes are sites of (a) macrophages, (b) proliferating B lymphocytes, (c) T lymphocytes, (d) all of these.

7. The red pulp areas of the spleen are sites of (a) venous sinuses and red blood cells, (b) clustered lymphocytes and macrophages, (c) connective tissue septa.

8. The lymphoid organ that functions primarily during youth and then begins to atrophy is the (a) spleen, (b) thymus gland, (c) palatine tonsils, (d) bone marrow.

9. Collections of lymphoid tissue that guard mucosal surfaces include all of the following except (a) GALT, (b) the tonsils, (c) Peyer's patches, (d) the thymus gland.

Short Answer Essay Questions

10. Compare and contrast blood, interstitial fluid, and lymph.

11. Compare the structure and functions of a lymph node to those of the spleen.

12. (a) What anatomical characteristic ensures that the flow of lymph through a lymph node is slow? (b) Why is this desirable?

Clinical Application Questions

13. A 59-year-old woman has undergone a left radical mastectomy (removal of the left breast and left axillary lymph nodes and vessels). Her left arm is severely swollen and painful, and she is unable to raise it more than shoulder height. (a) Explain her signs and symptoms. (b) Can she expect to have relief from these symptoms in time? How so?

14. A young woman arrives at the clinic complaining of pain and redness of her right ring finger. The finger and the dorsum of her hand are edematous, and red streaks (indicating inflammation of her lymphatics) are apparent on her right forearm. Antibiotics are prescribed, and the nurse applies a sling to the woman's right arm. Why was a sling ordered and the patient warned not to move the affected arm excessively?

CHAPTER 22

Nonspecific Defenses and the Immune System

Chapter Outline and Student Objectives

NONSPECIFIC BODY DEFENSES (pp. 674–681)

Surface Membrane Barriers (pp. 674–675)

1. Describe the surface membrane barriers and their protective functions.

Nonspecific Cellular and Chemical Defenses (pp. 675–681)

2. Explain the importance of phagocytosis and natural killer cells in nonspecific body defense.
3. Relate the events of the inflammatory process. Identify several inflammatory chemicals and note their specific roles.
4. Name the body's antimicrobial substances and describe their function.
5. Explain how fever helps protect the body against invading pathogens.

SPECIFIC BODY DEFENSES: THE IMMUNE SYSTEM (pp. 681–704)

Cells of the Immune System: An Overview (pp. 682–684)

6. Compare and contrast the origin, maturation process, and general function of B and T lymphocytes. Describe the role of macrophages in immunity.
7. Define immunocompetence and self-tolerance.

Antigens (p. 684)

8. Explain what an antigen is and how it affects the immune system.
9. Define complete antigen, hapten, and antigenic determinant.

Humoral Immune Response (pp. 684–692)

10. Define humoral immunity.
11. Describe the process of clonal selection of a B cell.
12. Note the roles of plasma cells and memory cells in humoral immunity.
13. Compare and contrast active and passive humoral immunity.
14. Describe the structure of an antibody monomer, and name the five classes of antibodies.
15. Explain the function(s) of antibodies.
16. Describe the development and clinical uses of monoclonal antibodies.

Cell-Mediated Immune Response (pp. 692–698)

17. Define cell-mediated immunity and describe the process of clonal selection of T cells.
18. Describe the functional roles of T cells in the body.
19. Indicate the tests ordered before an organ transplant is done, and note methods used to prevent transplant rejection.

Homeostatic Imbalances of Immunity (pp. 698–703)

20. Give examples of immune deficiency diseases and of hypersensitivity states.
21. Explain how tolerance of foreign antigens might develop.
22. Note factors involved in autoimmune disease.

Developmental Aspects of the Immune System (pp. 703–704)

23. Describe changes in immunity that occur with aging.
24. Briefly describe the role of the nervous system in regulating the immune response.

Preview of Selected Key Terms

Inflammatory (*inflamm* = setting on fire) **response** A nonspecific defensive process induced by physical, chemical, or biological assaults on the body. Indicated by redness, heat, swelling, and pain.

Complement A group of blood-borne proteins, which, when activated, enhance the inflammatory and immune responses and may lead to cell lysis.

Immune (*immun* = free) **response** Antigen-specific defenses mounted by activated lymphocytes (T cells and B cells).

Antigen (an′-tih-jin) A substance or part of a substance (living or nonliving) that is recognized as foreign by the immune system, activates the immune system, and reacts with immune cells or their products.

Antibody A protein molecule that is released by a plasma cell (a daughter cell of an activated B lymphocyte) and that binds specifically to an antigen.

Humoral (hyoo′-mer-ul) **immune response** Immunity conferred by antibodies present in blood plasma and other body fluids.

Cell-mediated immune response Immunity conferred by activated T cells, which directly lyse infected or cancerous body cells or cells of foreign grafts and release chemicals that regulate the immune response.

Most of us would find it wonderfully convenient if we could walk into a single clothing store and buy a complete wardrobe—hat to shoes—that fit us "to a T" regardless of any special figure problems. Our reason tells us that such a service would be next to impossible to find. And yet we take for granted our *immune system*, a built-in specific defense system, which stalks and eliminates with nearly equal precision almost any type of pathogen that intrudes into the body.

Although certain organs of the body (notably, lymphatic and cardiovascular organs) are intimately involved with the immune response, the immune system is a *functional system* rather than an organ system in an anatomical sense. Its "structures" are trillions of individual immune cells, which inhabit lymphatic tissues and circulate in body fluids, and a diverse array of molecules. The most important of the immune cells are *lymphocytes* and *macrophages*.

When the immune system is operating effectively, it swiftly protects the body from most infectious microorganisms, transplanted organs or grafts, and even the body's own cells that have turned against it. The immune system does this both by direct cell attack and by releasing protective antibody molecules. The resulting highly specific resistance to disease is called **immunity.**

The immune system has one major shortcoming: It must be primed by an initial exposure to each foreign substance (antigen) before it is capable of protecting the body against the substance. On the other hand, the less glamorous *nonspecific defenses* respond immediately to protect the body from all foreign substances whatever their nature. The nonspecific defenses, provided by intact skin and mucosae, the inflammatory response, and a number of proteins produced by body cells, effectively reduce the workload of the immune system by preventing entry of bacteria and viruses or by preventing their spread throughout the body. Although we will consider them separately, the specific and nonspecific defenses work together to intensify each other's effects. The inflammatory response, for example, sets the stage on which the immune response is played out, and many defensive chemicals are common to both defense systems.

NONSPECIFIC BODY DEFENSES

Some nonspecific resistance to disease is inherited—a so-called *species resistance*. Thus, there are certain diseases that humans never fall prey to, such as some forms of tuberculosis that affect birds. Birds, on the other hand, are not susceptible to the bacteria that cause sexually transmitted diseases in humans. Most often, however, the term **nonspecific body defense** refers to the membrane barriers that cover body surfaces and to a whole array of cells and chemical molecules that act on the initial battlefronts to protect the body from invading pathogens.

Surface Membrane Barriers

The body's *first line of defense* against the invasion of disease-causing microorganisms is the *skin* and the *mucous membranes*. As long as the epidermis is unbroken, this heavily keratinized epithelial membrane presents a formidable physical barrier to most microorganisms that swarm on the skin. Keratin is also resistant to most weak acids and bases and to bacterial enzymes and toxins. Intact mucosae provide similar mechanical barriers. These membranes are found on the outer surface of the eye and as the linings of all body cavities open to the exterior: the digestive, respiratory, urinary, and reproductive tracts. In addition to serving as physical barriers, epithelial membranes produce a variety of protective chemicals:

1. The acid pH of skin secretions inhibits bacterial growth; sebum, in particular, contains chemicals that are toxic to bacteria. Vaginal secretions of adult females are also very acidic.

2. The stomach mucosa secretes a concentrated hydrochloric acid solution and protein-digesting enzymes, both of which kill pathogens.

3. Saliva, which cleanses the oral cavity and teeth, and lacrimal fluid, which washes the external eye surfaces, contain *lysozyme*, an enzyme that destroys bacteria.

4. Sticky mucus produced by the gastrointestinal and respiratory mucosae traps many microorganisms that enter those passageways.

The respiratory tract mucosae also have structural modifications that counteract potential invaders. The mucus-coated hairs of the nasal mucosa trap inhaled particles, and the mucosa of the upper respiratory tract is ciliated. The cilia sweep dust- and bacteria-laden mucus superiorly toward the mouth, restraining it from entering the lower respiratory passages, where the warm, moist environment would provide an ideal site for bacterial growth.

Although the surface barriers are extremely effective, they are breached occasionally by small nicks and cuts resulting, for example, from brushing your teeth or shaving. When this happens and microorganisms infiltrate deeper tissues, other nonspecific mechanisms come into play to defend the body.

Nonspecific Cellular and Chemical Defenses

The body uses an enormous number of nonspecific defensive cellular and chemical devices to protect itself. Some of these rely on the destructive powers of phagocytes and natural killer cells. The inflammatory response enlists the help of macrophages, mast cells, and all types of white blood cells, as well as dozens of chemical substances that contribute to pathogen killing and tissue repair. Still other mechanisms involve antibacterial proteins present in blood (complement) or antiviral proteins released by virus-infected cells (interferon). Fever can also be considered a nonspecific protective response. Although these mechanisms are only a few of the possible protective responses, they are among the most significant. Table 22.2 on p. 680 provides a summary of the nonspecific defenses.

Phagocytosis

Phagocytosis (fa″-gō-sī-tō′-sis), the ingestion and destruction of particulate matter, is one of the most important nonspecific defense mechanisms. A phagocyte, such as a macrophage or neutrophil, engulfs a foreign particle, much the way an amoeba ingests a food particle. Flowing cytoplasmic extensions bind to the particle and then pull it inside, enclosed within a membrane-lined vacuole (Figure 22.1). The *phagosome* thus formed is then fused with a *lysosome*. Phagocytic attempts are not always successful. To accomplish ingestion, the phagocyte must first adhere to the particle; and the rougher the particle surface, the more easily the phagocyte adheres. Complement proteins and antibodies coat foreign particles, roughening their surfaces and making phagocytosis more efficient.

Pathogens that make it through epithelial barriers are confronted by macrophages, which are located in nearly every body organ. Though the macrophages are called Kupffer cells in the liver, Langerhans' cells in skin, and histiocytes in connective tissue, all are similar in structure and function and they act vigorously to protect the body.

The manner in which neutrophils and macrophages kill their ingested prey is interesting. It is much more than simple digestion of the engulfed microorganism by lysosomal enzymes. Once the phagosome is fused with the lysosome, specific lysosomal enzymes are activated that produce the *respiratory burst*, an event liberating large quantities of oxygen-free radicals (see page 79), which have potent cell-killing ability. In addition, neutrophils release to the extracellular space oxidizing substances and a chemical identical to household bleach, which cause more long-lasting and thorough killing activity in the local area. As a result neutrophils are destroyed in their defensive efforts, whereas macrophages, which rely only on intracellular killing, can go on to kill another day.

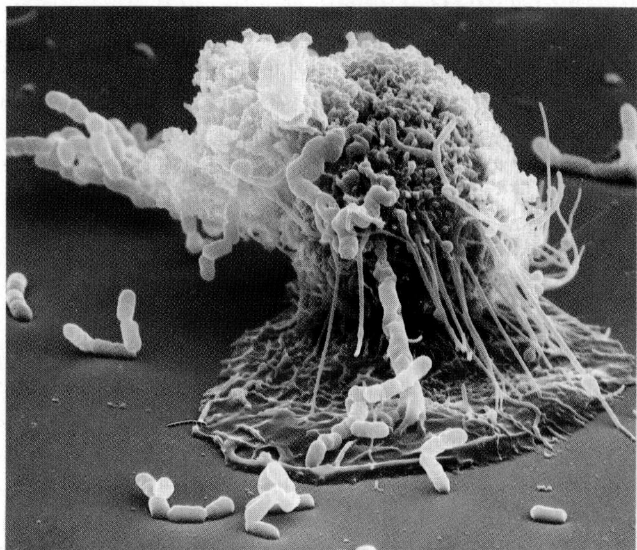

Figure 22.1 Phagocytosis by a macrophage. In this scanning electron micrograph (4300X), a macrophage is shown pulling sausage-shaped *E. coli* bacteria toward it with its long cytoplasmic extensions. Several bacteria on the macrophage's surface are being engulfed.

Natural Killer Cells

Natural killer (NK) cells, which roam the body in blood and lymph, are a unique group of defensive cells capable of killing (by causing lysis) cancer cells and virus-infected body cells before the immune system is enlisted in the fight. Originally, the NK cells were thought to be lymphocytes that help to mediate the immune response, but this has proven not to be true. The immune system lymphocytes can recognize and react only against specific virus-infected or tumor cells, whereas the natural killer cells can act spontaneously against any such target. This nonspecificity of their killing activity is reflected in their name, "natural" killer cells.

Tissue Response to Injury: Inflammation

When surface barriers are breached by pathogens or when tissues are injured by physical factors such as heat or cold, ultraviolet (UV) light (as from excessive exposure to sunlight), ionizing radiation (X-ray), or physical trauma, the **inflammatory response** is set in motion (Figure 22.2). The inflammatory response, which involves the interaction of cells, chemicals, and tissue fluid, is often referred to as the body's *second line of defense.* This response (1) prevents the spread of the injurious agent to nearby tissues, (2) disposes of cellular debris and pathogens, and (3) sets the stage for repair processes. Short-term, or *acute,* inflammation is always accompanied by four cardinal signs: *swelling, redness, heat,* and *pain.* We will see how these come about by describing the major events of the inflammatory process.

Vasodilation and Increased Vascular Permeability

The inflammatory process begins as a host of inflammatory chemicals are released into the extracellular fluid. Injured tissue cells, phagocytic leukocytes, lymphocytes, mast cells, and blood proteins all serve as sources of these inflammatory mediators, the most important of which are *histamine* (his'-tuh-mēn), *kinins* (kī'-ninz), *prostaglandins* (pros-tuh-glan'-dinz), *complement,* and *lymphokines* (limf'-ō-kīnz). Though histamine, kinins, and prostaglandins have individual inflammatory roles as well (see Table 22.1 on p. 678), they all promote local vasodilation. This allows more blood to flow into the area, causing a local *hyperemia* (congestion with blood), which accounts for the *redness* and the *heat* of an inflamed area.

The liberated chemicals also increase the permeability of local capillaries, and large amounts of exudate—fluid containing proteins such as clotting factors and antibodies—seep from the bloodstream into the tissue spaces. This exudate is the cause of the local edema, or *swelling.* If blood vessels have been injured, blood also escapes into the tissue spaces. The excessive fluid in the extracellular space presses on adjacent nerve endings, contributing to a sensation of *pain.* Pain may also result from the release of bacterial toxins and lack of nutrition to cells in the area, as well as from the sensitizing effects of released prostaglandins and kinins. Aspirin and some other anti-inflammatory drugs exert their effects by inhibiting prostaglandin synthesis.

If the swollen and painful area is a joint, joint movement may be hampered temporarily, permitting the injured part to rest and recover. Some authorities consider *impairment of function* to be the fifth cardinal sign of acute inflammation.

Although, at first glance, edema may seem to be detrimental, it isn't. The entry of protein-rich fluids into the tissue spaces (1) helps to dilute harmful substances that may be present, (2) brings in large quantities of oxygen and nutrients necessary for the repair process, and (3) allows the entry of clotting proteins, which begin a walling-off process (Figure 22.2). The gellike fibrin mesh that develops in the tissue space effectively isolates the injured area and prevents the spread of bacteria and other particulate matter into surrounding tissues. It also forms a scaffolding on which the permanent repair process can build.

Phagocyte Mobilization

Soon after inflammation begins, the damaged area is invaded by additional phagocytes—both neutrophils and macrophages. If the inflammation was provoked by pathogens, complement is activated and immune elements (lymphocytes and antibodies) also invade the injured site and mount an immune response.

Some of the chemicals released by injured cells act as *leukocytosis-inducing factors* that promote rapid release of stored neutrophils from the red bone marrow. Within a few hours the number of neutrophils in the bloodstream may increase four- to fivefold; this leukocytosis is a characteristic sign of inflammation. Neutrophils normally migrate randomly, but inflammatory chemicals seem to act as homing devices that attract them and other white blood cells to the site of the injury. This directional movement of cells in response to chemicals is called *chemotaxis.* Because of the tremendous outpouring of fluid from the blood into the injured area, blood flow in the region slows and the neutrophils begin to cling to the inner walls of the capillaries. This phenomenon is known as *margination.* The continued chemical signaling prompts

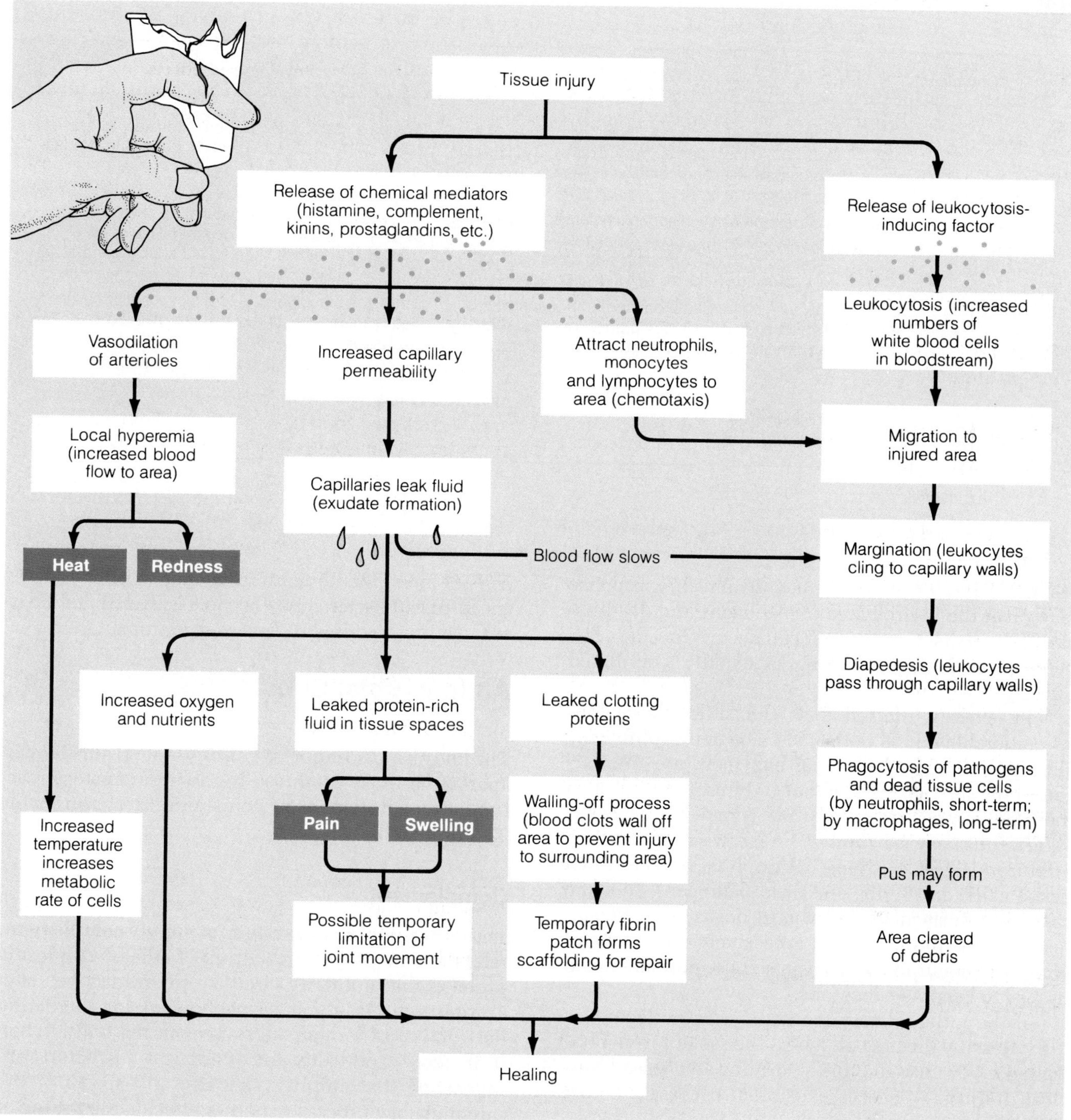

Figure 22.2 Flowchart of events in inflammation. The four cardinal signs of acute inflammation are shown in red boxes.

the neutrophils to squeeze through the capillary walls, a process called *diapedesis* (dī″-uh-puh-dē′-sis), and to make their way to the combat zone. Within an hour after the inflammatory response has begun, neutrophils have collected in the injury site and are actively phagocytizing bacteria, toxins, and dead tissue cells.

As the counterattack continues, neutrophils are gradually replaced by macrophages. These arise from monocytes entering the area from the bloodstream. Monocytes are fairly poor phagocytes, but after entering the tissues, they swell and develop large numbers of lysosomes, becoming macrophages within 8–12 hours. The relatively late-arriving macrophages are most important in the final disposal of cell debris as an acute inflammation subsides. Macrophages also predominate at sites of prolonged, or *chronic*, inflam-

Table 22.1 Inflammatory Chemicals

Chemical	Source	Physiological effects
Histamine	Granules of basophils and mast cells; released in response to mechanical injury, presence of certain microorganisms, and chemicals released by neutrophils	Promotes vasodilation of local arterioles; increases permeability of local capillaries, permitting exudate formation
Kinins (bradykinin and others)	A plasma protein, kininogen, is cleaved by the enzyme kallikrein found in plasma, urine, saliva, and lysosomes of neutrophils and other types of cells; cleavage releases active kinin peptides	Same as for histamine; also induce chemotaxis of leukocytes and stimulate neutrophils to release lysosomal enzymes, thereby enhancing generation of more kinins; induce pain
Prostaglandins	Fatty acid molecules produced from arachidonic acid; found in all cell membranes; generated by lysosomal enzymes of neutrophils and other cell types	Sensitizes blood vessels to effects of other inflammatory mediators; one of the intermediate steps of prostaglandin generation produces oxygen free radicals, which themselves can cause inflammation; induce pain
Complement	See Table 22.2	
Lymphokines	See Table 22.4	

mation. The ultimate goal of the inflammatory response is to clear the injured area of pathogens and dead tissue cells so that tissue repair can occur. Once this has been accomplished, healing is usually completed quickly.

In severely infected areas, the battle takes a considerable toll on both sides. The neutrophils may simply spill their lysosomal enzymes into the surrounding area, destroying not only bacteria but also themselves and surrounding cells. As a result, a creamy, yellow pus may be formed in the wound. Pus is a mixture of dead or dying neutrophils, broken-down tissue cells, and living and dead pathogens. In some cases the effects may be even more devastating. Recent research has indicated that excessively vigorous and prolonged neutrophil activity may cause normal human tissues to become cancerous.

If the inflammatory mechanism is ineffective in fully clearing the area of debris, the sac of pus may be walled off by collagen fibers, forming an *abscess*. Surgical drainage of abscesses is often necessary before healing can occur.

Some bacteria, such as the bacillus that causes tuberculosis, are resistant to digestion by macrophages. In fact, these bacteria are able to escape the effects of antibiotics by remaining snugly enclosed within their macrophage hosts. In such cases, granulomas form. A *granuloma* (gran-yoo-lō′-muh) contains a central region of infected macrophages, surrounded by uninfected macrophages and an outer fibrous capsule. A person may harbor pathogens walled off in granulomas for many years without displaying any symptoms of disease. However, if something happens to lower the person's resistance to infection, the encapsulated bacteria may become activated and break out, leading to clinical disease symptoms. ■

Antimicrobial Substances

The body's most important **antimicrobial substances,** apart from those produced in the inflammatory reaction, are complement proteins and interferon (Table 22.2 on p. 680).

Complement

The term **complement system,** or simply **complement,** refers to a heterogeneous group of at least 20 plasma proteins that normally circulate in an inactive state. Complement is a major mechanism for mediating destruction of foreign substances in the body. When it is fixed and activated, chemical mediators are unleashed that amplify virtually all aspects of the inflammatory process. Another effect is the killing of bacteria and certain other cell types by cell lysis. Although the complement attack is often directed against specific microorganisms that have been identified by antibody binding, complement itself is a nonspecific defensive mechanism that "complements" or enhances the effectiveness of both nonspecific and specific defenses.

Complement can be activated by either of two pathways (Figure 22.3): the classical pathway or the alternate pathway. Eleven proteins designated as C1 through C9 (3 of the 11 proteins form the C1 complex) act in the *classical pathway.* Initiation of this pathway

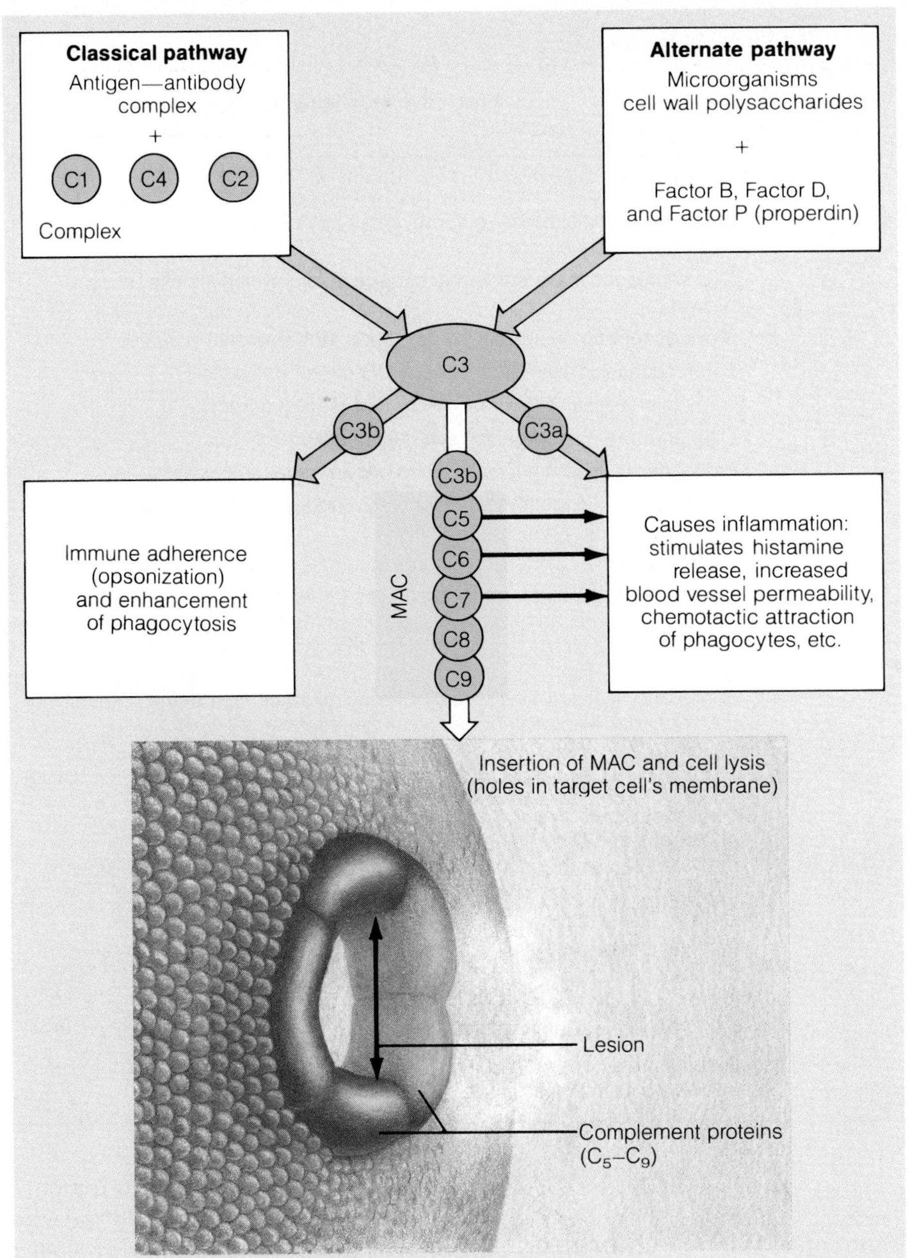

Figure 22.3 Events and results of complement activation. The complement system can be activated by either the classical pathway, which requires antigen–antibody interaction, or the alternate pathway, which occurs when factors B, D, and P interact with the cell wall polysaccharides of certain bacteria and fungi. The two pathways converge to activation of C3. Cleavage of C3 initiates a common terminal sequence that generates most of the biological activities of complement. Insertion of MAC, the membrane attack complex (components C5, C6, C7, C8, and C9) into the target cell's membrane creates a funnel-shaped lesion that causes cell lysis.

depends on the binding of antibodies to the invading organisms and the subsequent binding of C1 to the antigen–antibody complexes. The *alternate pathway,* which sidesteps the early phases of the classical pathway, is less well understood. Although it, too, can apparently be initiated by antibody binding, most commonly this pathway is triggered by an interaction between three plasma proteins—*factors B, D,* and *P (properdin)*—and polysaccharide molecules present on the surface of certain bacteria, parasites, and fungi.

Each pathway involves a cascade in which complement proteins are activated in an orderly sequence—the activated intermediates of each step causing catalysis of the next step. Both the classical and the alternate pathways lead to the cleavage of C3, and beyond this event the reaction sequence is the same in both pathways. The cleavage of C3 into two fragments, C3a and C3b, has important consequences because the terminal pathway beyond that point initiates cell lysis, promotes immune adherence, and enhances inflammation. Once C3b is bound to the target cell's surface, it enzymatically initiates the remaining steps of complement activation, which incorporate C5 through C9 into the foreign cell's membrane. These inserted proteins, appropriately called the *membrane attack complex (MAC),* form a hole in the target cell membrane. As a result, the membrane becomes leaky, allowing ions first and then larger cell solutes to pass freely

Table 22.2 Summary of Nonspecific Body Defenses

Category/associated elements	Protective mechanism
Surface membrane barriers	
Intact skin epidermis	Forms mechanical barrier that prevents entry of pathogens and other harmful substances into body
• Acid mantle	Skin secretions make epidermal surface acidic, which inhibits bacterial growth
• Keratin	Provides resistance against acids, alkalis, and bacterial enzymes
Intact mucous membranes	Form mechanical barrier that prevents entry of pathogens
• Mucus	Traps microorganisms in respiratory and digestive tracts
• Nasal hairs	Filter and trap microorganisms in nasal passages
• Cilia	Propel debris-laden mucus away from lower respiratory passages
• Gastric juice	Contains concentrated hydrochloric acid and protein-digesting enzymes that destroy pathogens in stomach
• Acid mantle of vagina	Inhibits growth of bacteria and fungi in female reproductive tract
• Lacrimal secretion (tears); saliva	Continuously lubricate and cleanse eyes (tears) and oral cavity (saliva); contain lysozyme, an enzyme that destroys microorganisms
Nonspecific cellular and chemical defenses	
Phagocytes	Engulf and destroy pathogens that breach surface membrane barriers; macrophages also contribute to immune response
Natural killer cells	Promote cell lysis by direct cell attack against virus-infected or cancerous body cells; do not depend on specific antigen recognition
Inflammatory response	Prevents spread of injurious agents to adjacent tissues, disposes of pathogens and dead tissue cells, and promotes tissue repair; chemical mediators released attract phagocytes (and immunocompetent cells) to the area
Antimicrobial substances	
• Interferons	Proteins released by virus-infected cells that protect uninfected tissue cells from viral takeover; mobilize immune system
• Complement	Lyses microorganisms, enhances phagocytosis by opsonization, and intensifies inflammatory response
Fever	Systemic response initiated by pyrogens; high body temperature inhibits microbial multiplication and enhances body repair processes

into the extracellular space. The membrane can repair itself until the very end of complement fixation, but the terminal step, insertion of C9, ensures lysis of the target cell by stabilizing the open lesion in its membrane (see Figure 22.3).

The C3b molecules that coat the microorganism roughen its surface, enabling macrophages and neutrophils to phagocytize the particle more rapidly. This phenomenon is called *opsonization* (*opsonize* = to make tasty) or *immune adherence*. C3a and other cleavage products produced during complement fixation have potent inflammatory actions. These products stimulate mast cells and basophils to release histamine, which increases vascular permeability, and they attract neutrophils and other inflammatory cells to the area.

Interferon

Viruses—essentially, nucleic acids surrounded by a protein coat—lack the cellular machinery required to generate ATP or synthesize proteins. They do their damage in the body by entering tissue cells and taking over the cellular metabolic machinery needed to reproduce themselves. Although the virus-infected cells can do little to save themselves, they help defend other cells by secreting small proteins called **interferons** (in-ter-fēr′-onz). The interferon molecules diffuse to nearby cells and bind to their membrane receptors, interfering with the ability of viruses to multiply within these cells. It is currently assumed that interferon stimulates the synthesis of proteins in the targeted tissue cells, which then act to inhibit production of the virus's genetic material and/or viral coat proteins.

The interferons are not virus-specific (that is, they are active against a variety of different viruses), but they are host-specific. Thus, interferons produced by mice or chicks have little or no antiviral activity in humans and vice versa.

Interferon is a family of related proteins, each with slightly different physiological effects. Most leukocytes produce alpha (α) interferon, fibroblasts produce beta (β) interferon, and lymphocytes secrete gamma (γ), or immune, interferon. The interferons have a number of effects on cell function in addition to their antiviral effects. For example, they all decrease the rate of cell division and activate macrophages, and the gamma interferons mobilize the natural killer cells. Because both macrophages and natural killer cells can act directly against malignant cells, the interferons are assumed to have some anticancer role.

When interferons were discovered in the 1950s, it was hoped that they could be isolated and used to treat human viral and neoplastic diseases, but their naturally occurring quantities have proved much too small to be clinically useful. Recently, genetic-engineering techniques have made it possible to produce and harvest usable quantities of interferons. These have been tested as anticancer agents, as the main defense against devastating viral infections in organ-trans- plant patients, and in last-ditch attempts to save AIDS victims. Thus far, success in cancer therapy has been mixed. Some patients have been helped dramatically, others not at all, and still others have suffered life-threatening neurological side effects. Although interferon has been used in a nasal spray against the common cold virus, it is unlikely that this will be a routine use. Like all proteins, interferon degrades in the body within a few days, making the nasal spray therapy much too expensive for use against a sniffle!

Fever

Fever, or abnormally high body temperature, represents a systemic response to invading microorganisms. Body temperature is regulated by a cluster of neurons in the hypothalamus, commonly referred to as the body's thermostat. Normally, the thermostat is set at approximately 37°C (98.6°F). However, the thermostat is reset upward in response to chemicals called *pyrogens* (*pyro* = fire), which are secreted chiefly by macrophages exposed to bacteria and other foreign substances in the body.

Although high fevers are dangerous because of the likelihood of heat-induced inactivation of enzymes, mild or moderate fever seems to benefit the body. Bacteria require large amounts of iron to proliferate, but during a fever the liver and spleen sequester iron, making it less available. Fever also increases the metabolic rate of tissue cells, speeding up defensive actions and repair processes.

SPECIFIC BODY DEFENSES: THE IMMUNE SYSTEM

The body's *third line of defense*, the immune response, tremendously amplifies the inflammatory response and is responsible for the bulk of complement activation. Further, it provides a unique protective service that is carefully targeted against specific antigens and is adaptive: The initial exposure to an antigen "primes" the body to react more vigorously to the same antigen at subsequent meetings.

The **immune system** is a functional system that recognizes foreign molecules (antigens) and acts to immobilize, neutralize, or destroy them. When it operates effectively, this system protects the body from a wide variety of infectious agents, as well as from abnormal body cells. When it fails, malfunctions, or is disabled, some of the most devastating diseases—such as cancer, rheumatoid arthritis, and AIDS—may result.

Although *immunology*, the study of immunity, is a fairly new science, the ancient Greeks knew that once someone had had a certain infectious disease, that person was unlikely to have the same disease again. About 2500 years ago, Thucydides of Athens, in describing an epidemic (probably of the plague), recorded that the sick and dying were tended by those who had recovered and who "were themselves free of apprehension. For no one was ever attacked a second time or with a fatal result."

The basis of this **immune response** was revealed in landmark experiments, done in 1890 by von Behring and Kitasato, which showed that animals surviving a serious bacterial infection have protective factors in their blood that protect them from future attacks by the same pathogen. Furthermore, if serum from the surviving animals (immune serum) was injected into animals that had not been exposed to the pathogen, those animals would also be protected. These experiments revealed three important aspects of the immune response:

1. It is *antigen-specific;* it recognizes and is directed against *particular* pathogens or foreign substances, that is, those antigens that incite the immune response.
2. It is *systemic;* immunity is not restricted to the initial infection site.
3. It has *"memory"*; it recognizes and mounts an enhanced attack on previously encountered pathogens.

Following these revelations, the protective factors in serum were identified as the highly reactive proteins we call *antibodies*. It was later found that injected serum did *not* always protect the recipient from diseases the donor had survived; however, in such cases, injection of the donor's lymphocytes did provide immunity. As the pieces of the puzzle began to fall into place, two separate but overlapping types of immunity were recognized, each using a variety of attack mechanisms that vary with the type of intruder.

Humoral immunity is provided by the antibodies in the body's "humors," or fluids. Though antibodies are produced by lymphocytes, they circulate freely in the blood and lymph, where they bind primarily to bacteria and their toxins and to free viruses, inactivating them temporarily and marking them for destruction by phagocytes or complement. **Cell-mediated immunity** is provided by nonantibody-producing lymphocytes, which (1) directly attack and lyse body cells infected by viruses or other intracellular parasites, cancer cells, and foreign grafts and (2) release chemical mediators that enhance the inflammatory response or help to activate lymphocytes or macrophages.

In summary, the immune system is a two-armed defensive system that uses lymphocytes, macrophages, and specific molecules (see Table 22.4) to identify and destroy all substances—both alive and inert (nonliving)—that are in the body but are not recognized as being part of the body, or as being *self*. Lacking any central controlling organ, the immune system's ability to respond to such threats depends on the ability of its individual cells to recognize foreign substances (antigens) in the body by binding to them, a capability called **immunocompetence,** and to communicate with one another so that the system as a whole mounts a response specific to those antigens. Before we describe the humoral and cell-mediated responses individually, we will consider the remarkable cells involved in these immune responses and the antigens that trigger their activity.

Cells of the Immune System: An Overview

The crucial cells of the immune system are two distinct populations of lymphocytes and macrophages. The **B lymphocytes** or **B cells** are those that oversee humoral immunity; the **T lymphocytes** or **T cells** are nonantibody-producing lymphocytes that constitute the cell-mediated arm of immunity. Unlike the two types of lymphocytes, macrophages do not respond to specific antigens but instead play essential auxiliary roles and, in the absence of macrophages, the immune system is severely impaired.

Like all blood cells, lymphocytes originate in red bone marrow from hemocytoblasts, and the immature lymphocytes released from the marrow are essentially identical (Figure 22.4). Whether a given lymphocyte matures into a B cell or a T cell depends on where in the body it becomes immunocompetent. T cells arise from lymphocytes that migrate to the thymus, where they undergo a maturation process of two to three days under the direction of thymic hormones. Within the thymus, the immature lymphocytes divide rapidly and their numbers increase enormously. However, most of the lymphocytes that enter the thymus die there. It appears that only those maturing T cells with the sharpest ability to identify foreign antigens are "selected" for survival. In any case, lymphocytes capable of binding strongly with *self-antigens* (cell-surface proteins of body cells) are vigorously weeded out and destroyed. Thus, the development of *tolerance* for self-antigens is an essential part of a lymphocyte's "education" or maturation process. (This is true not only for T cells but also for B cells.) B cells are believed to develop immunocompetence while still in the bone marrow, but very little is known about the factors that control B cell maturation in humans. B cells are so called because they were first identified in the *bursa of Fabricius*, a pouch of gut-associated lymphatic tissue in birds.

Acquisition of immunocompetence is signaled by the appearance of a single, unique type of cell-surface receptor protein on each T or B cell that enables the lymphocyte to recognize and bind to a specific antigen. Once immunocompetence has been acquired, a lymphocyte is *committed* to react to one distinct antigen and one only, because *all* of its antigen receptors have the same specificity. For example, the receptors of one lymphocyte can recognize only the hepatitis A virus, those of another lymphocyte can bind to pneumococcus bacteria, and so forth. Although the antigen receptors of T cells and B cells differ—those of B cells are actually bound antibody molecules, whereas those of T cells look more like "half-antibodies"—it appears that in many cases both types of lymphocytes are capable of responding to the same antigen variety.

While many of the details of lymphocyte processing are still unknown, we do know that lymphocytes become immunocompetent *before* meeting the antigens they may later attack; hence, development of immunocompetence is not antigen-driven. The fact that *each* lymphocyte bears unique antigen receptors allows the body to respond to an enormous number of antigens. Not all of the possible antigens lymphocytes are programmed to resist will invade the body. Consequently, of the total army of immunocompetent cells, only some will be mobilized during a person's lifetime; the others will be forever idle. Genes deter-

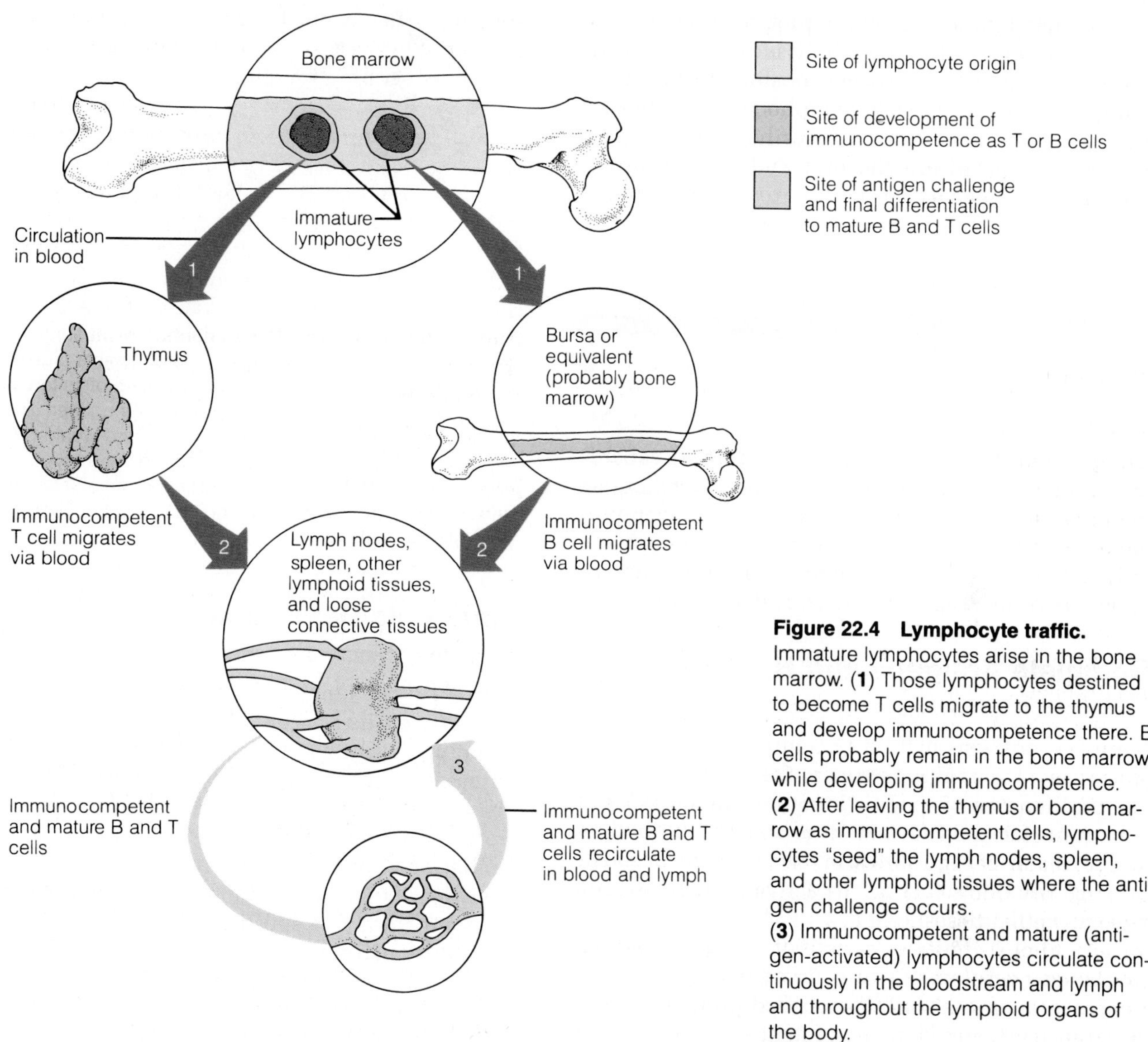

Figure 22.4 Lymphocyte traffic. Immature lymphocytes arise in the bone marrow. (**1**) Those lymphocytes destined to become T cells migrate to the thymus and develop immunocompetence there. B cells probably remain in the bone marrow while developing immunocompetence. (**2**) After leaving the thymus or bone marrow as immunocompetent cells, lymphocytes "seed" the lymph nodes, spleen, and other lymphoid tissues where the antigen challenge occurs. (**3**) Immunocompetent and mature (antigen-activated) lymphocytes circulate continuously in the bloodstream and lymph and throughout the lymphoid organs of the body.

mine which antigen a specific immunocompetent cell is able to attack by specifying the structure of its receptor proteins. The antigen determines only which *existing* B or T cells will proliferate and mount the attack against it. After becoming immunocompetent, both T cells and B cells disperse to the lymph nodes and spleen (and loose connective tissues), where the encounter with antigens occurs (Figure 22.4). When the lymphocytes bind with recognized antigens, they complete their differentiation into fully mature T cells and B cells.

Macrophages, which also become widely dispersed throughout the lymphoid organs and connective tissues, arise from monocytes formed in the bone marrow. A major role of macrophages is to engulf foreign particles and present critical portions of these antigens on their own surfaces for recognition by immunocompetent lymphocytes. Macrophages also secrete soluble proteins that activate T cells. Activated T cells, in turn, release chemicals that mobilize macrophages, causing them to become insatiable phagocytes, or *killer macrophages*. Although both macrophages and lymphocytes are found in each lymphoid organ, they have different localizations. The cortical area of lymph nodes houses regions of B and T cells, whereas macrophages are usually clustered around the sinuses. Macrophages tend to remain fixed in the lymphoid organs, as if waiting for antigens to come to them; but lymphocytes, especially the T cells (which account for 65% to 85% of blood-borne lymphocytes), are extremely mobile and circulate continuously throughout the body (see Figure 22.4). The circulatory pattern of immunocompetent cells greatly increases their chance of coming into contact with antigens located in different parts of the body, as well as with huge numbers of macrophages and other lymphocytes.

Because lymph capillaries pick up proteins and pathogens from nearly all body tissues, immune cells in lymph nodes are in a strategic position to encounter a large variety of antigens. Lymphocytes and macrophages in the tonsils act primarily against microorganisms that invade the oral and nasal cavities, whereas the spleen acts as a filter to trap blood-borne antigens.

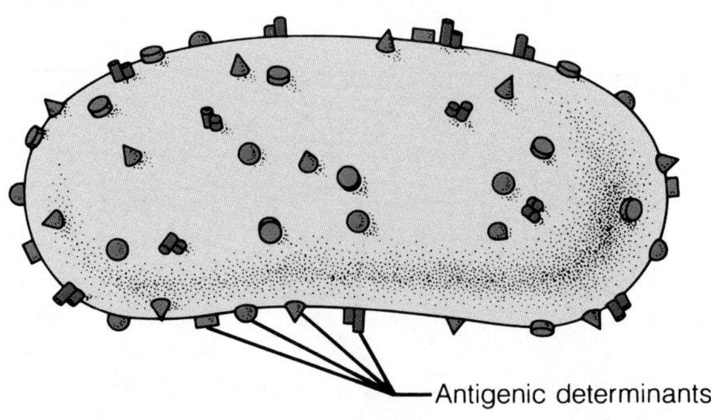

Figure 22.5 Antigenic determinants. Most antigens express several different antigenic determinants (specific cell surface molecules), which serve as markers of its foreign nature and elicit the immune response. The antigen illustrated has four different antigenic determinants and would be expected to elicit the production of four different types of antibodies, each antibody being specific for one of the antigenic determinants and capable of binding with the recognized antigenic determinant.

Antigens

Antigens (an′-tih-jinz) are substances that are capable of mobilizing the immune system and provoking an immune response. Most antigens are large, complex molecules (macromolecules) that are not normally present in the body. Consequently, as far as the immune system is concerned, they are intruders (or *nonself*).

Complete Antigens and Haptens

Complete antigens exhibit two important functional properties: (1) *immunogenicity*, the ability to stimulate proliferation of specific immunocompetent T and B lymphocytes and formation of specific antibodies; and (2) *reactivity*, the ability to react with the products of these reactions, that is, the activated lymphocytes and the antibodies released.

An almost limitless variety of foreign molecules can act as complete antigens, including virtually all foreign proteins as well as modified proteins (glycoproteins, nucleoproteins, lipoproteins) and many large polysaccharides. Of these, proteins are the most potent. Microorganisms—such as bacteria, fungi, and virus particles—and the cells of transplanted tissues are immunogenic because their surface membranes (or coats in the case of viruses) bear such foreign macromolecules. Most complete antigens have a molecular weight of 10,000 or greater.

Small molecules—such as peptides, nucleotides, and hormones—are not usually immunogenic by themselves; but when they link up with the body's own proteins, the immune system sometimes recognizes the combination as foreign and mounts an attack that is harmful rather than protective. (Such reactions, called *allergies*, are described later in the chapter.) The troublesome small molecules, called *haptens* (hap′-tenz), from *haptein* meaning to grasp, are *incomplete antigens:* They can react with antibodies or activated T cells, but they cannot initiate an immune response unless they are attached to protein carriers. Thus, haptens are reactive but not immunogenic.

Antigenic Determinants

The ability of a molecule to act as an antigen depends not only on its size but also on the complexity of its structure. Only small portions of an entire antigen are immunogenic, and typically these sites have unique three-dimensional structures, called *antigenic determinants* or *epitopes* (eh′-pih-tōps), to which free antibodies or activated B or T lymphocytes can bind. This binding occurs much in the same manner that an enzyme binds to a substrate.

Most naturally occurring antigens have several structurally different epitopes (Figure 22.5), each of which is immunogenic and reactive. The number of antigenic determinants on an antigen is called its *valence* (vā′-lints). Large proteins have hundreds of chemically different antigenic determinants, which accounts for their high immunogenicity and reactivity. Large, but simple molecules such as plastics, which have many identical, regularly repeating units and thus are not chemically complex, have little or no immunogenicity; such substances are used to make artificial implants (hip replacements and others) because they are not rejected by the body.

Humoral Immune Response

The *antigen challenge*, the first encounter between an immunocompetent lymphocyte and an invading antigen, usually takes place in the spleen or a lymph node. If the lymphocyte is a B cell, the challenging antigen

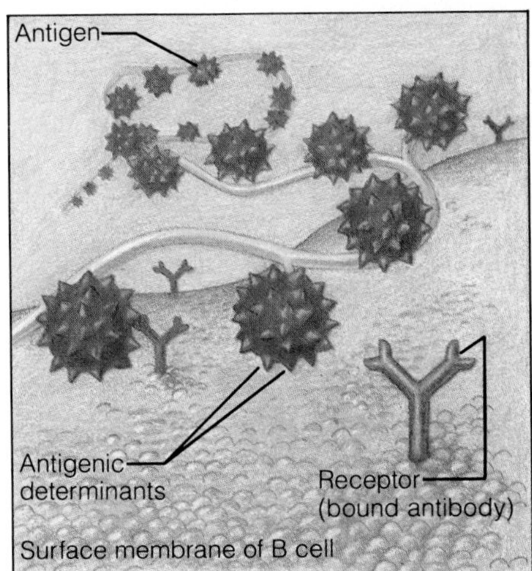

Figure 22.6 The capping theory of B cell activation. According to this theory, a B cell is activated when an antigen binds with its surface receptors (bound antibody molecules) and pulls them into a continuous cluster. For this to occur, the antigen must have multiple antigenic determinants of the same type (but not so many that they pull adjacent receptors into different clusters).

provokes the *humoral immune response,* in which antibodies are produced against the challenger.

Clonal Selection and Differentiation of B Cells

B cell activation occurs when antigen binds to its antigen-specific surface receptors and, in most cases, requires the collaboration of T cells (and macrophages). According to a recent theory, B cell mobilization requires that a multivalent antigen bind to several adjacent receptors and pull them into a continuous cluster, an event called *capping* (Figure 22.6). (This capping requirement explains why haptens alone cannot generate an immune response: They are too small and have insufficient antigenic determinants to cluster the cell surface receptors.) Capping is quickly followed by endocytosis of the cross-linked antigen–receptor complexes. These activating events trigger **clonal** (klō-nul) **selection;** that is, they stimulate the B cell into dramatic growth and then rapid multiplication to form a large number of cells all exactly like itself and bearing the same antigen-specific receptors (Figure 22.7). The resulting "family of identical cells," descended from the same ancestral cell, is called a **clone.** It is the antigen that does the selecting in clonal selection by "choosing" a lymphocyte with complementary receptors.

Shortly thereafter, the cloned cells begin to differentiate. Most of them become effector cells, called **plasma cells,** which develop the elaborate internal machinery (largely, rough endoplasmic reticulum) needed for antibody synthesis. The plasma cells produce and secrete nearly all the antibodies of this initial immune response. (The B cells themselves shed only very limited amounts of antibodies and are short-lived once activated.) Each plasma cell secretes the same kind of antibody at the unbelievable rate of about 2000 molecules per second for four to five days; then it dies. The secreted antibodies, each with the same antigen-binding properties as the receptor molecules on the surface of the parent B cell, circulate in the blood or lymph, where they bind to free antigens and mark them for destruction by other specific or nonspecific mechanisms. The clone cells which do not differentiate into plasma cells become long-lived *memory cells* capable of mounting an almost immediate humoral response if the same antigen is encountered again (see Figure 22.7).

Immunological Memory

The cellular proliferation and differentiation just described constitute the *primary humoral immune response,* which occurs when the body is first exposed to a particular antigen. The primary response is characterized by a lag period that persists for several days after the antigen challenge. This lag period mirrors the time required for the few B cells specific for that antigen to proliferate and for subsequent differentiation of the plasma cells into functional antibody-synthesizing factories. After the mobilization period,

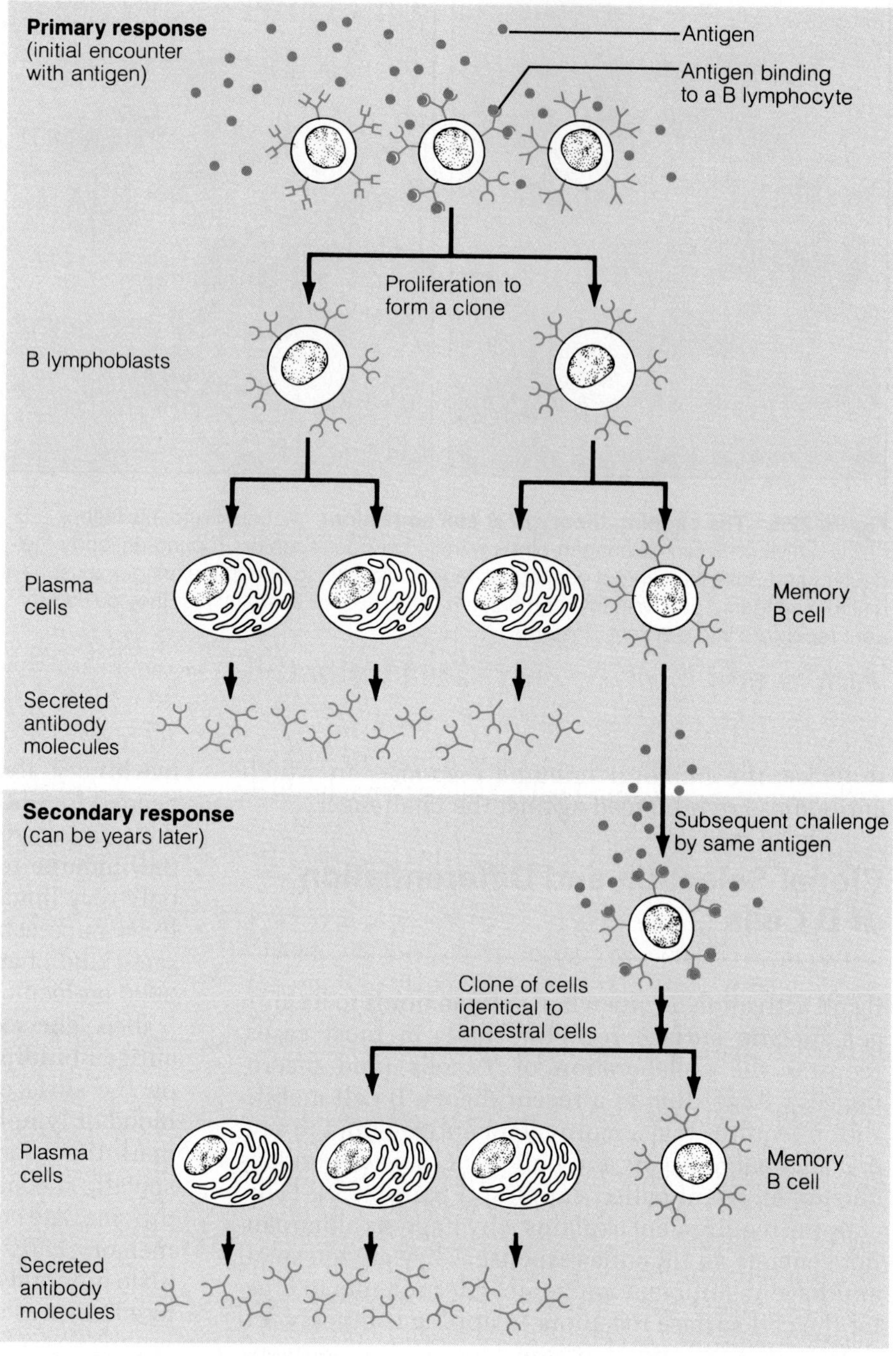

Figure 22.7 Simplified version of clonal selection of a B cell stimulated by antigen binding. The initial meeting stimulates the primary response in which the B cell proliferates rapidly, forming a clone of like cells most of which differentiate into antibody-producing plasma cells. (Though not indicated in the figure, the production of mature plasma cells takes about five days and eight cell generations.) Cells that do not differentiate into plasma cells become memory cells, which respond to subsequent meetings with the same antigen by cloning, again producing plasma cells and memory cells with the same antigen specificity. The responses generated by memory cells are called secondary responses.

plasma antibody levels slowly rise, reaching peak levels in about ten days, and then slowly begin to decline (Figure 22.8).

Subsequent challenges by the same antigen, whether the second or the twenty-second meeting, evoke secondary immune responses that are faster, more prolonged, and more effective. This is because the immune system has been primed to the antigen and sizable numbers of sensitized memory cells are already in place. Huge numbers of antibodies literally flood into the bloodstream within hours after recognition of the antigen. Even in the absence of antigenic stimulation, memory cells persist for long periods in humans and retain their capacity for producing powerful secondary humoral responses for as long as 20 years. The same phenomena occur in the cellular immune response: A primary response sets up a pool of activated lymphocytes (in this case, T cells) and generates memory cells that can then mount secondary responses.

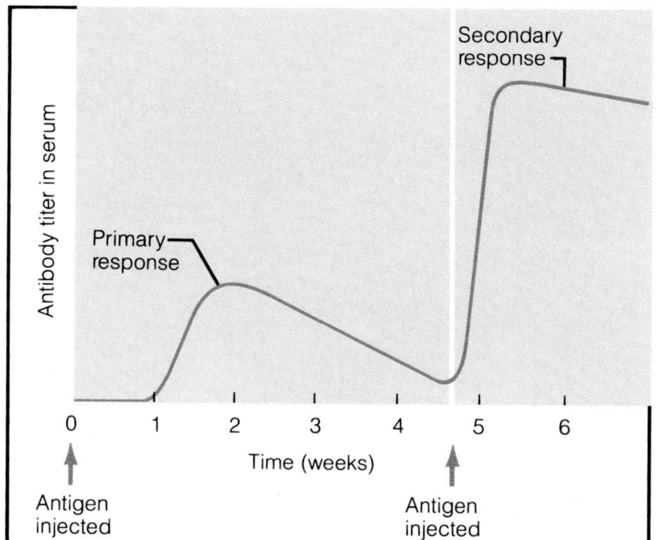

Figure 22.8 Primary and secondary humoral responses. Both a primary response and a secondary response to antigen A are depicted in the graph. In the primary response, there is a gradual rise and then a gradual decline in the level of antibodies in the blood. The secondary response is both more rapid and more intense. Additionally, antibody levels remain high for a much longer time.

Active and Passive Humoral Immunity

When B cells encounter antigens and their plasma-cell progeny produce antibodies against them, the body is exhibiting *active humoral immunity.* Active immunity is acquired naturally during bacterial and viral infections, but it can also be provoked artificially by means of a *vaccine.* Most vaccines contain dead or extremely weakened pathogens, which have the antigenic determinants necessary to stimulate the immune response, but are generally unable to cause disease. Active immunity acquired in this way is indistinguishable from immunity acquired in the course of an infection. Thus, vaccines bestow two benefits in addition to the protective antibodies formed: (1) The person is spared most of the signs and symptoms of the disease, which would otherwise occur during the primary response, and (2) *immunological memory* is established. Vaccines are currently available against microorganisms that cause pneumonia, smallpox, polio, tetanus, diphtheria, and many other diseases.

Passive humoral immunity, on the other hand, is different both in antibody source and in degree of protection. The antibodies that confer passive immunity are supplied by an immune human or animal donor. Because the body's own B cells are not challenged, neither active antibody production nor immunological memory is established and the protection provided by the "borrowed" antibodies ends when they naturally degrade in the body. Passive immunity is conferred naturally on a fetus when certain of the mother's antibodies cross the placenta into the fetal circulation. For several months after birth, the baby is protected from all the antigens to which the mother has been exposed. Passive immunity is artificially conferred when a person receives an infusion of immune serum (for example, gamma globulin). Gamma globulin is commonly administered after exposure to hepatitis. Other immune sera are used to treat poisonous snake bites, botulism, rabies, and tetanus because these diseases would kill a person before active immunity could be established. The infused antibodies provide immediate protection, but their effect is short-lived (two to three weeks). Meanwhile, however, the body's own defenses have taken over.

Antibodies

Antibodies, also called **immunoglobulins** (ih″-myoo-nō-glah′-byoo-linz) **(Igs),** constitute the *gamma globulin* part of blood proteins. As mentioned, antibodies are soluble proteins secreted by the plasma cell offspring of sensitized B cells in response to an antigen and are capable of binding specifically with that antigen.

Antibodies are formed in response to an incredible number of different antigens. However, despite their variety, all antibodies can be grouped into one of five Ig classes, each slightly different in structure and function. We will see how these Ig classes differ, but first let's take a look at how all antibodies are alike.

Basic Antibody Structure

Regardless of its class, each antibody has a basic structure consisting of four polypeptide chains linked together by disulfide (sulfur-to-sulfur) bonds (Figure 22.9). Two of these chains, the *heavy chains,* are identical to each other and contain approximately 400 amino acids each. The other two polypeptide chains,

Figure 22.9 Basic antibody structure. (**a**) Basic antibody structure consists of four polypeptide chains that are joined together by disulfide bonds. Two of the chains are short *light chains;* the other two are long *heavy chains.* Each chain has a variable region, which is different in different antibodies, and a constant region, which is essentially identical in different antibodies of the same class. The variable regions are the antigen-binding sites of the antibody; hence, each antibody monomer has two antigen-binding sites. (**b**) Computer-generated model of antibody structure. The spheres represent the individual amino acids of the polypeptide chains.

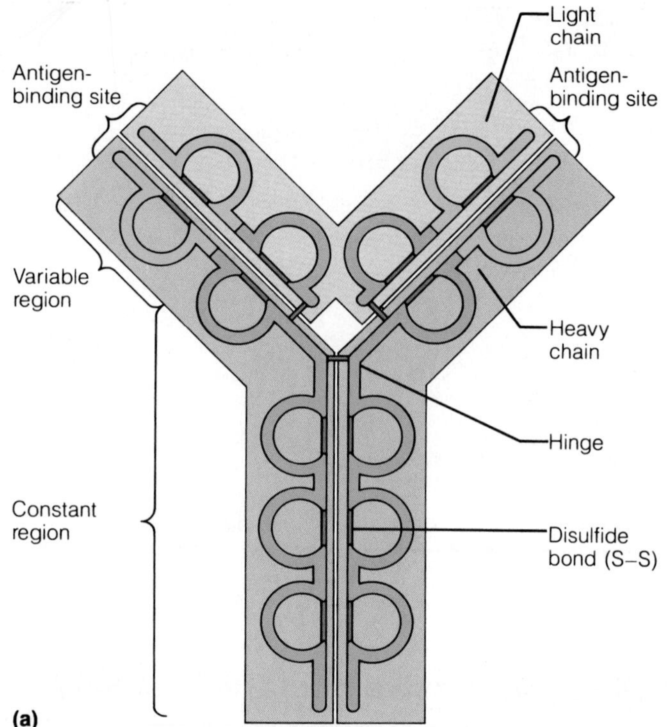

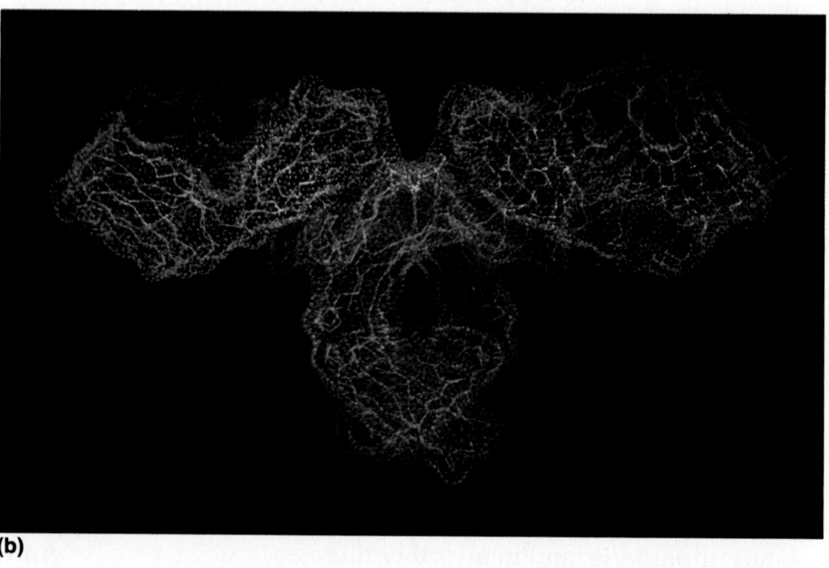

the *light chains,* are also identical to each other, but they contain only about 200 amino acids each. The antibody molecule formed, called a *monomer* (mah′-nō-mer), has two identical halves, each consisting of a heavy and a light chain. Free antibodies resemble the letter T; but after binding to an antigen, they appear smaller and Y-shaped. Antigen binding is believed to cause the antibody to pivot and change its three-dimensional shape—an event allowed by its flexible *hinge region.*

When researchers began investigating the structure of antibodies, they discovered something very peculiar. Each of the four chains forming an antibody has a *variable (V) region* confined entirely to one end of a much larger *constant (C) region.* The sequence of the amino acids in the variable regions is quite dissimilar in antibodies responding to different antigens, but their constant regions are the same (or nearly so) in all antibodies of a given class. This began to make sense when it was discovered that the variable regions of the heavy and light chains in each arm combine to form an *antigen-binding site* (see Figure 22.9) that is uniquely shaped to "fit" a specific antigenic determinant or an antigen. Because each antibody monomer has two such antigen-binding regions, it is said to be *bivalent.*

Table 22.3 Immunoglobulin Classes

Class	Generalized structure	Heavy chain constant region	Where found	Biological function
IgD	Monomer	Δ (delta)	Virtually always attached to B cell	Believed to be cell surface receptor of immunocompetent B cell; important in activation of B cell
IgM	J chain; Pentamer	μ (mu)	Attached to B cell; free in plasma	When bound to B cell membrane, serves as antigen receptor; first Ig class released to plasma by plasma cells during primary response; potent agglutinating agent; fixes complement
IgG	Monomer	γ (gamma)	Most abundant antibody in plasma; 75% to 85% of circulatory antibodies	Main antibody of both primary and secondary responses; crosses placenta and provides passive immunity to fetus; fixes complement
IgA	J chain; Monomer or dimer	α (alpha)	Some (monomer) in plasma; dimer in secretions such as saliva, tears, intestinal juice, and milk	Protects mucosal surfaces; prevents attachment of pathogens to epithelial cell surfaces
IgE	Monomer	ε (epsilon)	Secreted by plasma cells in skin, mucosae of gastrointestinal and respiratory tracts, and tonsils	Becomes bound to mast cells and basophils; when cross-linked by antigens, triggers release of histamine and other chemicals from mast cell or basophil that mediate inflammation and certain allergic responses

The constant regions of the polypeptide chains that form the stem of the antibody monomer can be compared with the handle of a key. A key handle has a common function in all keys: It allows you to hold the key and place its tumbler-moving portion in the lock. Likewise, the constant regions of antibody chains serve common functions in all antibodies: They determine what class of antibody will be formed, how the antibody class will carry out its immune roles in the body, and what cell types or chemicals the antibody can bind with. For example, some but not all antibodies can fix complement, some circulate in blood while others are found primarily in body secretions, some cross the placental barrier, and so on.

Antibody Classes

The five major immunoglobulin classes are designated as IgD, IgM, IgG, IgA, and IgE, on the basis of the C regions in their heavy chains. As illustrated in Table 22.3, IgD, IgG, and IgE have the same basic Y-shaped structure and are referred to as *monomers*. IgA occurs in both monomer and *dimer* (two linked monomers) forms. (Only the dimer is shown in Table 22.3.) Compared to the other antibody classes, IgM is a huge antibody, a *pentamer* (*penta* = five), constructed from five linked monomers. The antibodies of each class have a slightly different biological role and location in the body. For example, IgM, which accounts for about 10% of the gamma globulins in plasma, is the first antibody class that is *released* to the blood by plasma cells. IgG is the most abundant antibody in plasma and is the only immunoglobulin class that can cross the placental barrier. Hence, the passive immunity that a mother transfers to her fetus is "with the compliments" of her IgG antibodies. Only IgM and IgG can fix complement. The IgA dimer, also called *secretory IgA*, is found primarily in mucus and other secretions that bathe body surfaces and plays a major role in preventing pathogens from gaining entry into the body. Both IgM and IgD may be bound to the B cell surface and hence act as B cell receptors. IgE antibodies, almost never found in blood, are involved in allergies. These and other characteristics unique to each of the immunoglobulin classes are summarized in Table 22.3.

Mechanisms of Antibody Diversity

It has been estimated that the various plasma-cell clones can make over 1 billion different types of antibodies. Because a person probably encounters several thousand different antigens during the course of a lifetime, the immune system has provided for a wide margin of error. Because antibodies, like all proteins, are specified by genes, it would seem that an individual must have billions of genes. Not so; each body cell contains perhaps 100,000 genes that code for all the proteins the cells must make.

How can a limited number of genes generate an apparently limitless number of different antibodies? Recombinant-DNA studies have shown that the genes ultimately dictating the structure of each antibody are not present as such in embryonic cells. Rather than harboring a complete and active set of "antibody genes," embryonic cells contain about 300 genetic bits and pieces that can be thought of as an "erector set for antibody genes." These gene segments are shuffled and combined in different ways by each B cell as it becomes immunocompetent in a process called *somatic recombination*. This mixing and matching of antibody-specifying gene segments is random, and each B cell reassembles the parts to form its own unique antibody genes. The information of these reassembled genes is then expressed in the surface receptors of B cells and in the antibodies later released by the plasma cells of each clone.

In an immature B cell, the gene segments coding for the light and heavy chains are located on different chromosomes. The light and heavy chains are manufactured separately and then joined to form the antibody molecule. A simplified version of the somatic-recombination process used to produce one type of light antibody chain is described and illustrated in Figure 22.10. The sequence of events for forming the heavy chains is similar; however, the process is much more complex because of the larger variety of genes for the constant regions of the heavy chains (remember, the C regions of the heavy chains determine antibody class). A single plasma cell can make various kinds of heavy chains and does so at different stages of its development, producing two or more different antibody classes with the same antigen specificity. For example, the first antibody released in the primary response is IgM; then the plasma cell begins to secrete IgG. During secondary responses, almost all of the Ig protein is IgG.

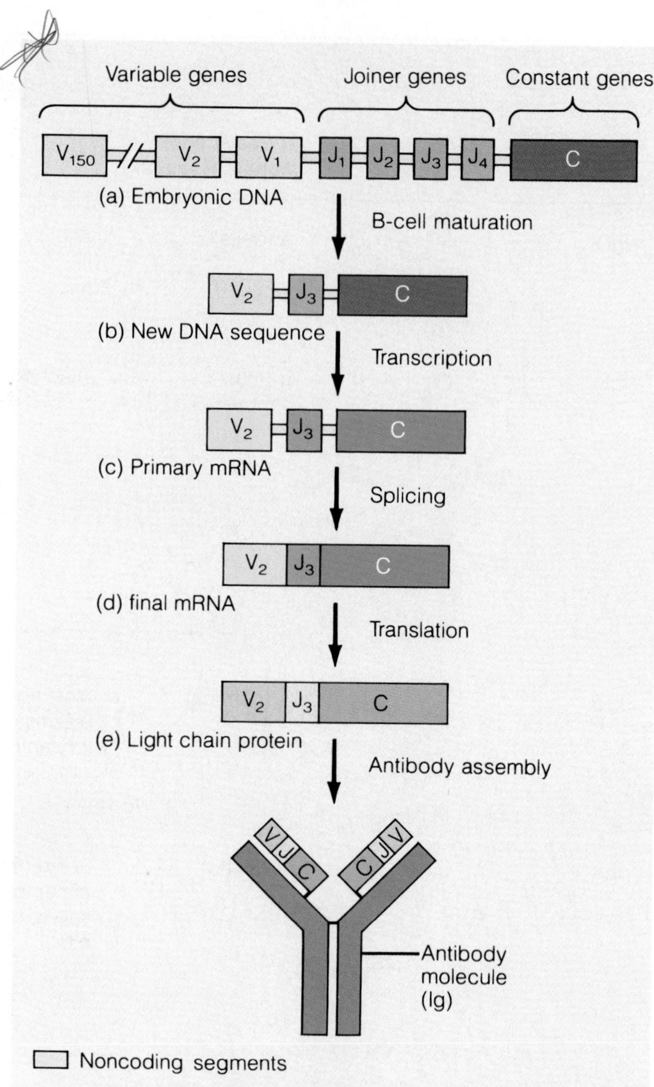

Figure 22.10 Somatic recombination and formation of a light antibody chain (kappa variety). (**a**) In an embryonic cell, there are one constant (C) gene, four joiner (J) genes, and at least 150 variable (V) genes that can be used to form the light antibody chain. These genes are separated by noncoding segments of DNA (gray). (**b**) During B cell maturation, one randomly selected V gene is combined with one of the J genes and the C gene to form the active gene for the light chain. The intervening genes are deleted (spliced out). (**c**) The new gene is transcribed to form primary messenger RNA (mRNA). (**d**) Noncoding segments are spliced out to form the final messenger RNA. (**e**) The final mRNA is translated into protein, forming the light chain with variable, joiner, and constant regions.

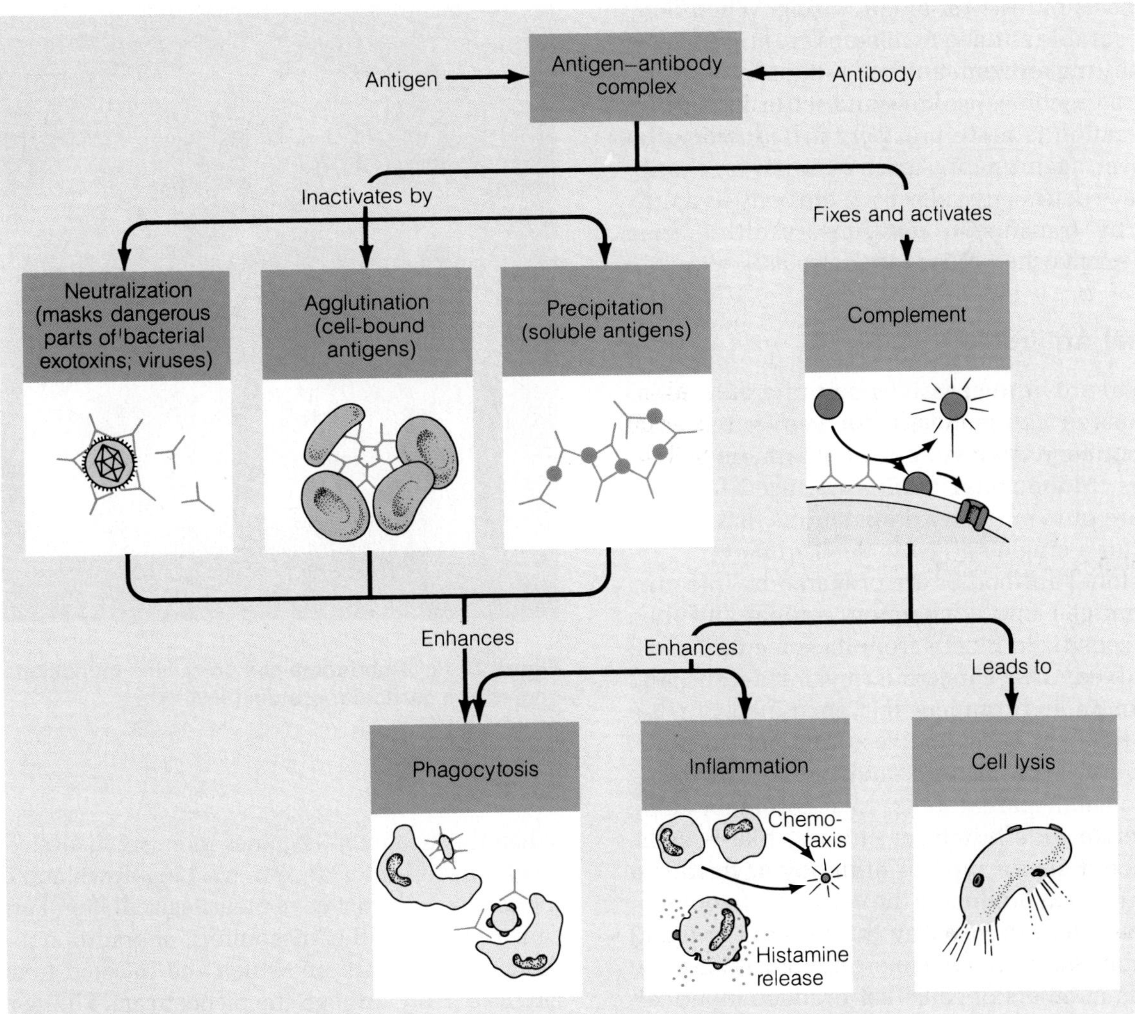

Figure 22.11 Mechanisms of antibody action.

Antibody Function

Soluble antibodies can inactivate antigens in various ways, depending on the nature of the antigen (Figure 22.11). The common thread is that all antibody–antigen interactions involve the formation of *antigen–antibody* (or immune) *complexes*. The defensive mechanisms employed by antibodies include complement fixation, neutralization, agglutination, and precipitation. Of these, complement fixation and neutralization are the most important in body protection.

Complement fixation is the chief antibody ammunition used against cellular antigens, such as bacteria or mismatched red blood cells. When antibodies bind to cellular targets, their shape changes such that complement-binding sites are exposed on their constant regions. This triggers, in rapid-fire sequence, a number of events that result in complement fixation into the antigenic cell's surface and subsequent cell lysis. Additionally, as described earlier, molecules released during complement activation enhance the inflammatory response and promote phagocytosis via opsonization.

Neutralization occurs when antibodies bind to specific sites on bacterial exotoxins (toxic chemicals secreted by bacteria) or viruses that can cause cell injury. Consequently, the exotoxin or virus cannot bind to receptor sites on tissue cells and is eventually destroyed by phagocytes.

Because antibodies have more than one antigen-binding site, they can bind to more than one antigen at a time; consequently, antigen–antibody complexes can be cross-linked into large lattices. When the cross linking involves cell-bound antigens, the process causes clumping, **agglutination** (uh-gloo"-tih-nā'-shun), of the foreign cells. IgM, with ten antigen-binding sites (see Table 22.3), is an especially potent agglutinating agent. This type of reaction occurs when mismatched blood is infused (the foreign red blood cells are clumped) and is the basis of tests for blood typing. When large numbers of soluble antigen molecules are cross-linked and the resulting antigen–antibody complexes are so large that they become insoluble and settle out of solution, the reaction is more precisely called **precipitation.** However, agglutination and precipitation of antigens may have detrimental effects on the body, as amply illustrated by transfusion reactions resulting from infusion of mismatched blood (see page 589).

Monoclonal Antibodies

In addition to providing passive immunity, antibodies can be prepared commercially for use in research, clinical laboratory testing, and the treatment of certain cancers. **Monoclonal antibodies,** used for such purposes, are pure antibody preparations that exhibit specificity for a single antigenic determinant.

Monoclonal antibodies are prepared by injecting a specific antigen into a laboratory animal and then harvesting sensitized B cells from its spleen. These B cells are mixed with cancerous myeloma cells and incubated in an environment that encourages fusion of the two types of cells. The resulting cell hybrids, called *hybridoma* (hī″-brih-dō′-muh) *cells,* have desirable traits of both parent cells. Like myeloma cells, they proliferate indefinitely in culture; like B cells, they produce a single type of antibody. Hybridoma cells (Figure 22.12) producing the desired monoclonal antibody (the one provoked by the injected antigen) are then isolated and allowed to proliferate, giving rise to enormous numbers of cells that produce identical, or monoclonal, antibodies.

Monoclonal antibodies are used for diagnosis of pregnancy, certain sexually transmitted diseases, hepatitis, and rabies. Compared with the conventional diagnostic tests for these conditions, monoclonal antibody tests are much more specific, sensitive, and rapid. Monoclonal antibody tests are being used for early diagnosis of certain types of cancer, such as colon cancer, before clinical disease symptoms appear, but the most exciting use of monoclonal antibodies is for cancer therapy. It is not known exactly how many cancers manage to escape detection by the immune system. Undoubtedly, part of the problem is that cancers arise from normal tissue cells and have many "self" cell markers. Since 1980, monoclonal antibodies have been used in isolated cases to treat cancers such as leukemia and lymphomas, which are present in the circulation and thus are easily accessible to injected antibodies. Many of these treatments have had positive results, and partial remissions have been reported in many cases thought to be hopeless. Monoclonal antibodies, attached to anti-cancer drugs, act as homing devices to deliver the drug only to the cancerous tissue.

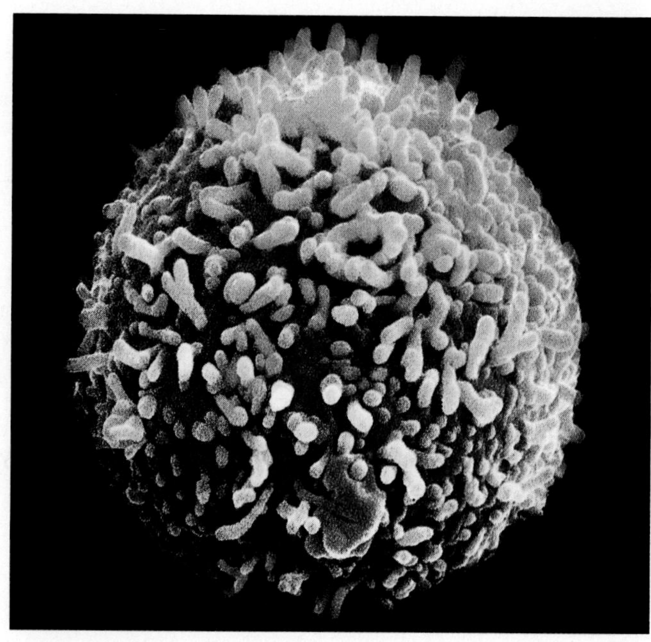

Figure 22.12 Hybridoma cell producing monoclonal antibodies to a particular protein (5600X).

When this treatment becomes more sophisticated, it will be of crucial importance in tracking down and destroying cancer cells that have metastasized. It will also spare cancer patients the discomfort of traditional chemotherapy treatments, in which the injected toxic drugs circulate freely through the bloodstream, killing not only cancer cells but also rapidly dividing normal tissue cells.

Cell-Mediated Immune Response

The T cells that mediate cellular immunity are a diverse lot, much more complex than B cells both in classification and function. The four major populations of T cells include two varieties of effector cells, the *cytotoxic* and *delayed hypersensitivity T cells,* and two varieties of regulatory cells, called *helper* and *suppressor T cells.* Each population is distinguished by antigen-specific receptors, which resemble the light chains of antibodies responding to the same antigen.

Before going into the specific details of the cell-mediated immune response, however, we will recap some introductory information and compare and contrast the major importance of the humoral and cellular responses in the overall scheme of immunity. Although both arms of the immune response respond to virtually the same antigens, they do so in substantially different ways.

Soluble antibodies produced by B cell progeny (the effector plasma cells) react with infectious microorganisms, soluble antigens, and foreign cells that circulate freely in body fluids. The formation of antigen–antibody complexes does not destroy the antigen; instead, it prepares the antigen for destruction by nonspecific defense mechanisms and cell-mediated responses mounted by activated T cells.

In contrast, T cells are unable to recognize free antigens carried in blood or lymph. Current evidence indicates that, for the most part, T cells can respond only to antigens attached to or displayed on surfaces of the body's own cells, and then only under specific circumstances. Consequently, T cells are best suited for cell-to-cell interactions, and most of their direct attacks on antigens (mediated by the cytotoxic T cells) are mounted against body cells infected by viruses, bacteria, or intracellular parasites; abnormal or cancerous body cells; and cells of infused or transplanted foreign tissues. As in the inflammatory response, chemical mediators enhance the response; but the chemical mediators involved in cellular immunity are the hormonelike **lymphokines,** soluble glycoproteins released by activated T cells themselves. Lymphokines enhance the defensive activity not only of T cells but also of B cells and macrophages as well. Additionally, macrophages secrete *monokines*, which stimulate T cells (see Table 22.4 on p. 697).

Clonal Selection and Differentiation of T Cells

Activation of immunocompetent T cells, like that of B cells, is triggered by antigen binding to the lymphocyte's surface receptors. However, the T cell receptor consists of only two amino acid chains (α and β), each with constant and variable regions; and T cell clonal selection involves a double recognition: a simultaneous recognition of nonself (the antigen) and self (a membrane glycoprotein of a body cell). Cell surface proteins that mark every body cell as self are coded for by genes of the **major histocompatibility complex (MHC).** Because there are millions of different combinations of these genes possible, it is unlikely that any two people, except identical twins, have the same MHC-encoded proteins.

Although several different proteins important in the immune response, including complement, are specified by MHC, only two of these protein classes are important to T cell activation: *class I proteins*, found on virtually all body cells; and *class II proteins*, found only on the surfaces of mature B lymphocytes and macrophages.

The cell that appears to play the major role in presenting the double activation signal to T cells is the macrophage. Hypothetically, antigens engulfed by macrophages are digested and a very small part of the antigen (perhaps a single antigenic determinant) is displayed on the macrophage's surface, attached to one of the two classes of MHC proteins. A T cell whose receptor binds to the *self–antiself complex* (like a "double handshake") begins to enlarge and then forms a clone of cells that differentiate and perform functions according to their T cell class. When the antigen is complexed with a class I protein, the members of the T cell clone differentiate into cytotoxic T cells. Immunocompetent T cells that bind class II antigen complexes produce progeny that differentiate into one of the other T cell populations, depending on the nature of the antigen. This primary response peaks within a week after a single exposure and is then gradually shut down by suppressor cells. In every case, some of the clone members become memory T cells; they persist, perhaps for life, and can mediate secondary responses to the same antigen in that T cell class. Clonal selection resulting in the production of a cytotoxic and a helper T cell clone is illustrated in Figure 22.13.

T cell proliferation is enhanced by two *lymphokines:* interleukin 1 and 2. Interleukin 1, released by macrophages, stimulates bound T cells to liberate interleukin 2. Interleukin 2 sets up a positive feedback cycle that encourages activated T cells to divide further or even more rapidly. (Genetically engineered interleukin 2 is undergoing clinical trials to test its usefulness in enhancing the activity of the body's own cytotoxic T cells against cancer.)

Additionally, all activated T cells secrete one or more other lymphokines that help to amplify and regulate a variety of immune and nonspecific responses. Many lymphokines mobilize other lymphocytes; others act primarily on macrophages. For example, *macrophage activating factor (MAF)* enhances the killing power of macrophages by prompting them to form more lysosomes, secrete enzymes that enhance inflammation, and form abundant amounts of toxic free radicals. Lymphokines that enhance antibody production or activate other T cells are termed *helper factors* and are secreted only by helper T cells. Lymphokines and their effects on target cells are summarized in Table 22.4 on p. 697.

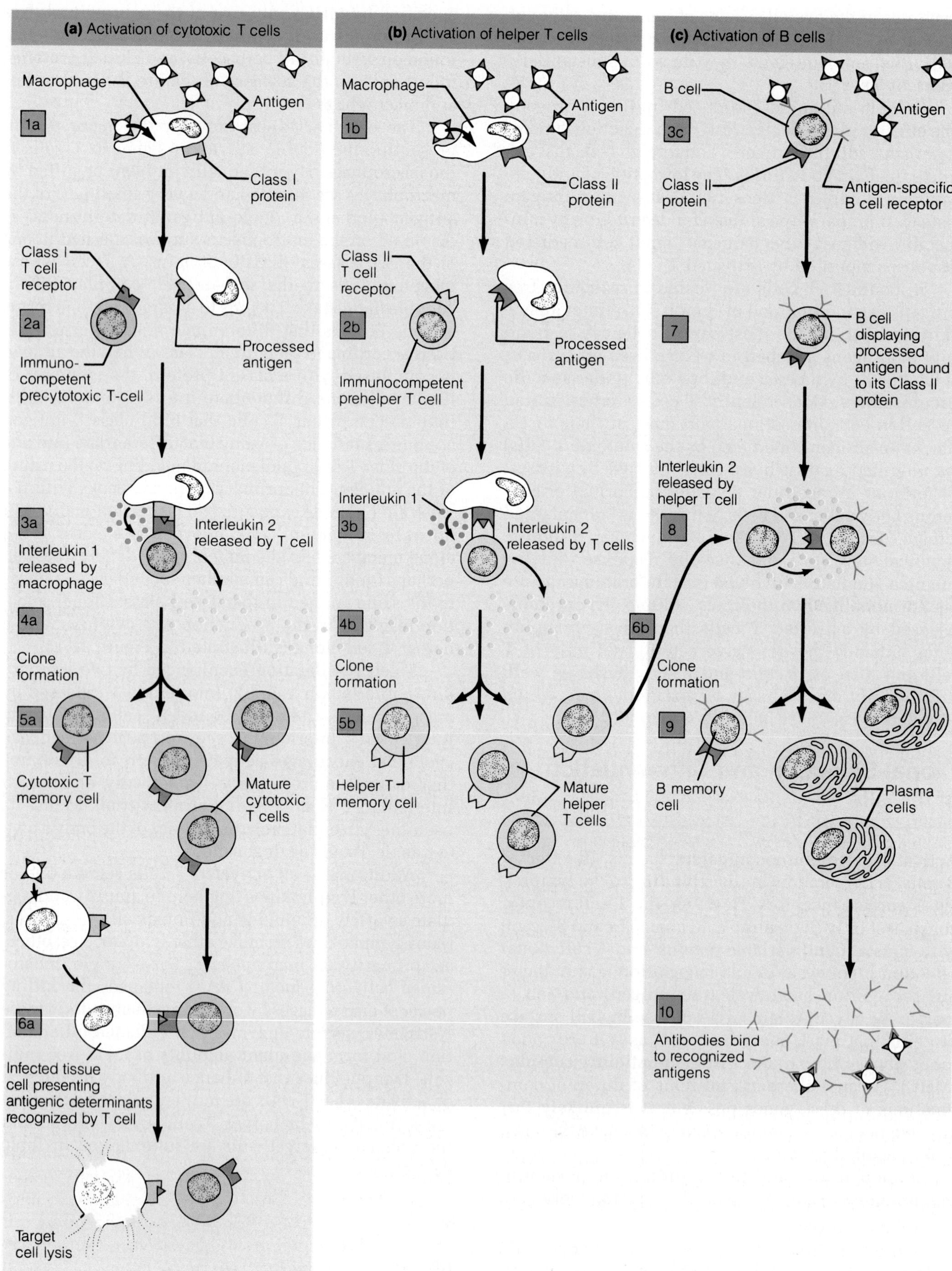
(a) Activation of cytotoxic T cells
Macrophage
Antigen
1a
Class I protein
Class I T cell receptor
2a
Processed antigen
Immuno-competent precytotoxic T-cell
3a
Interleukin 2 released by T cell
Interleukin 1 released by macrophage
4a
Clone formation
5a
Cytotoxic T memory cell
Mature cytotoxic T cells
6a
Infected tissue cell presenting antigenic determinants recognized by T cell
Target cell lysis
(b) Activation of helper T cells
Macrophage
Antigen
1b
Class II protein
Class II T cell receptor
2b
Processed antigen
Immunocompetent prehelper T cell
Interleukin 1
3b
Interleukin 2 released by T cells
4b
Clone formation
5b
Helper T memory cell
Mature helper T cells
6b
(c) Activation of B cells
B cell
Antigen
3c
Class II protein
Antigen-specific B cell receptor
7
B cell displaying processed antigen bound to its Class II protein
Interleukin 2 released by helper T cell
8
Clone formation
9
B memory cell
Plasma cells
10
Antibodies bind to recognized antigens

Figure 22.13 Schematic flowchart depicting theoretical events of clonal selection of (a) cytotoxic T cells, (b) helper T cells, and (c) B cells. (**1a and b**) An antigen is phagocytized by a macrophage and is digested. (**2a and b**) The antigen is then presented in partially digested form on the macrophage's surface, complexed to a class I or II MHC-encoded protein (a self-protein). (**3a and b**) A T cell whose antigen receptor "fits" and binds with the antigen–self-protein complex is selected for propagation by the complex. (**3a**) T cells that bind to antigens complexed with class I proteins will become cytotoxic T cells. (**3b**) Those T cells bound to antigens complexed with class II proteins will become helper T cells (in this example). (**3c**) In a similar manner, free antigen "selects" a B cell when it is able to bind to the B cell's surface receptors.

While the T cell is bound to the antigen–self-protein on the macrophage's surface, the macrophage releases interleukin 1. (**4a and b**) Interleukin 1 stimulates the T cell to proliferate and form a clone of like cells. The T cell in turn releases interleukin 2 as it differentiates and becomes mature, which stimulates other T cells into more vigorous proliferative activity. (The proliferative activity continues as long as the T cells are stimulated by antigen-presenting macrophages.)

(**5 a and b**) Mature T cells can then carry out their functional roles. Some remain circulating in the blood and lymph as memory T cells. (**6a**) If it is a cytotoxic T cell, it can bind to an infected body cell that presents the recognized antigen and cause its lysis. (**6b**) If it is a mature helper T cell, it also functions to stimulate the proliferation of antigen-activated B cells.

(**7**) After a B cell binds to antigen, it engulfs the antigen–receptor complexes and then processes the antigen. It presents a bit of the antigen on its surface in combination with a class II protein. (**8**) The mature helper T cell can then bind to the antigen–self-protein complex on the B cell and release interleukin 2, which stimulates the B cell to divide and differentiate into (**9**) memory cells or effector plasma cells. (**10**) Mature plasma cells release antibodies into the circulation that bind to recognized antigens and mark them for destruction.

Specific T Cell Roles

Cytotoxic T Cells

Cytotoxic T cells, also called **killer** or **cytolytic T cells,** are the only T cell population that can directly attack and lyse cells. Activated cytotoxic T cells roam the body, circulating in and out of the blood and lymph and through lymphatic organs in search of body cells displaying antigens to which they have been sensitized. Although the main target of the cytotoxic cells is virus-infected cells, they also mount the attack against tissue cells infected by certain intracellular bacteria (such as the tubercle bacillus) or parasites and against cancer cells and foreign human cells introduced into the body by blood transfusions or organ transplants. The continuous body search conducted by cytotoxic T cells for antigens of cancerous body cells is called *immune surveillance.*

Before the onslaught can begin, a double recognition must occur once more: The cytotoxic cell must identify both a class I self-protein and the foreign antigen displayed in combination. Remember, all body cells display class I antigens, so all infected or abnormal body cells can be destroyed by these T cells as long as the appropriate antigenic stimulus is also present. The attack on foreign human cells is more difficult to explain because *all* of the antigens are nonself; however, the cytotoxic cells apparently "see" the foreign class I antigens as a complex of a class I protein bound to an antigen.

For a long time, the mechanism of the cytotoxic cell's *lethal hit,* which leads to cell lysis, was unclear. Recently, it was revealed that the following events occur and account for at least some cases of lysis induced by cytotoxic cells: (1) The cytotoxic T cell binds tightly to the target cell and, during this period, a lymphotoxic chemical, named *perforin,* is released from its granules and becomes inserted into the plasma membrane of the target cell (Figure 22.14). (2) The cytotoxic cell then detaches and continues the search for another prey. Sometime later, the perforin molecules in the target-cell membrane begin to polymerize, forming cylindrical or porelike lesions very much like those produced by complement activation. This polymerization is dependent on adequate levels of Ca^{2+}. Consequently, the target cell becomes leaky and within 2 hours is dead.

Besides its direct role in cell lysis, the cytotoxic T cell is also the source of many lymphokines, particularly those that activate macrophages (MIF) and inhibit their migration (MIF); interferon; transfer factor, which causes nonsensitized lymphocytes in the area to take on the characteristics of the activated cytotoxic cells; and many others (see Table 22.4).

Helper T Cells

Helper T cells are regulatory cells that do not participate in direct cell killing. Once primed by macrophage presentation of antigen, their major function is to chemically or directly stimulate the proliferation of other T cells and of B cells that have already become bound to antigen. In fact, without the "director" helper T cells, there is no immune response. After an antigen–receptor complex has been endocytosed by a B cell, part of the antigen is displayed (in a macrophage-like manner) on the B cell surface in combination with

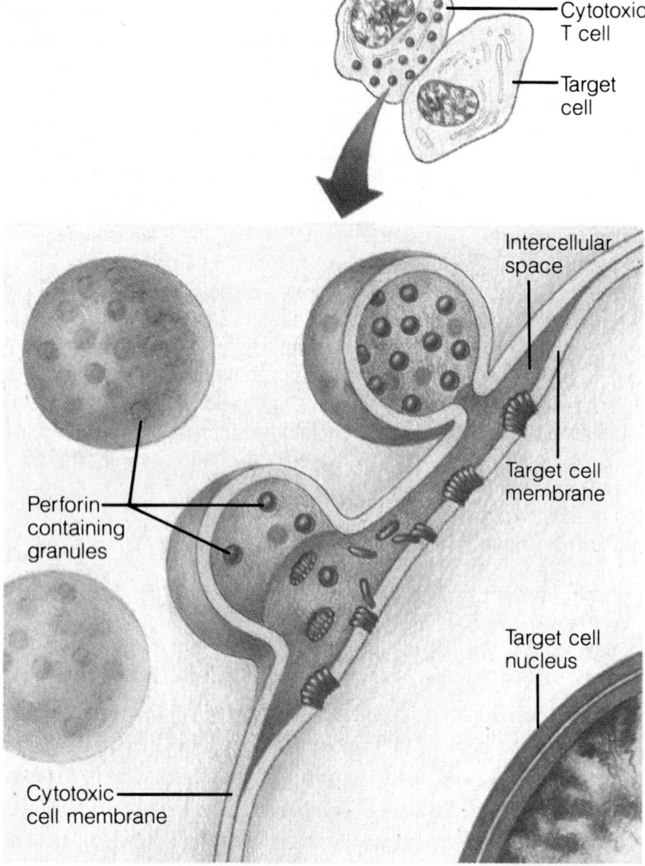

Figure 22.14 Proposed mechanism of target cell lysis by cytotoxic T cells. The initial event is the tight binding of the cytotoxic cell to the target cell. During this binding period, granules within the cytotoxic cell fuse with its plasma membrane and release their perforin contents by exocytosis. The perforin molecules insert into the target cell membrane and, in the presence of ionic calcium, they polymerize, forming cylindrical holes that allow free exchange of ions and water. Target cell lysis occurs long after the cytotoxic cell has detached.

a class II protein (Figure 22.13c). When a helper T cell capable of recognizing the antigen binds, it releases interleukin 2, which stimulates the B cell to divide and differentiate. The B cell division process continues as long as it is stimulated by the helper T cell. Thus, the helper T cells function to help unleash the protective potential of the B cells. Additionally, interleukin 2, released by the helper cells, also prompts the proliferation and differentiation of other activated helper and cytotoxic T cells.

Suppressor T Cells

Like the helper cells, **suppressor T cells** are regulatory cells. However, their action is to dampen the activity of both T cells and B cells by releasing lymphokines that suppress their activity. Hence, their function is to temper the normal immune response. As the antigenic stimulus wanes, suppressor T cell activity increases, bringing an end to the immune responses that are no longer needed. The suppressor cells are also important in controlling excessive antibody production and in preventing autoimmune reactions. It is thought that exceptionally strong immune responses and certain types of immunodeficiencies may result from deficits in suppressor T cell activity.

Delayed Hypersensitivity T Cells

The **delayed hypersensitivity T cells** are effector cells that play a major role in cell-mediated allergies and long-term, or chronic, inflammations. They are activated not only by macrophages in lymphatic tissues, but also by the Langerhans' cells of the skin. During subsequent meetings with the same antigen, they release a deluge of lymphokines such as MAF, MIF, and chemotactic factors that enhance migration of inflammatory cells into the area and promote an intense inflammatory response.

Organ Transplants and Prevention of Rejection

Organ transplants, a viable treatment option for many patients with end-stage cardiac or renal disease, have been done with mixed success for over 30 years. The breadth of immune system control presents a particular problem when the goal is to provide a person who has a failing heart or kidneys with functional organs from a living or recently deceased donor. Essentially, there are four major varieties of grafts:

1. **Autografts** are tissue grafts taken from one body site and transplanted in another on the same person.
2. **Isografts** are grafts donated by genetically identical individuals, the only example being identical twins.
3. **Allografts** are grafts transplanted from individuals that are not genetically identical but belong to the same species.
4. **Xenografts** are grafts taken from another animal species, such as the transplantation of a baboon heart into a human being.

The success of a transplant declines with the dissimilarity of the tissue because cytotoxic T cells will act vigorously to destroy foreign tissue that is in the body. Although cytotoxic T cells represent the major threat, macrophages are important in rejection phenomena as well. Given an adequate blood supply and no infection, autografts and isografts are the ideal donor tissues and are always successful because the MHC-

Table 22.4 Summary of Functions of Cells and Molecules Involved in the Immune Response

Element	Function in immune response
B cell	Lymphocyte that resides in lymph nodes, spleen or other lymphoid tissues, where it is induced to replicate by antigen binding and helper T cell interactions; its progeny (clone members) form memory cells or plasma cells
Plasma cell	Antibody-producing "machine"; produces huge numbers of antibodies (immunoglobulins) with the same antigen specificity; represents further specialization of B cell clone descendants.
Helper T cell	A regulatory T cell that binds with a specific antigen presented by a macrophage; upon circulating into spleen and lymph nodes, it stimulates production of other cells (killer T cells and B cells) to help fight invader; acts both directly and indirectly by releasing lymphokines
Cytotoxic T cell	Also called cytolytic or killer T cell; activated by antigen presented by a macrophage; recruited and activity enhanced by helper T cells; its specialty is killing virus-invaded body cells and cancer cells; it is involved in rejection of foreign tissue grafts
Suppressor T cell	Activated by free antigen or antigen presented by a macrophage; slows or stops activity of B and T cells once infection (or onslaught by foreign cells) has been conquered
Memory cell	Descendant of activated B cell or any class of T cell; generated during initial immune response (primary response); may exist in body for years after, enabling it to respond quickly and efficiently to subsequent infections or meetings with same antigen
Macrophage	Engulfs and digests antigens that it encounters and presents parts of them on its plasma membrane for recognition by T cells bearing receptors for same antigen; this function, antigen presentation, is essential for normal cell-mediated responses; also releases chemicals that activate T cells and prevent viral multiplication
Antibody (immunoglobulin)	Protein produced by B cell or by plasma cell; antibodies produced by plasma cells are released into body fluids (blood, lymph, saliva, mucus, etc.), where they attach to antigens, causing neutralization, precipitation, or agglutination, which "mark" the antigens for destruction by phagocytes or complement
Lymphokines	Chemicals released by sensitized T cells: • *Migration inhibitory factor (MIF)* inhibits macrophage migration and keeps them in the area of antigen deposition • *Macrophage-activating factor (MAF)* activates macrophages to become killer macrophages (MIF and MAF may be the same chemical) • *Interleukin 2* stimulates T cells to proliferate • *Helper factors* enhance antibody formation by plasma cells and stimulate other T cell activities • *Suppressor factors* suppress antibody formation or immune responses mediated by T cells • *Chemotactic factors* attract leukocytes (neutrophils, eosinophils, and basophils) to inflamed area • *Lymphotoxin (LT),* a cell toxin (perforin), causes cell lysis • *Transfer factor,* released by cytotoxic T cells, mobilizes nonsensitized lymphocytes, causing them to take on same characteristics as activated cytotoxic T cells • *Gamma interferon* renders tissue cells resistant to viral infection; also released by macrophages
Monokines	Chemicals released by activated macrophages: • *Interleukin 1* stimulates T cells to proliferate and causes fever (believed to be the pyrogen that resets thermostat of hypothalamus) • *Interferon* helps protect virus-invaded tissue cells by preventing replication of virus particles
Complement	Group of blood-borne proteins activated after binding to antibody-covered antigens; causes lysis of microorganism and enhances inflammatory response

encoded proteins are identical. Xenografts are essentially never successful in humans. Consequently, the problematic graft type, and the type that is also most frequently used, is the allograft, which is usually obtained from a cadaver.

Several steps must be taken before an allograft is even attempted. First, the donor and recipient ABO and other blood-group antigens must be determined. They must match to prevent rejection because of attacks on the blood-group antigens, which are also present on most body cells. Second, the recipient's serum must be checked for the presence of antibodies that might react to the donor's MHC-encoded proteins; this is a possibility if the recipient has been previously exposed to the donor's cells via blood transfusion, graft, or other avenues. If both these tests are compatible, the recipient and donor tissues are typed to determine the extent to which their MHC antigens match. Because the MHC genes are exhibited in at least 80 different forms, variation among human tissues is tremendous. Thus, good tissue matches between unrelated individuals are hard to obtain. However, the donor and recipient tissues are matched as closely as possible (at least a 75% match is essential).

Following surgery the patient is treated with immunosuppressive therapy involving one or more of the following: (1) corticosteroid drugs, such as prednisone, to suppress inflammation; (2) cytotoxic drugs, such as azathioprine; (3) radiation (X-ray) therapy; (4) antilymphocyte globulins; and (5) cyclosporine. All these drugs have severe side effects, but cyclosporine has markedly improved the survival of kidney, liver, heart, and heart–lung transplant patients. Approved in 1983, cyclosporine specifically inhibits interleukin 2 production and suppresses helper and cytotoxic T cells without affecting antibody production by presensitized lymphocytes or impairing marrow function.

The major problem with immunosuppressive therapy is that the patient's immune system cannot protect the body against other foreign agents. Explosive bacterial and viral infection still remains the most frequent cause of death in immunosuppressed patients. The key to successful graft survival is to provide enough immunosuppression to prevent graft rejection, but not enough to be toxic, and to use antibiotics to keep infection under control.

Although most instances of graft rejection are caused by cell-mediated immune responses, early acute rejection crises may result from surgical complications such as vascular obstructions leading to graft ischemia or from the onslaught of already circulating antibodies. Late rejection crises, occurring after a well-matched graft has remained intact for a considerable period, may occur because of uncontrolled bursts of antibody production, but the reason for this is unknown.

Homeostatic Imbalances of Immunity

Under certain circumstances, the immune system is depressed or exhibits activities that damage the body itself. Most such problems can be classified as immunodeficiencies, hypersensitivities, immunological tolerances, or autoimmune diseases.

Immunodeficiencies

The **immunodeficiencies** comprise a wide spectrum of congenital and acquired abnormalities in the production or function of immunocompetent cells, phagocytes, or complement. The congenital immunodeficiencies are described here. AIDS (acquired immunodeficiency syndrome) is spotlighted in the box on pp. 700 and 701.

Congenital thymic aplasia (DiGeorge's syndrome) is a devastating congenital disorder in which the thymus fails to develop. Because T cells mediate both arms of the immune response and enhance nonspecific defenses, afflicted children have virtually no protection against many pathogens and even common infections are lethal to them. Transplants of fetal thymic tissue help such children.

In *severe combined immunodeficiency disease (SCID)*, there is a marked depletion of both B and T cells, and the victims become deathly ill and wasted by infections that begin at about 3 months of age. Like DiGeorge's syndrome, SCID is fatal if untreated but some affected children are kept alive and healthy by bone marrow transplants, which provide normal stem cells. In the past, bone marrow transplants could be done only if the marrow donor was an identical twin; otherwise, the transplanted T cells would mount a potentially fatal attack on the recipient's tissues—the so-called *graft versus host (GVH) response*. New techniques use monoclonal antibodies to remove mature T cells, leaving only the marrow stem cells. This *makes* the donor's bone marrow more compatible with the recipient's even if their tissue types are substantially different. The stem cells are then transplanted to the recipient, where they set up a whole new immune system.

Hypersensitivities

At first, the immune response was thought to be purely protective; however, it was not long before its dangerous potentials were discovered. *Allergy* ("altered reaction") and *hypersensitivity* are synonyms for an antigen-induced state that results in abnormally intense immune responses to an antigen. In an allergic reaction, the immune system itself causes tissue damage as it fights off an antigen "threat" that is normally harmless. The term *allergen* is sometimes used to distinguish this type of antigen from that producing an essentially protective immune response.

There are different types of hypersensitivity reactions, which are distinguished by (1) their time course, and (2) whether antibodies or T cells are the principal immune elements involved. Allergies mediated by antibodies are referred to as *immediate hypersensitivities;* the most important cell-mediated allergic state is *delayed hypersensitivity.*

Immediate Hypersensitivities

The **immediate hypersensitivities** include anaphylaxis and cytotoxic and immune-complex hypersensitivities. All of these involve antibodies; the cytotoxic and immune-complex forms involve complement as well. These allergic reactions begin within minutes to hours after exposure to the allergen and can be passively transferred from a hypersensitive person to a nonallergic one via blood transfusion or in serum. However, the sensitivity is temporary and ends when the transferred antibodies are degraded in the body.

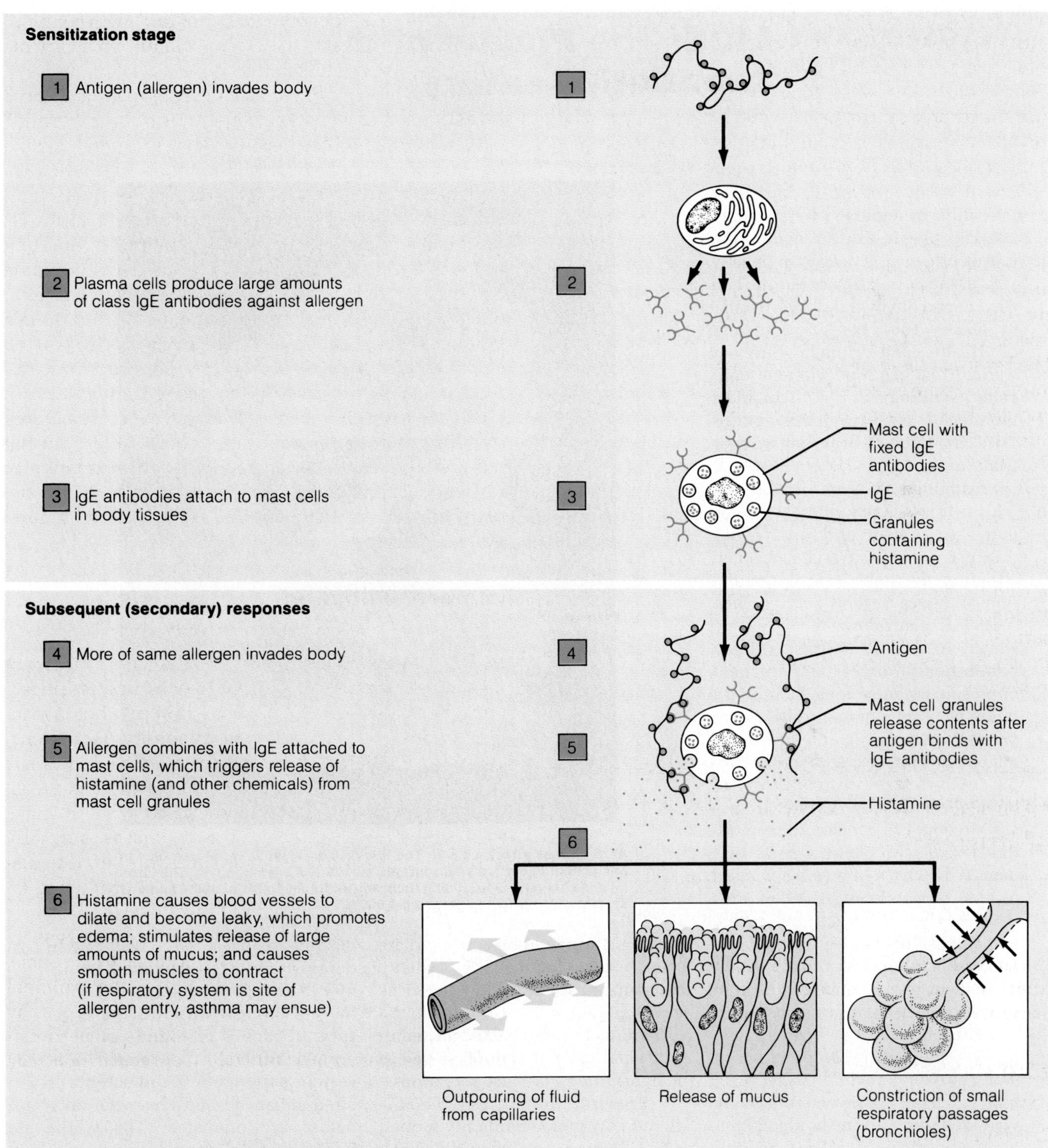
Sensitization stage
1 Antigen (allergen) invades body
2 Plasma cells produce large amounts of class IgE antibodies against allergen
3 IgE antibodies attach to mast cells in body tissues
Mast cell with fixed IgE antibodies
IgE
Granules containing histamine
Subsequent (secondary) responses
4 More of same allergen invades body
Antigen
5 Allergen combines with IgE attached to mast cells, which triggers release of histamine (and other chemicals) from mast cell granules
Mast cell granules release contents after antigen binds with IgE antibodies
Histamine
6 Histamine causes blood vessels to dilate and become leaky, which promotes edema; stimulates release of large amounts of mucus; and causes smooth muscles to contract (if respiratory system is site of allergen entry, asthma may ensue)
Outpouring of fluid from capillaries
Release of mucus
Constriction of small respiratory passages (bronchioles)

Figure 22.15 Mechanism of an acute allergic (immediate hypersensitivity) response.

Anaphylaxis. *Anaphylaxis* (a″-nuh-fih-lak′-sis), literally "against protection," is triggered when allergen molecules cross-link to IgE antibodies attached to mast cells or basophils. Basophils circulate in the bloodstream; mast cells are found throughout the body but are particularly abundant in connective tissues associated with the skin, respiratory system, and blood vessels. Binding of the allergen to the IgE antibodies stimulates the mast cells and basophils to degranulate, that is, to release the contents of their secretory granules into the extracellular space (Figure 22.15). The chemicals released include histamine, heparin, platelet-activating factors, serotonin, and chemotactic fac-

A CLOSER LOOK AIDS: The Plague of the Twentieth Century

In October 1347, several ships made port in Sicily and within days all of the sailors they carried were dead of bubonic plague. By the end of the fourteenth century, approximately 70% of the population of Europe had been wiped out by this "black death." In January 1987, the U.S. Secretary of Health and Human Services warned that *acquired immune deficiency syndrome* or (*AIDS*) might be the plague of the twentieth century. These are strong words, but are they true?

Although AIDS was first identified in this country in 1981 among homosexual men and intravenous drug users of both sexes, it had begun afflicting the heterosexual population of Africa several years earlier. AIDS is characterized by severe weight loss, night sweats, swollen lymph nodes, increasingly frequent opportunistic infections, including a rare type of pneumonia called *Pneumocystis pneumonia* and the bizarre malignancy *Kaposi's sarcoma* in people with no previous history of immune deficiency disease. (Kaposi's sarcoma is a cancer of the endothelial cells of blood vessels evidenced by purple-to-brown splotches of the skin.) Some AIDS victims develop slurred speech and severe dementia. The course of AIDS is inexorable, finally ending in complete debilitation and death from cancer or overwhelming infection.

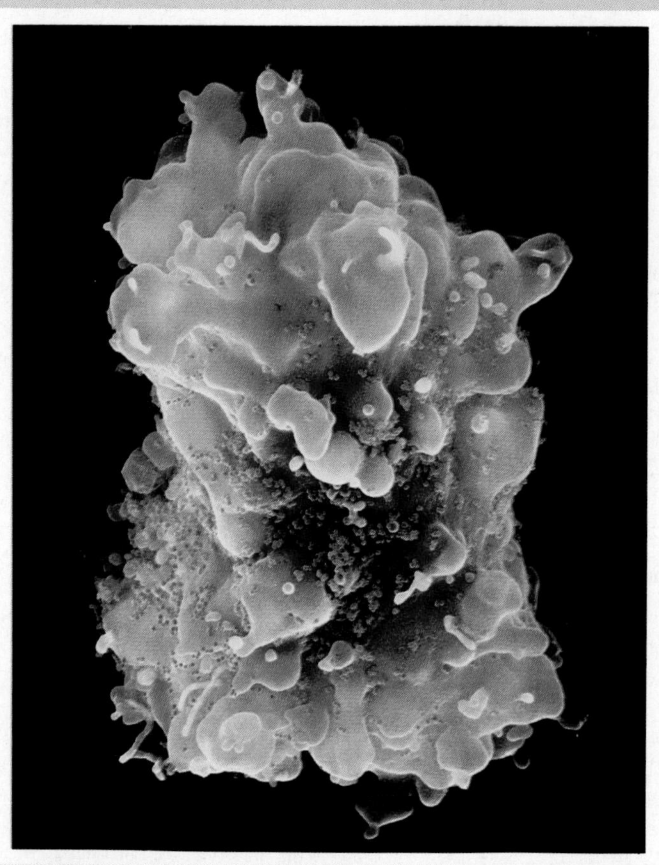

AIDS viruses attack a T cell: The HIV viruses, artificially colored blue in this scanning electron micrograph, surround a T lymphocyte. The virus invades helper T cells, killing them and seriously weakening the immune system of the infected person (14,000 X).

AIDS is caused by a virus transmitted principally in blood and semen (and possibly vaginal secretions). Most commonly, AIDS enters the body via blood transfusions or blood-contaminated needles and during intimate sexual contact in which the mucosa is torn (and bleeds) or where open lesions induced by sexually transmitted diseases allow the virus access to the blood. Although the AIDS virus has also been detected in saliva and tears, it is not believed to be transmitted via those secretions.

The virus, initially called HTLV-III, but now more commonly termed HIV (human immunodeficiency virus), specifically targets and destroys helper T cells, resulting in depression of cell-mediated immunity. Although antibody levels initially rise in response to viral exposure, in time a profound deficit of normal antibodies develops. Conversely, suppressor T cell activity is often enhanced and abnormal antibodies are produced. The whole immune system is turned topsy-turvy. The virus also invades the brain, per-

tors that attract eosinophils into the area. These induce the inflammatory response typical of anaphylaxis.

Anaphylactic reactions may be local or systemic. All local reactions are characterized by edema; other signs and symptoms are related to the specific tissues affected. Redness and hives are typical of skin reactions; excessive mucus and exudate production occurs when the nasal mucosa is involved, as in allergic rhinitis (hay fever). Allergen-initiated asthma attacks are local reactions involving severe bronchoconstriction. When the allergen is ingested (a drug or certain food chemicals), gastrointestinal discomfort (smooth muscle cramping, vomiting, or diarrhea) occurs. Antihistamines help relieve the distress of hay fever and gastrointestinal hypersensitivities because most of the symptoms are caused by histamine; antihistamines are

haps via infected macrophages, which accounts for the neurological symptoms (dementia) of some AIDS patients. Although there are exceptions, most AIDS victims die within a few months to six years after diagnosis.

The years since 1981 have witnessed an AIDS epidemic in the United States. By the end of 1987, over 53,000 cases had been diagnosed (29,000 had already died), and 1.5 million Americans were demonstrated to be carrying the anti-AIDS antibodies in their blood. The grim prophecy is that there will be 270,000 diagnosed cases nationwide by 1991 and 179,000 deaths will have occurred within the decade since the disease was first recognized in this country. Furthermore, there are probably 100 asymptomatic carriers of the virus for every diagnosed case because the virus may lurk undetectable in macrophages and the disease has a long incubation period (from a few months to ten years) between exposure and the appearance of clinically recognizable symptoms. Not only has the number of identified cases jumped exponentially in the at-risk populations, but victims have also begun to include people who do not belong to the original high-risk groups. Before reliable testing of donated blood was available, some people contracted the virus from blood transfusions. The virus can also be transmitted from an infected mother to her fetus. Although homosexual men still account for the bulk of cases where the virus is transmitted by sexual contact, more and more heterosexuals are contracting this disease.

Large city hospitals are seeing more AIDS patients daily, and statistics reporting the number of cases in inner-city ghettos, the corridors of poverty where intravenous drug use is the chief means of AIDS transmission, are alarming. One of every 61 babies born in New York City in December 1987, had anti-AIDS antibodies in its blood. Although some of these babies are exhibiting passive humoral immunity transmitted from the mother, many, if not most, have actually been infected with the AIDS virus. This shocking revelation prompted New York City to consider providing syringes to intravenous drug users in an attempt to suppress that means of transmission. Similar discussions have been initiated in other cities.

Though the disease is not spread by casual contact, fearful communities are demanding that AIDS-infected children be kept out of the schools, employers are beginning to demand AIDS testing before hiring, and insurance companies are denying insurance to those who will not submit to a test for the AIDS antibody. People have stopped donating blood, fearful that they might become infected (they cannot), so that blood supplies needed for surgical patients are deficient.

Although diagnostic tests to identify carriers of the AIDS virus are available, no cure has yet been found. Over 20 drugs, mostly antiviral drugs or immunity stimulants, are being developed to combat AIDS, and vaccines are undergoing clinical trials; but it is unlikely that an approved vaccine will be available before 1995. However, azidothymidine (AZT), the only drug currently approved for the treatment of AIDS in this country, has proven to be helpful in prolonging life by inhibiting replication of the virus. Although past medical practice has reserved experimental treatment only for those suffering from full-blown AIDS, treatment of extremely ill AIDS patients has yielded few positive returns. Currently, treatment focus is shifting to still-healthy, asymptomatic individuals with positive AIDS-antibody tests and ARC (AIDS-related complex), a condition marked by many of the same symptoms as AIDS that frequently leads to AIDS. Physicians are urging people to monitor their immune system status through the HIV-antibody test, which checks for the presence of anti-HIV antibodies, and many are ordering the P-24 antigen test, which measures serum levels of particular proteins thought to be present when the AIDS virus is active. Although federal guidelines approve AZT only in severe AIDS disease, physicians are beginning to prescribe the drug for those with ARC or positive antibody tests in an attempt to prevent the immunological devastation that precedes the AIDS diagnosis. Transfer factor, one of the lymphokines released by cytotoxic T cells, has shown promise in ARC patients, and an aerosol version of the antibiotic pentamidine, commonly used to treat the deadly *Pneumocystis* pneumonia, is being prescribed to prevent contraction of the pneumonia.

A number of nonapproved drugs are being prescribed or independently sought out and purchased in other countries by activist groups of AIDS patients. These include the antiviral drugs dextran sulfate and ribavirin, and Antabuse, a drug long used to treat alcoholism, which is similar to the experimental immune-boosting drug DTC. Some of these will prove useless and others will prove to have unforeseen harmful effects; but when life itself is at stake, this may be a small gamble to take. Perhaps, as urged in the media, the best defense is to practice "safe sex" by using condoms and knowing one's sexual partner; the alternative is abstinence.

relatively ineffective for treating other types of local anaphylactic reactions, in which other mediators are more important. There is little question that there is a familial predisposition for anaphylactic reactions, particularly in asthma and hay fever.

The systemic response, known as *anaphylactic shock*, is fairly rare. It typically occurs when an allergen—for example, bee venom or penicillin (which act as haptens)—enters the bloodstream of a susceptible person. The mechanism is essentially the same as that of a local response; however, in anaphylactic shock, mast cells and basophils throughout the body are enlisted and the flood of histamine and other chemical mediators produces an intense, widespread reaction. The bronchioles constrict, impairing breathing, and vasodilation and fluid loss from the bloodstream may precipitate circulatory collapse and physiological shock within minutes. Unless epinephrine is administered immediately to reverse these effects, the victim may die. The penicillin reaction is the most common example of drug-induced anaphylactic shock.

Cytotoxic Hypersensitivity. *Cytotoxic hypersensitivity* is equivalent to transfusion reactions, in which the antibody targets are foreign erythrocytes. Antigen–antibody complexes are formed following transfusion of mismatched ABO blood and agglutinate the foreign red blood cells. Binding of the antibodies to the red blood cells also results in complement activation and subsequent lysis of the target cells.

Immune-Complex Hypersensitivity. Antibodies and complement fixation mediate the *immune-complex hypersensitivities*. Activation of complement by the immune complexes is followed by an intensive inflammatory reaction in the area. Large numbers of neutrophils invade the area; as they release their lysosomal enzymes and huge amounts of oxidizing agents, local tissue damage occurs. Examples of localized responses include the pulmonary disorders known as farmer's lung, induced by inhaling moldy hay, and mushroom grower's lung, induced by inhaling mushroom spores. The systemic response, called *serum sickness*, may be triggered by certain drugs, streptococcal infection, persistent viral infection, and malaria and is characterized by fever, enlarged lymph nodes, skin rash, and widespread vascular lesions. This type of hypersensitivity is also implicated in certain autoimmune diseases. In such cases, the antibodies participating in the immune complexes are autoantibodies formed against the person's own tissues.

Delayed Hypersensitivities

As the name implies, **delayed hypersensitivity reactions** mediated by T cells are slower to appear (one to three days) than the immediate hypersensitivity reactions produced by antibodies. Their mechanism is basically that of a cell-mediated immune response. However, the usual reaction to most pathogens depends largely on cytotoxic T cells directed against specific antigens, whereas delayed hypersensitivity reactions involve both cytotoxic and delayed hypersensitivity T cells and depend to a greater extent on lymphokine-activated macrophages and nonspecific killing. Delayed hypersensitivity to antigens can be passively transferred from one person to another by whole blood transfusions.

The hallmark of delayed hypersensitivity is fibrin deposition at the inflamed site and infiltration by large numbers of macrophages and lymphocytes. Lymphokines released by T cells are the major mediators of these inflammatory responses, as opposed to histamine, which accounts for most symptoms of immediate hypersensitivity reactions. Thus, antihistamine drugs are not helpful against delayed hypersensitivity reactions. Corticosteroid drugs are used to provide relief.

The most familiar examples of delayed hypersensitivity reactions are those that follow skin contact with poison ivy, some heavy metals, and certain chemicals in cosmetics and deodorants. These types of agents act as haptens. After diffusing through the skin and attaching to self-proteins, they are perceived as foreign and the hypersensitivity response specifically known as *allergic contact dermatitis* occurs. Skin tests for tuberculosis depend on delayed hypersensitivity reactions. When the tubercle antigens are injected just under the skin, the area becomes red, swollen, and indurated (hardened) if the person has been previously sensitized to the antigen. The redness and swelling quickly disappear, but the induration may persist for days.

Delayed hypersensitivity reactions are effective against *facultative intracellular pathogens (FIP)*, which include malaria protozoans, salmonella bacteria, and some yeasts. FIP are readily phagocytized by macrophages, but they are not killed and may even multiply within their hosts unless the macrophages are activated to killer status by lymphokines. Thus, the enhanced release of lymphokines during delayed hypersensitivity reactions plays an important protective role.

Immunological Tolerances

Tolerance is defined as failure to mount an immune response, or unresponsiveness to a previously met antigen. Tolerance for self-antigens is normal and desirable; tolerance for foreign antigens is neither and may be dangerous. Although the immune system can respond to virtually every foreign antigen, it usually does not respond destructively to self-antigens. However, both arms of the immune system do produce antiself cells during the course of lymphocyte development, and B cell or T cell clones bearing receptors against self-antigens are suppressed or destroyed during the formative stages of the immune system. As a result of this *clonal inactivation* and *clonal elimination*, the immune system normally develops **self-tolerance** to body molecules during fetal life and fails to react against them thereafter.

In certain instances, tolerance to foreign antigens also occurs. For example, tolerance is easily established in the fetus. If foreign tissues are introduced into the fetal body before self-recognition and self-tolerance have been accomplished, they will be accepted and recognized as self. However, if the antigenic stimulus happens after clonal elimination has occurred, the foreign tissues will be rejected. Immunological tolerance may also be induced in adults under certain conditions. Very large antigen doses and antigens with many similar, repeating determinants can blockade the B cell receptors because each antigen can bind to a different B cell receptor; thus, clustering of receptors is prevented. As a result, a B cell clone that can recognize the antigen is effectively suppressed. Perhaps more importantly, T cells also become unresponsive under the same conditions. Extremely low antigen doses can also induce T cell tolerance in some cases, but this mechanism is less well understood.

An unresponsive, or tolerant, state may also reflect the route by which the antigen enters the body. For example, antigens that enter the blood directly and bypass the regional lymph nodes are more often tolerated than those that have been processed by lymph nodes.

Autoimmune Diseases

On occasion, the immune system's ability to tolerate self-antigens while still recognizing and attacking foreign antigens fails. When this happens, the body produces antibodies and sensitized effector T cells against its own tissues, causing their destruction. This phenomenon is called **autoimmune disease.**

How does the state of self-tolerance break down? Although there are probably hundreds of triggers for autoimmune disease and some, no doubt, are mediated by genetic factors, most seem to reflect one or more of the following events:

1. Changes in the structure of self-antigens. Proteins previously met and tolerated by immunocompetent cells may suddenly be recognized as intruders if they are distorted or altered in some way. If self-antigens are complexed with haptens, for example, they may be recognized as foreign. *Autoimmune thrombocytopenia* (throm″-bō-sī-tō-pē′-nē-uh), a condition in which platelets are destroyed, can be induced in susceptible people by drugs such as aspirin and antihistamines, which become bound to the platelet membranes, thus producing "new" antigens.

Antibodies directed against foreign antigens may be damaged or partially degraded by enzymes released during the immune response. As a result, antigenic sites on the body's own antibody molecules may appear that are not recognized as self. This is thought to be the cause of some forms of rheumatoid arthritis, in which an initial inflammatory response to viruses or streptococcal bacteria in a joint cavity leads to degradation of IgG antibodies formed against the pathogen. The altered IgG is then attacked by IgM antibodies. The immune complexes formed by the binding of IgM to the altered IgG antibodies activates complement, leading to local tissue destruction. This type of autoimmunity is actually an example of immune-complex hypersensitivity. In such cases, the IgM antibodies are referred to as *rheumatoid* (roo′-muh-toyd) *factor.*

Self-antigens on tissue cells damaged by disease or infection may become vulnerable to attack as well. For example, viral infections can lead to autoimmune reactions and play an important role in multiple sclerosis and juvenile diabetes. A common result of gene mutation in a tissue cell is the production of entirely new self-antigens. If the new protein is bound to the external cell surface, it will be recognized as foreign.

2. Appearance of self-proteins in the circulation that have not been previously exposed to the immune system. Not all self-antigens are exposed to immune system scrutiny during fetal development; consequently, they escape being recognized as self. Such hidden antigens are found in sperm cells, the lens of the eye, nervous tissue, and the thyroid gland. For example, the large storage form of thyroid hormone from which thyroxine is cleaved is retained entirely within the thyroid follicles and is not usually released to the blood. However, in cases of thyroid trauma, thyroglobulin may enter the bloodstream and be identified as foreign. The result is *Hashimoto's thyroiditis,* caused by a cell-mediated immune attack on the thyroid gland. Similarly, developing sperm cells are shielded from the immune system by the blood–testis barrier and, because sperm cells are not formed until puberty, the immune system is not exposed to their antigens. If the testes become inflamed, however, sperm antigens may escape into the circulation; the testes then become a target of the immune response, which causes sterility.

3. Cross-reaction of antibodies produced against foreign antigens with self-antigens. Antibodies made against a foreign antigen sometimes cross-react with a self-antigen that has very similar determinants. Generally, the self-antigens are attacked less vigorously because the two sets of determinants are not completely identical; but if the pathogen causes recurrent infections, the damage to body tissues is enhanced. Recurring infections by streptococcal bacteria may lead to *rheumatic fever,* an autoimmune sequel that causes swollen joints, rashes, and damage to the kidneys and heart long after the bacterial infection has subsided. ■

Developmental Aspects of the Immune System

The development of immunity relies on an orderly development of lymphoid organs and cells. Because the development of lymphatic system organs is described in Chapter 21, we will focus on the cells of the immune system here. Stem cells of the immune system originate from mesoderm during very early embryonic development. Soon thereafter, blood vessels appear and the stem cells migrate to temporary depots such as the liver and spleen, where they proliferate and become specialized as the various leukocyte lines. Later in fetal development, the bone marrow becomes the predominant source of stem cells (hemocytoblasts), and it persists in this role into adult life. In late fetal life and shortly after birth, the young lymphocytes develop self-tolerance and immunocompetence and then populate the lymphoid tissue. Upon antigen challenge, the T and B populations in these lymphoid tissues complete their development to fully mature effector and regulatory cells.

The ability of the immune system to recognize foreign substances is genetically determined. However, there is little doubt that the nervous system plays an important role in both the control and the activity of the immune response, and this intriguing topic deserves more than a passing mention. The sheer power of the mind on the body as a whole and on the general state of health is amazing. Scientists have confirmed that a witch doctor can cause death simply by telling those believing in his powers that they are going to die! In recent years, the magazine publisher Norman Cousins has maintained that sometimes the mind can cure what the doctor cannot, and has written a book detailing how he overcame an often-fatal neurological disease by using humor as a therapeutic tool. How did he do it? Can we think or will ourselves well? The highly complex interactions of the brain and the immune system have been largely unexplored. However, studies of neuroimmunomodulation—a term coined to describe links between the brain and the immune system—are beginning to reveal some answers.

The observation that left-handed people are more likely than right-handed people to suffer immune system disorders led scientists to study the relative roles of the two cerebral hemispheres on immunity. The conclusions indicate that the left hemisphere most directly controls the immune system (through the T cells), but the right hemisphere enhances and modulates the response. Other investigations have shown that while the nervous and immune systems have unique "languages," they seem to share a few of the same "words," namely, some of the opiate neuropeptides. Like neurons, many cells involved with the immune response produce or have receptors for the opiate neuropeptides, which have long been known to influence mood and behavior. Now it seems that some of these same neuropeptides have a specific immune system function as well: They attract and rally macrophages. Many scientists are convinced that macrophages have receptors for these neuropeptides, which are released by pain-sensing neurons and by lymphocytes. Why these cells of the immune system should release or react to chemicals used by the nervous system to deal with pain is not known, but it seems that they may be an important link in communication between the brain and the body.

Hormones such as corticosteroids and epinephrine may also provide chemical links between the two systems. For example, T cell responsiveness is depressed during stress, and severe, incapacitating depression unquestionably weakens the immune system, increasing susceptibility to physical illness.

The immune system normally serves us very well until late in life, when its efficiency begins to wane and the ability to fight infection declines. Old age is also accompanied by greater susceptibility to both autoimmune and immune deficiency diseases. The greater incidence of cancer in the elderly is assumed to be another example of the progressive failure of the immune system. Just why this decline of the immune system occurs is not really known, but "genetic aging" and its consequences are probably partly at fault. The immune system provides remarkable defenses against disease: Not only are the defenses amazing in their diversity, but also they are very precisely regulated both by cellular interactions and by a flood of chemicals. The nonspecific defensive mechanisms exhibit a different arsenal for body defense, more simple perhaps and more easily understood to be sure. However, the specific and nonspecific defenses are tightly interlocked, each providing what the other cannot and amplifying each other's effects. As summarized in Figure 22.16, all body systems benefit from the defensive mechanisms considered in this chapter.

Related Clinical Terms

Atopic dermatitis (ā-tah′-pik der″-muh-tī′-tis) Also called *eczema* (eg′-zih-muh). A type of anaphylactic hypersensitivity response that results in "weeping" skin lesions and intense itching. Onset is within the first five years of life in 90% of cases. The allergen is uncertain, but familial effects are strong.

Immunopathology Disease of the immune system.

Immunization The process of rendering a subject immune (by vaccination or injection of antiserum) or of becoming immune.

Systemic lupus erythematosus (uh-rih″-thuh-muh-tō′-sis) **(SLE)** A systemic autoimmune disorder that occurs mainly in young females. Diagnosis is helped by finding antinuclear (anti-DNA) antibodies in the patient's serum. DNA–anti-DNA complexes localize in the kidneys (the capillary filters, or glomeruli), in blood vessels, and in the synovial membranes of joints, resulting in glomerulonephritis, vascular problems, and painful arthritis. Red skin lesions, particularly a "butterfly rash" (the sign of the red wolf, or *lupus*) on the face are common.

Immune system

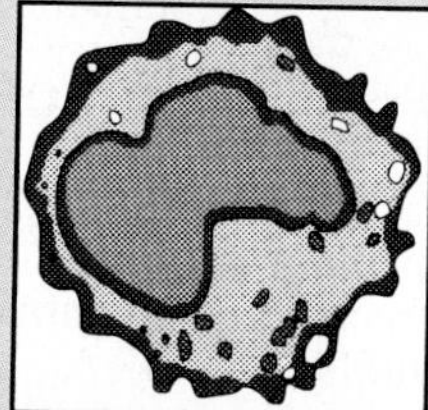

Integumentary system

▶ Keratinized epithelial cells provide mechanical barrier to pathogens; Langerhans' cells and dermal macrophages act as antigen presenters

◀ Enhances protective roles of skin by providing specific defense mechanisms

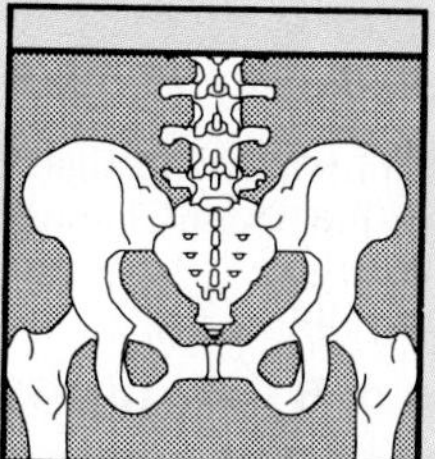

Skeletal system

▶ Lymphocytes and macrophages arise from stem cells in bone marrow

◀ Protects from specific pathogens, cancer cells, bacterial toxins

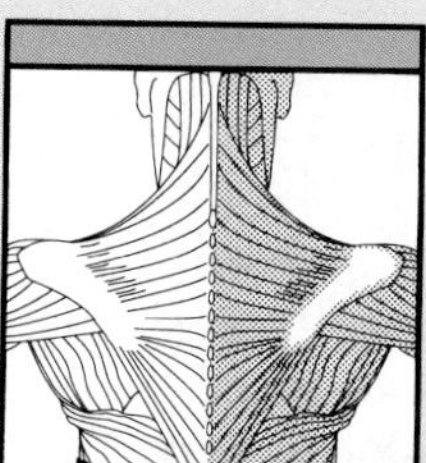

Muscular system

▶ Heat from muscle activity initiates fever-like effects

◀ Protects from specific pathogens, cancer cells, bacterial toxins

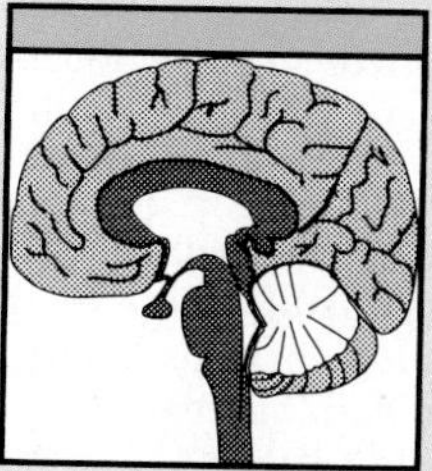

Nervous system

▶ Opiate neuropeptides may influence immune function; helps modulate and regulate immune response; emotional state influences T cell activity

◀ Provides protection of peripheral nervous system structures only as immune elements do not enter CNS

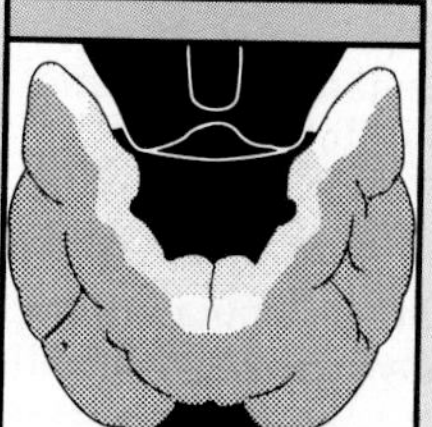

Endocrine system

▶ Hormones released during stress (epinephrine, glucocorticoids) depress immune activity; thymic hormones "program" T lymphocytes

◀ Protects from specific pathogens, cancer cells, bacterial toxins

Cardiovascular system

▶ Blood provides route for circulation of immune elements, complement, etc.; spleen is lymphoid organ as well as blood "filter"

◀ Protects from specific pathogens, cancer cells, bacterial toxins

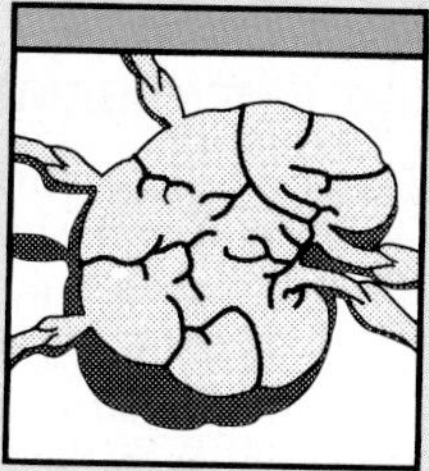

Lymphatic system

◀ Lymphoid organs house lymphocytes, macrophages and are site for antigen challenge, lymphocyte proliferation; lymph circulates antigens

▶ Protects from specific pathogens, cancer cells, bacterial toxins

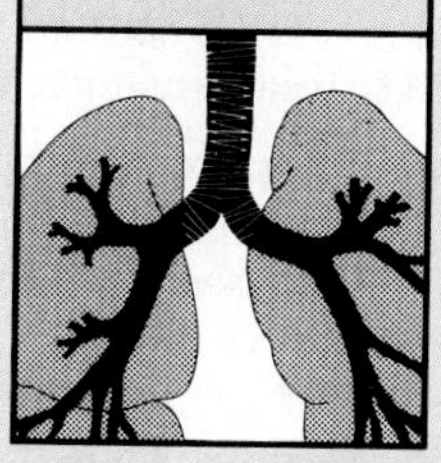

Respiratory system

◀ Provides oxygen needed by immune cells; eliminates carbon dioxide; pharynx houses tonsils

▶ Lymphocytes "seed" tonsils and secrete IgA to protect respiratory mucosa; asthma is allergic response

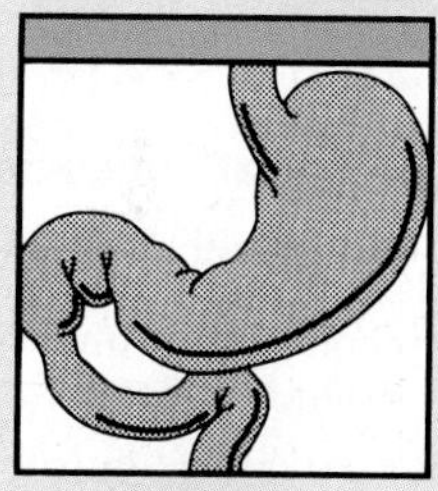

Digestive system

◀ Provides nutrients to immune cells; houses GALT; acid secretions inhibit entry of pathogens to blood

▶ Lymphocytes "seed" intestinal wall GALT; protects from specific pathogens, cancer cells

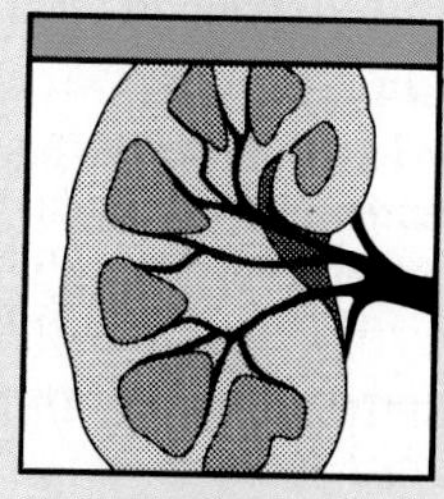

Urinary system

◀ Eliminates nitrogenous wastes; maintains homeostatic water/acid-base/electrolyte balance for immune cell functioning

▶ Protects from specific pathogens, cancer cells

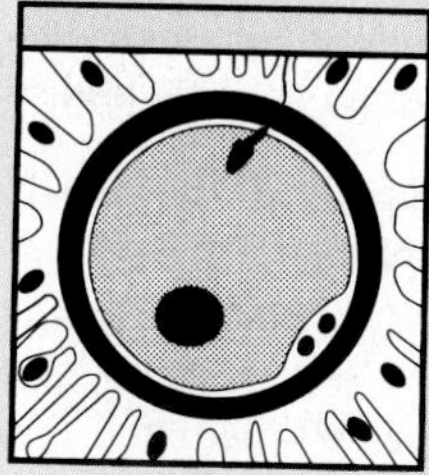

Reproductive system

◀ Reproductive hormones may influence immune functioning; vaginal secretion acidity prevents bacterial multiplication

▶ Protects from specific pathogens, cancer cells

Figure 22.16 Homeostatic interrelationships between the immune system and other body systems.

Chapter Summary

NONSPECIFIC BODY DEFENSES (pp. 674–681)

SURFACE MEMBRANE BARRIERS (pp. 674–675)

1. The role of skin and mucous membranes is to prevent pathogens from entering the body. Protective membranes line all body cavities and organs exposed to the exterior.

2. Surface membranes provide mechanical barriers to pathogens. Some have structural modifications and produce secretions that enhance their defensive effects: The skin's acidity, lysozyme, mucus, keratin, and ciliated cells are examples.

NONSPECIFIC CELLULAR AND CHEMICAL DEFENSES (pp. 675–681)

Phagocytosis (p. 675)

1. Phagocytes (macrophages and neutrophils) engulf and destroy pathogens that breach epithelial barriers. This process is enhanced when the pathogen's surface is altered by the attachment of antibodies and/or complement. Cell killing is enhanced by the respiratory burst.

Natural Killer Cells (p. 676)

2. Natural killer cells are nonimmune cells that act nonspecifically to lyse virus-infected and malignant cells.

Tissue Response to Injury: Inflammation (pp. 676–678)

3. The inflammatory response prevents the spread of injurious agents, disposes of pathogens and dead tissue cells, and promotes healing. Exudate is formed; protective leukocytes enter the area; the area is walled off by fibrin; and tissue repair occurs.

4. The cardinal signs of inflammation are swelling, redness, heat, and pain. These result from vasodilation and increased permeability of blood vessels induced by inflammatory chemicals.

Antimicrobial Substances (pp. 678–681)

5. When complement (a group of plasma proteins) becomes fixed on a foreign cell's membrane, lysis of the target cell occurs. Complement also enhances phagocytosis and the inflammatory and immune responses.

6. Interferon is a group of related proteins synthesized by virus-infected cells and certain immune cells that prevents viruses from multiplying in other body cells.

Fever (p. 681)

7. Fever enhances the body's fight against invading pathogens by increasing metabolism, which speeds up defensive actions and repair processes, and by prompting the liver and spleen to sequester iron, which is needed for bacterial multiplication.

SPECIFIC BODY DEFENSES: THE IMMUNE SYSTEM (pp. 681–704)

1. The immune system recognizes something as foreign and acts to immobilize, neutralize, or remove it. The immune response is antigen-specific and systemic and has memory.

CELLS OF THE IMMUNE SYSTEM: AN OVERVIEW (pp. 682–684)

1. Lymphocytes arise from the hemocytoblasts of the bone marrow. T cells develop immunocompetence in the thymus and confer cell-mediated immunity. B cells develop immunocompetence in the bone marrow and provide humoral immunity. Immunocompetent lymphocytes seed the lymphoid organs, where the antigen challenge occurs, and circulate between the blood, lymph, and lymphoid organs.

2. Immunocompetence is signaled by the appearance of antigen-specific receptors on the surfaces of the lymphocytes.

3. Macrophages arise from monocytes produced in bone marrow. They phagocytize pathogens and present antigenic determinants on their surfaces, for recognition by T cells.

ANTIGENS (p. 684)

1. Antigens are substances capable of generating an immune response.

Complete Antigens and Haptens (p. 684)

2. Complete antigens have both immunogenicity and reactivity. Incomplete antigens or haptens must combine with a body protein before becoming immunogenic.

Antigenic Determinants (p. 684)

3. Antigenic determinants are the portions of the antigen that are recognized as foreign. Most antigens have many such sites.

HUMORAL IMMUNE RESPONSE (pp. 684–692)

Clonal Selection and Differentiation of B Cells (p. 685)

1. Clonal selection and differentiation of B cells occur when antigens bind to their receptors, causing them to proliferate. Most of the clone members become plasma cells, which secrete antibodies. This is the primary immune response.

Immunological Memory (pp. 685–686)

2. Other clone members become memory B cells, capable of mounting a rapid attack against the same antigen, in subsequent encounters (secondary immune responses). The memory cells provide humoral immunological memory.

Active and Passive Humoral Immunity (p. 687)

3. Active humoral immunity is acquired during an infection or via vaccination. Passive immunity is conferred when a donor's antibodies are injected into the bloodstream, or when the mother's antibodies cross the placenta.

Antibodies (pp. 687–692)

4. The antibody monomer consists of four polypeptide chains, two heavy and two light, connected by disulfide bonds. Each chain has both a constant and a variable region. Constant regions determine antibody function and class. Variable regions enable the antibody to recognize its appropriate antigen.

5. Five classes of antibodies exist: IgA, IgG, IgM, IgD, and IgE. They differ structurally and functionally.

6. Antibody functions include complement fixation and antigen neutralization, precipitation, and agglutination.

7. Monoclonal antibodies are pure preparations of a single antibody type useful in diagnostic tests and cancer treatment. They are prepared by injecting an antigen into a laboratory animal, harvesting its B cells, and fusing them with myeloma cells.

CELL-MEDIATED IMMUNE RESPONSE (pp. 692–698)

Clonal Selection and Differentiation of T Cells (p. 693)

1. T cells are sensitized by binding simultaneously to an antigen and a major histocompatibility complex (MHC) protein displayed on the surface of a macrophage. Clonal selection occurs and the clone members differentiate into regulatory or effector T cells that mount the primary immune response. Certain of the clone members become memory T cells.

Specific T Cell Roles (pp. 695–696)

2. Cytotoxic T cells directly attack and lyse infected cells and cancer cells. Helper T cells release helper factors and interact directly with B cells and other T cells bound to antigen. Sup-

pressor T cells terminate normal immune responses by releasing suppressor factors that dampen the activity of helper T cells. Delayed hypersensitivity T cells release chemical mediators that enhance the inflammatory response.

3. The immune response is enhanced by interleukin 1, released by macrophages, and by lymphokines (interleukin 2, MIF, MAF, etc.), released by activated T cells.

Organ Transplants and Prevention of Rejection (pp. 696–698)

4. Grafts or foreign organ transplants will be rejected by cell-mediated responses unless the patient is immunosuppressed. Infections are major complications in such patients.

HOMEOSTATIC IMBALANCES OF IMMUNITY (pp. 698–703)

Immunodeficiencies (p. 698)

1. Immunodeficiency diseases include congenital thymic aplasia, severe combined immunodeficiency disease, and acquired immune deficiency syndrome (AIDS). Overwhelming infections cause death because the immune system is unable to combat them.

Hypersensitivities (pp. 698–702)

2. Hypersensitivity, or allergy, is an abnormally intense reaction to an allergen following the initial immune response. Immediate hypersensitivities mounted by antibodies include anaphylaxis and cytoxic and immune-complex hypersensitivities. Cell-mediated hypersensitivity is called delayed hypersensitivity.

Immunological Tolerances (pp. 702–703)

3. Tolerance occurs when the body does not react to an antigen. Self-tolerance results from clonal inactivation or clonal elimination during development of the immune system. Foreign antigen tolerance may reflect B cell blockade or excessive antigen concentrations.

Autoimmune Diseases (p. 703)

4. Autoimmune disease occurs when the body regards its own tissues as foreign and mounts an immune attack against them. Examples include rheumatoid arthritis and multiple sclerosis.

DEVELOPMENTAL ASPECTS OF THE IMMUNE SYSTEM (pp. 703–704)

1. Development of the immune response occurs around the time of birth. The ability of the immune system to recognize foreign substances is genetically determined.

2. The nervous system plays an important role in regulating immune responses, possibly through common mediators (neuropeptides). Depression impairs immune function.

3. With aging, the immune system becomes less responsive. The elderly more often suffer from immune deficiency, autoimmune diseases and cancer.

Review Questions

Multiple Choice/Matching

1. All of the following are considered nonspecific body defenses *except* (a) complement, (b) phagocytosis, (c) antibodies, (d) lysozyme, (e) inflammation.

2. The process by which neutrophils squeeze through capillary walls in response to inflammatory signals is called (a) diapedesis, (b) chemotaxis, (c) margination, (d) opsonization.

3. Antibodies released by plasma cells are involved in (a) humoral immunity, (b) immediate hypersensitivity reactions, (c) autoimmune disorders, (d) all of the above.

4. Which of the following antibodies can fix complement? (a) IgA, (b) IgD, (c) IgE, (d) IgG, (e) IgM.

5. Which antibody class is abundant in body secretions? (Use the choices from Question 4.)

6. Small molecules that must combine with large proteins to become immunogenic are called (a) complete antigens, (b) reagins, (c) idiotypes, (d) haptens.

7. Lymphocytes that develop immunocompetence in the thymus are (a) B lymphocytes, (b) T lymphocytes.

8. Cells that can directly lyse target cells include all of the following *except* (a) macrophages, (b) cytotoxic T cells, (c) helper T cells, (d) natural killer cells.

9. Which of the following is involved in the activation of a B cell? (a) antigen, (b) helper T cell, (c) lymphokine, (d) all of the above.

Short Answer Essay Questions

10. Besides acting as mechanical barriers, the skin epidermis and mucosae of the body have other attributes that contribute to their protective roles. Cite the common body locations and the importance of mucus, lysozyme, keratin, acid pH, and cilia.

11. Explain why attempts at phagocytosis are not always successful; note factors that increase the likelihood of success.

12. What is complement? How does it cause bacterial lysis? What are some of the other roles of complement?

13. Interferons are referred to as antiviral proteins. What stimulates their production, and how do they protect uninfected cells? What cells of the body are known to form interferons?

14. Differentiate between humoral and cell-mediated immunity.

15. Although the immune system has two arms, it has been said that "no T cells, no immunity." Explain.

16. Define immunocompetence. What event (or observation) signals that a B or a T cell has developed immunocompetence?

17. Describe the process of clonal selection of a helper T cell.

18. Differentiate between a primary and a secondary immune response. Which is more rapid and why?

19. Define antibody. Using an appropriately labeled diagram, describe the structure of an antibody monomer. Indicate and label: variable and constant regions, heavy and light chains.

20. What is the role of the variable regions of an antibody? Of the constant regions?

21. Name the five antibody classes and describe where each is most likely to be found in the body.

22. How do antibodies help defend the body?

23. Do vaccines produce active or passive humoral immunity? Explain your answer. Why is passive immunity less satisfactory?

24. Describe the specific roles of helper, suppressor, cytotoxic, and delayed hypersensitivity T cells in normal cell-mediated immunity.

25. Name several lymphokines and describe their role in the immune response.

26. Define hypersensitivity. List three types of hypersensitivity reactions. For each, note whether antibodies or T cells are involved and provide two examples.

27. What events can result in autoimmune disease?

28. Aging and declining efficiency of the immune system are thought to reflect what event?

Clinical Application Questions

29. Some people with a deficit of IgA exhibit recurrent paranasal sinus and respiratory tract infections. Explain these symptoms.

30. Explain the underlying mechanisms responsible for the cardinal signs of acute inflammation: heat, pain, redness, swelling.

CHAPTER 23

The Respiratory System

Chapter Outline and Student Objectives

Functional Anatomy of the Respiratory System (pp. 709–721)

1. Identify the organs forming the respiratory passageway(s) in descending order until the alveoli are reached. Distinguish between conducting and respiratory zone structures.
2. List and describe several protective mechanisms of the respiratory system.
3. Describe the makeup of the respiratory membrane, and relate structure to function.
4. Describe the structure and function of the lungs and pleural coverings.

Mechanics of Breathing (pp. 721–728)

5. Relate Boyle's law to the events of inspiration and expiration.
6. Explain the relative roles of the respiratory muscles and lung elasticity in effecting volume changes that cause airflow into and out of the lungs.
7. Explain the functional importance of the partial vacuum that exists in the intrapleural space.
8. Describe several physical factors that influence pulmonary ventilation.
9. Explain and compare the various lung volumes and capacities. Discuss types of information that can be gained from pulmonary function tests.
10. Define dead space.

Gas Exchanges in the Body (pp. 728–732)

11. Describe, in general terms, differences in composition of atmospheric and alveolar air, and attempt to explain these differences.
12. State Dalton's law of partial pressures and Henry's law. Relate each to events of external and internal respiration.

Transport of Respiratory Gases by the Blood (pp. 733–737)

13. Describe how oxygen is transported in the blood, and explain how oxygen loading and unloading is affected by temperature, pH, 2-3-DPG, and P_{CO_2}.
14. Describe carbon dioxide transport in the blood.

Control of Respiration (pp. 737–742)

15. Describe the neural controls of respiration.
16. Compare and contrast the influences of the following factors on respiratory rate and depth: lung reflexes, volition, emotions, arterial pH, and partial pressures of oxygen and carbon dioxide in arterial blood.

Respiratory Adjustments During Exercise and at High Altitudes (pp. 742–744)

17. Compare and contrast the hyperpnea of exercise with involuntary hyperventilation.
18. Describe the process and effects of acclimatization to high altitude.

Homeostatic Imbalances of the Respiratory System (pp. 744–745)

19. Compare the causes and consequences of chronic bronchitis, emphysema and lung cancer.

Developmental Aspects of the Respiratory System (pp. 745–746)

20. Follow briefly the development of the respiratory system in an embryo.
21. Note normal changes that occur in respiratory system functioning from infancy to old age.

Preview of Selected Key Terms

Respiratory system Organ system that includes the nose, pharynx, larynx, trachea, bronchi, and lungs and that carries out gas exchange.

Alveolus (al-vē′-uh-lus) (*alveol* = small cavity) One of the microscopic air sacs of the lungs.

Pleura (pler′-uh) The two-layered serous membrane that lines the thoracic cavity and covers the external surface of the lung.

Pulmonary ventilation Breathing; consists of inspiration and expiration.

Surfactant (ser-fak′-tent) Secretion produced by certain cells of the alveoli that reduces the surface tension of water molecules.

Vital capacity (VC) Total amount of exchangeable air.

Oxyhemoglobin (ok″-sē-hē′-muh-glō-bin) Oxygen-bound form of hemoglobin.

When John Donne, the seventeenth-century English poet and preacher, wrote "No man is an island, entire of itself," he was referring to the mind and spirit. Nonetheless, this classic statement applies to the workings of the body as well. Far from self-sustaining, the body is critically dependent on the external environment, both as a source for obtaining substances it needs to survive and as a catch basin into which it can cast its wastes.

The trillions of cells making up the body require a continuous supply of oxygen to carry out their vital functions. We cannot do without oxygen for even a little while, as we can without food or water. Further, as cells use oxygen, they give off carbon dioxide, a waste product the body must get rid of. The major function of the **respiratory system** is to fulfill these needs, that is, to supply the body with oxygen and dispose of carbon dioxide. To accomplish this function, at least four distinct events, collectively called **respiration**, have to happen:

1. **Pulmonary ventilation**. Air must be moved into and out of the lungs so that the gas in the air sacs (alveoli) of the lungs is continuously changed and refreshed. This is commonly called **ventilation**, or breathing.
2. **External respiration**. Gas exchange (oxygen loading and carbon dioxide unloading) between the blood and alveoli must occur.
3. **Respiratory gas transport**. Oxygen and carbon dioxide must be transported to and from the lungs and tissue cells of the body. This is accomplished by the cardiovascular system, which uses blood as the transporting fluid.
4. **Internal respiration**. Gas exchanges must be made between the blood and tissue cells.*

All of these processes are highlighted in the present chapter. Although only the first two processes are the special responsibility of the respiratory system, it cannot accomplish its primary goal of obtaining oxygen and eliminating carbon dioxide unless the third and fourth processes also occur. Thus, the respiratory and circulatory systems are inextricably coupled, and if either system fails, the body's cells begin to die from oxygen starvation.

*The actual use of oxygen and production of carbon dioxide by tissue cells, that is, *cellular respiration*, is the cornerstone of all energy-producing chemical reactions in the body. Cellular respiration, which occurs in all body cells, is discussed more appropriately in Chapter 25.

Functional Anatomy of the Respiratory System

The organs of the respiratory system include the nasal cavities, the pharynx, the larynx, the trachea, the bronchi and their smaller branches, and the lungs, which contain the terminal air sacs, or alveoli (Figure 23.1). Functionally, the respiratory system consists of respiratory and conducting zones. The **respiratory zone**, the actual site of gas exchange, is made up of the respiratory bronchioles, alveolar ducts, and alveoli. The **conducting zone**, sometimes called the dead air space, includes all other respiratory passageways, which serve as fairly rigid conduits to allow air to reach the gas exchange sites. The conducting zone organs also purify, humidify, and warm the incoming air. Thus, the air finally reaching the lungs has many fewer irritants than when it entered the system, and it resembles the warm, damp air of the tropics. The functions of the major organs of the respiratory system are summarized briefly in Table 23.1 on p. 720.

The Nose

The nose is the only externally visible part of the respiratory system. Unlike the often poetic references to other facial features, such as the eyes or lips, the nose is usually an irreverent target: We are urged to keep our nose to the grindstone and to keep it out of someone else's business. However, considering its important functions, the nose deserves more esteem. The nose (1) provides an airway for respiration, (2) moistens and warms entering air, (3) filters inspired air and cleanses it of foreign matter, (4) serves as a resonating chamber for speech, and (5) is the seat of the olfactory (smell) receptors.

The structures of the nose are divided into the external nose and the internal cavities for ease of consideration. The framework of the **external nose** is fashioned by the nasal and frontal bones superiorly (forming the bridge and root, respectively), the maxillary bones laterally, and movable plates of hyaline cartilage (the lateral and alar cartilages) inferiorly (Figure

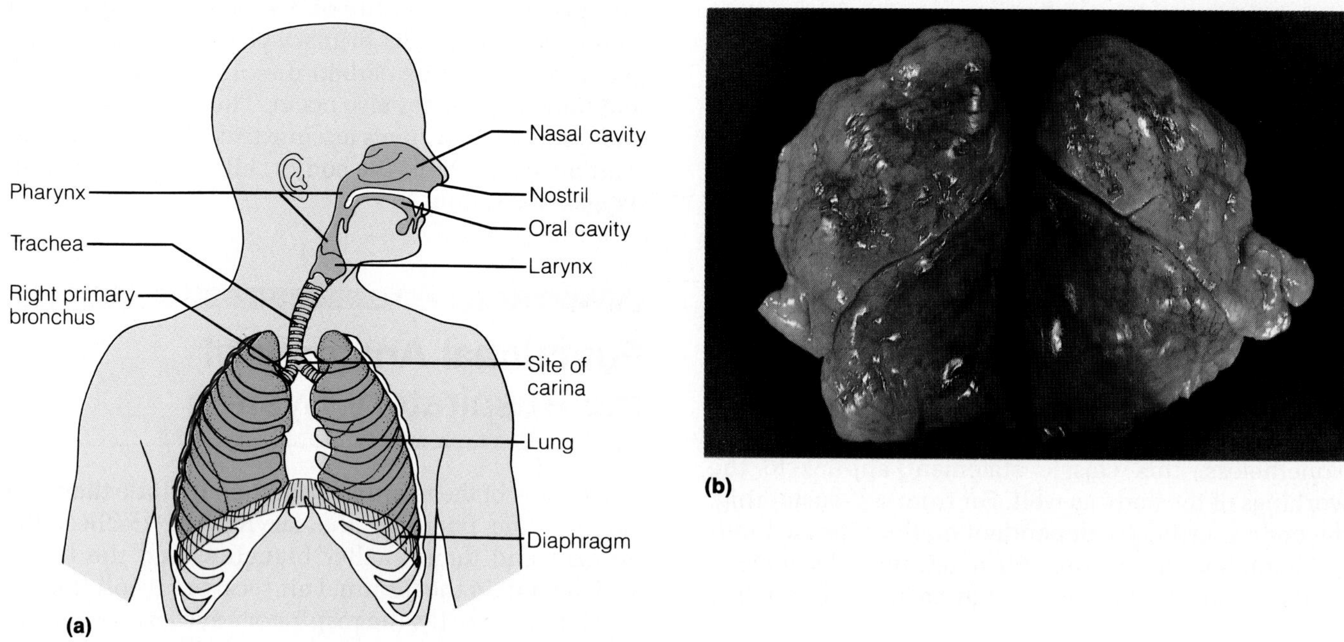

Figure 23.1 Organs of the respiratory system. (**a**) The major respiratory organs shown in relation to surrounding structures. (**b**) Photograph of human lungs.

23.2a). Noses vary a great deal in size and shape, largely because of differences in the nasal cartilages. The skin covering the nose's dorsal and lateral aspects is thin and contains many sebaceous glands.

During breathing, air enters the **nasal cavities** by passing through the **external nares** (nār′-ēz), or **nostrils** (Figure 23.2b). The two nasal cavities are separated by a midline **nasal septum**, formed anteriorly by hyaline cartilage and posteriorly by the vomer bone and perpendicular plate of the ethmoid bone (see Figure 7.10b, p. 184). The nasal cavities are continuous posteriorly with the nasal portion of the pharynx through the **internal nares,** or **choanae** (kō-a′-nē).

The roof of the nasal cavity is formed by the ethmoid and sphenoid bones of the skull; its floor is formed by the *palate* (pa′-lut), which separates it from the oral cavity below. Anteriorly, where the palate is supported by the maxillary processes and the palatine bones, it is called the **hard palate**. The unsupported posterior portion is the **soft palate**.

The portion of the nasal cavities just behind the nostrils, called the *vestibule*, is lined with skin containing sebaceous and sweat glands and numerous hair follicles. The hairs, or *vibrissae* (vī-brih′-sē) (*vibro* = to quiver), filter large particles from inspired air. The remainder of the nasal cavity is lined with two types of mucous membrane. The *olfactory mucosa*, lining the superior slitlike region of the nasal cavity, contains the special olfactory neurons that function in the sense of smell. The balance of the mucosa, the *respiratory mucosa*, is a ciliated columnar epithelium, containing goblet cells, that rests on a lamina propria richly supplied with mucous and serous glands. Each day, these glands secrete about a quart of sticky mucus containing lysozyme, an antibacterial enzyme. The mucus traps inspired dust, bacteria, and other debris; lysozyme attacks and destroys bacteria chemically.

The ciliated cells of the respiratory mucosa create a gentle current that moves contaminated mucus posteriorly toward the throat (pharynx), where it is swallowed. We are usually unaware of this important ciliary action, but when exposed to extremely cold environmental temperatures, the cilia become sluggish, allowing mucus to accumulate in the nasal cavity and to dribble outward through the nostrils. This helps explain why you might have a "runny" nose on a crisp, wintry day.

A rich vascular plexus underlies the mucosa and warms incoming air as it flows across the mucosal surface. When the temperature of inspired air drops, the vascular plexuses become engorged with blood, intensifying the air-heating process. Projecting medially from the lateral wall of each nasal cavity are three mucosa-covered lobes, the *superior, middle*, and *inferior conchae* (kon′-kē). The depression, or groove, below each concha is referred to as a *meatus* (mē-ā-tus). The conchae (also called the *turbinates*) enhance the air turbulence in the nasal cavity and greatly increase the mucosal surface area exposed to the air. The gases in inhaled air swirl through the twists and turns, but the heavier, nongaseous particles tend to be

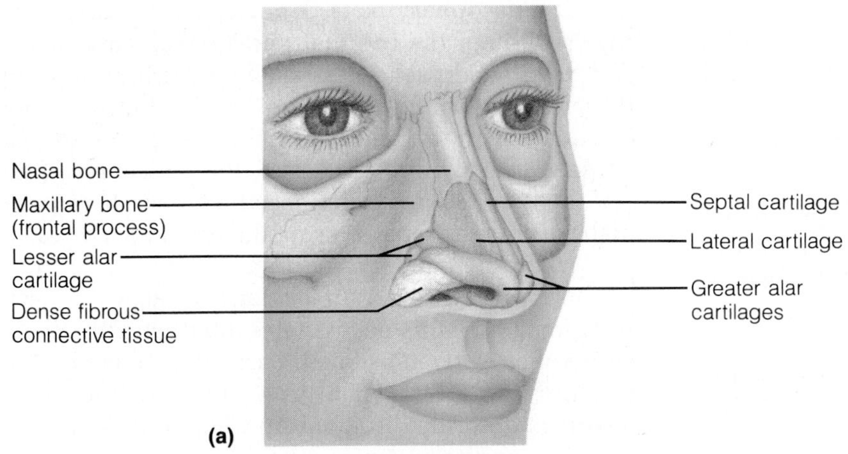

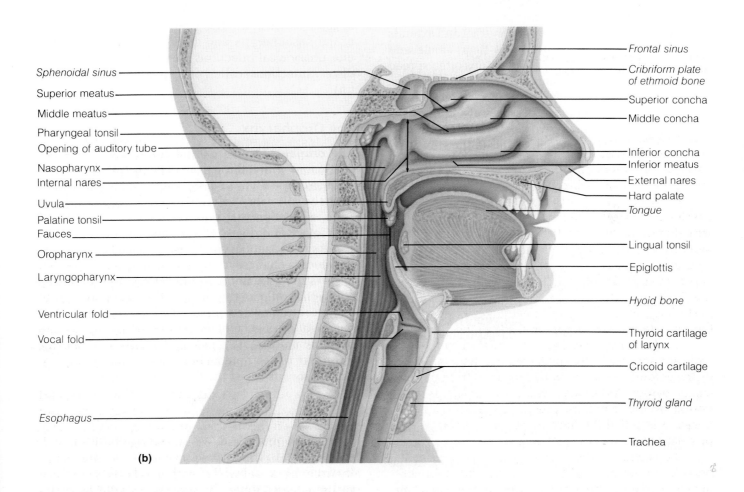

Sphenoidal sinus
Superior meatus
Middle meatus
Pharyngeal tonsil
Opening of auditory tube
Nasopharynx
Internal nares
Uvula
Palatine tonsil
Fauces
Oropharynx
Laryngopharynx
Ventricular fold
Vocal fold
Esophagus
Frontal sinus
Cribriform plate of ethmoid bone
Superior concha
Middle concha
Inferior concha
Inferior meatus
External nares
Hard palate
Tongue
Lingual tonsil
Epiglottis
Hyoid bone
Thyroid cartilage of larynx
Cricoid cartilage
Thyroid gland
Trachea
(b)

Figure 23.2 Basic anatomy of the upper respiratory tract. (**a**) External framework of the nose. (**b**) Sagittal section.

deflected and to strike the mucus-coated surfaces, where they become trapped. As a result, few particles larger than 4 μm are able to make it past the nasal cavities.

The Paranasal Sinuses

The nasal cavity is surrounded by a ring of **paranasal sinuses** located in the frontal, sphenoid, ethmoid, and maxillary bones (see Figure 7.11). As described in Chapter 7, the sinuses lighten the skull and act as resonance chambers in speech. They also produce mucus that drains into the nasal cavities and helps trap debris. The suctioning effect created by nose blowing helps to drain the sinuses.

Cold viruses, streptococcal bacteria, and various allergens can cause *rhinitis* (rī-nī′-tis), inflammation of the nasal mucosa, and excessive mucus production, which result in nasal congestion and postnasal drip. Because the nasal mucosa is continuous with that of the rest of the respiratory tract and extends tentacle-like into the nasolacrimal (tear) ducts and paranasal sinuses, nasal cavity infections often spread to those regions as well. *Sinusitis*, or inflamed sinuses, is difficult to treat and can result in marked changes in voice quality. When the passageways connecting the sinuses to the nasal cavity become blocked with mucus or infectious material, the air in the sinus cavities is absorbed. The result is a partial vacuum and a *sinus headache* localized over the inflamed area. ■

The Pharynx

The **pharynx** (fayr′-inks) is a funnel-shaped passageway about 13 cm (5 inches) long that extends from the base of the skull to the level of the sixth cervical vertebra (Figures 23.1 and 23.2b); it serves as a common passageway for food and air. Commonly called the *throat*, the pharynx vaguely resembles a short length of red garden hose.

On the basis of location and function, the pharynx is divided into three regions; from superior to inferior, they are the nasopharynx, oropharynx, and laryngopharynx. The muscular wall of the pharynx is composed of skeletal muscle throughout its length (see Table 10.3, p. 290–291), but the cellular composition and architecture of the mucosal lining varies in the three pharyngeal regions.

The Nasopharynx

The **nasopharynx** is located beneath the sphenoid bone and above the level of the soft palate. Because it lies above the point of food entry into the body, it serves only as an air passageway. During swallowing, the soft palate and its pendulous uvula (yoo′-vyoo-lah) are reflected upward, an action that effectively closes off the nasopharynx and prevents food from entering it.

The nasopharynx is continuous with the nasal cavity through the internal nares (see Figure 23.2b), and its ciliated pseudostratified epithelium picks up the mucus-propelling job where the nasal mucosa leaves off. Located in the mucosa high on its posterior wall are masses of lymphatic tissue, the *pharyngeal* (fahrin′-jē-al) *tonsils*, or *adenoids*, which serve to trap and dispose of pathogens entering the nasopharynx in air.

When the adenoids become inflamed and swollen, they obstruct air passage through the nasopharynx. Since this necessitates mouth breathing, the air is not properly moistened, warmed, or filtered before reaching the lungs. (The protective function of the tonsils is described in detail in Chapter 21.) ■

The *auditory*, or *eustachian* (yoo-stā′-shē-an), *tubes*, which drain the middle ear cavities, open into the lateral walls of the nasopharynx (see Figure 23.2b). These tubes provide an avenue for equalizing middle ear pressure with atmospheric pressure via the flow of gases between the two cavities. Since the mucosae of the nasopharynx and auditory tubes are continuous, pharyngeal infections are often followed by middle ear infections, called *otitis media* (ō-tī′-tis mē-dē-uh).

The Oropharynx

The **oropharynx** lies behind the oral cavity and is continuous with it through an opening called the *fauces* (fah′-sēz) (see Figure 23.2b). The oropharynx extends from the soft palate inferiorly to the level of the hyoid bone, a location that requires it to service both the respiratory and digestive systems.

As the nasopharynx blends into the oropharynx, the mucosa changes from pseudostratified epithelium to a stratified squamous epithelium, a structural adaptation that reflects the increased friction and greater chemical trauma accompanying food passage.

Two pairs of tonsils lie embedded in the oropharynx mucosa. The *palatine tonsils* are located at the end of the soft palate in its anterolateral walls; the *lingual tonsils* are at the base of the tongue.

The Laryngopharynx

Like the oropharynx above, the **laryngopharynx** (luhrin″-gō-fayr′-inks) serves as a common passageway for food and air and is lined with a stratified squamous epithelium. It extends from the hyoid bone to the larynx, where the respiratory and digestive pathways diverge and the laryngopharynx becomes continuous with the posteriorly positioned esophagus. The esophagus conducts food and fluids to the stomach; air enters the larynx anteriorly. During swallowing, food has the "right of way," and air passage is temporarily arrested.

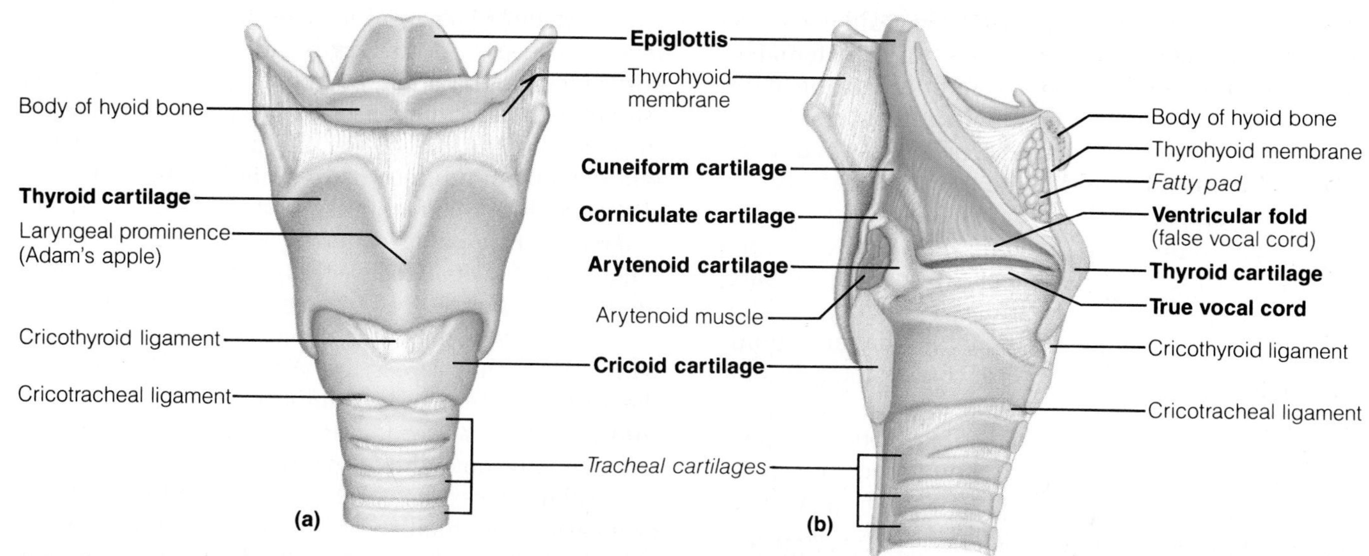

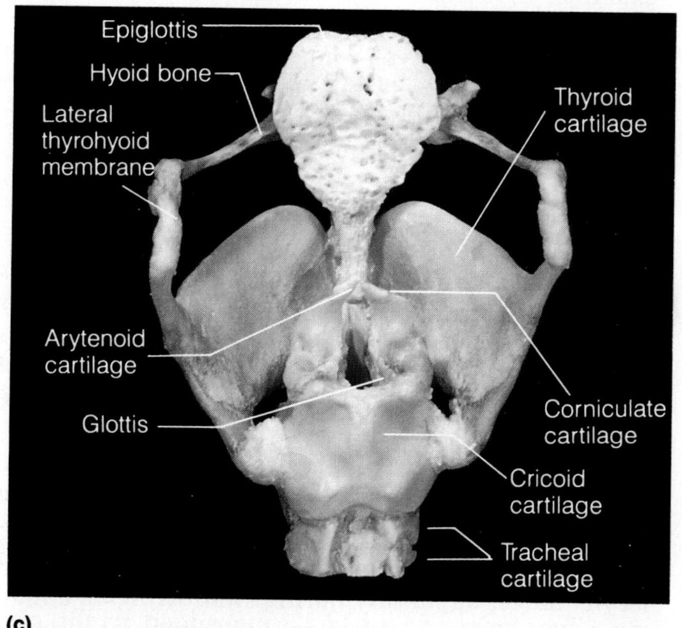

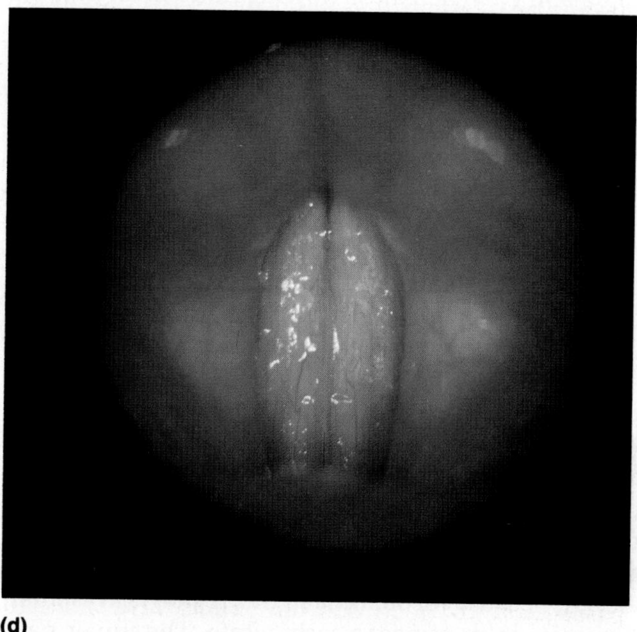

Figure 23.3 Anatomy of the larynx. (**a**) Anterior superficial view of the larynx. (**b**) Sagittal view of the larynx. (**c**) Photograph of the cartilaginous framework of the larynx, posterior view. (**d**) Photograph of the larynx with glottis closed and vocal cords stretched, superior superficial view.

The Larynx

The **larynx** (layr′-inks), or voicebox, extends for about 5 cm (2 inches) from the level of the fourth to sixth cervical vertebra. Superiorly it is attached to the hyoid bone and opens into the laryngopharynx; inferiorly it connects to the trachea (see Figure 23.2b).

The highly specialized larynx has three important functions. Its two principal tasks are to provide a patent (permanently open) airway and to act as a switching mechanism to route air and food into the proper channels. When food is being propelled through the pharynx, the inlet to the larynx is closed; when only air is flowing through the pharynx, the inlet to the larynx is open wide, allowing air to pass into the lower portions of the respiratory tract. Because the larynx houses the vocal cords, its third function is voice production.

The framework of the larynx is formed by an intricate arrangement of nine cartilages connected by membranes and ligaments (Figure 23.3). Except for the epiglottis, all laryngeal cartilages are hyaline cartilages. The large, shield-shaped **thyroid cartilage** is formed by the anteromedial fusion of two curving cartilage plates. The ridgelike *laryngeal* (luh-rin′-jē-ul) *prominence*, which marks the fusion point, is obvious externally as the *Adam's apple*. The thyroid cartilage is typically much larger in males than in females because of the growth-stimulating influence of male sex hormones during puberty. Below the thyroid cartilage is the signet ring–shaped **cricoid** (krī-koyd) **cartilage**, which is anchored to the trachea inferiorly.

Three pairs of small cartilages, the *arytenoid* (ar″-ih-tē′-noyd), *cuneiform* (kyoo-nē′-ih-form), and *corniculate cartilages* (see Figure 23.3b and c), form part of the lateral and posterior walls of the larynx. The most important of these are the pyramidal arytenoid cartilages, which anchor the vocal cords located within the larynx.

The ninth cartilage, the flexible, spoon-shaped **epiglottis** (eh″-pih-glah′-tis), is composed of elastic cartilage and is almost entirely covered by a taste bud–containing mucosa. The epiglottic stalk is attached to the anterior rim of the thyroid cartilage (see Figure 23.3b and c). When air is flowing into the larynx, the free edge of the epiglottis projects upward to the base of the tongue. But during swallowing, the situation changes dramatically; the larynx is pulled upward, and the epiglottis is tipped to cover the laryngeal inlet. Because this action routes food or fluid into the esophagus, the epiglottis has been referred to as the "guardian of the airways." If anything other than air enters the larynx, the cough reflex is initiated, which acts to expel the substance and prevent it from continuing down into the lungs. Because this protective reflex does not work when we are unconscious, it is never a good idea to try to give fluids to an unconscious person when attempting to revive the individual.

Lying under the laryngeal mucosa on each side are the *vocal ligaments*, which attach the arytenoid cartilages to the thyroid cartilage. These ligaments, composed largely of elastic fibers, produce two pairs of mucosal folds called *vocal folds* (see Figure 23.3b and d). Because the mucosa in this region is avascular, the vocal folds appear pearly white. The upper pair, called the **ventricular folds** or **false vocal cords**, play no part in sound production. The lower pair, the **true vocal cords**, vibrate and produce sound as air rushes upward from the lungs. The mucosa lining the superior portion of the larynx, an area subject to food contact, is stratified squamous epithelium, but below the vocal folds the epithelium changes to pseudostratified ciliated columnar. The power stroke of the cilia is exerted in an upward direction toward the throat so that mucus is continually moved away from the lungs.

Voice Production

Speech involves the intermittent release of expired air and the opening and closing of the true vocal cords. The length of the cords and the size of the **glottis** (the medial opening between them) is altered by the action of the intrinsic laryngeal muscles attached to the arytenoid cartilages. As the length and tension of the vocal cords change, the pitch, or frequency, of the sound is altered. Generally, the tenser the vocal cords, the faster they vibrate and the higher the pitch. The glottis is wide when we produce deep tones and narrows to a slit for high-pitched sounds. As a young boy's larynx enlarges during puberty, his vocal cords become both longer and thicker. Because this causes the vocal cords to vibrate more slowly, the voice becomes deeper.

Loudness of the voice depends on the force with which air rushes across the vocal cords. The greater the force, the stronger the vibration and the louder the sound. The vocal cords do not move at all when we whisper, but they vibrate vigorously when we yell.

Although the true vocal cords have the job of sound production, the quality of the voice depends on the coordinated activity of many other structures. For example, the entire length of the pharynx acts as a resonating chamber, or pipe organ, to amplify and enhance the quality of sound. The oral, nasal, and sinus cavities also contribute to this function. In addition, normal speech and good enunciation depend on the "shaping" of sound into recognizable consonants and vowels by the action of muscles in the pharynx, tongue, soft palate, and lips.

Inflammation of the vocal cords, or *laryngitis*, results in hoarseness or an inability to speak above a whisper. Overuse of the voice, bacterial infections, and inhalation of irritating chemicals can all cause laryngitis. ■

Sphincter Functions of the Larynx

There are two points at which the larynx can be closed by muscle action. As previously described, its inlet is closed by the flapping over of the epiglottis when the larynx moves superiorly during swallowing. In addition to the opening and closing of the glottis for speech, the vocal folds can perform a sphincter function under certain other conditions, such as when we swallow, cough, sneeze, or strain to have a bowel movement. The mechanisms involved in coughs and sneezes are summarized in Table 23.3 (p. 728). In abdominal straining (associated with defecation or urination), inhaled air is held temporarily in the lower respiratory tract by closing the glottis. The abdominal muscles then contract, and the intra-abdominal pressure rises, which helps to evacuate the rectum or bladder. These events, collectively known as the *Valsalva maneuver*, can also act to splint the body trunk when one lifts a heavy load.

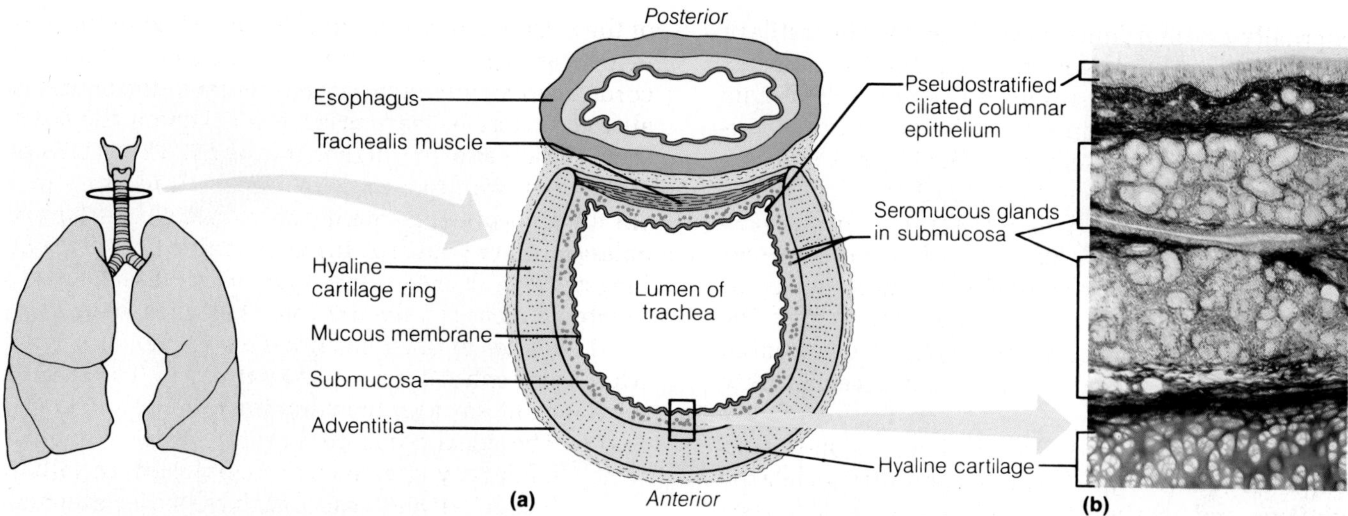

Figure 23.4 Tissue composition of the tracheal wall. (**a**) Cross-sectional view of the trachea, illustrating its relationship to the esophagus, the position of the supporting hyaline cartilage rings in its wall, and the trachealis muscle connecting the free ends of the cartilage rings. (**b**) Photomicrograph of a portion of the tracheal wall (cross-sectional view) (50X).

The Trachea

The **trachea** (trā′-kē-uh), or *windpipe*, begins in the neck, where it is continuous with the inferior larynx (see Figure 23.1). From there, it descends in front of the esophagus to enter the mediastinum, where it divides into the right and left primary bronchi at the level of the fifth thoracic vertebra. In humans, the trachea is 12–15 cm (4–5 inches) long and 2.5 cm (1 inch) in diameter. Unlike most other organs in the neck, the trachea is very mobile. It stretches and descends during inspiration and recoils during expiration, and it can be moved from side to side by probing fingers.

The tracheal mucosa, descriptively called the *mucociliary* (myoo′-kō-sil″-ē-ayr″-ē) *escalator*, is pseudostratified ciliated columnar epithelium containing abundant seromucous glands that produce a thick mucus-containing secretion. The cilia continuously propel the mucus, loaded with dust particles and other debris, to the throat, where it can be swallowed or expelled.

Smoking diminishes ciliary activity and ultimately destroys the cilia. When their function is lost, coughing is the only means of preventing mucus from accumulating in the lungs. For this reason, smokers with respiratory congestion should never be given medications that inhibit the cough reflex. ■

The balance of the tracheal wall is composed of smooth muscle and elastic connective tissue and is reinforced internally by 16 to 20 C-shaped rings of hyaline cartilage (Figure 23.4). Consequently, the trachea is flexible enough to permit twisting and elongation, but the cartilage rings prevent it from collapsing and keep the airway patent despite the pressure changes that occur during breathing. The open parts of the cartilage rings, which abut the esophagus (see Figure 23.4a), are connected by smooth muscle fibers of the *trachealis muscles* and elastic tissue, allowing the esophagus to expand anteriorly during swallowing. When the trachealis muscles contract, the diameter of the trachea decreases, causing expired air to rush upward from the lungs with greater force; this helps to expel mucus from the trachea when we cough. The last tracheal cartilage is expanded (see Figure 23.1), and a spar of cartilage, called the *carina* (kuh-rī′-nuh), projects posteriorly from its inner face, marking the tracheal bifurcation point.

Tracheal obstruction is life threatening. Many people have suffocated after choking on a piece of food that suddenly closed off the trachea. The Heimlich maneuver, a procedure in which the air in a person's own lungs is used to "pop out," or expel, an obstructing piece of food, has saved many people from becoming victims of such "café coronaries." ■

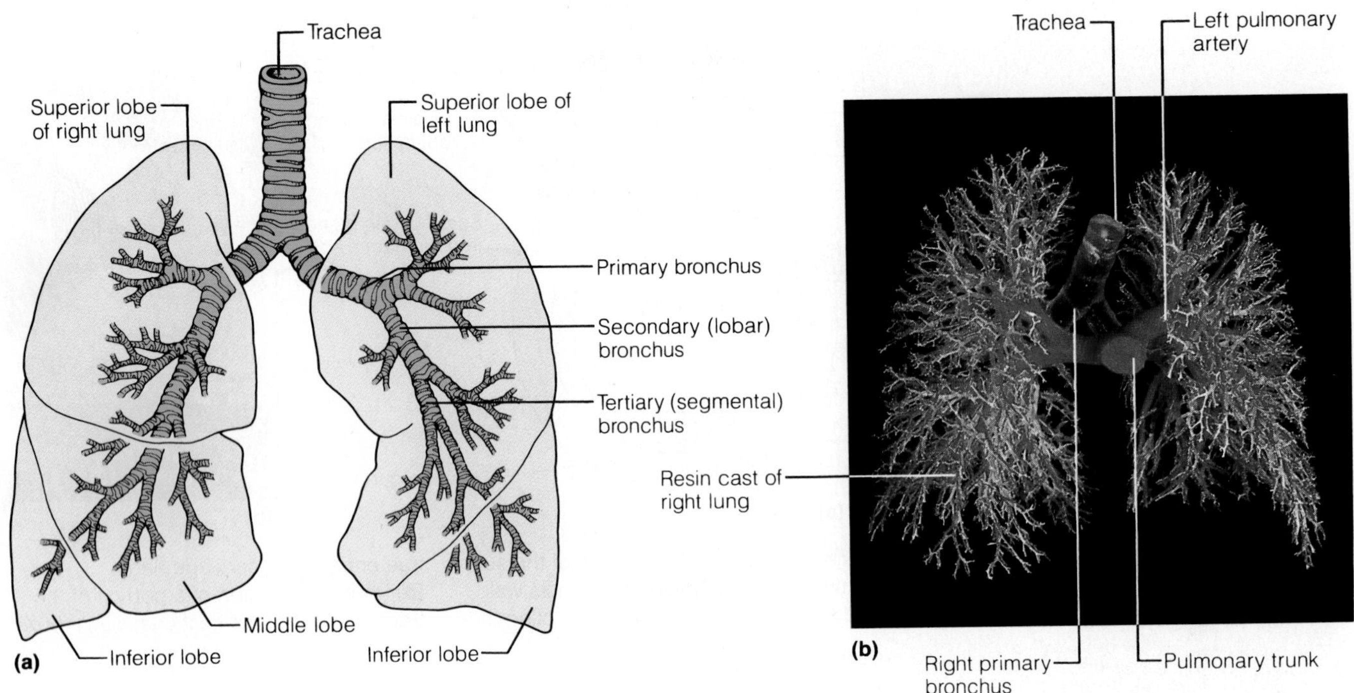

Figure 23.5 Conducting respiratory passages. (**a**) The air pathway inferior to the larynx consists of the trachea and primary, secondary, and tertiary bronchi, which in turn branch into smaller and smaller bronchioles until the terminal bronchioles of the lungs are reached. (**b**) Resin cast of the bronchial tree and associated pulmonary arterial supply of the lungs, viewed from the front. Clear resin fills the airways; red resin fills the pulmonary arteries and their branches.

The Bronchi and Subdivisions: The Bronchial Tree

Conducting Zone Structures

The **right** and **left primary (main) bronchi** (bron′-kī) are formed by the division of the trachea at the level of the sternal angle (Figure 23.5a). Each bronchus runs obliquely in the mediastinum before plunging into the medial depression (hilus) of the lung on its own side. The right primary bronchus is wider, shorter, and more vertical than the left and is the more common site for an inhaled foreign object to become lodged. By the time incoming air reaches the bronchi, it is warm, cleansed of most impurities, and saturated with water vapor.

Once inside the lungs, each main bronchus subdivides into **secondary (lobar) bronchi**—three on the right and two on the left, each of which enters and supplies one lung lobe. The secondary bronchi branch to form **tertiary (segmental) bronchi**, which in turn divide repeatedly into smaller and smaller **bronchioles** (bron′-kē-ōls). The most minute of the conducting passages are the **terminal bronchioles**, which are less than 0.5 mm in diameter. Because of this branching and rebranching pattern of the bronchi and bronchioles, the conducting network within the lungs is often called the *bronchial* or *respiratory tree*. A resin cast showing the extensive branching of the bronchial tree and its accompanying pulmonary arterial supply is illustrated in Figure 23.5b.

The tissue composition of the walls of the primary bronchi mimics that of the trachea, but as the conducting tubes become smaller, a number of structural changes occur:

1. The cartilage rings are gradually replaced by irregular plates of cartilage, and by the time the bronchioles are reached, supportive cartilage is no longer present in the tube walls.
2. The mucosal epithelium thins as it changes progressively from pseudostratified ciliated columnar epithelium to ciliated columnar epithelium and then nonciliated cuboidal epithelium in the terminal bronchioles. Any debris found at or below the level of the terminal bronchioles is normally removed by macrophages.
3. The relative amount of smooth muscle in the tube walls increases as the passageways become smaller. The complete layer of circularly arranged smooth muscle in the bronchioles makes them capable of providing substantial resistance to air passage under certain conditions (as described later).

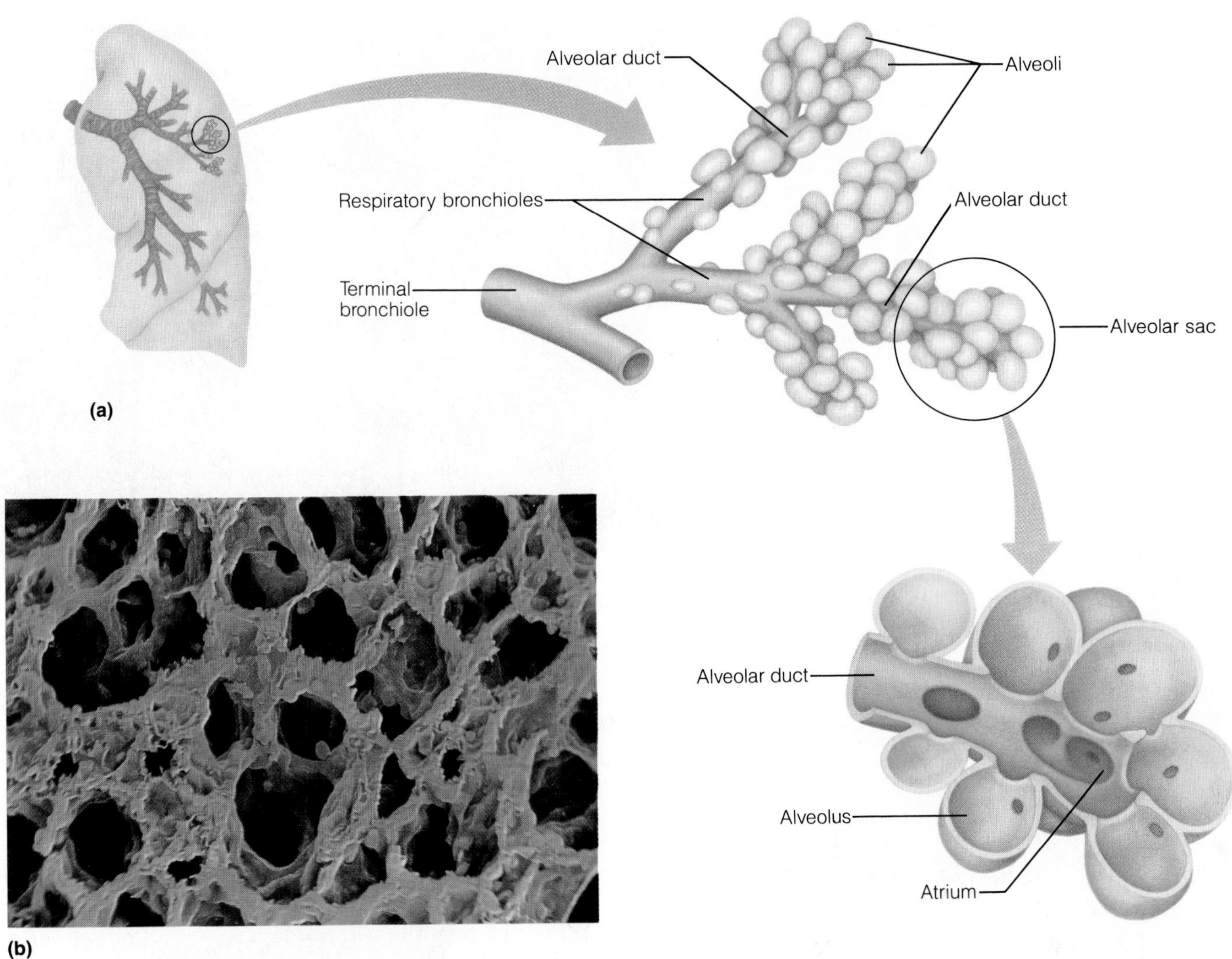

Figure 23.6 Respiratory zone structures. (**a**) Diagrammatic view of the respiratory unit (respiratory bronchiole, alveolar ducts, alveolar sacs, and alveoli). (**b**) Scanning electron micrograph (SEM) of a section of the human lung, showing the network of alveolar ducts and alveoli that form the final divisions of the bronchial tree (475X). Notice the thinness of the alveolar walls.

Respiratory Zone Structures

The respiratory zone begins as the terminal bronchioles feed into the **respiratory bronchioles**. The respiratory bronchioles exhibit scattered air sac outpocketings from their walls through which gas exchange can occur—hence, the name *respiratory bronchioles*. Respiratory bronchioles branch into less distinct **alveolar ducts**, which lead directly into **alveolar sacs** and **alveoli** (al-vē′-o-lī), the chambers where the bulk of gas exchange occurs. These relationships are illustrated in Figure 23.6a.

The alveoli, minute depressions along the walls of the alveolar sacs, are densely clustered together. As a result, the alveolar sacs resemble bunches of grapes opening into a common chamber, the *atrium*, at the terminus of the alveolar duct. As shown in Figure 23.6b, the millions of gas-filled alveoli in each lung account for the largest portion of lung volume and provide a tremendous surface area for gas exchange.

The Respiratory Membrane. The walls of the alveoli are composed of a single layer of squamous epithelium (*type I cells*) underlain by a flimsy basement membrane. The thinness of their walls is hard to imagine, but a sheet of notepaper is much thicker. The external surfaces of the alveoli are densely covered with a "cobweb" of pulmonary capillaries (Figure 23.7a). Together, the alveolar and capillary walls and their fused basal laminas form the **respiratory membrane**, which has gas on one side and blood flowing past on the other (Figure 23.7c). Gas exchanges occur

Figure 23.7 Anatomy of the respiratory membrane. (**a**) Scanning electron micrograph of casts of alveoli and associated pulmonary capillaries (255X). (**b, c**) Detailed anatomy of the respiratory membrane composed of the alveolar squamous epithelial cells (type I cells), the capillary endothelium, and the scant basement membranes (fused basal laminas) intervening between. Type II (surfactant-secreting) alveolar cells are also shown, as are the alveolar pores that connect adjacent alveoli. Diffusion of oxygen occurs from the alveolar air into the pulmonary capillary blood; carbon dioxide diffuses from the pulmonary blood into the alveolus.

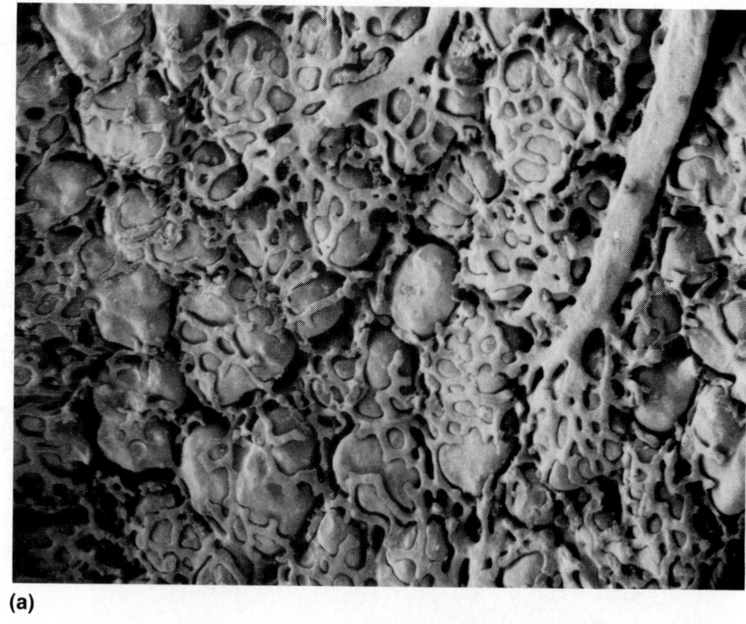

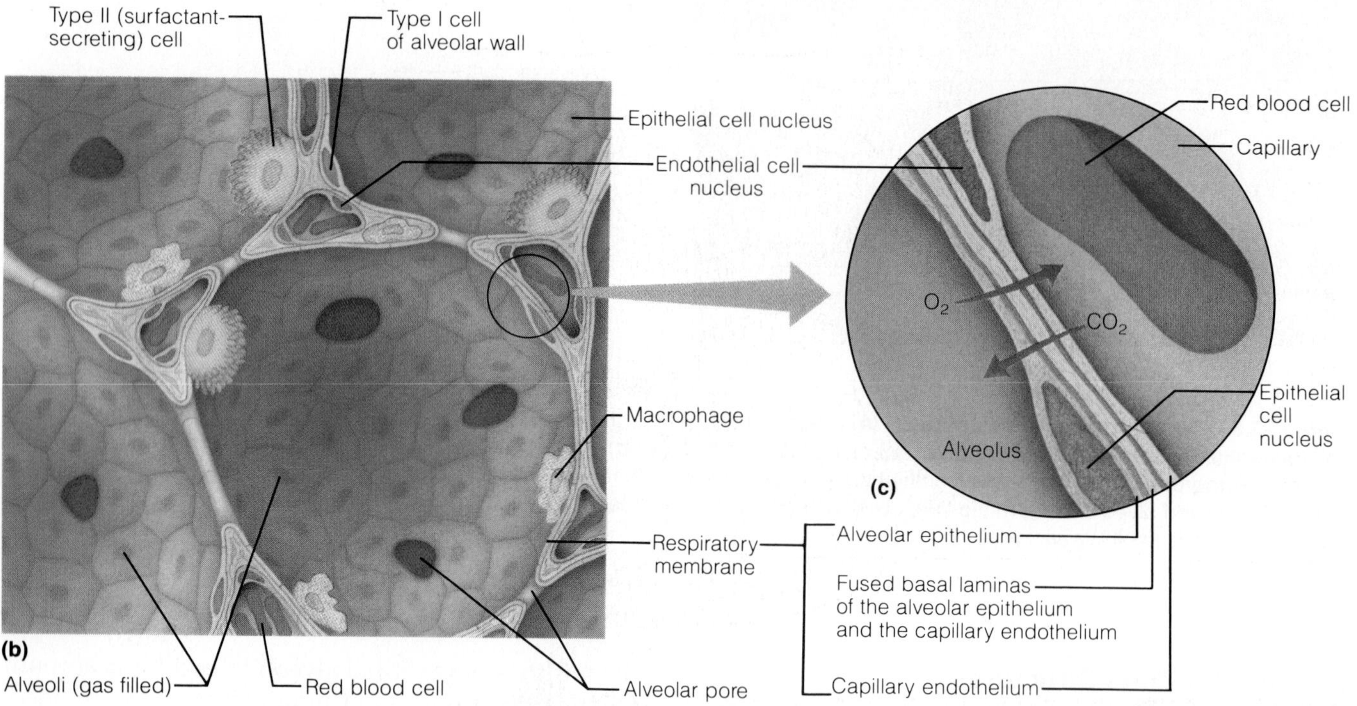

readily by simple diffusion across the respiratory membrane—the oxygen passing from the alveolus into the blood, and carbon dioxide leaving the blood to enter the gas-filled alveolus.

Scattered amid the type I squamous cells forming the major part of the alveolar walls are cuboidal *type II cells* (Figure 23.7b). Efficient gas exchange between air and blood requires a moist membrane, and the type II cells secrete a surfactant-containing fluid that coats the gas-exposed alveolar surfaces. (Surfactant's role in reducing the surface tension of the alveolar fluid is described later in this chapter.)

The Lungs and Pleural Coverings

Gross Anatomy of the Lungs

The paired **lungs** occupy all of the thoracic cavity except for the most central area, the *mediastinum*, which houses the heart, great blood vessels, bronchi, esophagus, and other organs (Figure 23.8). Each lung is suspended in its own pleural cavity by vascular and bronchial attachments, collectively called its *root*. The anterior, lateral, and posterior lung surfaces lie in close contact with the ribs and form a continuously curving surface referred to as the *costal surface*. Just deep to

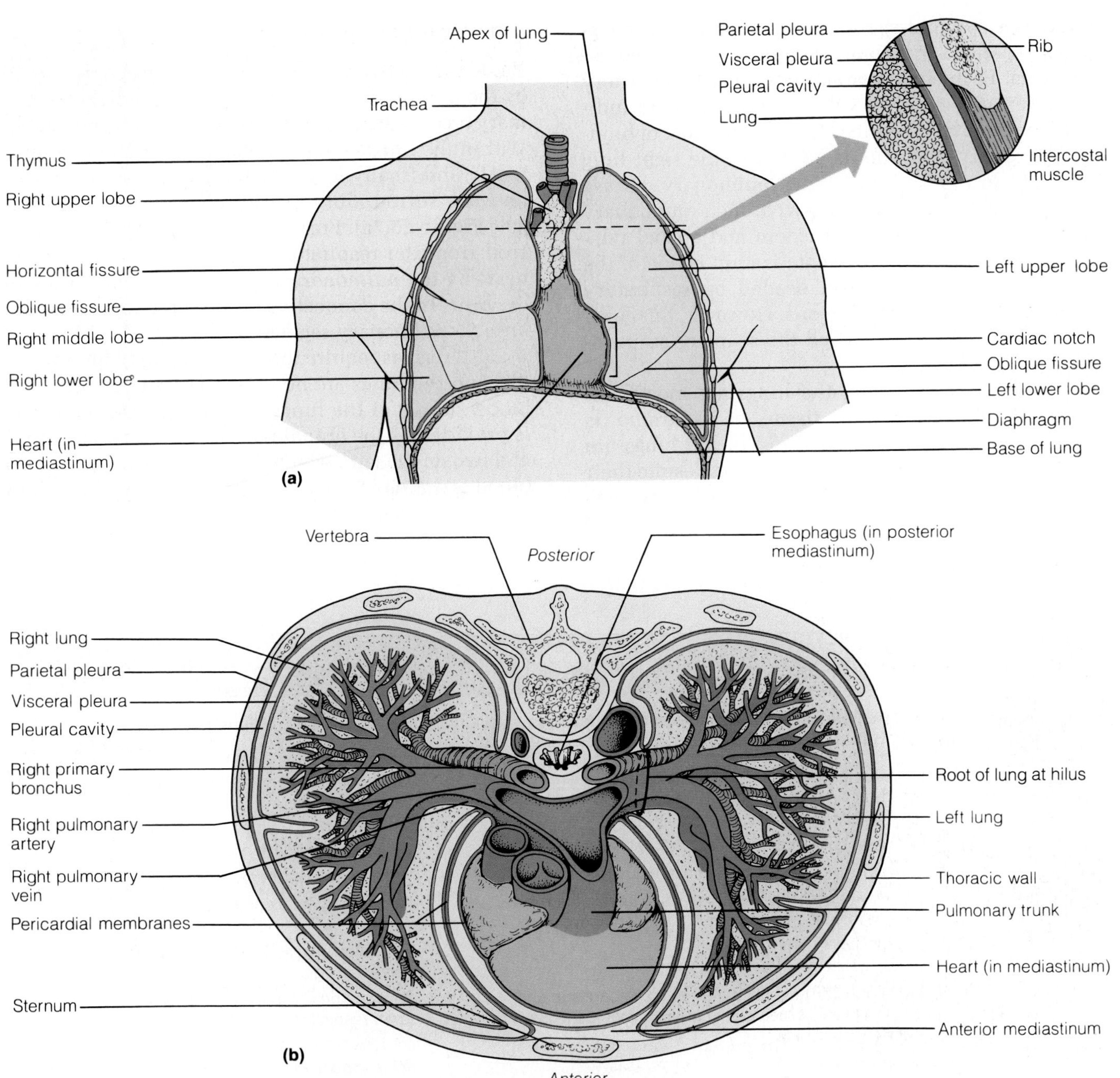

Figure 23.8 Anatomical relationships of organs in the thoracic cavity. (**a**) Anterior view of the thoracic cavity organs, showing the position of the lungs, which flank mediastinal structures laterally. (**b**) Transverse section through the thorax, showing the lungs, the pleural membranes, and the major organs present in the mediastinum. (The plane of the section is shown by the dotted line in **a**; the thymus has been omitted for clarity.)

the clavicle is the *apex*, the narrow superiormost portion of the lung; the somewhat concave, inferior surface that rests on the diaphragm is the *base*. On the medial (mediastinal) surface of each lung is an indentation, the *hilus*, through which blood vessels of the pulmonary and systemic circulations enter and leave the lungs. The primary bronchi also plunge into the hilus and begin to branch almost immediately. All conducting and respiratory passageways distal to the primary bronchi are found within the lung substance.

Because the heart is located slightly to the left of the median thorax plane, the two lungs are somewhat different in shape and size. The left lung is smaller than the right, and the **cardiac notch**—a concavity in its medial aspect—is molded to and accommodates the heart (see Figure 23.8a). The left lung is subdivided into two (upper and lower) **lobes** by the *oblique fissure*, whereas the right lung is partitioned into three lobes (upper, middle, and lower) by the *oblique* and *horizontal fissures*. Each of the lung lobes contains a number of pyramid-shaped **bronchopulmonary segments** separated from one another by connective tissue septa. Each such segment is served by an artery and a vein and receives its air supply from an individual segmental bronchus. The total number of bronchopulmonary segments is 18: ten in the right lung and eight in the left lung. Since pulmonary disease may be confined to one (or a few) bronchopulmonary segments, knowledge of the extent and spatial relationships of lung compartmentalization is one of the main pieces of information needed by respiratory therapists and thoracic surgeons. However, this level of anatomical detail is beyond the scope of an introductory textbook.

As mentioned earlier, the lungs are mostly air spaces; the balance of lung tissue, or its *stroma*, is elastic connective tissue. As a result, the lungs are light, soft, spongy, and elastic organs that despite their size together weigh only about 2.5 pounds. The elasticity of healthy lungs helps to reduce the work of breathing, as described shortly.

Blood Supply of the Lungs

Blood that is to be oxygenated in the lungs is delivered by the *pulmonary arteries*, which accompany the primary bronchi into the lungs (Figures 23.5b and 23.8b). Within the lungs, the pulmonary arteries branch profusely and finally feed into the *pulmonary capillary network* surrounding the alveolar ducts and alveoli (see Figure 23.7a). Freshly oxygenated blood is delivered from the respiratory zones of the lungs to the heart by the *pulmonary veins*, which course back to the hilus in the connective tissue septa separating the bronchopulmonary segments.

All tissues require their own nutritive blood supply, and the lungs are no exception. The oxygen-rich blood supply of the lungs that nourishes lung tissue is provided by the *bronchial arteries* (one on the right and two on the left), which arise from the aorta. Venous blood is drained from the lungs by both *bronchial* and *pulmonary veins.*

Table 23.1 Principal Organs of the Respiratory System

Structure	Description, general and distinctive features	Function
Nose	Jutting external portion supported by bone and cartilage; paired internal nasal cavities separated by midline nasal septum and lined with mucosa	Produces mucus; filters, warms, and moistens incoming air
	Roof of nasal cavities contain olfactory epithelium	Receptors for sense of smell
	Nasal cavity surrounded by paranasal sinuses	Resonance chambers for speech
Pharynx	Passageway connecting nasal cavities to larynx; three subdivisions: nasopharynx, oropharynx, and laryngopharynx	Passageway for air and food
	Houses tonsils	Facilitates exposure of immune system to inhaled antigens
Larynx	Connects pharynx to trachea; framework of cartilage and dense connective tissue; opening (glottis) can be closed by epiglottis or vocal folds	Air passageway; prevents food from entering lower respiratory tract
	Houses true vocal cords	Voice production
Trachea	Flexible tube running from larynx and dividing inferiorly into two primary bronchi; walls contain C-shaped cartilages that are incomplete posteriorly and connected by smooth muscle	Air passageway; filters, warms, and moistens incoming air
Bronchial tree	Consists of right and left primary bronchi, which subdivide within the lungs to form secondary and tertiary bronchi and bronchioles; bronchiolar walls contain complete layer of smooth muscle; constriction of this muscle impedes expiration	Air passageways connecting trachea with alveoli; warms and moistens incoming air
Alveoli	Microscopic chambers at termini of bronchial tree; walls of simple squamous epithelium underlain by thin basement membrane; external surfaces intimately associated with pulmonary capillaries	Main sites of gas exchange
	Special alveolar cells produce surfactant	Reduces surface tension; helps prevent lung collapse
Lungs	Paired composite organs located within pleural cavities of thorax; composed primarily of alveoli and respiratory passageways; stroma is fibrous elastic connective tissue, allowing lungs to recoil passively during expiration	House passageways smaller than primary bronchi
Pleurae	Serous membranes; parietal pleura lines thoracic cavities; visceral pleura covers external lung surfaces	Produce lubricating fluid and compartmentalize lungs

The Pleurae

The **pleura** (pler'-uh) is a thin, double-layered serosa (see Figure 23.8). The **parietal pleura** lines the thoracic wall and mediastinum. It continues around the heart and between the lungs, forming the mediastinal enclosure and a cufflike extension on each side (the *pulmonary ligament*) that helps support the lung and snugly encloses its root. From here, the pleura extends as the **visceral**, or **pulmonary, pleura** to cover the external lung surface, dipping into and lining its fissures.

The pleurae produce **pleural fluid**, a lubricating serous secretion that allows the lungs to glide easily over the thorax wall during breathing movements and causes the parietal and visceral layers to cling tightly together. The pleurae can slide easily from side to side across one another, but their separation is strongly resisted. Consequently, the lungs are held tightly to the thorax wall, and the pleural space is more of a potential space than an actual one. This factor contributes to the formation of a slight vacuum (or negative pressure) in the pleural space that is absolutely essential to normal lung function. The pleurae also help divide the thoracic cavity into three separate chambers—a central one containing mediastinal structures and two lateral compartments, each containing a lung. This compartmentalization helps prevent one mobile functioning organ from interfering with another; it also limits the extent of local infection or traumatic injury.

In *pleurisy* (pler'-ih-sē), inflammation of the pleura, secretion of pleural fluid often declines, which results in stabbing pain with each breath. Conversely, the pleurae may produce excessive amounts of fluid, which exerts pressure on the lungs. This type of pleurisy hinders breathing movements, but it is much less painful than the dry type. ■

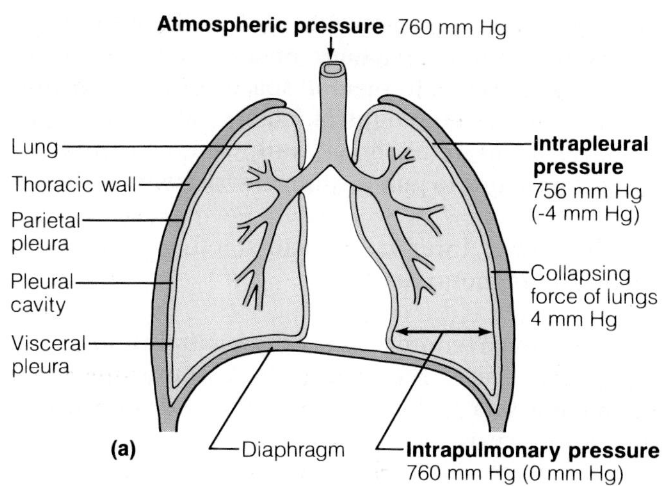

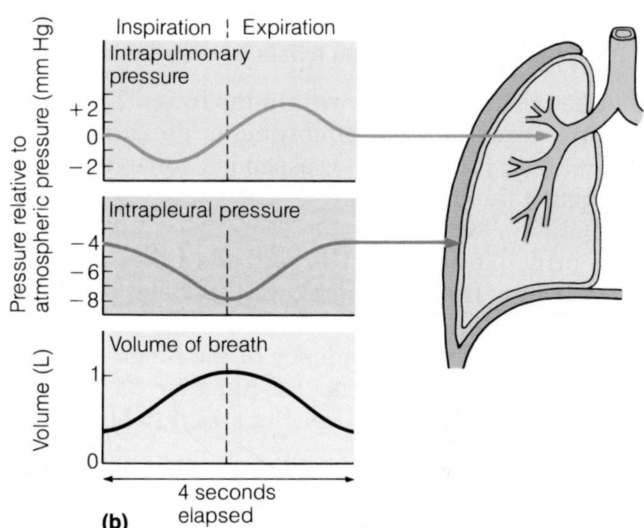

Figure 23.9 Intrapulmonary and intrapleural pressure relationships. (**a**) Intrapulmonary and intrapleural pressures in the resting position. Differences in pressure relative to atmospheric pressure are given in parentheses. (**b**) Changes in intrapulmonary and intrapleural pressures during inspiration and expiration. Notice that normal atmospheric pressure (760 mm Hg) is given a value of 0 on the scale.

Mechanics of Breathing

Breathing, or **pulmonary ventilation**, consists of two phases: **inspiration**, the period when air is flowing into the lungs, and **expiration**, the period when gases are leaving the lungs. The mechanical factors that promote these gas flows are the topics of this section.

Pressure Relationships in the Thoracic Cavity

Before we can begin to describe the breathing process, some groundwork must be laid—most importantly, the pressure relationships that normally exist in the thoracic cavity.

The **intrapulmonary pressure**, the pressure within the alveoli of the lungs, rises and falls with the phases of breathing, but it always equalizes itself with the atmospheric pressure outside the body (Figure 23.9). Atmospheric pressure is the pressure exerted by gases of the atmosphere; at sea level this pressure is equal to 760 mm Hg, the pressure exerted by a column of mercury 760 mm high.

The pressure within the intrapleural space, the **intrapleural pressure**, also fluctuates with breathing phases. However, the intrapleural pressure is always about 4 mm Hg less than the pressure in the alveoli and is said to be negative relative to both the intrapulmonary and atmospheric pressures. The negative pressure in the intrapleural space results from the interaction of two groups of factors: factors acting to hold the lungs to the thorax wall, which are opposed by factors acting to pull the lungs away from the thorax wall.

There are three main factors acting to hold the lungs to the thorax wall:

1. The adhesive force created by pleural fluid in the pleural space. The pleural fluid secures the pleurae together in the same way that a drop of water holds two glass slides to one another. They can slide from side to side easily, but they can never separate unless extreme force is applied.
2. Absorption of gases in the pleural space into the capillary blood. Since this absorption process reduces the intrapleural volume, it creates much of the partial vacuum that exists in the pleural space.
3. The positive pressure within the lungs. This pressure, equal to the atmospheric pressure when breathing is temporarily suspended, forces the lungs against the thorax wall.

Two principle forces act to pull the lungs (visceral pleura) away from the thorax wall (parietal pleura):

1. The natural recoil tendency of the lungs. Because of their highly elastic nature, lungs have a constant tendency to assume a smaller size if (when) able.
2. The surface tension of the fluid film in the alveoli, which constantly acts to collapse the alveoli or draw them to their smallest possible dimension.

The importance of the negative pressure in the intrapleural space and the tight coupling of the lungs to the thorax wall cannot be overemphasized. *Any condition that equalizes the intrapleural pressure with the intrapulmonary (or atmospheric) pressure causes immediate lung collapse.*

Atelectasis (a″-teh-lek′-tuh-sis), or lung collapse, renders the lung useless for ventilation. This phenomenon commonly occurs when air enters the pleural space through a chest wound, but may result from a rupture of the visceral pleura, which allows air to enter the pleural space from the respiratory tract. The presence of air in the intrapleural space is referred to as a *pneumothorax* (noo″-mō-thor′-aks). A pneumothorax is reversed by drawing air out of the intrapleural space with chest tubes, which allows the lung to reinflate and resume its normal function. ■

Pulmonary Ventilation: Inspiration and Expiration

Pulmonary ventilation, or breathing, is a completely mechanical process that depends on volume changes occurring in the thoracic cavity. A rule to keep in mind throughout the following discussion is: *Volume changes lead to pressure changes, which lead to the flow of gases to equalize the pressure.*

The relationship between the pressure and volume of gases is given by *Boyle's Law*, which states that under conditions of constant temperature, the pressure of a gas varies inversely with its volume; that is, $P_1V_1 = P_2V_2$, where P is the pressure of the gas in millimeters of mercury, V is its volume in cubic millimeters, and subscripts 1 and 2 represent the initial and resulting conditions respectively. Gases, like liquids, always conform to the shape of their container; however, unlike liquids, gases *fill* their container. Therefore, in a large volume, the gas molecules will be far apart and the pressure will be low. But, if the volume is reduced, the gas molecules will be closer together and the pressure will rise. Let us see how this relates to the phases of inspiration and expiration.

Inspiration

The process of inspiration is easy to understand if you can visualize the thoracic cavity as a box with a single entrance at the top: the tubelike trachea. The volume of this box is changeable and can be increased by enlarging all of its diameters, thereby decreasing the gas pressure within it. This, in turn, will result in air rushing into the box from the atmosphere, since gases always flow along their pressure gradients.

Identical relationships exist during normal quiet inspiration, when the **inspiratory muscles**—the diaphragm and external intercostal muscles—are activated:

1. When the dome-shaped diaphragm contracts, it moves inferiorly and flattens out (Figure 23.10a and d). As a result, the vertical diameter (height) of the thoracic cavity increases.
2. Contraction of the external intercostal muscles results in an elevation of the rib cage and in a thrusting forward of the sternum (Figure 23.10b and c). Since the ribs curve downward as well as forward around the chest wall, the broadest lateral and anteroposterior dimensions of the rib cage are normally directed obliquely downward. But when the ribs are raised and drawn together, they also swing outward (much like the action that occurs when a curved bucket handle is raised), expanding the diameter of the thorax both laterally and in the anteroposterior plane.

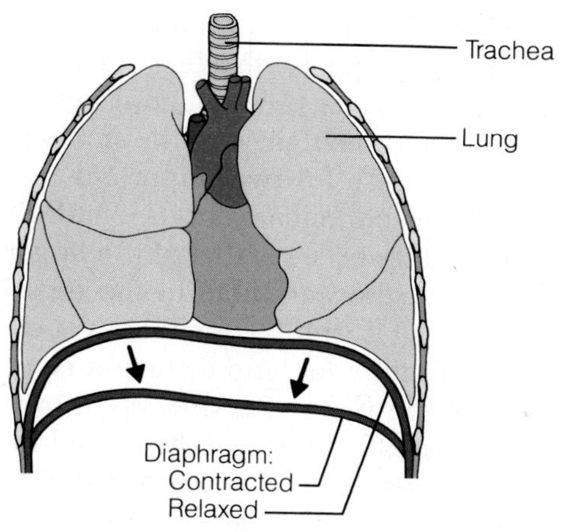

(a) Superoinferior expansion

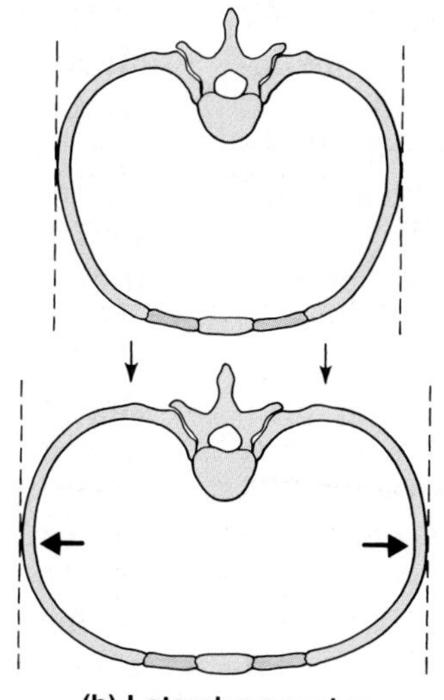

(b) Lateral expansion

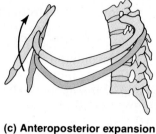

(c) Anteroposterior expansion

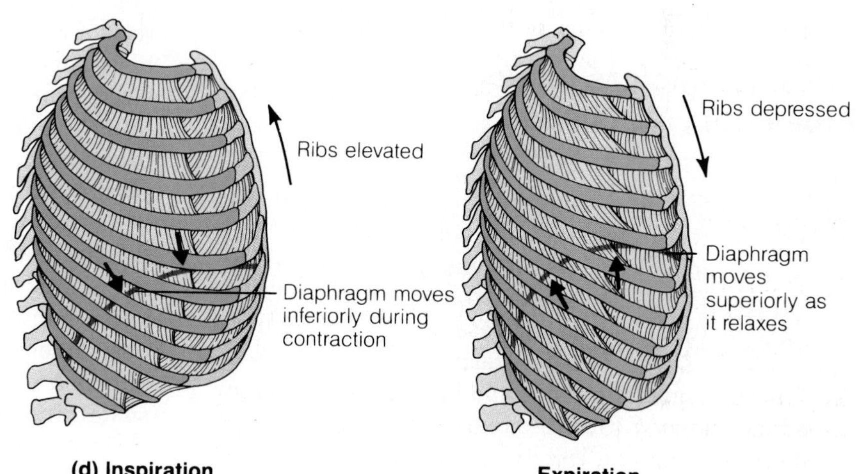

(d) Inspiration

Expiration

Figure 23.10 Changes in thoracic volume during breathing. (**a–c**) Ways in which the volume of the thorax is increased during inspiration. The diaphragm descends as it contracts, increasing the superoinferior dimension (**a**); the contraction of the external intercostal muscles results in a "bucket-handle" movement of the ribs, so that as the ribs are elevated, the thorax expands laterally (**b**) and in an anteroposterior plane (**c**). (**d**) Lateral view of the thorax during inspiration and expiration.

Although the thoracic dimensions are increased by only a few millimeters along each plane, this is sufficient to increase the volume of the thoracic cavity by almost half a liter—the usual volume of air that enters the lungs during a normal quiet inspiration. Of the two types of inspiratory muscles, the diaphragm is far more important in bringing about the volume changes that lead to normal quiet inspiration.

Because the lungs adhere tightly to the thorax wall, they are stretched and the intrapulmonary volume also increases as the thoracic dimensions increase. As a result, intrapulmonary pressure drops 1 to 3 mm Hg relative to atmospheric pressure, and air rushes into the lungs along this pressure gradient until the intrapulmonary and atmospheric pressures become equal. During the same period, the intrapleural pressure declines to about −8 mm Hg relative to atmospheric pressure (see Figure 23.9b).

During *deep* or *forced inspiration*, the capacity of the thorax is increased maximally by the activation of accessory muscles. Several muscles capable of raising the ribs even more than occurs during quiet inspiration are activated, including the anterior and medial scalenes, the sternocleidomastoid muscles of the neck, and the pectoral muscles of the chest. Additionally, the back is extended as the thoracic curvature is straightened by contraction of the erector spinae muscles.

Expiration

Quiet expiration in healthy individuals is chiefly a passive process that depends more on the natural elasticity of the lungs than on muscular contraction. As the inspiratory muscles relax and resume their initial resting length, the lungs recoil. Thus, both the thoracic and intrapulmonary volumes decrease simultaneously. This compresses the alveoli, and the intrapulmonary pressure rises to 1–3 mm Hg above atmospheric pressure (see Figure 23.9b), which results in an outflow of gases from the lungs.

In contrast, *forced expiration* is an active process produced by contraction of abdominal wall muscles, primarily the oblique and transversus muscles. These contractions (1) increase the intra-abdominal pressure, which forces the abdominal organs superiorly against the diaphragm, and (2) depress the rib cage. The internal intercostal muscles and latissimus dorsi muscles of the back may also contract to enhance rib cage depression.

It is interesting that the ability of a trained vocalist to hold a particular note depends on the coordinated activity of several muscles normally used in forced expiration. Thus, muscle control of expiration is very important during speaking and singing, when a precise regulation of airflow from the lungs is desired.

Physical Factors Influencing Pulmonary Ventilation

As we have seen, the lungs are stretched during inspiration and recoil passively during expiration. The inspiratory muscles consume energy to enlarge the thorax. Energy is also used to overcome various types of resistance that hinder air passage and pulmonary ventilation. These factors—respiratory passageway resistance, lung compliance and elasticity, and alveolar surface tension forces—are examined next.

Respiratory Passageway Resistance

The major nonelastic source of resistance to gas flow is friction, or drag, encountered in the respiratory passageways. The relationship between gas flow, pressure, and resistance is given by the following equation:

$$\text{Gas flow} = \frac{\text{pressure gradient}}{\text{resistance}}$$

Notice that the factors determining blood flow in the cardiovascular system and gas flow in the respiratory passages are equivalent. The volume of gas flow to and from the alveoli is directly proportional to the difference in pressure (pressure gradient) between the external atmosphere and the alveoli, and very small differences in pressure are sufficient to produce large volumes of gas flow. The average pressure gradient during normal quiet breathing is 3 mm Hg or less (see Figure 23.9b), and yet it moves 500 ml of air in and out of the lungs with each breath.

But as the equation also indicates, gas flow changes inversely with resistance; that is, the flow of gases decreases as resistance increases. Although it would seem that resistance would be highest in the smallest conduits, this is not the case for two important reasons: (1) The total cross-sectional area of the smallest respiratory passageways far exceeds that of the larger conducting passages that precede them in the respiratory pathway (Figure 23.11); and (2) gas flow, as such, stops at the terminal bronchioles, and diffusion takes over as the major driving force effecting gas movements. However, the smooth muscle of the bronchiolar walls is exquisitely sensitive to parasympathetic stimulation and inflammatory chemicals, both of which promote vigorous constriction of the bronchioles and a dramatic reduction in air passage. Indeed, the strong bronchoconstriction occurring during an acute asthma attack can stop pulmonary ventilation almost completely, regardless of the pressure gradient. Conversely, sympathetic nervous system activity and epinephrine dilate the bronchioles and reduce airway resistance. Under normal circumstances, the greatest resistance to gas flow is encountered in the medium-sized bronchi. Local accumulations of mucus, infectious material, or solid tumors in the passageways become important sources of airway resistance in respiratory disease.

Any factor that amplifies airway resistance demands that breathing movements become more strenuous. However, such compensation has its limits, and when the bronchioles are severely constricted or obstructed, even the most magnificent respiratory efforts are ineffective in restoring ventilation to life-sustaining levels.

Lung Compliance

Healthy lungs are unbelievably stretchy, or distensible. The ease with which the lungs can be expanded—that is, their distensibility—is referred to as **lung compliance**. Lung compliance depends not only on the elasticity of lung tissue itself, but also on that of the thoracic cage. Compliance is diminished by any factor that (1) reduces the natural resilience of the lungs, such as fibrosis, (2) blocks the bronchi or smaller respiratory passageways, since this reduces alveolar ventilation, or (3) impairs the flexibility of the thoracic cage. Lung compliance is assessed by measuring the increase in volume resulting from an increase in intrapulmonary pressure: The greater the volume increase for a given rise in pressure, the greater the compliance.

Deformities of the thorax, ossification of the costal cartilages (common during old age), or paralysis of the intercostal muscles all reduce the compliance of the lungs because they hinder thoracic expansion. ■

$CO_2 + H_2O =$

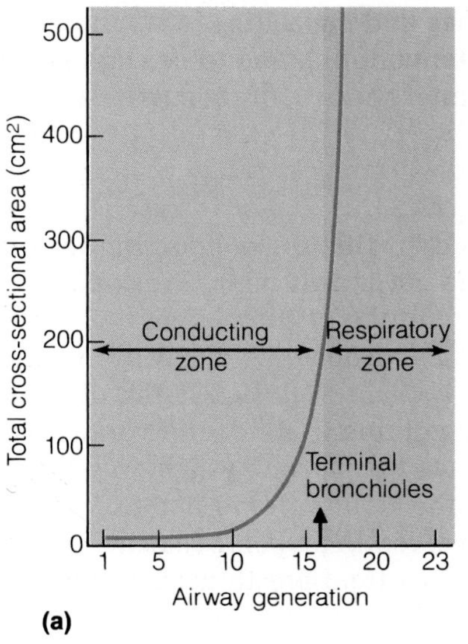

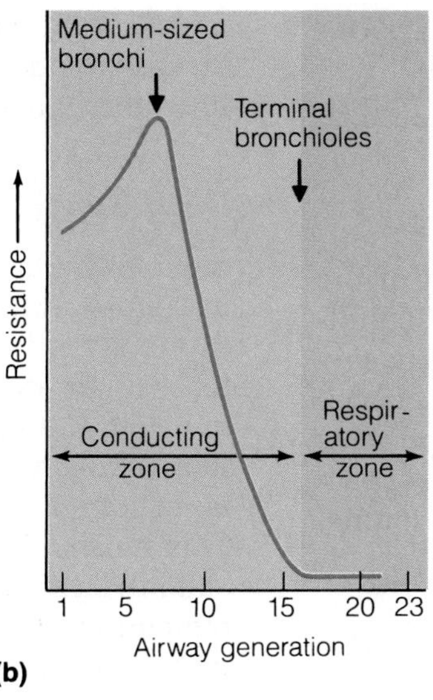

Figure 23.11 Cross-sectional area and resistance of the various segments of the respiratory passageways. (**a**) Of the 23 generations of respiratory passageways beginning with the trachea, the first 10 make up the conducting zone. Total cross-sectional area increases very little until the medium and smaller bronchi are encountered. The total cross-sectional area increases tremendously in the respiratory zone beyond the terminal bronchioles. This leads to a reduction in the velocity of airflow, which facilitates gas exchange across the respiratory membrane. (**b**) The peak of airway resistance occurs in the medium-sized bronchi and then declines rapidly as the total cross-sectional area of the airway increases rapidly.

Lung Elasticity

Whereas lung distension is required for normal inspiration, lung recoil is essential for normal expiration. These two functional attributes of fibroelastic lung tissue represent opposite faces of the same coin, *lung elasticity*. The importance of lung elasticity becomes glaringly clear when it declines, as in emphysema.

In sufferers of *emphysema* (em″-fuh-sē′-muh), the lungs become progressively less elastic and more fibrous, which hinders both inspiration and expiration. However, the major problem is the loss of recoil properties, which means that muscular activity must be enlisted to expire, a process that normally requires little or no energy consumption. Emphysema victims are sometimes referred to as "pink puffers," because their breathing is labored, but they do not become cyanotic (blue) because gas exchange remains adequate until late in the disease. They are perpetually exhausted because they must expend 15% to 20% of their total body energy supply just to breathe. This is quite a contrast to the 5% energy requirement for the work of breathing in healthy individuals. ■

Alveolar Surface Tension Forces

At any gas-liquid boundary, the molecules of the liquid are more strongly attracted to each other than to the gas. This phenomenon produces a state of tension at the liquid surface, called *surface tension*, that draws the liquid molecules ever more closely together and reduces their overall contact with the dissimilar gas molecules.

A watery film coats the alveolar walls, and water, composed of highly polar molecules, has a very high surface tension that perpetually acts to reduce the alveoli to their smallest possible size. If the alveolar film was pure water, the alveoli would, in fact, collapse between each breath; but the alveolar film contains surfactant. **Surfactant** (ser-fak′-tent), a detergent-like phospholipoprotein produced by the type II alveolar cells, interferes with the cohesiveness of water molecules, much the way a laundry detergent reduces the attraction of water for water, allowing water to interact with and pass through fabrics. As a result, the surface tension of alveolar fluid is reduced, and less energy is needed to overcome surface tension forces to expand the lungs.

When insufficient amounts of surfactant are present, surface tension forces become adequate to collapse the alveoli. The alveoli must be completely reinflated during each inspiration, which requires tremendous energy expenditure. This phenomenon accounts for many of the signs and symptoms of *infant respiratory distress syndrome (IRDS)*, a condition peculiar to premature babies. Since pulmonary surfactant is produced toward the end of fetal life, babies born prematurely often have too little to maintain patent alveoli between breaths. This condition is treated by using positive-pressure respirators to maintain a positive gas pressure in the alveoli and keep the alveoli open continuously. ■

Respiratory Volumes and Pulmonary Function Tests

Respiratory Volumes and Capacities

The amount of air that is flushed in and out of the lungs varies substantially depending on the conditions of inspiration and expiration. Consequently, several different respiratory volumes are described. Additionally, information about a person's respiratory status can be gained by measuring various lung capacities, which consist of differing combinations (sums) of the respiratory volumes.

Respiratory Volumes. The **respiratory**, or **lung, volumes** include tidal, inspiratory reserve, expiratory reserve, and residual volumes. The values recorded in Figure 23.12 represent normal values for a healthy 20-year-old male weighing about 70 kg (155 pounds).

During normal quiet breathing, about 500 ml of air moves into and out of the lungs with each breath. This respiratory volume is referred to as the **tidal volume (TV)**. The amount of air that can be inspired forcibly beyond the tidal volume is called the **inspiratory reserve volume (IRV)**; IRV is normally 2100–3200 ml.

The **expiratory reserve volume (ERV)** is the amount of air—normally 1000–1200 ml—that can be evacuated from the lungs over and above a tidal expiration. Even after the most strenuous expiration, about 1200 ml of air still remains in the lungs; this is the **residual volume (RV)**. Residual volume air is important because it helps to maintain the patency of the alveoli and prevents lung collapse.

Respiratory Capacities. The **respiratory capacities**, measured for diagnostic purposes, include inspiratory capacity, functional residual capacity, vital capacity, and total lung capacity (see Figure 23.12). As noted, the respiratory capacities always consist of two or more lung volumes.

The **inspiratory capacity (IC)** is the total amount of air that can be inspired after a tidal expiration; thus, it is the sum of the tidal and inspiratory reserve volumes. The **functional residual capacity (FRC)** is the combined residual and expiratory reserve volumes and represents the amount of air remaining in the lungs after a tidal expiration.

Vital capacity (VC) is the total amount of exchangeable air. It represents the sum of the tidal, inspiratory reserve, and expiratory reserve volumes. In healthy young males, the VC is approximately 4800 ml. The **total lung capacity (TLC)** is the sum of all lung volumes and is normally around 6000 ml in males. Lung volumes and capacities (with the possible exception of tidal volume) tend to be slightly less in women than in men because of their smaller size.

Dead Space

Some of the inspired air fills the conducting respiratory passageways and never contributes to gas exchange in the alveoli. The volume of these conduits, which make up the *anatomical dead space*, typically amounts to about 150 ml. (The rule of thumb is that the anatomical dead space volume in a healthy young adult is equal in milliliters to the person's weight in pounds.) This means that if the tidal volume is 500 ml, only 350 ml of this is involved in alveolar ventilation. The remaining 150 ml of the tidal breath is in the anatomical dead space.

If some of the alveoli become nonfunctional and cease to act in gas exchange (due to alveolar collapse or obstruction by mucus, for example), the *alveolar dead space* is added to the anatomical dead space, and the sum of the nonuseful volumes is referred to as **total dead space**.

Pulmonary Function Tests

Because the various lung volumes and capacities are often abnormal in people with pulmonary disorders, they are routinely measured. A *spirometer* (spī-rom′-eh-ter) is a simple instrument containing a hollow bell, inverted over water, that is displaced as the patient breathes into a connecting mouthpiece. As the test is conducted, a graphic recording is made on a rotating drum. Spirometry testing is most useful for evaluating losses in respiratory function and for following the course of certain respiratory diseases. Although it cannot provide a specific diagnosis, spirometry can distinguish between disease patterns interfering with expiration (obstructive disorders) and those interfering with inspiration (restrictive disorders). For example, increases in TLC, FRC, and RV may occur as a result of hyperinflation of the lungs in obstructive disease, whereas VC, TLC, FRC, and RV are reduced in restrictive diseases, which limit lung expansion.

Much more information can be obtained about a patient's ventilation status if the rate of gas movement in and out of the lungs is assessed. The **minute respiratory volume (MRV)** is the total amount of gas that flows into and out of the respiratory tract in 1 minute. During normal quiet breathing, the MRV in healthy people is about 6 L/min (500 ml per breath multiplied by 12 breaths per minute). During vigorous exercise, it is not uncommon for the MRV to reach 200 L/min, as both the depth and rapidity of breathing increase.

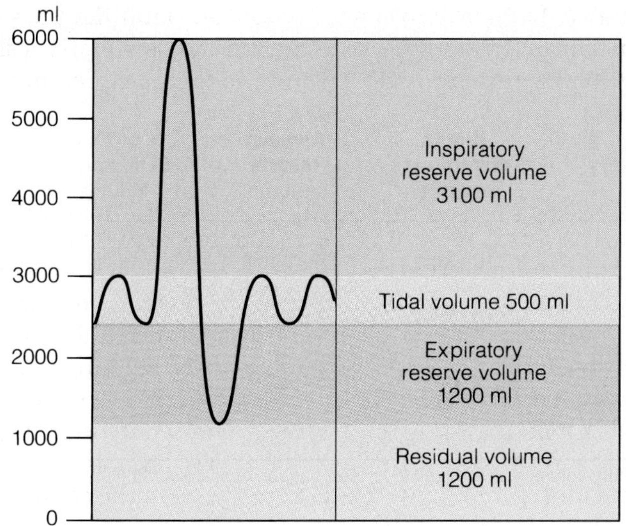

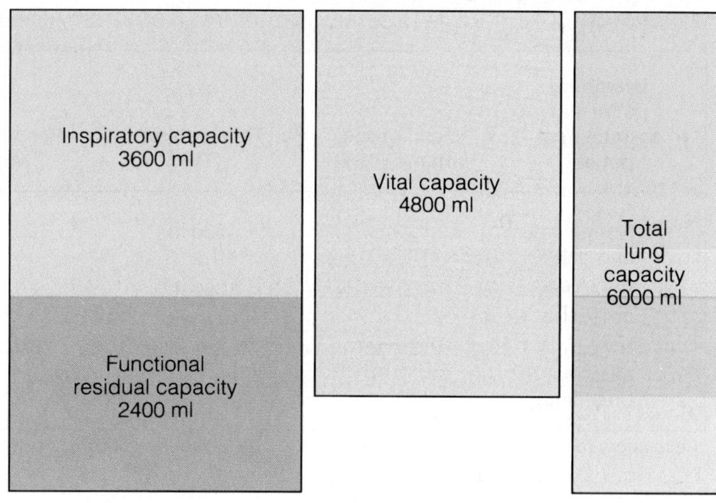

(a) Spirographic record

	Measurement	Adult male average value	Description
Respiratory volumes	Tidal volume (TV)	500 ml	Amount of air inhaled or exhaled with each breath under resting conditions
	Inspiratory reserve volume (IRV)	3100 ml	Amount of air that can be forcefully inhaled after a normal tidal volume inhalation
	Expiratory reserve volume (ERV)	1200 ml	Amount of air that can be forcefully exhaled after a normal tidal volume exhalation
	Residual volume (RV)	1200 ml	Amount of air remaining in the lungs after a forced exhalation
Respiratory capacities	Total lung capacity (TLC)	6000 ml	Maximum mount of air contained in lungs after a maximum inspiratory effort: TLC = TV + IRV + ERV + RV
	Vital capacity (VC)	4800 ml	Maximum amount of air that can be expired after a maximum inspiratory effort: VC = TV + IRV + ERV (should be 80% TLC)
	Inspiratory capacity (IC)	3600 ml	Maximum amount of air that can be inspired after a normal expiration: IC = TV + IRV
	Functional residual capacity (FRC)	2400 ml	Volume of air remaining in the lungs after a normal tidal volume expiration: FRC = ERV + RV

(b) Summary of respiratory volumes and capacities

Figure 23.12 Respiratory volumes and capacities. (**a**) Idealized spirographic record of respiratory volumes. (**b**) Summary of respiratory volumes and capacities in a healthy young male weighing approximately 70 kg.

Alveolar Ventilation Rate (AVR)

While the MRV provides a rough-and-ready yardstick for assessing respiratory efficiency, the **alveolar ventilation rate (AVR)** is a better index of effective ventilation. The AVR takes into account the volume of air wasted in dead space areas and provides a measurement of the concentration of fresh gases in the alveoli at a particular time. AVR may be computed with the following equation:

$$\underset{(\text{ml/min})}{\text{AVR}} = \underset{(\text{breaths/min})}{\text{frequency}} \times \underset{(\text{ml/breath})}{(\text{TV} - \text{dead space})}$$

In the absence of respiratory impairments, the AVR is usually about 12 breaths per minute multiplied by the difference of 500 − 150 ml per breath, or 4200 ml/min. Increasing the volume of each inspiration is more effective in enhancing alveolar ventilation and gas exchange than increasing the respiratory rate because anatomical dead space is a constant in any particular individual. When breathing is rapid and shallow, alveolar ventilation drops dramatically, since most of the inspired air never reaches the exchange sites. Furthermore, as tidal volume approaches the dead space

Table 23.2 Effect of Breathing Rate and Depth on Alveolar Ventilation of Three Hypothetical Patients

Breathing pattern of hypothetical patient	Dead space volume (DSV)	Tidal volume (TV)	Respiratory rate*	Minute respiratory volume	Alveolar minute volume	% of TV = dead space volume
I—Normal rate and depth	150 ml	500 ml	20/min	10 L/min	7 L/min	30%
II—Slow, deep breathing	150 ml	1000 ml	10/min	10 L/min	8.5 L/min	15%
III—Rapid, shallow breathing	150 ml	250 ml	40/min	10 L/min	4 L/min	60%

*Respiratory rate values are artificially adjusted to provide equivalent minute respiratory volumes as a baseline for comparison of alveolar ventilation.

value, effective ventilation approaches zero, regardless of the rapidity of breathing. The effects of rate and depth of breathing on effective alveolar ventilation are summarized for three hypothetical patients in Table 23.2.

Nonrespiratory Air Movements

There are many processes other than breathing that move air into or out of the lungs and that may modify the normal respiratory rhythm. Coughs and sneezes function to clear the air passages of debris or collected mucus; laughing and crying reflect our emotions. For the most part, these **nonrespiratory air movements** are a result of reflex activity, but some are produced voluntarily. Examples of the most common of these movements are given in Table 23.3.

Gas Exchanges in the Body

Basic Properties of Gases

Gas exchange in the body occurs by bulk flow of gases (and solutions of gases) and by diffusion of gases through tissues. To fully understand these processes, it is necessary to examine some of the physical properties of gases and their behavior in liquids. Two of the *ideal gas laws*—Dalton's law of partial pressures and Henry's law—provide most of the information we need.

Table 23.3 Nonrespiratory Air (Gas) Movements

Movement	Mechanism and result
Cough	Taking a deep breath, closing glottis, and forcing air superiorly from lungs against glottis; glottis opens suddenly and a blast of air rushes upward; can dislodge foreign particles or mucus from lower respiratory tract and propel such substances superiorly
Sneeze	Similar to a cough, except that expelled air is directed through nasal cavities instead of through oral cavity; depressed uvula closes oral cavity off from pharynx and routes air upward through nasal cavities; sneezes clear upper respiratory passages
Crying	Inspiration followed by release of air in a number of short expirations; primarily an emotionally induced mechanism
Laughing	Essentially same as crying in terms of air movements produced; also an emotionally induced response
Hiccups	Sudden inspirations resulting from spasms of diaphragm; believed to be initiated by irritation of diaphragm or phrenic nerves, which serve diaphragm; sound occurs when inspired air hits vocal folds of closed glottis
Yawn	Very deep inspiration, taken with jaws wide open; formerly believed to be triggered by need to increase amount of oxygen in blood but this theory is now being questioned; ventilates all alveoli (not the case in normal quiet breathing)

Table 23.4 Comparison of Gas Partial Pressures and Approximate Percentages in the Atmosphere and in the Alveoli

	Atmosphere (sea level)		Alveoli	
Gas	Approximate percentage	Partial pressure (mm Hg)	Approximate percentage	Partial pressure (mm Hg)
N_2	78.6	597	74.9	569
O_2	20.9	159	13.7	104
CO_2	0.04	0.3	5.2	40
H_2O	0.46	3.7	6.2	47
	100.0%	760	100.0%	760

Dalton's Law of Partial Pressures

Dalton's law of partial pressures, or simply **Dalton's law**, states that the total pressure exerted by a mixture of gases is the sum of the pressures exerted independently by each gas in the mixture. Further, the pressure exerted by each gas—its **partial pressure**—is directly proportional to its percentage in the total gas mixture.

As noted earlier, the pressure exerted by ambient air (that is, atmospheric pressure) is approximately 760 mm Hg at sea level. (It varies slightly with the weather.) Since air is a mixture of gases, we can examine it more closely in terms of the gases that comprise it. As indicated in Table 23.4, nitrogen makes up about 79% of air, and thus, the partial pressure of nitrogen (written as P_{N_2}) is 78.6 × 760 mm Hg, or 597 mm Hg. Oxygen, which accounts for nearly 21% of the atmosphere, has a partial pressure (P_{O_2}) of 159 mm Hg (20.9 × 760 mm Hg). As you can see, nitrogen and oxygen together contribute about 99% of the total atmospheric pressure. Air also contains 0.04% carbon dioxide, up to 0.5% water vapor, and insignificant amounts of inert gases such as argon and helium.

At high altitudes, where the atmosphere is less influenced by gravitational pull, all partial pressure values decline in direct proportion to the decline in atmospheric pressure. For example, at 10,000 feet above sea level, where the atmospheric pressure is 563 mm Hg, the partial pressure of oxygen is 110 mm Hg. Below sea level (beneath the sea or in deep tunnels), atmospheric pressure increases by 1 atmosphere (760 mm Hg) for each 33 feet of descent. Thus, at 99 feet below sea level, the total pressure on the body is equivalent to 4 atmospheres, or 3040 mm Hg, and the partial pressure exerted by each of the component gases is also tripled.

Henry's Law

Gases diffuse from high-pressure areas to low-pressure areas. The rate of gas diffusion depends on (1) the steepness of the partial pressure gradient and (2) the nature of the barrier between the two areas. According to **Henry's law**, when a mixture of gases is in contact with a liquid, each gas will dissolve in the liquid in proportion to its partial pressure and its solubility in the liquid. When equilibrium is achieved, the gas partial pressures in both phases are equal.

The various gases present in air have very different solubilities in water or plasma. Carbon dioxide is most soluble, with a *solubility coefficient* of 0.57. Oxygen, with a solubility coefficient of 0.024, is relatively insoluble. Nitrogen is least soluble, with a solubility coefficient of 0.012. Beyond this specific condition, the solubility of *any* gas in water increases with increasing partial pressure and decreases with increasing temperature. To understand this idea, think of club soda, which is produced by forcing carbon dioxide gas into water under high-pressure conditions. If you take the cap off and allow the bottle to stand at room temperature, within just a few minutes you will have plain water—that is, all the carbon dioxide gas will have escaped from solution.

Influence of Hyperbaric Conditions

The effect of hyperbaric (hī″-per-bayr′-ik), or high-pressure, conditions on gas exchange in the body is easily explained by the ideal gas laws. The use of *hyperbaric oxygen chambers* provides a clinical application of Henry's law. These chambers contain oxygen at pressures higher than 1 atmosphere and are used to force greater than normal amounts of oxygen into a patient's blood in cases of carbon monoxide poisoning, circulatory shock, and asphyxiation. The technique is also used to treat individuals with gas gangrene or tetanus poisoning, because the anaerobic bacteria causing these infections cannot live in the presence of high levels of oxygen.

Although breathing oxygen at 2 atmospheres presents no problems for short periods of time, *oxygen toxicity* develops rapidly when the P_{O_2} is greater than 2.5–3 atmospheres. Excessive oxygen concentrations generate large amounts of harmful free radicals; the result is profound central nervous system disturbances, culminating in coma and death.

Although nitrogen accounts for nearly 80% of the gases in air, it ordinarily has little effect on body functions because it is so insoluble in blood. However, when a deep-sea diver breathes under hyperbaric conditions, the partial pressures of all gases in the mixture increase dramatically. As a result, much more nitrogen gas is forced into solution in the blood and causes *nitrogen narcosis*, evidenced by dizziness, drowsiness, giddiness, and other symptoms similar to alcohol intoxication. For this reason, the condition is sometimes called "rapture of the deep."

As long as a diver ascends to the surface gradually, dissolved nitrogen gas can be eliminated by the lungs without problems. But if ascent is rapid, the P_{N_2} decreases abruptly, and the poorly soluble nitrogen gas appears to boil out of solution in the body fluids, forming large gas bubbles. Gas bubbles in blood represent life-threatening emboli, and those formed within tissues can cause excruciating pain and may impair the normal function of any organ, including the brain. These phenomena are referred to collectively as *the bends*, or more accurately, *decompression sickness*. ■

Composition of Alveolar Gas

As shown in Table 23.4, the gaseous composition of the atmosphere is quite different from that within the alveoli. Whereas the atmosphere consists almost entirely of oxygen and nitrogen, the alveoli contain more carbon dioxide and water vapor and considerably less oxygen. These differences reflect the effects of the following processes: (1) gas exchanges occurring in the lungs (oxygen diffuses from the alveoli into the pulmonary blood and carbon dioxide diffuses in the opposite direction), (2) humidification of air by the conducting respiratory passageways, and (3) the mixing of alveolar gas that occurs with each breath. Because only 500 ml of air is inspired with each tidal inspiration, alveolar gas is actually a mixture of newly inspired gases and gases remaining in the respiratory passageways between breaths.

The partial pressures of oxygen and carbon dioxide are easily changed by increasing the depth and rate of breathing. High alveolar ventilation flushes more oxygen into the alveoli, increasing the alveolar P_{O_2}, and rapidly eliminates carbon dioxide from the lungs.

Gas Exchanges Between the Blood, Lungs, and Tissues

As described earlier, oxygen enters and carbon dioxide leaves the blood in the lungs (external respiration). These same gases move in opposite directions by the same mechanism (diffusion) at the body tissues (internal respiration). External and internal respiration will be considered consecutively to emphasize similarities in these two functional processes, but keep in mind that gases must be transported by the blood between these two exchange sites.

External Respiration: Pulmonary Gas Exchange

During the course of external respiration, dark red venous blood moving through the pulmonary circulation is transformed into the scarlet river that is returned to the heart for distribution by systemic arteries to all body tissues. Although it is the oxygen exchange process and binding of oxygen to hemoglobin in red blood cells that accounts for the color change of blood, carbon dioxide exchanges are occurring with equal rapidity. A number of factors influence the movement of these two gases across the respiratory membrane. These include (1) partial pressure gradients and gas solubilities, (2) anatomical characteristics of the respiratory membrane, and (3) functional considerations, such as the matching of alveolar airflow and pulmonary capillary circulation.

Partial Pressure Gradients and Gas Solubilities. Since the P_{O_2} of pulmonary blood is only 40 mm Hg, as opposed to a P_{O_2} of 104 mm Hg in the alveoli, a steep oxygen partial pressure gradient exists, and oxygen diffuses rapidly from the alveoli into the pulmonary capillary blood (Figure 23.13). Equilibrium—that is, a P_{O_2} of 104 mm Hg on both sides of the respiratory membrane—has usually occurred in 0.25 second, which is about one-third the time a red blood cell is in a pulmonary capillary (Figure 23.14). The lesson to take away here is that the time of blood flow through the pulmonary capillaries can be reduced by two-thirds and still provide adequate oxygenation. Carbon dioxide moves in the opposite direction along a much less steep partial pressure gradient of about 5 mm Hg (45 mm Hg to 40 mm Hg) until equilibrium occurs at 40 mm Hg (see Figure 23.13). Carbon dioxide is then evacuated gradually from the alveoli during expiration. Notice that even though the pressure gradient for oxygen is much steeper than that for carbon dioxide, equal amounts of these gases are exchanged because carbon dioxide solubility in plasma and alveolar fluid is 20 times greater than that of oxygen.

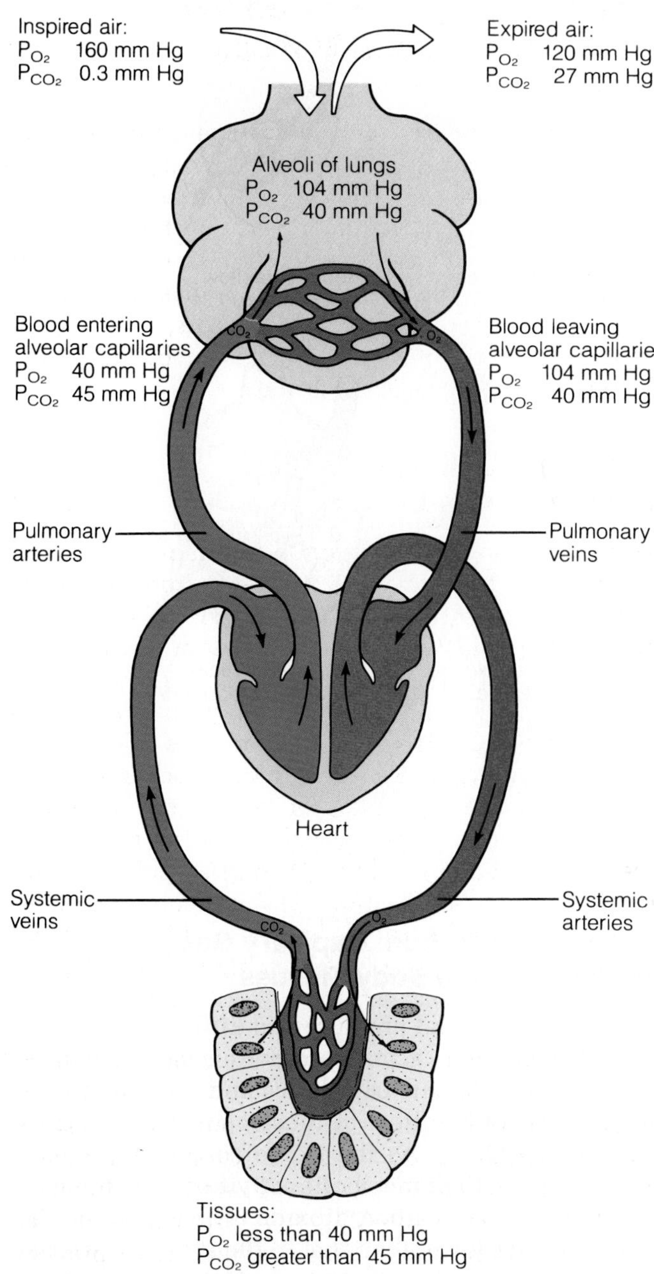

Figure 23.13 Partial pressure gradients promoting gas movements in the body. Gradients promoting oxygen and carbon dioxide exchanges across the respiratory membrane in the lungs (external respiration) are shown in the top part of the figure. Gradients promoting gas movements across systemic capillary membranes in the body tissues (internal respiration) are indicated in the bottom part of the figure.

Thickness of the Respiratory Membrane. In healthy lungs, the respiratory membrane is only 0.5–1 μm thick, and as noted, gas exchange is usually very efficient.

If the lungs become edematous (as during pulmonary congestion and pneumonia), the thickness of the exchange membrane may increase dramatically, restricting gas exchange. Under such conditions, even the total time (0.75 second) that the red blood cells are in transit through the pulmonary capillaries may be insufficient, and body tissues begin to suffer from oxygen deprivation, or *hypoxia* (hī-pok′-sē-uh). ■

Surface Area for Gas Exchange. The greater the surface area of the respiratory membrane, the greater the amount of gas that can diffuse across it in a given time period. The alveolar surface area is enormous in healthy lungs. It has been estimated that the total gas exchange surface provided by the alveolar walls of a man's lungs is 70–80 m^2, or approximately equal to the area of a racquetball court!

In certain pulmonary diseases, the alveolar surface area actually functioning in gas exchange may be drastically reduced. This occurs in emphysema, when the walls of adjacent alveoli break through and the alveolar chambers become more spacious; it also occurs when tumors, mucus, or inflammatory material block gas flow into the alveoli ■

Alveolar Airflow–Blood Flow Coupling. For gas exchange to be most efficient, there must be a precise match, or coupling, between the amount of gas reaching the alveoli and the blood flow in pulmonary capillaries. As explained in Chapter 20 and illustrated in Figure 23.15, local autoregulatory mechanisms acting in the respiratory zone continuously accommodate alveolar conditions. In regions where alveolar ventilation is inadequate, the P_{O_2} is low. As a result, the pulmonary vessels constrict, and blood is redirected to respiratory areas where oxygen pickup can be more efficient. On the other hand, when alveolar ventilation is maximal, the pulmonary arterioles dilate, increasing blood

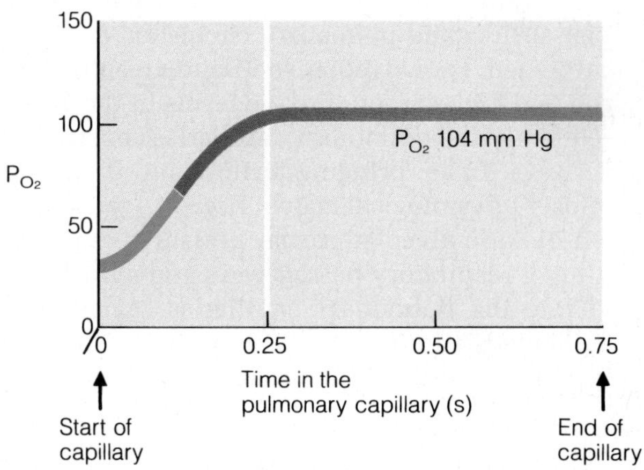

Figure 23.14 Oxygenation of blood in the pulmonary capillaries. Note that the time expired from the time blood enters the pulmonary capillaries (indicated by 0) until the P_{O_2} is 104 mm Hg is approximately 0.25 second.

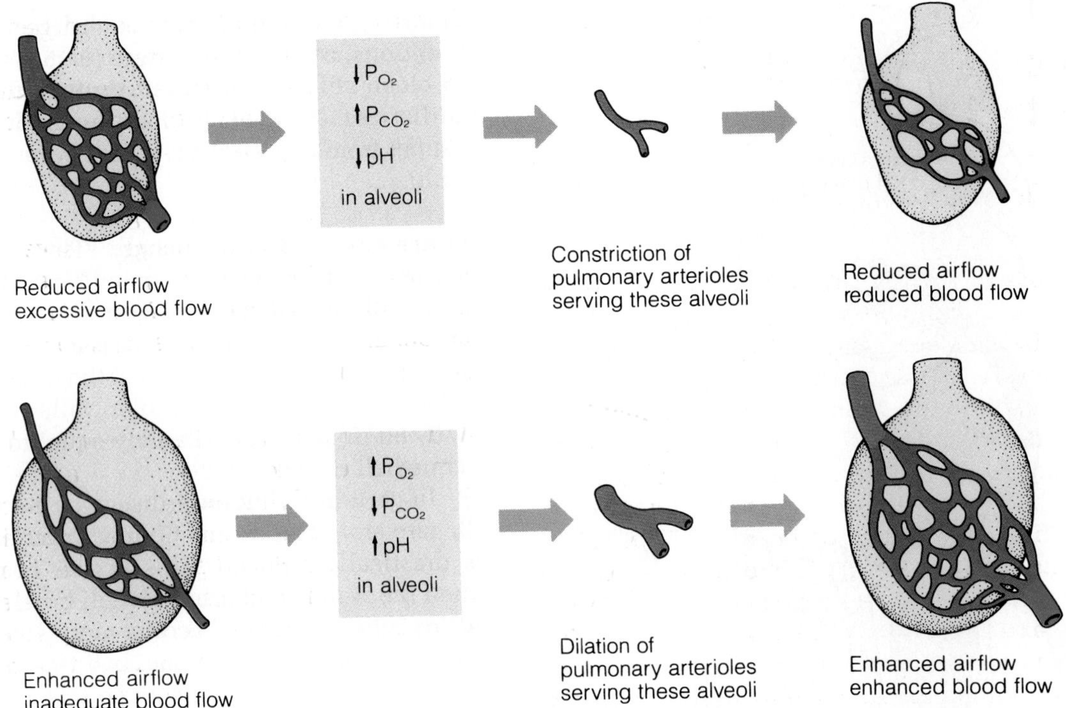

Figure 23.15 Alveolar airflow–blood flow coupling. The flowcharts follow the autoregulatory events that result in a local matching of alveolar ventilation and blood flow through the pulmonary capillaries.

flow into the pulmonary capillaries. Notice that the autoregulatory mechanism controlling pulmonary vascular muscle is exactly opposite that controlling most arterioles in the systemic circulation.

Changes in the P_{CO_2} within the alveoli cause changes in the diameters of the bronchioles. Passageways servicing areas where alveolar carbon dioxide levels are high dilate, allowing carbon dioxide to be eliminated from the body more rapidly; those serving areas where the P_{CO_2} is low constrict.

As a result of modifications in these two systems, alveolar airflow and pulmonary circulation are always synchronized. Poor alveolar ventilation results in low oxygen and high carbon dioxide levels in the alveoli; consequently, the pulmonary capillaries constrict and the airways dilate, bringing airflow and blood flow into closer physiological match. High oxygen and low carbon dioxide alveolar partial pressures cause constriction of respiratory passageways and a flushing of blood into the pulmonary capillaries. At all times, these homeostatic mechanisms provide the most appropriate conditions for efficient gas exchange.

Internal Respiration: Capillary Gas Exchange in the Body Tissues

Although the partial pressure and diffusion gradients are reversed, the factors promoting gas exchanges between the systemic capillaries and the tissue cells are essentially identical to those acting in the lungs (see Figure 23.13). As the tissue cells continuously use oxygen for their metabolic activities, they produce equal amounts of carbon dioxide. The P_{O_2} in the tissues is always lower than that in the systemic arterial blood (40 mm Hg versus 104 mm Hg). Consequently, oxygen moves rapidly from the blood into the tissues until equilibrium is reached, and carbon dioxide moves quickly along its partial pressure gradient into the blood. As a result, venous blood draining the capillary beds of the tissues and returning to the heart has a P_{O_2} of 40 mm Hg and a P_{CO_2} of 45 mm Hg.

In summary, the gas exchanges that occur between the blood and the alveoli on the one hand and between the blood and the tissue cells on the other take place by simple diffusion and depend on partial pressure gradients of oxygen and carbon dioxide that exist on the opposite sides of the exchange membranes.

Transport of Respiratory Gases by Blood

Oxygen Transport

Molecular oxygen is carried in blood in two ways: bound to hemoglobin within red blood cells and dissolved in plasma (see Figure 23.18, p. 736). Because oxygen is poorly soluble in water, only about 1.5% is carried in the dissolved form. Assuming normal lung function and arterial blood P_{O_2}, this amounts to about 0.3 ml of oxygen dissolved in each 100 ml of plasma. Indeed, if this were the only means of oxygen transport, a P_{O_2} of 3 atmospheres or a cardiac output of 15 times normal would be required to provide normal oxygen levels needed by body tissues! This problem, of course, has been solved by hemoglobin, and 98.5% of the oxygen ferried from the lungs to the tissues is carried in a loose chemical combination with hemoglobin.

Association and Dissociation of Oxygen and Hemoglobin

As described in Chapter 18, hemoglobin is composed of four polypeptide chains, each bound to an iron-containing heme group (see Figure 18.4). Since the iron groups are oxygen-binding sites, each hemoglobin molecule can combine with four molecules of oxygen.

The process of oxygen binding is rapid and reversible. The hemoglobin-oxygen combination, called *oxyhemoglobin* (ok″-sē-hē-muh-glō-bin), is customarily written as HbO_2. Hemoglobin that has released oxygen is called *reduced hemoglobin,* or *deoxyhemoglobin* and written as HHb. Both the loading and unloading of oxygen can be indicated by a single reversible equation:

$$HHb + O_2 \underset{\text{Tissues}}{\overset{\text{Lungs}}{\rightleftharpoons}} HbO_2$$

There is a kind of cooperation between the four polypeptides of the hemoglobin molecule. The binding of one O_2 molecule causes the remainder of the hemoglobin molecule to undergo a conformational change. As a result, hemoglobin more readily takes up another oxygen molecule and uptake of the fourth oxygen molecule is extremely facilitated. When all four of its heme groups are bound to oxygen, a hemoglobin molecule is said to be *fully saturated;* when one, two, or three oxygen molecules are bound, hemoglobin is *partially saturated.* By the same token, unloading of one oxygen molecule enhances the unloading of the next, and so on. Thus, the *affinity* of hemoglobin for oxygen changes with its state of oxygen saturation, and both loading and unloading of oxygen are very efficient.

The rate at which hemoglobin reversibly binds or releases oxygen is regulated by several factors: P_{O_2}, temperature, blood pH, P_{CO_2}, and the concentration of an organic chemical called 2,3-DPG in the blood. These factors interact to ensure adequate deliveries of oxygen to tissue cells.

Hemoglobin and P_{O_2}. The extent to which oxygen is bound to hemoglobin depends to a large degree on the P_{O_2} of the blood, but the relationship is not a precisely linear one. When hemoglobin saturation is plotted against oxygen partial pressure, the S-shaped oxygen-hemoglobin dissociation curve produced has a steep slope between 10 and 50 mm Hg P_{O_2} and then flattens out (plateaus) between 70 and 100 mm Hg (Figure 23.16).

Under normal resting conditions, each 100 ml of blood contains 19.7 ml of oxygen (approximately 20% by volume). A small amount of this oxygen is dissolved in plasma; the rest is combined with hemoglobin. Under these conditions, the hemoglobin is 97.4% saturated with oxygen. As arterial blood flows through

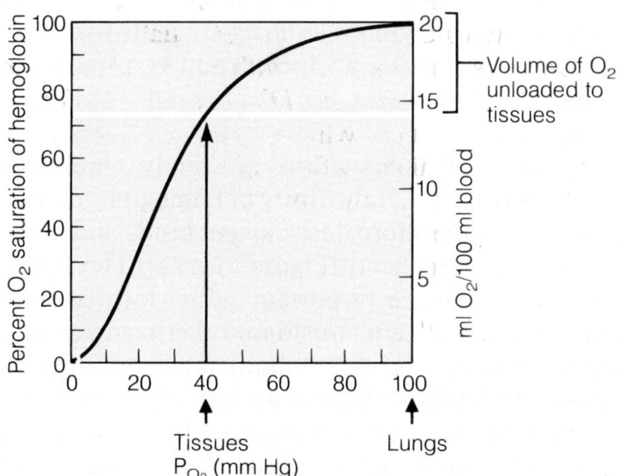

Figure 23.16 Oxygen-hemoglobin dissociation curve. Percent O_2 saturation of hemoglobin and blood oxygen content are shown at different oxygen partial pressures (P_{O_2}). Notice that hemoglobin is almost completely saturated with oxygen at a P_{O_2} of 70 mm Hg. Rapid loading and unloading of oxygen to and from hemoglobin occurs at the oxygen partial pressures corresponding to the steep portion of the curve. During the systemic circuit, approximately 25% of the hemoglobin-bound oxygen is unloaded to the tissues (that is, approximately 5 ml of the 20 ml of O_2 per 100 ml arterial blood is released). Thus, hemoglobin of venous blood is still about 75% saturated with oxygen.

the systemic capillaries, about 5 ml O_2 per 100 ml of blood is unloaded, resulting in a hemoglobin saturation of 75% in venous blood. Recognizing that hemoglobin is almost completely saturated in arterial blood is important, because it explains why breathing deeply (which increases both the alveolar and arterial blood P_{O_2} above 104 mm Hg) has little effect on increasing the oxygen saturation of hemoglobin. Remember, P_{O_2} measurements indicate only the amount of oxygen dissolved in plasma, not the amount bound to hemoglobin. However, P_{O_2} values provide a good index of lung function, and when arterial P_{O_2} is not equal to alveolar air P_{O_2}, some degree of respiratory impairment exists.

Two other important pieces of information can be gained from looking at a hemoglobin saturation curve. First, hemoglobin is almost completely saturated at a P_{O_2} of 70 mm Hg, and further increases in P_{O_2} cause only small increases in oxygen binding. The adaptive value of this is that adequate oxygen loading and delivery to the tissues can occur when the P_{O_2} of inspired air is substantially below its usual levels, a situation common at higher altitudes and in cardiopulmonary disease. Moreover, because most oxygen unloading also occurs on the steep portion of the curve, where the partial pressure changes very little, only 20% to 25% of bound oxygen is unloaded during one systemic circuit (see Figure 23.16), and substantial amounts of oxygen are still available in venous blood (the "venous reserve"). Thus, if oxygen tension drops to very low levels in the tissues, as might occur during vigorous muscle activity, much more oxygen can dissociate from hemoglobin to be used by the tissue cells without necessitating an increase in respiratory rate or cardiac output.

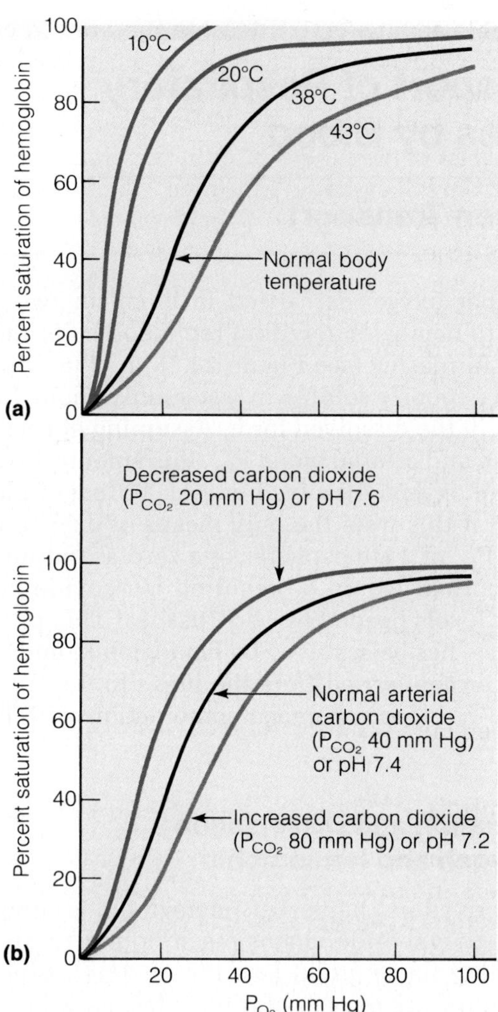

Figure 23.17 Effects of temperature, P_{CO_2}, and blood pH on the oxygen-hemoglobin dissociation curve. Oxygen unloading is accelerated at conditions of (**a**) increased temperature and (**b**) increased P_{CO_2} and/or decreased pH, resulting in a shift to the right of the dissociation curve.

Hemoglobin and Temperature. As body temperature rises above normal, the affinity of hemoglobin for oxygen declines; therefore, less oxygen binds and oxygen unloading is enhanced (Figure 23.17a). The temperature effect is more important when localized temperature increases are considered. Because one of the by-products of cellular metabolism is heat, active tissues are typically warmer than less active ones; consequently, more oxygen is released from hemoglobin in the vicinity of hard-working tissues. Conversely, when the tissues are chilled, oxygen unloading is inhibited and, although your exposed cheeks and ears become bright red (an indication of a high rate of blood flow through those areas), little oxygen is released to those tissues.

Hemoglobin and pH. Increases in hydrogen ion (H^+) concentration weaken the strength of the hemoglobin-oxygen bond. This phenomenon is called the *Bohr effect*. As blood pH declines (acidosis), the affinity of hemoglobin for oxygen declines as well (Figure 23.17b), and oxygen unloading to the tissues is enhanced. The Bohr effect is less pronounced at higher P_{O_2} levels, such as in the lungs, and only slightly less oxygen is loaded into the blood as blood pH declines. However, the P_{O_2} is always much lower in the tissues, and as H^+ rises, much more oxygen is made available to the tissues. The same relative effect occurs when the partial pressure of carbon dioxide rises because this leads to a fall in blood pH (see Figure 23.17b). These effects are beneficial because active tissues, such as exercising muscles, release more carbon dioxide and generate more hydrogen ions, both of which accelerate the release of oxygen from oxyhemoglobin to that tissue.

Hemoglobin and 2,3-DPG. Because mature red blood cells lack mitochondria, their metabolism of glucose is entirely anaerobic. The compound *2,3-diphosphoglycerate* (dī-fos″-fō-glih′-ser-āt) *(2,3-DPG)*, a unique intermediate compound that enhances the release of oxygen from hemoglobin, is formed in red blood cells in the pathway called anaerobic glycolysis.

The major effect of 2,3-DPG on oxyhemoglobin levels can be summarized as follows: Whenever the metabolic rate of red blood cells (RBCs) increases (as might occur during fever), more 2,3-DPG is produced. As the 2,3-DPG level rises, it exerts feedback inhibition on the enzyme that catalyzes its formation; consequently, RBC metabolism can continue only if 2,3-DPG levels decline. This is accomplished by the binding of 2,3-DPG to hemoglobin, an event that alters the structure of hemoglobin and, subsequently, enhances oxygen release and unloading to the tissues.

Certain hormones, such as thyroxine, testosterone, growth hormones, and catecholamines (epinephrine and norepinephrine), increase RBC metabolic rate and 2,3-DPG formation. As a result, these hormones directly enhance oxygen delivery to the tissues.

Impairments of Oxygen Transport

Whatever the cause, any state in which an inadequate amount of oxygen is delivered to body tissues is called *hypoxia*. Hypoxia can often be recognized simply by observing a person's skin color, because the skin and mucosae take on a blue cast (become cyanotic) when the blood is oxygen-poor.

Anemic hypoxia reflects poor oxygen delivery resulting from too few RBCs or from RBCs that contain abnormal or inadequate amounts of hemoglobin.

Stagnant hypoxia ensues when blood circulation is impaired or blocked. Congestive heart failure may cause body-wide stagnant hypoxia, whereas emboli or thrombi restrict oxygen delivery to tissues distal to the obstruction.

Hypoxemic hypoxia reflects some interference with gas exchange in the lungs. Possible causes include pulmonary diseases that impair lung ventilation or the breathing in of air that contains scanty amounts of oxygen.

Carbon monoxide poisoning represents a unique type of hypoxemic hypoxia. As mentioned in Chapter 18, carbon monoxide (CO) is an odorless, colorless gas that competes vigorously with oxygen for heme binding sites. Moreover, because the affinity of hemoglobin for carbon monoxide is over 200 times greater than that for oxygen, carbon monoxide is a highly successful competitor—so much so that even at minuscule partial pressures, carbon monoxide is able to crowd out or displace oxygen. For example, blood exposed to a mixture of gases containing 21% oxygen and only 0.1% carbon monoxide will bind equal amounts of the two gases.

Carbon monoxide poisoning is particularly dangerous because it kills its victims softly and quietly; it is the leading cause of death from fire. It does not produce the characteristic signs of hypoxia—cyanosis and respiratory distress. Instead, the skin becomes cherry red (the color of the hemoglobin–carbon monoxide complex), which the eye of the beholder interprets as a healthy "blush," and the person becomes confused. Those with CO poisoning are given 100% oxygen until the carbon monoxide has been cleared from the body. ■

Carbon Dioxide Transport

Normally active body cells produce about 200 ml of carbon dioxide each minute, which is the exact amount of carbon dioxide excreted by the lungs per minute. For carbon dioxide to be eliminated from the blood, it must be transported from the tissue cells to the lungs. This transport process occurs in three forms: as a gas dissolved in plasma, chemically bound to hemoglobin, and in plasma as the bicarbonate ion (Figure 23.18).

1. *Dissolved in plasma.* From 7% to 8% of carbon dioxide transported is simply dissolved in the plasma. Most of the remainder of the carbon dioxide molecules entering the plasma quickly enter the red blood cells, where most of the chemical reactions that prepare carbon dioxide for transport occur.

2. *Chemically bound to hemoglobin in red blood cells.* Approximately 20% of transported carbon dioxide is carried within RBCs as *carbaminohemoglobin* (kar″-ba-min-ō hē′-muh-glō-bin):

$$CO_2 + \text{hemoglobin} \rightleftharpoons HbCO_2 \text{ (carbaminohemoglobin)}$$

This reaction is rapid and does not require a catalyst. Since carbon dioxide binds directly with the amino acids of globin (and not to the heme), carbon dioxide transport within RBCs does not compete with the oxyhemoglobin transport mechanism.

The loading and unloading of carbon dioxide to and from hemoglobin is directly influenced by the P_{CO_2} and the degree of oxygenation of hemoglobin. Dissociation of carbon dioxide from hemoglobin occurs rapidly in the lungs, where the P_{CO_2} of alveolar air is quite low; carbon dioxide is loaded in the tissues, where the P_{CO_2} is higher than that of the blood. As a general rule, deoxygenated hemoglobin can combine with more carbon dioxide than oxygenated hemoglobin (see the discussion of the Haldane effect shortly). However, recent studies suggest that 2,3-DPG competes with carbon dioxide for hemoglobin binding sites,

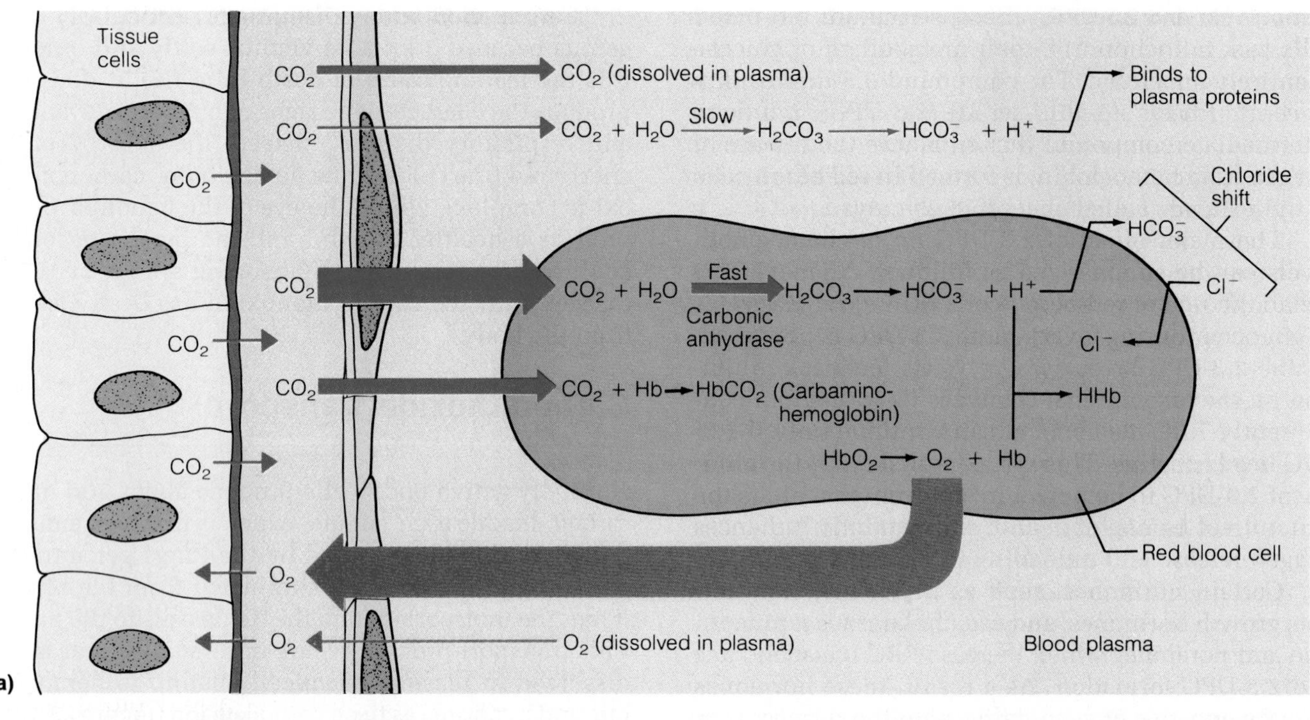

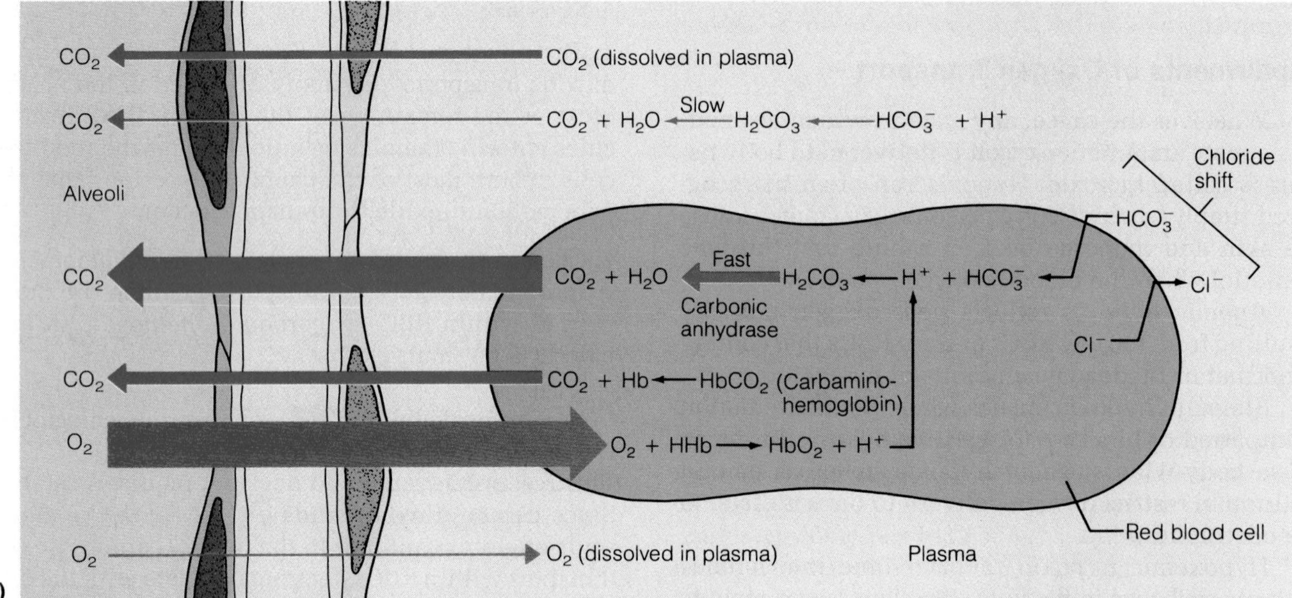

Figure 23.18 Transport and exchange of carbon dioxide and oxygen. Gas exchanges occurring (**a**) at the tissues and (**b**) in the lungs. Carbon dioxide is transported primarily as bicarbonate ion (HCO_3^-) in plasma (70%); smaller amounts are transported bound to hemoglobin (indicated as $HbCO_2$) in the red blood cells (22%) or in physical solution in plasma (7%). Nearly all the oxygen transported in blood is bound to hemoglobin as oxyhemoglobin (HbO_2) in red blood cells. Very small amounts (about 1.5%) are carried dissolved in plasma. (Reduced hemoglobin is indicated as HHb.)

and the amount of carbon dioxide transported by this mechanism is probably substantially less than the 20% originally assumed.

3. *As bicarbonate ion in plasma.* The largest fraction of carbon dioxide (70%) is converted to bicarbonate ions (HCO_3^-) and transported in plasma. As illustrated in Figure 23.18a, when carbon dioxide diffuses from the tissue cells into the RBCs, it combines with water, forming carbonic acid (H_2CO_3), which, being unstable, quickly dissociates into hydrogen ions and bicarbonate ions:

$$CO_2 + H_2O \rightleftharpoons H_2CO_3 \rightleftharpoons H^+ + HCO_3^-$$

Although this same reaction also occurs in plasma, it is thousands of times faster in erythrocytes because they (and not plasma) contain *carbonic anhydrase* (kar-bon'-ik an-hī-drās), an enzyme that reversibly catalyzes the conversion of carbon dioxide and water to carbonic acid. The hydrogen ions released during the reaction bind to hemoglobin; thus, oxygen release is enhanced by carbon dioxide loading (as HCO_3^-). Because of the buffering effect of hemoglobin, the liberated hydrogen ions cause little change in pH under resting conditions, and blood becomes only slightly more acidic (that is, the pH declines from 7.4 to 7.34) as it passes through the tissues.

Once generated, bicarbonate ions diffuse quickly from the RBCs into the plasma, where they are carried to the lungs. To counterbalance the net positive charge left within the RBCs by the rapid outrush of negative bicarbonate ions, chloride ions (Cl^-) move from the plasma into the erythrocytes. This ionic exchange process is called the **chloride shift** (see Figure 23.18).

At the lungs, the process is reversed (see Figure 23.18b). As blood is pumped through the pulmonary capillaries, the P_{CO_2} of the blood declines from 45 mm Hg to 40 mm Hg. But for this to occur, carbon dioxide must first be released from its "bicarbonate housing." Bicarbonate ions reenter the RBCs (and chloride ions move into the plasma) and bind with hydrogen ions to form carbonic acid, which is then split by carbonic anhydrase to release carbon dioxide and water. Carbon dioxide then diffuses along its partial pressure gradient from the blood into the alveoli.

The Haldane Effect

The amount of carbon dioxide transported in blood is markedly influenced by the degree of oxygenation of the blood. The lower the P_{O_2} and the less the degree of hemoglobin saturation with oxygen, the greater the amount of carbon dioxide that can be carried in the blood. This phenomenon, called the *Haldane effect*, reflects the enhanced ability of reduced hemoglobin to buffer H^+ by combining with it (see Figure 23.18a) and to form carbamino hemoglobin. As carbon dioxide enters the systemic bloodstream, it causes more oxygen to dissociate from hemoglobin (the Bohr effect), which in turn allows more carbon dioxide to combine with hemoglobin and more bicarbonate ions to be generated (the Haldane effect). In the pulmonary circulation (Figure 23.18b), the situation is reversed: The uptake of oxygen facilitates the release of carbon dioxide. As the hemoglobin becomes saturated with oxygen, the hydrogen ions released combine with HCO_3^-, facilitating the elimination of CO_2 from the pulmonary blood. The Haldane effect is particularly important because it encourages carbon dioxide uptake from the tissues and enhances its elimination from the blood in the lungs.

Influence of Carbon Dioxide on Blood pH

In most cases, the hydrogen ions released during carbonic acid dissociation are buffered by hemoglobin or other proteins within the red blood cells. The bicarbonate ions generated diffuse out of the red blood cells into the plasma, where they act as the alkaline reserve part of the carbonic acid–bicarbonate buffer system of the blood. The carbonic acid–bicarbonate buffer system is very important in resisting shifts in blood pH. For example, if the hydrogen ion concentration in blood begins to rise, H^+ is removed by combination with bicarbonate ions to form carbonic acid (a weak acid that does not dissociate at either physiological or acidic pH). If the blood concentration of H^+ drops below desirable levels, carbonic acid dissociates releasing hydrogen ions, thus lowering the pH again. Changes in respiratory rate or depth can cause dramatic changes in blood pH by modifying the amount of carbonic acid in the blood. For example, slow, shallow respirations—*hypoventilation*—result in the accumulation of carbon dioxide in the blood, causing carbonic acid levels to increase and blood pH to decrease. Conversely, *hyperventilation*, or deep breathing, which flushes carbon dioxide rapidly out of the blood and reduces carbonic acid levels, increases blood pH. Thus, respiratory ventilation can provide a fast-acting system for adjusting blood pH (and P_{CO_2}) in cases where it is disturbed by metabolic factors. (Acid-base balance of the blood is discussed in detail in Chapter 27.)

Control of Respiration

Neural Mechanisms and Generation of Breathing Rhythm

Although our tidelike breathing seems so beautifully simple, its control is fairly complex. The basic pattern of breathing is generated by the activity of neurons located in the reticular formation of the medulla and pons. We will begin our discussion of neural mechanisms controlling breathing by describing the role of the medulla, which sets the respiratory rhythm, and then move on to the postulated roles of the pons nuclei.

Medullary Respiratory Centers

The pace-setting nucleus within the medulla oblongata is called the **inspiratory center**, or, more precisely, the *dorsal respiratory group (DRG)* (Figure 23.19). Neurons of this center appear to have an intrinsic ability

Figure 23.19 Neural pathways involved in the control of respiratory rhythm. During normal quiet breathing, the self-exciting neurons of the medullary inspiratory center, or dorsal respiratory group (DRG), set the pace by alternately (1) depolarizing and sending impulses to the inspiratory muscles and then (2) becoming quiescent, which allows passive expiration to occur. The neurons of the ventral respiratory group (VRG) appear to be largely inactive during normal quiet breathing. When forced breathing is required, the inspiratory center enlists the activity of accessory respiratory muscles and stimulates the expiratory center to activate the muscles of forced expiration. The pons centers interact with the medullary centers in such a way that the breathing pattern is smooth. The apneustic center continuously stimulates the medullary inspiratory center unless inhibited by the pneumotaxic center. The pneumotaxic center normally limits the inspiratory phase and promotes expiration by inhibiting both the apneustic and inspiratory center neurons.

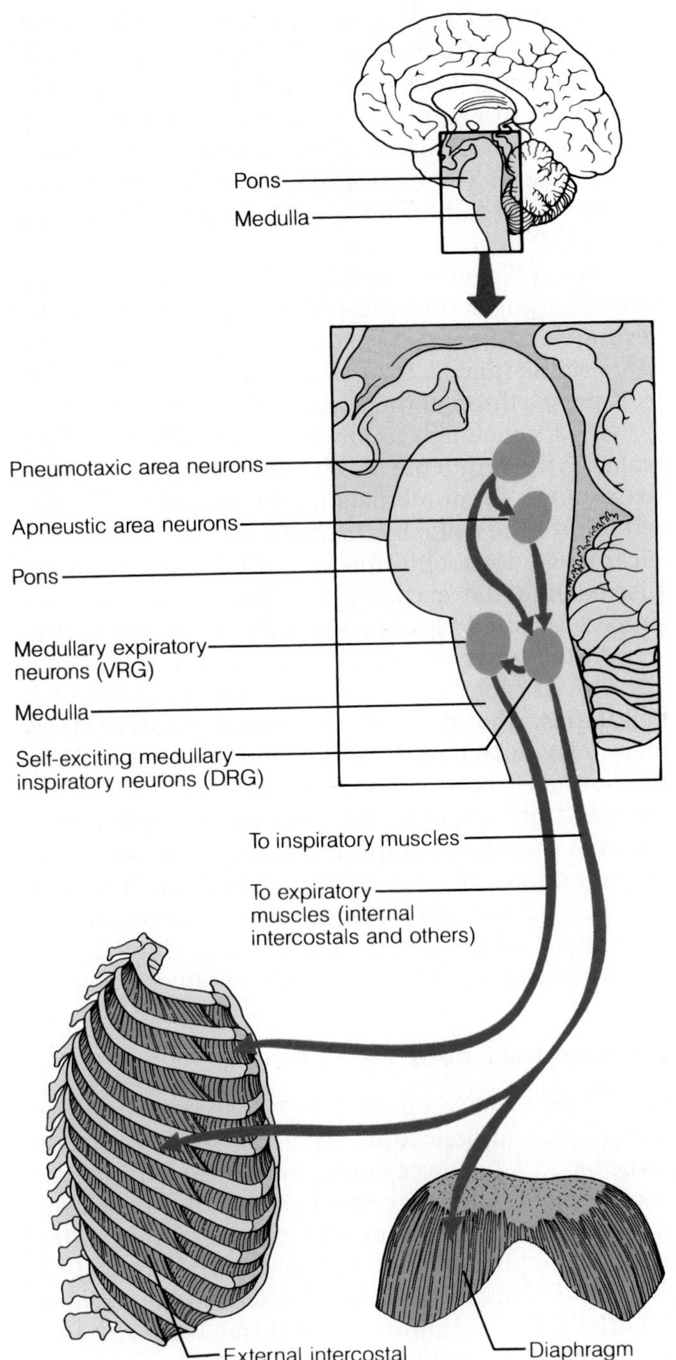

to depolarize in a spontaneous and rhythmic way, so that the oscillating pattern of breathing is produced.

When the inspiratory neurons are firing, nerve impulses travel along the *phrenic* and *intercostal nerves* to excite the diaphragm and external intercostal muscles, respectively, which then contract. The inspiratory center then becomes dormant, and expiration occurs passively as the inspiratory muscles are allowed to relax and the lungs recoil. This cyclic on-off activity of the inspiratory neurons is repeated continuously and produces a respiratory rate of 12–15 breaths per minute, with inspiratory phases lasting for about 2 seconds followed by expiratory phases lasting about 3 seconds. This normal respiratory rate and rhythm is referred to as **eupnea** (yoop′-nē-uh).

Although the pons respiratory centers and stretch receptors in the lungs can modify DRG activity, the inspiratory center can independently set the respiratory rhythm. Rhythmic breathing will still occur even when all afferent connections to the inspiratory center are severed. When the medulla itself is completely suppressed, as by an overdose of sleeping pills, morphine, or alcohol, respiration stops completely.

The function of the second medullary center, the **expiratory center**, or *ventral respiratory group (VRG)*, is not as well understood. Like the inspiratory center, it appears to stimulate the inspiratory muscles, but in a completely different manner. The VRG neurons appear to maintain the respiratory muscles in a state of slight contraction; that is, their activity is *tonic* rather than cyclic. However, that activity is interrupted by impulses from the inspiratory center when more forceful breathing becomes necessary. When this occurs, the expiratory center sends efferent impulses to the muscles of *expiration*—the internal intercostal and abdominal muscles—which cause vigorous depression of the rib cage and more strenuous expiratory movements. Notice that the expiratory center is not active in promoting expiration during normal quiet breathing.

Pons Respiratory Centers

Although the medullary inspiratory center generates the basic respiratory rhythm, breathing becomes abnormal when connections between the pons and the medulla are cut. Breathing is still rhythmic, but it occurs in gasps. Pons centers are believed to be important in producing smooth transitions from inspiration to expiration, and vice versa.

The **pneumotaxic** (noo″-mō-tak′-sik) **center**, the more superior pons center (see Figure 23.19), continuously transmits inhibitory impulses to the inspiratory center of the medulla. When its signals are particularly strong, the period of inspiration is shortened. The most important role of the pneumotaxic center is to fine-tune the breathing rhythm and prevent lung overinflation.

The **apneustic** (ap-noo′-stik) **center** provides inspiratory drive by continuously stimulating the medullary inspiratory center. Unless inhibited by the pneumotaxic center or by afferent impulses from the lungs, its effect is to prolong inspiration or to cause breath holding in the inspiratory phase. Breathing becomes deep and slow when the pneumotaxic efferents are cut, which indicates that the apneustic center is usually inhibited by pneumotaxic center neurons.

Factors Influencing the Rate and Depth of Breathing

Both the depth and rate of breathing can be modified in response to changing body demands. Depth of inspiration is determined by the frequency with which the respiratory center stimulates the respiratory muscles; the greater the frequency, the more motor units are excited and the stronger the muscle contractions. Rate of respiration is determined by how long the inspiratory center is active or, conversely, how quickly it is switched off.

The respiratory centers in the medulla and pons are sensitive to many different stimuli—some excitatory, some inhibitory. These influencing factors are summarized in Figure 23.20.

Pulmonary Irritant Reflexes

The lungs contain receptors that respond to an enormous variety of irritating factors. When activated, these receptors communicate with the respiratory centers via vagal nerve afferents. Irritation of the tracheal or bronchial mucosa by dust, noxious fumes, or accumulated mucus promotes a cough; the same irritants in the nasal cavity cause sneezing. Both reflexes help keep the airway clear. Air pollutants, cigarette smoke, or other inhaled debris in the lungs stimulate receptors in the bronchioles and lead to reflex constriction of those conduits.

The Hering-Breuer Reflex

The visceral pleurae and conducting passages in the lungs also contain numerous stretch receptors that are vigorously stimulated when the lungs become overinflated and excessively stretched (potentially damaging events). Inhibitory impulses are then dis-

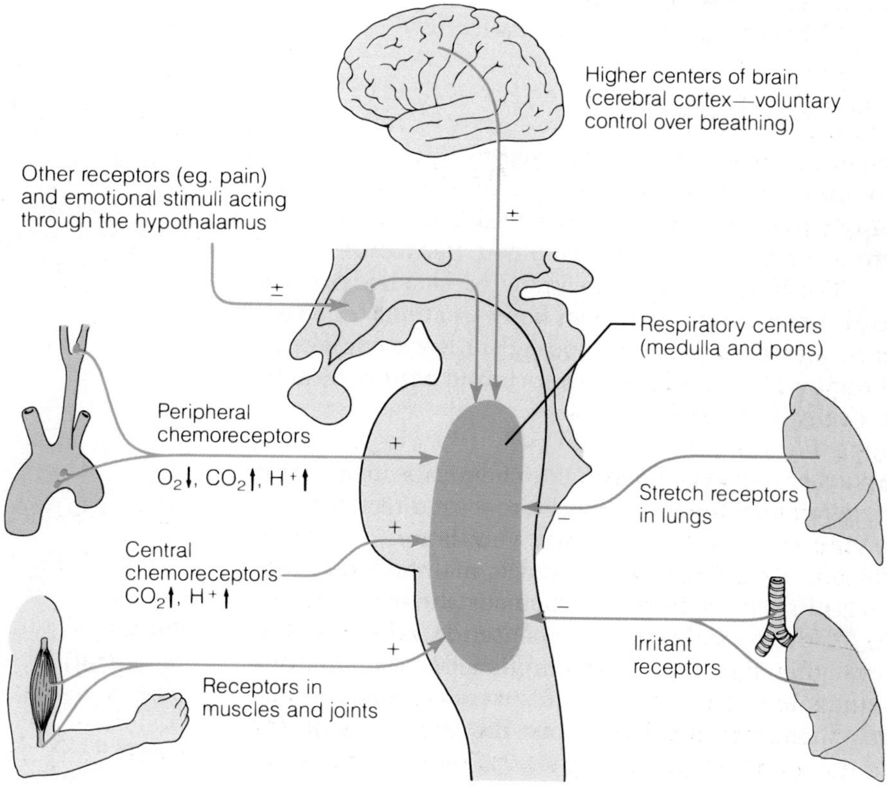

Figure 23.20 Neural and chemical influences on the respiratory centers in the medulla. Excitatory influences (+) increase the frequency of impulses sent to the muscles of respiration and result in deeper, faster breathing. Inhibition of the medullary center (−) has the reverse effect. In some cases, the impulses may be excitatory or inhibitory (indicated by ±), depending on the precise receptors or brain regions activated.

patched to the medullary inspiratory center and to the apneustic center of the pons, allowing expiration to occur. As the lungs recoil, the stretch receptors become quiet, and inspiration can be initiated once again. This reflex, called the **inflation** or **Hering-Breuer** (her′-ing broy′-er) **reflex**, is thought to be more a protective response rather than a mechanism important in the regulation of normal respiration. Although receptors responsive to extreme lung deflation are also known to exist, their activity seems to be unimportant in human lung function.

Influence of Higher Brain Centers

Hypothalamic Controls. Strong emotions and pain acting through the limbic region of the brain activate hypothalamic sympathetic nervous system centers, which can influence respiratory rate and depth by sending modulating signals to the respiratory centers. For example, have you ever touched something cold and clammy and gasped? That response was mediated through the hypothalamus.

Cortical Controls (Volition). Although breathing is normally regulated involuntarily by brain stem centers, connections between the cortical centers and the brain stem allow us to consciously alter our pattern of respiration. During singing and swallowing, breath control is extremely important. However, voluntary controls of breathing are limited, and the respiratory centers will simply ignore messages from the cortex and will escape from voluntary controls when the P_{CO_2} in the blood reaches certain critical levels. As a result, breathing resumes automatically.

Chemical Factors

Although many factors can modify the baseline respiratory rate and depth, the most important factors are changing levels of carbon dioxide, oxygen, and hydrogen ions in the arterial blood. Sensors responding to such chemical fluctuations, called **chemoreceptors**, are found in two major body locations: The **central chemoreceptors** are located in the medulla; the **peripheral chemoreceptors** are found within the great vessels of the neck.

Influence of P_{CO_2}. Of all the chemicals influencing respiration, arterial CO_2 levels are most directly related to the efficiency of alveolar ventilation. Normally, arterial P_{CO_2} is 40 mm Hg and is maintained within ± 3 mm Hg of this level by homeostatic mechanisms.

How is respiratory compensation in response to changing P_{CO_2} levels accomplished? Actually, this exquisitely sensitive mechanism is mediated through the influence of rising carbon dioxide levels on the pH of the cerebrospinal fluid surrounding the brain stem respiratory centers. Carbon dioxide diffuses easily from the blood into the cerebrospinal fluid, where it is hydrated and forms carbonic acid. As the acid dissociates, hydrogen ions are liberated. (This is the same reaction that occurs when carbon dioxide enters red blood cells, discussed on p. 736.) However, unlike RBCs or plasma, cerebrospinal fluid contains virtually no proteins that can buffer these free hydrogen ions. Thus, as P_{CO_2} levels rise, a condition referred to as **hypercapnia** (hī″-per-kap′-nē-uh), the pH of cerebrospinal fluid drops, exciting the central chemoreceptors, which make abundant synapses with the respiratory regulatory centers (Figure 23.21). As a result, the depth (and perhaps rate) of breathing is increased. This breathing pattern is called **hyperventilation**. Notice that although rising carbon dioxide levels act as the initial stimulus, it is the rising hydrogen ion levels that prod the central chemoreceptors into activity. In the final analysis, *control of breathing during rest is aimed primarily at regulating the H^+ concentration in the brain.*

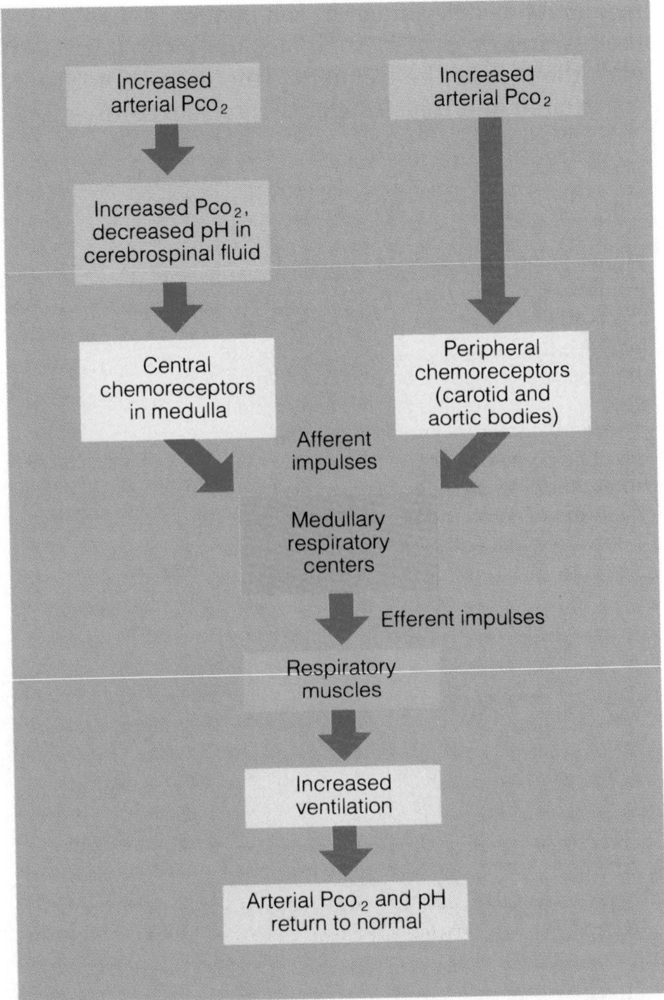

Figure 23.21 Mechanism for controlling the pH of cerebrospinal fluid by regulation of ventilation.

Since hyperventilation enhances alveolar ventilation and quickly flushes carbon dioxide out of the blood, increasing blood pH, sometimes competing swimmers will voluntarily hyperventilate to prolong their breath-holding period during swim meets. This is a dangerous practice, because hypoxia can lead to drowning before the swimmer has the urge to breathe. ■

People experiencing anxiety attacks may hyperventilate involuntarily, to the point where they become dizzy and may faint. This reflects the fact that low carbon dioxide levels in the blood (**hypocapnia**) cause vasoconstriction of the cerebral blood vessels, producing cerebral ischemia. Such attacks can be averted by breathing into a paper bag. Since the person is rebreathing expired air, which is enriched with carbon dioxide, carbon dioxide is retained in the blood. ■

An elevation of only 5 mm Hg in arterial P_{CO_2} results in a 100% increase in alveolar ventilation, even when arterial oxygen content and pH are unchanged. When P_{O_2} and pH are less than normal, the response to elevated P_{CO_2} is enhanced still further. On the other hand, when P_{CO_2} levels are abnormally low, respiration is inhibited and becomes slow and shallow, and periods of **apnea** (breathing cessation) may occur until the arterial P_{CO_2} rises and again provides the respiratory stimulus.

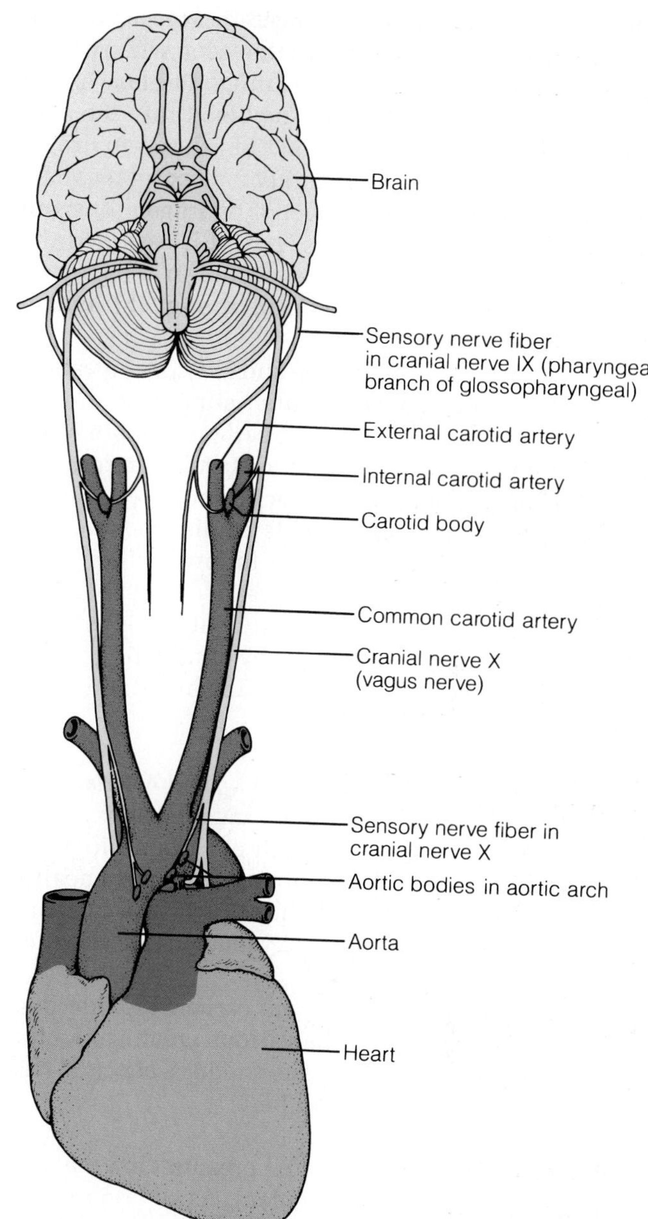

Figure 23.22 Location of the peripheral chemoreceptors in the carotid and aortic bodies. Indicated here is the afferent pathway from these receptors via cranial nerves IX and X to the respiratory centers of the medulla.

Influence of P_{O_2}. Cells sensitive to arterial oxygen levels are found in the peripheral chemoreceptors, that is, in the aortic bodies of the aortic arch and in the carotid bodies found at the bifurcation of the common carotid arteries (Figure 23.22). Of these peripheral chemoreceptors, those in the carotid bodies are the main oxygen sensors. Under normal conditions, the effect of declining P_{O_2} on alveolar ventilation is very slight and mostly limited to enhancing the sensitivity of central receptors to increased P_{CO_2}. Arterial P_{O_2} must drop substantially before oxygen levels become a major stimulus for increased ventilation. This is not as strange as it may appear; remember, there is a huge reservoir of oxygen bound to hemoglobin, and hemoglobin remains almost entirely saturated with oxygen unless or until the P_{O_2} of alveolar gas and arterial blood falls to 60 mm Hg or below. The central chemoreceptors then begin to suffer from oxygen starvation, and their activity is depressed. At the same time, the peripheral chemoreceptors become excited and stimulate the respiratory centers to initiate a reflexive increase in ventilation, even if P_{CO_2} is normal. Thus, the peripheral chemoreceptor reflex system can maintain ventilation in those whose alveolar oxygen levels are low even though brain stem sensors are depressed by hypoxia.

In people who retain carbon dioxide because of pulmonary disease (e.g., emphysema and chronic bronchitis victims), arterial P_{CO_2} is chronically elevated and chemoreceptors become unresponsive to this chemical stimulus. In such cases, it is the effect of declining P_{O_2} levels on the oxygen-sensitive peripheral chemoreceptors that provides the principal respiratory stimulus, or the so-called *hypoxic drive.* Gas mixtures administered to such patients during respiratory distress are only slightly enriched with oxygen. If pure oxygen were given, these individuals would stop breathing, because their respiratory stimulus (low P_{O_2} levels) would be removed. ■

Influence of Arterial pH. Changes in arterial pH can modify respiratory rate and rhythm even when carbon dioxide and oxygen levels are normal. Because hydrogen ions diffuse poorly from the blood into the cerebrospinal fluid, the direct effect of arterial hydrogen ion concentration on central chemoreceptors is insignificant compared to the effect of hydrogen ions generated by elevations in P_{CO_2}. Thus, it is assumed that the increased ventilation that occurs in response to decreased arterial pH is mediated through the peripheral chemoreceptors.

Although changes in P_{CO_2} and H^+ concentrations are interrelated, they are distinct stimuli. A drop in blood pH (acidosis) can reflect carbon dioxide retention, but it may also result from metabolic causes, such as the accumulation of lactic acid during exercise or of fatty (or other organic) acids in diabetes mellitus patients who are careless about following their insulin and/or diet regimen. Regardless of cause, as arterial pH declines, respiratory system controls attempt to compensate and raise blood pH by eliminating carbon dioxide (and carbonic acid) from the blood; thus, respiratory rate and depth increase.

Summary of Interactions of P_{CO_2}, P_{O_2}, and Arterial pH. Although every cell in the body must have oxygen to live, the body's need to rid itself of carbon dioxide is the most important stimulus for breathing in a healthy person. This reflects the fact that this whole system is "overengineered" to acquire oxygen, but just adequate for eliminating CO_2.

However, as we have seen, carbon dioxide does not act in isolation, and the various chemical factors interact to enforce or inhibit one another's effects. These interactions are summarized here:

1. Rising carbon dioxide partial pressures are the most powerful respiratory stimulant. As carbon dioxide is hydrated in cerebrospinal fluid, liberated H^+ acts directly on the central chemoreceptors, causing a reflexive increase in breathing rate and depth. Low P_{CO_2} levels depress respiration.
2. Under normal conditions, blood P_{O_2} affects breathing only indirectly by influencing chemoreceptor sensitivity to changes in P_{CO_2}. Low oxygen tensions augment P_{CO_2} effects; high P_{O_2} levels diminish the effectiveness of carbon dioxide stimulation.
3. When arterial P_{O_2} drops below 60 mm Hg, it becomes the major stimulus for respiration, and ventilation is increased via reflexes initiated by peripheral chemoreceptor activation. Depending on the precise conditions, this may increase oxygen loading into the blood, but it also causes hypocapnia (low P_{CO_2} blood levels) and an increase in blood pH, both of which inhibit respiration.
4. Changes in arterial pH resulting from carbon dioxide retention or metabolic factors act indirectly through the peripheral chemoreceptors to promote changes in ventilation, which in turn modify arterial P_{CO_2} and pH. Arterial pH does not appear to influence the central chemoreceptors directly.

Respiratory Adjustments During Exercise and at High Altitudes

Effects of Exercise

Respiratory adjustments during exercise are geared to both the intensity and duration of the exercise. Working muscles consume tremendous amounts of oxygen and evolve large amounts of carbon dioxide (Figure 23.23); thus, it is not surprising that ventilation can increase 10- to 20-fold during vigorous exercise. Breathing becomes deeper and more vigorous, but the respiratory rate is not significantly changed. This breathing pattern is called **hyperpnea** (hī″-perp-nē′-uh) to distinguish it from the deep and often rapid pattern of hyperventilation. Although it is tempting to suppose that the exercise-enhanced ventilation reflects and is prompted by rising P_{CO_2} blood levels and a declining P_{O_2}, this does not appear to be the case for two reasons. First, there is an abrupt increase in ventilation at or just prior to the onset of exercise, followed by a more gradual increase, and finally (within 4–6 minutes) a steady state of ventilation. Similarly, there is a sudden decline in ventilation when exercise is stopped, followed by a more gradual decrease to pre-exercise values. Second, arterial P_{CO_2} and P_{O_2} levels do not change significantly during exercise. In fact, P_{CO_2} levels may decline below normal levels, and P_{O_2} levels may rise very slightly because of the efficiency of the respiratory adjustments. How can these observations be explained? At present, our understanding of the mechanisms involved is sketchy, but the most accepted explanation is as follows.

The abrupt increase in ventilation that occurs as exercise is initiated reflects the interaction of the following neural factors: (1) psychic stimuli (that is, our conscious anticipation of exercise), (2) simultaneous cortical motor activation of the skeletal muscles and of the respiratory centers, and (3) proprioceptors in moving muscles, tendons, and joints, which send excitatory impulses to the respiratory centers. The subsequent gradual increase in respiration reflects chemical factors. Because oxygen and carbon dioxide levels in arterial blood remain surprisingly constant during exercise, researchers believe that the chemical

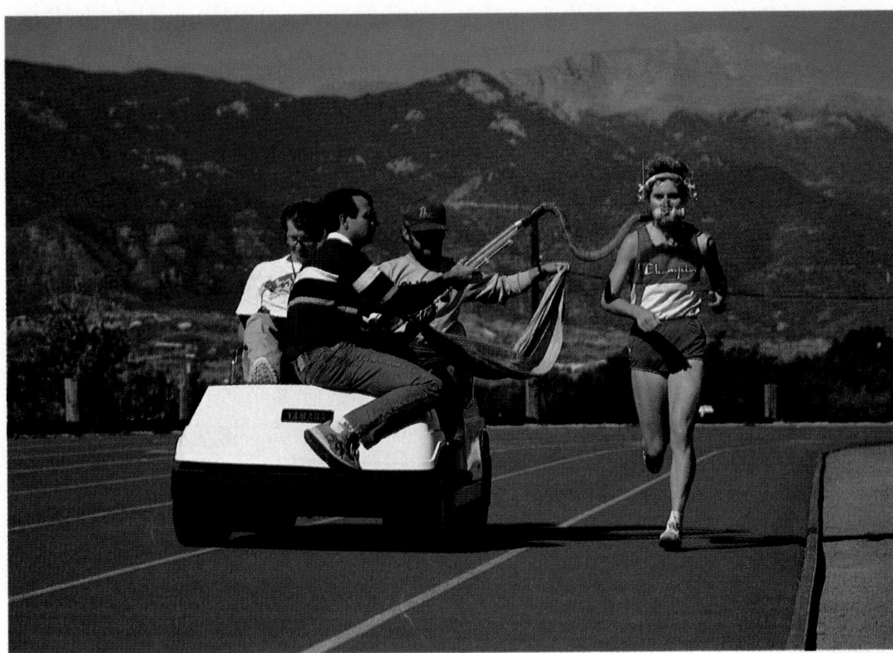

Figure 23.23 Scientists at the U.S. Olympic Training Center measure O_2 and CO_2 levels in an athlete's expired air. Blood samples are also taken. By analyzing this information, these scientists will be able to determine the athlete's oxygen use and lactic acid levels, which in turn will determine that athlete's training pace.

conditions in the respiratory centers of the brain may differ from those in the general circulation or that the sensitivity of the respiratory centers to minute changes in P_{CO_2} is vastly increased by the warmer body temperatures present during vigorous exercise.

The abrupt decrease in ventilation that occurs as exercise ends reflects termination of the neural regulatory mechanisms. The subsequent gradual decline in ventilation to baseline levels occurs as the oxygen debt is being repaid and respiration continues to be stimulated by low arterial pH due to lactic acid accumulation.

It is important to understand that the rise in lactic acid levels during vigorous exercise is not a result of inadequate respiratory function, because the alveolar airflow and pulmonary blood flow are matched just as exquisitely during exercise as during resting conditions. Rather, it reflects limitations of cardiac output or an inability of the skeletal muscles to increase their oxygen consumption any further (see Chapter 9). In light of this fact, the practice of inhaling pure oxygen by mask, commonly used by football players to replenish their "oxygen-starved" body as quickly as possible, is useless. The panting athlete does have an oxygen deficit. But extra oxygen will not help, because the shortage is in the muscles—not in the lungs. ■

Effects of High Altitude

Barometric pressure decreases with increasing altitude, and when you move from a sea-level region to the mountains, where air density and P_{O_2} are lower, several respiratory and hematopoietic adjustments must be made. This multistep adaptive response of the body is called *acclimatization*.

As already explained, decreases in arterial P_{O_2} cause the central chemoreceptors to become more responsive to increases in P_{CO_2}, and a substantial decline in P_{O_2} directly stimulates the peripheral chemoreceptors. As a result, hyperventilation occurs as the brain attempts to restore gas exchange to previous levels. Within a few days, the minute respiratory volume becomes stabilized at a level 2 to 3 L/min higher than that seen at sea level. Since hyperventilation also reduces arterial carbon dioxide levels, the P_{CO_2} of individuals living at high altitudes is typically below 40 mm Hg (its value at sea level).

Because less oxygen is available to be loaded, high altitude conditions always result in lower than normal hemoglobin saturation levels. For example, at 19,000 feet above sea level, the oxygen saturation of arterial blood is only 67% (compared to nearly 98% at sea level). But remember, hemoglobin unloads only 20% to 25% of its oxygen at sea level pressures; thus, even at the reduced hemoglobin saturations seen at high altitudes, oxygen needs of the tissues are still met adequately under resting conditions. As a partial compensation for declining hemoglobin saturations at high altitudes, hemoglobin's affinity for oxygen is also reduced, and enhanced amounts of oxygen are unloaded to the tissues during each circulatory period. This occurs because the low oxyhemoglobin content in erythrocytes stimulates the production of 2,3-DPG, which, as explained previously, interferes with hemoglobin's affinity for oxygen.

Although the tissues receive adequate oxygen under normal conditions, problems arise when all-out efforts are demanded of the cardiovascular and respiratory systems to supply vigorously working muscles (as discovered by U.S. athletes competing in Olympic events in the high mesa of Mexico City). Unless one has become fully acclimatized, such conditions almost guarantee that body tissues will become severely hypoxic. ■

When blood oxygen tensions decline, the kidneys intervene by accelerating the production of erythropoietin, which stimulates bone marrow production of red blood cells (see Chapter 18). This phase of acclimatization occurs slowly and is a long-term compensatory mechanism for dealing with oxygen-deficient high-altitude areas.

Homeostatic Imbalances of the Respiratory System

The respiratory system is particularly vulnerable to infectious diseases because it is open to airborne pathogens. Because many of these inflammatory conditions, such as rhinitis and laryngitis, were considered in conjunction with the specific organs afflicted, we will turn our attention to the most disabling respiratory disorders, the group of diseases collectively referred to as *chronic obstructive pulmonary disease (COPD)* and *lung cancer*. These disorders are "living proof" of cigarette smoking's devastating effects on the body. Long known to promote cardiovascular disease, cigarettes are perhaps even more effective at destroying the lungs.

Chronic Obstructive Pulmonary Disease (COPD)

The chronic obstructive pulmonary diseases, exemplified by chronic bronchitis and obstructive emphysema, are a major cause of death and disability in the United States and are becoming increasingly more prevalent. These diseases have certain features in common: (1) Patients almost invariably have a history of smoking; (2) *dyspnea* (disp-nē-uh), difficult or labored breathing often referred to as "air hunger", occurs and becomes progressively more severe; (3) coughing and frequent pulmonary infections are common; and (4) most COPD victims ultimately develop respiratory failure, accompanied by hypoxemia, carbon dioxide retention, and respiratory acidosis.

Obstructive emphysema is distinguished by permanent enlargement of the alveoli (the walls of adjacent alveoli break through), followed by their destruction. Pulmonary capillaries are lost, and chronic inflammation leads to lung fibrosis. Invariably, the lungs become "floppy" and lose their elasticity. Arterial oxygen and carbon dioxide levels remain essentially normal until late in the disease. As the lungs become less elastic, the airways collapse during expiration and obstruct the outflow of air. As a result, air is retained in the lungs, which promotes more effective oxygen exchange than would be expected; however, this hyperinflation leads to the development of a permanently expanded "barrel chest." Cigarette smoking is known to be a causative factor in emphysema, but hereditary factors seem important in some patients.

Chronic bronchitis is characterized by chronic excessive mucus production by the mucosae of the lower respiratory passageways. This severely impairs lung ventilation and, subsequently, gas exchange. The important pathology is reduction of airway diameters by inflammation of the mucosae and fibrosis. Chronic bronchitis patients are sometimes called "blue bloaters" because hypoxia and carbon dioxide retention occur early in the disease and cyanosis is common. However, the degree of dyspnea is usually moderate when compared to that of emphysema sufferers. So significant is cigarette smoking as a causative factor of chronic bronchitis that it has been hypothesized that this disease would be an insignificant health problem if cigarettes were unavailable. Environmental pollution is also a causative factor, but a minor one.

Lung Cancer

Lung cancer accounts for fully one-third of cancer deaths in the United States. Its incidence, strongly associated with cigarette smoking (over 90% of lung cancer patients were smokers), is increasing daily. In men, lung cancer is the most prevalent type of malignancy; in women, it is exceeded only by breast cancer. Lung cancer has a notoriously low cure rate; overall five-year survival of those with lung cancer is 7%. This reflects the fact that lung cancer is tremendously aggressive and metastasizes rapidly and widely, and most cases are not diagnosed until they are well advanced.

Lung cancer appears to follow closely the stepwise progression of oncogene-activating events outlined in the box in Chapter 3 (pp. 94–95). Ordinarily, nasal hairs, sticky mucus, and the action of cilia do a fine job of protecting the lungs from chemical and biological irritants, but when one smokes, these devices are overwhelmed and eventually become nonfunctional. Continuous irritation prompts the production of more mucus, but smoking paralyzes the cilia and depresses the activity of lung macrophages, leading to mucus pooling and increasingly frequent pulmonary infections, including pneumonia, and COPD. However, it is the irritant effects of the 15 or so carcinogens present in tobacco smoke that eventually translates into lung cancer, causing the mucosal cells to proliferate wildly and lose their characteristic histological nature.

The three most common types of lung cancer are (1) *squamous cell carcinoma*, which arises in the major bronchi or their larger subdivisions and tends to form masses that cavitate and bleed, (2) *adenocarcinoma* (a″-dih-nō-kar-sih-nō′-muh), which originates as solitary nodules primarily in peripheral lung areas from bronchial glands and alveolar cells, and (3) *small cell carcinoma*, also called *oat cell carcinoma*, which consists of lymphocyte-like cells that typically grow in cords or small grapelike clusters within the mediastinum. Some small cell carcinomas produce metabolic alterations beyond their effects on the lungs, because they become ectopic sites of hormone production. For example, some produce the hormone ADH. Others secrete ACTH (leading to Cushing's disease) or calcitonin (which results in low blood levels of ionic calcium, or hypocalcemia).

Whenever possible, the most effective treatment for lung cancer is complete resection (removal) of the diseased lung, because it has the greatest potential for prolonging life and cure. However, resection is a choice that is open to very few lung cancer patients; in most cases, radiation therapy and chemotherapy are the only options. ■

Developmental Aspects of the Respiratory System

Since embryos develop in a cephalocaudal (head-to-tail) direction, the upper respiratory structures materialize first. By the fourth week of development, a thickened plate of ectoderm, the *olfactory placode* (plak′-ōd), has appeared on the anterior aspect of the head (Figure 23.24); this quickly invaginates to form a pit that extends posteriorly to connect with the developing foregut, which is forming simultaneously from the endodermal germ layer.

The respiratory epithelium of the lower respiratory organs develops as an outpocketing of the foregut endoderm, which differentiates into the pharyngeal mucosa. This protrusion, called the *laryngotracheal bud*, is present by the fifth week of development. The proximal part of the bud forms the tracheal lining, and its distal end splits and forms the mucosae of the bronchi and all their subdivisions, including the terminal alveoli. By the eighth week of development, mesoderm clothes these ectoderm- and endoderm-derived linings and forms the walls of the respiratory passageways and the stroma of the lungs. By 28 weeks, the respiratory system has developed sufficiently to allow a baby born prematurely to survive birth and to breathe on its own. As noted earlier, infants born before this time are likely to exhibit infant res-

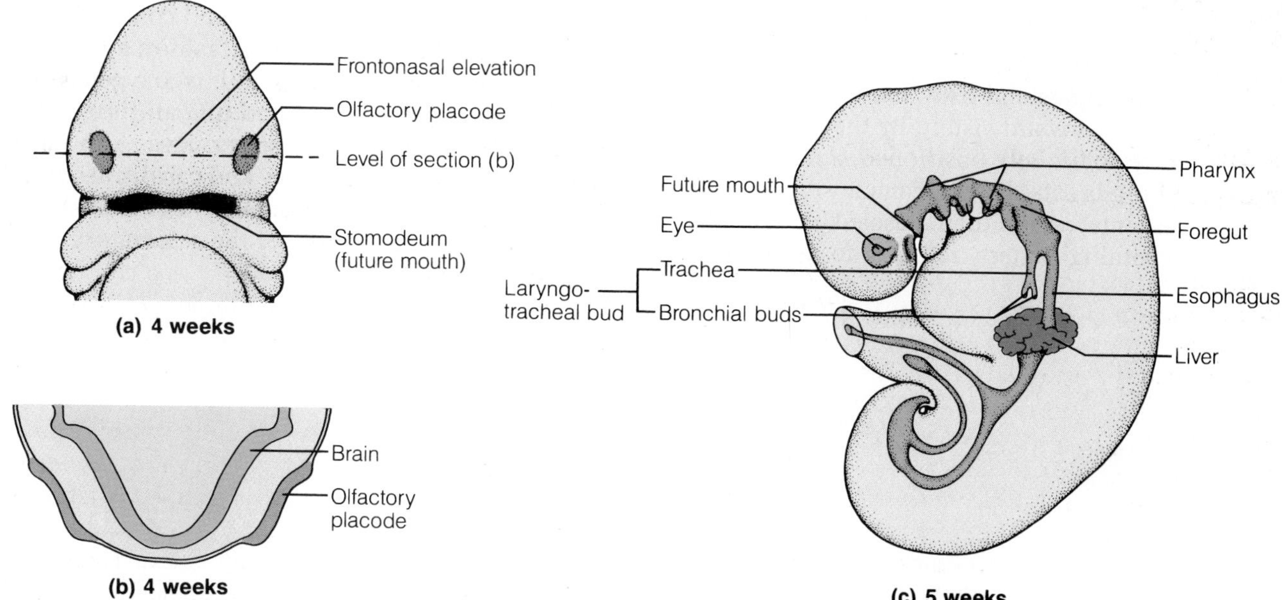

Figure 23.24 Embryonic development of the respiratory system. (**a**) Anterior superficial view of the embryo's head, showing the ectodermal olfactory placodes. (**b**) Transverse section of the embryo's head at four to five weeks, showing the location of the olfactory placodes [the level of transection is shown in (**a**)]. (**c**) Development of the lower respiratory passageway mucosae. The laryngotracheal bud is shown outpocketing from the foregut (endodermal) mucosa.

piratory distress syndrome resulting from inadequate surfactant production.

During fetal life, the lungs are filled with fluid, and all respiratory exchanges are made by the placenta. Vascular shunts (ductus arteriosus and foramen ovale) largely bypass the lungs (see Chapter 29, p. 972). At birth, the fluid-filled pathway is drained, and the respiratory passageways become filled with air. As the P_{CO_2} in the baby's blood rises, the inspiratory center becomes excited and causes the baby to take its first breath. Consequently, the alveoli inflate and begin to function in gas exchange. However, it takes nearly two weeks for the lungs to become fully inflated.

Important birth defects of the respiratory system include cleft palate and *cystic fibrosis*, a genetic disease that causes excess mucus production that clogs the respiratory passageways and predisposes the child to respiratory system infections. Cystic fibrosis—which accounts for about 5% of childhood deaths—affects other body systems as well; most importantly, it impairs the digestive process.

The respiratory rate is highest in newborn infants (about 40 respirations per minute). In infants, it is around 30 per minute; at 5 years it is around 25 per minute; and in adults it is between 12 and 15 per minute. In old age, the rate often increases again. The lungs continue to mature throughout childhood, and more alveoli are formed until young adulthood. Research has revealed that when smoking is begun during the early teens, complete maturation of the lungs never occurs, and those additional alveoli are lost forever.

The patterns of breathing movements depend largely on age and sex. In infants, the ribs take a nearly horizontal course; thus, infants rely almost entirely on descent of the diaphragm to increase thoracic volume for inspiration. This type of breathing is called *abdominal*, or *diaphragmatic, breathing* because as the diaphragm descends and rises, the abdominal wall moves outward and inward visibly. By the second year, the ribs are more obliquely positioned, and the adult form of breathing is established. Females rely mainly on rib movements rather than descent of the diaphragm, a breathing pattern referred to as *thoracic breathing*. Males use both methods, but abdominal breathing predominates.

Under normal conditions, the respiratory system works so efficiently and smoothly that we are not even aware of it. Most problems that do occur are the result of external factors—for example, viral or bacterial infections or the obstruction of the trachea by a piece of food. However, some unfortunate individuals are plagued with asthma, a condition frequently caused by allergy that causes them to wheeze and gasp for air as their bronchioles constrict. For many years, tuberculosis and pneumonia were the worst killers in the United States. Antibiotics have decreased their lethality to a large extent, but they are still dangerous diseases. By far the most problematic diseases at present are those described earlier, COPD and lung cancer.

As we age, the chest wall becomes more rigid and the lungs gradually begin to lose their elasticity. Both of these factors result in a slowly decreasing ability to ventilate the lungs. Vital capacity decreases by about one-third by the age of 70 years. Accompanying these changes is a decrease in blood oxygen levels and a reduced sensitivity to the stimulating effects of carbon dioxide, particularly in a reclining or supine position. As a result, many old people tend to become hypoxic during sleep and exhibit sleep apnea.

Additionally, just as the overall efficiency of the immune response declines with age, many of the respiratory system's protective mechanisms become less effective. Ciliary activity of its mucosa decreases, and the phagocytes in the lungs become sluggish. The net result is that the elderly are more at risk for respiratory tract infections, particularly pneumonia.

* * *

Lungs, bronchial tree, heart, and connecting blood vessels—together, these organs fashion a remarkable system that ensures that blood is oxygenated and relieved of carbon dioxide and that all tissue cells of the body have access to these services. Although the cooperation of the respiratory and cardiovascular systems is quite obvious, all organ systems of the body depend on the functioning of the respiratory system, as summarized in Figure 23.25.

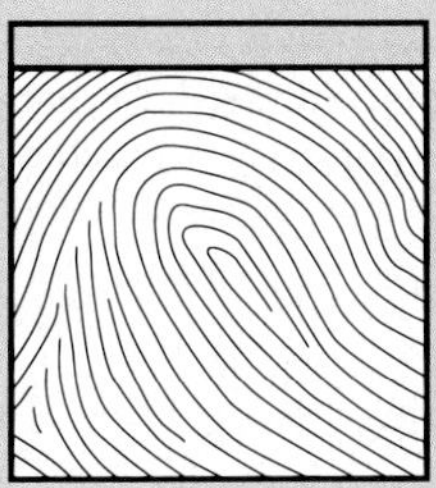

Integumentary system

Protects respiratory system organs by forming surface barriers ▶

Provides oxygen; disposes of carbon dioxide ◀

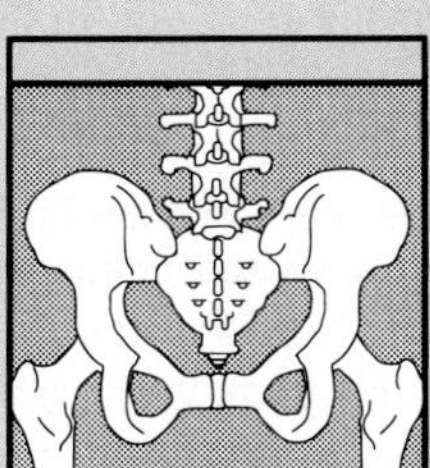

Skeletal system

Protects lungs and bronchi by enclosure by bone ▶

Provides oxygen; disposes of carbon dioxide ◀

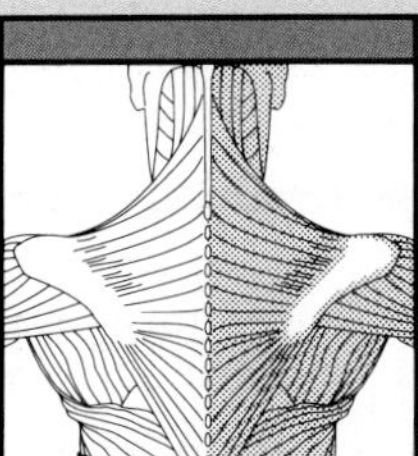

Muscular system

Regular exercise increases respiratory efficiency ▶

Provides oxygen needed for muscle activity; disposes of carbon dioxide ◀

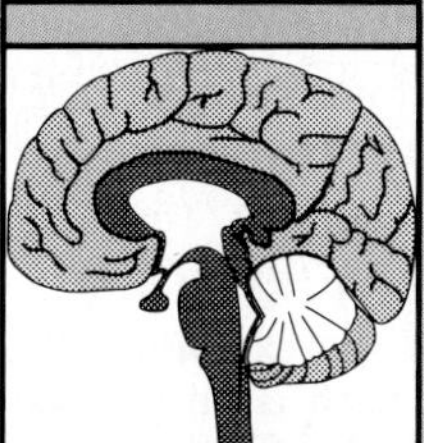

Nervous system

Medullary and pons centers regulate respiratory rate/depth; stretch receptors in lungs provide feedback ▶

Provides oxygen needed for normal neuronal activity; disposes of carbon dioxide ◀

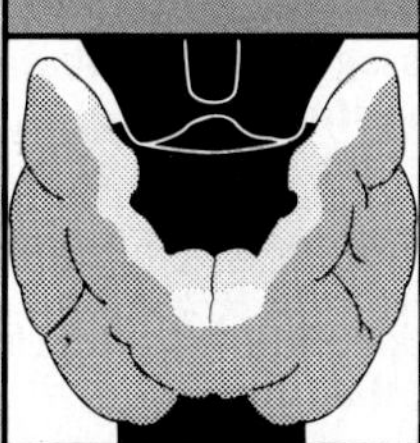

Endocrine system

Epinephrine dilates the bronchioles ▶

Provides oxygen; disposes of carbon dioxide ◀

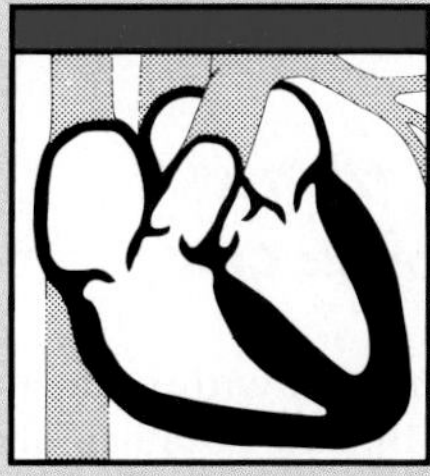

Cardiovascular system

Blood is the transport medium for respiratory gases ▶

Provides oxygen; disposes of carbon dioxide; carbon dioxide present in blood as HCO_3^- and H_2CO_3 contributes fo blood buffering ◀

Respiratory system

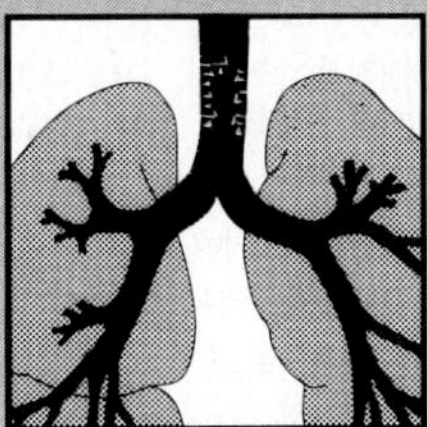

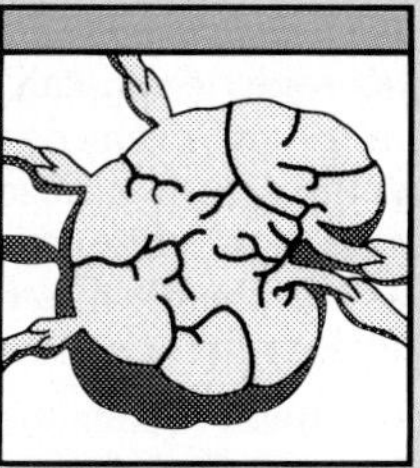

Lymphatic system

◀ Helps to maintain blood volume required for respiratory gas transport

▶ Provides oxygen; disposes of carbon dioxide

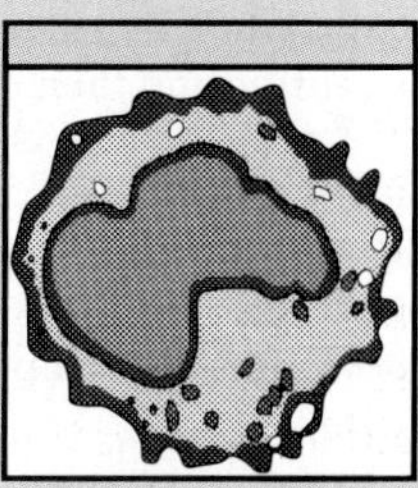

Immune system

◀ Protects repiratory organs from bacteria, bacterial toxins, viruses and cancer

▶ Provides oxygen; disposes of carbon dioxide; tonsils in pharynx house immune cells

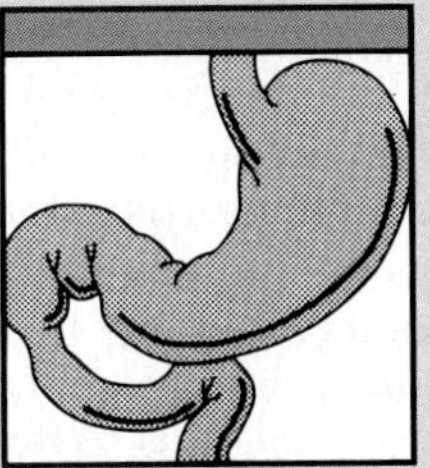

Digestive system

◀ Provides nutrients needed by respiratory system

▶ Provides oxygen; disposes of carbon dioxide

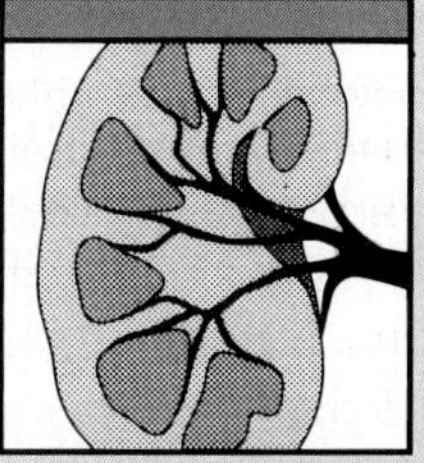

Urinary system

◀ Disposes of metabolic wastes of respiratory system organs (other than carbon dioxide

▶ Provides oxygen; disposes of carbon dioxide

Reproductive system

▶ Provides oxygen; disposes of carbon dioxide

Figure 23.25 Homeostatic interrelationships between the respiratory system and other body systems.

Related Clinical Terms

Bronchography (bron-kah′-gruh-fē) Diagnostic technique used to visualize and examine the bronchial tree; catheter is passed through the mouth or nose and then through the glottis into the trachea; then, a radiopaque chemical is infused and X-rays are taken as the chemical is distributed through the passageways; this method allows obstructions to be detected.

Cheyne–Stokes (chān-stōks) **breathing** A very abnormal breathing pattern sometimes seen just before death (the so-called "death rattle") and in people with combined neurological and cardiac disorders; consists of bursts of tidal volume breaths (first increasing and then decreasing in depth) alternating with periods of apnea; cause is poorly understood, but trauma and hypoxia of the brain stem centers, as well as imbalances between the P_{CO_2} levels of arterial blood and cerebrospinal fluid, may be factors.

Deglutition apnea Temporary cessation of breathing during swallowing.

Deviated septum Condition in which the nasal septum takes a more lateral course than usual and may obstruct breathing; often manifests in old age, but may also result from nose trauma.

Epistaxis (ep″-eh-stak′-sis) (*epistazo* = to bleed at the nose) Nosebleed; nasal hemorrhage; commonly follows trauma to the nose or excessive nose blowing; most nasal bleeding is from the highly vascularized anterior septum and can be stopped by pinching the nostrils closed or packing them with cotton.

Nasal polyps Mushroomlike growths (benign neoplasms) of the nasal mucosa; may occur in response to continual nasal irritation and may block airflow.

Sudden infant death syndrome (SIDS) Unexpected death of an apparently healthy infant during sleep; commonly called crib death, SIDS is one of the most frequent causes of death in infants under one year old; cause is unknown, but it is believed to be a problem of immaturity of the respiratory control centers.

Stuttering Condition in which the vocal cords of the larynx are out of control and the first syllable of words is repeated in "machine-gun" fashion; cause is undetermined, but problems with neuromuscular control of the larynx and emotional factors are suspect; many stutterers become fluent when whispering or singing, both of which involve a change in the manner of vocalization.

Tracheotomy (trā″-kē-ah′-tō-mē) Surgical opening of the trachea; done to provide an alternate route for air to reach the lungs when more superior respiratory passageways are obstructed (as by food or a crushed larynx).

CHAPTER SUMMARY

1. Respiration involves four processes: ventilation, external respiration, internal respiration, and transport of respiratory gases in the blood. Both the respiratory system and the cardiovascular system are involved in respiration.

FUNCTIONAL ANATOMY OF THE RESPIRATORY SYSTEM (pp. 709–721)

2. Respiratory system organs are divided functionally into conducting zone structures (nose to bronchioles), which filter, warm, and moisten incoming air, and respiratory zone structures (respiratory bronchioles to alveoli), where gas exchanges occur.

The Nose (pp. 709–712)

3. The nose provides an airway for respiration and houses the olfactory receptors.

4. The external nose is shaped by bone and cartilage plates. The nasal cavities, which open to the exterior, are separated by the nasal septum. Paranasal sinuses and nasolacrimal ducts drain into the nasal cavities.

The Pharynx (p. 712)

5. The pharynx extends from the base of the skull to the level of the sixth vertebra. The nasopharynx is an air passageway; the oropharynx and laryngopharynx serve as common passageways for food and air. Pairs of tonsils are found in the oropharynx and nasopharynx.

The Larynx (pp. 713–714)

6. The larynx, or voice box, contains the vocal cords. It also provides a patent airway and serves as a switching mechanism to route food and air into the proper channels.

7. The epiglottis prevents food or liquids from entering the respiratory channels during swallowing.

The Trachea (p. 715)

8. The trachea extends from the larynx to the primary bronchi. The trachea is reinforced by C-shaped cartilage rings, which keep the trachea patent, and its mucosa is ciliated.

The Bronchi and Subdivisions: The Bronchial Tree (pp. 716–718)

9. The right and left main bronchi run into their respective lungs, within which they continue to subdivide into smaller and smaller passageways.

10. The terminal bronchioles lead into respiratory zone structures: alveolar ducts, alveolar sacs, and finally alveoli. Gas exchange occurs in the alveoli, across the respiratory membrane.

11. As the respiratory conduits become smaller, cartilage is reduced in amount and finally lost; the mucosa thins, and smooth muscle in the walls increases.

The Lungs and Pleural Coverings (pp. 718–721)

12. The lungs, the paired organs of gas exchange, flank the mediastinum in the thoracic cavity. Each is suspended in its own pleural cavity via its root and has a base, an apex, and medial and costal surfaces. The right lung has three lobes; the left has two.

13. The lungs are primarily air passageways/chambers, supported by an elastic connective tissue stroma.

14. The pulmonary arteries carry blood returned from the systemic circulation to the lungs where gas exchange occurs. The pulmonary veins return newly oxygenated blood back to the heart to be distributed throughout the body. The bronchial arteries provide the nutrient blood supply of the lungs.

15. The parietal pleura lines the thoracic wall and mediastinum; the pulmonary pleura covers external lung surfaces. Pleural fluid produced by the pleurae reduces friction during breathing movements.

MECHANICS OF BREATHING (pp. 721–728)

Pressure Relationships in the Thoracic Cavity (pp. 721–722)

1. Intrapulmonary pressure is the pressure within the alveoli. Interpleural pressure is the pressure within the intrapleural space; it is always negative relative to intrapulmonary and atmospheric pressure.

Pulmonary Ventilation: Inspiration and Expiration (pp. 722–724)

2. Gases travel from an area of higher pressure to an area of lower pressure.

3. Inspiration occurs when the diaphragm and intercostal muscles contract, increasing the dimensions (and volume) of the thorax. As the intrapulmonary pressure drops, air rushes into the lungs until the intrapulmonary and atmospheric pressures are equalized.

4. Expiration is largely passive, occurring as the inspiratory muscles relax and the lungs recoil. When intrapulmonary pressure exceeds atmospheric pressure, gases flow from the lungs.

Physical Factors Influencing Pulmonary Ventilation (pp. 724–725)

5. Friction in the air passageways causes resistance, which decreases air passage and causes breathing movements to become more strenuous. The greatest resistance to air flow occurs in the midsize bronchi.

6. Lung compliance depends on elasticity of lung tissue and flexibility of the bony thorax. When either is impaired, expiration becomes an active process, requiring energy expenditure.

7. Surface tension of alveolar fluid acts to reduce alveolar size and collapse the alveoli. This tendency is resisted in part by surfactant.

Respiratory Volumes and Pulmonary Function Tests (pp. 726–728)

8. The four respiratory volumes are tidal, inspiratory reserve, expiratory reserve, and residual. The four respiratory capacities are tidal, functional residual, inspiratory, and total lung. Respiratory volumes and capacities may be measured by spirometry.

9. Anatomical dead space is the air-filled volume (about 150 ml) of the conducting passageways. If alveoli become nonfunctional in gas exchange, their volume is added to the anatomical dead space, and the sum is the total dead space.

10. Alveolar ventilation rate (AVR) is the best index of ventilation efficiency, because it accounts for anatomical dead space.

$$\text{AVR} = \left(\text{TV} - \underset{(\text{ml/breath})}{\text{anatomical dead space}}\right) \times \text{respiratory rate}$$

Nonrespiratory Air Movements (p. 728)

11. Nonrespiratory air movements are voluntary or reflex actions that clear the respiratory passageways or express emotions.

GAS EXCHANGES IN THE BODY (pp. 728–732)

Basic Properties of Gases (pp. 728–730)

1. Gaseous movements in the body occur by bulk flow and by diffusion.

2. Dalton's law states that each gas in a mixture of gases exerts pressure in proportion to its percentage in the total mixture.

3. Henry's law states that the amount of gas that will dissolve in a liquid is proportional to the partial pressure of the gas and its solubility coefficient.

Composition of Alveolar Gas (p. 730)

4. Alveolar gas contains more carbon dioxide and water vapor and considerably less oxygen than atmospheric air.

Gas Exchanges Between the Blood, Lungs, and Tissues (pp. 730–732)

5. External respiration is the process of gas exchange that occurs in the lungs. Oxygen enters the pulmonary capillaries; carbon dioxide leaves the blood and enters the alveoli. Factors influencing this process include the partial pressure gradients, the thickness of the respiratory membrane, surface area available, and the matching of alveolar airflow and pulmonary blood flow.

6. Internal respiration is the gas exchange that occurs between the systemic capillaries and the tissues. Carbon dioxide enters the blood, and oxygen leaves the blood and enters the tissues.

TRANSPORT OF RESPIRATORY GASES BY THE BLOOD (pp. 733–737)

Oxygen Transport (pp. 733–735)

1. Molecular oxygen is carried in blood bound to hemoglobin in the red blood cells. The amount of oxygen bound to hemoglobin depends on the P_{O_2} and P_{CO_2} of blood, blood pH, the presence of 2,3-DPG, and temperature. A very small amount of oxygen gas is transported dissolved in plasma.

2. Hypoxia occurs when inadequate amounts of oxygen are delivered to body tissues. When this occurs, the skin and mucosae may become cyanotic.

Carbon Dioxide Transport (pp. 735–737)

3. Carbon dioxide is transported in the blood dissolved in plasma, chemically bound to hemoglobin, and (primarily) as the bicarbonate ion in plasma. The loading and unloading of O_2 and CO_2 are mutually beneficial.

4. Accumulation of CO_2 leads to acidosis; depletion of CO_2 from blood leads to respiratory alkalosis.

CONTROL OF RESPIRATION (pp. 737–742)

Neural Mechanisms and Generation of Breathing Rhythm (pp. 737–739)

1. Medullary respiratory centers are the inspiratory center (dorsal respiratory group) and the expiratory center. The inspiratory center is responsible for the rhythmicity of breathing.

2. Respiratory centers in the pons, the pneumotaxic and apneustic centers, influence the activity of the medullary inspiratory center.

Factors Influencing the Rate and Depth of Breathing (pp. 739–742)

3. Pulmonary irritant reflexes are initiated by dust, mucus, fumes, and pollutants.

4. The Hering-Breuer reflex is a protective reflex initiated by extreme overinflation of the lungs; it acts to initiate expiration.

5. Emotions, pain, and other stressors can also alter respiration by acting through hypothalamic centers. Respiration can also be controlled voluntarily for short periods of time.

6. The most important chemical factors modifying baseline respiratory rate and depth are arterial levels of carbon dioxide, H^+, and oxygen.

7. An increasing level of carbon dioxide is the most powerful respiratory stimulant. It acts (via the release of H^+ in cerebrospinal fluid) on the central chemoreceptors to cause a reflexive increase in the rate and depth of breathing. Hypocapnia depresses respiration and results in hypoventilation and, possibly, apnea.

8. Acidosis and a decline in blood P_{O_2} act on peripheral chemoreceptors and enhance the response to carbon dioxide.

9. Arterial oxygen levels below 60 mm Hg constitute the hypoxic drive.

RESPIRATORY ADJUSTMENTS DURING EXERCISE AND AT HIGH ALTITUDES (pp. 742–744)

Effects of Exercise (pp. 742–743)

1. As exercise begins, there is an abrupt increase in ventilation (hyperpnea) followed by a more gradual increase in ventilation. When exercise stops, there is an abrupt decrease in ventilation followed by a gradual decline to baseline values. Continued stimulation of respiration is due to lactic acid accumulation.

2. P_{O_2}, P_{CO_2}, and blood pH remain quite constant during exercise and hence do not appear to account for changes in ventilation. Psychological factors and proprioceptor inputs may.

Effects of High Altitude (pp. 743–744)

3. At high altitudes, there is a decrease in arterial P_{O_2} and hemoglobin saturation levels because of the decrease in barometric pressure compared to sea level. Hyperventilation helps restore gas exchange to physiological levels.

4. Long-term acclimatization involves increased erythropoiesis.

HOMEOSTATIC IMBALANCES OF THE RESPIRATORY SYSTEM (pp. 744–745)

1. The major respiratory disorders are COPD (emphysema and chronic bronchitis) and lung cancer; a significant cause is cigarette smoking.

Chronic Obstructive Pulmonary Disease (COPD) (p. 744)

2. Emphysema is characterized by permanent enlargement and destruction of alveoli. The lungs lose their elasticity, and expiration becomes an active process.

3. Chronic bronchitis is characterized by excessive mucus production in the lower respiratory passageways, which severely impairs ventilation and gas exchange. Patients may become cyanotic as a result of chronic hypoxia.

Lung Cancer (pp. 744–745)

4. Lung cancer is extremely aggressive and metastasizes rapidly.

DEVELOPMENTAL ASPECTS OF THE RESPIRATORY SYSTEM (pp. 745–746)

1. The superior respiratory system mucosa develops from the invagination of the ectodermal olfactory placode; that of the inferior passageways develops from an outpocketing of the endodermal foregut lining. Mesoderm becomes associated with these mucosae and forms the walls of the respiratory conduits and the lung stroma.

2. Premature infants have problems keeping their lungs inflated owing to the lack of surfactant in their alveoli, resulting in infant respiratory distress syndrome (IRDS). Surfactant is formed late in fetal development.

3. With age, the thorax becomes more rigid, the lungs become less elastic, and vital capacity declines. In addition, the stimulatory effect of increased arterial levels of carbon dioxide is less apparent, and respiratory system protective mechanisms are less effective in old age.

Review Questions

Multiple Choice/Matching

1. Cutting the phrenic nerves will result in (a) air entering the pleural cavity, (b) paralysis of the diaphragm, (c) stimulation of the diaphragmatic reflex, (d) paralysis of the epiglottis.

2. Following the removal of his larynx, an individual would (a) be unable to speak, (b) be unable to cough, (c) have difficulty when he tries to swallow, (d) be in respiratory difficulty or arrest.

3. Under ordinary circumstances, the Hering-Breuer reflex is initiated by (a) the inspiratory center, (b) the apneustic center, (c) overinflation of the alveoli and bronchioles, (d) the pneumotaxic center.

4. The detergent-like molecule that keeps the alveoli from collapsing between breaths because it reduces the surface tension of the water film in the alveoli is called (a) lecithin, (b) bile, (c) surfactant, (d) reluctant.

5. Which of the following determines the direction of gas movement? (a) solubility in water, (b) partial pressure gradient, (c) diffusion coefficient, (d) molecular weight and size of the gas molecule.

6. When the inspiratory muscles contract, (a) the size of the thoracic cavity is increased in diameter, (b) the size of the thoracic cavity is increased in length, (c) the volume of the thoracic cavity is decreased, (d) the size of the thoracic cavity is increased in both length and diameter.

7. The nutrient blood supply of the lungs is provided by (a) the pulmonary arteries, (b) the aorta, (c) the pulmonary veins, (d) the bronchial arteries.

8. Oxygen and carbon dioxide are exchanged in the lungs and through all cell membranes by (a) active transport, (b) diffusion, (c) filtration, (d) osmosis.

9. Which of the following would not normally be treated by 100% oxygen therapy? (Choose all that apply.) (a) anoxia, (b) carbon monoxide poisoning, (c) respiratory crisis in an emphysema patient, (d) eupnea.

10. Most oxygen carried in the blood is (a) in solution in the plasma, (b) combined with plasma proteins, (c) chemically combined with the heme in red blood cells, (d) in solution in the red blood cells.

11. Which of the following has the greatest stimulating effect on the respiratory center in the brain? (a) oxygen, (b) carbon dioxide, (c) calcium, (d) willpower.

12. In mouth-to-mouth artificial respiration, the rescuer blows air from his or her own respiratory system into that of the victim. Which of the following statements are correct?

1. Expansion of the victim's lungs is brought about by blowing air in at higher than atmospheric pressure (positive-pressure breathing).
2. During inflation of the lungs, the intrapleural pressure increases.
3. This technique will not work if the victim has a hole in the chest wall, even if the lungs are intact.
4. Expiration during this procedure depends on the elasticity of the alevolar and thoracic walls.

(a) all of these, (b) 1, 2, 4, (c) 1, 2, 3, (d) 2, 4.

13. A baby holding its breath will (a) have brain cells damaged because of low blood oxygen levels, (b) automatically start to breathe again when the carbon dioxide levels in the blood reach a high enough value, (c) suffer heart damage because of increased pressure in the carotid sinus and aortic arch areas, (d) be termed a "blue baby."

14. Under ordinary circumstances, which of the following blood components is of no physiological significance? (a) bicarbonate ions, (b) carbaminohemoglobin, (c) nitrogen, (d) chloride.

15. Damage to which of the following would result in cessation of breathing? (a) the pneumotaxic center, (b) the medulla, (c) the stretch receptors in the lungs, (d) the apneustic center.

16. The bulk of carbon dioxide is carried (a) chemically combined with the amino acids of hemoglobin as carbamino hemoglobin in the red blood cells, (b) as the ion HCO_3^- in the plasma after first entering the red blood cell, (c) as carbonic acid in the plasma, (d) chemically combined with the heme portion of Hb.

Short Answer Essay Questions

17. Trace the route of air from the external nares to an alveolus. Name subdivisions of organs where applicable, and differentiate between conducting and respiratory zone structures.

18. (a) Why is it important that the trachea is reinforced with cartilage rings? (b) Of what advantage is it that the rings are incomplete posteriorly?

19. What is the mucociliary escalator, and where is it found?

20. The lungs are mostly passageways and elastic tissue. (a) What is the role of the elastic tissue? (b) Of the passageways?

21. Describe the functional relationships between volume changes and gas flow into and out of the lungs.

22. What is it about the structure of the respiratory membrane that makes the alveoli ideal sites for gas exchange?

23. Discuss how respiratory passageway resistance, lung compliance and elasticity, and alveolar surface tension influence pulmonary ventilation.

24. (a) Differentiate clearly between minute respiratory volume and alveolar ventilation rate. (b) Which provides a more accurate measure of ventilatory efficiency, and why?

25. State Dalton's law of partial pressures and Henry's law.

26. Define partial pressure of a gas.

27. (a) Define hyperventilation. (b) If you hyperventilate, do you retain or expel more carbon dioxide? (c) What affect does hyperventilation have on blood pH?

28. Describe age-related changes in respiratory function.

Clinical Application Questions

29. While diapering her two-year-old boy (who puts virtually everything in his mouth), a mother failed to find one of the small safety pins previously used. Two days later, the boy developed a cough and became feverish. What is likely to have happened to the safety pin, and where (anatomically) would you expect it to be found?

30. Mr. Jackson was driving to work when he was hit broadside by a car that jumped the red light. When freed from the wreckage, Mr. Jackson was deeply cyanotic and his respiration had ceased. His heart was still beating, but his pulse was fast and thready. His head was cocked at a peculiar angle, and the emergency personnel stated that it looked like he had a fracture at the level of the C_2 vertebra. (a) How might this explain the cessation of breathing? (b) What procedures should the emergency personnel initiate immediately? (c) Why is Mr. Jackson cyanotic? Define cyanosis.

CHAPTER 24

The Digestive System

Chapter Outline and Student Objectives

Overview of the Digestive System (pp. 753–757)

1. Identify the overall function of the digestive system, and differentiate between organs of the alimentary canal and accessory digestive organs.
2. List and define briefly the major processes occurring during digestive system activity.
3. Describe the location and function of the peritoneum and its extensions.
4. Describe the tissue composition of the four layers of the alimentary tube wall, and note the general function of each layer.

Anatomy of the Digestive System (pp. 758–777)

5. Describe the gross and microscopic anatomy of each digestive organ.
6. Describe the anatomy of the generalized tooth, and differentiate clearly between deciduous and permanent teeth.
7. Describe structural modifications of the wall of the stomach and small intestine that enhance the digestive process in these regions.

Physiology of Digestion (pp. 777–798)

8. Describe the mechanisms of ingestion, chewing, and swallowing.
9. Describe the composition and functions of saliva, and explain how salivation is regulated.
10. Describe the composition of gastric juice, name the cell types responsible for secreting its various components, and describe the importance of each component in stomach activity.
11. Explain how gastric secretion and mechanical digestion and propulsion in the stomach are regulated.
12. Describe the function of local hormones produced by the small intestine.
13. List the major enzymes or enzyme groups involved in chemical digestion; name the foodstuffs on which they act; and name the end products of protein, fat, carbohydrate, and nucleic acid digestion.
14. Describe the process of absorption of digested foodstuffs that occurs in the small intestine.
15. State the roles of bile and of bicarbonate ions in pancreatic juice in the digestive process.
16. Describe the processes regulating the entry of pancreatic juice and bile into the small intestine.
17. List the major functions of the large intestine, and describe the regulation of defecation.

Developmental Aspects of the Digestive System (pp. 798–799)

18. Note the role of endoderm and mesoderm in the embryonic development of the digestive system.
19. Describe important abnormalities of the gastrointestinal tract at different times during the life span.

Preview of Selected Key Terms

Alimentary canal (al″-ih-men′-tar-ē) (*aliment* = nourish) The continuous hollow tube extending from the mouth to the anus; its walls are constructed by the oral cavity, pharynx, esophagus, stomach, and small and large intestines.

Accessory digestive organs Organs that contribute to the digestive process but are not part of the alimentary canal; include the tongue, teeth, salivary glands, pancreas, liver, and gallbladder.

Digestion (*digest* = dissolved) Chemical or mechanical process of breaking down foodstuffs to substances that can be absorbed.

Absorption Process by which the products of digestion pass through the alimentary tube mucosa into the blood or lymph.

Chyme (kīm) (*chym* = juice) Semifluid, creamy mass consisting of partially digested food and gastric juice.

Peristalsis (payr″-ih-stal′-sis) (*peri* = around; *stalsis* = constriction, compression) Progressive, wavelike contractions that move foodstuffs through the alimentary tube organs (or that move other substances through other hollow body organs).

Villus (*villus* = shaggy hair) Fingerlike projections of the small intestinal mucosa that tremendously increase its surface area for absorption.

After each meal, we perform the remarkable feat of processing plant and animal tissues into the blood and bone of our bodies and into the energy needed to power our body processes. So—when we eat that hamburger, its nutrients may become part of us or may be converted into the energy for our daily jog. The digestive system provides the body with the nutrients, or raw materials, for these cellular conversions: It takes in food, changes it to a form that the cells can use, and then rids the body of the indigestible remains. The treatment that food undergoes, once it has been eaten, is a rough one. And once food is swallowed, it passes totally out of our voluntary control; our only digestive decisions are whether and what to eat.

Overview of the Digestive System

Digestive System Organs

The organs of the digestive system can be separated into two main groups: the *alimentary canal,* also called the *gastrointestinal (GI) tract,* and the *accessory digestive organs* (Figure 24.1). The alimentary canal **digests** food—breaks it down into smaller fragments—and **absorbs** the digested fragments through its lining into the blood. The accessory organs assist in the process of digestive breakdown.

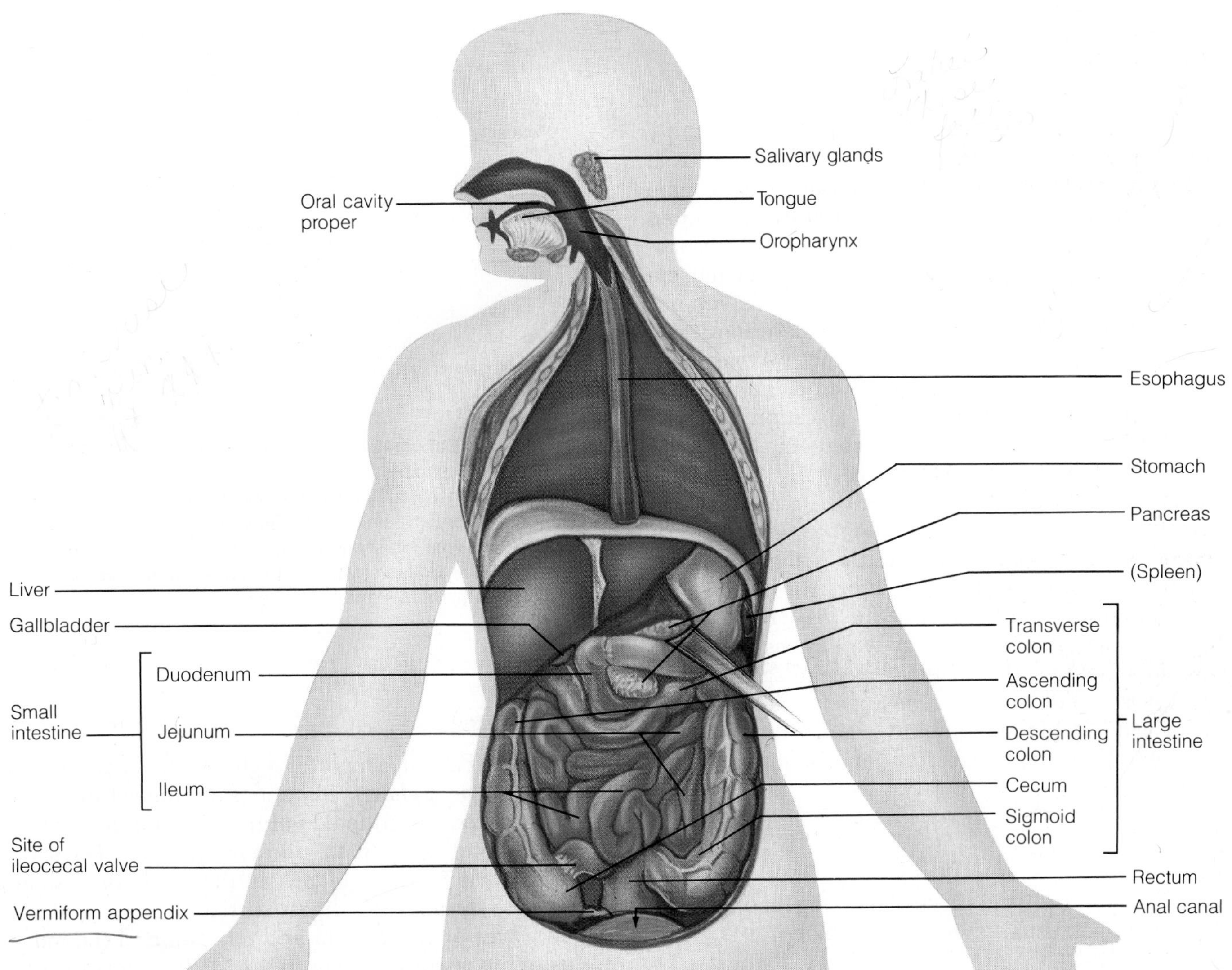

Figure 24.1 Organs of the alimentary canal and related accessory digestive organs.

The **alimentary canal** is a continuous, coiled, hollow, muscular tube that winds through the ventral body cavity and is open to the external environment at both ends. Its organs are the *mouth, pharynx, esophagus, stomach, small intestine,* and *large intestine;* the large intestine includes the terminal opening, or *anus.* In a cadaver, the alimentary canal is approximately 9 m (about 30 feet) long, but in a living person, it is considerably shorter because of its relatively constant muscle tone. Food material within this tube is technically outside the body, because it has contact only with cells lining the tract.

The **accessory digestive** organs are the *teeth, tongue, gallbladder;* and a number of digestive glands—the *salivary glands, liver,* and *pancreas.* The teeth and tongue protrude into the mouth, or oral cavity; the digestive glands and gallbladder lie outside the GI tract and are connected to it by ducts. The accessory digestive glands produce saliva, bile, and enzymes—secretions that contribute in specific ways to the breakdown of various food molecules.

Digestive Processes

The digestive tract can be viewed as a "disassembly line" in which food becomes less and less complex at each step of its processing and its nutrients become available to the body. Although this processing is often summarized in two words—digestion and absorption—many specific activities, such as smooth muscle contraction, secretion, and regulatory events, are not really covered by either term. Thus, to describe digestive system processes more completely, we must consider six essential activities: ingestion, propulsion, mechanical digestion, chemical digestion, absorption, and defecation. Some of these processes are the job of a single organ; for example, only the mouth ingests and only the large intestine defecates. But most digestive system activities require the cooperation of several organs and occur bit by bit as food moves along the tract (Figure 24.2). Later, when we discuss the functioning of each GI tract organ, we will consider the processes specific to it, as well as the neural or hormonal factors that regulate and integrate these processes. In the meantime, you need to know a little about each of the six functions, since they will be mentioned in the discussion of digestive system anatomy.

Ingestion

Ingestion is simply the process of taking food into the digestive tract, usually via the mouth.

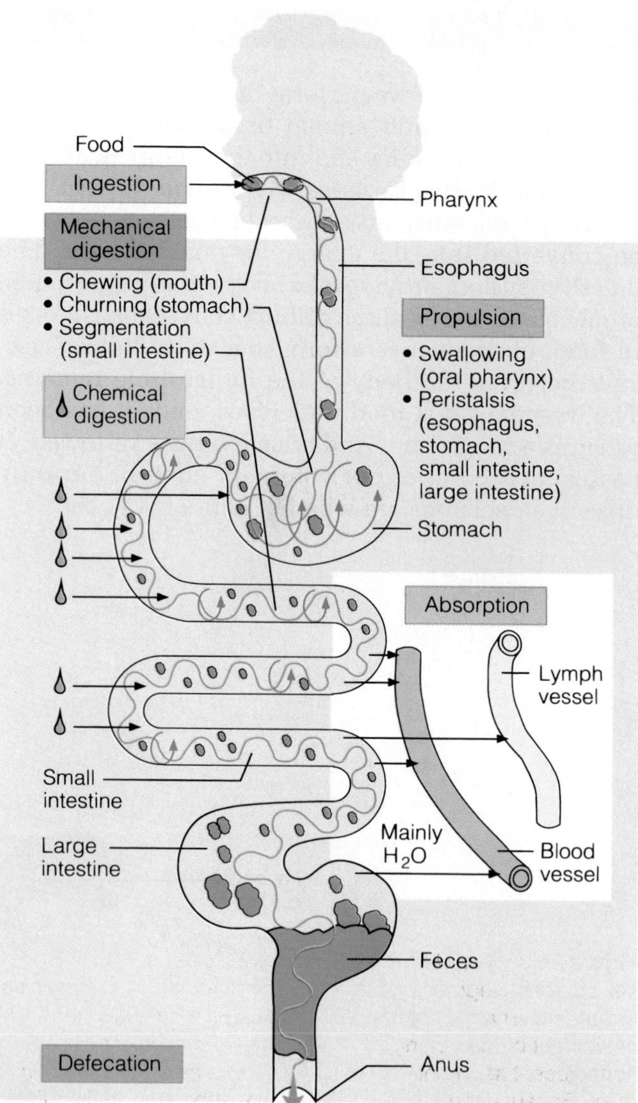

Figure 24.2 Schematic summary of gastrointestinal tract activities. Gastrointestinal tract activities include ingestion, mechanical digestion, chemical (enzymatic) digestion, propulsion, absorption, and defecation. Sites of chemical digestion are also sites where enzyme products are produced and secreted. The mucosa of virtually the entire GI tract secretes mucus, which protects and lubricates.

Propulsion

Processes that move food through the alimentary canal constitute **propulsion.** They include the act of *swallowing,* which is initiated voluntarily, and *peristalsis* (payr″-ih-stal′-sis), an involuntary process. **Peristalsis,** the major means of propelling foods (and food residues) through the digestive tract, involves alternate waves of contraction and relaxation of the muscles in the organ wall (Figure 24.3a). The net effect is to squeeze food along the pathway from one organ to the next, but some mixing occurs as well. From the

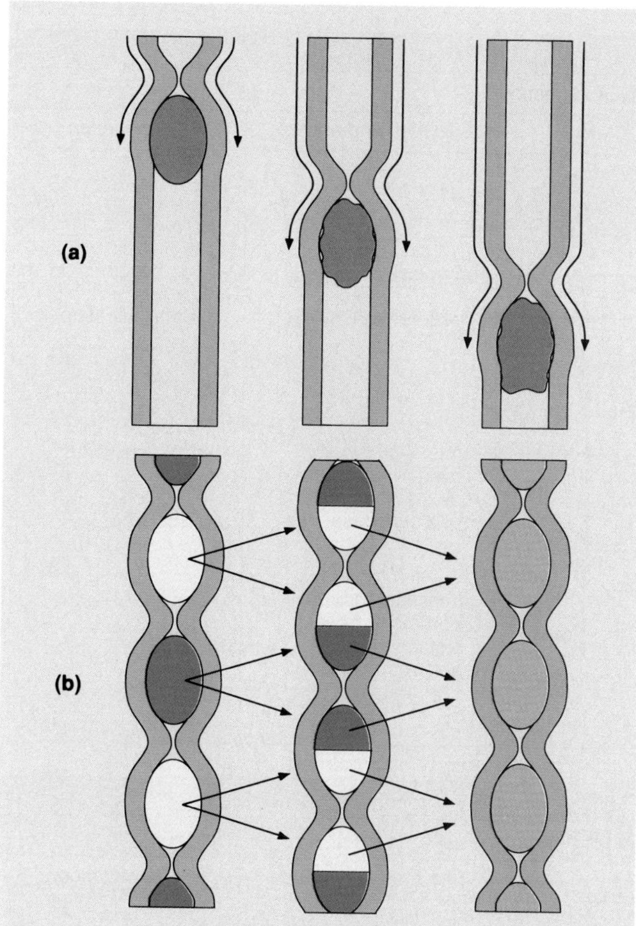

Figure 24.3 Peristalsis and segmentation. (**a**) In peristalsis, adjacent segments of the intestine (or other alimentary tract organs) alternately contract and relax, which results in the movement of food along the tract distally. (**b**) In segmentation, nonadjacent segments of the intestine alternately contract and relax. Because the active segments are separated by inactive regions, the food is moved forward and then backward; this results in mixing of the food, rather than food propulsion.

time food enters the pharynx, its movement is completely reflexive. In fact, the peristaltic waves are so powerful that food and fluids will reach your stomach even if you stand on your head. It is precisely because gravity plays little or no part in food transport that astronauts experiencing zero gravity in outer space can still swallow and get nourishment.

Mechanical Digestion

Mechanical digestion physically prepares food for chemical digestion by enzymes. Mechanical processes include chewing (mastication), mixing of food with saliva by the tongue, churning and mixing of food in the stomach, and **segmentation,** or rhythmic local constrictions of the intestine (Figure 24.3b). Segmentation mixes food with digestive juices and increases the rate of absorption by moving different parts of the food over the intestinal wall. In the small intestine, segmentation also appears to be important in food propulsion.

Chemical Digestion

Chemical digestion is a catabolic process in which large food molecules are broken down to their monomers (chemical building blocks), which are small enough to be absorbed by the GI tract lining (Figure 24.4). Chemical digestion is overseen by enzymes secreted by various glands into the lumen of the alimentary canal. The enzymatic breakdown of any type of food molecule is called **hydrolysis** (hī-drol′-ih-sis), since it involves the addition of a water molecule to each molecular bond to be broken (lysed).

The monomers, or building blocks, of carbohydrates are *monosaccharides* (simple sugars), which are absorbed immediately without further ado. Only three of these are common in our diet: *glucose, fructose,* and *galactose*. The only more complex carbohydrates that our digestive system is able to break down to monosaccharides are the disaccharides *sucrose* (table sugar), *lactose* (milk sugar), and *maltose* (grain sugar) and the polysaccharides *glycogen* and *starch*. In an average American diet, most digestible carbohydrates are in the form of starch, with smaller amounts of disaccharides and monosaccharides. Humans do not have enzymes capable of breaking down other polysaccharides, such as cellulose. Consequently, indigestible polysaccharides do not provide us with any nutrients, but they help move the food along the GI tract by providing bulk, or fiber.

Proteins are ultimately digested to their monomers, *amino acids*. Intermediate products of protein digestion are *polypeptides* and *peptides*. When lipids (fats) are completely digested, they yield two different types of building blocks, *fatty acids* and the sugar alcohol *glycerol*. *Nucleotides* are the units of nucleic acids, but nucleic acids are broken down even further by digestive system enzymes. The enzymatic breakdown of foodstuffs in the GI tract is summarized in the flowchart in Figure 24.4 and described in more detail later in the chapter.

Absorption

Absorption is the transport of digested end products from the lumen of the GI tract to the blood or lymph. For absorption to occur, the monomers must first enter the mucosal cells by active or passive transport processes. The small intestine is the major absorptive site.

Defecation

Defecation is the elimination of indigestible substances from the body via the anus in the form of feces.

(a) Carbohydrate digestion

Example of hydrolysis	Sequence and sites of chemical digestion		
	Foodstuff	Enzymes and source	Site of action
Maltose (HO-CH 2 ... two linked rings)	Lactose Starch Sucrose		
		Salivary amylase	Mouth
Carbohydrate-digesting enzymes; H_2O		Pancreatic amylase	Small intestine
	Dextrins and oligosaccharides		
		Intestinal (brush border) enzymes: dextrinase and glucoamylase	Small intestine
Glucose (2)	Lactose Maltose Sucrose		
	Galactose Glucose Fructose	Intestinal (brush border) enzymes: lactase, maltase, and sucrase	

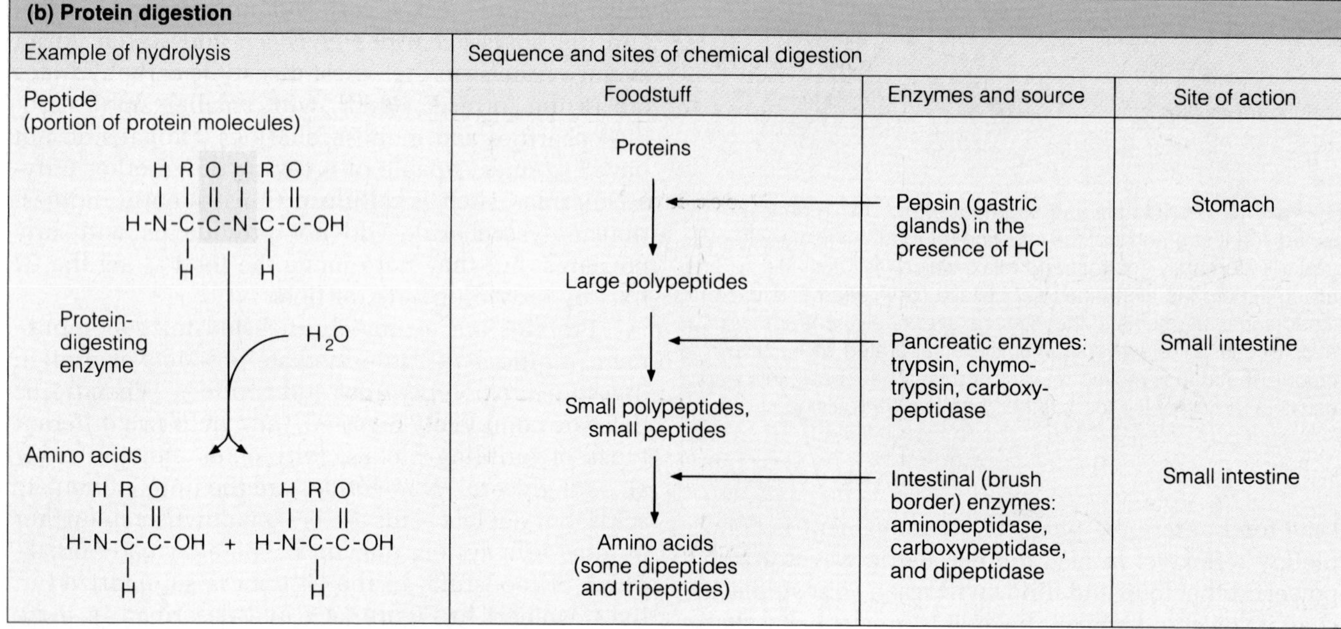

Figure 24.4 Chemical digestion of carbohydrates, proteins, lipids, and nucleic acids. In most cases, chemical (enzymatic) digestion of food molecules within the gastrointestinal tract involves hydrolysis of the food molecules to their monomers.

(c) Lipid digestion

Example of hydrolysis	Sequence and sites of chemical digestion		
	Foodstuff	Enzymes and source	Site of action
Fat: $C_{17}H_{35}C(=O)-O-CH_2$, $C_{17}H_{35}C(=O)-O-CH$, $C_{17}H_{35}C(=O)-O-CH_2$	Unemulsified fats		
Fat-digesting enzyme, $3H_2O$	↓	Emulsified by detergent action of bile salts from liver	Small intestine
	↓	Pancreatic lipase	Small intestine
Fatty acids: $3\ C_{17}H_{35}COOH$ + Glycerol: $HO-CH_2$, $HO-CH$, $HO-CH_2$	Monoglycerides and fatty acids; Glycerol and fatty acids		

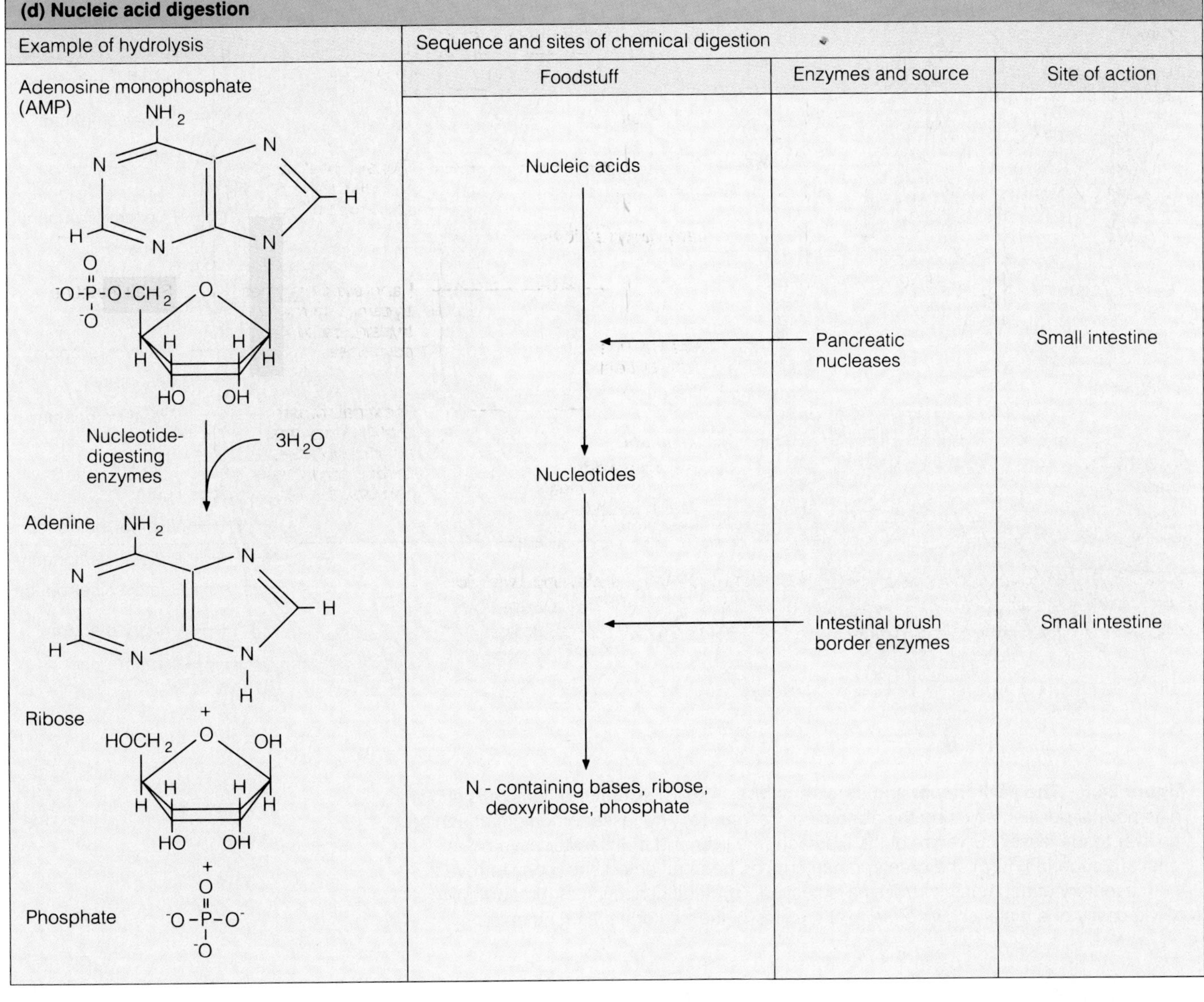

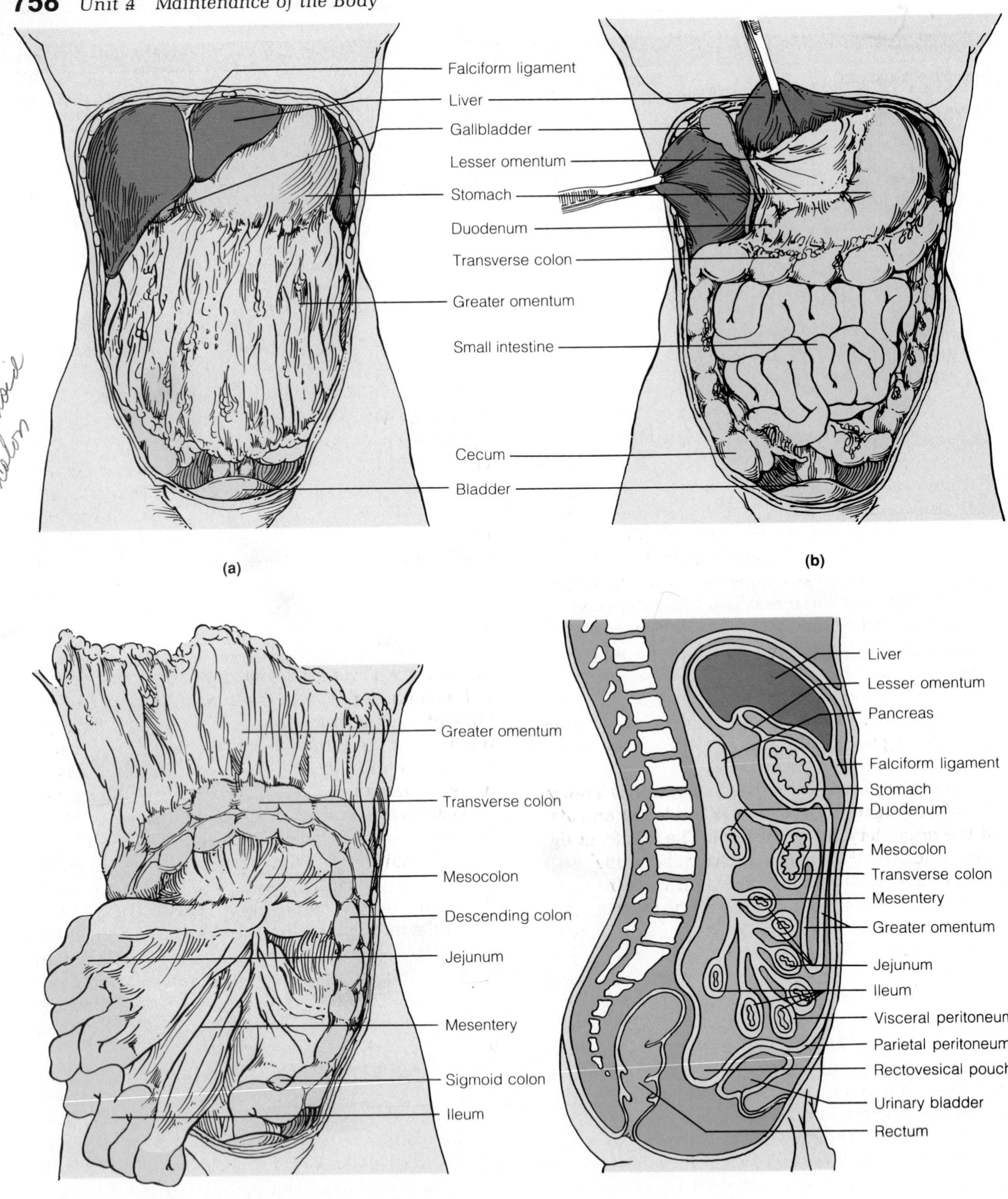

Figure 24.5 The peritoneum and its extensions. (**a**) The greater omentum is shown in its normal position covering the abdominal viscera. (**b**) The lesser omentum attaching the liver to the lesser curvature of the stomach (the liver and gallbladder have been reflected superiorly). (**c**) The greater omentum has been reflected superiorly to reveal the mesentery attachments of the small intestine. (**d**) Sagittal section of the abdominopelvic cavity of a male, showing the relationships of the peritoneal attachments.

Anatomy of the Digestive System

Because most digestive organs only have a "share" in the work of digestion, it is necessary to consider the structural characteristics of virtually all the digestive organs before we can describe digestive functions in any cohesive way. We will now describe this "anatomical landscape."

Relationships of the Digestive Organs and the Peritoneum

As described in Chapter 1, serous membranes are thin, friction-reducing membranes that cover the walls of the ventral body cavities and the surfaces of the organs within those cavities (see Figure 1.9). The *peritoneum* (payr″-ih-tō-nē′-um) of the abdominopelvic cavity is the most extensive serous membrane in the body. The **visceral peritoneum,** also referred to as the *serosa,* covers the external surfaces of most digestive organs. The visceral peritoneum is continuous with the slick **parietal peritoneum** that lines the walls of the abdominopelvic cavity by way of double-layered extensions or folds that suspend or anchor the digestive organs within the peritoneal cavity. These folds serve as routes for conducting blood vessels, lymphatics, and nerves to the digestive organs (Figure 24.5).

Important peritoneal folds include the **falciform** (fal′-si-form) **ligament,** the **greater** and **lesser omenta,** and the **mesentery** (Figure 24.5d). The falciform ligament binds the liver to the anterior abdominal wall and diaphragm. The lesser omentum runs from the liver to the anteromedial surface of the stomach (the lesser curvature), where it becomes continuous with the visceral peritoneum covering the stomach. The greater omentum drapes inferiorly from the lateral stomach surface (the greater curvature) to cover the coils of the small intestine and then runs dorsally and superiorly to wrap the transverse portion of the large intestine before forming the **mesocolon** (meh″-zō-kō′-lun). The mesocolon secures the transverse colon to the parietal peritoneum of the posterior abdominal wall. From there, the peritoneum runs to cover the anterior surface of the pancreas and extends to form the mesentery, which suspends the small intestine from the posterior abdominal wall.

The greater omentum is riddled with fat deposits that give it the appearance of a lacy apron covering and protecting the abdominal contents (see Figure 24.5a). It also contains large collections of lymph nodules. When the peritoneum is infected, a condition called *peritonitis* (payr″-ih-tuh-nī′-tis), the peritoneal coverings tend to stick together around the infection site. In this way, many intraperitoneal infections are sealed off and localized (at least initially), providing time for macrophages in the lymphatic tissue to mount the attack to prevent the inflammation from spreading to other regions of the peritoneal cavity.

Between the visceral and parietal peritoneums is the **peritoneal cavity,** a potential space that contains pale yellow serous fluid secreted by these serous membranes. The serous fluid lubricates the mobile digestive organs, allowing them to glide easily upon one another as they squeeze and churn the foodstuffs or propel them along the tract.

Histology of the Alimentary Tract Wall

Beginning with the esophagus, the walls of the GI tract organs have a basic tissue pattern that reflects their common functions. The basic alimentary canal wall structure, shown in Figure 24.6, has four distinct layers, or *tunics:* the *mucosa, submucosa, muscularis,* and *serosa.* Each layer contains a predominant tissue type that has a specific role to play in the food breakdown process.

The Mucosa

The **mucosa,** or **mucous membrane,** is a moist epithelial membrane that lines the lumen, or cavity, of the alimentary canal from the mouth to the anus. Its major functions are *secretion* (of mucus, digestive enzymes, and hormones), *absorption* of the end products of digestion into the blood, and *protection* against infectious disease. The mucosa in a particular region of the GI tract may express one or all three of these capabilities.

The mucosa consists of a surface epithelium underlain by a small amount of connective tissue called the lamina propria and a scanty smooth muscle layer called the muscularis mucosae. Typically, the **epithelium** of the mucosa is a *simple columnar epithelium* that is rich in mucus-secreting *goblet cells* and other types of glands. The slippery mucus it produces protects certain digestive organs from being digested themselves by the enzymes working within their cavities and eases the passage of foods along the tract. In some digestive organs, such as the stomach and small intestine, the mucosa contains both enzyme-secreting and hormone-secreting cells. Thus, in such sites, the mucosa is a diffuse kind of endocrine organ as well as part of the digestive organ.

The **lamina propria** (la′-mih-nuh prō′-prē-uh) (*proprius* = one's own), consisting of areolar connective tissue and containing lymph nodules, is important in the defense against bacteria and other pathogens, which have rather free access to the digestive tract. Particularly large collections of lymph nodules

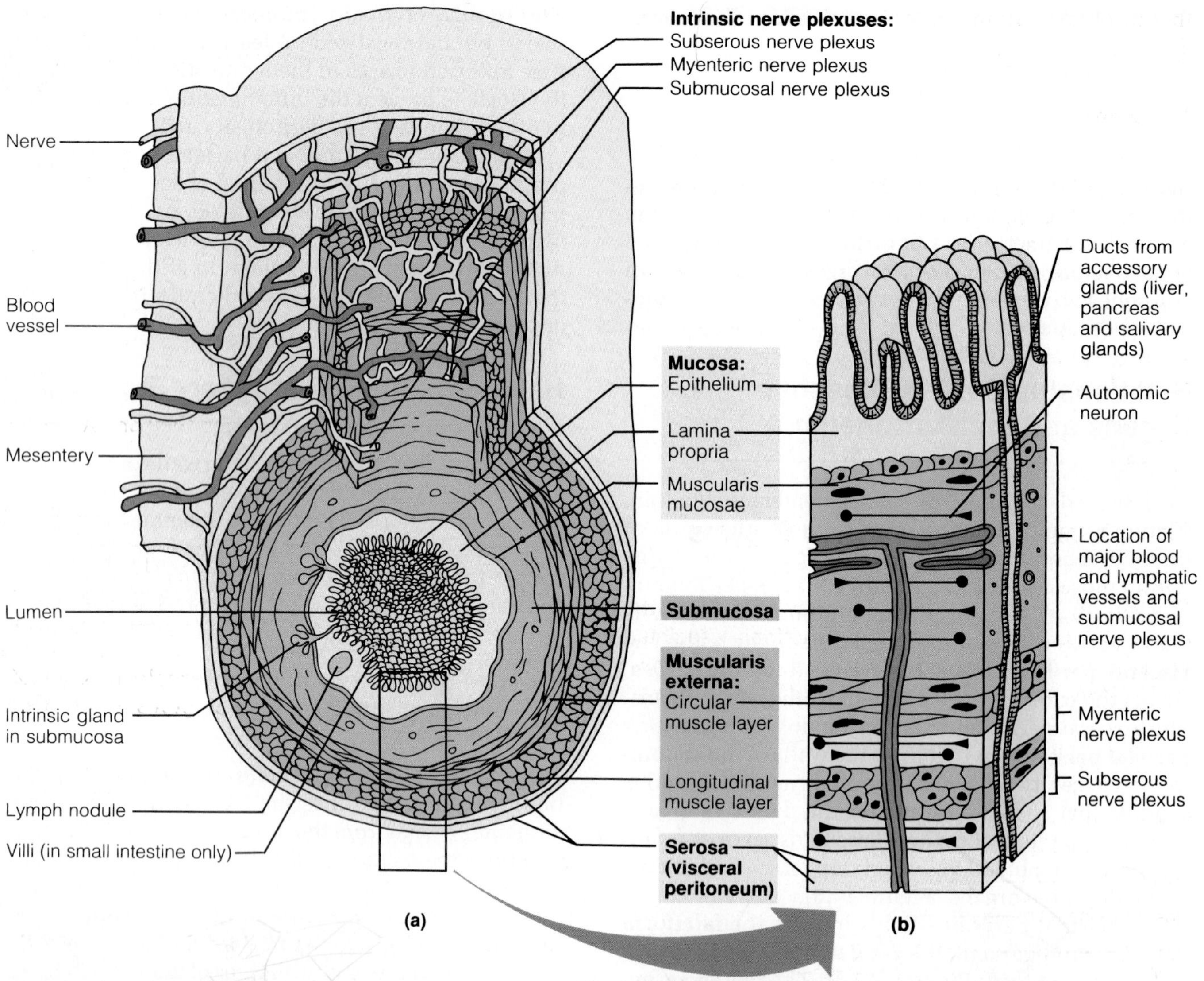

Intrinsic nerve plexuses:
Subserous nerve plexus
Myenteric nerve plexus
Submucosal nerve plexus
Nerve
Blood vessel
Mesentery
Lumen
Intrinsic gland in submucosa
Lymph nodule
Villi (in small intestine only)
Mucosa:
Epithelium
Lamina propria
Muscularis mucosae
Submucosa
Muscularis externa:
Circular muscle layer
Longitudinal muscle layer
Serosa (visceral peritoneum)
(a)
Ducts from accessory glands (liver, pancreas and salivary glands)
Autonomic neuron
Location of major blood and lymphatic vessels and submucosal nerve plexus
Myenteric nerve plexus
Subserous nerve plexus
(b)

Figure 24.6 The basic structure of the wall of the alimentary canal consists of four layers: the mucosa, submucosa, muscularis externa, and serosa. (**a**) Cross-sectional view with part of the wall removed to show the intrinsic nerve plexuses. (**b**) Schematic view of a portion of a cross section of the wall, indicating its major components and their arrangements.

are found at strategic locations, such as within the pharynx (as the tonsils), at the distal end of the small intestine (Peyer's patches), and at the appendix. In the small intestine, the **muscularis mucosae** throws the mucosal tunic into a series of small folds, which vastly increases its surface area for secretion and absorption.

The Submucosa

The **submucosa,** just deep to the mucosa, is areolar connective tissue containing blood vessels, lymphatic vessels, nerve endings, and epithelial glands. Its luxuriant vascular network supplies surrounding tissues and carries away absorbed nutrients. Its nerve supply, called the **submucosal plexus,** is part of the *intrinsic nerve supply* of the GI tract wall and is actually part of the autonomic nervous system.

The Muscularis

The **muscularis,** also called the **muscularis externa,** mixes and propels foodstuffs along the digestive tract. This muscular tunic usually has an inner circular layer and an outer longitudinal layer of smooth muscle cells. However, as described shortly, there are variations in this fundamental pattern and, at several points along the GI tract, the circular muscle layer becomes thickened to form *sphincters* that act as valves to prevent backflow and control food passage from one organ to the next.

The **myenteric** (mī″-en-tayr′-ik) **plexus,** another autonomic nerve plexus, is found between the two smooth muscle layers. It provides the major nerve supply to the GI tract wall and controls GI motility. In general, activity of parasympathetic fibers enhances GI motility and secretory activity, whereas sympathetic fibers inhibit motility, thereby depressing digestion.

The Serosa

The **serosa,** the protective outermost layer of most of the alimentary canal wall, is the visceral peritoneum. It is formed of areolar connective tissue covered with *mesothelium,* a single layer of squamous epithelial cells. The **subserous plexus,** a third autonomic plexus, is associated with the serosa.

In the esophagus, the serosa is replaced by an **adventitia** (ad″-ven-tih′-shuh), consisting of a denser, more fibrous areolar connective tissue that binds the esophagus to surrounding structures. The same is true of surfaces of abdominal organs that are retroperitoneal (behind the peritoneum).

The Mouth and Associated Organs

The mouth is simply the site of food entry into the alimentary canal. Most digestive functions associated with the mouth reflect the activity of associated accessory organs, such as the tongue, salivary glands, and teeth.

The Mouth

The **mouth,** a mucous membrane–lined cavity, is also called the **oral cavity,** or **buccal** (buk′-ul) **cavity.** Its boundaries are the lips, which protect its anterior opening (the *oral orifice*), the cheeks laterally, the palate superiorly, and the tongue inferiorly (Figure 24.7). Posteriorly, the oral cavity is continuous with the *oropharynx* at the junction called the *fauces* (faw′-sēz). The walls of the mouth have to withstand considerable wear and are lined with a stratified squamous epithelium instead of the more typical simple columnar epithelium. Although not keratinized elsewhere in the oral cavity, this epithelium *is* keratinized on the gums, hard palate, and dorsum of the tongue, surfaces that require more fortification because they are subjected to abrasion during eating.

The Lips and Cheeks. The **lips (labia)** and the **cheeks** have a core of skeletal muscle covered externally by skin. The orbicularis oris muscle forms the bulk of the fleshy lips; the cheeks are formed largely by the buccinators. The lips and the cheeks help keep food between the teeth when we chew and also play a role in speech. The recess bounded externally by the labia and cheeks and internally by the gums and teeth is

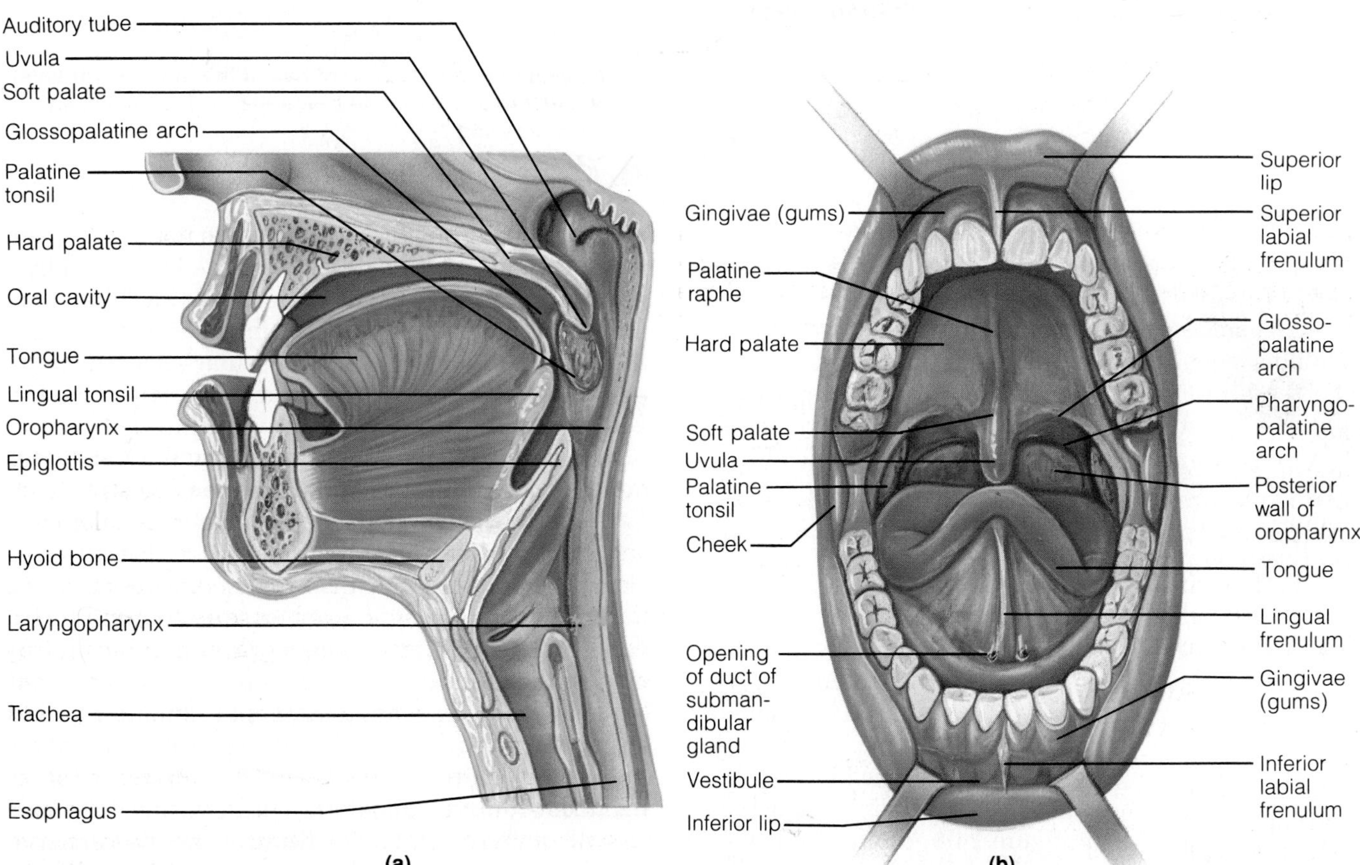

Figure 24.7 Anatomy of the oral cavity (mouth). (**a**) Sagittal section of the oral cavity and pharynx. (**b**) Anterior view of the oral cavity.

called the **vestibule;** the area that lies within the teeth and gums is the **oral cavity proper.**

At the free margin of each lip is a transitional zone where the keratinized membrane (skin) and nonkeratinized membranes (mucosa) meet. This region, called the *red margin,* is poorly keratinized and translucent, allowing the red color of blood in the underlying capillaries to flush through. Because there are no sweat or sebaceous glands on the red margin, it must be moistened with saliva periodically to prevent it from becoming dry and cracked.

The Palate. The **palate** (pal′-at), forming the roof of the mouth, has two distinct parts: the hard palate anteriorly and the soft palate posteriorly (see Figure 24.7). The **hard palate** is underlain by bone (the palatine bones and the palatine processes of the maxillae), and it forms a rigid surface against which the tongue can force food during chewing. The mucosa on either side of its *raphe* (rā′-fē), a midline ridge, is slightly corrugated, which helps to create friction.

The **soft palate** is a mobile fold formed mostly of skeletal muscle and lacking bony reinforcement. Dipping downward from its free edge is a fingerlike projection called the **uvula** (yoo-vyoo′-luh). The soft palate rises reflexly to close off the nasopharynx when we swallow food. It is anchored laterally to the tongue by the **glossopalatine arches** and connected to the wall of the oropharynx by the **pharyngopalatine arches.** Located in the recess between these arches on each side are the **palatine tonsils,** lymphatic tissue structures that are part of the body's defense system (see Chapter 21).

The Tongue

The **tongue** occupies the floor of the mouth (see Figure 24.7). Constructed of interlacing masses of skeletal muscle held together by connective tissue, the tongue contains mucous and serous glands, many small pockets of adipose cells, and the sensory receptors that allow us to taste our food. Besides making movements required for speech, the versatile tongue mixes food with saliva during chewing, forms the food into a compact mass called a *bolus* (bō′-lus), and initiates swallowing.

Both intrinsic and extrinsic skeletal muscle fibers form the tongue. The *intrinsic muscles* are confined to the tongue and are unattached to bone. They include bundles of fibers that run in three planes: longitudinal, transverse, and vertical (Figure 24.8). These muscles allow the tongue to change shape, becoming thicker, thinner, longer, or shorter, as needed for speech and swallowing. The *extrinsic muscles,* including the genioglossus shown in Figure 24.8, extend from the tongue to their points of origin on bones of the skull or the soft palate, as described in Chapter 10 (see p. 288 and Figure 10.6.) The extrinsic muscles alter the

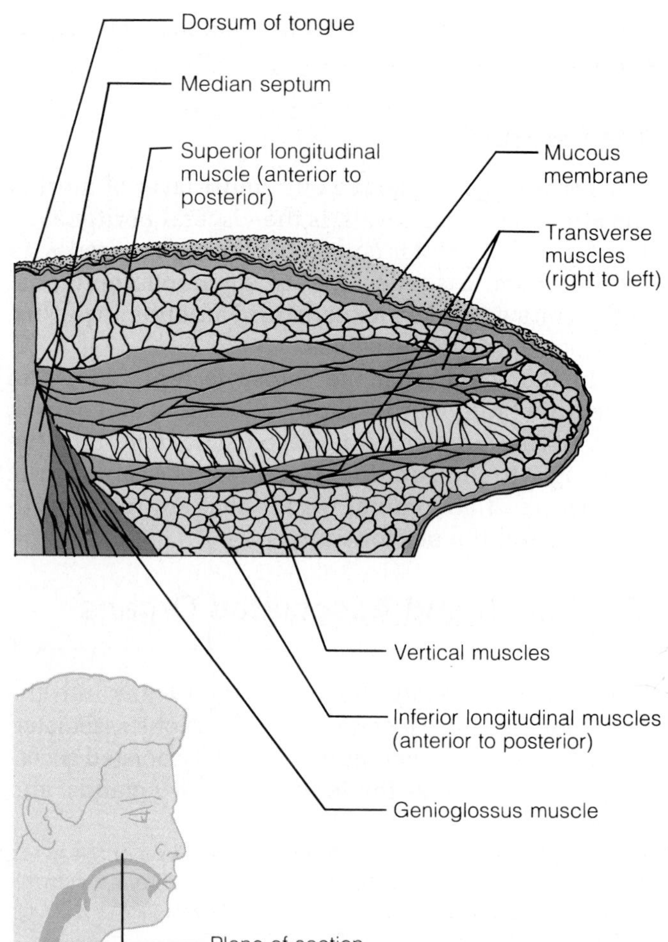

Figure 24.8 Frontal section of half of the tongue (left side) showing its intrinsic muscles, and the genioglossus, an extrinsic muscle.

position of the tongue: They protrude it, retract it, and move it from side to side. The tongue is divided by a midline connective tissue septum, and each half contains identical groupings of the intrinsic and extrinsic muscles. In addition to its muscular attachments to bone, the tongue has a fold of mucosa called the **lingual frenulum** (lin′-gwul fren′-yoo-lum) that secures it to the floor of the mouth (see Figure 24.7b). This frenulum also limits posterior movements of the tongue.

Children born with an extremely short lingual frenulum are often referred to as "tongue-tied" because of the speech distortions that result when movement of the tongue is abnormally restricted. This congenital condition can be corrected surgically by cutting the frenulum. ■

The mucosa covering most of the superior tongue surface bears **papillae** (puh-pih′-lē), projections of the underlying lamina propria. The three types of papillae—filiform, fungiform, and circumvallate—are shown in Figure 24.9. The **filiform papillae** are conical and in many animals (such as cats) have a sharp tip, making the tongue raspy. Human filiform papillae

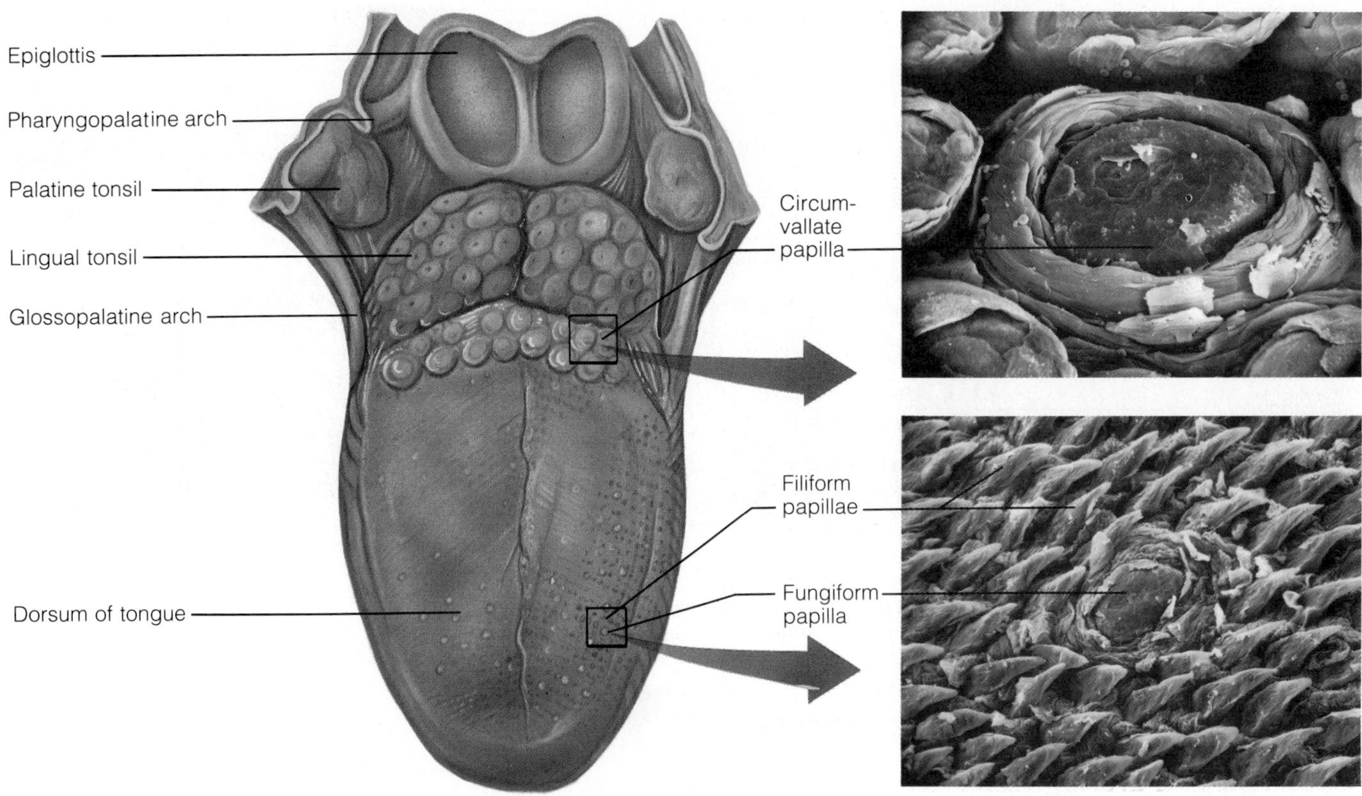

Figure 24.9 Surface view of the dorsal surface of the tongue. Also shown are the locations and detailed structures of the circumvallate, fungiform, and filiform papillae. The tonsils closely associated with the oral cavity are also illustrated. (Top right, 300×; bottom right, 140×.)

are less sharp, but they do give the tongue surface a roughness that aids in licking semisolid foods (such as ice cream) and provide friction for manipulating foods in the mouth. The filiform papillae, the most numerous type, cover the anterior two-thirds of the tongue dorsum. These papillae are heavily keratinized and give the tongue its whitish appearance.

The fungiform and circumvallate papillae house taste buds, the special sense organs of taste described in Chapter 16. The mushroom-shaped **fungiform papillae** are located primarily on the sides of the tongue and are interspersed among the filiform papillae on the tongue apex. Each has a vascular core that gives it a reddish hue. Ten to twelve large **circumvallate,** or **vallate, papillae** are located in a V-shaped row at the back of the tongue. They resemble the fungiform papillae but have an additional surrounding furrow, or trough.

The mucosa covering the posteriormost aspect and root of the tongue lacks papillae, but it is still bumpy because of the nodular **lingual tonsil,** which lies just deep to its mucosa (see Figure 24.9).

The Salivary Glands

A number of glands both inside and outside the oral cavity produce and secrete **saliva.** Saliva has many functions. It (1) cleanses the teeth; (2) dissolves food chemicals so that they can be tasted; (3) moistens food and aids in its compaction into a bolus; and (4) contains enzymes that begin the chemical breakdown of starchy foods.

Most saliva is produced by three pairs of major **salivary glands** that lie outside the oral cavity and empty their secretions into it. Their output is augmented slightly by small glands, called *buccal glands,* forming part of the oral cavity mucosa.

The salivary glands include the parotid, submandibular, and sublingual glands (Figure 24.10), paired compound tubuloalveolar glands that develop from the mucosa of the oral cavity and remain connected to it by ducts. The large **parotid** (puh-rah′-tid) **glands** lie anterior and inferior to the ears, where each is encased in a connective tissue capsule that lies between the masseter muscle and the skin. Their prominent ducts (*parotid ducts*) pierce the buccinator muscles and open into the vestibule next to the second molars of the upper jaw.

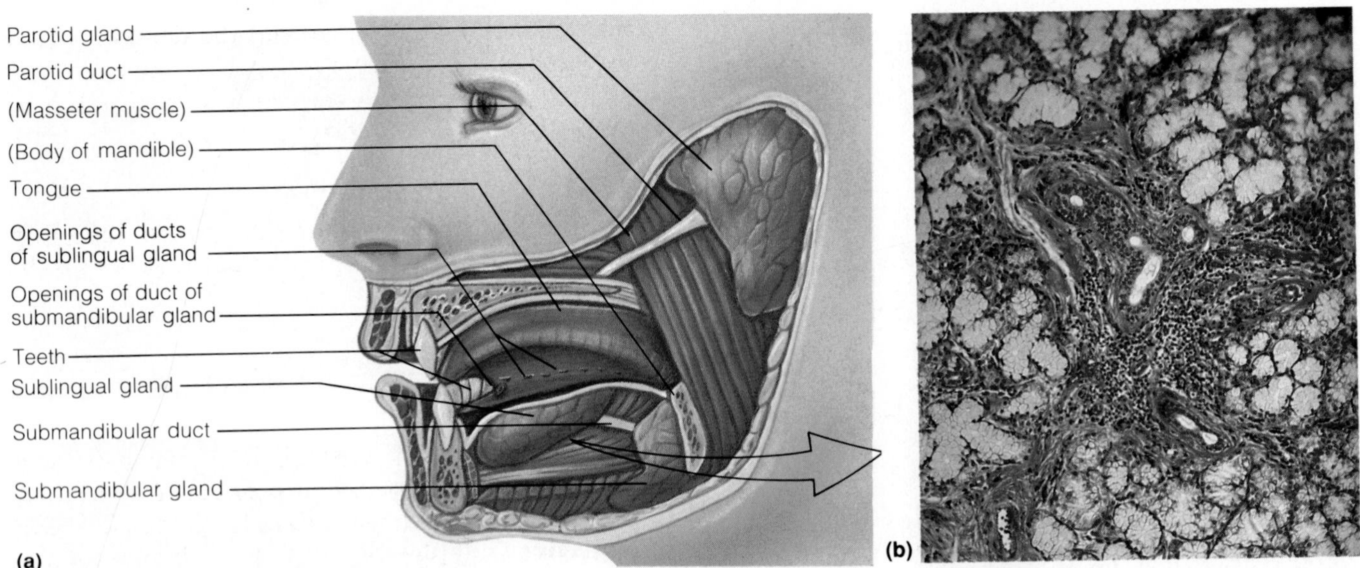

Figure 24.10 The major salivary glands. (**a**) The parotid, submandibular, and sublingual salivary glands associated with the left aspect of the oral cavity. (**b**) Photomicrograph of the sublingual salivary gland (75×), which is a mixed salivary gland consisting predominantly of mucin-producing cells (white) and a few serous-secreting units (purple). The serous cells sometimes form demilunes (caps) around the bases of the mucous cells.

Mumps, a common children's disease, is an inflammation of the parotid glands. If you check the location of the parotid glands in Figure 24.10a, you can readily understand why people with mumps complain that it hurts to open their mouth or chew. In males, the testes may also become inflamed, and result in sterility, particularly when the inflammation occurs during adulthood. ■

The **submandibular glands** lie bilaterally along the medial aspect of the mandibular angle. Their ducts, the *submandibular ducts,* run beneath the mucosa of the oral cavity floor and open at the base of the lingual frenulum (see also Figure 24.7). The small **sublingual glands** lie anterior to the submandibular glands under the tongue, and their multiple ducts open into the floor of the mouth.

To a greater or lesser degree, the salivary glands are composed of two major types of secretory cells: mucous and serous. The **serous cells** produce a watery secretion containing salivary amylase, whereas the **mucous cells** produce **mucus,** a stringy, viscous solution rich in the lubricating glycoprotein **mucin** (myoo′-sin). The parotid glands contain mostly serous cells. The submandibular glands are mixed glands containing approximately equal numbers of serous and mucous cells. The sublingual glands (see Figure 24.10b) and buccal glands contain mostly mucous cells and produce a saliva thick with mucus.

The Teeth

The role of the teeth in food processing needs little introduction. We **masticate,** or **chew,** by opening and closing our jaws and moving them from side to side while continually moving the food between our teeth and tongue to mix it with saliva. In the process, the teeth tear and grind the food, breaking it down mechanically into smaller fragments.

Dentition and the Dental Formula. Ordinarily, by age 21, two sets of teeth, or **dentitions,** have been formed, called the primary and permanent dentitions (Figure 24.11). The *primary dentition* consists of the **deciduous teeth** (*decid* = falling off), also called **milk teeth** or **baby teeth.** The first teeth to appear, at about age six months, are the lower central incisors. An additional pair of teeth erupts at one- to two-month intervals until all 20 milk teeth are through the gums by 20 to 24 months.

As the deeper **permanent teeth** enlarge and develop, the roots of the milk teeth are resorbed, causing the first set of teeth to loosen and fall out between the ages of 6 and 12 years. Generally, all the permanent teeth but the third molars have erupted by the end of adolescence; the third molars, also called the wisdom teeth, emerge later, between the ages of 17 and 25 years. There are 32 permanent teeth in a full set, but the wisdom teeth often fail to erupt, and sometimes they are completely absent.

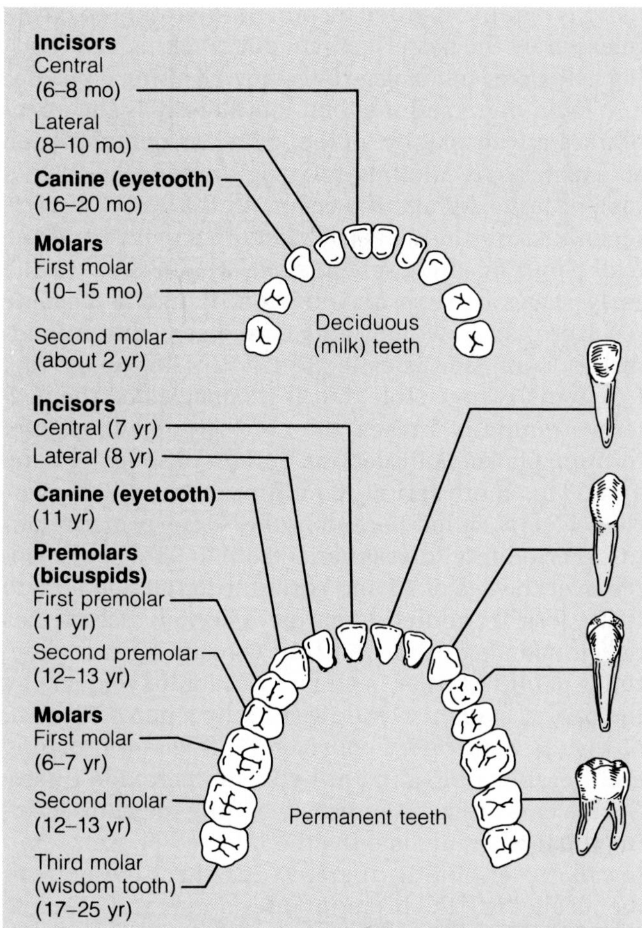

Figure 24.11 Human deciduous and permanent teeth. Approximate time of tooth eruption is shown in parentheses. Since the same number and arrangement of teeth exist in both upper and lower jaws, only the lower jaw is shown in each case. The shapes of individual teeth are shown on the right.

When teeth remain embedded in the jawbone, they are said to be *impacted*. Impacted teeth cause a good deal of pressure and pain and must be removed surgically. Wisdom teeth are most commonly impacted. ■

The teeth are classified according to shape and function as incisors, canines, premolars, and molars (see Figure 24.11). The chisel-shaped **incisors** are adapted for cutting, the conical or fanglike **canines** (eyeteeth) for tearing or piercing. The **premolars** (bicuspids) and **molars** have broad crowns with rounded cusps (tips) and are best suited for grinding or crushing.

The **dental formula** indicates the types, number, and position of teeth in one-half of the mouth. (Since the other side is a mirror image, the total dentition is simply obtained by multiplying the dental formula by 2.) The primary dentition consists of two incisors (I), one canine (C), and two molars (M) on each side of each jaw (see Figure 24.11), and the dental formula for the milk teeth is written as

$$\frac{\text{2I, 1C, 2M (upper jaw)}}{\text{2I, 1C, 2M (lower jaw)}} \times 2 \quad \text{(20 teeth)}$$

Similarly, the permanent dentition (two incisors, one canine, two premolars [PM], and three molars) is indicated by

$$\frac{\text{2I, 1C, 2PM, 3M}}{\text{2I, 1C, 2PM, 3M}} \times 2 \quad \text{(32 teeth)}$$

Tooth Structure. Each **tooth** has two major regions: the crown and the root (Figure 24.12). The enamel-covered **crown** is the part of the tooth visible above the **gingiva** (jin′-jih-vuh), or **gum**, which surrounds the tooth like a tight collar. **Enamel** is an acellular, brittle material that persists throughout life; heavily mineralized with calcium salts, it is the hardest substance in the body. The enamel is the only part of the tooth produced by ectoderm-derived cells; the rest of the tooth is formed by mesoderm. The cells that pro-

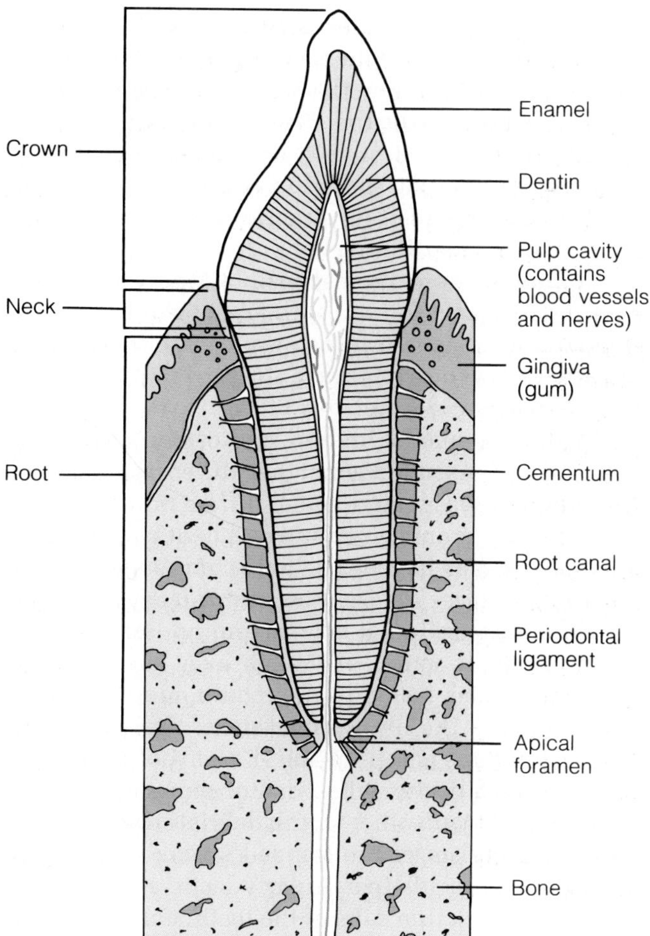

Figure 24.12 Longitudinal section of a canine tooth within its bony alveolus.

duce the enamel degenerate at the time of tooth eruption, leaving none for repair purposes. Consequently, any breaches of the enamel caused by decay or cracks must be artificially filled.

The portion of the tooth embedded in the jawbone is the **root.** Canine teeth and incisors have one root, premolars have two, and molars have two or three roots. The crown and root are connected by a constricted tooth region called the *neck*. The outer surface of the root is covered by a calcified connective tissue called **cementum,** which attaches the tooth to the thin **periodontal** (payr″-ē-ō-don′-tul) **ligament**. This ligament anchors the tooth in place in the bony alveolus of the jaw.

Where the gingiva borders on a tooth, it dips downward to form a shallow groove called the *gingival sulcus*. In youth, the gingival margin adheres tenaciously to the enamel covering the crown; but with age, the gums begin to recede, and then the gingiva adheres to the more sensitive cementum covering the superior region of the root. As a result, the teeth *appear* to get longer in old age, hence, the expression "long in the tooth" sometimes applied to elderly people.

Dentin, a bonelike material, forms the bulk of a tooth. It surrounds a central **pulp cavity,** which contains a number of soft tissue structures (connective tissue, blood vessels, and nerve fibers) collectively called **pulp**. Pulp supplies nutrients to the tooth tissues and provides for tooth sensation. Where the pulp cavity extends into the root, it becomes the **root canal.** At the proximal end of each root canal is an *apical foramen*, an opening that provides a route for blood vessels, nerves, and other structures to enter the pulp cavity of the tooth. Lining the pulp cavity is a layer of **odontoblasts,** the cells responsible for secreting the dentin. Dentin is formed throughout adult life and gradually encroaches on the pulp cavity. New dentin can also be laid down fairly rapidly to compensate for tooth damage or decay.

Although enamel, dentin, and cementum are all calcified and resemble bone, they differ from bone in that they are avascular. Enamel differs not only from bone, but also from cementum and dentin, because it lacks collagen as its main organic component.

Dental caries, or cavities, result from a gradual demineralization of enamel and underlying dentin by bacterial action. Although over 200 different bacterial species reside in the mouth and contribute to tooth decay, the most important of these is *Streptococcus mutans* (strep″-tō-kah′-kus myoo′-tanz). The decay process begins when *dental plaque* (a film of sugar, bacteria, and other mouth debris) adheres to the teeth. Metabolism of the trapped sugars by the bacteria produces acids (particularly lactic acid), which dissolve the calcium salts of the teeth. Once the salts are leached out, the remaining organic matrix of the tooth is readily digested by protein-digesting enzymes released by the bacteria. Frequent brushing and flossing helps prevent damage by removing forming plaque.

Even more serious than tooth decay is the effect of unremoved plaque on the gums. As dental plaque accumulates, it calcifies, forming *calculus* (kal′-kyoo-lus), or tartar. When this occurs in the gingival sulci, it can disrupt the seals between the gingivae and the teeth, putting the gums at risk for infection. In the early stages of such an infection, the gums become red, sore, and swollen and may bleed; this inflammatory condition is called *gingivitis* (jin″-jih-vī′-tis). Gingivitis is reversible, but if it is neglected the bacteria eventually invade the bone around the teeth, forming pockets of infection and dissolving the bone away. This more serious condition, once called pyorrhea, is now called *periodontal disease*, or *periodontitis*. Periodontal disease affects up to 95% of all people over the age of 35 and accounts for 80% to 90% of tooth loss in adults. Once periodontal disease has developed, its treatment is time consuming and sometimes painful. Initial treatment is calculus removal by the dental hygienist, followed up by a home regimen of plaque removal by consistently frequent brushing and flossing and saline or hydrogen peroxide rinses. Severe cases require surgery—cutting the gums, scaling the teeth, and cleaning the infected pockets—followed by antibiotic therapy. Together, these treatments alleviate the bacterial infestations and promote reattachment of the surrounding tissues to the teeth and bone. Nutrition plays an important role in the health of periodontal tissue: A diet heavy in carbohydrates, refined sugars, alcohol, and junk food encourages gingivitis. ■

The Pharynx

From the mouth, food passes posteriorly into first the **oropharynx** and then the **laryngopharynx** (see Figure 24.7), both common passageways for food, fluids, and air. (The nasopharynx, posterior to the nasal cavities, has no digestive role.)

The mucosa of the pharynx, like that of the oral cavity and esophagus, is stratified squamous epithelium amply provided with mucus-producing glands. The balance of the wall tissue lacks the pattern that typifies most other alimentary canal organs. The pharyngeal walls contain two *skeletal muscle* layers: The cells of the inner layer run longitudinally; those of the outer layer (the constrictor muscles) run around the wall in a circular fashion. Alternating contractions (peristalsis) of these two muscle layers propel food through the pharynx into the esophagus below.

The Esophagus

The **esophagus** (ē-sof′-uh-gus), or **gullet**, is a muscular tube approximately 25 cm (10 inches) long that is collapsed when not involved in food propulsion. After being transported through the laryngopharynx, food is routed into the esophagus posteriorly as the larynx is closed off to food entry by action of the epiglottis (see pp. 781–782).

As shown in Figure 24.1, the esophagus takes a fairly straight course through the mediastinum of the thorax and then pierces the diaphragm at the *esophageal hiatus* to join the stomach at the **cardiac orifice.** The cardiac orifice (see Figure 24.14) is surrounded by the **gastroesophageal sphincter** (gas′-trō-ē-sof″-uh-jē″-ul sfink′-ter), which is a physiological sphincter; that is, it acts as a valve, but only a slight thickening of the circular smooth muscle is seen at that point. The muscular diaphragm, which surrounds this sphincter, helps keep it closed when food is not being swallowed.

Unlike the walls of the mouth and pharynx, the esophagus wall has the basic alimentary tract structure described on p. 759, with the following exceptions:

1. The esophageal mucosa is nonkeratinized stratified squamous epithelium containing abundant mucus-secreting esophageal glands. At the esophageal-stomach junction, this abrasion-resistant mucosa changes abruptly to the simple columnar epithelium of the stomach, which is specialized for secretory activity.
2. The muscularis is skeletal muscle in the superior third of the esophagus, changing as it descends to finally become entirely smooth muscle in its inferior third.
3. Instead of a serosa, the esophagus has a fibrous adventitia composed entirely of connective tissue, which blends with surrounding structures along its route.

When the esophagus is empty, its submucosa is thrown into folds (Figure 24.13), but when food is in transit, these longitudinal folds are "ironed out."

A *hiatal hernia* is a structural abnormality in which the superior part of the stomach protrudes slightly above the diaphragm into the thoracic cavity. Since the diaphragm no longer reinforces the action of the gastroesophageal sphincter, gastric juice (which is extremely acidic) may regurgitate or flow back into

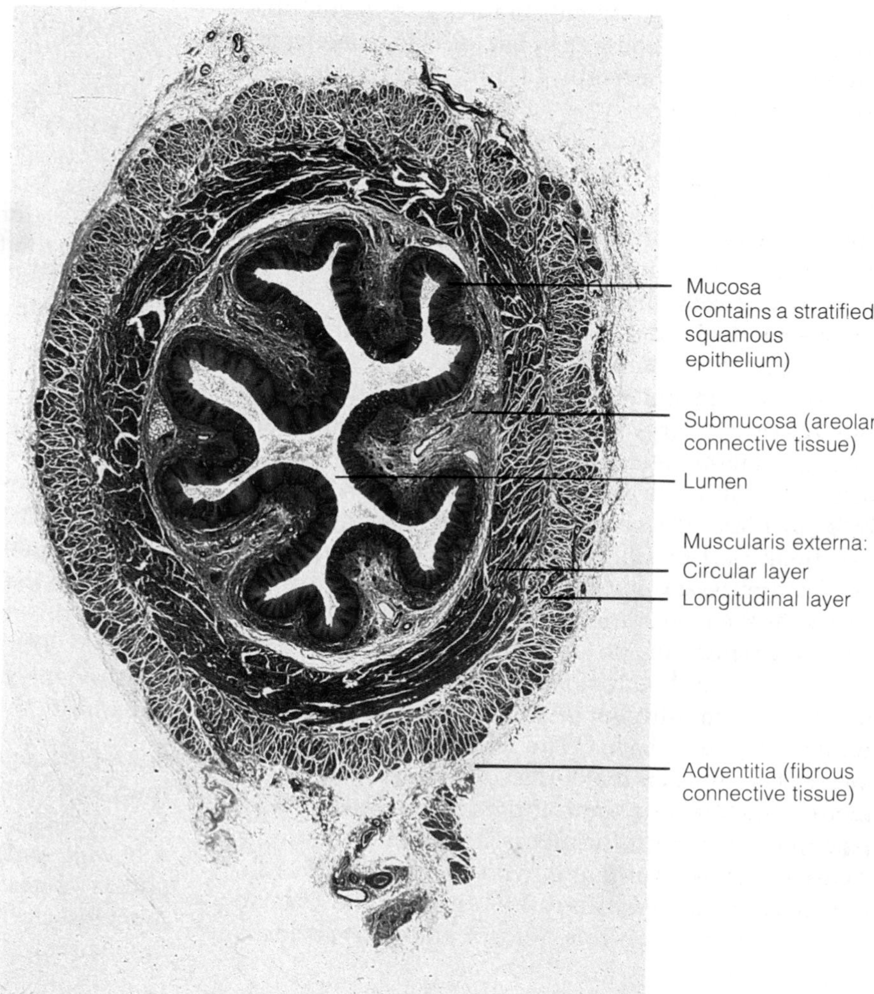

Figure 24.13 Microscopic structure of the wall of the esophagus, cross-sectional view (5X). The section shown is taken from the region close to the stomach junction, so the muscularis is composed of smooth muscle. (Superiorly, at the pharyngeal junction, the esophageal muscularis is composed of skeletal muscle.)

the esophagus. This is most likely to happen when one has eaten or drunk in excess and is lying down and in conditions that force abdominal contents superiorly, such as extreme obesity and pregnancy. Although the esophageal mucosa does produce mucus, its mucus protection is in no way equal to that of the stomach, and prolonged presence of gastric juice in the esophagus results in a burning, radiating substernal pain called *heartburn*. If the episodes are frequent and prolonged, *esophagitis* (inflammation of the esophagus) and *esophageal ulcers* may result. However, the condition can usually be managed by avoiding late-night snacks and by using antacid preparations. ■

The Stomach

Below the esophagus, the GI tract expands to form the **stomach** (see Figure 24.1), a temporary "storage tank" for food and an important site for mechanical and chemical breakdown of proteins. By the time ingested food leaves the stomach, it has been converted to a creamy, semifluid mass called *chyme* (kīm). In its normal position in the left side of the abdominal cavity, the stomach is nearly hidden by the liver and diaphragm. Though the stomach is relatively fixed at both (esophageal and small intestinal) ends, it is quite movable in between. It tends to lie high and run horizontally in short, obese people (steer-horn stomach) and is elongated vertically in tall, thin people (a J-shaped stomach).

Gross Anatomy

The stomach is approximately 25 cm (10 inches) long, but its diameter depends on how much food it contains. It can easily hold over 1.5 L of food and, when it is really distended, it can contain as much as 4 L. When empty, the stomach appears to collapse inward on itself, and its mucosa (and underlying submucosa) is thrown into large longitudinal folds called **rugae** (roo′-gē) (Figure 24.14a and c). The convex lateral surface of the stomach is its **greater curvature;** its concave medial surface is the **lesser curvature.** As described earlier, double layers of peritoneum, the greater and lesser omenta, respectively, extend from each of these curvatures (see Figure 24.5).

The regions of the stomach are shown in Figure 24.14a. The *cardiac region,* or *cardia,* surrounds the cardiac orifice, through which food enters the stomach from the esophagus. The *fundus* is the expanded part of the stomach that bulges superolaterally to the cardia. The *body* is the midportion of the stomach, and the *pylorus* (pī-lor′-us) is its inferior terminus. The pylorus is continuous with the duodenum (the first part of the small intestine) through the **pyloric sphincter,** which controls stomach emptying (*pylorus* = gatekeeper).

Microscopic Anatomy

The stomach wall exhibits the four basic tunics typical of most of the alimentary canal, but two of the tunics—the muscularis and the mucosa—exhibit modifications specifically related to the special roles of the stomach. Besides the usual circular and longitudinal layers of smooth muscle, the stomach muscularis has an innermost smooth muscle layer that runs obliquely (Figure 24.14a and b). This arrangement allows the stomach not only to move food along the tract, but also to churn, mix, and pummel the food, physically breaking it down into smaller fragments. It also contributes to the stomach's ability to be distended greatly.

The surface epithelium of the stomach mucosa is composed entirely of columnar cells, which produce large amounts of mucus. This otherwise smooth lining is dotted with millions of deep **gastric pits,** which lead into the **gastric glands** that collectively produce the stomach secretion called **gastric juice** (Figure 24.14b and d). The cells forming the walls of the gastric pits are generally like those of the lining mucosa, but those composing the gastric glands vary in cellular composition in different regions of the stomach. For example, the cells present in the glands of the cardia are primarily mucus secreting. Those of the fundus and body are substantially larger and secrete almost all of the hydrochloric acid, enzymes, and intrinsic factor (discussed shortly) produced by the stomach, as well as some of the mucus. The pylorus contains an abundance of cells that secrete the hormone gastrin, and mucus-producing glands. Most chemical digestion in the stomach occurs in the fundus and body; the glands in these regions contain a variety of secretory cells, including the four types of cells described here:

1. **Zymogenic** (zī″-mō-jeh′-nik), or **chief, cells** produce *pepsinogen* (pip-sin′-ō-jen), the inactive form of the protein-digesting enzyme *pepsin,* and insignificant amounts of other digestive enzymes. They are usually found at the bases of the gastric glands.

2. **Parietal (oxyntic) cells** secrete *hydrochloric acid (HCl)* and intrinsic factor. HCl makes the stomach contents extremely acidic (pH 1.5–3.5), a condition necessary for the activation and optimal activity of the proteolytic enzymes, and it kills many of the bacteria ingested with foods. **Intrinsic factor** is a glycoprotein required for the absorption of vitamin B_{12} in the small intestine.

3. **Mucous neck cells,** found in the "neck" regions of the glands interspersed among the parietal cells, produce a sticky, alkaline mucus that clings to the stomach wall and helps to shield it from the damaging effects of gastric juice. They are very similar to the mucus-secreting cells of the surface epithelium. The mucus-secreting cells are very sensitive to irritants

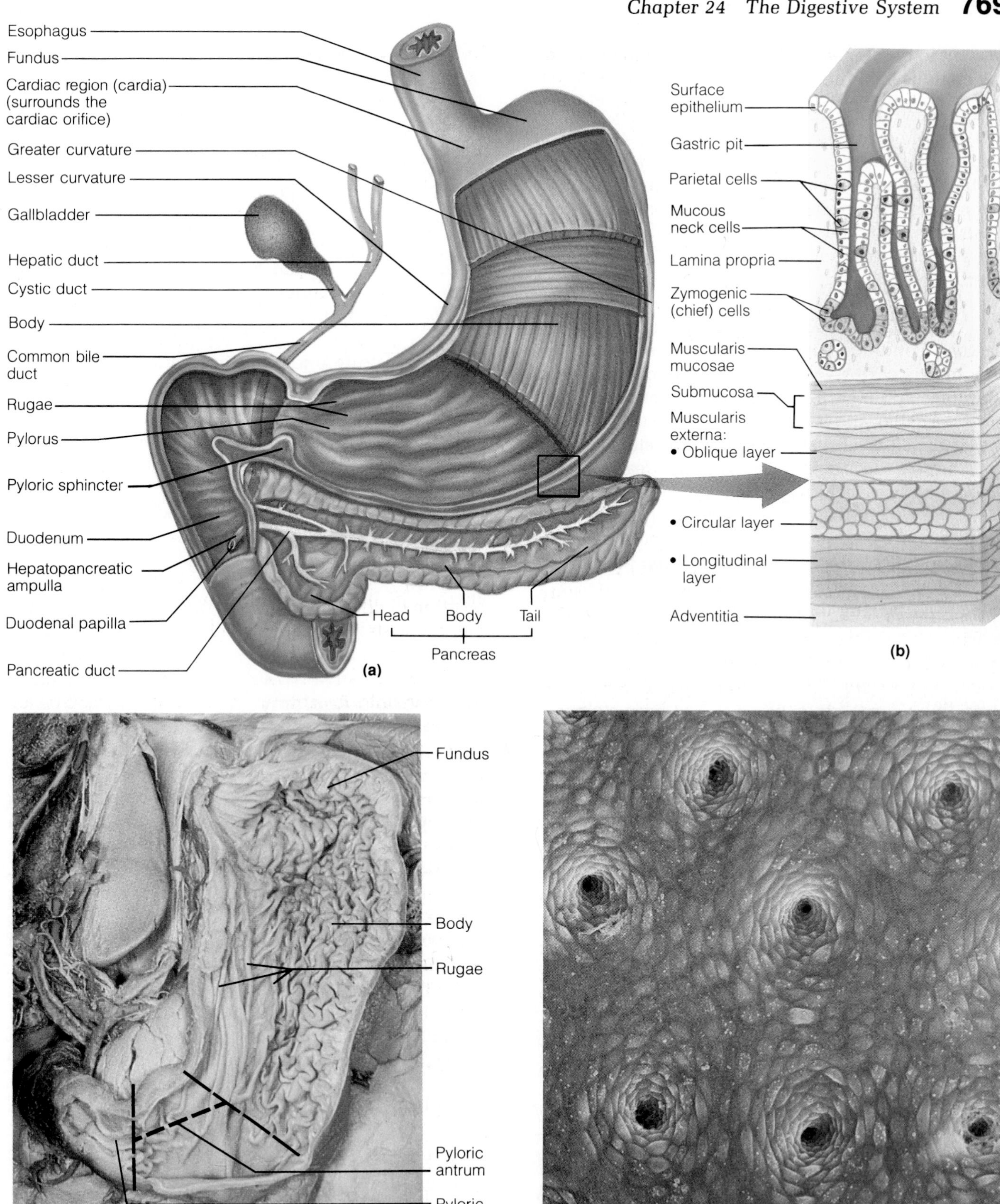

Figure 24.14 Anatomy of the stomach. (**a**) Gross internal anatomy (frontal section). Pancreatic, cystic, and hepatic ducts are also shown. (**b**) Enlarged view of the stomach wall (longitudinal section) showing gastric pits and glands. (**c**) Photograph of internal aspect of stomach; note the dense rugae. (**d**) Scanning electron micrograph of the stomach mucosa, showing gastric pits leading into the gastric glands (800X).

present in the stomach and quickly respond by secreting copious amounts of mucus.

4. Enteroendocrine (en'-ter-ō-en'-dō-krin) **cells** secrete a variety of hormones or hormonelike products directly into the lamina propria. These products, including **gastrin, histamine, endorphins** (natural opiates), **serotonin,** and **somatostatin,** then diffuse into the blood. Gastrin, in particular, plays essential roles in regulating stomach secretion and contraction, as described later in this chapter.

The stomach mucosa is exposed to some of the harshest conditions in the entire digestive tract. Gastric juice is corrosively acidic (H^+ concentrations in the stomach can be 100,000 times that found in blood or tissue cells), and its protein-digesting enzymes can digest the stomach itself. Just how the stomach wall resists these destructive agents is not really understood, but three factors appear to be important in creating the protective *mucosal barrier:* (1) The thick coating of alkaline mucus provides some measure of protection; (2) the epithelial cells of the mucosa are joined together by tight junctions that prevent gastric juice from leaking into the underlying tissue layers; and (3) damaged epithelial mucosal cells are shed and quickly replaced by cell division, and the stomach epithelium is completely renewed approximately every three days.

Anything that breaches the mucosal barrier results in inflammation of the underlying tissues, a condition called *gastritis*. Persistent damage to the underlying tissues can promote *gastric ulcers,* erosions of the stomach wall. The most distressing symptom of gastric ulcers is boring or gnawing pain (referred to the epigastric region) occurring 1–2 hours after eating and often relieved by eating again. Common causes of ulcer formation include hypersecretion of pepsin or hydrochloric acid and/or hyposecretion of mucus. Smoking, alcohol, coffee, and stress all appear to be predisposing factors. The danger of ulcers, regardless of where they occur (esophagus, stomach, or duodenum), is perforation of the organ wall followed by peritonitis and, perhaps, massive hemorrhage. ■

The Small Intestine

The **small intestine** is the body's major digestive organ. Within its twisted passageways, usable food is finally prepared for its journey into the cells of the body. Digestion is completed with the aid of secretions produced by the liver, the pancreas, and the small intestine itself.

Gross Anatomy

The small intestine is a convoluted tube extending from the pyloric sphincter to the **ileocecal** (il"-ē-ō-sē'-kul) **valve** (see Figure 24.1). It is the longest section of the alimentary tube, with an average length (in a living person) of 6 m (20 feet), but its diameter is only about 2.5 cm (1 inch). It hangs in sausagelike coils in the abdominal cavity, suspended from the posterior abdominal wall by the fan-shaped mesentery (see Figure 24.5) and encircled and framed by the large intestine.

The small intestine has three subdivisions: the duodenum, the jejunum, and the ileum. The relatively immovable **duodenum** (doo"-ō-dē'-num) (see Figure 24.14a), which curves around the head of the pancreas, is approximately 25 cm (10 inches) long. The ducts delivering bile and pancreatic juice from the liver and pancreas, respectively, join close to the duodenum at a common point called the **hepatopancreatic ampulla** (hep"-uh-tō-pan"-krē-a'-tak am'-pyoo-luh), which empties into the duodenum via the **duodenal papilla** (see Figure 24.14). The entry of bile and pancreatic juice is controlled by a muscular valve called the **sphincter of Oddi.**

The **jejunum** (jeh-joo'-num) is about 2.5 m (8 feet) long and extends from the duodenum to the terminal ileum. The **ileum** (il'-ē-um), approximately 3.6 m (12 feet) in length, joins the large intestine at the ileocecal valve.

Microscopic Anatomy

Modifications for Absorption. As already noted, nearly all nutrient absorption occurs in the small intestine, an organ highly adapted for its function. Its length alone provides a huge surface area, and its wall has three structural modifications that amplify this absorptive surface enormously: plicae circulares, villi, and microvilli (Figure 24.15). Most absorption occurs in the proximal part of the organ, and all of these modifications decrease in number toward the end of the small intestine.

The **plicae circulares** (plī'-kē ser"-kyoo-layr'-ēz), literally, "circular folds," are deep, permanent folds of the mucosa and submucosa (Figure 24.15a). Some of these folds extend entirely around the circumference of the small intestine; others only partway around. The plicae force the chyme to move spirally through the lumen, which continually mixes the chyme with intestinal juice and slows its movement through the lumen.

Villi (vih'-lī) are fingerlike projections of the mucosa, approximately 1 mm high, that give it a velvety appearance and feel, much like the soft nap of a Turkish towel (Figure 24.15b). The epithelial cells of the villi are chiefly absorptive columnar cells. Within the lamina propria core of each villus is a dense capillary bed and a modified lymphatic capillary called a **lacteal** (lak'-tē-ul). Digested foodstuffs are absorbed

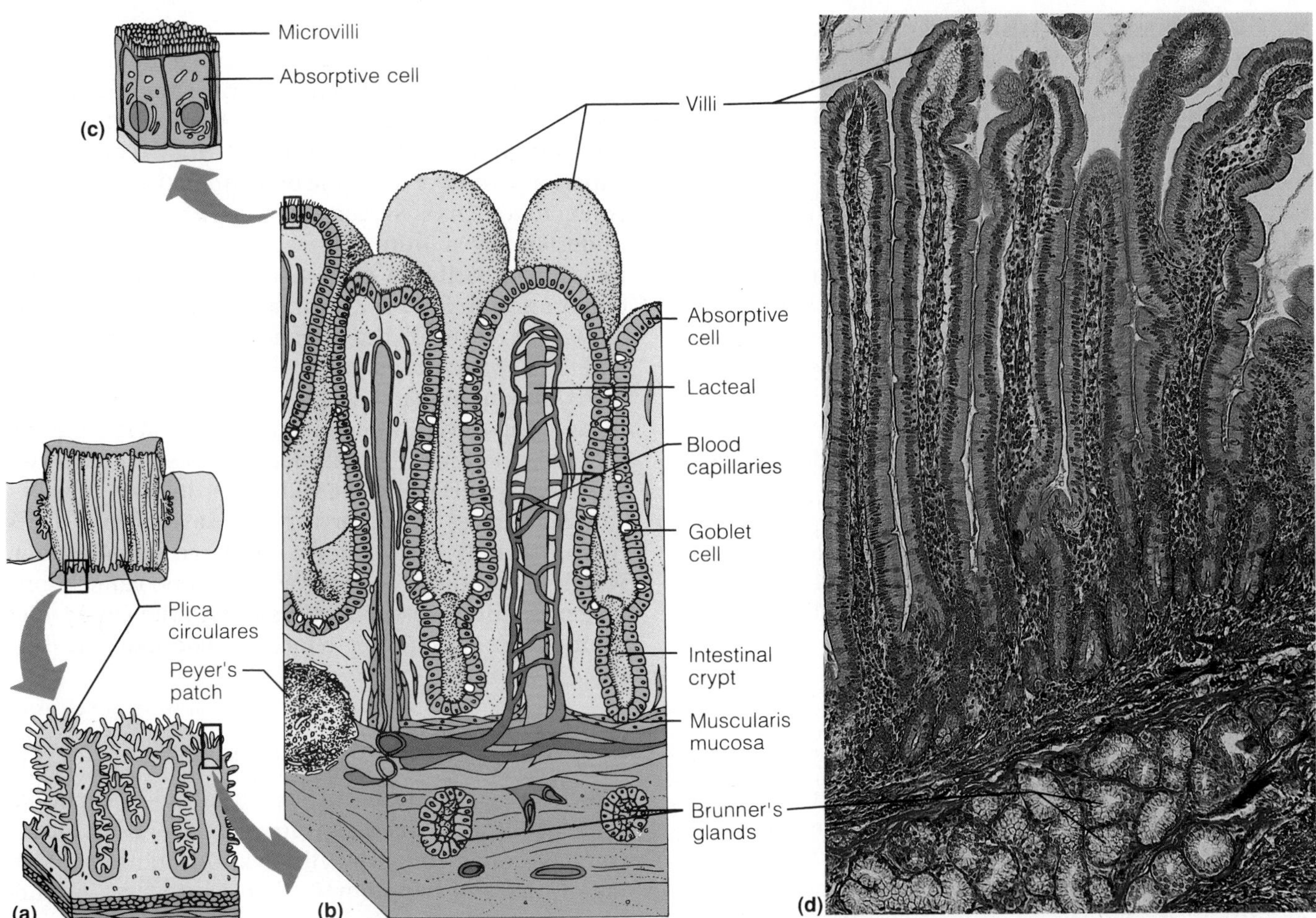

Figure 24.15 Structural modifications of the small intestine that increase its surface area for digestion and absorption. (**a**) Enlargement of a few plicae circulares, showing associated fingerlike villi. (**b**) Diagrammatic view of the structure of a villus. (**c**) Two absorptive cells that exhibit microvilli on their free (luminal) surface. (**d**) Photomicrograph of the small intestinal mucosa, showing villi. A small portion of the submucosa containing Brunner's glands is also shown (150×).

through the mucosal cells into both the capillary blood and the lacteal. The villi are large and leaflike in the duodenum (the site of most absorption) and gradually become narrower and shorter along the length of the small intestine.

Microvilli, tiny projections of the plasma membrane of the absorptive mucosal cells, give the mucosal surface a fuzzy appearance that is sometimes referred to as the **brush border** (Figure 25.15c). In addition to enhancing absorption, the plasma membrane of the microvilli bears intestinal digestive enzymes, referred to collectively as **brush border enzymes.**

Histology of the Wall. Although the subdivisions of the small intestine appear to be nearly identical in external anatomy, their internal and microscopic anatomies reveal some important differences. The same four tunics seen in the rest of the GI tract are also seen here, but the mucosa and submucosa have modifications that reflect the relative location of the intestinal region in the digestive pathway (see Figure 24.15b, d).

The mucosa is formed largely of simple columnar epithelial cells (absorptive cells) bound by tight junctions and richly endowed with microvilli. It also has many mucus-secreting goblet cells and scattered enteroendocrine cells. Between the villi, the mucosa is studded with pits that lead into tubular glands called **intestinal crypts,** or **crypts of Lieberkühn** (lē′-ber-koon). The cells of these glands secrete intestinal juice, a watery mixture of mucus and digestive enzymes, as well as some hormones. The various epithelial cells (absorptive columnar, goblet, and enteroendocrine cells) apparently arise from stem cells at the base of the crypts and gradually spread up the villi, where they are shed from the villus tips. In this way, the villus epithelium is renewed approximately every five to six days. The crypts decrease in number along the length of the small intestine, but the goblet cells become more abundant.

The submucosa is typical areolar connective tissue, but it has some very obvious local collections of lymphatic tissue (nodules) called **Peyer's** (pī′-erz) **patches.** Unlike the structural modifications that serve absorption, Peyer's patches increase in abundance toward the end of the small intestine, reflecting the fact that the undigested food residue contains huge numbers of bacteria that must be prevented from entering the bloodstream. A set of elaborated mucous glands is found in the submucosa of the duodenum only. These glands, called **Brunner's** (brun′-erz) **glands** (see Figure 24.15d), produce an alkaline mucus that helps neutralize the acidic chyme moving in from the stomach. When this protective mucous barrier is inadequate, the intestinal wall is eroded and duodenal ulcers result.

The muscularis is typically bilayered. Except for the bulk of the duodenum, which is retroperitoneal (behind the peritoneum) and has an adventitia, the outer intestinal surface is covered with the visceral peritoneum (serosa).

The Liver and Gallbladder

Like the pancreas, the liver and gallbladder are accessory organs associated with the small intestine. The liver has many metabolic and regulatory roles, and there is little question that it is one of the body's most important organs. However, its digestive function is to produce bile for export to the duodenum. Bile is a fat emulsifier; that is, it acts as a detergent to distribute fat, as tiny particles, throughout an aqueous medium so that it is accessible to digestive enzymes. We will describe bile and the emulsification process when we discuss digestive functions of the small intestine. Although the liver also processes nutrient-laden venous blood delivered to it from the digestive organs, this function concerns its metabolic rather than its digestive role. (The metabolic functions of the liver are described in Chapter 25, p. 839.) The gallbladder is chiefly a storage organ for bile.

Gross Anatomy of the Liver

The ruddy, blood-rich **liver** is the largest gland in the body, weighing about 1.4 kg (3 pounds) in the average adult. It is located under the diaphragm, extending farther to the right of the body midline than to the left, and it almost completely obscures the stomach (see Figure 24.1).

The liver has four lobes: the large *right* and *left lobes* and the smaller *caudate* and *quadrate* lobes. The caudate lobe is posterior to the right lobe; the quadrate lobe is inferior to the left lobe (Figure 24.16b). A delicate mesentery cord, the *falciform ligament*, separates the right and left lobes and suspends the liver from the diaphragm and anterior abdominal wall (see also Figure 24.5a and d). Running along the free edge of the falciform ligament is the **round ligament,** or **ligamentum teres** (tayr′-ēz), a fibrous remnant of the fetal umbilical vein (Figure 24.16). Except for the superiormost liver area (the *bare area*), the entire liver is enclosed in a serosa (the visceral peritoneum) underlain by a thin connective tissue capsule. As mentioned earlier, a peritoneal extension, the lesser omentum (see Figure 24.5b), anchors the liver to the lesser curvature of the stomach. The hepatic artery and the hepatic portal vein, which enter the liver, and the common bile duct and hepatic veins, which leave the liver, all travel through the lesser omentum to reach their destinations. The gallbladder rests in a recess on the inferior surface of the right liver lobe.

Bile leaves the liver through several bile ducts that ultimately fuse to form the large **hepatic duct,** which travels downward toward the duodenum. Along its course, the hepatic duct fuses with the **cystic duct** draining the gallbladder to form the **common bile duct** (Figure 24.14a).

Microscopic Anatomy of the Liver

The liver is composed of structural and functional units called **liver lobules** (Figure 24.17). Each lobule is a hexagonal, roughly cylindrical structure consisting of plates of **hepatocytes** (heh-pa′-tō-sitz), epithelial *parenchymal cells.* The hepatocyte plates radiate outward from a **central vein** running upward within the longitudinal axis of the lobule. At each of the six corners of a lobule is a **triad** (*portal tract*) region, so named because three basic structures are always present there: a branch of the *hepatic artery* (the functional blood supply of the liver), a branch of the *hepatic portal vein* (carrying nutrient-rich venous blood from the digestive viscera), and a *bile duct.* Between the hepatocyte plates are blood-filled spaces, or *sinusoids.* Blood from both the hepatic portal vein and the hepatic artery percolates from the triad regions through these sinusoids and empties into the central vein. The sinusoids are lined with special phagocytic cells called **Kupffer** (koop′-fer) **cells** (see Figure 24.17d), which remove debris, such as bacteria, from the blood as it flows past.

The hepatocytes have several major functions:

1. They produce bile.
2. They pick up nutrients from the blood passing through the sinusoids and process them in various ways. For example, they store glucose as glycogen for later use and use amino acids to make plasma proteins.
3. They store fat-soluble vitamins, such as vitamins A, E, D, and K.
4. They play important roles in detoxification, as described in Chapter 25.

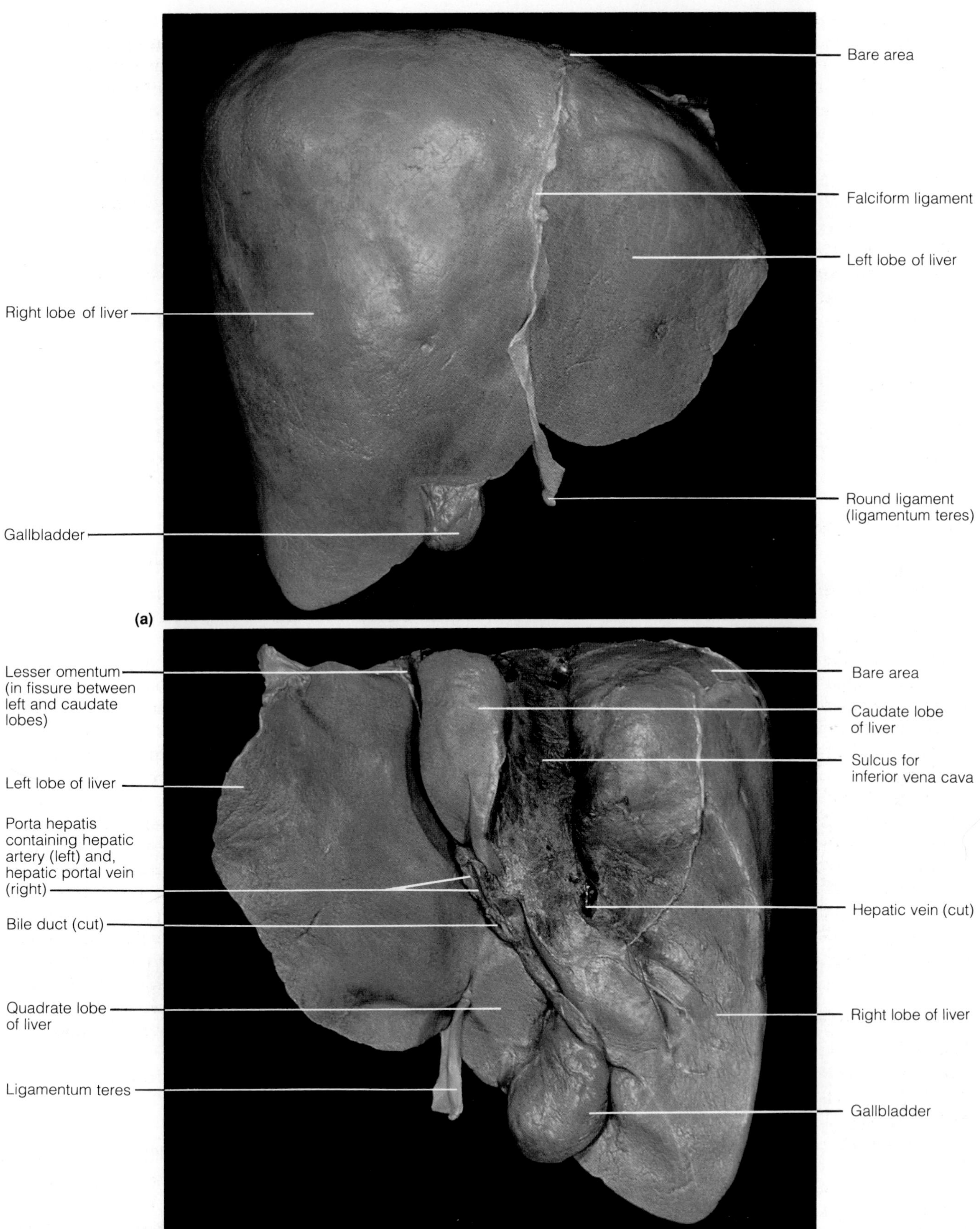

Figure 24.16 Gross anatomy of the human liver. (**a**) Anterior view of the liver. (**b**) Posteroinferior aspect (visceral surface) of the liver. The four liver lobes are separated by an H-shaped group of fissures in this view. The crossbar of the H is the *porta hepatis*, a deep fissure that contains the hepatic portal vein, hepatic artery, hepatic duct, and lymphatics.

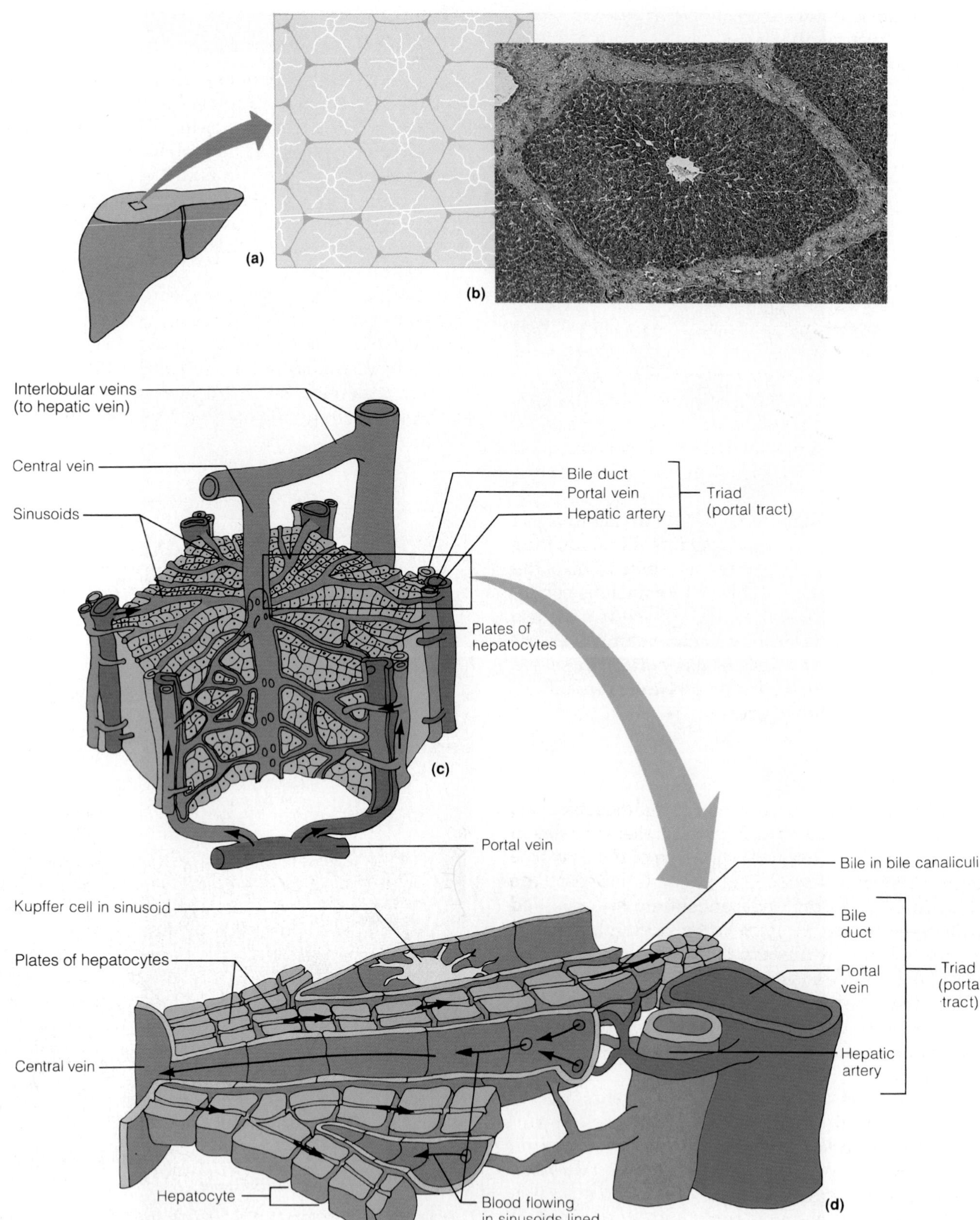

Figure 24.17 Microscopic anatomy of the liver. (**a**) Schematic view of the cut surface of the liver, showing the hexagonal nature of its lobules. (**b**) Photomicrograph of one liver lobule (50×). (**c**) Schematic representation of one liver lobule, showing its blood supply and the platelike arrangement of the hepatocytes. (The arrows indicate the direction of blood flow.) (**d**) Enlarged schematic view of a small portion of one liver lobule, illustrating the components of the triad (portal tract) region (bile duct and branch of the hepatic portal vein and hepatic artery), the positioning of the bile canaliculi, and the Kupffer cells in the sinusoids. The direction of blood and bile flow are also indicated.

Owing to the activities of the hepatocytes, the blood leaving the liver via the hepatic veins contains fewer nutrients and waste materials than the blood that entered it.

Although the hepatocytes make bile continuously (at the rate of approximately 500–1000 ml per day) bile production is enhanced by the intestinal hormone secretin and high levels of bile salts in the blood. Secreted bile flows through tiny canals, called **bile canaliculi** (kan″-ah-lik′-you-lī), that run between adjacent hepatocytes toward the bile duct branches in the triad regions (see Figure 24.17d). Notice that the directions of blood and bile flow in the liver lobule are exactly opposite. Bile entering the bile ducts eventually leaves the liver via the hepatic duct to travel toward the duodenum. When digestion is not occurring, the released bile backs up the cystic duct into the gallbladder, where it is stored until needed. Release of bile from the gallbladder is discussed in connection with the digestive functions of the small intestine (p. 790).

Hepatitis (hep″-uh-tī′-tis), or inflammation of the liver, is most often due to viral infection following ingestion of contaminated water or transmission of the virus in blood (by means of blood transfusions or contaminated needles). *Cirrhosis* (ser-ō′-sis) is a diffuse and progressive chronic inflammation of the liver that typically results from chronic alcoholism or severe hepatitis. The liver becomes fibrous and shrinks, and all of its functions are depressed. ■

The Gallbladder

The **gallbladder** is a thin-walled, green muscular sac, approximately 10 cm (4 inches) long, that snuggles in a shallow fossa on the ventral surface of the liver (see Figures 24.1 and 24.16). The major function of the gallbladder is to store and concentrate bile secreted by the liver. When empty, or when storing only small amounts of bile, its mucosa is thrown into folds that resemble the rugae of the stomach. When its muscular wall contracts, bile is expelled into the cystic duct and then flows into the common bile duct. The gallbladder, like most of the liver, is covered by the visceral peritoneum.

Although the liver continuously secretes bile, the sphincter of Oddi (guarding the entry of bile and pancreatic juice into the duodenum) is closed when bile is not needed for digestion. At these times, bile backs up into the gallbladder and is concentrated as the gallbladder mucosa absorbs some of its water and ions. In some cases, bile released from the gallbladder is ten times as concentrated as that entering it.

The Pancreas

The **pancreas** (pan′-krē-us), an accessory digestive organ associated with the small intestine, is the principal enzyme-producing organ of the digestive system. A soft, pink, triangular gland, it extends across the abdomen from its *tail*, abutting the spleen, to its *head* (the region closest to the duodenum) (see Figures 24.1 and 24.14a). The *body* and tail of the pancreas are retroperitoneal and lie behind the greater curvature of the stomach.

Within the pancreas are exocrine secretory units called *acini* (a′-sih-nī) (singular, acinus), which are essentially clusters of deeply staining secretory (acinar) cells surrounding ducts (Figure 24.18). These cells secrete **pancreatic juice,** an alkaline fluid containing a broad spectrum of enzymes. The acini are drained

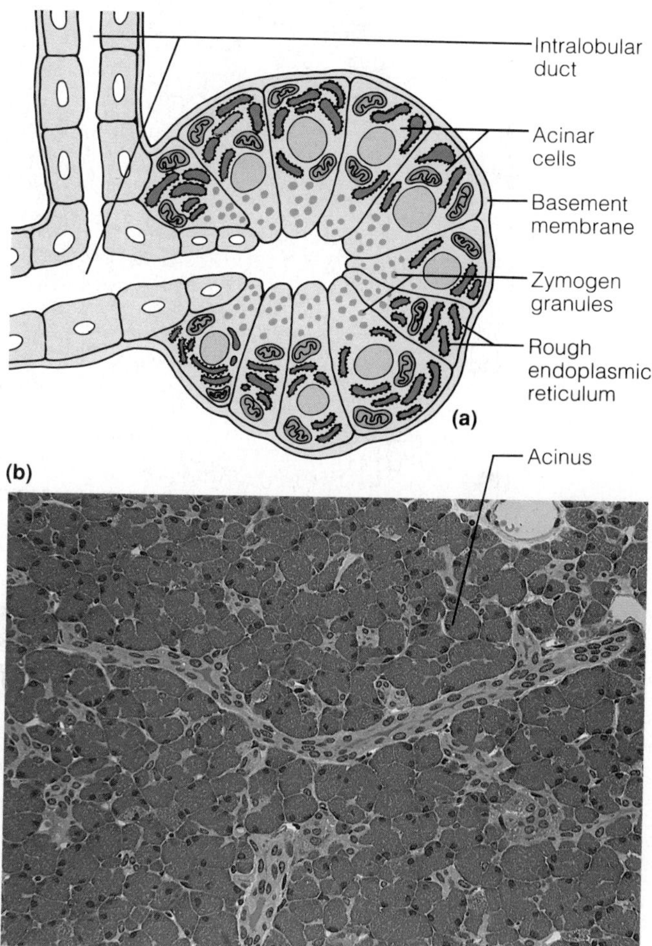

Figure 24.18 Structure of the acinar (enzyme-producing) tissue of the pancreas. (**a**) One acinus (a secretory unit). The acinar cells have abundant amounts of rough endoplasmic reticulum (reflecting their high level of protein synthesis) in their basal regions. The cell apices contain abundant zymogen (enzyme-containing) granules. (**b**) Photomicrograph of pancreatic acinar tissue (200×). Notice that the basal regions of the acinar cells stain deeply where ribosomes are abundant. A single acinus is pointed out.

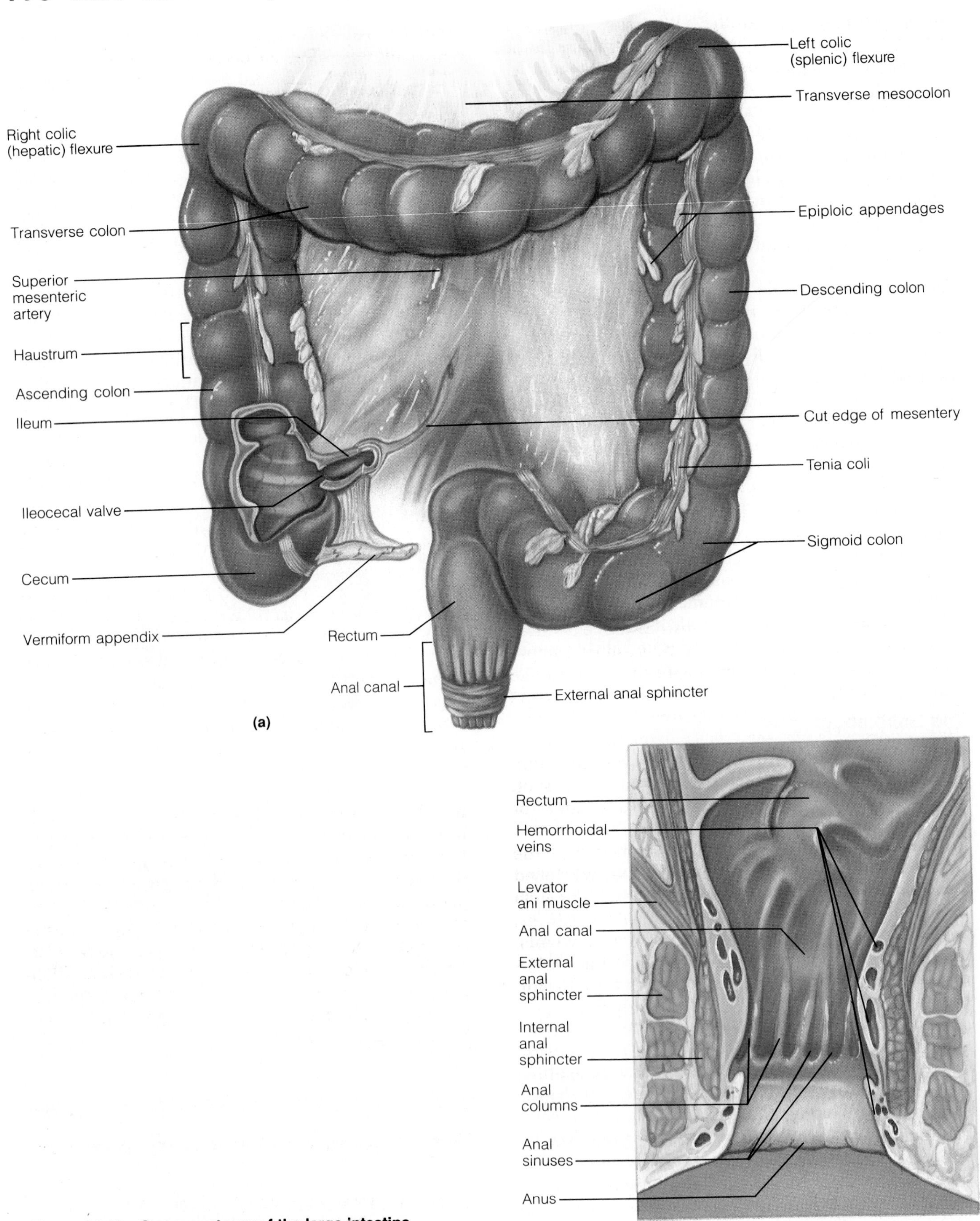

Figure 24.19 Gross anatomy of the large intestine. (**a**) Diagrammatic view. (**b**) Structure of the anal canal.

by ducts that feed into the centrally located **pancreatic duct.** The pancreatic duct usually fuses with the *common bile duct,* which conveys bile from the liver and gallbladder, just before it enters the duodenum. Hence, bile and pancreatic juice enter the duodenum together. The specific roles of the enzyme-rich pancreatic juice will be described later.

The pancreas also has an endocrine function: It produces two chief hormones, insulin and glucagon, both of which help to regulate carbohydrate metabolism. Scattered here and there in the acinar tissue are the more lightly staining *islets of Langerhans* (lon′-ger-honz), which contain the hormone-producing cells. The endocrine function of the pancreas is discussed in Chapter 17.

The Large Intestine

The **large intestine** (see Figure 24.1) is about 1.5 m (5 feet) long and extends from the ileocecal valve to the anus. Its major function is to dry out indigestible food residues by absorbing water and to eliminate these residues from the body as feces.

Gross Anatomy

The large intestine frames the small intestine on three sides and has the following subdivisions: cecum, appendix, colon, rectum, and anal canal. The saclike **cecum** (sē′-kum), which lies below the ileocecal valve, is the first part of the large intestine (Figure 24.19a). Hanging from its inferior surface is the blind, worm-like **vermiform appendix.** Although the appendix contains lymphatic tissue, it is a potential trouble spot; since it is usually twisted, it is an ideal location for enteric bacteria to accumulate and multiply.

Depending on its severity, inflammation of the appendix, or *appendicitis,* may be evidenced simply by edema (swelling) or may progress to ischemia or gangrene of the appendix. If the appendix ruptures, bacteria are sprayed over the abdominal contents, resulting in *peritonitis* (inflammation of the peritoneum), which may be followed by abscess or adhesion formation. The first symptom of appendicitis is usually pain in the umbilical region. Loss of appetite, nausea and vomiting, and relocalization of pain to the lower right abdominal quadrant follow. Immediate surgical removal of the appendix is the accepted treatment for suspected appendicitis. ■

The **colon** has several distinct regions. As the **ascending colon,** it travels up the right side of the abdominal cavity and makes a right-angle turn—the **right colic,** or **hepatic, flexure** —and travels across the abdominal cavity as the **transverse colon.** It then turns acutely at the **left colic (splenic), flexure,** continues down the left side as the **descending colon,** and enters the pelvis, where it becomes the S-shaped **sigmoid colon.** At the level of the third sacral vertebra, the sigmoid colon becomes the **rectum,** which runs inferiorly in front of the sacrum. The rectal terminus, called the **anal canal,** opens to the body exterior at the **anus.** The anus has two sphincters, an internal (smooth muscle) involuntary sphincter and an outer (skeletal muscle) voluntary sphincter (Figure 24.19b). The sphincters, which act rather like purse strings to open and close the anus, are ordinarily closed except during defecation.

Microscopic Anatomy

The wall of the large intestine differs in several ways from that of the small intestine. The colon mucosa is simple columnar epithelium except in the anal canal, but because most food is absorbed before reaching the large intestine, there is no brush border and no villi. However, the colon has its own version of plicae (the plicae semilunares), and there are tremendous numbers of goblet cells in its crypts (Figure 24.20). The mucus produced by the large intestinal cells plays an important role in easing the passage of feces to the end of the digestive tract. The mucosa of the anal canal is somewhat different. It hangs in long ridges or folds (*anal columns*) and is composed of stratified squamous epithelium, reflecting the greater abrasion that this region receives. Recesses between the anal columns, called *anal sinuses,* exude mucus when compressed by feces, which aids in emptying the anal canal.

In most of the large intestine, the longitudinal muscle layer of the muscularis is reduced to three bands of muscle referred to as the **teniae coli** (tē′-nē-ī kō′-lī). Since they usually maintain some degree of tone, they cause the intestinal wall to pucker into small pocketlike sacs, called **haustra** (see Figure 24.19a). The muscle layers in the rectum are complete, so haustra are absent in that region. Another unique, and obvious, feature of the colon is the presence of small fat-filled pouches on its visceral peritoneum, called *epiploic* (eh″-pih-ployk) *appendages* (see Figure 24.19a), but their significance is not known.

Physiology of Digestion

Basic Functional Concepts

A theme stressed throughout this book has been the body's thrust to maintain the constancy of its internal environment, particularly of the blood, which is in

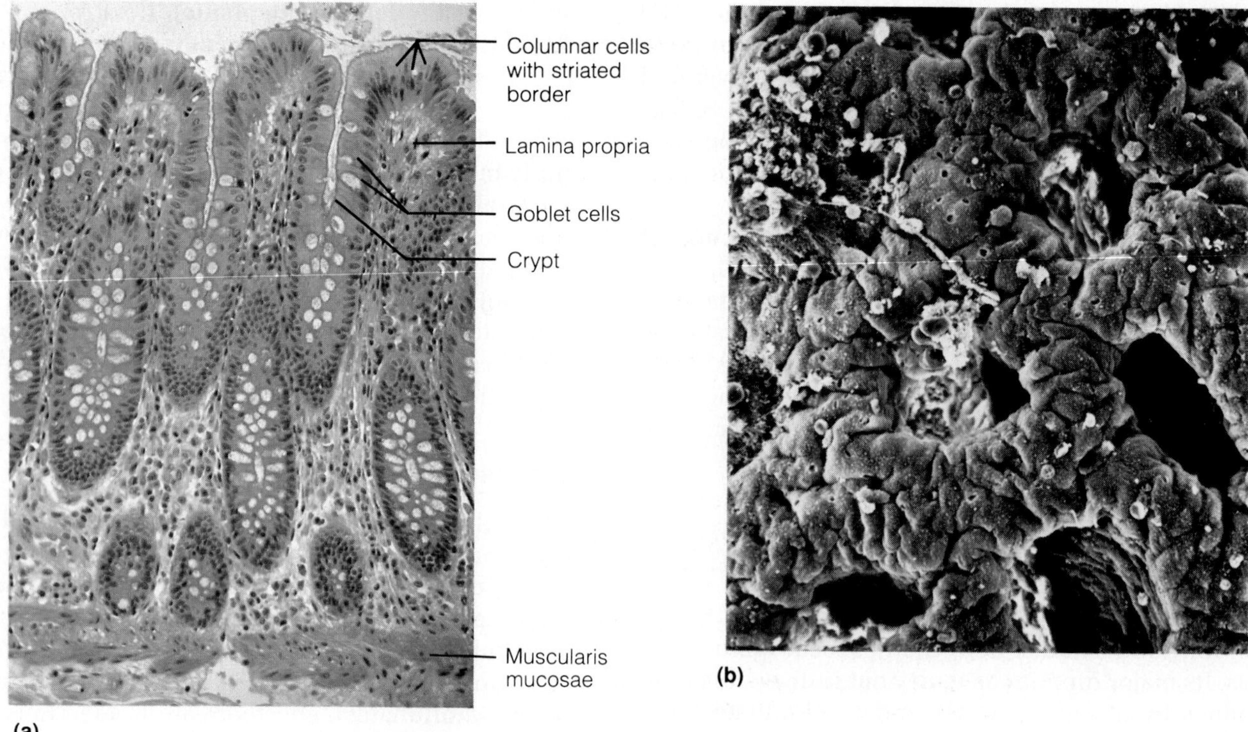

Figure 24.20 The mucosa of the large intestine. (**a**) Light photomicrograph of the mucosa of the large intestine. Notice the abundant mucus-secreting goblet cells (550×). (**b**) Scanning electron micrograph of the mucosal surface of the human colon, showing the entrances to the crypts (525×).

close contact with virtually all body cells. Most organ systems respond to changes in that environment either by attempting to restore some plasma variable to its former levels or by changing their own function. The digestive system, however, creates an optimal environment for its functioning in the lumen of the GI tract, an area that is actually outside the body, and essentially all digestive tract regulatory mechanisms act to control luminal conditions so that digestion and absorption can occur there as effectively as possible.

Sensors (mechanoreceptors, osmoreceptors, and chemoreceptors) involved in controls of GI tract activity are located in the walls of the tract organs and respond to a limited number of stimuli, the most important being (1) stretch of the organ wall by food in the lumen, (2) osmolarity (solute concentration) and pH of the chyme, and (3) the presence of end products of digestion. When activated by appropriate stimuli, these receptors initiate reflexes that activate or inhibit effectors that secrete digestive juices into the lumen or hormones into the blood (glandular cells) or mix lumen contents and move it along the length of the tract (smooth muscle of the muscularis). Another novel trait of the digestive tract is that many of its controlling systems are themselves intrinsic—a product of local hormone-producing cells or "in-house" nerve plexuses. As noted earlier, the wall of the alimentary canal exhibits three nerve plexuses. The most important, the submucosa and myenteric, extend essentially the entire length of the GI tract and influence each other both in the same and in different digestive organs. As a result, two kinds of reflex activity occur: *short reflexes*, mediated by the local plexuses in response to GI tract stimuli, and *long reflexes*, initiated by stimuli arising within or outside the GI tract and involving CNS centers and extrinsic autonomic nerves.

The stomach and small intestinal mucosae also contain individual hormone-producing cells that, when stimulated by local chemical factors, vagal nerve fibers, or, perhaps, local stretch, release their products to the extracellular space. These hormones then circulate in the blood and are distributed to their target cells within the same or different digestive tract organs, which they prod into secretory or contractile activity.

Now that we have summarized some points that unify the functioning digestive organs, we are ready to consider the special capabilities of each part of the system. Table 24.1 provides an overview of this information.

Table 24.1 Overview of the Functions of the Gastrointestinal Organs

Organ	Major functions	Comments/additional functions
Mouth and associated accessory organs	● Ingestion: food is voluntarily placed into oral cavity ○ Propulsion: voluntary (buccal) phase of deglutition (swallowing) initiated by tongue; propels food into pharynx ● Mechanical digestion: mastication (chewing) by teeth and mixing movements by tongue ○ Chemical digestion: chemical breakdown of starch is begun by salivary amylase present in saliva produced by salivary glands	Mouth serves as a receptacle; most functions performed by associated accessory organs; mucus present in saliva helps dissolve foods so they can be tasted and moistens food so that tongue can compact it into a bolus that can be swallowed; oral cavity and teeth cleansed and lubricated by saliva
Pharynx and esophagus	○ Propulsion: peristaltic waves move food bolus to stomach, thus accomplishing involuntary (pharyngeal-esophageal) phase of deglutition	Primarily food chutes; mucus produced helps to lubricate food passageways
Stomach	● Mechanical digestion and propulsion: peristaltic waves mix food with gastric juice and propel it into the duodenum ○ Chemical digestion: digestion of proteins begun by pepsin ○ Absorption: absorbs a few fat-soluble substances (aspirin, alcohol, some drugs)	Also serves as storage site for food until it can be moved into the duodenum; hydrochloric acid produced is a bacteriostatic agent and activates protein-digesting enzymes; mucus produced helps lubricate and protect stomach from self-digestion; intrinsic factor produced is required for intestinal absorption of vitamin B_{12}
Small intestine and associated accessory organs	● Mechanical digestion and propulsion: segmentation by smooth muscle of the small intestine continually mixes contents with digestive juices and moves food along tract and through ileocecal valve at a slow rate, allowing sufficient time for digestion and absorption ○ Chemical digestion: digestive enzymes conveyed in from pancreas and brush border enzymes attached to villi membranes complete digestion of all classes of foods ○ Absorption: breakdown products of carbohydrate, protein, fat, and nucleic acid digestion; vitamins; electrolytes; and water are absorbed by active and passive mechanisms	Small intestine is highly modified for digestion and absorption (plicae circulares, villi, and microvilli); alkaline mucus produced by intestinal glands and bicarbonate-rich juice ducted in from pancreas help neutralize acidic chyme and provide proper environment for enzymatic activity; bile produced by liver emulsifies fats and enhances (1) fat digestion and (2) absorption of fatty acids, monoglycerides, cholesterol, phospholipids, and fat-soluble vitamins; gallbladder stores and concentrates bile; bile is released to small intestine in response to hormonal signals
Large intestine	○ Chemical digestion: some remaining food residues are digested by enteric bacteria (which also produce vitamin K and some B vitamins) ○ Absorption: absorbs most remaining water, electrolytes (largely NaCl), and vitamins produced by bacteria ○ Propulsion: propels feces toward rectum by peristalsis, haustral churning, and mass movements ● Defecation: reflex triggered by rectal distention; eliminates feces from body	Temporarily stores and concentrates residues until defecation can occur; copious mucus produced by goblet cells eases passage of feces through colon

*The colored circles beside the functions correspond to the color coding of digestive functions illustrated in Figure 24.2.

Digestive Processes Occurring in the Mouth, Pharynx, and Esophagus

The mouth and its associated accessory digestive organs are involved in most digestive processes: ingestion, mechanical and chemical digestion, and propulsion. By contrast, the pharynx and esophagus merely provide a conduit from the mouth to the stomach, and their single digestive function is food propulsion, accomplished by the role they play in swallowing.

Since swallowing is initiated by the tongue within the mouth, we will discuss the functioning of the mouth, pharynx, and esophagus together. With the exception of a few drugs that are absorbed through the oral mucosa (for example, nitroglycerine), virtually no absorption occurs in the mouth, pharynx, or esophagus.

Ingestion

The mouth is the only part of the alimentary canal normally involved in the entry of food into the body. Although severely ill patients might well receive nutrients via an intravenous or gastric tube feeding, neither avenue is usually described as "ingestion."

Mechanical Digestion

As food enters the mouth, its mechanical breakdown is begun by **mastication,** or **chewing.** The cheeks and closed lips hold the food between the teeth, the tongue mixes the food with saliva to soften it, and the teeth cut and grind solid foods into smaller morsels. Mastication is partly voluntary and partly reflexive. We initiate it by voluntarily putting food into our mouth and contracting the muscles that close our jaws, but continued jaw movements are controlled by stretch reflexes and in response to pressure inputs from receptors in the cheeks, gums, and tongue.

Chemical Digestion

Chemical digestion of starch begins in the mouth. Salivary amylase (am′-ih-lās), an enzyme present in saliva, splits starch into smaller fragments of linked glucose molecules and the disaccharide maltose (see Figure 24.4a). If you chew a piece of bread for a few minutes, it will begin to taste sweet as some of the sugars are released. Salivary amylase works best in the slightly acid environment (pH 6.35–6.85) maintained by the buffering effects of bicarbonate and phosphate ions present in saliva. Starch digestion continues in the stomach until amylase is inactivated by stomach acid and broken apart by the stomach's protein-digesting enzymes.

Because several of the substances found in saliva have important physiological functions that enhance the activity of salivary amylase or that serve to protect the oral cavity, it is worthwhile to consider the composition of saliva in a bit more detail.

Composition of Saliva. Saliva is largely water—97% to 99.5%, depending on the precise glands that are active and the nature of the stimulus for salivation. Its solutes include electrolytes (sodium, potassium, chloride, phosphate, and bicarbonate ions); salivary amylase; the proteins mucin (myoo′-sin), lysozyme (lī-sō-zīm), IgA, and kallikrein (ka″-lih-krē′-in); and metabolic wastes. When dissolved in water, the glycoprotein *mucin* forms thick mucus that lubricates the oral cavity and hydrates foodstuffs. Protection against microorganisms is provided by (1) IgA antibodies; (2) lysozyme, a bacteriostatic enzyme that inhibits bacterial growth in the mouth and may help to prevent tooth decay; and (3) some as-yet unidentified fluid product of the sublingual and submandibular glands that blocks the AIDS virus from infecting human lymphocytes. Additionally, *growth factor,* found in the saliva of many animals, may aid in the healing of licked wounds and has recently been identified in human saliva. (Kallikrein, which helps regulate saliva production, is discussed in the next section.)

Control of Salivation. Saliva is secreted continuously in amounts just sufficient to keep the mouth moist. But when food enters the mouth, copious amounts of saliva pour out. The average output of saliva is 1000–1500 ml per day.

Salivation is controlled primarily by the autonomic nervous system, particularly the parasympathetic division. Parasympathetic fibers regulate saliva output both in the absence and presence of food stimuli. But when we ingest food, chemoreceptors and pressoreceptors in the mouth send signals to the salivatory nuclei in the brain stem (pons and medulla), and as a result, parasympathetic nervous system activity increases and motor fibers in the facial (V) and glossopharyngeal (IX) nerves trigger a dramatically increased output of watery (serous), enzyme-rich saliva. The pressoreceptors are stimulated by virtually any mechanical stimulus in the mouth—even rubber bands; the chemoreceptors are activated most strongly by acidic substances such as vinegar and citrus juice.

The sight or smell of food is often enough to get the juices flowing; in fact, the mere thought of hot fudge sauce on peppermint stick ice cream will make many a mouth water! Irritation of the lower regions of the GI tract by bacterial toxins, spicy foods, or hyperacidity—particularly when accompanied by a feeling of nausea—also increase salivation. This response may help wash away or neutralize the irritants.

Kallikrein, a solute present in saliva, enhances salivation chemically by converting a certain blood protein to bradykinin, which acts as a vasodilator. Since the salivary glands obtain their raw materials (including water) from the bloodstream, saliva secretion is enhanced by the local increase in blood flow that results from bradykinin-induced vasodilation. What happens here is a kind of positive feedback cycle in which saliva secretion results in more saliva secretion because of the effect of kallikrein.

In contrast to parasympathetic controls, the sympathetic division causes release of a mucin-rich viscous saliva. Extremely strong mobilization of the sympathetic division causes the blood vessels serving the salivary glands to constrict and almost completely

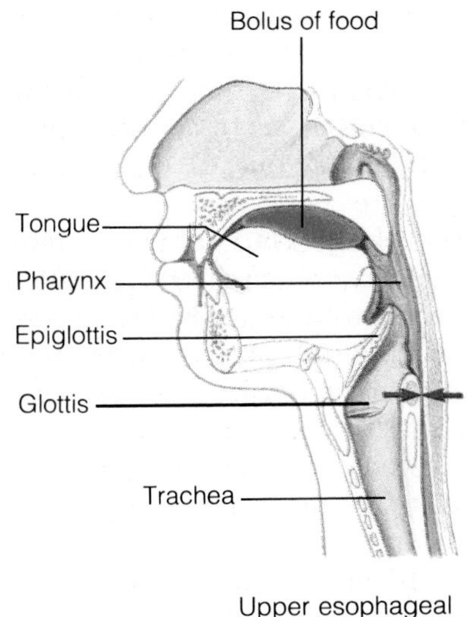

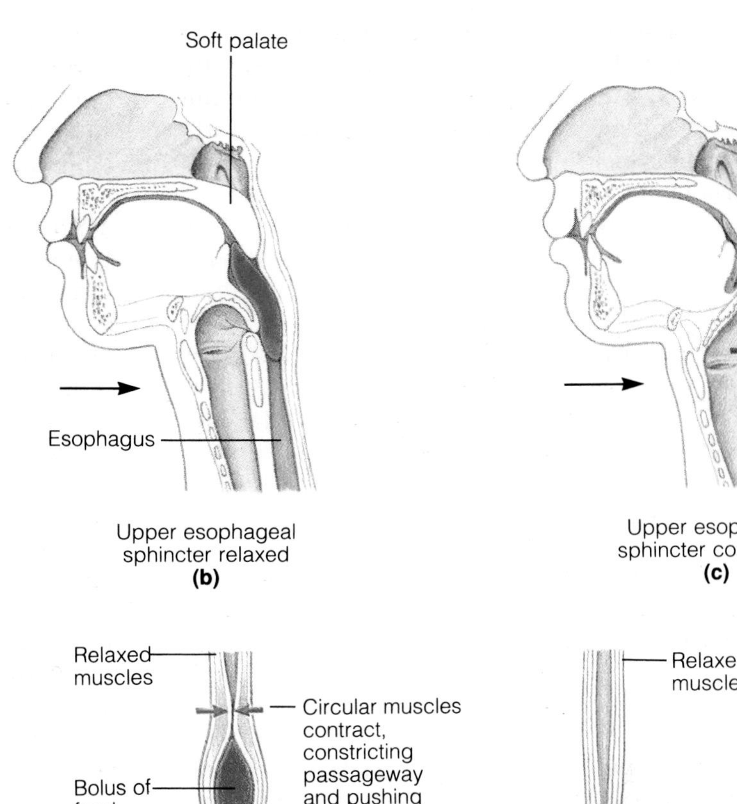

Figure 24.21 Deglutition (swallowing). The process of swallowing consists of both voluntary (buccal) and involuntary (pharyngeal-esophageal) phases. (**a**) During the buccal phase, the tongue rises and presses against the hard palate; in so doing, it forces the food bolus into the oropharynx. Once food enters the pharynx, the involuntary phase of swallowing begins. (**b**) Food passage into respiratory passageways is prevented by the rising of the uvula and larynx and relaxation of the upper esophageal sphincter to allow food entry into the esophagus. (**c**) The constrictor muscles of the pharynx contract, forcing food into the esophagus inferiorly, and the upper esophageal sphincter contracts once food entry has occurred. (**d**) Food is conducted along the length of the esophagus to the stomach by peristaltic waves. (**e**) The gastroesophageal sphincter opens, and food enters the stomach.

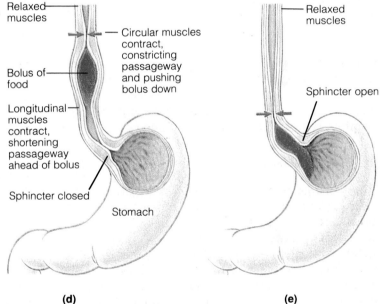

inhibits the release of saliva, causing the mouth to become dry. Dehydration also inhibits salivation because low blood volume results in reduced filtration pressure at capillary beds.

Any disease process that inhibits saliva secretion results in *halitosis* (hal″-ih-tō-sis) (bad breath), because decomposing food particles are allowed to accumulate, and bacteria flourish. ■

Propulsion

Propulsion of food through the digestive tract begins with **deglutition** (dē″-gloo-tish′-un), or **swallowing,** a complicated process that involves the coordinated activity of the tongue, soft palate, pharynx, esophagus, and over 22 separate muscle groups (Figure 24.21). Before swallowing can begin, the tongue must compact the food into an easily swallowed mass, or *bolus.* Once this is accomplished, we are ready to initiate the *buccal phase* of swallowing, which occurs in the mouth and is voluntary. In the buccal phase, we place the tip of the tongue against the hard palate and then, via tongue contractions, force the bolus into the oropharynx, where it stimulates tactile receptors and passes out of our voluntary control. The involuntary phase of swallowing, called the *pharyngeal-esophageal phase,* is controlled by the swallowing center located in the medulla and lower pons. Motor impulses from the swallowing center are transmitted via various cranial nerves (most importantly the vagus nerve) to the muscles of the pharynx and esophagus.

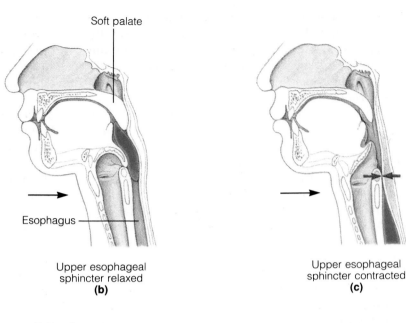

Once food enters the pharynx, all routes that it might take, except for the desired one, distally into the digestive tract, are blocked off. The tongue blocks off the mouth, the soft palate rises to close off the nasopharynx, and the larynx rises so that its opening into the respiratory passageways is covered by the flaplike epiglottis.

If we try to talk or inhale while swallowing, the various protective mechanisms may be short-circuited, and food may enter the respiratory passageways, triggering the protective coughing reflex in an attempt to expel the food. ■

Food is squeezed through the pharynx and then into the esophagus below by wavelike peristaltic contractions. The entire time for solid food passage from the oropharynx to the stomach is 4–8 seconds; fluids pass within 1–2 seconds. Just before the peristaltic wave (and food) reaches the end of the esophagus, the gastroesophageal sphincter relaxes reflexively to allow food to enter the stomach.

Digestive Processes Occurring in the Stomach

Besides serving as a holding area for ingested food, the stomach continues the demolition job begun in the oral cavity by the teeth and salivary amylase, further degrading food both physically and chemically. It then delivers chyme, the product of its activity, into the small intestine at a rate that best promotes the digestive and absorptive activities of the latter organ.

Despite the obvious benefits of preparing food to enter the intestine, the only stomach function essential to life is secretion of intrinsic factor. Intrinsic factor is required for intestinal absorption of vitamin B_{12}, which is involved in the production of mature erythrocytes. However, if vitamin B_{12} is administered by injection, individuals can survive with minimal digestive problems even after total gastrectomy (stomach removal). (The stomach's activities are summarized in Table 24.1.)

Regulation of Gastric Secretion

Gastric secretion is controlled by both neural and hormonal mechanisms (Figure 24.22), and under normal conditions, the gastric mucosa pours out as much as 2–3 L of gastric juice each day. Since gastric secretions are involved not only in chemical digestion, but also in the regulation of stomach contractions, we will discuss them before taking up the mechanical aspects of digestion.

Stimulation of Secretion. Nervous control is provided by the long and short nerve reflexes. When the vagus nerves actively stimulate the stomach, secretory activity of virtually all of its glands increases dramatically. (By contrast, activation of sympathetic nerves depresses secretory activity.) Hormonal control of gastric secretion is largely the province of gastrin, which stimulates the secretion of both enzymes and HCl.

The stimulation of gastric secretion is related to three distinct phases of gastric secretory activity: the cephalic, gastric, and intestinal phases (Figure 24.22).

1. The cephalic (reflex) phase. The cephalic, or reflex, phase of gastric secretion occurs before food has entered the stomach and is triggered by the sight, aroma, taste, or thought of food. During this phase, the brain gets the stomach ready for its upcoming digestive chore. Inputs from activated olfactory receptors and taste buds are relayed to the hypothalamus, which in turn stimulates the vagal nuclei of the medulla oblongata, causing motor impulses to be transmitted via vagal nerve fibers to the stomach glands. The enhanced secretory activity resulting from the sight or thought of food is believed to represent a *conditioned reflex* and occurs only when we like or want the food. If we are depressed or have no appetite, this particular part of the reflex is dampened.

2. The gastric phase. When food reaches the stomach, neural and hormonal mechanisms initiate the gastric phase, which accounts for about two-thirds of the gastric juice released. Stomach distention activates stretch receptors and initiates both local (myenteric) reflexes and the long vagovagal reflexes. In the latter type of reflex, impulses are transmitted to the medulla along vagal afferents and then back to the stomach via vagal efferents. Both types of neural activity lead to acetylcholine (ACh) release, which stimulates the output of more gastric juice; however, the hormone gastrin is probably more important than neural influences in stimulating the stomach glands during the gastric phase.

Chemical stimuli provided by foods, especially partially digested proteins and caffeine (present in colas, coffee, and tea), directly activate gastrin-secreting cells. Gastrin, in turn, stimulates the gastric glands to spew out even more gastric juice. Although gastrin stimulates the release of both enzymes and HCl, its most pronounced effects are exerted on the HCl-secreting parietal cells. Gastrin secretion is inhibited by high acidity, and when the pH of the gastric contents drops below 2, its secretion is damped. As a general rule, when protein foods are in the stomach, the pH of the gastric contents rises because proteins act as buffers to tie up H^+. The rise in pH removes the inhibition of gastrin and stimulates release of HCl, which in turn provides the acidic conditions needed for protein digestion. The more protein in a meal, the greater the amount of gastrin and HCl released. As proteins are digested, the gastric contents become more acidic, which again inhibits the enteroendocrine (gastrin-secreting) cells. This negative feedback mechanism helps maintain an optimal pH for the functioning of the gastric enzymes.

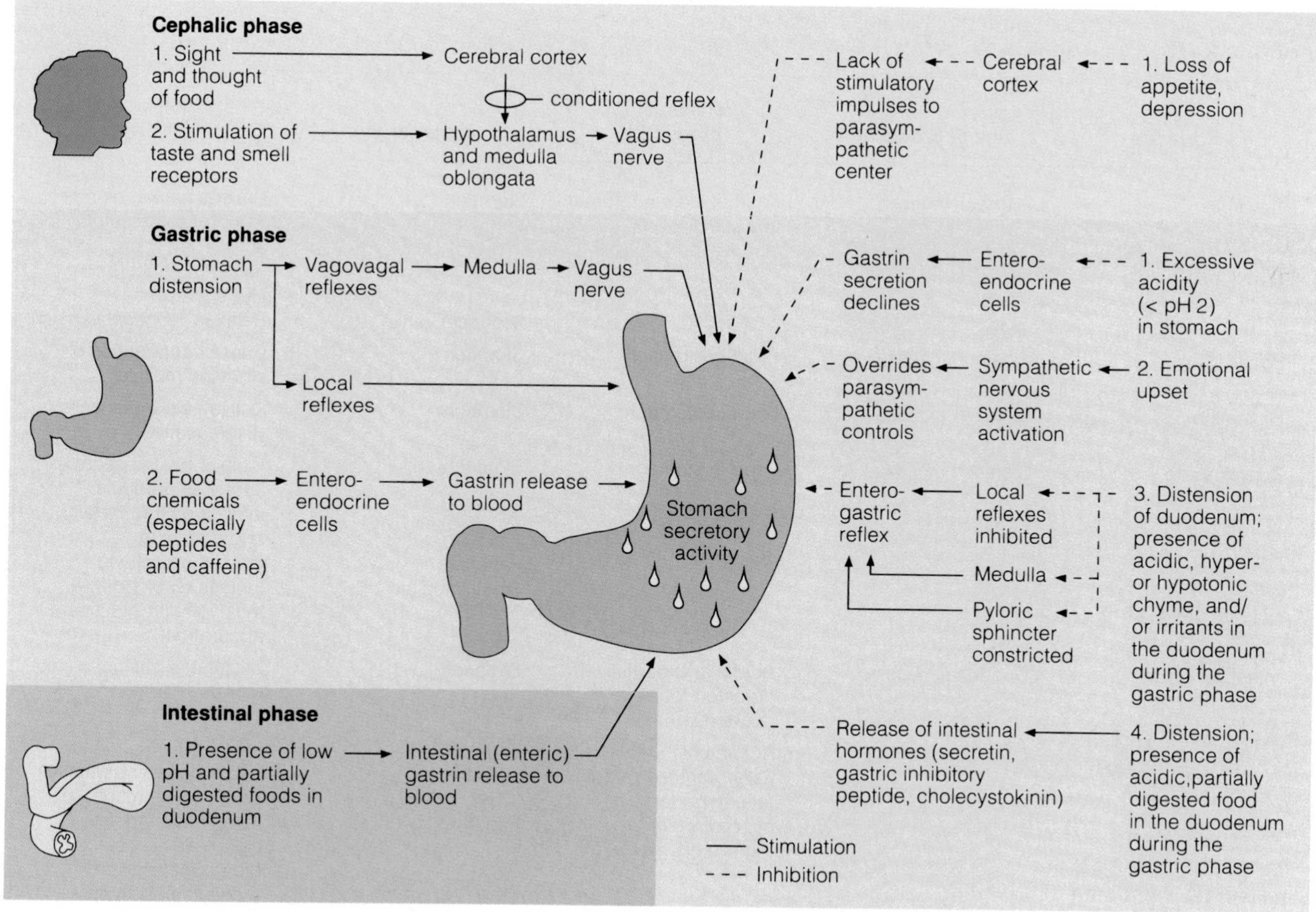

Figure 24.22 Neural and hormonal mechanisms that regulate the release of gastric juice. Stimulatory factors are shown on the left; inhibitory factors are shown on the right.

In addition to responding directly to chemical stimuli, the gastrin-secreting cells are activated by the neural reflexes already described. Emotional upsets, fear, anxiety, or other states that trigger sympathetic nervous system activity inhibit gastric secretion, because (during such times) the sympathetic division overrides parasympathetic controls of digestion (see Figure 24.22).

3. The intestinal phase. The intestinal phase of gastric secretion is set into motion when partially digested food begins to fill the initial part of the small intestine. This filling stimulates intestinal mucosal cells to release a hormone that encourages the gastric glands to continue their secretory activity. Although the precise nature of this hormone is not known, its effects mirror those of gastrin, and it has been named **intestinal (enteric) gastrin.** Many researchers believe that the intestinal phase involves other hormones or reflex mechanisms or both, acting to hasten digestion in the stomach before stomach emptying is completed.

Inhibition of Secretion. Although the presence of food in the small intestine stimulates gastric gland activity during the intestinal phase when the stomach is almost emptied, it *inhibits* gastric secretion during the gastric phase when the stomach contains food and its secretory activity is being regulated by gastric phase mechanisms (see Figure 24.22). During the gastric phase, factors such as distention of the duodenum, low pH, presence of partially digested proteins, hypertonic or hypotonic fluids, and irritants in the small intestine all trigger the **enterogastric reflex.**

The enterogastric reflex is actually a trio of reflexes that (1) inhibits the vagal nuclei in the medulla, (2) inhibits local reflexes, and (3) activates sympathetic fibers that cause the pyloric sphincter to tighten and prevent further food entry into the small intestine. As a result, gastric secretory activity declines.

In addition, the factors just named trigger the release of several intestinal hormones, including **secretin** (suh-krē′-tin), **cholecystokinin** (kō″-lē-sis″-tō-kī′-nin) **(CCK),** and **gastric inhibitory peptide (GIP).** All of these hormones inhibit gastric secretion during the gastric phase, when the stomach is actively secret-

Table 24.2 Hormones and Hormone-like Products That Act in Digestion

Hormone	Site of production	Stimulus for production	Target organ	Activity
Gastrin	Stomach mucosa	Food in stomach (chemical stimulation)	Stomach	Causes gastric glands to increase secretory activity; most pronounced effect is on HCl secretion
			Ileocecal valve	Relaxes ileocecal valve
Serotonin	Stomach mucosa	Food in stomach	Stomach	Causes contraction of stomach muscle
Histamine	Stomach mucosa	Food in stomach	Stomach	Activates parietal cells to release HCl
Somatostatin	Stomach mucosa	Food in stomach	Stomach	Inhibits gastric secretion of all products; inhibits gastric motility and emptying
			Pancreas	Inhibits secretion
			Small intestine	Inhibits GI blood flow; thus, inhibits intestinal absorption
Intestinal gastrin	Duodenal mucosa	Acidic and partially digested foods in duodenum	Stomach	Stimulates gastric glands
Secretin	Duodenal mucosa	Acidic chyme (also partially digested proteins, fats, hypertonic or hypotonic fluids, or irritants in chyme)	Stomach	Inhibits gastric gland secretion during gastric phase of secretion
			Pancreas	Increases output of pancreatic juice rich in bicarbonate ions
			Liver	Increases bile output
Cholecystokinin (CCK)	Duodenal mucosa	Fatty chyme, in particular, but also partially digested proteins	Stomach	Inhibits gastric gland secretion during gastric phase of secretory activity
			Pancreas	Increases output of enzyme-rich pancreatic juice
			Gallbladder	Stimulates organ to contract and expel stored bile
			Sphincter of Oddi	Relaxes to allow entry of bile and pancreatic juice into duodenum
Gastric inhibitory peptide (GIP)	Duodenal mucosa	Fatty chyme	Stomach	Inhibits gastric gland secretion during gastric phase
Enterogastrone	Duodenal mucosa	Fatty foods; hyperosmolar solutions	Stomach	Inhibits motility of stomach smooth muscle

ing gastrin. These hormones also play other roles, which are summarized in Table 24.2.

Production of HCl. So far, we have mentioned gastrin as a regulator of HCl secretion, but actually, HCl secretion by the parietal cells is stimulated by three chemicals. Besides responding to neural controls mediated by *acetylcholine* and hormonal controls exerted by *gastrin*, secretion of gastric juice is also responsive to *histamine* released by enteroendocrine cells. Apparently, parietal cells bear membrane receptors for all three chemicals, and their activity increases in direct proportion to the number of receptors activated. When only one of the three chemicals is bound, HCl secretion is scanty, but when all three bind, volumes of HCl pour out. The process of HCl formation within the parietal cells is complicated and as yet poorly understood. A postulated mechanism is illustrated in Figure 24.23. It is presumed that all three regulatory chemicals act in some manner to stimulate the active hydrogen pumps that extrude H^+ into the stomach lumen.

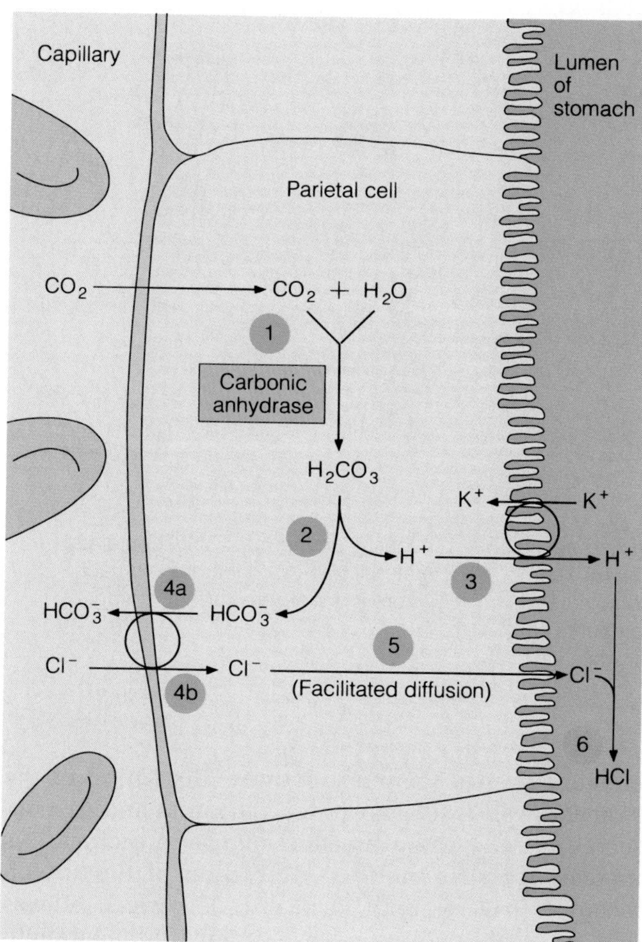

Figure 24.23 Postulated mechanism of HCl secretion by parietal cells of the stomach. Since both H^+ and Cl^- are moved into the stomach lumen against their electrochemical gradients, metabolic energy (ATP) is required for the transport of both ions (indicated by [symbol] in the figure). HCl secretion into the stomach lumen is believed to occur as follows. (1) Carbon dioxide (CO_2), produced as a product of metabolism or passing into the cells from blood plasma, is hydrated (combined with water), forming carbonic acid (H_2CO_3) in a reaction catalyzed by carbonic anhydrase. (2) Carbonic acid splits into H^+ and HCO_3^-. (3) H^+ obtained from the cleavage of H_2CO_3 is actively extruded from the parietal cell by an H^+-K^+ ATPase located at the cell apex, which transports one K^+ into the cell for each H^+ ejected from the cell. (4) (a) As H^+ is pumped from the cell, the amount of HCO_3^- within the cell becomes excessive, and HCO_3^- is ejected (along its electrochemical gradient) through the basal cell membrane and into the capillary blood. As a result, blood draining from the stomach is more alkaline than the blood serving it; this phenomenon is called the *alkaline tide.* (b) The proteins that mediate HCO_3^- transport out of the parietal cell simultaneously transport Cl^- into the cell along its electrochemical gradient; thus, the transport protein acts as an anion exchanger. (Some believe that there is also an Na^+-Cl^- cotransport system.) (5) Cl^- moves into the stomach lumen by facilitated diffusion. (6) Within the stomach lumen, H^+ and Cl^- form HCl.

Mechanical Digestion and Propulsion

Stomach contractions, accomplished by the trilayered muscularis, cause not only its emptying, but also compress, knead, twist, and continually mix the food with gastric secretions so that the semifluid chyme is formed. Since the mixing movements are accomplished by a unique type of peristalsis (for example, in the pylorus it is bidirectional rather than unidirectional), the processes of mechanical digestion and propulsion are inseparable in the stomach.

Response of the Stomach to Filling. As food enters the stomach, its wall is stretched to accommodate the increasing volume; however, internal stomach pressure remains constant until about 1 L of food has been ingested. (Thereafter, the pressure rises.) The relatively unchanging pressure in a filling stomach reflects two phenomena: (1) the plasticity of visceral smooth muscle and (2) reflex-mediated relaxation of the stomach musculature.

Plasticity is the intrinsic ability of visceral smooth muscle to exhibit the stress-relaxation-response, i.e., to be stretched without greatly increasing its tension and contracting expulsively. As described in Chapter 9 (p. 267), plasticity is very important in hollow organs, like the stomach, that must serve as temporary reservoirs. Reflexive relaxation of stomach smooth muscle occurs both in anticipation of and in response to food entry into the stomach. When we swallow food, the muscles of the stomach relax—this occurring while food is still traveling through the esophagus. This *receptive relaxation* response is coordinated by the swallowing center of the brain stem. The stomach also relaxes when its walls are actually distended by food (that is, by gastric filling), which activates mechanoreceptors in the wall. Both of these reflex regulation responses are mediated by the vagus nerves.

Gastric Contractile Activity. After a meal, peristalsis begins near the gastroesophageal sphincter. Contractions produced in this region produce only gentle rippling movements of the stomach wall, but as the peristaltic contractions approach the pylorus, they become much more powerful and their speed of transmission increases dramatically. These differences can be explained by structural factors: Muscle layers in the fundus and body are quite thin, but stomach musculature increases in thickness distally. Consequently, the contents of the fundus remain relatively undisturbed, while foodstuffs close to the pylorus receive a truly lively pummeling and mixing. (Starch digestion by salivary amylase may continue for an hour or more in the superior region of the stomach, where foods are poorly mixed with gastric juice, particularly if a large meal has been ingested.)

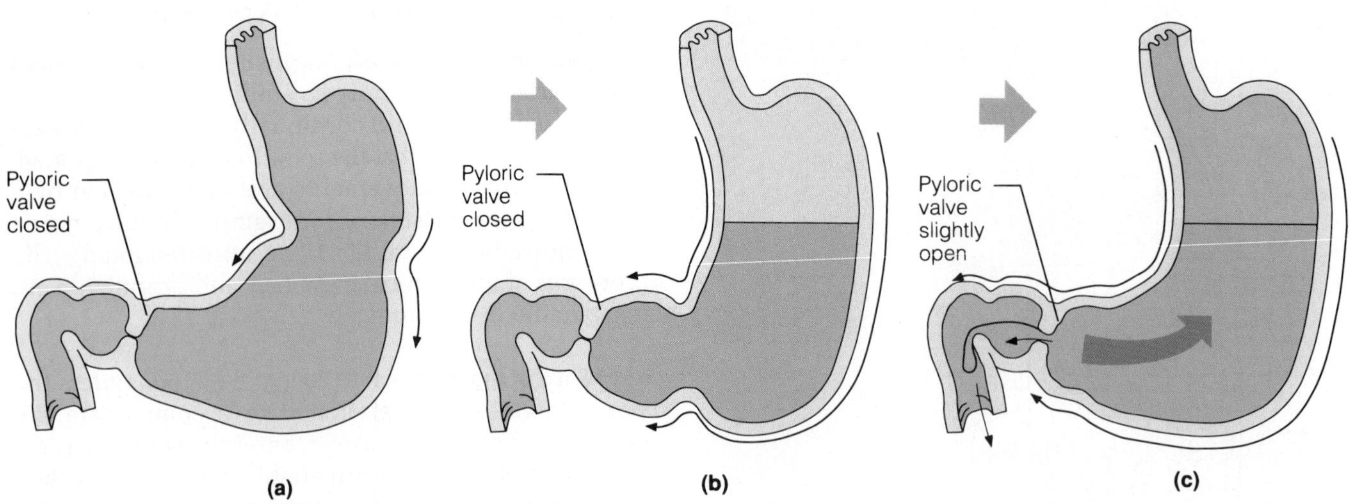

Figure 24.24 Peristaltic waves act primarily in the inferior portion of the stomach to mix and move chyme through the pyloric valve. (**a**) Peristaltic waves move toward the pylorus. (**b**) Most vigorous peristalsis and mixing action occurs close to the pylorus. (**c**) The pyloric end of the stomach acts as a pump that delivers small amounts of chyme into the duodenum, while simultaneously forcing most of its contained material backward into the stomach, where it undergoes further mixing.

The pyloric region of the stomach, which holds about 30 ml of chyme, acts as a dynamic filter that allows only liquids and small particles to pass through the barely open pyloric valve during the digestive period. Normally, each contraction of the pyloric muscle "spits" or squirts, 3 ml or less of chyme into the small intestine. Because the contraction also closes the valve, the rest (about 27 ml) is propelled backward into the stomach, where it undergoes further mixing (Figure 24.24). This backward pumping action, called *retropulsion*, provides an effective means of breaking up solids in the gastric contents.

Although the intensity of stomach peristaltic waves can be enhanced a good deal, the rate of these contractions is always around three contractions per minute. How can this be explained? It appears that the stomach's contractile rhythm is set by the spontaneous activity of *pacemaker cells* located within the longitudinal smooth muscle layer in the stomach's greater curvature. The pacemaker cells depolarize and repolarize spontaneously three times each minute, establishing the so-called *slow-wave* electrical activity of the stomach. Since these pacemakers are connected to the rest of the smooth muscle sheet by gap junctions (electrically coupled), their "beat" is transmitted efficiently and quickly to all of the stomach musculature. Although the pacemakers set the maximum rate of contraction, they do not initiate the contractions. Basically, the pacemakers generate subthreshold depolarization waves, which are then "ignited" (enhanced by further depolarization and brought to threshold) by extrinsic factors.

Factors that modulate and increase the strength of stomach contractions are the same factors that enhance gastric secretory activity. Distention of the stomach wall activates stretch receptors and gastrin-secreting cells; the ultimate response in each case is increased gastric motility. Activation of the stretch receptors triggers both local and vagovagal reflexes that, via acetylcholine release, intensify the strength of peristaltic contractions and elicit gastrin release. Gastrin enters the blood and eventually recirculates to the stomach, where it stimulates gastric smooth muscle. Thus, the more food there is in the stomach, the more vigorous the stomach mixing and emptying movements will be—within certain limits—as described next.

Regulation of Gastric Emptying. The stomach usually empties completely within 4 hours after a meal. The more liquefied its contained food, the faster the stomach empties. Fluids pass through the stomach very quickly; solids remain until they are well mixed with gastric juice and converted to the liquid state.

The rate of gastric emptying depends as much—and perhaps more—on the contents of the duodenum as on the state of food digestion in the stomach. The stomach and duodenum act in tandem and can be thought of as a "coupled meter" that functions at less than full capacity. As chyme enters the duodenum, receptors in its wall respond to chemical signals and to stretch, initiating the enterogastric reflex and the hormone mechanisms described earlier. These mechanisms prevent further duodenal filling by inhibiting gastric secretory activity and by reducing the vigor of pyloric contractions (Figure 24.25).

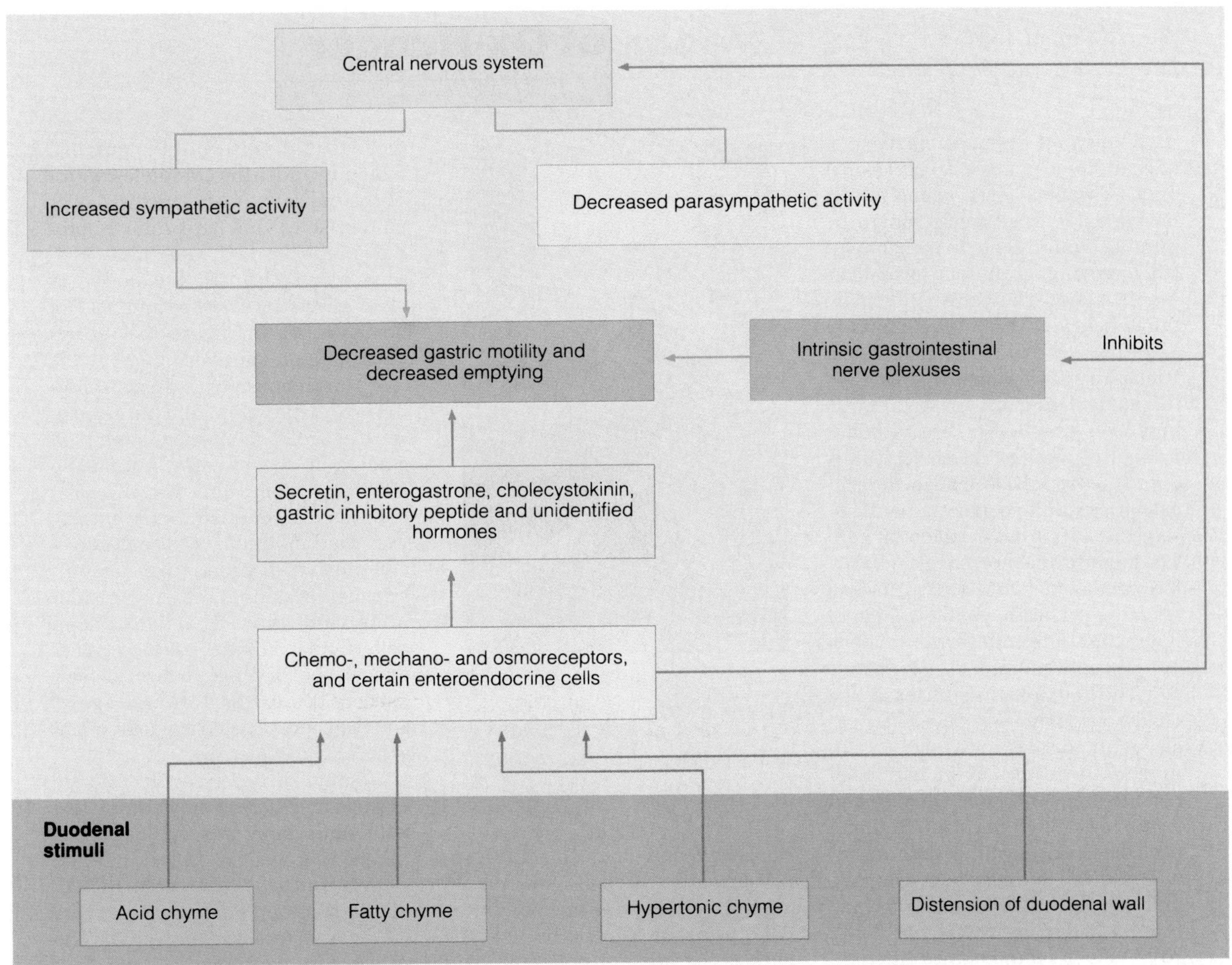

Figure 24.25 Neural and hormonal factors inhibiting gastric emptying. These controls ensure that the food will be well liquefied in the stomach and prevent the small intestine from being overwhelmed.

As a rule, a meal rich in carbohydrates moves through the duodenum rapidly, but fats form an oily layer at the top of the chyme and are digested more slowly by intestinal enzymes. Thus, when chyme entering the duodenum is high in fats, food may remain in the stomach for as long as 6 hours or more before gastric emptying is completed. Additionally, the presence of fats elicits the release of an intestinal hormone called **enterogastrone** (en″-ter-ō-gas′-trōn), which, like secretin elicited by low pH, also inhibits gastric motility.

After one eats an ordinary, balanced meal, the volume of the stomach contents remain the same during the first hour. This does not imply that gastric emptying is delayed; in fact, the rate of emptying is maximal shortly after eating a meal. However, the volume of food being emptied is counterbalanced by the volume of gastric secretions spewed out (secreted) during that time.

Vomiting is an unpleasant experience that causes reflex stomach emptying by a different route. It is usually initiated by extreme stretching of the stomach or intestine or by the presence of irritants such as bacterial toxins, excessive alcohol, spicy foods, and certain drugs in those organs. Sensory impulses from the irritated sites reach the *emetic* (ē-meh′-tik) *center* of the medulla and initiate a number of motor responses. The skeletal muscles of the abdominal wall and diaphragm contract and increase the intra-abdominal pressure, the smooth muscles of the GI tract contract forcibly, the gastroesophageal sphincter relaxes, and the soft palate rises to close off the nasal passages. As a result, the duodenal and stomach contents are forced upward through the esophagus and pharynx and out the mouth. Before vomiting, an individual typically is pale-faced, feels nauseated, and salivates. Excessive vomiting can lead to severe disturbances in water, salt,

A CLOSER LOOK The Paradox of Ox-Hunger

The more fortunate citizens of ancient Rome enjoyed an occasional orgy—a time of feasting and sexual merriment, interrupted only by purging (vomiting) to make room for still more gorging. *Bulimia* (byoo-lim′-ē-uh), a modern version of binge-purge behavior, has become a daily affair for ever-increasing numbers of young American women. But whereas the ancient Romans prized plumpness as a standard of beauty and a symbol of wealth, our society finds obesity so repulsive that some teenage girls would sooner starve themselves than harbor a single fat cell. The American abhorrence of obesity has produced two eating disorders affecting primarily young women of high school and college age: anorexia nervosa and bulimia.

The better known of these two conditions, *anorexia nervosa* (an″-or-eks′-ē-uh ner-vō′-suh), is a self-induced wasting away in which patients frequently overexercise and refuse to eat anything at all. *Bulimia*, from the Greek meaning "ox-hunger," or insatiable hunger, is a complex, multifaceted disease that is just beginning to come out of the closet. Although its symptoms are different, it is the other side of the same coin: Both disorders are rooted in a pathological fear of becoming fat and a passionate need to control eating.

For some bulimics, the syndrome evolves into a \$50- to \$100-a-day food habit. As if starving, they eat huge amounts of food, often ingesting up to 55,000 calories in a single hour. Then, they induce vomiting. This paradoxical behavior may be repeated three to four times daily. Others take 200 or more laxatives each week to prevent their bodies from retaining the enormous amounts of food consumed. Surprisingly, most bulimics have a normal body weight and look healthy. However, most bulimics are prone to such ailments as swollen salivary glands, pancreatitis, and liver and gallbladder problems. The constant vomiting damages both the stomach and the esophagus, and stomach acid flushing upward through the mouth severely erodes tooth enamel. In the worst cases, bulimia kills. Purging empties the body of potassium and causes severe electrolyte disturbances, which make the heart irritable and place it at risk to fail. Stomach rupture may cause death from massive hemorrhage. Kidney dysfunction is also a problem.

Chances are that this young woman, or one of her close college-student friends is a bulimic.

Although the disorder knows no race or class, bulimics are typically single, white women from affluent families who have had some college education. Despite their obvious accomplishments, they have little self-regard and are scared of success or feel that they do not deserve it. In many cases, bulimics have problems within their family in establishing their independence or personal growth. Some researchers believe that eating helps to relieve the anger and frustration that evolves as these young women strive for perfection. The eating itself is not a pleasant experience, but instead is associated with feelings of disgust. But the purging, which is under the complete control of the bulimic, is her "victory" over all obstacles and stressors. It provides relief from anxiety and disciplines and cleanses the body—in effect providing a new beginning. The bonus is a slim figure that brings attention and approval.

Many bulimics exhibit symptoms of depression, and up to one-third of such patients have attempted suicide. In 6% of cases, bulimia so fills the person's life that all social contacts are shunned. Some authorities view bulimia as an obsessive-compulsive disorder (like compulsive handwashing); others consider it an addiction. The latter camp believes that bulimics use food, rather than alcohol or drugs, to regulate their tensions because they are "good girls" and food abuse has no apparent moral or legal consequences.

Eating disorders evolve slowly and become increasingly entrenched with time. They are not just attention-getting devices; in fact, bulimia victims tend to be very secretive about their binges and most wait five years or more before seeking help. The seriousness of bulimia is all too often underestimated, and its treatment is extremely difficult. Techniques that are often used simultaneously include hospitalization (to control food intake or outgo or to isolate the patient from family), behavior modification (learning to eat three meals daily), education in nutrition, and antidepressant drugs for severely depressed patients. The treatment is usually prolonged and, when hospitalization is involved, always expensive. Moreover, treatment succeeds only in suppressing the symptoms of bulimia rather than in curing it. Like addicts who must constantly avoid their drug or alcohol habit, bulimics must fight the urge to resume their binge-purge behavior for the rest of their lives. Bulimia is not only a paradox, it is a nightmare of frustration and despair for those who suffer from it.

and acid-base balance of the body. Since large amounts of hydrochloric acid are lost in vomitus, the blood becomes alkaline as the stomach attempts to replace its lost acid. ■

Chemical Digestion

Protein digestion is initiated in the stomach and is essentially the only type of enzymatic digestion that occurs there. Pepsinogen secreted by the zymogenic cells is quickly activated to **pepsin** (actually a group of proteolytic, or protein-digesting, enzymes) in the lumen of the stomach. The first pepsinogen molecules are activated by HCl, but once pepsin is present, it also acts in the activation process. Thus, the activation of pepsinogen to pepsin is an example of a positive feedback process and is limited only by the amount of pepsinogen present. This activation process is accomplished by the removal of a small peptide fragment from the pepsinogen molecule, an event that causes it to change shape, exposing its active site. In general, the same factors that stimulate or inhibit HCl secretion exert similar effects on pepsinogen secretion.

Pepsin functions optimally in the highly acid pH range found in the stomach: 1.5–3.5. It preferentially cleaves bonds involving the amino acids tyrosine and phenylalanine, so that proteins are broken into polypeptides and small numbers of free amino acids (see Figure 24.4). The pepsins, responsible for hydrolyzing about 15% of ingested protein, are inactivated by the high pH in the duodenum, so their proteolytic activity is restricted to the stomach.

In infants, the gastric glands also produce large amounts of **rennin.** This enzyme works primarily on the milk protein casein, converting it to a curdy substance that looks like sour milk. Rennin is apparently not produced in adults. Although the gastric glands are also thought to secrete small amounts of lipases, lipases function optimally at a pH of 5–6, and little fat digestion occurs in the stomach.

Absorption

The only commonly ingested substances known to pass easily through the stomach mucosa into the blood are two lipid-soluble substances, aspirin and alcohol; but some other lipid-soluble drugs also pass. The movement of aspirin through the gastric mucosa has been shown to cause gastric bleeding; thus, aspirin should be avoided by those with gastric ulcers.

Digestive Processes Occurring in the Small Intestine

Although food reaching the small intestine is unrecognizable, it is far from digested chemically. Carbohydrates and proteins are partially digested, but virtually no fat digestion has occurred up to this point. The process of food digestion is accelerated during the chyme's tortuous 3- to 6-hour journey through the small intestine, and it is here that virtually all absorption of nutrients takes place.

The intestinal glands, located over the entire surface of the small intestine, normally secrete 2–3 L of intestinal juice per day. The major stimulus for production of this fluid is distention or irritation of the intestinal mucosa by hypertonic or acidic chyme. Normally, the pH range of intestinal juice is 6.5–7.8, and it is isotonic with blood plasma. Largely water, intestinal juice contains a good deal of mucus, which is secreted both by Brunner's glands of the duodenum and by goblet cells in the epithelium. It is relatively enzyme-poor because the bulk of intestinal enzymes are bound (brush border) enzymes, and most of the enzymes that do appear in intestinal juice are derived from the disintegration of sloughed-off mucosal cells that bear them.

Mechanical Digestion and Propulsion

The small intestine is able to process only small amounts of chyme at one time because the entering chyme is usually hypertonic. If large amounts of chyme were rushed into the small intestine, the osmotic water loss from the blood across the intestinal wall into the lumen would result in dangerously low blood volume. Additionally, the pH of the entering (acidic) chyme must be adjusted upward and the chyme must be mixed with bile and pancreatic juice before digestion can continue. Thus, food movement into the small intestine is carefully controlled by the pumping action of the stomach pylorus, which prevents the duodenum from being overwhelmed.

The major functions of intestinal smooth muscle are to mix chyme thoroughly with bile, pancreatic juice, and intestinal juice and to move food residues through the ileocecal valve into the large intestine. Although many authorities assume that segmentation does the mixing and peristalsis does the propelling (see Figure 24.3), there is some disagreement on this point and the more widely held view is discussed here. Peristaltic contractions in the small intestine are weak and listless—nothing like the forceful peristaltic waves seen in the esophagus. Examination of the small intestine by X-ray fluoroscopy after it is "loaded" with a meal reveals that the intestinal contents are simply moved backward and forward in the lumen a few centimeters at a time by alternating contraction and relaxation of rings of smooth muscle. The segmenting movements of the intestine, like the peristalsis of the stomach, are initiated by intrinsic pacemaker cells in the longitudinal smooth muscle. However, since the rate of spontaneous depolarization of the pacemakers decreases from about 12 contractions per minute in the duodenum to 8 or 9 contractions per minute in the ileum, the final result of this *segmentation* is a slow progress toward the ileocecal valve. The intensity of segmentation is controlled by long and short reflexes (parasympathetic activity enhances and sympathetic activity decreases it) and hormones. True peristalsis occurs only after most of the nutrients have been absorbed. At this point, segmenting movements wane, and peristaltic waves initiated at the duodenal end begin to sweep along the length of the intestine, moving 10–70 cm before dying out. Whatever the mechanism, the result is that food is properly mixed and propelled at a rate slow enough to allow ample time to complete digestion and absorption, and usually the small intestine is emptied of food by the time the next meal is eaten.

Most of the time, the ileocecal sphincter is constricted and closed. However, two mechanisms—one neural and the other hormonal—cause it to relax and allow food residues to enter the cecum. Enhanced secretory and contractile activity of the stomach initiates the **gastroileal** (gas″-trō-il′-ē-ul) **reflex,** which enhances ileal contractions. In addition, gastrin released by the stomach increases the motility of the ileum and causes relaxation of the ileocecal sphincter. Once the chyme has passed through, the valve's flaps are closed by the backward pressure of the material in the cecum, preventing regurgitation into the ileum.

Chemical Digestion

Chemical digestion begins in earnest in the small intestine. Bile conveyed into the duodenum prepares lipids for enzymatic degradation, and pancreatic and intestinal enzymes complete the breakdown of the various foodstuffs into absorbable fragments. We will begin by focusing on the composition and general roles of bile and pancreatic juice and by examining factors that regulate their secretion into the small intestine. Then we will move on to describe the enzymatic digestion of each category of foodstuffs.

Bile. Bile is a yellow-to-green, watery solution containing bile salts, bile pigments, cholesterol, neutral fats, phospholipids, and a variety of electrolytes. Of these components, only the bile salts and phospholipids aid the digestive process. Although many substances secreted in bile leave the body in feces, the bile salts are conserved by means of a recycling mechanism called the **extrahepatic circulation;** they are reabsorbed into the blood by the intestinal mucosa and returned to the liver via the hepatic portal blood, then resecreted in newly formed bile.

Bile salts, primarily cholic acid and chenodeoxycholic acids, are cholesterol derivatives. Their role is to *emulsify* fats—that is, to distribute them throughout the watery intestinal contents. (Another example of emulsification is homogenization, which distributes cream throughout the watery phase of milk.) The result is that the large fat globules entering the small intestine are physically separated into suspensions of millions of small fatty droplets that provide large surface areas for the fat-digesting enzymes (lipases) to work on. Bile salts also facilitate fat and cholesterol absorption (discussed later) and act with lecithin (a phospholipid) to solubilize cholesterol, both that contained in bile and that entering the small intestine via the diet.

The chief bile pigment is **bilirubin** (bih′-lih-roo-bin), a breakdown product of the heme of hemoglobin formed during the disposal of worn-out erythrocytes. The bilirubin is absorbed from the blood by the liver cells and secreted into the bile. Most of the bilirubin in bile is metabolized in the small intestine by resident bacteria, and one of its breakdown products, *urobilinogen* (yer″-ō-buh-lin′-ō-jen), gives feces its brown color. In the absence of bile, feces is gray-white in color.

Although bile is made continuously (under the influence of blood-borne secretin and bile salts), it does not usually enter the small intestine until the gallbladder contracts and releases its stores of concentrated bile. Parasympathetic impulses delivered by the vagus nerves have a minor impact on stimulating gallbladder contraction; the major stimulus is cholecystokinin (CCK), an intestinal hormone (Figure 24.26). CCK is released to the blood when acidic, fatty chyme enters the duodenum. Besides stimulating gallbladder contraction, CCK has two other important effects: It stimulates the secretion of pancreatic juice, and it relaxes the sphincter of Oddi so that bile and pancreatic juice can enter the duodenum. CCK and the other hormones that aid in the digestive process are summarized according to site of secretion and function in Table 24.2.

Bile is the major route for cholesterol excretion from the body. When the amount of bile salts or lecithin in bile is inadequate to solubilize its contained cholesterol (or when cholesterol levels are excessive), cholesterol may crystallize, forming *gallstones,* or *biliary calculi* (bil′-ē-ayr-ē kal-kyoo-lī), which obstruct the flow of bile from the gallbladder or through the cystic duct or common bile duct. The sharp crystals can cause agonizing pain when the gallbladder or its duct contracts. Blockage of the common bile duct prevents both bile salts and bile pigments from entering the intestine. As a result, the yellow bile pigments accumulate in the blood and eventually are deposited in the skin, causing it to become yellowish, or *jaundiced.* Jaundice caused by blocked ducts is called *obstructive jaundice,* but jaundice may also reflect other conditions. For example, in *hepatitis* or *fibrous cirrhosis,* the liver cannot carry out its normal metabolic duties, and undegraded bile pigments are deposited in the skin. ■

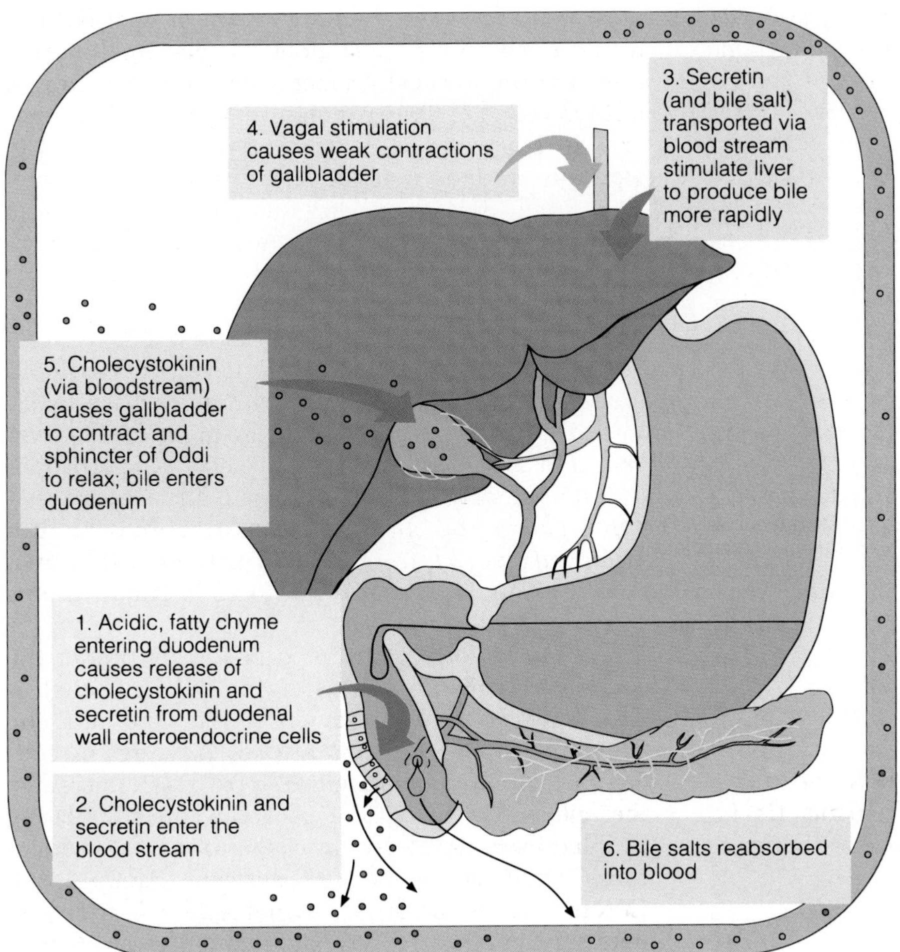
4. Vagal stimulation causes weak contractions of gallbladder
3. Secretin (and bile salt) transported via blood stream stimulate liver to produce bile more rapidly
5. Cholecystokinin (via bloodstream) causes gallbladder to contract and sphincter of Oddi to relax; bile enters duodenum
1. Acidic, fatty chyme entering duodenum causes release of cholecystokinin and secretin from duodenal wall enteroendocrine cells
2. Cholecystokinin and secretin enter the blood stream
6. Bile salts reabsorbed into blood

Figure 24.26 Mechanisms promoting the secretion of bile and its entry into the duodenum. When digestion is not occurring, bile is stored and concentrated in the gallbladder. However, when acidic chyme enters the small intestine, both neural and hormonal mechanisms are initiated that accelerate the liver's output of bile, cause the gallbladder to contract, and promote the relaxation of the sphincter of Oddi, which allows bile (and pancreatic juice) to enter the small intestine, where it plays a role in fat emulsification.

Pancreatic Juice. The pancreas produces 1200–1500 ml per day of pancreatic juice, a clear fluid consisting primarily of water. In addition to digestive enzymes, pancreatic juice has a high sodium bicarbonate content that makes it very alkaline (pH about 8). Pancreatic juice entering the duodenum neutralizes the acidic chyme, providing the optimal pH for the activity of both intestinal and pancreatic enzymes.

Like bile release from the gallbladder, secretion of pancreatic juice is regulated both by the parasympathetic nervous system and by local hormones (Figure 24.27). Vagal stimulation releases enzyme-rich pancreatic juice primarily during the cephalic and gastric phases of gastric secretion. Hormonal control is exerted by two intestinal hormones: secretin, released in response to the presence of HCl in the intestine, and cholecystokinin, released in response to the entry of proteins and fats. Both hormones target the pancreas, but secretin prompts the release of a watery bicarbonate-rich pancreatic juice, whereas CCK causes release of enzyme-rich pancreatic juice. Normally, the amount of HCl produced in the stomach is exactly balanced by the amount of bicarbonate (HCO_3^-) actively secreted by the pancreas, and as HCO_3^- is secreted into the pancreatic juice, H^+ enters the blood. However, the pH of venous blood returning to the heart remains relatively unchanged because the alkaline blood draining from the stomach mixes with the acidic blood draining the pancreas, and they neutralize each other's effects.

As described shortly, pancreatic juice contains enzymes capable of digesting all categories of foods (see Figure 24.4). Just as the pepsins of the stomach are produced in an inactive form, most of the pancreatic protein-digesting enzymes are released in inactive forms that are activated in the duodenum by other enzymes. This prevents the pancreas from self-digestion. Trypsinogen, for example, is activated to trypsin (trip′-sin) by enterokinase, an intestinal brush border enzyme. Trypsin, in turn, activates the other protease precursors to their active forms (Figure 24.28). Other pancreatic enzymes, like amylase and lipase, are secreted in active form, but require the presence of ions or bile in the intestinal lumen for optimal activity.

Intestinal Enzymes. The brush border enzymes of the intestinal microvilli include mostly disaccharases and peptidases (see Figure 24.4), which complete the digestion of carbohydrates and proteins, respectively.

Figure 24.27 Regulation of pancreatic juice secretion by neural and hormonal factors. Neural control is mediated by parasympathetic fibers of the vagus nerves, primarily during the cephalic and gastric phases of gastric secretory activity. Hormonal controls, exerted by secretin and cholecystokinin (steps 1–3), are the more important regulatory factors.

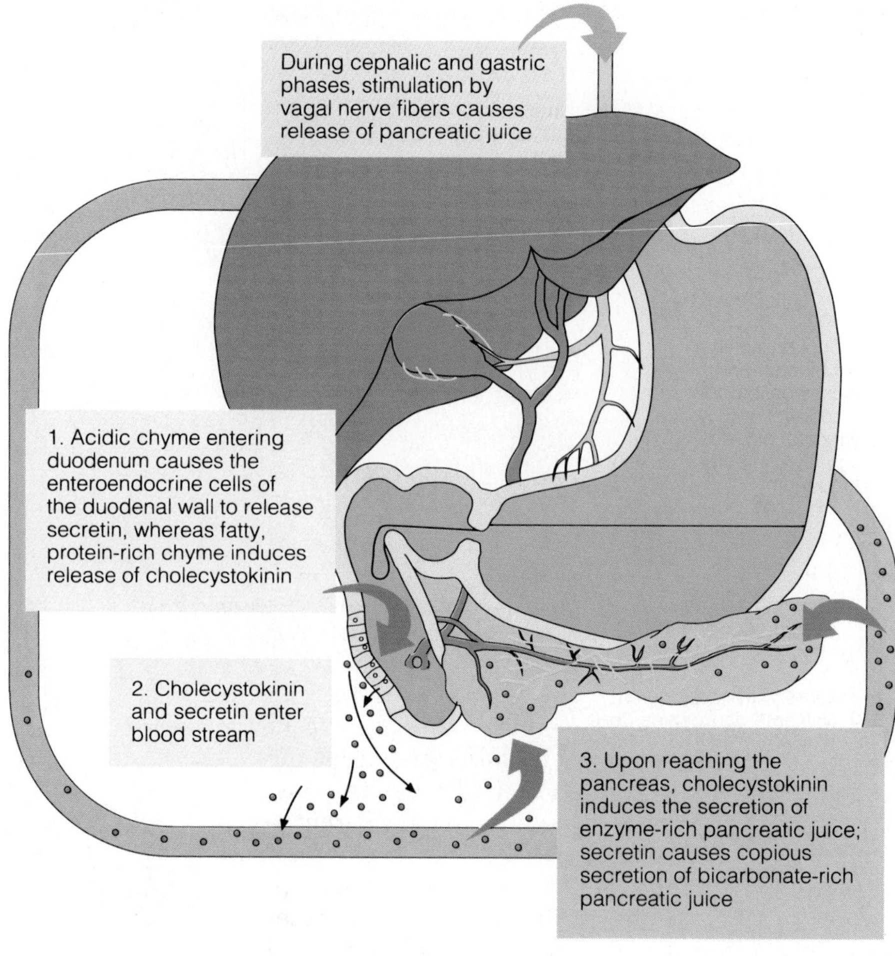
During cephalic and gastric phases, stimulation by vagal nerve fibers causes release of pancreatic juice
1. Acidic chyme entering duodenum causes the enteroendocrine cells of the duodenal wall to release secretin, whereas fatty, protein-rich chyme induces release of cholecystokinin
2. Cholecystokinin and secretin enter blood stream
3. Upon reaching the pancreas, cholecystokinin induces the secretion of enzyme-rich pancreatic juice; secretin causes copious secretion of bicarbonate-rich pancreatic juice

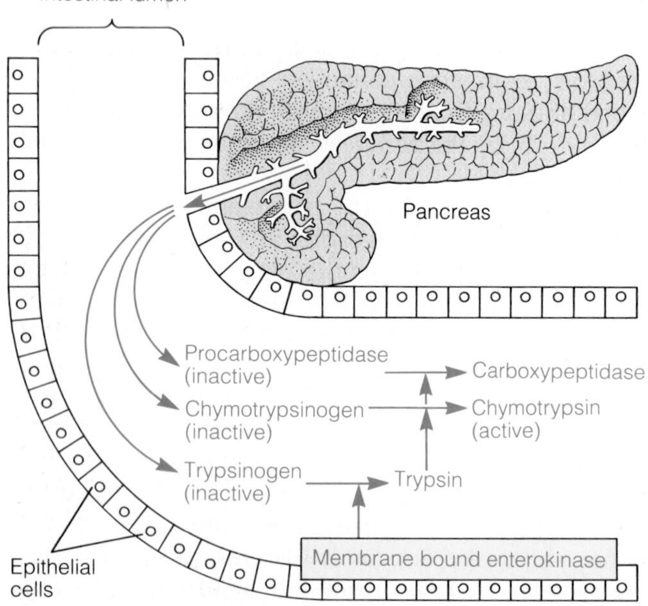
Intestinal lumen
Pancreas
Procarboxypeptidase (inactive)
Carboxypeptidase
Chymotrypsinogen (inactive)
Chymotrypsin (active)
Trypsinogen (inactive)
Trypsin
Membrane bound enterokinase
Epithelial cells

Figure 24.28 Activation of pancreatic proteases in the small intestine. Pancreatic proteases are secreted in an inactive form and are activated in the duodenum. Enterokinase, a membrane-bound (brush border) intestinal enzyme, activates trypsinogen to the active trypsin form. Trypsin, itself a proteolytic enzyme, then activates procarboxypeptidase and chymotrypsinogen.

We have already mentioned enterokinase, which initiates activation of the pancreatic proteases. Many of the others will be described in the next section.

Digestion of Specific Food Groups. Now we are ready to follow the enzymatic breakdown of carbohydrates, proteins, lipids, and nucleic acids that occurs in the small intestine.

1. Carbohydrates. Starchy foods that escape being digested in the mouth by salivary amylase are acted on by the highly active **pancreatic amylase** in the small intestine (see Figure 24.4). Within about 10 minutes after entering the small intestine, starch has been entirely converted to maltose and short chains of four to nine glucose molecules called dextrins and oligosaccharides. The further digestion of these products (and sucrose and lactose) to monosaccharides is accomplished by the intestinal brush border enzymes. The most important of these are **dextrinase, glucoamylase,** and the disaccharase enzymes **maltase, sucrase,** and **lactase.**

In some people, intestinal lactase is present at birth, but then declines in concentration and becomes deficient. In such cases, the person becomes intolerant of milk (the source of lactose), which causes bouts of bloating, flatulence (gaseousness), and diarrhea. Undigested solutes (in this case, lactose sugar) cannot be absorbed and create osmotic gradients that not only prevent water from being absorbed in the small and large intestines, but also pull water from the interstitial space into the intestines. Consequently, a watery stool is produced. The bloating and flatulence are caused by bacterial metabolism of the undigested solutes, which produces large amounts of gas in the large intestine. The gas distends the colon and produces cramping pain. The solution to this problem, which is believed to be largely genetically determined, is simple—don't drink milk! ■

2. Proteins. Proteins digested in the small intestine include not only dietary proteins (typically about 125 g per day), but also 15–25 g of enzyme proteins secreted into the GI tract by its various glands and (probably) an equal amount of protein derived from sloughed and disintegrating mucosal cells. In normal individuals, virtually all of this protein is digested and absorbed.

Protein fragments entering the small intestine from the stomach are cleaved to smaller peptides by **trypsin** and **chymotrypsin** (kī-mō-trip″-sin) secreted by the pancreas. These peptides, in turn, become the grist for the pancreatic and brush border enzyme **carboxypeptidase** (kar-box″-ē-pep′-tih-dās), which splits off one amino acid at a time from the end of the polypeptide chain that bears the carboxyl group, and other brush border enzymes such as **aminopeptidase** and **dipeptidase,** which liberate the final amino acid products (Figure 24.29). Aminopeptidase works to digest a protein, one amino acid at a time, by working from the amine end. Both carboxypeptidase and aminopeptidase are independently capable of completely dismantling a protein, but the teamwork between these enzymes and between trypsin and chymotrypsin, which attack the more internal parts of the protein, speeds up the process tremendously.

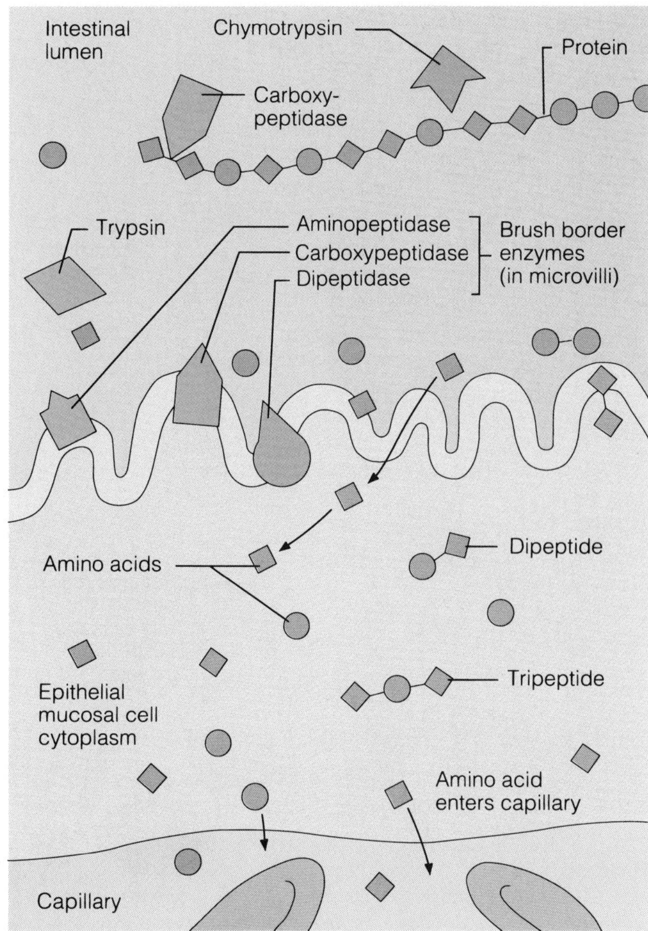

Figure 24.29 Protein digestion and absorption in the small intestine. Proteins and protein fragments are digested to amino acids by the action of pancreatic proteases (trypsin, chymotrypsin, carboxypeptidase, and aminopeptidases) and the tripeptidases and dipeptidases within the intestinal mucosal cells. The amino acids are then absorbed by active transport mechanisms into the capillary blood of the villi.

3. Lipids. The small intestine is the sole site of lipid digestion because the pancreas is essentially the only source of fat-digesting enzymes, or **lipases.** Neutral fats (triglycerides) are the most abundant fats in the diet. They enter the small intestine clumped together as large fat globules, but are quickly coated by bile salts on entry into the small intestine (Figure 24.30a). Bile salts have both nonpolar and polar regions; their nonpolar parts cling to the fat molecules, and their polar (ionized) parts allow them to repel each other and to interact with water. As a result, fatty droplets are pulled off the large fat globules, and an emulsion—an aqueous suspension of fatty droplets, each about 1 μm in diameter—is formed. It is important to understand that the process of emulsification does not break chemical bonds; it just reduces the attraction between fat molecules so that they become associated in minute amounts with the bile salts and are more widely dispersed. Since the water-soluble lipases can work only on the (aqueous) surface of fat droplets, lipids would be incompletely digested (in the time food remains in the small intestine) without the help of bile.

The pancreatic lipases catalyze the breakdown of fats by cleaving off two of the fatty acid chains, thus yielding free fatty acids and monoglycerides (glycerol with one fatty acid chain attached). Fat-soluble vitamins that ride along with the fats require no digestion.

4. Nucleic acids. Both DNA and RNA, present in small amounts in foods, are hydrolyzed to their nucleotide monomers by **pancreatic nucleases** present in pancreatic juice. The nucleotides are then broken apart by intestinal brush border enzymes, which release their free bases, pentose sugars, and phosphate ions.

A summary of digestive enzymes relative to source, substrate, and end products appears in Figure 24.4.

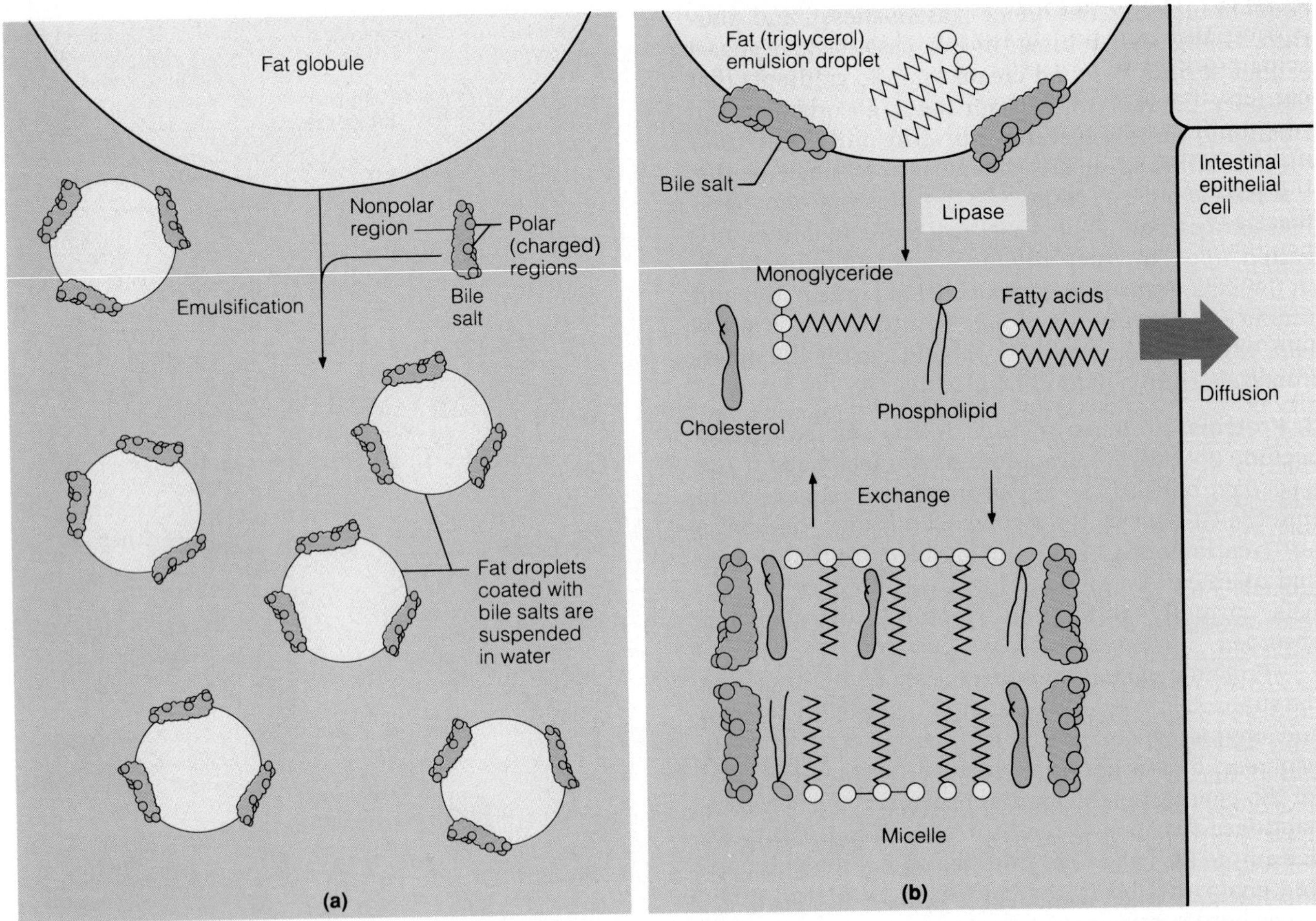

Figure 24.30 Role of bile salts in fat emulsification and absorption. (**a**) As large aggregates of fats enter the small intestine, bile salts cling, via their nonpolar parts, to the fat molecules. Their polar parts, facing the aqueous phase, interact with water and repel each other, causing the fatty globule to be physically broken up into smaller fat droplets. (**b**) Products of fat digestion (monoglycerides and fatty acids), cholesterol, and phospholipids combine with bile salts, forming water-soluble micelles that maintain an equilibrium between the fatty substances in the micelles and those in the free state available for diffusion into the intestinal cell.

Absorption

Up to 10 L of food, drink, and GI secretions enter the alimentary canal daily, but only 0.5–1 L reaches the large intestine. Virtually all of the foodstuffs, 80% of the electrolytes, and most of the water are absorbed in the small intestine. Although absorption occurs all along the length of the small intestine, most of it is completed by the time chyme reaches the ileum. At the end of the ileum, all that remains is some water, indigestible food materials (largely plant fibers such as cellulose), and millions of bacteria. This debris is passed on to the large intestine.

Most nutrients are absorbed through the mucosa of the intestinal villi by the process of *active transport*, driven by metabolic energy (ATP). They then enter the capillary blood in the villus to be transported in the hepatic portal vein to the liver. The exception is some of the lipids, which are absorbed passively by diffusion and then enter the lacteal in the villus to be carried to the blood via lymphatic fluid. The absorption of each nutrient class is described next.

Carbohydrates. The monosaccharides glucose and galactose, liberated by the breakdown of starch and disaccharides, are transported across the villus epithelium by common protein carriers and then move passively into the capillary blood. The carriers are located very close to the disaccharase enzymes on the microvilli and combine with the monosaccharides as soon as the disaccharides are broken down. The transport process for these sugars is coupled to the active transport of sodium ions; Na^+ and the sugar are ferried across the membrane at the same time (cotransport). The precise mechanism of fructose absorption is unclear, but it is thought to move passively by *facilitated diffusion*.

Proteins. There appear to be several specific types of carriers that transport the different classes of amino acids resulting from protein digestion. Some of these carriers, like those for glucose and galactose, are coupled to the active transport of sodium. Short chains of two or three amino acids (dipeptides and tripeptides, respectively) are also actively absorbed, but are digested to their amino acids within the epithelial cells before entering the capillary blood.

In rare cases, intact proteins are taken up by endocytosis and released on the reverse side of the epithelial cell by exocytosis. This process is most common in newborn infants, reflecting the immaturity of the intestinal mucosa, and accounts for many early food allergies. The immune system recognizes the intact proteins as antigenic and mounts an attack. These food allergies usually disappear as the mucosa matures. ■

Lipids. Just as bile salts accelerate lipid digestion, they are essential for absorption of the end products of fat digestion. Since cholesterol and the products of fat digestion—the monoglycerides and free fatty acids—are water-insoluble, they quickly become associated with bile salts to form micelles. **Micelles** (mī-selz) are collections of fatty elements (see Figure 24.30b) clustered together with bile salts in such a way that the polar ends of each molecule face the water and the nonpolar portions form the micelle core. Although micelles are similar to emulsion droplets, they are much smaller and easily diffuse between microvilli to come into very close contact with the mucosal cell surface. It is in these micelle "vehicles" that the fatty substances to be absorbed contact the microvilli of the epithelium (Figure 24.31). The fatty substances and cholesterol then leave the micelles, and because of their high lipid solubility, move across the lipid phase of the plasma membrane by simple diffusion. The micelles provide a source of solubilized fat breakdown products that can be released and absorbed as fat digestion continues. Generally, fat absorption is completed in the ileum; but in the absence of bile (and micelles), as might occur when a gallstone blocks the cystic duct, it occurs so slowly that most of the fat passes into the large intestine and is lost in the feces.

Once inside the epithelial cells, the free fatty acids and monoglycerides are resynthesized into triglycerides. The triglycerides are then combined with small amounts of phospholipids, cholesterol, free fatty acids, and some protein to form tiny fatty droplets called **chylomicrons,** which are then processed by the Golgi apparatus for extrusion from the cell. This series of events is quite different from the absorption of amino acids and simple sugars, which pass through the epithelial cells unchanged.

While a few free fatty acids enter the capillary blood, the chylomicrons are too large to pass through the basement membranes of the blood capillaries and instead enter the more permeable lacteals. Thus, most fat enters the lymphatic stream and is eventually emptied into the venous blood in the neck region via the thoracic duct, which drains the digestive viscera. While in the bloodstream, the triglycerides of the chylomicrons are hydrolyzed to free fatty acids and glycerol by *lipoprotein lipase,* an enzyme associated with the capillary endothelium. The fatty acids and glycerol can then pass through the capillary walls to be used by tissue cells for energy or stored as fats in adipose tissue. The residual chylomicron material is combined with proteins by the liver cells, and these lipoproteins are used as vehicles to transport cholesterol in the blood.

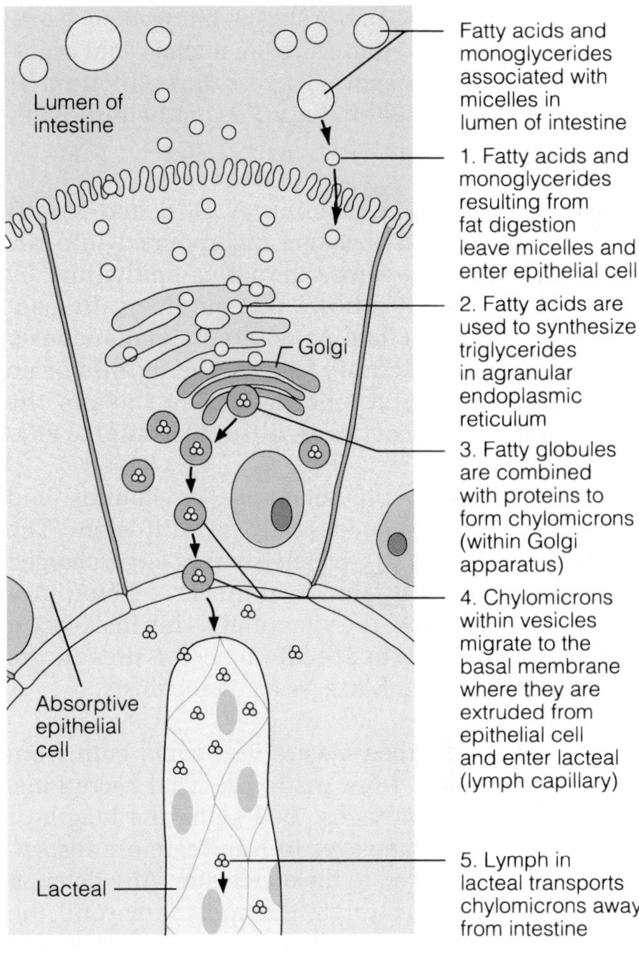
Lumen of intestine
Fatty acids and monoglycerides associated with micelles in lumen of intestine
1. Fatty acids and monoglycerides resulting from fat digestion leave micelles and enter epithelial cell
Golgi
2. Fatty acids are used to synthesize triglycerides in agranular endoplasmic reticulum
3. Fatty globules are combined with proteins to form chylomicrons (within Golgi apparatus)
4. Chylomicrons within vesicles migrate to the basal membrane where they are extruded from epithelial cell and enter lacteal (lymph capillary)
Absorptive epithelial cell
Lacteal
5. Lymph in lacteal transports chylomicrons away from intestine

Figure 24.31 Fatty acid absorption. Digestion products of fat breakdown include glycerol, fatty acids, and monoglycerides. The monoglycerides, free fatty acids, phospholipids, and cholesterol associate with bile salt micelles, which serve to "ferry" them to the intestinal mucosa. They then dissociate and enter the mucosal cells by diffusion. Within the mucosal epithelial cells, they are recombined to lipids and packaged with other lipoid substances (phospholipids and cholesterol) and protein to form chylomicrons. The chylomicrons are extruded from the epithelial cells and enter the lacteal for distribution in the lymph. Free fatty acids and monoglycerides enter the capillary bed (not illustrated).

Nucleic Acids. The pentose sugars, nitrogenous bases, and phosphate ions resulting from nucleic acid digestion are actively transported across the epithelium by special carriers present in the villus epithelium. They then enter the blood.

Vitamins. Fat-soluble vitamins (A, D, E, and K) dissolve in dietary fats, become incorporated into the micelles, and move across the villus epithelium by passive diffusion. It follows that gulping pills containing fat-soluble vitamins without simultaneously eating some fat-containing food results in little or no vitamin absorption. A bile deficiency impairs the absorption, not only of fats, but also of fat-soluble vitamins as well.

Most water-soluble vitamins (B vitamins and vitamin C) are absorbed easily by diffusion. The exception is vitamin B_{12}, which is a very large, charged molecule. Vitamin B_{12} binds to intrinsic factor, produced in the stomach; the vitamin B_{12}–intrinsic factor complex then binds to specific mucosal sites in the terminal ileum, which trigger its endocytosis.

Electrolytes. Absorbed electrolytes come both from ingested foods and from gastrointestinal secretions. Most ions are actively absorbed along the length of the small intestine; however, iron and calcium absorption are largely limited to the duodenum. Any detailed consideration of electrolyte transport is beyond the scope of this text, and only a few points about that process will be made here.

As noted, the absorption of sodium ions in the small intestine is coupled to the active absorption of glucose and amino acids. For the most part, anions passively follow the electrical potential established by sodium transport. However, chloride ions are also transported actively, and in the small intestine terminus, HCO_3^- is actively secreted into the lumen in exchange for Cl^-.

Potassium ions move across the intestinal mucosa by simple diffusion in response to osmotic gradients. As water is absorbed from the lumen, the resulting rise in potassium levels in chyme produces a concentration gradient for its absorption. Thus, anything that interferes with water absorption (such as diarrhea) not only reduces potassium absorption but also "pulls" K^+ from the interstitial space into the intestinal lumen.

For most substances, particularly the organic nutrients, the amount reaching the intestine is the amount absorbed, regardless of the nutritional state of the body. However, the story for iron and calcium is quite different: Their absorption is intimately related to the body's need for them at the time.

Ionic iron is actively transported into the mucosal cells, where it binds to the protein *ferritin* (fayr′-ih-tin). This phenomenon is called the mucosal iron barrier. The iron-ferritin complex then serves as an intracellular storehouse for iron. When body reserves of iron are adequate, very little is allowed to pass into the portal blood, and in such cases, most of the stored iron is lost as the epithelial cells later slough off. On the other hand, when iron reserves are depleted (as occurs during acute or chronic hemorrhage), iron uptake from the intestine and its release to the blood is accelerated. In the blood, iron binds to transferrin, a plasma protein that acts as its transport vehicle in the circulation. The absorption of iron is similar to that of many other trace minerals; that is, it involves cell storage proteins and plasma transport proteins, and control of absorption is the major means of maintaining the homeostatic levels of the mineral in the body.

Calcium absorption is closely related to blood levels of ionic calcium. It is locally regulated by the active form of vitamin D, which acts as a cofactor to facilitate calcium absorption. Decreased blood levels of ionic calcium prompt parathyroid hormone (PTH) release from the parathyroid glands. Besides facilitating the release of calcium ions from bone matrix and enhancing the reabsorption of calcium by the kidneys, PTH enhances levels of active vitamin D, which in turn accelerates calcium ion absorption in the small intestine.

Water. Water is the most abundant substance in chyme, and 95% of it is absorbed in the small intestine by osmosis. The normal rate of water absorption is about 200–400 ml per hour. Water moves freely in both directions across the intestinal mucosa, but *net osmosis* occurs whenever a concentration gradient is established by the active transport of solutes into the mucosal cells. Thus, water uptake is effectively coupled to solute uptake. Net water absorption, in turn, affects the rate of absorption of substances that normally pass by diffusion, and as water moves into the mucosal cells, these substances follow along their concentration gradients.

Digestive Processes Occurring in the Large Intestine

What is finally delivered to the large intestine contains few nutrients, but the residue still has 12–24 hours more to spend in the GI tract. With the exception of a small amount of digestion of the residue by the enteric bacteria, no further food breakdown occurs in the large intestine.

The large intestine is primarily concerned with propulsion, absorption (chiefly of sodium ions and water), and defecation. While it is undeniably essential for our comfort, it is not essential for life. If the colon is removed, as is often necessitated by colon cancer, the terminal ileum is brought out to the abdominal wall in a procedure called an *ileostomy* (il″-ē-os′-tuh-mē), and food residues are eliminated from there into a sac attached to the abdominal wall.

Propulsion

As described earlier, the passage of chyme from the ileum into the cecum is controlled by neural and hormonal mechanisms activated when a meal is eaten; at other times, the ileocecal valve remains tonically contracted. Once food enters the large intestine, three types of propulsive movements occur: peristalsis, haustral churning, and mass movements. These movements are intermingled and, with the exception of the mass movements, are barely perceptible.

Colon peristalsis is very sluggish and most likely contributes very little to propulsion. **Haustral** (hos′-trul) **churning** reflects local activation of the smooth muscle within the walls of the individual haustra. As a haustrum becomes filled with food, the distention activates its muscle to contract, which propels the luminal contents into the next haustrum. These movements also mix the residue, which aids in the absorption of water.

Mass movements (mass peristalsis) are long, slow-moving, but powerful contractile waves that move over large areas of the colon three or four times daily and force the contents toward the rectum. Typically, they occur during or just after eating, which indicates that the presence of food in the stomach and small intestine activates propulsive reflexes (*gastrocolic* and *duodenocolic* reflexes, respectively). Bulk, or fiber, in the diet increases the strength of colon contractions and softens the stool, allowing the colon to act like a well-oiled machine.

When the diet lacks bulk and the volume of residues in the colon is small, the colon narrows, and contraction of its circular muscles becomes more powerful, increasing the pressure on its walls. This promotes the formation of *diverticula* (dī″-ver-tik′-yoo-luh), small herniations of the mucosa through the colon walls, a condition called *diverticulosis*. *Diverticulitis*, a condition in which the diverticula become inflamed, can be life threatening if they rupture. ■

Absorption

The colon itself produces no digestive enzymes and has no chemical digestive function. However, the bacteria that live within its lumen do metabolize any remaining carbohydrates and amino acids, releasing hydrogen, carbon dioxide, methane, and hydrogen sulfide gases, which contribute to the odor of feces. The bacteria also make some vitamins, mostly vitamin K and some B vitamins.

Absorption by the large intestine is limited to the absorption of these vitamins, some electrolytes (largely Na^+ and Cl^-), and most of the remaining water, which amounts to 300–400 ml daily. The semisolid product delivered to the rectum is **feces** (fē′-sēz), or the stool, and contains undigested food residues, mucus, sloughed-off epithelial cells, millions of bacteria, and just enough water to allow its smooth passage.

Defecation

The rectum is generally empty, but when feces is forced into it by mass movements, stretching of the rectal wall initiates the **defecation reflex** (Figure 24.32). This is a spinal cord–mediated (sacral region) reflex that causes the walls of the sigmoid colon and the rectum to contract and the anal sphincters to relax. As feces is forced into the anal canal, messages reach the brain allowing us to decide whether the external (voluntary) sphincter should remain open or be constricted to stop the passage of feces temporarily. If the decision is to delay defecation, the reflex contractions end within a few seconds, and the rectal walls relax. With the next mass movement, the defecation reflex is initiated again.

We ordinarily aid defecation voluntarily by closing the glottis (to keep air in the lungs) and contracting

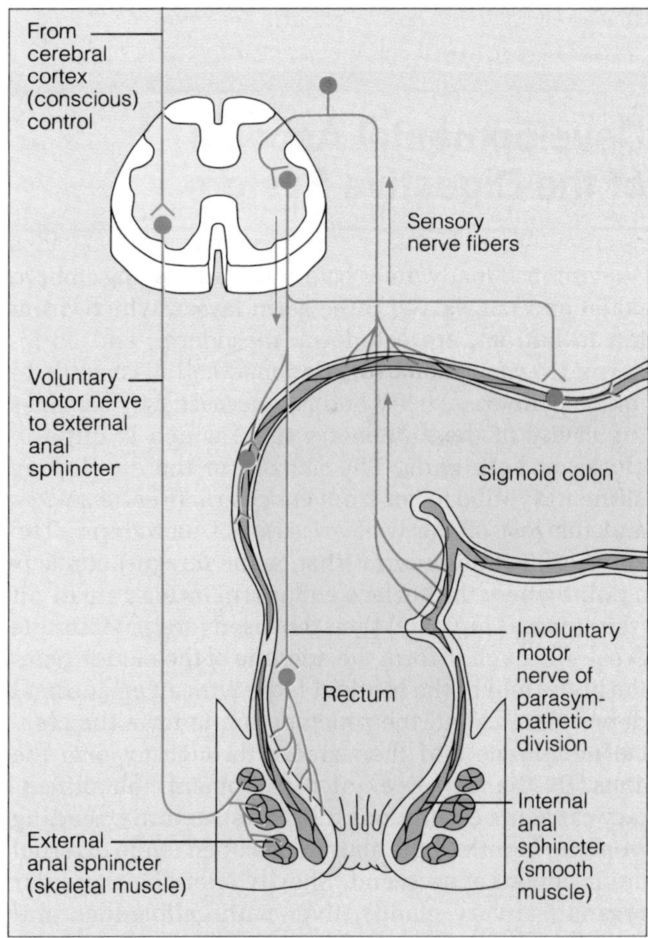

Figure 24.32 Defecation reflex. Distention, or stretch, of the rectal walls caused by movement of food residues into the rectum triggers the depolarization of afferent fibers, which synapse with spinal cord neurons. Parasympathetic efferent fibers, in turn, trigger contraction of the rectal walls and defecation. Defecation may be delayed temporarily by conscious controls descending from the cerebral cortex, which allow voluntary constriction of the external sphincter.

the diaphragm and abdominal wall muscles (to increase the intra-abdominal pressure). This breath-holding maneuver, called the *Valsalva* (val-sal′-vuh) *maneuver*, effectively forces the feces distally. Involuntary or automatic defecation (incontinence of feces) normally occurs in infants, because they have not yet acquired voluntary control of their external anal sphincter. It also occurs in those with spinal cord transections.

Watery stools, or *diarrhea*, result from any condition that rushes food residue through the large intestine before that organ has had sufficient time to absorb the remaining water (as in irritation of the colon by bacteria). Prolonged diarrhea may result in dehydration and electrolyte imbalance. Conversely, when food remains in the colon for extended periods, too much water is absorbed, and the stool becomes hard and difficult to pass. This condition, called *constipation*, may ensue from lack of fiber in the diet, improper bowel habits (failing to heed the "call"), lack of exercise, emotional upset, or laxative abuse. ■

Developmental Aspects of the Digestive System

As described many times before, the very young embryo is flat and consists of three germ layers, which, from top to bottom, are ectoderm, mesoderm, and endoderm. However, soon this flattened cell mass folds to form a cylindrical body, and its internal cavity becomes the cavity of the alimentary tube, which is initially closed at both ends. The mucosa of the developing alimentary tube forms from endoderm (Figure 24.33), and the rest of the wall arises from mesoderm. The anteriormost endoderm (that of the foregut) contacts a point where the surface endoderm forms a small pit (the *stomodeum*), and then the fused (oral) membrane breaks through to form the opening of the mouth. Similarly, the end of the hindgut fuses with an ectodermal depression, called the *proctodeum*, to form the cloacal membrane and then breaks through to form the anus. By the fifth week of development, the alimentary canal is a continuous tubelike structure extending from the mouth to the anus and is open to the external environment at each end. Shortly after, the glandular organs (salivary glands, liver with gallbladder, and pancreas) bud out from the mucosa at various points along its length (Figure 24.33b). These glands retain their connections, which become ducts leading into the digestive tract.

The digestive system is susceptible to many congenital defects that interfere with feeding. The most common congenital defects are *cleft palate* and *cleft lip* (see p. 746), which often occur together. Of the two, cleft palate is much more serious because the child is unable to suck properly. Another relatively common defect is *tracheoesophageal fistula*. In this condition, there is an opening between the esophagus and the trachea, and the esophagus often ends in a blind sac and lacks a connection to the stomach. The baby chokes, drools, and becomes cyanotic during feedings because food enters the respiratory passageways. These defects are usually corrected surgically.

Cystic fibrosis primarily affects the lungs, but it also significantly impairs the activity of the pancreas. In this genetic disease, the mucous glands produce huge amounts of mucus, which blocks the ducts or passageways of involved organs. Blockage of the pancreatic duct prevents pancreatic juice from reaching the small intestine. As a result, most fats and fat-soluble vitamins are not digested or absorbed, and the stools are bulky and fat-laden. Although cystic fibrosis is a severe and ultimately fatal disease, the pancreatic problems can be handled by administering pancreatic enzymes with meals.

During fetal life, the developing infant receives all of its nutrients through the placenta; obtaining and processing nutrients is no problem if the mother is adequately nourished. However, feeding is a newborn baby's most important activity, and several reflexes present in the newborn serve this purpose. The rooting reflex helps the infant find the nipple, and the sucking reflex helps the baby hold onto the nipple and swallow. The stomach of a newborn infant is very small, so feeding must be frequent (every 3–4 hours). Peristalsis is inefficient, and vomiting of the feeding is not at all unusual. As the teeth break through the gums, the infant progresses to more and more solid foods and is usually eating an adult diet by the age of two years. Appetite decreases in the school-age child and then increases again during the rapid growth of adolescence.

All through childhood and into adulthood, the digestive system operates with relatively few problems unless there are abnormal interferences, such as contaminated food or extremely spicy or irritating foods, which may cause inflammation of the GI tract, called *gastroenteritis* (gas″-trō-en″-ter-ī′-tis). Appendicitis is particularly common in teenagers and declines in frequency with age because the opening into the appendix narrows. Ulcers and gallbladder problems —inflammation, or *cholecystitis* (kō-lē-sis-tī′-tis), and gallstones—are problems of middle age. Ulcers seem to occur more frequently in those who are constantly rushing and under pressure.

During old age, gastrointestinal tract activity declines. Fewer digestive juices are produced, absorption becomes less efficient, and peristalsis slows. The result is less frequent bowel movements and, often, constipation. Taste and smell become less acute, and periodontal disease becomes endemic. Many elderly people live alone or on a reduced income. These fac-

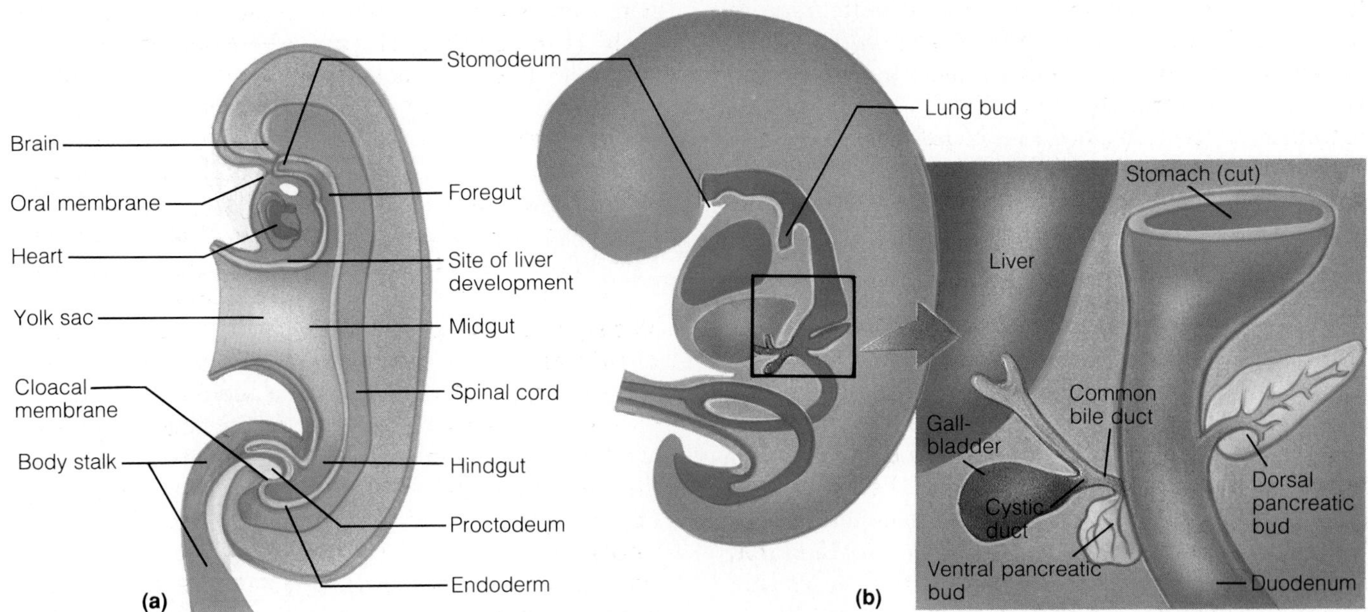

Figure 24.33 Embryonic development of the digestive system. (**a**) Three-week-old embryo. The endoderm has folded, and the foregut and hindgut have formed. (The midgut is still open and continuous with the yolk sac.) The anterior and posterior points of ectodermal-endodermal fusion (the oral and cloacal membranes, respectively) will shortly break through to form the mouth and anal openings. (**b**) By five weeks of development, the accessory organs are budding out from the endodermal layer, as shown in the enlargement.

tors, along with increasing physical disability, tend to make eating less appealing, and many of our elderly citizens are inadequately nourished.

Diverticulosis and cancer of the GI tract are fairly common problems of the aged. Cancer of the stomach and colon rarely have early signs, and the disease often progresses to an inoperable stage (that is, it has metastasized to distant sites) before a person seeks medical attention. However, when detected early, both diseases are treatable. Perhaps the best advice that can be given is to have regular dental and medical checkups. Most oral cancers are detected during routine dental examinations, 50% of all rectal cancers can be palpated digitally, and nearly 80% of colon cancers can be seen and removed during a colonoscopy. It has been suggested that diets high in plant fiber may decrease the incidence of colon cancer, but this has not been indisputably proved.

* * *

As summarized in Figure 24.34 on p. 800, the digestive system operates to keep the blood well supplied with the nutrients needed by all body tissues and organ systems to fuel their energy needs and to synthesize new proteins for growth and maintenance of health. Now we are ready to examine how these nutrients are utilized by body cells, the topic of Chapter 25.

Related Clinical Terms

Ascites (ah-sī′-tēz) (*asci* = bag, bladder) Abnormal accumulation of fluid within the peritoneal cavity; if excessive, causes visible bloating of the abdomen; may be caused by liver cirrhosis, heart disease, and kidney disease.

Colitis (kō-lī′-tis) (*itis* = inflammation) Inflammation of the large intestine, or colon.

Dysphagia (dis-fā′-juh) (*dys* = difficult, abnormal; *phag* = eat) Difficult swallowing; usually due to obstruction or physical trauma to the esophagus.

Enteritis (*enteron* = intestine) Inflammation of the intestine, especially the small intestine.

Gastrectomy (gas-trek′-tuh-mē) (*gastro* = stomach; *tomy* = cut, excise) Removal of all or part of the stomach; common in severe cases of gastric ulcer.

Hemochromatosis (hē″-mō-krō″-muh-tō′-sis) (*hemo* = blood; *chroma* = color; *osis* = condition of) A syndrome caused by a disorder in iron metabolism, owing to excessive/prolonged iron intake or a breakdown of the mucosal iron barrier; excess iron

Digestive system

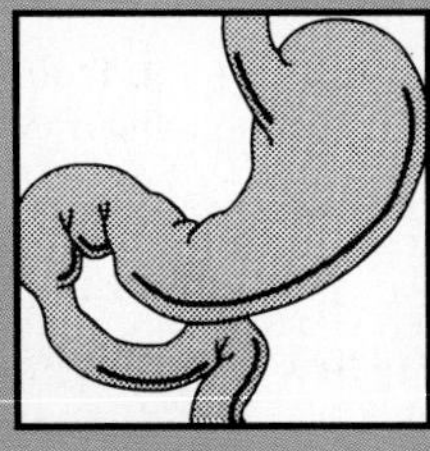

Integumentary system

Synthesizes vitamin D needed for calcium absorption from cholesterol; protects by enclosure

Provides nutrients needed for energy fuel, growth, and repair

Skeletal system

Protects some digestive organs by bone; cavities store some nutrients (e.g., calcium, fats)

Provides nutrients needed for energy fuel, growth, and repair

Muscular system

Activity increases motility of GI tract

Provides nutrients needed for energy fuel, growth, and repair

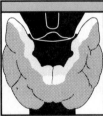

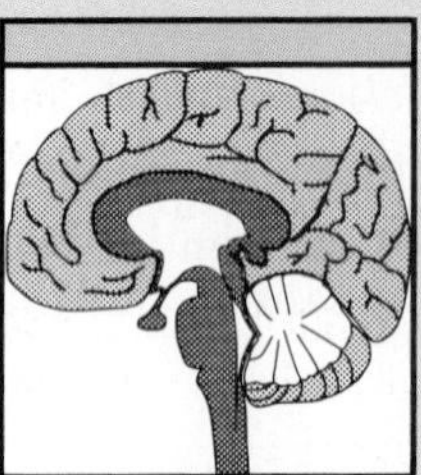

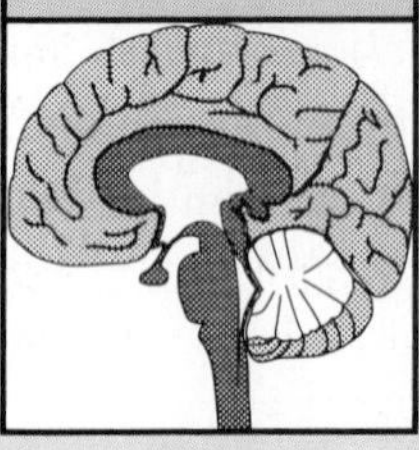

Nervous system

Neural controls of digestive function; in general, parasympathetic fibers accelerate and sympathetic fibers inhibit digestive activity

Provides nutrients needed for normal neural functioning

Endocrine system

Local hormones help regulate digestive function

Liver removes hormones from blood ending their activity; provides nutrients needed for energy fuel, growth, and repair

Cardiovascular system

Transports nutrients absorbed by alimentary canal to all tissues of body

Provides nutrients to heart and blood vessels; absorbs iron needed for hemoglobin synthesis

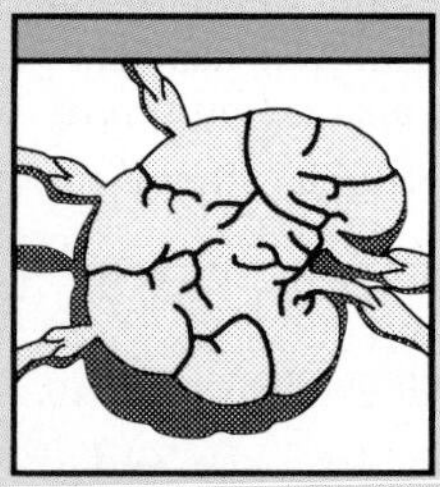

Lymphatic system

Lacteals drain fatty lymph from digestive tract organs and convey it to blood; lymphatic tissue in intestine and mesentery houses protective WBCs

Provides nutrients for normal functioning

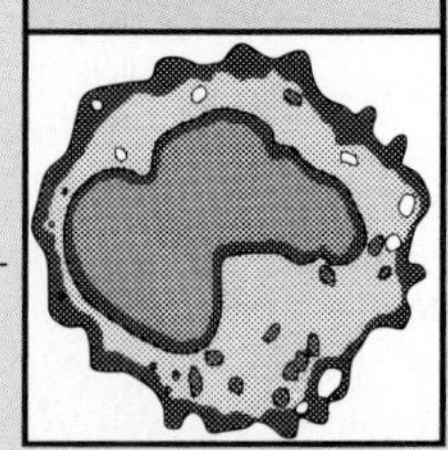

Immune system

Peyer's patches and lymphoid tissue in mesentery houses macrophages and immune cells that protect digestive tract organs against infection

Provides nutrients for normal functioning

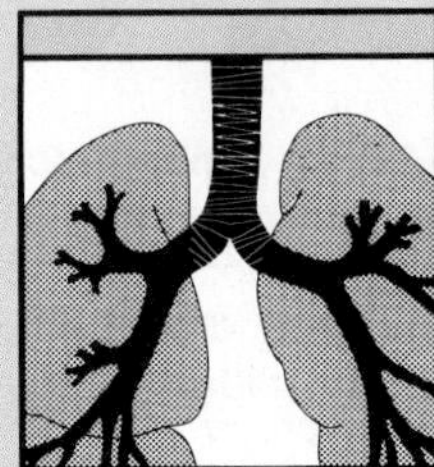

Respiratory system

Provides oxygen and carries away carbon dioxide produced by digestive system organs

Provides nutrients needed for energy metabolism, growth, and repair

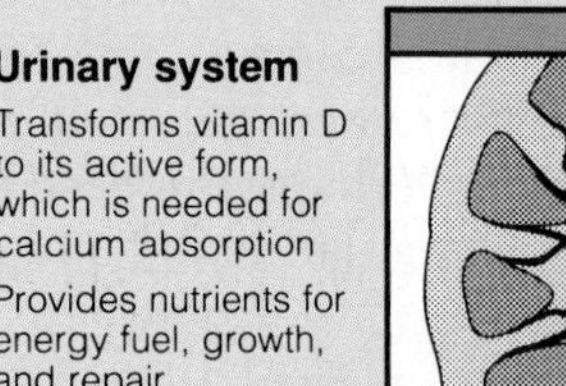

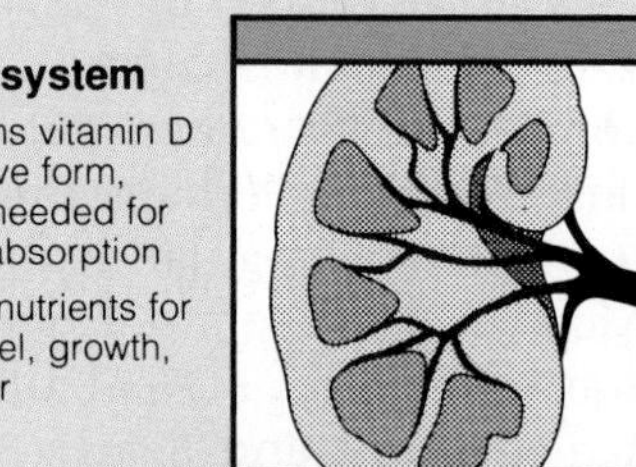

Urinary system

Transforms vitamin D to its active form, which is needed for calcium absorption

Provides nutrients for energy fuel, growth, and repair

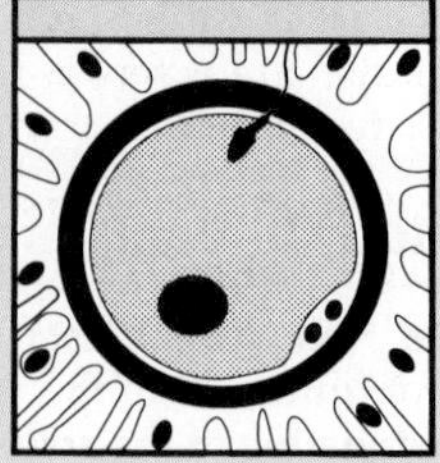

Reproductive system

Provides nutrients for energy fuel, growth, and repair

Figure 24.34 Homeostatic interrelationships between the digestive system and other body systems.

is deposited in the tissues, resulting in increased skin pigmentation and increased incidence of hepatic cancer and liver cirrhosis; also called *bronze diabetes* and *iron storage disease*.

Hemorrhoids (hem'-uh-roydz) Venous inflammation and weakening of the valves of veins in the anal canal (varicose veins), causing them to itch and bleed when hard stools are passed; risk factors include obesity, multiple pregnancies, and frequent constipation; also called *piles*.

Pancreatitis (pan"-krē-uh-tī-tis) A rare but extremely serious inflammation of the pancreas; most often results from activation of pancreatic enzymes in the pancreatic duct, causing the pancreatic tissue and duct to be digested; this painful condition can lead to nutritional deficiencies, because pancreatic enzymes are essential to food digestion in the small intestine.

Pyloric stenosis (pī-lor'-ik steh-nō'-sis) (*stenosis* = narrowing, constriction) Congenital abnormality in which the pyloric sphincter is abnormally constricted; there is usually no problem until the baby begins to take solid food, and then projectile vomiting begins; usually corrected surgically.

Chapter Summary

OVERVIEW OF THE DIGESTIVE SYSTEM (pp. 753–757)

Digestive System Organs (pp. 753–754)

1. The digestive system includes organs of the alimentary canal (mouth, pharynx, esophagus, stomach, small and large intestines) and accessory digestive system organs (teeth, tongue, salivary glands, liver, gallbladder, and pancreas).

Digestive Processes (pp. 754–757)

2. Digestive system activities include six functional processes: ingestion, or food intake; propulsion, or movement of food through the tract; mechanical digestion, or processes that physically mix or break foods down into smaller fragments; chemical digestion, or food breakdown by enzymatic action; absorption, or transport of products of digestion through the intestinal mucosa into the blood; and defecation, or elimination of the undigested residues (feces) from the body.

ANATOMY OF THE DIGESTIVE SYSTEM (pp. 758–777)

Relationships of the Digestive Organs and the Peritoneum (pp. 758–759)

1. The parietal and visceral layers of the peritoneum are continuous with one another via several extensions (mesentery, falciform ligament, lesser and greater omenta), and are separated by a potential space containing serous fluid, which decreases friction during organ activity.

Histology of the Alimentary Tract Wall (pp. 759–761)

2. All organs of the GI tract have the same basic pattern of tissue layers in their walls; i.e., all have a mucosa, a submucosa, a muscularis, and a serosa (or adventitia). Intrinsic nerve plexuses are found within the wall.

The Mouth and Associated Organs (pp. 761–766)

3. Food enters the GI tract via the mouth, which is continuous with the oropharynx posteriorly. The boundaries of the mouth are the lips and cheeks, the palate, and the tongue.

4. The oral mucosa is stratified squamous epithelium, an adaptation seen where abrasion occurs.

5. The tongue is mucosa-covered skeletal muscle. Its intrinsic muscles allow it to change shape; its extrinsic muscles allow it to change position.

6. Saliva is produced by many small buccal glands and three pairs of major salivary glands—parotid, submandibular, and sublingual—that secrete their product into the mouth via ducts.

7. The 20 deciduous teeth begin to be shed at the age of six and are gradually replaced during childhood and adolescence by the 32 permanent teeth.

8. Teeth are classed as incisors, canines, premolars, and molars. Each tooth has an enamel-covered crown and a cementum-covered root. The bulk of the tooth is dentin, which surrounds the central pulp cavity. A periodontal ligament secures the tooth to the bony alveolus.

The Pharynx (p. 766)

9. Food propelled from the mouth passes through the oropharynx and laryngopharynx. The mucosa of the pharynx is stratified squamous epithelium; the skeletal muscles in its wall (constrictor muscles) move food toward the esophagus.

The Esophagus (pp. 767–768)

10. The esophagus extends from the laryngopharynx and joins the stomach at the cardiac orifice, which is surrounded by the gastroesophageal sphincter.

11. The esophageal mucosa is stratified squamous epithelium. Its muscularis is skeletal muscle superiorly and changes to smooth muscle inferiorly. It has an adventitia rather than a serosa.

The Stomach (pp. 768–770)

12. The C-shaped stomach lies in the upper left quadrant of the abdomen. Its major regions are the cardia, fundus, body, and pylorus. When empty, its internal surface exhibits rugae.

13. The stomach mucosa is simple columnar epithelium dotted with gastric pits that lead into gastric glands. Secretory cells in the gastric glands include pepsinogen-producing zymogenic cells, parietal cells, which secrete hydrochloric acid and intrinsic factor; mucous neck cells, which produce mucus; and enteroendocrine cells, which secrete a variety of hormones.

14. The mucosal barrier, which protects the stomach from self-digestion and HCl, reflects the fact that the mucosal cells are connected by tight junctions, secrete a thick, alkaline mucus, and are quickly replaced when damaged.

15. The stomach muscularis contains a third layer of obliquely oriented smooth muscle that allows it to churn and mix food.

The Small Intestine (pp. 770–772)

16. The small intestine extends from the pyloric sphincter to the ileocecal valve. Its three subdivisions are the duodenum, jejunum, and ileum. The common bile duct and pancreatic duct join to form the hepatopancreatic ampulla and empty their secretions into the duodenum through the sphincter of Oddi.

17. The duodenal submucosa contains elaborate mucosal glands (Brunner's glands); that of the ileum contains Peyer's patches (lymph nodules). The duodenum is covered with an adventitia rather than a serosa.

18. Plicae circulares, villi, and microvilli increase the intestinal surface area for digestion and absorption.

The Liver and Gallbladder (pp. 772–775)

19. The liver is a four-lobed organ overlying the stomach. Its digestive role is to produce bile, which it secretes into the hepatic (and common bile) duct.

20. The structural and functional units of the liver are the hepatic lobules. Blood flowing to the liver via the hepatic artery and hepatic portal vein flows into its sinusoids from which Kupffer cells remove debris and the hepatocytes remove nutrients. The hepatocytes store glucose as glycogen, use amino acids to make plasma proteins, and detoxify metabolic wastes and drugs.

21. Bile is made continuously by the hepatocytes. Vagal stimulation, secretin, and bile salts in the blood stimulate bile production.

22. The gallbladder, a muscular sac that lies beneath the right liver lobe, stores and concentrates bile.

The Pancreas (pp. 775–777)

23. The pancreas is retroperitoneal between the spleen and small intestine. Its exocrine product, pancreatic juice, is carried to the duodenum via the pancreatic duct.

The Large Intestine (p. 777)

24. The subdivisions of the large intestine are the cecum (and appendix), colon (ascending, transverse, descending, and sigmoid portions), rectum, and anal canal. It opens to the body exterior at the anus.

25. The mucosa of most of the large intestine is simple columnar epithelium containing abundant goblet cells. The longitudinal muscle in the muscularis is reduced to three bands (teniae coli), which pucker its wall, producing haustra.

PHYSIOLOGY OF DIGESTION (pp. 777–798)

Basic Functional Concepts (pp. 777–779)

1. The digestive system controls the environment within its lumen to ensure optimal conditions for digestion and absorption of foodstuffs.

2. Receptors and hormone-secreting cells within the alimentary canal wall respond to stretch and chemical signals that result in stimulation or inhibition of GI secretory activity or motility. The alimentary canal also has a local (intrinsic) nerve supply.

Digestive Processes Occurring in the Mouth, Pharynx, and Esophagus (pp. 779–782)

3. The mouth and associated accessory organs accomplish food ingestion and mechanical digestion (chewing and mixing), initiate the chemical digestion of starch, and propel food into the pharynx (buccal phase of swallowing).

4. Teeth function to masticate food. Chewing is initiated voluntarily and is then controlled reflexively.

5. Largely water, saliva also contains ions, proteins, metabolic wastes, lysozyme, IgA, salivary amylase, and mucin.

6. Saliva moistens and cleanses the mouth; moistens foods, aiding their compaction; dissolves food chemicals to allow for taste; and begins chemical digestion of starch (salivary amylase).

7. Saliva output is increased by parasympathetic reflexes initiated by activation of chemical and pressure receptors in the mouth and by conditioned reflexes. The sympathetic nervous system depresses salivation.

8. The tongue mixes food with saliva, compacts it into a bolus, and initiates swallowing (the voluntary phase).

9. The pharynx and esophagus are primarily food conduits that conduct food to the stomach by peristalsis. The swallowing center in the medulla and pons controls this phase reflexively.

10. When the peristaltic wave approaches the gastroesophageal sphincter, the sphincter relaxes to allow food entry to the stomach.

Digestive Processes Occurring in the Stomach (pp. 782–789)

11. Gastric secretory activity is controlled by both nervous and hormonal factors.

12. The three phases of gastric secretion are the cephalic phase (reflex phase), gastric phase, and intestinal phase.

13. The presence of chyme in the small intestine during the gastric phase triggers the enterogastric reflex and release of secretin, CCK, and GIP, all of which inhibit gastric secretory activity. Sympathetic activity also inhibits gastric secretion.

14. Mechanical digestion in the stomach is triggered by stomach distention and coupled to food propulsion and stomach emptying. Food movement into the duodenum is controlled by the pylorus and feedback signals from the small intestine. Pacemaker cells in the smooth muscle sheet set the maximal rate of peristaltic waves.

15. Protein digestion is initiated by activated pepsins and requires acidic conditions (provided by HCl). Few substances are absorbed.

Digestive Processes Occurring in the Small Intestine (pp. 789–796)

16. The small intestine is the major digestive and absorptive organ. Intestinal juice is largely water and is relatively enzyme-poor. The major stimuli for its release are stretch and chemical stimuli.

17. Bile contains electrolytes, a variety of fatty substances, bile salts, and bile pigments in an aqueous medium. Bile salts are emulsifying agents; they disperse fats and form water-soluble micelles, which solubilize the products of fat digestion.

18. Cholecystokinin released by the small intestine stimulates the gallbladder to contract and the sphincter of Oddi to relax, allowing bile (and pancreatic juice) to enter the duodenum.

19. Pancreatic juice is a HCO_3^--rich fluid containing enzymes that digest all categories of foods. Secretion of pancreatic juice is controlled by the vagus nerves and intestinal hormones.

20. Mechanical digestion and propulsion in the small intestine mix chyme with digestive juices and bile and force the residues, largely by segmentation, through the ileocecal valve. Pacemaker cells set the rate of segmentation. Ileocecal valve opening is controlled by the gastroileal reflex and gastrin.

21. Chemical digestion is completed in the small intestine by intestinal (brush border) enzymes and, more importantly, by pancreatic enzymes. Alkaline pancreatic juice neutralizes the acidic chyme and provides the proper environment for the operation of the enzymes. Both pancreatic juice (the only source of lipases) and bile are necessary for normal fat breakdown.

22. Virtually all of the foodstuffs and most of the water and electrolytes are absorbed in the small intestine. Except for fat digestion products, fat-soluble vitamins, and most water-soluble vitamins (which are absorbed by diffusion), most nutrients are absorbed by active transport processes.

23. Fat breakdown products are resynthesized to triglycerides in the intestinal mucosal cells and combined with other lipids and protein; the resulting chylomicrons enter the lacteals. Other absorbed substances enter the villus blood capillaries and are transported to the liver via the hepatic portal vein.

Digestive Processes Occurring in the Large Intestine (pp. 796–798)

24. The major functions of the large intestine are absorption of water, electrolytes, and vitamins made by enteric bacteria; and defecation (evacuation of food residues from the body).

25. The defecation reflex is triggered when feces enters the rectum. It involves sacral reflexes leading to the contraction of the rectal walls and is aided by the Valsalva maneuver.

DEVELOPMENTAL ASPECTS OF THE DIGESTIVE SYSTEM (pp. 798–799)

1. The mucosa of the alimentary canal develops from the endoderm, which folds to form a tube. The glandular accessory organs (salivary glands, liver, pancreas, and gallbladder) form from outpocketings of the foregut endoderm. The remaining three tunics of the alimentary canal wall are formed by mesoderm.

2. Important congenital abnormalities of the digestive tract include cleft palate/lip, tracheoesophageal fistula, and cystic fibrosis. All interfere with normal nutrition.

3. Various inflammations plague the digestive system throughout life. Appendicitis is common in adolescents, gastroenteritis and food poisoning can occur at any time (given the proper irritating factors), ulcers and gallbladder problems increase in middle age.

4. The efficiency of all digestive system processes declines in the elderly, and periodontal disease is common. Diverticulosis and GI tract cancers such as stomach and colon cancer appear with increasing frequency in an aging population.

Review Questions

Multiple Choice/Matching

1. Obstruction of the sphincter of Oddi impairs digestion by reducing the availability of (a) bile and HCl, (b) HCl and intestinal juice, (c) pancreatic juice and intestinal juice, (d) pancreatic juice and bile.

2. The action of an enzyme is influenced by (a) its chemical surroundings, (b) its specific substrate, (c) its presence of needed cofactors or coenzymes, (d) all of these.

3. Carbohydrates are acted on by (a) peptidases, trypsin, and chymotrypsin, (b) amylase, maltase, and sucrase, (c) lipases, (d) peptidases, lipases, and galactase.

4. The parasympathetic nervous system influences digestion by (a) relaxing smooth muscle, (b) stimulating peristalsis and secretory activity, (c) constricting sphincters, (d) none of these.

5. The digestive juice product containing enzymes capable of digesting all three major foodstuff categories is (a) pancreatic, (b) gastric, (c) salivary, (d) biliary.

6. The vitamin associated with calcium absorption is (a) A, (b) K, (c) C, (d) D.

7. Someone has eaten a meal of buttered toast, cream, and eggs. Which of the following would you expect to happen? (a) Compared to the period shortly after the meal, gastric motility and secretion of HCl decrease when the food reaches the duodenum; (b) gastric motility increases even while the person is chewing the food (before any swallowing); (c) fat will be emulsified in the duodenum by the action of bile; (d) all of these.

8. The site of production of GIP and cholecystokinin is (a) the stomach, (b) the small intestine, (c) the pancreas, (d) the large intestine.

9. Which of the following is not characteristic of the large intestine? (a) It is divided into ascending, transverse, and descending portions; (b) it contains abundant bacteria, some of which synthesize certain vitamins; (c) is the main absorptive site; (d) it absorbs much of the water and salts remaining in the wastes.

10. The gallbladder (a) produces bile, (b) is attached to the pancreas, (c) stores and concentrates bile, (d) produces secretin.

11. The sphincter between the stomach and duodenum is (a) the pyloric sphincter, (b) the cardiac sphincter, (c) the sphincter of Oddi, (d) the ileocecal sphincter.

In items 12–16, trace the path of a single protein molecule that has been ingested.

12. The protein molecule will be digested by enzymes secreted by (a) the mouth, stomach, and colon, (b) the stomach, liver, and small intestine, (c) the small intestine, mouth, and liver, (d) the pancreas, small intestine, and stomach.

13. The protein molecule must be digested before it can be transported to and utilized by the cells because (a) protein is only useful directly, (b) protein has a low pH, (c) proteins in the circulating blood produce an adverse osmotic pressure, (d) the protein is too large to be readily absorbed.

14. The products of protein digestion enter the bloodstream largely through cells lining (a) the stomach, (b) the small intestine, (c) the large intestine, (d) the bile duct.

15. Before the blood carrying the products of protein digestion reaches the heart, it first passes through capillary networks in (a) the spleen, (b) the lungs, (c) the liver, (d) the brain.

16. Having passed through the regulatory organ selected above, the products of protein digestion are circulated throughout the body. They will enter individual body cells as a result of (a) active transport, (b) diffusion, (c) osmosis, (d) pinocytosis.

Short Answer Essay Questions

17. Make a simple line drawing of the organs of the alimentary tube and label each organ. Then add three labels to your drawing—salivary glands, liver, and pancreas—and use arrows to show where each of these organs empties its secretion into the alimentary tube.

18. Name the layers of the alimentary tube wall. Note the tissue composition and major function of each tunic.

19. What is the mesentery? Mesocolon? Greater omentum?

20. Name the six functional activities of the digestive system.

21. (a) Describe the boundaries of the oral cavity. (b) Why do you suppose its mucosa is stratified squamous epithelium rather than the more typical simple columnar epithelium?

22. (a) What is the normal number of permanent teeth? Of deciduous teeth? (b) What substance covers the tooth crown? Its root? (c) What substance makes up the bulk of a tooth? (d) What and where is pulp?

23. Describe the two phases of swallowing, noting the organs involved and the activities that occur.

24. Describe the role of the following cell types found in the gastric glands: parietal, chief, mucous neck, and enteroendocrine.

25. Describe the regulation of the cephalic, gastric, and intestinal phases of gastric secretion.

26. (a) What is the relationship between the cystic, hepatic, common bile, and pancreatic ducts? (b) What name is given to the point of fusion of the common bile and pancreatic ducts?

27. Explain why fatty stools result from the absence of bile and/or pancreatic juice.

28. Note the function of the Kupffer cells and the hepatocytes of the liver.

29. What are (a) brush border enzymes? (b) chylomicrons?

30. Name one inflammatory condition particularly common to adolescents, two common in middle age, and one common in old age.

31. What are the effects of aging on digestive system activity?

Clinical Application Questions

32. A baby is admitted to the hospital with a history of diarrhea and watery feces occurring over the last three days. The baby has sunken fontanels, indicating extreme dehydration. On examination, it is found that the baby has a bacterium-induced colitis, and antibiotics are prescribed. Because of the baby's loss of intestinal juices, what do you think that his blood would indicate, acidosis or alkalosis? Explain your reasoning.

33. A young attorney complains of a burning pain in the "pit of his stomach," usually beginning about 2 hours after eating and abating after drinking a glass of milk. When asked to indicate the site, he points to his epigastric region. The GI tract is examined by X-ray fluoroscopy. A gastric ulcer is visualized, and a selective vagotomy (cutting the vagal fibers serving the stomach) is recommended. (a) Why is the vagotomy suggested? (b) What are the possible consequences of nontreatment?

CHAPTER 25

Nutrition, Metabolism, and Body Temperature Regulation

Chapter Outline and Student Objectives

Nutrition (pp. 805–818)

1. Define nutrient, essential nutrient, and calorie.
2. List the six major nutrient categories. Note important dietary sources and the principal cellular uses of each.
3. Distinguish between nutritionally complete and incomplete proteins.
4. Define nitrogen balance and note possible causes of positive and negative nitrogen balance.
5. Distinguish between fat- and water-soluble vitamins, and list the vitamins belonging to each group.
6. For each vitamin, list its important sources and functions in the body, and describe consequences of its deficit or excess.
7. List minerals essential for health; note important dietary sources and describe how each is used in the body.

Metabolism (pp. 818–842)

8. Define metabolism. Explain how catabolism and anabolism differ.
9. Define oxidation and reduction and note the importance of these reactions in metabolism. Explain the role of coenzymes used in cellular oxidation reactions.
10. Follow the oxidation of glucose in body cells, noting the important events and products of glycolysis, Krebs cycle, and the electron transport chain.
11. Define glycogenesis, glycogenolysis, and gluconeogenesis.
12. Explain the process by which fatty acids are oxidized for energy.
13. Define ketone bodies, and note the stimulus for their formation.
14. Describe how amino acids are prepared for oxidation for energy.
15. Describe the need for protein synthesis in body cells.
16. Explain the concept of amino acid or carbohydrate/fat pools and describe pathways by which substances in these pools can be interconverted.
17. List important goals and events of the absorptive and postabsorptive states, and explain how these events are regulated.
18. List and describe several metabolic functions of the liver.
19. Differentiate between LDLs and HDLs relative to their structures and major functional roles.

Body Energy Balance (pp. 842–850)

20. Explain what is meant by body energy balance.
21. Describe the current theories of food intake regulation.
22. Define basal metabolic rate and total metabolic rate. Name several factors that influence metabolic rate.
23. Explain how body temperature is regulated, and describe the common mechanisms regulating heat production/retention and heat loss from the body.

Developmental Aspects of Nutrition and Metabolism (pp. 850–851)

24. Describe the effects of protein insufficiency on the fetal nervous system.
25. Describe the cause and consequences of the low metabolic rate typical of the elderly.
26. List ways in which medications commonly used by an aged population may influence their nutrition and health.

Preview of Selected Key Terms

Aerobic (ayr-ō′-bik) (*aero* = air) Oxygen-requiring.

Anaerobic (an″-ayr-ō′-bik) (*an* = without) Not requiring oxygen.

Calorie Abbreviated cal, the amount of energy needed to raise the temperature of 1 gram of water 1° Celsius. Energy exchanges associated with biochemical reactions are usually reported in kilocalories (1 kcal = 1000 cal) or large calories (Cal).

Metabolism Sum total of all the chemical reactions occurring in body cells.

Glycolysis (glī-kol′-ih-sis) (*glyco* = sugar; *lysis* = a loosening) Breakdown of glucose to pyruvic acid—an anaerobic process.

Krebs cycle Aerobic metabolic pathway occurring within mitochondria in which food metabolites are oxidized and CO_2 is liberated.

Oxidative phosphorylation (ok′-sih-dā-tiv fos″-for-ih-lā′-shun) Process of ATP synthesis during which an inorganic phosphate group is attached to ADP; occurs within the mitochondria.

Cytochromes (sī′-tō-krōmz) (*cyto* = cell; *chrom* = color) Brightly colored iron-containing proteins that form part of the inner mitochondrial membrane and function as electron acceptors in oxidative phosphorylation.

Although it seems at times that people can be divided into two camps according to their reactions to food—those that live to eat and those that eat to live—we all recognize the vital importance of food for life. Indeed, it has been said that "you are what you eat," and this is true in that part of the food we eat is converted to our living flesh. In other words, a certain fraction of nutrients is used to build cell structures, replace worn-out parts, and synthesize functional molecules. However, most foods are used as metabolic fuels; that is, they are oxidized and transformed into the chemical energy (ATP) needed by cells to drive their many activities. The energy value of foods is measured in units called *kilocalories* (kcal) or "large calories (C)." One kilocalorie is the amount of heat energy needed to raise the temperature of 1 kilogram of water 1°C (1.8°F) and is the unit scrupulously counted by dieters.

In Chapter 24, we considered how foods are digested and absorbed. But what happens to these foods once they have gained entry into the blood? Why do we need bread, meat, and fresh vegetables? Why does everything we eat seem to turn to fat? This chapter will try to answer these questions as it explains both the nature of nutrients and the "flameless metabolic furnace" that is stoked by absorbed foodstuffs.

Nutrition

A **nutrient** is a substance present in food that is used by the body to promote normal growth, maintenance, and repair. The nutrients needed for health divide neatly into six categories. The major nutrients—carbohydrates, lipids, and proteins—make up the bulk of what we eat. Vitamins and minerals, while equally crucial for health, are required in minute amounts. In the strict sense, water, which accounts for about 60% of the volume of the food we eat, is also considered to be a major nutrient. However, since its importance as a dissolving medium (solvent), as a reactant in many chemical reactions, and in many other aspects of body functioning is described in Chapter 2 (pp. 41–42), only the five nutrient classes listed above will be considered here.

Most foods offer a combination of nutrients. For example, a bowl of cream of mushroom soup contains all of the major nutrients plus some vitamins and minerals. A diet consisting of foods selected from each of the four food groups, that is, grains, fruits and vegetables, meats and fish, and milk products, normally guarantees adequate amounts of all of the needed nutrients.

The ability of cells, especially those of the liver, to convert one type of molecule into another is truly remarkable. These interconversions allow the body to use the wide range of chemicals found in different foods and to adjust to varying food intakes. But there are limits to this ability to conjure up new molecules from old. At least 45 and possibly 50 molecules, called **essential nutrients,** cannot be made by such interconversions and must be provided by the diet. As long as all of the essential nutrients are ingested, the body is capable of synthesizing the hundreds of additional molecules required for life and good health. The use of the word "essential" to describe those chemicals that must be obtained from outside sources is unfortunate and misleading, to say the least, because both the essential and the nonessential nutrients are equally vital (essential) for normal functioning.

In the remainder of this initial section, we will review the dietary sources and recommended requirements of each of the nutrient categories and note their general importance and uses in the body. Figure 25.1 illustrates the fates of the major nutrient categories in a very brief form.

Carbohydrates

Dietary Sources

With the exception of milk sugar (lactose) and small amounts of glycogen in meats, all the carbohydrates we ingest are derived from plants. Sugars (monosaccharides and disaccharides) come from fruits, sugar cane, sugar beets, honey, and milk; the polysaccharide starch is found in grains, legumes, and root vegetables. The polysaccharide cellulose, plentiful in most vegetables, is not digested by humans, but provides roughage, or fiber, which increases the bulk of the stool and facilitates defecation. A typical American ingests approximately one-third of the supply of dietary carbohydrates in starchy foods, another third in sweets, and the remaining third in vegetable and milk products.

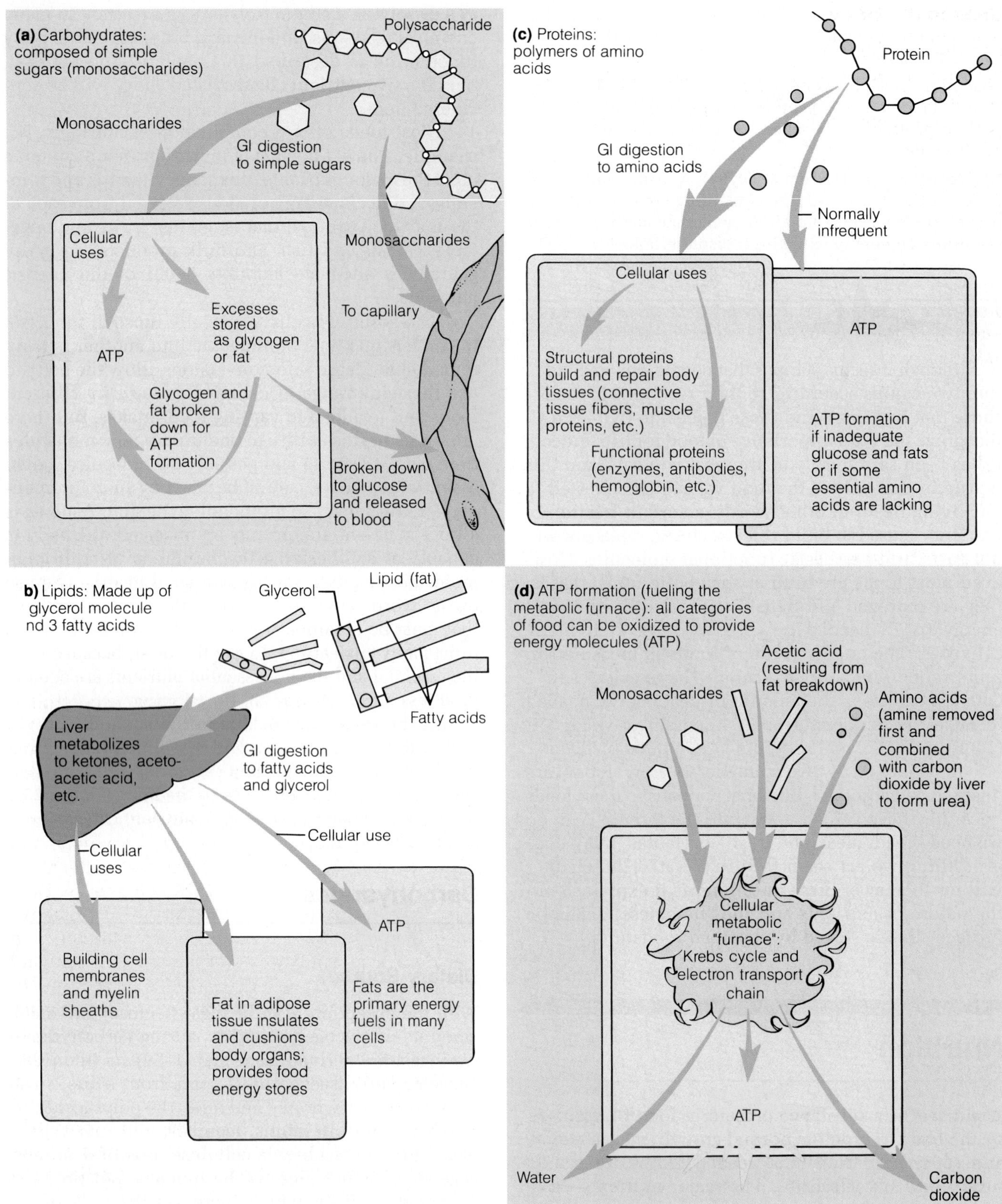

Figure 25.1 Overview of nutrient use by body cells. (**a**) Carbohydrates. (**b**) Lipids. (**c**) Proteins. (**d**) ATP formation.

Uses in the Body

Glucose is the carbohydrate molecule ultimately delivered to and used by body cells. Although fructose and galactose also result from carbohydrate digestion, these monosaccharides are converted to glucose by the liver before they enter the general circulation. Glucose is a major body fuel and is readily used to make ATP. Although many body cells can and do use fats as energy sources, certain cells, notably those of the brain and red blood cells, rely almost entirely on glucose to supply their energy needs. Because even a temporary shortage of blood glucose can severely depress brain function and lead to the death of neurons, the body scrupulously monitors and regulates blood sugar (glucose) levels.

Other uses of monosaccharides are meager, but small amounts of pentose sugars are used to synthesize nucleic acids and a variety of sugars are needed to glycosylate plasma membrane proteins and lipids. When blood glucose is present in excess of the body's needs, it is converted to glycogen or fat and stored.

Dietary Requirements

The low-carbohydrate diet of Eskimos and the high-carbohydrate diet of peoples in the Far East indicate that humans can be healthy even with wide variations in carbohydrate intake. This, no doubt, reflects the body's ability to use fats and amino acids as fuels. The minimum requirement for carbohydrates is not known, but 100 grams per day is assumed to be the minimum amount adequate to maintain adequate blood glucose levels. However, current recommendations are 125 to 175 grams carbohydrate per day with the emphasis on complex carbohydrates. When less than 50 grams per day is consumed, tissue proteins and fats will be broken down for glucose production and energy fuel.

American adults consume 200–300 grams of carbohydrate each day, accounting for about 46% of dietary food energy. Because starchy foods are less expensive than meat and other high-protein foods, carbohydrates constitute an even greater percentage of the diet in low-income groups. Starchy foods and milk offer many valuable nutrients, such as vitamins and minerals, but highly refined carbohydrate foods such as candy and soft drinks provide energy sources (calories) only. The term "empty calories" is commonly used to describe such foods. When people eat refined, sugary foods instead of more nutritionally complex carbohydrates (for instance, replacing milk with soft drinks and whole grain breads with cookies), they may develop nutritional deficiencies as well as becoming obese. Other possible consequences of excessive intake of simple carbohydrates are listed in Table 25.1 on p. 809.

Lipids

Dietary Sources

Although we ingest cholesterol and phospholipids, the most abundant dietary lipids are neutral fats (also called triglycerides [trī-glih′-ser-īdz] or triacylglycerols [trī-ah″-cyl-glih′-ser-olz]). We eat saturated fats in animal products such as meat and dairy foods and in a few plant products such as coconut. Unsaturated fats are present in seeds, nuts, and most vegetable oils. (In saturated fats, all carbon–carbon bonds are single; unsaturated fats have two or more double bonds.) Major sources of cholesterol are egg yolk, meats, particularly organ meats like liver, and milk products.

Fats (but not cholesterol) are digested to fatty acids and monoglycerides and then reconverted to triglycerides for transport in the lymph. Although the liver is very adept at converting one fatty acid to another, it cannot synthesize *linoleic* (lin″-ō-lē′-ik) *acid* (a fatty acid component of *lecithin* [les′-ih-thin]). Thus, linoleic acid is an *essential fatty acid* that must be ingested. Fortunately, linoleic acid is found in most vegetable oils and is the most common polyunsaturated fatty acid in foods.

Uses of Lipids in the Body

The use of triglycerides and cholesterol is controlled mainly by the liver and adipose (fatty) tissue, and is fairly complex. Like sugary carbohydrates, fats have fallen into disfavor, particularly in affluent groups, where food is plentiful and the "battle of the bulge" is a constant concern. But fats *are* essential for several reasons. Phospholipids are an integral component of *all* cellular membranes, and triglycerides are the major energy fuel of hepatocytes and skeletal muscle. Fatty deposits in adipose tissue provide (1) a protective cushion around body organs, such as the kidneys and eyeballs; (2) an insulating layer beneath the skin; and (3) a concentrated source of energy fuel. Additionally, dietary fats help the body absorb fat-soluble vitamins. Lecithin is used to build plasma membranes and to construct myelin sheaths around nerve fibers. When transformed into arachidonic (ar″-ak-ih-don′-ik) acid, linoleic acid is used to form an important class of regulatory molecules called *prostaglandins* (pros″-tah-glan′-dinz), which play a role in smooth muscle contraction, control of blood pressure, and mediation of inflammatory responses.

Unlike the neutral fats, cholesterol is not used for energy. It is important as the structural basis of bile salts, steroid hormones, and other essential functional molecules.

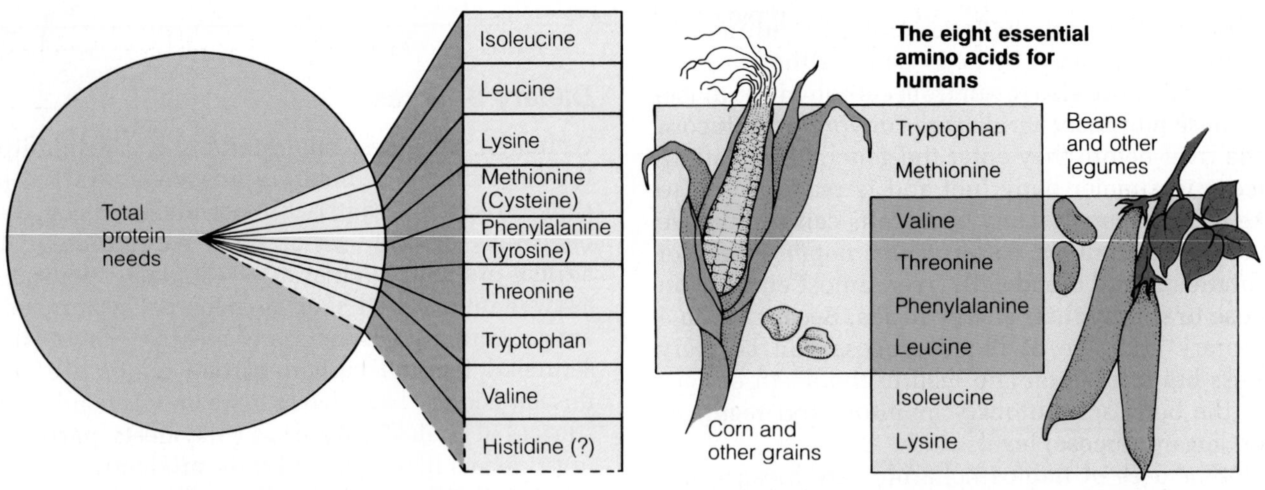

(a) Essential amino acids

(b) Vegetarian diets providing the eight essential amino acids for humans

Figure 25.2 Essential amino acids. Eight amino acids must be available simultaneously and in the correct relative amounts for protein synthesis to occur. **(a)** The relative amounts of the essential amino acids and total proteins needed by adults. Notice that the essential amino acids represent only a small part of the total recommended amino acid (protein) intake. Histidine, known to be an essential amino acid in children, is graphed with dotted lines because its requirement in adults has not been established. The amino acids shown in parentheses are *not* essential, but can substitute in part for methionine and phenylalanine. **(b)** Vegetarian diets must be carefully constructed to provide all essential amino acids. As shown, corn has little isoleucine and lysine. Beans have ample isoleucine and lysine, but little tryptophan and methionine. All essential amino acids can be obtained by consuming a meal of corn and beans.

Dietary Requirements

Fats represent roughly 10% of the calories in Oriental diets and over 40% of the calories in American diets. There are no precise recommendations on amount or type of dietary fats, but the American Heart Association suggests that (1) fats should represent 30% or less of total caloric intake, (2) saturated fats should be limited to 10% or less of total fat intake, and (3) daily cholesterol intake should be no more than 250 mg (the amount in one egg yolk). Because a diet high in saturated fats and cholesterol may contribute to cardiovascular disease, these are wise guidelines. Sources of the various lipid classes and consequences of their deficient or excessive intake are summarized in Table 25.1.

Proteins

Dietary Sources

Animal products contain the highest-quality proteins; that is, those with the greatest amount and best ratios of essential amino acids (Figure 25.2). Eggs, milk, and most meat proteins (Table 25.1) are *complete proteins* that meet all of the body's amino acid requirements for tissue maintenance and growth. Legumes (beans and peas), nuts, and cereals are also protein-rich, but their proteins are nutritionally incomplete because they are low in one or more of the essential amino acids. Cereals are generally low in lysine (lī′-sēn), while legumes, which are good sources of lysine, are low in methionine (meh-thī′-uh-nēn). Leafy green vegetables are well balanced in all essential amino acids except methionine, but contain only small amounts of protein. As you can see, strict vegetarians must carefully plan their diets to obtain all the essential amino acids and prevent protein malnutrition. For instance, cereal grains and legumes when ingested together provide all of the essential amino acids (Figure 25.2b), and some variety of this combination is found in the diets of all cultures (most obviously in the rice and beans seen on nearly every plate in a Mexican restaurant). For nonvegetarians, grains and legumes are useful as partial substitutes for more expensive animal proteins.

Uses in the Body

Proteins are important structural materials of the body, including, for example, keratin of skin, collagen and elastin of connective tissues, muscle proteins, and others. In addition, functional proteins such as enzymes, hemoglobin, some hormones, and plasma proteins regulate an incredible variety of body functions. Even so, whether amino acids are used for synthesis of new proteins or are burned for energy depends on a number of factors:

Table 25.1 Summary of Carbohydrate, Lipid, and Protein Nutrients

Food sources	Recommend daily amounts (RDA) for adults	Problems: Excesses	Problems: Deficits
Carbohydrates			
• *Complex carbohydrates (starches):* bread, cereal, crackers, flour, pasta, nuts, rice, potatoes • *Simple carbohydrates (sugars):* carbonated drinks, candy, fruit, ice cream, pudding, young (immature) vegetables • Foods such as pies, cookies, cakes contain both complex and simple carbohydrates	125 to 175 g; 55%–60% of total caloric intake	Obesity; nutritional deficits; dental caries; gastrointestinal irritation; elevated triglycerides in plasma	Tissue wasting (in extreme deprivation); metabolic acidosis resulting from accelerated fat use for energy
Lipids			
• *Animal sources:* lard, meat, poultry, eggs, milk, milk products • *Plant sources:* chocolate, corn, soy, cottonseed, olive oils; coconut; corn; peanuts	80 to 100 g; 30% or less of total caloric intake	Obesity (and saturated fats as for cholesterol)	Weight loss; fat stores and tissue proteins catabolized to provide metabolic energy; problems controlling heat loss (due to depletion of subcutaneous fat)
• *Essential fatty acids:* corn, cottonseed, soy oils; wheat germ; vegetable shortenings	6000 mg	Not known	Poor growth; skin lesions (eczemalike)
• *Cholesterol:* organ meats (liver, kidneys, brains), egg yolk, fish roe; smaller concentrations in milk products and meat	250 mg or less	Increased levels of plasma cholesterol and LDL lipoproteins; correlated with increased risk of cardiovascular disease	Rare because cholesterol is made by liver
Proteins			
• *Complete proteins:* eggs, milk, milk products, meat (fish, poultry, pork, beef, lamb) • *Incomplete proteins:* legumes (soybeans, lima beans, kidney beans, lentils); nuts and seeds; grains and cereals; vegetables	0.8 g/kg body weight	Obesity; possible aggravation of chronic disease states	Profound weight loss and tissue wasting; growth retardation in children; anemia; edema (due to deficits of plasma proteins) During pregnancy: miscarriage or premature birth

1. The all-or-none law. All amino acids needed to make a particular protein must be present in a cell at the same time and in sufficient amounts. If a single essential amino acid is missing, the protein cannot be made. Because essential amino acids cannot be stored for long-term use, those not used immediately to build proteins are oxidized for energy or converted to carbohydrates or fats.

2. Adequacy of caloric intake. For optimal protein synthesis, the diet must supply sufficient carbohydrate or fat calories for ATP production. If it does not, dietary and tissue proteins will be used for energy.

3. Nitrogen balance of the body. In healthy adults with adequate caloric intake, the rate of protein synthesis equals the rate of protein breakdown and loss. This homeostatic state is reflected in the **nitrogen balance** of the body, which can be determined by chemical analysis based on the fact that the nitrogen content of protein averages 16%. The body is in nitrogen balance when the amount of nitrogen ingested in proteins equals the nitrogen amount excreted in urine and feces.

The body is in *positive nitrogen balance* when the rate of protein synthesis is higher than the rate of its breakdown and loss—the normal situation in growing children and pregnant women. A positive balance also occurs when tissues are being rebuilt or repaired following illness or injury. All cases of positive nitrogen balance indicate that the amount of protein being incorporated into tissue is greater than the amount being broken down and used for energy.

In *negative nitrogen balance,* protein breakdown exceeds the use of protein for building structural or functional molecules. This occurs during physical and emotional stress (for example, infection, injury, burns, depression, fear, or anxiety), when the quality of dietary protein is poor, and during starvation. In such cases, the tissues lose protein faster than it is replaced—an undesirable state of affairs.

4. Hormonal controls. Certain hormones, called *anabolic hormones,* accelerate protein synthesis and growth. The effects of these hormones vary continually throughout life. For example, pituitary growth hormone stimulates tissue growth during childhood and conserves protein in adults; the sex hormones trigger the growth spurt of adolescence. Other hormones, such as the adrenal glucocorticoids, released during stress, enhance protein breakdown and conversion of amino acids to glucose.

Dietary Requirements

Besides supplying the essential amino acids, dietary proteins also furnish the raw materials for forming nonessential amino acids and various nonprotein nitrogen-containing substances. The amount of protein a person requires reflects his or her age, size, metabolic rate, and state of nitrogen balance at that particular time. However, as a rule of thumb, nutritionists recommend a daily protein intake of 0.8 g per kilogram of body weight—approximately 56 g for a 70-kg (154-pound) man and 48 g for a 58-kg (128-pound) woman. This amount of protein (almost 2 ounces) would be supplied by a generous serving of fish and a glass of milk daily. For pregnant and nursing women, an increase of 30 and 20 g/day respectively is recommended to take care of fetal growth or milk production.

Vitamins

The name **vitamin** (*vita* = life giving) is given to a group of potent organic compounds that are needed in minute amounts for growth and good health. Unlike other organic nutrients, vitamins are not used for energy and do not serve as building blocks, but they are crucial in helping the body to use those nutrients that do. Without vitamins, all of the carbohydrates, proteins, and fats we eat would be useless.

Most vitamins function as *coenzymes* (or parts of coenzymes); that is, they act with an enzyme to accomplish a particular type of catalysis. For example, the B vitamins riboflavin and niacin act as coenzymes (FAD and NAD, respectively) in the oxidation of glucose for energy. We will describe the roles of some vitamins in the discussion of metabolism shortly.

Except for vitamins D and K, vitamins are not made in the body and must be taken in via foods or vitamin supplements. Vitamin D is made by ultraviolet irradiation of modified cholesterol molecules in the skin, and vitamin K is synthesized by enteric bacteria. In addition, the body can convert *carotene* (kayr′-uh-tēn), the orange pigment in carrots and other foods, to vitamin A. (Thus, carotene and substances like it are called *provitamins.*)

Vitamins were discovered only in this century, and we still have a good deal to learn about them. Initially, they were named according to their ability to cure certain diseases and given a convenient letter designation that indicated the order of their discovery. For example, ascorbic (as-kor′-bik) acid was also called vitamin C and antiscorbutic (an″-tī-skor-byoo′-tik) acid because it prevents scurvy. This earlier terminology is still used, but chemically descriptive terms are now becoming more commonplace.

Vitamins are classed as either **fat-soluble** or **water-soluble.** The water-soluble vitamins, which include the B-complex vitamins and vitamin C, are absorbed along with water from the gastrointestinal tract. (The exception is vitamin B_{12}, which must become attached to intrinsic factor to be absorbed.) The water-soluble vitamins are not stored in significant amounts in the body and, unless used, are excreted in urine; consequently, few conditions resulting from excessive levels of these vitamins (*hypervitaminosis*) are known. The content (thus the benefits) of the water-soluble vitamins in a food may be reduced during food processing and/or by heating.

Fat-soluble vitamins (vitamins A, D, E, and K) bind to ingested lipids and are absorbed along with their digestion products. Anything that interferes with fat absorption, such as the absence of bile, interferes with the uptake of fat-soluble vitamins as well. Except for vitamin K, fat-soluble vitamins are stored in the body, and pathologies due to fat-soluble vitamin toxicity, particularly vitamin A hypervitaminosis, have been well documented clinically.

Vitamins are found in all major food groups, but no one food contains all required vitamins. Thus, a balanced diet is the safe and sane way to ensure a full vitamin complement. This is particularly important in the light of recent evidence that certain vitamins (A, C, and E) appear to have tangible effects as anticancer agents. Diets rich in broccoli, cabbage, and Brussels sprouts (all good sources of vitamins A and C) appear to reduce cancer risk. However, misconceptions about the ability of vitamins to work wonders abound, such as the idea that huge doses (megadoses) of vitamin C will prevent colds. The notion that taking megadoses of vitamin supplements is the road to eternal youth and glowing health is useless at best and, at worst, may cause serious health problems, particularly in the case of fat-soluble vitamins. Table 25.2 lists the vitamins, their sources, their importance to body function, and the consequences of their deficiency or excess.

Table 25.2 Vitamins

Vitamin	Description/ comments	Sources/ recommended daily amounts (RDA) for adults	Importance in body	Problems: Excesses	Problems: Deficits
Fat-soluble vitamins					
• A (retinol)	Group of compounds including retinol and retinal; 90% is stored in liver, which can supply body needs for a year; stable to heat, acids, alkalis; easily oxidized; rapidly destroyed by exposure to light	Formed from provitamin carotene in intestine, liver, kidneys; carotene found in deep-yellow and deep-green leafy vegetables; vitamin A found in fish liver oils, egg yolk, liver, fortified foods (milk, margarine) RDA: males 5000 IU females 4000 IU	Required for synthesis of photoreceptor pigments of rods and cones, integrity of skin and mucosae, normal tooth and bone development; normal reproductive capabilities; acts with vitamin E to stabilize cell membranes	Toxic when ingested in excess of 50,000 IU daily for months; symptoms: nausea, vomiting, anorexia, headache, hair loss, bone and joint pain, bone fragility, enlargement of liver and spleen	Night blindness; epithelial changes: dry skin and hair, skin sores; increased respiratory, digestive, urogenital infections; drying of conjunctiva; clouding of cornea; most prevalent vitamin deficiency in world
• D (antirachitic factor)	Group of chemically distinct sterols; concentrated in liver and to lesser extent in skin, kidneys, spleen, other tissues; stable to heat, light, acids, alkalis, oxidation	Vitamin D_3 (cholecalciferol) is chief form in body cells; produced in skin by irradiation of 7-dehydrocholesterol by UV light; active form (1,25-dehydroxy-vitamin D_3) produced by chemical modification of vitamin D_3 in liver, then kidneys; major food sources: fish liver oils, egg yolk, fortified milk RDA: 400 IU	Functionally a hormone; increases calcium blood levels by enhancing absorption of calcium; in conjunction with PTH, mobilizes calcium from bones; both mechanisms serve calcium homeostasis of blood (essential for normal neuromuscular function, blood clotting, bone and tooth formation)	1800 IU/day may be toxic to children, massive doses induce toxicity in adults; symptoms: vomiting, diarrhea, weight loss, calcification of soft tissues, renal damage	Faulty mineralization of bones and teeth; rickets in children, osteomalacia in adults; poor muscle tone, restlessness, irritability
• E (antisterility factor)	Group of related compounds called tocopherols, chemically related to sex hormones; stored primarily in muscle and adipose tissues; resistant to heat, light, acids; unstable in presence of oxygen	Vegetable oils, margarine, whole grains, dark-green leafy vegetables RDA: 30 IU	Appears to be an antioxidant; may help prevent oxidation of vitamins A and C in intestine; in tissues, decreases oxidation of unsaturated fatty acids, thus helps maintain integrity of cell membranes	Thrombophlebitis, hypertension; slow wound healing	Extremely rare, precise effects uncertain: possible hemolysis of RBCs, macrocytic anemia; fragile capillaries
• K (coagulation vitamin)	Number of related compounds known as quinones; small amounts stored in liver; heat resistant; destroyed by acids, alkalis, light, oxidizing agents; activity antagonized by certain anticoagulants and antibiotics that interfere with synthetic activity of enteric bacteria	Most synthesized by coliform bacteria in large intestine; food sources: leafy green vegetables, cabbage, cauliflower, pork liver RDA: not known, but obtaining adequate amount is not a problem	Essential for formation of clotting proteins and some other proteins made by liver; as intermediate in electron transport chain, participates in oxidative phosphorylation in all body cells	None known, not stored in appreciable amounts	Easy bruising and bleeding (prolonged clotting time)

(Table continues)

Table 25.2 (continued)

Vitamin	Description/ comments	Sources/ recommended daily amounts (RDA) for adults	Importance in body	Problems: Excesses	Problems: Deficits
Water-soluble vitamins:					
• C (ascorbic acid)	Simple 6-carbon crystalline compound derived from glucose; rapidly destroyed by heat, light, alkalis; about 1500 mg is stored in body, particularly in adrenal gland, retina, intestine, pituitary; when tissues are saturated, excess excreted by kidneys	Fruits and vegetables, particularly citrus, cantaloupe, strawberries, tomatoes, fresh potatoes, leafy green vegetables RDA: 60 mg	Acts in hydroxylation reactions in formation of nearly all connective tissues; in conversion of tryptophan to serotonin (vasoconstrictor); in conversion of cholesterol to bile salts; helps protect vitamins A and E and dietary fats from oxidation; enhances iron absorption and use; required for conversion of folacin (a B vitamin) to its active form	Result of megadoses (10 or more times RDA); enhanced mobilization of bone minerals and blood coagulation; exacerbation of gout, kidney stone formation	Defective formation of intercellular cement; fleeting joint pains, poor tooth and bone growth; poor wound healing, increased susceptibility to infection; extreme deficit causes scurvy (bleeding gums, anemia, cutaneous hemorrhage, degeneration of muscle and cartilage, weight loss)
• B_1 (thiamin)	Rapidly destroyed by heat; very limited amount stored in body; excess eliminated in urine	Lean meats, liver, eggs, whole grains, leafy green vegetables, legumes RDA: 1.5 mg	Part of coenzyme cocarboxylase, which acts in carbohydrate metabolism; required for transformation of pyruvic acid to acetyl CoA, for synthesis of pentose sugars and acetylcholine; for oxidation of alcohol	None known	Beriberi: decreased appetite; gastrointestinal disturbances; peripheral nerve changes indicated by weakness of legs, cramping of calf muscles, numbness of feet; heart enlarges, tachycardia
• B_2 (riboflavin)	Named for its similarity to ribose sugar: has green-yellow fluorescence; quickly decomposed by UV, visible light, alkalis; body stores are carefully guarded; excess eliminated in urine	Widely varying sources such as liver, yeast, egg white, whole grains, meat, poultry, fish, legumes; major source is milk RDA: 1.7 mg	Present in body as coenzymes FAD and FMN (flavin mononucleotide), both of which act as hydrogen acceptors in body; also is component of amino acid oxidases	None known	Dermatitis; cracking of lips at corners (cheilosis); lips and tongue become purple-red and shiny; ocular problems: light sensitivity, blurred vision; one of the most common vitamin deficiencies.
• Niacin (nicotinamide)	Simple organic compounds stable to acids, alkalis, heat, light, oxidation (even boiling does not decrease potency); very limited amount stored in body, day-to-day supply is desirable; excess secreted in urine	Diets that provide adequate protein usually provide adequate niacin because amino acid tryptophan is easily converted to niacin; preformed niacin provided by poultry, meat, fish; less important sources: liver, yeast, peanuts, potatoes, leafy green vegetables RDA: 20 mg	Constituent of NAD and NADP (nicotinamide adenine dinucleotide phosphate), coenzymes involved in glycolysis, oxidative phosphorylation, fat breakdown; inhibits cholesterol synthesis	Result of megadoses; hyperglycemia; vasodilation leading to flushing of skin, tingling sensations; possible liver damage; gout	Pellagra after months of deprivation (rare in U.S.); early signs are vague: listlessness, headache, weight loss, loss of appetite; progresses to soreness and redness of tongue and lips; nausea, vomiting, diarrhea; photosensitive dermatitis: skin becomes rough, cracked, may ulcerate; neu-

Table 25.2 (continued)

Vitamin	Description/ comments	Sources/ recommended daily amounts (RDA) for adults	Importance in body	Problems: Excesses	Problems: Deficits
					rological symptoms also occur: characterized by the 4 Ds: dermatitis, diarrhea, dementia, and death (in final stage)
• B_6 (pyridoxine)	Group of three pyridines occurring in both free and phosphorylated forms in body; stable to heat, acids; destroyed by alkalis, light; body stores very limited	Meat, poultry, fish; less important sources: potatoes, sweet potatoes, tomatoes, spinach RDA: 2 mg	Active form is coenzyme pyridoxal phosphate, which functions in several enzyme systems involved in amino acid metabolism; also required for conversion of tryptophan to niacin, for glycogenolysis, for formation of antibodies and hemoglobin	Depressed deep tendon reflexes, numbness, loss of sensation in extremities	Infants: nervous irritability, convulsions, anemia, vomiting, weakness, abdominal pain; adults: seborrhea lesions around eyes and mouth
• Pantothenic acid	Quite stable, little loss of activity with cooking except in acidic or alkaline solutions; liver, kidney, brain, adrenal, heart tissues contain large amounts	Name derived from Greek *panthos* meaning everywhere; widely distributed in animal foods, whole grains, legumes; liver, yeast, egg yolk, meat especially good sources; some produced by enteric bacteria RDA: 10 mg	Functions in form of coenzyme A in reactions that remove or transfer acetyl group, e.g., formation of acetyl CoA from pyruvic acid, oxidation and synthesis of fatty acids; also involved in synthesis of steroids and heme of hemoglobin	None known	Symtoms vague: loss of appetite, abdominal pain, mental depression, pains in arms and legs, muscle spasms, neuromuscular degeneration (neuropathy of alcoholics is thought to be related to deficits)
• Biotin	Urea derivative containing sulfur; crystalline in its free form; stable to heat, light, acids; in tissues, is usually combined with protein; stored in minute amounts, particularly in liver, kidneys, brain, adrenal glands	Liver, egg yolk, legumes, nuts; some synthesized by bacteria in gastrointestinal tract RDA: not established, probably about 0.3 mg (because biotin content of feces and urine is greater than dietary intake, it is assumed that formation of the vitamin by enteric bacteria provides far more than is needed)	Functions as coenzyme for a number of enzymes that catalyze carboxylation, decarboxylation, deamination reactions; essential for reactions of Kreb's cycle, for formation of purines and nonessential amino acids, for use of amino acids for energy	None known	Scaly skin, muscular pains, pallor, anorexia, nausea, fatigue; elevated blood cholesterol levels

(Table continues)

Table 25.2 (continued)

Vitamin	Description/ comments	Sources/ recommended daily amounts (RDA) for adults	Importance in body	Problems: Excesses	Problems: Deficits
• B_{12} (cyanocobalamin)	Most complex vitamin; contains cobalt; stable to heat; inactivated by light and strongly acidic or basic solutions; intrinsic factor required for transport across intestinal membrane; stored principally in liver; liver stores of 2000–3000 μg sufficient to provide for body needs for 3–5 years	Liver, meat, poultry, fish, dairy foods except butter, eggs; not found in plant foods RDA: 6 μg	Functions as coenzyme in all cells, particularly in gastrointestinal tract, nervous system, and bone marrow; in bone marrow, acts in synthesis of DNA; when absent, erythrocytes do not divide but continue to increase in size; essential for synthesis of methionine and choline	None known	Pernicious anemia, signified by pallor, anorexia, dyspnea, weight loss, neurological disturbances; most cases reflect impaired absorption rather than actual deficit
• Folic acid (folacin)	Pure vitamin is bright-yellow crystalline compound; stable to heat; easily oxidized in acidic solutions and light; stored mainly in liver	Liver, deep-green vegetables, yeast, lean beef, eggs, veal, whole grains; synthesized by enteric bacteria RDA: 0.4 mg	Basis of coenzymes that act in synthesis of methionine and certain other amino acids, choline, DNA; essential for formation of red blood cells	None known	Macrocytic or megaloblastic anemia; gastrointestinal disturbances; diarrhea

Notes

1. Each vitamin has specific functions; one vitamin cannot substitute for another. Many reactions in body require several vitamins, and lack of one can interfere with activity of others.

2. Diet that includes recommended amounts of the four food groups will furnish adequate amounts of all vitamins except vitamin D. Each food group makes a special vitamin contribution. For example, fruits and vegetables are principal sources of vitamin C; dark-green leafy vegetables and deep-yellow vegetables and fruits are primary sources of vitamin A (carotene); milk is principal source of riboflavin; meat, poultry, and fish are outstanding sources of niacin, thiamin, and vitamins B_6 and B_{12}. Whole grains are also important sources of niacin and thiamin.

3. Because vitamin D is present in natural foods in only very small amounts, infants, pregnant and lactating women, and people who have little exposure to sunlight should use vitamin D supplements (or supplemented foods, e.g., fortified milk).

4. Vitamins A and D are toxic in excessive amounts and should be used in supplementary form only when prescribed by a physician. Because most water-soluble vitamins are excreted in urine when ingested in excess, taking supplements is probably a waste of money.

5. Many vitamin deficiencies are secondary to disease (including anorexia, vomiting, diarrhea, or malabsorption diseases) or reflect increased metabolic requirements due to fever or stress factors. Specific vitamin deficiencies require therapy with the vitamins that are lacking.

Minerals

The body requires adequate supplies of seven **minerals** (calcium, phosphorus, potassium, sulfur, sodium, chloride, and magnesium) and trace amounts of about a dozen others (Table 25.3). Minerals make up about 4% of the body by weight, with calcium and phosphorus (as bone salts) accounting for about three-quarters of this amount.

Minerals, like vitamins, are not used for fuel, but work in tandem with other nutrients to ensure a smoothly functioning body. Sometimes minerals are incorporated into structures that must be strong. For example, calcium, phosphorus, and magnesium salts harden the teeth and strengthen the skeleton. However, most minerals are found ionized in body fluids or are bound to organic compounds to form molecules such as phospholipids, hormones, enzymes, and other functional proteins. For example, iron is essential to the oxygen-binding protein of hemoglobin and the electron-accepting part of cytochromes (as described later in the chapter). Sodium and chloride ions are the major electrolytes in blood and other extracellular fluids where they help to maintain (1) normal osmolarity and water balance of body fluids and (2) responsiveness of neurons and muscle cells to stimuli. The amount of a particular mineral present in the body gives little clue to its importance in body function. For example, a few milligrams of iodine can make a critical difference to health.

Table 25.3 Minerals

Mineral	Distribution in body/comments	Sources/ recommended daily allowances (RDA) for adults	Importance in body	Problems: Excesses	Problems: Deficits
Calcium (Ca)	Most stored in salt form in bones; most abundant cation in body; absorbed from intestine in presence of vitamin D; excess excreted in feces; blood levels regulated by PTH and calcitonin	Milk, milk products, leafy green vegetables, egg yolk, shellfish RDA: 800 mg	In salt form, required for hardness of bones, teeth; ionic calcium in blood and cells essential for transmission of nerve impulses, normal heart rhythm, blood clotting; activates certain enzymes; necessary for normal membrane permeability	Depressed neural function; calcium salt deposit in soft tissues; kidney stones	Children: retarded growth; rickets; adults: osteomalacia, osteoporosis; convulsions
Phosphorus (P)	About 80% found in inorganic salts of bones, teeth; remainder in muscle, nervous tissue, blood; absorption aided by vitamin D; about 1/3 dietary intake excreted in feces; metabolic by-products excreted in urine	Diets rich in proteins are usually rich in phosphorus; plentiful in milk, eggs, meat, fish, poultry, legumes, nuts, whole grains RDA: 800 mg	Component of bones and teeth, nucleic acids, proteins, phospholipids, ATP, phosphates (buffers) of body fluids; thus, important for energy storage and transfer, muscle and nerve activity, cell permeability	Not known, but excess in diet may depress absorption of iron and manganese	Rickets, poor growth
Magnesium (Mg)	In all cells, particularly abundant in bones; absorption parallels that of calcium; excreted chiefly in urine	Milk, dairy products, whole-grain cereals, nuts, legumes, leafy green vegetables RDA: 300–350 mg	Constituent of many coenzymes that play a role in conversion of ATP to ADP; required for normal muscle and nerve irritability	Diarrhea	Neuromuscular problems, tremors; seen in alcoholism and severe renal disease
Potassium (K)	Principal intracellular cation, 97% within cells; fixed proportion of K is bound to proteins, and measurements of body K are used to determine lean body mass; K^+ leaves cells during protein catabolism, dehydration, glycogenolysis; most excreted in urine	Widely distributed in foods; normal diet provides 2–6 g/day; avocados, dried apricots, meat, fish, fowl, cereals RDA: not established; diet adequate in calories provides ample amount, i.e., 2500 mg	Helps maintain intracellular osmotic pressure; needed for normal nerve impulse conduction, muscle contraction, glycogenesis, protein synthesis	Usually a complication of renal failure or severe dehydration, but may result from severe alcoholism; paresthesias, muscular weakness, cardiac abnormalities	Rare but may result from severe diarrhea or vomiting; muscular weakness, paralysis, nausea, vomiting, tachycardia, heart failure
Sulfur (S)	Widely distributed: particularly abundant in hair, skin, nails; excreted in urine	Meat, milk, eggs, legumes (all rich in sulfur-containing amino acids) RDA: not established; diet adequate in proteins meets body's needs	Essential constituent of many proteins (insulin), some vitamins (thiamin and biotin); found in mucopolysaccharides present in cartilage, tendons, bone	Not known	Not known

(Table continues)

Table 25.3 (continued)

Mineral	Distribution in body/comments	Sources/ recommended daily allowances (RDA) for adults	Importance in body	Problems: Excesses	Problems: Deficits
Sodium (Na)	Widely distributed: 50% found in extracellular fluid, 40% in bone salts, 10% within cells; absorption is rapid and almost complete; excretion, chiefly in urine, controlled by aldosterone	Table salt (1 tsp = 2000 mg); cured meats (ham, etc.), sauerkraut, cheese; diet supplies substantial excess RDA: not established; probably about 500 mg	Most abundant cation in extracellular fluid: principal electrolyte maintaining osmotic pressure of extracellular fluids and water balance; as part of bicarbonate buffer system, aids in acid–base balance of blood; needed for normal neuromuscular function; part of pump for transport of glucose and other nutrients	Hypertension; edema	Rare but can occur with excessive vomiting, diarrhea, sweating, or poor dietary intake; nausea, abdominal and muscle cramping, convulsions
Chlorine (Cl)	Exists in body almost entirely as chloride ion; principal anion of extracellular fluid; highest concentrations in cerebrospinal fluid and gastric juice; excreted in urine	Table salt (as for sodium); usually ingested in excess RDA: not established; normal diet contains 3–9 g	With sodium, helps maintain osmotic pressure and pH of extracellular fluid; required for HCl formation by stomach glands; activates salivary amylase; aids in transport of CO_2 by blood (chloride shift)	Vomiting	Severe vomiting or diarrhea leads to Cl loss and alkalosis; muscle cramps; apathy
Trace minerals*					
Iron (Fe)	60% to 70% in hemoglobin; remainder in skeletal muscle, liver, spleen, bone marrow bound to ferritin; only 2% to 10% of dietary iron is absorbed because of mucosal barrier; lost from body in urine, perspiration, menstrual flow, shed hair, sloughed skin and mucosal cells	Best sources: meat, liver; also in shellfish, egg yolk, dried fruit, nuts, legumes, molasses RDA: males 10 mg females 18 mg	Part of heme of hemoglobin, which binds bulk of oxygen transported in blood; component of cytochromes, which function in oxidative phosphorylation	Damage to liver (cirrhosis), heart, pancreas	Iron-deficiency anemia; pallor, gastrointestinal problems, flatulence, anorexia, constipation, paresthesias
Iodine (I)	Found in all tissues, but in high concentrations only in thyroid gland; absorption controlled by blood levels of protein-bound iodine; excreted in urine	Cod liver oil, iodized salt, shellfish, vegetables grown in iodine-rich soil RDA: 100–130 mg	Required to form thyroid hormones (T_3 and T_4), which are important n regulating cellular metabolic rate	Depresses synthesis of thyroid hormones	Hypothyroidism: cretinism in infants, myxedema in adults (if less severe, simple goiter)

Table 25.3 (continued)

Mineral	Distribution in body/comments	Sources/ recommended daily allowances (RDA) for adults	Importance in body	Problems: Excesses	Problems: Deficits
Manganese (Mn)	Most concentrated in liver, kidneys, spleen; excreted largely in feces	Nuts, legumes, whole grains, leafy green vegetables, fruit RDA: not established	Acts with enzymes catalyzing synthesis of fatty acids, cholesterol, urea, hemoglobin; needed for normal neural function, lactation, oxidation of carbohydrates, protein hydrolysis	Not known	Not known
Copper (Cu)	Concentrated in liver, heart, brain, spleen; excreted in feces	Liver, shellfish, whole grains, legumes, meat; typical diet provides 2–5 mg daily RDA: 2 mg	Required for synthesis of hemoglobin; essential for manufacture of melanin, myelin, some intermediates of electron transport chain	Rare; abnormal storage results in Wilson's disease	Rare
Zinc (Zn)	Concentrated in liver, kidneys, brain; excreted in feces	Seafood, meat, cereals, legumes, nuts, wheat germ, yeast RDA: 15 mg	Constituent of several enzymes; as component of carbonic anhydrase, important in CO_2 metabolism; required for normal growth, wound healing, taste, smell, sperm production	Difficulty in walking; slurred speech; masklike facial expression; tremors	Loss of taste and smell; growth retardation
Cobalt (Co)	Found in all cells; larger amounts in bone marrow	Liver, lean meat, poultry, fish, milk RDA: not established	A constituent of vitamin B_{12}, which is needed for normal maturation of red blood cells	Goiter, polycythemia; heart disease	See deficit of vitamin B_{12} (Table 25.2)
Fluorine (F)	Component of bones, teeth, other body tissues; excreted in urine	Flouridated water RDA: not established	Important for tooth structure; may help prevent dental caries (particularly in children) and osteoporosis in adults	Mottling of teeth	Not known
Selenium (Se)	Stored in liver, kidneys	Meat, seafood, cereals RDA: Not known	Antioxidant; constituent of certain enzymes; spares vitamin E	Nausea, vomiting, irritability, fatigue	Not known
Chromium (Cr)	Widely distributed	Liver, meat, cheese, whole grains, Brewer's yeast, wine RDA: not established	Essential for proper use of carbohydrates in body; enhances effectiveness of insulin on carbohydrate metabolism; may increase blood levels of HDLs while decreasing levels of LDLs	Not known	Not known

*Trace minerals together account for less than 0.005% of body weight.

A fine balance between uptake and excretion is crucial for retaining needed amounts of minerals while preventing toxic overload. For example, sodium, present in virtually all natural foods (and added to processed foods), may contribute to high blood pressure if ingested in excess; so too may deficiencies of calcium and potassium.

Fats and sugars are practically devoid of minerals, and highly refined cereals and grains are poor sources. The most mineral-rich foods are vegetables, legumes, milk, and some meats. Specific minerals and their roles in the body are summarized in Table 25.3.

METABOLISM

Overview

Once inside body cells, nutrients become involved in an incredible variety of biochemical reactions. These reactions are known collectively as **metabolism,** a broad term referring to all chemical reactions necessary to maintain life. During metabolism, there is a constant building up and tearing down of substances. Cells use energy to extract more energy from foods, and then use some of this energy to drive their activities. Even at rest, the body uses energy on a grand scale.

Anabolism and Catabolism

Metabolic processes are either *anabolic* (synthetic) or *catabolic* (degradative). **Anabolism** (ah-nab′-ō-lizm) includes reactions in which larger molecules or structures are built from smaller ones; for example, the bonding of amino acids to make proteins and of proteins and lipids to form cell membranes. **Catabolism** (kah-tab′-ō-lizm) involves processes that break down complex structures to simpler ones. The hydrolysis of foods in the digestive tract to reduce them to their absorbable subunits is catabolic. So too is the group of reactions collectively called **cellular respiration,** during which glucose, fatty acids, and amino acids are broken down within cells and some of the energy released as their bonds are broken is captured to form ATP.

Figure 25.3 is a flowchart of the metabolism of energy-containing nutrients in the body. As you can see, metabolism occurs in three major stages. Stage 1 is digestion, which occurs in the gastrointestinal tract, as described in Chapter 24. In stage 2, which occurs in the cytoplasm, absorbed nutrient molecules are transported in blood to the tissue cells, where they are (1) built into cellular molecules (lipids, proteins, and glycogen) by anabolic pathways or (2) broken down by catabolic pathways to pyruvic (pī-roo′-vik) acid and acetyl CoA (ah-sē-til kō-ā′). Stage 3, which is entirely catabolic, occurs within the mitochondria of cells, requires oxygen, and completes the breakdown of newly digested or stored foodstuffs, producing carbon dioxide and water and harvesting large amounts of ATP. (ATP, as you will recall from Chapter 2, is used as the cells' energy "currency.") The primary function of cellular respiration, consisting of glycolysis of stage 2 and all of the events of stage 3, is to generate ATP. ATP traps the chemical energy of the original food molecules in its own (high-energy) bonds and then, usually within a fraction of a minute, transfers that energy to power the various cell functions. Thus, food fuels, such as glycogen and fats, *store* energy in the body, and ATP *mobilizes that energy for cellular use.* It is not necessary at this point to try to memorize this schematic, but as you will soon see, it provides a cohesive summary of metabolic processes.

Oxidation-Reduction Reactions and the Role of Coenzymes

The energy-yielding (ATP-yielding) reactions within cells are **oxidation reactions,** a term that requires some explanation to prevent misunderstanding of the sections that follow. Originally, *oxidation* was defined as an event resulting from the combination of oxygen with other elements. Examples are the rusting of iron (the slow formation of iron oxide) and the burning of wood. In burning, oxygen combines rapidly with carbon, producing (in addition to carbon dioxide and water) an enormous amount of energy, which is liberated as heat and light. But then it was discovered that oxidation *also* occurs when hydrogen atoms are *removed* from compounds, so the definition was expanded to its current form: Oxidation occurs via the gain of oxygen or the loss of hydrogen. Whichever way oxidation occurs, the common event is that the oxidized substance *loses electrons.*

This process can be explained by reviewing some chemistry principles and understanding the consequences of different electron-attracting abilities of atoms (see p. 37). Because hydrogen is very electropositive, its lone electron usually spends more time orbiting other atoms of molecules that hydrogen helps to form. Thus, when a hydrogen *atom* is removed, its electron goes with it, and the molecule as a whole loses that electron. Conversely, oxygen is very electron-hungry (electronegative) and strongly attracts electrons. So when oxygen binds with other atoms, the shared electrons spend a disproportionate amount of time in oxygen's vicinity. Again, the molecule as a whole loses electrons. As you will soon see, essentially all oxidation

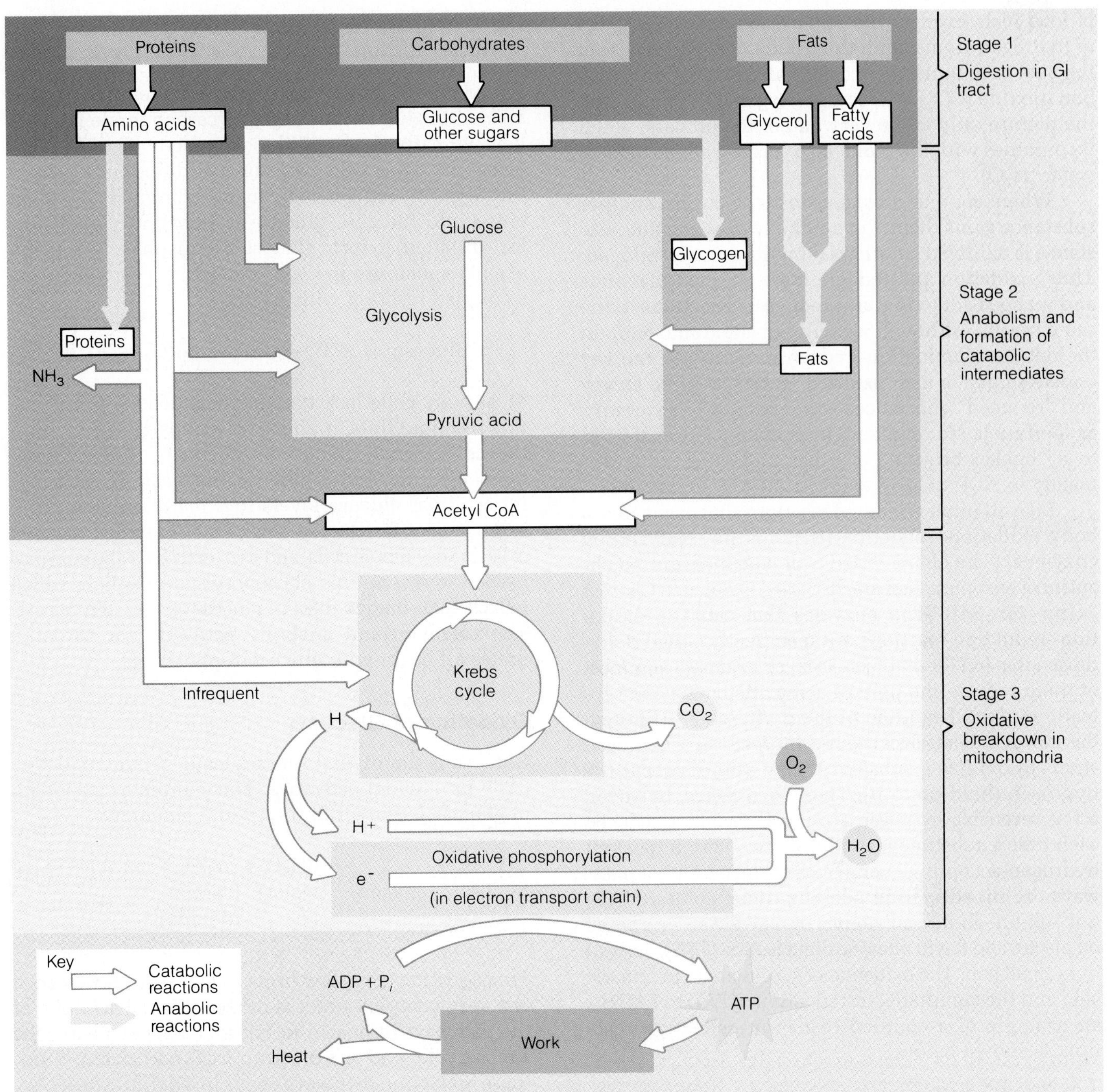

Figure 25.3 The three stages of metabolism of energy-containing nutrients in the body. In stage 1, foods are degraded to their absorbable forms by digestive enzymes within the GI tract. In stage 2, the absorbed nutrients are transported in blood to the body cells, where they may be incorporated into cellular molecules (anabolism) or funneled into catabolic pathways (stage 3) via glycolysis and intermediate breakdown products such as pyruvic acid or acetyl CoA. Stage 3 consists of the pathways of the Krebs cycle and oxidative phosphorylation, which go on within the mitochondria. During the Krebs cycle, acetyl CoA is broken apart: Its carbon atoms are liberated as carbon dioxide (CO_2), and the hydrogen atoms removed are delivered to a chain of receptors (the electron transport chain), which ultimately deliver them to molecular oxygen so that water is formed. Because the hydrogen atoms are split into protons and free electrons in the electron transport chain, an electrical current is generated and some of this electrical energy is used to form the bonds of high-energy ATP molecules. Once generated, ATP provides the energy needed by cells to power their many activities. The reactions of glycolysis (stage 2) and the Krebs cycle and oxidative phosphorylation (stage 3) collectively constitute cellular respiration.

of food fuels involves the step-by-step removal of pairs of hydrogen atoms (and, thus, pairs of electrons) from the substrate molecules, eventually leaving only carbon dioxide (CO_2). Molecular oxygen (O_2) comes into the picture only at the very end of the process, when it combines with the removed hydrogen atoms to form water (H_2O).

Whenever one substance loses electrons, another substance gains them. Put another way, when one substance is oxidized, another is simultaneously *reduced*. Thus, oxidation and reduction are coupled reactions and we speak of **oxidation–reduction reactions** in the same breath. Although we will not attempt to explain the details of oxidation–reduction reactions, the key understanding is that "oxidized" substances *lose* energy and "reduced" substances *gain* energy. Consequently, as food fuels are oxidized, their energy is transferred to a "bucket brigade" of other molecules and ultimately to ADP to form energy-rich ATP molecules.

Like all other chemical reactions that occur in the body, oxidation–reduction reactions are catalyzed by enzymes. (The characteristics of enzymes and mechanism of enzyme action are discussed in detail in Chapter 2 [pp. 52–54]). The enzymes that catalyze oxidation–reduction reactions are specifically called *dehydrogenases* (dē″-hī-drah′-jen-ās-iz) or *oxidases*, and most of them require the help of a specific coenzyme typically derived from one of the B vitamins. Although the enzymes themselves catalyze the removal of hydrogen atoms to oxidize a substance, they cannot *accept* the hydrogen (hold on to it). Their *coenzymes*, however, act as reversible hydrogen acceptors, becoming reduced each time a substrate is oxidized. Two very important hydrogen-accepting coenzymes of the oxidative pathways are **nicotinamide adenine dinucleotide (NAD)** (nik″-uh-tin′-ah-mīd ad′-eh-nēn dī-noo′-klē-ah-tīd), based on niacin, and **flavin adenine dinucleotide (FAD),** derived from riboflavin. The oxidation of succinic acid to fumaric acid and the simultaneous reduction of FAD to $FADH_2$, an example of a coupled oxidation–reduction reaction, is shown below.

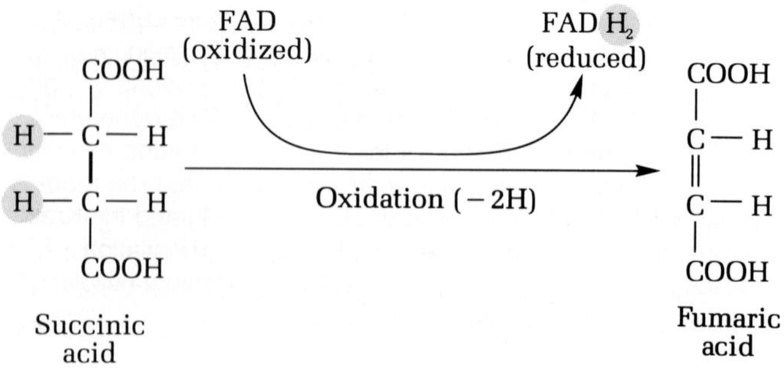

FAD (oxidized) → $FADH_2$ (reduced)

Succinic acid (COOH–CH₂–CH₂–COOH: $COOH-\underset{H}{\overset{H}{C}}-\underset{H}{\overset{H}{C}}-COOH$) — Oxidation (−2H) → Fumaric acid ($COOH-CH=CH-COOH$)

Carbohydrate Metabolism

Because all food carbohydrates are eventually transformed to glucose, the story of carbohydrate metabolism is really a tale of glucose metabolism. Glucose enters the tissue cells by facilitated diffusion, a process that is greatly enhanced by insulin. Immediately upon entry into the cell, glucose is *phosphorylated* (fos-for′-ih-lā-ted) to form glucose-6-phosphate by transfer of a phosphate group (PO_4^{3-}) to its sixth carbon during a coupled reaction with ATP:

$$\text{Glucose} + \text{ATP} \rightarrow \text{glucose-6-PO}_4 + \text{ADP}$$

Most body cells lack the enzymes needed to reverse this reaction; thus, it effectively traps glucose within the cells. Since glucose-6-phosphate is a *different* molecule from simple glucose, the reaction also keeps intracellular glucose levels low, maintaining a diffusion gradient for glucose entry. Only intestinal mucosal cells, kidney tubule cells, and liver cells have the enzymes needed to reverse this phosphorylation reaction, which reflects their unique roles in glucose uptake *and* release. The catabolic and anabolic pathways for carbohydrates all begin with glucose-6-phosphate.

Oxidation of Glucose

Glucose is the pivotal fuel molecule in the oxidative (ATP-producing) pathways. The complete catabolism of glucose is shown in the overall equation:

$$\underset{\text{(glucose)}}{C_6H_{12}O_6} + \underset{\text{(oxygen)}}{6O_2} \rightarrow \underset{\text{(water)}}{6H_2O} + \underset{\text{(carbon dioxide)}}{6CO_2} + 38\ \text{ATP} + \text{heat}$$

This equation gives few hints that glucose breakdown is a very complex process or that it involves three of the pathways indicated in Figure 25.3: glycolysis, the Krebs cycle, and the electron transport chain. Since these metabolic pathways occur in a definite order, we will consider them sequentially.

Glycolysis. **Glycolysis** (glī-kol′-ih-sis), literally translated as sugar breakdown, is a series of ten *reversible* chemical steps by which glucose is converted into two pyruvic acid molecules, yielding small amounts of ATP (2 ATP per glucose molecule). Glycolysis occurs in the cytoplasm of cells, where its steps are catalyzed by specific soluble enzymes. It is an *anaerobic* (an″-ayr-ō′-bik) *process*. Although this is sometimes (mistakenly) interpreted to mean that the pathway occurs only in the *absence* of oxygen, this is *not* the case. The correct interpretation is that glycolysis does not *use* oxygen and occurs *whether or not oxygen is present.* Figure 25.4 (pp. 822–823) shows the glycolytic pathway, in both its complete and its abbreviated forms. For simplicity, we will consider the three major phases shown in the abbreviated scheme on the lower right side of the figure.

1. **Sugar activation.** In phase 1, glucose is phosphorylated and converted to fructose, which is then phosphorylated again. These three steps yield fructose-1,6-diphosphate and a net deficit of two ATP molecules. The two separate reactions of the sugar with ATP provide the *activation energy* needed to prime the later stages of the pathway. (The importance of activation energy is described in Chapter 2.)

2. **Sugar cleavage.** During phase 2, fructose-1,6-diphosphate is split into two 3-carbon fragments that exist (reversibly) as one of two isomers: glyceraldehyde (glih″-ser-al′-deh-hīd) 3-phosphate or dihydroxyacetone (dī″-hī-drok″-sē-a′-suh-tōn) phosphate.

3. **Oxidation and ATP formation.** In phase 3, two major events happen. First, the two 3-carbon molecules are oxidized by the removal of hydrogen, which is picked up by NAD. Hence, some of glucose's energy is transferred to NAD. Second, inorganic phosphate groups (P_i) are attached to each oxidized fragment by high-energy bonds. Then, as these terminal phosphates are cleaved off later, enough energy is captured to form a total of four ATP molecules. The formation of ATP by this method, that is, by transferring phosphate groups from phosphorylated substrates (metabolic intermediates) to ADP, is called **substrate-level phosphorylation.**

The final products of glycolysis are two molecules of **pyruvic acid** and two molecules of reduced NAD (NADH + H^+)*, with a net gain of two ATP molecules per glucose molecule. (Four ATPs are produced, but two are used in phase 1 to "prime the pump.") Notice that each of the two pyruvic acid molecules has the formula $C_3H_4O_3$. Since glucose is $C_6H_{12}O_6$, you can see that between them the pyruvic acid products have lost 4 hydrogen atoms, which are now bound to two molecules of NAD. Although a small amount of ATP has been harvested, we have yet to see either of the other two end products (H_2O and CO_2) of glucose oxidation.

The fate of pyruvic acid, which still contains the bulk of glucose's chemical energy, depends on the availability of oxygen at the time it is produced. If glycolysis is to continue and the resulting pyruvic acid is to enter the Krebs cycle and be completely broken down, the reduced coenzymes (NADH + H^+) must be relieved of the accepted hydrogen so that they can continue to act as hydrogen acceptors. When oxygen is readily available, this is no problem. The NADH + H^+ simply deliver their burden of hydrogen atoms to the enzymes of the electron transport chain in the mitochondria, which in turn deliver them to the molecular oxygen, forming water. However, when oxygen is not present in sufficient amounts, as might occur during strenuous exercise, the NADH + H^+ produced during glycolysis begins to unload its hydrogen baggage *back onto pyruvic acid,* thus reducing it. This addition of 2 hydrogen atoms to pyruvic acid results in the production of **lactic acid** (see bottom of Figure 25.4). Some of the lactic acid produced diffuses out of the cells and is transported to the liver. When oxygen is again available, lactic acid is oxidized back to pyruvic acid, which then enters the **aerobic pathways** (the oxygen-requiring Krebs cycle and electron transport chain within the mitochondria) that completely oxidize it to water and carbon dioxide. The liver may also convert lactic acid all the way back to glucose-6-phosphate (reverse glycolysis) and then store it as glycogen or free it of its phosphate and release it to the blood.

Although glycolysis generates ATP very rapidly, glycolysis alone produces only 2 ATP/glucose molecule, as compared to the 38 ATP/glucose harvested during complete oxidation of glucose. Prolonged anaerobic respiration ultimately results in acid-base problems; and except for red blood cells (which typically carry out *only* glycolysis), *totally* anaerobic respiration involving lactic acid formation provides only a temporary (or emergency) means of rapid ATP production during conditions of oxygen deficit. It can go on without harm (tissue damage) for the longest periods in skeletal muscle, for much shorter periods in cardiac muscle, and almost not at all in the brain.

Krebs Cycle. The **Krebs cycle,** named after its discoverer Hans Krebs, is the next stage of glucose oxidation. The Krebs cycle occurs in the mitochondrial matrix and is fueled largely by pyruvic acid produced during glycolysis and by fatty acids resulting from fat breakdown.

Pyruvic acid enters the mitochondria; but before it can enter the Krebs cycle pathway, it must first be converted to acetyl CoA. This involves three events (Figure 25.5, p. 824):

1. Decarboxylation, in which one of pyruvic acid's carbons is removed and released as carbon dioxide gas (a waste product of metabolism), which diffuses out of the cells into the blood to be expelled by the lungs.
2. Oxidation by the removal of hydrogen atoms. The removed hydrogens are picked up by NAD^+

*Although it is sometimes not indicated, NAD carries a positive charge (NAD^+). Thus, when it accepts the hydrogen pair, NADH + H^+ is the resulting reduced product.

Step 1 Glucose enters the cell and is phosphorylated by the enzyme hexokinase, which catalyzes the transfer of a phosphate group, indicated as P, from ATP to the number six carbon of the sugar producing glucose-6-phosphate. The electrical charge of the phosphate group traps the sugar in the cell because of the impermeability of the plasma membrane to ions. Phosphorylation of glucose also makes the molecule more chemically reactive. Although glycolysis is supposed to *produce* ATP, in step 1 ATP is actually consumed—an energy investment that will be repaid with dividends later in glycolysis.

Step 2 Glucose-6-phosphate is rearranged and converted to its isomer, fructose-6-phosphate. Isomers, remember, have the same number and types of atoms but in different structural arrangements.

Step 3 In this step, still another molecule of ATP is used to add a second phosphate group to the sugar, producing fructose 1,6-diphosphate. So far, the ATP ledger shows a debit of −2. With phosphate groups on its opposite ends, the sugar is now ready to be split in half.

Step 4 This is the reaction from which glycolysis gets its name. An enzyme cleaves the sugar molecule into two different 3-carbon sugars: glyceraldehyde 3-phosphate and dihydroxyacetone phosphate. These two sugars are isomers of one another.

Step 5 The isomerase enzyme interconverts the 3-carbon sugars, and if left alone in a test tube, the reaction reaches equilibrium. This does not happen in the cell, however, because the next enzyme in glycolysis uses only glyceraldehyde phosphate as its substrate and is unreceptive to dihydroxyacetone phosphate. This pulls the equilibrium between the two 3-carbon sugars in the direction of glyceraldehyde phosphate, which is removed as fast as it forms. Thus, the net result of steps 4 and 5 is cleavage of a 6-carbon sugar into two molecules of glyceraldehyde phosphate; each will progress through the remaining steps of glycolysis.

Step 6 An enzyme now catalyzes two sequential reactions while it holds glyceraldehyde phosphate in its active site. First, the sugar is oxidized by the transfer of H from the number one carbon of the sugar to NAD+, forming NADH. Here we see in metabolic context the oxidation-reduction reaction described on page 820. This reaction releases substantial amounts of energy, and the enzyme capitalizes on this by coupling the reaction to the creation of a high-energy phosphate bond at the number one carbon of the oxidized substrate. The source of the phosphate is inorganic phosphate (P_i) always present in the cytosol. As products, the enzyme releases NADH and 1,3-diphosphoglyceric acid. Notice in the figure that the new phosphate bond is symbolized with a squiggle (~), which indicates that the bond is at least as energetic as the high-energy phosphate bonds of ATP.

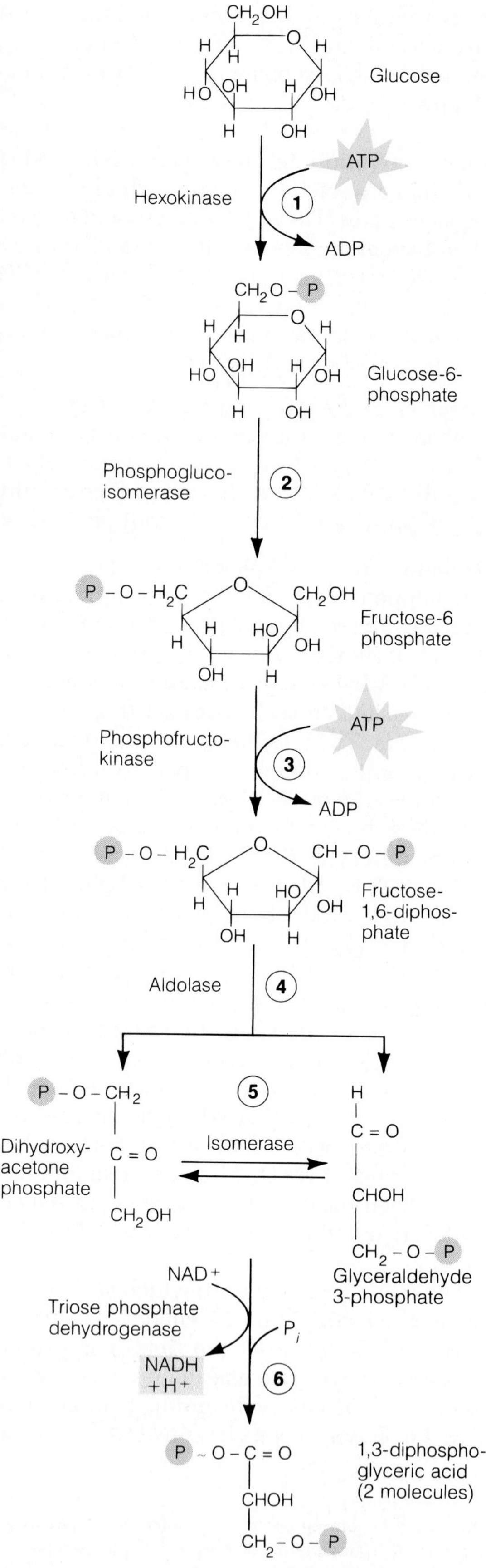

Figure 25.4 The ten steps of glycolysis. Each of the ten steps of glycolysis is catalyzed by a specific enzyme found dissolved in the cytoplasm. All steps are reversible. An abbreviated version of the three major phases of glycolysis appears in the lower right hand corner of the figure.

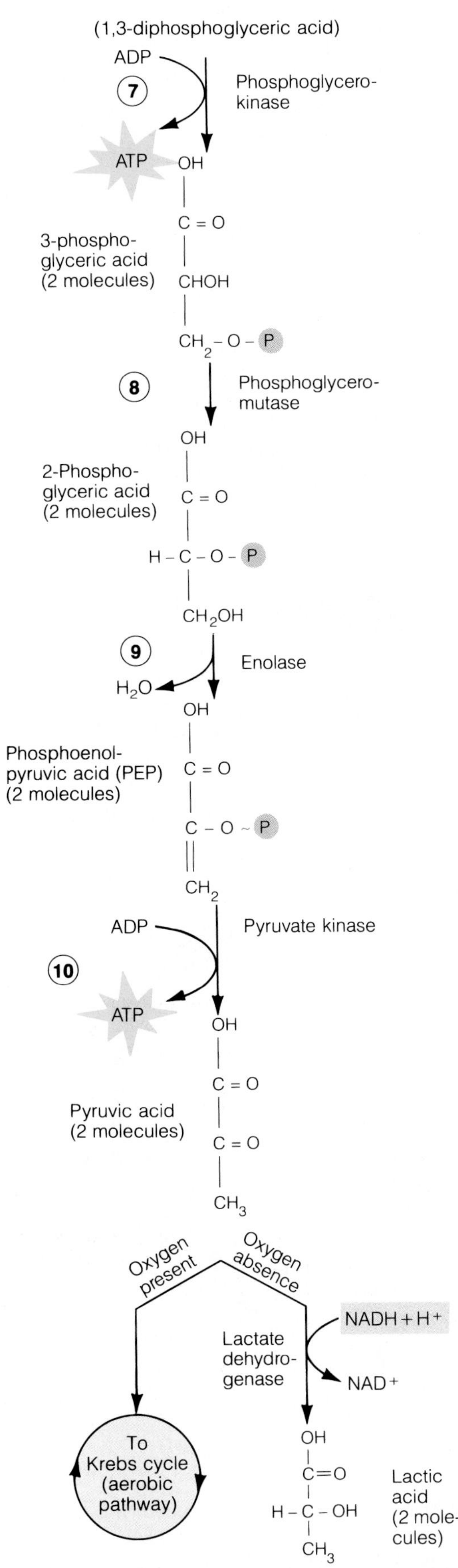

Step 7 Finally, glycolysis produces ATP. The phosphate group, with its high-energy bond, is transferred from 1,3-diphosphoglyceric acid to ADP. For each glucose molecule that began glycolysis, step 7 produces two molecules of ATP, because every product after the sugar-splitting step (step 4) is doubled. Of course, two ATPs were invested to get sugar ready for splitting. The ATP ledger now stands at zero. By the end of step 7, glucose has been converted to two molecules of 3-phosphoglyceric acid. This compound is not a sugar. The sugar was oxidized to an organic acid back in step 6, and now the energy made available by that oxidation has been used to make ATP.

Step 8 Next, an enzyme relocates the remaining phosphate group of 3-phosphoglyceric acid to form 2-phosphoglyceric acid. This prepares the substrate for the next reaction.

Step 9 An enzyme forms a double bond in the substrate by removing a water molecule from 2-phosphoglyceric acid to form phosphoenolpyruvic acid, or PEP. This results in the electrons of the substrate being rearranged in such a way that the remaining phosphate bond becomes very unstable; it has been upgraded to high-energy status.

Step 10 The last reaction of glycosis produces another molecule of ATP by transferring the phosphate group from PEP to ADP. Because this step occurs twice for each glucose molecule, the ATP ledger now shows a net gain of two ATPs. Steps 7 and 10 each produce two ATPs for a total credit of four, but a debt of two ATPs was incurred from steps 1 and 3. Glycolysis has repaid the ATP investment with 100% interest. In the meantime, glucose has been broken down and oxidized to two molecules of pyruvic acid, the compound produced from PEP in step 10.

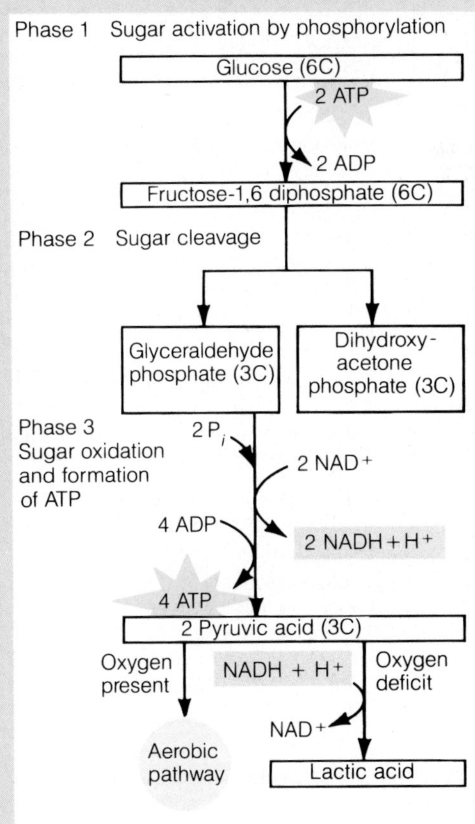

Step 1 Two-carbon acetyl CoA is combined with oxaloacetic acid, a 4-carbon compound. The unstable bond between the acetyl group and CoA is broken as oxaloacetic acid binds and CoA is freed to prime another 2-carbon fragment derived from pyruvic acid. The product is the 6-carbon citric acid, for which the cycle is named.

Step 2 A molecule of water is removed, and another is added back. The net result is the conversion of citric acid to its isomer, isocitric acid.

Step 3 The substrate loses a CO_2 molecule, and the remaining 5-carbon compound is oxidized, forming α ketoglutaric acid and reducing NAD^+.

Step 4 This step is catalyzed by a multi-enzyme complex very similar to the one that converts pyruvic acid to acetyl CoA. CO_2 is lost; the remaining 4-carbon compound is oxidized by the transfer of electrons to NAD^+ to form NADH and is then attached to CoA by an unstable bond. The product is succinyl CoA.

Step 5 Substrate-level phosphorylation occurs in this step. CoA is displaced by a phosphate group, which is then transferred to GDP to form guanosine-triphosphate (GTP). GTP is similar to ATP, which is formed when GTP donates a phosphate group to ADP. The products of this step are succinic acid and ATP.

Step 6 In another oxidative step, two hydrogens are removed from succinic acid, forming fumaric acid, and transferred to FAD to form $FADH_2$. The function of this coenzyme is similar to that of NADH, but $FADH_2$ stores less energy. The enzyme that catalyzes this oxidation-reduction reaction is the only enzyme of the cycle that is embedded in the mitochondrial membrane. All other enzymes of the citric acid cycle are dissolved in the mitochondrial matrix.

Step 7 Bonds in the substrate are rearranged in this step by the addition of a water molecule. The product is malic acid.

Step 8 The last oxidative step reduces another NAD^+ and regenerates oxaloacetic acid, which accepts a 2-carbon fragment from acetyl CoA for another turn of the cycle.

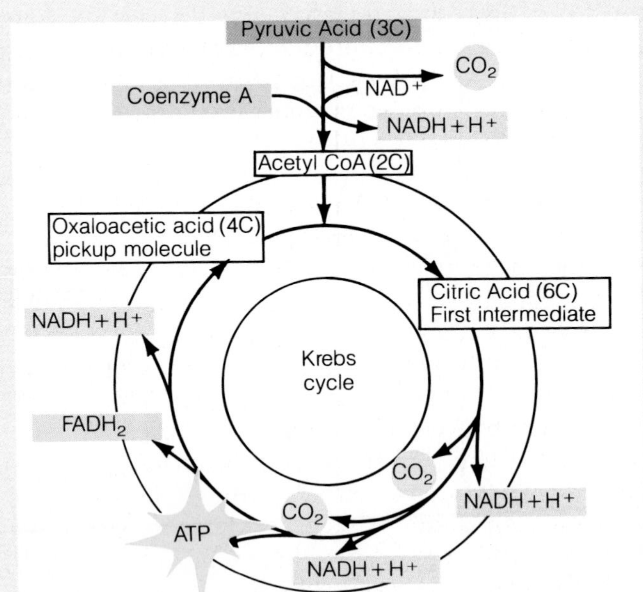

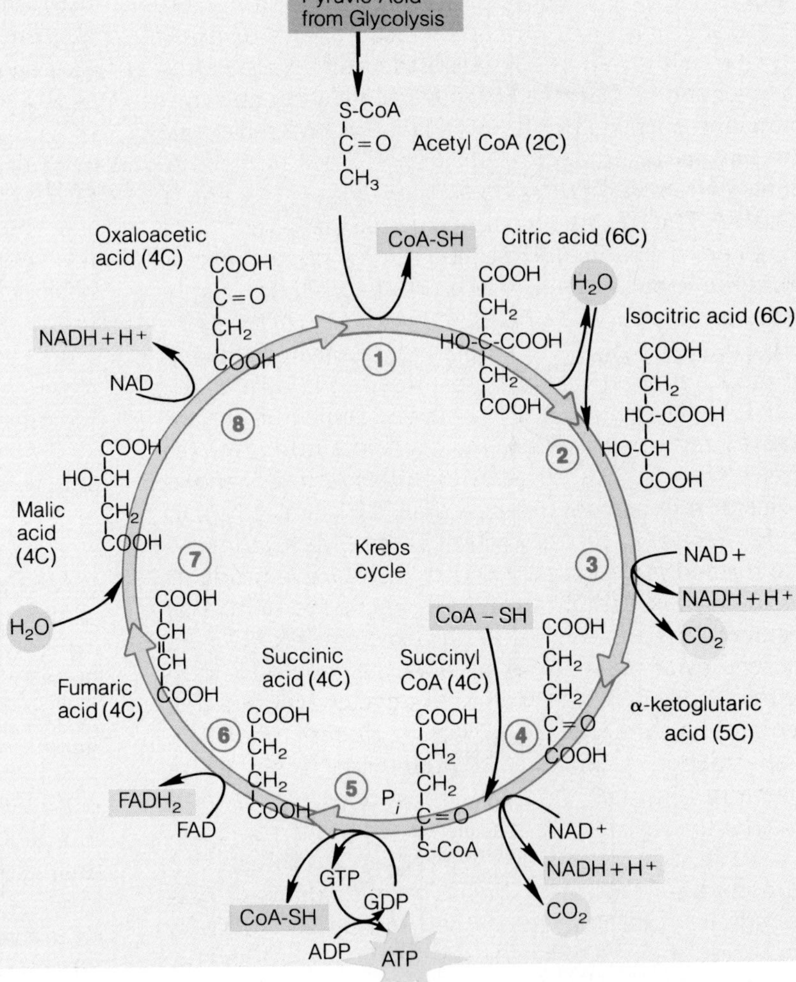

Figure 25.5 Events of the Krebs cycle. Abbreviated version shown at top; detailed pathway shown at bottom. All but one of the steps (step 6) occur in the mitochondrial matrix. The preparation of pyruvic acid (by oxidation, decarboxylation, and reaction with coenzyme A) to enter the cycle as acetyl CoA is shown above the cycle. Acetyl CoA is picked up by oxaloacetic acid to form citric acid; and as it passes through the cycle, it is oxidized four more times (forming three molecules of reduced NAD [$NADH + H^+$] and one of reduced FAD [$FADH_2$]) and decarboxylated twice (releasing 2 CO_2). Energy is captured in the bonds of GTP, which then acts in a coupled reaction with ADP to generate one molecule of ATP.

3. Combination of the resulting acetic acid with *coenzyme A* to produce the final product, **acetyl coenzyme A,** or simply **acetyl CoA.** Coenzyme A (CoA-SH) is a sulfur-containing coenzyme derived from the B vitamin pantothenic acid.

Acetyl CoA is now ready to be broken down completely by the mitochondrial coenzymes. Coenzyme A acts to shuttle the 2-carbon acetic acid to the proper enzyme that can condense it with a 4-carbon acid called *oxaloacetic acid* to produce the 6-carbon *citric acid.* Citric acid is the first intermediate of the cycle, and the Krebs cycle is also called the *citric acid cycle.* The basic events of the Krebs cycle, as summarized in the simplified schematic at the top of Figure 25.5, are fairly easy to remember. As the cycle grinds through eight successive steps, the atoms of citric acid are rearranged to produce different intermediate molecules, each called a **keto acid.** The acetic acid, which enters the cycle, is broken apart carbon by carbon (decarboxylated) and is oxidized, simultaneously generating NADH + H^+ and $FADH_2$. At the end of the cycle, acetic acid has been totally disposed of and oxaloacetic (ok-sah″-lō-ah-sē′-tik) acid, the pickup molecule, is regenerated. Because two decarboxylation and four oxidation events occur, the products of the Krebs cycle are two molecules of carbon dioxide and four molecules of reduced coenzymes (3 NADH + H^+ and 1 $FADH_2$). The addition of water at certain steps accounts for some of the released hydrogen. The energy harvest from the Krebs cycle is one molecule of ATP (via substrate-level phosphorylation) with each turn of the cycle (Figure 25.5, step 5). The detailed events of each step of the Krebs cycle are also described in Figure 25.5.

If we return to the pyruvic acid molecules entering the mitochondrion to take an accounting, we see that a total of three carbon dioxides and five molecules of reduced coenzymes—1 $FADH_2$ and 4 NADH + H^+ (equal to the removal of 10 hydrogen atoms)—are formed for each pyruvic acid. Thus, the products of glucose oxidation in the Krebs cycle are twice that (remember 1 glucose = 2 pyruvic acids): six CO_2, ten molecules of reduced coenzymes, and two ATP molecules. Notice that it is the Krebs cycle reactions that account for the CO_2 evolved during glucose oxidation. The reduced coenzymes, which carry their "extra electrons" in high-energy linkages, must now be oxidized if the Krebs cycle (and glycolysis) is to continue.

Although the glycolytic pathway is exclusive to carbohydrate oxidation, breakdown products of carbohydrates, fats, and proteins can be oxidized in Krebs cycle reactions. Fatty acids destined to be oxidized for energy can feed directly into the cycle as acetyl CoA. Likewise, acetyl CoA can be siphoned off to be used to synthesize fatty acids. Thus, fat and glucose metabolism are tightly interwoven. Amino acids to be used for ATP synthesis are deaminated (their amine groups are removed), and they are converted to keto acids, which can enter the citric acid cycle. In the reverse process, keto acids can be drawn off to be converted to nonessential amino acids as needed. Thus, the Krebs cycle, in addition to serving as the final common pathway for the oxidation of food fuels, is also a source of building materials for anabolic (biosynthetic) reactions.

Electron Transport Chain and Oxidative Phosphorylation. Although the citric acid cycle is considered to be part of *aerobic metabolism,* none of its reactions *directly* uses oxygen. This is the exclusive function of the **electron transport chain,** which oversees the final catabolic reactions that take place on the inner mitochondrial membranes. However, because NAD^+ and FAD are essential for both pathways, the citric acid cycle and electron transport chain reactions are coupled, and both phases are considered to be oxygen requiring.

In the electron transport chain, the hydrogen picked up by coenzymes when food fuels are oxidized is finally combined with molecular oxygen. The energy released during these reactions is harnessed to attach inorganic phosphate groups (P_i) to ADP. The term **oxidative phosphorylation** is used to distinguish this type of phosphorylation process from substrate level phosphorylation in which phosphate groups are transferred from phosphorylated substrates to ADP.

Hydrogen atoms delivered to the electron transport chain by the reduced coenzymes are split into protons and electrons. The protons escape into the fluid matrix between the mitochondrial membranes, and the free electrons are shuttled along the mitochondrial membrane from one electron acceptor to the next. Thus, the ultimate reaction

$$2H\ (2H^+ + 2e^-) + \tfrac{1}{2}O_2 \rightarrow H_2O$$

is forced to occur in many small steps so that much of the energy transferred to the reduced coenzymes can be stored in ATP bonds. If hydrogen combined directly with oxygen in the cell, energy would be released in one big burst and would be largely lost to the environment as heat.

The electron acceptors of the chain include 2 copper atoms and about 15 different proteins, each bound to a metal ion. The proteins form part of the membrane structure of the mitochondrial cristae, that is, the inner mitochondrial membrane. Two of the proteins are flavin proteins, six are iron- (heme) and sulfur-containing complexes, and five are brightly colored iron-containing pigments called **cytochromes** (sī′-tō-krōmz). Several neighboring carriers are clustered together to form three major **respiratory enzyme complexes** (Figure 25.6) that are alternately reduced and oxidized by picking up electrons and then passing them on to the next complex in the sequence. The first

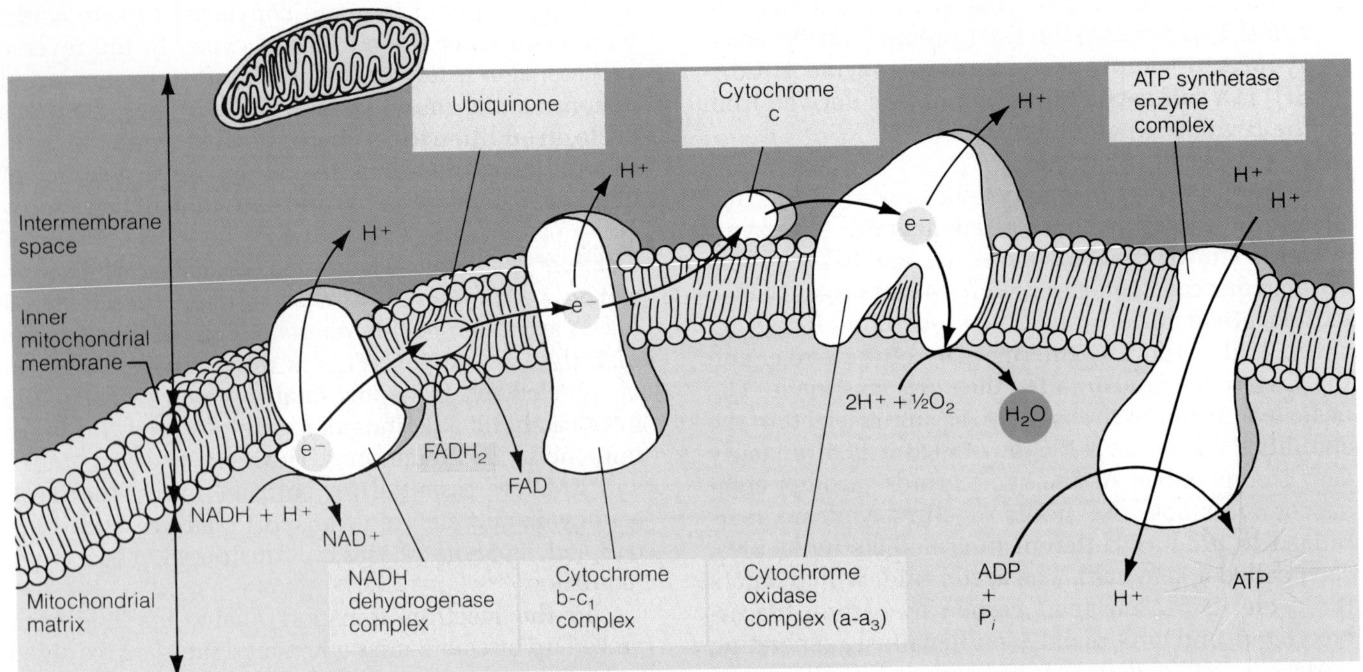

Figure 25.6 Mechanism of oxidative phosphorylation. Schematic diagram showing the flow of electrons through the three major respiratory enzyme complexes of the electron transport chain during the transfer of two electrons from reduced NAD to oxygen. Ubiquinone and cytochrome c are mobile and act as carriers between the major complexes. As the high-energy electrons flow along their energy gradient, some of the energy is used by each complex to pump protons into the intermembrane space. These protons create an electrochemical proton gradient that drives them back across the inner membrane through the ATP synthetase complex. ATP synthetase uses the energy of proton flow (electrical energy) to synthesize ATP from ADP and P_i present in the matrix. Oxidation of each NADH + H^+ to NAD^+ yields 3 ATP. Because $FADH_2$ unloads its hydrogen atoms to ubiquinone, which is beyond the first respiratory complex, less energy (2 ATP) is captured as a result of the oxidation of each $FADH_2$ to FAD.

of these complexes accepts electrons from NADH + H^+, oxidizing it to NAD^+. Because each successive group of carriers has a greater affinity for electrons than those preceding it, the electrons cascade from NADH + H^+ to progressively lower energy levels until they are finally delivered to oxygen, which has the greatest affinity of all for electrons.

The inner mitochondrial membrane functions as an energy conversion machine, converting the energy trapped in the reduced coenzymes into the high-energy phosphate bonds of ATP molecules. At the same time, the membrane regenerates the oxidized coenzymes so that they can continue to shuttle electrons from the citric acid cycle to the electron transport chain. As electrons stream down the chain from one carrier metal ion to the next, the energy released is used by each complex to pump (or *translocate*) the protons that were released into the matrix across the inner mitochondrial membrane into the intermembrane space (Figure 25.6). This creates an **electrochemical proton gradient** across the inner membrane that has two major consequences: (1) It creates a pH gradient, with the hydrogen ion concentration in the matrix much lower than that in the intermembrane space; and (2) it generates a voltage across the membrane that is negative (−) on the matrix side and positive (+) between the mitochondrial membranes. Both conditions strongly attract the protons back into the matrix; and as the protons flow back across the membrane through **ATP synthetase** (a large enzyme–protein complex), they create an electrical current. ATP synthetase harnesses this electrical energy to catalyze the attachment of a phosphate group to ADP to form ATP, thus completing the process of oxidative phosphorylation. Additionally, at the chain's terminus, hydrogen's protons and electrons are combined with molecular oxygen to form water. Virtually all the water resulting from glucose oxidation is formed during oxidative phosphorylation.

The electrochemical proton gradient also supplies energy to actively pump needed metabolites (ADP, pyruvic acid, inorganic phosphate) and calcium ions across the relatively impermeable inner mitochondrial membrane. Indeed, the calcium-sequestering mechanism of mitochondria is an important emergency backup mechanism for eliminating Ca^{2+} overloads in the cytoplasm. (The outer membrane is quite freely permeable to these substances, so no "help" is needed there.) However, the supply of energy from oxidation is not limitless; so when more of the electrochemical gradient energy is used to drive these transport processes, less is available to make ATP. For example, if large amounts of calcium enter the cells, the mitochondria stop producing ATP entirely and

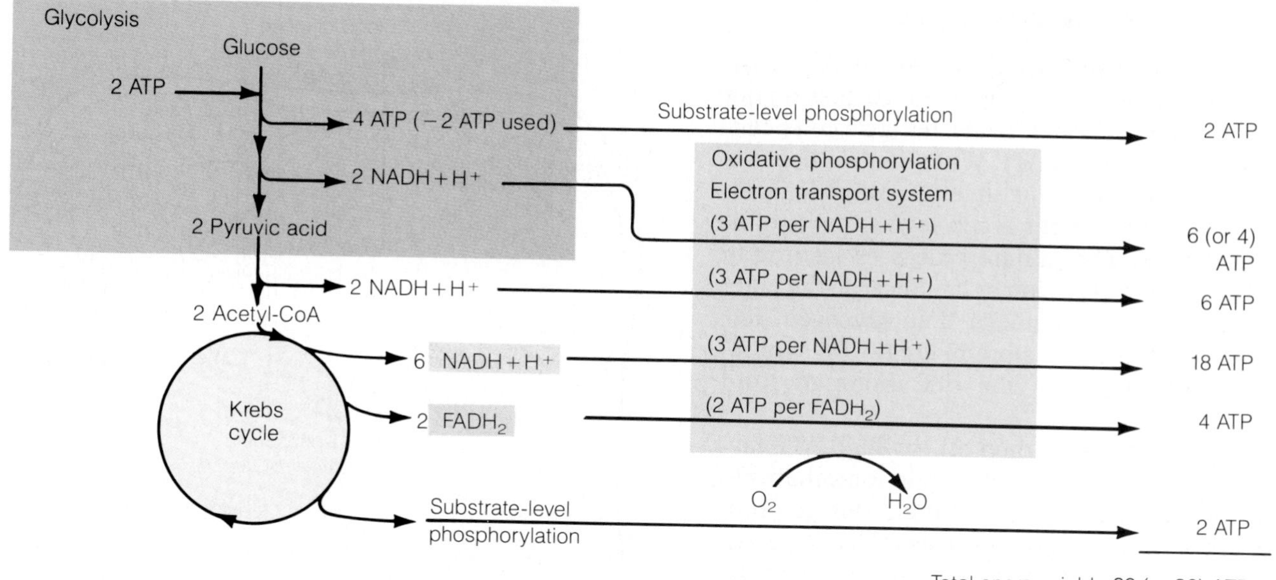

Figure 25.7 ATP yield from the aerobic oxidation of one molecule of glucose yields a net gain of 36 to 38 ATP molecules. A net gain of four ATP molecules is produced by substrate-level phosphorylation events occurring during glycolysis and the citric acid cycle. The balance is produced by oxidative phosphorylation events occurring in the electron transport chain.

devote all of the gradient energy to the uptake (active pumping) of calcium ions.

Summary of ATP production. In the presence of oxygen, cellular respiration is remarkably efficient. Of the total of 686 kilocalories (kcal) of energy present in 1 mole of glucose, as much as 263 to 278 kcal can be captured in the bonds of ATP molecules; the rest is liberated as heat. (The mole, a unit of concentration, is explained on p. 31.) This corresponds to an energy capture of 38% to 40.5%, which is far more efficient than any man-made machines (which use only 10% to 30% of the energy actually available to them). As ADP is transported in, providing the impetus for energy harvest, ATP is moved out in a coupled transport process. The rapid conversion of ADP to ATP in mitochondria keeps the cytoplasm well stocked with ATP reserves; typically, the amount of ATP in the cytoplasm is five to ten times greater than the ADP concentration.

Each time NADH + H^+ is oxidized (relieved of its picked-up electrons), *three* ATP molecules are formed during oxidative phosphorylation (see Figure 25.7). The oxidation of reduced FAD is somewhat less efficient: Each time it delivers 2 hydrogen atoms to the electron transport chain, *two* molecules of ATP are produced. Overall, the complete oxidation of one glucose molecule to carbon dioxide and water yields 36 or 38 molecules of ATP. This energy yield is depicted schematically in Figure 25.7 and can be accounted for as follows:

Substrate-level phosphorylation	
Net ATP gain during glycolysis	2 ATP
Net ATP gain for 2 pyruvic acid molecules during citric acid cycle	2 ATP
Oxidative phosphorylation via electron transport	
Oxidation of 8 molecules of NADH + H^+ formed during the citric acid cycle (3 ATP per molecule)	24 ATP
Oxidation of 2 molecules of $FADH_2$ formed during the citric acid cycle (2 ATP per molecule)	4 ATP
Oxidation of 2 molecules of reduced NAD formed during glycolysis (3 or 2 ATP per molecule)	6 ATP (or 4)
Total energy harvest	38 ATP (or 36)

The alternative figures given represent the present lack of certainty about the energy yield when the reduced NAD is generated outside the mitochondria. The inner mitochondrial membrane is *not* permeable to reduced NAD generated in the cytoplasm, so NADH + H^+ formed during glycolysis utilizes one of several *shuttle molecules* to deliver its extra electron pair to the electron transport chain. One shuttle molecule, *malate*, unloads the electrons at the very beginning of the electron transport chain, yielding three ATP; another, *glycerol phosphate*, unloads later and therefore yields only two ATP. It is not known which of these shuttle systems predominates in other kinds of human cells, but heart muscle and liver cells are known to use the malate system.

Glycogenesis and Glycogenolysis

Although most glucose is used to generate ATP molecules, unlimited amounts of available glucose do *not* result in unlimited ATP synthesis, because cells cannot store large amounts of ATP. When more glucose is available than can be immediately oxidized, intracellular ATP concentrations reach levels that inhibit glucose catabolism and initiate processes leading to storage of glucose as glycogen or fat. Because the body can store a good deal more fat than glycogen, fats account for 80% to 85% of stored energy. (Fat synthesis will be considered in the discussion of lipid metabolism.)

When glycolysis is "turned off" by high ATP levels, glucose molecules are combined in long chains to form glycogen, the animal carbohydrate storage product. This process, called **glycogenesis** (*glyco* = sugar; *genesis* = origin), begins when glucose entering cells is phosphorylated to form glucose-6-phosphate and then converted into its isomer, *glucose-1-phosphate*. The enzyme *glycogen synthetase* then cleaves off the terminal phosphate group as it catalyzes the polymerization of glucose to form glycogen (Figure 25.8). Liver and skeletal muscle cells are most active in glycogen synthesis and storage.

When blood levels of glucose drop, the reverse process of **glycogenolysis** (glī″-kō-jeh-nol′-ih-sis), literally, glycogen lysis or splitting, occurs. The enzyme *glycogen phosphorylase* oversees the phosphorylation and cleavage of glycogen to release glucose-1-phosphate, which is then converted to glucose-6-phosphate, a form that can enter the glycolytic pathway to be oxidized for energy.

In most cells, glucose-6-phosphate resulting from glycogenolysis is trapped because it cannot cross the cell membrane. However, hepatocytes (and some kidney and intestinal cells) contain a unique enzyme called *glucose-6-phosphatase*, which removes the terminal phosphate, producing free glucose. Because glucose *can* readily diffuse from the cell and enter the blood, the liver can use its large glycogen supply to provide blood sugar for the benefit of other organs when blood glucose levels drop. Liver glycogen is an important energy source for skeletal muscles that have depleted their own glycogen reserves.

A common misconception is that athletes need to eat more protein to improve their athletic performance and maintain their greater muscle mass. Actually, a diet rich in *complex* carbohydrates, which lays in muscle glycogen, is much more effective in sustaining intense muscle activity than are high-protein meals. Notice that the emphasis is on complex carbohydrates. Eating a candy bar before an athletic event to provide "quick" energy is more harmful than helpful because it stimulates insulin secretion, which favors glucose use and retards fat use at a time when fat use should be maximal. Furthermore, building muscle protein or avoiding its loss requires intake not only of extra protein, but also of extra (protein-sparing) calories to meet the greater energy needs of the increasingly massive muscles.

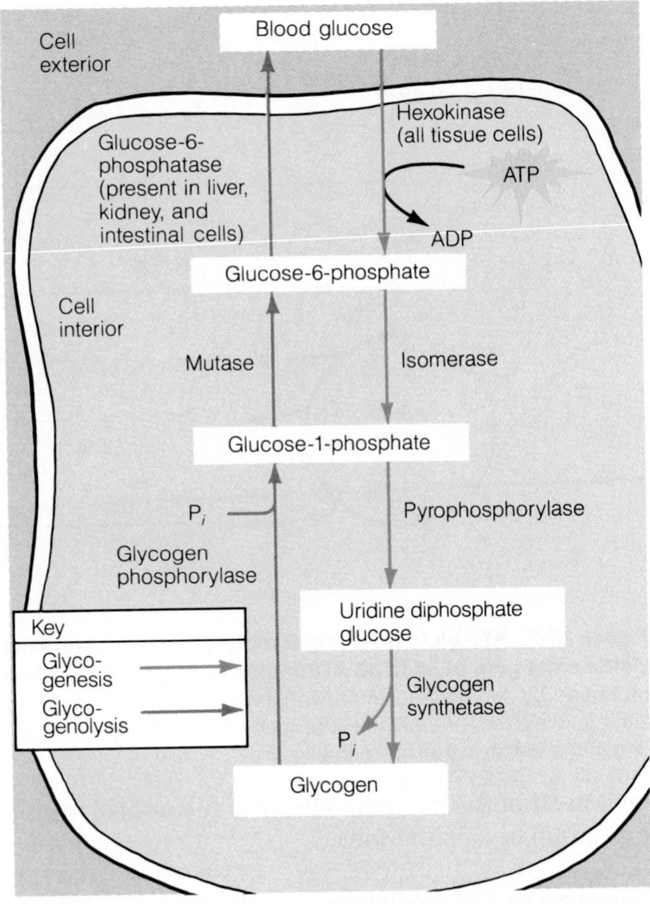

Figure 25.8 Glycogenesis and glycogenolysis. When glucose supplies exceed cellular need for ATP synthesis, glycogenesis, the conversion of glucose into glycogen for storage occurs. Glycogenolysis, the breakdown of glycogen to release glucose, is stimulated by falling blood sugar levels. Notice that glycogen is synthesized and degraded by different enzymatic pathways.

Long-distance runners in particular are well aware of the practice called glycogen loading, popularly called "carbo loading," for endurance events. Glycogen loading, which "tricks" the muscles into storing more glycogen than they normally have the capacity for, involves (1) eating meals high in protein and fat while exercising heavily over a period of several days (this depletes muscle glycogen stores) and then (2) decreasing exercise intensity and switching abruptly to a high-carbohydrate diet (this causes a rebound in muscle glycogen stores of two to four times the normal amount). However, this practice is not without risk. Glycogen retains water and the resulting weight increase and fluid retention may hamper the ability of muscle cells to obtain adequate oxygen, and some athletes using the procedure have complained of cardiac and skeletal muscle pain. Consequently, its use should be cautioned against, particularly in older athletes. ■

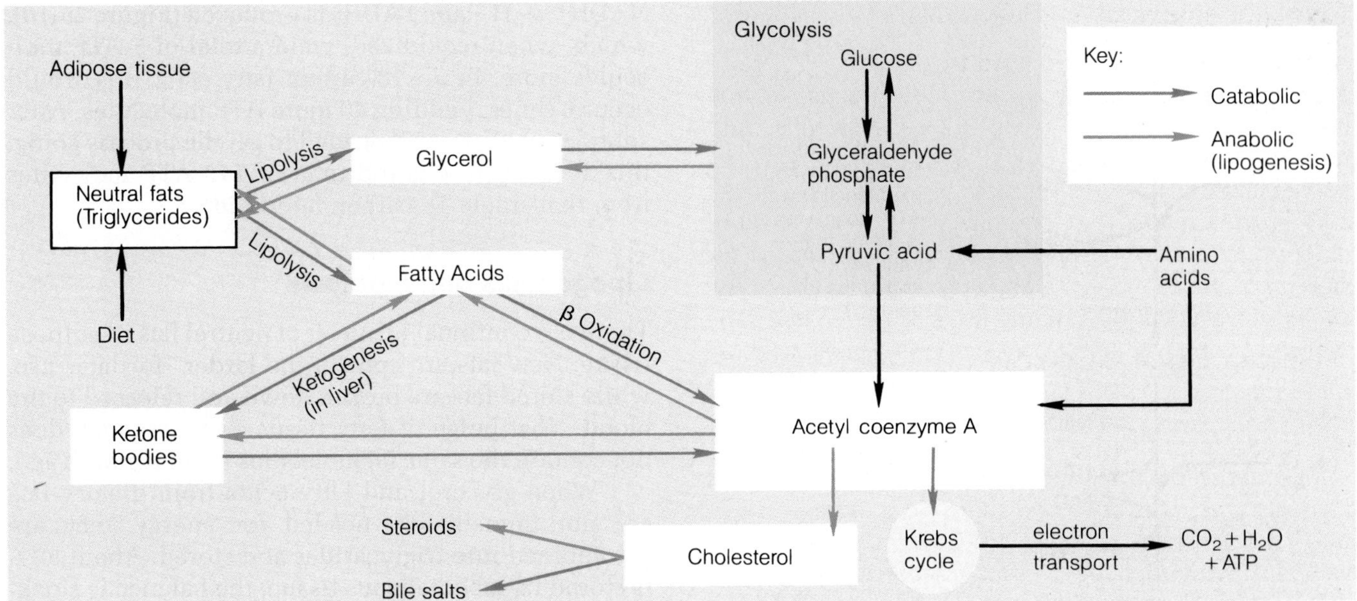

Figure 25.9 Metabolism of triglycerides. When needed for energy, dietary or stored fats enter the catabolic pathways. Glycerol enters the glycolytic pathway (as glyceraldehyde phosphate), and the fatty acids are broken down by beta oxidation to acetyl CoA, which enters the Krebs cycle. When fats are to be synthesized (lipogenesis) and stored in fat depots, the intermediates are drawn from glycolysis and the Krebs cycle in a reversal of the processes noted above. Likewise, excess dietary fats are stored in adipose tissues. When triglycerides are in excess or are the primary energy source, the liver releases their breakdown products (fatty acids → acetyl CoA) in the form of ketone bodies. Excessive amounts of carbohydrates and amino acids are also converted to triglycerides (lipogenesis).

Gluconeogenesis

When insufficient glucose is available to stoke the metabolic furnace, fatty acids and amino acids are converted to glucose. **Gluconeogenesis,** the formation of new (*neo*) sugar from *noncarbohydrate* molecules, occurs in the liver. It takes place when dietary sources and glucose reserves have been depleted and blood glucose levels are beginning to decline. Gluconeogenesis protects the body, the nervous system in particular, from the damaging effects of low blood sugar (*hypoglycemia*) by ensuring that ATP synthesis can continue. The hormones triggering this process and the way fats and proteins are siphoned into these pathways are considered in the discussions of fat and protein metabolism.

Lipid Metabolism

Fats are the body's most concentrated source of energy. They contain very little water, and the energy yield from the catabolism of fats is approximately twice that gained from either glucose or protein breakdown; that is, 9 kcal per g fat in contrast to 4 kcal per g from carbohydrate or protein. Most fat absorption products are transported in the lymph in the form of fatty-protein droplets called chylomicrons (see Chapter 24). Eventually, the lipids in the chylomicrons are hydrolyzed by plasma enzymes, and the resulting fatty acids and glycerol are taken up by body cells where they are processed in various ways. In this section we focus on the oxidation of fats for energy and on some of their anabolic uses (Figure 25.9).

Oxidation of Glycerol and Fatty Acids

Of the various classes of lipids, only the neutral fats are routinely oxidized for energy. Catabolism of neutral fats involves the separate oxidation of the two different building blocks: glycerol and fatty acid chains. Most body cells easily convert glycerol to one of the glycolysis intermediates, glyceraldehyde-phosphate, which then flows into the Krebs cycle. Because glyceraldehyde is equivalent to half a glucose molecule, ATP energy harvest from its complete oxidation is approximately half that of glucose (18–19 ATP/glycerol).

The initial phase of fatty acid oxidation, called **beta oxidation,** occurs in the mitochondria. Although many different types of reactions (oxidation, dehydration, and others) are involved, the *net result* of beta oxidation is that the fatty acid chains are broken apart into two carbon *acetic acid* fragments, and reduced coenzymes are produced (Figure 25.10). Each acetic acid molecule is then fused to coenzyme A, forming acetyl CoA. The term "beta oxidation" reflects the fact that cleavage occurs at every *second* (beta) carbon; it is as if the sequential carbons forming the fatty acid

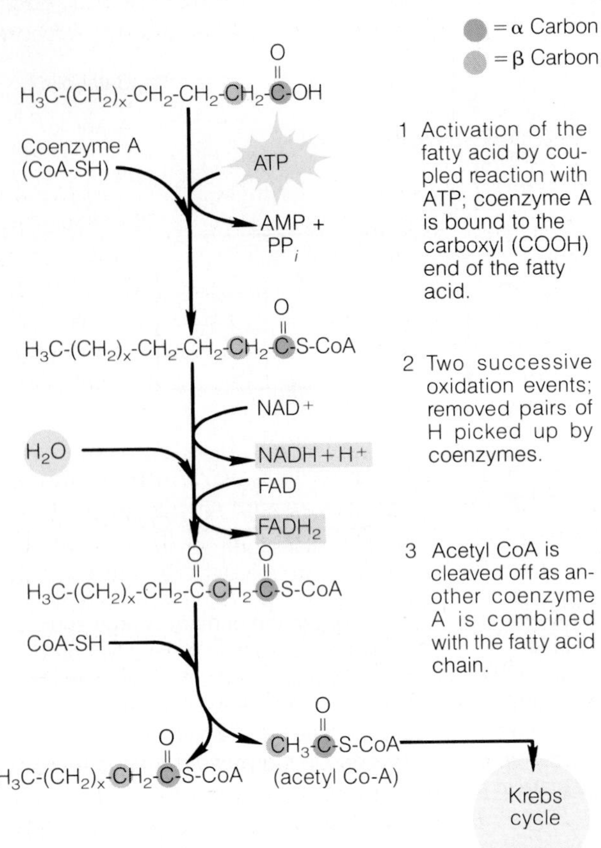
= α Carbon
= β Carbon
$H_3C-(CH_2)_x-CH_2-CH_2-CH_2-C(=O)-OH$
Coenzyme A (CoA-SH)
ATP
AMP + PP_i
1 Activation of the fatty acid by coupled reaction with ATP; coenzyme A is bound to the carboxyl (COOH) end of the fatty acid.
$H_3C-(CH_2)_x-CH_2-CH_2-CH_2-C(=O)-S-CoA$
NAD+
H_2O
NADH+H+
FAD
$FADH_2$
2 Two successive oxidation events; removed pairs of H picked up by coenzymes.
$H_3C-(CH_2)_x-CH_2-C(=O)-CH_2-C(=O)-S-CoA$
CoA-SH
3 Acetyl CoA is cleaved off as another coenzyme A is combined with the fatty acid chain.
$CH_3-C(=O)-S-CoA$ (acetyl Co-A)
$H_3C-(CH_2)_x-CH_2-C(=O)-S-CoA$
Krebs cycle

Figure 25.10 Events of beta oxidation. The process begins as the fatty acid is activated in a coupled reaction with ATP and coenzyme A is attached to the first carbon (the alpha or α carbon) of the fatty acid chain. Each successive 2-carbon (acetyl) segment is oxidized twice and then cleaved off as another coenzyme A is added to the remaining portion of the fatty acid. Acetyl CoA released with each round enters the Krebs cycle. The reduced coenzymes resulting from beta oxidation (like those resulting from the oxidation of acetyl CoA in the Krebs cycle) are oxidized via the electron transport chain. PPi = pyrophosphate (2 linked phosphate groups).

chain were labeled A, B, A, B, and so on (or alpha, beta, alpha, beta, and so on). Acetyl CoA is then picked up by oxaloacetic acid and enters the aerobic pathways to be completely oxidized to carbon dioxide and water.

The number of ATP molecules resulting from the oxidation of a particular fatty acid can be calculated easily by counting the number of carbon atoms in the fatty acid and dividing by 2 to determine the number of acetyl CoA molecules produced. For example, an 18-carbon fatty acid yields 9 acetyl CoA molecules. Because each of these yields 12 ATP molecules per turn of the Krebs cycle, a total of 108 ATP molecules is provided from the oxidative pathways: 9 from electron transport oxidation of 3 NADH + H^+, 2 from oxidation of 1 $FADH_2$, and a net yield of 1 ATP during the Krebs cycle. Also, for every acetyl CoA released during beta oxidation, an additional molecule each of $NADH^+$ + H^+ and $FADH_2$ is produced (Figure 25.10), which, when reoxidized, yield a total of 5 ATP molecules more. In an 18-carbon fatty acid, this would occur 8 times, yielding 40 more ATP molecules. After subtracting the 1 ATP needed to get the process going, this adds up to a grand total of 147 ATP molecules from that single 18-carbon fatty acid!

Lipogenesis and Lipolysis

There is a continual turnover of neutral fats in adipose tissue: New fats are "put in the larder" for later use, while stored fats are broken down and released to the blood. That bulge of fatty tissue you see today does not contain the same fat molecules it did a month ago.

When glycerol and fatty acids from dietary fats are not immediately needed for energy, they are recombined into triglycerides and stored. About 50% is stored in subcutaneous tissue; the balance is stockpiled in other fat depots of the body. Triglyceride synthesis, or **lipogenesis** (see Figure 25.9), occurs when ATP reaches high levels in the cells. Excess ATP leads not only to increased cellular levels of glucose (which is converted into glycogen) but also to an accumulation of intermediates of glucose metabolism, including acetyl CoA and glyceraldehyde-PO_4, which would otherwise feed into the krebs cycle. But, when present in excess, these two metabolites are channeled into pathways leading to triglyceride synthesis. Acetyl CoA molecules are condensed together, forming a fatty acid chain that grows 2 carbons at a time. (This accounts for the fact that almost all fatty acids in the body contain an even number of carbon atoms.) The fact that acetyl CoA, an intermediate in glucose catabolism, is also the *starting point* for synthesis of fatty acids (see Figure 25.9) explains the ease with which glucose is converted into fat. Glyceraldehyde-PO_4 is converted to glycerol, which is condensed with the fatty acids to form triglycerides. Thus, even if fats are restricted in the diet, an excessive carbohydrate intake provides all the raw materials needed to form neutral fats. When blood sugar levels are high, lipogenesis is the major activity occurring in adipose tissues and is also an important liver function.

Lipolysis (lih-pol'-ih-sis), the breakdown of stored fats into glycerol and fatty acids, is essentially the reverse of lipogenesis. The fatty acids and glycerol products are released to the blood, helping to ensure that body organs normally have continuous access to fat fuels for aerobic respiration. (The liver and resting skeletal muscles actually prefer fatty acids as an energy fuel.) The meaning of the adage "fats burn in the flame of carbohydrates" becomes clear when carbohydrate intake is inadequate. Under such conditions, lipolysis is accelerated as the body attempts to fill the fuel gap with fats. However, the entry of acetyl CoA into the Krebs cycle depends on the availability of oxaloacetic acid to act as a pickup molecule to form citric acid.

When carbohydrates are deficient, oxaloacetic acid is converted to glucose (to fuel the brain). Under such conditions, fat oxidation is incomplete and the liver converts acetyl CoA molecules into metabolites called **ketones,** or **ketone bodies,** which are released into the blood. This conversion process is called **ketogenesis** (Figure 25.9). Ketones include the 4-carbon acid acetoacetic acid, as well as the 4-carbon beta-hydroxybutric acid and the 3-carbon molecule acetone, which are formed from acetoacetic acid. (The *keto acids* cycling through the krebs cycle and the *ketone bodies* resulting from fat metabolism are quite different and should not be confused.)

When the rate of lipolysis and beta oxidation is so speeded up that ketones accumulate in the blood faster than they can be used by tissue cells for fuel and large amounts are being excreted from the body in urine, *ketosis* results. Ketosis is a common consequence of starvation, unwise dieting (in which inadequate carbohydrates are eaten), and diabetes mellitus. Because most ketone bodies (like fatty acids) are potent organic acids, a consequence of ketosis is *metabolic acidosis.* The body's buffer systems cannot tie up the acids fast enough, and blood pH drops to dangerously low levels. The person's breath smells fruity as acetone vaporizes from the lungs, and breathing becomes more rapid as the respiratory system tries to blow off blood carbonic acid as carbon dioxide to force the blood pH upward. In severe cases of untreated metabolic acidosis, the person may become comatose or even die as the acid pH depresses the nervous system. ■

Synthesis of Structural Materials

All cells of the body utilize neutral fats, phospholipids, and cholesterol to structure their membranes; phospholipids are also important components of the myelin sheaths of neurons. In addition, the liver (1) synthesizes lipoproteins for transport of cholesterol, fats, and other substances in the blood; (2) makes thromboplastin, a clotting factor; (3) synthesizes cholesterol from acetyl CoA; and (4) uses cholesterol for forming bile salts. Steroid hormones produced in certain endocrine organs (ovaries, testes, and adrenal cortex) use the cholesterol nucleus as the basis for the synthesis of their hormones.

Protein Metabolism

Like all biological molecules, proteins have a finite life span and must be broken down to their amino acids and replaced before they begin to deteriorate. Newly ingested amino acids transported in the blood are taken up by cells by active transport processes and used to *replace* tissue proteins at the rate of about 100 grams each day. When more protein is ingested than is needed to meet these constructive (or anabolic) purposes, amino acids are oxidized for energy, or converted to fat.

Oxidation of Amino Acids

Before amino acids can be oxidized for energy, their amine group (NH_2) must be removed. The resulting molecule is then converted to pyruvic acid or one of the keto acids that serves as an intermediate in the krebs cycle. The key molecule in these interconversions is *glutamic* (gloo-ta'-mik) *acid,* one of the common nonessential amino acids. The events that occur appear to be as follows:

1. **Transamination** (tranz"-am-ih-nā'-shun). A number of amino acids can transfer their amine group to α-ketoglutaric acid (a krebs cycle keto acid), to form glutamic acid (Figure 25.11). The transferring amino acid becomes a keto acid (that is, it has an oxygen atom where the amine group formerly was), and the keto acid (α-ketoglutaric acid) becomes an amino acid (glutamic acid). This reaction is fully reversible.

2. **Oxidative deamination.** In the liver, the amine group of glutamic acid is removed as *ammonia* (NH_3), and α-ketoglutaric acid is regenerated (Figure 25.11). The liberated ammonia molecules are combined with carbon dioxide, yielding **urea** and water. The urea is released to the blood and removed from the body in urine. Because ammonia is toxic to body cells, particularly the delicate brain cells, the ease with which glutamic acid funnels amine groups into the urea cycle is extremely important. This mechanism rids the body not only of the ammonia produced during oxidative deamination of amino acids, but also of blood-borne ammonia produced by intestinal bacteria.

3. **Keto acid modification.** The strategy of amino acid degradation is to produce molecules that can be oxidized in the Krebs cycle or converted to glucose. Hence, keto acids resulting from transamination are altered as necessary to produce one of the metabolites of the Krebs cycle that can then enter the cycle to be oxidized. The most important of these metabolites are pyruvic acid (actually an intermediate before krebs cycle pathways), acetyl CoA, and the keto acids beyond α-ketoglutaric acid in the Krebs cycle pathway (see Figure 25.5). Since the reactions of glycolysis are reversible, deaminated amino acids converted to pyruvic acid can also be reconverted to glucose and contribute to gluconeogenesis.

Synthesis of Structural and Functional Proteins

Amino acids are the most important anabolic nutrients. Not only do they form all protein structures, but they form the bulk of the body's functional molecules as

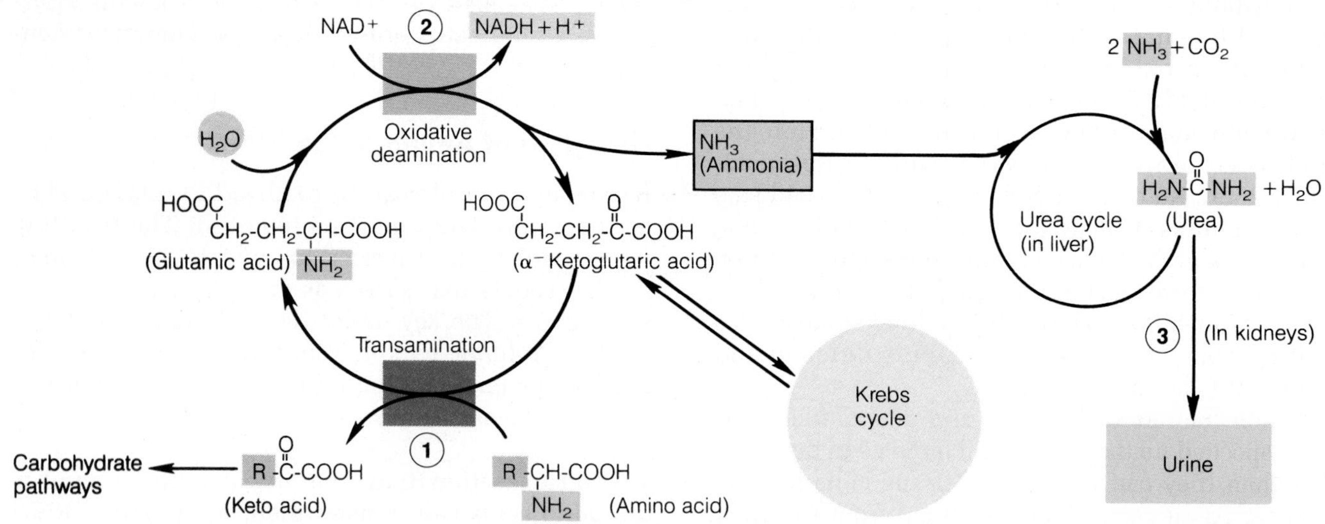

Figure 25.11 Transamination and oxidative deamination of amino acids. (1) Transamination occurs when an amino acid transfers its amine group to α-ketoglutaric acid (one of the krebs cycle intermediates). As α-ketoglutaric acid accepts the amine group, it becomes glutamic acid (a nonessential amino acid), which (2) easily unloads its amine group (as ammonia) to the urea cycle in the liver. (3) Although several steps are involved, urea is formed in the liver when ammonia is combined with CO_2. Urea is excreted by the kidneys in urine. Deamination of glutamic acid regenerates α-ketoglutaric acid.

well. As described in Chapter 3, protein synthesis (directed by messenger RNA molecules) occurs on ribosomes, where cytoplasmic enzymes oversee the formation of peptide bonds linking amino acids into protein polymers. The amount and type of protein synthesized is precisely controlled by hormones (growth hormone, thyroxine, sex hormones, insulin, and others), so protein anabolism reflects the hormonal balance at each stage in the life cycle.

During the course of your lifetime, your cells synthesize 225–450 kg (about 500–1000 pounds) of proteins, depending on your size. However, you need not consume anywhere near that amount because nonessential amino acids are easily formed by siphoning keto acids from the Krebs cycle and transferring amine groups to them by transamination. Most of these transformations occur in the liver, which provides nearly all the nonessential amino acids needed to produce the relatively small amount of protein that the body synthesizes each day. Keep in mind, however, that a complete set of amino acids must be present for protein synthesis to take place, so all of the essential amino acids must be provided by the diet. If some are not, the rest are oxidized for energy even though they may be needed for anabolism; negative nitrogen balance is always the result.

Catabolic–Anabolic Steady State of the Body

The body exists in a *dynamic catabolic–anabolic state.* With few exceptions (notably DNA), organic molecules are continuously broken down and rebuilt—frequently at a head-spinning rate.

The blood is the common transport pool for all body cells, and it is a pool that contains many kinds of nutrients—glucose, ketone bodies, fatty acids, glycerol, and lactic acid. Some organs routinely use blood energy sources other than glucose, thus sparing (saving) glucose for tissues with stricter glucose requirements.

The body's total supply of nutrients constitutes its **nutrient pools**—amino acid, carbohydrate, and fat stores—which can be drawn on to meet its varying needs (Figure 25.12a). These pools are interconvertible; that is, their pathways are linked by key intermediates (see Figure 25.12b). The liver, adipose tissue, and skeletal muscles are the primary effector organs determining the amount and direction of the molecular conversions shown in the figure.

The amino acid pool (Figure 25.12a) consists of the body's total supply of free amino acids. A small amount of amino acids and proteins is lost daily in urine and in sloughed hairs or skin cells and must be replaced via the diet; otherwise, amino acids arising

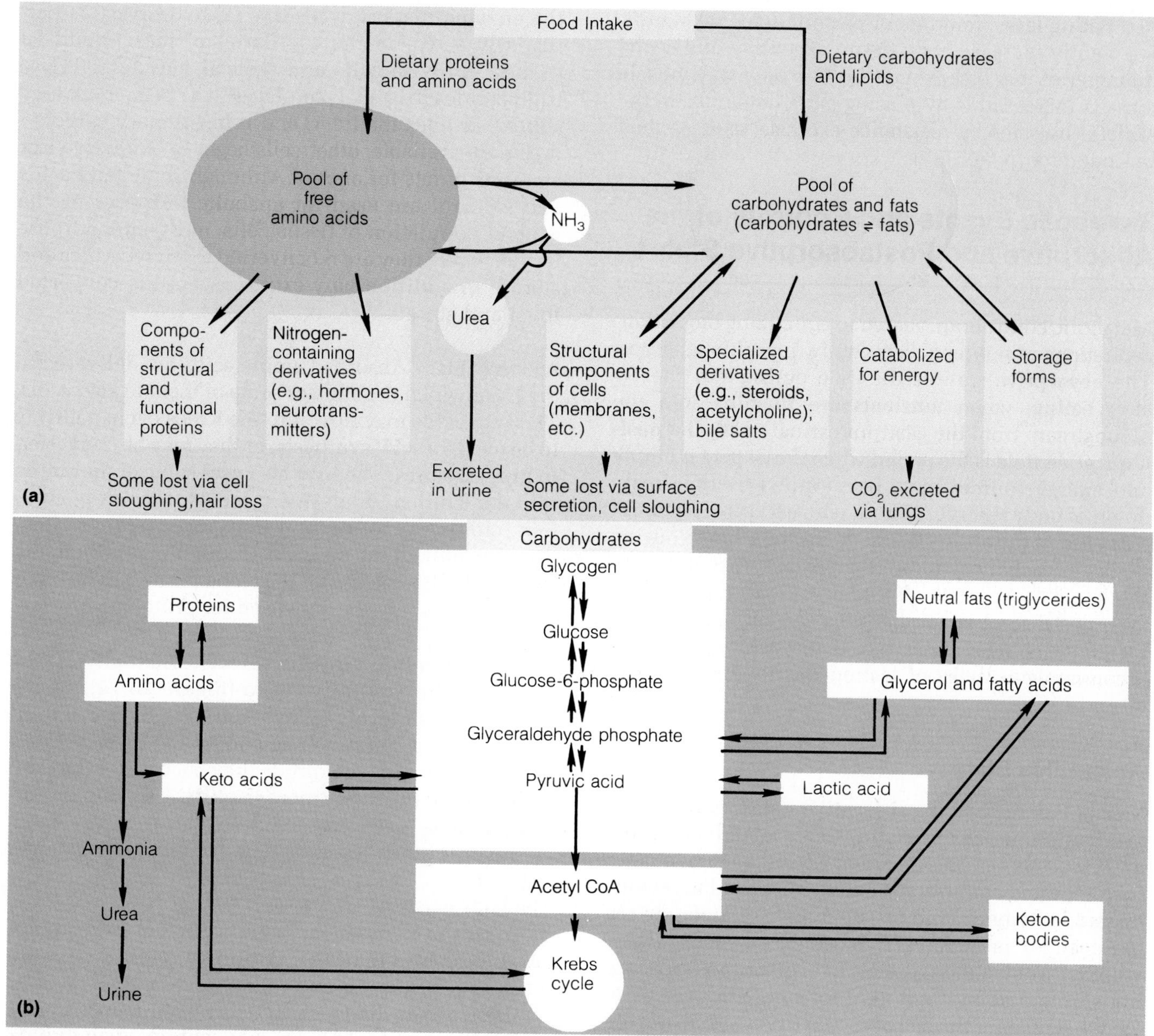

Figure 25.12 Carbohydrate, fat, and amino acid pools and pathways of interconversion. (**a**) Carbohydrate–fat and amino acid pools. Sources contributing to the pools and fates of intermediates drawn from the pools are indicated. (**b**) Simplified flowchart showing the intermediates through which carbohydrates, fats, and proteins can be interconverted.

from tissue breakdown return to the pool. This pool is the source of amino acids for resynthesizing body proteins and for forming a number of amino acid derivatives. In addition, deaminated amino acids can participate in the synthesis of carbohydrates and fats. Not all events of amino acid metabolism occur in all cells. For example, *only* the liver forms urea, and excreting urea from the body is a major role of the kidneys (and to a small extent the skin). Nonetheless, the concept of a common amino acid pool is valid because all cells are connected by the blood.

Because carbohydrates are easily and frequently converted to fats, the carbohydrate and fat pools are usually considered together (Figure 25.12). There are two major differences between this pool and the amino acid pool: (1) Fats and carbohydrates are oxidized directly to produce cellular energy, whereas amino acids supply energy *only after they have been converted to a carbohydrate intermediate.* (2) Excess carbohydrate and fat can be stored as such, whereas excess amino acids are *not* stored as protein, but are oxidized for energy or converted to fat or glycogen for storage.

Eating large amounts of protein does not significantly increase body tissue protein or add to the diameter of that biceps muscle! The only way muscle mass is increased is by placing extra demands on the skeletal muscles by resistance exercise, as described in Chapter 9 (p. 265). ■

Metabolic Events and Controls of the Absorptive and Postabsorptive States

Metabolic controls are geared to equalizing blood concentrations of nutrients between two nutritional states. The **absorptive state** is the time during and shortly after eating, when nutrients are flushing into the bloodstream from the gastrointestinal tract. The **postabsorptive state** is the period when the GI tract is empty and energy sources must be supplied by the breakdown of body reserves. Those who eat "three squares" a day are in the absorptive state for the 4 hours during and after each meal and in the postabsorptive state in the late morning, late afternoon, and all night. However, postabsorptive mechanisms can sustain the body for much longer intervals if necessary—even to accommodate weeks of fasting—as long as water is taken in.

Absorptive State

Metabolic events of the absorptive state (Figure 25.13) can be summarized by stating that anabolic processes exceed catabolic ones and glucose is the major energy fuel. Ingested amino acids and fats are used to remake degraded body protein or fat, and small amounts are oxidized to provide ATP. Most excess metabolites, whether from carbohydrates, fats, or amino acids, are transformed to fat if not used for synthetic purposes. We will consider the fate of each nutrient group during this phase and the hormonal control of its pathways.

Carbohydrates. Absorbed monosaccharides are delivered directly to the liver, where fructose and galactose are converted to glucose. Glucose, in turn, is released to the blood or converted to glycogen and fat. The glycogen formed in the liver is stored there, but most of the fat synthesized by the hepatocytes is released to the blood and picked up by and stored in adipose tissues. Blood-borne glucose not sequestered by the liver enters body cells to be metabolized for energy. Any excess is stored in skeletal muscle cells as glycogen or in adipose cells as fat.

Triglycerides. Nearly all products of fat digestion enter the lymph in the form of chylomicrons, which are hydrolyzed to fatty acids and glycerol before they can pass through the capillary walls. Lipoprotein lipase, the enzyme that catalyzes this fat hydrolysis, is particularly active in the capillaries of muscle and fat tissues. Adipose cells and skeletal muscle cells use triglycerides (triacylglycerols) as their primary energy source, as does the liver; and if few dietary carbohydrates are available, other cells begin to oxidize greater amounts of fats for energy. Although some fatty acids and glycerol are used for anabolic purposes by the general population of tissue cells, most enter adipose tissue, where they are reconverted to triglycerides and stored. In addition, any excess glucose is converted to fat and stored.

Amino Acids. Absorbed amino acids are delivered to the liver, which deaminates some of them to keto acids. The keto acids may flow into the Krebs cycle pathway to be used for ATP synthesis, or they may be converted to liver fat stores. The liver also uses some of the amino acids to synthesize plasma proteins, including albumin, clotting proteins, and transport proteins. However, the bulk of the amino acids flushing through the liver sinusoids remains in the blood for uptake by other body cells, where they are used for protein synthesis.

Hormonal Control. **Insulin** directs essentially all the events of the absorptive state (Figure 25.14). Rising blood glucose levels (above 100 mg glucose/100 ml blood) after a carbohydrate-containing meal act as a humoral stimulus that prods the beta cells of the pancreatic islets to secrete more insulin. A second important stimulus is elevated amino acid levels in the blood. As insulin binds to membrane receptors of its target cells, it

1. Activates carrier-mediated facilitated diffusion of glucose into the cells. Within seconds to minutes, the rate of glucose entry into tissue cells, especially muscle and adipose cells, is increased 15 to 20 times. (The exception is brain cells, which actively take up glucose whether or not insulin is present.)
2. Enhances glucose oxidation for energy.
3. Stimulates the conversion of glucose to glycogen (while inhibiting enzymes that catalyze glycogenolysis).
4. Increases the conversion of glucose to triglycerides in adipose tissues (while inhibiting lipolysis).
5. Stimulates the active transport of amino acids into tissue cells and promotes protein synthesis (while inhibiting enzymes that promote protein catabolism).
6. Inhibits virtually all liver enzymes that promote gluconeogenesis.

Insulin has a *hypoglycemic* (hī″-pō-glī-sē′-mik) *effect:* It sweeps glucose from the blood into the tissue

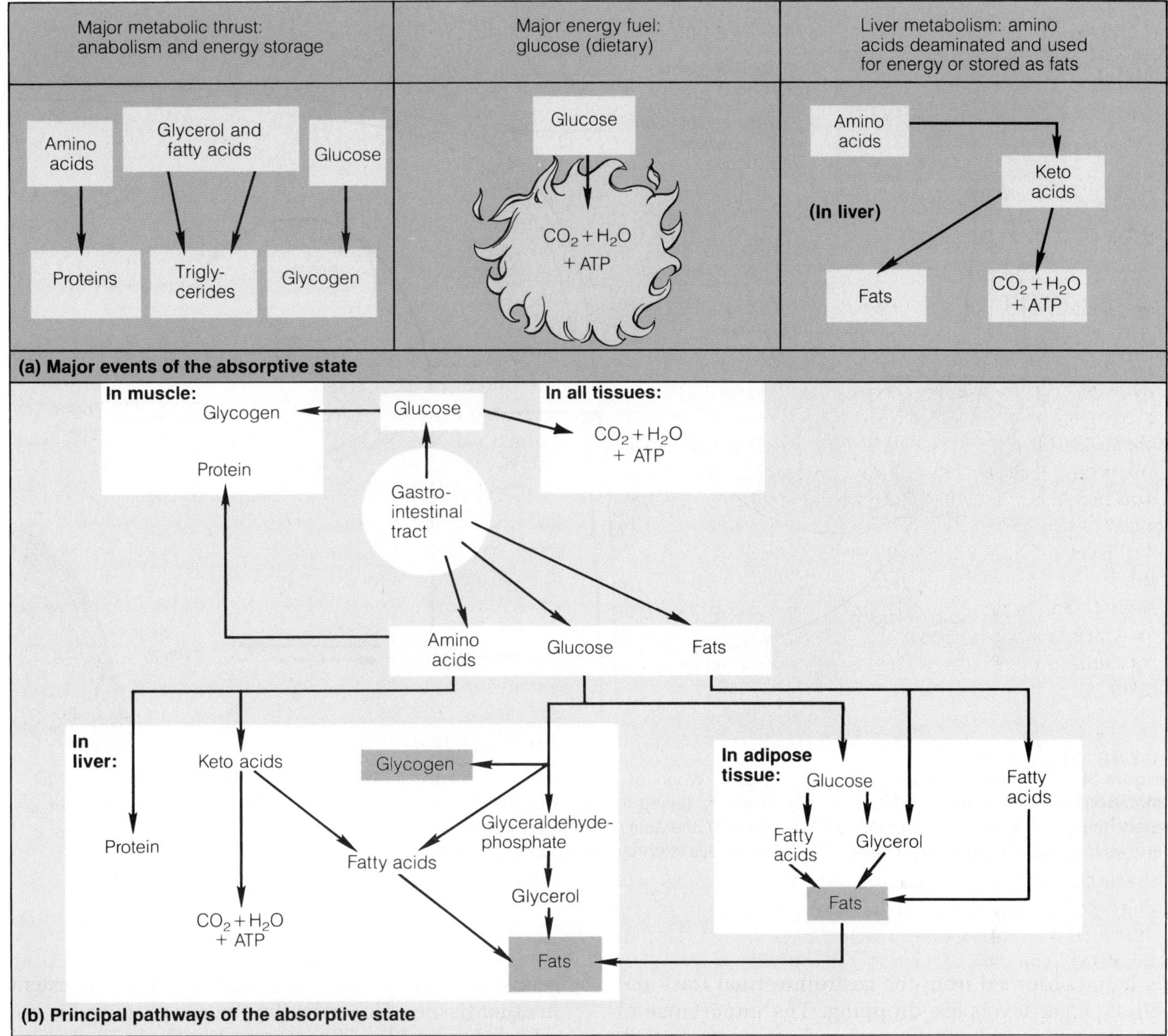

Figure 25.13 Summary of the major events and principal metabolic pathways of the absorptive state. (Although not indicated in part (**b**), amino acids are also taken up by tissue cells and used for protein synthesis, and fats are the primary energy fuel of muscle, liver cells, and adipose tissue.)

cells, thereby lowering blood sugar levels. Additionally, it enhances activities that accelerate oxidation of glucose or its conversion to glycogen or fat while simultaneously inhibiting any process that might increase blood sugar levels.

Diabetes mellitus, a consequence of inadequate insulin production or abnormal insulin receptors, is a metabolic disorder with far-reaching consequences. Without insulin (or receptors that "recognize" it), glucose becomes unavailable to most body cells; thus, blood glucose levels remain high, and large amounts of glucose are excreted in urine. Metabolic acidosis, protein wasting, and weight loss occur as large amounts of fats and tissue proteins are used for energy. (Diabetes mellitus is described in more detail in Chapter 17.) ■

Postabsorptive State

The primary goal during the postabsorptive state is to maintain blood glucose levels within the homeostatic range (90–100 mg glucose/100 ml) when no glucose

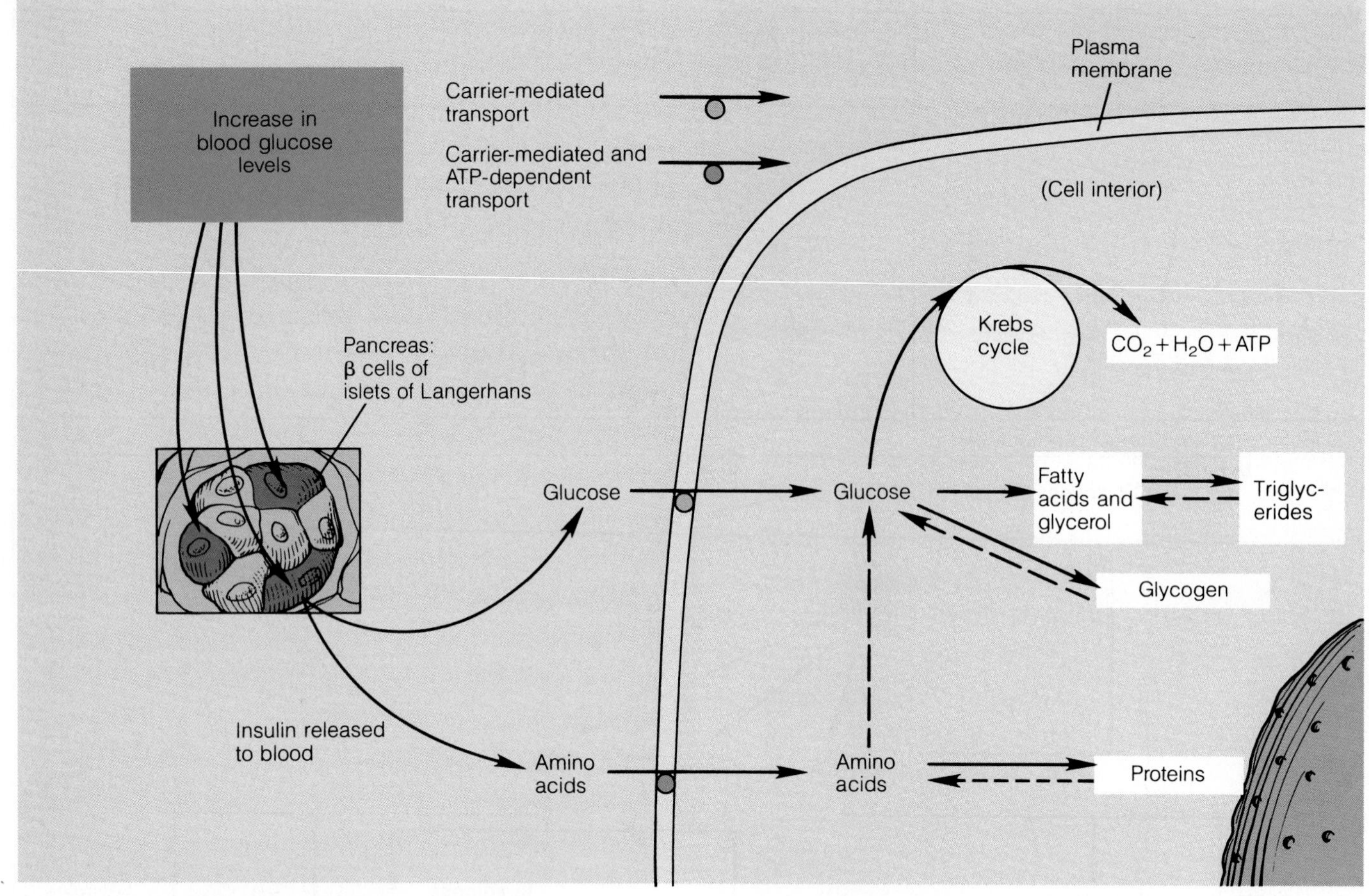

Figure 25.14 Effects of insulin on metabolism. Solid arrows indicate processes enhanced by insulin; dashed arrows indicate reactions inhibited by insulin. When insulin concentrations are low (i.e., during the postabsorptive period), the activity of the dashed-arrow pathways is enhanced and normal effects of insulin are inhibited. (Note that not all the effects indicated occur in all tissue cells.)

is being absorbed from the gastrointestinal tract and blood sugar levels are dropping. The importance of constant blood glucose has already been explained: The brain *must* use glucose as its energy source. Most events of the postabsorptive state either make glucose available to the blood or spare available glucose for the organs that most need it (Figure 25.15).

Sources of Blood Glucose. Glucose can be obtained from stored glycogen and fats and from tissue proteins by several routes:

1. Glycogenolysis in the liver. The liver's stores of glycogen (about 100 g) are the first line of glucose reserves. They are mobilized quickly and efficiently and can maintain blood sugar levels for about 4 hours during the postabsorptive period.

2. Glycogenolysis in skeletal muscle. Glycogen stores in skeletal muscle are approximately equal to liver reserves. Before liver glycogen is exhausted, glycogenolysis begins in skeletal muscle and to a lesser extent in other tissues. However, the glucose produced is not released to the blood because, unlike the liver, skeletal muscle does not have the enzymes needed to dephosphorylate glucose. Instead, glucose is partly oxidized to pyruvic acid (or, during anaerobic conditions, lactic acid), which enters the blood, is reconverted to glucose by the liver, and is released to the blood again. Thus, skeletal muscle contributes to blood sugar homeostasis indirectly, via liver mechanisms.

3. Lipolysis in adipose tissues and the liver. Adipose and liver cells produce glycerol by lipolysis, and the liver converts it to glucose (gluconeogenesis), which is released to the blood. Because acetyl CoA, the beta oxidation product of fatty acids, is produced beyond the *reversible* steps of glycolysis, fatty acids cannot be used to bolster blood sugar levels.

4. Catabolism of cellular protein. Tissue proteins become the major source of blood glucose when fasting is prolonged and glycogen and fat stores are close to exhaustion (and during stress accompanied by

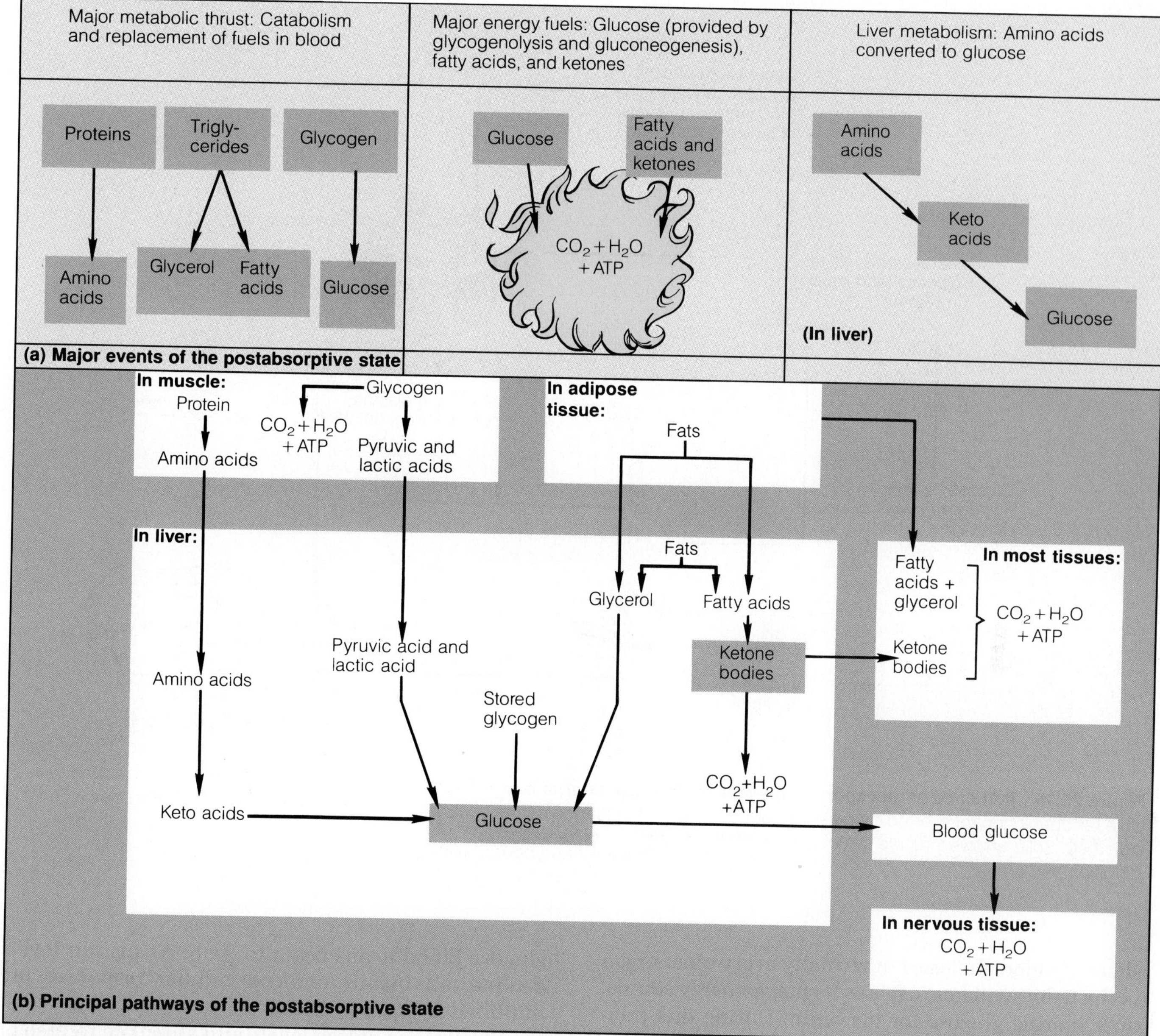

Figure 25.15 Summary of the major events and principal metabolic pathways of the postabsorptive state.

enhanced glucocorticoid secretion). The deamination and conversion of cellular amino acids to glucose (gluconeogenesis) occur mostly in the liver at first; but during extremely long fasts (several weeks), the kidneys also carry out gluconeogenesis and contribute fully as much glucose to the blood as does the liver.

Even in the dire situation of prolonged fasting, the body sets priorities. Muscle tissue is the first to go: Movement is not nearly as important as maintaining wound healing and the immune response. But as long as life continues, so does the body's production of ATP, which is needed to drive life processes. Obviously, there are limits to the amount of tissue protein breakdown that can occur before the body ceases to function. The heart is almost entirely muscle protein, and when it is severely cannibalized, the result is death.

Glucose Sparing. Even collectively, all of the manipulations to increase blood glucose are inadequate to provide energy supplies for prolonged periods. Luckily, the body can adapt to burn more fats and proteins, which enter the same metabolic pathways as glucose. The increased use of noncarbohydrate molecules (especially triacylglycerols) to conserve glucose is called **glucose sparing.**

As the body progresses from the absorptive to the postabsorptive state, the brain continues to take its

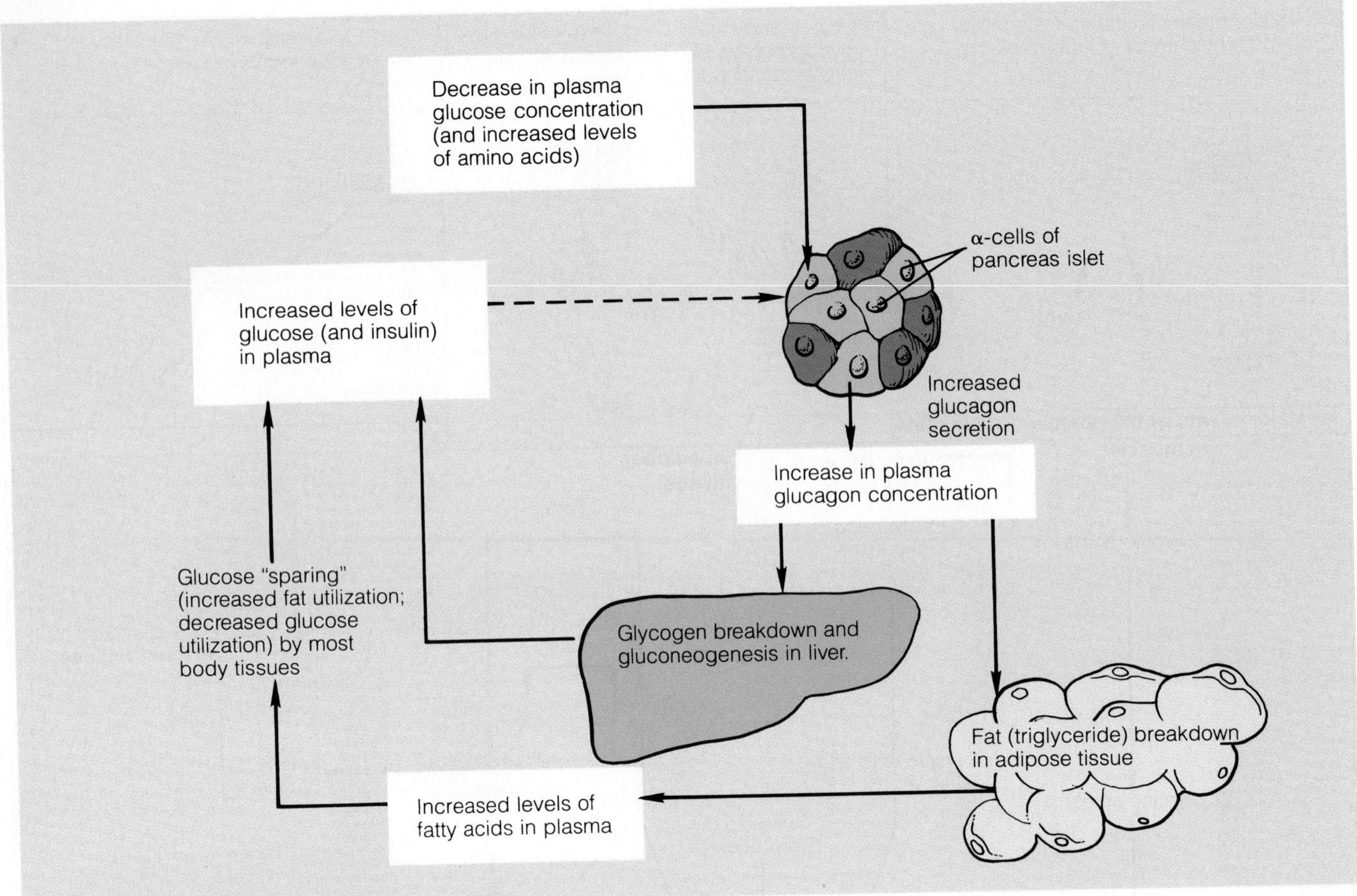

Figure 25.16 Influence of glucagon on plasma glucose concentrations. The negative-feedback control exerted by plasma glucose levels on glucagon secretion is also indicated. Solid arrows indicate stimulatory effects; the dashed arrow indicates inhibition.

share of blood glucose; but virtually every other organ in the body switches to fats as its major energy source, thus sparing glucose for the brain. During this transition phase, lipolysis begins in adipose tissues throughout the body and released fatty acids are picked up by tissue cells and oxidized for energy. In addition, the liver oxidizes fats to ketone bodies and releases them into the blood for use by tissue cells. If fasting continues for longer than four or five days, the brain too begins to use large quantities of ketone bodies as well as glucose as its energy fuel. The survival value of the brain's ability to use an alternative fuel source is obvious: Much less tissue protein has to be ravaged to form glucose.

Hormonal and Neural Controls. The sympathetic nervous system and several hormones interact to control events of the postabsorptive state. Consequently, regulation of this phase is much more complex than that of the absorptive state when a single hormone, insulin, holds sway.

An important trigger for initiating postabsorptive events is damping of insulin release, which occurs as glucose blood levels begin to drop. As insulin levels decline, all insulin-induced cellular responses are inhibited as well.

Declining glucose levels also stimulate release of the insulin antagonist **glucagon** by the alpha cells of the pancreatic islets. Like other hormones acting during the postabsorptive state, glucagon is a *hyperglycemic hormone,* that is, it promotes a rise in blood glucose. Glucagon targets the liver and adipose tissue (Figure 25.16). The hepatocytes respond by accelerating glycogenolysis and gluconeogenesis, whereas adipose cells mobilize their fatty stores (lipolysis) and release fatty acids and glycerol to the blood. Thus, glucagon acts to "refurbish" blood energy fuels by enhancing both glucose and fatty acid levels. Glucagon release is inhibited after the next meal or whenever blood sugar levels rise and insulin secretion is turned on once again.

Thus far, the picture is pretty straightforward. Increasing blood sugar levels trigger the release of insulin, which "pushes" glucose into the cells, and blood sugar levels decline. This stimulates the release of glucagon, which "pulls" glucose from the cells into the blood. However, there is more than a push–pull mechanism here, because *both* insulin and glucagon release are strongly stimulated by rising levels of amino acids in the blood. This effect is insignificant when we eat a normally balanced meal, but it has an important adaptive role when we eat a high-protein, low-carbohydrate meal. In this instance, the stimulus for insulin release is strong; and if it were not counterbalanced, the brain might be damaged by the resulting abrupt onset of hypoglycemia as glucose is rushed out of the blood. However, simultaneous release of glucagon modulates the effects of insulin and helps stabilize blood glucose levels.

The sympathetic nervous system plays a crucial role in supplying fuel quickly when blood sugar levels drop suddenly (Figure 25.17). Adipose tissue is well supplied with sympathetic fibers, and epinephrine released by the adrenal medulla in response to sympathetic activation acts on the liver, skeletal muscle, and adipose tissues. Together, these stimuli promote fat mobilization and glycogenolysis—essentially the same effects prompted by glucagon. Bodily injury, anxiety, anger, or any other stressor that mobilizes the fight-or-flight response (including declining blood glucose levels) will trigger this control pathway.

In addition to glucagon and epinephrine, a number of other hormones—including growth hormone, thyroxine, sex hormones, and corticosteroids—have important effects on metabolism and nutrient flow. Growth hormone secretion is enhanced by prolonged fasting or rapid declines in plasma glucose levels, and it exerts important anti-insulin effects (see Chapter 17, p. 539). However, the release and activity of most of these hormones are not specifically related to absorptive or postabsorptive events of metabolism. A checklist of the general metabolic effects of normal levels of these various hormones is provided in Table 25.4.

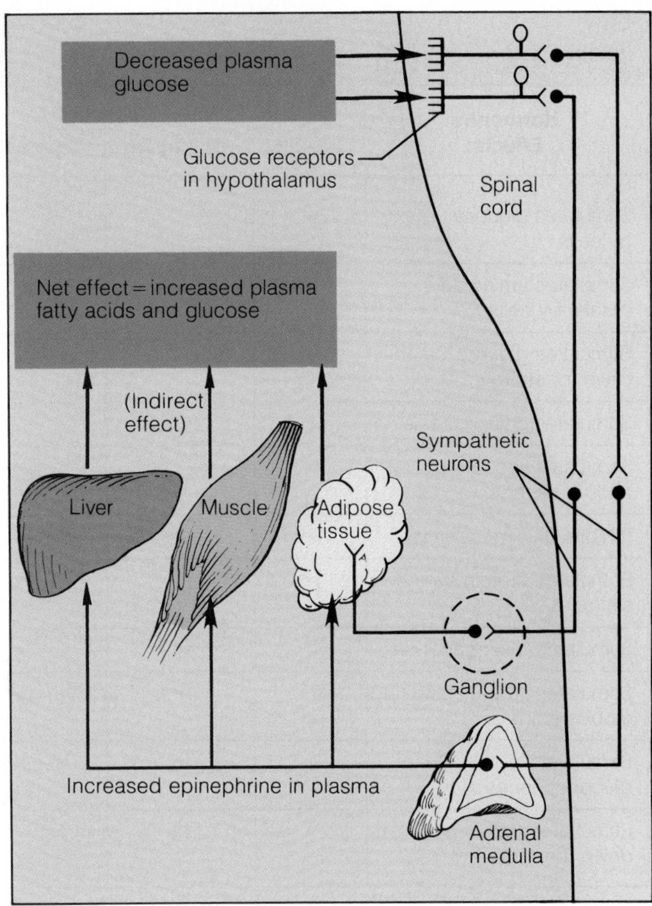

Figure 25.17 Sympathetic nervous system–epinephrine mechanism for providing glucose to the blood during the postabsorptive state. A decrease in blood glucose levels stimulates hypothalamic glucose receptors, and, via a reflex mechanism (assumed to be as illustrated), blood levels of glucose and fatty acids are increased.

Role of the Liver in Metabolism

The anatomy of the liver and its role in bile formation and digestive activities was described in Chapter 24. Here, we will focus on the liver's metabolic functions. Essentially, the liver processes nearly every class of nutrients absorbed from the digestive tract. In addition, it plays a major role in regulating plasma cholesterol levels.

General Metabolic Functions

The hepatocytes carry out some 500 or more intricate metabolic functions, including some already mentioned. The detailed credits that the liver deserves are well beyond the scope of this text, but a brief summary is provided next.

The liver (1) packages fatty acids into forms that can be stored or transported to be used for cellular fuel; (2) builds amino acids into plasma proteins, performs transamination to form nonessential amino acids, and converts the ammonia resulting from their deamination to urea, a less toxic excretory product; (3) stores glucose as glycogen and, via its central role in glycogenolysis and gluconeogenesis, regulates blood sugar homeostasis. In addition to these roles in fat, protein, and glucose metabolism, liver cells conserve iron salvaged from worn-out red blood cells and detoxify substances such as alcohol and drugs. These major metabolic functions are detailed in Table 25.5.

Cholesterol Metabolism and Regulation of Plasma Cholesterol Levels

Cholesterol, though a very important dietary lipid, has received little mention thus far, primarily because it is not used as an energy fuel. It serves instead as the structural basis of bile salts, steroid hormones,

Table 25.4 Summary of Normal Hormonal Influences on Metabolism

Hormone's Effects:	Insulin	Glucagon	Epinephrine	Growth hormone	Thyroxine	Cortisol	Testosterone
Stimulates glucose uptake by cells	✓				✓		
Stimulates amino acid uptake by cells	✓			✓			
Stimulates glucose catabolism for energy	✓				✓		
Stimulates glycogenesis	✓						
Stimulates lipogenesis and fat storage	✓						
Inhibits gluconeogenesis	✓						
Stimulates protein synthesis (anabolic)	✓			✓	✓		✓
Stimulates glycogenolysis		✓	✓				
Stimulates lipolysis and fat mobilization		✓	✓	✓	✓	✓	
Stimulates gluconeogenesis		✓	✓	✓		✓	
Stimulates protein breakdown (catabolic)						✓	

and vitamin D and as a major component of plasma membranes. Cholesterol is provided by the intake of animal foods and is also made from acetyl CoA by the liver and to a lesser extent by other body cells, particularly intestinal cells. Cholesterol is lost from the body when it is catabolized and secreted in bile salts, which are eventually excreted in feces.

Lipoproteins and Cholesterol Transport. Because fatty acids, triglycerides, and cholesterol are insoluble in water, they do not circulate free in the bloodstream. Instead, they are transported in body fluids bound to small lipid–protein complexes called **lipoproteins.** These complexes solubilize the hydrophobic lipids, and their protein parts contain signals that regulate the entry and exit of particular lipids at specific target cells.

The lipoproteins vary considerably in their relative fat–protein compositions, but they all contain triglycerides, phospholipids, and cholesterol in addition to protein (Figure 25.18). In general, the higher the percentage of lipid in the lipoprotein, the lower its density; and the greater the proportion of protein, the higher its density. On this basis, there are **high-density lipoproteins (HDLs), low-density lipopro-**

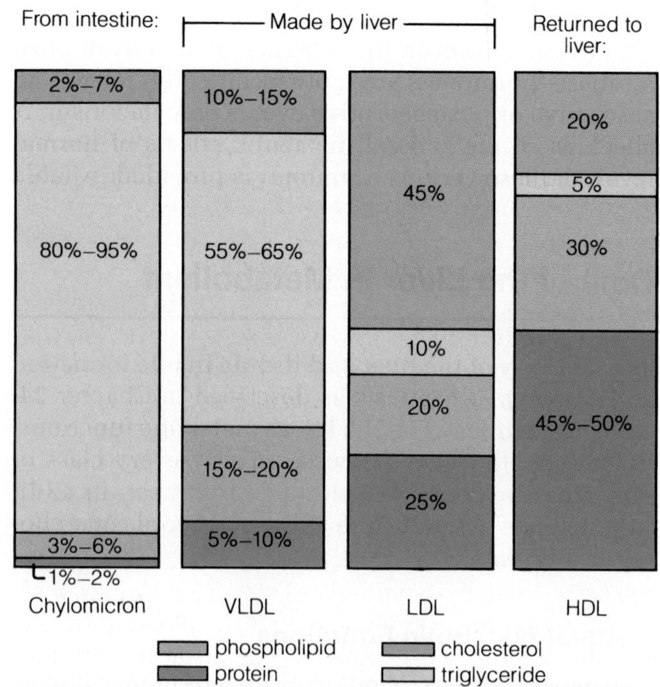

Figure 25.18 Composition of lipoproteins that transport lipids in body fluids. VLDL = very low density lipoprotein; LDL = low density lipoprotein; HDL = high density lipoprotein.

Table 25.5 Summary of Metabolic Functions of the Liver

Metabolic processes targeted	Functions
Carbohydrate metabolism Particularly important in maintaining blood glucose homeostasis	1. Converts galactose and fructose to glucose 2. Glucose buffer function: stores glucose as glycogen when blood glucose levels are high; under hormonal controls, performs glycogenolysis and releases glucose to blood 3. Gluconeogenesis: converts amino acids and glycerol to glucose when glycogen stores are exhausted and blood glucose levels are falling 4. Converts glucose to fats for storage
Fat metabolism Although most cells are capable of some fat metabolism, liver bears the major responsibility	1. Primary body site of beta oxidation (breakdown of fatty acids to acetyl CoA) 2. Converts excess acetyl CoA to ketone bodies for release to tissue cells 3. Stores fats 4. Forms lipoproteins for transport of fatty acids, fats, and cholesterol to and from tissues 5. Synthesizes cholesterol from acetyl CoA; catabolizes cholesterol to bile salts, which are secreted in bile
Protein metabolism Other metabolic functions of the liver could be dispensed with and body could still survive; however, without liver metabolism of proteins, severe survival problems ensue: many essential clotting proteins would not be made, ammonia would not be disposed of, and so on	1. Deamination of amino acids (required for their conversion to glucose or use for ATP synthesis); amount of deamination that occurs outside the liver is unimportant 2. Formation of urea for removal of ammonia from body; inability to perform this function (e.g., in cirrhosis or hepatitis) results in accumulation of ammonia in blood and depression of brain, a condition called *hepatic coma* 3. Forms most plasma proteins (exceptions are gamma globulins and some hormones and enzymes); plasma protein depletion causes rapid mitosis of the hepatocytes and actual growth of liver, which is coupled with increase in synthesis of plasma proteins until blood values are again normal 4. Transamination: intraconversion of nonessential amino acids; amount that occurs outside liver is inconsequential
Miscellaneous Vitamin/mineral storage	1. Stores a 1–2-year supply of vitamin A 2. Stores sizable amounts of vitamin D and B_{12} (1–4-month supply) 3. Stores iron; other than iron bound to hemoglobin, most of body's supply is stored in liver as ferritin until needed; releases iron to blood as blood levels drop
Biotransformation functions	1. Relative to drug metabolism, performs synthetic reactions which yield inactive products that can be secreted by the kidneys and nonsynthetic reactions which may result in products that are more active, changed in activity, or less active 2. Processes bilirubin resulting from RBC breakdown and excretes bile pigments in bile 3. Metabolizes blood-borne hormones to forms that can be excreted in urine

teins (LDLs), and **very low-density lipoproteins (VLDLs).** Chylomicrons, which transport dietary triglycerides, cholesterol, and other lipids from the gastrointestinal tract, are considered a separate class and have the lowest density of all.

The liver is the primary source of VLDLs, which transport triglycerides synthesized or processed in the liver *to adipose tissues.* Once the triglycerides are unloaded to the adipocytes, the residues of these VLDLs are converted into LDLs, which are cholesterol-rich. The role of the LDLs is to transport cholesterol *to the peripheral tissues,* making it available to the tissue cells for membrane or hormone synthesis and for storage for later use. The LDLs also regulate cholesterol synthesis in the tissue cells. The major function of HDLs, which are particularly rich in phospholipids and cholesterol, is to transport cholesterol from the peripheral tissues *to the liver,* where it is broken down and becomes part of bile.

High plasma levels of cholesterol (above 175 mg cholesterol/100 ml blood) have been linked to risk of arteriosclerosis, which clogs the arteries and causes strokes and heart attacks. However, it is not enough to simply measure total cholesterol count; clearly, the form in which cholesterol is transported in the blood is of major clinical importance. As a rule, lipid profiles revealing high levels of HDLs are considered good because the transported cholesterol is destined for degradation. High LDL levels are considered bad, because when LDLs are excessive, potentially lethal cholesterol deposits are laid down in the artery walls.

Factors Regulating Plasma Cholesterol Levels. There is a negative feedback loop that partially adjusts the amount of cholesterol produced by the liver according to cholesterol intake in the diet. In other words, a high cholesterol intake inhibits its synthesis by the liver, but it is not a one-to-one relationship because the liver produces a certain basal amount of cholesterol even when dietary intake is excessive. For this reason, severe restriction of dietary cholesterol, although helpful, does not lead to a correspondingly steep reduction in plasma cholesterol levels.

The relative proportion of saturated and unsaturated fatty acids in the diet has an important effect on plasma cholesterol levels. Saturated fatty acids *stimulate hepatic synthesis* of cholesterol while *inhibiting its excretion* from the body. Thus, moderate decreases in intake of saturated fats (found chiefly in animal fats and coconut oil) can reduce plasma cholesterol levels as much as 15% to 20%. In contrast, unsaturated fatty acids (found in most vegetable oils) *enhance excretion* of cholesterol and its catabolism to bile salts, thereby reducing total plasma cholesterol levels. In addition, certain unsaturated fatty acids (omega-3 fatty acids) found in fish oils lower the proportions of both saturated triglycerides and LDLs. These omega-3 fatty acids also seem to make the blood platelets less sticky, thus helping to prevent spontaneous clotting that can block blood vessels.

Factors other than diet also seem to influence plasma cholesterol levels. For example, cigarette smoking, coffee drinking, and stress have been implicated in increased LDL levels, whereas regular aerobic exercise has been shown to lower LDL levels and increase levels of HDLs.

Most cells other than liver and intestinal cells obtain the bulk of the cholesterol they need for membrane synthesis from the plasma. When a cell needs cholesterol, it makes receptor proteins for LDL and inserts them in its plasma membrane. LDL binds to the receptors and is engulfed in coated pits by endocytosis; within 10 to 15 minutes the endocytotic vesicles are fused with lysosomes, where the cholesterol is freed for use. When excessive cholesterol accumulates within a cell, it inhibits both the cell's own cholesterol synthesis and its synthesis of LDL receptors.

Body Energy Balance

When any fuel is burned, it consumes oxygen and liberates heat. The "burning" of food fuels by our cells is no exception. As described in Chapter 2, energy is finite. It can be neither created nor destroyed—only converted from one form to another. This principle (actually the first law of thermodynamics), applied to cell metabolism, means that bond energy released as foods are catabolized must be precisely balanced by the total energy output of the body. Thus, a dynamic balance exists between the body's energy intake and its energy output:

$$\text{Energy intake} = \text{total energy output} \\ (\text{heat} + \text{work} + \text{energy storage})$$

Energy intake is considered equivalent to the energy liberated during food oxidation. Undigested foods are not considered in the equation because they contribute no energy. Energy output includes the energy immediately lost as heat (about 60% of the total), plus that used to do work (driven by ATP), plus energy stored in the form of fat or glycogen. (Because losses of organic molecules in urine, feces, and perspiration are negligible in healthy people, they are usually ignored.) A close look at this situation reveals that *nearly all the energy derived from foodstuffs is eventually converted to heat:* Heat is lost during virtually every cellular activity—when ATP bonds are formed and when they are cleaved to do work, during muscle contraction, and through friction as blood flowing through blood vessels encounters resistance. Though the cells cannot use this energy for doing work, the heat warms the tissues and blood, thereby maintaining the homeostatic body temperature that allows metabolic reactions to occur at efficient levels. Energy storage becomes an important part of the equation only during periods of growth and during net fat deposit.

Regulation of Food Intake

When energy intake and energy outflow are in balance, body weight remains stable; when they are not, weight is either gained or lost. Body weight in most people is surprisingly stable, indicating the existence of physiological mechanisms that control food intake (and hence, the magnitude of food oxidation) or heat production or both.

The control of food intake presents difficult questions to researchers. For example, what type of receptor could possibly sense the total body content of calories and alert one to start or stop eating? Despite heroic research efforts, no such signal has been found. Current theories of how feeding behavior is regulated focus on one or more of four factors: nutrient signals related to total energy stores, hormones, body temperature, and psychological factors. All these factors appear to exert their effects through feedback signals to the feeding centers of the brain. Brain receptors are thought to include thermoreceptors and a variety of chemoreceptors (for glucose and probably glycerol, insulin, and others). For years, it was assumed that two hypothalamic nuclei acted in tandem to regulate hunger and satiety. However, recent evidence indicates that many other brain areas also play roles in regulating feeding behavior.

Nutrient Signals Related to Energy Stores

At any time, plasma levels of glucose, amino acids, fatty acids, and glycerol provide a good deal of information to the brain that may be helpful in adjusting energy intake to energy output. For example:

A CLOSER LOOK Obesity: Magical Solution Wanted

How fat is too fat? What distinguishes an obese person from one who is merely overweight? The bathroom scale is an inaccurate guide because body weight tells nothing of body composition. A skilled dancer with dense bones and well-developed muscles may weigh several pounds more than an inactive person of the same size. Most experts agree that a person is obese when he or she is 20% heavier than the "ideal weight" published in insurance company tables (which, by the way, are artificially low for the most healthy state). What is really needed is a measure of body fat, because the most common view of obesity is that it is a condition of excessive triglyceride storage. Although we bewail our inability to get rid of fat from our genetically determined fat depots, the real problem is that we keep refilling the storehouses by consuming too many calories. A body fat content of 18% to 22% of body weight (males and females, respectively) is considered normal for adults; having fat stores that exceed these amounts is defined as obesity.

However it's defined, obesity is a perplexing and poorly understood disease. The term "disease" is appropriate because all forms of obesity involve some imbalance in food intake control mechanisms. Despite its well-known adverse effects on health (the obese have a higher incidence of arteriosclerosis, hypertension, coronary artery disease, and diabetes mellitus), it is the most common health problem in the United States. About 20% of teenagers and 50% of adults are obese. In addition to the health problems mentioned, the obese may store excessive levels of toxic chemicals in their bodies. The greater the amount of fat in the body, the greater the potential for fat-soluble toxins, such as the potent insecticide DDT and PCB (a cancer-causing chemical), to collect and persist in the system for long periods of time. Because DDT interferes with the liver's ability to rid the body of other toxins, these effects may be very far-reaching.

As if this were not enough, the social stigma and economic disad-

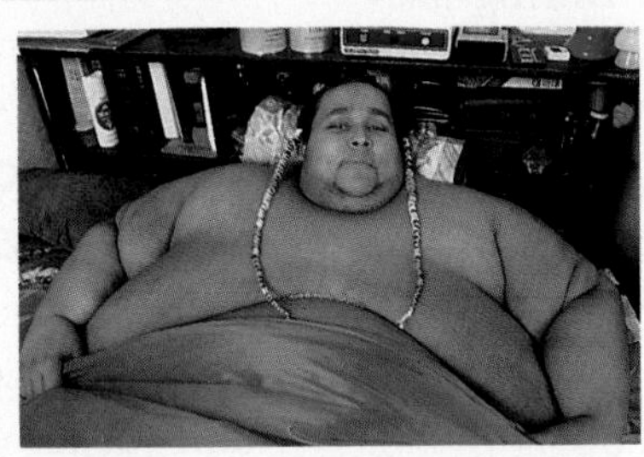

vantages of obesity are legendary. A fat person pays higher insurance premiums, is discriminated against in the job market, has fewer clothing choices, and is frequently humiliated both during childhood and adulthood.

With all its attendant problems, it's a pretty fair bet that few people choose to become obese. So what causes obesity? Some believe that overeating behaviors develop early in life (the "clean your plate" syndrome) and set the stage for adult obesity by increasing the number of fat cells formed during childhood. The number of adipose cells stabilizes during early adulthood. Thereafter, increases in adipose tissue mass occur by depositing more fat in the existing cells. Thus, the more cells there are, the more fat that can be stored. Further, when there are armies of incompletely filled fat cells, basal plasma levels of fatty acids and glycerol are lower and the stimulus for eating is greater, leading to abnormal hunger and a vicious self-perpetuating cycle for filling the fat depots. Possible causes of obesity cited by other theories include insensitivity of brain "monitors" to glucose or insulin, a satiety set point that is abnormally high, the use of food as a substitute for unfilled emotional needs, and genetic factors.

Although it is often assumed that the obese eat more than other people, this is not always true. In fact, many of the obese consume substantially fewer calories than people of normal weight. The "tickler" here seems to be activity level, which is typically lower in the obese than in people of more normal weight. This stresses the often-forgotten fact that there are two sides to the energy balance equation: If caloric intake is not balanced by energy output, surplus calories will be deposited in fat. Further, low activity levels actually stimulate eating while physical exercise depresses food intake (by increasing blood levels of glycerol) and increases the metabolic rate of the muscles not only during activity but also for some time after it.

Rumors and poor choices for dealing with obesity abound. Some of the most unfortunate strategies used for coping with obesity are listed here.

1. **"Water pills."** Diuretics, which prompt the kidneys to excrete more water, are sought as a means of losing weight. At best, these may cause a few pounds of weight loss for a few hours; they can also cause severe electrolyte imbalance and dehydration.

2. **Diet pills.** Some obese use amphetamines (best known as dexedrine and benzedrine) to reduce appetite. These work, but only temporarily (until tolerance develops) and can cause a dangerous dependency. Then the user has the problem of trying to dispense with the drug habit.

3. **Fad diets.** Many magazines print at least one new diet regimen yearly, and books on dieting are best-sellers. However, many of these diets are nutritionally unsound, particularly if they limit certain groups of nutrients.

4. **Surgery.** Sometimes sheer desperation prompts surgical solutions, such as having the jaws wired shut, intestinal bypass surgery, and lipectomy (removal of fat by suction). Any type of surgery should be carefully considered. Intestinal bypass surgery removes or disconnects part of the small intestine so that fewer nutrients can be absorbed. It is major surgery, accompanied by all the usual risks plus the risk of liver failure, intestinal infection, malnutrition, and massive bouts of diarrhea.

Unfortunately, there is no magical solution for obesity. The only way to lose weight is to take in fewer calories and increase physical activity. The only way to keep the weight off is to make these dietary and exercise changes lifelong habits.

1. During eating, plasma glucose levels rise and cellular metabolism of glucose increases. Detection of rising blood sugar levels by special glucose receptors in the brain alerts brain areas controlling feeding and ultimately depresses eating. During fasting, this signal would be absent and result in hunger and a "turning on" of food-seeking behaviors.
2. Elevated plasma levels of amino acids depress eating, whereas low amino acid levels in blood stimulate it. Neither the precise mechanism nor the receptor type mediating these effects is known.
3. Blood concentrations of fatty acids and glycerol may serve as indicators of the body's total energy stores (in adipose tissue) and act as a long-term mechanism for controlling hunger. According to this theory, the greater the fat reserves, the greater the basal amounts of fatty acids and glycerol released to the blood and the more eating behavior is inhibited.

Hormones

Blood levels of hormones that regulate plasma nutrient levels during the absorptive and postabsorptive states may also serve as negative feedback signals to the brain. Insulin, which is released during food absorption, is known to depress hunger and is presumed to be the most important satiety signal. In contrast, glucagon levels rise during fasting and stimulate hunger. Other possible hormonal controls include epinephrine (released during fasting) and cholecystokinin, an intestinal hormone secreted during food digestion; if these hormones are involved, it is assumed that epinephrine triggers hunger while cholecystokinin depresses it.

Body Temperature

Increased body temperature, related to ingestion and liver processing of foods, may inhibit eating behavior. Operation of such a thermal signal would help to explain why people in cold climates normally eat more than those in more temperate or warm regions. If a mechanism of this sort does exist, the brain's thermoregulation centers must interact with the feeding control centers.

Psychological Factors

All mechanisms described thus far are totally reflexive, but their final outcome may be strongly reinforced or inhibited by psychological factors that have little to do with caloric balance, such as the sight, taste, smell, or even thought of food. Psychological factors are thought to be very important in the obese. However, even when psychological factors are the underlying cause of obesity, these individuals do *not* continue to gain weight endlessly. Their feeding controls still operate, but have a higher "set point," which maintains total energy content at higher-than-normal levels.

Metabolic Rate and Body Heat Production

The body's energy expenditure or energy use per time unit (usually per hour) is called the **metabolic rate.** Metabolic rate, which is the sum of heat produced by all the chemical reactions and mechanical work of the body, can be measured directly or indirectly. In the *direct method,* the person enters a chamber called a *calorimeter* and heat liberated by the body is absorbed by water circulating around the chamber. The rise in the water's temperature is directly related to the heat produced by the person's body. The *indirect method* uses a *respirometer* (Figure 25.19) to measure oxygen consumption, which is directly proportional to heat production. For each liter of oxygen used, the body produces about 4.8 kcal of heat.

Because many factors influence metabolic rate, it is usually measured under standardized conditions. The person is in a postabsorptive state (has not eaten for at least 12 hours), is reclining, and is mentally and physically relaxed. The temperature of the room is maintained at a comfortable 20°–25°C. The measurement obtained under these circumstances, called the **basal metabolic rate (BMR),** reflects the energy the body needs to perform only its most essential activities, such as breathing and maintaining cellular membrane potentials and resting levels of neural, cardiac, liver, and kidney function. The BMR, often referred to as the "energy cost of living," is reported in kilocalories per square meter of body surface per hour (kcal/m^2/h). Although named the *basal* metabolic rate, this measurement is not the lowest metabolic state of the body. That situation occurs during sleep, when the muscles are completely relaxed.

An average 70-kg (154-pound) adult has a BMR of approximately 60–72 kcal/h. A quick approximation of one's BMR can be calculated by multiplying weight in kilograms (note that 2.2 pounds = 1 kg) times the factor 1 for males or the factor 0.9 for females.

Many factors influence BMR, including surface area, age, gender, stress, and hormones. Although BMR is related to overall body weight and size, the critical factor is surface area rather than weight itself. This reflects the fact that as the ratio of body surface area to body volume increases, heat loss to the environment increases and the metabolic rate must be higher to replace the lost heat. Hence, if two people weigh the same, the taller or thinner person will have a higher BMR than the shorter or fatter person.

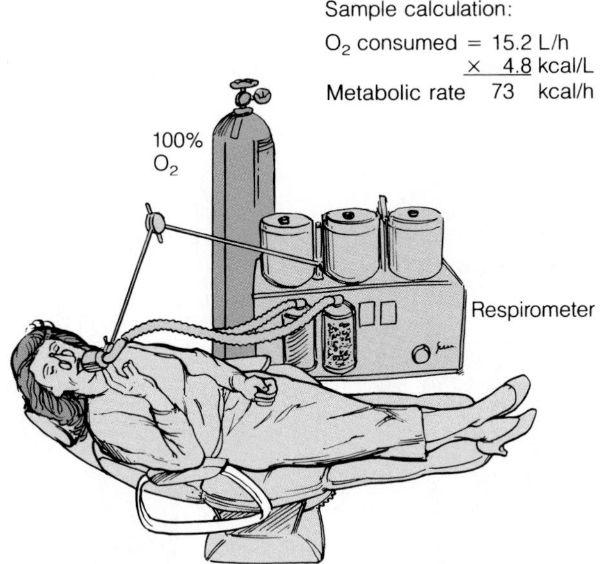

Figure 25.19 Indirect measurement of basal metabolic rate by respirometry. The indirect method for measuring metabolic rate is based on the fact that oxygen usage and heat liberation during food oxidation are directly proportional. The person breathes into a respirometer, and the total amount of oxygen consumed during testing is measured. The average amount of oxygen consumed per hour (L/h) is multiplied by 4.8 (the amount of energy liberated when 1 L of oxygen is consumed in the oxidation of protein, carbohydrate, or fat) to calculate the metabolic rate.

In general, the younger a person, the higher the BMR. Children and adolescents require large amounts of energy for growth and have relatively higher BMRs. In old age, BMR declines dramatically as skeletal muscles begin to atrophy. (This helps explain why those elderly who fail to decrease their caloric intake become obese.) Gender also plays a role, metabolic rate being disproportionately higher in males than in females. Males typically have greater amounts of muscle tissue, which is very active metabolically even during rest; fatty tissue, which is in greater relative amounts in females, is metabolically sluggish in comparison with muscle tissue. Body temperature tends to rise and fall with metabolic rate. Fever (hyperthermia), resulting from infections and other factors, results in a markedly higher metabolic rate. Stress, whether physical or emotional, increases metabolic rate by mobilizing the sympathetic nervous system. Both norepinephrine (released by sympathetic fibers) and epinephrine (released by adrenal medullary cells) provoke an increase in metabolic rate primarily by stimulating fat catabolism.

The amount of **thyroxine** produced by the thyroid gland is probably the most important hormonal factor in determining BMR; hence, thryroxine has been dubbed the "metabolic hormone." Its direct effect on most body cells (with the exception of brain cells) is to enhance their oxygen consumption, presumably by accelerating the use of cellular ATP for operation of the sodium-potassium pump. As intracellular ATP reserves decline, cellular respiration is accelerated; thus, the more thyroxine produced, the higher the BMR. In the past, most BMR testing was done to investigate whether sufficient thyroxine was being made. Today, thyroid activity is more easily assessed by blood tests.

Hyperthyroidism causes a host of effects resulting from the excessive metabolic rate it produces. The body catabolizes stored fats and tissue proteins, and, despite increased hunger and food intake, the person often loses weight. Bones become weak and body muscles, including the heart, begin to atrophy. In contrast, *hypothyroidism* results in slowed metabolism, obesity, and diminished thought processes. ■

The term **total metabolic rate (TMR)** refers to the total rate of kilocalorie consumption to fuel *all* ongoing activities—involuntary and voluntary. The BMR accounts for a surprisingly large part of the TMR. For example, a woman whose total energy needs per day are about 2000 kcal may spend well over half of these calories (1400 kcal or so) supporting vital body activities. Skeletal muscle activity results in the most significant and dramatic short-term changes in TMR, reflecting the fact that skeletal muscles make up nearly half of body mass. Even slight increases in muscular work can cause remarkable leaps in metabolic rate and body heat production. For example, vigorous exercise by a well-trained athlete for several minutes at a time can increase the metabolic rate to 20 times normal, and it remains elevated for several hours afterward. Surprisingly, physical training has little real effect on BMR. Although it would appear that athletes, with their greatly enhanced muscle mass, should have substantially higher BMRs than do nonathletes, there is little difference between those of the same sex and surface area during resting conditions. Ingestion of food also increases the total metabolic rate. This effect, called the *specific dynamic action* or *dietary thermogenesis*, is greatest when proteins are eaten. Because intravenous injection of amino acids causes the same effect, the metabolism-increasing effects of amino acids appear to reflect much more than just the added energy needed to break down and absorb nutrients. The heightened metabolic activity of the liver during such periods probably accounts for the bulk of additional energy use. In contrast, fasting or very low caloric intake depresses the metabolic rate and results in a slower breakdown of body reserves.

Regulation of Body Temperature

As shown in Figure 25.20, body temperature reflects the balance between heat production and heat loss. Although all body tissues produce heat, those that are

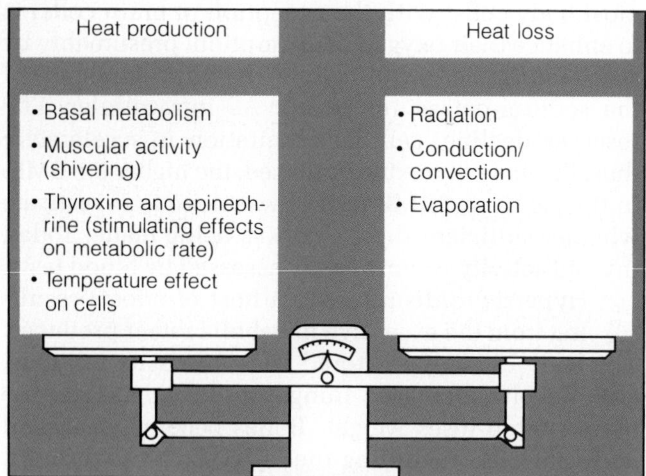

Figure 25.20 As long as heat production and heat loss are properly balanced, body temperature remains constant. Factors contributing to heat production (and temperature rise) are shown on the left side of the scale; those contributing to heat loss (and temperature fall) are shown on the right side of the scale.

most active metabolically produce the greatest amounts. When the body is at rest, the bulk of heat is generated by the liver, heart, brain, and endocrine organs; the inactive skeletal muscles account for 20% to 30% of body heat. However, this situation is changed by even slight alterations in muscle tone; and during vigorous exercise, the amount of heat produced by the skeletal muscles can be 30 to 40 times that produced by the rest of the body. For this reason, a change in muscle activity is one of the most important means of modifying body temperature.

Body temperature is usually maintained within a very narrow range of 36.1°–37.8°C (97°–100°F), regardless of the external temperature or how much heat the body is producing. The adaptive value of this precise temperature homeostasis relates to the effect of temperature on the rate of biochemical reactions, particularly enzyme activity. When body temperature is normal, conditions are optimal for enzymatic activity. As body temperature rises, catalysis is accelerated: With each rise of 1°C, the rate of chemical reactions becomes approximately 10% faster. As the temperature continues to spike upward and goes beyond the homeostatic range, neurons are depressed and proteins begin to degrade. Most people go into convulsions when their temperature reaches 41°C (106°F), and 43°C is the absolute limit for life. In contrast, most body tissues can withstand marked reductions in temperature as long as other conditions are carefully controlled. This fact underlies the use of body cooling during open heart surgery when the heart must be stopped. The low body temperature induced reduces metabolic rate (and consequently nutrient requirements of body tissues and the heart), allowing more time for surgery without incurring tissue damage.

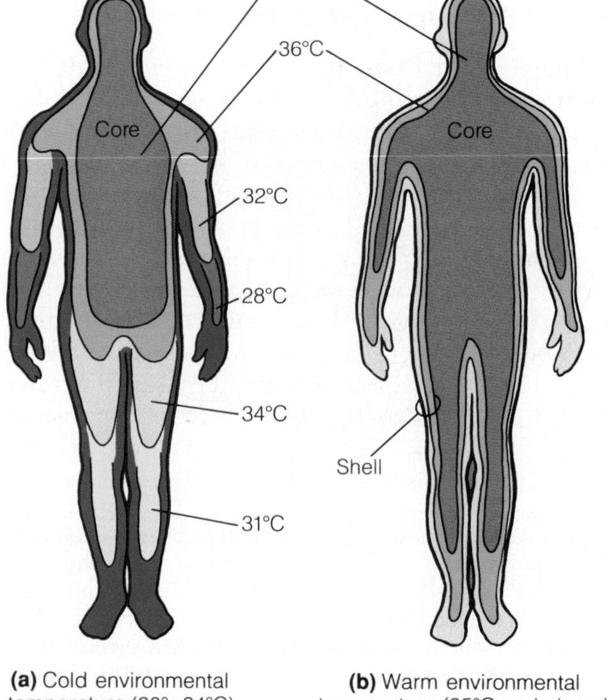

Figure 25.21 The temperature of the body core remains constant over a wide range of temperatures in the external environment. However, as illustrated in the diagrams, the thickness of the shell (heat exchange surface) (**a**) increases as the temperature of the external environment declines, and (**b**) decreases when the external environment is warm.

Core and Shell Temperatures

Different regions of the body have different temperatures at rest (Figure 25.21). The body's *core* (that is, organs within the skull and the thoracic and abdominal cavities) has the highest temperature; its *shell*, or heat loss surface—essentially, the skin—has the lowest temperature. Of the two body sites used routinely to obtain body temperature clinically, the rectum typically has a temperature about 0.5°C higher than the oral cavity and is considered more indicative of the core temperature.

Blood serves as the major *heat transfer* or *exchange agent* between the body core and shell. Whenever the shell is warmer than the external environment, heat is lost from the body; thus, when heat must be dissipated, warm blood is allowed to flush into the skin capillaries. On the other hand, when heat must be conserved, blood largely bypasses the skin capillaries, which reduces heat loss and allows the shell temperature to fall to that of the environment. Thus, the relative "thickness" of the shell changes with body activity and external temperature (Figure 25.21).

Role of the Hypothalamus

Although other brain regions contribute, the hypothalamus is the major integrating center for thermoregulation. The *heat-loss center* (a parasympathetic center) is located anteriorly in the preoptic area; the *heat-promoting center* (a sympathetic center) is in the posterior hypothalamus.

The hypothalamus receives afferent input both from **peripheral thermoreceptors,** which are located in the skin, and from **central thermoreceptors** (receptors sensitive to the temperature of blood), which are located in the body core, including the hypothalamus itself. Much like a thermostat, the hypothalamus responds to this input by initiating the appropriate heat-promoting or heat-loss reflex mechanisms via autonomic effector pathways (Figure 25.22). Although the central thermoreceptors are more critically located than the peripheral ones, changing inputs from the shell probably alert the hypothalamus that modifications must be made to prevent temperature changes in the core, i.e., they allow the hypothalamus to anticipate possible changes to be made.

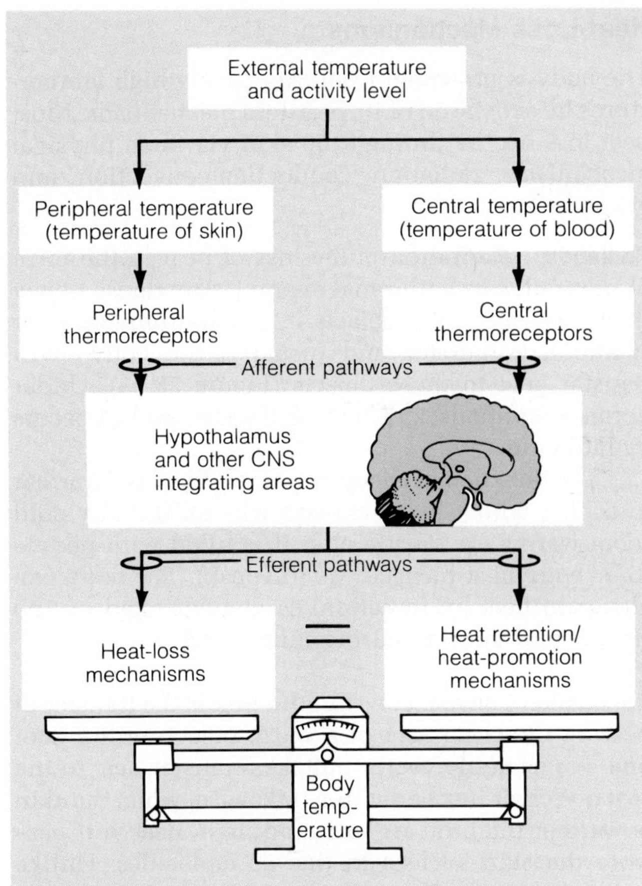

Figure 25.22 Central position of the hypothalamus and other CNS regulating centers in body temperature regulation.

Heat-Promoting Mechanisms

When the environmental temperature is cold (or the temperature of circulating blood falls), the hypothalamic heat-promoting center is activated. It, in turn, triggers one or more of the following mechanisms via sympathetic pathways to maintain or increase core body temperature (Figure 25.23).

1. Vasoconstriction. Activation of the sympathetic vasoconstrictor fibers serving the blood vessels of the skin causes them to be strongly constricted. As a result, blood is restricted to deep body areas and largely bypasses the skin. Because the skin is insulated by a layer of subcutaneous (fatty) tissue, heat loss from the shell is dramatically reduced and shell temperature drops toward that of the external environment.

Restriction of the circulation to the skin is not a problem for a brief period of time; but if it is extended (as during prolonged exposure to very cold weather), the skin cells, deprived of oxygen and nutrients, begin to die. This extremely serious condition is called *frostbite.* ■

2. Increase in metabolic rate. Cold stimulates the release of norepinephrine by sympathetic nerve fibers and of epinephrine by the adrenal medulla. These hormones elevate the rate of metabolic activity, which enhances heat production. This mechanism is referred to as **chemical thermogenesis.**

3. Shivering. If the mechanisms described above are not enough to handle the situation, shivering is triggered. Brain centers controlling muscle tone become more active (which in itself increases heat production); and when muscle tone has reached sufficient levels to alternately stimulate stretch receptors in antagonistic muscles, involuntary shuddering contractions of the skeletal muscles begin. Shivering is very effective in increasing body temperature, because muscle activity produces large amounts of heat.

4. Enhancement of thyroxine production. When environmental temperature decreases gradually, as in the transition from a hot to a cold season, the hypothalamus releases *thyrotropin releasing factor.* This activates the anterior pituitary to release *thyroid stimulating hormone,* which, in turn, induces the thyroid gland to liberate larger amounts of thyroxine to the blood. Because thyroxine increases metabolic rate, body heat production and the ability to maintain a constant body temperature at cold environmental temperatures increase.

Heat-Loss Mechanisms

The body is protected from excessively high temperatures by activation of its heat-loss mechanisms. Most heat loss occurs through the skin via three physical mechanisms: radiation, conduction/convection, and evaporation.

Radiation. **Radiation** is the loss of heat in the form of infrared waves (thermal energy). Any dense object that is warmer than objects in its environment—for example, a radiator and (usually) the body—will transfer heat to those objects (Figure 25.23). Under normal conditions, 25% to 40% of body heat loss occurs by radiation.

The direction of energy flow is always from warmer to cooler, which helps explain why an initially cold room warms up shortly after it is filled with people (the "body heat furnace," so to speak). The body can also gain heat by radiation, as demonstrated by the warming of the skin during sunbathing.

Conduction/Convection. **Conduction** is the transfer of heat by direct contact of a warm object with a cool one. For example, warm buttocks transfer heat to the seat of a chair by conduction. Likewise, when the skin is warmer than the air surrounding it, heat will pass from the skin surface to the air molecules. Unlike radiation, conduction requires molecule-to-molecule *contact* of objects; that is, the thermal energy must move through a material medium.

When the body shell transfers heat to the surrounding air, convection also comes into play. Because warm air tends to expand and rise, and cool air (being denser) falls, the warm air enveloping the body is continually replaced by cooler air molecules. This process, called **convection,** substantially enhances heat exchange from the body surface to the air, because the cooler air absorbs heat by conduction more rapidly than the already-warmed air. Together, conduction and convection account for 15% to 20% of heat loss to the environment. These processes are enhanced by anything that moves air more rapidly across the body surface, such as wind or a fan, that is, by *forced convection.*

Evaporation. Water evaporates because its molecules absorb heat from the environment and become energetic enough (that is, vibrate fast enough) to escape as gas. The heat absorbed by water during evaporation is called **heat of vaporization.** Because water has a high specific heat, it absorbs a great deal of heat before vaporizing; thus, its evaporation from body surfaces removes substantial amounts of body heat. For every gram of water evaporated, about one-half kilocalorie (0.58 kcal) of heat is removed from the body.

There is a basal level of body heat loss due to the continuous evaporation of water from the lungs, from the mucosa of the mouth, and through the skin. The unnoticeable water loss occurring via these routes is called *insensible water loss* and the accompanying heat loss is called *insensible heat loss*. Insensible heat loss dissipates about 10% of the basal heat production of the body and is a constant; that is, it is not controllable by body temperature regulatory mechanisms. However, the regulatory mechanisms do initiate heat-promoting activities to counterbalance this heat loss when necessary.

Evaporative heat loss becomes an active (sensible) process when body temperature rises and sweating provides increased amounts of water for vaporization. Extreme emotional states activate the sympathetic nervous system, causing body temperature to rise one degree or so, and vigorous exercise can thrust body temperature upward as much as 2°–3°C (5°–6°F). During vigorous muscular activity, when sweating is profuse, 1–2 L/h of perspiration can be produced and evaporated, causing 2000 kcal of heat to be removed from the body each hour. This is over 30 times the amount of heat lost via insensible heat loss!

When sweating is profuse, the loss of water and salt (NaCl) may cause painful spasms of the skeletal muscles called *heat cramps*. This situation is easily rectified by ingesting fluids. ■

Regulation of Heat-Loss Mechanisms. How do the heat-loss mechanisms fit into the temperature regulation scheme? The answer is quite simple. Whenever core body temperature rises above normal, the hypothalamic heat-promoting center is inhibited. At the same time, the heat-loss center is activated and damps the activity of the vasoconstrictor fibers controlling skin blood vessels. As the blood vessels dilate, the skin vasculature becomes swollen with warm blood (Figure 25.23), permitting heat loss from the shell by radiation, conduction, and convection. If the body is extremely overheated or if the environment is so hot that heat cannot be lost by conduction, increased evaporation becomes necessary. The sweat glands are strongly activated and spew out huge amounts of perspiration. Evaporation of perspiration is an efficient means of ridding the body of surplus heat as long as the air is dry. However, when the air is very humid, evaporation occurs much more slowly. In such cases, the heat-liberating mechanisms cannot work well, and we feel miserable and irritable.

When normal heat loss processes become ineffective, the *hyperthermia*, or elevated body temperature, that ensues depresses the hypothalamus. As a result, all heat-control mechanisms are suspended, creating a vicious positive-feedback cycle: Sharply increasing temperatures increase the metabolic rate, which, in turn, increases heat production. The skin becomes hot and dry; and as the temperature continues to spiral upward, permanent brain damage becomes a distinct possibility. This condition, called *heat stroke,*

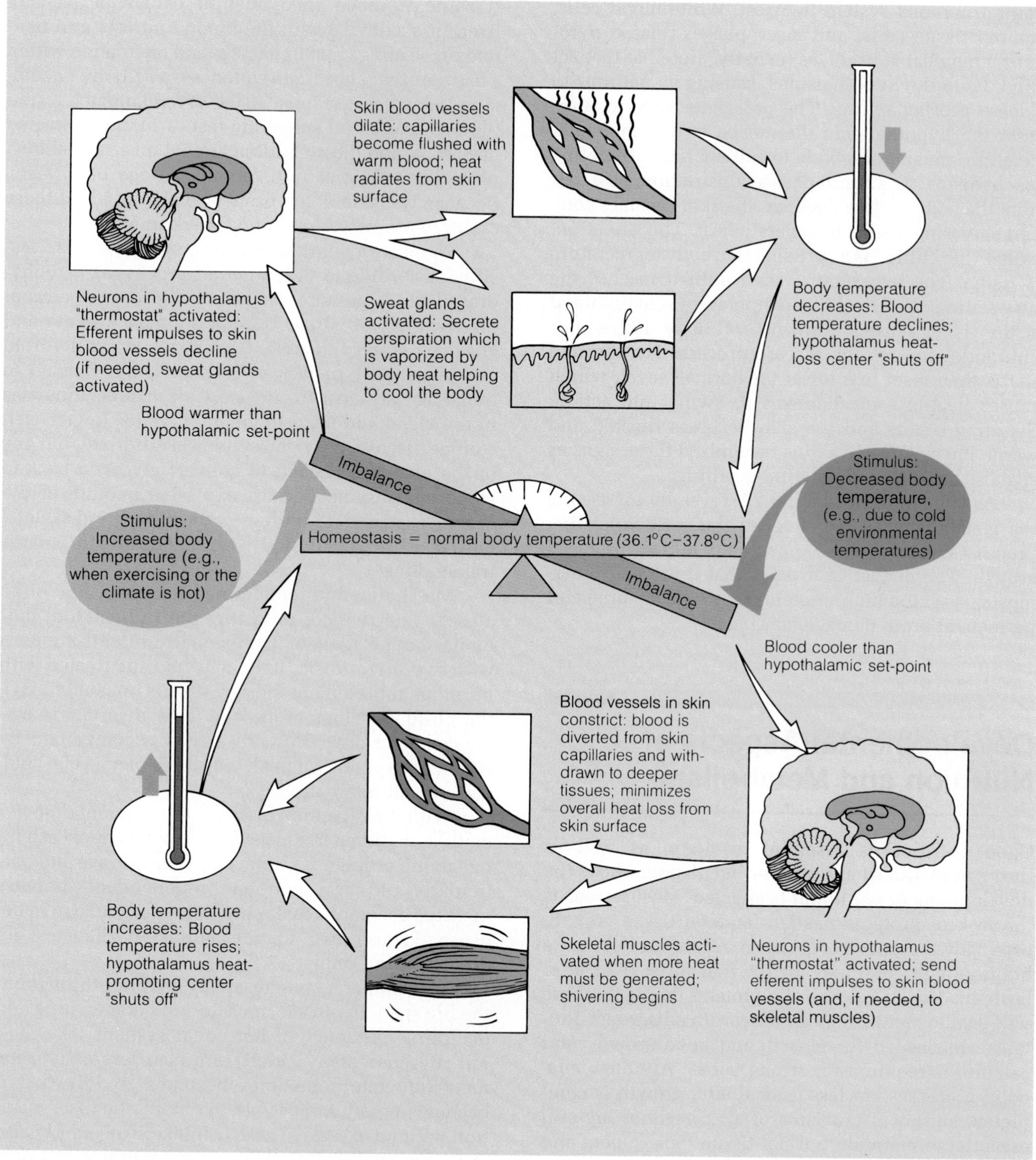
Skin blood vessels dilate: capillaries become flushed with warm blood; heat radiates from skin surface
Neurons in hypothalamus "thermostat" activated: Efferent impulses to skin blood vessels decline (if needed, sweat glands activated)
Sweat glands activated: Secrete perspiration which is vaporized by body heat helping to cool the body
Body temperature decreases: Blood temperature declines; hypothalamus heat-loss center "shuts off"
Blood warmer than hypothalamic set-point
Imbalance
Stimulus: Decreased body temperature, (e.g., due to cold environmental temperatures)
Stimulus: Increased body temperature (e.g., when exercising or the climate is hot)
Homeostasis = normal body temperature (36.1°C–37.8°C)
Imbalance
Blood cooler than hypothalamic set-point
Blood vessels in skin constrict: blood is diverted from skin capillaries and withdrawn to deeper tissues; minimizes overall heat loss from skin surface
Body temperature increases: Blood temperature rises; hypothalamus heat-promoting center "shuts off"
Skeletal muscles activated when more heat must be generated; shivering begins
Neurons in hypothalamus "thermostat" activated; send efferent impulses to skin blood vessels (and, if needed, to skeletal muscles)

Figure 25.23 Mechanisms of body temperature regulation.

can be fatal unless rapid corrective measures are initiated immediately (immersion in cool water and administration of fluids and electrolytes).

A much less serious consequence of extreme heat exposure is *heat exhaustion*. Heat exhaustion results from excessive loss of body fluids and electrolytes and is evidenced by low blood pressure and a cool, clammy skin. In contrast to heat stroke, heat-loss mechanisms are still functional in heat exhaustion. ■

Fever

Fever is *controlled hyperthermia*. Most often, it results from infection somewhere in the body, but it may be caused by other conditions (cancer, allergic reactions, central nervous system injuries). White blood cells, injured tissue cells, and macrophages release *pyrogens*, chemical substances (usually, proteins) that act directly on the hypothalamus, causing its neurons to release prostaglandins. The prostaglandins, in turn, reset the hypothalamic thermostat to a higher temperature, causing the body to initiate heat-promoting mechanisms. As a result of vasoconstriction, heat loss from the body surface declines, the skin becomes cool, and shivering begins to generate heat. This situation, called "the chills," is a sure sign that body temperature is rising. The temperature rises until it reaches the new setting; then the body temperature is maintained at the "fever setting" until natural body defenses or antibiotics reverse the disease process. The thermostat is then reset to a lower (or normal) level, which causes heat-loss mechanisms to swing into action: Sweating begins and the skin becomes flushed and warm. Physicians have long recognized these signs as signals that body temperature is falling.

As explained in Chapter 22, fever, by increasing the metabolic rate, helps speed the various healing processes, and it also appears to inhibit bacterial growth. The danger of fever is that if the body thermostat is set too high, proteins may be denatured and permanent brain damage may occur.

Developmental Aspects of Nutrition and Metabolism

Good nutrition is essential *in utero,* as well as throughout life. If the mother is ill nourished, the development of her infant is affected. Most serious is the lack of proteins needed for fetal tissue growth, especially, the growth of the brain. Additionally, because brain growth continues for the first three years after birth, inadequate calories and proteins during this time will lead to mental deficits or learning disorders. Proteins are needed for muscle and bone growth, and calcium is required for strong bones. Although anabolic processes are less critical after growth is completed, sufficient nutrients of all categories are still essential to maintain normal tissue replacement and metabolism.

There are many inborn errors of metabolism (or genetic disorders), but perhaps the two most common are *cystic fibrosis* and *phenylketonuria* (feh″-nul-kē″-tō-ner′-ē-uh) *(PKU)*. Cystic fibrosis is described in Chapter 24. In PKU, the tissue cells are unable to use one particular amino acid, phenylalanine (feh″-nul-al′-uh-nēn), which is present in all protein foods. The defect involves a deficiency of the enzyme phenylalanine hydroxylase, which converts phenylalanine to tyrosine. Because phenylalanine cannot be metabolized, it accumulates in the blood and acts as a neurotoxin, leading to brain damage and retardation within a few months. These consequences of PKU are uncommon today because most states have mandated a simple urine or blood screening test to identify affected newborns, and these children are put on a special low-phenylalanine diet until about the age of 6 years. Because tyrosine is the basis of melanin, all of these children are very blond and fair-skinned.

There are a number of other enzyme defects, classified according to the impaired process as carbohydrate, lipid, or mineral metabolic disorders. For example, two carbohydrate deficits are galactosemia and glycogen storage disease. *Galactosemia* results from an abnormality or lack of liver enzymes needed to transform galactose to glucose. Galactose accumulates in the blood and leads to mental deficits. In *glycogen storage disease*, glycogen synthesis is normal, but one of the enzymes needed to convert glycogen back to glucose is missing. Because excessive amounts of glycogen are stored, its storage organs (liver and skeletal muscles) become glutted with glycogen and enlarge tremendously.

Metabolic and endocrine disorders are often closely intertwined, reflecting the influence of hormones on metabolism. Occasionally, infants are born with hypothyroidism. If not detected and treated with hormone replacement therapy (this is unusual today), the child develops *cretinism*. This disorder is evidenced by a low BMR, dwarfism accompanied by skeletal deformities, a thick tongue, intolerance to cold, and mental retardation.

With the exception of *insulin-dependent diabetes mellitus*, children free of genetic disorders rarely exhibit metabolic problems. However, by middle age and particularly old age, *non-insulin-dependent diabetes mellitus* becomes a major problem, particularly in the obese. (These two manifestations of diabetes are described in Chapter 17.)

Metabolic rate progressively declines throughout the life span. In old age, muscle and bone wasting and declining efficiency of hormonal systems take their toll. Because many elderly are also less active, the metabolic rate sometimes becomes so low that it becomes nearly impossible to obtain adequate nutrition without excessive caloric intake. The elderly also use more medications than any other group, at a time of life when the liver has become less efficient in its detoxifying duties, and many of the agents prescribed for age-related medical problems influence nutrition. For example:

- Some diuretics prescribed for congestive heart failure or hypertension (to flush fluids out of the body) can cause severe hypokalemia by promoting the excessive loss of potassium.

- Some antibiotics affect absorption of nutrients. For example, sulfa drugs, tetracycline, and penicillin interfere with systems responsible for digestion and absorption. They may also cause diarrhea, thereby decreasing nutrient absorption.
- Mineral oil, a popular laxative with the elderly, interferes with the absorption of fat-soluble vitamins.
- Alcohol is used by about half the elderly population in this country. When alcohol is substituted for food, nutrient stores can be depleted; excessive alcohol intake leads to malabsorption problems and damage to the liver and pancreas. Alcoholics are particularly at risk for certain types of deficiencies, including protein, B complex, and the minerals magnesium, potassium, and zinc.

In short, the elderly are at risk not only from the declining efficiency of the metabolic processes themselves, but also from a huge variety of life-style and medication factors that have an impact on their nutrition.

* * *

Nutrition is one of the most overlooked areas in clinical medicine. Yet, what we eat influences nearly every phase of metabolism and plays a major role in our overall health. Now that we have examined the fates of nutrients within body cells, we are ready to study the urinary system, the organ system that works tirelessly to rid the body of nitrogen wastes resulting from metabolism and to maintain the purity of our internal fluids.

Related Clinical Terms

Appetite The desire to eat. A psychological phenomenon, as opposed to hunger, which is a physiological need to eat.

Hypercholesterolemia (hī″-per-kuh-les″-ter-ol-ē′-mē-uh) A genetically induced condition in which the LDL receptors are abnormal, the uptake of cholesterol by tissue cells is blocked, and the total concentration of cholesterol (and LDLs) in the blood is enormously elevated (e.g., 680 mg cholesterol/100 ml blood). Affected individuals develop arteriosclerosis at an early age, and most die during childhood or adolescence from coronary artery disease.

Hypothermia (hī″-pō-ther′-mē-uh) Low body temperature resulting from prolonged uncontrolled exposure to cold. Vital signs (respiratory rate, blood pressure, and heart rate) decrease as cellular enzymes become sluggish. Drowsiness sets in and, oddly, the person becomes comfortable even though previously he or she felt extremely cold. Uncorrected, the situation progresses to coma and finally death, when body temperatures approach 21°C (70°F).

Ideal weight Actually a misnomer. The average (but not necessarily the most desirable) weight given in insurance tables for persons of a given height and sex.

Kwashiorkor (kwash″-ē-or′-ker) Severe protein deficiency in children, resulting in mental retardation and failure to grow. Characterized by a bloated abdomen because the level of plasma proteins is inadequate to keep fluid in the bloodstream.

Marasmus (mah-raz′-mus) Protein–calorie malnutrition, accompanied by progressive wasting.

Pica (pī′-kuh) Craving for substances not normally considered nutrients, such as clay.

Skin-fold test Clinical test of body fatness. A skin fold in the back of the arm or below the scapula is measured with a caliper. A fold over 1 inch in thickness indicates excess fat. Also called the fat-fold test.

Chapter Summary

NUTRITION (pp. 805–818)

1. Nutrients include water, carbohydrates, lipids, proteins, vitamins, and minerals. The bulk of the organic nutrients is used as fuel to produce cellular energy (ATP). The energy value of foods is measured in kilocalories (kcal).

2. Essential nutrients cannot be synthesized by body cells and must be ingested in the diet.

Carbohydrates (pp. 805–807)

3. Carbohydrates are obtained primarily from plant products. Absorbed monosaccharides other than glucose are converted to glucose by the liver.

4. Monosaccharides are used primarily for cellular fuel. Small amounts are used for nucleic acid synthesis and to glycosylate plasma membranes.

5. Minimum carbohydrate requirement is 100 g/day.

Lipids (pp. 807–808)

6. Most dietary lipids are triglycerides. The primary sources of saturated fats are animal products; unsaturated fats are present in plant products. The major source of cholesterol is egg yolk.

7. Linoleic acid is an essential fatty acid.

8. Neutral fats provide reserve energy, cushion body organs, and insulate the body. Phospholipids are used to synthesize plasma membranes and myelin. Cholesterol is used in plasma membranes and is the structural basis of vitamin D, steroid hormones, and bile salts.

9. Fats should represent 30% or less of caloric intake, and saturated fats should be replaced by unsaturated fats if possible. Cholesterol intake should be restricted to 250 mg or less daily.

Proteins (pp. 808–810)

10. Animal products provide high-quality protein containing all (8) of the essential amino acids. Most plant products lack one or more of the essential amino acids.

11. Amino acids are the structural building blocks of the body and of important regulatory molecules.

12. Protein synthesis can/will occur if all essential amino acids are present and sufficient carbohydrate (or fat) calories are available for ATP production. Otherwise, amino acids will be burned for energy.

13. Nitrogen balance occurs when protein synthesis equals protein loss.

14. A dietary intake of 0.8 g of protein/kg body weight is recommended for adults.

Vitamins (pp. 810–814)

15. Vitamins are organic compounds needed in minute amounts. Most act as coenzymes.

16. Except for vitamin D, vitamins are not made in the body.

17. Water-soluble vitamins (B and C) are not stored in the body. Fat-soluble vitamins include vitamins A, D, E, and K; all but vitamin K are stored in the body and can accumulate to toxic amounts.

Minerals (pp. 814–818)

18. Besides calcium, phosphorus, potassium, sulfur, sodium, chloride, and magnesium, the body requires trace amounts of at least a dozen other minerals.

19. Minerals are not used for energy. Some are used to mineralize bone; others are bound to organic compounds or exist as ions in body fluids, where they play various roles in cell processes.

20. Mineral uptake and excretion are carefully balanced and regulated to prevent mineral toxicity. The richest sources of minerals are animal products, vegetables, and legumes.

METABOLISM (pp. 818–842)

Overview (pp. 818–820)

1. Metabolism encompasses all chemical reactions necessary to maintain life. Metabolic processes are either anabolic or catabolic.

2. Cellular respiration refers to catabolic processes during which energy is released and some is captured in ATP bonds.

3. Energy is released when organic compounds are oxidized. Cellular oxidation is accomplished primarily by the removal of hydrogen (electrons). When molecules are oxidized, others are simultaneously reduced by accepting hydrogen (or electrons).

4. Most enzymes catalyzing oxidation–reduction reactions require coenzymes as hydrogen acceptors. Two important coenzymes in these reactions are NAD^+ and FAD.

Carbohydrate Metabolism (pp. 820–829)

5. Carbohydrate metabolism is essentially glucose metabolism.

6. Phosphorylation of glucose on entry into cells effectively traps it in most tissue cells.

7. Glucose is oxidized completely to carbon dioxide and water via three successive pathways: glycolysis, Krebs cycle, and electron transport chain. Some ATP is harvested in each pathway, but the bulk is captured in the electron transport chain.

8. Glycolysis is a reversible pathway in which glucose is converted into two pyruvic acid molecules; two molecules of reduced NAD are formed, and there is a net gain of two ATPs. Under aerobic conditions, pyruvic acid enters the citric acid pathway; under anaerobic conditions, it is reduced to lactic acid.

9. The Krebs cycle is fueled by pyruvic acid (and fatty acids). To enter the cycle, pyruvic acid is converted to acetyl CoA. The acetyl CoA is then oxidized and decarboxylated. Complete oxidation of two pyruvic acid molecules yields 6 CO_2, 8 NADH + H^+, 2 $FADH_2$, and a net gain of 2 ATP. Much of the energy originally present in the bonds of pyruvic acid is now present in the reduced coenzymes.

10. In the electron transport chain, (a) reduced coenzymes are oxidized by delivering hydrogen to a series of oxidation–reduction acceptors; (b) hydrogen is split into hydrogen ions and electrons (as electrons run downhill from acceptor to acceptor, the energy released is used to pump H^+ into the mitochondrial intramembrane space, which creates an electrochemical proton gradient); (c) the electrochemical proton gradient drives H^+ back through ATP synthetase, which uses the energy to form ATP; (d) H^+ and electrons are combined with oxygen to form water.

11. For each glucose molecule oxidized to carbon dioxide and water, 36 or 38 ATP are formed: 4 ATP from substrate-level phosphorylation and 32 or 34 ATP from oxidative phosphorylation.

12. When cellular ATP reserves are high, glucose catabolism is inhibited and glucose is converted to glycogen (glycogenesis) or to fat (lipogenesis). Much more fat than glycogen is stored.

13. Gluconeogenesis is the formation of glucose from noncarbohydrate molecules. It occurs in the liver when blood glucose levels begin to fall.

Lipid Metabolism (pp. 829–831)

14. End products of lipid digestion (and cholesterol) are transported in blood in the form of chylomicrons.

15. Glycerol is converted to glyceraldehyde-PO_4 and enters the Krebs cycle or is converted to glucose.

16. Fatty acids are oxidized by beta oxidation into acetic acid fragments. These are bound to coenzyme A and enter the Krebs cycle as acetyl CoA. Dietary fats not needed for energy or structural materials are stored in adipose tissues.

17. There is a continual turnover of fats in fatty depots. Breakdown of fats to fatty acids and glycerol is called lipolysis.

18. When excessive amounts of fats are used, the liver converts acetyl CoA to ketone bodies and releases them to the blood. Excessive levels of ketone bodies (ketosis) leads to metabolic acidosis.

19. All cells use neutral fats, phospholipids, and cholesterol to build their plasma membranes. The liver forms many functional molecules (lipoproteins, thromboplastin, etc.) from lipids.

Protein Metabolism (pp. 831–832)

20. To be oxidized for energy, amino acids are converted to keto acids that can enter the Krebs cycle. This involves transamination, oxidative deamination, and keto acid modification.

21. Amine groups removed during deamination (as ammonia) are combined with carbon dioxide by the liver to form urea. Urea is excreted in urine.

22. Deaminated amino acids may also be converted to fatty acids and glucose.

23. Amino acids are the body's most important building blocks. Nonessential amino acids are made in the liver by transamination.

24. In adults, most protein synthesis serves to replace tissue proteins and to maintain nitrogen balance.

25. Protein synthesis requires the presence of the essential amino acids. If any are lacking, amino acids are used as energy fuels.

Catabolic–Anabolic Steady State of the Body (pp. 832–834)

26. The amino acid pool provides amino acids for synthesis of proteins and amino acid derivatives, ATP synthesis, and energy storage. To be stored, amino acids are first converted to fats or glycogen.

27. The carbohydrate–fat pool primarily provides fuels for ATP synthesis and forms that can be stored as energy reserves.

28. The nutrient pools are connected by the bloodstream; fats, carbohydrates, and proteins may be interconverted via common intermediates.

Metabolic Events and Controls of the Absorptive and Postabsorptive States (pp. 834–839)

29. During the absorptive state (during and after a meal), glucose is the major energy source; needed structures and functional molecules are made; excesses of carbohydrates, fats, and amino acids are stored as glycogen and fat.

30. Events of the absorptive state are controlled by insulin, which enhances the entry of glucose (and amino acids) into cells and accelerates its use for ATP synthesis or storage as glycogen or fat.

31. The postabsorptive (fasting) state is the period when blood-borne fuels are provided by breakdown of energy reserves. Glucose is made available to the blood by glycogenolysis, lipolysis, and gluconeogenesis.

32. Glucose sparing begins and, if fasting is prolonged (4–5 days), the brain too begins to metabolize ketone bodies.

33. Events of the postabsorptive state are controlled largely by glucagon and the sympathetic nervous system (and epinephrine), which mobilize glycogen and fat reserves and trigger gluconeogenesis.

Role of the Liver in Metabolism (pp. 839–842)

34. The liver is the body's main metabolic organ and plays a crucial role in processing (or storing) literally every nutrient group. It helps maintain blood energy sources, and it detoxifies drugs and other substances.

35. The liver synthesizes cholesterol, catabolizes cholesterol and secretes it in the form of bile salts, and makes lipoproteins.

36. LDLs transport triglycerides and cholesterol from the liver to the tissues, whereas HDLs transport cholesterol from the tissues to the liver (for catabolism).

37. Excessively high LDL levels are implicated in arteriosclerosis, cardiovascular disease, and strokes.

BODY ENERGY BALANCE (pp. 842–850)

1. Body energy intake (derived from food oxidation) is precisely balanced by energy output (heat, work, and energy storage). Eventually, all of the energy intake is converted to heat.

2. When energy balance is maintained, weight remains stable. When excess amounts of energy are stored, the result is obesity (condition of excessive fat storage, 20% or more above the norm).

Regulation of Food Intake (pp. 842–844)

3. The hypothalamus and other brain centers are involved in the regulation of eating behavior.

4. Factors thought to be involved in regulating food intake include (a) nutrient signals related to total energy storage (e.g., large fatty reserves and increased plasma levels of glucose and amino acids depress hunger and eating); (b) plasma concentrations of hormones that control events of the absorptive and postabsorptive states, providing feedback signals to brain feeding centers; (c) body temperature; and (d) psychological factors. Psychological factors are thought to underlie some forms of obesity.

Metabolic Rate and Body Heat Production (pp. 844–845)

5. Energy used by the body per hour is the metabolic rate.

6. Basal metabolic rate (BMR), reported in kcal/m^2/h, is the measurement obtained under basal conditions; i.e., the person is at comfortable room temperature, supine, relaxed, and in the postabsorptive state. BMR indicates energy needed to drive only the resting body processes.

7. Factors influencing metabolic rate include age, sex, size, body surface area, thyroxine levels, specific dynamic action of foods, and muscular activity.

Regulation of Body Temperature (pp. 845–850)

8. Body temperature reflects the balance between heat production and heat loss and is normally 36.1°–37.8°C, which is optimal for physiological activities.

9. At rest, most body heat is produced by the liver, heart, brain, kidney, and endocrine organs. Activation of skeletal muscles causes dramatic increases in body heat production.

10. The body core (organs within the skull and the thoracic and abdominal cavities) generally has the highest temperature; its shell (the skin) is coolest. The shell is the heat-exchange surface.

11. Blood serves as the major heat-exchange agent between the core and the shell. When blood flushes into skin capillaries and the skin is warmer than the environment, heat will be lost from the body. When blood is withdrawn to deep organs, heat loss from the shell is inhibited.

12. The hypothalamus acts as the body's thermostat. Its heat-promotion and heat-loss centers receive inputs from peripheral and central thermoreceptors, integrate these inputs, and initiate responses leading to heat loss or heat promotion.

13. Heat-promoting mechanisms include vasoconstriction of skin vasculature, increase in metabolic rate (via release of norepinephrine and epinephrine), and shivering. If environmental cold is prolonged, the thyroid gland is stimulated to release more thyroxine.

14. Heat-loss mechanisms include radiation, conduction/convection, and evaporation.

15. Evaporation, the conversion of water to water vapor, requires the absorption of heat. For each gram of water vaporized, about 0.5 kcal of heat is absorbed.

16. When heat must be removed from the body, dermal blood vessels are dilated, allowing heat loss through radiation, conduction, and convection. When greater heat loss is mandated (or the environmental temperature is so high that radiation and conduction are ineffective), sweating is initiated. Evaporation of perspiration is a very efficient means of heat loss as long as the humidity is low.

17. Profuse sweating can lead to heat exhaustion, indicated by a drop in blood pressure. When the body cannot rid itself of surplus heat, body temperature rises to the point where all thermoregulatory mechanisms become ineffective—a potentially lethal condition called heat stroke.

18. Fever is controlled hyperthermia, which follows thermostat resetting to higher levels by prostaglandins and initiation of heat-promotion mechanisms, evidenced by the chills. When the disease process is reversed, heat-loss mechanisms are initiated.

DEVELOPMENTAL ASPECTS OF NUTRITION AND METABOLISM (pp. 850–851)

1. Good nutrition is essential for normal fetal development and normal growth during childhood.

2. Inborn errors of metabolism include cystic fibrosis, PKU, glycogen storage disease, galactosemia, and many others. Hormonal disorders, e.g., lack of insulin or thyroxine, may also lead to metabolic abnormalities. Diabetes mellitus is the most significant metabolic disorder in both the young and the old.

3. In old age, metabolic rate declines as enzyme and endocrine systems become less efficient and skeletal muscles atrophy. Reduced caloric needs make it difficult to obtain adequate nutrition without incurring obesity.

4. Elderly individuals ingest more medications than any other age group, and many of these drugs negatively affect their nutrition.

Review Questions

Multiple Choice/Matching

1. Which of the following reactions would liberate the greatest amount of energy? (a) complete breakdown and oxidation of a molecule of sucrose to CO_2 and water, (b) conversion of a molecule of ADP to ATP, (c) respiration of a molecule of glucose to lactic acid, (d) conversion of a molecule of glucose to carbon dioxide and water.

2. The formation of glucose from glycogen is (a) gluconeogenesis, (b) glycogenesis, (c) glycogenolysis, (d) glycolysis.

3. The net gain of ATP from the complete metabolism (aerobic) of glucose is (a) 2, (b) 30, (c) 38, (d) 4.

4. Which of the following best defines cellular respiration? (a) intake of carbon dioxide and output of oxygen by cells, (b) excretion of waste products, (c) inhalation of oxygen and exhalation of carbon dioxide, (d) oxidation of substances by which energy is released in usable form to the cells.

5. During aerobic respiration, hydrogen and electrons are passed down the electron transport chain and ______ is formed. (a) oxygen, (b) water, (c) glucose, (d) NADH + H^+.

6. Metabolic rate is relatively low in (a) youth, (b) physical exercise, (c) old age, (d) fever.

7. In a temperate climate under ordinary conditions, the greatest loss of body heat occurs through (a) radiation, (b) conduction, (c) evaporation, (d) none of the above.

8. Which of the following is *not* a function of the liver? (a) glycogenolysis and gluconeogenesis, (b) synthesis of cholesterol, (c) detoxification of alcohol and drugs, (d) synthesis of glucagon, (e) deamination of amino acids.

9. Amino acids are essential (and important) to the body for all the following *except* (a) production of some hormones, (b) production of antibodies, (c) formation of most structural materials, (d) as a source of quick energy.

10. A person has been on a hunger strike for seven days. Compared to normal, he has (a) increased release of fatty acids from adipose tissue, ketosis, and ketonuria, (b) elevated glucose concentration in the blood, (c) increased plasma insulin concentration, (d) increased glycogen synthetase (enzyme) activity in the liver.

11. Transamination is a chemical process by which (a) protein is synthesized, (b) an amine group is transferred from an amino acid to a keto acid, (c) an amine group is cleaved from the amino acid, (d) amino acids are broken down for energy.

12. Three days after removal of the pancreas from an animal, the researcher finds a persistent increase in (a) acetoacetic acid concentration in the blood, (b) urine volume, (c) blood glucose, (d) all of the above.

13. Hunger, appetite, obesity, and physical activity are interrelated. Thus, (a) hunger sensations arise *primarily* from the stimulation of receptors in the stomach and intestinal tract in response to the absence of food in these organs; (b) obesity, in most cases, is a result of the abnormally high enzymatic activity of the fat-synthesizing enzymes in adipose tissue; (c) in all cases of obesity, the energy content of the ingested food has exceeded the energy expenditure of the body; (d) in a normal individual, increasing blood glucose concentration increases hunger sensations.

14. Body temperature regulation is (a) influenced by temperature receptors on the skin, (b) influenced by the temperature of the blood perfusing the heat regulation centers of the brain, (c) subject to both neural and hormonal control, (d) all of the above.

15. Which of the following yields the greatest caloric value per gram? (a) fats, (b) proteins, (c) carbohydrates.

Short Answer Essay Questions

16. What is cellular respiration? What is the common role of FAD and NAD^+ in cellular respiration?

17. Describe the site, major events, and outcomes of glycolysis.

18. Pyruvic acid is a product of glycolysis, but it is not the substance that enters the Krebs cycle. What is the substance, and what must occur if pyruvic acid is to be transformed into this molecule?

19. Define glycogenesis, glycogenolysis, gluconeogenesis, and lipogenesis. Which is (are) likely to be occurring (a) shortly after a carbohydrate-rich meal, (b) just before waking up in the morning?

20. What is the harmful result when excessive amounts of fats are burned for energy? Name two conditions that might lead to this result.

21. Make a flowchart that indicates the pivotal intermediates through which glucose can be converted to fat.

22. Distinguish between the role of HDLs and that of LDLs.

23. List some factors that influence plasma cholesterol levels. Also list the sources and fates of cholesterol in the body.

24. What is meant by "body energy balance," and what happens if it is not in precise balance?

25. Explain the effect of the following on metabolic rate: thyroxine levels, eating, body surface area, muscular exercise, emotional stress, starvation.

26. Explain the terms "core" and "shell" relative to body temperature balance. What serves as the heat-transfer agent from one to the other?

27. Compare and contrast mechanisms of heat loss with mechanisms of heat promotion, and explain how these mechanisms determine body temperature.

Clinical Application Questions

28. Every year dozens of elderly people are found dead in their unheated apartments and listed as victims of hypothermia. What is hypothermia, and how does it kill?

29. Fran has been diagnosed as having severe arteriosclerosis and high blood cholesterol levels. He is told that he is at risk for a stroke or a heart attack. First, what foods would you suggest that he avoid like the plague? What foods would you suggest he add or substitute? What activities would you recommend to him?

CHAPTER **26**

The Urinary System

Chapter Outline and Student Objectives

Kidney Anatomy (pp. 856–863)

1. Describe the gross anatomy of the kidney and its coverings.
2. Trace the blood supply through the kidney.
3. Describe the anatomy of a nephron.

Kidney Physiology: Mechanisms of Urine Formation (pp. 863–877)

4. List several kidney functions that help maintain body homeostasis.
5. Identify the parts of the nephron responsible for filtration, reabsorption, and secretion, and describe the mechanisms underlying each of these functional processes.
6. Explain the role of aldosterone and of atrial natriuretic factor in sodium and water balance.
7. Describe the mechanism that maintains the medullary osmotic gradient.
8. Explain the formation of dilute versus concentrated urine.
9. Describe the normal physical and chemical properties of urine.
10. List several abnormal urine components, and name the condition when each is present in detectable amounts.

Ureters (pp. 877–878)

11. Describe the general structure and function of the ureters.

Urinary Bladder (pp. 878–879)

12. Describe the general structure and function of the urinary bladder.

Urethra (pp. 879–880)

13. Describe the general structure and function of the urethra.
14. Compare the course, length, and functions of the male urethra with those of the female.

Micturition (p. 880)

15. Define micturition and describe the micturition reflex.

Developmental Aspects of the Urinary System (pp. 881–883)

16. Describe the embryonic development of the kidneys.
15. List several changes in urinary system anatomy and physiology that occur with age.

Preview of Selected Key Terms

Nephron (neh'-fron) (*nephr* = pertaining to the kidney) Structural and functional unit of the kidney; consists of the glomerulus and renal tubule.

Glomerulus (glō-mer'-yoo-lus) (*glom* = little ball) Cluster of capillaries forming part of the nephron; forms filtrate.

Juxtaglomerular apparatus (juks-tuh-glō-mer'-yoo-ler) (*juxta* = near to) Cells located close to the glomerulus that play a role in blood pressure regulation by releasing the enzyme renin.

Glomerular filtration rate (GFR) Rate of filtrate formation by the kidneys.

Aldosterone (al-dah'-ster-ōn) Hormone produced by the adrenal cortex that regulates sodium ion reabsorption.

Antidiuretic hormone (an''-tih-dī-yoo-ret'-ik) (*anti* = against; *diure* = urine flow) Hormone produced by the hypothalamus and released by the posterior pituitary; stimulates the kidneys to reabsorb more water, reducing urine volume.

Micturition (mik''-tū-rish'-un) (*mictur* = urinate) Urination, or voiding; emptying the bladder.

The kidneys, which maintain the purity and constancy of fluids in our internal environment, are perfect examples of homeostatic organs. Much like sanitation workers, who keep a city's water drinkable and who dispose of its waste, the kidneys are usually unappreciated until there is a malfunction and "internal garbage" piles up. Every day the kidneys filter gallons of fluid from the bloodstream, allowing toxins,

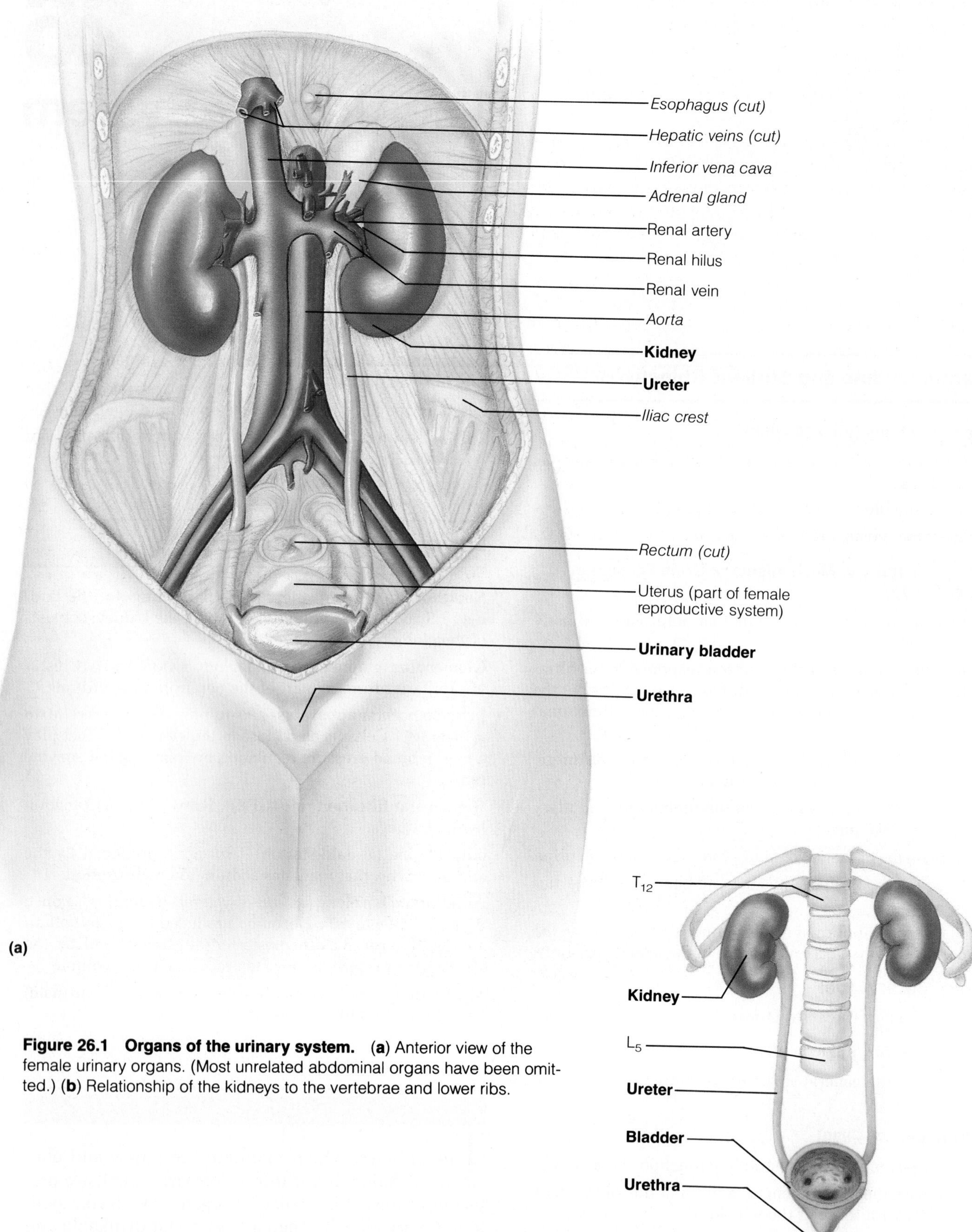

T_{12}
Kidney
L_5
Ureter
Bladder
Urethra
(b)

Figure 26.1 Organs of the urinary system. (**a**) Anterior view of the female urinary organs. (Most unrelated abdominal organs have been omitted.) (**b**) Relationship of the kidneys to the vertebrae and lower ribs.

metabolic wastes, and excess ions to leave the body in urine, while returning needed substances to the blood. Although the lungs and the skin also play roles in excretion, the kidneys bear the major responsibility for eliminating nitrogenous wastes, toxins, and drugs from the body.

Disposing of wastes and excess ions is only one aspect of the work of the kidneys. As they perform these excretory functions, they simultaneously regulate the volume and chemical makeup of the blood, maintaining continuously the proper balance between water and salts and between acids and bases. Frankly, this would be tricky work for a chemical engineer, but the kidneys do it efficiently most of the time.

The kidneys have other regulatory functions as well: They produce the enzyme *renin* (reh′-nin), which helps regulate blood pressure and kidney function, and the hormone *erythropoietin* (eh-rith″-rō-poy′-eh-tin), which stimulates red blood cell production in bone marrow (see Chapter 18). In addition, kidney cells metabolize vitamin D to its active form (see Chapter 17).

Besides the urine-forming kidneys, the organs of the **urinary system** are the paired *ureters* (ū-rē′-terz), the *urinary bladder*, and the *urethra* (ū-rē′-thrah) (Figure 26.1), which provide temporary storage reservoirs or transportation channels for urine.

Kidney Anatomy

Location and External Anatomy

The bean-shaped kidneys lie between the dorsal body wall and the parietal peritoneum in the *superior* lumbar region (Figure 26.2). Since they extend approximately from the level of the twelfth thoracic vertebra to the third lumbar vertebra, the kidneys receive some protection from the lower part of the rib cage (see Figure 26.1). The right kidney is crowded by the liver and lies slightly lower than the left. An adult kidney weighs about 150 g (5 ounces) and is approximately 12.5 cm (5.0 inches) long, 7.5 cm (3.0 inches) wide, and 2.5 cm (1.0 inch) thick. The lateral surface of the kidney is convex. Its medial surface is concave and has a vertical cleft called the *hilus* that leads into a space called the *renal sinus*. Several structures, including the ureters, the renal blood vessels, lymphatics, and nerves, enter or exit the kidney at the hilus and occupy the sinus. Atop each kidney is an *adrenal* (or *suprarenal*) *gland*, which is distinctly separate functionally (see Figure 26.1a).

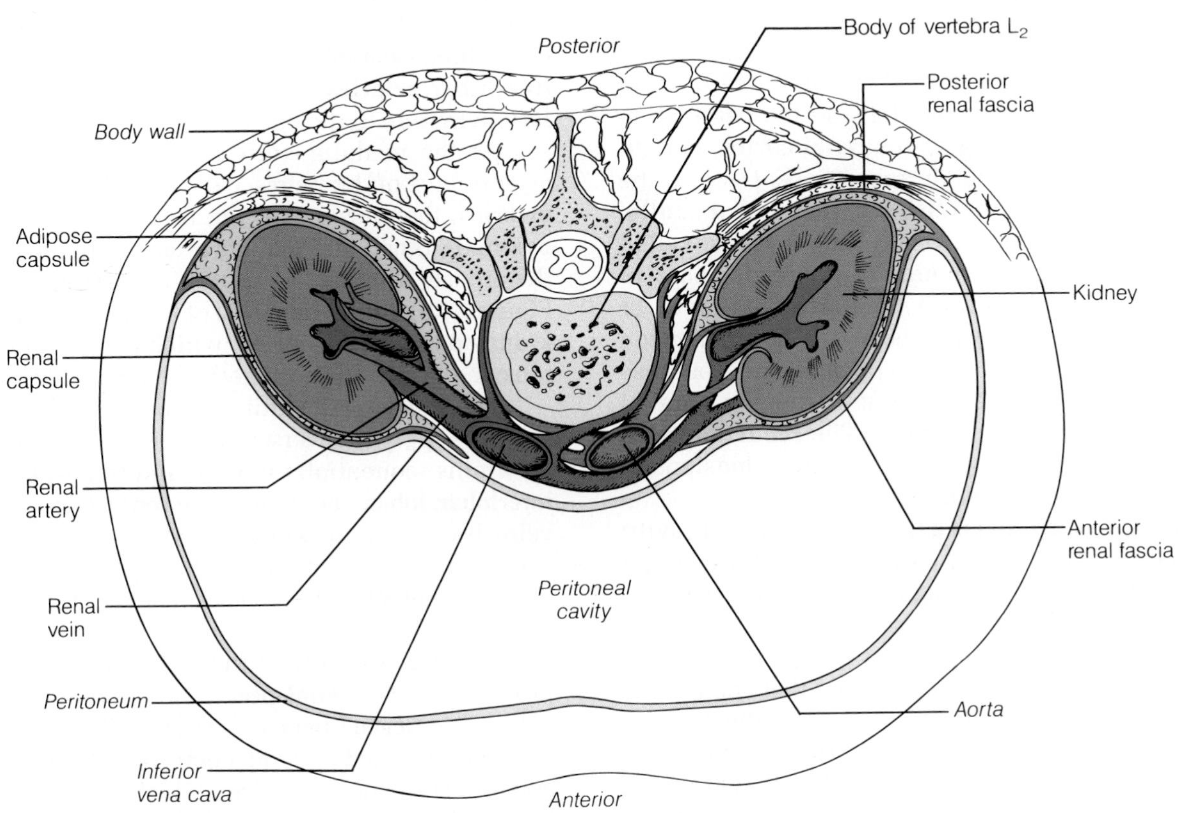

Figure 26.2 Position of the kidneys against the posterior body wall. The kidneys are retroperitoneal and surrounded by three membranes: the renal capsule, the adipose capsule, and the renal fascia.

Three layers of supportive tissue surround each kidney (Figure 26.2). The layer closest to the kidney is the fibrous, transparent **renal capsule**, which gives a fresh kidney its glistening appearance. This capsule provides a strong barrier that prevents infections in surrounding regions from spreading to the kidneys and plays a minor role in protecting kidney tissue from physical trauma. The middle layer is a fatty mass called the **adipose capsule**, which helps to hold the kidney in place against the posterior trunk muscles and cushions it against blows. The outermost layer, the **renal fascia**, is dense, fibroareolar connective tissue. It surrounds not only the kidney and its attendant membranes, but also the adrenal gland, and anchors these organs to surrounding structures.

The fatty encasement of the kidneys is extremely important in holding the kidneys in their normal body position. If the amount of fatty tissue dwindles (as with rapid weight loss), the kidneys may drop to a lower position, a condition called *ptosis* (tō′-sis), from the Greek word meaning a fall. Ptosis creates problems when a ureter becomes kinked because the urine, prevented from draining, backs up into the kidney and exerts pressure on its tissue. This condition, called *hydronephrosis* (hī-drō-neh-frō′-sis), can severely damage the kidney, leading to necrosis (tissue death) and renal failure. ■

Internal Anatomy

A coronal section of a kidney reveals three distinct regions: the cortex, the medulla, and the pelvis (Figure 26.3). The outer region, the **renal cortex**, is light in color and has a granular appearance. Deep to the cortex is the darker, reddish-brown **renal medulla**, which exhibits cone-shaped tissue masses called **medullary** or **renal pyramids**. The broader *base* of each pyramid faces toward the cortex; its apex, or *papilla* (literally, nipple), points toward the inner region of the kidney. The pyramids appear striped because they are formed almost entirely of roughly parallel bundles of microscopic tubules. The **renal columns**, areas of lighter-staining tissue that are extensions of cortical tissue, separate the pyramids.

Medial to the hilus is a flat, funnel-shaped cavity, the **renal pelvis**, which is continuous with the ureter leaving the hilus. Branching extensions of the pelvis form two or three **major calyces** (kā′-lih-sēz) (singular, *calyx*), each of which subdivides further to form several **minor calyces**, cup-shaped areas that enclose the papillae of the pyramids. The calyces collect urine, which drains continuously from the papillae, and empty it into the renal pelvis. Urine then flows through the renal pelvis and into the ureter, which transports it to the bladder to be stored. The walls of the calyces, pelvis, and ureter contain smooth muscle, which contracts rhythmically and propels urine along its course by peristalsis.

Kidney inflammations include *pyelitis* (pī-eh-lī′-tis), involving the renal pelvis and calyces, and *pyelonephritis* (pī″-eh-lō-neh-frī′-tis), which affects the entire kidney. These usually result from infections spreading upward fom the lower urinary tract or, less often, from infections elsewhere in the body that are carried to the kidney by blood-borne bacteria. In severe cases of pyelonephritis, the kidney becomes swollen, abscesses form, and the pelvis fills with pus. Untreated, the kidneys may be severely damaged, but antibiotic therapy usually achieves total remission. ■

Blood and Nerve Supply

Since the kidneys must continuously cleanse the blood and adjust its composition, it is not surprising that they have a very rich blood supply (Figure 26.3). Under resting conditions, the large **renal arteries** deliver approximately one-fourth of the total cardiac output (about 1200 ml) to the kidneys each minute. The renal arteries issue at right angles from the abdominal aorta between the first and second lumbar vertebra. Because the aorta lies to the left of the midline, the right renal artery is typically longer than the left. As each renal artery approaches a kidney, it divides into five **lobar**, or **segmental, arteries**, which enter the hilus. Within the kidney, each lobar artery branches to form several **interlobar arteries**, which pass between the medullary pyramids to the cortex.

At the medulla–cortex junction, the interlobar arteries give off branches, called **arcuate** (ar′-kyoo-āt) **arteries**, that curve over the bases of the medullary pyramids. Small **interlobular arteries** run upward from the arcuate arteries to supply the cortical tissue (see Figure 26.5). More than 90% of the blood entering the kidney perfuses the cortex, which contains the nephrons, the structural and functional units of the kidneys.

Veins leaving the kidney trace the pathway of the arterial supply in reverse: Blood leaving the renal cortex drains sequentially into the **interlobular, arcuate, interlobar, lobar**, and, finally, the **renal veins**. The renal veins issue from the kidneys and empty into the inferior vena cava, which lies to the right of the vertebral column; consequently, the left renal vein is about twice as long as the right.

The nerve supply of the kidney and its ureter is provided by the **renal plexus**, a variable network of autonomic nerve fibers and ganglia. The renal plexus is largely supplied by sympathetic fibers from the lesser and lowest splanchnic nerves, which course along with the renal artery to reach the kidney.

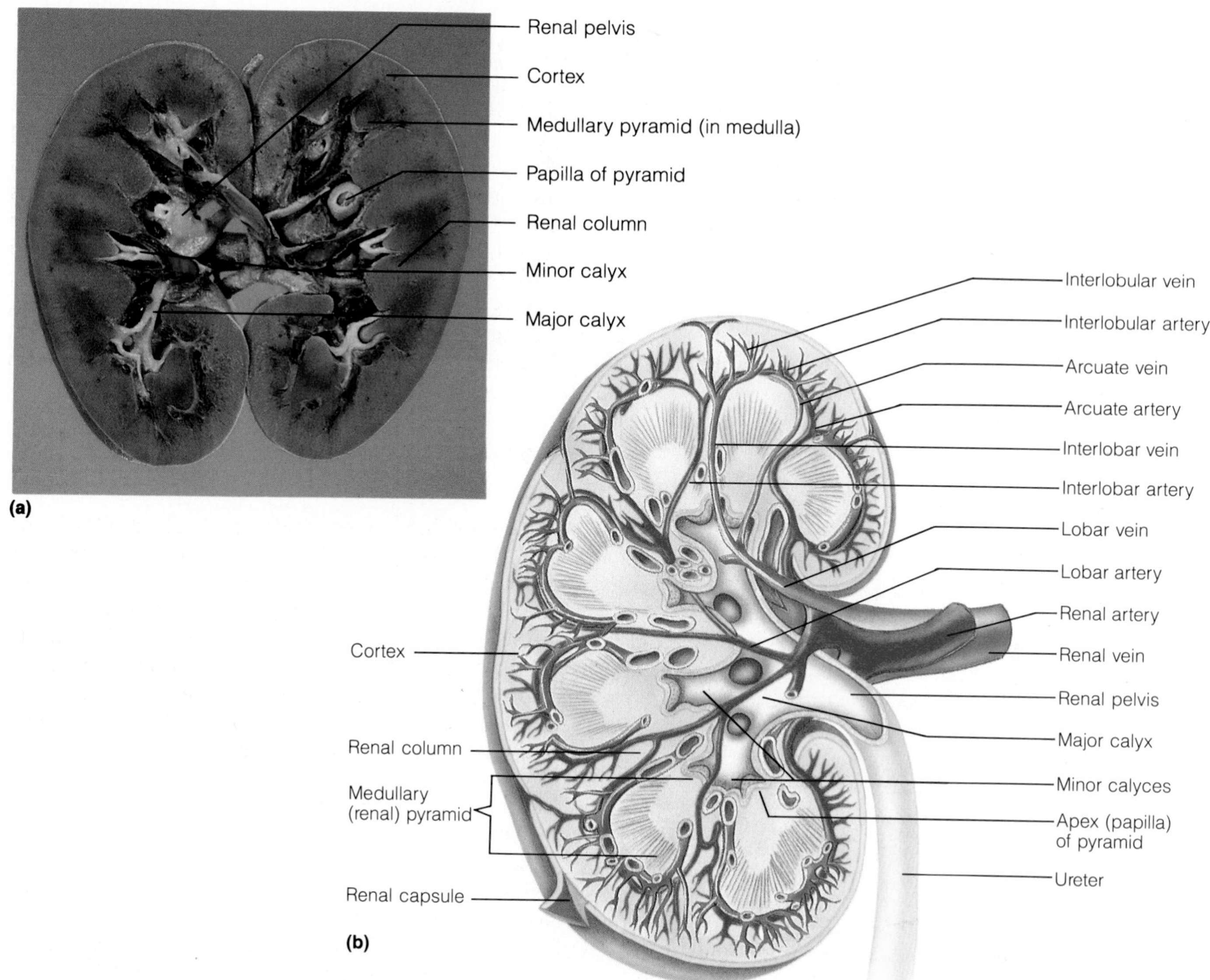

Figure 26.3 Internal anatomy of the kidney. (**a**) Photograph of a coronally sectioned kidney. (**b**) Diagrammatic view of a coronally sectioned kidney, illustrating major blood vessels.

Nephrons

Each kidney contains over 1 million tiny blood-processing units called **nephrons** (neh′-fronz), which carry out the processes that form urine (Figure 26.4). Each nephron consists of a **glomerulus** (glō-mer′-yoo-lus), a tuft of capillaries, associated with a **renal tubule.** The end of the renal tubule, called the **glomerular (Bowman's) capsule,** is blind, enlarged, and cup-shaped and completely surrounds the glomerulus. Collectively, the glomerular capsule and the enclosed glomerulus are referred to as the **renal corpuscle.** The glomerular endothelium is *fenestrated* (penetrated by many small pores), which makes these capillaries exceptionally porous: They allow virtually all substances, except blood cells and most plasma proteins, to pass from the blood into the glomerular capsule. This solute-rich fluid that filters from the bloodstream into the glomerular capsule is called *filtrate* and is the raw material that is processed by the renal tubules.

The *parietal,* or outer, *layer* of the glomerular capsule consists of simple squamous epithelium. This layer plays no role in filtrate formation. The *visceral layer,* which lies adjacent to the glomerulus (Figure 26.4b), is made up of highly modified, branching epithelial cells called **podocytes** (pō′-dō-sīts), which form part of the filtration membrane. The branches of the octopuslike podocytes terminate in *foot processes* (*podo* = foot) or *pedicles,* which intertwine with one another and cling to the basement membrane of the glomerulus. Clefts or openings between the foot processes, called **filtration slits** or **slit pores** (see Fig-

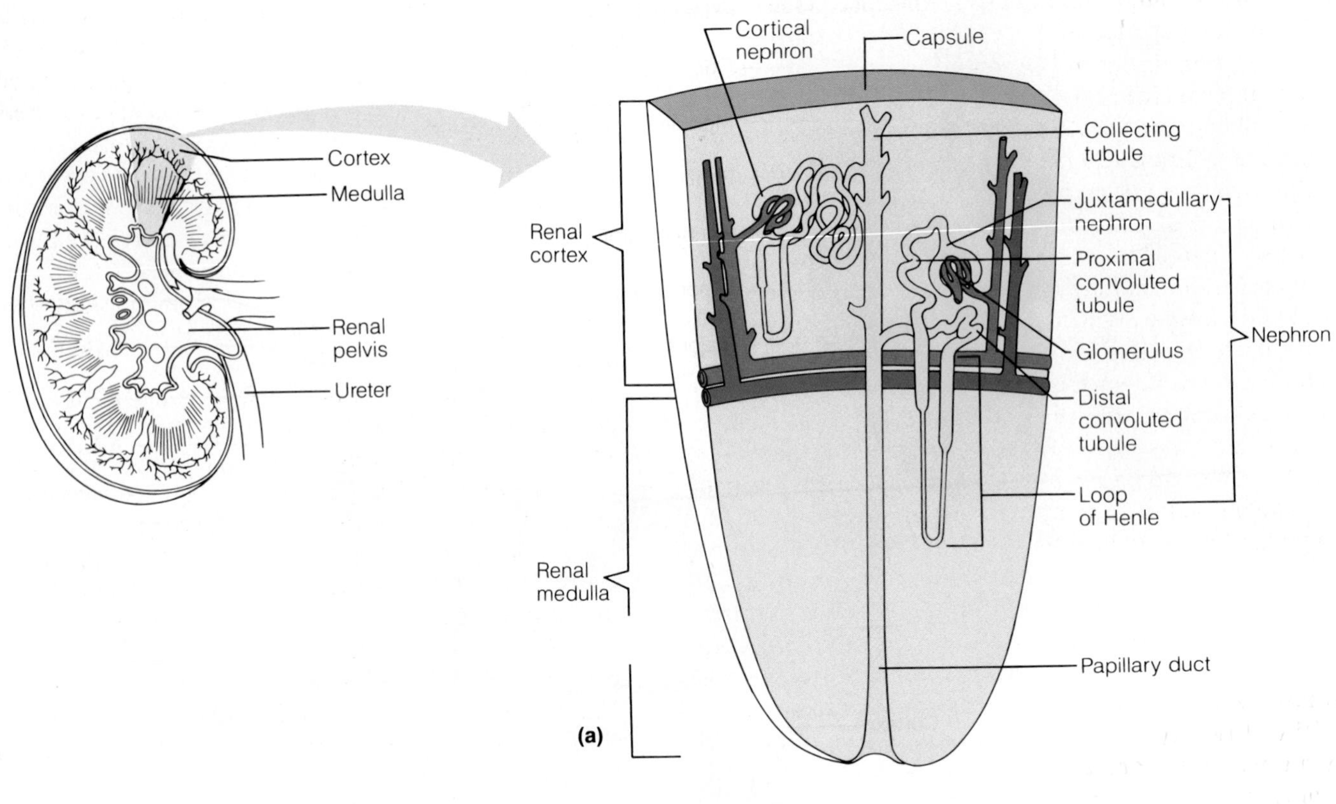

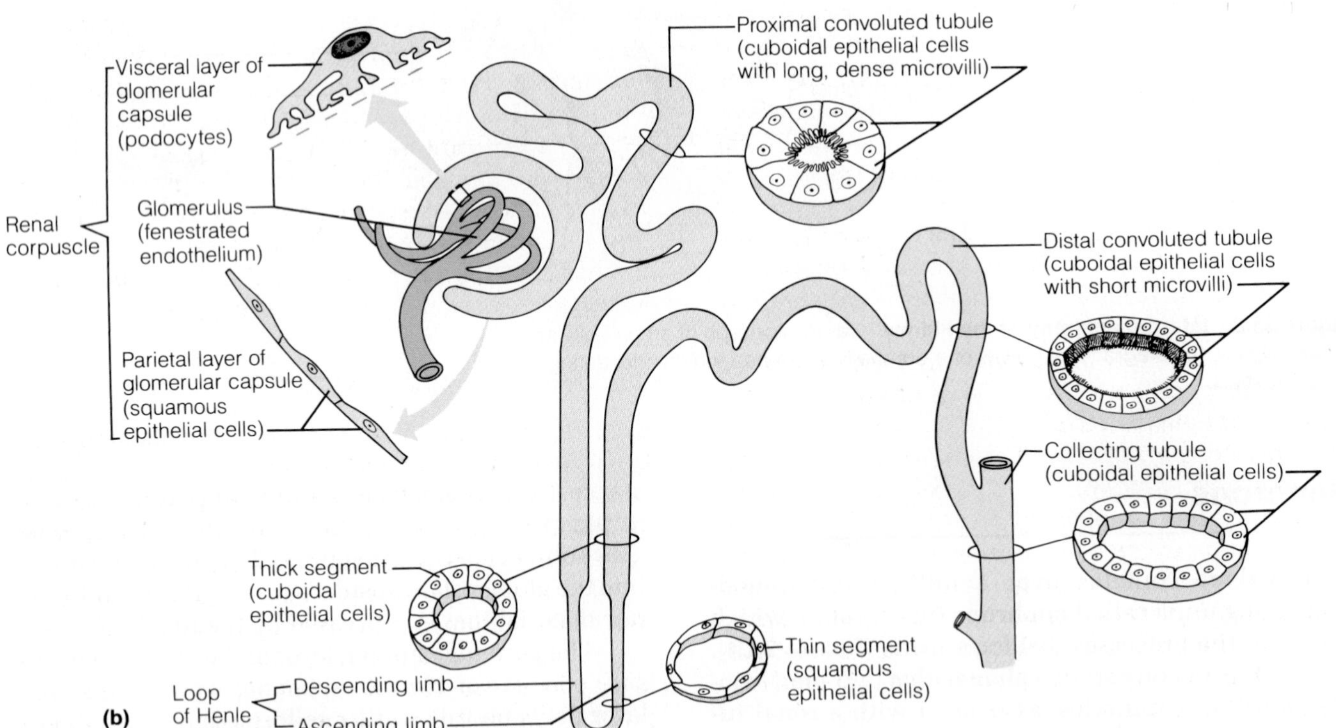

Figure 26.4 Location and structure of nephrons. (**a**) Wedge-shaped section (lobule) of kidney tissue, indicating the location of nephrons in the kidney. (**b**) Schematic view of a nephron depicting the structural characteristics of epithelial cells forming its various regions.

ure 26.8), are small enough to keep most plasma proteins in the capillaries.

The remainder of the renal tubule (Figure 26.4b) is about 3 cm (approximately 1.25 inches) long. It leaves the glomerular capsule as the elaborately coiled **proximal convoluted tubule (PCT),** makes a hairpin loop called the **loop of Henle,** and then winds and twists again as the **distal convoluted tubule (DCT)** before it enters a urine-collecting duct, the **collecting tubule**. This meandering nature of the renal tubule increases its length and enhances its capabilities for processing the filtrate it contains. The collecting tubules, each of which receives urine from many nephrons, run through the medullary pyramids and give them their striped appearance. As the collecting tubules approach the renal pelvis, they fuse to form larger **papillary ducts,** which deliver urine into the minor calyces via the papillae of the pyramids.

Each region of the renal tubule has a unique cellular anatomy, which reflects its specific filtrate-processing function. The walls of the proximal convoluted tubule are formed by cuboidal epithelial cells whose luminal (exposed) surfaces have dense microvilli that nearly fill the tubule lumen. The microvilli tremendously increase the surface area of the proximal tubule cells, and, thereby, their capacity for reabsorbing water and solutes. The *thin segment* of the loop of Henle is lined with squamous epithelium, but the epithelium gradually becomes cuboidal or even low columnar in the ascending part of the loop, which therefore becomes the *thick segment.* In some nephrons, the thin segment is found only in the descending limb of the loop of Henle; in others, it extends into the ascending limb as well. The epithelial cells of the distal tubule, like those of the proximal tubule, are cuboidal, but are somewhat thinner and almost entirely lack microvilli.

Most nephrons are called **cortical nephrons** because they either are located entirely in the cortex or have only small portions of their loops of Henle dipping into the medulla. The remaining nephrons are somewhat different in structure and play an important role in the kidney's ability to adjust urine concentration. These **juxtamedullary** (juks″-tuh-meh′-dyoo-layr-ē) **nephrons** are located close to the cortex–medulla junction. Their loops of Henle deeply invade the medulla, and their thin segments are much more extensive than those of cortical nephrons (Figures 26.4a and 26.5b).

Capillary Beds of the Nephron

Every nephron is closely associated with two capillary beds: the glomerulus and the **peritubular capillary bed** (Figure 26.5). The glomerulus, specialized for filtration, is unlike any other capillary bed in the body in that it is both fed and drained by arterioles—the *afferent arteriole* and the *efferent arteriole*, respectively. The afferent arterioles arise from the interlobular arteries that penetrate through the renal cortex (see also Figure 26.3b). Because (1) arterioles are high-resistance vessels and (2) the afferent arteriole is larger than the efferent, the blood pressure in the glomerulus is extraordinarily high for a capillary bed and easily forces fluid and solutes out of the blood into the glomerular capsule along nearly its entire length. Most of this filtrate is eventually reabsorbed by the renal tubule cells and returned to the blood in the peritubular capillary bed.

The peritubular capillary bed consists of capillaries that arise from the efferent arteriole draining the glomerulus. These capillaries cling closely to the renal tubule and empty into the renal venous system. The peritubular capillaries are adapted for absorption rather than filtrate formation. They are low-pressure, porous capillaries that readily absorb solutes and water from the tubule cells as these substances are reabsorbed from the filtrate. The juxtamedullary nephrons have additional, thin-walled looping vessels called the **vasa recta** (va′-suh rek′-tuh), literally, straight vessels, which parallel the course of their longer loops of Henle in the medulla (Figure 26.5).

As blood flows through the renal circulation, it encounters high resistance, first in the afferent and then in the efferent arterioles. As a result, renal blood pressure declines from approximately 95 mm Hg in the renal arteries to about 8 mm Hg (or less) in the renal veins (Figure 26.6). The resistance of the afferent arterioles assures that the glomeruli are protected from large fluctuations in systemic blood pressure. Resistance encountered in the efferent arterioles reinforces the high glomerular pressure and reduces the hydrostatic pressure in the peritubular capillaries.

Juxtaglomerular Apparatus

Each nephron has a region called the **juxtaglomerular** (juks-tuh-glō-mer′-yoo-ler) **apparatus (JGA),** where its coiling distal tubule lies against the afferent arteriole, feeding the glomerulus (Figure 26.5a). At the point of contact, both structures are modified.

The afferent arteriole wall has **juxtaglomerular (JG) cells**—enlarged, smooth muscle cells with prominent renin-containing granules. These cells appear to be mechanoreceptors that directly sense the blood pressure in the afferent arteriole. The **macula densa** (ma′-kyoo-luh den′-suh) is a group of tall, closely packed, distal tubule cells that lie adjacent to the JG cells. The macula densa cells are thought to be chemoreceptors or osmoreceptors that respond to changes in the solute concentration of the filtrate. These two distinct cell populations play important roles in regulating the rate of filtrate formation.

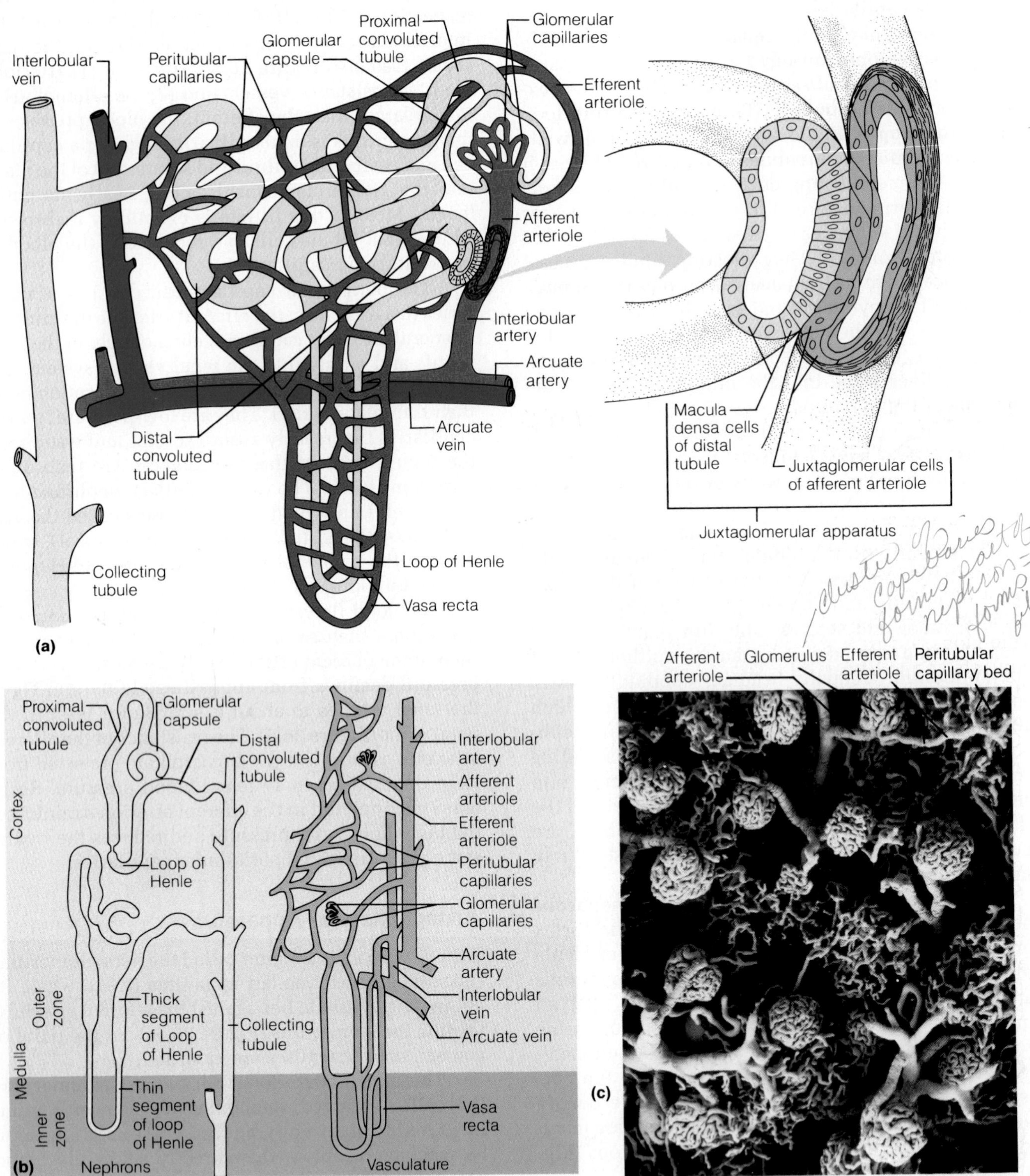

Figure 26.5 Detailed anatomy of a nephron and its vasculature.
(a) Structure of a juxtamedullary nephron and its associated capillaries. **(b)** Comparison of the tubular and vascular anatomy of cortical and juxtamedullary nephrons drawn to the same scale. **(c)** Scanning electron micrograph of a cast of blood vessels associated with nephrons (65X).

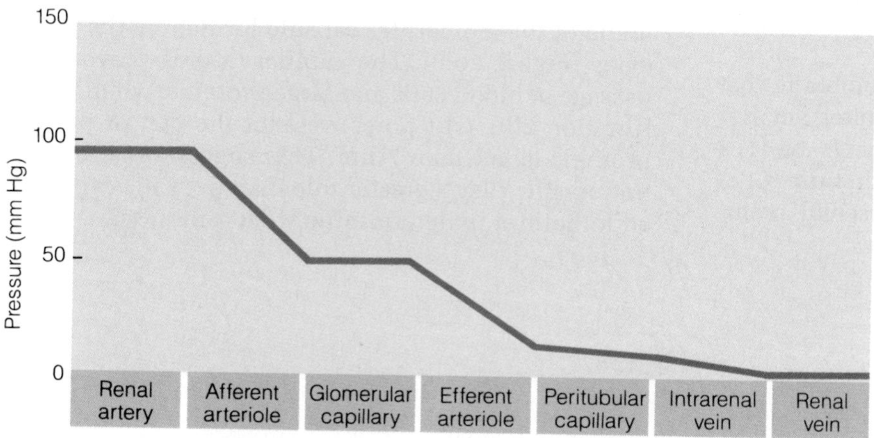

Figure 26.6 Relative blood pressures in the renal circulation. The two areas of highest resistance (thus, points of greatest decline in blood pressure) are the afferent and efferent arterioles, respectively, feeding and draining the glomeruli.

Kidney Physiology: Mechanisms of Urine Formation

The marvelously complex kidneys process about 180 L (45 gallons) of blood-derived fluid daily. Of this amount, only about 1% (1.5 L) actually leaves the body as urine; the rest is returned to the circulation. Of the approximately 1000–1200 ml of blood passing through the glomeruli each minute, about 650 ml is plasma, and about one-fifth of this (120–125 ml) is forced into the renal tubules. This is equivalent to filtering out your entire plasma volume about 60 times each day! Considering the magnitude of their task, it is not surprising that the kidneys (which account for only about 0.4% of body weight) consume about 10% of all the oxygen used by the body at rest.

The entire responsibility for urine formation falls on the nephrons. Essentially, urine formation and the simultaneous adjustment of blood composition involve three processes (Figure 26.7). *Glomerular filtration* is mainly the job of the glomeruli, which act largely as nonselective filters. The renal tubules carry out *tubular reabsorption* and *secretion*, both of which are carefully regulated by renal and hormonal controls.

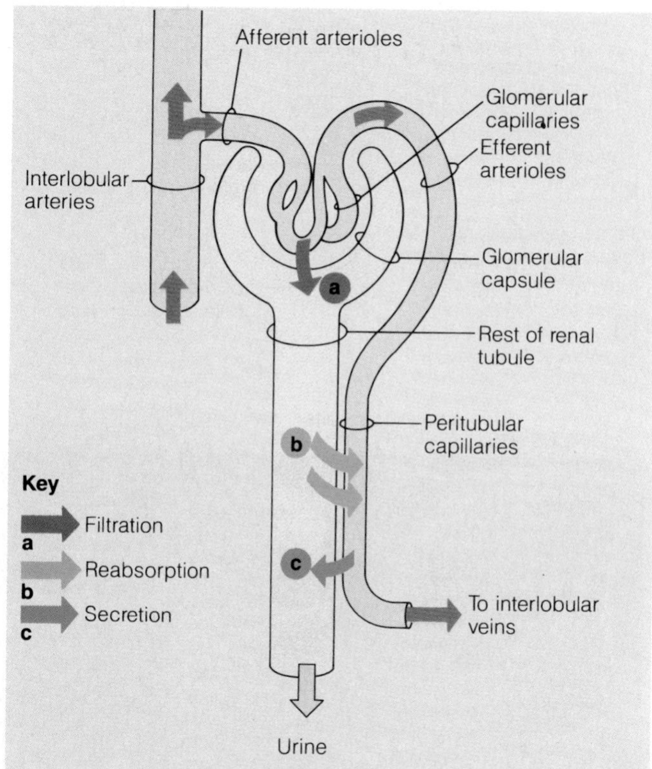

Figure 26.7 The kidney depicted as a single, large nephron. A kidney actually has millions of nephrons acting in parallel. The three major mechanisms by which the kidneys adjust the composition of plasma are (**a**) glomerular filtration, (**b**) tubular reabsorption, and (**c**) tubular secretion.

Glomerular Filtration

Urine formation begins with **glomerular filtration.** Filtration is a nonselective process in which fluids and solutes are forced through a membrane by hydrostatic pressure (see Chapter 3). Thus, filtrate formation is a passive process, driven by pressure gradients, which does not require the use of metabolic energy, and the glomeruli are simple mechanical filters. The principles of fluid dynamics that account for the formation of tissue fluid at all capillary beds (see Chapter 20) apply to the formation of filtrate at the glomerulus as well. However, the glomerulus is a much more efficient filter than are other capillary beds because (1) its *filtration membrane* is thousands of times more permeable to water and solutes than are other capillary membranes, and (2) glomerular blood pressure is much higher than that of other capillary beds (approximately 55 mm Hg as opposed to 15–20 mm Hg), resulting in a much higher *net filtration pressure.* As a result of these differences, the kidneys produce about 180 L of filtrate daily, in contrast to the 3 L formed daily by all other capillary beds of the body.

The Filtration Membrane

The **filtration membrane** is a porous membrane that allows free passage of water and all solutes smaller than plasma proteins. Its anatomical elements are (1) a fenestrated capillary (glomerular) endothelium, (2) a thin basement membrane, and (3) a visceral membrane of the glomerular capsule formed by the podocytes (Figure 26.8). The capillary pores prevent the passage of blood cells and large proteins, whereas the filtration slits (slit pores) restrict the exit of plasma proteins larger than 7 nm. This suggests that the filtration slits play a greater role than does the capillary endothelium in determining what will pass.

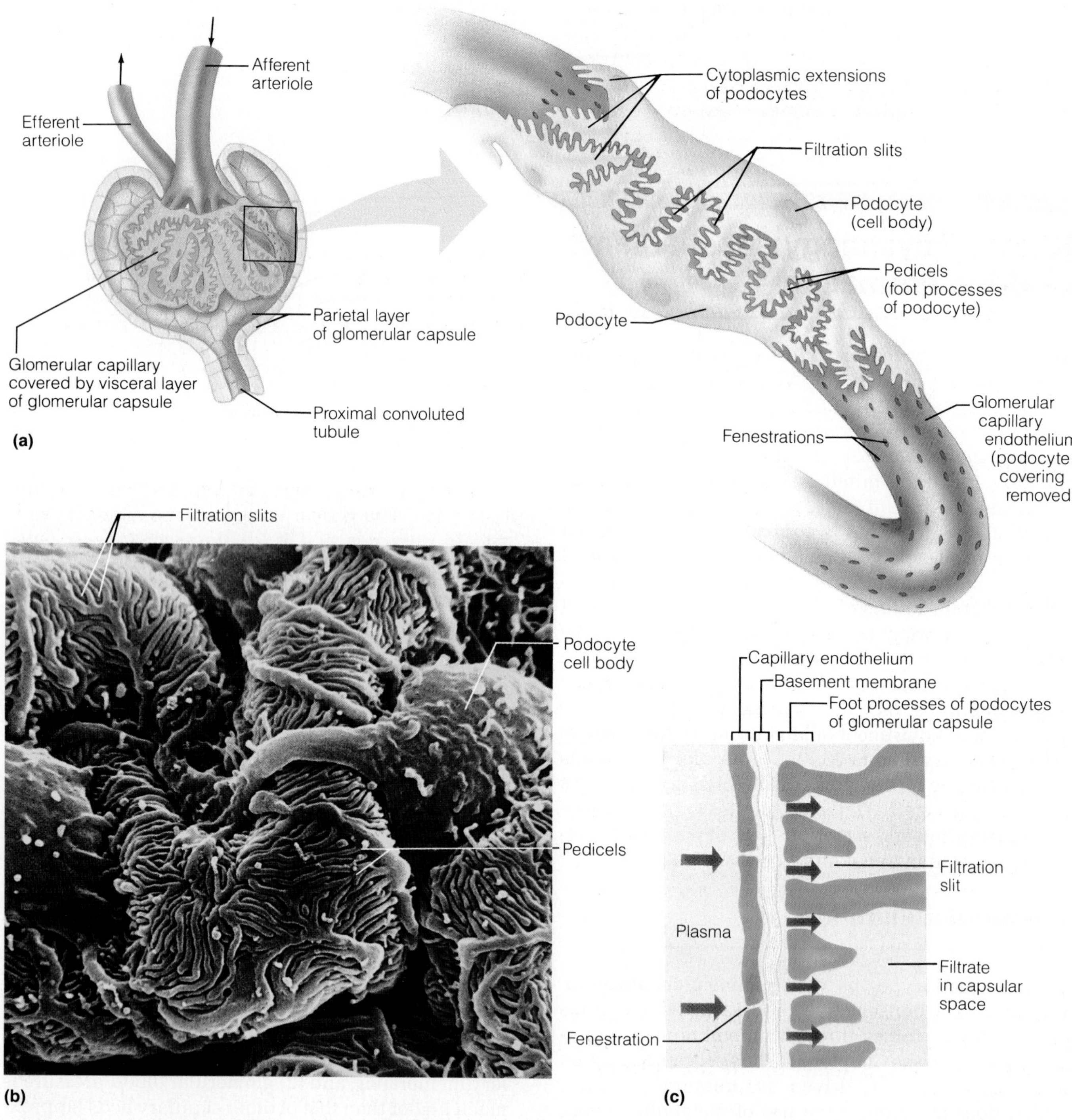

Figure 26.8 The filtration membrane. The filtration membrane is composed of three layers: the glomerular endothelium, the podocyte-containing visceral layer of the glomerular capsule, and the intervening basement membrane. (**a**) Diagrammatic three-dimensional view of the relationship of the visceral layer of the glomerular capsule to the glomerular capillaries. The visceral epithelium is drawn incompletely to show the fenestrations in the underlying capillary wall. (**b**) Scanning electron micrograph of the visceral layer. Filtration slits between the podocyte foot processes are evident (39,000X). (**c**) Diagrammatic view of a section taken through the filtration membrane showing all three structural elements.

In general, the porosity of the filtration membrane is such that molecules smaller than 3 nm in diameter—such as water, glucose, amino acids, and nitrogenous wastes—pass freely from the glomerulus into the renal tubule and are usually present in equal concentrations in both the glomerular blood and the filtrate. Molecules larger than 3 nm pass with greater difficulty, and those larger than 7–9 nm are usually completely barred from entering the tubule. Albumin, a fairly small blood protein (diameter of about 7.5 nm), is rarely seen in the filtrate of healthy individuals, and larger proteins and blood cells are totally absent. (Because erythrocytes have a diameter of about 7 μm, about 250 times that of an albumin molecule, this is easy to understand.) Keeping the plasma proteins in the capillaries maintains the colloid osmotic pressure of the glomerular blood, preventing the loss of all of its water to the renal tubules. The presence of proteins or blood cells in the urine usually indicates some problem with the filtration membrane.

There is a bit more to the story of filtration than simple porosity of the membrane. The structural makeup of the basement membrane also appears to confer some selectivity on the filtration process. Most basement membrane proteins are anionic (negatively charged) glycoproteins. Because these proteins repel other anionic molecules and hinder their passage into the tubule, the filtrate contains relatively larger amounts of cationic (positively charged) and uncharged molecules.

Net Filtration Pressure

To determine the **net filtration pressure (NFP)** responsible for filtrate formation, we must consider the various forces acting at the glomerular bed (Figure 26.9). *Glomerular hydrostatic pressure* (essentially, glomerular blood pressure) is the chief force pushing water and solutes across the filtration membrane. Although theoretically colloid osmotic pressure in the intracapsular space of the glomerular capsule "pulls" the filtrate into the tubule, this pressure is essentially zero because normally no proteins enter the capsule. Glomerular hydrostatic pressure (55 mm Hg) is opposed by forces that drive fluids back into the glomerular capillaries: (1) *glomerular osmotic pressure*, or colloid osmotic pressure of plasma proteins in the glomerular blood (about 30 mm Hg), and (2) *capsular hydrostatic pressure* (about 15 mm Hg), exerted by fluids in the glomerular capsule. Thus, the NFP responsible for forming renal filtrate from plasma is 10 mm Hg:

NFP = glomerular hydrostatic pressure − (glomerular osmotic pressure + capsular hydrostatic pressure)
NFP = 55 mm Hg − (30 mm Hg + 15 mm Hg)
NFP = 10 mm Hg

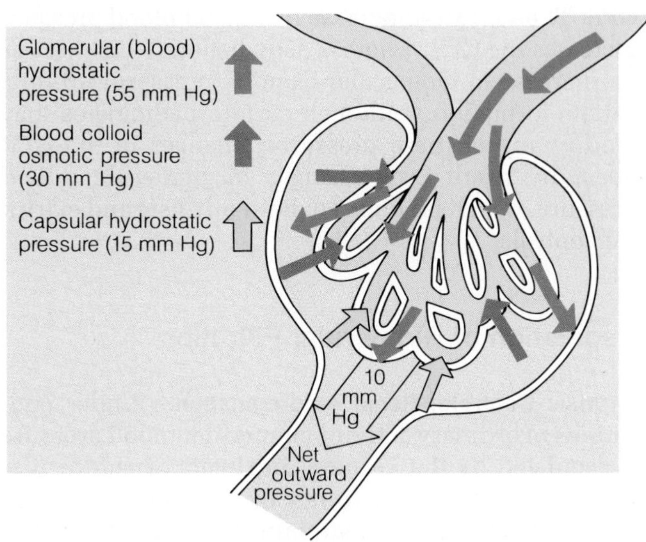

Figure 26.9 Forces that determine glomerular filtration and the effective filtration pressure. The glomerular hydrostatic (blood) pressure is the major factor forcing fluids and solutes out of the blood. This is opposed by the colloid osmotic pressure of the blood and the hydrostatic pressure that exists within the glomerular capsule. The pressure values cited in the diagram are approximate.

The NFP is *not* constant along the length of the glomerulus. The glomerular capillaries are short and wide, so glomerular hydrostatic pressure declines only slightly toward the end of the capillary bed. However, capillary osmotic pressure increases dramatically in the same direction, because as fluid is lost, the plasma proteins represent an increasingly large proportion of the plasma. Thus, at some point before the end of the glomerulus, the osmotic pressure gradient opposing filtration is exactly balanced by the hydrostatic pressure gradient favoring filtration. Consequently, the distal portions of the glomerular capillaries do not usually form filtrate.

Filtration Rate

Factors governing filtration at capillary beds are (1) total surface area available for filtration, (2) filtration membrane permeability, and (3) net filtration pressure. The rate at which fluid filters from the blood into the glomerular capsule, the **glomerular filtration rate** or **GFR,** is directly proportional to the net filtration pressure. Because glomerular capillaries are exceptionally permeable and have a huge surface area (collectively equal to the surface area of the skin), large amounts of filtrate can be produced even with the usual modest net filtration pressure of 10 mm Hg.

The normal GFR in both kidneys in adults is 120–125 ml/min (7.5 L/hour, or 180 L/day). A change in any of the pressures acting at the filtration membrane (see Figure 26.9) changes the NFP and thus the

GFR. Therefore, an increase in arterial blood pressure increases the GFR, whereas dehydration (which causes an increase in glomerular osmotic pressure) inhibits filtrate formation. Although certain pathologies may modify any of these pressures, changes in the GFR *normally* result from changes in glomerular blood pressure, which is subject to both intrinsic and extrinsic controls.

Regulation of Glomerular Filtration

Intrinsic Controls: Renal Autoregulation. Under conditions of ordinary activity, glomerular blood pressure is regulated by the kidneys' intrinsic, or autoregulatory, system. Thus, the kidney, by determining its own rate of blood flow, can maintain the glomerular filtration rate despite fluctuations in systemic arterial blood pressure. The importance of **renal autoregulation** becomes clear when one realizes that reabsorption of water, nutrients, and other substances from the filtrate depends to some extent on the *rate* at which the filtrate flows through the renal tubules. Although the GFR can increase by 30% without impairing the balance between glomerular filtration and tubular reabsorption, when large amounts of filtrate are formed and flow is very rapid, needed substances are inadequately reabsorbed and are lost to the body in urine. On the other hand, when filtrate is scant and flows slowly, nearly all of it is reabsorbed, including most of the wastes that are normally disposed of. Thus, for optimal filtrate processing, the intrinsic control system must precisely regulate the glomerular filtration rate of each nephron. It does this by directly regulating the diameter of the afferent (and perhaps the efferent) arterioles. Although the mechanism of renal autoregulation is still something of a mystery, there is little question that a myogenic mechanism is involved. Also probable is the involvement of the juxtaglomerular apparatus, which regulates glomerular pressure both directly and indirectly by adjusting afferent arteriole resistance and systemic blood pressure.

The **myogenic** (mī″-ō-jeh′-nik) **mechanism** reflects the tendency of vascular smooth muscle to contract when it is stretched. An increase in systemic blood pressure causes afferent arterioles to constrict, which impedes blood flow into the glomerulus and prevents glomerular blood pressure from rising to damaging levels. Conversely, a decline in systemic blood pressure causes dilation of afferent arterioles and an increase in glomerular hydrostatic pressure. Both responses help maintain a normal GFR.

Autoregulation by the **juxtaglomerular apparatus** reflects the activity of its juxtaglomerular cells (modified smooth muscle cells in the wall of an afferent arteriole) or its macula densa cells (chemoreceptors in the wall of the distal tubule) or both.

The **renin–angiotensin mechanism,** triggered by the juxtaglomerular cells, reflects their ability to release the enzyme renin in response to various stimuli. Renin acts on *angiotensinogen*, a plasma globulin made by the liver, to form *angiotensin I*, which is in turn converted to *angiotensin II* by enzymes present in plasma and in various body tissues, particularly the lungs. Angiotensin II, the most potent vasoconstrictor produced by the body, activates vascular smooth muscle throughout the body, causing systemic blood pressure to rise, and stimulates the adrenal cortex to release aldosterone, which causes the renal tubules to reclaim more sodium ions from the filtrate. Because water follows sodium osmotically, blood volume and blood pressure rise (Figure 26.10). Angiotensin II may also cause vasoconstriction of the efferent arteriole, thereby increasing glomerular blood pressure and the GFR. However, whether this occurs or not is still a matter of dispute.

Renin release is triggered by several factors acting independently or collectively:

1. Reduced stretch of the juxtaglomerular (JG) cells, caused by a drop in systemic blood pressure below 70 mm Hg, directly stimulates them to release renin.
2. Stimulation of the JG cells by macula densa cells, which have been activated by some unknown osmotic signal (perhaps filtrate with a low chloride and sodium content or low osmolarity in general) or by slowly flowing filtrate.
3. Stimulation of the JG cells by the sympathetic nervous system. Although the renin–angiotensin mechanism contributes to renal autoregulation, its main thrust is to stabilize systemic blood pressure and extracellular fluid volume—processes considered in more detail in Chapter 27.

As just noted, the macula densa cells can contribute to the renin–angiotensin mechanism, but they can independently influence the glomerular filtration rate as well. Under conditions in which they prompt the JG cells to release renin, they also promote an intense vasodilation of the afferent arteriole, which increases glomerular filtration pressure. On the other hand, when the macula densa cells are exposed to rapidly flowing filtrate or filtrate with high osmolarity, they promote vasoconstriction of the afferent arteriole. This hinders blood flow into the glomerulus, decreasing the GFR and allowing more time for filtrate processing. The direct role of the macula densa cells in renal autoregulation is referred to as the *tubuloglomerular feedback mechanism*. This mechanism is believed to be very important in balancing filtration and tubular reabsorption.

Renal autoregulation maintains a relatively constant kidney perfusion over an arterial pressure range

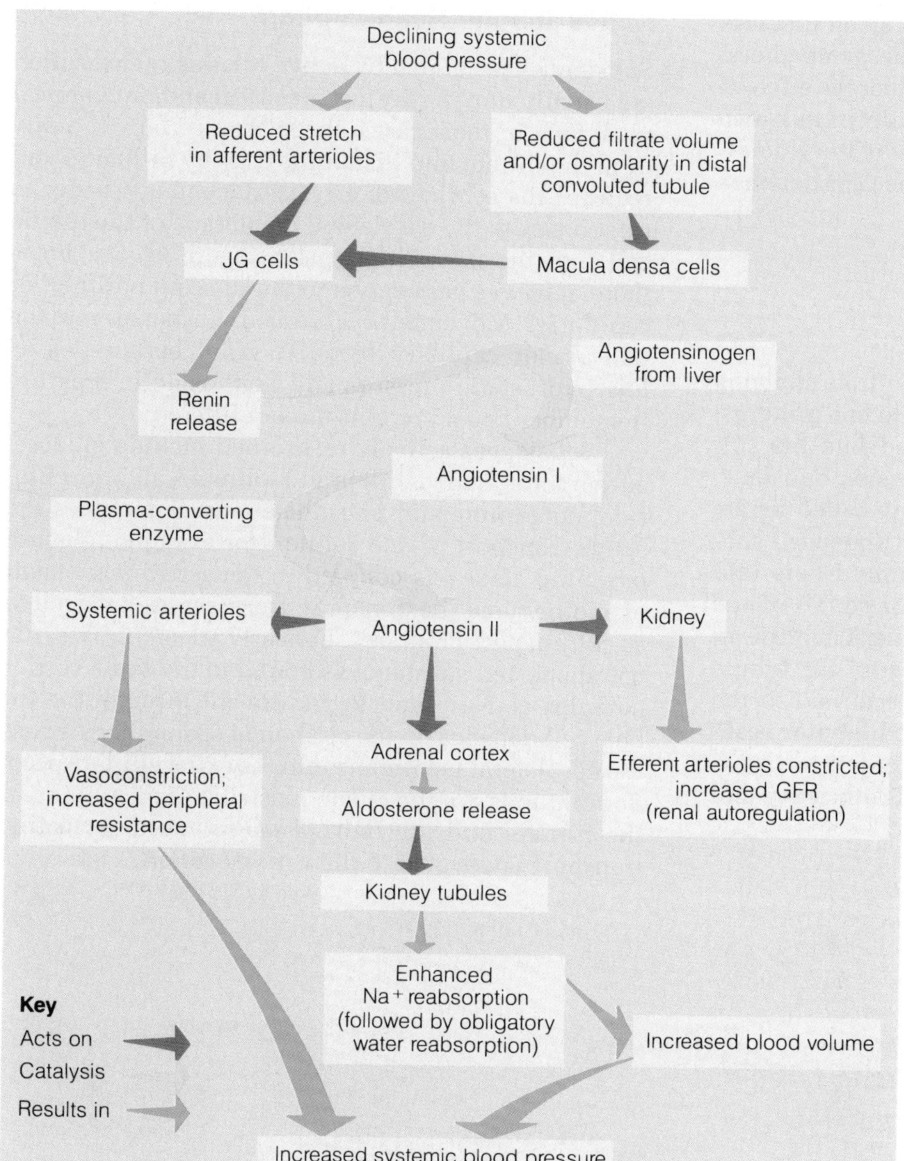

Figure 26.10 Flowchart illustrating the role of renin in regulating blood pressure under conditions of declining systemic blood pressure. Under conditions of normal or high blood pressure, less renin is released by the juxtaglomerular (JG) cells. Consequently, less angiotensin II is generated and less sodium and water are conserved. Thus, blood volume declines and blood pressure decreases.

from about 80 to 180 mm Hg, preventing large changes in water and solute excretion. However, this intrinsic control system cannot handle extremely low systemic blood pressure, such as might result from serious hemorrhage (*hypovolemic shock*). Once systemic blood pressure has dropped below 50 mm Hg (the point at which glomerular filtration pressure exactly balances the opposing forces), filtration stops.

An abnormally low urinary output (less than 50 ml/day) is called *anuria* (ah-noo′rē-uh). It usually indicates that glomerular blood pressure is too low to cause filtration but anuria may also result from acute nephritis, transfusion reactions, or crush injuries. ■

Even within its effective range, the autoregulatory system is not 100% efficient. When systemic blood pressure is *substantially* increased—for example, by drinking enough fluids to increase blood volume dramatically—the GFR increases slightly. However, such large volumes of filtrate are normally formed that even a small relative increase in the GFR can cause much more filtrate to be produced and more urine to be excreted.

Extrinsic Controls: Sympathetic Nervous System Stimulation. Neural controls serve the needs of the body as a whole, and renal autoregulatory mechanisms may be overcome during periods of extreme stress or emergency when it is necessary to shunt blood to the heart, brain, and skeletal muscles at the expense of the kidneys. At such times, stimulation by sympathetic nerve fibers and release of epinephrine by the adrenal medulla cause strong constriction of the afferent arterioles and inhibit filtrate formation. The sympathetic nervous system also stimulates the JG cells (via the binding of norepinephrine to beta-adrenergic receptors) to release renin, bringing about an increase in systemic blood pressure. When the sympathetic nervous system is less intensely activated, the efferent and afferent arterioles are constricted to about the same extent; because blood flow into and out of the glomerulus is hindered to the same degree, the GFR declines only slightly.

Tubular Reabsorption

Filtrate and urine are quite different. Filtrate contains everything that blood plasma does except proteins; but by the time filtrate has percolated into the collecting ducts, it has lost most of its water, nutrients, and essential ions. What remains, now called **urine,** contains mostly metabolic wastes and unneeded substances. Because the total blood volume filters into the renal tubules every 45 minutes or so, all of the plasma would be drained away as urine within an hour were it not for the fact that most of the tubule contents are quickly reclaimed and returned to the blood. This reclamation process, called **tubular reabsorption,** begins as soon as the filtrate enters the proximal convoluted tubules and involves both active and passive transport.

Active Tubular Reabsorption

Substances reclaimed by **active tubular reabsorption** are usually moving against electrical and/or chemical gradients. In most cases, the substance moves easily from the filtrate into the tubule cells by diffusion, but requires the services of an ATP-dependent carrier to eject it across the basolateral membrane of the tubule cell into the interstitial space (Figure 26.11). From there, it moves passively into the adjacent peritubular capillaries. Movement of absorbed substances into the peritubular capillary blood is rapid because of its low hydrostatic pressure and high osmotic pressure (remember, most proteins are not filtered).

Substances actively reabsorbed include glucose, amino acids, and vitamins and sodium, calcium, chloride, potassium, and phosphate ions. Sometimes the active transport of one solute—for example, glucose or amino acids—is *coupled* to the active movement of another; and the transport processes are referred to as *cotransport processes.* In nearly all such cases, the cotransported substance is bound to the same carrier complex as Na^+, and its movement through the tubule cell depends on the ejection of sodium ions across the basolateral membrane into the extracellular space. Sodium ions are the single most abundant cations in the filtrate, and the bulk of energy used for active transport is devoted to their reabsorption.

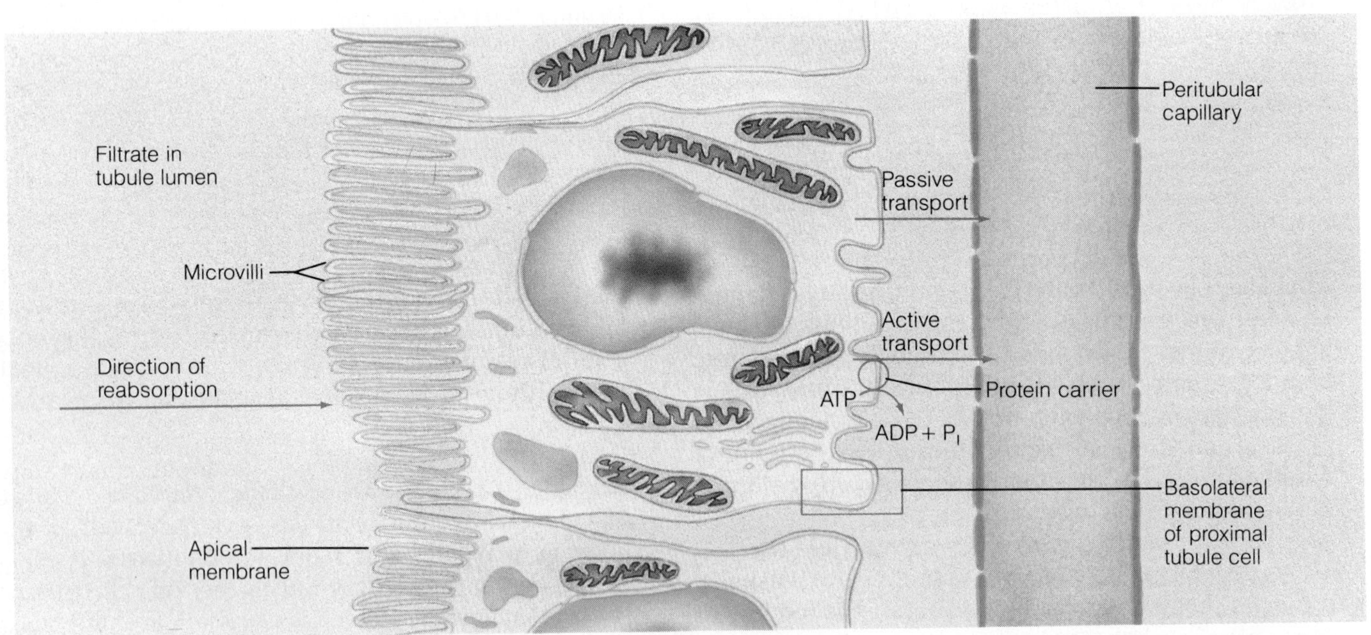

Figure 26.11 Directional movement of reabsorbed substances. Most substances reabsorbed by renal tubule cells pass by diffusion from the filtrate through the tubule cell's apical membrane and then through the tubule cell to its basolateral membrane facing the peritubular capillaries. From there the substances move through the basolateral membrane into the interstitial fluid and then into the peritubular capillary blood. The ATP-dependent carriers that transport Na^+ and other substances that must be transported actively are thought to be located in the basolateral membranes of the tubule cells. Most such carriers transport Na^+ along with another solute.

There is some overlap of carriers; for example, fructose and galactose compete with glucose for the same Na^+-associated carrier. Even so, the transport systems for the various solutes are quite specific *and limited*. There is a **transport maximum (T_m)** for nearly every substance that is actively reabsorbed, reflecting the number of carriers in the renal tubules available to "ferry" that particular substance. In general, there are abundant carriers and high transport maximums for substances such as glucose that need to be retained and few or no carriers for substances of no use to the body. When the carriers are saturated—that is, all bound to the substance they transport—excesses of that substance are excreted in urine. This is exactly what happens in individuals who become hyperglycemic because of uncontrolled diabetes mellitus. Large amounts of glucose are lost in the urine simply because the T_m for glucose has been exceeded. The renal tubules are still functioning normally.

Any plasma proteins that squeeze through the filtration membrane are completely removed from the filtrate in the proximal tubule by pinocytosis, which also requires ATP. Once inside the tubule cells, the proteins are digested to their amino acid monomers, which then move into the peritubular blood.

Passive Tubular Reabsorption

In **passive tubular reabsorption,** which encompasses diffusion, facilitated diffusion, and osmosis, substances move along their electrochemical gradient without the use of metabolic energy (see Chapter 3). In the renal tubules, the electrochemical gradient that drives passive transport of water and many solutes is established by the active reabsorption of sodium ions from the filtrate. As positively charged sodium ions move through the tubule cells into the peritubular capillary blood, they establish an electrical gradient that favors the passive diffusion of anions (either HCO_3^- or Cl^-) into the peritubular capillaries to restore electrical neutrality in the filtrate and plasma. Just which anion is absorbed depends on the pH of the blood at the time. Sodium movement also establishes a strong osmotic gradient, and water moves by osmosis into peritubular capillaries. Because water is "obliged" to follow salt, this sodium-linked water flow is referred to as **obligatory water reabsorption**. As water leaves the tubules, the relative concentrations of substances still present in the filtrate increase dramatically, and they too begin to follow their concentration gradients into the tubule cells. This phenomenon of solutes following the movement of the solvent is called **solvent drag;** it explains the passive reabsorption of some of the urea present in the filtrate and of a number of lipid-soluble substances such as fatty acids. Although the gradient provides the driving force, remember that solute size and lipid solubility may be limiting factors. Solvent drag also explains in part why many drugs and environmental toxins, which are nonpolar and hence highly lipid-soluble, are difficult to excrete.

Nonreabsorbed Substances

Some substances are not reabsorbed or are reabsorbed incompletely because they lack carriers, are not lipid-soluble, or are too large to pass through plasma membrane pores of the tubular cells. The most important of these substances are the nitrogenous end products of protein and nucleic acid metabolism: urea, creatinine, and uric acid. Urea is small enough to diffuse through the membrane pores, and about 40% of the urea present in filtrate is reclaimed. Creatinine, a large, non-lipid-soluble molecule, is not reabsorbed at all—a characteristic that makes it useful for measurement of the GFR and glomerular function (see p. 875).

Absorptive Capabilities of Different Regions of the Renal Tubules

Although the entire length of the renal tubule is involved in reabsorption to some degree (Table 26.1), the proximal tubule cells are by far the most active reabsorbers and the events just described occur mainly in this tubular segment. Normally, all of the glucose and amino acids are reabsorbed by the time the filtrate reaches the loop of Henle. About 80% of the Na^+ present in filtrate is absorbed in the proximal tubule; and because water follows in the same proportion, approximately 80% of the water is also reabsorbed there. Hence, the filtrate entering the loop of Henle has the same relative concentrations of sodium ions and water as that entering the glomerular capsule. Likewise, the bulk of the *selective* or active-transport-dependent reabsorption of electrolytes has been accomplished by the time the filtrate reaches the loop of Henle. For the most part, the reabsorption of electrolytes is hormonally controlled and intimately related to the need to regulate their plasma levels. (These controls are described in Chapter 27.) Nearly all of the potassium ions and uric acid are reabsorbed in the proximal tubule, but both are secreted back into the filtrate.

Beyond the proximal convoluted tubule, the permeability characteristics of the tubule epithelium change dramatically (Figure 26.12). Water can leave the descending limb of the loop of Henle but not the ascending limb, which, however, is quite permeable to sodium chloride. For reasons that will be explained shortly, the permeability differences in the different regions of the loop of Henle play a vital role in the kidneys' ability to form dilute and concentrated urine. Reabsorption of the remainder of the filtrate (about 20% of the total volume filtered) in the distal and collecting tubule depends on the needs of the body at the time and is regulated by hormones. If necessary, nearly all of the water and Na^+ reaching these regions

Table 26.1 Reabsorption Capabilities of Different Segments of the Renal Tubules and Collecting Tubules

Tubule segment	Substance reabsorbed	Mechanism
Proximal convoluted tubule	Sodium ions (Na^+)	Active transport via ATP- dependent carrier; sets up electrochemical gradient for passive solute diffusion and osmosis
	Virtually all nutrients (glucose, amino acids, vitamins)	Active transport; cotransport with Na^+
	Cations (K^+, Mg^{2+}, Ca^{2+}, and others)	Active transport; cotransport with Na^+
	Anions (Cl^-, HCO_3^-)	Passive transport; driven by electrochemical gradient
	Water	Osmosis; driven by solute reabsorption (obligatory)
	Urea and lipid-soluble solutes	Passive diffusion via solvent drag created by osmotic movement of water
	Small proteins	Pinocytosed by tubule cells and digested to amino acids within tubule cells
Loop of Henle		
Descending loop	Water	Osmosis
Ascending loop		
Thin segment	Na^+ and Cl^-	Passive diffusion; driven by concentration gradient
Thick segment	Na^+ and Cl^-	Active transport of Cl^-; Na^+ follows passively by diffusion (Na^+ may be contransported with Cl^-)
Distal convoluted tubule	Na^+	Active transport; requires aldosterone
	Anions	Diffusion; follow electrochemical gradient created by Na^+ reabsorption
Collecting tubule	Water	Osmosis; facultative water reabsorption; depends on ADH to increase porosity of tubule epithelium
	Urea	Diffusion in response to concentration gradient; most remains in medullary interstitial space

can be reclaimed. In the absence of regulatory hormones, the distal tubule and collecting duct are relatively impermeable to both water and Na^+. The reabsorption of more water depends on the presence of antidiuretic hormone, which makes the collecting tubules more permeable to water. This mechanism is described shortly.

Reabsorption of the remaining Na^+ is under the control of the hormone aldosterone. Several conditions—for example, decreased blood volume or blood pressure, low Na^+ concentration (hyponatremia) or increased K^+ concentration (hyperkalemia) in the extracellular fluid—cause the adrenal cortex to release aldosterone to the blood. Except for hyperkalemia, just about all of these conditions promote the renin–angiotensin mechanism, which, in turn, stimulates the release of aldosterone (see Figure 26.10). Aldosterone targets the renal tubules, particularly their distal segments, and enhances Na^+ reabsorption, so that very little leaves the body in urine. In the absence of aldosterone, essentially no sodium is reabsorbed by these segments, resulting in catastrophic Na^+ losses that are incompatible with life.

A second effect of aldosterone is to increase water absorption because, as sodium is reabsorbed, water follows it back into the blood. Aldosterone also regulates (reduces) potassium ion concentrations in the blood because aldosterone-induced reabsorption of sodium is coupled to potassium ion secretion via the sodium-potassium pump.

In contrast to aldosterone, which acts to conserve Na^+, *atrial natriuretic factor (ANF)*, a hormone released by the atrial cardiac cells when blood volume and/or blood pressure is elevated, inhibits Na^+ reabsorption and thereby reduces water reabsorption and blood volume.

Tubular Secretion

The failure of tubule cells to reabsorb some or all of certain filtered solutes is an important means of clearing plasma of unwanted substances. A second such mechanism is **tubular secretion**—essentially, reabsorption in reverse. Substances such as hydrogen and potassium ions, urea, creatinine, ammonia, and certain organic acids move from the blood of the peritubular capillaries through the tubule cells or from

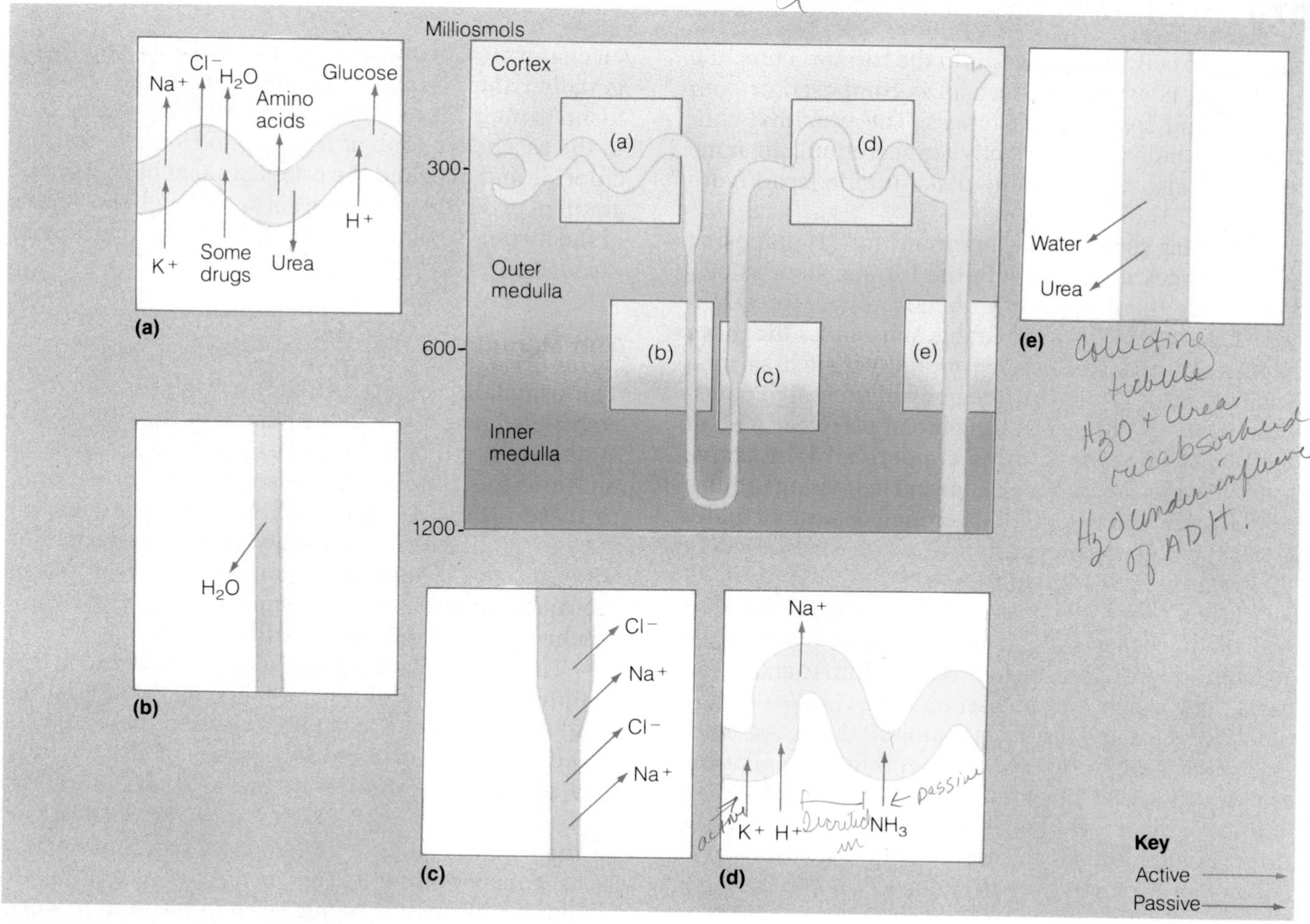

Figure 26.12 Summary of nephron functions. The different regions of the nephron tubule carry out reabsorption, secretion, and maintain a gradient of osmolality within the medullary interstitial fluid. Urea is the most important solute in this gradient, with salt (Na^+ and Cl^-) making a minor contribution. Varying osmolality at different points in the tubule and interstitial fluid are symbolized in the figure by gradients of color. The inserts that accompany the central figure describe the transport functions of the four regions of the nephron tubule and of the collecting tubule. Blue arrows indicate transport processes believed to be passive; red arrows represent active transport. Blood vessels are not shown.

(**a**) **Proximal tubule.** Filtrate that enters the proximal tubule from the glomerular capsule has about the same osmolality as blood plasma. The main activity of the transport epithelium of the proximal tubule is reabsorption of certain solutes from the filtrate back into the blood. Nearly all nutrients, such as glucose and amino acids, and about 80% of Na^+ are actively transported out of the tubule and enter the peritubular capillaries; Cl^- and water follow passively. The proximal tubule cells also secrete specific substances, such as K^+, and nitrogenous wastes into the filtrate and help maintain a constant pH in blood and interstitial fluid by the controlled secretion of H^+. By the end of the proximal tubule, the filtrate volume has been reduced by 80%.

(**b**) **Descending limb.** The wall of the descending limb of the loop of Henle is freely permeable to water, but not very permeable to salt or urea. As filtrate in the limb descends into the medulla, it loses water by osmosis to the interstitial fluid, which is increasingly hypertonic in that direction. Consequently, salt and other solutes become more concentrated in the filtrate.

(**c**) **Ascending limb.** The filtrate rounds the bend and heads up the ascending limb of the loop of Henle, which has a thin segment and a thick segment. The epithelium of the thin ascending limb is not permeable to water, but is permeable to Na^+ and Cl^-. These ions, which become concentrated within the descending limb, diffuse out of the thin segment of the ascending limb and contribute to the high osmolality of interstitial fluid in the inner medulla (however, most of this salt reenters the blood in the peritubular capillaries). The thick segment of the ascending limb contributes to the osmolality of the outer medulla by expelling salt from the tubule, but this time by active transport. Cl^- is pumped out of the tubule, with Na^+ following passively (or both ions may be actively transported). Because the epithelium here is not permeable to water, the filtrate becomes more and more dilute.

(**d**) **Distal tubule.** The distal tubule, like the proximal tubule, is specialized for selective secretion and reabsorption. K^+ and H^+ are secreted and, in the presence of aldosterone, more Na^+ is reabsorbed.

(**e**) **Collecting duct.** The urine, normally quite dilute at this point, begins its journey via the collecting duct back into the medulla, with its increasing osmolality gradient. The wall of the collecting duct is permeable to urea, and a certain amount of this nitrogenous waste diffuses out of the duct. Some of the urea that leaks out reenters the nephron by diffusing into the ascending limb of the loop of Henle. However, most urea remains in the interstitial space. Hence, urea accounts for most of the high osmolality of the inner medulla. In the absence of antidiuretic hormone, the collecting tubules are nearly impermeable to water, and the dilute urine passes out of the kidney. In the presence of antidiuretic hormone, the pores of the collecting tubule wall enlarge, and the filtrate loses water by osmosis as it passes through medullary regions of increasing osmolality. Consquently, water is conserved, and concentrated urine is excreted.

the tubule cells themselves into the filtrate. Thus, the urine that is eventually formed is composed of both *filtered and secreted substances*. The proximal convoluted tubules are most active in secretion, but some secretion also occurs in the distal tubule (see Figure 26.12).

Tubular secretion is important for (1) disposing of substances not already in the filtrate, such as certain drugs (penicillin and phenobarbital, for example); (2) eliminating undesirable substances that have been reabsorbed by passive processes, such as urea and uric acid; (3) ridding the body of excessive potassium ions; and (4) controlling blood pH. Nearly all of the potassium ions in urine are derived from active tubular secretion into the distal and collecting tubules under the influence of aldosterone because virtually all of the K^+ present in the filtrate is reabsorbed in the proximal convoluted tubule.

When blood pH begins to drop toward the acid end of its homeostatic range, the renal tubule cells actively secrete hydrogen ions into the filtrate and retain more bicarbonate and potassium ions. As a result, the blood pH rises and the urine drains off the excess acid. Conversely, when blood pH approaches the alkaline end of its range, Cl^- rather than HCO_3^- is reabsorbed and more bicarbonate is allowed to leave the body in urine. Renal mechanisms for controlling the acid–base balance of the blood are described in more detail in Chapter 27.

Regulation of Urine Concentration and Volume

Because we will make frequent use of the term *osmolality* (oz″-mō-lal′-ih-tē) in this discussion, it is important that this concept be clearly understood. A solution's *osmolality*, or *osmotic concentration*, is the number of solute particles dissolved in 1000 g of water and is reflected in the solution's ability to cause osmosis. This capability, called *osmotic activity*, is determined only by the concentration of nonpenetrating solute particles and is independent of their type and nature: Ten sodium ions have the same osmotic activity as ten glucose molecules or ten amino acids in the same volume of solution. Since the *osmol* (equivalent to 1 mole of a nonionizing substance in 1000 g of water) is a fairly large concentration unit, the **milliosmol** (mih′-lē-oz-mōl) **(mOsm)**, equal to 0.001 osmol, is used to describe the solute concentration of body fluids.

One of the crucial functions of the kidneys is to keep the solute concentration of body fluids constant at about 300 mOsm by regulating urine concentration and volume. Just how the kidneys accomplish this feat is still a matter of controversy (as is the name of the mechanism). The most current hypothesis, whether it is called the *counter current* or *two-solute mechanism*, explains it in terms of differential permeability of the ascending limb of the loop of Henle to sodium chloride and urea and the establishment of an osmotic gradient extending from the cortex through the depths of the medulla. Hence, this is where we will begin our discussion.

The Medullary Osmotic Gradient

The osmolality of the filtrate entering the proximal convoluted tubule is identical to that of both the plasma and the interstitial fluid of the renal cortex, about 300 mOsm. Thus, the filtrate is *iso-osmotic*, or *isotonic*, to these fluids. As described earlier, approximately 80% of the filtrate is reabsorbed by the proximal tubule cells; and though the amount and flow of filtrate are much reduced by the time the loop of Henle is reached, 20% of the filtrate still remains.

The stage is set for conserving most of this remaining filtrate as the osmolality of the interstitial fluid and the filtrate becomes more and more concentrated from the cortex to the depths of the medulla. The solute concentration increases from 300 mOsm in the cortex to about 1200 mOsm in the deepest part of the medulla (Figure 26.13). How can this tremendous concentration of the medullary fluids be explained? The answer seems to lie in the unique workings of the loops of Henle of the juxtamedullary nephrons and their closely associated vasa recta, which penetrate deep into the renal medulla. Let us follow filtrate processing through the loop of Henle, as portrayed in Figure 26.13, to see how this concentration gradient occurs and is maintained.

1. The descending limb of the loop of Henle is relatively impermeable to solutes and freely permeable to water. Because the osmolality of the medullary interstitial fluid increases along the course of the descending limb, water passes out of the filtrate osmotically all along its course. The water enters the vasa recta and is carried away; thus, the osmolality of the filtrate and medullary interstitial fluid is the same (1200 mOsm) at the lowest point of the loop of Henle.

2. The thin segment of the ascending limb of the loop of Henle is freely permeable to sodium and chloride, poorly permeable to urea, and impermeable to water. As the filtrate rounds the corner into the thin segment of the ascending limb, the tubule permeability changes, becoming impermeable to water and nearly impermeable to urea. Because sodium and chloride ions are in high concentration in the filtrate (and interstitial fluid concentrations are lower), these ions diffuse along their concentration gradients into the interstitial fluid and most are picked up by the vasa recta.

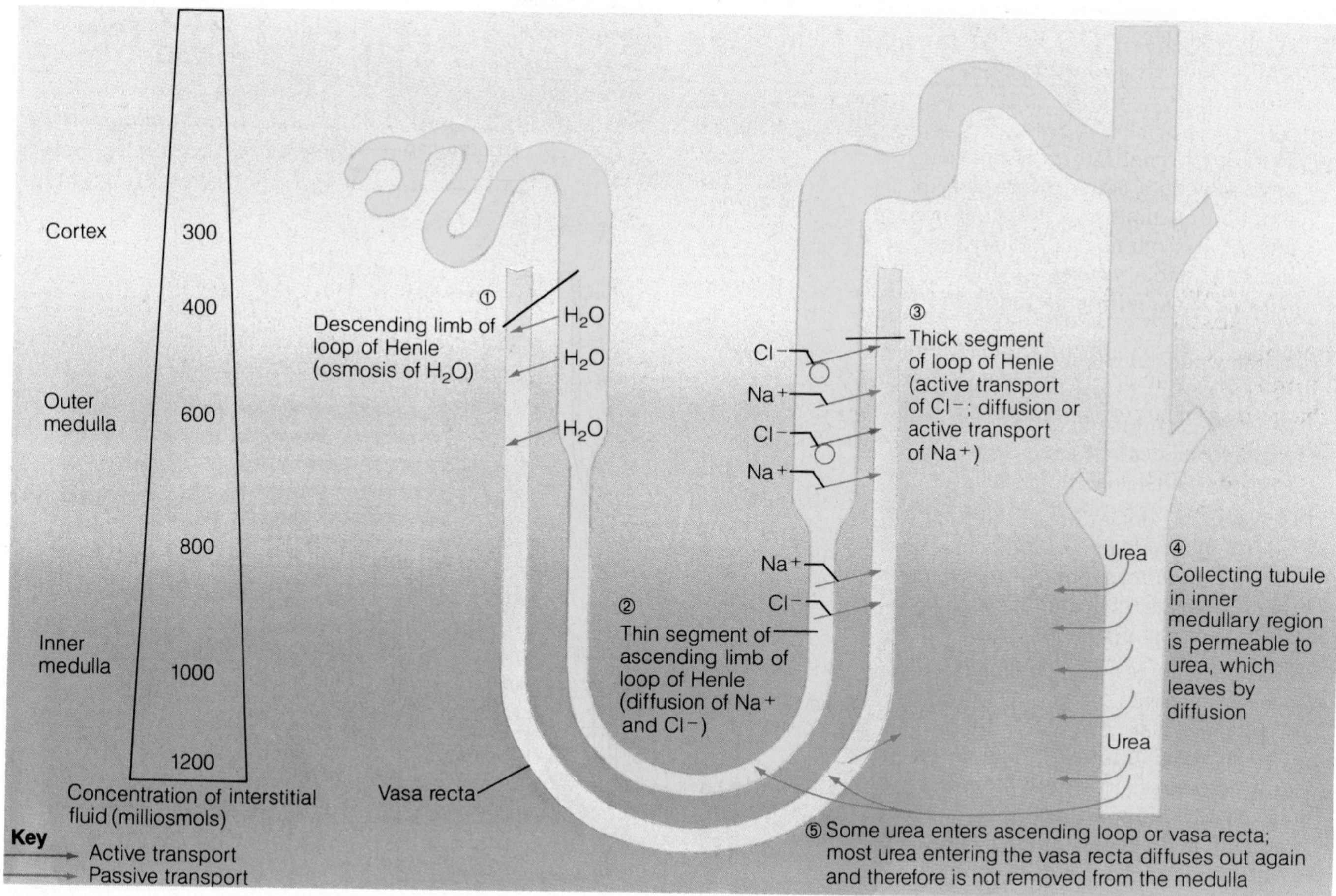

Figure 26.13 The mechanism for establishing and maintaining the osmotic gradient. (The numbered events correspond to the numbered sequence of the text discussion beginning on page 872).

3. The thick segment of the ascending limb of the loop of Henle has an active transport mechanism for chloride (and probably sodium) ions. Exactly what happens in the thick segment of the loop of Henle has been difficult to determine. One idea is that an active chloride ion pump transports Cl^- out of the filtrate into the interstitial space and that, because this leads to a deficit of negative charge in the filtrate, sodium follows passively. More current evidence suggests that two Cl^-, one Na^+, and one K^+ are cotransported simultaneously by an ATP-driven carrier and that the positive electrical potential in the lumen results because Na^+ leaks back into the filtrate more rapidly than Cl^-. Whichever the case, the movement of Na^+ and Cl^- out of the filtrate establishes a strong osmotic gradient for water reabsorption. However, the ascending limb of the loop of Henle is nearly impermeable to both water and urea, which, therefore, remain in the filtrate. Because the volume of the filtrate moving through the ascending limb does not change, removal of Na^+ and Cl^- dilutes the filtrate until it is hypo-osmotic, or hypotonic, to blood plasma and cortical fluids. Additionally, because most of the Na^+ and Cl^- enters the vasa recta, the relative osmolality of the filtrate and interstitial fluid remains the same, as if in lockstep.

4. The collecting tubules in the deep medullary regions are permeable to urea. In a normally hydrated person, the dilute urine passes through the distal convoluted tubule and the superior portions of the collecting ducts without modification. However, as urine passes through the deep medullary regions, urea begins to diffuse out of the collecting tubule into the interstitial space. It is this loss of urea to the medullary interstitial space that accounts for its buildup and establishes the gradient of osmolality that exists in the medulla. Even though the ascending limb of the loop of Henle is poorly permeable to urea, when urea concentrations in the medullary interstitial space are very high, some urea does enter the limb. However, it simply gets recycled back to the collecting duct, where it diffuses out again.

5. The vasa recta removes very little urea, thus maintaining the osmotic gradient. The characteristics of the vasa recta, which parallel the loops of Henle, are important in maintaining the osmotic gradient estab-

A CLOSER LOOK **Renal Failure and the Artificial Kidney**

Although renal failure is not common, it occurs when the number of functioning nephrons becomes too few to carry out normal kidney functions. Possible causes of acute or chronic renal failure include:

1. Repeated damaging infections of the kidneys.
2. Physical injury to the kidneys or major physical trauma (crushing injury) to other parts of the body.
3. Chemical poisoning of the tubule cells by heavy metals (mercury or lead) or organic solvents (dry-cleaning fluids, acetone, and others).
4. Inadequate blood delivery to the tubule cells (as may happen with arteriosclerosis).

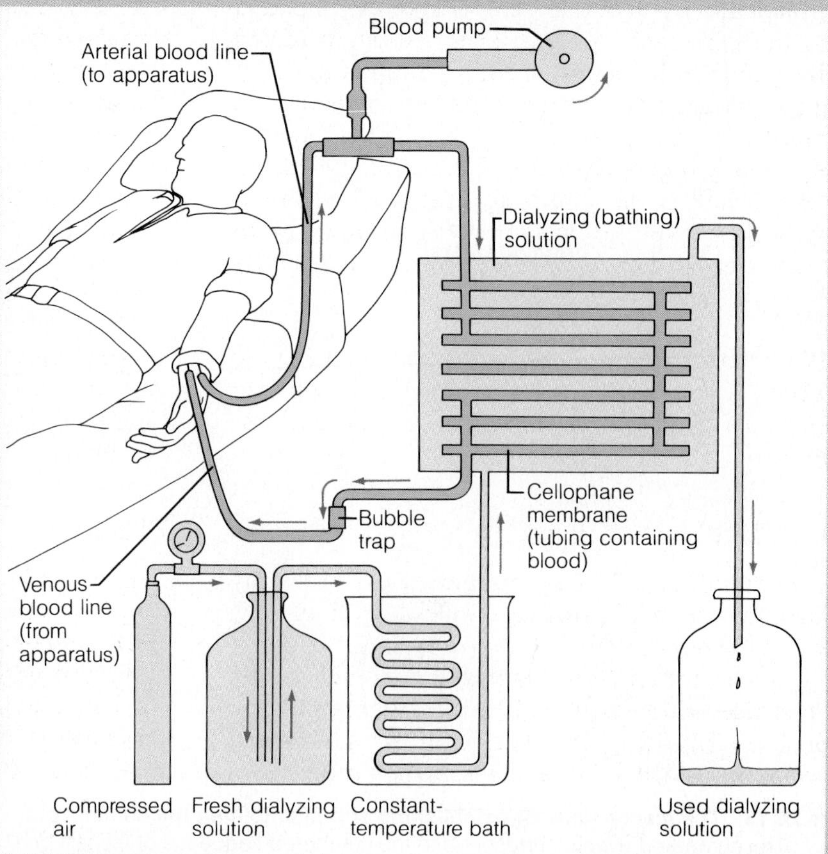

Apparatus for renal dialysis.

In renal failure, filtrate formation decreases or stops completely. Nitrogenous wastes accumulate quickly in the blood when the tubule cells are not working, a condition called *azotemia* (ā″-zō-tē′-mē-uh), and blood pH tumbles toward the acidic range. *Uremia* (yer-ē′-mē-uh), the bodywide (or systemic) response to azotemia, and electrolyte imbalance set in shortly after and totally disrupt life processes. Signs and symptoms of uncontrolled uremia include diarrhea, vomiting, edema (due to Na^+ retention), labored breathing, cardiac irregularities (due to excess K^+), convulsions, coma, and finally death.

To prevent these consequences of azotemia, *dialysis* by an artificial kidney must be performed to cleanse and modify the composition of the blood while the kidneys are shut down. In dialysis, the patient's blood is passed through a membrane tubing (see the diagram) that is permeable only to selected substances, and the tubing is immersed in a bathing solution that differs slightly from normal cleansed plasma. As blood circulates through the tubing, substances such as nitrogenous wastes and K^+ present in the blood (but not in the bath) diffuse out of the blood into the surrounding solution and substances to be added to the blood (mainly buffers for H^+) move from the bathing solution into the blood. In this way, needed substances are retained in the blood or added to it, while substances such as nitrogenous wastes and ion excesses are removed.

When renal damage is irreversible, as in chronic, slowly progressing renal failure, the kidneys become totally incapable of processing plasma or concentrating urine, and a kidney transplant is the only answer. Unhappily, the signs and symptoms of this life-threatening kidney problem become obvious only after about 75% of renal function has been lost. (This large margin of safety reflects the fact that individual nephrons hypertrophy and begin to function even more effectively as others are progressively lost.) At this point, the signs of *renal insufficiency* become obvious. Glomerular filtration declines and azotemia occurs as the kidneys lose their ability to form both concentrated and dilute urine, and the urine becomes isotonic with blood plasma. The final or end stage of renal failure, uremia, occurs when about 90% of the nephrons have been lost.

lished by urea. Because these capillaries receive only about 10% of the renal blood supply, blood flow through the vasa recta is very sluggish. It removes only limited amounts of Na^+ and Cl^- from the interstitial space and practically none of the urea. Most of the urea that enters the vasa recta diffuses back out again as the capillary loop follows the ascending limb of the loop of Henle toward the distal tubule. As a result, the osmolality of the vasa recta blood leaving the medulla is only slightly higher than that entering (Figure 26.13). The inefficiency of the vasa recta directly reflects their sluggish blood flow, because if the blood flowed rapidly, the urea would be carried out of the medulla before it could leak back out.

Now that the basic mechanisms establishing the medullary gradient have been explained, we can explore how the kidneys excrete dilute and concentrated urine.

Formation of Dilute Urine

Because tubular filtrate is diluted while traveling through the ascending limb of the loop of Henle, all the kidney needs to do to secrete dilute urine is allow the filtrate to continue its course into the renal pelvis. When antidiuretic hormone is not being released by the posterior pituitary, that is basically what happens. The collecting tubules remain essentially impermeable to water, and no further water reabsorption occurs. Moreover, added sodium and selected other ions can be moved from the filtrate by active or passive mechanisms so that the urine becomes even more dilute. The osmolality of urine can become as low as 50 mOsm, less than one-fifth as concentrated as glomerular filtrate or blood plasma.

Formation of Concentrated Urine

As its name reveals, **antidiuretic** (an″-tīh-dī″-yoo-ret′-ik) **hormone (ADH)**, inhibits *diuresis* (dī″-yoo-ē′-sis), or urine flow. ADH accomplishes this by causing the pores in the collecting tubules to enlarge so that water passes easily into the interstitial space. As a result, the osmolality of the filtrate tends to become equal to that of the interstitial fluid. In the distal tubules in the cortex, the osmolality of the filtrate is approximately 100 mOsm; but as the filtrate flows through the collecting tubules and is subjected to the hyperosmolar conditions in the medulla, water rapidly leaves the filtrate and exits the collecting tubules. Depending on the amount of ADH released (which is keyed to the level of body hydration), urine concentration may rise as high as 1200 mOsm, the concentration of the interstitial fluid in the deepest part of the medulla.

Although ADH opens the door (pores) for enhanced water reabsorption, it is the hypertonicity of the medullary fluids that establishes the gradient for osmotic reabsorption of water. ADH also affects this hypertonicity by reducing blood flow through the vasa recta. Consequently, removal of Na^+ and Cl^- (and urea) from the medulla becomes even more sluggish than usual and the osmolality of medullary fluids may reach levels as high as 1400 mOsm. In such cases, 99% of the water in the filtrate is reabsorbed and returned to the blood and only a very small amount of highly concentrated urine is excreted. Water reabsorption that depends on the presence of ADH is called **facultative water reabsorption**.

ADH is released more or less continuously unless the solute concentration of the blood drops too low; but its release is enhanced by any event that increases plasma osmolality to above 300 mOsm, such as excessive water loss through sweating or diarrhea, or reduced blood volume or blood pressure (hemorrhage). These mechanisms are considered in Chapter 27.

Diuretics

Diuretics are chemicals that enhance urinary output. Any substance that is not reabsorbed or that exceeds the ability of the renal tubules to reabsorb it will carry water out with it, acting as an *osmotic diuretic*. The high blood glucose levels of diabetes mellitus patients and the high blood urea levels created by high-protein diets both act as osmotic diuretics. Alcohol encourages diuresis by inhibiting release of ADH. Caffeine (found in coffee, tea, and colas) and many of the diuretic drugs prescribed for hypertension or edema cause diuresis by increasing the glomerular filtration rate (GFR) and inhibiting sodium ion reabsorption and, thereby, the obligatory water reabsorption that normally follows it.

Renal Clearance

The term **renal clearance** refers to the kidneys' ability to clear or cleanse a given volume of plasma of a particular substance in a given time, usually 1 minute. Renal clearance tests are used to determine the GFR, which provides information about the amount of functioning renal tissue. Renal clearance tests are also used to detect glomerular damage and to follow the progress of diagnosed renal disease.

The renal clearance (RC) of any substance, in ml/min, is calculated from the equation $RC = UV/P$, where U is the concentration (mg/ml) of the substance in urine, P is the concentration of the same substance in plasma, and V is the flow rate of urine formation (ml/min). *Inulin* (in′-yoo-lin), a polysaccharide with a molecular weight of approximately 5000, is frequently used as the standard to determine the GFR by renal clearance because it is neither reabsorbed nor secreted to any significant extent. Because infused inulin is excreted in the urine, its renal clearance value is equal to the GFR. Measured values for inulin are $U = 125$ mg/ml, $V = 1$ ml/min, and $P = 1$ mg/ml. Therefore, the computation of its renal clearance is RC = $(125 \times 1)/1 = 125$ ml/min, meaning that, in 1 minute, the kidneys have removed all the inulin present in 125 ml of plasma.

A clearance value less than that of inulin means that the substance is partially reabsorbed. For example, urea has a clearance value of 70 ml/min, meaning that, of the 125 ml of glomerular filtrate formed each minute, approximately 70 ml is completely cleared of urea, while the urea in the remaining 55 ml is recovered and returned to the plasma. If the renal clearance is zero, reabsorption is complete; glucose has a clearance value of zero in normal individuals. If clearance is greater than that of inulin, the tubule cells must be secreting the substance into the filtrate; this is the case with creatinine, which has an RC of 140 ml/min. The renal clearance values of several substances are provided in Table 26.2.

Table 26.2 Relative Renal Clearance Values of Selected Substances

Substance	Renal clearance (ml/min)
Glucose	0.0
Urea	70.0
Uric acid	14.0
Creatinine	140.0
Inulin	125.0
PAH	585.0
SO_4^{2-}	47.0
Na^+	0.9
K^+	12.0
Ca^{2+}	1.2
Mg^{2+}	5.0
Cl^-	1.3
HCO_3^-	0.5
$H_2PO_4^-$ & HPO_4^{2-}	25.0

Characteristics and Composition of Urine

Physical Characteristics

Color and Transparency. Freshly voided urine is generally clear and pale to deep yellow in color. The yellow color is due to the presence of *urochrome* (yoo′-rō-krōm), a pigment that results from the body's destruction of hemoglobin (via bilirubin or bile pigments). The more concentrated the urine, the deeper the yellow color. An abnormal color such as pink or brown, or a smoky tinge, may result from certain foods (beets, rhubarb) or may reflect the presence of bile or blood in the urine. Additionally, some commonly prescribed drugs and vitamin supplements alter the color of urine. Occasionally, fresh urine is slightly turbid (cloudy) due to mucus produced by the lining of the urinary tract. However, excessive or persistent turbidity usually indicates blood or pus in the urine and hints of an infection in some part of the urinary tract. Most urine samples become turbid on standing as a result of bacterial growth.

Odor. Fresh urine is slightly aromatic; but if it is allowed to stand, it develops an ammonia odor due to bacterial metabolism of its urea solutes. Some drugs and vegetables (asparagus) alter the usual odor of urine, as do some diseases. For example, in diabetes mellitus the urine smells fruity because of its acetone content.

pH. Urine is usually slightly acidic (around pH 6), but changes in body metabolism or diet may cause it to vary from about 4.5 to 8.0. A diet that contains large amounts of protein and whole wheat products, called an *acid–ash* diet, produces acidic urine; a vegetarian diet is called an *alkaline–ash* diet because it causes urine to become quite basic. Prolonged vomiting and bacterial infection of the urinary tract also may cause the urine to become quite alkaline. Although the solutes of urine are not usually visible, insoluble phosphate and urate sediments are likely to form when urine is alkaline.

Specific Gravity. Because urine is water plus solutes, it weighs more (is more dense) than distilled water. The term used to compare how *much* heavier urine is than an equal volume of distilled water is **specific gravity**. The specific gravity of distilled water is 1.0; that of urine can range from 1.001 to 1.035, depending on its solute concentration. When urine becomes extremely concentrated, solutes begin to precipitate out of solution.

Chemical Composition

Water accounts for about 95% of urine volume; the remaining 5% consists of solutes. The largest component of urine by weight, apart from water, is urea, which is derived from the breakdown of amino acids. Other nitrogenous wastes found in urine include uric

Table 26.3 Abnormal Urinary Constituents

Substance	Name of condition	Possible causes
Glucose	Glycosuria	Nonpathological: excessive intake of sugary foods Pathological: diabetes mellitus
Proteins	Proteinuria, or albuminuria	Nonpathological: excessive physical exertion; pregnancy; high-protein diet Pathological (over 250 mg/day): heart failure, severe hypertension; glomerulonephritis; often initial sign of asymptomatic renal disease
Ketone bodies	Ketonuria	Excessive formation and accumulation of ketone bodies, as in starvation and untreated diabetes mellitus
Hemoglobin	Hemoglobinuria	Various: transfusion reaction, hemolytic anemia, severe burns, etc.
Bile pigments	Bilirubinuria	Liver disease (hepatitis, cirrhosis) or obstruction of bile ducts from liver or gallbladder
Erythrocytes	Hematuria	Bleeding in urinary tract (due to trauma, kidney stones, infection, or neoplasm)
Leukocytes (pus)	Pyuria	Urinary tract infection

acid (a metabolite of nucleic acid metabolism) and creatinine (an end product of creatine phosphate, found in large amounts in skeletal muscle tissue). Normal solute constituents of urine, in order of decreasing concentration, include urea, sodium, potassium, phosphate, and sulfate ions, creatinine, and uric acid. Much smaller but highly variable amounts of calcium, magnesium, and bicarbonate ions are also found in urine. Abnormally high concentrations of any of these urinary constituents may indicate pathology.

With certain diseases urine composition changes dramatically and may include, for example, glucose, blood proteins, red blood cells, hemoglobin, white blood cells (pus), or bile. The presence of such abnormal substances in urine is an important warning of illness and may be helpful in diagnosis (Table 26.3). A urinalysis should be a part of every complete physical examination.

Ureters

The **ureters** are slender tubes, each about 25–30 cm (10–12 inches) long and 6 mm (0.25 inch) in diameter, that convey urine from the kidneys to the bladder (see Figure 26.1). Each ureter runs behind the peritoneum from the renal hilus to the level of the bladder, makes a medial turn, and courses obliquely through the posterior bladder wall before opening into the bladder cavity. This arrangement prevents backflow of urine into the ureters during bladder filling or emptying, because at such times the bladder compresses and closes the distal ends of the ureters.

Histologically, the ureter wall is trilayered. Its inner mucosa of transitional epithelium is continuous with that of the kidney pelvis superiorly and the bladder medially. Its middle portion is composed chiefly of two smooth muscle layers: the inner longitudinal and the outer circular. Its outer surface is covered with fibrous connective tissue adventitia (Figure 26.14).

The ureters play an active role in urine transport. As urine enters the renal pelvis, peristaltic waves initiated there force urine into the ureter, distending it. Distension of the ureter stimulates its contraction, which propels urine into the bladder. The vigor and frequency of the peristaltic waves are adjusted to the rate of urine formation; thus, the waves may occur several times each minute or may be separated in time by several minutes. Although the ureters are innervated by both sympathetic and parasympathetic fibers, neural control of their peristalsis appears to be insignificant compared to the local response of ureteral smooth muscle to stretch.

On occasion, calcium, magnesium, or uric acid salts in urine may crystallize and precipitate in the renal pelvis, forming *renal calculi* (kal′-kyoo-lī), or kidney stones. Renal calculi can cause excruciating pain, which radiates to the flank on the same side, when they obstruct a ureter, blocking urine drainage and increasing pressure within the kidney. Pain also occurs when the ureter walls close in on the sharp calculi as they are being eased through the ureter by peristalsis. Predisposing conditions are frequent bacterial infections of the urinary tract, urinary retention, and alkaline urine. Surgery has been the treatment of choice, but a newer, noninvasive procedure that uses ultrasound waves to shatter the calculi is becoming more popular. The pulverized, sandlike remnants of the calculi are eliminated in the urine. ■

Figure 26.14 Structure of the ureter wall (cross-sectional view, 10X). The wall of the ureter contains two smooth muscle layers: an outer, circular layer; and an inner, longitudinal layer. Its lumen is lined by a transitional epithelial membrane. The external surface of the ureter is covered by a loose, connective tissue adventitia.

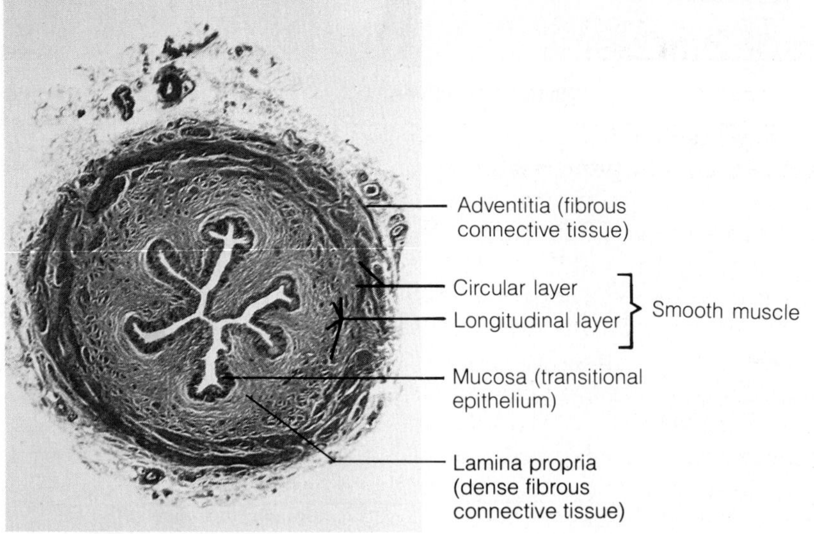

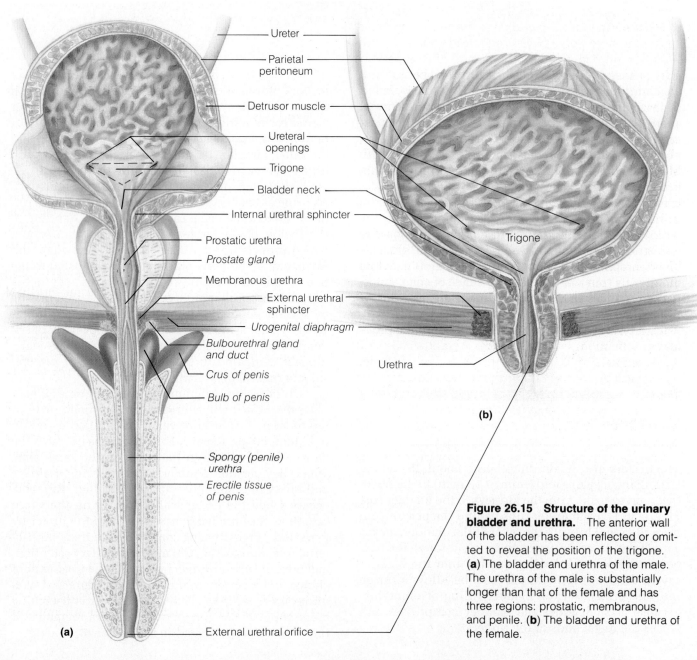

Ureter
Parietal peritoneum
Detrusor muscle
Ureteral openings
Trigone
Bladder neck
Internal urethral sphincter
Prostatic urethra
Prostate gland
Membranous urethra
External urethral sphincter
Urogenital diaphragm
Bulbourethral gland and duct
Crus of penis
Bulb of penis
Spongy (penile) urethra
Erectile tissue of penis
External urethral orifice
(a)
Trigone
Urethra
(b)

Figure 26.15 Structure of the urinary bladder and urethra. The anterior wall of the bladder has been reflected or omitted to reveal the position of the trigone. (**a**) The bladder and urethra of the male. The urethra of the male is substantially longer than that of the female and has three regions: prostatic, membranous, and penile. (**b**) The bladder and urethra of the female.

Urinary Bladder

The **urinary bladder** is a smooth, collapsible, muscular sac located retroperitoneally on the pelvic floor just posterior to the pubic symphysis. In males, the bladder lies immediately in front of the rectum, and the prostate gland (part of the male reproductive system) surrounds its neck where it empties into the urethra. In females, the bladder abuts the vagina and uterus anteriorly.

The interior of the bladder has openings for the ureters and the urethra (Figure 26.15). The smooth, triangular region of the bladder base outlined by these three openings is the **trigone** (trī′gōn). The trigone is important clinically because infections tend to persist in this bladder region.

The bladder wall has four layers: a mucosa of transitional epithelium, a connective tissue submucosa, a muscular layer, and a fibrous adventitia (except on its superior surface, where it is covered by the parietal peritoneum). The muscular layer, called the **detrusor** (deh-troo′-zer) **muscle** (from *detrudere*, meaning to thrust out), consists of highly intermingled smooth muscle fibers arranged in inner and outer longitudinal layers and a middle circular layer. The bladder is very distensible and uniquely suited for its function of urine storage. When there is little or no urine, the bladder collapses to 5.0–7.5 cm (2–3 inches) long and its walls are thick and thrown into folds (*rugae*). But when urine accumulates, the bladder expands and becomes pear-shaped as it rises superiorly in the pelvic cavity (Figure 26.16). The muscular wall stretches and thins, and the transitional epithelial cells slide over one another, increasing the internal volume of the bladder by decreasing the thickness of its wall. These changes allow the bladder to store larger amounts of urine without a significant rise in internal pressure (at least until 300 ml has accumulated). A moderately full bladder is about 12.5 cm (5.0 inches) long and holds approximately 500 ml (1 pint) of urine. However, the bladder can hold more than double that amount if necessary; when it is quite distended with urine, it becomes firm and can be palpated well above the pubic symphysis.

Although urine is formed continuously by the kidneys, it is usually stored in the bladder until its release is convenient. Mucus produced by the bladder mucosa helps protect the wall from the urine; this protection is particularly important when the urine is especially acidic or alkaline.

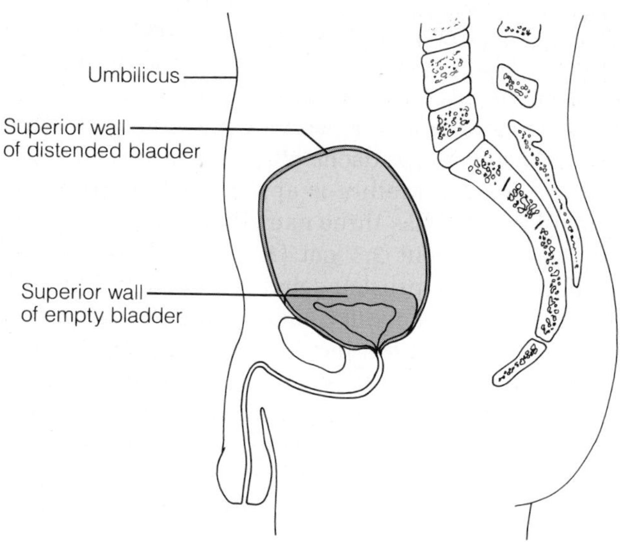

Figure 26.16 Relative positioning and shape of a distended and an empty urinary bladder.

Urethra

The **urethra** is a thin-walled muscular tube that drains urine from the bladder floor and conveys it out of the body. Its mucosal lining is mostly stratified columnar epithelium. However, near the bladder it becomes transitional epithelium, and near its external opening the lining membrane changes to a stratified squamous epithelium.

At the bladder–urethra junction a thickening of the smooth muscle forms the **internal urethral sphincter**, an involuntary sphincter that keeps the urethra closed when urine is not being passed. A second sphincter, the **external urethral sphincter**, is fashioned by skeletal muscle as the urethra passes through the pelvic floor. This sphincter is voluntarily controlled.

The length and functions of the urethra differ in the two sexes (see Figure 26.15). In females the urethra is 3–4 cm (1.5 inches) long and is tightly bound to the anterior vaginal wall by fibrous connective tissue. Its external opening, the *external urethral orifice (meatus)*, lies anterior to the vaginal opening and posterior to the clitoris (see Figure 28.14).

Because the female's urethra is very short and its external orifice is close to the anal opening, improper toilet habits (wiping back to front) can easily carry fecal bacteria into the urethra. Since the mucosa of the urethra is continuous with that of the rest of the urinary tract, an inflammation of the urethra (*urethritis*) can ascend the tract to cause bladder inflammation (*cystitis*) or even kidney inflammation (*pyelitis* or *pyelonephritis*). Symptoms of urinary tract infection include *dysuria* (painful urination), urinary *urgency* and *frequency*, fever, and sometimes cloudy or blood-tinged urine. When the kidneys are involved, back pain and a severe headache often occur as well. ■

In males the urethra is approximately 20 cm (8 inches) long and has three named regions. The *prostatic urethra*, about 2.5 cm (1.0 inch) long, is surrounded by the prostate gland. The *membranous urethra* extends for about 2 cm from the prostate gland to the beginning of the penis. The *spongy*, or *penile*, *urethra*, about 15 cm long, passes through the penis and opens at its tip via the external urethral orifice. The male urethra has a double function: It carries semen as well as urine out of the body. The reproductive function of the male urethra is discussed in Chapter 28.

Micturition

Micturition (mik″-tū-rish′-un), also called *voiding* or *urination*, is the act of emptying the bladder. Ordinarily, when about 200 ml of urine has accumulated, distension of the bladder wall activates stretch receptors, triggering a reflex arc. Afferent impulses are transmitted to the sacral region of the spinal cord, and efferent impulses return to the bladder via the parasympathetic *pelvic nerves*, causing the detrusor muscle to contract and the internal sphincter to relax (Figure 26.17). As the contractions increase in intensity, stored urine is forced through the internal sphincter into the upper part of the urethra. Afferent impulses are also transmitted to the brain, so one feels the urge to void at this point. Because both the lower external sphincter and the muscles of the pelvic diaphragm that surround it are skeletal muscles and are voluntarily controlled, this sphincter can be kept closed and bladder emptying postponed temporarily. On the other hand, if the time is convenient, the muscles of the pelvic diaphragm can be relaxed voluntarily by inhibiting impulses delivered to it by the *pudendal* (pyoo-den′-dul) *nerves* and urine is expelled from the bladder. (The external sphincter itself is believed to be relatively unimportant in voluntary control, but it contracts and relaxes in concert with the pelvic diaphragm.)

When one chooses not to micturate, reflex contractions of the bladder subside within a minute or so and urine continues to accumulate. After 200–300 ml more has collected, the micturition reflex occurs again and, if urination is delayed again, is damped once more. However, the urge to void eventually becomes irresistible and micturition occurs, whether one wills it or not. Although the sympathetic nervous system also innervates the bladder, its influence is uncertain.

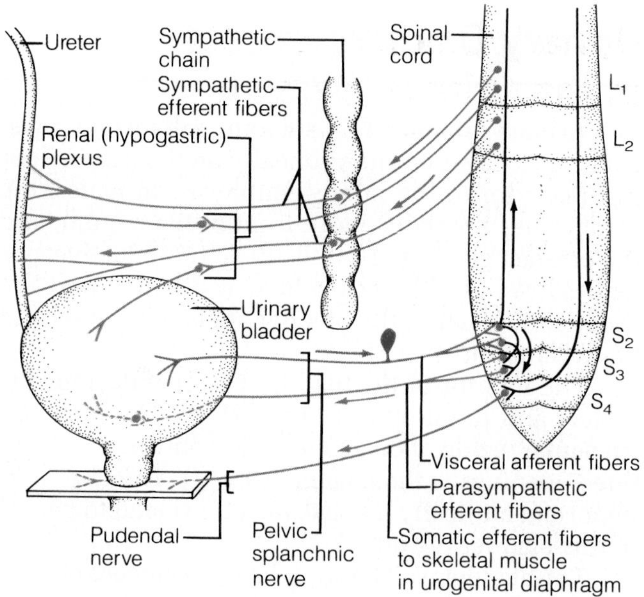

Figure 26.17 Micturition reflex arc pathway. Stretching of the bladder wall by urine causes afferent impulses to be transmitted to the sacral region of the spinal cord. Efferent impulses are delivered to the bladder detrusor muscle and the internal sphincter via parasympathetic fibers of the pelvic nerves. The pudendal nerve serves the skeletal muscle fibers of the voluntary external sphincter (largely in the pelvic diaphragm). The role of sympathetic efferents is controversial.

One suggestion is that it decreases the tone of the detrusor muscle of the trigone region.

Incontinence, the inability to control micturition voluntarily, is normal in very young children. Reflex voiding occurs each time a baby's bladder fills enough to activate the stretch receptors, but normal sphincter activity prevents dribbling of urine between voidings just as it does in adults. After the toddler years, incontinence is usually a result of emotional problems, physical pressure during pregnancy, or nervous system problems (stroke or spinal cord lesions). In *stress incontinence*, a sudden increase in intra-abdominal pressure (as during laughing and coughing) forces urine through the external sphincter. This condition is a common consequence of stretching the muscles of the pelvic floor and the urogenital diaphragm that support the external sphincter. Such stretching often occurs toward the end of pregnancy.

In *urinary retention*, the bladder is unable to expel its contained urine. Urinary retention is common after general anesthesia has been given (it seems that it takes a little time for the smooth muscles to regain their activity). Urinary retention in men often reflects hypertrophy of the prostate gland, which narrows the urethra, making it very difficult to void. When urinary retention is prolonged, a slender rubber drainage tube called a *catheter* (kath′-ih-ter) must be inserted through the urethra to drain the urine and prevent bladder trauma from excessive stretching. ■

Developmental Aspects of the Urinary System

As the kidneys develop in a young embryo, it almost seems as if they are unable to "make up their mind" how to go about it. As illustrated in Figure 26.18, three different pairs of kidneys develop from the *urogenital ridges*, paired elevations of the dorsal intermediate mesoderm that give rise to both the urinary and the reproductive organs. Only the last set persists to become the adult kidneys. During the fourth week of development, the first tubule system, the *pronephros* (prō-nef-rus), forms and then begins to degenerate as a second, lower set appears. Although the pronephros is never functional and is entirely gone by the sixth week, the *pronephric duct* that connects it to the cloaca persists and is used by the later-developing kidneys. (The cloaca is the terminal part of the gut that opens to the body exterior.) As the second renal system, the *mesonephros* (meh′-zō-nef″-rus), claims the pronephric duct, it comes to be called the *mesonephric duct*. The mesonephric kidneys, in turn, degenerate once the third set, the *metanephros* (meh′-tuh-nef″-rus), makes its appearance.

The metanephros starts to develop at about five weeks as hollow *ureteric buds* that push superiorly from the mesonephric duct. The expanded distal ends of the ureteric buds form the renal pelves and associated calyces and collecting tubules; their unexpanded proximal parts, now called the *metanephric ducts*, become the ureters. Nephrons develop from a mass of intermediate mesoderm that clusters around the top of each ureteric bud. Because the embryonic

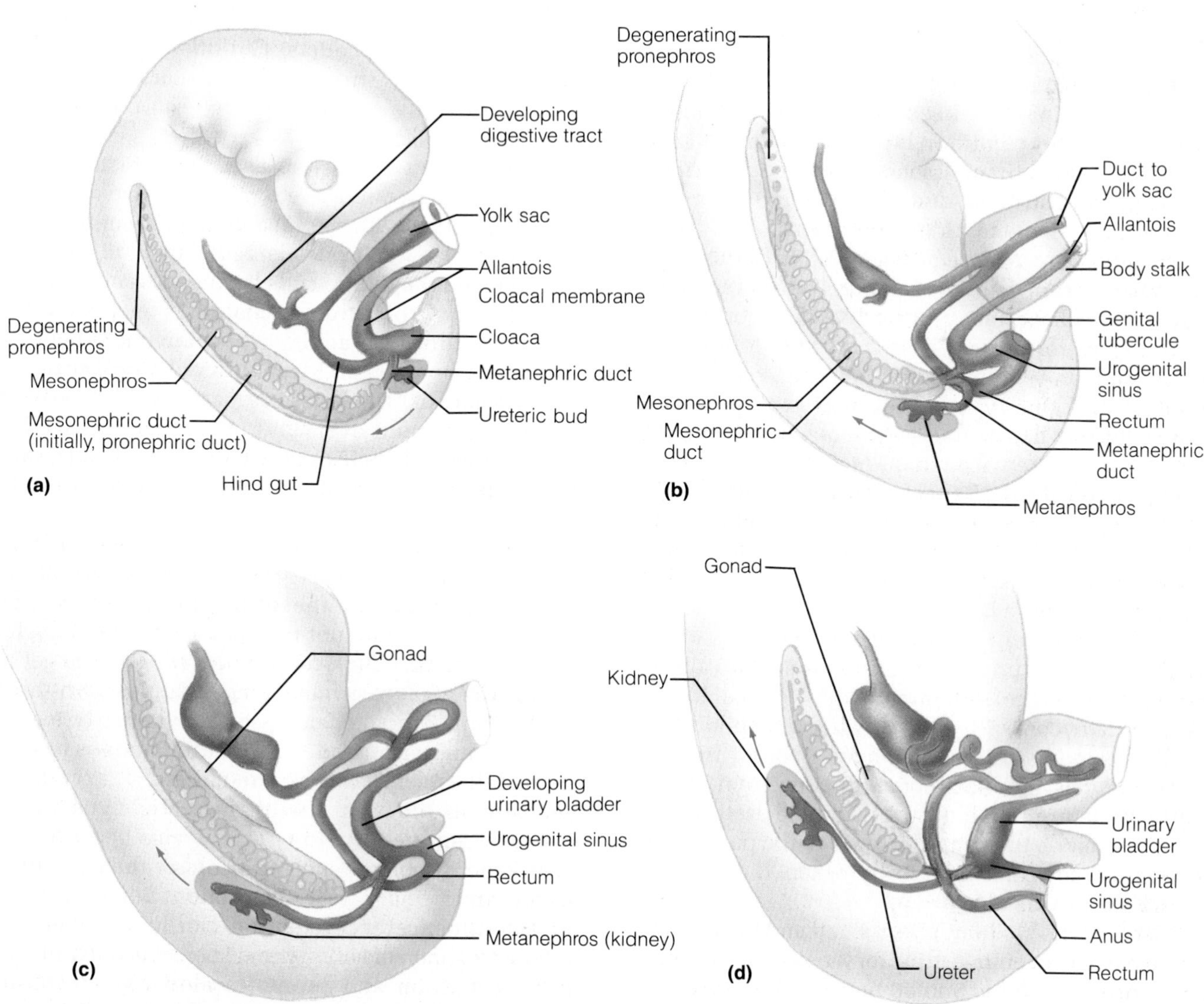

Figure 26.18 Development of the urinary system in the embryo. (**a**) Fifth week. (**b**) Sixth week. (**c**) Seventh week. (**d**) Eighth week.

kidneys develop in the pelvis and then ascend to their final position in the abdomen, they receive their blood supply from successively higher sources. Usually, the lower blood vessels degenerate as the upper ones appear. However, failure of some of these vessels to deteriorate is common and results in multiple renal arteries. The metanephric kidneys are excreting urine by the third month of fetal life, but the fetal kidneys do not work nearly as hard as they will after birth because exchange through the placenta allows the mother's urinary system to clear most of the undesirable substances from the fetal blood.

As the metanephros is developing, the cloacal chamber subdivides to form the *urogenital sinus*, into which the urinary and genital ducts empty, and the *rectum*, which empties into the anal canal. The urinary bladder develops from the urogenital sinus, and the urethra forms from a lengthening of the short duct that extends from the bladder to the urogenital sinus.

There are many congenital abnormalities of the urinary system, but four of the most common are horseshoe kidney, polycystic kidney, exstrophy of the bladder, and hypospadias.

When forming in the pelvis the kidneys are close together, and in 1 out of 600 people, they become fused across the midline, forming a single, U-shaped *horseshoe kidney*. This condition is usually asymptomatic, but there are sometimes associated kidney abnormalities, such as obstructed drainage, which place a person at risk for frequent kidney infections.

In *polycystic disease*, a degenerative disease that appears to run in families, one or both kidneys are enlarged and have many blisterlike sacs containing blood, mucus, or urine. These cysts tend to increase in size and, in extreme cases, the kidneys become "knobby" and grossly enlarged, increasing in weight from about one-half pound to as much as 30 pounds. The cysts interfere with renal function, but the damage progresses slowly. Renal failure as a result of polycystic disease is rare before the mid-40s, and many of its victims live without problems until their 60s. Little can be done for this condition except to prevent further kidney damage by avoiding infection.

In *exstrophy of the bladder*, the bladder and the distal ends of the ureters protrude through an abnormal opening in the abdominal wall. This complication is usually corrected surgically in early childhood.

Hypospadias (hī″-pō-spā′-dē-us), found in male infants only, is the most common congenital abnormality of the urethra. It occurs when the urethral orifice is located on the ventral surface of the penis. This problem is corrected surgically when the child is around 12 months old.

Because the bladder is very small and the kidneys are unable to concentrate urine for the first two months, a newborn baby voids 5 to 40 times daily, depending on the volume of fluids taken in. By 2 months of age, the infant is voiding approximately 400 ml/day, and the amount steadily increases until about the age of 8, when urine output reaches 1000 ml/day. By adolescence, adult urine output (about 1500 ml/day) is achieved.

Control of the voluntary urethral sphincter goes hand in hand with nervous system development. By 15 months, most toddlers are aware when they have voided. By 18 months, they can usually hold urine in their bladder for about two hours, which is the first sign that toilet training for micturition can begin. Daytime control usually occurs well before nighttime control is achieved. As a rule, it is unrealistic to expect that complete nighttime control will occur before the child is 4 years old.

From childhood through late middle age, most urinary system problems are infectious conditions. *Escherichia coli* (eh″-sher-ē′-shē-uh kō′-li′) are normal residents of the digestive tract and generally cause no problems there, but these bacteria account for 80% of all urinary tract infections. *Sexually transmitted diseases*, which are chiefly reproductive tract infections (discussed in Chapter 28), may also cause inflammation of the urinary tract, leading to the clogging of some of its ducts. Childhood streptococcal infections such as strep throat and scarlet fever, if not treated promptly, may cause long-term inflammatory damage to the kidneys.

Only about 3% of the aged have histologically normal kidneys, and there is a progressive decline in kidney function with age. The kidneys become smaller (atrophy), the nephrons decrease in both size and number, and the tubule cells become less efficient. By age 70, the rate of filtrate formation is only about half that of middle-aged adults. This slowing is believed to result from impaired renal circulation caused by arteriosclerosis. Diabetics are particularly at risk for renal disease, and over 50% of those who have had demonstrable diabetes mellitus for 20 years (regardless of age) are in renal failure owing to the vascular ravages of this disease.

The bladder of an aged person is shrunken, with less than half the capacity of a young adult (250 ml versus 600 ml), and the desire to urinate is often delayed. Loss of bladder tone causes an annoying frequency. *Nocturia* (nok-too′-rē-uh), the need to get up during the night to urinate, plagues almost two-thirds of this population. Many people eventually experience incontinence, and with it a tremendous blow to their self-esteem. Urinary retention is a common problem in older men with prostatic hypertrophy. However, an inability to void may also be associated with other problems such as a cold or the flu, certain drugs, and urinary tract obstruction. Some problems of incontinence and retention can be avoided by a regular regimen of activity that keeps the body in optimum condition and promotes alertness to voiding signals.

* * *

The ureters, urinary bladder, and urethra have important roles to play in transporting, storing, and eliminating urine from the body; but when the term "urinary system" is used, it is the kidneys, which produce the urine, that capture center stage. As summarized in Figure 26.19 on p. 884, the other organ systems of the body contribute to the well-being of the urinary system in many ways; but without continuous kidney function, internal body fluids quickly become contaminated with nitrogenous wastes, and the electrolyte and fluid balance of the blood is dangerously disturbed. No body cell can escape the harmful effects of such imbalances.

Now that renal mechanisms have been described, we are ready to integrate kidney function into the larger topic of fluid and electrolyte balance of the body as a whole—the focus of Chapter 27.

Related Clinical Terms

Acute glomerulonephritis (glō-mer″-yoo-lō″-neh-frī′-tis) **(GN)** Inflammation of the glomeruli, often arising from some type of immune response. In some cases, circulating immune complexes (antibodies bound to foreign substances, such as streptococcal bacteria), become trapped in the glomerular membranes. In other cases, immune responses are mounted against one's own kidney tissue, leading to glomerular damage. In either case, the inflammatory response that follows damages the filtration membrane, increasing its permeability. Blood proteins and even blood cells begin to pass into the renal tubules and are lost in urine. As the osmotic pressure of the blood drops, fluid seeps from the bloodstream into the tissue spaces, causing bodywide edema. Renal shutdown (requiring dialysis; see the box on page 874) may occur temporarily, but normal renal function usually returns within a few months. However, if permanent glomerular damage occurs, chronic GN and, ultimately, renal failure result.

Cystocele (sis′-tō-sēl) (*cysto* = a sac, the bladder; *cele* = cyst) Herniation of the urinary bladder into the vagina.

Diabetes insipidus (in-sih′-pih-dus) (*insipid* = tasteless, bland) Condition in which large amounts (up to 40 L/day) of dilute urine flush from the body. Results from little or no ADH release due to injury to, or a tumor in, the hypothalamus or posterior pituitary. Can lead to severe dehydration and electrolyte imbalances unless the individual drinks large volumes of liquids to maintain normal fluid balance.

Intravenous pyelogram (IVP) (pī′-uh-lō-gram″) (*pyelo* = kidney pelvis; *gram* = written) An X-ray of the kidney and ureter obtained after intravenous injection of a contrast medium. Allows assessment of renal blood vessels for obstructions, viewing of renal anatomy (pelvis and calyces), and determination of rate of excretion of the contrast medium.

Nephrotoxin A substance (heavy metal, organic solvent, or bacterial toxin) that is toxic to the kidney.

Renal infarct Area of dead, or necrotic, renal tissue. May result from inflammation, hydronephrosis, or blockage of the vascular supply to the kidney. A common cause of localized renal infarct is obstruction of an interlobar artery. Because interlobar arteries are end arteries, their obstruction leads to ischemic necrosis of the portions of the kidney they supply.

Chapter Summary

KIDNEY ANATOMY (pp. 856–863)

Location and External Anatomy (pp. 857–858)

1. The paired kidneys are retroperitoneal in the superior lumbar region.

2. A renal capsule, an adipose capsule, and renal fascia surround each kidney. The fatty adipose capsule helps hold the kidneys in position.

Internal Anatomy (p. 858)

3. A kidney has an outer cortex, a deeper medulla consisting mainly of medullary pyramids, and a medial pelvis. Extensions of the pelvis (calyces) surround and collect urine draining from the apices of the medullary pyramids.

Blood and Nerve Supply (p. 858)

4. The kidneys receive 25% of the total cardiac output/minute.

5. The vascular pathway through a kidney is as follows: renal artery → lobar arteries → interlobar arteries → arcuate arteries → interlobular arteries → afferent arterioles → glomeruli → efferent arterioles → peritubular capillary beds → interlobular veins → arcuate veins → interlobar veins → lobar veins → renal vein.

6. The nerve supply of the kidneys is derived from the renal plexus.

Nephrons (pp. 859–862)

7. Nephrons are the structural and functional units of the kidneys.

8. Each nephron consists of a glomerulus (a high-pressure capillary bed) and a renal tubule. Subdivisions of the renal tubule (from the glomerulus) are the glomerular capsule, proximal tubule, loop of Henle, and distal tubule. A second capillary bed, the low-pressure peritubular capillary bed, is closely associated with the renal tubule of each nephron.

9. The more numerous cortical nephrons are located entirely in the cortex. Juxtamedullary nephrons are located at the cortex–medulla junction, and their loops of Henle dip deeply into the medulla. Juxtamedullary nephrons play an important role in establishing the medullary osmotic gradient.

10. Collecting tubules receive urine from many nephrons and form the medullary pyramids.

11. The juxtaglomerular apparatus is at the point of contact between the afferent arteriole and the distal convoluted tubule. It consists of the juxtaglomerular cells and the macula densa cells.

Urinary system

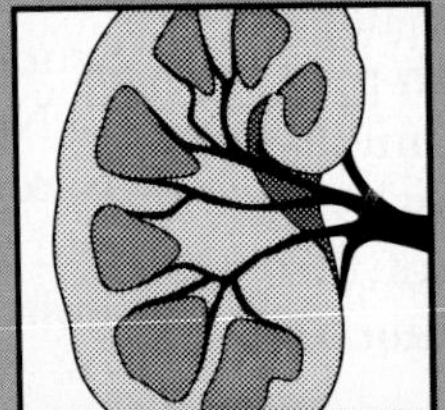

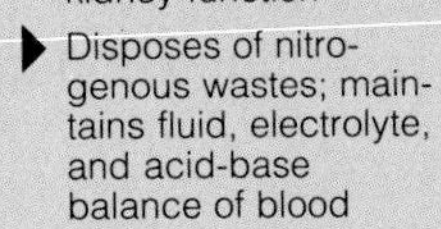

Integumentary system

Provides external protective barrier; serves as site for water loss (via perspiration); vitamin D synthesis site ▶

◀ Disposes of nitrogenous wastes; maintains fluid, electrolyte, and acid-base balance of blood

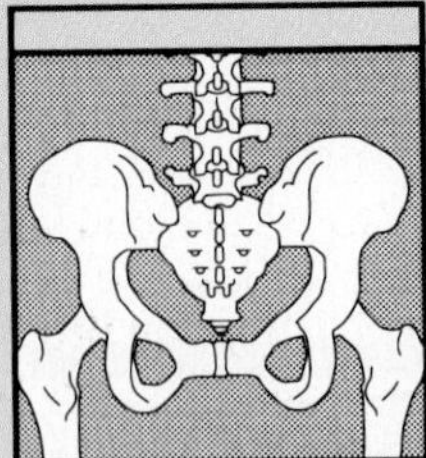

Skeletal system

Bones of rib cage provide some protection to kidneys; ▶

◀ Disposes of nitrogenous wastes; maintains fluid, electrolyte, and acid-base balance of blood

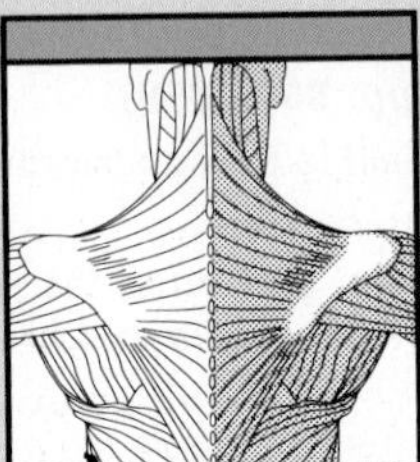

Muscular system

Muscles of pelvic diaphragm and external urethral sphincter function in voluntary control of micturition ▶

◀ Disposes of nitrogenous wastes; maintains fluid, electrolyte, and acid-base balance of blood

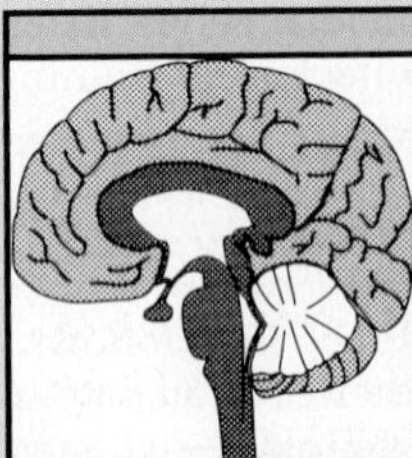

Nervous system

Neural controls involved in micturition; sympathetic nervous system activity reduces GFR during extreme stress ▶

◀ Disposes of nitrogenous wastes; maintains fluid, electrolyte, and acid-base balance of blood

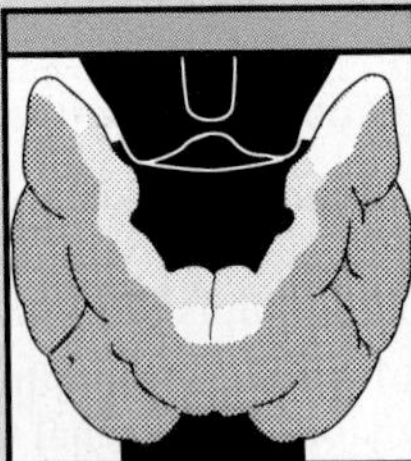

Endocrine system

ADH, aldosterone, ANF, and other hormones help regulate renal reabsorption of water and electrolytes ▶

◀ Same as integumentary; also, kidneys produce the hormone erythropoietin

Cardiovascular system

Systemic arterial blood pressure is the driving force for glomerular filtration; heart secretes atrial natriuretic factor; transports nutrients, oxygen, etc. to urinary organs ▶

◀ Same as integumentary

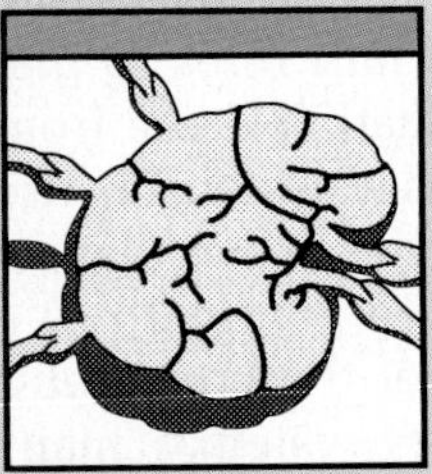

Lymphatic system

◀ By returning leaked plasma fluid to cardiovascular system, helps assure normal systemic arterial pressure required for kidney function

▶ Disposes of nitrogenous wastes; maintains fluid, electrolyte, and acid-base balance of blood

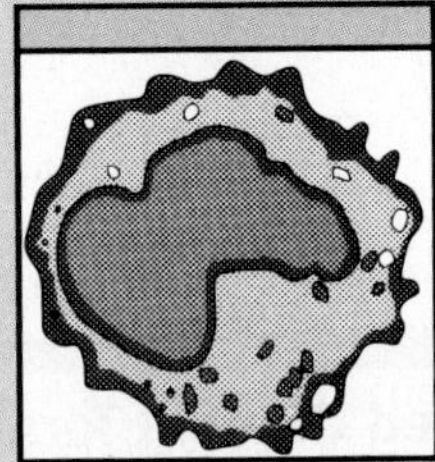

Immune system

◀ Protects urinary organs from infection, cancer, and other foreign substances

▶ Disposes of nitrogenous wastes; maintains fluid, electrolyte, and acid-base balance of blood

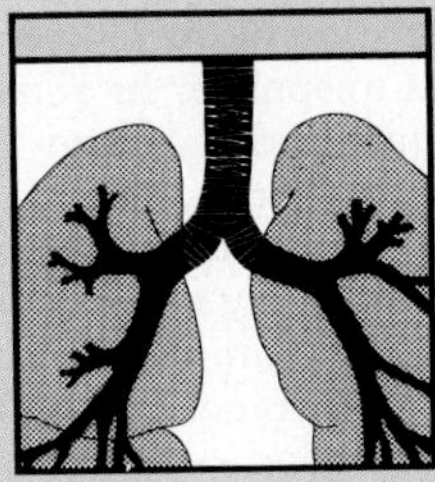

Respiratory system

◀ Provides oxygen required by kidney cells for their high metabolic activity; disposes of carbon dioxide

▶ Disposes of nitrogenous wastes; maintains fluid, electrolyte, and acid-base balance of blood

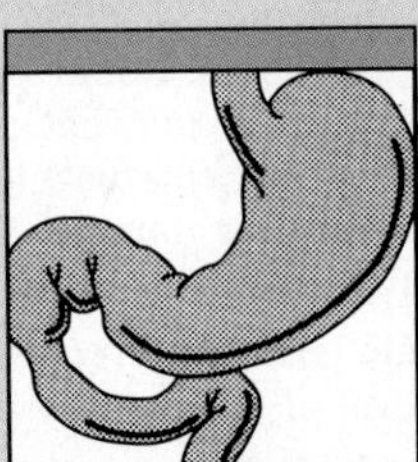

Digestive system

◀ Provides nutrients needed for kidney cell health

▶ Same as integumentary; also, metabolizes vitamin D to the active form needed for calcium absorption

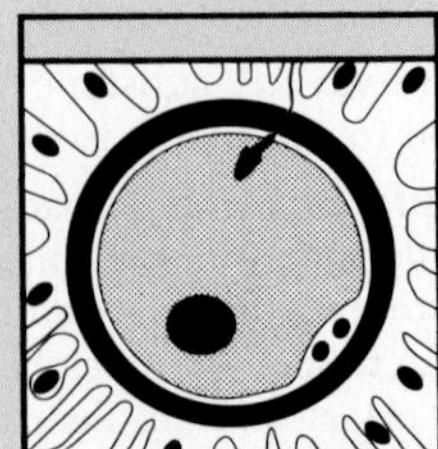

Reproductive system

▶ Disposes of nitrogenous wastes; maintains fluid, electrolyte, and acid-base balance of blood

Figure 26.19 Homeostatic interrelationships between the urinary system and other body systems.

KIDNEY PHYSIOLOGY: MECHANISMS OF URINE FORMATION (pp. 863–877)

1. Functions of the nephrons include filtration, tubular reabsorption, and tubular secretion. Via these functional processes, the kidneys eliminate nitrogenous metabolic wastes, and regulate the volume, composition, and pH of the blood.

Glomerular Filtration (pp. 863–868)

2. The glomeruli function as filters. High glomerular blood pressure (55 mm Hg) occurs because the glomeruli are fed and drained by arterioles, and the afferent arterioles are larger in diameter than the efferent arterioles.

3. About one-fifth of the plasma flowing through the kidneys is filtered from the glomeruli into the renal tubules.

4. The filtration membrane consists of the fenestrated glomerular endothelium, the basement membrane, and the podocyte-containing visceral membrane of the glomerular capsule. It permits free passage of substances smaller than (most) plasma proteins.

5. Net filtration pressure (usually about 10 mm Hg) is determined by the relationship between forces favoring filtration (glomerular hydrostatic pressure) and forces that oppose it (capsular hydrostatic pressure and blood colloid osmotic pressure).

6. The glomerular filtration rate is directly proportional to the net filtration pressure and is about 125 ml/min (180 L/day).

7. Renal autoregulation, which enables the kidneys to maintain a relatively constant renal blood pressure and glomerular filtration rate, is believed to involve a myogenic mechanism, the renin–angiotensin mechanism mediated by the juxtaglomerular (JG) cells, and a tubuloglomerular feedback mechanism mediated by the macula densa.

8. Sympathetic nervous system activation causes constriction of the afferent arterioles, which decreases filtrate formation and stimulates renin release by the JG cells.

Tubular Reabsorption (pp. 868–870)

9. During tubular reabsorption, needed substances are removed from the filtrate and returned to the peritubular capillary blood.

10. Active tubular reabsorption requires protein carriers and ATP. Transport of many substances is coupled to the active transport of sodium ions and is limited by the number of carriers available. Substances reabsorbed actively include nutrients and most ions.

11. Passive tubular reabsorption is driven by electrochemical gradients established by active reabsorption of sodium ions. Water, many anions, and various other substances (for example, urea) are reabsorbed passively.

12. Certain substances (creatinine, drug metabolites, etc.) are not reabsorbed or are reabsorbed incompletely because of the lack of carriers, their size, or nonlipid solubility.

13. The proximal tubule cells are most active in reabsorption. Most of the nutrients, 80% of the water and sodium ions, and the bulk of actively transported ions are reabsorbed in the proximal convoluted tubules.

14. Reabsorption of additional sodium ions and water occurs in the distal and collecting tubules and is hormonally controlled: Aldosterone increases the reabsorption of sodium (and obligatory water reabsorption); antidiuretic hormone enhances water reabsorption by the collecting tubules.

Tubular Secretion (pp. 870–872)

15. Tubular secretion, which may be active or passive, is a means of adding substances to the filtrate (from the blood or tubule cells). It is important in eliminating drugs, urea, and excess ions and in maintaining the acid–base balance of the blood.

Regulation of Urine Concentration and Volume (pp. 872–875)

16. The graduated hyperosmolality of the medullary fluids (largely due to urea) ensures that the filtrate reaching the distal convoluted tubule is dilute (hypo-osmolar). This allows urine with osmolalities ranging from 50 to 1400 mOsm to be formed.

  a. The descending limb of the loop of Henle is permeable to water, which leaves the filtrate and enters the vasa recta blood. The filtrate and medullary fluid at the tip of the loop of Henle is hyperosmolar.
  b. The ascending limb is impermeable to water and poorly permeable to urea. Na^+ and Cl^- diffuse freely from the thin segment. Events occurring in the thick segment are less clear: They may involve (1) active transport of Cl^- out of the filtrate, followed by passive diffusion of Na^+, or (2) active transport of both Cl^- and Na^+ out of the tubule. Na^+ and Cl^- enter the vasa recta blood, and the filtrate becomes more dilute as it continues to lose salt.
  c. As filtrate flows through the collecting tubules in the inner medulla, urea diffuses into the interstitial space. Some urea enters the ascending limb and is recycled.
  d. The blood flow in the vasa recta is sluggish. Most urea that enters the vasa recta diffuses out again; thus, the high solute (urea) concentration of the medulla is maintained.

17. In the absence of antidiuretic hormone, dilute urine is formed because the dilute filtrate reaching the distal tubule is simply allowed to pass from the kidneys.

18. As blood levels of antidiuretic hormone rise, the collecting tubules become more permeable to water, and water moves out of the filtrate as it flows through the hyperosmotic medullary areas. Consequently, smaller amounts of more concentrated urine are produced.

Renal Clearance (pp. 875–876)

19. Renal clearance is the rate at which the kidneys clear the plasma of a particular solute. Studies of renal clearance provide information about renal function or the course of renal disease.

Characteristics and Composition of Urine (pp. 876–877)

20. Urine is typically clear, yellow, aromatic, and slightly acidic. Its specific gravity ranges from 1.001 to 1.035.

21. Urine is 95% water; solutes include nitrogenous wastes and various ions (always sodium, potassium, sulfate, and phosphate).

22. Substances not normally found in urine include glucose, plasma proteins, erythrocytes, pus, hemoglobin, and bile.

23. Daily urinary volume is typically 1.5–1.8 L, but this depends on the state of hydration of the body.

URETERS (pp. 877–878)

1. The ureters are slender tubes running retroperitoneally from each kidney to the bladder. They conduct urine by peristalsis from the renal pelvis to the urinary bladder.

URINARY BLADDER (pp. 878–879)

1. The urinary bladder, which functions to store urine, is a distensible muscular sac that lies posteriorly to the pubic symphysis. It has two inlets (ureters) and one outlet (urethra) that outline the trigone. In males, the prostate gland surrounds its outlet.

2. The bladder wall consists of a transitional epithelium mucosa, a submucosa, a three-layered detrusor muscle, and an adventitia.

URETHRA (pp. 879–880)

1. The urethra is a muscular tube that conveys urine from the bladder to the body exterior.

2. Where the urethra leaves the bladder, it is surrounded by an internal urethral sphincter, an involuntary smooth muscle sphincter. Where it passes through the urogenital diaphragm, the voluntary external urethral sphincter is formed by skeletal muscle.

3. In females the urethra is 3–4 cm long and conducts only urine. In males it is 20 cm long and conducts both urine and semen.

MICTURITION (p. 880)

1. Micturition is emptying of the bladder.

2. Stretching of the bladder wall by accumulating urine initiates the micturition reflex, which causes the detrusor muscle to contract and the internal urethral sphincter to relax (open).

3. Because the external sphincter is voluntarily controlled, micturition can usually be delayed temporarily.

DEVELOPMENTAL ASPECTS OF THE URINARY SYSTEM (pp. 881–883)

1. Three sets of kidneys (pronephric, mesonephric, and metanephric) develop in sequence from the intermediate mesoderm. The metanephros is excreting urine by the third month of development.

2. The more common congenital abnormalities of this system include horseshoe kidney, polycystic kidney, exstrophy of the bladder, and hypospadias.

3. The kidneys of newborns cannot concentrate urine; their bladder is small and voiding is frequent. Neuromuscular maturation generally allows toilet training for micturition to begin by 18 months of age.

4. The most common urinary system problems in children and young to middle-aged adults are bacterial infections.

5. Renal failure is uncommon but extremely serious. The kidneys are unable to concentrate urine, nitrogenous wastes accumulate in the blood, and acid–base and electrolyte imbalances occur.

6. With age, nephrons are lost, the filtration rate decreases, and tubule cells become less efficient at concentrating urine.

7. Bladder capacity and tone decrease with age, leading to frequency and (often) incontinence. Urinary retention is a common problem of elderly men.

REVIEW QUESTIONS

Multiple Choice/Matching

1. The lowest blood concentration of nitrogenous waste occurs in the (a) hepatic vein, (b) inferior vena cava, (c) renal artery, (d) renal vein.

2. The capillaries of the glomerulus differ from other capillary networks in the body because they (a) have a larger area of anastomosis, (b) are derived from and drain into arterioles, (c) are not made of endothelium, (d) cause the filtrate to be forced from the blood.

3. Damage to the renal medulla would interfere *first* with the functioning of the (a) glomerular capsules, (b) distal convoluted tubules, (c) collecting tubules, (d) proximal convoluted tubules.

4. Which is reabsorbed by the proximal convoluted tubule? (a) Na^+, (b) K^+, (c) amino acids, (d) all of the above.

5. Glucose is not normally found in the urine because it (a) does not pass through the walls of the glomerulus, (b) is kept in the blood by colloid osmotic pressure, (c) is reabsorbed by the tubule cells, (d) is removed by the body cells before the blood reaches the kidney.

6. Filtration at the glomerulus is directly related to (a) water reabsorption, (b) arterial blood pressure, (c) capsular hydrostatic pressure, (d) acidity of the urine.

7. Renal reabsorption (a) of glucose and many other substances is a T_m-limited active transport process, (b) of chloride is always linked to the passive transport of Na^+, (c) is the movement of substances from the blood into the nephron, (d) of sodium occurs only in the proximal tubule.

8. If a freshly voided urine sample contains excessive amounts of urochrome, it has (a) an ammonia-like odor, (b) pH below normal, (c) a dark yellow color, (d) a pH above normal.

9. Conditions such as diabetes mellitus, starvation, and low-carbohydrate diets are closely linked to (a) ketosis, (b) pyuria, (c) albuminuria, (d) hematuria.

Short Answer Essay Questions

10. What is the importance of the adipose capsule that surrounds the kidney?

11. Trace the pathway a creatinine molecule takes from a glomerulus to the urethra. Name every microscopic or gross structure it passes through on its journey.

12. Explain the important differences between blood plasma and renal filtrate, and relate the differences to the structure of the filtration membrane.

13. Describe the mechanisms that contribute to renal autoregulation.

14. Describe what is involved in active and passive tubular reabsorption.

15. Explain how the peritubular capillaries are adapted for receiving reabsorbed substances.

16. Explain the process and purpose of tubular secretion.

17. How does aldosterone modify the chemical composition of urine?

18. Explain why the filtrate becomes hypotonic as it flows through the ascending limb of the loop of Henle. Also explain why the filtrate at the tip of the loop of Henle (and the interstitial fluid of the deep portions of the medulla) is hypertonic.

19. How does the bladder anatomy support its storage function?

20. Define micturition and describe the micturition reflex.

21. Describe the changes that occur in kidney and bladder anatomy and physiology in old age.

Clinical Application Questions

22. While repairing a frayed utility wire, an experienced lineman slips and falls to the ground. Medical examination reveals a fracture of his lower spine and transection of the lumbar region of the spinal cord. How will micturition in this man be controlled from this point on? Will he ever again feel the need to void? Will there be dribbling of urine between voidings? Explain the rationale for all your responses.

23. What is cystitis? Why are women more frequent cystitis sufferers than men?

24. A 55-year-old woman is awakened by an excruciating pain that radiates from her right abdomen to the loin and groin regions on the same side. The pain is not continuous, but recurs at intervals of 3–4 minutes. Diagnose this patient's problem and cite factors that might favor its occurrence. Also, attempt to explain why this woman's pain comes in "waves."

25. Patients who are suffering severe hemorrhage undergo a number of functional changes involving their kidneys. During such periods the GFR decreases and blood levels of renin, aldosterone, and ADH rise. Explain the cause and consequences of each of these changes.

CHAPTER 27

Fluid, Electrolyte, and Acid–Base Balance

Chapter Outline and Student Objectives

Body Fluids (pp. 888–891)

1. List the factors that determine body water content and describe the effect of each factor.
2. Note the relative fluid volume and solute composition of the three fluid compartments of the body.
3. Contrast the overall osmotic effects of electrolytes and nonelectrolytes.
4. Describe factors that determine fluid shifts in the body.

Water Balance (pp. 891–894)

5. List the routes by which water enters and leaves the body.
6. Describe the feedback mechanisms that regulate water ingestion and hormonal controls of water output in urine.
7. Explain the importance of obligatory water losses.
8. Describe possible causes and consequences of dehydration, hypotonic hydration, and edema.

Electrolyte Balance (pp. 894–901)

9. Describe the routes of electrolyte entry and loss from the body.
10. Describe the importance of ionic sodium in fluid and electrolyte balance of the body, and note its relationship to normal cardiovascular system functioning.
11. Briefly describe mechanisms involved in regulating sodium and water balance.
12. Explain how potassium, calcium, magnesium, and anion balance of plasma is regulated.

Acid–Base Balance (pp. 901–908)

13. List important sources of acids in the body.
14. Name the three major chemical buffer systems of the body and describe how they operate to resist pH changes.
15. Describe the influence of the respiratory system on acid–base balance.
16. Describe how the kidneys regulate hydrogen and bicarbonate ion concentrations in the blood.
17. Distinguish between acidosis and alkalosis resulting from respiratory and metabolic factors. Describe the importance of respiratory and renal compensations to acid–base balance.

Developmental Aspects of Fluid, Electrolyte, and Acid–Base Balance (pp. 908–909)

18. Explain why infants and the aged are at greater risk for fluid and electrolyte imbalances than are young adults.

Preview of Selected Key Terms

Extracellular fluid (*extra* = outside) Internal fluid located outside cells; includes plasma and interstitial fluid.

Intracellular fluid (*intra* = within) Internal fluid located within cells.

Electrolyte (*electro* = electricity; *lyt* = that which may be loosed) Chemical substances, such as salts, acids, and bases, that ionize and dissociate in water and are capable of conducting an electrical current.

Water balance Situation in which water intake and output are equal.

Dehydration (*de* = out; *hydra* = water) Condition of excessive water loss.

Electrolyte balance Situation in which electrolyte intake and output are equal.

Natrium Latin for sodium.

Acid Proton donor; substance capable of releasing hydrogen ions in solution.

Base Proton acceptor; substance capable of binding with hydrogen ions.

pH A measure of the hydrogen ion concentration of a solution.

Acid–base balance Situation in which the pH of the blood remains between 7.35 and 7.45.

Buffer Chemical substance or system that minimizes changes in pH by releasing or binding hydrogen ions.

Acidosis (a″-sih-dō′-sis) (*osis* = state or condition) State of abnormally high hydrogen ion concentration in the extracellular fluid.

Alkalosis (al″-kuh-lō′-sis) (*alka* = alkaline) State of abnormally low hydrogen ion concentration in the extracellular fluid.

Have you ever wondered why on certain days you don't urinate for hours at a time, while on others you void large amounts of urine, seemingly every few minutes? Or why on occasion you cannot seem to get enough to drink? These situations and many others reflect one of the body's most important functions: that of maintaining fluid, electrolyte, and acid–base balance.

Cell function depends not only on a continuous supply of nutrients and removal of metabolic wastes, but also on the physical and chemical homeostasis of the fluids that surround cells. This was recognized with great style in 1857 by the French physiologist Claude Bernard, who said, "It is the fixity of the internal environment which is the condition of free and independent life." In this chapter, we will examine the composition and distribution of fluids in the internal environment and then consider the roles of various body organs and functions in establishing, regulating, or altering this balance.

Body Fluids

Body Water Content

If you are a healthy young adult, water probably accounts for about half your body weight (mass). However, not all bodies contain the same amount of water, and total body water is a function not only of weight, age, and sex, but also of the relative amount of body fat. Because of their low body fat and low bone mass, infants are 73% or more water. But total water content declines throughout life, accounting for only about 45% of body weight in old age. A healthy young man is 57% to 63% water; a healthy young woman about 50%. This significant difference between the sexes reflects the relatively larger amount of body fat and smaller amount of skeletal muscle in females. Of all body tissues, adipose tissue is *least* hydrated; even bone contains more water than does fat. Thus, lean people have proportionately more body water than do obese people.

Fluid Compartments

Water occupies three main locations within the body, known as **fluid compartments** (Figure 27.1). A little less than two-thirds by volume is said to be in the **intracellular fluid (ICF) compartment**, which actually consists of trillions of tiny individual compartments: the cells. In an adult male of average size (70 kg, or 154 pounds), the ICF accounts for about 25 L of the 40 L of body water. The remaining one-third or so of body water is outside cells, in the **extracellular fluid (ECF) compartment.** The ECF constitutes the "internal environment" of the body referred to by Claude Bernard and the external environment of each body cell. The ECF compartment is, in turn, divisible into two important subcompartments: (1) *plasma*, the fluid portion of blood within the blood vessels, and (2) *interstitial fluid (IF)*, the fluid in the microscopic spaces between tissue cells. Additionally, there are numerous other examples of ECF—lymph, cerebrospinal fluid, humors of the eye, synovial fluid, serous fluid, and secretions of the gastrointestinal tract—that are distinct from both plasma and interstitial fluid. However, most are similar to interstitial fluid and they are usually considered to be part of it. In the 70-kg adult male, interstitial fluid accounts for approximately 12 L and plasma is about 3 L of the 15 L ECF volume (Figure 27.1).

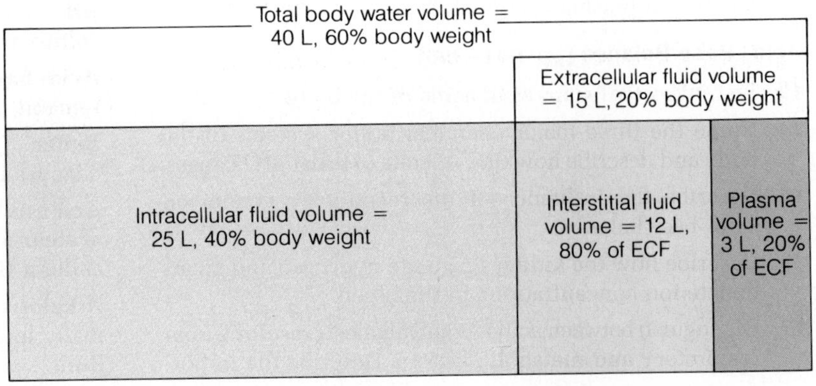

Figure 27.1 The major fluid compartments of the body. Approximate volume and percentage of body weight values are noted for a 70-kg (154-pound) male.

Composition of Body Fluids

Solutes: Electrolytes and Nonelectrolytes

Water serves as the universal solvent in which a variety of solutes are dissolved. The solutes may be classified broadly as *electrolytes* and *nonelectrolytes*. The nonelectrolytes have bonds (usually, covalent bonds) that do not permit them to dissociate in solution; therefore, they have no electrical charge. Most nonelectrolytes are organic molecules—glucose, lipids, creatinine, and urea, for example. In contrast, electrolytes are chemical compounds that do dissociate into ions (ionize) in water. (See Chapter 2, if necessary, to review these concepts of chemistry.) Because ions are charged particles, they can conduct an electrical current—hence the name *electrolyte*. Typically, electrolytes include inorganic salts, both inorganic and organic acids and bases, and many proteins.

Although all dissolved solutes contribute to the osmotic activity of a fluid, electrolytes have much greater osmotic power than nonelectrolytes do because each electrolyte molecule dissociates into at least two ions. For example, a molecule of sodium chloride (NaCl) contributes twice as many solute particles as the amino acid alanine (which remains undissociated), and a molecule of magnesium chloride ($MgCl_2$) contributes three times as many:

$$NaCl \rightarrow Na^{+} + Cl^{-} \quad \text{(two particles)}$$

$$MgCl_2 \rightarrow Mg^{2+} + 2Cl^{-} \quad \text{(three particles)}$$

$$\text{alanine} \rightarrow \text{alanine} \quad \text{(one particle)}$$

Because water moves according to osmotic gradients (from areas of lesser osmolality to those of greater osmolality), regardless of the type of solute particle, electrolytes have the greatest ability to cause fluid shifts.

Electrolyte concentrations of body fluids are usually expressed in *milliequivalents per liter (mEq/L)*, a measure of the number of electrical charges in 1 liter of solution. Because the total number of negative charges (anions) in a solution is always equal to the number of positive charges, the milliequivalent system of reporting electrolyte concentrations makes it easier to follow electrolyte shifts.

The concentration of any ion in solution in mEq/L can be computed using the equation

$$\text{mEq/L} = \frac{\text{concentration of ion (mg/L)}}{\text{atomic weight of ion}} \times \begin{matrix}\text{number of}\\ \text{electrical}\\ \text{charges on}\\ \text{one ion}\end{matrix}$$

Thus, to compute the mEq/L of sodium or calcium ions in solution in plasma, we would determine the normal concentration of these ions in plasma, look up their atomic weights in the periodic table, and plug these values into the equation:

$$Na^{+}: \frac{3300\ \text{mg/L}}{23} \times 1 = 143\ \text{mEq/L}$$

$$Ca^{2+}: \frac{100\ \text{mg/L}}{40} \times 2 = 5\ \text{mEq/L}$$

Notice that for ions with a single charge, 1 mEq is equal to 1 mOsm, whereas 1 mEq of bivalent ions (those, like calcium, with a double charge) is equal to ½ mOsm. In either case, 1 mEq provides the same degree of reactivity.

Comparison of Extracellular and Intracellular Fluids

Even a quick glance at the bar graphs in Figure 27.2 reveals that each fluid compartment has a distinctive pattern of electrolytes. But with the exception of the relatively high protein content in plasma, the extracellular fluids are very similar: Their chief cation is sodium, and their major anion is chloride. However, because the nonpenetrating plasma proteins normally exist as anions and plasma is electrally neutral, it contains somewhat fewer chloride ions than does interstitial fluid. In contrast to extracellular fluids, intracellular fluid contains only small amounts of Na^{+} and Cl^{-}. Its most abundant cation is potassium, and its major anion is phosphate (HPO_4^{2-}). Cells also contain moderate amounts of magnesium ions and substantial quantities of soluble proteins (equivalent to three times the concentration found in plasma).

Notice that sodium and potassium ion concentrations in extracellular and intracellular fluids are nearly opposite (Figure 27.2). The characteristic distribution of these ions on the two sides of cellular membranes reflects the activity of cellular ATP-dependent sodium-potassium pumps, which keep intracellular Na^{+} concentrations low while maintaining high intracellular K^{+} concentrations (see Chapter 2). Renal mechanisms can reinforce these ion distributions by secreting potassium as sodium is reabsorbed from the filtrate.

Electrolytes are the most *abundant* solutes in the various body fluids and determine most of their chemical and physical reactions, but they do not constitute the *bulk* of dissolved solutes in these fluids. Proteins and some of the nonelectrolyte molecules (phospholipids, cholesterol, and neutral fats) found in the ECF are much larger molecules, and they account for about 90% of the mass of dissolved solutes in plasma, 60% in interstitial fluid, and 97% in the intracellular compartment.

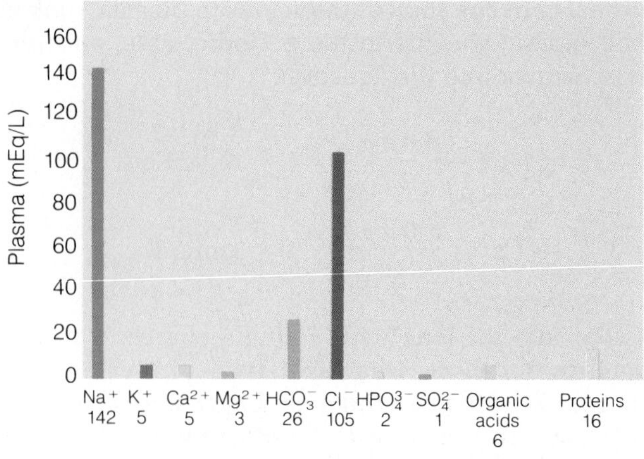

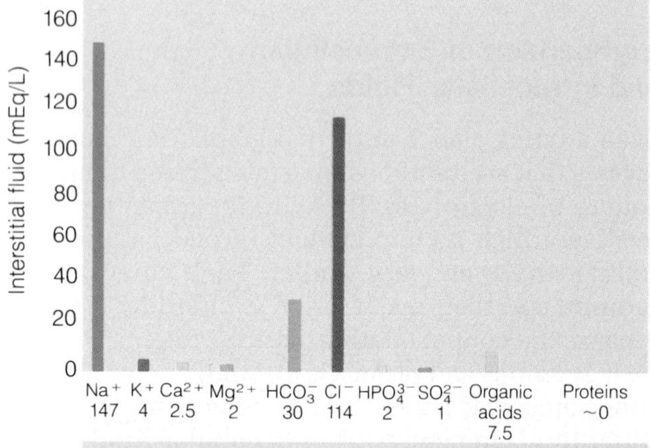

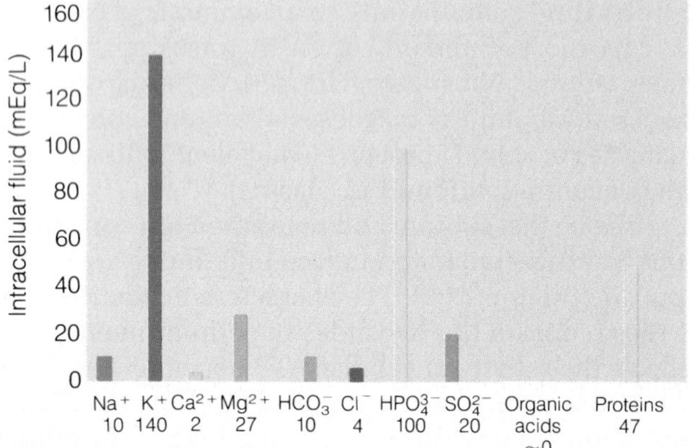

Figure 27.2 Comparison of the electrolyte composition of blood plasma, interstitial fluid, and intracellular fluid.

Movement of Fluids Between Compartments

The continuous exchange and mixing of body fluids are regulated by osmotic and/or hydrostatic pressures (Figure 27.3). Although water moves freely between the compartments along osmotic gradients, solutes are unequally distributed because of their molecular size, electrical charge, or dependence on active transport.

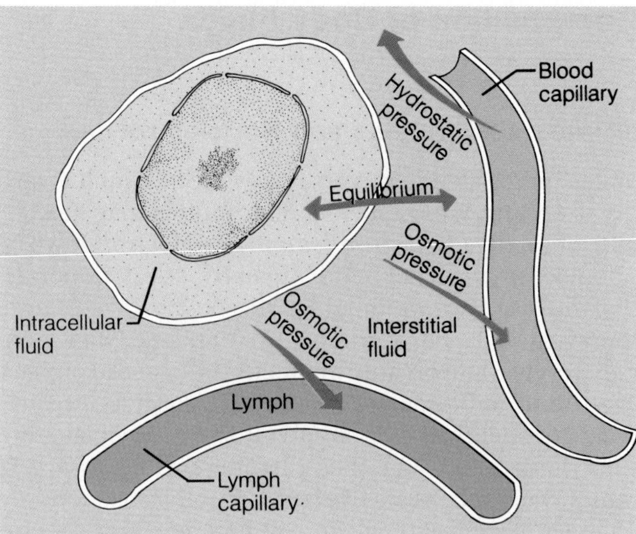

Figure 27.3 Fluid movement between compartments follows hydrostatic and osmotic pressure gradients.

Anything that changes the solute concentration in any compartment will lead to net water flows.

Exchanges between plasma and interstitial fluid occur across capillary membranes. The pressures driving these fluid movements are described in detail in Chapter 20 on pp. 638–639, and the major points are recapped in the discussion of glomerular filtration on pp. 863–865. We will just note the outcome of these mechanisms here. Driven by hydrostatic pressure of the blood, nearly protein-free plasma filters into the interstitial space. Its almost complete reabsorption into the bloodstream reflects the colloid osmotic pressure exerted by plasma proteins. Under normal circumstances, the small net leakage that occurs is picked up by lymphatic vessels and returned to the bloodstream.

Exchanges between the interstitial and intracellular fluids are more complex because of the selective permeability of cellular membranes. As a general rule, two-way osmotic flows of water are substantial, but ion fluxes are restricted and, in most cases, ions move selectively by active transport. Movements of nutrients, respiratory gases, and wastes are typically unidirectional. For example, following their concentration gradients, glucose and oxygen move into the cells passively and metabolic wastes move out of the cells and into the blood.

Of the various body fluids, only plasma circulates throughout the body and serves as the link between the external and internal environments (Figure 27.4). Although exchanges that occur almost continuously in the lungs, gastrointestinal tract, and kidneys do alter plasma composition and volume, these changes are quickly followed by compensating adjustments in the other two fluid compartments so that balance is restored.

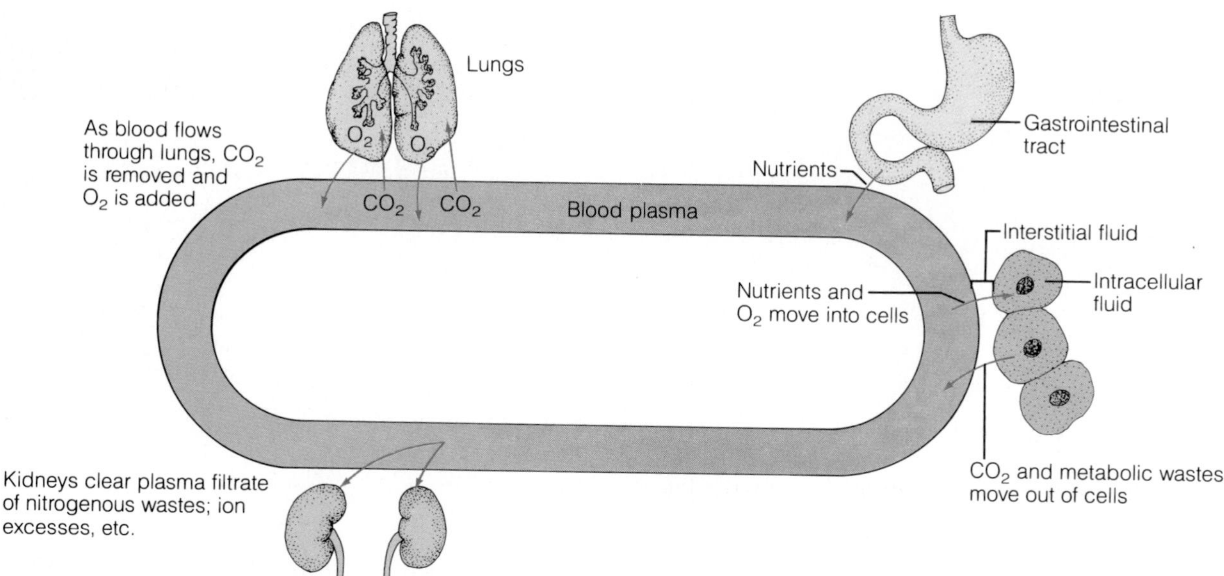

Figure 27.4 The continuous mixing of body fluids. Blood plasma serves as the communicating medium between the external and internal environments. Exchanges are made between the plasma and tissue cells (intracellular fluid) through the interstitial space.

Many factors can cause marked changes in ECF and ICF volumes; but since water moves freely between compartments, the osmolalities of all body fluids are equal, except for the first few minutes after a change in one of the fluids occurs. Increasing the osmolality of the ECF (most importantly, sodium chloride concentrations) can be expected to cause osmolality changes in the ICF—namely, a shift of water out of the cells. Conversely, decreasing the osmolality of the ECF will cause water to move into the cells. Thus, the volume of the ICF is a function of the solute concentration of the ECF. These concepts underlie all the events that control fluid balance in the body and should be understood thoroughly.

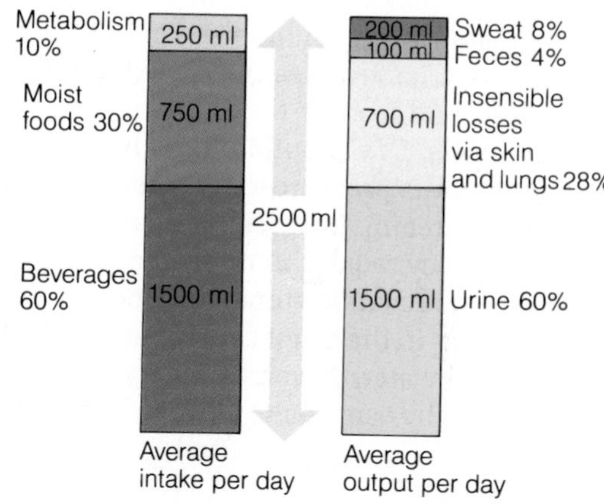

Figure 27.5 Major sources of body water (water intake) and routes of water loss from the body (water output). When intake and output are in balance, the body is adequately hydrated.

Water Balance

If one is to remain properly hydrated, water intake must equal water output. *Water intake* varies widely from person to person and is strongly influenced by habit, but it is typically about 2500 ml a day in adults (Figure 27.5). Most water enters the body through ingested fluids (about 60%) and moist foods (about 30%). About 10% of body water is produced by cellular metabolism; this is called *water of oxidation* or *metabolic water.* *Water output* occurs by several routes: Some water vaporizes out of the lungs in expired air or diffuses directly through the skin (28%); some is lost in perspiration (8%) and in feces (4%); and the balance (60%) is excreted by the kidneys in urine.

Healthy people have a remarkable ability to maintain the tonicity of their body fluids within very narrow limits (285–300 mOsm/L). A rise in plasma osmolality triggers thirst, which leads to water intake, and antidiuretic hormone (ADH) release, which causes the kidneys to excrete concentrated urine. Conversely, a decrease in osmolality is followed by inhibition of both thirst and ADH release, the latter followed by excretion of large volumes of dilute urine.

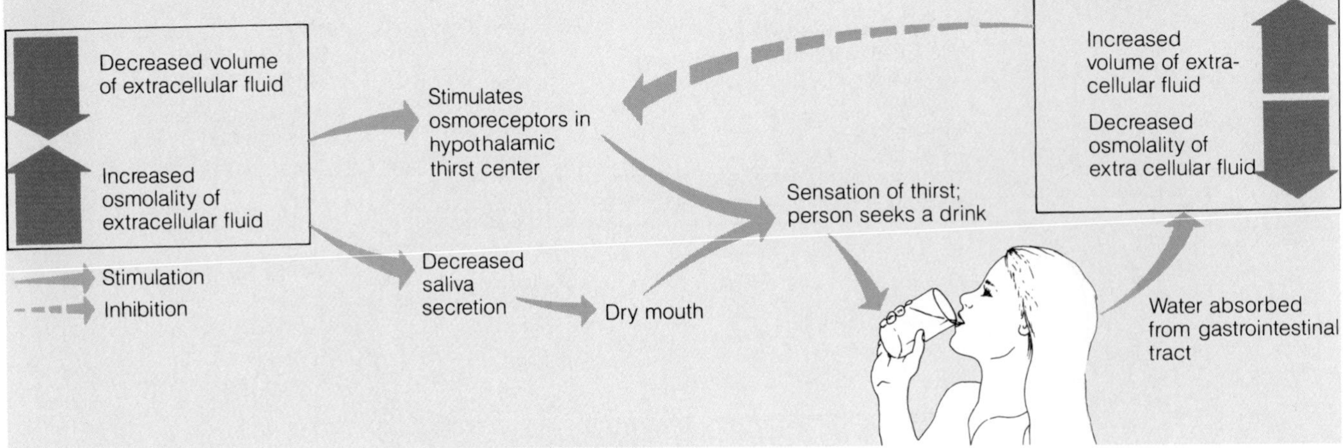

Figure 27.6 The thirst mechanism for regulating water intake.

Regulation of Water Intake: The Thirst Mechanism

Thirst is the driving force for water intake, but the **thirst mechanism** is poorly understood. It is believed that a decrease in plasma volume of 10% (or more, as from hemorrhage) and/or an increase in plasma osmolality of 1% to 2% (as from eating salty foods) results in a dry mouth and excites the hypothalamic *thirst*, or *drinking, center*. Dry mouth occurs because when the plasma osmotic pressure rises, less fluid filters from the bloodstream. Because the salivary glands obtain the water they require from the blood, less saliva is produced. Hypothalamic stimulation occurs because the osmoreceptors in the thirst center become irritable and depolarize as water, driven by the hypertonic ECF, moves out of them by osmosis. Collectively, these events cause a subjective sensation of thirst, which prods us to get a drink (Figure 27.6).

Curiously, thirst is quenched almost as soon as we drink the required amount of water, even though the water has yet to be absorbed into the blood. The quenching of thirst begins as the mucosa of the mouth and throat is moistened and continues as stretch receptors in the stomach and intestine are activated, providing feedback signals that inhibit the hypothalamic thirst center. This premature slaking of thirst has adaptive value because it prevents us from drinking far more than we need and overdiluting our body fluids.

As effective as thirst is, it is not always a reliable indicator of need. The elderly or confused may not recognize or heed it. Conversely, fluid-overloaded renal or cardiac patients may feel thirsty despite their condition. ■

Regulation of Water Output

Certain amounts of water output are uncontrolled and cannot be avoided. These *obligatory water losses* help to explain why we cannot survive long without drinking: Even the most heroic conservation efforts on the part of the kidneys cannot compensate for zero water intake. Obligatory water losses include *insensible water losses* from the lungs and through the skin, water that accompanies undigested food residues in feces, and a minimum daily sensible water loss in urine. Obligatory water loss in urine reflects the fact that when we eat an adequate diet, our kidneys must excrete about 900–1200 mOsm of solutes to maintain blood homeostasis. Unlike the kidneys of desert mice, which can excrete a dry, pasty urine, human kidneys must remove urine solutes from the body in water. Thus, even when urine is maximally concentrated, there is an absolute minimum of 500 ml of water lost in urine daily.

Beyond obligatory water loss, the concentration and volume of urine excreted depend on fluid intake, diet, and water loss via other avenues. For example, if you jog on a hot day and perspire profusely, much less urine than usual must be excreted to maintain water balance. Normally, the kidneys begin to eliminate excess water about 30 minutes after it is ingested. This delay mainly reflects the time required for inhibition of ADH release. Diuresis reaches its peak in 1 hour and then declines to its lowest level after 3 hours.

The body's water volume is closely tied to a powerful water "magnet," ionic sodium. Moreover, maintaining water balance through urinary output is really a problem of sodium *and* water balance because the two are always regulated in tandem by mechanisms that serve cardiovascular function and blood pressure.

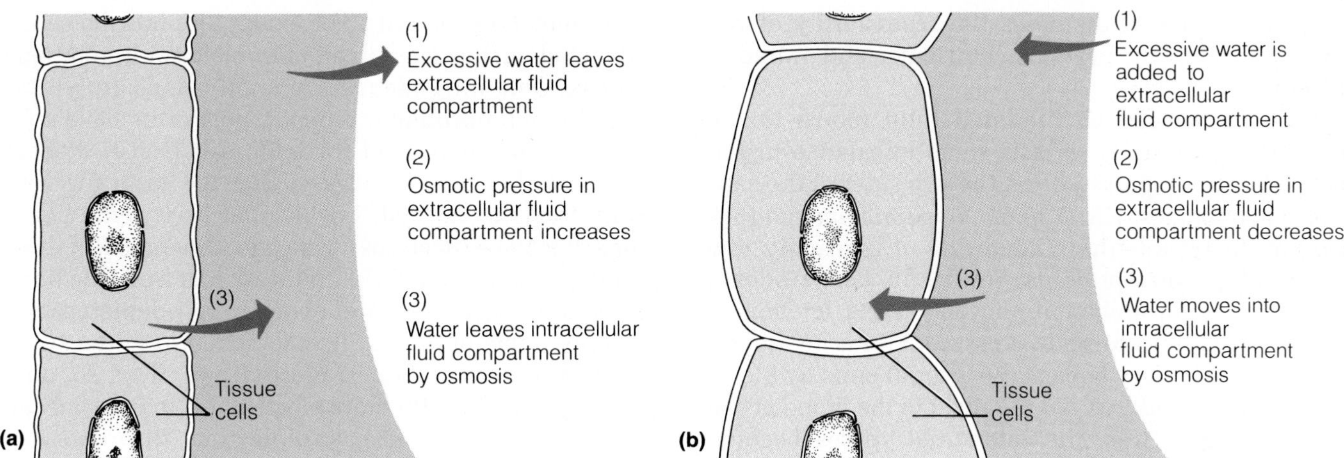

Figure 27.7 Disturbances in water balance. (**a**) Mechanism of dehydration. (**b**) Mechanism of hypotonic hydration (water intoxication).

Disorders of Water Balance

The principal abnormalities of water balance are dehydration, hypotonic hydration, and edema.

Dehydration

Dehydration occurs when water loss exceeds water intake over a period of time and the body is in negative water balance. Dehydration is a common sequel to hemorrhage, severe burns, prolonged bouts of vomiting or diarrhea, profuse sweating, water deprivation, and diuretic abuse. Excessive water loss leading to dehydration may also be caused by endocrine disturbances, such as diabetes mellitus or diabetes insipidus (see Chapter 17). Early signs and symptoms of dehydration include a "cottony" or sticky oral mucosa; thirst; dry, flushed skin; and decreased urine output (oliguria). If prolonged, dehydration may lead to weight loss, fever, and mental confusion. In all these situations, water is lost from the ECF (Figure 27.7a). This is followed by the movement of water (by osmosis) from the cells into the ECF, which keeps the osmolality of the extracellular and intracellular fluids equal even though the total fluid volume has been reduced. The overall effect is called dehydration, but it rarely involves only a deficit of water; most often, as water is lost, electrolytes are lost as well.

Hypotonic Hydration

Hypotonic hydration, also called *dilutional hyponatremia* or *water intoxication,* may result from renal insufficiency or from drinking extraordinary amounts of water very quickly. In either case, the ECF is diluted: The sodium content is normal, but the water content is excessive. Thus, the hallmark of this condition is *hyponatremia* (low ECF Na^+ level). This, in turn, promotes net osmosis into the tissue cells, causing them to swell as they become abnormally hydrated (Figure 27.7b). These events must be reversed quickly—for example, by intravenous administration of hypertonic saline or mannitol, which reverses the osmotic gradient and "pulls" water out of the cells. Otherwise, the resulting electrolyte dilution leads to severe metabolic disturbances evidenced by nausea, vomiting, muscular cramping, and cerebral edema. Cerebral edema quickly leads to disorientation, convulsions, coma, and death.

Edema

Edema (eh-dē′-muh) is an atypical accumulation of fluid in the interstitial space, leading to tissue swelling. Edema may be caused by any event that steps up fluid flow out of the bloodstream or hinders its return.

Factors that enhance the rate of fluid loss from the bloodstream include increased capillary hydrostatic pressure and enhanced capillary permeability. Increased capillary hydrostatic pressure can result from incompetent venous valves, localized blockage of a blood vessel, congestive heart failure, hypertension, or high blood volume resulting, for instance, during pregnancy or from abnormal retention of sodium (which promotes increased water retention). Whatever the cause, the abnormally high hydrostatic pressure of the blood intensifies the rate of filtration and, thus, the fluid shift out of the capillary beds. Increased capillary permeability usually reflects an ongoing inflammatory response. Chemical mediators released during an inflammatory response cause the local capillaries to become very porous, allowing large amounts of exudate to be formed. The importance of this response is that it increases the accessibility of the region to clotting proteins, nutrients, and immune elements.

Edema caused by hindered fluid return to the bloodstream usually reflects some imbalance in the colloid osmotic pressures on the two sides of the capillary membranes. For example, *hypoproteinemia* (hī″-pō-prō″-tēn-ē′-mē-uh), a condition of unusually low levels of plasma proteins, results in tissue edema because protein-deficient plasma exerts an abnormally low colloid osmotic pressure. Fluids are forced out of the capillary beds at the arterial ends by blood pressure as usual, but fail to return to the blood at the venous ends. Thus, the interstitial spaces become congested with fluid. Hypoproteinemia may result from protein malnutrition, liver disease (in which the liver fails to make its normal complement of plasma proteins), or glomerulonephritis (in which plasma proteins pass through "leaky" renal filtration membranes and are lost in urine). Although the cause differs, the result is the same when lymphatic vessels are blocked or have been surgically removed. The small amounts of plasma proteins that normally seep out of the bloodstream are not returned to the blood as usual. As the leaked plasma proteins accumulate, they exert an ever-increasing colloid osmotic pressure, which draws fluid from the blood and holds it in the interstitial space.

Because excess fluid in the interstitial space increases the distance nutrients and oxygen must diffuse between the blood and the cells, edema can impair tissue function. However, the most serious problems resulting from edema concern the operation of the cardiovascular system. As described in Chapter 20, when fluid leaves the bloodstream and accumulates in the interstitial space, both blood volume and blood pressure decline and efficiency of circulatory delivery is severely hampered. ■

Electrolyte Balance

Electrolytes include salts, acids, and bases, but the term **electrolyte balance** usually refers to the salt balance in the body. Salts provide minerals essential to neuromuscular excitability, secretory activity, membrane permeability, and many other cellular functions. Additionally, salts are dominant factors in controlling fluid movements. Although a number of electrolytes are crucial for cellular activity, we will specifically examine the regulation of only four of these; sodium, potassium, calcium, and magnesium. Acids and bases are considered in the next section.

Salts enter the body in foods and, to a limited extent, in water. Additionally, small amounts result from metabolic activity. For example, phosphates are liberated during catabolism of nucleic acids and bone matrix. Obtaining adequate amounts of electrolytes is usually not a problem; indeed, most of us have a far greater taste than need for table salt (NaCl). Recent nutritional research suggests that the taste for very salty foods is learned; however, at least some liking for salt may be innate, assuring adequate intake of these vital ions (Na^+ and Cl^-). Salts are lost from the body in perspiration, feces, and urine. In salt-depleted conditions, some of the salt is reabsorbed from the sweat ducts and perspiration is more dilute. Even so, on a hot day, significantly more electrolytes than usual can be lost in sweat, and gastrointestinal disorders can lead to large losses in feces or vomitus. Thus, the flexibility of renal mechanisms that regulate the electrolyte balance of the blood is a critical asset. The causes and consequences of electrolyte excesses and deficiencies are summarized in Table 27.1, p. 900.

The Central Role of Sodium in Fluid and Electrolyte Balance

Sodium is pivotal to fluid and electrolyte balance and to the homeostasis of all body systems. Regulating the balance between sodium input and output is one of the most important functions of the kidneys. Sodium salts ($NaHCO_3$ and NaCl) account for 90% to 95% of all solutes in the ECF, and they contribute about 280 mOsm of the total (300 mOsm/L) ECF solute concentration. At its normal plasma concentration of 142 mEq/L, sodium is the single most abundant cation in the ECF and is the only one exerting *significant* osmotic pressure. This, plus the fact that Na^+ does not easily cross cellular membranes, gives sodium the primary role in controlling ECF volume and water distribution in the body.

It is important to understand that *while the sodium content of the body may be altered, its concentration in the ECF remains stable because of immediate adjustments in water volume.* Further, because all body fluids are in osmotic equilibrium, a change in plasma sodium levels affects not only the plasma volume and blood pressure, but also the fluid volumes of the other two compartments. In addition, sodium ions continually move back and forth between the ECF and body secretions. For example, large volumes (about 8 L) of sodium-containing secretions (gastric, intestinal, and pancreatic juice; saliva; bile) are released into the digestive tract daily, only to be almost completely reabsorbed. Finally, the operation of renal acid–base control mechanisms is coupled to sodium ion transport and engineered to balance sodium ion concentrations.

Regulation of Sodium Balance

Despite the crucial importance of sodium ion concentration, sodium receptors that specifically monitor sodium ion levels in the body fluids have not been found. Regulation of sodium balance or, rather, sodium–water balance is inseparably linked to blood pressure and entails a variety of neural and hormonal controls.

Influence and Regulation of Aldosterone

The hormone aldosterone exerts a major influence on renal regulation of sodium ion concentrations in the ECF; but whether aldosterone is present or not, some 80% of the sodium in the renal filtrate is reabsorbed in the proximal tubules of the kidneys (see Chapter 26). When aldosterone concentrations are high, most of the remaining Na^+ (actually NaCl, because Cl^- follows) is actively reabsorbed in the distal convoluted and collecting tubules. Water follows if it is able, that is, if the tubule porosity has been increased by ADH. Thus, targeting of the kidneys by aldosterone usually promotes both sodium and water retention. When aldosterone release is inhibited, virtually no Na^+ reabsorption occurs beyond the proximal tubules. So, while urinary excretion of large amounts of sodium *always* results in the excretion of copious amounts of water as well, the reverse is *not* true. Substantial amounts of nearly sodium-free urine can be eliminated as needed to achieve water balance.

Recall that aldosterone is produced by adrenal cortical cells. Although elevated K^+ levels (and to a lesser extent low Na^+ levels) in the ECF directly stimulate the adrenal cells to release aldosterone, the most important trigger for aldosterone release is the renin–angiotensin mechanism mediated by the juxtaglomerular apparatus of the renal tubules. When juxtaglomerular (JG) cells in the afferent arterioles respond to decreased stretch (due to decreased blood pressure or volume and/or to sympathetic nervous sytem stimulation) and macula densa cells of the distal convoluted tubules respond to decreased filtrate osmolality, the JG cells release renin. Renin catalyzes the series of reactions that produce angiotensin II, which, in turn, prompts aldosterone release (Figure 27.8). Conversely, high renal blood pressure and high filtrate osmolality depress renin–angiotensin–aldosterone release.

People with Addison's disease (hypoaldosteronism) lose tremendous amounts of salt and water to urine. As long as adequate amounts of salt and fluids are ingested, people with this condition can avoid

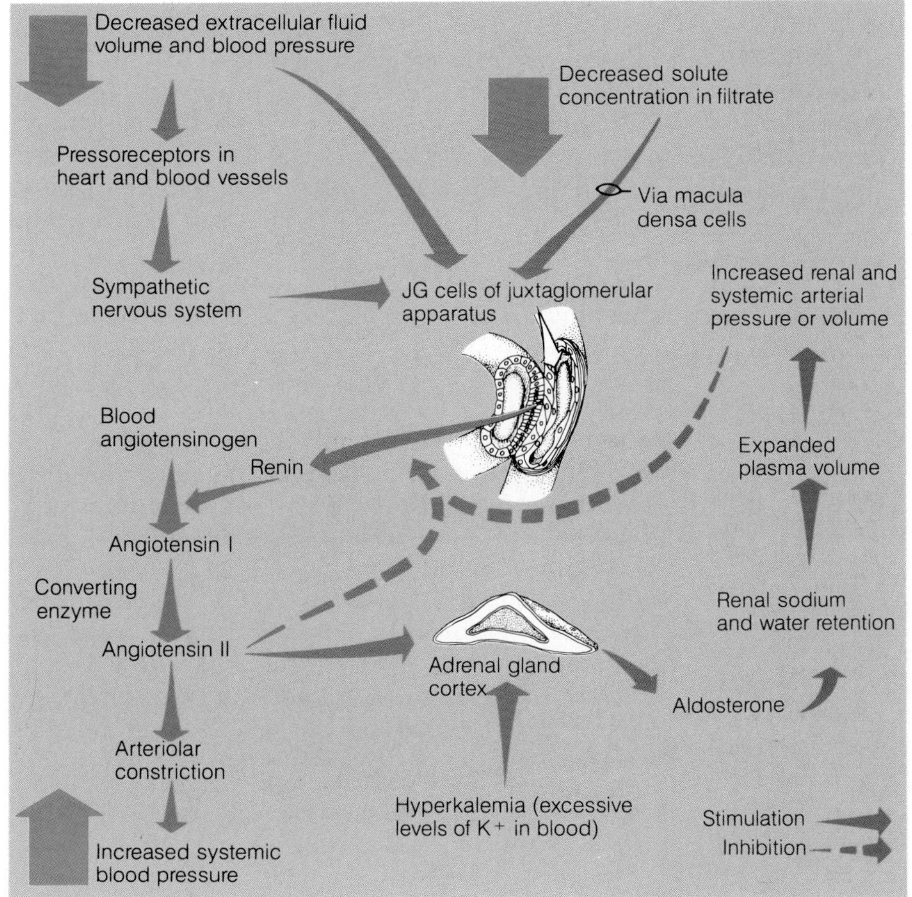

Figure 27.8 Mechanisms by which aldosterone release is controlled.

problems, but they are perpetually teetering on the brink of hypovolemia and dehydration. ■

Aldosterone brings about its effects slowly, over a period of hours to days; and although it has a dramatic effect on tubular reabsorption of sodium, it has a relatively small impact on blood volume. The principal effects of aldosterone on the renal tubules are to diminish urinary output and increase blood volume; but before these can change more than a few percent, basic feedback mechanisms for blood volume control come into play. Even people with Cushing's disease (hyperaldosteronism) rarely have ECF and blood volumes more than 5% to 10% above normal.

Cardiovascular System Pressoreceptors

Blood volume is carefully monitored and regulated to maintain blood pressure and cardiovascular function. As blood volume (hence pressure) rises, the pressoreceptors in the heart and in the large vessels of the neck and thorax (carotid arteries and aorta) communicate this information the hypothalamus. As a result, sympathetic nervous system impulses to the kidneys decrease, allowing the afferent arterioles to dilate. As the glomerular filtration rate rises, sodium and water output increases dramatically. This phenomenon, called *pressure diuresis,* reduces blood volume and, consequently, blood pressure. In contrast, drops in systemic blood pressure lead to constriction of the afferent arterioles, which reduces filtrate formation and urinary output and increases systemic blood pressure (see Figure 27.10). Thus, the pressoreceptors provide information on the "fullness" or volume of the circulation that is critical for maintaining cardiovascular homeostasis. Because sodium ion concentration determines fluid volume, the pressoreceptors might be regarded as sodium receptors.

Influence and Regulation of ADH

The amount of water reabsorbed in the distal segments of the kidney tubules is proportional to ADH release. When ADH levels are low, most of the water reaching the collecting tubules is allowed to pass through, diluting the urine and reducing the volume of body fluids. When ADH levels are high, the tubule pores are so widened that nearly all of the water is reabsorbed, resulting in the output of a small volume of highly concentrated urine.

Osmoreceptors of the hypothalamus sense the ECF solute concentration and trigger or inhibit ADH release from the posterior pituitary accordingly: A decrease in sodium ion concentration (which may be due to increased blood volume) inhibits ADH release and allows more water to be excreted in urine, restoring normal Na^+ levels in the blood. An increase in sodium

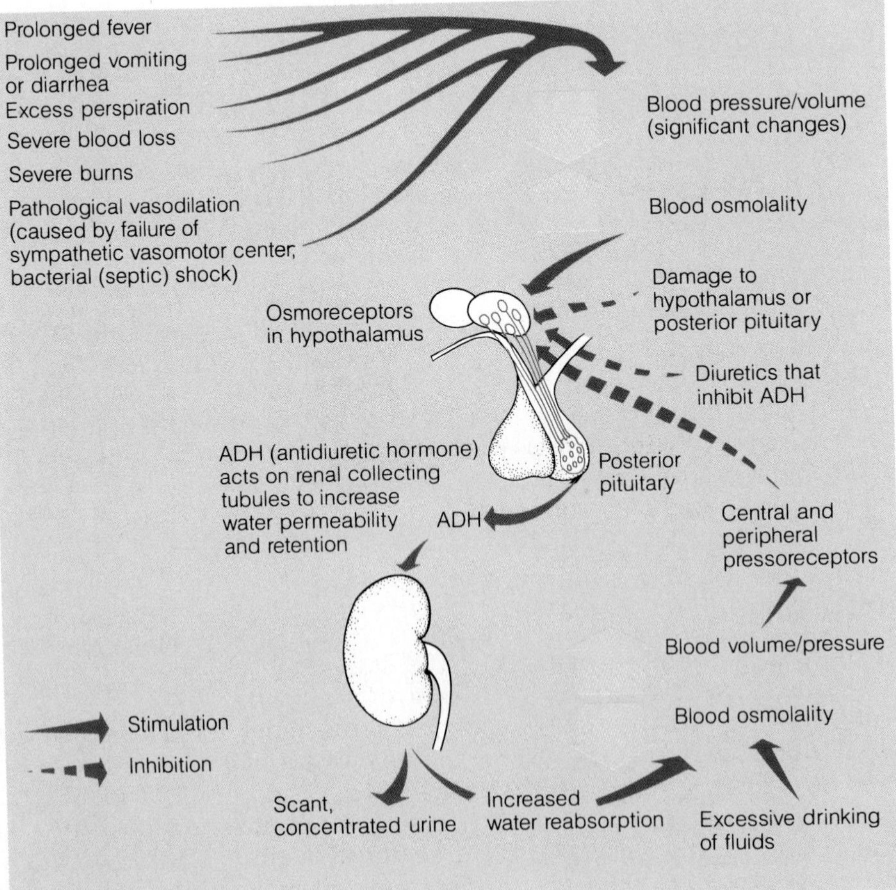

Figure 27.9 Mechanisms by which ADH release is controlled.

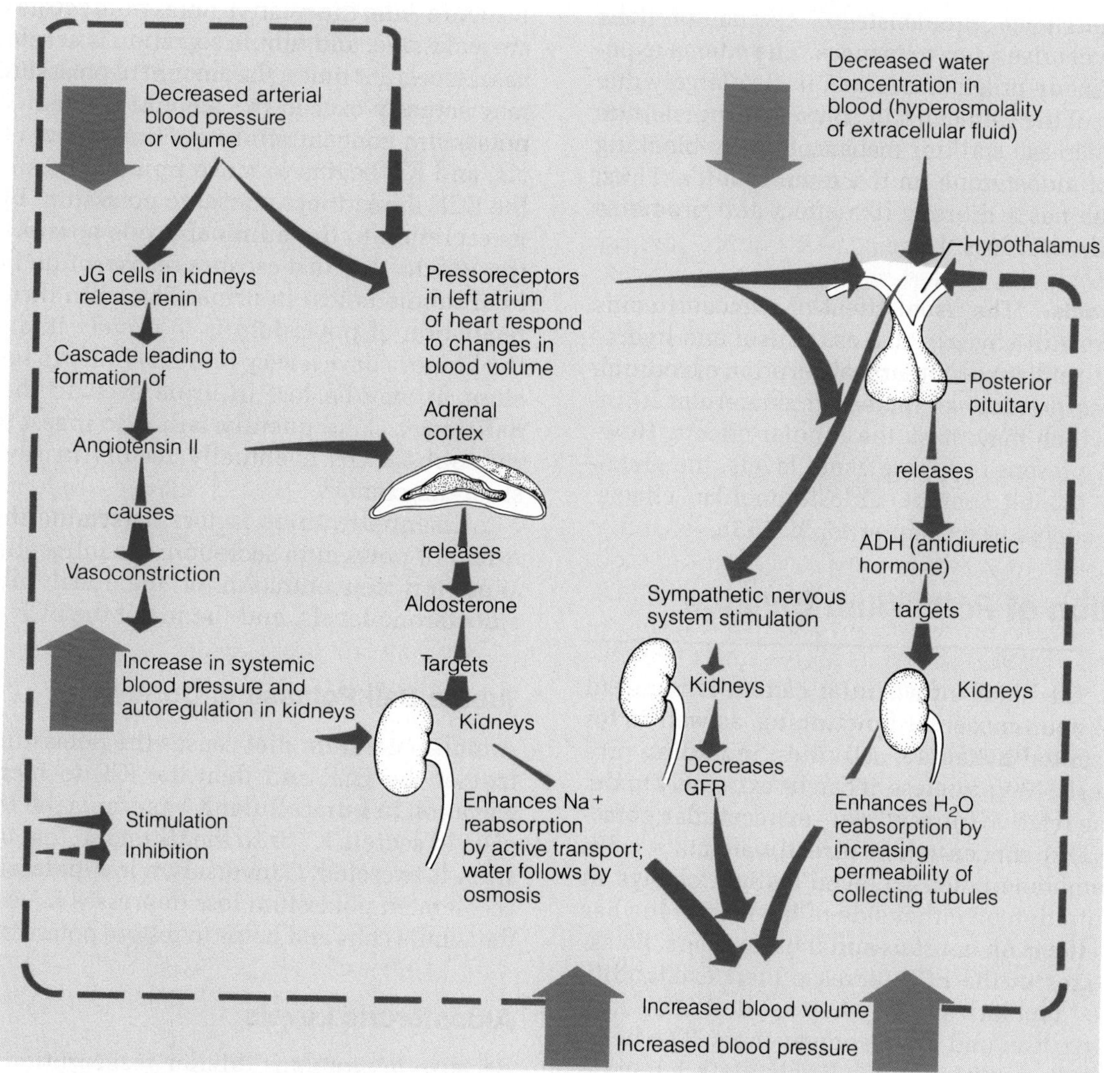

Figure 27.10 Flowchart illustrating the interaction of ADH, aldosterone, and blood pressure in the regulation of sodium and water balance mechanisms mediated through the kidneys.

levels (which may be due to decreased blood volume) stimulates ADH release. Figure 27.9 illustrates specific factors that alter the amount of ADH released. Figure 27.10 summarizes the manner in which renal mechanisms involving aldosterone, angiotensin II, and ADH are interrelated and tied into overall controls of blood volume and blood pressure.

Influence and Regulation of Atrial Natriuretic Factor

The influence of *atrial natriuretic* (na″-tri-yoo-reh′-tik) *factor (ANF)* can be fairly well summarized in one sentence: It reduces blood pressure and blood volume by inhibiting nearly all events that promote vasoconstriction and Na^+ and water retention. Atrial natriuretic factor is a hormone that is released by certain cells of the heart atria when they are stretched by the effects of elevated blood pressure; it has potent diuretic and natriuretic (literally, salt-excreting) effects. It promotes excretion of sodium and water by the kidneys, presumably by directly inhibiting the ability of the distal tubule cells to reabsorb Na^+ and by suppressing the release of ADH, renin, and aldosterone. Additionally, ANF acts both directly and indirectly (by inhibiting renin-induced generation of angiotensin II) to relax vascular smooth muscle; thus it causes vasodilation. Collectively, these effects reduce blood pressure, but the precise way in which this natriuretic hormone brings about these responses is still being investigated.

Influence of Other Hormones

Female Sex Hormones. Because of their chemical similarity to aldosterone, the *estrogens* exert a similar effect on the renal tubules: They enhance reabsorption of sodium chloride. Because water follows, many women experience *premenstrual edema* as their estrogen levels rise prior to menses. The edema experienced by many pregnant women is also largely due to the effect of the estrogens. In contrast, progesterone seems to decrease sodium reabsorption by blocking the effect of aldosterone on the renal tubules. Thus, progesterone has a diuretic-like effect and promotes sodium and water loss.

Glucocorticoids. The usual effect of *glucocorticoids* (gloo″-kō-kor′-tih-koydz), such as cortisol and hydrocortisol, is to enhance tubular reabsorption of sodium, but they also promote an increased glomerular filtration rate, which may mask the tubular effects. However, when present in high plasma levels, the glucocorticoids exhibit potent aldosteronelike effects, promoting edema (see Chapter 17, p. 553).

Regulation of Potassium Balance

Potassium, the chief intracellular cation, is required for normal neuromuscular functioning, as well as for several essential metabolic activities, including protein synthesis. Nevertheless, it can be extremely toxic. Because the relative intracellular–extracellular potassium ion (K^+) concentration directly affects a cell's resting membrane potential, even a slight change in K^+ concentration on either side of the membrane has profound effects on neurons and muscle fibers. Potassium excesses in the ECF increase their excitability and cause depolarization, whereas deficits cause hyperpolarization and nonresponsiveness. The heart is particularly sensitive to K^+ levels: Both hyperkalemia and hypokalemia can cause abnormalities of cardiac rhythm and even cardiac arrest (see Table 27.1).

Potassium is also part of the body's buffer system, which resists changes in the pH of body fluids. Shifts of hydrogen ions (H^+) into and out of cells are compensated for by corresponding shifts of potassium in the opposite direction to maintain cation balance. Thus, ECF potassium levels rise with acidosis, as K^+ leaves the cells, and fall with alkalosis, as K^+ moves into the cells. Although these pH-driven shifts do not change the total amount of potassium in the body, they can seriously interfere with the activity of excitable cells.

Like sodium balance, potassium balance is maintained chiefly by renal mechanisms. However, there are important differences in the way this balance is achieved. The amount of sodium reabsorbed in the tubules is precisely tailored to need, and sodium ions are never secreted into the filtrate. In contrast, potassium reabsorption from the filtrate is a constant, leaving 10% to 15% to be lost in urine regardless of need. Because K^+ content of the ECF (and, therefore, the filtrate) is normally very low compared to Na^+ content, K^+ balance is accomplished chiefly by changing the amount of potassium *secreted* into the filtrate.

As a rule, the relative potassium levels in the ECF are excessive, and tubule secretion is accelerated over basal levels. At times, the amount of potassium excreted may actually exceed the amount filtered. When ECF potassium concentrations decline below normal levels, and K^+ begins to move from the tissue cells into the ECF, the kidneys conserve potassium by reducing its secretion to the minimal levels possible. Even so, the 10% to 15% that escapes reabsorption in the proximal tubule is lost in urine. The main thrust of renal regulation of potassium is to *excrete* it; and because the kidneys have a very limited ability to retain potassium, it may be lost in urine even in the face of a deficiency. Consequently, failure to ingest potassium-rich substances eventually results in severe potassium deficiency.

Essentially, three factors determine the rate and extent of potassium secretion: the intracellular potassium ion concentration of the renal tubule cells, aldosterone levels, and the pH of the ECF.

Tubule Cell Potassium Content

A high-potassium diet causes the potassium ion content of the ECF, and then the ICF, to increase. This elevation in intracellular K^+ prompts the renal tubule cells to secrete K^+ into the filtrate so that more potassium is excreted. Conversely, a low-potassium diet or accelerated potassium loss depresses K^+ secretion by the tubule cells and helps to restore potassium balance.

Aldosterone Levels

Because it promotes tubular reabsorption of sodium ions, aldosterone released by the renin–angiotensin mechanism helps regulate the rate of potassium ion secretion as well. To maintain electrolyte balance there is a one-for-one exchange of Na^+ and K^+ in the distal tubules: For each Na^+ reabsorbed, a K^+ is secreted. Thus, as plasma levels of sodium increase, potassium levels decrease proportionately. (However, the collecting tubules are not obliged to exchange one K^+ for one Na^+.) Aldosterone also increases the passive movement of potassium into the filtrate by increasing the permeability of luminal membranes of the tubule cells to potassium.

The adrenal cortical cells are also *directly* sensitive to the potassium ion concentration of the ECF bathing them. When K^+ concentrations in the ECF increase even slightly, the adrenal cortex is strongly stimulated to release aldosterone, which increases potassium secretion by the exchange process just described. Thus, feedback regulation of aldosterone release provides a very potent system by which potassium controls its own concentrations in the ECF.

In an attempt to reduce NaCl intake, many people have turned to salt substitutes. However, heavy consumption of these substitutes, which are high in potassium, is safe only when adrenocortical function is normal. Otherwise, severe hyperkalemia, resulting in muscle weakness and, in extreme cases, heart block, is likely. In the absence of aldosterone, hyperkalemia is swift and lethal regardless of K^+ intake. Conversely, when a person has an adrenocortical tumor that pumps out tremendous amounts of aldosterone, ECF potassium levels become so depressed that neurons all over the body are hyperpolarized and paralysis occurs. ■

pH of the Extracellular Fluid

The excretion of both K^+ and H^+ is linked to the reabsorption of sodium ions; that is, K^+ and H^+ are cotransported ions and, thus, are competitors for secretion. When the pH of the blood begins to fall, H^+ secretion is accelerated and K^+ secretion is depressed.

Regulation of Calcium Balance

About 99% of the body's calcium is found in bones in the form of calcium phosphate salts, which provide strength and rigidity to the skeleton. Ionic calcium in the ECF is important for normal blood clotting, cell membrane permeability, and secretory behavior. Like sodium and potassium, ionic calcium has potent effects on neuromuscular excitability. Hypocalcemia increases excitability and causes muscle tetany; hypercalcemia is equally dangerous because it inhibits neurons and muscle cells and may cause cardiac arrhythmias (see Table 27.1).

Calcium balance is regulated primarily by the interaction of two hormones: parathyroid hormone and calcitonin. The bony skeleton provides a dynamic reservoir from which calcium and phosphorus can be withdrawn or deposited to maintain the balance of these electrolytes.

Effects of Parathyroid Hormone

The most important controls of Ca^{2+} homeostasis are exerted by *parathyroid hormone (PTH)*, released by the tiny parathyroid glands located on the posterior aspect of the thyroid gland in the throat. Declining plasma levels of Ca^{2+} directly stimulate the parathyroid glands to release PTH, which promotes an increase in calcium levels by targeting the following organs (see Figure 17.11, p. 548):

1. Bones. PTH activates osteoclasts (bone-digesting cells), which break down the bone matrix, resulting in the release of Ca^{2+} and PO_4^{2-} to the blood.

2. Small intestine. PTH enhances intestinal absorption of Ca^{2+} indirectly by stimulating the kidneys to transform vitamin D to its active form (1,25-dihydroxycholecalciferol), which is a necessary cofactor for calcium absorption by the small intestine.

3. Kidneys. PTH increases the reabsorption of calcium by the renal tubules while simultaneously decreasing phosphate ion (PO_4^{2-}) reabsorption. Thus, calcium conservation and phosphate excretion go hand in hand. The adaptive value of this effect is that the *product* of calcium and phosphate concentrations in the ECF remains constant, preventing the deposit of calcium phosphate salts in bones or soft body tissues (see Chapter 6).

As a rule, most filtered phosphate is reabsorbed in the proximal tubules by active transport. In the absence of PTH, phosphate reabsorption is regulated by the *overflow mechanism;* that is, its transport maximum provides for a certain amount of phosphate ion reabsorption, and amounts present in excess of that maximum simply flow out in urine. But when the PTH level rises, active transport of PO_4^{2-} is inhibited.

When calcium levels in the ECF are within normal limits (9–11 mg/100 ml blood) or high, PTH secretion is inhibited. Consequently, the release of calcium from bone is inhibited, larger amounts of calcium are lost in feces and urine, and more phosphate is retained.

Effect of Calcitonin

Calcitonin (kal″-sih-tō′-nin), a hormone produced by the parafollicular cells of the thyroid gland, is released in response to rising blood calcium levels. Calcitonin targets bone, where it encourages deposit of calcium salts and inhibits bone reabsorption. Calcitonin is an antagonist of PTH, but its contribution to calcium homeostasis is much smaller.

Regulation of Magnesium Balance

Magnesium activates the coenzymes needed for carbohydrate and protein metabolism and plays an essential role in neuromuscular functioning. Half of the magnesium in the body is in the skeleton; most of the remainder is found intracellularly—in heart and skeletal muscle and in the liver.

Control of magnesium balance is poorly understood, but a renal transport maximum for magnesium is known to exist. Because the handling of magnesium by many body cells is similar to that of K^+, it is likely that increased aldosterone levels promote increased Mg^{2+} excretion by the renal tubules.

Regulation of Anions

Chloride is the major anion accompanying sodium in the ECF and, like sodium, it helps maintain the osmotic pressure of the blood. When blood pH is within normal limits or slightly alkaline, about 99% of the fil-

Table 27.1 Causes and Consequences of Electrolyte Imbalances

Ion	Abnormality/serum value	Possible causes	Consequences
Sodium	Hypernatremia (Na^+ excess in ECF: >145 mEq/L)	Uncommon in healthy individuals; most often a result of excessive NaCl administration by intravenous route	Increased blood pressure; water leaves cells to enter ECF; edema; congestive heart failure in cardiac patients
	Hyponatremia (Na^+ deficit in ECF: <130 mEq/L)	Excessive Na^+ loss through burned skin, excessive sweating, vomiting, diarrhea, tubal drainage of stomach, and as a result of excessive use of diuretics; deficiency of aldosterone (Addison's disease); renal disease	Dehydration; decreased blood volume and blood pressure (shock); if sodium is lost and water is not, symptoms are the same as with water excess (mental confusion, giddiness, muscular twitching, convulsions, coma)
Potassium	Hyperkalemia (K^+ excess in ECF: >5.5 mEq/L)	Renal failure; deficit of aldosterone; rapid intravenous infusion of KCl; burns or severe tissue injuries, which cause K^+ to leave cells	Bradycardia; cardiac arrhythmias, depression, and arrest; skeletal muscle weakness, flaccid paralysis
	Hypokalemia (K^+ deficit in ECF: <3.5 mEq/L)	Gastrointestinal tract disturbances (vomiting, diarrhea), gastrointestinal suction; chronic stress; Cushing's disease; inadequate dietary intake (starvation); hyperaldosteronism; diuretic therapy	Cardiac arrhythmias, possible cardiac arrest; muscular weakness; alkalosis; hypoventilation
Magnesium	Hypermagnesemia (Mg^{2+} excess in ECF: >6 mEq/L)	Rare (occurs when Mg is not excreted normally); deficiency of aldosterone; excessive ingestion of Mg^{2+}-containing antacids	Lethargy; impaired CNS functioning, coma, respiratory depression
	Hypomagnesemia (Mg^{2+} deficit in ECF: <1.4 mEq/L)	Alcoholism; loss of intestinal contents, severe malnutrition; diuretic therapy	Tremors, increased neuromuscular excitability, convulsions
Chloride	Hyperchloremia (Cl^- excess in ECF: >105 mEq/L)	Increased retention or intake; hyperkalemia	Metabolic acidosis due to enhanced loss of bicarbonate; stupor; rapid, deep breathing; unconsciousness
	Hypochloremia (Cl^- deficit in ECF: <95 mEq/L)	Vomiting; hypokalemia; excessive ingestion of alkaline substances	Metabolic alkalosis due to bicarbonate retention
Calcium	Hypercalcemia (Ca^{2+} excess in ECF: >5.8 mEq/L or 11 mg%)	Hyperparathyroidism; excessive vitamin D; prolonged immobilization; renal disease (decreased excretion); malignancy; Paget's disease; Cushing's disease accompanied by osteoporosis	Tetany, bone wasting, pathological fractures; flank and deep thigh pain; kidney stones, nausea and vomiting, cardiac arrhythmias and arrest; depressed respiration, coma
	Hypocalcemia (Ca^{2+} deficit in ECF: <4.5 mEq/L or 9 mg%)	Burns (calcium trapped in damaged tissues); increased renal excretion in response to stress and increased protein intake; diarrhea; vitamin D deficiency; alkalosis	Tingling of fingers, tremors, tetany, convulsions; depressed excitability of the heart, bleeder's disease

tered chloride ions is reabsorbed by passive transport: They simply follow sodium ions out of the filtrate and into the peritubular capillary blood. However, when acidosis occurs, fewer chloride ions accompany sodium as bicarbonate ion reabsorption is stepped up to restore blood pH to its normal range. Thus, the choice between chloride and bicarbonate ions serves acid–base regulation.

Most other anions, such as sulfates and nitrates, have definite transport maximums; and when their concentrations in the filtrate exceed their renal thresholds, excesses begin to spill over into urine. Hence, the concentrations of most anions in plasma are regulated by the overflow mechanism.

Acid–Base Balance

Because all functional proteins (enzymes, hemoglobin, cytochromes, and others) are influenced by hydrogen ion concentration, it follows that nearly all biochemical reactions are influenced by the pH of their fluid environment. Accordingly, the acid–base balance of body fluids is critical and closely regulated. (For a review of the basic principles of acid–base reactions and pH, see Chapter 2.)

The optimal pH of various body fluids differs, but not very much. The normal pH of arterial blood is 7.4, that of venous blood and interstitial fluid is 7.35, and that of intracellular fluid averages 7.0. The lower pH of cells and venous blood reflects their greater amounts of acidic metabolites (such as lactic acid) and carbon dioxide, which combines with water to form carbonic acid.

Whenever the pH of arterial blood rises above 7.45, a person is said to have **alkalosis** (ak″-kuh-lō′-sis). A drop in arterial pH to below 7.35 results in **acidosis** (a″-sih-dō′-sis). Because a pH of 7.0 is neutral, 7.35 is not acidic, chemically speaking; however, it represents a higher-than-optimal H^+ concentration for the functioning of most cells. Therefore, any arterial pH between 7.35 and 7.0 is called *physiological acidosis*.

Although small amounts of acidic substances enter the body via ingested foods, most hydrogen ions originate as by-products or end products of cellular metabolism (Figure 27.11). The breakdown of phosphorus and sulfur-containing proteins (and certain other molecules) releases *phosphoric acid* and *sulfuric acid* into the ECF. Anaerobic respiration of glucose produces *lactic acid*, and fat metabolism yields other organic acids, such as fatty acids, and *ketone bodies*. Additionally, as described in Chapter 23, the loading and transport of carbon dioxide in the blood as bicarbonate liberates hydrogen ions. Finally, although

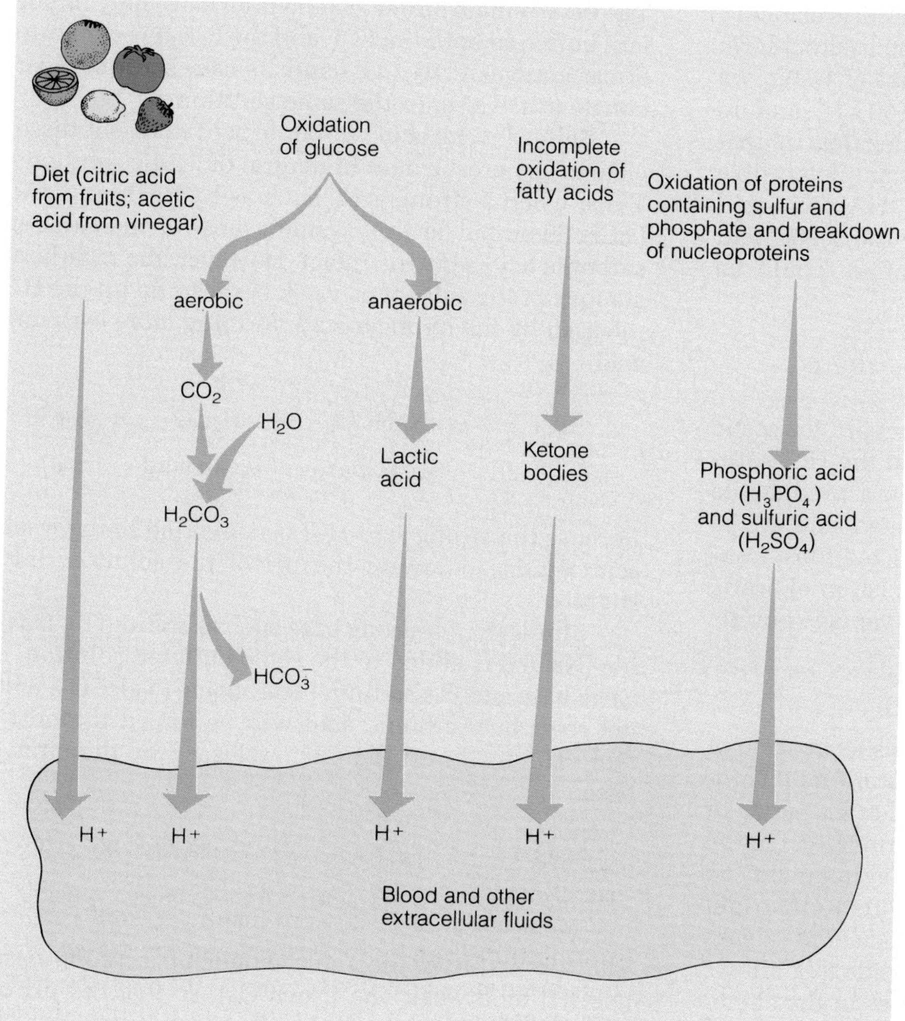

12 ½
- 8 ½
4 0

Figure 27.11 Sources of hydrogen ions (H^+) in the body.

hydrochloric acid produced by the stomach is not really *in* the body, it is a source of H^+ that must be buffered if digestion is to occur normally in the small intestine.

The concentration of H^+ in body fluids is regulated sequentially by (1) chemical buffer systems, (2) the respiratory center in the brain stem, and (3) renal mechanisms. Chemical buffers act within a fraction of a second to resist a pH change and are the first line of defense. Adjustments in respiratory rate and depth begin to compensate for acidosis and alkalosis within 1–3 minutes. The kidneys, although the most potent of the acid–base regulatory systems, ordinarily require hours to a day or more to effect changes in blood pH.

Chemical Buffer Systems

A comprehension of the way chemical buffer systems operate requires a thorough understanding of strong and weak acids and bases. The first important concept is that the acidity of a solution reflects only the free hydrogen ions, not those bound to anions. Consequently, acids that dissociate completely and liberate all their H^+ in water are called *strong acids,* because they have the capability of dramatically changing the pH of a solution. Examples are hydrochloric acid and sulfuric acid. If we could count out 100 hydrochloric acid molecules and place them in 1 ml of water, we could expect to end up with 100 H^+, 100 Cl^-, and no undissociated hydrochloric acid molecules in that solution. Acids that do not dissociate completely, like carbonic acid (H_2CO_3) and acetic acid (HAc), are said to be *weak acids.* If we were to place 100 acetic acid molecules in 1 ml of water, the reaction would be something like this:

$$100\ HAc \rightarrow 90\ HAc + 10\ H^+ + 10\ Ac^-$$

Because undissociated acids do not affect pH, the acetic acid solution is much less acidic than the HCl solution. Weak acids tend to dissociate in a predictable way, and molecules of the undissociated acid are in dynamic equilibrium with the dissociated ions. Consequently, we could look at the dissociation of acetic acid from the following more precise vantage point:

$$HAc \rightleftharpoons H^+ + Ac^-$$

Viewing it from this perspective allows us to see that if H^+ (a strong acid) is added to the acetic acid solution, the equilibrium will shift to the left and some of the H^+ and Ac^- will recombine to form HAc. On the other hand, if a base is added to the system and the pH begins to rise, the equilibrium shifts to the right and more of the HAc molecules dissociate to release H^+. It is this characteristic of weak acids that allows them to play extremely important roles in the chemical buffer systems of the body.

The concept of strong and weak bases is more easily explained. Remember that bases are proton acceptors. Thus, strong bases are those, like hydroxides, that dissociate easily in water and quickly tie up H^+. Ammonia (NH_3) ionizes very incompletely in water to form the ammonium ion (NH_4^+); because it accepts relatively few protons, it is considered a weak base.

Chemical acid–base buffers are systems of one or two molecules that act to prevent pronounced changes in H^+ concentration when a strong acid or base is added. They do this by binding to hydrogen ions whenever the pH of body fluids drops and releasing them when pH rises. The three major chemical buffer systems in the body are the bicarbonate, phosphate, and protein buffer systems, each of which is important in maintaining the pH of one or more fluid compartments. They all work together: When anything causes a shift in hydrogen ion concentration in any compartment, it simultaneously causes a change in the others. Thus, the buffer systems actually buffer one another, so that any drifts in pH are resisted by the *entire* buffering system.

Bicarbonate Buffer System

The **bicarbonate buffer system** is an extremely important buffer, in both the ECF and the ICF. It is a mixture of carbonic acid (H_2CO_3) and its salt, sodium bicarbonate ($NaHCO_3$), in the same solution.

Being a weak acid, carbonic acid does not dissociate to any great extent in neutral or acidic solutions. Thus, when a strong acid, such as hydrochloric acid (HCl), is added to this system, most of the existing carbonic acid remains intact. However, the bicarbonate ions of the salt act as weak bases to tie up the H^+ released by the stronger acid, forming more carbonic acid:

$$\underset{\text{strong acid}}{HCl} + \underset{\text{weak base}}{NaHCO_3} \rightarrow \underset{\text{weak acid}}{H_2CO_3} + \underset{\text{salt}}{NaCl}$$

Because the strong acid HCl is converted to the weak acid H_2CO_3, it lowers the pH of the solution only slightly.

Similarly, if a strong base such as sodium hydroxide (NaOH) is added to the same buffered solution, a weak base such as sodium bicarbonate ($NaHCO_3$) will not react but carbonic acid will be forced to donate more H^+ to tie up the OH^- released by the strong base:

$$\underset{\text{strong base}}{NaOH} + \underset{\text{weak acid}}{H_2CO_3} \rightarrow \underset{\text{weak base}}{NaHCO_3} + \underset{\text{water}}{H_2O}$$

The net result is the replacement of a strong base (NaOH) by a weak one ($NaHCO_3$), so that the pH of the solution rises very little.

Although the bicarbonate salt in the example is sodium bicarbonate, the specific cation dissociated is unimportant; other bicarbonate salts function in an identical way. Within cells, where little sodium is present, potassium and magnesium bicarbonates act as part of the bicarbonate buffer system.

The buffering power of this type of system is directly related to the concentrations of the buffering substances. Thus, if acids are poured into the blood at such a rate that all the available bicarbonate ions (often referred to as the *alkaline reserve*) are tied up, the buffer system becomes ineffective in resisting pH changes. The bicarbonate ion concentration in the ECF is normally around 25 mEq/L and is closely regulated by the kidneys. The concentration of carbonic acid is only about one-twentieth that of bicarbonate, but since the supply of carbonic acid (available from cellular respiration) is almost limitless, obtaining that member of the buffer pair is usually not a problem. Carbonic acid concentration is subject to respiratory controls.

Phosphate Buffer System

The operation of the **phosphate buffer system** is nearly identical to that of the bicarbonate buffer. Its components are the sodium salts of dihydrogen phosphate ($H_2PO_4^-$) and monohydrogen phosphate (HPO_4^{2-}). NaH_2PO_4 acts as a weak acid; Na_2HPO_4 acts as a weak base.

Again, hydrogen ions released by strong acids are tied up in weak acids:

$$\underset{\text{strong acid}}{HCl} + \underset{\text{weak base}}{Na_2HPO_4} \rightarrow \underset{\text{weak acid}}{NaH_2PO_4} + \underset{\text{salt}}{NaCl}$$

and strong bases are converted to weak bases:

$$\underset{\text{strong base}}{NaOH} + \underset{\text{weak acid}}{NaH_2PO_4} \rightarrow \underset{\text{weak base}}{Na_2HPO_4} + \underset{\text{water}}{H_2O}$$

Because the phosphate buffer system is present in low concentrations in the ECF (approximately one-sixth that of the bicarbonate buffer system), it is relatively unimportant for buffering blood plasma. However, it is a very effective buffer in both urine and intracellular fluid, where phosphate concentrations are usually quite high.

Protein Buffer System

Proteins in plasma and within cells are the body's most plentiful and powerful source of buffers. In fact, at least three-quarters of all the buffering power of body fluids resides within the cells, and most of this reflects the buffering activity of intracellular proteins that constitute the **protein buffer system.**

As described in Chapter 2, proteins are polymers of amino acids. Some of the linked amino acids have free, or exposed, groups of atoms called *organic acid (carboxyl) groups* (—COOH), which dissociate to release H^+ when the pH begins to rise:

$$R\text{—}COOH \rightarrow R\text{—}COO^- + H^+$$

Additionally, some amino acids have groups of atoms that can act as bases and accept protons. For example, an exposed $\text{—}NH_2$ group can bind with hydrogen ions, becoming $\text{—}NH_3^+$:

$$R\text{—}NH_2 + H^+ \rightarrow R\text{—}NH_3^+$$

Because this withdraws free hydrogen ions from solution, it prevents the solution from becoming too acidic. Thus, the same protein molecules can function as either acids or bases depending on the pH of their environment. Individual molecules that can act reversibly as acid–base buffers are called *amphoteric* (am″-fō-tayr′-ik) *molecules.*

Hemoglobin of red blood cells provides an excellent example of a protein that functions as an intracellular buffer. As explained earlier, carbon dioxide released from the tissues forms carbonic acid, which dissociates liberating hydrogen and bicarbonate ions in the blood. However, at the same time, hemoglobin is unloading oxygen, becoming reduced hemoglobin, which carries a negative charge. Because the hydrogen ions rapidly bind to the hemoglobin anions, pH changes are minimized (see Chapter 23, p. 736). In this case, carbonic acid, a weak acid, is being buffered by an even weaker acid, hemoglobin.

Respiratory System Regulation of Hydrogen Ion Concentration

As described in Chapter 23, the respiratory system eliminates carbon dioxide from the blood while replenishing its stores of oxygen. Carbon dioxide generated by cellular respiration enters erythrocytes in the circulation and is converted to bicarbonate ions for transport in the plasma as shown by the equation:

$$CO_2 + H_2O \underset{}{\overset{\text{carbonic anhydrase}}{\rightleftharpoons}} \underset{\text{carbonic acid}}{H_2CO_3} \rightleftharpoons H^+ + \underset{\text{bicarbonate ion}}{HCO_3^-}$$

There is a reversible equilibrium between dissolved carbon dioxide and water on the one hand and carbonic acid on the other, and also between carbonic acid and the hydrogen and bicarbonate ions. Consequently, an increase in any of these chemical species will push the reaction in the opposite direction. Notice also that the right side of the equation is equivalent to the bicarbonate buffer system.

In healthy individuals, carbon dioxide is expelled from the lungs at the same rate it is formed in the tissues. During carbon dioxide unloading, the reaction shifts to the left and the H^+ generated from carbonic acid is reincorporated into water. Thus, H^+ produced by CO_2 transport is not allowed to accumulate and has little or no effect on blood pH. However, the respiratory centers in the medulla are very sensitive to changes in blood pH resulting from virtually any metabolic process, and they respond by modifying respiratory rate and depth (see Figure 23.21).

A rising plasma H^+ concentration (acidosis) excites the respiratory center to stimulate deeper, more rapid respirations. As alveolar ventilation increases, more carbon dioxide is removed from the blood, pushing the reaction to the left and reducing the hydrogen ion concentration. On the other hand, when the H^+ concentration begins to fall (alkalosis), the respiratory center is depressed. As the respiratory rate drops and respirations become more shallow, carbon dioxide accumulates; the equilibrium is pushed to the right, causing H^+ concentrations to increase. Again the blood pH is restored to the normal range. Generally, these respiratory system–mediated corrections of blood pH (via regulation of the the CO_2 content of the blood) are accomplished within a minute or so.

Respiratory system regulation of acid–base balance provides a *physiological*, or *functional, buffering system*. Although it acts more slowly than the chemical buffers, it has one to two times the buffering power of all the chemical buffers combined. Modifications of alveolar ventilation can produce dramatic changes in blood pH—far more than is needed. For example, a doubling or halving of alveolar ventilation can raise or lower blood pH by about 0.2 pH unit. Because normal arterial pH is 7.4, a change of 0.2 pH unit would yield a blood pH of 7.6 or 7.2—both well beyond the normal limits of blood pH. Actually, alveolar ventilation can be increased about fifteenfold or reduced to zero. Thus, respiratory controls of blood pH have a tremendous reserve capacity.

Anything that impairs respiratory system functioning causes acid–base imbalances. For example, carbon dioxide retention leads to acidosis; hyperventilation can cause alkalosis. When the cause of the pH imbalance is due to respiratory system problems, the resulting condition is either **respiratory acidosis** or **respiratory alkalosis.**

Renal Mechanisms of Acid–Base Balance

Chemical buffers can tie up excess acids or bases temporarily, but they cannot eliminate these substances from the body. And while the lungs can dispose of the *volatile acid* carbonic acid by eliminating CO_2, only the kidneys can rid the body of the other acids generated by cellular metabolism: sulfuric, phosphoric, uric, and keto acids. These acids are sometimes referred to as *metabolic (fixed) acids,* and the acidosis resulting from their accumulation is called **metabolic acidosis.** This terminology is both unfortunate and incorrect because carbon dioxide (hence, carbonic acid) is also a product of metabolism. Additionally, only the kidneys have the power to regulate blood levels of alkaline substances and to renew the chemical buffers that are used up in regulating H^+ levels in the ECF. (HCO_3^-, which helps regulate H^+ levels, is lost to the body when CO_2 is eliminated by the lungs.) Thus, the ultimate acid–base regulatory organs are the kidneys, which act slowly but surely to compensate for acid–base imbalances resulting from enormous variations in diet or metabolism or from disease states. Renal mechanisms for regulating acid–base balance of the blood include secretion of H^+ and conservation or elimination of bicarbonate ions.

Regulation of Hydrogen Ion Secretion

Virtually all the H^+ that leaves the body in urine is *secreted* into the filtrate. The tubule cells (including those of the collecting ducts) appear to respond directly to the pH of the ECF and to alter their rate of H^+ secretion accordingly. The secreted hydrogen ions are obtained from the dissociation of carbonic acid within the tubule cells (Figure 27.12). For each H^+ actively secreted into the tubule lumen, one sodium ion is reabsorbed into the tubule cell from the filtrate, maintaining the electrochemical balance. The intracellular supplies of carbonic acid come from the combination of carbon dioxide and water within the cells.

The rate of H^+ secretion rises and falls with CO_2 levels in the ECF. The higher the content of CO_2 in the peritubular capillary blood, the faster the rate of H^+ secretion. Because carbon dioxide concentrations in the blood are directly related to blood pH, this system can respond to both increasing and decreasing H^+ concentrations. Also notice that excreted H^+ can combine with bicarbonate ions (HCO_3^-) in the filtrate, generating CO_2 and water within the tubular lumen. In this case, H^+ is bound in water, but the CO_2 enters the tubule cell, where it promotes still more H^+ secretion.

Conservation of Bicarbonate Ions

Bicarbonate ions (HCO_3^-) are an important part of the bicarbonate buffer system, the most important inorganic blood buffer. If this reservoir of base, or alkaline reserve, is to be maintained, the kidneys must be able to do more than just eliminate hydrogen ions to counter rising blood H^+ levels. This is somewhat more

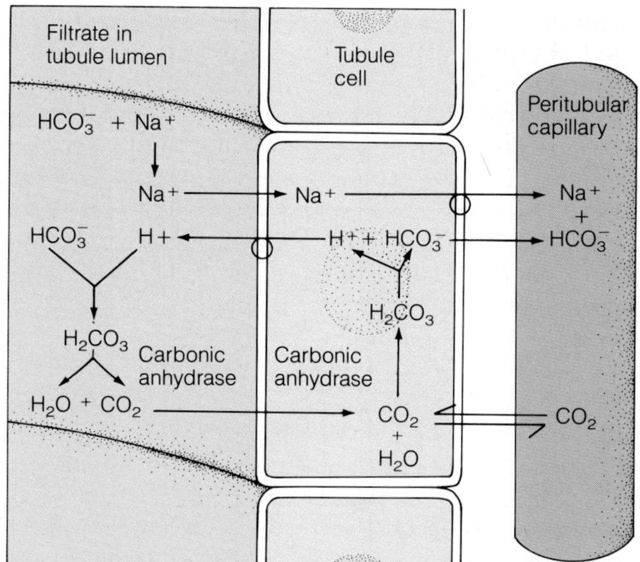

Figure 27.12 Events occurring during hydrogen ion secretion by the kidneys. For each H^+ secreted into the filtrate, a Na^+ is reabsorbed. Secreted H^+ ions can combine with bicarbonate ions (HCO_3^-) present in the tubular filtrate, forming carbonic acid, which dissociates to release carbon dioxide and water. Carbon dioxide then diffuses into the tubule cells, where it acts to trigger H^+ secretion (via the intracellular formation of carbonic acid and then its dissociation). Active transport processes are indicated by a circle beneath the reaction arrow.

complex than it may seem because the tubule cells are almost completely impermeable to bicarbonate ions present in the filtrate—they cannot reabsorb them. However, the kidneys are able to conserve bicarbonate by a rather roundabout mechanism, also illustrated in Figure 27.12. As you can see, the dissociation of carbonic acid within the tubule cells liberates HCO_3^- as well as H^+. While the tubule cells cannot reabsorb bicarbonate from the filtrate, they can, and do, shunt bicarbonate ions generated within them into the peritubular capillary blood. Thus, reabsorption of bicarbonate depends on the presence of H^+ in the filtrate; and when large amounts of hydrogen ions are secreted, bicarbonate ions are almost completely removed from the filtrate.

Buffering of Excreted Hydrogen Ions. Eliminating weak acids and replenishing depleted stores of HCO_3^- also depend on H^+ secretion. Although it might appear that once H^+ has been secreted into the filtrate, the kidneys are done with it, this is not the case. Only small amounts of excreted H^+ are bound in water. To remove the balance in urine, the hydrogen ions must bind with buffers in the filtrate. Otherwise, a urine pH of approximately 1.4 would result—a situation incompatible with life. (H^+ secretion ceases when urine pH falls to 4.5.) The most important urine buffers are the phosphate buffer system and the *ammonia–ammonium ion buffer system*.

The members of the phosphate buffer system filter freely into the tubules, but are not absorbed in appreciable amounts. Therefore, the buffer pair becomes more and more concentrated as the filtrate moves through the renal tubules. As shown in Figure 27.13a, secreted H^+ combines with Na_2HPO_4, forming NaH_2PO_4, which then flows out in urine. Sodium ions released during this reaction move into the cells and are reabsorbed along with bicarbonate ions (generated within the tubule cells) into the peritubular capillary blood.

The **ammonia–ammonium ion buffer system** utilizes the ammonia (NH_3) produced by metabolism of glutamine and other amino acids within the tubule cells. The ammonia diffuses into the filtrate, where it combines with H^+ to form ammonium ions (NH_4^+), which are excreted in urine in combination with chloride ions (Figure 27.13b). Each time an ammonia molecule combines with a hydrogen ion, another ammonia molecule is allowed to diffuse into the filtrate. Thus, the rate of ammonia secretion is coupled to the rate of H^+ secretion. (Although it might seem that hydrogen ions could simply be excreted in urine in combination with chloride ions, this is not possible because HCl is a strong acid.) As with the phosphate buffer system, bicarbonate production and sodium bicarbonate reabsorption also accompany the ammonia–ammonium buffering mechanism, replenishing the alkaline reserve of the blood.

Abnormalities of Acid–Base Balance

All cases of acidosis and alkalosis can be classed according to cause as either *respiratory* or *metabolic*. The causes and consequences of acid–base disorders are summarized in Table 27.2, p. 908.

Respiratory Acidosis or Alkalosis

Respiratory acidosis or alkalosis results from some failure of the respiratory system to perform its normal pH-balancing role. The partial pressure of carbon dioxide (P_{CO_2}) is the single most important indicator of the adequacy of respiratory function. When respiratory system functioning is normal, the P_{CO_2} fluctuates between 35 and 45 mm Hg. Values above 45 mm Hg indicate respiratory acidosis; values below 35 mm Hg indicate respiratory alkalosis.

Respiratory acidosis most often occurs when a person breathes shallowly or when gas exchange is hampered by diseases such as pneumonia, cystic fibrosis, or emphysema. Under such conditions, CO_2 accumulates in the blood. Thus, respiratory acidosis is predicted by a falling blood pH and a rising P_{CO_2}.

Respiratory alkalosis results when carbon dioxide is eliminated from the body faster than it is pro-

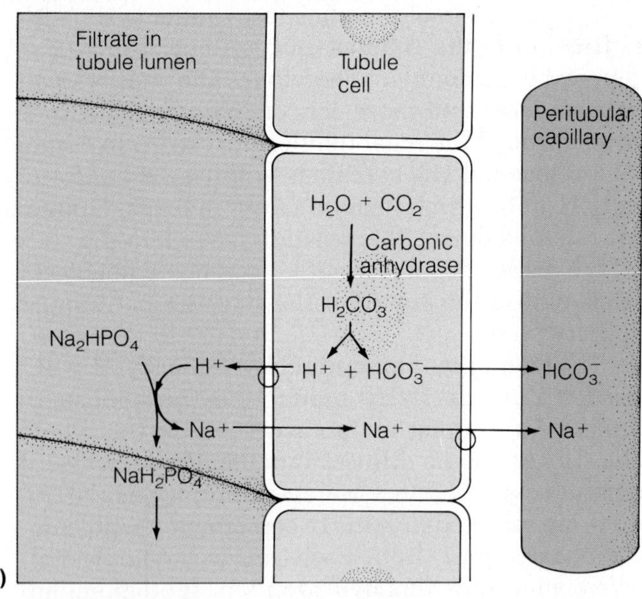

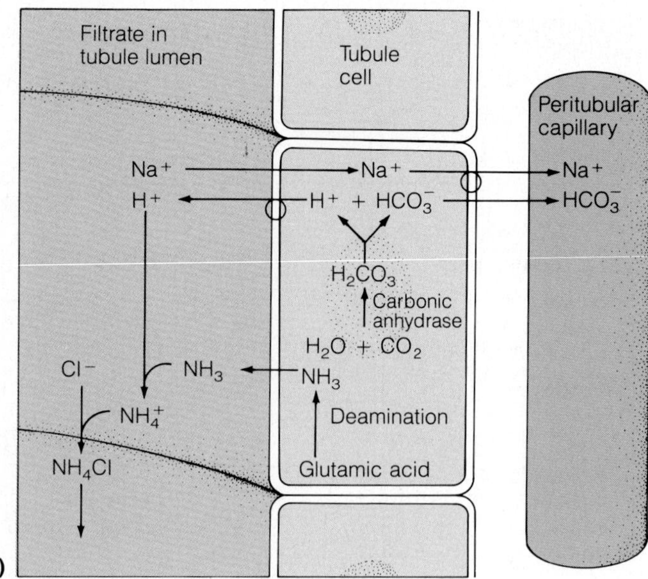

Figure 27.13 Buffering of hydrogen ions in urine and production of bicarbonate ions. (**a**) Buffering of secreted H^+ by the phosphate buffer system. Sodium ions released during the reaction are reabsorbed along with the generated bicarbonate ions into the peritubular capillary blood, which also contributes to a rise in blood pH. (**b**) Buffering of secreted H^+ by ammonia. Secreted H^+ binds to ammonia, forming ammonium ions, which take the place of sodium ions in the filtrate. Sodium is reabsorbed into the tubule cells and into the peritubular capillary blood along with produced bicarbonate ions. Note that active transport processes are indicated by a circle below the reaction arrow.

duced, causing the blood to become more alkaline. This is a common result of hyperventilation. Notice that respiratory acidosis is frequently associated with pathology, but respiratory alkalosis rarely is.

Metabolic Acidosis or Alkalosis

Metabolic acidosis or alkalosis includes all abnormalities of acid–base imbalance, *except* those caused by too much or too little carbon dioxide in the blood. Bicarbonate ion levels above or below the normal range of 22–28 mEq/L are indicative of a metabolic acid–base imbalance.

The most common causes of **metabolic acidosis** are ingestion of too much acid (drinking too much alcohol) and excessive loss of bicarbonate as might result from persistent diarrhea. It may also be caused by an accumulation of lactic acid during exercise or shock and by the ketosis that occurs in diabetic crisis or starvation. Kidney failure, in which excess H^+ is not eliminated in urine, is an infrequent cause. Metabolic acidosis is recognized by blood pH and bicarbonate levels below their homeostatic ranges.

Metabolic alkalosis, indicated by rising blood pH and bicarbonate levels, is much less common. Typical causes are vomiting of the acidic contents of the stomach (or loss of those secretions through gastrointestinal suctioning), intake of excess base (antacids, for example), and constipation, in which more than the usual amount of HCO_3^- is reabsorbed by the colon.

Effects of Acidosis and Alkalosis

The absolute blood pH limits for life are a low of 7.0 and a high of 7.8. When blood pH falls below 7.0, the central nervous system becomes so severely depressed that the person goes into coma and death soon follows. In contrast, alkalosis causes overexcitement of the nervous system. Characteristic signs include muscle tetany, extreme nervousness, and convulsions. Death often results from respiratory arrest.

Respiratory and Renal Compensation

When an acid–base imbalance arises due to inadequate functioning of one of the physiological buffer systems (the lungs or kidneys), the other system tries to compensate. The respiratory system will attempt to compensate for metabolic acid–base imbalances, and the kidneys (although much slower) will work to correct imbalances caused by respiratory disease. These **respiratory** and **renal compensations** can be recognized by changes in plasma P_{CO_2} and bicarbonate ion concentrations.

As a rule, changes in respiratory rate and depth are evident when the respiratory system is attempting to compensate for metabolic acid–base imbalances. In metabolic acidosis, the respiratory rate and depth are usually elevated—an indication that the respiratory centers are stimulated by the high H^+ levels. The blood pH is low (below 7.35), the bicarbonate level is low (below 22 mEq/L), and the P_{CO_2} falls below the normal

A CLOSER LOOK

Sleuthing: Using Blood Values to Determine the Cause of Acidosis or Alkalosis

Students, particularly nursing students, are often provided with blood values and asked to determine whether the patient is in acidosis or alkalosis, the cause of the condition (respiratory or metabolic), and whether or not the condition is being compensated. Such determinations are not nearly as difficult as they may appear if they are approached systematically. When attempting to analyze a person's acid–base balance, scrutinize the blood values in the following order:

1. Note the pH. This tells you whether the person is in acidosis or alkalosis, but does not tell you the cause.

2. Check the P_{CO_2} to see if this is the cause of the acid–base imbalance. Because the respiratory system is a fast-acting system, an excessively high or low P_{CO_2} may indicate that the condition is respiratory system–caused or that the respiratory system is compensating. For example, if the pH indicates acidosis and (a) the P_{CO_2} is over 45 mm Hg, the respiratory system *is the cause* of the problem and the condition is a respiratory acidosis; (b) the P_{CO_2} is below normal limits (below 35 mm Hg), the respiratory system is *not the cause but is compensating;* (c) the P_{CO_2} is within normal limits, the condition is *neither caused nor compensated* by the respiratory system.

3. Check the bicarbonate level. If step 2 proves that the respiratory system is not responsible for the imbalance, then the condition is metabolic and should be reflected in increased or decreased bicarbonate levels: Metabolic acidosis is indicated by HCO_3^- values below 22 mEq/L, and metabolic alkalosis by values over 28 mEq/L. Notice that whereas P_{CO_2} levels vary inversely with blood pH (P_{CO_2} rises as blood pH falls), HCO_3^- levels vary directly with blood pH (increased HCO_3^- results in increased pH).

Consider two examples of this approach:

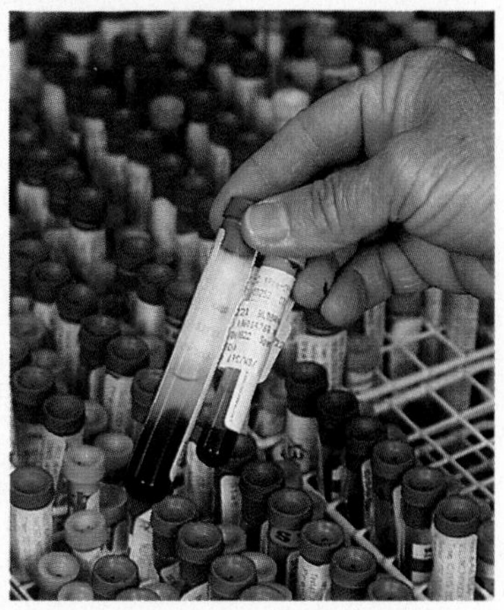

Problem 1

Blood values given: pH 7.5; P_{CO_2} 24 mm Hg; HCO_3^- 24 mEq/L.

Analysis:

1. The pH is elevated = alkalosis.

2. The P_{CO_2} is very low = the cause of the alkalosis.

3. The HCO_3^- value is within normal limits.

Conclusion: This is a respiratory alkalosis that is not compensated by renal mechanisms, as might occur during short-term hyperventilation.

Problem 2

Blood values given: pH 7.48; P_{CO_2} 46 mm Hg; HCO_3^- 32 mEq/L.

Analysis:

1. The pH is elevated = alkalosis.

2. The PCO_2 is elevated = the cause of *acidosis*, not alkalosis; thus, the respiratory system is compensating and is not the cause.

3. The HCO_3^- is elevated = the cause of the alkalosis.

Conclusion: This is a metabolic alkalosis being compensated by respiratory acidosis (retention of CO_2 to restore blood pH to the normal range).

A simple chart to help you in your future determinations is provided below.

	pH	HCO_3^-	PCO_2
Normal range in plasma	**7.35–7.45**	**22–28 mEq/L**	**35–45 mm Hg**
Acid–base disturbance			
Respiratory acidosis	↓	↑ if compensating	↑
Respiratory alkalosis	↑	↓ if compensating	↓
Metabolic acidosis	↓	↓	↓ if compensating
Metabolic alkalosis	↑	↑	↑ if compensating

Table 27.2 Causes and Consequences of Acid–Base Imbalances

Condition and hallmark	Possible causes; comments
Metabolic acidosis (HCO_3^- <22 mEq/L; pH <7.4)	*Severe diarrhea:* bicarbonate-rich intestinal (and pancreatic) secretions rushed through digestive tract before their solutes can be reabsorbed; bicarbonate ions are replaced by removal from blood *Renal disease:* failure of kidneys to rid body of acids formed by normal metabolic processes *Untreated diabetes mellitus:* lack of insulin or inability of tissue cells to respond to insulin, resulting in inability to use glucose; fats are used as primary energy fuel, and ketoacidosis occurs *Starvation:* lack of dietary nutrients for cellular fuels; body proteins and fat reserves are used for energy—both yield acidic metabolites as they are broken down for energy *Excess alcohol ingestion:* results in excess acids in blood *High ECF potassium concentrations:* potassium ions compete with H^+ for secretion in renal tubules; when ECF levels of K^+ are high, H^+ secretion is inhibited
Metabolic alkalosis (HCO_3^- >28 mEq/L; pH >7.4)	*Vomiting of chloride-containing gastric contents:* loss of stomach HCl requires that H^+ be withdrawn from blood to replace stomach acid; thus, H^+ decreases and HCO_3^- increases proportionately *Selected diuretics:* cause rapid reabsorption of Na^+ in distal tubule segments, leaving behind an excess of buffers in tubules that can combine with H^+ and thus remove excessive amounts *Ingestion of excessive amount of sodium bicarbonate:* bicarbonate moves easily into ECF, where it enhances natural alkaline reserve *Constipation:* prolonged retention of feces, resulting in increased amounts of HCO_3^- being reabsorbed. *Excess aldosterone (e.g., adrenal tumors):* promotes excessive reabsorption of Na^+, which pulls increased amount of H^+ into urine
Respiratory acidosis (P_{CO_2} >45 mm Hg; pH <7.4)	*Any condition that impairs gas exchange or lung ventilation (chronic bronchitis, cystic fibrosis, emphysema):* increased airway resistance and decreased expiratory air flow, leading to retention of carbon dioxide *Rapid, shallow breathing:* tidal volume markedly reduced *Narcotic or barbiturate overdose or injury to brain stem:* depression of respiratory centers, resulting in hypoventilation and respiratory arrest
Respiratory alkalosis (P_{CO_2} <35 mm Hg; pH >7.4)	*Direct cause is always hyperventilation:* hyperventilation in asthma, pneumonia, and at high altitude represents effort to raise P_{O_2} at the expense of excessive carbon dioxide excretion *Brain tumor or injury:* abnormality of respiratory controls

range of 35–45 mm Hg as the respiratory system "blows" off CO_2 to rid the blood of excess acid. (Conversely, in respiratory acidosis, the respiratory rate is typically depressed and *is the immediate cause of the acidosis.*)

In metabolic alkalosis, on the other hand, the respiratory compensation involves promoting slow, shallow breathing, which allows carbon dioxide to accumulate in the blood. A metabolic alkalosis that is being compensated by respiratory mechanisms is revealed by a high pH (over 7.45), elevated bicarbonate levels (over 28 mEq/L), and a rising P_{CO_2} (over 45 mm Hg).

When the imbalance is of respiratory origin, renal mechanisms are stepped up to correct the imbalance. For example, a hypoventilating individual will exhibit acidosis. When renal compensation is occurring, both the P_{CO_2} and the bicarbonate levels are high. The high P_{CO_2} reveals the cause of the acidosis; the rising HCO_3^- levels indicate that the kidneys are retaining bicarbonate to offset the acidosis. Conversely, a person with renal-compensated respiratory alkalosis will have a high blood pH and a low P_{CO_2}. Bicarbonate ion levels began to fall as the kidneys act to eliminate more bicarbonate from the body. ■

Developmental Aspects of Fluid, Electrolyte, and Acid–Base Balance

An embryo and a very young fetus are over 90% water, but as fetal development continues, solids accumulate and at birth an infant is 70% to 80% water. (The average value for adults is 58%.) Infants have proportionately more extracellular fluid than do adults and, consequently, a much higher sodium chloride content in relation to potassium, magnesium, and phosphates. The distribution of body water begins to change about two months after birth and achieves the adult pattern when the child is 2 years old. Although plasma electrolyte concentrations are not markedly different in infants and adults, potassium values are higher and magnesium, bicarbonate, and calcium levels are lower in the first few months of life than at any other time.

At puberty, sex differences in body water content become obvious as females develop relatively greater amounts of body fat.

Problems with fluid, electrolyte, and particularly acid–base balance are far more common in infancy than in later life, reflecting the following conditions:

1. The very low residual volume of the infant's lungs (equivalent to half that of adults relative to body weight). When respiration is altered, rapid and dramatic changes in the P_{CO_2} can result.

2. The high rate of fluid intake and output in infants (about seven times higher than that in adults). Infants may exchange fully half their ECF daily. Though infants have proportionately much more body water than adults, this does not protect them from excessive fluid shifts. Even slight alterations in fluid balance can cause serious problems. Further, although adults can live without water for about ten days, infants can survive for only three to four days.

3. The relatively high metabolic rate of infants (about twice that of adults). The higher metabolic rate leads to the generation of much larger amounts of metabolic wastes and acids that need to be excreted by the kidneys. This, along with buffer systems that are not yet fully effective, results in a tendency toward acidosis.

4. The high rate of insensible water loss in infants because of their larger surface area relative to body volume (about three times that of adults). Infants lose substantial amounts of water through their skin.

5. The inefficiency of the kidneys in infants (about half that of adults). At birth, the kidneys are functionally immature and only about half as proficient as adult kidneys at concentrating urine. The infant's kidneys are also notably inefficient in ridding the body of acids.

All these factors put newborns at risk for dehydration and acidosis, at least until their kidneys achieve reasonable proficiency by the end of the first month. Bouts of vomiting or prolonged diarrhea greatly amplify the risk. With such extreme variations in their internal environment, it is little wonder that mortality of premature infants in particular is so high.

In old age, there is often a decrease in total body water (largely from the intracellular compartment) because of a progressive decline in muscle mass and a rise in body fat. Few changes in solute concentrations of body fluids occur; hence, aging itself has little effect on the pH or osmotic pressure of the blood. However, the speed with which homeostasis of the internal environment is restored after the balance has been disrupted declines with age. Elders may also be unresponsive to thirst cues and thus are at risk for dehydration. Additionally, the aged are the most frequent prey of diseases that lead to severe fluid, electrolyte, or acid–base problems, such as congestive heart failure (and its attendant edema) and diabetes mellitus. Because most fluid, electrolyte, and acid–base imbalances occur when body water content is highest or lowest, the very young and the very old are the most frequent victims.

* * *

This chapter has examined the chemical and physiological mechanisms that provide, to the greatest extent possible, the optimal internal environment for survival. Although the kidneys are the "Rolex" among homeostatic organs in regulating water, electrolyte, and acid–base balance, they do not and cannot act alone. Rather, their activity is made possible by a host of hormones and enhanced both by blood-borne buffers, which give the kidneys time to react, and by the respiratory system, which shoulders a substantial responsibility for acid–base balance of the blood.

Related Clinical Terms

Acidemia (a″-sih-dē′-mē-uh) Arterial blood pH below 7.35.

Alkalemia (al″-kuh-lē′-mē-uh) Arterial blood pH above 7.45.

Renal tubular acidosis A metabolic acidosis resulting from impaired renal reabsorption of bicarbonate; the urine is alkaline.

Chapter Summary

BODY FLUIDS (pp. 888–991)

Body Water Content (p. 888)

1. Water accounts for 45% to 75% of body weight, depending on age, sex, and amount of body fat.

Fluid Compartments (p. 888)

2. About two-thirds (25 L) of body water is found within cells in the intracellular fluid (ICF) compartment; the balance (15 L) is in the extracellular fluid (ECF) compartment. The ECF includes plasma and interstitial fluid, considered to be in separate compartments.

Composition of Body Fluids (p. 889)

3. Solutes dissolved in body fluids include electrolytes and nonelectrolytes. Electrolyte concentration is expressed in mEq/L.

4. Plasma contains more proteins than does interstitial fluid; otherwise, extracellular fluids are similar. The most abundant ECF electrolytes are sodium, chloride, and bicarbonate ions.

5. Intracellular fluids contain large amounts of protein anions and potassium, magnesium, and phosphate ions.

Movement of Fluids Between Compartments (p. 890)

6. Fluid exchanges between compartments are regulated by osmotic and hydrostatic pressures: (a) Filtrate is forced out of the capillaries by hydrostatic pressure and pulled back in by osmotic pressure; (b) water moves freely between the ECF and the ICF by osmosis, but solute movements are restricted by size, charge, and dependence on active transport; (c) water flows always follow changes in ECF osmolality.

7. Plasma links the internal and external environments.

WATER BALANCE (pp. 891–894)

1. Sources of body water are ingested foods and fluids and metabolic water.

2. Water leaves the body via the lungs, skin, gastrointestinal tract, and kidneys.

Regulation of Water Intake: The Thirst Mechanism (p. 892)

3. Increased plasma osmolality (or decreased plasma volume) triggers the thirst mechanism, which is mediated by hypothalamic osmoreceptors. Thirst is inhibited first by distention of the gastrointestinal tract by ingested water and then by osmotic signals.

Regulation of Water Output (p. 892)

4. Obligatory water loss is unavoidable and includes insensible water losses from the lungs, the skin, in feces, and about 500 ml of urine output daily.

5. Beyond obligatory water loss, the volume of urinary output depends on water intake and loss via other routes and reflects the influence of antidiuretic hormone and aldosterone on the renal tubules.

Disorders of Water Balance (pp. 893–894)

6. Dehydration occurs when water loss exceeds water intake over time. It is evidenced by thirst, dry skin, and decreased urine output.

7. Hypotonic hydration occurs when body fluids are excessively diluted and cells become swollen by water entry. The most serious consequence is cerebral edema.

8. Edema is an abnormal accumulation of fluid in the interstitial space, which may impair blood circulation.

ELECTROLYTE BALANCE (pp. 894–901)

1. Most electrolytes (salts) are obtained from ingested foods and fluids. Salts, particularly NaCl, are often ingested in excess of need.

2. Electrolytes are lost in perspiration, feces, and urine. The kidneys are most important in regulating electrolyte balance.

The Central Role of Sodium in Fluid and Electrolyte Balance (p. 894)

3. Sodium salts are the most abundant solutes in ECF. They exert the bulk of ECF osmotic pressure and control water volume and distribution in the body.

4. Na^+ transport by the renal tubule cells is coupled to and helps regulate K^+, Cl^-, HCO_3^-, and H^+ concentrations in the ECF.

Regulation of Sodium Balance (pp. 895–898)

5. Sodium ion balance is linked to water balance and blood pressure regulation and involves both neural and hormonal controls.

6. Declining blood pressure and falling filtrate osmolality stimulate the juxtaglomerular cells to release renin. Renin, via angiotensin II, enhances systemic blood pressure and aldosterone release.

7. Cardiovascular system pressoreceptors sense changing arterial blood pressure, prompting changes in sympathetic vasomotor activity. Rising arterial pressure leads to vasodilation and enhanced Na^+ and water loss in urine. Falling arterial pressure promotes vasoconstriction and conserves Na^+ and water.

8. Atrial natriuretic factor, released by certain atrial cells in response to rising blood pressure (or blood volume), causes systemic vasodilation and inhibits renin, aldosterone, and ADH release. Hence, it enhances Na^+ and water excretion, reducing blood volume and blood pressure.

9. Estrogens and glucocorticoids increase renal retention of sodium. Progesterone promotes enhanced sodium and water excretion in urine.

Regulation of Potassium Balance (pp. 898–899)

10. From 85% to 90% of filtered potassium is reabsorbed in the proximal convoluted tubules; the balance is lost in urine.

11. The main thrust of renal regulation of K^+ is to excrete it. Potassium ion secretion is enhanced by (a) increased tubule cell K^+ content, (b) aldosterone, and (c) a low H^+ content in the blood or tubule filtrate.

Regulation of Calcium Balance (p. 899)

12. Calcium balance is regulated primarily by parathyroid hormone, which targets the bones, intestine, and kidneys, thereby enhancing blood Ca^{2+} levels.

13. Calcitonin accelerates the deposit of Ca^{2+} in bone and inhibits its release from bone matrix.

Regulation of Magnesium Balance (p. 899)

14. Magnesium levels are regulated by the influence of aldosterone on the kidneys. Aldosterone accelerates Mg^{2+} excretion.

Regulation of Anions (pp. 899–901)

15. When blood pH is normal or slightly high, chloride is the major anion accompanying sodium reabsorption. In acidosis, chloride is replaced by bicarbonate.

16. Most other anions appear to be regulated by renal overflow mechanisms.

ACID–BASE BALANCE (pp. 901–908)

1. Acids are proton (H^+) donors; bases are proton acceptors. Acids that dissociate completely in solution are strong acids; those that dissociate incompletely are weak acids.

2. The homeostatic pH range of arterial blood is 7.35 to 7.45. A higher pH represents alkalosis; a lower pH reflects acidosis.

3. Some acids enter the body in foods, but most are generated by the breakdown of phosphorus- and sulfur-containing proteins, incomplete oxidation of fats or glucose, and the loading and transport of carbon dioxide in the blood.

4. Acid–base balance is achieved by regulating the hydrogen ion concentration of body fluids.

Chemical Buffer Systems (pp. 902–903)

5. Chemical buffers are single or paired (a weak acid and its salt) sets of molecules that act rapidly to resist excessive shifts in pH by releasing or binding H^+.

6. Chemical buffers of the body include the bicarbonate, phosphate, protein, and ammonia buffer systems.

Respiratory System Regulation of Hydrogen Ion Concentration (pp. 903–904)

7. Respiratory regulation of acid–base balance of the blood utilizes the bicarbonate buffer system and the fact that carbon dioxide and water are in reversible equilibrium with H_2CO_3.

8. Acidosis activates the respiratory center to increase respiratory rate and depth, which eliminates more CO_2 and causes blood pH to rise. Alkalosis depresses the respiratory center, resulting in CO_2 retention and a fall in blood pH.

Renal Mechanisms of Acid–Base Balance (pp. 904–905)

9. The kidneys provide the major long-term mechanism for controlling acid–base balance by maintaining stable HCO_3^- levels in the ECF. Metabolic acids (organic acids other than carbonic acid) can be eliminated from the body only by the kidneys.

10. All hydrogen ions excreted in urine are secreted into the filtrate. For each H^+ secreted, a Na^+ is reabsorbed.

11. Secreted hydrogen ions come from the dissociation of carbonic acid generated within the tubule cells.

12. Tubule cells are impermeable to bicarbonate in the filtrate, but they can conserve bicarbonate ions generated within them by the dissociation of carbonic acid.

13. For secreted H^+ ions to be eliminated from the body in urine, they must be buffered. The major urine buffers are the phosphate and ammonia buffer systems.

Abnormalities of Acid–Base Balance (pp. 905–908)

14. Classification of acid–base imbalances as metabolic or respiratory indicates the cause of the acidosis or alkalosis.

15. Respiratory acidosis results from carbon dioxide retention; respiratory alkalosis occurs when carbon dioxide is eliminated faster than it is produced.

16. Metabolic acidosis occurs when fixed acids (lactic acid, ketone bodies, and others) accumulate in the blood or when bicarbonate is lost from the body; metabolic alkalosis occurs when bicarbonate levels are excessive.

17. Extremes of pH for life are 7.0 and 7.8.

18. Compensations occur when the respiratory system or kidneys act to reverse acid–base imbalances resulting from abnormal or inadequate functioning of the alternate system. Respiratory compensations involve changes in respiratory rate and depth. Renal compensations involve changes in HCO_3^- levels in the blood.

DEVELOPMENTAL ASPECTS OF FLUID, ELECTROLYTE, AND ACID–BASE BALANCE (pp. 908–909)

1. Factors that place infants at risk for dehydration and acidosis include low lung residual volume, high rate of fluid intake and output, high metabolic rate, relatively large body surface area, and functional immaturity of the kidneys at birth.

2. The elderly are at risk for dehydration because of their low percentage of body water and insensitivity to thirst cues. Diseases that promote fluid and acid–base imbalances (cardiovascular disease, diabetes mellitus, and others) are most common in the aged.

Review Questions

Multiple Choice/Matching

1. Body water content is greatest in (a) infants, (b) young adults, (c) elderly adults.

2. Potassium, magnesium, and phosphate ions are the predominant electrolytes in (a) plasma, (b) interstitial fluid, (c) intracellular fluid.

3. Sodium balance is regulated primarily by control of amount(s) (a) ingested, (b) excreted in urine, (c) lost in perspiration, (d) lost in feces.

4. Water balance is regulated primarily by control of amount(s) (use choices in question 3).

5 through 10. Answer questions 5 through 10 by choosing a response from the following: (a) ammonia, (b) bicarbonate, (c) calcium, (d) chloride, (e) hydrogen ions, (f) magnesium, (g) phosphate, (h) potassium, (i) sodium, (j) water.

5. Three ionic substances regulated (at least in part) by the influence of aldosterone on the kidney tubules.

6. Two substances regulated by parathyroid hormone.

7. Two substances secreted into the distal convoluted tubules in exchange for sodium ions.

8. Part of an important chemical buffer system in plasma.

9. When secreted by the tubule cells, it combines with H^+ to form a substance that cannot be reabsorbed and is, thus, excreted in urine.

10. Substance regulated by ADH's effects on the renal tubules.

11. Which of the following factors will enhance ADH release? (a) increase in ECF volume, (b) decrease in ECF volume, (c) decrease in ECF osmolality, (d) increase in ECF osmolality, (e) increase in blood pressure, (f) decrease in blood pressure.

12. The pH of blood varies directly with (a) HCO_3^-, (b) P_{CO_2}, (c) neither of the above.

13. In an individual with metabolic acidosis, a clue that the respiratory system is compensating is provided by (a) high blood bicarbonate levels, (b) low blood bicarbonate levels, (c) rapid, deep breathing, (d) slow, shallow breathing.

Short Answer Essay Questions

14. Name the body fluid compartments, noting their locations and the approximate fluid volume in each.

15. Describe the thirst mechanism, noting how it is triggered and terminated.

16. Explain why and how sodium and water balance are jointly regulated.

17. Describe the role of the respiratory system in controlling acid–base balance.

18. Explain how the chemical buffer systems resist changes in pH.

19. Explain the relationship of the following to renal secretion and excretion of hydrogen ions: (a) plasma carbon dioxide levels, (b) ammonia, and (c) sodium bicarbonate reabsorption.

20. Note several factors that place newborn babies at risk for acid–base imbalances.

Clinical Application Questions

21. For each of the following sets of blood values, name the acid–base imbalance (acidosis or alkalosis), determine its cause (metabolic or respiratory), decide whether the condition is being compensated, and cite at least one possible cause of the imbalance. *Problem 1:* pH 7.63; P_{CO_2} 19 mm Hg; HCO_3^- 19.5 mEq/L. *Problem 2:* pH 7.22; P_{CO_2} 30 mm Hg; HCO_3^- 12.0 mEq/L.

22. Explain how emphysema and congestive heart failure can lead to acid–base imbalance.

23. A 70-year-old woman is admitted to the hospital. Her history states that she has been suffering from diarrhea for three weeks. On admission, she complains of severe fatigue and muscle weakness. A blood chemistry study yields the following information: Na^+ 142 mEq/L; K^+ 1.5 mEq/L; Cl^- 92 mEq/L; P_{CO_2} 32 mm Hg. Which electrolytes are within normal limits? Which are so abnormal that the patient has a medical emergency? Which of the following represents the greatest danger to this patient? (a) a fall due to her muscular weakness, (b) edema, (c) cardiac arrhythmia and cardiac arrest.

The three chapters of this final unit survey the process and product of reproduction. Chapter 28 reviews the anatomy of the organs of the male and female reproductive systems and describes how they function and are regulated. It also considers how we become and act as sexual human beings. Chapter 29 examines the events of human development from conception to birth and considers the effects of pregnancy on the mother. The concluding chapter overviews essential concepts of genetics and the chromosomal interactions that make each of us unique.

Crystals of estrogen.

UNIT 5

CONTINUITY

CHAPTER 28

The Reproductive System

Chapter Outline and Student Objectives

1. Describe the common function of the male and female reproductive systems.

Anatomy of the Male Reproductive System (pp. 915–920)

2. Describe the structure and function of the testes, and explain the importance of their location in the scrotum.
3. Describe the location, structure, and function of the accessory ducts and glands of the male reproductive system.
4. Describe the structure of the penis, and note its role in the reproductive process.
5. Discuss the sources and functions of semen.

Physiology of the Male Reproductive System (pp. 920–929)

6. Define meiosis. Compare and contrast it to mitosis.
7. Outline the events of spermatogenesis.
8. Describe the phases of the male sexual response.
9. Discuss hormonal regulation of testicular function and the physiological effects of testosterone on male reproductive anatomy.

Anatomy of the Female Reproductive System (pp. 929–934)

10. Describe the location, structure, and function of the ovaries.
11. Describe the location, structure, and function of each of the organs of the female reproductive duct system.
12. Describe the anatomy of the female external genitalia.
13. Discuss the structure and function of the mammary glands.

Physiology of the Female Reproductive System (pp. 935–942)

14. Describe the process of oogenesis and compare it to spermatogenesis.
15. Describe the phases of the ovarian cycle, and relate them to events of oogenesis.
16. Describe the regulation of the ovarian and menstrual cycles.
17. Discuss the physiological effects of estrogen and progesterone.
18. Describe the phases of female sexual response.

Sexually Transmitted Diseases (pp. 942–943)

19. Describe the infectious agents and modes of transmission of gonorrhea, syphilis, chlamydia, and genital herpes.

Developmental Aspects of the Reproductive System: Chronology of Sexual Development (pp. 943–950)

20. Discuss the determination of genetic sex and prenatal development of male and female structures.
21. Note the significant events of puberty and menopause.

Preview of Selected Key Terms

Reproductive system Organ system that functions to produce offspring.

Gonad (gō′-nad) (*gono* = seed) Primary reproductive organ; i.e., the testis of the male or the ovary of the female.

Gamete (ga′-mēt) (*gam* = marriage) A sex cell; a sperm or ovum.

Seminiferous tubule Site of sperm formation in the testis.

Semen (sē′-min) Fluid mixture containing sperm and secretions of the male accessory reproductive glands.

Diploid (2*n*) chromosomal number (dih′-ployd) The number of chromosomes in most body cells; in humans, this number is 46.

Meiosis (mī-ō′-sis) (*meio* = less) Nuclear division process that reduces the chromosomal number by half and results in the formation of four **haploid (n)** cells; occurs only in certain reproductive organs.

Spermatogenesis (sper″-mah-tō-jeh′-nih-sis) (*spermat* = sperm; *genesis* = beginning) The process of sperm (male gamete) formation.

Oogenesis (ō″-ō-jeh′-nih-sis) (*oo* = ovum, egg) The process of ovum (female gamete) formation.

Follicle Ovarian structure consisting of a developing egg surrounded by one or more layers of follicle cells.

Ovulation Ejection of an immature egg (oocyte) from the female ovary.

Puberty Period of life when reproductive maturity is achieved.

Menopause Period of life when ovulation and menstruation cease, owing to hormonal changes.

Most organ systems of the body function almost continuously to maintain the well-being of the individual. The reproductive system, however, appears to "slumber" until puberty. Although male and female reproductive organs are quite different, their common purpose is to produce offspring.

The role of the male in the reproductive scheme is to manufacture male **gametes** (ga′-mēts), or sex cells, called spermatozoa and to deliver them to the female reproductive tract, where fertilization can occur. The mutually complementary role of the female is to produce female gametes, called ova or eggs. When these events are appropriately timed, a sperm and egg fuse to form a fertilized egg, the first cell of the new individual, from which all body cells will arise. The male and female reproductive systems are equal partners in events leading up to fertilization, but once fertilization has occurred, the female uterus provides a protective environment in which the embryo develops until birth. Additionally, the testes and ovaries secrete hormones that play vital roles both in the development and function of the reproductive organs and in sexual behavior and drives. The gonadal hormones also influence the growth and development of many other organs and tissues of the body.

Although the biological drive to reproduce is powerful in all animals, emotional, cultural, and social factors can enhance or restrain its expression in humans. Television, newspapers, and magazines sell products by using advertisements that play on our natural interest in sex. Why, for example, is a scantily clad young woman used to advertise automobile tires? This illogical emphasis on sex can make it difficult to view our reproductive abilities and drives as natural traits that enrich our lives.

Anatomy of the Male Reproductive System

The primary sex organs of the male are the **testes** (tes′-tēz), or **male gonads** (gō′-nadz), which produce sperm and male sex hormones (androgens). Males are males because they possess testes. All other male reproductive organs (scrotum, ducts, glands, and penis) are *accessory reproductive structures* that protect the sperm and aid in their delivery to the exterior of the body or to the female reproductive tract. The male reproductive system is shown in Figure 28.1.

The Scrotum and Testes

The paired, oval testes, or *testicles,* lie suspended in the saclike scrotum outside the abdominopelvic cavity. This seems a rather vulnerable position for a man's testes, which contain his entire genetic heritage. However, viable sperm cannot be produced at body temperature in humans. Hence, the scrotum, which provides a temperature about 3°C lower, is an essential adaptation.

The Scrotum

The **scrotum** (skrō′-tum) is a pouch of skin and superficial fascia that hangs at the root of the penis, just anterior to the anal opening (Figure 28.2). The superficial fascia forms an incomplete septum that divides the scrotum into right and left halves, one compartment for each testis. Covered with sparse hairs and with more heavily pigmented skin than elsewhere on the body, the scrotum varies in its appearance. For example, when it is cold or when a man is sexually aroused, the scrotum becomes shorter and heavily wrinkled as it is pulled up closer to the warmth of the body wall. When it is warm, the scrotal skin is flaccid and loose, and the testes hang lower. These changes in scrotal surface area help maintain a fairly constant intrascrotal temperature and reflect the activity of two sets of muscles. The **dartos** (dar′-tōs) **muscle,** a layer of smooth muscle in the superficial fascia, causes scrotal wrinkling. The **cremaster** (krē-mas′-ter) **muscles,** bands of skeletal muscle that arise from the internal oblique muscles of the trunk (and form part of the external spermatic fascia), elevate the testes.

The Testes

Each testis is approximately 4 cm (1.5 inches) long and 2.5 cm (1 inch) in diameter and is surrounded by two tunics. The outer tunic is the **tunica vaginalis**

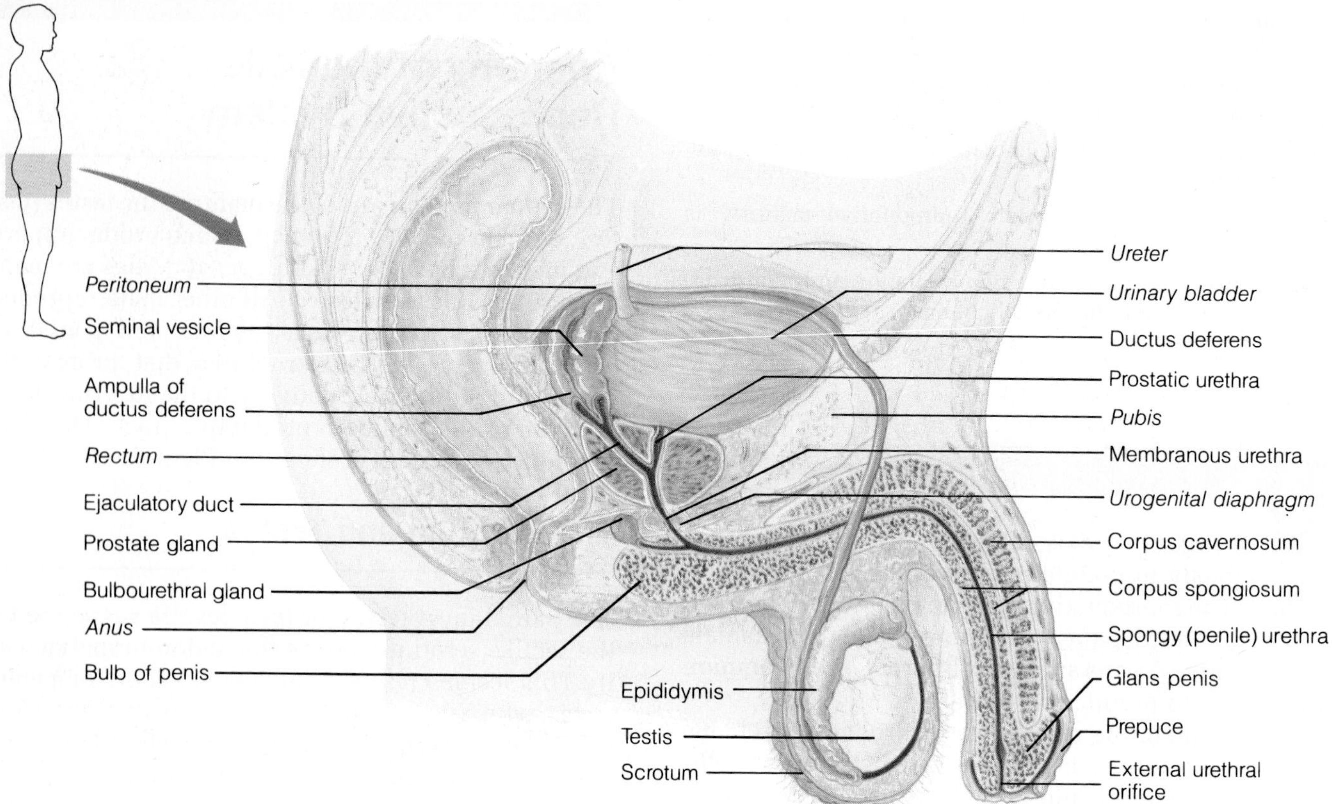

Figure 28.1 Reproductive organs of the male, sagittal view. A portion of the pubis of the os coxa is shown to illustrate the relationship of the ductus deferens to the bony pelvis.

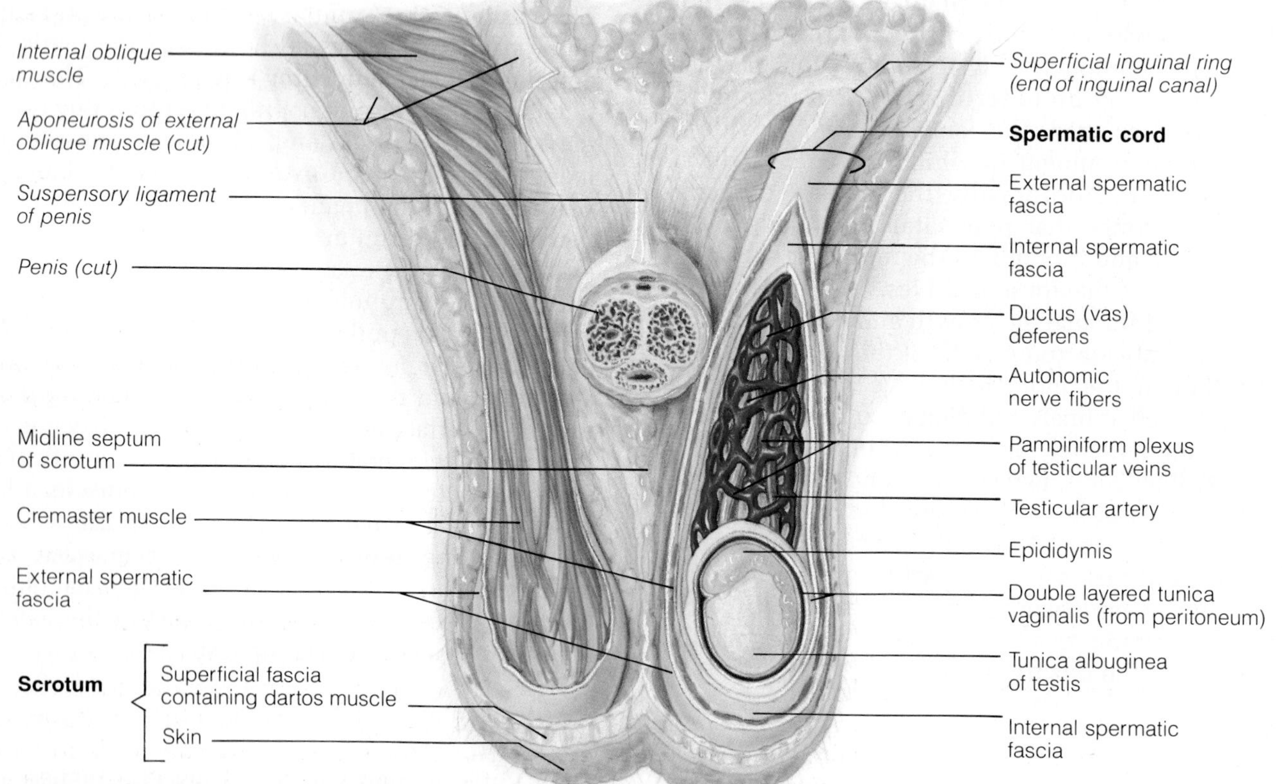

Figure 28.2 Relationships of the testis to the scrotum and spermatic cord. The scrotum has been opened and its anterior portion removed.

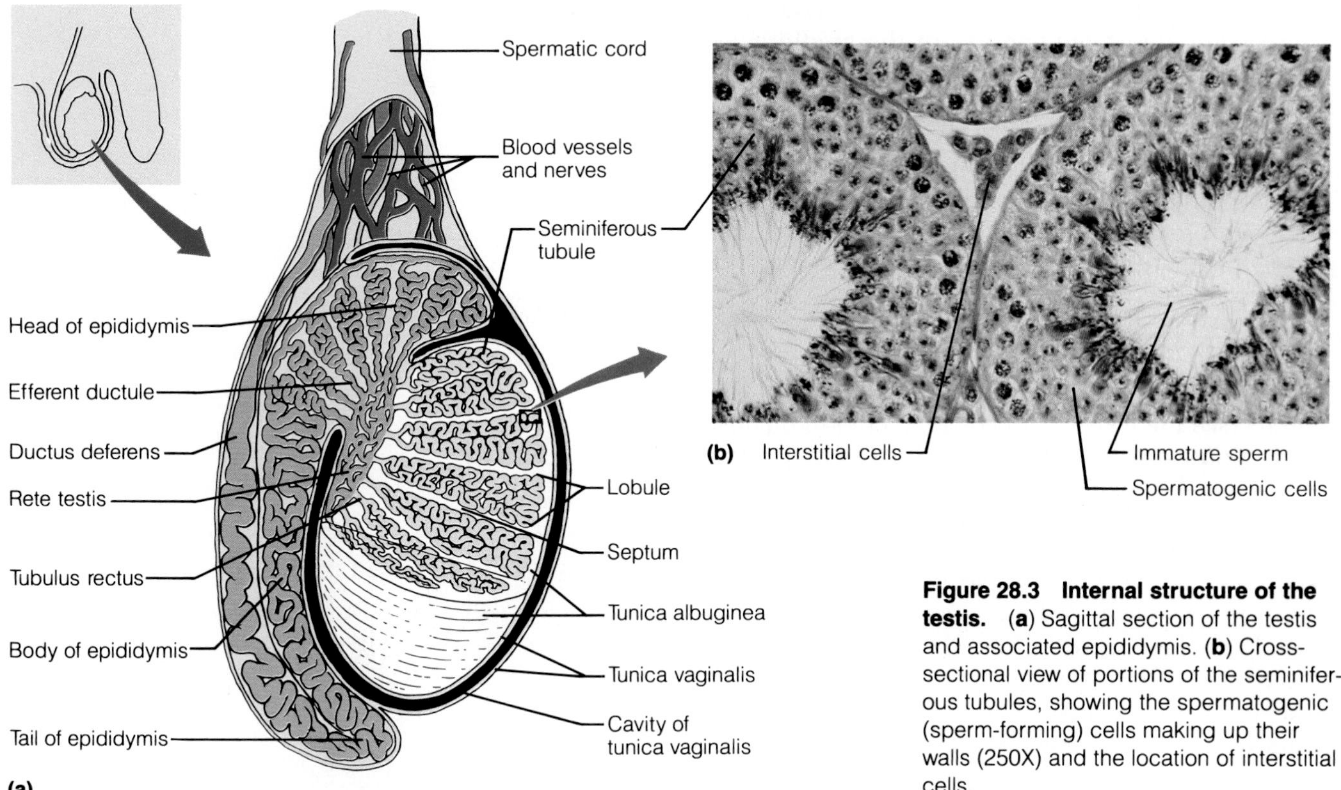

Figure 28.3 Internal structure of the testis. (**a**) Sagittal section of the testis and associated epididymis. (**b**) Cross-sectional view of portions of the seminiferous tubules, showing the spermatogenic (sperm-forming) cells making up their walls (250X) and the location of interstitial cells.

(va″-jih-na′-lis), derived from the peritoneum (see Figures 28.2 and 28.3). Deep to this is the **tunica albuginea** (al-byoo-jih′-nē-uh), literally "white coat," a fibrous connective tissue capsule. Septal extensions of the tunica albuginea divide the testis into 250 to 300 wedge-shaped compartments. or *lobules* (Figure 28.3). Each lobule contains one to four tightly coiled **seminiferous** (seh-mih-nih′-fer-us) **tubules**, the actual "sperm factories."

The seminiferous tubules of each lobule converge to form a *tubulus rectus,* a straight tubule that conveys sperm into the *rete* (rē′-tē) testis, a tubular network located on one side of the testis. Sperm leave the testis through the *efferent ductules* and enter the *epididymis* (eh″-pih-dih′-dih-mis), which hugs the external surface of the testis.

Lying in the soft connective tissue surrounding the seminiferous tubules are the **interstitial** (in″-ter-stih′-shul) **cells** or *Leydig cells,* functionally distinct cells that produce androgens (most importantly, *testosterone*). Thus, the sperm-producing and hormone-producing functions of the testis are carried out by completely different cell populations.

The testes receive their arterial blood supply from the *testicular arteries,* which arise from the abdominal aorta. The *testicular veins,* which drain the testes, form a vinelike network called the *pampiniform* (pam-pih′-nih-form) *plexus* around the testicular artery (see Figure 28.2). This plexus absorbs heat from the arterial blood, cooling it before it enters the testes, thereby providing an additional avenue for maintaining the testes at their homeostatic temperature. Scrotal structures are served by both divisions of the autonomic nervous system; associated sensory nerves transmit impulses that result in agonizing pain and nausea when the testes are hit forcefully. The nerve fibers are enclosed, along with the blood vessels and lymphatics, in a connective tissue sheath called the **spermatic cord** (see Figure 28.2).

The Duct System

The accessory organs forming the duct system, which transports sperm from the body, are the epididymis, the ductus deferens, and the urethra.

The Epididymis

The comma-shaped **epididymis** (see Figures 28.1 and 28.3a) is a coiled tube about 6 m (20 feet) long. Its *head,* which receives immature sperm from the efferent ductules, caps the superior aspect of the testis; its *body* and *tail* regions lie on the posterolateral aspect of the testis. Some of the cells of its pseudostratified epithelial mucosa exhibit long, nonmotile microvilli (stereocilia), which are thought to absorb fluid or to pass nutrients to the sperm in the lumen.

The immature, nearly nonmotile sperm that leave the testis are stored temporarily in the epididymis. During their transport along the tortuous course of the epididymis (a trip that takes about 20 days), the sperm become both motile and fertile. When a male is sexually stimulated and ejaculates, the smooth muscle in the walls of the epididymis contracts, expelling sperm from the tail portion into the next segment of the duct system, the ductus deferens.

The Ductus Deferens

The **ductus deferens** (duk'-tus deh'-fer-enz), or *vas deferens* (see Figure 28.1), is about 45 cm (18 inches) long. It runs upward from the epididymis through the inguinal canal into the pelvic cavity and is easily palpated as it passes anterior to the pubic bone. It then arches medially over the ureter and descends along the posterior aspect of the bladder. Its terminus expands to form the *ampulla* and joins with the duct of the seminal vesicle (a gland) to form the short **ejaculatory duct,** which passes through the prostate gland and merges with the urethra. The main function of the ductus deferens is to propel live sperm from their storage sites, the epididymis and distal part of the ductus deferens, into the urethra. At the moment of ejaculation, the thick layers of smooth muscle in its walls create peristaltic waves that rapidly squeeze the sperm forward.

As Figure 28.1 illustrates, part of the ductus deferens lies in the scrotal sac, from which it runs superiorly into the abdominopelvic cavity enclosed within the spermatic cord. Some men voluntarily opt to take full responsibility for birth control by having a *vasectomy* (va-sek'-tuh-mē). In this relatively minor operation, the physician makes a small incision into the scrotum and then cuts through or cauterizes the ductus deferens. Although sperm continue to be produced for the next several years, they can no longer reach the body exterior. Eventually, they deteriorate and are reabsorbed.

The Urethra

The **urethra,** the terminal portion of the male duct system, serves both the urinary and reproductive systems (see Figures 28.1 and 28.4); it conveys urine and semen (at different times) to the tip of the penis. Its three continuous regions are the *prostatic urethra* (prah-sta'-tik yoo-rē'-thruh), the portion immediately draining the bladder and surrounded by the prostate gland; the *membranous urethra,* the portion that passes through the urogenital diaphragm; and the *spongy (penile) urethra,* which runs through the penis and opens to the exterior at the *external urethral orifice.* The spongy urethra is about 15 cm (6 inches) long and accounts for 75% of urethral length.

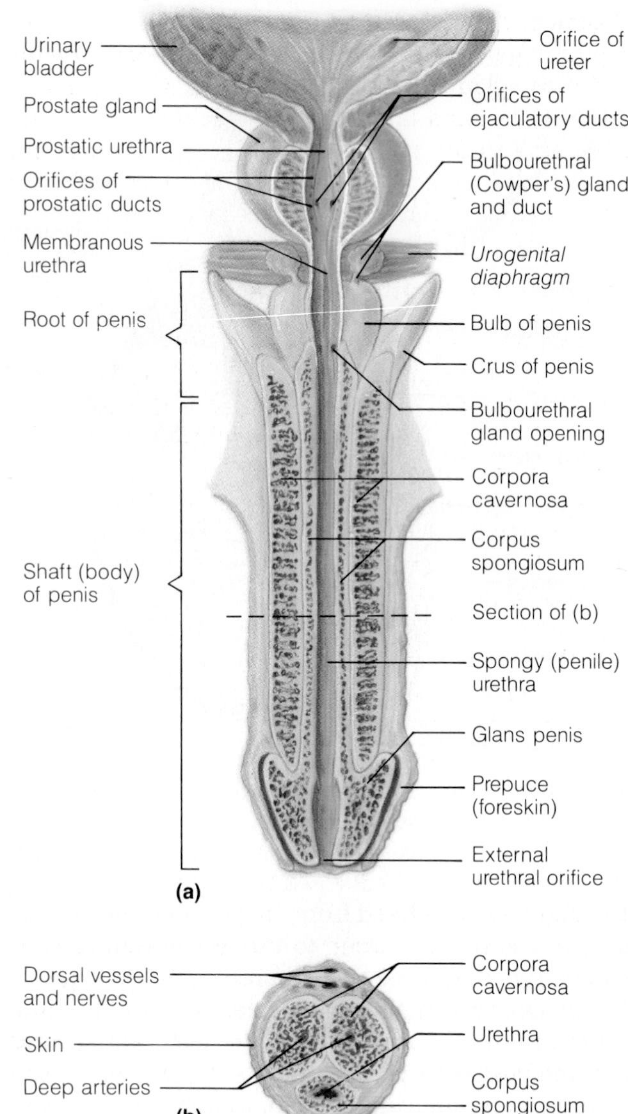

Figure 28.4 Structure of the penis. (**a**) Longitudinal (coronal) section of the penis. (**b**) Transverse section of the penis.

Accessory Glands

The accessory glands include the paired seminal vesicles and bulbourethral glands and the single prostate gland (Figure 28.1). These glands produce the bulk of semen (sperm plus secretions of accessory glands and ducts).

The Seminal Vesicles

The **seminal** (seh'-mih-nul) **vesicles,** located at the base of the bladder, produce about 60% of the fluid volume of semen. Their secretion is a yellowish viscous fluid containing fructose sugar, ascorbic acid, amino acids, and prostaglandins. As noted, the duct of each seminal vesicle joins that of the ductus deferens on the same side to form the ejaculatory duct. Sperm and seminal fluid mix in the ejaculatory duct and enter the prostatic urethra together during ejaculation.

The Prostate Gland

The **prostate** (prah′-stāt) **gland** is a single gland about the size of a walnut (see Figures 28.1 and 28.4); it encircles the upper part of the urethra just below the bladder. Enclosed by a thick connective tissue capsule, it is made up of 20 to 30 tubuloalveolar glands surrounded by smooth muscle and a fibroelastic stroma. The prostatic gland secretion, accounting for up to one-third of the semen volume, is a milky alkaline fluid that plays a role in activating sperm. It enters the prostatic urethra via several ducts when the prostatic smooth muscle contracts during ejaculation. Since the prostate is located immediately anterior to the rectum, its size and texture can be palpated by digital (finger) examination through the anterior rectal wall.

The prostate gland has a reputation as a health destroyer (perhaps reflected in the common mispronunciation "prostrate"). Hypertrophy of the prostate gland, which affects nearly every elderly male, strangles the urethra. This troublesome condition makes urination difficult and enhances the risk of bladder infections (cystitis) and kidney damage. Treatment is usually surgical. *Prostatitis* (prah″-stuh-tī′-tis), inflammation of the prostate, is the single most common reason for a man to consult a urologist, and prostatic cancer is the third most prevalent type of cancer in men. As a rule, prostatic cancer is a slow-growing, hidden condition, but it can also be a swift and deadly killer. ■

The Bulbourethral Glands

The **bulbourethral** (bul-bō-yoo-rē′-thrul) **glands,** or *Cowper's glands,* are tiny pea-sized glands situated inferior to the prostate gland (see Figures 28.1 and 28.4). They produce a thick, clear mucus, which drains into the spongy urethra. This secretion is released prior to ejaculation when a male first becomes sexually excited. It is believed to neutralize traces of acidic urine in the urethra and to serve as a lubricant during sexual intercourse.

The Penis

The **penis** is a copulatory organ, designed to deliver sperm into the female reproductive tract (see Figure 28.4). The penis and scrotum (described earlier) make up the external reproductive structures, or **external genitalia** (jeh-nih-tā′-lē-uh), of the male. The penis consists of an attached *root* and a free *shaft* or *body* that ends in an enlarged tip, the **glans penis.** The skin covering the penis is loose, and it slides distally to form a cuff of skin called the **prepuce** (prē′-pyoos), or **foreskin,** around the proximal end of the glans. Frequently, the foreskin is surgically removed shortly after birth, a procedure called *circumcision.*

Internally, the penis contains the spongy urethra and three columns of erectile tissue. *Erectile tissue* is a spongy network of connective tissue and smooth muscle riddled with vascular spaces. During sexual excitement, the vascular spaces fill with blood, causing the penis to enlarge and become rigid. This event, called *erection,* helps the penis serve as a penetrating organ. The slender midventral column of erectile tissue, the **corpus spongiosum** (spun-jē-ō′-sum), literally "spongy body", surrounds the urethra and expands distally to form the glans. Its proximal end is also enlarged, forming the portion of the root called the **bulb of the penis.** The bulb is covered externally by the sheetlike bulbospongiosus muscle and is secured to the urogenital diaphragm. The paired dorsal columns are called the **corpora cavernosa** (kor′-per-uh ka-ver-nō′-suh), meaning "cavernous bodies." Their proximal ends form the **crura** (kroo′-ruh) **of the penis** (singular, crus), each of which is surrounded by an ischiocavernosus muscle and anchored to the pubic arch of the bony pelvis.

The Male Perineum

The diamond-shaped region located between the pubic symphysis anteriorly, the coccyx posteriorly, and the ischial tuberosities laterally is referred to as the male *perineum.* The floor of the male perineum is formed by muscles described in Chapter 10 (pp. 300–301).

Semen

Semen (sē′-min) is a milky white, somewhat sticky fluid mixture of sperm and accessory gland secretions. The liquid provides a transport medium and nutrients and contains chemicals that protect the sperm and facilitate their movement. Mature sperm cells are streamlined cellular "missiles" containing very little cytoplasm or stored nutrients; the fructose in the seminal vesicle secretion provides essentially all of their energy fuel. The prostaglandins in semen are thought to decrease the viscosity of mucus guarding the entry (cervix) of the uterus and to cause reverse peristalsis in the uterus, facilitating the movement of sperm through the female reproductive tract. The presence of the hormone *relaxin* and certain enzymes in semen enhance sperm motility. The relative alkalinity of prostatic fluid causes semen as a whole to be slightly basic (pH 7.2–7.6) and helps neutralize the acid environment (pH 3.5–4) of the female's vagina, protecting the delicate sperm and enhancing their motility. Sperm are very sluggish under acidic conditions (below pH 6). Semen also contains a chemical called *seminalplasmin,* which has bacteriostatic properties.

Not the least of semen's functions is its sperm-diluting role; without such dilution, sperm motility is severely impaired. The amount of semen propelled out of the male duct system during ejaculation is relatively small, only 2–6 ml, but there are between 50 and 100 million sperm in each milliliter.

Male infertility may be caused by anatomical obstructions, hormonal imbalances, and many other factors. One of the first series of tests done when a couple has been unable to conceive is *semen analysis.* Factors analyzed that may give clues to particular abnormalities include sperm count, motility and morphology (shape and maturity), and semen volume, pH, and fructose content. For example, a low fructose content may indicate congenital absence or obstruction of the seminal vesicle ducts. A low sperm count accompanied by a high percentage of immature sperm may hint that a man has a *varicocele* (var'-ih-kō'-sēl), a condition that hinders drainage of the testicular vein, resulting in an elevated temperature in the scrotum that interferes with normal sperm development. A sperm count lower than 20 million per milliliter makes impregnation improbable. ■

Physiology of the Male Reproductive System

Spermatogenesis

Spermatogenesis (sper"-mah-tō-jeh'-nih-sis) is the sequence of events in the seminiferous tubules of the testes that leads to the production of male gametes, or sperm. The process begins during puberty, around the age of 14 years in males, and continues throughout life. Every day, a healthy adult male makes several hundred million sperm, and the cellular divisions that support this prodigious output of sperm are rapid and continuous. It seems that nature has made sure that the human species will not be endangered for lack of sperm.

Meiosis

Both spermatogenesis and its counterpart in the female, oogenesis, involve **meiosis** (mī-ō'-sis), the unique kind of nuclear division that, for the most part, occurs only within the gonads. Recall that *mitosis* (the process by which other body cells divide) distributes replicated chromosomes equally between the two daughter cells (see Chapter 3). Consequently, each daughter cell receives a set of chromosomes identical to that of the mother cell. Meiosis, on the other hand, consists of two consecutive nuclear divisions without a second replication of the chromosomal material between divisions. As a result, four daughter cells are produced instead of two, and each has half as many chromosomes as the mother cell or other body cells. (The root *mei* in *meiosis* means "less.") Mitosis and meiosis are compared in Figure 28.5.

The normal chromosome number in most body cells (the fertilized egg and all body cells derived from it by mitotic cell division) is referred to as the **diploid** (dih'-ployd) or **2*n* chromosomal number** of the organism. In humans, this number is 46, and such diploid cells contain 23 pairs of similar chromosomes called **homologous** (hō-mah'-luh-gus) **chromosomes** or **homologues.** One member of each pair is from the male parent (the paternal chromosome); the other is from the female parent (the maternal chromosome). The two homologues of each chromosome pair look alike and carry genes that code for the same traits, though not necessarily for identical expression of those traits. Consider, for example, the homologous genes controlling the expression of freckles; the paternal gene might code for an ample sprinkling of freckles and the maternal gene for no freckles. (In Chapter 30, we will consider how maternal and paternal genes interact to produce our visible traits.)

In contrast, the number of chromosomes present in human gametes is 23, referred to as the **haploid** (ha'-ployd) or ***n* chromosomal number,** and gametes contain only one member of each homologous pair. The importance of reducing the gamete chromosomal number by half, from $2n$ to n, is clear; when sperm and egg fuse, they form a fertilized egg, or *zygote,* once again containing the diploid chromosomal number characteristic of human cells. Moreover, if meiosis did not occur, the chromosomal number would double with each new generation, resulting in daughter cells packed with redundant genetic material, which would cause lethal developmental problems.

The two nuclear divisions of meiosis, called *meiosis I* and *meiosis II,* are divided into phases for convenience. Although these phases are given the same names as those of mitosis (prophase, metaphase, anaphase, and telophase), the events of meiosis I are quite different from those of mitosis, as detailed below and in Figure 28.6.

Before an ordinary mitotic division, all the chromosomes are replicated, and throughout prophase and during metaphase alignment, the identical copies remain together as *sister chromatids* connected by a centromere. Then, at anaphase, the centromeres split and the sister chromatids are separated from each other

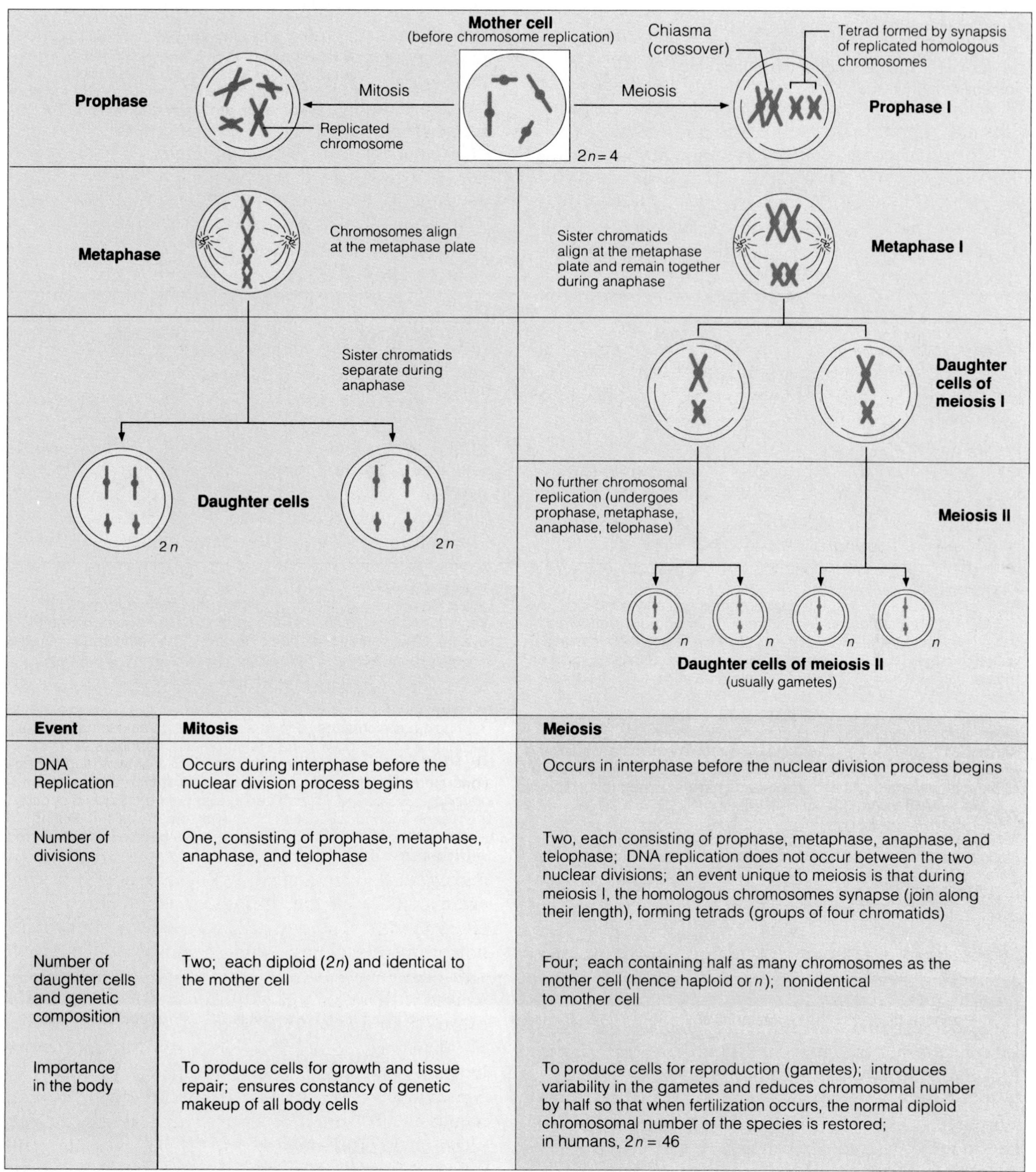

Event	Mitosis	Meiosis
DNA Replication	Occurs during interphase before the nuclear division process begins	Occurs in interphase before the nuclear division process begins
Number of divisions	One, consisting of prophase, metaphase, anaphase, and telophase	Two, each consisting of prophase, metaphase, anaphase, and telophase; DNA replication does not occur between the two nuclear divisions; an event unique to meiosis is that during meiosis I, the homologous chromosomes synapse (join along their length), forming tetrads (groups of four chromatids)
Number of daughter cells and genetic composition	Two; each diploid ($2n$) and identical to the mother cell	Four; each containing half as many chromosomes as the mother cell (hence haploid or n); nonidentical to mother cell
Importance in the body	To produce cells for growth and tissue repair; ensures constancy of genetic makeup of all body cells	To produce cells for reproduction (gametes); introduces variability in the gametes and reduces chromosomal number by half so that when fertilization occurs, the normal diploid chromosomal number of the species is restored; in humans, $2n = 46$

Figure 28.5 Comparison of mitosis and meiosis in a mother cell with a diploid number ($2n$) of 4. Mitosis is shown on the left, meiosis on the right. Not all phases of mitosis and meiosis are shown.

Figure 28.6 Meiotic cell division. This series of diagrams shows meiotic cell division for an animal cell with a diploid number (2*n*) of 4. The behavior of the chromosomes is emphasized.

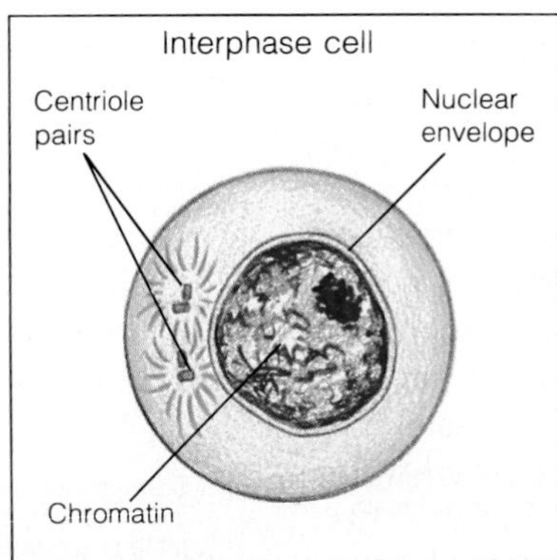

Interphase events
As in mitosis, meiosis is preceded by events occurring during interphase that lead to DNA replication and other preparations needed for the division process. Just before meiosis begins, the replicated chromatids, held together by centromeres, are ready and waiting.

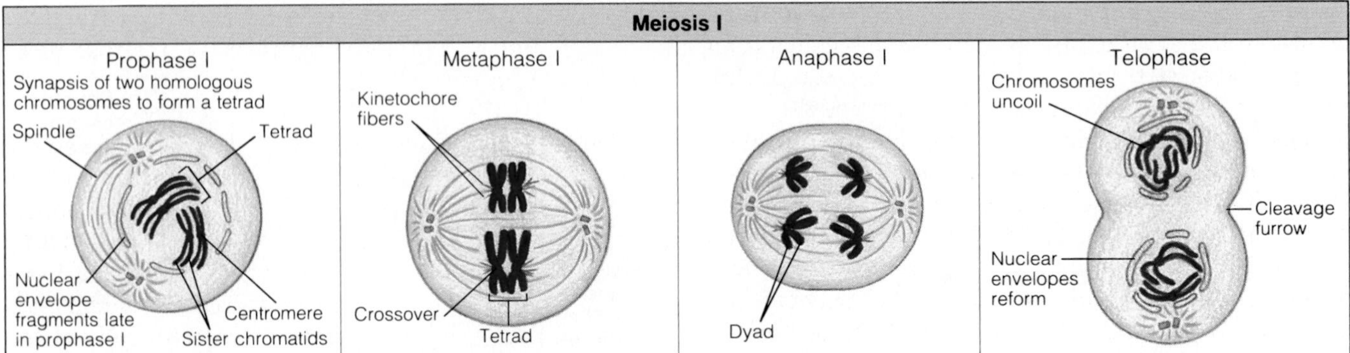

Prophase I
As in prophase of mitosis, the chromosomes coil and condense, the nuclear membrane and nucleolus break down and disappear, and the spindle is formed. However, a unique event not seen in mitosis, called synapsis, occurs in prophase I of meiosis. Synapsis involves the coming together of homologous chromosomes, to form tetrads, little packets of four chromatids. While in synapsis, the "arms" of adjacent homologous chromatids become wrapped around each other, forming several points of crossover or chiasmata. Generally speaking, the longer the chromatids, the more chiasmata are formed. Prophase I is the longest period of meiosis, accounting for about 90% of the total period; by its end, the tetrads have attached to the spindle and are moving toward the spindle equator.

Metaphase I
During metaphase I, the tetrads align on the spindle equator in preparation for anaphase.

Anaphase I
Unlike the anaphase events of mitosis, the centromeres do not break during anaphase I of meiosis, and so the sister chromatids (dyads) remain firmly attached. However, the homologous chromosomes do separate from each other, and the dyads are moved toward opposite poles of the cell.

Telophase I
The nuclear membranes re-form around the chromosomal masses, the spindle breaks down, and the chromatin reappears as telophase and cytokinesis are completed, so that two daughter cells are formed. The daughter cells (now haploid) then enter a second interphase-like period called interkinesis, before meiosis II occurs. There is no second replication of DNA before meiosis II. During human spermatogenesis, the daughter cells remain interconnected by cytoplasmic extensions.

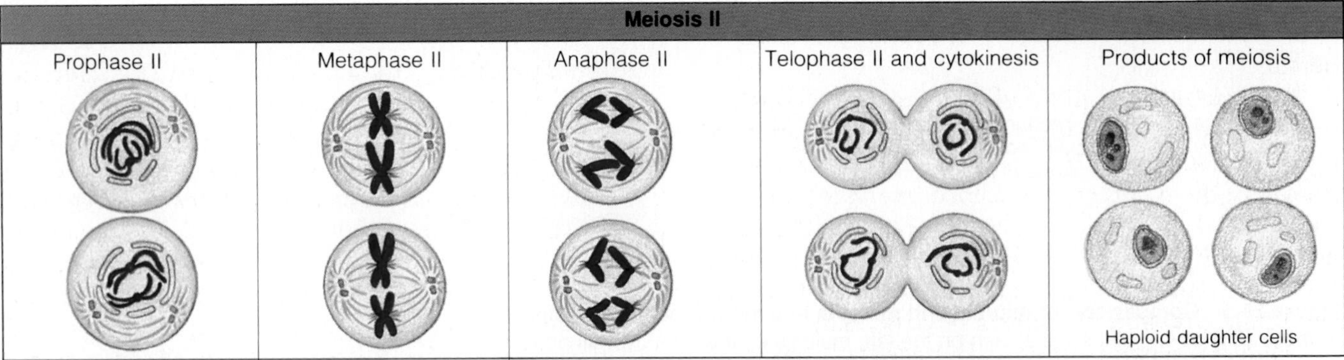

Meiosis II begins with the two daughter cells produced by meiosis I. All of the normal events of prophase recur. The nucleus and nucleolus dissipate, the chromatids recoil, and the spindle re-forms. During metaphase II, the dyads (left over from meiosis I) align on the spindle equator, and during anaphase II, their centromeres break and the individual chromatids are distributed to opposite ends of the cell. During telophase II, the spindles break down, nuclear membranes and nucleoli reappear, and cytokinesis is completed. Since each of the (two) daughter cells of meiosis I has now undergone meiosis II, the product of meiosis is four haploid daughter cells.

so that each daughter cell inherits a copy of *every* chromosome possessed by the mother cell (see Figure 28.5).

The requirement that gametes receive only one member of each pair of homologues places some extra demands on the meiotic process. As in mitosis, chromosomal replication precedes the onset of meiosis. But in prophase of meiosis I, an event never seen in mitosis occurs. The replicated chromosomes seek out their homologous partners and become aligned with them along their entire length (see Figure 28.6). This intimate pairing of the homologous chromosomes is called **synapsis,** and it is so precise that analogous genes lie side by side. Since each chromosome is composed of two sister chromatids, what we see during this phase is little groups of four chromatids called **tetrads.** During synapsis, a second unique event called crossover occurs. **Crossovers,** interactions within each tetrad of one maternal and one paternal chromatid, are formed as the free ends of the chromatids wrap around each other at one or more points. Crossover allows the exchange of genetic material between the maternal and paternal chromosomes. (Neither tetrads nor crossovers are seen during mitosis.)

During metaphase I, the tetrads line up at the spindle equator. This alignment is random; that is, either the paternal or maternal chromosome can be on a given side of the equator.

During anaphase I, the sister chromatids representing each homologue behave as a unit—almost as if replication had not occurred—and the *homologous chromosomes* (each still composed of two joined sister chromatids) are distributed to opposite ends of the cell. Thus, when meiosis I is completed, the following conditions exist: Each daughter cell has (1) *two* copies of one chromosome of each homologous pair (either the paternal or maternal) and none of the other, and (2) a diploid amount of DNA but a *haploid* chromosomal number because the still-united sister chromatids are considered to be a single chromosome. Since meiosis I reduces the chromosome number from $2n$ to n, it is sometimes called the *reduction division* of meiosis.

The second meiotic division, meiosis II, mirrors mitosis in every way, except that the chromosomes are not replicated before it begins. Instead, the chromatids present in the two daughter cells of meiosis I are simply parceled out among four cells. Because the chromatids are distributed equally to the daughter cells (as in mitosis), meiosis II is sometimes referred to as the *equational division* of meiosis (Figure 28.6).

Meiosis accomplishes two important tasks: It reduces the chromosomal number by half and introduces genetic variability. The random orientation of the homologous pairs during meiosis I provides tremendous variability in the resulting gametes by mixing up, or scrambling, genetic characteristics derived from the two parents in different combinations. Variability is increased further by the crossovers that occur between homologous chromosomes. As a result, no two gametes are exactly alike, and all are different from the original mother cells. Genetic variability arising from both these features is discussed in more detail in Chapter 30.

Summary of Events in the Seminiferous Tubules

Now that we have described meiosis, let us turn to the specific events of spermatogenesis, which take place in the seminiferous tubules.

Cell Divisions. If we scrutinize a histological section of an adult testis, we can see that the majority of the cells making up the walls of the seminiferous tubules are in various stages of cell division (Figure 28.7). Each seminiferous tubule is surrounded by a basement membrane, and the cells in direct contact with this membrane are undifferentiated stem cells called **spermatogonia** (sper″-mah-tō-gō′-nē-uh). The spermatogonia, literally "sperm seed," divide more or less continuously *by mitosis,* and until puberty, all the daughter cells become new spermatogonia. Spermatogenesis begins during puberty, and from then on, each mitotic division of a spermatogonium results in two distinctive daughter cells (see Figure 28.7). A *type A* daughter cell remains at the basement membrane to keep the germ cell line going; the other cell, *type B,* gets pushed toward the lumen, where it becomes a **primary spermatocyte** (sper-ma′-tō-sīt) destined to produce four sperm. The primary spermatocyte undergoes meiosis I, forming two smaller haploid cells called **secondary spermatocytes.** The secondary spermatocytes continue on rapidly into meiosis II, and their daughter cells, called **spermatids** (sper′-mah-tidz), can be seen as small round cells with large spherical nuclei closer to the lumen of the tubule.

Spermiogenesis. Although each spermatid has the correct chromosomal number for fertilization (n), it is a nonmotile cell. It still must undergo a streamlining process called **spermiogenesis** (Figure 28.8a), during which most of its superfluous cytoplasmic "baggage" is sloughed off and a tail is fashioned. The resulting sperm, or **spermatozoon** (sper′-mah-tō-zō′-on), has three major regions: a head, a midpiece, and a tail, which correspond roughly to genetic, metabolic, and locomotor regions, respectively. The head of the sperm consists almost entirely of the flattened spermatid nucleus, which contains the DNA. Adhering to the nucleus is a helmetlike **acrosome,** produced by the

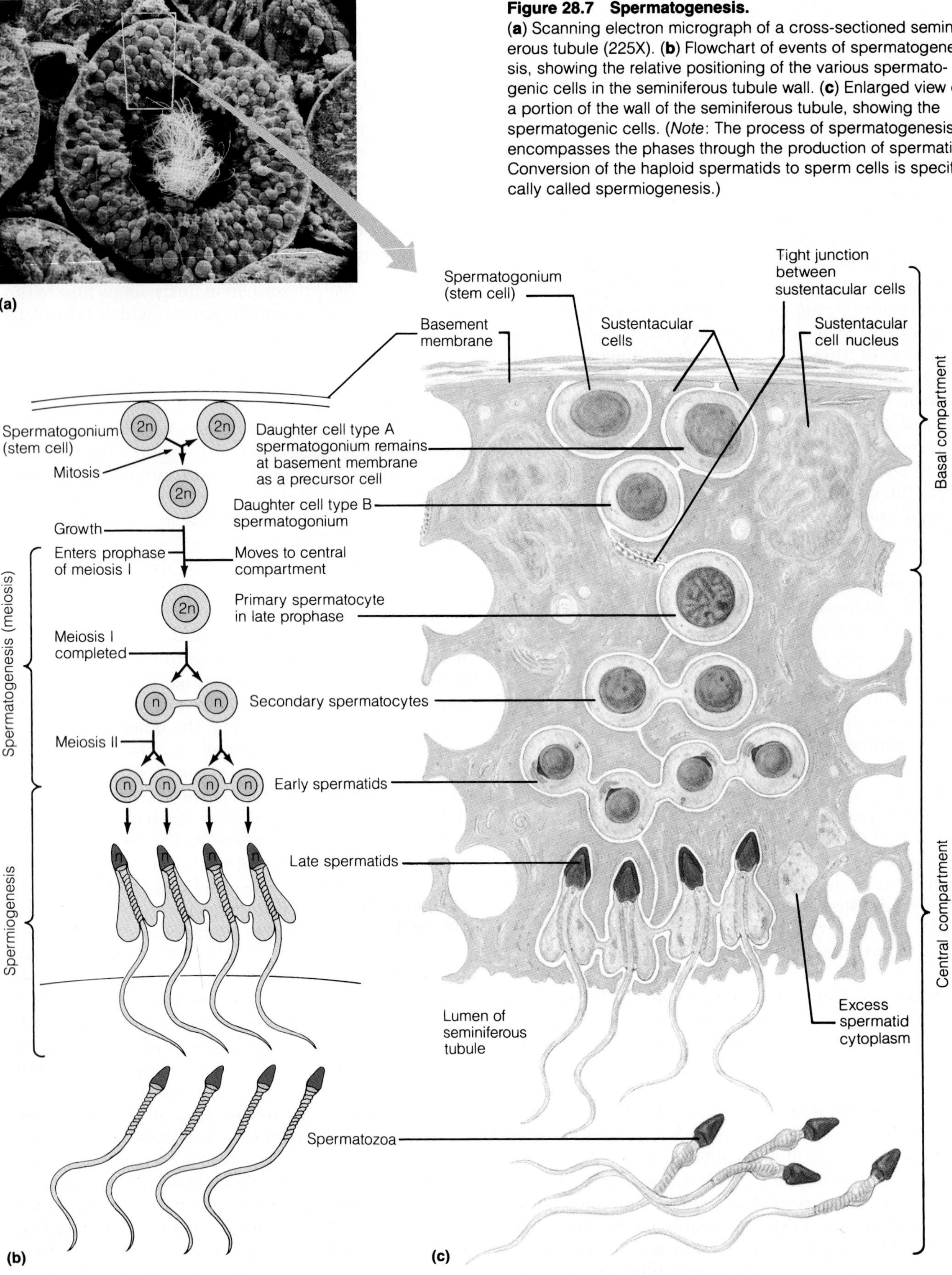

Figure 28.7 Spermatogenesis.
(**a**) Scanning electron micrograph of a cross-sectioned seminiferous tubule (225X). (**b**) Flowchart of events of spermatogenesis, showing the relative positioning of the various spermatogenic cells in the seminiferous tubule wall. (**c**) Enlarged view of a portion of the wall of the seminiferous tubule, showing the spermatogenic cells. (*Note*: The process of spermatogenesis encompasses the phases through the production of spermatids. Conversion of the haploid spermatids to sperm cells is specifically called spermiogenesis.)

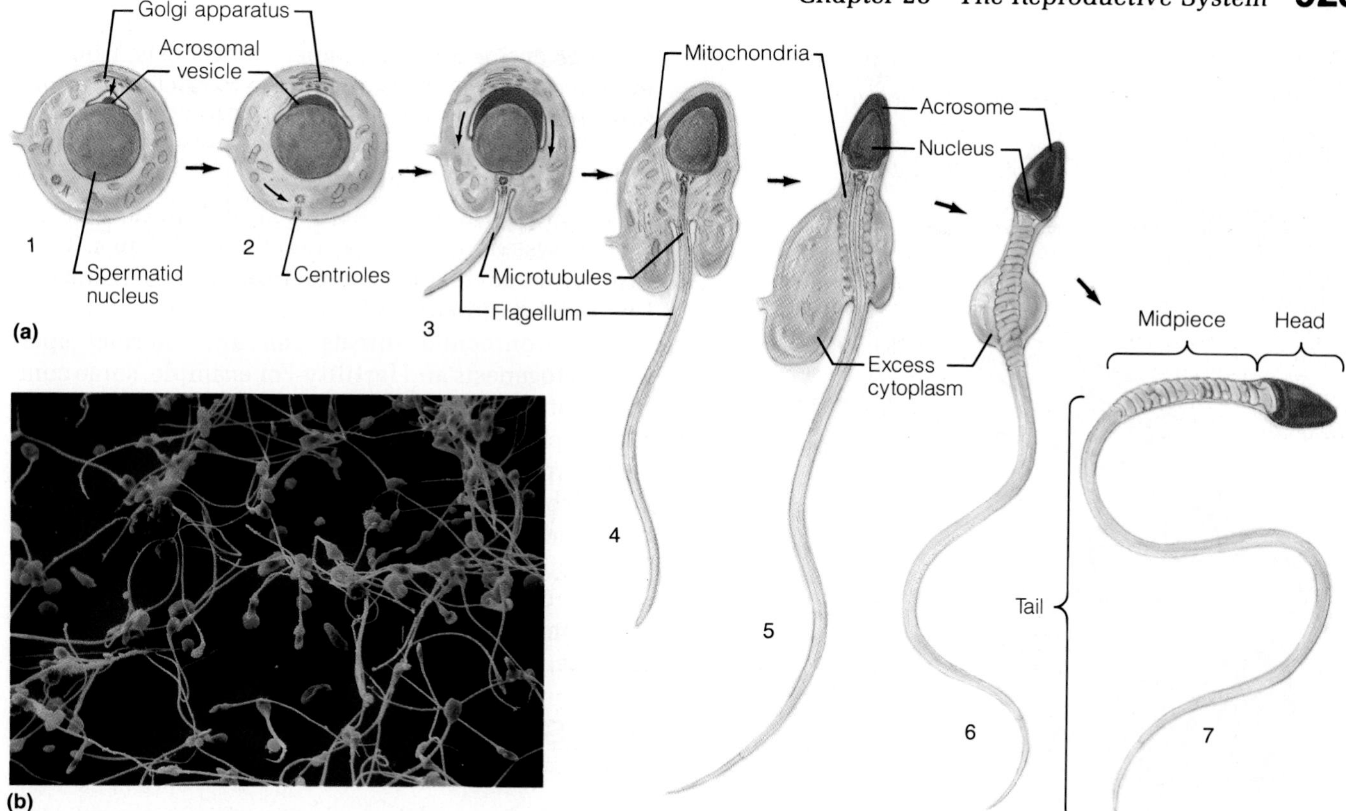

Figure 28.8 Spermiogenesis: transformation of a spermatid into a functional sperm. (**a**) The stepwise process of spermiogenesis consists of (1) activity of the Golgi apparatus to package the acrosomal enzymes, (2) positioning of the acrosome at the anterior end of the nucleus and of the centrioles at the opposite end of the nucleus, (3) elaboration of microtubules to form the flagellum of the tail, (4) multiplication of mitochondria and their positioning around the proximal portion of the flagellum, and (5) sloughing off of excess spermatid cytoplasm. (6) Structure of an immature sperm that has just been released from a sustentacular cell. (7) Structure of a fully mature sperm. (**b**) Scanning electron micrograph of mature sperm (430X).

Golgi apparatus and containing hydrolytic enzymes (hyaluronidase and others) that enable the sperm to penetrate and enter an egg. (Semen contains vesicles that hold cholesterol to the sperm's membrane, helping to keep the acrosome intact until after the sperm has entered the female reproductive tract. We will speak more of this in Chapter 29.) The sperm midpiece consists of mitochondria spiraled tightly around the contractile filaments of the tail, a typical flagellum elaborated by centrioles. The mitochondria provide the metabolic energy (ATP) needed for the whiplike movements of the tail that propel the sperm at a speed of 1–4 mm/min.

Role of the Sustentacular Cells. Throughout the spermatogenic process, the descendants of the same primary spermatocyte remain closely attached to one another by cytoplasmic bridges (see Figure 28.7). Additionally, they are surrounded by and connected to supporting cells of a special type, called *sustentacular cells* (*Sertoli cells*), which extend from the basement membrane to the lumen of the tubule (see Figure 28.7c). The sustentacular cells, bound to each other by tight junctions, form an unbroken layer within the seminiferous tubule, dividing it into two compartments (Figure 28.9). The *basal compartment* is the spermatogonia-containing compartment between the basement membrane and the tight junctions of the sustentacular cells; the *central compartment* includes the meiotically active cells and the tubule lumen.

The sustentacular cell barrier, often called the **blood-testis barrier,** prevents the membrane antigens (mostly proteins) of differentiating sperm from escaping through the basement membrane into the bloodstream. Since sperm are not formed until puberty, they are absent when the immune system is being programmed to recognize one's own tissues during early life. Moreover, all spermatogenic cells formed subsequent to meiosis I are genetically different from somatic body cells. Consequently, if the blood-testis barrier did not exist, sperm membrane antigens entering the blood might provoke an autoimmune response to one's own sperm. The spermatogonia, which are identical to other body cells, are outside the barrier and can be influenced by chemical messengers that prompt sper-

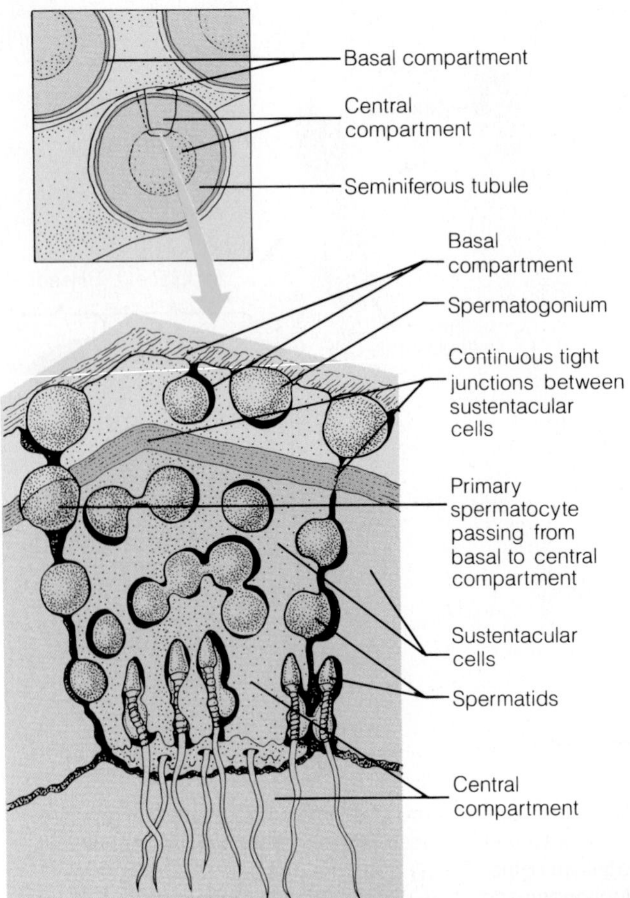

Figure 28.9 Schematic view of the relationship between a sustentacular cell and the spermatogenic cells and of the sustentacular cell's role in maintaining the blood-testis barrier. Continuous tight junctions, connecting sustentacular cells laterally at their bases, separate the basal compartment containing spermatogonia from the central compartment containing cells in various stages of spermatogenesis.

matogenesis. Following mitosis of the spermatogonia, the tight junctions of the sustentacular cells open to allow primary spermatocytes to pass through and enter the central compartment—much as locks in a canal are opened to allow a boat to pass.

In the central compartment, spermatocytes and spermatids are nearly enclosed in recesses in the sustentacular cells (see Figure 28.9), which seem to move them along toward the lumen. The sustentacular cells deliver nutrients to the dividing cells, secrete a fluid that provides the transport medium for sperm in the lumen, and dispose of the excess cytoplasm sloughed off the spermatids as they become transformed into sperm. As described shortly, the sustentacular cells also produce chemical mediators that play very active roles in the regulation of spermatogenesis.

The entire process of spermatogenesis, from formation of a primary spermatocyte to release of immature sperm into the lumen, takes 64 to 72 days. Sperm in the lumen show little motility and are incapable of fertilizing an egg. They are moved by peristalsis, and by the pressure of the sperm that follow, through the tubular system of the testes into the epididymis. There they undergo further maturation, which results in increased motility and fertilizing power.

Environmental threats can alter normal spermatogenesis and fertility. For example, some common antibiotics such as penicillin and tetracycline may suppress sperm formation; and radiation, lead, certain pesticides, marijuana, cigarettes, and excessive alcohol can cause abnormal (two-headed, multiple-tailed, etc.) sperm to be produced. Moreover, even if the sperm themselves are unaffected, toxic substances carried in a man's semen may be transferred to a woman, increasing the risk of birth defects, miscarriage, and stillbirths. ■

Male Sexual Response

The chief phases of the male sexual response are erection of the penis, which allows it to penetrate into the female vagina, and ejaculation, which deposits semen in the vagina.

Erection

Erection occurs when the erectile tissue of the penis, particularly the corpora cavernosa, becomes engorged with blood. When a man is not sexually aroused, the arterioles supplying the erectile tissue are constricted, and the penis is flaccid. However, during sexual excitement, a reflex is triggered during which parasympathetic nerve fibers stimulate these arterioles, causing them to dilate. In addition, vascular shunts between the blood supply of the ventral corpus spongiosum and dorsal corpora cavernosa constrict and reroute the blood destined for the bottom side of the penis to its top side. As a result, the vascular spaces of the corpora cavernosa fill with blood, causing the penis to become enlarged and rigidly erect. Expansion of the penis compresses its drainage veins, retarding the outflow of blood and enhancing the engorgement still further. Erection of the penis is one of the rare examples of parasympathetic control of arterioles. Another parasympathetic effect is stimulation of the bulbourethral glands, which causes lubrication of the glans penis.

The erection reflex can be initiated by a variety of sexual stimuli, such as touching the genital skin, mechanical stimulation of the pressure receptors in the head of the penis, and pleasurable sights, sounds, and smells. The CNS responds to such stimuli by discharging efferent impulses to the second through fourth sacral segments of the spinal cord, activating parasympathetic neurons that innervate the internal pudendal arteries serving the penis. In some instances, erection is induced solely by emotional or higher mental activity (the thought of a sexual encounter). Emotions and thoughts can also inhibit erection, causing vasoconstriction and resumption of the flaccid penile state.

Failure to attain erection is called *impotence*. Temporary periods of impotence are common among normal males owing to psychological factors. The decline in potency seen in inebriated individuals is thought to reflect alcohol's effects on higher brain centers, and the ability of many drugs (for example, certain antihypertensive agents) to cause impotence is well known. In some cases, impotence results from congenital absence of the shunt arterioles, or failure of the autonomic nervous system to constrict them. Various types of implantable penile devices are available to restore sexual activity to those whose impotence is nonreversible. ■

Ejaculation

Ejaculation (*ejac* = to shoot forth) is the propulsion of semen from the male duct system. When afferent impulses provoking erection reach a certain critical level, a spinal reflex is initiated, and a massive discharge of nerve impulses occurs over the sympathetic nerves (largely at the level of L_1 and L_2) serving the genital organs. As a result, (1) the reproductive ducts and accessory glands contract peristaltically, emptying their contents into the urethra; (2) the bladder sphincter muscle constricts, preventing expulsion of urine or reflux of semen into the bladder; and (3) the bulbospongiosus muscles of the penis undergo a rapid series of contractions, propelling semen from the urethra. These rhythmic muscle contractions are accompanied by intense pleasure and many systemic changes, such as generalized muscle contraction, rapid heartbeat, and elevated blood pressure. The entire event is referred to as *climax* or *orgasm*. Orgasm is quickly followed by muscular and psychological relaxation and vasoconstriction of the arterioles serving the penis, which allows the penis to become flaccid once again. After ejaculation, there is a latent period, ranging in time from minutes to hours, during which a man is unable to achieve another orgasm.

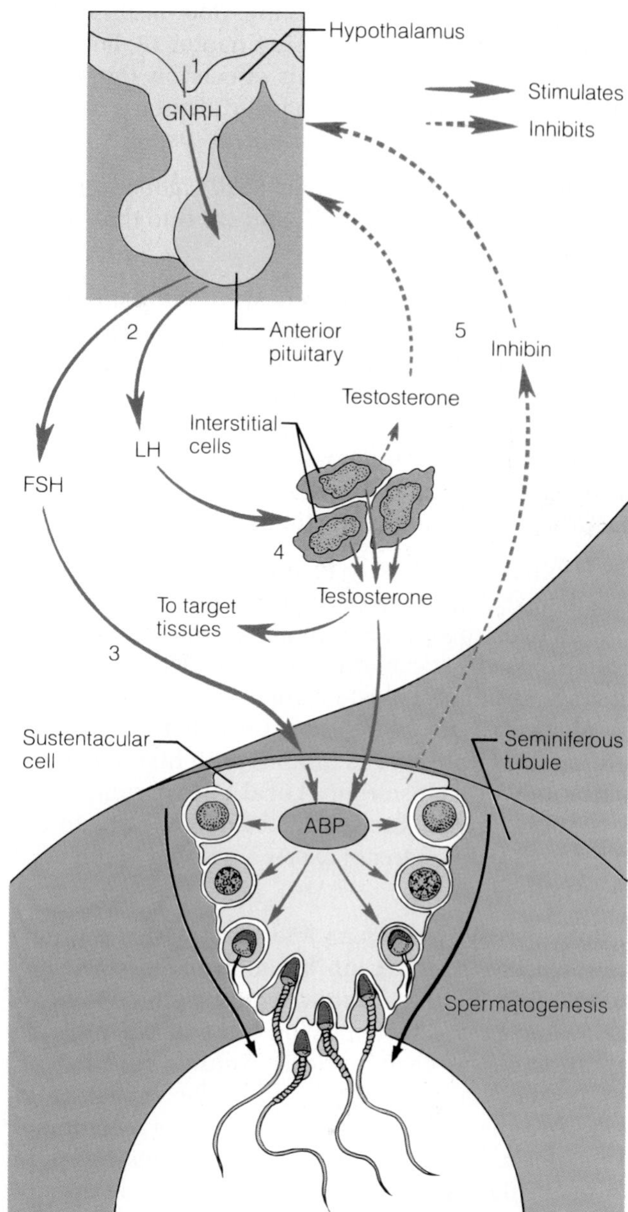

Figure 28.10 Hormonal regulation of testicular function, the brain-testicular axis. (1) The hypothalamus secretes gonadotropin-releasing hormone (GnRH). (2) GnRH stimulates the anterior pituitary to release gonadotropins—follicle-stimulating hormone (FSH) and luteinizing hormone (LH). (3) FSH acts on the sustentacular cells, causing them to release androgen-binding protein (ABP). (4) LH acts on the interstitial cells, enhancing their release of testosterone. ABP binding of testosterone then enhances spermatogenesis. (5) Rising levels of testosterone and inhibin (released by the sustentacular cells) exert feedback inhibition on the hypothalamus and pituitary.

Hormonal Regulation of Male Reproductive Function

The Brain-Testicular Axis

Hormonal regulation of spermatogenesis and testicular androgen production involves interactions between the hypothalamus, anterior pituitary gland, and testes, a relationship sometimes called the **brain-testicular axis.** The sequence of regulatory events, shown schematically in Figure 28.10, is as follows:

1. The hypothalamus releases *gonadotropin-releasing hormone (GnRH)*, which controls the release of the anterior pituitary gonadotropins, *follicle-stimulating* *hormone (FSH)* and *luteinizing* (loo′-tē-ih-nī-zing) *hormone (LH)*. (As described in Chapter 17, both FSH and LH were named for their effects on the female gonad.) GnRH reaches the anterior pituitary cells via the blood of the hypophyseal portal system.

2. Binding of GnRH to pituitary cells (gonadotrophs) prompts the secretion of FSH and LH into the general circulation.

3. FSH stimulates spermatogenesis in the testes. However, it does not act directly on the spermatogenic cells, as once thought. Instead, it stimulates the sustentacular cells to release **androgen-binding protein (ABP)**, which enables the spermatogenic cells to bind and concentrate testosterone; testosterone, in turn, stimulates spermatogenesis. Thus, FSH makes the cells receptive to testosterone's stimulatory effects.

4. LH binds specifically to the interstitial cells and stimulates them to secrete testosterone (and a small amount of estrogen). LH is therefore sometimes called interstitial cell–stimulating hormone (ICSH) in males. The interstitial cell clusters are heavily invested with lymphatic vessels, which may provide the means for bathing the seminiferous tubules with high local concentrations of testosterone. Local testosterone serves as the final trigger for spermatogenesis; testosterone entering the bloodstream exerts a number of effects at other body sites.

5. Both the hypothalamus and the anterior pituitary are subject to feedback inhibition by blood-borne hormones. Testosterone inhibits hypothalamic release of GnRH and may act directly on the anterior pituitary to inhibit gonadotropin release. **Inhibin** (in-hih′-bin) is a protein hormone produced by the sustentacular cells, and the amount released serves as a barometer of the normalcy of spermatogenesis. When the sperm count is high, inhibin release increases; it directly inhibits anterior pituitary release of FSH and probably inhibits GnRH release by the hypothalamus as well. When the sperm count falls below 20 million per milliliter, inhibin secretion declines steeply.

As you can see, the amount of testosterone and sperm produced by the testes reflects a balance among three sets of hormones: (1) GnRH; (2) gonadotropins, which stimulate the testes (GnRH indirectly via its effect on FSH and LH release); and (3) testicular hormones (testosterone and inhibin), which exert negative feedback controls on the hypothalamus and anterior pituitary. Since the hypothalamus is also influenced by input from other brain areas, it is safe to say that the whole axis is under CNS control. Both GnRH and gonadotropins are required for normal testicular function. In their absence, the testes atrophy, and for all practical purposes, sperm and testosterone production cease.

Development of male reproductive structures (discussed later in the chapter) depends on prenatal secretion of male hormones, and for a few months after birth, a male infant has plasma gonadotropin and testosterone levels nearly equal to that of a midpubertal boy. Shortly thereafter, blood levels of these hormones recede and they remain low throughout childhood. The means of this suppression of the hypothalamic-pituitary axis is not understood, but it is suggested that during childhood, the hypothalamus is inhibited by low levels of testosterone secreted by the boy's testes. As puberty nears, the threshold for hypothalamic inhibition rises, and much higher levels of testosterone are required to suppress hypothalamic release of GnRH. As more GnRH is released, more testosterone is secreted by the testes, but the threshold for hypothalamic inhibition keeps rising until the adult pattern of hormone interaction is achieved. Maturation of the brain-testicular axis takes about three years, and once established, the balance between the interacting hormones remains relatively constant. Consequently, an adult male's sperm and testosterone production also remain fairly stable, whereas in females, cyclic swings in gonadotropin and female sex hormone levels are the normal condition.

Mechanism and Effects of Testosterone Activity

Testosterone, like all steroid hormones, is synthesized from cholesterol. It exerts its effects by activating specific genes to transcribe messenger RNA molecules, which results in enhanced synthesis of certain proteins in the target cells. (Details are discussed in Chapter 17.)

In some target cells, testosterone must be transformed into another steroid before it can exert its effects. In cells of the prostate gland, for example, testosterone must be converted to *dihydrotestosterone* before it can bind within the nucleus. In certain neurons of the brain, testosterone is converted to *estrogen* (es′-trō-jin) to bring about its stimulatory effects; this seems strange—a "male" hormone that must be transformed into a "female" hormone to exert its masculinizing effects.

As puberty ensues, testosterone not only prompts spermatogenesis, but has multiple anabolic effects throughout the body. It targets all accessory reproductive organs—ducts, glands, and the penis—causing them to grow and assume adult functions. In adult males, normal plasma levels of testosterone are required to maintain the normal structure and function of these organs. When the hormone is deficient or absent, all accessory organs atrophy, the volume of semen declines markedly, and erection and ejaculation are impaired. Thus, a man becomes both sterile and impotent. This situation is easily rectified by testosterone replacement therapy.

Male **secondary sex characteristics** that make their appearance at puberty are testosterone-dependent. These include the appearance of pubic, axillary, and facial hair, enhanced hair growth on the chest or other body areas in some men, and a deepening of the voice as the larynx enlarges. The skin thickens and becomes oilier (which predisposes young men to acne), bones grow and increase in density, and skeletal muscles increase in size and mass. The last two effects are often referred to as the somatic effects of testosterone (*soma* = body). If present in high enough levels, testosterone will also enhance muscle mass in adult men, and with this in mind, many athletes dose themselves with testosterone-like anabolic steroids. However, as discussed in Chapter 10, there are more drawbacks than benefits in such a practice.

Testosterone also boosts basal metabolic rate and influences behavior. It is the basis of the sex drive (libido) in both males and females, be they heterosexual or homosexual. Thus, although testosterone is called a "male" sex hormone, it should not be specifically tagged as a promoter of male sexual activity. The masculinization of brain anatomy by testosterone is dealt with in Chapter 12.

The testes are not the only source of androgens; the adrenal glands of both sexes also release androgens. However, since adrenal androgens are unable to support normal testosterone-mediated functions when the testes fail to produce androgens, it is safe to say that it is testicular testosterone production that supports male reproductive function.

Anatomy of the Female Reproductive System

The reproductive role of the female is far more complex than that of a male. Not only must she produce gametes, but her body must prepare to nurture a developing embryo for a period of approximately nine months. **Ovaries** (ō′-ver-ēz) are the primary reproductive organs of a female, and like the male testes, ovaries serve a dual purpose: Besides their gametogenic functions, they produce the female sex hormones **estrogen** and **progesterone** (prō-jes′-ter-ōn). The accessory ducts (uterine tubes, uterus, and vagina) transport or otherwise serve the needs of the reproductive cells and/or the developing fetus.

The ovaries and duct systems of females are mostly located within the pelvic cavity and are collectively known as the *internal genitalia* (Figure 28.11). External organs, forming the vulva, are referred to as the *external genitalia*.

The Ovaries

The paired ovaries, which flank the uterus on each side (see Figure 28.11a), are about the size and shape of almonds. Each ovary is held in place by several ligaments: The **ovarian ligament** anchors the ovary medially to the uterus; the **suspensory ligament** anchors it laterally to the pelvic wall; and the **mesovarium** (meh-zō-vayr′-ē-um) suspends it in between. The mesovarium is part of the **broad ligament,** a peritoneal fold that also supports the uterine tubes, uterus, and vagina; the fibrous ovarian and suspensory ligaments are enclosed within the broad ligament.

The ovaries are served by the *ovarian arteries,* branches of the abdominal aorta, and by the ovarian branch of the uterine arteries. The ovaries are innervated by both divisions of the autonomic nervous system. The ovarian blood vessels and nerves reach the ovaries by travelling through the suspensory ligaments and mesovaria.

Like a testis, the external surface of an ovary is surrounded by a fibrous **tunica albuginea** (Figure 28.12). The tunica albuginea is, in turn, covered externally by a layer of cuboidal epithelial cells called the *germinal epithelium,* which is continuous with the peritoneum of the mesovarium. The term *germinal epithelium* is a misnomer because this layer does not give rise to ova. The ovary has an outer cortex and an inner medullary region, but the relative extent of each region is ill defined.

Within the ovary are many tiny saclike structures called **ovarian follicles,** embedded in a connective tissue stroma containing a rich blood supply. Each follicle consists of an immature egg, called an *oocyte* (ō-ō-sīt), surrounded by one or more layers of very different cells, called *follicle cells* if a single layer is present and *granulosa cells* when many layers are present. Follicles at different stages of maturation are distinguished by their structure. In a **primary follicle,** one layer of follicle cells surrounds an oocyte; a **growing follicle** has two or more layers of granulosa cells; and **mature,** or **Graafian** (grā′-fē-en), **follicles** contain a fluid-filled cavity called an *antrum*. Each month, in adult women, one of the ripening follicles ejects its oocyte from the ovary, an event called *ovulation*. After ovulation, the ruptured follicle is transformed into a very different looking structure called the **corpus luteum** (loo′-tē-um) (plural, corpora lutea), which eventually degenerates. As a rule, most of these structures can be seen within the same ovary. In older women, the surface of the ovaries are scarred and pitted, revealing that many oocytes have been released.

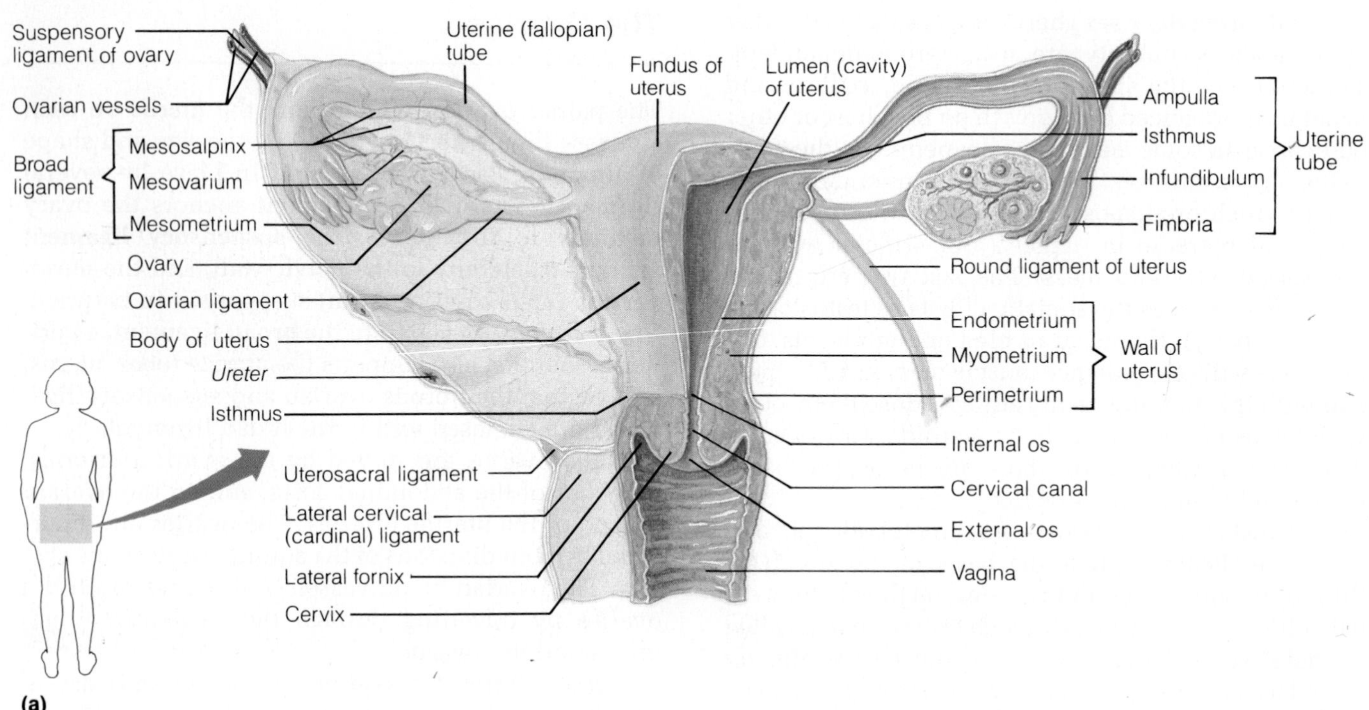

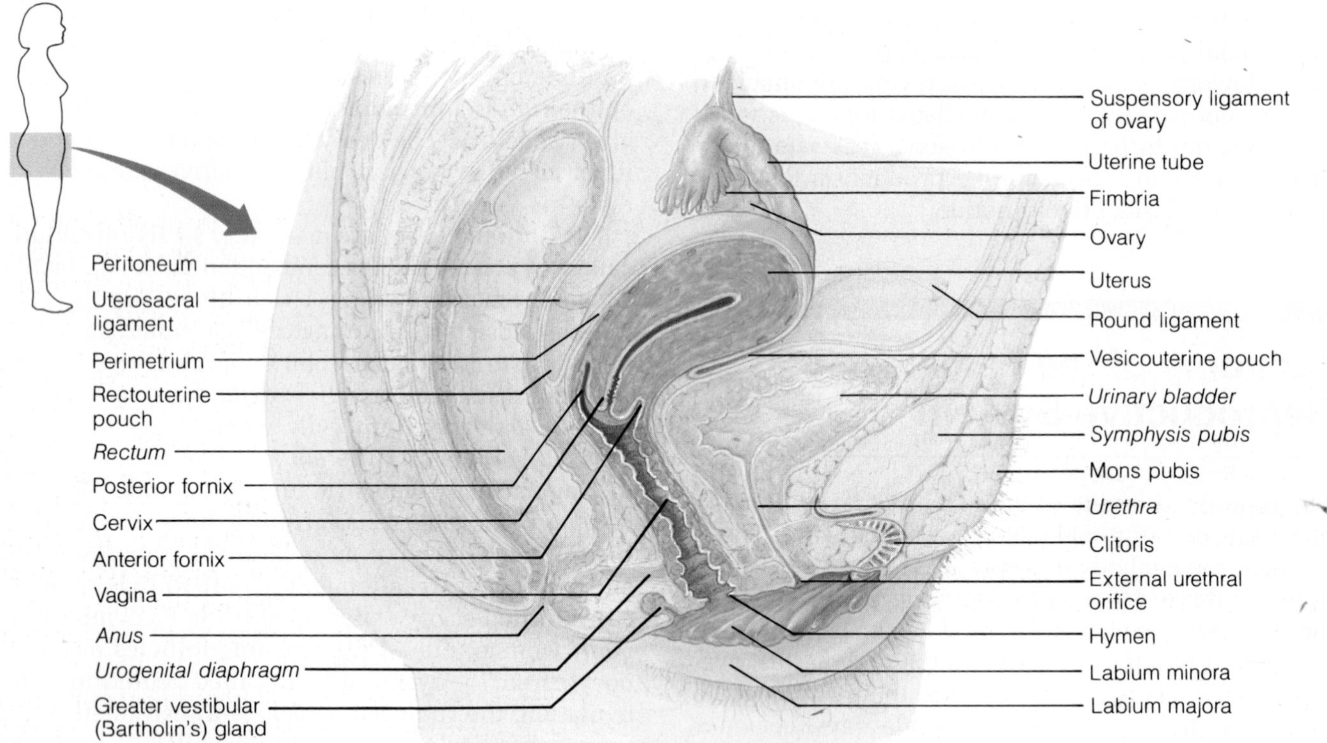

Figure 28.11 Internal organs of the female reproductive system.
(**a**) Posterior view of the female reproductive organs. The posterior walls of the vagina, uterus, and uterine tubes and the broad ligament have been removed on the right side to reveal the shapes of the lumens of these organs. (**b**) Midsagittal section of the female pelvis.

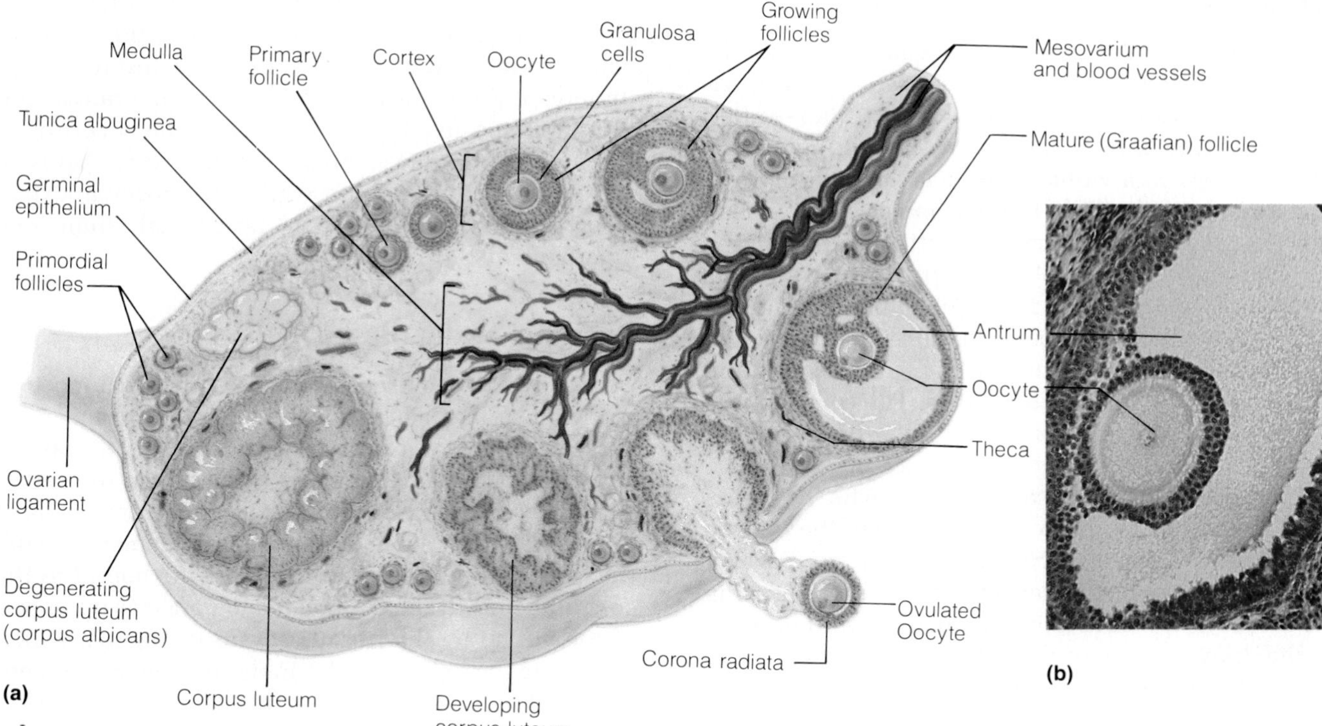

Figure 28.12 Structure of an ovary. (**a**) The ovary has been sectioned to reveal the follicles in its interior. Note that all of these structures would not appear in the ovary at the same time. (**b**) Photomicrograph of a mature (Graafian) follicle (250X).

Duct System

The Uterine Tubes

The **uterine** (yoo′-ter-in), or **fallopian, tubes** form the initial part of the female duct system (see Figure 28.11a). They receive the ovulated oocyte and provide a site where fertilization can occur. Each of the uterine tubes is about 10 cm (4 inches) long and 1 cm in diameter and extends medially from the region of an ovary to empty into the superolateral region of the uterus via a constricted region called the **isthmus** (is′-mus). The distal end of each uterine tube expands as it curves around the ovary, forming the **ampulla;** fertilization usually occurs in this region. The ampulla ends in the **infundibulum** (in-fun-dih′-byoo-lum), an open, funnel-shaped structure bearing ciliated, fingerlike projections called **fimbriae** (fim′-brē-ē) that drape over the ovary. Unlike the male duct system, which is continuous with the tubules of the testes, the uterine tubes have little or no actual contact with the ovaries. An ovulated oocyte is cast into the peritoneal cavity, and many oocytes are lost there. However, the fimbriae become very active close to the time of ovulation and undulate to create currents in the peritoneal fluid. These currents usually carry the oocyte into the uterine tube, where it begins its journey toward the uterus.

The structure of the uterine tube wall allows it to aid the progress of an oocyte. Its muscularis consists of circular and longitudinal layers of smooth muscle, and its thick, highly folded mucosa contains both ciliated and nonciliated cells. The oocyte is carried toward the uterus by a combination of peristalsis and the rhythmic beating of the cilia. Nonciliated cells of the mucosa have dense microvilli and produce a secretion that keeps the oocyte (and sperm, if present) moist and nourished. The uterine tubes are covered externally and supported along their length by the portion of the broad ligament called the **mesosalpinx** (meh″-zō-sal′-pinx), a term that indicates the broadening of this membrane (*salpin* = trumpet) as it approaches the ovary (see Figure 28.11a).

The fact that the uterine tubes are not continuous distally with the ovaries places women at risk for infections spreading into the peritoneal cavity from other parts of the reproductive tract. Gonorrhea bacteria sometimes infect the peritoneal cavity in this way, causing an extremely severe inflammation called *pelvic inflammatory disease (PID).* Unless treated promptly, PID can cause scarring of the narrow uterine tubes and of the ovaries and lead to sterility. Indeed, scarring and closure of the uterine tubes is one of the major causes of female infertility. ■

The Uterus

The **uterus** (Latin for "womb") is located in the pelvis, anterior to the rectum and posterosuperior to the bladder (see Figure 28.11b). It is a hollow, thick-walled organ that functions to receive, retain, and nourish a fertilized egg. In a woman who has never been pregnant, the uterus is about the size and shape of a pear, but it is usually somewhat larger after childbirth.

The major portion of the uterus is referred to as the **body** (see Figure 28.11a). The superior rounded region above the entrance of the uterine tubes is the **fundus,** and the slightly narrowed region between the body and the cervix is the **isthmus.** The **cervix** of the uterus is its narrow neck, or outlet, which projects into the vagina inferiorly. The cavity of the cervix, called the *cervical canal,* communicates with the vagina via the *external os* and with the cavity of the uterine body via the *internal os.* Normally, the uterus is flexed anteriorly between the cervix and the body, causing the uterus as a whole to be inclined forward or *anteverted;* however, the organ is frequently turned backward, or *retroverted,* in older women.

The Uterine Wall. The wall of the uterus is composed of three layers (see Figure 28.11a): the perimetrium, the myometrium, and the endometrium. The **perimetrium,** the outermost serous layer, is the visceral peritoneum. The **myometrium** (mī″-ō-mē′-trē-um) is the bulky middle layer of the uterus, composed of interlacing bundles of smooth muscle. The myometrium plays an active role during childbirth when it contracts rhythmically to force the baby out of the mother's body. The mucosal lining of the uterine cavity is the **endometrium** (Figure 28.13), a simple columnar epithelium underlain by a thick, extremely cellular connective tissue *stroma* (strō-muh). Should fertilization occur, the young embryo burrows into the endometrium (implants) and resides there for the rest of its development. The endometrium has two chief layers. The superficial **stratum functionalis** (funk-shuh-na′-lis) undergoes cyclic changes in response to blood levels of ovarian hormones and is shed during menstruation (approximately every 28 days). The thin, deeper **stratum basalis** (buh-sa′-lis) shows little response to ovarian hormones and is responsible for re-forming a new functionalis after menstruation ends. The endometrium has numerous glands that change in length as the endometrial thickness changes.

To understand the cyclic changes of the uterine endometrium (discussed later in the chapter), it is essential to understand the vascular supply of the uterus. As shown in Figure 28.13b, the uterine arteries break into several *arcuate* (ar′-kyoo-ut) *arteries* within the middle layer of the myometrium. These arteries send radial branches into the endometrium, where they, in turn, give off the straight and the spiral arteries. The *straight arteries* supply the stratum basalis; the *spiral,* or *coiled, arteries* supply the capillary beds of the stratum functionalis. The spiral arteries undergo repeated degeneration and regeneration, and it is their spasms that actually cause the shedding of the functionalis layer during menstruation. Veins in the endometrium are thin-walled and form an extensive venous network with occasional sinusoidal enlargements.

Supports of the Uterus

The uterus is enclosed by the **mesometrium** portion of the broad ligament, through which the uterine blood vessels and nerves pass (Figure 28.11). The **lateral cervical (cardinal) ligaments** extend from the cervix and superior part of the vagina to the lateral walls of the pelvis, and the paired **uterosacral ligaments** secure the uterus to the sacrum. The uterus is bound to the subcutaneous tissue of the labia majora (the outer lips of the vulva) by the fibrous **round ligaments,** which run through the inguinal canals to their attachment points. These various ligaments allow the uterus a good deal of mobility, and its position changes as the rectum and bladder fill and empty. The principal supports of the uterus are the muscles that form the pelvic floor, namely, the muscles of the urogenital diaphragm and the levators ani and coccygeal muscles (see Chapter 10). The undulating course of the peritoneum around and over the various pelvic structures results in cul-de-sacs, or blind-ended peritoneal pouches; the most important of these are the *vesicouterine* (veh″-sih-kō-yoo″-ter-in) *pouch* and the *rectouterine pouch* (see Figure 28.11b).

Cancer of the cervix is the third most common female cancer (after breast and lung cancer). It primarily strikes women between the ages of 30 and 50. Risk factors include frequent cervical inflammations, venereal disease, multiple pregnancies, and an active sex life with multiple partners. A yearly Pap smear is the single most important diagnostic technique for detecting this slow-growing cancer. ■

The Vagina

The **vagina** is a thin-walled fibromuscular tube, 8–10 cm (3 to 4 inches) long. It lies between the bladder and the rectum and extends from the body exterior to the cervix (see Figure 28.11). Often called the *birth canal,* the vagina provides a passageway for delivery of an infant and for menstrual flow. Since it receives the penis (and semen) during sexual intercourse, it is the female organ of copulation.

The wall of the vagina consists of three coats: an outer fibrous *adventitia,* a smooth muscle *muscularis,* and a *mucosa* marked by transverse ridges, or *rugae.* The mucosa is a stratified squamous epithelium adapted to stand up to friction; it is not normally keratinized, but it can become so in the event of vitamin

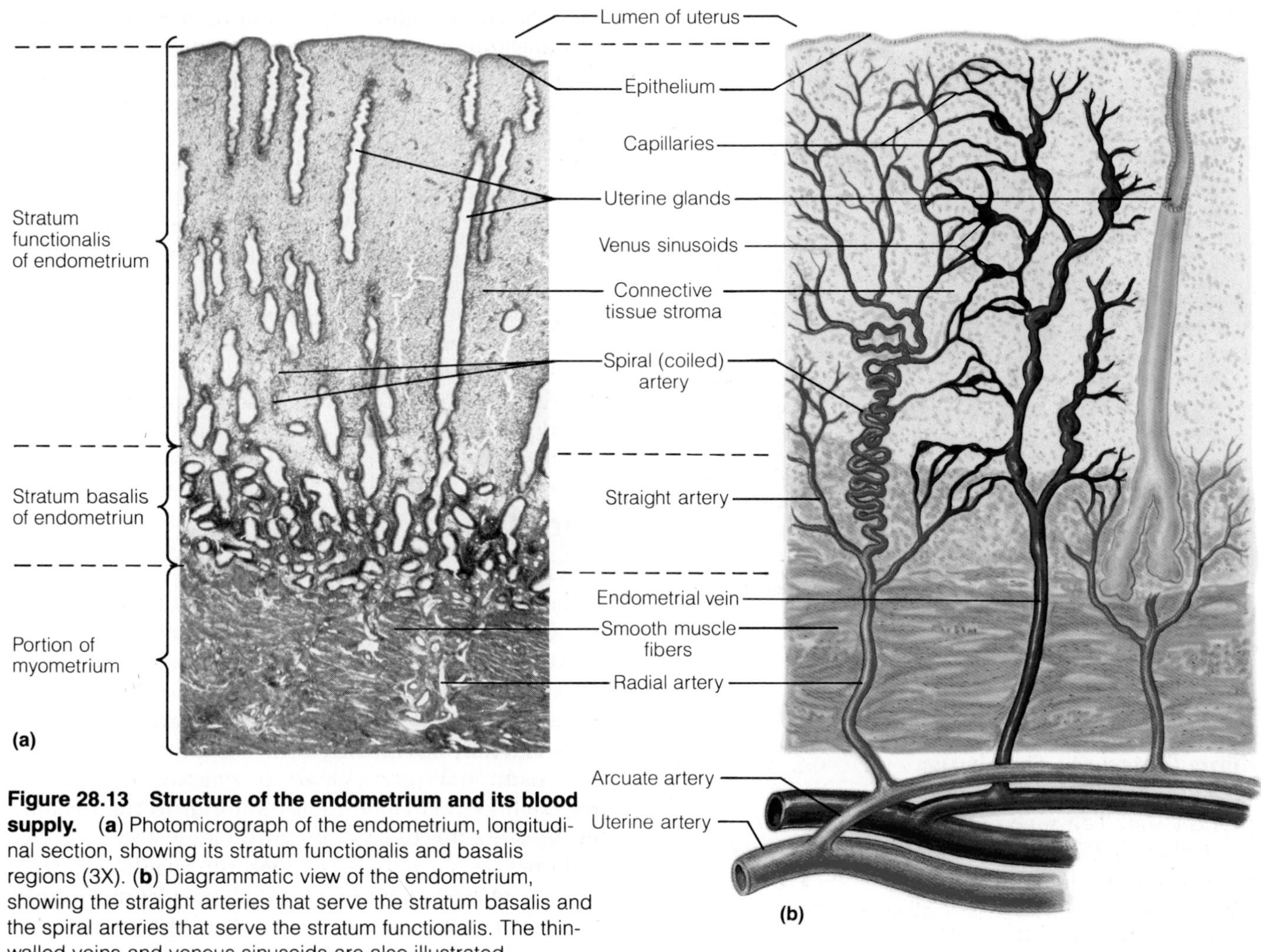

Figure 28.13 Structure of the endometrium and its blood supply. (**a**) Photomicrograph of the endometrium, longitudinal section, showing its stratum functionalis and basalis regions (3X). (**b**) Diagrammatic view of the endometrium, showing the straight arteries that serve the stratum basalis and the spiral arteries that serve the stratum functionalis. The thin-walled veins and venous sinusoids are also illustrated.

A deficiency. The vaginal mucosa has no glands; its lubrication is provided by cervical mucous glands or vestibular glands located outside the vagina. Its mucosal cells store large amounts of glycogen, which is anaerobically metabolized to lactic acid by resident bacteria; consequently, the pH of the vagina is normally quite acidic (pH 3.5–4). This acidity helps keep the vagina healthy and free of infection, but it is also hostile to sperm.

Near the distal *vaginal orifice* (in virgins), the mucosa is elaborated, forming an incomplete partition called the **hymen** (hī'-min) (Figure 28.14). The hymen is very vascular and tends to bleed when it is ruptured during the first coitus (sexual intercourse). However, its durability varies. In some females, it is ruptured during a sports activity, tampon insertion, or pelvic examination. Occasionally, it is so tough that it must be breached surgically if intercourse is to occur. The upper end of the vaginal canal loosely surrounds the cervix of the uterus, giving rise to a vaginal fold called the **fornix.** The posterior part of this recess, the *posterior fornix,* is much deeper than the *lateral* and *anterior fornices* (see Figure 28.11a and b). Generally, the lumen of the vagina is quite small, and except where it is held open by the cervix, its posterior and anterior walls are in contact with one another. The vagina stretches considerably during copulation and childbirth, but its lateral distention is limited by the ischial spines and the sacrospinous ligaments.

Because the uterus tilts away from the vagina, attempts by untrained persons to enter the uterus with an unsterilized instrument may result in puncturing of the posterior wall of the vagina, followed by hemorrhage and subsequent peritonitis. This has happened in the past during unskilled attempts to induce an abortion with a coat-hanger wire or other sharp objects.The legalization and ready availability of safe therapeutic abortions has all but eliminated this type of problem. ■

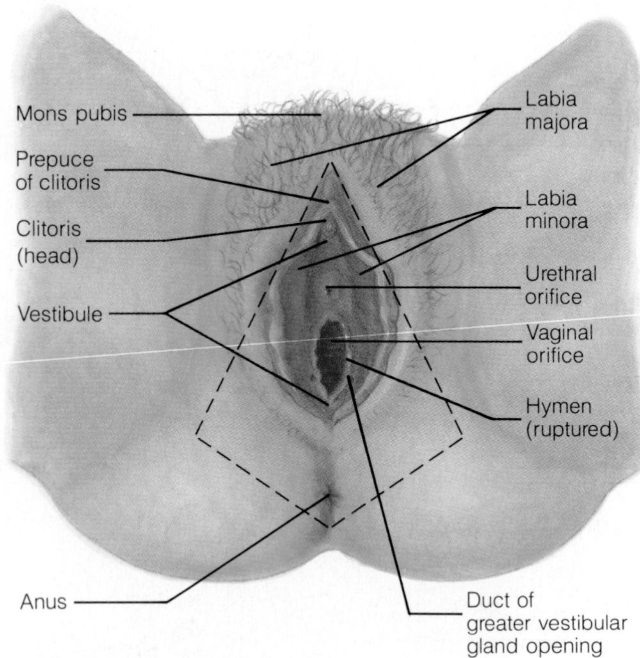

Figure 28.14 The external genitalia (vulva) of the female. The region enclosed by the dotted lines is the perineum.

The External Genitalia

The female reproductive structures located external to the vagina are called the **external genitalia** (see Figure 28.14). The external genitalia, also called the **vulva** (vul′-vuh), include the mons pubis, labia, clitoris, and structures associated with the vestibule (the urethral and vaginal orifices and vestibular glands).

The **mons pubis** (monz pyoo′-bis) is a fatty, rounded area overlying the pubic symphysis; after puberty, this area is covered with pubic hair. Running posteriorly from the mons pubis are two elongated, hair-covered fatty skin folds, the **labia majora** (lā′-bē-uh mah-jor′-uh), which are the female counterpart, or *homologue*, of the male scrotum (that is, they are derived from the same embryonic tissue). The more heavily pigmented labia majora enclose the **labia minora** (mih-nōr-uh), two thin, delicate, hair-free folds that are covered with mucosa and richly supplied with sebaceous glands. The labia minora enclose a region called the **vestibule,** which contains the clitoris, most anteriorly, followed by the external openings of the urethra and the vagina. Flanking the vaginal opening are a number of glands that release mucus into the vestibule and help to keep it moist and lubricated, facilitating intercourse. These include the **greater vestibular (Bartholin's) glands,** homologous to the bulbourethral glands of males, and the slightly more anterior **lesser vestibular glands.**

The **clitoris** (klit′-ō-ris) is a small, protruding structure, composed largely of erectile tissue, that is homologous mostly to the glans penis of the male. It is hooded by a skin fold called the **prepuce of the clitoris,** formed by the junction of the labia minora folds. The clitoris is richly innervated with sensory nerve endings sensitive to touch, and it becomes swollen with blood during tactile stimulation, contributing to a female's sexual arousal. Like the penis, the clitoris has dorsal erectile columns (corpora cavernosa); but it lacks a corpus spongiosum. In males, the urethra carries both urine and semen and runs through the penis, but the female urinary and reproductive tracts are completely separate, and neither of them runs through the clitoris.

The female **perineum** is a diamond-shaped region located between the anterior end of the labial folds, the anus posteriorly, and the ischial tuberosities laterally. The soft tissues of the perineum overlie the muscles of the pelvic outlet (see Chapter 10).

The Mammary Glands

The **mammary glands** are present, of course, in both sexes, but they normally become functional only in females (Figure 28.15). Since the biological role of the mammary glands is to produce milk to nourish a newborn baby, they are actually important only when reproduction has already been accomplished.

Each mammary gland is contained within a rounded skin-covered breast anterior to the pectoral muscles of the thorax. Slightly below the center of each breast is a pigmented area, the **areola** (uh-rē′-ō-luh), which surrounds a central protruding **nipple.** Numerous large sebaceous glands in the areola make it slightly bumpy and produce sebum that lubricates the areola and nipple during nursing. Autonomic nervous system controls of smooth muscle fibers in the areola and nipple cause the nipple to become erect when stimulated by tactile or sexual stimuli and when exposed to cold.

Internally, each mammary gland consists of 15 to 25 **lobes** that radiate around the nipple. The lobes are padded and separated from each other by fibrous connective tissue and fat. The interlobar connective tissue forms **suspensory ligaments** that attach the breast to the underlying muscle fascia and to the overlying dermis. Within the lobes are smaller chambers called **lobules,** which contain **alveolar** (al-vē′-ō-ler) **glands** that produce milk when a woman is lactating. These compound tubuloalveolar glands pass the milk into the **lactiferous** (lak-tih′-fer-us) **ducts,** which open to the outside at the nipple. Just deep to the areola, each lactiferous duct has a dilated region called a **lactiferous sinus** or **ampulla.** Milk accumulates in these sinuses during lactation.

Cancers of the breast are the leading cause of death in adult women. Since most breast lumps are discovered by women themselves in routine monthly self-exams, this simple examination should be a health maintenance priority in every woman's life. ■

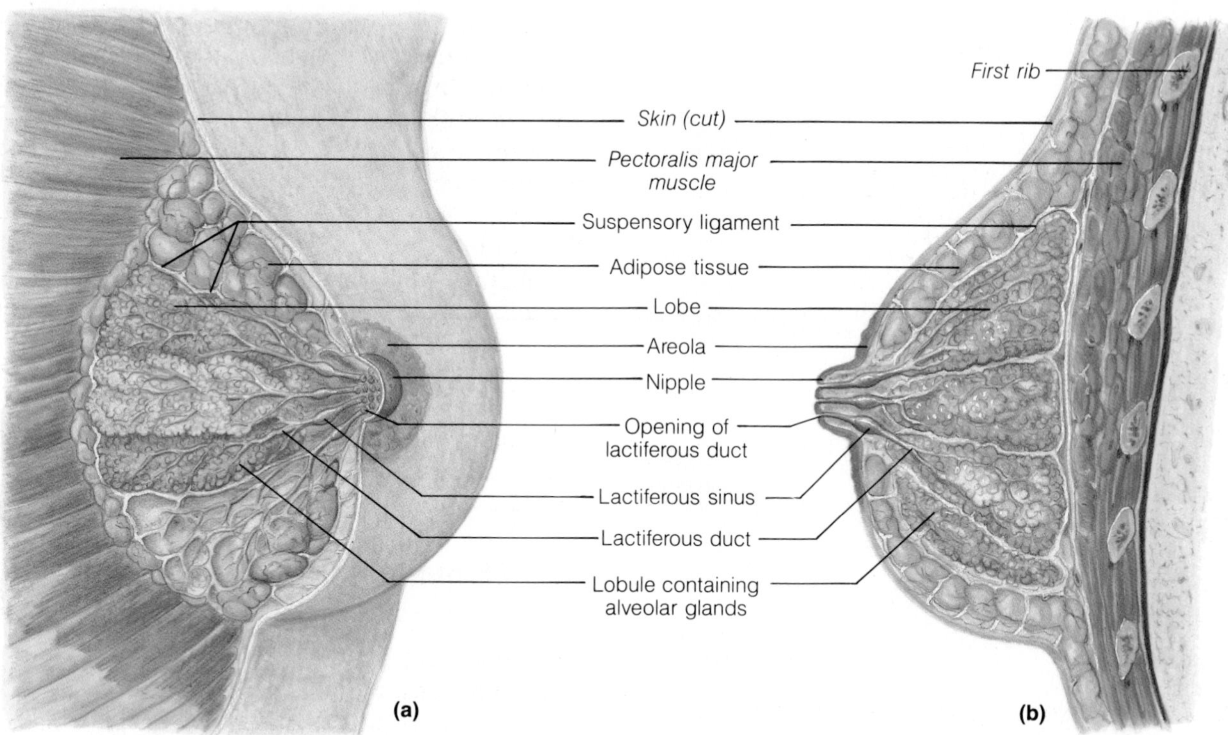

Figure 28.15 Structure of lactating mammary glands. (**a**) Anterior view of a partially dissected breast. (**b**) Sagittal section of a breast.

Physiology of the Female Reproductive System

Oogenesis

As noted earlier, sperm production in males begins at puberty and generally continues throughout life. The situation is quite different in females. The total supply of eggs that a female can release has already been determined by the time she is born, and the time span in which she releases them extends from puberty to about the age of 50.

Meiosis, the specialized type of nuclear division that occurs in the testes to produce sperm, also occurs in the ovaries. In this case, female sex cells are produced, and the process is called **oogenesis** (ō″-ō-jeh′-nih-sis), literally, "the beginning of an egg." The events of oogenesis, shown in Figure 28.16, begin during fetal development.

In a female fetus, **oogonia,** the diploid stem cells of the ovaries that correspond to the spermatogonia of the testes, multiply rapidly by mitosis. The oogonia then enter a growth phase and lay in nutrient reserves. Gradually, **primordial follicles** (see Figure 28.12a) begin to appear, and the oogonia are transformed into **primary oocytes,** which become surrounded by a single layer of flattened follicle cells. The primary oocytes replicate their DNA and begin the first meiotic division, but become stalled late in prophase I and do not complete it. Many primordial follicles deteriorate before birth, but those that remain occupy the cortical region of the immature ovary. By birth, a female's lifetime supply of primary oocytes, approximately 700,000 of them, is already in place in the primordial follicles awaiting the chance to complete meiosis and produce functional ova. Since they remain in their state of suspended animation all through childhood, the wait is a long one—10 to 14 years at least, and 60 years at most!

Beginning at puberty, a small number of primary oocytes are activated to grow each month, but only one is "selected" to continue meiosis I, producing two haploid cells (each with 23 replicated chromosomes), which are quite dissimilar in size. The smaller cell is called the **first polar body;** the larger cell, which contains nearly all the cytoplasm of the primary oocyte, is the **secondary oocyte.** The events that occur as the primary oocyte proceeds from its arrested state in prophase I to complete the first maturation division are interesting. A spindle forms at the very edge of the oocyte (see Figure 28.16, left) as the area is cleared of other organelles, and a little "nipple," into which the

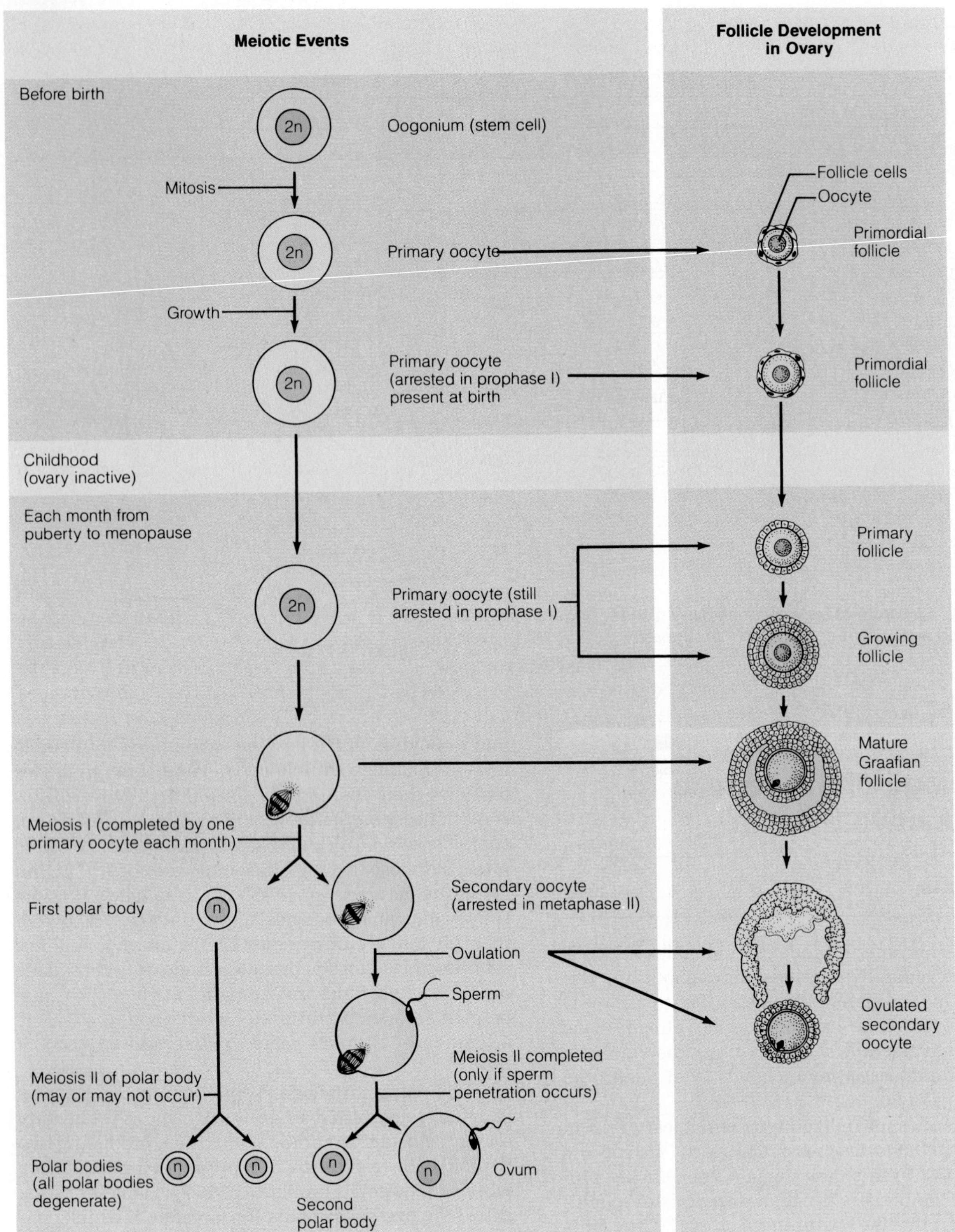

Figure 28.16 Events of oogenesis. Left, flowchart of meiotic events. Right, correlation with follicle development and ovulation in the ovary.

polar body chromosomes will be cast, appears at that edge. This sets up the polarity of the oocyte and ensures that the polar body receives almost no cytoplasm or organelles.

The first polar body usually undergoes meiosis II, producing two even smaller polar bodies. However, in humans, the secondary oocyte becomes arrested in metaphase II and is ovulated. If it is not penetrated by a sperm, the secondary oocyte deteriorates; but when sperm penetration does occur, the secondary oocyte quickly completes meiosis II, yielding one large **ovum** and a tiny **second polar body** (see Figure 28.16). Completion of meiosis II and the union of the egg and sperm nuclei will be described in Chapter 29. What you should realize now is that the products of oogenesis are usually three tiny polar bodies, nearly devoid of cytoplasm, and one large ovum. All of these cells are haploid, but only the ovum is a functional fertilizable gamete. This is quite different from spermatogenesis, where four viable gametes—spermatozoa—are produced during meiosis. An additional point must be made here. Although you commonly hear that females produce *ova*, or *eggs*, this really never happens in humans prior to sperm penetration. (However, it does happen in certain other species.) Consequently, even though the term *ovum* is used loosely to indicate an *ovulated oocyte*, this is not precise terminology when referring to the human species.

The unequal cytoplasmic divisions that occur during oogenesis assure that the fertilized egg has enough nutrient reserves to support it during its seven-day journey to the uterus. The polar bodies degenerate and die because they lack nutrient-containing cytoplasm. Since the reproductive life of a female is at best about 45 years (from the age of 11 to approximately 55) and there is typically only one ovulation event each month, only 400 to 500 oocytes out of her potential of 700,000 are released during a woman's lifetime. Again, nature has provided us with a generous oversupply of sex cells, even accounting for those that begin to develop but are never ovulated.

The Ovarian Cycle

The **ovarian cycle** can be described in terms of three consecutive phases. The **follicular phase** is the period of follicle growth from the first to the tenth day of the cycle; the **ovulatory phase** encompasses days 11–14 and culminates in ovulation; and the **luteal phase** is the period of corpus luteum activity, days 14–28. The "typical" ovarian cycle repeats at intervals of 28 days, but it may be longer or shorter. In any case, the luteal phase is constant: It is 14 days between the time of ovulation and the end of the cycle. The hormonal controls of these events will be described shortly; here we will concentrate on what happens each month within the ovary (Figure 28.17).

The Follicular Phase

Maturation of a primordial follicle to the mature state involves several events, shown in Figure 28.17, steps 1–6. When the primordial follicles are activated, the squamouslike cells surrounding the primary oocyte swell, becoming cuboidal cells, and the follicle comes to be called a primary follicle. Then the primary oocyte increases in size; the granulosa (follicle) cells proliferate, forming a multicellular capsule, the zona granulosa, around the oocyte; and specialized connective tissue cells of the ovarian stroma form a fibrous capsule around the follicle called the **theca folliculi** (thē'-kuh fah-lik'-you-lī). As the follicle increases in size, the thecal cells and the granulosa cells cooperate to produce estrogen, and a fluid-filled space, the

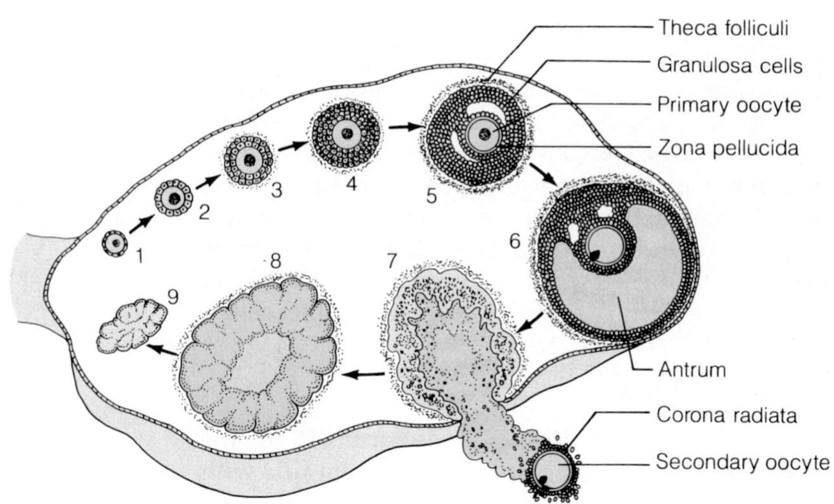

Figure 28.17 The ovarian cycle: development and fate of ovarian follicles. (1) Primary oocyte surrounded by flattened cells (primordial follicle). (2) A primary follicle containing a primary oocyte. (3–5) The growing follicle. This follicle is secreting estrogen as it continues to mature. (6) The Graafian (mature) follicle, ready to be ovulated. Meiosis I, producing the secondary oocyte and first polar body, occurs in the mature follicle. (7) The ruptured follicle and ovulated secondary oocyte surrounded by its corona radiata of granulosa cells. (8) The corpus luteum, formed from the ruptured follicle under the influence of LH, produces progesterone (and estrogen). (9) The corpus albicans.

antrum, begins to form in the midst of the granulosa cells. The granulosa cells also secrete a glycoprotein-rich substance that forms a thick transparent membrane, the **zona pellucida** (peh-loo'-sih-duh), around the oocyte. However, the granulosa cells remain connected to the oocyte membrane by cytoplasmic extensions, which may allow (1) estrogen to pass to the primary oocyte and/or (2) the granulosa cells to mediate the effects of LH, which prompts the primary oocyte to resume meiosis.

The Ovulatory Phase

After ten or more days of growth, the follicle is a fully mature Graafian follicle and bulges from the external ovarian surface. The large central antrum is tense with fluid, and the oocyte and its capsule of granulosa cells are pushed to one side. One of the final events of follicle maturation is completion of meiosis I by the primary oocyte to form the secondary oocyte and first polar body. The tiny first polar body is ejected and lies between the plasma membrane of the secondary oocyte and the zona pellucida. The stage is now set for ovulation, which occurs midcycle, typically on the fourteenth day.

Ovulation occurs when the ovary wall at the site of the ballooning ruptures and expels the secondary oocyte into the peritoneal cavity (see Figure 28.17, step 7). The ovulated secondary oocyte is still surrounded by its granulosa cell capsule, now called the *corona radiata* ("radiating crown"). Some women experience a twinge of abdominal pain in the lower abdomen when ovulation occurs. This phenomenon, called *mittelschmerz* (German for "middle pain"), is thought to be caused by the intense stretching of the ovarian wall during ovulation.

In the ovaries of adult females, there are always several follicles at different stages of maturation. As a rule, one follicle outstrips the others to become the *dominant follicle* and is at the peak stage of maturation when the hormonal stimulus is given for ovulation. How this follicle is selected, or selects itself, is still uncertain. The others undergo degeneration, or *atresia* (uh-trē'-zhuh), and are referred to as *atretic follicles*. In 1% to 2% of all ovulations, more than one oocyte is ovulated, which can result in multiple births. Since different oocytes are fertilized by different sperm, the siblings are *fraternal*, or nonidentical, twins. (Identical twins result from the fertilization of a single oocyte by a single sperm, followed by separation, in early development, of the fertilized egg's daughter cells.)

The Luteal Phase

After ovulation and discharge of antrum fluid, the ruptured follicle collapses, and the antrum fills with clotted blood. The granulosa cells increase in size and form a new, quite different endocrine gland, the *corpus luteum* ("yellow body") (see Figure 28.17, step 8). Once formed, the corpus luteum begins to secrete progesterone and some estrogen. Its fate depends on that of the ovulated oocyte. If pregnancy does not occur, the corpus luteum begins to degenerate in about ten days, and its hormonal output ceases. In this case, all that ultimately remains is a mass of scar tissue called the *corpus albicans* (al'-bih-kanz), literally the "white body" (see Figure 28.17, step 9). On the other hand, if the oocyte is fertilized and pregnancy ensues, the corpus luteum persists until the placenta is ready to take over its hormone-producing duties.

Hormonal Regulation of the Ovarian Cycle

Because there are two interacting, cyclically released hormones, ovarian events are much more complicated than those occurring within the testes. However, the hormonal controls set into motion at puberty are quite similar. Gonadotropin-releasing hormone (GnRH), the pituitary gonadotropins, and, in this case, ovarian estrogen and progesterone interact to produce the cycles of follicle growth and appearing and disappearing corpora lutea.

Establishment of the Ovarian Cycle

During childhood, the ovaries grow and continuously secrete small amounts of estrogen, which inhibit hypothalamic release of GnRH. But as puberty nears, the hypothalamus becomes less sensitive to estrogen and begins to release GnRH in a discontinuous, or pulselike, manner. GnRH, in turn, stimulates the anterior pituitary to release FSH and LH, which act on the ovaries.

Gonadotropin levels continue to increase for about four years, and during this time, the pubertal girl is anovulatory and incapable of becoming pregnant. Eventually, the adult cyclic pattern is achieved, and hormonal interactions are stabilized. Apparently, the ovaries have little or no role to play in initiating the cycling of anterior pituitary gonadotropins, because this pattern develops at puberty even if the ovaries have been removed. Achievement of the adult pattern of gonadotropin cycling is heralded by the young woman's first menstrual period, referred to as **menarche** (meh-nar'-kē). However, in the one to two years that follow menarche, many cycles are still anovulatory. Usually, it is not until the third year postmenarche that the cycle becomes regular and the luteal phase reaches its normal duration of 14 days.

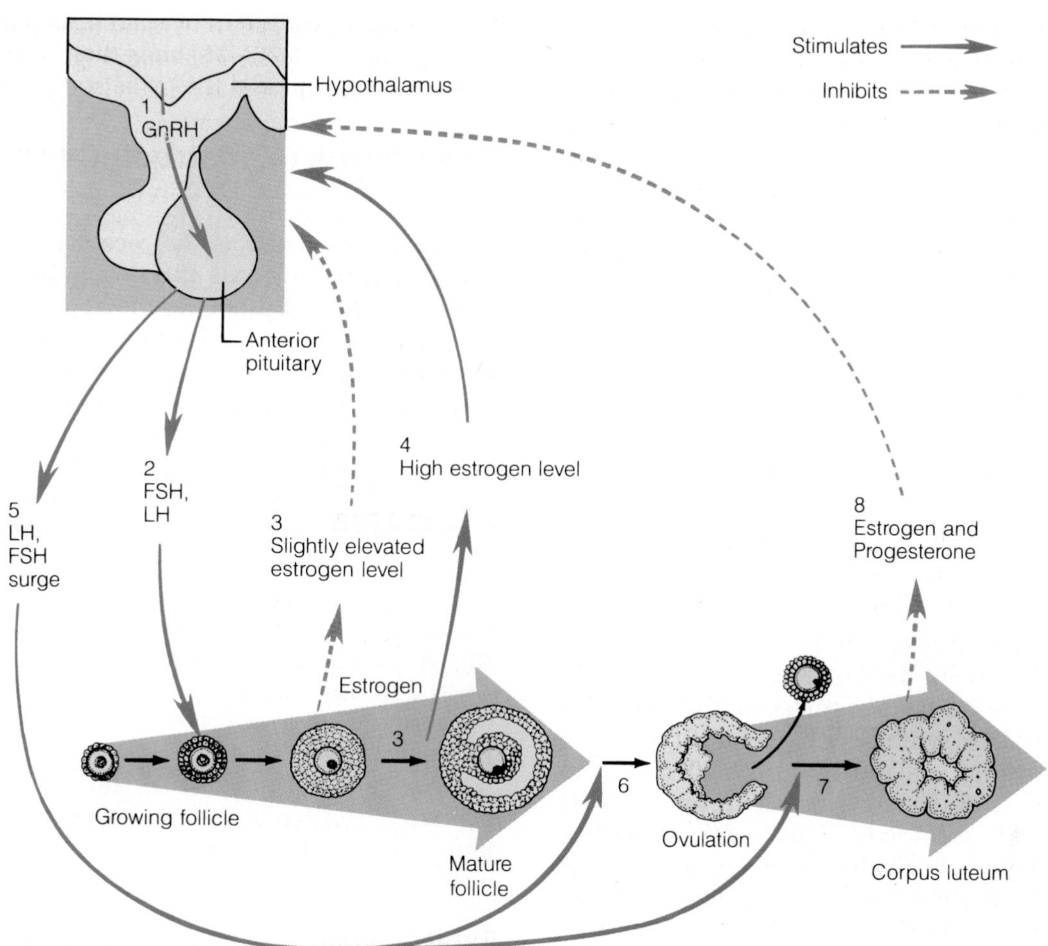

Figure 28.18 Feedback interactions involved in the regulation of ovarian function. Numbers refer to the events listed in the text. Events that follow step 8 (negative feedback inhibition of progesterone and estrogen on the hypothalamus and anterior pituitary) are not depicted, but involve a gradual deterioration of the corpus luteum and, therefore, a decline in ovarian hormone production. Ovarian hormones reach their lowest levels in the plasma around day 28.

Hormonal Events of the Ovarian Cycle

The waxing and waning of the anterior pituitary and ovarian hormones and the positive and negative feedback interactions that regulate ovarian function appear to be as described next. Note that events 1–8 in the following discussion directly correspond to the same numbered steps in Figure 28.18.

1. Follicle growth and oocyte maturation are controlled in tandem by the interaction of follicle-stimulating hormone (FSH), luteinizing hormone (LH), and estrogen. On day 1 of the cycle, rising levels of GnRH from the hypothalamus stimulate increased production of FSH and LH in the anterior pituitary.

2. FSH and LH stimulate follicular growth. (Why only *some* follicles are sensitive to these hormonal stimuli is still a mystery; however, there seems to be little doubt that their responsiveness reflects their development of increased numbers of gonadotropin receptors, which enhances their sensitivity to the hormonal signals.) As the follicles enlarge, estrogen secretion begins under the influence of both LH and FSH. LH targets the thecal cells to produce androgens, which diffuse through the basement membrane to the granulosa cells; there the androgens are converted to estrogens by the FSH-primed granulosa cells. Only minute amounts of ovarian androgens enter the blood, because they are almost completely converted to estrogen within the ovaries.

3. As estrogen levels slowly rise in the plasma, they exert negative feedback on the anterior pituitary and hypothalamus, causing the pituitary output of FSH and LH to fall.* But within the ovary, estrogen intensifies the effect of FSH on follicle growth and maturation, thereby increasing the output of estrogen.

4. While the initial small rise in estrogen blood levels inhibits the hypothalamic-pituitary axis as just described, high estrogen levels have the opposite effect. Once estrogen reaches a critical concentration in the blood, it exerts positive feedback on the brain and anterior pituitary.

5. The stimulatory effects of high estrogen levels set a cascade of events into motion. First there is a sudden burstlike release of LH (and, to a lesser extent, FSH) by the anterior pituitary. This occurs approximately midcycle (also see Figure 28.19a).

6. The sudden flush of LH stimulates resumption of meiosis in the primary oocyte of the mature follicle, causing it to complete the first meiotic division; the secondary oocyte formed continues on to metaphase II. LH also induces synthesis of enzymes that cause the bulging ovarian wall to rupture, and ovulation occurs at or around day 14. The role (if any) of FSH in this process is unknown. Shortly after ovulation, estrogen levels begin to decline; this probably reflects the damage done to the dominant estrogen-secreting follicle during ovulation.

7. The LH surge also promotes transformation of the ruptured follicle into a corpus luteum (hence the name luteinizing hormone). LH stimulates the corpus luteum to produce progesterone and smaller amounts of estrogen almost immediately after it is formed.

8. As progesterone and estrogen levels rise in the blood, the combination exerts a powerful negative feedback, or inhibitory, effect on the release of LH and FSH from the anterior pituitary. As gonadotropins decline, development of new follicles during the luteal phase is inhibited, and subsequent LH surges that might cause additional oocytes to be ovulated are prevented.

9. As LH blood levels decline slowly, the stimulus for corpus luteum activity ends, and the corpus luteum begins to degenerate. As goes the corpus luteum, so go the levels of ovarian hormones, and blood estrogen and progesterone levels drop sharply. (However, if implantation of an embryo has occurred, the activity of the corpus luteum is maintained by an LH-like hormone released by the developing embryo.)

10. A marked decline in ovarian hormones at the end of the cycle (days 26–28) ends their blockade of FSH and LH secretion, and the cycle starts anew.

The Uterine (Menstrual) Cycle

Although the uterus is the receptacle in which the young embryo implants and develops, it is receptive to implantation only for a very short period each month. Not surprisingly, this brief interval coincides exactly with the time when a developing embryo would normally reach the uterus, approximately seven days after ovulation. The **uterine,** or **menstrual** (men′-stroo-ul), **cycle** is a series of cyclic changes that the uterine endometrium goes through, month after month, as it responds to changes in levels of ovarian hormones in the blood.

Generally speaking, both the ovarian and uterine cycles are about 28 days long with ovulation occurring at the midpoint of the cycles, on day 14. Figure 28.19 describes the uterine cycle and illustrates how the hormonal and local events of the ovarian and uterine cycles fit together.

Notice that the menstrual and proliferative phases overlap the follicular and ovulatory stages of the ovarian cycle, and the secretory phase corresponds to the ovarian luteal phase. The essential point to keep in mind is that uterine changes are responses to the changing levels of estrogen and progesterone released during the ovarian cycle.

Exercise has long been touted as good for your bones, and so it is, in moderation. However, the extremely vigorous activity of trained athletes can delay onset of menarche in girls and can disrupt the normal menstrual cycle in adult women, even causing *amenorrhea* (ā″-meh-nor-ē′-uh), the cessation of menses. Part of the problem appears to be that female athletes have little body fat, and fat deposits aid in conversion of adrenal androgens to estrogens. In addition, hypothalamic controls appear to be blocked in some way by severe physical regimens. Although these effects are believed to be totally reversible when the athletic training is discontinued, a worrisome consequence of amenorrhea in young, healthy adult women is that they begin to suffer the dramatic loss in bone mass usually seen only in old age. Once estrogen levels drop and the menstrual cycle stops, whether because of disease, menopause, or exercise, bone loss begins. Currently, female athletes are encouraged to increase their daily calcium intake to 1.5 g, roughly the amount contained in a quart of milk. ■

Extrauterine Effects of Estrogen and Progesterone

Estrogen is analogous to the testosterone of males. As estrogen levels rise during puberty, it promotes oo-

*There is also some evidence that *inhibin*, released by the granulosa cells, also exerts negative feedback controls of FSH release during this period.

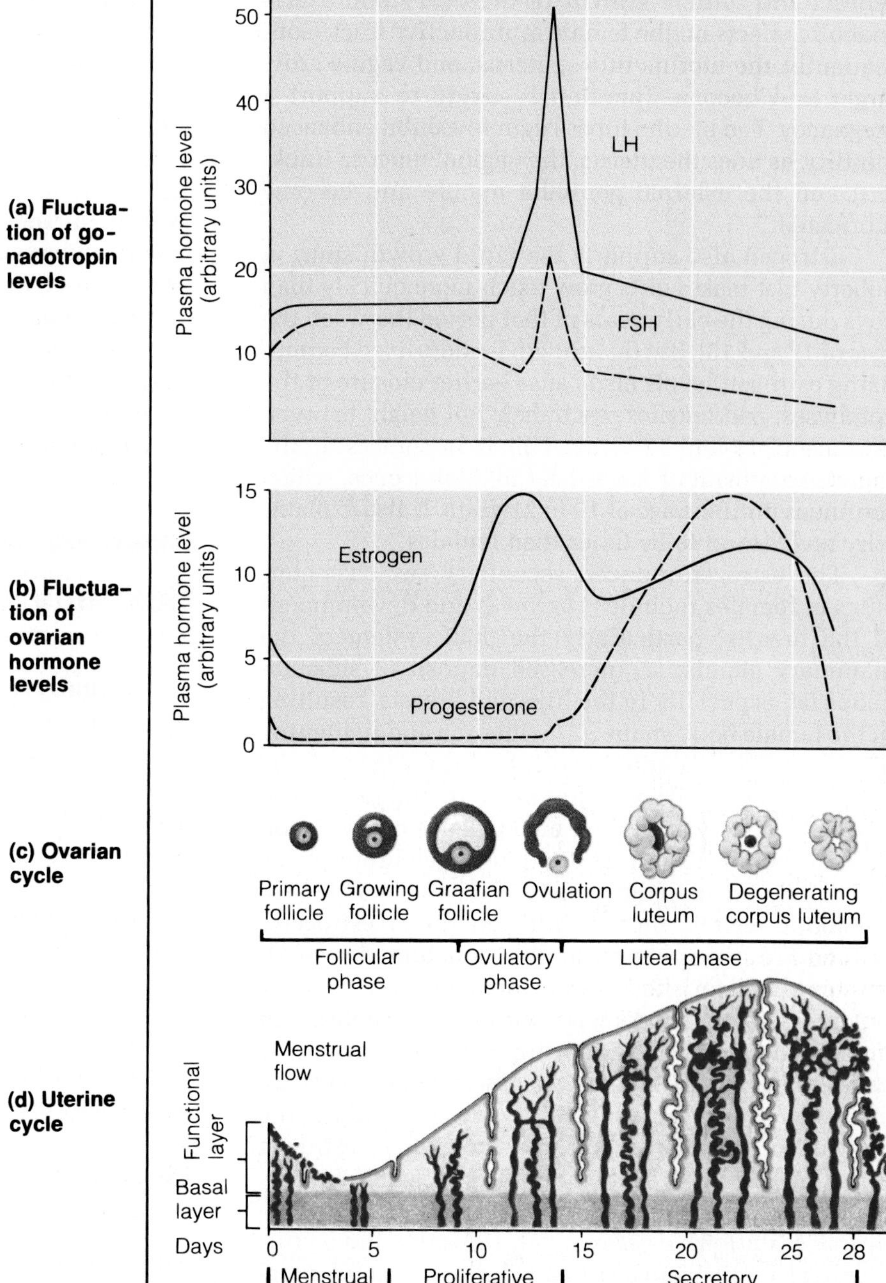

Figure 28.19 The uterine cycle and its correlation to the ovarian cycle. (**a**) The fluctuating blood levels of gonadotropins that help regulate the events of the ovarian cycle. (**b**) The fluctuating blood levels of ovarian hormones that cause the endometrial changes of the uterine cycle (and feed back to influence gonadotropin levels). (**c, d**) The correlation between local ovarian and uterine events during the 28-day cycle.

The events of the three stages of the uterine cycle, depicted in (**d**), are as follows:

1. Days 1–5, **menstrual phase.** The thick functionalis layer of the uterine endometrium becomes detached from the uterine wall, a process that is accompanied by bleeding for three to five days. The detached tissues and blood pass through the vagina as the menstrual flow; the average blood loss during this period is 50–150 ml. Meanwhile, the growing ovarian follicles are starting to produce estrogen. (see **b**).

2. Days 6–14, **proliferative phase.** Under the influence of rising blood levels of estrogen, the endometrium is repaired by the stratum basalis. The stratum functionalis thickens, tubular glands form, and the spiral arteries become more numerous. (See also Figure 28.13.) The mucosa once again becomes velvety, thick, and well vascularized. Estrogen also induces synthesis of progesterone receptors in the endometrial cells, readying them for interaction with progesterone.

Normally, the cervical mucus is thick and sticky, but rising estrogen levels cause it to thin and become crystalline, forming channels that facilitate the passage of sperm into the uterus. Ovulation occurs in the ovary at the end of this stage, in response to the burst-like release of more LH from the anterior pituitary.

3. Days 15–28, **secretory phase.** Rising levels of progesterone, produced by the corpus luteum, act on the estrogen-primed endometrium, causing the spiral arteries to elaborate and coil and converting the functionalis layer to a secretory mucosa. The glands enlarge and begin secreting glycogen into the uterine cavity. These nutrients will sustain a developing embryo until it has implanted in the blood-rich endometrial lining. Increasing progesterone levels also cause the cervical mucus to become viscous again, forming the *cervical plug,* which blocks sperm entry and plays an important role in keeping the uterus "private" in the event an embryo has begun to implant.

If fertilization has not occurred, the corpus luteum begins to degenerate toward the end of the secretory phase as LH blood levels decline. Progesterone levels fall, depriving the endometrium of hormonal support, and the spiral arteries kink and go into spasms. Denied oxygen and nutrients, the endometrial cells begin to die, and as their lysosomes rupture, the stratum functionalis begins to "self-digest," setting the stage for menses to begin on day 28. Menses is precipitated by the sudden dilation of arteries supplying the endometrium on day 28. As blood flushes into the weakened capillary beds, they fragment, causing the functionalis layer to slough off. The cycle starts again with the first day of menstrual flow.

genesis and follicle growth in the ovary and exerts anabolic effects on the female reproductive tract. Consequently, the uterine tubes, uterus, and vagina grow larger and become functional—ready to support a pregnancy. The uterine tubes begin to exhibit enhanced motility, as does the uterus; the vaginal mucosa thickens; and the external genitalia mature and become lubricated.

Estrogen also supports the rapid growth spurt at puberty that makes girls grow much more quickly than boys during the early years of that period (between the ages of 12 and 13). But this growth is short-lived because rising estrogen levels also cause earlier closure of the epiphyses, and females reach their full height between the ages of 15 and 17 years. This is in contrast to the aggressive growth of males later in adolescence, which continues until the age of 19 to 21 years. It also explains why males tend to be taller than females.

The estrogen-induced secondary sex characteristics of females include (1) growth and development of the breasts, particularly the duct system of the mammary glands; (2) increased deposit of subcutaneous fat, especially in the hips and breasts, resulting in the female body shape; (3) widening and lightening of the pelvis (adaptations for childbirth); (4) the appearance of axillary and pubic hair; and (5) several metabolic effects, including maintenance of low blood cholesterol levels and facilitation of calcium uptake, which helps sustain the density of the skeleton. (These metabolic effects, though initiated under estrogen's influence during puberty, are not considered secondary sex characteristics in the strict sense.)

Progesterone works with estrogen to establish and then help regulate the uterine cycle and promotes changes in cervical mucus. Its other major effects are exhibited during pregnancy, when it inhibits the motility of the uterus and takes up where estrogen leaves off in preparing the breasts for lactation. Indeed, progesterone is named for these important roles (*pro* = for; *gestation* = pregnancy). However, the source of progesterone and estrogen during most of pregnancy is the placenta, not the ovaries.

Female Sexual Response

The **female sexual response** is similar to that of males in most respects. Sexual excitement is accompanied by engorgement of the clitoris, vaginal mucosa, and breasts with blood; erection of the nipples; and increased secretory activity of the vestibular glands, which lubricates the vestibule and facilitates the entry of the penis. These events, though more widespread, are analogous to the *erection* phase in men. Sexual excitement is promoted by touch and psychological stimuli and is mediated along the same autonomic nerve pathways as in males.

The final phase of the female sexual response, *orgasm*, is not accompanied by ejaculation, but muscle tension increases throughout the body, pulse rate and blood pressure rise, and rhythmic contractions of the uterus occur. As in males, orgasm is accompanied by an intense sensation of pleasure and followed by general relaxation, but in addition, a sensation of warmth surges through the body. Unlike the situation in males, orgasm in females is not followed by a refractory period. Consequently, females may experience multiple orgasms during a single sexual experience. A man must achieve orgasm and ejaculate if fertilization is to occur, but female orgasm is not required for conception. Indeed, some women never experience orgasm, yet are perfectly able to conceive.

Sexually Transmitted Diseases

Sexually transmitted diseases (STDs), also called *venereal diseases (VD),* are infectious diseases spread through sexual contact; as a group, they are the single most important cause of reproductive disorders. Until recently, gonorrhea and syphilis, caused by bacterial infections, were the most common STDs, but within the past decade, viral diseases such as genital herpes and AIDS have taken center stage. AIDS, caused by the HIV virus that cripples the immune system, was described in Chapter 22. The important bacterial STDs (gonorrhea, syphilis, and chlamydia) and genital herpes are described here. Condoms are effective in helping to prevent the spread of STDs, and their use is strongly urged since the advent of AIDS.

Gonorrhea

The causative agent of **gonorrhea** (gah″-nor-ē′-uh), called "the clap" in common usage, is *Neisseria gonorrhoeae,* which invades the mucosae of the reproductive and urinary tracts and associated structures. These kidney bean–shaped bacteria are spread by genital contact, as well as by contact with the anal and pharyngeal mucosal surfaces.

The most common symptom of gonorrhea in males is *urethritis* (inflammation of the urethra), accompanied by painful urination and a discharge of pus from the penis (penile "drip"). In women, symptoms are varied, ranging from none (about 20% of cases) to lower abdominal discomfort, vaginal discharge, abnormal uterine bleeding, and, occasionally, urethral symptoms similar to those seen in males.

Untreated gonorrhea can cause urethral constriction and inflammation of the entire male duct system;

in women, it causes pelvic inflammatory disease and sterility. These consequences have been rare since the 1950s, because gonorrhea has been easily treated with penicillin and certain other antibiotics. However, strains resistant to those antibiotics are becoming increasingly prevalent, making this disease more difficult to cure.

Syphilis

Syphilis (sih'-fih-lis), caused by *Treponema pallidum,* a corkscrew-shaped bacterium, is usually transmitted sexually. However, it can be contracted congenitally from an infected mother. This bacterium easily penetrates intact mucosae and abraded skin and enters the local lymphatics and the bloodstream; within a few hours of exposure, an asymptomatic systemic infection is in progress. The incubation period is typically two to three weeks (but may be as long as 60 days), at the end of which a red, painless primary lesion called a *chancre* (shank'-er) appears at the site of bacterial invasion. In males, this is typically the penis, but in females the lesion is usually within the vagina or on the cervix and often goes undetected. The primary lesion persists for one to a few weeks during which time it ulcerates and becomes crusty; then it heals spontaneously and disappears.

If syphilis is untreated, its secondary signs appear several weeks later. (The time varies, but six weeks is typical.) Any organ may be involved. A pink skin rash that appears all over the body is one of the first symptoms. Fever, joint pain, and a general feeling of illness are common. Anemia may develop, and the hair may begin to fall out in clumps. Like primary syphilis, the secondary stage disappears spontaneously, usually within three to six weeks. Then the disease enters the *latent period* and is detectable only by a serological diagnostic test, such as the venereal disease research lab (VDRL) test. The latent stage may last a person's lifetime (or the bacteria may be killed by the immune system), or it may be followed by the signs of *tertiary syphilis* at any time. Tertiary syphilis is characterized by *gummas* (guh'-muz), destructive lesions of the CNS, blood vessels, bones, and skin. Penicillin, which interferes with the ability of dividing bacteria to synthesize new cell walls, is still the treatment of choice for all stages of syphilis, but prolonged therapy is necessary because the bacterium multiplies very slowly.

Chlamydia

Chlamydia (kluh-mih'-dē-uh) is a largely unrecognized, silent epidemic that infects perhaps 3 to 4 million people yearly, making it the most common sexually transmitted disease in the United States. Since chlamydia was not recognized as a health problem until the 1970s, and physicians are not required to report it, accurate statistics are hard to come by. However, chlamydia is responsible for 25% to 50% of all diagnosed cases of pelvic inflammatory disease, and each year, more than 150,000 infants are born to infected mothers. About 20% of men and 30% of women infected with gonorrhea are also infected by *Chlamydia trachomatis,* the causative agent of chlamydia.

Chlamydia is a bacterium with a viruslike dependence on host cells. Its incubation period is about one week. Symptoms, which often go unrecognized, include urethritis; vaginal discharge or burning; abdominal, rectal, or testicular pain; painful intercourse; and irregular menses. In men, this infection can cause arthritis as well as widespread urogenital tract infection; in women, its most severe consequence is sterility. Newborns infected in the birth canal tend to develop conjunctivitis and respiratory tract inflammations including pneumonia. The disease can be diagnosed by cell culture techniques and is easily treated with tetracycline.

Genital Herpes

Human herpesviruses (*herpes simplex, Epstein-Barr virus,* and others) have a worldwide distribution and are among the most difficult human pathogens to control. They remain silent for weeks or years and then suddenly flare up, causing a burst of blisterlike lesions. Direct transmission of herpes simplex virus type 2, the common cause of **genital herpes** (her'-pēz), is via infectious secretions. The resulting painful lesions that appear on the reproductive organs of the unlucky minority of infected adults are usually more of a nuisance than a threat to life. However, congenital herpes infections can cause severe malformations of a fetus, and herpes infections have been implicated as a cause of cervical cancer. Most people who have genital herpes do not know it, and it has been estimated that one-quarter to one-third of all adult Americans harbor the type 2 herpes simplex virus. ■

Developmental Aspects of the Reproductive System: Chronology of Sexual Development

Thus far, we have been describing the structure and function of the reproductive organs as they exist and operate in adults. Now we are ready to look at the timing of events that cause us to become sexual individuals. These events begin long before birth and end, at least in women, in late middle age.

Embryological and Fetal Events

Determination of Genetic Sex

Genetic sex is determined at the instant the genes of a sperm combine with those of an ovum. The determining factor is the **sex chromosomes** each gamete contains. Of the 46 chromosomes in the fertilized egg, two (one pair) are sex chromosomes; the other 44 are called **autosomes.** Two types of sex chromosomes, quite different in size, exist in humans: the large ***X* chromosome** and the much smaller ***Y* chromosome.** The body cells of females have two *X* chromosomes and are designated *XX*; the ovum resulting from normal meiosis in a female always contains an *X* chromosome. Males have one *X* chromosome and one *Y* in each body cell (*XY*); approximately half of the sperm produced by normal meiosis in males contain an *X* and the other half a *Y*. If the fertilizing sperm delivers an *X* chromosome, the fertilized egg and its daughter cells will contain the female (*XX*) composition, and the embryo will develop ovaries. If the sperm bears the *Y*, the offspring will be male (*XY*) and will develop testes. Thus, the father determines the genetic sex of the offspring, and all subsequent events of sexual differentiation depend on which gonads are formed during embryonic life.

When meiosis fails to distribute the sex chromosomes to the gametes properly (an event called *nondisjunction*), abnormal combinations of sex chromosomes can occur in the fertilized egg and lead to striking abnormalities in sexual and reproductive system development. For example, females with a single *X* chromosome (*XO*), a condition called *Turner's syndrome*, never develop ovaries. Males with no *X* chromosome (*YO*) die during embryonic development. *Metafemales*, or females carrying *XXX*, *XXXX*, or other multiples of the *X* chromosome, have underdeveloped ovaries and limited fertility and as a rule are mentally retarded. *Klinefelter's syndrome*, which affects one out of 500 live male births, is the most common sex chromosome abnormality. Affected individuals usually have a single *Y* chromosome and two or more *X* chromosomes and are sterile males. Although *XXY* males are most often normal (or only slightly below normal) intellectually, the incidence of mental retardation increases as the number of *X* sex chromosomes rises. ■

Sexual Differentiation of the Reproductive System

The gonads of both males and females begin their development during the fifth week of gestation from a mass of mesoderm called the **gonadal ridges** (Figure 28.20). The gonadal ridges appear as bulges just medial to the mesonephros (the developing kidney system that precedes the final metanephros-derived kidney in the abdominal cavity). The **paramesonephric** (peh-ruh-meh-zō-neh′-frik), or **Müllerian, ducts** (future female ducts) develop laterally to the **mesonephric ducts** (future male ducts), and both sets of ducts empty into a common chamber called the *cloaca*. At this stage of embryonic development, the embryo is said to be in the *indifferent stage*, because the gonadal ridge tissue can develop into either male or female gonads and both duct systems are present.

Shortly after the gonadal ridges make their appearance, *primordial germ cells* migrate to them from an embryonic structure called the *yolk sac* (see Figure 29.7) and seed the developing gonads with stem cells destined to become spermatogonia or oogonia. Once these cells are in residence, the gonadal ridges differentiate into testes or ovaries, depending on the genetic makeup of the embryo. The process begins about the seventh week in male (*XY*) embryos. Seminiferous tubules form in the internal part of the gonadal ridges and become continuous with the mesonephric duct via the efferent ductules. Continued development of the mesonephric duct produces the accessory reproductive organs of the male. The paramesonephric ducts play no part in the male's developmental process and soon degenerate.

If the embryo is a developing female (*XX*), the process begins slightly later, at about the eighth week. The outer, or cortical, part of the immature ovaries form follicles. Shortly thereafter, the paramesonephric ducts differentiate into the structures of the female duct system, and the mesonephric ducts degenerate.

Like the gonads, the external genitalia arise from the same structures in both sexes (Figure 28.21). During the indifferent stage, all embryos exhibit a small projection called the **genital tubercle** on the external body surface. The urogenital sinus, which develops from subdivision of the cloaca, is just deep to the tubercle, and the **urethral groove,** which serves as the external opening of the urogenital sinus, is on the tubercle's inferior surface. The urethral groove is flanked laterally by the **urethral folds** and then the **labioscrotal swellings.**

During the eighth week, the external genitalia begin to develop rapidly. In males, the genital tubercle enlarges, forming the penis. The urethral folds fuse around the urethral groove, forming the corpus spongiosum of the penis; only the tips of the folds remain unfused to form the *urethral orifice* at the tip of the penis. The labioscrotal swellings fuse externally to the urethral folds, forming the scrotum. In females, the genital tubercle gives rise to the clitoris; the unfused urethral folds become the labia minora, and the unfused labioscrotal folds become the labia majora.

Differentiation of accessory structures and the external genitalia as male or female structures is totally dependent on the presence or absence of testosterone. When testes are formed, they quickly begin to release

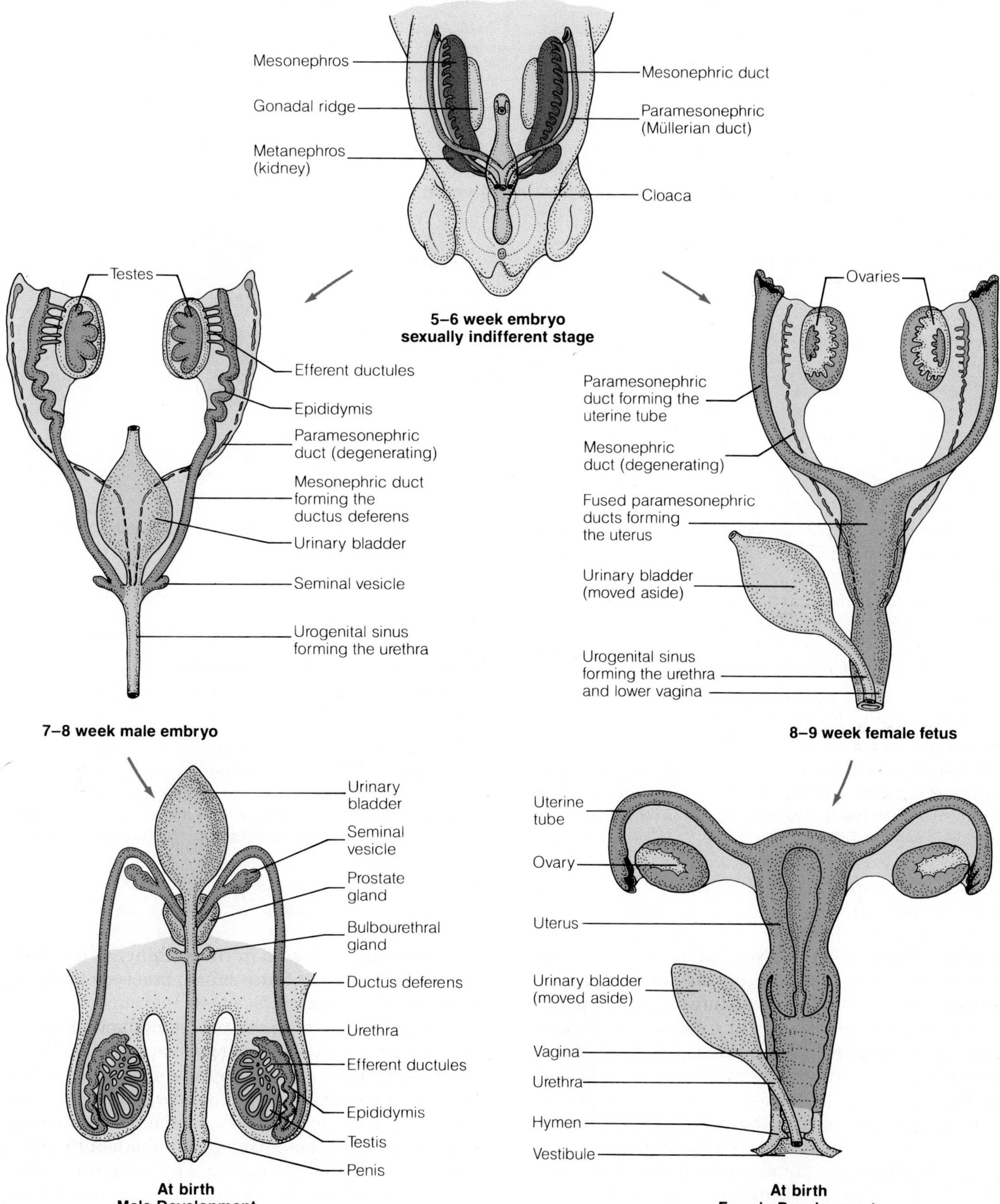
Mesonephros
Gonadal ridge
Metanephros (kidney)
Mesonephric duct
Paramesonephric (Müllerian duct)
Cloaca
5–6 week embryo sexually indifferent stage
Testes
Efferent ductules
Epididymis
Paramesonephric duct (degenerating)
Mesonephric duct forming the ductus deferens
Urinary bladder
Seminal vesicle
Urogenital sinus forming the urethra
7–8 week male embryo
Ovaries
Paramesonephric duct forming the uterine tube
Mesonephric duct (degenerating)
Fused paramesonephric ducts forming the uterus
Urinary bladder (moved aside)
Urogenital sinus forming the urethra and lower vagina
8–9 week female fetus
Urinary bladder
Seminal vesicle
Prostate gland
Bulbourethral gland
Ductus deferens
Urethra
Efferent ductules
Epididymis
Testis
Penis
At birth Male Development
Uterine tube
Ovary
Uterus
Urinary bladder (moved aside)
Vagina
Urethra
Hymen
Vestibule
At birth Female Development

Figure 28.20 Development of the internal reproductive organs.

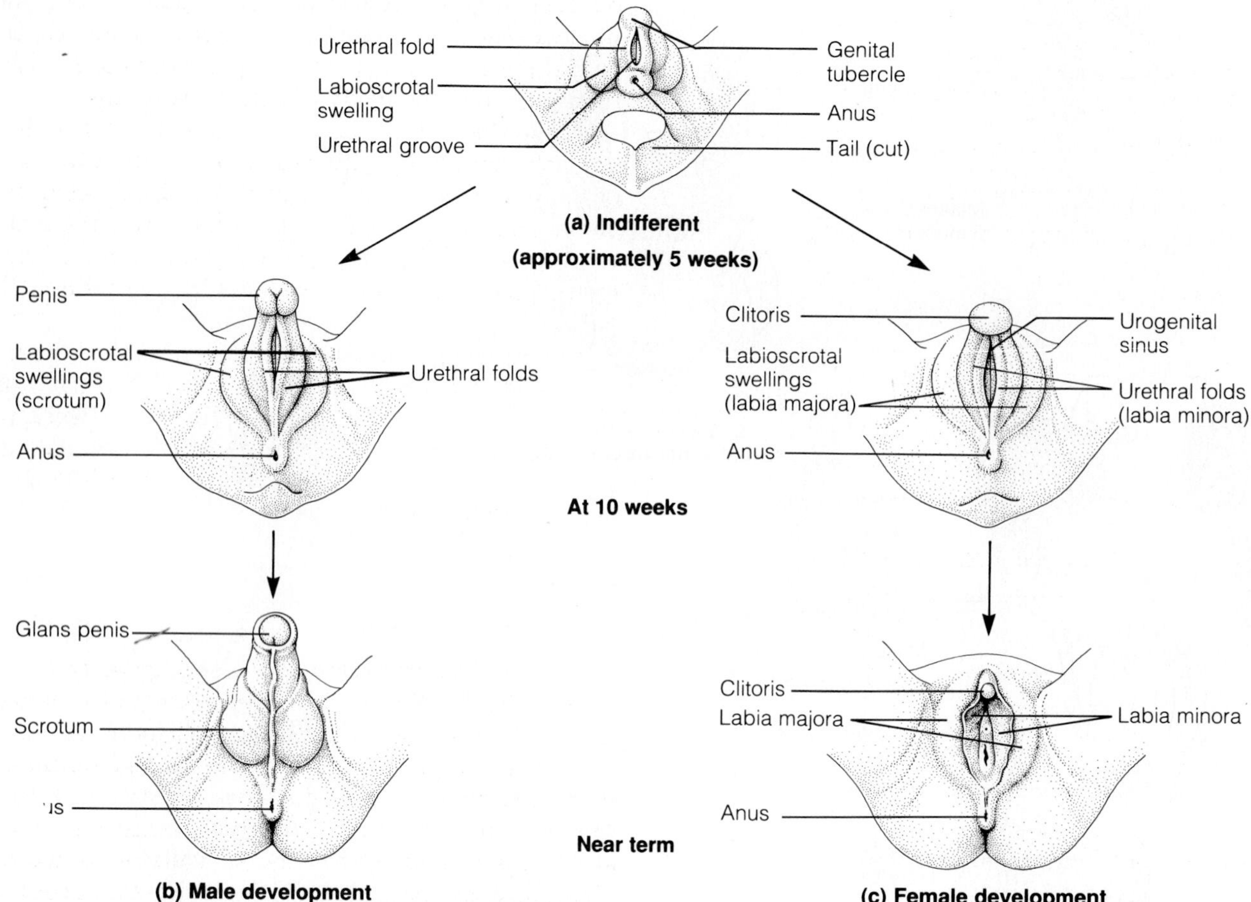

Figure 28.21 Development of external genitalia in humans. (**a**) The external reproductive organs are undifferentiated until about the eighth week of gestation. During the indifferent stage, all embryos have a conical elevation called the genital tubercle, which has a ventral opening called the urethral groove. The groove is surrounded by urethral folds, which in turn are bounded by the labioscrotal swellings. (**b**) In males, under the influence of testosterone, the urethral groove elongates and closes completely, and the urethral folds give rise to the shaft of the penis. The labioscrotal swellings develop into the scrotum. (**c**) In females (or in the absence of testosterone), the urethral groove remains open, and the urethral folds become the labia minora. The labioscrotal swellings develop into the labia majora.

testosterone, which causes the mesonephric ducts to develop into male accessory organs and the genital tubercle and its associated structures to form a penis and scrotum. In the absence of testosterone, female accessory organs arise from the paramesonephric ducts, and the genital tubercle structures develop in the female pattern.

Any interference with the normal pattern of sex hormone production in the embryo results in bizarre abnormalities. For example, if the embryonic testes fail to produce testosterone, a genetic male develops the female accessory structures and external genitalia. On the other hand, if a genetic female is exposed to testosterone (as might happen if the mother has an androgen-producing tumor of her adrenal gland), the embryo has ovaries but develops male accessory ducts and glands, as well as a penis and an empty scrotum. It appears that the female pattern of reproductive structures has an intrinsic ability to develop, and in the absence of testosterone, proceeds to do so, regardless of the embryo's genetic makeup.

Individuals with accessory reproductive structures that do not "match" their gonads are called *pseudohermaphrodites* (soo-dō-her-ma′-frō-dīts), to distinguish them from true *hermaphrodites,* rare individuals who possess both ovarian and testicular tissue. In recent years, many pseudohermaphrodites have sought sex-change operations to match their outer selves (external genitalia) with their inner selves (gonads). ■

Descent of the Gonads

About two months before birth, the testes begin their descent toward the scrotum, dragging their supplying blood vessels and nerves along behind them (Figure

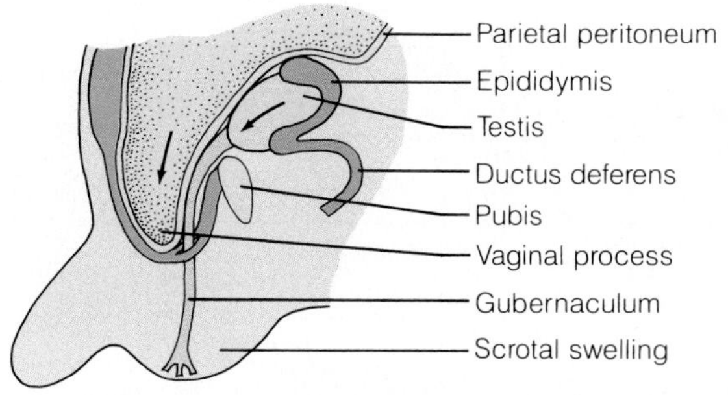

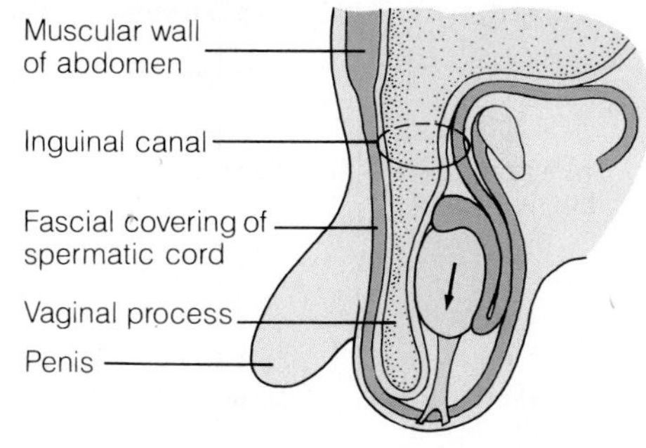

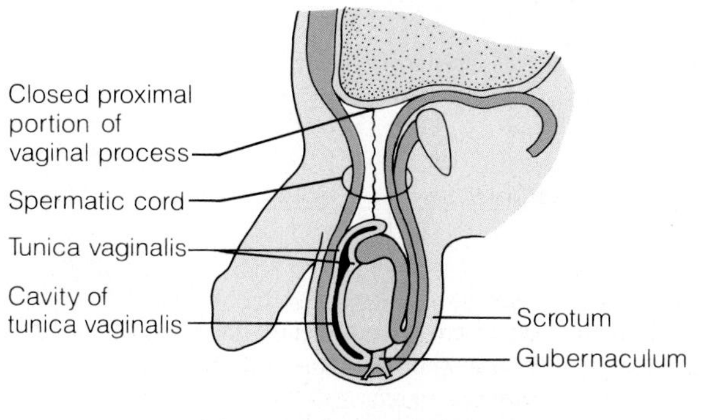

Figure 28.22 Descent of the testes. (**a**) A testis begins to descend behind the peritoneum about two months before birth, and the vaginal process (a peritoneal extension) protrudes into the tissue that will become the scrotum (scrotal swellings). (**b**) The testis, accompanied by the vaginal process, descends into the scrotum via the inguinal canal. (**c**) In the one-month-old infant, descent is complete. The lower part of the vaginal process becomes the tunica vaginalis covering of the testis. The proximal portion of the vaginal process is obliterated.

28.22). They finally exit from the pelvic cavity via the inguinal canals, oblique passages through the oblique and transverse muscles of the inferior abdominal wall, and enter the scrotum. This migration is stimulated by testosterone made by the male fetus's testes, but it appears to be guided mechanically by a strong fibromuscular cord called the **gubernaculum,** anchored both to the testis and to the floor of the scrotal sac. The *tunica vaginalis* covering of the testis that is pulled into the scrotum with the testis is derived from an outpocketing of the parietal peritoneum, the *vaginal process.* The accompanying blood vessels, nerves, and fascial layers form part of the *spermatic cord,* which helps suspend the testis within the scrotum.

Like the testes, the ovaries descend during fetal development, but in this case only to about the level of the pelvic brim. Each is guided in its descent by a gubernaculum (anchored in the labium majora) that later divides, becoming the ovarian and round ligaments that help support the internal genitalia in the pelvis.

Failure of the testes to make their normal descent leads to *cryptorchidism* (*crypt* = hidden, concealed; *orchi* = testicle). Because this condition causes sterility and increases the risk of testicular cancer, surgery is usually performed during early childhood to rectify this problem. ■

Puberty

Puberty is the period of life, generally between the ages of 10 and 15 years, when the reproductive organs grow to their adult size and become functional under the influence of rising levels of gonadal hormones (testosterone in males and estrogen in females). After this time, reproductive capability continues until old age in males and menopause in females. Since the secondary sex characteristics and regulatory events of puberty were described earlier, these details will not be repeated here. But it is important to remember that puberty represents the earliest period of reproductive system activity in the scheme of sexual development.

The events of puberty occur in the same sequence in all individuals, but the age at which they occur varies widely. In males, the event that signals puberty's onset is enlargement of the testes and scrotum between the ages of 13 and 14 years, which is followed by the appearance of pubic, axillary, and facial hair. Growth of the penis goes on over the next two years, and sexual maturation is evidenced by the presence of mature sperm in the semen. In the meantime, the young man has unexpected, and often embarrassing, erections and frequent nocturnal emissions ("wet dreams") as his hormones surge and the hormonal control axis struggles to achieve a normal balance.

The first sign of puberty in females is budding breasts, often apparent by the age of 11 years; menarche usually occurs about two years later. Dependable ovulation and fertility await the maturation of the hormonal controls, which takes nearly two more years.

Menopause

Most women reach the peak of their reproductive abilities in their late 20s. After that, ovarian function declines gradually, presumably because the ovaries

A CLOSER LOOK Contraception: To Be or Not To Be

In a society such as ours, where many women opt for professional careers or must work for economic reasons, contraception (*contra* = against; *cept* = taking), or birth control, is often seen as a necessity. Although scientists across the country are making some headway in the search for a male contraceptive, thus far the burden for birth control has fallen on women's shoulders, and most birth control products are female directed.

There is little doubt that many potential offspring die naturally before their presence is even recognized, but the key to birth control is dependability. As shown by the red arrows in the accompanying flowchart, the birth control techniques and products currently available have many sites of action for blocking the reproductive process. Let's examine the relative advantages of a few of these more closely.

The most used contraceptive product in the United States is the *birth control pill*, or simply "the pill," a preparation containing minute amounts of estrogens and progestins (progesterone-like hormones) that is taken daily, except for the last five days of the 28-day cycle. The pill tricks the hypothalamic-pituitary axis and "lulls it to sleep," because the relatively constant blood levels of ovarian hormones make it appear that the woman is pregnant (both estrogen and progesterone are produced throughout pregnancy). Ovarian follicles do not develop and ovulation ceases. The endometrium does proliferate slightly and is sloughed off monthly when the pill is discontinued, but menstrual flow is much reduced. However, since hormonal balance in the body is one of the most precisely controlled of all body functions, some women simply cannot

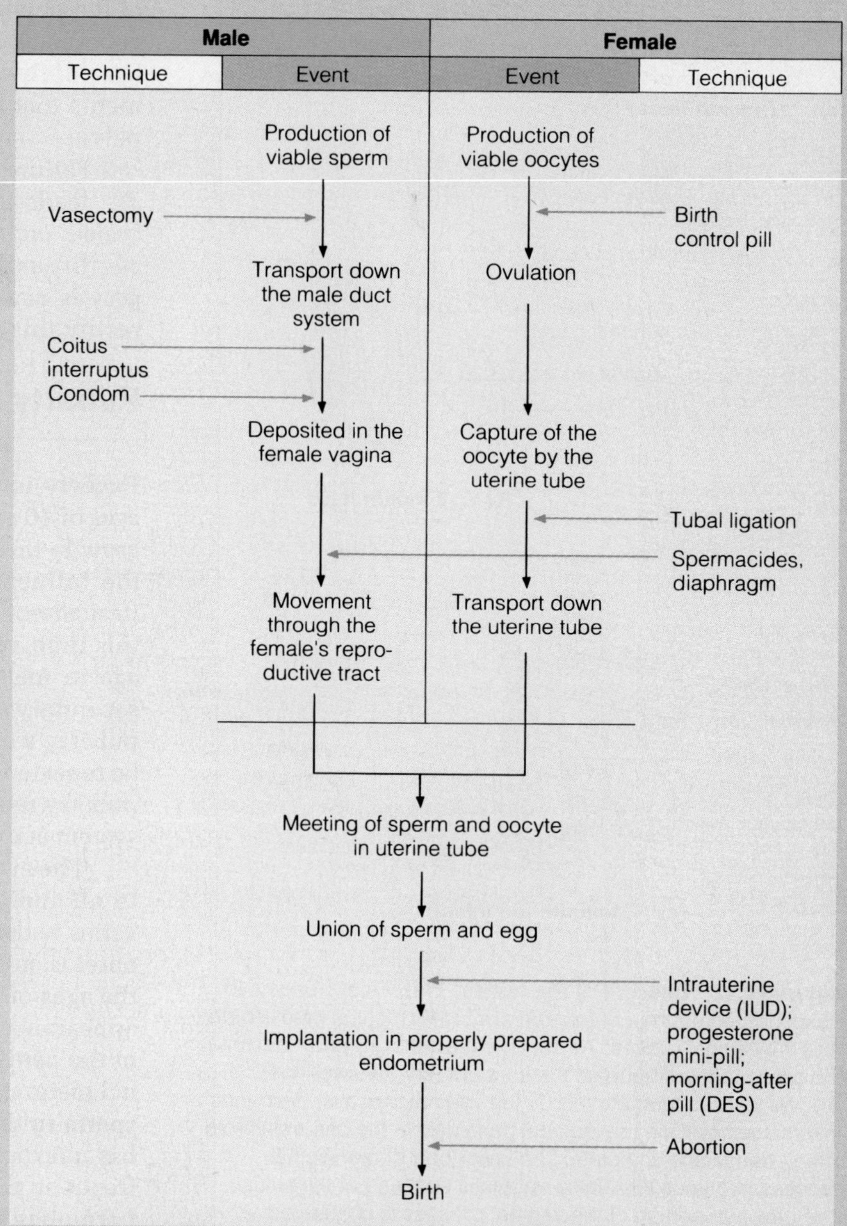

Flowchart of the events that must occur to produce a baby. Techniques or products that interfere with the process are indicated by colored arrows at the site of interference.

tolerate the changes caused by the pill—they become nauseated and/or hypertensive. For a while, the pill was suspected of increasing the incidence of breast and uterine cancer, but it now appears that the new very-low-dose preparations may actually help protect against some forms of cancer (ovarian and endometrial) and have also reduced the incidence of serious cardiovascular side effects, such as strokes, heart attacks, and blood clots, that occurred (rarely) with earlier forms of the pill. Presently, the pill is one of the most widely used drugs in the world; well over 50 million women use these drugs to prevent pregnancy. Its incidence of failure is less than one pregnancy per 100 women per year.

For several years, the second most used contraceptive method was the *intrauterine device (IUD)*, a plastic or metal device inserted into the uterus that prevented implantation of the fertilized egg in the endometrial lining. The failure rate of the IUD was nearly as low as that of the pill. However, with the exception of a single product, IUDs have been taken off the market in the United States because of occasional contraceptive failure, uterine perforation, or pelvic inflammatory disease (resulting in lawsuits).

Some methods, such as *tubal ligation* or *vasectomy* (cutting or cauterizing the uterine tubes or vas deferens, respectively) are nearly foolproof and are the choice of approximately 33% of couples of childbearing age in the United States. Both procedures can be done in the physician's office. However, these techniques are usually permanent, making them unpopular with individuals who still plan to have children but want to select the time.

Coitus interruptus, or withdrawal of the penis just before ejaculation, is simply against nature, and control of ejaculation is never assured. *Rhythm* or *fertility awareness methods* are based on recognizing the period of ovulation or fertility and avoiding intercourse during those intervals. This may be accomplished by (1) recording daily basal body temperatures (body temperature drops slightly [0.2°–0.6°F] immediately prior to ovulation and then rises slightly [0.2°–0.6°F] after ovulation) or (2) recording changes in the consistency of vaginal mucus (the mucus first becomes sticky and then clear and stringy, much like egg white, during the fertile period). Both of these rhythm techniques require accurate record keeping for several cycles before they can be used with confidence, but have a high success rate for those willing to take the trouble. *Barrier methods*, such as diaphragms, cervical caps, condoms, spermicidal foams, gels, and sponges, are quite effective, especially when used by both partners. But many avoid them because they can reduce the spontaneity of sexual encounters.

The *morning-after pill*, a high dose of synthetic estrogen called *diethylstilbestrol (DES)*, is a postcoital contraceptive. Its precise mechanism is not known, but it prevents implantation. It would seem that this would be the most desirable type of contraceptive—one that need be taken only when sexual relations have actually occurred—but its side effects (severe nausea and others) are so unpleasant that DES is routinely used only for victims of rape or incest.

Several new products are presently undergoing clinical trials, including the following:

1. *Subdermal implants*. These tiny capsules, which are inserted under the skin and slowly release progestins, seem to be very effective. Their advantage, like that of the IUD, is passivity; that is, they do not rely on the woman's remembering to take a daily pill or the couple's willingness to use a barrier every time they have intercourse. Another advantage is that the protection lasts for about five years.

2. *Inhibin*. Inhibin acts on the anterior pituitary gland to depress FSH secretion without changing levels of the other pituitary hormones that are often released in tandem (like LH). Some researchers believe that it can provide an ideal contraceptive for both women and men.

3. *Anti-HCG vaccine*. HCG, a hormone released shortly after fertilization occurs, is crucial to retention of the endometrium for implantation of the embryo. Without HCG, menses occurs instead of pregnancy. A vaccine that induces the formation of anti-HCG antibodies by the mother's body is now undergoing clinical trials. Thus far, there have been no observable side effects with this vaccine, and normal fertility is restored after 7 to 16 months.

This summary has barely touched the large number of experimental birth control drugs now awaiting clinical trials; and other methods are sure to be developed in the near future. In the final analysis, however, the only 100% effective means of birth control is the age-old one—*total abstinence*.

become less and less responsive to gonadotropin signals. As estrogen production declines, many ovarian cycles become anovulatory, and menstrual periods become erratic and shorter in length. Eventually, ovulation and menses cease entirely; this normally occurs between the ages of 46 and 54 years, an event called menopause. **Menopause** is considered to have occurred when a whole year has passed without menstruation.

Although estrogen production continues for a while after menopause, the ovaries finally become nonfunctional as endocrine organs. Deprived of the stimulatory effects of estrogen, the reproductive organs and breasts begin to atrophy. The vagina becomes dry; intercourse may become painful, particularly if infrequent; and vaginal infections become increasingly common. Other sequelae to cessation of estrogen release include irritability and other mood changes (depression in some); intense vasodilation of the skin's blood vessels (seen in 75% to 80% of menopausal women), which causes uncomfortable "hot flashes" and night sweats; gradual thinning of the skin and loss of bone mass; and slowly rising blood cholesterol levels, which place postmenopausal women at risk for cardiovascular disorders. Some physicians prescribe low-dose estrogen-progestin preparations to help women through this often difficult period and to prevent the skeletal and cardiovascular complications.

There is no equivalent of menopause in males. While aging men do exhibit a steady decline in testosterone secretion and a longer latent period after orgasm, a male's reproductive capability seems unending. Healthy elderly men are able to father offspring well into their eighth decade of life.

* * *

The reproductive system is unique among organ systems in at least two ways: (1) It is nonfunctional during the first 10 to 15 years of life, and (2) it is capable of interacting with the complementary system of another person—indeed, it *must* do so to carry out its biological function. The events of this complex interaction culminate in pregnancy and birth. To be sure, having a baby is not always what the interacting partners have in mind, and we humans have devised a variety of techniques for preventing this outcome (see the box on p. 948–949).

The major focus of the reproductive system is on ensuring the health and function of its own organs so that conditions are optimal for producing offspring. However, as illustrated in Figure 28.23, gonadal hormones do exert major effects on certain body organs, and like all organ systems, the reproductive system depends on other body organs for oxygen and nutrients and to carry away and dispose of its wastes.

Now that we know how the reproductive system functions to prepare itself for childbearing, we are ready to consider the events of pregnancy and prenatal development of a new living being, the topics of Chapter 29.

Related Clinical Terms

Dysmenorrhea (dis-menō-rē′-ah) (*dys* = bad; *meno* = menses, a month) Painful menstruation; may reflect abnormally high prostaglandin activity during menses.

Endometriosis (en-dō-me-trē-ō′-sis) (*endo* = within; *metrio* = the uterus, womb) An inflammatory condition in which endometrial tissue occurs and grows atypically in the pelvic cavity; characterized by abnormal uterine or rectal bleeding, dysmenorrhea, and pelvic pain; may cause sterility.

Gynecology (gī″-neh-kol′-lō-jē) (*gyneco* = woman; *ology* = study of) Specialized branch of medicine that deals with the diagnosis and treatment of female reproductive system disorders.

Hysterectomy (his″-teh-rek′-tō-mē) (*hyster* = uterus; *ectomy* = cut out) Surgical removal of the uterus.

Inguinal hernia Protrusion of part of the intestine into the scrotum or through a separation in the abdominal muscles in the groin region; since the inguinal canals represent weak points in the abdominal wall, inguinal hernia may be caused by heavy lifting or other activities that increase intra-abdominal pressure.

Oophorectomy (ō-ō-for-ek′-tō-mē) (*oophor* = ovary) Surgical removal of the ovary.

Oophoritis Inflammation of an ovary.

Salpingitis (sal″-pin-jī′-tis) (*salpingo* = uterine tube) Inflammation of the uterine tubes.

Uterine prolapse Protrusion of the uterus through the vaginal orifice due to weakening of uterine supports.

Vaginitis (va-jih-nī′-tis) (*vagin* = a sheath) Inflammation of the vagina.

Chapter Summary

The function of the reproductive system is to produce offspring. The gonads produce gametes (sperm or ova) and sex hormones. All other reproductive organs are accessory organs.

ANATOMY OF THE MALE REPRODUCTIVE SYSTEM (pp. 915–920)

The Scrotum and Testes (pp. 915–917)

1. The scrotum contains the testes. It provides a temperature slightly lower than that of the body, as required for viable sperm production.

2. Each testis is covered externally by a tunica albuginea that extends internally to divide the testis into many lobules. Each lobule contains sperm-producing seminiferous tubules and interstitial cells that produce androgens.

The Duct System (pp. 917–918)

3. The epididymis hugs the external surface of the testis and serves as a site for sperm maturation and storage.

Reproductive system

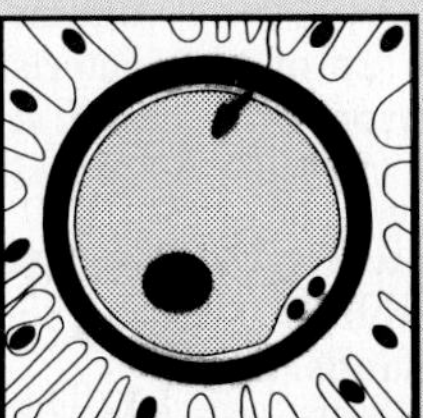

Integumentary system

▶ Protects all body organs by external enclosure

◀ Androgens activate oil glands which lubricate skin and hair; estrogen increases skin hydration; enhanced facial skin pigmentation during pregnancy

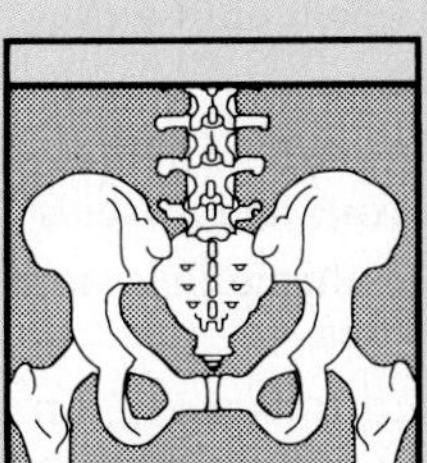

Skeletal system

▶ The bony pelvis encloses some reproductive organs

◀ Androgens masculinize the skeleton and increase bone density; estrogen feminizes skeleton and maintains bone mass

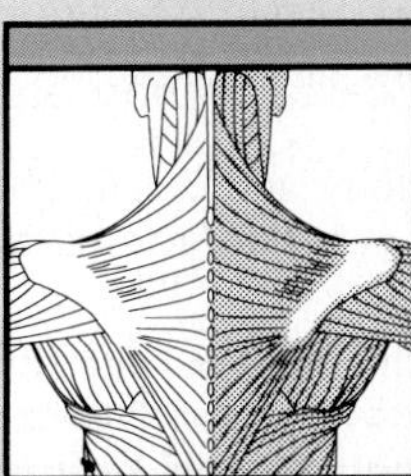

Muscular system

▶ Abdominal muscles active during childbirth; muscles of the pelvic floor support reproductive organs and aid erection of penis/clitoris

◀ Androgens promote increase in muscle mass

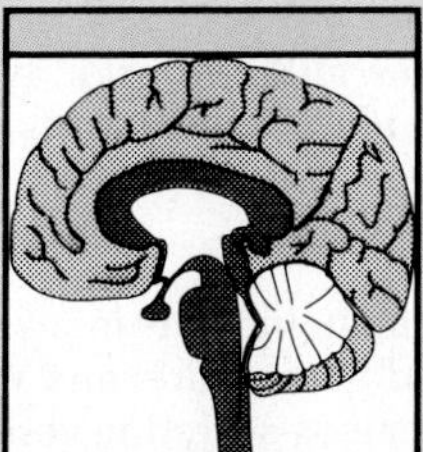

Nervous system

▶ Hypothalamus regulates timing of puberty; neural reflexes regulate events of sexual response

◀ Sex hormones masculinize or feminize the brain

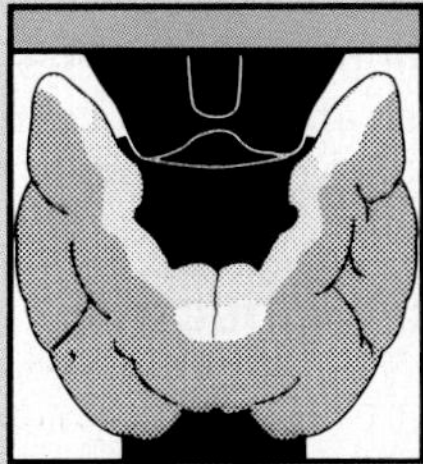

Endocrine system

▶ Gonadotropins (and GnRH) help regulate function of gonads

◀ Gonadal hormones exert feedback effects on hypothalamic pituitary axis; placental hormones promote maternal hypermetabolism

Cardiovascular system

▶ Transports needed substances to organs of reproductive system; vasodilation involved in erection; transports sex hormones

◀ Estrogen lowers blood cholesterol levels; pregnancy increases workload of the cardiovascular system

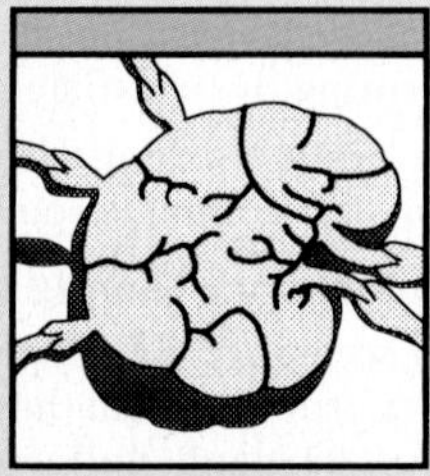

Lymphatic system

◀ Drains leaked tissue fluids; transports sex hormones

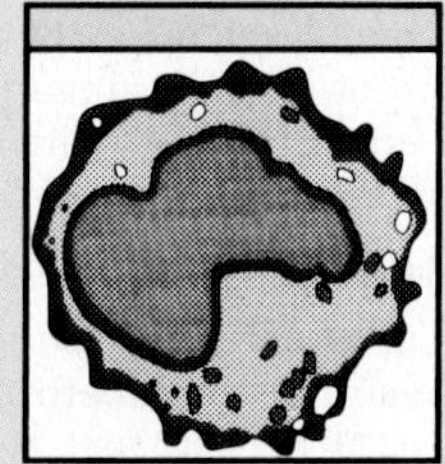

Immune system

◀ Protects reproductive organs from disease

▶ Developing embryo/fetus escapes immune surveillance (not rejected)

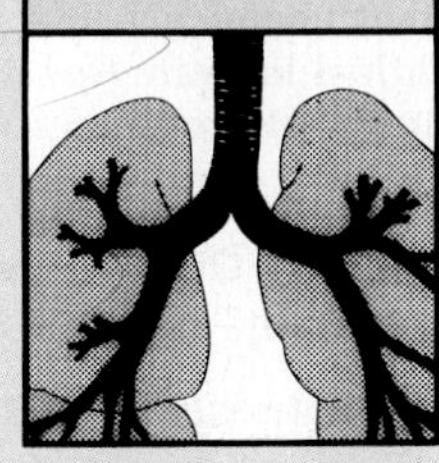

Respiratory system

◀ Provides oxygen; disposes of carbon dioxide; vital capacity and respiratory rate increase during pregnancy

▶ Pregnancy impairs descent of the diaphragm

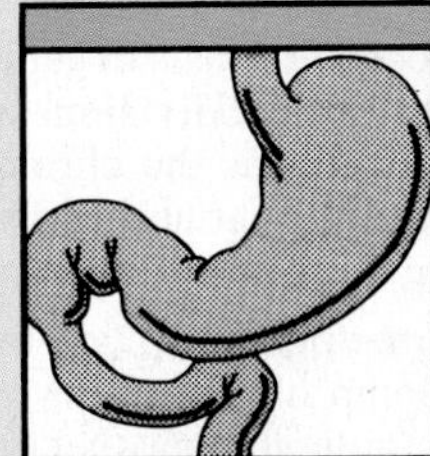

Digestive system

◀ Provides nutrients needed for health

▶ Digestive organs crowded by developing fetus; heart burn common

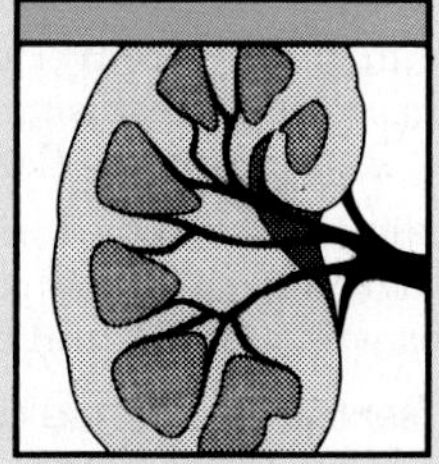

Urinary system

◀ Disposes of nitrogenous wastes and maintains acid–base balance of blood of mother and fetus; semen discharged through the urethra of the male

▶ Compression of bladder during pregnancy leads to urinary frequency

Figure 28.23 Homeostatic interrelationships between the reproductive system and other body systems.

4. The ductus (vas) deferens, extending from the epididymis to the urethra, propels sperm into the urethra by peristalsis during ejaculation. Its terminus fuses with the duct of the seminal vesicle, forming the ejaculatory duct.

5. The urethra extends from the urinary bladder to the tip of the penis. It conducts semen and urine to the body exterior.

The Accessory Glands (pp. 918–919)

6. The accessory glands produce the bulk of the semen, which contains fructose from the seminal vesicles, alkaline fluid from the prostate gland, and mucus from the bulbourethral glands.

The Penis (p. 919)

7. The penis, the male copulatory organ, is largely erectile tissue (corpus spongiosum and corpora cavernosa). Engorgement of the erectile tissue with blood causes the penis to become rigid, an event called erection.

The Male Perineum (p. 919)

8. The male perineum is the region encompassed by the pubic symphysis, ischial tuberosities, and coccyx.

Semen (pp. 919–920)

9. Semen is an alkaline fluid that dilutes and transports sperm. Important chemicals in semen are nutrients, prostaglandins, and seminalplasmin. An ejaculation contains 2–6 ml of semen, with 50 to 100 million sperm per milliliter in normal adult males.

PHYSIOLOGY OF THE MALE REPRODUCTIVE SYSTEM (pp. 920–929)

Spermatogenesis (pp. 920–926)

1. Spermatogenesis, production of male gametes in the seminiferous tubules, begins at puberty.

2. Meiosis, the basis of gamete production, consists of two consecutive nuclear divisions without DNA replication in between. Meiosis reduces the chromosomal number by half and introduces genetic variability. Events unique to meiosis include synapsis and crossover of homologous chromosomes.

3. Spermatogonia divide by mitosis to maintain the germ cell line. Some of their progeny become primary spermatocytes, which undergo meiosis I to produce secondary spermatocytes. Secondary spermatocytes undergo meiosis II, producing a total of four haploid (n) spermatids.

4. Spermatids are converted to functional sperm by spermiogenesis, during which superfluous cytoplasm is stripped away and an acrosome and a flagellum (tail) are produced.

5. Sustentacular cells form the blood-testis barrier, nourish spermatogenic cells and move them toward the lumen of the tubules, and secrete fluid for sperm transport.

Male Sexual Response (pp. 926–927)

6. Erection is controlled by parasympathetic reflexes.

7. Ejaculation is expulsion of semen from the male duct system, promoted by the sympathetic nervous system. Ejaculation is part of male orgasm, which also includes pleasurable sensations and increased pulse and blood pressure.

Hormonal Regulation of Male Reproductive Function (pp. 927–929)

8. GnRH, produced by the hypothalamus, stimulates the anterior pituitary gland to release FSH and LH. FSH stimulates spermatogenesis by causing sustentacular cells to produce androgen-binding protein (ABP). LH stimulates interstitial cells to release testosterone, which binds to ABP, stimulating spermatogenesis. Testosterone and inhibin (produced by sustentacular cells) feed back to inhibit the hypothalamus and anterior pituitary.

9. Maturation of hormonal controls occurs during puberty and takes about three years.

10. Testosterone stimulates maturation of the male reproductive organs and triggers the development of the secondary sex characteristics of the male. It exerts anabolic effects on the skeleton and skeletal muscles, stimulates spermatogenesis, and is responsible for sex drive.

ANATOMY OF THE FEMALE REPRODUCTIVE SYSTEM (pp. 929–934)

1. The female reproductive system produces gametes and sex hormones and houses a developing infant until birth.

The Ovaries (p. 929)

2. The ovaries flank the uterus laterally and are held in position by the ovarian and suspensory ligaments and mesovaria.

3. Within each ovary are oocyte-containing follicles at different stages of development and corpora lutea.

The Duct System (pp. 931–934)

4. The uterine tube, supported by the mesosalpinx, extends from near the ovary to the uterus. Its fibriated distal end creates currents that help move an ovulated oocyte into the uterine tube. Cilia of the uterine tube mucosa help propel the oocyte toward the uterus.

5. The pear-size uterus has fundus, body, and cervical regions. It is supported by the broad, lateral cervical, uterosacral, and round ligaments.

6. The uterine wall is composed of the outer perimetrium, the myometrium, and the inner endometrium. The endometrium consists of a stratum functionalis, which sloughs off periodically unless an embryo has implanted, and an underlying stratus basalis, which replenishes the functionalis.

7. The vagina extends from the uterus to the exterior. It is the copulatory organ and allows passage of the menstrual flow and a baby.

The External Genitalia (p. 934)

8. The female external genitalia (vulva) include the mons pubis, labia majora and minora, clitoris, and the urethral and vaginal orifices. The labia majora house the mucus-secreting vestibular glands.

The Mammary Glands (p. 934)

9. The mammary glands lie over the pectoral muscles of the chest and are surrounded by adipose and fibrous connective tissue. Each mammary gland consists of many lobules, which contain milk-producing alveolar glands.

PHYSIOLOGY OF THE FEMALE REPRODUCTIVE SYSTEM (pp. 935–942)

Oogenesis (pp. 935–937)

1. Oogenesis, the production of eggs, begins in the fetus. Oogonia, the diploid stem cells of female gametes, are converted to primary oocytes before birth. The infant female's ovaries contain about 700,000 primordial follicles, each containing a primary oocyte arrested in prophase of meiosis I.

2. At puberty, meiosis resumes. Each month, one primary oocyte completes meiosis I, producing a large secondary oocyte and a tiny first polar body. Meiosis II of the secondary oocyte produces a functional ovum and a second polar body, but does not occur unless the secondary oocyte is penetrated by a sperm.

3. The ovum contains most of the primary oocyte's cytoplasm. The polar bodies are nonfunctional and degenerate.

The Ovarian Cycle (pp. 937–938)

4. During the follicular phase (days 1–10), several primary follicles begin to mature. The follicle cells proliferate and produce

estrogen, and a connective tissue capsule (theca) is formed around the maturing follicle. Generally, only one follicle per month completes the maturation process and protrudes from the ovarian surface.

5. During the ovulatory phase (days 11–14), the primary oocyte in the dominant follicle completes meiosis I. Ovulation occurs, usually on the fourteenth day, releasing the secondary oocyte into the peritoneal cavity. Other developing follicles undergo atresia.

6. In the luteal phase (days 15–28), the ruptured follicle is converted to a corpus luteum, which produces progesterone and estrogen for the remainder of the cycle. If fertilization does not occur, the corpus luteum degenerates in about ten days.

Hormonal Regulation of the Ovarian Cycle (pp. 938–940)

7. Beginning at puberty, the hormones of the hypothalamus, anterior pituitary, and ovaries interact to establish and regulate the ovarian cycle. Establishment of the mature cyclic pattern, indicated by menarche, takes about four years. The hormonal events of each cycle are as follows: (1) Release of GnRH stimulates the anterior pituitary to release FSH and LH, which stimulate follicle maturation and follicular production of estrogen. (2) When blood estrogen reaches a certain level, positive feedback is exerted on the hypothalamic-pituitary axis, causing a sudden release of LH that stimulates the primary oocyte to continue meiosis and triggers ovulation. LH then causes conversion of the ruptured follicle to a corpus luteum and stimulates its secretory activity. (3) Rising levels of progesterone and estrogen inhibit the hypothalamic-pituitary axis, the corpus luteum deteriorates, ovarian hormones drop to their lowest levels, and the cycle begins anew.

The Uterine (Menstrual) Cycle (p. 940)

8. Varying levels of ovarian hormones in the blood trigger events of the uterine cycle.

9. During the menstrual phase of the uterine cycle (days 1–5), the stratum functionalis sloughs off in menses. During the proliferative phase (days 6–14), rising estrogen levels stimulate regeneration of the stratum functionalis, making the uterus receptive to implantation about one week after ovulation. During the secretory phase (days 15–28), the uterine glands secrete glycogen, and endometrial vascularity increases further.

10. Falling levels of ovarian hormones during the last few days of the ovarian cycle cause the spiral arteries to become spastic and cut off the blood supply to the stratum functionalis, and the uterine cycle begins again.

Extrauterine Effects of Estrogen and Progesterone (pp. 940–942)

11. Estrogen promotes oogenesis. At puberty, it stimulates the growth of the reproductive organs and the growth spurt and promotes the appearance of the secondary sex characteristics.

12. Progesterone cooperates with estrogen in breast maturation.

Female Sexual Response (p. 942)

13. The female sexual response is similar to that of males. Orgasm in females is not accompanied by ejaculation and is not necessary for conception.

SEXUALLY TRANSMITTED DISEASES (pp. 942–943)

1. Sexually transmitted diseases (STDs) are infectious diseases spread via sexual contact. Gonorrhea, syphilis, and chlamydia are bacterial diseases; if untreated, they can cause sterility. Genital herpes is a viral infection and is implicated in cervical cancer. Syphilis has more far-reaching consequences than the other venereal diseases, since it can infect organs throughout the body. (AIDS, a viral STD, is discussed in Chapter 22.)

DEVELOPMENTAL ASPECTS OF THE REPRODUCTIVE SYSTEM: CHRONOLOGY OF SEXUAL DEVELOPMENT (pp. 943–950)

Embryological and Fetal Events (pp. 944–947)

1. Genetic sex is determined by the sex chromosomes: an *X* from the mother, an *X* or a *Y* from the father. If the fertilized egg contains *XX*, it is a female and develops ovaries; if it contains *XY*, it is a male and develops testes.

2. Gonads of both sexes arise from the mesodermal gonadal ridges. The mesonephric ducts produce the male accessory ducts and glands. The paramesonephric ducts produce the female duct system.

3. The external genitalia arise from the genital tubercle and associated structures. The development of male accessory structures and external genitalia depends on the presence of testosterone produced by the embryonic testes. In its absence, female structures develop.

4. The testes form in the abdominal cavity and descend into the scrotum.

Puberty (p. 947)

5. Puberty is the interval when reproductive organs mature and become functional. It begins in males with penile and scrotal growth and in females with breast development.

Menopause (pp. 947–950)

6. During menopause, ovarian function declines, and ovulation and menstruation cease. Hot flashes and mood changes may occur. Postmenopausal events include atrophy of the reproductive organs, bone mass loss, and increasing risk for cardiovascular disease.

Review Questions

Multiple Choice/Matching

1. The structures that draw an ovulated oocyte into the female duct system are (a) cilia, (b) fimbriae, (c) microvilli, (d) stereocilia.

2. The usual site of embryo implantation is (a) the uterine tube, (b) the peritoneal cavity, (c) the vagina, (d) the uterus.

3. The male homologue of the female clitoris is (a) the penis, (b) the scrotum, (c) the penile urethra, (d) the testis.

4. Which of the following is correct relative to female anatomy? (a) The vaginal orifice is the most dorsal of the three openings in the vulva, (b) the urethra is between the vaginal orifice and the anus, (c) the anus is between the vaginal orifice and the urethra, (d) the urethra is the most ventral of the three orifices.

5. Secondary sex characteristics are (a) present in the embryo, (b) a result of male or female sex hormones increasing in amount at puberty, (c) the testis in the male and the ovary in the female, (d) not subject to withdrawal once established.

6. Which of the following produces the male sex hormones? (a) seminal vesicles, (b) corpus luteum, (c) developing follicles of the testes, (d) interstitial cells.

7. Which will occur as a result of nondescent of the testes? (a) Male sex hormones will not be circulated in the body, (b) sperm will have no means of exit from the body, (c) inadequate blood supply will retard the development of the testes, (d) viable sperm will not be produced.

8. The normal diploid number of human chromosomes is (a) 48, (b) 47, (c) 46, (d) 23, (e) 24.

9. Relative to differences between mitosis and meiosis, choose the statements that apply *only* to events of meiosis.

(a) tetrads present
(b) produces two daughter cells
(c) produces four daughter cells
(d) occurs throughout life
(e) reduces the chromosomal number by half
(f) synapsis and crossover of homologues occur

10. Match the items in column B with the phrases in column A.

Column A	Column B
(1) ——, —— hormones that directly regulate the ovarian cycle	**(a)** Androgen-binding protein
(2) ——, —— chemicals that inhibit the pituitary-testicular axis	**(b)** Estrogen
(3) —— hormone that makes the cervical mucus viscous	**(c)** FSH
(4) —— potentiates the activity of testosterone on spermatogenic cells	**(d)** GnRH
(5) ——, —— in females, exerts feedback inhibition on the hypothalamus and anterior pituitary	**(e)** Inhibin
(6) —— stimulates the secretion of testosterone	**(f)** LH
	(g) Progesterone
	(h) Testosterone

11. The menstrual cycle can be divided into three continuous phases. Starting from the first day of the cycle, their consecutive order is (a) menstrual, proliferative, secretory, (b) menstrual, secretory, proliferative, (c) secretory, menstrual, proliferative, (d) proliferative, menstrual, secretory, (e) secretory, proliferative, menstrual.

12. Spermatozoa are to seminiferous tubules as oocytes are to (a) fimbriae, (b) corpus albicans, (c) ovarian follicles, (d) corpora lutea.

13. Which of the following does not add a secretion that contributes to semen? (a) prostate, (b) bulbourethral glands, (c) testes, (d) vas deferens.

14. The corpus luteum is formed at the site of (a) fertilization, (b) ovulation, (c) menstruation, (d) implantation.

15. The sex of a child is determined by (a) the sex chromosome contained in the sperm, (b) the sex chromosome contained in the oocyte, (c) the number of sperm fertilizing the oocyte, (d) the position of the fetus in the uterus.

16. FSH is to estrogen as estrogen is to (a) progesterone, (b) LH, (c) FSH, (d) testosterone.

17. A drug that "reminds the pituitary" to produce gonadotropins might be useful as (a) a contraceptive, (b) a diuretic, (c) a fertility drug, (d) an abortion stimulant.

Short Answer Essay Questions

18. Why is the term *urogenital system* more applicable to males than to females?

19. The spermatid is haploid, but it is not a functional gamete. Name and describe the process during which a spermatid is converted to a motile sperm, and describe the major structural (and functional) regions of a sperm.

20. Oogenesis in the female results in one functional gamete—the egg, or ovum. What other cells are produced? What is the significance of this rather wasteful type of gamete production—that is, production of a single functional gamete instead of four, as seen in males?

21. List three secondary sex characteristics of females.

22. Describe the events and possible consequences of menopause.

23. Define menarche. What does it indicate?

24. For three popular types of birth control—the pill, the diaphragm, and coitus interruptus—note (a) the point of blockage of the reproductive process, (b) the mode of action, (c) the relative efficiency or effectiveness, and (d) the inherent problems in each method.

25. Trace the pathway of a sperm from the male testes to the uterine tube of a female.

Clinical Application Questions

26. A sexually active adolescent male appeared in the emergency room complaining of a penile "drip" and pain on urination. An account of his recent sexual behavior was requested and recorded. (a) What do you think his problem is? (b) What is the causative agent of this disorder? (c) How is the condition treated, and what may happen if it isn't treated?

27. A 36-year-old mother of four is considering tubal ligation as a means of ensuring that her family gets no larger. She asks the physician if she will become "menopausal" after the surgery. (a) How would you answer her question and explain away her concerns? (b) Explain what a tubal ligation is.

28. A 76-year-old gentleman, interested in a much younger woman and concerned about his age, asks his urologist if he will be able to father a child. What questions would a physician ask of this man, and what diagnostic tests would be ordered?

CHAPTER 29

Pregnancy and Human Development

Chapter Outline and Student Objectives

From Egg to Embryo (pp. 956–964)

1. Describe the importance of capacitation to the ability of sperm to penetrate an oocyte.
2. Explain the mechanism of the fast and slow blocks to polyspermy.
3. Define fertilization.
4. Explain the process and product of cleavage.
5. Describe the processes of implantation and placenta formation, and list placental functions.

Events of Embryonic Development (pp. 964–972)

6. Describe the process of gastrulation and its consequence.
7. Name and describe the formation, location, and function of the embryonic membranes.
8. Define organogenesis and note the important roles of the three primary germ layers in this process.
9. Describe the unique features of the fetal circulation.

Events of Fetal Development (p. 972)

10. Indicate the duration of the fetal period, and note the major events of fetal development.

Effects of Pregnancy on the Mother (pp. 972–975)

11. Describe changes in maternal reproductive organs and in cardiovascular, respiratory, and urinary system functioning during pregnancy.
12. Note the effects of pregnancy on maternal metabolism and posture.

Parturition (Birth) (pp. 975–978)

13. Explain how labor is initiated, and describe the three stages of labor.

Adjustments of the Infant to Extrauterine Life (pp. 978–979)

14. Outline the events leading to the first breath of a newborn.
15. Describe the changes that occur in the fetal circulation after birth.

Lactation (pp. 979–980)

16. Explain how the breasts are prepared for lactation and nursing.

Preview of Selected Key Terms

Capacitation Process during which a sperm becomes capable of the acrosomal reaction.

Fertilization Fusion of the sperm and egg nuclei.

Zygote (zī′-gōt) (*zygo* = yoke) A fertilized egg.

Cleavage Mitotic division process occurring during early development, which produces a blastocyst.

Gastrulation (*gaster* = the stomach, belly) Developmental process that produces the three primary germ layers (ectoderm, mesoderm, and endoderm).

Embryo Developmental stage extending from gastrulation to the end of the eighth week in humans.

Fetus Developmental stage extending from the ninth week of development to birth.

Placenta (pluh-sen′-tuh) (*plac* = plate, flat) Temporary organ formed from both fetal and maternal tissues that provides nutrients and oxygen to the developing fetus, carries away fetal metabolic wastes, and produces the hormones of pregnancy.

Neurulation The process during which the neural plate forms and becomes the neural tube.

Because the birth of a baby is such a familiar event, we tend to lose sight of the wonder of this accomplishment. In every instance, it begins with a single cell, the fertilized egg, and ends with an extremely complex human being consisting of trillions of cells. The details of this complicated process can fill a good-sized book. Our intention here is simply to outline the important events of gestation from both the developing infant's and the mother's vantage points. We will

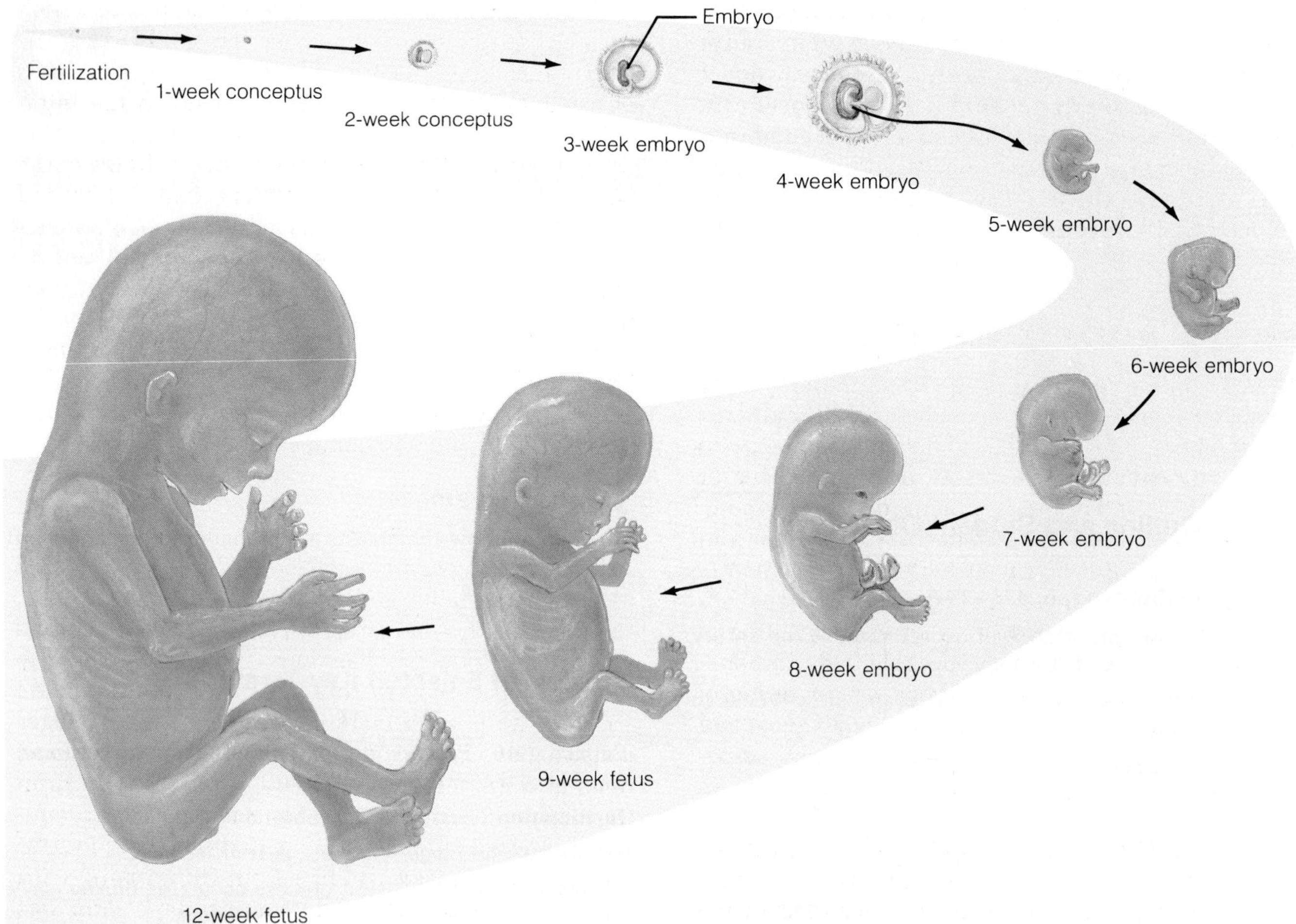

Figure 29.1 Selected diagrams showing the actual size of a human conceptus from fertilization to the early fetal stage. The embryonic stage begins in the third week after fertilization; the fetal stage begins in the ninth week.

also consider briefly the events that occur immediately after birth.

But to get started, we need to define some terms. The term **pregnancy** refers to events that occur from the time of fertilization until the infant is born. The woman is pregnant; her developing offspring, called the **conceptus** (kun-sep′-tus), undergoes preembryonic, embryonic, and fetal development in succession. The time during which development occurs is referred to as the **gestation period** and extends by convention from the last menstrual period until birth, approximately 280 days. Thus, at the moment of fertilization, the mother is officially two weeks pregnant! For two weeks following fertilization, the product of conception undergoes preembryonic development and is known simply as the conceptus. From the third through eighth weeks after fertilization, the conceptus is called an **embryo,** and from the ninth week through birth, a **fetus.** Figure 29.1 shows the changing size and shape of the conceptus as it progresses from fertilization to the early fetal stage.

From Egg to Embryo

Accomplishing Fertilization

Before fertilization can occur, the sperm must reach the ovulated secondary oocyte. The oocyte is viable for 12–24 hours after it is cast out of the ovary, and sperm generally retain their fertilizing power within the female reproductive tract for 12–48 hours after ejaculation. Some "super sperm," however, are viable for 72 hours. Consequently, for fertilization to occur, coitus must occur no more than 72 hours before ovulation and no later than 24 hours after, at which point the oocyte is approximately one-third of the way down the length of the uterine tube. **Fertilization** occurs at the moment the genetic material of a sperm combines with that of an ovum to form a fertilized egg, or **zygote,** which represents the first cell of the new individual. Let's look at the events that must occur for this to happen.

Sperm Transport and Capacitation

Although sperm are self-propelling, their journey through the female reproductive tract is a difficult one, to say the least. The acidity of the vagina is hostile to sperm, and millions of them leak from the vagina or die almost immediately after being deposited there. Survivors forge ahead through the cervical mucus and upward through the uterus and uterine tubes. However, this is presuming that the cervical mucus has been made fluid by estrogen; otherwise, millions of sperm fail to make it through that relatively impenetrable barrier. Within the uterus, thousands more sperm are destroyed by resident phagocytic leukocytes, and it is estimated that only a few thousand (and sometimes fewer than 500) sperm, out of the millions in the male ejaculate, finally reach the uterine tubes, where the oocyte may be drifting leisurely toward the uterus.

These difficulties aside, there is still another hurdle to be overcome. Sperm freshly deposited in the vagina are incapable of penetrating an oocyte. They must first be **capacitated**; that is, their membranes must become fragile so that the hydrolytic enzymes in their acrosomes can be released. As mentioned in Chapter 28 (p. 925), semen contains vesicles that donate cholesterol to the sperm membranes to keep them tough and intact. As sperm swim through the cervical mucus, uterus, and uterine tubes, the cholesterol is depleted, and the sperm undergo a gradual capacitation for the next few hours. Thus, even though the sperm may actually reach the oocyte within a few minutes, they must wait around (so to speak) for capacitation to occur. This is a pretty elaborate mechanism for preventing the spilling of acrosomal enzymes, but consider the alternative. If sperm had fragile acrosomal membranes while still within the male reproductive tract, they could rupture prematurely, causing some degree of autolysis (self-digestion) of the male reproductive system.

Acrosomal Reaction and Sperm Penetration

The **acrosomal** (a″-krō-sō′-mul) **reaction** is the release of acrosomal enzymes (hyaluronidase, acrosin, proteases, and others) that occurs in the immediate vicinity of the oocyte. The ovulated oocyte is encapsulated by the corona radiata and, deep to that, the zona pellucida, both of which must be breached before the oocyte membrane itself can be penetrated. Hundreds of acrosomes must rupture to break down the intercellular cement, rich in hyaluronic acid, that holds the granulosa cells together and to digest holes in the zona pellucida. But once a path has been cleared and a single sperm makes contact with the oocyte membrane receptors, its nucleus is pulled into the oocyte cytoplasm (Figure 29.2a). This is one case that does not bear out the old adage, "The early bird catches the worm." A sperm that comes along later, after virtually hundreds of sperm have undergone acrosomal reactions to expose the oocyte membrane, is in the best position to be *the* fertilizing sperm.

Blocks to Polyspermy

Although *polyspermy* (pah″-lē-sper′-mē), or the entry of several sperm into an egg, occurs in some animals, in humans only a single sperm is allowed to penetrate the oocyte. Two mechanisms ensure *monospermy* (mah″-nō-sper′-mē), or the one-sperm condition. As soon as the plasma membrane of one sperm makes contact with the oocyte membrane, sodium channels open and ionic sodium moves into the oocyte from the extracellular space, causing its membrane to depolarize. This electrical event, called the *fast block to polyspermy,* prevents other sperm from fusing with the oocyte membrane. Then, once the sperm has entered, the *cortical reaction* occurs (see Figure 29.2a). The depolarization that serves as the fast block to polyspermy also causes ionic calcium to be released into the oocyte cytoplasm. This surge in intracellular calcium levels activates the oocyte and prepares it for cell division. It also causes the cortical granules, located just deep to the plasma membrane, to spill their contents into the extracellular space beneath the zona pellucida. This spilled material binds water, and as it swells, it detaches all sperm still in contact with the oocyte membrane, accomplishing the permanent *slow block to polyspermy.*

Completion of Meiosis II and Fertilization

After a sperm has entered the oocyte, it remains temporarily in the peripheral cytoplasm. What happens next is shown in Figure 29.3. The secondary oocyte, activated by the ionic calcium signal, now completes meiosis II to form the ovum nucleus and ejects the second polar body. The ovum and sperm nuclei swell, becoming the female and male *pronuclei,* (prō-noo′-klē-ī), and approach each other as a mitotic spindle develops between them. The pronuclei membranes then

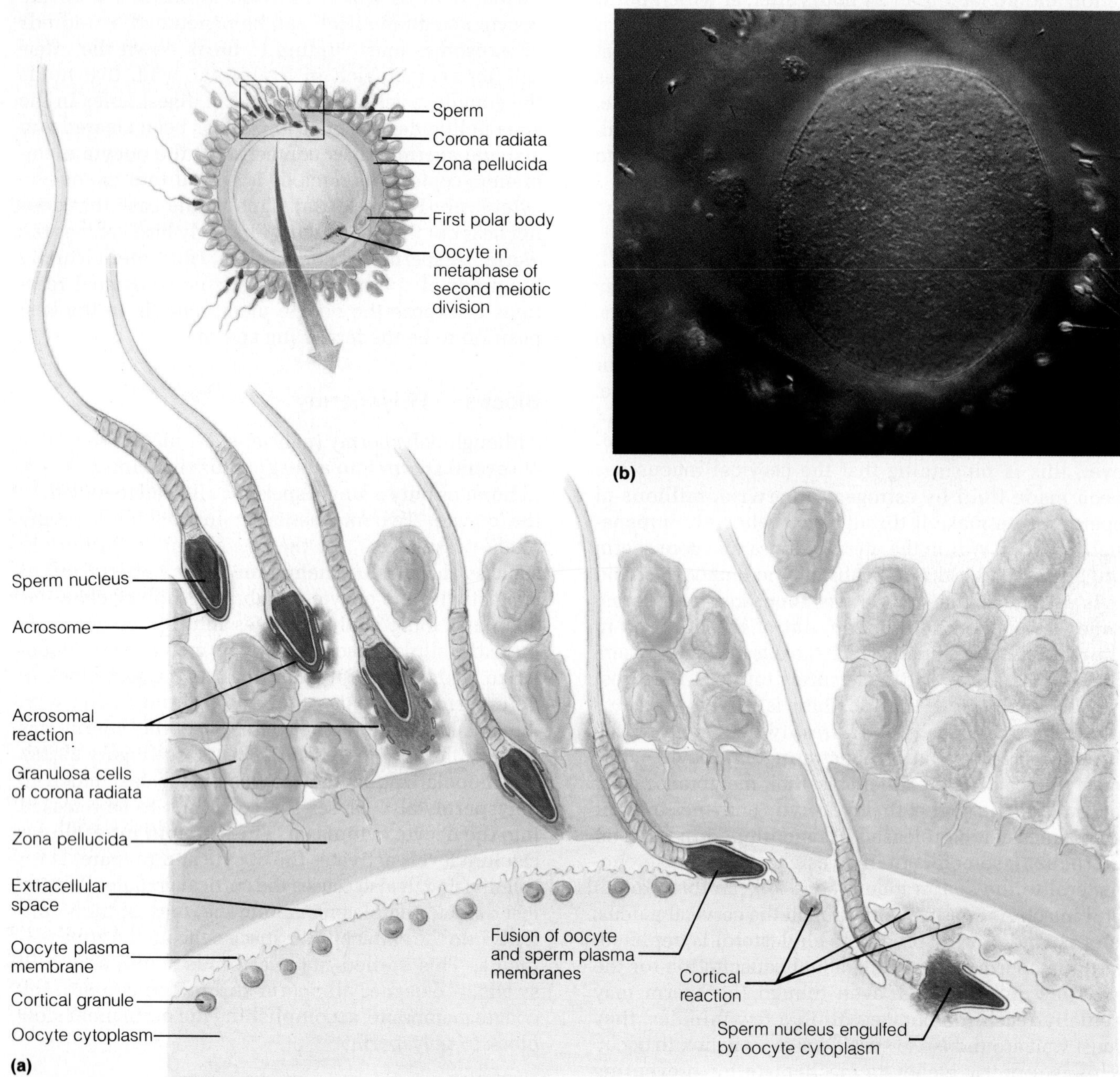

Figure 29.2 Sperm penetration and the cortical reaction (slow block to polyspermy). (**a**) The sequential steps of oocyte penetration by a sperm are depicted from left to right. Sperm penetration through the corona radiata and zona pellucida surrounding the oocyte is accomplished by release of acrosomal enzymes by many sperm. Fusion of the plasma membane of a single sperm with the oocyte membrane is followed by engulfment of the sperm by the oocyte cytoplasm and the cortical reaction. The cortical reaction involves the release of the contents of the oocyte's cortical granules into the extracellular space, preventing entry of more than one sperm. (**b**) Photomicrograph of human sperm surrounding a human oocyte (750x).

rupture, releasing their chromosomes into the immediate vicinity of the spindle. Combination of the maternal and paternal chromosomes constitutes the act of fertilization and produces the diploid zygote. Almost as soon as the male and female pronuclei come together, their chromosomes replicate. Then, the first mitotic division of the conceptus begins (Figure 29.3 d–f).

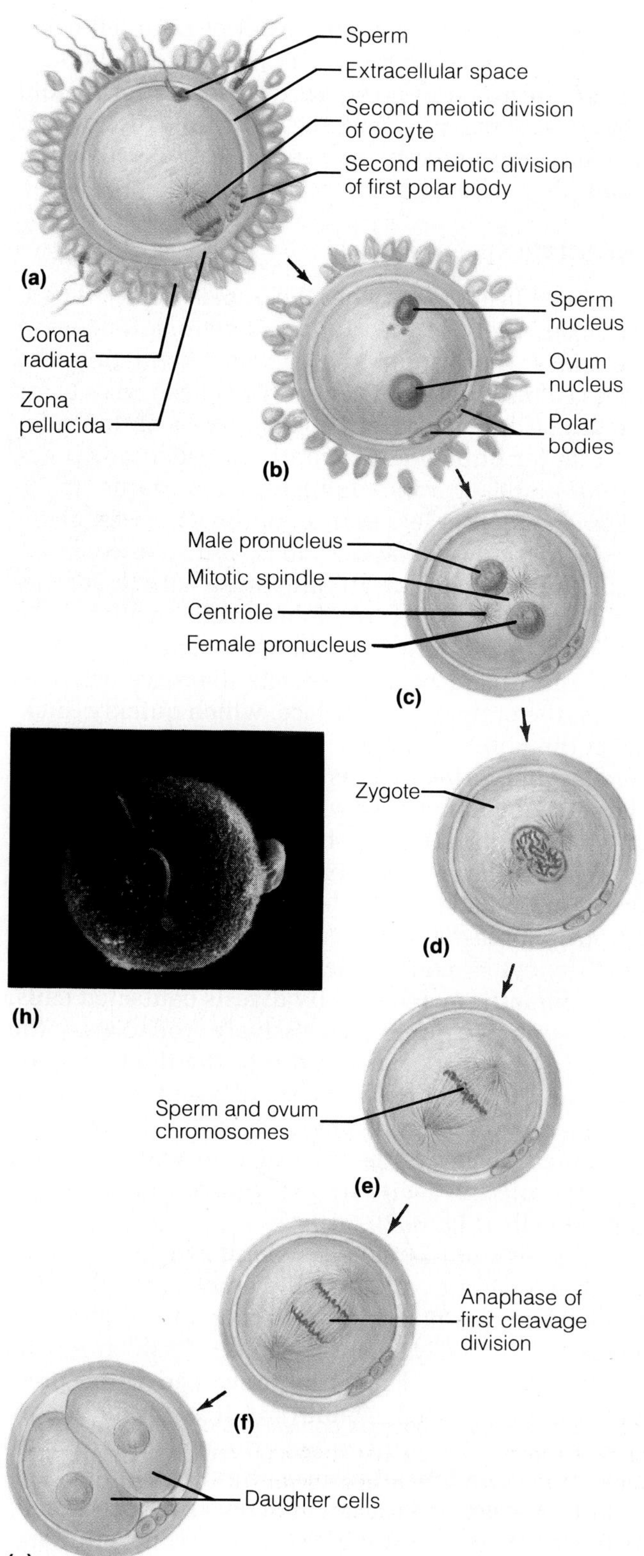

Figure 29.3 Events immediately following sperm penetration. (**a**) Once a sperm has entered the oocyte, the secondary oocyte completes meiosis II and casts out the second polar body. (**b–c**) The ovum nucleus formed during meiosis II begins to swell, and the sperm and egg nuclei approach each other. When fully swollen, the nuclei are called pronuclei. (**d–g**) As the male and female pronuclei come together (constituting the act of fertilization), they immediately replicate their DNA, and then form chromosomes which attach to the mitotic spindle and undergo a mitotic division. This division is the first cleavage division and results in two daughter cells. (**h**) Scanning electron micrograph of a fertilized egg (zygote). Notice the second polar body protruding at the right. A single sperm is seen against the zygote's surface (500x).

Preembryonic Development

Preembryonic development begins with fertilization and continues as the conceptus travels through the uterine tube, floats free in the uterine lumen, and finally implants in the uterine wall. Significant events of the preembryonic period are cleavage, which results in the formation of the structure called a blastocyst, and implantation of the blastocyst.

Cleavage and Blastocyst Formation

Cleavage is a period of fairly rapid mitotic divisions of the zygote following fertilization. Since there is not much time for cell growth between divisions, the daughter cells become smaller and smaller (Figure 29.4). Cleavage thus results in cells with a high surface-to-volume ratio, which enhances the uptake of nutrients and oxygen and the disposal of wastes. It also provides a large number of cells to serve as building blocks for constructing the embryo. Consider, for a moment, the difficulty of constructing a building from a huge block of granite. If you now consider how much easier your task would be if you could use hundreds of brick-size granite blocks, you will quickly grasp the importance of cleavage.

By about 36 hours after fertilization, the first cleavage division has produced two identical cells called *blastomeres;* the second division (60 hours after fertilization) produces a total of four blastomeres, and the third (72 hours after fertilization) a total of eight. Eventually, the solid mass forms a berry-shaped *morula* (mor′-uh-luh) ("little mulberry"); all the while, transport of the conceptus toward the uterus continues. By the fourth or fifth day after fertilization, the conceptus consists of about 100 cells and is floating freely in the uterus. The zona pellucida now starts to break down, and the inner structure, or blastocyst, escapes from it. The **blastocyst** (bla′-stō-sist) is a hollow sphere composed of a single layer of large, flattened cells called **trophoblast** (trō′-fō-blast) **cells** (*troph* = nourish; *blast*

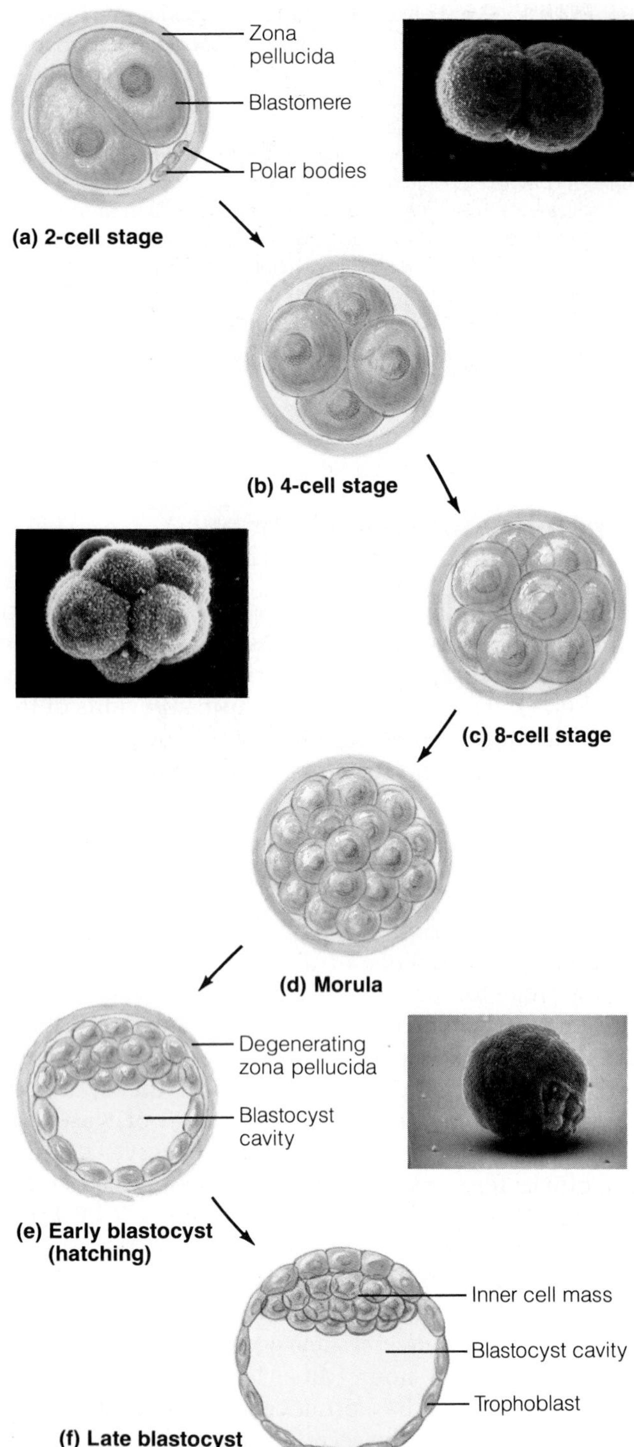

Figure 29.4 Cleavage is a rapid series of mitotic divisions that begins with the zygote and produces the preembryonic form called the blastocyst. (**a**) Two-cell stage, diagram and corresponding photomicrograph. (**b**) Four-cell stage. (**c**) Eight-cell stage, photomicrograph and corresponding diagram. (**d**) The morula, a solid ball of blastomeres. (**e**) Diagram and corresponding photomicrograph of an early blastocyst; the morula hollows out, becomes fluid-filled, and "hatches" from the zona pellucida. (**f**) Late blastocyst formed of an outer sphere of trophoblast cells and an eccentric cell cluster called the inner cell mass. Notice that the final blastocyst is only slightly larger than the two-cell stage, because little time is allowed for growth between successive divisions.

= sprout, bud) and a small cluster of rounded cells, called the **inner cell mass,** at one side. The trophoblast cells are destined to take part in placenta formation; the inner cell mass becomes the *embryonic disc,* which will form the embryo proper and associated membranes.

Implantation

When the blastocyst reaches the uterine cavity, it usually floats freely in the uterine secretions for two to three days; then, about six days after ovulation, the process of **implantation** begins. The blastocyst's outer layer, the trophoblast, tests the readiness of the endometrium for implantation, and if the mucosa is properly prepared, it implants high in the uterus. If the endometrium is at less than an optimal stage of maturity, the blastocyst detaches and floats to a lower level, finally implanting when it finds a site with the proper chemical signals. The trophoblast cells overlying the inner cell mass adhere to the endometrium (Figure 29.5a and b) and begin to secrete digestive enzymes against the endometrial surface, which quickly thickens at the point of contact. The trophoblast then proliferates and forms two distinct layers (Figure 29.5c). The cells of the inner layer, called the *cytotrophoblast* (sī″-tō-trō′-fō-blast), retain their cell boundaries; those in the outer layer lose their plasma membranes and form a multinuclear cytoplasmic mass called the *syncytiotrophoblast* (sin-sī″-tē-ō-trō′-fō-blast) (*syn* = together; *cyt* = cell), which projects invasively into the endometrium and rapidly digests contacted cells. As the endometrium is progressively eroded away, the blastocyst becomes burrowed into the thick, velvety lining and surrounded by a pool of blood leaked from degraded endometrial blood vessels. Shortly, the implanted blastocyst is covered over and sealed off from the uterine cavity by proliferation of the endometrial cells (Figure 29.5d).

Implantation is usually completed by the fourteenth day after ovulation—the exact time when the endometrium is normally beginning to slough off in menses. Obviously, menses would flush away the embryo as well and must be prevented if the pregnancy is to continue. Viability of the corpus luteum is maintained by an LH-like hormone called **human chorionic gonadotropin** (kor-ē-ah′-nik gō″-na-dō-trō′-pin) **(HCG),** which is secreted by the trophoblast cells of the blastocyst and which bypasses the pituitary-ovarian controls at this critical time. HCG prompts the corpus luteum to continue secreting progesterone and estrogen. The *chorion,* which develops from the trophoblast after implantation, continues this hormonal stimulus; thus, the developing conceptus takes over the hormonal control of the uterus during this early phase of development. HCG is usually detectable in the mother's blood by the third week of gestation. Blood levels of HCG rise until the end of the second month and then decline sharply by 4 months

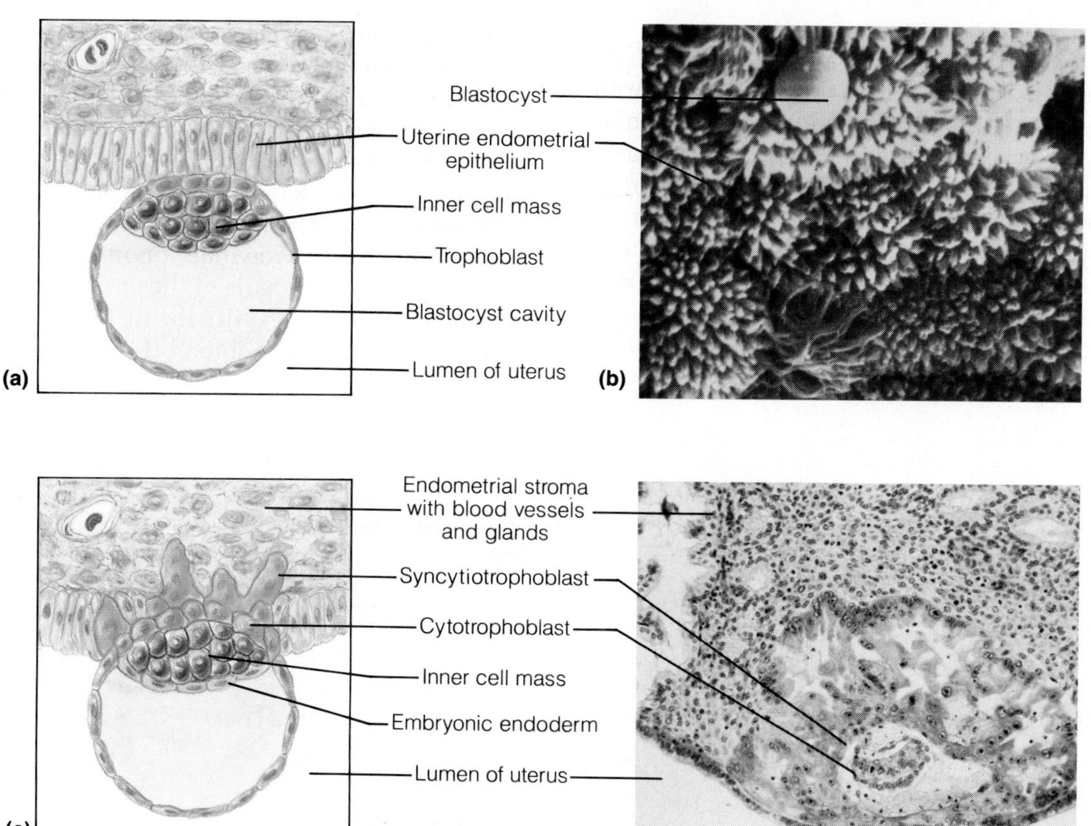

Figure 29.5 Implantation of the blastocyst. (**a**) Diagrammatic view of a blastocyst that has just adhered to the uterine endometrium. (**b**) Scanning electron micrograph (SEM) of a blastocyst as it begins to implant in the uterine endometrium. (**c**) Slightly later stage of an implanting embryo (approximately seven days after ovulation), depicting the cytotrophoblast and syncytiotrophoblast portions of the eroding trophoblast. (**d**) Light micrograph of an implanted blastocyst (approximately ten days after ovulation).

to reach a low value, which persists for the remainder of gestation (Figure 29.6). Between the second and third month, the placenta assumes the role of progesterone and estrogen production for the remainder of the pregnancy. The corpus luteum then degenerates and the ovaries remain inactive until after birth. All pregnancy tests used today are antibody or receptor protein tests that detect HCG in a woman's blood or urine.

Initially, the implanted embryo obtains its nutrition by digesting the endometrial cells, but by the second month, the placenta is providing nutrients and oxygen to the embryo and carrying away embryonic metabolic wastes. Since placenta formation is a continuation of the events of implantation, we will consider it next, although we will be getting ahead of ourselves a little as far as embryonic development is concerned.

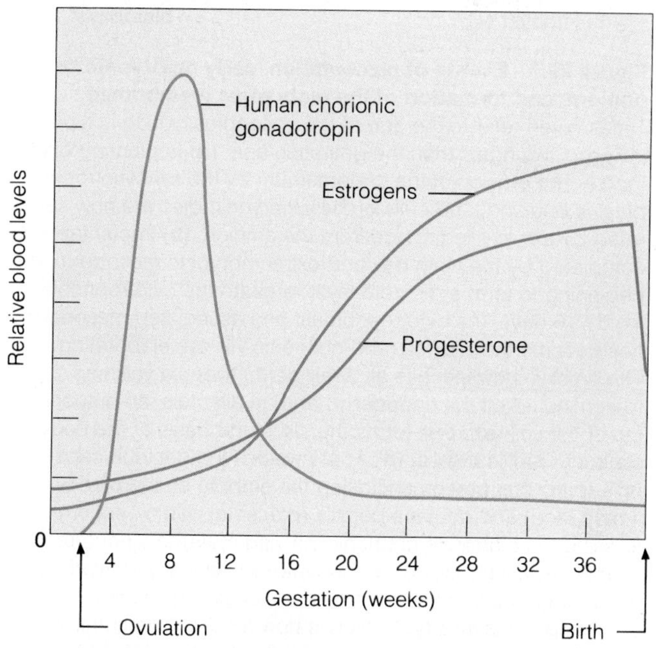

Figure 29.6 Relative change in maternal blood levels of human chorionic gonadotropin, estrogen, and progesterone during pregnancy. (Actual relative blood concentrations are not depicted.)

Placentation

Placentation (pla″-sen-tā′-shun) refers to the formation of a **placenta,** a temporary organ that originates from both embryonic (trophoblastic) and maternal (endometrial) tissues. When the proliferating trophoblast acquires a layer of extraembryonic mesoderm at its inner surface (Figure 29.7b), it becomes the **chorion.** (These extraembryonic mesodermal cells arise from the caudal part of the embryonic disc.) The chorion develops fingerlike *chorionic villi* (Figure 29.7c), which become elaborate where they are in contact with the maternal blood. Soon the mesodermal cores of the villi become richly vascularized by newly forming blood vessels, which extend to the embryo as the umbilical arteries and vein. Thus, the embryo's blood vessels are connected to the fetal (chorionic) portion of the placenta. The continuing erosion process forms large, blood-filled *lacunae,* or *intervillus spaces,* in the stratum functionalis of the endometrium (see Figure 28.13), and the villi lie in these spaces totally immersed in maternal blood (Figure 29.7d). The por-

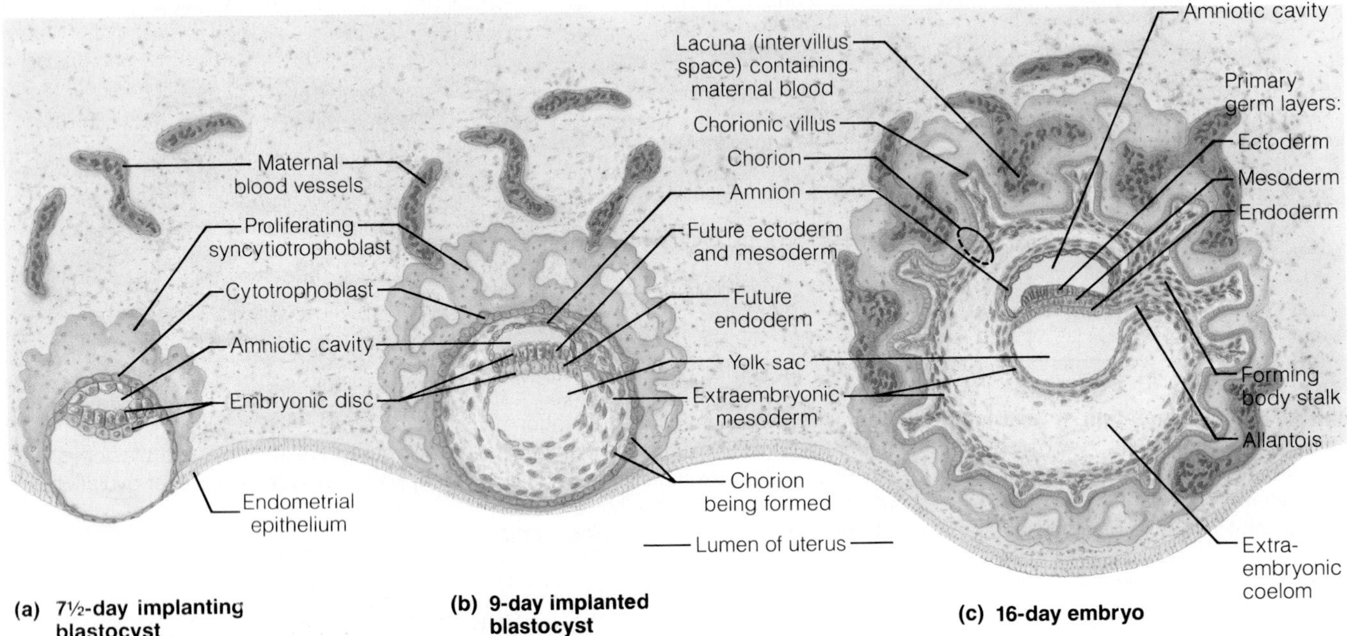

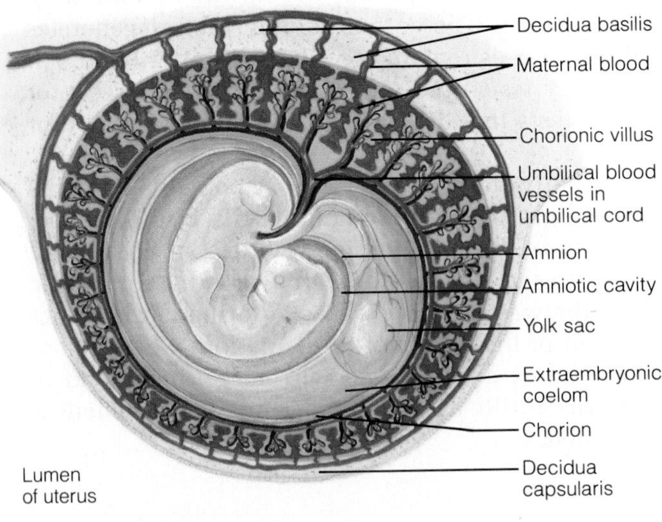

Figure 29.7 Events of placentation, early embryonic development, and formation of the embryonic membranes. Times given refer to the age of the conceptus, which is typically 2 weeks "younger" than the gestation age. (**a**) Implanting blastocyst. The erosion of the endometrium by the syncytiotrophoblast is ongoing, and cells of the embryonic disc are now separated by a fluid-filled space from the amnion. (**b**) Implantation is completed by the ninth day and extraembryonic mesoderm is beginning to form a discrete layer beneath the cytotrophoblast. (**c**) By 16 days, the cytotrophoblast and associated mesoderm has become the chorion, and chorionic villi are elaborating. The embryo now exhibits all three germ layers, a yolk sac (delaminated off the endoderm) and an allantois, an outpocketing of the yolk sac that forms the structural basis of the body stalk, or umbilical cord. (**d**) At 4½ weeks, the decidua capsularis (placental portion enclosing the embryo on the uterine lumen face) and decidua basalis (placental portion encompassing the elaborate chorionic villi and maternal endometrium) are well formed. The chorionic villi lie in blood-filled intervillus spaces within the endometrium. Organogenesis is occurring in the embryo, which is now receiving its nutrition via the umbilical vessels that connect it (via the body stalk) to the placenta. The amnion and yolk sac are well developed. (**e**) 13-week fetus. (**f**) Detailed anatomy of the vascular relationships in the mature decidua basalis. This state of development has been accomplished by the third month of development.

tion of the endometrium that lies between the chorionic villi and the stratum basalis becomes the **decidua basalis** (deh-sih″-dyoo-uh buh-sa′-lis); endometrial cells surrounding the uterine cavity face of the implanted embryo form the **decidua capsularis** (cap″-suh-layr′-is) (Figure 29.7d and e). Together, the chorionic villi and the decidua basalis form the pancake-shaped placenta. Since the placenta detaches and sloughs off after the infant is born, naming of the maternal portion as the *decidua* ("that which falls off") is very appropriate. The decidua capsularis expands to accommodate the growing embryo, which eventually fills and stretches the uterine cavity. The fetal side of the placenta is easily recognized because it is slick and smooth, and the umbilical cord projects from its surface (see Figure 29.18c on p. 977). In contrast, the maternal side is "bumpy," reflecting the shape of the chorionic villus masses.

The placenta is usually fully functional as a nutritive, respiratory, excretory, and endocrine organ by the end of the second month of pregnancy. However, well before this time, oxygen and nutrients are diffusing from maternal to embryonic blood, and embryonic metabolic wastes are passing in the opposite direction. The barriers to free passage of substances between the two blood supplies are embryonic barriers: the membranes of the chorionic villi and the endothelium of embryonic capillaries. Although the maternal and embryonic blood supplies are very close, they normally never intermix.

While the placenta secretes HCG from the very beginning, its ability to produce the steroid hormones of pregnancy (estrogen and progesterone) matures much more slowly. If, for some reason, the placenta is producing inadequate amounts of these hormones when HCG levels wane, the endometrium degenerates and the pregnancy is aborted. Throughout pregnancy, blood levels of estrogen continue to increase (see Figure 29.6), and the estrogen-progesterone ratio changes from 1:100 at the beginning of pregnancy to close to 1:1 by its end. The rising levels of estrogen and progesterone throughout the gestation period encourage growth and further differentiation of the mammary glands, readying them for lactation. The placenta also produces other hormones, such as *placental lactogen, human chorionic thyrotropin,* and *relaxin.* The effects of these hormones on the mother will be described shortly.

Since many potentially harmful substances can cross placental barriers and enter the fetal blood, a pregnant woman should be very much aware of what she is taking into her body. *Teratogens* (ter′-ah-tō-jinz), factors that may cause severe congenital abnormalities or even fetal death, include alcohol, nicotine, many drugs (anticoagulants, sedatives, antihypertensives, and some antibiotics), and maternal infections, particularly German measles. For example, when a

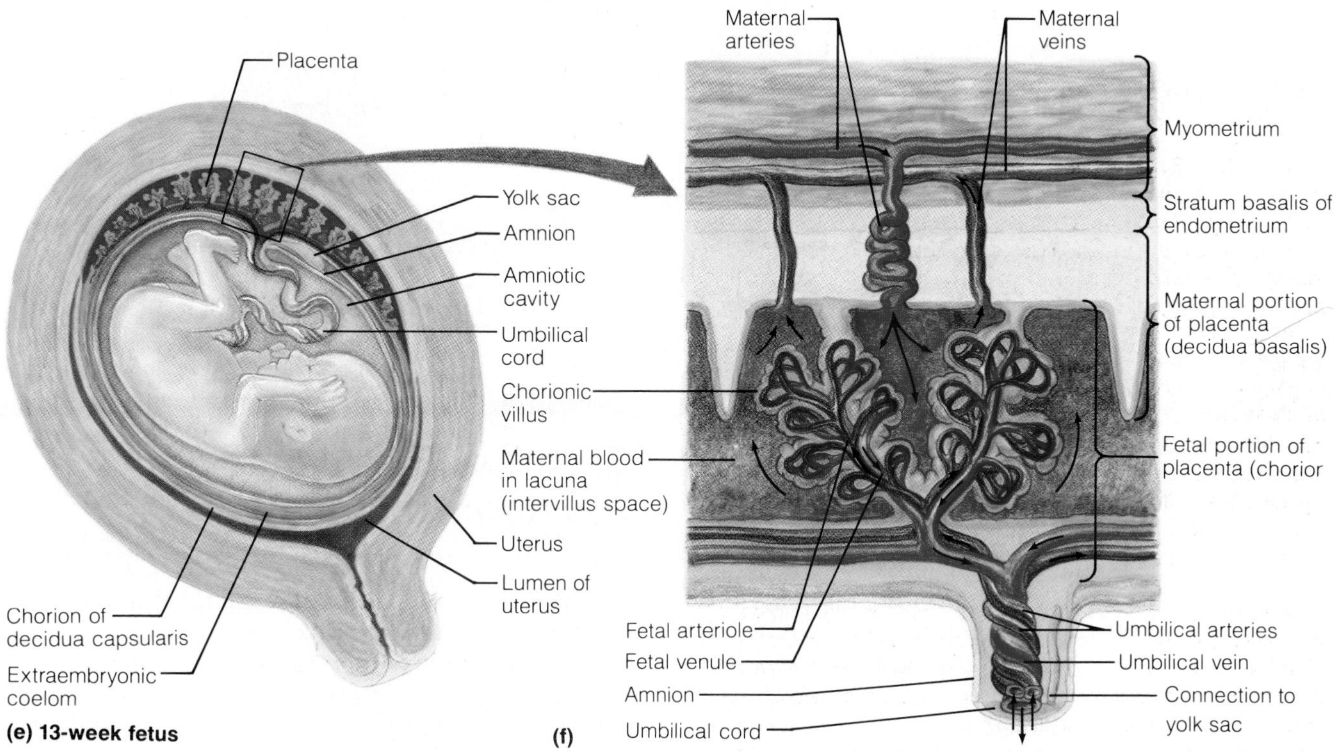

Figure 29.7 (continued)

A CLOSER LOOK Baby Making: Popsicles, GIFTs, and Surrogate Moms

In the summer of 1978, Louise Joy Brown was born. A cause for her parents to rejoice, for sure, but tiny Louise was unique. She was the first baby born who was conceived in a glass dish, or, more precisely, in an *in vitro* laboratory setting. Most who read of this event were overcome with amazement that a child could be conceived outside the body. For scientists in the field of reproductive physiology, Louise's birth was a stunning breakthrough that fueled the subsequent avalanche of new techniques to provide childless couples with offspring.

Making babies outside the privacy of the bedroom is quite a recent phenomenon. In the 1970s, the simplest and oldest of the so-called alternative forms of conception became available. This technique, *artificial insemination by donor (AID)* was directed at couples where the male was sterile, or nearly so, and at single women. Almost overnight, sperm banks that collected, froze, and stored sperm from anonymous donors began to spring up; today, in the United States, there are 17 sperm banks with at least 100,000 frozen ejaculates for sale. The technique is simple by today's standards: All that is required is sperm, a syringe, and a fertile woman. The sperm is deposited in the woman's vagina or cervix

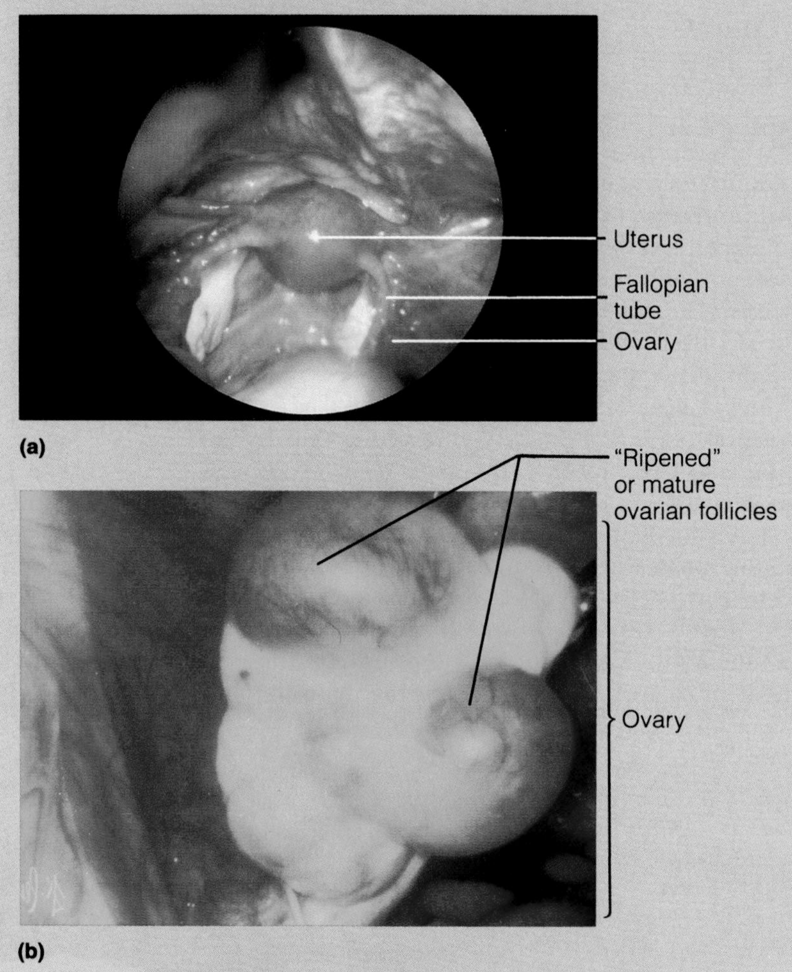

Photographs of (**a**) an unstimulated ovary and (**b**) an ovary that has been stimulated with hormones to develop multiple mature follicles, which can be seen protruding from the ovary surface.

woman imbibes too much alcohol, her fetus gets inebriated as well. However, the fetal consequences may be much more lasting and result in the *fetal alcohol syndrome (FAS)* typified by microcephaly (small head), mental retardation, and incomplete and abnormal growth. Nicotine hinders oxygen delivery to the fetus, impairing normal growth and development. *Thalidomide* (thuh-lih′-duh-mīd), a sedative used by thousands of pregnant women in Europe in the 1960s to alleviate morning sickness, was banned in the United States, but did get marketed by one U.S. company. When taken during the period of limb bud differentiation (approximately days 26–56), it sometimes resulted in tragically deformed infants with flipper-like legs and arms. ■

Events of Embryonic Development

Having followed placental development into the fetal stage, we will now backtrack and consider development of the embryo during and after implantation. Even while implantation is occurring, the blastocyst is being converted to the next developmental stage, the **gastrula** (gas′-truh-luh), in which the three primary germ layers are evident, and the embryonic membranes are developing. Since the embryonic

at the appropriate time of the month, and nature is allowed to take its course. Presently, AID accounts for over 20,000 American babies each year. It is an "oldie but goodie" that does what it claims to do.

Although the number of *test-tube babies* like Louise is swelling, they are still relatively few because only a few hundred test-tube clinics exist around the world. (The first U.S. test-tube clinic was opened in 1980.) *In vitro* (literally, "in glass") fertilization is an elegant, but tricky, procedure. The woman is primed with hormones to promote the development of several oocytes in her ovaries. (The photographs show an unstimulated and stimulated ovary.) The oocytes are then sucked from the ovary through a small opening in the abdomen into which a viewing instrument called a *laparoscope* is inserted. A few of these oocytes are fertilized with her partner's sperm in a glass receptacle, and after a few days of growth, all preembryos that show normal early development are implanted back into her uterus. The other oocytes are also fertilized and frozen "for spares" in case the first implantation is unsuccessful—hence, the catchy term *popsicle babies*. Since it is estimated that only 30% of natural fertilizations result in pregnancy and a healthy baby, the success rate of 15% to 20% reported by these clinics is actually very good, and the number in the popsicle baby club is beginning to become significant.

Meanwhile, the use of *surrogate* mothers had been growing by leaps and bounds. Surrogates (for a fee) provide ova and substitute uteri for women who have fertile husbands but who have been unsuccessful in becoming pregnant. The surrogate is artificially fertilized with the sponsor husband's sperm (as in AID), and the baby is eventually delivered to the sponsoring parents. Anyone who reads the newspapers is well aware that the use of surrogates is teeming with legal and ethical problems, primarily because some surrogate mothers have refused to relinquish the baby after birth. Whose baby is it?

Surrogates will no longer be needed when *embryo transfers* from the womb of one woman into that of another become more common. In this procedure, a conceptus is harvested from a donor, who has been artificially inseminated by the infertile recipient's mate, and then transferred into the infertile recipient's uterus. The harvesting procedure involves flushing the donor's uterus out during the first week of pregnancy through a soft catheter (plastic tube). As you can imagine, this procedure is even more rife with complications than *in vitro* fertilization, but the first successful transfer occurred in early 1984 with the birth of a healthy baby boy.

In 1985, another new technique was added to the roster, *gamete intrafallopian transfer (GIFT)*. GIFT is an attempt to bypass the hazards (high acidity, unsuitable mucus, phagocytes, and others) that sperm encounter in a female's reproductive tract by bringing oocytes and sperm together in a woman's uterine tubes. As with the *in vitro* procedure, the woman is given ovulatory stimulant drugs, and the oocytes are harvested by vacuum aspiration. The oocytes and the partner's sperm are mixed and then immediately inserted into the woman's uterine tubes via the same laparoscope tubing. GIFT has a cost advantage over *in vitro* fertilization because everything is done as a single procedure.

These new scientific breakthroughs are good news for childless couples, but for some people they have raised the fear of a "brave new world" where babies will gestate in laboratory bottles and only those with the right gender or selected genetic traits will be chosen to live. This is the nature of every major change—intense joy within a veil of tears.

membranes contribute to the well-being of the embryo or aid its development in various ways, we will consider their formation first.

Formation and Roles of the Embryonic Membranes

The **embryonic membranes** that form during the first two to three weeks of development include the amnion, yolk sac, chorion, and allantois (see Figure 29.7). The **amnion** (am′-nē-on) begins to develop when the surface cells of the inner cell mass detach from the embryonic disc and fashion themselves into a transparent membranous sac. This sac, the amnion, becomes filled with *amniotic fluid*, which bathes the cells of the embryonic disc. As the embryonic disc curves to form the tubular body (a process we will describe shortly), the amnion curves with it. Eventually, the sac extends all the way around the embryo, broken only by the umbilical cord (see Figure 29.7d).

Sometimes called the "bag of waters," the amnion provides a buoyant environment that protects the developing embryo against physical trauma (much as the cerebrospinal fluid protects the brain) and helps to maintain a constant homeostatic temperature. The fluid also helps to keep the rapidly growing embryonic parts from adhering and fusing together and allows the embryo considerable freedom of movement. Initially, amniotic fluid is formed by absorption from the maternal blood, but as the fetus develops and its kidneys become functional, urine is added to the amniotic fluid volume. The water portion of amniotic fluid turns over very rapidly and is completely changed every 3 hours. Sloughed-off fetal cells in the amniotic fluid can be aspirated and examined for genetic and chromosomal abnormalities, a procedure called *amniocentesis* (described in Chapter 30).

The **yolk sac** (see Figure 29.7b) forms from (endodermal) cells on the surface of the embryonic disc opposite the forming amnion. These cells proliferate and fashion themselves into a small sac that eventually hangs from the ventral surface of the embryo (see Figure 29.7d). The yolk sac is very important in birds and reptiles, where it surrounds and digests the yolk and dispatches the nutrients to the embryo. But human eggs contain very little yolk, and nutritive functions have been taken over by the placenta. The yolk sac is nevertheless important in humans as the site of earliest blood cell formation and the source of the primordial germ cells that migrate into the embryonic body to seed the gonads.

The *chorion,* which arises from the trophoblast and helps to form the placenta, has already been described (see Figure 29.7). Since it is the outermost membrane, the chorion encloses all other membranes and the embryonic body. As the fetus enlarges during gestation, the amniotic and chorionic membranes become tightly fused to form the composite *amniochorionic membrane.*

The **allantois** (uh-lan′-tō-is) (see Figure 29.7c) forms as a small outpocketing at the caudal end of the yolk sac. In animals that develop within shelled eggs, the allantois is a disposal site for solid metabolic wastes, but the placenta has done away with the need for this fetal excretory function. In humans, the allantois serves as a structural base for constructing the umbilical cord that links the embryo to the placenta, and becomes part of the urinary bladder. Shortly after the allantois appears, it becomes covered with extraembryonic mesoderm, which produces the umbilical arteries and vein. When the umbilical cord is fully formed, it contains a core of embryonic connective tissue (Wharton's jelly) and is covered externally by amniotic membrane.

Gastrulation: Germ Layer Formation

The process of gastrula formation, or **gastrulation,** entails cellular rearrangements and widespread cell migrations that transform the inner cell mass into a three-layered **embryo,** in which **ectoderm** (ek′-tō-derm), **mesoderm** (meh′-zō-derm), and **endoderm** (en′-dō-derm) are present. The surface cells of the inner cell mass split off to form the amnion, and the rest of the inner cell mass is referred to thereafter as the **embryonic disc** (see Figures 29.7a and 29.8a). The embryonic disc soon elongates and broadens anteriorly, becoming a pear-shaped plate of cells. A midline depression, called the **primitive streak,** develops along the median longitudinal plane of the embryonic disc (Figure 29.8a and b). The primitive streak establishes the longitudinal axis of the embryo and is the site through which some of the surface cells of the embryonic disc will invade the middle region.

The cell migrations that follow appear rather frantic when observed microscopically. Surface cells of the embryonic disc migrate medially across other cells, enter the primitive streak, and then push laterally between the cells at the upper and lower surfaces (see Figure 29.8a and b). When this migratory behavior ends (about the second week after fertilization), the cells at the superior boundary of the embryonic disc are ectodermal cells, those at the lower surface are endoderm, and the migratory cells that now occupy the middle of the "sandwich" form the mesoderm. The mesodermal cells immediately beneath the early primitive streak quickly aggregate, forming a rod of mesodermal cells called the **notochord** (nō′-tō-kord), which serves as the first axial support of the embryo (Figure 29.8c). The conceptus, now called an embryo, is about 2 mm long.

The three **primary germ layers** serve as the primitive tissues, or cell populations, from which all body organs will derive. The ectoderm goes on to fashion the structures of the nervous system and the skin epidermis. The endoderm forms the mucosal linings of the digestive and respiratory systems, and parts of the mucosae of the urinary and reproductive systems. The mesoderm forms virtually everything else. Table 29.1 lists the germ layer derivatives; some of the details of the differentiation processes are described next.

Organogenesis: Differentiation of the Germ Layers

Gastrulation lays down the basic structural framework of the embryo and sets the stage for **organogenesis,** (or″-gah-nō-jeh′-nih-sis), formation of body organs and organ systems. During organogenesis, the continuity of the germ layers is lost as their cells rearrange themselves, forming clusters, rods, or membranes that differentiate into definitive tissues and organs. By the end of the embryonic period, when the embryo is eight weeks old and about 22 mm (slightly less than 1 inch) long from head to buttocks (referred to as the crown-rump measurement), all the adult organ systems are recognizable. It is truly amazing how much organogenesis occurs in such a short time in such a small amount of living matter.

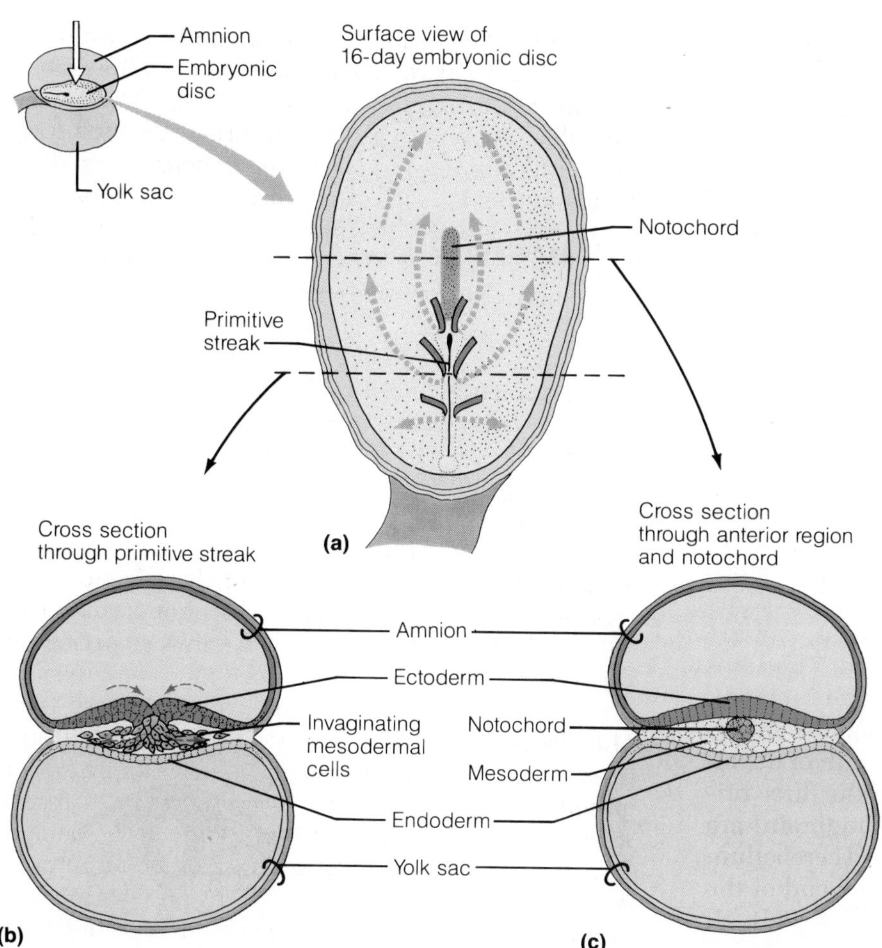

Figure 29.8 Gastrulation: formation of the three primary germ layers. (**a**) Surface view of an embryonic disc. The solid blue lines indicate the path taken by surface cells as they migrate toward the median primitive streak. The broken lines indicate the path of the same cells as they enter the primitive streak and migrate laterally just beneath the surface cells. (**b**) Cross-sectional view of the embryonic disc, showing the germ layers resulting from cell migration. Arrows show the migratory pathway of the cells that have invaded the primitive streak. (**c**) Migratory behavior is over; the cells at the surface of the embryonic disc are ectodermal cells; those at the inferior surface are endodermal cells; and the migratory cells that now occupy the middle position are mesodermal cells.

Table 29.1 Derivatives of the Primary Germ Layers

Ectoderm	Mesoderm	Endoderm
All nervous tissue	Skeletal, smooth, and cardiac muscle	Epithelium of digestive tract (except that of oral and anal cavities)
Epidermis of skin and epidermal derivatives (hairs, hair follicles, sebaceous and sweat glands, nails)	Cartilage, bone, and other connective tissues	Glandular derivatives of digestive tract (liver, pancreas)
Cornea and lens of eye	Blood, bone marrow, and lymphoid tissues	Epithelium of respiratory tract, auditory tube, and tonsils
Epithelium of oral and nasal cavities, of paranasal sinuses, and of anal canal	Endothelium of blood vessels and lymphatics	Thyroid, parathyroid, and thymus glands
Tooth enamel	Serosae of ventral body cavity	Epithelium of reproductive ducts and glands
Epithelium of pineal and pituitary glands and adrenal medulla	Fibrous and vascular tunics of eye	Epithelium of urethra and bladder
	Synovial membranes of joint cavities	
	Organs of urogenital system (ureters, kidneys, gonads, and reproductive ducts)	

Specialization of the Ectoderm

The first event in organogenesis is **neurulation,** the differentiation of ectoderm that gives rise to the brain and spinal cord (Figure 29.9). This process is induced (stimulated to happen) by chemical signals provided by an axial supporting rod of mesoderm, the notochord mentioned earlier. Although the notochord is eventually replaced by the vertebral column, parts of it persist later in life as the nucleus pulposus of the intervertebral discs. By the seventeenth day of embryonic development, the ectoderm overlying the notochord thickens, forming the *neural plate,* and by the twenty-first day, the raised edges of the neural plate form prominent *neural folds.* Two days later, the superior margins of the neural folds fuse, forming a *neural tube,* which detaches from the surface ectoderm to lie beneath it in the body midline. As described in Chapter 12, the anterior end of the neural tube develops into the brain and associated sensory organs, and the posterior portion becomes the spinal cord. The associated neural crest cells give rise to the cranial, spinal, and sympathetic ganglia and to the medulla of the adrenal gland. By the end of the first month of development, eye rudiments are present, and the three primary brain vesicles (fore-, mid-, and hindbrain) are obvious, but the cerebral hemispheres and cerebellum have yet to make their appearance. By the end of the second month (the end of the embryonic period), all brain flexures are evident, the cerebral hemispheres cover the top of the brain stem (see Figure 12.4), and brain waves can be recorded. Most of the remaining ectoderm forming the surface layer of the embryonic body differentiates into the epidermal layer of the skin. Other ectodermal derivatives are indicated in Table 29.1.

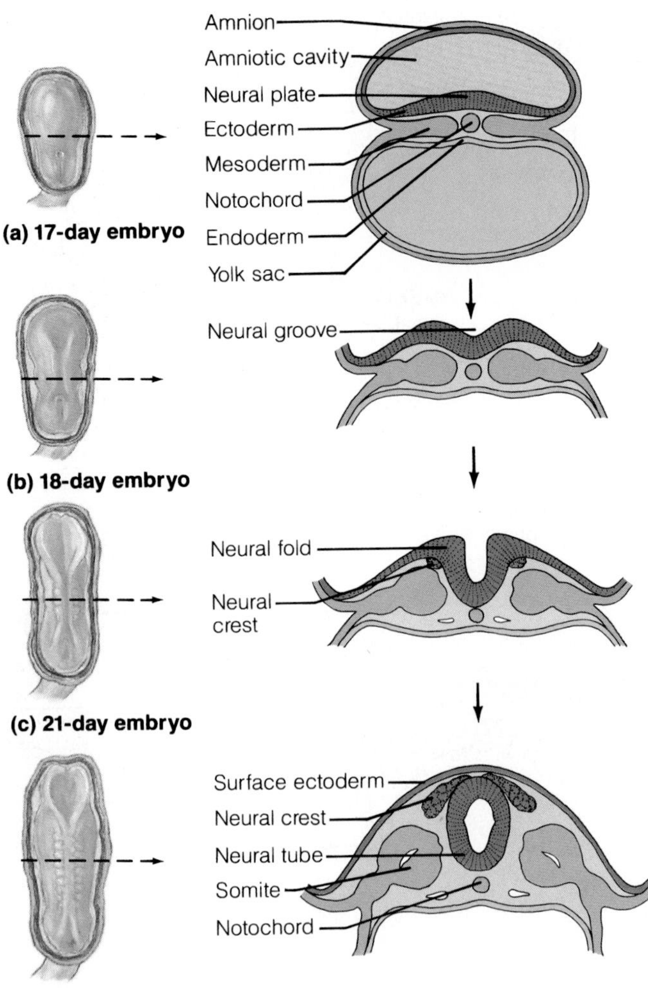

Figure 29.9 Events of neurulation. Dorsal surface views on the left; frontal sections on the right. (**a**) The embryonic disc, which has accomplished gastrulation. The notochord and neural plate have formed. (**b**) Formation of the neural folds by folding of the neural plate. (**c**) The neural folds begin to close. (**d**) The neural tube has detached from the surface ectoderm and lies between the surface ectoderm and the notochord; the neural crest is evident. While the neural tube is forming, the embryonic disc is folding to form the embryonic body, as shown in Figure 29.10.

Specialization of the Endoderm

As noted, the embryo starts off as a flat embryonic plate, but as it grows rapidly, it folds to form the more familiar tubular structure of the body trunk (Figure 29.10). The folding occurs simultaneously on both sides (lateral folds) and from the rostral (head fold) and caudal (tail fold) ends, and progresses toward the central part of the embryonic body, where the yolk sac and umbilical vessels protrude. As the endoderm undercuts the embryo and its edges come together and fuse, it forms the tubular epithelial lining, or mucosa, of the gastrointestinal tract (Figure 29.11). The specialized regions of the tract (pharynx, esophagus, stomach, intestines) quickly become apparent, and then the oral and anal openings perforate. The mucosal lining of the respiratory tract forms as an outpocketing from the foregut (pharyngeal endoderm), and the glands that derive from endoderm arise as endodermal outpocketings at various points along the tract. For example, the thyroid, parathyroids, and thymus form from the endoderm of the pharynx region, and the accessory digestive glands (liver and pancreas) arise from the intestinal mucosa (midgut) more inferiorly. All of these structures are present in rudimentary form by the fifth week of embryonic development. Notice that while the glands mentioned form entirely from endodermal cells, the endoderm forms only the mucosal lining of the hollow organs of the digestive and respiratory tracts. Mesoderm forms the rest of their wall structure.

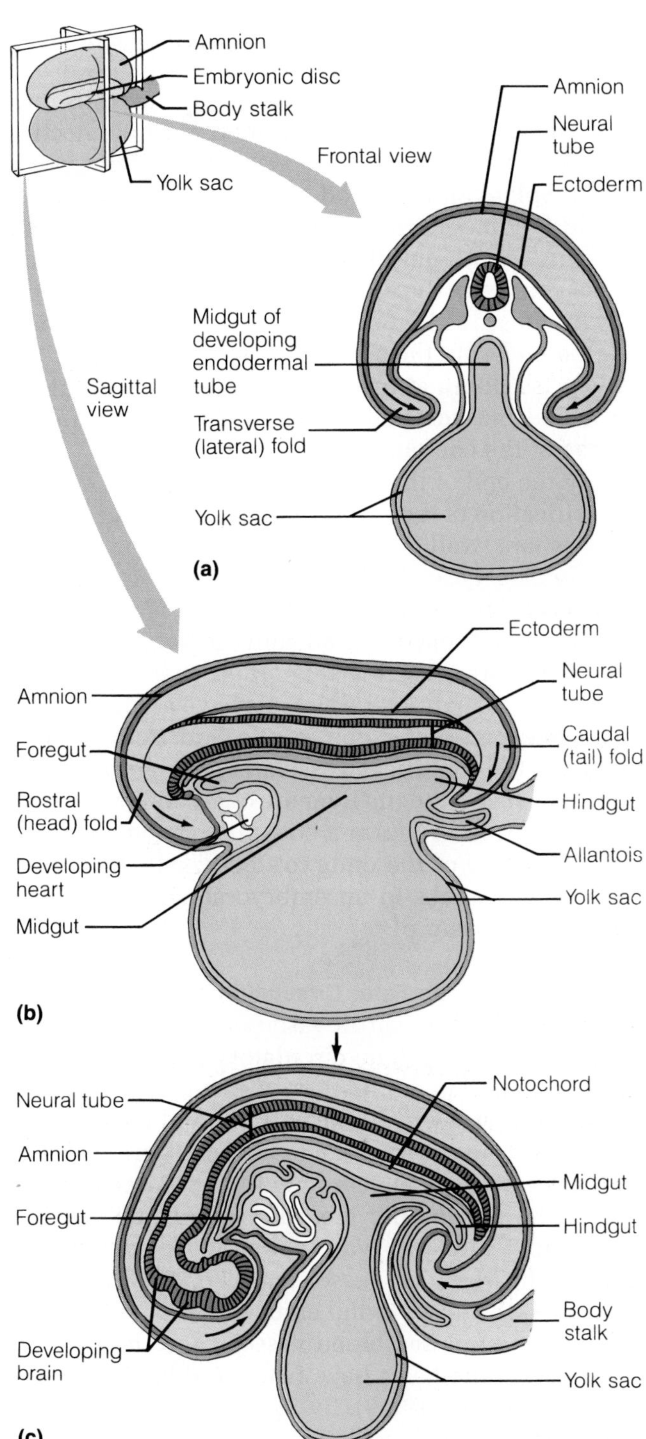

Figure 29.10 Folding of the embryonic body that produces the tubular body trunk. (**a**) Frontal view showing the lateral folding of the embryonic disc. Sagittal views of the (**b**) early and (**c**) late folding process, which begins simultaneously from the rostral and caudal ends of the embryo. Folding results in the formation of an internal endodermal tube that constitutes the embryonic foregut, midgut, and hindgut.

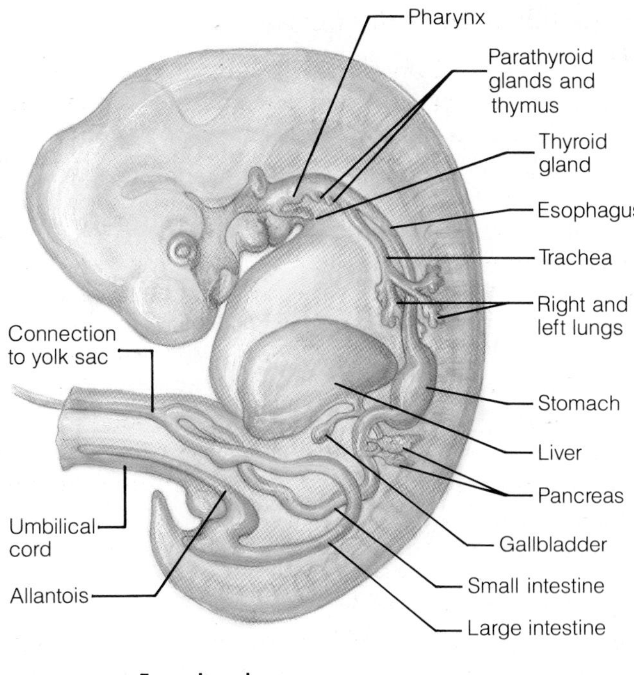

Figure 29.11 Endodermal differentiation to form the digestive and respiratory system mucosae and associated glands.

Specialization of the Mesoderm

Mesoderm forms all body parts except the nervous system, skin epithelium and derivatives, and mucosae of the digestive and respiratory organs. Since we have covered embryonic development of the various organ systems in earlier chapters, we will focus only on the very early events of mesodermal segregation and specialization here.

The first evidence of mesodermal differentiation is the appearance of the notochord in the embryonic disc (see Figure 29.8c). Shortly thereafter, mesodermal aggregates appear on either side of the notochord (Figure 29.12a). The largest of these, the *somites* (sō′-mīts), are a series of paired mesodermal blocks that hug the notochord on either side. All 40 pairs of somites are present by the end of the fourth week of development. Flanking the somites laterally are small clusters of mesoderm called the *intermediate mesoderm* and then double sheets of *lateral mesoderm*.

The cells of somite mesoderm on either side cooperate to form the early (cartilaginous) vertebrae, which are laid down in a corresponding segmental manner; this vertebrae-forming portion of each somite is called the *sclerotome* (skleh′-rō-tōm) (Figure 29.12b). The skeletal muscles of the body neck and trunk are also laid down in segments (much like the structural plan of an earthworm) from somite regions called

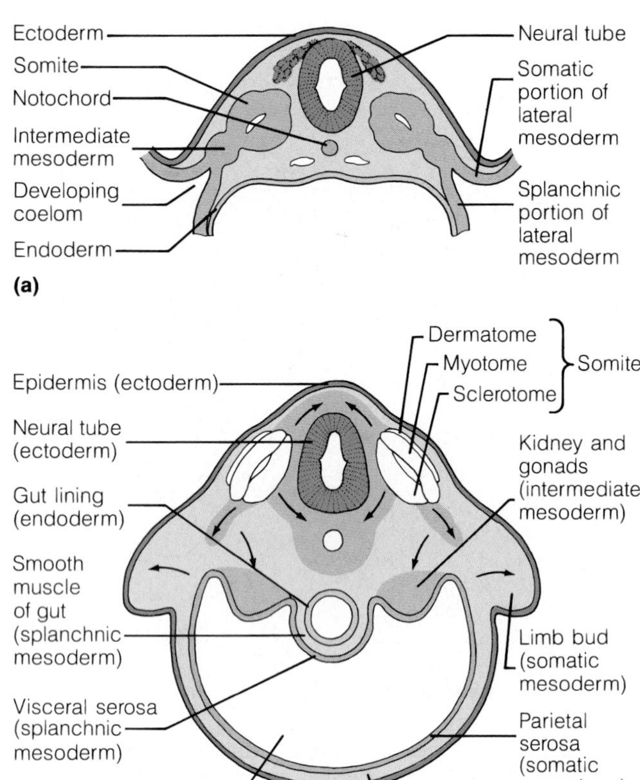

Figure 29.12 Early mesodermal differentiation. (**a**) Segregation of the mesoderm (with the exception of notochord formation, which occurs earlier) takes place as neurulation is occurring. The notochord becomes flanked bilaterally by a somite, intermediate mesoderm, and lateral mesoderm successively. (**b**) Schematic view of the mesodermal divisions and relationships in an undercut embryo. The different regions of each somite—the sclerotome, dermatome, and myotome—contribute to the formation of the vertebrae, skin dermis, and skeletal muscles, respectively. The intermediate mesoderm forms the kidneys, gonads, and associated structures. The somatic mesoderm portion of the lateral mesoderm forms the limb buds and parietal serosa and contributes to the dermis; the splanchnic mesoderm forms the muscle, connective tissues, and serosae of the walls of the visceral organs.

myotomes, which develop in conjunction with the vertebrae. Each myotome is quickly invaded by a spinal nerve that supplies both its sensory and motor supply. As a result, a segmented arrangement of muscle, bone, and nerves is produced along the length of the embryo. The *dermatome*, or outer wall of each somite, helps to form the dermis of the skin. Cells of the intermediate mesoderm produce the glands and ducts of the urogenital system and the adrenal cortex.

The lateral mesoderm consists of two mesodermal sheets: the *somatic mesoderm* and the *splanchnic* (splank′-nik) *mesoderm*. Cells of the somatic mesoderm have three major roles: (1) They migrate beneath the surface ectoderm and help to form the dermis of the skin; (2) they form the parietal serosa that lines the ventral body cavity; and (3) they produce the *limb buds* that give rise to the muscles and bones of the limbs (see Figure 29.12b). Mesenchymal cells that form the cardiovascular system and most connective tissues of the body arise from the splanchnic mesoderm. Splanchnic mesodermal cells also collect around the endodermal mucosal linings, where they form the smooth muscle, connective tissues, and serosal coverings of the digestive and respiratory organs. As you can see in Figure 29.12b, folding of the embryonic body forms the ventral body cavity, called the *coelom* (sē′-lom), at this time. As already described, the lateral mesodermal layers cooperate to form the serosae of the coelom.

By the end of the embryonic period (end of week 8), ossification of the bones has begun and the skeletal muscles are well formed and contracting spontaneously. The mesonephric kidneys are at the height of their development, and the metanephros is developing. The gonads have formed, and the lungs and digestive organs are attaining their final shape and body position. The main blood vessels have assumed their final plan, and blood delivery to and from the placenta via the umbilical vessels is constant and efficient. The heart and the liver are competing with each other for space and form a conspicuous bulge on the ventral surface of the embryo's body. All this by the end of eight weeks in an embryo about 1 inch long from crown to rump!

Development of the Fetal Circulation. Embryonic development of the cardiovascular system lays the groundwork for the fetal circulatory pattern, which is converted to the adult pattern at birth. The primitive blood cells arise within the yolk sac and the extraembryonic mesoderm associated with the chorion. Before the third week of development, tiny spaces appear within the splanchnic mesoderm and link together into rapidly spreading vascular networks, destined to form the heart, blood vessels, and lymphatics. By the end of the third week, the embryo has a fairly elaborate system of paired blood vessels, and the two vessels forming the heart have fused and bent into an S shape (see Figure 19.19). By 3½ weeks, the miniature heart is pumping blood for an embryo less than a quarter inch long.

The **umbilical arteries** and **vein** and three *vascular shunts* are unique vascular modifications seen only during prenatal development (Figure 29.13); all of these structures are occluded at birth. The large umbilical vein carries oxygenated blood returning from the placenta into the embryonic body, where it is conveyed to the liver. From there blood flows into the inferior vena cava via two routes. The direct route is through the **ductus venosus** (duk′-tus veh-nō′-sus), a vascular shunt that allows most of the blood to bypass the liver circulation. The indirect route is through the

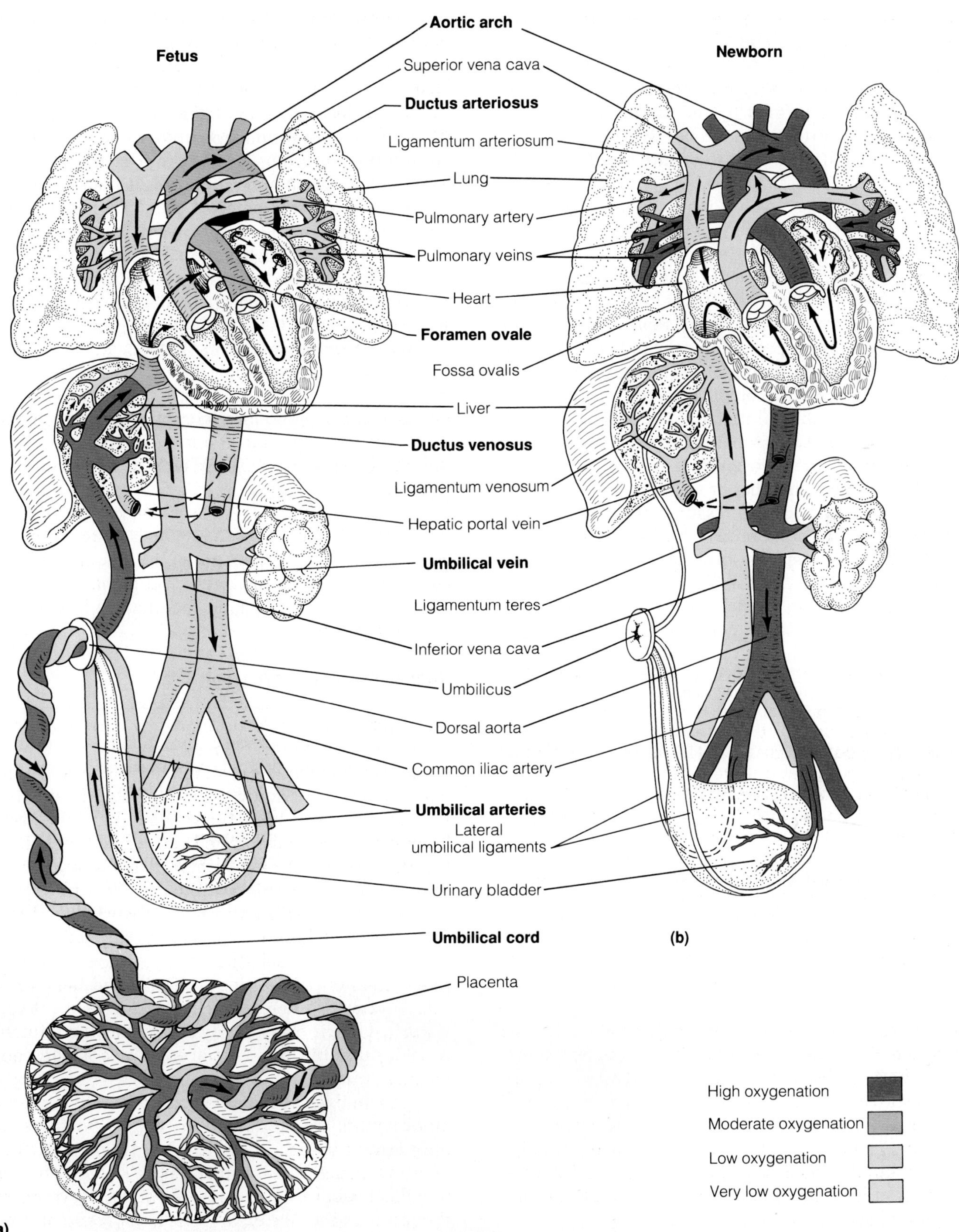

Figure 29.13 Circulation in fetus and newborn. Arrows indicate direction of blood flow. (**a**) Special adaptations for embryonic and fetal life. The umbilical vein carries oxygen- and nutrient-rich blood from the placenta to the fetus; the umbilical arteries carry waste-laden blood from the fetus to the placenta; the ductus arteriosus and foramen ovale bypass the nonfunctional lungs; and the ductus venosus allows blood to partially bypass the liver circulation. (**b**) Changes in the cardiovascular system at birth. The umbilical vessels are occluded, as are the liver and lung bypasses (ductus venosus and arteriosus, and the foramen ovale).

liver sinusoids and out the hepatic veins. The inferior vena cava conveys blood to the right atrium of the heart. After birth, the liver plays an important role in nutrient processing, but during embryonic life these functions are performed by the mother's liver. Consequently, blood flow through the liver during development is important only to ensure viability of the liver cells.

Blood entering and leaving the heart encounters two more shunt systems, each of which serves to bypass the nonfunctional lungs. Some of the blood entering the right atrium flows directly into the left side of the heart via the **foramen ovale** ("oval hole"), an opening in the interatrial septum. Blood that does enter the right ventricle is pumped out into the pulmonary trunk. However, the second shunt, the **ductus arteriosus,** transfers much of that blood directly into the aorta again, bypassing the pulmonary circuit. (The lungs do receive adequate blood to maintain their growth.) Blood tends to enter the two pulmonary bypass shunts because the heart chamber or vessel on the other side of each shunt is a lower-pressure area, owing to the low volume of venous return from the lungs. Blood flowing distally along the aorta eventually reaches the umbilical arteries, which are branches of the internal iliac arteries serving the pelvic structures. From here the largely deoxygenated blood, laden with metabolic wastes, is delivered back to the capillary circulation in the chorionic villi of the placenta.

Events of Fetal Development

Virtually all the groundwork for fetal development has been laid by the onset of the fetal period at the beginning of the ninth week. All body organ systems are present at least in rudimentary form, and some, such as the cardiovascular and urinary systems, are busily involved in their normal activities. Most changes that occur from this point on involve further tissue and organ specialization and growth, accompanied by changes in body proportions. During the fetal period, the developing fetus grows from a crown-to-rump length of about 22 mm (slightly less than 1 inch) and a weight of approximately 1 g (0.03 ounce) to about 360 mm (14 inches) and 2.7–4.1 kg (6–10 pounds) or more. (Total body length at birth is about 550 mm, or 22 inches.) As you might expect with such tremendous growth, the changes in fetal appearance are quite dramatic (Figure 29.14). The most significant of these changes are summarized in Table 29.2.

Effects of Pregnancy on the Mother

Pregnancy can be a difficult time for the mother. Not only are there obvious anatomical changes, but striking changes in her metabolism and physiology as well.

Anatomical Changes

As pregnancy progresses, the female reproductive organs become increasingly vascularized and engorged with blood, and the vagina develops a purplish hue (*Chadwick's sign*) that is diagnostic for the pregnant condition. This enhanced vascularity causes a marked increase in vaginal sensitivity and sexual intensity, and some women achieve orgasm for the first time when they are pregnant. The breasts, too, become

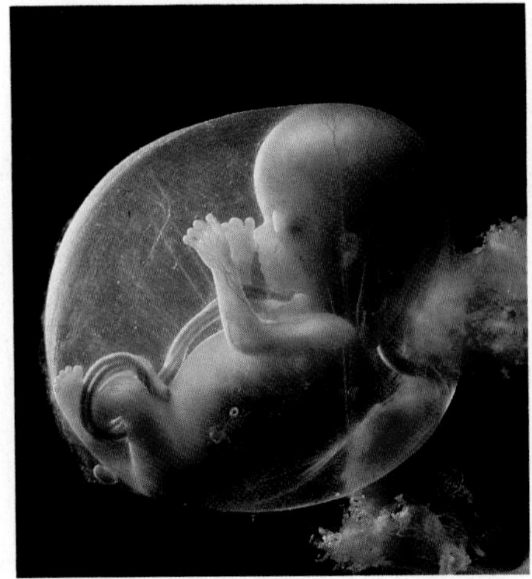

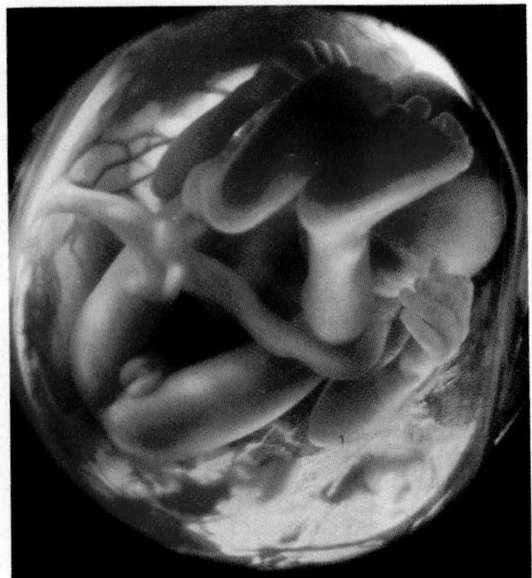

Figure 29.14 Photographs of developing fetuses. The major events of fetal development are growth and tissue specialization. All the organ systems are laid down, at least in rudimentary form, during the embryonic period of development. (**a**) Fetus of 14 weeks, approximately 6 cm long. (**b**) Fetus of 20 weeks, approximately 19 cm long. By birth the fetus is typically 35 cm in length crown to rump.

Table 29.2 Developmental Events of the Fetal Period

Time	Changes/Accomplishments
8 weeks (end of embryonic period)	Head nearly as large as body; all major brain regions present Liver disproportionately large Limbs present; ossification just begun; weak, spontaneous muscle contractions occur Cardiovascular system fully functional All body systems present in at least rudimentary form Approximate crown-to-rump length: 22 mm.
9–12 weeks	Head still dominant, but body elongating; brain continues to enlarge, shows its general structural features; cervical and lumbar enlargements apparent in spinal cord; retina of eye is present Skin epidermis and dermis obvious; nails seen on digits; facial features present in crude form Liver prominent and bile being secreted; palate is fusing; most glands of endodermal origin are developed; walls of hollow visceral organs gaining smooth muscle Blood formed in bone marrow Notochord degenerating and ossification accelerating; limbs nicely molded Sex readily detected Approximate crown-to-rump length at end of interval: 36 mm.
13–16 weeks	Cerebellum becoming prominent; neuroglia beginning to differentiate; general sensory organs differentiated; eyes and ears assume characteristic position and shape; blinking of eyes and sucking motions of lips occur Face looks human and body beginning to outgrow head Hard palate fused; glands developed in GI tract; meconium is collecting; elastic fibers apparent in lungs Kidneys attain typical structure Most bones now distinct and joint cavities apparent; quickening occurs (mother feels spontaneous muscular activity of fetus) Approximate crown-to-rump length at end of interval: 140 mm.
17–20 weeks	Eyelashes and eyebrows present; vernix caseosa (fatty secretions of sebaceous glands) covers body; lanugo (silklike hairs) covers skin Fetal position (anterior flexion) assumed because of space restrictions Limbs achieve final proportions Approximate crown-to-rump length at end of interval: 190 mm.
21–30 weeks	Substantial increase in weight; may survive if born prematurely at 27–28 weeks, but hypothalamic temperature regulation and lung production of surfactant are immature Myelination of cord begins; eyes are open Skin is wrinkled and red; tooth enamel is forming on deciduous teeth in gums Body is lean and well proportioned Mesenteric attachments completed; pulmonary branching is two-thirds complete Blood formation in bone marrow increasing; blood formation in liver is decreasing Distal limb bones beginning to ossify Approximate crown-to-rump length at end of interval: 280 mm.
30–40 weeks (term)	Skin whitish pink; fingernails and toenails present; fat laid down in subcutaneous tissue Testes in scrotum (in males) Approximate crown-to-rump length at end of interval: 350–360 mm.

engorged with blood, and prodded by rising levels of estrogen and progesterone, they enlarge and become heavy, and their areolae darken. Some women develop increased pigmentation of facial skin of the nose and cheeks, a condition called *chloasma* (klō-az′-muh) or the "mask of pregnancy."

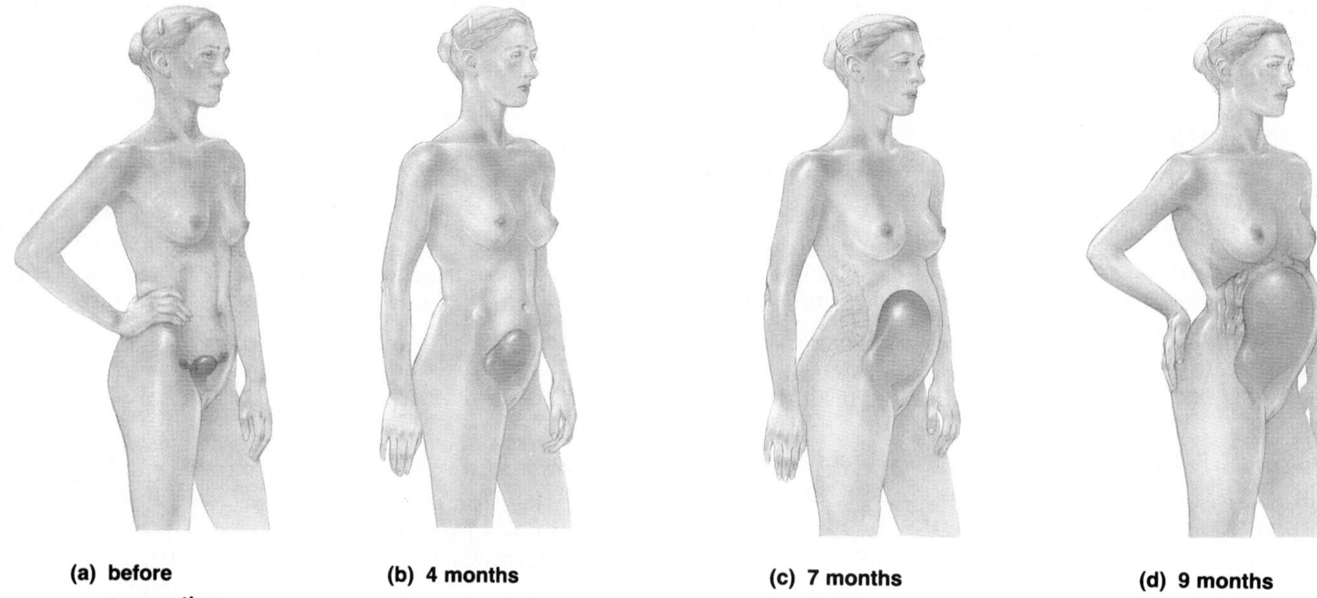

Figure 29.15 Relative size of the uterus before conception and during pregnancy. (**a**) Before conception, the uterus is the size of a fist and resides within the pelvis. (**b**) At 4 months, the fundus of the uterus is halfway between the pubic symphysis and the umbilicus. (**c**) At 7 months, the fundus is well above the umbilicus. (**d**) At 9 months, the fundus reaches the xiphoid process.

The degree of uterine enlargement during pregnancy is nothing less than remarkable. Starting as a fist-sized organ, the uterus occupies most of the pelvic cavity by 16 weeks (Figure 29.15b). The fetus itself is only about 120 mm long at this time, but the placenta is fully formed, the uterine muscle is hypertrophied, and the amniotic fluid is increasing in volume. As pregnancy continues, the uterus pushes higher and higher into the abdominal cavity, exerting increasing pressure on both abdominal and pelvic organs (Figure 29.15c). As birth nears, the uterus reaches the level of the xiphoid process and occupies the bulk of the abdominal cavity (Figure 29.15d). The crowded abdominal organs press superiorly against the diaphragm, which encroaches on the thoracic cavity. As a result, the ribs flare, causing the thorax to widen.

The increasing bulkiness of the abdomen changes the woman's center of gravity, and many women develop an accentuated lumbar curvature (lordosis), often accompanied by backaches, during the last few months of pregnancy. Placental production of the hormone **relaxin** causes pelvic ligaments and the pubic symphysis to relax, widen, and become more flexible. This increased motility eases birth passage, but it may also result in a waddling gait in the interim.

Normal pregnancy is responsible for considerable weight gain. Because some women are over- or underweight before pregnancy begins, it is almost impossible to state the ideal or desirable weight gain. However, summing up the weight increases resulting from fetal and placental growth, growth of the maternal reproductive organs and breasts, and increased blood volume during pregnancy, a weight gain of approximately 13 kg (about 29 pounds) usually occurs.

Obviously, good nutrition is necessary all through pregnancy if the developing fetus is to have all the building materials (especially proteins, calcium, and iron) necessary to form its tissues and organs. However, the popular belief that a pregnant woman is eating for two has encouraged many women to eat twice the amount of food actually needed during pregnancy, which of course leads to excessive weight gain. Actually, a pregnant woman only needs about 300 additional calories daily to sustain proper fetal growth. The emphasis should be on eating high-quality food, not just more food.

Metabolic Changes

As the placenta enlarges, it secretes increasing amounts of **human placental lactogen (HPL),** which works cooperatively with estrogen and progesterone to stimulate maturation of the breasts for lactation. In addition, HPL has anabolic effects (promotes growth) and a glucose-sparing effect in the mother. This last effect means that maternal cells metabolize more fatty acids and less glucose than usual, sparing glucose for use by the fetus. The placenta also releases **human chorionic thyrotropin (HCT),** a glycoprotein hormone similar to thyroid-stimulating hormone of the anterior pituitary. HCT activity increases the rate of maternal metabolism throughout the pregnancy, causing hypermetabolism.

Physiological Changes

Gastrointestinal System

Excessive salivation often occurs during pregnancy, and many women suffer nausea, commonly called *morning sickness*, during the first few months, until their system becomes adjusted to the elevated levels of progesterone and estrogen. (It is interesting to note that nausea is an important side effect of many birth control pills.) Regurgitation of stomach acid into the esophagus, causing *heartburn*, is common because the esophagus is displaced and the stomach is crowded by the growing uterus. Another problem is constipation, because motility of the digestive tract declines during pregnancy.

Urinary System

Since the kidneys have the additional burden of disposing of fetal metabolic wastes, they produce more urine during pregnancy. Because the uterus compresses the bladder, urination becomes more frequent, more urgent, and sometimes uncontrollable. (The last condition is called stress incontinence.)

Respiratory System

The nasal mucosa responds to estrogen by becoming edematous and congested; thus, nasal stuffiness and occasional nosebleeds may occur. Vital capacity increases during pregnancy, as does respiratory rate, but residual volume declines, and many women exhibit *dyspnea* (disp′-nē-uh), or difficult breathing, during the later stages of pregnancy.

Cardiovascular System

Perhaps the most dramatic physiological changes occur in the cardiovascular system. Total body water rises; this acts as a safeguard against blood loss during birth. Blood volume increases by 25% to 40% by the thirty-second week, owing to increases in both formed elements and plasma volume to accommodate the additional needs of the fetus. Blood pressure and pulse typically rise and increase cardiac output by 20% to 40% at various stages of pregnancy; this helps propel the greater circulatory volume around the body. Because the uterus presses on the pelvic blood vessels, venous return from the lower limbs may be impaired and result in varicose veins.

Parturition (Birth)

Parturition (par″-tū-rish′-un) is the culmination of gestation—giving birth to the baby. It usually occurs within 15 days of the calculated due date (the calculated due date is 280 days from the last menstrual period). The series of events that accompany birth are referred to collectively as **labor.**

Initiation of Labor

While the precise mechanism that triggers labor is not clear, several events and three hormones appear to be interlocked in this process. During the last few weeks of pregnancy, estrogen reaches its highest levels in the mother's blood. This has two important consequences: It causes abundant oxytocin receptors to form on the myometrial cells of the uterus, and it antagonizes progesterone's quieting influence on the uterine muscle. As a result, the myometrium becomes increasingly irritable, and weak, irregular uterine contractions begin to occur. These contractions, called *Braxton Hicks contractions*, have caused many women to go to the hospital, only to be told that they were in *false labor* and sent home.

As birth nears, two more chemical signals cooperate to convert these false labor pains into the real thing. Certain cells of the fetus begin to produce *oxytocin* (ok′-sē-tō″-sin), which in turn acts on the placenta, stimulating production and release of *prostaglandins* (pros-tuh-glan′-dinz). Both hormones are powerful uterine muscle stimulants, and since the myometrium is now highly sensitive to oxytocin, contractions become more frequent and more vigorous. At this point, the increasing emotional and physical stresses activate the hypothalamus, which signals for oxytocin release by the posterior pituitary. The combined effects of elevated levels of oxytocin and prostaglandins initiate the rhythmic expulsive contractions of true labor. Once the hypothalamus is involved, a positive feedback mechanism is propelled into action—greater contractile force causes the release of more oxytocin, which causes greater contractile force, and so on (Figure 29.16).

Since both oxytocin and prostaglandins are essential to initiate labor in humans, interference with production of either of these hormones will hinder the onset of labor. For example, antiprostaglandin drugs such as aspirin and ibuprofen can inhibit labor at the early stages, and such drugs are used occasionally to prevent preterm births.

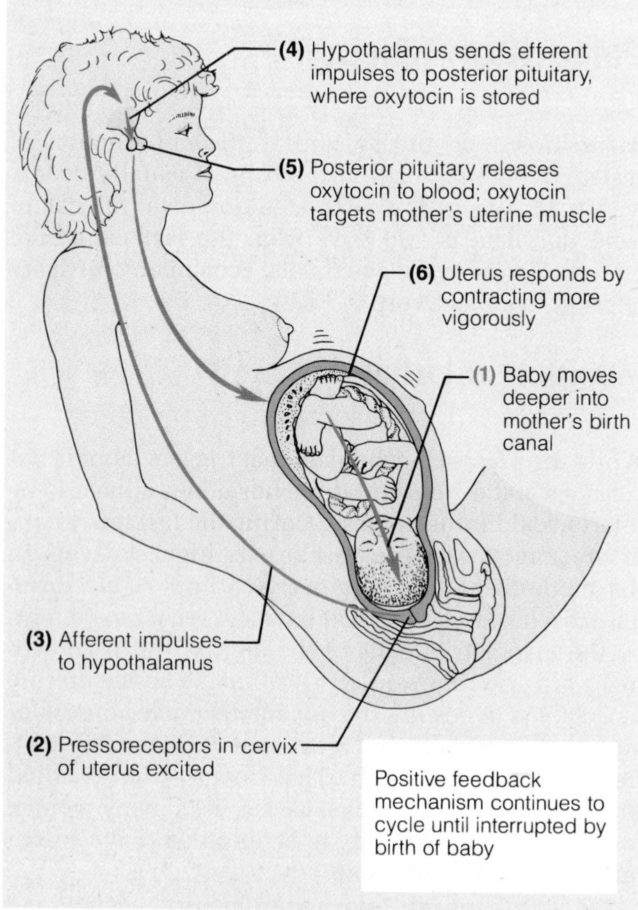
(4) Hypothalamus sends efferent impulses to posterior pituitary, where oxytocin is stored
(5) Posterior pituitary releases oxytocin to blood; oxytocin targets mother's uterine muscle
(6) Uterus responds by contracting more vigorously
(1) Baby moves deeper into mother's birth canal
(3) Afferent impulses to hypothalamus
(2) Pressoreceptors in cervix of uterus excited
Positive feedback mechanism continues to cycle until interrupted by birth of baby

Figure 29.16 The positive-feedback mechanism by which oxytocin promotes labor contractions during birth.

Stages of Labor

Stage 1: Dilation Stage

The **dilation stage** (Figure 29.17a and b) encompasses the time from labor's onset until the cervix is fully dilated (about 10 cm in diameter). As labor starts, regular uterine contractions (much like peristaltic contractions) begin in the upper part of the uterus and move downward toward the vagina. At first, the contractions are weak and only the superior uterine muscle is active. These initial contractions are 15–30 minutes apart and last for 10–30 seconds. As labor progresses, the contractions become more vigorous and more rapid, and the lower uterine segment gets involved. As the infant's head is forced against the cervix with each contraction, the cervix begins to soften, becomes thinner (effaces), and dilates. Eventually, the amnion ruptures, releasing the amniotic fluid, an event commonly called "breaking the water." The dilation stage is the longest part of labor and lasts for 6–12 hours (or considerably more), depending on the size of the infant and on whether the woman has previously given birth. This phase begins when the infant's head descends into the true pelvis, an event called *engagement*. As descent continues through the birth canal, the baby's head rotates so that its greatest dimension is in the anteroposterior line, which allows it to navigate the narrow dimensions of the pelvic outlet.

Stage 2: Expulsion Stage

The **expulsion stage** (Figure 29.17c) is the period from full dilation to delivery of the infant, or actual childbirth. Ordinarily, by the time the cervix is fully dilated, strong contractions are occurring every 2–3 minutes and lasting about 1 minute, and a mother experiencing natural childbirth (that is, undergoing labor without local anesthesia) has an increasing urge to push or bear down with the abdominal muscles. Although this phase can take as long as 2 hours, it is typically 50 minutes in a first birth and around 20 minutes in subsequent births.

When the largest dimension of the baby's head is distending the vulva, an event called *crowning* (Figure 29.18a), an *episiotomy* (eh-pih″-zē-ah′-tō-mē) may be performed to reduce tissue tearing. An episiotomy is a midline incision from the vaginal orifice nearly to the rectum. The baby's head extends as it exits from the perineum, and once the head has been delivered, the rest of the baby's body is delivered much more easily. After birth, the umbilical cord is clamped and cut.

When the infant is in the usual *vertex* (head-first) presentation, the skull (its largest diameter) acts as a wedge to dilate the cervix; the head-first presentation also allows the baby to be suctioned free of mucus and to breathe even before it has completely exited from the birth canal (Figure 29.18b). In *breech* (buttock-first) presentations and other nonvertex presentations, these advantages are lost and delivery is much more difficult, often requiring the use of forceps. Additionally, if the woman has a deformed or android (malelike) pelvis, labor may be prolonged and difficult. This condition is called *dystocia* (dĭs-tō′-shuh) (*dys* = difficult; *toc* = birth). Besides causing extreme maternal fatigue, another possible consequence of

Figure 29.17 ▶ Parturition. (**a**) Dilation stage (early). The baby's head has entered the true pelvis and is engaged. The widest head dimension is along the left-right axis. (**b**) Late dilation. The baby's head rotates so that its greatest dimension is in the anteroposterior axis as it moves through the pelvic outlet. Dilation of the cervix is nearly complete. (**c**) Expulsion stage. The baby's head extends as it reaches the perineum and is delivered. (**d**) Placental stage. After the baby has been delivered, the placenta is detached by the continuing uterine contractions and is delivered.

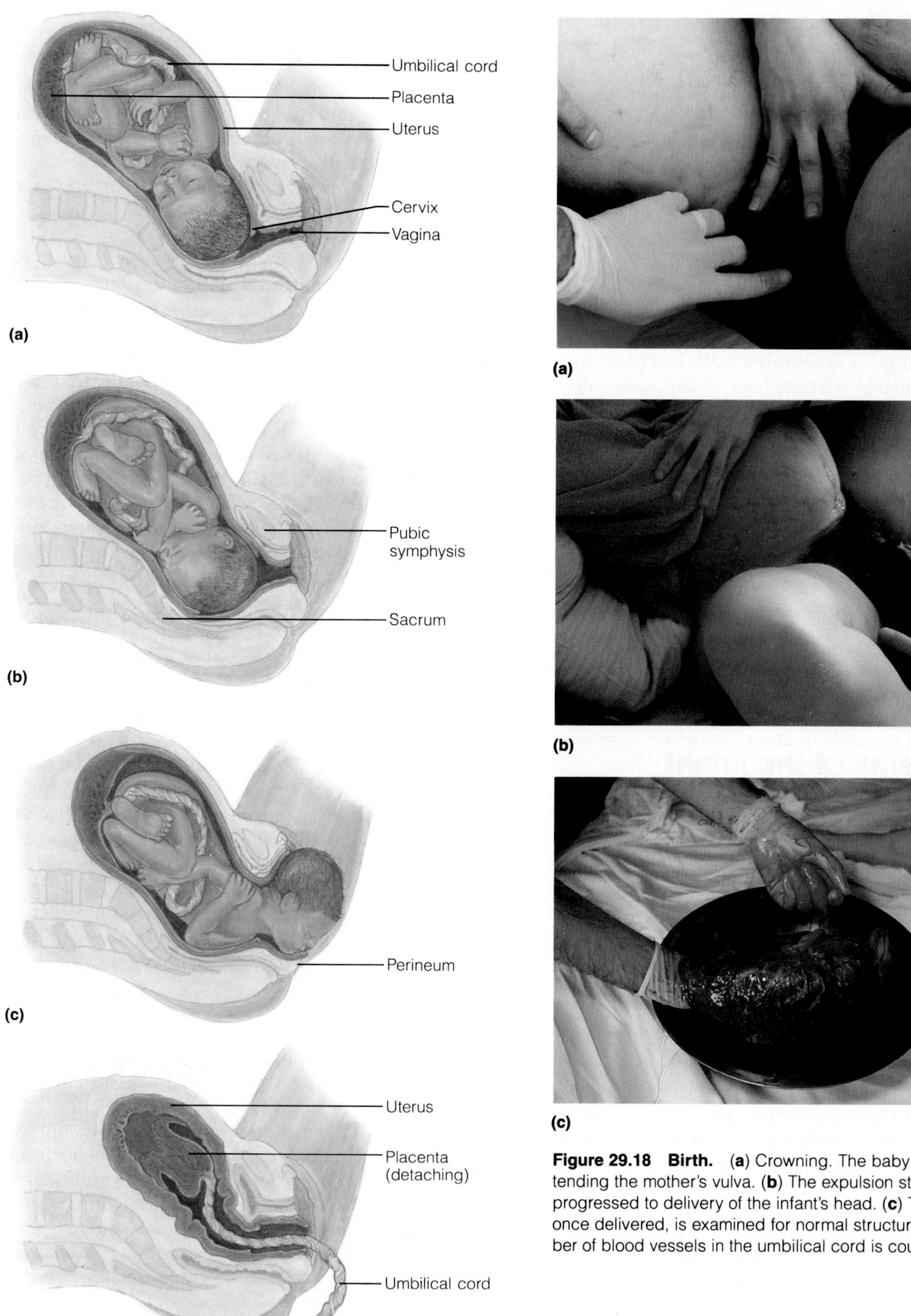

Figure 29.18 Birth. (**a**) Crowning. The baby's head is distending the mother's vulva. (**b**) The expulsion stage of labor has progressed to delivery of the infant's head. (**c**) The placenta, once delivered, is examined for normal structure, and the number of blood vessels in the umbilical cord is counted.

dystocia is fetal brain damage (resulting in cerebral palsy or epilepsy) and decreased viability of the infant. To prevent these outcomes, a *cesarean* (seh-zayr′-ē-un) *(C-) section* is performed in many such instances. A C-section is delivery of the infant through a surgical incision made through the abdominal and uterine walls.

Stage 3: Placental Stage

The **placental stage** (Figure 29.17d), or the delivery of the placenta, is accomplished within 15 minutes after birth of the infant. Forceful uterine contractions continue after birth and compress the uterine blood vessels, preventing hemorrhage and causing the placenta to detach from the uterine wall. Placental separation is signaled by a firmly contracting, rising uterine fundus and a lengthening of the umbilical cord. The placenta and its attached fetal membranes, collectively called the *afterbirth,* is then easily removed by slight tension on the umbilical cord. It is very important that all placental fragments be removed to prevent continued uterine bleeding after birth (*postpartum bleeding*). Since absence of one of the umbilical arteries is often associated with other cardiovascular disorders in the infant, the number of blood vessels in the severed umbilical cord is counted after the placenta has been delivered (Figure 29.18c).

Adjustments of the Infant to Extrauterine Life

The *neonatal period* is the four-week period immediately after birth. Here we will be concerned with the events of just the first few hours after birth in a normal infant. As you might suspect, birth represents quite a shock to the infant. Exposed to intense physical trauma during the birth process, it is suddenly cast out of its watery, warm environment into the cold, and its placental life supports are severed. Now it must do for itself all that the mother's body has been doing for it—respire, obtain nutrients, excrete, and maintain its body temperature.

One minute after birth, the infant's physical status is assessed based on five signs: heart rate, respiration, color, muscle tone, and reflexes (tested by slaps on the feet). Each observation is given a score of 0 to 2, and the total is called the *Apgar rating*. An Apgar rating of 8 to 10 indicates a healthy baby; lower scores reveal problems in one or more of the physiological functions assessed.

Taking the First Breath

The crucial first event is to breathe. Once carbon dioxide is no longer removed by the placenta, it begins to accumulate in the baby's blood, causing acidosis. This, along with mechanical stimulation and the cold environmental temperature, excites the respiratory control centers in the brain and triggers the first inspiration. The first breath requires a tremendous effort—the airways are tiny, and the lungs are collapsed. However, once the lungs have been inflated in full-term babies, lung surfactant in the alveolar fluid helps to reduce the surface tension in the alveoli, and breathing becomes easier. The rate of newborn respiration is rapid (about 45 respirations per minute) during the first two weeks and then gradually declines to normal levels.

Keeping the lungs inflated is much more difficult for premature infants (those weighing less than 2500 g, or about 5.5 pounds, at birth) because surfactant production occurs during the last months of prenatal life. Consequently, preemies are usually put on respiratory assistance (a ventilator) until their lungs are mature enough to function on their own.

The Occlusion of Special Fetal Blood Vessels and Vascular Shunts

The special umbilical blood vessels and fetal shunts, no longer necessary, are occluded shortly after birth (see Figure 29.13b). The umbilical arteries and vein constrict and become fibrosed. The proximal parts of the umbilical arteries persist as the *superior vesical arteries* that supply the urinary bladder, and their distal parts become the *lateral umbilical ligaments.* The umbilical vein becomes the *ligamentum teres* that attaches the umbilicus to the liver. The ductus venosus collapses as blood stops flowing through the umbilical vein and is eventually converted to the *ligamentum venosum* of the liver.

As the pulmonary circulation becomes functional, left heart pressure increases and right heart pressure decreases. These changing pressure relationships cause the pulmonary shunts to close. The flap of the foramen ovale moves to the closed position, and its edges fuse to the septal wall; ultimately, only a slight depression, the *fossa ovalis,* marks its position in adults. The ductus arteriosus constricts and is converted to the cordlike *ligamentum arteriosum,* which persists as a fibrous connection between the aorta and pulmonary trunk. Some investigators believe that bradykinin, a vasoconstrictor released by the newly functioning lungs, also promotes constriction of the ductus arteriosus.

Except for the foramen ovale, all of the special circulatory adaptations of the fetus are functionally occluded within 30 minutes after birth, although the fibrosis that converts them into the adult structures goes on for several weeks. Closure of the foramen ovale is not complete for nearly a year. As described in Chapter 19, failure of the ductus arteriosus or foramen ovale to close leads to congenital heart defects.

The Transitional Period

Infants pass through an unstable *transitional period* lasting 6–8 hours after birth, during which they adjust to extrauterine life. For the first 30 minutes, the baby is awake, alert, and active and may look hungry as it makes sucking motions with its mouth. Heart rate increases above the normal infant range of 120 to 160 beats per minute, respirations become more rapid and irregular, and body temperature falls. Then activity gradually diminishes, and the baby sleeps for 3 hours or so. A second period of activity then occurs, and the baby gags frequently as it regurgitates mucus and debris. After this, the infant sleeps again and then stabilizes, with waking periods (dictated by hunger) occurring every 3–4 hours.

Lactation

Lactation is the production of milk by the hormone-prepared mammary glands. Rising levels of (placental) estrogen, progesterone, and lactogen toward the end of pregnancy stimulate the hypothalamus to release prolactin-releasing hormone (PRH). The anterior pituitary gland responds to PRH by secreting **prolactin.** (This mechanism is described in more detail in Chapter 17, p. 541.) After an initial two- to three-day delay, true milk production begins. During this delay (and also during late gestation), a yellowish fluid called **colostrum** (kō-los′-trum) is secreted. Colostrum has a lower lactose content than milk and contains almost no fat, but it contains more protein, vitamin A, and minerals than true milk. Like milk, colostrum is rich in IgA antibodies. Since these antibodies are probably digested in the gastrointestinal tract, it is doubtful that they provide passive immunity to the nursing infant (as was once claimed); however, they may help somewhat to protect the infant's digestive tract against bacterial infection.

After birth, prolactin release gradually wanes toward prebirth levels, and continual milk production depends on mechanical stimulation of the nipples, normally provided by the sucking infant. Mechanoreceptors in the nipple send afferent nerve impulses to the hypothalamus, stimulating secretion of PRH. This results in a burstlike release of prolactin, which stimulates milk production for the next feeding.

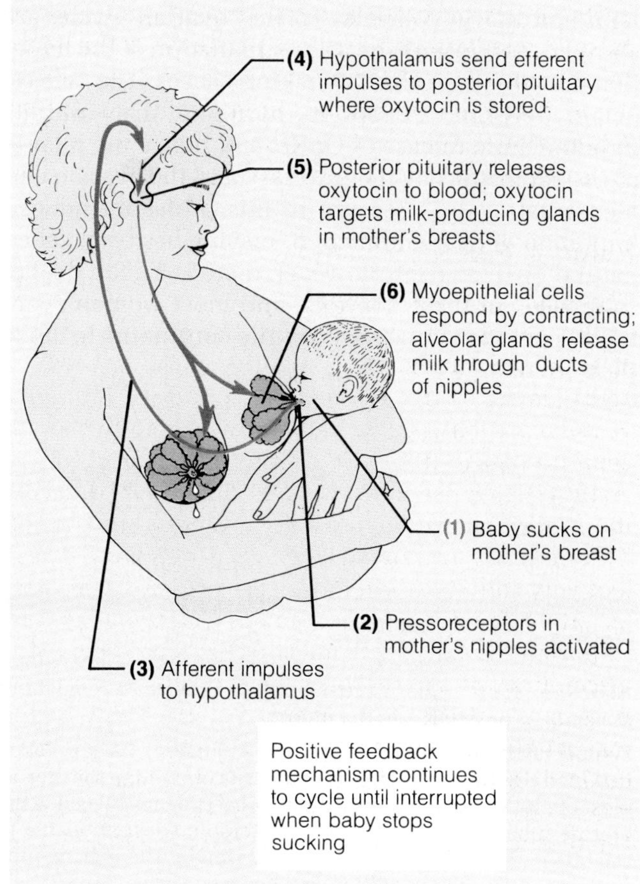
(4) Hypothalamus send efferent impulses to posterior pituitary where oxytocin is stored.
(5) Posterior pituitary releases oxytocin to blood; oxytocin targets milk-producing glands in mother's breasts
(6) Myoepithelial cells respond by contracting; alveolar glands release milk through ducts of nipples
(1) Baby sucks on mother's breast
(2) Pressoreceptors in mother's nipples activated
(3) Afferent impulses to hypothalamus
Positive feedback mechanism continues to cycle until interrupted when baby stops sucking

Figure 29.19 Positive feedback mechanism of the milk let-down reflex.

The same afferent impulses also prompt hypothalamic release of oxytocin from the posterior pituitary via a positive feedback mechanism. Oxytoxin causes the *milk let-down reflex,* the actual ejection of milk from the alveoli of the mammary glands (Figure 29.19). Let-down occurs when myoepithelial cells surrounding the glands are stimulated by binding of oxytocin to their membrane receptors, triggering milk ejection from both breasts, not just the suckled one. Oxytocin also targets the recently emptied uterus, causing it to contract, helping it to return to (nearly) its prepregnant size. For this reason, as well as the advantages of breast milk to the infant (the fats and iron in human milk are better absorbed, and the amino acids are more efficiently metabolized than those of cow's milk), many obstetricians recommend that mothers nurse their babies.

When nursing is discontinued, the stimulus for prolactin release, and thus milk production, ends and the ability of the mammary glands to produce milk declines. However, milk secretion can continue for several years if appropriate stimulation continues. While prolactin levels are high, the normal hypothalamic-pituitary controls of the ovarian cycle are damped, supposedly because stimulation of the hypothalamus by the infant's sucking causes it to release beta-endorphin, a peptide hormone that inhibits hypothalamic release of GnRH and hence the release of gonadotropins by the pituitary and the ovarian and menstrual cycles. Because of this prolactin-induced inhibition of ovarian function, nursing has been called natural birth control. However, there is a good deal of "slippage" in these controls, and most women eventually begin to ovulate even while continuing to nurse their infants.

* * *

In this chapter, we have followed the changes that occur during human *in utero* development. But the phenomenon of differentiation has barely been touched upon. How does an unspecialized cell that can become *anything* in the body develop into a specific *something* (a heart or bone cell, for example)? And what paces the developmental sequence, so that if a particular process fails to occur at a precise time, it never occurs at all? Scientists are beginning to believe that there are master switches in the genes. Chapter 30, the final chapter of this book, describes a small part of the "how" as it examines the interaction of genes that determine what we finally become.

Related Clinical Terms

Abortion (*abort* = born prematurely) Premature removal of the embryo or fetus from the uterus.

Ectopic pregnancy (ek-tah'-pik) (*ecto* = outside) A pregnancy in which the embryo implants in any site other than the uterus; most often the site is a uterine tube (tubal pregnancy); since the uterine tube (as well as most other ectopic sites) is unable to establish a placenta or accommodate growth, the uterine tube ruptures unless the condition is diagnosed early.

Eclampsia (ih-klamp'-sē-uh) (*eclamp* = shine) Dangerous condition in which the pregnant woman becomes edematous and hypertensive, and proteinuria and seizures occur; formerly called toxemia of pregnancy.

Hydatid (hydatidiform) mole (hī'-duh-tid) (*hydat* = watery) Developmental abnormality of the placenta; the conceptus degenerates and the chorionic villi convert into a mass of vesicles that resemble tapioca; signs include vaginal bleeding, which contains some of the grapelike vesicles.

Meconium (mih-kō'-nē-um) The greenish fecal matter discharged by a newborn shortly after birth; contains sloughed-off digestive epithelial cells and other debris.

Physiological jaundice (jon'-dis) Jaundice sometimes occurring in normal newborns within three to four days after birth; fetal erythrocytes are short-lived, and they break down rapidly after birth; the infant's liver may be unable to process the bilirubin (breakdown product of hemoglobin pigment) fast enough to prevent its accumulation in blood and subsequent deposit in body tissues.

Placenta abruptio (uh-brup'-tē-ō) (*abrupt* = broken away from) Premature separation of the placenta from the uterine wall; if this occurs before labor, it can result in fetal death due to anoxia.

Placenta previa (prē'-vē-uh) Placental formation adjacent to or across the internal os of the uterus; represents a problem because as the uterus and cervix stretch, tearing of the placenta may occur; additionally, the placenta precedes the infant during labor.

Ultrasonography (ul"-truh-suh-nah'-gruh-fē) Noninvasive technique that uses sound waves to visualize the position and size of the fetus and placenta (see p. 32).

Chapter Summary

1. The gestation period of approximately 280 days extends from the woman's last menstrual period to birth. The conceptus undergoes preembryonic development for about two weeks after fertilization, then embryonic development (weeks 3–8) and fetal development (week 9–birth).

FROM EGG TO EMBRYO (pp. 956–964)

Accomplishing Fertilization (pp. 956–959)

1. An oocyte is fertilizable for up to 24 hours; sperm are viable within the female reproductive tract for up to 72 hours.

2. Sperm must survive the hostile environment of the vagina and become capacitated.

3. Hundreds of sperm must release their acrosomes to break down the egg's corona radiata and zona pellucida.

4. When one sperm penetrates the egg, it triggers first the fast block to polyspermy (membrane depolarization) and then the slow block (release of cortical granules).

5. Following sperm penetration, the secondary oocyte completes meiosis II. Then the ovum and sperm pronuclei fuse (fertilization), forming a zygote.

Preembryonic Development (pp. 959–964)

6. Cleavage, a rapid series of mitotic divisions without intervening growth, begins with the zygote and ends with a blastocyst. The blastocyst consists of the trophoblast and an inner cell mass. Cleavage produces a large number of cells with a favorable surface-to-volume ratio.

7. The trophoblast adheres to, digests, and implants in the endometrium. Implantation is completed when the blastocyst is entirely surrounded by endometrial tissue, about 14 days after ovulation.

8. HCG released by the blastocyst maintains hormone production by the corpus luteum, preventing menses. HCG levels decline after four months. Typically, the placenta becomes functional as an endocrine organ by the third month.

9. The placenta acts as the respiratory, nutritive, and excretory organ of the fetus and produces the hormones of pregnancy; it is formed from embryonic (chorionic villi) and maternal (endometrial decidua) tissues. The chorion develops when the trophoblast becomes associated with extraembryonic mesoderm.

EVENTS OF EMBRYONIC DEVELOPMENT (pp. 964–972)

Formation and Roles of the Embryonic Membranes (pp. 965–966)

1. The fluid-filled amnion forms by detachment from the superior surface of the inner cell mass. It protects the embryo from physical trauma and adhesion formation, provides a constant temperature, and allows fetal movements.

2. The yolk sac forms from the endoderm; it is the source of primordial germ cells and early blood cells.

3. The chorion is the outermost membrane and takes part in placentation.

4. The allantois, a caudal outpocketing of the yolk sac, forms the structural basis of the umbilical cord. Its coating of extraembryonic mesoderm forms the umbilical vein and arteries.

Gastrulation: Germ Layer Formation (p. 966)

5. Gastrulation involves cellular rearrangements and migrations that transform the inner cell mass into a three-layered embryo (gastrula) containing ectoderm, mesoderm, and endoderm. Cells destined to become mesoderm migrate to the middle layer from the surface of the embryonic disc by moving through the midline primitive streak.

Organogenesis: Differentiation of the Germ Layers (pp. 966–972)

6. The ectoderm forms the nervous system and the epidermis of the skin and its derivatives. The first event of organogenesis is neurulation, which produces the brain and spinal cord. By the eighth week, all major brain regions are formed and brain waves can be recorded.

7. The endoderm forms the mucosa of the digestive and respiratory systems, as well as all associated glands (thyroid, parathyroids, thymus, liver, pancreas). It becomes a continuous tube when the embryonic body undercuts and fuses ventrally.

8. The mesoderm forms all other organ systems and tissues. It segregates early into (1) a dorsal superior notochord, (2) paired somites that form the vertebrae, skeletal trunk muscles, and part of the dermis, and (3) paired masses of intermediate and lateral mesoderm. The intermediate mesoderm forms the urogenital glands and ducts; the somatic layer of the lateral mesoderm forms the dermis of skin, parietal serosa, and bones and muscles of the limbs; the splanchnic layer of the lateral mesoderm forms the cardiovascular system and the visceral serosae.

9. The fetal cardiovascular system is formed in the embryonic period. The umbilical vein delivers nutrient- and oxygen-rich blood to the embryo; the paired umbilical arteries return oxygen-poor, waste-laden blood to the placenta. The ductus venosus allows most of the blood to bypass the liver; the foramen ovale and ductus arteriosus are pulmonary shunts.

EVENTS OF FETAL DEVELOPMENT (p. 972)

1. Since all organ systems have been laid down during the embryonic period, growth and tissue/organ specialization are the major events of the fetal period.

2. During the fetal period, fetal length increases from about 22 mm to 360 mm, and weight increases from less than an ounce to 6–10 pounds.

EFFECTS OF PREGNANCY ON THE MOTHER (pp. 972–975)

Anatomical Changes (pp. 972–974)

1. Maternal reproductive organs and breasts become increasingly vascularized during pregnancy, and the breasts enlarge.

2. The uterus eventually occupies nearly the entire abdominopelvic cavity. Abdominal organs are pushed superiorly and encroach on the thoracic cavity, causing the ribs to flare.

3. The increased abdominal mass changes the woman's center of gravity; lordosis and backache are common. A waddling gait occurs as pelvic ligaments and joints are loosened by placental relaxin.

4. A typical weight gain during pregnancy in a woman of normal weight is 29 pounds.

Metabolic Changes (pp. 974–975)

5. Human placental lactogen has anabolic effects and promotes glucose sparing in the mother. Human chorionic thyrotropin results in maternal hypermetabolism.

Physiological Changes (p. 975)

6. Many women suffer morning sickness, heartburn, and constipation during pregnancy.

7. The kidneys produce more urine, and pressure on the bladder may cause frequency, urgency, and stress incontinence.

8. Vital capacity and respiratory rate increase, but residual volume decreases. Dyspnea is common.

9. Total body water and blood volume increase dramatically. Heart rate and blood pressure rise, resulting in enhancement of cardiac output of 20% to 40% in the mother.

PARTURITION (BIRTH) (pp. 975–978)

1. Parturition encompasses a series of events called labor.

Initiation of Labor (p. 975)

2. When estrogen levels are sufficiently high, they induce oxytocin receptors on the myometrial cells and inhibit progesterone's quieting effect on uterine muscle. Weak, irregular contractions begin.

3. Fetal cells produce oxytocin, which stimulates prostaglandin production by the placenta; both hormones stimulate contraction of uterine muscle. Increasing stress activates the hypothalamus, causing oxytocin release from the posterior pituitary; this sets up a positive feedback loop resulting in true labor.

Stages of Labor (pp. 976–978)

4. The dilation stage is from the onset of rhythmic, strong contractions until the cervix is fully dilated (10 cm). The head of the fetus rotates as it descends through the pelvic outlet.

5. The expulsion stage extends from full cervical dilation until the birth of the infant.

6. The placental stage is the delivery of the afterbirth (the placenta and attached fetal membranes).

ADJUSTMENTS OF THE INFANT TO EXTRAUTERINE LIFE (pp. 978–979)

1. The infant's Apgar score is recorded immediately after birth.

Taking the First Breath (p. 978)

2. Once the umbilical cord is clamped, carbon dioxide accumulates in the infant's blood, causing respiratory centers in the brain to trigger the first inspiration. Mechanical and thermal stimuli are also involved.

3. Once the lungs are inflated, breathing is eased by the presence of surfactant, which decreases the surface tension of the alveolar fluid.

The Occlusion of Special Fetal Blood Vessels and Vascular Shunts (pp. 978–979)

4. Inflation of the lungs causes pressure changes in the circulation; as a result, the umbilical arteries and vein, ductus venosus, and ductus arteriosus collapse, and the foramen ovale closes. The occluded blood vessels are converted to fibrous cords; the site of the foramen ovale becomes the fossa ovalis.

The Transitional Period (p. 979)

5. For the first 8 hours after birth, the infant is physiologically unstable and adjusting. After stabilizing, the infant wakes approximately every 3–4 hours in response to hunger.

LACTATION (pp. 979–980)

1. The breasts are prepared for lactation during pregnancy by high blood levels of estrogen and progesterone and by placental lactogen.

2. Colostrum, a premilk fluid, is a fat-poor fluid that contains more protein, vitamin A, and minerals than true milk. It is produced toward the end of pregnancy and for the first two to three days after birth.

3. True milk is produced around the third day in response to suckling, which stimulates the hypothalamus to prompt anterior pituitary release of prolactin and posterior pituitary release of oxytocin. Prolactin stimulates milk production; oxytocin triggers milk let-down. Continued breast-feeding is required for continued milk production.

4. Ovulation and menses are absent or irregular at first during nursing, but in most women, the ovarian cycle is eventually reestablished while still nursing.

Review Questions

Multiple Choice/Matching

1. Indicate whether each of the following statements is describing (a) cleavage or (b) gastrulation.

_____ (1) period during which a morula is formed
_____ (2) period when vast amounts of cell migration occur
_____ (3) period when the three embryonic germ layers appear
_____ (4) period during which the blastocyst is formed

2. Most systems are operational in the fetus by four to six months. Which system is the exception to this generalization, impacting on premature infants? (a) the circulatory system, (b) the respiratory system, (c) the urinary system, (d) the digestive system.

3. The zygote contains chromosomes from (a) the mother only, (b) the father only, (c) both the mother and father, but half from each, (d) each parent and synthesizes others.

4. The outer layer of the blastocyst, which later attaches to the uterus, is the (a) decidua, (b) trophoblast, (c) amnion, (d) inner cell mass.

5. The fetal membrane that forms the basis of the umbilical cord is the (a) allantois, (b) amnion, (c) chorion, (d) yolk sac.

6. In the fetus, the ductus arteriosus carries blood from (a) the pulmonary artery to the pulmonary vein, (b) the liver to the inferior vena cava, (c) the right ventricle to the left ventricle, (d) the pulmonary trunk to the aorta.

7. Which of the following changes occur in the baby's cardiovascular system after birth? (a) blood clots in the umbilical vein, (b) the pulmonary vessels dilate as the lungs expand, (c) the ductus venosus becomes obliterated, as does the ductus arteriosus, (d) all of these.

8. Following delivery of the infant, the delivery of the afterbirth includes (a) the placenta only, (b) the placenta and decidua, (c) the placenta and attached (torn) fetal membranes, (d) the chorionic villi.

9. Identical twins result from the fertilization of (a) one ovum by one sperm, (b) one ovum by two sperm, (c) two ova by two sperm, (d) two ova by one sperm.

10. The umbilical vein carries (a) waste products to the placenta, (b) oxygen and food to the fetus, (c) oxygen and food to the placenta, (d) oxygen and waste products to the fetus.

11. The germ layer from which the skeletal muscles, heart, and skeleton are derived is the (a) ectoderm, (b) endoderm, (c) mesoderm.

12. Which of the following cannot pass through placental barriers? (a) blood cells, (b) glucose, (c) amino acids, (d) gases, (e) antibodies.

13. The most important hormone in initiating and maintaining lactation after birth is (a) estrogen, (b) FSH, (c) prolactin, (d) oxytocin.

14. The initial stage of labor, during which the neck of the uterus is stretched, is the (a) dilation stage, (b) expulsion stage, (c) placental stage.

Short Answer Essay Questions

15. Fertilization involves much more than a mere restoration of the diploid chromosome number. (a) What does the process of fertilization involve on the part of both the egg and sperm? (b) What are the effects of fertilization?

16. Cleavage is an embryonic event that mainly involves mitotic divisions. How does cleavage differ from mitosis occurring during life after birth, and what are its important functions?

17. The life span of the ovarian corpus luteum is extended for nearly three months after implantation, but otherwise it deteriorates. (a) Explain why this is so. (b) Explain why it is important that the corpus luteum remain functional following implantation.

18. The placenta is a marvelous, but temporary, organ. Starting with a description of its formation, show how it is an intimate part of both fetal and maternal anatomy and physiology during the gestation period.

19. Why is it that only one sperm out of the hundreds (or thousands) available enters the oocyte?

20. What is the function of the gastrulation process?

21. (a) What is a breech presentation? (b) Cite two problems with this type of presentation.

22. What factors are believed to bring about the uterine contractions at the termination of pregnancy?

Clinical Application Questions

23. During Mrs. Jones's labor, the obstetrician decided that it was necessary to perform an episiotomy. What is an episiotomy, and why is it done?

24. A woman in substantial pain called her doctor and explained (between sobs) that she was about to have her baby "right here." The doctor calmed her and asked how she had come to that conclusion. She said that her water had broken and that her husband could see the baby's head. (a) Was she right? If so, what stage of labor was she in? (b) Do you think that she had time to make it to the hospital 60 miles away? Why or why not?

25. Mary is a heavy smoker and has ignored a friend's advice to stop smoking during her pregnancy. On the basis of what you know about the effect of smoking on physiology, describe how Mary's smoking might affect her fetus.

CHAPTER 30

Heredity

Chapter Outline and Student Objectives

The Vocabulary of Genetics (pp. 984–985)

1. Define allele.
2. Differentiate clearly between genotype and phenotype.

Sexual Sources of Genetic Variation (pp. 985–987)

3. Describe events that lead to genetic variability of gametes.

Types of Inheritance (pp. 987–991)

4. Compare and contrast dominant-recessive inheritance with incomplete dominance and codominance.
5. Describe the mechanism of sex-linked inheritance.
6. Explain how polygene inheritance differs from that resulting from the action of a single pair of alleles.

The Influence of Environmental Factors on Gene Expression (p. 991)

7. Provide examples illustrating how gene expression may be modified by environmental factors.

Genetic Screening and Counseling (pp. 991–993)

8. List and explain several techniques used to determine or predict genetic diseases.

Preview of Selected Key Terms

Allele (ah-lēl′) (*all* = other, another) One of two or more alternate genes present at a given site on a chromosome.

Heterozygous (heh″-ter-ō-zī′-gus) (*hetero* = other; *zygo* = a yoke) Condition in which the members of an allele pair are not alike; e.g., *Aa*.

Homozygous (hō-mō-zī′-gus) (*homo* = same) Condition in which the members of an allele pair are identical; e.g., *AA* and *aa*.

Dominant allele An allele that is expressed whether present in the homozygous or heterozygous state, and which masks or suppresses the expression of the alternate allele.

Recessive allele An allele that is not expressed unless it is present in the homozygous state.

Genotype (jē′-nō-tīp) (*geno* = offspring; sex) The genes that a person has; one's "genetic formula."

Phenotype (fē′-nō-tīp) (*pheno* = show, seem, appear) The appearance or obvious nature of gene action.

Sex-linked gene A gene that is located on a sex chromosome, usually the *X* chromosome.

The wondrous growth and development of a new individual is guided by the codes of the gene-bearing chromosomes it receives from its parents in the egg and sperm. As described in Chapter 3, individual genes, or DNA segments, contain the genetic blueprints for proteins, many of which are enzymes that dictate the synthesis of virtually all of the body's molecules; thus, genes are ultimately expressed in your hair color, sex, blood type, and so on. However, genes do not act as free agents. As you will see, the ability of a gene to prompt the development of a specific trait is enhanced or inhibited by interactions with other genes, as well as by environmental factors.

Although the science of **genetics**, which studies the mechanism of heredity, is still relatively young, our understanding of how human genes interact to determine our characteristics has advanced considerably since the rules of gene transmission were first proposed by the monk Gregor Mendel in the mid-1800s. Part of the problem in unraveling the complexities of human genetics is that, unlike the pea plants that provided convenient subjects for Mendel's experiments, humans have long life spans, produce relatively few offspring, and cannot be mated experimentally just to examine the nature of their offspring. But the urge to understand human inheritance is a powerful one, and advances made in recent years have enabled geneticists to manipulate and engineer genes in order to examine their expression and treat or cure disease. Here, however, we will concentrate on the enlightening principles of heredity proposed by Mendel over a century ago.

The Vocabulary of Genetics

All human cells except gametes contain the diploid number of chromosomes (46), consisting of 23 pairs of *homologous* (hō-mah′-luh-gus) *chromosomes*. Two of this set are the *sex chromosomes* (*X* and *Y*) that determine our genetic sex; the other 44 are the 22 pairs of *autosomes*, which guide the expression of most other traits. The complete human *karyotype* (kayr′-ē-ō-tīp), or diploid chromosomal complement, is illustrated in Figure 30.1. The diploid *genome* (jē′-nōm), or genetic complement, actually represents two sets of genetic instructions—one from the egg and the other from the sperm.

Gene Pairs (Alleles)

Since chromosomes are paired, it follows that the genes within them are paired as well. Consequently, each of us receives two genes, one inherited from each parent, that interact to dictate a particular trait. These matched genes, which are at the same *locus* (location) on homologous chromosomes, are called **alleles** (ah-lēlz′). Alleles may code for the same or for alternative forms of a trait. For example, there are two alleles that dictate whether or not you have loose thumb ligaments. One variant codes for tight thumb ligaments; the other codes for loose ligaments (the so-called double-jointed thumb condition). When the two alleles controlling a single trait are exactly alike, a person is said to be **homozygous** (hō-mō-zī′-gus) for that particular trait. When the two alleles differ in their expression, the individual is **heterozygous** (heh″-ter-ō-zī′-gus) for the trait.

Sometimes, one allele masks or suppresses the expression of its partner. Such an allele is said to be **dominant,** whereas the allele that is masked is said to be **recessive.** By convention, a dominant allele is represented by a capital letter (for example, *J*), and a recessive allele by a lowercase letter (*j*). Dominant alleles are expressed, or make themselves "known," when they are present in either single or double dose; but for recessive alleles to be expressed, they must be present in double dose, the homozygous condition. Now let us return to our example of the double-jointed thumb. A person whose genetic makeup includes either the gene pair *JJ* (the homozygous dominant condition) or *Jj* (the heterozygous condition) will have double-jointed thumbs. The combination *jj* (the homozygous recessive condition) is required to produce tight thumb ligaments.

A common misunderstanding is that dominant traits are automatically seen more often because they are expressed whether the allele is present in single or double dose. However, dominance and recessiveness alone do not determine the frequency of a trait in a population; this also depends on the frequency or relative abundance of the respective dominant or recessive alleles in that population.

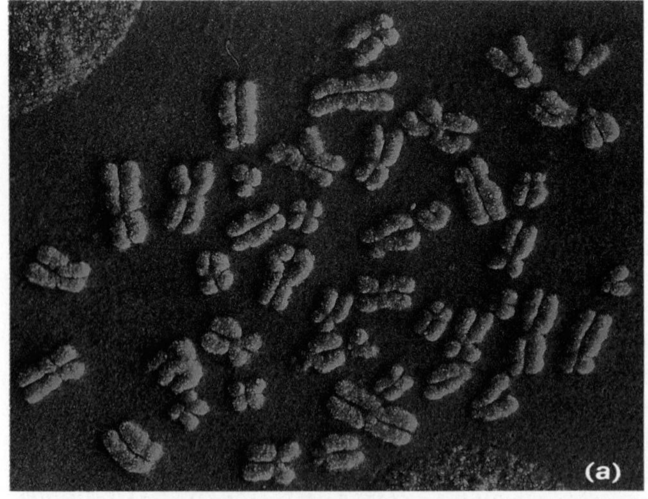

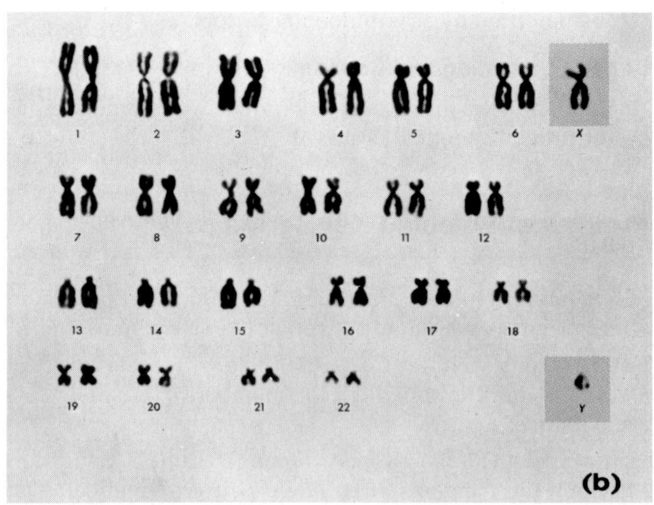

Figure 30.1 The human karyotype or chromosomal complement. (**a**) Scanning electron micrograph of human chromosomes (9300x). (**b**) A karyotype. The 46 human chromosomes are arranged in homologous pairs for examination of their structure and number. The twenty-third pair, the sex chromosomes (*X* and *Y*) are not truly homologous because the *X* chromosome bears many more genes than the much smaller *Y* chromosome.

Genotype and Phenotype

A person's genetic makeup (that is, whether one is homozygous or heterozygous for the various gene pairs) is referred to as his or her **genotype** (jē′-nō-tīp). The way that genotype is expressed in the body is called one's **phenotype** (fē-nō-tīp). Thus, the double-jointed condition is the phenotype produced by a genotype of *JJ* or *Jj*.

Sexual Sources of Genetic Variation

Before we examine how genes interact, let us consider why each of us is one of a kind, with a unique genotype and phenotype. This variability reflects three phenomena that occur before we are even a twinkle in our father's eye: independent assortment of chromosomes, crossover of homologues, and random fertilization of eggs by sperm.

Segregation and Independent Assortment of Chromosomes

As described in Chapter 28 (p. 923), each pair of homologous chromosomes synapses during meiosis I, forming a tetrad. This happens during both spermatogenesis and oogenesis. The alignment and orientation of the tetrads on the metaphase I spindle depends on chance, which ultimately means that maternal and paternal chromosomes are randomly distributed to the daughter nuclei. As illustrated in Figure 30.2, this simple event leads to an amazing amount of variation in the gametes that result. The cell in our example has a diploid chromosome number of 6; hence, three tetrads are formed. As you can see, the possible combinations of alignments of the three tetrads result in eight different types of gametes. Since the way each tetrad aligns is random, and many mother cells are undergoing meiosis simultaneously, each alignment and each type of gamete occurs with the same frequency as all others. Notice two important points here: (1) The members of the allele pair determining each trait are *segregated*, or distributed to different gametes, during meiosis; and (2) alleles on different pairs of homologous chromosomes are distributed independently of each other. The net result is that each gamete has a single allele for each trait, but that allele represents only one of the four possible parent alleles.

The number of different gamete types resulting from **independent assortment** of the homologues during meiosis I can be quickly calculated for any genome by the formula 2^n, where n is the number of homologous pairs. In our example, $2^n = 2^3$ (or $2 \times 2 \times 2$), for a total of eight different gamete types. The number of gamete types increases dramatically as the chromosome number increases. A cell with six pairs of homologues would produce 2^6, or 64, kinds of gametes. In a man's testes, the kinds of gametes that can be produced on the basis of independent assortment alone is 2^{23}, or about 8.5 million possibilities—an incredible variety. Since a human female's ovaries produce at most 500 completed reduction divisions in her lifetime, the number of different gamete types produced simultaneously is significantly less; still, each ovulated oocyte will most likely be novel genetically because of the completely random assortment of chromosomes.

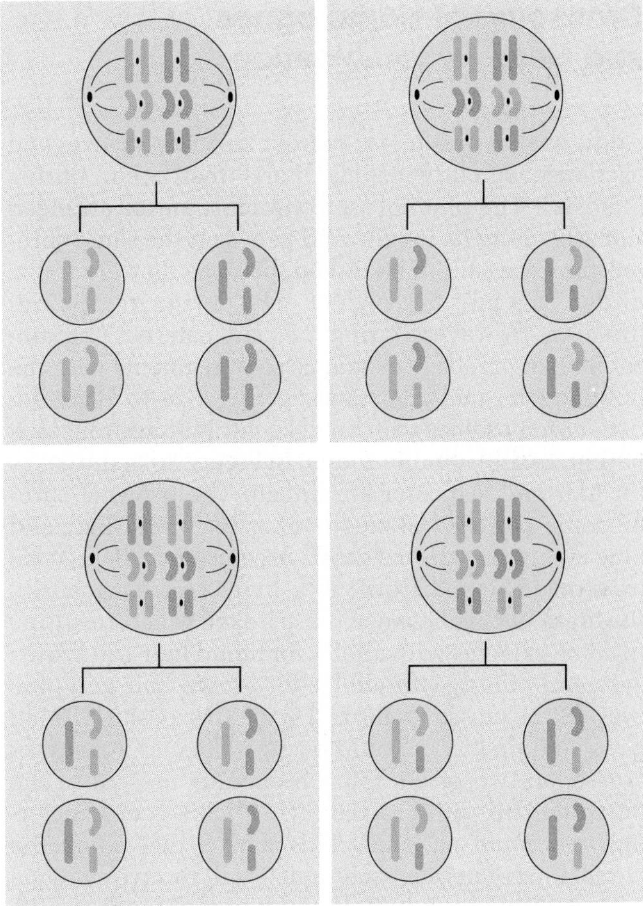

Figure 30.2 Production of variability in gametes by independent assortment of homologous chromosomes during metaphase of meiosis I. The large circles depict the possible metaphase I alignments of a mother cell with a diploid number of 6. The male homologues are depicted as the purple chromosomes; the maternal chromosomes are green. The small circles represent the genetic complements of the daughter cells (gametes) arising from each alignment. Some of the gametes contain all maternal or all paternal chromosomes; others have maternal and paternal chromosomes in varying combinations.

Crossover of Homologues and Gene Recombination

Additional variation results from the crossing over and exchange of chromosomal parts that occur during meiosis I. The genes of each chromosome are arranged linearly along its length, and genes on the same chromosome are said to be **linked,** because they are transmitted as a unit to daughter cells during mitotic cell division. However, during meiosis, paternal chromosomes can precisely exchange gene segments with the homologous maternal ones, giving rise to *recombinant chromosomes* with mixed contributions from each parent. In the example shown in Figure 30.3, the genes for hair and eye color are linked. The paternal chromosome contains alleles coding for blond hair and blue eyes, while the maternal chromosomal alleles code for brown hair and brown eyes. In the crossover shown, the break occurs between these linked genes, resulting in some gametes with alleles for blond hair and brown eyes and others with alleles for brown hair and blue eyes. (If the crossover formed at another position, other gene combinations would occur.) Thus, as a result of crossover, two of the four chromatids present in the tetrad end up with a mixed set of alleles—some maternal and some paternal. This means that when the chromatids segregate, each gamete will receive a unique and scrambled combination of parental genes.

It appears that only two of the four chromatids in a tetrad take part in crossing over and recombination, but these two may make many crossovers during synapsis. Furthermore, each crossover causes recombination of many genes, not just two as shown in the example. In general, the points at which crossovers are made are random (a sort of molecular lottery), and the longer the chromosome, the greater the number of possible crossovers. Since humans have 23 tetrads, with crossovers going on in most of them during meiosis I, the variability resulting from this factor alone is tremendous.

Random Fertilization

At any point in time, the process of gametogenesis is turning out gametes with all possible variations introduced by independent assortment and random crossovers. Compounding the variety still more is the fact that a single human egg will be fertilized by a single sperm on a totally haphazard basis. If we consider

Key
H = allele for brown hair
h = allele for blond hair
E = allele for brown eyes
e = allele for blue eyes

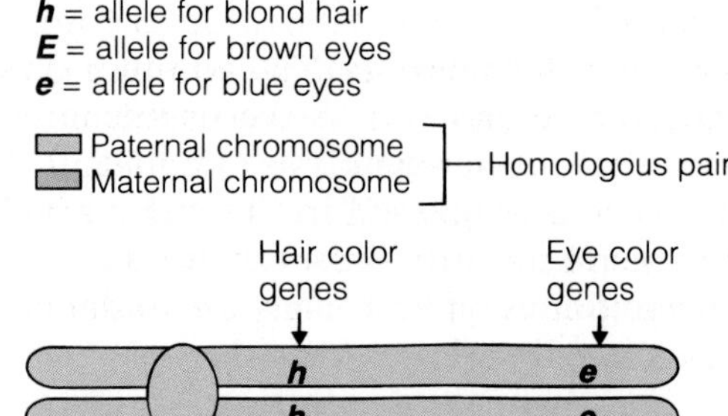

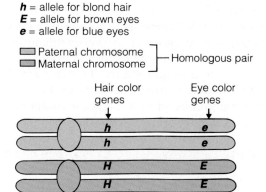

Homologous chromosomes synapse during prophase I of meiosis; each chromosome consists of two sister chromatids

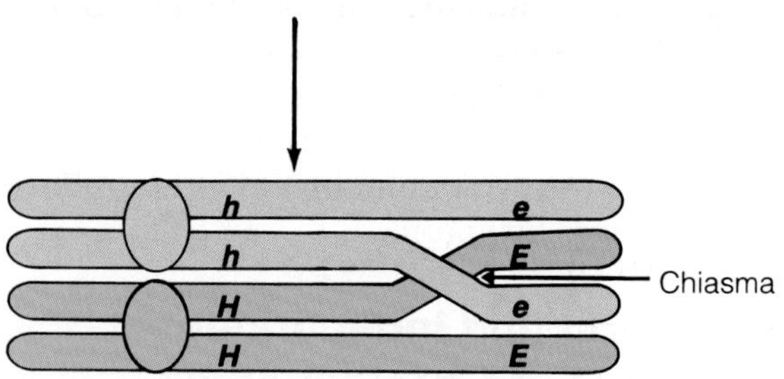

One chromatid segment exchanges positions with a homologous chromatid segment; crossing over occurs, with the formation of a chiasma

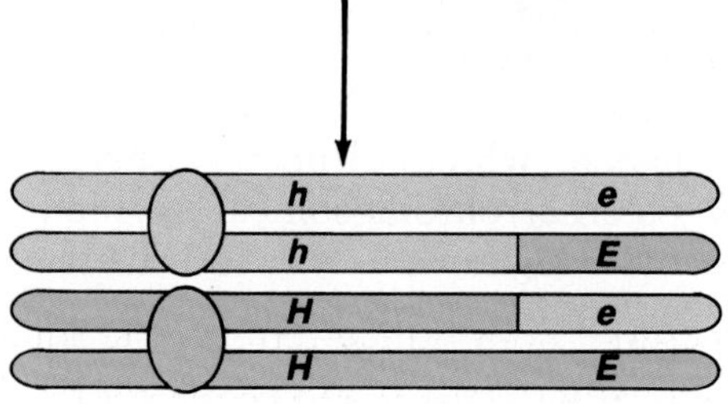

The chromatids forming the chiasma break and join their homologues

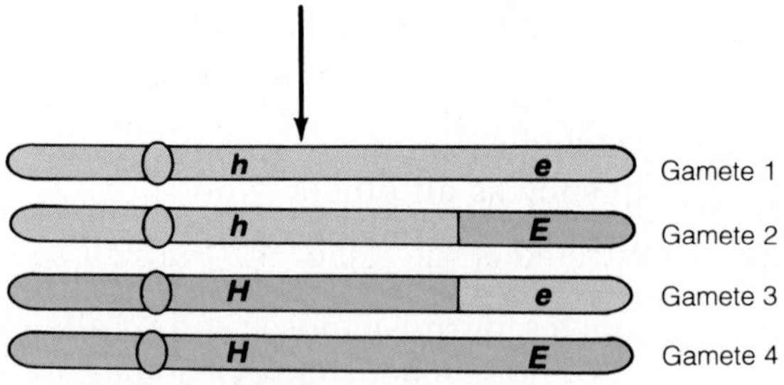

At the conclusion of meiosis, each haploid gamete has one of the four chromosomes shown; two of the chromosomes are recombinant; they carry new combinations of genes

Figure 30.3 Crossover and genetic recombination occurring during meiosis I introduces genetic variability in the gametes formed.

variation resulting only from independent assortment and random fertilization, any resulting offspring represents one out of the close to 72 trillion (8.5 million × 8.5 million) zygotes possible. The additional variation introduced by crossovers increases this number exponentially. Perhaps now you can understand why brothers and sisters are so different, and marvel at how they can also be so alike in many ways.

Once these sources of variability are understood, it is also easy to see that the concept that most people have of their inheritance is erroneous. We often hear something like "I'm one-half German, one-quarter Scotch, and one-quarter Italian," which reveals a belief that genes from both sides of the family are parceled out in mathematically precise combinations. While it is true that we receive half of our genes from each parent, they may not (indeed, do not) contain one-quarter from each grandparent, or one-eighth from each great grandparent, and there are no such things as German or Italian genes.

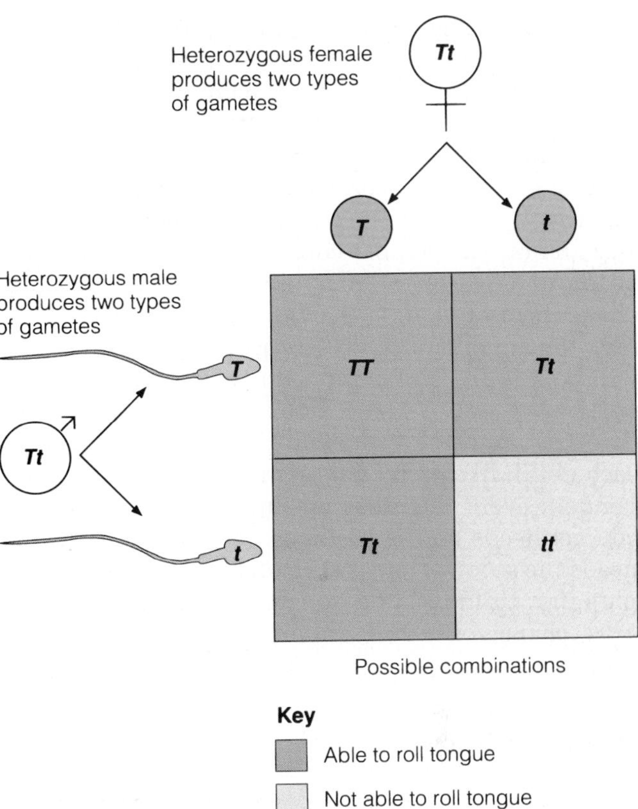

Figure 30.4 Use of a Punnett square to determine genotypes and phenotypes resulting from a mating of two heterozygous parents. The alleles carried by the mother are shown at the top; those carried by the father, along the side. The square shows all possible combinations of those alleles in the zygote. In this example the *T* allele is dominant and determines the tongue-rolling ability; the *t* allele is recessive. Homozygous recessive individuals (*tt*) are unable to exhibit tongue rolling. The *TT* and *Tt* offspring are tongue rollers.

Types of Inheritance

Although it is certainly possible to find examples in humans of visible traits, or phenotypes, that can be traced to a single gene pair (as described shortly), most such traits are likely to be very limited in nature, or reflect a variation in a single enzyme. Most human traits are determined by multiple alleles or by the interaction of several gene pairs.

Dominant-Recessive Inheritance

Dominant-recessive inheritance reflects the interaction of dominant and recessive alleles. A simple diagram, called the **Punnett square,** is used to figure out the possible combinations of genes for a single trait that would result from the mating of parents of known genotypes (Figure 30.4). In the example shown, both parents can roll their tongues into a U because both are heterozygous for the dominant allele (*T*) that determines this ability. In other words, each parent has the genotype *Tt*. The gamete types (alleles) of one parent are indicated at the top of the Punnett square, and the gamete types for the other parent are shown along one side. Then the alleles are combined down and across to determine the possible gene combinations (genotypes) and their expected frequency. As you can see from the completed Punnett square, the probability of these parents producing a homozygous dominant child (*TT*) is 25% (1 out of 4); of producing a heterozygous child (*Tt*), 50% (2 out of 4); and of producing a homozygous recessive child, 25% (1 out of 4). Since the *T* allele is dominant, the *TT* and *Tt* offspring will be tongue rollers; only the *tt* offspring will be unable to form a U shape with their tongues.

Although the Punnett square allows the results of simple genetic crosses to be predicted quickly, keep in mind that it only predicts the *probability* of having a certain percentage of offspring with a particular genotype (and phenotype). The larger the number of offspring, the greater the likelihood that the ratios will conform to the predictions—just as the chances of getting half heads and half tails when we toss a coin increase with the number of tosses. If we toss only twice, we may well get heads both times; likewise, if the couple in our example has only two children, it would not be surprising if both children had the genotype *Tt*.

What are the chances of having two children of the same genotype? To determine the probability of two events happening in succession, we must multiply together the probabilities of the separate events happening. The probability of getting heads for one coin toss is 1/2, so the probability of getting two heads in a row is $1/2 \times 1/2 = 1/4$. (In other words, out of a large number of double coin tosses, we will get two heads one-quarter of the time.) Now consider the chances of a couple having two children, both non–tongue rollers (*tt*). The probability that one child will be *tt* is 1/4, so the probability that both children will be *tt* is $1/4 \times 1/4 = 1/16$, or only slightly better than 6%. Remember though, that the production of each child, like each coin toss in a series, is an *independent event* that does not influence the others. If you get heads on the first toss, the chance of getting heads the second time is still 1/2; likewise, if our couple's first child is a non–tongue roller, they still have a 1/4 chance of getting a non–tongue roller the next time.

Dominant Traits

Human traits known to be dictated by dominant alleles include dimples, freckles, and the condition of detached earlobes, in which the lobe hangs inferior to the point of attachment of the ear.

Genetic disorders caused by dominant genes are fairly uncommon because *lethal dominant genes* are almost always expressed, resulting in the death of the embryo, fetus, or child. Thus, the deadly genes do not usually get passed along to successive generations. However, there are some dominant disorders where the person is less impaired or at least survives long enough to reproduce, such as achondroplasia (ā″-kon-drō-plā′-zhuh) and Huntington's disease. *Achondroplasia* is a rare type of dwarfism resulting from an impaired ability of the fetus to form cartilage bone. *Huntington's disease* is an unremitting, fatal nervous system disease involving degeneration of the basal nuclei. However, in this case, the gene is a *delayed-action gene* that is not expressed until the affected individual is in his or her late 30s or early 40s. Offspring of a parent with Huntington's disease have a 50% chance of inheriting the lethal gene. (The parent is inevitably heterozygous, since the dominant homozygous condition is lethal to the fetus.) Many informed offspring of such parents are opting not to become parents themselves. These and other dominant gene–determined traits are listed in Table 30.1.

Recessive Traits

Many examples of recessive inheritance cause no problems, and some are actually the more desirable genetic condition. For example, normal vision is dictated by recessive alleles, whereas farsightedness and astigmatism are dictated by dominant alleles. However, many, if not most, genetic disorders are inherited as simple recessive traits. These include conditions as different as *albinism* (lack of skin pigmentation); *cystic fibrosis*, a condition of excessive mucus production, which impairs lung and pancreatic function-

Table 30.1 Traits Determined by Simple Dominant-Recessive Inheritance

Phenotype due to expression of dominant genes (genotype *ZZ* or *Zz*)	Phenotype due to expression of recessive genes (genotype *zz*)
Tongue roller	Inability to roll tongue into a U shape
Free (unattached) earlobes	Attached earlobes
Farsightedness	Normal vision
Astigmatism	Normal vision
Freckles	Absence of freckles
Dimples in cheeks	Absence of dimples
Feet with normal arches	Flat feet
PTC taster	PTC nontaster
Widow's peak	Straight hairline
Double-jointed thumb	Tight thumb ligaments
Broad lips	Thin lips
Polydactyly (extra fingers and toes)	Normal number of fingers and toes
Syndactyly (webbed digits)	Normal digits
Achondroplasia (heterozygous: dwarfism; homozygous: lethal)	Normal cartilage bone formation
Huntington's disease	Absence of Huntington's disease
Normal skin pigmentation	Albinism
Absence of Tay-Sachs disease	Tay-Sachs disease
Absence of cystic fibrosis	Cystic fibrosis
Normal mentation	Schizophrenia

ing; and *Tay-Sachs disease,* a disorder of brain lipid metabolism that has a disproportionately high incidence among Jews of eastern European origin. Tay-Sachs disease, which is caused by an enzyme deficit, shows itself a few months after birth. The baby's brain becomes clogged with nonmetabolizable lipids (gangliosides) and deteriorates functionally. The infant suffers seizures and progressive blindness and dies within a few years.

The higher incidence of recessive genetic disorders, as opposed to those caused by dominant alleles, reflects the fact that those who carry a single recessive allele for a recessive genetic disorder do not themselves express the disease, but can pass the gene on to their offspring. For this reason, such people are called **carriers** of the disorder. This understanding underlies the laws that prohibit *consanguineous* (kon″-san-gwih′-nē-us) *marriages* (literally, "same blood" marriages)—those between close relatives such as siblings or first cousins. (The term *consanguineous* is a holdover from times when it was believed that inherited traits were carried by a person's blood.) The chance of inheriting two identical deleterious recessive genes, and thus expressing the disease they code for, is much higher in such matings than in matings between unrelated people. The greater incidence of stillbirth and severe genetic disorders seen in animals that are extremely inbred, such as show dogs, has amply proved the truth of this statement.

Incomplete Dominance (Intermediate Inheritance)

Not all genes comply with the rules of dominant-recessive inheritance, in which one allele variant completely masks the other. Some traits exhibit **incomplete dominance,** or **intermediate inheritance,** in which the heterozygote has a phenotype intermediate between that of homozygous dominant and homozygous recessive individuals. Incomplete dominance is very common in plants and some animals, but less so in humans.

Perhaps the best human example is inheritance of the *sickling gene (s),* which causes a substitution of one amino acid in the β chain of hemoglobin. Hemoglobin molecules containing the abnormal polypeptide chains crystallize and become sharp when the oxygen tension in blood is low, causing the erythrocytes to assume a sickle shape. The sickling gene is especially widespread among black people (those now living in the malaria belt of Africa or their descendants). It is also prevalent in other tropical regions, India, and eastern Mediterranean countries. Those with a double dose of the sickling allele (ss) have *sickle-cell anemia,* and any condition that lowers the oxygen level of the blood, such as respiratory difficulty or excessive exercise, can precipitate a *sickle-cell crisis*. The deformed erythrocytes jam up and fragment in small capillary channels, causing intense pain, and vital organs may suffer ischemic damage. The only treatment for sickle-cell anemia is blood transfusion to replace the sickling red blood cells with normal cells. Administration of oxygen and infusion of bicarbonate can be helpful.

Individuals heterozygous for the sickling gene (Ss) make both normal and sickling hemoglobin. As a rule, these individuals are quite healthy, but they can suffer a crisis if there is prolonged reduction of blood oxygen tension, as might happen when traveling in high-altitude areas. It is estimated that 10% of U.S. blacks have the heterozygous condition, known as *sickle-cell trait.* SS homozygotes are normal.

Multiple-Allele Inheritance

Although we inherit only two alleles for each gene, some genes actually exhibit more than two alternate forms, leading to a phenomenon called **multiple-allele inheritance.** Inheritance of the ABO blood types is such an example. There are three possible alleles that determine the ABO blood types in humans: I^A, I^B, and *i*; each of us receives two of these. The I^A and I^B alleles are *codominant,* and both are expressed when present; the *i* allele is recessive to the other two alleles. Genotypes determining the four possible ABO blood groups are indicated in Table 30.2.

Sex-Linked Inheritance

Inherited traits determined by genes on the sex chromosomes are said to be **sex-linked.** The *X* and *Y* sex chromosomes are not homologous in the true sense. The *Y* chromosome, which determines maleness, is only about one-third as large as the *X* chromosome and lacks many of the genes present on the *X* that code for nonsexual characteristics. For example, genes for production of certain clotting factors and cone pigments are present on the *X* chromosome but not the *Y* chromosome. A gene found only on the *X* chromosome is said to be **X-linked.**

Table 30.2 ABO Blood Groups

Blood group (phenotype)	Genotype	Frequency in white Americans
O	ii	45%
A	I^AI^A or I^Ai	41%
B	I^BI^B or I^Bi	10%
AB	I^AI^B	4%

When a male inherits an X-linked recessive allele (for example, for the blood-clotting disorder hemophilia or for red-green color blindness), its expression is never masked or damped, because there is no corresponding allele on his Y chromosome. Consequently, the recessive gene is always expressed, even when it is present only in single dose. In contrast, females must have two X-linked recessive alleles to express such a disorder; as a result, very few females exhibit any of these X-linked conditions. X-linked traits are typically passed from a mother to her son. (Since males receive no X chromosome from their fathers, X-linked traits are never passed from father to son.) Of course, the mother can also pass the recessive allele to her daughter, but unless the daughter receives another such allele from her father, she will not express the trait.

It should also be noted that some segments of the Y chromosome have no counterpart on the X chromosome. Traits coded for by genes in these segments (including hairy ear pinnae) appear only in males and are passed from fathers to sons. This type of sex-linked gene transmission is referred to as **Y-linked inheritance.**

Polygene Inheritance

So far, we have considered only those traits inherited by mechanisms of classical Mendelian genetics, which are fairly easily understood. Such traits are characterized by having two, or perhaps three, alternate forms. However, there are many phenotypes that depend on the action of several diffferent gene pairs located at different loci and acting in tandem. **Polygene inheritance** results in *continuous*, or *qualitative*, phenotypic variation between two extremes and explains many human characteristics. Skin color, for instance, is controlled by three separately inherited genes, each existing in two allelic forms: *A*, *a*; *B*, *b*; and *C*, *c*. The *A*, *B*, and *C* alleles confer dark skin pigment, and their effects are additive, whereas the *a*, *b*, and *c* alleles confer pale skin tone. Hence, an individual with an *AABBCC* genotype would be about as dark-skinned as a human can get, while an *aabbcc* person would be very fair. However, when individuals heterozygous for at least one of these gene pairs mate, a broad range of pigmentation is possible in their offspring. Representative gradations in skin tone according to genotype are illustrated in Figure 30.5a. Such polygene

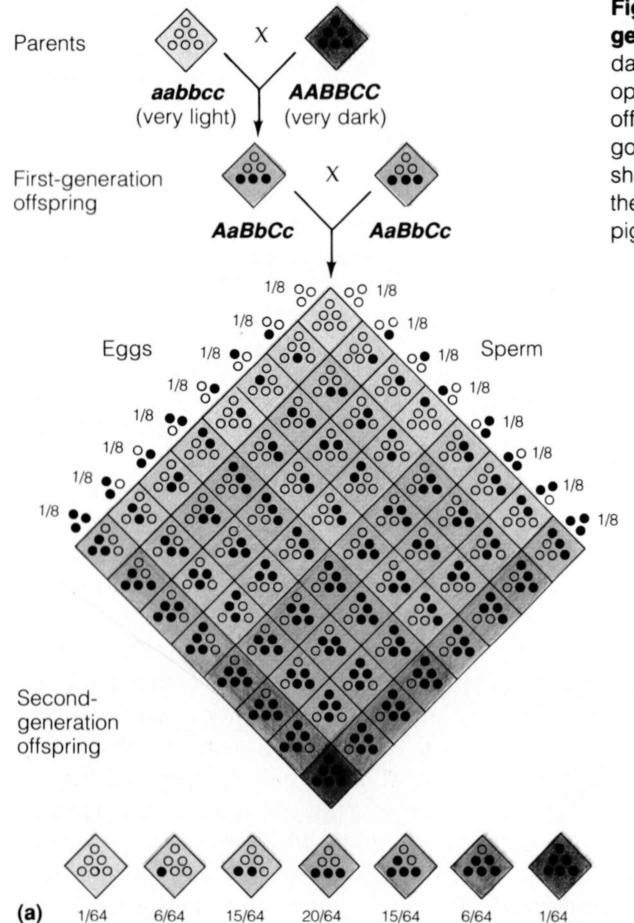

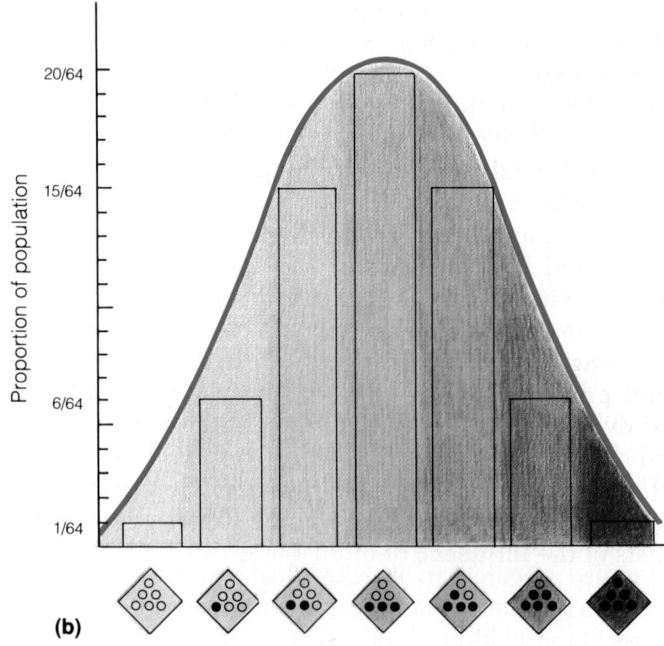

Figure 30.5 Polygene inheritance of skin pigmentation based on three gene pairs. (**a**) Each dominant gene (*A*, *B*, or *C*) contributes one unit of darkness to the phenotype. If, as in the example shown, the parents are at opposite extremes of the phenotype range, their children (first-generation offspring) will have intermediate pigmentation, since they will be heterozygotes. When heterozygotes mate, their offspring (second generation) may show a wide variation in pigmentation, owing to independent assortment of the three gene pairs. (**b**) The histogram shows the normal distribution of pigmentation types in the second-generation offspring.

inheritance results in a distribution of phenotypes that, when plotted, yields a bell-shaped curve (Figure 30.5b).

The amount of brown pigment in the iris (which determines eye color) is also regulated by polygenes, as are intelligence and height. Height is controlled by four gene pairs, and different combinations of the tallness or shortness alleles are reflected in differences in stature, much the same way as described for inheritance of skin tone. The phenomenon of polygene inheritance helps explain why parents of average height can produce very tall or very short offspring.

The Influence of Environmental Factors on Gene Expression

Although we have been talking about genes as though they function in a vacuum, there are many situations in which environmental factors override genetic ones or at least influence gene expression. It appears that while our genotype (discounting mutations) is as unchanging as the Rock of Gibraltar, our phenotype is more like clay. If this were not the case, we would never get a tan, the bulging muscles of trained athletes would never appear, and there would be no hope for treating genetic disorders.

Sometimes, maternal factors (drugs, pathogens, and others) alter normal gene expression during embryonic development. Take, for example, the case of the "thalidomide babies" (discussed on p. 964). As a result of drug ingestion by their mothers, the embryos developed phenotypes other than that directed by their genes (flipperlike appendages developed to various degrees and lengths). Such environmentally produced phenotypes that mimic conditions that may be caused by genetic mutations are called *phenocopies*.

Less dramatic but equally significant examples of environmental factors influencing genetic expression after birth include the effect of poor infant nutrition on subsequent brain growth, general body development, and height. There is little question but that a person with "tall genes" can be stunted by insufficient nutrition. Furthermore, hormonal deficits during the growing years can lead to abnormal skeletal growth and proportions as exemplified by cretinism, a type of dwarfism resulting from hypothyroidism. This is an example of genes (or genetic abnormalities), other than those dictating height, influencing the height phenotype. Hence, part of a gene's environment consists of the influence of other genes.

Genetic Screening and Counseling

Newborn infants are routinely screened for a number of anatomical disorders (congenital hip dysplasia, imperforate anus, and others), and phenylketonuria (PKU) testing is mandated by law in many states. These tests are performed much too late to allow any type of parental choice concerning whether or not to have the child, but they do alert the new parents that treatment is necessary to ensure the well-being of their infant. The anatomical defects are usually treated surgically, and PKU is managed by strict dietary measures that exclude most phenylalanine-containing foods during early childhood. However, for today's prospective parents, *genetic screening* and *genetic counseling* provide information and options not even dreamed of 100 years ago. Adult children of parents with Huntington's disease are obvious candidates for genetic screening and counseling, but there are many other genetic conditions that place babies at risk. For example, a woman pregnant for the first time at the age of 35 may wish to know if her baby has trisomy-21 (Down's syndrome), a chromosome abnormality with a high incidence in children of older mothers. Depending on the precise condition to be investigated, the screening process can occur before conception with carrier recognition or during fetal testing.

Carrier Recognition

When a prospective parent displays a recessive genetic disorder, but his or her partner does not, it is important to determine if the partner is heterozygous for the same recessive gene. If not, the offspring will receive only one such gene and will not express the trait. But if the other parent *is* a carrier, then the child's chance of having two deleterious genes is 50%.

There are two major avenues for detecting carriers: pedigrees and blood tests. Through **pedigree analysis,** a particular genetic trait is traced through several generations. A genetic counselor collects information on the phenotypes of as many family members as possible and uses it to construct the pedigree. Figure 30.6 uses the rare woolly hair trait of northern Europeans to illustrate how a pedigree is constructed and read. Woolly hair, resulting from the presence of a dominant allele (*W*), is fuzzy-textured hair that breaks easily; it is not the same as the tightly curled hair of blacks. Looking at the pedigree, we know that individuals with the woolly hair phenotype (the colored symbols) have at least one dominant gene (*WW*

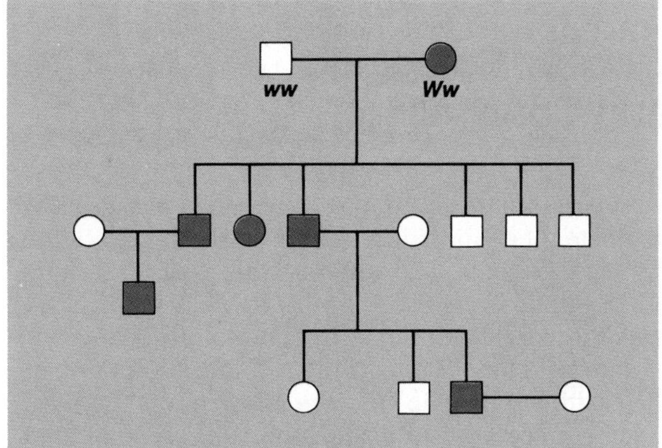

Figure 30.6 A pedigree of inheritance of the woolly hair gene in three generations of a family. Circles represent females; squares represent males. Horizontal lines indicate a mating of the connected individuals; vertical lines show their offspring. Colored symbols represent individuals expressing the dominant woolly hair trait. White (uncolored) symbols represent those not expressing the trait; these individuals are homozygous recessive (*ww*).

or *Ww*), while those with normal hair must be homozygous recessive (*ww*). By working backward and applying the rules of dominant-recessive inheritance, we can deduce the genotypes of the parents. Since three of their offspring have normal hair (*ww*), we know that each parent must have at least one recessive gene. The mother has woolly hair, determined by the *W* gene, so her genotype must be *Ww*, and the father with normal hair is homozygous recessive. You should be able to take it from here and figure out the most likely genotypes of the offspring. (Try it!) This pedigree represents one of the simplest cases, because many human traits are controlled by multiple alleles or polygenes and are much more difficult to unsnarl.

Simple blood tests are used to screen for the sickling gene in heterozygotes, and sophisticated blood chemistry tests and DNA probes (in which DNA hybridization techniques are used to identify aberrant genes) can identify the presence of other unexpressed recessive genes. At present, carriers of the Tay-Sachs and cystic fibrosis genes can be ascertained with such tests.

Fetal Testing

Fetal testing is used when there is a known risk of a genetic disorder. The most common type of fetal testing is *amniocentesis* (am″-nē-ō-sen-tē′-sis) (Figure 30.7a). In this fairly simple procedure, a wide-bore needle is inserted into the amniotic sac through the mother's abdominal wall, and a small sample (about 10 ml) of fluid is aspirated (withdrawn). Because there is a chance of injuring the fetus before ample amniotic fluid is present, this procedure is not normally done before the fourteenth week of pregnancy. Ultrasound visualization of fetal position and the amniotic sac has dramatically reduced the risk of this procedure. Cer-

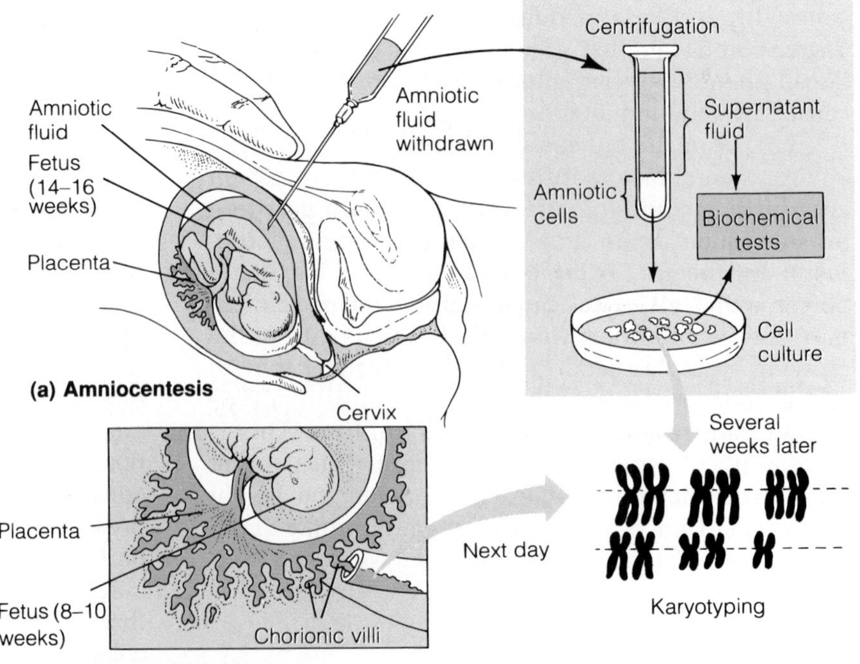

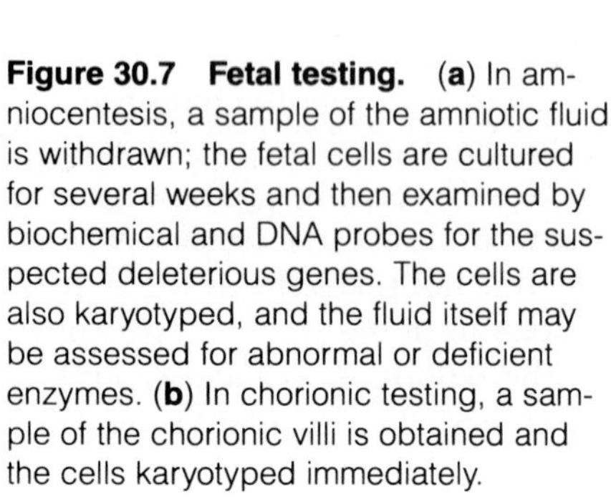

Figure 30.7 Fetal testing. (**a**) In amniocentesis, a sample of the amniotic fluid is withdrawn; the fetal cells are cultured for several weeks and then examined by biochemical and DNA probes for the suspected deleterious genes. The cells are also karyotyped, and the fluid itself may be assessed for abnormal or deficient enzymes. (**b**) In chorionic testing, a sample of the chorionic villi is obtained and the cells karyotyped immediately.

tain tests can be done on the fluid itself to determine the presence of chemicals (enzymes and others) that serve as markers for specific diseases, but most are done on the sloughed-off fetal cells in the fluid. These cells are isolated and cultured (grown and proliferated) in laboratory dishes over a period of several weeks and then examined for DNA markers of genetic disease. The cells are also karyotyped to check for chromosomal abnormalities such as Down's syndrome.

A newer procedure, *chorionic* (kor-ē-ah′-nik) *villi sampling*, snips off bits of the chorionic villi from the placenta (Figure 30.7b). A small tube is inserted through the vagina and cervical os and guided by ultrasound to an area where a piece of placental tissue can be removed. This procedure allows earlier testing (at eight weeks) and is much faster, but may be more risky for the fetus. However, karyotyping can be done almost immediately on the rapidly dividing chorionic cells.

Since both of these procedures are invasive, they carry with them an inherent risk to both fetus and mother. Although they are routinely ordered for pregnant women over the age of 35 (because of the enhanced risk of Down's syndrome), they are performed on younger women only when the probability of finding a severe fetal disorder is greater than the probability of doing harm via the procedure. If a serious genetic or congenital defect is detected in the developing fetus, the parents must decide whether or not to continue the pregnancy.

* * *

In this chapter, we have explored some of the most basic principles of genetics, the manner in which genes are expressed, and the means by which gene expression can be thwarted. Considering the precision required to make perfect copies of genes and chromosomes, and considering the incredible mechanical events of meiotic division, it is amazing that genetic defects are as rare as they are. Perhaps, after reading this chapter, you have a greater sense of wonder that you turned out as well as you did.

Related Clinical Terms

Deletion Chromosomal aberration in which part of a chromosome is lost.

Down's syndrome Also called mongolism, this condition usually reflects the presence of an extra autosome (trisomy of chromosome 21); the child has slightly slanted eyes, flattened facial features, a large tongue, and a tendency toward short stature and stubby fingers; some, but not all, offspring are mentally retarded; the most important risk factor appears to be advanced maternal age.

Karyotype (kayr′-ē-ō-tīp) (*karyon* = nucleus) The chromosomal content of a cell; the term is usually used to indicate diagrams made by lining up the homologous chromosomes and drawing them to true proportion (see Figure 30.1); karyotypes are used to judge whether chromosomes are normal in number and structure.

Mutation (*mutare* = change) A permanent structural change in a gene; the mutation may or may not affect function, depending on the precise site of the alteration.

Nondisjunction Abnormal segregation of chromosomes during meiosis, resulting in gametes receiving two or no copies of a particular parental chromosome; if the abnormal gamete participates in fertilization, the resulting zygote will have an abnormal chromosomal complement (monosomy or trisomy) for that particular chromosome (as in Down's syndrome).

Chapter Summary

Genetics is the study of heredity and mechanisms of gene transmission.

THE VOCABULARY OF GENETICS (pp. 984–985)

1. A complete diploid set of chromosomes is called the *karyotype* of an organism; the complete genetic complement is the *genome*. A person's genome consists of two sets of instructions, one from each parent.

Gene Pairs (Alleles) (p. 984)

2. Genes coding for the same trait and found at the same locus on homologous chromosomes are called alleles.

3. Alleles may be the same or different in expression. When the allele pair is identical, the person is homozygous for that trait; when the alleles differ, the person is heterozygous.

Genotype and Phenotype (p. 985)

4. The actual genetic makeup of cells is the genotype; phenotype is the manner in which those genes are expressed.

SEXUAL SOURCES OF GENETIC VARIATION (pp. 985–987)

Segregation and Independent Assortment of Chromosomes (p. 985)

1. During meiosis I of gametogenesis, tetrads align randomly on the metaphase plate, and chromatids are randomly distributed to the daughter cells. This is called independent assortment of the homologues. Each gamete receives only one allele of each gene pair.

2. Each different metaphase I alignment produces a different assortment of parental chromosomes in the gametes, and all combinations of maternal and paternal chromosomes are equally possible; in humans, this results in about 8.5 million gamete types.

Crossover of Homologues and Gene Recombination (p. 986)

3. During meiosis I, two of the four chromatids (one maternal and the other paternal) may cross over at one or more points and exchange corresponding gene segments. The recombinant chromosomes contain new gene combinations, adding to the variability arising from independent assortment.

Random Fertilization (pp. 986–987)

4. The third source of genetic variation is random fertilization of eggs by sperm.

TYPES OF INHERITANCE (pp. 987–991)

Dominant-Recessive Inheritance (pp. 987–989)

1. Dominant genes are expressed when present in single or double dose; recessive genes must be present in double dose to be expressed.

2. For traits following the dominant-recessive pattern, the laws of probability predict the outcome of a large number of matings.

3. Genetic disorders more often reflect the homozygous recessive condition than the homozygous dominant or heterozygous condition because dominant genes are expressed, and if they are lethal genes, the pregnancy is usually aborted. Genetic disorders caused by dominant alleles include achondroplasia and Huntington's disease; recessive disorders include cystic fibrosis and Tay-Sachs disease.

4. Carriers are heterozygotes who carry a deleterious recessive gene (but do not express the trait) and have the potential of passing it on to offspring. The chances of carriers of the same trait mating are higher in consanguineous marriages.

Incomplete Dominance (Intermediate Inheritance) (p. 989)

5. In incomplete dominance, the heterozygote exhibits a phenotype intermediate between that of the homozygous dominant and recessive individuals. Inheritance of sickle-cell trait is an example of incomplete dominance.

Multiple-Allele Inheritance (p. 989)

6. Multiple-allele inheritance involves genes that exist in more than two allelic forms in a population. Only two of the alleles are inherited, but on a random basis. Inheritance of ABO blood types is an example of multiple-allele inheritance in which the I^A and I^B alleles are codominant.

Sex-Linked Inheritance (pp. 989–990)

7. Traits determined by genes on the X and Y chromosomes are said to be sex-linked. The small Y chromosome lacks many genes present on the X chromosome. Recessive genes present only on the X chromosome are expressed in single dose in males. Examples of such X-linked conditions, passed from mother to son, include hemophilia and red-green color blindness. There are a few Y-linked genes, passed from father to son only.

Polygene Inheritance (pp. 990–991)

8. Polygene inheritance occurs when several gene pairs interact to produce phenotypes that vary qualitatively over a broad range. Height, eye color, and skin pigmentation are examples.

THE INFLUENCE OF ENVIRONMENTAL FACTORS ON GENE EXPRESSION (p. 991)

1. Environmental factors may influence the expression of genotype.

2. Maternal factors that cross the placenta may alter expression of fetal genes. Environmentally provoked phenotypes that mimic genetically determined ones are called phenocopies. Nutritional deficits or hormonal imbalances may alter anticipated growth and development during childhood.

GENETIC SCREENING AND COUNSELING (pp. 991–993)

Carrier Recognition (pp. 991–992)

1. The likelihood of an individual carrying a deleterious recessive gene may be assessed by constructing a pedigree. Some of these genes can be detected by various blood tests and DNA probes.

Fetal Testing (pp. 992–993)

2. Amniocentesis is fetal testing based on aspirated samples of amniotic fluid. Fetal cells in the fluid are cultured for several weeks, then examined for chromosomal defects (karyotyped) or for DNA markers of genetic disease. Amniocentesis cannot be performed until the fourteenth week of pregnancy.

3. Chorionic villi sampling is fetal testing based on a sample of the chorion. Because this tissue is rapidly mitotic, karyotyping can be done almost immediately. Samples may be obtained by the eighth week.

Review Questions

Multiple Choice/Matching

1. Match the terms in column B with the appropriate descriptions in column A.

Column A	Column B
(1) ______ genetic makeup	(a) alleles
(2) ______ how the genetic makeup is expressed	(b) autosomes
(3) ______ chromosomes that dictate most body characteristics	(c) dominant allele
(4) ______ alternate forms of the same gene	(d) genotype
(5) ______ an individual bearing two alleles that are the same for a particular trait	(e) heterozygote
(6) ______ an allele that is expressed, whether in single or double dose	(f) homologous
(7) ______ an individual bearing two alleles that differ for a particular trait	(g) homozygote
(8) ______ an allele that must be present in double dose to be expressed.	(h) phenotype
	(i) recessive allele
	(j) sex chromosomes

2. Match the types of inheritance listed in column B with the descriptions in column A.

Column A	Column B
(1) ______ only sons show the trait	(a) dominant-recessive
(2) ______ homozygotes and heterozygotes have the same phenotype	(b) incomplete dominance
(3) ______ heterozygotes exhibit a phenotype intermediate between that of the homozygotes	(c) multiple-allele
(4) ______ phenotypes of offspring may be more varied than those of the parents	(d) polygene
(5) ______ inheritance of ABO blood types	(e) sex-linked
(6) ______ inheritance of stature	

Short Answer Essay Questions

3. Describe the important mechanisms that lead to genetic variations in gametes.

4. The ability to taste PTC (phenylthiocarbamide) depends on the presence of a dominant gene *T*; nontasters are homozygous for the recessive gene *t*. This is a situation of classical dominant-recessive inheritance. (a) Consider a mating between heterozygous parents producing three offspring. What proportion of the offspring will be tasters? What is the chance that all three offspring will be tasters? Nontasters? What is the chance that two will be tasters and one will be a nontaster? (b) Consider a mating between *Tt* and *tt* parents. What is the anticipated percentage of tasters? Nontasters? What proportion can be expected to be homozygous recessive? Heterozygous? Homozygous dominant?

5. Most albino children are born to normally pigmented parents. Albinos are homozygous for the recessive gene (*aa*). What can you conclude about the genotypes of the nonalbino parents?

6. A woman with blood type A has two children. One has type O blood and the other has type B blood. What is the genotype of the mother? What is the genotype and phenotype of the father? What is the genotype of each child?

7. In skin color inheritance, what will be the relative range of pigmentation in offspring arising from the following parental matches: (a) AABBCC × aabbcc, (b) AABBCC × AaBbCc, AAbbcc × aabbcc?

8. A color-blind man marries a woman with normal vision. The woman's father was also color-blind. (a) What is the chance that their first child will be a color-blind son? A color-blind daughter? (b) If they have four children, what is the chance that two will be color-blind sons? (Be careful on this one.)

9. Compare and contrast amniocentesis and chorionic villi sampling as to the time at which they can be performed and the techniques used to obtain information on the fetus's genetic status.

Clinical Application Questions

10. Brian is a college student. His genetics assignment is to do a family pedigree for dimples in the cheeks. Absence of dimples is recessive; presence of dimples reflects a dominant allele. Brian has dimples, as do his three brothers. His mother and maternal grandmother are dimple-free, but his father and all other grandparents have dimples. Construct a pedigree spanning three generations for Brian's family. Show the phenotype and genotype for each person.

11. Mr. and Mrs. Lehman have sought genetic counseling. Mrs. Lehman is concerned because she is (unexpectedly) pregnant and her husband's brother died of Tay-Sachs disease. She can recall no incidence of Tay-Sachs disease in her own family. Do you think biochemical testing should be recommended to detect the deleterious gene in Mrs. Lehman? Explain your answer.

APPENDIX A

The Metric System

MEASUREMENT	UNIT AND ABBREVIATION	METRIC EQUIVALENT	METRIC TO ENGLISH CONVERSION FACTOR	ENGLISH TO METRIC CONVERSION FACTOR
Length	1 kilometer (km)	= 1000 (10^3) meters	1 km = 0.62 mile	1 mile = 1.61 km
	1 meter (m)	= 100 (10^2) centimeters = 1000 millimeters	1 m = 1.09 yards 1 m = 3.28 feet 1 m = 39.37 inches	1 yard = 0.914 m 1 foot = 0.305 m
	1 centimeter (cm)	= 0.01 (10^{-2}) meter	1 cm = 0.394 inch	1 foot = 30.5 cm 1 inch = 2.54 cm
	1 millimeter (mm)	= 0.001 (10^{-3}) meter	1 mm = 0.039 inch	
	1 micrometer (μm) [formerly micron (μ)]	= 0.000001 (10^{-6}) meter		
	1 nanometer (nm) [formerly millimicron (mμ)]	= 0.000000001 (10^{-9}) meter		
	1 angstrom (Å)	= 0.0000000001 (10^{-10}) meter		
Area	1 square meter (m^2)	= 10,000 square centimeters	1 m^2 = 1.1960 square yards 1 m^2 = 10.764 square feet	1 square yard = 0.8361 m^2 1 square foot = 0.0929 m^2
	1 square centimeter (cm^2)	= 100 square millimeters	1 cm^2 = 0.155 square inch	1 square inch = 6.4516 cm^2
Mass	1 metric ton (t)	= 1000 kilograms	1 t = 1.103 ton	1 ton = 0.907t
	1 kilogram (kg)	= 1000 grams	1 kg = 2.205 pounds	1 pound = 0.4536 kg
	1 gram (g)	= 1000 milligrams	1 g = 0.0353 ounce 1 g = 15.432 grains	1 ounce = 28.35 g
	1 milligram (mg)	= 0.001 gram	1 mg = approx. 0.015 grain	
	1 microgram (μg)	= 0.000001 gram		
Volume (solids)	1 cubic meter (m^3)	= 1,000,000 cubic centimeters	1 m^3 = 1.3080 cubic yards 1 m^3 = 35.315 cubic feet	1 cubic yard = 0.7646 m^3 1 cubic foot = 0.0283 m^3
	1 cubic centimeter (cm^3 or cc)	= 0.000001 cubic meter = 1 milliliter	1 cm^3 = 0.0610 cubic inch	1 cubic inch = 16.387 cm^3
	1 cubic millimeter (mm^3)	= 0.000000001 cubic meter		
Volume (liquids and gases)	1 kiloliter (kl or kL)	= 1000 liters	1 kL = 264.17 gallons	1 gallon = 3.785 L 1 quart = 0.946 L
	1 liter (l or L)	= 1000 milliliters	1 L = 0.264 gallons 1 L = 1.057 quarts	
	1 milliliter (ml or mL)	= 0.001 liter = 1 cubic centimeter	1 ml = 0.034 fluid ounce 1 ml = approx. $\frac{1}{4}$ teaspoon 1 ml = approx. 15–16 drops (gtt.)	1 quart = 946 ml 1 pint = 473 ml 1 fluid ounce = 29.57 ml 1 teaspoon = approx. 5 ml
	1 microliter (μl or μL)	= 0.000001 liter		
Time	1 second (s)	= $\frac{1}{60}$ minute		
	1 millisecond (ms)	= 0.001 second		
Temperature	Degrees Celsius (°C)		°F = $\frac{9}{5}$°C + 32	°C = $\frac{5}{9}$(°F − 32)

APPENDIX B

Answers to Multiple Choice and Matching Questions

Chapter 1
1. c
2. a
3. e
4. a, d
5. (a) wrist (b) hipbone (c) nose (d) toes (e) scalp
6. c, d
7. (a) dorsal (b) ventral (c) dorsal (d) ventral (e) ventral
8. b
9. b

Chapter 2
1. b, d
2. d
3. b
4. a
5. b
6. a
7. c, d
8. b
9. a
10. a
11. b
12. a, c
13. (1) a, (2) c
14. c
15. d
16. e
17. d
18. d
19. a
20. b
21. b
22. c

Chapter 3
1. d
2. a, c
3. b
4. d
5. b
6. e
7. c
8. d
9. a
10. a
11. b
12. d
13. c
14. b
15. d
16. a
17. b
18. d

Chapter 4
1. a, c, d, b
2. c, e
3. b, f, a, d, g
4. b
5. c

Chapter 5
1. a
2. c
3. e, b, a and e
4. d
5. d
6. b
7. b
8. c
9. c
10. b
11. b
12. a
13. d
14. b

Chapter 6
1. e
2. b
3. c
4. d
5. e
6. b
7. c
8. 3, 2, 4, 1, 5, 6
9. b
10. e
11. b
12. c
13. b
14. b

Chapter 7
1. (1) b, g; (2) h; (3) d; (4) d, f; (5) e; (6) c; (7) a, b, d, h; (8) i
2. (1) g, (2) f, (3) b, (4) a, (5) b, (6) c, (7) d
3. (1) b, (2) c, (3) e, (4) a, (5) h, (6) e

Chapter 8
1, (1) c, (2) a, (3) a, (4) b, (5) c, (6) b, (7) b, (8)a
2. b
3. d
4. d
5. b
6. d
7. d

Chapter 9
1. c
2. b
3. c
4. (1) b, (2) a, (3) a, (4) b
5. c
6. a
7. a
8. d
9. a
10. (1) a; (2) a, c; (3) b; (4) c; (5) b; (6) b
11. a
12. c
13. (1) c, d, e; (2) a, b
14. c
15. c
16. b

Chapter 10
1. c
2. (1) e, (2) c, (3) g, (4) f, (5) d
3. a, c
4. c
5. d
6. c
7. c
8. b
9. d
10. b
11. a
12. c

Chapter 11
1. b
2. (1) d, (2) b, (3) f, (4) c, (5) a
3. c
4. c
5. a
6. c
7. b
8. d
9. b,1
   a,1
   a,5, 3
   a,4
   a,2
   c,2
10. b
11. c
12. (1) d, (2) b, (3) a, (4) c

Chapter 12
1. a
2. (1) c, (2) f, (3) e, (4) g, (5) b, (6) f, (7) i, (8) a
3. d
4. c
5. a
6. b
7. c
8. a

Chapter 13
1. (1) f; (2) i; (3) b; (4) g, h, l; (5) e; (6) i; (7) c; (8) k; (9) l; (10) c, d, f, k
2. (1) b 6: (2) d 1, 8; (3) c 2; (4) c 5; (5) a 4; (6) a 3, 9; (7) a 7; (8) a 7; (9) d 1
3. b

Chapter 14
1. d
2. (1) S, (2) P, (3) P, (4) S, (5) S, (6) P, (7) P, (8) S, (9) P, (10) S, (11) P, (12) S

Chapter 15
1. b
2. c
3. c
4. e
5. b
6. (1) d, (2) e, (3) d, (4) a

Chapter 16
1. d
2. a
3. d
4. c
5. c
6. d
7. a
8. b
9. c
10. d
11. b
12. a
13. b
14. b
15. b
16. d
17. b
18. c
19. d
20. b
21. b
22. e
23. b
24. c

Chapter 17
1. b
2. (1) c, (2) b, (3) f, (4) d, (5) e, (6) g, (7) a
3. a
4. c
5. d
6. d
7. c
8. b
9. d
10. b
11. d
12. b

Chapter 18
1. c
2. c
3. d
4. b
5. d
6. a
7. a
8. b
9. c
10. d

Chapter 19
1. a
2. c
3. b
4. b
5. b
6. b
7. c
8. d
9. b

Chapter 20
1. d
2. b
3. d
4. c
5. e
6. d
7. c
8. b
9. b
10. a
11. b
12. c

Chapter 21
1. c
2. c
3. a, d
4. c
5. a
6. b
7. a
8. b
9. d

Chapter 22
1. c
2. a
3. d
4. d, e
5. a
6. d
7. b
8. c
9. d

Chapter 23
1. b
2. a
3. c
4. c
5. b
6. d
7. d
8. b
9. c, d
10. c
11. b
12. b
13. b
14. c
15. b
16. b

Chapter 24
1. d
2. d
3. b
4. b
5. a
6. d
7. d
8. b
9. c
10. c
11. a
12. d
13. d
14. b
15. c
16. a

Chapter 25
1. a
2. c
3. c
4. d
5. b
6. c
7. a
8. d
9. d
10. a
11. b
12. d
13. c
14. d
15. a

Chapter 26
1. d
2. b
3. c
4. d
5. c
6. b
7. a
8. c
9. a

Chapter 27
1. a
2. c
3. b
4. b
5. f, h, i
6. d, g
7. h, e
8. b
9. a
10. j
11. b, d, f
12. a
13. c

Chapter 28
1. b
2. d
3. a
4. d
5. b
6. d
7. d
8. c
9. a, c, e, f
10. (1) c, f; (2) e, h; (3) g; (4) a; (5) b, g; (6) f
11. a
12. c
13. d
14. b
15. a
16. b
17. c

Chapter 29
1. (1) a, (2) b, (3) b, (4) a
2. b
3. c
4. b
5. a
6. d
7. d
8. c
9. a
10. b
11. c
12. a
13. c
14. a

Chapter 30
1. (1) d, (2) h, (3) b, (4) a, (5) g, (6) c, (7) e, (8) i
2. (1) e, (2) a, (3) b, (4) d, (5) c, (6) d

APPENDIX C

Word Roots, Prefixes, Suffixes, and Combining Forms

PREFIXES AND COMBINING FORMS

a-, an- *absence or lack* acardia, lack of a heart; anaerobic, in the absence of oxygen
ab- *departing from; away from* abnormal, departing from normal
acou- *hearing* acoustics, the science of sound
acr-, acro- *extreme or extremity; peak* acrodermatitis, inflammation of the skin of the extremities
ad- *to or toward* adorbital, toward the orbit
aden-, adeno- *gland* adeniform, resembling a gland in shape
amphi- *on both sides; of both kinds* amphibian, an organism capable of living in water and on land
angi- *vessel* angiitis, inflammation of a lymph vessel or blood vessel
ant-, anti- *opposed to; preventing or inhibiting* anticoagulant, a substance that prevents blood coagulation
ante- *preceding; before* antecubital, in front of the elbow
arthr-, arthro- *joint* arthropathy, any joint disease
aut-, auto- *self* autogenous, self-generated
bi- *two* bicuspid, having two cusps
bio- *life* biology, the study of life and living organisms
blast- *bud or germ* blastocyte, undifferentiated embryonic cell
broncho- *bronchus* bronchospasm, spasmodic contraction of bronchial muscle
bucco- *cheek* buccolabial, pertaining to the cheek and lip
caput- *head* decapitate, remove the head
carcin- *cancer* carcinogen, a cancer-causing agent
cardi, cardio- *heart* cardiotoxic, harmful to the heart
cephal- *head* cephalometer, an instrument for measuring the head
cerebro- *brain, especially the cerebrum* cerebrospinal, pertaining to the brain and spinal cord
chondr- *cartilage* chondrogenic, giving rise to cartilage
circum- *around* circumnuclear, surrounding the nucleus
co-, con- *together* concentric, common center, together in the center
contra- *against* contraceptive, agent preventing conception
cost- *rib* intercostal, between the ribs
crani- *skull* craniotomy, a skull operation
crypt- *hidden* cryptomenorrhea, a condition in which menstrual symptoms are experienced but no external loss of blood occurs
cyt- *cell* cytology, the study of cells
de- *undoing, reversal, loss, removal* deactivation, becoming inactive
di- *twice, double* dimorphism, having two forms
dia- *through, between* diaphragm, the wall through or between two areas
dys- *difficult, faulty, painful* dyspepsia, disturbed digestion
ec, ex, ecto- *out, outside, away from* excrete, to remove materials from the body
en-, em- *in, inside* encysted, enclosed in a cyst or capsule
entero- *intestine* enterologist, one who specializes in the study of intestinal disorders
epi- *over, above* epidermis, outer layer of skin
eu- *well* euesthesia, a normal state of the senses
exo- *outside, outer layer* exophthalmos, an abnormal protrusion of the eye from the orbit
extra- *outside, beyond* extracellular, outside the body cells of an organism
gastr- *stomach* gastrin, a hormone that influences gastric acid secretion
glosso- *tongue* glossopathy, any disease of the tongue
hema-, hemato-, hemo- *blood* hematocyst, a cyst containing blood
hemi- *half* hemiglossal, pertaining to one half of the tongue
hepat- *liver* hepatitis, inflammation of the liver
hetero- *different or other* heterosexuality, sexual desire for a person of the opposite sex
hist- *tissue* histology, the study of tissues
hom-, homo- *same* homeoplasia, formation of tissue similar to normal tissue; homocentric, having the same center
hydr-, hydro- *water* dehydration, loss of body water
hyper- *excess* hypertension, excessive tension
hypno- *sleep* hypnosis, a sleeplike state
hypo- *below, deficient* hypodermic, beneath the skin; hypokalemia, deficiency of potassium
hyster-, hystero- *uterus or womb* hysterectomy, removal of the uterus; hysterodynia, pain in the womb
im- *not* impermeable, not permitting passage, not permeable
inter- *between* intercellular, between the cells
intra- *within, inside* intracellular, inside the cell
iso- *equal, same* isothermal, equal, or same, temperature
leuko- *white* leukocyte, white blood cell
lip-, lipo- *fat, lipid* lipophage, a cell that has taken up fat in its cytoplasm
macro- *large* macromolecule, large molecule
mal- *bad, abnormal* malfunction, abnormal functioning of an organ
mamm- *breast* mammary gland, breast
mast- *breast* mastectomy, removal of a mammary gland
meningo- *membrane* meningitis, inflammation of the membranes of the brain
meso- *middle* mesoderm, middle germ layer
meta- *beyond, between, transition* metatarsus, the part of the foot between the tarsus and the phalanges
metro- *uterus* metroscope, instrument for examining the uterus
micro- *small* microscope, an instrument used to make small objects appear larger

mito- *thread, filament* mitochondria, small, filamentlike structures located in cells
mono- *single* monospasm, spasm of a single limb
morpho- *form* morphology, the study of form and structure of organisms
multi- *many* multinuclear, having several nuclei
myelo- *spinal cord, marrow* myeloblasts, cells of the bone marrow
myo- *muscle* myocardium, heart muscle
narco- *numbness* narcotic, a drug producing stupor or numbed sensations
nephro- *kidney* nephritis, inflammation of the kidney
neuro- *nerve* neurophysiology, the physiology of the nervous system
ob- *before, against* obstruction, impeding or blocking up
oculo- *eye* monocular, pertaining to one eye
odonto- *teeth* orthodontist, one who specializes in proper positioning of the teeth in relation to each other
ophthalmo- *eye* ophthalmology, the study of the eyes and related disease
ortho- *straight, direct* orthopedic, correction of deformities of the musculoskeletal system
osteo- *bone* osteodermia, bony formations in the skin
oto- *ear* otoscope, a device for examining the ear
oxy- *oxygen* oxygenation, the saturation of a substance with oxygen
pan- *all, universal* panacea, a cure-all
para- *beside, near* paraphrenitis, inflammation of tissues adjacent to the diaphragm
peri- *around* perianal, situated around the anus
phago- *eat* phagocyte, a cell that engulfs and digests particles or cells
phleb- *vein* phlebitis, inflammation of the veins
pod- *foot* podiatry, the treatment of foot disorders
poly- *multiple* polymorphism, multiple forms
post- *after, behind* posterior, places behind (a specific) part
pre-, pro- *before, ahead of* prenatal, before birth
procto- *rectum, anus* proctoscope, an instrument for examining the rectum
pseudo- *false* pseudotumor, a false tumor
psycho- *mind, psyche* psychogram, a chart of personality traits
pyo- *pus* pyocyst, a cyst that contains pus
retro- *backward, behind* retrogression, to move backward in development
sclero- *hard* sclerodermatitis, inflammatory thickening and hardening of the skin
semi- *half* semicircular, having the form of half a circle
steno- *narrow* stenocoriasis, narrowing of the pupil
sub- *beneath, under* sublingual, beneath the tongue
super- *above, upon* superior, quality or state of being above others or a part
supra- *above, upon* supracondylar, above a condyle
sym-, syn- *together, with* synapse, the region of communication between two neurons
tachy- *rapid* tachycardia, abnormally rapid heartbeat
therm- *heat* thermometer, an instrument used to measure heat
tox- *poison* antitoxic, effective against poison
trans- *across, through* transpleural, through the pleura
tri- *three* trifurcation, division into three branches
viscero- *organ, viscera* visceroinhibitory, inhibiting the movements of the viscera

SUFFIXES

-able *able to, capable of* viable, ability to live or exist
-ac *referring to* cardiac, referring to the heart
-algia *pain in a certain part* neuralgia, pain along the course of a nerve
-ary *associated with, relating to* coronary, associated with the heart
-cide *destroy or kill* germicide, an agent that kills germs
-ectomy *cutting out, surgical removal* appendectomy, cutting out of the appendix
-emia *condition of the blood* anemia, deficiency of red blood cells
-ferent *carry* efferent nerves, nerves carrying impulses away from the CNS
-fuge *driving out* vermifuge, a substance that expels worms of the intestine
-gen *an agent that initiates* pathogen, any agent that produces disease
-gram *data that are systematically recorded, a record* electrocardiogram, a recording showing action of the heart
-graph *an instrument used for recording data or writing* electrocardiograph, an instrument used to make an electrocardiogram
-ia *condition* insomnia, condition of not being able to sleep
-iatrics *medical specialty* geriatrics, the branch of medicine dealing with disease associated with old age
-itis *inflammation* gastritis, inflammation of the stomach
-logy *the study of* pathology, the study of changes in structure and function brought on by disease
-lysis *loosening or breaking down* hydrolysis, chemical decomposition of a compound into other compounds as a result of taking up water
-malacia *soft* osteomalacia, a process leading to bone softening
-mania *obsession, compulsion* erotomania, exaggeration of the sexual passions
-odyn *pain* coccygodynia, pain in the region of the coccyx
-oid *like, resembling* cuboid, shaped as a cube
-oma *tumor* lymphoma, a tumor of the lymphatic tissues
-opia *defect of the eye* myopia, nearsightedness
-ory *referring to, of* auditory, referring to hearing
-pathy *disease* osteopathy, any disease of the bone
-phobia *fear* acrophobia, fear of heights
-plasty *reconstruction of a part, plastic surgery* rhinoplasty, reconstruction of the nose through surgery
-plegia *paralysis* paraplegia, paralysis of the lower half of the body or limbs
-rrhagia *abnormal or excessive discharge* metrorrhagia, uterine hemorrhage
-rrhea *flow or discharge* diarrhea, abnormal emptying of the bowels
-scope *instrument used for examination* stethoscope, instrument used to listen to sounds of various parts of the body
-stasis *arrest, fixation* hemostasis, arrest of bleeding
-stomy *establishment of an artificial opening* enterostomy, the formation of an artificial opening into the intestine through the abdominal wall
-tomy *to cut* appendectomy, surgical removal of the appendix
-ty *conditon of, state* immunity, condition of being resistant to infection or disease
-uria *urine* polyuria, passage of an excessive amount of urine

APPENDIX D

Cadaver Photos

Muscles

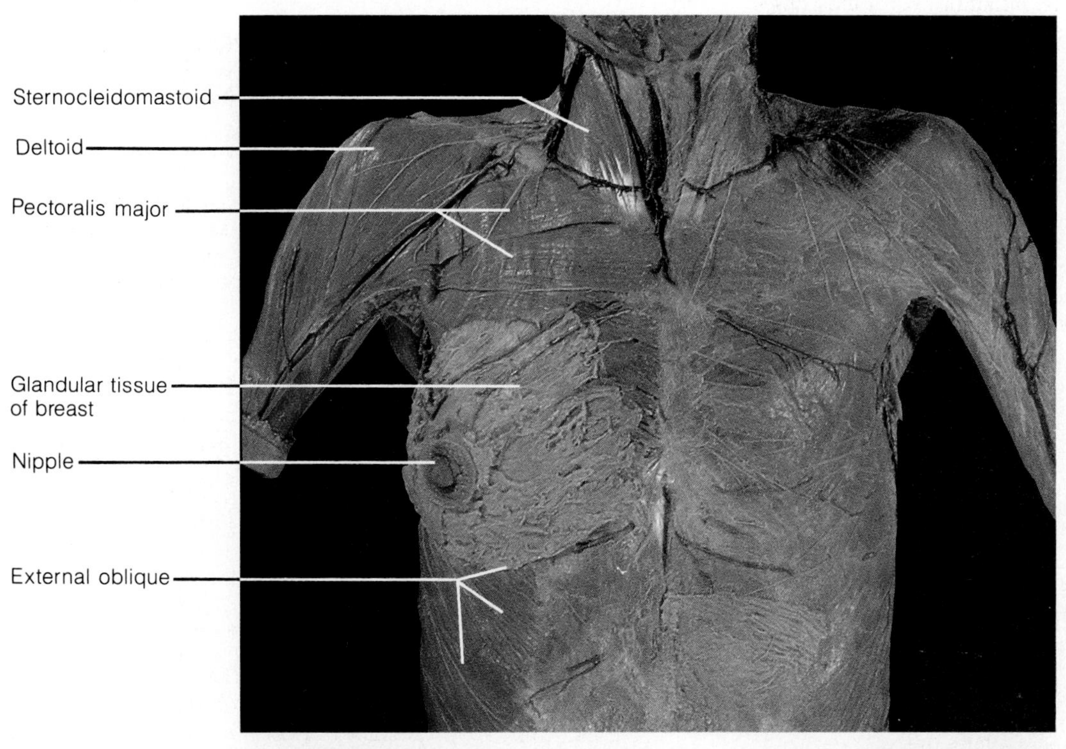

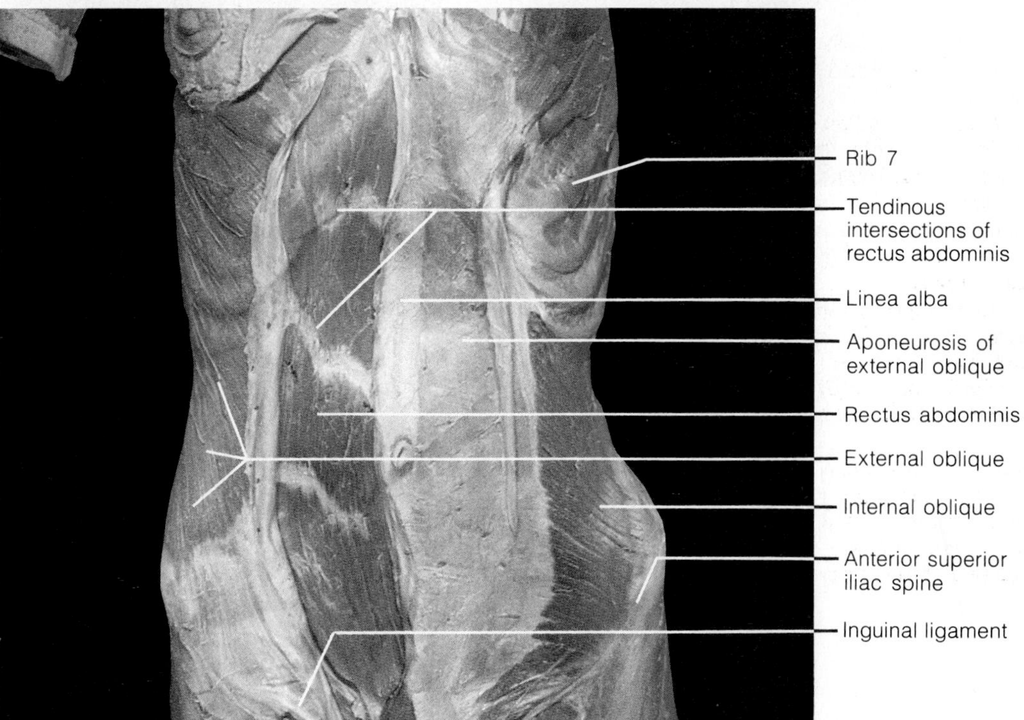

Muscles

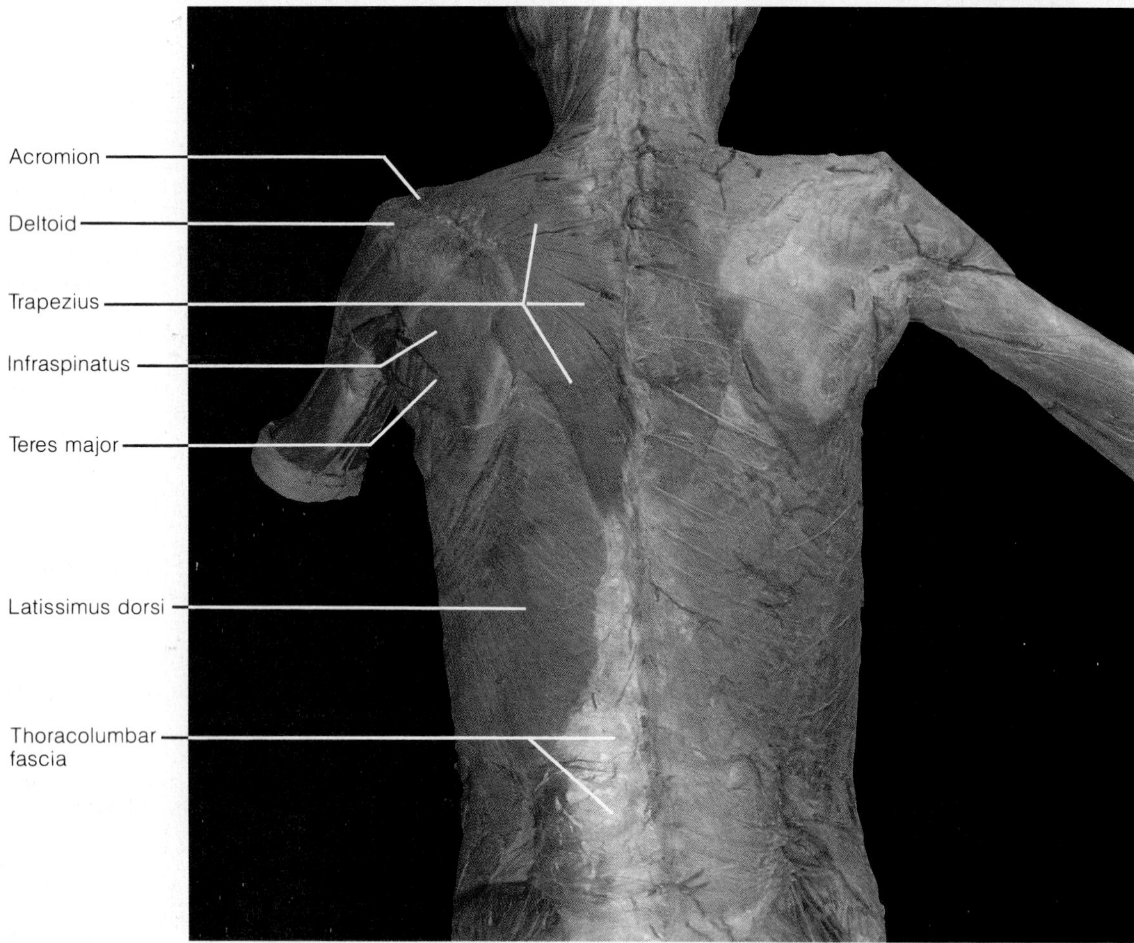
Acromion
Deltoid
Trapezius
Infraspinatus
Teres major
Latissimus dorsi
Thoracolumbar fascia

Cardiovascular

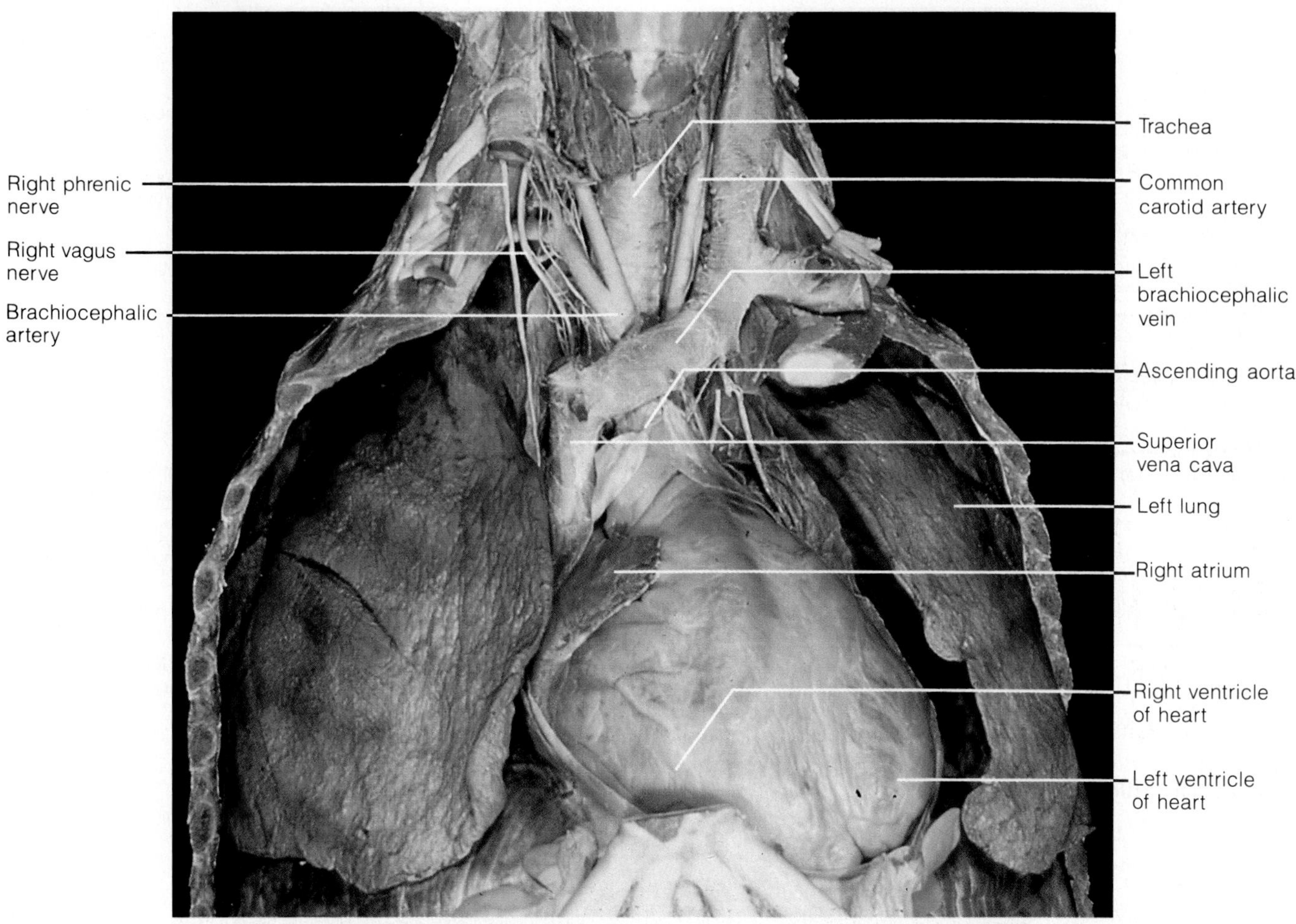
Right phrenic nerve
Right vagus nerve
Brachiocephalic artery
Trachea
Common carotid artery
Left brachiocephalic vein
Ascending aorta
Superior vena cava
Left lung
Right atrium
Right ventricle of heart
Left ventricle of heart

Respiratory

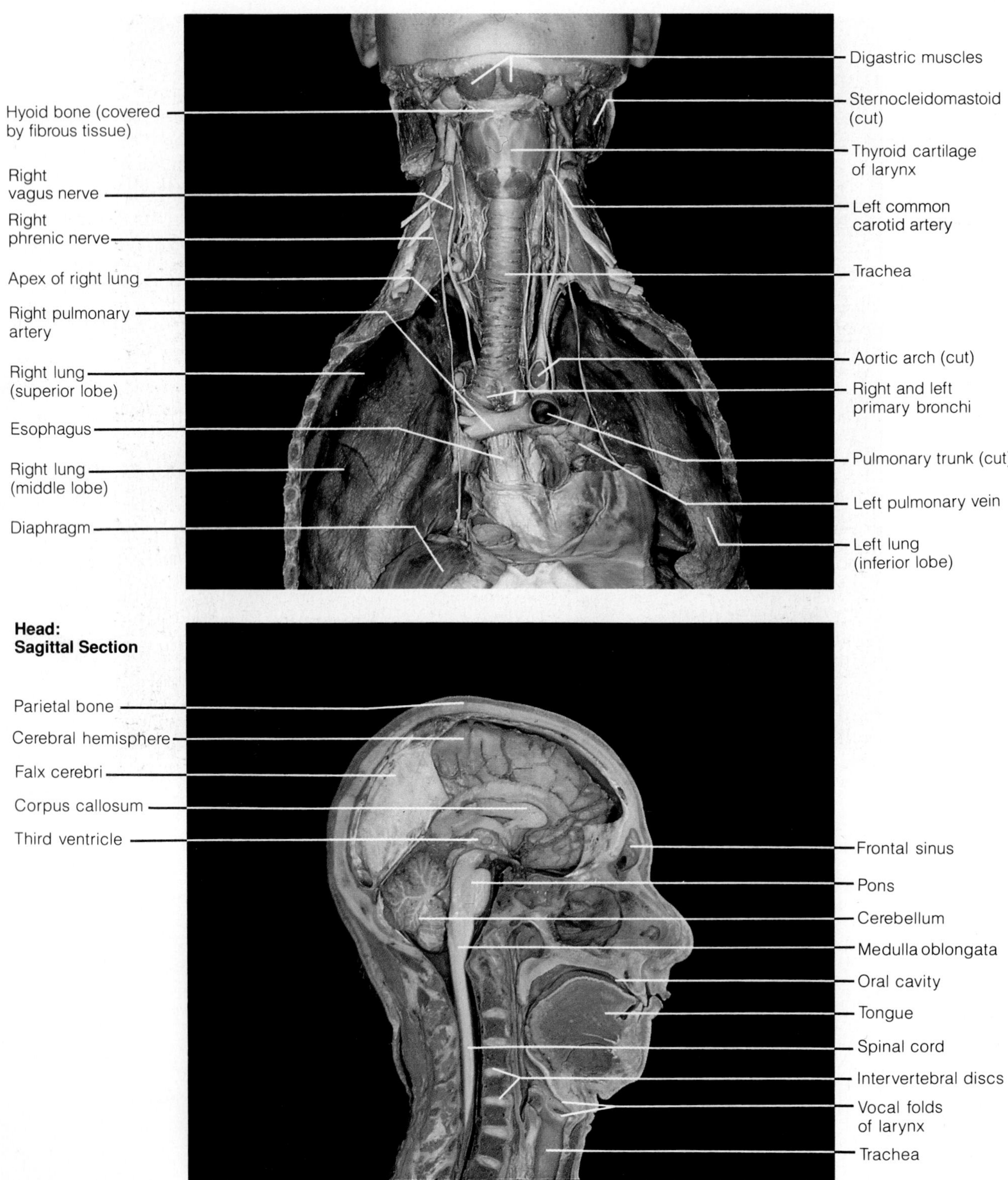

Hyoid bone (covered by fibrous tissue)
Right vagus nerve
Right phrenic nerve
Apex of right lung
Right pulmonary artery
Right lung (superior lobe)
Esophagus
Right lung (middle lobe)
Diaphragm
Digastric muscles
Sternocleidomastoid (cut)
Thyroid cartilage of larynx
Left common carotid artery
Trachea
Aortic arch (cut)
Right and left primary bronchi
Pulmonary trunk (cut)
Left pulmonary vein
Left lung (inferior lobe)
Head: Sagittal Section
Parietal bone
Cerebral hemisphere
Falx cerebri
Corpus callosum
Third ventricle
Frontal sinus
Pons
Cerebellum
Medulla oblongata
Oral cavity
Tongue
Spinal cord
Intervertebral discs
Vocal folds of larynx
Trachea

Digestive

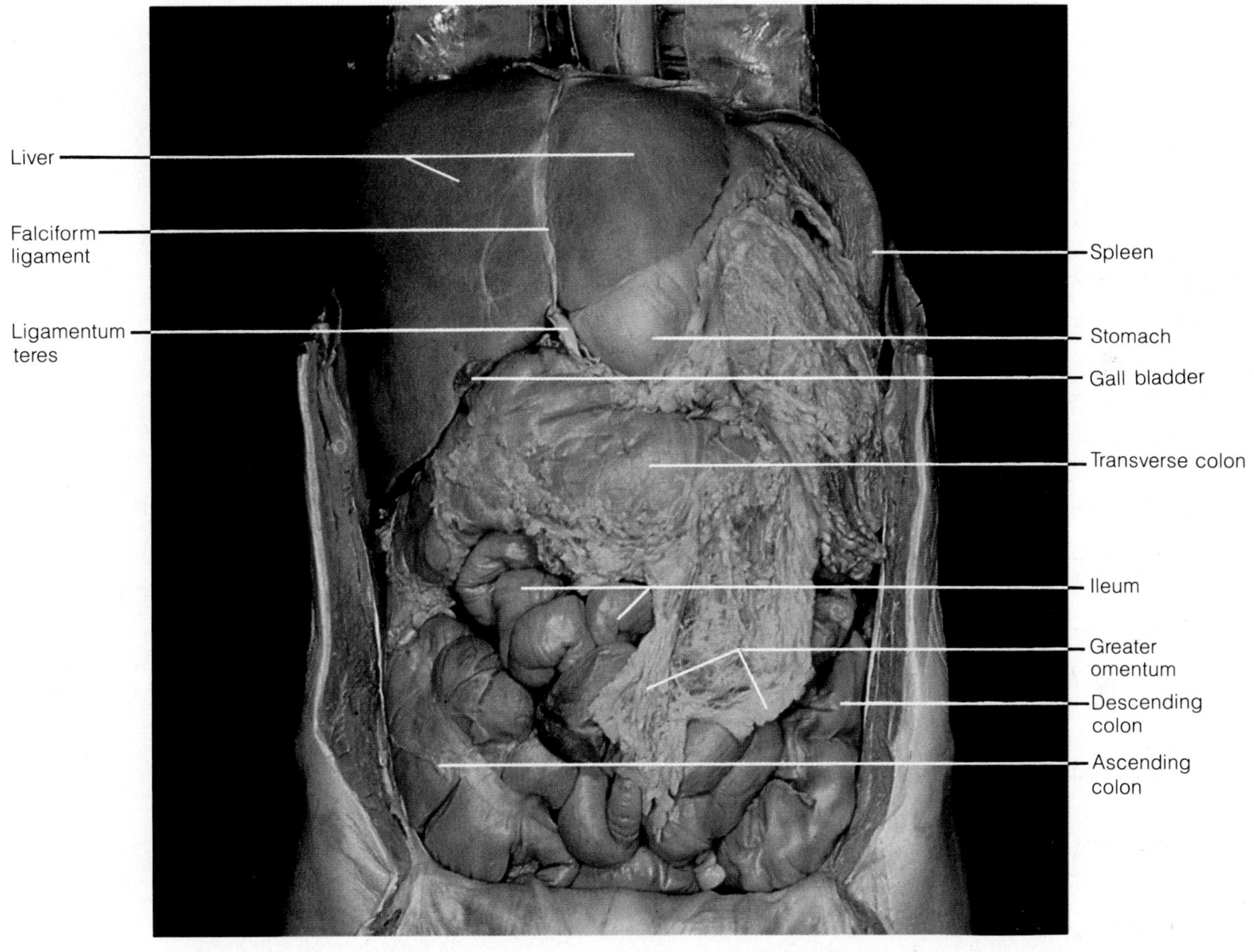
Liver
Falciform ligament
Ligamentum teres
Spleen
Stomach
Gall bladder
Transverse colon
Ileum
Greater omentum
Descending colon
Ascending colon

Urinary

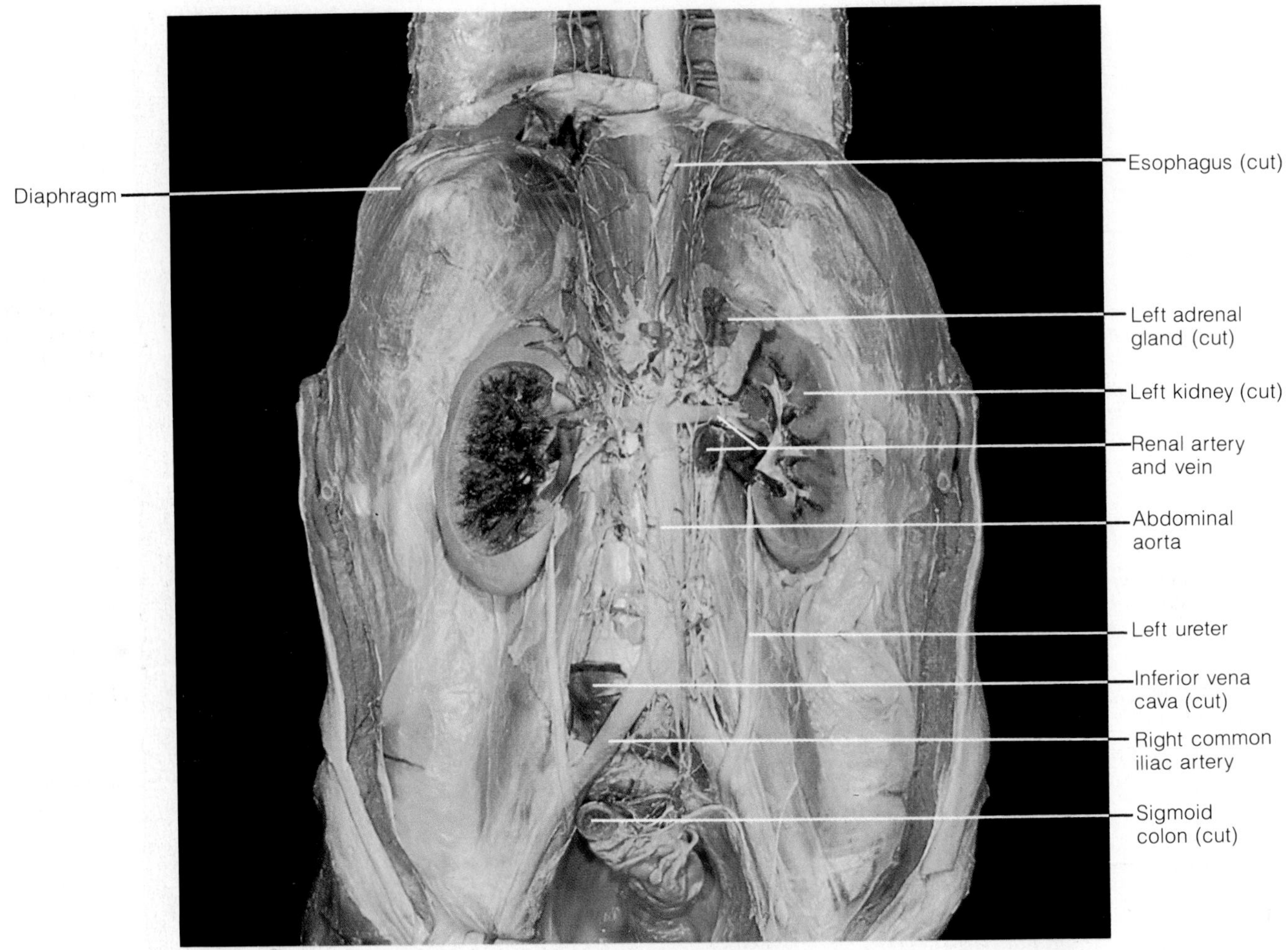
Diaphragm
Esophagus (cut)
Left adrenal gland (cut)
Left kidney (cut)
Renal artery and vein
Abdominal aorta
Left ureter
Inferior vena cava (cut)
Right common iliac artery
Sigmoid colon (cut)

Male Reproductive

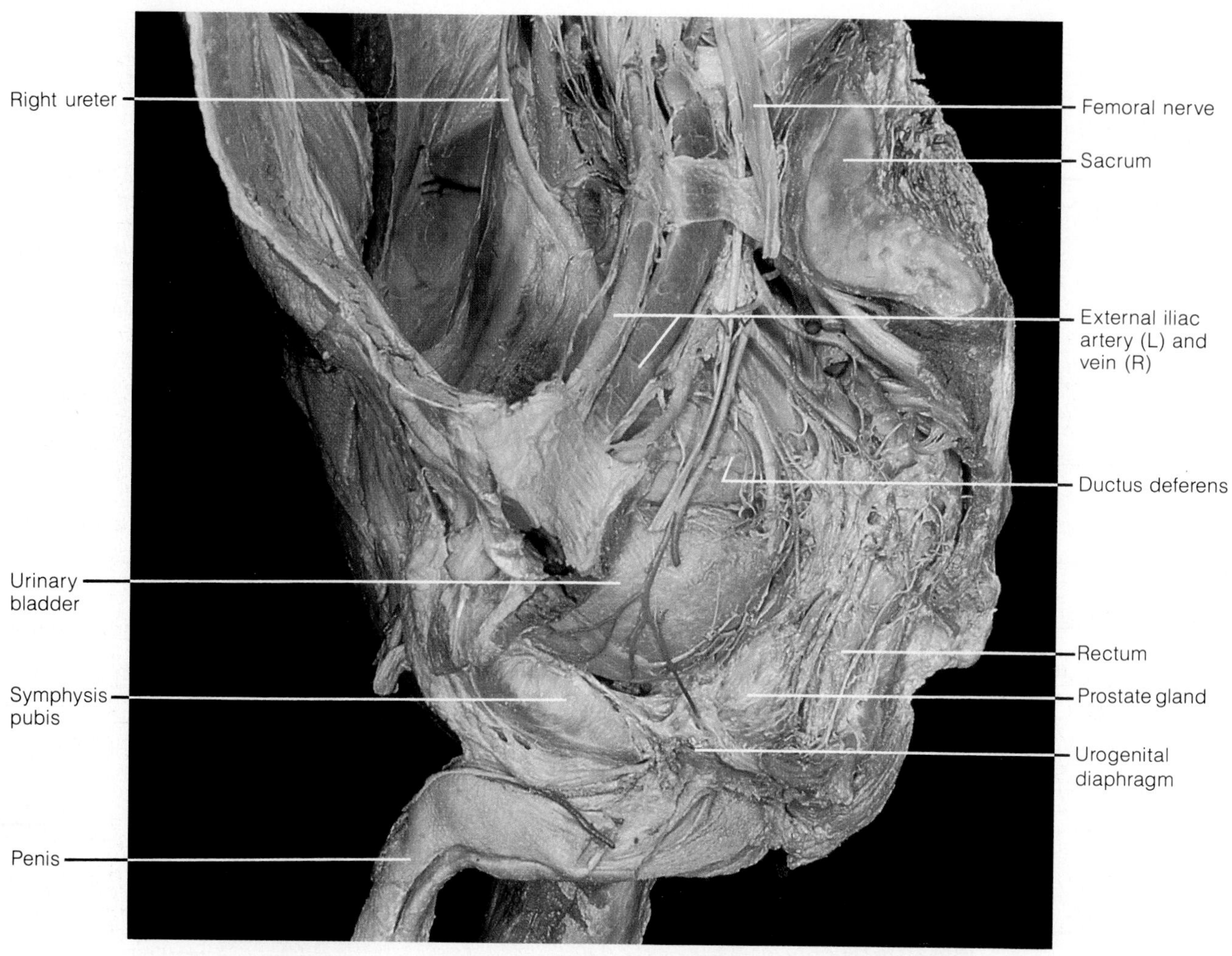

Female Reproductive

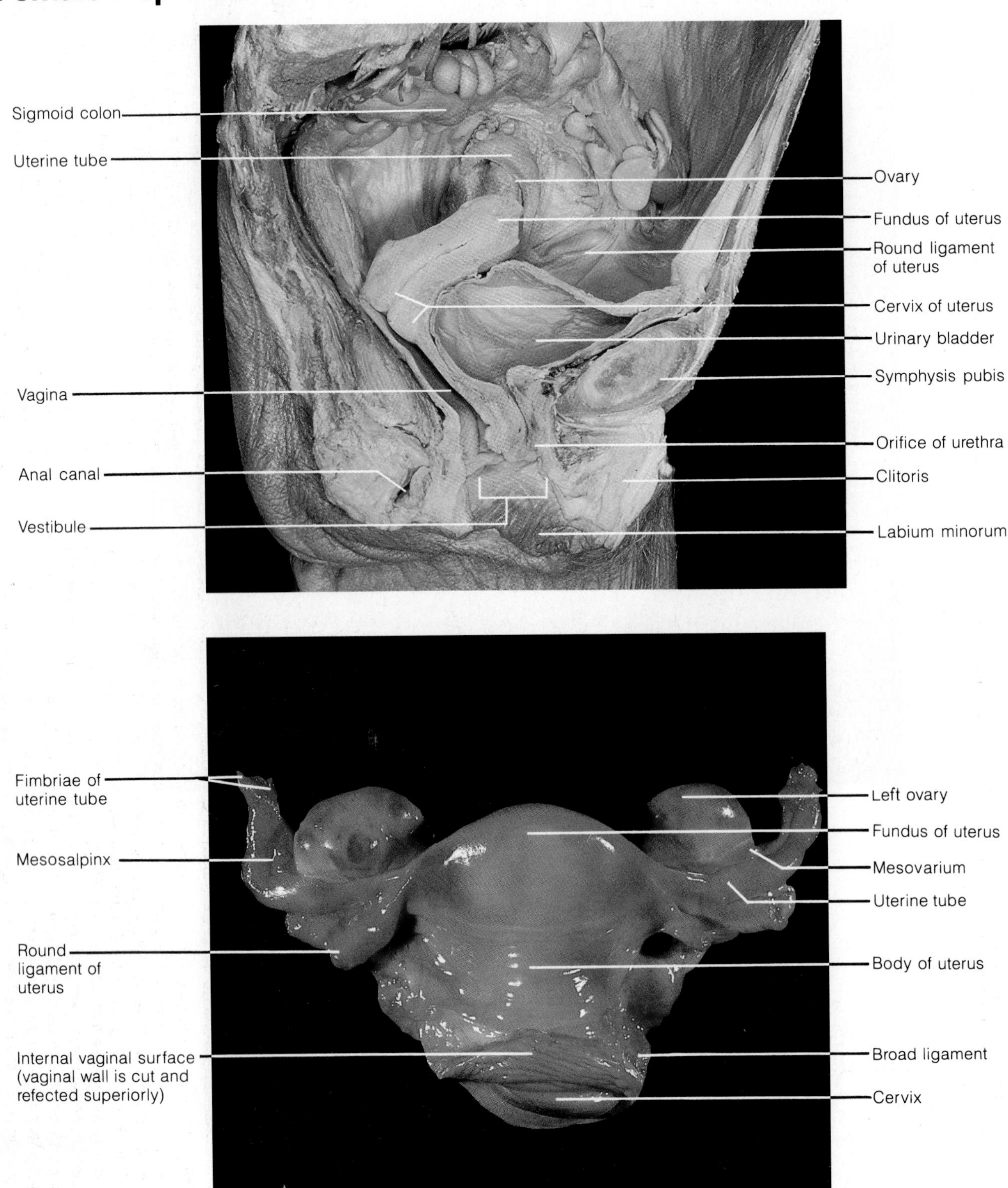
Sigmoid colon
Uterine tube
Ovary
Fundus of uterus
Round ligament of uterus
Cervix of uterus
Urinary bladder
Symphysis pubis
Vagina
Orifice of urethra
Anal canal
Clitoris
Vestibule
Labium minorum
Fimbriae of uterine tube
Left ovary
Fundus of uterus
Mesosalpinx
Mesovarium
Uterine tube
Round ligament of uterus
Body of uterus
Internal vaginal surface (vaginal wall is cut and refected superiorly)
Broad ligament
Cervix

APPENDIX **E**

PERIODIC TABLE OF THE ELEMENTS

1 H 1.00794	
1	Atomic number
H	Symbol
1.00794	Atomic weight

IA 1	IIA 2	IIIB 3	IVB 4	VB 5	VIB 6	VIIB 7	VIIIB 8	VIIIB 9	VIIIB 10	IB 11	IIB 12	IIIA 13	IVA 14	VA 15	VIA 16	VIIA 17	VIIIA 18
1 H 1.008																	**2 He** 4.003
3 Li 6.941	**4 Be** 9.012											**5 B** 10.81	**6 C** 12.01	**7 N** 14.01	**8 O** 16.00	**9 F** 19.00	**10 Ne** 20.18
11 Na 22.99	**12 Mg** 24.31											**13 Al** 26.98	**14 Si** 28.09	**15 P** 30.97	**16 S** 32.06	**17 Cl** 35.45	**18 Ar** 39.95
19 K 39.10	**20 Ca** 40.08	**21 Sc** 44.96	**22 Ti** 47.88	**23 V** 50.94	**24 Cr** 52.00	**25 Mn** 54.94	**26 Fe** 55.85	**27 Co** 58.93	**28 Ni** 58.70	**29 Cu** 63.55	**30 Zn** 65.38	**31 Ga** 69.72	**32 Ge** 72.59	**33 As** 74.92	**34 Se** 78.96	**35 Br** 79.90	**36 Kr** 83.80
37 Rb 85.47	**38 Sr** 87.62	**39 Y** 88.91	**40 Zr** 91.22	**41 Nb** 92.91	**42 Mo** 95.94	**43 Tc** (98)	**44 Ru** 101.1	**45 Rh** 102.9	**46 Pd** 106.4	**47 Ag** 107.9	**48 Cd** 112.4	**49 In** 114.8	**50 Sn** 118.7	**51 Sb** 121.8	**52 Te** 127.6	**53 I** 126.9	**54 Xe** 131.3
55 Cs 132.9	**56 Ba** 137.3	**57 La*** 138.9	**72 Hf** 178.5	**73 Ta** 180.9	**74 W** 183.9	**75 Re** 186.2	**76 Os** 190.2	**77 Ir** 192.2	**78 Pt** 195.1	**79 Au** 197.0	**80 Hg** 200.6	**81 Tl** 204.4	**82 Pb** 207.2	**83 Bi** 209.0	**84 Po** (209)	**85 At** (210)	**86 Rn** (222)
87 Fr (223)	**88 Ra** (226.0)	**89 Ac**** (227)	**104** (261)	**105** (262)	**106** (263)	**107**	**108**	**109**									

*Lanthanides	**58 Ce** 140.1	**59 Pr** 140.9	**60 Nd** 144.2	**61 Pm** (145)	**62 Sm** 150.4	**63 Eu** 152.0	**64 Gd** 157.3	**65 Tb** 158.9	**66 Dy** 162.5	**67 Ho** 164.9	**68 Er** 167.3	**69 Tm** 168.9	**70 Yb** 173.0	**71 Lu** 175.0
**Actinides	**90 Th** 232.0	**91 Pa** (231)	**92 U** 238.0	**93 Np** (237)	**94 Pu** (244)	**95 Am** (243)	**96 Cm** (247)	**97 Bk** (247)	**98 Cf** (251)	**99 Es** (252)	**100 Fm** (257)	**101 Md** (258)	**102 No** (259)	**103 Lr** (260)

The periodic table arranges elements according to atomic number and atomic weight into horizontal rows called *periods* and eighteen vertical columns called *groups* or *families*. The elements in the groups are classified as being in either A or B classes.

Elements of each group of the A series have similar chemical and physical properties. This reflects the fact that members of a particular group have the same number of valence shell electrons, which is indicated by the number of the group. For example, group IA elements have one valence shell electron, group IIA elements have two, and group VA elements have five. In contrast, as you progress across a period from left to right, the properties of the elements change in discrete steps, varying gradually from the very metallic properties of groups IA and IIA elements to the nonmetallic properties seen in group VIIA (chlorine and others), and finally to the inert elements (noble gases) in group VIIIA. This change reflects the continual increase in the number of valence shell electrons seen in elements (from left to right) within a period.

Class B elements are referred to as *transition elements*. All transition elements are metals, and in most cases they have one or two valence shell electrons. (In these elements, some electrons occupy more distant electron shells before the deeper shells are filled.)

The 24 elements found in the human body are listed on p. 27 with their atomic symbols, names, and functions.

Credits

Photograph Credits

Frontispiece and Contents

I: © Manfred Kage/Peter Arnold, Inc. V: © Cheryl Sheppard. XVI: © Howard Sochurek 1986. XVII: Courtesy of J.V. Small and G. Rinnerthaler, Austrian Academy of Sciences. XVIII: © Manfred Kage/Peter Arnold, Inc. IXX: A STEREOSCOPIC ATLAS OF HUMAN ANATOMY, by David L. Bassett, M.D. XX: © Science Photo Library/Photo Researchers, Inc. XXIII: © Manfred Kage/Peter Arnold, Inc. XXV: © Science Photo Library/Photo Researchers, Inc. XXVII: © Kris Erikson. XXVIII: A STEREOSCOPIC ATLAS OF HUMAN ANATOMY, by David L. Bassett, M.D. XXX: © Manfred Kage/Peter Arnold, Inc. XXXI: © David Scharf/Peter Arnold, Inc.

Table 7.4; Figures 7.5; 7.6a, b, and c; 8.10b; 8.11b; 8.12c; 12.1; 12.7c; 12.11a; 12.13b; 12.17a; 12.23b; 14.4; 16.5; 17.4b; 19.4b; 19.7b; 21.3a; 23.3c; 24.14c; 24.16a, b; 26.3a; and Appendix D: From *A Stereoscopic Atlas of Human Anatomy*, by David L. Bassett, M.D.

Unit Openers

Unit 1: Computer Graphics Laboratory, University of California, San Francisco. Units 2, 3, 4: © Manfred Kage/Peter Arnold, Inc. Unit 5: © Phillip A. Harrington/Peter Arnold, Inc.

Chapter 1

1.7a: © Science Source/Photo Researchers, Inc. 1.7b: © Howard Sochurek 1986. 1.7c: © Science Photo Library/Photo Researchers, Inc.

Chapter 2

2.21c: © Computer Graphics Laboratory, University of California, San Francisco. "A Closer Look": (a) © Howard Sochurek 1986. (b) top: © Maria Karras; inset: Henry N. Wagner, Jr., M.D., The Johns Hopkins Medical Institution, Baltimore, MD. (c) © Howard Sochurek 1986.

Chapter 3

3.3a: © L. Orci and A. Perrelet, *Freeze-Etch Histology.* Heidelberg: Springer-Verlag, 1975. 3.3b: © Douglas J. Kelly, *J. Cell Biol.* 28(196):51. Reproduced by Copyright permission of the Rockefeller University Press. 3.3c: © C. Peracchia and A.F. Dulhunty, *J. Cell Biol.* 70(1976):419. Reproduced by Copyright permission of the Rockefeller University Press. 3.12b: © Courtesy N. Simionescu. 3.13b: © J.F. King, University of California School of Medicine/BPS. 3.14b: © Barry King, University of California, Davis/BPS. 3.16: R. Rodewald, University of Virginia/BPS. 3.17b: Courtesy of J.V. Small and G. Rinnerthaler, Austrian Academy of Sciences. © Science Photo Library/Photo Researchers, Inc. 3.20b, c: W.L. Dentler, University of Kansas/BPS. 3.21b: L. Orci and A. Perrelet, *Freeze-Etch Histology.* Heidelberg: Springer-Verlag, 1975. 3.22a: © D.E. Olins and A.L. Olins, *Amer. Sci.* 66(1978):704. Reprinted by permission, *American Scientist*, Journal of Sigma Xi. Photo courtesy of Ada L. Olins and Donald E. Olins, University of Tennessee—Oak Ridge. 3.25 (6 photos): © Ed Reschke.

Chapter 4

4.2a–c: Marian Rice. 4.2d, e: © Ed Reschke. 4.2f: Marian Rice. 4.2g: © John D. Cunningham/Visuals Unlimited. 4.2h: © Ed Reschke. 4.4a: © Biophoto Associates/Photo Researchers, Inc. 4.8a: © Ed Reschke. 4.8b: Marian Rice. 4.8c, d: © Ed Reschke. 4.8e: Marian Rice. 4.8f–i: © Ed Reschke. 4.8j: © Biology Media/Photo Researchers, Inc. 4.8k, l: © Ed Reschke. 4.9a: © Eric Grave/Photo Researchers, Inc. 4.9b: Marian Rice. 4.9c: University of California, San Francisco. 4.10: © Ed Reschke. "A Closer Look": Caryn Levy © *Sports Illustrated*, Time Inc.

Chapter 5

5.2: © Ed Reschke. 5.3: © Science Photo Library/Photo Researchers, Inc. 5.4: Marian Rice. 5.7: © Manfred Kage/Peter Arnold, Inc. 5.9: © Dan McCoy/Black Star.

Chapter 6

6.3b, 6.6: © Ed Reschke. 6.8b: © Dr. Daniel Grande/Peter Arnold, Inc. "A Closer Look": Courtesy of Dr. Harold M. Dick, New York.

Chapter 7

7.24a, b, 7.25: University of California, San Francisco. 7.27a, b: William Thompson.

Chapter 8

8.3b: © Science Photo Library/Photo Researchers, Inc. 8.5: Manny Milan © *Sports Illustrated*, Time Inc. 8.14: © Science Photo Library/Photo Researchers, Inc. "A Closer Look": Courtesy of Cemax, Inc.

Chapter 9

9.1b: Marian Rice. 9.2: Courtesy of Thomas S. Leeson. Table 9.3: (left) © Eric Grave/Photo Researchers, Inc.; (middle) Marian Rice; (right) University of California, San Francisco. 9.4a: © Ed Reschke. 9.7: Courtesy of Dr. H.E. Huxley. 9.12b: © Eric Grave/Photo Researchers, Inc. 9.17: © Dave Black. "A Closer Look": © Bob Krist/Black Star.

Chapter 10

10.3a, 10.4a: William Thompson. "A Closer Look": © Mike Neveux.

Chapter 11

11.3a: © Manfred Kage/Peter Arnold, Inc. 11.4g: Courtesy G.L. Scott, J.A. Feilbach, and T.A. Duff. 11.8a: From *Tissues and Organs: A Text-Atlas of Scanning Electron Microscopy* by Richard G. Kessel and Randy H. Kardon. W.H. Freeman and Company. Copyright © 1979. "A Closer Look": © M.P.L. Fogden/Bruce Coleman, Inc.

Chapter 12

12.8b: Lester V. Bergman, NY. 12.22: © Howard Sochurek 1986. 12.25b: © Biophoto Associates/Photo Researchers, Inc.

Chapter 13

Table 13.1: William Thompson. 13.14a, b: Charles Bassin. "A Closer Look": Courtesy of Warren E. Collins, Inc.

Chapter 14

"A Closer Look": © Bobbie Kingsley/Photo Researchers, Inc.

Chapter 15

15.4a: © Alexander Tsiaras/Photo Researchers, Inc.

Chapter 16

16.10: © SIU/Photo Researchers, Inc. 16.14: © Kodansha/Imageback West.

Chapter 17

17.6: Clinical Pathological Conference, *American Journal of Medicine*, 20:133, 1956, with permission of Albert I. Mendeloff, M.D. 17.7b, 17.13b, 17.17b: © Ed Reschke. "A Closer Look": © Louie Psihoyos.

Chapter 18

18.2: © Biophoto Associates/Photo Researchers, Inc. 18.3b: © Science Photo Library/Photo Researchers, Inc. 18.3c: © Ed Reschke. 18.12: © Lennart Nilsson/Boehringer Ingelheim International.

Chapter 19

19.3a: Robinson, Cohen-Gould and Factor, from LABORATORY INVESTIGATION, Vol. 49 (4), pp. 482–498. Copyright © by U.S. & Canadian Academy of Pathology, 1983. 19.10: © Science Photo Library/Photo Researchers, Inc. 19.11a: Marian Rice.

Chapter 20

20.1b: From *Tissues and Organs: A Text-Atlas of Scanning Electron Microscopy* by Richard G. Kessel and Randy H. Kardon. W.H. Freeman and Company. Copyright © 1979. 20.4: From U.S. Ryan, J.W. Ryan, D.S. Smith and H. Winkler, TISSUE AND CELL 7:181, 1975. With permission from Longman Group, Ltd. 20.10a: © Ed Reschke. 20.10b: © F. Sloop and W. Ober/Visuals Unlimited. 20.15b: © Science Photo Library/Photo Researchers, Inc. "A Closer Look": © Howard Sochurek.

Chapter 21

21.3c, 21.5c: © Biophoto Associates/Photo Researchers, Inc. 21.6a, 21.6b, 21.7: © John Cunningham/Visuals Unlimited. 21.8: © Biophoto Associates/Photo Researchers, Inc.

Chapter 22

22.1: © Lennart Nilsson/Boehringer Ingelheim International. 22.9b: Photography and Molecular Modeling, Arthur J. Olson, Ph.D. © 1985. 22.12: © Science Photo Library/Photo Researchers, Inc. "A Closer Look": © Lennart Nilsson/Boehringer Ingelheim International.

Chapter 23

23.1b: © Martin M. Rotker/Taurus Photos. 23.3d: © Lennart Nilsson/Boehringer Ingelheim International. 23.4b: University of California, San Francisco. 23.5b, 23.6b: © Science Photo Library/Photo Researchers Inc. 23.7a: From *Tissues and Organs: A Text-Atlas of Scanning Electron*

Microscopy by Richard G. Kessel and Randy H. Kardon. W.H. Freeman and Company. Copyright © 1979. 23.23: © Kris Erikson.

Chapter 24
24.9: (top) From *Tissues and Organs: A Text-Atlas of Scanning Electron Microscopy*, by Richard G. Kessel and Randy H. Kardon. W.H. Freeman and Company. Copyright © 1979; (bottom) R.G. Kessel and C.Y. Shih, *Scanning Electron Microscopy in Biology*, p. 277 (Springer-Verlag, 1974). 24.10b: © Science Photo Library/Photo Researchers, Inc. 24.13: © Biophoto Associates/Photo Researchers, Inc. 24.14d: Courtesy of Peter M. Andrews. 24.15d: © Monfred Kage/Peter Arnold, Inc. 24.17b: © M.I. Walker/Photo Researchers, Inc. 24.18b: Tufts Medical School, Department of Anatomy. 24.20a: © Ed Reschke/Peter Arnold, Inc. 24.20b: © Biophoto Associates/Photo Researchers, Inc. "A Closer Look": © American Stock Photography.

Chapter 25
"A Closer Look": © Stephen Ferry/Gamma-Liaison.

Chapter 26
26.5c: From *Tissues and Organs: A Text-Atlas of Scanning Electron Microscopy*, by Richard G. Kessel and Randy H. Kardon. W.H. Freeman and Company. Copyright © 1979. 26.8b: © David M. Phillips/Visuals Unlimited. 26.14 © Biophoto Associates/Photo Researchers, Inc.

Chapter 27
"A Closer Look": © David York/Medichrome.

Chapter 28
28.3b: © Ed Reschke. 28.7a: From *Tissues and Organs: A Text-Atlas of Scanning Electron Microscopy*. W.H. Freeman and Company. Copyright © 1979. 28.8b: © Manfred Kage/Peter Arnold, Inc. 28.12b: © Ed Reschke. 28.13a: © Dr. Michael Ross.

Chapter 29
29.2b: © Lennart Nilsson/Boehringer Ingelheim International. 29.3h, 29.4 (3 photos): Courtesy of Dr. C.A. Ziomek, Worcester Foundation for Experimental Biology, Shrewsbury, MA. 29.5b: E.S.E. Hafez. 29.5d: Photographed from the Carnegie Collection by Dr. Allen C. Enders, University of California, Davis. 29.14a, b: © Lennart Nilsson/Boehringer Ingelheim International. 29.18a: © Michael Alexander. 29.18b, c: © Hal Stoezle. "A Closer Look": Courtesy IVF Clinic, Stanford Medical Center.

Chapter 30
30.1a: © Lennart Nilsson, *The Body Victorious*, Delacourt Press, Dell Publishing Co. 30.1b: E. de Harven, Rockefeller University.

Illustration Credits

Chapter 1
1.1, 1.2, 1.5, 1.9: Jeanne Koelling. 1.3: Darwen and Vally Hennings; modified from N.A. Campbell, *Biology*, (Menlo Park, CA: Benjamin/Cummings, 1987), p. 782. © The Benjamin/Cummings Publishing Company, Inc. 1.4, 1.6–1.8, 1.10, 1.11, Table 1.1: Joan Carol/Kenneth R. Miller.

Chapter 2
2.1–2.10, 2.12–2.22: Georg Klatt. 2.11: Cecile Duray-Bito (modified by Georg Klatt); from N.A. Campbell, *Biology*, (Menlo Park, CA: Benjamin/Cummings, 1987), p. 44. © The Benjamin/Cummings Publishing Company, Inc.

Chapter 3
3.1, 3.3–3.8, 3.11–3.16, 3.18, 3.21–3.23, "A Closer Look": Carla Simmons. 3.2: Barbara Cousins (modified by Carla Simmons); from N.A. Campbell, *Biology*, (Menlo Park, CA: Benjamin/Cummings, 1987), pp. 161. © The Benjamin/Cummings Publishing Company. 3.9, 3.10: Barbara Cousins; from N.A. Campbell, *Biology*, (Menlo Park, CA: Benjamin/Cummings, 1987), pp. 173, 174. © The Benjamin/Cummings Publishing Company, Inc. 3.17a: Carol Verbeeck; from "The Ground Substance of the Living Cell," by Keith R. Porter and Jonathan B. Tucker, Copyright © 1981 by Scientific American, Inc. All rights reserved. 3.17c, 3.25: Carla Simmons; from N.A. Campbell, *Biology*, (Menlo Park, CA: Benjamin/Cummings, 1987), pp. 144, 232–233. © The Benjamin/Cummings Publishing Company, Inc. 3.20: Carol Verbeeck; adapted from K.D. Johnson, D.L. Rayle, and H.L. Wedberg, *Biology: An Introduction*, (Menlo Park, CA: Benjamin/Cummings, 1984) p. 52. © The Benjamin/Cummings Publishing Company, Inc. 3.24, 3.26, 3.27: Georg Klatt.

Chapter 4
4.1–4.12: Linda McVay.

Chapter 5
5.1, 5.2, 5.4–5.6, 5.8: Linda McVay. 5.10: Kenneth R. Miller.

Chapter 6
6.1–6.12, Table 6.1, Table 6.2: Barbara Cousins. 6.13: Kenneth R. Miller.

Chapter 7
7.1–7.23, 7.25, 7.26, 7.28–7.32, Table 7.2–7.4: Nadine Sokol.

Chapter 8
8.1–8.4, 8.6–8.13, Table 8.1: Barbara Cousins.

Chapter 9
9.1, 9.3–9.16, 9.18–9.23, Table 9.1, Table 9.3: Raychel Ciemma. 9.24: Kenneth R. Miller.

Chapter 10
10.1–10.22: Raychel Ciemma.

Chapter 11
11.1–11.30: Charles W. Hoffman. 11.31: Kenneth R. Miller.

Chapter 12
12.2–12.21, 12.23–12.28: Stephanie McCann.

Chapter 13
13.1–13.13, Table 13.1: Stephanie McCann.

Chapter 14
14.1–14.3: Stephanie McCann.

Chapter 15
15.1–15.7, "A Closer Look": Charles W. Hoffman.

Chapter 16
16.1–16.4, 16.6–16.9, 16.11–16.13, 16.23–16.30, 16.32, 16.34, 16.35: Charles W. Hoffman. 16.14: Carla Simmons; from N.A. Campbell, *Biology*, (Menlo Park, CA: Benjamin/Cummings Publishing, 1987), p. 210. © The Benjamin/Cummings Publishing Company, Inc. 16.15–16.22, 16.31, 16.33: Kenneth R. Miller.

Chapter 17
17.1–17.4, 17.7, 17.9–17.11, 17.13, 17.14, 17.16, 17.17, Table 17.2: Charles W. Hoffman. 17.5, 17.8, 17.15, 17.18, 17.20, 17.21: Kenneth R. Miller. 17.12: Janet Hayes; from N.A. Campbell, *Biology*, (Menlo Park, CA: Benjamin/Cummings, 1987), p. 898. © The Benjamin/Cummings Publishing Company, Inc. 17.19: John and Judy Waller; from N.A. Campbell, *Biology*, (Menlo Park, CA: Benjamin/Cummings, 1987), p. 896. © The Benjamin/Cummings Publishing Company, Inc.

Chapter 18
18.1 (top): Darwen and Vally Hennings; from N.A. Campbell, *Biology*, (Menlo Park, CA: Benjamin/Cummings, 1987), p. 824. © The Benjamin/Cummings Publishing Company, Inc. 18.1 (bottom), 18.3–18.11, 18.13, Table 18.1, Table 18.4: Nadine Sokol.

Chapter 19
19.1–19.14, 19.16–19.19, "A Closer Look": Barbara Cousins. 19.15: From N.M. Holloway: *Nursing the Critically Ill Adult: Applying Nursing Diagnosis*, Second edition, (Menlo Park, CA: Addison-Wesley, 1984), pp. 277, 282, 298, 285.

Chapter 20
20.1–20.3, 20.7, 20.8, 20.11–20.26: Barbara Cousins. 20.5, 20.6, 20.9: Darwen and Vally Hennings; from N.A. Campbell, *Biology*, (Menlo Park, CA: Benjamin/Cummings, 1987), pp. 821, 8.20. © The Benjamin/Cummings Publishing Company, Inc. 20.27: Kenneth R. Miller.

Chapter 21
21.1–21.5: Nadine Sokol. 21.9: Kenneth R. Miller.

Chapter 22
22.2–22.11, 22.13–22.15, Table 22.3: Carla Simmons. 22.16: Kenneth R. Miller.

Chapter 23
23.1–23.11, 23.13–23.20, 23.22, 23.24: Raychel Ciemma. 23.12: Raychel Ciemma/Kenneth R. Miller. 23.21, 23.25: Kenneth R. Miller.

Chapter 24
24.1, 24.7, 24.9, 24.10, 24.14, 24.19, 24.33: Kenneth R. Miller/Nadine Sokol. 24.2–24.6, 24.8, 24.11, 24.12, 24.15, 24.17, 24.18, 24.21–24.27, 24.29–24.32, 24.34: Kenneth R. Miller. 24.28: Darwen and Vally Hennings; from N.A. Campbell, *Biology*, (Menlo Park, CA: Benjamin/Cummings, 1987), p. 806. © The Benjamin/Cummings Publishing Company, Inc.

Chapter 25
25.1, 25.2a, 25.3–25.23: Charles W. Hoffman. 25.2b: Elizabeth Morales; from N.A. Campbell, *Biology*, (Menlo Park, CA: Benjamin/Cummings, 1987), p. 793. © The Benjamin/Cummings Publishing Company, Inc.

Chapter 26
26.1–26.11, 26.13, 26.15–26.18, "A Closer Look": Linda McVay. 26.12: Linda McVay; adapted from N.A. Campbell, *Biology*, (Menlo Park, CA: Benjamin/Cummings, 1987), p. 868. © The Benjamin/Cummings Publishing Company, Inc. 26.19: Kenneth R. Miller.

Chapter 27
27.1–27.13: Linda McVay.

Chapter 28
28.1–28.4, 28.7–28.18, 28.20, 28.22: Martha Blake. 28.5, 28.6, 28.23, "A Closer Look": Kenneth R. Miller. 28.19, 28.21: Darwen and Vally Hennings; from N.A. Campbell, *Biology*, (Menlo Park, CA: Benjamin/Cummings, 1987), pp. 911, 913. © The Benjamin/Cummings Publishing Company, Inc.

Chapter 29
29.1–29.5, 29.7–29.12, 29.16, 29.17, 29.19: Martha Blake. 29.6: Kenneth R. Miller. 29.13: Martha Blake; from L.G. Mitchell, J.A. Mutchmor, and W.D. Dolphin, *Zoology*, (Menlo Park, CA: Benjamin/Cummings, 1988), p. 809. © The Benjamin/Cummings Publishing Company, Inc. 29.15: Charles W. Hoffman; from S.B. Olds, M.L. London, and P.A. Ladewig, *Maternal-Newborn Nursing: A Family-Centered Approach*, (Menlo Park, CA: Addison-Wesley, 1988), insert.

Chapter 30
30.2, 30.4: Kenneth R. Miller. 30.3: Georg Klatt; from N.A. Campbell, *Biology*, (Menlo Park, CA: Benjamin/Cummings, 1987), p. 255. © The Benjamin/Cummings Publishing Company, Inc. 30.5, 30.7: Barbara Cousins; from N.A. Campbell, *Biology*, (Menlo Park, CA: Benjamin/Cummings, 1987), pp. 273, 278. © The Benjamin/Cummings Publishing Company, Inc. 30.6: Carol Verbeeck; from N.A. Campbell, *Biology*, (Menlo Park, CA: Benjamin/Cummings, 1987), p. 274. © The Benjamin/Cummings Publishing Company, Inc.

Glossary

Pronunciation Key:
′ = Primary accent
″ = Secondary accent

Pronounce	as in
a	hat
ah	father
ar	tar
ā	say
ayr	air
e, eh	hen
ē	bee
ēr	deer
er	her
ew	new
g	go
i, ih	him
ī	kite
j	jet
ng	ring
o	frog
ō	no
oo	soon
or	for
ow	plow
oy	boy
sh	she
th	thick
u, uh	sun
z	zebra
zh	measure

Abdomen (ab′dō-men) Portion of the body between the diaphragm and the pelvis.

Abduct (ab-dukt′) To move away from the midline of the body.

Abscess (ab′-ses) Localized accumulation of pus and disintegrating tissue.

Absolute refractory period Period following stimulation during which no additional action potential can be evoked.

Absorption Process by which the products of digestion pass through the alimentary tube mucosa into the blood or lymph.

Accessory digestive organs Organs that contribute to the digestive process but are not part of the alimentary canal; include the tongue, teeth, salivary glands, pancreas, liver, and gallbladder.

Acetabulum (a″-sih-ta′-byoo-lum) Cuplike cavity on lateral surface of hipbone that receives the femur.

Acetylcholine (a″-seh-til-kō′-lēn) Chemical transmitter substance released by some nerve endings.

Achilles tendon Tendon that attaches the calf muscles to the heel bone.

Acid Proton donor; substance capable of releasing hydrogen ions in solution.

Acidosis (a″-sih-dō′-sis) State of abnormally high hydrogen ion concentration in the extracellular fluid.

Actin (ak′-tin) A contractile protein of muscle.

Action potential A large transient depolarization event, including polarity reversal, that is conducted along the membrane of a muscle cell or a nerve fiber.

Active immunity Immunity produced by an encounter with an antigen; provides immunologic memory.

Active transport Membrane transport processes for which ATP is provided; e.g., solute pumping and endocytosis.

Adaptation (1) Any change in structure or response to suit a new environment; (2) decline in the transmission of a sensory nerve when a receptor is stimulated continuously and without change in stimulus strength.

Adduct (a-dukt′) To move toward the midline of the body.

Adenohypophysis (a″-deh-nō-hī-pof′-ih-sis) Anterior pituitary; the glandular part of the pituitary gland.

Adenoids (a′-deh-noydz) Pharyngeal tonsils.

Adenosine triphosphate (ATP) (uh-den′-uh-sēn trī″-fos′-fāt) Organic molecule that stores and releases chemical energy for use in body cells.

Adipose (a′-dih-pōs) Fatty.

Adrenal glands (uh-drē′-nul) Hormone-producing glands located superior to the kidneys; each consists of medulla and cortex areas.

Adrenalin (uh-dreh′-nuh-lin) See Epinephrine.

Adrenergic fibers (ad″-ren-er′-jik) Nerve fibers that release norepinephrine.

Adrenocorticotropic hormone (ACTH) (uh-drē″-nō-kor″-tih-kō-trō′-pik) Anterior pituitary hormone that influences the activity of the adrenal cortex.

Adventitia (ad″-ven-tish′-uh) Outermost layer or covering of an organ.

Aerobic (ayr-ō′-bik) Oxygen-requiring.

Afferent (a′-fer-ent) Carrying to or toward a center.

Afferent neuron Nerve cell that carries impulses toward the central nervous system.

Agglutination (uh-gloo″-tih-nā′-shun) Clumping of (foreign) cells; induced by cross-linking of antigen–antibody complexes.

Agonist (a′-guh-nist) Muscle that bears primary responsibility for causing a certain movement.

AIDS Acquired immune deficiency syndrome; caused by human immunodeficiency virus (HIV); symptoms include severe weight loss, night sweats, swollen lymph nodes, opportunistic infections.

Albumin (al-byoo′-min) The most abundant plasma protein.

Aldosterone (al-dah′-ster-ōn) Hormone produced by the adrenal cortex that regulates sodium ion reabsorption.

Alimentary canal (al″-ih-men′-tuh-rē) The continuous hollow tube extending from the mouth to the anus; its walls are constructed by the oral cavity, pharynx, esophagus, stomach, and small and large intestines.

Alkalosis (al″-kuh-lō′-sis) State of abnormally low hydrogen ion concentration in the extracellular fluid.

Allantois (a-lan′-tō-is) Embryonic membrane; its blood vessels develop into blood vessels of the umbilical cord.

Allergy Overzealous immune response to an otherwise harmless antigen.

Alveolus (al-vē′-uh-lus) One of the microscopic air sacs of the lungs.

Amino acid (uh-mē′-nō) Organic compound containing nitrogen, carbon, hydrogen, and oxygen; building block of protein.

Amnion (am′-nē-on) Fetal membrane that forms a fluid-filled sac around the embryo.

Amphiarthrosis (am″-fē-ar-thrō′-sis) A slightly movable joint.

Anabolism (uh-na′-buh-lizm) Energy-requiring building phase of metabolism in which simpler substances are combined to form more complex substances.

Anaerobic (an″-ayr-ō′-bik) Not requiring oxygen.

Anastomosis (uh-nas″-tō-mō′-sis) A union or joining of blood vessels or other tubular structures.

Androgen (an′-drō-jen) A hormone that controls male secondary sex characteristics.

Anemia (uh-nē′-mē-uh) Reduced oxygen-carrying ability of blood resulting from too few erythrocytes or abnormal hemoglobin.

Aneurysm (an′-yer-ih-zim) Blood-filled sac in an artery wall caused by dilation or weakening of the wall.

Angina pectoris (an-jī′-nuh pek′-tōr-is) Severe suffocating chest pain caused by brief lack of oxygen supply to heart muscle.

Anion (an′-ī-on) An ion carrying one or more negative charges and therefore attracted to a positive pole.

Anorexia (an″-ō-rek′-sē-uh) Loss of appetite or desire for food.

Anoxia (uh-nok′-sē-uh) Deficiency of oxygen.

Antagonist (an-ta′-guh-nist) Muscle that reverses, or opposes, the action of another muscle.

Anterior (an-tēr′-ē-er) The front of an organism, organ, or part; the ventral surface.

Anterior pituitary See Adenohypophysis.

Antibody A protein molecule that is released by a plasma cell (a daughter cell of an activated B lymphocyte) and that binds specifically to an antigen.

Antidiuretic hormone (ADH) (an″-tī-dī″-yer-eh′-tik) Hormone produced by the hypothalamus and released by the posterior pituitary; stimulates the kidneys to reabsorb more water, reducing urine volume.

Antigen (an′-tih-jin) A substance or part of a substance (living or nonliving) that is recognized as foreign by the immune system, activates the immune system, and reacts with immune cells or their products.

Antithrombin (an″-tī-throm′-bin) Substance in blood plasma that neutralizes thrombin, thereby inhibiting coagulation.

Anus (ā′-nus) Distal end of the digestive tract; outlet of the rectum.

Aorta (ā-or′-tuh) Major systemic artery; arises from the left ventricle of the heart.

Aortic body Receptor in the aortic arch sensitive to changing oxygen, carbon dioxide, and pH levels of the blood.

Apocrine gland (a′-puh-krin) The less numerous type of sweat gland; produces a secretion containing water, salts, proteins, and fatty acids.

Aponeurosis (a″-pō-ner-ō′-sis) Fibrous or membranous sheet connecting a muscle and the part it moves.

Appendicular skeleton (a″-pen-dih′-kyoo-ler) Bones of the limbs and limb girdles that are attached to the axial skeleton.

Appendix (uh-pen′-diks) Wormlike sac attached to the cecum of the large intestine.

Aqueous humor (ā′-kwē-us) Watery fluid in the anterior chambers of the eye.

Arachnoid (uh-rak′-noyd) Weblike; specifically, the weblike middle layer of the three meninges.

Areola (uh-rē′-ō-luh) Circular, pigmented area surrounding the nipple.

Arrector pili (uh-rek′-ter pih′-lē) Tiny, smooth muscles attached to hair follicles; cause the hair to stand upright when activated.

Arteries Blood vessels that conduct blood away from the heart and into the circulation.

Arteriole (ar-tēr′-ē-ōl) A minute artery.

Arteriosclerosis (ar-tēr′-ē-ō-skler-ō′-sis) Any of a number of proliferative and degenerative changes in the arteries leading to their decreased elasticity.

Arthritis Inflammation of the joints.

Articular capsule Double-layered capsule composed of an outer fibrous capsule lined by synovial membrane; encloses the joint cavity of a synovial joint.

Articulation Joint; point where two bones meet.

Atherosclerosis (a″-ther-ō″-skler-ō′-sis) Changes in the walls of large arteries consisting of lipid deposits on the artery walls; the early stage of arteriosclerosis.

Atlas First cervical vertebra; articulates with the occipital bone of the skull and the second cervical vertebra (axis).

Atom Smallest particle of an elemental substance that exhibits the properties of that element; composed of protons, neutrons, and electrons.

Atomic mass number Sum of the number of protons and neutrons in the nucleus of an atom.

Atomic number Number of protons in an atom.

Atomic symbol A one- or two-letter symbol indicating a particular element.

Atomic weight Average of the mass numbers of all of the isotopes of an element.

Atria (ā′-trē-uh) Paired, superiorly located heart chambers that receive blood returning to the heart.

Atrioventricular bundle (AV bundle) (ā″-trē-ō-ven-trih′-kyoo-ler) Bundle of specialized fibers that conduct impulses from the AV node to the right and left ventricles; also called bundle of His.

Atrioventricular node (AV node) (ā″-trē-ō-ven-trih′-kyoo-ler) Specialized mass of conducting cells located at the atrioventricular junction in the heart.

Atrophy (a′-trō-fē) Reduction in size or wasting away of an organ or cell resulting from disease or lack of use.

Auditory ossicles (ah′-sih-kulz) The three tiny bones serving as transmitters of vibrations and located within the middle ear: the malleus, incus, and stapes.

Auditory tube Tube that connects the middle ear and the pharynx. Also called eustachian tube.

Autoimmune response Production of antibodies or effector T cells that attack a person's own tissue.

Autonomic nervous system Efferent division of the peripheral nervous system that innervates cardiac and smooth muscles and glands; also called the involuntary or visceral motor system.

Axial skeleton Portion of the skeleton that forms the central (longitudinal) axis of the body; includes the bones of the skull, the vertebral column, and the bony thorax.

Axilla (ak-sih′-luh) Armpit.

Axis (1) Second cervical vertebra; has a vertical projection called the dens around which the atlas rotates; (2) imaginary line about which a joint or structure revolves.

Axon Neuron process that carries impulses away from the nerve cell body; efferent process; the conducting portion of a nerve cell.

B cells Lymphocytes that oversee humoral immunity; their descendants differentiate into antibody-producing plasma cells. Also called B lymphocytes.

Baroreceptor (bayr″-ō-rē-sep′-ter) Pressoreceptor; receptor that is stimulated by pressure changes.

Basal metabolic rate (BMR) (bā′-sul meh″-tuh-bol′-ik) Rate at which energy is expended (heat produced) by the body per unit time under controlled (basal) conditions: 12 hours after a meal, at rest.

Basal (cerebral) nuclei Gray matter areas located deep within the white matter of the cerebral hemispheres. Also called basal ganglia.

Base Proton acceptor; substance capable of binding with hydrogen ions.

Basement membrane Extracellular material between epithelium and underlying connective tissue.

Basophil (bā′-zō-fil) White blood cell whose granules stain deep blue with basic dye; has a relatively pale nucleus and granular-appearing cytoplasm.

Benign (bē-nīn′) Not malignant.

Biceps (bī′-seps) Two-headed, especially applied to certain muscles.

Bile Greenish-yellow or brownish fluid produced in and secreted by the liver, stored in the gallbladder, and released into the small intestine.

Bilirubin (bil″-ih-roo′-bin) Red pigment of bile.

Bipolar neuron Neuron with axon and dendrite that extend from opposite sides of the cell body.

Blastocyst (blas′-tō-sist) Stage of early embryonic development; the product of cleavage.

Blood-brain barrier Mechanism that inhibits passage of materials from the blood into brain tissues; reflects relative impermeability of brain capillaries.

Blood flow Amount of blood flowing through a vessel or organ at a particular time.

Blood pressure Force exerted by blood against a unit area of the blood vessel walls; differences in blood pressure between different areas of the circulation provide the driving force for blood circulation.

Bolus (bō′-lus) A rounded mass of food prepared by the mouth for swallowing.

Bone remodeling Process involving bone formation and destruction in response to hormonal and mechanical factors.

Bony thorax Bones that form the framework of the thorax; includes sternum, ribs, and thoracic vertebrae.

Bowman's capsule (bō′-manz) Double-walled cup at the end of a renal tubule. Encloses a glomerulus.

Bradycardia (brā″-dē-kar′-dē-uh) A heart rate below 60 beats per minute.

Brain stem Collectively the midbrain, pons, and medulla of the brain.

Brain ventricle Fluid-filled cavity of the brain.

Bronchus (brong′-kus) One of the two large branches of the trachea that leads to the lungs.

Buffer Chemical substance or system that minimizes changes in pH by releasing or binding hydrogen ions.

Bursa A fibrous sac lined with synovial membrane and containing synovial fluid; occurs between bones and muscle tendons (or other structures), where it acts to decrease friction during movement.

Calcitonin (cal″-sih-tō′-nin) Hormone released by the thyroid that promotes a decrease in calcium levels of the blood.

Calculus (kal′-kyoo-lus) A stone formed within various body parts.

Calorie (cal.) Amount of energy needed to raise the temperature of 1 gram of water 1° Celsius. Energy exchanges associated with biochemical reactions are usually reported in kilocalories (1 kcal = 1000 cal) or large calories (Cal).

Calyx (kā′-liks) A cuplike extension of the pelvis of the kidney.

Canaliculus (ka″-nuh-lih′-kyoo-lus) Extremely small tubular passage or channel.

Cancer A malignant, invasive cellular neoplasm that has the capability of spreading throughout the body or body parts.

Capillaries (kah′-pih-layr″-ēs) The smallest of the blood vessels and the sites of exchange between the blood and tissue cells.

Carbohydrate (kar″-bō-hī′-drāt) Organic compound composed of carbon, hydrogen, and oxygen; includes starches, sugars, cellulose.

Carbonic anhydrase (kar-bah′-nik an-hī′-drās) Enzyme that facilitates the combination of carbon dioxide with water to form carbonic acid.

Carcinogen (kar-sin′-ō-jin) Cancer-causing agent.

Cardiac cycle Sequence of events encompassing one complete contraction and relaxation of the atria and ventricles of the heart.

Cardiac muscle Specialized muscle of the heart.

Cardiac output Amount of blood pumped out of a ventricle in one minute.

Carotid body (kuh-rah′-tid) A receptor in the common carotid artery sensitive to changing oxygen, carbon dioxide, and pH levels of the blood.

Carotid sinus (sī-nus) A dilation of a common carotid artery; involved in regulation of systemic blood pressure.

Cartilage (kar′-tih-lij) White, semiopaque connective tissue.

Catabolism (kuh-ta′-buh-lizm) Process in which living cells break down substances into simpler substances.

Catecholamines (ka″-tih-kō′-luh-mēnz) Epinephrine and norepinephrine.

Cation (kat′-ī-on) An ion with a positive charge.

Caudal (kah′-dul) Literally, toward the tail; in humans, the inferior portion of the anatomy.

Cecum (sē′-kum) The blind-end pouch at the beginning of the large intestine.

Cell membrane *See* Plasma membrane.

Cell-mediated immune response Immunity conferred by activated T cells, which directly lyse infected or cancerous body cells or cells of foreign grafts and release chemicals that regulate the immune response.

Cellular respiration Metabolic processes in which ATP is produced.

Cellulose (sel′-yoo-lōs) A fibrous carbohydrate that is the main structural component of plant tissues.

Central nervous system(CNS) Brain and spinal cord.

Centriole (sen′-trē-ōl) Minute body found near the nucleus of the cell; active in cell division.

Cerebellum (sayr″-ih-beh′-lum) Brain region most involved in producing smooth, coordinated skeletal muscle activity.

Cerebral aqueduct (suh-rē′-brul a′-kwih-dukt″) The slender cavity of the midbrain that connects the third and fourth ventricles; also called the aqueduct of Sylvius.

Cerebral cortex (suh-rē′-brul) The outer gray matter region of the cerebral hemispheres.

Cerebrospinal fluid (suh-rē″-brō-spī′-nul) Plasmalike fluid that fills the cavities of the CNS and surrounds the CNS externally; protects the brain and spinal cord.

Cerebrum (suh-rē′-brum) The cerebral hemispheres and the structures of the diencephalon.

Cervix (ser′-viks) The inferior necklike portion of the uterus.

Chemical bond An energy relationship holding atoms together; involves the interaction of electrons.

Chemical energy Energy stored in the bonds of chemicals.

Chemoreceptor (kē′-mō-rē-sep″-ter) Receptors sensitive to various chemicals in solution.

Chemotaxis (kē″-mō-tak′-sis) Movement of a cell, organism, or part of an organism toward or away from a chemical substance.

Cholecystokinin (CCK) (kō″-lē-sis″-tō-kī′-nin) An intestinal hormone that stimulates gallbladder contraction and pancreatic juice release.

Cholesterol (kuh-les′-ter-ol″) Steroid found in animal fats as well as in most body tissues.

Cholinergic fibers (kō″-lih-ner′-jik) Nerve endings that, upon stimulation, release acetylcholine.

Chondroblast (kon′-drō-blast) Actively mitotic cell form of cartilage.

Chondrocyte (kon′-drō-sīt) Mature cell form of cartilage.

Chorion (kor′-ē-on) Outermost fetal membrane; helps form the placenta.

Choroid plexus (kor′-oyd plek′-sus) A capillary knot that protrudes into a brain ventricle; involved in forming cerebrospinal fluid.

Chromatin (krō′-muh-tin) Structures in the nucleus that carry the hereditary factors (genes).

Chromosomes (krō′-muh-somz) Barlike bodies of tightly coiled chromatin; visible during cell division.

Chyme (kīm′) Semifluid, creamy mass consisting of partially digested food and gastric juice.

Cilia (sih′-lē-uh) Tiny; hairlike projections on cell surfaces that move in a wavelike manner.

Circle of Willis A union of arteries at the base of the brain.

Circumduction (ser″-kum-duk′-shun) Movement of a body part so that it outlines a cone in space.

Cirrhosis (ser-ō′-sis) Chronic disease, particularly of the liver, characterized by an overgrowth of connective tissue or fibrosis.

Cleavage An early embryonic phase consisting of rapid mitotic cell divisions without intervening growth periods; product is a blastocyst.

Clonal selection (klō′-nul) Process during which a B cell or T cell becomes sensitized through binding contact with an antigen.

Clone Descendants of a single cell.

Cochlea (kōk′-lē-ah) Snail-shaped chamber of the bony labyrinth that houses the receptor for hearing (the organ of Corti).

Coenzyme (kō-en′-zīm) Nonprotein substance associated with and activating an enzyme.

Colloid (kah′-loyd) Solute particles dispersed in a medium; particles do not settle out readily and do not pass through natural membranes.

Colloidal osmotic pressure (kah-loyd′-ul oz-mah′-tik) Pressure exerted on a membrane by the particles in a colloid; usually refers to the osmotic pressure of blood plasma and body fluids resulting from the presence of protein.

Command neuron An interneuron located in a brain stem extrapyramidal nucleus that helps regulate the spinal cord motor circuits.

Complement A group of blood-borne proteins, which, when activated, enhance the inflammatory and immune responses and may lead to cell lysis.

Compound Substance composed of two or more different elements, the atoms of which are chemically united.

Concentration gradient The difference in the concentration of a particular substance between two different areas.

Conductivity Ability to transmit an electrical impulse.

Condyle (kon′-dīl) A rounded projection at the end of a bone that articulates with another bone.

Cones One of the two types of photoreceptor cells in the retina of the eye; provide for color vision.

Congenital (kun-jeh′-nih-tul) Existing at birth.

Conjunctiva (kon″-junk-tī′-vuh) Thin, protective mucous membrane lining the eyelids and covering the anterior surface of the eye itself.

Connective tissue A primary tissue; form and function vary extensively. Function includes support, storage, and protection.

Contraception The prevention of conception; birth control.

Contraction To shorten or develop tension, an ability highly developed in muscle cells.

Contralateral Opposite; acting in unison with a similar part on the opposite side of the body.

Convergence Turning toward a common point from different directions.

Cornea (kor′-nē-uh) Transparent anterior portion of the eyeball.

Corona radiata (kor-ō′-nuh rā-dē-ah′-tuh) (1) Arrangement of elongated follicle cells around a mature ovum; (2) Crownlike arrangement of nerve fibers radiating from the internal capsule of the brain to every part of the cerebral cortex.

Cortex (kor′-teks) Outer surface layer of an organ.

Corticosteroids (kor″-tih-kō-stayr′-oydz) Steroid hormones released by the adrenal cortex.

Cortisol (kor′-tih-sol) Glucocorticoid produced by the adrenal cortex.

Covalent bond (kō-vā′-lent) Chemical bond created by electron sharing between atoms.

Cranial nerves The 12 nerve pairs that arise from the brain.

Creatine phosphate (krē′-uh-tin fos′-fāt) Compound that serves as an alternative energy source for muscle tissue.

Cutaneous (kyoo-tā′-nē-us) Pertaining to the skin.

Cyclic AMP Important intracellular second messenger that mediates hormonal effects; formed from ATP by the action of adenylate cyclase, an enzyme associated with the plasma membrane.

Cytokinesis (si″-tō-kuh-nē′-sis) Division of cytoplasm that occurs after the cell nucleus has divided.

Cytoplasm (sī′-tō-pla-zim) The cellular material surrounding the nucleus and enclosed by the plasma membrane.

Cytotoxic T cell Effector T cell that directly kills (lyses) foreign cells, cancer cells, or virus-infected body cells. Also called a killer T cell.

Deamination (de-am″-ih-nā′-shun) Removal of an amino group from an organic compound by reduction, hydrolysis, or oxidation.

Defecation (def″-ih-kā′-shun) Elimination of the contents of the bowels (feces).

Deglutition (deh″-gloo-tih′-shun) Swallowing.

Dehydration (dē″-hī-drā′-shun) Condition of excessive water loss.

Dendrite (den′-drīt) Branching neuron process that serves as a receptive, or input, region; transmits the nerve impulse to the cell body.

Deoxyribonucleic acid (DNA) (dē-ok″-sē-rī″-bō-noo-klē″-ik) A nucleic acid found in all living cells; carries the organism's hereditary information.

Depolarization (dē-pō″-ler-ih-zā′-shun) Loss of a state of polarity; loss of a negative charge inside the cell.

Dermis Layer of skin deep to the epidermis; composed of dense irregular connective tissue.

Desmosome (dez′-muh-sōm) Cell junction composed of thickened plasma membranes joined by filaments.

Diabetes insipidus (dī″-uh-bē′-tēz in-sih′-pih-dus) Disease characterized by passage of a large quantity of dilute urine plus intense thirst and dehydration caused by inadequate release of antidiuretic hormone (ADH).

Diabetes mellitus (meh-lī′-tus) Disease caused by deficient insulin release, leading to inability of the body cells to use carbohydrates at a normal rate.

Dialysis (dī-al′-ih-sis) Diffusion of solute(s) through a semipermeable membrane.

Diapedesis (dī″-uh-puh-dē′-sis) Passage of blood cells through intact vessel walls into tissue.

Diaphragm (dī′-uh-fram) (1) Any partition or wall separating one area from another; (2) a muscle that separates the thoracic cavity from the lower abdominopelvic cavity.

Diaphysis (di-a′-fih-sis) Elongated shaft of a long bone.

Diarthrosis (dī″-ar-thrō′-sis) Freely movable joint; also called a synovial joint.

Diastole (dī-as′-tuh-lē) Period when either the ventricles or the atria are relaxing.

Diencephalon (dī″-en-seh′-fuh-lon) That part of the forebrain between the cerebral hemispheres and the midbrain including the thalamus, the third ventricle, and the hypothalamus.

Diffusion (dih-fyoo′-zhun) The spreading of particles in a gas or solution with a movement toward uniform distribution of particles.

Digestion Chemical or mechanical process of breaking down foodstuffs to substances that can be absorbed.

Disaccharide (dī-sa′-kuh-rīd″) Literally, double sugar; e.g., sucrose, lactose.

Distal (dis′-tul) Away from the attached end of a limb or the origin of a structure.

Diverticulum (dī″-ver-tih′-kyoo-lum) A pouch or sac in the walls of a hollow organ or structure.

Dorsal (dor′-sul) Pertaining to the back; posterior.

Duct Canal or passageway.

Duodenum (doo″-ō-dē′-num) First part of the small intestine.

Dura mater (doo′-ruh ma′-ter) Outermost and toughest of the three membranes (meninges) covering the brain and spinal cord.

Dynamic equilibrium Sense that reports on angular or rotatory movements of the head in space.

Ectoderm (ek′-tō-derm) Embryonic germ layer; tissue forms the outer covering of the body and nervous tissues.

Edema (eh-dē′-muh) Abnormal accumulation of fluid in body parts or tissues; causes swelling.

Effector (eh-fek′-ter) Organ, gland, or muscle capable of being activated by nerve endings.

Efferent (eh′-fer-ent) Carrying away or away from, especially a nerve fiber that carries impulses away from the central nervous system.

Elastin (ē-las′-tin) Main protein in elastic fibers of connective tissues.

Electrocardiogram (ECG) (ih-lek″-trō-kar′-dē-ō-gram″) Graphic record of the electrical activity of the heart.

Electroencephalogram (EEG) (ih-lek″-trō-en-seh′-fuh-lō-gram″) Graphic record of the electrical activity of nerve cells in the brain.

Electrolyte (ih-lek′-trō-līt) Chemical substances, such as salts, acids, and bases, that ionize and dissociate in water and are capable of conducting an electrical current.

Electron Negatively charged subatomic particle; orbits the atomic nucleus.

Element One of a limited number of unique varieties of matter that composes substances of all kinds; e.g., carbon, hydrogen, oxygen.

Embolism (em′-bō-lizm) Obstruction of a blood vessel by a clot floating in the blood; may also be a bubble of air in the vessel (air embolism).

Embryo (em′-brē-ō) Developmental stage extending from gastrulation to the end of the eighth week.

Encephalitis (en″-seh-fuh-lī′-tis) Inflammation of the brain.

Endocarditis (en″-dō-kar-dī′-tis) Inflammation of the inner lining of the heart.

Endocardium (en″-dō-kar′-dē-um) Endothelial membrane that lines the interior of the heart.

Endocrine glands (en′-duh-krin) Ductless glands that empty their hormonal products directly into the blood.

Endocrine system Body system that includes internal organs that secrete hormones.

Endocytosis (en″-dō-sī-tō′-sis) Means by which substances enter cells; e.g., phagocytosis, pinocytosis, receptor-mediated endocytosis.

Endoderm (en′-dō-derm) Embryonic germ layer; forms the lining of the digestive tube and its associated structures.

Endometrium (en-dō-mē′-trē-um) Mucous membrane lining of the uterus.

Endomysium (en″-dō-mī′-sē-um) Thin connective tissue surrounding each muscle cell.

Endoplasmic reticulum (en″-dō-plaz′-mik rih-tih′-kyoo-lum) Membranous network of tubular or saclike channels in the cytoplasm of a cell.

Endothelium (en″-dō-thē′-lē-um) Single layer of simple squamous cells that line the walls of the heart, blood vessels, and lymphatic vessels.

Enzyme (en′-zīm) A protein that acts as a biological catalyst to speed up chemical reactions.

Eosinophil (ē″-ō-sin′-ō-fil) Granular white blood cell whose granules readily take up a stain called eosin.

Epidermis (eh″-peh-der′-mis) Superficial layer of the skin; composed of keratinized stratified squamous epithelium.

Epididymis (eh″-pih-dih′-dih-mis) That portion of the male duct system in which sperm mature. Empties into the ductus (or vas) deferens.

Epiglottis (eh″-puh-glah′-tis) Elastic cartilage at the back of the throat; covers the opening of the larynx (glottis) during swallowing.

Epimysium (eh″-pih-mī′-sē-um) Sheath of fibrous connective tissue surrounding a muscle.

Epinephrine (eh″-pih-neh′-frin) Chief hormone produced by the adrenal medulla. Also called adrenalin.

Epiphyseal plate (ih-pih″-fih-zē′-ul) Plate of hyaline cartilage at the junction of the diaphysis and epiphysis that provides for growth in length of a long bone.

Epiphysis (ih-pih′-fih-sis) The end of a long bone, attached to the shaft.

Epithelial (eh″-pih-thē′-lē-ul) Pertaining to a primary tissue that covers the body surface, lines its internal cavities, and forms glands.

Erythrocytes (eh-rih′-thrō-sīts) Red blood cells; when mature, erythrocytes are literally sacs of hemoglobin, the iron-containing pigment that transports oxygen in the blood.

Erythropoiesis (eh-rih″-thrō-poy-ē′-sis) Process of erythrocyte formation.

Erythropoietin (eh-rih″-thrō-poy′-eh-tin) Hormone that stimulates production of red blood cells.

Estrogen (es′-trō-jin) Hormone that stimulates female secondary sex characteristics; female sex hormones.

Eustachian tube (yoo-stā′-shun) *See* Auditory tube.

Excretion (ek-skrē′-shun) Elimination of waste products from the body.

Exocrine glands (ek′-sō-krin) Glands that have ducts through which their secretions are carried to a particular site.

Exocytosis (ek″-sō-sī-tō″-sis) Mechanism by which substances are moved from the cell interior to the extracellular space.

Exogenous (ek-sah′-jeh-nus) Developing or originating outside an organ or part.

Expiration Act of expelling air from the lungs.

Exteroceptor (ek″-ster-ō-sep′-ter) Sensory end organ that responds to stimuli from the external world.

Extracellular Outside a cell.

Extracellular fluid Internal fluid located outside cells; includes plasma and interstitial fluid.

Extrinsic (ek-strin′-sik) Originating outside an organ or part.

Exudate (ek′-syoo-dāt) Material including fluid, pus, or cells that has escaped from blood vessels and has been deposited in tissues.

Facet (fa′-set) A smooth, nearly flat surface on a bone for articulation.
Fallopian tube (fah-lō′-pē-un) *See* uterine tube.
Fascia (fa′-shē-uh) Layers of fibrous tissue covering and separating muscles.
Fascicle (fa′-sih-kul) Bundle of nerve or muscle fibers bound together by connective tissues.
Fatty acid Building block of fats.
Feces (fē′-sēz) Material discharged from the bowel; composed of food residue, secretions, bacteria.
Fenestrated (feh′-neh-strā″-tid) Pierced with one or more small openings.
Fertilization Fusion of the sperm and egg nuclei.
Fetus Developmental stage extending from the ninth week of development to birth.
Fibrin (fī′-brin) Fibrous insoluble protein formed during blood clotting.
Fibrinogen (fī-brin′-ō-jin) A blood protein that is converted to fibrin during blood clotting.
Fibroblast (fī′-brō-blast) Young, actively mitotic cell that forms the fibers of connective tissue.
Fibrocyte (fī′-brō-sīt) Mature fibroblast; maintains the matrix of fibrous types of connective tissue.
Filtration Passage of a solvent and dissolved substances through a membrane or filter.
Fissure (fih′-zher) (1) A groove or cleft; (2) the deepest depressions or inward folds on the brain.
Fixator (fix′-ā-ter) Muscle that immobilizes one or more bones, allowing other muscles to act from a stable base.
Flagella (fluh-jeh′-luh) Long, whiplike extensions of the plasma membrane of some bacteria and sperm; propel the cell.
Follicle (fah′-lih-kul) (1) Ovarian structure consisting of a developing egg surrounded by one or more layers of follicle cells; (2) colloid-containing structure of the thyroid gland.
Follicle-stimulating hormone (FSH) Hormone produced by the anterior pituitary that stimulates ovarian follicle production in females and sperm production in males.
Foramen (for-ā′-min) Hole or opening in a bone or between body cavities.
Forebrain Anterior portion of the brain consisting of the telencephalon and the diencephalon.
Formed elements Cellular portion of blood.
Fossa (fah′-suh) A depression, often an articular surface.
Fovea (fō′-vē-uh) A pit.
Frontal (coronal) plane Longitudinal plane that divides the body into anterior and posterior parts.
Fundus (fun′-dus) Base of an organ; that part farthest from the opening of the organ.
Funiculi (fuh-ni′-kyoo-lī) (1) Cordlike structures; (2) divisions of white matter in the spinal cord.

Gallbladder Sac beneath the right lobe of the liver used for bile storage.
Gamete (ga′-mēt) Sex cell; sperm or oocyte.
Gametogenesis (ga″-meh-tō-jeh′-neh-sis) Formation of gametes.
Ganglion (gan′-glē-on) Collection of nerve cell bodies outside the CNS.
Gap junction A passageway between two adjacent cells; formed by transmembrane proteins called connexons.
Gastrin (gas′-trin) Hormone that stimulates gastric secretion, especially hydrochloric acid release.
Gastroesophageal sphincter (gas″-trō-ē-sof″-uh-jē′-ul sfink′-ter) Valve between the esophagus and the stomach.
Gastrulation (gas″-troo-lā′-shun) Developmental process that produces the three primary germ layers (ectoderm, mesoderm, and endoderm).
Gene One of the biological units of heredity located in chromatin; transmits hereditary information.
Genetic code Information encoded in nucleotide base sequence.
Germ layers Three cellular layers (ectoderm, mesoderm, and endoderm) that represent the initial specialization of cells in the embryonic body and from which all body tissues arise.
Gestation (jes-tā′-shun) The period of pregnancy; about 280 days for humans.
Gland Organ specialized to secrete or excrete substances for further use in the body or for elimination.
Glia (glē′-uh) *See* Neuroglia.
Glomerular filtration rate (GFR) (glō-mer′-yoo-ler) Rate of filtrate formation by the kidneys.
Glomerulus (glō-mer′-yoo-lus) Cluster of capillaries forming part of the nephron; forms filtrate.
Glottis (glah′-tis) Opening between the vocal cords in the larynx.
Glucagon (gloo′-kuh-gon) Hormone formed by alpha cells of islets of Langerhans in the pancreas; raises the glucose level of blood.
Glucocorticoids (gloo″-kō-kor′-tih-koydz) Adrenal cortex hormones that increase blood glucose levels and aid the body in resisting long-term stressors.
Glucose (gloo′-kōs) Principal blood sugar.
Glycerol (glih′-ser-ol) A sugar alcohol; a building block of fats.
Glycogen (glī′-kō-jin) Main carbohydrate stored in animal cells; a polysaccharide.
Glycogenesis (glī″-kō-jeh′-nih-sis) Formation of glycogen from glucose.
Glycogenolysis (glī″-kō-jeh″-nō-lī′-sis) Breakdown of glycogen to glucose.
Glycolysis (glī-kol′-ih-sis) Breakdown of glucose to pyruvic acid—an anaerobic process.
Goblet cells Individual cells of the respiratory and digestive tracts that produce mucus.
Gonad (gō′-nad) Primary reproductive organ; i.e., the testis of the male or the ovary of the female.
Gonadotropins (gō″-na-dō-trō′-pinz) Gonad-stimulating hormones produced by the anterior pituitary.
Graafian follicle (grā′-fē-en) Mature ovarian follicle.
Graded potential A local change in membrane potential that varies directly with the strength of the stimulus, declines with distance, and is not conducted along the nerve fiber.
Gray matter Gray area of the central nervous system; contains cell bodies and unmyelinated fibers of neurons.
Growth hormone (GH) Hormone that stimulates growth in general; produced in the anterior pituitary; also called somatotropin (STH).
Gustation (guh-stā′-shun) Taste.
Gyrus (jī′-rus) An outward fold of the surface of the cerebral cortex.

Hair follicle The epithelial–connective tissue sheath within which a hair is formed by epithelial cells.
Haustra (hah′-struh) Pouches of the colon.
Haversian system (hah-ver′-zhen) System of interconnecting canals in the microscopic structure of adult compact bone; unit of bone. Also called osteon.
Heart block Impaired transmission of impulses from atrium to ventricle.
Heart murmur Abnormal heart sound (usually resulting from valve problems).
Helper T cell Type of T lymphocyte that orchestrates cellular immunity by direct contact with other immune cells and by releasing chemicals called lymphokines; also helps to mediate the humoral response by interacting with B cells.
Hematocrit (hē-mah′-tō-krit) The percentage of erythrocytes to total blood volume.
Hematopoiesis (hem″-ah-tō-poy-ē′-sis) Blood cell formation; hemopoiesis.

Heme (hēm) Iron-containing pigment that is essential to oxygen transport by hemoglobin.

Hemocytoblast (hē″-mō-sī′-tō-blast) Bone marrow stem cell that gives rise to all formed elements.

Hemoglobin (he′-muh-glō-bin) Oxygen-transporting component of erythrocytes.

Hemolysis (hē-mah′-lih-sis) Rupture of erythrocytes.

Hemopoiesis (hē″-mō-poy-ē′-sis) Blood cell formation; hematopoiesis.

Hemorrhage (hem′-or-ij) Loss of blood from the vessels by flow through ruptured walls; bleeding.

Hemostasis (hē″-mō-stā′-sis) Stoppage of bleeding or blood flow.

Hepatic portal system (heh-pa′-tik) Circulation in which the hepatic portal vein carries dissolved nutrients to the liver tissues for processing.

Hepatitis (hep″-uh-tī′-tis) Inflammation of the liver.

Hernia (her′-nē-uh) Abnormal protrusion of an organ or a body part through the containing wall of its cavity.

Histamine (his′-tuh-mēn) Substance that causes vasodilation and increased vascular permeability.

Histology (his-tah′-luh-jē) Branch of anatomy dealing with the microscopic structure of tissues.

Holocrine glands (hō′-luh-krin) Glands that accumulate their secretions within their cells; secretions are discharged only upon rupture and death of the cell.

Homeostasis (hō″-mē-ō-stā′-sis) State of body equilibrium, or the maintenance of a stable internal environment of the body.

Homologous (hō-mah′-luh-gus) Parts or organs corresponding in structure but not necessarily in function.

Hormones Steroidal or amino acid–based molecules released to the blood that act as chemical messengers to regulate specific body functions.

Humoral (hyoo′-mer-ul) **immune response** Immunity conferred by antibodies present in blood plasma and other body fluids.

Hydrochloric acid (HCl) (hī″-drō-klor′-ik) Acid that aids protein digestion in the stomach; produced by parietal cells.

Hydrogen bond Weak bond in which a hydrogen atom forms a bridge between two electron-hungry atoms. An important intramolecular bond.

Hydrolysis (hī-drah′-lih-sis) Process in which water is used to split a substance into smaller particles.

Hydrophilic (hī″-drō-fih′-lik) Refers to molecules, or portions of molecules, that interact with water and charged particles.

Hydrophobic (hī″-drō-fō′-bik) Refers to molecules, or portions of molecules, that interact only with nonpolar molecules.

Hydrostatic pressure (hī″-drō-sta′-tik) Pressure of fluid in a system.

Hypertension (hī″-per-ten′-shun) High blood pressure.

Hypertonic (hī″-per-tah′-nik) Excessive, above normal, tone or tension.

Hypertrophy (hī-per′-truh-fē) Increase in size of a tissue or organ independent of the body's general growth.

Hypodermis (hī″-pō-der′-mis) Subcutaneous connective tissue; also called superficial fascia.

Hypothalamus (hī″-pō-tha′-luh-mis) Region of the diencephalon forming the floor of the third ventricle of the brain.

Hypotonic (hī″-pō-tah′-nik) Below normal tone or tension.

Hypoxia (hī-pok′-sē-uh) Condition in which inadequate oxygen is available to tissues.

Ileum (il′-ē-um) Terminal part of the small intestine; between the jejunum and the cecum of the large intestine.

Immune response Antigen-specific defenses mounted by activated lymphocytes (T cells and B cells).

Immunity (ih-myoo′-nih-tē) Ability of the body to resist many agents (both living and nonliving) that can cause disease; resistance to disease.

Immunocompetence Ability of the body's immune cells to recognize (by binding) specific antigens; reflects the presence of plasma membrane–bound receptors.

Immunodeficiency disease Disease resulting from the deficient production or function of immune cells or certain molecules (complement, antibodies, and so on) required for normal immunity.

Immunoglobulin (ih″-myoo-nō- glah′-byoo-lin) Protein molecule released by plasma cells that mediates humoral immunity; an antibody.

In vitro (in vē′-trō) In a test tube, glass, or artificial environment.

In vivo (in vē′-vo) In the living body.

Infarct (in′-farkt) Region of dead, deteriorating tissue resulting from a lack of blood supply.

Inferior (caudal) Pertaining to a position near the tail end of the long axis of the body.

Inflammation (in″-fluh-mā′-shun) A physiological response of the body to tissue injury; includes dilation of blood vessels and an increase in vessel permeability; indicated by redness, heat, swelling, and pain.

Inguinal (in′-gwih-nul) Pertaining to the groin region.

Inner cell mass Accumulation of cells in the blastocyst from which the embryo develops.

Innervation (in″-er-vā′-shun) Supply of nerves to a body part.

Insertion Movable part or attachment of a muscle as opposed to origin.

Inspiration Drawing of air into the lungs.

Insulin (in′-suh-lin) Hypoglycemic hormone produced by beta cells of the pancreas affecting carbohydrate and fat metabolism, blood glucose levels, and other systemic processes.

Integumentary system (in-teh″-gyoo-men′-tah-rē) Skin (and its derivatives); provides the external protective covering of the body.

Intercalated discs (in-ter′-kuh-lā-tid) Gap junctions connecting muscle cells of the myocardium.

Intercellular (in″-ter-sel′-yoo-ler) Between body cells.

Intercellular matrix (mā′-triks) Material between adjoining cells; important in connective tissue.

Interferon (in-ter-fēr′-on) Chemical that is able to provide some protection against virus invasion of the body; inhibits viral growth.

Internal capsule Band of white matter between the basal ganglia and the thalamus.

Internal respiration Exchange of gases between blood and tissue fluid and between tissue fluid and cells.

Interoceptor (in″-ter-ō-sep′-ter) Nerve ending situated in the viscera, which is sensitive to changes and stimuli within the body's internal environment; also called visceroceptor.

Interstitial fluid (in″-ter-stih′-shul) Fluid between the cells.

Intervertebral disks (in″-ter-ver′-teh-brul) Disks of fibrocartilage between vertebrae.

Intervertebral foramina (for-a′-mih-nuh) Openings between the dorsal projections of adjacent vertebrae through which the spinal nerves pass.

Intracellular (in″-truh-sel′-yoo-ler) Within a cell.

Intracellular fluid Fluid within a cell.

Intrinsic factor (in-trin′-sik) Substance produced by the stomach that is required for vitamin B_{12} absorption.

Involuntary muscle Muscle not under control of the will; cardiac and smooth muscle.

Ion (ī′-on) Atom with a positive or negative electric charge.

Ionic bond (ī-ah′-nik) Chemical bond formed by electron transfer between atoms.

Ipsilateral (ip″-sih-la′-ter-ul) Situated on the same side.

Irritability Ability to respond to a stimulus.

Ischemia (is-kē′-mē-uh) Local decrease in blood supply.

Isometric contraction (ī″-sō-meh′-trik) Contraction in which the muscle does not shorten (the load is too heavy) but its internal tension increases.

Isotonic contraction (ī″-sō-tah′-nik) Contraction in which muscle tension remains constant and the muscle shortens.
Isotopes (ī″-suh-tōps) Different atomic forms of the same element. Isotopes vary only in the number of neutrons they contain.

Jejunum (juh-joo′-num) The part of the small intestine between the duodenum and the ileum.
Joint Junction of two or more bones; an articulation.
Juxtaglomerular apparatus (juks-tuh-glō-mer′-yoo-ler) Cells located close to the glomerulus that play a role in blood pressure regulation by releasing the enzyme renin.

Keratin (kayr′-uh-tin) Water-insoluble protein found in the epidermis, hair, and nails that makes those structures hard and water-repellent.
Ketosis (kē-tō′-sis) Abnormal condition during which an excess of ketone bodies is produced.
Killer T cell See Cytotoxic T cell.
Kinins (kī′-ninz) Group of polypeptides that dilate arterioles, increase vascular permeability, and induce pain.
Krebs cycle Aerobic metabolic pathway occurring within mitochondria, in which food metabolites are oxidized and CO_2 is liberated.

Labia (lā′-bē-uh) Lips.
Labyrinth (la′-bih-rinth″) Bony cavities and membranes of the inner ear.
Lacrimal (la′-krih-mul) Pertaining to tears.
Lactation (lak-tā′-shun) Production and secretion of milk.
Lacteal (lak′-tē-ul) Special lymphatic capillaries of the small intestine that take up lipids.
Lactic acid (lak′-tik) Product of anaerobic metabolism, especially in muscle.
Lacuna (luh-kyoo′-nuh) Little depression or space; the lacunae of bone or cartilage, lacunae are occupied by cells.
Lamina (la′-mih-nuh) (1) A thin layer or flat plate; (2) the portion of a vertebra between the transverse process and the spinous process.
Larynx (layr′-inks) Cartilaginous organ located between the trachea and the pharynx; voice box.
Latent period Period of time between stimulation and the onset of muscle contraction.
Lateral Away from the midline of the body.
Leukocytes (loo′-kō-sīts) White blood cells; formed elements involved in body protection that take part in inflammatory and immune responses.
Ligament (lih′-guh-ment) Band of fibrous tissue that connects bones.
Limbic system (lim′-bik) Functional brain system involved in emotional response.
Lipid (lih′-pid) Organic compound formed of carbon, hydrogen, and oxygen; examples are fats and cholesterol.
Lumbar (lum′-bar) Portion of the back between the thorax and the pelvis.
Lumen (loo′-min) Cavity inside a tube, blood vessel, or hollow organ.
Luteinizing hormone (LH) (loo′-tē-in-ī″-zing) Anterior pituitary hormone that aids maturation of cells in the ovary and triggers ovulation in females. In males, causes the interstitial cells of the testis to produce testosterone.
Lymph (limf′) Protein-containing fluid transported by lymphatic vessels.
Lymph node Small lymphatic organ that filters lymph; contains macrophages and lymphocytes.
Lymphatic system (lim′-fa-tik) System consisting of lymphatic vessels, lymph nodes, and other lymphoid organs and tissues; drains excess tissue fluid from the extracellular space and provides a site for immune surveillance.
Lymphatics General term used to designate lymphatic vessels.
Lymphocyte Agranular white blood cell that arises from bone marrow and becomes functionally mature in the lymphoid organs of the body.
Lymphokines (lim′-fō-kīnz) Substances involved in cell-mediated immune responses that enhance immune and inflammatory responses.
Lysosomes (lī′-sō-sōmz) Organelles that originate from the Golgi apparatus and contain strong digestive enzymes.
Lysozyme (lī′-sō-zīm) Enzyme in lacrimal gland secretions that is capable of destroying certain kinds of bacteria.

Macrophage (ma′-krō-fāj″) Protective cell type common in connective tissue, lymphatic tissue, and certain body organs that phagocytizes tissue cells, bacteria, and other foreign debris; important as an antigen-presenter to T cells and B cells in the immune response.
Malignant (muh-lig′-nent) Life threatening; pertains to neoplasms that spread and lead to death, such as cancer.
Mammary glands (mam′-er-ē) Milk-producing glands of the breasts.
Mastication (mas″-tih-kā′-shun) Chewing.
Matrix Specialized extracellular substance secreted by connective tissue cells that determines the specialized function of each connective tissue type; typically includes ground substance (fluid to hard) and fibers (collagen, elastic, and/or reticular).
Meatus (mē-ā′-tus) External opening of a canal.
Mechanoreceptor (meh″-kuh-nō-rē-sep′-ter) Receptor sensitive to mechanical pressures such as touch, sound, or contractions.
Medial (mē′-dē-ul) Toward the midline of the body.
Mediastinum (mē″-dē-uh-stī′-num) Region of the thoracic cavity between the lungs.
Medulla (meh-dul′-ah) Central portion of certain organs.
Medulla oblongata (meh-dul′-ah ob-lon-gah′-tuh) Inferior part of the brain stem.
Meiosis (mī-ō′-sis) Nuclear division process that reduces the chromosomal number by half and results in the formation of four haploid (n) cells; occurs only in certain reproductive organs.
Melanin (meh′-luh-nin) Dark pigment formed by cells called melanocytes; imparts color to the skin and hair.
Memory cells Members of T cell and B cell clones that provide for immunologic memory.
Meninges (meh-nin′-jēz) Protective coverings of the central nervous system; from the most external to the most internal, the dura mater, arachnoid, and pia mater.
Meningitis (meh″-nin-jī′-tis) Inflammation of the meninges.
Menstruation (men″-stroo-ā′-shun) The periodic, cyclic discharge of blood, secretions, tissue, and mucus from the mature female uterus in the absence of pregnancy.
Merocrine glands (mayr′-ō-krin) Glands that produce secretions intermittently; secretions do not accumulate in the gland.
Mesencephalon (meh″-zin-seh′-fuh-lon) Midbrain.
Mesenchyme (meh′-zin-kīm) Common embryonic tissue from which all connective tissues arise.
Mesenteries (meh′-zen-tayr″-ēz) Double-layered extensions of the peritoneum that support most organs in the abdominal cavity.
Mesoderm (meh′-zō-derm) Tissue that forms the skeleton and muscles of the body.
Metabolic rate (meh″-tuh-bah′-lik) Energy expended by the body per unit time.
Metabolism (meh-tah′-buh-lizm) Sum total of all the chemical reactions occurring in body cells, reactions that transform substances into energy or materials the body can use or store by means of anabolism or catabolism.

Metastasis (meh-tas′-tuh-sis) The spread of cancer from one body part or organ into another not directly connected to it.

Microvilli (mī″-krō-vih′-lē) Tiny projections on the free surfaces of some epithelial cells; increase surface area for absorption.

Micturition (mik″-ter-ih′-shin) Urination, or voiding; emptying the bladder.

Midbrain Region of the brain stem between the diencephalon and the pons.

Minerals Inorganic chemical compounds found in nature; salts.

Mineralocorticoid (min″-er-al″-ō-kor′-tih-koyd) Steroid hormone of the adrenal cortex that regulates mineral metabolism and fluid balance.

Mitochondria (mī″-tō-kon′-drē-uh) Cytoplasmic organelles responsible for ATP generation for cellular activities.

Mitosis (mī-tō′-sis) Process during which the chromosomes are redistributed to two daughter nuclei; nuclear division.

Mixed nerves Nerves containing the processes of motor and sensory neurons; their impulses travel to and from the central nervous system.

Mixture A substance consisting of particles dispersed together and in which the particles retain their unique properties; e.g., a solution, colloid, or suspension.

Molar (mō′-ler) A solution concentration determined by mass of solute—one liter of solution contains an amount of solute equal to its molecular weight in grams.

Molecule Particle consisting of two or more atoms joined by chemical bonds.

Monoclonal antibodies (mah″-nō-klō′-nul) Pure preparations of identical antibodies that exhibit specificity for a single antigen.

Monocyte (mah′-nō-sīt) Large single-nucleus white blood cell; agranular leukocyte.

Monosaccharide (mah″-nō-sa′-kuh-rīd) Literally, one sugar; building block of carbohydrates; e.g., glucose, fructose.

Motor nerves Nerves that carry impulses leaving the brain and spinal cord.

Motor unit A motor neuron and all the muscle cells it stimulates.

Mucous membranes Membranes that form the linings of body cavities open to the exterior (digestive, respiratory, urinary, and reproductive tracts).

Mucus (myoo′-kus) A sticky thick fluid, secreted by mucous glands and mucous membranes, that keeps the free surface of membranes moist.

Multipolar neuron Neuron that has one long axon and numerous dendrites.

Muscle fiber Muscle cell.

Muscle spindle Complex receptor found in skeletal muscle that is sensitive to stretch.

Muscle twitch A single rapid contraction of a muscle followed by relaxation.

Myelencephalon (mī″-uh-len-seh′-fuh-lon) Lower part of the hindbrain, especially the medulla oblongata.

Myelin sheath (mī′-uh-lin) Fatty insulating sheath that surrounds all but the smallest nerve fibers.

Myocardial infarction (mī″-ō-kar′-dē-ul in-fark′-shun) Condition characterized by dead tissue areas in the myocardium; caused by interruption of blood supply to the area.

Myocardium (mī″-ō-kar′-dē-um) Layer of the heart wall composed of cardiac muscle.

Myofibril (mī″-ō-fī′-bril) Rodlike bundle of contractile filaments (myofilaments) found in muscle cells.

Myofilament (mī″-ō-fil′-uh-ment) Filament that constitutes myofibrils. Of two types: actin and myosin.

Myogenic (mī″-ō-jeh′-nik) Having the potential to contract automatically without nervous system stimulation.

Myoglobin (mī″-ō-glō′-bin) Oxygen-binding pigment in muscle.

Myometrium (mī″-ō-mē′-trē-um) Thick uterine musculature.

Myosin (mī′-ō-sin) One of the principal proteins found in muscle.

Nares (nār′-ēz) The nostrils.

Necrosis (neh-krō′-sis) Death or disintegration of a cell or tissues caused by disease or injury.

Negative feedback Feedback that tends to cause the levels of a variable to change in a direction opposite to that of an initial change.

Nephron (neh′-fron) Structural and functional unit of the kidney; consists of the glomerulus and renal tubule.

Nerve fiber Axon or dendrite of a neuron.

Neuroglia (ner-ah′-glē-uh) Nonexcitable cells of neural tissue that support, protect, and insulate the neurons.

Neurohypophysis (ner″-ō-hī-pah′-fih-sis) Posterior pituitary; portion of the pituitary gland derived from the brain.

Neuromuscular junction Region where a motor neuron comes into close contact with a skeletal muscle cell.

Neuron (ner′-on) Cell of the nervous system specialized to generate and transmit nerve impulses.

Neurotransmitter Chemical released by neurons that may, upon binding to receptors of neurons or effector cells, stimulate or inhibit them.

Neutron (noo′-tron) Uncharged subatomic particle; found in the atomic nucleus.

Neutrophil (noo′-truh-fil) Most abundant type of white blood cell.

Nucleic acid (noo-klē′-ik) Class of organic molecules that includes DNA and RNA.

Nucleoli (noo-klē′-uh-lī) Small spherical bodies in the cell nucleus.

Nucleotide (noo′-klē-uh-tīd) Building block of nucleic acids; consists of a sugar, a nitrogen-containing base, and a phosphate group.

Nucleus (noo′-klē-is) Control center of a cell; contains genetic material.

Occipital (ok-sih′-pih-tul) Pertaining to the area at the back of the head.

Occlusion (ah-kloo′-zhun) Closure or obstruction.

Olfaction (ol-fak′-shun) Smell.

Olfactory epithelium Sensory receptor region in the nasal cavity containing olfactory neurons that respond to volatile chemicals that enter the nasal passages in air.

Oocyte (ō′-ō-sīt) Immature egg.

Oogenesis (ō″-ō-jeh′-neh-sis) Process of ovum (female gamete) formation.

Ophthalmic (op-thal′-mik) Pertaining to the eye.

Optic (op′-tik) Pertaining to the eye.

Optic chiasma (op′-tik ki-az′-muh) The partial crossover of fibers of the optic nerves.

Organ A part of the body formed of two or more tissues and adapted to carry out a specific function; e.g., the stomach.

Organ system A group of organs that work together to perform a vital body function; e.g., the nervous system.

Organelles (or″-guh-nelz′) Small cellular structures (ribosomes, mitochondria, and others) that perform specific metabolic functions for the cell as a whole.

Organic Pertaining to carbon-containing substances, such as proteins, fats, and carbohydrates.

Origin Attachment of a muscle that remains relatively fixed during muscular contraction.

Osmoreceptor (oz″-mō-rē-sep′-ter) Structure sensitive to osmotic pressure or concentration of a solution.

Osmosis (os-mō′-sis) Diffusion of a solvent through a membrane from a dilute solution into a more concentrated one.

Osmotic pressure Water-attracting ability of plasma proteins.

Ossicles (ah′-sih-kulz) The three bones of the middle ear: hammer, anvil, and stirrup.
Osteoblasts (ah′-stē-ō-blasts) Bone-forming cells.
Osteoclasts (ah′-stē-ō-klasts) Large cells that reabsorb or break down bone matrix.
Osteocyte (ah′-stē-ō-sīt) Mature bone cell found in each lacuna.
Osteogenesis (ah″-stē-ō-jeh′-neh-sis) Process of bone formation.
Osteon (ah′-stē-on) *See* Haversian system.
Osteoporosis (os″-tē-ō-por-ō′-sis) Increased softening of the bone resulting from a gradual decrease in rate of bone formation.
Ovarian cycle (ō-vayr′-ē-un) Monthly cycle of follicle development, ovulation, and corpus luteum formation in an ovary.
Ovary (ō′-ver-ē) Female sex organ in which ova (eggs) are produced; female gonad.
Ovulation (ah″-vyoo-lā-shun) Ejection of an immature egg (oocyte) from the ovary.
Ovum (ō′-vum) Female gamete (germ cell); egg.
Oxidation (ok″-sih-dā′-shun) Process of substances combining with oxygen or the removal of hydrogen.
Oxidative phosphorylation (ok′-sih-dā′-tiv fos″-for-ih-lā′-shun) Process of ATP synthesis during which an inorganic phosphate group is attached to ADP; occurs within the mitochondria.
Oxygen debt The volume of oxygen required after exercise to oxidize the lactic acid formed during exercise.
Oxyhemoglobin (ok″-sē-hē′-muh-glō-bin) Oxygen-bound form of hemoglobin.
Oxytocin (ok″-sih-tō′-sin) Hormone released by the posterior pituitary that stimulates contraction of the uterus during childbirth and the ejection of milk during nursing.

Pacemaker Region of specialized cardiac tissue (the SA node) that has the fastest rate of spontaneous depolarization, and hence sets the rate of contraction for the heart as a whole.
Palate (pa′-lut) Roof of the mouth.
Pancreas (pan′-krē-us) Gland located behind the stomach, between the spleen and the duodenum; produces both endocrine and exocrine secretions.
Pancreatic juice (pan″-krē-a′-tik) Bicarbonate-rich secretion of the pancreas containing enzymes for digestion of all food categories.
Parasympathetic (payr″-uh-sim″-puh-theh′-tik) The division of the autonomic nervous system that oversees digestion, elimination, and glandular function; the resting and digesting subdivision.
Parathyroid glands (payr″-uh-thī′-royd) Small endocrine glands located on the posterior aspect of the thyroid gland.
Parathyroid hormone (PTH) Hormone released by the parathyroid glands that regulates blood calcium level.
Parietal (puh-rī′-eh-tul) Pertaining to the walls of a cavity.
Passive immunity Short-lived immunity resulting from the introduction of "borrowed antibodies" obtained from an immune animal or human donor; immunological memory is not established.
Passive transport Membrane transport processes that do not require cellular energy (ATP); e.g., diffusion, which is driven by kinetic energy.
Pectoral (pek′-ter-ul) Pertaining to the chest.
Pectoral girdle Bones that attach the upper limbs to the axial skeleton; includes the clavicle and scapula.
Pelvic girdle Consists of the paired coxal bones that attach the lower limbs to the axial skeleton.
Pelvis (pel′-vis) A basin-shaped bony structure; lower portion of body trunk.
Penis (pē′-nis) Male organ of copulation and urination.
Pepsin Enzyme capable of digesting proteins in an acid pH.
Peptide bond (pep′-tīd) Bond joining the amine group of one amino acid to the acid carboxyl group of a second amino acid with the loss of a water molecule.
Pericardium (payr″-ih-kar′-dē-um) Double-layered serosa enclosing the heart and forming its superficial layer.
Perichondrium (payr″-ih-kon′-drē-um) Fibrous, connective-tissue membrane covering the external surface of cartilaginous structures.
Perimysium (payr″-ih-mī′-sē-um) Connective tissue enveloping bundles of muscle fibers.
Perineum (payr″-ih-nē′-um) That region of the body spanning the region between the ischial tuberosities and extending from the anus to the scrotum in males and from the anus to the vulva in females.
Periosteum (payr″-ē-ah′-stē-um) Double-layered connective tissue that covers and nourishes the bone.
Peripheral nervous system (PNS) Portion of the nervous system consisting of nerves and ganglia that lie outside of the brain and spinal cord.
Peripheral resistance A measure of the amount of friction encountered by blood as it flows through the blood vessels.
Peristalsis (payr″-ih-stal′-sis) Progressive, wavelike contractions that move foodstuffs through the alimentary tube organs (or that move other substances through other hollow body organs).
Peritoneum (payr″-ih-tuh-nē′-um) Serous membrane lining the interior of the abdominal cavity and covering the surfaces of abdominal organs.
Peritonitis (payr″-ih-tuh-nī′-tis) Inflammation of the peritoneum.
Permeability That property of membranes that permits passage of molecules and ions.
pH (pē-āch) The measure of the relative acidity or alkalinity of a solution: hydrogen ion concentration in moles per liter.
Phagocyte (fa′-gō-sīt) Cell capable of engulfing and digesting particles or cells harmful to the body.
Phagocytosis (fa″-gō-sī-tō′-sis) Engulfing of foreign solids by cells.
Pharynx (fayr′-inks) Muscular tube extending from the region posterior to the nasal cavities to the esophagus.
Phospholipid (fos″-fō-lip′-id) Modified lipid containing phosphorus.
Photoreceptor (fō″-tō-rē-sep′-ter) Specialized receptor cells that respond to light energy.
Pinocytosis (pē″-nō-sī-tō′-sis) Engulfing of extracellular fluid by cells.
Pituitary gland (pih-too′-ih-tayr″-ē) Neuroendocrine gland located beneath the brain that serves a variety of functions including regulation of gonads, thyroid, adrenal cortex, lactation, and water balance.
Placenta (pluh-sen′-tuh) Temporary organ formed from both fetal and maternal tissues that provides nutrients and oxygen to the developing fetus, carries away fetal metabolic wastes, and produces the hormones of pregnancy.
Plasma The nonliving fluid component of blood within which formed elements and various solutes are suspended and circulated.
Plasma cells Members of a B cell clone; specialized to produce and release antibodies.
Plasma membrane Membrane that encloses cell contents; outer limiting cell membrane.
Platelet (plāt′-let) Cell fragment found in blood; involved in clotting.
Pleura (pler′-uh) Two-layered serous membrane that lines the thoracic cavity and covers the external surface of the lung.
Plexus (plek′-sus) A network of converging and diverging nerve fibers.
Plica (plī′-kuh) A fold.
Podocyte (pod′-ō-sīt) Epithelial cell located on the basement membrane of the glomerulus, spreading thin cytoplasmic projections over the membrane. Forms capsular part of the filtration membrane.

Polarized State of the plasma membrane of an unstimulated neuron or muscle cell in which the inside of the cell is relatively negative in comparison to the outside; the resting state.

Polypeptide (pah″-lē-pep′-tīd) A chain of amino acids.

Polysaccharide (pah″-lē-sa′-kuh-rīd) Literally, many sugars, a polymer of linked monosaccharides; e.g., starch, glycogen.

Pons (1) Any bridgelike structure or part; (2) the part of the brain stem connecting the medulla with the midbrain, providing linkage between upper and lower levels of the central nervous system.

Positive feedback Feedback that tends to cause the level of a variable to change in the same direction as an initial change.

Posterior (pōs-tēr′-ē-er) The back of an organism, organ, or part; the dorsal surface.

Posterior pituitary See Neurohypophysis.

Postganglionic neuron (pōst″-gan-glē-ah′-nik) Autonomic motor neuron that has its cell body in a peripheral ganglion and projects its axon to an effector.

Preganglionic neuron Autonomic motor neuron that has its cell body in the central nervous system and projects its axon to a peripheral ganglion.

Pressoreceptor A nerve ending in the wall of the carotid sinus and aortic arch sensitive to vessel stretching.

Primary (immune) response Initial response of the immune system to an antigen; involves clonal selection and establishes immunological memory.

Prime mover Muscle that bears the major responsibility for effecting a particular movement; agonist.

Process (1) Prominence or projection; (2) series of actions for a specific purpose.

Progesterone (prō-jes′-ter-ōn) Hormone responsible for preparing the uterus for the fertilized ovum.

Pronation (prō-nā′-shun) Inward rotation of the forearm causing the radius to cross diagonally over the ulna—palms face posteriorly.

Prone Refers to a body lying horizontally with the face downward.

Proprioceptor (prō″-prē-ō-sep′-ter) Receptor located in a muscle or tendon; concerned with locomotion and posture.

Protein (prō′-tēn) Complex nitrogenous substance; main building material of cells.

Proton (prō′-ton) Subatomic particle that bears a positive charge; located in the atomic nucleus.

Proximal (prok′-sih-mul) Toward the attached end of a limb or the origin of a structure.

Pseudostratified (soo″-dō-stra′-tih-fīd) Pertaining to epithelium that appears to be stratified (consisting of many layers) but is not.

Puberty Period of life when reproductive maturity is achieved.

Pulmonary (pul′-muh-nayr-ē) Pertaining to the lungs.

Pulmonary circuit System of blood vessels that serves gas exchange in the lungs; i.e., pulmonary arteries, capillaries, and veins.

Pulmonary edema (eh-dē′-muh) Leakage of fluid into the air sacs and tissue of the lungs.

Pulmonary ventilation Breathing; consists of inspiration and expiration.

Pulse Rhythmic expansion and recoil of arteries resulting from heart contraction; can be felt from outside the body.

Pupil Opening in the center of the iris through which light enters the eye.

Purkinje fibers (per-kin′-jē) Modified cardiac muscle fibers of the conduction system of the heart.

Pus Fluid product of inflammation composed of white blood cells, the debris of dead cells, and a thin fluid.

Pyloric region (pī-lor′-ik) The distal portion of the stomach; joins with the duodenum.

Radioactivity The process of spontaneous decay seen in some of the heavier isotopes, during which particles or energy is emitted from the atomic nucleus; results in the atom becoming more stable.

Radioisotope (rā″-dē-ō-ī′-suh-tōp) Isotope that exhibits radioactive behavior.

Ramus (rā′-mus) Branch of a nerve, artery, vein, or bone.

Receptor (rē-sep′-ter) Peripheral nerve ending specialized for response to particular types of stimuli.

Receptor potential A graded potential that occurs at a sensory receptor membrane.

Reduction Restoring broken bone ends (or a dislocated bone) to their (its) original position.

Reflex Automatic reaction to stimuli.

Relative refractory period Period following stimulation during which only a stronger than usual stimulus can evoke an action potential.

Renal (rē′-nul) Pertaining to the kidney.

Renin (rē′-nin) Substance released by the kidneys that is involved with raising blood pressure.

Respiratory system Organ system that carries out gas exchange; includes the nose, pharynx, larynx, trachea, bronchi, lungs.

Rete (rē′-tē) A network; often composed of nerve fibers or blood vessels.

Reticulocyte (rih-tih′-kyoo-lō-sīt) Immature or young erythrocyte.

Reticular formation (rih-tih′-kyoo-ler) Functional system that spans the brain stem; involved in regulating sensory input to the cerebral cortex, cortical arousal, and control of motor behavior.

Retina (reh′-tih-nuh) Neural tunic of the eyeball; contains photoreceptors (rods, cones).

Rhinencephalon (rīn″-en-seh′-fuh-lon) That portion of the cerebrum concerned with reception and integration of olfactory impulses.

Ribonucleic acid (RNA) (rī″-bō-noo-klē′-ik) Nucleic acid that contains ribose; acts in protein synthesis.

Ribosomes (rī′-buh-sōmz) Cytoplasmic organelles at which proteins are synthesized.

Rods One of the two types of photosensitive cells in the retina.

Rugae (roo′-gē) Elevations or ridges, as in the mucosa of the stomach.

Sacral (sa′-krul) Lower portion of the back, just superior to the buttocks.

Sagittal plane (sa′-jih-tul) A longitudinal plane that divides the body or any of its parts into right and left portions.

Saltatory conduction Transmission of an action potential along a myelinated fiber in which the nerve impulse appears to leap from node to node.

Salt Ionic compound that dissociates into charged particles (other than hydrogen or hydroxyl ions) when dissolved in water.

Sarcomere (sar′-kō-mēr) The smallest contractile unit of muscle; contains myofilaments composed mainly of contractile proteins (actin, myosin).

Sarcoplasmic reticulum (sar″-kō-plaz′-mik rih-tik′-yoo-lum) Specialized endoplasmic reticulum of muscle cells.

Sclera (sklayr′-uh) Outer fibrous tunic of the eyeball.

Scrotum (skrō′-tum) External sac enclosing the testes.

Sebaceous gland (suh-bā′-shus) Epidermal gland that produces an oily secretion called sebum.

Sebum (sē′-bum) Oily secretion of sebaceous glands.

Second messenger Intracellular molecule generated by the binding of a chemical (hormone or neurotransmitter) to a plasma membrane receptor; mediates intracellular responses to the chemical messenger.

Secondary (immune) response Second and subsequent responses of the immune system to a previously met antigen; more rapid and more vigorous than the primary response.

Secretion (suh-krē′-shun) (1) The passage of material formed by a cell to its exterior; (2) Cell product that is transported to the exterior of the cell.

Section A cut through the body (or an organ) that is made along a particular plane; a thin slice of tissue prepared for microscopic study.

Semen (sē′-min) Fluid mixture containing sperm and secretions of the male accessory reproductive glands.

Semilunar valves (seh″-mē-loo′-ner) Valves that prevent blood return to the ventricles after contraction.

Seminiferous tubules (seh″-mih-nih′-fer-us too′-byulz) Highly convoluted tubes within the testes; form sperm.

Sensory nerve Nerve that contains processes of sensory neurons and carries nerve impulses to the central nervous system.

Sensory neuron Neuron that initiates nerve impulses following receptor stimulation.

Sensory receptor Dendritic end organs, or parts of other cell types, specialized to respond to a stimulus.

Sensory transduction Conversion of stimulus energy into a nerve impulse.

Serous fluid (sēr′-us) Clear, watery fluid secreted by cells of a serous membrane.

Serum (sēr′-um) Amber-colored fluid that exudes from clotted blood as the clot shrinks and then no longer contains fibrinogen.

Sinoatrial (SA) node (sī″-nō-ā′-trē-ul) Specialized myocardial cells in the wall of the right atrium; pacemaker of the heart.

Sinus (sī′-nus) (1) Mucous-membrane-lined, air-filled cavity in certain cranial bones; (2) dilated channel for the passage of blood or lymph.

Smooth muscle Muscle consisting of spindle-shaped, unstriped (nonstriated) muscle cells; involuntary muscles.

Solute (sol′-yoot) The substance that is dissolved in a solution.

Somatic nervous system (sō-ma′-tik) Division of the peripheral nervous system that provides the motor innervation of skeletal muscles.

Somite (sō′-mīt) A mesodermal segment of the body of an embryo.

Special senses The senses of taste, smell, vision, hearing, and equilibrium.

Spermatogenesis (sper″-muh-tō-jeh′-nih-sis) The process of sperm (male gamete) formation; involves meiosis.

Sphincter (sfink′-ter) A muscle surrounding an opening; acts as a valve.

Spinal nerves The 31 nerve pairs that arise from the spinal cord.

Spinal reflex A somatic reflex mediated through the spinal cord.

Squamous (skwā′-mus) Pertaining to flat, thin cells that form the free surface of some epithelial tissues.

Static equilibrium Sense of head position in space with respect to gravity.

Stenosis (steh-nō′-sis) Constriction or narrowing.

Steroids (stayr′-oydz) Group of chemical substances including certain hormones and cholesterol.

Stimulus (stim′-yoo-lus) An excitant or irritant; a change in the environment that evokes a response.

Stressor Any stimulus that directly or indirectly causes the hypothalamus to initiate stress-reducing responses, such as the fight or flight response.

Striated muscle (strī′-ā-tid) Muscle consisting of cross-striated (cross-striped) muscle fibers; cardiac and skeletal muscle.

Stroke Condition in which brain tissue is deprived of a blood supply, as in blockage of a cerebral blood vessel.

Stroke volume Amount of blood pumped out of a ventricle during one contraction.

Subcutaneous (sub″-kyoo-tā′-nē-us) Beneath the skin.

Sudoriferous gland (soo-duh-rih′-fer-us) Epidermal gland that produces sweat.

Sulcus (sul′-kus) A furrow on the brain, less deep than a fissure.

Summation Accumulation of effects, especially those of muscular, sensory, or mental stimuli.

Superficial Located close to or on the body surface.

Superior Refers to the head or upper body regions.

Supination (soo″-pih-nā′-shun) The outward rotation of the forearm causing palms to face anteriorly.

Supine (soo′-pīn) Refers to a body lying horizontally with the face upward.

Suppressor T cells Regulatory T lymphocytes that suppress the immune response.

Surfactant (ser-fak′-tent) Secretion produced by certain cells of the alveoli that reduces the surface tension of water molecules, thus preventing the collapse of the alveoli after each expiration.

Suspension (sus-pen′-shun) A dispersing of particles throughout a body of liquid.

Suture (soo′-cher) An immovable joint; with one exception, all bones of the skull are united by sutures.

Sweat gland *See* Sudoriferous gland.

Sympathetic division Division of the autonomic nervous system that activates the body to cope with some stressor (danger, excitement, etc.); the fight, fright, and flight subdivision.

Sympathetic tone State of partial vasoconstriction of the blood vessels maintained by sympathetic fibers.

Symphysis (simp′-fih-sis) A joint in which the bones are connected by fibrocartilage.

Synapse (sih′-naps) Functional junction or point of close contact between two neurons or between a neuron and an effector cell.

Synaptic cleft (sih-nap′-tik) Fluid-filled space at a synapse between neurons.

Synaptic delay Time required for an impulse to cross a synapse between two neurons.

Synarthrosis (sin″-ar-thrō′-sis) Immovable joint.

Synchondrosis (sin″-kon-drō′-sis) A joint in which the bones are united by hyaline cartilage.

Syndesmosis (sin″-dez-mō′-sis) A joint in which the bones are united by a ligament or a sheet of fibrous tissue.

Synergist (sih′-ner-jist) Muscle that aids the action of a prime mover by effecting the same movement or by stabilizing joints across which the prime mover acts to prevent undesirable movements.

Synergy (sih′-ner-jē) Coordinated activity of agonist and antagonist muscles that results in smooth, well-controlled movements.

Synostosis (sin″-os-tō′-sis) A completely ossified joint; a fused joint.

Synovial fluid (sih-nō′-vē-ul) Fluid secreted by the synovial membrane; lubricates joint surfaces and nourishes articular cartilages.

Synovial joint (sih-nō′-vē-ul) Freely movable joint; also called a diarthrosis.

Systemic (sis-teh′-mik) Pertaining to the whole body.

Systemic circuit System of blood vessels that serves gas exchange in the body tissues.

Systole (sis′-tō-lē) Period when either the ventricles or the atria are contracting.

Systolic pressure (sis-tah′-lik) Pressure generated by the left ventricle during systole.

T tubule Extension of the muscle cell plasma membrane (sarcolemma) that protrudes deeply into the muscle cell.

T cells Lymphocytes that mediate cellular immunity; include helper, killer, suppressor, and memory cells. Also called T lymphocytes.

Target cell A cell that is capable of responding to a hormone because it bears receptors to which the hormone can bind.
Taste buds Sensory receptor organs that house gustatory cells, which respond to dissolved food chemicals.
Telencephalon (tel″-en-seh′-fuh-lon) Anterior subdivision of the primary forebrain that develops into olfactory lobes, cerebral cortex, and corporal striata.
Tendon (ten′-dun) Cord of dense fibrous tissue attaching muscle to bone.
Testis (tes′-tis) Male primary sex organ that produces sperm; male gonad.
Tetanus (tet′-nus) (1) A smooth, sustained muscle contraction resulting from high-frequency stimulation; (2) an infectious disease caused by an anaerobic bacterium.
Thalamus (tha′-luh-mis) A mass of gray matter in the diencephalon of the brain.
Thermoreceptor (ther″-mō-rē-sep′-ter) Receptor sensitive to temperature changes.
Thoracic (thor-a′-sik) Refers to the chest.
Thoracic duct Large duct that receives lymph drained from the entire lower body, the left upper extremity, and the left side of the head and thorax.
Thorax (thor′-aks) That portion of the body trunk above the diaphragm and below the neck.
Threshold Weakest stimulus capable of producing a response in an irritable tissue.
Thrombin (throm′-bin) Enzyme that induces clotting by converting fibrinogen to fibrin.
Thrombocytes (throm′-bō-sīts) Platelets; cell fragments that participate in blood coagulation.
Thymus gland (thī′-mus) Endocrine gland active in immune response.
Thyroid gland (thī′-royd) One of the largest of the body's endocrine glands.
Tight junction Area where plasma membranes of adjacent cells are fused.
Tissue A group of similar cells (and their intercellular substance) specialized to perform a specific function; primary tissue types of the body are epithelial, connective, muscle, and nervous tissue.
Tolerance Failure of the body to mount a specific immune response against a particular antigen.
Tone Refers to the state of continuous muscular or neuron activity.
Tonicity (tuh-nih′-sih-tē) A measure of the ability of a solution to cause a change in cell shape or tone by promoting osmotic flows of water.
Trabecula (truh-beh′-kyoo-luh) Any one of the fibrous bands extending from the capsule into the interior of an organ.
Trachea (trā′-kē-uh) Windpipe; cartilage-reinforced tube extending from larynx to bronchi.
Tract A collection of nerve fibers in the central nervous system having the same origin, termination, and function.
Transverse process One pair of projections that extend laterally from each neural arch of a vertebra.
Trochanter (trō-kan′-ter) A large, somewhat blunt process.
Trophoblast (trō′-fō-blast) Outer sphere of cells of the blastocyst.
Tropic hormone (trō′-pik) A hormone that regulates the function of another endocrine organ.
Trypsin (trip′-sin) An active enzyme that splits proteins.
Tubercle (too′-ber-kul) A nodule or small rounded process.
Tuberosity (too″-ber-ah′-sih-tē) A broad process, larger than a tubercle.
Tumor An abnormal growth of cells; a swelling; cancerous at times.
Tunica (too′-nih-kuh) A covering or tissue coat; membrane layer.
Twitch A brief contraction of muscle in response to a stimulus.
Tympanic membrane (tim-pa′-nik) Eardrum.
Ulcer (ul′-ser) Lesion or erosion of the mucous membrane, such as gastric ulcer of stomach.
Umbilical cord (um-bih′-lih-kul) Structure bearing arteries and veins connecting the placenta and the fetus.
Umbilicus (um-bih′-lih-kus) Navel; marks site where umbilical cord was attached in fetal stage.
Unipolar neuron Neuron in which embryological fusion of the two processes leaves only one process extending from the cell body.
Urea (yer-ē-uh) Main nitrogen-containing waste excreted in urine.
Ureter (yer′-eh-ter) Tube that carries urine from kidney to bladder.
Urethra (yer-ē′-thruh) Canal through which urine passes from the bladder to outside the body.
Uterine tube (yoo′-ter-in) Tube through which the ovum is transported to the uterus. Also called fallopian tube.
Uterus (yoo′-ter-us) Hollow, thick-walled organ that receives, retains, and nourishes fertilized egg; site where embryo/fetus develops.
Uvula (yoo′-vyoo-luh) Tissue tag hanging from soft palate.

Valence shell (vā′-lints) Outermost energy level of an atom that contains electrons.
Vas (vaz′) A duct; vessel.
Vasa recta (vā′-zuh rek′-tuh) Capillary branches that supply loops of Henle and collecting ducts.
Vasoconstriction (va″-zō-kon-strik′-shun) Narrowing of blood vessels.
Vasodilation (va″-zō-dī-lā′-shun) Relaxation of the smooth muscles of the blood vessels producing dilation.
Vasomotion (va″-zō-mō′-shun) An increase or decrease in caliber of a blood vessel.
Vasomotor center (va″-zō-mō′-ter) Brain area concerned with regulation of blood vessel resistance.
Vasomotor fibers (va″-zō-mō′-ter) Sympathetic nerve fibers that regulate the contraction of smooth muscle in the walls of blood vessels, thereby regulating blood vessel diameter.
Vasopressin (va″-zō-preh′-sin) Antidiuretic hormone.
Veins (vānz′) Blood vessels that return blood toward the heart from the circulation.
Ventral Pertaining to the front; anterior.
Ventricles (ven′-trih-kulz) Paired, inferiorly located heart chambers that function as the major blood pumps.
Venule (ven′-ūl) A small vein.
Vertebral column (ver′-tih-brul) The spine, formed of a number of individual bones called vertebrae and two composite bones (sacrum and coccyx).
Vesicle (veh′-sih-kul) A small liquid-filled sac or bladder.
Villus (vih′-lus) Fingerlike projections of the small intestinal mucosa that tremendously increase its surface area for absorption.
Visceral (vih′-ser-ul) pertaining to an internal organ of the body or the inner part of a structure.
Viscosity (vis-kah′-sih-tē) State of being sticky or thick.
Vital capacity The volume of air that can be expelled from the lungs by forcible expiration after the deepest inspiration; total exchangeable air.
Vitamins Organic compounds required by the body in minute amounts.
Voluntary muscle Muscle under control of the will; skeletal muscle.
Vulva (vul′-vuh) Female external genitalia.

White matter White substance of the central nervous system; myelinated nerve fibers.

Yolk sac (yōk) Endodermal sac that serves as the source for primordial germ cells.

Zygote (zī′-gōt) Fertilized egg.

Index

Note: **Boldface** page numbers indicate definitions or major discussions; *italic* page numbers refer to tables or illustrations.